P9-EED-161

FOR REFERENCE

Do Not Take From This Room

Second Edition

EDITOR-IN-CHIEF

Renato Dulbecco
The Salk Institute

EDITORIAL ADVISORY BOARD

John Abelson
California Institute of Technology

Peter Andrews
Natural History Museum, London

John A. Barranger
University of Pittsburgh

R. J. Berry
University College, London

Konrad Bloch
Harvard University

Floyd Bloom
The Scripps Research Institute

Norman E. Borlaug
Texas A&M University

Charles L. Bowden
University of Texas Health Science Center at San Antonio

Ernesto Carafoli
ETH-Zentrum

Stephen K. Carter
Bristol-Myers Squibb Corporation

Edward C. H. Carterette
University of California, Los Angeles

David Carver
University of Medicine and Dentistry of New Jersey

Joel E. Cohen
Rockefeller University and Columbia University

Michael E. DeBakey
Baylor College of Medicine

Eric Delson
American Museum of Natural History

W. Richard Dukelow
Michigan State University

Myron Essex
Harvard University

Robert C. Gallo
Institute for Human Virology

Joseph L. Goldstein
University of Texas Southwestern Medical Center

I. C. Gunsalus
University of Illinois, Urbana–Champaign

Osamu Hayaishi
Osaka Bioscience Institute

Leonard A. Herzenberg
Stanford University Medical Center

Kazutomo Imahori
Mitsubishi-Kasei Institute of Life Science, Tokyo

Richard T. Johnson
Johns Hopkins University Medical School

Yuet Wai Kan
University of California, San Francisco

Bernard Katz
University College, London

Seymour Kaufman
National Institutes of Health

Ernst Knobil
University of Texas Health Science Center at Houston

Glenn Langer
University of California Medical Center, Los Angeles

Robert S. Lawrence
Johns Hopkins University

James McGaugh
University of California, Irvine

Henry M. McHenry
University of California, Davis

Philip W. Majerus
Washington University School of Medicine

W. Walter Menninger
Menninger Foundation

Terry M. Mikiten
University of Texas Health Science Center at San Antonio

Beatrice Mintz
Institute for Cancer Research, Fox Chase Cancer Center

Harold A. Mooney
Stanford University

Arno G. Motulsky
University of Washington School of Medicine

Marshall W. Nirenberg
National Institutes of Health

G. J. V. Nossal
Walter and Eliza Hall Institute of Medical Research

Mary Osborn
Max Planck Institute for Biophysical Chemistry

George E. Palade
University of California, San Diego

Mary Lou Pardue
Massachusetts Institute of Technology

Ira H. Pastan
National Institutes of Health

David Patterson
University of Colorado Medical Center

Philip Reilly
Shriver Center for Mental Retardation

Arthur W. Rowe
New York University Medical Center

Ruth Sager
Dana-Farber Cancer Institute

Alan C. Sartorelli
Yale University School of Medicine

Neena B. Schwartz
Northwestern University

Bernard A. Schwetz
National Center for Toxicological Research, FDA

Nevin S. Scrimshaw
United Nations University

Michael Sela
Weizmann Institute of Science, Israel

Satimaru Seno
Shigei Medical Research Institute, Japan

Phillip Sharp
Massachusetts Institute of Technology

E. R. Stadtman
National Institutes of Health

P. K. Stumpf
University of California, Davis (emeritus)

William Trager
Rockefeller University (emeritus)

Arthur C. Upton
University of Medicine and Dentistry of New Jersey, Robert Wood Johnson Center

Itaru Watanabe
Kansas City VA Medical Center

David John Weatherall
Oxford University, John Radcliffe Hospital

Klaus Weber
Max Planck Institute for Biophysical Chemistry

Thomas H. Weller
Harvard School of Public Health

Harry A. Whitaker
Université du Québec á Montréal

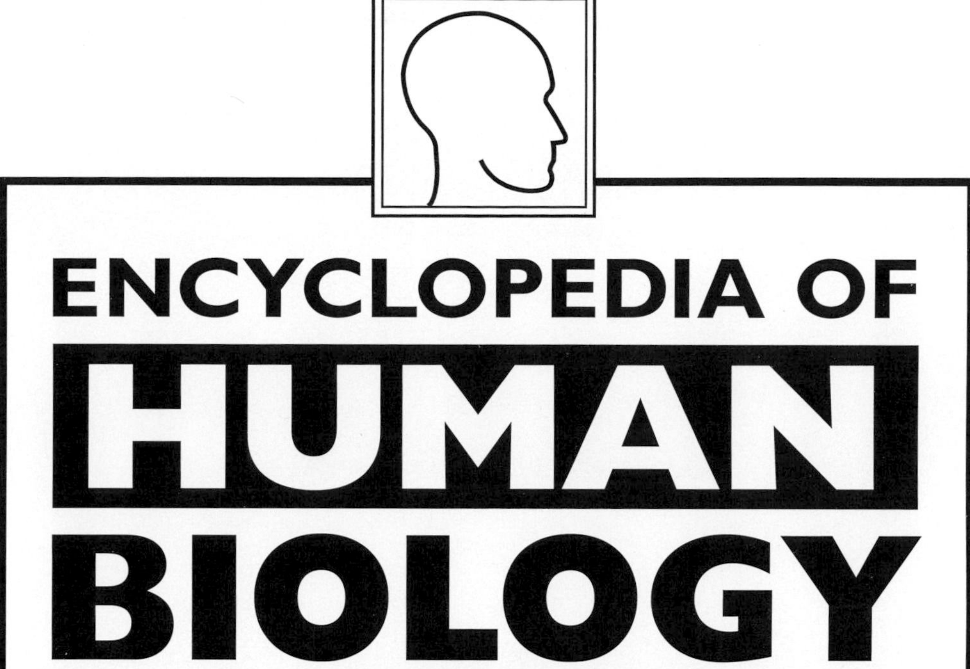

ENCYCLOPEDIA OF HUMAN BIOLOGY

Volume 2 Bl–Com

Second Edition

Editor-in-Chief

RENATO DULBECCO

The Salk Institute

La Jolla, California

Riverside Community College
'98 Library
4800 Magnolia Avenue
Riverside, California 92506

ACADEMIC PRESS

San Diego London Boston New York Sydney Tokyo Toronto

REF QP 11 .E53 1997 v.2

Encyclopedia of human biology

This book is printed on acid-free paper. ♾

Copyright © 1997, 1991 by ACADEMIC PRESS

All Rights Reserved.
No part of this publication may be reproduced or transmitted in any form or by any means, electronic or mechanical, including photocopy, recording, or any information storage and retrieval system, without permission in writing from the publisher.

Academic Press
a division of Harcourt Brace & Company
525 B Street, Suite 1900, San Diego, California 92101-4495, USA
http://www.apnet.com

Academic Press Limited
24-28 Oval Road, London NW1 7DX, UK
http://www.hbuk.co.uk/ap/

Library of Congress Cataloging-in-Publication Data

Encyclopedia of human biology / edited by Renato Dulbecco. -- 2nd ed.
p. cm.
Includes bibliographical references and index.
ISBN 0-12-226970-5 (alk. paper: set). -- ISBN 0-12-226971-3 (alk. paper: v. 1). -- ISBN 0-12-226972-1 (alk. paper: v. 2). -- ISBN 0-12-226973-X (alk. paper: v. 3). -- ISBN 0-12-226974-8 (alk. paper: v. 4). -- ISBN 0-12-226975-6 (alk. paper: v. 5). -- ISBN 0-12-226976-4 (alk. paper: v. 6). -- ISBN 0-12-226977-2 (alk. paper : v. 7). -- ISBN 0-12-226978-0 (alk. paper: v. 8). -- ISBN 0-12-226979-9 (alk. paper: v. 9)
1. Human biology--Encyclopedias. I. Dulbecco, Renato, date.
[DNLM: 1. Biology--encyclopedias. 2. Physiology--encyclopedias. QH 302.5 E56 1997]
QP11.E53 1997
612'.003-dc21
DNLM/DLC
for Library of Congress 97-8627
CIP

PRINTED IN THE UNITED STATES OF AMERICA
97 98 99 00 01 02 EB 9 8 7 6 5 4 3 2 1

CONTENTS OF VOLUME 2

Contents for each volume of the Encyclopedia appears in Volume 9.

C

PREFACE TO THE FIRST EDITION

We are in the midst of a period of tremendous progress in the field of human biology. New information appears daily at such an astounding rate that it is clearly impossible for any one person to absorb all this material. The *Encyclopedia of Human Biology* was conceived as a solution: an informative yet easy-to-use reference. The Encyclopedia strives to present a complete overview of the current state of knowledge of contemporary human biology, organized to serve as a solid base on which subsequent information can be readily integrated. The Encyclopedia is intended for a wide audience, from the general reader with a background in science to undergraduates, graduate students, practicing researchers, and scientists.

Why human biology? The study of biology began as a correlate of medicine with the human, therefore, as the object. During the Renaissance, the usefulness of studying the properties of simpler organisms began to be recognized and, in time, developed into the biology we know today, which is fundamentally experimental and mainly involves nonhuman subjects. In recent years, however, the identification of the human as an autonomous biological entity has emerged again—stronger than ever. Even in areas where humans and other animals share a certain number of characteristics, a large component is recognized only in humans. Such components include, for example, the complexity of the brain and its role in behavior or its pathology. Of course, even in these studies, humans and other animals share a certain number of characteristics. The biological properties shared with other species are reflected in the Encyclopedia in sections of articles where results obtained in nonhuman species are evaluated. Such experimentation with nonhuman organisms affords evidence that is much more difficult or impossible to obtain in humans but is clearly applicable to us.

Guidance in fields with which the reader has limited familiarity is supplied by the detailed index volume. The articles are written so as to make the material accessible to the uninitiated; special terminology either is avoided or, when used, is clearly explained in a glossary at the beginning of each article. Only a general knowledge of biology is expected of the reader; if specific information is needed, it is reviewed in the same section in simple terms. The amount of detail is kept within limits sufficient to convey background information. In many cases, the more sophisticated reader will want additional information; this will be found in the bibliography at the end of each article. To enhance the long-term validity of the material, untested issues have been avoided or are indicated as controversial.

The material presented in the Encyclopedia was produced by well-recognized specialists of experience and competence and chosen by a roster of outstanding scientists including ten Nobel laureates. The material was then carefully reviewed by outside experts. I have reviewed all the articles and evaluated their contents in my areas of competence, but my major effort has been to ensure uniformity in matters of presentation, organization of material, amount of detail, and degree of documentation, with the goal of presenting in each subject the most advanced information available in easily accessible form.

Renato Dulbecco

PREFACE TO THE SECOND EDITION

The first edition of the *Encyclopedia of Human Biology* has been very successful. It was well received and highly appreciated by those who used it. So one may ask: Why publish a second edition? In fact, the word "encyclopedia" conveys the meaning of an opus that contains immutable information, forever valid. But this depends on the subject. Information about historical subjects and about certain branches of science is essentially immutable. However, in a field such as human biology, great changes occur all the time. This is a field that progresses rapidly; what seemed to be true yesterday may not be true today. The new discoveries constantly being made open new horizons and have practical consequences that were not even considered previously. This change applies to all fields of human biology, from genetics to structural biology and from the intricate mechanisms that control the activation of genes to the biochemical and medical consequences of these processes.

These are the reasons for publishing a second edition. Although much of the first edition is still valid, it lacks the information gained in the six years since its preparation. This new edition updates the information to what we know today, so the reader can be confident of its full validity. All articles have been reread by their authors, who modified them when necessary to bring them up-to-date. Many new articles have also been added to include new information.

The principles followed in preparing the first edition also apply to the second edition. All new articles were contributed by specialists well known in their respective fields. Expositional clarity has been maintained without affecting the completeness of the information. I am convinced that anyone who needs the information presented in this encyclopedia will find it easily, will find it accessible, and, at the same time, will find it complete.

Renato Dulbecco

A GUIDE TO USING THE ENCYCLOPEDIA

The *Encyclopedia of Human Biology, Second Edition* is a complete source of information on the human organism, contained within the covers of a single unified work. It consists of nine volumes and includes 670 separate articles ranging from genetics and cell biology to public health, pediatrics, and gerontology. Each article provides a comprehensive overview of the selected topic to inform a broad spectrum of readers from research professionals to students to the interested general public.

In order that you, the reader, derive maximum benefit from your use of the Encyclopedia, we have provided this Guide. It explains how the Encyclopedia is organized and how the information within it can be located.

ORGANIZATION

The *Encyclopedia of Human Biology, Second Edition* is organized to provide the maximum ease of use for its readers. All of the articles are arranged in a single alphabetical sequence by title. Articles whose titles begin with the letters A to Bi are in volume 1, articles with titles from Bl to Com are in Volume 2, and so on through Volume 8, which contains the articles from Si to Z.

Volume 9 is a separate reference volume providing a Subject Index for the entire work. It also includes a complete Table of Contents for all nine volumes, an alphabetical list of contributors to the Encyclopedia, and an Index of Related Titles. Thus Volume 9 is the best starting point for a search for information on a given topic, via either the Subject Index or Table of Contents.

So that they can be easily located, article titles generally begin with the key word or phrase indicating the topic, with any descriptive terms following. For example, "Calcium, Biochemistry" is the article title rather than "Biochemistry of Calcium" because the specific term *calcium* is the key word rather than the more general term *biochemistry*. Similarly "Protein Targeting, Basic Concepts" is the article title rather than "Basic Concepts of Protein Targeting."

TABLE OF CONTENTS

A complete Table of Contents for the *Encyclopedia of Human Biology, Second Edition* appears in Volume 9. This list of article titles represents topics that have been carefully selected by the Editor-in-Chief, Dr. Renato Dulbecco, and the members of the Editorial Advisory Board (see p. ii for a list of the Board members). The Encyclopedia provides coverage of 35 specific subject areas within the overall field of human biology, ranging alphabetically from Behavior to Virology.

In addition to the complete Table of Contents found in Volume 9, the Encyclopedia also provides an individual table of contents at the front of each volume. This lists the articles included within that particular volume.

INDEX

The Subject Index in Volume 9 contains more than 4200 entries. The subjects are listed alphabetically and indicate the volume and page number where information on this topic can be found.

ARTICLE FORMAT

Articles in the *Encyclopedia of Human Biology, Second Edition* are arranged in a single alphabetical list by title. Each new article begins at the top of a right-hand page, so that it may be quickly located. The author's name and affiliation are displayed at the beginning of the article. The article is organized according to a standard format, as follows:

- Title and author
- Outline
- Glossary
- Defining statement
- Body of the article
- Bibliography

OUTLINE

Each article in the Encyclopedia begins with an outline that indicates the general content of the article. This outline serves two functions. First, it provides a brief preview of the article, so that the reader can get a sense of what is contained there without having to leaf through the pages. Second, it serves to highlight important subtopics that will be discussed within the article. For example, the article "Gene Mapping" includes the subtopic "DNA Sequence and the Human Genome Project."

The outline is intended as an overview and thus it lists only the major headings of the article. In addition, extensive second-level and third-level headings will be found within the article.

GLOSSARY

The Glossary contains terms that are important to an understanding of the article and that may be unfamiliar to the reader. Each term is defined in the context of the particular article in which it is used. Thus the same term may appear as a Glossary entry in two or more articles, with the details of the definition varying slightly from one article to another. The Encyclopedia includes approximately 5000 glossary entries.

DEFINING STATEMENT

The text of each article in the Encyclopedia begins with a single introductory paragraph that defines the topic under discussion and summarizes the content of the article. For example, the article "Free Radicals and Disease" begins with the following statement:

A FREE RADICAL is any species that has one or more unpaired electrons. The most important free radicals in a biological system are oxygen- and nitrogen-derived radicals. Free radicals are generally produced in cells by electron transfer reactions. The major sources of free radical production are inflammation, ischemia/reperfusion, and mitochondrial injury. These three sources constitute the basic components of a wide variety of diseases. . . .

CROSS-REFERENCES

Many of the articles in the Encyclopedia have cross-references to other articles. These cross-references appear within the text of the article, at the end of a paragraph containing relevant material. The cross-references indicate related articles that can be consulted for further information on the same topic, or for other information on a related topic. For example, the article "Brain Evolution" contains a cross reference to the article "Cerebral Specialization."

BIBLIOGRAPHY

The Bibliography appears as the last element in an article. It lists recent secondary sources to aid the reader in locating more detailed or technical information. Review articles and research papers that are important to an understanding of the topic are also listed.

The bibliographies in this Encyclopedia are for the benefit of the reader, to provide references for further reading or research on the given topic. Thus they typically consist of no more than ten to twelve entries. They are not intended to represent a complete listing of all materials consulted by the author or authors in preparing the article.

COMPANION WORKS

The *Encyclopedia of Human Biology, Second Edition* is one of an extensive series of multivolume reference works in the life sciences published by Academic Press. Other such works include the *Encyclopedia of Cancer, Encyclopedia of Virology, Encyclopedia of Immunology,* and *Encyclopedia of Microbiology,* as well as the forthcoming *Encyclopedia of Reproduction.*

Blood Binding of Drugs

JEAN-PAUL TILLEMENT
ROLAND ZINI
JÉRÔME BARRÉ
FRANÇOISE HERVÉ
Faculté de Médecine de Paris

GLOSSARY

Acceptor Protein that binds the drug; the relevant interaction induces no pharmacological effect (synonymous with silent receptor)

Apparent volume of distribution Expressed in liter per kilogram of body weight, a theoretical concept that relates the amount of drug in the body to a specified concentration: $V = A/C$, where V is plasma-apparent volume of distribution, A is amount of drug in the body, and C is plasma concentration

Drug monitoring Strategy whereby the dosing regimen for a patient is guided by repeated measurements of plasma, serum, or blood concentration

Drug targeting Methodology designed to deliver a drug to its site(s) of action

Extraction ratio Fraction of drug entering a clearing organ (liver or kidney), which is removed during transit

Pharmacokinetics Study of the time course of drug and metabolite concentrations in biological fluids, tissues, or excreta

Unbound fraction Ratio of unbound to total drug concentration

SEVERAL BINDERS MAY BE INVOLVED IN BLOOD binding of drugs, including the main plasma proteins, red cells, specific carriers (e.g., transcortin), and sometimes specific circulating receptors. All of them act as drug carriers from blood to tissues. According to the parameters of the relevant interaction, each bond brings its own pharmacokinetic consequences, either a permissive or a restrictive effect on tissue transfer. Moreover, these two effects may coexist for the same drug–protein complex depending on the tissue or organ involved. These particularities have varying implications in pharmacokinetics, monitoring, and the design of new drugs.

I. BACKGROUND

A. Possible Implications of Plasma-Bound Drugs

In blood, a drug can be either almost fully free or largely bound to proteins. Free drugs are generally small molecules, sufficiently lipid soluble to be quickly transferred across the membranes by a passive mechanism and, thus, available to act directly in and on the tissues. In contrast, bound drugs form a high molecular weight complex too big to be easily transferred into the tissues. Assuming that the effects of the drug are located in tissues, the plasma-free fraction is thus considered as active because it is diffusible; the bound part is considered inactive because it is nondiffusible. However, as this interaction is reversible, the binding is only a temporary inactivation process, whose duration and intensity are related to the stability of the drug–protein complex and to the amount of drug retained in the blood.

The concept of inactivation by binding is a general one that is applicable to tissue as well as plasma binding. For example, a high degree of nonspecific binding

ENCYCLOPEDIA OF HUMAN BIOLOGY, Second Edition, VOLUME 2. Copyright © 1997 by Academic Press. All rights of reproduction in any form reserved.

in the tissues also removes a drug from the site of action.

B. Relevant Binding Parameters of Drug–Protein Complexes

The drug–protein interaction may be considered as a reversible equilibrium, following the mass action law, which requires verified fixation in each case. The association and dissociation rates of drug–protein complexes are very high. At equilibrium, the bound (B) and free (F) drug concentrations can be measured, thus allowing the calculation of the following: (1) the plasma-binding percentage (B/T; T = total drug concentration) or the bound fraction (Fb); (2) the affinity constant (K_a) or the dissociation constant (K_d) of the binding process; and (3) the number of binding sites per mole (n) or the plasma-binding sites concentration (N), where P is the protein concentration and $N = n \cdot P$.

Numerous techniques can be used to determine either directly or indirectly these B and F concentrations. Among the techniques that measure the free and/or bound concentrations, equilibrium dialysis and ultrafiltration are most widely used.

Equilibrium dialysis is useful for measuring the binding of low molecular weight ligands ($<1000\ M_r$) to macromolecules. This method is widely used to determine the parameters of the binding of drugs to plasma proteins. Its accuracy relies on the fact that the acceptor (i.e., binding protein) concentration is often higher than the total drug concentration in therapeutic conditions. Ultrafiltration separating drug-bound from free drug is also extremely popular because of the ease of handling with a large number of samples as well as the commercial availability of a variety of filtration devices. Equilibrium dialysis and ultrafiltration can be considered as thermodynamically identical and their respective errors are similar. It seems appropriate to choose ultrafiltration to evaluate the binding percentage of a drug for clinical purposes even though a slight but significant difference can be observed between data obtained from ultrafiltration and equilibrium dialysis.

C. Multiple Binders of Drugs in Blood

Human plasma contains about 100 proteins of which 13 have a concentration higher than 1 g/liter. Among these, five macromolecules can bind noticeable amounts of drugs: serum albumin (HSA): α_1-acid glycoprotein (AAG); and very low-density, low-density, and high-density lipoproteins (VLDL, LDL, and HDL, respectively). HSA is quantitatively the most important protein, corresponding to 60% of total plasma proteins. Albumin exists in the tissues as well as in plasma, and accordingly drug binding to albumin occurs both in tissues and in plasma. The conformational structure of HSA is flexible due to its 17 disulfide bonds, which cause the formation of six large loops. This property may explain why HSA is able to bind many drugs. Endogenous substances such as bilirubin, oleate, palmitate, linoleate, and L-thyroxin are carried by HSA with a high affinity.

HSA binds a wide variety of both acidic and basic drugs. Most acidic drugs bind almost exclusively to HSA with a high affinity and a low capacity. These drugs can bind to two well-defined binding sites, termed site I or II. However, inhibition studies performed with HSA show that saturation is often difficult to demonstrate, suggesting that clear-cut sites do not exist for the majority of drugs. They bind to a region of HSA molecule rather than to a well-defined site, so several areas can coexist and overlap, thus sharing the binding of other drugs or variably excluding the simultaneous binding of two drugs.

HSA often contributes significantly to the total binding of basic drugs in plasma. However, the percentages bound in plasma are too high to be accounted for by binding to HSA alone. This protein, in fact, tends to show a low affinity and a high binding capacity for basic drugs. Thus, a slight variation in HSA concentration does not result in marked changes in their overall plasma–protein binding. The flexibility of the HSA molecule also explains the possible induction of conformational changes by drugs that may induce variations of its initial binding characteristics, capacity, and/or affinity.

AAG (or orosomucoid) is a glycoprotein that shows a high concentration of carbohydrates (45%), notably sialic acids, with a very high acidic isoelectric point (pH = 2.7). AAG (40 kDa) is highly soluble in water and shows a wide polymorphism due to 21 interchangeable amino acids and sialyl residues, which cause a heterogeneous electrophoretic running (seven to eight bands). AAG is present in plasma at a concentration that normally is 100 times lower than that of HSA. Some acidic drugs such as warfarin and meloxicam are able to bind to a single site on AAG, perhaps in the protein part of the macromolecule. It is unlikely that AAG contributes significantly to the plasma binding of such acidic drugs because the affinity and the capacity of HSA for these drugs are too high. In contrast, AAG is the main binding protein for a large number of basic or cationic drugs such as β-adrenoceptor blockers, antiarrhythmics, opiates, antidepressants, neuroleptics, local anesthetics, calcium channel blockers, and macrolides, for which it exhibits a high-affinity and low-binding site capacity. Binding of drugs to AAG seems to take place at a single hydrophobic pocket in the protein, which accepts both acidic and basic drugs. Changes in plasma AAG concentrations in certain disease states (inflammation, infection, trauma, or malignancy) may be responsible for changes in the extent of protein binding of a large number of basic drugs. Recent studies also indicate that some isoforms of physiological AAG bind the same drug to different degrees. Thus, according to their genetic characteristics, they may lead to differences in plasma and tissue distribution of the drugs they bind.

Plasma lipoproteins, which carry fatty acids, triglycerides, phospholipids, and cholesterol, may also be responsible for the binding of certain drugs such as chlorpromazine and imipramine. However, nearly all types of drugs are able to bind to lipoproteins provided they exhibit a certain degree of lipophilicity. Probucol, cyclosporin, and nicardipine are at least 50% if not totally bound to lipoproteins in plasma. [*See* Plasma Lipoproteins.]

Drugs may also bind to red blood cells (RBC). High plasma–protein binding may result in low distribution to RBC, even if the affinity of the drug to RBC is high. Propranolol enters RBC according to its lipid solubility. About 25% of the amount of phenytoin in blood is distributed in RBC. Three major components in RBC are able to bind drugs: hemoglobin (phenothiazines and phenobarbital), carbonic anhydrase inhibitors (acetazolamide and chlorthalidone), and RBC membrane (chlorpromazine and imipramine). Cyclosporin A binds to the intraerythrocytic protein cyclophilin according to a saturable process. [*See* Hemoglobin.]

Some drugs bind exclusively to one plasma protein, generally to HSA, but many drugs bind to several proteins in blood. Moreover, at the range of therapeutic concentration of drugs, the proteins carry much less drug than they could, i.e., they are not saturated. On this premise, it is possible to derive a mathematical expression of the total bound drug fraction to all protein. From this estimation, the importance of each plasma protein can then be determined and quantified and the relative role of erythrocytes in the drug binding can be estimated (Fig. 1). The conclusion is that bound and free drug fractions in blood depend on the number of proteins involved, the number of binding sites on each protein concentration, and the affinity of the drug for the various proteins.

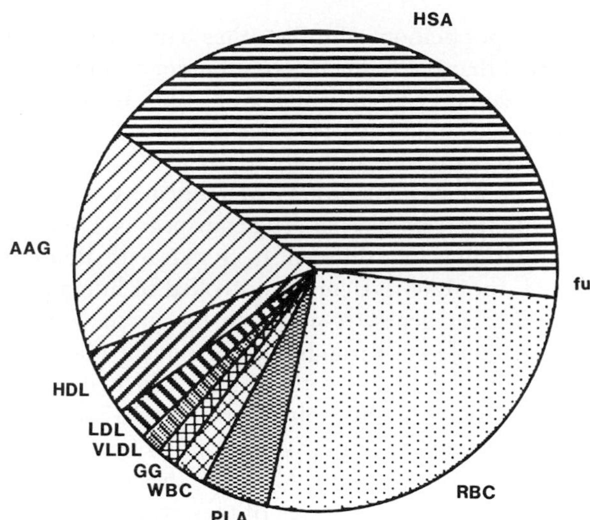

FIGURE 1 Simulation of blood distribution of bepridil in humans from Albengres *et al.* (1984). Concentration of bepridil, 10 μM; physiological components of blood (from volunteers) in normal range; hematocrit = 0.47. AAG, α_1-acid glycoprotein; GG, γ-globulins; HDL, high-density lipoprotein; LDL, low-density lipoprotein; PLA, platelets; RBC, red blood cells; VLDL, very low-density lipoprotein; WBC, white blood cells; fu, unbound fraction.

D. Individual Variations of Drug Binding in Plasma

More than 50 different variants have been characterized for albumin; 47 of these reflect single-point mutations in the single albumin gene and 4 are chain-termination mutants, with altered and shortened polypeptide chains. A reduction in the binding of various drugs (e.g., warfarin, salicylate, diazepam) has been found for a number of albumin variants compared to normal albumin. Furthermore, drug binding is affected differently by the mutations, agreeing with the presence of different binding sites for drugs on the albumin molecule. In about half of the cases of albumin mutants with diminished drug binding, the primary association constant was reduced by one order of magnitude, giving rise to an increase of the unbound fraction of a therapeutic dose of a drug. However, in heterozygous individuals, normal and mutant albumin will each make up 50% of the total albumin level, which remains in the normal range.

The incidence of heterozygote carriers in most populations is relatively low (between 1 : 1000 and 1 : 10,000), although it can be 20% or more in some isolated populations. On the other hand, it has been reported that the frequency of albumin and proalbumin variants is probably higher (1 : 400). If so, alloalbuminemia could have a greater pharmacological importance than hitherto believed.

The other case of protein heterogeneity with significant effects on plasma drug transport is AAG, but the situation is very different from that with albumin. Indeed, the AAG system is controlled in humans by two different genes, resulting in the presence of at least two different variants for AAG in the plasma of all the individuals. The two genes which code for human AAG are highly polymorphic, and 57 different genetic variants have been identified so far for AAG in the human population. Although the AAG phenotypes have only two or three of these variants, all the others account for less than 1% of the population. The three main AAG variants are revealed by sensitive isoelectric focusing techniques and are designated F1, S, and A according to their electrophoretic migration. Depending on the presence of two or three of the F1, S, and A variants in plasma, three main phenotypes are observed for AAG in a general population, F1S/A, F1/A, and S/A, the respective frequencies of which being approximately 50, 35, and 15%. Genetic characterization of the AAG variants has shown that variants F1 and S are encoded by two alleles of the same gene and differ by only a few amino acids in the peptide chain. The A variant is encoded by the other gene of AAG and differs from variants F1 and S by at least 22 amino acids.

The A variant was found to play a significant role in the binding of tricyclic antidepressant (e.g., imipramine and amitriptyline), analgesic (e.g., methadone), and antiarrhythmic (e.g., disopyramide and lidocaine) drugs in contrast to the negligible role of the F1 and S variants in the binding of these drugs. Conversely, a preferential binding of antiaggregant and anticoagulant drugs (e.g., warfarin and dipyridamole) and of the steroid antagonist, mifepristone, to the F1 and S variants rather than to the A variant was shown. These findings indicate the specialization of each AAG variant with respect to its separate genetic origin in the plasma transport of determined drug categories and may explain the interindividual variations that have been observed in the plasma binding of drugs, such as methadone and amitriptyline, to AAG. Finally, measurements of plasma drug binding to AAG (or to albumin) do not usually take into account the individual variant-bound species of the drug, but these should be distinguished because each may have its own specific effect on the transfer of drugs to tissues.

II. CURRENT STATUS

A. Different Drug Distributions between Plasma Proteins and RBC

The distribution among blood components is governed by two main characteristics. The first one deals with the ionization or the absence of ionization of the drug, and the nature of their ionization (i.e., acidic or basic) should be considered. The second one deals with the lipophilicity of their nonionized form.

Tables I and II indicate the corresponding classification that shows the drugs that may bind to satura-

TABLE I
Protein Interactions of Acidic Drugs in Human Plasma

	Types of Drugs			
	I: Subtype a	I: Subtype b	II	III
Reference drug	Warfarin Apazone	Diazepam	Indomethacin	Phenytoin
pK_a	5.05	3.4	4.5	8.33
Binding proteins	HSA AAG	HSA	HSA	HSA
Binding	Saturable	Saturable	Saturable and nonsaturable	Nonsaturable
Association constants (K_a; M^{-1})	10^4–10^6	10^5	10^3–10^5	10^2–10^3
Binding sites per mole of protein	1–3	1	6	Many (>30)
Drug-HSA saturation	Possible	Possible	No	No
Free fatty acid-induced drug displacement	Possible	Possible	Possible	Possible

TABLE II
Protein Interactions of Nonionizable and Basic Drugs in Human Plasma

	Types of Drugs		
	IV	V	VI
Reference drug	Digitoxin	Erythromycin	Imipramine
pK_a	Nonionizable	8.8	9.5
Binding proteins	HSA: ns	HSA: ns AAG: s	HSA: ns AAG: s Lipo: ns
Drug plasma-binding saturation	No	Possible	No
Drug-induced fu increase	No	Possible	No
Free fatty acid-induced fu increase	Possible	No	No

Note. Lipo, lipoproteins (VLDL, LDL, HDL); ns, nonsaturable; s, saturable; fu, unbound fraction.

tion; generally, neither basic nor lipophilic drugs can saturate their respective proteins. Specific bindings may be observed for drugs that have circulating receptors, e.g., cyclosporin and T lymphocytes, β-adrenergic agonists (or antagonists) and RBC β-receptors, colchicine, and leucocytes.

B. Pharmacokinetic Consequences of Drug Binding in Blood

Blood binding may modify the apparent distribution volume, facilitate or prevent cell penetration, and govern the clearance of the relevant drug. Experimental data show that the volume in which the drug is distributed varies in relation to the value of fraction of free drug. For drugs with large volumes of distribution relative to the plasma volume, the apparent volume increases virtually in proportion to the value of the free fraction of drug. [*See* Pharmacokinetics.]

Depending on the protein-binding characteristics, two distribution patterns—restrictive versus nonrestrictive distribution—can be distinguished. Restrictive distribution means that the binding to plasma proteins greatly impairs diffusion into tissues. This is the case with acidic drugs, which are exclusively and tightly bound to a few sites on HSA. As a consequence, the distribution of these drugs is superimposed with that of albumin within the body.

In contrast, the distribution of basic and neutral drugs, which have a much greater affinity to tissues than to plasma proteins, is classified as nonrestrictive. These drugs are either bound to AAG alone or to various other proteins. Because the fraction of drug in plasma is much smaller than that in the extravascular compartments, plasma binding does not greatly affect the overall distribution (Table III). The influence of plasma binding on hepatic or renal clearance may be best understood in terms of the extraction ratio of the drug by these organs. The elimination of drugs, with extraction ratios close to 1, is insensitive to the extent of plasma binding because both the free and the bound drug are extracted by the organ. On the other hand, when the extraction ratio of a drug is low, the percentage of serum drug available for extraction is less than or equal to the free drug. Therefore, in this situation, protein binding governs the fraction of drug available for enzymatic metabolism as well as renal glomerular filtration and secretory processes.

The time course of a drug in the body reflected by its elimination half-life ($T_{1/2}$) can be affected by changes in binding because this parameter is related to the volume of distribution and clearance. Therefore, differences in the half-life can be anticipated from the dependence of these two parameters on plasma–protein binding.

Historically, only the free drug was assumed to be the pharmacologically active species, based on animal data in which pharmacodynamic effects showed a strong correlation with an unbound drug concentration. However, for many drugs, a pharmacodynamic effect cannot be correlated with a plasma concentration. There is a considerable lack of data establishing the correlation between pharmacodynamic (i.e., therapeutic or toxic responses) and unbound drug, although some attempts were made with antiepileptic and antiarythmic drugs. So far the results of these studies do not show that the unbound concentration is a much better correlate of response than the total plasma concentration. At present, one must acknowledge that our capability to interpret correctly the free drug level data is limited, despite some studies that have demonstrated improved pharmacokinetic–pharmacodynamic correlations using unbound rather than total drug concentrations.

C. Blood Binding and Drug Transport

Brodie's hypothesis stated that, considering the two states of a drug, free and protein bound, in the blood, only the first was available for tissue transfer. Two

TABLE III
Effect of Plasma Protein on Drug Distribution and Elimination

	Free drug in plasma (%)	Apparent volume of distribution (liter · kg^{-1})	Clearance (ml · min^{-1})
Acidic drugs			
Carbenoxolone	1	0.10	5.8
Ibuprofen	1	0.14	57.0
Phenylbutazone	1	0.10	8.0
Naproxen	2	0.09	5.2
Clofibric acid	3	0.09	6.1
Fusidic acid	3	0.15	12.0
Warfarin	3	0.10	3.8
Bumetamide	4	0.18	48.5
Dicloxacillin	4	0.29	114.0
Furosemide	4	0.20	190.0
Tolbutamide	4	0.14	16.2
Cloxacillin	5	0.34	160.0
Fluorophenindione	5	0.09	2.3
Nalidixic acid	5	0.35	162.0
Sulfaphenazole	5	0.29	23.0
Chlorpropamide	8	0.20	6.8
Oxacillin	8	0.44	210.0
Nafcillin	10	0.63	160.0
Basic and neutral drugs			
Nifedipine	2	1.2	70
Chlorpromazine	2–5	21	602
Digitoxin	3	0.54	3.9
Amiodarone	4	66	
Prazosin	5	0.6	210
Propranolol	7	3.9	840
Oxprenolol	8	0.86	600
Methadone	11	3.9	98
Erythromycin	15–30	0.80	630
Lidocaine	30	1.1	644
Disopyramide	32–70	0.59	80
Digoxin	75	0.54	35

Note. The apparent volume of distribution of acidic drugs is mainly governed by that of its main binding protein (i.e., HSA): the lower the free fraction, the smaller the apparent volume of distribution (≃ close to that of HSA). On the contrary, no clear relationship between free fraction and volume of distribution is observed with basic and neutral drugs. For these drugs, the plasma-binding capacity is low as compared with those of the tissues. Because clearance depends on blood flow for drugs with high-extraction ratio and on binding and intrinsic clearance for compounds with low-extraction ratio, no clear relationship can be established between clearance and plasma-free drug.

main findings supported this view: (1) the drug can be found free in various organs, without noticeable amounts of its plasma-binding proteins, and (2) when a drug binds to a plasma protein, the complex is too big to easily cross the membranes by a passive diffusion mechanism. The stability of the drug–protein complex is the main factor impairing blood to tissue transfer.

A significant breakthrough was made when it became possible to assess the effect of each plasma protein on its bound drug transfer for a single organ. This was done by injecting a volume of the tested drug into an artery and measuring the amount transferred to the corresponding organ after a single capillary pass. Different measurements are made using the same amount of labeled drug to which are added increasing amounts of the test protein, chosen to significantly modify the unbound fraction of the drug when entering the blood vasculature of the tissue. Theoretically, the available fraction is the free drug, and it is possible to calculate the tissue uptake according to the imposed unbound fraction. This theoretical curve is afterwards compared with the measured one (Fig. 2). The drug uptake is expressed by comparison with those of a readily diffusible tracer (tritiated water or [^{14}C]butanol), supposedly representing the maximal extraction.

Used on rat brain, this method shows that both curves of propranolol uptake are superimposable when HSA was used but that they markedly differed when AAG was the binding protein. In both cases, brain uptake indices decreased when the protein concentrations increased, but when propranolol and AAG were associated, the relevant uptakes were higher than expected from the unbound fraction measured *in vitro*. This result was attributed to a dissocia-

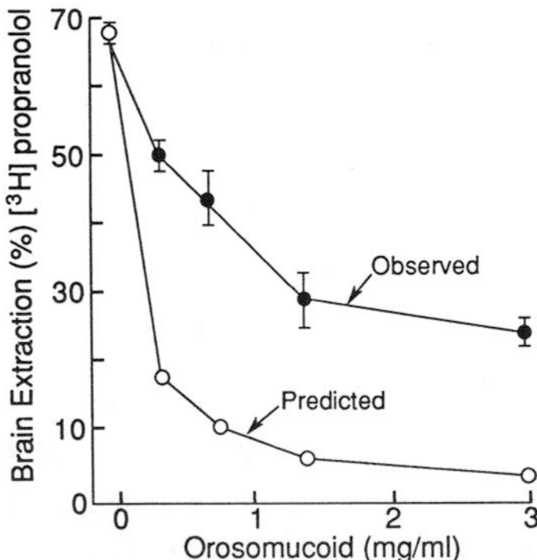

FIGURE 2 *In vivo* effects of orosomucoid binding on propranolol brain extraction. Brain extraction of [^{3}H]propranolol is plotted against orosomucoid concentration. The predicted value is calculated assuming that only the free fraction of the drug entering the brain vasculature is available for extraction and that the propranolol–orosomucoid complex is stable during brain transit. The observed values are in fact higher, suggesting the free intermediate drug hypothesis.

tion of the drug–protein complex during the blood transit in the brain and, as a consequence, the intermediate free drug hypothesis was raised (Fig. 3). Assuming a quick transfer rate of the free drug across the capillary walls, a high dissociation rate of the drug–protein complex in the capillaries, and a sufficiently long-lasting transit time of blood, it is conceivable that more than the initial free fraction may be available for tissue transfer.

Two other observations are interesting. First, HSA–propranolol complexes apparently do not dissociate in the brain capillaries, although the affinity of the drug for HSA is lower than for AAG, which forms dissociable complexes. Thus, the transferability of protein-bound propranolol apparently is not directly related, as expected, to its affinity but probably to the structure of the protein in the capillaries and, more precisely, to conformation and accessibility of its relevant binding site. Nevertheless, although nondeterminant, the affinity of the drug for the protein can partially affect factor to transfer tissue. This was shown with the warfarin–HSA complex: Increasing affinity by the addition of palmitic acid decreased the amount of transferred warfarin. Second, it was also observed that neither HSA nor AAG in liver, protect even partly bound propranolol from liver uptake. This may be explained by the anatomical organization of liver capillary, which allows direct contact between plasma, interstitial fluid, and the hepatocyte membrane, whereas in the brain, the endothelial tight junctions prevent such a contact and limit drug transfer. Moreover, the blood mean transit time favors liver uptake, as it is longer than the corresponding brain time (5 sec versus 1 sec in the rat).

The main interest of these data is to show for the first time a dual specificity in blood to tissue drug transfer: a specific role of the plasma-binding protein, an organ specificity based on the anatomical disposition of its capillaries, their endothelium organization, and their size, and on its blood flow rate as the corresponding physiological characteristic. The enhanced dissociation of the AAG–propranolol complex in brain capillaries was explained by the hypothesis of a local binding inhibitor, released from the endothelium and able to dissociate the complex locally. This hypothesis, however, was not proved. The dissociation-limited model was proposed to explain the high liver uptake of endogenous compounds and drugs, following the HSA–receptor model. Many results suggest, although indirectly, that the drug–HSA complex may be bound to endothelium receptors on the lumen side, where either the drug is partly released or the complex is internalized. A valid argument for drug release is the recent description of such HSA receptors, but these data need to be confirmed. It is also important to keep in mind that HSA is a flexible molecule, thus able to exhibit quick conformational structure changes leading perhaps to a decrease of the drug–protein association constant.

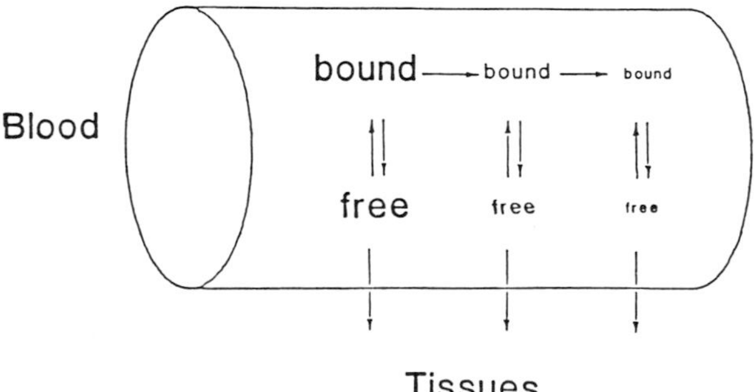

FIGURE 3 The free intermediate hypothesis. This hypothesis assumes that the initial free drug entering the brain vasculature is rapidly transferred and, as a result, the equilibrium bound-free drug is disrupted and subsequently readjusted, with additional release of free drug, which is then available for brain transfer.

III. FUTURE DIRECTIONS

A. Drug Monitoring in Pathological States

Theoretically, drugs showing a high binding percentage associated with a low apparent volume of distribution are worth being monitored provided they have low therapeutic indices. In these circumstances, the amount of drug available for a pharmacological effect is closely related to the concentration of unbound drug. Cu stands for unbound or free concentration. Since the early 1980s, progress in pharmacokinetics has enabled us to gain more insight into the mechanism regulating the unbound fraction and Cu. There are many situations where unbound drug may be altered, whereas Cu remains unchanged. This obviously implies a change in total concentration of the drug. These observations suggest that Cu instead of unbound fraction should be monitored. Currently, therapeutic and/or toxic ranges of free concentrations have not been established. Controlled clinical investigations have to be implemented to fully evaluate the correlation between free drug levels and clinical and toxic effects to warrant the monitoring of unbound concentration.

Probably in many circumstances, modifications of drug effects are not only related to plasma binding but also to deep modifications of drug biotransformations and elimination as has been shown for erythromycin in cirrhotic patients. A good strategy in elucidating the effects of drugs in these complicated states would be a global pharmacokinetic study including protein binding as one pathological factor of variation.

B. Drug Targeting

The principle of drug targeting is to increase the amount of drug transferred from blood to a specific target organ and/or to decrease the drug transfer in other nonpertinent tissues. This is theoretically possible considering that drug plasma binding may be restrictive for some tissues or permissive for others, depending on the protein carrier. For example, HSA restricts brain transfer but permits liver transfer of propranolol. In contrast, when prazosin bound to AAG is not taken up by the liver, but when desialylated AAG is used, both drug and protein are taken up. Another way may be to use physiological transport mechanisms of proteins to which drugs are bound and cotransported. This receptor-mediated transport is, for instance, applied to brain penetration using specific peptides known to be transferred by their own endothelial receptors, insulin, transferrin, and insulin-like growth factors 1 and 2. Another direction is to try to modify the protein structure to render it transferable across appropriate endotheliums by an endocytosis process. Cationized albumin is, for instance, transferred through the blood–brain barrier by an absorptive-mediated endocytosis and it seems that this may carry drugs. Lactosylated albumin is also another possible carrier for increasing hepatic drug transfer.

C. Individual Variations of the Response to Drugs

It is well known that responses of individuals to the same dose of a drug may vary and that these differences are related to genetic characteristics. For drugs which plasma binding is restrictive to proteins exhibiting genetic polymorphism, i.e., albumin and AAG, differences in the amount of drug bound in the plasma and thus of tissue distribution may at least partly explain the variability of the response.

BIBLIOGRAPHY

Albengres, E., *et al.* (1984). Multiple binding of bepridil in human blood. *Pharmacology* **28**, 139.

Baumann, P., *et al.* (1989). Alpha 1-Acid Glycoprotein, Genetics, Biochemistry, Physiological Functions and Pharmacology." A. R. Liss, New York.

Birkett, D. J., and Wanwimolruk, S. (1985). Albumin as a specific binding protein for drugs and endogenous compounds. *In* "Protein Binding and Drug Transport" (J. P. Tillement and E. Lindenlaub, eds.). Schattauer, Stuttgart/New York.

Hervé, F., Gomas, E., Duché, J.-C., and Tillement, J.-P. (1993). Evidence for differences in the binding of drugs between the two main genetic variants of human alpha-1-acid glycoprotein. *Br. J. Clin. Pharamcol.* **36**, 241–249.

Kragh-Hansen, U. (1993). Genetic variants of human serum albumin and proalbumin: Identification and possible pharmacological significance. *In* "Blood Binding and Drug Transfer." (Tillement *et al.*, eds.), pp. 31–48. Editions Fort et Clair, Paris.

Kremer, J. M. H., *et al.* (1988). Drug binding to human α_1 acid glycoprotein in health and disease. *Pharmacol. Rev.* **40**, 1.

Pardridge, W. M. (1981). Transport of protein-bound hormones into tissues in vivo. *Endocr. Rev.* **2**, 103.

Tillement, J. P., Houin, G., Zini, R., Urien, S., Albengres, E., Barré, J., Lecomte, M., D'Athis, P., and Sébille, B. (1984). The binding of drugs to blood plasma macromolecules: Recent advances and therapeutic significance. *Adv. Drug Res.* **13**, 59.

Blood–Brain Barrier

PETER J. ROBINSON
The Procter & Gamble Company

STANLEY I. RAPOPORT
National Institutes of Health

GLOSSARY

Active transport Energy-dependent uphill movement of substances across cell membranes, stochiometrically coupled to an energy-producing chemical reaction; reduced by cold, lack of oxygen, or metabolic inhibitors

Capillary heterogeneity Differences in functionally important properties, such as body flow and surface area, from capillary to capillary in the capillary bed of an organ. Leads to variations in extraction fraction among capillaries and to an overall reduction in extraction from the capillary bed as a whole

Extraction fraction Proportion of the arterial (input) concentration, c_a, of a substance passing through an organ that is taken up by the organ: extraction $(E) = (c_a - c_v)/c_a$, where c_v is the venous (output) concentration

Facilitated diffusion Enhanced transfer of specific substrates across plasma membranes, which is downhill (i.e., from a higher to a lower concentration or electrochemical potential), stereospecific, and saturable

Osmotic pressure gradient Results from a difference in concentration of molecules or ions across a membrane which has a different permeability to water and to solute; leads to a movement of water toward the side with the greater concentration; osmotic pressure is equal to the hydrostatic pressure required to prevent this transfer of water

Partition coefficient Ratio of steady-state concentrations of a solute between two solvents. The degree to which a solute will dissolve in a lipid membrane (membrane–water partition coefficient) is approximated by its octanol/water or olive oil/water partition coefficients

Permeability Ratio of net solute flux across membrane to driving force, either the concentration difference across the membrane (nonelectrolytes) or the difference in the electrochemical potential across the membrane (ions or electrolytes)

Reflection coefficient Determines the osmotic pressure gradient across a membrane, generated by a particular concentration gradient of solute; related inversely to the permeability of the membrane to the solute. For a membrane permeable to water, but not to the solute (semipermeable membrane), the coefficient $(\sigma) = 1$; for a membrane equally permeable to water and solute, $\sigma = 0$ and no osmotic pressure gradient is generated

Sodium pump Active transport of sodium ions across a cell membrane, frequently mediated by Na^+,K^+-ATPase

Tight junctions (zonulae occludens) Complete bands of close membrane apposition surrounding cells as continuous belts, restricting intercellular diffusion. Membranes from adjacent cells are closely linked by common fibrillar strands so that the overall width of both apposing membranes is less than twice that of each plasma membrane (i.e., about 130 Å). They are found connecting many epithelial cells, including those of the choroid plexus, and the endothelial cells of cerebral capillaries, arterioles, and venules

THE BLOOD–BRAIN BARRIER SEPARATES TWO OF the major compartments of the central nervous system—the brain and the cerebrospinal fluid—from the third compartment—the blood. The sites of the barrier are the interfaces between the blood and the other two compartments: the choroid plexus, the arachnoid membrane that overlies the subarachnoid space, and the blood vessels of the brain and the subarachnoid space (see Fig. 2). All barrier sites are characterized by cells connected by tight junctions that restrict inter-

ENCYCLOPEDIA OF HUMAN BIOLOGY, Second Edition, VOLUME 2. Copyright © 1997 by Academic Press. All rights of reproduction in any form reserved.

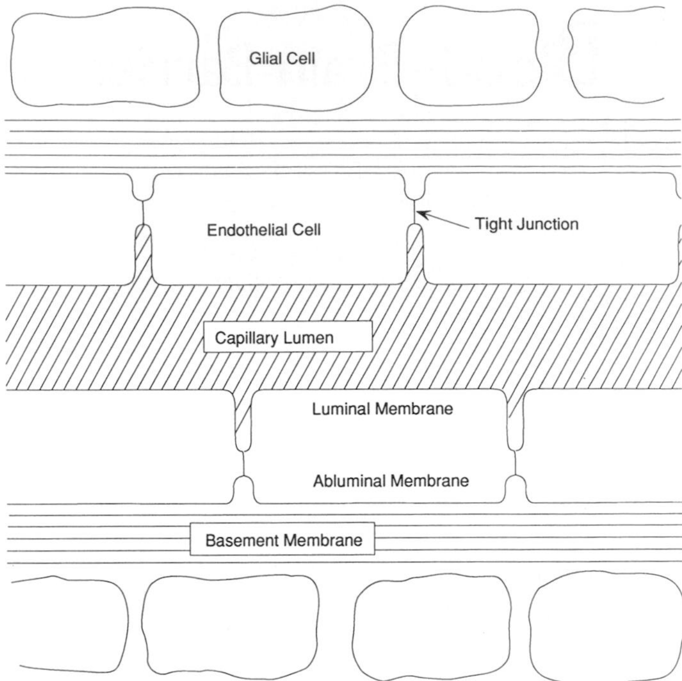

FIGURE 1 Schematic section through a cerebral capillary. Endothelial cells are connected by continuous belts of tight junctions, which restrict intercellular diffusion. The capillary is surrounded by a basement membrane and by glial cells with spaces between them.

cellular diffusion so that the cells act almost like a continuous cell layer (see Fig. 1). Lipid-soluble substances easily penetrate the membrane surrounding each cell (i.e., the plasma membrane), and so equilibrate rapidly between the blood and the brain. On the other hand, lipid-insoluble nonelectrolytes and proteins (and substances highly and tightly bound to plasma proteins) enter the brain much more slowly than they enter other tissues, with the result that the brain is substantially protected from inevitable and potentially disruptive fluctuations of such substances in the bloodstream. However, many substrates essential for brain metabolism, for syntheses of brain proteins and neurotransmitters, and some inorganic ions are not limited by the passive restrictions of the blood–brain barrier. They can make use of specific transport systems that facilitate their entry into the brain. The role of the blood–brain barrier in regulating the chemical environment of the brain is thus achieved by combining low passive permeability with highly selective transport between the blood and the brain.

I. STRUCTURE AND FUNCTION OF THE BLOOD–BRAIN BARRIER

The existence of a functional barrier between the brain and the blood was inferred at the beginning of the century from studies in which certain dyes and pharmacologically active substances were injected either into the bloodstream or into the cerebrospinal fluid of experimental animals. Although the brain was stained or pharmacologically affected after injection of a dye or drug into the cerebrospinal fluid (CSF), it often remained unaffected after an intravenous injection (although other tissues and organs were often stained or modified). Early electron micrographs did not initially support the notion of a physical barrier, but showed that cerebral capillaries were similar to those elsewhere. However, by the late 1960s, when various macromolecules, such as horseradish peroxidase and microperoxidase, could be visualized in electron micrographs, these compounds were found to be confined to the capillary lumen when injected intravenously.

Cerebral capillaries have a continuous lining of endothelial cells held together by tight junctions, which form a continuous belt surrounding each of them. Intravascular protein markers penetrate between the overlapping endothelial cells until blocked by a tight junction so that the remainder of the interendothelial cleft remains free of the marker. Similarly, when injected on the other side of the barrier into the CSF, the markers readily diffuse within the brain interstitial fluid and penetrate between the processes of the glial sheath surrounding the capillaries and through the basement membranes to reach the interendothelial clefts from the other side, where they are again stopped by the tight junctions. It has become clear that the blood–brain barrier within the brain is located at the endothelial cell layer of the cerebral capillary (Fig. 1) and that this cell layer determines the permeability properties of the barrier. Tight junctions are found between many epithelial cells in the body, but not generally between capillary endothelial cells, except in the brain, nerve, and eye.

In addition to the diffusion barrier located at the cerebral capillary, there is also an interface between blood and brain at the choroid plexus and at the arachnoid membrane that overlies the subarachnoid space and separates it from the subdural space (see Fig. 2). These additional sites share the property that access to the brain is again restricted by intercellular tight junctions. Specifically, they form a barrier between the blood and the CSF. This "blood–CSF barrier" is often distinguished from the blood brain barrier (situated at the cerebral capillary endothelium) to explain why intravascular substances sometimes enter the brain and the CSF at different rates. However, these kinetic differences can often be interpreted by taking into account gross anatomical relationships among brain, blood, and CSF. There is no "subbarrier" between the brain and the CSF to impede the transfer of solute.

A. Choroid Plexus

The choroid plexus is located on the roofs of the third and fourth ventricles and in the walls of the lateral ventricles. It is a convoluted structure with minute protrusions (villi) extending from the brain surface "like coral fronds into a sea of CSF" and is richly endowed with capillaries. Passive exchange of dissolved substances between the blood and the CSF is restricted at the choroid plexus, not by the layer of endothelial cells lining these choroidal capillaries, which is not fully interconnected and porous (fenestrated), but by choroidal epithelial cells, which are connected by tight junctions. Nevertheless, choroidal epithelium is classified as "leaky" in that it discriminates more poorly between diffusing ions than other "tighter" epithelia.

The choroid plexus regulates, to a large extent, the production and composition of the CSF. Active transport of sodium (Na^+) and potassium (K^+) ions across the choroid plexus is regulated by a Na^+,K^+-ATPase-catalyzed pump, which drives Na^+ toward the ventricular surface of the plexus into the CSF [accompanied by chloride (Cl^-) and bicarbonate (HCO_3^-) ions] and drives K^+ [and hydrogen (H^+)] ions in the opposite direction. The resulting standing osmotic gradient leads to movement of water across the choroidal epithelium into the CSF. Water movement into the CSF also results from the hydrostatic pressure difference between capillary blood and CSF. In all, some 25% of the water in the blood flowing to the choroid plexus normally goes to the CSF, accounting for about 70% of the net CSF production (see Section I,B). Also, the choroid plexus can rapidly remove many drugs and other agents from the CSF, and has been likened to an "extrarenal" kidney.

B. Cerebrospinal Fluid

The total volume of CSF in humans is about 140 cm^3, of which about 23 cm^3 resides in the brain ventricles. The net production of CSF is about 0.35 cm^3/min, replacing about 0.25% of the CSF every minute; complete replacement occurs every 8 hr. About 70% of the CSF is produced by secretion at the choroid plexus (see Section I,A), about 12% results from metabolic production of water in the brain, and the remainder is from other extrachoroidal sources, such as filtering through the capillary cells (capillary ultrafiltrate). The major pathways for the bulk flow of CSF are shown in Fig. 2. CSF is formed at a hydrostatic pressure of 15 cm of H_2O, which drives it through the ventricular system and into the subarachnoid space and the spinal canal. It is returned to the blood by the hydrostatic pressure through one-way valves (arachnoid villi) which protrude from the subarachnoid space (see Fig. 2). Many unwanted metabolic products, drugs, and other substances are removed from the brain through the CSF; a stream of CSF carries these solutes away through the arachnoid villi into the venous blood.

The chemical composition of the CSF has a pro-

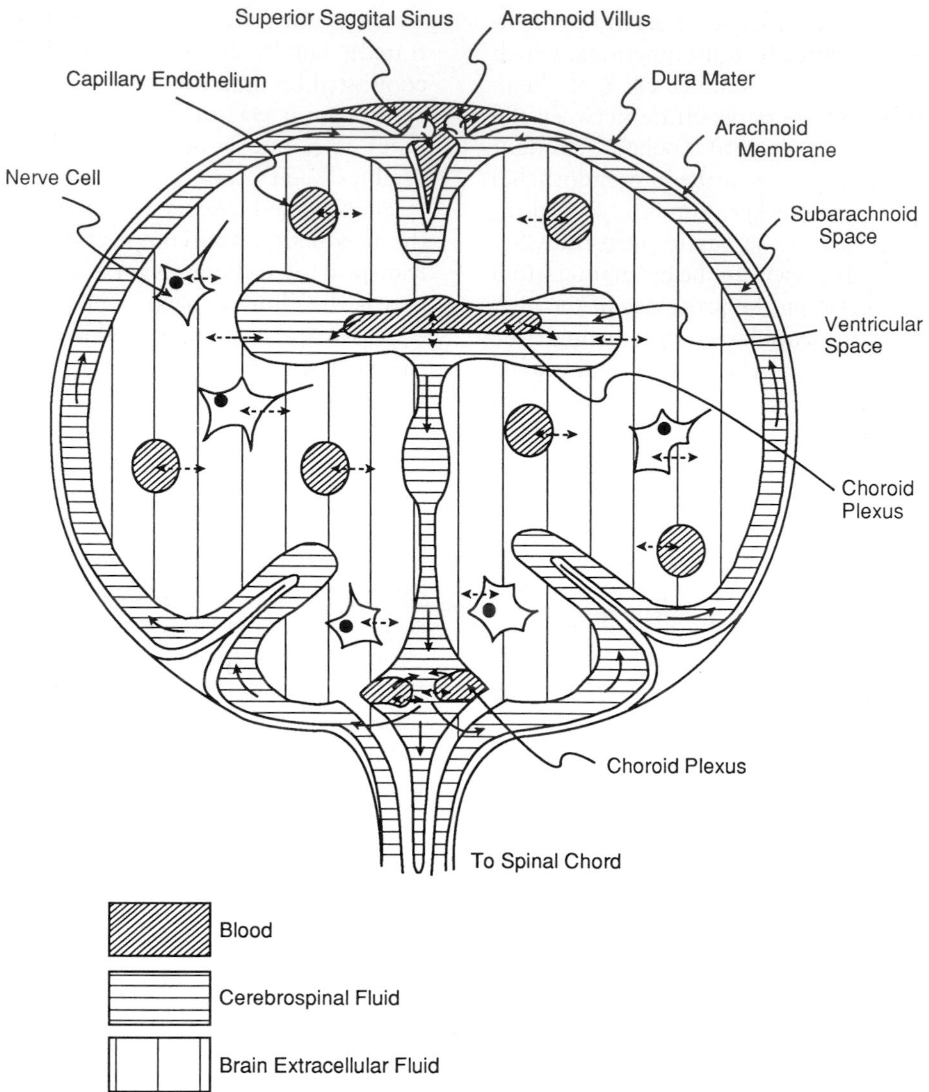

FIGURE 2 The major fluid compartments of the central nervous system, showing major pathways for solute exchange (dashed arrows) and bulk circulation of fluid (solid arrows).

found influence on the functional integrity of the central nervous system because there is no diffusion barrier between CSF and the neurons and glia of the brain. In humans, the distance for free diffusion from the CSF to any cerebral site through the spaces between brain cells is less than about 15 mm. Thus, changes in the concentrations of calcium, potassium, and magnesium in the CSF, especially if they are maintained for some time, may profoundly affect many functions controlled by the brain, such as blood pressure, heart rate, vasomotor reflexes, respiration, gastric motility, muscle tone, and emotional state, while the degree of acidity (i.e., pH) governs respiration, autoregulation of the cerebral blood flow, and brain metabolism. The composition of the CSF is determined by a large number of factors, including metabolism, production, or uptake of solutes by cells; restriction of intercellular diffusion, coupled with special transport mechanisms at the choroid plexus and at the endothelium of the brain capillaries; and rates of CSF production and excretion. The combined effect of these factors helps create a chemical environment for neurons and glia that differs from, and is more stable than, that of the blood.

II. PERMEABILITY OF THE BLOOD–BRAIN BARRIER

A. General Considerations

The experimental determination of blood–brain barrier permeability shares many characteristics with measurements in other organs, such as the liver or lung, in that a local membrane property (the permeability, or P), is inferred from the rate of uptake of a substance from an entire network of capillaries (whether within a region or within the organ as a whole). In general, such a capillary bed is heterogeneous with respect to capillary blood flow, surface area and length, so that the uptake of the substance is the sum total of (possibly widely differing) individual capillary uptakes. How such capillary heterogeneities influence the experimental determination of blood–brain barrier permeabilities will be discussed later. In many cases, however, it is sufficient to assume that the capillary bed can be described adequately as a (large) number of functionally identical capillaries in parallel so that the uptake characteristics of a single representative capillary reflect the uptake of the brain or brain region as a whole.

In such a single-capillary model, the concentration of the solute of interest (usually a radiolabeled tracer) decreases as it passes from the blood across the blood–brain barrier into the brain extracellular fluid. If we assume negligible backflux from the brain to the blood, then the depletion of solute from a small volume of blood passing through the capillary is proportional to the concentration of solute in the blood and the permeability of the blood–brain barrier. This gives a characteristic exponential concentration profile of the solute (relative to the capillary surface area passed by the volume element) from the arterial to the venous end of the capillary. The venous and arterial solute concentrations (c_v and c_a, respectively) are related by

$$c_v = c_a e^{-PA/F}, \quad (1)$$

where A is the capillary surface area and F is the capillary blood flow. For a number of identical capillaries in parallel sharing a common c_a, all venous concentrations, c_v, are the same, and Eq. (1) applies to the whole capillary bed, A now being the total capillary surface area and F being the total blood flow. Total (or regional) brain extraction, E, of solute (i.e., the proportion of solute taken up by the brain) is given by

$$E = (c_a - c_v)/c_a = 1 - e^{-PA/F} \quad (2)$$

so that

$$PA = -F\ln(1 - E). \quad (3)$$

The permeability–surface area product, PA, for the brain or brain region is not measured directly, but is calculated from measurements of the extraction fraction, E, by means of a model-dependent relationship, one of the simplest of which is given in Eq. (3).

For slowly extracted solutes ($PA/F \ll 1$), the exponential in Eq. (2) can be expanded so that

$$E \approx PA/F \quad (PA/F \ll 1). \quad (4)$$

Defining the unidirectional transfer constant, K_i (cm^3/sec/g of brain), as

$$K_i = FE, \quad (5)$$

then in this limit we have, from Eqs. (4) and (5),

$$K_i \approx PA \quad (PA/F \ll 1). \quad (6)$$

In other words, K_i is independent of blood flow, F, in this extreme of membrane- or permeability-limited uptake. If, on the other hand, the solute is highly extracted ($PA/F \gg 1$), then E approaches 1 and K_i approaches F, and we have the special case of flow-limited uptake, in which the flow alone to a brain region determines uptake. As we pass from more slowly to more highly penetrating solutes, it becomes progressively more important to have an accurate determination of regional blood flow in order to determine reliable values for PA from experimental measurements of K_i.

In many cases, the single-capillary model for brain uptake is adequate to describe the experimental findings. In some cases, however, the inevitable variation in functional properties of capillaries, such as surface area and blood flow, needs to be taken into account. If the degree of variability of such functional properties in a capillary bed is characterized by the coefficient of variation, ε (the standard deviation divided by the mean), then Eq. (2) again holds, provided PA is replaced with an "effective" PA, $(PA)_{\text{eff}}$:

$$E = 1 - e^{-(PA)_{\text{eff}}/F}, \quad (7)$$

where, to a first approximation for small values of ε,

$$(PA)_{\text{eff}} = PA(1 - \tfrac{1}{2}\varepsilon^2 PA/F) \leq PA. \quad (8)$$

It is clear from this expression that the larger the degree of functional heterogeneity, ε, the lower the value of $(PA)_{\text{eff}}$ and hence the lower the extraction for the same value of PA/F. Alternatively, if E is determined experimentally and PA is calculated on the basis of the single-capillary model [$\varepsilon = 0$; Eq. (3)], then this value is an underestimate of the true PA, which is given by Eqs. (7) and (8), if in fact there is some degree of functional heterogeneity ($\varepsilon > 0$). However, the smaller the value of PA/F (i.e., the lower the extraction), the less important the term $\frac{1}{2}\varepsilon^2 PA/F$ becomes in Eq. (8) and the less effect a particular degree of capillary heterogeneity has on the determination of PA. At the other extreme, for $PA/F \gg 1$, the extraction, E, approaches 1 and is also insensitive to the degree of functional heterogeneity.

B. Protein Binding

Analysis of brain uptake in terms of capillary permeability is further complicated by the fact that many drugs and other substances are reversibly bound to circulating plasma proteins such as albumin. Such large protein molecules do not normally pass through the intact blood–brain barrier, so that only the locally free or unbound fractions of such substances are available for entry into the brain. The uptake of a protein-bound substance is thus determined (in addition to the factors already discussed) by the equilibrium free fraction, f, in plasma, plus what is generated by the release (and subsequent uptake) of initially bound drug during its passage through the cerebral capillary. [*See* Blood Binding of Drugs.]

An exact analysis of the effects of protein binding on uptake is quite complex; however, a useful lower bound to the PA is obtained by assuming that the dissociation (and association) rate constants are large so that bound and free forms of a substance equilibrate rapidly following uptake at each point along the capillary (i.e., "instantaneous equilibration"). Under these circumstances, PA is given by a modified form of Eq. (3):

$$(PA)_{\text{lower}} = -(F/f)\ln(1 - E). \qquad (9)$$

In reality, it takes a finite time for a new equilibrium to be established between bound and free forms to compensate for the uptake of free drug along the capillary. Thus, less free drug is available for uptake than the assumption of instantaneous equilibrium would allow, and the actual PA is somewhat higher than that given by Eq. (9). For many substances, however, the difference is negligible.

C. Cerebral Blood Flow

The amount of a substance taken up by the brain depends on the cerebral blood flow, F, which is, in turn, strongly dependent on the experimental conditions. In the mammalian brain, flow (per gram of tissue) is at least two to six times higher in gray matter than in white matter. Variations also exist between gray matter regions of as much as two to three times. Such differences are attributed to regional differences in the number of perfused capillaries per unit mass of tissue, which is two to five times larger in gray matter, rather than to differences in flow velocities. [*See* Central Gray Area, Brain.]

A number of factors may alter regional and total cerebral blood flow, including arterial carbon dioxide pressure, anesthesia, age, and extracerebral nervous and humoral stimuli. Such changes in blood flow may be brought about either by alterations in the total number of open or perfused capillaries (capillary recruitment) or by changing the flow velocity through a fixed capillary bed. It is likely that both mechanisms are involved, to a greater or lesser extent, as changes in blood flow are brought about by various means.

It is, however, no simple matter to distinguish between such alternative mechanisms on the basis of uptake studies alone. Among the possible complications are, first, that capillaries may be perfused intermittently. Under such a condition, uptake would be highly dependent on how it is measured, whether by depletion of the substance from plasma or by its accumulation in the brain; it would also depend on the duration of the experiment compared with the average cycling time for a capillary between successive periods of perfusion. Second, capillaries may change their diameters in response to flow changes, possibly leading to an increase in PA with increasing flow that could simulate the recruitment of additional capillaries. Third, the recruitment of capillaries could also be simulated by increases in flow velocity through a fixed heterogeneous capillary bed [see Eq. (8)]. In order to unambiguously show that capillary recruitment is taking place from its effects on uptake, it is necessary to look at a substance taken up by a saturable mechanism (see Section III). At saturation, such a substance would have a maximum uptake rate proportional to the total number of perfused capillaries.

D. Experimental Methods for Determining Permeability

A number of experimental methods to characterize and quantify the permeability and transport properties of the blood–brain barrier have been developed since the mid-1960s. Table I summarizes many of the methods that currently use whole brain preparations. (The permeability characteristics of isolated brain capillaries have also been studied.) These methods can be broadly classified into two main groups: those that look at the influx of tracer into the brain (measured either by accumulation of tracer in brain tissue or by depletion of tracer from the plasma) and those that look at efflux of the tracer back to the plasma from an initially loaded brain. The tracer influx studies again form two groups: those that measure the extraction fraction, E, of a solute during a single pass following a bolus injection in the carotid artery and those that measure the unidirectional transfer constant, K_i, from multiple passes of the tracer through the brain following either an intravenous injection or a continuous infusion. The single-pass studies include the indicator dilution method and the brain uptake index (BUI) method, both of which are relatively quick and easy to perform. Because the perfusate goes directly to the brain, there is relatively little mixing with the blood, and a high degree of control of the chemical composition of the perfusate in the brain capillaries is possible.

In the indicator dilution method, the venous outflow is measured as the bolus leaves the brain. At each time point E is estimated from the difference in the outflow between the concentrations of the indicator and of an appropriately matched intravascular tracer (which is not extracted, but otherwise behaves as closely as possible like the test solute). The extraction measured in this way may vary with time due to the capillary heterogeneities of extraction. At earlier times, E is contributed by capillaries with shorter transit times, whereas at later times there may be back-diffusion of already extracted indicator. It is still possible, however, to obtain a secure lower bound to whole brain PA under these circumstances. The indicator dilution method has the advantage that it allows multiple consecutive measurements in the same brain, although it does not allow estimates of regional values for permeability.

The BUI method also uses the intracarotid injection of a solution containing both a test and a reference solute. However, the reference substance is a highly diffusible flow-limited compound, such as butanol, and the brain is sampled to determine extraction. This procedure allows a comparison of some regional permeabilities (e.g., by means of autoradiography), but extraction in certain parts of the brain cannot be studied because the injected bolus does not reach the entire brain. An additional limitation is that no consecutive measurements are possible in the same animal.

All single-pass methods share the limitation that they can accurately determine the PA values only for moderate to highly penetrating substances (with E values between about 0.05 and 0.9, or PA values between about 5×10^{-4} and 0.03 cm^3/sec/g of brain). For more slowly penetrating substances, it is necessary to allow the tracer to accumulate over a number of passes through the brain, as in continuous uptake studies.

Continuous uptake techniques are of two types. In the intravenous administration technique, an indicator is given intravenously starting at time $t = 0$ (either

TABLE I
Commonly Used Experimental Methods to Estimate Regional or Whole Brain Permeability to Tracers

Method		Extraction range[a]	Regional values obtained	Repeat experiments in same animal	Control of perfusate
Tracer influx studies					
Single pass	Indicator dilution	Moderate–high	No	Yes	Yes
	Brain uptake index	Moderate–high	Yes	No	Yes
	Single injection–external registration	Moderate–high	No	Yes	Yes
Continuous uptake	Intravenous administration	Low–high	Yes	No	No
	In situ brain perfusion	Low–high	Yes	No	Yes
Tracer efflux studies	Brain washout	Moderate–high	Yes	No	—
	Ventriculocisternal perfusion	Moderate–high	Only near perfused surfaces	No	—

[a]Low extraction, $E \leq 0.05$ (corresponding to permeability $10^{-8} \leq P \leq 10^{-6}$ cm/sec); moderate extraction, $E \geq 0.05$, $P \geq 10^{-6}$ cm/sec; high extraction, $P \geq 10^{-4}$ cm/sec.

as a bolus injection or as an infusion); the arterial plasma concentration is then monitored until time T, which can vary between 1 min and several hours, when the concentration of the indicator in the brain (or brain region) is determined. The unidirectional uptake constant, K_i, is calculated as the ratio of the brain concentration of the indicator c_{br} [corrected for residual intravascular radioactivity, $V_{br}c_a(T)$, where V_{br} is the intravascular volume and c_a is the concentration of the indicator in the arterial plasma] divided by the integral of the plasma concentration during the time of brain exposure to the solute:

$$K_i = \frac{c_{br} - V_{br}c_a(T)}{\int_0^T c_a\,dt}. \tag{10}$$

If brain uptake is determined (in separate animals) over several time intervals, T, then a plot of c_{br} against the plasma concentration integral, $\int_0^T c_a\,dt$, will give a straight line, with slope K_i and intercept $V_{br}c_a(T)$, provided that uptake is indeed unidirectional so there is negligible back-diffusion from the brain to the plasma. (If back-diffusion is significant, the results may be analyzed by means of a multicompartment model, which, however, requires additional information on the fate of the solute in the brain.) Thus, two time points suffice to determine K_i and V_{br} for any solute; more time points may be useful to check unidirectionality. If V_{br} is known independently (e.g., from a suitably matched reference tracer), then a single time point may be sufficient to determine K_i. Once K_i is known, the PA can be determined, for example, from the single capillary model by means of Eqs. (5) and (3).

The second type of continuous uptake study is the *in situ* brain perfusion technique. In this method, one cerebral hemisphere of an anesthetized rat is perfused completely for 5–300 sec via a catheter in the appropriate external carotid artery. This method shares many of the advantages of the intravenous administration technique, particularly its sensitivity to compounds of low permeabilities (down to about 10^{-8} cm/sec). However, it has the great advantage that it allows almost total control of perfusate composition so that saturation, inhibition, and competition in carrier-mediated transport can be examined in detail (see Section III). Also, pH, osmolality, ionic content, and protein concentrations can be varied over a much greater range than would be tolerated systemically.

E. Water Transfer

Although the diffusional exchange of water across the blood–brain barrier (measured with radiolabeled water under conditions in which bulk flow of water across the barrier is negligible) is quite rapid, E is less than 0.95 and PA is less than about 0.03 cm^3/sec/g of brain. This indicates that water flux is partly membrane limited, except at low blood flow rates (less than 0.008 ml/sec/g of brain), when E approaches 1. This differs, for example, from skeletal muscle, in which the distribution of water is flow limited over a broad range of tissue perfusion rates. Furthermore, the octanol–water partition coefficient for water is about 0.07. It is therefore concluded that there are few, if any, water-filled channels between blood and brain extracellular fluid when the blood–brain barrier is intact: Water passes from the blood to the brain by dissolving in, and diffusing through, the membranes and the cytoplasm of the cerebral capillary endothelial cells. The transfer coefficients for diffusion of water across the blood–brain barrier therefore apply to a transcellular double-membrane system.

Under a pressure gradient (either hydrostatic or osmotic), water also moves across the blood–brain barrier by means of bulk flow. The hydraulic conductivity, L_p, of the barrier is the proportionality coefficient which relates the volume flow, J_v, to the driving pressure force:

$$J_v = -L_p\left(\Delta P - \sum_s \sigma_s \Delta\Pi_s\right), \tag{11}$$

where ΔP is the hydrostatic pressure difference across the barrier and $\Delta\Pi_s$ is the osmotic pressure difference for solute s, which has a reflection coefficient σ_s, and $\sum_s$ means that contributions from all solutes s are added. σ_s is a measure of the permeability of a membrane to a particular solute, relative to that of water.

The convective flow of water across a membrane can produce a coupled movement of solute by solute drag. The composition of the fluid (the filtrate) that actually moves across the membrane depends on the σ_s of each solute, which is a direct indicator of its osmotic effectiveness. A completely impermeant solute has a σ_s of 1, develops the full osmotic pressure across the membrane (given by the van't Hoff equation), and is not carried by solute drag in bulk fluid flow. On the other hand, a highly permeant solute has a σ_s close to zero, exerts little or no osmotic pressure across the membrane, and is readily carried across the membrane by the bulk movement of water [see Eq. (11)].

The reflection coefficients for small molecules and ions at the blood–brain barrier are often close to unity. For example, σ_s for sucrose and NaCl have been estimated to be about 0.91 and 1.00, respectively. (For comparison purposes, σ_s for sucrose at the heart capillary is only about 0.3.) The filtrate would be very dilute in such solutes and would tend to lower the osmolality of the fluid into which it flows, thus setting up a countering osmotic gradient. This restriction to solute movement when fluid flows across the intact blood–brain barrier is a major part of the volume-regulating role played by the normal blood–brain barrier and is critical in protecting the brain against edema.

F. Other Compounds and Drugs

For organic molecules, other than those confined to the extracellular space, transfer across the cerebral capillaries takes place mainly or exclusively through the endothelial cells. There are two main processes for this movement: (1) diffusing into, and through, the endothelial membranes and cytoplasm, and (2) combined carrier-mediated and diffusional transport across these structures. Carrier-mediated transport is discussed in Section III. Some general features of diffusional transfer are discussed here.

Most drugs and other organic compounds probably exchange between the blood and the brain extracellular fluid by passing through the lipid membranes of the capillary endothelial cells because (1) the capillary wall lacks water-filled channels of suitable size for entry by aqueous diffusion, (2) it is unlikely that cytoplasmic vesicles contribute significantly to pinocytotic transfer, and (3) the endothelial cell membranes usually lack appropriate carrier systems for facilitated diffusion. Movement of a drug across the entire capillary therefore usually entails entry into and release from a lipid phase at both the luminal and abluminal membranes, together with aqueous diffusion through the cytoplasm. Transfer through the blood–brain barrier should therefore depend strongly on the lipid solubility of the drug.

The lipid/water partition hypothesis of drug entry into the central nervous system postulates that blood–brain barrier permeability is proportional to the ability of the free (unbound) and unionized drug to partition between lipid and aqueous media (the usual fluid combination is octanol–water or olive oil–water). Both the ionized form of the drug and the fraction bound to plasma proteins are assumed to be excluded from the membrane and are therefore blocked from movement across the brain capillaries. Figure 3 shows a plot of measured blood–brain barrier permeability against the octanol/water partition coefficient for a number of test solutes. The relatively good fit of the linear regression over several orders of magnitude of permeability confirms, in general terms, the lipid/water partition hypothesis for drug entry into the brain. However, high molecular weight compounds (of 1000 or more), in which the hydrophobic regions may be distinct from the hydrophilic regions, have *PA* values lower than expected on the basis of their lipid/water partition coefficients. These compounds are presumably able to insert their hydrophobic regions into the membrane, but are too large to enter totally. [*See* Plasma Lipoproteins.]

III. TRANSPORT AT THE BLOOD–BRAIN BARRIER

It has been known since the early 1960s that some lipid-insoluble substances traverse the blood–brain barrier by carrier-mediated processes. These systems display the characteristics of stereospecificity, self-saturation, and competitive inhibition. Two forms of carrier-mediated transport are possible at the blood–brain barrier: (1) facilitated diffusion, in which the carrier expedites the movement of material in either direction, according to existing electrochemical potential gradients; and (2) active transport, in which solute is driven, with the expenditure of energy, against the electrochemical potential gradient either from the blood to the brain (influx pump) or vice versa (efflux pump). Evidence shows that specific facilitated carrier systems exist at the blood–brain barrier for each of the following classes of compounds: hexoses (such as glucose), monocarboxylic acids, large neutral amino acids, acidic amino acids, basic amino acids, and nucleic acid precursors.

In characterizing the transport of a substance across the blood–brain barrier, it is customary to determine the maximum rate of unidirectional transport, V_{max}, and the half-saturating concentration, K_m, usually by measuring the effect of various concentrations of the substance of interest in the blood (or vascular fluid) on the uptake of the same radiolabeled substance in a tracer amount. For a single membrane that separates plasma from brain extracellular fluid, and assuming a simple mobile carrier model (plus passive diffusion), the measured *PA* has the Michaelis–Menten form

$$PA = V_{max}/(c_{plasma} + K_m) + K_d, \tag{12}$$

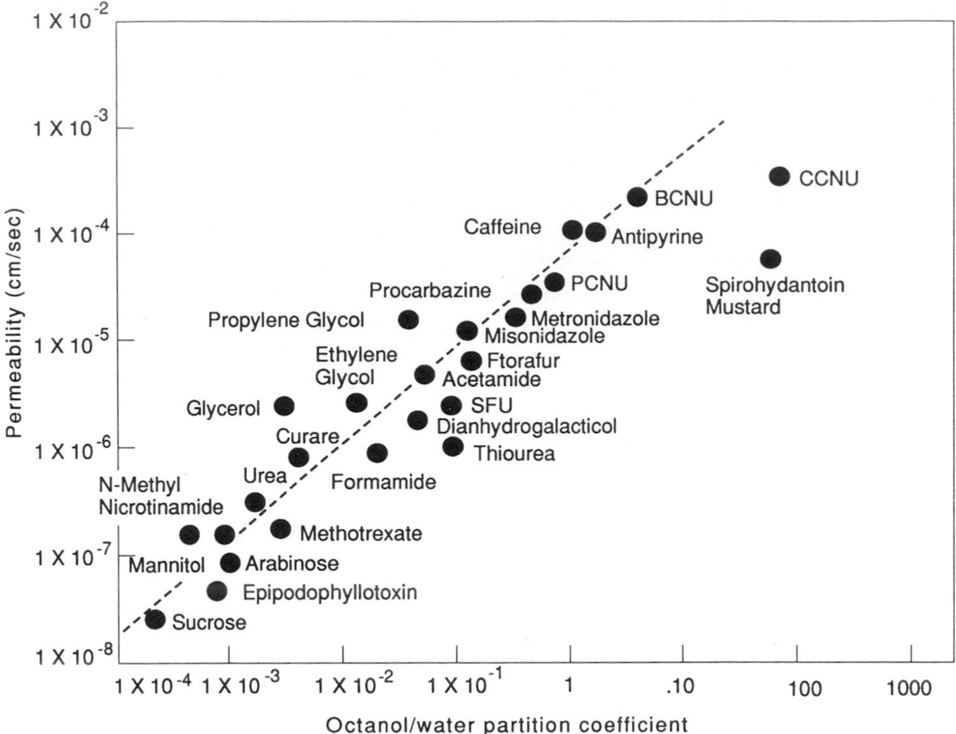

FIGURE 3 Relationship between cerebrovascular permeability and the octanol–water partition coefficient. (The surface area of brain capillaries was taken as 240 cm^2/g of brain.) CCNU, 1-(2-chloroethyl)-3-cyclohexyl-1-nitrosourea, Lomustine; BCNU, 1,3-bis(2-chloroethyl)-1-nitrosourea, carmustine; PCNU, 1-(2-chloroethyl)-3-(2,6-dioxo-3-piperidinyl)-1-nitrosourea; 5FU, 5-fluorouracil.

where c_{plasma} is the concentration of the unlabeled substance in the capillaries and K_d is the constant for nonsaturable transport. Such a simple model has a number of difficulties. First, transfer of the solutes across the blood–brain barrier actually involves serial passage across both luminal and abluminal membranes (which have been shown in a number of cases to have different transport properties) and across the cytoplasm of the endothelial cells. In addition, the Michaelis–Menten form of Eq. (12) is strictly appropriate only locally, with c_{plasma} referring to the concentration of unlabeled substrate at the carrier site. Clearly, if there is appreciable uptake, c_{plasma} will decrease significantly along the capillaries. In such cases, c_{plasma} in Eq. (12) is often replaced with its spatial average over the capillaries. Such an approximation reduces the accuracy of PA determinations as the extraction fraction becomes large.

A related problem of interpretation of measured PA (or K_i) values has already been mentioned in relation to first-order (i.e., nonsaturable) uptake, in which the single capillary model for uptake was modified to take into account the effects of capillary functional heterogeneity. In the case of saturable uptake, the effects of such heterogeneity are negligible at high concentrations ($c_{plasma} \gg K_m$) because the transport sites are already saturated so it makes no difference how they are distributed among capillaries. At lower substrate concentrations, however, effects of heterogeneity become more and more apparent, until in the first-order limit ($c_{plasma} \ll K_m$), an effective PA, given by Eq. (8), again applies. In general, the effects of capillary heterogeneity are of importance only for substances with fairly high extractions. Further complexities arise when two or more substrates share the same transport system.

Because of these theoretical limitations, and the relevant experimental shortcomings discussed in Section II, calculations of V_{max} and K_m are often, at best, semiquantitative and operational. The rest of this section discusses some general results relating to carrier-mediated transport at the blood–brain barrier.

A. Monosaccharides

Glucose moves across the blood–brain barrier mainly by a carrier system. This system exhibits stereospecificity, transport saturation, and competition and does not seem to involve active transfer in either direction. In the steady state, the glucose transfer system can be approximated by a single-membrane carrier, with an apparent unidirectional V_{max} of 1–3 μmol/min/g of brain and an apparent K_m of 5–6 mM. At normal glucose concentrations, net glucose uptake into the brain seems to be determined mainly by metabolic demand and is relatively independent of fluctuations in plasma glucose levels. At lower glucose concentrations, however, the initial step in the metabolism of glucose (its phosphorylation by brain hexokinase) becomes progressively less saturated, and transport across the blood–brain barrier becomes a progressively more important factor in determining uptake.

Transport of D-glucose is markedly inhibited by other hexoses. The affinities ($1/K_m$ values) of various hexoses for the carrier can be ranked as follows: α-D-glucose > 2-deoxy-D-glucose ≥ 3-*O*-methyl-D-glucose > β-D-glucose > D-mannose ≥ D-galactose ≥ D-xylose > L-glucose = D-fructose.

B. Amino Acids

Most amino acids cross the blood–brain barrier by carrier processes that are stereospecific, saturable, and competitive. Only the small neutral amino acids—glycine and proline—do not seem to enter the brain in this way, and instead cross the blood–brain barrier from the blood to the brain at a very low rate. Neutral amino acids, such as phenylalanine and leucine, compete with each other for entry into the brain, as do the basic amino acids arginine, lysine, and ornithine. The order of affinity among neutral amino acids for the cerebrovascular transport system appears to be phenylalanine > tryptophan > leucine > methionine > isoleucine > tyrosine > histidine > valine > threonine > cysteine > glutamine > asparagine > serine > alanine. This is similar to the so-called L-system for neutral amino acid transport, demonstrated in many tissues throughout the body. However, the magnitudes of the affinities ($1/K_m$) are 10- to 100-fold higher at the blood–brain barrier, with the consequence that the transport system is nearly saturated with neutral amino acids at normal plasma concentrations.

Because of transport saturation, each amino acid must compete for available transport sites according to its affinity relative to that of the other amino acids. Phenylalanine and leucine together occupy more than 50% of the capillary transport sites at normal plasma amino acid concentrations. Transport saturation and competition have a significant homeostatic role in the regulation of brain neutral amino acid concentrations. Plasma concentrations of neutral amino acids show characteristic fluctuations related to food consumption, protein synthesis, and hormonal activity. However, because these plasma concentrations tend to increase or decrease for all neutral amino acids as a group, alterations in brain influx are minimized by transport competition.

C. Other Substances

Among the other organic compounds transported across the blood–brain barrier by a facilitated diffusion mechanism are monocarboxylic acids, such as acetate and pyruvate, and nucleic acid precursors, such as L-dopa (3,4-dihydroxyphenylalanine). In addition, it appears that at least one exogenous anticancer drug (melphalan, an alkylating phenylalanine derivative) can make use of an existing carrier system (in this case the large neutral amino acid system) for facilitated entry into the brain (albeit with a significantly lower affinity than the endogenous large neutral amino acids).

Finally, it appears that transferrin (an iron transport protein) may cross the blood–brain barrier by a receptor-mediated mechanism, suggesting that the brain may derive its iron through the transfer of iron-loaded transferrin across the brain microvasculature.

IV. MODIFICATION OF THE BLOOD–BRAIN BARRIER

Irreversible disruption of the blood–brain barrier may be caused by trauma, heavy metal poisoning, freeze lesions, irradiation, some tumors, and oxygen deprivation. Reversible increases in barrier permeability, followed by the restoration of normal permeability within 30 min to a few hours, can be produced by acute and large osmotic imbalances. Hypertension, seizures, and elevations in the carbon dioxide pressure of the blood (i.e., hypercapnia) may also cause reversible increases in barrier permeability, although some brain damage may occur. Under most circumstances, such increases in permeability are disruptive to normal brain function, but there are circumstances, such as in the treatment of some brain tumors and in enzyme replacement therapies, in which reversible disruption of the blood–brain barrier is desirable to allow entry of water-soluble drugs or high molecular weight compounds into the brain. The osmotic method for enhancing blood–brain barrier permeability has been shown to have therapeutic effectiveness in the treatment of primary lymphomas and glioblastomas in humans.

A. Osmotic Opening of the Blood–Brain Barrier

The permeability of the capillary wall may be increased by applying hyperosmotic solutions to either side of the blood–brain barrier. Experiments in which such solutions are applied topically to pial vessels within the subarachnoid space show that the more lipid soluble a substance, the higher the concentration needed to induce opening. This result is consistent with the concept that barrier opening is induced by shrinkage of the endothelial cells, with loosening of the intercellular tight junctions, due to the inverse relationship between cell shrinkage and lipid solubility observed *in vitro* (see Fig. 4). Highly lipophilic solutes can open the barrier irreversibly, probably by destroying endothelial cell membranes.

Similar effects are obtained by infusing the appropriate hypertonic solution directly into the internal carotid artery. By infusing hypertonic arabinose in the rat, it can be shown that there is a threshold concentration for barrier opening at a fixed infusion duration, as well as a minimal duration at a fixed concentration. These results suggest that, in a more general way, a threshold product of infusion concentration multiplied by duration is required to produce barrier opening. This threshold product is not constant, but appears to increase as the duration of exposure is lengthened.

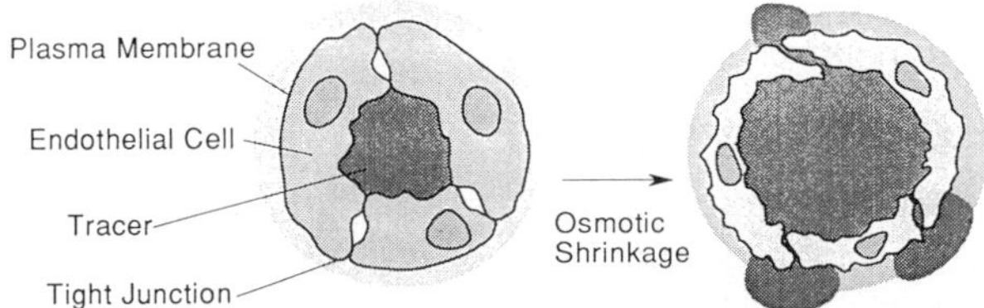

FIGURE 4 Model for a normal barrier (left) and barrier opening by widening of tight junctions between endothelial cells. When endothelial cells shrink in a hypertonic environment (right), their membranes stress the tight junctions and make them permeable to intravascular tracer.

Blood–brain permeability following a hyperosmotic infusion into the carotid can be determined by a modification of the intravenous administration technique described in Section II. After a certain interval of time following the infusion, a solution of the test radiotracer of interest is injected intravenously. Plasma samples are taken, and the brain is removed for analysis after sufficient radioactivity has accumulated for a determination of the (regional) unidirectional transfer constant, K_i, according to Eq. (10). Values for PA are then determined by Eqs. (3) and (5). These experiments show that, at early times after the barrier is opened, PA for water-soluble substances is increased markedly. For example, 1–6 min after a 30-sec infusion of 1.8 M arabinose, regional PA values for sucrose increased by a factor of about 20. PA values for higher molecular weight compounds, such as inulin and dextran, increased even more dramatically. Thereafter, PA declines, until, after about 2–4 hr, the barrier has reclosed.

In addition to osmotically induced loosening of tight junctions, it has been postulated that enhanced vesicular transport and passage through large channels formed by the fusion of vesicles play a major role in the transport of solutes following osmotic opening of the blood–brain barrier. Although hypertonic infusions do seem to induce vesicle formation in the cerebrovascular endothelium, the weight of the evidence supports modification of the tight junctions as the basis for osmotic opening of the blood–brain barrier and suggests that vesicular transport is, at best, a minor component of enhanced barrier permeability.

Barrier reclosure following osmotic opening is dependent on molecular size in a manner consistent with a pore mechanism with bulk fluid flow (and solute drag) from the blood to the brain. Thus, data on the uptake of sucrose (molecular radius, 5 Å), inulin (15 Å), and various molecular weight dextrans of up to 100 Å radius suggest an effective pore radius of about 200 Å and bulk flow from the blood to the brain of about 1.6×10^{-4} cm^3/sec/g of brain within 6 min of barrier opening, falling to about 7×10^{-6} cm^3/sec/g of brain at 55 min. Such bulk flow values are consistent with observed increases in brain water content during the first 10 min after the opening of the barrier. This reduction in bulk flow seems to be the major factor in the reclosure of the barrier within the first hour or so following osmotic opening. The effective pore density seems to decline only slightly, if at all, during this period: It has been calculated to be about 5×10^{-3} pores per μm^2 at 5 and 35 min, corresponding to less than 0.001% of the total capillary surface area.

Clinically, osmotic opening of the blood–brain barrier has been used to enhance the chemotherapy of brain tumors, using minimally toxic water-soluble drugs. Although blood vessels within metastatic cerebral tumors and primary gliomas may sometimes have a fenestrated, discontinuous endothelium and an ineffective blood–brain barrier, it is likely that the barrier is completely or partially intact at the proliferating edge of a tumor, and in some cases within the entire tumor, because systemically administered drugs are generally ineffective for central nervous system tumors. Indeed, in some patients, previously unrecognized tumors with intact barriers can take up intravascular water-soluble markers after osmotic treatment. Regression of a number of primary or metastatic brain tumors has been observed following osmotic treatment combined with appropriate chemotherapy.

Entry of high molecular weight enzymes used in enzyme replacement therapies has also been enhanced by osmotic blood–brain barrier disruption. For example, delivery of hexosaminidase A, the enzyme deficient in Tay–Sachs disease, is markedly increased over control values following osmotic barrier opening in rats.

B. Other Methods for Enhancing Drug Entry into Brain

At present, osmotic blood–brain barrier opening is perhaps the most highly developed method for reversible barrier modification to enhance drug entry into brain. However, other techniques have been studied, including Metrazol-induced seizure activity, acute hypertension, and X-irradiation.

Other methods for enhancing drug entry into the brain that do not involve barrier modification include intracerebrospinal fluid drug administration, utilization of carrier-mediated transport systems at the barrier, and chemical modification of the drugs themselves to increase their brain uptake (e.g., by increasing their lipid solubility or by enabling them to make use of existing transport systems). Using a drug such as a neurotransmitter agonist or antagonist, or an anesthetic agent, to increase regional metabolic rates, and hence regional blood flows, may also increase the regional delivery of that drug (or another drug) to the brain. Finally, for a drug extensively bound to circulating plasma proteins, only the free fraction is normally available for entry into a brain with an intact blood–brain barrier. Increasing this free fraction (e.g., by introducing a high-affinity competitor for the protein–binding sites) may significantly enhance its entry.

BIBLIOGRAPHY

Bradbury, M. W. B. (1979). "The Concept of a Blood–Brain Barrier." Wiley, Chichester, England.

Davson, H., Welch, K., and Segal, M. B. (1987). "Physiology and Pathophysiology of the Cerebrospinal Fluid." Churchill-Livingstone, Edinburgh.

Fenstermacher, J. D., and Rapoport, S. I. (1984). Blood–brain barrier. *In* "Handbook of Physiology—The Cardiovascular System IV," Ch. 21, pp. 969–1000. American Physiological Society, Bethesda.

Flannagan, T. R., Emerich, D. F., and Winn, S. R. (eds.) (1994). Providing pharmacological access to the brain: Alternate approaches. *In* "Methods in Neurosciences," Vol. 21. Academic Press, San Diego.

Neuwelt, E. A. (1989). Implications of the Blood–Brain Barrier and Its Manipulation," Vols. 1 and 2. Plenum, New York.

Rapoport, S. I. (1976). "Blood–Brain Barrier in Physiology and Medicine." Raven, New York.

Blood Coagulation, Hemostasis

CRAIG M. JACKSON
Del Mar, California

GLOSSARY

Activation Conversion from a form without biological (functional) activity to a fully biologically (functionally) active participant in a biochemical reaction

α-2 Antiplasmin Plasma protein inhibitor that inactivates plasmin

Antithrombin III (heparin cofactor) Plasma protein inhibitor that inactivates many of the proteinases of blood clotting; it is aided by heparin

C1 inactivator Plasma protein inhibitor that can inactivate factor XIIa

Cofactor protein Protein that participates in a biochemical reaction in the blood clotting system but that is not directly involved in the chemical reaction (proteolysis)

Factor V (proaccelerin) Circulating precursor of factor Va, the protein cofactor that acts with factor Xa during activation of prothrombin

Factor VII (proconvertin) Proteinase that acts with tissue factor to activate factor X

Factor VIII (antihemophilic factor, AHF, AHG) Circulating precursor of factor VIIIa, protein cofactor that acts with factor IXa during activation of factor X

Factor IX (antihemophilic factor B, Christmas factor) Precursor of proteinase (factor IXa) that, with factor VIIIa, activates factor X

Factor X (Stuart-Prower factor) Precursor of proteinase (factor Xa) that, with factor Va, activates prothrombin

Factor XI (plasma thromboplastin antecedent) Precursor of proteinase (factor XIa) that activates factor IX in the "intrinsic pathway"

Factor XII (Hageman factor) Precursor of proteinase (factor XIa) that activates factor XI and prekallikrein

Factor XIII (fibrin-stabilizing factor) Precursor to transamidase that stabilizes fibrin by cross-linking

Fibrinogen (factor I) Soluble plasma protein that after proteolytic cleavage by thrombin polymerizes to form the fibrin clot

Heparin Glycosaminoglycan that accelerates proteinase inactivation by antithrombin III and heparin cofactor II

Heparin cofactor II Plasma protein inhibitor that inactivates thrombin

High-molecular-weight kininogen Protein cofactor of factor XIIa and kallikrein in the activation of factor XII and prekallikrein in the "contact phase" of the coagulation cascade

Inactivation Destruction or elimination by active site blockage of the biological (functional) activity of a clotting system component

α-2 Macroglobulin General inhibitor of hemostatic and fibrinolytic proteinases

Plasminogen Precursor of proteinase (plasmin) that dissolves the fibrin clot

Plasminogen activator inhibitor, types I, II, III Inhibitors of tissue plasminogen activators in plasma; type III also inhibits activated protein C

Platelet Cell (megakaryocyte) fragment that adheres to tissue exposed as a result of injury; it forms the primary hemostatic plug

Prekallikrein Precursor of kallikrein, the proteinase that activates factor XII

Proenzyme Inactive precursor to an enzymatically active proteolytic enzyme

Protein C Precursor of proteinase (activated protein C) that inactivates factor Va and factor VIIIa

Protein S Protein cofactor to activated protein C

Prothrombin (factor II) Circulating precursor (proenzyme) to the proteinase thrombin

Thrombomodulin Membrane protein that acts as a protein cofactor in the activation of protein C by thrombin

ENCYCLOPEDIA OF HUMAN BIOLOGY, Second Edition, VOLUME 2. Copyright © 1997 by Academic Press. All rights of reproduction in any form reserved.

Tissue factor (thromboplastin, factor III) Cellular protein from ruptured cells that is responsible for initiation of clotting; in a mixture with phospholipids it is called tissue thromboplastin

Tissue plasminogen activator (tissue-type plasminogen activator, TPA) Proteinase of endothelial cell origin that activates plasminogen

Urokinase-type plasminogen activator (prourokinase, scu-PA) Proteinase of cellular origin that activates plasminogen

von Willebrand factor Plasma protein required for platelet adhesion to cells that acts as a carrier for factor VIII in the circulating blood

HEMOSTASIS IS THE SPONTANEOUS ARREST OF bleeding from ruptured blood vessels. Normal hemostatic response to vessel injury includes blood vessel constriction, adhesion of platelets to blood vessel cells not normally exposed to blood, and a complex series of proteolytic reactions that culminate in the formation of a fibrin network that reinforces and consolidates the platelet plug at the injury site. The hemostatic response is finely balanced under normal physiological conditions. Through the synergistic actions of the components that promote the proteolytic reactions of the blood clotting system and the antagonistic actions of the inactivators of these components, the risk of excessive bleeding (hemorrhage) is balanced against the risk of formation of unwanted blood clots (thrombosis). Decreased numbers of functional platelets or deficiencies of clotting factors predispose to hemorrhage; deficiencies in inhibitors of the clotting proteinases or in the enzymes that digest the blood clot after tissue repair is completed predispose to thrombosis.

The principal chemical reactions of the blood clotting cascade are transformations of proenzymes of proteinases into active proteolytic enzymes. Physiologically adequate proenzyme activation occurs when an activating proteinase, a protein cofactor in a proteolytically modified form, and the proenzyme that is the substrate in the next stage of the clotting cascade bind to a cell membrane surface to form a complex. In the complex, proenzyme activation is several hundred-thousand times faster than it is when only the activating proteinase is causing the proteolytic cleavage. The formation of the complex on the surface of the damaged cell membranes localizes rapid activation to the site of injury. Activator complexes are characteristic of the proenzyme activation reactions in blood clotting.

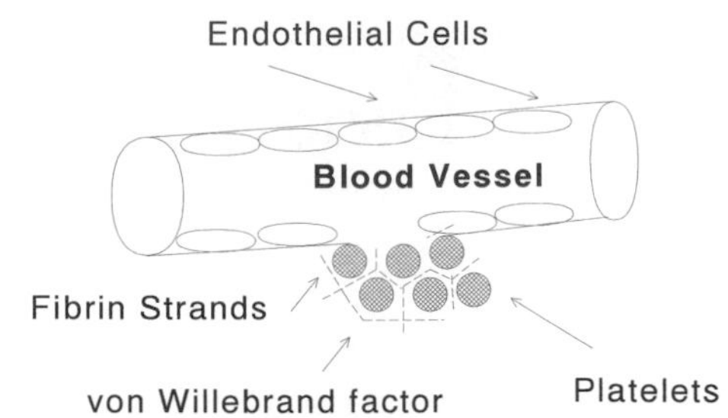

FIGURE 1 An endothelial cell-lined blood vessel with platelets released into the extravascular space via a rupture in the vessel wall.

I. PHYSIOLOGY OF HEMOSTASIS AND BLOOD CLOTTING

Rupture of blood vessels initiates the normal hemostatic response. The ruptured blood vessels, particularly arteries, constrict, thus narrowing the opening through which blood may escape. When blood proteins and platelets come into contact with cells or cell components underneath the endothelial cells that form the normal lining of the blood vessel, they adhere to them. A platelet plug forms that occludes the opening in the ruptured vessel. Adhesion of platelets to the smooth muscle cells and exposed collagen fibrils beneath the endothelium requires the plasma protein, von Willebrand factor. Von Willebrand factor binds to specific receptors on the outer membrane of the platelet and thus anchors the platelets to the subendothelial cells and the extracellular matrix. Deficiency in von Willebrand factor is among the more common of the blood clotting disorders and leads to prolonged bleeding in affected individuals. A schematic depiction of the primary hemostatic response is shown in Fig. 1.

Simultaneously with platelet adhesion, the sequence of blood clotting reactions commences with the formation of a complex between a membrane protein, tissue factor, and factor VII.[1] The formation of the tissue factor–factor VII complex initiates the sequence of reactions that culminates in formation of the proteolytic enzyme thrombin. Thrombin then catalyzes the conversion of soluble fibrinogen into fibrin monomer, the protein that polymerizes to produce the fibrin meshwork of the hemostatic plug or blood clot. Thrombin also stimulates platelets to release several

[1]The Roman numeral designations are those given in 1958 [I. Wright, *J. Am. Med. Assoc.* **170**, 135–138 (1959)] to eliminate the use of patient names for coagulation factors. The letter "a" following the numeral indicates the activated derivative.

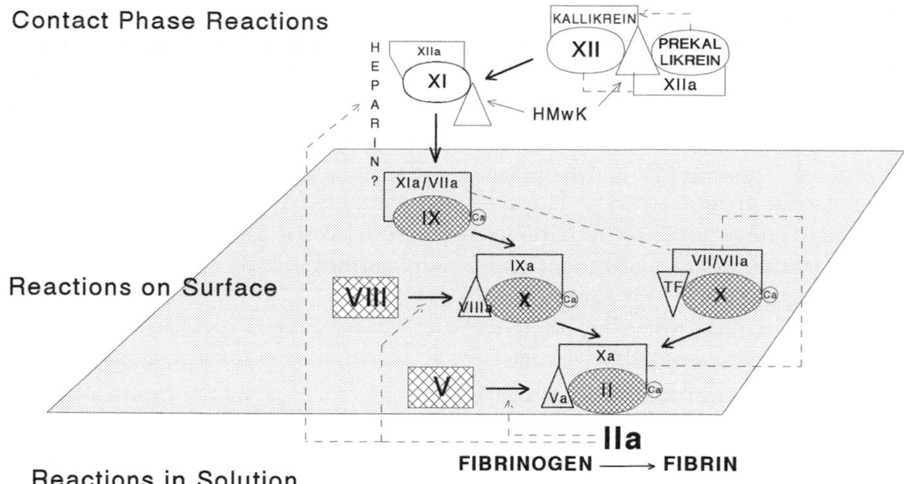

FIGURE 2 The blood clotting cascade. Components are enclosed by geometric figures that indicate the type of component as follows: proenzymes are shown as ellipses; proteinases are shown as polygons; protein cofactors are shown as rectangles; activated protein cofactors are shown as triangles. The tissue factor triangle (TF) is inverted to indicate that it "sticks into" the membrane. The contact phase reactions are shown away from the cell surface because the surfaces to which they bind are only known *in vitro* and are not of biological origin. The pathway that is initiated through exposure of tissue factor is called the extrinsic pathway; that initiated through contact of factor XII and prekallikrein with an artificial surface is the intrinsic pathway. HMwK, high-molecular-weight kininogen. [Reproduced, with permission, from C. M. Jackson and Y. Nemerson (1980). Blood coagulation. *Annu. Rev. Biochem.* **49**, 765. © 1980 by Annual Reviews Inc.]

clotting factors and aggregating agents such as adenosine diphosphate from storage granules within the platelet. These platelet granule contents promote further development of the platelet plug. The hemostatic response that is initiated by tissue injury, or *in vitro* by the addition of a tissue homogenate (tissue thromboplastin) to a blood sample, is called the *extrinsic pathway of blood clotting*. The alternate pathway of blood clotting that most likely begins with factor XI, rather than tissue factor, is called the *intrinsic pathway* (Fig. 2). [*See* Blood Coagulation, Tissue Factor.]

II. THE BIOCHEMISTRY OF THE HEMOSTATIC SYSTEM COMPONENTS

The blood clotting cascade consists of a series of proteolytic reactions in which circulating, inactive blood clotting factors are converted into catalytically active forms. The proenzymes are converted to active proteinases. The protein cofactors, which have latent binding sites for a proteinase and proenzyme, have these sites unmasked by proteolysis. The transformations of proenzymes into active proteinases are the basic reactions of the blood clotting cascade; however, they occur rapidly only when a proteinase is assisted by an activated protein cofactor and when all components, including the proenzyme substrate, are bound to a membrane phospholipid surface. Such assembled activator complexes catalyze activation of their respective proenzymes at more than 100,000 times the rate observed with a proteinase alone.

A. Clot Formation: The Polymerization of Fibrin

The only protein required to form the gel-like blood clot is the colorless protein fibrin, which comes from fibrinogen. A blood clot formed in a test tube, or a thrombus formed in a vein in which there is stagnant flow, is red because it entraps red blood cells. In a hemostatic plug, fibrin fibers are attached to platelets via specific receptor molecules in the platelet membrane. The fibrin forms a mesh that envelopes the platelets and trapped cells. Fibrin is not present in blood in appreciable concentration prior to injury and initiation of the reactions of blood clotting; fibrinogen, the precursor of fibrin, is the normally circulating form. The presence of fibrin without an injury may indicate ongoing pathological thrombus formation.

Fibrinogen molecules, because of their large size, can be visualized from electron microscopic pictures. They consist of three globular domains (Fig. 3); two are at the ends of the molecule and are attached to a central domain by long, thin protein strands. Thrombin action on fibrinogen to convert it to fibrin removes four short polypeptides from the central domain. This small change, however, exposes sites that enable noncovalent association of individual fibrin monomers, leading to formation of macroscopic fibrin strands. Under electron microscopy the fibrin strands appear to be similar to a frayed and tangled rope with small fibrils extending from the large fibers.

The fibrin clot formed only by noncovalent association of fibrin monomers is not stable in flowing blood. The blood that flows by the hemostatic plug permits fibrin monomer molecules to dissociate from the fibrin polymer mesh and the fibrin redissolves and bleeding occurs. Such dissolution is prevented by the plasma transamidase, factor XIIIa. Factor XIIIa catalyzes the formation of covalent cross-links between fibrin monomer molecules within the polymer. Factor XIII is a necessary component in the normal hemostatic system, and when it is defective or deficient, poor wound healing frequently occurs. The most dramatic example of the importance of factor XIII is that of a woman who suffered many spontaneously aborted pregnancies prior to diagnosis of factor XIII deficiency. After transfusion to replenish the factor XIII, she delivered a normal infant without significant complications.

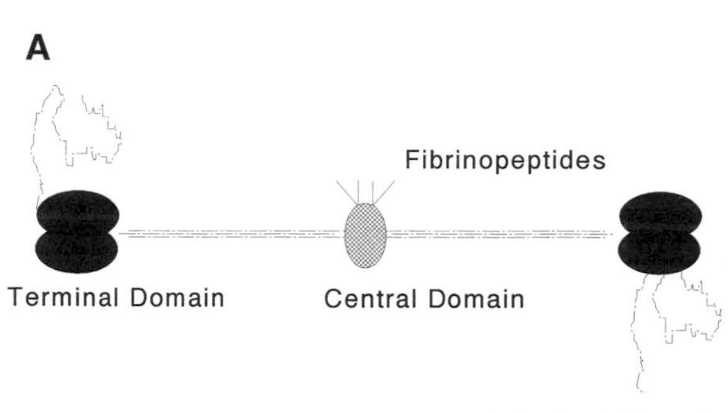

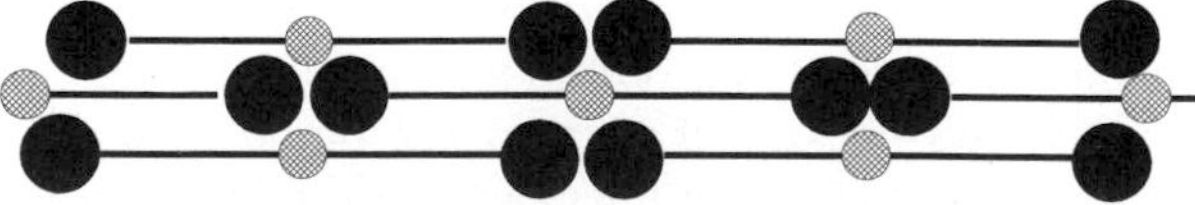

FIGURE 3 (A) Fibrinogen is shown with its central domain, attached fibrinopeptides, and the two terminal domains with the free additional polypeptide chains that are not folded into the terminal domains. (B) Polymerized fibrin monomers are shown, with central and terminal domains of adjacent molecules making contact. The free polypeptide chain portions of the terminal domains have been omitted.

B. Activation of the Circulating Proenzymes

1. Initiation of Blood Clotting and the Extrinsic Activator of Factor X

The blood clotting factors circulate as inactive precursors, until injury to the blood vessel occurs. Upon injury, clotting factors accumulate at the injury site and are converted to their biochemically (proteolytically) active forms by proteolysis at one or two specific positions along their amino acid sequences. A possible exception to this generalization is factor VII. Factor VII might express limited activity without proteolytic cleavage. However, factor VII's efficiency as a proteolytic enzyme is extremely low and becomes significant only after proteolytic cleavage and when it is associated with its protein cofactor, tissue factor.

It seems almost certain, although not proven conclusively, that exposure of tissue factor as a result of endothelial injury initiates the extrinsic pathway of blood clotting. The exposed tissue factor binds factor VIIa and VIIa in the presence of negatively charged phospholipids that are exposed to the blood as a result of the cell rupture and cell membrane lipid rearrangement. The binding of factor VIIa to the phospholipid molecules requires calcium ions; this is a primary way in which calcium is involved in blood clotting. Factor VIIa is more than 100,000 times more efficient as a proteinase in the complex with tissue factor than in its absence. Consequently, significant factor VIIa proteolytic action is effectively restricted to the site where tissue factor is exposed. The complexed factor VIIa catalyzes the cleavage of a single peptide bond in factor X, transforming it from proenzyme to active proteolytic enzyme, factor Xa. In turn, factor Xa acts on circulating factor VII producing VIIa which creates a maximally active complex with tissue factor. Physiologically, this mechanism of activity enhancement as a result of complex formation with a protein cofactor and binding to membrane lipids localizes both the initiation of clotting and the subsequent reactions to the site of injury. Factor VIIa also activates factor IX and thereby provides a link between the extrinsic and the intrinsic pathways of blood clotting.

An *in vitro,* routine laboratory test of the function of the clotting system uses a mixture of tissue factor and lipid vesicles in the form of an extract of lung or brain, which is called thromboplastin. Historically, because this mixture contains tissue factor, a material not normally present in the circulating blood, the resulting clotting reactions were described as occurring by the extrinsic pathway of blood clotting.

2. Prothrombinase and the Activation of Prothrombin

The prototype activator complex of the coagulation system is prothrombinase, shown schematically in Fig. 4. This complex contains the substrate prothrombin, the proteinase, factor Xa, the activated protein cofactor, factor Va, the requisite lipids from cell or platelet membranes, and calcium ions. Activation of prothrombin to form thrombin can be catalyzed by factor Xa alone, but in the absence of factor Va and a membrane surface it is extremely slow. Factor Xa in the complex converts prothrombin to thrombin more than 300,000 times faster than factor Xa by itself. Expressed in terms of time required to form a given amount of thrombin, a 300,000-fold increase in reaction rate is equivalent to reducing the time for thrombin formation from 6 months to little more than a minute.

Binding of prothrombin and factor Xa to a cell membrane (in the absence of factor Va) increases the concentrations of these components, thereby increasing the rate of prothrombin activation about 100 times. Such activation of prothrombin in the absence

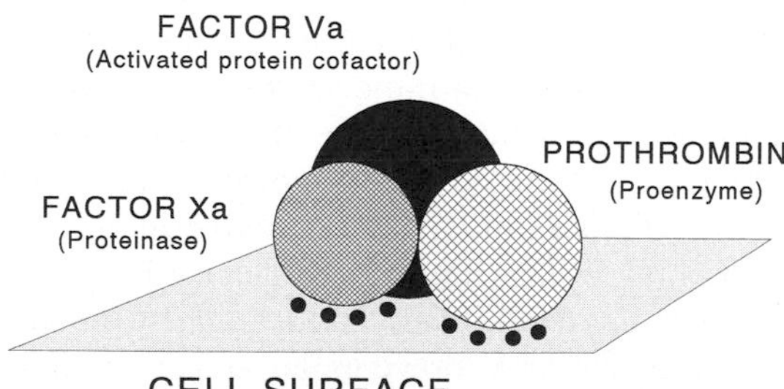

FIGURE 4 The association of factor Va (the activated protein cofactor), factor Xa (the proteinase), prothrombin (the proenzyme substrate), and calcium ions to create the prothrombinase complex on a membrane surface is illustrated. The overlapping circles represent the contacts made among all of the individual components of the complex.

of the activated cofactor is probably sufficient to produce enough thrombin for activation of factor V, and thus unmask factor V's binding sites for factor Xa and prothrombin. This cofactor activation enables formation of the complete complex, which optimally converts prothrombin into thrombin. Because factor V and factor Va bind to the cell membrane surface, along with prothrombin and factor Xa, the activation of prothrombin, like that of factor X by factor VIIa, is localized to the injury site. When the membrane surface is provided by platelets stimulated by thrombin, factor Va may also bind to a specific platelet receptor molecule. The physiological importance of complex formation here, as for factor X activation, is localization and rapidity of response at the site at which risk of blood loss exists.

Prothrombin and factor Xa binding to factor Va and to the membrane phospholipids in the presence of calcium ions in the prothrombinase complex depends on specialized domains present at the amino-terminal ends of the protein molecules. Structural and functional properties of these domains are related to vitamin K action in blood clotting (see Section II,F,3).

The prothrombinase complex, and other complexes as well, differ in one important respect from the factor VII–tissue factor complex. Whereas the tissue factor in the latter is always anchored to the cell membrane, the individual components of prothrombinase can spontaneously dissociate because they are held together by weaker noncovalent interactions. Dissociation of factor Xa and prothrombin from the complex makes factor Va very susceptible to inactivation by activated protein C, a component of the anticoagulant pathway. Furthermore, factor Xa separated from factor Va and the membrane surface is readily inactivated by the proteinase inhibitor, antithrombin III. These inactivation reactions, which shut off the clotting cascade, are discussed in Section II,D.

C. Other Activator Complexes of the Clotting Cascade

1. The Intrinsic Activator of Factor X

Factor X, in addition to being activated by the tissue factor–factor VIIa complex, is activated by a complex in which factor IXa is the proteinase, factor VIIIa is the activated protein cofactor, and either damaged cells or platelets provide the membrane surface. Similar to the activation of prothrombin, the complex activator increases the rate of activation of factor X by nearly 500,000 times. Factor VIII, the protein cofactor precursor, is activated by thrombin, similarly to the activation of factor V by thrombin. Factor VIII is the clotting factor that is missing or defective in classic hemophilia, a life-threatening disease characterized by bleeding into joints. Hemophilia, or more specifically hemophilia A, is perhaps best known because of its prevalence within royal families of Europe. Factor VIII circulates in the blood bound to von Willebrand factor, although von Willebrand factor is not involved in factor X activation. Factor IX deficiency, hemophilia B, gives rise to a bleeding disorder that is similar to hemophilia A, not surprising because both of these factors are components of the same complex activator. The genes for both factor VIII and factor IX are on the X chromosome; thus hemophilias are diseases of males, and females are carriers of the disease but unaffected. For historical reasons, this activator of factor X is sometimes referred to as the intrinsic factor X activator. [*See* Hemophilia, Molecular Genetics.]

2. The Contact Phase of Blood Clotting: An *in Vitro* Phenomenon

Several reactions that occur when blood clots in a test tube may not be important for *in vivo* hemostasis. These reactions are labeled in Fig. 2 as the contact phase reactions of blood clotting. These apparently nonphysiological reactions take place upon contact of blood with the surface of the test tube or, in one of the *in vitro* clotting tests, when kaolin, a type of clay, is added to the blood plasma. As in the previous cases, the contact phase reactions occur most efficiently in complexes that contain a protease and a protein cofactor. The contact phase clotting factors are factor XII, prekallikrein, and high-molecular-weight kininogen. Individuals deficient in the contact system factors do not have an evident bleeding disorder, which indicates that they are not important for physiological hemostasis.

Factor XI, usually included among the contact phase factors, is important for hemostasis because factor XI-deficient individuals suffer excessive bleeding under some circumstances. Although one activator of factor XI is factor XIIa, thrombin can activate factor XI when glycosaminoglycans such as those found on endothelial cells are present in *in vitro* experiments. Thombin is likely to be the more important physiological activator of factor XI.

D. Regulation of Blood Clotting Reactions

Two distinctly different mechanisms exist for terminating the reactions of the blood clotting cascade. The first mechanism by which clotting is stopped is reaction of the proteinases with irreversible inhibitors, most importantly antithrombin III and heparin cofactor II. Antithrombin III reacts with all of the proteinases, although slowly with VIIa, to form a covalent complex that renders the proteinase inactive. Reaction of these proteinases with antithrombin III is catalyzed by heparin, a sulfated polysaccharide, or glycosaminoglycan. Heparin increases the rate of inactivation of thrombin nearly a million times, and that of factor Xa more than a thousand times. Heparin is found in small amounts on the surface of endothelial cells, where it may have a role in preventing clotting on the normal uninjured vessel lining; it is primarily present in blood only when added as a therapeutic agent to impede the formation of thrombi. The second heparin-related inhibitor, heparin cofactor II, is limited to inactivating thrombin. However, heparin cofactor II is as efficient as antithrombin III in inactivating thrombin and can be aided by glycosaminoglycans more common than heparin. Other inhibitors of clotting proteinases are α-1 proteinase inhibitor, α-2 macroglobulin, and C1 inactivator. These inhibitors are believed to be less important than antithrombin III and heparin cofactor II.

The second mechanism for stopping clotting is proteolytic inactivation of the protein cofactors, factor Va and factor VIIIa, by activated protein C. Activated protein C is optimally effective only when assisted by protein S acting as a cofactor and in the presence of a membrane phospholipid surface and calcium ions; again the importance of activation complexes, rather than single enzymes, is evident. The reaction sequence that leads to factor Va and VIIIa inactivation has been designated the *anticoagulant pathway of hemostasis.* Protein C is activated by thrombin and, like the other proenzyme activations, requires a protein cofactor, the integral membrane protein thrombomodulin, for an optimal activation rate. The physiological importance of the anticoagulant pathway is evident from the increased risk of thrombosis (unwanted blood clotting that occludes blood vessels and thus starves the surrounding tissues of nutrients) in individuals with defects in this pathway. Risk for thrombosis is particularly great if a mutation in factor V prevents factor Va from being inactivated by activated protein C.

Close and efficient regulation of thrombin is extremely important because of thrombin's multiple actions in the hemostatic process. Thrombin action on fibrinogen produces clotting and promotes its own formation through its activation of factors V and VIII, as well as through stimulation of platelets.

On the other hand, thrombin retards its further formation by activating protein C and, as a consequence, inactivating factors Va and VIIIa. Optimal hemostatic response must prevent excessive bleeding when injury occurs, but without causing thrombosis. The intricacy of the reactions in the clotting cascade and the synergistic interactions between plasma proteins and cellular elements in the hemostatic response are a reflection of the importance of the control that is necessary for balancing hemostatic response and thrombotic risk.

E. Fibrinolysis

The fibrin strands that are intertwined within the hemostatic plug are transient structures and are digested by the fibrinolytic enzyme, plasmin. The proenzyme, plasminogen, is activated by plasminogen activators, proenzymes, and proteinases that are released from cells into the circulating blood. Two types of plasminogen activator are known, tissue-type plasminogen activators and urokinase-type plasminogen activators. Activation of plasminogen and the action of plasmin on fibrin are facilitated by their binding to the fibrin strands. Therefore, fibrin, acts like a protein cofactor for plasminogen activation and localizes normal fibrinolysis to fibrin. Fibrinogenolysis is an abnormal process, or is a consequence of therapeutic administration of fibrinolytic enzymes for the purpose of digesting a thrombus. The association of the tissue-type plasminogen activators and plasminogen with fibrin for maximally efficient activation of plasminogen and fibrinolytic activity provides the rationale for the development of recombinant tissue-type plasminogen activators as targeted therapeutic agents for dissolving thrombi in coronary arteries and other blood vessels. Irreversible inhibitors of plasmin and of plasminogen activators, similar to the inhibitors of the clotting proteinases, are present in blood and act to restrict the action of these proteinases to the fibrin clot. Deficiencies of these inhibitors, α-2 antiplasmin and the plasminogen activator inhibitors, produce excessive fibrinolytic action and result in bleeding similar to that associated with clotting factor deficiencies.

F. Protein Structural Domains and the Functional Properties of the Coagulation and Fibrinolytic System Proenzymes

The ability of the clotting proteins to form specific activator complexes with high catalytic efficiencies is a direct consequence of unique structural domains within these molecules. These domains are found primarily in the amino-terminal portions of the proenzymes and are approximately one-half of the total amino acid sequence of the proenzyme. The proteinase is derived from the remaining portion of the proenzyme molecule. Whereas the proteinase domains are structurally very similar, the amino-terminal regions of the individual molecules are different in each proenzyme. These parts of the proenzymes contain several common structural features, but in different combinations that are related to the specific interactions between proenzyme, protein cofactors, and proteinases in the activation complexes. Schematic, two-dimensional structures are shown in Fig. 5 that represent the six distinct domain structures of the proenzyme molecules.

1. The Proteinase Domains: The Carboxy-Terminal Regions

The proteinase domains of the proenzymes, identified by the similarity of their amino acid sequences with that of trypsin, possess all the functional groups necessary for cleaving the peptide bonds in their protein substrates. Several particular amino acid residues, common to all of the proteinases, are directly involved in the hydrolysis of the peptide bond. These residues, the catalytic triad or active site residues, consist of one histidine, one serine, and one aspartate residue. All of the hemostatic system proteinases hydrolyze peptide bonds at the carboxyl group of argininyl or lysyl residues. The complementarity between the sequences of the amino acid residues that precede the arginine or lysine residues in the proenzyme substrates and the active site regions of the proteinases is responsible for much of the specificity of a particular proteinase for its substrate. The remainder of the specificity is determined by the amino acid residues that are adjacent to the active site residues in the three-dimensional structures of the proteinases and the residues in the proenzyme or protein cofactor molecules that are adjacent to the peptide bonds that are cleaved.

In all of the coagulation and fibrinolytic proenzymes, except prothrombin, the proteinase domain is covalently attached to the domains of the amino-terminal region through a single disulfide bond. This simple structural feature maintains all proteinases, except thrombin, covalently bound to their amino-terminal regions, and thus to the other components of the activator complex. Because thrombin is not so bound it is able to diffuse to the several proteins of the clotting system on which it acts (see Fig. 2).

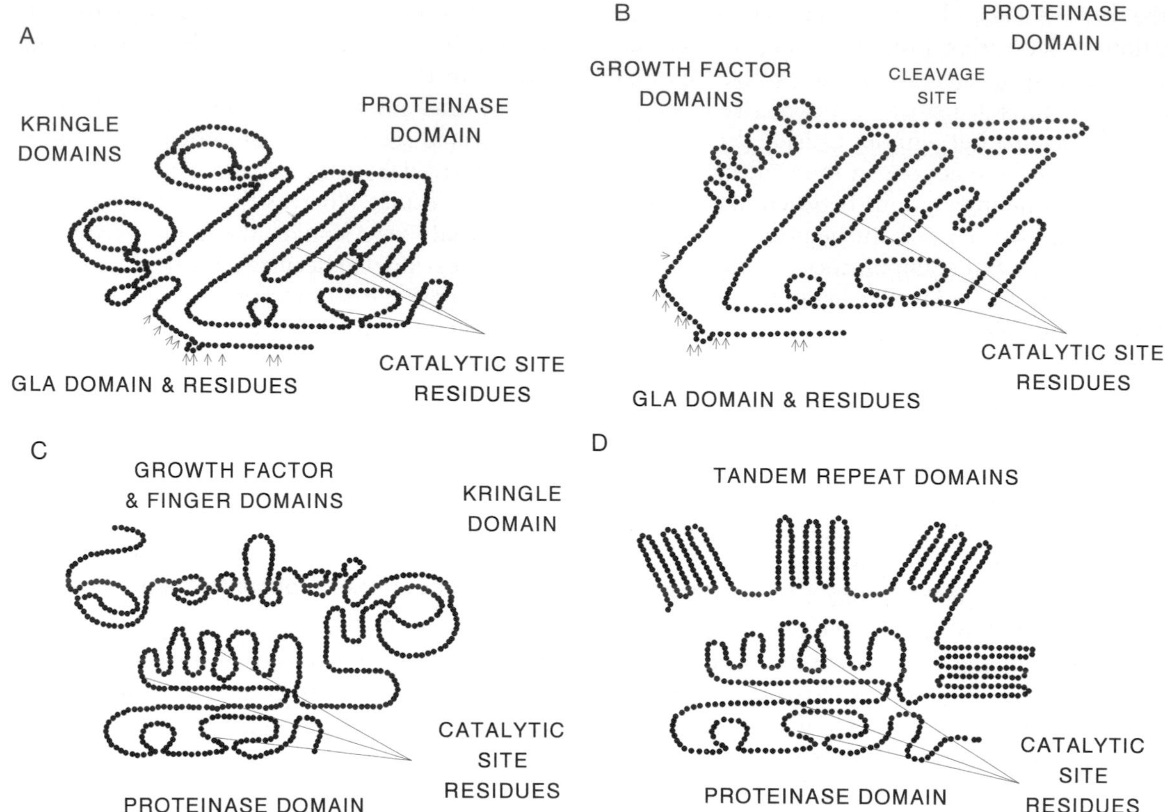

FIGURE 5 Schematic structures for four proenzyme molecules: (A) prothrombin, (B) factor X, (C) factor XII, and (D) prekallikrein. The various domain structures as represented in two dimensions are shown to illustrate the varieties of structure that make up the proenzyme molecules. Gla, γ-carboxyglutamic acid. [Modified from E. W. Davie (1987). *In* "Hemostasis and Thrombosis: Basic Principles and Clinical Practice" (R. W. Colman, J. Hirsh, V. J. Marder, and E. W. Salzman, eds.), 2nd Ed., pp. 242–267. Lippincott, Philadelphia.]

2. The Amino-Terminal Regions of the Clotting Proenzymes

In contrast to the great similarities in structure and enzymatic function of the proteinase domains, the amino-terminal regions possess a variety of structural domains with specific biological functions. The domain structure called a kringle (after a form of Danish pretzel) is illustrated in Fig. 5A. This structure is found in at least five of the clotting or fibrinolytic system proenzymes. Prothrombin and the tissue-type plasminogen activator each contain two kringles; factor XII and the urinary-type plasminogen activator each contain one kringle. Plasminogen contains five kringles.

Specific functions have been identified with the kringle domains in prothrombin and plasminogen. in prothrombin the first kringle has the Gla domain attached to it and interacts with it (see the following). The second kringle domain in prothrombin is necessary for the binding of prothrombin to activated factor V. Sites in two of the kringles of plasminogen and plasmin are responsible for plasminogen and plasmin binding to fibrin.

3. Gla Domains in the Vitamin K-Dependent Proteins of Clotting

The clotting factor precursors and active proteinases that depend on the action of vitamin K for full activity contain several residues of γ-carboxyglutamic acid, abbreviated Gla. The Gla residues are located in a 33- to 45-amino-acid domain at the amino-terminal end of the protein molecule. Gla domains are found in prothrombin (Fig. 5A), factor VII, factor IX, factor X (Fig. 5B), protein C, and protein S. Ten to 11 Gla residues are present in each of the human vitamin K-related proteins and are found in three pairs plus four to five separate residues. The presence of Gla residues confers unique calcium ion binding properties on the vitamin K-dependent proteins.

4. Vitamin K Action and the Formation and Function of Gla Residues

The vitamin K-dependent clotting proteins undergo postribosomal modification to add an extra carboxyl group to 10–11 glutamic acid residues within the Gla domain, changing them to γ-carboxyglutamic acid. The carboxylation depends directly on vitamin K. These proteins are synthesized and carboxylated in the liver, and then secreted into the blood. Calcium ions bind to the Gla residues, enabling the vitamin K-dependent proteins to bind to the negatively charged lipids that are exposed on cell membranes at the site of injury. Several drugs (oral anticoagulants) block the formation of Gla residues and are used therapeutically to reduce the occurrence of unwanted blood clotting, particularly after deep vein thrombosis and heart attacks. Anticoagulation by these drugs is achieved because the non-γ-carboxylated forms of the vitamin K-dependent proteins do not bind to the membranes of platelets and at injury sites. Without the ability to bind to the membranes, the activator complexes form very poorly and the large reaction rate enhancements that normally occur as a result of activator complex formation do not occur. The unbound proteinases are more readily inactivated by the proteinase inhibitors also. The extent of interference with normal hemostatic response by vitamin K anticoagulant drugs is sufficiently great that patients administered such drugs are monitored to ensure that the reduction in γ-carboxylation is not so great that the individual becomes susceptible to hemorrhage.

5. Growth Factor Domains

A domain first identified in polypeptide growth factors is found in several clotting and fibrinolytic system proteins. In the vitamin K-dependent proteins, growth factor domains are found in factor VII, factor IX, factor X (Fig. 5B), protein C, and factor XII (Fig. 5C). The tissue-type plasminogen activator and the urinary-type plasminogen activator each have a single growth factor domain. A second modified amino acid, β-hydroxyaspartic acid, is found in the growth factor domains and is involved in calcium ion binding; this binding is distinct from Ca binding by Gla residues. These domains all appear to be important in the protein–protein interactions that underlie the formation of the specific activation complexes in which the proteins are involved.

6. "Finger" Domains

Another structural domain, the "finger" domain, is found in factor XII, which contains two such domains separated by a growth factor domain (Fig. 5C). The second finger domain is followed by a second growth factor domain, which in turn is followed by a kringle domain. Factor XII is thus the most varied of the proteins with respect to structural domains within the amino-terminal region.

7. Tandem Repeat, "Apple" Domains

Two of the clotting factors, prekallikrein (Fig. 5D) and factor XI, possess unique domain structures that in two dimensions resemble tandem repeated amino acid sequences. These too appear to participate in the interactions between proenzyme and protein cofactor recognition in activation complexes.

G. Protein Cofactor Structural and Functional Domains

Amino acid sequences have been established from the cDNAs for the protein cofactors factor V and factor VIII. The two cofactor proteins show extensive sequence homology. The cofactor activity of factor Va and factor VIIIa in the activation complex is derived from portions of the precursor molecules located at the two ends of each cofactor precursor polypeptide chain. An ion links the two polypeptide chains of the activated protein cofactors. No function has been identified for the very large internal region of the protein cofactor molecules. The carboxy-terminal polypeptide of the activated cofactors is involved in binding to membranes and, in factor VIIIa, to von Willebrand factor. Specific peptide bond cleavages convert the precursor forms of the protein cofactors to their functionally competent forms. Other, but equally important, specific peptide bonds are cleaved by activated protein C in the anticoagulant pathway.

III. SUMMARY AND CONCLUSIONS

Although the hemostatic system comprises many components and reactions, the structural and functional similarities among the components make it possible, although not simple, to understand how this system works. Similar structures express similar functions. Interactions between and among the various components of the activator complexes produce similar consequences. Extremely large increases in activation reaction rates occur in the activation complexes, which normally form only at the site of an injury where blood loss is threatened. Irreversible inhibitors inactivate the proteinases after completion of hemostatic response and proteolytic destruction of the activated protein cofactors eliminates their contributions to the large rate enhancements that occur only when the activation complexes exist. Digestive processes remove the fibrin after it is no longer required, and thus blood flow to the tissues to which it transports nutrients and other substances is restored.

BIBLIOGRAPHY

Bloom, A. L., Forbes, C. D., Thomas, D. P., and Tuddenham, E. G. D. (eds.) (1994). "Haemostasis and Thrombosis," 3rd Ed. Churchill Livingstone, New York.

Jackson, C. M., and Nemerson, Y. (1980). Blood coagulation. *Annu. Rev. Biochem.* **49**, 765.

Miletich, J. P., Prescott, S. M., White, R., Majerus, P. W., and Bovill, E. G. (1993). Inherited predisposition to thrombosis. *Cell* **72**, 477.

Blood Coagulation, Tissue Factor

JAMES H. MORRISSEY
Oklahoma Medical Research Foundation

GLOSSARY

Cofactor Substance required by an enzyme for activity; for the coagulation proteases, this can include metal ions (Ca^{2+}), phospholipids, or even other proteins; tissue factor is the protein cofactor for coagulation factor VIIa

Plasma Portion of the blood remaining when the cells are removed

Protease Enzyme that catalyzes the cleavage of peptide bonds in proteins (proteolysis); in general, proteases of the coagulation cascade are highly specific in which proteins they will attack, cleaving only one or two specific peptide bonds of the proteins that they recognize

Thromboplastin Depending on usage, a synonym of tissue factor, or can refer to relatively crude tissue extracts that possess tissue factor activity; the latter is in reference to commercial preparations employed in a clinical test of the clotting time of plasma, the Prothrombin Time test

Tissue factor Also known as tissue thromboplastin and, infrequently, as coagulation factor III

Vascular endothelial cells Cells that line the blood vessels

Zymogen Inert precursor of an enzyme; zymogens of the coagulation protease cascade are inert precursor forms of proteases; they are converted into active proteases by highly specific cleavage of one or two peptide bonds

TISSUE FACTOR IS THE CELL-SURFACE PROTEIN that triggers the coagulation protease cascade, leading to blood clotting, an essential component of hemostasis (the process whereby bleeding stops after tissue injury). Tissue factor is also thought to be the protein responsible for triggering blood clotting in a variety of life-threatening diseases, such as heart attacks and certain types of stroke. Under normal circumstances, tissue factor is present only on cells located outside the vasculature, or bloodstream. Physical damage to blood vessels severe enough to result in leakage of blood permits plasma proteins to come into contact with cells that have tissue factor on their surfaces. A key triggering event of blood clotting occurs when one of these soluble plasma proteins, coagulation factor VII, binds to tissue factor. The complex of tissue factor and factor VIIa is a protease that converts coagulation factors IX and X to their active forms by highly specific, limited proteolysis. This conversion initiates a cascade of subsequent proteolytic activation events that ultimately results in a cross-linked fibrin gel, or clot. In some forms of inflammatory and cellular immune responses, tissue factor is induced on cells located inside the vasculature (specifically on monocytes and possibly on endothelial cells). Induced expression of tissue factor during inflammation is thought to be beneficial in rejecting tumors or foreign tissues, in limiting the spread of infection, and perhaps in wound healing. In other settings, such as septic shock, intravascular expression of tissue factor may contribute to the pathogenesis of life-threatening disease states.

I. THE TRIGGERING OF BLOOD COAGULATION

A. The Coagulation Protease Cascade

Tissue factor is the initiator of the extrinsic pathway of blood coagulation, so named because it is triggered by a substance that is absent from the plasma and

ENCYCLOPEDIA OF HUMAN BIOLOGY, Second Edition, VOLUME 2. Copyright © 1997 by Academic Press. All rights of reproduction in any form reserved.

blood cells (including platelets). The existence of tissue factor was initially inferred from old observations that tissue extracts, particularly from brain or lung, could stimulate blood coagulation in the test tube. We now know that tissue factor is a protein embedded in the surface membrane of a variety of different cell types located outside the vasculature. In this respect, tissue factor is very different from all the other procoagulant, or clot-promoting, proteins of the coagulation protease cascade, which are components of the blood.

Another mechanism of triggering blood coagulation, known as the intrinsic pathway (so named because it appears to be an intrinsic property of the plasma to form clots), is completely independent of tissue factor. Both intrinsic and tissue factor pathways ultimately activate factors IX and X, and all subsequent events in the protease cascade are in common. The intrinsic pathway is actually initiated by the surface of the vessel in which the blood is collected (glass is a good initiator of this pathway). Consequently, the intrinsic pathway of blood coagulation is often referred to as the contact pathway. Although intensively studied, the contact pathway may not play a direct role in normal hemostasis. This idea is supported, in part, by the existence of individuals who genetically lack components that initiate the intrinsic pathway and yet do not have any detectable bleeding tendency. The single exception is individuals who, with a deficiency in factor XI, may have a mild bleeding disorder. Thus, factor XI is implicated as probably undergoing activation in hemostasis by a mechanism separate from the contact pathway of blood coagulation. The contact pathway will not be considered further in this article.

Initiation of the tissue factor pathway of blood coagulation begins when tissue factor binds factor VII, a protease zymogen composed of a single polypeptide chain that circulates in the plasma. A schematic diagram of the initiation of blood clotting by tissue factor is given in Fig. 1. Factor VII bound to tissue factor becomes much more susceptible to proteolytic conversion to the active form, factor VIIa. Termed a serine protease because serine forms an essential part of the active site, factor VIIa belongs to a group known as the vitamin K-dependent serine proteases, meaning that their synthesis by the liver requires vitamin K. This vitamin is required for the addition of extra carboxyl groups to several glutamic acid residues located near the N terminus of factor VII.

The tissue factor–factor VIIa complex activates factor X by specific peptide bond cleavage to form the active protease, factor Xa. (Factor Xa is another vitamin K-dependent serine protease related to factor VIIa.) Although factor Xa has small amounts of enzyme activity on its own, its activity is greatly enhanced when it binds to its cofactor (factor Va) on cell surfaces or phospholipid vesicles. The association of factors Xa and Va on phospholipids is known as the prothrombinase complex because it activates prothrombin to thrombin. (Prothrombin is yet another vitamin K-dependent serine protease zymogen related to factors VII and X.) Thrombin, the activated form of prothrombin, is a soluble enzyme that cleaves specific peptide bonds on fibrinogen to form fibrin. Once fibrin is generated, it polymerizes to form a gel, or clot. The sequential activation of coagulation proteases has been likened to a cascade, and because each step generates an enzyme or enzyme complex capable of many rounds of catalysis, the initial triggering stimulus is tremendously amplified. In this manner, relatively tiny amounts of tissue factor can initiate the coagulation of large amounts of plasma.

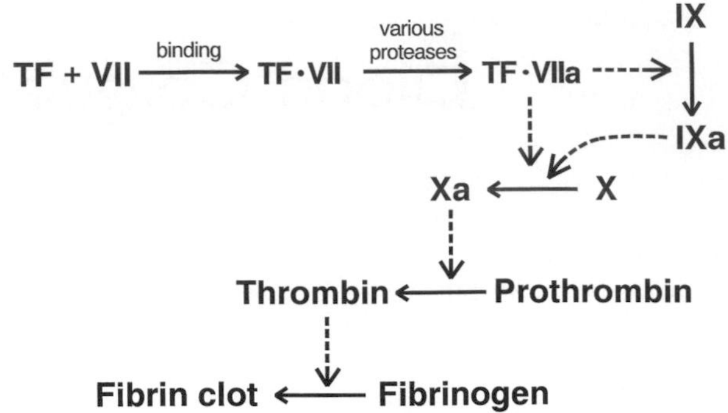

FIGURE 1 Schematic diagram of the initiation of clotting by tissue factor. Tissue factor (TF), which is embedded in cell membranes at all times, binds plasma factor VII (VII) to form the cell-surface TF · VII complex. In turn, bound factor VII is activated to factor VIIa by limited proteolysis, a process that can be catalyzed by serine proteases such as factors VIIa, Xa, IXa, and thrombin. Once the TF · VIIa complex is formed, it activates two serine protease zymogens, factors IX and X, by limited proteolysis. Factor IXa, in complex with its protein cofactor, factor VIIIa (not shown for clarity), also activates factor X to Xa by limited proteolysis, providing an additional amplification loop under some circumstances. Factor Xa is the final common enzyme in the clotting system. In complex with its protein cofactor, factor Va (not shown), factor Xa activates prothrombin to thrombin by limited proteolysis. Thrombin converts fibrinogen to fibrin, which immediately polymerizes to form a gel, or clot. Not shown is the cross-linking of fibrin by factor XIIIa, to form a highly stable, cross-linked clot.

Factor X is the preferred substrate for the tissue factor–factor VIIa complex *in vitro;* however, this complex will also activate factor IX, another vitamin

K-dependent serine protease zymogen of the plasma. When complexed with its protein cofactor (factor VIIIa) on cell surfaces or phospholipid vesicles, factor IXa can activate factor X to Xa. This is an alternative way in which factor Xa can be generated by the tissue factor–factor VIIa complex, thought to constitute an important additional amplification step when the amounts of tissue factor are low.

B. The Cofactor Function of Tissue Factor

Tissue factor is the essential cofactor for factor VIIa, which has almost no enzyme activity in its absence. Tissue factor binds factor VIIa in the presence of calcium ions and causes it to undergo a change in shape, allowing it to function as an active enzyme. Tissue factor is fully active only when it is incorporated into phospholipid membranes. This is in part because tissue factor is a protein spanning the outer membrane of the cell, which is made up of the phospholipid bilayer. Furthermore, factor VII, like other vitamin K-dependent serine proteases, has an appreciable affinity for phospholipids, which may be important in stabilizing the binding of factor VII or VIIa to tissue factor.

As with all other coagulation proteases, the vast majority of factor VII circulates in plasma in the single-chain (zymogen) form. An open question for some years has been how factor VII is converted to factor VIIa upon binding to tissue factor. This has been a puzzle because tissue factor is not an enzyme and factor VII, as a zymogen, should be enzymatically inert. Furthermore, serine proteases are inactivated so rapidly by plasma protease inhibitors that the known activators of factor VII should be effectively unable to circulate in the plasma. However, factor VIIa is unusually resistant to plasma protease inhibitors and, when injected, will circulate for hours in the blood. This led to the proposal that trace levels of circulating factor VIIa may be present in the plasma at all times, and that these trace levels of preexisting factor VIIa are responsible for priming the clotting system. In this view, the initially small amounts of tissue factor–factor VIIa complexes formed owing to preexisting factor VIIa would give rise to small amounts of factor Xa, which could then back-activate bound factor VII to VIIa. Another way for preexisting factor VIIa to prime the clotting system is for the tissue factor–factor VIIa complex to directly activate factor VII through a process termed autoactivation. In either case, once the rest of the bound factor VII is converted to factor VIIa, the tissue factor–factor VIIa complex triggers clotting through activation of factors IX and X. As described in the following section, a new test has been developed for measuring trace levels of plasma factor VIIa. It has shown that about 1% of the total factor VII in normal individuals is in the form of factor VIIa, confirming earlier predictions.

C. Clinical Use of Tissue Factor (Thromboplastin)

Laboratory tests on the ability of plasma to form clots are an important means by which the physician can infer if certain specific problems in hemostasis are present. One test, known as the Prothrombin Time (PT) test, consists of mixing citrated plasma with calcium chloride and tissue factor in a test tube. In actuality, the "tissue factor" is usually a relatively crude saline extract of brain, lung, and/or placental tissue, often of rabbit origin. Such preparations are known as thromboplastins and are used as a convenient source of active tissue factor to initiate the extrinsic coagulation protease cascade. Cloning of the tissue factor gene has enabled production of recombinant tissue factor, which can be purified and readily reconstituted into phospholipid vesicles for full clotting activity. Thromboplastin reagents based on recombinant human tissue factor are now being marketed for use in clotting assays like the PT test. These synthetic thromboplastin reagents have the advantage over crude tissue extracts in that their composition, and therefore their sensitivity, can be carefully controlled.

Citrate is an anticoagulant by virtue of its ability to trap the calcium ions required for most of the steps of the coagulation protease cascade. Adding excess calcium ions overcomes the anticoagulant activity and allows the cascade to initiate and propagate. If a reproducible amount of tissue factor activity is added to each sample, then the time between adding calcium ions and clot formation is inversely related (in a rather complex way) to the content of coagulation factors participating in the extrinsic pathway. This particular test is useful for discovering genetic or acquired deficiencies in factors VII, X, or prothrombin, which could prolong the clotting time.

The PT test is used to measure the effect of anticoagulants administered to individuals diagnosed as having an increased risk of developing recurrent thromboses (clots within blood vessels). Warfarin and its relatives are often used as oral anticoagulants in such cases because they are antagonists of vitamin K. Their addition results in the synthesis, by the liver, of partially active vitamin K-dependent serine proteases that have reduced carboxylation of their glutamic acid residues. The PT test is sensitive to the plasma levels of three of these vitamin K-dependent proteases and, therefore, reflects the level of anticoagulation achieved by warfarin administration.

A mutant form of recombinant tissue factor has been produced in which the membrane-anchoring domain has been deleted. Unlike wild-type tissue factor, the resulting protein is highly water soluble. This soluble, mutant tissue factor has selectively lost one of the two functions of tissue factor: it still acts as the cofactor for factor VIIa enzymatic activity, but it no longer supports the conversion of factor VII to VIIa. Therefore, when soluble tissue factor is used in place of the wild-type protein in clotting tests, the clotting time of the plasma depends on its content of preexisting factor VIIa, while zymogen factor VII remains inert. This unique property of soluble tissue factor has been exploited to create a clotting test to measure trace levels of preexisting factor VIIa in plasma, without interference from the vast excess of zymogen factor VII. From the results of epidemiologic studies on risk factors for heart disease, some investigators have proposed that elevated factor VIIa may be a risk factor for heart disease and other potentially fatal thrombotic disorders. The soluble tissue factor-based test for plasma factor VIIa is being used in clinical studies to examine this idea.

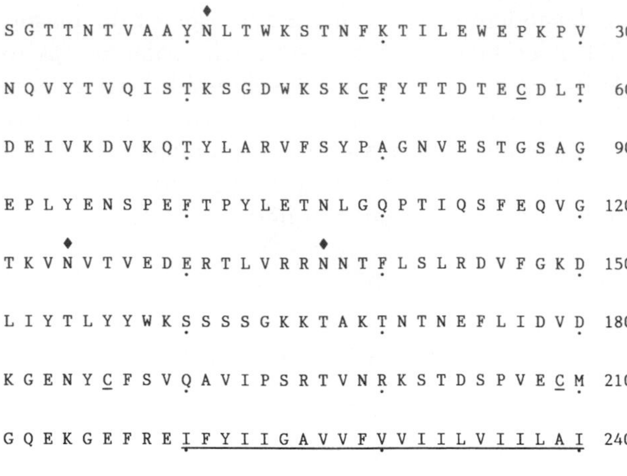

FIGURE 2 Amino acid sequence of the human tissue factor protein. Numbering is from the mature N terminus, and amino acids are abbreviated with the standard one-letter code. Cysteine residues (C) and the transmembrane domain are underlined. Asparagine residues (N) predicted to be linked to carbohydrate are indicated with a diamond. The cysteines at positions 49 and 57 are disulfide-bonded, as are the cysteines at positions 186 and 209. The cysteine at position 245 is attached to a fatty acyl group by a thioester linkage. Amino acids 1–219 constitute the predicted extracellular domain, whereas amino acids 243–263 constitute the predicted cytoplasmic domain. [From Morrissey *et al.* (1987). *Cell* **50**, 129–135. Reprinted with permission of the copyright holder, Cell Press.]

II. TISSUE FACTOR PROTEIN STRUCTURE AND FUNCTION

A. The Primary Structure of Tissue Factor

The sequence of the complete tissue factor protein has been deduced from the sequence of its messenger RNA through isolation of cDNA clones. It is a single polypeptide chain of 263 amino acids (Fig. 2). The sequence information suggests that most of the protein (219 amino acids) protrudes outside the cell; a hydrophobic 23-amino acid sequence spans the cell membrane, and the remainder (21 amino acids) forms a short "tail" that sticks into the cytoplasm. Tissue factor also has a fatty acid connected to the cysteine residue located in the cytoplasmic tail, which may serve as an additional membrane anchor, although its purpose is really not known. The four cysteine residues located in the extracellular portion of the molecule form two disulfide-bonded loops. At least one of these loops is required for the function of tissue factor, because the molecule is irreversibly inactivated by reducing agents. Tissue factor is also predicted to have three carbohydrate groups attached to asparagine residues in the extracellular domain.

B. The Three-Dimensional Structure of Tissue Factor

The three-dimensional structure of the soluble, extracellular domain of tissue factor has been determined to a resolution of 2.2 Å by X-ray crystallography (Fig. 3). This is the part of tissue factor located outside the cell; it is also the part that binds factors VII and VIIa. Structurally, tissue factor is a member of the cytokine/growth factor receptor family of proteins, which can be divided into two classes based on structural similarities that include the pattern of disulfide bonds. The members of Class 1 include the receptors for growth hormone, hematopoietin, and several interleukins. Tissue factor is a member of Class 2, which also includes the interferon and interleukin-10 receptors.

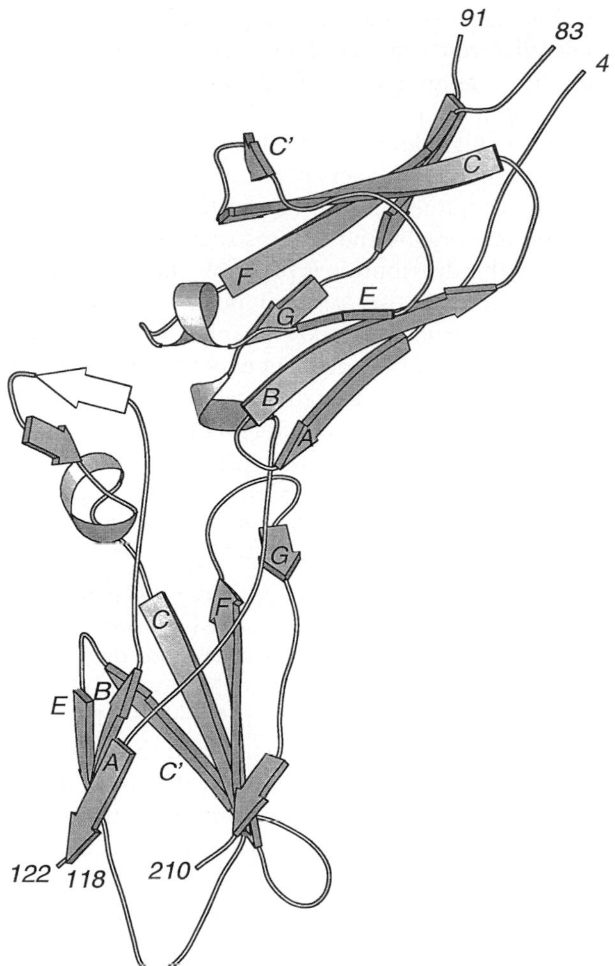

FIGURE 3 Ribbon diagram of the three-dimensional structure of the extracellular domain of tissue factor. In this form of recombinant tissue factor, the membrane-anchoring and cytoplasmic domains have been deleted. The extracellular portion of tissue factor is composed of two fibronectin type III domains, which are each in turn composed of seven beta strands (indicated by the flat arrows labeled A, B, C, C′, E, F, and G). The two domains are joined together by a relatively rigid interdomain region with a hydrophobic core. Near the interdomain interface are two short regions of alpha helix. The N terminus of the protein is near the top of the diagram, whereas the C terminus is located near the bottom. The C terminus of this part of tissue factor would normally be joined to the membrane-anchoring domain in wild-type tissue factor. [From Muller *et al.* (1994). *Biochemistry* 33, 10864–10870. Reprinted with permission from the copyright holder, The American Chemical Society.]

The extracellular portion of tissue factor, like other members of the cytokine/growth factor receptor family, is made up of two domains that are structurally homologous to the fibronectin type III portion of the immunoglobulin superfamily. Each fibronectin type III domain contains seven beta strands. In the case of tissue factor, there are also short stretches of alpha helix, and the two domains are joined together at a relatively rigid interface. Alanine-scanning mutagenesis has been used to map the surface on tissue factor responsible for binding to factor VIIa. These studies have implicated a number of amino acid side chains located on beta strands of both modules, as well as the interface region, as making up the putative factor VIIa binding site. The mechanism by which tissue factor alters the enzymatic activity of factor VIIa is still unknown.

C. Tissue Factor as a Cell-Surface Receptor

Most cell-surface receptors are thought either to transduce signals into the cell (i.e., from hormones or cytokines) or to bring nutrients or other substances into the cell. The main function of tissue factor, however, is to alter the outside environment of the cell. In hemostasis, it triggers clot formation and, as described later, in some settings it may trigger the coagulation protease cascade for other purposes. [*See* Cell Receptors.] Recent evidence suggests that binding of factor VIIa to tissue factor causes transient changes in the concentration of intracellular calcium ions. Whether or not this represents classical signal transduction as seen with other cellular receptors remains to be determined.

The calculated molecular weight for the polypeptide chain of tissue factor is approximately 30,000 daltons. This makes tissue factor a remarkably small cell-surface receptor protein as well as a remarkably small protein cofactor of the coagulation protease cascade. Factors V and VIII, for example, are over eight times the size of tissue factor. Another difference between tissue factor and factors V and VIII is that proteolysis is not required to activate tissue factor.

D. Molecular Biology of Tissue Factor

The messenger RNA (mRNA) encoding tissue factor in humans is approximately 2300 nucleotides long, including the poly(A) tail. Less than half of the message actually encodes the tissue factor protein. Most of the balance consists of a relatively long 3′ noncoding sequence, including a region very rich in adenine and uridine that probably causes rapid degradation of the tissue factor message.

The gene for tissue factor is located on the short arm

of chromosome 1, in segment 1p21–22. It contains approximately 12,400 base pairs and is split into six exons interrupted by five introns. Each of the six exons encodes a portion of the tissue factor protein: exon 1 encodes the signal peptide responsible for targeting the protein to the endoplasmic reticulum; exons 2–5 encode the extracellular domain; and exon 6 encodes the transmembrane domain, the cytoplasmic domain, and the extensive 3′ noncoding region of the tissue factor mRNA.

The first exon and the probable control region (promoter) of the tissue factor gene are located within a CpG island. CpG islands, also known as HTF islands, are found associated with the 5′ ends of many genes. They have a very high content of guanine and cytosine, do not have suppression of CpG dinucleotides, and generally lack DNA methylation. These features are present in the CpG island of the tissue factor gene. Often, but not always, CpG islands are located at the 5′ ends of housekeeping genes (i.e., genes expressed by every cell); however, tissue factor is decidedly a tissue-specific gene.

The promoter region of the tissue factor gene is being studied in order to understand how transcription of the tissue factor gene is controlled. Binding sites for transcription factors AP-1 and members of the κB family have been identified 5′ to the start site of transcription. They have been implicated as being important in regulating transcription of the tissue factor gene in monocytes and endothelial cells in response to inducers such as bacterial lipopolysaccharide.

Minor sequence polymorphisms (i.e., difference between individuals) have been described for the tissue factor gene, including a polymorphism consisting of a single nucleotide difference in the coding region. The effect is to replace an amino acid with one of similar properties, which probably does not affect function. No known genetic deficiencies of tissue factor have been reported, probably because a total lack of tissue factor would be fatal *in utero*. Indeed, transgenic mice have been produced in which the tissue factor gene has been inactivated. Mice homozygous for the inactivated tissue factor gene die before birth.

III. REGULATION OF TISSUE FACTOR

A. Expression of Tissue Factor by Cells

Although tissue factor, as its name implies, is found in all tissues of the body, its expression on a cellular level is highly specific and regulated. Tissue factor is generally present in very low levels in connective tissues but is abundant in the adventitial cell layer surrounding most blood vessels. It is also expressed at relatively high levels in the renal glomerulus, the outer, precornified layers of the epidermis (and other squamous epithelia), and myoepithelial cells, which encapsulate organs and organ structures. The significance of this distribution is probably that tissue factor constitutes a hemostatic envelope around blood vessels, organ structures, and the entire body. Furthermore, tissue factor is safely out of contact with plasma proteins as long as blood vessels are intact but is available to initiate blood coagulation following injury. The lack of tissue factor expression in endothelial cells and most vascular smooth muscle cells is likely to be important in decreasing the risk of thrombosis (blood clotting inside a blood vessel).

Tissue factor is expressed on the outer surface of cells. Nevertheless, when cells are damaged or lysed, they are more efficient at initiating blood coagulation. This effect is probably beneficial because the need for tissue factor activity in hemostasis is probably greater in heavily damaged tissues than when the tissues are relatively intact. Part of the reason that tissue factor is not fully active in intact cells may be that acidic phospholipids, which are known to enhance the catalytic activity of the tissue factor–factor VII complex, are selectively present on the inner leaflet of the plasma membrane. Lysis or damage of cells can destroy this polarity and allow acidic phospholipids to become located on both faces of the plasma membrane. Cells that have been examined so far do not have an intracellular store of active tissue factor, so liberation of intracellular tissue factor is not likely to be the basis of the increased procoagulant activity of lysed and damaged cells.

Although tissue factor is normally not expressed by any blood cell or by the cells lining the blood vessels, its expression can be induced in circulating monocytes by a variety of inflammatory agents. In addition, endothelial cells cultured *in vitro* have been shown to express tissue factor after stimulation with various inflammatory agents and cytokines. Intravascular expression of tissue factor may lead to the formation of a thrombus and consequent blockage of the vessel in which it is expressed. Such processes may be beneficial in graft and tumor rejection and for walling off infection to prevent dissemination of pathogens. Activation of the protease cascade by vascular cells may be important in cell–cell interactions and remodeling of the extracellular matrix. Thrombin, one of the end products of the coagulation protease cascade, has many biological effects in addition to its ability to cleave fibrinogen. Thrombin is a potent mitogen, chemotactic agent, and activator of cell types such as platelets, monocytes, neutrophils, fibroblasts, and endothelial cells.

Inducers of tissue factor expression by monocytes include certain viruses, antigen–antibody complexes, activated complement, and bacterial endotoxin. The monocyte response to bacterial endotoxin in particular is substantially augmented in the presence of T lymphocytes, and the response to some inducers, such as antigens in cell-mediated immunity, shows an absolute requirement for lymphocyte collaboration. The basis for this collaboration is not completely understood. Expression of tissue factor is not merely a general consequence of monocyte stimulation, because it is not caused by other known inducers of monocyte gene expression such as interleukin-3 or interferon-gamma. [*See* Lymphocytes.]

Vascular endothelial cells normally have an anticoagulant surface owing to the lack of tissue factor as well as the active expression of natural anticoagulant molecules such as thrombomodulin, heparin-like polysaccharides, and inhibitors of platelet function. However, studies of endothelial cells cultured *in vitro* have shown that bacterial endotoxin or cytokines such as interleukin-1 or tumor necrosis factor-alpha cause the endothelial cell surface to become procoagulant. One aspect of this change is the transient expression of tissue factor. Because interleukin-1 and tumor necrosis factor-alpha are products of activated monocytes, their effect may represent an additional means of amplifying the procoagulant aspect of cellular immune responses. [*See* Cytokines and the Immune Response.]

B. Inhibitor of Tissue Factor Function

In vivo, coagulation reactions cannot be allowed to propagate indefinitely. Otherwise, clot formation would be so extensive that blood vessels would be in danger of becoming occluded every time the clotting system was triggered. Accordingly, every activated protease of the clotting cascade (including the tissue factor–factor VIIa complex) has a corresponding inhibitor to control it. A specific inhibitor of the tissue factor pathway of coagulation is tissue factor pathway inhibitor (TFPI), a plasma protein that is a member of the Kunitz-type family of protease inhibitors. TFPI has a rather complex mechanism of action against tissue factor. By itself, TFPI is a poor inhibitor of either free factor VIIa or the tissue factor–factor VIIa complex. TFPI is, however, a rapid inhibitor of factor Xa. Once the complex of TFPI and factor Xa forms, it becomes an efficient inhibitor of the tissue factor–factor VIIa complex. This results in an enzymatically inactivate complex composed of four proteins.

The other major plasma inhibitor of tissue factor–factor VIIa is antithrombin (formerly known as antithrombin III). This protein inhibits many of the activated proteases of the clotting system. Antithrombin is essentially inactive against free factor VIIa, but it does inactivate the tissue factor–factor VIIa complex and this ability is greatly stimulated by heparin. It is currently unknown which of these two proteins is the more important physiologic inhibitor of the tissue factor–factor VIIa complex *in vivo*.

IV. TISSUE FACTOR IN DISEASE

A. Hemophilia

Individuals suffering from the hereditary bleeding disorder hemophilia A or B are deficient in coagulation factor VIII or IX, respectively. Both forms of hemophilia exhibit the same types of clinical bleeding, particularly a tendency for bleeding into muscles and joints. Interestingly, joint tissues and skeletal muscle are locations where tissue factor is expressed in very low amounts. The tissue factor–factor VII complex appears to depend preferentially on activation of factor IX when tissue factor is limiting. This relationship may explain why deficiencies of factor IX (or its cofactor, factor VIII) result in bleeding tendencies of these anatomic locations. Skin is particularly abundant in tissue factor; these high levels permit the tissue factor–factor VII complex to activate factor X directly, so that the clinical test known as bleeding time (in which a small superficial wound is made in the skin) does not depend on factors VIII or IX.

B. Atherosclerosis and Thrombosis

Myocardial infarction (heart attack) and some types of stroke are due to the formation of a thrombus, or clot, that obstructs blood flow. A major predisposing factor in thrombosis is atherosclerosis, and it is now thought that fissure or rupture of atherosclerotic plaques, present under the endothelium in arteries, is often the direct cause of such thrombi. Tissue factor expressed by cells within atherosclerotic lesions is, thus, probably responsible for triggering the coagulation protease cascade. Not only would its activation cause the formation of a fibrin clot within the lumen of the vessel, but it would also activate platelets owing to the generation of thrombin, causing extension of the clot and local inflammation. [*See* Atherosclerosis.]

C. Disseminated Intravascular Coagulation

Disseminated intravascular coagulation (DIC) is a disease in which extensive activation of the coagulation protease cascade results in the consumption of the coagulation factors of both the procoagulant and anticoagulant pathways. This condition can arise from a variety of different causes. The consequences of DIC can be a bleeding disorder, thrombosis, or both. DIC is associated with many different diseases, some of which may be due to extensive intravascular expression of tissue factor. Examples are septic shock caused by gram-negative bacteria, cancer, certain viral diseases, obstetric complications, and brain injuries.

As the likely triggering agent of the clotting system in most, if not all, thrombotic disorders, tissue factor is responsible for initiating the underlying fatal events in heart attack, certain types of stroke, deep-vein thrombosis, and DIC. Therefore, the development of specific inhibitors of tissue factor or factor VIIa function may prove to be beneficial in treating or preventing such life-threatening thrombotic diseases.

BIBLIOGRAPHY

Bach, R. R. (1988). Initiation of coagulation by tissue factor. *CRC Crit. Rev. Biochem.* **23**, 339–368.

Carson, S. D., and Brozna, J. P. (1993). The role of tissue factor in the production of thrombin. *Blood Coag. Fibrinol.* **4**, 281–292.

Drake, T. A., Morrissey, J. H., and Edgington, T. S. (1989). Selective cellular expression of tissue factor in human tissues: Implications for disorders of hemostasis and thrombosis. *Am. J. Pathol.* **134**, 1087–1097.

Furie, B., and Furie, B. C. (1988). The molecular basis of blood coagulation. *Cell* **53**, 505–518.

Hoffman, R., Benz, E. J., Jr., Shattil, S. J., Furie, B., Cohen, H. J., and Silberstein, L. E. (1995). "Hematology: Basic Principles and Practice," 2nd Ed. Churchill Livingstone, New York.

Mackman, N. (1995). Regulation of the tissue factor gene. *FASEB J.* **9**, 883–889.

Mann, K., Jenny, R. J., and Krishnaswamy, S. (1988). Cofactor proteins in the assembly and expression of blood clotting enzyme complexes. *Annu. Rev. Biochem.* **57**, 915–956.

Martin, D. M. A., Boys, W. G., and Ruf, W. (1995). Tissue factor: Molecular recognition and cofactor function. *FASEB J.* **9**, 852–859.

Nemerson, Y., and Bach, R. (1982). Tissue factor revisited. *Prog. Hemost. Thromb.* **6**, 237–261.

Wilcox, J. N., Smith, K. M., Schwartz, S. M., and Gordon, D. (1989). Localization of tissue factor in the normal vessel wall and in the atherosclerotic plaque. *Proc. Natl. Acad. Sci. USA* **86**, 2839–2843.

Blood Platelets and Their Receptors

G. A. JAMIESON
American Red Cross

GLOSSARY

Adhesion Attachment of a monolayer of platelets to adhesive substrates such as collagen, laminin, or immobilized von Willebrand factor

Aggregation Commonly, the fibrinogen-dependent platelet–platelet interaction induced by soluble mediators such as thrombin or adenosine diphosphate; alternatively, the piling up of layers of platelets on the adherent or anchored monolayers

Anchorage Subsequent state involving the activation and spreading of adherent platelets resulting in a more tenacious interaction with the substrate

M_r Molecular mass based on electrophoretic mobility

PLATELETS ARE THE SMALLEST CELLS IN THE blood, having a diameter of ~3 μm. In the blood, platelets normally circulate in a unique discoid shape, being protected from activation by the endothelial cells lining the vasculature. However, if subendothelial components are exposed owing to a wound or other injury to the wall of the blood vessel, platelets are activated and undergo a remarkable series of morphological transformations. Within seconds, they adhere to the exposed surfaces, put out pseudopodia, undergo changes in the expression of surface receptors, recruit other platelets to form an aggregate on the adherent platelets, and secrete various molecules from specific internal compartments. These changes in platelets are closely interdigitated with the activation of the intrinsic and extrinsic clotting systems of the plasma. This leads to the conversion of fibrinogen to fibrin, whose fibers act like cables binding the activated platelets and other blood cells to form a thrombus, which serves to stop the bleeding and prepares the way for healing. Methods have been developed to store and transfuse platelets obtained from healthy donors, since patients can have severe bleeding problems as a result of low numbers of platelets in their blood because of acquired and hereditary diseases, or as the result of chemotherapy or radiation.

I. PLATELET FORMATION

Except for platelets, blood cells develop by the maturation of a precursor cell. Platelets are not true cells since they lack a nucleus and they arise by the fragmentation of their precursor, the megakaryocyte, in the bone marrow or during passage through the microcirculation of the lung. [*See* Hemopoietic System.] Because of this, platelets lack the capacity to divide or to synthesize proteins. However, they carry out metabolic processes such as glycolysis and respiration like most cells in the body. Platelets are the shortest-lived blood cell with a life span in the circulation of 8–10 days, requiring the formation of approximately 2×10^{11} new platelets per day in the normal adult. The main regulator of platelet formation is a cytokine termed thrombopoietin, whose identity had remained unknown despite heroic efforts by numerous investigators over several years. Recently, however, the cDNA for thrombopoietin has been cloned and shown to express a protein with a molecular weight of 35,000, injection of which into mice caused a fourfold increase in circulating platelet counts within 7 days.

ENCYCLOPEDIA OF HUMAN BIOLOGY, Second Edition, VOLUME 2. Copyright © 1997 by Academic Press. All rights of reproduction in any form reserved.

II. PLATELET ULTRASTRUCTURE

Nonactivated platelets have a major axis of ~3 μm and a minor axis of ~1.5 μm and a discoid shape that is maintained by several circumferential bands of microtubules lying just under the plasma membrane (Fig. 1). Invaginations of the surface membrane form an interconnecting, internal, three-dimensional meshwork called the open canalicular system. The outer surface of the plasma membrane, the glycocalyx, and the internal surfaces of the open canalicular system, which are extensions of it, are covered with a dense coat of glycoproteins and receptors that are important for platelet function. Other structures discernible in platelets are primitive mitochondria with poorly developed cristae, as well as dense granules, alpha granules, and lysosomes, which are important in platelet secretion, the dense tubular system, which is involved in Ca^{2+} sequestration, and glycogen particles, which constitute energy stores of the resting platelet. The

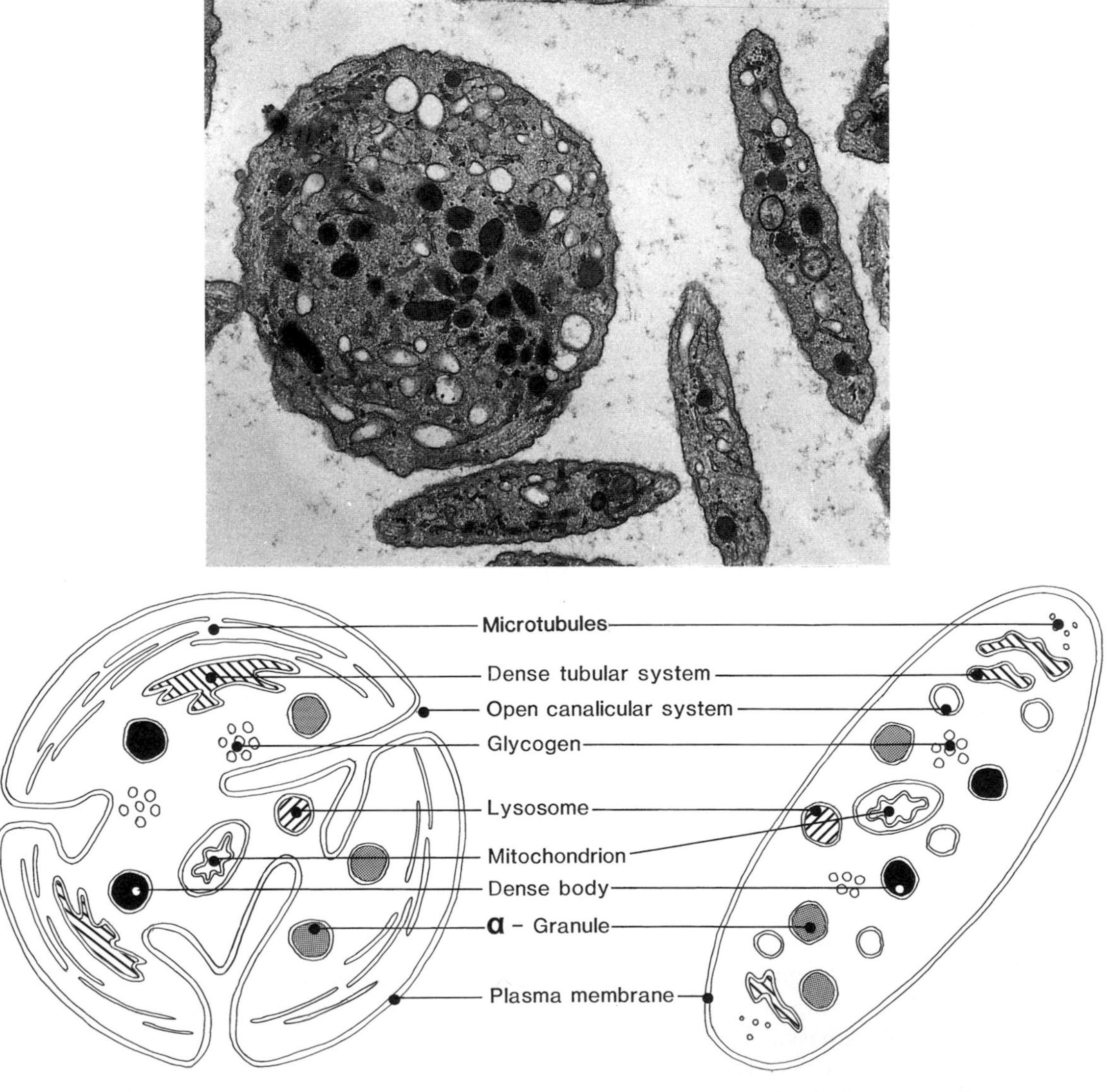

FIGURE 1 The electron photomicrograph (top) shows equatorial and longitudinal cross sections of resting platelets. (Courtesy Dr. James White, University of Minnesota.) The drawing (bottom) identifies ultrastructural features. [Reprinted with permission from N. N. Tandon, J. T. Harmon, and G. A. Jamieson (1988). *In* "Lipid Domains and the Relationship to Membrane Function" (R. C. Aloia, C. C. Curtain, and L. N. Gordon, eds.), pp. 83–100. Liss, New York.]

changes in platelet morphology that occur on platelet activation are accompanied by the centralization of these dispersed organelles within a ring of contracting microtubules and the secretion of their contents into the channels of the open canalicular system.

III. ADHESION AND AGGREGATION

Rheological forces in flowing blood drive red and white blood cells to the central portion of the lumen of the blood vessel, whereas the platelets, being smaller, become more concentrated in the plasma nearest to the vessel wall. As the blood circulates these platelets may make contact with the nonthrombogenic endothelial cell surface, but they become adherent only where they encounter damaged or denuded endothelium (Fig. 2).

The blood vessel wall consists of several layers of different structure and composition and contains many substances to which platelets can adhere once the integrity of the endothelial cell layer is compromised: these include collagen, laminin, fibronectin, and von Willebrand factor. Fibrillar collagen is generally considered to be the most thrombogenic of these components, but the relative contributions of these adhesive substrates appear to differ depending on the shear rate in different blood vessels; for example, at high shear rates (>800 sec^{-1}), which occur in the microvasculature, where blood vessel diameters are smaller than about 25 μm (arterioles, capillaries, and venules), adhesion to collagen is dependent on the presence of von Willebrand factor. However, von Willebrand factor is not required for adhesion at the lower shear rates found in arteries and veins, suggesting that other adhesive mechanisms are operational under these conditions. Studies on these phenomena have been facilitated by the use of *in vitro* perfusion devices in which blood is passed over a surface coated with the adhesive substrate of interest, or a segment of denuded blood vessel such as rabbit aorta. Attachment of platelets to the surface is then evaluated under a variety of shear conditions or in the presence of different inhibitors. [*See* Collagen, Structure and Function.]

On initial attachment to collagen in the absence of Mg^{2+}, platelets remain discoid or spherical in shape and this morphology is quite stable if Mg^{2+} ions continue to be excluded from the medium in which the platelets are suspended. However, in the presence of Mg^{2+}, which would be the case under physiological conditions, there is a very rapid (within seconds) change in platelet morphology with the development of pseudopodia, the formation of spiny spheres, and the spreading of the platelets across the adhesive surface. This is accompanied by signal transduction events within the platelet, including protein phosphorylation and dephosphorylation, the mobilization of Ca^{2+} from internal stores, and its uptake from the surrounding medium through membrane channels. The changes that occur in the membrane result in the formation of new receptors that then allow these adherent platelets to capture platelets from the blood to form aggregates or thrombi, involving the formation of fibrinogen bridges between adjacent platelets.

Platelet aggregation also occurs if platelet-activating agents, for example, thrombin, collagen, or adenosine diphosphate, at appropriate concentrations are added to platelets suspended in plasma or in a physiological medium with added fibrinogen in the case of the last two agonists. The occurrence of shape change and the formation of aggregates cause changes in the optical density of the suspension, which can be measured to evaluate platelet reactivity (Fig. 3). The metabolic basis of these platelet activation processes involves an initial rapid increase in both glycolytic and oxidative adenosine triphosphate (ATP) production, changes in cytoplasmic pH, elevation of cytoplasmic Ca^{2+} levels, and the stimulation of signal transduction pathways involving phospholipase A_2 and phospholipase C. A countervailing inhibitory effect on platelet activation is the stimulation of a Ca^{2+}/Mg^{2+}-ATP-dependent pump by cyclic adenosine monophosphate, leading to the reduction of cytoplasmic Ca^{2+} levels.

IV. PLATELET SECRETION

The formation of aggregates is accompanied by secretion from specialized platelet granules that occurs over 10–120 sec depending on the nature and concentration of the aggregating agent. The substances secreted in this "platelet release reaction" can augment thrombus formation, modulate the responses of other cell types in the thrombus or in surrounding tissues, and begin to prepare the thrombus for its own dissolution. [*See* Blood Coagulation, Hemostasis.] Two types of granule are recognized in the electron microscope; the dense granule is so called because of its electron density and it is revealed as a dark disk in electron microscopy (see Fig. 1). α-Granules are much less electron-dense and appear to be heterogeneous, so they have been subclassified into α-granules and lysosomes.

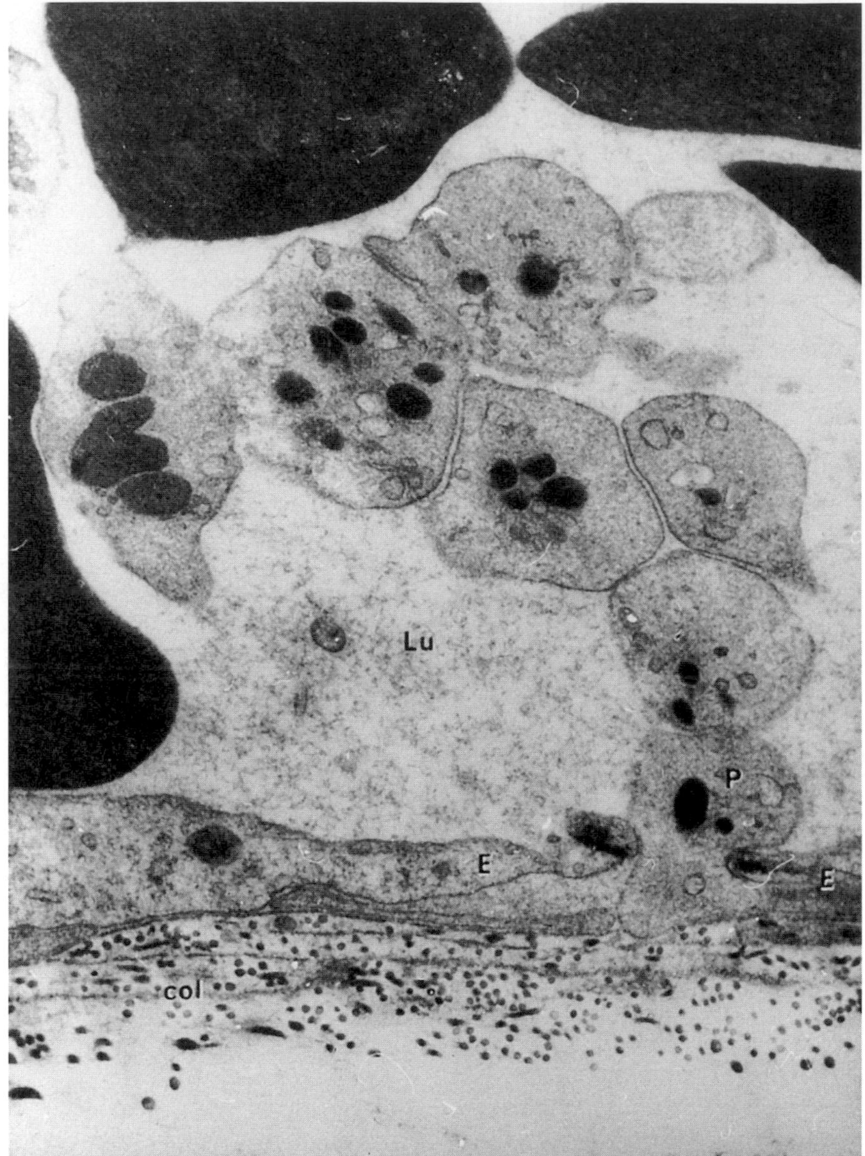

FIGURE 2 A platelet (P) in a small vein has insinuated itself into a gap between two endothelial cells (E) and is adhering to collagen fibers (col) in the vessel wall. This causes changes in the membrane of the adherent platelet, which allows the recruitment of other platelets to form a small aggregate in the lumen (Lu). This electron photomicrograph provides a unique visualization of a microthrombus. [From J. E. French, R. G. MacFarlane, and A. G. Saunders (1964). *Br. J. Exp. Pathol.* **45**, 467–474. Reprinted by permission of John Wiley & Sons, Inc.]

Each of these compartments contains distinct classes of chemical entities. Dense granules contain low-molecular-weight compounds such as adenine nucleotides (ADP,ATP,GDP,GTP), divalent cations (Ca^{2+}, Mg^{2+}), and serotonin. Secretion of these components, particularly ADP, by platelet agonists can lead to a second wave of aggregation resulting from the interaction of ADP with its own receptors on the platelet surface (Fig. 3). Interestingly, the secreted nucleotides are not part of the platelets' metabolic pool, suggesting that they may be packaged into the dense granules as they are being formed in the megakaryocyte. However, radiolabeled serotonin outside the platelet can be taken up into the dense granules, and its subsequent release induced by platelet agonists provides a useful measure of platelet activation.

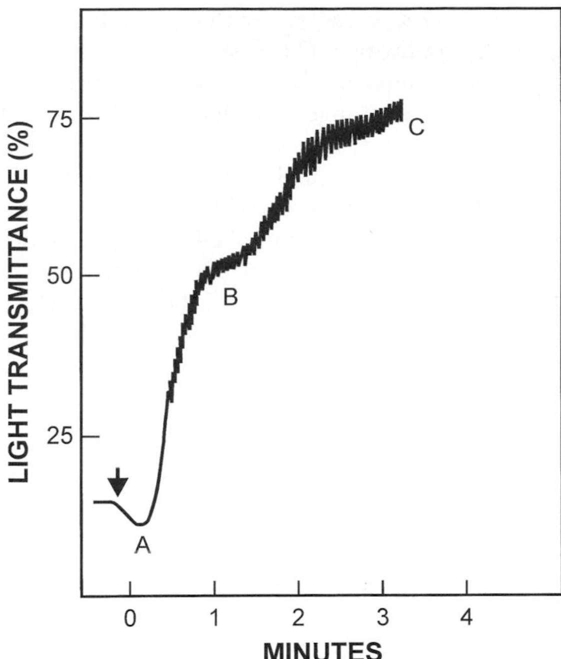

FIGURE 3 A tracing of aggregation of platelets in plasma induced by collagen (added at the arrow) showing the effects on light transmittance of shape change (A), the primary wave of aggregation (A–B), and the secondary wave of aggregation (B–C) owing to the effects of ADP secreted from the platelets. The oscillations of the pen are due to the optical effects of the increasingly large size of the platelet aggregates.

The α granules contain a wide range of proteins, some of which are specific to platelets (β-thromboglobulin), some of which also occur in plasma (albumin, thrombospondin, fibronectin, and the coagulation factors Va, VIII, and fibrinogen), and cationic proteins such as platelet-derived growth factor (PDGF), which play roles in cell permeability, chemotaxis, and mitogenesis. [*See* Blood Coagulation, Tissue Factor.]

The lysosomal secretion products (β-hexosaminidase, β-glucuronidase, β-galactosidase, and α-arabinosidase) are termed acid hydrolases because of the low pH required for their optimum activity (pH 3.5–5.5), and they can degrade the complex carbohydrate structures that play a role in wound healing and tissue remodeling.

V. RECEPTORS

Since application of a stimulus (e.g., thrombin, collagen, ADP) to platelets rapidly results in a response (e.g., aggregation or secretion), platelets have been studied as models of stimulus–response coupling, which is of importance in many cell types although it seldom occurs with the speed with which it occurs in platelets. Platelets can also be activated by cultured tumor cells *in vitro* and may also play a role in metastasis *in vivo*. In this case, *in vitro* activation appears to depend on the ability of the tumor cells to produce the platelet agonist ADP or to elaborate microvesicles expressing tissue factor, which then leads to the production of α-thrombin.

A large number of membrane proteins and glycoproteins have been identified as components of the platelet membrane by the use of polyacrylamide gel electrophoresis and isoelectric focusing combined with sensitive techniques for protein and glycoprotein detection. Several of these proteins have covalently linked fatty acids that help to further stabilize them in the membrane beyond the effects of the hydrophobic amino acid sequences of the membrane-spanning domains. Nomenclature can be confusing since several different systems have been used. The most common system uses terms such as GPIIb or GPIIIa to identify individual glycoproteins; a second system uses the nomenclature based on leukocyte differentiation clusters so that GPIIb/IIIa would be termed CD41/61; a third system, which is applicable only to the α and β subunits of integrins, would term it $\alpha_{IIb}\beta_3$. Caution is required, however, since the terms α and β can be used to describe the components in a single glycoprotein; for example, GPIIb contains α and β subunits, which together constitute the α component of the $\alpha_{IIb}\beta_3$ integrin.

In a number of cases, these glycoproteins have been shown to function as receptors mediating platelet activation processes, but the receptors for certain agonists have not yet been identified and the function of large numbers of the glycoproteins is not known. Where they have been cloned, these receptors have been found to be members of gene families widely dispersed in the animal kingdom.

Platelets express both human leukocyte antigens (HLA antigens) and major red cell antigens, particularly ABH and Le^a. There are also platelet-specific antigens associated with specific platelet membrane glycoproteins, and structural polymorphisms in these antigens are detectable in a small number of individuals. All of these antigens are of clinical significance since transfusion of incompatible platelets can lead to immune-mediated platelet destruction.

Each of the receptors for which reasonably detailed information is available will be discussed in the alphabetical order of the agonists with which they interact.

A. Adenosine

Adenosine receptors are coupled to adenylate cyclase, the enzyme that in cells produces cyclic adenosine monophosphate, a secondary messenger in signal transduction. These receptors are of two main types: the A1 receptors, which inhibit the cyclase, and the A2 receptors, which stimulate it. Both of these types have been cloned and shown to be of the seven-transmembrane domain type. The adenosine receptors in platelets are exclusively of the A2 subtype. This conclusion is based on the relative ability of a series of adenosine analogues to stimulate the cyclase. The most potent of these bind to isolated platelet membranes with high affinity [dissociation constant (K_d) of 12 nM].

B. Adenosine Diphosphate

Adenosine diphosphate has an important role as a platelet-aggregating agent. Exogenous ADP from red cells or other cellular sources can cause both platelet activation and the inhibition of stimulated adenylate cyclase. This activation, whether induced by ADP itself or by other agonists, then induces the secretion from platelets of ADP packaged in the dense granules. This secreted ADP can then promote further aggregation and recruitment of platelets to the thrombus.

The ADP receptor on platelets is termed a P_{2T} purinergic receptor but it does not closely resemble adenine nucleotide (P_2) receptors in other cells. In other tissues, including endothelial cells, both ADP and ATP are agonists, and there are no known reversible antagonists. In platelets, adenosine triphosphate is a powerful antagonist and is effective at the same concentration as ADP itself. Nucleotide receptors in other cells also generally show much lower stereoselectivity than does the ADP receptor of platelets, which has an absolute requirement for the natural form of the nucleotide. Essentially, any modification in the structure of ADP, whether in the purine, ribose, or phosphate moieties, results in the loss of platelet-aggregating activity. The only exception is ADP analogues substituted in the C2 position of the adenine ring, which are also able to induce full platelet activation. Many other nucleotides or structural analogues are partial or complete antagonists and block the interaction of ADP with its receptors. It has not been resolved whether platelet activation is mediated by a single type of receptor coupled to two pathways, one inducing activation and the other affecting adenylate cyclase, or by two different receptors, each separately affecting one of the pathways. The first, or single-receptor, hypothesis is supported by the constant ratio in the activities of a wide range of structurally diverse ADP analogues in their effects as agonists or antagonists of platelet activation and inhibition of stimulated adenylate cyclase. The two-receptor hypothesis is supported by the fact that certain ADP analogues induce aggregation but do not inhibit adenylyl cyclase.

Attempts to affinity-label the ADP receptor(s) of platelets have been largely unsuccessful. 5′-Fluorosulfonylbenzoyl adenosine, which has been used to label nucleotide-binding sites in a number of purified enzyme systems, labels a membrane component (M_r 100,000) that has been termed aggregin, but whether this is the P_{2T} receptor and what the mechanism is by which this protein may mediate ADP-induced platelet activation have not been determined. Photoaffinity labeling of the α-chain of GPIIb (M_r 120,000) by ADP analogues suggests that it may be part of a platelet–ADP–receptor complex but is not itself the receptor since platelets from patients with Glanzmann's thrombasthenia, which lack GPIIb, are activated by ADP. Recent studies with a new photoaffinity analogue, AzPET-ADP, have identified a membrane molecule of M_r 43,000 that is the size that would be expected of a seven-transmembrane domain receptor, which appears to be the gene family to which the P_{2T} receptor belongs. The recent identification of two unrelated patients exhibiting defects in ADP-induced platelet activation may facilitate the characterization of this important receptor.

C. Collagen

Direct binding of platelets to collagen without the interpolation of other adhesive proteins appears to occur in certain regions of the vasculature. Numerous candidates have been proposed as the platelet collagen receptor, but few have been adequately characterized. Three distinct membrane glycoproteins are currently being investigated for their roles in platelet–collagen interaction. Available data are consistent with the interpretation that GPIV (CD36) and/or GPVI mediate the initial interaction of platelets with collagen and convert the integrin GPIa/IIa ($\alpha_2\beta_1$) into a binding-competent form, but this hypothesis remains to be proven.

CD36 (M_r 88,000) has been isolated, cloned, and shown to be a member of a new gene family that may be involved in various adhesive processes. CD36 itself has been shown to bind to collagen, and antibodies to one of its epitopes reduce the rate of platelet adhesion at physiological shear rates. On the other hand, antibodies to another epitope of CD36 block the attachment of red cells infected with *Plasmodium falciparum* malaria, as well as the binding of thrombospondin and oxidized low-density lipoprotein, suggesting that CD36 may play a role in atherosclerosis. CD36 is not detectable on the platelets of individuals of the Nak^a-negative phenotype, which constitutes about 3% of East Asian populations but only about 0.2% of the blood donor population in the United States. However, it may occur with higher frequency in African-Americans. Individuals lacking CD36 do not suffer from bleeding problems, although their platelets adhere to collagen slightly more slowly than do control platelets.

The absence of GPVI (M_r 62,000) in some patients, or the presence of antibodies to it, has also been shown to be associated with defective platelet adhesion to collagen and bleeding problems. Platelets from patients lacking GPIa/IIa ($\alpha_2\beta_1$) show reduced adhesion to collagen and those platelets that do adhere do not spread or form thrombi. Monoclonal antibodies to GPIa/IIa block platelet adhesion to collagen and lipid vesicles (liposomes) containing purified GPIa/IIa adhere to collagen.

The way in which these three membrane components may interact to induce platelet adhesion to collagen is under active investigation in several laboratories.

D. Epinephrine

Epinephrine is unique as a platelet agonist in that it induces aggregation without the platelets undergoing shape change. It has not yet been unequivocally established whether epinephrine is a platelet agonist in its own right or whether it merely potentiates the activation induced by other agonists. The platelet α_2-adrenergic receptor has been isolated, and the corresponding cDNA has been cloned. The deduced primary amino acid structure shows that it is a member of the family of G-protein-coupled receptors. The peptide backbone contains seven regions of 20–25 predominantly hydrophobic amino acids in an α-helical conformation, which loop back and forth through the plasma membrane and are numbered 1–7 from the amino-terminal domain. There are a total of six loops of hydrophilic amino acids connecting the membrane-spanning regions, three projecting into the cytoplasm and three into the extracellular space. The fourth membrane-spanning domain is involved in ligand binding to human platelet α_2-adrenergic receptors.

E. Fibrinogen

The binding of fibrinogen to platelets following activation is a prerequisite for their aggregation. Extensive studies in numerous laboratories have shown that the binding site for fibrinogen becomes expressed in the GPIIb/IIIa complex after platelet activation. This conclusion is based on the ability to bind fibrinogen covalently to GPIIb/IIIa using cross-linking agents, the inhibition of fibrinogen binding by monoclonal antibodies to GPIIb/IIIa, and the lack of aggregation seen with platelets from patients with Glanzmann's thrombasthenia, who have bleeding problems because of the lack of GPIIb/IIIa. An Arg-Gly-Asp sequence in fibrinogen and certain other adhesive molecules is the counter-ligand for activated GPIIb/IIIa.

Structural features of GPIIb and GPIIIa have been deduced from biochemical and molecular studies and have established that the two molecules are members of the integrin superfamily of adhesive binding proteins. GPIIb is a 135-kDa integral membrane protein composed of 100-kDa α-chain disulfide-linked to a 23-kDa β-chain and contains four sequences of amino acids homologous to the Ca^{2+}-binding domains of calmodulin. GPIIIa is a single chain of 95 kDa containing 21 intrachain disulfide bonds in four cysteine-rich repeats. GPIIb is synthesized as a larger precursor that undergoes posttranslational processing prior to forming a complex with GPIIIa, which is then inserted into the plasma membrane. Both GPIIb and GPIIIa contain a single transmembrane-spanning region and a short cytoplasmic domain.

The molecular basis of fibrinogen receptor exposure following platelet activation remains unknown but extracellular divalent cations play an important role in the stability and function of GPIIb/IIIa, and the complex may serve as a channel for the influx of extracellular Ca^{2+}. The majority of high-affinity binding sites for Ca^{2+} on the platelet surface have been localized to GPIIb/IIIa, possibly within the calmodulin-like domains of GPIIb. Ca^{2+} is required to maintain GPIIb and GPIIIa as a complex in solution. Although it is possible that Ca^{2+} functions as a bridge between GPIIb and GPIIIa, it seems more likely that it functions to maintain one or both of these glycoproteins in a conformation required for their stable association. Full fibrinogen-binding function of the complex has been observed only when it is present in its native membrane environment within the intact platelet.

F. Laminin

Platelets adhere to laminin, a constituent of the basal lamina that lies directly under the endothelial cells lining blood vessels. In contradistinction to their attachment to other adhesive proteins such as collagen and von Willebrand factor (vWF), platelets adherent to laminin under static conditions do not spread or undergo other morphological changes characteristic of activation. Platelets will also adhere directly to the laminin-derived peptide Tyr-Ile-Gly-Ser-Arg. Antilaminin antibodies inhibit the adhesion of platelets to segments of vascular subendothelium in flowing blood, indicating the physiological relevance of this interaction and, under these conditions, platelets do become activated, possibly because of the added effects of fluid flow.

A 67-kDa laminin-binding protein, identical to that found in human breast carcinoma tissue and a variety of other normal and transformed cells, has been affinity-isolated from platelets. Monoclonal antibodies to this receptor inhibit the binding of platelets to laminin. Monoclonal antibodies to the VLA-6 integrin $\alpha_6\beta_2$, corresponding to the GPIc/IIa complex, also inhibit platelet adhesion to laminin. The relationship between adhesion mediated by the 67-kDa receptor and that mediated by GPIc/IIa is not clear, but this may be another example, like that seen with collagen and vWF, of different receptors being involved in the adhesion and anchorage phases of platelet attachment.

G. Leukocyte Adhesion

P-selectin mediates the interaction of platelets with neutrophils and monocytes during inflammation, wound healing, and thrombogenesis. P-selectin is an unusual surface receptor since it is packaged in the α-granules in resting platelets but is expressed on the plasma membrane within 15–30 sec when the two membranes fuse following platelet activation. P-selectin (M_r 140,000) is a heavily glycosylated, single-chain protein that is homologous with E-selectin of endothelial cells and L-selectin of lymphocytes, and it is characterized by the presence of nine complement-related domains, an N-terminal lectin domain, and an endothelial cell growth factor-related domain: the two latter domains interact in the binding of P-selectin to its counter-ligand PSGL-1 (P-selectin glycoprotein ligand-1) on neutrophils and monocytes. PSGL-1 (M_r 110,000 reduced) is a mucin-like molecule containing 50% carbohydrate and its binding to selectin appears to require both its protein component and the heterosaccharide sialyl Lewis X antigen.

H. Platelet-Activating Factor

The structure of platelet-activating factor (PAF) is 1-*O*-alkyl-2-acetyl-*SN*-glycero-3-phosphocholine, with hexadecyl and octadecyl moieties comprising the major alkyl constituents, and it is the paradigm of a class of biologically active phospholipids of the general structure alkylacylglycerophosphocholine. It causes aggregation of washed human platelets at low concentrations (1–10 n*M*) but may have a much wider range of biological functions, as it is also a potent activator of a variety of cells, including neutrophils, monocytes, endothelial cells, smooth muscle cells, mesangial cells, and neuronal cells. PAF on human platelets binds with high affinity (K_d 0.5–5 n*M*) to a specific receptor that has been cloned. PAF-binding proteins of M_r 160,000–180,000 have been identified that may, in fact, be on the inner surface of the membrane but may be accessible to PAF since its structural similarity to membrane phospholipids may enable it to cross cell membranes.

I. Thrombin

Thrombin is the most potent platelet agonist and can cause full aggregation at very low concentrations (~0.3 n*M* or 30 milliunits/ml). Several lines of evidence indicate that α-thrombin activates platelets by two distinct pathways corresponding to the high-affinity (K_d 0.3 n*M*, 50 sites/platelet) and moderate-affinity (K_d 10 n*M*, 1700 sites/platelet) binding sites that have been identified. The two pathways differ in their requirements for receptor occupancy and in the role of guanine nucleotide regulatory proteins. Both receptors are required to ensure the maximum rate and extent of platelet activation, and a model has been proposed in which the high-affinity receptor activates a phospholipase A_2-dependent pathway and the moderate-affinity receptor activates a phospholipase C-dependent pathway.

GPIb has been identified as the high-affinity thrombin receptor, as demonstrated most directly by the effect of monoclonal antiGPIb antibodies that decrease binding of, and activation by, low concentrations of thrombin (<0.4 nM). GPIb exists in a complex comprising the disulfide-link d subunits GPIbα (M_r 140,000) and GPIbβ (M_r 27,000) and the noncovalently bound GPIX (M_r 22,000); this GPIb–IX complex is associated with GPV (M_r 82,000) in a ratio of 2 to 1. However, of the approximately 30,000 copies of the GPIb complex present on the platelet surface, only about 50 that are present in a supercomplex with a functional molecular weight of ~900,000 as determined by radiation inactivation and target analysis are able to bind thrombin with high affinity. The thrombin binding domain of GPIbα has been localized to the sequence Gly^{271}–Glu^{285} in the carbohydrate-poor amino-terminal domain and sulfation of three tyrosine residues in this region appears to modulate thrombin binding.

All of the components of the GPIb–IX complex have been cloned and shown to belong to a gene family characterized by repeated leucine-rich sequences, but the mechanism by which the binding of α-thrombin to the supercomplex initiates platelet activation through the high-affinity pathway is not known. Patients suffering from Bernard-Soulier syndrome lack the components of the GPIb–IX complex and do not respond to low concentrations of α-thrombin.

The moderate-affinity thrombin receptor (M_r 66,000) has recently been cloned and shown to be a member of the seven-transmembrane family of G-protein-coupled receptors. Thrombin activates this receptor by an unusual mechanism involving cleavage of the Arg^{41}/Ser^{42} peptide bond in the amino-terminal extracellular sequence of the receptor, giving rise to a new amino terminus for the receptor that has the initial sequence S^{42}FLLRN and that then reacts intramolecularly with the receptor itself to effect platelet activation: a synthetic peptide with this sequence can itself induce platelet activation. A second protease-activated seven-transmembrane domain G-protein-coupled receptor has recently been characterized and termed protein-activated receptor 2 (PAR-2), suggesting that the moderate-affinity thrombin receptor may be similarly termed PAR-1. Platelets from "knockout" mice lacking the gene for the moderate affinity receptor have recently been shown to have a normal hemostatic response indicating that this receptor is not obligatory for platelet activation by α-thrombin.

J. von Willebrand Factor

In the regions of the microvasculature where the blood moves with high velocity, the adhesion of platelets to subendothelium appears to be mediated by vWF bound to collagen and, possibly, to other subendothelial components. The initial interaction with platelets is dependent on the presence of GPIb on the platelet surface, and the vWF-binding domain of GPIb has been localized as being in the sequence Ser^{251}–Tyr^{279}, which overlaps the binding site for thrombin. Thus patients with Bernard-Soulier syndrome, whose platelets lack the GPIb–IX complex, have deficient platelet adhesion as well as a deficient response to α-thrombin. Similar deficient platelet adhesion is seen in Glanzmann's thrombasthenia owing to the absence of vWF from these patients' plasma. Agglutination of formalin-fixed platelets by vWF in the presence of the antibiotic ristocetin has been a useful laboratory test for this GPIb-dependent interaction.

This initial attachment of platelets to the vessel wall is mediated by the interaction of GPIb with the sequence V^{449}–Q^{460} in the amino-terminal fourth of the vWF polypeptide chain. This results in the activation of a phospholipase A_2, which leads to the expression of GPIIb/IIIa in a form that binds to the Arg^{1794}-Gly-Asp^{1796} sequence present in the carboxy-terminal third of the vWF molecule. These steps provide the clearest evidence in platelets of different receptors mediating adhesion and anchorage to the same adhesive substrate.

BIBLIOGRAPHY

Greco, N. J., and Jamieson, G. A. (1991). High and moderate affinity pathways for α-thrombin-induced platelet activation. *Proc. Soc. Exp. Biol. Med.* **198**, 792–799.

Greenwalt, D. E., Lipsky, R. H., Ockenhouse, C. F., Ikeda, H., Tandon, N. N., and Jamieson, G. A. (1992). Membrane glycoprotein CD36: A review of its roles in aherence, signal transduction and transfusion medicine. *Blood* **80**, 1105–1115.

Kaushansky, K. (1995). Thrombopoietin: The primary regulator of platelet production. *Blood* **86**, 419–456.

Lopez, J. A. (1994). The platelet glycoprotein Ib–IX complex. *Blood Coag. Fibrinolys.* **5**, 97–119.

Mills, D. C. B. (1996). ADP receptors on platelets. *Thromb. Haemostas.* **76**, 835–856.

State of the Art Lectures (1995). *Thromb. Haemostas.* **74**, 1–579.

Blood Sampling, Fetal

FERNAND DAFFOS
Institut de Puericulture de Paris

GLOSSARY

Alloimmunization Antibody synthesis in an individual who receives an antigen from another individual of the same species

Amniotic fluid Fluid surrounding the fetus in the uterus

Anoxemia Decrease of oxygen in the blood

Buffy coat Coat of white blood cells and platelets after sedimentation

PLA_1, Zwa platelet system Antigens fixed on platelets (e.g., A and B antigens fixed on red blood cells)

Semiology Specialty of medicine that concerns the signs of diseases

JUST AS BLOOD TESTS HAVE BECOME A *SINE QUA non* factor of current adult medicine, the possibility of sampling fetal blood easily and without risks during pregnancy has modified radically the treatment of *in utero* diseases and has opened the vascular compartment of the fetus to diagnosis and therapy, turning prenatal medicine into a primary care specialty in its own right—fetal medicine. This specialty is still young, and its relevant semiology is yet to be discovered; however, some major syndromes such as infection, anemia, and hypoxia are already revealed, and some fetal pathologies can be treated through the venous system.

The specificity of fetal medicine is twofold: The fetus is a fast-growing individual whose biological data will evolve throughout the pregnancy. Although the fetus is totally mother-dependent through the placenta for a number of biological parameters, it is biologically different from her.

I. TECHNIQUE AND RESULTS OF FETAL BLOOD SAMPLING

A. Technique

The technique of fetal blood sampling (FBS) from the umbilical vein by means of an ultrasonically guided needle has finally replaced the former techniques of fetoscopy or placentesis. The principle is simple: A real-time ultrasound transducer locates the umbilical cord. While this transducer is maintained perfectly still, in conditions of surgical asepsy, a long 20- or 22-gauge needle is introduced near the transducer into the mother's abdominal wall, in the plane of the ultrasonic image. The tip of the needle sends an echo, which remains clearly visible, and the needle is guided, through permanent monitoring, toward the cord's vein, which is punctured 1 cm away from its insertion on the placenta, used as a fixed point (Fig. 1). According to the stage of pregnancy, 2–5 ml of fetal blood is taken, that is, <5% of the total feto-placental blood volume.

There are many advantages to this technique: (1) It can be performed on outpatients without any premedication. Nothing more than local anesthesia at the point of puncture is needed to ensure the patient's comfort and stillness. (2) It can be carried out from Week 18 after the last menstruation until the end of pregnancy. (3) It can be repeated several times in the course of a pregnancy, which allows a follow-up of the development of a pathology or the efficiency of a therapy.

ENCYCLOPEDIA OF HUMAN BIOLOGY, Second Edition, VOLUME 2. Copyright © 1997 by Academic Press. All rights of reproduction in any form reserved.

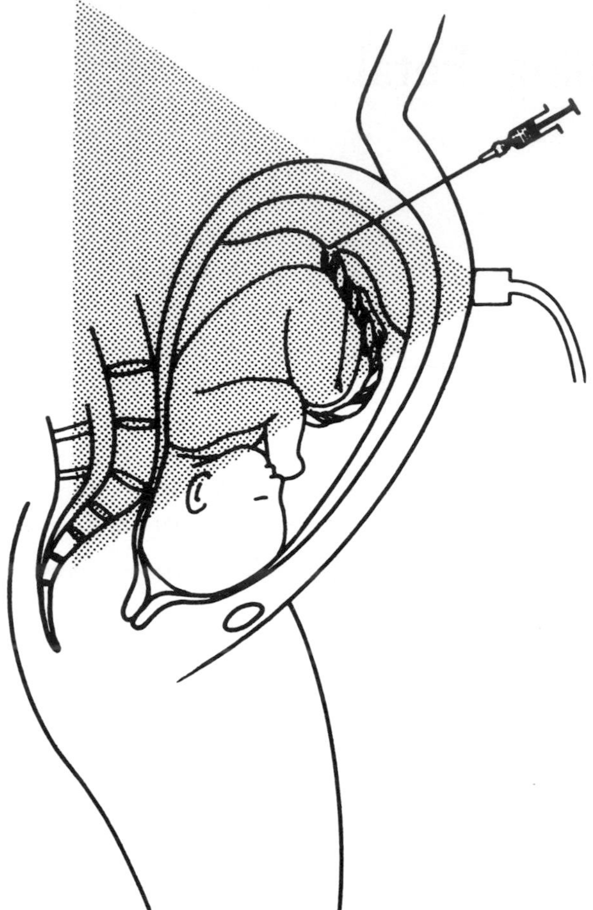

FIGURE 1 Umbilical cord blood sampling using a needle guided with ultrasound.

B. Results

Over 10,000 FBSs have been carried out to this day in the world. At the 37th Annual Meeting of the American College of Obstetricians and Gynecologists, held in May 1989, American and European teams obtained the following results. In Europe, 6336 samplings were carried out. The total fetal loss was 2.59% (ranging from 0.7 to 6.46%) and the procedure-related risk was evaluated at 1.05% (ranging from 0.4 to 2.98%). In the United States, 1610 samplings were performed with a procedure-related risk of 1.56% (ranging from 0 to 6%). The complications most frequently encountered were infection, membrane rupture, hemorrhage at the point of the puncture on the umbilical vein, severe slowdown of the fetal heart, and blood clotting.

C. Specific Technical Problems

When practiced by a well-trained team, FBS usually takes little time (<10 min in 92% of cases), but there are exceptions to the rule. When the mother is obese, it is necessary to use longer needles, which are consequently more difficult to guide, and the quality of the ultrasound image may be poor. High-resolution ultrasonography is indispensable, quality being an essential factor to simplify the technique and make it successful. In cases of fetal mobility, excessive fetal movements may occasionally block access to the cord. The problem can usually be overcome with a little patience. In rare instances, an adequtae view of the cord insertion may not be possible, and the needle must be introduced near the abdominal insertion (the other fixation point) or, in exceptional circumstances, on a loop of the cord.

D. Purity of the Blood Samples

Controlling the quality of the sample is an absolute prerequisite before the fetal biological results can be interpreted. It has to be done immediately after sampling through different tests. The sample may be contaminated with maternal blood or diluted with amniotic fluid. For instance, a 1/1000 contamination with maternal blood may lead mistakenly to conclude the presence of IgM in the fetal blood and incorrectly to diagnose congenital infection, or substantial dilution by amniotic fluid may suggest the presence of fetal anemia; on the other hand, a slight dilution of amniotic fluid could induce the activation of coagulation and bring about false information about the blood-clotting apparatus of the fetus.

II. PRENATAL DIAGNOSIS

Prenatal diagnosis of fetal pathology (i.e., affections) was first reduced to simple diagnoses such as the study of the presence or absence of a given protein or enzyme in fetal blood when a risk of hereditary diseases such as hemophilia or hemoglobinopathies exists. The knowledge of normal fetal biology now makes it possible to recognize more complex illnesses contracted in the course of a pregnancy: infections, anoxemia, and feto-maternal immunological conflict (Table I). [*See* Embryofetopathology.]

A. Fetal Infection

Affection of a fetus by an infection contracted by the mother during pregnancy may have very serious

TABLE I
Indications of Fetal Blood Samplings

Prenatal diagnosis	• Rapid chromosome analysis	Late pregnancy Amniotic fluid mosaicism Ultrasound detected pregnancy abnormality
	• Genetic disorders	Fragile X chromosome Hemoglobinopathies Coagulation disorders Miscellaneous
	• Congenital infection	Toxoplasmosis Rubella CMV—varicella Other viral infection
Assessment of fetal welfare	• Alloimmunizations	Rhesus system Platelet systems
	• Idiopathic thrombocytopenic purpura	
	• Growth retardation	Anoxemia Nutritional assessment
Fetal therapy	• Intravascular transfusions	Red blood cells Platelets
	• Drugs	Curare Digitoxin Thyroxin
	• Assessment of maternal therapy on fetal state	Steroids IgG Antibiotics Oxygen
Prenatal pharmacology	• Placental transfer of drugs • Fetal effects	

consequences for the fetus. Roughly, the earlier in the course of pregnancy the infection occurs, the more serious it will be. However, from Weeks 18–20 of pregnancy, the fetus is able to produce its own specific antibodies in sufficient quantities, and the sequelae will therefore be less serious after that period.

Many congenital infections can be diagnosed *in utero* (e.g., toxoplasmosis, rubella, cytomegalo-virus (CMV), varicella, parvovirus, chlamydiae]. As for adults, a diagnosis of fetal infection relies on two types of biological indices.

1. The Specific Indices of an Infection

a. Identifying the Pathogenic Organism in Fetal Blood or in the Amniotic Fluid

Parasites and viruses, and any live pathogenic organism, can be revealed by means of cell culture or by inoculating an animal with the buffy coat of the fetal blood or with the residue from centrifugation of the amniotic fluid. According to the agent considered, the identification is usually easily obtained and more or less constant.

A culture of a rubella virus is almost always positive in fetal blood when the fetus is infected, irrespective of the period when the fetal blood was sampled, whereas the cytomegalovirus is not always found in fetal blood in a case of congenital infection. For *Toxoplasma gondii,* the parasite is not always found in fetal blood or in amniotic fluid (Table II). The parasite is more frequently found in fetal blood when the congenital infection took place shortly before the sampling of fetal blood. Apparently, fetal parasitemia ceases rapidly, and the parasite is later eliminated through fetal urine and can be found in the amniotic fluid.

b. The Presence of Specific IgM Antibodies

The possibility for the fetus to synthesize IgM antibodies appears progressively in the course of fetal life. Total IgM can exceptionally be found in fetal blood before 17 weeks of pregnancy, its rate increasing rapidly between 17 and 22 weeks, and fetuses are capable of antigenic response of good quality from Week 22 of amenorrhea onward.

TABLE II
Presence of *Toxoplasma gondii* in Cases of Congenital Toxoplasmosis

Time of maternal infection	Time of fetal blood sampling (weeks gestation ± SD)	Inoculation into mice (% positive results)		
		With fetal blood alone	With amniotic fluid alone	With both
6–16 weeks	23.03 ± 2.5	16	16	50
16–25 weeks	26.54 ± 2.4	45	20	25

Great care should be taken in interpreting this specific biological test to assess the presence or absence of congenital infection, for the slightest trace of contamination with maternal blood containing specific IgM might mistakenly suggest a fetal infection. On the other hand, a fetal synthesis of IgM, when tested, is always very low and may go undetected if the biological techniques used are not sensitive enough. Generally speaking, one should always wait at least until Week 22 of pregnancy to expect a reliable result.

Finally, the fetal synthesis of specific IgM antibodies also depends on the pathogenic agent, the blocking effect of maternal IgG antibodies, and the treatment received by the mother. For instance, whereas specific IgM is found in 100% of congenital rubella, the same is found in only 20% of congenital toxoplasmosis.

2. Nonspecific Biological Indices of Infection

In the same way that a general syndrome of infection exists in an adult, whatever the cause may be, so does the fetus present general indices such as leukocytosis, thrombocytopenia, and liver damage. But the different tests performed, whether isolated or combined, may, in a very specific context of prenatal diagnosis of infection, prove extremely valuable in predicting a very high specificity (Table III), which helps form a diagnosis.

B. Fetal Anoxemia

Intrauterine growth retardation in nonmalformed fetuses is the most frequent cause of fetal and neonatal suffering in the whole obstetrical population. Whatever the etiologies of such suffering, the factor common to them all is a placental dysfunctioning resulting in a reduction of feto-maternal exchanges.

FBS makes it possible to evaluate the degree of fetal anoxemia objectively. At present, relatively little is known of fetal anoxemia, but some objective parameters are already being used to evaluate the degree of fetal suffing and to decide on what practical steps to take (Table IV). The biological signs of chronic fetal suffering, which appear progressively, are more constant and reliable than those of acute decompensation, which may vary extremely rapidly according to maternal oxygenation or to the presence of uterine contractions.

C. Fetal Thrombocytopenia (Deficiency of Blood Platelets)

The causes of fetal thrombocytopenia are many and varied (Table V). These affections represent 5.9% of the total number of the FBSs performed; however, severe fetal thrombocytopenias (<50,000 platelets/mm^3), which may cause intracerebral hemorrhagic accidents in the fetus during delivery, or even during pregnancy, are principally of immunological origin. The three main causes are alloimmunization in the PLA$_1$ (or Zwa) platelet system, idiopathic thrombocy-

TABLE III
No Specific Biological Signs of Congenital Rubella

Test	Biological value	Predictive value of positive test
White blood cell count	>6.34 × 10^9/liter	89%
Platelet count	<147 × 10^9/liter	100%
Interferon acid labile	>2 IU[a]	100%
Erythroblasts count	>27%	73%
Total IgM	>5.6 mg/100 ml	85%
Gamma glutamyl transferase	>72 IU	75%

[a]IU, international units.

TABLE IV

Biological Patterns in Cases of Intrauterine Growth Retardation

Long-term stress	Acute stress
• Stimulation of erythropoiesis	Acidosis
• Red blood cell destruction	Decrease pO_2 concentration Increase pCO_2 concentration
• Shortened red blood cell life span —Normal or decreased RBC count	
• Liver damage —Increase gamma glutamyl transferase	
• Decreased amino acid levels —Arginine, leucine, valine —Glutamic acid + glutamine, tyrosine	
• Decreased growth factor activities —Thymidine —Somatomedin C —Insulin	

topenic purpuras (ITP), or asymptomatic thrombocytopenia (AST) toward the end of pregnancy.

FBS has confirmed that no correlation exists between the rate of maternal antiplatelet antibodies and the fetal platelet count, nor between the fetal and the maternal platelet counts in cases of idiopathic thrombocytopenic purpura. The risks of fetal thrombocytopenia in cases of ITP are 38%, whereas they are only 12% in cases of AST.

To conclude, the very early appearance (as early as Week 20 of amenorrhea) of severe fetal thrombocytopenia (<10,000 platelets/mm^3) in cases of anti-PLA_1 alloimmunization has actually been confirmed by FBS, which in turn explains why 10% of fetal intracerebral hemorrhages occur *in utero* before delivery.

TABLE V

Etiologies of Fetal Thrombocytopenia in Our Series[a]

Etiology	Percentage of samplings
Platelet alloimmunizations (PLA_1, Bak systems)	16.6
Fetal malformations	11.8
Intrauterine growth retardation	11.1
Idiopathic thrombocytopenic purpura	7.7
Rhesus isoimmunization	3.5
Infections (toxoplasmosis, rubella)	46.5
Miscellaneous	2.8

[a]Mean platelet count = 85 ± 51 10^3/mm^3 (from 3 to 150).

III. NORMAL FETAL BIOLOGY

Because of the unique opportunity of carrying out a large number of FBSs for a prenatal diagnosis of congenital toxoplasmosis among a population in which only 5% of the fetuses were infected, we have been able to establish, retrospectively, the normal fetal biological parameters from the remaining 95% of fetuses born without infection. Though we cannot enumerate all of the fetal biological standards that have already been studied in hematology, hemostasis, biochemistry, immunology, and so on in this chapter, two basic principles must preside over the study: understanding and interpretation of fetal biological parameters. [*See* Fetus.]

A. The Gradual Maturation of the Fetus's Different Systems

Though fetal development has already been described with regard to the immunity system, it also applies to other systems. For instance, hematopoiesis (production of blood cells) is first achieved in the thymus, then in the liver, and then in bone marrow. And the biological values of the different fetal hematological parameters (according to the stage of pregnancy) must be known before one can correctly interpret a case of fetal anemia or of erythroblastosis (Fig. 2). The same is true for the fetal blood gases (Fig. 3). [*See* Hemopoietic System.]

B. The Transplacental Passage of the Different Molecules

Although it is biologically different, the fetus lives in constant symbiosis with its mother, and some molecules easily permeate through the placental barrier and are, therefore, in equal proportions on either side of the barrier, whereas other molecules cannot go through. It is well known that antibodies belonging to the IgG type found in fetal blood are of maternal origin and can cross the placenta, whereas IgM does not cross the placental barrier. Therefore, a correct interpretation of fetal biochemical and immunological parameters is possible only if a study of their transpla-

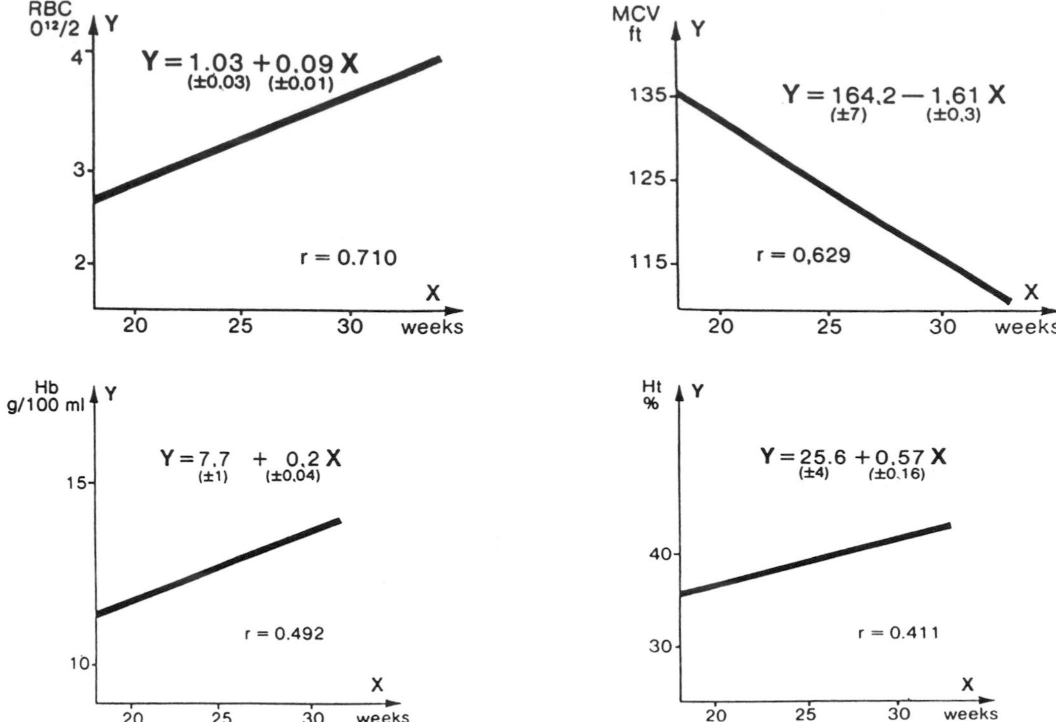

FIGURE 2 Evolution of some hematological parameters throughout the pregnancy. RBC, red blood cell count; MCV, mean corpuscular volume; Hb, hemoglobin; Ht, hematocrit.

cental passage is made beforehand. Figure 4 shows that fetal creatininemia is of no diagnostic or prognostic interest, because fetal and maternal creatininemias are always evenly balanced, whereas no correlation exists between the lactic dehydrogenase (LDH) activities of the mother and her fetus; therefore, this parameter can be considered as a reliable reflection of the fetus's condition.

IV. FETAL THERAPY AND PRENATAL PHARMACOLOGY

The fetus can be treated either through its mother or directly by intravenous infusion. In either case, FBS is essential. When the fetus is treated through its mother, repeated fetal sampling makes it possible to check the effectiveness of the treatment on the fetus.

Injecting drugs directly into the fetal vascular compartment is also possible. At present, the most frequent intravenous therapies are transfusion of red blood cell concentrates in cases of Rhesus immunization and transfusions of platelet concentrates in cases of PLA_1 alloimmunizations. By evaluating the amount of red blood cells in the fetal blood before and at the end of transfusion, we can almost perfectly monitor the exact count and frequency of transfusions to avoid severe fetal anemia, which might entail irreversible damage.

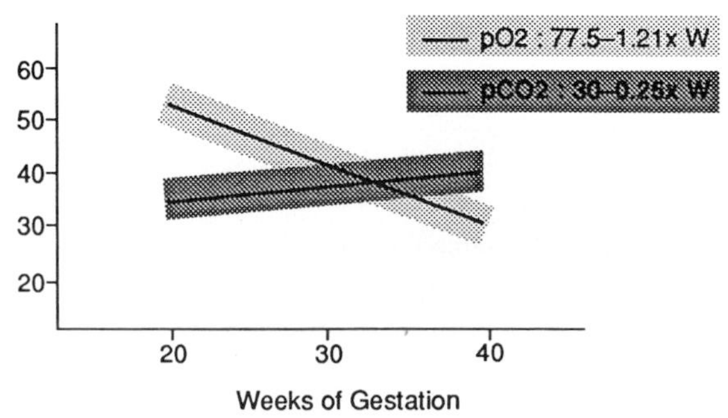

FIGURE 3 Evolution of fetal blood gases in umbilical vein according to the stage of the pregnancy.

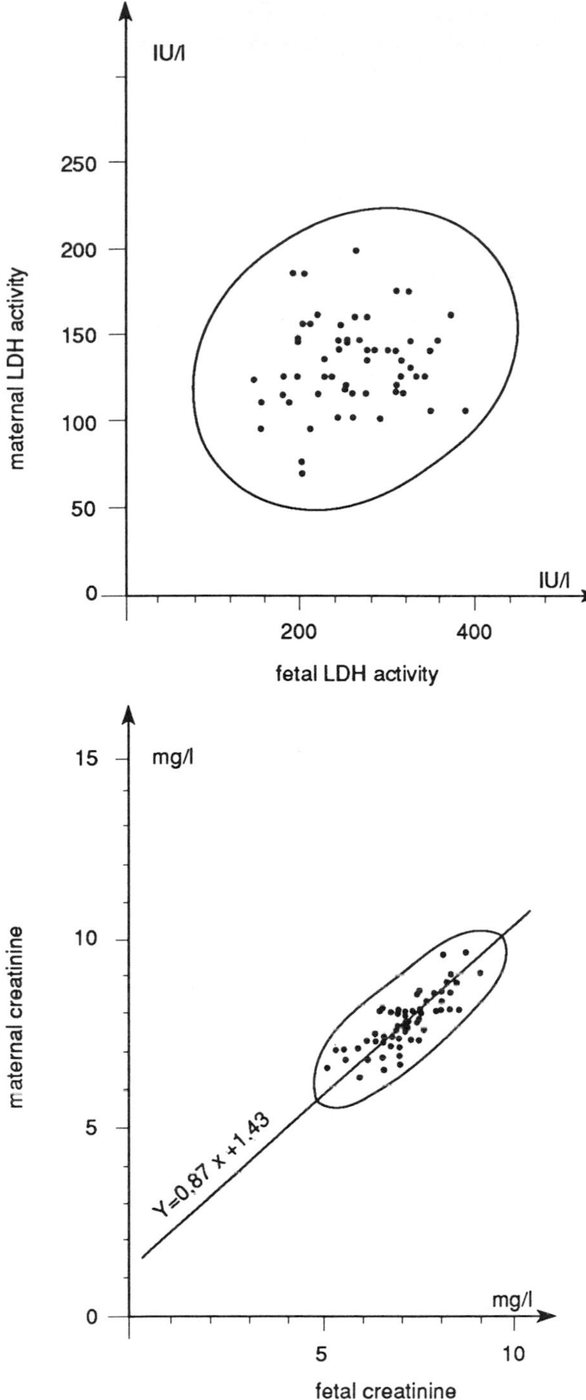

FIGURE 4 Spearman coefficient between maternal and fetal values. Lactic dehydrogenase does not cross the placenta, whereas creatinine does.

The possibility of having simultaneous access to either side of the placental barrier, that is, to fetal and maternal blood, has also opened exceptional prospects in prenatal pharmacology, provided these studies are carried out in accordance with appropriate ethical conditions. The study of the transplacental passages of vitamin K_1 is a suitable example.

Two questions arise in matters of prenatal pharmacology, and the answers to such questions are not necessarily univocal:

- Does the drug permeate through the placental barrier?
- Does it have some effect on the physiology of the fetus?

Tables VI and VII show that if vitamin K_1 easily crosses through the placental barrier, it has no effect on the synthesis of the coagulation factors that depend on vitamin K_1; they remain very low in the fetus, not because of a lack of vitamin K_1, as was formally believed, but because of the immaturity of the fetal synthesis in the liver.

TABLE VI
Transplacental Passage of Vitamin K_1 during 2nd and 3rd Trimesters of Pregnancy after Maternal Supplementation with Vitamin K[a]

	Not supplemented (pg/ml)	Supplemented (pg/ml)
2nd trimester		
Mother	565	81,130
Fetus	30	765
Birth		
Mother	395	45,190
Fetus	21	783

[a]Supplemented daily with 20 mg 3–7 days before sampling.

TABLE VII
Prothrombin Activity[a] in Fetal Plasma during 2nd and 3rd Trimesters of Pregnancy According to Maternal Supplementation

	Not supplemented	Supplemented
2nd trimester	16.5	16.9
Birth	49.5	47.8

[a]Expressed as percentage of that in normal adult plasma.

BIBLIOGRAPHY

Berkowitz, R. L., Chitkara, U., Wilkins, I. A. Lynch, L., *et al.* (1988). Intravascular monitoring and management of erythroblastosis fetalis. *Am. J. Obstet. Gynecol.* **158,** 783–795.

Cox, W. L., Daffos, F., Forestier, F., *et al.* (1988). Physiology and management of intrauterine growth retardation: A biologic approach with fetal blood sampling. *Am. J. Obstet. Gynecol.* **159,** 36–41.

Daffos, F., and Forestier, F. (1988). "Médecine et biologie du foetus humain." S. A. Maloine, Paris.

Daffos, F., Capella-Pavlovsky, M., and Forestier, F. (1985). Fetal blood sampling during pregnancy with use of a needle guided by ultrasound. A study of 606 consecutive cases. *Am. J. Obstet. Gynecol.* **153,** 655–660.

Daffos, F., Forestier, F., Capella-Pavlovsky, M., *et al.* (1988). Prenatal management of 746 pregnancies at risk for congenital toxoplasmosis. *N. Engl. J. Med.* **318,** 271–275.

Forestier, F., Daffos, F., Rainaut, M., and Cox, W. L. (1988). The assessment of fetal blood samples. *Am. J. Obstet. Gynecol.* **158,** 1184–1188.

Haig, D. (1993). Genetic conflicts in human pregnancy. *Quart. Rev. Biol.* **68,** 495.

Nicolaides, K. H. (1989). Studies on fetal physiology and pathophysiology in Rhesus disease. *Semin. Perinat.* **13,** 328–337.

Blood Supply Testing for Infectious Diseases

IRA A. SHULMAN
University of Southern California, Los Angeles

GLOSSARY

Sensitivity Percentage of positive tests obtained in a cohort known to have the condition the test was designed to detect

Seroconversion Change in antibody status of an individual from antibody negative to antibody positive

Specificity Percentage of negative tests obtained in a cohort known not to have the condition the test was designed to detect

Surrogate test Substitute test used to determine viral exposure when a specific test for the virus is not available

THE NATION'S BLOOD SUPPLY IS A LIMITED resource principally maintained by voluntary blood donations. Although blood products are lifesaving, they can also transmit infectious diseases, some of which are deadly. The goal of blood bank professionals is to provide the safest possible blood products for patients who need transfusions. The testing methods in use to screen donated blood for infectious diseases are highly sensitive, but are not 100% sensitive. Therefore, the potential for transfusion-transmitted infectious diseases continues to exist.

I. SCOPE OF TESTING IN 1996

In 1996, the following laboratory tests were routinely performed on *all* blood products in an effort to reduce the risk of transfusion transmitted infectious diseases. These tests were as follows:

1. Human immunodeficiency viruses (HIV-1/2) antibody test
2. Human T-cell lymphotrophic virus-1 (HTLV-1) antibody test
3. Hepatitis B surface antigen (HBsAg) test
4. Hepatitis B core antibody (HBcAb) test
5. Hepatitis C virus antibody (HCV) test
6. Serologic test for syphilis (STS)
7. Human immunodeficiency virus p24 antigen

Tests performed on some (but not all) donated blood products included:

1. Serum alanine aminotransferase (ALT)
2. Cytomegalovirus antibody (CMV) test

A. Human Immunodeficiency Virus (HIV-1/2) Antibody Test

The HIV-1/2 antibody test is used to screen donors for exposure to both of the known acquired immunodeficiency syndrome (AIDS) viruses, HIV-1 and HIV-2. The AIDS viruses are part of a family of viruses known as retroviruses. These RNA viruses, through transcription, form DNA in the presence of an enzyme known as reverse transcriptase. It is this synthesized DNA that integrates itself into the host genome. Because the DNA is formed from an RNA template rather than a DNA template, the class of viruses is designated as "retro" (meaning DNA is synthesized from RNA rather than the reverse). [*See* Acquired Immunodeficiency Syndrome, Virology.]

ENCYCLOPEDIA OF HUMAN BIOLOGY, Second Edition, VOLUME 2. Copyright © 1997 by Academic Press. All rights of reproduction in any form reserved.

Once infectious with HIV, it usually takes at least 22 days before enough antibody is formed to be detected. The HIV antigen test (see below) detects evidence of HIV-1 infection at 16 days following infectivity with the virus. It is during this 16-day "window" period that HIV-infected blood donations can slip through the safety net and be transfused to a patient. In addition, errors in specimen handling, error in testing, or results reporting may allow an unacceptable unit of blood to be used for transfusion. At the time of this writing, the risk per unit of contracting HIV from blood transfusion was about 1 in 750,000 compared to a risk estimate in 1989 of 1 in 38,000 to 1 in 150,000.

B. HIV Antigen Testing

The testing of donor blood for HIV-related proteins, such as an antigen designated HIV p24, shortens the "window" period between infectivity and a positive test for infection to about 16 days.

In March of 1996, the FDA licensed the use of donor screening tests for the HIV p24 antigen.

C. HTLV-I Antibody Test

This antibody test is used by blood banks to detect exposure to another retrovirus known as HTLV-1. Rarely, infection with HTLV-1 may result in leukemia or neurologic disease. It is estimated that approximately 1% of HTLV-1-infected individuals develop leukemia following a 20- to 30-year latency period. With the advent of the HTLV-1 antibody test, the per unit risk of contracting HTLV-1 from blood transfusion is about 1 in 50,000 to 1 in 70,000.

D. Hepatitis Tests

Viral hepatitis is the most frequent transfusion-transmitted infectious disease. Because the hepatitis A virus is rarely transmitted by blood transfusion (fewer than five cases have been reported since 1980), blood banks do not test for this virus. Hepatitis B (HBV) and hepatitis C (HCV) viruses both cause transfusion-transmitted disease, with most cases being caused by HCV (formerly called non-A, non-B hepatitis agent).

Hepatitis B virus infection can cause a serious disease that is occasionally fatal. The per unit risk is about 1 in 100,000. Although 90% of adults infected with this virus recover completely, almost 10% remain chronically infected. Individuals with chronic HBV infection lasting 20–30 years are at increased risk of developing cirrhosis and liver cancer (hepatocellular carcinoma). A child infected with hepatitis B has a 90% chance of becoming a chronic carrier of HBV and is likely to suffer the sequelae of this disease, including chronic active hepatitis, cirrhosis, and hepatocellular carcinoma. The testing of donated blood by the HBsAg test and anti-HBc test helps prevent posttransfusion HBV infection.

Hepatitis C virus infection accounts for the vast majority of posttransfusion hepatitis cases. The per unit risk is about 1 in 100,000. This virus is associated with a chronic carrier state in approximately 50% of infected patients. Furthermore, 10% of these patients develop cirrhosis, and deaths secondary to this virus have been reported. The testing of donated blood with the anti-HCV test helps prevent posttransfusion HCV infection.

E. Serologic Test for Syphilis

Transfusion-transmitted syphilis is extremely rare, partly because a serologic test for syphilis is routinely used by blood bank laboratories to screen donated blood and partly because the spirochete which causes syphilis (*Treponema pallidum*) dies when cooled to 4°C (the temperature at which blood is stored).

F. Cytomegalovirus Antibody Testing

Symptomatic cytomegalovirus virus infection is an unusual transfusion risk when cellular blood products are administered to individuals who have a normal immune system. However, some patients (such as very low birth weight neonates and immunocompromised adults) may suffer severe complications if infected by CMV following a blood transfusion. Consequently, blood banks do not routinely test all blood donors for exposure to cytomegalovirus (CMV). Rather, when a patient at high risk of severe CMV infection needs a blood transfusion, donated blood is tested to select CMV antibody negative units or, if the antibody status of the blood product cannot be determined, the blood products may be depleted of white blood cells, which removes CMV virus. CMV resides within white blood cells, and the removal of white blood cells from blood products renders those products essentially free of CMV infection risk. Patients who benefit from CMV risk-free transfusions include neonates weighing less than 1200 g at birth, pregnant women who are CMV negative, HIV-infected individuals who are CMV antibody negative, and CMV-negative transplant patients who receive organs or tissues from CMV-negative donors. CMV-negative blood products are of no proven benefit once a patient is known to be CMV antibody positive.

G. Alanine Aminotransferase Testing

ALT testing is a surrogate or indirect test for infection with a hepatitis-causing virus. This test used to be performed on all donated blood products, but was no longer required when data showed that specific tests for hepatitis virus infection were adequate in preventing posttransfusion hepatitis.

II. DONOR HISTORY AS A SCREENING TEST

Since it is not possible to eliminate the risk of infectious disease transmission from blood transfusion by testing alone, attempts are made to increase the safety of transfusion by using a careful predonation history to weed out unacceptable donors, by efforts to recruit volunteer rather than paid blood donors, and by avoiding improper donor incentives and coercion, which could alter the truthfulness of some donors.

Each individual blood donor is required to read information about blood safety and is encouraged to leave, without explanation, if he or she recognizes that giving blood would be inappropriate. Potential donors are also asked a series of questions about their health and life-style (including direct questions on sexual behavior designed to identify high-risk activities) and undergo a miniphysical before being allowed to donate. The questions and examinations are designed to prevent individuals who are at high risk for HIV, hepatitis, and other infectious diseases from donating blood. This process is continually refined in order to ensure that blood is drawn from the most appropriate individuals.

Blood donors may be offered a confidential opportunity to exclude their blood from use in transfusion by attaching stickers to the paperwork identifying the collected unit for use or withdrawal. If a donor knows of any reason why his or her blood should not be used for transfusion, he or she places the sticker indicating that the unit should not be transfused on the label. This is done to ensure that no pressure is exerted on the donor to give blood.

Every donation is checked against existing records. If a donor was indefinitely deferred, the collected unit is withdrawn from circulation and potential use. This process acts as a barrier to prevent the release of any blood from a donor who was previously judged to be indefinitely unacceptable. When a specific test for a disease does not exist, a surrogate test for a similar disease entity is sometimes used as an alternative screening method.

III. TESTING IN 1997 AND BEYOND

A. Babesiosis

Babesiosis is a malaria-like illness caused by *Babesia microti*. The organism is an intraerythrocytic parasite transmitted by the tick *Ixodes dammini*. A few cases of babesiosis transmitted by blood have been reported. At this time the blood supply is not tested for babesiosis.

B. Chagas' Disease

Chagas' disease is caused by a parasite (*Trypanosoma cruzi*) endemic in parts of Mexico, and South and Central American countries. Some authorities believe that transfusion-transmitted Chagas' disease could become a problem in certain areas within the United States in which there are large number of immigrants from endemic regions. At this time, reliable tests for this blood parasite have been developed, but none of these have been licensed by the FDA for blood donor screening.

C. Hepatitis G Viruses

A new hepatitis virus has recently been described. Research on this virus is ongoing, including what risk, if any, the virus presents to the U.S. blood supply.

BIBLIOGRAPHY

Busch, M. P., and Alter, H. J. (1995). Will human immunodeficiency virus p24 antigen screening increase the safety of the blood supply and, if so, at what cost? *Transfusion* **35**(7), 536–539.

Busch, M. P., Lee, L. L. L., Satten, G. A., *et al.* (1995). Time course of detection of viral and serologic markers preceding human immunodeficiency virus type 1 seroconversion: Implications for screening of blood and tissues donors. *Transfusion* **35**(2), 91–97.

Cumming, P. D., Wallace, E. L., Schorr, J. B., and Dodd, R. Y. (1989). Exposure of patients to human immunodeficiency virus through the transfusion of blood components that test antibody-negative. *N. Engl. J. Med.* **321**, 941–946.

Dodd, R. Y. (1992). The risk of transfusion transmitted infection. *N. Engl. J. Med.* **327**, 419–420.

FDA memorandum (1995). Recommendations on donor screening with a licensed test for HIV antigen.

Lackritz, E. M. Sattem, G. A., Raimondi, C. P., *et al.* (1995). "Estimated Risk of HIV Transmission by Screened Blood in the U.S." 2nd International Conference on Human Retroviruses and Related Infections, Washington, D.C. [Updated at the 48th Meeting of the Blood Products Advisory Committee, Bethesda, MD, June 23, 1995.]

NIH Consensus Development Conference on Infectious Disease Testing for Blood Transfusions (1995). Program and Abstracts. Bethesda, Maryland.

Schlauder, G. C., Dawson, G. J., Simons, J. N., *et al.* (1995). Molecular and serologic analysis in the transmission of the GB hepatitis agents. *Med. Virol.* **46,** 81–90.

Ward, J. W., Holmberg, S. D., Allen, J. R., *et al.* (1988). Transmission of human immunodeficiency virus (HIV) by blood transfusions screened as negative for HIV antibody. *N. Engl. J. Med.* **318,** 473–478.

Body Fat, Menarche, and Fertility

ROSE E. FRISCH
Harvard Center for Population Studies

GLOSSARY

Adipose tissue Body fat consisting of the fat cell, the adipocyte, and connective tissue, the stroma

Amenorrhea Absence of menstrual cycles, hence infertility

Androgens Steroid hormones secreted by the testis and the adrenal glands

Anovulatory Absence of ovulation in a menstrual cycle, hence infertility

Dyzygotic twins Nonidentical twins resulting from two separate ovulations

Estrogen Steroid hormone (particularly estradiol) secreted by the ovary, which regulates the development of the uterus and the breasts and other tissues of the female reproductive tract. Estrogens also regulate the characteristic deposition of female fat in the hips, thighs, and breasts

Follicular phase Period of the menstrual cycle up to ovulation during which there is growth of a follicle containing an egg (the ovum) and rapid growth of the lining of the uterus (endometrium); normally lasts about 14 days

Follicular-stimulating hormone (FSH) Hormone that promotes ovarian follicular growth

Gonadotropin-releasing hormone Hormone secreted by a part of the brain (hypothalamus) that controls the release of follicle-stimulating hormone and luteinizing hormone by the pituitary gland

Hypothalamus Part of the brain (the diencephalon) that controls reproduction and other basic functions such as food intake, temperature, and control of the emotions. Among other functions, the hypothalamus produces and secretes releasing hormones that control release of pituitary hormones

Lactation Breast feeding

Luteinizing hormone (LH) Hormone that together with FSH stimulates estrogen secretion; the surge of LH in the middle of the menstrual cycle stimulates ovulation. LH then controls the transformation of the ruptured follicle to become the corpus luteum (yellow body), which secretes both estrogen and progesterone

Luteal phase Period of the menstrual cycle after ovulation; the lining of the uterus (endometrium) prepares for implantation of the fertilized egg. If this does not occur, there is shedding of the lining of the uterus and menstruation occurs, normally in about 14 days after ovulation

Menarche First menstrual cycle. This cycle can often be without ovulation, and the cycles following may be irregular for 1 or 2 years

Monozygotic twins Identical twins resulting from one egg splitting into two parts early in development

Pituitary gland Secretions of the pituitary gland, controlled by the releasing factors of the hypothalamus, regulate other endocrine organs of the body, including the ovary and testis

ENCYCLOPEDIA OF HUMAN BIOLOGY, Second Edition, VOLUME 2. Copyright © 1997 by Academic Press. All rights of reproduction in any form reserved.

Puberty General term covering the period of time of the rapid growth of the adolescent growth spurt and development of the secondary sex characteristics before menarche in girls

Testosterone Androgen produced by the testis. Testosterone regulates the development of male genitalia and male characteristics of the skeleton and muscular system

TOO LITTLE BODY FAT AND TOO MUCH BODY FAT are both associated with the disruption or delay of the reproductive ability of women. Evidence gathered from women with moderate or extreme weight loss caused by injudicious dieting, intensive exercise, or both indicates that this association is causal and that the large amount of body fat (26–28% of body weight) stored by the human female at maturity influences reproduction directly.

Many young girls who diet excessively or who are well-trained athletes or dancers have a delayed menarche (the first menstrual cycle). Menarche may be delayed until as late as 19 or 20 years. If their athletic training begins after menarche, girls may have anovulatory menstrual cycles, irregular cycles, or a complete absence of cycles (secondary amenorrhea).

In addition to these extreme effects of weight loss and athletic training on the menstrual cycle, women who train moderately or who are regaining weight into the normal range may have a menstrual cycle that appears to be normal, but that actually has a shortened luteal phase or is anovulatory.

All these disruptions of reproductive ability are usually reversible after varying periods of time after weight gain, decreased athletic training, or both.

Secondary amenorrhea occurs in dieting girls and women or in athletes and dancers when weight loss is 10–15% of normal weight for height, which is equivalent to a loss of about one-third of body fat. Primary amenorrhea (absence of menarche at age 16 or older) also occurs in association with excessive thinness. These data suggest that a minimum level of body fat (i.e., stored, easily mobilized energy), in relation to the lean body mass, is necessary for the onset and maintenance of regular ovulatory menstrual cycles. Both the absolute and the relative amounts of fat are important because lean mass and fat must be in a particular absolute range, as well as a relative range (i.e., the individual must be big enough to reproduce successfully).

Data on obese women show that excessive fatness is also associated with infertility, as was described for cattle and mares almost a century ago. Loss of weight restores fertility of the women and of the animals. Too little or too much fat are thus both associated with infertility.

I. WHY FAT?

Reproduction costs calories; a pregnancy requires about 50,000 calories above normal metabolic requirements, and lactation requires about 1000 calories a day. In premodern times, lactation was an essential part of reproduction; storage of fat when food supplies are uncertain, as they were in our prehistory, would therefore be of selective advantage to the female.

Whatever the reason, during the adolescent growth spurt that precedes menarche, girls increase their body fat by 120% compared with a 44% increase in lean body mass. Changes in body composition can be monitored by direct measurements of body water. Because fat contains only about 10% water compared with about 80% water in muscle and viscera, an increase in fatness results in a decrease in body water as a percentage of total body weight. Direct measurements of body water of girls from birth to completion of growth show a continuous decline in the proportion of body water because of the large relative increase in body fat (Fig. 1). At menarche, girls average about 24% of their body weight as fat (11 kg, 24 lb). At the completion of growth, between ages 16 and 18

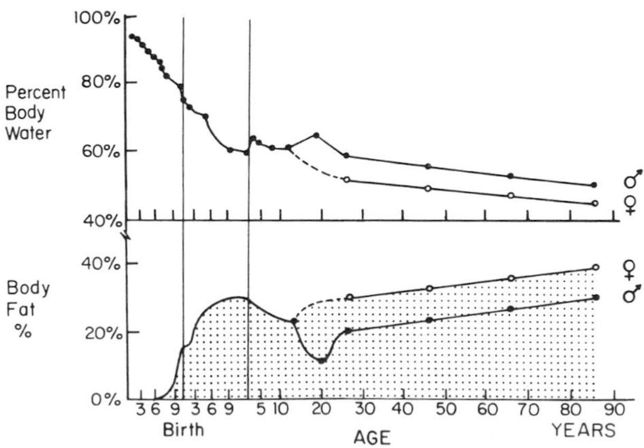

FIGURE 1 Changes in body water as percentage of body weight throughout the life span and corresponding changes in the percentage of body fat. [Adapted from Friis-Hansen, B. (1965). Hydrometry of growth and aging. *In* "Human Body Composition" (J. Brožek, ed.), Vol. VII, pp. 191–209. Pergamon Press, Oxford. Reprinted with permission.]

TABLE I
Total Body Water as Percentage of Body Weight, an Index of Fatness: Comparison of an 18-Year-Old Girl and a 15-Year-Old Boy of the Same Height and Weight

Variable	Girl	Boy
Height (cm)	165.0	165.0
Weight (kg)	57.0	57.0
Total body water (liter)	29.5	36.0
Lean body weight (kg)[a]	41.0	50.0
Fat (kg)	16.0	7.0
Fat/body weight (%)[b]	28.0	12.0
Total body water/body weight (%)	51.8	63.0

[a]Lean body weight = total body water/0.72.
[b]Fat/body weight % = 100 − [(total body water/body wt %)/0.72].

years, the body of a well-nourished young girl in the United States contains on average about 26–28% fat (16 kg, 35 lb) and about 52% water. In contrast, the body of a boy at the same height and weight contains about 12% fat and 63% water (Table I). At the completion of growth, men are about 15% fat and about 61% water. The main function of the 16 kg of stored female fat, which is equivalent to 144,000 calories, may be to provide energy for a pregnancy and for about 3 months lactation.

A. Is the Amenorrhea of Athletes and Underweight Girls Adaptive?

Infant survival is correlated with birth weight, and birth weight is correlated with the prepregnancy weight of the mother and, independently, with her weight gain during pregnancy. An underweight woman is therefore at high risk of an unsuccessful pregnancy. As Dr. J. M. Duncan observed more than a century ago, if a seriously undernourished woman could get pregnant, the chance of her giving birth to a viable infant, or herself surviving the pregnancy, is small. Therefore, the amenorrhea of underweight girls and women can be considered adaptive.

II. HOW ADIPOSE TISSUE MAY REGULATE FEMALE REPRODUCTION

There are at least four mechanisms by which body fat (adipose tissue) may directly affect ovulation and the menstrual cycle, hence fertility: (1) Conversion of androgen to estrogen takes place in adipose tissue of the breast and abdomen, the omentum, and the fatty marrow of the long bones. Adipose tissue therefore is a significant extragonadal source of estrogen. (2) Body weight, hence fatness, influences the direction of estrogen metabolism to the most potent or least potent forms. (3) In fatter girls and women, the capacity of serum sex hormone-binding globulin to bind estradiol is diminished, resulting in an elevated percentage of free serum estradiol. (4) Adipose tissue stores steroid hormones.

Changes in relative fatness might also affect reproductive ability indirectly through disturbance of the regulation of body temperature and energy balance by the hypothalamus. Lean amenorrheic women, both anorectic and nonanorectic, display abnormalities of temperature regulation at the same time that they have delayed response, or lack of response, to exogenous gonadotropin-releasing hormone (GnRH).

III. HYPOTHALAMIC DYSFUNCTION, GONADOTROPIN SECRETION, AND WEIGHT LOSS

It is now known that the amenorrhea of underweight and excessively lean women, including athletes, is due to hypothalamic dysfunction. Consistent with the view that this type of amenorrhea (hence infertility) is adaptive, the pituitary–ovarian axis is apparently intact and functions when exogenous GnRH is given. [*See* Hypothalamus.]

Girls and women with this type of hypothalamic amenorrhea have both quantitative and qualitative changes in the secretion of gonadotropins, luteinizing hormone (LH), and follicle-stimulating hormone (FSH): (1) LH and FSH are low as are estradiol levels. (2) The secretion of LH and the response to GnRH are reduced in direct correlation with the amount of weight loss. (3) Underweight patients respond to exogenous GnRH with a pattern of secretion similar to that of prepubertal children; the FSH response is greater than the LH response. The return of LH responsiveness is correlated with weight gain. (4) The maturity of the 24-hr LH secretory pattern and body weight are related; weight loss results in an age-inappropriate secretory pattern resembling that of prepubertal or early pubertal children. Weight gain restores the postmenarcheal secretory pattern.

IV. WEIGHT AT PUBERTY

The idea that relative fatness is important for female reproductive ability followed from the findings that the events of the adolescent growth spurt, particularly menarche in girls, were closely related to an average critical body weight. The mean weight at menarche for a sample of middle class U.S. girls was 47.8 ± 0.5 kg, at the mean height of 158.5 ± 0.5 cm at the mean age of 12.9 ± 0.1 years. This mean age included girls from Denver, who had a slightly later age of menarche than the sea-level populations because of the slowing effect of altitude on the rate of weight growth. The average age of menarche of U.S. girls is now 12.6–12.8 years.

A. Secular (Long-Term) Trend to an Earlier Age of Menarche

Even before analyzing the meaning of the average critical weight for an *individual* girl, the idea that menarche is associated with a critical weight for a population explained many observations associated with early or late menarche. Observations of earlier menarche are associated with attaining the critical weight more quickly. The most important example is the secular (long-term) trend to an earlier menarche of about 3 or 4 months per decade in Europe in the past 100 years (Fig. 2). Our explanation is that children now are bigger sooner; therefore, girls on the average reach 46–47 kg, the average weight at menarche of U.S. and many European populations, more quickly. Also, the secular trend should end when the rate of weight growth of children of successive cohorts remains the same because of the attainment of maximum nutrition and child care; this now has happened in the United States.

Conversely, a late menarche is associated with body weight growth that is slower prenatally, postnatally, or both so that the average critical weight is reached at a later age: undernutrition delays menarche, twins have later menarche than do singletons of the same population, and high altitude delays menarche. Diseases such as cystic fibrosis, juvenile onset diabetes, Crohn's disease, and sickle cell anemia also delay menarche.

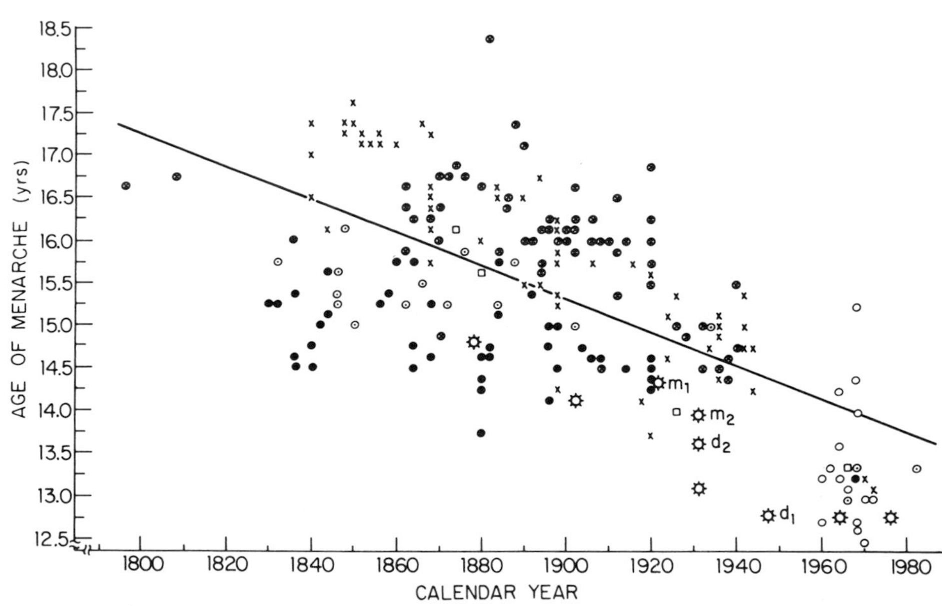

FIGURE 2 Mean or median age of menarche as a function of calendar year from 1790 to 1980. The symbols refer to England (⊙), France (●), Germany (⊗), Holland (□), Scandinavia (×) (Denmark, Finland, Norway, and Sweden); Belgium, Czechoslovakia, Hungary, Italy, Poland (rural), Romania (urban and rural), Russia (15.2 years at an altitude of 2500 m and 14.4 years at 700 m), Spain, and Switzerland (all labeled ○); and the United States (✲) (data not included in the regression line). Twenty-seven points were identical and do not appear on the graph. The regression line cannot be extended indefinitely. The age of menarche has already leveled off in some European countries, as it has in the United States (see text). [From Wyshak, G., and Frisch, R. E. (1982). *N. Engl. J. Med.* **306,** 1033–1035.]

B. Components of Weight at Menarche

Individual girls have menarche at varied weights and heights. The components of body weight at menarche, total body water (TW), lean body weight (LBW; TW/0.72), and fat (body weight minus lean body weight) showed that the predictive factor for weight of an individual girl was total water as the percentage of body weight, an indicator of relative fatness. Body fat has little water (about 10%) compared with muscle and viscera (about 80%). Therefore as the body increases in fatness, the percentage of total water in the body decreases (Fig. 1). TW and LBW are more closely correlated with metabolic rate than is body weight because they represent the metabolic mass as a first approximation. Metabolic rate and relative fatness are considered important clues since the late G. C. Kennedy hypothesized a food intake–metabolic signal related to stored fat to explain his findings on weight and puberty in the rat.

The greatest change in body composition of both early and late maturing girls during the adolescent growth spurt was a large increase in body fat, from 5 to 11 kg, an increase of 120%, compared with a 44% increase in LBW. Thus there was a change in the ratio of LBW to fat from 5 : 1 at initiation of the spurt to 3 : 1 at menarche.

V. FATNESS AS A DETERMINANT OF MINIMAL WEIGHTS FOR MENARCHE AND THE RESTORATION OF MENSTRUAL CYCLES

Figures 3 and 4 are nomograms that have been found useful clinically in the evaluation and treatment of young girls and women with primary or secondary amenorrhea caused by weight loss. The nomograms were developed from data on relative fatness of normal girls at menarche and at the completion of growth of the same girls at age 18 years. The diagonal lines on the nomograms are percentiles of total water at percent of body weight, indicating levels of relative fatness.

The *minimum* weight necessary for a particular height for *restoration* of menstrual cycles is indicated on the weight scale of Fig. 3, at the point where the diagonal 10th percentile line of total water as percent of body weight (56.1%) crosses the appropriate vertical height line. This percentile of TW/body weight percentage is equivalent to about 22% fat of body weight. For example, a 20-year-old woman whose height is 160 cm (63 in.) should weigh at least 46.3 kg (102 lb) before menstrual cycles would be expected to resume.

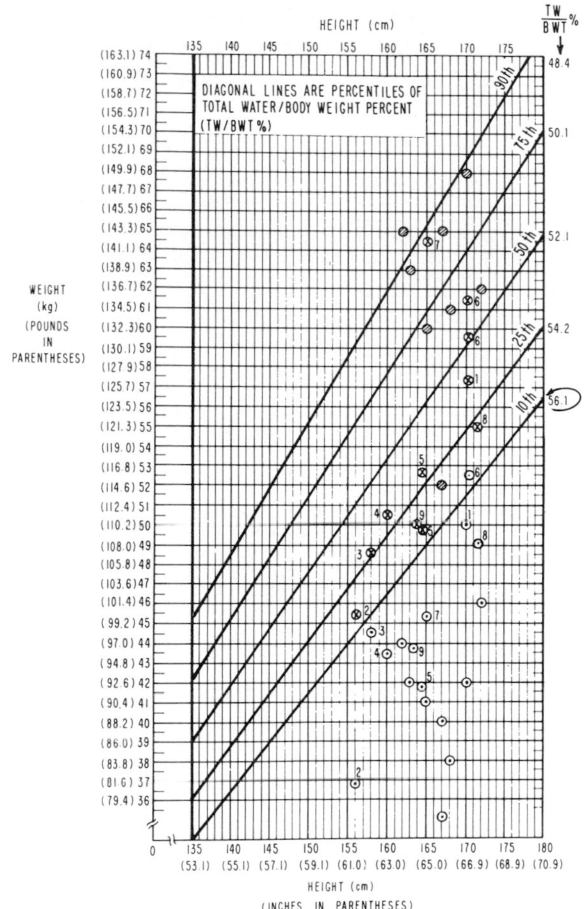

FIGURE 3 Minimal weight necessary for a particular height for *restoration* of menstrual cycles is indicated on the weight scale by the 10th percentile diagonal line of total water (TW)/body weight (BWT) percent, 56.1%, as it crosses the vertical height line. For example, a 20-year-old woman whose height is 160 cm should weigh at least 46.3 kg (102 lb) before menstrual cycles would be expected to resume. [From Frisch, R. E., and McArthur, J. W. (1974). *Science* **185**, 949–951.]

The *minimum* weight necessary for a particular height for the *onset* of menstrual cycles is indicated on the weight scale of Fig. 4 at the point where the diagonal 10th percentile, representing the 10th percentile of TW/body weight percentage at menarche (59.8%), crosses the appropriate vertical height line. This percentile is equivalent to about 17% fat of body weight. The height growth of girls must be completed, or nearing completion, for use of the nomogram for

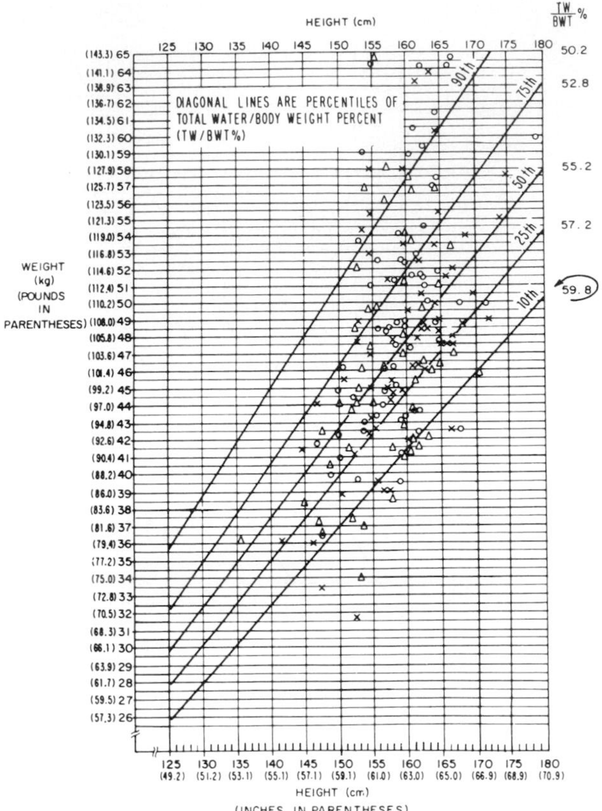

FIGURE 4 Minimal weight necessary for a particular height for *onset* of menstrual cycles is indicated on the weight scale by the 10th percentile diagonal line of total water/body weight percentage, 59.8%, as it crosses the vertical height lines. Height growth of girls must be completed or approaching completion. For example, a 15-year-old girl whose completed height is 160 cm (63 in.) should weigh at least 41.4 kg (91 lb) before menstrual cycles can be expected to start. [From Frisch, R. E., and McArthur, J. W. (1974). *Science* 185, 949–951.]

menarche. For example, a 15-year-old girl whose completed height is 160 cm (63 in.) should weigh at least 41.4 kg (91 lb) before menstrual cycles can be expected to start. The minimum weights for menarche would also be used for girls who become amenorrheic as a result of weight loss shortly after menarche, as is often found in cases of anorexia nervosa.

The weights at which menstrual cycles ceased or resumed in postmenarcheal women were 10% heavier than the minimal weights for the same height observed at menarche. In accord with this finding, the body composition data showed that both early and late maturing girls gained an additional 4.5 kg (10 lb) of fat from menarche to age 18 years. Almost all this gain was achieved by age 16 years, when mean fat is 15.7 kg (35 lb), 27% of body weight. At age 18 years, mean fat is 16.0 kg, 28% of the mean body weight of 57 kg (125 lb). Reflecting this increased fatness, the total water/body weight percentage decreases from 55.1% at menarche to 52.1% at age 18 years.

The prediction of the minimum weights for height is from total water as percent of body weight, not fat/percent of body weight, indicating that the size of the lean mass is important in relation to the amount of fat (i.e., the prediction is based on a lean : fat ratio).

Other factors, such as emotional stress, affect the maintenance or onset of menstrual cycles. Therefore, menstrual cycles may cease without weight loss and may not resume in some subjects even though the minimum weight for height has been achieved. Also, these minimal weight standards apply thus far only to Caucasian women in the United States and Europe. Different races have different critical weights at menarche, and it is not yet known whether the different critical weights represent the same critical body composition of fatness.

Some amenorrheic athletes, such as shot-putters, oarswomen, and some swimmers, are not lightweight because they are very muscular; muscles are heavy (80% water) compared with fat (about 10% water). The cause of their amenorrhea is, nevertheless, most probably an increased lean mass and a reduced fat content of their bodies; gaining body fat or ceasing exercise usually restores menstrual cycles. In relation to athletic amenorrhea, it is important to note that body composition may change without any change in body weight. A woman may increase muscle mass by increasing training and at the same time lose fat, without a perceptible change in body weight. Direct measurements of body composition with magnetic resonance imaging showed that well-trained athletes had 30–40% less fat than comparable sedentary women, although the body weights of the two groups of women did not differ.

VI. PHYSICAL EXERCISE, DELAYED MENARCHE, AND AMENORRHEA

Does intense exercise cause delayed menarche of dancers and athletes or do late maturers choose to be dancers and athletes? In accord with many reports, the mean age of menarche of college swimmers and runners was significantly later (13.9 ± 0.3 years) than that of the general population (12.8 ± 0.05 years). However, the mean menarcheal age of the athletes whose training began *before* menarche was 15.1 ± 0.5 years, significantly later than the mean menarcheal age (12.8 ± 0.2 years) of the athletes whose training began after their menarche. The latter mean age was similar to that of the college controls (12.7 ± 0.4 years) and the general population. Therefore, training, not preselection, is the delaying factor. Each year of premenarcheal training delayed menarche by 5 months (0.4 year). The training also disrupted the regularity of the menstrual cycles in both pre- and postmenarcheal trained athletes. Athletes with irregular cycles or amenorrhea had hormonal levels confirming lack of ovulation, hence infertility.

VII. LONG-TERM REGULAR EXERCISE DECREASES THE RISK OF SEX HORMONE–SENSITIVE CANCERS

The amenorrhea and delayed menarche of the college athletes raised the question: Are there differences in the long-term reproductive health of moderately trained athletes compared with nonathletes?

A study of 5398 college alumnae aged 20–80 years, 2622 of whom were former athletes and 2776 nonathletes, showed that the former athletes had a significantly lower lifetime occurrence (prevalence rate) of breast cancer and cancers of the reproductive system (Fig. 5) compared with nonathletes; 82.4% of the former college athletes began their training in high school or earlier compared with 24.9% of the nonathletes. The analysis controlled for potential confounding factors including age, age of menarche, age of first birth, smoking, and family history of cancer. The former college athletes were leaner in every age group compared with the nonathletes. [*See* Breast Cancer Biology.]

Although we can only speculate at present as to the reasons for the lower risk of sex hormone–sensitive cancers among the former athletes, the most likely explanation is that the former athletes had lower levels of estrogen because they were leaner and more of the estrogen was metabolized to a nonpotent form of estrogen (catechol estrogens). The former college athletes also had a lower prevalence of benign tumors of the breast and reproductive system and a lower prevalence of diabetes, particularly after age 40, compared with the nonathletes.

VIII. DOUBLE-MUSCLED CATTLE AND OTHER PERTINENT ANIMAL DATA

Carcass analyses of meat animals and of experimental animals provide important data unattainable from

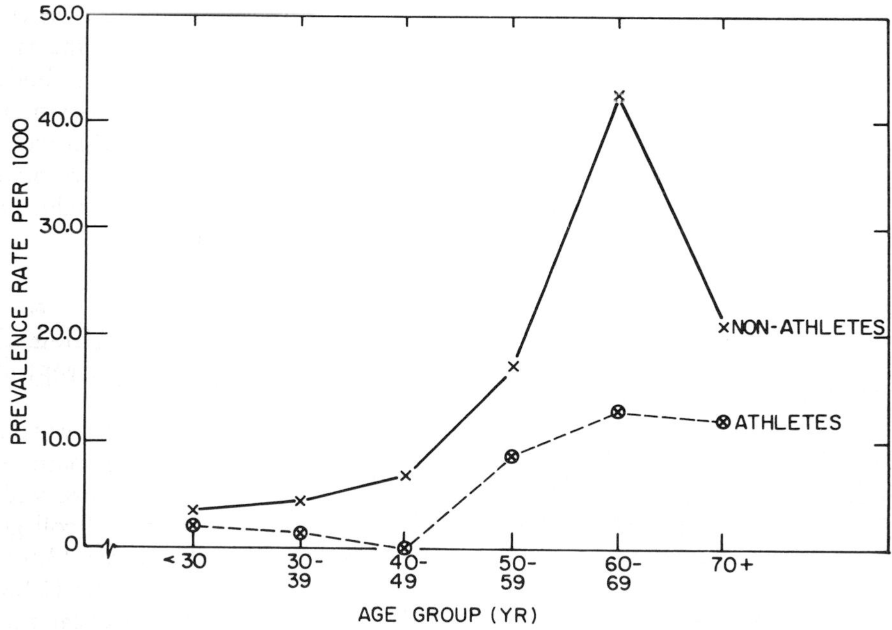

FIGURE 5 Prevalence rate (lifetime occurrence rate) of cancers of the reproductive system for athletes and nonathletes by age group. [From Frisch *et al.* (1985). *Br. J. Cancer* **52**, 885.]

human beings. That loss of fat and extreme leanness, not weight loss per se, are important factors in infertility can be deduced from the relative infertility of the breed of double-muscled Charolais cattle bred for lean beef. These female cattle bred for increased muscular mass have physiological troubles with their puberty, their fertility, and their sexuality in general. The Charolais bulls are also not very fertile.

Rats fed a high-fat (HF) diet, the fat being substituted isocalorically for carbohydrate, had first estrus (equivalent to menarche) significantly earlier than did rats fed a low-fat (LF) diet. The caloric intake per 100 g of body weight of the HF and LF diet rats did not differ at vaginal opening or at first estrus, whereas the two groups significantly differed at both events in age, absolute food intake, relative food intake, and absolute caloric intake. Direct carcass analysis data showed that the HF and LF diet rats had similar body compositions at first estrus, although the HF diet rats had estrus at a lighter body weight than the LF diet rats.

A. Link between Food Intake and Ovulation: "Flushing"

"Flushing" is the increase in the rate of twinning in sheep resulting from short-term (e.g., 1 week) high-caloric feeding of the ewe before mating to the ram. The well-nourished human female fortunately does not normally have multiple ovulations in response to a high-caloric intake, such as a large steak dinner; however, there is evidence for some residual flushing effect even in human beings. The rate of human dyzygotic twinning, which results from two independent ovulations, but not monozygotic twinning, fell during wartime restrictions of nutrition in Holland; the dyzygotic rate returned to normal after the return of a normal food supply.

IX. NUTRITION AND MALE REPRODUCTION

Undernutrition delays the onset of sexual maturation in boys, similar to the delaying effect of undernutrition on a girl's age of menarche. Undernutrition and weight loss in men also affect their reproductive ability. Loss of libido is the first effect of a decrease in caloric intake and subsequent weight loss. Continued weight loss to 25% of normal body weight results in a loss of prostate fluid and decreases of sperm motility and sperm longevity, in that order. Refeeding results in a restoration of function in the reverse order of loss. [*See* Malnutrition.]

A. Effects of Exercise on Males

Male marathon runners have a decreased hypothalamic GnRH secretion similar to women runners. Also reported are changes in serum testosterone levels with weight loss in wrestlers and a reduction in serum testosterone and prolactin levels in male distance runners. Hypothalamic dysfunction and hormonal changes in the male are apparently associated with more intense training and higher levels of physical activity than are observed in hypothalamic dysfunction in female athletes.

X. NUTRITION, PHYSICAL WORK, AND NATURAL FERTILITY: HUMAN REPRODUCTION RECONSIDERED?

The effects of hard physical work and nutrition on reproductive ability suggest that differences in the fertility of populations, historically and today, may be explained by a direct pathway from food intake to fertility, in addition to the classic Malthusian pathway through mortality. Charles Darwin described this common sense direct relation, observing that (1) domestic animals, which have a regular, plentiful food supply without working to get it, are more fertile than the corresponding wild animals; (2) "hard living retards the period at which animals conceive"; (3) the amount of food affects the fertility of the same individual; and (4) it is difficult to fatten a cow that is lactating. All of Darwin's dicta apply to human beings.

A. Paradox of Rapid Population Growth in Undernourished Populations

In many historical populations with slow population growth, poor couples living together to the end of their reproductive lives had only six to seven living births. Most poor couples in many developing countries today also only have six to seven living births during their reproductive life span. But, six children per couple today in developing countries results in a rapid rate of population growth because of decreased mortality rates, resulting from the necessary introduction of modern public health procedures. The difference between the birth rate/1000 and the death rate/1000 (which gives the growth rate) is now as high as 2, 3, and 4%. Populations growing at these rates double in 35, 23, and 18 years, respectively.

A total fertility rate of 6 or 7 births, typical of many developing countries today, is far below the human maximum of 11 or 12 children found among noncontracepting, well-nourished populations such as the Hutterites. The usual explanation of the lower than maximum fertility observed in both historical and contemporary societies is that it is due to the use of "folk" methods of contraception, abortion, or venereal disease, in combination with social customs that can affect fertility, such as late age of marriage or a taboo on intercourse during lactation. Because food intake and high energy outputs can directly affect fecundity, undernutrition and hard physical work are an alternate explanation of the observed submaximum fertility (Fig. 6).

British mid-19th century data on growth rates, food intake, age-specific fertility, sterility, and ages of menarche and menopause show that females who grew relatively slowly to maturity, completing height growth at ages 20–21 years (instead of 16–18 years as in well-nourished contemporary populations), differ from well-nourished females in each event of the reproductive span: Menarche is later (e.g., 15.0–16.0 years, compared with 12.8 years); adolescent sterility is longer; the age of peak nubility is later; the levels of specific fertility are lower; pregnancy wastage is higher; the duration of lactational amenorrhea is longer; the birth interval is therefore longer; and the age of menopause is earlier, preceded by a more rapid period of perimenopausal decline (Fig. 7).

A shortened, less-efficient reproductive span is also observed among the poor populations of many developing countries today, when data on age of menarche, age of menopause, length of birth intervals, and pregnancy wastage are available. Natural fertility therefore would be expected to rise as desired higher socioeconomic levels of a population are attained. For example, data show an increase in natural fertility in

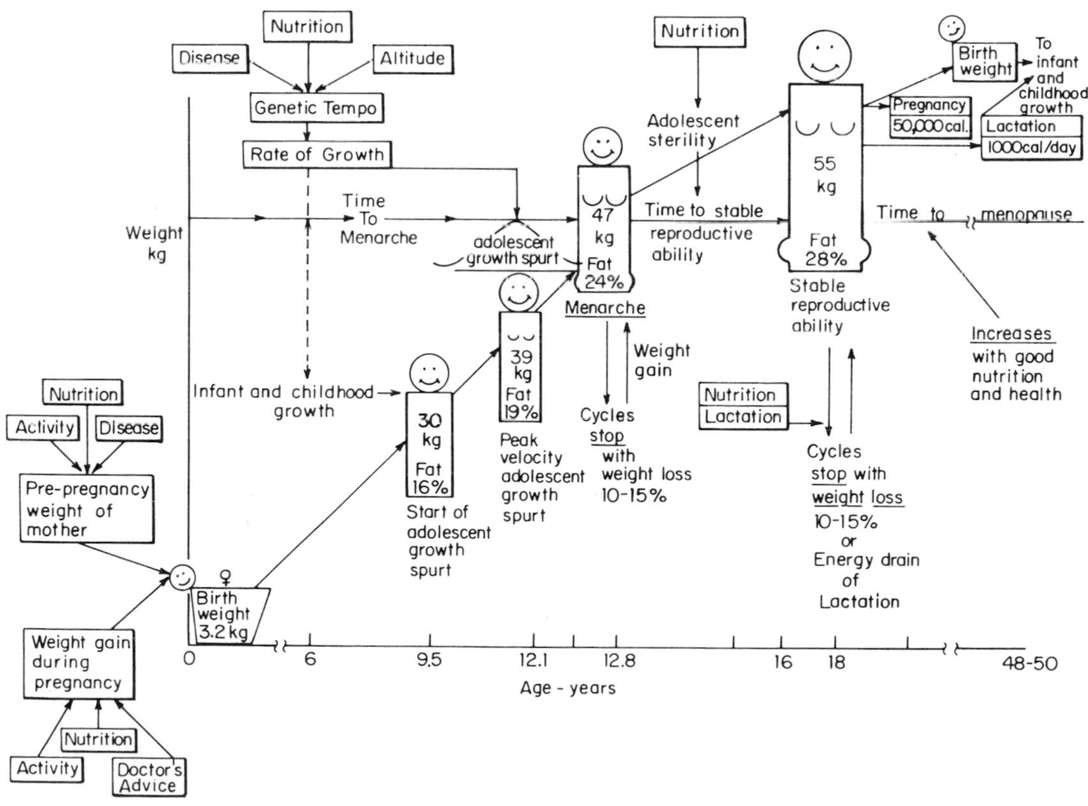

FIGURE 6 Biological determinants of female fecundity. Each reproductive milestone can be affected by environmental factors, as shown. The maintenance of regular ovulatory cycles is related to a minimum level of fat storage and is thus directly affected by undernutrition and energy-draining activities, such as lactation. [From Frisch, R. E. (1975). *Soc. Biol.* 22, 17–22.]

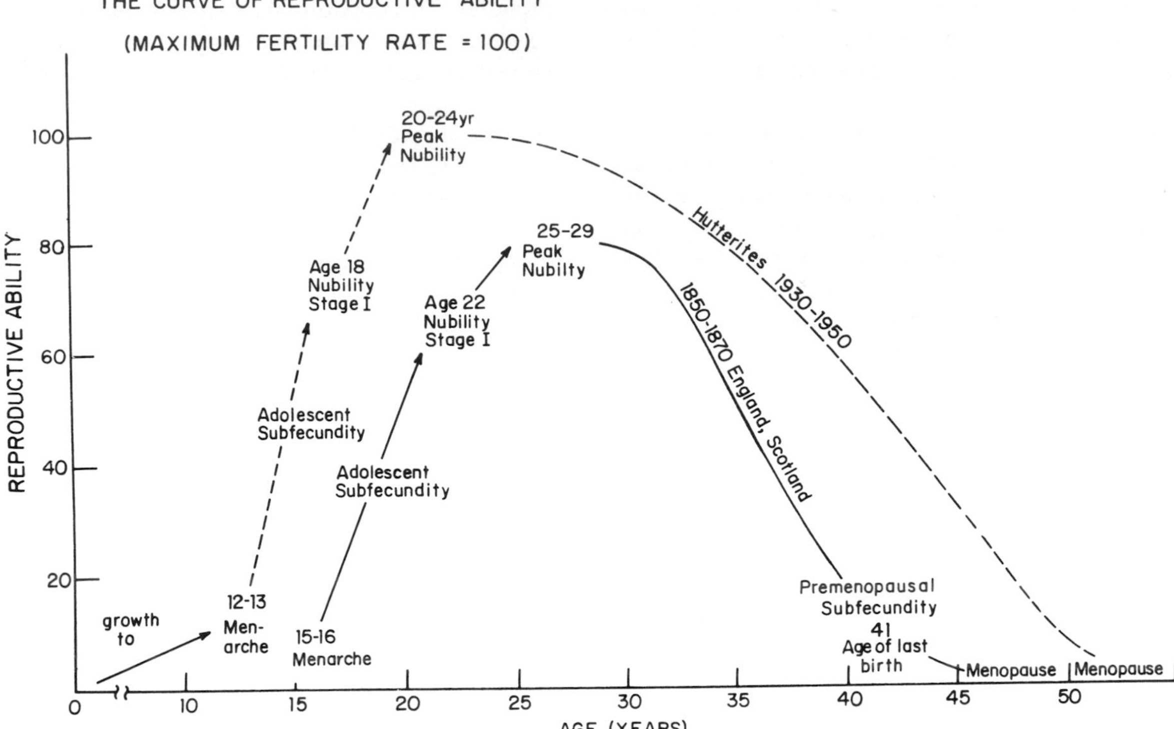

FIGURE 7 Mid-19th century curve of female reproductive ability (variation of the rate of child bearing with age) compared with that of the well-nourished, noncontracepting modern Hutterites. Whereas the Hutterite fertility curve results in an average of 10–12 children, the 1850–1870 fertility curve results in about 6 to 8 children. [From Frisch, R. E. (1978). *Science* **199**, 22–30.]

Taiwan, which accompanied recent improvements in health and nutrition. Therefore, the need for family planning programs, when there is the hoped-for progress in standards of living, may be much greater than realized heretofore during the transitional period to a desired lower family size.

BIBLIOGRAPHY

Friis-Hansen, B. (1965). Hydrometry of growth and aging. *In* "Human Body Composition" (J. Brožek, ed.), Vol. VII, pp. 191–209. Pergamon, Oxford.

Frisch, R. E. (1975). Demographic implication of the biological determinants of female fecundity. *Soc. Biol.* **22**, 17–22.

Frisch, R. E. (1978). Population, food intake and fertility. *Science* **199**, 22–30.

Frisch, R. E. (1985). Fatness, menarche and female fertility. *Perspect. Biol. Med.* **28**, 611–633.

Frisch, R. E. (1988). Fatness and fertility. *Sci. Am.* **258**, 88–95.

Frisch, R. E., and McArthur, J. W. (1974). Menstrual cycles: Fatness as a determinant of minimum weight for height necessary for their maintenance or onset. *Science* **185**, 949–951.

Frisch, R. E., Snow, R. C., Johnson, L. A., *et al.* (1993). Magnetic resonance imaging of overall and regional body fat, estrogen metabolism, and ovulation of athletes compared to controls. *J. Clin. Endo. Metab.* **77**, 471–477.

Frisch, R. E., Wyshak, G., and Vincent, L. (1980). Delayed menarche and amenorrhea of ballet dancers. *N. Engl. J. Med.* **303**, 17–19.

Siiteri, P. K. (1981). Extraglandular oestrogen formation and serum binding of estradiol: Relationship to cancer. *J. Endocrinol.* **89**, 119p–129p.

Vigersky, R. A., Anderson, A. E., Thompson, R. H., and Loriaux, D. L. (1977). Hypothalamic dysfunction in secondary amenorrhea associated with simple weight loss. *N. Engl. J. Med.* **297**, 1141–1145.

Wyshak, G., and Frisch, R. E. (1982). Evidence for a secular trend in age of menarche. *N. Engl. J. Med.* **306**, 1033–1035.

Body Temperature and Its Regulation

K. E. COOPER
University of Calgary

GLOSSARY

Antipyretic Substance that when administered to a febrile person reduces the body temperature

Fever Raised body temperature caused by a pathological process such as infection or diseased tissues

Hyperthermia Raised body temperature such as may be caused by an imbalance between heat dissipation and heat load (e.g., during the running of a race)

Hypothermia Condition in which the core temperature is below the normal regulated set point range

Metabolic rate Total energy production of an individual in unit time

Pyrogen Substance derived from materials outside the body (exogenous) (e.g., bacterial endotoxins) or from tissues within the body (endogenous) (e.g., interleukin-1), which can be produced by the body's macrophages and causes fever

Set point Temperature or range of temperatures of the body core, which is maintained and defended in the face of change in the thermal environment

Temperature regulation, autonomic Regulation of body temperature by involuntary processes such as sweating, circulatory adjustments, or shivering

Temperature regulation, behavioral Regulation of body temperature by voluntary means such as selection of different environmental temperatures or adjusting clothing

IT IS REMARKABLE THAT THE TEMPERATURE OF the human body, a large mass containing multiple sources of metabolic heat, which can vary its heat producing activity so widely and which can be exposed to environmental temperatures varying from at least +40°C to −40°C, can be regulated to within ±1–2°C over the long term and with only a few degrees alteration in severe physical activity. However, it is important to realize that all internal parts of the body are not always at the same temperature. It is customary to divide the body theoretically into two regions, the "core" and the "shell" (Fig. 1). The core consists of the contents of the skull, the thorax, and the abdomen, and the shell is made up of the skin, subcutaneous tissues, muscles, and all the limbs. The core temperature is usually kept within ±1°C of 37°C, whereas the temperature of the shell can widely vary before the core temperature changes significantly. The shell can thus act as a locus for heat "storage" and thus cushion the effects of changing environmental temperature on the body core. The core temperature is what we refer to when we use the term "body temperature." There are some variations of temperature within the core, with the rectal temperature being higher and slower to respond to heat or cold exposure than that measured in the esophagus or other upper body sites. For clinical purposes it is usual to measure the body temperature with a thermometer in the closed mouth under the tongue or in the rectum. For scientific studies, other measuring loci may be used, including the tympanic membrane, the external auditory canal, and the esophagus. The locus of the measurement is determined by the information and the speed of response needed to answer the problem under study. Depending on the location of measurement and the information required, the measurement of core temperature is made with mercury-in-glass thermometers, thermocouples, thermistor devices, resistance thermometers, and, on the skin surface, infrared thermography.

The human belongs to a group of animals known as endotherms, which maintain their core temperatures

ENCYCLOPEDIA OF HUMAN BIOLOGY, Second Edition, VOLUME 2. Copyright © 1997 by Academic Press. All rights of reproduction in any form reserved.

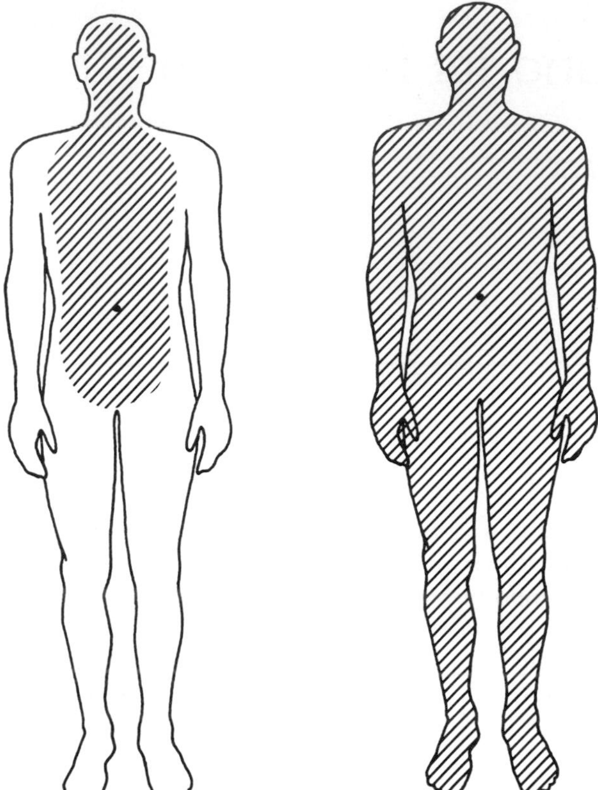

FIGURE 1 The shaded area in the figure on the left shows the region known as the body "core" and the unshaded area is the body "shell." Temperature of the core is maintained relatively constant and is defended, whereas that of the shell can widely vary under different thermal conditions of the environment. The individual on the left could have been sitting in a cold room for some time. The individual on the right had been resting in a hot environment, and the shell temperature is at the core temperature. The individual on the right has much more "stored" heat in the body at the same core temperature as compared with the one on the left.

constant in widely altering environmental conditions by both physiological and behavioral means using a neural control system. Other animals (e.g., desert lizards) are able to maintain their body temperatures fairly constant above environmental temperature by the behavioral device of shuttling between sunlight and shade whereas others seek certain warmer or cooler waters (i.e., having thermal preferenda); such creatures are known as ectotherms.

In the human, there is a circadian rhythm of core temperature. Body temperature is at its peak in the early evening and falls to a minimum in the early morning. The extent of this fluctuation varies from one person to another and may be close to 1°C. The human female also has a temperature change during the menstrual cycle. The body temperature taken on waking rises about 0.5°C on the morning after ovulation and remains at this elevated level until the start of the next menstrual period. This change is often used to pinpoint the time of ovulation in the management of infertility.

It is of interest that most mammals have core temperatures in the range of 36–39°C whereas birds have a slightly higher body temperature. The precise reason, if there is a single one, for this selected temperature range is still not defined. The endotherm does have the advantage of being active and potentially alert at all times, the most do not become torpid as the environmental temperature falls. It is presumed that the body temperature range is favorable for the best performance of the full range of enzyme systems and biochemical processes in the mammal. However, whether the selection of a constancy of core temperature preceded the selection of the biochemical processes that operate optimally at the higher temperature or vice versa or, more likely, whether there was parallel evolution is a matter for speculation.

I. PHYSICAL MEANS OF HEAT EXCHANGE

Heat may be exchanged between the body and the environment by four physical processes: convection, conduction, radiation, and evaporation. The equations for heat exchange are given, using °C as the unit of temperature. In much of the thermodynamic literature, °K (absolute) is used.

A. Convection

Convection of heat from the body surface involves transfer of heat to the air at the surface and the movement of the heated air away from the surface. If there is no wind at the body surface and the body is still, the convection is termed *free convection.* Air close to the skin is heated, becomes less dense or more buoyant, and rises. Thus an upward stream of air flows over the body surface and carries heat away. Such movement of heated air over the body can be visualized by Schlieren photography, which makes use of the variations in the refractive index of the layers of air of different densities. If there is air movement (i.e., wind) playing onto the body surface or if the body is moving, the convection is increased and the term *forced convection* is used. There may be a thin layer of still air at the body surface, the boundary layer, which can act as an insulating layer. There are also points at which there are pockets of still air, the so-called stagnation points (e.g., the perineum and axilla). The rate of heat exchange per unit area is given by the equation $C = h_c(T_1 - T_2)$, where h_c (W/m²/°C) is the convection heat transfer coefficient, T_1 and T_2 are the temperatures of the skin surface (°C) and the adjacent air, respectively, and C is the heat exchange in W/m². The coefficient h_c depends on the pattern of flow of air and on the thermal properties of the air (or on immersion in a fluid on the properties of the fluid). It is of interest that there can be no free convection in space travel in zero-gravity conditions, adding to the problems of thermoregulation.

B. Conduction

This process involves the transfer of thermal energy from molecule to molecule through a substance, either solids or, under some circumstances, liquids. Heat flows from the region of higher temperature to that of lower temperature. The heat transfer, H (in watts), is determined by the following relation: $H = (T_1 - T_2) \cdot R^{-1}$, where T_1 and T_2 are the temperatures of the sites through which the heat flows and R is the thermal resistance between them and is defined as °C/m²/W. The inverse of this resistance is termed *conductance* (C), and its units are W/m²/°C (i.e., heat flux per unit area per unit of temperature). Conductive heat loss occurs from the body when it is in contact with a cooler solid (i.e., lying on a thin mattress on the snow), it takes place through clothing, and it is combined with convection when the body is immersed in air or water. The thermal conductivity of a substance, k, is the heat flow through unit thickness of the substance per unit area per temperature difference across the unit thickness, and it has the units W/m/°C. A metal object having a high thermal conductivity will extract heat rapidly from skin in contact with it, leading to serious frostbite at low environmental temperatures.

C. Radiation

Heat will be exchanged between two facing surfaces by radiation even across empty space. The amount of heat exchanged is proportional to the difference in the fourth powers of the absolute temperatures of the surfaces. The radiation heat exchange also depends on the properties of the radiating surfaces and their emissivities (or ability to radiate or absorb radiant energy at different wavelengths) and can be expressed mathematically as $H_r = \sigma e_1 e_2(T_1^4 - T_2^4)$ in W/m², where T_1 and T_2 are the temperatures of the radiating surfaces in °K (°C + 273), σ is the Stefan–Boltzmann constant ($5 \cdot 67 \cdot 10^{-8}$ W/m²/K⁴), and e_1 and e_2 are the emissivities of the radiating and receiving surfaces. When the temperature difference between the body surface and the environment to which radiation takes place is small (i.e., <15°C), the radiant heat loss, R, in W/m², is approximately $k_r(T_s - T_e)$, where $k_r = \sigma T_s^3$ (T_s is mean skin temperature and T_e is radiation receiving environment temperature). Radiation heat losses can be large (e.g., from a cross country skier radiating heat to a snowbank, or radiation to a clear night sky), and heat gain from solar radiation can also be large. Clothing can be used both to minimize radiation heat loss and also to reflect radiation from a hot source.

D. Evaporation

The evaporation of water from the skin surface and from the respiratory tract is an important means of heat loss. It is of vital importance to body temperature regulation in hot climates and can be a major cause of hypothermia in a cold, damp exposure. As water evaporates, heat is required to effect the change of state, and this heat in the case of the body is extracted from the body. This *latent heat of vaporization* (L) is partly dependent on the temperature and is given by the equation $L = 2490.9 - 2.34\ T(J_g)$, where T is the water temperature in °C. At 30°C, evaporation of 1 g of water from the body surface would extract 2421 J (or 576 gcal) of body heat. Water reaches the skin surface in two physiological ways: insensible and sensible perspiration. In the former case, water diffuses through the skin and evaporates from the skin surface at a rate entirely dependent on the water vapor pressure at the skin surface. This process is without physiological control. The other mode of water reaching the skin surface is active excretion via the sweat glands, the mechanism of which will be considered below.

E. Insulation of Clothes

Thermal insulation, also known as thermal resistance, is the inverse of thermal conductivity. It is often measured and expressed in an arbitrary but convenient unit, the "clo" (1 clo = 0.155°C/m²/W). This provides enough insulation to maintain a seated person comfortable at 21°C, in air having an air movement of

0.1 m/sec. A person in a light suit would be wearing about 0.6 clo insulation and, in a winter parka with appropriate winter trousers and underwear, about 3–4 clo.

F. Heat Balance Equation

The exchange of heat between the body and its environment may be expressed in a simple form, namely, $S = M - E \pm W \pm C \pm R \pm K$, where S is heat storage (W or W/m^2 or W/kg), M is metabolic heat production, E is evaporative heat loss, W is positive or negative work accomplished or absorbed by the body, C is convective heat transfer from or to the body, R is radiant heat exchange from or to the body, and K is conductive heat transfer from or to the body. At perfect body thermal equilibrium, $S = 0$.

II. PHYSIOLOGICAL MECHANISMS USED IN BODY TEMPERATURE REGULATION

A. Heat Balance

1. Behavior

The mechanism of thermoregulation that has been the most important in enabling the human race to spread to all climatic regions of the earth is behavioral thermoregulation. The ability to adjust the insulation of clothing, to construct shelters, and to use fire and other energy sources to heat or other means to cool the immediate environment are of paramount importance in human body temperature control. The use of such a mechanism implies good sensing of the environmental temperature and the appropriate response to it. Some elderly people who have seriously dulled perception of cold may fail to heat their residences or to wear appropriate clothes in winter and become hypothermic. Similar perception problems may lead them to suffer heat illness in heat waves. Apart from adjusting the environmental temperature or clothing, the person may increase voluntary activity to make more heat and may adopt a curled-up (fetal) posture to reduce the surface area and so minimize the heat loss. Huddling together of a group is also an example of behavioral thermoregulation.

2. Skin Circulation

Blood flowing through the skin carries heat from the body core to the body surface from which, in cool environments, it is transferred to the environment. The skin contains complex networks, or plexuses, of capillaries, and the blood flow into these is regulated by muscular precapillary sphincters, which are under the control of the autonomic nervous system, which is connected to the thermoregulatory system. In some regions of the body, specialized capillary loops are close to the skin surface at the bases of which are larger diameter shunts with muscular walls. When the muscular walls relax, a huge blood flow occurs through the shunts, and although they are deeper than the capillaries, the heat loss from such a large blood flow is much greater than could be obtained from a slower flow in the superficial capillaries. These shunts are found in the fingers and the toes and to a lesser extent in the more proximal parts of the limbs. They are valuable in that during, for example, exposure of the digits to cold water, they open up in a cyclical fashion and heat the local tissues, so helping to avert cold injury. This high blood flow in cold digits is called "cold vasodilatation," or the "Lewis's hunting response." At neutral environmental temperatures, the blood vessels in the skin of the distal parts of the limbs are moderately *constricted* by tonic impulses in the sympathetic nerves to the blood vessels, but as the body warms up, less tonic impulses flow to the blood vessels and the vessels open up to allow more blood to flow. This process is often termed *release of sympathetic tone.* In contrast to this, blood vessels in the more proximal parts of the limbs and much of the trunk are *dilated* by tonic impulses in the sympathetic nerves going to them, a process known as *active vasodilatation.* Some hold that part of this active vasodilatation results from release of substances from activated sweat glands, thereby linking the blood flow to the secretory activity of the sweat glands, but this mechanism is still regarded as controversial. [*See* Skin.]

The blood vessels in the skin constrict or dilate in response to two sets of circumstances: (1) a rise or fall in core temperature or (2) reflexes from warming or cooling the skin. Thermally sensitive nervous receptors in the central nervous system can appropriately respond by modulating the sympathetic nervous tone to skin blood vessels to limit the extent of the changing deep body temperature; these receptors are discussed below. Short-lived vasoconstriction can be induced by suddenly cooling a relatively small area of skin, and profound vasodilatation can be induced in the skin by exposure of a larger area of the rest of the body surface to, for example, radiant heat, and these responses depend on stimulation of thermally sensitive nerve endings in the skin. Presumably they represent rapid first lines of defense of the body temperature in response to large and rapid changes in the thermal environment.

Another important avenue of heat conservation lies with the distribution of the veins in the limbs. There is an extensive network of superficial veins in the limbs and, in addition, there are veins that run in close apposition to the main arteries. These are called the *venae comitantes*. In a hot environment, blood returning from the periphery of the limbs courses back to the core via the superficial veins, and thus blood arriving in the digits is maximally warm, thus leading to maximal heat loss. During cold exposure, however, blood returns from the digits to the core mostly by way of the venae comitantes and, in so doing, precools the blood going to the digits. The result is reduced heat loss from the digits and some warming of the cool blood returning to the core. This process is called the *countercurrent heat exchange mechanism*, a mechanism that has been copied by engineers in efforts to conserve heat.

3. Evaporative Heat Loss

This takes place in three ways. First, there is diffusion of water, which takes place through the skin from the surface of which it evaporates. The water evaporated in this way is called *insensible perspiration*, and its extent is governed by the water vapor pressure prevailing at the skin surface. A resting individual at an average comfortable room temperature might evaporate some 16 g/m^2 hr of water by insensible perspiration. It is not under physiological control other than behavioral adjustment to the environmental humidity. Second, an important route of evaporative heat loss is from the respiratory passages. Air breathed in becomes saturated with water vapor, which comes from the lungs and the respiratory passages, and this saturated gas is exhaled. Water loss by this route can be large (e.g., during vigorous exercise during mountain climbing at high altitudes). In such circumstances the large loss of water can lead to dehydration, and the heat loss from evaporation can contribute to subsequent hypothermia. Respiratory evaporative heat loss is a major avenue of heat loss and body temperature control in animals that have no sweat glands, such as the dog, and the variation in this heat loss is brought about by panting. Breathing dry air, the human could at high levels of pulmonary ventilation lose up to 1.5 g water per minute, giving an evaporative heat loss of 60 W. To the evaporative heat loss so caused must be added the heat lost in heating the inspired air, or have subtracted the heat gained in cooling it. Third and of supreme importance in the human is sweating.

Distributed throughout the skin are glands consisting of coiled ducts with their bases in the corium and their outlet into sweat pores (Fig. 2), which are funnel-shaped cavities in the corium, or outer layer of skin. Sweat glands are divided into two types: the apocrine and eccrine glands. The eccrine glands are distributed all over the body surface of the human, but there are some racial variations in their density, whereas the apocrine glands occur in the axillae, the groin regions, the genitalia, and in other regions in which there are also racial variations. The eccrine glands subserve mainly the function of providing a film of water over the body surface, which is the main source of evaporative heat loss, and the apocrine glands secrete in response to emotional stimuli, fear, and pain and may have some sexual function. Sweat is a watery fluid that contains sodium chloride (salt) and also some lactate, urea, potassium, and other

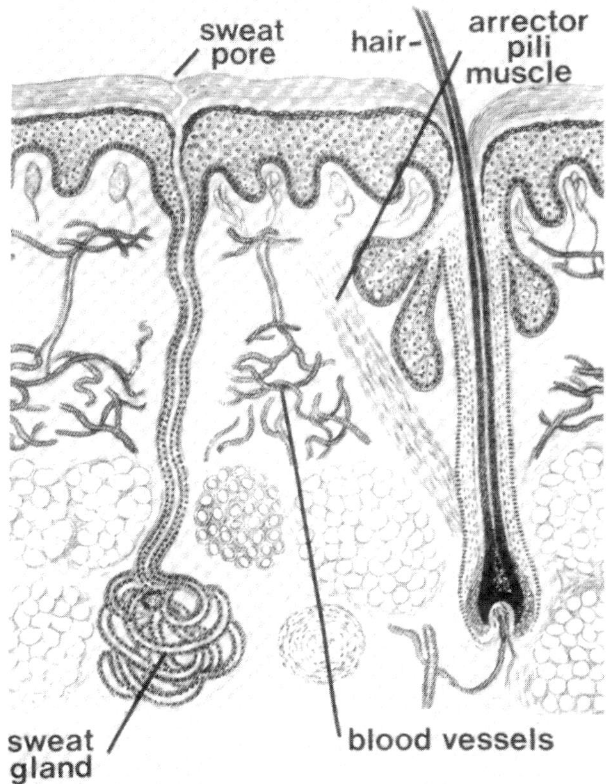

FIGURE 2 Section of the skin showing a sweat gland with its duct and pore, a hair follicle and its arrector pili muscle, and skin blood vessels.

inorganic substances that would be expected in an ultrafiltrate of plasma as well as small organic molecules of the same origin. Apart from water, sweating can result in significant loss of salt and also of potassium. On a hot day, performing light work, a man could sweat some 125 g/hr; doing moderate work in the desert, sweat loss could range from 150 to 700 g/hr and, under more severe conditions, can range from 1.2 kg/hr in unacclimatized men to 3.5 kg/hr in acclimatized men. The efficacy of heat loss by evaporation of sweat is determined by a number of factors: (1) the area of sweating skin exposed for evaporation (thus a person in a tight impermeable suit would lose most of the advantage of sweating); (2) the evenness of the distribution of the sweat over the skin surface, which is often called the *effective wettedness*; and (3) the water vapor pressure at the skin surface and the salt concentration in the sweat. This latter has two effects on the sweat evaporation, namely, by reducing the vapor pressure, it diminishes the evaporative potential, but by increasing the ability of the sweat to wet the skin as a slimy film, it enlarges the surface area available for evaporation.

An important process that occurs on repeated daily exposure to a hot climate, particularly if physical work is performed, is known as *acclimatization*. There is a progressive reduction in the rise of core temperature for a given work load in the same hot environment, and much of this alteration runs parallel to an increased sweat rate and more efficient wetting of the skin. The increased volume of sweat produced is accompanied by a reduced salt concentration. Given sweat rates (at below maximum) are achieved at lower skin temperatures after acclimatization. It is possible, by daily raising the body temperature passively to about 38°C for 1–2 hr, to achieve good acclimation to heat. The term *acclimation* is often used to denote experimentally induced changes characteristic of naturally induced acclimatization. Although the evidence is not as complete as for sweating, there is also some cardiovascular acclimatization, although in many studies it is difficult to separate the effects of exposure to the heat from the effect of exercise training.

4. Piloerection

Furry animals are able to raise the hairs over the body in the cold and trap a deeper layer of still air over the skin, increasing the thermal insulation of the furry coat and reducing heat loss. The human has the apparatus to erect the skin hairs, but not having a thick furry coat, it is of little thermoregulatory value. The result of contraction of the little muscle attached to the hair follicle, the arrector pili, is known colloquially as a "goose bump."

5. Heat Production

The chemical reactions that accompany the processes of life are overall exothermic (i.e., they release energy in the form of heat). The minimum release of this energy in a subject at complete rest, fasting, awake, and in a thermoneutral environment is called the basal metabolic rate (BMR). Because these conditions are often difficult to achieve, conditions can be standardized to be as close as possible to those used for BMR, and thus the standard metabolic rate (SMR) is measured. Its units are W, W/m^2, W/kg, or W/kg$^{3/4}$. Other measures of metabolic rate that are used are *maximum* metabolic rate (i.e., the highest aerobic work sustainable over a standard period), *peak* metabolic rate (i.e., the maximum achieved in cold exposure), *resting* metabolic rate (i.e., that of a person at rest in a thermoneutral environment but not fasting), and many others. The BMR of men is slightly higher than that for women. Expressed as a function of body surface area, it is highest in young children of both sexes and diminishes throughout life from about the age of 6 years. It is raised in conditions in which excess thyroid hormones are secreted and diminished in states of low thyroid hormone release. When food is eaten, the two effects on metabolic heat production are, namely, an immediate fairly short-lived rise in metabolism (i.e., the diet-induced thermogenesis, which is maximum with eating the most palatable and attractive foods) and a later release of energy associated with the breakdown of the foods, particularly proteins, and the synthesis of these products into new compounds (sometimes called specific dynamic action of food).

Three other avenues of increased heat production are physical activity, shivering, and nonshivering thermogenesis. An awake resting man could have an energy expenditure of 90 W, it may be 200 W, doing light exercise and 550 W in severe exercise (77, 172, and 473 kg/cal/hr). So it is clear that physical activity has a profound effect on the body's heat production and that it can aid resistance to a cold environment, providing that an adequate food intake is maintained to provide metabolic substrates.

Shivering is uncoordinated widespread muscle contractions that perform no useful work but that cause the liberation of heat. Shivering can be triggered by a substantial drop in skin temperature and is more intense as the core temperature falls. It can be an important element in defense against external cooling because it can increase the metabolic rate by a factor of 50–100%, and this can be maintained over several hours. Greater increases can occur over a few minutes only.

Nonshivering thermogenesis is an important element in cold resistance in many mammals and is important in newborn babies. The evidence for its occurrence in adult humans is, at present, incomplete. In nonshivering thermogenesis, heat is produced, in response to body cooling, partly in the liver and also in specialized fat depots known as "brown adipose tissue." In the latter, the fat cells are smaller than in other fat depots, they have a rich blood supply under the control of the sympathetic nervous system, and they are packed with mitochondria. Sympathetic nerve stimulation releases norepinephrine at the surface of these fat cells, which, acting through the adrenergic β receptors, triggers chemical processes that release heat. The heat generation is correlated to the oxidation of fatty acids, and a polypeptide (P 32000 polypeptide) present in the inner mitochondrial membrane plays an important role in this by causing a proton shunt that bypasses oxidative phosphorylation. The brown adipose tissue is principally found in the newborn between the shoulder blades, in the back of the neck, and in the perirenal regions. Veins draining these fat pads carry heated blood directly into the body core. The release of heat from brown fat is rapidly induced on cold exposure.

Some think that there are widely dispersed brown fat cells in the adult human and that they play an important role in thermoregulation, but more evidence is needed to substantiate this view.

B. Sensing Temperature

For the temperature control mechanism to be effective, it must be supplied with information about the temperature of the body surface and about that of the body core.

There are nerve endings in the skin that are sensitive to temperature and to temperature change. The information about temperature is transmitted to the central nervous system in the form of a frequency code. Two types of receptor are found in the skin, one of which responds to heating and the other to cooling. They are known as "warm-sensitive" and "cold-sensitive" receptors, respectively. The pattern of firing of these receptors is in the form of a bell-shaped curve. For a warm receptor, the firing rate increases as its temperature rises up to a peak after which the rate falls. There are numerous such receptors in the skin having a range of peak firing rates. The cold receptors are similar, but the firing rate increases as the temperature falls. In addition to these relations that pertain to steady temperatures, the skin nerve endings have what is called *dynamic responses* to a sudden change in temperature. When the temperature alters, the firing rate of the ending is rapid and it then slows to a plateau rate, or static firing rate characteristic of that ending at that temperature. These phenomena are illustrated in Fig. 3. These characteristics enable the central nervous system to be aware of rapid changes in skin temperature as well as of the absolute skin temperature. It is also possible that by having

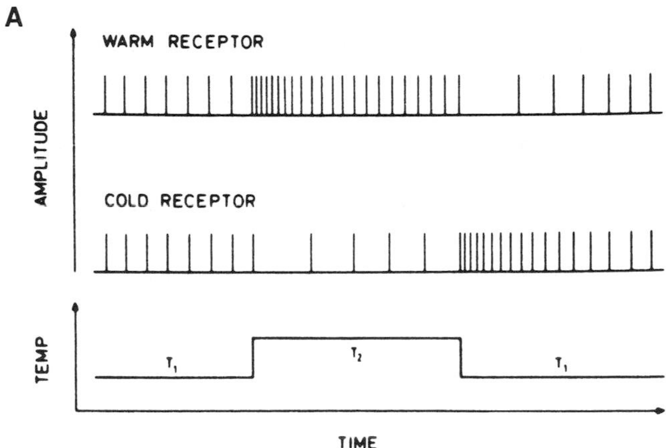

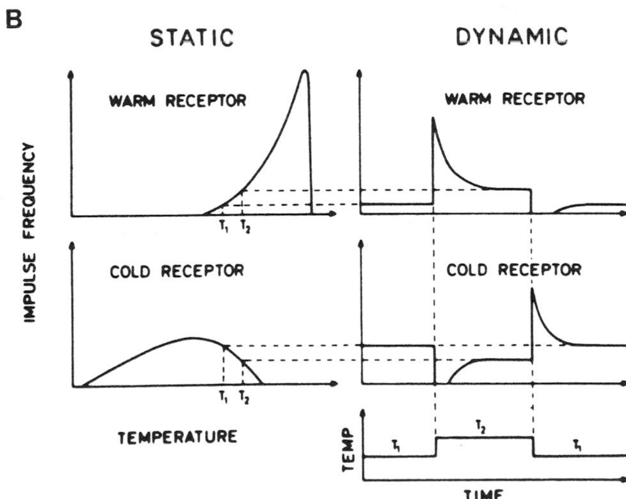

FIGURE 3 Properties of thermoreceptors. (A) Nerve impulses from single warm and cold receptors in response to sudden changes in temperature. (B) Static and dynamic responses of these receptors in response to steady temperatures or to changes in temperature.

thermoreceptors at different skin depths, the brain could be informed about the rates of heat flux through the skin.

Temperature-sensitive neuronal units within the brain appear to be important in thermoregulation. They are found in the anterior hypothalamus, the preoptic area, the septum pellucidum, the posterior hypothalamus, and throughout the brain stem. The greatest density of highly thermosensitive neurons is in the anterior hypothalamus and preoptic areas. Here there are more warm-sensitive than cold-sensitive neurons. The characteristics of the brain thermosensitive neurons are shown in Fig. 4. They fire at rates linearly related to their temperatures over their effective ranges, with warm-sensitive units increasing their firing rates as their temperatures rise and vice versa. These units are positioned so as to monitor the temperature of the blood perfusing the base of the brain. Thus, in addition to the information about the thermal condition of the body surface, the temperature controller is also informed about the blood, or core, temperature. There are also less sensitive thermoreceptors in the spinal cord, and there is evidence for others in some of the viscera. Thus the controller has good information about the integrated state of the thermal condition of the body. The mechanism of transduction of temperature into impulses by thermosensitive units is not properly understood. There are also highly thermosensitive neurons in parts of the brain not thought to play a role in thermoregulation, and hence it is important to demonstrate that role before assigning such units to the thermoregulatory system. [*See* Hypothalamus.]

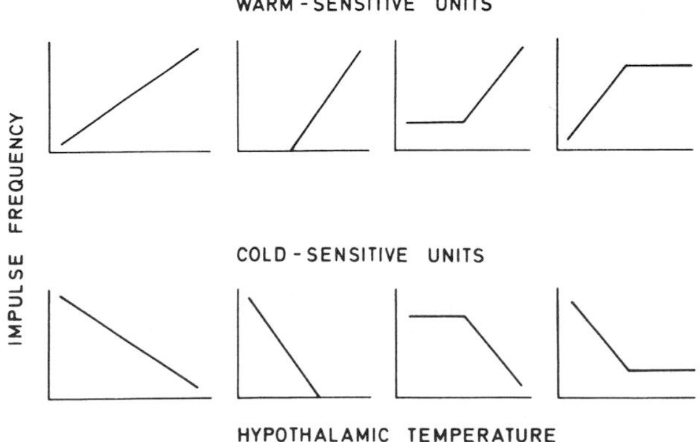

FIGURE 4 Frequency/temperature curves for thermoreceptors found in the hypothalamus.

C. Body Temperature Set Point

If the body temperature is raised or lowered by nonpathological processes (e.g., exercise or immersion in cold water), it returns to the original level when the altering activity ceases. Furthermore, attempts to alter the core temperature are resisted, and the appropriate heat loss or heat conservation measures are invoked to prevent a displacement of the body temperature. The defended temperature behaves as though there is a set point of core temperature, analogous to the setting of a thermostat. Some have argued that the apparent set point is no more than a dynamic result of the sum of the activities of heat gain and heat loss mechanisms, but the evidence is heavily against such a view. The brain mechanism for the set point is not understood. Some regard it as a comparator using nonthermally sensitive neurons as reference signals against which the output of the thermosensitive units is gauged. Mathematical models can be used in which, for example, the intersection of the activity curves of cold and warm sensors provides the set point temperature. Others use the relative secretion of neurotransmitters used in heat loss and heat gain pathways to determine the apparent set point. One set of experiments has suggested the hypothesis that the set point is determined by the relative concentrations of Na^+ and Ca^{2+} in some part of the posterior hypothalamic thermoregulatory neuron pools. We do not yet know for certain how the set point system works, what transmitters are used, or where it may be located. We do know that the set point fluctuates on a daily cycle and that in women it rises after ovulation by up to 0.5°C.

D. Temperature Controllers

Between the temperature sensors and the effector mechanisms for the induction of heat gain or heat loss, there must be a regulating system or systems. The nature and location of these are not known at present. It is likely that the final common path for the effectors used in thermoregulation is in the posterior hypothalamus. It is also likely that the old, but often quoted, statement that there is a center for heat loss in the anterior hypothalamus and one for heat conservation in the posterior hypothalamus is erroneous. It is clear that if the heat conservation pathways are driven by, for example, cold exposure, the heat loss activity pathways will be inhibited and vice versa. Evidence suggests that whereas an essential part of the control of autonomic thermoregulation lies in the rostral hypothalamus and preoptic region, the lateral hypothalamus is essential to behavioral thermoregulation.

III. DISTURBANCES OF THERMOREGULATION

A. Fever

One of the oldest recognized symptoms of disease is fever. In fever there is a rise in body temperature which is part of a coordinated host defense response known as the "acute phase response"; this includes a reduction in the plasma level of iron and of zinc as well as alterations in plasma levels of several trace metals. The fall in plasma iron may help reduce the rate of bacterial replication for which iron seems necessary, but the value of changes in the levels of other metallic ions is far from clear. Changes in plasma levels of immunoglobulins occur as well as other changes due to immune system stimulation.

The major breakthrough in our understanding of fever came in the 1940s with the discovery by Dr. Paul Beeson that pyrogen-free, sterile suspensions of rabbit white cells could release a fever-producing substance on warm incubation. This substance was called *leukocyte* pyrogen or later *endogenous* pyrogen. In the mid-1980s, this was shown to be a low molecular weight (15–18 kD) peptide now known to be the cytokine interleukin-1 (IL-1). This cytokine, present in several forms, is one of a family of peptide cytokines of importance to the immune system in the host defense response. Other cytokines released in febrile states include tumor necrosis factor (TNF), interferons-α and -β (elaborated during viral infections), and interleukin-6 (IL-6). These and others are mediators of fever.

IL-1 is released from monocytic cells and from fixed macrophages of the reticuloendothelial system when these are stimulated by bacterial pyrogens. Bacterial pyrogens, often called endotoxins, are lipopolysaccharides (10^6 D) derived from bacterial (particularly gram-negative organisms) cell walls. IL-1 acts on the brain (and some say also peripherally) to cause the release of prostaglandin E_2 (PGE_2) which acts within the anterior hypothalamus and preoptic areas (AH/POA) and the septal area of the brain to stimulate heat conservation and heat production and to diminish heat loss. Thus the body temperature rises to a new level which is defended, that is, to a new level of the set point for body temperature. The locus of action in the brain is still the subject of much research. If injected into the anterior hypothalamus/preoptic area it causes fever, but there is no evidence that it can penetrate the blood–brain barrier to diffuse into this area. It may penetrate into one or more of the circumventricular organs which are in communication with the inner surface of the brain and which have less tight cell junctions in their blood vessels. One such organ lying close to the septum, and a very likely candidate for the site of action of pyrogenic cytokines, is the organum vasculosum laminae terminalis (OVLT). IL-1 could act within the OVLT to cause local release of PGE_2 which could diffuse to the AH/POA; it could act there to stimulate nerve fibers going to the AH/POA which then release PGE_2 there; or it could, and this is less likely, diffuse into the brain from the OVLT. There are IL-1-containing neurons in the AH/POA and neurons there which can release PGE_2. Maybe these could be part of a system for inducing fever in infections within the brain, and the OVLT route can induce fever in peripheral infections. If IL-1 were to be released within the hypothalamus, the effect of thermosensitive neurons there is to increase the firing rate of cold-sensitive units and to diminish the firing rate of warm-sensitive neurons. Such activity would be expected to diminish body heat loss and enhance heat production, thus raising the body temperature until the new temperature set point is reached. The effect of PGE_2 on thermosensitive neurons is less clear. However, the net effect of the action of the prostaglandin release within the hypothalamus is to cause shivering and, in some animals, there is an increase in heat production in brown adipose tissue. There is intense vasoconstriction in the skin. The skin vasoconstriction leads to a sense of cold, and behavioral responses such as covering the body with blankets and turning on bed heaters take place. When the effects of the hypothalamically released mediators of fever wear off, the body temperature set point returns to normal. Sweating occurs and vasodilatation takes place in the skin, the heat production falls and the behavioral responses also alter to increase heat loss.

Some work done on ectotherms and small mammals suggests that fever, especially in the first few hours after an infection, may be beneficial in fighting the disease. However, this should not prevent the physician having modern medicines from preventing dangerous rises in core temperature of risking febrile convulsions in young children.

If fever has been selected during evolution as an important defense mechanism against infection, then it would not be surprising to find that another mechanism may prevent excessive and harmful fever may be available to the body. This process, known as *endogenous antipyresis,* exists in six mammalian species. At present, two peptides, namely arginine vasopressin (AVP) and α-melanocyte-stimulating hormone (α-MSH), have been shown to exist in neuronal pathways which connect to thermoregulatory regions of the brain. These substances, or fragments of them, have been shown to reduce the extent of fever when infused into septal region. For example, AVP is secreted by neurons derived from the bed nucleus of the stria terminalis or the hypothalamic paraventricular nuclei into the ventral septal area close to the diagonal band of Broca where it reduces fever but does not alter normal body temperature. α-MSH in neurons derived from the arcuate nucleus acts similarly in the lateral septum, and its fragment also have a peripheral antipyretic action. The antipyretic action of these peptides within the brain has been demonstrated in several species, but for obvious ethical reasons, it has not been possible to make similar studies in humans. Thus the presence or actions of endogenous antipyretic peptides in the brain have not been demonstrated in the human, and it may take many years to obtain evidence for or against such a fever feedback control system in the human.

The action of AVP within the septum has been shown in some species to be associated with the refractoriness to fever that occurs in the mothers at the time of birth and for a few hours after birth in the newborn. It might be associated with the absence of fever often seen during severe infections in the newborn human infant. A team in South Africa has shown that the blood concentrations of naturally occurring IL-1 receptor antagonists rise in the maternal circulation close to term in the human and that these antagonists cross the placenta into the fetal circulation. These high levels of circulating IL-1 receptor antagonists may be of major importance in the suppression of fever in the newborn human infant.

B. Hyperthermia

This term refers to a high body temperature, usually not associated with a rise in set point, which occurs as a result of an imbalance between heat production and gain and heat loss. For example, forest or brush fire-fighting crews in Australia may be exposed to intense heat for long periods while dressed in relatively impervious protective garments. The result of the high energy expenditure of the fire fighting combined with restricted heat loss can lead to heat stroke, a catastrophic rise in body temperature which often causes death. Other factors involved in the induction of heat stroke under conditions in which heat loss by sweating would be impaired include restricted water intake and inadequate salt intake. The old notion that taking salt tablets would make up for the salt loss in sweat is no longer accepted because the tablets tend to lie undissolved in the stomach. The usual daily intake of salt in the diet should be adequate to maintain sweating. Minor infection may also lead to an abrupt cessation of sweating with a consequent severe rise in body temperature. The use of hot whirlpool tubs has also led to death from overheating, particularly when alcohol is consumed during the immersion in the hot pool. It is worth noting that the elderly are particularly at risk in severe heat waves. This is partly due to (1) diminished thermoregulatory control in some, but not all, of the elderly (2) socioeconomic factors such as the inability to afford air conditioning.

A deadly form of hyperthermia, *malignant hyperthermia,* occurs rarely during surgical procedures in response to certain anesthetics and some muscle relaxants. The disorder is a failure of the normal membrane calcium actions in muscle fibers determined by the anesthetic or muscle relaxant. It results in massive muscle contractions and enormous heat production in the muscles, which in a few minutes can lead to death from hyperthermia. The condition has a familial component and can often be aborted, if it occurs, by prompt treatment with the drug *dantrolene*. No preventative drug therapy is yet known. A similar malignant hyperthermia can be induced in some animals that are stressed. This can be of importance in the moving of wildlife from one location to another.

C. Hypothermia

This is a condition in which the body temperature falls below the normal range. Usually we refer to a body core temperature less than 35°C as hypothermia. As the core temperature falls, there is progressive impairment of tissue and organ function. As shivering ceases, the metabolic rate falls, the heart rate falls, breathing is slow and shallow, and consciousness is impaired or lost.

There are two main causes of hypothermia. (a) Exposure to environmental conditions such that more heat is extracted from the body than is generated by the body over a substantial period. Immersion in cold water, which because of its high thermal conductivity and capacity, can reduce body temperature rapidly. However, the old belief that immersion in water at, say 2°C, would lead to fatal hypothermia within a few minutes is no longer tenable. Swimming in cold water, because of the increase of blood flow in the limbs and the stirring of the water in contact with the skin, accelerates the rate of body cooling. The choice between attempting to swim to shore and staying curled up attached to an upturned boat can be difficult, but at water temperatures below 10°C, swimming over 1–2 km will usually lead to severe hypothermia. (b) Any process that impairs normal thermoregulation may lead to hypothermia in a cold environment. A major cause of such thermoregulatory impairment in cities in winter is alcohol abuse. Other causes include coma (e.g., caused by diabetes), accidental brain injury, drugs that impair hypothalamic thermoregulation such as barbiturates or some mood-altering drugs, low levels of thyroid hormone in the circulation, i.e., myxedema, and the special case of those elderly persons with impaired thermoregulation or reduced perception of cold. At body temperatures below 30°C, the body may lose the ability to produce enough heat to rewarm itself or to prevent further falls in core temperature.

The first aid principles of managing hypothermia include the addition of thermal insulation to reduce further heat loss and reduction in core temperature and, where possible, the gentle addition of heat such as body to body contact with a warm individual. Warm drinks are not to be given unless the victim is fully conscious, and then only in small amounts liberally laced with sugar or glucose. Some mountain rescue teams have recorded beneficial effects from supplying very warm air saturated with water vapor to breath, provided that the correct air-warming apparatus is used. In hospital the victim should be managed under full physiological control, such as can be achieved in an intensive care unit, where the blood pressure, electrocardiograph, blood gases, and pH can all be monitored and where means varying from heat supplied from an extracorporeal circulation or from warm peritoneal lavage can be supplied to the body. The danger in rewarming is arrest of the heart in the condition of ventricular fibrilation; this requires special equipment for its management. With proper management, even in the apparent absence of life, full recovery is possible from very low body temperatures. A recent hypothermic victim, a little child, was revived in Saskatoon, Canada, from a body core temperature that fell to 14°C.

A special case of immersion hypothermia is that which occurs when the body is completely immersed with the head under water in ice-cold water—submersion hypothermia. Even if the victim has no vital signs when pulled out of the water, if cardiopulmonary resuscitation is started immediately and the patient is transferred to a hospital without delay, there is a good chance of complete recovery in the case of submersions of 20 min or even longer.

Further details of the causes, the first aid, and the hospital management of hypothermia can be obtained from specialized medical texts on the subject.

BIBLIOGRAPHY

Clark, R. P., and Edholm, O. G. (1985). "Man and His Thermal Environment." Edward Arnold, London.

Cooper, K. E. (1987). The neurobiology of fever: Thoughts on recent developments. *Annu. Rev. Neurosci.* **10**, 297–324.

Cooper, K. E. (1995). "Fever and Antipyresis: The Role of the Nervous System." Cambridge Univ. Press, Cambridge, UK.

Hensel, H. H. (1981). "Thermoreception and Temperature Regulation." Academic Press, New York.

Kerslake, D. M. K. (1972). "The Stress of Hot Environments." Cambridge Univ. Press, Cambridge, UK.

Bone Cell Genes, Molecular Analysis

MICHAEL HORTON
Imperial Cancer Research Fund and University College, London

MIEP HELFRICH
University of Aberdeen, Aberdeen

GLOSSARY

Cluster of differentiation Definition of hemopoietic antigens by monoclonal antibodies that is widely used for classification

Colony-forming unit (CFU) Progenitor cell committed (or if uncommitted, and thus multipotent, termed a stem cell) to differentiation toward a particular hemopoietic (blood cell) lineage under the influence of a colony-stimulating factor (a regulator of cell growth and differentiation)

Hemopoiesis (hematopoiesis) Process by which mature blood cells develop from immature precursor cells (and ultimately stem cells) in the bone marrow; it is regulated by interaction with other blood cells and colony-stimulating factors

Integrin (receptor) Glycoprotein transmembrane receptor capable of binding to extracellular matrix and linking extracellular components to the cytoskeleton; they play a major role in cell adhesion to matrix or other cells

Noncollagenous bone proteins Protein components of mineralized bone matrix that exclude fibrillar (largely Type I) collagen

Osteoblast Cell that synthesizes osteoid during formation of bone. It mediates key hormonal interactions that regulate osteoclastic bone resorption and is derived from immature precursor cells (fibroblastic CFU) in common with stromal/fibroblastic cells in other tissues. Quiescent osteoblasts are termed bone lining cells

Osteoclast Multinucleated cell formed by fusion of postmitotic mononuclear precursors of hemopoietic origin that mediates bone resorption under the control of systemic and osteoblast-derived signals

Osteocyte Osteoblast that has ceased osteoid production and become incorporated into bone; its function is not known

Osteoid Unmineralized extracellular matrix of bone consisting largely of Type I collagen

Vitronectin receptor Specific type of integrin receptor capable of binding a range of extracellular proteins, including vitronectin, and expressed at high level by osteoclasts

THE SKELETON CAN BE CONSIDERED AS BEING composed of two major phases. The first is bone itself. Largely acellular, apart from entrapped osteocytes, bone consists of a characteristic set of proteins (predominantly Type I collagen) that are together termed *osteoid.* When mineralized they impart mechanical strength to the mammalian skeleton. The biochemical composition of bone reflects the activity of the second phase, the major cellular components of bone—osteoblasts (OB), osteocytes, and osteoclasts (OC) (Fig. 1). OCs, of hemopoietic origin, are the main cells carrying out resorption of bone. In contrast, OBs, which are derived from local mesenchyme, synthesize osteoid; they also provide an important regulatory influence on OC activity. The overall balance in skeletal mass—that is, whether a negative (OC bone resorption) or positive (OB synthetic activity) effect predominates—depends on a complex interplay between these two cell types. Furthermore, it reflects the influence of both systemic and locally produced calciotropic hormones and factors and a series of complex

ENCYCLOPEDIA OF HUMAN BIOLOGY, Second Edition, VOLUME 2. Copyright © 1997 by Academic Press. All rights of reproduction in any form reserved.

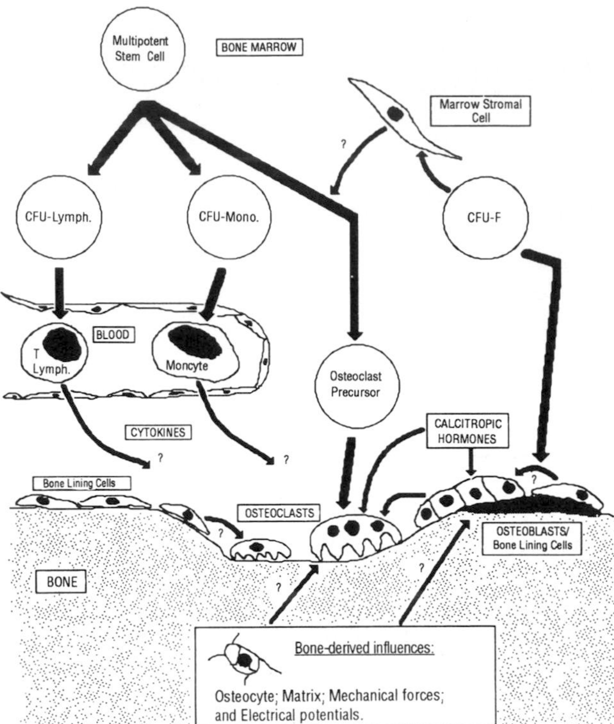

FIGURE 1 Developmental pathways of bone and hemopoietic cells, including osteoblasts and osteoclasts, and the regulatory effects modifying bone cell function. →, precursor to progeny relationship; ?→, proven or putative regulatory effects on osteoclast or osteoblast activity or development; CFU, colony-forming unit; CFU-F, fibroblast CFU; Lymph., lymphocyte; Mono., monocyte.

interactions with other cell types of the bone/bone marrow compartment. This article concentrates on the biochemical features that are specific for, or at least characteristic of, OBs and OCs and attempts to relate them to the maintenance of skeletal homeostasis. As bone protein is largely synthesized by OBs, its overall biochemical composition will also be summarized.

I. OVERVIEW OF THE CELL BIOLOGY OF BONE

The mammalian skeleton is frequently perceived as a rigid and unchanging structure. In truth, the opposite holds, with continuing bone turnover: a small percentage of the total bone is demolished and replaced with new bone each year. The percentage varies according to age and site. Moreover, every time a mechanical force is imparted upon the skeleton—such as standing or running or a force severe enough to fracture bone—a series of changes takes place, the extent of which relates to the strength of the stimulus. These include reorientation of preexisting collagen fibers, generation of electrical potentials within bone, and alteration in the behavior of OBs. Changes in osteoblastic activity include alteration in the pattern of gene expression, cell proliferation, and transition to a form with a different enzyme profile which actively secretes osteoid. OBs also communicate locally with OCs to alter the extent of bone resorption (Fig. 1). Increased OC activity may be regulated by the production by OBs of short-range protein "factor(s)" that are at present ill-defined, or by direct cell-to-cell contact. Stoppage of resorption occurs via the inhibitory effects of local mediators such as prostaglandins (PG) and transforming growth factor-β (TGF-β). At a higher level of control, various calciotropic hormones and factors regulate overall skeletal homeostasis (Fig. 1)—these include parathyroid hormone (PTH), regulating bone resorption by an action on OCs that is mediated by OBs, and calcitonin (CT), which is the only hormone proven to act directly on OCs (and results in cessation of resorption). Many other factors—either local "growth factors" present in bone such as TGF-β or systemically active cytokines including interleukin-1 (IL-1)—also impart profound and complex actions on bone. A detailed exposition on the regulation of bone cell function is outside the scope of this review. [*See* Transforming Growth Factor-β.]

This article will concentrate on the biochemical features (and hence gene function) of OBs and OCs, the two cell types in bone that have important actions on the skeleton and that have been studied the most extensively. Because the proteinaceous component of bone is largely derived from osteoid synthesized by OBs, the structural and functional aspects of collagen and the main noncollagenous proteins (NCPs) will also be summarized.

The molecular biology of "minor" cell types of bone, such as bone lining cells (quiescent osteoblasts) and osteocytes (osteoblasts entrapped in mineralized bone), will not be covered in detail because of lack of data—this does not imply, however, that they may not have important actions on the skeleton. Likewise, it is reasonable to assume, because bone and bone cells are in intimate contact with cells of the hemopoietic system in the bone marrow cavity and with the endothelium of the extensive bone microvasculature and the neural network of bone, that important effects are transmitted between bone and these nonskeletal compartments. Indeed, some workers consider that cytokines released from lymphocytes play a pivotal role in the regulation of OC function and differentiation from hemopoietic precursors. Also, regulation of OC function by endothelial cell products, such as nitric oxide and reactive oxygen species, is suggested and is the subject of intense study. Again, these important areas will not be covered in detail in this article. [*See* Cytokines in the Immune Response; Lymphocytes.]

II. OSTEOBLAST-SPECIFIC AND RELATED GENES AND THEIR PRODUCTS

A. Methodology

To a certain extent, OBs can be considered specialized fibroblastic cells sharing many functions, and hence biochemical features, with stromal cells at other locations. This apparent lack of specialized features presents a major obstacle to the application of the usual methodologies of molecular biology for investigating cellular or tissue-specific genes and their products. Reliance on the use of immunological or cloning methodologies to analyze bone proteins and OB molecules that have already been defined in other ways is, however, beginning to yield a greater understanding of bone physiology. In particular, information is being accrued on the mechanism by which the expression of genes encoding known bone proteins is regulated in osteoblastic cells.

A further problem that has delayed progress centers on the heterogeneity of the "osteoblastic" cells used for functional or biochemical studies. The major source of information comes from *in vitro* analysis of clonal rodent (usually rat) osteosarcoma cell lines or primary, and hence mixed, cultures of OBs from normal long bone or calvaria. The data from rodent systems cannot be translated directly to the human situation owing to interspecies variation and heterogeneity between cell populations.

With these limitations, a variety of OB activities can be studied experimentally in laboratory animals and, increasingly, in humans. Animal studies involve either the whole animal *in vivo* or *ex vivo* organs (whole mouse calvaria, rat radius). In humans, OBs grown out of trabecular bone chips or isolated from osteosarcoma tumors can be studied; they seem to exhibit osteoblastic features. OB activities that can then be measured in humans include: assay of collagens and noncollagenous proteins either directly or by immunohistochemical methods, study of the expression of the genes involved at the level of mRNA and regulation of specific gene transcription, especially by the powerful *in situ* mRNA hybridization technique that identifies the mRNAs in individual cells, and the determination of the ability of the synthesized extracellular matrix to mineralize. These features can be considered "bone specific" only if the OB function measured responds appropriately to regulation by calciotropic hormones and factors.

B. Proteins of Osteoblasts and Bone

1. Overview

The principal proteins found in bone are summarized in Fig. 2 and detailed in Table I. From a biochemical standpoint, bone protein can be divided into collagens and NCPs. They are mainly derived from the synthetic and secretory activity of OBs, however. A significant minority of the noncollagenous component is derived from plasma proteins entrapped in bone by the binding capacity of hydroxyapatite. These include serum albumin and α2HS glycoprotein, which is enriched severalfold in bone over circulating levels. The function of the majority of plasma-derived proteins in bone is largely unknown. However, several of the growth factors present within bone (e.g., IL-1, TGF-β, FGFs, and IGFs) are totally or partially derived from blood plasma. Though biologically highly active, they cannot be considered quantitatively "major" bone proteins. Osteocytes also retain the capacity to synthesize a similar range of bone proteins as OBs

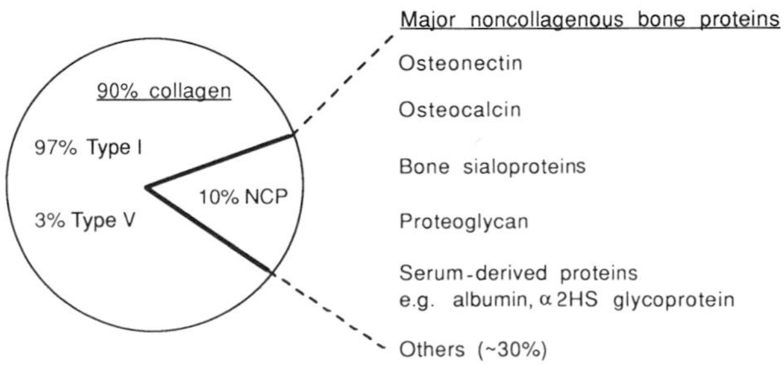

FIGURE 2 Distribution of collagens and noncollagenous proteins in bone.

TABLE I
Bone Proteins, Their Structure, and Function

Protein	Percentage	Molecular mass (kDa)	Sequence motif[b]	Possible functions
Collagens	≈90			
Collagen, Type I	(97)	High; aldehyde cross-linked		Fibrillar collagen—mechanical strength
Collagen, Type V	(3)			
Noncollagenous proteins	≈10			
Osteonectin	(15)	45 [32][a]	EF hand	Collagen, hydroxyapatite, and Ca^{2+} binding; mineralization
Osteocalcin	(15)	10 [6]	γ-carboxy Glu	Hydroxyapatite binding
Osteopontin	(3)	80 [32]	RGD, phosphorylation	Hydroxyapatite binding; cell adhesion via RGD sequence; mineralization
Bone sialoprotein	(12)	80 [34]	RGD, phosphorylation	Hydroxyapatite binding; cell adhesion via RGD sequence
Proteoglycan I (-biglycan)	(5)	240 [36]	Leu-rich motif, GAG	?
Proteoglycan II (-decorin)	(5)	120 [36]	Leu-rich motif, GAG	Interaction with collagen; fibril formation
Matrix Gla protein	(2)	14 [9]	γ-carboxy Glu	? Role in mineralization
Collagen N-terminal propeptide		24		?
BAG75	(1)	75	Poly Asp	? mineralization; ? cell adhesion
Bone morphogenetic protein(s)[c]	—	—		OB growth/differentiation
(Bone) growth factors[d]	—	—		OB growth/differentiation/function
Thrombospondin	—	450	RGD	? Cell adhesion; ? antiadhesive
Serum proteins	(≈15)			
Albumin	(3)	67		? ↓ mineralization
α2HS glycoprotein	(10)	60		? Cell attachment
Others	≈30%	—	—	—

[a]Unmodified core protein size (kDa).
[b]GAG, glycosaminoglycan; RGD, Arg-Gly-Asp tripeptide sequence; EF hand, calcium binding motif.
[c]For example, BMP, TGF-β homologues.
[d]For example, TGF-β, PDGF, FGF, IGF I and II.

and their contribution may be important on a local level.

2. Bone Collagens

Collagen accounts for approximately 90% of protein and 30% of total weight of compact bone (the remainder being accounted for by the mineral phase, calcium phosphate/hydroxyapatite). Bone collagen is principally Type I collagen, a large (300 nm in length) triple-helical protein formed from two α1 (1) chains and one α2 (1) chain; the two chain types are products of distinct genetic loci. Each unit associates longitudinally to form characteristic banded fibrils. In bone there is extensive cross-linking between fibers (at hydroxylysine residues) to form the rigid conformation of bone collagen required for structural support in the skeleton. Type I collagen is not bone specific. It is, however, an important OB product owing to its abundance and extensive cross-linking, its association with Type V collagen and NCPs that are enriched in bone, and its ability to mineralize. [*See* Collagen, Structural and Function.]

About 3% of bone collagen comprises Type V collagen, a further class I collagen that is capable of forming longitudinal fibrils in extracellular matrix. Its function is not known, but it is of interest that its synthesis is developmentally regulated.

3. Noncollagenous Proteins of Bone

The main noncollagenous proteins of bone (and hence largely OB derived) are listed in Table I with an estimate of their relative abundance, molecular mass, some of their putative functions, and indication whether the gene has been cloned.

Two-dimensional electrophoretic analysis, which affords good separation of individual proteins, suggests that several hundred proteins, and their degraded or modified derivatives, may exist in the noncollagen protein fraction of bone. However, only a dozen or so of these proteins have been cloned and studied in sufficient detail to suggest a role in bone, and even for these the role is often ambiguous. Thus, for example, osteonectin is an abundant 45-kDa phosphoprotein that is conserved in bone across all species tested (though at different levels), suggesting an important function in the skeleton. Osteonectin binds calcium and hydroxyapatite at high affinity (10^{-7} to 10^{-8} *M*) and also binds to Type I collagen. In solution osteonectin can inhibit hydroxyapatite crystal growth, but when associated with collagen it appears to have the opposite action. These data support an important role in bone. However, osteonectin has been shown to be present in several extraskeletal sites (platelets, mouse endoderm, endothelia, BM-40 from extracellular matrix from certain tumor cell lines). Osteonectin thus may be involved in cell proliferation or differentiation, perhaps related to calcium binding without having a bone-specific action.

C. Functions of Bone Proteins

The structural (mechanical) features of Type I collagen of bone in its mature fibrillar, cross-linked, and calcified form are well established. Less clear is why this collagen mineralizes in bone but not at other sites where it is equally abundant.

In contrast, the functions of NCPs (collagen binding, nucleation of mineralization, cell adhesion) have largely been gleaned from their study in various *in vitro* systems or by analysis of their distribution *within* the skeleton and other tissues. Their true role *in vivo* is usually unknown or is inferred from such experimental, and hence probably simplistic, systems. For instance, the functions of osteocalcin, an abundant NCP with a characteristic vitamin K-dependent γ-carboxyl modification, is not clearly understood. In fact, blockade of vitamin K with warfarin in experimental animals, and hence inhibition of osteocalcin modification, fails to produce major biochemical or morphological changes in bone.

Despite the foregoing, OBs and their products are clearly involved in the mineralization process (Table II). Cell membrane "vesicles," probably derived from OBs, are found at the site of mineralization. Many NCPs (e.g., osteonectin, osteopontin, and bone proteoglycan II) are known to interact with collagen and others (e.g., osteonectin, osteocalcin, and bone sialoproteins) with the calcium phosphate/hydroxyapatite mineral phase of bone (Table I). They are thus implicated in mineralization but by undefined mechanisms. Less clear is the role of alkaline (and other) phosphatases in bone—it has been suggested that they may stimulate mineralization by removing inhibitors of mineral nucleation, such as pyrophosphate. Finally, it is possible that OBs, with their active calcium transport mechanisms, might generate cation gradients in bone, thus favoring mineralization. [*See* Proteoglycans.]

TABLE II

Osteoblast "Products" Implicated in the Mineralization of Bone

- Osteoblast-derived matrix vesicles (site of induction of mineralization)
- Ca^{2+} binding proteins, e.g., annexins
- Synthesis of noncollagenous proteins of osteoid (e.g., osteonectin, sialoproteins): induction of mineralization process (nucleation)
- Alkaline (and other) phosphatases: elimination of nucleation inhibitors (e.g., pyrophosphate)
- ? Generation of local Ca^{2+} gradient favoring mineralization

Bone contains locally and systematically derived growth factors, the majority of which have been demonstrated to have effects on bone cells *in vitro* and hence might be expected to affect bone *in vivo*. Some of the available information about them is contradictory. Table III summarizes the important and noncontroversial features of a range of local mediators that have effects on osteoblastic function (bone formation) and compares them with systematically active calciotropic hormones.

Several bone proteins have a proven activity in promoting cell adhesion. Type I collagen, the bone sialoproteins (osteopontin and BSPII), and thrombospondin are capable of binding to cell-surface receptors of the integrin type (discussed in some detail in Section III,C); collagen binds to a receptor (VLA-2) of the fibronectin receptor family, and other proteins bind to the "vitronectin" receptor. The serum protein, α2HS glycoprotein, and bone proteoglycans have also been shown to promote cell attachment *in vitro*. Their role in bone—whether adhesion of bone cells to osteoid or some other function—remains to be clarified.

D. Regulation of Osteoblast Genes: A Cautionary Note

Key to our understanding of OB gene function is how the expression of bone proteins is regulated. Molecu-

TABLE III
Hormones, Cytokines, and Growth Factors Implicated in the Regulation of Bone Cell Function

	Major effects[a,b]	
	Resorption	Formation
(a) Systemic hormones, vitamins		
Parathyroid hormone/PTHrP	↑	↑/↓
Calcitonin	↓	(↑)
Sex steroids	↓	↑
Glucocorticoids	↑	↓
Growth hormone		↑
Thyroid hormone	↑	↑/↓
1, 25-$(OH)_2$ vitamin D3	↑	↑/↓
(b) Locally acting agents—hemopoietic origin		
Interleukin-1	↑	↑/↓
Tumor necrosis factor α and β	↑	↓
γ Interferon	↓	↓
CSFs (M-, GM-, IL-3)	↑ (↓ ,M-CSF)	
(c) Locally acting—bone derived, etc.		
Prostaglandins	↑/↓	↑
Transforming growth factor-β	↑/↓	↑/↓
PDGF	↑	
Bone morphogenetic protein(s)		↑
IGF I and II		↑
(d) Ionic		
Ca^{2+} ↑	↓	
pH ↓	↑	
(e) Physical		
Immobilization	↑	↓
Exercise	↓	↑

[a]Effects on OC resorption and OB bone formation; ↑ = stimulation; ↓ = inhibition.
[b] ↑/↓: discordant effects between *in vivo* and *in vitro* analysis (e.g., prostaglandins and bone resorption) or due to biphasic responses (e.g., PTH on bone formation).

lar probes are now available for the majority of bone proteins (Table I). By far the greatest amount of information is available for the Type I procollagen genes encoding the pro $\alpha1$ (1) and pro $\alpha2$ (1) chains of collagen. The molecular approach has provided the structural information required to define the mutations in the collagen gene that give rise to osteogenesis imperfecta, an inherited disorder leading to brittle bones. These studies have not, however, enabled scientists to discover how the expression of the two structural genes is regulated in tissues—such as bone—that mineralize compared with those—such as tendon—that do not. Perhaps bone mineralization involves subtle changes of the gene products occurring posttranslationally or complex interactions between large numbers of proteins. Similarly, we are a long way from discovering the molecular basis for the skeletal disease osteoporosis, which is of major social and economic impact in the aging population.

III. GENE PRODUCTS EXPRESSED BY OSTEOCLASTS

A. Methodology

OCs can be operationally defined *in vivo* as multinucleated cells attached to and resorbing the surface of bone. Once removed from bone, the distinction between OCs (which may be mononuclear and still functional) and other hemopoietic cells becomes problematic. It is generally agreed that multinucleated cells, if freshly isolated from embryonic bone and particularly if adhering to and spreading on bone or other solid substrate, are likely to be true OCs. Cultured cells—either isolated from bone in short-term culture or derived from hemopoietic tissues such as marrow—can be incorrectly identified as OCs unless strict criteria are followed. Often identification is based on multinuclearity or the presence of tartrate-resistant acid phosphatase (TRAP), but these properties are not discriminatory, as macrophages (and tumor cells) can be multinucleate but are incapable of resorbing bone. Cells can be classified as OCs only when they resorb bone in a three-dimensional manner and bear receptors for, and respond to, the hormone calcitonin.

Studies can be carried out with isolated OCs, or long bone or calvarial explant cultures (or whole animals), which have the advantage of retaining regulatory contact with OBs and maintaining the overall control mechanisms that are lost in isolated cell preparations. The use of different systems has led to divergent results on the role of, for example, prostaglandins on bone resorption—they are inhibitory on isolated OCs but stimulatory in organ culture. Functional criteria of OC activity include OC shape change, bone resorption, enzymatic activity (such as TRAP), and acid production. Over the last few years, bone marrow culture and hemopoietic stem cell isolation techniques have been used increasingly to analyze OC differentiation. Likewise, the proton and ion channel pumping mechanisms operative in the functioning OC have begun to be studied using electrophysiological and biochemical techniques. Because of the difficulty

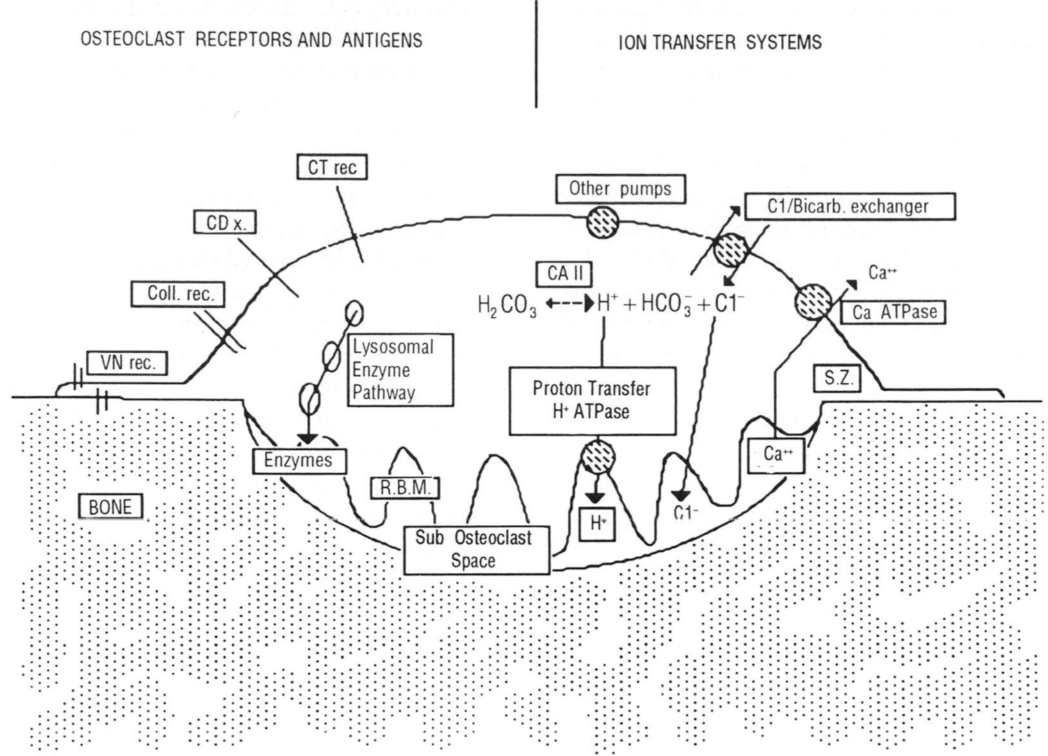

FIGURE 3 Diagrammatic view of the major molecules and enzymatic functions of osteoclasts. VN rec., vitronectin receptor; CT rec., calcitonin receptor; CD x., hemopoietic antigens by CD number; R.B.M., ruffled border membrane; S.Z., sealing zone; CA II, carbonic anhydrase II.

of obtaining mammalian (and especially human) OCs, immunological and cloning methodologies have been relatively underused when compared with OBs—however, immunocytochemical and *in situ* mRNA hybridization techniques, which are applicable to whole tissues, are eminently suited to the study of this rare cell type.

B. The Biochemical Basis for Osteoclast-Mediated Bone Resorption

To resorb bone the OC must carry out two prime functions: first, it must produce an extracellular environment of low pH that favors dissolution of calcium bound to skeletal protein; second, the demineralized collagen and NCPs are enzymatically degraded. A schematic view of these two functions is depicted in Fig. 3; the major enzyme systems involved are listed in Table IV.

The current view (established from electrophysiological measurement and immunohistochemical studies of *in situ* or isolated OCs) of the mechanism by

TABLE IV
Prominent Enzymes and Other Proteins of Osteoclasts, Demonstrated Immunologically or Structurally

Function	Enzyme, etc.
Acidification mechanism	Carbonic anhydrase II, vacuolar H^+-ATPase (kidney type)
Compensatory mechanism for resorption process	HCO_3/Cl^- exchanger, Na^+, K^+-ATPase, Ca^{2+}-ATPase
Lysomal enzymes in exocytic pathway	Tartrate-resistant acid phosphatase (TRAP), cysteine proteinases(s) (cathespins B, L, C), arylsulfatase, β-glycerophosphatase
Secretory vesicles	Metalloproteinases (collagenase, etc.)
Protease inhibitors	TIMPs
Other proteins	Osteopontin, TGF-β, IL-6, NADPH-diaphorase (nitric oxide synthase)

which OCs produce acid is as follows. Intracellular carbonic acid produces protons under the action of the enzyme carbonic anhydrase (type II isoenzyme), which is abundant in OCs. The protons are then pumped into the extracellular space by an electrogenic proton pump, kidney type H^+-ATPase, which is present in the correct configuration in the ruffled membrane of the OC. Protons are cotransported with chloride ions, which enter the OC to balance the exit of bicarbonate ions probably via a dorsal membrane anion exchanger. The ionic and electrical integrity of the osteoclast intracellular environment is thus maintained by a series of exchangers and ion channels listed in Table IV and Fig. 3. Dissolved calcium is thought to exit from beneath the OC either by transcellular movement via a dorsal membrane Ca^{2+}-ATPase or on disruption of the sealing zone once resorption ceases. The role of membrane-bound acid phosphatase in dissolution of bone mineral is unclear.

Enzymatic digestion of bone collagen by OCs is carried out by lysosomal cysteine proteinase of the cathepsin family—this is capable of collagenolytic activity at the acidic pH expected in the OC environment. The role of mammalian collagenase is unclear. This enzyme is produced by OBs (as well as OCs). It is optimally active at neutral pH, secreted in latent proenzyme form in association with tissue inhibitor of metalloproteinase (TIMP) and activated by tissue plasminogen activator (TPA). This enzyme may "prepare" the bone surface, allowing optimal OC resorption, or remove partially degraded collagen following alkalinization in the low pH subosteoclast space that occurs upon cessation of resorption and movement away of the osteoclast.

C. Morphological and Biochemical Features of Osteoclast Cell Adhesion: Vitronectin and Other Adhesion Receptors

The main, and possibly the only, role of the OC is to provide an environment suited to degradation of bone matrix and removal of mineral. It does this by forming, when actively resorbing bone, an "extracellular vacuole" in which the acidic, enzyme-rich subosteoclast space is bounded at its periphery by a zone of tight cell adhesion to the bone surface (the sealing zone) and roofed by specialized cell membrane (the ruffled border membrane), which is highly folded and enriched with proton pump, acid phosphatase, and other enzyme systems (see Fig. 3 and Section III,B).

Characteristic structures (podosomes) have been identified by electron microscopy associated with the zone of contact between bone and the ventral OC cell membrane. They are enriched in actin filaments. Immunohistological studies have suggested that cell membrane extracellular matrix receptors (of integrin type) and proteins (vinculin, talin) linking these receptors to the actin cytoskeleton are also concentrated at these sites. In isolated chicken OCs the appearance of these adhesion structures seems to be regulated by external pH and intracellular calcium levels, which may provide a mechanism of feedback regulation of OC adhesion to bone. Together these structural and biochemical features might operate to form and regulate the zone of tight membrane—bone apposition that is required to maintain the extracellular acid environment under the OC.

Two main cell adhesion molecules, both of the integrin extracellular matrix receptor superfamily, have been demonstrated in OCs. One is a collagen receptor, termed VLA-2 (chain structure $\alpha2\beta1$), whose expression is widespread in bone (and other tissues). OC adhesion to collagen has recently been demonstrated. The second receptor is the vitronectin receptor, $\alpha v\beta3$. It is highly expressed by OCs (and generally not other bone or hemopoietic cells *in vivo*), is conserved in all species tested, localizes to adhesion sites *in vivo* or *in vitro,* and is known to bind bone proteins, osteopontin and bone sialoprotein, in addition to serum proteins including vitronectin. Moreover, antibodies and peptide inhibitors of integrin–ligand binding block bone resorption both *in vitro* and *in vivo*. Together these data support a specific role for vitronectin receptor in bone resorption. It is likely that this involves cell adhesion and hence maintenance of resorption, homing, and fusion of differentiating OC precursors, but effects on integrin-induced signaling cannot be excluded. Currently this inhibitory effect is being investigated as a potential route for treatment of osteoporosis.

D. Calcitonin and Other Regulatory Receptors of Osteoclast

OC function, and hence the level of bone resorption, is controlled both by the direct effects of hormones and locally acting factors and by indirect regulatory signals from OBs and, in all probability, other cell types in the skeleton (summarized in Table III). The final common pathway for molecules affecting OCs indirectly via OBs (such as IL-1, PTH, vitamin D3) suggests that a local factor is secreted by OBs that

TABLE V
Functional Receptors of Osteoclasts[a]

Calcitonin receptor:	High affinity, $>10^6$ sites per cell; molecular mass of 90 kDa demonstrated for OC receptor; 7-membrane-spanning, G-protein-linked receptor
Macrophage CSF (CSF-1) receptor:	c-fms protooncogene
Integrins:	$\alpha v\beta 3$ (vitronectin receptor), $\alpha 2\beta 1$
Functionally defined receptors: (no structural information in osteoclasts)	Prostaglandin receptors; receptor for OB-derived stimulatory factor

[a]Note that receptors for PTH, IL-1, and vitamin D3 are absent from OCs as these compounds act indirectly via OBs. Also, it is disputed whether OCs possess the estrogen receptor in addition to expression in OBs.

would act via an OC-specific receptor; this remains uncharacterized.

Three compounds are known to act on OCs directly; calcitonin, prostaglandins, and macrophage colony-stimulating factor (M-CSF). The receptor for calcitonin has been structurally defined as a member of the seven-membrane-spanning, G-protein-linked family of receptors (Table V). The calcitonin receptor is a well-defined 90-kDa membrane protein exhibiting high affinity ligand binding. Mammalian OCs express remarkably high numbers of receptors (several million per cell), which are not expressed by other marrow or bone cells; as such they make a valuable marker of the OC phenotype. Avian OCs, which are frequently used in bone research, oddly express neither detectable calcitonin receptors nor respond to the hormone.

The direct effects of prostaglandins are presumably receptor mediated, but the structure is unknown. Recent observations have shown that M-CSF stimulates OC precursor proliferation and differentiation and inhibits bone resorption by a direct action on OCs. The M-CSF receptor of OCs has been shown to be identical to that expressed by macrophages, the c-fms protooncogene product.

E. Immunologically Defined Molecules of Osteoclasts

Many cell-surface molecules of OCs have been identified from the study of the expression of antigens defined by antibodies to antigens of hemopoietic cells (OCs are of hemopoietic origin, Fig. 1). Few molecules are shared between OCs and their lineage neighbors, myeloid cells. They are summarized in Table VI. Some information is available on the function of these molecules; the collagen and "vitronectin" adhesion receptors have already been discussed. It is of interest that OCs express high levels of the enzyme aminopeptidase N (the CD13 antigen), which is known to preferentially cleave short peptides and thus may play a role in modifying the action of regulatory

TABLE VI
Immunologically Defined Hemopoietic Antigens Expressed by Osteoclasts

	Antigen[a]	Myeloid cell cross-reactivity[b]	Molecular mass (kDa)
—	CD9	+ (all)	25
Amino-peptidase N	CD13	++ (all)	150
Collagen receptor (integrin)	CD29 (VLA-β1)	++ (all)	130 (red.)
	CD49b (VLA-α2)	+ (subpop.)	170 (red.)
Vitronectin receptor (integrin)	CD61 (gpIIIa)	± (subpop.)	110 (red.)
	CD51 (VNRα)	± (subpop.)	125 (red.)
Leukocyte common	CD45	+++ (all)	"200"
iCAM-1	CD54	++ (mono.)	85
—	CD68	+++ (mono., mac.)	110

[a]Antigen definition using CD nomenclature. red., reduced form.
[b]± to +++, range of cellular reactivity; subpop., subpopulation of myeloid cells reactive; mono., monocyte; mac., macrophage.

factors on OC bone resorption and CD9, which might act as an anion channel. [*See* Hematopoietic System.]

IV. SUMMARY

Bone is a prime example of an "organ" that relies on complex cellular activities—which need to be regulated and integrated—for its ultimate function, the structural endoskeleton. Our current knowledge of how OBs interact with OCs is at an early stage. A few of the important questions are: Are OBs a heterogeneous population of cells (e.g., functionally or by cell lineage)? Do hemopoietic or immunoregulatory cells play a role in normal or pathological bone remodeling, and how do the vascular or nervous systems affect bone? These cell biological questions are likely to be the center of study in the future.

The study of collagen mutations in osteogenesis imperfecta has amply demonstrated the truism that to study disease can profoundly affect our understanding of normal physiological processes. A similar approach to other inherited disease (such as osteopetrosis, a defect resulting in increased bone density due to reduced bone resorption), transgenic mice carrying deleted c-src and c-fos oncogenes (leading to altered bone turnover), and acquired diseases (e.g., osteoporosis) is providing important information on normal skeletal homeostasis.

BIBLIOGRAPHY

Aarden, E. M., Burger, E. H., and Nijweide, P. J. (1994). Functions of osteocytes in bone. *J. Cellular Biochem.* **55**, 286.

Horton, M. A., and Rodan, G. A. (1996). Integrins as therapeutic targets in bone disease. *In* "Adhesion Receptors as Therapeutic Targets" (M. A. Horton, ed.), pp. 223–245. CRC Press, Boca Raton, FL.

Kelly, P. J., and Eisman, J. A. (1989). Hypercalcaemia of malignancy. *Cancer Metastasis Rev.* **8**, 23.

Marcus, R., Feldman, D., and Kelsey, J. (eds.) (1996). "Osteoporosis." Academic Press, San Diego.

Nesbitt, S., Nesbit, A., Helfrich, M., and Horton, M. (1993). Biochemical characterisation of human osteoclast integrins. *J. Biol. Chem.* **268**, 16737.

Noda, M. (ed.) (1993). "Cellular and Molecular Biology of Bone." Academic Press, San Diego.

Peck, W. A. (ed.) "Bone and Mineral Research" (series). Elsevier Science, New York.

Rifkin, B. R., and Gay, C. V. (eds.) (1992). "Biology and Physiology of the Osteoclast." CRC Press, Boca Raton, FL.

Bone Density and Fragility, Age-Related Changes

A. MICHAEL PARFITT
University of Arkansas for Medical Sciences

GLOSSARY

Activation The initiating event in the process of bone remodeling

Bone (as a substance) A mineralized (and consequently hard) connective tissue

Bone (as an organ) An individual component of the skeleton

Bone density Mass (mainly mineral) per unit volume of bone, either as a substance (true density) or as an organ (apparent density)

Bone formation Production of bone by osteoblasts

Bone resorption Destruction of bone by osteoclasts

Fatigue Damage, initially at a submicroscopic level, to a structural material as a result of frequently repeated cyclical loading

Modeling Manner in which bone resorption and formation are coordinated to accomplish changes in the size and shape of the bones during growth

Remodeling Manner in which bone resorption and bone formation are coordinated to accomplish the replacement and turnover of bone, with little or no change in the size or shape of the bones

Strain Relative deformation of a structural material in response to mechanical loading

THE INCREASING RISK OF FRACTURE WITH AGE IS partly the result of increasing bone fragility, which in turn is partly the result of continued loss of bone with age. These aspects of aging are sometimes collectively referred to as osteoporosis, a convenient term for denoting a region of interest within the universe of biological knowledge, but one which causes much confusion when used as the name of a disease that a person can be said to have since the relationship between fracture risk and any property of bone is continuous, not dichotomous. In addition to mechanical support and protection for the soft tissues, the skeleton provides a reservoir of calcium and other minerals and a suitable environment for hemopoiesis. The main focus of this article is on the first of these functions, which depends on the three-dimensional organization of a hard tissue (bone) into individual organs (the bones). [For a detailed description of the skeleton as a whole and of its constituent bones, *see* Skeleton.]

I. BONE AS A TISSUE AND BONES AS ORGANS

Mammalian bone is a specialized connective tissue consisting of a mineralized organic matrix in which living cells are dispersed. The matrix is composed of collagen fibers embedded in a ground substance containing abundant protein–polysaccharide complexes. The fibrous protein collagen forms about 90% by weight of bone matrix and about 60% of total

ENCYCLOPEDIA OF HUMAN BIOLOGY, Second Edition, VOLUME 2. Copyright © 1997 by Academic Press. All rights of reproduction in any form reserved.

body collagen is in bone. The ground substance of bone, as of other connective tissues, is a gel with two phases differing in water content whose consistency depends on the degree of polymerization of its main structural components, proteoglycans and glycoproteins. Bone matrix contains small amounts of a large number of noncollagenous proteins, of which the most abundant is osteocalcin, a vitamin K-dependent protein containing the calcium-binding amino acid γ-carboxyglutamic acid. The mineral of bone contains mainly calcium, phosphate, and carbonate ions, in an approximate molar ratio of 10 : 6 : 1. The mineral becomes progressively more crystalline with time and the crystals increase in size with age. The lattice structure of the crystals conforms to that of hydroxyapatite, with the empirical formula $3(Ca_3(PO_4)_2)Ca\ OH_2$. [*See* Collagen, Structure and Function.]

Each bone has an outer and an inner envelope. The outer (periosteal) envelope includes the surfaces of the layers of cartilage adjacent to joints as well as the fibrous sheath known as the periosteum to which the muscles are attached. The inner (endosteal) envelope separates the bone substance from, and encloses, the soft tissue of the bone marrow and the intracortical canals. It is useful to classify bones as members of either the appendicular or the axial skeleton (Table I). Anatomists classify the pelvis as appendicular, but functionally it is more appropriate to classify it as axial. At the microscopic level, two distinct structural types of bone can be discerned (Table II). Cortical bone is solid and forms the outer wall of all bones immediately beneath the periosteum, whereas cancellous bone is a network of plates and bars (trabeculae) that fills most of the bones of the axial skeleton and occupies the ends of the long bones of the appendicular skeleton. The cancellous bone surface forms one subdivision of the endosteal envelope and is in continuity with the endocortical and intracortical subdivisions. The interstices of cancellous bone are filled with marrow, consisting of blood-forming tissue (red marrow) and fat (yellow marrow) in varying proportions, with the former predominating in the axial and the latter in the appendicular skeleton. Because of their geometric differences, cortical bone provides most of the volume in the skeleton, whereas cancellous bone provides most of the surface.

TABLE I
Subdivisions of the Skeleton

Feature	Appendicular	Axial
Main bone tissue	Cortical	Cancellous
Main soft tissue	Muscle	Viscera
Main joint type	Synovial	Various
Cortices	Thick	Thin
With age, cancellous tissue	Contracts[a]	Expands[b]
Marrow	Fatty	Hematopoietic
Turnover	Low	High

[a]Caused by retreat of boundary with diaphysis toward metaphysis.
[b]Caused by cancellization of inner region of cortex.

TABLE II
Some Differences between Main Structural Types of Bone

	Cortical	Cancellous
Envelopes	Two[a]	One
Surfaces	Three[b]	One
Proximity to marrow	Variable	Close
Surface/volume	Low	High
Porosity	Low	High
Turnover	Low	Variable[c]
Response	Slow	Rapid
Mineral reservoir	Yes[d]	No[e](!)

[a]Periosteal and endosteal.
[b]Periosteal, Haversian (intracortical), and endocortical.
[c]Related to proportion of hemopoietic tissue in marrow.
[d]Sustained physiological demands for bone mineral, such as during growth, pregnancy, lactation, and antler production, are met mainly by temporary increases in cortical porosity.
[e]Except for transient net loss of calcium ions to subserve the homeostatic function of the bone surface.

At the microscopic level, bone in the adult skeleton is of a lamellar (laminated) structure like plywood, and the regularity of collagen fiber orientation imparts a characteristic grain. Between the lamellae are small lacunae in which lie osteocytes, the most numerous type of cell in bone. Osteocytes make contact with each other and with the cells on the surface by long cell processes that lie within fine canaliculi running through the bone. During early embryonic and neonatal growth, the bone formed is of irregular fiber orientation; the differences between such woven bone and lamellar bone are summarized in Table III. Woven bone is also formed in fracture callus, in osseous metaplasia in various diseases, and wherever ectopic bone is laid down in muscle or skin. Lamellar bone com-

TABLE III
Some differences between Main Microscopic Types of Bone

	Lamellar bone[a]	Woven bone[a]
Osteoblast organization	Monolayer Coordinated	Isolated Uncoordinated
Osteoblast activity	Polarized	Nonpolarized
Collagen fiber orientation	Regular	Irregular
Lacunar size	Small	Large
Osteocyte density	Low	High
Mineral density	High	Low
Production rate	Slow	Rapid

[a]Whether the osteoblasts that make the different types of bone have undergone different pathways of differentiation or are responding to differences in local environment is not known.

prises many small structural units made at different times, held together by a different form of bone matrix known as cement substance, which in a two-dimensional histologic section is visible as the cement lines that demarcate the boundaries of individual structural units. In cortical bone, the characteristic structural unit is the Haversian system, or osteon, a cylinder about 250 μm in diameter, roughly parallel to the long axis of the bone, with a central canal containing blood vessels, lymphatics, nerves, and loose connective tissue. In cancellous bone, the structural units are thin irregular plates that lie parallel to the surface and adjacent to the bone marrow. In histological sections, the structural units appear as annuli in cortical bone and as crescents in cancellous bone. [*See* Bone, Embryonic Development.]

II. TURNOVER OF BONE AND ITS MINERALS: REMODELING AND HOMEOSTASIS

Like many other tissues, bone is continually replaced and renewed throughout life, but (unlike other tissues) at a much slower rate than the cells responsible for the renewal. The rate of turnover is generally lower in cortical than in cancellous bone because of its lower surface-volume ratio and is generally lower in the appendicular than in the axial skeleton, partly because of its higher proportion of cortical bone and partly because turnover is higher adjacent to red than to yellow marrow. The cellular basis of bone turnover is that old bone is resorbed by osteoclasts and new bone is formed in its place by osteoblasts. Osteoclasts are multinucleated giant cells (polykaryons) derived from the hemopoietic stem cells of the bone marrow. They attach to the bone surface by a tight seal at the periphery of the cell, within which are secreted protons which dissolve mineral, as well as hydrolytic enzymes which depolymerize proteoglycans and digest collagen and other matrix proteins. Osteoblasts are derived from the stromal stem cells of the bone marrow and synthesize all the proteins of bone matrix. Unlike resorption, bone formation occurs in two stages; matrix is first deposited in a layer referred to as an osteoid seam, which undergoes mineralization several weeks later. Because the mineral crystals displace water, matrix synthesis determines the volume of bone formed but not its density, whereas mineralization increases the density but does not alter the volume. [*See* Bone Cell Genes, Molecular Analysis.]

Osteoclasts and osteoblasts can be studied separately *in vitro*, but their recruitment and activity are closely coordinated *in vivo*. Bone is remodeled in temporally discrete episodes that occur at spatially discrete sites, each episode comprising the erosion and refilling of a cavity on a bone surface by an organized group of cells, the basic multicellular unit (BMU) that includes, as well as successive teams of osteoclasts and osteoblasts, new blood vessels, nerves, and connective tissue. The BMUs dig tunnels through cortical bone or trenches across the surface of cancellous bone for about 6 months, during which the spatial and temporal relationships between the constituent cells are maintained. Depending on whether the cavity (tunnel or trench) is overfilled, exactly filled, or underfilled, the outcome of each episode is a net gain, no change, or a net loss of bone, but once completed, the transaction is irrevocable. The summation of all such transactions determines the direction and magnitude of bone balance on a particular surface over a particular period of time. All physiologic, pathologic, and pharmacologic influences on bone balance are mediated by the cellular mechanisms that determine the net result of each remodeling episode and the frequency with which the episodes occur. Every point on a bone surface goes through successive stages of quiescence, activation, resorption, reversal, formation, and back to quiescence; in cancellous bone, one such cycle occurs about once every 3 years and the active phase of the cycle lasts for 3–4 months. [*See* Bone Remodeling.]

During the quiescent stage, the bone surface is covered by a layer of thin flat lining cells that separate the bone from the marrow and represent one mode of terminal differentiation of osteoblasts. The remodeling cycle is initiated by the event of activation that involves recruitment of mononuclear osteoclast precursor cells, proliferation of new blood vessels to carry them to the chosen site, and preparation of the surface to allow them access to the mineralized bone, where they undergo fusion into osteoclasts. A group of new osteoclasts form a team that erodes a cavity of characteristic size and shape; progression of the BMU through or across the surface of bone is sustained by continued recruitment of new osteoclasts. When erosion at any location is completed, the osteoclasts disappear, probably as the result of apoptosis, and are replaced by mononuclear cells of uncertain nature that clear the floor of the cavity of debris, make its surface smoother, and deposit the cement substance. A team of new osteoblasts then assembles in response to growth factors released either from osteoclasts or from resorbed bone; its task is to restore the bone surface as closely as possible to its original location. Both matrix synthesis and mineralization proceed most rapidly when the osteoblasts are young and they gradually slow down as they get older. Some osteoblasts become immured in the new bone as osteocytes and some die. Those that remain become progressively thinner and flatter and eventually become the cells that will cover the new region of bone surface until the next cycle of remodeling occurs.

The most obvious difference between the bone removed and the new bone laid down in its place is in their age. The mean age of cortical bone is about 10 years, whereas the mean age of cancellous bone is about 2 years, but bone in the middle of thick trabeculae may be as old as cortical bone. As bone becomes older, it becomes more highly mineralized, and so more dense, less able to exchange ions with the extracellular fluid, more brittle, more prone to fatigue damage because it is subjected to more loading cycles, and more likely to undergo death of its osteocytes. Prevention of these adverse effects of excessive aging is presumably the principal function of remodeling, and the rate of renewal is controlled mainly by the frequency of remodeling activation. This is influenced by the deformation consequent on biomechanical loading (strain), geometric factors such as curvature, and both local and systemic humoral agents. Most remodeling is stochastic, but to remove bone that has undergone fatigue damage or has become hypermineralized, remodeling must be directed to specific locations; how this is accomplished is unknown. Stochastic remodeling is influenced by abnormally low or high levels of secretion of thyroid and parathyroid hormones, but whether these participate in physiologic regulation in normal conditions is unknown. Once activation has occurred, the outcome depends mainly on the number of osteoclasts and osteoblasts that are recruited rather than on the function of the differentiated cells. The mechanisms involved are complex and poorly understood but, like activation, probably involve the interaction of biomechanical signals, locally produced growth factors and cytokines, and systemic hormones. [*See* Bone Regulatory Factors.]

The principal homeostatic function of bone is the regulation of plasma calcium. Deviations from the normal level lead to reciprocal changes in the secretion of parathyroid hormone (PTH), which acts on bone and kidney to restore the previous level, thus closing a feedback loop. [*See* Parathyroid Gland and Hormones.] Plasma calcium regulation is relatively independent both of the whole body rate of turnover and of the balance between total resorption and total formation. PTH regulates plasma calcium in two ways. First, it controls the solubility of the surface mineral that participates in the reversible exchange of calcium ions with the extracellular fluid at quiescent bone surfaces, such that inward and outward fluxes equilibrate at a particular level. Second, it regulates the reabsorption of calcium in the renal tubules, and so determines the plasma level needed to ensure excretion of the surplus calcium derived from net bone resorption and net intestinal absorption. A second feedback loop involves the stimulation by PTH of the biosynthesis of calcitriol, the active metabolite of vitamin D made in the kidney, which in turn directly inhibits both the synthesis and the secretion of PTH. In a third feedback loop, calcitriol raises plasma calcium both by increasing the intestinal absorption of calcium and by potentiating the effects of PTH on bone, and a rise in plasma calcium directly inhibits calcitriol production in the kidney by some mechanism independent of PTH. [*See* Vitamin D.]

III. BONE DENSITY AND OTHER DETERMINANTS OF BONE STRENGTH

The strength of a bone and its ability to withstand injury depend on the material properties of the tissue, the amount of structural material present in the bone, the manner in which the material is arranged, distributed, and connected in space, and the ability of bone, unlike man-made structural materials, to repair itself.

The material properties reflect the toughness of collagen fibers, the hardness and rigidity imparted by the mineral, and the laminated mode of construction. The amount of material depends on the thickness of the cortex, how far cancellous bone extends along the shafts of the long bones, and the thicknesses of individual trabeculae and the average distance between them. The three-dimensional architecture of each bone, both external and internal, is determined by interaction between the genetic program of growth and the biomechanical demands to which the growing bone must adapt. Cortical bone acquires the cross-sectional geometry at each level that is most efficient for load bearing, and individual trabeculae become aligned in directions that enable them to preferentially resist either tension or compression and are interconnected to reduce unsupported length and to minimize buckling. Like all structural materials that undergo repetitive cyclical loading, bone is subject to fatigue damage but can repair itself at the microscopic level by the remodeling process just described, which serves to arrest crack propagation before it accumulates into overt fracture. Once fracture has occurred, repair at the macroscopic level is accomplished by the rapid deposition of woven bone around the fracture site, whether of a whole bone or of an individual trabecula.

The mineral content of bone is its most distinctive feature, conferring the greatest density of any tissue. Density is mass/unit volume, but the volume to which mass is referred must be carefully defined. For true bone density, or bone substance density, the volume is of mineralized bone excluding the voids that result from osteocyte lacunae and canaliculi. True bone density mainly depends on bone age, increasing rapidly to about 1.4 g/cm^3 within a few days after the onset of mineralization, and then more slowly to 1.8–2.0 g/cm^3 over the next 6 months, continuing to increase at an exponentially decreasing rate for several more years. Normally, >95% of all bone is >6 months old, so that the total amount of mineral provides a reasonable estimate of the total amount of mineralized bone. This relationship is the basis for using ash weight as an *in vitro* measure and bone mineral content as an *in vivo* measure of bone quantity. For bone organ or apparent density, the volume referent is that of the whole bone as would be measured by Archimedes' principle. Apparent bone density depends on the volumetric proportions of cortical bone tissue, cancellous bone tissue, and fatty marrow in the shaft and relates the total amount of bone to the size of the bone. The distinction between apparent bone density and true bone density underlies the paradox that with increasing age the density of the *bones* declines, but the density of *bone* shows little change and may even increase as a consequence of increasing bone age.

IV. ACCUMULATION AND LOSS OF BONE WITH AGE

The amount of bone in the skeleton increases from zero at conception to a peak value at around 25–30 years, followed, after a plateau of variable duration, by a progressive decline. Most of the increase is associated with growth. Cortical bone is formed by net subperiosteal apposition and so is linked to growth in width, whereas cancellous bone is formed by endochondral ossification and so is linked to growth in length. After epiphysial closure and cessation of growth, net periosteal apposition continues very slowly throughout life, but no new trabeculae can be formed. While increasing in size, the shape of the bones is preserved by continuous resorption and apposition at appropriate locations, processes collectively referred to as modeling (Table IV). The most rapid gain in bone mass occurs during the prepubertal growth spurt, when there is a very high rate of bone turnover, and cortical bone becomes more porous (and consequently weaker) because sites of intracorti-

TABLE IV
Some Differences between Remodeling and Modeling

	Modeling	Remodeling
Purpose	Redistribution	Replacement
Timing	Continuous	Cyclical
Location of resorption and formation	Different surfaces	Same surface
Extent	100% of surface	20% of surface
Coupling	Systemic	Local
Balance	Net gain	Net loss[a]
Activation[b]	Not needed	Needed
Apposition rate	2 to 20 μm/day	0.3 to 1.0 μm/day
Need for osteoblast		
Stimulus	Growth	Resorption
Location	Periosteal	Endosteal
Effect of strain	Large	Small

[a]Except for a brief period between completion of growth and consolidation, and onset of age-related loss.

[b]In sense of transformation of quiescent to active surface.

cal remodeling are more numerous. Temporary internal removal of cortical bone to meet an increased demand for minerals occurs also in deer during the antler growth cycle (Table II). When growth slows down, the rate of remodeling falls and porosity declines to the adult level of 2–5%. At about the same time, for a few years there is a net gain of bone on the inner surface of cortical bone, in contrast to the net loss of bone from this surface at all other times of life. These processes of consolidation continue slowly for some years after cessation of growth in height and contribute about 5–10% to peak adult bone mass.

Total body calcium, measured by whole body counting after neutron activation, provides an estimate of peak total bone mass, otherwise measurable directly only in cadavers. In both sexes, the values are higher in blacks than in whites, and in both races the values are higher in men than in women. Approximate mean values for peak total body calcium are 1400 g for black men, 1200 g for white men, 1000 g for black women, and 800 g for white women; the risk of fracture in each demographic group is in the reverse order. Within each group, large differences exist between individuals—about 1 woman out of 40 has a peak adult bone mass that is less than the average value for a woman aged 65 years. Such variability makes a substantial contribution to individual differences in fracture risk within groups; it is partly genetic in origin, but other factors can jeopardize attainment of the full genetic potential for bone accumulation. Appropriately timed changes in pituitary, thyroid, and gonadal hormone secretion are necessary for accelerated growth, and adequate physical activity and nutrition are essential, particularly during the growth spurt. In each species, the individual grows the bones and muscles that will be needed to support its customary ranges of physical activity, with respect to both duration and intensity, provided these levels are repeatedly attained during growth. During adolescence, skeletal growth requires a net calcium retention of 300–500 mg/day; to allow for relatively inefficient intestinal absorption (despite increased serum levels of calcitriol) and obligatory renal excretion, dietary calcium intake needs to be 1.5–2.5 g/day, more than at any other time of life. [*See* Growth, Anatomical.]

Some years after peak adult bone mass is attained (about 5–10 years sooner in women than in men), age-related bone loss begins, a universal phenomenon of human biology that occurs regardless of gender, race, occupation, habitual physical activity, dietary habits, economic development, or geographic location. Loss of bone with age can even be inferred from the remains of prehistoric populations. In its universality, the age-related loss of bone differs from the age-related increase in blood pressure, which although ubiquitous in developed countries, does not occur in many communities remote from Western civilization. Because most of the loss takes place after cessation of reproduction, its absence would confer little evolutionary advantage. Bone is lost from every skeletal site amenable to measurement, but only from the endocortical and cancellous subdivisions of the endosteal envelope in contact with the bone marrow; as previously mentioned, the periosteal envelope continues to gain bone slowly throughout life, and in most bones there is little change in cortical porosity. Few serial measurements have been reported in blacks, but data suggest that they differ from whites less in rate of bone loss than in peak adult bone mass, so that the absolute differences established at maturity tend to persist throughout life.

In males, average cortical bone loss is about 0.3% of peak adult bone mass per year whereas cancellous bone loss is slightly faster. In females, the average loss is about 1% of peak adult bone mass per year for both cortical and cancellous bone. As a result of estrogen deficiency, there is an approximately twofold increase in the frequency of remodeling activation and bone turnover and an increase in the depth of resorption cavities throughout the skeleton. These changes lead to an approximately threefold acceleration in the rate of loss for about 5 years after menopause and a slower rate at both earlier and later times. The sex difference is greater for the femoral shaft than for the rib or the spine. After about age 90 years, the rate of endocortical loss may fall below the rate of periosteal gain so that cortical thickness, having declined progressively for 40–50 years, very slowly increases. In both genders those with most bone tend to lose most quickly, implying an exponential component to the loss. There are large individual differences in rate of loss that conform to a Gaussian distribution and that tend to be maintained over time. However, current bone mass is the best predictor of future bone mass, at least for about the next 20 years. [*See* Estrogen and Osteoporosis.]

V. BONE LOSS AND FRAGILITY: MICROARCHITECTURE AND CELLULAR MECHANISMS

Noninvasive methods of measuring bone mass and density that can be repeated over time in the same person have provided a much more detailed description of age-related bone loss than was previously possible, but the observations must be related to the underlying structural changes in bone and to the cellular mechanisms that are responsible. In cortical bone, porosity increases with distance from the periosteum. In the inner one-third to one-half of the cortices, the cavities are larger and more numerous, and they communicate with the bone marrow. Because cortical bone loss is initiated by the enlargement of subendocortical cavities, both porosity and surface-volume ratio, the most important geometric properties of bone, increase from the low values characteristic of cortical bone to the high values characteristic of cancellous bone, a process referred to as cancellization (Fig. 1). The walls separating the cavities from each other and from the marrow become thinner and eventually break down so that the enlarged subendocortical cavities communicate with each other and with the marrow cavity. The newly created trabeculae resemble those originally present in cancellous bone but are usually thicker and more irregular in alignment. After this process is complete, the less porous outer one-half to two-thirds of the original cortex continues to lose bone diffusely from its inner surface and the cortices continue slowly to get thinner despite net periosteal gain, at least until the 10th decade.

The architectural changes with age are more complex in cancellous than in cortical bone and are less evident from two-dimensional histologic sections. Although an oversimplification, it is a useful first step to consider the amount of bone present as made up of some number per unit length or area of individual structural elements, conveniently thought of as plates, of a particular mean thickness. Application of this simple model indicates that a major component of age-related loss of cancellous bone is the complete removal of some plates, so that those remaining are more widely separated and less well connected. The marrow spaces are increased in size as the result of complete removal of part of their walls so that previously separated spaces coalesce in a manner conceptually similar to the enlargement of alveoli in an emphysematous lung. The process begins with the focal perforation of a plate, continues by enlargement of the perforation to a window, and may progress by conversion of the plates to a lattice of rods, which in turn may be transected or completely removed (Fig. 2). During this sequence of events, and continuing after it has run its course, the remaining trabeculae slowly get thinner throughout life. In the spine, horizontal trabeculae are removed first so that vertical

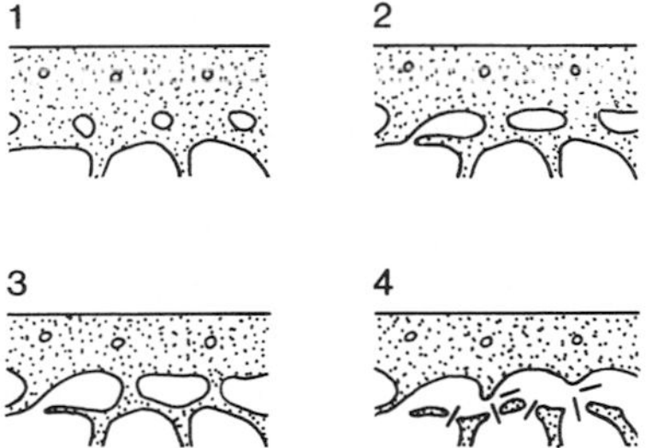

FIGURE 1 Microarchitecture of cortical bone loss. (1) Normal adult cortex with larger Haversian canals closer to the endocortical surface. (2) Enlargement of subendocortical spaces and communication with the marrow cavity. (3) Further enlargement of the spaces and conversion of the inner third of the original cortex to a structure intermediate in geometric properties between cortical and cancellous bone and consequent expansion of the marrow cavity. (4) Perforation and disconnection of the new trabeculae. [Reprinted with permission from Parfitt, A. M. (1984). Age-related structural changes in trabecular and cortical bone: Cellular mechanisms and biomechanical consequences. a) Differences between rapid and slow bone loss. b) Localized bone gain. *Calc. Tissue Int.* **36,** S123–128. Springer-Verlag, Heidelberg.]

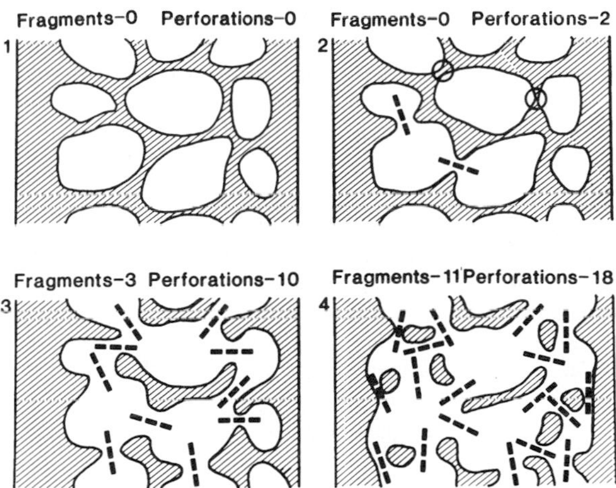

FIGURE 2 Microarchitecture of cancellous bone loss. Successive stages in the age-related conversion of a continuous to a discontinuous trabecular network. Fragments are isolated profiles seen in a two-dimensional section. They do not lie free in a marrow space but are connected in the third dimension. Perforations are the focal breaks in continuity that initiate rapid cancellous bone loss; they bring adjacent but separated marrow cavities into communication. The same changes occur in the cancellous bone that arises from the inner third of the cortex, as shown in Fig. 1. (Upper right) The circles highlight regions of local thinning that probably precede perforation. [Reprinted with permission from Parfitt, A. M. (1984). Age-related structural changes in trabecular and cortical bone: Cellular mechanisms and biomechanical consequences. a) Differences between rapid and slow bone loss. b) Localized bone gain. *Calc. Tissue Int.* **36,** S123–128. Springer-Verlag, Heidelberg.]

trabeculae are more liable to buckle under load bearing, but the compensatory thickening of vertical trabeculae, inferred from misinterpretation of radiographs and predicted from biomechanical theory, does not occur.

It is both a necessary and a sufficient condition for bone to be lost from a surface that the depth of a resorption cavity exceeds the thickness of new bone formed within the cavity. Based on this simple proposition, bone loss can either be osteoclast mediated, with increased resorption depth, or be osteoblast mediated, with decreased formation thickness. Increased resorption depth leads to subendocortical cavitation and cancellization of inner cortical bone (Fig. 1) and to trabecular plate perforation and architectural discontinuity (Fig. 2). Decreased formation thickness leads to slow diffuse thinning of both cortices and trabeculae. The principal features of these two forms of bone loss are contrasted in Table V. Osteoclast-mediated loss is characteristic of early menopause; it is a consequence of estrogen deficiency and is preventable by estrogen replacement. Osteoblast-mediated loss is characteristic of aging and is mainly the result of defective osteoblast recruitment; possible mechanisms include depletion of stem cells as a result of replacement of hemopoietic by fatty marrow and impaired generation, transmission, or reception of the signal that normally couples formation to resorption. The rate at which the loss occurs is amplified by an increased frequency of remodeling activation as a result of age-related declines in renal function, calcitriol biosynthesis, and intestinal calcium absorption, augmented in some persons by insufficient dietary calcium intake and, particularly in elderly nursing home residents, by nutritional vitamin D deficiency, all leading to an increased secretion of PTH.

TABLE V
Comparison of Two Morphologic Types of Bone Loss

Characteristic	Osteoclast mediated	Osteoblast mediated
Cellular defect	Longer life span	Lack of number
Remodeling mechanism	Deeper resorption	Shallower formation
Usual cause	Estrogen deficiency	Aging
Structure		
Cancellous	Perforation and discontinuity	Simple thinning
Cortical	Subendocortical cavitation	Simple thinning
Reduction in strength	More than predicted[a]	As predicted[a]
Timing	Early	Late
Rate	Rapid	Slow
Magnitude	Usually greater	Usually lesser
Frequency of activation	Often increased	Often decreased

[a]From reduction in mass or density based on mineral content.

VI. NONTRAUMATIC FRACTURES: FRAGILE BONES IN FRAIL PEOPLE

Fragility of a bone is manifested by fracture. Any fracture results from the interplay of three factors: the mechanical energy resulting from bodily movement or from a fall or injury, the capacity of soft tissue absorption and muscle contraction to dissipate energy, and the ability of the bone to withstand the force resulting from the residual undissipated energy. Each of these three factors can vary continuously over a wide range, and their relative contributions to any particular fracture differ markedly in different circumstances, but it is a convenient simplification to classify fractures as either traumatic, in which the energy available is excessive, or as nontraumatic, in which the energy available is no greater than expected for the normal activities of daily living. Sharing features of both types are stress fractures, which occur as a result of frequently repeated cyclical loading during activity that may be unusual for the individual but is not abnormal. They represent the accumulation of microscopic cracks due to fatigue into macroscopic fracture because the repair mechanism is overwhelmed.

With the same residual undissipated energy, the risk of fracture increases with age. In this sense, fracture risk can be regarded as an inverse function of bone strength or as a direct function of bone fragility. The concept of fragility leaves open the possibility that some aspects of fracture susceptibility may not be captured by conventional biomechanical tests of strength. The increase in bone fragility with age is to a large extent the result of the loss of bone with age, a relationship that is clearer if bone loss is expressed as a change in apparent density rather than in mass. In persons who have sustained one of the age-related fractures, average values for bone density and mass are lower than in persons of the same age, gender, and race who have not sustained such a fracture. This difference is usually evident regardless of site or method of measurement, but it is greatest for measurements made at the site that is most representative of the site of fracture. Such measurements also provide an index of future fracture risk, and if combined with an estimate of life expectancy, the probability of fracture during the remainder of life can be calculated.

Bone fragility and fracture risk are also influenced by qualitative factors that are partly independent of changes in bone density. Although mean values differ, there is invariably a substantial overlap in individual values between subjects with and without fracture. Stochastic models that describe individual fracture risk, in terms both of individual bone deficits and of the change in frequency of falls with age, match observed changes in fracture incidence with age most accurately if an additional age-related factor is included in the model, such as a change in fall severity or bone quality. *In vitro* testing invariably shows a high correlation between bone density and bone strength, but the unexplained variance is usually in the range of 20–40%, and there is considerable uncertainty in relating loads and resulting strains experienced by bone *in vivo* to those studied *in vitro*. For many bones there is a large margin of safety between instantaneous breaking strain and the peak strain attainable during maximum physical activity, such that even a reduction in density by half would not by itself result in nontraumatic fracture. Consequently, the loss of bone with age is a necessary, but not always a sufficient, condition for the occurrence of age-related fractures.

Fractures at many sites are more common in the elderly, but three sites are of particular importance: lower forearm (wrist), vertebral body, and upper femur (hip). Wrist fractures are very common in both genders during adolescence because the increase in cortical porosity associated with rapid growth has a disproportionate effect on strength where the cortices are thin, as in the distal radius. After completion of growth and consolidation, the incidence of wrist fractures falls to a very low level; it increases sharply in women around the time of menopause, but increases only slightly in men. This difference reflects the nonspecific effect of estrogen deficiency to increase the rates of remodeling activation and bone turnover, and thus make cortical bone more porous, as well as the specific effect to increase resorption depth. The pathogenesis of vertebral fracture is more complex. These fractures frequently occur without symptoms; unlike other fractures, there is no displacement or separation of fragments but compaction leading to altered vertebral body shape. Minor degrees of vertebral height loss can result from remodeling rather than fracture and contribute to the age-related loss of stature. The architectural disruption of cancellous bone mentioned earlier increases the risk of vertebral collapse, but thin vertical trabeculae may buckle as a result of the temporary weakness induced by remodeling, even without perforation. In some women, for unknown reasons remodeling activation falls and bone age increases in axial cancellous bone. There is no evidence that this predisposes to fatigue damage accumulation. Rather, it may be a response to a defect in the completion of secondary mineralization. [*See* Vertebra.]

One in six elderly persons who suffers a hip fracture will die within 6 months and many of those that survive will never again be able to walk. The incidence of hip fracture doubles for every 5-year increase in age after 50 years; the point at which the care of hip fractures consumes all available medical resources may be the most intractable constraint on human longevity. Many hip fractures follow a fall, and the incidence of falls increases with age in a similar manner to the incidence of hip fractures, but only a minority of falls result in fracture. The site of fracture is influenced by the type of fall: a fall forward tends to result in wrist fracture, a fall backward in vertebral fracture, and a fall sideways in hip fracture. The energy released by a fall from standing height, if undissipated by soft tissue, is sufficient to break even the strongest bone. The elderly are not only more likely to fall but, because of impaired neuromuscular coordination, are less able to protect themselves during a fall or to absorb the energy released by the fall by appropriately timed muscle contraction. Older persons also frequently have less subcutaneous soft tissue to cushion the effect of the fall. Some of the neuromuscular factors that make falls both more likely and more dangerous can be detected by abnormalities of gait, posture, or balance.

Frailty of the body is one of the main reasons why the elderly are especially prone to hip fracture, but fragility of the bones is also important. Some hip fracture subjects recall hearing or feeling the break before they fall; stress fractures can occur in the femoral neck even in young people, and such apparently spontaneous fractures in the elderly may similarly result from the accumulation of microscopic fatigue damage. Differences between persons with and without fracture lessen progressively with increasing age, but the risk of hip fracture increases as local bone density declines, independent of age or neurologic status. This relationship is usually attributed to a reduction in instantaneous breaking strength, but could also reflect increased susceptibility to fatigue damage because of greater loads on the remaining bone. In the young, >95% of bone lacunae in the upper femur contain living osteocytes, but this proportion falls with increasing age. Much of the bone at the fracture site is dead, but this could be the result of the fracture rather than one of its causes. Nevertheless, replacement of hemopoietic tissue by much less vascular fatty tissue, as occurs in the head and neck of the femur in the elderly, would be expected to reduce the viability of the bone as well as increasing its age because of lower turnover. These factors are more important for hip than for vertebral fracture because there is a lower margin of safety for bone turnover in the appendicular than in the axial skeleton.

An unexplained but important feature of the biologic significance of hip fractures is that the most robust predictor of incidence in different countries is the gross national product. Hip fractures are much less common in South African blacks than in either South African whites or in North American blacks, differences that cannot be accounted for by ethnicity alone or by differences in bone density. Evidently, some consequence of economic development promotes either increased frailty or increased fragility of the bones. The mechanism responsible is unknown, but decreased physical activity because of a sedentary life-style, accumulation of an unidentified osteoblast toxin as a result of environmental deterioration, deficiency of some nutrient not yet recognized as essential for bone health, and increased renal wasting of calcium due to excessive intake of salt or protein have each been proposed. Whatever the factor (or combination of factors) turns out to be, it is sufficient to offset any benefit conferred by increased dietary calcium intake, which correlates positively with fracture incidence among different countries, even though it may correlate negatively with fracture incidence within the same country. Traumatic fractures can be studied as a problem in applied engineering, but nontraumatic fractures cannot be fully understood except as an expression of the biology of human aging, conceived in the broadest sense, untrammelled by geographic or cultural parochialism.

BIBLIOGRAPHY

Avioli, L. V., and Krane, S. M. (eds.) (1990). "Metabolic Bone Disease," 2nd Ed. Saunders, Philadelphia.

Coe, F. L., and Favus, M. J. (eds.) (1992). "Disorders of Bone and Mineral Metabolism." Raven Press, New York.

Frost, H. M. (1986). "Intermediary Organization of the Skeleton." CRC Press, Boca Raton, FL.

Hall, B. K. (ed). (1990–1993). "Bone: A Treatise." Telford Press, Caldwell, NJ.

Mundy, G. R., and Martin, T. J. (eds.) (1993). "Physiology and Pharmacology of Bone." Springer-Verlag, Berlin.

Parfitt, A. M. (1994). Osteonal and hemi-osteonal remodeling: The spatial and temporal framework for signal traffic in adult human bone. *J. Cell. Biochem.* **55**, 273–286.

Parfitt, A. M. (1994). The two faces of growth—benefits and risks to bone integrity. *Osteo. Int.* **4**, 382–398.

Parfitt, A. M. (1996). Skeletal heterogeneity and the purposes of bone remodeling: Implications for the understanding of osteoporosis. *In* "Osteoporosis" (R. Marcus, D. Feldman, and J. Kelsey, eds.). Academic Press, San Diego.

Riggs, B. L., and Melton, L. J. (eds.) (1995). "Osteoporosis, Etiology, Diagnosis and Management," 2nd Ed. Raven Press, New York.

Bone, Embryonic Development

BRIAN K. HALL
Dalhousie University

GLOSSARY

Bone Vascularized calcified skeletal tissue consisting of cells (osteoblasts and osteocytes) embedded in a biochemically and structurally distinct extracellular matrix

Cartilage Mostly avascular, usually calcified, skeletal tissue consisting of cells (chondroblasts and chondrocytes) embedded in a biochemically and structurally distinct extracellular matrix

Chondroblasts/chondrocytes Progressive stages of cartilage-forming cells (chondroblasts differentiate into chondrocytes)

Differentiation Progressive specialization of cells during embryonic development, in wound repair, fracture healing, and regeneration

Endochondral ossification Development of bone within a cartilage model

Intramembranous ossification Direct development of bone without a cartilaginous model

Morphogenesis Progressive development of the shape of a cell, tissue, organ, or individual

Osteoblast/osteocyte Progressive stages of bone-forming cells, found either on the surfaces (osteoblasts) or embedded within (osteocytes) bone; osteoblasts differentiate into osteocytes

Skeleton Organ system of cartilages and bones that supports and protects the body, provides attachment sites for muscles, and houses the marrow from which blood-forming cells arise

BONE, A SUPPORTING SKELETAL TISSUE, IS FOUND in the earliest known vertebrates. Bony plates can be identified in fossil vertebrates that are over 500 million years old. That such ancient bone looks like the bone of modern-day vertebrates tells us that the fundamental organization and structure of bone have changed little since the origins of the vertebrates themselves. Bone begins to form in the embryo through the action of bone-forming cells (i.e., preosteoblasts, osteoblasts, and osteocytes) which arise in one of two fundamental embryonic layers: from the mesoderm that also forms the muscles, heart, and connective tissues in the trunk, and from the ectoderm that also forms the brain and the outer layer of the skin in the head.

Bone can develop in one of two major ways: either directly in a cellular condensation (intramembranous ossification) or by replacement of a preexisting tissue, as in endochondral, tendinous, or ligamentous ossification, when cartilage, tendon, or ligament, respectively, is the tissue being replaced. Both in the trunk, where bone forms vertebrae, ribs, and long bones, and in the head, where bone forms the skull and the jaws, future bone-forming cells have to migrate within the developing embryo to the final positions where bones will develop. Once at their final site, these cells condense and begin to differentiate into osteoblasts. In both of the major types of ossification, bone-forming cells differentiate only as a result of interactions with other developing cells and tissues. These cell and tissue interactions allow osteoblasts to differentiate and to deposit the biochemically and structurally distinct bony extracellular matrix that creates bones of

ENCYCLOPEDIA OF HUMAN BIOLOGY, Second Edition, VOLUME 2. Copyright © 1997 by Academic Press. All rights of reproduction in any form reserved.

specific shapes with joints at precise locations. Subsequent growth of bone and of bones, even after birth, can be traced back to the same processes responsible for the embryonic development of bone.

I. WHAT IS BONE?

It may seem odd to begin by asking this question, for, of course, we all know what bone is. However, not all may realize that bone falls into at least three levels of biological organizations, for bone is, at the same time, a tissue, an organ, and an organ system.

A. Bone as a Tissue

As a tissue, bone is made up of cells embedded in a matrix, rather like raisins embedded in a muffin mix. The cells of bone are (1) preosteoblasts, which are stem cells that act as a reserve for the production of future bone-forming cells (Figs. 1 and 2); (2) osteoblasts, which arise from preosteoblasts, line both the inner (endosteal) and outer (periosteal) surfaces of all bones and deposit the bony matrix; and (3) osteocytes, which develop from osteoblasts once the latter have completely surrounded themselves with bony material (Fig. 3). A fourth category of cells also exists, the osteoclasts and their precursors, the preosteoclasts. Osteoclasts function to break down, or resorb, bone. As this article is concerned with how bone develops, not with how bone is resorbed or remodeled, no more is said about osteoclasts.

Preosteoblasts are identified on the basis of their ability to become osteoblasts, usually using cell, tissue, or organ culture techniques (Fig. 4). Osteoblasts can be identified from their location on bone surfaces and by the production of specific bone matrix products that show that they are depositing bone. The osteocyte is identified by being completely encased in bone and by its long processes (i.e., canaliculi) (Fig. 3) running through channels in the bone. (In fact, all cells of a bone are in communication with one another through cellular connections. This is why signals such as a hormone pulse or an electrical impulse can travel so rapidly throughout a bone.)

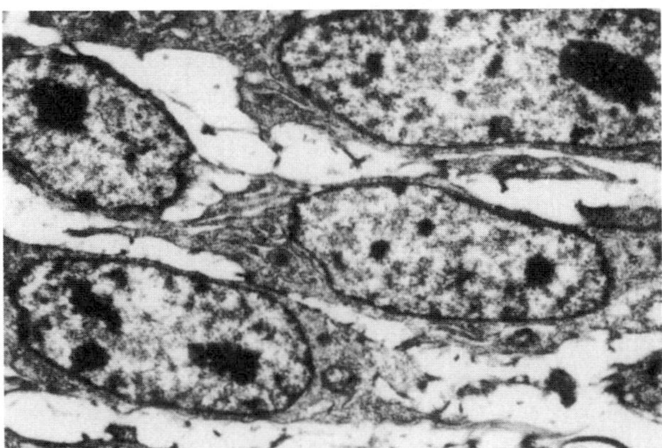

FIGURE 1 Preosteoblasts from the periosteum of a direct developing bone, illustrating their elongate shape and the paucity of extracellular matrix between them. ×16,600.

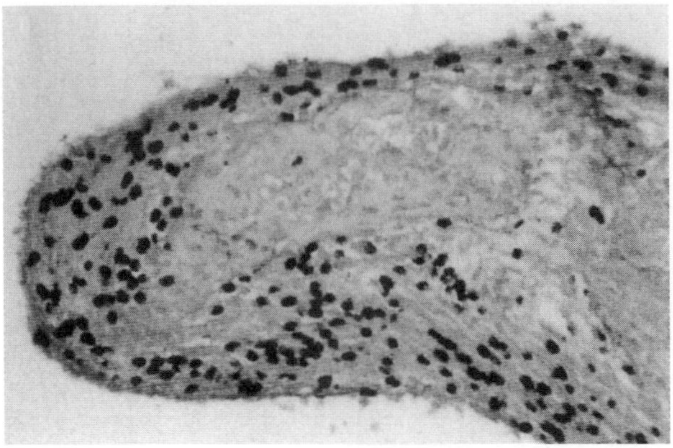

FIGURE 2 An autoradiograph following labeling with radioactive thymidine (a precursor of DNA) to demonstrate the intense mitotic activity of periosteal preosteoblasts (black circles; the periosteum is the layer of cells covering a bone). The bone in the center is unlabeled.

The matrix of bone (the "muffin mix" around the "raisins") is an extracellular matrix, i.e., it surrounds the osteocytes, its organic phase having been secreted by the osteoblasts/osteocytes. Bone matrix consists of two phases: (1) an organic matrix, of which 70–80% is the structural protein collagen and 1–5% is bone-specific proteins such as osteocalcin, osteonectin, and bone sialoprotein; and (2) an inorganic mineral phase consisting of calcium phosphate salts deposited into the organic matrix (Fig. 3b). The latter is, of course, the mineral that makes bone a hard mineralized tissue, enabling it to serve as a supporting system in living vertebrates and to be fossilized as the remains of ancient vertebrates. [*See* Extracellular Matrix.]

When first deposited, the extracellular matrix of bone is unmineralized and is known as osteoid; a rim of osteoid can be found along the surfaces of all growing and developing bones. How mineral is deposited into this organic matrix is not fully under-

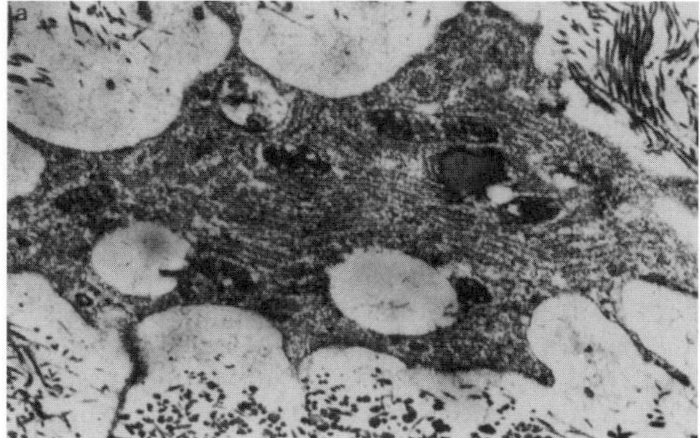

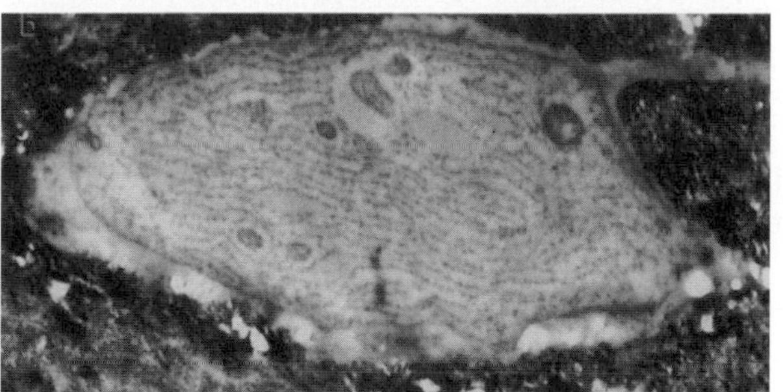

FIGURE 3 (a) A newly formed osteocyte, surrounded by extracellular collagen fibers (some in cross-section and some in longitudinal section). Note the numerous cell processes (canaliculi) extending from the cell into the extracellular matrix. ×31,800. (b) A mature osteocyte embedded in a fully mineralized bony extracellular matrix. Two canalaculi can be seen at the right. ×18,500.

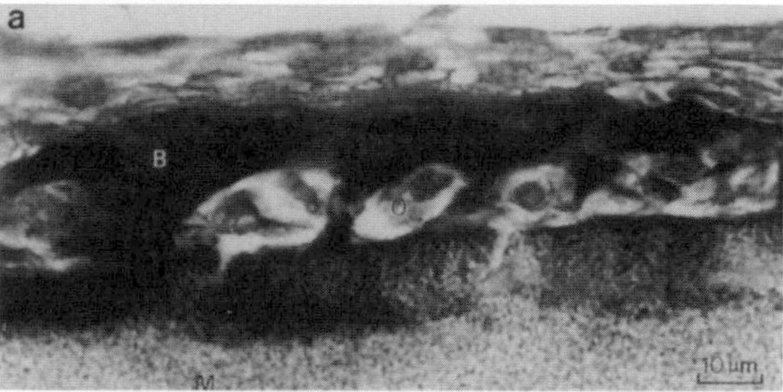

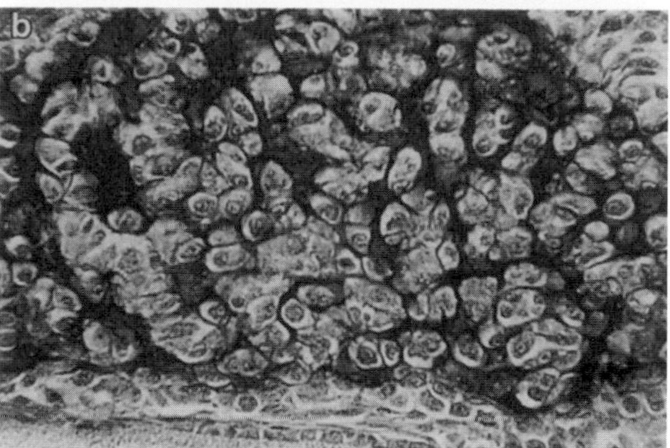

FIGURE 4 (a) Embryonic mesenchyme placed on a Millipore filter substrate (M) and cultured for several days forms bone (B). Osteoblast (O) depositing the bone matrix, some of which is accumulating in the filter. (b) More extensive osteogenesis occurs with increasing time in culture. Many osteocytes have differentiated and deposited bony matrix (black).

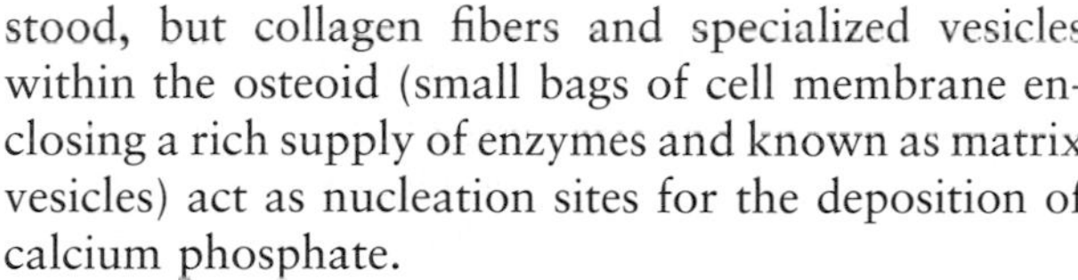

stood, but collagen fibers and specialized vesicles within the osteoid (small bags of cell membrane enclosing a rich supply of enzymes and known as matrix vesicles) act as nucleation sites for the deposition of calcium phosphate.

B. Bone as an Organ

When we use the term "bone" in everyday life, we are usually referring to bone as an organ. The wishbone of the Thanksgiving turkey is an example of bone as an organ (actually, it is two bones—the fused clavicles or collar bones—known in birds as the furcula). The drumstick has a bony organ at its meaty center. Thus, bone as an organ represents the organization of bone cells and matrix into a specific shape that is both recognizable and reproducible from individual to individual (Fig. 5). Individual bones, unless at the end of an extremity, such as the last bones in the fingers, are usually joined to other bones by joints.

C. Bones as an Organ System

The skeletal system, one of the major organ systems of the body, consists of all of the individual bones assembled into their proper organization and arrangement—the dinosaur skeleton in the museum, the human skeleton in medical school, and the fossil embedded in the rocks of ancient times. [*See* Skeleton.]

II. HOW ANCIENT IS BONE?

That we are able to find fossilized bones obviously shows that bone is an ancient tissue and organ. Bone is found only in vertebrates, i.e., animals with a backbone and a dorsal nervous system, such as humans. Therefore, the origins of bone have to be sought among fossil vertebrates. In fact, the earliest evidence we have for the presence of vertebrates on the earth consists of scraps of bone in rocks from the Upper

FIGURE 5 The normal shape of an embyronic long bone, in this case, the femur of an embryonic chick grown as a graft from the condensation stage on. The bone develops its normal shape as an elongate rod with expanded ends (i.e., the condyles) where the joint would form.

Cambrian period over 500 million years ago. These flat plates of bone are associated with, and apparently support, a layer of dentine, which is the tissue that coats the inner portion of teeth. The bone and dentine formed dermal denticles or ossicles embedded in the skin. Such denticles were initially isolated, but later came to fuse to form an external (dermal) skeleton around our vertebrate ancestors. It is thought that the dentine of these denticles initially served a sensory function (because it is well supplied with nerves), with the bone acting as a support.

Thus, the first known bone was ancient indeed, was associated with a tissue now confined to teeth (except for body denticles in some fishes), and formed a primitive external skeleton. Only later in vertebrate evolution did bone produce the now familiar internal skeletal system, which replaced the external dermal skeleton.

What is amazing is that the bone of these 500-million-year-old fossils can be so readily recognized as bone; that is, bone, when it first arose, was so like modern-day bone in terms of its cells and matrix that we had no difficulty in assigning it to this category of skeletal tissue. Consequently, specialized fields such as paleopathology exist in which the diseases of ancient animals can be determined, studied, and compared with those of modern-day animals. We can tell that dinosaurs suffered from arthritis or when syphilis first appeared just from the study of fossil bones.

III. MODES OF BONE DEVELOPMENT

The two basic ways in which bone develops can be most simply classified as direct or indirect osteogenesis.

A. Direct Osteogenesis

In direct osteogenesis, osteoblasts differentiate (i.e., progressively specialize from precursor cells) directly, without involvement or replacement of any other tissue. Cells aggregate, begin to synthesize and deposit the specialized proteins of bone, and gradually separate from one another as the newly deposited extacellular matrix accumulates (Fig. 4). Such osteogenesis is usually termed "intramembranous ossification" because the bone develops directly in what early microscopists and histologists thought looked like a membrane—a loose collection of cells. Intramembranous ossification produces membrane or dermal bones, such as the flat plates of bone of the skull, the rod-like collar bones, and the first bone that surrounds the shafts of long bones, such as the femur or the tibia of the leg (Fig. 6).

B. Indirect Osteogenesis

In this mode of ossification, the bone that is deposited replaces a preexisting tissue that acts as a model or

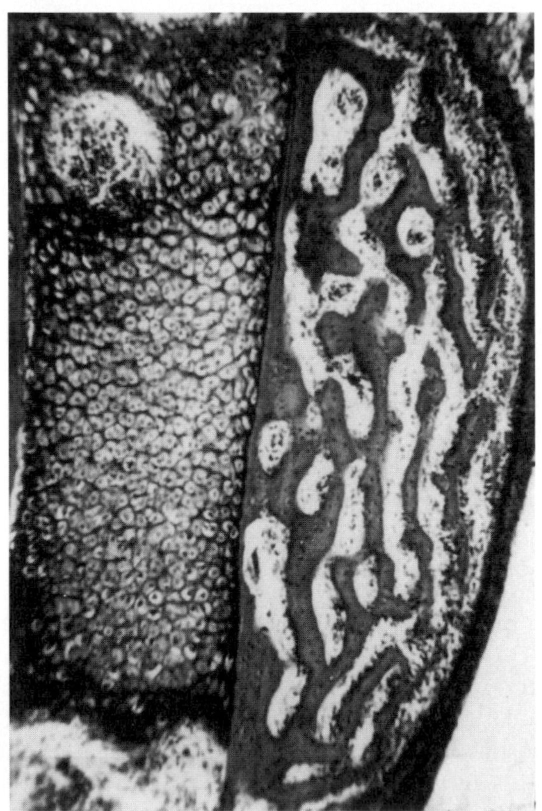

FIGURE 6 The cartilaginous model of a long bone (left) on which an extensive amount of intramembranous bone has been deposited (right), forming an exostosis. Note the sharp boundary between cartilage and bone, indicative of the development of bone on cartilage rather than by replacing cartilage. Marrow can be seen replacing the cartilage at the top and the bottom.

template for bone formation. In humans, that template is usually cartilage that is laid down early in embryonic life. The cartilaginous precursor is gradually removed and is replaced by bone in a process known as endochondral ossification (literally, bone formation in cartilage).

Within a single bone, such as the femur, intramembranous and endochondral ossifications take place simultaneously, the former laying down bone on the surface of the cartilage, while the latter is depositing bone within, and replacing, the cartilaginous model. Intramembranous ossification occurs from preosteoblasts and osteoblasts in the perichondrium (i.e., the cell layer surrounding the cartilage) as it is transformed into a periosteum (i.e., the cell layer surrounding bone) (Fig. 6). Endochondral ossification occurs from cells that arise in the marrow within the cartilage core or from cells that are brought into the eroding cartilage by invading blood vessels. Thus, the femur is not an endochondral bone; it is a long bone that forms by both endochondral and intramembranous ossifications.

We now have considerable knowledge concerning the cellular and, to some extent, the molecular and biochemical steps involved in deposition of the cartilaginous model and its replacement by bone. Initially, a rod of small-celled cartilage is laid down by chondroblasts and chondrocytes. The chondrocytes at the center of this rod then enlarge in a process known as chondrocyte hypertrophy, which is associated with production of at least two molecules that are specific for, or at least present in greatest amounts in, hypertrophic chondrocytes, type X collagen, and alkaline phosphatase. Mineralization of the matrix surrounding hypertrophic chondrocytes follows the acquisition of alkaline phosphatase activity, although the precise role of this enzyme in mineralization remains obscure. [*See* Cartilage.]

Chondrocyte hypertrophy and matrix mineralization gradually spread along the cartilage model as intramembranous ossification begins in the perichondrium/periosteum adjacent to the hypertrophic chondrocyte zone, producing a collar of bone around the cartilage model.

The mineralized cartilage begins to break down as blood vessles invade it, creating both space and an environment conducive for marrow and endochondral bone development. The cartilage is then replaced, partly by marrow (Fig. 6) and partly by bone. Endochondral ossification progressively spreads away from the center of the shaft until only the cartilages at the ends of the bone remain unreplaced; these serve as the articular cartilages at joints with adjacent bones.

Indirect osteogenesis can also occur by the replacement of tissues other than cartilage, the prime examples being replacement of ligaments (ligamentous ossification) or of tendons (tendinous ossification). Although such modes of indirect ossification are relatively common in nonmammalian vertebrates (the bones that connect the processes on the neck vertebrae of many large dinosaurs developed by ligamentous ossification), they are relatively uncommon in humans, in which they are often associated with pathological conditions.

Bone can also develop outside the skeleton, when it is known as ectopic or heterotopic bone (the process, then, is known as ectopic osteogenesis). Ectopic bone can develop by either direct or indirect osteogenesis, often arising after injury or trauma or in scar tissue following surgery.

IV. ORIGINS OF BONE-FORMING CELLS

The osteoblasts and osteocytes that form bone of the head and the trunk have quite separate and distinctive embryological origins. Some of the skull and all of the facial bones (e.g., jaws, cheeks, and ear bones) have an embryonic origin that reflects the ancient origin of bone in the vertebrates, for they arise from cells of the neural ectoderm that otherwise go to form nerve cells of the developing brain. The cells that form these bones (and the cartilages and connective tissues of the head) are known as neural crest cells because of their origin in the crests of neural folds that are the precursors of the brain. Because they arise in the head, they are known as cranial neural crest cells.

Although neural crest cells are found along the portion of the neural tube that will form the spinal cord, these trunk neural crest cells cannot form bone or cartilage. Consequently, the skeleton of the trunk, including the ribs, vertebrae, and limb skeletons, cannot arise from neural crest cells, but instead arises from mesoderm, the embryonic germ layer that produces muscle, organs such as the heart and kidneys, and connective tissue (Fig. 7). A few of the most posterior and roofing bones of the skull also arise from mesoderm.

If we attempt to relate the development of bone to its evolutionary origins, we come to the conclusion that the earliest vertebrate bone (the supporting bone of the denticles) was of neural crest origin. Evidence shows that (1) the dentine only forms from neural

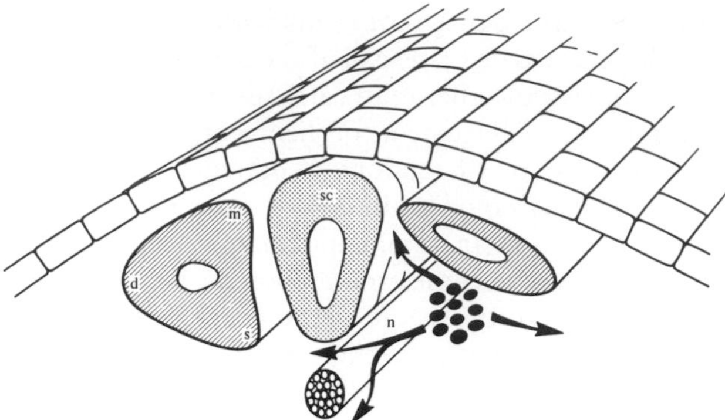

FIGURE 7 Cells from a somite (black circles) are migrating toward the spinal cord (sc) and the notochord (n), where they differentiate as the chondroblasts of the vertebrae of the back bone. d, m, and s dermis-, muscle-, and skeletal-forming regions of the somite, respectively.

crest cells and (2) the bone that supports dentine in the teeth of modern vertebrates differentiates from the same cell population as does dentine. Therefore, give the conservation of developmental processes over evolutionary time, we can conclude that the first bone was neural crest in origin. Only later, in vertebrate evolution, did a postcranial nonneural crest and mesodermally derived skeleton arise. Thus, knowledge of embryonic development of bone allows us to, in effect, perform "evolutionary experiments" on the origin of bone.

V. CELL MIGRATION

Knowing the embryological origin of the bone-forming cells, we can begin to explore how those cells progressively differentiate and deposit the extracellular matrix that makes bone and bones.

The first obvious conclusion is that the cells that become osteoblasts, especially in the head, do not arise where bone will form. Cranial neural crest cells arise in the developing brain, not a place where one wants bone to develop. It is one of the wonders of embryonic development that, as Hans Spemann says, "We are standing and walking with parts of our body which we could have used for thinking if they had been developed in another position in the embryo." Neural crest cells must therefore migrate to the proper location in the embryo for bone to develop in its proper place.

Similarly, the cells that form the vertebrae do not arise around the spinal cord and notochord, but rather arise in segmented blocks of mesoderm (i.e., somites) on either side of the developing spinal cord. These cells, too, must migrate away from the somites to surround the spinal cord before development of the vertebral column can begin (Fig. 7).

Cell migration is therefore an important initial step in the embryonic development of bone. The fact that this step and the neural origin of many bone-forming cells seem so far removed from bone formation illustrates the hierarchical organization of embryonic development and the need to search early in development for mechanisms that control events that occur much later in development.

The waves of neural crest cells, which migrate away from the developing brain, or of somitic cells, which move away from the developing spinal cord, migrate by utilizing molecules in the extracellular spaces through which they move. Those molecules (fibronectin, for example, is one of considerable importance to the migratory process) either lie free in spaces within the embryo or are associated with the basement membranes on which sheets of cells (embryonic epithelia, the future skin cells) reside. They are used as the guidance substrates along which future skeleton-forming cells move and with which they interact. Although it is still not clear how directionality is imposed on these migrating cells or how they know when to cease their migration and to settle down at specific locations, cell migration and localization are central to the embryonic mechanisms that determine where individual bones will develop. In fact, many developmental abnormalities that involve defective skeletal development—such as Treacher–Collins syndrome and other craniofacial abnormalities, in which jaw or skull development is abnormal in position, duration, amount, or conformation—can be traced to defective migration of the neural crest cells and can be modeled with the injection of chemical agents that disrupt cell migration, such as retinoic acid, a close chemical relative of vitamin A.

VI. CELL CONDENSATION (AGGREGATION)

One of the consequences of cell migration is that groups of cells are brought together so that they can interact. Although cells divide as they migrate, it is localization at the final site that triggers cell division and allows the first visible primordium of each bone to arise. This primordium consists of a condensation (i.e., an aggregation, or blastema) of cells. The great geneticist Hans Grüneberg called this stage the "membranous skeleton" to emphasize that it was an important, although often neglected, stage in the development of all cartilages and bones. Defects at this stage of skeletal formation can have far-reaching consequences. Small condensations often produce abnormally small bones. In some mutants, bone formation will not begin unless a condensation reaches a critical size; that is, an abnormally small condensation means that the corresponding bone will not form.

The size of a condensation depends on a number of factors, chiefly the number of cells that migrate to the site, local (paracrine) factors for the promotion of cell division at the condensation site, production of (autocrine) factors that promote cell division by cells in the condensation itself, and the length of time between the initiation of the condensation and the deposition of bone (i.e., the time available for a condensation to grow). Variation in any of these factors will affect the final size of the bone. In fact, many evolutionary biologists believe that variation in processes that regulate the timing of developmental events (heterochrony) accounts for variations in the location, number, size, and shape of bones among species.

Evidence has accumulated that the formation of the condensation triggers the next stage of differentiation of cartilage- or bone-forming cells. The cells in an early condensation are not yet osteoblasts or chondroblasts. Unless we knew that a bone or carilage was to form at that site, we could not even identify these cells as preosteoblasts or prochondroblasts, for they do not demonstrate any of the cellular, molecular, or biochemical characteristics that we associate with bone- or cartilage-forming cells. However, once the condensation starts to form, there is a tremendous amplification of specific gene products for the particular skeletal tissue: cartilage if the bone develops by endochondral ossification or bone if it develops by intramembranous ossification. Thus, for a condensation in which cartilage will develop, the synthesis of mRNA for type II collagen (the type characteristic of cartilage) increases 100-fold, while the synthesis of mRNA for type I collagen (the type characteristic of bone) remains at basal rates. Condensation formation, therefore, triggers the selective synthesis of cartilage- or bone-specific products, even before the cells can be identified as chondroblasts or osteoblasts.

It is thought by some, although by no means yet proven, that the process of bringing like cells into close association in a condensation permits those cells to begin to synthesize local factors, such as autocrine growth/differentiation factors, that play a role in cellular differentiation. The signal for this process may lie in the changed shape of the cells in a developing condensation; cells become rounded, and a rounded shape is associated with selective expression of cell type-specific products. The scenario would be that a rounded cell shape alters the configuration of the cytoskeleton–cell membrane connections and/or alters calcium flux into the cells, leading to the synthesis of cartilage- or bone-specific products.

VII. SIGNALS FOR BONE CELL DIFFERENTIATION

As discussed earlier, one of the signals for bone cell differentiation, especially the initial synthesis of bone cell-specific products, is the formation of a condensation of cells. Growth/differentiation factors may be produced by the cells as a consequence of the condensation process and, indeed, growth factors have emerged over the past few years as major candidates for differentiation signals in osteogenesis.

A. Growth Factors

Growth factors are small peptides with specific actions on responsive cells, i.e., on cells with receptors for the growth factor(s). Many growth factors have now been identified, the major ones with actions on bone being platelet-derived growth factor, epidermal growth factor, fibroblast growth factor, insulin-like growth factor (IGF), and transforming growth factor β (TGF-β). Such growth factors do not act alone, but act in concert with other growth factors, often having their action modified, or even suppressed, by another growth factor. That more than one growth factor can use a single receptor and/or modify the action of other factors by receptor modulation considerably complicates the analysis of growth factor effects on bone development. [*See* Transforming Growth Factor-β.]

Growth factors have two major functions: they control cell division (hence, their name) and they play a role in the control of cellular differentiation (they would probably be better termed "growth and differentiation factors"). Therefore, we would expect different growth factors or combinations of growth factors to be active at different stages of osteogenesis or chondrogenesis—those that promote cell division being active in early stages of condensation formation and those that promote differentiation only being active once condensation has begun. A recent model for cartilage formation, based on cell and tissue culture experiments, proposes that TGF-β, which is present in prechondroblasts before condensation begins, stimulates synthesis of fibronectin, which, in turn, functions to adhere cells to one another and so initiates condensation formation. The condensation event then triggers changes in second messengers such as cyclic AMP or Ca^{2+} flux, which, in turn, act directly or indirectly on prechondrocyte DNA to initiate the synthesis of cartilage-specific products.

Growth factors can act as paracrine or autocrine factors. A paracrine factor is one that is produced by a cell type/population different from that which it acts on; an autocrine factor is produced by the same cell that it acts on. Within a condensation, both autocrine and paracrine factors could be operating. As all of the cells are undergoing the same differentiation (i.e., becoming chondroblasts or osteoblasts), it might be difficult to distinguish autocrine from paracrine effects. When a cell or tissue outside the future bone-forming cells acts on those cells, it could be doing so by release of a paracrine growth factor.

B. Interactions with Other Developing Tissues

Many examples of cells outside the bone-forming cells now play a role in the initiation of bone cell differentiation.

Interactions between two unlike cells that lead to the differentiation of a third cell type are known as inductions; when they occur in the embryo, as most do, they are known as embryonic inductions. Sufficient inductive interactions have been identified in the differentiation of bone and cartilage cells that we can regard them as necessary signals for cellular differentiation. Many such interactions are between an epithelium, which provides the necessary inductive signal, and a responding population of mesenchymal cells (i.e., cells organized in a loose network, or aggregation, as in the condensations).

Epithelial–mesenchymal interactions have now been identified as necessary steps along the differentiation pathway for the vast majority of cartilages and bones investigated in vertebrates and may turn out to be universal after more skeletal elements, species, and/or groups have been investigated experimentally. For several individual bones or regions of bones, these interactions have been shown to initiate condensation formation. Such interactions may take place at the site where the individual bone will differentiate, involving an epithelium at that final site (as in the differentiation of bones in the beak of the embryonic chick or in the lower jaw of the mouse), or they may occur earlier in the history of the cells, either during their migration (as in development of the cartilage of the lower jaw in amphibians) or even before the onset of migration (as in cartilage development in the lower jaw of birds). More than one such interaction may be required for osteoblasts to differentiate, leading to the concept of cascades of interactions, each playing, in turn, its role in the initiation or promotion of cellular differentiation.

Although we know much about these interactions, unfortunately, we do not yet understand their molecular basis. The local production of growth factors would clearly be an attractive mechanism for interactions that promote both cell division and cellular differentiation, and several laboratories are using such a model as their working hypothesis for designing experiments on embryonic bone development. Bone morphogenetic protein, a member of the TGF-β superfamily of growth factors, is a strong candidate as an embryonic inducer of bone.

C. Hormones and Mechanical Factors

Other classes of signals also play a role in embryonic development, but are less critical for differentiation than are the inductions and growth factors described. The two other major signals are hormones and mechanical factors.

Hormones play vital roles in skeletal growth, in maintaining the bulk of the skeleton in the adult, and in modulating calcium mobilization from the skeleton to maintain the calcium balance of the body. Their role in direct bone development is secondary; bone can develop quite well in chemically defined culture media lacking hormones, although it is difficult to get such bone to mineralize in the absence of hormones. Hormones, especially growth hormone and somatomedin C (i.e., IGF-1), do play an important role in regulating both the division and the differentiation of chondrocytes in the growth plate and are therefore important regulators of development and growth of those bones that form by endochondral ossification.

Mechanical factors also play an important role in maintaining the adult skeleton by adjusting bone formation and resorption so that skeletal mass, structure, and shape are optimal to resist the stresses imposed on the skeleton. Except for several specialized instances, these mechanical factors do little to control differentiation of bone or cartilage, although they do play an important role in producing a skeletal system of normal size and shape, with fully formed and functional joints. Thus, if an embryo, such as that of the domestic chicken, is paralyzed during skeletal development, all cartilages and bones develop, except for one specialized type of cartilage (i.e., secondary cartilage) that is specifically adapted to only differentiate in response to embryonic movement. However, skeletal growth is greatly reduced, and many of the joints either fail to form or, having formed, secondarily fuse.

VIII. SHAPES OF BONES AND JOINTS

The process by which shape is generated is morphogenesis. There are various levels at which morphogenesis of bone can be considered.

A. First-Order Morphogenesis

The first order of morphogenesis is the fundamental shape of the bone, be it a rod (long bone), a flat plate (skull), or a cuboidal block (vertebra). This level of morphogenesis is intrinsically controlled within the developing bone itself, either as a property of groups of cells or, indeed, as a property of individual cells, albeit acting in concert with other like cells. Thus, if an embryonic rudiment of a bone at the mesenchymal condensation stage is placed into tissue culture or grafted, it will develop a predictable shape, often even in the face of considerable forces that would be expected to deform its shape. Thus, primordia of long bones develop as rods with the expanded ends (i.e., condyles) characteristic of the future joints (Fig. 5). We therefore speak of the fundamental shape as being intrinsic, which does not tell us how that shape comes about, except that factors outside the condensation are not required. Differential growth is one important factor.

That first-order morphogenesis is a property of individual cells can be amply illustrated by experiments in which cells of one future bone or cartilage are separated from another (usually by enzymatic digestion) and are then reassociated (usually by centrifugation) before being cultured. Such cultured cells produce shapes that could have been predicted from knowledge of the normal shape of the intact bone or cartilage.

B. Second-Order Morphogenesis

The second order of morphogenesis is the shape of the "minor" architectural features of bone: the knobs, bumps, holes, and crevices associated with the attachment of muscles, ligaments, or tendons or where nerves or blood vessels course through bone. These features do not develop in the tissue-cultured primordium, for they are functionally induced. The additional bony knob that normally forms where a muscle attaches does not form if that muscle is not attached (as in culture) or if it is attached, but not functional (as in paralysis). Thus, this second-order morphogenesis depends on bones developing in a normal functional environment.

Maintenance of joints falls into this second category; in the absence of normal function, joints will fuse as the bones on either side of the joint ankylose. Similarly, if a bone is fractured and the fracture site is kept mobile, a false joint (pseudoarthrosis) can, and often does, form at the mobile fractured site. Here, morphogenesis is modeled by function.

IX. GROWTH OF BONE

Although most bone growth occurs after birth as the individual grows and matures, considerable growth does occur during development. In fact, it is not easy to separate the growth of bone from differentiation and morphogenesis, for all three processes occur simultaneously. It has been argued that the only way to study the mechanism of bone growth is to study bone before it is growing, the notion being that controls for growth operate early in development. [*See* Growth, Anatomical.]

Growth of bone is influenced by virtually every factor that we have discussed so far: the number of cells that migrate away from the initial embryonic source; how many times those cells divide as they migrate; the rate of cell division in the condensation; the number of cells in the condensation that become osteoblasts (some will die whereas others will fail to differentiate or will remain as stem cells); the amount of extracellular matrix that is synthesized and deposited; and the amount of hormonal, muscular, nervous, and vascular stimulation of the developing bone.

Three aspects of growth need to be considered: when growth starts, rate of growth, and when growth stops. Different factors or combinations of factors influence each of these aspects of growth. Only the second and third aspects operate after birth; only the first and second operate in the embryo.

Many vertebrates show determinate growth, i.e., an individual reaches a final size and stops growing. Final height in humans is set by the duration and extent of growth in the growth plates of the long bones; once all of the stem cells in the growth plate are used up, growth in height ceases. Other vertebrates, such as many fishes and tortoises, exhibit indeterminate growth, in which growth continues, albeit slowly, throughout life.

Even in animals with determinate growth, such as humans, growth can be reinitiated during the repair of fractured bone and even in the regeneration of selected regions of the skeleton, such as the tips of the fingers in young children. In both of these circumstances, growth is only reinitiated in the context of, and probably as a consequence of, cellular differentiation. In this, we see that embryonic and developmental processes hold the key to understanding not only the embryonic development of bone, but also bone repair, regeneration, and transplantation.

BIBLIOGRAPHY

Hall, B. K. (1983). "Cartilage," Vols. 1–3. Academic Press, New York.

Hall, B. K. (1987). Earliest evidence of cartilage and bone development in embryonic life. *Clin. Orthop. Relat. Res.* **225**, 255.

Hall, B. K. (1988). The embryonic development of bone. *Am. Sci.* **76**, 174.

Hall, B. K. (1988). "The Neural Crest." Oxford Univ. Press, Oxford, England.

Hall, B. K. (1990–94). "Bone," Vols. 1–9. CRC Press, Boca Raton, FL.

Hall, B. K. (1991). Closing address: What is bone growth. *In* "Fundamentals of Bone Growth: Methodology and Applications" (A. D. Dixon, B. G. Sarnat, and D. A. N. Hoyte, eds.), p. 605. CRC Press, Boca Raton, FL.

Hall, B. K., and Miyake, T. (1995). Divide, accumulate, differentiate: Cell condensation in skeletal development revisited. *Int. J. Dev. Biol.* **39**, 881.

Ham, A. W., and Cormack, D. H. (1979). "Histophysiology of Cartilage, Bone and Joints." Lippincott, Philadelphia.

Johnson, D. R. (1986). "The Genetics of the Skeleton." Oxford Univ. Press, Oxford, England.

Shipman, P., Walker, A., and Bichell, D. (1985). "The Human Skeleton." Harvard Univ. Press, Cambridge, MA.

Smith, M. M., and Hall, B. K. (1990). Development and evolutionary origins of vertebrate skeletogenic and odontogenic tissues. *Biol. Rev.* **65**, 277.

Spemann, H. (1943). "Forschung und Leben," (F. W. Spemann, ed.) p. 167. J. Engelhorn, Stuttgart.

Bone Marrow Transplantation

ROBERTSON PARKMAN
Children's Hospital, Los Angeles, and University of Southern California School of Medicine

Revised by
JOEL M. RAPPEPORT
Yale University School of Medicine

GLOSSARY

Chemotherapy Administration of drugs to transplant recipients to eliminate either normal or abnormal cells (e.g., normal stem cells and residual neoplastic cells) before transplantation

Cryopreservation Viable freezing of cells in liquid nitrogen (−120 to −180°C)

E-rosette formation Binding of T lymphocytes to sheep red blood cells ("E" for erythrocyte)

Graft-versus-host disease Disease caused by the attack of immunocompetent donor T lymphocytes against recipient histocompatibility antigens not present in the donor

Histocompatibility Degree of genetic identity between the donor and the recipient; histocompatibility genes are found on the short arm of chromosome 6 at the human leukocyte antigen loci

Human leukocyte antigens Principal antigens controlling transplantation immunity in humans

Stem cell Cell capable of both self-replication and differentiation; bone marrow stem cells differentiate into lymphoid and hematopoietic cells

BONE MARROW TRANSPLANTATION IS THE treatment of choice for some patients with immunological, genetic, hematological or oncological diseases. Bone marrow can be obtained either from another individual (allogeneic transplantation) or from the individual himself (autologous transplantation). Patients receive chemotherapy and/or irradiation prior to transplantation to eliminate the abnormal cells in the patient's body and/or to provide adequate immunosuppression; the exact therapy depends on the patient's disease. Following transplantation there is a recapitulation of donor immunological ontogeny, resulting in varying degrees of clinical immunodeficiency. Clinincal results in children are superior to those in adults. Patients with immunological, genetic, or hematological diseases have a greater likelihood of cure than do patients with oncological diseases.

I. NORMAL BONE MARROW DIFFERENTIATION

In prenatal life a pluripotent stem cell is present within the fetal liver and bone marrow and gives rise to the cellular elements of both the lymphoid and hematopoietic systems. In postnatal life a pluripotent stem cell might no longer exist, and two differentiated stem cells, one giving rise to the elements of the hematopoietic system (hematopoietic stem cell) and one giving rise to the immune system (lymphoid stem cell), are present. Bone marrow transplantation, therefore, can result in the engraftment of donor lymphoid stem cells and/or donor hematopoietic stem cells, depending on the nature of the disease for which the transplant is performed. [*See* Hemopoietic System.]

Lymphoid stem cells give rise to T and B lymphocytes. Hematopoietic stem cells produce the circulating hematopoietic elements (e.g., red blood cells, gran-

ENCYCLOPEDIA OF HUMAN BIOLOGY, Second Edition, VOLUME 2. Copyright © 1997 by Academic Press. All rights of reproduction in any form reserved.

ulocytes, monocytes, and platelets) and the fixed tissue macrophages (e.g., Kupffer cells in the liver, alveolar macrophages in the lungs, osteoclasts in the bone, and most likely the microglial cells in the brain). The hematopoietic stem cells normally undergo differentiation and proliferation within the bone marrow before they exit as mature cells. The hematopoietic stem cells first differentiate into progenitor cells, which are committed to red blood cell, granulocyte–monocyte, or platelet differentiation. The committed progenitors can be grown *in vitro* and permit the quantitation of the intramedullary pool of committed hematopoietic progenitors. The time for intramedullary bone marrow differentiation is between 5 and 9 days. [*See* Lymphocytes; Macrophages.]

II. HISTOCOMPATIBILITY

In all species there is a series of genetic loci (the major histocompatibility region) that controls transplantation immunity. In humans the major histocompatibility region is called human leukocyte antigen (HLA) and is found on the short arm of chromosome 6. HLA is composed of several loci. Loci that control histocompatibility antigens expressed on most nucleated cells are called class I antigens, three of which exist: HLA-A, -B, and -C. Loci that control histocompatibility antigens initially identified by their ability to stimulate foreign lymphocytes to divide in a mixed-lymphocyte culture are called class II antigens. Three class II loci have been identified: HLA-DR, -DP, and -DQ. These antigens are identified both serologically and by molecular techniques. Transplantation antigens outside the HLA loci are also involved in transplantation immunity. Such antigens are referred to as minor histocompatibility, or non-HLA, antigens. Genetic identity at the major histocompatibility region results in a low rate of solid-organ (e.g., skin or kidney) graft rejection.

Within a family, siblings who have inherited the same chromosome 6 pair and who therefore express the same HLA antigens are referred to as being genotypically identical (i.e., HLA identical). HLA-identical siblings have the same class I and class II histocompatibility antigens and are nonstimulatory in mixed-lymphocyte culture. Since children inherit one chromosome from each parent, children have one chromosome 6 in common with each parent; children and their parents are referred to as being haploidentical (i.e., having one haplotype or one chromosome 6 in common).

When an individual receives a bone marrow transplant from an HLA-identical sibling, this is referred to as a histocompatible transplant. Transplants into a patient from a haploidentical parent donor are referred to as haploidentical. Donor–recipient combinations that are not histocompatible are also referred to as histoincompatible, or nonhistocompatible. For some neoplastic diseases a person can receive a transplant of his own bone marrow, which has been previously removed and cryopreserved. Transplantation with a patient's own bone marrow is referred to as autologous bone marrow transplantation.

Because of the heterogeneity of the HLA system, the probability of finding a random individual who is HLA identical to a patient is low. The probability has been increased by the development of computer-based lists of potential transplant donors. With a database of 100,000 individuals, there is a 70% probability of finding an HLA-identical unrelated donor for most individuals, although the probability is variable depending on the ethnic background of the recipient and the representation of that ethnicity within the potential donor pool. This variability results from differing frequency of HLA antigens within different genetic propulations. A transplant with such a donor is called a matched unrelated donor transplant.

III. CLINICAL BONE MARROW TRANSPLANTATION

In histocompatible bone marrow transplantation, bone marrow is carefully aspirated from the pelvic bones of the donor under anesthesia. The bone marrow is collected in a tissue culture medium containing heparin to prevent clotting. Approximately 10 ml of bone marrow is collected per kilogram of recipient body weight, resulting in the transplantation of an average of 3×10^8 nucleated bone marrow cells per kilogram of recipient weight. An alternative source of allogeneic stem cells is the collection of cord blood, which has a uniquely high concentration of circulating hematopoietic and lymphoid stem cells. This source of stem cells has been used successfully to reconstitute both histocompatible family members as well as unrelated recipients. The donor bone marrow is infused intravenously into the recipient. For most transplants the histocompatible bone marrow is not treated *in vitro* before it is given to the recipient. For a haploidentical or histoincompatible bone marrow transplant, the donor bone marrow is treated prior to its infusion into the recipient to remove the donor T lymphocytes capable of causing graft-versus-host disease.

Graft-versus-host disease is due to an attack against recipient tissues by donor T lymphocytes that recognize the HLA antigen differences between the donor and the recipient. The attack is directed primarily against the skin, liver, and gastrointestinal tract and is usually fatal between histoincompatible donor–recipient pairs. The methods presently used to eliminate the donor T lymphocytes include their lysis by monoclonal antibodies to *T* lymphocyte differentiation antigens and their physical removal by agglutination with soybean lectin and E-rosette formation. The utility of positive selection of stem cells from the bone marrow is currently under active investigation.

In some patients with neoplastic disease, the marrow contains tumor cells. If these cells can be killed *in vitro* by treatment with drugs and/or monoclonal antibodiees, the patient's bone marrow is harvested and the tumor cells are removed. Autologous stem cells may be obtained by bone marrow harvest and more recently and with increasing frequency by peripheral blood stem cell harvest. The small number of circulating stem cells is increased by the rebound normally seen following the administration of moderately intense chemotherapy and/or administration of growth factors such as granulocyte colony-stimulating factor. Adequate stem cells are then collected by intensive pheresis of large volumes of blood. The major advantages to this technique are less contamination of the stem cells by malignant cells in the marrow as well as more rapid reinstitution of hematopoietic functions. The stem cells are then cryopreserved in liquid nitrogen until it is reinfused into the patient following high-dose chemo/radiation therapy. [*See* Tumors of Bone.]

Prior to transplantation the patient receives chemotherapy, the nature of which depends on the patient's disease. In patients with neoplastic disease, the purpose of the chemotherapy is to destroy residual leukemic or tumor cells, whereas in patients with aplastic anemia (i.e., failure of the bone marrow) the chemotherapy is needed to eliminate the patient's immune system to permit the engraftment of donor bone marrow cells. Patients with genetic diseases need to have their abnormal lymphoid and /or hematopoietic stem cells eliminated to permit the engraftment of normal donor lymphoid and hematopoietic stem cells. In most transplant centers, patients undergoing bone marrow transplants are kept in isolation to protect them from infections that might occur during the transplant period. Following the administration of chemotherapy, there are reductions in the patient's circulating white blood cell, platelet, and red blood cell counts that require the transfusion of red blood cells and platelets and aggressive prevention and treatment of infections. Following infusion of the donor bone marrow, signs of donor hematopoiesis can be detected as early as 8–10 days posttransplantation. By 30 days following transplantation, patients have self-supporting white blood cell, red blood cell, and platelet counts, although normal levels might not be obtained for several months.

Following transplantation the donor lymphoid stem cells undergo normal development, generating the various cell types found in the normal immune system. Committed T lymphocytes do not engraft and therefore the recipient does not derive donor immunity to infectious agents (e.g., chicken pox, measles, or mumps). However, donor-derived T lymphocytes are capable of producing graft-versus-host disease; to prevent it, histocompatible recipients routinely are immunosuppressed with drugs such as methotrexate, cyclosporine, and/or steroids. The prophylactic administration of these drugs has reduced the incidence of significant acute graft-versus-host disease to 20% in histocompatible bone marrow transplant recipient. The likelihood of acute graft-versus-host disease is increased by a number of factors, including an increased recipient age and a sex difference between the donor and the host. Significant acute graft-versus-host disease may require treatment with additional immunosuppression, including high-dose corticosteroids and/or antithymocyte globulin.

Some patients also develop chronic graft-versus-host disease, which has many characteristics of an autoimmune disease, particularly scleroderma. These patients develop fibrotic skin changes and abnormalities of their liver, lungs, and gastrointestinal tract. Chronic graft-versus-host disease can be treated by immunosuppression, with clinical improvement in many, but not all, patients.

Since the donor immunological system is not yet developed immediately after transplantation, bone marrow transplant patients are, like newborn infants, at increased risk for viral, fungal, and bacterial infections during the early posttransplant period and must be reimmunized with the standard vaccines. Patients are particularly at increased risk for infection with encapsulated respiratory bacteria and therefore are routinely placed on antibiotic prophylaxis until they acquire a normal ability to make anticarbohydrate antibodies.

IV. PRESENT STATUS

A. Severe Combined Immunodeficiency

Severe combined immunodeficiency (SCID) is a genetic disorder involving the lymphoid stem cell. Affected infants have an absence of normal T and B lymphocyte immunity and therefore are at an increased risk of infections by viral, bacterial, fungal, and protozoan organisms. Although a variety of primary defects have been identified (e.g., deficiency of the enzyme adenosine deaminase, absence of lymphoid stem cells and absence of interleukin-2 production), all forms of SCID are potentially correctable by histocompatible bone marrow transplantation. Because of the small size of infants, only small amounts of bone marrow are needed for successful transplantation. Since patients with SCID have normal hematopoiesis, engraftment of only the donor lymphoid stem cells is required for correction of the patient's primary disease. Successful transplantation in infants with SCID has been achieved with bone marrow doses as low as 50×10^6 nucleated cells per kilogram. Seventy to 80% of children with SCID who received transplants from histocompatible donors achieve lymphoid engraftment, normal immunological function, and long-term survival.

Patients with SCID have been in the forefront for many advances in bone marrow transplantation, especially the use of haploidentical parental bone marrow transplantation. Infants with SCID who received transplants of T lymphocyte-depleted haploidentical bone marrow have a survival rate of 70%, with immune reconstitution, a figure similar to that achieved with histocompatible bone marrow transplantation. Thus, SCID in children can be successfully corrected with either histocompatible or haploidentical bone marrow.

B. Other Immunodeficiencies

Histocompatible bone marrow transplantation is the treatment of choice for children with other immunodeficiency states. Immunodeficiency states can be divided into those that primarily affect the lymphoid stem cell (e.g., Wiskott–Aldrich syndrome) and those affecting the hematopoietic stem cell (e.g., chronic granulomatous disease, granulocyte actin deficiency, and congenital agranulocytosis). Children with Wiskott–Aldrich syndrome also have abnormalities of platelets and, therefore, for successful correction both the abnormal lymphoid and hematopoietic stem cells must be eliminated by pretransplant chemotherapy. Cyclophosphamide eliminates the lymphoid stem cells, and busulfan eliminates the hematopoietic stem cells. After combined chemotherapy, 90% of the children with Wiskott–Aldrich syndrome can be cured by histocompatible bone marrow transplantation.

Other children have immune dysfunction involving only the progency of hematopoietic stem cells. Because of the clinical heterogeneity of children with such granulocyte disorders, only those with the most severe forms require bone marrow transplantation as potentially curative therapy. The affected children can receive successful transplants following preparation with cyclophosphamide and busulfan.

C. Aplastic Anemia

Aplastic anemia is due to an acquired absence of hematopoietic stem cells, resulting in the decreased production of red blood cells, granulocytes, and platelets. If a patient with aplastic anemia has an identical twin, bone marrow transplantation without pretransplant chemotherapy is possible and restores normal hematopoietic function in 80% of the patients. The remaining 20% can receive successful transplants following intense immunosuppression with cyclophosphamide alone, indicating that the aplastic anemia in such patients might be of immunological origin. It is not possible to transplant into histocompatible siblings without prior immunosuppression, because the recipient's normal immune system rejects the donor cells owing to non-HLA antigen differences. Thus, immunosuppression with cyclophosphamide is required to eliminate the normal immune system of the recipient.

Following pretransplant chemotherapy, engraftment of both donor lymphoid and hematopoietic stem cells is achieved. Patients with aplastic anemia who receive their transplants before receiving any blood transfusions have a cure rate of 85–90%, with patients now having survived in excess of two decades. Patients who have received transfusions have an increased incidence of marrow graft rejection resulting in a lower (i.e., 75%) survival rate. The latter patients may require additional immunosuppression. The major long-term sequela found in patients transplanted for aplastic anemia is chronic grafts-versus-host disease.

D. Leukemia

1. Acute Lymphoblastic Leukemia

Acute lymphoblastic leukemia is the most common form of leukemia in children, but is less frequent in adults. Therapy with chemotherapy alone has had poor results in adult patients, and therefore histocompatible bone marrow transplantation is used to treat adult patients in first remission. Acute lymphoblastic leukemia in children has a good response to chemotherapy, and therefore bone marrow transplantation is reserved for patients who have relapsed or are in second remission, although children with poor prognostic features may be transplanted in first remission. Patients are prepared for transplantation with high-dose chemotherapy (cyclophosphamide or cytosine arabinoside) and total body irradiation. Following the pretransplant radiochemotherapy the patients are infused with histocompatible bone marrow. Forty percent of second-remission patients who have received histocompatible bone marrow are long-term disease free survivors. The principal causes of death are leukemia relapse (30–40%) and transplant-related problems (20–30%). For patients without a matched donor for an allogeneic bone marrow transplant, the use of autologous cryopreserved stem cells has demonstrated some utility. The patient's bone marrow is collected in either first or second remission and usually undergoes *in vitro* purging to remove residual leukemic cells. In acute lymphoblastic leukemia the marrow is most commonly purged with monoclonal antibodies that recognize leukemic-specific antigens and complement. The treated marrow is then cryopreserved and reinfused following completion of the preparative regimen. When utilized in first remission, disease-free survival of 20–60% has been reported, and when applied in second or subsequent remission, long-term results of 20–40% have been noted. The major cause of failure is leukemia relapse. [*See* Leukemia.]

2. Acute Myelogenous Leukemia

Acute myelogenous leukemia is the most common form of leukemia in adults and is less frequent in children. Only 25–30% of the patients can be cured by chemotherapy alone. The usual strategy for bone marrow transplantation is again, to perform transplants in patients in remission, when the tumor burden is minimal, by giving them high-dose radiochemotherapy to destroy residual leukemic cells. In the past, patients with acute myelogenous leukemia were prepared for transplantation with a combination of cyclophosphamide and total body irradiation; presently, a combination of cyclophosphamide and busulfan is often used. Both combinations result in elimination of the patient's residual leukemic cells but also in destruction of the patient's normal lymphoid and hematopoietic stem cells. Overall, a 50% disease-free survival rate is found in patients who received their transplants while in first remission using histocompatible donors.

Recently, the transplantation of autologous bone marrow has been used to treat patients with acute myelogenous leukemia in first or second remission for whom a histocompatible donor is not available. The patient's bone marrow is collected in either first or second remission. Whether or not a "remission" marrow requires *in vitro* purging in acute myelogenous leukemia remains controversial. Purging techniques utilize activated chemotherapeutic agents that relatively spare the hematopoietic stem cell or monoclonal antibodies, which recognize specific leukemic antigens. Autologous peripheral blood stem cells have also been utilized for marrow reconstitution in an attempt to decrease the burden of infused leukemic cells. After pretransplant conditioning, which is similar to that used for histocompatible bone marrow transplantation, the cyropreserved purged autologous bone marrow is infused. From 35 to 60% disease-free survival has been reported in recipients of autologous marrow transplanted in first remission and 25–50% in patients transplanted in second and subsequent remission. In the latter setting, there is no long-term survival for patients receiving chemotherapy alone.

3. Chronic Myelogenous Leukemia

Chronic myelogenous leukemmia is primarily a disease of older adults. In the past, bone marrow transplantation was restricted to patients less than 40 years of age, but recently successful transplants have been performed in patients in good health up to 55 years of age. Most patients are Philadelphia chromosome positive. The Philadelphia chromosome is found in multiple cell types of both lymphoid and hematopoietic lineage, suggesting that the abnormality occurs in stem cells; therefore, at the present time, autologous bone marrow transplantation is not a potential treatment. Chronic myelogenous leukemia in its chronic phase may be an indolent disease. Some patients survive many years before their disease develops into an acute leukemia called a "blast crisis," ultimately leading to the patient's death. The question faced by patients with chronic myelogenous leukemia who have a histocompatible donor is when to receive a bone marrow transplant.

On an average, patients have a 3-year interval between the time of diagnosis and the development of blast crisis, although the time course is extremely variable. Thus, the physician and the patient have to decide whether to transplant early, with the risk of death from transplant-related causes, or to wait until the patient develops a blast crisis, at which time bone marrow transplantation is much less successful. Bone marrow transplantation with histocompatible bone marrow can result in a 40–50% disease-free survival rate when patients receive their transplants while in the stable phase. Once patients develop blast crisis, less than 10% of the patients can be cured by transplantation.

Another group of hematopoietic stem cell disorders successfully treated with allogeneic bone marrow transplantation is myelodysplasia. Again these disorders are seen predominantly in older patients and therefore the number of patients for whom transplantation is appropriate is limited. These disorders usually arise *de novo* or may result from prior chemoradiation therapy for the treatment of a malignancy. Patients with these disorders are at risk of infection, hemorrhage, of ultimately the evolution into an acute leukemia. These secondary leukemias would not be good candidates for autologous marrow transplantation given the underlying stem cell disorder.

E. Lymphoma

Lymphomas occur primarily in adult patients and can be divided into Hodgkin's disease and non-Hodgkin's lymphoma. Subcategories of lymphoma respond well to chemotherapy, and bone marrow transplantation has no role in the initial stages in these cases. However, once patients relapse, there is little likelihood of cure with conventional chemotherapy for many patients. Bone marrow transplantation allows higher doses of radiochemotherapy, capable of destroying lymphoma cells resistant to normal doses of chemotherapy. The high-dose radiochemotherapy also results in the destruction of the patient's normal bone marrow, requiring the transplantation of either histocompatible or cryopreserved autologous bone marrow. Some forms of non-Hodgkin's lymphoma have such a poor response to chemotherapy that bone marrow transplantation is used as treatment during first remission. [*See* Lymphoma.]

Patients with lymphoma can receive transplants of either histocompatible sibling bone marrow or autologous bone marrow, from which potential lymphoma cells have been removed by *in vitro* treatment. In some lymphomas the patient's bone marrow has no morphological or immunological evidence of involvement. Overall, the survival rates with both tratments are equivalent, although the causes of death are different. Patients receiving histocompatible bone marrow are more likely to die from complications of transplantation, including acute graft-versus-host disease and infections, whereas patients receiving autologous bone marrow are more likely to die from recurrent lymphoma.

The decreased incidence of lymphoma and acute lymphoblastic leukemia relapse following histocompatible bone marrow transplantation suggests that there is a graft-versus-leukemia effect as well as graft-versus-host disease following transplantation. Such a graft-versus-leukemia effect is not seen when syngeneic (i.e., identical twin) or autologous bone marrow is used.

F. Other Oncological Diseases

Both allogeneic and autologous transplants have been successful in selected patients with multiple myeloma and chronic lymphocytic leukemia. Autologous bone marrow transplantation has been used increasingly in the treatment of adult patients with a variety of solid tumor malignancies including carcinomas of the breast, brain and testicle. The theory behind autologous transplantation is that the bone marrow in such patients is rarely involved with tumor and that, for many chemotherapeutic regimens, bone marrow toxicity limits the dose of chemotherapy. In autologous transplantation, patients receive high-dose chemotherapy without regard for potential marrow toxicity. The reinfusion of cryopreserved autologous bone marrow then "rescues" the patients from the toxicity associated with the high-dose chemotherapy.

Pediatric patients with neuroblastoma have received successful transplants with the use of autologous bone marrow, from which neuroblastoma cells have been removed *in vitro* by treatment with monoclonal antibodies to the neuroblastoma cells. The results of transplantation with autologous bone marrow are equivalent to those seen with histocompatible bone marrow (i.e., a 40% survival rate in patients who receive their transplants before disease progression).

G. Genetic Diseases

An area of increasing interest is the use of bone marrow transplantation to treat genetic diseases. SCID and genetic disorders of granulocyte function in children were the first genetic diseases to be treated successfully by transplantation. More recently, patients with disorders of erythroid function (e.g., thalassemia and sickle-cell disease) or with enzyme deficiencies (e.g., Hurler's syndrome, Hunter's disease, and Gaucher's disease) have received successful transplants. Patients with disorders of erythroid function can receive successful transplants if their abnormal hematopoietic stem cells and normal lymphoid stem cells are eliminated by pretransplant chemotherapy, usually cyclophosphamide and busulfan. The subsequent engraftment of normal donor hematopoietic and lymphoid stem cells results in patients with normal erythroid function. Overall, a disease-free survival rate of 84% has been seen in patients with thalassemia who received transplants in Italy. The major causes of failure are graft-versus-host disease, graft rejection, and hepatic complication resulting from iron overload. Transplantation in younger patients might result in decreases in these complications. [*See* Genetic Diseases.]

An area of particular potential challenge is the treatment of enzyme deficiency states. If normal bone marrow-derived cells contain the enzyme that the patient is missing, then the ablation of the patient's lymphoid and hematopoietic stem cells and replacement with normal donor lymphoid and hematopoietic stem cells would result in circulatory donor-derived cells that contain the normal enzyme. Patients with Hurler's syndrome (i.e., α-L-iduronidase deficiency) who received transplants of histocompatible bone marrow have had decreases in corneal clouding and hepatomegaly. Circulating cells of donor origin with normal enzyme levels have been demonstrated.

An area of continuing controversy is whether cells of lymphohematopoietic origin are a normal component of the central nervous system. The origin of brain microglial cells is important, since many enzyme deficiencies are characterized by abnormalities of central nervous system function. The correction of the non-central nervous system aspects of a genetic disease would be of no significant clinical benefit if central nervous system deterioration continued. Long-term follow-up of patients with enzyme deficiencies, involving the evaluation of central nervous system function, is necessary before definitive statements about the role of bone marrow transplantation in the treatment of genetic diseases involving the central nervous system can be made. Bone marrow transplantation, however, represents curative therapy for some genetic disease (e.g., Wiskott–Aldrich syndrome, sickle-cell anemia, and thalassemia) for which only supportive care was previously available.

V. FUTURE PROSPECTS

A. Gene Therapy

Although some genetic diseases characterized by enzyme deficiency can be successfully treated by histocompatible bone marrow transplantation, only 20% of potential patients have histocompatible donors. Much research, therefore, has centered on the question of whether a normal gene could be inserted into the patient's lymphoid and/or hematopoietic stem cells *ex vivo* and the treated bone marrow reinfused into the patient. If the transplanted bone marrow expressed the product of the inserted gene, then the patients would derive the benefits of histocompatible bone marrow transplantation without the concomitant risk of graft-versus-host disease. Most attempts at gene therapy are centered on inserting the normal gene into a virus (i.e., retrovirus) and using the modified virus as a vector to introduce the normal gene into recipient cells.

The present difficulties are (1) determining systems by which the retroviral vector can selectively infect the hematopoietic stem cells, which occur at a frequency of approximately one per 10^5 nucleated bone marrow cells, and (2) determining the conditions under which the inserted gene can be appropriately expressed, giving rise to the product it specifies. Little difficulty has been encountered in achieving the expression of the inserted gene in mature cells (e.g., T lymphocytes and fibroblasts); however, difficulty has been observed when bone marrow stem cells are infected with the normal genes, suggesting that the regulation of gene expression in the progeny of stem cells differs from that of mature cells.

Gene therapy can be divided into attempts to inset the noraml gene into somatic cells (i.e., bone marrow, keratinocytes, and liver cells) and those to insert the gene into germ-line cells. At the moment for both medical and ethical reasons, researchers do not propose germ-line gene therapy. However, investigators plan to study the role of somatic cell gene therapy for diseases such as adenosine deaminase deficiency (i.e., SCID) and glucocerebrosidase deficiency (i.e., Gaucher's disease).

B. New Diseases

Although bone marrow transplantation has been used for a wide range of immunological hematological, oncological, and genetic diseases, new applications of bone marrow transplantation still exist. One potential future use is in the treatment of patients infected with human immunodeficiency virus type I (HIV-1). The first attempts to use bone marrow transplantation to treat patients with clinical acquired immunodeficiency syndrome (AIDS) were unsuccessful. In part, these failures might be due to the fact that the lessons learned from the treatment of patients with oncological and immunological diseases were not adhered to. The optimal use of bone marrow transplantation to treat patients infected with HIV-1 would include the use of (1) histocompatible, as opposed to identical twin, donors to introduce genetic factors that could retard disease progression, (2) pretransplant chemotherapy to reduce the number of HIV-1-infected cells, (3) antiviral chemotherapy after transplantation to block viral replication until the transplanted immune system is established, (4) drugs such as cyclosporine to retard lymphocyte activation after transplantation, thus reducing the likelihood of HIV-1 infection of the newly engrafted lymphoid cells, and (5) the transplantation of patients at stages of their disease when opportunistic infections have not occurred. Use of bone marrow transplantation for the treatment of HIV-1 infected patients using these parameters could produce better results than those previously achieved. [*See* Acquired Immunodeficiency Syndrome (Virology).]

C. Solid-Organ Transplantation

The major problem associated with solid-organ transplantation is graft rejection. In the case of patients receiving kidney and liver transplants, posttransplant immunosuppression with cyclosporine, antithymocyte globulin, and steroids has reduced the incidence of graft rejection. In renal transplantation, however, the presence of preexisting anti-HLA antibodies results in an acute antibody-mediated rejection and immediate graft failure. Such antibody-positive patients cannot receive successful transplants, regardless of the immunosuppression used posttransplantation. It might be possible in the future to perform autologous bone marrow transplants on highly sensitized patients requiring renal transplants. The autologous bone marrow would be purged of T and B lymphocytes, eliminating the preexisting antibody-producing cells, making kidney transplantation possible.

BIBLIOGRAPHY

Cunningham, I. (1995). Allogeneic bone marrow transplantation. *In* "Hematology" (R. Hoffman, E. J. Benz, S. J. Shattil, B. Furie, H. J. Cohen, and L. E. Silberstein eds.), 2nd Ed., pp. 400–414. Churchill Livingston, New York.

Gale, R. P., and Champlin, R. E. (eds.) (1989). "Bone Marrow Transplantation: Current Controversies." Liss, New York.

Kernan, N.A., Bartsch, G., Ash, R. C., *et al.* (1993). Analysis of 462 transplantations from unrelated donors facilitated by the National Marrow Donor Program. *N. Engl. J. Med.* **328,** 593–602.

Kessinger, A. (1993). Utilization of peripheral blood stem cells in autotransplantation. *Hematol. Oncol. Clin. NA.* **7,** 535–547.

Krivit, W., and Paul, N. W. (eds.) (1986). "Bone Marrow Transplantation for Treatment of Lysosomal Storage Diseases." Liss, New York.

Parkman, R. (1986). The application of bone marrow transplantation to the treatment of genetic diseases. *Science* **232,** 1373–1378.

Reed, E.C., and Kessinger, A. (1995). Autologous bone marrow transplantation. *In* "Hematology." (R. Hoffman, E. J. Benz, S. J. Shattil, B. Furie, H. J. Cohen, and L. E. Silberstein, eds.), 2nd Ed., pp. 400–414.

Vogelsang, G. B., and Hess, A. D. (1994). Graft-versus-host disease: New directions for a persistent problem. *Blood* **84,** 2061–2067.

Wagner, J. E. (1993). Umbilical cord blood stem cell transplantation. *Am. J. Pediatr. Hematol. Oncol.* **15,** 169–174.

Weinstein, H. J., Rappeport, J. M., and Ferrara, J. L. M. (1993). Bone marrow transplantation. *In* "Hematology of Infancy and Childhood" (D. G. Nathan and F. A. Oski, eds.), 4th Ed., pp. 317–344. Saunders, Philadelphia.

Bone Regulatory Factors

ANTONIO PECILE
University of Milan

GLOSSARY

Analog Chemical compound with a structure similar to that of another but differing from it with respect to a certain component; it may have a similar or different action

Mesenchyma Meshwork of embryonic connective tissue in the mesoderm from which are formed the connective tissue of the body and the blood vessels and lymphatic vessels

Osteoblasts Cells that arise from fibroblasts and that are associated with the production of bone

Osteoclasts Large multinuclear cells associated with the absorption and removal of bone

Osteocytes Osteoblast that has become embedded within the bone matrix, occupying a flat oval cavity (lacuna) and sending through apertures in its walls thin cytoplasmic processes that directly connect with other osteocytes in bone

BONE IS A COMPLEX TISSUE MADE UP OF CELLS and extracellular material with organic and mineral components. The cells have a common origin in the mesenchymal tissue. The principal cell types are the (a) chondrocytes or cartilage cells, which secrete the collagen matrix of the cartilage region; (b) osteoblasts or bone-forming cells; and (c) osteoclasts, which are multinucleated cells involved in bone resorption.

The formation of functional bone tissue can be divided into two phases, one concerned with production and secretion of the extracellular bone matrix and the other with the deposition of the calcium hydroxyapatite crystals in the matrix.

Bone is the essential component of the vertebrate skeleton, which protects vulnerable internal structures, is a rigid framework facilitating locomotion, and is also a vast reservoir of Ca^{2+} and phosphate ions. It is important to appreciate that both the metabolic activity of the various cell populations present and the kinetics of mineral flux into and out of the various bone compartments are dynamic. Bone undergoes continuous, although generally slow, changes that enable it to be remodeled and to adapt the skeleton to load-elicited variations. This means there must be bone resorption and new bone formation. Bone is also a bank for Ca^{2+} and phosphate ions, which move into and out of the skeleton rapidly when storage or release are required. The formation or resorption of tissue and deposit or release of mineral must take place accurately and precisely.

There are different "compartments" of bone mineral. These include the "readily exchangeable" bone mineral compartment that contains the rapidly and slowly mobilizable calcium component. The term "rapid" implies the ability to exchange on a minute-by-minute basis. The "stable" bone mineral compartment is the pool of calcium that participates in the bone remodeling processes associated with bone mineralization, accretion (as in growth), or resorption.

Some of the regulatory processes of bone formation and bone resorption and of calcium and phosphate storage or release are well known. Others are emerging in current research. The well-known factors are the classic hormonal regulatory factors of bone and calcium metabolism: parathyroid hormone, calcitonin, and vitamin D. In addition to these, other

ENCYCLOPEDIA OF HUMAN BIOLOGY, Second Edition, VOLUME 2. Copyright © 1997 by Academic Press. All rights of reproduction in any form reserved.

hormones directly or indirectly influence bone metabolism so that they too must be considered bone regulatory factors. Local bone regulatory factors are becoming known from more recent research. To highlight the importance of bone regulatory factors, clinical aspects of the hormonal factor excesses or defects will also be illustrated briefly.

I. CLASSIC HORMONAL REGULATORY FACTORS OF BONE AND CALCIUM METABOLISM

A. Chemistry and Biochemistry

1. Parathyroid Hormone

Parathyroid hormone (PTH) is a polypeptide secreted by the chief cells of the parathyroid gland. PTH is a single-chain polypeptide of 84 amino acids (molecular weight about 9300). The structures of the native 84-amino-acid polypeptide hormones from the bovine, porcine, rat, and human species are similar and have many amino acid residues in common. [*See* Parathyroid Gland and Hormones]

PTH is biosynthesized as a larger precursor molecule by the parathyroid gland. A preproparathyroid hormone (Pre-ProPTH) is the initial product of synthesis on the ribosomes. This polypeptide, which has 31 additional amino acids present at the N-terminal region of PTH, is converted into proparathyroid hormone (ProPTH) by removal of 24 amino acids from the NH_2-terminal end. The ProPTH then reaches the Golgi region, where it is stored in vesicles and is converted into PTH by removal of the NH_2-terminal hexapeptide. PTH is stored in the secretory granules until released into the circulation in response to a fall in blood calcium concentration.

The NH_2-terminal third of the molecule is critical for binding of the hormone to specific receptors on cells, for activation of intracellular adenylate cyclase, and for biological activity. Removal of the two NH_2-terminal amino acids destroys biological activity but not receptor-binding activity. Binding to the receptor depends on two regions within the molecule (residues 10–27 and 25–34). The sequence 1–27 is the minimum required for detectable biological activity. Synthetic polypeptides corresponding to the first 34-amino-acid sequence of PTH generally are fully as active (and some analogs are more active) as the entire 1–84 sequence. A synthetic peptide lacking two amino acids (Ala-Ser) at the amino terminus is a competitive inhibitor of PTH action *in vitro* and a peptide containing amino acids 7–34 is a low-affinity blocker of PTH action *in vivo*.

The circulating levels of PTH have been studied by radioimmunoassay techniques that have made it possible to identify the intact 1–84 PTH and two PTH fragments the importance of which is not yet clear. Although some investigators believe that the biologically active hormone is the intact 1–84 molecule, others think a smaller sequence could be the major biologically active form. PTH is destroyed by the liver and the half-life of the 1–84 molecule in human plasma is only 20 min.

For clinical use, specific carboxy-terminal, midregion, and amino-terminal radioimmunoassays have been developed. The most promising for clinical use are the amino-terminal assays and the midregion assays. The midregion assays are preferable to C-terminal assays because they are more sensitive.

2. Calcitonin

Calcitonin (CT) is a small polypeptide hormone secreted by the ultimobranchial glands of fish, amphibians, reptiles, and birds. In mammals, calcitonin is produced by the parafollicular or "C" cells of the thyroid gland. These cells are also present in small numbers outside the thyroid. The primordial cells that give rise to the parafollicular cells are derived from ectodermal neural crest precursors that migrate ventrally into the branchial pouch. Localization of the hormone with specific antibodies using cytological methods provides direct evidence that parafollicular cells synthesize and secrete CT. [*See* Peptide Hormones and Their Convertases.]

Calcitonins from different species all consist of a 32-amino-acid polypeptide with an amino-terminal seven-membered disulfide ring and a C terminus of prolineamide. Six of the 7 amino-terminal residues are identical in all calcitonins, but as many as 19 of the 32 amino acids differ in the most diverse (human versus ovine) forms of the polypeptide. The nonmammalian hormones are the most potent molecules—from 10 to 50 times more potent than mammalian hormones.

Modifications of the seven-membered amino-terminal dicysteine ring structure by deletion of serine or other substitutions are compatible with significant biological activity.

Calcitonin is synthesized from a large CT precursor that is later processed. This precursor is formed by differential splicing of the primary transcript of the calcitonin/calcitonin gene-related peptide (CGRP) gene (Fig. 1). As a consequence of alternative pro-

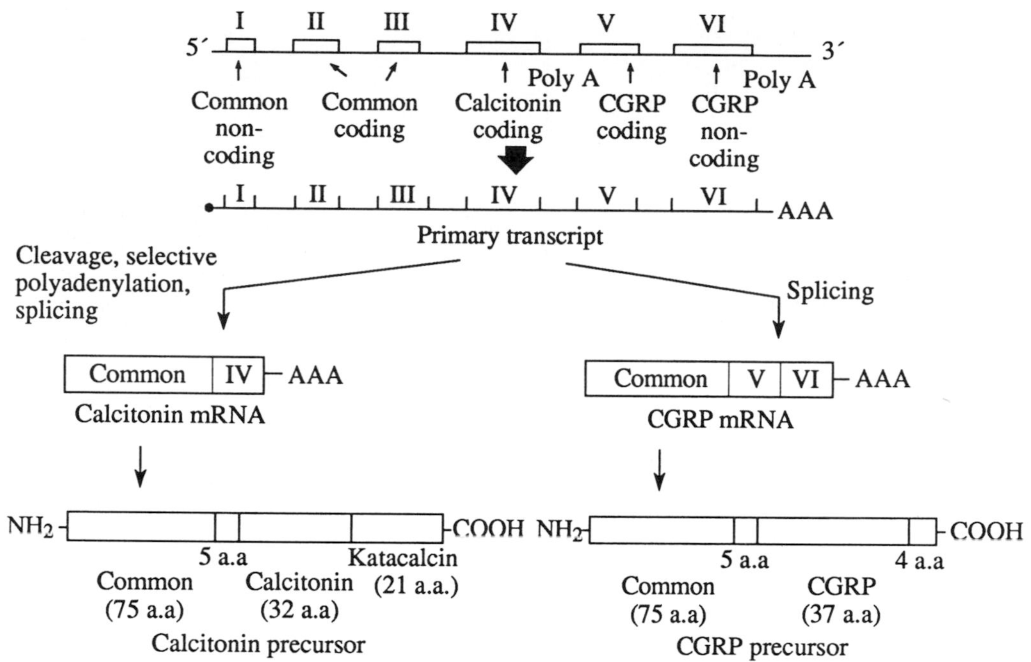

FIGURE 1 Organization of the human α-calcitonin/CGRP gene. The same primary transcript is alternatively processed to form mRNA encoding either calcitonin or CGRP. [Reproduced by permission from M. Zaidi, L. H. Breimer, and I. MacIntyre (1987). Biology of peptides from the calcitonin genes. *Quart. J. Exp. Physiol.* 72, 375.]

cessing of RNA transcripts, this gene causes the production of the hormone calcitonin in the parafollicular cells of the thyroid, and of CGRP in the neural tissue.

Plasma CT levels increase when calcium rises and decrease when it falls. There is a negative feedback between the concentration of calcitonin in the circulation and the content of CT in the thyroid gland. Immunoreactive CT exists in human plasma in multiple forms, including calcitonin monomer, oxidized monomer, a dimer, and possibly a precursor of CT—the calcitonin–katacalcin complex (see Fig. 1). Thus a mixture of different forms of calcitonin appears to be secreted by the C cells (or is formed peripherally). Because some of these multiple forms are inactive, for example, the dimer (which, however, can be converted to the monomer with restoration of biological activity), it is particularly difficult to interpret radioimmunoassay data in terms of biological activity of calcitonin itself.

3. Vitamin D Endocrine System

The D vitamins (calciferols) are a family of seco-steroids, that is, steroid-related molecules in which one of the rings of the cyclopentanoperhydrophenanthrene ring structure has been opened. In the vitamin D family, the bond between C_9 and C_{10} of the B ring has been cleaved, giving a conjugated triene structure (Fig. 2). [*See* Vitamin D.]

D vitamins are synthesized *in vivo* from precursors (ergosterol produced in plants, and 7-dehydrocholesterol produced in animals). Vitamin D_2 (or ergocalciferol) and D_3 (or cholecalciferol), which differ only in the side-chain structure, are generated by radiation of the precursors with ultraviolet light (230–313 nm wavelength). Collectively, vitamin D_3 and vitamin D_2 are called calciferols. In humans, the UV radiation penetrates as far as the upper portion of the dermis, generating several sterols, including vitamin D. The calciferols are inactive and are activated by a sequence of hydroxylations. Vitamin D_3 is first transformed by a liver microsomal enzyme into 25-hydroxy vitamin D_3. This circulating form of seco-steroid then serves as a substrate for either the renal 25(OH)D-1-hydroxylase or the 25(OH)D-24-hydroxylase. The resulting $1,25(OH)_2D_3$ form of the vitamin is the major biologically active form, particularly in the intestine. This compound induces renal 24-hydroxylase to coproduce $24,25(OH)_2D_3$, which appears to be an important factor in normal bone development and parathyroid function. Thus $1,25(OH)_2D_3$ and 24,

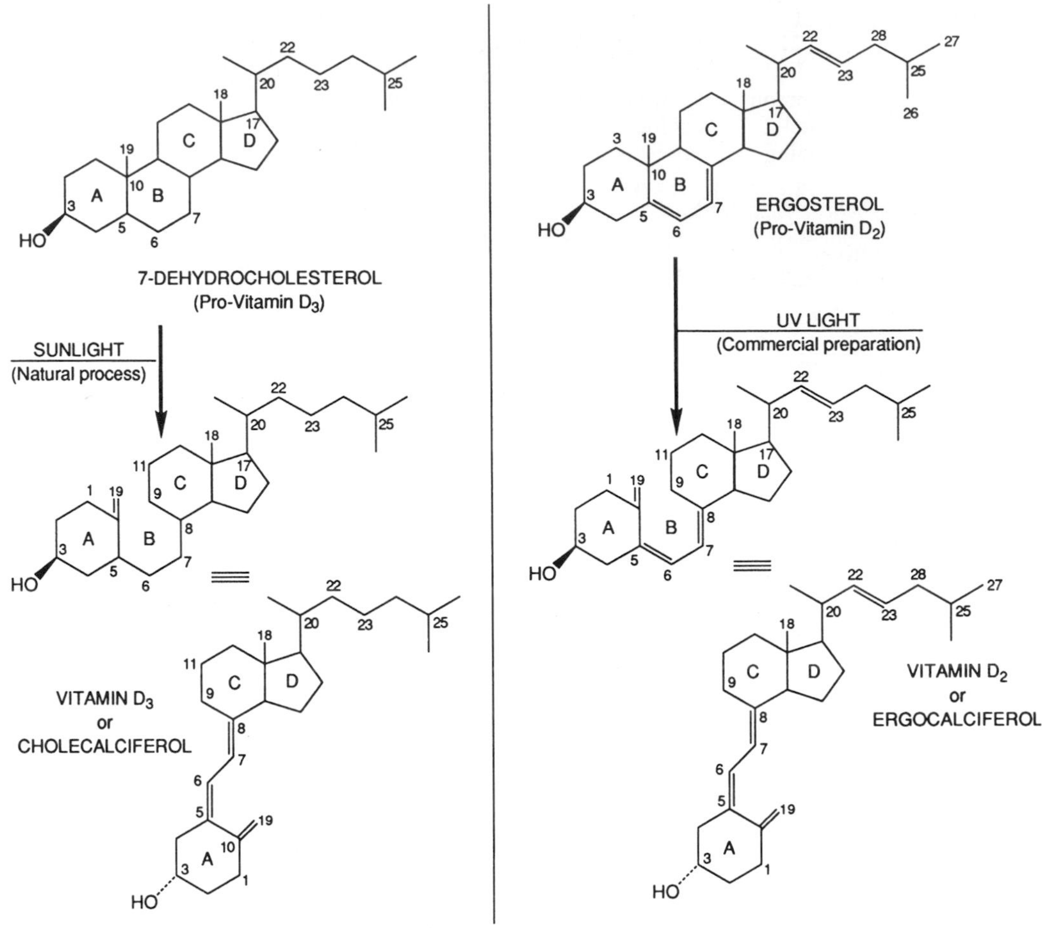

FIGURE 2 Structural relationship of vitamin D_3 (cholecalciferol) and vitamin D_2 (ergocalciferol) to their respective provitamins. The two structural representations presented at the bottom for both vitamin D_3 and vitamin D_2 are equivalent; these are simply different ways of drawing the same molecule. [Reproduced by permission from A. W. Norman and G. Litwack (1987). "Hormones," p. 375. Academic Press, New York.]

$25(OH)_2D_3$ are the two principal forms of vitamin D_3. Vitamin D_2 (irradiated ergosterol) or ergocalciferol, which has the side chain of ergosterol, is not a naturally occurring form of the vitamin but a commercial preparation. Vitamin D_2 is metabolized as vitamin D_3 and has equivalent biological potency.

B. Biology and Physiological Significance

1. Parathyroid Hormone

The primary function of PTH is to control the concentration of calcium in the extracellular fluids by acting on the kidneys and bones. In the kidneys, PTH stimulates the rate of reabsorption of calcium from the glomerular filtrate. In addition, through its influence on the renal formation of active vitamin D metabolites, PTH influences the rate of absorption of calcium from the gastrointestinal tract. It also increases the excretion of phosphate by the kidney. In bone, PTH enhances the rate of calcium absorption.

PTH increases osteolysis of bone for bone resorption and remodeling but also appears to promote bone formation, with results that vary depending on time and the concentration of the hormone. These effects are complex and not yet fully understood. It is clear that PTH can stimulate bone resorption directly by increasing the activity and number of osteoclasts in bone organ cultures. However, no specific membrane receptors for PTH have been identified in bone. The major effect appears to be an increase in osteoclast number, probably by stimulating the fusion of existing precursors into active osteoclasts. [*See* Bone Remodeling.]

PTH also acts directly on osteoblasts, the cells that generate bone. Because osteoblasts can produce substances that activate isolated osteoclasts, this phenomenon could be another pathway to bone resorption. However, there is as yet no evidence that the increase in osteoclast number is mediated by an osteoblast product. Another possible mechanism is the secretion of collagenase and plasminogen activator by osteoblasts at high concentrations of PTH because it may initiate the bone resorptive process, perhaps by preparing the bone surface for the osteoclasts.

PTH stimulates cyclic AMP (cAMP) production in osteoblasts, but its role in bone resorption is unclear. Thus agents that increase cAMP production can stimulate bone resorption, but analogs of PTH that block PTH stimulation of adenylate cyclase do not inhibit bone resorption and PTH analogs that have relatively little effect on cAMP can stimulate bone resorption. Synthetic 1–34 PTH, a related peptide, is more potent than PTH in stimulating cAMP production in cells of the osteoblast lineage, but less potent in stimulating osteoclastic bone resorption in organ cultures.

Regarding the effects of PTH on bone formation, intermittent exposure to low concentrations of PTH results both *in vitro* and *in vivo* (e.g., in rats treated with subcutaneous injection of PTH) in an increase in the number as well as the activity of the osteoblasts. The extent of bone-forming surfaces and the rate of mineral and matrix apposition on these surfaces are increased. PTH could produce this effect by stimulating the production of a local growth factor. The insulin-like growth factor (IGF-I) is a good candidate as mediator, because it is produced in bone and because growth hormone deficiency, which presumably impairs local production of IGF-I, also impairs the bone-forming response to PTH. PTH also stimulates the production of prostaglandins (PGE_2) in bone and PGE_2 may be an important stimulator of bone formation. Recent data have shown an increase in bone mass in patients with osteoporosis treated intermittently with low doses of human synthetic 1–34 PTH given in conjunction with small amounts of $1,25(OH)_2D_3$. It is therefore possible that PTH, which is normally secreted in small amounts and intermittently, is a physiological stimulator of skeletal growth. Increased bone resorption and decreased formation would occur only when there was a marked and sustained increase of PTH concentration.

In kidney, specific membrane receptors for PTH have been characterized. It has been proposed that PTH stimulates calcium reabsorption from the kidney tubule by a Na^+–Ca^+ exchange mechanism.

The effect of PTH on gastrointestinal absorption of calcium is indirect, through regulation of synthesis of $1,25(OH)_2D_3$ in the kidney, which enhances absorption of calcium from the gastrointestinal tract. The mechanism will be discussed in Section I,B,3.

2. Calcitonin

The main action of calcitonin, and certainly the one of greatest physiological importance, is on bone. Calcitonin lowers serum calcium by an acute inhibition of osteoclasts. It has been demonstrated that osteoclasts incubated with calcitonin rapidly lose the ruffled borders that are characteristic of resorbing bone. Calcitonin directly inhibits osteoclastic motility and spreading. Plasma calcium falls because the inhibition of osteoclastic activity markedly diminishes the flow of calcium from bone to blood. Calcitonin also decreases phosphates in plasma as well as alkaline phosphatase and hydroxyproline production. These effects reflect inhibition of bone resorption, as well as stimulated urinary excretion of both calcium and phosphate.

Calcitonin receptors on isolated osteoclasts (the physiological target cells of the hormone) have been characterized. Whether cyclic AMP is the intracellular second messenger involved in the calcitonin effect is not clear.

Plasma calcitonin levels increase when calcium rises and decrease when it falls. It has been proposed that calcitonin has a specific role after eating: its secretion induced by gastrointestinal peptides results in greater skeletal retention of calcium than would normally occur. Calcitonin also acts on the kidney to enhance the production of $1,25(OH_2)D_3$ and this might be a physiological effect of importance during pregnancy and childhood.

3. Vitamin D Endocrine System

All the biological activities attributed to vitamin D itself are due to the biologically active derivatives $1,25(OH)_2D_3$ and $24,25(OH)_2D_3$.

$1,25(OH)_2D_3$ acts with parathyroid hormone and calcitonin on the classic target tissues: bone, intestine, kidney, and parathyroid glands. It stimulates bone mineralization indirectly by providing calcium and phosphorus for incorporation into bone matrix through increased intestinal absorption.

$1,25(OH)_2D_3$ plays a part in the regulation of the function of osteoblasts, which possess $1,25(OH)_2D_3$ receptors. $1,25(OH)_2D_3$ modulates their proliferation, their alkaline phosphatase production, and their synthesis of γ-carboxyglutamic acid protein (osteocalcin) and of matrix γ-carboxyglutamic acid protein. It reduces the production by osteoblasts of type I collagen (in fetal rat calvaria), and it increases the receptors of epidermal growth factor in osteoblasts and the activity of transforming growth factor. The physiological importance of $1,25(OH)_2D_3$ in osteoblast-mediated processes of bone remodeling awaits better clarification.

Bone resorption is also influenced by $1,25(OH)_2D_3$. In organ cultures of bone, it causes calcium release after several hours, and *in vivo* it causes increased formation of osteoclasts over a period of several days. Since osteoclasts originates from cells of early macrophage lineage, $1,25(OH)_2D_3$ may have a maturational effect on myeloid precursor cells, inducing them to differentiate toward functional osteoclasts. The function of mature osteoclasts, however, is not directly influenced by $1,25(OH)_2D_3$.

One of the major contributions of $1,25(OH)_2D_3$ is the provision of mineral for bone formation through stimulation of the intestinal lumen–plasma flux of calcium and phosphate by acting on an intestinal receptor. There is evidence that it increases the production of a calcium-binding protein known as calbindin-D, the levels of which in the intestinal mucosa are positively correlated with the rate of calcium transport or absorption. The exact role of calbindin-D in this process has not yet been defined.

Though parathyroid hormone is an important tropic stimulator of the renal synthesis of $1,25(OH)_2D_3$, serum levels of $1,25(OH)_2D_3$ decrease the release of parathyroid hormone through two different mechanisms. In a short feedback loop, $1,25(OH)_2D_3$ directly inhibits the synthesis of parathyroid hormone by acting on the preproparathyroid hormone gene. In a long feedback loop, the increased serum concentrations of ionized calcium are an inhibitory signal for secretion and production of parathyroid hormone.

II. OTHER HORMONAL BONE REGULATORY FACTORS

A. Gonadal Steroids

The most important source of information about the effects of sex hormones on bone metabolism in mammals has been clinical studies of estrogen withdrawal and lifelong replacement or premature deficiency of the principal estrogens in women (or the principal androgens in men), which is associated with an increased incidence of osteoporosis. [*See* Estrogens and Osteoporosis.]

At menopause there is an increase in bone resorption accompanied by a smaller increase in bone formation, so that bone mass decreases. The rates of those changes vary considerably in different skeletal sites, but all sites are involved.

The administration of estrogens to menopausal women in "physiological replacement" amounts decreases serum calcium and phosphate and urinary calcium, with retention of total body calcium; hydroxyproline excretion (signaling bone resorption) is decreased; PTH and $1,25(OH)_2D_3$ concentrations rise in the serum. All of these changes are taken as evidence of decreased bone resorption. Cultured bone cells and osteosarcoma cells with an osteoblast phenotype have been shown to contain estrogen receptors and to show metabolic changes when these receptors are activated. Estrogen receptors are present in bone cells of the osteoblast lineage but there may also be estrogen receptors in osteoclasts.

A direct action of estrogens on bone cells is at present only a speculation. A reasonable alternative is that estrogens act indirectly to alter the production of one or more of the local regulators of bone metabolism. In fact, calvarial bones after ovary removal from rats produce substantially more prostaglandin PGE_2 than bones from sham-operated controls; pretreatment with estrogen reduces PGE_2 production. Because PGE_2 can stimulate both bone resorption and formation, an increase in its production could explain the bone changes observed after menopause. Estrogens may inhibit production of cytokines by cells of the osteoblast lineage: this is the most attractive and convincing possibility emerging from the recent data.

B. Glucocorticoids

States of glucocorticoid excess are associated with accelerated loss of skeletal mass, particularly from regions of trabecular bone such as vertebrae. The histological appearance of the skeleton suggests decreased bone formation and increased bone resorption.

Glucocorticoid excess lowers serum Ca^{2+} concentration. There is decreased calcium absorption in the intestine due to both direct interference of glucocorticoids with intestinal calcium absorption and an indirect effect of reduced plasma concentrations of 25(OH)D and $1,25(OH)_2D_3$. Lower calcium concentrations lead to secondary hyperparathyroidism, with increased bone resorption and a direct inhibitory effect on bone formation.

C. Thyroid Hormones

Conditions characterized by excess of thyroid hormones, especially thyrotoxicosis, are associated with mild increase of calcium serum concentration and elimination of Ca^{2+} through urine, as well as decreased bone mass. The histological features of bone in thyrotoxicosis suggest increased skeletal resorption and formation. Elevation of plasma calcium levels decreases circulating PTH, which accounts for the excessive elimination of Ca^{2+} through the urine. *In vitro*, thyroid hormone directly stimulates bone resorption.

The effects of excess thyroid hormone contrast with those of normal concentrations, which stimulate maturation of cartilage and normal bone formation. [*See* Thyroid Gland and Its Hormones.]

D. Insulin

Insulin is a regulator of cartilage and bone growth, with both direct and indirect stimulatory effects. In untreated diabetic patients, as in animals with experimental diabetes, bone and cartilage formation and bone mineralization may be impaired, with a high incidence of osteoporosis. Insulin directly stimulates cartilage and bone formation. It stimulates amino acid transport and RNA, collagen, and noncollagen protein synthesis in bone cultures. An indirect effect of insulin derives from its regulation of release of somatomedin (which stimulates bone DNA synthesis) by the liver.

Insulin may also be required for the normal mineralization of bone. Its effect on calcium absorption and bone mineralization is probably mediated by $1,25(OH)_2D_3$, since insulin is required for its synthesis by the kidney. [*See* Insulin and Glucagon.]

E. Growth Hormone and Somatomedins

Growth hormone has no direct effects on cartilage or bone formation but has indirect effects mediated by somatomedins, a family of insulin-like peptides with growth-promoting activity in a variety of tissues. Somatomedins are regulated by growth hormone and are believed to mediate its effects on skeletal growth. They have a direct growth-promoting activity on bone and cartilage, but apparently have no effect on bone resorption or bone mineralization. The major somatomedins are insulin-like growth factor I (IGF-I) and insulin-like growth factor II (IGF-II), which act like insulin on chondrocytes and bone cells. They are synthesized by the liver and also by skeletal cells. Thus IGFs may be considered both systemic and local factors that increase cellular replication and/or enhance the formation of matrix. Their contribution to the maintenance of bone mass also seems to be caused by a decrease of collagen degradation [*See* Growth, Anatomical; Insulin-like Growth Factors and Fetal Growth.]

III. LOCAL ENDOGENOUS BONE REGULATORY FACTORS

Local regulators of bone remodeling have been identified, usually in *in vitro* systems and under experimental conditions that are not easily comparable for various reasons. Therefore, the exact role of each factor is somewhat unclear. Effects on cell replication and differentiation are, however, pronounced and are probably of physiological significance. Local factors may also mediate the effects of systemic hormones, which may modify the synthesis or effects of the local factors. Local regulators of skeletal growth may be classified into (a) growth factors synthesized by skeletal cells, (b) growth factors isolated from bone matrix, (c) growth factors synthesized by cells from adjoining tissues, and (d) prostaglandins (as shown in the following list). [*See* Growth Factors and Tissue Repair.]

LOCAL REGULATORS OF BONE GROWTH

I. Polypeptides
   A. Synthesized by skeletal cells[a] and/or from bone matrix[b]
      1) Transforming growth factor (TGF-β)[a,b]
      2) Bone-derived growth factor (BDGF) or beta_2 microglobulin (β_2m)[a,b]
      3) Somatomedin (Sm) C or insulin-like growth factor (IGF-I)[a,b]
      4) Platelet-derived growth factor (PDGF)[a,b]
      5) Fibroblast growth factor (FGF) a ("acidic")[b]
      6) Fibroblast growth factor (FGF) b ("basic")[b]

B. Synthesized by adjoining tissue from cartilage[a] or blood cells[b]
   1) Somatomedin C (IGF-I)[a]
   2) Fibroblast growth factor (FGF) b (basic)[a]
   3) Interleukin I (IL-I)[b]
   4) Tumor necrosis factor α (TNF-α)[b]
   5) Macrophage-derived growth factor[b]
   6) Platelet-derived growth factor[b]
   7) Lymphotoxin or tumor necrosis factor β (TNF-β)[b]
   8) Interferon γ (IFN-γ)[b]

II. Endogenous prostanoids
   A. Prostaglandins (PGs)

A. Growth Factors Synthesized by Skeletal Cells

Insulin-like growth factor IGF-I (also known as somatomedin C) has already been discussed. It is a growth hormone-dependent polypeptide that stimulates cellular replication in cartilage and enhances the formation of matrix. Local IGF-I seems to be a major regulator of growth. Growth hormone, which stimulates IGF-I synthesis in liver and other tissues, may directly stimulate its synthesis in bone also. Anabolic effects of parathyroid hormone on bone appear to be mediated in part by IGF-I. The inhibitory effect of corticoids on bone formation also involves a decrease of IGF-I synthesis.

Bone-derived growth factor (BDGF) is a beta$_2$ microglobulin (β_2m) isolated from matrix and bone cultures. It stimulates bone collagen and DNA synthesis in bone cultures.

Platelet-derived growth factor (PDGF) is a polypeptide originally isolated from platelets but subsequently isolated from normal and neoplastic tissues, including bone matrix and osteosarcoma cells. PDGF stimulates bone DNA and protein synthesis. When released by platelets during aggregation, it may have important effects in fracture healing. When synthesized and released by bone cells, it may interact locally with other hormones and growth factors in bone formation. PDGF has also been shown to stimulate bone resorption, probably via prostaglandin-mediated mechanisms.

Transforming growth factor β (TGF-β) is a peptide belonging to a large gene family of many related regulatory proteins. This peptide is now known to act on nearly every tissue and cell type, including mesenchymal cells. It plays essential roles in embryogenesis (particularly during periods of morphogenesis) and also, in the adult, during the normal processes of tissue remodeling and repair. It also plays aberrant roles in various pathological processes.

Although platelets are the most concentrated source of TGF-β in the body, the largest amount of TGF-β in the body is found in bone. The presence of TGF-β in bone matrix and the observation that in healing fractures the levels of TGF-β mRNA are greatly elevated at the time of endochondral ossification suggest that the mechanisms of TGF-β action in bone formation in the embryo are reiterated in the process of bone remodeling and repair of bone injury. TGF-β has been shown to activate gene transcription and increase synthesis and secretion of bone matrix proteins; to decrease synthesis of proteolytic enzymes that degrade matrix proteins; to increase synthesis of protein inhibitors that block the activity of these enzymes; and to increase the transcription, translation, and processing of cellular receptors for matrix proteins.

TGF-β is mitogenic for osteoblasts in culture. It may stimulate collagen synthesis in osteoblasts, but it is unclear whether this is a direct action or secondary to its effects on the proliferation of the cells.

TGF-β activity is increased in cultures of fetal rat calvaria incubated with agents that stimulate bone resorption such as parathyroid hormone or 1,25$(OH)_2D_3$. Moreover, other data show that TGF-β inhibits the formation of osteoclasts in bone marrow cultures and this further suggests that it may act as a local regulator of bone remodeling. TGF-β might be involved in the strict coupling of the processes of bone resorption and bone formation characteristic of the remodeling process in adult bone. It has been proposed that the local acidic proteolytic environment created by osteoclastic activity would result in activation of matrix-associated latent TGF-β. The activated TGF-β would then inhibit new osteoclast formation and expand the local population of osteoblasts, resulting in deposition of matrix for mineralization. The physiological mechanisms of activation of TGF-β, signaling pathways, mechanisms of activation of target genes, and control of TGF-β expression are all important areas for future research.

The intricate mechanisms by which TGF-β regulates bone formation are likely to be fundamental to understanding the processes of skeletal growth during development, maintenance of bone mass in adult life, and healing of bone fractures. Acidic fibroblast growth factor (aFGF) and basic fibroblast growth factor (bFGF) belong to the polypeptide family of heparin-binding growth factors (HBGFs). They have been shown to be mitogenic for bone cells in bone cultures, but their specific function in bone cell biology has not been clarified [*See* Transforming Growth Factor, Beta.]

B. Growth Factors Isolated from Bone Matrix

In bone matrix extracts, a number of growth factors have been found: transforming growth factors $\beta 1$ and $\beta 2$, β_2 microglobulin, somatomedin, platelet-derived growth factor, bone-derived growth factor, and acid and basic fibroblast growth factors. They may be synthesized by skeletal cells or by cells from extraskeletal tissue or only trapped by bone matrix.

Recently a bone morphogenetic protein (BMP) family has been isolated from bone matrix. BMPs act by initiating bone formation and controlling other steps in the osteoinduction sequence. A cooperative interaction between BMPs and collagenous substratum is required for BMP bone induction *in vivo*.

C. Growth Factors Synthesized by Cells from Adjoining Tissues

Cartilage and bone marrow may synthesize factors that act directly on the skeleton. Somatomedin and fibroblast growth-like factors have been isolated from cartilage. Blood cell-derived factors that have been shown to modulate bone remodeling are monokines and lymphokines. [*See* Cartilage.]

The monokines are secreted by the monocytes/macrophages and include:

1. Interleukin I (IL-I), a polypeptide that stimulates bone resorption and is now known to be a component of the activity termed "osteoclast activating factor" (OAF). IL-I causes a decrease in collagen production in high doses but at low doses stimulates bone collagen synthesis. In addition, IL-I stimulates cell replication in bone cultures. Since IL-I stimulates bone resorption and bone formation, it may be an important link between the two processes.
2. Tumor necrosis factor α (TNF-α) is a monokine produced by activated necrophages known for its cytostatic, cytolytic, and antiviral effects. TNF-α stimulates bone resorption and cell replication in the osteoblastic lineage. On the contrary, TNF-α directly inhibits osteoblastic collagen synthesis.

The lymphokines are secreted by the activated lymphocyte and include:

1. Lymphotoxin or tumor necrosis factor β (TNF-β), whose effects on bone formation and resorption are identical to those of tumor necrosis factor α (TNF-α) (see preceding).
2. Interferon γ (IFN-γ), a glycoprotein with antiviral and antiproliferative activities. IFN-γ inhibits collagen synthesis and resorption in bone cultures, although the inhibitory effects of IFN-γ are not limited to bone cultures. [*See* Interferons.]

Monokines and lymphokines stimulate bone resorption and preosteoblast cell replication, suggesting a possible role in coupling.

Osteotropic cytokine production is impaired by estrogens and removal of estrogens may lead to enhanced production that may cause an increased bone resorption.

D. Prostaglandins

Prostaglandins (PGs), complex endogenous prostanoids, are derived from phospholipid breakdown and are potent stimulators of bone resorption. PGE_2 increases cAMP production in bone. Although there is a close parallel between the ability to increase cAMP production, presumably in cells of the osteoblast lineage, and the resorptive response to PGE_2, the association between these two responses is not as clear as it is for PTH. Prostaglandins may serve as local mediators of skeletal resorption. This may be relevant to the bone loss associated with skeletal inflammatory processes, such as rheumatoid arthritis or periodontal disease. Unlike PTH and OAF, PGE_2 does not inhibit bone formation, although in concentrations higher than those that stimulate bone resorption it may, like PTH, inhibit collagen synthesis in bone selectively. Low concentrations of PGE_2, which still stimulate bone resorption, also appear to enhance bone formation. The effect on bone formation is relatively slow and appears to involve replication and differentiation of osteoblast progenitors. PGE_2 effects on bone formation may be indirect, through activation of latent bone formation stimulatory factors in the bone matrix during the resorptive process.

Prostaglandin production in bone appears to be increased by mechanical forces that promote remodeling, with initial resorption and subsequent formation of new bone. There is also evidence that prostaglandins may mediate both the rapid increase in resorption that occurs with immobilization and the pathological ectopic bone formation that occurs with implantation of demineralized bone matrix.

The osteoinductive activity of demineralized bone matrix may well involve prostaglandin, but its main cause may be osteogenin, a bone-inductive glycoprotein recently isolated and purified from bovine bone and rat tooth matrix.

IV. CLINICAL ASPECTS

A. Parathyroid Hormone

Hypoparathyroidism is a clinical disorder induced by decreased secretion or peripheral action of PTH. Secretion may be inadequate, the hormone may be biologically ineffective, or organ sensitivity to PTH may be defective. The most prominent clinical feature of hypoparathyroidism is hypocalcemia and hyperphosphatemia, which are consequences of diminution of PTH action on kidney and bone.

Hypoparathyroidism due to target organ unresponsiveness to PTH rather than to PTH deficiency is called pseudohypoparathyroidism and represents a true hormone resistance syndrome. It is a rare genetic disorder involving bone and mineral metabolism. Vitamin D and calcium supplements are the therapeutic tools of all forms of hypoparathyroidism.

Persistent hypersecretion of PTH produces extensive resorption of the skeleton with generalized demineralization. The single parathyroid adenoma (about 80% of cases) or the hyperplastic glands (about 15% of cases) may be surgically removed and that leads to the cure of hyperparathyroidism. Permanent hypoparathyroidism is a serious but unusual complication of parathyroid surgery that requires life-long treatment with vitamin D and supplemental calcium.

B. Calcitonin

Calcitonin is secreted at abnormally high rates, inducing a recognized hypercalcitonism, when a tumor called medullary thyroid cancer, which arises in the parafollicular or C cells of the thyroid (those that synthesize CT), develops in the gland. This tumor accounts for less than 10% of thyroid cancers so that the disease is considered rare. Unlike many other hormones when secreted at abnormally high rates, calcitonin does not induce obvious and specific metabolic abnormalities.

There is no disease state known to be related to hyposecretion of calcitonin. Some authors claim a CT hyposecretion during menopause that could be in part responsible for postmenopausal osteoporosis.

C. Vitamin D Endocrine System

The classic disease state consequent to vitamin D deficiency is a bone disease called rickets in children or osteomalacia in adults. The characteristic feature of this disease is the failure of the organic matrix of bone (osteoid) to calcify.

The uncalcified bone is prone to deformities, especially in bones exposed to high load like knees, ankles, and wrists. Deformities occur mainly in children, who have a growing and developing skeleton; adults experience few bone deformities and mainly bone pain and muscular weakness.

Administration of vitamin D has practically eliminated rickets, formerly a major medical problem, but is of no benefit in the rare cases of vitamin D-resistant rickets.

The responsibility of a vitamin D metabolite deficiency due, for instance, to reduced synthesis of $1,25(OH)_2D_3$ by the kidneys in the most common generalized disorder of bone (called osteoporosis, in which there is a thinning of the bones with a concomitant increase in the number of fractures) is debatable. In some patients with osteoporomalacia, vitamin D treatment is successful.

It is important to underline that other causes of osteoporoses are well recognized, such as prolonged calcium deficiency, prolonged immobilization, reduced estrogens as in postmenopausal women, adrenal cortical hyperfunction, or prolonged high-dosage administration of corticoids. It is not yet clear if osteoporoses that are not due to recognized condition(s) are attributable to increased bone resorption or decreased rates of new bone formation or both.

D. Local Endogenous Bone Regulatory Factors

Our knowledge about local regulators of bone formation and bone resorption is still limited, and no specific clinical aspects due to lack or excess of one factor or to an imbalance among factors have been demonstrated, although it is strongly suspected that the local factors are important, particularly in the coupling of bone formation to bone resorption.

The morphogenetic proteins are expected to be used clinically for their bone-inductive capacity, mainly in surgical procedures and in dentistry.

V. CONCLUSIONS

Bone formation and bone resorption are complex processes that are regulated by many humoral and local factors with direct and indirect effects on bone. In addition to the main calcium regulatory hormones (parathyroid hormone, calcitonin, and calciferols), glycocorticoids, thyroid hormones, growth hormone, insulin, and sex steroids participate in the many events controlling bone formation and resorption.

The recent spectacular advances in biological research have identified an array of local regulators of bone. Some of them are polypeptides synthesized and released by skeletal cells or released by bone matrix, where they are trapped and storaged in latent form. Various polypeptides are also synthesized by cells from adjoining tissues (e.g., cartilage, bone marrow cells, monocytes, macrophages, lymphocytes) that are in close proximity to bone. The exact role of local bone regulatory factors has not yet been defined.

BIBLIOGRAPHY

Canalis, E., McCarthy, T., and Centrella, M. (1993). Factors that regulate bone formation. *In* "Physiology and Pharmacology of Bone" (G. R. Mundy and T. J. Martin, eds.), pp. 249–266. Springer-Verlag, Berlin/Heidelberg.

Mundy, G. R. (1993). Cytokines of bone. *In* "Physiology and Pharmacology of Bone" (G. R. Mundy and T. J. Martin, eds.), pp. 185–214. Springer-Verlag, Berlin/Heidelberg.

Norman, A. W., and Litwach, G. (1987). "Hormones." Academic Press, New York.

Pecile, A., and De Bernard, B. (eds.) (1990). "Advances in Bone Regulatory Factors." Plenum, New York.

Wozney, J. M., and Rosen, V. (1993). Bone morphogenetic proteins. *In* "Physiology and Pharmacology of Bone" (G. R. Mundy and T. J. Martin, eds.), pp. 725–748. Springer-Verlag, Berlin/Heidelberg.

Zaidi, M., Breimer, L. H., and MacIntyre, I. (1987). Biology of peptides from the calcitonin genes. *Quart. J. Exp. Physiol.* 72, 371–408.

Bone Remodeling

JAMES A. GALLAGHER
JANE P. DILLON
University of Liverpool

GLOSSARY

Basic multicellular unit Assemblage of cells responsible for a distinct focus of remodeling

Bone structural unit Quantum of bone produced by one basic multicellular unit

Diaphysis Shaft of a long bone

Endosteum Tissue that lines the marrow cavity of bone

Epiphysis Expanded region at the ends of long bones, separated from the shaft in immature bones by growth-plate cartilage

Lamellar bone Mature bone organized in layers or lamellae; each lamella is composed of collagen fibers oriented in one plane, which is distinct from that of the collagen in adjacent lamellae

Modeling Process by which the shapes of bones are defined by cellular activity during growth

Parathyroid hormone Important hormone regulating calcium and bone homeostasis; it is secreted from the parathyroid glands

Periosteum Membranous tissue surrounding a bone

Remodeling Turnover of bone in which discrete microscopic packets of tissue are removed by resorption and subsequently replaced by formation

Renewed modeling Process by which the shapes of bones are altered by cellular activity after growth has ceased

WHEN THE LONGITUDINAL GROWTH OF BONE IS arrested at epiphyseal fusion, the activity of bone-forming and bone-resorbing cells does not cease. Instead, bone is continually turned over by cellular activity, a phenomenon referred to as remodeling. Bone remodeling plays a role in the maintenance of calcium homeostasis and is also important in the mechanical function of bone, providing a mechanism for adaptive reconstruction. Bone remodeling involves the removal of discrete microscopic packets of bone by bone-resorbing osteoclasts, followed by replacement of an equivalent amount of bone by bone-forming osteoblasts. This complex sequence of cellular activities is regulated by a combination of local and systemic factors. Remodeling is normally a replacement phenomenon with little change in the architecture of bone. In some situations, such as when a new pattern of loading is imposed on the skeleton, cellular activity brings about radical changes, which involve redesigning the architecture of individual bones. In this type of turnover and in contrast to normal remodeling, a resting bone surface often directly transforms to one of active formation. This phenomenon is described as renewed modeling to distinguish it from normal remodeling.

I. INTRODUCTION

A. The Functions of Bone

We can distinguish two distinct functions of bone. The first one is mechanical—bone is the material of which the skeleton is mainly composed. The skeleton provides a framework that supports the soft tissues of the body, allows movement through a system of

ENCYCLOPEDIA OF HUMAN BIOLOGY, Second Edition, VOLUME 2. Copyright © 1997 by Academic Press. All rights of reproduction in any form reserved.

levers, and provides protective casing around delicate organs. Thus, the skull protects the brain, the rib cage protects the lungs and heart, and the best protected of all is the hematopoietic tissue, which is encased deep inside some bones. [*See* Skeleton.]

The second function of bone is in mineral homeostasis. The skeleton acts as a reservoir for calcium and other minerals. Calcium is essential in the regulation of many cellular activities and, reflecting this, the calcium concentration ([Ca]) of extracellular fluid (ECF) is maintained in a narrow range around 2.5 m*M*. Calcium is exchanged continuously between bone and ECF, and this exchange together with renal tubular reabsorption regulates the steady-state ECF [Ca] and plasma [Ca]. Furthermore, in times of calcium stress, when the demand cannot be met by dietary intake, large amounts of calcium are mobilized from the skeleton. Examples of this mobilization are seen during egg laying, antler formation, and pregnancy and lactation. It is within the context of these two functions, mechanical support and mineral homeostasis, that we must consider bone remodeling.

B. Growth, Modeling, and Remodeling

Bone matrix is made by cells described as osteoblasts. The special composition of bone, particularly its high mineral content, restricts the manner in which osteoblasts can lay down bone. Thus bone, unlike other tissues, cannot increase in size by interstitial growth (i.e., expansion from within) but only by apposition (i.e., addition to the surface). Osteoblastic apposition is not sufficient to sustain rapid longitudinal growth of bone. For this reason, most of the bones of the skeleton start life as cartilaginous models. Cartilage undergoes interstitial growth and is replaced by (not transformed into) bone by the process of endochondral ossification. The capacity for longitudinal growth is maintained until skeletal maturity by the persistence of growth cartilages usually sited between the epiphyses and diaphyses of long bones (Fig. 1). [*See* Bone, Embryonic Development; Cartilage.]

Accompanying the growth of bone is the process of modeling by which the ultimate shape of individual bone is formed. In addition to new bone formation, growth and modeling require the removal or resorption of bone. For example, increase in length of the femur is accompanied by an increase in the diameter of the medullary or marrow cavity. The only way that this increase could be achieved in a mineralized structure is by net removal at the inner surface, called endosteal resorption, accompanied by net accretion at the outer surface, called periosteal formation. Removal or resorption of bone is achieved by the activity of osteoclasts, which are multinucleate cells with a spectacular microscopic appearance. Thus the development, growth, and modeling of bone require the coordinated activity of at least two types of bone cell.

Longitudinal growth of an individual bone ceases at epiphyseal fusion when the last remnant of the growing cartilage, the epiphyseal growth plate, degenerates. However, with this cessation of growth, the activity of bone-forming and bone-resorbing cells does not cease. Instead, some bone material is continually removed and replaced. This turnover is usually described as remodeling. [*See* Growth, Anatomical.]

C. Remodeling and Renewed Modeling

Some cellular activity, postepiphyseal fusion, is directed toward significant changes in the architecture of bone, such as in fracture repair or in the correction of malformed bone. Similarly, bone can respond to new types of loading regimes, such as those imposed by the adoption of new exercise routines, by reorganization of its internal and external structure. These load-induced changes are clearly adaptive, as bones are responding to increased strain by changing their architecture, involving deposition and resorption, until the new shape of the bone is more suited to withstand the applied loads. In this type of reconstruction the initial cellular activity can be either resorption or formation.

In situations where the pattern of loading is relatively consistent, previous reconstruction is likely to have adapted the architecture of individual bones to a near-optimal design, yet turnover of bone still continues. In this type of turnover, which appears to be simply a replacement phenomenon, resorption almost invariably precedes formation.

The term "bone remodeling" has been used by different authors to describe various phenomena. Some have reserved this term to describe the turnover of bone that occurs exclusively in the adult skeleton and in which there is no discernible change in architecture. Others have used the term more loosely to include modeling of immature bones and adaptive responses in adult bones in which there is structural alteration.

The term "modeling" should be used to describe that process by which the shapes of bones are defined by cellular activity during growth. The term "remodeling" should be reserved for reconstruction, in which there is little or no discernible change in architecture and which follows a sequence in which resorption precedes formation. Reconstruction in adult bone that

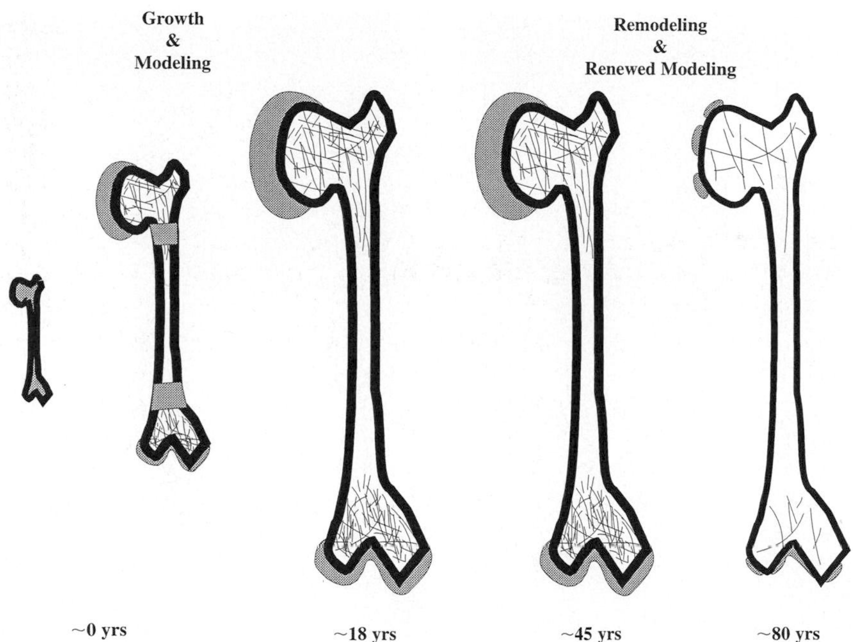

FIGURE 1 Stages in the lifetime of a human long bone. Cartilage is indicated by gray shading and bone by thick black borders. The bone develops from a cartilage model laid down during embryogenesis. Bone replaces cartilage by the process of endochondral ossification, beginning in the primary ossification center in the middle of the bone and spreading outward. Secondary ossification centers develop in the ends of the bones, but a thin plate of cartilage, the epiphyseal growth plate, persists to allow longitudinal growth. Growth is accompanied by modeling, in which the shapes of the bones are fashioned by the coordinated activity of bone-forming and bone-resorbing cells. Bone shape is determined partly by genetic blueprint and partly by adaptive response, in which the architecture of the bone is modified to withstand the stresses of load bearing and muscular contraction. When the growth-plate cartilage degenerates at epiphyseal fusion, longitudinal growth ceases, and the only cartilage remaining is the thin covering at the articular surfaces. Throughout life, bone undergoes continual reconstruction through remodeling and in certain situations through renewed modeling. Initially, the rate at which bone is removed is equal to the rate of replacement, but after middle age the balance is shifted in favor of removal and bone is lost, resulting in cortical thinning and loss of trabeculae. Bone loss is particularly severe in females after menopause, leading to osteoporosis in which the mechanical integrity of the skeleton is compromised and fractures may ensue. Aging is also associated with a loss of articular cartilage.

results in a new architecture should be described as "renewed modeling." In this type of turnover, formation is not usually preceded by resorption, instead there is direct transition from a resting surface to a forming surface. The distinction between remodeling and renewed modeling may be artificial; these may simply represent two ends of the spectrum of adaptive reconstruction in bone.

D. Where Does Remodeling Occur?

Bone remodeling occurs in compact and in cancellous bone and involves periosteal, endosteal, and trabecular surfaces (Fig. 2). Compact bone is subjected to Haversian remodeling (see Section II,B), in which cores of bone are removed by resorption and then refilled with new bone constituting a secondary osteon. Haversian remodeling is sometimes referred to as internal remodeling. In cancellous bone the sequence of events is essentially the same, but the excavations are along the surface of the trabeculae leaving trench-like resorption cavities rather than the cylindrical cavities formed in compact bone. These resorption cavities are subsequently filled in with new bone to form osteons, which are analogous to the secondary osteons formed in Haversian remodeling, but are

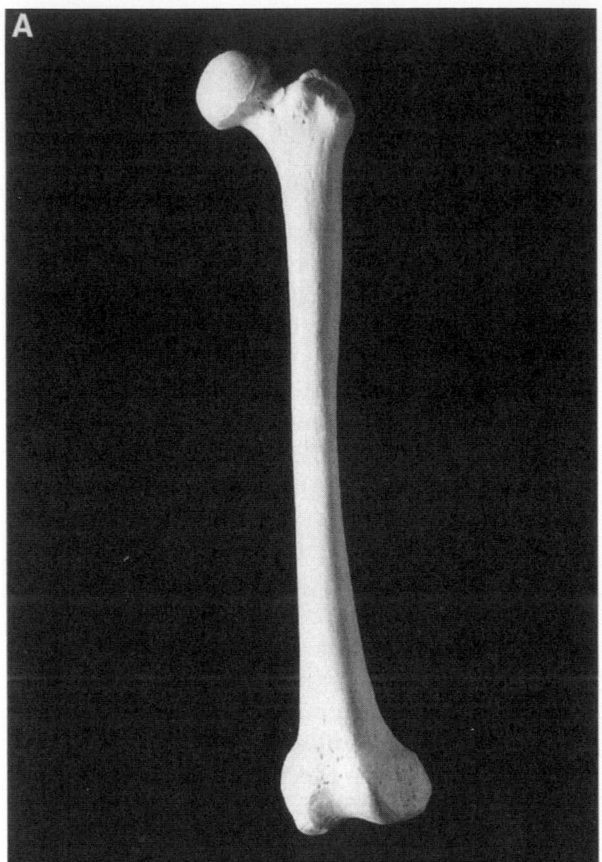

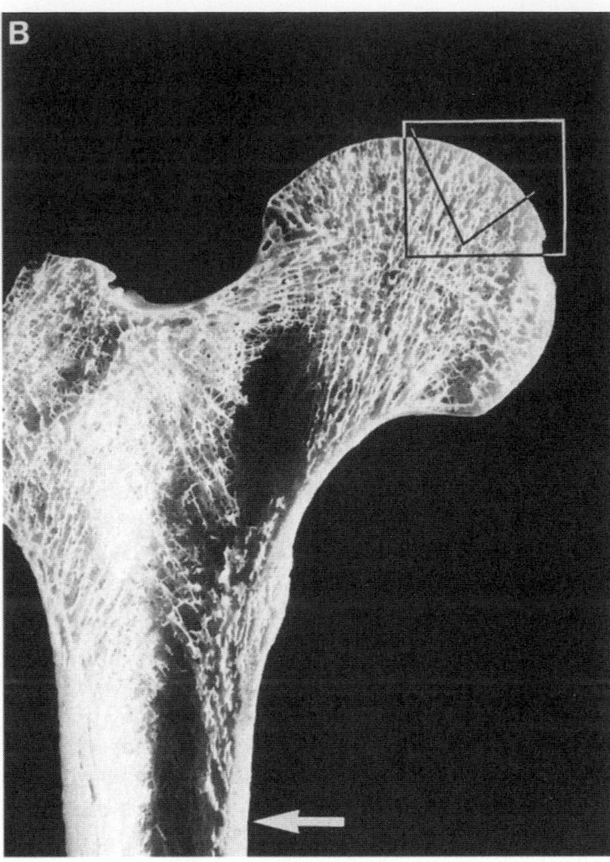

FIGURE 2 (A) The intact human femur. (B) A longitudinal section through the upper part of a human femur showing the organization of the cancellous bone, otherwise known as spongy bone or trabecular bone, covered by a thin shell of compact bone, which becomes thicker in the region of the shaft or diaphysis. A transverse section through the shaft at the level of the white arrow is shown in Fig. 4, and the area enclosed by the box is shown at higher power in Fig. 3.

sometimes described as hemiosteons because of their geometric shape.

II. WHY REMODEL?

A. Functions of Remodeling

What are the reasons for the continuous removal and replacement of bone that constitutes remodeling? As yet our understanding of this question is incomplete, but we can consider the roles that remodeling may play in the two principal functions of bone, in calcium homeostasis and in mechanical support.

B. Remodeling and Calcium Homeostasis

If we first consider calcium homeostasis, it is widely believed that remodeling is important in the maintenance of plasma [Ca]. However, most of the calcium exchange bewteen bone and blood does not occur at surfaces where there is active formation and resorption, but rather at surfaces associated with lining cells or osteocytes. These cells (described in Section IV,B) are found at quiescent bone surfaces not involved in active remodeling. The osteocytes and lining cells can be regarded as a functional syncytium regulating the equilibrium between bone and blood calcium and the short-term transfer of calcium in and out of the skeleton, responding to the important demands of plasma calcium homeostasis. Thus, remodeling does not appear to be a major factor in the minute-to-minute maintenance of plasma [Ca], yet clearly there are situations in which remodeling is accelerated to mobilize calcium from the skeleton. Laying hens resorb medullary bone, a specialized form of trabecular bone, to meet the demands for calcium during eggshell formation, and then they lay down new medullary bone when calcium is available in excess. This avian example is not identical to mammalian remodeling, but a similar tax on the skeleton can be observed in stags during antler growth. This period sees a large increase in resorption surfaces as remodeling is accelerated and calcium is mobilized from bone and shunted to the growing antlers. When antler formation is complete, it leaves the stag's bones with as much as a fivefold increase in porosity due to the abundance of resorption cavities. If the remodeling cycle is allowed to progress normally, with formation following resorption, all of the extra resorption cavities eventually will be refilled, creating an increased demand for calcium during the phase of formation. In effect, calcium is borrowed from the skeleton as the phase of enhanced remodeling gets under way and is then normally redeposited as the cycle is completed. The calcium laid down in the antlers is ultimately derived from the diet; the bone calcium pool is called on so that the acquisition of dietary calcium can be spread over a year rather than the much shorter period of rapid antler growth. Increased bone turnover is also used to provide calcium during pregnancy and lactation. In normal circumstances, this calcium is returned to the skeleton and, in fact, bone mass is often increased by multiparity.

These examples clearly illustrate that even though remodeling is not a major factor in responding to the short-term daily demands of calcium homeostasis, individuals can respond to calcium stress by enhancing their remodeling rate, thereby gaining a temporary loan of calcium from the skeleton. But is this the exclusive function of remodeling? Almost certainly not, for extensive remodeling occurs in populations that are likely never to have suffered from calcium stress, for example young adults in developed countries.

C. Mechanical Advantages of Remodeling

What is the involvement of remodeling in the mechanical functions of bone? Renewed modeling is clearly advantageous mechanically, matching bone structure to applied loads, but what about the continuous reconstruction of remodeling? Reconstruction can offer two distinct but interrelated advantages; the first is that it allows adaptive change. Haversian remodeling can change the grain of bone making it more suitable to resist loading, and remodeling at bone surfaces can produce subtle changes in the alignment of trabeculae.

The second advantage of reconstruction is simply the replacement of new material for old, which can be independent of any architectural reorganization. This replacement may be targeted; for example, one proposed function of remodeling is to remove microcracks from bone and/or to remove worn-out bone. Although some evidence supports this proposition, clearly not all remodeling activity is directed to these ends, because not only dead bone is removed and remodeling is not always greatest in those regions subjected to the greatest strains and presumably most likely to bear microfractures. Remodeling may be a preemptive response; that is, some detection system in bone determines and records its strain history and targets specific sites for replacement before microcracks even develop.

III. REMODELING IN COMPACT AND TRABECULAR BONE

A. Basic Structure of Bone

In any individual bone, the bony material is organized in two distinct forms: compact bone comprising the outer shell and cancellous bone forming an internal network of interconnecting bars and plates. The internal structure of the human femur is demonstrated in Fig. 2. Cancellous bone, seen in cross section in Fig. 3, is also referred to as spongy or trabecular bone

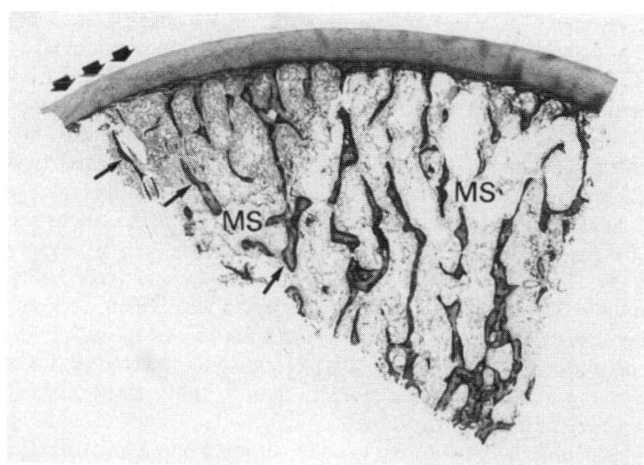

FIGURE 3 A 7-μm-thick section through an area similar to that outlined in Fig. 2. A layer of articular cartilage (thick arrows) is present on the surface. Underlying that is a thin shell of compact, subchondral bone. The interior of the bone is composed of a network of interconnecting rods or plates (arrows), described as trabeculae. The spaces between the trabeculae are occupied by the marrow space (MS), which contains hematopoietic tissue, a rich vascular supply, osteogenic precursor cells, and fibrous tissue.

and the individual plate-like or rod-like elements are described as trabeculae, derived from the Latin word for little beams. Like beams, their function is one of support and they are generally arranged along the lines of principal stress. Trabecular bone presents a large surface, continuous with the endosteum, and all of which is lined with cells and is accessible for remodeling. Compact bone is essentially composed of the same material organized in a different way. It consists of a solid mass of bone perforated by vascular canals (seen in Fig. 4 and at higher magnification in Fig. 5). The cylinders of bone surrounding each vascular canal are osteons and can be of two types—primary or secondary. Primary osteons are formed during growth, by a special form of osteoblastic apposition at the periosteal surface, in which bone is laid down and envelops capillaries, which run up and down the periosteal surface. Primary osteons of compact bone are replaced by secondary osteons during Haversian remodeling.

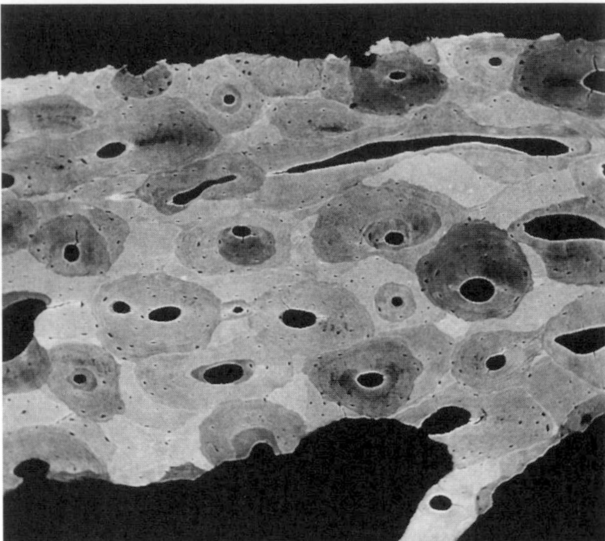

FIGURE 5 Transverse section of human cortical bone studied by backscattered electron imaging. This technique can be used to analyze mineral density distribution (light areas have high density and darker areas have low density) and is an excellent method to illustrate the osteonal structure of compact bone produced by Haversian remodeling. The osteons, seen here in cross section, are cylinders of bone containing a central Haversian canal surrounded by concentric lamellae of bone. The darkest-staining osteons are the most recently formed and some have not yet been completed. A schematic representation of the formation of secondary osteons is shown in Fig. 10.

B. Quantum Concept of Remodeling

The structure of compact bone and the inability of bone to grow interstitially dictate that, during remodeling, resorption must precede formation. In other words, new bone cannot be formed until a space has been made for it. This constraint does not apply at trabecular or endosteal surfaces; however, even at these sites, resorption normally precedes formation. This cycle of activity, in which resorption generally precedes formation, led to a quantum concept of remodeling, applying both to compact and to cancellous bone, in which bone is turned over in discrete microscopical foci following the sequence activation–resorption–formation. Osteoclasts are recruited to the bone surface and excavate a cavity, which is then refilled by osteoblasts. The assemblage of cells, both bone-resorbing and bone-forming, responsible for a distinct foci of remodeling is described as the basic multicellular unit (BMU). The term bone structural unit (BSU) describes the quantum or packet of bone produced by one BMU. BSUs are demarcated from older bones by histologically recognizable boundaries called cement lines (see Section III,C). In compact bone, the BSU is synonymous with the secondary osteon.

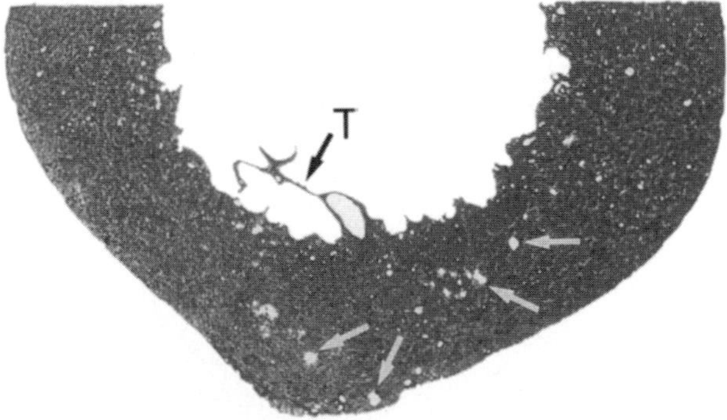

FIGURE 4 A 100-μm-thick section through a devitalized split femur. This section was prepared from a region similar to that identified by the large white arrow in Fig. 2B. Trabecular bone (T) can be seen projecting from the concave endosteal surface. The largest holes in the compact bone are new resorption cavities (white arrows). These will eventually be filled in with concentric lamellae of new bone surrounding a Haversian canal, usually carrying a single capillary. In this section, the Haversian canals of completely filled osteons are not visible as holes but rather as small dark spots.

Throughout the average skeleton, there are approximately 1.5 million sites of remodeling or BMUs. At each of those sites, the amount of bone recplaced should equal the amount removed. As the activity of BMUs is not synchronized, throughout the skeleton the amount of bone being resorbed generally equals the amount being formed, with skeletal mass remaining constant. Acceleration of the remodeling rate by increasing the birth of new BMUs will tip the balance temporarily in favor of resorption, as increased numbers of cavities are excavated. This is the mechanism by which calcium can be borrowed from the skeleton in times of calcium stress (see Section II,B). With the consequent increase in formation as these cavities are filled, balance returns. If the remodeling rate then slows down to its original rate, there will be a phase during which formation outstrips resorption and the calcium borrowed from the skeleton will be returned. The duration of the remodeling sequence at any individual site is approximately 3 months for Haversian remodeling and 2 months for trabecular and endosteal remodeling. The lifetime of a BSU is between 3 and 20 years depending on the local remodeling rate. Remodeling demands that the activity of the osteoblasts must be temporally and spatially coordinated with that of the osteoclasts. This mechanism has been described as coupling (it is described in detail in Section IV,C).

C. Cement Lines

Microscopic examination of stained sections of human bone reveals deeply stained lines at the boundaries of BSUs in compact and trabecular bone. These cement lines mark the farthest extent of prior osteoclastic excavation. As resorption cavities are refilled, the cement line forms a three-dimensional sheath separating the new BSU from the older bone. They are collagen-free regions, rich in glycoprotein and highly mineralized. Cement lines are also observed when a quiescent surface is transformed directly into a bone-forming surface as in renewed modeling. These are referred to as resting lines, whereas those formed during normal remodeling sequence are described as reversal lines. These two types are easily distinguishable in stained sections because of the scalloped appearance of the latter.

IV. CELLULAR AND BIOCHEMICAL EVENTS

A. Cellular Heterogeneity in Bone

If we compare bone with cartilage, the other major tissue involved in load bearing, the most striking feature is the heterogeneity of the former tissue. Whereas cartilage is composed of one major cell type and a fairly uniform matrix, devoid even of blood capillaries, bone is a complex tissue with many constituent cell types, bone-forming and bone-resorbing cells, blood cells, endothelial cells, and the wide spectrum of cells resident in bone marrow (Fig. 6). The different cell types are not found in discrete sites; instead, each microscopic locus is a microcosm of the heterogeneity of the tissue. Therefore, there is a wide range of potential cell–cell and cell–matrix interactions regulating bone cell function and bone remodeling. An involvement in bone turnover has been proposed for many of these cell types (e.g., endothelial cells and macrophages secrete factors that have been shown to modulate the activities of osteoclasts and osteoblasts *in vitro*). The elucidation of these cellular interactions is the subject of much current research. In this article, we consider in detail only the roles of the four major cell types.

B. Bone-Forming and Bone-Resorbing Cells

1. The Osteoclast

The osteoclast is the cell responsible for removal or resorption of bone; it is distinctive because of its multinuclearity, containing up to 20 nuclei in normal bone

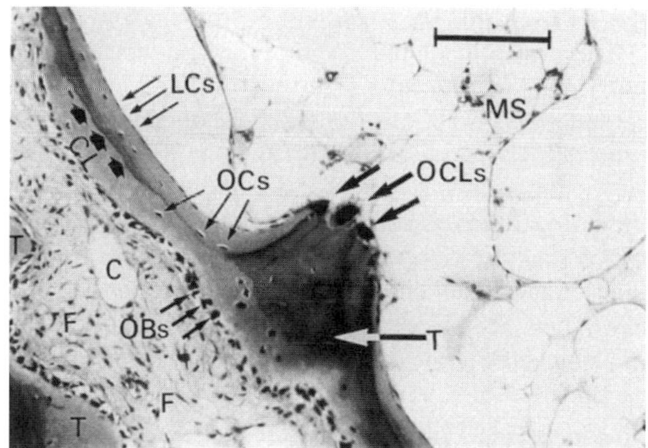

FIGURE 6 A photomicrograph of a trabecula (T) from a sample of bone from Fig. 3. This is a pathological specimen from a 30-year-old female with accelerated turnover due to secondary hyperparathyroidism. Flattened lining cells (LCs) can be seen forming a continuous layer on one surface, whereas on the other surface active cuboidal osteoblasts (OBs) are engaged in apposition. Three large multinucleate osteoclasts (OCLs) can be seen in Howship's lacunae. Osteocytes (OCs) are present entombed in bone matrix in spaces referred to as lacunae. Cement lines, marking the extent of previous osteoclastic resorption, are clearly visible (CL). The bone marrow space (MS) to the right of the trabecula is occupied by fat cells and hematopoietic and vascular tissue, whereas on the left side there has been a proliferation of fibrous tissue (F) in which capillaries (C) can be identified. Bar = 100 μm.

and 100 nuclei in some pathologies (Fig. 7). It is found in association with Howship's lacunae, irregular indentations in the bone surface excavated by the resorbing cell itself. Osteoclasts are formed by fusion of mononuclear precursors, probably derived from hematopoietic stem cells. Osteoclasts are incapable of division; expansion of the population can be achieved only by recruitment of new precursors. Resorption of bone is a formidable task requiring dissolution of the mineral phase and subsequent degradation of the matrix, which is mainly composed of the tough fibrous and highly cross-linked protein type I collagen. Demineralization is brought about by a low-pH environment, created by proton pumps on the osteoclast membranes. Degradation of bone matrix is probably achieved by the concerted activity of several types of proteolytic enzymes, including the neutral metalloproteases, which are involved in the degradation of extracellular matrices in other tissue. The osteoclast also utilizes lysosomal enzymes, including cathepsin K, which is highly concentrated in the resorbing cell. To contain the enzymes and the protons, the osteoclast has a special region, the clear zone, which is an annular ring devoid of organelles but rich in actin. The clear zone acts as a sealant ring enclosing the site of active resorption, known as the ruffled border. There the cell membrane is thrown into many folds, presumably to expand the surface area of active membrane involved in proton production and enzyme secretion. The interface between the ruffled border and the bone thus has the characteristics of a gian extracellular lysosome.

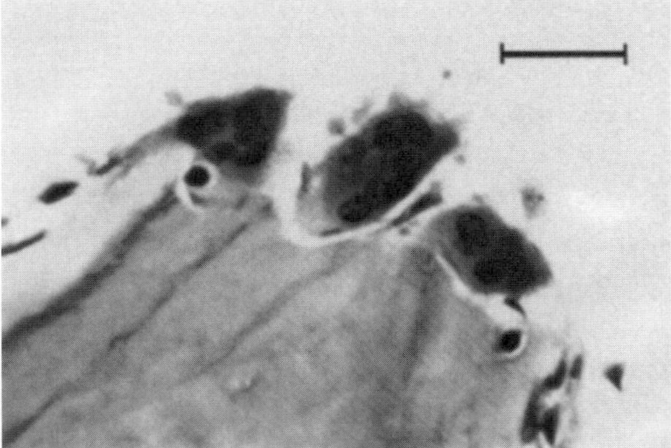

FIGURE 7 A higher-power view of the osteoclasts from Fig. 7. Note the extent of the multinucleation. The characteristic ultrastructural features of osteoclasts, the ruffled border, and the clear zone can be clearly identified only by electron microscopy. Bar = 20 μm.

2. The Osteoblast

The principal function of the osteoblast is the synthesis of bone matrix and its priming for subsequent mineralization. In cross section, osteoblasts appear as large cuboidal cells with abundant secretory apparatus. Viewed from above by scanning electron microscopy, these cells are seen to form an epithelioid layer on bone surfaces. Osteoblasts are polarized, that is, they secrete matrix onto the underlying substrate, which thus grows by apposition. The matrix is composed mainly of type I collagen and of noncollagenous proteins, some of which are specific to bone such as osteocalcin. The uncalcified matrix, referred to as osteoid, is generally mineralized so rapidly that little is found in the sections of bone. However, in phases of rapid growth or in situations where there is a mineralization defect (e.g., the deficiency of vitamin D), a seam of osteoid is seen at bone-forming surfaces. The origin of osteoblasts is distinct from that of osteoclasts. The bone-forming cell is part of a family including chondrocytes, marrow fibroblasts, and adipocytes, which are the progeny of stromal stem cells resident in the marrow.

3. The Osteocyte

During osteoblastic apposition, some cells become engulfed in bone matrix and eventually entombed within lacunae. These cells are described as osteocytes. Cells located in lacunae in other tissues, such as chondrocytes, can continue to divide and synthesize extracellular matrix and thus the tissues undergo expansion from within, or interstitial growth. In contrast, osteocytes are surrounded by a rigid mineralized matrix and so their capacity to divide and/or synthesize new matrix is extremely limited. In the lacunae they undergo a change in shape, involving the loss of secretory organelles. The mineralized matrix also prevents diffusion of nutrients, oxygen, and metabolites and so the osteocyte must maintain junctions with other entombed cells via little channels or cannaliculi that are ultimately connected to cells on the bone surface and hence to the vascular supply. Why do these cells stay alive and what is their function? Together with the surface lining cells, they present a large surface for calcium exchange (estimated to be between 1000 and 1500 m^2 in human bone) and they are probably the major

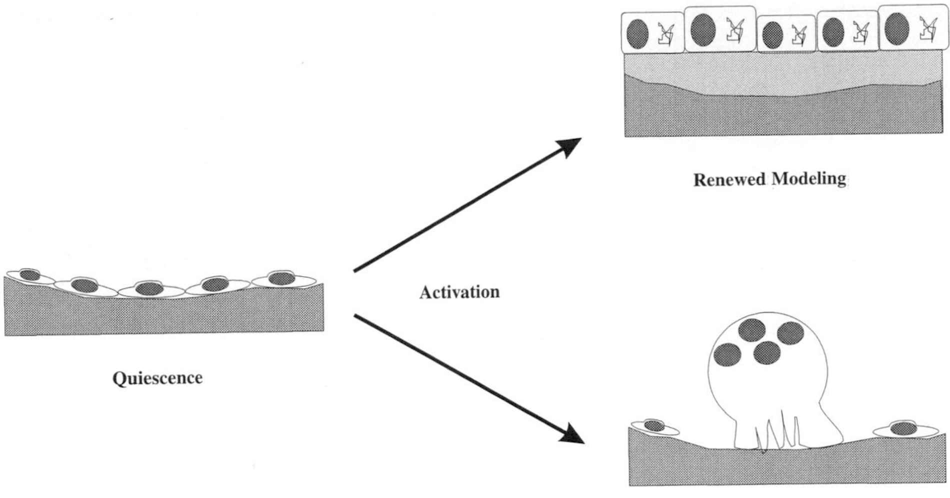

FIGURE 8 Role of bone lining cells in renewed modeling and remodeling. Quiescent bone surfaces can be activated by redifferentiation of lining cells into active osteoblasts or retraction of lining cells from the bone surface followed by the arrival of bone-resorbing osteoclasts. The former type of activation is normally seen in renewed modeling in which the architecture of the bone is adapted to new loading regimes, whereas the latter type of activation is associated with remodeling in which there is reconstruction without distinct modification. At least three mechanisms have been proposed whereby lining cells may regulate resorption: (1) retraction of the cells to expose the underlying surface to osteoclastic resorption, with subsequent release of matrix-derived chemotactic factors to recruit osteoclastic precursors; (2) direct proteolysis, particularly of a fringe of uncalcified matrix refractory to osteoclastic attach thick arrows; and (3) release of local regulatory factors in response to stimulation by systemic hormones, including PTH.

regulators of calcium fluxes in and out of bone. They are target cells for hormones involved in calcium homeostasis, including parathyroid hormone (PTH).

Another function proposed for these cells is that they act like strain gauges, monitoring and recording the extent of physical loading. These cells are ideally positioned to monitor strain and there is much evidence that indicates rapid changes in osteocyte metabolism in response to loading. They can communicate with other osteocytes and with bone-lining cells via their junction contacts.

4. The Bone Lining Cells

Bone lining cells are found on resting or quiescent bone surfaces, that is, those that are not involved in active remodeling. They are sometimes described as resting osteoblasts because they are derived from bone-forming cells, which have completed their synthetic activity, and they can be reactivated into matrix secretion as the prelude to renewed modeling (Fig. 8). Nevertheless, these should be considered as a specific cell type with functions distinct from those of their parent cell. In addition to cooperation with the osteocyte in regulating calcium exchange from bone surfaces, the lining cell may have a paramount role in the activation of the remodeling sequence (see Fig. 8). This cell type is responsive to PTH and other calcium-regulating hormones.

C. The Sequence of Cellular Events

Remodeling is brought about by the coordinated activity of the foregoing cell types. The sequence of cellular events is outlined schematically in Fig. 9. The sequence described is that observed at the trabecular surface, but the pattern is essentially similar to that observed during remodeling of compact bone as shown in Fig. 10. The following phases can be identified: activation, resorption, reversal, and formation.

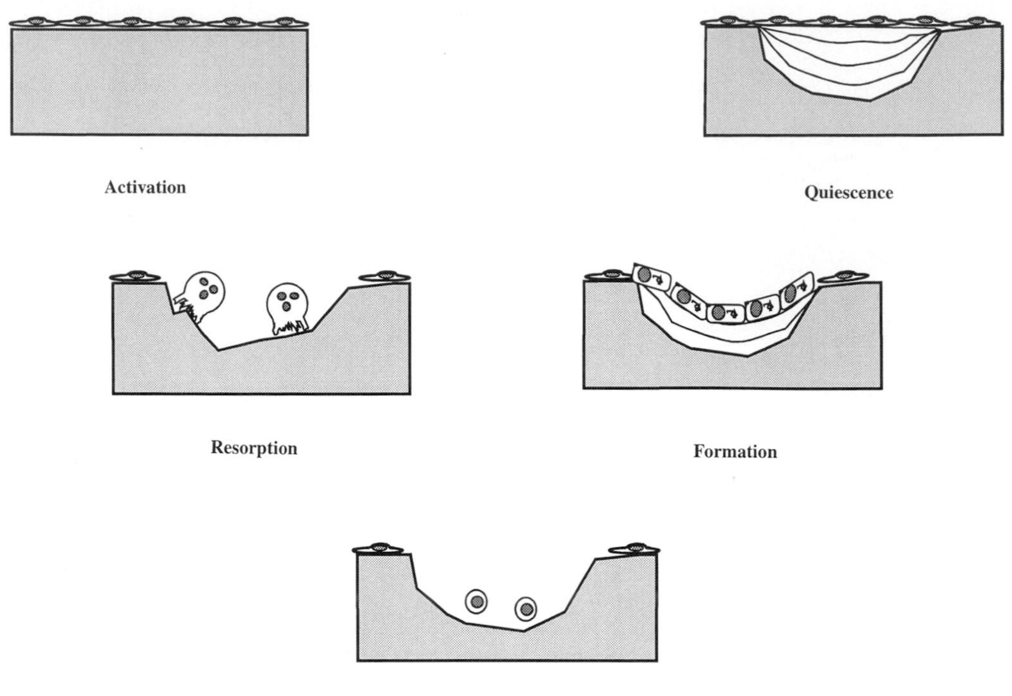

FIGURE 9 Schematic representation of the sequence of cellular events in bone remodeling. See text for information.

1. Activation

Activation applies to the phase in which a resting bone surface covered by lining cells is transformed into an active surface. Lining cells retract from the surface exposing the underlying substrate to resorption, and osteoclast precursor cells are recruited to the remodeling locus. It is now widely believed that it is the lining cell that initiates the remodeling sequence, because resorption cannot proceed until the bone surface has been exposed and also because the lining cells rather than osteoclasts express receptors for hormones, which stimulate bone resorption. Exposure of the bone surface could result in release of chemotactic and other active factors targeting the site for resorption. Lining cells may also contribute to activation and progression of remodeling by production of local hormones or cytokines, which stimulate the birth and activity of osteoclasts. Such factors may be produced in response to systemic hormones including PTH, which cannot stimulate the osteoclasts directly. It has also been proposed that lining cells may assist resorption by direct proteolysis. Some, if not all, bone surfaces are lined with a fringe of unmineralized tissue sometimes referred to as the lamina limitans. This fringe may act as a barrier to osteoclasts, which normally prefer to absorb fully mineralized tissue. Protease production by lining cells could affect the removal of this layer. This hypothesis is supported by the findings in tissue culture, in which cells of the osteoblastic lineage synthesize and secrete proteolytic enzymes.

2. Resorption

In resorption, a team of osteoclasts proceeds to excavate a quantum of bone. The regulatory signals involved in the progression and cessation of this activity are not known. The extent of resorption determines the size of the subsequent BSU.

3. Reversal

Reversal describes the intermediate phase between cessation of resorption and initiation of formation. This phase is associated with the formation of the cement lines, which mark the extent of osteoclastic resorption and eventually form the boundary of the new BSU. The cell type responsible for laying down the cement line has not been firmly identified; osteoclasts and uncharacterized mononuclear cells have been proposed, but it should be borne in mind that

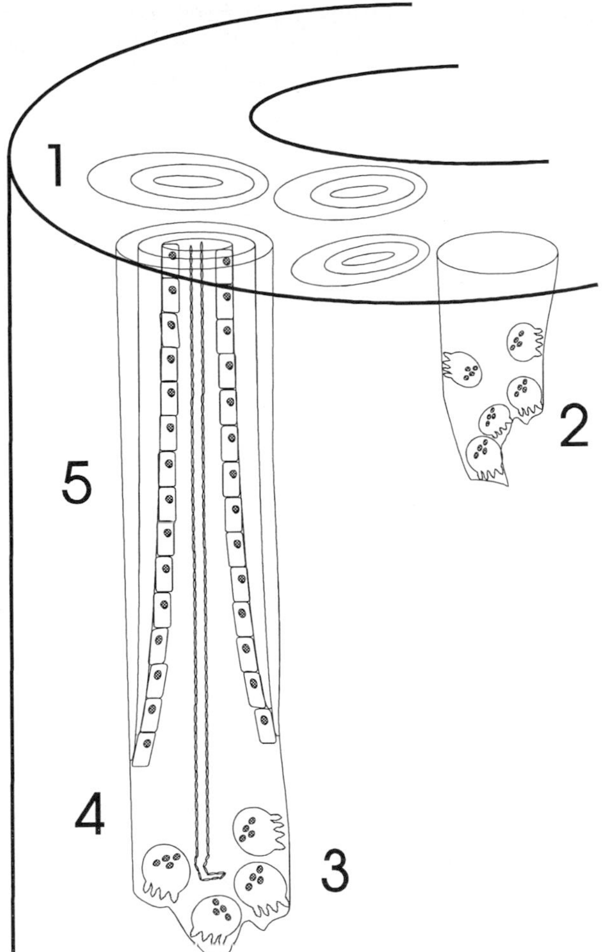

FIGURE 10 Remodeling in compact bone depicted by a schematic representation of the formation of a secondary osteon. The typical appearance of the osteons in transverse section is shown (1). In longitudinal section, a team of osteoclasts, which constitutes a cutting cone, can be seen excavating a tunnel running parallel to the long axis of the bone and thereby increasing the porosity of the compact bone (2). Another cutting cone is shown in (3); here the team of osteoclasts is followed by a migration of endothelial cells forming a new capillary and behind this is a zone of reversal (4). Eventually, osteoblasts, which make up the closing cone, start to fill in the excavation by laying down concentric lamellae of bone (5). When the osteoblasts have completed the synthetic phase they differentiate into lining cells on the quiescent surface around the Haversian canal.

resting lines, another form of cement lines, are almost certainly the product of cells of the osteoblastic lineage.

4. Formation

Osteoblasts are recruited to the remodeling locus and fill in the resorption cavity, usually by laying down lamellar bone. Some osteoblasts are engulfed in matrix and are transformed into osteocytes. When the cavity has been refilled and synthesis of matrix ceases, the osteoblasts differentiate into bone lining cells. A photomicrograph of a completed BSU is shown in Fig. 11.

D. How Are Resorption and Formation Coupled?

The remodeling sequence outlined here requires that the activities of the osteoclasts and osteoblasts are spatially and temporally coordinated. In other words, the appropriate number of osteoblasts must be recruited to the site of remodeling, and there they must lay down approximately the same amount of bone as was previously removed by the osteoclasts. The mechanism by which this coordination is achieved is described as coupling. Some authors prefer to reserve this term to describe only the recruitment of osteoblasts to the remodeling locus, whereas others include the activity of the osteoblasts postrecruitment. Considerable research has concentrated on the identification of the factor(s) responsible for coupling. A soluble mediator has been suggested, and a number of candidates have been proposed, either derived from one or many cell types in close proximity to the re-

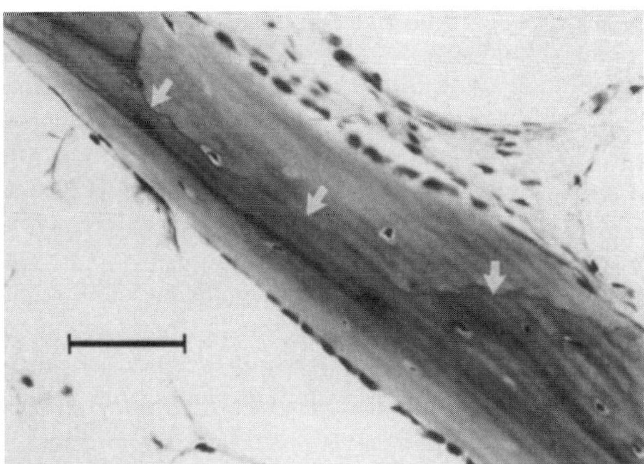

FIGURE 11 Photomicrograph of a completed bone structural unit (BSU) in human trabecular bone. A quantum of bone was removed by osteoclastic resorption to be replaced subsequently by osteoblastic apposition. The cement line (white arrows) marks the farthest extent of osteoclastic progress and is the boundary of the new BSU. Note that this section was from a pathological section in which there had been trabecular thinning. In trabeculae of normal dimensions, the average depth of the BSU (referred to as the mean wall thickness) does not generally exceed one-third of the trabecular depth. Bar = 50 μm.

modeling locus or alternatively liberated from the resorbed matrix. How a soluble mediator could regulate cellular activities separated in time but highly specific in location is difficult to envisage. This objection has led other authors to propose that the coupling factor may be a solid-phase factor associated with the cement line. This is an attractive hypothesis, especially when one considers the potential of glycosaminoglycans, in which the cement line is rich, to bind and present growth factors to cells. However, until the cement line is shown to provide an osteogenic stimulus, this hypothesis remains no more than a speculation.

V. REGULATION OF REMODELING

A. Effects of Mechanical Loading

Bone remodeling occurs throughout the skeleton, but at some sites the rate is much greater than at others. This variation is probably related to the amount of reconstruction required to adapt bone to its load-bearing function. It is often stated that the remodeling rate is increased by loading, but this is a misinterpretation, because high rates of remodeling can be seen both in bones bearing heavy loads and in bones in disuse. Despite this, much evidence indicates that loading affects bone shape and mass during growth (i.e., modeling) and also after epiphyseal fusion (i.e., remodeling). Tennis players have larger mass of bone in their playing arms than in their nonplaying arms, presumably an adaptive response to the greater loads imposed. Conversely, astronauts freed from gravitational force lose bone, which is mechanically superfluous, until they return to earth. These adaptive responses may involve remodeling and renewed modeling. In the case of remodeling, it is not just the rate that is regulated, but also the balance between formation and resorption that is affected. Thus, when bones are freed from normal physical loading such as in immobilization and bedrest, teams of osteoclasts continue to excavate tunnels through compact bone but osteoblastic activity is reduced, leaving cavities that are not completely refilled, resulting in increased porosity in bone. A shift in the balance in favor of resorption may also be observed in remodeling at trabecular surfaces. These examples demonstrate that the formation phase of remodeling does not progress simply as a consequence of previous resorption but is influenced by the functional demands of bone.

How does bone detect and record loading? When loads are applied to bone they produce strains. Numerous mechanisms have been postulated by which strains in bone may be measured. Much interest has centered on load-induced electrical effects in which current flows in response to deformation. Electrical currents have been shown to affect the behavior of bone cells *in vitro* and *in vivo*. Potential differences may be generated in bone by piezoelectricity, when crystals are strained, or by streaming potentials, when a polar liquid flows past a solid surface. Electrical and magnetic fields have been used to promote bone formation and fracture healing *in vivo,* with varying claims of success. It has been proposed that strain can induce changes in the orientation of glycosaminoglycans in bone proteoglycans. The attraction of this hypothesis is that it provides a basis for "strain memory" and, furthermore, it is not difficult to envisage a cellular response to conformational changes in proteoglycan structure, because it is clear that members of this family of molecules have regulatory functions and can interact with receptors on cell membranes. The junctional contacts between osteocytes and bone lining cells may be important in transducing the response to strain. It has also been proposed that osteocytes and bone lining cells may release ATP in response to loading. Bone cells are known to express receptors for extracellular nucleotides, and binding of ATP to these purinoceptors enhances responsiveness of cells to PTH. This may be one of the mechanisms determining how PTH activates remodeling at one site but not another.

B. Effects of Systemic Factors

In the average skeleton at any moment, there may be 1.5 million different sites of remodeling distributed throughout the remodeling cycle, some in the resorptive phase, some in reversal, and some in formation. The lack of synchrony demands a local level of regulation. Nevertheless, systemic hormones can influence the rate of remodeling and the balance within the remodeling cycle between resorption and formation. Some of the important systemic factors that regulate bone remodeling are outlined in Fig. 12. PTH enhances bone turnover by stimulating the birth of new BMUs. The hormone acts at the level of activation where one of the target cells is the bone lining cell. It may also stimulate the recruitment of osteoclasts and osteoblasts. Calcitonin has a direct inhibitory effect on osteoclast activity, but its physiological role in regulating remodeling is not clear.

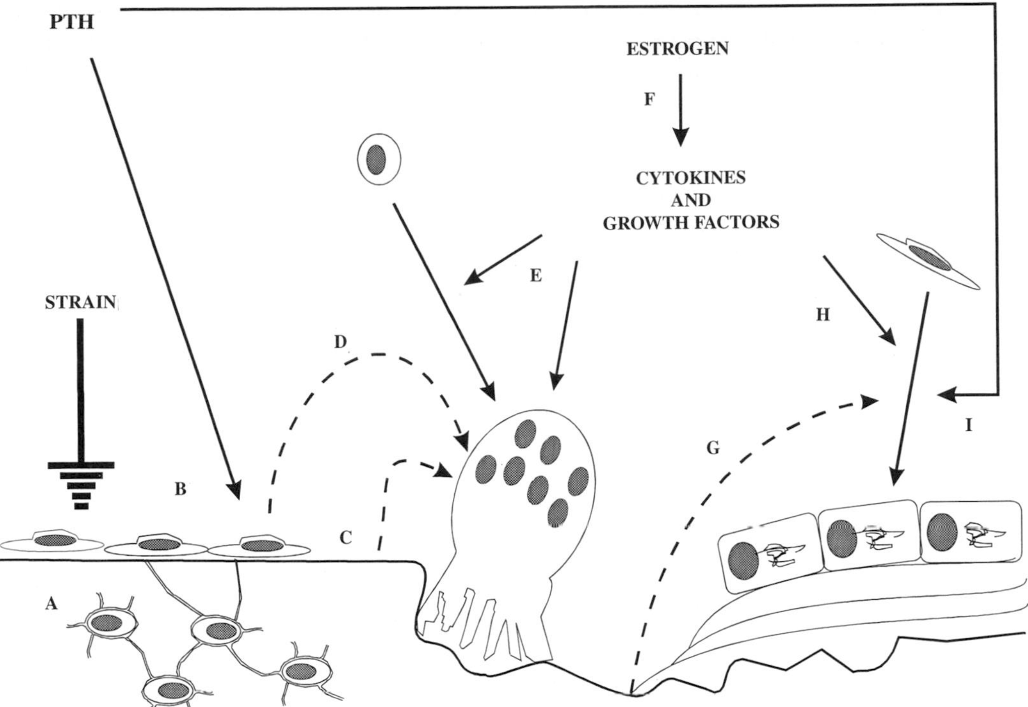

FIGURE 12 Scheme of the potential sites of action of systemic and local factors in the regulation of bone remodeling. Osteocytes detect and record mechanical strain (A) and release locally active effector molecules, possibly cytokines or extracellular ATP or other neurotransmitters, which increase the responsiveness of bone lining cells at that specific locus to parathyroid hormone (B). PTH, acting on the lining cells, causes them to retract from the bone surface and thus initiates the remodeling sequence. Mononuclear osteoclast precursors are attracted to the exposed bone surface, possibly by chemotaxis, where they fuse (C). PTH also induces bone lining cells to secrete paracrine factors that stimulate osteoclastic activity (D). The recruitment and activity of osteoclasts are promoted by cytokines, including interleukin-1, interleukin-6, and tumor necrosis factor, which could potentially be derived from numerous cell types present in bone and marrow (E). Normally the secretion of these cytokines is kept in check by estrogen, but when the circulating concentration of this hormone falls after menopause, local cytokine production may become elevated, leading to enhanced bone resorption (F). Individual osteoclasts undergo apoptosis and when no further precursors cells are recruited to the remodeling locus the resorptive phase of remodeling ends. Signals released from the resorption cavity, possibly the release of bone morphogenic proteins and transforming growth factor β, promote the recruitment and differentiation of osteoblastic precursor cells (G). Osteoblastic differentiation and subsequent bone formation are also regulated by cytokines and growth factors derived from adjacent cells (H) and by systemic hormones, including PTH (I).

$1,25(OH)_2D_3$, the hormonal form of vitamin D, is required for the calcification of newly synthesized osteoid. Reflecting this, the most obvious skeletal manifestation of vitamin D deficiency is a mineralization defect; however, this hormone is widely believed to play a major role in the regulation of skeletal growth and turnover. *In vitro,* $1,25(OH)_2D_3$ is a potent stimulator of bone resorption, yet when this hormone is administered *in vivo,* enhanced resorption does not always ensue. [*See* Vitamin D.]

The role of estrogens in bone remodeling has been the subject of intensive research. The remodeling imbalance observed in postmenopausal osteoporosis is associated with deficiency of the hormone. Until recently, it was widely thought that bone cells were not directly responsive to estrogens. With the introduction of techniques for culturing cells from human bone, evidence now indicates that human osteoblasts do have receptors for sex steroids. This hormone may promote the proliferation of osteoblast precursors and the matrix-synthesizing capacity of mature osteoblasts. It has also been suggested that estrogens suppress the local secretion of bone-resorbing cytokines from cells in the bone and marrow environment. [*See* Estrogen and Osteoporosis.]

C. Effects of Local Factors

The local regulation of bone turnover is the focus of much current research aimed particularly at the identification of "coupling factors" responsible for coordinating the activity of bone-resorbing and bone-forming cells. The numerous cell types present in the vicinity of the remodeling locus all serve as potential sources of effector molecules. A wide spectrum of factors, including cytokines, growth factors, prostanoids, nitric oxide, and neurotransmitters, including extracellular ATP, have been shown to influence bone resorption and bone formation in experimental situations. In addition to these factors, it has recently been demonstrated that bone cells secrete parathyroid hormone-related protein (PTHrP), a protein that has some homology with PTH. PTHrP shares many biological activities with PTH but also has some unique activities. The cytokines, interleukin-1 (IL-1), tumor necrosis factor, and IL-6 are believed to be important in regulating osteoclast recruitment. These factors are probably involved in the complex regulation of bone turnover in health and in disease. Some of them are undoubtedly responsible for local destruction of bone, which can occur in malignancy and also in rheumatoid arthritis close to the degenerating joint.

Bone matrix is a rich source of growth factors, including transforming growth factor β, insulin-like growth factors, and the bone morphogenic proteins, a family of proteins originally identified by their capacity to induce osteogenesis in nonosseous sites. These factors may be released from the resorbed surface of the matrix and have an informational role in directing the activity of newly recruited osteoblasts. [*See* Bone Regulatory Factors.]

VI. BONE REMODELING AND BONE DISEASE

A. Paget's Disease

The normal sequence of cellular activity in bone remodeling is often maintained even in pathological conditions in which the rate of remodeling is greatly accelerated. Perhaps the best example is in Paget's disease, which is characterized by aggressive resorption of bone. In this condition, osteoclast number is increased and the number of nuclei per cell is elevated. Despite massive increases in resorption, the activity of the osteoblast remains coupled and formation is subsequently accelerated. The excessive turnover of bone can lead to severe malformations in skeletal architecture, and the disease has lytic and sclerotic phases in which resorption or formation predominate. Despite this, the coupling phenomenon ensures that the skeletal mass remains approximately in balance. The primary defect underlying Paget's disease has not yet been identified. Some researchers have suggested that there is localized infection with an unidentified slow virus, but conclusive evidence is still lacking.

B. The Concept of Bone Remodeling Applied to Bone Loss

With increasing age, the amount of bone deposited by osteoblasts in the resorption cavity becomes less than that removed by osteoclasts. Therefore, each BMU is responsible for a small deficit of bone, which, when accumulated over the whole skeleton, leads to a reduction in bone mass. This imbalance in remodeling is particularly severe in females after menopause, leading to bone loss and fracture, a condition described as osteoporosis. If the remodeling rate is increased, for example, in response to raised concentrations of PTH, the number of new BMUs is increased and the deficit becomes even greater. The rate of bone loss is influenced by hormonal status, nutritional factors, and exercise. There is accumulating evidence of the importance of genetic predisposition. However, at the cellular level we still do not know what is the primary cellular defect in osteoporosis. It may be a failure to recruit sufficient numbers of osteoblasts to the remodeling locus or a reduction in the synthetic activity per cell. A reduction in the capacity to synthesize extracellular matrix has been noted in other cell types as a consequence of aging. Until we understand the mechanisms regulating remodeling, bone loss will remain a major problem in aging populations. [*See* bone Density and Fragility, Age-Related Changes.]

ACKNOWLEDGMENTS

We are grateful to Professors Sheila Jones and Alan Boyde for Fig. 5.

BIBLIOGRAPHY

Baron, R. (1996). Anatomy and ultrastructure of bone. *In* "Primer on the Metabolic Bone Disease and Disorders of Mineral Metabolism," 3rd Ed., pp. 3–11. Lippincott–Raven, New York.

Currey, J. C. (1984). "The Mechanical Adaptations of Bone." Princeton Univ. Press, Princeton, New Jersey.

Frost, H. M. (1964). Dynamics of bone remodeling. *In* "Bone Biodynamics" (H. M. Frost, ed.). Little, Brown, Boston.

Parfitt, A. M., Mundy, G. R., Roodman, G. D., Hughes, D. E., and Boyce, B. F. (1996). A new model for the regulation of bone resorption, with particular reference to the effects of bisphosphonates. *J. Bone Min. Res.* **11**, 150–159.

Pead, M., Skerry, T. M., and Lanyon, L. E. (1988). Direct transformation from quiescence to formation in the adult periosteum following a single brief period of bone loading. *J. Bone Min. Res.* **3**, 647–656.

Rodan, G. A., and Martin, T. J. (1981). Role of osteoblasts in hormonal control of bone resorption: An hypothesis. *Calcif. Tissue Int.* **33**, 349–351.

Shipman, P., Walker, A., and Bichell, D. (1985). "The Human Skeleton." Harvard Univ. Press, Cambridge, Massachusetts.

Bone Tumors

HARLAN J. SPJUT
Baylor College of Medicine

GLOSSARY

Chondroblast Cell that produces cartilage

Diaphysis Shaft of a long bone; it has a lesser diameter than the metaphysis or epiphysis

Epiphysis Subarticular end of a long bone formed from an ossification center from cartilage

Fine needle aspiration Process in which a fine-bore needle is inserted into the tumor through the skin; samples of cells are obtained with continuous suction on a syringe attached to the needle

Growth plate Also known as epiphyseal cartilage; it is the zone of active growth of bone that separates the epiphysis from the metaphysis anatomically

Metaphysis Region of the shaft of a long bone adjacent to the growth plate; in the young it is the site of active growth

Osteoblast Uninuclear cell that produces bone; it is closely associated with bone

Osteoclast Multinucleated cell that is capable of resorbing bone

A NEOPLASM OF BONE IS MORE OR LESS uncontrolled growth of an element or elements of bone. For example, osteoblasts of fibrous components may proliferate to become a mass that replaces bone and/or provokes bone formation. Thus, with destruction of bone or neoformation of bone the bone is radiographically visualized. Growth is not uncontrolled necessarily in that there are benign as well as malignant skeletal neoplasms. Those that are benign may take many years to develop to the point at which they would become detectable, whereas those that are malignant may develop rather rapidly and in the course of a year may become detectable clinically and radiographically. A simple subclassification of tumors of bone is into benign and malignant categories. However, a more detailed classification is given that has clinical relevance. This classification is based on the pathological, particularly the histological, features of neoplasms (Table I). It has been observed that these features correlate quite well with expected biological behavior, which is an important component of a classification of diseases. There are approximately 60 different histologically identified neoplasms, both benign and malignant, of the skeleton. A few of these represent diseases that mimic neoplasms of bone and are not true skeletal neoplasms.

Regardless of the number of lesions identifiable, the incidence of neoplasms of the skeleton is low. A true estimate of the number of benign neoplasms is difficult to attain as many are never brought to medical attention, being asymptomatic. However, it is estimated that approximately one new malignant tumor of the skeleton arises for every one million people in the United States per year. This amounts to approximately 2500 to 3000 new malignant tumors per year in the United States. Of the malignant tumors, osteosarcoma is the most common followed by chondrosar-

ENCYCLOPEDIA OF HUMAN BIOLOGY, Second Edition, VOLUME 2. Copyright © 1997 by Academic Press. All rights of reproduction in any form reserved.

TABLE I
Classification of Bone Tumors

- I. Bone-forming tumors
  - A. Benign
    - 1. Osteoma
    - 2. Osteoid osteoma
    - 3. Osteoblastoma (benign osteoblastoma)
  - B. Indeterminate
    - Aggressive osteoblastoma
  - C. Malignant
    - 1. Osteosarcoma (osteogenic sarcoma)
    - 2. Juxtacortical osteosarcoma (parosteal osteosarcoma)
    - 3. Perioseal osteosarcoma
    - 4. Telangiectatic osteosarcoma
    - 5. Well-differentiated intraosseus osteosarcoma
    - 6. Multicentric osteosarcoma
    - 7. Osteosarcoma of the jaws
    - 8. Small cell osteosarcoma
    - 9. High-grade surface osteosarcoma
    - 10. Osteosarcoma arising in Paget's disease of bone and irradiated bone
- II. Cartilage-forming tumors
  - A. Benign
    - 1. Chondroma (enchondroma)
    - 2. Osteochondroma (osteocartilaginous exostosis)
    - 3. Periosteal chondroma
    - 4. Chondroblastoma (benign chondroblastoma)
    - 5. Chondromyxoid fibroma
  - B. Malignant
    - 1. Chondrosarcoma
    - 2. Mesenchymal chondrosarcoma
    - 3. Dedifferentiated chondrosarcoma
    - 4. Clear cell chondrosarcoma
- III. Giant cell tumor (osteoclastoma)
- IV. Marrow tumors
  - A. Ewing's sarcoma
  - B. Malignant lymphoma
  - C. Myeloma
- V. Vascular tumors
  - A. Benign
    - 1. Hemangioma
    - 2. Lymphangioma
    - 3. Glomus tumor (glomangioma)
  - B. Intermediate or indeterminant
    - Hemangiopericytoma
  - C. Malignant
    - Angiosarcoma (hemangioendothelioma)
- VI. Other connective tissue tumors
  - A. Benign
    - 1. Desmoplastic fibroma
    - 2. Lipoma
    - 3. Fibrous histiocytoma
  - B. Malignant
    - 1. Fibrosarcoma
    - 2. Malignant fibrous histiocytoma
    - 3. Lipoma
    - 4. Liposarcoma
    - 5. Leiomyosarcoma
    - 6. Malignant mesenchyma
    - 7. Undifferentiated sarcoma
- VII. Other tumors
  - A. Chordoma
  - B. Parachordoma
  - C. "Adamantinoma" of long bones
  - D. Neurilemoma (schwannoma, neurinoma)
  - E. Neurofibroma
- VIII. Unclassified tumors
- IX. Tumor-like lesions
  - A. Solitary bone cyst (simple or unicameral bone cyst)
  - B. Aneurysmal bone cyst
  - C. Juxtaarticular bone cyst (intraosseous ganglion)
  - D. Metaphyseal fibrous defect (nonossifying fibroma)
  - E. Eosinophilic granuloma (Langerhan's cell granuloma)
  - F. Fibrous dysplasia and ossifying fibroma
  - G. "Myositis ossificans"
  - H. "Brown tumor" of hyperparathyroidism
  - I. Giant cell reaction
  - J. Invasive benign lesions, e.g., pigmented villoglandular synovitis
  - K. Bizarre parosteal osteochondromatous proliferation
- X. Metastatic malignant neoplasms

coma. Of the benign tumors, osteochondroma is the most common.

The diagnosis of bone tumors depends on the clinical presentation, the radiographic and imaging findings, and the pathological examination of a biopsy of the lesion or the resected specimen. The clinical presentation of most neoplasms of the skeleton is one of a mass, visible or palpable in the region of the tumor, associated with some degree of discomfort or pain. A pathological fracture may occur in any of the lesions that are benign or malignant and this may be the presenting feature. Radiographic and imaging findings are particularly important in that they give considerable information in regard to the localization of the tumor, what part of the bone is involved, and the probability of their being benign or malignant (Figs. 1 and 2). In addition, the imaging techniques now available, such as computerized tomography and magnetic resonance imaging, give detailed information pertinent to the extent of the tumor and this in turn is important in determining the biopsy site and/or surgical techniques that might be needed to extirpate the tumor.

There are several forms of biopsies: the incisional biopsy, excisional biopsy, needle biopsy, and fine nee-

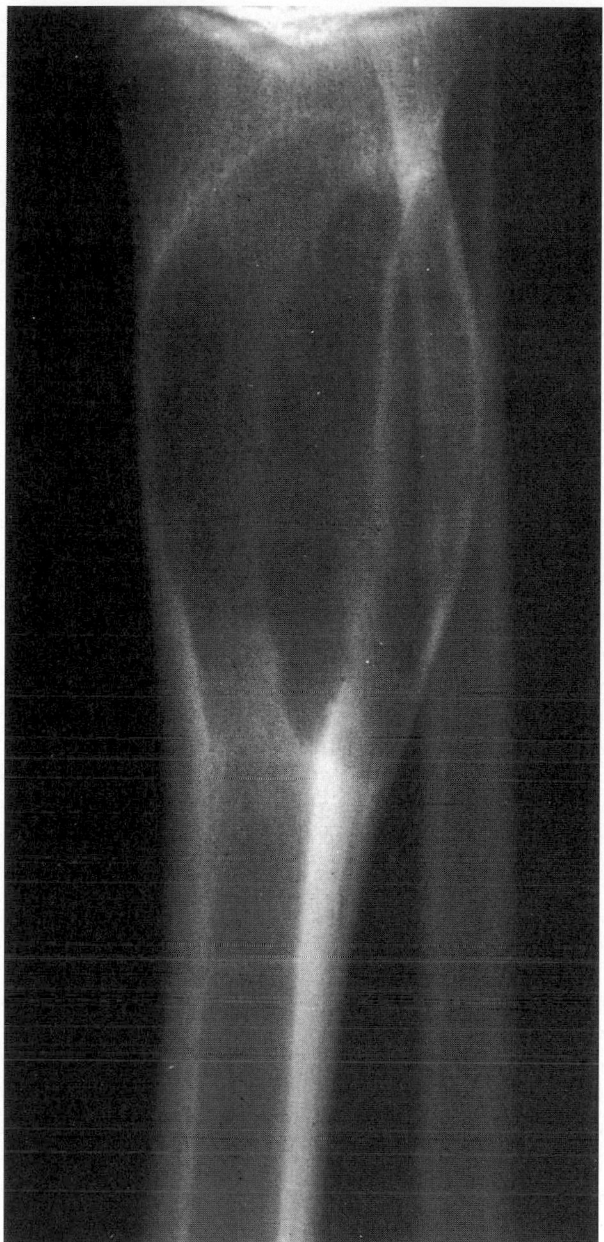

FIGURE 1 An aneurysmal bone cyst in a child. It is well demarcated and expansile. The cortices, though thin, are intact. The lesion is benign.

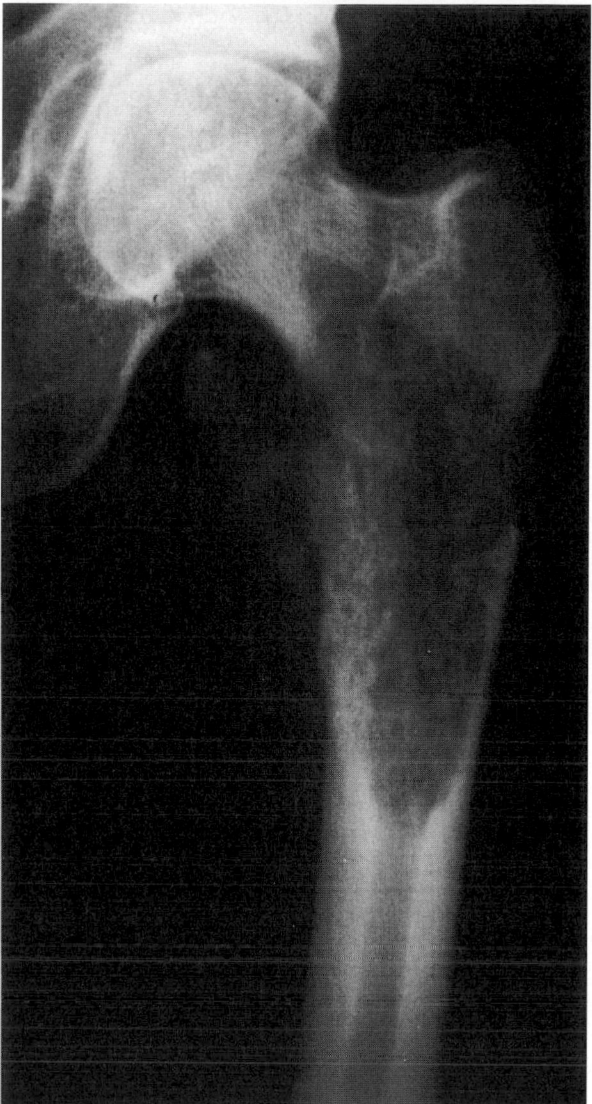

FIGURE 2 A malignant tumor, lymphoma of bone, that destroys cortex and except for its lower margin permeates the femur. This tumor occurred in a 60-year-old man.

dle aspiration. Each of these has proved to be successful and has a role in the diagnosis of tumors of bone. At present, many suspected malignant skeletal neoplasms are subjected to needle biopsy or fine needle aspiration. It has been observed that both are successful in that they obtain tissue that is quite representative of the underlying lesion. The amount of tissue obtained is less than that by an incisional or excisional biopsy; this leads to uneasiness on the part of some surgeons and pathologists. Fine needle aspiration relies upon studying tiny bits of tissue and individual cells. Excisional biopsies are satisfactory in the sense that they are useful in removing small lesions, particularly those considered to be radiographically benign. Thus, the excision generally serves as a means of identifying the neoplasm and also its treatment. Incisional biopsies are those in which a wedge of tissue or fragments of tissue are obtained for pathological examination. In all forms of biopsies, tissue should be set aside for special staining techniques, electron microscopy, or flow cytometry, should these be indicated, particularly in those lesions difficult to classify histologically.

For the majority of neoplasms of the skeleton the etiological factors are unknown. However, there are skeletal neoplasms in which an etiological factor is known, for example, those osteosarcomas or other malignant neoplasms that arise from irradiation and those that arise on a basis of preexistent benign diseases such as Paget's disease, infarcts, or fibrous dysplasia. It has been noted that, experimentally and occasionally in the human tumor, virus-like particles are identified and, in experimental animals, virus particles are capable of inducing osteosarcomas. Thus, virus may well play a role in the histogenesis of this neoplasm. Genetic effects may well be present since some benign tumors such as osteochondromas have a familial or hereditary pattern, particularly those that involve multiple bones. [*See* Neoplasms, Etiology.]

Description of examples of the different types of benign and malignant neoplasms arising in the skeleton follow.

I. OSTEOID OSTEOMA

One of the common benign tumors of osseous origin is osteoid osteoma. It has a tendency to occur in children and adolescents, being more frequent in boys than in girls. It has a widespread skeletal distribution but is most commonly seen in the long bones, particularly those of the lower extremities. It is one of the few skeletal tumors that has a symptom complex that is recognizable in some of the patients. The symptoms include pain during the night that is readily relieved by aspirin. Pain is not severe enough to be of any great note during the waking hours. This lesion has radiographic findings that again are somewhat specific in that it provokes considerable bone production resulting in visibly increased radiodensity and a central, less dense, area known as the nidus. This combination is characteristic of osteoid osteoma with few exceptions. The nidus is less than 2 cm and most frequently less than 1 cm in greatest dimension. Histologically, osteoid osteoma is composed of irregular trabeculae of bone and osteoid associated with numerous osteoblasts and osteoclasts. The intervening stroma is fine and vascular. This lesion is considered to be benign and apparently does not have a malignant counterpart.

II. OSTEOSARCOMA

The most common primary malignant neoplasm arising in bone is osteosarcoma. Its definition is based on histological features that include a malignant undifferentiated stroma associated with bone and osteoid production from osteoblasts that are cytologically malignant. In some, if not most, osteosarcomas, one may find combinations of patterns, some with dominance of fibroblasts, cartilage, or osteoid. These components are malignant. In other words, a mixture of patterns and tissues is a common finding in this tumor. Osteosarcoma occurs more frequently in boys or men than in girls or women. It is a lesion that dominates in childhood and adolescence but may be seen throughout life. Osteosarcoma arises in almost any part of the skeleton but particularly in the lower extremities. Here it is most common at the ends of long bones, for example, the distal end of the femur or the proximal end of the tibia. The symptoms of osteosarcoma are usually pain or swelling in the region of the tumor. In other words, the symptoms are often nonspecific, merely calling attention to an underlying disease. Radiographically, there is evidence of a malignant tumor in that there is destruction of bone, medulla, and cortex, and an irregular periosteal reaction. Frequently, an extraosseous component of the sarcoma is visible radiographically. The growth plate generally serves as a barrier to progression of tumor into the epiphysis or knee joint but at times the osteosarcoma will destroy or penetrate it.

As can be seen from the classification (Table II), osteosarcoma has been subclassified into 10 different categories. This is the result of observations made by many pathologists that with certain histological and radiographic patterns the osteosarcomas behave in a different fashion. For example, parosteal osteosarcoma has a great tendency to occur at the lower end of the femur posteriorly and has certain radiographic features that are recognizable. It grows slowly and, if properly treated, the 5- and 10-year survival rates are excellent. Some of the lesions in the subclassification, on the other hand, are unusually aggressive; for example, patients having small cell osteosarcoma are likely to die of the tumor.

III. OSTEOCHONDROMA

Benign tumors of cartilaginous origin include the most commonly found primary tumor of the skeleton, osteochondroma. It may occur as single or as multiple

TABLE II
Classification of Osteosarcoma[a]

Conventional
Osteoblastic
Chondroblastic
Fibroblastic
Clinical variants
Osteosarcoma of the jaws
Postradiation osteosarcoma
Osteosarcoma in Paget's disease of bone
Multifocal osteosarcoma
Osteosarcoma arising in benign conditions
Morphological variants
Intraosseous well-differentiated osteosarcoma
Osteosarcoma resembling osteoblastoma
Telangiectatic osteosarcoma
Small cell osteosarcoma
Dedifferentiated chondrosarcoma (?)
Malignant fibrous histiocytoma (?)
Surface variants
Parosteal osteosarcoma
Dedifferentiated parosteal osteosarcoma
Periosteal osteosarcoma
High-grade surface osteosarcoma

[a] From S. G. Silverberg (1989). "Principles and Practice of Surgical Pathology," 2nd Ed. Churchill Livingstone, New York.

lesions; the latter have a hereditary background. Osteochondroma commonly occurs at distal ends of long bones and may involve other tubular and flat bones as well. It has characteristic radiographic findings that allow its identification. These include a sweep of the cortex up into the osteochondroma and then the more or less cauliflower configuration of the body of the tumor. It has a lobular pattern and may be lightly to heavily calcified. Since many osteochondromas occur at the distal end of the long bones near the growth plate, it is felt that osteochondromas, at least in this location, represent an error in formation of the growth plate that allows growth of a lesion that comes to have a configuration of an osteochondroma. This same configuration has been induced experimentally by transposing fragments of growth plate near the metaphysis of the long bone. It is felt by many observers that osteochondromas are capable of transforming into chondrosarcomas. This must be an uncommon if not rare event. Some chondrosarcomas do have a configuration that would suggest an origin from an osteochondroma. Patients having multiple osteochondromas are at greater risk for developing a chondrosarcoma than is the population having solitary osteochondromas.

IV. CHONDROMA

Benign cartilaginous tumors that arise in and are confined to the medullary cavity of bone, particularly in the metaphysis or diaphysis, usually are known as chondromas or enchondromas. The hyalin cartilage that comprises the tumor is often more cellular than normal hyalin cartilage. Chondromas are common in the small bones of the hands and feet and present as radiolucent defects in a phalanx with some scalloping of the inner surface of the cortex. Evidence of calcification may be present. The lesions usually are not discovered unless a pathological fracture has occurred or if the lesion slowly expands to cause swelling of the phalanx. There are two syndromes, Maffucci's and Ollier's, associated with multiple chondromas. The first is associated with angiomatous lesions of the soft and visceral tissues and the latter is simply multiple enchondromas. The multiple skeletal lesions often cause severe deformities of the affected parts. Patients with both of these syndromes are at risk for developing chondrosarcomas.

V. CHONDROBLASTOMA

An uncommon but important benign cartilaginous tumor is chondroblastoma. Some of its importance is related to its rich cellularity that may lead to a mistaken diagnosis of a malignant tumor by an unwary pathologist. This lesion has a strong propensity to occur in the epiphysis or at least have a major epiphyseal component with extension in the metaphysis. Chondroblastomas occur in young persons, usually teenagers or younger, who often complain of pain in the adjacent joint. Radiographically chondroblastoma is a benign-appearing lesion that involves the epiphysis or the epiphysis and metaphysis. At times, calcification is visible and aids in the radiographic diagnosis. Histologically, this lesion is richly cellular, composed of uniformly rounded cells with fairly distinct cytoplasmic boundaries. The cells have been demonstrated to be chondroblasts, hence the name of the lesion. There may be associated hyalin cartilage formation; multinucleated giant cells are a constant feature. Calcification seen histologically has a pattern that is quite specific in that the calcium is laid down to surround individual cells. The major method of treatment for chondroblastoma is curettage and packing of the defect with bone chips. Consequently, there is a risk of local recurrence. There are only rare examples of lesions designated as malignant chondroblastomas as metastases have been demonstrated. But for all practical purposes a chondroblastoma is a benign cartilaginous tumor.

VI. CHONDROSARCOMA

Chondrosarcomas are malignant tumors of cartilaginous origin. They are generally classified into two categories: a peripheral form apparently arising from the surface of a bone and a central chondrosarcoma that arises in the medullary cavity. Whether there has been a preexisting benign lesion in either one is often difficult to prove, but there is radiographic and sometimes histological evidence suggesting that there may be a relationship between benign cartilaginous tumors and chondrosarcomas. Chondrosarcomas have a tendency to occur in the flat bones, particularly the pelvis and the upper end of the femur. However, there is a wide skeletal distribution of chondrosarcomas. They have a tendency to occur later in life compared to osteosarcoma, being more common beyond the fourth decade of life. In contrast to osteosarcoma, it is rare for chrondrosarcoma to occur in the pediatric age group. A patient with a peripheral chondrosarcoma usually presents to the physician because of a slowly growing mass, say, in the region of the pelvis. A patient who has a central chondrosarcoma commonly found in the upper end of the femur and upper end of the humerous presents because of pain in the region. A number of central chondrosarcomas are found on chest films that include the upper end of the humerus. The histological diagnosis of chondrosarcoma is dependent on nuclear alterations. This includes binucleate chondrocytes and the enlargement of the nuclei associated with nuclear pleomorphism. As the grade of the lesion becomes higher, mitotic figures become visible. A lesion in which no mitoses are found and the nuclear cellularity is not great is a grade I chondrosarcoma; one in which there is hypercellularity associated with considerable nuclear pleomorphism and easily found mitoses would be a grade III chondrosarcoma. The grade is closely related to the expected prognosis in that patients with grade I chondrosarcoma have an excellent prognosis and those with grade III chondrosarcomas have a poor prognosis. Fortunately, the majority of chondrosarcomas are low-grade tumors.

Two subtypes of chondrosarcoma deserve mentioning: mesenchymal chondrosarcoma and dedifferentiated chondrosarcoma. The former is of interest in that it is a combination of histologically malignant cartilage with an undifferentiated malignant component that at times may have patterns resembling fibrosarcoma or even hemangiopericytoma. This lesion is slowly progressive and over a period of years results in the death of the patient. Another point of interest is that about one-third of these lesions occur extraskeletally. Dedifferentiated chondrosarcoma is a recognizable, usually central chondrosarcoma in which there is histologically an undifferentiated mesenchymal component often resembling a malignant fibrous histiocytoma. The outcome for these patients is particularly bleak.

VII. EWING'S SARCOMA

Ewing's sarcoma, often referred to as a small round cell tumor, is of disputable histogenesis. It was considered at one time to be endothelial in origin and then of primitive mesenchymal origin. Most recently, electron microscopic and immunohistochemical evidence lend support to the observation that Ewing's sarcoma may be of neuroectodermal origin. This malignant tumor occurs most frequently in the first and second decades of life and at least 80% of the patients are younger than 30 years of age. It has a wide distribution in the skeleton with a dominance in the flat bones and the long bones. Patients with this tumor often present because of pain and swelling in the region of the tumor. A few patients may have fever associated with tenderness and redness in the area of the tumor. Radiographically, Ewing's sarcoma is infiltrative, causing ill-defined destruction of bone. Periosteal reaction is prominent and one expects to find an extraosseous component. Grossly the tumor appears to arise in the medullary cavity, permeate this cavity, and eventually penetrate the cortex. This correlates well with the radiographic findings. Pathologically, one notes a richly cellular tumor that has the appearance of lymphoid tissue. Widespread necrosis or hemorrhagic destruction may be present. Histologically, the tumor is composed of small round cells of a uniform appearance. The nuclei are rather bland and nucleoli, when present, are tiny. There is not much chromatin alteration. Mitotic figures may be sparse. Necrosis and hemorrhage may be present. Occasionally, one might find foci of bone and cartilage. Ewing's sarcoma histologically resembles a malignant lymphoma, leukemia, metastatic neuroblastoma, and other undifferentiated malignant tumors of the small cell type. Thus, it is important that if a specific diagnosis cannot be made that electron microscopic and immunohistochemical studies be done on the biopsy specimen. A stain that is helpful is the periodic acid Schiff (PAS) stain, which is positive in about three-quarters of Ewing's sarcoma; it demonstrates glycogen in the cytoplasm of the tumor cells. This tumor has a propensity to metastasize to the lungs and to other bones. In about 20% of patients, nodal metastases may be demonstrated.

At one time the expected survival of a patient with Ewing's sarcoma varied from 0 to 10% in a 5-year period; however, with the use of various chemotherapeutic protocols, associated at times with radiation therapy and with limb-sparing surgical procedures, the survival rate now is at least 50% at 5 years.

VIII. LYMPHOMA OF BONE

A tumor that is important in the differential diagnosis of Ewing's sarcoma is primary lymphoma of bone. This tumor, however, occurs in an older age group, usually beyond the age of 20, and is rare in the pediatric age group. It too may present radiographically as a permeative, destructive, ill-defined lesion of bone. It is more likely to be lytic than sclerotic but may be a combination of the two. Histologically, these tumors are identical with those that occur in the hematopoietic system. The subclassification of malignant lymphomas of the skeleton is similar to that used for lymph nodes. The differential diagnosis can often be made on the basis of a histological examination in that the nuclei tend to be larger and nucleoli are frequently prominent in contrast to the infrequent enlargement of nucleoli in Ewing's sarcoma. Lymphomas are rarely PAS positive, another histological feature that serves to differentiate the two lesions. With a diagnosis of skeletal lymphoma, it becomes important for the clinicians to exclude the possibility that there is secondary lymphomatous involvement of the skeleton; thus a complete immunological workup is in order for these patients. For those who are found to have primary lymphomas of the bone, the survival rate is quite good, that is, one can expect a 40% 5-year survival. [*See* Lymphoma.]

IX. PRIMARY FIBROSARCOMA

Primary fibrosarcoma of bone is a malignant tumor composed of spindle cells that do not produce osteoid, bone, or cartilage. Since the appearance of malignant fibrous histiocytoma of bone, this lesion has become rare. Some authors believe it is the low-grade part of the malignant fibrous histocytoma spectrum. These spindle cells have been demonstrated to be fibroblasts. This tumor occurs equally in men and women and is more commonly found beyond the third decade of life. It occurs in almost all bones of the skeleton but has a tendency to arise in the long bones, usually at the distal ends. The presenting signs and symptoms are not specific but swelling and pain are common. Radiographically, the lesion is permeative with destruction of the medullary portion of bone and the cortex with extension into the soft tissues. Grossly, the tumor is a firm, gray white, fibrous appearing mass that is generally quite well circumscribed. There may be areas of necrosis and hemorrhagic destruction. Histologically, the tumor resembles that of fibrosarcoma of the soft tissues. Many sections must be studied to rule out the possibility of a fibroblastic type of osteosarcoma. Tumor osteoid bone and cartilage should not be seen in a fibrosarcoma. The prognosis for patients treated for fibrosarcoma of bone is somewhat dependent on the location of the lesion. For those that are dominantly in the medullary cavity the survivial rate is approximately 25–30% at 5 years, whereas those that are periosteal or surface in location have a better prognosis, that is, approaching 50% 5-year survival.

X. MALIGNANT FIBROUS HISTIOCYTOMA

Malignant fibrous histiocytoma, a tumor composed of fibroblastic and histiocytic components, has been somewhat controversial in that it is not certain that it differs from fibrosarcoma. However, it does have a different histological pattern in that it generally has considerable nuclear pleomorphism and a storiform pattern to the stroma that tends to set it off from a conventional form of fibrosarcoma. In addition, one may find foci of osteoid in malignant fibrous histiocytoma whereas this would not be acceptable for fibrosarcoma. The lesion has been fairly recently described and thus large numbers of patients are not available for review. In general, the radiographic findings are those of malignant tumor but not necessarily specific for a malignant fibrous histiocytoma. All ages may be involved, but there is a tendency for the majority of the lesions to occur beyond the third decade of life. It has widespread skeletal involvement. The lesion grossly may have findings that are similar to those of fibrosarcoma with necrotic areas. There may be a yellowish brown to tan discoloration of the tumor.

Histologically, the findings are those that have been mentioned previously. The survival rates after treatment are similar to those of fibrosarcoma.

There are several rare malignant tumors of the skeleton that need to be mentioned. One is adamantinoma. This tumor occurs dominantly in the tibia and results in multiple lytic and sclerotic areas when seen radiographically. The name derives from the fact that it has a resemblance to adamantinoma of the jaws. There has been considerable controversy as to the histogenesis of this lesion, varying from angiosarcomatous to squamous epithelium to sweat duct origin. This lesion is slowly progressive and somewhat indolent but is capable of metastasis. Primary liposarcomas arise in the skeleton and bear a close resemblance to those seen in the soft tissues. Leiomyosarcomas have also been described and have a resemblance histologically to those that occur in the soft tissues. Angiosarcomas, which are malignant tumors of endothelial origin, may appear as solitary lesions or as multiple lesions of the skeleton. Their histological pattern is similar to that of angiosarcomas in the skin and soft tissues. The survival rate is dependent on the differentiation of the neoplasm. For example, with a well-differentiated angiosarcoma, the 5-year survival rate is near 90–95%. For those who have poorly differentiaged (grade III) tumors, the survival rate at 5 years is 20%.

XI. GIANT CELL TUMOR

Giant cell tumor generally behaves as a benign lesion. It is one of the few neoplasms of the skeleton that is dominant in girls and women. It has a wide skeletal distribution but arises frequently in the lower end of the femur and upper end of the tibia. Histologically it consists of multinucleated giant cells that are believed to be of osteoclastic or stromal origin. A stromal component is also present. The majority of patients with giant cell tumors are skeletally mature, thus the lesion is distinctly rare in the first decade and in the early part of the second decade. The age dominance is in the late second, third, and fourth decades. The lesion often presents with a swelling of the affected part, sometimes associated with pain. Pathological fractures may occur. Radiographically, the lesion has an epiphyseal and metaphyseal component. Without the two components, it is difficult to diagnose giant cell tumor radiographically. It is usually a fairly well delineated lesion with attenuation of the cortex and at times breakout through the cortex into the soft tissues. Usually little periosteal reaction is present. It is dominantly a radiolucent lesion but at times may have dense areas within it. Grossly, the tumor is gray red to yellowish brown and friable. It may be partly fibrous. A few giant cell tumors may be dominantly cystic. Histologically, one finds myriads of multinucleated giant cells associated with a stroma that tends to be rounded but may be quite spindled. Hemosiderin (brown yellow pigment granules) and areas of degeneration are common. Osteoid formation, often the result of small fractures, or major fractures may be present.

Giant cell tumors may metastasize even though histologically they are benign. Most of the metastases occur after the tumor has been treated, usually by curettage, thus it is felt that these metastases are the result of the procedure. Nevertheless, it is recognized that the so-called benign giant cell tumors are capable of producing metastases, but this is an uncommon phenomenon. It shows the same location and radiographic appearance of a conventional giant cell tumor but differs in that it contains a high-grade sarcoma. The major method of treatment of giant cell tumors has changed from time to time and at present thorough curettage with cement packing of the cavity seems to be the favored method of treatment. Curettage obviously has a risk of local recurrence. However, curettage does afford the preservation of joints and this is important since giant cell tumors are commonly located near major joints.

XII. TUMOR-LIKE LESIONS

An important category of tumors of bone are those designated as tumor-like lesions. These are important in that radiographically, clinically, and pathologically they often mimic primary skeletal neoplasms, benign and malignant. Common among these are the reactive lesions such as localized myositis ossificans and periostitis. These lesions are often but not always commonly associated with trauma. Radiographically, at times they resemble an osteosarcoma or a surface tumor such as periosteal osteosarcoma or parosteal sarcoma. Histologically, because of the abundance of osteoid and bone formation, they may be mistaken for osteosarcoma unless the pathologist is aware of the features of reactive lesions. Aneurysmal bone cysts are benign lesions that cause defects, usually in the distal ends of bones. However, they may be large and quite destructive radiographically. Histologically, they are composed of large blood-filled spaces with the trabeculae lined by osteoid and a fibrous stroma and multinucleated giant cells. Their main importance histologically lies in differentiating them from osteosarcoma, particularly telangiectatic osteosarcoma. This is done by carefully studying the stroma, the osteoid, and bone to determine whether there are malignant features or not. Solitary cysts of the skeleton are most commonly seen in children in the upper end of the humerus and the upper end of the femur. The cyst is considered to be an error in bone formation resulting in accumulation of fluid. These lesions, in the young person, occur near the epiphyseal plate; as the patient matures skeletally the lesions appear to migrate away from the growth plate. At present, they are commonly treated by the injection of steroids, which has been noted to be successful in many. If not successful then the cyst is curetted and packed with bone chips.

There are a number of skeletal lesions that contain multinucleated giant cells that can be mistaken for giant cell tumor. Among these are the so-called giant cell reparative granuloma, which is commonly found in the jaw bones but may be seen in the small bones of the hands and feet. Another important lesion is the brown tumor associated with hyperparathyroidism, which is so similar to a giant cell tumor that without clinical history and radiographic studies one may not be able to differentiate the two lesions histologically. Eosinophilic granuloma, also known as Langerhan's cell granulomatosis, is a lesion composed of histiocytic cells, which are the Langerhan's cells, and inflammatory cells, usually lymphocytes and a few plasma cells and eosinophils; the latter may be abundant. There may be associated necrosis. Eosinophilic granulomas occur in children and in adolescence dominantly and present radiographically as a destructive lesion of bone with periosteal reaction, closely mimicking a Ewing's sarcoma, for example. The diagnosis of these lesions can be made with needle biopsy or curettage. They may be treated with injection of steroids.

Metastases from carcinomas and sarcomas and the involvement of the skeleton by a myeloma are the commonest forms of malignant tumor to involve the skeleton. However, these are considered to be secondary. They may cause destructive areas, pathological fractures, and pain, and at times resemble primary tumors radiographically. Often metastases and myeloma can be readily diagnosed radiographically. For definitive diagnosis or for confirmation, these lesions lend themselves to fine needle aspiration or needle biopsy. Soft tissue tumors that invade bone are also capable of destroying a portion of bone or an entire bone. The soft tissue tumors need not be malignant, as benign lesions also impress themselves upon the bone.

BIBLIOGRAPHY

Campanacci, M., Baldini, N., Boriani, S., and Sudanese, A. (1987). Giant-cell tumor of bone. *J. Bone Jt. Surg.* **69**(A), 106.

Dahlin, D. C., and Unni, K. K. (1986). "Bone Tumors," 4th Ed. Charles C. Thomas, Springfield, Illinois.

Fechner, R. E., and Mills, S. E. (1993). "Tumors of the Bones and Joints. Atlas of Tumor Pathology," 3rd Series, Fascicle 8. Armed Forces Institute of Pathology, Washington, D.C.

Huvos, A. G., Heilweil, M., and Bretsky, S. S. (1985). The pathology of malignant fibrous histiocytoma of bone. *Am. J. Surg. Pathol.* **9**, 853.

Ostrowski, M. L., Unni, K. K., Banks, P. M., Shives, T. C., Evans, R. G., O'Connell, M. J., and Taylor, W. F. (1986). Malignant lymphoma of bone. *Cancer* **58**, 2646.

Raymond, A. K., Simms, W., and Ayala, A. G. (1995). Osteosarcoma. Specimen management following primary chemotherapy. *Hematol./Oncol. Clinics of North America* **9**(4), 841–867.

Taconis, W. K., and Mudler, J. D. (1984). Fibrosarcoma and malignant fibrous histiocytoma of long bones: Radiographic features and grading. *Skel. Radiol.* **11**, 237.

Ushigome, S., Takakuwa, T., Shinagawa, T., Takagi, M., Kishimoto, H., and Mori, N. (1984). Ultrastructures of cartilaginous tumors and S-1000 protein in the tumors. *Acta Pathol. Jpn.* **34**, 1285.

Ushigome, S., Shimoda, T., Takaki, K., Nikaido, T., Takakuwa, T., Ishikawa, E., and Spjut, H. J. (1989). Immunocytochemical and ultrastructural studies of the histogenesis of Ewing's sarcoma and putatively relative tumors. *Cancer* **64**, 52.

Variend, S. (1985). Small cell tumours in childhood: A review. *J. Pathol.* **145**, 1.

Bony Pelvis of Archaic *Homo sapiens*

LORI D. HAGER
University of California, Berkeley

GLOSSARY

Greater sciatic notch Large notch found on the lateral aspect of the os coxae that is composed of ischial and iliac elements. In life, the greater and lesser sciatic notches are divided by ligaments that form foramina through which muscles, nerves, and vessels pass out of the pelvis

Hominid Any human or immediate human ancestor of the family Hominidae; there are two extinct genera, *Ardipithecus* and *Australopithecus*, and one extant genus, *Homo*

Neandertal Upper Pleistocene form of *Homo* that differs morphologically from anatomically modern humans, *Homo sapiens sapiens*

Os coxae Large, irregularly shaped bones forming the two lateral halves of the pelvis. Each os coxa is composed of three separate bones that fuse together in the walls of the acetabulum (the hip socket) at puberty: the ilium, ischium, and pubis

Sexual dimorphism Presence of distinct morphologies between members of the same species due to secondary sexual characteristics

THE PELVIS IS COMPOSED OF THE TWO HIPBONES (os coxae), the sacrum, and the coccyx, which are held together during life by cartilaginous and ligamentous attachments. The pelvic girdle supports and transmits the weight of the body during locomotion and is intrinsically involved with reproduction in females (i.e., labor and delivery of the infant). Because one of the hallmarks of being human is the ability to habitually walk upright, the pelvis is exceedingly important in human evolutionary studies. In addition, the human female pelvis must meet both locomotory and reproductive requirements. Because the human neonate is large-headed with broad shoulders, the human female pelvis requires a birth canal of adequate capacity for its successful birth. [*See* Skeleton.]

The fossils referred to as archaic *Homo sapiens* fall between the earlier *Homo erectus* forms and the later modern *Homo sapiens sapiens*. The pelvis of archaic *Homo sapiens* more closely resembles the modern human condition than do the earlier hominids, even though the differences between males and females are not as fully developed as in their modern counterparts.

I. THE PELVIS IN HUMAN EVOLUTIONARY STUDIES

A. Bipedalism

The study of the pelvis in human evolution has focused primarily on bipedalism. It was once thought that the enlargement of the brain came first in the evolutionary history of humans, followed by, or even precipitating, upright walking. That bipedalism actually preceded the enlargement of the brain in the evolution of humans was first made evident in 1947 with the discovery of a nearly complete pelvis from a site in South Africa called Sterkfontein. Although the exact nature of its locomotory capabilities has been disputed, the fossil pelvis was unequivocally from a bipedal individual, whereas all associated cranial material from the site indicated a small brain size. The antiquity of bipedalism has been reaffirmed by find-

ENCYCLOPEDIA OF HUMAN BIOLOGY, Second Edition, VOLUME 2. Copyright © 1997 by Academic Press. All rights of reproduction in any form reserved.

ings of other pelvic and lower limb elements at many other sites in East and South Africa, and has been extended back in time to over 4 million years ago (myr) with the important finds from the Middle Awash in Ethiopia and Kanapoi in Kenya. Studies of fossil pelvic and lower limb remains have confirmed the notion that the adaptation to bipedalism is one of the hallmarks of humankind.

B. Reproduction

Because the ability to walk upright is such a critical adaptation in the evolution of hominids, the intensity of study regarding the pelvis in terms of locomotion has overshadowed most substantive work on the reproductive function of the hominid female pelvis. However, recently there has been an increased interest in the hominid pelvis as it concerns reproduction rather than locomotion. The birth of the modern human neonate is often a difficult and lengthy affair due to the tight constraints placed on the female pelvis for the successful birth of the relatively large infant. Several studies have looked at the fossil pelvic remains to determine if these same obstetrical constraints existed for the early hominids. Some believe the birth of hominid infants 2–4 myr would have been quick and easy, as it is in our closest living relatives, the great apes. Others suggest that obstetrical trauma began with the earliest hominids or that, at the very least, the mechanism of birth was different than what is evident for modern human females and their infants.

Although these studies have provided intriguing and much needed models on "paleo-obstetrics" and the hominid pelvis as a reproductive unit, the models are founded on the basic premise that the fossils are sexed correctly as male and female. Sex determinations of hominid fossils have been based on two main sets of criteria. First, when pelvic elements are present, the sex of the fossils has been determined based on the well-documented sex differences in the modern human pelvis. Even though the relevance of using modern sexing criteria on the fossil material has been called into question, if a pelvis is present, modern standards of sex differences are invariably used to determine the sex of the fossil. Second, differences in overall size, whether it be teeth or long bones, have played a significant role in diagnosing the sex of the fossil material since among dimorphic nonhuman primates males tend to be larger and more robust than females. The nature and extent of sexual dimorphisms, such as body size, in the hominid fossil record are currently disputed. Some believe that our earliest ancestors were highly sexually dimorphic in body size, with the males being twice as large as females, as in modern gorillas. Others believe that the earliest hominids were only moderately dimorphic, more like what we find in ourselves and in our closest living relatives, the chimpanzees. Variation between specimens would therefore represent interspecific variation rather than intraspecific variation as in the single-species, highly dimorphic model. Determination of sex is more difficult when moderate dimorphism in body size exists than it is in highly dimorphic species.

It is well known that pelvic dimorphisms exist in modern human populations. The majority of differences are in the lower portion of the pelvis, the "true" pelvis or birth canal, because in females it serves as the bony ring through which the fetus must pass during parturition. The upper portion of the pelvis is the "false" pelvis whose main function is to support the abdominal viscera and the growing fetus in the pregnant female. In general, the female pelvis is characterized by a complex of traits that renders a more spacious pelvic capacity to the true pelvis. Large neonatal heads and shoulders need a relatively large maternal pelvis. There are, of course, biomechanical constraints as to how large the female pelvis can be and still be efficient at bipedal locomotion. The female pelvis exhibits morphologies mutually beneficial for both locomotory *and* reproductive functions. Thus in modern humans, the true pelvis of females is larger than in males.

These same sex differences do not exist in the earliest hominids nor in the great apes. In the australopithecines, for example, all the pelves look more like those of modern human females than modern human males. Thus, the basal hominid pelvic morphology was more female-like than male-like. Sex is therefore difficult to determine in these hominids when examining the pelvis.

Among the great apes, it is difficult to sex chimpanzee pelves as male or female because there is a great deal of overlap in the pelvic traits without much difference in size. Gorilla and orang pelves, on the other hand, are relatively easy to sort by size, but not necessarily by shape, because the male is larger than the female. Whether the fossil hominids were more like our closest living relatives, the chimpanzees, more like the size-dimorphic gorillas and orangs, or more like their modern human counterparts is an intriguing question.

An increased cranial capacity in hominids is suggested by the fossil record to have occurred with the emergence of the genus *Homo* approximately 2 myr.

If this increase in adult cranial capacity is correlated to an increase in neonatal head and/or body size, one would expect commensurate selective pressures on the maternal pelvis to accommodate the relatively larger neonate. This could be accomplished by an increase in female body size since the shape of the basic hominid pelvis is female-like already. However, the male pelvis would not be under any selective pressures related to reproduction and could thus vary more substantially than could females with regard to their pelvic morphology. Differences between males and females might therefore first occur in the early *Homo* material and then continue to develop until the level evident in modern humans was reached. [*See* Comparative Anatomy; Evolving Hominid Strategies.]

II. WHO WERE THE "ARCHAIC *HOMO SAPIENS*"?

Evolving in Africa from an earlier *Homo* stock, the first hominids to migrate outside of Africa belong to the taxon *Homo erectus*. Until approximately 500 kyr (thousand years ago), *Homo erectus* lived throughout much of the Old World (with the possible exception of Europe). We refer generally to the fossils that mark the end of *Homo erectus* and the beginning of *Homo sapiens* as "archaic *Homo sapiens*." Morphologically, the archaic *Homo sapiens* specimens do not fully resemble those of the earlier *Homo erectus,* and yet they do not fully resemble later anatomically modern *Homo sapiens sapiens* either. The archaic *Homo sapiens* can be divided into early non-Neandertal forms, chronologically ranging from approximately 350–400 kyr to 100 kyr, and later forms, the Neandertals, which lived from approximately 200 to 30 kyr.

The ability to establish a chronological sequence for the sites yielding fossils between 400 and 50 kyr has proven to be particularly difficult since this time period is outside the range of conventional absolute dating techniques. For example, absolute dating by ^{14}C cannot be extended back in time far enough to apply to these sites, whereas techniques for older sites, such as potassium/argon or argon/argon (K/Ar or Ar/Ar), cannot be applied to sites in many parts of the Old World. Chronologies based on relative dating methods remain uncertain and imprecise. Material from archaic *Homo sapiens* sites are constantly being subjected to new techniques with the hopes of arriving at a more certain dating scheme in which to place these important fossils. Recent attempts include thermoluminescence dating of burnt flints and electron spin resonance determinations on bovid teeth from hominid-bearing sites in southwestern Asia.

The evolutionary history of humans is not a fully understood history, and at no point is this more apparent than in the time frame concerning the origin of modern humans. Whether or not the fossils representing archaic *Homo sapiens* actually gave rise to modern *Homo sapiens sapiens* is a matter of considerable debate. Although the role of archaic *Homo sapiens,* especially the Neandertals, in the evolution of modern humans has always been a controversial issue, current arguments focus on two main opposing viewpoints. First, the regional continuity hypothesis states that modern *Homo sapiens* arose from local populations of archaic *Homo sapiens* already established throughout the Old World. Proponents for this "regional continuity theory" cite the similarities in morphological features, particularly cranial ones, in specimens of archaic *Homo sapiens* and the later modern Upper Pleistocene fossils in each geographical region. For example, it has been suggested that the anatomically modern *Homo sapiens* fossils from Mladeč, Czechoslovakia, which date to approximately 35 kyr, share morphological similarities to the earlier Neandertals of the same region, and that Asian archaic *Homo sapiens* fossils such as those found at Dali in China share similarities with modern Asian populations.

By contrast, the "out of Africa" hypothesis proposes a single source of modern *Homo sapiens* in Africa that then radiated out of its homeland, replacing other populations of hominids in the different regions throughout the Old World. In addition to the fossil evidence, current proponents of this theory have the support of genetic data on modern humans that trace the inheritance of mitochondrial DNA (mtDNA). Analysis of the degree of variation in the mtDNA in various modern human populations suggests that all modern humans can point to a single African source for their mtDNA. The amount of variation is consistent with an origin of approximately 150–250 kyr, which suggests that one African line of archaic *Homo sapiens* evolved into modern humans, then migrated to other areas of the world, replacing the existing populations. Whether there was interbreeding with the local populations or not remains uncertain, although the extreme view states that there was complete replacement of the archaic *Homo sapiens* in these other regions by the modern forms. These genetic data are currently a hotly disputed topic in human evolution.

The paleontological evidence for the "out of Africa" theory centers on Africa and southwest Asia. In these areas, there is evidence for a more ancient origin of modern humans than was previously thought. For example, at Klasies River Mouth in South Africa, remains of modern humans have been uncovered that date to approximately 100 kyr, not only much older than previously thought but also older than some of the archaic *Homo sapiens*. In addition, modern humans from sites in Israel, such as Qafzeh and Skhul, predate Neandertals by several thousand years at nearby sites such as Kebara and Amud. Thus, populations of modern humans predate and then coexisted with Neandertal populations for thousands of years rather than evolved from them and, according to some researchers, ultimately replaced them. [*See* Human Evolution.]

III. FOSSIL PELVIC MATERIAL OF ARCHAIC *HOMO SAPIENS*

A. Background

The fossil remains of the early archaic *Homo sapiens* are relatively scarce but the number of available specimens increases by Neandertal times. The pelvis as a whole does not lend itself to good preservation due to the fact that it actually consists of several bones (two hipbones, sacrum, and coccyx) held together during life by cartilaginous and ligamentous attachments. The irregular shape of each os coxae and the thin pubic and ischial rami in the anterior aspect contribute to the fragmentary nature of the fossil pelvic remains discovered thus far. The iliac portion of the hipbone is more often preserved than are the ischial or pubic portions. Nonetheless, several pelvic specimens of archaic *Homo sapiens* do exist.

B. Early Archaic *Homo sapiens*

The earliest hominid pelvic remains come from the continent of Africa. Primarily found in South and East Africa, these specimens include members of the species *Australopithecus* and early *Homo*. Two of these early *Homo* fossils, KNM-ER-3228 from East Lake Turkana and OH.28 from Olduvai Gorge, are considered to be morphologically similar to a later specimen from Europe, Arago 44 (Fig. 1). Only Arago 44 has been considered a possible representative of archaic *Homo sapiens*. These three hipbones are separated temporally with KNM-ER-3228 dated to 1.9 mya, OH.28 to 500 kyr, and Arago 44 less precisely dated to 250–400 kyr.

KNM-ER-3228 is a well-preserved right hipbone consisting of much of the ilium and ischium but missing the pubis. It is robust with a greater sciatic notch that is extremely narrow, suggesting that this specimen was male. OH.28 is a left hipbone also preserving much of the ilium and ischium and lacking the pubis. Also robust, OH.28 has been sexed as female, but the presence of both male and female morphologies has made sex determination less certain for this fossil. KNM-ER-3228 is taxonomically defined as early *Homo,* perhaps *H. habilis* or *H. erectus,* whereas OH.28 has consistently been regarded as a *Homo erectus* specimen.

Arago 44 is a left hipbone that was badly crushed and then reconstructed from the many available fragments. It consists primarily of the ilium and the supe-

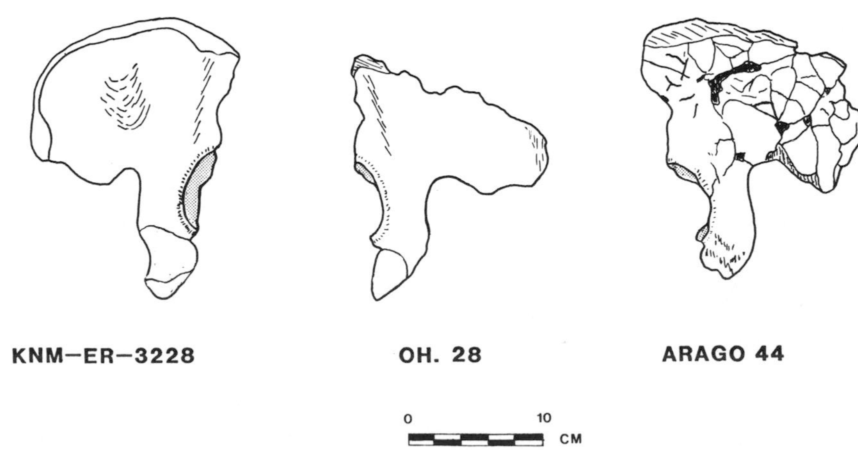

FIGURE 1 Lateral views of KNM-ER-3228, OH.28, and Arago 44 os coxae.

rior ischium but is missing the inferior ischium and the entire pubis. The wide-open sciatic notch has consistently sexed this specimen as female.

Two specimens from the Broken Hill mines in Kabwe, Zambia, may also fall into the taxonomic category of archaic *Homo sapiens*. These specimens, E719 and E720, are unprovenienced remains; the dating provisionally places these specimens at 130–250 kyr. Broken Hill E719 is a right os coxa that preserves part of the ilium and ischium but is missing the pubis. Broken Hill 720 is a partial left ilium that is missing the entire posterior iliac portion, the ischium, and the pubis.

The La Grotte du Prince right hipbone fragment is also considered to be a member of early archaic *Homo sapiens*. This specimen from the Grimaldi site in Italy is dated approximately to the Riss glaciation (125–250 kyr).

C. Late Archaic *Homo sapiens* (Neandertals)

The later forms of archaic *Homo sapiens* are generally referred to as Neandertals. These are more numerous and better preserved than the earlier archaic *Homo sapiens*. The remains of Neandertals found in Europe and Southwest Asia have figured prominently in the discussion on the origins of modern humans. The Neandertals share a suite of cranial and postcranial characters that hold them together as a group. The pelvic remains of the Neandertals are distinctive in that the superior pubic ramus is mediolaterally elongated in all Neandertal specimens found thus far.

Southwestern Asia sites such as Kebara, Tabun, Amud, and Shanidar have yielded important Neandertal specimens with pelvic remains present. The Kebara 2 skeleton comes from a cave site in Israel. The dating of burnt flints from the fossil layers at the site has recently yielded a date of 48–60 kyr. The skeleton was complete except for the cranium, with the pelvis nearly intact. The right os coxa and sacrum are in relatively good condition, whereas the left os coxa has been flattened due to postdepositional processes. Through mirror-imaging of the right hipbone and sacrum, a reconstruction of the complete pelvis has been accomplished. From this complete pelvis, it has been suggested that the long pubis bone in Neandertals resulted in an externally rotated pelvis, possibly related to locomotory and postural adaptations.

Mugharet-et-Tabun is a cave site in Israel that was excavated from 1929 to 1934. New dates for Tabun suggest that the site is of greater antiquity than previously thought, perhaps as old as 100 kyr. The Tabun C1 specimen is associated with cranial and postcranial elements. The hipbones were both badly crushed and fragmented. The left os coxa consists of the reconstructed ilium and the partially reconstructed pubis. The right pubis includes portions of the superior pubic ramus and the pubic body. This individual has consistently been sexed as a female due to the overall gracile nature of the bones and the morphology of the pubis bone.

Amud 1 is also from Israel and has cranial and postcranial elements. The left hipbone is primarily of the ilium but a separate left pubis fragment is also present. This fossil has been sexed as male and is often referred to as the "Amud Man."

The remains of several individuals come from the cave site of Shanidar in Iraq. Of the nine individuals discovered, only the pelvic elements of Shanidar 1, 3, and 4 are adequately preserved. Shanidar 1 has preserved the right and left hipbones with portions of the ilium, ischium, and pubis intact. Shanidar 3 consists of a right hipbone with parts of the ilium, ischium, and pubis preserved, and a less intact left hipbone. This left hipbone consists of fragments of the pubis and the ischium. Shanidar 4 has fragments of the ischium and pubis of the right hipbone. The more complete left hipbone has iliac, ischial, and pubic portions preserved.

In Europe, sites such as Krapina, La Ferrassie, and La Chapelle-aux-Saints have yielded Neandertal specimens that include pelvic remains. The first remains discovered of a human fossil came from the Neander Valley near Dusseldorf, Germany, in 1856. These remains, Neandertal 1, represent a nearly complete skeleton dating to approximately 40–70 kyr. Of the pelvis, the left os coxa is preserved but is broken at the ilium and ischium and is missing the pubis.

The Krapina material from the former Yugoslavia represents some of the older Neandertal specimens found in Europe. Tentatively dated to approximately 70–100 kyr, the remains of numerous individuals, both adult and immature, were discovered in a rock shelter in the early 1900s. The pelvic material comes from nine individuals, with three specimens, Krapina 207, 208, and 209/212, being the best preserved (Fig. 2). Krapina 207 is a left hipbone fragment that is missing the posterior superior iliac and pubic elements. Krapina 208 is a right hipbone preserving the inferior portion of the ilium at the acetabulum, the ischial body, and the superior ramus of the publis. Krapina 209/212 is a partial right hipbone consisting

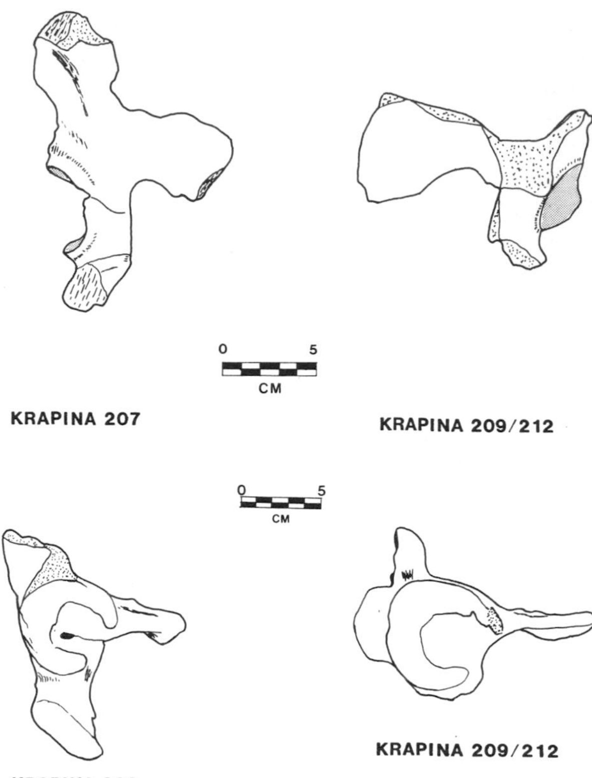

FIGURE 2 Lateral and frontal views of hipbones from the Krapina site in Yugoslavia: Krapina 207, Krapina 208, and Krapine 209/212.

of the inferior ilium, the acetabulum, and the superior pubic ramus. The preservation of so many individuals from one site is somewhat unusual. Further examination of these specimens will help answer questions concerning the nature and extent of morphological variation, even though the chronological sequence of the site remains uncertain.

The site of La Ferrassie produced two adult skeletons and several children that are dated to approximately 40–70 kyr. La Ferrassie 1 includes right and left os coxae fragments although the left is in better condition than the right. Both hipbones have been reconstructed in several areas. La Ferrassie 2 consists of two iliac fragments, right and left, both preserving the superior aspect of the sciatic notches.

The Neandertal skeleton from the site of La Chapelle-aux-Saints received considerable attention in the early part of this century. This specimen served as the model for the earlier reconstructions of Neandertals that portrayed them as brutish, stooped-over, and unwitting. Recognition that this particular individual was older and possibly pathological helped to dispel this disparaging image of these hominids. The pelvic remains of La Chapelle 1 include a partial left hipbone that is missing the pubis entirely.

IV. SEX DIFFERENCES

The pelvic remains of archaic *Homo sapiens* offer an opportunity to test the hypothesis that modern pelvic dimorphisms exist as a response to selective pressures exerted on the female pelvis for the successful delivery of an infant that was becoming increasingly larger in head and body size. The pelvic morphology of the earliest hominids, the australopithecines, can be principally explained in terms of the adaptation to bipedalism, such that by modern standards sex differences are poorly developed. For the early *Homo* material, the presence of a clearly "male" sciatic notch in KNM-ER-3228 suggests that pelvic dimorphisms may have been at least partially present at 1.9 mya. This is in contrast to the earlier hominids whose sciatic notch morphologies are dominated by a female appearance that does not appear to be based on sex differences. In fact, the sciatic notch appears to be one of the first of the modern pelvic dimorphisms to develop in hominids that is not dependent on males being larger than females. In addition, the posterior iliac dimension is small in KNM-ER-3228 as in modern human males, but relatively large in the earlier australopithecines as in modern human females.

Fossils like OH.28 and the early archaic *Homo sapiens* such as Arago 44, La Grotte du Prince, and Broken Hill E719 have a series of "contradictory features" in their pelves, that is, some pelvic variables sex these fossils as male whereas other variables sex them as females. For example, in each of these fossils, the sciatic notch sexes these specimens as females whereas the large acetabulum suggests that these fossils are more like modern human males. In addition, the posterior aspect of the ilium is small in OH.28, Broken Hill E719, and Broken Hill E720, like in modern human males, whereas in Arago 44 and La Grotte du Prince the posterior iliac dimension is as large as in modern human females. This suggests that the posterior iliac dimension is making a large contribution to the size of the pelvic inlet in Arago 44 and La Grotte du Prince. Functionally, a large posterior iliac dimension increases the capacity of the posterior pelvic inlet through which the infant must pass during childbirth.

By the time of the Neandertals, there is a more equal representation of male and female sciatic notch morphologies. For example, Krapina 207 has a sciatic notch that resembles that of modern human males whereas that of Krapina 209/212 resembles that of modern human females. On the other hand, like the earlier *Homo* material, against modern criteria the ischium and acetabulum suggests that these fossils are always male. In other words, the ischium and acetabulum of Neandertal males and females closely resemble those of modern human males in being rather large.

In hominids, the ischium and acetabulum increase in size from the earliest australopithecines due to their relationship with body size, which also increases through time in hominids. In fact, the acetabulum and ischium remain relatively large even in the Upper Pleistocene anatomically modern humans. As sex discriminators, the ischium and acetabulum are difficult to use on the fossils because of this relationship with this overall trend toward greater body size. Therefore, these sexing criteria, which are based on size-related differences, should be used intraspecifically since comparisons to earlier or later hominid species may not yield accurate results.

Another feature of the pelvis that appears to be a relatively good sex discriminator in modern humans is the length of the pubis bone. In the modern human female, the pubis tends to be longer with a thinner and more narrow superior pubic ramus than in modern human males. A longer pubis bone is generally consistent with increased proportions of the birth canal. The only available fossil hominids preserving the pubis are two early australopithecines, some later Neandertals, and a few Upper Pleistocene anatomically modern humans. The earlier australopithecines have relatively long pubis bones, as do the Neandertals. Since some of these Neandertals have been sexed as males from cranial attributes and/or postcranial robusticity, the presence of a long pubis bone in both sexes of Neandertals has raised considerable debate in recent years. Since the two available australopithecine specimens have relatively long pubis bones, as do the pongids, it is possible that a long, thin pubis bone is a feature that the early hominids share with the pongids owing to common ancestry. The pubis bone in Neandertals may therefore represent the persistence of this feature in the hominid line rather than the "lengthening" of this bone in Neandertals. It is quite likely that this morphology posed unique locomotory and postural adaptations in the larger-bodied Neandertals than in the smaller-bodied australopithecines, but it is unlikely that it affected reproductive patterns. The pubis bone does not appear to become a sexually dimorphic trait in hominids until the advent of anatomically modern humans.

By early anatomically modern humans, there is a more equal representation of male and female pelvic morphologies, although the acetabulum continues to be relatively large. Modern sexing criteria are appropriate to these essentially modern humans for all other aspects of the pelvis.

V. ARCHAIC *HOMO SAPIENS* AND THE EVOLUTION OF THE MODERN HUMAN BONY PELVIS

The hominid pelvis diverges greatly from the basic primate pelvis in being adapted for upright posture and locomotion. For example, in hominids the ilium shortened and expanded dorsally to increase the area for the attachment of muscles that maintain erect posture and render stability to the pelvis during locomotion. These changes are apparent on the earliest australopithecine pelves. It has been suggested that later hominids improved on these locomotory adaptations and were better bipeds than the earlier hominids, but there is little agreement over this issue. Nonetheless, by archaic *Homo sapiens*, the pelvis is well adapted to bipedalism. The further evolution of the bony pelvis mainly involves the development of sex differences.

It can be argued that modern pelvic dimorphisms exist as a response to distinct selective pressures acting on the male and female pelvis. The birth of the relatively small-headed australopithecine, though possibly more difficult than in the great apes, was most likely not as tightly constrained by the maternal pelvis as in modern humans. It is more plausible that pelvic dimorphisms developed as cranial capacity increased, beginning at approximately 2 myr, due to the reproductive demands being placed on the maternal pelvis for the birth of a relatively larger neonate. In females, strong selection pressures maintained a true pelvis of large capacity in females. In males, since the pelvis is functionally related only to locomotion, the true pelvis could be small and narrow without evolutionary consequences. Thus pelvic dimorphisms would result. Fossils like OH.28, Arago 44, and Broken Hill E719 have "contradictory features" with regard to sex because modern pelvic dimorphisms are not fully developed at this point in human evolution. Even for the later Neandertals, the pelvis has not achieved the modern condition of sexual differentiation. It is not until the Upper Pleistocene *Homo sapiens sapiens* that modern patterns of sex differences in the pelvis are evident.

The hominid pelvis has changed since the transition to bipedality, perhaps toward a more efficient bipedal gait, but certainly in the development of sex differences. Like the transition to upright walking, the development of sex differences was a complex issue. Even though pelvic dimorphisms are not as developed as in modern humans, the archaic *Homo sapiens* more closely resemble modern humans in their pelves than do the earlier hominids.

BIBLIOGRAPHY

Aiello, L. C. (1993). The fossil evidence for modern human origins in Africa: A revised view. *Am. Anthropologist* **95**, 73–96.

Frayer, D. W., Wolpoff, M. H., Thorne, A. G., Smith, F. H., and Pope, G. G. (1993). Theories on modern human origins: The paleontological test. *Am. Anthropologist* **95**, 14–50.

Hager, L. D. (1997). Sex and gender in paleoanthropology. *In* "Women in Human Evolution" (L. D. Hager, ed.), pp. 1–28. Routledge, London.

Rosenberg, K. R. (1992). The evolution of modern human childbirth. *Yearbook Phys. Anthropol.* **35**, 89–124.

Stringer, C. B. (1990). The emergence of modern humans. *Sci. Am.*, December, 33–37.

Stringer, C. B., and Gamble, C. (1993). "In Search of the Neanderthals." Thames & Hudson, New York.

Trinkaus, E., and Shipman, P. (1992). "The Neandertals." Knopf, New York.

Wolpoff, M. (1995). "Paleoanthropology." McGraw–Hill, New York.

Brain

BRYAN KOLB
IAN Q. WHISHAW
University of Lethbridge, Canada

JAN CIOE
Okanagan University-College, Canada

GLOSSARY

Action potential Brief electrical impulse by which information is conducted along an axon. It results from short-lived changes in the membrane's permeability to sodium

Amnesia Partial or total loss of memory

Aphasia Defect or loss of power of expression by speech, writing, or signs or of comprehending spoken or written language; caused by injury or disease of the brain

Cerebral cortex Layer of gray matter on the surface of the cerebral hemispheres composed of neurons and their synaptic connections that form four to six sublayers

Hippocampus Primitive cortical structure lying along the medial region of the temporal lobe; named after its shape, which is similar to a sea horse, or hippocampus

Neuron Basic unit of the nervous system; the nerve cell. Its function is to transmit and store information; it includes the cell body (soma), many processes called dendrites, and an axon

Neurotransmitter Chemical released from a synapse in response to an action potential and acting on postsynaptic receptors to change the resting potential of the receiving cell; chemically transmits information from one neuron to another

Synapse Functional junction between one neuron and another

THE BRAIN IS THAT PART OF THE CENTRAL NERvous system that is contained in the skull. It weighs approximately 1450 g at maturity and is composed of brain cells (neurons) and their processes, as well as support cells, that are organized into hundreds of functionally distinct regions. Neurons communicate both chemically and electrically so that different brain regions form functional systems to control behavior. Measurement of brain structure, activity, and behavior has allowed neuroscientists to reach inferences regarding the mechanisms of the basic functions, including (1) the body's interactions with the environment through the sensory systems (e.g., vision, audition, touch) and motor systems, (2) internal activities of the body (e.g., breathing, temperature, blood pressure), and (3) mental activities (e.g., thought, language, affect). By studying people with brain injuries it is possible to propose brain circuits that underly human behavior.

I. ANATOMICAL AND PHYSIOLOGICAL ORGANIZATION OF THE HUMAN BRAIN

A. Cellular Composition

The brain is composed of two general classes of cells: neurons and glial cells. Neurons are the functional units of the nervous system, whereas glial cells are support cells. Estimates of the numbers of cells in the human brain usually run around 10^{10} neurons and 10^{12} glial cells, although the numbers could be even higher. Only about 2–3 million cells (motor neurons) send their connections out of the brain to animate

ENCYCLOPEDIA OF HUMAN BIOLOGY, Second Edition, VOLUME 2. Copyright © 1997 by Academic Press. All rights of reproduction in any form reserved.

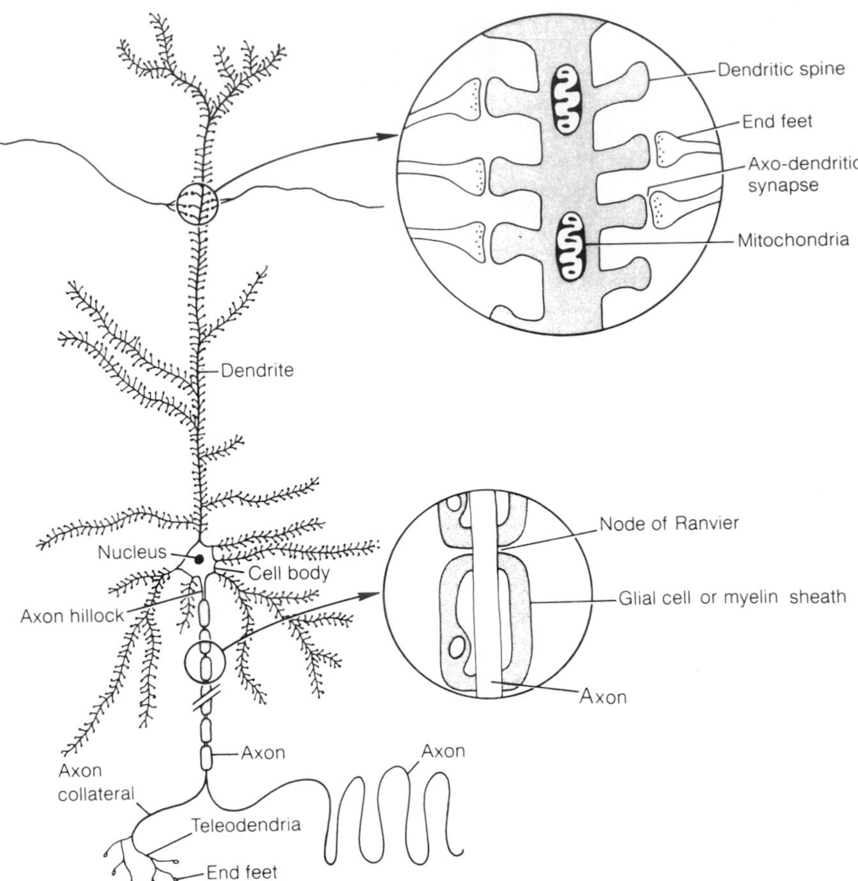

FIGURE 1 Summary of the major parts of a stylized neuron. Enlargement at the top right shows the gross structure of the synapse between axons and the spines on dendrites. Enlargement on the bottom right shows the myelin sheath that surrounds the axon and acts as insulation. [From Kolb, B., and Whishaw, I. Q. (1996). "Fundamentals of Human Neuropsychology," 4th Ed. Freeman, New York.]

muscle fibers, leaving an enormous number of cells with other functions.

There are numerous types of neurons (e.g., pyramidal, granule, Purkinje, Golgi I, motoneurons), but they share several features in common. First, they have a cell body, which like most cells contains a variety of substances that determine the function of the cell; processes called *dendrites*, which function primarily to increase the surface area on which a cell can receive information from other cells; and a process called an *axon*, which normally originates in the cell body and transmits information to other cells (Fig. 1). Different types of neurons are morphologically distinct, reflecting differences in function. The different types are distributed differentially to different regions of the brain reflecting regional differences in brain function.

Neurons are connected with one another via their axons; any given neuron may have as many as 15,000 connections with other neurons. These connections are highly organized so that certain regions of the brain are more closely connected to one another than they are to others. As a result, these closely associated regions form functional systems in the brain, which control certain types of behavior.

B. Gross Anatomical Organization of the Brain

The most obvious feature of the human brain is that there are two large hemispheres, which sit on a stem, known as the brain stem. Both structures are composed of hundreds of regions; nearly all of them are found bilaterally. Traditionally, the brain is described by the gross divisions observed phylogenetically and

TABLE I
Divisions of the Central Nervous System

Primitive divisions	Mammalian divisions	Major structures
Prosencephalon (forebrain)	Telencephalon (endbrain)	Neocortex Basal ganglia Limbic system Olfactory bulb Lateral ventricles
	Diencephalon (between brain)	Thalamus Hypothalamus Epithalamus Third ventricle
Mesencephalon (midbrain)	Mesencephalon (midbrain)	Tectum Tegmentum Cerebral aqueduct
Rhombencephalon (hindbrain)	Metencephalon (across brain)	Cerebellum Pons Fourth ventricle
	Myelencephalon (spinal brain)	Medulla oblongata Fourth ventricle

embryologically as summarized in Table I. The most primitive region is the hindbrain, whose principal structures include the cerebellum, pons, and medulla. The cerebellum was originally specialized for sensory-motor coordination, which remains its major function. The pons and medulla also contribute to equilibrium, balance, and the control of gross movements (including breathing). The midbrain consists of two main structures, the tectum and the tegmentum. The tectum consists primarily of two sets of nuclei, the superior and inferior colliculi, which mediate whole body movements to visual and auditory stimuli, respectively. The tegmentum contains various structures including regions associated with (1) the nerves of the head (so-called cranial nerves), (2) sensory nerves from the body, (3) connections from higher structures that function to control movement, and (4) a number of structures involved in movement (substantia nigra, red nucleus), as well as a region known as the reticular formation. The latter system plays a major role in the control of sleep and waking.

The forebrain is conventionally divided into five anatomical areas: (1) the neocortex, (2) the basal ganglia, (3) the limbic system, (4) the thalamus, and (5) the olfactory bulbs and tract. Each of these regions can be dissociated into numerous smaller regions on the basis of neuronal type, physiological and chemical properties, and connections with other brain regions.

The neocortex, which is usually called the *cortex*, is composed of approximately six layers, each of which have distinct neuronal populations, comprises 80% of the human forebrain by volume, and is grossly divided into four regions, which are named by the cranial bones lying above them (Fig. 2). It is wrinkled, which is nature's solution to the problem of confining a large surface area into a shell that is still small enough to pass through the birth canal. The cortex has a thickness of only 1.5–3.0 mm but has a total area of about 2500 cm^2. The cortex can be subdivided into dozens of subregions on the basis of the distribution of neuron types, their chemical and physiological characteristics, and their connections. These subregions can be shown to be functionally distinct. [*See* Neocortex.]

The basal ganglia are a collection of nuclei lying beneath the neocortex. They include the putamen, caudate nucleus, globus pallidus, and amygdala. These nuclei have intimate connections with the neocortex as well as having major connections with midbrain structures. The basal ganglia have principally a motor function, as damage to different regions can produce changes in posture or muscle tone and abnormal movements such as twitches, jerks, and tremors.

The limbic system is not really a unitary system but refers to a number of structures that were once believed to function together to produce emotion. These include the hippocampus, septum, cingulate cortex, and hypothalamus, each of which have different functions (Fig. 3). [*See* Hippocampal Formation; Hypothalamus; Limbic Motor System.]

The thalamus provides the major route of information to the neocortex, and different neocortical regions are associated with inputs from distinct thalamic regions connected with the neocortex. The different thalamic areas receive information from sensory and motor regions in the brain stem as well as the limbic system. [*See* Thalamus.]

C. Physiological Organization of the Brain

Like other cells in the body, the neuron has an electrical voltage (potential) across its membrane, which results from the differential distribution of different ions on the two sides of the membrane. In contrast to other body cells, however, this electrical potential is used to transmit information from one neuron to another in the nervous system, which is accomplished in the following way. Neurons have a resting potential across the membrane of the dendrites, cell body, and

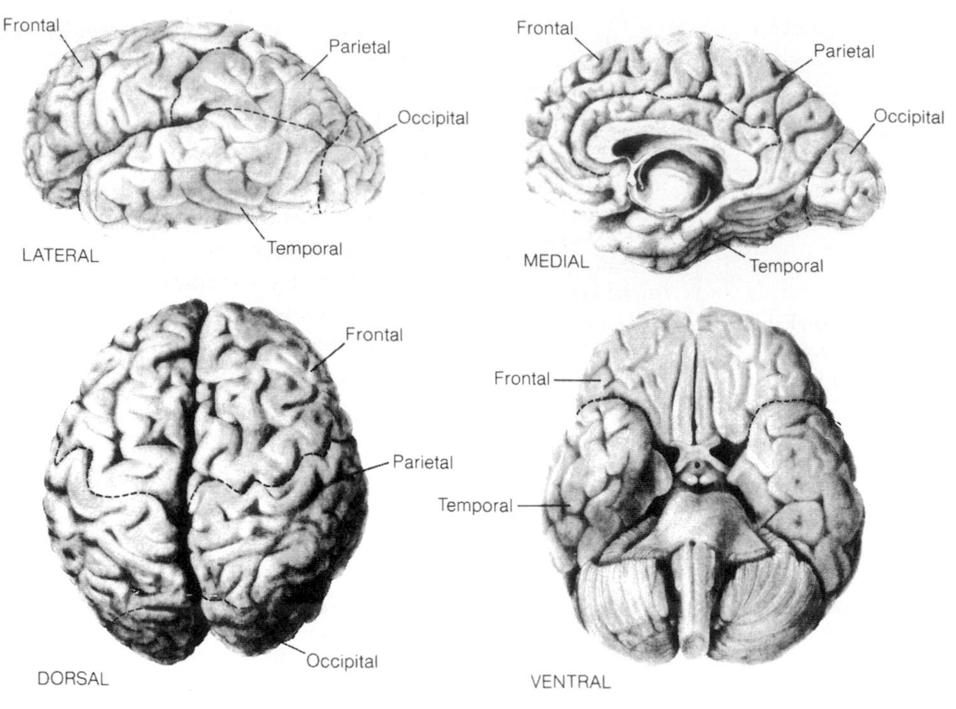

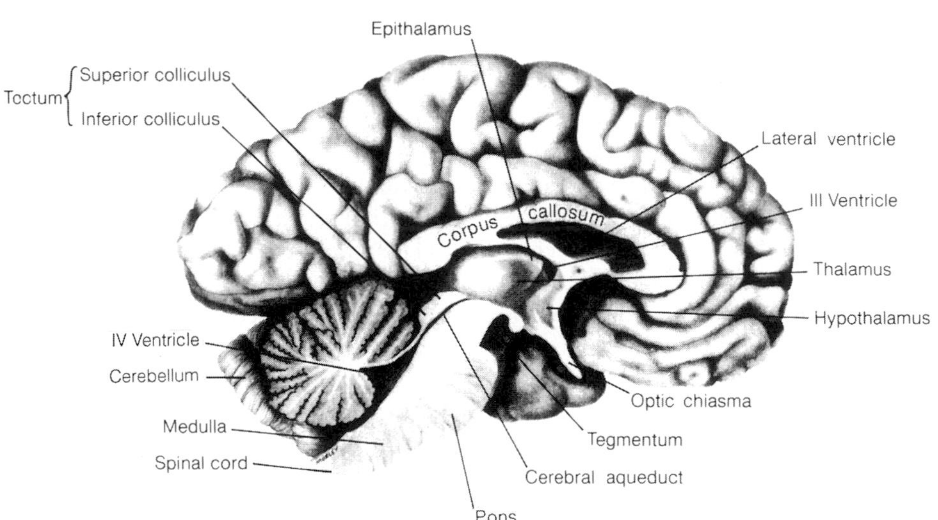

FIGURE 2 (Top) Summary of gross regions of the neocortex of the human brain. (Bottom) View of major structures of the brain. [From Kolb, B., and Whishaw, I. Q. (1996). "Fundamentals of Human Neuropsychology," 4th Ed. Freeman, New York.]

axon, which remains relatively constant at about −70 mV. If the membrane permeability for different ions changes, the electrical potential will also change. If it becomes more negative the cell is said to be hyperpolarized, and if it becomes less negative (i.e., more positive) it is said to be depolarized. When the membrane of a neuron is perturbed by the signals coming from other neurons or by certain external agents (e.g., chemicals), the voltage across the membrane changes, becoming either hyperpolarized or depolarized. These

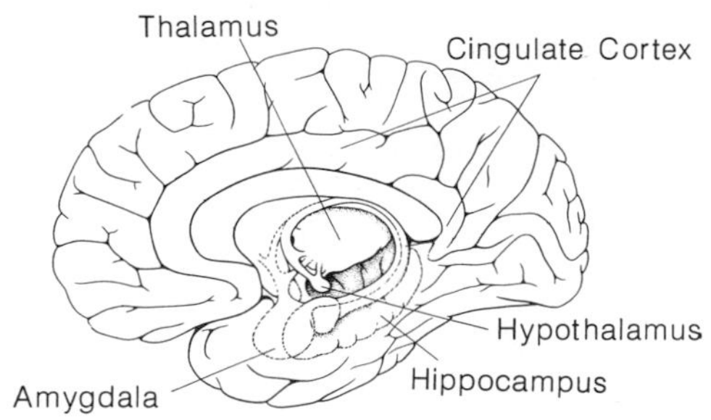

FIGURE 3 Medial view of the right hemisphere illustrating positions of limbic structures. Anterior is to the left.

changes are normally restricted to the area of membrane stimulated, but if there are numerous signals through many synapses to the same cell, they will summate, altering the membrane potential of a larger region of the cell. If the excitation is sufficient to reduce the membrane potential to about −50 mV, the membrane permeability for positive Na^+ ions changes. The influence of ions raises the potential until it becomes positive (e.g., +40 mV). This change in membrane potential spreads across the cell and, if it reaches the axon, it travels down the axon, producing a propagating signal. This change in permeability is quickly reversed by the cell in about 0.5 msec, allowing the cell to send repeated signals during a short period of time. The signal that travels down the axon is known as an action potential (or nerve impulse), and when it occurs, the cell is said to have fired. The rate at which the impulse travels along the axon varies from 1 to 100 m/sec and can occur as frequently as 1000 times/sec, depending on the diameter of the axon; the most common rate is about 100 times/sec.

D. Chemical Organization

Once the nerve impulse reaches the end of the axon (the axon terminal), it initiates biochemical changes that result in the release of a chemical known as a neurotransmitter into the synapse. Although the action of transmitter is complex, their effect is either to raise or to lower the membrane potential of the postsynaptic cell, with the effect of making it more or less likely to transmit a nerve impulse.

Dozens of chemicals are known to be neurotransmitters, including a variety of amino acids (e.g., glutamic acid, glycine, aspartate, and γ-aminobutyric acid), monoamines (e.g., dopamine, norepinephrine, serotonin), and peptides [e.g., substance P, β-endorphin, corticotrophin (ACTH)]. Any neuron can receive signals from neurotransmitters through different synapses. The distribution of different transmitters is not homogeneous in the brain as different regions are dominated by different types. Because drugs that affect the brain act by either mimicking certain transmitters or interfering with the normal function of particular transmitters, different drugs alter different regions of the brain and subsequently have different behavioral effects. [*See* Neurotransmitter and Neuropeptide Receptors in the Brain.]

II. FUNCTIONAL ORGANIZATION OF THE BRAIN

A. Principles of Brain Organization

The fundamental principle of brain organization is that it is organized hierarchically such that the same behavior is represented at several levels in the nervous system. The function of each level can be inferred from studies in which the outer levels have been removed, as is summarized in Fig. 4. The principal idea is that the basic units of behavior are produced by the lowest level, the spinal cord, and at each successive level there is the addition of greater control over these simple behavioral units. At the highest level the neocortex allows the addition of flexibility to the relatively stereotyped movement sequences generated by lower levels, as well as allowing greater control of behavior by complex concepts such as space and time. Although new abilities are added at each level in the hierarchy, it remains difficult to localize any process to a particular level because any behavior requires the activity of all regions for its successful execution. Nonetheless, brain injury at different levels will produce different symptoms, depending on what functions are added at that level. [*See* Spinal Cord.]

B. Principles of Neocortical Organization

The cortex can be divided into three general types of areas: (1) sensory areas, (2) motor areas, and (3) association areas. The sensory areas are the regions that function to identify and to interpret information coming from the receptor structures in the eyes, ears, nose, mouth, and skin. The motor complex is the region involved in the direct control of movement. The association cortex is the cortex that is not ascribed specific sensory or motor functions.

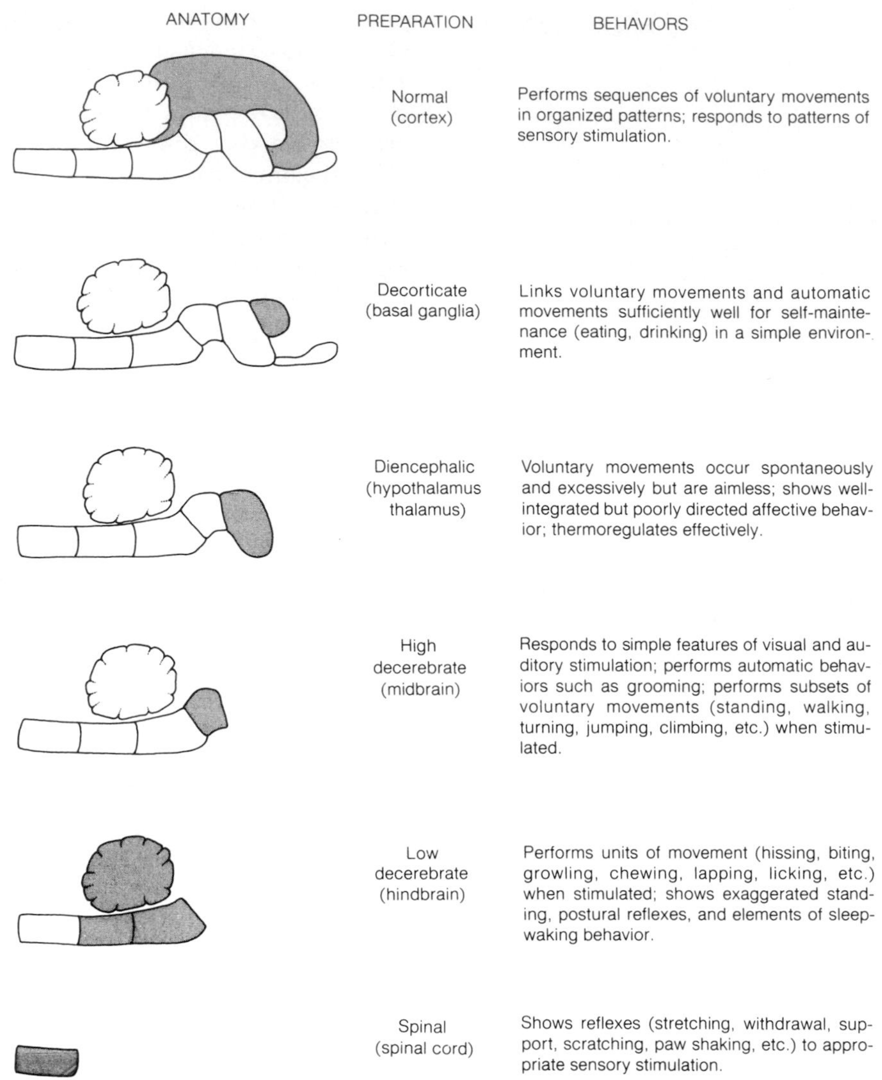

FIGURE 4 Summary of the behavior that can be supported by different levels of the nervous system. Shading indicates higher remaining functional area in each preparation.

Distinct sensory cortex is associated with each of these *sensory systems*, and each is made up of numerous subregions, each of which functions to process a distinct type of sensory information. For example, in the visual system, separate regions are devoted to the analysis of form, color, size, movement, etc. Damage to each of these regions will produce a distinct loss of sensory experience. In each system there is a region of sensory cortex that produces an apparent inability to detect sensory information such that a person will, for instance, appear to be blind, unable to taste, or will have numbness of the skin, etc. In each case, however, it can be shown that other aspects of sensory function are intact. Thus, a person who is unable to "see" an object may be able to indicate its color and position! Similarly, a person may be able to locate a place on the body where they were touched while being unable to "feel" the touch. Regions of sensory cortex producing such symptoms are referred to as primary sensory cortex. The other regions are known as unimodal association regions, and damage to them is associated with various other symptoms. Thus, in every sensory system there are regions of cortex that, when damaged, result in an inability to understand the significance of sensory events. Such symptoms are known as agnosias. For example, although able to perceive an object (e.g., toothbrush) and to pick it up, a person may be unable to name it or to identify its use. Similarly, a person may be able to perceive a sound (e.g., that of an insect) but be unable to indicate what the sound is from. Some agnosias are relatively specific (e.g., an inability to recognize faces or colors). [*See* Visual System.]

The motor cortex represents a relatively small region of the cortex that controls all voluntary movements and is specialized to produce fine movements such as independent finger movements and complex tongue movements. Damage to this region prevents certain movements (e.g., of the fingers), although others (e.g., arm and body movements) may be relatively normal. Although the motor cortex is a relatively small region of the cortex, a much larger region contributes to motor functions, including some of the sensory regions as well as the association cortex. For example, a person may be unable to organize behaviors such as those required for dressing or for using objects. Such disorders are known as apraxias, which refers to the inability to make voluntary movements in the absence of any damage to the motor cortex. [*See* Motor Control.]

The regions of the neocortex not specialized as sensory or motor regions are referred to as association cortex. This cortex receives information from one or more of the sensory systems and functions to organize complex behaviors such as the three-dimensional control of movement, the comprehension of written language, and the making of plans of action. The principal regions of association cortex include the prefrontal cortex in the frontal lobe, the posterior parietal cortex, and regions of the temporal cortex. Although not neocortex, it is convenient to consider the medial temporal structures, including the hippocampus, amygdala, and associated cortical regions, as a type of association cortex. Taken together, damage to the association regions causes a puzzling array of behavioral symptoms that include changes in affect and personality, memory, and language. [*See* Cortex.]

III. CEREBRAL ASYMMETRY

One of the most distinctive features of the human brain is that the two cerebral hemispheres are both anatomically and functionally different, a property referred to as cerebral asymmetry.

A. Functional Asymmetry

The clearest functional difference between the two sides of the brain is that structures of the left hemisphere are involved in language functions whereas those of the right hemisphere are involved in nonlanguage functions such as the control of spatial abilities. This asymmetry can be demonstrated in each of the sensory systems, with differences in left-handed and right-handed persons. We will first consider the common case, the right-handed individual. In the visual system the left hemisphere is specialized to recognize printed words or numbers, whereas the right hemisphere is specialized to process complex nonverbal material such as is seen in geometric figures, faces, and route maps. Similarly, in the auditory system the left hemisphere analyzes words whereas the right hemisphere analyzes tone of voice (prosody) and certain aspects of music. Asymmetrical functions go beyond sensation, however, to include memory and affect. In the control of movement, the left hemisphere is specialized for the production of certain types of complex movement sequences, as in meaningful gestures (e.g., salute, wave) or writing. The right hemisphere has a complementary role in the production of other movements such as in drawing, dressing, or constructing objects. Similarly, the left hemisphere has a favored role in memory functions related to language (e.g., written and spoken words), whereas the right hemisphere plays a major role in the memory of places and nonverbal information such as music and faces. [*See* Language.]

B. Anatomical Asymmetry

The functional asymmetry of the human brain is correlated with various asymmetries in gross brain morphology, cell structure, neurochemical distribution, and blood flow. Differences in gross morphology and cell structure are most easily seen in the regions specialized for language, including the anterior (Broca's) and posterior speech areas (Fig. 5). For example, one region in the posterior speech area, the planum temporale, is twice as large on the left hemisphere in most brains, whereas a region involved in the processing of musical notes, Heschl's gyrus, is larger in the right temporal lobe than in the left.

C. Variations in Cerebral Asymmetry: Handedness and Sex

There is considerable variation in the details of both functional and anatomical asymmetry in different

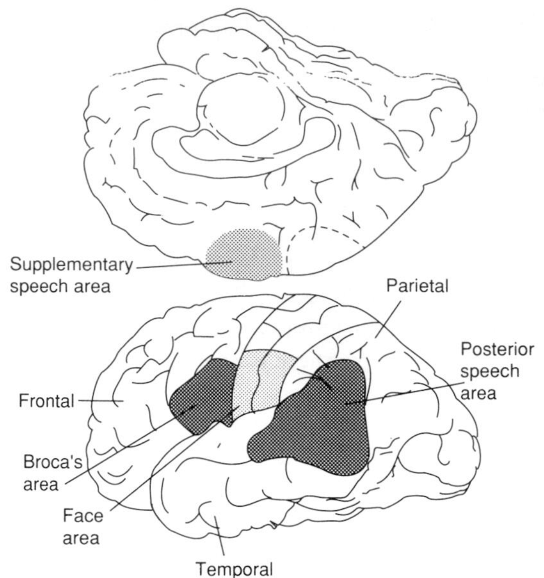

FIGURE 5 Schematic diagram showing the anterior speech area (Broca's area) in the frontal lobe and the posterior speech area (including Wernicke's area) in posterior temporal and parietal regions. The shaded region between speech areas is the motor and sensory cortex controlling the face and tongue. [From Kolb, B., and Whisihaw, I. Q. (1996). "Fundamentals of Human Neuropsychology," 4th Ed. Freeman, New York.

people. Two factors, handedness and sex, appear to account for much of this variation. First, left-handers have a different pattern of anatomical organization than do right-handers. For example, they appear to have a larger bundle of fibers connecting the two cerebral hemispheres, the corpus callosum, which implies that the nature of hemispheric interaction differs in left- and right-handers. Similarly, left-handers are less likely to show the large asymmetries in the structure of the language-related areas than are right-handers. Functionally, the organization of the left-handed brain shows considerable variation: Language is located primarily in the left hemisphere in about two-thirds of left-handers, in the right hemisphere in about one-sixth, and in both hemispheres in about one-sixth. Left-handers with different speech organization than right-handers do not simply have a reversal of brain organization, however, although the nature of their cerebral organization is still poorly understood. Second, males and females also differ in functional and structural organization. For example, the corpus callosum of females is larger relative to brain size than that of males, and females are less likely to show gross asymmetries or to have reversed asymmetries. Animal studies have shown a clear relation between anatomical organization and the presence of the perinatal gonadal hormones present at about the time of birth, suggesting that these hormones differentially organize the brain of males and females. Functionally, the effect of brain damage in males and females differs as well, although in complex ways that are poorly understood. It does appear, however, that frontal lobe injuries in both human and nonhuman subjects have differential effects in the two sexes, with larger behavioral effects of frontal lobe injury observed in females. Other factors are also believed to influence the nature of cerebral asymmetry, especially experience, interacting with both sex and handedness. [*See* Hemispheric Interactions.]

IV. ORGANIZATION OF HIGHER FUNCTIONS

Complex functions such as memory or emotion are not easily localized in the brain, as the circuits involved include vast areas of both the cerebral hemispheres and other forebrain structures. Part of the difficulty in localizing such functions is that they are not unitary things but are inferred from behavior, which in turn results from numerous processes. Nonetheless, it is possible to reach some generalizations regarding such functions.

A. Memory

Memory is an inferred process that results in a relatively permanent change in behavior, which presumably results from a change in the brain. Psychologists distinguish many types of memory, each of which may have a distinct neural basis. These include, among others, (1) long-term memory, which is the recall of information over hours, days, weeks, and years; (2) short-term memory, which is the recall of information over seconds or minutes; (3) declarative memory, which is the recall of facts that are accessible to conscious recollection; (4) procedural memory, which is the ability to perform skills that are "automatic" and that are not stored with respect to specific times or places (e.g., the movements required to drive); (5) verbal memory, which is memory of language-related material; and (6) spatial memory, which is the recall of places or locations.

The neural basis of human memory can be considered at two levels: cellular and neural location. Thus, changes in cell activity and structure are associated with processes like memory, which may occur extremely rapidly in the brain, possibly in the order of seconds or at least minutes. Further, there is a variety of candidate regions for memory processes, the region varying with the nature of memory process. One structure that plays a major role in various forms of memory is the hippocampus. Bilateral damage to this structure leads to a condition of anterograde amnesia, which refers to the inability to recall, after a few minutes, any new material that is experienced after the damage. There is only a brief period of retrograde amnesia, which refers to the inability to recall material before the injury. The relatively selective effects of hippocampal injuries in producing anterograde but not retrograde amnesia suggest that different brain regions are involved in the initial learning of information and their later retrieval from memory. As a generalization, it appears that the temporal lobe is involved in various types of long-term memory processes, whereas the frontal and parietal lobes play a role in certain short-term memory processes. Damage to these regions thus produces different forms of memory loss, which are further complicated by whether the injury is to the left or right hemisphere. [*See* Learning and Memory; Neurology of Memory.]

B. Language

Damage to either of the major speech areas (Broca's or the posterior speech zone, which is sometimes referred to as Wernicke's area) will produce a variety of dissociable syndromes, including aphasia, an inability to comprehend language; alexia, an inability to read; and agraphia, an inability to write. There are various forms of each of these syndromes that relate to the precise details of the brain injury. Various other forms of language disturbance result from damage outside the speech areas, including changes in speech fluency [i.e., the ability to generate words according to certain criteria (e.g., write down words starting with "D"; give the name of objects)], spontaneous talking in conversation, the ability to categorize words (e.g., apple and banana are fruits), and so on. It appears that nearly any left hemisphere injury will affect some aspect of these language functions, as does damage to some regions of the right hemisphere. [*See* Speech and Language Pathology.]

C. Emotional Processes

Like memory processes, emotional processes are inferred from behavior and include many different functions, including autonomic nervous system activity, "feelings," facial expression, and tone of voice. Certain subcortical regions (hypothalamus, amygdala) play a major role in the generation of affective behaviors, especially the autonomic components such as blood pressure, respiration, and heart rate. In addition, damage nearly anywhere in the cortex will alter some aspect of cognitive function, which in turn will alter personality and emotional behavior, but damage to the right hemisphere produces a greater effect on emotional behavior than similar damage to the left. Moreover, the frontal lobe plays a special role as well, possibly because it has direct control of autonomic function as well as of spontaneous facial expression and other nonverbal aspects of personality. Thus, damage to the right hemisphere, or the frontal lobe of either hemisphere, is likely to lead to complaints from relatives regarding a change in "personality" or "affect." The control of emotional behavior may not only be relatively localized to different regions of the brain, but also related to specific neurotransmitter systems in the brain. For example, one dominant theory of the cause of schizophrenia proposes that there is an overactivity of the dopaminergic neurons (i.e., neurons that use dopamine as neurotransmitter) in the forebrain, likely in the frontal cortex (hence, the dopamine hypothesis of schizophrenia). Similarly, the dominant theory of depression is that it is related to low activity in systems that employ norepinephrine or serotonin as neurotransmitters. Because there are asymmetries in the cerebral distribution of these transmitters, it is reasonable to expect that depression may be more related to the right than left hemisphere, which is consistent with the dominant role of the right hemisphere in emotion. [*See* Depression, Neurotransmitters and Receptors; Schizophrenic Disorders.]

D. Space

The concept of space has many different interpretations, which are not equivalent. Objects (and bodies) occupy space, move through space, and interact with other things in space; we can form mental representations of space, and we have memories for the location of things. It is difficult therefore to define space or to know how the brain codes spatial information. Damage to different parts of the cerebral hemispheres can produce a wide variety of spatial disturbances, including the inability to appreciate the location of one's body or even the location of one's body parts relative to one another. It is generally accepted that the right hemisphere plays the major role in spatial behavior, but there can also be spatial disruptions from left hemisphere damage, especially if the behavior involves verbalizing space. The major region in the control of spatial behavior is the parietal cortex, although the hippocampus is also involved. [*See* Binocular Vision and Space Perception.]

E. Attention and Consciousness

It is often claimed that the central question in brain science is the relationship between mind and brain. The question is easy to state, yet it is not so easy to grasp what it is we need to explain. One thing to explain is the process by which organisms select information to act upon. Another is the process of selecting behaviors. As sensory and cognitive processes increase in complex brains, so does the process of selection. Electrophysiological evidence shows that there are three different attentional systems for selecting information: one in the parietal cortex that enhances attention to places (so-called spatial attention); one in the visual cortex that selects features of objects for analysis; and one in the temporal cortex that selects the objects themselves. In addition, there are two systems in the frontal lobe for selecting behaviors: one for selecting behaviors on the basis of internal information (such as knowledge or emotional state) and one for selecting behaviors on the basis of external information that is processed by the three information-selecting systems just described. Thus, the process of attention allows the brain to focus on aspects of the world with specific behaviors.

Consciousness is also not a unitary process. Most authors perceive it to be composed of several independent processes. Most definitions of consciousness exclude the conditions of simply being responsive to sensory information or simply being able to produce movement. Consciousness is usually assumed to be composed of a number of relatively separate functions that together form what we know as consciousness. Brain damage can limit, change, or reduce consciousness for specific types of information, but people with restricted brain damage do not lose their consciousness. For example, people who have damage to the visual areas may lose conscious awareness of certain types of visual stimulation but they are still aware, and can comprehend, auditory information. People who are aphasic may no longer be conscious of language but they are still perceived as being conscious and can maintain other forms of cognitive activity. Thus, the reason that no one structure can be equated with consciousness is that consciousness is a composite or a product of all cortical areas and their connections. In sum, the processes of attention and consciousness are properties of the brain that have evolved simply because the brain is complex and requires such processes to analyze and use the information available.

F. Conclusions

One of the key features of the brain is the localization of function. We have seen that certain brain structures are critical for a number of the higher human functions and that damage in specific locations on one particular side of the brain can result in the loss of behaviors that are not seen when the other side is injured. This localization of function results from the unique inputs and outputs of the subregions of the brain. Nevertheless, the hierarchical organization of the brain guarantees that control of a specific behavior is seldom restricted to a single area. There are multiple representations of the different cognitive functions in the different levels of neural organization but they are not simply backup circuits. Each level adds its own unique contribution that ultimately produces a degree of complexity that makes it possible for humans to behave flexibly in a changing environment.

BIBLIOGRAPHY

Kolb, B., and Whishaw, I. Q. (1996). "Fundamentals of Human Neuropsychology," 4th Ed. Freeman, New York.

Nauta, W. J. H., and Feirtage, M. (1986). "Fundamental Neuroanatomy." Freeman, New York.

Brain, Central Gray Area

ALVIN J. BEITZ
University of Minnesota

GLOSSARY

Afferent Fibers or impulses leading to a predetermined site in the nervous system

Cytoarchitecture Arrangement of nerve cell bodies in the central nervous system

Efferent Fibers or impulses leading away from a predetermined site in the nervous system

Neuromodulator Substance released from neurons that modifies presynaptic or postsynaptic function

Neurotransmitter Substance synthesized in neurons and usually released from an axon terminal, which acts on a receptor site to produce either excitation or inhibition of the target cell

Nociception Appreciation or transmission of a sensation produced by damage to body tissue

Nucleus Well-defined cluster of nerve cell bodies

Receptor Protein entity existing within the cell cytoplasm or in conjunction with the cell membrane that specifically binds neurotransmitters and other mediators of cell signaling to ultimately produce a biological effect

THE CENTRAL GRAY AREA, ALSO KNOWN AS the periaqueductal gray, is a midline region that encircles the mesencephalic aqueduct of the human midbrain. It comprises several types of neurons (i.e., nerve cells) and neuroglial cells and can be separated into four anatomical subdivisions and several longitudinal columns based on its cytoarchitecture, histochemical staining, and functions.

The ventrolateral subdivision contains the largest neurons and plays a major role in the involvement of the central gray in pain modulation and analgesia. The dorsolateral subdivision contains the greatest number of neurons per unit area and appears to be involved in vocalization and behavioral reactions. The medial subdivision comprises elongated neurons that form a rim around the mesencephalic aqueduct. These neurons are oriented predominantly parallel to the edge of the aqueduct and are purported to participate in aversive behavior (i.e., escape responses are induced by stimulation of this region in animals). The dorsal subdivision is anatomically the most distinct subregion. It has the lowest number of neurons per unit area, the smallest neurons, and the largest glia (i.e., nonneuron) to neuron ratio.

The neurons of the human central gray possess a plethora of neurochemicals (i.e., neurotransmitters and neuromodulators) and receptors. The relationship of these components to the numerous functions of the central gray is only beginning to be defined.

I. ANATOMICAL ORGANIZATION

A. Location, Subdivisions, and Longitudinal Columns

The human central gray consists of a mass of cells and fibers that surrounds the mesencephalic aqueduct of Sylvius and traverses the entire longitudinal extent of the midbrain. It is bordered anteriorly by the posterior commissure, a structure that constitutes a fibrous arch at the transition of the midbrain into the diencephalon, and posteriorly it extends to a point immediately anterior to the decussation of the trochlear nerve. As indicated in Fig. 1, the central gray is bounded dorsally and dorsolaterally by the superior and inferior colliculi, ventrally by the oculomotor and

ENCYCLOPEDIA OF HUMAN BIOLOGY, Second Edition, VOLUME 2. Copyright © 1997 by Academic Press. All rights of reproduction in any form reserved.

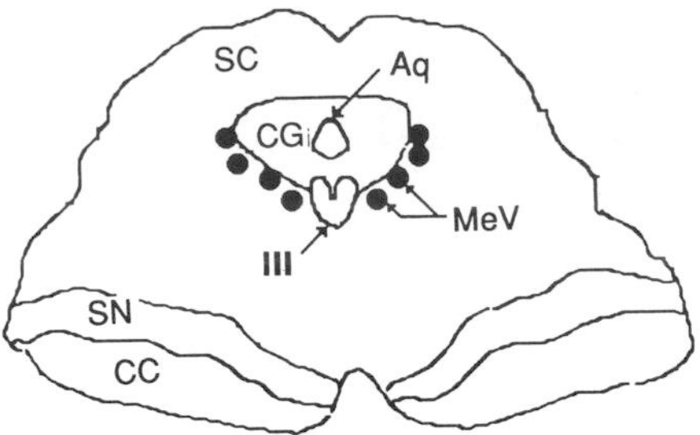

FIGURE 1 Transverse (coronal) section through the caudal midbrain, illustrating the location of the central gray (CG). SC, superior colliculus; III, oculomotor nucleus; MeV, mesencephalic trigeminal nucleus; CC, crus cerebri; SN, substantia nigra; Aq, mesencephalic aqueduct.

trochlear nuclei, and laterally by the mesencephalic trigeminal nucleus and several reticular nuclei.

Although the question of anatomical subdivisions within the central gray has been the source of considerable controversy, the evidence indicates that the human central gray is divisible into four distinct regions. The dorsal subdivision, illustrated in Fig. 2, is the most distinct subdivision observed in stained sections. As indicated in Table I, this subdivision contains the smallest neurons (mean neuronal area, 128.17 μm^2), has the lowest neuronal packing density, and exhibits the highest glia–neuron ratio. Thus, in contrast to other subdivisions, the dorsal subdivision is characterized by a marked accumulation of glial nuclei, among which occasional small neurons are found.

The medial subdivision forms a ring around the mesencephalic aqueduct and, as noted in Table I, contains a larger number of neurons per cubic millimeter than the dorsal subdivision. The hallmark of the medial subdivision is the distinctive orientation of its neurons. The majority of the neurons in this subdivision exhibit an orientation preference between 70° and 130°; that is, their long axis is arranged parallel to the edge of the aqueduct. By contrast, the majority of the neurons in the ventrolateral and dorsolateral subdivisions are oriented at obtuse or acute angles to the edge of the aqueduct (Table I).

The ventrolateral and dorsolateral subdivisions comprise the lateral regions of the central gray, as illustrated in Fig. 2. These two divisions have the greatest neuronal density and contain the largest neurons of the four subregions (Table I). Although these two subdivisions are difficult to distinguish in routine stained midbrain sections, several histochemical and immunocytochemical staining procedures, as well as

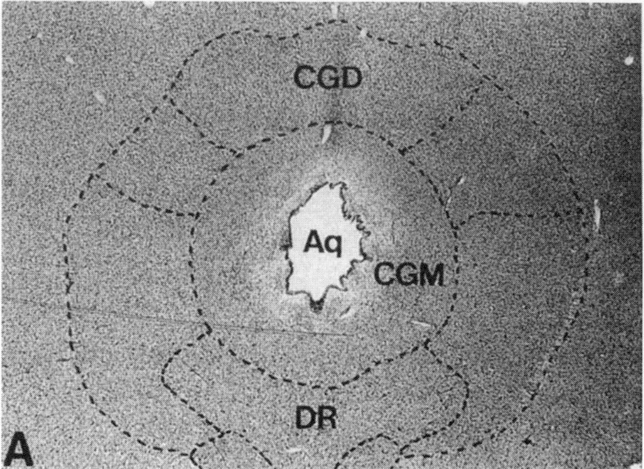

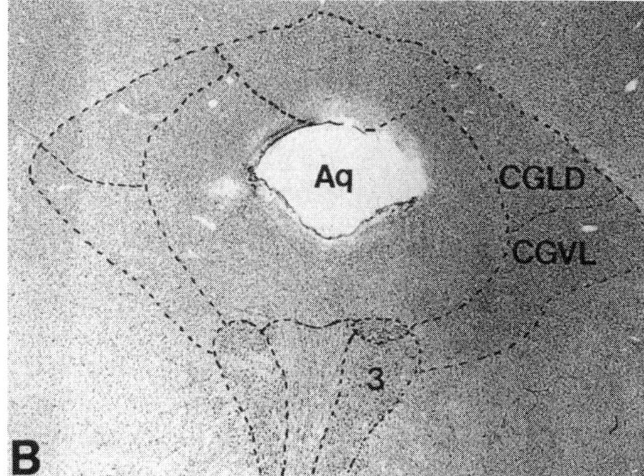

FIGURE 2 Coronal sections through the human midbrain central gray at the levels of the trochlear (A) and oculomotor (B) nuclei. The dorsal (CGD), medial (CGM), dorsolateral (CGLD), and ventrolateral (CGVL) subdivisions are illustrated. DR, dorsal raphe nucleus; 3, oculomotor nucleus; Aq, mesencephalic aqueduct. [Reproduced with permission from A. J. Beitz (1990). Central gray. *In* "The Human Nervous System" (G. Paxinos, ed.), pp. 307–320. Academic Press, San Diego.]

certain *in vitro* receptor binding assays, provide additional evidence to support their existence in the central gray.

Although the question of distinct anatomical subdivisions within the central gray appears to have been settled, recent data from several laboratories now suggest that the central gray also contains a previously unsuspected organizational design that appears to be superimposed on the anatomical subdivisions just described. This organization takes the form of longitudinal columns that display both functional and anatomical specificity. The anatomical specificity of these columns stems from distinct longitudinal arrangements of afferent inputs, output neurons, and intrinsic

TABLE I
Neuronal Data for the Four Human Central Gray Subdivisions[a]

Central gray subdivision	Mean neuronal packing density (cells/mm^3)	Mean neuronal area (μm^2)	Mean neuronal diameter (μm)	Axial ratio[b]	Orientation preference[c]	Glia–neuron ratio
Medial	6801.1 (307.6)	147.41 (1.21)	9.31 (0.05)	1.95 (0.03)	70–130°	6 : 1
Dorsal	4079.8 (234.2)	127.17 (0.96)	9.78 (0.07)	1.76 (0.04)	50–110°	19 : 1
Dorsolateral	8894.1 (393.3)	142.28 (0.51)	9.90 (0.02)	1.71 (0.01)	10–50° and 150–170°	6 : 1
Ventrolateral	7123.5 (181.6)	166.54 (0.87)	10.81 (0.04)	1.73 (0.01)	130–170°	8 : 1

[a]Data are based on the measurement of over 20,000 neurons. [Reproduced with permission from A. J. Beitz (1990). Central gray. *In* "The Human Nervous System" (G. Paxinos, ed.), pp. 307–320. Academic Press, San Diego.

[b]Axial ratio is derived from the ratio of cell body length to width and provides an indication of shape, since round cells have a ratio of approximately 1, whereas more elongate neurons have a ratio greater than 1.

[c]Orientation is relative to a vertical axis through the center of the mesencephalic aqueduct.

interneurons. Many of the important functions that are classically associated with the central gray, including analgesia, autonomic reactions, and rage and fear reactions, are integrated by these overlapping longitudinal zones of neurons. Five discrete longitudinal columns (dorsomedial, dorsolateral, lateral, ventrolateral, and juxtaaqueductal) have been proposed. In many respects these newly defined longitudinal columns appear to represent inferior to superior three-dimensional arrangements of the anatomical subdivisions. Thus the dorsomedial neuronal column is almost identical to the dorsal subdivision, the juxtaaqueductal column is homologous to the medial subdivision, the dorsolateral neuronal column is comparable to the dorsolateral subdivision, and the ventrolateral subdivision contains the lateral and ventrolateral neuronal columns. Previous studies of the anatomical circuitry and function of the central gray focused on individual subdivisions from a rather narrow two-dimensional perspective and basically neglected the longitudinal, three-dimensional organization of this region, which the columns clearly define. It is currently believed that different classes of stressful, threatening, or painful stimuli trigger distinct coordinated patterns of skeletal, autonomic, and antinociceptive adjustments by selectively targeting specific central gray columnar circuits.

B. Neuronal Types and Cytoarchitecture

The neurons of the human central gray can be classified into four broad categories: fusiform, stellate, multipolar, and pyramidal. The fusiform cells have an elliptical cell body, with at least one process emerging from each pole. Although they predominate in the medial subdivision, they are present throughout the central gray and exhibit rather distinctive orientations in the four subdivisions. Fusiform cells have a vertical orientation in the dorsal subdivision, are arranged parallel to the edge of the aqueduct in the medial subdivision, and are arranged either vertically or horizontally in the two lateral subdivisions.

Both stellate and multipolar cells are distributed throughout the central gray. Stellate cells are characterized by an oval soma that gives rise to four to six randomly oriented primary dendrites. The multipolar neurons are distinguished by an extensive dendrite arbor, spreading preferentially in the transverse plane.

The fourth neuronal type, the pyramidal cell, is most numerous in the lateral subdivisions of the central gray. It has a pyramidal cell body and is characterized by an extensive dendritic arborization, which often penetrates well into the overlying superior colliculus or into the adjacent cuneiform nucleus.

Although neurons in the central gray can be classified into these four broad neuronal categories, it should be emphasized that the neurons of this region constitute a heterogeneous population, and some neurons are not easily classified.

C. Connections with Other Brain Regions

1. Afferent Connections

The central gray receives neural inputs from a large number of central nervous system regions, which allow this midbrain area to be influenced by motor, sensory, autonomic, and limbic system structures. The major inputs to the central gray are summarized in Fig. 3 and are based on the results of studies of the monkey central gray.

The hypothalamus provides the greatest descending input to the primate central gray. Major projections arise from the anterior, posterior, and lateral hypothalamus and probably allow the central gray to play a key integrative role in descending limbic (i.e., those portions of the brain involved in learning, emotions, and behavior) and autonomic (i.e., those portions of

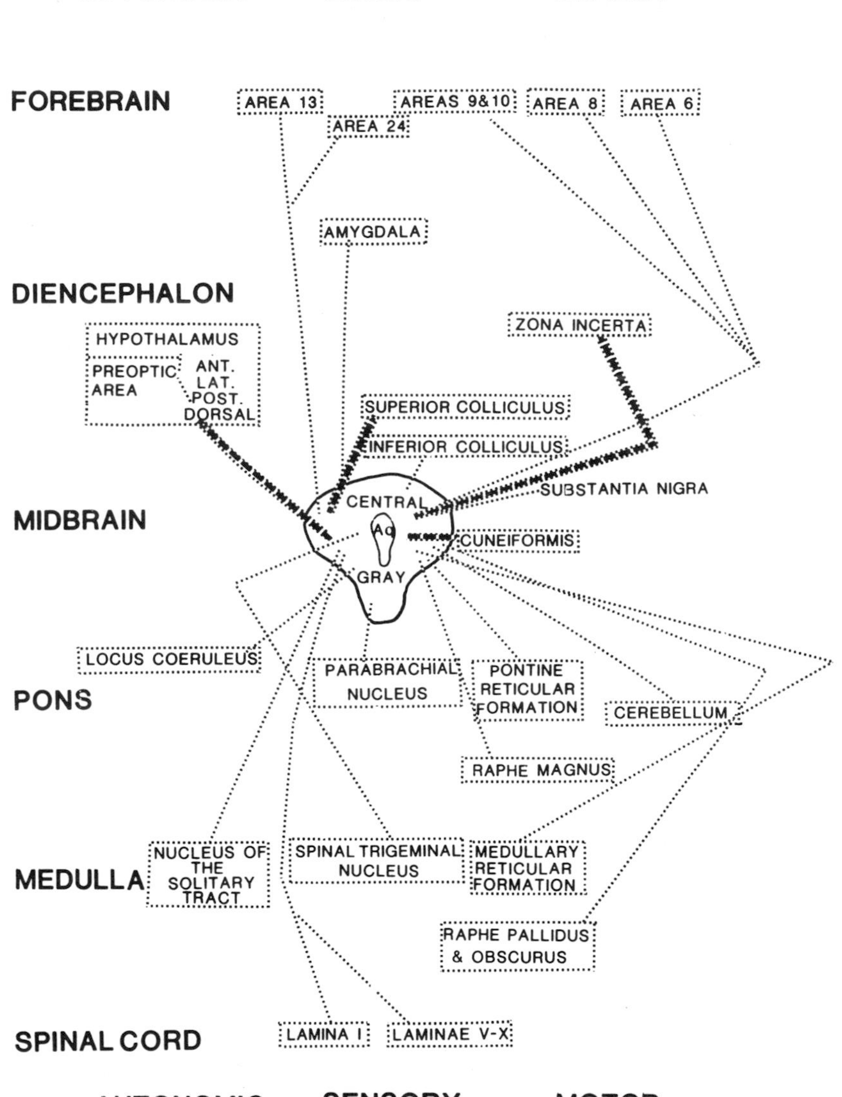

FIGURE 3 Origin of the major afferent projections to the central gray from different levels of the central nervous system. The largest inputs are indicated by heavy dashed lines and include the hypothalamus, zona incerta, and cuneiform nucleus. The various regions are also broadly classified under motor, limbic, autonomic, and sensory systems. Obviously, some regions (e.g., the reticular formation and the superior colliculus) can be grouped under more than one functional system. [Reproduced with permission from A. J. Beitz (1990). Central gray. *In* "The Human Nervous System" (G. Paxinos, ed.), pp. 307–320. Academic Press, San Diego.]

the brain and the spinal cord controlling heart rate, respiration rate, and other visceral functions) multisynaptic pathways. [*See* Hypothalamus.]

In addition to the hypothalamus, the importance of other descending limbic system input to the primate central gray should be emphasized. Anatomical studies have shown direct projections from the central and basolateral amygdala, bed nucleus of the stria terminalis, and cingulate gyrus (a component of the limbic system) to this midbrain region, whereas electrophysiological studies have demonstrated that over 50% of the neurons in the central gray are influenced by stimulation of the hippocampus and the amygdala. These diverse limbic system inputs to the central gray underscore the involvement of this region in certain aspects of human behavior (see Section III,E). [*See* Hippocampal Formation.]

The central gray also receives a substantial projection from areas 6, 8, 9, and 10 of the cerebral cortex, as well as from the insular and prefrontal cortices. The prefrontal cortex has a rather selective projection, sending fibers predominantly to the dorsolateral subdivision of the central gray. It is likely that this projection allows neocortical regulation of our perception of noxious stimuli. [*See* Cortex.]

The central gray receives an important descending projection from the zona incerta, as well as inputs from the mesencephalic reticular formation and the deep layers of the superior colliculus. In addition, the primate central gray receives a large number of ascending projections, including afferent connections from the locus coeruleus, parabrachial nuclei, raphe magnus, raphe pallidus, and a variety of brain stem reticular nuclei. Neurons in the spinal trigeminal nucleus and the spinal cord also project to this midbrain region.

It is likely that these ascending projections convey autonomic, nociceptive, and other somatosensory information to the central gray. This is supported by electrophysiological data demonstrating the presence of neurons in the primate central gray that respond to genital, rectal, innocuous somatosensory, or various forms of noxious stimulation.

2. Efferents

A review of the literature indicates that the central gray is reciprocally linked with most brain regions. Thus, the primate central gray projects to the anterior, dorsal, lateral, and posterior hypothalamic nuclei. All of these nuclei have been demonstrated to provide reciprocal projections back to the central gray region, as indicated earlier. The central gray also sends efferent fibers to the zona incerta; mesencephalic, pontine, and medullary reticular formation; the superior colliculus; the lateral parabrachial nucleus; the raphae magnus and pallidus; the spinal trigeminal nucleus; and the spinal cord. Each of these regions sends reciprocal projections back to the central gray. This two-way communication system between the central gray and numerous other brain structures underlies its unique role in integrating numerous functional systems (see Section III).

II. NEUROTRANSMITTERS AND NEUROMODULATORS

The central gray contains a myriad of neurotransmitters and neuromodulators. Unfortunately, our knowledge of the presence and distribution of these neurochemicals within the central gray far exceeds our understanding of their relationship to the neural circuitry of this region or their functional roles within this midbrain area. The four following sections summarize the existing data concerning the localization of the neurochemicals in the human central gray.

A. Monoamines

The catecholamines dopamine, norepinephrine, and epinephrine and the indoleamine serotonin are neurotransmitters in the central nervous system. Immunocytochemical studies using antibodies against tyrosine hydroxylase or phenylethanolamine *N*-methyltransferase, catecholamine-synthesizing enzymes, have demonstrated catecholamine-containing neurons and fibers in the ventrolateral subdivision of the central gray in both fetal and adult human midbrains. Biochemical studies have shown that epinephrine and norepinephrine are the key neurotransmitters of the catecholamine-containing cell bodies and fibers in this area.

This is further supported by *in vitro* receptor binding data demonstrating the presence of relatively high densities of receptors for these neurotransmitters (called adrenergic receptors) in the human central gray, whereas both D_1 and D_2 dopamine receptors are present in low densities. Although catecholamines could play a role both in behavioral responses related to central gray stimulation and in the production of analgesia in lower mammals, their role in the human central gray remains to be elucidated. [*See* Adrenergic and Related G Protein-Coupled Receptors; Catecholamines and Behavior.]

Serotonin-containing neuronal cell bodies are confined to the ventrolateral central gray, whereas serotoninergic fibers are most numerous in the medial subdivision of this midbrain structure. Serotoninergic neurons are round or fusiform, with their longitudinal axis oriented dorsoventrally. Both serotonin-1 and serotonin-2 receptors have been demonstrated in the human central gray, and there is some evidence suggesting that the serotonin system in this midbrain region could play a role in human behavior.

B. Acetylcholine

In addition to catecholamines and indoleamines, acetylcholine probably also plays an important role in the normal function of the human central gray. The human brain contains all of the elements of the cholinergic synapse (which releases the transmitter acetylcholine): the biosynthesizing enzyme choline acetyltransferase, the catabolizing enzyme acetylcholinesterase, uptake and storage mechanisms, and both postsynaptic and presynaptic receptors.

The human central gray contains a high density of one type of acetylcholine receptor, the muscarinic receptor, especially in the dorsolateral and ventrolateral subdivisions. Acetylcholinesterase is also present in the central gray, where it appears to be more prominent in the dorsolateral and ventrolateral subdivisions. This is consistent with the receptor binding data and suggests that acetylcholine might play an important role in these two subdivisions. In fact, injection of cholinergic agonists (i.e., drugs that stimulate acetylcholine action) into the ventrolateral central gray of animals produces a dose-dependent bradycardia (i.e., slowing of the heart beat), implicating this transmitter in autonomic regulation by the central gray.

C. Amino Acid Transmitters

Glutamate and possibly aspartate represent two of the most important excitatory neurotransmitters in the central nervous system, whereas γ-aminobutyric acid (GABA) and glycine represent important inhibitory transmitters in the brain. Based on their widespread distribution throughout the central nervous system, it is not surprising that these four simple amino acid transmitters are present in the human central gray.

Using immunocytochemical procedures, glutamate-immunoreactive cell bodies are found predominantly in the ventrolateral and dorsal subdivisions. Aspartate-immunoreactive cell bodies are typically fusiform or triangular and are localized to the dorsolateral and ventrolateral subdivisions. The presence of glutamate- and aspartate-immunoreactive neurons in the human central gray could be functionally significant in light of the results of animal studies demonstrating that the central gray projection to the raphe magnus utilizes excitatory amino acids as transmitters. This projection plays an important role in the descending pain modulation system discussed in Section III,A. The recent cloning of multiple glutamate receptor subtypes and the subsequent localization of several of these subtypes to both the central gray and the raphe magnus emphasize the importance of excitatory amino acid transmitters in these two regions.

GABA-immunoreactive neurons are scattered throughout the rostrocaudal extent of the central gray and are concentrated in the ventrolateral and dorsal subdivisions. These neurons are predominantly fusiform and display distinct orientations dependent on their location. Although GABA receptors have not been studied in the human central gray, benzodiazepine binding sites, which are related to GABA receptors, are found in relatively high amounts in this midbrain region.

Analysis of glycine immunoreactivity in the central gray reveals predominantly fiber and terminal staining. The dorsal and dorsolateral subdivisions contain the highest density of both fibers and glycine receptors.

D. Neuropeptides

The large class of neuropeptides is structurally quite diverse and plays important neurotransmitter or neuromodulator functions in the mammalian central nervous system. Although a majority of these neuropeptides have now been detected in the human brain, only recently have immunohistochemical techniques been used to localize peptides within the human brain stem. Opioid peptides were among the first detected in the human midbrain. Both β-endorphin and Met-enkephalin are present in the central gray, as well as several peptides derived from proenkephalin A. The μ, δ, and κ opiate receptor binding sites have also been identified in this midbrain area, reinforcing the hypothesis that opioid peptides play an important role in the human central gray, especially with regard to the activation of descending pain modulatory systems (see Section III,A).

Substance P- and somatostatin-immunoreactive cells and fibers are found throughout the central gray. Substance P fibers are concentrated in the dorsal half of the central gray, whereas somatostatin fibers are distributed homogeneously. Cholecystokinin- and vasoactive intestinal polypeptide-immunoreactive fibers have also been described throughout the human central gray. The functional roles of these four peptides in the human central gray remain to be elucidated.

III. FUNCTIONS AND RECEPTORS

A. Pain Modulation and Opiate Receptors

The initial demonstration by Reynolds in 1969 that electrical stimulation of the central gray induces a profound analgesia (i.e., loss of pain sensation), coupled with the discovery of an opioid system that is dependent on the central gray for inducing pain relief, resulted in an explosion of interest in the role played by this region in both nociception and analgesia. It is now well established that an endogenous pain modulation system exists in the central nervous system and that the central gray, together with the nucleus raphe magnus and several other brain stem nuclei, is a key component of this system. The original studies by Reynolds and others demonstrating behavioral antinociception (i.e., inhibition of pain sensation) in animals following central gray stimulation laid the foundation for determining the therapeutic effect of such stimulation in humans. [*See* Pain.]

Clinical evidence has now established that stimulation of this region in humans is effective in the control of pain syndromes that are responsive to opiates (e.g., morphine). Initial studies indicated that stimulation-produced analgesia was reserved by the opiate antagonist naloxone, which blocks the interaction of opiates with their receptors. It is not clear, however, whether pain relief from electrical stimulation of the human central gray involves an opioid mechanism or whether the response to opiates is predictive of the effectiveness of central gray stimulation in a given patient.

Studies in animals have indicated that the μ opiate receptor is responsible for descending pain inhibition originating in the central gray. This is not surprising since both the mRNA and protein for the μ opioid receptor are more prevalent in the central gray than the message or protein for the δ or κ opioid receptor. Moreover, the endogenous opioid peptides, B-endorphin and enkephalin, are both present in the central gray and have been shown to be potent ligands of the μ receptor. Opioid receptors in this region clearly play a role in the analgesic effect produced by exogenously administered opiate drugs, like morphine, but the endogenous opioid system present in this region may also play a critical role in stimulation-produced analgesia and in acupuncture-induced analgesia. However, this does not exclude the possibility that central gray stimulation might also operate via activation of nonopioid systems. Consistent with this possibility are studies in human patients indicating that pain relief obtained by stimulation of the dorsal central gray is not reversed by naloxone and is not accompanied by alterations in the endorphin level in ventricular cerebral spinal fluid.

These data imply a possible heterogeneity of the analgesic system existing at the level of the central gray in humans. Although further work is necessary to clarify the role of opioid and nonopioid systems in analgesia induced by electrical stimulation in the central gray, stimulation of this region provides a viable alternative means of long-term pain control.

B. Vocalization and Amino Acid Receptors

Evidence over the past two decades has indicated that excitatory amino acids are the most prevalent excitatory transmitters in the central nervous system. Glutamate- and aspartate-immunoreactive cell bodies, fibers, and terminals, as well as the three major classes of excitatory amino acid receptors (i.e., *N*-methyl-D-aspartate, quisqualate, and kainate), have been localized to the central gray. Based on their extensive distribution in this region, it is not surprising that excitatory amino acids play a role in several proposed functions of the central gray, including its involvement in vocalization.

It has been known since the late 1930s that stimulation of the primate central gray yields species-specific calls in rhesus monkeys. Vocalization, the nonverbal production of sound, can be elicited in many vertebrates by stimulation in several regions of the brain, but most easily in the central gray. Furthermore, recent data indicate that the posteriolateral central gray, but not the anterior part, is involved in vocal motor control. Vocalization can be induced by microinjections of excitatory amino acid agonists into the dorsolateral primate central gray, implicating excitatory amino acid receptors in the vocalization process. Anatomical and physiological data indicate that the central gray projects to the nucleus retroambiguus, which, in turn, projects to vocalization motor neurons, and that this projection system forms the final common pathway in vocalization.

C. Eye Movements

The central gray has also been implicated in eye movements, and the dorsolateral subdivision appears to be especially important in regulating presaccadic activity (i.e., neural activity occurring immediately before the quick jump of the eyes from one fixation point to another). Bilateral lesions of the human dorsolateral central gray have been reported to cause selective paralysis of the downward gaze, whereas lesions of the dorsal subdivision have been reported to cause upward gaze paralysis. However, because the central gray is located close to several nuclear groups and fiber bundles known to be directly involved in eye movement, such reports must be interpreted with some caution.

D. Autonomic Function and Opiate Receptors

Several autonomic reactions can be elicited by stimulation or lesions of the central gray, including temperature regulation and changes in stomach and bladder tone, respiratory pattern, pressor responses, and pupillary dilation. Recent work demonstrating that morphine microinjected into the ventral central gray produces hyperthermia (i.e., increased body temperature) provides further evidence for a role in temperature regulation. Microinjection of morphine into the central gray has also been reported to inhibit intestinal propulsion. In addition to their role in analgesia, it appears that the opioids could play an important role in the involvement of the central gray in autonomic function. [*See* Autonomic Nervous System; Brain Regulation of Gastrointestinal Function.]

With regard to effects on respiration and blood pressure, stimulation of the central gray causes increases in respiratory rate and blood pressure. Hyperventilation and apneic periods (i.e., without breathing) have also been reported in humans following stimulation of this region. Finally, it should be noted that excitatory amino acids microinjected into the central gray of animals cause changes in both blood pressure and heart rate whereas injections of cholinergic agonists produce a dose-dependent bradycardia, as indicated in Section III,B. This suggests that excitatory amino acid receptors and cholinergic receptors might also be involved in autonomic functions associated with the central gray. Finally, it should be noted that one of the earliest and strongest indications of functional columns in the central gray stems from investigations demonstrating that discrete activation of the lateral column elicited pressor responses, whereas activation of the ventrolateral column produced depressor responses. Moreover, recent studies have now shown that peripheral changes in resting arterial blood pressure activate discrete columns of central gray neurons. The relationship between these "pressor- and depressor-activated" input columns and the central gray output neurons, whose activation produces selective changes in arterial pressure, remains to be elucidated.

E. Role in Behavior

A possible role for the central gray in human behavior can be predicted from the rich interconnections that exist between this midbrain area and limbic structures. Thus, it is not surprising that the central gray has been implicated in sexual behavior, rage and fear reactions, and memory storage. Interestingly, stimulation of the human central gray produces feelings of intense fear. Similarly, stimulation of the dorsal central gray induces aversive behavior in animals, and it has been suggested that aversive stimulation of this region might serve as a model for human anxiety. Indeed, benzodiazepines (which are commonly used to treat human anxiety) have been shown to diminish the aversive behavior produced by stimulation of the central gray in animals. With regard to anxiety, it is possible that the autonomic responses produced by central gray stimulation, as described in the previous section, reflect an "anxiety" component common to both fear and pain.

Recent evidence suggests that there may be a "sexual" longitudinal column in the central gray. This is based in part on the selective projection from the medial preoptic area of the hypothalamus to the central gray. The medial preoptic area is complex region located just lateral to the walls of the third ventricle. This region has been strongly implicated in sexual behavior and neuroendocrine regulation and the projections from this region to the central gray exhibit a striking degree of columnar specificity, targeting the dorsomedial column along its entire length as well as the dorsolateral column superiorly and the lateral column inferiorly. Moreover, neurons expressing estrogen receptors are preferentially located in brain structures implicated in reproductive behavior and both the central gray and medial preoptic area contain abundant populations of neurons that express estrogen receptors. Recent evidence indicates a remarkable overlap of medial preoptic axon terminals in the central gray with neurons expressing estrogen receptors. It is currently hypothesized that a functional medial preoptic–central gray–medullary circuit exists that engages preautonomic and somatic premotor neurons involved in spinal sexual reflexes and sexual behaviors.

The central gray transmitters and receptors that mediate the involvement of this region in behavior are only beginning to be elucidated, but it appears likely that a large number of transmitters and their receptors are involved. Microinjection of GABA antagonists into the dorsal central gray produces escape-like behavior in animals, and this behavioral effect is modulated (i.e., modified) by preinjection of morphine into the same central gray site. Similarly, affective defensive behavior produced by electrical stimulation of the dorsal central gray is suppressed by microinjection of the enkephalin analog D-Ala^2-Met^2-enkephalinamide into the site of stimulation.

Excitatory amino acids and serotonin have also been implicated in the modulation of defensive behavior. Thus, GABA benzodiazepine, opiate, excitatory amino acid, and serotonin receptors appear to be involved in the mediation of certain types of behavior by the central gray, indicating the chemical complexity of this particular aspect of central gray function.

BIBLIOGRAPHY

Bandler, R., and Shipley, M. T. (1994). Columnar organization in the midbrain periaqueductal gray: Modules for emotional expression? *Trends Neurosci.* **17,** 379.

Beitz, A. J. (1990). Central gray. *In* "The Human Nervous System" (G. Paxinos, ed.), pp. 307–320. Academic Press, San Diego.

Dennis, B. J., and Meller, S. T. (1993). Investigations of the periaqueductal gray (PAG) of the rabbit: With consideration of experimental procedures and functional roles of the PAG. *Prog. Neurobiol.* **41,** 403.

Depaulis, A., and Bandler, R. (eds.) (1991). "The Midbrain Periaqueductal Gray Matter: Functional, Anatomical and Neurochemical Organization." Plenum, New York.

Depaulis, A., Penchnick, R. N., and Liebeskind, J. C. (1988). Relationship between analgesia and cardiovascular changes induced by electrical stimulation of the mesencephalic periaqueductal gray matter in the rat. *Brain Res.* **451,** 326.

Fields, H. L., and Besson, J. M. (eds.) (1988). Pain modulation. *Prog. Brain Res.* **77,** 141–312.

Jurgens, U., and Richter, K. (1986). Glutamate-induced vocalization in the squirrel monkey. *Brain Res.* **373,** 349.

Reynolds, D. V. (1969). Surgery in the rat during electrical analgesia induced by focal brain stimulation. *Science* **164,** 444.

Young, R. F., and Chambi, V. I. (1987). Pain relief by electrical stimulation of the periaqueductal and periventricular gray matter. *J. Neurosurg.* **66,** 364.

Brain Evolution

RALPH L. HOLLOWAY
Columbia University

GLOSSARY

Allometry Mathematical relationship between the size of bodily components and some larger unit of body size, such as body weight

Encephalization quotient Measure of the amount of brain tissue that an animal possesses beyond that expected for its body weight

Endocast Cast, either natural or man-made, of the inside of a skull, which may or may not show the cerebral convolutions of the brain

Lunate sulcus Sulcus that usually defines the anterior limit of primary visual striate cortex in primates

Organization In the context of brain studies, the relationships between nuclear masses and the fiber tracts that interconnect them, as well as the patterns by which these relations emerge during growth and development

Paleoneurology Study of brain endocasts of past living animals

Petalias Small extensions of parts of the cerebral cortex into the internal table of bone of the skull; they are used to assess cerebral asymmetries

Reorganization Quantitative shifts in the relationships between neural nuclei and fiber tracts and the different regions of the cerebral cortex that take place when new species evolve

HOW DID THE HUMAN BRAIN EVOLVE? WOULD A full knowledge of its evolutionary development be anymore interesting and informative than a full knowledge of the evolution of animal brains from groups ranging from aardvarks to zebras? And just what does it mean that neither aardvarks nor zebras are asking these questions? Clearly, the human animal has embarked upon an evolutionary trajectory that has capitalized on intelligence and rationalization based on arbitrary symbol systems as a keystone of its adaptive prowess, and yet it is a prowess with ever-present attending folly and colossal potential for self-annihilation. Despite the wonders of our brains and what we can (and don't) do with them, it is well to remember that this adaptation has had a fairly short evolutionary duration compared to most genera of other animals. The human genus *Homo,* partly defined by a brain size above the limits of any known hominoid brain and also with a large relative size compared to its body weight, has had a duration of about 2 million years. What we regard as typical of ourselves as a genus and a species that is, *Homo sapiens sapiens,* has had a duration of about 100,000 years or less. Many, if not most, animal genera have a recognizable duration in the fossil record of roughly 5–10 million years. If duration is any measure of evolutionary wisdom or truly successful adaptation, we certainly have a long way to go. This article focuses on how that evolution occurred. What lines of evidence are available to examine this evolutionary odyssey?

I. APPROACHES

There are basically two important approaches to such an understanding, regardless of what species or groups of species we are studying. The first approach

ENCYCLOPEDIA OF HUMAN BIOLOGY, Second Edition, VOLUME 2. Copyright © 1997 by Academic Press. All rights of reproduction in any form reserved.

considers the following: comparative neuroscientific information on extant, living species, which provides data on how large the brain is absolutely and relatively; quantitative information on how the brain is divided into units at the gross to microscopic level, and data regarding how these nuclei, cortical regions, and fiber tracts are interconnected; how the brain develops ontogenetically; and the relationships between neurological structural variation and behavioral variation.

The second approach deals with the actual fossil evidence for evolutionary changes within the taxa being studied. This is often called *paleoneurology*. Paleoneurological data consist of brain endocasts taken from the interior of the skulls of once-living animals. Sometimes these are natural, having been formed through permineralization of sediments that fill the skull after the brain tissue rots and is replaced. Other endocasts are made using rubber or silicone molding agents to make a cast of the internal bony surface of the skull. The endocasts yield valuable data about the size of the brain and its shape, and rough ideas about some of its parts, cortical asymmetries, and, occasionally, convolutional details regarding the size and disposition of the cerebral gyri and sulci.

At least two very important points need to be made about these approaches. The first is that the comparative approach is incredibly data-rich whereas paleoneurology is incredibly data-poor. The second is that extant species, however assembled or arranged, are only the terminal end points of their own lines of evolution. They do not represent an evolutionary sequence. This fact has largely been lost on most comparative and clinically practicing neuroscientists. I think of these two approaches as the *indirect* and *direct* evidence, respectively. The obvious goal, of course, is to blend these two approaches together in both complemental and supplemental fashion.

Perhaps in the future, a third approach can be taken, which might be called neurogenetics (or "paleoneurogenetic"). Here the goal would be to unravel some of the 40,000 or so genes that control the development and operation of the human brain and to understand what each one's parts is in the whole, and then to compare this information for closely related taxa. In other words, if we knew which genes turned "on" and "off" the growth and development of different neural nuclei, fiber tracts, and cortical regions, we might have a better understanding of how the human brain evolved. But this is currently in the realm of science fiction.

II. ORGANIZATION OF THE HUMAN BRAIN

The human brain, like that of many other animals, is composed of thousands if not billions of "parts" depending on how microscopic the analysis might be. Naming all the various nuclei of the brain and the fiber pathways that connect them involves hundreds of names. To those beginning their first neuroscience course, it seems as if everything might be connected to everything else. Fortunately, that is not the case, but there is more than enough connectivity to make the study of the brain a humbling experience.

At a more microscopic level, the brain is composed of tens of billions of units called neurons, which are nerve cells in either one of two states, firing or not firing. If each "state" depended on some particular combination of nerves firing or not firing, the possible permutations for a human brain are somewhere on the order of 2 raised to the ten-billionth power. This is a fair amount considering that if there are 100 billion galaxies, each with 100 billion stars, that is only 10 raised to the 22nd power. It is likely that chimpanzees (*Pan troglodytes*) have more states than stars in the universe, and a chimpanzee brain is not that much different from our own.

There is also a hierarchical organization to the brain, which refers to relationships between the cerebral cortex, the basal ganglia, the limbic system, and the olfactory bulbs; together they compose the *telencephalon* or forebrain. These structures surround the thalamus, epithalamus, hypothalamus, and pineal gland, which together make up the *diencephalon*. Moving further "downward" is the brain stem, which contains the colliculi (superior and inferior, visual and auditory, respectively); it is known as the *mesencephalon* or midbrain. Finally, structures such as the cerebellum, the pons, the medulla, and the third and fourth ventricles are integrated with the spinal chord. This is called the *metencephalon* and *myelencephalon*. This organization represents a case of extraordinary neurogenetic conservatism, as it is basic to almost all vertebrate brains, and the ontogenetic unfolding of these structures is also highly conserved. [*See* Brain.]

This structural conservatism underlines another important fact, which is that the total amount of neurogenetic change between ourselves and our earliest hominid ancestors of 4–5 million years ago has unquestionably been very minor. There are no "new" structures known in the human brain that are not present in primates such as the macaque and chimpanzee. Even the fiber tracts, which appear in part to determine our language capabilities, are present in these nonhuman primates. This leads us to several questions. How do our brains differ from those of our closest relatives, and how did it evolve? First we will consider the size of the brain, and then its organization.

III. SIZE OF THE HUMAN BRAIN

The average size of the human brain is about 1330 g, varying between roughly 900 to 2000 g at its extremes. It is the largest brain in the primate order. Elephant and whale (and dolphin) brains are absolutely larger, some attaining weights of 7500 g and greater. Relatively speaking, the human brain is about 2% of total body weight, a ratio higher than that among elephants and whales, but not the highest in mammals. Squirrel monkeys have higher ratios than humans. It is really the *combination* of absolute and relative brain size that makes the human animal unique. Biologists have shown this very clearly by plotting the log (base 10) of brain weight against the log of body weight. Depending on the totality of animals plotted, a fairly tight regression line is found that usually can be described by a formula in which brain weight is a function of a constant multiplied by a body weight raised to a particular exponent, such as 0.66 or 0.75, depending on whether the relationship is geometrical or metabolic. This is often called an allometric equation, because it indicates that there is some lawful constraint about brain size that relates to body size:

$$\text{brain weight} = \text{constant} \times \text{body weight}^N$$

Figure 1 shows such an allometric plot for the primate order. It is clear that the value for the human (*Homo*) point is well above the trend of points. Indeed, when the human value for brain weight is calculated from the equation, the real human value is more than three times as great as that calculated. In other words, the human brain weight is more than three times what would be expected for a primate of our body weight. This also yields a ratio known as the *encephalization quotient,* or EQ, which will vary depending on the equation used and the animal data set used to generate the equation. Table I illustrates the use of several different basic allometric equations to calculate EQ's. These EQ's provide a ratio between the actual brain weight of an animal to that expected

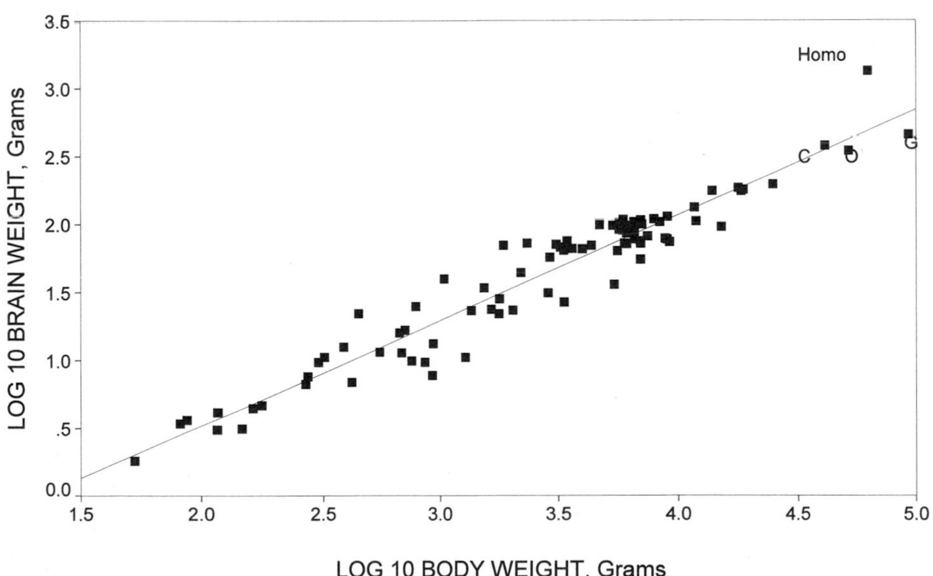

FIGURE 1 Log-log plot of brain weight against body weight for 85 species, resulting in a linear regression line with a slope of roughly 0.76 and a correlation coefficient of about 0.98. Note that the *Homo* value is a major outlying value, as the human brain is about three times larger than would be predicted for a primate of its body weight. C, chimp; G, gorilla; O, orangutan. (Data from H. Stephan, personal communication.)

TABLE I
Encephalization Quotients for Various Primates Based on Different Formulas[a]

Species	Brain weight (g)	Body weight (g)	EQ Homo	EQ Jerison	EQ Primates	EQ Stephan
Lemur	23.3	1,400	21	1.56 (22.6)[b]	0.94 (32.7)	5.66 (19.6))
Baboon	201	25,000	28	1.97 (28.5)	0.90 (31.3)	7.94 (27.5)
Gorilla	465	165,000	23	1.56 (22.5)	0.61 (21.2)	6.67 (23.2)
Orang	370	55,000	31	2.15 (31.1)	0.91 (31.7)	8.90 (30.9)
Chimp	420	46,000	39	2.63 (38.1)	1.18 (41.1)	11.3 (39.3)
Homo	1,330	65,000	100	6.91 (100)	2.87 (100)	28.80 (100)

EQ Homo = Brain wgt/1.0 Body wgt$^{0.64906}$

EQ Jerison = Brain wgt/0.12 Body wgt$^{0.66}$

EQ Primates = Brain wgt/0.0991 Body wgt$^{0.76237}$

EQ = Stephan = Brain wgt/0.0429 Body wgt$^{0.63}$

[a]These formulas are themselves based on different data sets. The EQ Homo equation is based simply on the mean value of both the brain and body weights for modern *Homo sapiens.* The EQ Jerison equation is based on almost 200 mammalian species. The EQ Primates equation is for all of the known primates, whereas the EQ Stephan equation is based on an equation for insectivores only. As can be seen, while the percentages of the *Homo sapiens* values are essentially similar in each case, except for the lemur using the EQ Stephen equation, there is a relativity to these relative quotients.

[b]Number in parentheses is percentage of *Homo sapiens* value.

based on its particular body weight, where the nominator is the animal's actual brain weight and the denominator is the allometric equation being used. As Table I indicates, using the "homocentric" equation provides an instant value in terms of the human value.

It is interesting to reflect that within our species, *Homo sapiens,* the variation between upper and lower brain weights is about 1000 g (e.g., from 1000 to 2000 g), and that is roughly the same as the total amount of evolutionary brain size change from our earliest australopithecine ancestors of 4 million years ago to the present, that is, from about 400 g to 1350–1400 g.

The change in brain size throughout hominid evolution seems to have been a process of both punctuated equilibria and gradualism. That is, at some times the increase in brain size was rapid, appearing almost as a discontinuity with the past, whereas at other times the change was slow and seemingly continuous. How one views these changes depends very much on how representative our samples of brain volumes are for our fossil ancestors, and whether we actually have an adequate fossil record to distinguish between the two kinds of rates. Put simply, we don't. Another confounding factor regarding the change of brain size with time is whether it is related to an accompanying change in body size or is purely related to an increase in brain size without any concomitant increase in body size. The known fossil record demonstrates that it is a mixture of both of these processes. [*See* Human Evolution.]

For example, the earliest known australopithecines, *Australopithecus afarensis,* living 3–4 million years ago (mya), had cranial capacities well within present-day chimpanzee ranges, about 350–450 cc (Table II). The body size of these fossil hominids is roughly the same as that of modern chimps. The next fossil hominid group is *Australopithecus africanus,* known from South Africa and dating at about 2–3 mya, which may have been intermediate between *A. afarensis* and the earliest *Homo,* the latter known from both East and South Africa. The brain size is perhaps slightly higher in *A. africanus,* on average, but the body weight appears to be the same as in *A. afarensis.* One offshoot from *A. afarensis* was a more robust line known as *A. aethiopicus,* which led to both *Zinjanthropus boisei* and *Australopithecus* (some prefer *Paranthropus*) *robustus,* which existed up until roughly 1.5 mya. Their brain sizes were larger (about 500–550 cc), but then so were their bodies, and it seems most plausible to attribute the increase in robust australopithecine brain size to simple allometry.

The earliest examples of our own genus *Homo* appear at about 1.9 mya and have been designated *Homo habilis.* The evidence from the cranial fragments from East Africa strongly suggests brain sizes

TABLE II
Hominid Endocranial Brain Volumes[a]

Taxon	N	Region	Mean endocranial volume (cc)	Range (cc)	Dating (million years ago)
Australopithecus afarensis	3	E. Africa	435	320–500	3–4
A. africanus	8	E. Africa	440	420–500	2–3
A. aethiopicus	1	E. Africa	410	410	2.5
A. robustus	6	E. Africa	512	475–530	1.6–2
Homo rudolphensis?	2	E. Africa	775	752–800	1.8
Homo habilis	6		612	510–687	1.7–2.0
Homo ergaster?	2		826	804–848	1.6
Homo erectus	2	E. Africa	980	900–1067	1.0–1.6
Homo erectus	8	Indonesia	925	780–1059	1
Homo erectus	8	China	1029	850–1225	0.6
Archaic Homo sapiens	6	Indonesia	1148	1013–1250	0.13
Archaic Homo sapiens	6	Africa	1190	880–1367	0.125
Archaic Homo sapiens	3	Asia	1237	1120–1390	0.25–0.075
Archaic Homo sapiens	7	Europe	1315	1200–1450	0.5–0.25
Homo sapiens neanderthalensis	25	Europe, Middle East	1415	1125–1740	0.09–0.03
Homo sapiens sapiens	11	Europe, Middle East, Africa	1506	1250–1600	0.025 to 0.01

[a]The volumes are always somewhat larger than the actual brain weights, as they include the meningeal coverings and cerebrospinal fluid. The difference is seldom more than 10%, and most paleoanthropologists use brain volume and brain weight interchangeably.

of between 575 and 700 cc, but it is uncertain how large their bodies were. One intriguing specimen, OH 62, from Olduvai Gorge, Tanzania, suggests a very small and gracile body, as does the foot from OH 8. If so, and this species of *Homo* is derived from *A. africanus,* the change in brain size was not a function of body size, but an autonomous increase in brain size itself. These two patterns, one related to body size and the other to the brain alone, suggest different selection pressures.

Unfortunately, the picture becomes somewhat more confusing because there appears to be a mini-adaptive radiation of early *Homo* species at roughly the time period of 1.6–2.0 mya. The famous KNM-ER 1470 cranium, which yielded a brain size of 752 cc according to my reconstruction, is now known as *Homo rudolphensis,* dated at about 1.8 mya, and from associated limb bones it was clearly larger than earlier habilines. Two additional crania, dated at about 1.6 mya, are now designated *Homo ergaster,* with brain sizes of about 800–850 cc. No limb bones were recovered, however, so their body sizes are a moot point. When first discovered, they were thought to be the very familiar genus and species *Homo erectus,* well known from Asia. Indeed, one other specimen, KNM-WT 15,000, from the Nariokotome site on the western side of Lake Turkana, Kenya, is called *Homo erectus* and lived at roughly the same time. He was a youth in his teens, but his brain size was about 900 cc and his body size much like that of our own species. There are other fossil crania known from Lake Turkana on the east side, namely, KNM-ER 1805 and KNM-ER 1813, that are smaller in brain size (about 585 and 513, respectively), and we frankly do not know whether these were truly early *Homo* or more advanced australopithecines.

If these hominids are derived from something akin to OH 62, then there clearly were increases in both body size and brain size, but how much of the increase was allometric and how much the brain alone cannot be untangled at this time. It is very clear, however, from the Nariokotome youth that at about 1.6 mya the body size of *Homo* was already at essentially modern levels, and that the subsequent increase in brain size was only partly allometric. Stone tools made to standardized patterns appear at roughly 2.5 mya, predating any evidence we have for early *Homo.* At about 1.8–2.0 mya, the tool patterns such as Oldowan and Acheulean become much more common and widespread. Many researchers believe there must be some link between the brain size attaining minimal modern human standards and the behavioral adaptations that allowed for stone tool making. Some of us, this author included, would even be willing to assert that primitive language abilities were present by about 1.8 mya. This, of course, is outright speculation.

From about 1.5 mya to the present, there is much variation in hominid brain size, but the trend is inexorably upward, culminating in Neandertals of about 50,000 years ago, with an average brain size slightly larger than our own but also accompanied by a musculoskeletal system that had more lean body mass, possibly as part of a cold weather adaptation. Hence, there is the possibility that that increase was basically an allometric one also. Since the Upper Paleolithic, there has been a slight decline of brain size to our modern selves, as well as a slight diminution of musculoskeletal ruggedness. But brain size is only part of the story (see table II for brain sizes).

IV. REORGANIZATION OF THE HUMAN BRAIN

Earlier it was noted that there is no evidence for any new or different neuroanatomical structures in the human brain. The same neural nuclei and fiber tracts prevail in almost all mammal brains, and certainly so in the primates. Obviously, each species has some specific repertoire of behaviors that reflects the organization of the underlying nervous system. At what levels of neurological organization do the differences exist and what exactly are they? We do not know the answers to these questions, except to note that in addition to size, the quantitative relationships that exist between various nuclei and fiber tracts vary in different animals, and both comparative neuroanatomy and paleoneurology suggest that some important *reorganizational* changes took place between early hominids and their more ape-like hominoid ancestors. Reorganization refers to mean changes in quantitative relationships between various nuclei, fiber systems, and the distribution of cortical tissue types in the cerebral cortex.

In discussing allometry in brain–body size relationships, it was noted that log-log regressions showed a linear relationship between brain and body weight for different taxa of primates (indeed, vertebrates). Similarly, there are allometrical relationships between parts of the brain and the brain as a whole. If we plot the log of the cerebral cortex against the log of brain weight for all primates, we find a very strong linear relationship, with a correlation coefficient of roughly 0.98, as seen in Fig. 2. The human value falls almost exactly on the regression line, indicating that the cerebral cortex of the human primate is as large as would be expected for a primate of its brain weight. The catch is that the cerebral cortex accounts for 76% of the weight of the brain in the human animal and decreases to about 60% in other primates (i.e., monkeys). Thus the measurement of the cortex is in fact a large part of the whole, and the correlation can be expected to be almost perfect. The human cerebellum takes up much less of the brain weight, and the correlation is still quite high, but the cerebellum is roughly 10% more than would be expected for a primate of such a brain weight. These regression exercises can be done for many of the neural nuclei making up the brain, and for the human animal, different nuclei will deviate from the expected allometric pattern to different degrees. Differences on the order of up to 25–50% might be attributable to error, as most of the data are based on a single brain for each species. But some deviations are considerable, as the following two examples show (refer to Table III).

If the volume of primary visual striate cortex (Brodmann's area 17) of the brain is plotted against the total volume of the brain, the human point falls some 121% below the value expected for a primate of its brain weight (Fig. 3). The next relay area from the visual striate cortex is in the thalamus, and that structure is the lateral geniculate nucleus. If that volume is plotted against the volume of the brain (again, these are log-log plots), the human point falls some 146% below the expected value (Fig. 4). There is nothing deficient in human visual capabilities compared to those of other primates. What these figures indicate is that there has been a positive relative increase in posterior parietal cerebral cortex in humans, that which we used to call the "association" cortex. This is the best example of reorganization in the human brain, indicating that at some point in the course of hominid evolution there occurred a reduction in primary visual striate cortex and an increase in parietal association cortex. Table III provides some examples of different structures in the human brain and how much they deviate from expected values. Not all regions of the brain have been adequately measured, such as particular nuclei within the hypothalamus or different parts of the amygdala or septal nuclei, and many of the fiber tracts such as the arcuate tract defy precise measurement. Nor have different dendritic branching patterns in different parts of the primate cerebral cortex been adequately quantified as yet. A

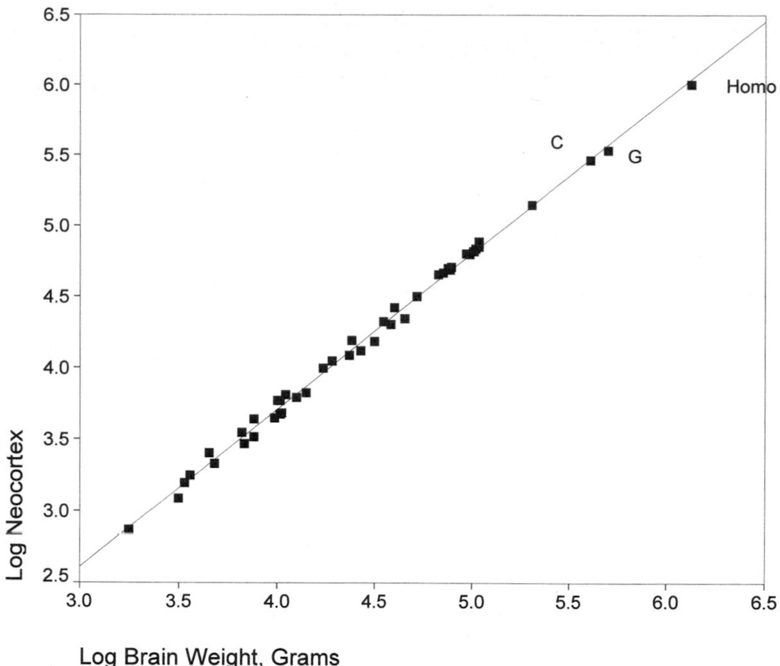

FIGURE 2 When volumes of neocortex are plotted against brain volume, the human point lies exactly on the regression line, and the overall dispersion shows few if any significant outliers. It must be remembered that in the human case, the neocortex accounts for about 76% of total brain volume, so the correlations are expectedly high. C, chimp; G, gorilla; $R^2 = 0.9963$. [Data from H. Stephan, *et al.* (1981).]

full quantitative neuromorphology of the primate brain is a long way off.

Perhaps the most interesting question is when in the course of human evolution did some of these "reorganizations" take place? The answer is quite controversial, because the meninges covering the brain during life hide the full details of the convolutions of the underlying brain, and consequently the endocasts show only imperfect and incomplete traces of those details.

Two fossil australopithecine fragments in particular have left endocranial brain traces of their underlying cerebral organization. These are the Taung child specimen described by Raymond Dart in 1925 and a specimen from *A. afarensis* known as AL 162-28. In all apes, the primary visual striate cortex known as Brodmann's area 17 is delineated from the adjacent posterior parietal cortex by a large sulcus called the lunate sulcus. Its position in chimpanzees, gorillas, and orangutans is relatively invariate and is placed far more anterior than in any human brain described to date. In our species, this sulcus, when it appears in its lunate shape, is situated in a much more posterior position on the cortex, a position fully consonant with the allometric relationship described earlier.

The question is: Can the lunate sulcus be seen on these early hominid endocasts and, if so, where was the lunate sulcus situated? Dart believed that the lunate sulcus on the Taung brain endocast (a natural one) was located in a posterior position, quite close to the lambdoid suture. Sir Arthur Keith, a Dart antagonist, believed that it was in a pongid-like anterior position, defined by a tiny dimple on the endocast surface. When Sir LeGros Clark published his observations on the anatomy of the autralopithecine remains in 1947, he stated that Dart was correct, because he could find no other anterior location with the requisite anatomy.

Then in 1981, D. Falk concurred with Keith's placement, whereas my own research led me to concur with LeGros Clark and Dart. Falk's placement of the lunate sulcus in the dimple is roughly 2.5 standard deviations *anterior* to the position expected for any ape, based on a sample of several chimpanzee brains

TABLE III
Residuals for Different Brain Structures Based on Allometric Relationships[a]

Dependent variable	Independent variable	Number of species	Correlation coefficient (*R*)	Actual value (*A*)	Expected value (*E*)	*A/E* ratio	% difference *A/E Homo*
Striate cortex	Brain weight (C)[b]	37	0.971	22,866	50,598	0.45	−121.30
		19	0.977		38,097	0.60	−66.60
Lateral geniculate	Brain weight (C)	37	0.978	416	1,026	0.41	−146.60
		19	0.982		857	0.49	−106.00
Cerebellum	Brain weight (C)	44	0.990	137,421	128,932	1.07	6.20
		26	0.994		150,535	0.91	−9.50
Diencephalon	Brain weight (C)	44	0.995	33,319	51,512	0.65	−54.60
		26	0.998		47,899	0.70	−43.70
Septum	Brain weight (C)	44	0.983	2,610	2,085	1.25	20.10
		26	0.991		2,201	1.19	15.70
Amygdala	Brain weight (C)	16	0.990	3,015	4,633	0.65	−53.70
		7	0.985		3,753	0.80	−24.50
Lateral geniculate	Thalamus	21	0.979	416	731	0.57	−75.72
		10	0.988	416	636	0.65	−52.88
Neocortex	Brain weight (C)	44	0.998	1,006,530	1,120,341	0.90	−11.33

[a]The residuals are calculated by regressing the structure in question against either brain or body size, without the *Homo* data. The resulting equation is then used to predict what the *Homo* value would be if it were a primate of the respective brain weight. This table shows the actual human value as well as that expected based on the derived equations, which are linear equations as derived from regressions similar to those seen in Figs. 1–3 when log-log data are plotted. It is unlikely that residuals less than about 25% in either direction are meaningful. Deviations of more than 100% require explanation as they are unlikely to be simply statistical artefacts. Note the strong negative departures for the visual striate cortex, lateral geniculate nucleus, and the lateral geniculate versus the thalamus. Compare these to the cerebellum or neocortex. Based on Stephan *et al.* (1981) data.

[b]The "(C)" means corrected (i.e., The structure is subtracted from the total brain weight.).

that I have measured. The controversy continues to this day.

The AL 162-28 specimen offers a somewhat different set of observations. Falk and I agree on the placement of a key sulcus called the intraparietal sulcus, which divides superior and inferior parietal lobules. In apes, the posterior end of that sulcus abuts directly on the lunate sulcus. The distance of the posterior end of that sulcus to the occipital pole measures roughly one-half the distance for the 18 chimpanzee brains that I have measured, most of them of a smaller volume than the brain of AL 128-68 would have been. If so, then the lunate would have had to have been in an essentially human rather than ape position. This must mean that by 3–4 mya, the brains of early hominids as represented by *A. afarensis and A. africanus* were reorganized to an essentially human cerebral pattern.

Unfortunately, both the Taung and AL 162-28 specimens are incomplete, and thus we can say nothing else about other regions of the cortex such as the inferior frontal convolution, or whether or not there was a true Broca's region there. It is not until 1.8 mya, with the brain endocast of KNM-ER 1470, now known as *Homo rudolphensis,* that we are certain its frontal lobe was constructed in a human-like pattern. However, the paleoneurological evidence of succeeding species of *Homo* is not distinct enough to trace in any satisfactory detail the evolutionary changes of the cerebral cortex to our own modern status. As much as some paleontologists and prehistorians like to believe that the Neandertals of our past were less intelligent than modern anatomical humans, the brain endocasts do not show any truly primitive characteristics to support this. Table IV provides a listing of the major cortical regions that were probably affected in hominid brain evolution.

V. ASYMMETRIES OF THE HUMAN BRAIN

One other important source of data provided by brain endocasts relates to asymmetries of the cerebral hemispheres. Human brains are generally characterized as being asymmetrical and apparently grow and develop

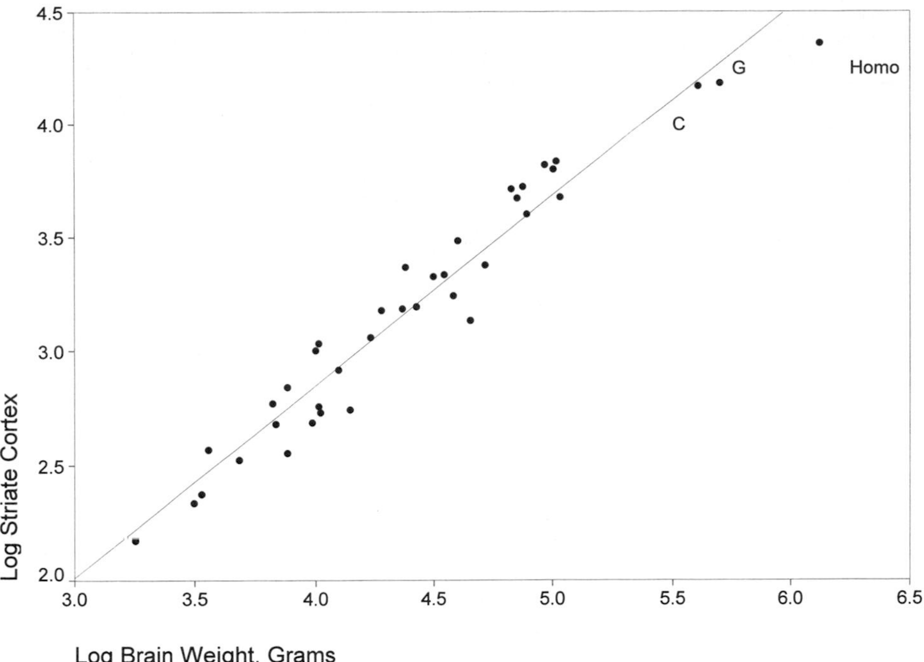

FIGURE 3 When primary visual striate cortex (area 17) is plotted against brain weight, the value for *Homo* is 121% less than predicted for a primate of that brain volume. This is not an indication that vision became less important, but that there was a relative expansion of parietal association area cortex (e.g., peri- and para-striate cortex, areas 18 and 19), lying anterior to area 17. C, chimp; G, gorilla. [Data from H. Stephan, *et al.* (1981).]

that way. We know that the two cerebral hemispheres are seldom equipotential for all tasks, and the hemispheres specialize for particular tasks or modes of processing information. Thus we commonly encounter suggestions that our right hemisphere is more competent at integrating visuospatial information, whereas our left hemisphere is more adept at language processing. The right hemisphere is sometimes characterized as being "gestalt" and "intuitive" while our left hemisphere is "analytical." These designations are crude, but basically this paradigm holds for right-handed individuals and many ambidextrals. Functional magnetic resonance imaging and positron emission tomography scanning shows these processes more clearly.

We do know that there are gross correlations between handedness and hemispheric specialization and what have been called *petalias*. Petalias are small extensions of parts of the cerebral cortex that show asymmetry in growth and extend slightly more in one direction than in the opposite hemisphere. For example, most right-handers show a petalial pattern in which the left occipital lobe protrudes farther back than the right occipital lobe, whereas the breadth of the right frontal lobe is greater than that of the left frontal lobe. In left-handers, the opposite condition holds, and most ambidextrals tend to follow the right-handed pattern. The petalial patterns are easily discernible on radiographs and brain endocasts. My colleagues and I (see Holloway and de Lacoste-Lareymondie, 1982) studied our entire collection of ape and fossil hominid endocasts and found that while the gorilla was the most asymmetrical of the apes, it seldom showed a combination often found in the human animal of a torsional petalial pattern of left-occipital and right-frontal cerebral cortex.

In hominids, on the other hand, these patterns of asymmetries are more apparent than in apes, and some of the australopithecines show such petalias. This, coupled with findings by the archaeologist Nick Toth that early stone tool makers almost 2 mya were right-handed, clinches the argument that the hominid brain had undergone some reorganizational changes and may have had incipient modes of cognitive functioning most similar to our own. [*See* Cerebral Specialization.]

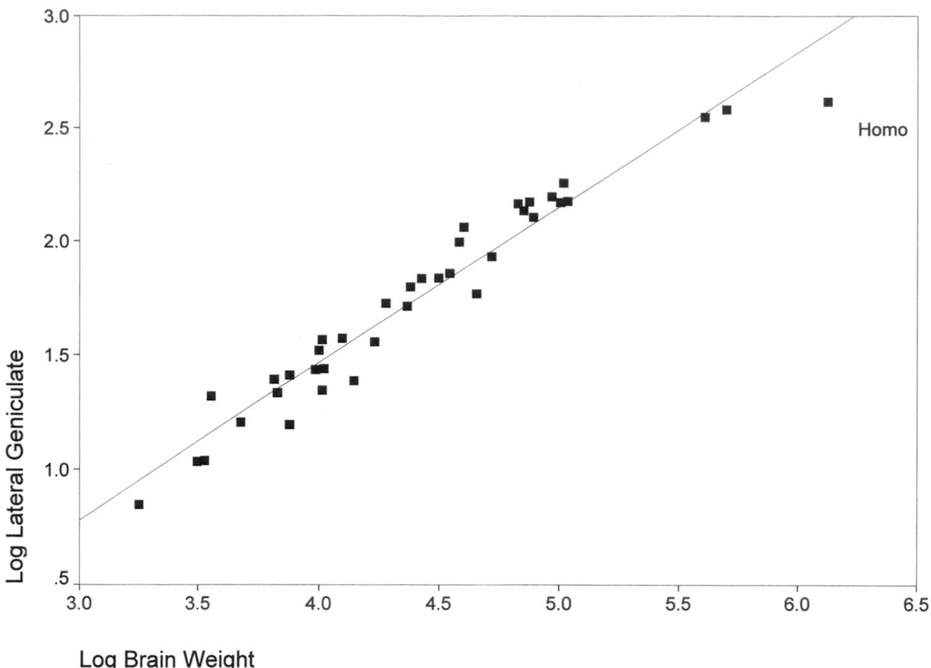

FIGURE 4 This figure shows the same pattern as in Fig. 3. The lateral geniculate body in the thalamus is the first way station for optic nerve fibers, which then radiate out to the visual cortex. In this plot, the human value is 146% less than expected. This, together with primary visual striate cortex (see Fig. 3), confirms a reduction in visual cortex with a concomitant increase in posterior association cortex. [Data from H. Stephan, *et al.* (1981).]

VI. SYNTHESIS OF SIZE AND ORGANIZATIONAL CHANGES

Tables V and VI suggest that the evolution of the human brain was a reticulated process of two sorts: (1) genetic changes that affected the organization of the brain and (2) genetic changes that led to increased size of the brain. Because these changes would have been either hyperplastic (increasing numbers of units) or hypertropic (increasing size of the

TABLE IV
Major Cortical Regions in Early Hominid Evolution

Cortical regions	Brodmann areas[a]	Functions
Posterior occipital striate cortex	17	Primary visual
Posterior parietal and anterior occipital (peri- and parastriate) cortex	18, 19	Secondary and tertiary visual integration with area 17
Posterior parietal, superior lobule	5, 7	Secondary somatosensory
Posterior parietal, inferior lobule (mostly right side; left side processes are symbolic-analytical)	39	Angular gyrus, perception of spatial relations among objects
Posterior parietal, inferior lobule (mostly right side; see above)	40	Supramarginal gyrus, spatial ability
Posterior superior temporal cortex	22	Wernicke's area, posterior superior temporal gyrus, comprehension of language
Posterior inferior temporal	37	Polymodal integration, visual, auditory; perception and memory of objects' qualities

[a]Area 17, the primary visual cortex, underwent a relative reduction in size, whereas areas 18, 19, 39, 40, and possibly 37 underwent a relative increase in volume. There is no solid evidence for an increase in size of the anterior prefrontal areas.

TABLE V
Reorganizational Changes of the Human Brain

Brain changes (reorganization)	Taxon
1. Reduction of primary visual striate cortex, area 17, and relative increase in posterior parietal cortex	*A. afarensis* *A. africanus*
2. Reorganization of frontal lobe (third inferior frontal convolution, Broca's area)	*Homo habilis*
3. Cerebral asymmetries, left occipital, right-frontal petalias	? *Homo habilis*
4. Refinements in cortical organization to a modern *Homo* pattern	? *Homo erectus* to present?

units), it seems most probable that the genetic changes would have mainly affected *regulatory* genes. That is, the rates of mitotic divisions in different brain areas and their durations probably varied at different times during the course of human evolution. The same patterns could also affect the hypertrophic changes, leading to longer duration of growth periods to achieve larger sizes of the nerve cells, the nuclei where they are found, and the overall volume of the brain.

In fact, this was most likely an incredibly complex skein of interactive processes, with different rates at different times, which emphasizes the mosaic nature of this organ's evolution. It is unlikely that we will fully understand just how this evolutionary mosaicism occurred, without some incredibly more sophisticated advances in our knowledge of neurogenetics.

The fossil record of human evolution also includes evidence for behavior and some of the cognitive processes that may have occurred in our past. We know that stone tools were made from about 2.5 mya, and at about 2.0 mya they were made in standardized forms. With time, these stone tools became more diverse and numerous and were clearly a major part of human behavioral and ecological adaptation. We can assume that other forms of perishable tools and items were also used, such as digging sticks, baskets, containers, and possibly clothing, as well as protective structures such as rocks and animal skins. From Toth's work, right-handedness seems apparent by at least 1.8 mya, suggesting some aspect of cerebral hemispheric dominance, which correlates well with the petalias of the brain endocasts.

All of this evidence indicates a strong dependence on the brain, meaning both its size and organization, during human adaptation and evolution. The human brain did not emerge instantaneously, but through a long process of myriad changes taking place at different tempos at different times.

A particularly exciting aspect is that the fossil record of the brain endocasts does suggest (but does not prove) that one of the earliest changes was a reorganization of the cerebral cortex that involved a relative decrease of primary visual striate cortex and a concomitant enlargement of posterior parietal cerebral cortex. This most likely occurred some 3–4 mya in *A. afarensis* and most probably was related to their growing adaptation to a mixed habitat, including both savannah and riverine forest econiches, based on bipedal locomotion in which an enhanced visuospatial competence would have been enormously valuable in finding patchy food and water resources. It is reasonable that an expansion of the parietal lobe would also have involved increased competence in social communication, providing for increasing com-

TABLE VI
Brain Size Change in Human Evolution

Brain changes (brain size related)	Taxon	Time (mya)	Evidence
1. Small increase, allometric[a]	*A. afarensis* to *A. africanus*	3.0 to 2.5	Brain endocasts increase from 400 to 450 ml
2. Major increase, rapid, both allometric and nonallometric	*A. africanus* to *Homo habilis*	2.5 to 1.8	KNM-1470, 752 ml (ca. 300 ml increase)
3. Modest allometric increase in brain size to 800–1000 ml (assumes *Homo habilis* is KNM-ER 1470-like)	*Homo habilis* to *Homo erectus*	1.8 to 0.5	*Homo erectus* brain endocasts and postcranial bones, e.g., KNM-ER 17000
4. Gradual and modest size increase to archaic *Homo sapiens*, nonallometric	*Homo erectus* to *Homo sapiens neanderthalensis*	0.5 to 0.075	Archaic *Homo* and Neandertal endocasts 1200 to 1700+ ml
5. Small reduction in brain size among modern *Homo sapiens*	*Homo s. sapiens*	0.015 to the present	Modern endocranial capacities

[a]Related to increase in body size only.

plexity of social behavior, including a greater degree of child care than before.

In sum, the major underlying selectional pressures for the evolution of the human brain were mostly social. It was an extraordinary evolutionary "decision" to go with an animal that would take longer to mature, reach sexual maturity later, and be dependent for its food and safety upon its caretakers (parents?) for a longer period of time. The benefits for the animal were many, including a longer learning period, a more advanced, larger, and longer-growing brain, and an increasing dependence on social cohesion and tool making and tool using to cope with the environments that they encountered. Needless to say, language abilities using arbitrary symbol systems were an important ingredient in this evolution.

The fossil record shows us that there was a feedback between the complexity of stone tools (which must be seen as a part of social behavior) and increasing brain size and the expansion of ecological niches. The "initial kick," however, the process that got the ball rolling, was a neuroendocrinological change affecting regulatory genes and target tissue–hormonal interactions that caused delayed maturation of the brain and a longer growing period, during which learning became one of our most important adaptations.

BIBLIOGRAPHY

Allman, J. (1990). Evolution of neocortex. *In* "Cerebral Cortex" (E. G. Jones and A. Peters, eds.), Vol. 8A, pp. 269–283. Plenum, New York.

Connolly, C. J. (1950). "The External Morphology of the Primate Brain." C.C. Thomas, Springfield, Illinois.

Falk, D. (1980). A reanalysis of the South African australopithecine natural endocasts. *Am. J. Phys. Anthropol.* **53**, 529–539.

Geschwind, N., and Galaburda, A. M. (1984). "Cerebral Dominance: The Biological Foundation." Harvard Univ. Press, Cambridge, Massachusetts.

Holloway, R. L. (1967). The evolution of the human brain: Some notes toward a synthesis between neural structure and the evolution of complex behavior. *General Systems,* **12**, 3–19.

Holloway, R. L. (1968). The evolution of the primate brain: Some aspects of quantitative relationships. *Brain Res.* **7**, 121–172.

Holloway, R. L. (1969). Culture: A human domain. *Curr. Anthropol.* **10**, 359–412.

Holloway, R. L. (1975). "The Role of Human Social Behavior in the Evolution of the Brain," 43rd James Arthur Lecture (1973). American Museum of Natural History, New York.

Holloway, R. L. (1979). Brain size, allometry, and reorganization: Toward a synthesis. *In* "Development and Evolution of Brain Size: Behavioral Implications" (M. Hahn, C. Jemsen, and B. Dudek, eds.), pp. 59–88. Academic Press, New York.

Holloway, R. L. (1988). The brain. *In* "Encyclopedia of Human Evolution and Prehistory" (I. Tattersall, E. Delson, and J. Van Couvering, eds.), pp. 98–105. Garland, New York.

Holloway, R. L. (1995). Toward a synthetic theory of human brain evolution. *In* "Origins of the Human Brain" (J. P. Changeux and J. Chavaillon, eds.), pp. 42–54. Oxford Univ. Press, New York.

Holloway, R. L., and de Lacoste-Lareymondie, M. C. (1982). Brain endocast asymmetry in pongids and hominids: Some preliminary findings on the paleontology of cerebral dominance. *Am. J. Phys. Anthropol.* **58**, 101–110.

Jerison, H. J. (1973). "Evolution of Brain and Intelligence." Academic Press, New York.

Kolb, K., and Whishaw, I. Q. (1985). "Fundamentals of Human Neurophysiology." Freeman, New York.

LeMay, M. (1976). Morphological cerebral asymmetries of modern man, fossil man, and nonhuman primates. *Ann. N.Y. Acad. Sci.* **280**, 349–366.

Martin, R. D. (1983). "Human Evolution in an Ecological Context," James Arthur Lecture (1982). American Museum of Natural History, New York.

Radinsky, L. B. (1972). Endocasts and studies of primate brain evolution. *In* "The Functional and Evolutionary Biology of the Primates" (R. Tuttle, ed.), pp. 175–184. Aldine, Chicago.

Stephan, H., Frahm, H., and Baron, G. (1981). New and revised data on volumes of brain structures in insectivores and primates. *Folia Primatologica,* **35**, 1–29.

Toth, N. (1985). Archaeological evidence for preferential right-handedness in the lower and middle Pleistocene, and its behavioral implications. *J. Human Evol.* **14**, 607–614.

Brain, Higher Function

STEVEN E. PETERSEN
Washington University School of Medicine

GLOSSARY

Cytoarchitecture Arrangement of cells in the brain, used in defining functional areas

Equiluminance When two (or more) visual stimuli have the same perceived brightness but can vary in other characteristics (such as color)

Functional area Brain region that is specialized to perform a set of related information-processing operations

MT Primate brain region that appears to be specialized for the processing of visual motion

Myeloarchitecture Distribution of myelin in the brain, used to define functional areas

Neuron Specialized cells in the nervous system that are electrically excitable and perform information processing in the brain

Parallel distributed processing models Computer simulations that use many simple elements arranged in parallel (working at the same time) to perform complex computations, based on simplified concepts of how the brain processes information

Positron emission tomography Method for measuring the distribution of low-level positron radiation in tissues, used to measure activity in the brain related to how hard different regions are working

THE DOMAIN OF NEUROSCIENCE EXTENDS FROM the biological basis of individuality to the molecular biology of nerve cells; higher brain function focuses on a subdiscipline within this larger field. The study of higher brain function in its largest sense is the study of the brain as an information-processing device. Studies of higher brain function generally emphasize aspects of information processing that are neither simple input processing (sensation and sensory transduction) nor output processing (final implementation of movement), and also do not focus on the implementation of organism maintenance such as breathing and thermoregulation. Examples of the information-processing problems that are commonly approached include studies of language, attention, perception, learning, and memory. Although higher brain functions have been studied for more than 100 years, recent technological, methodological, and conceptual advances allow for a useful interface among several disciplines including neurobiology, experimental psychology, and computer science.

I. INTRODUCTION

Because studies of higher brain function are at the interface of several disciplines, these studies produce many different kinds of information. For example, a study in computer vision, designed to simulate the way that a person recognizes a particular shape, produces very different information from neurobiological studies in which electrical signals are recorded from single nerve cells in the parts of a monkey's brain that are related to vision. Because of these differences, debates are often waged over which type of information is most appropriate for understanding human visual perception. Most profitably, however, the information from these different studies should be viewed cooperatively rather than competitively; bits of information from both computer science and neurobiology can provide clues as to how the brain implements the complex function of visual perception.

ENCYCLOPEDIA OF HUMAN BIOLOGY, Second Edition, VOLUME 2. Copyright © 1997 by Academic Press. All rights of reproduction in any form reserved.

To understand how kinds of information could profitably be integrated, perhaps it would be useful to think of how a reverse engineer (i.e., an engineer trying to understand an existing machine's design instead of trying to design a currently nonexistent machine) might approach any information-processing machine (machine X).

The engineer might begin by taking an outside-in approach, observing *what* machine X is capable of doing when given many different types of input, under many different output demands, and with different settings of any controls on the machine. Our engineer might well be able to make preliminary guesses about several ways that machine X might have been designed.

Armed with the knowledge of what machine X's capabilities are, the reverse engineer might take the cover off and attempt to get a general view of the *where* some of the possible design components are. He or she might attempt to trace different input and output lines from the outside-in and to determine how different subcomponents are used by the machine. By understanding the general physical layout of the machine's components, the engineer can begin to test ideas about how the machine works. Measurements of the electrical activities might be made in different regions of the machine when the machine is performing in different conditions. He or she might remove some components to see how they affect the performance of the machine during particular performance tests.

The reverse engineer might now turn to the question of exactly *how* machine X was designed to perform its varous functions. As any engineer knows, many different sets of chips, circuit boards, and physical layouts of other electrical components can be used to realize a similar end product. If our interest is in how machine X works in particular, our reverse engineer must now explore how the design is implemented in the special case of this machine, looking at exactly what components are used, and in what exact configurations.

A reverse engineering approach can be directed at answering three fundamental questions about a machine. *What* are the performance characteristics of the machine with different inputs, or output demands, or in different states? *Where* in the machine are components located that contribute to the different capabilities of the machine (i.e., what is the general physical layout of the machine)? *How* does the machine implement the performance of its different capabilities in terms of particular circuit designs and problem-solving techniques?

This same reverse engineering approach can be applied to the study of the brain. The many disciplines that study aspects of higher brain function tend to produce information that will give partial answers to one or the other of these questions. From these partial answers, fundamental concepts about how the brain functions as an information processor have emerged. The next section will describe some of these fundamental concepts. In the final section, two very different functions—the processing of visual motion and the processing of visually presented words—will be used as examples of how higher brain function is studied. It is hoped that these examples will show how integrating information from very different approaches can advance our understanding of how the brain supports human performance.

II. FUNDAMENTAL CONCEPTS IN THE STUDY OF HIGHER BRAIN FUNCTION

A. Elementary Mental Operations (What)

When a human performs a task such as answering the question "Is the red object moving to the right?", the task can be thought of as consisting of many component subtasks: understanding the question, looking at the object, appreciating its color, determining its direction of movement, deciding if the object is red, if it is moving to the right, programming a response, and making the response. Each of these component jobs may or may not be broken into further components. Each of the component jobs is a candidate for an elementary mental operation.

An elementary mental operation is difficult to define precisely, but such an operation has several characteristics: like a subroutine in a computer program, it is composed of a set of related computations; it is not easily divided into more component parts; and when one of these operations is performed, information is present in a form that was not available before the process took place.

Many elementary mental operations will be orchestrated in the performance of any task. Different processing operations will be used to deal with the input presented, the instructions or training needed to perform the task, in organizing further processing, and in programming output.

The methods used to determine elementary operations measure the performance characteristics of the brain (what the brain is capable of doing). The studies usually measure reaction time (how long it takes a person to perform a task), percent correct (how often a person can correctly perform a task), and other, more specialized measurements of behavior. The subdisciplines that most often address performance studies include psychophysics and cognitive psychology.

Psychophysics is the study of human performance characteristics when stimuli are manipulated along different physical dimensions such as color, size, or speed of movement in vision, and pitch or loudness in hearing. By carefully observing the responses to different stimuli, insight might be gained into the nature of perception. For example, color psychophysics studies the ability of people to distinguish between different colors, or how bright different colored objects must be before they can be seen. The manipulations of the different stimuli can get quite complicated, so that the effect of color changes in one part of the visual field on the person's judgment of color in another part of the visual field might be tested. [*See* Vision, Physiology.]

Not only the physical nature of stimuli can be manipulated in performance experiments, but other processes such as the ability to remember, to learn, to assign meaning, and to attend to different stimuli when instructed can also be studied. The study of these processes is the turf of the cognitive psychologist.

Section III includes one example from each of the foregoing subdisciplines.

B. Functional Area (Where)

Where in the brain are the different elementary mental operations taking place? Perhaps each elementary mental operation is distributed all over the brain, so that studying the physical layout becomes a useless level of analysis. Many lines of evidence argue, however, for functional localization; that is, different parts of the brain perform different functions.

Strong evidence that different functions are localized in the brain comes from the study of deficits following brain injury. Injuries to the back of the brain produce disturbances in vision. Injuries to other regions of the brain may produce disturbances in touch or hearing, or some more complex function (Fig. 1).

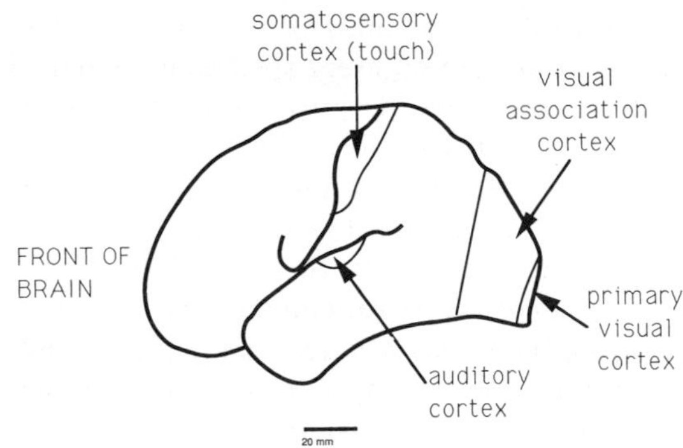

FIGURE 1 A diagrammatic, lateral (side) view of a human brain showing regions related to touch, vision, and hearing. Extrastriate or nonprimary visual areas such as MT (see text) are located in the visual association cortex.

Given the observation that different locations of brain injury cause different deficits, can different functional units, or functional areas, of the brain be described in more precise terms? A brief history about this question may prove illuminating.

In animals, the precise anatomy and physiology of the brain can be studied with a variety of techniques that have shaped the understanding of functional areas. In the nineteenth century, it was observed that movements could be elicited by electrically stimulating a region of a dog's cerebral cortex. If movements of the paw were elicited at one site, then an adjacent site of stimulation would produce movements in the same leg. It was as if a little "map" of one-half of the dog's body was on this small piece of cortex. The left half of the dog was mapped to the right hemisphere, and the right half of the dog was mapped to the left hemisphere.

Around the same time, people were beginning to look at the microscopic anatomy of slices of the brain in several species, including humans. Six layers of cells were observed in mammalian cerebral cortex, and the thicknesses of the layers, the types of cells in the layers, and the arrangement of cells within the layers were not uniform across the cortex but seemed to organize into a series of distinct zones. Agreement as to the number of zones or to their exact boundaries was unclear, but the presence of regional differences in what was called the architectonics of the brain was well accepted. [*See* Cortex.]

In the twentieth century, these two observations were pursued. Topographic mapping of the cortex, not just for electrically elicited movements but for responses to different kinds of sensory stimulation, was done for several different species of animals. For example, microscopically fine electrodes could be placed in the cortex of an anesthetized animal, and different kinds of visual stimuli could be presented in different parts of the visual field, or different types of sounds could be played to the ears. When this was done, for each of the sensory modalities, as well as for movements, there was not just one map, but several separate and complete maps. For the most completely studied system, the visual system, it was shown that for cats, several species of monkeys, and presumably for humans as well, there are some two dozen or more visually responsive "maps."

The twentieth century also saw much progress in anatomical techniques. New staining techniques made architectonic regional analysis clearer. Several techniques were also developed that allowed the tracing of connections between these different zones, both in the forward direction (so that all of the regions to which one region sends messages could be labeled) and in the backward direction (so that all of the regions from which a region receives messages could be labeled).

Finally, in the second half of the twentieth century, the exact responses of the cells in these different areas began to be studied. As mentioned earlier, though there were several electrically defined maps of the visual field, the cells in these different regions do not all seem to be doing the same type of processing. One visual region has cells that signal the direction and speed of moving visual stimuli, and other regions carry more information about the color and shape of visual stimuli. These analyses of the information carried by different cells are studies of the functional properties of neurons. [*See* Visual System.]

As the studies of these different methodologies progressed, particularly in the visual system, the different types of studies began to agree with one another. For example, when the borders of an electrically recorded map of the visual field were compared with the cytoarchitectonic or myeloarchitectonic borders, there was agreement. An architectonic area seemed also to have a coherent set of connections with other areas (i.e., in most cases, if one part of area A connects with areas B and C, but not with area D, then all parts of area A will connect with areas B and C, and not D).

Because the distinctions between different areas on any one measure can be very subtle, the current way of defining functional areas in animals is to find agreement between these different methodologies:

1. distribution of functional properties of cells;
2. distributions of anatomical cell and fiber distributions;
3. uniformity of connections with other areas;
4. presence of an electrically recorded (or electrically stimulated) topographic map.

Only a small number of areas to date have been studied with all four methodologies, but those that have produce very encouraging results; the study of visual motion processing area MT, presented later, is an example of such a study.

Although the use of these four criteria is not possible for directly studying human functional areas, the animal studies provide a firm conceptual basis for understanding functional localization in humans.

C. Neural Implementation (How)

A third concept important in understanding the brain as an information processor is in the way a functional area actually might implement the elementary operation, that is, the way in which the processing is specified on the hardware of the brain. At this point, there is no complete description of this level for any elementary operation, but there is information about how the nervous system carries and encodes information and, in some specific cases, how processing is carried out in quite specific ways at the level of single nerve cells.

To understand the idea of a neural implementation, it is necessary to know something about how nerve cells work. Neurons (nerve cells) are usually made up of a cell body, dendrites, and an axon. The cell body, which contains the nucleus, performs many of the same functions that all other types of cells must perform in terms of cell maintenance and so on. Mostly, the axon and dendrites distinguish neurons from many other cells. Neurons are electrically excitable cells and can produce spikes of electrical activity. These spikes often travel along the length of the axon, a long hair-like process that extends away from the cell body. The electrical message usually travels from the cell body along the axon, where it makes connections to other cells. These connections, or synapses, are often made on the dendrites of the receiving cell, the dendrites being a network of processes extending from the cell body. The connections between an axon and its receiving cell are not electrical, however, but chemical. When the electrical message comes to the end of the axon, it induces the release of a minute amount of a chemical (called a neurotransmitter) that moves across the synapse to find a home (receptor) on the receiving cell. This in turn induces electrical changes in the receiving cell. These changes can either encourage the cell to send an electrical message further on in the system, in which case we call the synapse an excitatory synapse, or they can discourage a message, in which case we call the connection inhibitory. Each cell of the brain receives messages from many others and must total up all of its inputs in deciding whether or not to send a message in terms of electrical spikes, and how many spikes to send. [*See* Neurotransmitter and Neuropeptide Receptors in the Brain.]

The electrochemical nature of the synapse makes the transmission of information from one neuron to the next slower than if information was transmitted by direct electrical connections. The time course of these steps in information processing contributes to our ability to measure reaction time differences in different tasks. If each step of a task took only microseconds (instead of the milliseconds that it normally takes), our reaction time in any task would be so quick that measuring differences would be very unlikely. Fortunately, for generations of psychophysicists and cognitive psychologists, this is not the case.

Most often, studies of neural implementation take advantage of the spiking nature of the electrical message that neurons use to communicate with one another. Neurobiologists can measure the spikes that a cell is producing by placing microscopically fine wires (electrodes) next to a cell body. The functional properties of a cell are most often assessed by comparing the number of spikes in one condition with the spikes produced under another set of conditions. If the number of spikes is affected by some manipulation, the cell is assumed to be coding information important to that manipulation. In the example presented in the next section, this idea is used to assess the information encoded in the directionally selective cells of visual area MT.

A second approach to the study of neural implementation is to take what is known about how neural systems are organized and to make computer simulation using, in essence, very simplified model versions of little pieces of the brain. Sometimes computers are used to simulate human performance, and different types of simulations include control systems, or connectionist (parallel distributed processing or PDP) models. The hope is that such simplifying models will help give insight into how the complicated system of the brain might be organized. The most explicitly neural of these types of models have been called neural network, connectionist, or PDP models.

Both of these approaches to neural implementation, direct measures of the information-processing capacities of neurons, and computer simulation are present in the following examples.

III. EXAMPLE STUDIES OF HIGHER BRAIN FUNCTION

The way in which information about elementary operations, functional areas, and neural implementation can help in understanding higher brain function can be seen in a very simple framework. Any task performance consists of the orchestration of several elementary operations. Each of these operations might be localized in a separate functional area of the brain. Within that area, the computations are implemented through the specific actions of single neurons, which carry and transform the information necessary to compute the elementary operation. The examples below are intended to clarify these ideas.

A. Visual Motion Processing

In the early 1970s, several workers described a region in several different primate species that had a high concentration of cells that coded for the direction of movement of a visual stimulus. Remember that the way neuronal coding is assessed is by the frequency of spikes they make. In the case of direction-selective cells, many more spikes are counted when an object is moved in one direction (e.g., upward) than when it is moved in the opposite direction (downward). This area is called MT.

MT has been the focus of studies utilizing several criteria for the definition of a functional area. Area MT had a distinctive architectonic appearance in brain sections; the area was heavily stained by myelin-staining techniques. The area defined by this heavily myelinated zone contained a complete representation of the visual half-field as measured by electrical mapping. The heavy concentration of directionally selective cells agreed with borders defined by the myelin and by the representation of the visual field assessed by electrical mapping.

MT also has a unique set of connections when compared with nearby cortex, particularly a set of two-way connections with primary visual cortex, the first way station for visual information in the cerebral cortex.

Along with the functional property of direction selectivity, further evidence that MT related to processing of visual motion information came from a series of microlesion experiments. In these studies, a monkey was trained to find and track with its eyes a moving spot of light that was presented in different locations of the visual field. After training, a microscopically small amount of a chemical that inactivates brain cells was applied to a small part of MT. Because MT has a map of the visual field, inactivating part of MT inactivates processing of only a part of the visual field. When this was done, the animal could find and move its eyes to a stationary spot with the same accuracy as prior to the inactivation. But when the stimulus was moving, the accuracy of the eye movement to the stimulus and the ability to track the movement of the stimulus was poorer than before the inactivation. The animal could still see a stimulus and could still move its eyes accurately to it if it was stationary, but it acted as if it could not appreciate the motion and speed of the stimulus. Thus, the monkey was inaccurate in movements to a moving stimulus and in matching the velocity of its eye to the velocity of the stimulus. The sum of evidence at this point was that area MT was part of a pathway processing visual motion information.

Further study of the functional properties of these cells also showed velocity tuning; the cells fired much more vigorously at some velocities than at others. These cells were also less sensitive to changes in the color of visual stimuli than they were sensitive to changes in brightness, irrespective of the color of the stimulus.

Another study further delved into the neural implementation of motion processing in MT. In these studies, monkeys were trained to perform a discrimination of the motion of a field of randomly placed moving dots. The field contained a large number of dots, and the direction of movement of each dot could be independently determined. The direction information in the field could be totally random, or a percentage of the dots could be moved in the same direction, while the remaining dots were randomized. The percentage of dots moving in the same direction varied, and the monkey was to report one of two directions that the correlated dots were moving. When there was near 0% correlation (each dot moving in a random direction), the monkey would guess right about 50% of the time by chance, and when all of the dots were moving in the same direction (100% correlation), the monkeys rarely, if ever, reported the incorrect direction. As the amount of correlation increased from 0 to 100%, the number of correct responses also increased.

Single cells were recorded in area MT during the performance of this task. If the cell was directionally selective for movement in the upward direction, the two directions chosen for the monkey's task were up and down. When all of the dots were moving in an upward (preferred) direction, the cell's response was very vigorous, and the animal's response was about 100% "up." When the dots were all moving down, the cell's response was very low, and the animal's response was about 100% "down." For these extremes, the response of the cells would be a good predictor of the response of the animal: when the cell fires vigorously, the animal's response will be "up"; when the cell fires a little, the response will be "down." But this was known before the test was performed. What about for the intermediate correlations for up and down dot percentages? Would the intermediate firing rates of the cells predict the intermediate percentages of up and down responses? The answer was a startling yes. Single cells in MT seemed to carry enough information to predict the percentage of up and down responses of the whole monkey. In other words, if the decision-making apparatus of the monkey could "look" at only this one cell in area MT, the cell would be just as good at this task as the whole monkey. It appears as if the information for *perceiving* the coherent motion of dots in a noisy background is precisely coded at the level of single MT neurons. Much of the raw information used for visual motion *perception* itself is probably encoded in the visual area MT.

How does knowledge about this motion processing area map onto studies of human performance? One observation about MT was that although much information about visual movement is computed, MT cells do not appear to carry much information about the color of the stimulus. MT cells are much more sensitive to levels of brightness than to changes in color. It is as if the motion-processing pathway in the brain is somewhat color-blind.

If this were the case, then what would happen if a person was asked to judge the movement of objects that were different only in color, not in the level of brightness? There is a long history of studies on this question, but a recent review of these studies illuminates a very interesting aspect of this question. When colors are equiluminant (same brightness), a person's appreciation of the velocity of the stimulus is much less accurate.

For example, imagine that a movie is made of a black cylinder on a black background. Nothing can be seen. Now paste small white circles to both the cylinder and the background. A pattern of white spots can be seen, but the cylinder blends into the background; it is effectively camouflaged. If the cylinder is rotated, the cylindrical surface becomes apparent, the only clue being the relative motions of the dots of light with one another. Even if the dots in the background are moved randomly, the rotating cylinder will still stand out.

A computer can be made to simulate this effect, which is called three-dimensional structure from motion. In this case, rather than white dots on a black background, red dots can be placed on a green background. If the red dots are much brighter than the green background, and are moved with the same relative motions as the lights glued to the can, again a cylinder is "seen" by an observer. Now the observer turns down the brightness on the red dots. When the red dots are the same perceived brightness as the green background, the cylinder disappears! The red dots are still visible against the green background, but there is no perception of the cylinder. The observer continues to turn down the brightness of the red dots. When the red dots are sufficiently dimmer than the green background, the cylinder reappears.

When the moving dots have a brightness signal, the motion pathway can "see" them and can compute the cylinder from the relative motion of the dots. When the only difference between the dots and the background is the color, then the color-blind motion pathway cannot see them, but some other color-sensitive pathway can. So in both the same- and different-brightness conditions, the dots can be seen, but only in the different-brightness condition can the cylinder be seen. This psychophysical observation concurs with the neurobiological evidence that motion and color information are computed by separate pathways at some point in the visual system.

B. Visual Processing of Words

Similar approaches can be made to a different kind of visual "understanding," how visual words are processed during reading.

Many different elementary operations must be performed when reading. The lines and curves that make up the words must be visually processed, recognized as words, the meaning of the word must be "looked up," and a series of words must be integrated to understand the full meaning of a sentence. This example focuses on the visual processing of single words. [*See* Reading Processes and Comprehension.]

The visual processing of a word is often thought to consist of three levels of processing: (1) visual features, the lines and squiggles that make up letters; (2) letters, the single characters that make up words; and (3) words, a group of letters that make up a functional unit.

Studies in cognitive psychology have supported the idea that there are complex computations made on words and word-like strings of letters that unify the visual input into a chunk. For example, each letter of a word can be perceived at dimmer contrasts than when that letter is presented alone or as part of a nonsense string of letters. This effect—that letters within words are more visible than the same letters in random letter strings—is known as the word superiority effect. The word superiority effect extends to meaningless letter strings that are visually similar to words (e.g., POLT), suggesting that this effect does not involve the meaning of the string but its regularity (i.e., similarity to strings of letters that could be words). It seems reasonable from these results to conclude that some level of visual processing involves coding at the level of visual words.

These results have led to some models of visual word processing that implement separate levels of feature, letter, and word form analysis. One of these models takes the form of a connectionist, or neural network, model run on a computer.

Each level is made up of a set of elements. Each of the elements is thought of as a neuron or set of neurons. Each element at the feature level represents one of the lines or squiggles that make up a letter; each element at the letter level represents a letter; each element at the word level, a single word. Elements at a single level inhibit one another, and elements between levels excite one another. All of the connections between elements are reciprocal (they go both ways). As visual information enters at the feature level, a particular set of features might be activated that make up the letters DOG. Elements at this letter level coding these letters become active and inhibit other letters at its level. Elements at this letter level in turn excite elements at the word level containing the letters DOG. The excited elements also excite appropriate features at the feature level through backward connections. The same thing takes place at the word level with the element representing the word DOG inhibiting the other words and adding further excitation to the constituent letters at the letter level. A random collection of letters does not get this extra feedback from the word level, so the input of a word or near word produces a word superiority effect. For sets of letters that are close to words, several elements at the word level are partially activated and still produce added activation at the letter level. In terms of a quasi-neuronal model of implementation, a multilayer processing network is a possible implementation that simulates human processing of visual words.

Two different types of studies provide evidence that separate functional areas in the human brain are related to these different levels of processing.

Studies of patients with damage to particular parts of visual association cortex provide evidence that visual word form codes might be present in areas near the back of the cortex. Such lesions sometimes cause pure alexia (i.e., the inability to read words without other language deficits). Some people with pure alexia can read the letters that make up words and, by saying the letters aloud, can assemble the words, thus allowing them to read in a very laborious way. Letter-by-letter reading could be attributed to a deficit in the visual word form processing, but it could also involve earlier stages of the network. These results are certainly consistent with an important role of the extrastriate cortex in processing visually presented words. [*See* Speech and Language Pathology.]

A second set of results that reinforces this localization comes from positron emission tomographic (PET) studies of normal individuals. Methods have been developed that allow the PET scanner to localize areas of the human brain that are active during the performance of a task. Very low doses of a short-lived radioisotope, oxygen-15, are administered to a subject in the form of radioactive water to trace changes in blood flow during task performance. Forty seconds are needed to collect information to develop a scan of the activity that takes place. The scan is made up of pictures of the brain at seven horizontal levels. By placing these pictures in exact registry, it is possible to subtract the activity in one task from another, or to average the activity in the same task when performed by different people. Use of a standard brain atlas allows identification of the actual brain regions activated.

When subjects were presented with visual words, areas in visual association cortex were activated. Some of these areas were activated only by words but were not activated by other types of visual stimuli. The area activated in the left hemisphere is at a similar location to that found damaged in pure alexics.

From the examples presented in the previous section, it can be seen that very different kinds of information can influence the understanding of how the brain supports different mental processes. Only through the integration of these different kinds of information can progress be made toward understanding the mysteries of how we perceive, think, and communicate.

BIBLIOGRAPHY

Damasio, H., and Damasio, A. R. (1989). "Lesion Analysis in Neuropsychology." Oxford Univ. Press, New York.

Newsome, W. T., Shadlen, M. N., Zohary, E., Britten, K. H., and Movshon, J. A. (1994). Visual motion: Linking neuronal activity to psychophysical performance. *In* "The Cognitive Neurosciences" (M. S. Gazzaniga, ed.), pp. 401–414. MIT Press, Cambridge, MA.

Petersen, S. E., and Fiez, J. A. (1993). The processing of single words studied with positron emission tomography (PET). *Annu. Rev. Neurosci.* **16**, 509–522.

Posner, M. I., Petersen, S. E., Fox, P. T., and Raichle, M. E. (1988). Localization of cognitive functions in the human brain. *Science* **240**, 1627–1631.

Rumelhart, D. E., and McClelland, J. L. (1986). "Parallel Distributed Processing," Vols. 1 and 2. MIT Press, Cambridge, MA.

van Essen, D. C., and DeYoe, E. A. (1994). Concurrent processing in the primate visual cortex. *In* "The Cognitive Neurosciences" (M. S. Gazzaniga, ed.), pp. 383–400. MIT Press, Cambridge, MA.

Brain Messengers and the Pituitary Gland

EUGENIO E. MÜLLER
University of Milan

GLOSSARY

Hypophysiotropic regulatory hormone Low- to medium-molecular-weight compound, especially synthesized in mediobasal hypothalamic nuclei, released episodically from nerve terminals into the hypophyseal portal capillaries and transported to the anterior pituitary cells, where it interacts with specific receptor sites

Median eminence Neurohemal structure consisting of nerve cells and blood vessels located at the base of the hypothalamus, where hypophysiotropic regulatory hormones are released into portal capillaries

Neuroactive drug Drug that acts by mimicking or opposing via different modalities the action of one or more neurotransmitters, thus affecting the secretion of anterior pituitary hormones

Neuropeptide Peptide generated from a large pre-pro-peptide molecule in the rough endoplasmic reticulum, packaged as pre-peptide in the Golgi complex, released by exocytosis at the nerve terminal into the synaptic cleft, and inactivated by proteolytic enzymes

Neurotransmitter Low-molecular-weight water-soluble compound, mostly monoamine or amino acid, that exerts localized, short-lived responses at the synapse; it is rapidly inactivated following the completion of the signal

ONCE THOUGHT TO BE THE MASTER GLAND OF the body, it is now clear that the anterior pituitary gland (AP) is under the influence of hypothalamic and extrahypothalamic structures. A host of chemical messenger substances are released from neurons located in the hypothalamus and conveyed to the AP via a portal system of capillaries. The functional activity of hypothalamic neurosecretory neurons, which elaborate and deliver specific hypophysiotropic regulatory hormones into the portal system, is in turn regulated by a host of neurotransmitters and neuropeptides. These substances, via typical or atypical synaptic connections, relay to the hypophysiotropic neurosecretory neurons of the hypothalamus neural or neurohormonal influences, which are translated into hormonal responses to be conveyed to the AP. As a corollary, pharmacologically induced suppression or activation of this neurotransmitter–neuropeptide system of control induces profound changes in the secretion of AP hormones, and regulatory hormones and central nervous system (CNS)-acting compounds can be used as probes of pituitary or hypothalamic function, respectively, and in humans in the diagnosis and therapy of neuroendocrine disorders.

I. BRAIN NEUROTRANSMITTERS AND NEUROPEPTIDES

Neurotransmitter and peptidergic systems in the endocrine hypothalamus and extrahypothalamic-related areas interact functionally to ensure the proper control of AP function. The widely held distinction between the two systems is now blurred. It is now generally recognized that hypothalamic peptidergic neurons are widely distributed to extrahypothalamic CNS areas, which is compatible with their additional behavioral role, that few classic neurotransmitters [e.g., dopamine (DA) and norepinephrine (NE)] mediate a neurohormonal type (e.g., diffuse and slow) of synaptic transmission, and in addition that neuro-

ENCYCLOPEDIA OF HUMAN BIOLOGY, Second Edition, VOLUME 2. Copyright © 1997 by Academic Press. All rights of reproduction in any form reserved.

transmitters [e.g., DA, γ-aminobutyric acid (GABA), adrenaline] may be delivered into the portal capillaries and vehicled to the AP, where they act as neurohormones.

Another reason for breakdown of demarcations between these messenger substances is the recognition that they may coexist in the same neuron in different CNS areas as well as in the mediobasal hypothalamus. The functional role of these costored neurotransmitters and neuropeptides is unsettled because demonstrating costorage also within nerve terminals of these neuronal systems at the median eminence level has not yet been possible. However, evidence for interaction between release mechanisms for the transmission lines (corelease), between decoding mechanisms (codecoding), and between transduction mechanisms (cotransduction) does exist. Figure 1 is a diagram illustrating the principal neural afferents involved in pituitary regulation.

Description of the influences exerted by brain messengers on the neuroendocrine control of AP secretion requires a sketch of the mechanisms of biosynthesis, release, metabolic disposal of principal neurotransmitters, nonhypophysiotropic and hypophysiotropic neuropeptides, of the CNS-acting compounds capable of interfering with the different metabolic steps or with pre-postsynaptic receptors, and, finally, mention of the regional distribution of the main neuronal systems in the CNS. [*See* Neuroendocrinology.]

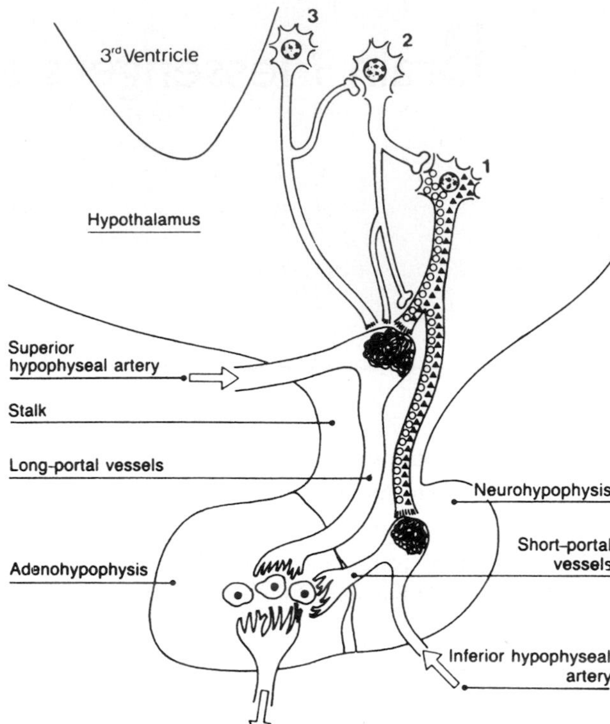

FIGURE 1 Neurotransmitter–neurohormonal control of AP secretion. The diagram illustrates the principal neural afferents involved in pituitary regulation. Neuron 1 denotes a tubero-hypophyseal peptidergic neuron that manufactures hypothalamic hormones, which are then released into the hypophyseal portal vessels and are relayed to the AP. Different symbols in the neuron (▲, ○) refer to the possibility of peptide–amine or peptide–peptide colocalization, respectively. Neuron 2 denotes a neurotransmitter nerve ending in relation to a peptidergic neuron (axosomatic or axoaxonic contact). The possibility is also depicted that a neurotransmitter neuron (DA, epinephrine, GABA, etc.) ends directly in relation with the primary plexus of blood capillaries at the median eminence level. Neuron 3 is a peptidergic neuron that modulates the activity of a neurotransmitter neuron (axosomatic or axoaxonic contact). This neuron may end in direct contact with hypophyseal portal vessels. For the sake of clarity, peptidergic neurons of the supraoptic and paraventricular-hypophyseal systems, with cell bodies in the hypothalamus and nerve endings in the neurohypophysis, have been omitted. [Reproduced, with permission, from E. E. Müller (1986). Brain neurotransmitters and the secretion of growth hormone. *Growth and Growth Factors* **1**, 65–74.]

A. Principal Neurotransmitters

1. Catecholamines

Synthesis of catecholamines (DA, NE, and epinephrine) occurs *in vivo* from the precursor amino acids phenylalanine and tyrosine. The first limiting step in catecholamine biosynthesis is the transformation of L-tyrosine to L-dopa, a reaction catalyzed by the enzyme tyrosine hydroxylase. Dopamine formed from L-dopa by L-aromatic amino acid decarboxylase is subsequently oxidized to NE. There are neurons in the CNS that take up NE and methylate it to epinephrine.

Catecholamines synthesized and stored in specific granules in nerve terminals are released into the extracellular space following neuronal depolarization and interact with specific receptor sites located on the postsynaptic or presynaptic membranes. Following release, most of the neurotransmitter is captured into the presynaptic nerve terminal, where it undergoes deamination by monoamine oxidases or is stored again in secretory granules. The uptake process is the principal mechanism for termination of catecholamine effects at postsynaptic sites.

Many drugs are capable of inhibiting the functional activity of catecholamine neurons. They comprise biosynthesis inhibitors at the level of different enzymatic steps: reserpine, which depletes the granular pool, and neurotoxic agents, such as 6-hydroxydopamine, which destroy catecholamine terminals. Conversely, catecholamine precursors or inhibitors of metabolic degradation or uptake potentiate catecholamine mechanisms.

NE has high affinity for both α and β receptors, whereas epinephrine has higher affinity for β than for α receptors. These receptors are now divided into α_{1A}, α_{1B}, α_{1C}, α_{2A}, α_{2B}, and α_{2C} subclasses and β_1, β_2, and β_3 receptors. α_1-Receptors are located postsynaptically and are functionally related to phosphoinositide metabolism; α_2-receptors are located pre- and postsynaptically and are functionally coupled to inhibition of adenylate cyclase. All β-adrenergic receptors stimulate adenylate cyclase. DA receptors are divided into an increasing number of classes, including D_1, D_2, D_3, D_4, and D_5. There are now specific agonists and antagonists for the different DA receptor subtypes.

2. Serotonin

The synthesis of serotonin (5-HT) involves several mechanisms: uptake of the precursor amino acid L-tryptophan (Trp) into the terminals, hydroxylation of Trp to 5-hydroxytryptophan (5-HTP), and decarboxylation to 5-HT, by the same enzyme that forms DA from L-dopa. Administration of either Trp or 5-HTP increases 5-HT neurotransmission, although the specificity of 5-HTP is poor. At high doses, in fact, this compound leads to substantial 5-HT accumulation in cells that do not ordinarily contain this indoleamine and also interferes with catecholamines and their precursors for transport, storage, and metabolsim within the CNS.

5-HT synthesis can also be controlled pharmacologically, either via inhibition of Trp-hydroxlase or by destroying 5-HT nerve terminals with specific neurotoxic agents.

It would seem now that 5-HT receptors belong essentially to two superfamilies: 5-HT_{1A}, 5-HT_{1B}, 5-HT_{1C}, 5-HT_{1D}, 5-HT_2, and 5-HT_4, which are coupled to G proteins and cAMP or phosphoinositide transduction, and 5-HT_3, which is coupled directly to cationic channels, though controversies still exist on the classification of these receptor subtypes.

Many of the 5-HT receptor agonists and antagonists have poor neuropharmacological specificity on the different receptor subtypes. Drugs that seem to be most effective in influencing 5-HT neurotransmission comprise a series of amine-uptake inhibitors (chloroimipramine, fluoxetine, fluvoxamine) and 5-HT releasers (fenfluramine).

3. Acetylcholine

In general, the biochemical mechanisms on which cholinergic [acetylcholine (ACh)-mediated] neurotransmission depends are very similar to those of catecholamines and 5-HT. The biosynthesis and metabolic degradation of ACh are controlled by two enzymatic activities, that is, choline acetyltransferase (CAT) and acetylcholinesterase (ACh-ase). Choline, the precursor of ACh, is taken up by the extracellular fluid into the axoplasm. ACh is stored in synaptic vesicles and, following its release, is hydrolyzed almost immediately to choline and acetic acid. Activation of cholinergic neurotransmission ensues inhibition of ACh-ase by drugs. There are two classes of ACh receptors: nicotinic and muscarinic, subdivided into M_1, M_2, M_3, M_4, and M_5. They have distinct anatomical distributions and physiological functions. Muscarinic receptors are also present in the AP, whereas both receptor types are present in the CNS.

4. Histamine

Histamine formation is made by decarboxylation of histidine. The existence of three types of histamine receptors (H_1, H_2, and H_3) is well established.

5. GABA

GABA, the main inhibitory amino acid neurotransmitter, is formed in the nervous tissue by the decarboxylation of L-glutamate. The catabolism of GABA involves a transamination. Several drugs have been found capable of inhibiting transamination, thus increasing brain GABA concentrations. GABA, once released into the synaptic cleft, acts on specific receptors to produce hyperpolarization of the postsynaptic membrane and a selective increase in Cl^- permeability. GABA receptors are part of a supramolecular assembly in which there is a GABA recognition site operationally coupled with a Cl^- ionophore and a benzodiazepine receptor, whose functional activation leads to an enhancement of GABA binding to its high-affinity binding sites.

Multiple sites exist for GABA action: $GABA_A$-receptors, which are bicuculline-sensitive, and $GABA_B$-receptors, which are bicuculline-sensitive, and $GABA_B$-receptors, which are bicuculline-insensitive, present in both the CNS and the periphery, and mainly at the presynaptic level, at least peripherally.

6. Glutamate

A large body of evidence now exists in favor of a transmitter role in the CNS for L-glutamate and L-aspartate, though the widespread distributon of these two dicarboxylic amino acids in the CNS and their role in the intermediary metabolism have tended to obscure the action that they might have as transmit-

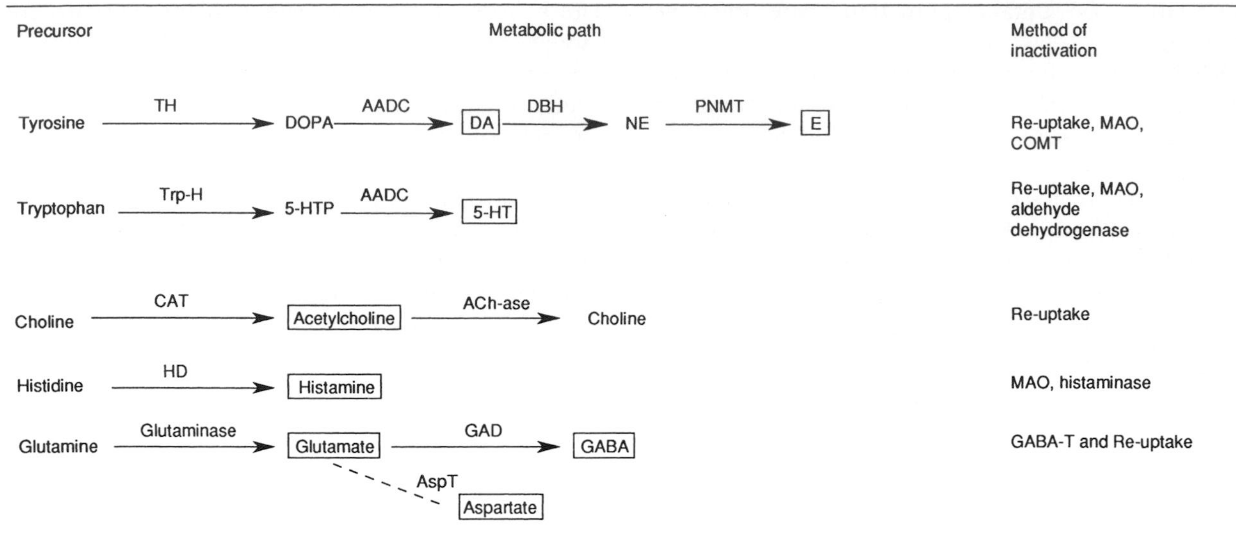

FIGURE 2 Steps in the formation of classic neurotransmitters. AADC, aromatic amino acid decarboxylase; ACh, acetylcholine-sterase, AspT, mitochondrial aspartate transaminase; CAT, choline acetyltransferase; COMT, catechol-O-methyltransferase; DA, dopamine; DβH, dopamine β-hydroxylase; DOPA, dihydroxyphenylalanine; E, epinephrine; GABA-T, γ-aminobutyric acid transaminase; GAD, glutamic acid decarboxylase; HD, histidine decarboxylase; 5-HT, 5-hydroxytryptamine; 5-HTP, 5-hydroxytryptophan; MAO, monoamine oxidase; NE, norepinephrine; PNMT, phenylethanolamine-*N*-methyltransferase; TH, tyrosine hydroxylase; Trp-H, tryptophan hydroxylase.

ters. It is now universally acknowledged that both amino acids exert extremely powerful excitatory effects on neurons in virtually every region of the CNS. Based on the relative potencies of synthetic agonists and the discovery of selective antagonists, many subtypes of receptors for excitatory amino acids have been characterized pharmacologically. Glutamate receptors can be categorized into two principal groups: ionotropic and metabotropic. Ionotropic receptors can be subdivided into *N*-methyl-D-aspartate (NMDA), kainate, and DL-α-amino-3-hydroxy-5-methyl-4-isoxazole propionic acid (AMPA); metabotropic receptors, which are coupled to G proteins and modulate the production of second messengers, are activated most potently by quisqualate.

Figure 2 shows the principal steps involved in the formation of the classic neurotransmitters here reviewed, and Table I lists a series of drugs that act as agonists or antagonists at neurotransmitter receptors or that alter different aspects of neurotransmitter function.

B. Topographical Localization of Neurotransmitter Systems of Neuroendocrine Relevance in the CNS

Progress in the topographical localization of principal neurotransmitter systems has been made possible by many technical advances, especially immunohistochemical methodology, which is suitable for analyzing multiple antigens in one and the same neuron.

Most studies of catecholamine neurons have been carried out on the rat brain, but tissues from other species, including primates, have also been analyzed. There are about 50,000 catecholamine neurons in the rat brain stem, of which DA neurons constitute 80% and NE neurons 20%. The number of epinephrine neurons has so far not been calculated. About 70% of all catecholamine neurons are found in the mesencephalon. The ascending NE pathway comprises (1) a dorsal bundle, which originates in the locus coeruleus and innervates the cortex, the hippocampus, and the cerebellum, and (2) a ventral bundle, which originates in the pons and medulla and gives rise to pathways innervating the lower brain stem, the hypothalamus, and the limbic system. The DA pathways comprise three main systems of long DA neurons—the nigrostriatal, the mesolimbic, and the mesocortical systems—and two systems composed of short intrahypothalamic DA axons, which originate in the arcuate nucleus and project to the external layer of the median eminence [tuberoinfundibular DA (TIDA) neurons] and in the rostral periventricular hypothalamus and project to anterior and dorsal hypothalamic areas and to the zona incerta (incertohypothalamic system). Of particular relevance for neuroendocrine

TABLE I
Drugs Altering Neurotransmitter Function by Various Mechanisms[a]

Drug	Mechanisms of action	Observations
Apomorphine	Direct stimulation of pre- and postsynaptic receptors	Increase in brain 5-HT and 5-HIAA
Bromocriptine (2-Br-α-ergocriptine)	Direct stimulation of pre- and postsynaptic DA receptors	Action depending in part on brain CA stores
Lisuride	Direct stimulation of pre- and postsynaptic DA receptors	Peripheral antiserotoninergic activity
Piribedil	Direct and indirect stimulation of DA receptors	—
Amantadine	Direct and indirect stimulation of DA receptors	Blockade of DA reuptake
Nomifensine	Blockade of DA and NE reuptake	—
Clonidine	Direct stimulation of central and peripheral α_2-adrenoceptors	Stimulation of central histamine H_2-receptors
L-Dopa	Increase in synthesis of DA and NE	Displacement of 5-HT by DA formed in serotoninergic neurons
α-MpT (α-methyl-paratyrosine)	Blockade of TH and of DA, NE, and E synthesis	Conversion to α-methyl DA and α-methyl NE endowed with intrinsic receptor stimulating activity
FLA-63 bis(4-methyl-1-homo-piperazinyl-thiocarbonyl) disulfide	Blockade of the conversion of DA to NE	—
6-OHDA (6-hydroxydopamine)	Neurotoxic for CA neurons	Does not cross the BBB
Reserpine	Release of intragranular pool of CAs and inhibition of granular uptake; same effect on 5-HT neurons	—
Carbidopa	Selective inhibition of peripheral aromatic L-amino acid decarboxylase	—
Chlorpromazine	Blockade of DA and NE receptors	Hypersensitivity to adrenergic stimulation
Haloperidol	Blockade of DA receptors	—
Pimozide	Blockade of DA receptors	At high doses blockade of NE receptors
Phentolamine	Blockade of α_1- and α_2-adrenoceptors	Long-lasting action
Phenoxybenzamine	Blockade of α_1- and α_2-adrenoceptors	Long-lasting action; blockade also of 5-HT, H, and ACh receptors
Prazosin	Blockade of α_1-adrenoceptors	—
Yohimbine	Blockade of α_2-adrenoceptors	—
Propranolol	Blockade of β_1- and β_2-adrenoceptors	—
L-Tryptophan	Selective increase in 5-HT synthesis, storage, and metabolism	—
5-Hydroxytryptophan	Increase in 5-HT synthesis	Decrease (by displacement) of brain DA and NE levels
pCPA (*p*-chlorophenyl-alanine)	Blockade of Trp-H and of 5-HT synthesis	Release of CAs
Methysergide	Blockade of 5-HT receptors	Potential dopaminergic, antidopaminergic, and antiserotoninergic activity
Cyproheptadine	Blockade of 5-HT receptors	Antihistamine, anticholinergic, and anticatecholaminergic activity
Metergoline	Blockade of 5-HT receptors	Potential dopaminergic activity
5,6-DHT (5,6-dihydroxytryptamine)	Neurotoxic for serotoninergic neurons	Release of CAs
5,7-DHT (5,7-dihydroxytryptamine)	Neurotoxic for serotoninergic neurons	Release of CAs
Fluoxetine	Selective blockade of 5-HT reuptake	—
Acetylcholine	Direct stimulation of muscarinic and nicotinic receptors	—

(*continues*)

TABLE I *(Continued)*

Drug	Mechanisms of action	Observations
Pyridostigmine	Reversible inhibition of acetylcholinesterase	Does not cross the BBB
Atropine	Antagonism at muscarinic M_1- and M_2-receptors	—
Pirenzepine	Antagonism at muscarinic M_1-receptors	Poor penetration of the BBB
Aminooxyacetic acid	Inhibition of GABA catabolism	—
Sodium valproate	Inhibition of GABA catabolism	—
Muscimol	Direct stimulation of $GABA_A$-receptors	Poor penetration of the BBB
Baclofen	Direct stimulation of $GABA_B$-receptors	Action not antagonized by bicuculline
Bicuculline	Antagonism at GABA receptors	—
2-Methylhistamine	H_1-receptor agonist	—
Dimaprit	H_2-receptor agonist	—
Diphenhydramine	H_1-receptor antagonist	Anticholinergic activity
Meclastine	H_1-receptor antagonist	No anticholinergic activity
Cimetidine	H_2-receptor antagonist	Poor penetration of the BBB
Ranitidine	H_2-receptor antagonist	Poor penetration of the BBB

[a]CA, catecholamine; E, epinephrine; H,histamine; 5-HIAA, 5-hydroxy-indoleacetic acid; Trp-H, tryptophan hydroxylase; BBB, blood–brain barrier; for other abbreviations see text. (Reproduced, with permission, from E. E. Müller (1986). Brain neurotransmitters and the secretion of growth hormone. *Growth and Growth Factors* **1**, 65–74.

control is the TIDA system, which innervates the median eminence, where regulatory hormones, neurotransmitters, and neuropeptides are released into the portal capillaries to be conveyed to the AP.

The 5-HT system arises from cell bodies situated in the mesencephalic and pontine raphe and sending axons especially to the hypothalamus, median eminence, preoptic area, limbic system, septal area, striatum, and cerebral cortex. There are also 5-HT neurons in the hypothalamus.

Availability of antibodies specific for CAT has made it possible to outline cholinergic neurons and to study their projections. The enzymatic activity is present in most hypothalamic nuclei, including the arcuate nucleus and the median eminence. Since only small changes have been detected in the concentrations of CAT in the mediobasal hypothalamus following mechanical separation of this area, the existence of a TI-cholinergic pathway, similar to the TIDA pathway, has been envisaged. This may play an important role in mediating most of the neuroendocrine effects of cholinergic drugs.

The regional localization of histamine in the brain and also in individual nuclei of the hypothalamus has been studied in rodents and primates. In the monkey hypothalamus, the highest concentrations are found in the mammillary bodies, the supraoptic nucleus, the ventromedial nucleus, the ventrolateral nucleus, and the median eminence. A similar pattern of distribution is present for histamine in the human brain; the highest levels are found in the mammillary bodies and the mid-hypothalamus. After deafferentation of the mediobasal hypothalamus in rats, levels of histamine do not decrease significantly from the control value in the arcuate nucleus, ventromedial nucleus, dorsomedial nucleus, and median eminence, which suggests that histamine is present in the posterior two-thirds of the hypothalamus in cells that are intrinsic to this area.

Specific antibodies raised against glutamate decarboxylase (GAD-I) have made precise localization of GABAergic neurons in the hypothalamus possible. A dense network of GAD-positive nerve fibers is present in different hypothalamic nuclei of rodent and cat brains. In the median eminence, a dense immunofluorescent plexus is found in the external layer, extending across the entire mediolateral axis from the rostral part to the pituitary gland. In the median eminence, GAD activity remains unaffected by a total deafferentation of the mediobasal hypothalamus but decreases about 50% following neurochemical lesioning of the arcuate nucleus, by monosodium glutamate. This suggests the existence of a TI-GABAergic system, a hypothesis confirmed by the presence of GAD-positive immunoreactive cell bodies in the arcuate nucleus and periventricular nuclei.

NMDA receptor is expressed throughout the brain,

TABLE II
Neuropeptides[a]

Pituitary peptides	Opioid peptides
Adrenocorticotropin hormone	Enkephalins
α-Melanocyte-stimulating hormone	β-Endorphin
Growth hormone	Dynorphin
Luteinizing hormone	Kyotorphin
Prolactin	Dermorphin
Thyroid-stimulating hormone	Hypothalamic regulatory hormones
Oxytocin	Corticotropin-releasing hormone
Vasopressin	Gonadotropin-releasing hormone
Circulating hormones	Thyrotropin-releasing hormone
Angiotensin	Growth hormone-releasing hormone
Calcitonin	Somatostatin
Glucagon	Miscellaneous peptides
Insulin	Bombesin
IGF-I; IGF-II	Gastrin-releasing peptide
Atrial natriuretic factor	Bradykinin
Gut hormones	Neuropeptide Y
Avian pancreatic polypeptide	Histidine-isoleucine peptide
Pancreatic polypeptide	Neurotensin
Cholecystokinin	Carnosine
Gastrin	Proctolin
Motilin	PACAP
Galanin	Cytokines
Secretin	Interleukins (ILs); TNF-α, IFN-γ
Substance P	
Vasoactive intestinal peptide	

[a]These peptides have been described in mammalian CNS neurons and nerve terminals other than those related to endocrine or neuroendocrine functions.

with the greatest expression in the hippocampus, hypothalamus, and olfactory bulb. In addition, elegant ultrastructural studies have demonstrated that presynaptic boutons in the suprachiasmatic nucleus, ventromedial nucleus, arcuate nucleus, and parvocellular and magnocellular paraventricular nucleus in the rat hypothalamus show strong immunoreactivity for glutamate.

C. Neuropeptides

During the last 15 years, it has become increasingly apparent that large numbers of neuropeptides are present in CNS neurons (Table II). Availability of synthetic neuropeptides to be used for the production of specific antibodies for radioimmunoassay and immunocytochemistry has served as a powerful tool for localization; application of rDNA or rRNA technologies has permitted the study of neuropeptide biosynthesis and its regulation. Detailed analysis of biosynthesis, metabolism, and localization in the CNS of principal neuropeptides is beyond the scope of this article. Only some generalities will be considered here.

Neurosecretory neurons synthesize, transport, process, and secrete neuropeptides by mechanisms similar to those in peripheral hormone-producing tissues. Peptides are generated from large precursor molecules produced in the rough endoplasmic reticulum and packaged in secretory granules or vesicles in the Golgi complex. The granules are transported out from cell bodies to the terminals (axonal transport), where they release their content by exocytosis upon neuronal depolarization. Neuropeptide release can also be stimulated or inhibited by application of neurotransmitters or other neuropeptides, indicating the existence of neurotransmitter–neuropeptide and neuropeptide–neuropeptide interactions, an important step in the control of neurohormonal or neuromodulator function of the peptide. [*See* Peptides.]

Proteolytic enzymes are not only important in the processing of pro-hormones to active component forms but also in terminating the action of active neuropeptides upon their release. In general, neuropeptides are present in the brain at concentrations much lower than those of the classic amine neurotransmitters, and their concentrations are even lower

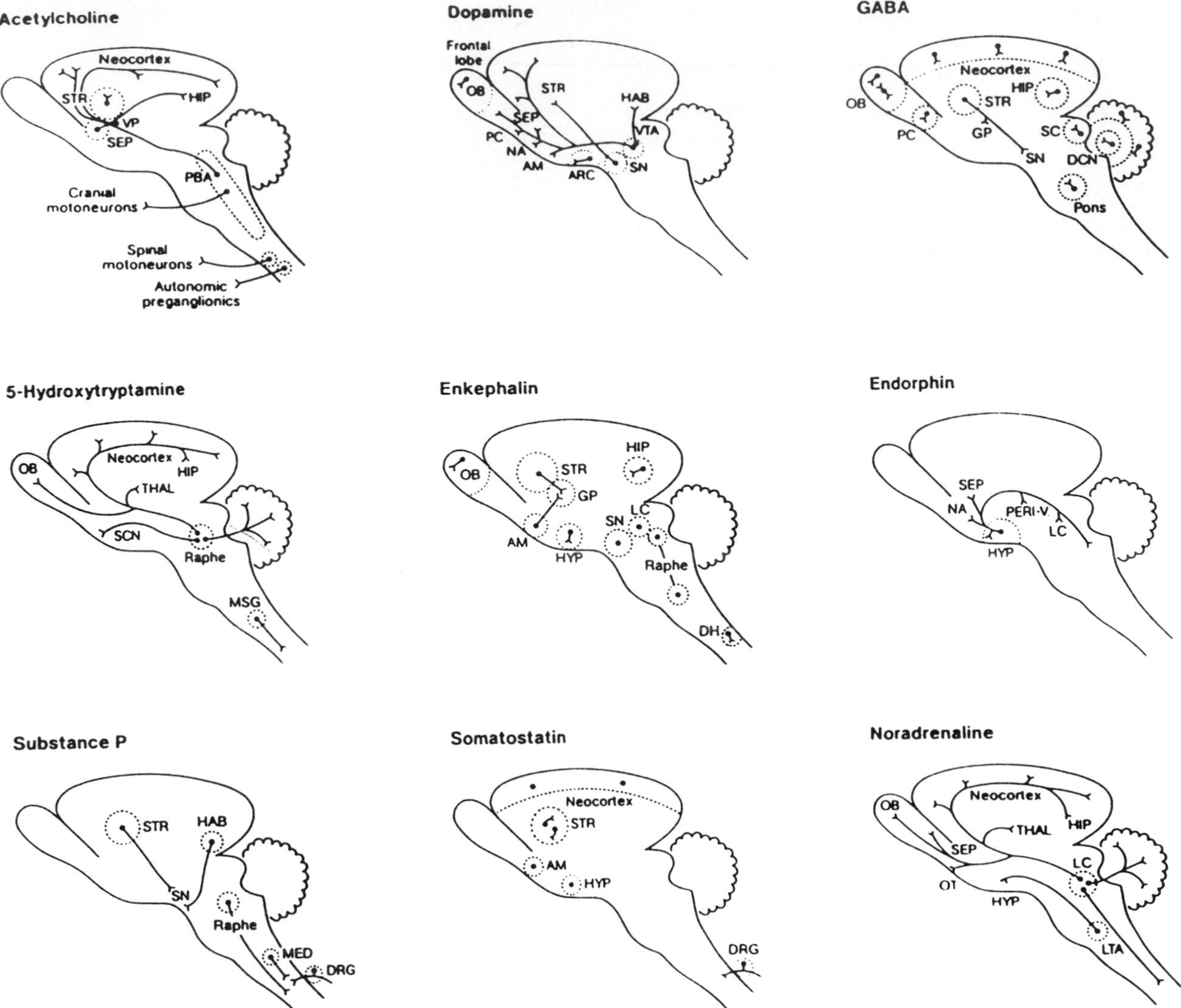

FIGURE 3 Principal locations of neuronal pathways. AM, amygdala; ARC, arcuate nucleus; DCN, deep cerebellar nuclei; DH, dorsal horn; DRG, dorsal root ganglion; GP, globus pallidus; HAB, habenula; HIP, hippocampus; HYP, hypothalamus; LC, locus coeruleus; LTA, lateral tegmental area; MED, medulla; MSG, medullary serotonin group; NA, nucleus accumbens; OB, olfactory bulb; OT, olfactory tubercle; PBA, parabrachial area; PC, pyriform cortex; PERI-V, periventricular gray; SC, superior colliculus; SCN, suprachiasmatic nucleus; SEP, septum; SN, substantia nigra; STR, striatum; THAL, thalamus; VP, ventral pallidum; VTA, ventral tegmental area. [Reproduced, with permission, from J. K. McQueen (1987). Classical transmitters and neuromodulators. *In* "Basic and Clinical Aspects of Neuroscience" (E. Flückiger, E. E. Müller, and M. O. Thorner, eds.), Vol. 2, pp. 7–16. Springer Sandoz Advanced Texts, Springer-Verlag, Heidelberg.]

when compared with those of amino acid neurotransmitters such as GABA and glycine. Once released into the synaptic cleft or transported via the extracellular fluid for a short or long distance (volume transmission), neuropeptides act on specific receptors to produce an alteration in the level of one or another intracellular second messenger.

Detailed maps of the distribution of various types of neuropeptides are available, and some generalities can be expressed on the basis of these distribution studies. Thus, there are some brain areas that are rich in both peptide-immunoreactive cell bodies and terminals. Other areas, such as the cerebellum, have low levels of most neuropeptides; the thalamus is also poor in neuropeptides. Cortical areas are particularly rich in some neuropeptides, such as vasoactive intestinal peptide, cholecystokinin, somatostatin, and corticotropin-releasing hormone (CRH). Peptides are present mainly in small interneurons, a notable exception being the long projections of endorphinergic neurons, whose cell bodies in the arcuate nucleus send terminals to innervate various nuclei of the brain stem.

In many peptidergic neuronal systems, a neuropeptide–receptor mismatch exists (e.g., neuropeptide stores), but no receptors are present (topological mismatch), or the amount of neuropeptide stores does not correlate across several brain regions with the density of the receptors (functional mismatch). Factors contributing to the mismatch phenomenon include the existence of volume transmission, breaking down of the parent peptide after release into active fragments that are then recognized by specific receptors, nonfunctional or spare receptors, and unrecognized low-affinity or occupied receptors, which impede proper evaluation of the density of the peptide receptor (i.e., functional mismatch). Figure 3 depicts schematically the topography of some neurotransmitters and neuropeptides in the rat brain.

II. REGULATORY HORMONES FOR AP CONTROL

In mammals, the existence of at least six hypothalamic regulators of the AP is now reasonably well established: five of them have been chemically characterized (Table III) and therefore conform to the designation of regulatory hormones. Although this designation, in view of the expanding list of neuropeptides, which are also endowed with intrinsic hydrophysiotropic actions, now appears to be a misnomer, it has historical and heuristic value and will be used here.

A. Generalities

Regulatory hormones are low- to medium-molecular-weight compounds of about 800–4000 (Table III), which are especially synthesized in the mediobasal hypothalamus nuclei and in the anterior hypothalamus, although they can be elaborated ubiquitously in the CNS. Each peptide is synthesized as a pre-pro-hormone in the ribosomes; the pre-pro-hormone enters the rough endoplasmic reticulum, where the leader sequence is removed; and the pre-hormone accumulates in secretory granules of the Golgi apparatus, which are then transported by axoplasmic flow to the nerve terminals abutting on the median eminence and released into the hypophyseal portal system. Regulatory hormone secretion is not continuous but occurs in episodic fashion (one major pulse each 1–3 hr), likely due to the rhythmic, spontaneous discharge of neurosecretory neurons. The pulsatile release of regulatory hormones into the portal capillaries is a *sine qua non* condition to allow proper sensitivity of pituitary receptor sites, which otherwise will undergo a process of down-regulation (diminution of receptor number). The interaction between regulatory hormones and their receptors prompts the formation or inhibits the functional activation of one or more intracellular messengers (cAMP, Ca^{2+}, hydrolysis products of phosphoinositides, e.g., diacylglycerol, inositol-1,4,5-triphosphate), which mediate the hormonal response of the AP.

A peculiar feature of regulatory hormones is provided by their abundant extrahypothalamic distribution, particularly in the most ancient phylogenetic species (some even lacking the AP), which suggests a primitive role as "regulators" of CNS function and more recent cooptation during the phylogenesis to act as controllers of the pituitary function. The ubiquitous distribution in the CNS and its ability to evoke neurochemical and behavioral effects in animals and humans, even in the absence of the pituitary, denote the neurotransmitter or neuromodulator role of these substances. It is also noteworthy that, contrary to previous belief, regulatory hormones do not present with a rigid neurohormonal specificity; in fact, some of them affect the release of more than one AP hormone. Thyrotropin-releasing hormone (TRH), for instance, in addition to thyroid-stimulating hormone (TSH), also affects prolactin secretion and, in some pathologic conditions, growth hormone secretion; somatostatin inhibits growth hormone but also TSH and adrenocorticotropin hormone (ACTH) secretion.

1. Corticotropin-Releasing Hormone

It is now evident that the secretion of ACTH is under regulation by a specific 41-residue peptide (CRF-41) first identified in sheep hypothalami. CRF-41 residue peptides with identical biological activities were isolated from the rat and the pig, and the sequence of the human peptide was deduced from cloned human genomic DNA. The human and rat CRF-41 molecules proved to be identical in sequence, differing in seven residues from the sheep peptide. CRF alone does not account for the ACTH-releasing activity of the hypothalamus; the bioactivity of CRF may be modulated in fact by other hypothalamic substances, which show a synergistic interaction with CRF (e.g., vasopressin, oxytocin, angiotensin). Within the hypothalamus, CRF is localized in the parvocellular neurons of the

TABLE III
Hypothalamic Regulatory Hormones[a]

Hypothalamic hormone or factor	Structure or candidate
Corticotropin-releasing hormone (CRH or CRF)[b]	H-Ser-Glu-Glu-Pro-Pro-Ile-Ser-Leu-Asp-Leu-Thr-Phe-His-Leu-Leu-Arg-Glu-Val-Leu-Glu-Met-Ala-Arg-Ala-Glu-Gln-Leu-Ala-Gin-Gin-Ala-His-Ser-Asn-Arg-Lys-Leu-Met-Glu-Ile-Ile-NH_2
Thyrotropin-releasing hormone (TRH)	pGlu-His-Pro-NH_2
Luteinizing hormone–follicle-stimulating hormone-releasing hormone (LHRH, or GnRH)	pGlu-His-Trp-Ser-Tyr-Gly-Leu-Arg-Pro-Gly-NH_2
Growth hormone-releasing hormone (GHRH, or GRF)[c]	H-Tyr-Ala-Asp-Ala-Ile-Phe-Thr-Asn-Ser-Tyr-Arg-Lys-Val-Leu-Gly-Gln-Leu-Ser-Ala-Arg-Lys-Leu-Leu-Gln-Asp-Ile-Met-Ser-Arg-Gln-Gln-Gly-Glu-Ser-Asn-Gln-Glu-Arg-Gly-Ala-Arg-Ala-Arg-Leu-NH_2
Growth hormone-releasing–inhibiting hormone (GH-RIH, or somatostatin)[d]	S-S (Cys–Cys) H-Al-Gly-Cys-Lys-Asn-Phe-Phe-Trp-Lys-Lys-Thr-Phe-Thr-Ser-Cys-OH
Prolactin-inhibiting factor (PIF)[e]	Dopamine
Prolactin-releasing factor (PRF)[f]	

[a]Reproduced, with permission, from E. E. Müller and G. Nisticò (1989). "Brain Messengers and the Pituitary." Academic Press, San Diego.
[b]The sequence of human CRF is shown.
[c]The 44-amino-acid structure.
[d]The 14-amino-acid molecular form of somatostatin is indicated.
[e]Dopamine is the most likely candidate to be the PIF, although the sequence of a 56-amino-acid peptide named GAP (GnRH-associated peptide) and endowed with PIF activity has also been reported.
[f]Vasoactive intestinal peptide may be a PRF.

paraventricular nucleus with the axons projecting to the external palisade zone of the median eminence, in both rats and humans. From the same source arise vasopressin-containing fibers present in the median eminence in about 50% of which CRH and vasopressin coexist in the same neurosecretory granules.

Acute administration of ovine CRH triggers in normal healthy volunteers a prompt, consistent, and long-lasting rise in circulating levels of ACTH, followed by a gradual increase of cortisol levels; after CRH doses of 0.3–30 μg kg, i.v., plasma ACTH reaches a peak at 10–15 min, declines until 90 min, and rises to a second peak at 3 hr. Plasma ACTH and cortisol levels are elevated 7 and 10 hr, respectively, after the highest dose. The hypercortisolemia that follows CRH injection blunts the responsiveness of the pituitary gland to subsequent stimuli, and with high CRH doses this effect may last for >12 hr. In the doses commonly used (approximately 100 μg), the only side effects observed are facial flushing, which occurs in about one-third of subjects, and an occasional transient tachycardia. CRH, although of no therapeutic use, has diagnostic power in conditions of hyperdysfunction of the hypothalamo–pituitary adrenal axis and in differentiating primary from secondary insufficiency of the adrenal gland.

2. Thyrotrophin-Releasing Hormone

TRH, the first regulatory hormone to be isolated and characterized, is the dominant regulator of the secretion of TSH through a tonic stimulating action. In mammals, TRH is synthesized in the thyrotropic area of the hypothalamus, principally the parvocellular division of the paraventricular nucleus, and then transported to the AP via portal capillaries. TRH interacts with high-affinity receptors on the pituitary thyrotrophs, whose number is regulated in part by thyroid hormones, and activates the phosphatidylinositol pathway. TRH stimulates the synthesis of TSH, in addition to its release.

In humans, i.v. doses of as little as 15 μg TRH induce significant increases in plasma TSH; the effect is dose-dependent up to a concentration of 400 μg, is higher in females than in males (a reflection of estrogen's ability to increase pituitary TRH-binding sites), and decreases in aged subjects. Administration of TRH causes not only a prompt rise in plasma TSH concentration, 15–30 min later, but also a significant increase in circulating plasma triiodothyronine (T_3) (the thyroid hormone), which occurs 20 min after the elevation in TSH. Administration of TRH provides a valuable tool for the assessment of pituitary or thyroid ability to respond to respective therapies and for distinguishing whether idiopathic hypopituitarism is due to a pituitary disease (impaired TSH response) or hypothalamic disease (normal or exaggerated response). In primary hypothyroidism, the response of TSH to TRH is exaggerated in >70% of cases; conversely, in hyperthyroidism the response is usually absent, to indicate that TSH secretion is controlled by the negative feedback of thyroid hormones on the AP.

TRH is also a potent prolactin releaser *in vivo* and from *in vitro* APs in both animals and humans. In humans, the minimum effective dose that releases prolactin is similar to the dose that releases TRH.

3. Gonadotropin-Releasing Hormone

Unequivocal demonstration for the existence of a neurohormonal control of gonadotropin secretion was given with the isolation and chemical identification of a peptide that had both luteinizing hormone (LH)- and follicle-stimulating hormone (FSH)-releasing activity, termed LHRH or, more recently, GnRH (gonadotropin-releasing hormone). GnRH induces FSH and LH release in many animal species, including primates.

GnRH is a linear decapeptide (Table III), which in the mammalian hypothalamus and the placenta is synthesized as part of a 92-amino-acid precursor protein.

The structure of the GnRH precursor comprises the decapeptide preceded by a signal sequence of 23 amino acids and followed by a Gly-Lys-Arg sequence necessary for enzymatic cleavage of the precursor and C-terminal amidation of GnRH. A sequence of 56 amino acid residues occupies the C-terminal region of the precursor and constitutes the GnRH-associated peptide, GAP. GAP is also capable of stimulating gonadotropin secretion and inhibiting prolactin secretion from rat AP cells in culture; although GAP has been suggested to be a candidate prolactin-inhibiting factor, unequivocal evidence for this role is still lacking. GnRH acts in the pituitary on specific receptors to foster release of LH and also FSH from the gland. Peripheral factors (e.g., gonadal steroids and inhibin, a peptide produced by the gonads, which selectively suppresses FSH release) probably exert a crucial role in dictating preferential FSH or LH release. Also, the modalities of pulsatile GnRH secretion contribute to the differential secretion of gonadotropins.

In mammals, the hypothalamo-infundibular GnRH tract constitutes the major axonal route for GnRH neurons. GnRH is present in high concentrations in the median eminence region (lateral wings of the external layer) but also in the suprachiasmatic nucleus, in the preoptic area and the vascular organ of the lamina terminalis, and in other circumventricular organs. In primates, GnRH cell bodies are mainly localized in the mediobasal hypothalamus, and GnRH neurons present in the preoptic area exert likely extrapituitary actions. In the mediobasal hypothalamus, the arcuate nucleus is an oscillator that generates signals (period 1/hr in the monkey or 1/1–2 hr in the human) that result in the release of an amount of GnRH into the stalk blood and, consequently, a pulse of LH and FSH from the gonadotrophs. Continuous infusion of or frequently repeated GnRH amounts or, conversely, reduction in the discharge frequency inhibits the secretion of both or only one (LH) gonadotropin, respectively.

During each menstrual cycle, the maturation of the ovarian follicle is largely due to FSH secretion, while secretion of estrogen from the maturing follicle is FSH- and LH-dependent. Persistence of plasma estradiol levels exceeding a threshold of approximately 200 pg/ml for at least 2 days at mid-cycle stops the negative feedback action of the steroid and triggers the preovulatory gonadotropin surge (positive feedback).

The secretion of FSH and LH is maximal after 15–30 min from i.v. or subcutaneous (s.c.) administration of a GnRH amount (50–150 μg), whereas chronic, pulsatile delivery of GnRH via minipumps effects successful pituitary and gonadal stimulation in men and women with hypothalamic hypogonadotropic hypogonadism. Superactive analogues of GnRH, after an initial phase of pituitary stimulation, lead to an impaired gonadotropin secretion by persistently occupying and down-regulating pituitary receptors. The use of these compounds allows a selective chemical castration therapy in conditions requiring temporary, reversible suppression of gonadotropin secretion (precocious puberty, endometriosis, prostate cancer, breast cancer, etc.).

4. Growth Hormone-Releasing Hormone and Somatostatin

The secretion of growth hormone is regulated by the CNS via two specific regulatory hormones, a stimulatory growth hormone-releasing hormone (GHRH) and an inhibitory somatostatin, both of which have been isolated, chemically identified, and synthesized.

GHRH-containing neurons are located mainly in the arcuate nucleus, the medial perifornical region of the lateral hypothalamus, paraventricular nucleus, dorsomedial nucleus, and lateral and medial borders of the ventromedial nucleus. Somatostatin-containing neurons are, instead, mainly localized in the periventricular-anterior region, from where they send axons directed caudally through the hypothalamus to enter the median eminence at the level of the ventromedial nucleus.

GHRH and somatostatin are rhythmically secreted from median eminence nerve terminals into the portal circulation with a periodicity of about 3–4 hr but 180° out of phase. The asynchronous secretion of GHRH and somatostatin allows a pulsatile pattern of growth hormone secretion, which is crucial for maintaining optimal sensitivity of growth hormone receptors in target tissues.

The action of GHRH appears to be very specific because both *in vivo* and *in vitro* it stimulates the pituitary to secrete only growth hormone. At the dose commonly used to study pituitary function (1 μg/kg, i.v.), the peptide does not induce side effects, with the exception sometimes of facial flushing. GHRH, at high doses and in subjects affected by a GH-secreting adenoma (acromegaly), may trigger the release of prolactin.

In addition to GHRH and somatostatin, a novel GH regulatory system has been described. GH-releasing peptide-6 (GHRP-6), His-D-Trp-Ala-Trp-D-Phe-Lys-NH_2, is a synthetic enkephalin-like peptide that specifically releases GH *in vitro* and *in vivo* without causing the concomitant release of other AP hormones. The endogenous nature of this or related peptides has been suggested by demonstration and cloning of specific functional receptor sites in the pituitary and hypothalamus. GHRP-6 or its congeners release GH in different animal species with an efficacy greater than that of GHRH, and are effective in humans after acute oral or intranasal administration, in both normal men and short-stature children.

Somatostatin decreases either basal or stimulated growth hormone secretion (i.e., that occurring after insulin hypoglycemia, arginine, L-dopa, physical exercise, or sleep) and suppresses growth hormone secretion also in acromegalics, although its effect is evanescent. The peptide also inhibits basal and stimulated TSH secretion, prolactin secretion under certain conditions, and ACTH secretion from tumor tissues.

Growth hormone autoregulates its own secretion through a feedback mechanism operating at either the hypothalamic or AP level. Intracerebroventricular administration of growth hormone in the rat increases somatostatin concentrations in the portal capillaries and suppresses pulsatile growth hormone release and the growth hormone-releasing effect of many secretagogues, including GHRH.

Growth hormone may also influence its own secretion via the formation of peripheral peptides, that is, somatomedins secreted by many tissues, particularly liver. Somatomedins, and especially somatomedin-C, induce, like growth hormone, release of somatostatin from the hypothalamus but, in addition, are potent inhibitors of growth hormone synthesis and GHRH-stimulated release. GHRH and somatostatin interact functionally: GHRH increases somatostatin secretion and gene expression in the hypothalamus; conversely, somatostatin inhibits release of GHRH acting on specific receptors located in the arcuate nucleus.

5. Prolactin-Inhibiting and -Releasing Factors

The secretion of prolactin, an AP hormone with multiple roles and sites of action, is under dual hypothalamic control, with a predominant dopaminergic inhibition and sometimes overlapping stimulatory input. DA, a classic neurotransmitter produced by neurons with cell bodies in the arcuate nucleus, is secreted as a neurohormone and transported via the portal circulation to the AP, where it interacts with high-affinity, specific DA D_2-type receptors situated in the cell membrane of the lactotroph. DA inhibits the secretion of prolactin so that, under most circumstances, blockade of DA receptors results in an elevation of prolactin levels. Fluctuations in DA levels in portal blood occur, and this may contribute to the changing levels of prolactin in animals and humans of both sexes during different physiological situations. To date, DA remains the single most important inhibitory neural signal in the regulation of prolactin secretion, although other candidates appear as potential prolactin-inhibiting factors; they include GABA and the previously alluded to GAP.

It is also apparent that the neural lobe of the pituitary provides both inhibitory and stimulatory signals for prolactin release, although the exact nature of these factors is, as yet, unknown. In addition to the inhibitory dopaminergic tone, certain aspects of prolactin secretion are mediated by substances with prolactin-releasing factor activity. Factors that have been found to participate in the stimulation of prolactin release under different physiological conditions include TRH, vasoactive intestinal peptide, peptide histidine isoleucine amide, and oxytocin. These substances are selectively involved in the stimulation of prolactin secretion in specific situations but do not entirely fulfill the criteria necessary to identify them as a physiological prolactin-releasing factor, because they do not play the same role as other regulatory hormones when regulating specific pituitary hormones.

III. NEUROTRANSMITTER REGULATION—AP

A. ACTH

1. Catecholamines

The effect of catecholamines on the hypothalamo–pituitary adrenal axis in experimental animals and in humans has received much attention. There seems to be a central α_1-adrenergic mechanism that exerts a stimulant effect on ACTH secretion in humans. Methoxamine (Methox), a highly selective α_1-adrenergic agonist, stimulates ACTH secretion in a dose-dependent manner, when infused into normal subjects. The site of action of Methox is within the blood–brain barrier (BBB), presumably on the paraventricular nucleus, and not at the pituitary. Administration of β_1- and β_2-adrenergic agonists had no effect on the secretion of ACTH and cortisol, suggesting that in humans, in contrast to rodents, circulating catecholamines do not play an important physiological role on ACTH and have ready access to the pituitary level, an area lying outside the BBB. [*See* Catecholamines and Behavior.]

2. Serotonin

Available results on the effect of the 5-HT system on the hypothalamo–pituitary adrenal axis are sometimes confusing, and many points still need elucidation. Nonspecificity of effects of some of the drugs at the doses used, existence of multiple 5-HT receptor subtypes, uncertainty about the site(s) of action, and the likelihood of multiple sites of action, differences in species, or experimental design are among the contributing factors.

5-HT precursors, direct 5-HT receptor signals, 5-HT uptake inhibitors, or releasing drugs induce an elevation in plasma corticosteroids when injected into rodents or members of other species, including humans. In particular, 5-HT stimulates bioactive CRH secretion from rat hypothalamic fragments *in vitro*. Oral administration of quipazine, or other direct 5-HT agonists, to normal volunteers reliably increases plasma cortisol levels. Administration of 5-HTP was found to be followed by an increase of plasma ACTH and cortisol, but the ambiguous pharmacologic effects of this compound dictate caution in interpreting these effects.

Studies in rats on the nature of the 5-HT receptors involved in the activation of the hypothalamo–pituitary adrenal axis indicate that they may be of the 5-HT_{1A}- and 5-HT_2-receptor subtypes. Also, animal studies indicate that 5-HT plays a modulatory role in the regulation of the time of ACTH secretion. Lesions of 5-HT pathways or blockade of 5-HT synthesis abolished the circadian rhythm of plasma corticosterone in rats.

Like ACTH, β-endorphin and related peptides are under a stimulatory serotoninergic control in both animals and humans.

3. ACh

It is now reasonably well established that in several subprimate species an increase in cholinergic transmission is often accompanied by hypothalamo–pituitary adrenal axis activation. The mechanism underlying this effect appears to be direct stimulation of CRH release from the hypothalamus, likely the paraventricular nucleus, as suggested by *in vitro* experiments. Nicotinic receptors would be involved in this effect. Cholinergic agonists do not have a direct stimulant effect on ACTH secretion by the AP *in vivo* or *in vitro*.

In humans, stimulation of ACTH secretion is observed after administration of the ACh-ase inhibitor physostigmine, an effect which is more evident in older subjects. Physostigmine administered to normal adults also induces an escape from suppression of the hypothalamo–pituitary adrenal axis induced by dexamethasone.

4. Histamine

Histamine, given systemically, enhances plasma ACTH and corticosteroids in animals and humans. It is unclear whether a peripheral or a central site is involved. Failure of histamine to stimulate ACTH release from rat hypothalamus *in vitro* suggests an indirect mechanism, perhaps mediated by vasopressin or catecholamine. Similar to 5-HT, histamine seems to participate in the circadian rhythmic regulation of hypothalamo–pituitary adrenal axis function. The role of histamine in regulation of ACTH release in response to stress has not yet been clarified. In humans, meclastine, a rather specific H_1 antagonist, causes a decrease of plasma ACTH concentrations in hypoadrenal patients and blunts the ACTH response to insulin hypoglycemia and metyrapone.

5. GABA

It is now sufficiently well established that GABA exerts an inhibitory role on ACTH secretion, although some *in vivo* studies also point to a stimulatory function. These data include the involvement of both $GABA_A$- and $GABA_B$-receptors, which can be shown as present on corticotrophs. Similar to GABA, benzodiazepines suppress *in vitro* 5-HT-stimulated CRH-like immunoreactivity secretion, whereas potent inverse agonists of the benzodiazepine receptors induce a significant release of CRH-like immunoreactivity.

In humans, some results, in part consistent with those of animal studies, have been presented. In healthy subjects, baclofen or y-vinyl-GABA, an irreversible inhibitor of GABA-T, decreases baseline plasma cortisol levels and the response to insulin hypoglycemia. These findings agree with the ability of sodium valproate to lower high plasma ACTH concentrations in some subjects with Nelson's syndrome (see Section IV,C).

B. TSH

1. Catecholamines

A series of studies in rodents demonstrates that central NE transmission increases secretion of TSH by acting on the α_2-adrenoceptors. The experimental evidence suggests that a facilitatory effect of α_2-adrenergic stimulation on TSH secretion is exerted through modifications in the activity of TRH-containing neurons within the paraventricular nucleus and dorsomedial nucleus areas.

In contrast to the facilitatory effect exerted on TSH secretion by NE transmission is the inhibitory influence that the dopaminergic system exerts in both animals and humans. In subjects with primary hypothyroidism, a single L-dopa dose lowers the elevated TSH levels, although it does not alter the response to TRH. Bromocriptine and DA are instead effective in lowering baseline levels and inhibiting the TSH response to TRH in hypothyroid subjects, and DA also in euthyroid subjects. The site of action of DA is unknown, although a median eminence or pituitary site of action is suggested by many findings.

2. Histamine

Evidence from a few *in vitro* studies indicates that H_2-receptors may be involved in the stimulatory control of TSH secretion. Specific H_2 antagonists counteracted TRH release from rat mediobasal hypothalamus slices or hypothalamic synaptosomes induced by histamine.

C. Gonadotropins

1. Catecholamines

Based on studies in rodents, central NE neurons clearly exert a facilitatory role for allowing GnRH neurons to produce and discharge their products on the afternoon of proestrus, the phase of the estrous cycle during which ovulation occurs. This activation, however, is not an absolute requirement, because in rats the estrous cycle is reestablished a few days after severance of the ascending NE bundle. In ovariectomized rats or monkeys, pulsatile secretion is inhibited by the use of blockers of NE synthesis or of α-adrenergic receptors, suggesting that NE is an essential neurotransmitter in the regulation of pulsatile LH secretion in both species.

In humans, the role played by catecholamines seems less prominent. DA might exert an inhibitory action on GnRH-producing neurons, as suggested by the existence of significant overlapping between GnRH and DA nerve terminals in the lateral wings of the external layer of the median eminence. Inhibition of pulsatile and phasic gonadotropin secretion by DA might be one of the mechanisms through which reproductive function is impaired in hyperprolactinemic states.

2. ACh

Under certain circumstances, the cholinergic system may exert a braking action on LH release. No evidence has been given for a role of ACh on gonadotropin secretion in humans.

3. Histamine

Numerous observations suggest that histamine may be involved in the control of gonadotropin release, although the sites and mechanism of action are uncertain. The observations indicate that the effect of histamine on gonadotropins is sex-related both in humans and in rodents.

4. GABA

The predominant effect of GABA appears to be inhibition of LH release. Data presented from animal studies suggest a direct action of GABA at the gonadotrophs and are consistent with some of the human findings.

D. Growth Hormone

1. Catecholamines

α-Adrenergic mechanisms are important in the regulation of growth hormone secretion in both subprimate and primate species. Blockade of catecholamine synthesis, or depletion of catecholamine storage granules by drugs, almost completely suppresses episodic growth hormone secretion in conscious rats. Selective inhibition of NE and epinephrine synthesis also suppress spontaneous bursts of growth hormone secretion; human growth hormone antagonizes this effect. Apparently, α_2-adrenergic influences act by stimulating GHRH release from the α_2-hypothalamus, whereas α_1-adrenergic stimulation, which is inhibitory to growth hormone release in rats and dogs, occurs via stimulation of somatostatin release. Dopaminergic pathways are also involved in the control of growth hormone secretion, although their role appears to be largely ancillary. It is generally recognized that the DA system may stimulate growth hormone release in humans: both directly and indirectly acting DA agonists elevate human growth hormone levels; however, DA and its agonists inhibit human growth hormone release in patients with acromegaly and also inhibit stimulated human growth hormone release. These effects are due to direct activation of DA receptors located on the tumoral somatotrophs (acromegaly) and the demonstration that DA agonists trigger not only GHRH but also somatostatin release from the hypothalamus, respectively.

2. ACh

Cholinergic neurotransmission is an important modulator of growth hormone secretion. Studies in rats and dogs have shown that the central or peripheral administration of muscarinic agonists or antagonists stimulates or suppresses, respectively, growth hormone release. In humans, administration of ACh precursors or agonists is invariably associated with a small, but unequivocal, rise in plasma growth hormone. Conversely, atropine or its congeners, regardless of whether or not they cross the BBB, proves effective in blocking most of the physiological or neurotransmitter stimuli to growth hormone release, except insulin hypoglycemia. Cholinergic modulation appears to affect growth hormone release via stimulation or inhibition of hypothalamic somatostatin release.

3. Histamine

Studies in both dogs and humans point to a facilitatory role of brain histamine system on growth hormone release.

4. GABA

Activation of GABAergic function induces in rats a dual effect on growth hormone secretion. Both central administration of GABA and systemic administration of a GABA-T inhibitor that freely crosses the BBB, amino-oxyacetic acid, elicit a dose-related, prompt increase in serum growth hormone, an effect blocked by bicuculline. Opposing, growth-hormone lowering effects are obtained instead by systemic administration of GABAergic drugs unable to cross the BBB. These findings are best explained by the ability of GABAergic neurotransmission to inhibit somatostatin neurons located within the BBB, and to act similarly on terminals of GHRH neurons located outside the BBB (median eminence). In humans, studies point to a stimulatory role for GABAergic neurotransmission on baseline growth hormone secretion.

E. Prolactin

1. Catecholamines

The role of DA in the inhibitory control of prolactin secretion has already been mentioned (Section II,B,5). Drugs mimicking at the receptor level the action of the amine (directly or indirectly acting DA agonists) are potent inhibitors of prolactin secretion and can be used on different paradigms of prolactin hypersecretion. Conversely, blockers of DA biosynthesis or receptor antagonists do increase plasma prolactin levels. Both NE and epinephrine exert only an ancillary role in the control of prolactin secretion.

2. Serotonin

The 5-HT system exerts in rodents and monkeys an important facilitatory role on the stimulated prolactin secretion (stress or suckling in lactating dams). The action of 5-HT is not direct on the pituitary but is mediated by a prolactin-releasing factor, most likely vasoactive intestinal peptide. Baseline prolactin secretion is instead barely affected in the rat by functional alterations of 5-HT neurons. In humans, the evidence for a facilitatory role of 5-HT is not so clear-cut.

3. ACh

ACh, via muscarinic receptors, may play an inhibitory role on phasic prolactin secretion. Cholinergic nicotinic receptors located in the hypothalamus also seem to be involved in the inhibitory control.

4. GABA

GABA exerts in rodents an inhibitory control over prolactin secretion. Neurons of the TI-GABA system directly secrete the amino acid into the hypophyseal capillaries from where it is transported to the lactotroph cells to interact with specific binding sites. Pituitary GABA receptors are not a homogeneous population, and both high-affinity, low-capacity and low-affinity, high-capacity sites have been detected. GABA receptors are also present in human APs. The high-affinity GABA-binding site would be the one responsible for the GABA-induced inhibition of prolactin secretion in the rat.

In addition to this inhibitory component of GABA action on prolactin secretion, which is preferentially activated by systemically administered GABA or GABA mimetic drugs, a central, stimulatory component exerts its action via inhibition of TIDA neuronal function. This stimulatory component is also evident from studies in humans, following intracisternal administration of GABAergic compounds to psychiatric patients. Intravenous administration of GABA to normal subjects elicits a biphasic response, because a transient increase in plasma prolactin levels is then followed by a sustained inhibition of prolactin secretion.

5. Glutamatergic Neurotransmission

Evidence gathered in a variety of species indicates that excitatory amino acid (EAA) neurotransmitters also play a role in neuroendocrine regulation. The neuroendocrine functions of EAAs have been most extensively studied and better characterized on the reproductive system. EAAs are potent stimulators of the secretion of LH and to some extent FSH, through a suprapituitary mechanism involving enhanced GnRH secretion. Activation of the specific glutamate receptor subtypes NMDA, kainate, and AMPA, but not metabotropic receptors, triggers enhanced release of GnRH from the hypothalamus. NMDA effects are most prominent on GnRH cell bodies in the preoptic area of the rat, whereas kainate and AMPA effects appear to be most prominent in the arcuate nucleus/median eminence, where GnRH nerve terminals are located. *In vitro* studies using immortalized GnRH cells argue for a direct EAA regulation of GnRH neurons, in which mediation of nitric oxide is involved. Use of specific NMDA and non-NMDA antagonists has disclosed the role of EAAs in the steroid-induced and preovulatory LH surges, as well in the regulation and synchronization of GnRH secretory pulses. This may explain the ability of EAAs to advance puberty in both male and female rodents and primates.

EAA administration also induces GH release in rats. In these studies, subcutaneous injections of NMDA or kainate, but not quisqualate, significantly elevated serum GH levels. In prepubertal male monkeys, systemic administration of NMDA for up to 14 weeks increased, progressively, body weight, crown–rump length, and testicular volume. The mechanism of NMDA regulation of GH release appears to involve stimulation of GHRH release from the ARC in the hypothalamus.

Administration of NMDA stimulates prolactin secretion in male and female monkeys, intact and castrate male rats, and cycling female rats. Evidence has been provided that both NMDA and non-NMDA receptors play a physiologically important role in the regulation of prolactin secretion, for example, proestrus prolactin surge in the female rat or in the pregnant mare serum gonadotropin-primed immature rats. The mechanism(s) whereby EAAs stimulate prolactin release are at present unclear. NMDA induces c-Fos immunoreactivity in the arcuate nucleus, the site of the DA cell bodies in the hypothalamus; the latter are the more likely site of EAA regulation in the control of prolactin release.

NMDA, kainate, and quisqualate are all capable of increasing plasma ACTH levels, implying that all three types of EAA receptors could be involved in the regulation of ACTH secretion. The site of action of EAAs in the control of ACTH release is most likely the hypothalamus, namely, the paraventricular nucleus, where CRH neuronal cell bodies reside. EAAs would also stimulate the release of another ACTH secretagogue in the paraventricular nucleus, that is, vasopressin. In addition to a direct hypothalamic site of action of EAAs on ACTH release, EAAs may also act at extrahypothalamic sites, for example, the locus coeruleus, a major site of NE cell bodies, which project to the hypothalamus, including the paraventricular nucleus. Thus, it is possible that NE plays a role in glutamate-induced ACTH release.

Table IV lists the stimulatory or inhibitory effects on AP secretion of brain neurotransmitters in humans, as derived from the reviewed experimental evidence.

F. Brain Peptides and AP Hormone Secretion

In addition to the known regulatory hormones, a host of CNS neuropeptides (Section I,C) exerts profound neuroendocrine effects. The presence of some of these compounds in specific hypothalamic nuclei in relatively high concentrations, their secretion from hypothalamic nerve endings, and their detection in hypo-

TABLE IV

Neurotransmitters and Anterior Pituitary Hormone Secretion[a,b]

Hypothalamic/ pituitary axis	Subject	E	α_1	α_2	β_1	β_2	NE	DA	5-HT	ACh	H_1	H	H_2	GAGA	Glutamate
CRF/ACTH	Animals	↑	(↑)	(→)	—	(↑)	↑	↑	↑[c]	↑	(↑)	↑	(⇅)[d]	↓	↑[j]
	Humans	—	(↑)	(→)[e]	(→)	(→)	—	↑[f]	↑	↑[?]	(↑)	—	—	↓	—
GHRH-Somato-statin/GH	Animals	↑	(↓)	(↑)	(↑)[p]	(↓)	↑	⇅	⇅[?,d]	↑	(⇵)	↑	?	⇅	↑[j]
	Humans	↑[g]	—	(↑)	(↓)	—	↑	↑[h]	⇅[?]	↑	(↑)	—	—	↑	—
LHRH/FSH-LH	Animals	↑	(↓)[i]	(⇅)[i]	(↓)[i](→)[j]	—	↑	⇅[?]	⇅[k]	↓	(↑)	↑[l]	(↑)	↓[m]	↑
	Humans	—	(→)	—	—	—	—	↓→	—	⇅[?]		—	?	↓[n]	—
PIF-PRF/Prolactin	Animals	↓[o]	—	(↓)	—	(↑→)[p]	↓[o]	⇵[q]	↑	⇅[?]	(→)	↑	(↑)	⇅	↑[j]
	Humans	—	—	(→↓)[r]	—	—	→	↓	↑	—	—	—	(→)	⇅	—
TRH/TSH	Animals	↑	(⇵)	(↑→)	—	—	↑	↓	⇵	→	(→)	→	(↑)[s]	⇅	—
	Humans	—	—	—	(→)	(→)	—	↓[t]	↓[u]→	—	(→)	—	(→)	↓	—

[a]Key to symbols: →, no effect; ↑, stimulation; ↓, inhibition; —, action not ascertained; ?, action still questionable.
[b]The effect of activation of receptor subtypes is indicated with parentheses.
[c]Modulation of circadian periodicity.
[d]Inhibition in the dog.
[e]Inhibition in depressed patients.
[f]Obtained with L-dopa.
[g]In combination with propranolol.
[h]Inhibitory in acromegaly and *in vitro* AP.
[i]Prolonged administration.
[j]In monkeys.
[k]Modulatory role.
[l]In ovariectomized, steroid-primed rats.
[m]Modulation of the LHRH-induced LH rise *in vitro*.
[n]Inhibition of the LHRH-induced LH rise *in vitro*.
[o]*In vitro* at high doses.
[p]*In vitro*.
[q]Through D_1 receptors.
[r]Inhibitory in children.
[s]TRH release *in vitro*.
[t]TRH-induced rise; hypothyroid subjects.
[u]Hypothyroid subjects.

physeal portal blood suggest a neurohormonal role and a hypophysiotropic function. However, these peptides can alter endocrine function not only as hormones but also as neurotransmitters or neuromodulators; moreover, their physiological role in the control of AP hormone secretion has yet to be unequivocally established. A separate, succinct description of the neuroendocrine effects of some of these compounds, therefore, seems proper. Discussion of the neuroendocrine effects of products of the immune system or growth factors is beyond the scope of this article.

1. Opioid Peptides

Among neuropeptides, a major role in the neural mechanisms underlying the control of pituitary function must be credited to opioid peptides. These peptides used so far stimulate the secretion of growth hormone in either subprimates or primates acting on the hypothalamus to release GHRH. The physiological significance of opioid peptides for growth hormone secretion, however, is not clear, because naloxone, the antagonist of opioid peptide receptors, fails to alter basal growth hormone secretion. Opioid peptides also stimulate prolactin secretion via a suprapituitary site of action, that is, by inhibiting TIDA neuronal function, but, at least in humans, do not exert a tonic stimulatory action on basal prolactin secretion. There is unequivocal evidence that opioid peptides play an important role in the control of gonadotropin secretion and, hence, reproductive function in both animals and humans. They act mainly by inhibiting GnRH release from the hypothalamus via decrease of excitatory adrenergic influences and also play an important role in the inhibitory feedback effects of gonadal steroids on LH release, making hypothalamic neurons hyperresponsive to the steroids. Concerning their role on the hypothalamo–pituitary adrenal axis, their major effect is to tonically inhibit the secretion of ACTH. Naloxone, although at high doses, increases ACTH and corticosterone secretion in rats and cortisol secretion in humans, and chronic administration of morphine inhibits stress-induced activation of the hypothalamo–pituitary adrenal axis.

2. Other Neuropeptides

An interesting feature of the neuroendocrine activity of CNS neuropeptides is the diversity of effects they can exert on the secretion of distinct hormones, according to their target site of action, a fact that compounds interpretation of their actual, physiological role. Thus, experimental evidence suggests for substance P in the rat a dual stimulatory and inhibitory role on the secretion of gonadotropins, exerted at the hypothalamic and pituitary levels, respectively.

It is also apparent that different organismic variables may greatly influence the effects of neuropeptides. For instance, neurotensin injected into the medial preoptic area of ovariectomized, estrogen-primed rats significantly facilitates the circadian afternoon rise of LH secretion, although in unprimed rats it does not affect the existing LH secretion; vasoactive intestinal peptides, essentially ineffective in eliciting growth hormone release from superfused rat pituitary cell reaggregates, strongly stimulate growth hormone release in the presence of dexamethasone in the culture medium. In addition, vasoactive intestinal peptide effects direct stimulation of growth hormone release from human somatotropinomas *in vitro* and is an active prolactin releaser in healthy humans but not in subjects bearing prolactin-secreting adenomas. Also, species-related differences are present.

3. Cytokines and the Neuroendocrine System

In recent years, evidence has accumulated for the existence of common peptide signals, receptors, and functions in cells of the immune and the neuroendocrine systems, and this provides the logical basis for a regulatory loop between the two systems. Concerning the different pathways for immune modulation of the neuroendocrine function, though one possibility is that endotoxin and other bacterial or viral products might directly stimulate pituitary cells or hypothalamic neurons, the current understanding is instead that most of the physiological effects of immune stimulation are brought about by the host of effector molecules produced by activated immune and vascular endothelial cells. Circulating cytokines produced from lymphoid, reticuloendothelial, or other cells may act in the periphery to stimulate primary endocrine organs in such a way to create indirect effects on the pituitary via altered blood levels of target hormones. Similarly, direct cytokine effects on the gastrointestinal system could influence amino acid or glucose metabolism with ensuing effects on the pituitary. Cytokines could act to influence afferent nerves leading to alterations in neural inputs to the hypothalamus and final modulation of hypothalamic regulatory hormones. It is unlikely that circulating cytokines cross the BBB in significant amounts, but they may access the CNS via the circumventricular organs. Finally, cytokines produced within the hypothalamus or pituitary may act as intrinsic modulators of secretion.

To summarize from these and other data, the interactions either within different brain messengers or between them and pituitary and target gland hormones, whose ultimate result is that of assuring proper functioning of the pituitary and the endocrine system, are still poorly understood. The effects on the hypothalamus and/or the pituitary of some neuropeptides are reported in Tables V and VI.

IV. NEUROACTIVE COMPOUNDS IN THE DIAGNOSIS AND TREATMENT OF NEUROENDOCRINE DISORDERS

The notion gained from the preceding sections that in the CNS a host of neuropeptides and neurotransmitters interact functionally to ensure the physiological secretion of AP hormones leads to the ultimate conclusion that their dysfunction may be the trigger for specific neuroendocrine disorders. The principal diagnostic and therapeutic uses of CNS-acting compounds will be succinctly reviewed here. Mention of the clinical applications of regulatory hormones has already been made in the specific subsections.

A. Growth Hormone Deficiency and Excess

The ability of GHRH to affect somatotroph function cannot be used as a test for identifying patients with inadequate spontaneous growth hormone secretion, owing to the poor inter- and intraindividual reproducibility of the growth hormone responses to GHRH. Because the major factors that plague evaluation of the growth hormone response to GHRH are fluctuations in the hypothalamic function and release of somatostatin, compounds such as pyridostigmine, which deprive the pituitary from inhibitory somatostatin inputs, given in advance to GHRH, appear to be useful for a full evaluation of pituitary somatotroph function and, thus, differentiation from a primary hypothalamic origin of the disease.

The finding that most adults and children with growth hormone deficiency show variable but unequivocal rises in plasma growth hormone levels after administration of GHRH suggests that growth hor-

TABLE V
Neuropeptides and Pituitary Hormone Release: Action on the CNS[a]

	Hormone[b]					
Peptide (dosage)	ACTH	Prolactin	Growth hormone	TSH	FSH	LH
Substance P (μg)	NT	+?	−?	0	0	+
Neurotensin (μg)	NT	−, +	+	0	0	+?
Vasoactive intestinal peptide (ng)	NT	+	+	0	0	+
Gastric inhibitory polypeptide (μg)	NT	0	+	0	−	0
Motilin (μg)	NT	NT	−[c]	NT	NT	NT
Galanin (ng)	0[d]	+[d]	+	0[d]	0[d]	0[d]
Cholecystokinin (ng)	+	+	+	−	0	−
Angiotensin II (μg)	+	−	−	0	NT	+
Neuropeptide Y (ng)	+	0[e]	−	0[e]	0	−?
Bombesin (ng)	NT	+[f]	+	0	0[d]	0[d]
Calcitonin (νg)	+[d]	−?	−	NT	NT	NT
IL-1β (ng)	+	0	−	−	NT	−
TNF-α (ng)	0	NT	NT	−	NT	NT
IFN-γ (ng)	+	0	NT	−	NT	NT

[a]Modified from E. E. Müller and G. Nisticò (1989). "Brain Messengers and the Pituitary." Academic Press, San Diego.
[b]Key to symbols: NT, not tested; +, stimulation; −, inhibition; 0, no effect; ?, controversial findings.
[c]Given intracerebroventricularly.
[d]Human data.
[e]Data derived from the effect of bovine and avian pancreatic polypeptides.
[f]Blockade of stress-induced prolactin release.

mone deficiency is rarely due to a functional impairment of somatotrophs but is more likely to be due to hypothalamic dysfunction. Pituitary growth hormone is present but not secreted, probably from a lack of GHRH synthesis and/or release. The presence of low but detectable GHRH levels in the cerebrospinal fluid of children with idiopathic growth hormone deficiency is suggestive of dysfunction of neurons regulating the release of GHRH from GHRH containing neurons. In this context, recent attempts to stimulate growth hormone release in children with idiopathic growth hormone deficiency, intrauterine growth retardation, constitutional growth delay by treatment with DA agonists (L-dopa, bromocriptine), or α_2-adrenergic agonists (Clo) must be considered. In all, the results obtained seem to be promising, although broadening and confirmation of these findings are awaited. Finally, administration of pyridostigmine alone or combined with GHRH has been considered for the treatment of short stature, but results obtained so far are elusive.

Evidence that in acromegalic patients direct DA agonists induce a consistent suppression of the elevated growth hormone levels in about 60% of patients for a direct action on DA receptors located on the somatotrophs was the rationale for the use of ergot drugs, bromocriptine, lisuride, pergolide, and cabergoline. Apart from lowering plasma human growth hormone levels, clinical and metabolic improvements have been reported following the institution of medical therapy, although apparently patients benefit from but are not cured by chronic treatment with ergot drugs. The introduction of potent analogues of somatostatin capable of long-lasting reductions in plasma growth hormone levels will limit the therapeutic use of ergot derivatives in acromegaly.

B. Hypogonadotropic Hypogonadism

Secondary amenorrhea is by far the most common symptom attributable to pituitary function in women. It is usually transient and unaccompanied by structural abnormalities of hypothalamus, pituitary, or ovary (hypothalamic amenorrhea). Supporting evidence for a role of opioid peptides in the etiology of amenorrhea is derived from studies of amenorrheic women treated with naloxone. A clear

TABLE VI
Neuropeptides and Pituitary Hormone Release: Action on the Pituitary[a]

Peptide (dosage)	Hormone[b]					
	ACTH	Prolactin	Growth hormone	TSH	FSH	LH
Substance P (ng)	NT	+	0	0	–[c]	–[c]
Neurotensin (ng)	NT	+	0	+?	0	0
Vasoactive intestinal peptide (μg)	+[d]	+	+[e,f]	0	0	0
Peptide histidine isoleucine amide (μg)	NT	+	+[e,f]	NT	NT	NT
Gastric inhibitory polypeptide (μg)	NT	NT	–	NT	+	+
Motilin (μg)	NT	NT	+	NT	NT	NT
Galanin (μg)	NT	0	0	NT	NT	NT
Cholecystokinin (μg)	0	+[g]	0	0	0	0
Angiotensin II (ng)	+	+	0	0	NT	0
Neuropeptide Y (μg)	NT	NT	–[f]	NT	+	+
Bombesin (ng)	NT	+[d]	+[d]	NT	NT	NT
Calcitonin (μg)	NT	+	0	–	NT	–
IL-1β (ng)	+	0	0	–	NT	0
TNF-α (ng)	+	0	0	–	NT	NT
IFN-γ (ng)	0	NT	NT	NT	NT	NT

[a]Modified from E. E. Müller and G. Nisticò (1989). "Brain Messengers and the Pituitary." Academic Press, San Diego.
[b]Same symbols as in Table V.
[c]Inhibition of GnRH-stimulated release.
[d]Only on tumor cells.
[e]In the presence of dexamethasone.
[f]On human somatotropinomas.
[g]At huge doses on the rat AP.

increment in LH levels was observed in women with amenorrhea and/or hyperprolactinemia, suggesting that the acyclicity was due, at least in part, to the effect of an increased opioid peptide tone on GnRH and gonadotropin secretion. Thus, opioid antagonists may represent a useful therapeutic approach in these cases.

C. Cushing's Disease

The awareness that a host of neurotransmitters and neuropeptides are involved in the regulation of ACTH secretion, and the evidence that some neuroactive drugs initially thought to act on the hypothalamus to decrease CRH activity actually may act at the pituitary level, account for a medical approach to therapy in Cushing's disease. Thus far, drugs used in this context encompass cyproheptadine, whose use relied on the known stimulatory action of the 5-HT neuronal system on the hypothalamo–pituitary adrenal axis, bromocriptine, and sodium valproate, ultimately capable of stimulating DA and GABA receptors located on the corticotrophs. Although on distinct cases some clinical and biochemical remission is evident, the role of the medical therapy in Cushing's disease is ancillary to transsphenoidal microsurgery.

D. Prolactinomas

Prolactin-secreting tumors, either ≤10 mm diameter (microprolactinomas) or higher (macroprolactinomas), are the most frequently occurring neoplasms in the human pituitary. Clinically, hyperprolactinemia is associated with amenorrhea, galactorrhea, infertility, decreased libido, impotence, and, in macroprolactinomas, visual disturbances. In patients with microprolactinomas, DA agonist–ergot-related drugs represent a primary medical therapy. Administration of these drugs causes immediate and sustained prolactin suppression with restoration of fertility in women and normalization of hyperprolactinemic hypogonadism in men. Usually within 2 months of the return of menstruation, ovulation and adequacy of the luteal phase are achieved and galactorrhea disappears. In men, libido and potency return to normal and, if reduced, the seminal volume also is normalized. When treatment is started, bromocriptine, the drug most commonly used, and other ergolines may cause different neurovegetative symptoms owing to activation of central and peripheral DA receptors. Thus, low doses of drugs that are increased slowly and taken during a meal rather than after food are mandatory to minimize the side effects. Two long-acting injectable preparations of bromocriptine are available whose injection in patients with prolactinoma is followed by a prompt and steep prolactin decrease lasting for weeks or months. In many cases, shrinkage of the pituitary tumor can also be documented. Only transient and mild to moderate side effects are noticed.

The effects of medical therapy are particularly noteworthy in macroprolactinomas because these tumors are rarely cured by surgery and with radiotherapy subsequent hypopituitarism is common. It is now evident that reduction of tumor size as documented by tomographic scan and amelioration of visual disturbances can be anticipated in about 75% of the patients. Interestingly, a tumor shrinks only when prolactin secretion is inhibited by dopaminergic stimulation, but tumor size may remain unaltered despite the reduction of prolactin secretion. From the foregoing evidence, it appears that medical treatment of macroprolactinomas is more appropriate than neurosurgical transsphenoidal exploration for the primary treatment of the disease. The critical issue is whether or not this therapy effects a real cure of the disease. Overall, it would seem that only in a minority of patients does long-term medical treatment result in a persisting correction of the underlying cause of the adenoma. Possibly, more prolonged drug regimens, different drug doses, or newer DA agonists will prove more effective in this context.

BIBLIOGRAPHY

Bloom, F. E. (1996). Neurotransmission and the central nervous system. *In* "Goodman & Gilman's. The Pharmacological Basis of Therapeutics" (J. G. Hardman, L. E. Limbird, P. B. Molinoff, R. W. Ruddon, A. Goodman Gilman, eds.), pp. 267–293. McGraw-Hill, New York.

Brann, D. W., Mahesh, V. B. (eds) (1995). Excitatory aminoacids: Their role in neuroendocrine function. CRC Press, Boca Raton.

Hökfelt, T., Meister, B., Everitt, B., Staines, W., Melander, T., Schalling, M. Mutt, V., Hulting, A.-N., Werner, S., Bartfai, T., Nordström, O., Fahrenkrug, J., and Goldstein, M. (1986). Chemical neuroanatomy of the hypothalamo-pituitary axis: Focus on multimessenger systems. *In* "Integrative Neuroendocrinology: Molecular, Cellular and Clinical Aspects" (S. M. McCann and R. I. Weiner, eds.), pp. 1–34. Karger, Basel.

Korbonotis, M., Grossman, A. B. (1995). Growth hormone-releasing peptide and its analogues. *Tem.* **6**, 43–49.

Martin, J. B., and Reichlin, S. (1987). "Clinical Neuroendocrinology." F. A. Davis Company, Philadelphia.

Müller, E. E., and Nisticò, G. (1989). "Brain Messengers and the Pituitary." Academic Press, San Diego.

Reichlin, S. (1995). Neuroendocrinology. *In* "Williams Textbook of Endocrinology" (J. D. Wilson and D. W. Foster, eds.), 8th ed., pp. 135–219. Saunders, Philadelphia.

Brain Regulation of Gastrointestinal Function

YVETTE TACHÉ
University of California, Los Angeles

ERIK BARQUIST
University of Rochester

GLOSSARY

Dorsal vagal complex Association of two medullary nuclei: the dorsal motor nucleus of the vagus (which contains neurons projecting to the gut through the vagus) and the nucleus tractus solitarius (which contains terminals of afferent vagal neurons from the gut)

Enteric nervous system Neuronal network embedded within the gut wall that serves as relay for signals from and to the brain or spinal cord but also can, independently from the brain, receive information from various kinds of sensory receptors and generate neural outflow

Hypothalamus Nuclei in the forebrain that are involved in the regulation of pituitary hormone secretion and visceral function and are subdivided in lateral, ventromedial, and paraventricular parts

Migrating myoelectrical complex A cyclically occurring phenomenon that begins in the stomach and duodenum and is propagated to the ileum. It is composed of three well-defined phases: phase I, noncontractile activity; II, intermittent and irregular contractions; III or activity front, period of intense spikes and contractile activity

Monosynaptic vago-vagal reflex Transmission of the information along vagal sensory neurons directly to vagal motoneurons in the dorsal motor nucleus of the vagus

Peptide Molecules formed of a small number (below 100) of amino acids

MAMMALIAN GASTROINTESTINAL FUNCTIONS ARE subjected to a diversity of regulatory controls exerted at multiple levels including the brain, spinal cord, peripheral autonomic ganglia, and enteric plexuses (Fig. 1) and by hormones acting through endocrine or paracrine mechanisms. The term *brain–gut axis* refers to the control of gut functions including secretion, motility absorption, and growth exerted by the brain or spinal cord through the autonomous nervous system. The extent of the central nervous system (CNS) influence on the gut ranges from the esophagus to the colon and includes the liver, pancreas, and gallbladder. Brain–gut interactions encompass knowledge on localization of the specific brain nuclei involved in modulating gut function, the anatomical, chemical, and electrophysiological characterization of the connections between the brain and the gut (efferent pathways) and from the gut to the brain (afferent pathways), the identification of chemicals in the brain, and the gut coding the neuronal transmission and physiological stimuli that use these regulatory mechanisms.

I. NEUROANATOMICAL BASIS FOR BRAIN–GUT INTERACTIONS

A. Brain Sites Influencing Gastrointestinal Function

1. Medullary Nuclei

Primary nuclei in the brain stem involved in the control of gastrointestinal function are the dorsal motor

ENCYCLOPEDIA OF HUMAN BIOLOGY, Second Edition, VOLUME 2. Copyright © 1997 by Academic Press. All rights of reproduction in any form reserved.

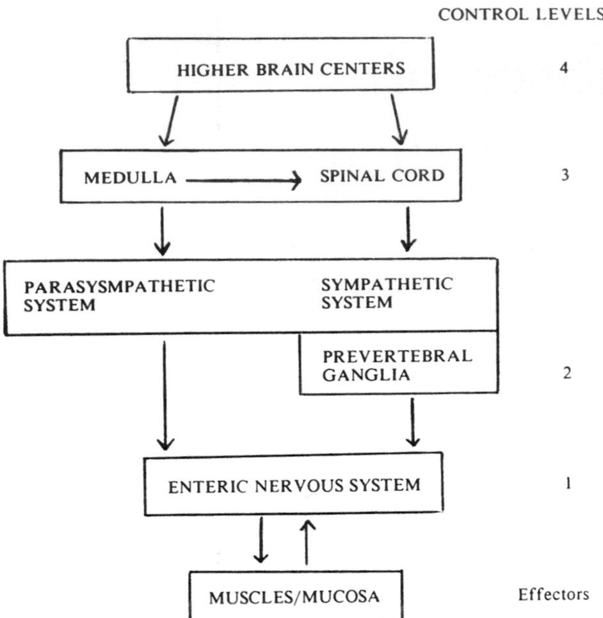

FIGURE 1 Various levels of neural control of gastrointestinal function.

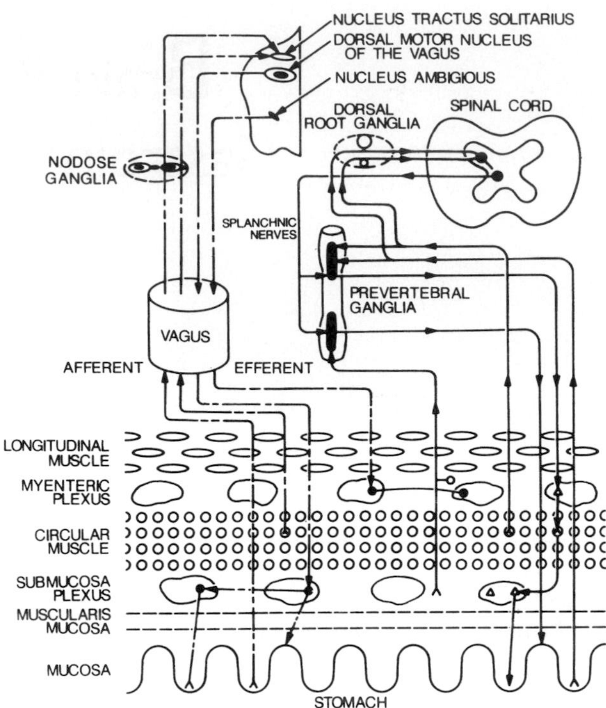

FIGURE 2 Schematic diagram illustrating medullary afferent and efferent pathways to the stomach.

nucleus of the vagus and the nucleus tractus solitarius, referred to as the dorsal motor vagal complex. Other medullary nuclei eliciting a gastrointestinal response are the nucleus ambiguus, raphe nucleus, reticular substance, and medial longitudinal fascicle (Fig. 2). Preganglionic gastric motor neurons located in the dorsal motor nucleus of the vagus have a dendritic arborization with direct synaptic contacts on axon terminals from vagal gastric sensory neurons arising from the nucleus tractus solitarius.These axodendritic contacts provide the anatomical basis for monosynaptic gastric vago-vagal reflexes in the brain stem.

2. Forebrain Nuclei

In the forebrain, several nuclei can influence gastrointestinal function: the cingulate cortex, hypothalamus (lateral, ventromedial, posterior, and paraventricular nuclei), locus coeruleus, bed nucleus of the stria terminalis, parabrachial nucleus, and central nucleus of the amygdala. These nuclei, the hypothalamus and locus coeruleus in particular, receive and process interoceptive and exteroceptive afferent information through poly- or monosynaptic reciprocal connections with medullary nuclei and provide a major input to the brain stem circuitry responsible for coordinating autonomic reflexes. The influence of these higher centers can be exerted either on the efferents of the dorsal motor nucleus of the vagus and/or the afferents to the nucleus tractus solitarius before the sensory signals are relayed to neurons in the dorsal motor nucleus of the vagus. [*See* Hypothalamus.]

B. Extrinsic Innervation of the Gut

The gastrointestinal tract is innervated by the parasympathetic and sympathetic divisions of the autonomic nervous system. The efferent autonomic innervation is involved in transmitting the information from the brain to the gut, whereas the autonomic afferent innervation allows peripheral information to reach the brain. The afferent fibers outnumber the efferent fibers by a ratio of 9 : 1 in the vagus and 3 : 1 in the splanchnic nerve.

1. Efferent Innervation

The autonomic efferent pathways convey input from the CNS to an intrinsic neuronal circuitry present in the gastrointestinal wall [enteric nervous system (ENS)], which then relays the information to gastrointestinal effectors such as the mucosal cells, smooth muscles, or blood vessels. The ENS serves not only as a relay station to the autonomic nervous system, but also acts as a local integrative system, particularly in the small intestine.

a. Parasympathetic

The efferent parasympathetic preganglionic neurons originate in the dorsal motor nucleus of the vagus and to a smaller extent in the nucleus ambiguus (Fig. 2). These neurons send axons in the vagus nerve of the same side and end in the gastrointestinal tract from the esophagus to proximal colon. The parasympathetic innervation to the distal colon and rectum comes from the pelvic nerves. Cell bodies arise in the spinal cord at the sacral levels 2–4, and terminals synapse with the pelvis plexus. The vagal efferent terminals are proportionally more numerous in the stomach than other bowel segments. Electrophysiological studies demonstrated that vagal efferent fibers display continuous spontaneous discharges generated centrally independently from vagal input to the brain stem. This ongoing activity provides a background vagal tone. In the gastrointestinal wall, terminals of preganglionic vagal fibers synapse with neurons in the intrinsic plexuses. These postganglionic neurons innervate smooth muscles, mucosal secretory and immune cells, and blood vessels. Neuroeffector transmission of excitatory vagal pathways use acetylcholine as transmitter and inhibitory vagal pathways use peptidergic [vasoactive intestinal peptide (VIP)], nitric oxide (NO), or purinergic transmitter (serotonin).

b. Sympathetic

The sympathetic efferent pathways are organized hierarchically between spinal and supraspinal levels. The preganglionic neurons of the efferent sympathetic nerve fibers are cholinergic. They are located in the intermediolateral column of the spinal cord mainly between the 5th to the 11th thoracic and lumbar 1–3 segments, although species variation exists. They send processes, which synapse with the postganglionic fibers arising in the prevertebral ganglia (celiac and mesenteric); these postganglionic noradrenergic fibers project to the gut where they further synapse within the intrinsic plexuses (Fig. 2). The noradrenergic terminals are densely represented, particularly at the levels of sphincters and blood vessels and to a lower extent in the mucosa and muscular layers.

2. Afferent Pathways

a. Parasympathetic

The majority of gastric vagal afferents are composed of unmyelinated fibers, which arise from the nodose ganglia. Axons project centrally almost exclusively to the medial subnucleus of the caudal areas of the nucleus tractus solitarius and peripherally to the vagus nerve, where terminals innervate the various mucosal, muscular, and vascular components of the gastrointestinal wall (Fig. 2). Vagal sensory fibers transmit nonconsciously perceived signals generated by food ingestion and digestions, which are encoded by mechano-, chemo-, osmo-, and thermoreceptors located in the gastrointestinal tract. Evidence suggests that vagal afferent inputs arising from the stomach and liver can converge onto the same neurons in the nucleus tractus solitarius. However, there is also a somatotopic organization of vagal afferents innervating the gut because they project along the rostrocaudal axis of the nucleus tractus solitarius.

b. Sympathetic

The sympathetic afferents have their cell bodies in the thoracolumbar and sacral dorsal root ganglia. They terminate centrally in the dorsal horn of the spinal cord at the level of the laminae I, II, V–VII, and X. Hence, terminals synapse onto ascending spinal pathways. These spinal pathways play an important role in conveying nociceptive signals and CNS representation of visceral pain. Sympathetic afferents also initiate enterogastric, intestino-intestino and intestino-colonic reflexes in response to physiological stimulation of enteroreceptors. These reflexes are mediated through projections of the afferents to preganglionic sympathetic neurons located in the intermediolateral column of the spinal cord or to postganglionic neurons located in the prevertebral ganglia. The chemical coding of afferent projections from the gut to the prevertebral ganglia is cholinergic and contains a variety of peptides, including calcitonin gene-related peptide (CGRP), substance P, cholecystokinin, enkephalin, dynorphin, gastrin-releasing peptide, and VIP.

II. BRAIN REGULATION OF GASTRIC FUNCTION

The main function of the stomach can be divided into motor, secretory, either exocrine (acid, pepsin, bicarbonate, mucus, prostaglandin) or endocrine (gastrin, somatostatin, serotonin, histamine), and vascular (blood flow). It is well documented that the CNS can modulate gastric function and that the vagus plays an important role in mediating this influence. Yet, as we will see for other areas of the gastrointestinal tract, only recent developments in research techniques have allowed elucidation of the chemical messengers that make up this control. [*See* Digestive System, Anatomy and Physiology.]

A. Gastric Secretion

1. Centrally Mediated Stimulation of Gastric Secretion

Gastric exocrine and endocrine secretions are well established to be under a vagal stimulatory control expressed through peripheral muscarinic receptors. Experimental evidence indicates that vagal activation can be triggered centrally by hypothalamic, limbic, and/or medullary input to preganglionic neurons in the dorsal motor nucleus of the vagus, leading to an increase in parasympathetic outflow and stimulation of gastric secretion. Convergent neuropharmacological, electrophysiological, and neuroanatomical data indicate that the tripeptide thyrotropin-releasing hormone (TRH) and TRH receptors located in the dorsal vagal complex are involved in initiating the activation of the parasympathetic outflow, leading to the stimulation of gastric secretion (acid, pepsin, bicarbonate, serotonin, histamine, prostaglandin, NO). Other transmitters, γ-aminobutyric acid (GABA) acting on $GABA_B$ receptors, acetylcholine acting on muscarinic receptors, or peptides such as oxytocin and somatostatin can also stimulate gastric secretion through central activation of vagal pathways. However, less substantial evidence has accumulated to assign them a physiological role.

Several physiological stimuli increase gastric acid and bicarbonate secretion through an action on preganglionic neurons in the dorsal motor nucleus of the vagus. Stimuli can originate in the brain [e.g., the sight, smell, or chewing of tasty food and the suggestion or anticipation of eating (cephalic phase of gastric secretion)]. The exact neural circuitries involved in the various afferent components of the cephalic phase are not completely elucidated. Processing of gustatory signals takes place in the hypothalamus where the sensory information projects by ways of the parabranchial nucleus. Then hypothalamic efferents converge on the brain stem nuclei responsible for coordinating vagal outflow. The cephalic phase of acid secretion accounts for one-third of the total acid response to eating in healthy subjects and is prolonged by gastric distention (Fig. 3). Other stimuli can originate in the periphery (e.g., gastric distention or peptide release, somatovisceral and stimulation). The acid response to gastric distention is initiated through mechanoreceptors located in the gastric wall. They trigger

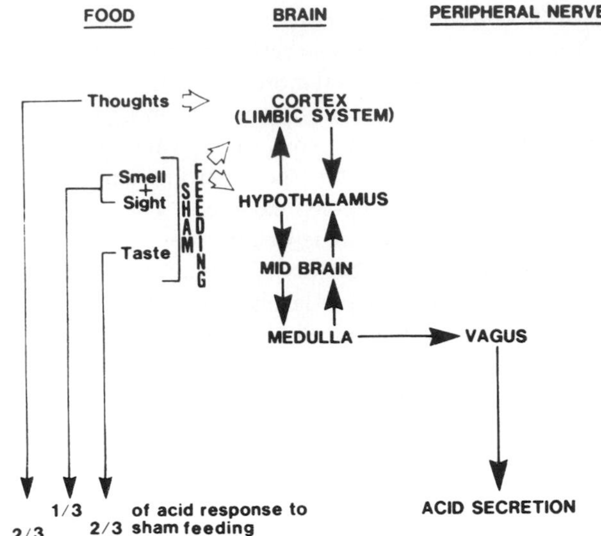

FIGURE 3 Schematic representation of pathways involved in shamfeeding–induced gastric acid secretion.

a flow of vagal afferent impulses to discrete neurons in the brain stem and hypothalamus, which feed back to excite preganglionic neurons in the dorsal motor nucleus of the vagus. Hypoglycemia induced by injection of nonmetabolized glucose analogues or insulin has been established for several decades to be potent vagal stimulants of gastric secretion through a hypothalamic site of action. However, the physiologic importance of this activation is uncertain because the levels of hypoglycemia required to stimulate gastric secretion are not commonly seen even under a fasting state. The centrally mediated gastric secretory response to all these stimuli does not return to baseline immediately after withdrawing the stimuli. For instance, acid secretion induced by sham feeding outlasts the duration of chewing by 60–90 min. This sustained effect required the integrity of the vagus; however, its underlying central mechanisms are unknown. The chemicals encoding the afferent and efferent transmission evoked by these stimuli are also yet to be established. However, circumstantial evidence indicates that TRH located in the dorsal vagal complex, nucleus ambiguus, and raphe nucleus is a likely candidate involved in initiating vagal stimulation.

2. CNS-Mediated Inhibition of Gastric Secretion

Knowledge has accumulated on the inhibitory influence exerted by the brain on gastric secretion. Many peptides, specifically bombesin, calcitonin, CGRP, corticotropin-releasing factor (CRF), interleukin 1β, neurotensin, and opioid peptides have been shown to act in the brain to inhibit gastric acid and pepsin secretion. Their sites of action are in the hypothalamus (lateral, ventromedial, or paraventricular nuclei), nucleus ambiguus, and/or the dorsal vagal complex. Thus, it seems that the same nuclei can be involved in both stimulation or inhibition of gastric secretion, depending on the type of receptors activated. Pharmacological studies further demonstrated that an interaction exists in the dorsal vagal complex between centrally acting stimulatory (TRH) and inhibitory (bombesin, CRF, CGRP, interleukin-1β) peptides.

The centrally mediated inhibition of gastric acid secretion can be achieved by inhibiting the vagal stimulatory pathways, by stimulating sympathetic noradrenergic inhibitory pathways, or by both. For instance, the central inhibition of gastric acid secretion induced by CGRP or opioid peptides is vagally dependent, whereas that of bombesin is mediated by sympathetic activation. CRF involves both pathways.

Stimuli inhibiting gastric acid secretion through cephalic influence are mostly related to stress exposure. CRF, which is centrally released during stress, and central CRF receptors play a physiological role in the inhibition of gastric acid secretion induced by various stressors in experimental animals. Removal of sympathetic pathways stimulates gastric acid secretion. To what extent this represents a CNS tonic inhibitory control exerted by these peptides is still to be addressed.

B. Gastric Motility

The motor function of the stomach varies strikingly in relation to the prandial state. During the interdigestive period, the motility in the corpus and antrum is characterized by a cyclic recurring period of high-amplitude contractions known as phase III of the interdigestive migrating motor complex (MMC). Phase III usually is initiated in the antrum and propagated to the duodenum. On food ingestion, the fasted pattern is disrupted and replaced by a postprandial pattern consisting of continuous irregular contractions. Also the fundus must accommodate to incoming nutrients by the process of receptive relaxation while antral and fundic motor activity are increased. The stomach also regulates the amount of food entering the intestine by modulating gastric emptying. Neural mechanisms, either local or long reflexes as well as hormonal factors, regulate these processes.

1. Central Vagal Regulation of Gastric Motility

Vagal activation conveys both excitatory and inhibitory inputs to gastric intrinsic plexus projecting to smooth muscle effectors. The anatomical substrate of such a dual response is the existence of two kinds of preganglionic fibers in the vagus: excitatory and inhibitory. The ones that convey an excitatory response on motility have a lower threshold of excitability than the inhibitory fibers. They are composed of preganglionic cholinergic fibers, which synapse with postganglionic cholinergic neurons. Their stimulation leads to an increase in the amplitude of corpus, antral, and pyloric phasic contractions and, to a lesser extent, their frequency.

Based on experimental studies, brain nuclei that can induce a vagally mediated excitation of gastric contractility are the dorsal motor nucleus of the vagus, nucleus ambiguus, raphe nucleus, and hypothalamus (lateral and paraventricular parts). TRH or acetylcholine in these medullary nuclei serve as mediators to activate the vagal excitatory pathways. Preliminary evidence, still to be further substantiated, indicates that somatostatin acting in the dorsal vagal complex may also play a role in such a process. Physiological stimuli influencing gastric motor function through the vagal excitatory pathways are sham feeding or distention of the corpus and antrum. The latter increases antral motility through vago-vagal reflexes well characterized electrophysiologically.

The vagal inhibitory pathway contains vagal efferent fibers that fire spontaneously at low frequency. The preganglionic vagal fibers synapse directly or through a nicotinic intermediate on intramural inhibitory neurons. These postganglionic inhibitory neurons are nonadrenergic and noncholinergic and most likely use VIP and NO as their neurotransmitter. Excitation of vagal inhibitory pathways leads to the relaxation of the gastric wall predominantly at the proximal stomach (fundus). Preliminary evidence suggests that oxytocin in the dorsal motor nucleus of the vagus may be the chemical signal activating this efferent inhibitory pathway, particularly during conditions that provoke satiety and nausea. Several physiological stimuli (e.g., swallowing, distention of the esophagus or antrum, or strong antral contractions) induce relaxation of the fundus through the long vago-vagal reflex, using the inhibitory pathways as the efferent limb of the reflex. Sham feeding has also been reported to inhibit gastric motility. This may be related to activation of vagal inhibitory pathways and/or subsequent inhibitory reflexes caused by the increased acid secretion and arrival of acid in the proximal duodenum.

The role of the vagus in the control of interdigestive patterns of gastric motility remains controversial. There are two schools of thoughts as to whether the extrinsic neural control is required to initiate gastric MMC. Some studies indicate that the cyclic interdigestive motor patterns of the stomach can occur in the absence of extrinsic neural control. Other more recent data, based on total extrinsic denervation or vagal cooling, favor a vagal tonic modulation of the gastric MMC. This is further supported by electrophysiological studies demonstrating that the majority of vagal efferent fibers display continuous activity. Moreover, the spontaneous discharge rate of excitatory fibers fluctuates in relation with the various phases of the MMC.

Similarly, the part that central vagal input plays in the initiation of the fed motility pattern is not clearly established. In favor of such a role are the facts that as soon as food is offered, there is an increase in the discharge rate of vagal excitatory fibers, which is maintained for several hours if the food is ingested. Moreover, an increase in parasympathetic activity triggers a postprandial pattern of motor activity. Peptides such as cholecystokinin (CCK) and neurotensin have been proposed to be putative peptides acting in the ventromedial hypothalamus to initiate (neurotensin) and maintain (CCK) the fed pattern of motility.

2. Central Sympathetic Regulation of Gastric Motility

Activation of the splanchnic sympathetic activity leads to a decreased tone in the proximal stomach and peristaltic activity in the antrum. This inhibitory effect is adrenergic in nature and is exerted at the myenteric nervous plexus by inhibiting acetylcholine release from intramural cholinergic neurons through an action on α-adrenergic receptors. Some studies also indicate a direct effect on the smooth muscles exerted by β-adrenergic receptors. Increased sympathetic outflow caused by removal of the inhibitory tone exerted by supraspinal centers or by blockade of vagal input prevents the initiation of phase III of the interdigestive MMC.

3. Central Regulation of Gastric Emptying

Various psychological, physical, or chemical stressors alter gastric contractility and emptying, producing mainly an inhibitory effect on motor function. In addition, delayed gastric emptying has been described in a variety of neurological disorders including brain tumors, bulbar poliomyelitis, diabetic gastroparesis, paraplegia, and high cord transection. Neural pathways initiating changes in the rate of gastric transit in response to these physiological or pathological conditions are not fully characterized, although it has been ascribed to an increased sympathetic outflow. Regarding the central chemical coding, growing experimental evidence indicates that CRF plays a physiological role in mediating the delay in gastric emptying related to stress exposure. Other peptides (e.g., bombesin, calcitonin, CGRP, neurotensin, neuropeptide Y, and μ- or δ-opioid peptides) act centrally to delay gastric emptying of a nonnutritive solution. The inhibitory effect of these peptides, except calcitonin, is also exerted through the vagus. However, their physiological role in the control of gastric transit is not clearly established because of the lack of specific antagonists for most of them and the inability to monitor their release in the brain in response to centrally acting stimuli.

III. BRAIN REGULATION OF INTESTINAL FUNCTION

After nutrients leave the stomach, they are further digested in the duodenum. It is here that gastric acid must be neutralized and nutrients mixed with pancreatic enzymes. The duodenum, as the recipient of gastric acid, has extensive feedback pathways to the stomach at local, hormonal, and neural levels to limit the amount of liquid entering the duodenum. Local protective mechanisms also exist (e.g., a thin layer of bicarbonate rich in mucus, which titrates incoming acid). Proper control of bicarbonate secretion and mucus production is essential to good duodenal function.

Although the autonomic nervous system, whether through central or spinal input, plays a major role in the control of gastric, colonic, and pancreatic function, it appears less involved in the overall control of small intestinal function. In the small intestine, the enteric nervous system is reciprocally connected with the prevertebral ganglia. For instance, in the prevertebral ganglia, cell bodies of sympathetic secretomotor inhibitory and motility inhibitory neurons receive direct input from sensory pathways originating from the intestine. These connections allow an array of reflex activity initiated by enteroreceptors and processed in the prevertebral ganglions or the ENS with a large degree of autonomy from the CNS.

A. Duodenal Bicarbonate Secretion

1. Vagal Control

It has long been held that vagal stimulation causes increased duodenal bicarbonate secretion. The peripheral transmitters involved in this effect are unknown, although pharmacological studies indicate a nicotinic receptor intermediate and a nonmuscarinic postganglionic transmission. Knowledge on central control of the vagally mediated alkaline secretion is still fragmentary. Several peptides (e.g., bombesin, CRF, somatostatin, TRH) cause a centrally mediated increase in duodenal bicarbonate secretion through vagal efferent pathways. The end neurotransmitter is most likely VIP and not cholinergic, as atropine does not block either TRH- or sham feeding–induced duodenal bicarbonate secretion increases.

2. Splanchnic Control

The local splanchnic nerves, normally quiescent, inhibit bicarbonate secretion when activated by stressful stimuli. For instance, hypovolemic stress induced by controlled blood loss is known to cause decreased bicarbonte secretion through splanchnic adrenergic pathways. Because the response is blocked by α_2-adrenergic antagonists, it implies an effect mediated by endogenous α_2-adrenergic receptors. Stimulation of the medial hypothalamus can trigger inhibition of duodenal alkaline secretion through spinal adrenergic pathways.

B. Intestinal Absorption/Secretion

Although CNS control of the duodenum is primarily directed toward protecting against gastric acid and enzyme effect and, to a lesser extent, motility, the more distal duodenum and remaining small bowel require coordination of nutrient absorption, fluid and electrolyte absorption/secretion, and motility. Intestinal absorption takes place simultaneously with intestinal secretion, making studies of either mechanism difficult. Most studies have measured either net absorption or net secretion.

1. Influence of the Vagus

Central activation of vagal efferent outflow causes net secretion. Neurotransmitters involved in this effect are both cholinergic and peptidergic. Atropine blocked the centrally stimulated effect of increased secretion. More distal stimulation of the vagus in atropinized rats still caused net secretion. A likely candidate for the more distal transmitter is VIP, which is released during electrical stimulation and causes net secretion when administered locally. However, in two studies in which vagal stimulation was produced by sham feeding, the increased net secretion in response to this stimulus was inconclusive. More recent studies performed in dogs with isolated jejunal segments concluded to the absence of a cephalic phase in the intestinal absorption/secretion.

2. Influence of the Sympathetic Nerve

Direct stimulation of sympathetic outflow generally causes intestinal absorption. This effect is blocked by the peripheral use of adrenergic antagonists. One center in particular, the lateral nucleus of the hypothalamus, causes absorption when stimulated with microelectrodes in rats. Several centrally acting agents (e.g., δ- and μ-opioid peptides, as well as angiotensin II and III) cause centrally mediated increases in intestinal secretion. Part of this effect may be mediated in hypothalamic centers.

The tonic nature of sympathetic control of intestinal absorption can be shown by the profuse hypersecretion and diarrhea that results from chemical or surgical sympathectomy. With the removal of the sympathetic efferent arm, the bowel tends to secrete more fluid and electrolytes than it absorbs, leading to diarrhea. Inputs from higher centers probably play a role in this effect but are not mandatory for its tonic functioning. Norepinephrine and somatostatin act as neurotransmitters in this system, and fibers containing these compounds project to the villi and crypts and originate in the prevertebral ganglia.

C. CNS Control of Intestinal Motility

In the fasting state, the small intestine exhibits a cyclic burst of contractions, which lasts a few minutes, and propagates from the proximal duodenum to ileocolonic junction with a velocity of 2–8 cm/min. The intestinal MMC is controlled by neural and hormonal mechanisms. There is clear evidence that the initiation and the propagation of the MMC take place at the level of the ENS and do not require the CNS. However, the brain can modulate the characteristics of the MMC by reducing their time intervals, disrupting the pattern, or initiating MMC in a postprandial state. For instance, cephalic influence produced by sham feeding can increase duodenal motor activity. Experimental studies demonstrated that several peptides present in the brain (e.g., calcitonin, CGRP, neurotensin, neuropeptide Y, and μ-opioid peptides) act centrally to induce a fasted MMC pattern of intestinal motility in fed animals, whereas growth hormone–releasing factor (GRF) and substance P shorten the duration of the fed pattern. TRH acts centrally to stimulate intestinal motility and, microinjected into the medial septum, medial, or lateral hypothalamus, increases intestinal transit in fasted animals. Central injection of CRF bombesin, calcitonin, CGRP, CRF, neurotensin, and μ-opioid peptides inhibit intestinal transit through vagal (CGRP, neurotensin, CRF) or nonvagal pathways. The periaqueductal gray is a site of action for the neurotensin and opioid peptide antitransit effect.

It is well known that the sympathetic nervous system inhibits intestinal motility. In the normal, unstressed state, however, these neurons are silent and are not required to assume normal digestive function. When activated, they cause contraction of sphincters, inhibit intestinal motility, and delay the incidence of MMC. The action is exerted presynaptically through inhibition of neural signals within the myenteric ganglia and not by a direct effect on the smooth muscles. The activation of this sympathetic efferent arm can be caused either by stimulation of intestinal receptors (e.g., by intestinal distention, intraperitoneal administration of irritants, or peritonitis) or by nonenteric stimuli. Nonenteric stimuli include those associated with severe pain or stress (e.g., surgery to nonintraperitoneal structures, systemic hypotension, or psychological stressors). It appears that this sympathetic effect is mediated at both the CNS and spinal levels.

IV. BRAIN REGULATION OF EXOCRINE PANCREATIC SECRETION

The CNS control of the pancreas is less well studied than that of the stomach, yet the two organs are closely interrelated by neural connections to allow integrated functioning of the foregut. Both organs mix their outputs in the proximal duodenum, where alkalization of gastric content allows further enzymatic degradation of nutrients. Information from the duodenum is extensively used as feedback on pancreatic function, through both vagal and sympathetic pathways.

A. Central Vagal Stimulation of Pancreatic Secretion

As with acid secretion, the vagus represents the main pathway that stimulates pancreatic exocrine secretion. This is achieved by the activation of cholinergic and noncholinergic secretomotor pathways using serotonin as well as peptides, VIP, bombesin, GRP, CCK, substance P, enkephalins, neuropeptide Y, and CGRP acting as neurotransmitters or hormones. The central transmitters implicated in modulating vagal pancreatic secretion are less well known. Based on pharmacological studies, TRH is a likely candidate because the peptide acts centrally to increase pancreatic secretion. Its action is expressed through noncholinergic, VIP postganglionic neurons. [*See* Peptide Hormones of the Gut; Peptides.]

A physiological stimuli activating this central vagal pathway is sham feeding. In humans, this response occurs within 1 min of food ingestion. This rapid response is unlikely to be related to gastric acid contamination of the duodenum, but rather represents a direct cephalic influence. In dogs, a sham-feeding pancreatic response may approximate 60 to 70% of maximal response to CCK and is similar to the effect of chemical vagal stimulation produced by glucose analogs (e.g., 2-deoxy glucose or 5-thio glucose) or electrical stimulation of the vagus. Stimuli originating peripherally such as stimulation of intestinal chemoreceptors by amino acids, acid, or fatty acids induced marked pancreatic secretory response, 50% of which is mediated by enteropancreatic long vago-vagal cholinergic reflexes.

Pancreatic secretion can be influenced by nuclei in the brain stem, hypothalamus, and medial amygdala. Stimulation of the anterior hypothalamus tends to exert the parasympathetic effect and increases basal pancreatic output without affecting periodicity.

B. Central Sympathetic Regulation of Pancreatic Secretion

The influence of sympathetic stimulation on pancreatic exocrine secretion is more controversial. Stimulation of the posterior hypothalamus has a sympathetic effect to increase the periodicity of basal pancreatic output. Much less is known about centrally acting agents that might stimulate the sympathetic arm of

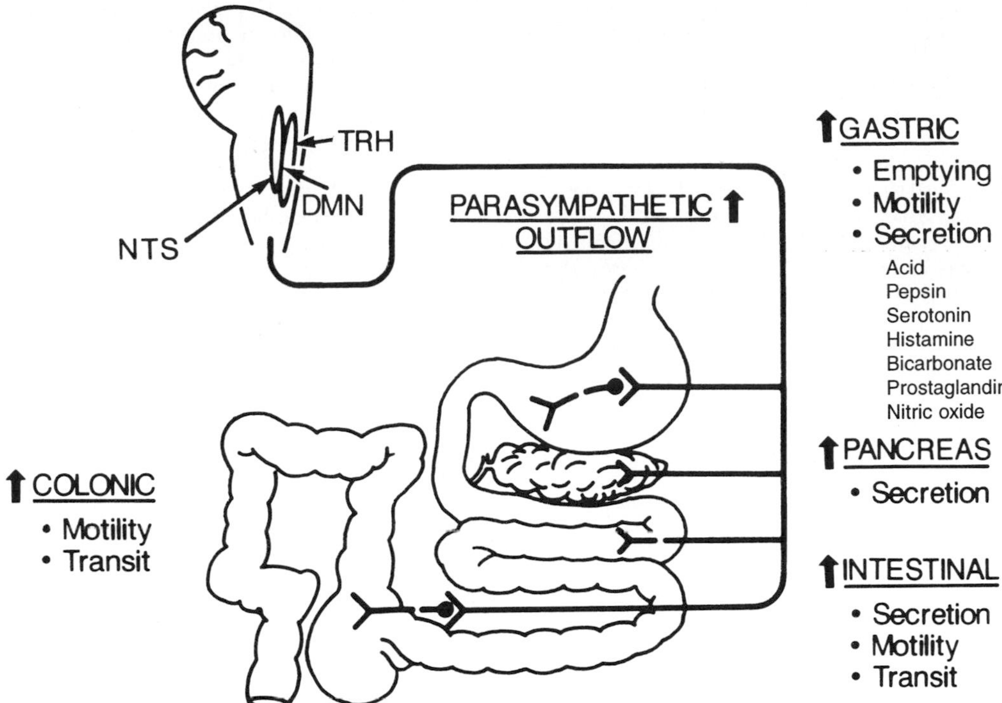

FIGURE 4 Summary of the centrally mediated effect of tripeptide TRH on gastrointestinal secretory and motor function exerted through activation of the vagal outflow in the dorsal vagal complex.

pancreatic innervation. Opioid peptides acting on μ receptors or α_2 agonists such as clonidine decrease pancreatic outflow when administered in the brain.

V. CONCLUSIONS

There has been a resurgence of interest in investigating the interactions between the CNS and the gastrointestinal tract. Positive impacts came from the characterization of better tools to probe these interactions, such as selective peptides able to influence gastrointestinal secretory and motor functions through the CNS. The development of sophisticated electrophysiological and neuroanatomical techniques for recording neural activity and tracing neural circuitry has also contributed to this knowledge. This has led to an increase in the understanding of CNS control of gastrointestinal motility and secretion, particularly in relation to neuroanatomy and neurochemistry. Functional studies have established that vagal efferents and afferents play an important tonic or phasic role in the overall regulation of gastrointestinal secretory and motor function, particularly to initiate the cephalic phase or food-induced receptive fundic relaxation. Vagal control is exerted through the dorsal vagal complex, along with input to these nuclei from higher hypothalamic or limbic centers. Convincing experimental evidence suggests that the tripeptide TRH and its receptors, localized in medullary nuclei, play a physiological role in mediating the increased parasympathetic outflow to the gut, leading to increased gastrointestinal secretory and motor activity (Fig. 4). The sympathetic nervous system through modulation of parasympathetic outflow and spinal or prevertebral reflexes can also markedly influence gastrointestinal function. Sympathetic influence can be tonic (intestinal absorption) but is mostly recruited in relation with a stress situation. Growing experimental data support a role of central CRF in mediating the inhibitory effect of stress on gastric function and stimulation of colonic activity.

BIBLIOGRAPHY

Gillis, R. A., Quest, J. A., Pagani, F. D., and Norman, W. P. (1989). Control centers in the central nervous system for regulating gastrointestinal motility. *In* "Handbook of Physiology: The Gastrointestinal System," pp. 621–683. American Physiological Society.

Taché, Y. (1992). Inhibition of gastric acid secretion and ulcers by calcitonin gene-related peptide. *Ann. N.Y. Acad. Sci.* **657**, 240–247.

Taché, Y., Garrick, T., and Raybould, H. (1990). Central nervous system action of peptides to influence gastrointestinal motor function. *Gastroenterology* **98**, 517–528.

Taché, Y., Mönnikes, H., Bonaz, B., and Rivier, J. (1993). Role of CRF in stress-related alterations of gastric and colonic motor function. *Ann. N.Y. Acad. Sci.* **697**, 233–243.

Taché, Y., and Saperas, E. (1992). Potent inhibition of gastric acid secretion and ulcer formation by centrally and peripherally administered interleukin-1. *Ann. N.Y. Acad. Sci.* **659**, 353–368.

Taché, Y., and Yang, H. (1990). Brain regulation of gastric acid secretion by peptides: Sites and mechanisms of action. *Ann. N.Y. Acad. Sci.* **597**, 128–145.

Taché, Y., Yang, H., and Yoneda, M. (1993). Vagal regulation of gastric function involves thyrotropin-releasing hormone in the medulary raphe nuclei and dorsal vagal complex. *Digestion* **54**, 65–72.

Taché, Y., and Wingate, D. (1991). "Brain–Gut Interactions," pp. 3–361. CRC Press, Boca Raton, FL.

Taché, Y., Wingate D., and Burks, T. F. (1994). "Innervation of the Gut: Pathophysiological Implications," pp. 3–343. CRC Press, Boca Raton, FL.

Brain Spectrin

WARREN E. ZIMMER
STEVEN R. GOODMAN
University of South Alabama

GLOSSARY

Actin A 43-kDa protein found in eukaryotic cells that has the capability to form a thin helical filament

Alternate splicing Mechanism occurring in the nucleus that allows multiple proteins to be produced from a single gene

Amelin A family of protein 4.1-related spectrin binding proteins found in neurons

Ankyrin Family of proteins that binds with spectrin and anchors the spectrin membrane skeleton to the membrane via association with integral membrane proteins

Calmodulin Highly conserved calcium-binding protein that appears to be ubiquitous among eukaryotes and thought to be involved in the transmission of intracellular calcium signals by its calcium-dependent interactions with proteins

Complementary DNA Synthesized from a messenger RNA

Fodrin Original name given to a brain protein (Grk. *fodros*, lining) that has been demonstrated to be a structural analog of erythrocyte spectrin and is now referred to as brain spectrin (αSpII/βSpII).

Indirect immunofluorescence Procedure used to find the location of a protein–antibody complex within a cell by binding a secondary antibody containing a fluorescent dye to the complex, and detection of the fluorescent antibody using a fluorescence microscope

Peptide mapping Technique for examining the structure of a protein by partial digestion of the protein with proteases and analysis of the cleavage products

Spectrin Class of related proteins that contains two high molecular mass subunits and forms the structural framework of the spectrin membrane skeleton

Spectrin membrane skeleton Protein complex that contains spectrin and actin as its major components and forms a two-dimensional meshwork lining the cytoplasmic surface of eukaryotic cell membranes

SPECTRIN IS THE MAJOR CONSTITUENT OF THE skeletal protein mesh work that is closely associated with the cortical cytoplasm of most, if not all, eukaryotic cells. This membrane protein structure was first identified in red blood cells, and it was considered to be a cellular structure peculiar to these cells until spectrin and spectrin-associated proteins were detected in many cell types. The complex has been termed the spectrin membrane skeleton because spectrin appears to form the structural backbone of the complex, and this nomenclature distinguishes the membrane skeletal structure from cytoskeletal structures, which transverse the cytoplasm of nucleated cells. The best-characterized nonerythroid spectrin molecules are those found in the brain. While brain spectrin appears to be similar to its erythrocyte counterpart with respect to physical and morphological parameters, structural analyses of isolated brain spectrin molecules demonstrate that they represent a diverse class of proteins that are related but clearly not identical to red blood cell spectrin. The diversity of brain spectrin is manifested not only at the structural level, but includes expression of multiple spectrin isoforms exhibiting distinct localization within a single neuronal or glial cell. The diversity of brain spectrin molecules must be taken into account when assigning functional roles to the spectrin membrane skeleton in

ENCYCLOPEDIA OF HUMAN BIOLOGY, Second Edition, VOLUME 2. Copyright © 1997 by Academic Press. All rights of reproduction in any form reserved.

brain cells. The close association of brain spectrin and spectrin-binding proteins with the plasma membrane and membranes of various organelles suggests that the membrane skeleton may play a role in the maintenance and modifications of membrane behavior. The best-characterized examples of the involvement of spectrum in neuronal processes are axonal transport, regulation of the lateral mobility of the 180-kDa neural cell adhesion molecule (N-CAM_{180}) in the plasma membrane, and control of synaptic transmission. [*See* Brain.]

I. NONERYTHROCYTE SPECTRIN

A. Discovery

Spectrin was originally described as the major constituent detected in low ionic strength extracts of human erythrocyte membranes. The structure and the function of erythrocyte spectrin have been subjects of intensive investigations beginning with its description in the late 1960s. During the course of these studies, the prevailing thought was that spectrin existed only in erythrocytes.

The first demonstration of spectrin or spectrin-like molecules in nonerythroid cells came in the early 1980s. Using an antibody that reacted with human erythrocyte spectrin, the presence of spectrin was detected in a diverse set of nonerythroid cells, including embryonic chicken heart cells, mouse fibroblast cells, and rat hepatoma (liver) cells, by indirect immunofluorescence techniques. While these experiments suggested that these very different cell types contained proteins that shared antigenetic determinants with erythroid spectrin, they did not address whether these molecules were close relatives of the erythrocyte protein. Thus, further experiments were performed, demonstrating that the erythrocyte spectrin antibodies precipitated two polypeptides from lysates of the chicken heart cells that not only represented spectrin in size [240 and 230 kDa apparent molecular masses by sodium dodecyl sulfate–polyacrylamide gel analysis (SDS-PAGE)] but were present in equal numbers (i.e., at 1 : 1 mol/mol stoichiometry). Comparison of the immunoprecipitated polypeptides with spectrin isolated from chicken erythrocytes by proteolytic digestion with the enzyme chymotrypsin documented that the proteins immunoprecipitated from chicken heart cells were similar in structure to the human erythrocyte spectrin α and β subunits. These experiments established that nonerythroid cells contain polypeptides that are related to erythrocyte spectrin subunits, not only in size and stoichiometry but also antigenically and structurally. These pioneering experiments opened the door for investigations of spectrin molecules and the spectrin membrane skeleton in nonerythroid cells.

B. Early Experiments with Brain Spectrin

Studies appeared in the literature about the same time as the discovery of nonerythroid spectrins, which, in retrospect, provided insights about the spectrin molecules found in brain cells. A high molecular mass protein, which bound the calcium-binding protein calmodulin, contained subunits of 235 and 230 kDa, and had the ability to bind F-actin, was purified from bovine brain tissues. Although this complex resembled spectrin in subunit molecular mass, it did not appear to react with erythrocyte spectrin antibodies and was termed calmodulin-binding protein I (CBP-I). Additionally, a 240-kDa polypeptide, which was able to bind F-actin and stimulate actomyosin Mg^{2+}-adenosine triphosphatase (ATPase) activity, was isolated from pig brain tissue by calmodulin affinity chromatography and was called brain actin-binding protein. Finally, two axonally transported actin-binding proteins of 250 and 240 kDa molecular mass were found to be concentrated at the internal periphery of neurons, Schwann cells, and a variety of nonneural cells. This protein complex was termed fodrin because of its lining of the cell membrane. Subsequent to these results, each of these proteins was shown to be related to erythrocyte spectrin. Thus, these experiments provided the knowledge that brain spectrin bound calmodulin to its 240-kDa α subunit, associated with F-actin and was axonally transported, and that related molecules were present in a variety of nonneural cells. These experiments also provided some confusion because of the various names that had been given to the same protein. Moreover, this confusion has been exacerbated with experiments demonstrating the extreme complexity of spectrins in brain, including the identification of unique β subunits using monoclonal antibody technology and the elucidation of spectrin sequences using molecular cloning technology. To ease the confusion, a unifying spectrin nomenclature has been adopted based on molecular structural information as summarized in Table I. In this nomenclature, spectrin subunits of known protein isoforms are classified as α and β (αSp or βSp) and are then given Roman numerals in the order of their discovery; with erythrocyte spectrin first (αSpI

TABLE I
Spectrin Nomenclature

Spectrin isoform	Previous names	Molecular mass (kDa)	mRNA size (kb)	Genomic locus	Mouse	Human
αSpIΣ1	α erythrocyte spectrin	280	8.0	*Spna1*	1	1
αSpIIΣ1	Nonerythroid α spectrin α fodrin	280	7.8	*Spna2*	2	9
βSpIΣ1	β erythrocyte spectrin	246	6.0/8.0	*Spnb1*	12	14
βSpIΣ2	Spectrin 235E, βSpIb	268	11.0	*Spnb1*	12	14
βSpIIΣ1	Nonerythroid β spectrin β fodrin	275	9.0	*Spnb2*	11	2

or βSpI) and nonerythroid spectrin second (αSpII or βSpII). Isoforms of each spectrin subunit generated by alternate splicing of premessenger RNAs (mRNAs) are given symbols Σ1, Σ2, etc., in the order in which they are described. If the relationship of a particular molecule to known isoforms has not been established by sequence criteria, then the symbol Σ* is used. This nomenclature greatly simplifies the confusing spectrin molecule terminology (both erythrocyte and nonerythroid) and will be utilized throughout this article.

II. BRAIN SPECTRIN

A. Structure

Spectrin has been isolated from brain tissue of a variety of species, including chicken, pig, mouse, rat, bovine, and human. In general, the brain spectrin from this variety of species contains high molecular mass subunits of 240 kDa (α) and 235 kDa (β by SDS-PAGE), which are present in a 1 : 1 mol/mol stoichiometry. Examination of the physical properties of brain spectrin demonstrates that it is a highly asymmetrical molecule with a calculated molecular weight of ~970,000. These experiments indicate that, similar to erythrocyte spectrin, the spectrin isolated from brain tissue exhibits properties of an $(\alpha\beta)_2$ tetrameric unit.

Isolated brain spectrin takes the shape of a long, flexible rod of an ~200-nm contour length when visualized by rotary shadowing and electron microscopy. The morphology of the brain spectrin revealed by these studies is that of two αβ heterodimers found in a head-to-head interaction with the two strands of the spectrin interwoven into a tight double helix structure containing few gaps. In this regard, brain spectrin is nearly identical to the erythrocyte spectrin, although the latter forms a helical structure that is more loosely woven than that of the brain molecule. The head-to-head alignment of the heterodimers is indicated by the bilateral symmetry of binding sites for monoclonal antibodies against brain spectrin, F-actin, ankyrin, calmodulin, and synapsin I. These studies showed that when two copies of any one of these molecules are bound to brain spectrin, the binding sites appear on opposite strands, equidistant from the center of spectrin tetramer. Thus, this bivalent, symmetrical binding of proteins along the brain spectrin tetramers demonstrates the head-to-head association of the αβ spectrin heterodimers. A model of the brain spectrin tetramer summarizing its structure and protein-binding sites is presented in Fig. 1.

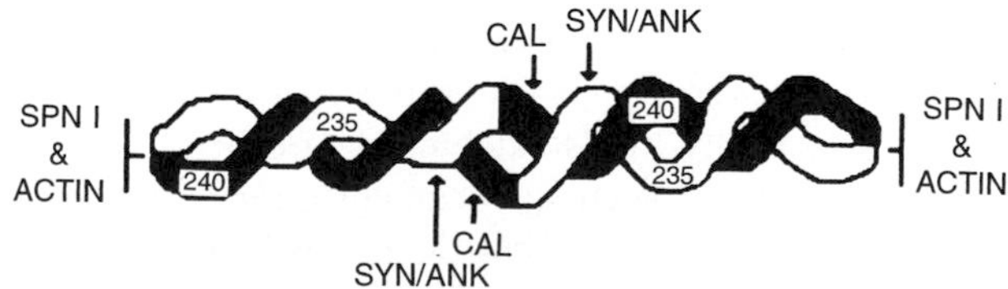

FIGURE 1 Model of the brain spectrin molecule. The morphology of the 200-nm brain spectrin tetramer is shown, along with the bivalent binding sites for calmodulin (CAL), ankyrin (ANK), actin, and synapsin I (SPN I). [Reproduced with permission from S. R. Goodman, B. M. Riederer, and I. S. Zagon (1986). *Bioessays* 5, 25.]

Examination of the primary sequence of brain spectrin subunits has relied on analyses of nucleic acid sequence from cloned DNA molecules representative of the mRNA [complementary DNAs (cDNAs)]. The complete structures of known erythroid and nonerythroid spectrin subunits have been recently elucidated by these experiments. The deduced amino acid sequence from cDNA analyses has demonstrated a 106-amino-acid-repeating motif, which is characteristic of the internal structure first described for erythrocyte spectrin. These 106-amino-acid repeats may each form a triple helical structure, and each spectrin molecule contains multiples of these internal repeating structures. Additionally, each of the spectrin polypeptides contains regions that do not conform to the 106-amino-acid motif and these are perhaps important functional regions of these molecules.

Erythrocyte α spectrin, αSplΣ1, is a polypeptide of 2429 amino acids with a calculated molecular mass of 280 kDa. The calculated molecular mass differs by 40 kDa from the molecular mass estimate of this subunit generated by electrophoretic mobilities (280 kDa compared with 240 kDa). There are 22 106-amino-acid repeat units within the αSplΣ1 molecule, beginning at the NH_2 terminus and extending through to the COOH terminus. There are, however, three regions of the molecule that do not conform to this motif: the extreme NH_2 and COOH termini and a sequence near the center of the molecule referred to as segment/repeat 10. The first ~40 amino acids of αSplΣ1 form the NH_2-terminal domain, which appears to be similar to the third helix of a canonical spectrin motif. Because of the antiparallel nature of the spectrin dimer, it is thought that the αSplΣ1 NH_2-terminal domain participates in important interactions for the dimer associations that form the native spectrin tetramer. The αSplΣ1 COOH terminus does not resemble a spectrin motif and is homologous with the carboxyl-terminal domain of the F-actin-binding protein α-actinin. This domain represents two E-F hand domains, which are important for calcium binding. While there is some thought that calcium has an influence on membrane skeleton structure, there is no direct evidence that the αSplΣ1 COOH-terminal domain actually binds calcium. The central portion of the αSplΣ1 molecule, a segment referred to as repeat 10, represents another deviation from the spectrin 106-amino-acid motif. This region shares extensive homology with a domain found in the modulatory nonkinase family of src proteins called the SH3 domain. Thus, this region of the molecule may be important for interactions of the membrane skeleton and cellular signaling pathways. Interestingly, an αSplΣ* protein and mRNA are expressed in mammalian brain in addition to the αSpllΣ1 products. However, there is a differential localization of these molecules within a single neuron (see Section II,B).

The structure of nonerythroid α spectrin, αSpllΣ1, demonstrates a high degree of similarity with αSplΣ1, ~58% identity in primary sequence. However, there are some striking differences between these molecules. There are two segments of the αSpll molecules, 20 amino acids at the junction of repeat 10 and 11, and 6 amino acids within segment 21 that are variable within the proteins formed from this gene, even among molecules expressed in the same cell. This variability occurs through an alternate splicing mechanism operating on the primary mRNA transcribed from the gene; thus, the products from this gene are referred to as αSpllΣ1, Σ2, etc. An important difference between the αSplΣ1 and the αSpllΣ1 molecules is the presence of a 36-amino-acid domain at the end of segment/repeat 11 within the αSpll domain which creates a high-affinity calmodulin-binding site within the polypeptide. It is interesting that, unlike αSpl, the COOH-terminal E-F hand structures of αSpll have been shown to bind calcium. Therefore, calcium may play an important role in the function of αSpll molecules, both through calcium–calmodulin interactions and through direct calcium binding via the COOH-terminal domain.

Molecular cloning of the erythrocyte β spectrin, βSplΣ1, has been accomplished for a variety of species, including mouse and human. The deduced primary sequence of βSplΣ1 from full-length cDNA clones indicates that it is a 2137-amino-acid polypeptide with a calculated molecular mass of 246 kDa. This molecule contains 17 spectrin repeats and there are two domains within βSplΣ1 that do not resemble the 106-amino-acid motifs. The extreme NH_2 terminus (the first ~276 amino acids) exhibits strong homology to other actin-binding proteins, including α-actinin, dystrophin, and filamin. Moreover, this segment of the βSplΣ1 molecule has been shown to bind with actin filaments, confirming that this region of the β subunit is the major binding site for actin in the membrane skeleton. The COOH-terminal domain of βSplΣ1 also does not conform to the spectrin repeat motif and appears to be involved, together with the NH_2-terminal domain of the α subunit, in spectrin dimer association. Interestingly, this COOH-terminal domain is replaced by a different amino acid sequence in skeletal muscle and brain tissues through alternate splicing of the *Spnb1* gene primary transcript in nuclei of nonerythroid cells. This nonerythroid-specific COOH-terminal domain in βSpIΣ2 adds 21.4 kDa to the βSpIΣ1 molecule which accounts for the difference in apparent molecular mass between these polypeptides (220 kDa for βSpIΣ1 and 235 kDa for βSpIΣ2 as determined by SDS-PAGE).

The complete sequence for the nonerythroid β subunit, βSpIIΣ1, has been recently elucidated. This molecule is highly conserved between mouse and human (~90% identical); however, βSpII and βSpI molecules within the same species are on the order of 59% identical. Similar to βSpIΣ1/Σ2, the βSpIIΣ1 molecule is composed of 17 spectrin repeat motifs. Moreover, the NH_2-terminal domain of βSpIIΣ1 shares extreme homology with that of βSpIΣ1/Σ2, and this region of βSpIIΣ1 is capable of binding F-actin. The COOH-terminal domains of βSpIIΣ1 and βSpIΣ2 are homologous and may account for the observed difference in spectrin tetramer formation and stability between nonerythroid and erythrocyte subunits. The most interesting functional insight generated from the examination of the βSpIIΣ1 sequence was the finding of a 142-amino-acid domain within spectrin repeat motifs 11 and 12 that shares significant homology with the heme-binding domain of various proteins such as α and β globins. The demonstration that brain spectrin, αSpIIΣ1/βSpIIΣ1, is capable of binding heme *in vitro* is a seminal finding in the brain spectrin field and suggests that brain spectrin in the neuron may bind heme and serve to transfer electrons, scavenge free heme, or bind oxygen/nitric oxide/carbon dioxide/carbon monoxide through a heme iron. A model of the βSpIIΣ1 polypeptide illustrating the structure of the molecule and the various binding domains along the protein is shown in Fig. 2.

B. Isoforms

The discovery of mammalian brain spectrin isoforms actually resulted from a reconciliation of seemingly contradictory data. On the one hand, immunohistochemistry experiments used antibodies, which recognized axonally transported spectrin (the antibody reacting primarily with the 240-kDa α subunit), and localized the protein in the cortical cytoplasm of guinea pig neuronal cell bodies, dendrites, and axons in the peripheral nervous system. On the other hand, an antibody made against mouse red blood cell spectrin that reacted with the 240- and 235-kDa spectrin subunits among total mouse brain proteins localized brain spectrin to mouse cell neuronal cell bodies and dendrites, but no reactivity was detected in axons. These data presented the rather interesting question of how two antibodies, both of which had been demonstrated to react specifically with the brain spectrin subunits, could demonstrate a very different localization of spectrin in the brain cells. The answer to this question came from careful preparation of antibodies against isolated mouse brain spectrin from a fraction enriched in synaptic-axonal membranes and mouse erythrocyte spectrin, and use of these antibodies in parallel studies to localize spectrin within brain tissue. After cleaning these antibodies through an affinity column to which spectrin from the opposite tissue was attached, the antibody made from the synaptic-axonal spectrin reacted with the 240-kDa α subunit of the synaptic-axonal spectrin, but did not cross-react with erythrocyte spectrin. In contrast, the eryth-

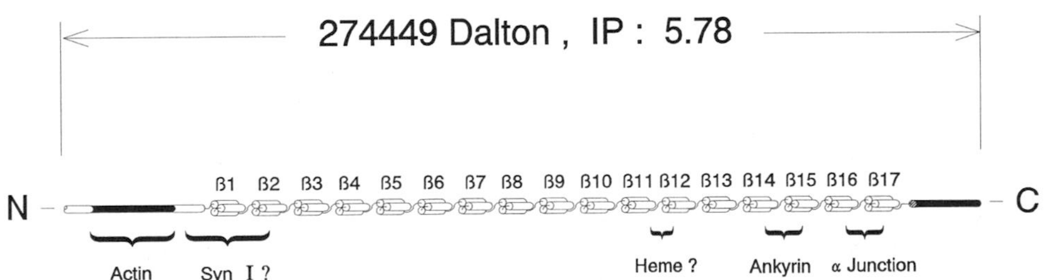

FIGURE 2 Model of the βSpIIΣ1 molecule based on cDNA sequencing analyses. This diagram represents a model of the mouse βSpIIΣ1 protein based on sequences deduced from cDNA(s). The central region of this molecule is composed of 17 triple helical segments of ~106 amino acids in length, which are given arabic numbers β1 through β17. The extreme amino and carboxyl termini which do not share this repeating motif are shown as filled bars. The domains of the βSpIIΣ1 molecule for actin, synapsin I (SynI), heme, ankyrin, and α junctional binding are shown by brackets.

rocyte spectrin antibody did not show reaction with the synaptic-axonal spectrin while retaining strong reaction with the erythrocyte spectrin. These two antibodies revealed two different localizations of spectrin when used to stain mouse brain tissue in parallel immunohistochemistry experiments. Brain spectrin that reacted with the synaptic-axonal spectrin was found to be enriched in mouse neuronal axons and to a lesser extent in cell bodies, but did not stain glial cells. Conversely, brain spectrin that demonstrated a reaction with the erythrocyte antibody was present in mouse neuronal cell bodies and dendrites and in glial cell types. These findings demonstrated that at least two immunologically distinct spectrin isoforms exist in mouse brain. Recently, using similar techniques, a third isoform named brain spectrin (240/235A) has been found exclusively in astrocyte soma and processes. Although sequence criteria have not conclusively determined the presence of a third distinct isoform in brain, it is clear that the synaptic-axonal-reacting spectrin contains αSpllΣ*/βSpllΣ1 and the erythroid-like spectrin contains αSplΣ*/βSplΣ2 polypeptides. The asterisk designation of α subunits is necessary because the *αSpll* gene creates many proteins via alternate splicing and because the αSpl polypeptide determined to be in brain by molecular hybridization experiments has yet to be completely sequenced. Significantly, these isoforms exhibit the same localizations in all mammalian brain tissues examined to date, including human. A summary of the location of αSpllΣ*/βSpllΣ1 and αSplΣ*/BSplΣ2 within a single neuron is presented in Fig. 3.

A detailed examination of the spectrin isoform distribution in brain tissue by immunohistochemistry techniques using electron microscopy revealed that brain spectrin αSplΣ*/BSplΣ2 was associated with the cytoplasmic surface of the plasma membrane and with organelle membranes, including mitochondria, endoplasmic reticulum, and the nuclear envelope. This isoform was located in the synaptic spines of dendrites exhibiting strong association with postsynaptic densities. Brain spectrin αSpllΣ*/βSpllΣ1 was found in axons and presynaptic elements, where it was associated with the cytoplasmic surface of the plasma membrane, organelle membranes, synaptic vesicles, and cytoskeletal structures. These findings are consistent with the observation that this isoform is transported down the axon, with specific neural cell structures. Interestingly, both of the brain spectrin isoforms were found to be associated with microtubules, neurofilaments, and actin filaments throughout the neuronal cytoplasm, indicating that although

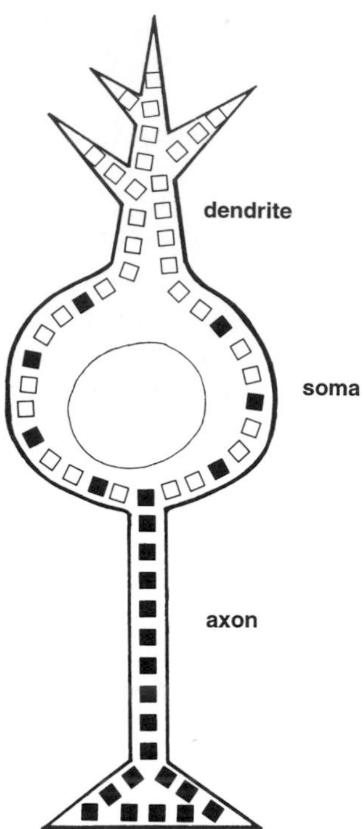

FIGURE 3 Summary of the distribution of spectrin isoforms within the mammalian neuronal cell. Brain spectrin αSpIΣ*/βSpIΣ2, detected with a red blood cell spectrin antibody, is localized throughout the dendrites and cell body (□). Brain spectrin αSpIIΣ*/βSpIIΣ1 is found primarily in the axon and to a lesser extent in the cell body (■). [Reprinted with permission from S. R. Goodman, B. M. Riederer, and I. S. Zagon (1986). *Bioessays* 5, 25.]

these isoforms exhibit very different localizations within the cell, there is a continuum of contacts between the spectrin membrane skeletal apparatus and the cytoplasmic organization within the cell. Moreover, the discovery that two spectrin isoforms exhibit specific compartmentalization within a single neuronal cell would suggest that brain spectrin and perhaps the membrane skeletal complex are more structurally and functionally versatile than that ascribed for erythroid cells. This is supported by the finding that brain spectrin may play a pivotal role in cellular oxygen and oxide metabolism through heme iron binding. The location and associations of brain spectrin revealed by electron microscopy are summarized in Fig. 4.

There is a differential expression of the spectrin isoforms during brain development. Using specific antibodies to quantitate the amount of spectrin isoform, brain spectrin αSpllΣ*/BSpllΣ1 was detected in fetal mouse brain tissues and increased twofold in content to the levels measured in adult tissues. Northern hy-

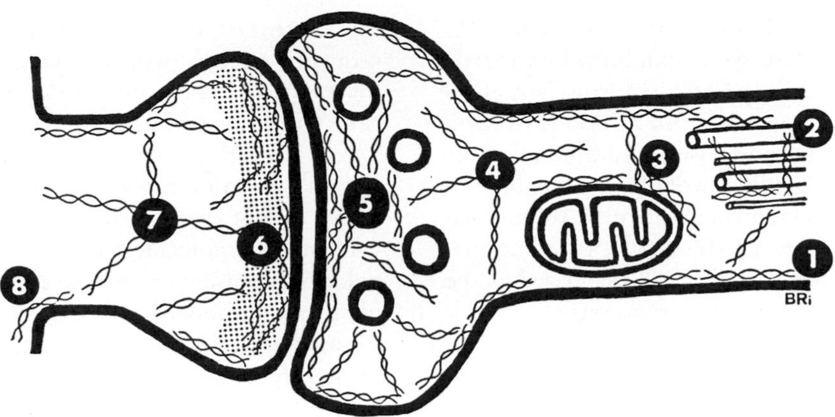

FIGURE 4 Summary of spectrin isoform location at an axodendritic synapse. This model summarizes the location and interactions of spectrin isoforms at an axo-dendrite synapse based on immunoelectron microscopy. The locations of brain spectrin αSpIIΣ*/βSpIIΣ1 and brain spectrin αSpIΣ*/βSpIΣ2 are demonstrated at positions 1–5 and positions 6–8, respectively. 1, axonal plasma membrane; 2, neurofilaments and microtubules; 3, mitochondria; 4, cytoplasmic cytoskeleton; 5, vesicles and presynaptic membrane; 6, postsynaptic densities and postsynaptic membrane; 7, dendritic spine shaft; 8, dendritic plasma membrane. [Reprinted with permission from Goodman *et al.* (1988). Copyright CRC Press, Inc., Boca Raton, Florida.]

bridization experiments confirmed these results. The 7.8-kb αSpIIΣ* mRNA(s) and the 9.0-kb βSpIIΣ1 mRNA were both detected in early fetal tissues and increased coordinately to the levels measured in adult tissue. This isoform was enriched in the cortical cytoplasm of primary and secondary germinative neural cells and was detected within fibers resembling axons in fetal tissues. The erythroid-like spectrin αSpIΣ*/βSpIΣ2 was not detected in mouse fetal and neonatal brain tissues, but exhibited a rapid increase in concentration during the second postnatal week. Moreover, molecular hybridization experiments showed a rapid increase in the 11-kb βSpIΣ2 mRNA during the second postnatal week, indicating that the developmental increase in this isoform is controlled, at least in part, by the concentration of the mRNA encoding this polypeptide. This isoform was detected in the cell body and dendrites of differentiating neurons and glial cells but was not found in mitotic cells. The differential expression of these brain spectrins might indicate that early neural differentiation events rely on the functional attributes of the axonal brain spectrin αSpIIΣ*/βSpIIΣ2, which is present at the origin of the neuronal tissues. Additionally, the observation that the erythroid-like brain spectrin isoform is expressed at later stages of brain development, at a time of cellular specialization within the neural tissue, implies that this isoform may impart a specialized role of the associated membrane skeleton within the cell, such as the development of cerebellar glomeruli (synaptic contacts involving granule cells, Golgi cells, and mossy fibers) and/or the maturation of sensory motor systems, both of which are initiated during the second postnatal week in the mouse.

Taken together, these results have demonstrated the presence of spectrin isoforms within brain cells. These isoforms not only exhibit discrete localizations within a brain cell, but are differentially expressed during development and specialization of the brain. Clearly, the organization of brain spectrin and perhaps the spectrin-based membrane skeleton is much more diverse, or complicated, than that of the erythrocyte. Whether this diversity of the spectrin in brain tissue is due to a requirement for specialized or regional functions within a single neuronal cell is, at present, not well understood. However, it is evident from these observations that experiments aimed at understanding the role of "brain spectrin" must take the diverse nature of the spectrin isoforms into consideration in assigning the functional capabilities of spectrin within brain tissues and cells.

III. PROTEIN INTERACTIONS

This section reviews the current knowledge of proteins that interact directly and indirectly with brain spectrin. These proteins include calmodulin, ankyrin, adducin, amelin, synapsin I, and actin. This section discusses experiments that were conducted primarily with comixtures of brain spectrin isoforms. The affinity of a spectrin-binding protein to a specific isoform has only begun to be addressed. Additionally, the list of brain spectrin-binding proteins is probably not exhaustive, as additional proteins are probably associated with specific neuronal structures, or additional specific spectrin isoforms will be characterized as the examination of brain spectrins progresses.

A. Calmodulin

As described previously, the interaction of brain spectrin with calmodulin was first demonstrated by the isolation of the 240-kDa subunit by calmodulin affinity chromatography. Subsequent studies have demonstrated that spectrin is a major calmodulin-binding protein in brain cells and that calmodulin binds with the 240-kDa α subunit in the presence of physiological concentrations of calcium. Analysis of calmodulin binding by electron microscopy revealed that the calmodulin-binding sites on the α subunits occur at a distance of ~15 nm from the junction of the heterodimers. The exact placement of calmodulin binding along the α subunit has been defined as within the region of sequence between α11 and α12 domains of the molecule by examination of proteins derived from bacteria, which had been induced to express this region of the α subunit from isolated cDNA sequences. These studies localized the calmodulin-binding domain to an extra arm of 36 amino acids that do not exhibit the 106-repeat consensus motif of the spectrin molecule, which is located just after the α11 repeat unit. Interestingly, inspection of the deduced sequence from cDNA clones isolated from chicken brain tissue implicated a sequence near the carboxyl-terminal region of the molecule as a calmodulin-binding site by homology with calmodulin-binding sites found in other proteins; however, the ability of this sequence to bind calmodulin has not been investigated. In mammals, erythrocyte spectrin does not contain a high-affinity calmodulin-binding site, although a weak interaction between the calmodulin and the erythrocyte β subunit has been reported. The exact role for the calmodulin–spectrin interaction remains unclear.

B. Ankyrin

The presence of an immunoreactive analog of erythrocyte ankyrin in brain tissue was demonstrated before the discovery of brain spectrin. There are three sequence-related proteins with apparent molecular masses of 220, 210, and 150 kDa, which exhibit reaction with an antibody prepared to erythrocyte ankyrin. Recently, it has been demonstrated that there are ankyrin molecules present in nonerythroid cells which do not share antigenetic sites with the erythrocyte ankyrin molecules. Thus, similar to the spectrins, ankyrin represents a diverse class of related, but not identical, proteins. In brain, the erythroid ankyrin (termed ankyrin$_R$) is the product of the ANK1 gene located on chromosome I in human and mouse. It is located in cell bodies and dendrites of neurons in spinal cord and cerebellum and arises late in development; it is highly suggestive that this ankyrin isoform is a likely candidate for the attachment of the erythroid brain spectrin (αSpIΣ*/βSpIΣ2) to membrane within the soma and dendrites on neurons. Ankyrin$_B$ encoded by the ANK2 gene located on human chromosome 4 is the major isoform of ankyrin in brain, localized on the plasma membranes of neurons as well as on glial cells throughout neural tissue, and is likely the protein for attaching brain spectrin (αSpIIΣ*,/ BSpIIΣ1) to plasma membranes. Ankyrin$_{node}$ is a specialized isoform which localizes to the nodes of Ranvier and axon node initial segments, most likely playing an important role for the attachment of brain spectrin at these specialized regions of the axonal plasma membrane.

Brain ankyrins bind to the β subunit of brain spectrin at sites that are ~20 nm from the junction of the spectrin heterodimers. Similar to erythrocyte ankyrin, brain ankyrin can be divided into a 72-kDa spectrin-binding domain and a 93-kDa membrane-bound domain by digestion with the proteolytic enzyme chymotrypsin. It is known that ankyrin binds the spectrin β subunit with the binding domain, as defined through analyses of erythrocyte β-spectrin βSpIΣ1, being the region of the molecule containing amino acids 1768–1898. Brain spectrin βSpIIΣ1 contains a stretch of amino acids at the same position (repeat number 15 of the 17 spectrin 106 motifs found in the β subunit) that is 91% identical to the erythrocyte β subunit; it is highly suggestive that this represents the βSpIIΣ1 ankyrin-binding domain. Although the exact role of the ankyrin–spectrin interaction in brain is not well defined, it is thought that ankyrins link spectrins to a number of physiologically important transmembrane proteins, including band 3 (anion transporter), membrane sodium/potassium–ATPase, and the voltage-dependent sodium channels. These interactions attach spectrin to the neural plasma membranes; through these interactions, spectrin may limit the motility of proteins within the membrane bilayer.

C. Actin

In erythrocytes, spectrin tetramers are capable of cross-linking F-actin filaments (protofilaments of ~13 actin monomers) into a two-dimensional mesh work. Actin binding occurs at the tails of the spectrin tetramer and it is the amino-terminal domain of the β subunit that confers this binding capacity. The actin-binding domain of the erythrocyte β spectrin has been defined as a 140-amino-acid stretch beginning at residue number alanine^{47} and extending to amino acid 186, lysine^{186}. This region is identical to the erythroid β spectrin found in brain βSplΣ2 and is 89% identical to the amino-terminal domain of βSpllΣ1, suggesting that the specific amino acids necessary for actin binding are well conserved among the brain spectrin isoforms.

D. Amelin

Initial characterizations of brain spectrin included experiments that demonstrated that this protein could bind with F-actin and that F-actin binding was stimulated or strengthened with the addition of erythrocyte protein 4.1. These experiments suggested that brain cells might contain a protein that could stimulate the F-actin–spectrin association analogous to the erythroid protein 4.1. Using antibodies that reacted with the erythrocyte protein 4.1, an immunoreactive and structural analog of the erythrocyte protein has been identified in mammalian brain tissue. The brain 4.1 analog, termed amelin (Greek, *Amelew,* overlook), is a single polypeptide exhibiting a molecular mass of 93 kDa as determined by SDS-PAGE. This protein has been demonstrated to be structurally related to the erythrocyte protein 4.1 by peptide mapping studies. Recent experiments have demonstrated that amelin is a family of proteins, each having a distinct localization within neurons. Amelin$_E$ is localized in the cell bodies and dendrites of neurons as well as in certain glial cells, which is the same localization observed for the spectrin αSplΣ*/βSplΣ2 isoform. Amelin$_{Axon}$ is located in axons and in the soma of neurons in the cerebellum, a pattern identical to brain spectrin αSpllΣ*/βSpllΣ1. This colocalization of amelin and brain spectrin might reflect a difference in the affinity of specific brain spectrin isoforms for amelin polypeptides and may provide a specialization for the spectrin–F-actin interactions.

E. Synapsin I

Synapsin I is a neuron-specific phosphoprotein that is associated with the cytoplasmic surface of small synaptic vesicles. This protein has the capacity to bind spectrin (αSpllΣ*/βSpllΣ1) and thereby attaches small spherical synaptic vesicles to the tail ends of the spectrin tetramer. Originally, synapsin I was suggested as a structural analog in brain of erythrocyte protein 4.1; however, subsequent analyses of this protein demonstrated that although synapsin I is a functional analog of the erythrocyte protein 4.1, it bears little structural resemblance to the erythroid 4.1.

Synapsin I is composed of two polypeptides of 76 and 70 kDa (termed Ia and Ib) molecular mass, which give nearly identical peptide maps independent of which mammalian species is the source of the protein. Examination of synapsin I by molecular cloning techniques has demonstrated that the synapsin I proteins belong to a family of at least four distinct proteins (termed synapsin Ia, Ib, IIa, and IIb), all of which exhibit localization to synaptic nerve terminals. These proteins are the product of two distinct genes (synapsin I and synapsin II) and the isoforms of each are formed through differential splicing mechanisms.

As indicated earlier, synapsin I is a phosphoprotein, and it has been demonstrated that the protein is differentially phosphorylated by calcium–calmodulin-dependent protein kinase II and cAMP-dependent protein kinase. Phosphorylation of synapsin I by calcium–calmodulin-dependent protein kinase II and cAMP-dependent protein kinase does not affect its interaction with brain spectrin (αSPIIΣ*/βSpIIΣ1).

F. Adducin

Adducin is a recently characterized calmodulin-binding phosphoprotein, which in erythrocytes is a 205-kDa heterodimer complex. Adducin stimulates the spectrin–F-actin interaction and has been demonstrated to cause F-actin bundling *in vitro.* Although these functions suggest that this protein is similar to protein 4.1, adducin and protein 4.1 appear to be independent and do not compete for spectrin–actin binding but are additive in their activities. Whether or not adducin cross-links spectrin with F-actin or creates new (altered) binding sites on the actin filament is unknown; however, the activities of adducin are downregulated by calmodulin and calcium. It has been recently demonstrated that three sequence-related polypeptides purified from brain tissue share properties with erythrocyte adducin, including the ability to bind calmodulin, to stimulate brain spectrin–actin interactions, and to react with antibodies to the erythrocyte adducin molecule. It remains to be demonstrated whether or not these molecules exhibit a localization similar to that of the brain spectrin isoforms. However, these observations suggest that calmodulin may regulate several aspects of the spectrin membrane skeletal complex.

IV. FUNCTIONS

We are in the early stages of experimentation aimed at elucidating the function(s) of spectrin in brain cells. The static views of brain spectrin isoforms within individual neuronal and glial cells supplied by immunoelectron microscopy as well as the studies of its protein and membrane interactions *in vitro* have led to an initial understanding of brain spectrin function. Based on the detailed knowledge of the erythrocyte spectrin membrane skeleton, we can predict that brain spectrin isoforms, which are bound to the cytoplasmic surface of the plasma membrane and to organelle membranes, will exhibit functions of (1) giving stability to these membranes, (2) regulating their contour, (3) controlling the flip-flop of phospholipids across the membrane bilayer, and (4) limiting the lateral mobility of integral membrane proteins through the bilayer. While it is clear that a single spectrin species manifests multiple functions in erythrocytes, the finding of multiple spectrins within a single brain cell might indicate that the individual spectrins each impart some specialized function(s) within the cell. This suggestion is strengthened by the observations that the brain spectrin isoforms show distinct compartmentalization within neural cells and that certain spectrin-binding proteins exhibit a colocalization with a specific spectrin isoform. Thus, as the examinations of the functions of spectrin in brain cells proceed, it will certainly be interesting to analyze whether an individual spectrin isoform can adequately perform all the tasks required by the brain cells or if individual spectrin isoforms have evolved to provide a specific functional capability to the cells. Although the preponderance of data supports the latter, the former cannot be totally excluded until experiments are designed to assay the individual brain spectrin isoforms.

A. Axonal Transport

Brain spectrin is synthesized in the neuronal cell body, and, thus, brain spectrin, which is found in axons and presynaptic terminals, must be transported from the cell body to the synapse. Experiments examining the movement of proteins through the axon have shown that five different populations of proteins, referred to as I–V, travel through the axon at different velocities. Brain spectrin (presumably brain spectrin αSpIIΣ*/βSpIIΣ1) travels down the axon at various rates, suggesting that distinct populations of brain spectrin are associated with neuronal structures. The brain spectrin traveling down the axon with the group of proteins exhibiting the slowest rate of migration, termed group V (at a rate of $\sim$1 mm/day), has been demonstrated to be associated with large, complex structures. The predominant protein components of group V include the neurofilament proteins and tubulin, indicating that this group may include movement of the cytoskeleton. Studies have demonstrated that a crude particulate fraction of brain tissue, when added to purified polymerized tubulin, caused ATP-dependent gelation–contraction *in vitro*. This particulate fraction had microtubule-stimulated ATPase activity and moved slowly ($\sim$1 μm/min) along microtubule walls in the presence of ATP. This fraction was essential for the gelation–contraction, which is thought to be the equivalent of slow axonal transport (group V), and contained brain spectrin as a major protein component. These observations suggest an important role of brain spectrin (αSpIIΣ*/βSpIIΣ1) in slow axonal transport.

B. Regulation of the Lateral Mobility of N-CAM_{180}

N-CAM has been implicated in the morphogenesis of neural and nonneural tissues. In the mouse, N-CAM consists of three integral membrane proteins of 180, 140, and 120 kDa molecular mass. The extracellular amino-terminal domain of these molecules shares a common sequence, whereas the cytoplasmic carboxyl-terminal regions exhibit differences in length and sequence. N-CAM_{180} contains the largest cytoplasmic domain and is accumulated at sites of cell–cell contact. This molecule has a restricted lateral mobility within the neuronal plasma membrane compared with the other N-CAMs, suggesting that the cytoplasmic domain may play a role in the restrictive movement of the molecule. The observation that ankyrin and brain spectrin coisolates with N-CAM_{180}, but not N-CAM_{140}, suggests that the membrane skeleton may play a role in regulating the lateral mobility of this molecule in the neuronal plasma membrane. Brain spectrin binds directly with N-CAM_{180} and does not bind with N-CAM_{140} or N-CAM_{120}. This interaction is presumably mediated by the cytoplasmic domain of the N-CAM_{180} and may be specific for the brain spectrin isoform based on colocalization studies. Taken together, these results suggest that brain spectrin plays a role in limiting the lateral mobility of this essential adhesion molecule, restricting it to regions of cell–cell contact. Although the role of ankyrin in this process is unclear, these studies demonstrate a possible function of brain spectrin that is analogous to the role of spectrin on the erythrocyte membrane.

C. Synaptic Transmission

The transmission of information through the neural system occurs via the specific release of neurotransmitters from the axon of one cell, which stimulates a second cell. This neurotransmitter release occurs at a region where the two cells are in close apposition, termed the synaptic cleft, which has a width of 10–20 nm. A large number of vesicles ranging in size from 10 to 140 nm in diameter are present in the presynaptic cytoplasm, closely associated with the presynaptic membrane. These vesicles are thought to contain the neurotransmitters that are released from the presynaptic terminal in response to nerve stimulation, with the resultant stimulation of the postsynaptic cell. At the cellular level, the first steps in the release of the neurotransmitter would include (1) release of the vesicle from actin filaments, (2) vesicle translocation, (3) vesicle attachment to the presynaptic plasma membrane, and (4) fusion of the vesicle membrane with the presynaptic membrane releasing the neurotransmitter into the synaptic cleft. The demonstration that brain spectrin is in contact with the axonal plasma membrane as well as vesicle membranes suggests that spectrin may play a role in the neurotransmitter release from the synaptic vesicles. [*See* Synaptic Physiology of the Brain.]

Characterizing the role of brain spectrin in synaptic transmission has advanced with the identification of proteins associated with the membrane of the synaptic vesicle and the characterization of the neural spectrin cytoskeleton in the presynaptic terminal. Electron microscopy experiments have shown that spectrin αSpIIΣ*/βSpIIΣ1 is associated with the cytoplasmic surface of synaptic vesicles and the plasma membrane of neurons. Furthermore, using specialized electron microscopic techniques, spectrin has been shown to form ~100-nm-long fibrous strands which interconnect synaptic vesicles with the presynaptic plasma membrane. Thus, spectrin αSpIIΣ*/βSpIIΣ1 appears to link synaptic vesicles to fusion sites on the presynaptic membrane and is therefore an integral component of the machinery that docks the synaptic vesicles near the membrane, making them ready for calcium-stimulated release. It is clear that the synapsins play an important role in the attachment of the synaptic vesicles to the membrane-bound spectrin. Interestingly, binding experiments using purified synaptic vesicles and brain spectrin showed that the docking of vesicles to membrane-bound spectrin occurs independent of synapsin I and II protein phosphorylation, which is consistent with the finding that mutant mice lacking synapsin I expression but containing a full complement of Synapsin II are not lethal with regard to synaptic function.

If spectrin tetramers are rod-shaped molecules extending ~200 nm parallel to the plasma membrane (as found in the red blood cell), how can they be the docking sites for synaptic vesicles which appear to be ~100-nm fibers that are perpendicular to the membrane? One model explaining this phenomenon is referred to as the "casting the line" hypothesis. In this hypothesis, as a synaptic vesicle approaches the plasma membrane it binds to an attachment site on the β spectrin subunit (likely to reside within residues 207–445; a conserved region of the molecule juxtaposed to the actin-binding domain), releasing that spectrin tail from its interaction with actin. Because spectrin tetramers are linked near their junctional end to ankyrin, the release of one tail of a spectrin tetramer from actin protofilaments in the membrane skeleton would result in a ~100-nm filament (equivalent to heterodimer in length) with a synaptic vesicle binding end on that is now perpendicular to the plasma membrane. A result of this model is a localized dissociation of the membrane skeleton at the site where the plasma membrane will invaginate to form a dimple leading to the fusion pore complex with the vesicle membrane. It has been demonstrated that the proteins which form the fusion pore complex (and thus cause the release of neurotransmitters) are sensitive to calcium ion fluctions; thus a local pertubation in the plasma membrane would bring it into close proximity with the

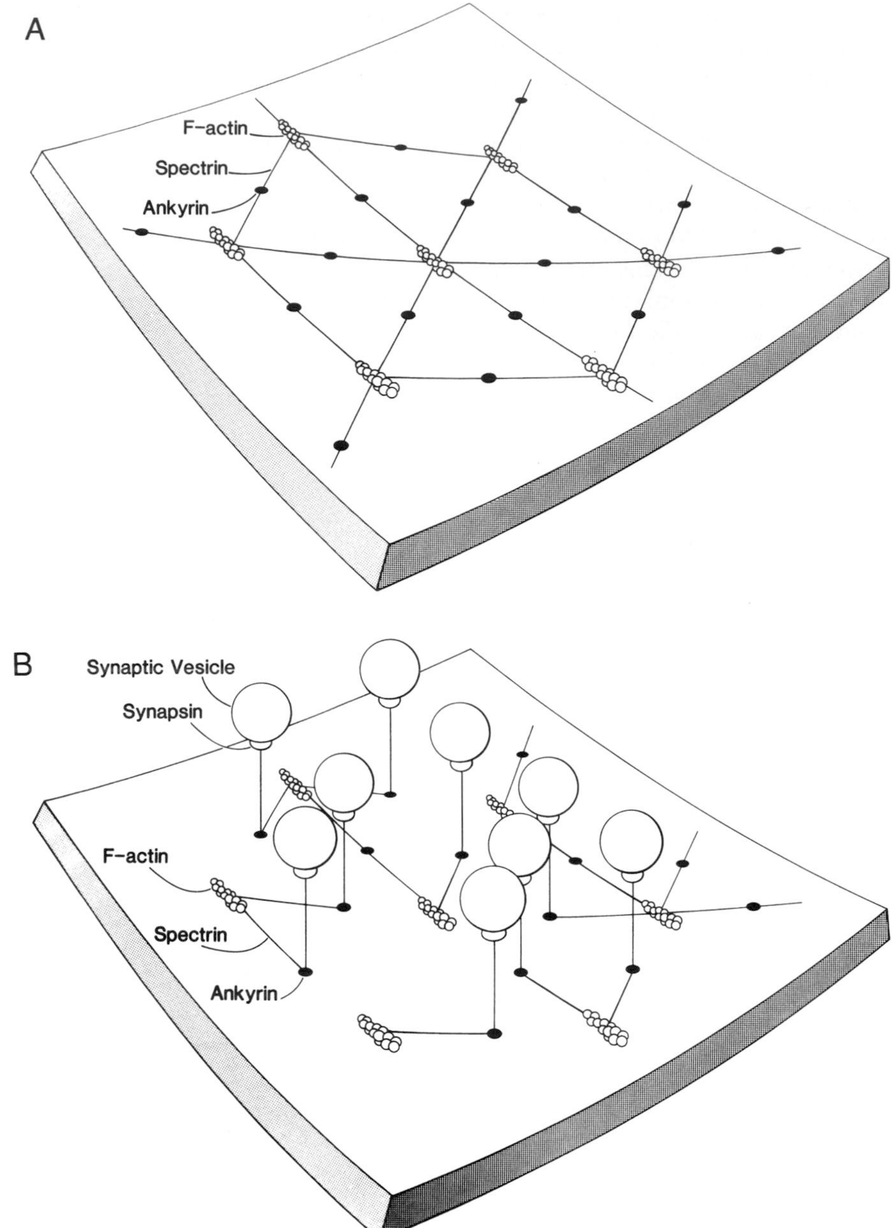

FIGURE 5 Model of the "casting the line" hypothesis. This figure represents diagrams for the role of spectrin in synaptic transmission as suggested by the casting the line hypothesis. (A) The appearance of a segment of the membrane skeleton covering a small patch of the active zone of the presynaptic plasma membrane. This model of the membrane skeleton assumes a configuration of proteins similar to that found in the erythrocyte membrane skeleton. When synaptic vesicles approach the plasma membrane, they bind to the tail region of a spectrin tetramer, releasing this tail from its interactions with membrane-associated actin filaments (B). The results would be the attachment of synaptic vesicles, via synapsin I or II, to ~100-nm fiber representing approximately one-half of the spectrin tetramer. This synaptic vesicle docking suggests a local disruption of the plasma membrane, causing a dimple that juxtaposes the membrane of the synaptic vesicle with that of the plasma membrane, such that increases in calcium ion concentrations would allow for rapid membrane fusion events and neurotransmitter release.

synaptic vesicles and the pore complex proteins such that a wave of calcium ions (as that which occurs in a stimulated neuron) would allow for rapid membrane fusion events and neurotransmitter release. In addition, brain spectrin contains high-affinity calcium ion-binding sites at or near the synapsin I-binding domain (in both the α and β subunits), and synaptic vesicles cannot bind with brain spectrin αSpIIΣ*/βSpIIΣ1 in conditions of elevated calcium. Therefore, a second effect of increasing calcium ion concentrations during nerve stimulation would be the facilitated release of the synaptic vesicles from their spectrin tether as they are undergoing fusion with the plasma membrane and releasing their contents into the synaptic cleft. The key points of the "casting the line" model are illustrated in Fig. 5.

BIBLIOGRAPHY

Bennett, V. (1992). Ankyrins. *J. Biol. Chem.* **267**, 8703.

Elferink, L. A., Peterson, M. S., and Scheller, R. H. (1993). A role for synaptotagmin (p65) in regulated exocytosis. *Cell* **72**, 153.

Goodman, S. R., Krebs, K. E., Whitfield, C. F., Riederer, B. M., and Zagon, I. S. (1988). Spectrin and related molecules. *CRC Crit. Rev. Biochem.* **23**, 171.

Goodman, S. R., Zimmer, W. E., Clark, M. B., Zagon, I. S., Barker, J. E., and Bloom, M. L. (1995). Brain spectrin: Of mice and men. *Brain Res. Bull.* **36**, 593.

Goodman, S. R., Zagon, I. S., and Kulikowski, R. R. (1981). Identification of a spectrin-like molecule in non-erythroid cells. *Proc. Natl. Acad. Sci. USA* **78**, 7570.

Monck, J. R., and Fernandez, J. M. (1994). The exocytotic fusion pore and neurotransmitter release. *Neuron* **12**, 707.

Winkelmann, J. C., and Forget, B. G. (1993). Erythroid and nonerythroid spectrins. *Blood* **81**, 3173.

Brain Tumors

BRIAN COPELAND
Scripps Clinic and Research Foundation

GLOSSARY

Astrocytoma Tumor that results from the abnormal growth of astrocytes, the supporting cells of the brain

Computerized axial tomography (CAT) Method of analyzing thousands of small X-rays with a computer to create an image of the whole structure (same thing as a CT scan)

Glioma Same thing as an astrocytoma (astrocytes are sometimes called glial cells)

Meningioma Tumor that results from the abnormal growth of the meninges, the membranes that cover the brain

Magnetic resonance imaging Process similar to CAT scanning that substitutes a large magnetic field for the X-rays and produces images of exquisite anatomic detail

BRAIN TUMORS ARE ABNORMAL GROWTHS OF cells that occur inside the cranium. Like tumors elsewhere in the body, they can be benign or malignant. They can occur in infants, children, and adults of any age. Although some tumors have a predilection for men or women, both sexes are equally effected overall. About 25,000 new cases are diagnosed each year in the United States. The impact of a brain tumor on the affected individual is extraordinarily variable. Some cases are completely benign and do not require any treatment whereas others are rapidly fatal. Most fall somewhere in between.

I. SYMPTOMS

The symptoms caused by brain tumors are extremely variable. They fall into three categories: those symptoms that are related to increased pressure inside the head, those related to the particular area of the brain affected by the tumor, and those that are related to epileptic seizures.

It is important to understand that the human skull forms a close-fitting rigid container for the brain. When a tumor begins to grow, there is no way that the skull can expand to accommodate the enlarging mass. Therefore, the space occupied by a brain tumor comes at the expense of the surrounding normal brain. Although there is a certain tolerance of the brain to compression, eventually the tumor growth will lead to increased pressure inside the calvarium. This increased intracranial pressure manifests itself in several ways. Headache is the most obvious symptom. This headache is often worse in the morning, after a patient has been recumbent for several hours, and better later in the day, after the patient has been in the upright position. Increased pressure also causes nausea and vomiting. Often, the vomiting is sudden and projectile. Pressure on the optic nerve can cause diminution of vision and creates swelling of the nerve called papiledema, which can be seen when the eye is viewed through an ophthalmoscope. When intracranial pressure increases greatly, it can cause a decreased level of consciousness and eventually lead to coma and death.

There is another way that brain tumors can cause increased intracranial pressure, by blocking the normal flow of cerebral spinal fluid (CSF). All of us manu-

ENCYCLOPEDIA OF HUMAN BIOLOGY, Second Edition, VOLUME 2. Copyright © 1997 by Academic Press. All rights of reproduction in any form reserved.

facture a clear, water-like fluid that constantly bathes the brain. This is called cerebrospinal fluid and it is formed in hollow chambers in the center of the brain, called ventricles. Under normal circumstances this fluid flows out of the center of the brain through specific channels and then up over the top of the brain where it is reabsorbed into the bloodstream. A tumor can block this normal pattern of CSF flow, leading to an accumulation of fluid and increased pressure. This condition is called hydrocephalus, commonly known as water on the brain. [*See* Cerebral Ventricular System and Cerebrospinal Fluid.]

The symptoms created by direct effect of the tumor on the surrounding brain are legion and are dependent on which area of the brain that is affected. Tumors pressing against the motor area can cause a partial paralysis, similar to that seen with a stroke. Tumors near the speech area can cause aphasias, whereas tumors in the cerebellum can cause tremors and incoordination. Frequently, family and friends note changes in personality or functional abilities, of which the patient himself is unaware. Shy, reserved individuals can become loud and uninhibited, whereas the outgoing and gregarious can become sullen and withdrawn. Tumors around the pituitary gland can lead to a hormonal imbalance that can mimic endocrinological disorders. Virtually any neurologic dysfunction can be related to a brain tumor, if it is located in the appropriate area.

Many brain tumors come to light after they have caused an epileptic seizure. In fact, if an adult presents with the first seizure of his life, brain tumor is the most likely diagnosis. Seizures can be full-blown convulsions known as "grand mal" or much less dramatic events such as "absence attacks," where a patient develops a blank stare and loses contact with his surroundings for several seconds. Although grand mal epilepsy is easy to recognize, some seizure activity can be subtle behavioral aberrations, the epileptic nature of which can only be determined by an electroencephalogram (brain wave test or EEG).

II. DIAGNOSIS

In the past, the diagnosis of brain tumors was difficult, depending on a combination of detailed neurologic examinations and cumbersome and invasive radiographic procedures, such as angiography (looking at the blood vessels of the brain on X-rays) and pneumoencephalography (taking X-rays of the skull after air has been injected through a spinal tap or a hole drilled in the skull).

Today, the diagnosis is made by a computerized axial tomography (CAT) scan or a magnetic resonance imaging (MRI) scan. The first of these tests was introduced in the early 1970s and the quality of the images produced has continuously improved. Not only has the quality increased, but so has their availability. Almost every hospital in the United States has a CAT scanner, and there are few Americans who do not live within a short drive of a MRI scan.

The CAT scanner works by taking thousands of small X-rays and using a computer to create a picture of the whole brain. This creates discrete "slices" of the brain that can be viewed from top to bottom. Intravenous contrast material is given to the patient to help delineate a tumor. Normally there is a blood–brain barrier that prevents molecules the size of the contrast material from passing from the blood vessels into the brain, but brain tumors break down this blood–brain barrier and allow the contrast to pass through into the tumor, creating a very bright spot on the scan (Fig. 1).

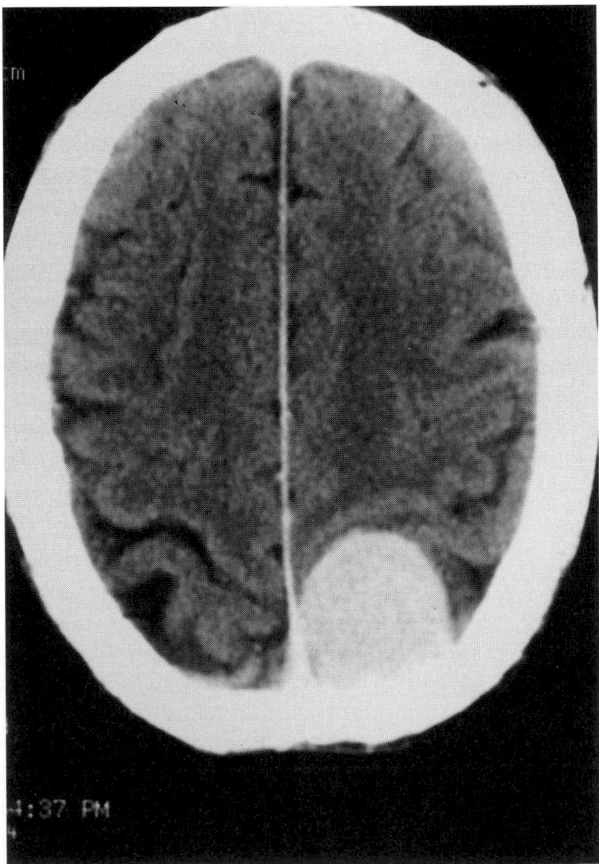

FIGURE 1 CAT scan of patient with meningioma.

The MRI scan synthesizes information in a similar way, but uses a large magnetic field rather than small X-rays. Originally known as nuclear magnetic resonance (NMR), the technique is familiar to many who have used it in a college chemistry course to identify unknown compounds. The name was changed to magnetic resonance imaging because physicians worried that patients would be afraid of a procedure that contained the word "nuclear," but the principle remains the same. The images created by this technology have exquisite detail, rivaling that of an anatomic atlas (Fig. 2).

Anyone suspected of having a brain tumor is now given one of these tests, and the question is answered in a few minutes. Scanning technology can detect tumors as small as 1 or 2 mm anywhere in the brain. Not only can brain tumors be diagnosed this way, but serial scans are used as a way to follow a patient and judge their response to treatment. [*See* Magnetic Resonance Imaging.]

III. TYPES

The diagnosis of the exact type of tumor can only be made histologically. A piece of tumor is obtained by surgical biopsy, and a pathologist examines the cellular pattern under a microscope. There have been a large number of different tumor types described in the brain, but for the purpose of this discussion they will be divided into three categories: astrocytomas, meningiomas, and all the others. The first two tissue types encompass the vast majority of all patients who develop brain tumors. It should be noted that this article deals with primary brain tumors and does not describe metastatic brain tumors, those tumors that come to the brain from cancer elsewhere in the body. These are best considered as part of the systemic manifestations of cancer. Primary brain tumors are confined to the central nervous system and, with exceedingly rare exception, never spread to elsewhere in the body.

A. Astrocytoma (Glioma)

Astrocytes (glial cells) are the support cells of the brain. They supply the structural and nutritional support that the working cells of the brain, the neurons, require to function properly. Tumors that develop from these cells are called astrocytomas or gliomas. These two terms are absolutely equivalent. As men-

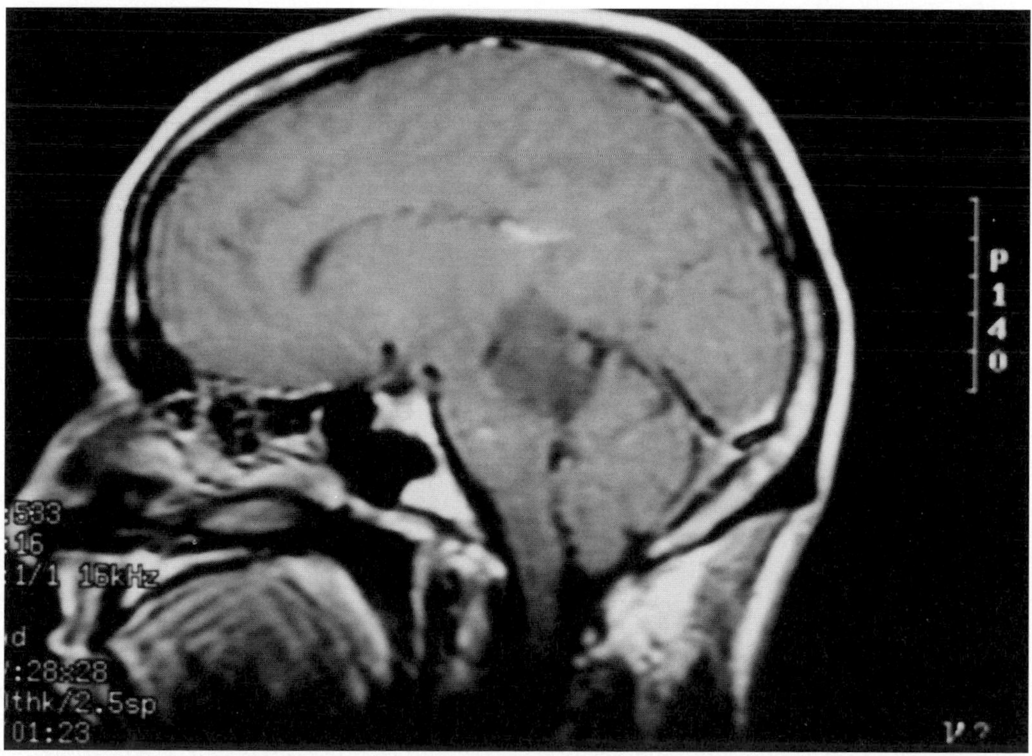

FIGURE 2 MRI scan of patient with anaplastic astrocytoma.

tioned earlier, the symptoms that a brain tumor creates are a function of its size and location, not its tissue type. There is no specific symptom characteristic of a glioma. These tumors are in the brain itself and there is no discrete border between the tumor and the surrounding brain tissue. Frequently the diagnosis can be suspected on the basis of the appearance of the tumor on the scan, but one can never be certain until the tissue is examined microscopically.

Astrocytomas are graded on a scale 1 to 3, depending on the degree of malignancy. A grade 1 astrocytoma is known as "low grade astrocytoma." It is composed of benign appearing cells with low mitotic activity. Although these tumors are indolent appearing and slow growing, they have the potential for evolving into higher grade lesions in time. Grade 2 astrocytomas are called "anaplastic astrocytomas." They are intermediate in their appearance and malignant potential. A grade 3 astrocytoma is referred to as glioblastoma multiforme. This tumor has highly malignant features and is very aggressive.

Oligodendroglial cells are a specific type of astrocytes whose neoplastic manifestation is an oligodendroglioma. Although this type of brain tumor often behaves like a low grade astrocytoma, it can undergo malignant degeneration.

B. Meningioma

The meninges is the term given to the multilayered membrane that covers the brain and spinal cord. Meningiomas are tumors that arise from the cells that comprise the outer and most substantial of these membranes, the dura mater. These tumors can grow into the surrounding brain and deform it, but a discrete boundary remains between these tumors and the surrounding brain. A neurosurgeon can exploit this anatomic fact to remove the tumor while avoiding damage to the surrounding normal brain. These tumors are most always benign, and successful surgical removal usually results in a cure.

C. Other

The overwhelming majority of brain tumors fall into the just-described two categories. There are, however, many other rarer types of brain tumors. The following list describes many of the most often encountered, but is, of necessity, incomplete. Tumors of the pituitary gland are most often benign. They can cause visual problems by compressing the optic nerves or disturbances of other endocrine glands by secreting hormones. Tumors can arise from the pineal gland, in the very center of the brain. They are called pinealomas or pinealcytomas and can be either benign or malignant. Hemangiomas are vascular tumors that have a predilection for the cerebellum.

In children, a whole different set of brain tumors exist. The most common of these is juvenile pilocytic astrocytoma. This is different from the astrocytoma discussed earlier both in its appearance and in its behavior. It is usually discrete and cured by surgical removal. The second major type of pediatric brain tumor is meduloblastoma, referred to as a type of primitive neural ectodermal tumor (PNET). This is a malignant tumor related to similar appearing tumors in other locations, such as retinoblastoma and neuroblastoma. Along with ependymomas, which arise from the cells that line the channels along which the CSF flows, these tumors form the bulk of pediatric brain cancer.

Any specialized cell type in the central nervous system can give rise to a tumor. There are choroid plexus papilomas that come from the cells that manufacture CSF, craniopharyngiomas that come from embryonic rest cells at the base of the brain, gangliogliomas that come from the cells that form neurons, and teratomas and dysgerminomas that evolve near the pituitary and pineal gland. Brain tumors can form along nerves, the most common of which is acoustic neuroma, which is a tumor of the supporting cells of the eighth cranial nerve, the nerve that is responsible for hearing. This tumor is associated with neurofibromatosis (von Recklinghausen's disease).

IV. TREATMENT

Most brain tumor patients do receive medical therapy. This usually consists of a steroid medication, the most common being dexamethasone. The steroids affect the junction between the capillaries and the brain surrounding the tumor, and they have been very effective at reducing the swelling associated with brain tumors. This swelling is called cerebral edema, and its alleviation often results in clinical improvement. However, this is temporary and does not obviate the need for definitive treatment. The amount of time that a patient stays on steroids, after surgery or radiation, ranges from a few days to several months or even years. A second type of medicine used in brain tumor patients is anticonvulsant medication. Any patient who has had a seizure and those that are felt to be at risk for developing epilepsy are given drugs such as diphenylhydantoin (Dilantin), carbemazapine (Tegretol), and phenobarbitol.

The cornerstone of treatment for brain tumors is surgery. There are several goals of surgical therapy. First, one must establish a tissue diagnosis. This means that a sufficient amount of tumor must be given to a pathologist so that it can be examined under a microscope. In some cases this is accomplished by a procedure known as stereotactic biopsy, wherein a rigid frame or some sort of fiducial marker is fixed to the patient's skull. A CAT or MRI scan is then obtained which localizes the tumor and establishes its relationship to the frame. Since the brain cannot move in relation to the frame, coordinates can be established that define the exact location of the tumor. The patient can then be taken to the operating room where a small hole is drilled in the skull, often under local anesthesia, and a biopsy forceps placed at the precise location where one wishes to obtain a tissue sample. These techniques can attain accuracy of less then 1 mm.

More often, an attempt is made to remove the tumor. Usually, this is accomplished under general anesthesia, but can be performed, in certain settings, with the patient awake. An armamentarium of technological advances aid the neurosurgeon in this regard. The first step in any brain surgery is to incise the scalp and make a hole in the skull. Once the hole is created, a power saw, much like a jigsaw, can be used to create a bone flap of any desired shape and size. Magnification using an operating microscope improves the visualization of the tumor and the surrounding brain, and is used to facilitate the removal of tumors from small delicate structures. Forceps that carry electrical current are extremely helpful in cutting tissue and stopping bleeding. Ultrasonic tumor aspirators and lasers also assist the surgeon. The mortality for modern brain tumor surgery is well below 5%.

The extent to which a tumor can be removed is determined by what type it is and its anatomic location. For example, meningiomas are much more amenable to complete surgical removal than astrocytomas, and tumors on the surface of the brain's cortex are much easier to remove than tumors wrapped around the optic nerve.

Another type of surgery that is sometimes performed in brain tumor patients is the placement of a shunt. As mentioned earlier, brain tumors can block the normal flow of CSF and create hydrocephalus. When this happens, a plastic tube can be implanted in the brain ventricle and the fluid can be shunted, usually through a valve placed under the scalp, into the abdominal cavity or a large vein near the heart.

In patients where surgery is not curative, adjunct therapies are available. Radiation therapy has proved to be highly effective in some tumors, such as dysgerminoma, and is somewhat effective in astrocytomas. There is a cumulative life time effect of radiation on the surrounding normal brain, which limits its use. Most patients with astrocytoma undergo radiation therapy at some point. Although radiation therapy is usually not curative, it can often cause tumor regression and delay tumor recurrence. Most patients tolerate radiation therapy well; the major side effect is hair loss.

A newer technique used to deliver radiation to tumors is called stereotactic radiosurgery, sometimes known by the trade name Gamma Knife. In this instance, a stereotactic frame is used as it is in the biopsy, and a scan is performed. Instead of taking tissue, a computer-planned, highly focused radiation beam is delivered without making an incision. A precisely defined area can be treated with limited spread to surrounding tissues. Although it is effective in some tumors, such as acoustic neuromas, its value in other types is less certain.

Chemotherapy has been used for many years in brain tumors. The most common class of drugs used are the nitrosoureas such as CCNU and BCNU. The efficacy of these drugs in extending life has been well established, but they are rarely curative. [*See* Chemotherapy, Cancer.]

A number of experimental therapies exist: immune therapy, gene alteration, and a host of newer drugs that are currently in the testing stage.

V. OUTCOME

Meningiomas are almost always benign. Once they are surgically removed, most patients are cured. However, there are some patients who experience recurrence of their tumors, sometimes years after successful surgery. Therefore, many physicians recommend continuing follow-up with CAT scans or MRI scans. These tumors rarely turn malignant or recur multiple times. In these cases, radiation therapy is often given.

Astrocytomas are a very different situation. Despite the best efforts of neurosurgeons, oncologists, radiation therapists, and others, patients with this diagnosis are very rarely cured of their tumor. The average survival of low grade astrocytoma patients is in the range of 7 to 10 years, that for anaplastic astrocytoma is about 3 years, and that for glioblastoma multiforme is about 1 year. Of course, one must remember that these averages are created by studying some patients who have died immediately after surgery and some who have lived for 20 years. There have been reported cases of patients who have been cured, but these are rarities. Patients and families must come to terms with the fact that this diagnosis usually means a limited life expectancy.

As far as the other tumors mentioned, the following is a brief description of their prognosis. Pituitary tumors are usually cured or controlled by surgery and/or radiation, although many patients require hormone replacement. Pineal gland tumors are too variable in their prognosis to categorize. Childhood juvenile pilocytic astrocytoma is usually cured with surgery alone. The PNET type of tumors requires surgery and chemotherapy and sometimes radiation as well. Today, as many as 50% of the patients with pediatric brain cancer may be tumor free 5 years after diagnosis. Choroid plexus papilomas, hemangiomas, and craniopharyngiomas are often cured with surgery alone. Teratomas and dysgerminomas are often controlled with surgery and/or radiation therapy. Acoustic tumors can be cured with surgery or controlled with stereotactic radiosurgery.

BIBLIOGRAPHY

Ausman, J. I., French, L. A., and Baker, A. B. (1995). Intracranial neoplasms. *In* "Clinical Neurology" (R. J. Joynt, ed.), Vol. 2, Chapter 14. Lippincott-Raven, Philadelphia/New York.

Deveaux, B., O'Fallon, J., and Kelly, P. (1993). Resection, biopsy and survival in malignant glioma neoplasms. *J. Neurosurg.* **78**, 767–775.

Lunsford, L. D. (1994). Contemporary management of meningiomas. *J. Neurosurg.* **80**, 187–190.

Piepmeir, J., *et al.* (1996). Variations in the natural history and survival in patients with supratentorial low grade astrocytomas. *Neurosurgery* **38**, 872–879.

Schiller, F. (1996). Early approaches to brain tumors. *Neurosurgery* **38**, 1023–1030.

Breast Cancer Biology

SAM C. BROOKS
ROBERT J. PAULEY
Wayne State University School of Medicine and Karmanos Cancer Institute

GLOSSARY

Allelic loss or loss of heterozygosity Absence of one of two distinguishable alleles at a heterozygous locus in tumor DNA as compared to nontumor DNA

Antioncogene, recessive oncogene, or tumor-suppressor gene Gene whose functional product diminishes the likelihood of transformation of normal cells to tumor cells

cDNA Complementary DNA synthesized *in vitro* from an mRNA template rather than a DNA template

Epigenetic Describes changes in the genome that do not alter the structure of the genome per se, but affect its expression

Growth factor Molecule, usually a polypeptide, that influences cellular proliferation and differentiation through binding to a specific high-affinity cell membrane receptor

Hormone-responsive element DNA element that is recognized by the receptor DNA binding domain and is a 15-base pair (bp) palindrome comprising two 6-bp arms separated by a 3-bp spacer

Oncogene Gene whose product potentiates the transformation of a normal cell to a tumor cell

Protooncogene Normal gene that, by alteration, can become an oncogene

Zinc finger The DNA binding domain of steroid receptors is rich in cysteines and basic amino acids. Four cysteine residues are coordinated with a zinc ion, forming a positively charged loop (i.e., a finger) with the intervening amino acids. Two such fingers are formed in the DNA binding domain, one of which is positively charged (basic amino acids)

BREAST CANCER IS A TUMOR OF THE MAMMARY gland epithelium occurring almost exclusively in women. The tumor arises from a hyperplastic growth of the epithelium, comprising the ducts, terminal ducts, or ductules. Although the exact causes of the human disease remain unknown, breast cancer has been shown to be related to endocrine factors, environmental agents, genetic anomalies, and socioeconomic discrepancies. These neoplasias are complex, displaying heterogeneity in cellular composition, hormone dependence, karyotype, genotype, metastatic potential, and prognosis. Unfortunately, the rate of death due to this disease in populations of the Western Civilizations has been unchanged over the past three decades.

I. INTRODUCTION

Breast cancer is a major disease of women in the Western Hemisphere. More than 130,000 women in the United States develop this disease each year, which represents approximately 30% of all cancer in

ENCYCLOPEDIA OF HUMAN BIOLOGY, Second Edition, VOLUME 2. Copyright © 1997 by Academic Press. All rights of reproduction in any form reserved.

women. Most disturbing is the fact that the incidence of this disease has not decreased in the past 35 years.

Several risk factors associated with increased incidence of primary breast cancer appear to be endocrine related, including (1) first full-term pregnancy after the age of 30, (2) early menarche or late menopause, and (3) nulliparity (i.e., the condition of having never given birth to a viable infant). On the other hand, oophorectomy before the age of 35 has been associated with a lower incidence of breast cancer. In addition, a genetic factor might be involved in a small percentage of these patients, as evidenced by the existence of high-risk families with a history of breast cancer. Furthermore, rates of breast cancer vary strikingly among countries, and migrants moving from low- to high-risk countries adopt the rates of the new habitat. Thus, the environment appears to have a role in this disease. Neither the use of birth control pills nor the use of estrogens in postmenopausal women has been reliably associated with increased risk for the development of breast cancer.

The hormones that regulate breast tissue have been identified. In humans it is believed that estrogen is essential for the growth of the ductal epithelium of the breast and that progesterone is required for the development of acini. Maintenance of human mammary glands also involves growth hormone and prolactin. Additionally, tissue culture investigations have implicated cortisol, insulin, and certain growth factors in the proliferation of human breast epithelium.

It would appear that the hormonal milieu or, possibly, the past exposure of breast tissue to hormonal influences could play a role in the initiation of breast tumors. Though the exact nature of hormonal involvement in the human disease is unknown, certain extrapolations might be made from experiments with laboratory animals and the culture of breast cancer cells. In the Sprague–Dawley rat, for example, estrogens must be present for the initiation and growth of mammary tumors induced by chemical carcinogens. More direct evidence has been gained from experiments with cell cultures of neoplastic human breast epithelium (MCF-7 and ZR75-1). The growth of these cells *in vitro* is enhanced by estrogen, and their *in vivo* growth, when transplanted into athymic mice, depends on the presence of estrogens, as well as prolactin.

Like carcinomas of other endocrine target tissues, the growth and progression of breast tumors to more malignant stages can be influenced by hormones and antihormones. Moreover, there is direct and indirect evidence that hormones have a role in the etiology of breast cancer. Finally, recent evidence implicates prolonged dietary animal fat consumption in the incidence of breast cancer. Although not conclusive, these studies show a relationship between the occurrence of breast tumors and the level of saturated fatty acids in the diet, as well as the total calories consumed.

II. DIFFERENTIATION OF THE MAMMARY GLAND AND SUSCEPTIBILITY TO CARCINOGENESIS

A. Differentiation

The human mammary gland is a complex organ. Under the influence of body growth and hormonal stimulation, its glandular structure undergoes continuous changes from birth to senescence. At birth the mammary gland contains primitive lobular structures composed of ducts and ductules lined by a single layer each of epithelial and myoepithelial cells. Throughout childhood there is little change in these structures other than growth commensurate with general body growth.

Puberty initiates the most dramatic changes in the mammary gland (Fig. 1), beginning with the growth of glandular tissue and the surrounding stroma. At this time the glandular tissue forms bundles of primary and secondary ducts undergoing repeated bifurcation. Ducts ultimately grow into terminal end buds, which proceed to form alveolar buds, a primitive form of the mature resting acinus. The alveolar buds cluster around the terminal duct in a structure termed a type 1 lobule, which contains approximately 11 alveolar buds. These buds are lined with two layers of epithelial cells, whereas the terminal end buds are lined with four layers of cells. [*See* Puberty.]

In the adult woman, normal breast epithelium undergoes cyclic variations in proliferation, which can be measured by DNA labeling with tritiated thymidine [termed the DNA-labeling index (DNA-LI)]. Diminished DNA-LI has been observed during the follicular phase of the menstrual cycle, followed by a significant increase in the luteal phase. Cellular proliferation and cell death appear to balance in the resting breast. However, the mammary development induced by ovarian hormones throughout the menstrual cycle never fully returns to the level of the preceding cycle. Accordingly, each cycle slightly fosters mammary development, with new budding of structures occurring continuously until about age 35. This development is

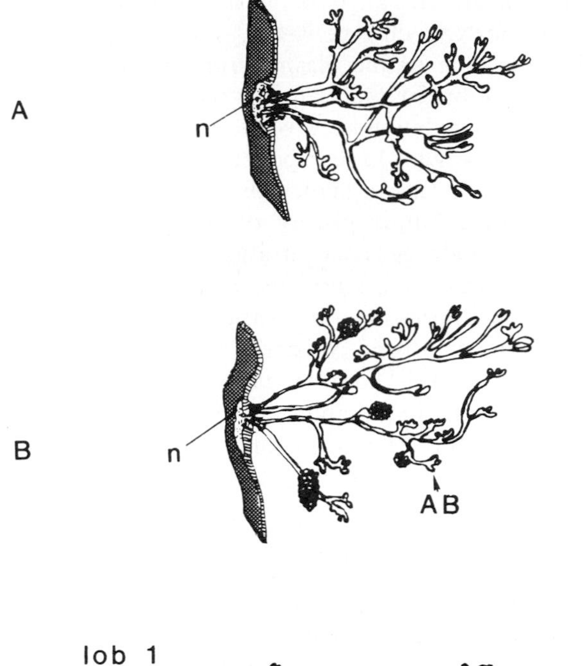

FIGURE 1 Breast development. (A) At puberty or during its onset, the ducts grow and divide in a dichotomous sympodial basis, ending in terminal end buds. (B) After the first menstruation, the initial lobular structures appear (lobule type 1); these are composed of alveolar buds (AB). Some branches end in terminal end buds or terminal ducts. (C) The number of lobules increases with age, and in the adult nulliparous female breast three types of lobules can be found (lobule types 1, 2, and 3). n, nipple; lob, lobule. [From M. C. Neville and C. W. Daniel (eds.) (1987). "The Mammary Gland," by permission of Plenum Publishing Corporation, New York.]

expressed in the appearance of two additional types of lobules in the breast of adult women. These are designated types 2 and 3 (see Fig. 1).

The gradual development of type 1 lobules into type 2, followed by further differentiation into type 3, is a process of sprouting of new alveolar buds that reach approximately 47 in number (type 2 lobules) and as many as 80 in type 3 lobules. Type 1 lobules are predominantly found in the breasts of nulliparous young women, whereas types 2 and 3 lobules are more frequent in the breasts of parous women. Measurement of the DNA-LI of these structures indicates that the proliferative activity of the breast epithelium decreases appreciably from the most active terminal end bud through each lobule type.

The systemic hormonal patterns that accompany pregnancy stimulate the breast to attain its maximum development. Growth is initiated that is characterized by proliferation of the distal elements of the ductal tree, resulting in the formation of new ductules and bringing about the development of type 3 lobules to a level of budding and degree of lobule formation beyond that seen in a virginal breast. Following pregnancy the parous organ contains more glandular tissue than if pregnancy had never occurred. At menopause, involution of the glandular tissue occurs.

B. Susceptibility to Carcinogenesis

The development of experimental systems, principally the rat 7,12-dimethylbenz[*a*]anthracene (DMBA)-induced mammary tumor model, has enabled investigators to pinpoint the site of tumor induction in the breast. The similarity of mammary gland growth and differentiation between the rat and human mammary glands is significant; therefore, it is reasonable to assume that a comparable state exists during tumor initiation in these species. Furthermore, as has been demonstrated in the human disease, there is an endocrine (i.e., estrogen, progesterone, and prolactin), dietary (i.e., fat), parity, and genetic influence on the initiation of mammary neoplasia in the rat model.

Treatment of virgin female rats with the carcinogen DMBA at the age when the terminal end buds begin to differentiate into alveolar buds (i.e., 35–42 days) ultimately results in the greatest number of mammary tumors. The observation that mammary carcinomas arise from undifferentiated structures such as the terminal end bud indicates that the carcinogen requires an adequate structural target for the induction of neoplastic lesions. Benign lesions have been associated with the more differentiated mammary structures (e.g., the alveolar buds) following carcinogen administration. The higher susceptibility of terminal end buds to neoplastic transformation is attributed to the fact that this structure is composed of actively proliferating epithelium. Furthermore, autoradiographic studies show that the greatest uptake of tritiated DMBA occurs in the nucleus of epithelial cells of the terminal end buds, indicating that the highest DMBA–DNA interaction is associated with the structure with the highest proliferative rate. This observation has been corroborated by *in vitro* experiments using human breast tissue.

Terminal end buds have also been demonstrated to metabolically produce more polar metabolites of the carcinogen than the differentiated lobular cells. Such polar metabolites are required for the carcinogenic action, since they are associated with the formation of DNA adducts during carcinogenesis (see Section III,C). Furthermore, the removal of these adducts (i.e., DNA repair) is less efficient in the terminal end buds, thereby facilitating tumor induction.

III. STEROID HORMONE METABOLISM

A. Androgens

More than three decades ago, investigators were attempting to relate the steroid content of a breast cancer patient's urine to the prognosis of her disease. The outcome of this extensive effort was the discovery that, as a consequence of nonspecific illness or exposure to therapeutic drugs, the metabolic fate of endogenous steroids in many patients is altered, resulting in diminished adrenal androgen excretion. A prospective study of 5000 apparently normal women on the island of Guernsey in the English Channel has showed the prediagnosis excretion of low levels of the urinary metabolite of adrenal androgens, etiocholanolone, to be correlated with the patient's eventual presentation with breast cancer. This well-controlled comprehensive investigation, which spanned some 20 years, established that urinary androgens are indeed decreased in the breast cancer population, both before and after diagnosis. This finding might reflect a diminution of the urinary 17-ketosteroid (etiocholanolone) levels via an increased hepatic steroid hydroxylation, principally at position 16α. [*See* Steroids.]

This specific metabolic pathway could result from the induction of certain hepatic mixed-function oxidases (e.g., cytochrome P-450) by stress or drugs, bringing to mind the earlier studies of chronically ill patients. Nevertheless, the discovery, in laboratory animals, that the inducibility of certain cytochrome P-450 systems is genetically determined allows speculation that certain women destined to develop breast cancer are predisposed to elevated levels of hepatic mixed-function oxidase activity over an extended period. The continuous bathing of breast tissue with an altered pattern of hepatically influenced plasma steroid metabolites might be causally related to breast cancer. [*See* Cytochrome P-450.]

Examination of the plasma from women with a high risk of breast cancer (e.g., those displaying low urinary etiocholanolone 5 years before detection of disease, those who have experienced early menarche or a first full-term pregnancy above the age of 30, or those with a family history of breast cancer) has shown low androgen concentrations to be characteristic of the individuals who later present with this disease. This interesting observation, carried out on white women from Western Europe or North America, has not been seen in African or Asian women. Thus, the reliability of this prognostic discriminant for breast cancer could vary among populations.

B. Estrogens

Of all the steroid hormones examined for a possible role in breast cancer, the estrogens have received the most attention. The obvious relationship between estrogens and mammary gland growth and function has prompted numerous laboratories to investigate the influence of this active steroid in neoplastic breast disease. Initially, these studies were limited to the examination of the urinary levels of estrone, estradiol, or estriol in breast cancer patients. The results from these investigations were offset by the discovery of a number of other estrogen metabolites. For the most part, these metabolites were composed of estrogens that had been hydroxylated at various positions on the molecule (e.g., 2- or 4-hydroxyestrogens and 16α-hydroxyestrone). With a more complete knowledge of the metabolic fate of estrogens in the women, it has recently become possible to attempt to relate estrogen metabolism to the disease process.

As discussed in the previous section with respect to the urinary androgen metabolites, hepatic hydroxylating enzymes might influence the pattern of estrogens in the urine and the plasma, a pattern that has been related to breast cancer occurrence. Increased 2-hydroxylation, whether of estrone, estradiol, or estriol, has been associated with a lowered risk of breast cancer. On the other hand, enhanced 16α-hydroxylation of estrone has been linked to the propensity of certain women to develop breast cancer and is elevated in the urine of breast cancer patients. In addition, this reactive metabolite of estrone is formed in mice infected with the mammary tumor virus, which induces breast cancer in these animals. It has been proposed that heightened 16α-hydroxylation of estrogens at the expense of 2-hydroxylation might be a metabolic pattern of women with breast cancer or at high risk for breast cancer.

Although the levels of numerous other estrogen metabolites have been examined in the urine and the plasma, only the elevation of 16α-hydroxylated estrogens in systemic fluids remains related to neoplasia of the breast. It is known that the reactive α-hydroxyketone structure on carbons 16 and 17 of 16α-hydroxyestrone will form covalent linkages with amines, sulfhydryl groups, and the guanine moiety. In view of the fact that this metabolite binds efficiently to the nuclear estrogen receptor, it is postulated that 16α-hydroxyestrone might react with informational macromolecules within the nucleus, resulting in transformation of the breast epithelial target cell.

Steroids are also metabolized by the tissues of breast tumors. For the most part this metabolism serves to deactivate the hormones or their precursors that enter the tumor from the plasma. Once within the neoplastic target tissue, estradiol might be bound by its specific nuclear receptor, oxidized to estrone, or esterified to sulfate at the 3-phenolic hydroxyl. Both of these metabolic products are considered to be attenuated or inactive estrogens, since they do not bind to receptor at physiological concentrations. Possibly of greater importance is the observation that certain breast tumors contain the enzyme aromatase and are therefore capable of converting common plasma steroids such as dehydroepiandrosterone and androstenediol into estradiol. It appears that aromatase in tumor stromal cells contributes the major portion of estrogen synthesized in breast tumors and that this local synthesis may increase estradiol levels and affect growth rates.

C. Steroid Hormones as Carcinogens

To date, extensive investigations of a possible role of steroid hormones in the etiology of breast cancer have produced considerable data regarding the association of these hormones in a process that ultimately results in neoplasia. However, the available data do not support steroid hormones as causative. Present theories of carcinogenesis stipulate that the responsible agents form DNA adducts or strand breaks in order to promote transformation. Steroid hormones or their metabolites have not been shown to carry out these prerequisite functions. Furthermore, epidemiological data compiled from populations of women using oral contraceptives do not indicate a direct association between estrogens and the appearance of breast tumors. Finally, estrogen's stimulation of the pituitary gland to secrete prolactin, although implicated in rodent mammary tumorigenesis, has not been shown to be involved in the human disease. It should be noted, however, that metabolites of the antiestrogen tamoxifen have been shown to form DNA adducts in animals and possibly in humans, a contingency that must be considered when using this compound in therapy or prevention of breast cancer (see Section IV,B,4).

IV. STEROID RECEPTORS

A. Receptor Theory

Breast epithelium, like all target tissues for steroid hormones, contains nuclear proteins that bind specific steroids with finite capacity and high affinity. These proteins, termed receptors, are essential to the proliferative activity or the promotion of differentiation brought about by steroid hormones in responsive cells. Distinct receptors have been characterized, each of which is bound with high affinity by a particular steroid hormone. The structural characteristics of each hormone are recognized by the binding site of its receptor. Therefore, a cell that is responsive to a given hormone must contain the receptor specific for that steroid hormone.

The cDNA of each receptor has been cloned, and the makeup of these important gene-regulatory proteins has been determined. A superfamily of nuclear receptors has thus been identified, each containing a ligand binding domain at the carboxyl end (domain E in Fig. 2), a highly conserved DNA binding domain made up of two zinc fingers near the center (domain C in Fig. 2), and a more diverse immunogenic amino end (domain B in Fig. 2).

It is presently envisioned that steroid hormones act on normal or neoplastic breast epithelium by spontaneously diffusing into the epithelial cell from the plasma (see Fig. 2). Upon their appearance in the nucleus, these hormones bind to the steroid binding domain of their specific receptors, thereby derepressing the DNA binding domain, which in turn binds to precise areas of the DNA known as the hormone-responsive elements. This interaction of the receptor complex with DNA results in initiation of transcription of the responsive genes. Specific regions of the receptor, other than the DNA binding domain, are known to be required for the activation of gene transcription. Generally these are subdomains of 14 to 47 amino acids in the B region (transactivation

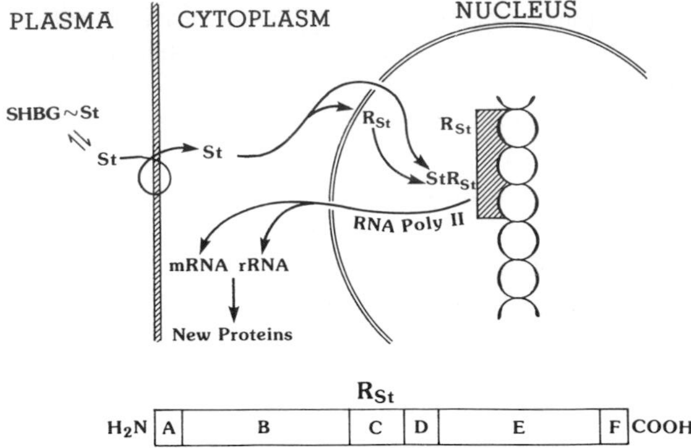

FIGURE 2 The mechanism of steroid hormone action in target tissues. Plasma free sex steroids ($\overline{S_t}$), which are in equilibrium with those bound to sex hormone-binding globulin (SHBG), enter the target cell by diffusion. The specific nuclear binding protein of each steroid is represented by R_{St}, chromatin is indicated by a helix, and specific genes are shown by boxed material. RNA polymerase II (RNA poly II), mRNA, and rRNA are also indicated. One of the "new proteins" specifically induced by the interaction of the estradiol–receptor complex with chromatin is the progesterone receptor. The structural domains of the R_{St} are indicated at the bottom.

function-1) and in the E region (transactivation function-2). The new proteins induced by this process are responsible for proliferation and differentiation of the target breast epithelial cell. [*See* DNA and Gene Transcription.]

Peptide hormones function in breast epithelial cells by binding to the extracellular portion of plasma membrane receptors and initiating the production of secondary messengers such as cAMP, inositol triphosphate, diacylglycerol, or the tyrosine kinase cascade. Prolactin is an example of a hormone that causes production of cAMP. Although the role of prolactin in human breast cancer is not clear, there is ample evidence that the peptide growth factors [i.e., epidermal growth factor (EGF), transforming growth factor-α (TGF-α), and insulin-like growth factor I (IGF-I)] do contribute to the process of neoplastic transformation. Many of these factors, or their receptors, are the products of protooncogenes, which are growth-promoting genes (discussed in Section VII). At this point it will suffice to mention that estrogens have been shown to induce certain of these growth factors (e.g., TGF-α) during their stimulation of breast epithelium, particularly in neoplastic tissue. [*See* Peptide Hormones and Their Convertases; Transforming Growth Factor-α.]

B. Steroid Receptors in Therapy and Prognosis

1. Therapy

The knowledge of the precise mechanisms by which particular agents regulate the growth of breast epithelium has been important to recent developments in the understanding of breast tumor growth, resulting in improved diagnosis and treatment of breast cancer. Examination of a large number of breast tumor biopsies has shown that 75% of the patients bear neoplasias that contain significant levels of the estrogen receptor (Table I). Clinical experience has demonstrated that approximately one-half of the breast tumors that contain a level of estrogen receptor greater than 10 fmol/mg of soluble protein are hormone (i.e., estrogen) dependent for growth. Patients with higher concentrations of estrogen receptors in their tumors (some are greater than 1000 fmol/mg of protein) more often respond favorably to hormonal therapy.

Although the reasons underlying these variations are unknown, there is evidence that each breast tumor is made up of both estrogen receptor-positive and -negative cells. Thus, a higher receptor value could represent a greater percentage of estrogen receptor-containing cells in the biopsy. The heterogeneity of the cellular makeup of breast neoplasias is discussed in Section V.

Tumors devoid of estrogen receptor do not respond to estrogens or the various protocols for hormone therapy and are considered to be hormone-independent neoplasias. Soon after the clinical importance of the level of estrogen receptors in breast tumor biopsies was recognized, it became accepted practice to initiate

TABLE I
Estrogen and Progesterone Receptor Status of Primary Breast Tumors

	Receptor status[a]			
Biopsies	ER^+, PgR^+	ER^+, PgR^-	ER^-, PgR^+	ER^-, PgR^-
Number	112	111	6	74
Percentage	37%	37%	2%	24%

[a]Estrogen receptor-positive (ER^+) tumors contained more than 3 fmol of estrogen binding capacity per 10 mg of tumor tissue. Progesterone receptor-positive (PgR^+) tumors contained more than 10 fmol of progesterone binding capacity per 10 mg of tumor tissue. ER^- and PgR^- indicate breast tumors with binding capacity below the defined amounts.

treatment of patients with estrogen receptor-negative tumors by administering chemotherapeutic agents or radiation.

The empirical experience from decades of patient observation has ascertained that one-third of breast cancer patients respond to hormonal therapy. Furthermore, as pointed out earlier, only one-half of the estrogen receptor-positive tumors respond when the patient is treated by hormonal manipulations. The most likely explanation for this phenomenon is that the neoplastic process in certain estrogen receptor-positive tumors prevents the estrogen receptor complex from its normal interaction with the genes. Indeed, several variant estrogen receptor mRNAs have been detected in both normal and cancerous breast tissues. Exon 3- and exon 7-deleted variants may act as dominant negative regulators of the wild-type receptor, whereas exon 5-deleted estrogen receptor has displayed ligand-independent transcriptional activity. It is clinically important to determine which estrogen receptor-containing tumors will respond to hormonal therapy.

Such a determination can be achieved by recognizing that estrogen target tissues display, as one parameter of response, the induction of progesterone receptor (see Fig. 2). In fact, this finding probably explains the requirement for estrogen priming to render uterine endometrium capable of responding to progesterone. With this in mind, numerous breast tumor biopsies were examined for progesterone receptor, showing that approximately one-half of the estrogen receptor-positive tumors are progesterone receptor-positive (see Table I). Furthermore, the clinical experience has been that breast cancers that contain both estrogen and progesterone receptors respond more often (i.e., 75%) to hormonal therapy. Patients with tumors devoid of both receptors usually do not respond to hormone therapy.

Although this dual-receptor assay is helpful in predicting the response of breast tumors, it must be kept in mind that in many postmenopausal women breast tumors might contain estrogen receptor but no progesterone receptor, because endogenous estrogen is insufficient in amount to bring about induction of the progesterone receptor within the tumor. Such patients, when primed with physiological levels of estradiol, display the induction of progesterone receptors in tumors previously devoid of this protein. Finally, a small fraction (i.e., 2%) of breast cancers has frequently been shown to contain progesterone receptor in the absence of wild-type estrogen receptor (see Table I). This anomaly appears to be associated with the presence of an exon 5-deleted receptor that is constitutively active, that is, that can induce progesterone receptor without ligand.

2. Prognosis

The retention of estrogen receptor in breast tumor cells suggests a more differentiated state of the neoplasia, better prognosis, and longer survival. This, indeed, proves to be true. In fact, the prognosis is even more favorable when estrogen's capability to induce progesterone receptor is retained. More important prognostically is the ability of breast tumor cells to invade regional lymph nodes, which can be determined when the tumor is removed by surgery. Consideration of this metastatic potential of a breast cancer, together with its receptor status, yields a discriminant with the greatest prognostic value (Fig. 3).

The presence of steroid receptors might, however, simply indicate slower growth of these more differentiated tumors. Indeed, retrospective studies and studies carried out for longer periods (i.e., longer than 54 months) indicate that the prognostic advantage of patients with estrogen receptor-positive tumors eventually disappears. These findings could reflect the tendency of breast tumor cells to dedifferentiate and hence lose their capacity to generate the estrogen and progesterone receptors. Such tumors usually display a higher growth rate.

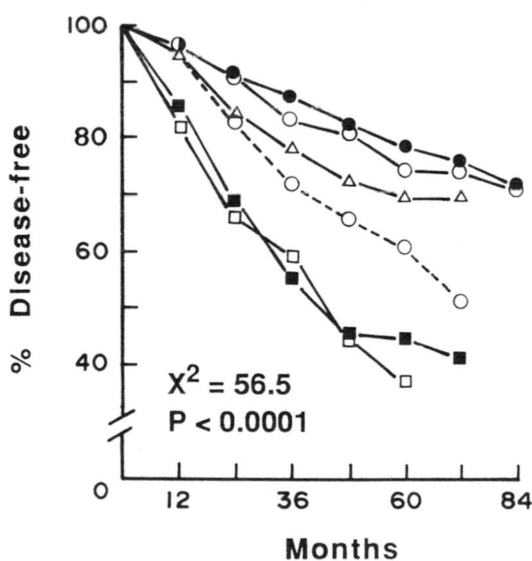

FIGURE 3 Survival and disease-free survival of breast cancer patients based on progesterone receptor (PgR), estrogen receptor (ER), and lymph node (LN) status. (○) PgR^+, ER^+, LN^-; (●) PgR^-, ER^+, LN^-; (△) PgR^-, ER^-, LN^-; (—○—) PgR^+, ER^+, LN^+; (■) PgR^-, ER^+, LN^+; (□) PgR^-, ER^-, LN^+.

3. Metastasis

The receptor and lymph node status of primary breast tumors has proved to be of considerable value to the clinician in designing protocols for the treatment of recurrent disease. Regardless of receptor content, lymph node involvement indicates dissemination of the tumor and calls for systemic hormonal therapy and/or chemotherapy. Virtually all primary neoplasias that are initially steroid receptor-negative remain negative upon recurrence. Receptor-positive cancers usually, although not always, recur as receptor-positive metastatic lesions.

Different patterns of metastatic sites are associated with the estrogen receptor status of neoplastic cells. Estrogen receptor-positive breast tumors predominantly metastasize to bone, whereas recurrent receptor-negative breast tumors most commonly involve the viscera. Metastasis to other soft tissues is the same for receptor-positive or -negative tumors. [*See* Metastasis.]

4. Antiestrogens

Knowledge of the receptor-mediated mechanism of activity of steroid hormones has stimulated the design of compounds capable of interfering with the binding of natural hormones to receptor. Once bound, these agents form a complex that is ineffective in gene activation. Such "antihormones" have been used successfully in the treatment of a number of neoplasias of the endocrine system. The most effective antiestrogens have the structure of triphenylethylenes. They are relatively nontoxic, although not without risk (see Section III,C). They have less affinity for the estrogen receptor than natural estrogens, but at high doses they are capable of binding to the receptor and preventing the binding of estrogen. Typical of the antiestrogens are clomiphene and tamoxifen. The latter compound has been used most extensively, particularly in the treatment of hormone-dependent breast cancer. Tamoxifen is metabolically converted *in vivo* to an even more potent antiestrogen via hepatic hydroxylation at position 4. Importantly, this metabolite is a partial agonist in that it performs as an estrogen in certain tissues. Postmenopausal patients maintained on a tamoxifen regimen benefit by displaying premenopausal levels of plasma calcium and cholesterol, both constituents normally regulated in part by estradiol.

Following the binding of 4-hydroxytamoxifen by the estrogen receptor, the cells are unable to initiate growth or differentiation, which are induced by the natural hormone. Under the influence of this antihormone, estrogen-dependent cells remain in a dormant or nondividing state for long periods. Hormone-dependent tumors will actually regress. Laboratory experiments have shown hormone-dependent neoplastic mammary cells, which have been implanted in mice treated with tamoxifen, to enter an active state of growth after removal of the antiestrogen. Although antihormones are growth inhibitory, they do not appear to be lethal.

V. CELLULAR HETEROGENEITY AND BREAST CANCER PROGRESSION

A. Cellular Heterogeneity

A salient property of breast cancer is the extensive heterogeneity of the cells that compose a single tumor. Some of this heterogeneity is only phenotypic and may be related to discontinuities within the tumor microenvironment. However, genotypic heterogeneity may also be responsible, as shown by the ability to establish multiple distinct, clonal lines from single cancers. Distinct subpopulations of neoplastic cells can be observed in rodent tumors induced by hormones, viruses, or chemicals, and also in human breast cancers. Variations in the cellular characteristics within a single tumor span a wide range of histological and biochemical properties (e.g., ultrastructural features) and biochemical markers (e.g., the expression of casein, keratin, type IV collagen, and a variety of tumor-associated antigens). Of importance to therapy is the heterogeneity in the cellular distribution of hormone receptors, as well as differences in intrinsic sensitivity to drugs and irradiation. Cells composing a single tumor might also show differences in ploidy (i.e., the number of chromosomes) and molecular differences in the genome.

Diversity in the cellular subpopulations of breast cancers can be reflected in the behavior of the neoplasm. Thus, a single rodent tumor can yield cellular clones that differ in their ability to metastasize. Tumor subpopulations can also differ in growth rate, immunogenicity, and, as mentioned, hormone dependency. Of particular clinical significance are the reported differences among tumor subpopulations in response to chemotherapy and irradiation.

Curiously, breast tumors do not usually display the extreme behavior seen among the individual subpopulations isolated from biopsies, suggesting that characteristics of subpopulations can be modified by the other cell types that make up the environment within the tumor. Indeed, subpopulation interactions that modify growth rate, genetic stability, drug sensitivity, and ability to metastasize have been demonstrated directly in experiments with mixed tumor subpopulations. The mechanisms of such interactions include host-mediated events such as the immune response and drug metabolism, as well as tumor-mediated events such as the secretion by one subpopulation of growth factors that affect other subpopulations. The behavior of a neoplasm is not just the sum of the component parts, but rather a reflection of an interlocking network of cellular subpopulations, a cellular society.

B. Tumor Progression

It has been hypothesized that cancer cells are "genetically unstable" in comparison to their normal counterparts. It is therefore conceivable that, within a breast tumor, genetic instability might underlie the emergence of cellular heterogeneity, even if the tumor was of clonal origin. Over time, evolution of new phenotypic characteristics and selection of dominent clones is thought to be responsible for a cancer's becoming refractory to a therapeutic protocol or exhibiting a new property (e.g., metastasis). This change in a tumor's characteristics over time is termed "progression."

Understanding the underlying mechanisms of genetic instability is clearly important to the eventual control of tumor growth and progression. Possibly contributing to this phenomenon are genomic alterations such as point mutations, base pair shifts, gene amplification, chromosomal rearrangements, and ploidy alterations. The appearance of new cellular variants might also arise from epigenetic changes, including normal differentiational processes. Regardless of the mechanism, tumor heterogeneity forces one to view breast cancer as a dynamic interacting tissue, the behavior of which is the result of a multiplicity of factors rather than a series of linear responses or a hierarchy of cause–effect relationships.

VI. GROWTH FACTORS AND GROWTH FACTOR RECEPTORS

The activity of growth factors (GF) and their receptors (GF-R) in normal and neoplastic breast cells is relevant because altered expression and function are associated with breast cancer development and progression. The importance of growth factors and their receptors in breast cancer resides in their control of cell growth, their alteration by mutation to an oncogene, and their potential as targets for therapy. Numerous growth factors have been described in normal and neoplastic breast tissue and cells. Only the best understood and most widely studied are reviewed here.

A. EGF and TGF-α

Although EGF and TGF-α are distinct protein molecules, they are discussed together since both bind to the EGF receptor with the same affinity. TGF-α and EGF are antigenically distinct and have significant, but not complete, sequence homology. EGF receptor, a product of the *EGFR/ERBB1/HER1* gene, is a 180-kDa transmembrane protein characterized by an extracellular ligand binding domain, as well as transmembrane and cytoplasmic tyrosine kinase domains. The cytoplasmic domain contains sites for tyrosine phosphorylation. [*See* Transforming Growth Factor-α.]

EGF and TGF-α both have important roles in normal and neoplastic mammary gland development in rodents and humans. Short-term cultures and established cell lines of normal breast cells, when cultured in a defined medium, require EGF for cellular growth. Biologically active TGF-α has been detected in chemical carcinogen-induced rat mammary tumors, in most primary human breast tumors, and in several human breast cancer cell lines. Murine mammary cells transfected with the TGF-α gene acquire a transformed phenotype *in vitro* (colony formation in soft agar) and tumor formation *in vivo,* indicating that TGF-α is an important determinant of the neoplastic phenotype.

The concept that EGF or TGF-α and the EGF receptor have a role in breast cancer is further supported by observations of EGF production and an increased quantity of EGF receptor in some human breast cancer cell lines. In addition, elevated levels of the EGF receptor in human breast tumors has been correlated with increased rates of proliferation and with a clinically more aggressive breast tumor phenotype. Recently, it has been postulated that TGF-α is an autocrine modulator of human breast cancer cell growth.

B. TGF-β

TGF-β can either stimulate or inhibit proliferation, depending on the cell type. Originally described in retrovirus-transformed rodent cells, TGF-β is distinct from TGF-α in its structure, the biological response it elicits, and the receptor to which it binds. A unique high-affinity receptor for TGF-β is expressed on nearly all cell types, including epithelial cells. [*See* Transforming Growth Factor-β.]

Expression of TGF-β has been demonstrated in normal as well as malignant human breast cells. It has been postulated to have a role in hormone-dependent breast cancer growth because antiestrogens that inhibit the growth of MCF-7 cultures also increase TGF-β secretion. No correlation, however, has been demonstrated among TGF-β expression and hormone receptor status, cellular growth rate, or tumorigenicity.

C. IGF-I and IGF-II

IGF-I and -II correspond to human somatomedins C and A, respectively. Both IGF-I and -II have been postulated to stimulate cellular growth and metabolism in an autocrine fashion. The receptors for IGF-I and -II are structurally and functionally different from each other and from the insulin receptor. [*See* Cell Receptors; Cell Signaling.]

The insulin requirement for the growth of normal murine and human mammary cells *in vitro* is replaced by IGF-I, indicating that IGF-I is acting through the IGF-I receptor. IGF-I receptors are elevated in breast cancer as compared to in normal or benign breast tissue. Human breast cancer cell lines have IGF-I or -II receptor, or both. Furthermore, these cell lines are capable of producing IGF-I, as well as IGF-II, mRNA, and the secreted protein. It is likely, therefore, that the growth characterization and the response of normal and neoplastic breast cells to these growth factors are dependent on the amount of each of the interacting components, IGFs, receptors, and IGF-binding proteins, in the cell itself and in its environment.

D. Other Growth Factors

Although breast carcinoma is derived from epithelial cells, these tumors also contain stromal tissue, which is composed of fibroblasts, endothelial cells, and infiltrating lymphoid cells. The excessive proliferation of stromal fibroblasts, resulting in collagen formation and desmoplasia, might be the result of growth factor production by the transformed epithelial component of the tumor. In fact, breast cancer cell lines produce a growth factor with mitogenic activity that is similar, if not identical, to the platelet-derived growth factor (PDGF). This is a potent growth enhancer for connective tissue-derived cells, which possess high-affinity PDGF receptors. It is possible that fibroblast growth factors (FGFs) or heparin-binding growth factors are important in breast cancer, because a member of the FGF gene family, *FGF3/INT2,* is a murine and a putative human breast cancer oncogene. There is no direct evidence to support a role for FGFs in breast cancer growth, but FGF action on tumor stroma might, in turn, influence breast tumor growth.

VII. GENETIC BASIS OF BREAST CANCER

Breast cancer develops and progresses in large part as a result of multiple somatic alterations to the genome; inherited germ line mutations contribute to 10% or less of breast cancer. The framework in which any mutation, or change in DNA sequence, must be viewed is the effect of the mutation upon the function of a DNA sequence and most importantly upon the amount and activity of the gene product. Chromosomal alterations detected at the level of chromosome number and structure by karyotyping imply genetic change. Alterations detected as amplification, an increase in gene copy number from the normal diploid $2n$ amount, imply increased gene product amount, which along with direct measures of gene product amount and activity suggest function of a dominant-acting oncogene. Alterations detected as allelic loss or loss of heterozygosity and as sequence mutations demonstrate that regions of the genome are altered, implying that gene function is lost and the location of a putative tumor suppressor gene. Additionally, inheritance of a specific chromosomal region containing a mutated gene and exhibiting linkage to breast cancer occurrence in members of families with hereditary breast cancer syndromes indicates that some individuals may inherit susceptibility/increased risk for breast cancer.

VIII. CHROMOSOMAL ABNORMALITIES

Karyotypic characterization of most human breast cancers has been limited to enumeration of the chromosome complement. In general, the modal chromosome number is abnormal (i.e., aneuploid), with few diploid cells. Identification of chromosomal alterations by chromosomal banding has been limited generally to established breast cancer cell lines, because direct karyotyping of tumor biopsies is technically difficult. Karyotypes observed in established cell lines might differ from the primary tumor karyotype. Nevertheless, recurring sites of chromosomal changes in breast cancer have been reported. For example, chromosome 1 has frequently been observed to be overrepresented, and the q arm to undergo translocations. Chromosome 11 is structurally altered in many human breast cancer cells, however, the translocation breakpoints and the chromosomal partner are variable. Chromosomes 6 and 7 and, to a lesser extent, chromosomes 3 and 9, also show frequent structural alterations. Karyotypic changes in breast cancer cells have not consistently implicated specific gene alterations, in contrast to chromosomal alterations in cancers such as Burkitt's lymphoma and retinoblastoma. [*See* Chromosome Patterns in Human Cancer and Leukemia.]

IX. ONCOGENES

A. *HRAS* Gene

Several lines of evidence suggest that alteration of the *HRAS* protooncogene to an oncogene contributes to the neoplastic development of mammary cells. In carcinogen-induced murine mammary tumors, mutational activation to a transforming form of H-*ras* is frequently observed. Also, mouse mammary cells are transformed when infected with the activated viral H-*ras* oncogene, and in transgenic mice containing either the v-*H-ras* or a mutationally activated human *HRASI* oncogene. However, this transformation probably requires cooperating phenotypic changes (e.g., *Myc* expression). [*See* Oncogene Amplification in Human Cancer.]

Human breast cancer cell lines and human breast tumors have been examined for mutationally activated members of the *RAS* gene family (e.g., *HRAS, KRAS2,* and *NRAS*), for amplification of the protooncogenes, and for overexpression of mRNA and p21 protein. Mutational activation has been rarely observed in cell lines and primary breast tumors, and it is thought not to be involved in human breast tumorigenesis. Conflicting reports concerning overexpression of the *RAS* protooncogene at the level of either mRNA or p21 antigen do not allow any conclusion regarding expression of *HRAS1* and *KRAS2* or *NRAS,* and the development of human breast tumors or their phenotypic properties. The observation that immortalized human breast cell lines are weakly transformed or exhibit features of proliferative breast disease by the *H-ras/HRAS* oncogenes suggests that this oncogene might require other concurring alterations for manifestation of the transformed phenotype.

B. *MYC* Gene

A role for the *MYC* oncogene in mammary neoplasia is implied from the development of mammary tumors in transgenic mice containing activated myc constructs. Approximately 30% of human breast tumors contain amplified *MYC* protooncogenes, whereas about 15% have *MYC* gene rearrangements. Neither of these markers has consistently correlated with the behavior of the primary tumor in terms of short-term prognosis for the patient. A high proportion, about 75% of human breast tumors, have higher levels of Myc RNA expression than benign breast tissue, and in many cases gene amplification and rearrangement correlated with enhanced expression. The precise role for Myc expression during breast tumorigenesis has not been elucidated. Elevated Myc expression could reflect the proliferative status of tumor cells and the presence of lymphoid cell infiltrates.

C. *ERBB2/HER2/neu* Gene

The transforming avian v-*erbB* oncogene, the EGF receptor gene (*EGFR*), which is the human counterpart, and the rat *neu* oncogene are closely homologous. The *neu* oncogene encodes a 185-kDa transmembrane domain that differs from the nontransforming c-*erbB2* gene by a single mutation. The observation that transgenic mice with the rat *neu* oncogene have a high incidence of early-appearing mammary tumors has implicated *neu* and its protooncogene counterpart, c-*erbB2,* in the initiation and progression of mammary tumors. The *erbB2/neu* receptor does not bind EGF, and the ligand for this receptor has not been conclusively identified.

Several studies have demonstrated amplification of the chromosome 17q21 *ERBB2* locus in primary tumors and in numerous human breast cancer cell lines. The amplified segment generally includes the *ERBA1/THRA* locus at chromosome 17q22, which is homologous to the viral *erbA* oncogene and related to the thyroid hormone receptor gene *THR.* Amplification and increased ErbB2 mRNA and cell-surface antigen expression occur in approximately 20% of primary breast tumors. *ERBB2* amplification and overexpression might be correlated with shorter disease-free survival in a patient population, but the evidence remains controversial.

D. Other Candidate Oncogenes

Mammary tumors occurring in mouse mammary tumor virus (MMTV)-infected cells contain acquired MMTV proviral DNA that acts as an insertional mutagen activating adjacent *int2* and *int1* protooncogenes. An activated oncogene identified in human stomach tumor DNA transfected into NIH3T3 cells, designated *HSTFI/FGF4,* is proximal to, by about 1 megabase, *INT2/FGF3* on human chromosome 11. These genes are members of the FGF gene family. Human *INT2* and *HSTF1* genes are components of an amplified region of contiguous DNA in approximately 15% of breast tumors. The significance of DNA amplification remains to be determined, because there is only limited evidence for Int2 or Hstf1 RNA expression. Some evidence suggests that *INT2* amplification correlates with poor prognosis in terms of local recurrence or distal metastasis. Also within the 11q13 amplicon is the cyclin D1 (*CCND1/PRAD1*) gene, whose increased expression correlates with amplification. Cyclin D function in the cell cycle is regulated by cyclin-dependent kinases (cdk's) and cdk inhibitors. The precise consequences of increased cyclin expression remain to be proven, but altered cell-cycle regulation is possibly a common feature of transformed breast cells. [*See* Cell Cycle.]

X. TUMOR SUPPRESSOR GENES

The concept that tumor suppressor genes have a role in human breast cancer stems from several lines of evidence. [*See* Tumor Suppressor Genes.] For example, deletions of specific regions and loss of whole chromosomes implies loss of gene function. Mothers of children with osteosarcoma, a second primary tumor observed in individuals with bilateral retinoblastoma and caused by alteration of the *RB1* antioncogene, are at increased risk for breast cancer. Furthermore, there is an increased risk for breast cancer development in first-degree relatives of women with breast cancer, particularly in families with at least two individuals with premenopausal breast cancer (see Section XI). The strategy to identify antioncogenes has involved identification of genes for which there is loss of an allele in tumor DNA as compared to normal cell DNA. The allelic loss is usually detected by the loss of a region on a chromosome, which can be mapped using multiple markers, or mutational analysis of specific tumor suppressor genes and gene products.

Recent studies have shown, in breast tumors and breast cancer cell lines, gene loss at numerous loci, including the *RB1* locus on chromosome 13q14, the p arm of chromosome 11 proximal to *HRAS* and the β-globin locus at p15.5, the q arm of chromosome 1 in the region q23–q32, and *TP53* on the p arm of chromosome 17 in the region of p13.3. Variable frequencies of loss of heterozygosity reported are due in part to long separation between markers and the antioncogene, and to tumors' being mixed populations of cells.

Analysis of a few primary breast tumors and breast cancer cell lines indicates specific chromosome 13 alterations to the *RB1* locus in approximately 15% of the samples examined, based on analysis for Rb expression. In several breast cancer cell lines (e.g., MDA-MB-468, MDA-MB-436, BT-549, and DU4475) there were homozygous internal or 3′ deletions, homozygous total deletions, or duplication of a portion of the *RB1* gene that altered the transcript and protein products. Many human breast cancer cell lines (e.g., MCF-7, BT20, T47D, and ZR-75) lacked *RB1* locus alteration by restriction mapping.

Genetic alteration of *TP53* or *p53* is the most common specific gene altered in breast cancer, implying that it has a major role in tumorigenesis. The wild-type p53 gene product inhibits the transformed phenotype with mutation, resulting in loss of function or for certain mutations function as a dominant acting oncogene product. The p53 protein has a complex role in regulating transcription of growth-regulatory genes and in cell-cycle progression, acting as a checkpoint to prevent cell-cycle progression in cells with damaged DNA and directing cells to apoptosis/cell death. About 40% of primary breast tumors have mutational alteration of *TP53* detected by mutational analysis or increased p53 protein accumulation. However, p53 accumulation is not synonymous with *TP53* mutation, because of complications pertaining to the antibody used, localization of intracellular p53, and interpretation. Nevertheless, p53 accumulation appears to be reliably associated with poor breast cancer prognosis and more aggressive biological features of the tumor, such as higher histological grade, signifying a less differentiated carcinoma, and there are suggestions of poorer response to chemotherapy.

Together, these observations suggest that mutational alteration of the *TP53* and *RB1* antioncogenes is involved in breast tumorigenesis. Importantly, reexpression of normal Rb1 and p53 products in human breast cancer cells with mutationally inactivated *RB1* and *TP53,* respectively, suppresses tumorigenicity on the basis of several criteria. This demonstration and the negative implications of mutated TP53 function for breast cancer make restoration of normal p53 function a goal in breast and other cancers.

XI. INHERITED BREAST CANCER SUSCEPTIBILITY

Family history is a strong risk factor for breast cancer, because about 5–10% of breast cancers occur in individuals with a family history, implicating an inherited genetic susceptibility. Investigators have sought to identify genes linked to family history in these families using the tools of classic and molecular genetics.

Categorization of breast cancer families as familial or hereditary requires rigorous criteria because breast cancer can be coincident by chance, not genetically inherited, in a family due to the 1 in 8 lifetime risk for breast cancer in women. Familial breast cancer indicates families with two or more first- or second-degree relatives with breast cancer, but because breast cancer occurrence in these families lacks clear Mendelian segregation or genetic inheritance, it is probable that breast cancer is related to common environmental experiences (e.g., dietary factors), biological factors (e.g., reproductive history), or coincidence. Hereditary breast cancer indicates families with two or more first- or second-degree relatives with breast cancer, whose breast cancer segregates as an autosomal dominant trait; additional important features are early age of cancer onset, an excess of bilateral and multifocal breast cancer, and sometimes other related cancers. The hereditary breast cancer syndromes include hereditary breast-ovarian, hereditary breast, and Li-Fraumeni syndromes; in Li-Fraumeni, inherited mutations of the *TP53* gene are usually linked to cancer occurrence. However, as illustrated in the 1990 study implicating a chromosome 17q locus in hereditary breast cancer, variation within category of hereditary breast cancer exists because linkage was most strongly associated with breast cancer in families having a median age of breast cancer diagnosis of 45 years or less.

To date, chromosome 17q and 13q loci have been linked to hereditary breast-ovarian and breast cancer syndromes, respectively. This led to identification of two inherited breast cancer susceptibility genes, *BRCA1* and *BRCA2,* respectively. Both *BRCA1* and *BRCA2* are very large genes, encoding proteins of 1863 and 3418 amino acids, respectively. Because they were identified only in 1994 and 1995, respectively, there are more questions than answers about gene product function and how altered function relates to breast cancer susceptibility. Humans are diploid for these genes. Sequence comparisons suggest that the Brca1 protein may interact with DNA and/or proteins to effect transcription, but evidence for the nuclear localization, cytoplasmic localization as a granin, and multiple protein products complicates interpretation and necessitates definitive biological experiments to prove function. Several lines of evidence implicate normal function as tumor suppressor genes, with germ line inheritance of a mutated allele meaning that total loss of function requires somatic mutation of the normal allele.

Mutations at numerous sites in each gene, for *BRCA1* over 200 to date, have been identified in families with hereditary breast cancer syndromes, with most having been identified in only one or two families. Although rigorous criteria have generally been used to restrict these mutations to those causing loss of function, not all inherited mutations appear to increase breast cancer risk. About two-thirds of the occurrences of breast cancer in some hereditary breast cancer families are associated with inheritance of a mutated allele of *BRCA1* and *BRCA2* genes. Of importance is the *BRCA1* exon 2 codon 185 AG deletion (185delAG) identified in more than 20 Jewish families with breast or ovarian cancer family history. In Ashkenazi Jews evaluated without regard to breast cancer family history, the prevalence of this mutation is about 1%, a frequency about 10-fold higher than the frequency of inherited *BRCA1* mutations in young women with breast cancer in the population. Evidence that *BRCA1* mutation is associated with an estimated lifelong risk of about 85% and 40% for breast and ovarian cancer, respectively, implies that *BRCA1* mutation is associated with a significant but not absolute lifelong risk. Whether these lifelong risk estimates established from families with multiple early-onset members will apply to those without clear early-onset hereditary breast cancer remains to be established. At this point, the presence of a *BRCA1* mutation in multiple hereditary breast cancer family members may be informative of risk, but other factors, such as reproductive and hormonal history, may affect the actual occurrence of breast cancer in an individual with a *BRCA1* mutation.

There is considerable impetus to provide, and controversy about, presymptomatic testing for *BRCA* mutations. The ultimate guideline for testing should be that the benefits must exceed the hazards; most benefits appear to be potentially substantial but perhaps not all hazards are recognized. Profound uncertainty exists about the benefits realized by women with detected inherited breast cancer susceptibility gene mutations. For example, genetic counseling between caregiver and recipient is complex and requires increased education. Intensified breast cancer screening, by self-examination and mammography, are advised but are of unclarified benefit for women less than 50 years of age. Prophylactic mastectomy is an option considered and taken by some, whose risk benefit for cancer must be weighed against the fact that susceptible epithelial tissue remains after surgery. Further, treatment regimens, including chemoprevention and postcancer therapies, need to be identified that may be optimally beneficial for this group. Weighing against the possible advantage of better risk estimate are concerns about health insurance and employment discrimination, the psychological and functional health consequences on the person and family, and the limitations of test results because some mutations are not associated with breast cancer and some implied mutations are not detectable. Current evidence that inherited *BRCA* mutations are rare in women with sporadic/nonhereditary breast cancer, who represent the majority of women with a 12% breast cancer lifetime risk, indicates that testing for inherited *BRCA* mutations in most women will not be informative. Testing currently is recommended only to members of high-risk, human breast-ovarian cancer families involved in research protocols. As benefits and hazards become clarified through investigation, guidelines for testing will evolve to permit women in families with hereditary breast cancer to make major decisions with greater and more reliable information.

BIBLIOGRAPHY

Brooks, S. C., and Singhakowinta, A. (1982). Steroid hormones in breast cancer. *In* "Special Topics in Endocrinology and Metabolism" (M. Cohen and P. Foa, eds.), Vol. 4, p. 29. Liss, New York.

Brunner, N., Zugmaier, G., Bano, M., Ennis, B. W., Clarke, R., Cullen, K. J., Kem, F. D., Dickson, R. B., and Lippmann, M. E. (1989). Endocrine therapy of human breast cancer cells: The role of secreted polypeptide growth factors. *Cancer Cells* **1**, 81.

Callahan, R., and Campbell, G. (1989). Mutations in human breast cancer: An overview. *J. Natl. Cancer Inst.* **81**, 1780.

Collins, F. (1996). *BRCA1*—Lots of mutations, lots of dilemmas. *N. Engl. J. Med.* **334**, 186.

Dickson, R. (1995). The molecular basis of breast cancer. *In* "Molecular Biology in Cancer Medicine" (R. Kurzrock and M. Talpaz, eds.), p. 241. Oxford Univ. Press, New York.

Ethier, S. P., and Heppner, G. H. (1987). Biology of breast cancer *in vivo* and *in vitro*. *In* "Breast Diseases" (J. R. Harris, S. Hellman, I. C. Henderson, and D. W. Kime, eds.), p. 135. Lippincott, Philadelphia.

Hall, J. M., Lee, M. K., Newman, B., *et al.* (1990). Linkage of early-onset familial cancer to chromosome 17q21. *Science* **250**, 1684.

Harris, J. R., Lippman, M. E., Veronesi, U., and Willett, W. (1992). Breast cancer. *N. Engl. J. Med.* **327**, 319, 390, 473.

Hines, N., and Dickson, R. (1996). Molecular aspects of breast cancer. *J. Mammary Gland Biol. Neoplasia* **1**, 137.

Kidwell, W. R., Monaham, S., and Salomon, D. S. (1987). Growth factor production by mammary tumor cells. *In* "Cellular and Molecular Biology of Mammary Cancer" (D. Medina, W. Kidwell, G. Heppner, and E. Anderson, eds.), p. 239. Plenum, New York.

Miki, Y., Swensen, J., Shattuck-Eidens, D., *et al.* (1994). A strong candidate for the breast and ovarian cancer susceptibility gene *BRCA1*. *Science* **266**, 66.

Miller, F. R., Soule, H. D., Tait, L., Pauley, R. J., Wolman, S. R., Dawson, P. J., and Heppner, G. H. (1993). Xenograft model of progressive human proliferative breast disease. *J. Natl. Cancer Inst.* **85**, 1725.

Russo, J., and Russo, I. (1987). Development of the human mammary gland. *In* "The Mammary Gland" (M. C. Neville and C. W. Daniel, eds.), p. 67. Plenum, New York.

Russo, J., and Russo, I. (1987). Biological and molecular basis of mammary carcinogenesis. *Lab. Invest.* **57**, 112.

Slocum, H. K., Heppner, G. H., and Rustum, Y. (1985). Cellular heterogeneity of human tumors. *In* "Biological Responses in Cancer" (E. Mihich, ed.), Vol. 4, p. 183. Plenum, New York.

Wolman, S. R. (1987). Chromosomes in breast cancer. *In* "Cellular and Molecular Biology of Mammary Cancer" (D. Medina, W. Kidwell, G. Heppner, and E. Anderson, eds.), p. 47. Plenum, New York.

Wooster, R., Bigness, G., Lancaster, J., *et al.* (1995). Identification of the breast cancer susceptibility gene *BRCA2*. *Nature* **378**, 789.

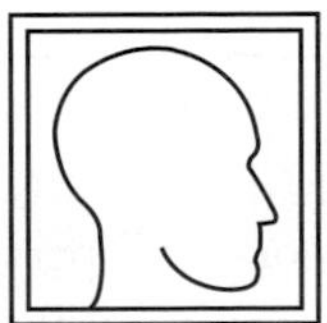

Caffeine

JOSEPH P. BLOUNT
Park College

W. MILES COX
University of Wales, Bangor

GLOSSARY

Caffeinism Another term for "caffeine intoxication," the official diagnostic category for excessive caffeine consumption to a point of intoxication

Double blind Type of research design in which neither the subject nor the experimenter (but only some third party) knows whether the subject has received a drug (e.g., caffeinated coffee) or an inert substance (e.g., decaffeinated coffee). The advantage of this design is that it allows the investigator to isolate the pharmacological effects of the drug from the psychological effects

DSM-IV The current diagnostic manual of the American Psychiatric Association

Epidemiological study Study that deals with the incidence, distribution, and control of a disease in a population

Ergogenic aid Substance that helps an athlete generate increased force or endurance

Etiologic science Science that deals with the causes of a disease or other abnormal condition

Mutagenic risk Factor that increases the chances that a mutation (a permanent change in chromosomes or genes) will occur

Protopathic bias Bias in an epidemiological study resulting from analyses based on nonrepresentative early measures of factors contributing to a disease without recognizing that early manifestations of the disease might have triggered the patient or physician to alter those very factors

State-dependent learning Drug-induced effect on memory that occurs when information that is learned in one drug state is later best recalled when the individual is in the same drug state rather than a different one

Statistical significance Assurance, through a statistical test, that an observed data pattern is genuine rather than having occurred by chance

Teratogen Exogenous agent, such as a drug ingested by the mother, that produces structural changes in the embryo or fetus

Theobromine Xanthine closely related to caffeine that is found especially in cacao beans and chocolate

Theophylline Xanthine closely related to caffeine that is found in tea leaves

Withdrawal symptoms Symptoms that occur when a person suddenly stops taking a drug than he or she is accustomed to taking. Headache and fatigue are the most common symptoms associated with withdrawal from caffeine

CAFFEINE IS A PSYCHOACTIVE DRUG, A METHYLated xanthine, that occurs naturally in many foods and beverages (e.g., coffee, tea, cocoa) and is added to many commercial products (e.g., soft drinks, analgesics). It is used very widely, often in high doses, and it has been the subject of much research. Four major questions have been raised regarding caffeine and its effects: (1) What are its physical and mental benefits?

ENCYCLOPEDIA OF HUMAN BIOLOGY, Second Edition, VOLUME 2. Copyright © 1997 by Academic Press. All rights of reproduction in any form reserved.

(2) What are the possible harmful effects of caffeine from single doses or chronic use? (3) Is caffeine consumption a social problem? (4) What are the short- and long-term mechanisms that cause its physiological and behavioral effects? Scientific studies have given us a definitive understanding of the chemical reactions, by-products, and biological cycles involved when caffeine is metabolized by the body. Although other research results are less conclusive, they have greatly increased our understanding of the complex ways in which caffeine can produce beneficial effects in one individual but harmful effects in another, or different effects in the same individual at different times. Researchers have also shown the apparent absence of certain suspected health risks and the seriousness of others. A book-length evaluative review of such results (James, 1991) is listed in the bibliography. Given the current state of knowledge, mild caution about excessive use of caffeine is prudent for the general population, reduced intake is indicated for certain clinical populations, and further research is needed in particular areas.

I. INTRODUCTION

Caffeine is almost certainly the most widely used psychoactive substance. It is present in coffee, tea, chocolate, soft drinks, analgesics, and other medications. Despite recent concerns about health risks, more than a hundred billion doses of caffeine are consumed annually in the United States. This popularity is probably attributable to the stimulant action of caffeine. Caffeine has a long history: written records mention a legendary Chinese emperor drinking tea in 2737 BC. Because it occurs naturally in more than 60 plant species, many of which have been used as food sources, people may have consumed it as early as the Paleolithic period. So prevalent a substance deserves the attention that caffeine has received. In the last 20 years, caffeine research has increased, perhaps stimulated by media attention to possible health risks. The purpose of this article is to present some basic facts about caffeine as they relate to human biology, to survey the major methodologies that have been used in caffeine research, and to discuss the implications of this research for health and social policy. However, because empirical knowledge about caffeine is constantly expanding, the summaries of research that we present here should be taken as the current state of the field and should be interpreted with caution.

Caffeine intake can be calculated by multiplying a person's consumption of food, beverages, or other products containing caffeine by the caffeine content of these products. Such data can be obtained by having people keep diaries of what they consume, by retrospectively reporting what they remember consuming, or by having a third party record what they consume. There is substantial agreement among these different methods of collecting data. Furthermore, there is close agreement between these methods and certain measures of caffeine production, such as the amount of coffee imported per capita. These intake data reveal interesting patterns. For instance, the most prominent source of caffeine in Britain is tea, but in the United States it is coffee. The shift in preference from tea to coffee in the United States has been attributed to the British tax on tea in the 1700s. The U.S. peak of more than three cups of coffee per day per capita in 1962 fell to less than two cups in 1983, a shift that has been variously attributed to poor advertising and/or a decline in the quality of coffees in the 1960s. Consumption has remained relatively constant across the 1980s and into the 1990s. Currently in the United States, females drink more coffee than males, and the younger generation gets its caffeine primarily from soft drinks. Caffeine-consuming habits also change with age. For example, 35- to 64-year-olds drink more coffee than do younger or older age groups; on the other hand, tea consumption shows no difference across ages, and soft drink consumption declines with increasing age.

The second factor necessary to calculate caffeine intake is the amount of caffeine in a source. The amount of caffeine depends on both methods of commercial production and personal methods of preparation. Tea is especially variable, ranging from 8 to 91 mg per serving. Because of such variation and differences in the definitions of serving size (and sometimes definitions are not reported), various authors report different data on caffeine content. Nonetheless, representative values for the caffeine content of major dietary sources are useful (Tables I and II). Many people believe that the dark color of soft drinks indicates which ones contain caffeine. In actuality, caffeine forms a white powder, a yellow residue (caffeine is a xanthine, Greek for yellow), or a clear solution. Note that root beer is dark and contains no caffeine, whereas Mountain Dew is clear and contains caffeine.

Despite this variability, a number of authors have used the figures for caffeine content to calculate the average person's daily caffeine intake. For American

TABLE I
Levels of Caffeine in Common Beverages and Foods

Beverage/food	Serving size (oz)	Approximate mg caffeine/serving
Coffee		
Drip	5	150
Percolated	5	110
Instant, regular	5	50–100
Instant, flavored mix	5	25–75
Decaffeinated coffee	5	1–6
Black tea		
1-min brew	5	20–35
3-min brew	5	35–45
5-min brew	5	40–50
Green tea		
1-min brew	5	10–20
3-min brew	5	20–35
5-min brew	5	25–35
Instant tea	5	30–60
Cocoa beverage	5	2–20
Soft drinks		
Jolt	12	70
Caffeinated cola drinks	12	30–65
Mountain Dew, Mello Yello, Sunkist Orange	12	40–50
7-Up, Sprite, RC-100, Fanta Orange, Hires Root Beer	12	0
Chocolate		
Cake	1/16 of 9-inch cake	14
Ice cream	2/3 cup	5
Mr. Goodbar	1.65	6
Special dark, Hershey	1.02	23

adults (consumers and nonconsumers), a representative average is 200 mg (3 mg/kg body weight) per day. Adults who would be considered heavy consumers ingest 500 mg (7 mg/kg) per day or more. Some individuals who consumed 2000 mg/day have sought professional help to reduce their intake. Note that several segments of society consume less than these figures: pregnant women appear to consume about 2.1 mg/kg, although the data are limited and pregnancy-related weight gains have not always been taken into account. Children less than 18 years old (both consumers and nonconsumers) ingest about 37 mg (1 mg/kg) per day. The health code of the Mormons, as well as that of certain other religious groups, prohibits all use of stimulants.

TABLE II
Levels of Caffeine in Common Drugs

Drug	Standard adult dose	Approximate mg caffeine/standard dose
Prescription painkillers		
Darvon compound capsule	1	32
Cafergot tablet (migraine)	1	100
Nonprescription (over-the-counter) painkillers		
Anacin, Midol, Vanquish	2	65
Plain aspirin	2	0
Cold/allergy medicine		
Dristan	2	30
Coryban-D, Sinarest, Triaminicin	1	30
Stimulants		
No-Doz	2	200
Vivarin	1	200

II. BIOCHEMISTRY

This section addresses the chemistry of caffeine, the processes by which caffeine is taken into the body, how it is distributed in fluids and tissues, its immediate effects on body systems, how tolerance to caffeine develops, and how it is cleared from the body.

A. Chemistry

Caffeine is an alkaline compound in the family of naturally occurring derivatives of xanthine. The three primary xanthines are caffeine, theophylline, and theobromine. All three are structurally similar with only minor variations in the number and position of methylated sites. Thus, they have similar effects on the body. Specifically, caffeine is 1,3,7-trimethylxanthine. Because it was one of the first pharmacological agents to be chemically isolated (in Germany in 1820), it is the most extensively studied xanthine.

B. Distribution in the Body

Although most other alkaloids are insoluble in water, caffeine is slightly soluble and becomes still more so in the form of complex double salts. Of greatest importance to humans is that caffeine is readily absorbed by the gastrointestinal tract after oral ingestion (99% complete within 45 min), and it is distributed into all body fluids and tissues, including the brain, testes, fetus, and mother's milk. Rates of caffeine absorption have been observed to be slower for soft drinks than for coffee, a surprising finding because of the fact that the rate of absorption of alcohol is increased by carbonation. Peak blood plasma levels of caffeine are reached within 15–45 min after ingestion.

C. Effects

Caffeine has a number of effects on the nervous system, cardiovascular system, and other body systems. It is a powerful central nervous system stimulant. Moderate doses of 200 mg (about two cups of coffee taken close together) activate the cortex of the brain enough to slow changes in a person's electroencephalogram. However, considerably higher doses or injection are required to show effects in the medulla, spinal cord, or autonomic nervous system. All such effects begin about 0.5 hr after intake and maximal effects are reached about 2 hr after intake.

Caffeine has several effects on the cardiovascular system, although the effects of caffeine are weaker and not as clinically useful as the effects of theophylline. The major effects of caffeine are on heart rate, blood pressure, blood flow, and serum cholesterol. Caffeine stimulates the heart directly by acting on the myocardium to increase the force of muscle contraction, heart rate, and cardiac output (stroke index). Such effects may be masked at low dosage levels because caffeine also stimulates the medullary vagal nuclei, which in turn tends to produce a decrease in heart rate. [*See* Cardiovascular System Physiology and Biochemistry.] Single caffeine doses decrease heart rate during the first hour after administration and increase it during the following 2 hr. Chronic use appears to elevate basal heart rate so that clinically significant reductions in heart rate are obtained upon abstention from caffeine. Previous contradictory reports about effects on blood pressure have now been reconciled. Single doses can increase blood pressure among both users and nonusers; however, the size of the increase is significantly larger for nonusers. Doses of from 100 to 300 mg of caffeine are enough to cause noticeable increases in nonusers, but higher doses are needed to cause such increases in users. Caffeine dilates systemic blood vessels, but constricts cerebral blood vessels. The accompanying change in peripheral blood flow is too short-lived and of too small a magnitude to be of therapeutic value. On the other hand, the reduced cerebral blood flow has been used in the treatment of headache. Both acute and chronic caffeine intake seem partially responsible for increased blood levels of cholesterol. Some researchers have disagreed; however, repeated replications, demonstrations that discontinuing coffee reduces serum cholesterol, and at least one finding of a dose–response relationship tend to outweigh such reservations. [*See* Cholesterol.]

In addition to effects on the nervous system and circulatory system, caffeine has a number of other effects. Caffeine causes a slight increase in basal metabolic rate (10%); the increase occurs as a sharp rise in rate during the first several hours after ingestion. Caffeine increases the secretion of stomach acids, increases the respiratory rate, and slightly increases the production of urine. It acts as a bronchodilator by relaxing the smooth muscles. On the other hand, caffeine strengthens the contraction of skeletal muscles.

It has been suggested that the basis for the effects of caffeine is multifactorial. The primary mechanism seems to be that caffeine blocks receptors for adenosine. Caffeine is also supposed to block receptors for benzodiazepines and to alter the movement of calcium within cells via the cyclic nucleotides. The molecular basis for the effects of caffeine remains unclear.

D. Tolerance and Individual Sensitivity

When an individual has developed tolerance to a drug, larger dosages are required than previously to achieve a given effect. This pattern of decreasing effects can cause drug users to increase dosages to compensate for the decrease. With chronic caffeine use, it takes larger dosages to achieve the same (mild) diuretic and salivary effects. As mentioned earlier, caffeine users show smaller blood pressure increases than nonusers for the same caffeine dosage. Some evidence suggests that these different levels of tolerance develop after only a few days of chronic use and are lost within a day, although one would like to see this research replicated. On the other hand, less tolerance seems to develop for the stimulating effects of caffeine on the central nervous system. A recent conflicting report claims complete tolerance development to a CNS effect among moderate coffee drinkers, but it involves subjective reports, not direct measures of CNS effects, and it has other weaknesses. Furthermore, for many individuals, increases in caffeine dosage would be self-limiting because higher doses exacerbate undesirable symptoms (nervousness, anxiety, restlessness, insomnia, tremors, gastrointestinal disturbances, and feelings of uneasiness). (Withdrawal symptoms will be discussed in Section VII.)

E. Toxicity

In contrast to the high toxicity of theophylline, caffeine is not very toxic. Nevertheless, caffeine can produce symptoms that require medical consultation. For example, sudden increases in consumption have been associated with a variety of adverse effects, such as delirium, abdominal cramps, vomiting, high anxiety and hostility, and psychosis. More gradual increases may not show toxicity because tolerance develops. Higher doses of caffeine can cause convulsions and still higher doses can cause death from respiratory failure. A lethal dose in adults appears to be 5–10 g, the amount of caffeine in approximately 200 colas. There is little concern that death could occur from beverage consumption because gastric distress and vomiting would prevent concentrations from reaching life-threatening levels. Although similar principles would seem to apply to over-the-counter caffeinated drugs, at least seven deaths from ingested caffeinated medications were reported between 1959 and 1980. One death after injection of 3.2 g has also been reported.

F. Clearance

About 95% of a dose of caffeine is eliminated from the body by being metabolized to other products in the liver; the remainder is excreted unchanged by the kidneys. The exact metabolites may vary among races, for example, Asians have different metabolites than Caucasians. Rates of clearance also vary greatly, with clearance being much slower in infants because they lack certain enzymes. Even among healthy adults of the same race, there is wide variation. For example, it takes between 2.5 and 7.5 hr to remove 50% of a dose of caffeine from the body, and 97% is eliminated in 15–30 hr. Many factors may cause the rate of clearance to vary: smoking, liver disease, pregnancy, and the use of oral contraceptives have all been found to decrease clearance. Other medications can increase or decrease clearance. When a chronic user abruptly discontinues all use of caffeine, complete removal of caffeine from the body can take up to 7 days.

In summary, a comprehensive metabolic pathway has been established for caffeine. Future work may produce minor modifications in our understanding of this pathway, but it is more likely to focus on individual differences and on the mechanisms by which caffeine achieves its various effects.

III. RESEARCH METHODOLOGY

Much of the research on caffeine has produced seemingly contradictory findings. In point of fact, the differences can be attributed to insufficient attention to methodological details. The most prominent methodological issues involve measurement, biased self-report, confounding factors, and improper interpretation of results.

At the most basic level of measurement of quantity, there is a problem with caffeine research. Is a "cup" of coffee 5, 6, or 8 oz? Often researchers do not report their reference volumes. Furthermore, when researchers ask members of the general public to report their consumption habits, those people may not realize that their "mugs" of coffee contain two of the researcher's reference volumes—and thus they may be consuming twice the caffeine they think they are.

There are also problems at the chemical level: measured caffeine content on beverages or foods can vary owing to the analytical method employed. Caffeine has no recommended clinical chemistry because it is not routinely measured. However, a number of methods could be used, including ultraviolet spectrophotometry, liquid chromatography, thin-layer chromatography, immunoassays, and gas chromatography, as well as other methods if they too provide appropriate sensitivity, specificity, and feasibility. Such techniques unanimously show that plant variety, growing conditions, and method of preparation (for coffee and tea) dramatically affect caffeine content.

Even when measured quantities of caffeine are administered during a laboratory experiment, specifying the pharmacologically active dose is difficult because many factors affect responses to caffeine. In other words, what is a relatively small dose for one person can be a relatively large dose for another person. As a dramatic example, consider that because of individual sensitivity and body size, a single candy bar could have the same effect on a young child as five cups of instant coffee would have on an adult. The important factors to take into consideration are differences in body weights, different rates of stomach emptying and intestinal absorption, and inherited caffeine sensitivities. Furthermore, differences in chronic, occasional, or no-caffeine use; recent intake; length of abstention; and blood caffeine levels can cause differences in response to caffeine. Psychological factors, such as stress or being told (deceptively) that one was administered caffeine when one was not, also cause differences in response. A common way to rule out psychological factors and study pharmacological effects in isolation is to use the "double-blind" design. In this design, caffeinated and uncaffeinated sources are identical in appearance and a third party schedules which will be given to each participant. Neither the participant nor the researcher who has direct contact with the participant knows whether the caffeinated or uncaffeinated treatment is being used. They are both blind to the kind of treatment administered. In some studies it has been found that participants who are "blind" and receive caffeine have different responses from those who know that they are receiving caffeine. Keeping participants blind to the nature of their treatment condition is difficult in some situations because people can taste and feel differences in treatments of 200 mg of caffeine or larger. Differences of 50 mg or smaller are not detectable by most individuals. The traditional balanced-placebo design involves four conditions and allows the experimenter to separate psychological and pharmacological effects. Knowledge that one is receiving a drug is important, as well as one's beliefs that its effects should be positive or negative. Some caffeine researchers recommend an expanded balanced-placebo design involving six groups (three larger groups, those told they received the drug, those told they did not, and those told they are being kept blind as to what they received; each larger group is subdivided into those receiving the drug versus the placebo).

Among all of these difficulties in accuracy of measurement, there is a reassuring fact: a number of biochemical and other validations have shown that self-reports of caffeine habits are for the most part reliable. Such findings allay concerns about recall bias in retrospective studies.

Important research that cannot be done in the laboratory includes studying whether caffeine contributes to the occurrence or nonoccurrence of certain diseases in human populations. No such epidemiological studies have been conducted with caffeine itself, but there are many with coffee drinking. Caffeine is only one of many active ingredients in coffee, thus it would be remarkable if caffeine proved to be the only substance of significance. Furthermore, those who are coffee drinkers bring confounding characteristics with them: they smoke more, consume more alcohol, are more extroverted, and have different levels of education, dietary habits, and life-styles (e.g., they exercise more). Controlling for these factors has often made apparent links between coffee consumption and health disappear.

The critical reader of an epidemiological study should watch for a number of potential difficulties. One problem is insufficient objectivity of recorded data when the researchers did not use "blinding" (explained earlier) or did not seek corroboration of self-reports through spouses or other sources. Another common problem is insufficient attention to protopathic bias, that is, when analyses are based on nonrepresentative early measures of factors contributing to a disease without recognizing that early manifestations of the disease triggered the patient or physician to make alterations in those very factors. Before accepting the results of these studies, the critical reader will require the researcher to have appropriately tested the statistical significance of the findings. Chance factors might always cause a pattern to appear in data that would not be observed if the study were repeated. Thus, researchers should take statistical precautions not to read meanings into patterns that are not really there. To do so would be analogous to reading meaning into tea leaves or "seeing" objects in the random shapes of clouds or a man in the moon. For researchers to say the pattern of data is compatible with their theory is not enough; they must also demonstrate that the results are statistically significant, that is, that they are very unlikely to have occurred only by chance. Moreover, although a simple test of statistical significance might be appropriate for research that tests only one hypothesis, performing the same statistical test repeatedly on different portions of the data is improper. Instead, the researcher must use a different statistical technique designed for so-called "multiple comparisons."

When a statistically significant association is found between coffee intake and some health risk, this result by itself is not sufficient to conclude that coffee intake causes the health effect. Other explanations are possible. For example, the disease symptoms may cause the increase in coffee intake, or a third factor may cause both. Consequently, when researchers find an association, they should look further to see if the size of a dose is related to the size of a response. The lack of a dose–response relationship is usually taken as lack of causality and as a clue that other factors are involved. Another possibility is that caffeine has no direct effects, but modulates the effects of other agents, either antagonistically or synergistically. While epidemiological studies must take into account one set of factors, laboratory studies of performance effects must take into account a different set of factors. For instance, caffeine has been shown to have different effects on the performance of introverted and extroverted people. Factors affecting a person's level of arousal are also important. For example, caffeine may have a different effect on the same person early in the day than later, because of the person's diurnal rhythm of bodily arousal. Finally, caffeine may have different effects depending on whether or not a person is under stress. In summary, caffeine research involves many complexities, and there are rigorous scientific methods for handling them.

IV. PHYSICAL PERFORMANCE

People report that caffeine increases their physiological arousal. Assuming that these subjective impressions reflect underlying physiological changes, one would expect effects on fine motor coordination, spontaneous gross motor activity, and athletic endurance. In fact, hand steadiness has been shown to be about 25% worse after consuming 200 mg of caffeine (about two cups of strong coffee). Furthermore, there have been a number of empirical reports of caffeine impairing motor skills that involve delicate muscular coordination and accurate timing. Hand steadiness is a very sensitive measure and there is great consistency in the findings.

Animal studies have consistently found that caffeine increases activity without producing the "locomotor stereotypy" or persistent repetitive movements produced by amphetamine. There are only a few studies with humans; those using high dosages have found increases in gross motor activity for both children and adults. Some studies have found no increases in activity, but only decreases for high consumers who abstain in order to take part in the study and then are in a no-caffeine condition. Careful naturalistic observation studies with doses equivalent to two-thirds of a soft drink have failed to find any changes in activity for 5-year-olds.

Many people believe that caffeine is an ergogenic aid, a substance that helps an athlete generate increased force or endurance. Witness the coffee-drinking rituals preceding marathons or the use of coffee to get through the daily grind of training. Empirical studies have shown improved work production in trained cyclists, runners, and cross-country skiers, for example, extending mean cycling time to exhaustion by 20% when cycling at 80% of maximal capacity. Such ergogenic effects occur only during prolonged work and not during short-term work episodes. When the work conditions have been varied or when dosage or caffeine habits are not sufficiently accounted for, effects have been unclear and equivocal (although none has been in the reverse direction).

Caffeine seems to delay deterioration in performance due to fatigue through both psychological and physical effects. It decreases perceived exertion, perception of fatigue, and drowsiness; it increases self-reported alertness and motivation. The mechanisms for these psychological effects may involve reduced neuronal thresholds in the central nervous system, influences on catecholamine receptors, and/or direct effects on the adrenal medulla. On the other hand, caffeine may produce its physical effects through increased fuel availability or increased contractile activity of skeletal muscles. In turn, several mechanisms have been suggested for the increase in contractile activity: enhanced transmission of nerve impulses or potentiated twitch responses in both rested and fatigued muscles. Similarly, several mechanisms have been suggested for the increase in fuel availability. For example, one proposal involves enhanced fat metabolism: caffeine may stimulate adipocyte lipolysis via the activation of lipase. Unfortunately, research has failed to clearly demonstrate such enhanced fat metabolism and competitive athletes should have plenty of epinephrine for lipolysis, even if they do not ingest caffeine. This example demonstrates the intricacies to consider in any viable explanation of the effects of caffeine.

An athlete concerned about the use of caffeine should realize that responses to caffeine vary greatly. For example, in contrast to others, more sensitive individuals may become overstimulated and show performance decrements. For some activities, the diuretic effects of caffeine may be a problem if maintaining hydration is important and difficult. The athlete who decides to use caffeine needs to consider how to obtain a beneficial effect while avoiding acquiring tolerance. Regular, heavy use throughout training may reduce benefits during competitive performance and may increase blood cholesterol and the risk of heart attack or other medical problems (see Section VI.) Finally, the athlete should realize that the International Olympic Committee has banned caffeine when its values exceed more than 15 μg/ml in a urine test.

V. PERCEPTUAL AND COGNITIVE PERFORMANCE

In addition to physical arousal, people report that caffeine increases their mental arousal. Many people believe this arousal leads to improvements in actual performance. Researchers have attempted to objectively assess the effects of caffeine on perception, speed of reaction, memory, and/or the flow of thoughts.

At the perceptual level, we can ask whether or not the taste of caffeine is noticeable. It has been claimed to enhance the taste of colas. However, humans in experimental settings have been unable to detect its presence or absence at common cola levels of concentration. Caffeine intensifies the taste of certain sweeteners, lowers visual luminance threshold, and improves auditory vigilance. On visual vigilance tasks, caffeine increases alertness and speeds up responding while reducing attention to detail. More definitive studies that show effects of caffeine on vision, hearing, and skin conductance would be desirable.

Regarding reaction time, small to moderate amounts of caffeine (i.e., 32–200 mg) help speed reactions to simple, routinized tasks, such as indicating whether an even or odd digit was presented, pressing buttons corresponding to bulbs lit in a circular pattern, or watching for strings of three even numbers. Conversely, when habitual caffeine users abstain for 2 days, their reactions are slowed and their attention is impaired. (The physical indicator of impaired attention is less anticipatory heart-rate deceleration.) For novel or slightly more complex tasks, whether caffeine will improve or impair performance is difficult to predict. An example of such a task is watching a random sequence of letters and responding each time the letter A is preceded by the letter X. Because to a large extent driving a car is routinized, the improvements in auditory vigilance and visual reaction time would seem to imply benefits for late-night driving if not counteracted by loss of fine motor coordination. No researcher seems to have carefully tested the net effects.

It is noteworthy, though, that several researchers have tested the common belief that caffeine counteracts the effects of alcohol. Contrary to the common belief, initial research showed that coffee further impairs rather than improves performance. For example, a person who has consumed enough alcohol to be close to the legal level of intoxication and then drinks a cup and a half of coffee (150 mg caffeine) has even slower reaction times than if only the alcohol had been consumed. In short, the extra coffee may make one more prone to accidents rather than less so. On the other hand, three more recent studies have shown that caffeine can partially offset the debilitating effects of alcohol on certain driving performance components. In summary, although the common belief is not completely wrong, it is dangerously misleading. Caffeine combined with alcohol does provide some benefits, but for other components it provides decrements or no benefits, and where it does provide benefits it does not fully restore normal functioning. The practical importance of this issue calls for careful research that uses doses of caffeine typical of social use (instead of only excessive doses), pairing wider ranges of alcohol and caffeine doses, adjusting both kinds of doses to subjects' body weights, taking into account subjects' typical patterns of alcohol and caffeine use, and including as part of the experiment a demonstration that caffeine alone actually improves performance.

Many people believe that arousal produced by caffeine is beneficial to learning and retention. At the simplest level of habituation, caffeine reduces the rate of habituation. At the level of learning and memory, some research has, in fact, shown beneficial effects; however, many studies involving short-term memory have shown no effect or impaired performance due to caffeine. It has been proposed and partially demonstrated that this lack of consistency can be resolved by disentangling the complex interactions among personality type (extrovert–introvert and/or high–low impulsivity), diurnal rhythm of arousal, task requirements (e.g., sustained information transfer, short-term memory), dosages matched to body weights, and the curvilinear relationship with arousal level (moderate arousal enhances performance, whereas excessive arousal hinders it). For example, coffee may improve an extrovert's performance on a particular task in the morning, but impair the same person's performance on the same task in the afternoon. Conversely, coffee may hinder an introvert in the morning, but facilitate the introvert in the afternoon.

Another kind of drug-induced effect on memory is state-dependent learning. This occurs when information learned in one drug state is later recalled better when the individual is in the same drug state rather than a different one. An example of state-dependent learning is the alcohol drinker who while sober forgets what he or she did while intoxicated but recalls it again when next intoxicated. In the experimental laboratory, state-dependent learning with alcohol has been demonstrated with social drinkers; however, several attempts to demonstrate state-dependent learning in the laboratory with caffeine have been unsuccessful. One possible cause for the latter negative results is that the "drug" and "nondrug" states that were supposed to be different actually were not. The experimenters assumed that the drug state involved a high level of arousal because caffeine was consumed, and that the other state involved a low level of arousal because a placebo (i.e., no caffeine) was consumed.

However, subjects in the later condition may have also been aroused because, for example, they had been challenged to perform well on the experimental task, or they may have been excited by being in an unfamiliar setting and in the presence of a stranger. The first explanation is especially plausible because many people find that having to take any kind of test causes them to be anxious. Assuming that alcohol reduces test anxiety would account for the state-dependent learning that was found in one study that, from the learning to the test phase of the experiment, shifted subjects from a combined alcohol/caffeine state to either (1) the same state (which produced no decrement in performance), (2) an alcohol-only state (which produced a performance decrement), (3) a caffeine-only state (which produced a small decrement), or (4) a no-drug state (which produced maximal detriment). In short, at the present time, whether caffeine does or does not produce state-dependent learning is not entirely clear. There seems to be a good chance that caffeine does produce state-dependent learning, but that this effect will be difficult to isolate.

Many people report that caffeine helps them think more clearly and creatively. In the public stereotype, creative persons use drugs to excess. Survey research has shown that, contrary to the stereotype, writers, artists, and musicians do not use drugs, not even caffeine, to excess. In fact, most creative people say they learned early in their career that drugs interfere with the creativity process. When they do use caffeine, it is only to counteract the effects of lack of sleep.

Whether or not ordinary people's more mundane cognitive performance improves while they are under the influence of common levels of caffeine has been tested using a variety of tasks, ranging from simple subtraction, to identifying errors in written passages, to taking the Graduate Record Examination (among many others). A very wide range of positive results has been found. A few studies have reported negative findings that could be due to lack of control for variables such as the personality characteristics of subjects or the time of day of testing. These variables are important because other research has demonstrated that they modulate the effects of caffeine. For example, several studies that controlled for personality characteristics and time of day found that caffeine both improved performance and showed dose–response effects. On the other hand, one study reported that high caffeine intake among college students was associated with low academic grades; however, it is impossible to conclude that coffee drinking is the cause of the low grades. Because this report is frequently cited, it should be replicated and cause and effect relationships should be investigated.

Furthermore, it would be important to have more studies of other "real-world" forms of thinking, such as reading comprehension, creativity, and problem solving. Studies that compared fatigued or bored subjects with alert ones would help resolve the issue of whether caffeine actually improves performance generally or is largely confined to restorative effects. Do the benefits of caffeine occur at the expense of impaired performance later in the day? Is the period of increased stimulation–metabolism followed by a restorative period of decreased stimulation–metabolism? Researchers do not seem to have addressed these simple, practical questions. The design problems described here are a classic example of the complex contextual nature of social science research in contrast to the searches for universal laws with a limited number of variables that are common in the physical sciences.

VI. PHYSICAL HEALTH

Physicians have long been interested in caffeine for its therapeutic uses. Caffeine has been beneficially used as a cardiac stimulant, to reduce bronchial asthma, therapeutically with infant apnea, to treat acne and other skin disorders, to reduce migraine headaches, to enhance the effects of analgesics, and to counteract the depressant side effects of various medications. There are two reports of anticancer effects of caffeine, but unfortunately they are not convincing.

There have been sporadic reports of evidence and counterevidence for health risks due to caffeine. As a result, public interest in caffeine has mushroomed in the last 25 years. The major debates have involved birth-related problems, cancer, ailments of the gastrointestinal system, and diseases of the cardiovascular system. In the following sections, we will evaluate the evidence related to each of these possible effects.

A. Birth-Related Risks

There is reason to be concerned about whether caffeine increases birth-related risks. In general, embryonic and fetal growth is a time of great risk. With regard to caffeine in particular, animal studies have demonstrated that caffeine is a teratogen among mammals, which compelled the Food and Drug Administration to remove caffeine from its list of drugs "generally regarded as safe" (GRAS) and in 1978 to release a warning concerning the ingestion of caffeine during pregnancy. Moreover, several better-designed studies with humans have shown a relationship between coffee consumption and miscarriage or spontaneous abortion. Other studies have found negative results, but were less well designed. Future studies should control for the confounding factor of alcohol abuse and are important to clarify the exact nature of the relationship. Research has shown no association between coffee consumption and preterm birth, preterm labor, use of Caesarean sections, breech births, or premature births. Whether or not coffee consumption affects ease of conception is an open question that is difficult to study because many additional life-style variables must be taken into account.

In addition to the effects on the mother and her child's birth, a pregnant woman's caffeine consumption may have effects on both her fetus and the infant after it is born. Certainly many fetuses are exposed to caffeine. In fact, more than 80% of pregnant women report consuming caffeine during their gestational period. Although many report reducing the quantity of their caffeine intake, few report that they totally abstain. In fact, pregnancy is a time of low energy, and women desiring a "lift" may actually increase their intake of coffee or other beverages containing caffeine. Furthermore, some women may unknowingly ingest caffeine through medications. These possibilities are of concern because caffeine is known to cross the placenta and diffuse into breast milk. Moreover, the concentrations of caffeine in the fetus or newborn may be considerably higher than in the mother because the former lack the enzymes necessary to metabolize caffeine. In fact, the slower clearance of caffeine in young infants has been empirically verified.

When the possibility of an association between coffee drinking and congenital malformation was first suggested, suspecting detrimental effects of caffeine was partially justified, because studies with microorganisms had shown that caffeine causes mutations. However, closer examination of the original rationale and of the studies with humans currently indicates that there is no mutagenic risk. The extrapolation of the results from microorganisms to humans does not hold because of biological differences between these organisms and because the experiments used caffeine concentrations 40 to 4000 times those achieved by heavy coffee drinkers. Most large-scale epidemiological studies with humans have found no association between coffee intake and congenital malformations, although a few recent studies with small numbers of subjects have found associations. Based on the large-scale studies, mutagenic risk is not presently a major concern, but further epidemiological research is needed.

The relationship between caffeine intake and low birthweight is closely related to concerns about mutagenic risk. With regard to birthweight, the evidence is mixed. Although some early studies found that increased caffeine consumption by pregnant women was associated with slightly greater frequency of low birthweights, the magnitude of the associations was small and, because of a lack of statistical significance, they were discounted. Another early study did find a significant association, but it too was discounted because the study had not controlled for smoking and gestational age (and researchers assumed that controlling for such extraneous factors would give a nonsignificant pattern of results). At the time that this study was conducted, the cumulative pattern of evidence seemed to indicate no association between coffee intake and low birthweights. However, several more recent studies, including one carefully controlled one, have found a significant association. The import of the association is strengthened by two additional statistically significant results: maternal caffeine consumption has been found to be associated with smalled head circumferences and neuromuscular immaturity of the newborn. This shift in the balance of evidence led researchers to reanalyze the early study whose significant results had been discounted. They found that when appropriate corrections were made for smoking and gestational age, a significant association was still present, indicating that the original results should not have been discounted. In summary, caffeine intake does seem to be related to low birthweights.

The research summarized thus far involved chronic coffee use. However, short-term intake of caffeine during pregnancy has also been associated with detrimental effects, albeit behavioral rather than physical. For example, maternal caffeine consumption 3 days before delivery has been shown to increase general arousal level, depress muscle tone, and make it harder to console the infant when it was upset. There is also some suggestion that spontaneous sleep states are affected.

In short, many of the suspected birth-related risks of caffeine have been disproved, and evidence regarding other effects is mixed. However, the current evidence regarding spontaneous abortion, low birthweight, and behavioral effects is convincing enough to advise pregnant mothers to limit, or discontinue entirely, their caffeine consumption, especially because safe levels of caffeine use during pregnancy have not been established. In fact, pregnant mothers should avoid taking any unnecessary drugs during pregnancy, or should take them only under the supervision of a physician.

B. Caffeine and Cancer

As just described, caffeine has been suspected of being a mutagen. Many mutagens also cause cancer. Caffeine has been suspected of both directly initiating cancer and interacting with other carcinogenic agents. In studies with animal tissue, caffeine has been shown to either increase or decrease the growth of tumors at the cellular level, depending on which carcinogen was used. Epidemiological studies in humans have focused on coffee rather than caffeine specifically and have been fraught with all of the potential difficulties of etiologic science (as discussed in Section II). In fact, initial reports of a link between coffee consumption and pancreatic cancer have been retracted by the original authors (as well as criticized by others). Many epidemiological studies show no link with cancer of the urinary bladder, although one study attributes 25% of all bladder tumors to drinking more than 1 cup of coffee a day. Many, but not all, reports show no link between coffee drinking and cancer of the breast through fibrocystic breast disease. Only a small amount of research exists on associations between coffee drinking and cancer of the ovary, and the results of these studies are mixed. On the other hand, there have been isolated reports of beneficial effects of caffeine: higher coffee consumption has been linked to lower carcinoma of the renal parenchyma, the skin, and the colon. In summary, the evidence does not allow us to conclude that coffee intake causes cancer. At the same time, more research is needed using all the rigorous controls of etiologic science.

C. Gastrointestinal Risks

There are several ways in which caffeine could lead to gastrointestinal problems. Caffeinated beverages can increase the secretion of stomach acids and could thereby exacerbate an ulcer that is already present or could contribute to the formation of a new one. Caffeinated beverages also lower esophageal sphincter pressure, perhaps allowing or adding to gastric acid reflux. One piece of evidence that tends to support this reasoning is the fact that a reduction in caffeine intake has been associated with reduced abdominal complaints. In particular, patients taking the drug cimetidine for gastrointestinal ulcers may find that it slows caffeine clearance by the body. In summary, coffee drinkers with ulcer problems may want to reduce their intake of caffeine, but switching to decaffeinated coffee is not a solution because it also stimulates the secretion of stomach acids.

D. Cardiovascular Risks

The effects of caffeine on the cardiovascular system have been extensively studied, yet remain unresolved. This research is especially difficult to interpret because of numerous interactions of caffeine with other lifestyle habits. Because of its effects on heart rate, blood pressure, and serum cholesterol, caffeine is implicated in arrhythmias, hypertension, and ischemic heart disease. Caffeine is contraindicated for individuals with cardiac arrhythmias. Perhaps because this dictum is so widely believed, there is disappointingly little published evidence of a link between caffeine intake and arrhythmias. More research is needed for an understanding of the real risks and the mechanisms through which caffeine operates. For example, it has been suggested that caffeine causes irregularity by affecting the heart's contractility through alterations in the movement of calcium in and out of the cells.

The relation of caffeine to hypertension has also been questioned. Although the effects of caffeine may be small, they may be clinically significant if a patient is exposed to stress or other factors associated with cardiovascular disease. Clinically significant reductions in blood pressure have been observed in chronic users who abstain. Future research should also investigate whether or not caffeine interacts with antihypertensive medication and negates the effect of the medication. Until more evidence has been gathered, individuals worried about blood pressure may consider reducing or eliminating caffeine from their diet.

During the 1970s, sporadic correlations between coffee intake and heart attackes were reported, but researchers tended to believe that these results did not indicate that caffeine consumption was one of the casual factors in ischemic heart disease. However, two reports since 1985 have shown a link between heavy coffee use and heart attacks, thus raising the issue again. Actually, two issues may be distinguished: Does chronic use contribute to the development of the disease? Does a single dose produce the stimulation that triggers a heart attack? Future research should distinguish these issues and should include careful controls for smoking, diet, exercise, and stress. Researchers should attempt to obtain coffee consumption habits both immediately preceding the coronary event and distant in time from it.

E. Other Health Problems

Research on sleep disturbances has shown that coffee consumed shortly before bedtime increases the time needed to fall asleep and the number of spontaneous awakenings during the night. It also decreases total sleep time, amount of deep sleep (time in stage 3 sleep and stage 4 sleep), and perceived quality of sleep. Habitual caffeine use has been shown to be correlated to habitual sleep duration. Single doses >300 mg can produce temporary insomnia. Reducing caffeine intake has been shown to reduce sleep disturbances, but with wide individual variations. For example, eliminating an evening soft drink may help some children with problems getting to sleep, but not others. General conclusions are difficult to draw because the effects of caffeine on sleep can be modified by factors such as the time when caffeine is ingested, an individual's habitual pattern of caffeine intake, and individual differences in sensitivity to caffeine. Moreover, caffeine can have adverse consequences on sleep without the individual being aware of such effects.

Several other health problems have been suggested: because it affects calcium flow, caffeine may accelerate the development of osteoporosis. Habitual use of caffeine has been associated with subclinical (i.e., premorbid) symptoms of poor somatic and psychological health. However, because this research was based on self-reports, one must be cautious about whether the symptoms are more imagined than real.

VII. MENTAL HEALTH

Many years ago, the well-known psychologist Harry Stack Sullivan observed "incipient depression and neurasthenic states" in a client after "unwitting denial of the accustomed caffeine dosage." He surmised that "there might be times when a cup of coffee would delay the outcropping of a mental disorder." Recent concerns reflect the opposite point of view, namely, that consumption of caffeine might lead to caffeine intoxication, that it might exacerbate other psychological disorders, or that its symptoms might be misdiagnosed as another disorder.

A. Addiction and Withdrawal

Is caffeine really a drug of addiction? By the criteria used in the 1970s, a drug of addiction must have psychoactive properties, must have reinforcing properties, and must result in withdrawal symptoms when its use is abruptly discontinued. Caffeine is psychoactive: witness the stimulating effects that people report. Caffeine is reinforcing according to people's reports as well as carefully controlled experimental studies. Does removing caffeine produce withdrawal symptoms? The adverse effects are well documented. When caffeine users abstain, they experience symptoms such as dysphoria, drowsiness, yawning, poor concentration, disinterest in work, runny nose, facial flushing, headache, fatigue, irritability, and anxiety. The symptoms typically begin between 12 and 24 hr after the person discontinues use of caffeine. The symptoms vary from individual to individual; they can be mild to extreme, peak within 20–48 hr, and can last for a week. Headache, for example, is reported by about one quarter of the heavy users who abstain. In a few cases the symptoms have been reported to appear when caffeine intake was gradually reduced over several weeks rather than abruptly. Even someone who is a relatively light user, habitually consuming as little as 200 mg of caffeine per day, may experience withdrawal symptoms. Nonusers can become quickly addicted—within as little as 6–15 days—if high doses are consumed.

Drugs of addiction tend to upset the homeostasis of the body. Addicts would, in fact, be in constant disequilibrium except for the fact that they develop compensatory responses, physiological changes opposite of those induced by the drug. Furthermore, these compensatory responses become conditioned to the cues that precede drug use. In short, the body prepares itself for the drug assault. Caffeine users, like users of other addicting drugs, develop compensatory responses that become conditioned to the stimulus cues associated with caffeine consumption. For example, the sight of coffee inhibits salivation in chronic users of caffeine, a response that compensates for the increase in salivation produced by caffeine. Note that decaffeinated coffee provides the same visual and gustatory cues as caffeinated coffee, thereby also inhibiting salivation. Clearly, then, caffeine is a drug of addiction by these standards—in spite of the fact that the general public does not generally regard it as such.

More recent discussions of the "abuse liability" of drugs emphasize drug use properties that maintain self-administration despite other drug-use properties that interfere with fulfilling life's responsibilities or with healthy bodily functions. Additional clinical criteria include unsuccessful efforts to control use, continued use despite knowledge of a problem caused by use, and tolerance to the behavioral effects of the drug. With regards to these criteria, caffeine has not maintained self-administration behavior as reliably as classic drugs of abuse such as cocaine or *d*-amphetamine. Because caffeine generally has weaker functions or has effects only in more limited situations, some authors have suggested (reasonably) that it has less dependence potential than, say, *d*-amphetamine. Others have gone even further and (unreasonably) suggested that caffeine is not capable of promoting dependence at all.

B. Caffeine Intoxication

The Diagnostic and Statistical Manual (DSM-IV) lists four criteria for "caffeine intoxication": (1) recent "excess" consumption of caffeine, (2) the presence of five or more symptoms that developed from the caffeine use (e. g., restlessness, insomnia, flushed face), (3) that the symptoms reach a clinically significant level of distress or impairment, and (4) that the symptoms not be accounted for by another medical condition. Because health professionals have traditionally ignored patterns of caffeine consumption (cf. the earlier anecdote about Sullivan), this diagnostic category performs the useful function of drawing attention to caffeine and helping prevent misdiagnosis. DSM-IV also lists "caffeine-induced anxiety disorder with onset during intoxication" and "caffeine-induced sleep disorder with onset during intoxication."

"Caffeine withdrawal" is not an official category but is included in DSM-IV as a research category for possible addition once its utility has been documented. "Caffeine dependence" and "caffeine abuse" are not included as official categories for lack of sufficient evidence at this time. [*See* Nonnarcotic Drug Use and Abuse.]

C. Anxiety Disorders

The relationship between caffeine and emotional health has been most often studied in terms of generalized anxiety disorder (GAD) and panic disorder (PD). At low doses, some people interpret the stimulation provided by caffeine as a pleasant, general elevation in mood, whereas others find it unpleasant. After consuming moderate amounts of caffeine (200 mg), many people report increased feelings of restlessness, tension, and anxiety. Larger doses (300 mg) can lead to further anxiety, hostility, and depression. Although the data indicating these effects were obtained from self-reports, they have been corroborated by objective observers. In double-blind experiments, observers were able to reliably see the increasd restlessness and "drug effect" of caffeine on users. In even larger doses, the symptoms may be indistinguishable from anxiety disorders. Reductions in daily caffeine level can reduce anxiety, although sudden withdrawal can increase it.

Patients with panic disorders are more sensitive than normals to the anxiogenic (anxiety-producing) effects of caffeine. In addition to the usual symptoms, they show palpitations, nervousness, fear, nausea, and tremors. They show clear dose–response effects. A 480-mg dose is enough to create a panic attack in these patients, although a much larger dose would be required to create panic in a normal person. Several biologically plausible mechanisms for anxiogenic effects of caffeine have been proposed: the effects are mediated by autonomic nervous system activity, plasma adenosine levels, blocking the actions of adenosine, or increased lactate. Definitive results have not yet been obtained, and additional studies are needed. In the meantime, all the available data reinforce the clinical wisdom that patients with anxiety disorders should avoid caffeine-containing foods and beverages. [*See* Mental Disorders.]

D. Depressive Disorders

Some laboratory studies have found that caffeine increases feelings of depression and exacerbates manic–depressive symptoms; furthermore, reducing daily caffeine intake can improve mood. Caution must be used in applying this finding clinically. Several other studies have found null results. In two cases of bipolar affective disorder who were on lithium treatment, reduction of caffeine intake increased lithium tremors. A contrasting theory suggests that during depressive episodes, patients self-medicate with caffeine to raise themselves out of their depression. At this time, the relationship between caffeine and depression, if any, is unclear. [*See* Depression; Mood Disorders.]

The role of low-dose oral or intravenous caffeine in electroconvulsive therapy (ECT) presents a much clearer picture. When ECT is indicated for major depression patients, it is often difficult to achieve desired seizure duration and to hold settings within desired ranges. Pretreatment with doses of 100 to 125 mg of caffeine can maintain duration and settings and achieve equivalent therapeutic outcome with no cardiac complications, cognitive side effects, or additional complications. One note of caution: patients can respond differently on different trials, hence careful monitoring on every trial is necessary. One study found the lengthening of seizure duration, but not the lowering of convulsive threshold. As chemists invent new xanthine derivatives, there is some chance that they will find one without unwanted side effects and with more selectivity for adenosine receptors, thus allowing more circumscribed pharmacological intervention than is currently possible with caffeine.

E. Schizophrenic Disorders

Although many schizophrenic patients drink no coffee, others have been observed to drink 20 cups of coffee a day, wear coffee-brown "mustaches," and snort instant-coffee crystals. Caffeine ingestion has been shown to exacerbate schizophrenic symptoms and caffeine removal has been shown to reduce them. Other studies have failed to find any effect of caffeine removal, but this may be explained by different subgroups of schizophrenics reacting differently to caffeine. More research is needed. [*See* Schizophrenic Disorders.]

F. Clinical Precautions and Other Issues

Clinicians working with anorexics may want to monitor their patients' caffeine intake. In their striving to be thin, anorexics have been observed to consume large quantities of diet colas or coffee, apparently because these beverages have few calories and suppress the appetite.

Clinicians working with patients who are taking psychotropic medication should note that diazepam has antagonistic and synergistic interactions with caffeine, although the exact nature of these interactions is controversial. Presumably other benzodiazapines have similar interactions.

The purported link between caffeine consumption and severity of premenstrual syndrome is not well established at this time. The use of caffeine in the treatment of headache or obesity is debatable. Expanded coverage of these issues and other psychosocial effects of caffeine may be found in Blount and Cox (1994).

VIII. CONCLUSIONS

Research to date has provided a definitive metabolic pathway for caffeine in humans and a number of well-established immediate effects of caffeine on the bodily systems. Caffeine has both beneficial and adverse effects on physical and mental performance, but these are less well established because of the intricacies of the research methodology. The data suggest, but do not conclusively prove, that caffeine has adverse effects on physical and mental health. Specifically, there appear to be links between caffeine intake and hypercholesterolemia, miscarriages, low birthweights, hypertension, heart disease, anxiety, and psychiatric disorders. There is much more to be learned about the mechanisms by which caffeine produces its effects. Researchers have come to opposite conclusions on many of the issues, and in the process have revealed extraneous variables that must be taken into account. It is important that future research resolve these issues, and that the studies be well designed and well controlled, taking these new variables into account. Caffeine deserves appropriate respect as a biologically and psychologically active drug. Because the objective results have remained mixed regarding the more troublesome effects, the medical community has not yet put caffeine in the same category as alcohol and nicotine.

Some organizations have used the research on caffeine to advocate that caffeine consumption be considered a social problem. This level of alarm is inappropriate; however, more education of the general public regarding caffeine is warranted. Educational efforts should cover known and probable health risks, misconceptions that caffeine counteracts the effects of alcohol, misconceptions about sources of caffeine, becoming aware of one's actual intake, and ways to reduce one's intake. Those who want to reduce or eliminate caffeine intake may do so on their own by reducing the concentration of caffeine in the foods or beverages that they consume (e.g., by steeping tea for 1 min instead of 5 min, mixing caffeinated and decaffeinated coffee), substituting noncaffeinated products for caffeinated ones (e.g., carob for chocolate, fruit juice during "coffee" breaks, caffeine-free for caffeinated over-the-counter medications), gradually eliminating occasions on which caffeine is consumed (e.g., coffee with the evening meal), and organizing a support group of friends or coworkers. Some individuals may find it hard to reduce their intake of caffeine because they lack motivation, because of social pressure to consume, or because they do not want to give up the stimulatory effects of caffeine. Such individuals may want to seek the help of health-care professionals who use systematic multicomponent interventions proven to be successful. In short, for individuals to decide to continue or change their caffeine habits, they must be informed.

BIBLIOGRAPHY

American Psychiatric Association. (1994). "Diagnostic and Statistical Manual of Mental Disorders," 4th Ed. APA, Washington, D.C.

Ashton, C. H. (1987). Caffeine and health. *Br. Med. J.* **295,** 1293–1294.

Blount, J. P., and Cox, W. M. (1985). Perception of caffeine and its effects: Laboratory and everyday abilities. *Percep. Psychophys.* **38,** 55–62.

Blount, J. P., and Cox, W. M. (1994). Caffeine: Psychosocial effects. *In* "Encyclopedia of Human Behavior" (V. S. Ramachandran), Vol. I. Academic Press, San Diego.

Bruce, M. S., and Lader, M. H. (1986). Caffeine: Clinical and experimental effects in humans. *Hum. Psychopharmacol.* **1,** 63–82.

Dews, P. B. (ed.). (1984). "Caffeine: Perspectives from Recent Research." Springer-Verlag, Berlin.

Garattini, S. (1993). "Caffeine, Coffee, and Health" Raven, New York.

Gilbert, R. (1992). "Caffeine: The Most Popular Stimulant." Chelsea House, New York.

Goldstein, A. (1994). "Addiction: From Biology to Drug Policy." Freeman, New York.

Griffiths, R. R., and Woodson, P. P. (1988). Caffeine physical dependence: A review of human and laboratory animal studies. *Psychopharmacol.* **94,** 437–451.

Heller, J. (1987). What do we know about the risks of caffeine consumption in pregnancy? *Br. J. Addiction* **82,** 885–889.

Humphreys, M. S., and Revelle, W. (1984). Personality, motivation, and performance: A theory of the relationship between individual differences and information processing. *Psychol. Rev.* **91,** 153–184.

James, J. E. (1991). "Caffeine and Health." Academic Press, San Diego.

Linde, L. (1995). Mental effects of caffeine in fatigued and non-fatigued female and male subjects. *Ergonomics* **38,** 864–885.

Pincomb, G. A. (1996). Acute blood pressure elevations with caffeine in men with boarderline hypertension. *Am. J. Cardiology* **77,** 270–274.

Revelle, W., Humphreys, M. S., Simon, L., and Gilliland, K. (1980). The interactive effect of personality, time of day, and caffeine: A test of the arousal model. *J. Exp. Psychol. General* **109,** 1–31.

Troyer, R. J., and Markle, G. E. (1984). Coffee drinking: An emerging social problem? *Social Problems* **31,** 403–416.

Watson, R. R. (1988). Caffeine: Is it dangerous to health? *Am. J. Health Promotion* **2**(4), 13–22.

Calcium Antagonists

WINIFRED G. NAYLER
University of Melbourne

GLOSSARY

Ca^{2+} Calcium in its ionized form

Extracellular fluid Fluid bathing the outer surface of the cells

Myocytes Single muscle cells

Na^+ Sodium in its ionized form

CALCIUM ANTAGONISTS ARE A NEWLY DISCOVered group of drugs that are now being used in the management of patients with a wide variety of disorders, including angina and hypertension (high blood pressure). The prototypes of the group were initially developed because of their ability to dilate the coronary blood vessels of the heart, but almost by chance it was discovered that these drugs also depress the pumping activity of the heart. Moreover, it was noted that this depressant effect on the heart could be counteracted or reversed simply by adding more calcium. This discovery, coupled with the fact that calcium ions (Ca^{2+}) are needed for muscle contraction, resulted in the conclusion that these drugs limit the availability of calcium to the muscle cells. Hence the term "calcium antagonist" was introduced to define them, and although subsequent investigations have shown that these drugs interact with only one of the possible ways in which calcium ions can be made available for muscle contraction, the term remains in common usage. Alternatives include "calcium channel blockers" and "calcium entry blockers."

I. DEFINITION OF CALCIUM ANTAGONISM

A. Background

This article deals mainly with the synthetic calcium antagonists that are now available. The discussion will concentrate on the chemistry of these drugs, how they modulate calcium transport, and the consequences of that modulation in terms of their current and prospective clinical use. It may be useful to know, however, that calcium antagonists are not the perogative of Western medicine. They form an active ingredient in some of the traditional Chinese medicines used in the treatment of cardiac disorders. Two such compounds are Tanshinone and Tetrandrine. As far as Western medicine is concerned, however, a wide variety of calcium antagonists are now available, with the enormity of the range reflecting the need to have drugs to target a particular end organ. For example, one of the calcium antagonists, nimodipine, has a high degree of specificity for the cerebral blood vessels. Another, nisoldipine, has a high degree of specificity for the blood vessels that regulate the supply of blood to the heart (the coronary vasculature). Others, including the prototype of the group, verapamil, are most effective in modulating calcium availability in the tissues that are responsible for regulating the distribution of the excitatory stimulus from its site of origin in the "pacemaker" area of the heart to the other chambers of the heart. This makes this particular calcium antagonist useful for treating patients with irregular heart

ENCYCLOPEDIA OF HUMAN BIOLOGY, Second Edition, VOLUME 2. Copyright © 1997 by Academic Press. All rights of reproduction in any form reserved.

beats (arrhythmias) caused by excessive or irregular pacemaker activity.

B. Site of Action

All muscle and excitable cells, including those of the brain and nervous tissue, are surrounded by a complex membrane that contains thousands of "pore-like" channels. When closed (inactive), these channels are impermeable to ions, but when open (active), they allow millions of ions to traverse the membrane each second (Fig. 1). The direction of movement of the ions through these channels is governed by the concentration gradients of the ions across the membrane. This means that sodium ions (Na^+) move in an inward direction into the cells, whereas potassium ions (K^+) move in the opposite direction, to accumulate in the fluid (extracellular) that surrounds the cells. Calcium ions (Ca^{2+}) resemble Na^+ ions in moving from the extracellular fluid into the cell. The movement of these ions (Ca^{2+}, Na^+, and K^+) does not occur in a haphazard fashion but rather through ion-specific channels. Thus, there are channels that selectively admit Na^+ into the cell. Other channels provide the route of K^+ exit, whereas others selectively admit Ca^{2+}. The calcium antagonists exert their effect by modulating the movement of Ca^{2+} through these Ca^{2+}-selective channels. [*See* Ion Pumps.]

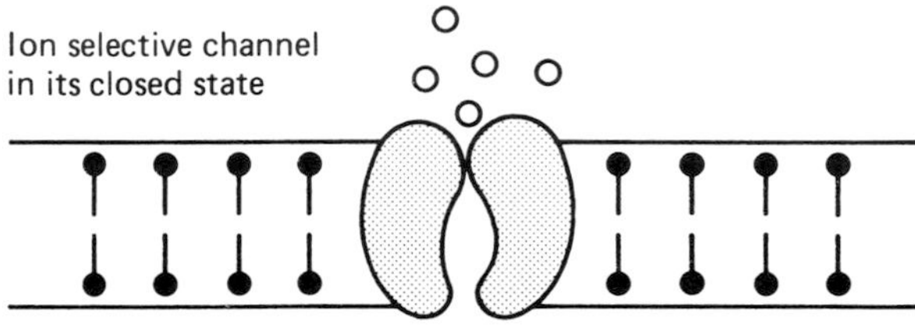

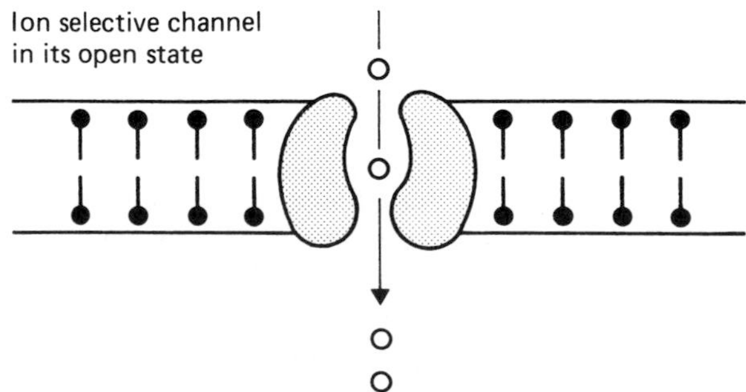

FIGURE 1 Schematic diagram of an ion-selective channel. The opening of each channel has a diameter large enough to admit Ca^{2+}. Larger ions are excluded.

In most excitable tissues the Na^+-carrying channels open within milliseconds of the cells being stimulated, whereas the Ca^{2+}-conducting channels open more slowly. It was for this reason that the terminology of "fast" Na^+ and "slow" Ca^{2+} channels was introduced. The currents that are generated by the inward movement of the ions through their respective channels accordingly are referred to as "fast inward Na^+" and "slow inward Ca^{2+}" currents.

When the Na^+ channels are open, Na^+ ions file through them at a rate of approximately 50 million ions every 10 sec. The Ca^{2+}-conducting channels selectively admit Ca^{2+} ions at about the same rate. Their ability to exclude Na^+ ions has nothing to do with the actual size of the channels because the diameter of a Na^+ ion is smaller than that of a Ca^{2+} ion. Instead the specificity for Ca^{2+} revolves around the chemical rejection of Na^+ from the lumen of the Ca^{2+} channel.

The importance of the Ca^{2+}-conducting channels and of the Ca^{2+} ions that move through them cannot be underestimated because these Ca^{2+} ions play a critical role in a wide variety of physiological processes, including muscle contraction, impulse propagation in the heart, and neuronal activity. These channels provide the primary site of action of the calcium antagonist drugs.

The overall effect of the calcium antagonists is to restrict Ca^{2+} ion entry, so that over a given period of time fewer Ca^{2+} ions become available for participation in the various intracellular events with which they are involved. Hence the alternative terminology: "calcium channel blockers," "slow calcium channel blockers," or "Ca^{2+} entry blockers."

II. MODE OF ACTION OF CALCIUM ANTAGONISTS

A. Techniques Used to Study the Functioning of Ca^{2+} Channels

The techniques that have been used to study the electrical activity associated with the opening and closing of Ca^{2+} channels include

1. Insertion of a conducting electrode into a single cell to record the electrical activity associated with the opening and closing of the channels in the membrane. By using appropriate bathing solutions it is possible to preset the potential difference across the membrane to a point at which only the Ca^{2+}-conducting channels are operative.

2. The electrical activity associated with the opening and closing of a single Ca^{2+}-conducting channel can be recorded from a small "patch" of the membrane that has been sucked into the orifice of a pipette. This is often called "patch clamping."

3. In addition, with modern technology it is now possible to harvest a single Ca^{2+} channel and insert it into an artificial lipid bilayer so that its properties can be studied without possible interference from surrounding tissue.

As soon as investigators started using these last two techniques it became apparent that the membranes of many excitable tissues, including those of the heart and the blood vessels, contain several different types of Ca^{2+} channels.

B. Heterogeneity of the Ca^{2+} Channels

There are at least three different types of Ca^{2+} channels, and by convention they are designated as L, T, and N types. These channels can be differentiated from one another in terms of (1) the transmembrane potential difference at which they become operational, (2) their Ca^{2+} ion-carrying capacity, or conductance, (3) their stability in isolated membrane patches, (4) their sensitivity to calcium antagonists, (5) the rate at which they open and close, and (6) their sensitivity to naturally occurring toxins.

The L-type channels were originally given this designation because once activated they remain open for a relatively long period of time and have a large Ca^{2+} ion-carrying capacity. In contrast, the T channels were so designated because of their brief opening time. The N channels were originally called N because their characteristics were neither of the L nor T type. However, N now is commonly assumed to mean neuronal because these channels assume prevalence in neuronal tissue. Only the L-type channels are sensitive to the drugs that are now classed as calcium antagonists. The Ca^{2+} conducting T-type channels, however, can be blocked by certain inorganic divalent cations, including cobalt and nickel. The sensitivity of the L-type channels to the organic-based calcium antagonist drugs can be explained simply in terms of the fact that sites with which these drugs interact actually form part of the L-type calcium channel.

C. Size and Structure of the Calcium Antagonist-Binding Sites

Biochemists who are interested in establishing how the calcium antagonists modulate Ca^{2+} ion entry through the L-type Ca^{2+} ion-conducting channels have identified the complex that contains the binding sites for these drugs. The molecular weight is around 170 kDa and, as such, it accounts for approximately half of the calcium channel. The binding site can be subdivided into four polypeptide subunits designed as α_1, α_2, β, and γ. The α_1 subunit contains the binding sites for the calcium antagonist drugs.

Until recently the calcium antagonist-binding site was thought to have the same structure and composition, irrespective of its tissue location. Quite recently, however, Schwartz and colleagues presented evidence that there are tissue-dependent differences in these binding sites. Thus it might help to explain why some of the antagonists act preferentially on certain tissues.

D. Effect of Calcium Antagonists on the Functioning of the Ca^{2+} Channels

Each Ca^{2+}-conducting channels has a half-life of about 40 hr, and it admits as many as 4 million Ca^{2+} ions per second when it is in its open or "activated" state. The calcium antagonists modulate the Ca^{2+}-carrying capacity of these channels not by providing a physical block or "plug" at the channel orifice, but rather by reducing the likelihood of the channels being open. Each channel can exist in three modes: open, shut (and hence unavailable for activation), or resting (and hence available for activation). The calcium antagonist drugs either prolong the closed time for each channel or reduce the likelihood of its being available for activation. Only those tissues that contain L-type channels are sensitive to the calcium antagonists (organic). They include the heart, the blood vessels, the electrical conducting pathways within the heart, the uterus, and the gastrointestinal tract.

III. CHEMISTRY OF THE CALCIUM ANTAGONISTS

The calcium antagonists can be subdivided into two major groups according to their chemistry: inorganic and organic. Cobalt (Co^{2+}), nickel (Ni^{2+}), manganese (Mn^{2+}), and lanthanum (La^{3+}) belong to the inorganic group, and although they are useful laboratory tools, they cannot be used clinically because of their toxicity and because they do not discriminate betwen the L- and T-types of calcium channels. In contrast, the organic calcium antagonists are relatively specific for the L-type channel. Nevertheless, they exhibit a large

TABLE I
Calcium Antagonists Subgroups

Subgroup		Molecular weight
Phenylalkylamines		
Prototype	Verapamil	454.5
Derivatives	Gallopamil (D600)	485.6
	Desmethoxyverapmil (D888)	461.0
	Animamil	520.8
	Ronipamil	460.7
	Devapamil	424.6
	Tiapamil	592.1
	Fendiline	315.5
Dihydropyridines		
Prototype	Nifedipine	346.3
Derivatives	Nisoldipine	388.4
	Nitrendipine	490.6
	Nimodipine	418.5
	Niludipine	490.5
	Nicardipine	388.4
	Felodipine	384.3
	Isradipine	371.4
	Amoldipine	408.9
Benzothiazepines		
Prototype	Diltiazem	414.5
Piperazines		
Prototype	Lidoflazine	490.6
Derivatives	Cinnarizine	360.5
	Flunarizine	406.5
Others		
	Bepridil	366.6
	Perhexiline	277.5
	Prenylamine	329.5

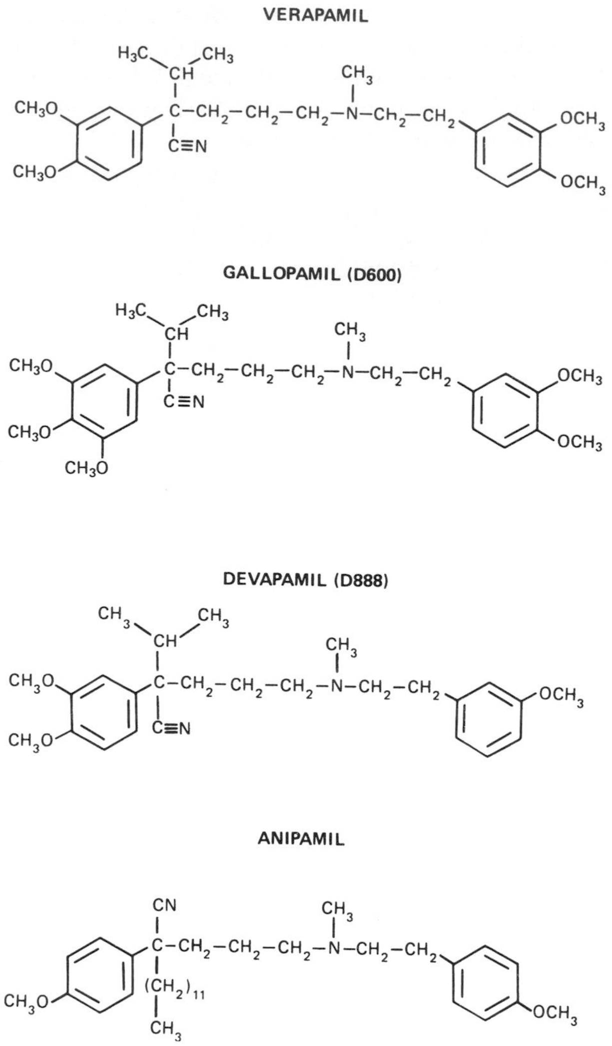

VERAPAMIL

GALLOPAMIL (D600)

DEVAPAMIL (D888)

ANIPAMIL

FIGURE 2 Chemical structure of verapamil and its derivatives: gallopamil, devapamil, and anipamil.

amount of chemical heterogeneity. Thus (Table I) some are phenylalkylamines, some are dihydropyridines, and others are either benzothiazepines or piperazines.

Calcium antagonists differ from one another in terms of their potency, tissue selectivity, and duration of action.

A. Phenylalkylamines

The prototype of the calcium antagonists is the phenylalkylamine verapamil (Fig. 2). Compounds that have a similar chemistry and that are now classed as calcium antagonists include (Fig. 2) gallopamil, verapamil, and anipamil.

Verapamil was originally selected for development as a therapeutic agent because of its potency as a coronary blood vessel dilator, but relatively early studies on this compound showed that it also slowed the heart rate and depressed the foce of cardiac contractions (negative inotropy). When the drug reached clinical trial status, it became apparent that it would be useful in the management of patients with hypertension because it has a blood pressure-lowering effect. Gallopamil resembles verapamil in its pharmacology, but its potency as a calcium antagonist exceeds that of the parent compound by a factor of 10. Anipamil matches gallopamil in terms of potency, but because it only slowly dissociates from its binding sites, it has a long duration of action.

The essential features of the phenylalkylamine-based calcium antagonists are (1) the presence of the two benzene rings, and (2) a tertiary-amino nitrogen in the chain linking the two benzene rings (Fig. 2). As a rule the drugs exist in either the d or l isomeric state, but it is the l isomer that has the greatest calcium antagonist activity.

B. Dihydropyridines

Nifedipine is the prototype of this group. Structurally it is quite dissimilar from verapamil (Fig. 3) and, unlike verapamil, it is rapidly degraded and inactivated when exposed to either daylight or ultraviolet light.

Various substitutions within the two rings of nifedipine have provided a wide range of dihydropyridine-based calcium antagonists, some of which are listed in Table I. The compounds differ from one another not only in terms of their potency and duration of action, but also in terms of their tissue specificity. For example, amlodipine is a long-acting calcium antagonist, relative to nifedipine, but like nifedipine it does not discriminate between the various vascular beds. In contrast, nisodipine, another dihydropyridine-based calcium antagonist, acts preferentially on the coronary blood vessels, whereas nimodipine is highly selective for the cerebral blood vessels.

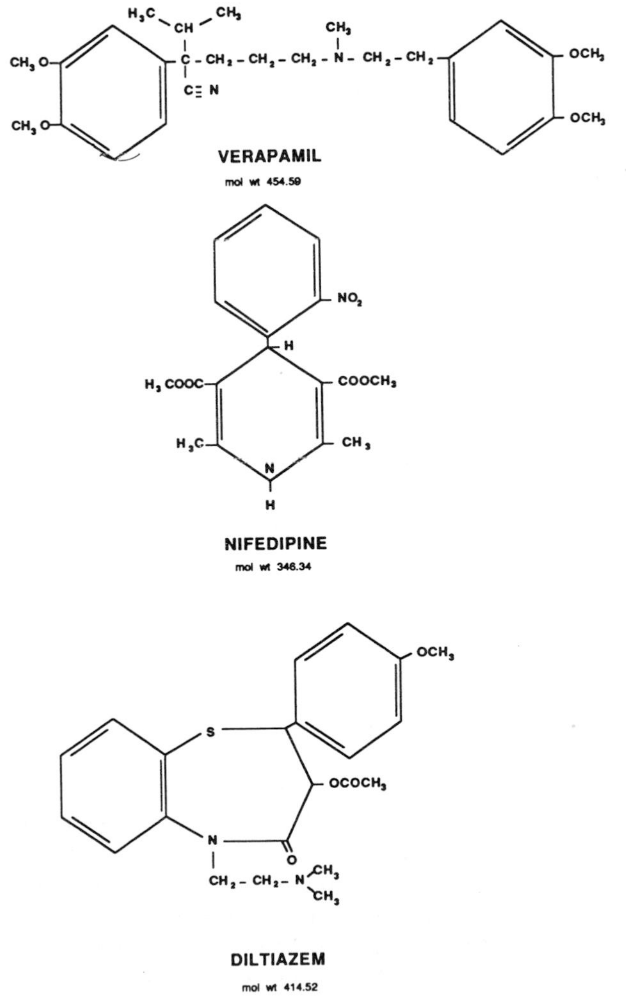

C. Benzothiazepines

Diltiazem (Fig. 3) is the prototype of this group. It resembles the phenylalkylamines in lacking vascular selectivity, but within the vasculature, it has a powerful dilator effect on the coronary arteries. It is not light sensitive and, like verapamil but in contrast to nifedipine, it is water soluble. Derivatives of diltiazem are being developed, but they are not yet available for clinical use.

D. Piperazines

Lidoflazine, cinnarizine, and flunnarizine are examples of this group. However, because these drugs also inhibit Na^+ ion entry through the Na^+-selective channels, some investigators argue that they should not be classed simply as calcium antagonists. Nevertheless they exert a potent inhibitory effect on the slow calcium channels in the vasculature.

VI. CRITERIA FOR CLASSIFYING A DRUG AS A CALCIUM ANTAGONIST

Two criteria at least must be satisfied: (1) The drug must exert a dose-dependent inhibitory effect on the slow inward Ca^{2+} current (reflecting Ca^{2+} ion entry through the Ca^{2+}-selective channels), and (2) the drug must interact specifically with the known binding sites associated with the Ca^{2+} channels.

V. TISSUE SELECTIVITY OF CALCIUM ANTAGONISTS

There are many examples of the tissue selectivity of these drugs. For example, although all the calcium antagonists have a dose-dependent inhibitory effect on cardiac and vascular smooth muscle, they have

VERAPAMIL

mol wt 454.59

NIFEDIPINE

mol wt 346.34

DILTIAZEM

mol wt 414.52

FIGURE 3 Structural formulas of the three calcium antagonist prototypes: verapamil (a phenylalkylamine), nifedipine (a dehydropyridine), and diltiazem (a benzothiazepine).

little effect on skeletal muscle. In addition they have relatively little effect on antigen- or spasmogen-induced contractions in tracheal and bronchial smooth muscle, although they relax vascular smooth muscle. As a rule they do not interfere with the excitation–secretion process, although many of these are Ca^{2+} dependent.

Various factors contribute to the tissue selectivity of these drugs. They include (1) differences in the source of Ca^{2+} ions needed for the physiological response; (2) differences in the chemical profiles of the drugs, which will determine how closely they can approach the receptors in a particular tissue; (3) differences in the density, distribution, and type (L, T, or N) of Ca^{2+}-selective channel; and (4) differences in the requirements of the binding sites and other ancillary properties of the drugs, including their interaction with other receptors, their effect on other ion-conducting channels, and, in some cass, possible intracellular effects.

The insensitivity of skeletal muscle is readily explicable because in that tissue the Ca^{2+} ions required to activate the contractile proteins originate from within the muscle cell. This contrasts with the situation that exists in cardiac and smooth muscle, where some extracellular Ca^{2+} must move inward across the cell membrane before contraction occurs.

Some calcium antagonists exhibit "use dependence." This is probably associated with the fact that the drug only interacts with the relevant receptor when the channel is in the "open" state. This applies particularly to the calcium antagonists that are ionized when in solution (verapamil and diltiazem). As a corollary, these drugs are particularly effective in those tissues in which the Ca^{2+}-conducting channels are being repeatedly activated, as at the atrioventricular node. This property of "use" dependence probably explains why drugs like verapamil, and to a lesser extent diltiazem, are particularly useful in controlling certain cardiac arrhythmias.

As far as the vascular selectivity of the dihydropyridines is concerned (Table II), the most likely explanation is that the transmembrane potential difference across vascular smooth muscle cells, which differs from that across the membranes of cardiac muscle cells, favors the binding of the dihydropyridine-based molecule. Different sensitivities within the vascular system involve a host of possibilities, including (1) differences in T-, L-, and N-type channel distribution, (2) relevance of reflex control mechanisms, (3) different stimuli for contraction, and (4) regional differences in membrane excitability and lipid composition, the latter being of importance because the lipid-soluble dihydropyridines may approach their receptors by way of the lipid bilayer of the membrane.

TABLE II
Tissue Selectivity of the Calcium Antagonists[a]

Drug	Myocardium	Vasculature	Conducting tissue	Skeletal muscle
Verapamil	+	+	+	−
Gallopamil	+	+	+	−
Diltiazem	+	+	+	−
Nifedipine	+	++	−	−
Nisoldipine	+	++++	−	−
Amlodipine	+	++++	−	−
Felodipine	=	++++	−	−

[a] + denotes the presence and − the absence of an effect on the activity of that tissue at therapeutically achievable concentrations.

The relative insensitivity of neuronal tissue is not altogether surprising because although the Ca^{2+}-selective channels that are of importance for transmitter release do occur in the dendrites, most of them are N-type channels, which are insensitive to the calcium antagonists.

The fact that the calcium antagonists have little or no effect on neuronal tissues is clinically important, and it explains why neuronally mediated processes are not affected by these drugs, which therefore lack mental side effects. In general, therefore, the calcium antagonists will only affect the functioning of those tissues in which (1) the L-type Ca^{2+} channels predominate, (2) the Ca^{2+} ions that mediate the functional response originate largely from the extracellular fluid, and (3) the Ca^{2+}-selective L-type channels are in an operational state that favors the binding of the drugs to the receptors associated with the Ca^{2+} channels.

VI. CLINICAL RELEVANCE

When first introduced into clinical practice, the calcium antagonists were considered to be potentially useful for the management of patients with ischemic (inadequate blood supply to the wall of the heart) heart disease, including angina pectoris. Experience has shown, however, that their usefulness includes a wide variety of cardiovascular disorders (Table III). Hypertension, certain forms of arrhythmias, and even, under certain circumstances, heart failure can provide areas of use. The drugs also have other potential uses,

TABLE III
Clinical Uses of Calcium Antagonists

Cardiovascular
Arrhythmias
Atrial flutter
Paroxysmal supraventricular tachycardia
Angina
Chronic, stable, and vasospastic
Hypertension
Peripheral vascular disease
Raynaud's disease
Cerebral spasm
Congestive heart failure
Hypertrophic cardiomyopathy
Stroke
Migraine headache
Others
Achalasia
Premature labor
Adjunct to immunosuppressive therapy
Potentiation on oncotic drug activity
Tissue protection
Heart, kidney, and brain
Cocaine intoxication
Prevention of cardiac toxicity
Morphine withdrawal symptoms
Ethanol withdrawal symptoms
Atherosclerosis

including the management of premature labor and Raynaud's disease (i.e., a disease of the peripheral blood vessels that is caused by their intense constriction so that the fingers and toes receive an inadequate supply of blood). The drugs are also being used to treat patients with esophageal achalasia (a disorder characterized by sustained constriction of the esophagus, which makes swallowing difficult or impossible). Other areas include the protection of muscle and brain tissue from secondary damage caused by inadequate perfusion, such as may occur if a thrombus (blood clot) disturbs blood flow or the vessels suddenly develop a sustained contraction (spasm). There is even evidence that suggests that these drugs might provide a useful way of slowing the growth of atherogenic plaques (calcified fatty deposits), which form in the lumen of blood vessels, including those of the heart.

A. Calcium Antagonists and Angina

The pathology of angina is complex. The condition can arise for a variety of reasons, including (1) a sudden reduction in the lumen of a major coronary artery caused by contraction of the muscle in the wall of the artery; (2) progressive reduction in blood flow through the coronary arteries because of platelet aggregations (blood clots) or atherosclerotic lesions; or (3) a sudden increase in the oxygen requirements of the heart, caused by an increase in workload on the heart.

The calcium antagonists, particularly the vascular selective calcium antagonists, can be useful for the management of these conditions because by restricting the entry of Ca^{2+} ions into the muscle cells of the coronary vasculature, they cause the vessels to dilate, thereby increasing the effective blood flow. However, if the inadequate flow of blood is due to a fixed lesion (e.g., an adherent clot or large atherosclerotic plaque), the drugs will be of little benefit.

B. Calcium Antagonists and Arrhythmias

Some calcium antagonists can be used to manage certain arrhythmias caused by excessive or irregular conduction of impulses through the conducting pathways in the heart.The dihydropyridines (nifedipine), however, cannot be used for this purpose because of their potency as vasodilators, but verapamil and, to a lesser extent, diltiazem are effective. Their use depends on their ability (1) to depress the electrical activity of the sinus node (i.e., the region of the heart where the "heart beat" actually originates), and (2) to slow down conduction through the conducting tissues so that even if the sinus node does trigger a rapid volley of excitatory stimuli, the excess pulses are prevented from spreading to the other areas of the heart.

C. Calcium Antagonists and Hypertension

An essential characteristic of hypertension is excessive constriction of the peripheral vasculature. This excessive constriction is caused by abnormalities in the mechanisms that are responsible for maintaining homeostasis with respect to Ca^{2+}. Because the force of muscle contraction is directly proportional to the amount of Ca^{2+} that is available, it is logical to use drugs that limit Ca^{2+} availability to treat a disorder that reflects either an excess availability of or excessive sensitivity to Ca^{2+} ions. The calcium antagonists are particularly useful for this purpose and, in contrast to many other blood pressure-lowering agents, they neither alter the plasma lipid profile (ratio of high-density to low-density lipoproteins) nor cause the retention of sodium ions (Na^{+}). To the contrary, because of their effect on the glomerular filtration rate in the kidney, they cause Na^+ and water loss, thereby preventing the development of edema (water retention). [*See* Hypertension.]

D. Calcium Antagonists and the Slowing of Atherogenesis

Laboratory studies on cholesterol-fed rabbits and monkeys and one study in humans that extended over a 3-year period indicate that the calcium antagonists may slow the rate of growth of atherogenic plaques. Some evidence suggests that this effect is due to the ability of these drugs to slow the rate of growth of smooth muscle cells, an event that is Ca^{2+} dependent and fundamental to the development of atherogenic lesions. [*See* Atherosclerosis.]

E. Calcium Antagonists and Cell Protection

In most organs the immediate cause of cell death appears to be an uncontrolled entry to Ca^{2+} ions. This applies to the heart that has been deprived of blood flow through its coronary vessels for a prolonged period of time and, likewise, to the brain. Laboratory evidence shows that calcium antagonists can have a beneficial effect under these conditions, but only if they are present before blood flow to the affected tissue is restored. In clinical practice, this presents a dilemma because of the difficulty of predicting which patients are at risk.

F. Other Uses and Future Developments

There are other conditions in which the calcium antagonists have been found to be of use, including the management of patients receiving immunosuppressive therapy associated with transplantation procedures.

Research on new drugs is continuing, with the aim of producing drugs with greater tissue specificity, prolonged duration of action, and reduced side effects. The newly developed drugs, however, like their prototypes, must retain the basic property of modulating Ca^{2+} ion movement through the Ca^{2+}-selective channels.

BIBLIOGRAPHY

Epstein, M. (ed.) (1992). "Calcium Antagonists in Clinical Medicine." Hanley and Belfus.

Finkel, M. S., Patterson, R. E., Roberts, W. G., Smith, T. D., and Keiser, H. R. (1988). Calcium channel binding characteristics in the human heart. *Am. J. Cardiol.* **62,** 1281–1284.

Fleckenstein, A. (ed.) (1983). "Calcium Antagonists in the Heart and Smooth Muscle." Wiley, New York.

Godfraind, T., *et al.* (eds.) (1993). "Calcium Antagonists: Pharmacology and Clinical Research." Kluwer Academic.

Mikam, A., Imotoa, K., Tanabe, T., Niidome,T., Mori, Y., Takeshima, H., Narumiya, S., and Numa, S. (1989). Primary structure and functional expression of the dihydropyridine-sensitive calcium channel. *Nature* **340,** 230–236.

Nayler, W. G. (ed.) (1988). "Calcium Antagonists." Academic Press, London.

Stone, P. H., and Antman, E. M. (eds.) (1983). "Calcium Channel Blocking Agents in the Treatment of Cardiovascular Disorders." Futura Publishing, New York.

Vaghy, P. L., Itagaki, K., Miwa, K., McKenna, E., and Schwartz, A. (1988). Mechanism of action of calcium channel modulator drugs: Identification of a unique, labile, drug-binding polypeptide in a purified calcium channel preparation. *Ann. N.Y. Acad. Sci.* **522,** 176–186.

Vaghy, P. L., McKenna, E., Itagki, K., and Schwartz, A. (1988). Resolution of the identity of the Ca^{2+}-antagonist receptor in skeletal muscle. *Trends Pharmacol. Sci.* **9,** 398–402.

Weiss, W. F., and Simic, M. G. (1988). Calcium antagonists in tissue protection. *Pharmacol. Ther.* **39,** 385–388.

Calcium, Biochemistry

JOACHIM KREBS
Swiss Federal Institute of Technology

GLOSSARY

Calmodulin Intracellular Ca^{2+}-binding modulator protein involved in triggering Ca^{2+} signals in the cell. It is a highly conserved, acidic protein containing four specific Ca^{2+}-binding domains (EF-hands). It belongs to a superfamily of highly homologous Ca^{2+}-binding proteins

Channel Transmembrane proteins responsible for ion-specific transport across the membrane down the ion's gradient

Coordination number Definition of the number of ligands bound to a central atom, which gives it a geometric configuration (e.g., tetraeder, octaeder)

EF-hand motif Expression coined by R. H. Kretsinger to describe Ca^{2+}-binding domains of specific proteins. It is composed of two helices and the Ca^{2+}-binding loop enclosed by the former, therefore it is also called the "helix–loop–helix" motif. The motif can be modeled by the forefinger and the thumb of the right hand (representing the two helices) and by the bent midfinger (representing the loop)

EF-hand type proteins Ca^{2+}-binding proteins containing "EF-hand" or "helix–loop–helix" motifs (e.g., calmodulin, parvalbumin, and troponin C)

Electrogenic transport Net transport of charges across a membrane, usually facilitated by an integral membrane protein

G proteins Specific GTP-binding proteins, consisting of three subunits (α, β, γ), involved in the signal-transducing process, transferring the message from a specific, hormonally activated receptor to the enzyme producing the second messenger

Growth factors Small peptide hormones [e.g., epidermal growth factor (EGF), platelet-derived growth factor (PDGF), insulin-like growth factor (IGF)] involved in cell proliferation. By binding to specific receptors they induce cell proliferation as a result of specific signal-transducing pathways

Ion exchanger Transmembrane protein involved in ion transport using the downhill gradient of one ion as an energy source to transport the other against its gradient (e.g., Na^+/Ca^{2+} exchanger)

Ion pump Ion-transporting enzyme that transports the ion (e.g., Ca^{2+}) across the membrane against its gradient by using ATP as an energy source, thereby forming an acylphosphate intermediate

Reticulum Intracellular organellar system (e.g., sarcoplasmic reticulum, endoplasmic reticulum), important for the control of Ca^{2+} homeostasis in the cell

Second messenger Either small molecules (e.g., cyclic nucleotides or inositolpolyphosphates) or ions (e.g., Ca^{2+}) that are pivotal for the transduction of extracellular stimuli (e.g., hormonal signals = primary messengers) into intracellular events

CALCIUM PLAYS A PIVOTAL ROLE IN BIOLOGICAL systems. It occurs predominantly in a complexed form, fulfilling either a static, structure-stabilizing function or a dynamic, signal-transducing role. Because of large gradients of Ca^{2+} across cellular membranes, the cell can use even small changes of its membrane permeability for the ion to cause significant changes in the intracellular free Ca^{2+} concentration, thereby transmitting and amplifying a signal from the extracellular space. Therefore the concentration of

ENCYCLOPEDIA OF HUMAN BIOLOGY, Second Edition, VOLUME 2. Copyright © 1997 by Academic Press. All rights of reproduction in any form reserved.

ionized calcium within the cell has to be carefully controlled, which involves a number of different systems, to use Ca^{2+} as an intracellular second messenger. This also involves specific intracellular Ca^{2+}-binding proteins with characteristic "helix–loop–helix" motifs as Ca^{2+}-binding domains important for the modulation of the Ca^{2+} message.

I. INTRODUCTION

Calcium is one of the most common elements found on earth, making up about 3% of the crust's composition. It was discovered by Humphry Davy as a chemical element in 1808. The biological importance of calcium minerals and their stabilizing function in shells, bones, and teeth soon became apparent, and today it is known that calcium is a very old component of organisms, the record (blue-green algae) being about 2×10^9 years. Therefore, it has long been recognized that calcium plays an important role in biology. It occurs predominantly in a complexed form, both in minerals and in solution. In most higher organisms, more than 99% of the total calcium is precipitated as hydroxyapatite $[Ca_{10}(PO_4)_6(OH)_2]$ in the skeleton, thereby fulfilling a rather static, structure-stabilizing function.

At the end of the nineteenth century, experiments on muscle contraction carried out by Sidney Ringer showed that calcium also played an important dynamic function in the regulation of cellular events. However, Earl Sutherland's discovery of cyclic AMP as an intracellular regulatory constituent in 1956 and the subsequent development of the "second messenger" concept prepared the ground to recognize the pivotal role of Ca^{2+} as an intracellular regulator.

Compared to the amount of calcium precipitated in the skeleton, the Ca^{2+} found in the extracellular fluid and intracellularly in the cytosol and in other intracellular compartments is rather minute. The concentration of calcium in the extracellular fluid or in the intracellular reticular system is in the millimolar range (2–5 mM, of which about 50% is ionized, i.e., unbound), whereas the free intracellular Ca^{2+} concentration in the cytosol of a resting cell is in the submicromolar range (200–500 nM). This results in a 10,000-fold concentration gradient of ionized Ca^{2+} across the membrane, and the resulting large electrochemical force permits that minor changes of the free Ca^{2+} concentration due to membrane permeability changes as the result of an external signal cause significant oscillations of its concentration in the cytosol. Those fluctuations provide the possibility of transmitting signals to a large variety of different biochemical activities, such as intracellular metabolic events, muscle contraction, neurotransmitter release, cell growth, proliferation and fertilization, stimulus–secretion coupling, and mineralization. Many of these activities are brought about by the interaction of Ca^{2+} with specific proteins, thereby stabilizing, activating, and modulating them, resulting in a conformational change and subsequent specific modulations of protein–protein interactions.

The dynamic role of Ca^{2+} as a signal transducer in the form of a so-called second messenger became the center of intense research activities in the last two decades. In this article, the different properties of calcium as a structure-stabilizing and signal-transducing factor will be described in detail, and the interrelationship of the different roles and functions will be discussed.

II. Ca^{2+} LIGATION

Ca^{2+} ligation in complexes usually occurs via carboxylates (mono- or bidentate) or neutral oxygen donors. The number of donor centers and the geometry of their arrangement have a great influence on the binding strength (e.g., in the range between 10^3 and 10^{12}). The superior binding properties of Ca^{2+} over Mg^{2+} or other cations like Na^+ or K^+, which are present at much higher concentrations, enable Ca^{2+} to fulfill its function as a signal transducer. Furthermore, besides its ability to choose oxygen donors as ligands, it has great flexibility in coordination (coordination numbers 6 to 8) with a largely irregular geometry in both bond length and bond angles. This geometry is suitable for binding to proteins, which usually provide irregularly shaped cavities. Since Mg^{2+} (0.64 Å) is smaller in diameter than Ca^{2+} (0.97 Å), it requires a regular octahedron with six coordinating oxygen atoms, whereas Ca^{2+} puts much less constraint on the complexing protein; thus its greater versatility in coordinating ligands and its much higher rate of exchanging them. The exchange rate, which is about 3 orders of magnitude faster for Ca^{2+} than for Mg^{2+} owing to a significantly slower dehydration rate for the latter, makes Ca^{2+} ideally suited for a signal-transducing factor. Since for Ca^{2+} the exchange rate with water in the inner coordination sphere is close to the diffusion limit, the on-rate in Ca^{2+} binding to proteins is often diffusion-limited, whereas the off-rates are dependent on the binding strength of the proteins.

These advantages of Ca^{2+} over other cations make it possible for Ca^{2+} to bind to cavities of proteins accepting a greater variation of interatomic distances. This ability in turn permits selective binding owing to a favorable charge-to-size ratio.

III. CALCIUM IN THE EXTRACELLULAR SPACE

Calcium in the extracellular space is tightly controlled to maintain its concentration in a range between 2 and 5 m*M*. It has long been known that certain specialized cells, such as the parathyroid cells, can recognize changes of Ca^{2+} levels in the extracellular space, but the mechanism remained unknown. Recently, E. M. Brown and his colleagues from the Harvard Medical School identified by expression cloning a 120-kDa protein from bovine parathyroid cells containing seven transmembrane domains similar to G-protein-coupled receptors, which can sense calcium changes in the extracellular space. The protein contains a large extracellular domain rich in clusters of acidic amino acids, which are possibly important for calcium binding. It is of interest to note that mutations in the human homologous receptor can cause familial hypercalcemia and neonatal severe hyperparathyroidism. [*See* Extracellular Matrix.]

It has long been recognized that one of the important roles of Ca^{2+} in the extracellular space is to stabilize the structure of proteins. Several such proteins have been crystallized and the Ca^{2+}-binding sites have been determined. By comparing their features with those of the intracellular trigger proteins (e.g., calmodulin) discussed in detail later, several important differences could be noted. Usually the Ca^{2+}-binding site of extracellular proteins, unlike those of the intracellular proteins, is preformed and relatively fixed. Therefore the on-rate for Ca^{2+} can be relatively slow, that is, much slower than the diffusion limit. Since the off-rate often is fairly slow, the affinity of Ca^{2+} for the protein is relatively low. However, since the Ca^{2+} concentration of the extracellular fluid is adequate for this affinity, these proteins usually occur in their Ca^{2+}-bound form, which is necessary to protect against proteolytic cleavage. The binding of Ca^{2+} to the protein reduces the probability of unfolding, which in turn would increase the probability of proteolytic cleavage.

The preformed cavity of Ca^{2+}-binding sites of extracellular proteins and their relatively high degree of rigidity are reflected in the arrangement of the ligands contributing to the binding site. In contrast to the intracellular Ca^{2+}-binding proteins (see the following), the amino acids participating in Ca^{2+} binding usually are located at distant positions in the amino acid sequence and are not sequential. Furthermore, these extracellular Ca^{2+}-binding proteins are not homologous to each other nor to one of the intracellular, so-called EF-hand type proteins (see Glossary).

Bone-forming cells, that is, osteoblasts, synthesize and secrete a number of noncollagenous Ca^{2+}-binding proteins such as osteocalcin, osteopontin, and osteonectin, which bind to bone minerals such as hydroxyapatite in a calcium-dependent manner. Osteocalcin belongs to a class of Ca^{2+}-binding proteins rich in γ–carboxyl glutamic acids mediating the Ca^{2+}-binding property, whereas osteopontin, a glycoprotein, is rich in sequences of glutamic and aspartic acid residues. In addition, both proteins contain arginine–glycine–aspartic acid (RGD) sequence domains known to be important for interaction with cell-surface receptors of the integrin type. Osteonectin or fibronectin, on the other hand, belongs to the family of EF-type Ca^{2+}-binding proteins typical for intracellular Ca^{2+}-trigger proteins. They are involved in bone formation or in blood clotting, respectively.

IV. SIGNAL TRANSDUCTION PRINCIPLES

Extracellular signals recognized by cell-surface receptors are transmitted into a limited number of intracellular signal transducers, so-called second messengers, to multiply the incoming signals using a cascade of intracellular events such as phosphorylation/dephosphorylation of a variety of different proteins or enzymes. It has long been known that the action of hormones such as adrenaline might be accompanied by changes of intracellular Ca^{2+} concentration, but it was not until the discovery of the intracellular formation of cyclic nucleotides upon an extracellular stimulus (e.g., β-adrenergic receptor stimulation) and the subsequent development of the second messenger hypothesis that the role of Ca^{2+} as an intracellular second messenger was recognized. To date four classes of intracellular messengers are known:

1. cyclic nucleotides (cyclic adenosine 3′-5′ monophosphate or cyclic AMP; cyclic guanosine 3′-5′ monophosphate or cyclic GMP);

2. derivatives of phosphatidylinositol [inositolpolyphosphates, e.g., inositol-1,4,5-triphosphate (IP_3), and diacylglycerol (DAG), stemming from the same precursor phosphatidylinositol-4,5-diphosphate (PIP_2)];
3. free Ca^{2+} ions; and
4. gases such as nitric oxide (NO) and probably carbon monoxide (CO).

Several steps are common to most signal-transducing pathways. After the receptor in the plasma membrane (the main barrier that information coming from outside has to overcome) receives the primary signal (e.g., a hormone), the information is transferred either indirectly via so-called G proteins (proteins that are modulated in their function because of the binding of GTP or GDP) or directly to signal-transducing enzymes responsible for the production of the second messengers. These are usually small phosphorylated molecules that originate either by converting energy-rich nucleotides (ATP, GTP) into cyclic nucleotides with the use of specific enzymes (e.g., adenylate or guanylate cyclases) or by the cleavage of phosphorylated forms of phosphatidylinositol located in the plasma membrane. Thus PIP_2 is cleaved into inositol-1,4,5-triphosphate and diacylglycerol by the specific, membrane-bound phospholipase C. IP_3 is closely connected to Ca^{2+} as second messenger, since it can release Ca^{2+} from intracellular stores, probably the endoplasmatic reticulum. Diacylglycerol amplifies its signal owing to the activation of protein kinase C, a Ca^{2+}- and phospholipid-dependent enzyme.

The way calcium performs its second messenger function is quite different. The essential difference is that for Ca^{2+}, fluctuations on a steep concentration gradient across the cell membrane modulate its messenger function, in contrast to metabolic synthesis and degradation. Upon receiving a signal from outside the cell (e.g., a nerve pulse), the voltage across the cell membrane changes (depolarization), thus opening voltage-sensitive Ca^{2+} channels that permit Ca^{2+} to enter the cell down its concentration gradient. The increased intracellular Ca^{2+} binds to specific proteins, causing a conformational change. These proteins provide the triggering device to multiply the signal. The occurrence of these specific Ca^{2+}-binding intracellular trigger proteins was first described in muscle tissues (i.e., troponin C), paving the way for the detection of a whole class of homologous proteins (see Section V).

A striking observation was made in 1986 by P. H. Cobbold and coworkers, who reported that upon addition of the calcium-mobilizing hormone vasopressin to hepatocytes, the intracellular Ca^{2+} did not rise to a sustained level, but instead increased and decreased repetitively with a certain frequency, that is, in calcium spikes. The frequency, but not the amplitude, of spiking increased proportional to the rise in concentration of the added hormone, that is, an extracellular *analog* signal was converted to an intracellular *digital* signal. This observation was later confirmed on a variety of different cells. Two models describing these phenomena in an attempt to rationalize the information transfer in time and space by calcium "spikes" or "waves" have been recently developed:

1. The IP_3–Ca^{2+} cross-coupling (ICC) model by T. Meyer and L. Stryer.
2. The calcium-induced calcium release (CICR) model by M. J. Berridge and coworkers.

The intracellular interplay between the different second messengers can be manifold and very complex. Figure 1 shows how the different second messengers not only are central components of intracellular control mechanisms, but are connected in their action by a complex network of feedback relationships. Signal transduction pathways are also connected with the interaction of polypeptide growth factors and their specific, high-affinity receptors on target cells. One of the earliest responses (within seconds) upon binding of growth factors (e.g., EGF, PDGF) to their receptors is an increase in ion fluxes (Na^+, K^+, H^+) across plasma membranes and the release of Ca^{2+} from intracellular stores, leading to a rapid increase of cytosolic free Ca^{2+}. It is proposed that upon binding of the growth factor to its receptor and subsequent autophosphorylation of the latter due to its tyr-specific kinase activity, the receptor activates the membrane-bound phospholipase C, which specifically cleaves PIP_2 to release IP_3 and DAG. IP_3 in turn releases Ca^{2+} from intracellular stores, whereas DAG activates the Ca^{2+} and phospholipid-dependent protein kinase C.

V. INTRACELLULAR Ca^{2+}-BINDING PROTEINS

In contrast to extracellular Ca^{2+}-binding proteins, which display a rather low affinity for calcium, intracellular Ca^{2+}-binding proteins bind Ca^{2+} with high affinity and specificity. In these proteins the amino acids that bind Ca^{2+} usually are arranged in a sequential manner often following a characteristic pattern,

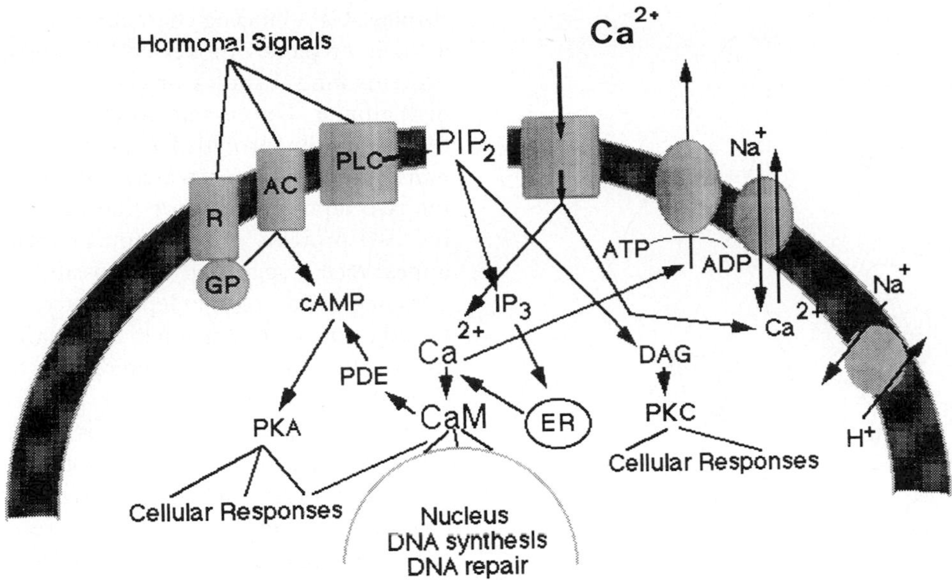

FIGURE 1 Schematic view of signal transduction pathways. AC, adenylate cyclase; CaM, calmodulin; DAG, diacylglycerol; ER, endoplasmic reticulum; GP, GTP-binding protein; IP_3, inositol-1,4,5-triphosphate; PDE, cyclic nucleotide phosphodiesterase; PIP_2, phosphatidylinositol-4,5-diphosphate; PKA, cAMP-dependent protein kinase (=protein kinase A); PKC, protein kinase C; PLC, phospholipase C; R, receptor.

typical for this class of proteins. Here Ca^{2+} has a triggering rather than a structure-stabilizing function. There may exist an additional class of Ca^{2+}-binding proteins, that is, the exchange proteins, which transport Ca^{2+} across membranes, using either ATP or an appropriate ion gradient as an energy source. Little is known, however, about the structural properties of these proteins.

Intracellular Ca^{2+}-binding proteins (e.g., calmodulin, troponin C) are nonenzymatic effector molecules transducing the Ca^{2+} signal into cellular responses. Calmodulin is the pivotal intracellular Ca^{2+}-dependent modulator protein that controls a great variety of cellular events, including phosphorylation and dephosphorylation of proteins, synthesis and degradation of cyclic nucleotides, and cell growth and differentiation. It is ubiquitous among eukaryotic organisms. By contrast, troponin C is specific in its function as the Ca^{2+}-dependent regulator of the troponin/tropomyosin system in muscle cells, and is limited in its distribution. This difference in their functional specificity is also reflected in the conservation of the primary structure of these proteins. Calmodulin is one of the most conserved proteins known to date (e.g., sequences known from different vertebrates, including humans, are identical). By contrast, troponin C displays a much greater degree of variation in its amino acid sequences, including tissue-specific differences.

In 1973, Kretsinger and Nockolds reported the first crystal structure of an intracellular Ca^{2+}-binding protein, parvalbumin. It consisted of three homologous domains, each of which contained two α-helices perpendicular to each other and enclosing a loop responsible for the specific Ca^{2+} binding. In parvalbumin the six helices were named A through F. The protein contains only two Ca^{2+}-binding sites, named CD and EF sites (Fig. 2). Kretsinger coined the term "EF-hand" to describe the binding domain of a Ca^{2+}-binding protein because it can be modeled by the forefinger (helix E), the thumb (helix F), and the bent middle finger (Ca^{2+}-binding loop) of the right hand. It is now often called the "helix–loop–helix" model. The helices of the homologous Ca^{2+}-binding proteins are usually 10 amino acids in length, often start with glutamic acid, and have hydrophobic amino acids in positions 2, 5, 6, and 9. The loop domain usually contains 12 amino acids and has carboxylate-containing amino acids in positions 1, 3, 5, and 12, the latter ligating calcium via both oxygens of the carboxylate side chain. Since Ca^{2+} is complexed in these proteins in a pentagonal-bipyrimidal fashion, two further ligands

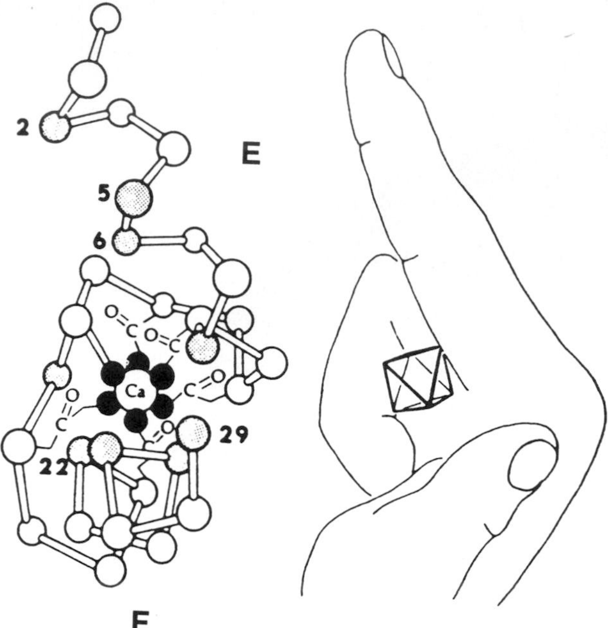

FIGURE 2 The EF-hand model. The model is built according to suggestions made by R. H. Kretsinger. [Reprinted with permission from R. H. Kretsinger *et al.* (1988). *In* "The Calcium Channel: Structure, Function and Implications" (M. Morad *et al.*, eds.), pp. 16–35. Copyright Springer-Verlag, Berlin.]

are in positions 7 and 9. The ligands complexing Ca^{2+} can be vertexed according to a Cartesian system with X nominating the first position. On the basis of the similarity of "EF-hand" domains of different Ca^{2+}-binding proteins, the following general sequence principle of "helix–loop–helix" motifs of these proteins emerged:

X Y Z -Y -X -Z
Eh**hh**hD*D*DG*hD**Eh**hh**h

(D = aspartic acid, E = glutamic acid, G = glycine, h = hydrophobic, * = variable).

The original EF-hand model (Fig. 2), designed on the basis of the parvalbumin crystal structure, appears to be valid for other Ca^{2+}-binding proteins like troponin C or calmodulin, which have a high degree of sequence homology. This generalization is supported by the determination of the crystal structure of the intestinal calcium-binding protein (ICaBP), of troponin C, and of calmodulin. But in contrast to parvalbumin and ICaBP, which consist of only two Ca^{2+}-binding domains as pairs of EF-hands, calmodulin—as well as troponin C (from skeletal muscle)—displays Ca^{2+}-binding characteristics compatible with a "pair of pairs" model of EF-hands, that is, both proteins bind 4 moles of Ca^{2+}/mole of protein with high affinity. The crystal structures of troponin C and of calmodulin provided evidence that the two EF-hand pairs are connected by a long helix, separating the two domains by more than 60 Å. This provided the two proteins with an unusual dumbbell-shaped appearance, suggesting the possibility that the two domains bind ions independently, as was indeed found using nuclear magnetic resonance (NMR) techniques. In addition, it is interesting to note that the central part of this long helix has a high degree of conformational mobility. This is probably of importance considering the interaction of these proteins with their targets, which can involve bending of the central helix. This has been shown recently by solving the three-dimensional structures of complexes between peptides corresponding to the calmodulin-binding domains of either skeletal or smooth muscle myosin light chain kinase or calmodulin kinase II by either X-ray crystallography or NMR, respectively. These structures provided evidence that binding of those peptides to calmodulin bended the latter to a more globular structure by gripping the peptide with the two N- and C-terminal domains like two hands capturing a rope and forming a hydrophobic channel enclosing the peptide helix.

To date more than 300 different intracellular Ca^{2+}-binding proteins of the EF-hand type have been identified by sequence determination. These can be grouped into about 10 subfamilies. Figure 3 shows an ancestral tree with the possible evolutionary relationship of these subfamilies. Proteins within a subfamily of this tree are more related to each other than to members of another subfamily. The closeness of such a relationship between two subfamilies is indicated by the high degree of sequence homology between calmodulin and skeletal troponin C, which share about 45% of identity and more than 70% homology. This high degree of homology in the primary structure is somewhat reflected functionally since calmodulin and troponin C can replace each other to a certain extent.

The so-called "S100" proteins (named by B. W. Moore according to their peculiar property of being soluble in 100% ammonium sulfate solution) are closely related to ICaBP. No specific function has yet been ascribed to these proteins, containing some 17 different members to date, which are localized mainly in the nervous tissue of vertebrates and invertebrates. These proteins often exist as dimers that can bind 4 moles of Ca^{2+}. Two of these, S100A6 or calcyclin and

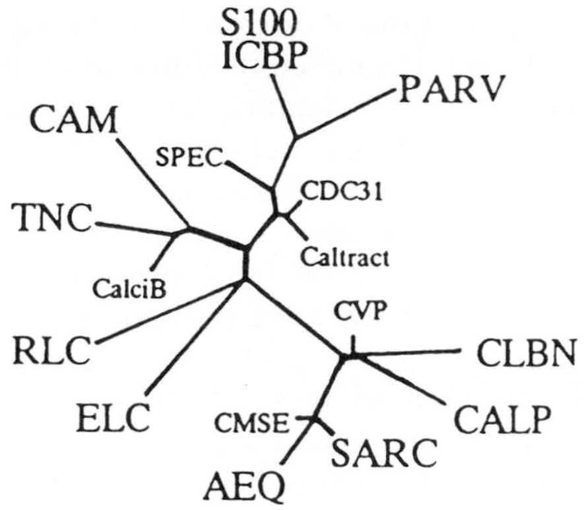

FIGURE 3 The evolutionary tree of the relationship among the 10 subfamilies of EF-hand type proteins. Caltract, caltractin from *Chlamydomonas reinhardtii;* CDC31, the cell division cycle gene 31 product from *Saccharomyces cerevisiae;* SPEC, the group of ectodermal proteins from *Strongylocentrotus purpuratus;* PARV, parvalbumin; S100/ICBP, two EF-hand domain families, including the intestinal calcium-binding protein 9K; CAM, calmodulin; TNC, troponin C; CalciB, B subunit of calcineurin; RLC, regulatory light chain of myosin; ELC, enzymatic light chain of myosin; CMSE, calcium-binding protein from *Streptomyces erythraeus;* AEQ, aequorin; SARC, sarcoplasmic calcium-binding protein; CALP, calpain; CLBN, calbindin and calretinin; CVP, calcium vector protein from *Branchiostoma lanceolatum.* [Reprinted with permission from R. H. Kretsinger *et al.* (1988). *In* "The Calcium Channel: Structure, Function and Implications" (M. Morad *et al.*, eds.), pp. 16–35. Copyright Springer-Verlag, Berlin.]

S100A10 or p11, form complexes with members of another calcium-binding protein family, the annexins, namely, annexin XI and annexin II, respectively. An interesting observation has been made recently by C. W. Heizmann and his coworkers. They reported that at least nine different S100 proteins are located as a gene cluster on human chromosome 1 (1q21) within approximately 500 kilobase-pairs. This region is reported to show a high frequency of chromosomal rearrangement during breast cancer development.

Parvalbumin, the subfamily of proteins that were crystallized first, were originally found only in fish and in reptiles, but were subsequently identified in higher organisms. No specific function has been assigned to this class of Ca^{2+}-binding proteins, but it is interesting to note that their highest concentrations occur in muscle cells of fast-twitch (i.e., fast-relaxing) fibers. In brain, parvalbumin is found in certain types of neurons.

In recent years another family of intracellular Ca^{2+}-binding proteins has been discovered. These soluble, amphipathic proteins bind to membranes containing negatively charged phospholipids in a Ca^{2+}-dependent manner and are therefore called annexins. They are widespread in the animal and plant kingdom and have been claimed to be involved in a variety of cellular functions, such as interaction with the cytoskeleton, signal transduction, phospholipase inhibition, membrane fusion, and anticoagulation. On the basis of the crystal structure of the most-studied annexin, it was proposed that annexin V may function as a calcium channel, and some experiments using a reconstituted system seem to support a voltage-gated mechanism.

VI. SYSTEMS CONTROLLING INTRACELLULAR Ca^{2+} CONCENTRATION: STRUCTURAL AND FUNCTIONAL PROPERTIES

As discussed in the previous sections, the ionized Ca^{2+} is kept in a narrow concentration range (100–500 n*M*) in the resting cell since this is crucial for its signaling function. Several transmembrane Ca^{2+}-transporting systems participate in controlling the free Ca^{2+} concentration in the cell. These systems are located either in the plasma membrane (Ca^{2+} channel, ATP-dependent transporting system, Na^{+}/Ca^{2+} exchanger), in the sarco(endo)plasmic reticular system (ATPase, Ca^{2+}-release channel), or in mitochondria (an electrophoretic uptake system, a Na^{+}-dependent Ca^{2+} exchanger). It is likely that the Golgi membranes and lysosomes also contain ATP-driven Ca^{2+} pumps, which have not yet been characterized. Some important structural features of these systems will be summarized in the following sections.

A. The Calcium Channel

As indicated earlier, transient changes of free Ca^{2+} inside cells trigger many different cellular functions. These Ca^{2+} ions can derive either from intracellular stores or from the extracellular fluid by passing through the plasma membrane down their electrochemical gradient using specifically regulated (i.e., gated) channels. The calcium channels located in the plasma membranes in cells of different tissues mediate calcium influx during depolarization. They play an important role in excitation–contraction coupling in muscle tissues, or couple changes in membrane potential at nerve terminals to the release of neurotransmitters. However, in this context one should emphasize that Ca^{2+} channels also exist in nonexcitable cells, since no cell can afford diffusion of Ca^{2+} into the cell down its concentration gradient without a tight control, which only selectively gated channels (i.e., proteins) can provide. The selectivity of these gates gives rise to different permeability rates for various cations, decreasing in the sequence $Ba^{2+} > Sr^{2+} > Ca^{2+} > Mg^{2+}$, whereas monovalent cations are much less permeable. On the other hand, cations such as La^{3+}, Cd^{2+}, Co^{2+}, Ni^{2+}, or Mn^{2+} can bind to the Ca^{2+}-binding sites with high affinity without being transported, that is, these cations block the channel. Excitable cells provide great advantages for studying the properties of Ca^{2+} channels and therefore most of the studies have been performed with those cell types.

On the basis of their known physiological characteristics, one can distinguish between four different types of channels: the L-type, the T-type, the N-type, and the P-type. Using the now well-established patch-clamp technique for studying single channels, it can be shown that these channels differ by their opening kinetics and their conductance. Channels of the T-type produce a transient inward current with relatively small conductance, whereas the L-type can be characterized by its long-lasting inward current and an approximately threefold stronger conductance. The third channel, the N-type, has been found so far only in sensory neurons and has properties found neither in L- nor in T-channels. Characteristics of the N-channel include an activation by rather strong depolarizations and an intermediate conductance. P-type channels are particularly abundant in cerebellar Purkinje and granule cells. They belong to the class of high voltage activated Ca^{2+} channels. Pharmacologically they differ from N-type and L-type channels, but they are sensitive to ω-AgaIVA, a peptide from the venom of a web spider. Ion flux through Ca^{2+} channels is on the order of 10^6 Ca^{2+} ions per second. The channel density in membranes can be up to 1 per μm^2 in heart cells, but is well above this value in membranes like the T-tubules of skeletal muscle. The four channel types differ in their tissue distribution and also in their response to neurotransmitters or pharmacological agents. It is now well established that the sensitivity of Ca^{2+} channels to adrenergic neurotransmitters is due to cyclic AMP-dependent increase in channel-opening probability, that is, increase in inward calcium current. The same effect can be obtained by intracellular injection of the catalytic subunit of the cAMP-dependent protein kinase, indicating that protein phosphorylation mediates this effect.

Pharmacological agents can block the channel in a well-defined conformational state, for example, in either the open (agonist) or closed conformation (antagonist). Quite a number of different compounds interacting with Ca^{2+} channels have been described in recent years, among them tertiary amines (e.g., verapamil, D600), benzothiazepines (e.g., diltiazem), and 1,4-dihydropyridines (e.g., nifedipine, felodipine). The latter selectively block the L-type channel, which seems to be the most common type of Ca^{2+} channel.

The L-type channel is a voltage-dependent Ca^{2+} channel with distinct modes of gating, that is, control of opening. The open-state probability increases with increasing depolarization of the membrane. It can be modulated by β-adrenergic agonists. As a consequence of this stimulation, one of the subunits of the channel proteins can be phosphorylated in a cAMP-dependent fashion, thus increasing its opening probability.

The selective interaction of 1,4-dihydropyridines and their derivatives (e.g., nifedipine) with the L-type channel allowed their identification and isolation in a pure form from brain cells, from rabbit skeletal muscles, and from guinea pig skeletal muscles. According to these studies, the L-type channel is a glycoprotein of 170–220 kDa with a subunit composition of either 140 kDa (α), 50 kDa (β), and 30 kDa (γ) or 140 and 30 kDa. Subunits α and β can be phosphorylated in a cAMP-dependent manner, but only phosphorylation of the α subunit seems to be of functional relevance.

Cloning of the cDNA for the larger subunit of the dihydropyridine receptor from skeletal muscle (1873 amino acids = 212 kDa) suggested that it is similar to two other channel proteins, a voltage-sensitive Na^+ channel and a K^+ channel from *Drosophila*. The similarity is especially significant in several internal repeat units that are thought to serve as voltage sensors.

B. Ca^{2+} Pump of Plasma Membranes

The ATP-dependent Ca^{2+}-transport system of plasma membranes is a protein of low abundance (e.g., 0.1% of the total membrane proteins of human erythrocytes). It has a very high affinity and specificity for Ca^{2+} and extrudes the ions from the cell against its concentration gradient using ATP as an energy source. Therefore it belongs to the class of so-called E_1–E_2 ATPases, that is, it has at least two different conformational states (E_1 and E_2) during the reaction cycle. The energy provided by ATP is conserved intramolecularly during the cycle by forming an aspartylphosphate intermediate [*See* Adenosine Triphosphate (ATP).]

The pump can be activated by calmodulin. As a result, the Ca^{2+} affinity of the ATPase and its transport rate are increased. The direct, strictly Ca^{2+}-dependent interaction between the two proteins has been used to isolate the Ca^{2+} pump from solubilized erythrocyte membranes by applying calmodulin affinity chromatography. The purified enzyme consists of a single polypeptide of 1220 amino acids (135 kDa). A number of functional domains that are highly conserved among ion-transporting pumps could be identified: Asp 465 is the residue of the aforementioned acylphosphate and Lys 591 is part of the ATP-binding site. These two residues are well separated within the sequence, but are thought to be brought close to each other in space during the reaction cycle with the help of another highly conserved domain, the so-called "hinge region." Residues 1100–1127 compose the calmodulin-binding domain, which was identified using a cross-linking reagent. Some isoforms of the Ca^{2+} pump can also be regulated by cAMP-dependent phosphorylation (Ser 1178), which increases the affinity of the enzyme for Ca^{2+}. The pump has 10 membrane-spanning domains that appear in comparable areas of the sequence of other ion-transporting pumps. The enzyme has a stoichiometry of transported Ca^{2+}/hydrolyzed ATP close to 1, and is an electroneutral Ca^{2+}/H^+ antiporter, that is, it transports Ca^{2+} and H^+ in opposite directions and in equivalent amounts.

The stimulation of the Ca^{2+}-ATPase by calmodulin (CaM) has been studied using calmodulin fragments obtained by controlled proteolysis. The structural properties of some of these fragments are very similar to those of the native protein. These studies showed that the C-terminal half of CaM (i.e., fragment 78–148) can fully activate the ATPase in contrast to the N-terminal half (i.e., fragment 1–77). This finding is further corroborated by the recent observation that a synthetic peptide corresponding to the CaM-binding domain of the Ca^{2+}-ATPase binds to the C-terminal half of CaM but not to the N-terminal half. Depending on the length of the peptide, binding to calmodulin can either leave the extended dumbbell-shape structure of calmodulin intact or bend it to a more globular form as demonstrated by small-angle X-ray spectroscopy.

Activation of the plasma membrane Ca^{2+} pump alternative to calmodulin can be achieved either by exposing the enzyme to negatively charged phospholipids or to fatty acids, by phosphorylation with protein kinase A or C, by dimerization, or by controlled proteolysis. The latter was important for the understanding of structure–function relationships of the enzyme. Recently, it became evident that the calmodulin-binding domain also had autoinhibitory functions similar to those of other calmodulin-dependent enzymes. Using cross-linking techniques it could be demonstrated that the calmodulin-binding domain had two receptor sites within the enzyme: one located between the phosphorylation site (Asp 465) and the ATP-binding site (Lys 591) and the other one within the so-called transduction domain near the N terminus of the pump. These results suggest that the calmodulin-binding domain exhibits its inhibitory function within the Ca^{2+} pump by "bridging" two functionally important domains of the enzyme in the absence of calmodulin and thereby limiting the access of the substrate Ca^{2+} to the catalytic site.

DNA cloning techniques and peptide sequencing identified at least four different gene families of the plasma membrane Ca^{2+} pump in both humans and rats. The human genes PMCA1-4 have been mapped to their chromosomal localizations: PMCA1 to 12q21-23, PMCA2 to 3p25-26 close to the Van Hippel–Lindau syndrome, PMCA3 to chromosome X, and PMCA4 to 1q25-32. Because of alternative splicing that occurs mainly near the phospholipid-binding domain and within the calmodulin-binding domain close to the C-terminal region of the protein, each of these gene products could subdivide into several isoforms. In fact, more than 30 different spliced isoforms of the four different genes have been identified to date, and especially those of PMCA2 and PMCA3 demonstrate distinct tissue distributions with the possibility of being differentially regulated in their expression depending on the special needs of the tissue.

C. Sarco(endo)plasmic Reticulum

Skeletal, heart, and smooth muscle cells contain a reticular system that is important for the control of the free Ca^{2+} concentration in the cell. Most studies have been carried out on the sarcoplasmic reticulum (SR) of striated muscles, but recently interest in studying the endoplasmic reticular (ER) system grew rapidly since it became evident that Ca^{2+} stored in this system can be released into the cytosol by inositol-1,4,5-triphosphate. The corresponding receptor has been purified (260 kDa) and localized in the ER membrane.

The SR is an important element in the excitation–contraction coupling of muscles. In striated muscles the SR forms longitudinal tubules. In the vicinity of the transverse tubules (T-tubules), which are periodic inflections from the plasma membranes at the level of the Z-lines, the SR forms compartments called the terminal cisternae that contain "feet-like" projections connecting the former to the plasma membrane and to the T-tubules (diadic or triadic junctions).

Recently, one of these "feet-like" structures, that is, of the triadic junction, has been identified as the calcium release channel of SR. It was identified as the ryanodine receptor (ryanodine is a plant alkaloid that potentiates Ca^{2+} release in SR) in skeletal as well as in heart muscle cells. The functional unit of the channel is a tetramer. The amino acid sequence of the monomer, deduced from the cDNA, consists of 5037 amino acids (molecular weight of 565,223). There are only four potential transmembrane domains situated right at the C-terminal end, suggesting that 90% of this protein protrudes into the cytosol. This is in line with the morphological observation that the ryanodine receptor spans the 150-Å gap between the SR and the T-tubules, forming the basis for the very attractive hypothesis that there exists a physical interaction between the Ca^{2+}-release channel and the voltage-dependent Ca^{2+} channel of the T-tubules. This would provide a rational explanation for a direct triggering of Ca^{2+} release by the Ca^{2+} channel located in the T-tubules as observed by a number of laboratories.

A major component of the SR (less abundant in ER) is an ATPase that pumps Ca^{2+} out of the cytosol into the lumen of the reticular system. It can represent up to 90% of the membrane protein (in SR of skeletal muscles), but even in less favorable cases (e.g., heart cells) it is still 50%. The SR Ca^{2+} pump also belongs to the E_1–E_2 type of ion-transporting ATPases, similar to the plasma membrane Ca^{2+} pump. The SR enzyme is smaller than its counterpart from the plasma membrane (1001 amino acids as compared to 1220 amino acids of the plasma membrane Ca^{2+} pump), reflecting a major difference in the regulation of the two enzymes: the SR Ca^{2+}-ATPase does not directly interact with calmodulin and therefore lacks the corresponding regulatory domain.

The Ca^{2+} pump of cardiac or smooth muscle SR (but not of fast skeletal muscles) is regulated via a highly hydrophobic, phosphorylatable protein called phospholamban, a pentamer of a subunit with a mass of 5–6 kDa. The protein serves as a substrate of several protein kinases, such as cAMP-dependent protein kinase, a calmodulin-dependent kinase, and protein kinase C. The recently determined primary structure of the monomer (52 amino acids = 6080 Da) showed that the amino acids phosphorylated by the various kinases are located next to the N terminus. They are a serine residue that becomes phosphorylated by a protein kinase A and next to it a threonine residue that can be phosphorylated by a calmodulin-dependent kinase.

A direct interaction between phospholamban and the SR Ca^{2+} pump exists, and its binding site has been recently identified by cross-linking experiments. It is interesting that it could be localized in a region of the enzyme similar to the calmodulin-binding domain receptor site within the plasma membrane Ca^{2+} pump. Therefore it is not surprising that phospholamban also influences the Ca^{2+}-transport rate across the membrane. Transport is strongly influenced by the phosphorylation state of the phospholamban: phosphorylation induces an increase in the enzyme's affinity for Ca^{2+}. The action of phospholamban on the SR Ca^{2+} pump could be compared with an inhibitor of the enzyme, which becomes ineffective by phosphorylation and thereby permits the pump to express its full activity.

D. Na^+/Ca^{2+} Exchanger of Plasma Membranes

It has been pointed out before that the fine and rapid tuning of the Ca^{2+} signal is performed by a membrane Ca^{2+} pump since only this enzyme possesses a high Ca^{2+} affinity. On the other hand, these enzymes have low capacity and are not able to transport Ca^{2+} in bulk quantities across the membrane. In contrast to these, there exists in the plasma membrane a Na^+/Ca^{2+} exchanger that has a rather low affinity but high transport capacity for Ca^{2+}. It is an electrogenic system, that is, it transports 3 Na^+ for 1 Ca^{2+}, in directions that depend on the magnitude and direction of the transmembrane gradients of Na^+ and Ca^{2+} or the transmembrane electrical potential.

The Na^+/Ca^{2+} exchanger is particularly active in plasma membranes of cells from excitable tissues. Recently, the exchanger has been cloned from cardiac tissues. The deduced amino acid sequence corresponds to a molecular mass of 120 kDa, or possibly even higher due to glycosylation. The rather polar protein has a large cytosolic extruding unit and 12 proposed transmembrane domains. A number of consensus phosphorylation sites could be identified within the sequence, explaining the interesting observation that ATP could activate the Na^+/Ca^{2+} exchange process. Indeed it was observed that the Ca^{2+} affinity of the Na^+/Ca^{2+} exchanger can be influenced by a phosphorylation process under the control of a CaM-dependent protein kinase that can be reversed by a CaM-dependent phosphatase.

E. Ca^{2+}-Transporting Systems in Mitochondria

Over the years, the Ca^{2+}-transport systems of isolated mitochondria have been studied most extensively since for a long time these systems have been regarded as extremely important in buffering the free cytosolic Ca^{2+} concentration. However, more detailed kinetic studies revealed that the affinity of mitochondria for Ca^{2+} is not sufficiently sensitive to play a role in the regulation of reactions in the submicromolar range of Ca^{2+}. In this context, it is relevant that the mitochondrial Ca^{2+}-uptake rate is about 10-fold slower than that of the SR. On the other hand, mitochondria possess by far the largest buffering capacity within the cell, which might be especially significant when the Ca^{2+} concentration in the cytosol increases pathologically. Mitochondria are able to accumulate Ca^{2+} at a high rate in the presence of the permeant anion inorganic phosphate by producing insoluble calcium phosphate precipitation, probably as amorphous hydroxyapatite.

It has been extensively documented that under the influence of toxic agents interfering with the Ca^{2+} permeability of plasma membranes, mitochondria accumulate substantial amounts of Ca^{2+}, which become visible under the electron microscope as electron-opaque granules. Similar observations have been made with hormonally induced hypercalcemias, resulting in similar dense mitochondrial granules, especially in tissues where the hormones specifically increase Ca^{2+} transport (e.g., kidney, bone). This safety device is of utmost importance for the cells, since such an excess of Ca^{2+} entry into cells is an early and frequent observable phenomenon in cell injury ("calcium overload") and can be secured owing to the significant Ca^{2+}-storing capacity of mitochondria. The electrogenic Ca^{2+}-uptake system of mitochondria operates under normal physiological conditions only at a marginal rate. Because of their large membrane potential (negative inside), mitochondria would continuously take up Ca^{2+} if there were not a mechanism for exporting Ca^{2+} against the large membrane potential. Indeed, such a system has been identified as an electroneutral Na^+/Ca^{2+} exchanger that operates independently of the electrophoretic uptake route.

Besides the Na^+-promoted Ca^{2+}-release route, there also exists a Na^+-independent route that has highest activity in mitochondria of nonexcitable tissues (e.g., liver, kidney), where the Na^+/Ca^{2+} exchanger has its lowest activity. Whether this route is a Ca^{2+}/H^+ antiporter or is due to changes of the permeability properties of the inner mitochondrial membrane is still controversial. It is possible that a Ca^{2+}-release route is linked to the hydrolysis of pyridine nucleotides concomitant with the ADP-ribosylation of a protein from the inner mitochondrial membrane.

VII. CONCLUSIONS

The following general points can be made on the basis of the material presented here:

1. Calcium plays a pivotal role in biological systems. It occurs predominantly in a complexed form, fulfilling either a static, structure-stabilizing function or a dynamic, signal-transducing role.
2. Cells are exposed to a large gradient of Ca^{2+} across their membranes. This is the condition that permits the cell to use even small changes of membrane permeability to cause significant changes in the free Ca^{2+} concentration inside the cell, which can be used to transmit and amplify a signal transduced from the extracellular space. On the other hand, it is clear that the concentration of ionized calcium within the cell has to be carefully controlled in order to use Ca^{2+} as an intracellular second messenger.
3. Intracellular Ca^{2+}-binding proteins have developed during evolution that complex Ca^{2+} with high affinity and specificity. Their characteristic structural principle is a common motif—the helix–loop–helix or EF-hand principle—which permits these proteins to bind Ca^{2+} with high selectivity in the presence of high concentrations of other ions. By contrast, Ca^{2+}-binding properties of extracellular proteins normally are of lesser specificity and affinity. In the latter proteins, calcium often plays a structure-stabilizing function.
4. Signal transformation from outside the cell changing the free Ca^{2+} concentration in the cytosol can occur either directly by opening calcium channels in the plasma membrane or indirectly by opening Ca^{2+} pathways from intracellular stores using other second messengers, that is, inositolpolyphosphates. In this way the signal is transferred from the extracellular space to intracellular receptors and amplified using different second messenger pathways.
5. The intracellular interplay between different second messenger systems can be manifold and very

complex as outlined in Fig. 1. In this respect, it is interesting to note that under many conditions signal transduction is transferred not only from the plasma membrane to the cytosol but also to the nucleus. Thus the synthesis of several calmodulin-binding proteins is induced in the nucleus. Future research will dissect in more detail the complexity of signal transduction pathways, and especially will identify not only cytosolic but also nuclear responses of the cell upon receiving extracellular signals.

BIBLIOGRAPHY

Carafoli, E. (1987). Intracellular calcium homeostasis. *Annu. Rev. Biochem.* **56**, 395–433.

Carafoli, E., Krebs, J., and Chiesi, M. (1988). Calmodulin in the transport of calcium across biomembranes. *In* "Calmodulin" (C. B. Klee and P. Cohen, eds.), pp. 297–312. Elsevier, Amsterdam.

Cheung, W. Y. (ed.). (1980). "Calcium and Cell Function," a series of monographs. Academic Press, New York.

Clore, G. M., Bax, A., Ikura, M., and Gronenborn, A. M. (1993). Structure of calmodulin–target peptide complexes. *Curr. Op. Struct. Biol.* **3**, 838–845.

Kretsinger, R. H. (1987). Calcium coordination and the calmodulin fold: Divergent versus convergent evolution. *Cold Spring Harbor Sympos. Quant. Biol.* **52**, 499–510.

Meyer, T., and Stryer, L. (1991). Calcium spiking. *Annu. Rev. Biophys. Chem.* **20**, 153–174.

Schaefer, B. W., and Heizmann, C. W. (1996). The S100 family of EF-hand calcium-binding proteins: Functions and pathology. *Trends Biochem. Sci.* **21**, 134–140.

Strynadka, N. C. J., and James, M. N. G. (1989). Crystal structures of the helix–loop–helix calcium-binding proteins. *Annu. Rev. Biochem.* **58**, 951–998.

Swairjo, M. A., and Seaton, B. A. (1994). Annexin structure and membrane interactions: A molecular perspective. *Annu. Rev. Biophys. Biomol. Struct.* **23**, 193–213.

Cancer Genetics

KATHERINE A. SCHNEIDER
Dana-Farber Cancer Institute

GLOSSARY

Cancer genetic counseling Discussion of an individual's risk of developing inherited forms of cancer

Carcinogenesis Process by which normal cells become transformed into a malignant (cancer) cell

Dominant inheritance Genetic condition caused by the presence of a single mutated gene. Individuals with dominant condition have one mutated gene and one normal gene. The mutated gene can be inherited by a parent or occur as a new event

Germline mutation Mutation that has been inherited by a parent or represents a new mutation from the egg or sperm

Mismatch repair gene Gene which identifies and corrects replication errors within the cell

Oncogene Activated form of a proto-oncogene, which can stimulate excess growth and lead to a malignant tumor population

Somatic mutation Mutation that originates within a single cell of the body

Tumor suppressor gene Gene that plays a role in growth replication of a cell. If mutated, it can lead to cellular transformation through release of the normal controls on cell growth

CANCER SUSCEPTIBILITY GENES FOR BREAST cancer and colon cancer have recently been identified, thus transforming cancer genetics from an esoteric discussion of rare syndromes to front page news. Because sporadic cases of cancer arise from genetic mechanisms similar to that for inherited forms of cancer, all molecular insights are potentially relevant to our overall understanding of cancer development. As the genetic changes instrumental in carcinogenesis are unraveled, the closer we move toward being able to offer more effective cancer prevention, surveillance, and treatment options. The incidence of cancer in the general population is one in three. The number of individuals with an inherited predisposition to cancer may be significant, and genetic testing is now possible for a handful of hereditary cancer syndromes. Although some people welcome the ability to learn their genetic risks of cancer, others wish wholeheartedly that these options did not exist. Research is ongoing to learn more about the impact of genetic testing on the lives of those being tested and their relatives.

I. GENERAL FEATURES OF HEREDITARY CANCER SYNDROMES

Over 100 hereditary cancer syndromes have now been identified, although most are quite rare. In hereditary cancer syndromes, malignancy is the central and sometimes only feature. The risks of cancer in a person known to carry an inherited mutation in a cancer susceptibility gene are substantial, not absolute. Associated risks may vary, ranging from a 60% risk of medullary thyroid carcinoma in multiple endocrine neoplasia, type 2A (MEN,2A) to the 100% risk of colon cancer in familial adenomatous polyposis (FAP). Most hereditary cancer syndromes have lifetime risks of cancer approaching 90%. A wide spectrum of disease also exists in hereditary cancer syndromes. Even among members of the same family, the development of cancer may differ greatly in terms

ENCYCLOPEDIA OF HUMAN BIOLOGY, Second Edition, VOLUME 2. Copyright © 1997 by Academic Press. All rights of reproduction in any form reserved.

of age of onset, tumor grading, clinical staging, survival rates, and occurrence of multiple primaries.

The majority of hereditary cancer syndromes follow an autosomal dominant pattern of inheritance. The underlying genetic defect in a hereditary cancer syndrome can be a mutation in a growth regulatory gene (oncogene or tumor suppressor gene) or in a gene which somehow facilitates the development of neoplasm (mismatch repair gene). Although every cell of the body contains the mutation, cancer susceptibility genes appear to confer increased risks for specific cancers. For most of the hereditary cancer syndromes, the underlying genetic defects remain unknown.

II. MODELS OF CARCINOGENESIS

A. General Process of Carcinogenesis

The transformation of a normal cell to a malignant cell requires multiple genetic changes to take place. The development of a tumor begins with a single cancer cell somewhere in the body which multiplies and creates a small colony of cells within the same tissue. Further genetic changes (such as oncogene activation) occur within this colony of abnormal cells, leading to a fully malignant tumor. This stochastic process can be completed within a few months or may take a decade or more to complete.

The phases of carcinogenesis are termed initiation, promotion, progression, and metastasis (see Fig. 1). Initiation is the first stage of carcinogenesis and is the result of a genetic alteration leading to the abnormal proliferation of a single cell. This alteration may be inherited as a germline mutation, but the majority occur somatically due to an error during mitosis or exposure to specific carcinogens, such as tobacco or radiation. Initiated or precancerous cells can spontaneously revert back to normal states, remain in precancerous states indefinitely, or progress toward malignancy.

Promotion follows initiation. In this phase, cells have acquired a selective growth advantage, which

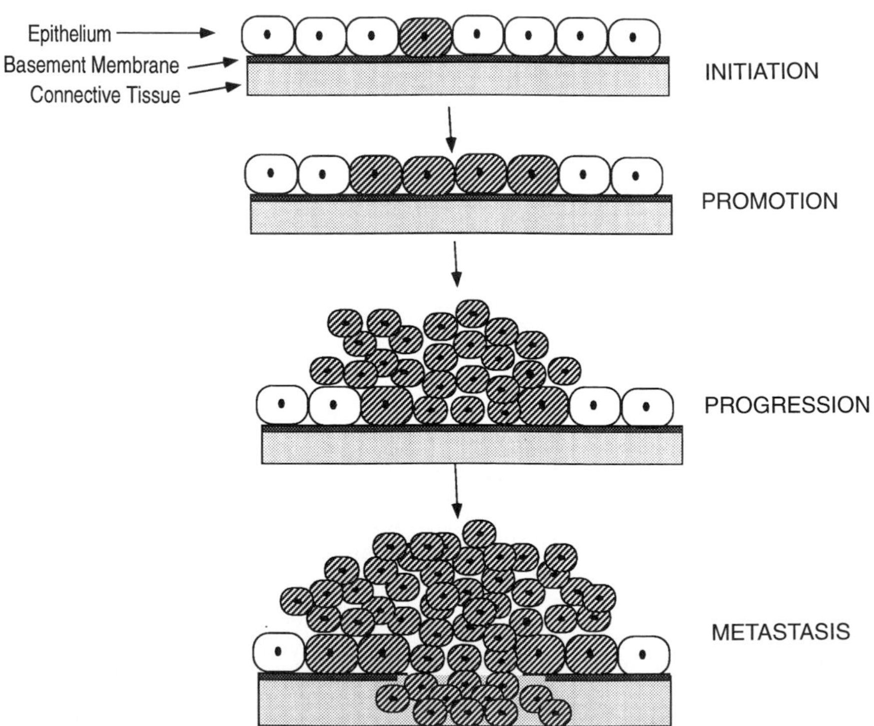

FIGURE 1 The four phases of carcinogenesis in colon cells. Initiation involves inheriting or acquiring a single mutation which leads to a precancerous cell. In promotion, the cell acquires additional mutations and undergoes cell division. Further genetic changes during progression result in a fully malignant tumor population, which will eventually spread to other sites of the body (metastasis).

leads to rapid growth and to the formation of a small, benign tumor population. Promotion occurs because of a random error during cell division or exposure to specific carcinogens, such as hormones or dietary fat. To date, all genetic changes leading to promotion have been somatically acquired.

In the progression phase, further genetic alterations lead to a larger colony of cells with enhanced growth potential and other special features (such as increased mobility and creation of circulatory system within the tumor, termed angiogenesis). Genetic alterations continue to occur as a result of the rapid, continuous cell divisions, leading to the stage of progression. At this stage the population of tumor cells are fully malignant. Malignant cells must undergo further genetic changes in order to gain metastatic properties.

Metastasis involves several separate steps, including the separation of the cancer cell from the primary tumor, entry into the circulatory or lymphatic system, and attachment to the surface of the new tissue site. The most common sites for metastases are the lung, liver, bone, and central nervous system. However, cancer cells can potentially be carried to any site of the body.

Two separate models developed by Drs. Knudson and Vogelstein have been instrumental in furthering our understanding of carcinogenesis.

B. Knudson's Model of Retinoblastoma

In 1971, Dr. Alfred Knudson sought to describe cancer development in a way that explained both inherited and sporadic forms of retinoblastoma. As illustrated in Fig. 2, Dr. Knudson theorized that both retinoblastoma (RB) alleles must be nonfunctional in order for an eye tumor to develop. In the inherited form, one copy of the retinoblastoma gene contains a germline mutation whereas the second copy of the gene is lost or damaged at the somatic level. The sporadic form of retinoblastoma is caused by two separate events in a single retinal cell, which renders both copies of the retinoblastoma genes nonfunctional. [*See* Tumor Suppressor Genes, Retinoblastoma.]

The Knudson two-hit hypothesis described cancer development as a process rather than the result of a single event or exposure, thus revolutionizing the understanding of carcinogenesis. Until that point, it was difficult to conceive how certain cancers could have both inherited and sporadic forms. The two-hit model also illustrates that the gene mutation can be passed to offspring in a dominant manner, yet behaves recessively at the cellular level. This is true of germline mutations in either a tumor suppressor or a mismatch repair gene.

C. Vogelstein's Model of Colon Cancer

The model of carcinogenesis in colorectal cancers, first described by Dr. Bert Vogelstein, nicely expands Knudson's model by demonstrating that a combination of genetic events leads to malignancy. As shown in Fig. 3, the initial event is a germline (or somatic) mutation in the *APC* tumor suppressor gene. Addi-

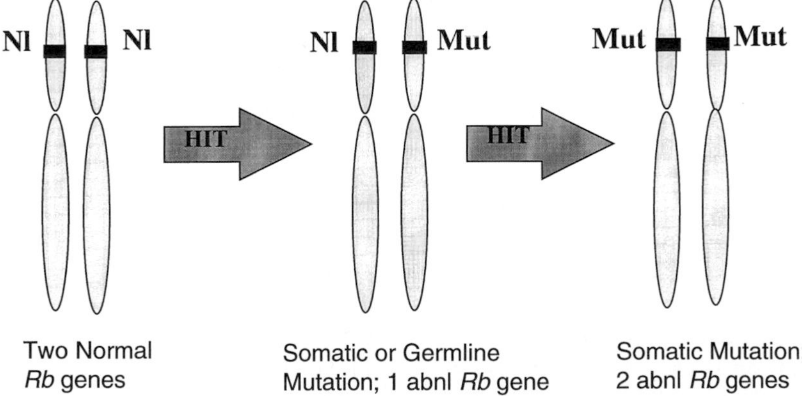

FIGURE 2 Knudson's two-hit hypothesis. In this model, retinoblastoma (an eye tumor) develops only if both copies of the *Rb* gene become mutated. The first mutation "hit" can occur at the germline or somatic level. The second mutation "hit" always occurs somatically. Other cancers that can be hereditary in some cases and sporadic in others fit this model as well.

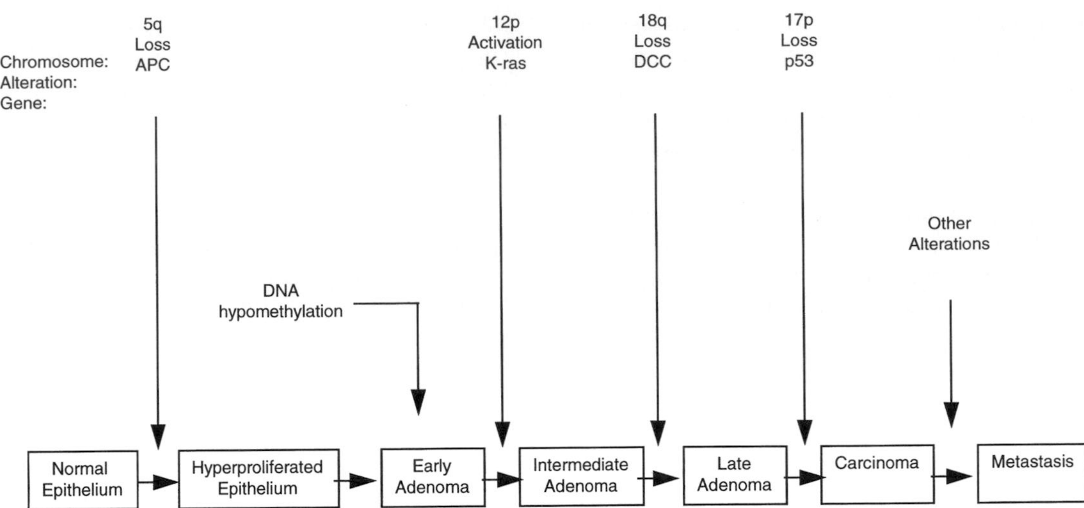

FIGURE 3 Vogelstein's model correlates the sequence of genetic changes that occur during carcinogenesis with the changes occurring in an epithelial colon cell, beginning with a normal cell, leading to a noncancerous tumor (adenoma), and ending with a fully malignant cell (carcinoma) with metastatic capabilities.

tional genetic mutations then lead to the activation of the *K-ras* oncogene and to loss of the *DCC* and *p53* tumor suppressor genes. The accumulation of genetic events is what is important. The exact order of genetic changes can vary, although in specific tumors, some genetic changes occur more frequently as early events and others typically happen later in the process. The series of genetic changes in colorectal cells has been shown to correspond to the progression of adenomas to metastatic carcinoma.

Similar models of carcinogenesis have been developed for other types of tumors. Thus, the inheritance of a germline mutation in a cancer susceptibility gene is not enough to lead to tumor development. Other genetic changes at the cellular level must occur in tandem. Unfortunately, the likelihood that these additional genetic changes will occur is exceedingly high.

III. CANCER SUSCEPTIBILITY GENES

A malignant tumor is caused by a series of DNA mutations occurring within specific genes. These genetic changes can occur at the germline or somatic level. A germline mutation is present in every cell of a person's body and either has been inherited by a parent or has occurred as a new mutation from a parent's egg or sperm. In contrast, a somatic mutation occurs during a person's lifetime in a single cell of his or her body. It is estimated that 80–90% of all cancers are due to somatic, not germline, mutations. Germline mutations in oncogenes, tumor suppressor genes, and mismatch repair genes have been shown to lead to increased risks of cancer.

A. Oncogenes

Oncogenes were identified in the 1960s and provided the first evidence that specific genetic changes within a cell nucleus could induce cancer. Oncogenes originate from normal cellular genes termed proto-oncogenes, which play a central role in the regulation of normal cellular growth and signal transduction. [*See* Oncogenes and Proto-oncogenes.]

The activation of an oncogene from a proto-oncogene typically occurs at the somatic level. An example of somatic oncogene activation is the Philadelphia chromosome. The Philadelphia chromosome consists of the *abl* proto-oncogene, originally on chromosome 9, which has fused to the central region of a gene on chromosome 22 called *bcr*. This rearrangement has caused the activation of the *abl* oncogene, which makes a hybrid protein with growth-promoting properties that predisposes cells to becoming leukemic. The Philadelphia chromosome is present in about 90% of individuals with chronic myelogenous leukemias.

Currently, there is only one known example of a germline mutation leading to oncogene activation. The *ret* gene is located on the lower portion (long arm) of chromosome 10 (10q11). The *ret* gene has been identified as the underlying genetic defect in MEN,2A. Individuals with MEN,2A have a higher chance of developing medullary thyroid carcinoma, parathyroid tumors, and pheochromocytomas. The *ret* mutation is dominantly inherited, and the activation of just one of the two copies of the *ret* oncogenes is sufficient for cancer to occur. [*See* Oncogene Amplification in Human Cancer.]

B. Tumor Suppressor Genes

A tumor suppressor gene is a growth regulatory gene whose function is to send signals to the cell that the replication process should cease. The presence of a single functional tumor suppressor gene appears sufficient to provide the appropriate signals to a cell. Loss of both copies of a tumor suppressor gene results in deregulated growth and the potential for tumor growth. [*See* Tumor Suppressor Genes.]

Inactivation of the first tumor suppressor allele can occur at either the germline or the somatic level. The second tumor suppressor allele always becomes inactivated at the somatic level. Most hereditary cancer syndromes identified to date are due to germline mutations in tumor suppressor genes, including Li–Fraumeni syndrome (LFS) (the *p53* gene), retinoblastoma (the *RB* gene), familial Wilms tumor (the *WT1* gene), and breast–ovarian cancer syndrome (the *BRCA1* gene). Although tumor suppressor genes are dominantly inherited, cancer does not occur until the loss of the second copy of the gene within a cell.

Carrying a germline mutation in a tumor suppressor gene can dramatically increase an individual's cancer risks. For example, the risk of retinoblastoma in children carrying a germline mutation in the *RB* gene is about 90% versus a risk of 1 in 30,000 for children in the general population. In addition, children carrying a germline *RB* mutation have increased risks of additional malignancies, including osteosarcoma, soft tissue sarcomas, malignant melanoma, and cancers of the lung and bladder.

C. Mismatch Repair Genes

The latest cancer susceptibility gene to be identified is the mismatch repair gene. First described in 1993, the function of the mismatch repair gene is to correct DNA mismatches that occur routinely during the replication process. The absence of a functioning mismatch repair gene leads to an increased incidence of replication errors (RER) within a cell. It is the accumulation of these errors within a single cell that can cause the formation of a malignant tumor. Because of the magnitude of RERs that occur, the likelihood of cancer is high. Similar to tumor suppressor genes, the mutation in the first copy of a mismatch repair gene can occur at the germline or somatic level and the mutation or loss of the second copy always occurs somatically. Only when a cell has lost both copies of a mismatch repair gene will there be microsatellite instability and the potential for tumor development.

The majority of RER positive tumors are due to somatic mutations of mismatch repair genes. However, most families with hereditary nonpolyposis colon cancer (HNPCC) have a germline mutation in one of the four mismatch repair genes identified to date: *MSH2, MLH1, PMS1,* or *PMS2*. Individuals with a germline mutation in one of these four genes appear to have 80–90% lifetime risks of developing colorectal cancer.

IV. GENES PREDISPOSING TO BREAST OR COLON CANCER

A. Cancer Susceptibility Genes Predisposing to Breast Cancer

The lifetime risk of breast cancer in the general population is about 11% (one in nine). It is estimated that 5–10% of total breast cancer cases are due to a dominantly inherited gene. There are currently three known dominant genes associated with inherited forms of breast cancer. [*See* Breast Cancer Biology.]

1. *BRCA1* Gene

BRCA1, mapped to chromosome 17q21, is associated with increased risks of breast and ovarian cancer. *BRCA1* is estimated to account for 50% of families with inherited forms of breast cancer and more than 80% of families with both breast and ovarian cancer. Risks of breast cancer may be as high as 85%, with 50% of the women developing breast cancer prior to age 50. Ovarian cancer risks are between 40 and 60%. Risks of prostate cancer are slightly increased in men carrying a *BRCA1* germline mutation, and risks of colon cancer may be increased two- to threefold for both men and women. The increased risks of prostate and colon cancer seem to occur after the age of 50, which is when these cancers typically occur in the general population.

The *BRCA1* gene consists of 22 exons and is approximately 100 kb in length, making it a very large gene to study. The *BRCA1* gene is known to be a tumor suppressor gene, although its exact function within the cell is unknown. There is conjecture that its function is important for breast growth during puberty.

More than 60 germline mutations have been reported scattered throughout the gene. Two specific mutations, termed 185delAG and 5382InsC, appear to be more common in Jewish families of Eastern European, or Ashkenazi, origin. Initial studies place the frequency of the 185delAG mutation among the Ashkenazi Jewish population at about 1%, which in terms of genetic traits is a high frequency. Further studies are needed to verify that the cancer risks associated with a *BRCA1* mutation in families with striking histories of cancer are also relevant for a person with the 185delAG or 5382InsC mutation who does not have any family history of cancer.

2. *BRCA2* Gene

BRCA2, recently mapped to chromosome 13q12, leads to increased lifetime risks of breast cancer for both women and men who carry a gene alteration. Lifetime breast cancer risks for women who carry a gene alteration are 80–90%. For men, the lifetime risk of breast cancer is about 6%. Because men in the general population have extremely low risks of breast cancer risks, this represents a substantial increased risk. Other forms of cancer associated with a *BRCA2* mutation include cancer of the pancreas, prostate, and Fallopian tube. Germline *BRCA2* mutations are thought to account for about 40% of families with inherited forms of breast cancer.

The *BRCA2* gene is 70 kb in length, with 27 exons. The properties of the *BRCA2* gene suggest that it is a tumor suppressor gene, although this has not yet been proven. The function of the *BRCA2* gene at the cellular level is currently unknown.

3. *p53* Gene

p53 is mapped to the short arm (upper portion) of chromosome 17 (17p21). The *p53* gene is the underlying genetic defect in Li–Fraumeni syndrome. Multiple pediatric and adult-onset cancers are associated with LFS. The most commonly associated cancers are breast cancer, leukemia, osteosarcomas, brain tumors, soft tissue sarcomas, and carcinomas of the adrenal gland. Several other types of cancers (including colon and prostate) have also been noted in these families. Individuals with LFS appear to have a 50% risk of developing cancer by age 30 and a lifetime cancer risk of about 90%. Premenopausal breast cancer is the most commonly occurring cancer in women with LFS. However, because LFS is such a rare cancer syndrome, *p53* germline mutations are thought to account for less than 1% of inherited forms of breast cancer. [*See* Tumor Suppressor Genes, p53]

The *p53* gene is one of the smaller cancer susceptibility genes, consisting of 11 exons. The *p53* gene normally encodes a 53-kDa nuclear phosphoprotein that is involved in the control of cell growth. The *p53* gene has been highly conserved across species, which is evidence for its importance in controlling cell growth. Chromosome studies performed on a variety of tumors have noted that the loss of the short arm of chromosome 17, the region containing the *p53* gene, is one of the most frequent genetic changes to occur in sporadic tumors. This may explain the high cancer risks and wide spectrum of tumors associated with the inheritance of a *p53* mutation.

B. Cancer Susceptibility Genes Predisposing to Colon Cancer

The risk of colon cancer in the general population to age 70 is about 6%. It is estimated that 5–10% of total colon cancer cases are due to a dominantly inherited gene mutation. Inherited forms of colon cancer are categorized as polyposis or nonpolyposis syndromes, based on the total number of polyps present in the colon at the time of diagnosis. In the polyposis syndromes, the colon may be carpeted with hundreds of polyps, whereas the hereditary nonpolyposis colon cancer syndromes have significantly fewer; less than 100 by definition. About 95% of inherited colon cancer cases are classified as HNPCC. Familial adenomatous polyposis has been mapped to a single gene and HNPCC has been mapped to four separate genes. [*See* Colon Cancer Biology.]

1. *APC* Gene

APC lies on chromosome 5q21 and is the underlying defect in FAP. FAP has an incidence of about 1 in 8000, and an *APC* germline mutation accounts for about 1% of all colon cancer cases. The *APC* gene is a tumor suppressor gene. Individuals with an inherited *APC* gene mutation typically develop greater than 100 benign tumors (adenomas) in the colon and rectum during the first 20 years of life. Progression to malignancy approaches 100% by age 50. A less severe form of the syndrome, termed attenuated FAP, is associated with fewer colonic polyps and a later age of onset. Individuals with FAP may also develop sebaceous cysts, osteomas, and supernumery cysts. This constellation of features is also known as Gardner syndrome.

Attenuated FAP and Gardner syndrome are also associated with mutations within the *APC* gene. The *APC* gene consists of 15 exons and, as stated earlier, is a tumor suppressor gene.

2. Mismatch Repair Genes

A germline mutation in one of the mismatch repair genes, *MLH1, MSH2, PMS1,* or *PMS2,* is associated with HNPCC. Individuals with a germline mutation in one of these genes have 80–90% lifetime risks of developing cancer of the colon or rectum. Some individuals may have increased risks of developing other forms of cancer, including cancer of the uterus, ovary, pancreas, stomach, and urinary tract. This constellation of cancers may also be referred to as Lynch syndrome, type II.

The *MLH1* gene, consisting of 19 exons, is located on chromosome 3p21, and the *MSH2* gene, which has 16 exons, is located on chromosome 2p22. The *PMS1* and *PMS2* genes are located on chromosomes 2q31 and 7p22, respectively. These four genes are responsible for repairing DNA mismatches that occur during cell replication. *MLH1* and *MSH2* mutations appear to be much more common, accounting for about 90% of HNPCC families.

V. CANCER GENETIC COUNSELING

The recognition that cancer susceptibility can be an inherited condition has created a new clinical genetics specialty, termed cancer genetic counseling. Cancer genetic counseling can be provided by a master's level trained genetic counselor or geneticist (physician trained in genetics) or another health care professional, such as a nurse or oncologist, who has received special training. Cancer genetic counseling, also referred to as cancer risk counseling, is a communication process regarding an individual's risk of developing inherited forms of cancer. This includes a discussion of the personal and family history of cancer, risks of developing inherited forms of cancer, psychosocial issues, possible early detection and risk reduction strategies, and option of DNA testing. This section focuses on the identification of a hereditary cancer syndrome and the option of DNA testing.

A. Identification of a Hereditary Cancer Syndrome

A cancer genetic counseling session begins with the construction of a family tree, termed a pedigree. As illustrated in the "colon cancer family" shown in Fig. 4, a pedigree indicates the relationships of family members and types of cancer that have developed. The pedigree ideally includes information about first-degree (parents, siblings, offspring), second-degree (grandparents, aunts, uncles, nieces, and nephews), and third-degree relatives (first cousins and great-aunts and great-uncles). A meaningful cancer risk assessment may not be possible in cases where little or no family history information is known or family size is extremely small. The specific information collected includes the exact site of cancer diagnosis, age and date of cancer diagnosis, and current age or age at death. The pedigree also needs to include information about individuals who have never had cancer, including current age or age at and cause of death. Written confirmation of each cancer diagnosis is extremely important because individuals may have incomplete or inaccurate information regarding their relatives' diagnoses. Confirmation of the cancer diagnosis is ideally a pathology report, but can also be a summary note in a medical record or a death certificate.

Distinguishing families with inherited forms of cancer from those with sporadic forms can be difficult for two reasons. First, the histology and morphology of inherited tumors resemble that of sporadic tumors. Second, in most cancer syndromes, at-risk individuals do not have any distinctive physical features.

Because the incidence of cancer in the general population is substantial, it is not uncommon for an individual to have one or more relatives who have developed some type of cancer. Therefore, it is important to consider additional features of cancer histories which raise the likelihood that cancers in the family are due to inherited gene mutations (summarized in Table I).

1. Several Relatives with Same or Related Cancers

The family history should include at least two first-degree relatives (or one first-degree relative and two or more second-degree relatives) affected with the same or related cancers. The more family members affected with cancer, the stronger the likelihood there is an inherited predisposition. However, the presence of the same type of cancer in a few relatives is more striking than several family members with unrelated cancers. This is because a cancer susceptibility gene leads to a specific pattern of cancer.

2. Unusually Young Ages of Onset

Inherited forms of cancer typically have earlier ages of onset than sporadic tumors. This is true for pediat-

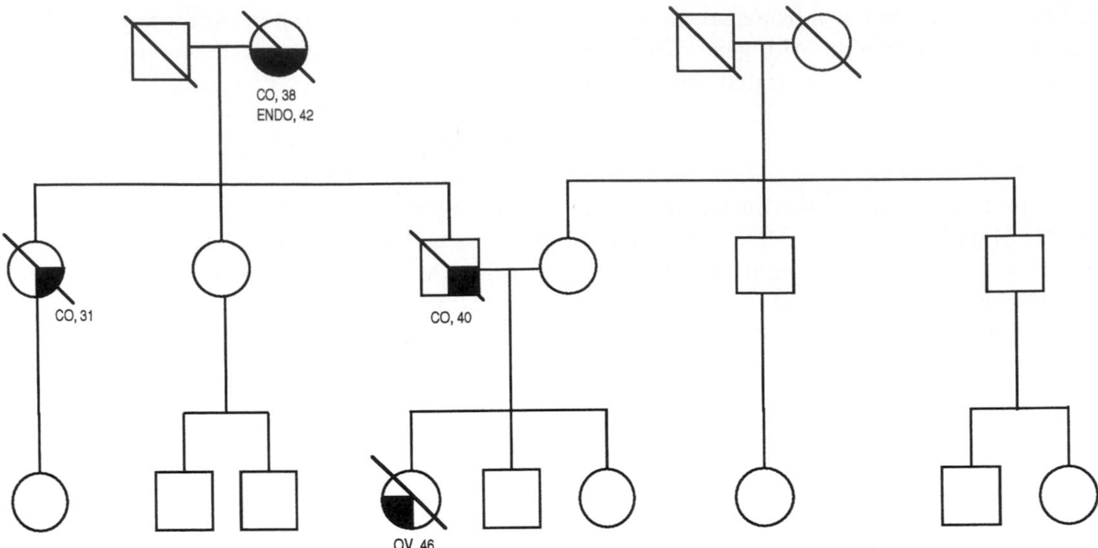

FIGURE 4 Pedigree of a family with a form of hereditary nonpolyposis colon cancer, termed Lynch syndrome, type II. ○, female; □, male; CO,38, diagnosis, age of onset; OV, ovarian cancer; CO, colon cancer; ENDO, endometrial cancer.

ric as well as adult forms of cancer. Tumors occurring during childhood may occur several months earlier than is usual whereas adult onset inherited tumors may occur decades earlier than their sporadic counterparts. If one or more family members have developed cancer at unusually young ages, it is strongly suggestive of an inherited form of cancer. However, cancer occurring at later ages may also be due, in part, to inherited factors.

3. Autosomal Dominant Pattern of Cancer

The hereditary cancer syndromes identified to date follow autosomal dominant inheritance patterns. This means that the incidences of cancer within a family should be present in at least two generations.

TABLE I
Family History Features Suggestive of a Hereditary Cancer Syndrome

Several relatives with same or related cancers
Unusually young ages of onset
Autosomal dominant pattern of cancer
Presence of rare cancers
More than one cancer in an individual
Absence of other risk factors
Physical findings suggestive of hereditary syndrome

4. Presence of Rare Cancers

With cancer occurring frequently in the general population it is important to rule out sporadic cases. The presence of rare forms of cancer in one or more family members provides evidence for an inherited etiology. For example, MEN,2A includes medullary thyroid carcinoma and pheochromocytomas, two uncommon forms of endocrine tumors. Cancer histories suggestive of inherited forms of cancer can also include malignancies in people less susceptible to developing that specific form of cancer. Examples include breast cancer in men and lung cancer in nonsmokers less than 50 years old.

5. More than One Cancer in an Individual

Most tumors are monoclonal, meaning that the population of malignant cells has arisen from a single cancer cell. Less commonly, individuals present with malignancies that are multifocal or bilateral. Cancer survivors who have an inherited susceptibility to cancer are at substantially increased risks for developing subsequent malignancies. It is important to distinguish between new primaries and cancer that has spread to neighboring or distant sites of the body. For example, an individual with breast cancer that spreads to the liver or bones has only one primary cancer; however, an individual with renal cell carcinoma (kidney cancer) and hemangioblastoma (brain tumor) has two primary tumors. The most conclusive way to distinguish primary from secondary tumors is to obtain pathology reports describing the features of the tumor cells.

6. Absence of Other Risk Factors

It is also important to consider potential environmental causes of tumors. This is especially true for cancers commonly associated with nongenetic risk factors. Certain forms of cancer are associated with preexisting medical conditions. Examples include Kaposi or non-Hodgkin lymphoma in persons with acquired immune deficiency syndrome (AIDS), colon cancer in those with inflammatory bowel disease, and testicular cancer in men with undescended testes. Malignancies are also associated with exposures to specific carcinogens, such as the various tumors linked with smoking cigarettes or heavy use of alcohol.

7. Physical Findings Suggestive of Hereditary Syndrome

In some of the hereditary cancer syndromes, malignancy may be preceded by the development of benign tumors. Examples include multiple nevi in individuals at risk for familial melanoma and multiple colonic polyps in those at risk for hereditary polyposis cancer syndromes. In addition, a few cancer disorders are associated with congenital defects. For example, children born with aniridia may be at increased risk for developing Wilms' tumor.

Cancer histories are classified according to the amount of cancer in the family and the presence of other cancer risk factors. Cancer within a family can be classified as

1. sporadic: no other cancer in family, no known risk factors
2. environmental: exposure to a known carcinogen
3. familial: inherited factors and shared environment
4. hereditary: an inherited single-gene mutation

For families with a hereditary cancer syndrome, individuals are quoted risks according to their placement within the pedigree. An individual has a 50% risk of carrying an inherited gene mutation if a parent is known to have the mutation and a 25% risk if a grandparent has the mutation and the parent's gene status is unknown. For some of the known hereditary cancer syndromes, genetic testing is available to clarify the risks of cancer for individual family members.

B. Option of Genetic Testing

Genetic testing is currently available for several hereditary cancer syndromes and will almost certainly become even more readily available in the future. Table II lists hereditary cancer syndromes for which genes have been identified. Because the DNA analysis is time-consuming and difficult, testing typically follows rather than precedes the identification of a family with a hereditary cancer syndrome. Once a germline mutation has been identified within a family, other family members also have the option of being tested. This type of testing, termed predisposition testing, is technically much easier because it involves looking for the presence of a specific mutation; however, deciding whether or not to have such testing can be extremely difficult. Genetic testing is just beginning to be made available to people who have not had a cancer susceptibility gene; the potential impact of such testing is still being evaluated.

Roughly half of the people who are eligible for predisposition testing elect to be tested. As listed in Table III, there are several factors why individuals decline or accept testing. Motivations for being tested include the desire to clarify the cancer risks for themselves or their children, hope that this information will lead to better cancer surveillance or prevention options, and wanting to end the uncertainty about whether they carry the gene mutation. For the many people who decide against being tested, the major

TABLE II
Dominant Cancer Susceptibility Genes Identified to Date, Their Chromosomal Location, and Associated Hereditary Cancer Syndrome

Gene name	Gene location	Cancer syndrome
APC	5q21	Familial adenomatous polyposis
BRCA1	17q21	Breast–ovarian cancer syndrome
BRCA2	13q12	Breast–ovarian cancer syndrome
MLH1	3p21	Hereditary nonpolyposis colon cancer
MSH2	2p22	Hereditary nonpolyposis colon cancer
PMS1	2q31	Hereditary nonpolyposis colon cancer
PMS2	7p22	Hereditary nonpolyposis colon cancer
p53	17p13	Li–Fraumeni syndrome
RB	13q14	Retinoblastoma (familial form)
RET	10q11	Multiple endocrine neoplasia, type 2A
VHL	3p24	von Hippel Lindau syndrome
WT1	11p13	Wilms' tumor (familial form)

TABLE III
Reasons Why People Who Have Never Had Cancer Decline or Accept Predisposition Testing for a Cancer Susceptibility Gene

Reasons for declining predisposition testing
Would rather not know if cancer risks are increased
Results may lead to increased depression or anxiety
No proven surveillance strategies for people with inherited gene mutation
No known strategies to prevent cancer in people with inherited gene mutation
Potential loss of insurance if inherited gene mutation is present
Reasons for accepting predisposition testing
To better clarify cancer risks of cancer for self
To better clarify cancer risks for offspring
To end uncertainty about whether mutation is present
Chance of learning that mutation is not present
May motivate toward more vigilant surveillance or healthier life-style.

reasons cited are a lack of interest in learning the results, fear that they do carry the gene mutation, perceived lack of medical benefits from learning information, and concerns about possible insurance discrimination.

Commercial laboratories are poised to perform and market DNA testing for cancer susceptibility genes and ordering such tests may be simple in the future. Currently, though, testing options are limited and usually occur within research programs at academic medical centers. These research programs typically offer testing within a framework of genetic counseling, psychological support, and follow-up. A predisposition testing program may include three discussions with a genetic counselor and a medical oncologist. The initial visit includes a discussion about the individual's motivation for testing, basic information about cancer genetics, possible results and implications, accuracy and limitation of results, risks and benefits, assessment of coping resources and support network, confidentiality issues, and program specifics. Individuals who present with clinical levels of depression or anxiety are referred to a psychologist for additional assessment and support. At the second visit, results are disclosed and discussed in a supportive environment. The third visit provides a review of the information already covered in the previous visits and involves a more detailed discussion about options for cancer surveillance and risk reduction.

BIBLIOGRAPHY

Bellacosa, A., Genuardi, M., Anti, M., *et al.* (1996). Hereditary nonpolyposis colorectal cancer: Review of clinical, molecular genetics, and counseling aspects. *Am. J. Med. Genet.* **62**, 353–364.

Bishop, J. M. (1991). Molecular themes in oncogenesis. *Cell* **64**, 235–248.

Hodgson, S. V., and Maher, E. R. (1993). "A Practical Guide to Human Cancer Genetics." Cambridge University Press, Cambridge.

Hoskins, K. F., Stopfer, J. E., Calzone, K. A., *et al.* (1995). Assessment and counseling for women with a family history of breast cancer. *J. Am. Med. Assoc.* **273**(7), 577–585.

Offit, K., and Brown, K. (1994). Quantitating familial cancer risk: A resource for clinical oncologists. *J. Clin. Oncol.* **12**, 1724–1736.

Varmus, H., and Weinberg, R. A. (1993). "Genes and the Biology of Cancer." Scientific American Library, New York.

Cancer Immunology

ARNOLD E. REIF
Mallory Institute of Pathology

GLOSSARY

Antibody Protein secreted by an immune cell that can specifically bind to an antigen

Antigen Substance that elicits a specific immune response when introduced into the body

BCG Bacillus Calmette-Guérin, a viable mycobacterium used for immunotherapy

Biomodulator Abbreviated term for biological response modifiers: agents that improve the biological response of the host to diseases such as cancer

C. parvum *Corynebacterium parvum,* a killed bacterium used for immunotherapy

Cytokines Tissue activators secreted by cells of the lymphoid system

Immune system Organs, cells, and secretions that participate in immune reactions

Immunogenic Capable of inducing an immune response

Immunotherapy Treatment by stimulation of the immune system or by transfer of immune cells or factors produced extraneously

Immunosurveillance Surveillance (guarding) of the body by the immune system

Lymphocytes White cells; spherical cells that are the chief constituents of the lymphoid tissues

Monoclonal antibody Antibody produced by immortalized cells derived from one clone

Natural killer (NK) cells Lymphoid cells that can attack tumor cells in a host that has not been immunized, but that do not attack normal cells

Tumor-specific antigen Antigen found on tumor cells, not on normal tissues

CANCER IMMUNOLOGY IS THE SUBJECT THAT deals with all aspects of the interaction between cancer cells and cells of the immune system. The title "Immunity to Cancer" was chosen as an alternative, because it hints at a therapeutic benefit. It is also the title I chose for a series of conferences on cancer immunology, which were held in 1984, 1987, and 1989. Since then, they have been renamed the Williamsburg Conference and have continued under the auspices of the Society for Biological Therapy.

Tumor immunology is a young science, in which some of the most relevant discoveries may still lie ahead. The body maintains a precise control over the numbers and types of its cells. Depending on the kind and concentration of the agent that causes cancer, the body is able to mount a defense against the cancer cells that ranges from very weak to strong. This defense employs cells of the immune system and their secretions, namely, antibodies and lymphokines. Uncertainties still underlie such central problems as the potentialities and limitations of immunosurveillance against cancer in human beings and the degree to which the various types of human cancer are immunogenic. The answer to this latter question holds the key to successful immunotherapy of cancer, because no strong, active (ongoing) immune response is possible unless a cancer is immunogenic.

Immunotherapy uses a variety of strategies to intensify the natural defenses against cancer. However, the key to potent treatment of cancer with immunotherapy has proven elusive. Even in work in experimental

ENCYCLOPEDIA OF HUMAN BIOLOGY, Second Edition, VOLUME 2. Copyright © 1997 by Academic Press. All rights of reproduction in any form reserved.

animals, we know only the immunogenic strengths of tumors induced by high doses of a carcinogen (cancer-causing agent) but do not understand the extent to which immunogenicity is reduced as the dose of carcinogen is reduced. Nor do we understand as yet how to boost the immune response maximally by injecting appropriate types and doses of biomodulators during appropriate phases of the immune response against tumors. On the bright side, the development of the use of monoclonal antibodies makes it possible to produce antibodies that react with specific antigenic portions of cancer cells. Trials are presently in progress on vaccination of patients with a minimal load of cancer cells against their very own tumor. Also under trial is vaccination against hepatitis B virus for prevention of liver cancer in Southeast Asia. This may be the prototype for the ultimate answer to those types of human cancer caused by viruses: prevention through vaccination. Unfortunately, most of such viruses have yet to be identified and purified.

I. DESTRUCTION OF "FOREIGN" CELLS

A. Cells from a Different Individual

In 1883, the Russian scientist Elie Metchnikoff performed the experiment that launched immunology. Intrigued with the idea that host defense cells might attack a foreign body, he stuck rose thorns under the skin of some transparent starfish larvae. When he viewed them under the microscope the next day, the thorns were surrounded by mobile host defense cells, which he later called "macrophages"—scavenger cells. Subsequent work showed that cells from a genetically different organism are destroyed (rejected).

Cancer cells are also foreign intruders, in the sense that they have constituents not specified in the genetic blueprint for the body, or else appear in the body at a time far removed from their normal appearance—for instance, during embryogenesis or during regeneration of injured cells. Research focused on the transplantation of tissues has shown that the rejection of normal tissues depends on the presence of genetic disparity between donor of the tissue and the host injected with it. In general, the greater the disparity, the stronger the rejection reaction, which is immunological in nature. When the tissue in question is cancer, then there is the additional problem for the host defense cells that they must marshall against an invader whose numbers are increasing through cell division. If the rate of growth of the cancer cells overwhelms the rate at which the defending cells can be produced, the cancer cells will win the battle, even if they are antigenically quite distinct. In addition, cancer cells can elaborate immunosuppressive factors that may disrupt host defenses.

Nevertheless, work in animal systems has shown that if there is a considerable genetic (and therefore antigenic) difference between the donor of the cancer cells and the host, most frequently the cancer cells are completely eradicated. The wider the genetic difference between cancer cell and host, the more powerful is this rejection response. Since this response depends on *normal* antigens on the tumor cells, such a rejection response tells us nothing about *tumor-specific* antigens. [*See* Immunobiology of Transplantation.]

B. Cells from a Genetically Identical Person

Transplantation research has shown that tissues grafted between identical twins are successful: the tissues are not rejected. For this reason, experiments directed at detecting a tumor-specific antigen are carried out in strains in which all animals are genetically as similar as identical twins. These strains are produced by continual "inbreeding"—matings between brothers and sisters of a single litter. After 21 successive generations of inbreeding, uniformity in genetic makeup is attained, and animals can be called "isogeneic" (Greek "iso" = equal, "−geneic" = genetically). Grafts of normal tissues performed between isogeneic animals are accepted, just as with grafts between identical twins.

To avoid confusing *normal* antigens with *tumor-specific* antigens, for the last 50 years most experiments in tumor immunology have been carried out in isogenic strains of animals. Tumors can be either induced *de novo* (meaning "new" in each animal) by use of a suitable carcinogen or transplanted from animal to animal. Because there are no genetic differences in an isogenic strain, tumor transplants between animals are successful. But if transplanted many times, tumor cells can undergo mutations that give rise to new antigens on their cell surface. This danger is almost certainly avoided if a given tumor is not transplanted more than 10 times. But many types of tumor can be suspended in a medium containing antifreeze and frozen for months before being thawed, transplanted, and grown again. Thus, a good strategy for a researcher who wishes to perform many immunological experiments with a given tumor, yet avoid the changes in tumor characteristics that occur with continual transplantation, is to store frozen multiple portions of a given tumor.

C. How Does Tumor Rejection Depend on the Cancer-Causing Agent?

When a known oncogenic (cancer-causing) agent is used in an animal of an inbred strain at appropriate dose and application, a tumor characteristic of that oncogenic agent is produced. Tumors induced by different types of oncogenic agents have widely varying immunogenicities (Table I). With the exception of some of the UV-induced tumors, all tumors are accepted if injected into an isogenic "naive" animal—one that has not been immunized (vaccinated) before this "challenge" by injection of viable tumor cells. To immunize a naive animal, the tumor cells must be inactivated by a method that prevents further cell division, such as irradiation with a high does (12,000 rad) of X rays. Naive animals are first immunized by injection of inactivated tumor cells. After a suitable time interval such as 1 or 2 weeks, animals are challenged by injection of a known number of viable tumor cells.

The results of such immunization experiments are recorded in Table I. Thus, "++++" for UV-induced tumors indicates that some of the tumors are so highly immunogenic that they may be rejected even without prior immunization. The notation "+++" indicates that the tumor will be rejected if only a single vaccination is given, whereas "+" shows that three or four vaccinations may be necessary before a challenge with viable tumor cells can be rejected. In contrast, several injections of "inactivated" normal tissues from isogeneic animals will not prevent acceptance of a graft of the corresponding viable normal tissue. Thus, rejection of the graft of a tumor by an isogeneic animal preimmunized with inactivated tumor indicates the presence of an antigen or antigens not found in normal tissue, which we call "tumor-specific."

Animal experiments (Table I) indicate that several classes of tumors have very low immunogenicities: those that appear spontaneously, those induced by ionizing agents (X rays, radioisotopes), and perhaps also those induced by very low doses of chemical carcinogens or by vertical transmission of some types of oncogenic viruses. Unfortunately, we lack adequate data for these latter two categories. In contrast, some classes of tumors are highly immunogenic: many of the tumors induced by high doses of chemical carcinogens, or by injection at birth of oncogenic viruses, or by UV radiation.

Two tumors are said to be cross-reacting when an animal immunized against one tumor exhibits changes (usually, increases) in immunity to the other tumor. The data on cross-reactivity (Table I) are important for work with oncogenic viruses, by showing that only tumor cells produced by a particular oncogenic virus possess the same tumor-specific antigen that is present in *all* tumors induced by that virus—even those induced in different individuals. Thus, it is possible to immunize one animal with inactivated cells of a given virally induced tumor taken from another animal. The cross-reactivity of virally induced tumors

TABLE I
Relation between Etiology of a Tumor and Its Immunogenicity

Etiological agent	Dose of agent	Approximate immunogenicity	Cross-reactivity with other tumors caused by that agent[a]
None: tumors are spontaneous	—	−, ±	−, ±
X rays, radioisotopes	High	−, ±	−, ±
Chemical carcinogens	High	++, +	±
	Medium	+, ±	(±)
Oncogenic viruses			
Injected at birth	High	+++, ++	+++, ++
Vertical transmission[b]	—	+, −	+, −
UV radiation	High	++++, +++	±

[a]Immune cells or antibody obtained by immunization with one tumor also react positively with other tumors initiated by the same carcinogenic agent.

[b]Vertical transmission means passage from the mother to the fetus in her uterus.

in animal systems suggests that human tumors caused by known oncogenic viruses can be successfully prevented by vaccination; indeed this is the assumption behind the ongoing trial of vaccination with hepatitis B virus to prevent liver cancer.

In contrast, for the other agents producing strongly immunogenic tumors (namely, chemical carcinogens and UV irradiation), only a very slight degree (±) of cross-reactivity exists. Thus, to obtain significant protection against challenge with viable tumor cells, it is necessary to immunize with the identical tumor to which immunity is desired. This requirement makes the process of vaccination against a tumor difficult, because the vaccine must be prepared from the patient's own tumor cells.

II. CELLS IMPLICATED IN TUMOR REJECTION

Many different types of immune cells have been described that take part in a positive or negative way in the rejection of tumor cells in naive or immunized animals.

A. Natural Killer Cells

Natural killer (NK) cells possess *spontaneous* cytotoxicity (cell-directed toxicity) against tumor cells and against a limited number of normal cell subpopulations, for instance, virus-infected fibroblasts. Unlike T (thymus-derived) cells, NK cells require neither sensitization nor the presence of major histocompatibility (tissue-compatibility) antigens, without which T cells cannot act. NK cells can kill target cells directly or through antibody-dependent cellular cytotoxicity (ADCC), which consists in the killing by NK cells of tumor cells that have been coated by an antibody that facilitates this NK cell action. In human beings, NK cells represent about 5–8% of circulating white cells. They can be distinguished from T cells because they do not express T-cell-antigen receptors nor have rearranged their T-cell-receptor genes. In animal experiments, NK cells have been shown to play an important role in immunosurveillance against cancer. The journal *Science* of April 1995 (p. 367) reports that human NK cells have genes that produce proteins on their cell surface that act as receptors for signals that tell NK cells not to kill. This seems to be the mechanism by which the body prevents indiscriminate killing by NK cells. T cells (see Section II,D) appear to have a similar mechanism that enables them to be disarmed. [*See* Natural Killer and Other Effector Cells.]

B. Lymphokine-Activated Killer Cells

The cytolytic (cell-killing) activity of NK cells can be increased by interleukin-2 (IL-2) both *in vivo* and *in vitro*. IL-2 can greatly potentiate the natural antitumor activity of NK cells and recruit cytotoxic cells of T-cell origin. NK cells treated with IL-2 can kill types of tumor cells that untreated NK cells cannot kill. Clinical trials with oft-repeated large doses of IL-2, supplemented by infusing cells obtained from the patient's blood and expanded through tissue culture in the presence of IL-2 [which turns these cells into lymphokine-activated killer (LAK) cells], have produced durable complete responses in 15–30% of patients with renal-cell carcinoma or melanoma. These responses sometimes include disappearance of the tumor. However, IL-2 is highly toxic at the doses employed, and the results are far less positive for more common types of cancer. A possible explanation is that renal-cell carcinoma and melanoma are highly immunogenic and that the positive results for these tumors result because LAK cell-killing of tumor cells is followed by conventional T-cell-mediated cytotoxicity (see Section II,D) that requires the presence of strongly immunogenic tumor antigens. Most recently, Rosenberg has separated tumor-infiltrating lymphocytes (TILs) and expanded these cells *in vitro* before injection into melanoma patients: much improved results have been obtained by this technique.

Science of March 1995 (p. 1391) reports that S. A. Rosenberg *et al.* have tried to insert the gene that produces the tumor necrosis factor (TNF), which is toxic to tumor cells, into TIL cells from cancer patients, prior to infusing the genetically altered TIL back into the same patient from which they were taken. In this very first experiment on gene therapy for cancer, it was hoped that the TIL cells would home in on the tumors in the body and there produce TNF, which would kill the tumors. But Rosenberg's interpretation of the clinical data was far more enthusiastic than that of the scientific advisory committee that evaluated the results. Thus, the jury is still out on the first experiment in which gene therapy was combined with the immunotherapy of human cancer. [*See* Interleukin-2 and the IL-2 Receptor.]

C. Macrophages

Macrophages are named for their ability to *swallow* bacteria or foreign particles. They are mononuclear (single-nucleus) cells present in both blood and body tissues. Macrophages can kill tumor cells in two ways. They can be "activated" nonspecifically, to become directly cytotoxic to tumor cells. Alternatively, they can kill tumor cells coated with specific antitumor antibody by means of ADCC (described in Section II,A). Macrophages also initiate the response to a strongly immunogenic tumor by presenting tumor antigen to T-helper lymphocytes (see Section II,D). Further, macrophages can secrete many different factors involved in host defense. In animal models, macrophages have been shown to prevent or cure cancer by both nonspecific and specific mechanisms. Despite their impressive potentialities and positive animal experiments, the evidence for the involvement of macrophages in immune surveillance and in the suppression of metastases is suggestive rather than definitive. Indeed, macrophages also have suppressive functions. The identification of macrophages is made possible because they cling to surfaces and possess specific cell-surface markers. Work in human beings has focused mainly on probing their involvement with primary tumors and metastases (extension of cancer to distant organs). [*See* Macrophages.]

D. B and T Lymphocytes

As mentioned earlier, strongly immunogenic tumors can elicit a classic immune response in which macrophages act as antigen-presenting cells for helper/inducer (CD4-positive) T lymphocytes. These activated helper cells begin a sequence of cellular interactions that results either in B (bursa-derived)-cell activation and/or activation of cytotoxic (CD8-positive) T lymphocytes to cytolytic cells (CTLs). The B-cell activation can lead to the secretion of specific antibodies to the tumor antigen involved, whereas the cytotoxic T cells can kill tumor cells bearing the antigen to which they have been specifically sensitized. Indeed, serum antibodies have been found in persons with certain types of tumors, and specifically cytotoxic T lymphocytes have been described in patients with melanomas, with UV-induced skin tumors, and with certain other types of solid tumors. Although these findings suggest that the respective tumors are highly immunogenic, the presence of antibodies or of cytotoxic T cells unfortunately does not imply that a given tumor will regress. [*See* Lymphocytes.]

Antibodies can attack foreign antigens only if they are present on the surface of a target cell. Thierry Boon credits Alain Townsend for the discovery that cytolytic T lymphocytes in some instances can detect viral proteins hidden within a cell. All the proteins in the cell's cytoplasm are cut into small fragments called peptides, which can attach to proteins called class I major histocompatibility (MHC) molecules. Then this complex can move to the cell surface, where cytolytic T cells can detect the viral peptides as foreign and destroy the cell that displays them. Subsequently, Boon has found that some tumor cell-surface antigens can function as tumor rejection antigens only if they are bound to a particular type of MHC molecule. Only then can they be recognized by cytolytic T cells.

E. Suppressor Cells and Factors

Suppressor cells down-regulate both humoral (B-cell-mediated) and cellular (T-cell-regulated) immunity. In human beings, T suppressor cells (Ts) bear the CD8 marker (also displayed by cytotoxic T cells) and also the CD11b marker, which distinguishes them from cytotoxic T cells. Suppressor cells act through soluble factors, which directly or indirectly inhibit the activity of effector cells. Also, antigen-nonspecific suppressor T cells exist, which have a similar effect. Unfortunately, many types of immunogenic tumors stimulate host production of Ts, which depress the immune response to those tumors. To inhibit Ts, low doses of certain biologic agents as well as X rays have been used. At high doses, cyclophosphamide inhibits the growth of rapidly dividing cells, but at low doses it inhibits Ts in cancer patients and thereby potentiates tumor immunity. Therefore, in current clinical immunotherapy trials, low doses of cyclophosphamide are frequently used to inhibit Ts cells.

Several suppressive factors have been found in cancer patients. Some of these factors represent tumor antigens produced by (or through breakdown of) the tumor. Others represent secretions of suppressor T cells or of suppressor macrophages, whereas others still are of unknown origin; all depress the immune response to immunogenic tumors.

III. TUMOR-ASSOCIATED AND TUMOR-SPECIFIC ANTIGENS

Tumors often arise from the basement layer of cells that—even in adult life—maintain active cell division and provide replacements for cells that have died. As has been illustrated in the white cell series, such replacement cells may go through a series of differentiation steps (changes to more specialized function), at each of which tumors can arise that correspond to their normal counterparts at that particular step of differentiation. Because the number of normal cells seen at that step may be quite small, the corresponding tumor cells that are often present in large numbers may be judged incorrectly to possess a new "tumor-specific" antigen. In reality, the antigen will be a normal tissue antigen previously unrecognized, because it is confined to a small subpopulation of normal cells. Such antigens are "tumor-associated" antigens—normal antigens associated with the tumor.

A. Oncofetal Antigens

Carcinoembryonic antigen (CEA) was named thus because it is found in large amounts only in human embryonic tissues and on the part of the tumor cell surface (the glycocalix) from which it is shed. Another such antigen is α-fetoprotein (AFP), which is secreted by human fetal liver and by most human liver tumors. Because small amounts of both CEA and AFP are present in normal adults, it is possible that both CEA- and AFP-secreting tumor developed from small subpopulations of embryonic-type stem cells retained even in adulthood to serve as basement layers for cell replacement. Because such embryonic cells possess a relatively unstable gene capable of causing the cell to differentiate into various types of more specialized cells, embryo-type cells are particularly capable of being transformed into cancer cells. Because oncofetal antigens are normal antigens, if found on tumors they are called "tumor-associated."

B. Oncogene Product Antigens

During the past decade, a virtual revolution has taken place in our understanding of the molecular mechanism of carcinogenesis. Today we believe that many types of cancer cells develop because one or two normal genes—called "oncogenes" because they can endow a cell with cancer-like properties—are switched on at a time inconsistent with the time sequence of their activation set out in the genetic blueprint of the body. The most vital cancer-like property is the ability to continue cell division, irrespective of host attempts to limit it or to destroy the cells. Also, the body's ability to limit the proliferation of embryonic-type cells becomes gradually impaired in old age. To date, dozens of human oncogenes have been discovered, and their molecular mode of action is now partly understood, according to Theodore Krontiris, writing in the *New England Journal of Medicine* in 1995 (p. 303). Oncogenes encode molecules that are normal body constituents and would therefore be only "tumor-associated" and not "tumor-specific" antigens if located at the cell surface—unless mutations occurred in the oncogenes, leading to the encoding of tumor-specific antigens. Because oncogene-encoded molecules may be present in large amounts, they—like oncofetal antigens—may serve a valuable role as tumor markers and, to a lesser extent, also as potential targets for immunotherapy. [*See* Oncogene Amplification in Human Cancer.]

C. Virally Induced Antigens

Hepatitis B virus (HBV), herpes viruses, retroviruses, and papovaviruses all can cause tumors under natural conditions. They can cause tumors not only in some species of mammals, but also in nonhuman primates and in human beings. In terms of the oncogene theory, oncogenic viruses can produce tumors in several ways. Most directly, once a DNA virus has inserted its DNA into the cellular genome, the viral DNA can function as an activated oncogene. For retroviruses, the viral RNA first must be transcribed into DNA before that DNA can be inserted. Other oncogenic viruses act by switching on the host cell's own oncogenes or else by inactivating oncogene repressor genes.

Oncogenic viruses can generate tumor-specific cell-surface antigens by several different mechanisms. Most simply, and probably most frequently, they can produce copious new virus particles or viral components; these appear on the cell surface, thereby conferring the antigenicity of the oncogenic virus onto its virally induced tumor cell. Less directly, the viral DNA within the genome of the transformed cell can specify novel cell-surface antigenicities that are distinct from those on the native virus. Finally, genetic recombinations between two or more viruses can occur that may specify novel cell-surface antigens. As already noted, most tumor-specific antigenicities induced by one particular oncogenic virus are identical, irrespective of the individual or even of the species in which the tumor has occurred. This means that such tumors cross-react and that vaccination against such tumors—most of which are highly immunogenic—is entirely feasible.

Liver cancer is far more common in hot, wet areas of the world than in Western nations or in the United States, where it ranks 25th. Presently, experts believe that 80–90% of cases of liver carcinoma are caused by HBV. The mode of HBV carcinogenesis is unusual: the surface antigen of HBV acts as a chemical carcinogen for liver, whereas integration of the viral DNA into the cellular genome is merely a random event that has no effect on induction of liver cancer. Principal among the other agents that cause liver carcinoma or act as cofactors for HBV are aflatoxin B, a toxin produced by a fungus that thrives on cereals and nuts stored under hot and moist conditions, and long-continued heavy use of alcohol, which produces a 10% incidence of cirrhosis. In the United States, most cases of primary liver cancer develop in livers cirrhotic because of a heavy consumption of alcohol. Secondary liver cancer, meaning cancer that has spread to the liver from its origin at a primary site in another organ, is unrelated to the causes of primary liver cancer.

Chronic HBV infection results in a relentless pathologic process that, barring death from other causes, leads to eventual death from either cirrhosis and/or liver cancer. Effective vaccines against HBV are now available. In China, where liver cancer is the most common form of cancer, a program of vaccinating all newborn children has begun. In the United States, a cancer prevention center based in Philadelphia provides hepatitis B vaccinations for high-risk groups (i.e., the families of HBV carriers, drug addicts, and others). In the great majority of HBV-negative persons who have been vaccinated, complete protection against HBV infection—and therefore against HBV-induced liver cancer—is expected. In any case, vaccination against HBV is a valuable preventive measure, since one can obtain an HBV infection from eating food contaminated by an infected food handler, and HBV can cause liver failure and death even when it does not cause liver cancer.

Many types of leukemogenic viruses have been discovered in animals. Among the most interesting is feline leukemia virus (FeLV), a retrovirus (RNA virus) that infects about 3% of cats in the United States. Of cats exposed to FeLV through contact with infected cats, about one-third receive insufficient virus to become either infected or immune, one-third become immune to the virus, and one-third become infected. Within 4 years of infection, more than 80% of infected cats die of leukemia. A vaccine is now commercially available to immunize cats against FeLV. The vaccine contains the FeLV envelope gp70 protein, which is produced by recombinant DNA techniques. [*See* Retroviral Vaccines.]

In the case of human leukemias, viral causation is certain only for one rare type: T-cell leukemias caused by the human T-cell leukemia virus-I (HTLV-I). This virus is gradually spreading around the world. Although it shares only 2% of its genes with the AIDS virus, like that virus, HTLV-I is spread by blood transfusions, sex, and mother's milk. Although the virus has been discovered in a significant proportion of people in Japan, the Caribbean, and the southernmost United States, no efforts to prepare a vaccine against it seem to have been made as yet. [*See* Leukemia.]

A study by Margaret Davis published in the *Journal of the National Cancer Institute* in 1989 showed that children who are breast-fed for more than six months are partially protected against childhood lymphoma, which is a rare disease. These children develop lymphomas only one-half as frequently as children who were breast-fed for a shorter period. Because breast-feeding allows the baby to absorb from the mother's milk her antibodies that protect against many types of bacteria and viruses, this finding suggests that some types of childhood lymphomas are caused by one or more cancer-causing viruses. When these have been identified, vaccination of babies against lymphomas will be possible. We have explored such vaccination in a model system in our laboratory. We found that vaccination of newborn mice against a known mouse leukemia virus is very effective in one but not in another strain of mice. This suggests that a future human leukemia virus vaccine will be effective in many but not in all infants. [*See* Lymphoma.]

The most common childhood tumor in hot and wet areas of Africa and southeast Asia is Burkitt's lymphoma (BL). In the same areas, nasopharyngeal carcinoma (NC) is also common. In those areas in Africa, around 99% of patients with BL and a high percentage of patients with NC have tumor cells bearing markers of the Epstein–Barr virus (EBV), a herpes virus. (In contrast, BL is rare in the United States, and only 10% of U.S. patients with BL have cells that bear the EBV antigen.) All patients have a translocation involving the oncogene *myc*. Possible cofactors include immune depression through injury by diseases such as malaria or schistosomiasis, whereas possible causes other than EBV include carcinogens produced by fungi common in the same hot and humid climates and other, as yet unrecognized viruses. Thus, three different factors combine to cause BL: EBV, the *myc* oncogene, and additional cofactors. [*See* Epstein–Barr Virus.]

Although the precise role of EBV is uncertain, it may not be premature to use a genetically engineered vaccine to EBV in areas where BL and NC are common, to determine whether immunity to EBV protects against these tumors. Because more than 80% of adults the world over have been exposed to EBV, immunization would be done in infants.

Potentially oncogenic venereal infections include herpes simplex virus type 2, cytomegalovirus, human papillomavirus (HPV), and the AIDS virus (HIV-I). Recently, interest has focused on HPV. A strong correlation exists between a silent HPV infection of the lower genital tract of women and the occurrence of cervical cancer. Analysis of tissue culture lines of human cervical cancer indicate the presence of the viral genome and suggest a functional role for HPV in malignant (cancer) transformation. Further, for men who develop relatively rare penile cancers, these cancers contain DNA sequences of HPV in one-third of those whose female partners had cervical cancer, but only in 1% of those whose partners had normal cervical tissues. Because as many as one-third of German women who are sexually active are infected with HPV virus, and a similar situation may prevail in other Western nations, it is to be hoped that a genetically engineered vaccine against HPV will be prepared and tested soon. [*See* Acquired Immunodeficiency Syndrome (Virology); Herpesviruses; Papillomaviruses and Neoplastic Transformation.]

Proof of the viral nature of animal tumors is largely based on Ludwig Gross' discovery of 1951 that tumor viruses will produce tumors when injected in high doses into newborn animals, because newborns have immature immunological defenses. In newborns, even the outgrowth of antigenic tumor cells usually fails to elicit a successful rejection reaction.

Proof of the viral causes of human cancers is difficult to obtain, because it is unethical to infect human beings deliberately with suspected viral agents. Most of our present information comes from epidemiological studies and depends on questionable correlations between the incidence of a certain type of tumor and the presence of a particular virus (or viruses) in that tumor, as compared with the incidence of that virus in the general population. Therefore, for all but a very few types of human cancers, a viral etiology presently is questionable. However, the new techniques of molecular biology are greatly helping to speed this quest. Such work is vital, because success in the identification of a particular virus as the cause of one type of human cancer brings with it the probability that a vaccine against the virus could be engineered, which could be used eventually to eradicate that type of cancer. Because prevention of cancer is far superior and incomparably less traumatic than cancer therapy, the preparation of effective and safe vaccines against various types of oncogenic viruses is by far the most promising approach to eliminating the types of cancers that otherwise would develop as a result of infection with those viruses.

D. Chemically Induced Antigens

Tumors induced by high doses of chemical carcinogens in laboratory animals tend to be immunogenic, although often not as highly as virally induced tumors. Such chemically induced tumors possess cell-surface antigens that are stable and heritable. These antigens can be used to immunize animals both prophylactically (i.e., animals are first immunized, then challenged with viable tumor cells) or else immunotherapeutically (i.e., the immunizations are performed only *after* challenge with viable tumor cells).

Work with tumors induced by different types of chemical carcinogens has shown that for each individual tumor, its tumor-specific antigen is unique—even in the situation in which two tumors arise on the same animal. Tumors induced by a single chemical may show cross-reactivity, but this is often of such low degree that it is difficult to detect. This lack of cross-reactivity between different tumors induced by the same chemical carcinogen poses a nearly insurmountable problem for use of such tumor-specific antigens for diagnosis, vaccination, or immunotherapy, except in the situation in which these are personalized to deal with a single patient's tumor. However, if a given chemical carcinogen produces a type of tumor cell that is the neoplastic (cancer) counterpart of a small, antigenically distinct subclass of normal cells, its tumor-associated (normal) antigens can be used for all tumors of that type to accomplish the same purpose, although perhaps not as easily. For diagnosis, the presence of an antigenically identical subclass of normal cells must pose only a surmountable problem; for therapy, eradication of this subclass of normal cells must not threaten the patient's life.

The dose of the chemical carcinogen used to produce tumors in experimental animals is often orders of magnitude higher than the doses of carcinogens to which human beings are exposed. What little information we have suggests that the immunogenicity of chemically induced tumors decreases as the dose of the carcinogen is decreased. We completely lack information about the immunogenicity of tumors induced at the very low doses at which we encounter chemical carcinogens in our daily lives.

E. Radiation-Induced Tumors

Radiation with a low wavelength and high frequency, such as UV light, atomic radiation, or X rays, is sufficiently energetic to cause mutations in the DNA of any and all cells of our bodies, especially at high-dose levels. However, some body tissues are exceptionally sensitive to lower-dose levels of radiation, for instance, white blood cells. The cancer that develops from these cells is leukemia. Brilliant experimental work by radiologist Henry Kaplan proved that radiation-induced leukemias of mice were seldom caused directly, due to radiation-induced mutations in DNA of white cells. More commonly, the effect of the radiation was permissive rather than direct: "it was to kill many of the cells that control white cell growth, thereby permitting their outgrowth as leukemia cells initiated by leukemogenic viruses carried by the mice."

A similar situation may well obtain in mice for carcinogenesis by radioisotopes. For instance, the bone-seeking radioisotope strontium-90 appears to act primarily in a permissive role for the initiation of bone sarcomas. The primary oncogenic agents are C-type oncogenic viruses carried by the host, viruses that remain latent and only occasionally produce tumors in unirradiated hosts. In that case, as above, the main effect of the radiation appears to be the killing of surveillance cells that normally guard against the outgrowth of cells mutated by the virus.

To answer the question "How immunogenic are radiation-induced tumors?" we need to take into account the effect of passaging and concentrating oncogenic viruses in the laboratory before injecting them into newborn animals. In direct contrast, oncogenic viruses are present in their native form and at low concentration when activated by the permissive action of radiation. Thus, we might expect radiation-induced tumors to be far less immunogenic than those produced by virus injection into newborn, and this seems to be the situation (Table I).

F. Tumors Induced by Other Causes—or None

Tumors may also arise by "physical" carcinogenesis: as when a piece of plastic film is introduced into the peritoneal cavity; or when asbestos fibers, too long to be swallowed and disposed of by macrophages, are inspired into the lung. As in the case of radiation carcinogenesis, physical carcinogens may merely permit latent oncogenic viruses to initiate tumor cells—which are initially out of reach of attack by host immune cells, being physically protected by film or fibers. Such tumors should have immunogenicities similar to those of radiation-induced tumors.

When tumors appear spontaneously, they may do so because they are genetically fated to arise, because of promotion by hormones or other cofactors, or because of the action of various carcinogens—viruses, chemicals, radiations, and fibers—on cells. Even when the primary oncogenic agent is a virus, the same dose concentration effects mentioned earlier may apply.

In human beings, much work has been done on melanomas, because they appear to be highly immunogenic. Unfortunately, animal studies suggest that UV-induced tumors (which include some melanomas) possess an extraordinary escape mechanism. A single tumor induced in a mouse by UV light may display a number of antigens on its surface; we shall call three of these A, B, and C. The host responds with a rejection reaction directed at the strongest antigen, A. Meanwhile, clones that possess only antigens B and C can grow unrestrained. Then the host eradicates all tumor cells displaying the next strongest antigen, B. However, in the meanwhile, clones displaying only C have grown unmolested; when the host finally switches to react against C, it finds itself overwhelmed by the large number of cells bearing antigen C. At least some human melanomas behave in a way that suggests that they use this escape mechanism. [*See* Melanoma Antigens and Antibodies.]

Other types of human tumors worthy of special attention because of suggestive evidence of immunogenicity include transitional cell carcinomas of the bladder, renal carcinoma, skin carcinomas, Burkitt's lymphoma, neuroblastoma type IVS, childhood lymphomas and leukemias, bone sarcoma, and choriocarcinoma.

The cell surfaces of human tumors contain complex carbohydrate molecules that sometimes display deletions or changes at their ends. Such changes are responsible for the appearance of new antigenic specificities. For instance, a gastric adenocarcinoma arising in a patient whose blood type is O may display the "illegitimate" blood group antigen A. Further, tumor cells sometimes display the embryonic rather than the adult form characteristic of a given cell-surface glycoprotein; this suggests that they originated from embryonic-type stem cells or else were converted to such cell types through activation of an oncogene. If identical antigenic changes were present on *all* tumor cells, their antigenic uniqueness could be used for immunoprevention or immunotherapy.

IV. IMMUNOSURVEILLANCE IN LIGHT OF AIDS

The basic thesis of immunosurveillance is that the immune system prevents the emergence of many more tumors than actually appear. Unfortunately, this thesis sometimes has been debated as by a lawyer seeking to make a winning case for a client, rather than as by a scientist seeking a fair exposition of pros and cons, to arrive at present truth dependent on present evidence. Here are some of the pros and cons:

1. *Pro:* the age-related incidence for some human cancers shows a peak for children 2–3 years old, expected from oncogenic viral infections when immunity is immature.
   Con: The cause of these tumors may be genetic, not viral.
2. *Pro:* The incidence of most human cancers increases greatly with age. This increase corresponds to a gradual decrease in immune competence beyond early adulthood.
   Con: The increase in tumor incidence is caused by the cumulative effects of exposure to carcinogens and promotors and reflects the long latent period of cancer.
3. *Pro:* In the many types of human immune deficiency diseases, the incidence of solid tumors is only about 25% of the incidence of leukemias and lymphomas.
   Con: Survival of persons with these diseases is not as brief as is claimed; if immunosurveillance was a reality, more carcinomas and sarcomas should appear.
4. *Pro:* Transplant recipients and others receiving immunosuppressive treatment have a greatly increased incidence of leukemias and lymphomas; epithelial tumors and sarcomas are increased 50-fold, and sun-related skin cancer in sunny areas by 100-fold.
   Con: The immunosuppressive agents used are carcinogenic.
5. *Pro:* Athymic (nude) mice that have severe deficiencies in rejecting grafted tissues do develop malignancies when prevented from dying when young.
   Con: Many more malignancies might be expected if immunosurveillance is real.
6. *Pro:* Natural antibodies directed against tumor are found in some types of cancer.
   Con: Because the tumor continues to grow, these antibodies must be ineffective.
7. *Pro:* Specifically sensitized host cells have been demonstrated in many types of tumor.
   Con: Some such cells have been shown to enhance rather than inhibit the tumor.
8. *Pro:* Failure to reject an immunogenic tumor is due to presence of suppressor cells.
   Con: This merely explains why immunosurveillance is ineffective.
9. *Pro:* Tumor cells but not normal cells are attacked by NK cells and by macrophages.
   Con: On occasion, macrophages can stimulate rather than retard growth.

Vaccination for prevention of cancer and cancer immunotherapy are often perceived as potentiating the natural propensity for surveillance against tumors—and as impractical if surveillance does not exist. This perception is partly wrong, for artificial means for inducing immunity can be more potent than natural immunity. Because cancer strikes more than 20% of the U.S. population, it is obvious that immunosurveillance frequently fails. No wonder it is hotly debated. [*See* Immune Surveillance.]

Because the AIDS virus (HIV-I) specifically destroys human helper/inducer (CD4-positive) T lymphocytes, the spread of this virus allows us to see the outcome of this insult to the human immune system. Because helper cells are implicated in the T-cell response that only takes place against immunogenic tumors, we should expect AIDS patients to develop a higher than normal incidence of the type of tumors induced by oncogenic viruses, UV radiation, or high doses of chemical carcinogens (see Table I). Once the deficiency of helper T cells is sufficient for AIDS to develop, at present survival is rarely more than 3 years. Given this short survival, it is all the more remarkable that the incidence of Kaposi's sarcoma in male homosexuals has been reported as 2000 times higher than that in age-matched controls and that the incidence of B-cell and other types of lymphomas is increased severalfold. This limited and specific outgrowth of tumors after destruction of helper T cells suggests that immunosurveillance by the host employs many different cell types and mechanisms and that the destruction of only one type of host defense cells will lead only to a limited outgrowth of tumors. However, transgenic mouse studies suggest that Kaposi's sarcoma is directly induced by HIV-I and not the result of immunodeficiency. In fact, it usually begins before much immunodeficiency is evident.

However, there are many other arguments in favor of immunosurveillance: for instance, the experimental success in initiating tumors by injecting oncogenic viruses into newborn animals (which are immunologically immature) and the failure of most such viruses to induce tumors in adults. The fact that the body obeys nearly all the time the dictates of its genetic blueprint suggests that there are precise mechanisms for maintaining the genetically designed integrity of each organ, which of necessity implies destruction of cells not designed to appear. Although we understand little about how this control is maintained, it argues for surveillance.

But there is one aspect of immunosurveillance that is so simple to understand and so strongly supported by evidence that it can be claimed as established. That is the situation in which the body carries an oncogenic virus that cannot produce tumors when the immune system is intact, but can do so when it is deficient. Since many solid organs possess their own particular type of resident white cell (for instance, the microglial cells of the brain), it is only when such resident surveillance cells are selectively injured that we can expect the outgrowth of tumors induced by oncogenic viruses. Thus, the reason why immunodeficiency produces solid tumors less frequently than leukemias is that immune deficiency means a deficiency in *circulating* lymphocytes. But it is organ-resident lymphocytes that would be expected to provide most of the surveillance function in solid organs. Thus, a high incidence of solid tumors should occur only when *organ-resident* lymphocytes as well as circulating lymphocytes have been depleted.

V. IMMUNODIAGNOSIS OF CANCER

The chief aim of immunodiagnosis is the detection of tumor antigens, of viral antigens, and/or of antibodies related to either one or both of these (see Section III). A second aim of immunodiagnosis is the evaluation of the state of patients to determine whether the tumor has harmed their general immune competence and whether they possess specific reactivity (either humoral or cellular) against their tumor (see Section II). Here, we shall deal with the first aim.

A. Monoclonal Antibodies

Monoclonal antibodies have revolutionized the diagnosis of cancer at the cellular level. Their name reflects their secretion by cells expanded from a single clone of cells. This clone is constructed by fusing a cell secreting the desired antibody (a splenic lymphocyte from an animal immunized with tumor cells or purified tumor antigen) with a myeloma cell (a malignant tumor cell derived from the type of normal lymphocytes that secrete antibodies). The myelomas used for fusion are selected because they do not secrete a complete immunoglobulin. These hybrid cells (hybridomas) can be grown indefinitely to produce large amounts of the antibodies made by the splenic lymphocyte used for the fusion. Their main advantage over conventional antibodies is their specificity. Because many hybridomas can be screened for specificity, the investigator has a good chance of finding one hybridoma secreting an antibody of (or close to) the desired specificity and binding strength. The usual quest is for an antibody that reacts only with one type of cancer but not with its benign precursor cells, nor with other types of tumors or normal tissues.

It was only in 1966 that the first diagnosis of leukemic lymphocytes using a conventional antibody was performed in the writer's laboratory. Since the advent of monoclonal antibodies, the precise differentiation between different subsets of neoplastic cells of a given type is now possible for leukemias and lymphomas. Even for routine histological diagnosis using tissue sections, staining techniques employing monoclonal antibodies with precise specificities are now a valuable adjunct to conventional staining techniques, some of which may soon be obsolete.

Despite commercial availability, many investigators need to prepare their own monoclonals for special projects. For instance, monoclonals can reveal variations in antigenicity within a given tumor, information that is valuable if immunotherapy is planned. Other uses are to determine antigenic differences between tumors of the same histologic type taken from different patients. Also, changes in antigenicity can be detected as tumors of a single type progress through stages from benign to malignant. [*See* Monoclonal Antibody Technology.]

VI. IMMUNOTHERAPY OF CANCER

In "active" immunotherapy, the patient's own immune system is stimulated to become the "effector"—the active agent—for rejection of his or her tumor. In "passive" immunotherapy, injected substances or cells are the effectors. The distinction between these two kinds of immunotherapy sometimes blurs, because effector materials may stimulate the immune system as a secondary effect.

A. Active Immunotherapy

Early in this century, Coley's toxin, a mixture of killed bacilli that causes fever rather like a bacterial infection, was used to obtain partial or complete regression of cancer in a number of patients. At least part of the explanation for its effects lies in the nonspecific "recruitment" (mobilization) of host immune cells as part of the fever process, with immune cells attacking the tumor cells as if they were "innocent bystanders," without specific immunization. However, later work in my laboratory showed that if this attack was to be powerful, it had to be followed by an immune response specific for tumor antigens, and that this specific response could occur only if the tumor was immunogenic. As a result of work in many laboratories on BCG, on *C. parvum* (a killed bacillus), and on many other biomodulators (biological response modifiers), the following precepts have emerged:

1. Genetic background greatly affects the host's response to any type of antigen—including tumor antigens.
2. Life history, including the extent of exposure to and immunity against different antigens, age, and current status of immune responsiveness are also important.
3. Restriction of immunotherapy to minimal tumor loads is an essential requirement for success, as illustrated by animal experiments. Thus, the best time for immunotherapy is soon after the primary tumor and accessible metastatic tissue have been surgically removed and any remaining tumor further reduced in mass by chemotherapy or radiotherapy.
4. A multifaceted approach should be used, in which immune stimulators (tumor vaccines and biomodulators) and inhibitors (cyclophosphamide, X rays) are used sequentially at times chosen to maximize the tumor rejection response and to minimize the emergence of suppressors.
5. Immunotherapy with tumor vaccines, or with BCG or *C. parvum* in animals is ineffective unless the tumor is strongly immunogenic. A similar conclusion is suggested by recent trials of immunotherapy with IL-2 or cancer vaccines in human beings.
6. Lacking a method for determining the immunogenicity of human tumors, only a rough guide can be obtained from indirect indications (last paragraph, Section III,C).
7. Because some animal models employ tumors with strong immunogenicities—often without quantifying them—the need for conservative extrapolation from the results of immunotherapy in animals to immunotherapy in patients is evident.
8. Methods that increase the immunogenicity of tumor cells should be explored further.
9. When indicated, restorative immunotherapy—to redress the immune imbalance created by the tumor in the patient—might well be chosen to precede immunotherapy.

The time has not yet arrived when immunotherapy can take its place beside surgery, radiation therapy, and chemotherapy as a fourth method of cancer treatment. The startling advances in chemotherapy during the past 30 years were based on a consensus of what constituted meaningful animal model systems and how the results of experiments should be expressed: namely, in terms of the increase in tumor load (expressed in logarithms of tumor cell kill) that was rejected in animals receiving therapy as compared with untreated controls.

But reaching a similar consensus in immunotherapy is far more difficult. With regard to the precepts evolved from previous experiments (see the foregoing), serious deficiencies exist that prevent broad-based advances in immunotherapy. Because of consensus conclusion number 5, that only immunogenic tumors respond to immunotherapy, a direct method of determining the relative immunogenicity of various types of human tumors is urgently needed. Though the author has described such a method for animal systems, it is labor-intensive and requires the use of many animals. Therefore, it needs to be replaced by a tissue culture method that gives parallel results. If such a method were developed, it could also be used to determine the relative immunogenicity of human tumors. At present, few research workers make an effort to determine the relative immunogenicities of the tumors that they use for their experiments in immunotherapy. But without such data, a meaningful comparison of their results with those of other workers who use different tumors is difficult. In addition, the extrapolation of the results of animal immunotherapy to immunotherapy in human beings is highly questionable. Additional complications exist, for instance, whether a standard regimen should be used to eliminate any suppressor cells that are generated during the test.

In immunotherapy, one standout success is the use of "adjuvant contact immunotherapy" in the case of bladder cancer. In this approach, BCG is injected directly into the tumor, where it can induce an intense delayed-type hypersensitivity reaction (a rejection reaction performed by immune cells). In a recent clinical trial that compared BCG immunotherapy with doxorubin chemotherapy for the treatment of bladder cancer, immunotherapy succeeded in producing 1.5-fold fewer recurrences and 8-fold longer durations of response than did chemotherapy. Bladder cancer has an incidence of almost 50,000 cases per year and comprises 2.5% of all cancer deaths in the United States.

Use of the lymphokine IL-2 deservedly has attracted much attention for immunotherapy of patients with renal-cell carcinoma or melanoma (see Section II,B). Equally promising—and far less toxic—is immunotherapy with tumor antigen vaccines, which is under way mostly in patients with the same types of tumors. Many different types of vaccine are being investigated. The protocol of one of the most exciting ongoing trials is based on extensive animal experiments. In this trial, the tumor of a patient with colon or rectal cancer is surgically removed (see 3 above), then immediately dissociated and stored frozen. When needed, tumor cells are thawed, irradiated to make them nontumorigenic, admixed with BCG, and used to vaccinate the patient. The initial result has been a 50% reduction in recurrences. It is still too early to know whether 5-year survival rates (the gold standard for meaningful cancer therapy) will be improved).

Many new concepts in active immunotherapy arise from our increasing understanding of how T cells recognize antigens; some of these were reviewed by Antonio Lanzavecchia in 1993. In the same year, John Travis commented on the finding by three groups of investigators that to stage an effective attack on tumor cells, T cells require not only presentation to them of a tumor antigen, but also "costimulation" of their cell-surface molecule known as CD2; this required a second tumor cell-surface antigen named B7 that was unrelated to the tumor antigen. By immunizing mice with tumor cells into which the B7 gene had been introduced by genetic engineering, an effective immune response was elicited not only against the genetically altered tumor cells, but also against the main mass of unaltered tumor cells.

Many other types of immunotherapy are presently in different phases of clinical trial. Particularly interesting is work with tumor necrosis factor and with interferon (INF). TNF is a product of stimulated macrophages that can induce hemorrhagic tumor necrosis directly or can activate macrophages to destroy tumor cells. Because TNF is toxic at the high doses required for effectiveness, clinical trials are being conducted in which TNF at lower total doses is introduced directly into the tumor. Endotoxin (a bacterial toxin that is also the active principal in Coley's toxin) is highly effective in stimulating macrophages to produce TNF and also induces secretion of IL-1 and INF. This explains the efficacy of Coley's toxin.

Interferons (there are several kinds) have many biological activities, including the enhancement of tumor-associated antigens such as CEA; they can act in synergy with TNF. INF-α can induce the remission and alter the natural history of hairy-cell leukemia and chronic myelogenous leukemia. INFs have shown lesser activity against several other types of tumors.

Renal-cell carcinoma is eventually fatal in two-thirds of those who contract it. In 1993, G. P. Haas and coworkers reported that both IL-2 and INF-α yield promising results in the treatment of this cancer; but A. Manseck and M. Wirth concluded that, because not a single cure with these agents had been reported as yet and only about 20% of patients experienced an objective remission, these and other immunotherapeutic approaches for renal-cell carcinoma must be regarded as experimental. [*See* Interferons.]

Hematopoietic growth factors (HGFs) constitute another family of cytokines that promises to show clinical usefulness. They act on the bone marrow to stimulate rapid growth and maturation of bone marrow cells. HGFs may prove capable of countering bone marrow failure resulting from chemotherapy.

B. Passive Immunotherapy

Because cytokines such as TNF and HGF have *direct* as well as indirect effects on tumor cells, they could be classified equally well as passive immunotherapy agents. Modes of passive immunotherapy include the expansion in tissue culture of host lymphocytes in the presence of IL-2, followed by reinfusion of a large number of these expanded cells (LAK or TIL cells) into the tumor-bearing host. Several different clinical trials employing either LAK or TIL cells are in progress presently (see Section II,B).

Monoclonal antibodies possess the greatest potentiality for tumor therapy. Their production and use for diagnosis have already been described (Section V,A). Because substances from a different species provoke a powerful immune response, repeated use of mouse monoclonals in human beings leads to their rapid elimination. Instead, human monoclonals are coming into use. However, when these antibodies are chimeric, for instance, when composed of the mouse variable region and the human constant region, as described by Y.-R. Zou and coworkers in 1993, then because these chimeric antibodies still contain a portion of a mouse protein, their immunogenicity in human beings is only reduced, not eliminated. Monoclonals have been used for therapy mainly in three different ways.

First, they can act as cytotoxic agents: either for direct killing of tumor cells with the help of host complement (a family of proteins that facilitates cell killing), or indirectly by acting as binding agents between host effector and tumor cells, thereby mediating the destruction of tumor cells. Presently, this type of therapy has benefited only a few of the patients enrolled in clinical trials.

Second, monoclonals can act as carriers for cytotoxic drugs, natural toxins, or radioactive isotopes. Their mission is to act as bullets, delivering their attached lethal charge to each tumor cell in the body. The difficulty lies not in attaching a lethal substance to an antibody molecule, which is easy, but in achieving antibody localization in a tumor, which is very difficult. There are many problems: presence of tumor antigens in the circulation; lack of sufficient strength in the binding of antibodies to tumor cells; slow and uneven uptake of antibody by the tumor due to variations in blood supply and in antigen content between different portions of the tumor; loss of antibodies through specific or nonspecific binding to cells of the reticulo-endothelial system and to other types of tissues; and immunization of the patient against injections of the monoclonal antibody, when it contains foreign antigens (see earlier). Despite such basic problems, radiolabeled antitumor monoclonals already have attained some limited success in the diagnosis of metastases.

Third, anti-idiotypic antibodies (antibodies against the idiotype, the active binding portion of another antibody) have been used for therapy in two different ways. The first type of therapy takes the same approach as the body uses to regulate the numbers of cells in its normal B-cell clones: it produces anti-idiotypic antibodies against the idiotype of the antibodies attached to the cell surface of a particular B-cell clone, thereby suppressing it. In early work from Ronald Levy's laboratory, monoclonal mouse antibodies were prepared against the idiotypes of each of the immunoglobulins produced by patients with B-cell lymphomas. Just 1 of 11 patients injected with monoclonal antibodies tailor-made for their own lymphoma immunoglobulin developed a long-lasting remission. A more successful approach has been to stimulate the immune system of each patient to produce its own anti-idiotypic antibodies. Using this concept, the same team reported in 1992 that 2 of 9 patients, when injected with the specific immunoglobulin protein derived from their very own B-cell lymphoma cells, had complete regressions of their tumor. To increase the body's immune response to its own lymphoma immunoglobulin, each patient's immunoglobulin was injected attached to a protein carrier and mixed with an immunologic stimulant.

A very different approach is to use idiotypes to mimic tumor antigens. If the tumor idiotype is shaped like a hand and the anti-idiotypic antibody as a glove that fits over the hand, a second antibody prepared against the first anti-idiotypic antibody (the glove) again assumes the shape of part of a hand—thereby mimicking the original tumor idiotype, which can act as a tumor-specific antigen. Thus, it is possible to use such second anti-idiotypic antibodies instead of the actual tumor antigen for immunization. So far, this approach has proven only mildly effective. An even more complex method was used in 1995 in Heinz Kohler's laboratory to induce breast cancer-specific antibody responses in monkeys. Instead of an antigenic "hand," they used a monoclonal anti-idiotypic antibody to mimic antigenic sites on that "hand." The method appears to be too intricate to be useful.

C. Tumors: Resourceful Adversaries

Finally, it must not be forgotten that tumors have at their disposal a variety of mechanisms that can frustrate or even co-opt host defenses. One such mechanism is "antigenic modulation," the outgrowth of tumor cells with antigenicities different from those of earlier generations of the same tumor. A second mechanism is the release of blocking factors that can be either specifically or nonspecifically immunosuppressive. A third mechanism is the copious shedding of tumor-specific or tumor-associated cell-surface antigens, thereby providing sham targets for host effector cells. Because of the existence of these mechanisms, immunotherapy is most effective when the tumor is still small. By the same token, the most effective strategy is to mobilize immunity at a time when the tumor load is zero, by vaccination against likely future tumors.

VII. SUMMARY

Although the body can strongly reject cells that are genetically foreign, the mechanisms for regulating the numbers of its own cells are delicate. Tumors are the body's own cells but have escaped its finely tuned mechanism that impels normal cells to conform to the genetic blueprint. Depending on the tumor's immunogenicity, different types of cells react against it. NK cells can kill tumor cells spontaneously and can be activated by IL-2 to kill more effectively. Macrophages can be activated to kill tumor cells both nonspecifically and specifically. Strongly immunogenic tumors induce a host response of cytotoxic T lymphocytes, but this can be inhibited by suppressor T cells and by suppressor factors. Some tumors express normal antigens in abnormal amounts, whereas others display tumor-specific antigens on their cell surface. Tumors induced by UV light, by viruses, and by high doses of chemical carcinogens fall into the latter class. Because tumors caused by a single type of virus share common antigens, vaccination against oncogenic viruses eventually should greatly reduce the incidence of virus-caused cancers in humans. Because prevention is vastly less painful, less threatening, and less costly than cure, vaccination against the different types of viruses that cause human cancer is a vital goal to aim for in the future.

Immunosurveillance predicts that the immune system prevents the emergence of many more tumors than actually appear. For virally induced tumors, the evidence in favor of immunosurveillance seems overwhelming. Immunodiagnosis of cancer has been revolutionized by the introduction of monoclonal antibodies. Their use for therapy of human cancer—whether as cytotoxic agents, as carriers of toxic molecules, or as anti-idiotypic antibodies—is only in its beginning stages. In "active" immunotherapy, the patient's own immune system is stimulated to reject the tumor. For success, immunotherapy requires attention to many conditions, which include restriction of this therapy to strongly immunogenic tumors, reduction to minimal tumor mass before therapy, and inhibition of the emergence of suppressor cells. Some promising results have been obtained with tumor vaccines containing the patient's own tumor antigens and with lymphokine-activated killer cells. Work with other lymphokines, such as tumor necrosis factor and interferon, suggests higher effectiveness for multimodal therapy.

Finally, for specific types of cancer for which an effective vaccine can be made, vaccination prior to the incidence of cancer is vastly more effective than immunotherapy after cancer has developed.

BIBLIOGRAPHY

Awwad, M., and North, R. J. (1989). Cyclophosphamide-induced immunologically mediated regression of a cyclophosphamide-resistant murine tumor: A consequence of eliminating precursor L3T4+ suppressor T-cells. *Cancer Res.* **49,** 1649.

Badger, C. C., and Bernstein, I. D. (1989). Treatment of murine lymphomas with anti-Thy-1.1 antibodies. *In* "Cell Surface Antigen Thy-1" (A. E. Reif and M. Schlesinger, eds.). Marcel Dekker, New York.

Boon, T. (1993). Teaching the immune system to fight cancer. *Sci. Am.* **266** (March 3), 82.

Chakraborty, M., Mukerjee, S., Foon, K. A., Kohler, H., Ceriani, R. L., and Bhattacharya-Chatterjee, M. (1995). Induction of human breast cancer-specific antibody responses in cynomolgus monkeys by a murine monoclonal anti-idiotypic antibody. *Cancer Res.* **55,** 1525.

Haas, G. P., Hillmann, G. G., Redman, B. G., and Pontes, J. E. (1993). Immunotherapy of renal cell carcinoma. *CA, A Cancer Journal for Clinicians* **43,** 177.

Ioachim, H. L. (1990). The opportunistic tumors of immune deficiency. *Adv. Cancer Res.* **54,** 301.

Johnson, H. M., Bazer, F. W., Szente, B. E., and Jarpe, M. A. (1994). How interferons fight disease. *Sci. Am.* **270**(5), 68.

Kagan, J. M., and Fahey, J. L. (1987). Tumor immunology. *JAMA* **258,** 2988.

Kwak, L. W., Campbell, M. J., Czerwinski, D. K., Hart, S., Miller, R. A., and Levy, R. (1992). Induction of immune responses in patients with B-cell lymphoma against the surface-immunoglobulin idiotype expressed by their tumors. *New Engl. J. Med.* **327,** 1209.

Lanzavecchia, A. (1993). Identifying strategies for immune intervention. *Science* **260,** 937.

Law, L. W. (ed.) (1985). Tumour antigens in experimental and human systems. *Cancer Surveys* **4,** number 1.

Manseck, A., and Wirth, M. (1993). Immunotherapy of metastatic renal cell cancer. *Urologe A* **32,** 360.

Metzgar, R. S., and Mitchell, M. S. (1989). "Human Tumor Antigens and Specific Tumor Therapy." Alan R. Liss, New York.

Mitchell, M. S. (ed.) (1989). "Immunity to Cancer, II." Alan R. Liss, New York.

Reif, A. E. (1978). Evidence for organ specificity of defenses against tumors. *In* "The Handbook of Cancer Immunology" (H. Waters, ed.), Vol. 1. Garland STPM Press, New York.

Reif, A. E. (1982). Antigenicity of tumors: A comprehensive system of measurement. *Methods Cancer Res.* **20,** 3.

Reif, A. E. (1985). Vaccination of adult and newborn mice of a resistant strain (C57BL/6J) against challenge with leukemias induced by Moloney murine leukemia virus. *Cancer Res.* **45,** 25.

Reif, A. E. (1986). Relationship of success in classical immunotherapy to the relative immunorejective strength of the tumor. *J. Natl. Cancer Inst.* **77,** 899.

Reif, A. E. (1987). Immunosurveillance reevaluated in light of AIDS. *In* "Lectures and Symposia of the 14th International Cancer Congress," Vol. 5. Akademiai Kiado, Budapest.

Reif, A. E., and Mitchell, M. S. (eds.) (1985). "Immunity to Cancer." Academic Press, Orlando, Fla.

Reif, A. E., Curtis, L. E., Duffield, R., and Shauffer, I. A. (1974). Trial of radiolabeled antibody localization in metastases of a patient with a tumor containing carcinoembryonic antigen (CEA). *J. Surg. Oncol.* **6**, 133.

Rosenberg, S. A., *et al.* (1990). Gene transfer into humans—Immunotherapy of patients with advanced melanoma, using tumor-infiltrating lymphocytes modified by retroviral gene transduction. *New Engl. J. Med.* **323**, 570.

Schwartz, R. S. (1995). Therapeutic clonotypic vaccines. *New Engl. J. Med.* **327**, 1236.

Socie, G., *et al.* (1993). Malignant tumors occurring after treatment of aplastic anemia. *New Engl. J. Med.* **329**, 1152.

Travis, J. (1993). A stimulating new approach to cancer treatment. *Science* **259**, 310.

Zou, Y.-R., Gu, H., and Rajewsky, K. (1993). Generation of a mouse strain that produces immunoglobulin κ chains with human constant regions. *Science* **262**, 1271.

Cancer Prevention

WALTER TROLL
JONG S. LIM
New York University Medical Center

GLOSSARY

Biomarkers Biological indicators that serve to detect exposure to the carcinogenic process. Biomarkers are prognostic or diagnostic indicators or intermediate endpoints used during clinical intervention trials. The biomarker of carcinogenic processes serve to greatly enhance the chemoprevention study design and efficiency

Cancer chemoprevention Systemic administration of specific natural or synthetic chemical agents to prevent the carcinogenic process of cells by suppressing or reversing the progression of a premalignancy to invasive cancer

Carcinogenesis A disease process that can lead to invasive cancer. The process proceeds in a series of sequential stages: initiation, involving changes at genetic levels, following by promotion, and progression to malignancies

Oncogene A gene that, when expressed at abnormal levels or in a mutated state, can contribute to carcinogenesis through the process of deregulation of cell proliferation and differentiation

Reactive oxygen species Hydrogen peroxide, superoxide anions, and hydroxyl radicals that are generated within cells as intermediates of biological processes. Under uncontrolled circumstances, they can cause reversible or irreversible damage to DNA, proteins, or lipids in the cells and can increase oxidative stress in the cells, causing altered gene expression. Cumulatively, these processes result in cancer formation.

Tumor suppressor gene A gene that under normal conditions functions coordinately as a physiologic inhibitor of abnormal cell division, preventing the outgrowth of cells that have undergone mutations

CANCER DEATH RATES HAVE REMAINED VERY HIGH despite some advancements. Oncologists and scientists have only begun to come to grips with the innermost secrets of malignant cells. It is now realized that cancer is a disease with somatic alterations of the cell's genes, a process known as cancer initiation. The availability of sophisticated biochemical tools in molecular epidemiology and cell biology has provided a better understanding of the process of cancer formation or carcinogenesis at its gene level. The process of cancer formation is caused by a series of multiple mutations in the genes that control cell division. These genes have been called oncogenes, which stimulate cells to divide, and tumor-suppressor genes, whose function is to inhibit cell division. These two types of genes in their normal form work together in a coordinated fashion to enable our bodies to replace dead cells and repair damaged cells. When mutations in these genes occur, either inherited or acquired from the environment, the cells undergo a process that would cause them to divide and grow out of control, leading to cancer. [*See* Cancer Genetics; Oncogenes and Proto-oncogenes; Tumor Suppressor Genes.]

It is a characteristic of cancer to occur in one part of the human body, e.g., the lungs, brain, cells, breasts, or prostate. Single cells with mutated oncogenes or tumor-suppressor genes divide to become two, then four, then eight, etc. This is the clonal theory of cancer formation. During the early stages of "mutated cells" they are able to elute from the body's detection and eradication. Among one of the most effective strategies is that the cells pile mutation upon mutation by silencing gene products that are critical

ENCYCLOPEDIA OF HUMAN BIOLOGY, Second Edition, VOLUME 2. Copyright © 1997 by Academic Press. All rights of reproduction in any form reserved.

in monitoring DNA replication for chemical errors. These cells also acquire the ability to stimulate the formation of new blood vessels known as angiogenesis in order to nourish their own growth. These processes of extracellular and cellular mechanisms involved in the modulation of initiated cells of the carcinogenesis process are known as tumor promotion. As the cancer formation process continues, a malignant cell begins to grow into a deadly mass that invades the surrounding tissue and spreads by metastasis to organs located far from its original site. These mechanisms are collectively known as tumor progression, invasion, and metastasis. [*See* Metastasis.]

I. CANCER AS A PREVENTABLE DISEASE

The recent understanding of the biology of cancer has enabled enough accumulated data to allow doctors and scientists to attack the disease when it occurs and to develop strategies to define and modulate the risk by means of agents that could alter critical stages in the multiple carcinogenesis process to prevent the disease from occurring (Fig. 1). Among the cutting-edge strategies being employed are interrupting the cell cycle of the mutated cells using agents that minimize the already mutated cells from piling up more mutations or those that can activate messages in the cell to commit suicide by destroying its DNA. All of these concepts of cancer prevention could be accomplished by chemoprevention, which refers to the use of pharmacologic agents that interfere with or block the process of carcinogenesis, or by changing one's life-style. Chemoprevention agents are available both in synthetic and in naturally occurring (from plants) forms. [*See* Cell Cycle.]

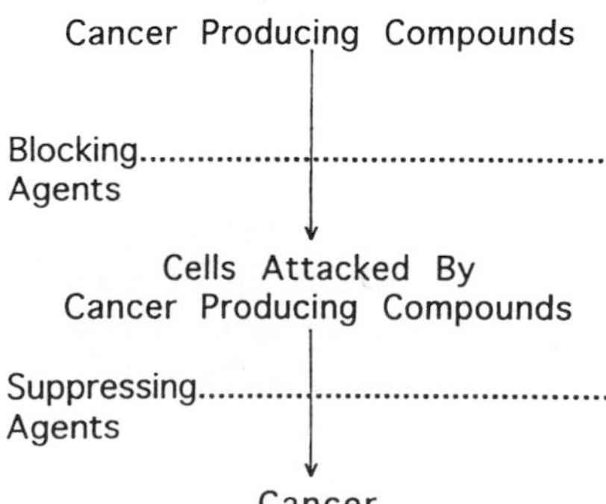

FIGURE 1 Classification of chemopreventive agents on the basis of the time at which they exert their protective effects. Reprinted with permission from *Cancer Res.* 53, 5890 (1993).

Cancer can be prevented by changing one's life-style as seen from epidemiologic observations of the low occurrence of cancer among vegetarian populations such as the Seventh-Day Adventists. This group of people has a significantly lower occurrence of virtually all human cancers, including lung cancer. The low lung cancer rate is attributed to the fact that as part of their religious observations they do not smoke. Epidemiologic studies in Asian countries (e.g., China, Japan, and Thailand) whose diets contain large amounts of fiber and soybeans to substitute for meat showed that they have a lower occurrence of breast, prostate, and colon cancer compared to the United States, the Netherlands, and Sweden. The higher consumption of meat also equates to a higher intake of saturated fat and calories, which are major contributors to high breast and colon cancer rates in Western countries. [*See* Dietary Factors of the Cancer Process.]

II. CHEMOPREVENTION AGENTS

Dr. Wattenberg identified several different classes of chemoprevention agents. The first class has antimutagenic actions that interfere with the formation of carcinogens by extracellular mechanisms. They act by preventing the endogenous production of nitrosamine from precursor amines or amides and nitrites. The second class of agents interferes with metabolic activation of procarcinogens by accelerating the detoxification of chemicals or by scavenging reactive electrophiles formed by carcinogens. In this way the preventive agents can stop the reactive products from interacting with critical cell macromolecules such as DNA and proteins or lipids. The third class of chemoprevention agents acts by inhibiting tumor promotion, progression, invasion, and metastases. This class has antioxidant defenses, anti-inflammatory effects, and is antiproliferative or inducers of cell differentiation. A new class of chemoprevention agents is being investigated. This new class of compounds acts by modulating DNA replication and repair in the control of gene expression.

The identification of these potential cancer preventive agents is being facilitated by our understanding of carcinogenesis in animal models in the preclinical stages before using them in human trials. It is through the animal models that biomarkers are identified and introduced for monitoring during human trials. These biomarkers serve to provide an indicator that appropriate doses and dose schedules are being followed in the trial subjects. The biomarkers used are generally classified into three major categories: (1) genomic

TABLE I

Screening Chemopreventive Agents Using Biomarkers[a]

Compound	Assay response of *in vitro* biomarker assays summary results					
	TK[b]	ODC[c]	GSH[d]	FR[e]	PADPR[f]	DNA binding[g]
N-Acetyl-L-cysteine[h]	+++	++	+++	+	++	−
Anethole trithione[i]	−	+	++	−	+++	−
Ascorbyl palmitate[i]	+++	+++	+	+	++	−
Aspirin[h]	+	+++	−	++	+++	+++
BASF 47851[i]	++	+++	−	−	−	+++
Benzyl isothiocyanate[i]	++	+	+	−	−	+++
Bismuththiol I[i]	+++	+++	+	+++	+++	+++
Butyrate, sodium[h]	+++	+++	−	+++	++	+
Caffeic acid[h]	−	+++	+++	++	+	++
Calcium-D-glucarate[i]	+++	+++	−	++	+++	++
Carbenoxolone[j]	−	+++	−	−	+++	+++
L-Carnosine[h]	+	+++	+++	−	+++	+
β-Carotene, *trans*[k]	+++	+++	−	+++	+++	+++
Catechin[i]	+++	+	++	−	+++	+++
Chlorogenic acid[h]	−	++	+++	−	+	−
Chlorophyll[h]	+++	−	−	−	−	−
Curcumin[i]	−	−	+++	−	+++	+
Difluoromethyl ornithine[h]	+++	+++	++	−	+++	++
Dehydroepiandrosterone analog 8354[i]	−	+++	−	++	−	−
Dehydroepiandrosterone[i]	−	−	+++	+++	+	−
Diallyl disulfide[i]	++	−	++	++	++	−
Diallyl sulfide[i]	+	+++	+++	++	−	−
Dimethyl prostaglandin E_2[k]	−	+	++	++	+++	+++
Ellagic acid[h]	++	−	++	−	++	+++
Esculetin[i]	+	+++	+	+	++	+++
Ethyl vanillin[h]	+++	−	−	−	++	+++
Etoperidone[i]	+++	+++	+	++	+++	+++
Fluocinolone acetonide[h]	+++	−	−	−	++	+++
Folic acid[h]	+++	+++	++	++	+++	++
Fumaric acid[h]	+++	+++	−	−	+++	−
Glucaric acid[h]	+++	+++	+	+++	+++	−
Glucaro-1,4-lactone[h]	+++	−	++	+	+++	−
18-*β*-Glycyrrhetinic acid[i]	++	++	−	+++	+++	+
N-(4-Hydroxyphenyl) retinamide[i]	+++	+++	+	+++	+++	−
Hydrocortisone[i]	+++	+++	+	+++	+++	+
Ibuprofen[h]	+++	−	−	−	++	++
Indole-3-carbinol[j]	+++	+	++	+++	+++	+++
Indomethacin[i]	+++	+	+++	−	++	+
Levamisole[h]	−		−	−	+++	−
D-Limonene[i]	++	++	−	++	+++	−
Lovastatin[i]	++	+	−	+++	+	+
2-Mercaptoethane sulfonic acid[h]	−	+++	+	++	+++	++
Miconazole[i]	−	+++	−	++	++	++
Molybdate, sodium[h]	++	+++	−	+++	+++	+
Nicotinic acid (vitamin B_3)[h]	+++	+++	+++	++	+	−
Nordihydroguaiaretic acid[i]	+++	+++	++	+	−	−
Oltipraz[i]	+++	−	+++	+	−	++
2-Oxothiazolidine-4-carboxylate[h]	+	+++	++	+++	+++	−
Palmitoylcarnitine HCl[h]	+++	+++	++	−	+++	−
Phenethyl isothiocyanate[i]	−	++	−	−	+	−
Phenidone[i]	+++	++	−	−	+++	+
Piroxicam[i]	+++	++	−	−	+++	+
Potassium glucarate[h]	+++	+++	−	−	+++	−

(*continues*)

TABLE I (*Continued*)

Compound	Assay response of *in vitro* biomarker assays summary results					
	TK[b]	ODC[c]	GSH[d]	FR[e]	PADPR[f]	DNA binding[g]
Praziquantel[i]	+++	+++	−	−	+++	−
Prednisone[i]	+++	++	−	+++	++	−
Progesterone	+	+++	+++	++	++	−
Promethazine[i]	+	+++	++	−	+++	+
Propyl gallate[i]	++	++	++	+++	++	−
Purpurin[i]	+++	+++	−	+++	++	−
Quercetin[i]	+++	+	−	−	++	−
Retinoic acid, all trans[i]	+++	++	−	++	++	+++
Retinol	+++	+++	++	+++	+++	−
Rhodamine B[b]	−	+++	−	−	+++	−
Riboflavin-5-phosphate[b]	−	++	+	−	+++	−
RO 16-9100[i]	+++	+++	−	++	+	+++
RO 19-2968[i]	+++	+++	−	++	+++	+
Rutin[b]	++	+++	−	−	+++	−
Selenate, sodium[b]	+++	+++	−	++	+++	−
Selenite, sodium[b]	++	+++	+++	++	−	+
D,L-Selenomethionine[b]	+++	++	−	−	++	−
L-Selenomethionine[b]	++	+++	+	−	+++	−
Silymarin[i]	+	+	−	−	−	−
β-Sitosterol	+++	+++	−	−	+++	−
Suramin, sodium[b]	+++	++	++	−	+	+++
Tamoxifen[i]	+++	+++	+++	−	+++	−
Taurine[b]	+++	++	++	−	+++	−
Temaroten[i]	+++	++	+++	+++	−	−
Transforming growth factor-β[b]	+	+++	+++	+	+	−
Thioctic acid	+++	−	++	++	+	−
Thiolutin[i]	+	+++	++	+++	−	+++
Thiosulfate, sodium[b]	+++	+++	+++	−	+++	−
α-Tocopherol acetate[i]	+	+++	−	+++	+	+
α-Tocopherol succinate PEG 1000[i]	++	+++	−	+++	+++	+++
α-Tocopherol succinate[i]	+	++	++	++	+++	+++
Vanillin[b]	+++	++	+++	−	++	++
Verapamil[i]	+	++	+++	−	++	−
Vitamin C[b]	+	+++	++	+	++	+
Vitamin K_3[i]	+++	+++	−	+	+++	++
Vitamin D_3[i]	+++	+++	+++	−	+++	−
(*N*-6-Aminohexyl)-5-chloro-1-napthalenesulfonamide[i]	+++	+++	−	++	+++	−

[a]Reprinted with permission from *Cancer Res.* **54,** 5848 (1994).

[b]Inhibition of TK: +++, dose-dependent inhibition or all doses inhibitory or two doses show 100% inhibition or three doses show greater than 50% inhibition; ++, two or more doses inhibitory; +, one dose only inhibitory and showing 20% or more inhibition of TPA-induced activity; and −, no inhibition or inhibition less than 20% of TPA-induced enzyme activity.

[c]Inhibition of ODC: +++, dose-dependent inhibition or all doses inhibitory or two doses show inhibition between 60 and 100%; ++, two or more doses show between 30 and 60% inhibition; +, one or more doses show between 15 and 30% inhibition; and −, all doses show less than 15% inhibition.

[d]Induction of reduced glutathione: +++, all doses induce or one dose induces above 30% of media control or three doses induce 30% above media control; ++, two doses induce 20% above media control; +, one dose induces at least 10% above media control; and −, no induction.

[e]Free radical inhibition: +++, all doses inhibitory or two or more doses show greater than 30% inhibition; ++, two or more doses inhibitory; +, one dose inhibitory ([mt]10%); and −, no inhibition or less than 10% of TPA-induced free radical formation.

[f]Inhibition of PADPR: +++, all doses inhibitory or three doses show between 60 and 100% inhibition; ++, two or more doses show between 40 and 60% inhibition; +, one or more doses show between 20 and 40% inhibition; and −, no inhibition or inhibition less than 20% of propane sultone-induced activity.

TABLE II
Examples of Chemopreventive Agents in Clinical Trials by the National Cancer Institute

Compounds	Source	Mechanisms of action
1. Aspirin	Synthetic, nonsteroidal anti-inflammatory drug	Inhibits prostaglandin synthesis by inhibiting the cyclooxygenase activity of PGH_2 synthase. Primarily considered for the chemoprevention of colon carcinogenesis.
2. Bowman–Birk inhibitor	Soybeans	Trypsin and chymotrypsin inhibitor, which can normalize carcinogen-induced effects of increased proteolytic activities. Inhibition of reactive oxygen species formation, gene amplification, and c-*myc* expression. Clinical trials in preventing colon and lung cancer.
3. Tamoxifen	Synthetic, nonsteroidal triphenylethylene	A competitive antagonist/agonist of estrogen. Acts by altering gene expression and cell proliferation and by inhibiting reactive oxygen species and lipid peroxidation. Modulates growth factor secretions such as TGF-β, IGF-1, and EGF. Induces apoptosis. Phase III trial for prevention of breast cancer in high-risk women.
4. Finesteride	Synthetic	Competitive inhibition of the Δ^3-3-ketosteroid 5α-reductase enzyme that acts by converting testosterone 5α-dihydrotestosterone in the prostate.
5. L-(4-Hydroxyphenyl)-retinamide	Synthetic all-*trans*-retinoic acid analog	Antiproliferation and differentiation inducing. Inhibition of ornithine decarboxylase induction, actions of reactive oxygen species prostaglandin synthesis. Modulates activities of tyrosine kinase and induces apoptosis. Trials in prevention of breast, oral, and prostate cancer.
6. Vitamin D_3	Synthetic	Decreases estrogen receptors of mammary gland cells. Decreases cell proliferation. Clinical trials in patients with a high risk for breast and colon cancer.
7. Vitamin E	Soybeans, lettuce	Active compound in other naturally occurring compounds known as *d*-α-tocopherol. Antioxidant by reacting with reactive oxygen species. Stimulates immune system, induces cell differentiation, and inhibits arachidonic metabolism, cell proliferation, and ornithine decarboxylase activities. Clinical trials in preventing esophageal and oral cancer.

markers, (2) indices of cell proliferation, and (3) phenotypic changes related to tumor differentiation. Table I summarizes some of the chemoprevention agents that are currently undergoing complex screening and experiments using some biomarkers as monitoring indicators.

Currently the National Cancer Institute (NCI) has a list of cancer prevention agents that includes approximately 200 agents being tested by *in vitro* bioassays. More than 100 agents are now being tested in animal models. Some of these chemoprevention agents are included by the NCI in human clinical trials. Table II lists some of the chemoprevention agents that are currently undergoing clinical trials.

III. CHANGES IN LIFE-STYLE

Epidemiologic studies in countries like China, Japan, and Thailand, where the general populations have a

[g] Inhibition of carcinogen-DNA binding: +++, all doses inhibitory or one dose above 25% or three doses show greater than 20% inhibition or two doses show greater than 40% inhibition; ++, two doses inhibitory; +, one dose inhibitory and inhibition greater than 15%; and −, no inhibition or inhibition less than 15% of B(*a*) P-DNA binding.
[h] Compound dissolved in media.
[i] Compound dissolved in dimethyl sulfoxide.
[j] Compound dissolved in ethanol.
[k] Compound dissolved in tetrahydrofuran.

high fiber diet, have shown that their diets contribute to the lower incidence of colon cancer. In these countries, where soybeans are substituted for meat as the major protein source, there also is a lower incidence of breast and prostate cancers. The protective mechanisms of dietary fiber, especially insoluble fiber that acts by increasing stool size, dilute tumor initiators and promoters of either an exogenous or an endogenous source. The soluble fibers that are metabolized by bacterial enzymes in the intestine can inhibit colon carcinogenesis by releasing protease inhibitors. The increased intake of soybeans, which have a high content of protease inhibitors and phytoestrogens, showed a low incidence of breast and prostatic cancers in these countries. Protease inhibitors act by normalizing the carcinogen-induced effects of proteolytic activities and by inhibiting reactive-oxygen species formation and gene amplification. The phytoestrogens act as inhibitors of estrogen's actions and modulators of hepatic metabolism. [*See* Colon Cancer Biology.]

Drinking tea as the main beverage has been shown to decrease the risk of cancer. One of the main components found in tea is the catechin class of polyphenols widely known as (−)·epigallocatechin-3-gallate and the aflavins class of polyphenols. These polyphenols are antimutagenic and have antioxidant properties that inhibit carcinogenic processes.

Lung cancer, which results from chronic exposure of the lung epithelium to critical levels of carcinogens from cigarette smoke, initiates a sequence of molecular events that cause cellular transformation. The chronic levels of carcinogens proved to be from cigarette smoking were shown to cause mutation in the tumor suppressor gene p53. The mutation of this gene, which under normal conditions acts coordinately to control cell growth, initiates a series of uncontrolled cell division that ultimately results in cancer formation. The cessation of cigarette smoking will significantly lower the risk of getting lung cancer. [*See* Lung Cancer; Tumor Suppressor Genes, p53.]

Melanoma and two other forms of nonpigmented skin cancer, squamous cell carcinoma and basal cell carcinoma, are the major forms of malignancy caused by high exposure to sunlight. The ultraviolet radiation from the sun can damage lipids, proteins, and DNA by direct crosslinking or by the induction of the cells to produce free radicals. These would cause cell mutations and, together with the lowered efficiency of the body's immunosurveillance caused by ultraviolet radiation, would significantly increase the rate of cancer formation. In order to reduce the risk of these skin cancers, effective use of a sunscreen in prolonged sunlight and reducing the amount of exposure to direct sunlight are the best preventive measures.

IV. CONCLUSION

Our present knowledge indicates that cancer can be prevented. Chemoprevention and life-style changes, such as dietary modifications and cessation of cigarette smoking, are complementary approaches to cancer prevention. With the availability of specific chemoprevention compounds that can be used as oral supplements, especially micronutrients like vitamins or other dietary components alone or in complex combinations, cancer incidence can be lowered. Similarly, it also is true for macrodietary components such as fibers.

The lower cancer incidences observed in populations consuming different foods are most likely the result of a complex interaction of life-style as well as multiple dietary components. Cancer chemoprevention and dietary modification research to identify more compounds is the goal of improving cancer prevention.

BIBLIOGRAPHY

Bertino, J. R. (ed.) (1996). "Encyclopedia of Cancer," Vol. 1. Academic Press, San Diego.

Ho, C.-T., Osawa, T., Huang, M.-T., and Rosen, R. T. (eds.) (1994). "Food Phytochemicals for Cancer Prevention II," American Chemical Society, Washington, D.C.

Troll, W., and Kennedy, A. R. (eds.) (1993). "Protease Inhibitors as Cancer Chemopreventive Agents." Plenum, New York.

Wattenberg, L. W. (1990). Inhibition of carcinogenesis by naturally-occurring and synthetic compounds. *Basic Life Sci.* **53**, 155.

Wattenberg, L., Lipkin, M., Boone,C. N., and Kelloff, G. J. (eds.) (1992). "Cancer Prevention." CRC Press, Boca Raton, FL.

Cardiac Muscle

C. E. CHALLICE
The University of Calgary

ALEXANDRE FABIATO
Medical College of Virginia

GLOSSARY

Action potential Sequence of changes in transmembrane potential responsible for triggering contraction of the myocardial cells

Myofibril Component of cardiac cells responsible for providing contraction

Sarcomere Regular repeating element along the length of a myofibril consisting of interdigitating actin and myosin filaments. In a relaxed state it is approximately 2.5 μm in length and in a contracted state it is approximately 1.5 μm.

Sarcolemma Outer membrane of the cardiac (or other types of muscular) cell composed of several layers among which the innermost is the lipid bilayer generating the action potential

Sarcoplasmic reticulum Tubular structure generally surrounding the myofibrillar elements; responsible for controlling the concentration of calcium ions, which, in turn, modulates contractile activity and contractile force

Transverse tubules Invaginations of the sarcolemma

CARDIAC MUSCLE IS THE MUSCLE UNIQUE TO THE heart. Although it is similar in many ways to skeletal muscle, it differs from it in a number of important respects, and because of this it is designated as a third type of muscle, the others being smooth muscle and skeletal muscle (Table I). Cellular structure and properties vary with their position in the heart. Although the majority of heart muscle cells have as their purpose the generation of a contractile force and the resultant production of a periodic contraction, all cardiac muscle cells conduct the electrical event that triggers contraction, and some cardiac muscle cells have conduction as their main purpose. Such muscle cells are often termed "specialized" cells and they include the muscle cells responsible for the initiation of the contractile impulse, the so-called pacemaker.

I. INTRODUCTION

A. Historical Background

Older light-microscopic studies showed a great deal of structural similarity between cardiac muscle and skeletal muscle, but these studies were unable to distinguish individual cellular elements or fibers. Thus, it was thought that the heart consisted of a single multinucleated cell or syncytium. The advent of electron microscopy made it possible to distinguish individual cellular elements, in which cellular junctions crossed the direction of contraction, and where the junction included intercalated disks. Later it was shown that these junctions included regions in which cellular contact was so intimate as to permit the passage not only of the electrical impulse, but also of molecules of substantial size (through so-called intercellular gap junctions). Thus, although later work demonstrated the multicellular nature of the heart, it

ENCYCLOPEDIA OF HUMAN BIOLOGY, Second Edition, VOLUME 2. Copyright © 1997 by Academic Press. All rights of reproduction in any form reserved.

TABLE I
Principal Differences between Skeletal and Cardiac Muscle

Skeletal muscle	Cardiac muscle
Fibrillar elements continuous in single fiber cell from end to end	Fibrillar continuity is interrupted by the intercellular abuttments at intercalated disks
Fibrils discrete along length	Fibrillar structures branch and interconnect, such that a single fibril cannot be precisely defined
Sarcoplasmic reticulum forms well-defined sheath around fibrils, forming triads at Z line level with T system	Sarcoplasmic reticulum less abundant and presenting different types of junctions with not only the transverse tubules, but also the superficial sarcolemma, and even extended junctional sarcoplasmic reticulum that is not apposed to either transverse tubules or sarcolemma
Activation neurogenic; controlled by central nervous system	Activation myogenic; production and conduction of activating action potential occur by the muscle cells themselves
Can maintain sustained contraction or tetanus	Cannot be tetanized (except in the presence or certain drugs)
Contraction irregular and widely variable in force, usually produced by voluntary action. Force of contraction modulated by increase in recruitment of fibers	Contraction regularly periodic, with frequency and force modulated in response to demands of body metabolism

showed that in many respects it could be regarded as a functional syncytium. [*See* Electron Microscopy.]

B. Origin of Information

The available information on the cardiac cell comes from microscopic and experimental studies. For both types of analysis, it is necessary to have healthy tissue, removed from the host and processed within seconds of its normal *in vivo* existence. These requirements make it impossible to have more than a very occasional observation on human tissue. Thus, most of our information is derived from studies of laboratory animals. However, occasional studies on primate hearts have shown that the information obtained from laboratory mammals can reliably be taken as closely indicative of the structure and function of the human cardiac cell.

C. Function of Cardiac Muscle

The essential function of the heart is that of a pump, operating by a regular periodic decrease of volume, which, in combination with unidirectional valves, generates a regular unidirectional flow of blood through the cavity, or lumen. Thus, the essential function of the cells of the heart is to produce a regular periodic contraction, and in describing the cardiac cell it is perhaps simplest to focus on this function and describe first the element responsible for the contraction. The other elements can then be seen to serve this function. Following this description the intercellular relationships important in the electrical activation of the cardiac cells are examined.

II. STRUCTURE AND FUNCTION OF THE CARDIAC CELLS

In this section on the structure and function of the cardiac cell, emphasis will be placed on the cell of the working ventricular myocardial cell of mammalian species.

A. Contractile Fibrils

The contractile element is, as with skeletal muscle, made up of fibrillar elements which, in turn, consist of sarcomeres (Fig. 1). These vary in length between about 2.2 μm when relaxed and about 1.7 μm when contracted. In cross section the sarcomeres are seen to consist of interdigitating thin (actin) and thick (myosin) filaments. This much smaller functional range of sarcomere length than the skeletal muscle is limited by the abundant and stiff elastic tissue around and within the cardiac cell. Electron microscopy of longitudinal sections reveals them to interdigitate in a regular symmetrical fashion, producing bands of width dependent on the degree of contraction (Fig. 1). Dense (by electron microscopy) Z bands join adjacent sarcomeres by joining their respective abutting actin filament ends. The interdigitating part of the sarcomeres (i.e., the middle section) behaves anisotropically in examination by polarized light (i.e., it causes polarization), and is thus known as the anisotropic, or A, band and the remainder—the region between the A bands, which has the Z band in the middle—behaves isotropically in examination by polarized light, and is thus the isotropic, or I, band. A fine structure can often be seen at the middle of the A band in the form of a thin dark band (the M line), each side of which is a thin pale band.

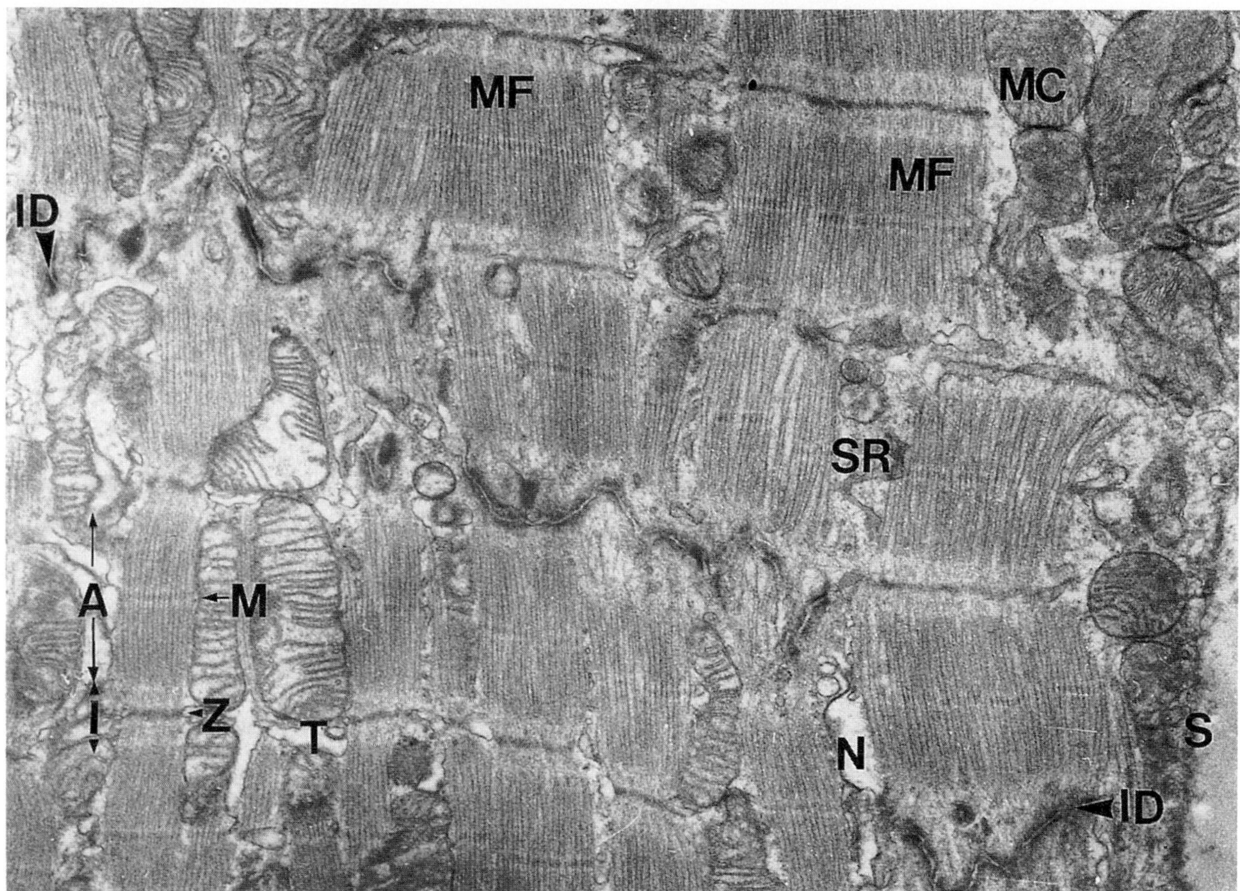

FIGURE 1 Transmission electron micrograph of a section from the ventricle of a rat heart in which the myofibrils (MF) are cut longitudinally. The bands of the sarcomeres are labeled (lower left): A, anisotropic band; I, isotropic band; Z, Z disk or line; M, M line. An intercellular junction traverses the micrograph from top left to lower right, showing the structure of the intercalated disks (ID). A nexus, or gap junction (N), is present. The way in which the myofibrillar elements divide and link up can be seen by following the two central elements from the top of the micrograph to the bottom. The discretenes of the elements seen at the top is not maintained. SR, sarcoplasmic reticulum; T, transverse tubule; S, plasmalemma. Original magnification ×117,000.

Whereas in skeletal muscle the fibrils are longitudinally continuous and transversely discrete, in cardiac muscle they branch and link neighboring elements, making the fibrillar contents of a cell essentially a single, branching, interconnected contractile element (Fig. 1).

B. Mechanism of Contraction

In the sarcomeres of the fibrils, energy is used (by hydrolysis of ATP to ADP) to generate a mechanical force and consequent contraction. The energy is actually used up only during relaxation. Thus, the interruption of energy delivery at the time of death causes the skeletal muscle or cardiac cell to be immobilized in a rigid state (the rigidity of a cadaver). The following represents a simplified summary of the process and of the microstructures involved.

Detailed analyses have shown that the thick myosin filament is made up of a bundle of molecules. Each molecule is a long double-helical coil, one end of which terminates abruptly, while at the other each strand terminates in a globular head (Fig. 2). The bundle (i.e., the thick filament) is assembled symmetrically about the middle, with the molecular heads farthest from the middle, equal numbers in each direction, and regularly distributed along the length (Fig. 3). As they interdigitate in the fibril with the actin filaments (each of which is a double strand of connected globular actin molecules), the myosin heads produce regularly arrayed cross-bridges, approximately at right angles with the filaments, connecting them with the actin filaments (Fig. 4).

Under specific conditions, notably of Ca^{2+} and ATP concentration, the myosin heads go through a motion similar to rowing. The myosin heads (the "oars")

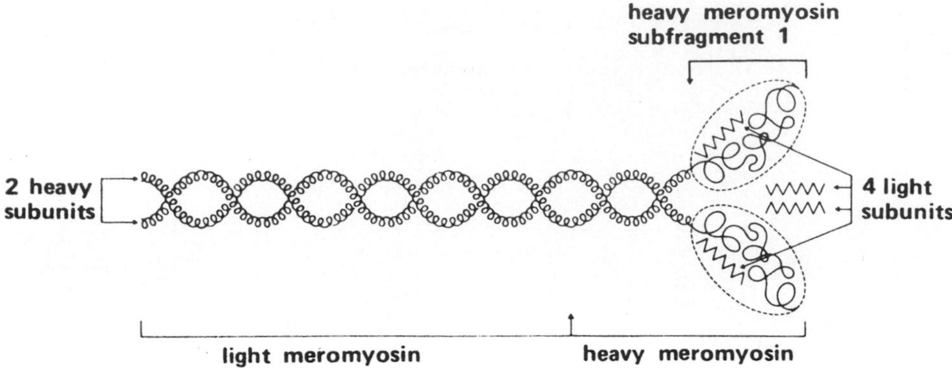

FIGURE 2 Structure of a myosin molecule showing the long double-helical tail and the double-globular head. The length of the tail is about 134 nm. It consists of two identical heavy chains, each of about 200,000 Da. One end of the helical coil ends abruptly (i.e., the carboxy terminus), whereas at the other globular end (i.e., the amino terminus) each molecule has two light chains (each about 20,000 Da) attached. [From Katz, A. M. (1977). "Physiology of the Heart." Raven Press. Reproduced by permission.]

contact the actin, bend in such a manner as to produce a force on the filaments (and thus movement in the direction of contraction), disconnect, return to their original condition, and then repeat the whole process. The control factor in this mechanism is the concentration of Ca^{2+} adjacent to the fibrils, which determines the degree of contractile activity and, hence, the degree of contraction.

C. Sarcolemma and Transverse Tubules

The sarcolemma, or outer membrane of the cardiac cell, is in fact made of four layers. The innermost layer is the lipid bilayer containing ionic channels that generate the action potential. Then comes a layer termed glycocalix or basement membrane containing sialic residues that buffer the free calcium concentration at the outer surface of the bilayer and which is much more developed in cardiac than in skeletal muscle. Finally, there are two superficial elastic layers.

In the working ventricular cell of mammalian species, the surface membrane invaginates, generally at the level of the Z bands of the sarcomeres (Fig. 1). These invaginations are termed transverse (T) tubules as is skeletal muscle but they differ from those observed in skeletal muscle in several important aspects: they are much broader; they branch; and they are constituted of invaginations of both the lipid bilayer and the glycocalix whereas only the lipid bilayer invaginates in skeletal muscle.

The most important difference between cardiac and skeletal is that the transverse tubules are generally missing in all cardiac cells other than the mammalian working ventricular cells. They are infre-

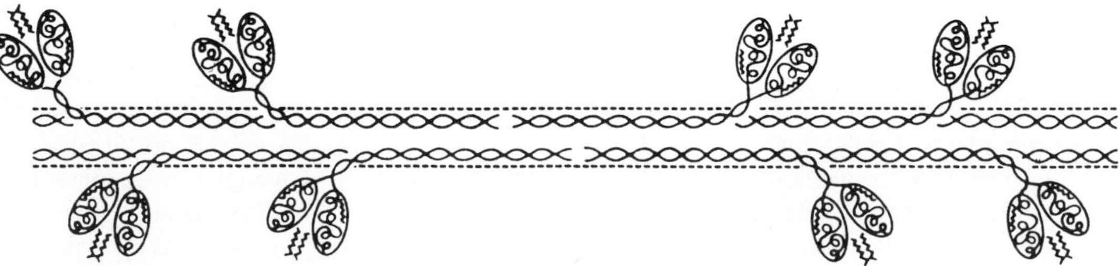

FIGURE 3 Myosin molecules associating to form a myosin filament. The long chains associate end to end at the carboxy termini in an overlapping manner to create a "bare" zone in the middle and a length at each end with regularly distributed globular heads. [From Katz, A. M. (1977). "Physiology of the Heart." Raven Press. Reproduced by permission.]

FIGURE 4 A single sarcomere showing the manner in which the thin (actin) and thick (myosin) filaments interdigitate in a sarcomere, indicating the way in which the myosin heads form the cross-bridges between the actin and the myosin.

quently observed in the working atrial cells and are always lacking in the specialized cells (pacemaker and conducting tissues) of mammalian species. They are absent in all the cardiac cells, including the working myocardial ventricular cells, of birds, reptiles, and batracians.

D. Sarcoplasmic Reticulum

The control of the Ca^{2+} concentration in the vicinity of the myofibrils is affected by the sarcoplasmic reticulum, which is composed of a longitudinal sarcoplasmic reticulum that reaccumulates calcium at the time of the relaxation following the contraction, and of terminal cisternae (plural latin word) that are believed to release the calcium turning on the contraction of the myofibrils (Fig. 5).

The sarcoplasmic reticulum is less abundant in the cardiac than in the skeletal muscle cell and differs in its relationship with transverse tubules and sarcolemma. In skeletal muscle, the terminal cisternae almost exclusively come in close apposition with the transverse tubules. In cardiac muscle, they can come in apposition with the transverse tubules, but may also form peripheral couplings with the superficial sarcolemma. In addition, there are terminal cisternae of the cardiac sarcoplasmic reticulum that are free within the cell, not apposed to any sarcolemma or transverse tubules.

The mechanism or excitation–contraction coupling, i.e., the mechanism whereby the action potential traveling in the sarcolemma causes calcium release from the sarcoplasmic reticulum, remains unknown but is the subject of an intense research activity. It is likely to differ in skeletal and cardiac muscle. In skeletal muscle, the action potential propagates from the sarcolemma down the transverse tubules to the center of the fiber. The transverse tubule sends an electrical signal to the terminal cisternae of the sarcoplasmic reticulum, probably through the movement of charged particles located within the transverse tubular membrane and capable of influencing calcium channels across the sarcoplasmic reticular membrane.

Such charge-movement-induced transmission of the signal of excitation–contraction coupling is unlikely to play a major role in cardiac excitation–contraction coupling, where a chemical transmission using calcium as a second messenger is the most likely mechanism. Some arguments for this hypothesis are the lack

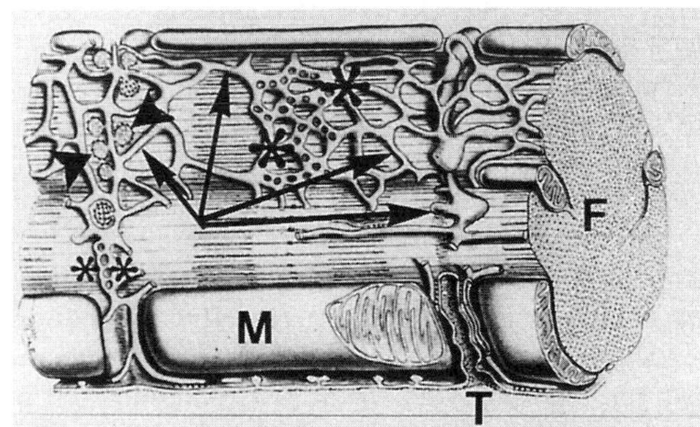

FIGURE 5 A sarcomere in a cardiac ventricular cell. The invagination of the sarcolemma forming a T tubule (T) is seen (lower right). The detailed structure of the sarcoplasmic reticulum (SR) is seen as it relates to the sarcomere. Blebs on the SR ("corbular" SR) are seen at the Z line level (arrowheads). SR consists generally of narrow tubules (arrowheads). SR consists generally of narrow tubules (arrows) with a tendency to form flattened lamellae with fenestrations at the M line level (large asterisk) and the same occurs to a lesser extent at the Z line level (small asterisk). F, myofibril; M, mitochondrion. (Modified from Bossen, E., Sommer, J. R., and Waugh, R. A. (1978). *Tissue Cell* **10**, 773–784. Reproduced by permission.)

of transverse tubules in most cardiac cells, the lack of any apposition to sarcolemma or transverse tubules of some terminal cisternae of the sarcoplasmic reticulum precluding any direct transmission of the signal of an electrical type, and the well-established dependence of cardiac excitation–contraction coupling on extracellular calcium, especially on the transsarcolemmal calcium current (see Section V,B) that is almost nonexistent in skeletal muscle.

It is of interest that in smooth muscle a chemical transmission of the signal of excitation–contraction coupling is also favored, but here the second messenger would be inositol-1,4,5-trisphosphate instead of calcium.

E. Cytoskeleton

As in other cells, filaments are present that have the functions of both stabilizing the distribution of cell components and controlling the size and shape of the cells. A measure of rigidity is important in many types of cells, particularly in contractile cells where the action of contraction introduces a strong tendency to change the shape of the cell.

While various filaments serve as the cytoskeleton, the principal element in adult cardiac muscle is in the so-called "intermediate" class (approximately 10 nm in diameter) and is primarily desmin (Fig. 6). It is aligned both parallel to and at right angles with the myofibrils, attaching to many types of organelles. They are sometimes referred to as "anchor fibers," present in the intercalated disks (see Section III,B) also attached to the invaginations of the T system and connecting Z discs to each other and to the plasmalemma.

F. Other Structures

Mitochondria, the site of ATP production in cells, are abundant (i.e., approximately 30% of the total volume of cardiac muscle cells), reflecting an available high level of energy conversion. They assume a location and shape largely dictated by the arrangement of the myofibrils, but on average are distributed uniformly within the sarcoplasm, again reflecting the use of energy throughout the cell (Fig. 1). The Golgi apparatus or complex is usually prominent, again reflecting the consistently high level of metabolic activity. The nucleus is located centrally, and frequently more than one can be seen, although it has not been possible to establish whether a consistent number of nuclei per cell exists. [*See* Golgi Apparatus; Mitochondria; Mitochondrial Respiratory Chain.]

Other cytoplasmic components common to most cells are found, including liposomes, multivesicular bodies, autophagic vacuoles, lipofuchsin granules, peroxisomes, glycogen granules, and lipid droplets.

III. CELL MEMBRANE AND INTERCELLULAR RELATIONSHIPS

The innermost layer of the cardiac sarcolemma also fulfills an essential electrical function involving the transmembrane movement of ions and, hence, electrical charge. The cycle of ionic movement that generates an action potential is described later (see Section V,B).

A. Lipid Bilayer

In the lipid bilayer of the cardiac cell, as is in that of the other types of muscle cells, each layer is made up of phospholipids in which the hydophobic lipids from the apposing layers face each other. Associated with this are various so-called membrane proteins which sometimes sit on one surface leaving the bilayer undisturbed, sometimes penetrate must one of the layers, and sometimes completely traverse the membrane. These proteins give the membrane its specific properties. Some form the structural support for the membrane, whereas others include antigens, receptors, enzymes, ion channels, and pumps. The membrane structure has a high degree of fluidity, permitting the protein molecules to move around.

The cells of the myocardial walls are, in most places, separated by a space of varying width. However, there are a number of specialized regions of intercellular abutment.

B. Intercalated Disks

The intercalated disk is the structure present where the continuity of the myofibrillar structure crosses from one cell to another (Figs. 6 and 7). It is usually found to be convoluted and is rarely at right angles to the myofibrillar orientation. It interdigitates, forming almost a two-dimensional dovetail junction, which arises presumably by virtue of the intercellular adhesion necessary to hold the cells together when subjected to the tension generated by the myofibrillar contraction. Within the intercalated disk are also found the specific specialized functions referred to in the next section.

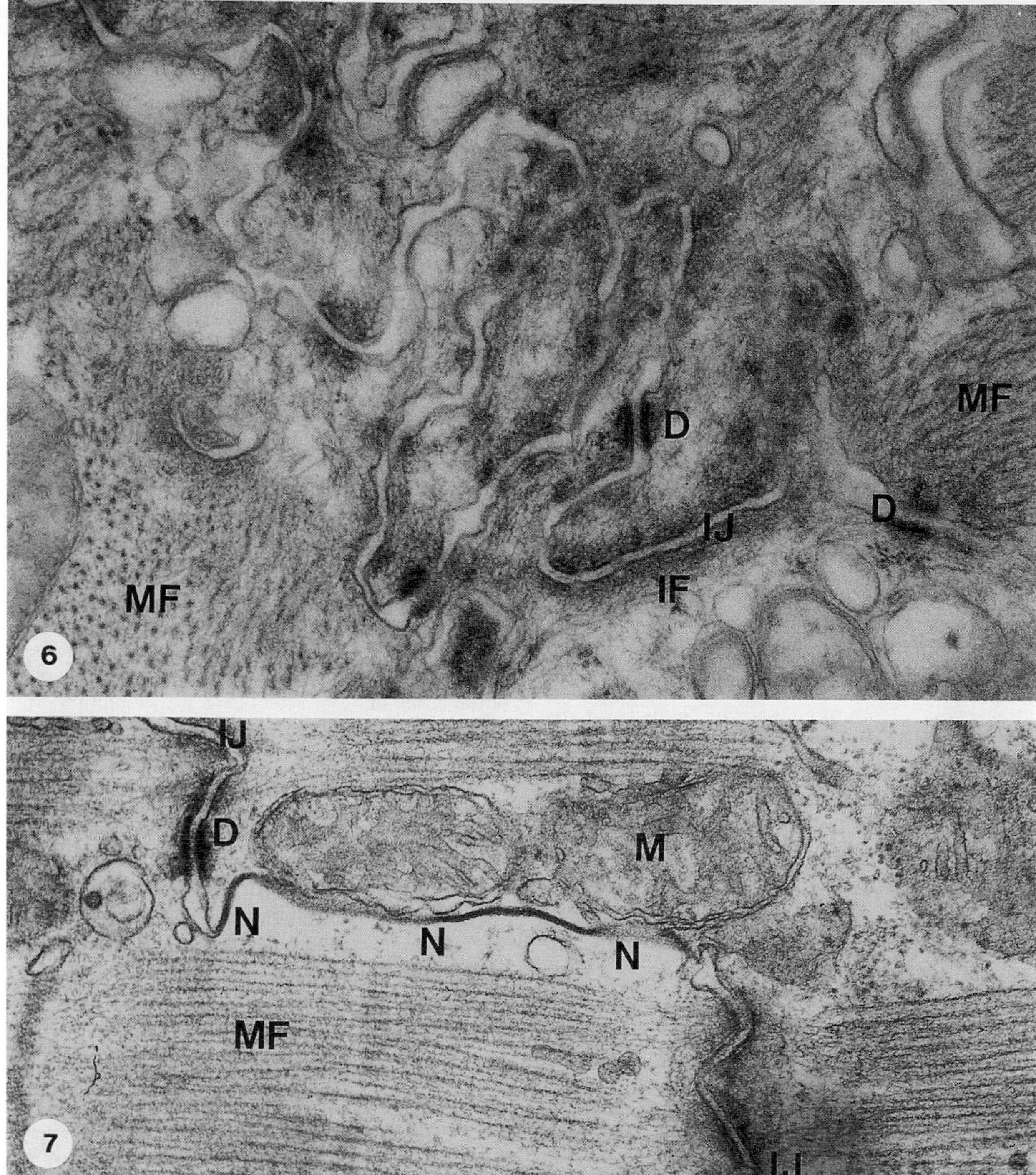

FIGURE 6 Electron micrograph of rat myocardium cut approximately at right angles to the myofibrils, at the level of an intercalated disk. The convoluted geometry is readily seen; the two cells interdigitate extensively. (Lower left) The cells are cut almost transversely where the fairly regular array of thick (myosin) and thin (actin) filaments can be seen, whereas at the right the myofibril is cut obliquely, showing that in these regions the myofibrillar orientation is not entirely constant. D, desmosomes; IJ, intermediate junction; IF, intermediate filaments; MF, myofibrillar elements. Original magnification ×60,000.

FIGURE 7 Electron micrograph of rat myocardium cut parallel with the myofibrils (MF) at the level of an intercalated disk. A desmosome (D) and intermediate junctions (IJ) are seen. A long nexus (Ns) is present, whose orientation with respect to the section varies along its length, showing the general irregularity of arrangement. M, mitochondrion. Original magnification ×60,000.

C. Nexuses, or Gap Junctions

In gap junctions the two apposing membranes contact each other to form essentially a single structure (Fig. 7). Freeze-fracture electron microscopic studies have shown pores to exist in these junctions which communicate between the respective sarcoplasms. The pores (i.e., connexons) are formed by a hexagonal array of six parallel protein molecules (i.e., a connexin) which form a tube. It has been suggested that the opening and closing are affected by a rotation of one end of the tube with respect to the other. Gap junctions permit the intercellular passage of not only ions but molecules and are the structures which render cardiac muscle a functional syncytium. [*See* Cell Junctions.]

D. Desmosomes and Intermediate Junctions

Desmosomes and intermediate junctions are believed to serve mainly to prevent the cells from being pulled apart. Desmosomes have a discrete substructure (Fig. 6). Intermediate junctions contain vinculin and α actin, which are both absent in desmosomes. Desmin is associated with desmosomes, but not actually within the structures.

IV. ATRIAL CELLS AND SPECIALIZED TISSUE CELLS

Most of what has been said until now refers to the myocardial ventricular cells of mammalian species. It has already been mentioned that the other myocardial cells generally lack transverse tubules. The following describes some of the other characteristics of the atrial muscle cells and of the specialized tissues of the heart.

A. Atrial Muscle Cells

During the embryonic developmental process in the ventricle, contractile trabeculae consolidate to form the ventricular wall, which, at maturity, consists of layers within which the fibrillar alignment is parallel. The atrial wall is much thinner, made up to smaller cells (i.e., 6–8 μm in diameter), and even at maturity is formed of bundles rather than layers. Historically, these bundles have been given names and proposed as specific conduction pathways, but this is no longer accepted. The smaller cells often lack transverse tubules, and atrial cells commonly contain dense "specific atrial granules," which have been shown to contain a natiuretic factor. [*See* Atrial Natriuretic Factor.]

B. "Specialized" Tissues

Skeletal muscle is activated by nervous stimulation; the nerves receive their signals from the central nervous system. Although nerves are present in the heart, their function is only to modify the contractile functioning of the heart cells, not to initiate it. Both the initiation and the conduction of the electrical stimulus, which leads to contraction in the heart, are performed by cardiac myocytes, many of which are specialized in their development to perform that function.

The heart beat is initiated by a small group of cells located in the right atrial wall close to its function with the sinus venosus. These form the pacemaker, or sinoatrial node. From here the impulse spreads over the atrium, generating atrial contraction, and then arrives at another small group of cells situated on the dorsal side of the function of the atria and the ventricles. Here the impulse is carried for a short distance quite slowly (about one-hundredth of the speed elsewhere) through the atrioventricular node to the ventricular septum, to connect with the trunk (i.e., common, or His, bundle) of a tree-like conduction system (i.e., the Purkinje system which is insulated from the surrounding myocardium by blood vessels and connective tissue. This system conveys the impulse to the various parts of the ventricular musculature, causing it to contract in a systematic manner beginning at the periphery and moving toward the aorta.

The most obvious feature of these cells is that the contractile material is sparse and largely unaligned, with a substantial part of the sarcoplasm containing apparently undifferentiated material. Nodal cells are small (Figs. 8 and 9), whereas Purkinje cells are often

FIGURE 8 A comparatively low magnification transmission electron micrograph of a section through the sinoatrial (i.e., pacemaker) region of a ferret heart, showing the small size of the nodal cells and both wide and very narrow spacings between them. Nervous elements (NE) and capillaries (C) are present. Original magnification ×4250.

FIGURE 9 A higher magnification electron micrograph from a region similar to that of Fig. 8. The nucleus (NU) is rounded, and the sarcoplasm has comparatively large areas without definable organelles. A Golgi complex (G) is present. Myofibrils (MF) are sparse and inconsistently oriented. Nervous elements (NE) are also present. M, mitochondrion. Original magnification ×12,750.

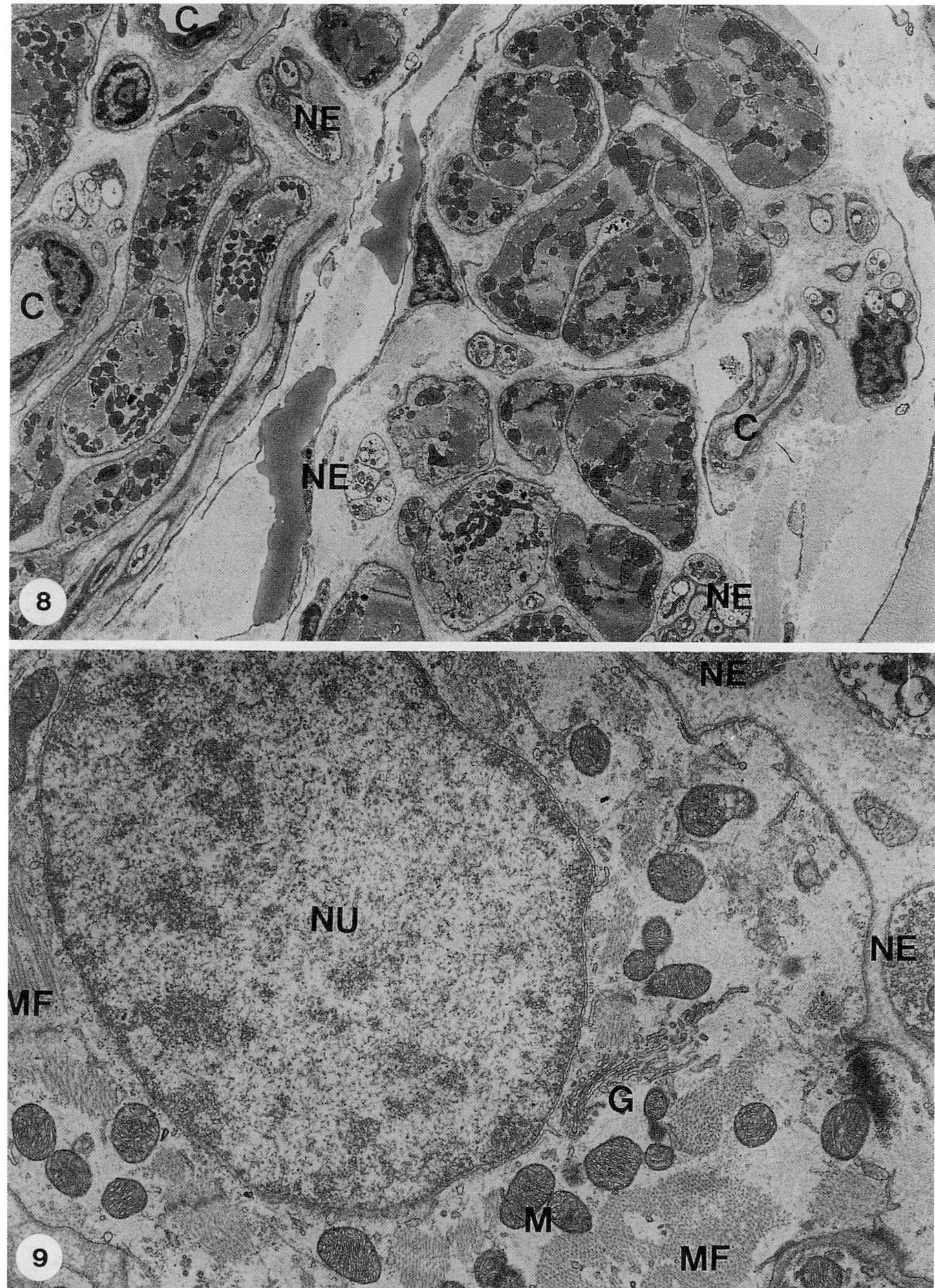
C
NE
C
C
NE
NE
8
NE
NE
NU
MF
G
M
MF
9

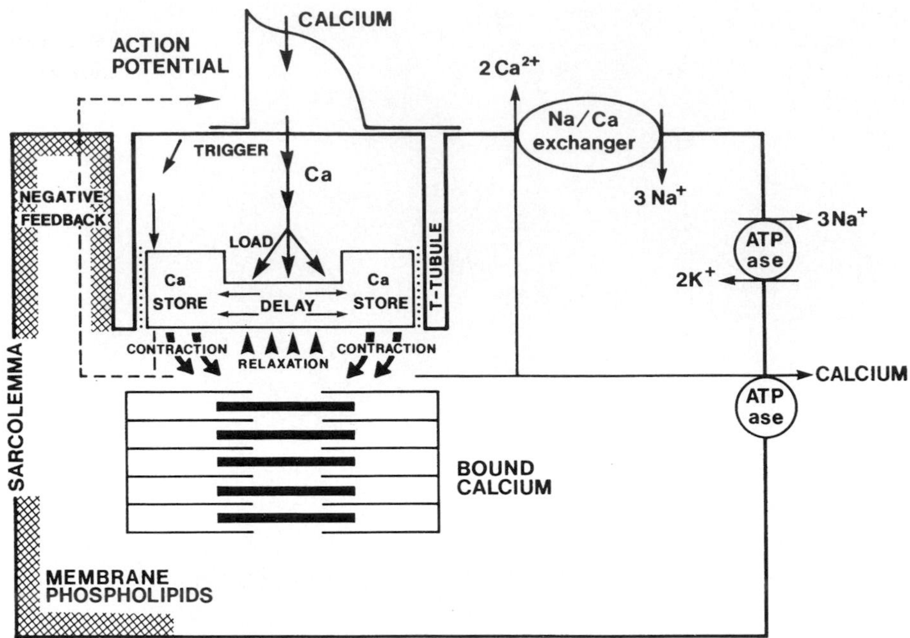

FIGURE 10 Ionic movement in the cardiac cell. The triggering cyclic action of calcium ions is indicated by the heavy arrows. During an action potential (top left), together with transmembrane movement of sodium and potassium ions (see Fig. 12), calcium ions enter the cell. This, in turn, stimulates the release of more calcium ions from the lateral (or junctional) regions of the sarcoplasmic reticulum (SR) by the mechanism of "calcium-induced calcium release" enunciated by Fabiato, and also produces an increase in cytoplasmic calcium, permitting the SR to be "reloaded." The sharply increased Ca^{2+} concentration in the immediate proximity of the contractile fibers triggers the contraction mechanisms, and the subsequent drop in Ca^{2+} concentration is brought about by Ca^{2+} uptake by the central (or longitudinal) region of the SR, which in turn brings about relaxation. This Ca^{2+} release–sequestration sequence is part of the regular cycle of events associated with the contraction–relaxation periodicity in the heart. There is a delay before the sequestered calcium is available for release by the SR. It has been speculated that this delay represents the control process preventing sustained contraction by cardiac muscle and thereby contributing to stabilization of the heartbeat. The thinner arrows indicate the method by which the overall increase in cytoplasmic Ca^{2+} is remedied—by a sarcolemmal Na/Ca exchange mechanism and by ATPase activity. ATPase activity also brings about the resting potential sarcolemmal balance of Na^+ and K^+ ions. (Modified from H. Banijamali, M.Sc. Thesis, The University of Calgary, 1989, with permission.)

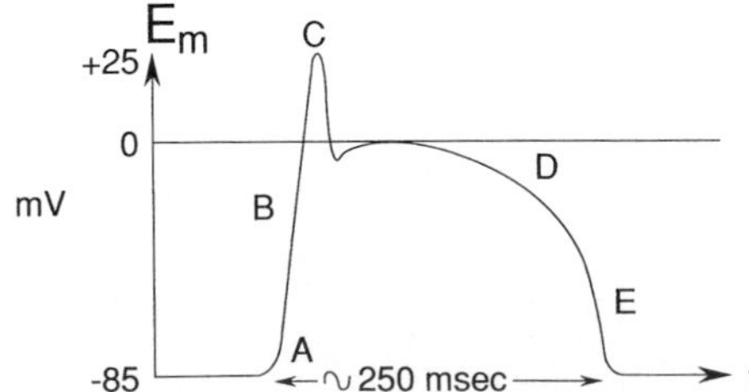

FIGURE 11 Sequence of transmembrane potential (E_m) as a function of time (t) in an action potential of a cardiac cell. (A) The point at which the depolarization is applied, (B) the period of rapid inward flow of Na^+ (fast inward current), (C) the point at which sodium current is "switched off," (D) the plateau, and (E) the period when inward current ceases.

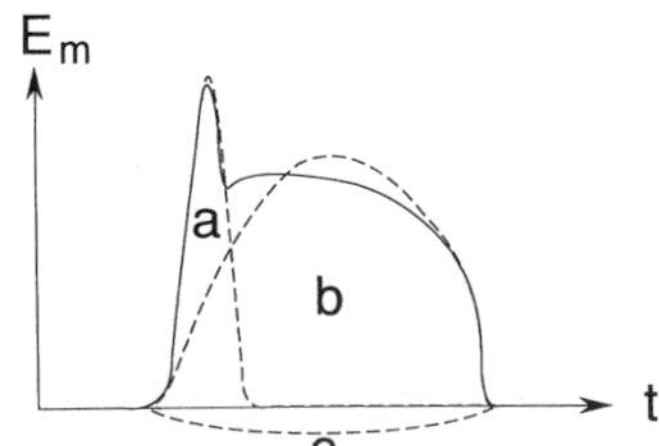

FIGURE 12 The principal current ingredients which produce the transmembrane potential changes. (a) The inward fast sodium current, (b) the slow inward calcium-containing current, and (c) the outward potassium current. The dotted lines refer to individual ionic currents; the solid line to the total electrical effect of the sum of these currents.

large. Specialized functions (i.e., nexuses, or gap functions) between cells are numerous. As one progresses along the Purkinje arborizations from the His bundle toward the myocardial wall, the cell structure is observed to change steadily toward that of the ordinary contractile myocardial cells.

V. ELECTRICAL ACTIVITY OF CARDIAC MUSCLE

A. Resting Membrane Potential

By penetrating inside the cell across the sarcolemma with a very fine glass micropipette (i.e., approximately 0.5 μm in diameter) that is filled with a 3*M* KCl solution; using this as a electrode, the transmembrane electrical potential of a cardiac cell can be measured. In the resting state it is about −85 mV with respect to its surroundings. This resting membrane potential reflects the difference in composition between the intracellular and the extracellular fluids, which includes a difference in ionic content. This difference arises as a result of the selective permeability of the membrane combined with an active transmembrane ion pumping process. A steady state is reached in which transmembrane ionic diffusion is prevented from producing equal ionic concentrations on either side of the membrane, or equal charge distribution, and hence a transmembrane potential difference is produced. In the resting state the $K+$ concentration is greater within the cell than outside it, whereas Na^+ is more concentrated outside than inside, as is Ca^{2+}. Figure 10 is a schematic representation of the way the resting potential is maintained, along with the way an action potential leads to contraction.

B. Action Potential

Although all living cells maintain a measure of resting potential, nerve and muscle are peculiar in developing an action potential. Reduction of the resting potential to about −60 mV (i.e., depolarization) causes the transmembrane potential to go through a spontaneous cycle, quickly becoming approximately +20 mV, then returning more slowly to its resting potential. For cardiac muscle the cycle is shown in Fig. 11; this cycle and the associated transmembrane ionic currents trigger the contractile process (Fig. 10).

Following depolarization of the membrane to −60 mV (the threshold potential), ionic channels in the membrane open, permitting the inward flow of Na^+ and Ca^{2+} and the outward flow of K^+. The sharp reversal of the potential is produced by a fast inward flow of Na^+, resulting in a closure of the channel and reestablishment of the resting transmembrane balance of Na^+. The same process occurs (in the opposite direction) with K^+, but the mechanism is slower. Peculiar to the cardiac muscle cell is the inward transmembrane current which carries Ca^{2+}. This is a slower process than the Na^+ current and is responsible for the overall transmembrane potential's demonstrating a "plateau" (Figs. 11 and 12).

When this process takes place locally in a cardiac cell, the depolarization then causes depolarization of the neighboring area to the threshold potential, with repetition of the described events. Thus, the action potential travels along the length of the cell to be conveyed also to neighboring cells.

An important feature of the cardiac action potential is that immediately following an action potential there is a brief refractory period during which it cannot be triggered; this helps secure the regular periodicity of the heart.

C. Origin of the Action Potential

In the heart some cells spontaneously self-depolarize, thereby initiating the contraction of the whole heart (i.e., pacemaker). In such cells, notably those of the sinus node (Figs. 8 and 9), the transmembrane potential spontaneously decreases until it reaches the threshold, when the action potential occurs (Fig. 13). While some other cells of the heart (notably in the atrium and parts of the conduction system) have their property, the periodicity is slower, and as a result they

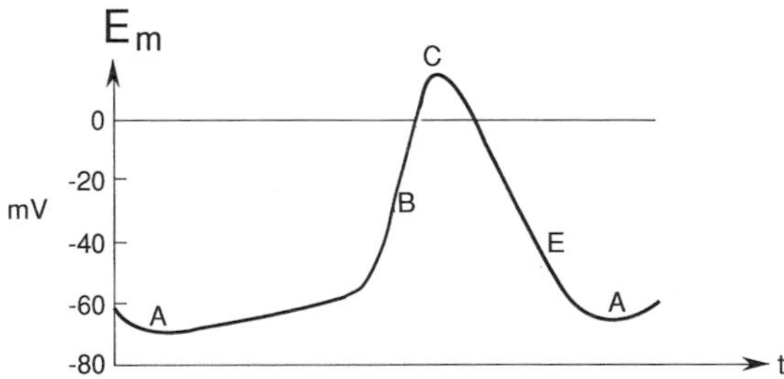

FIGURE 13 Spontaneous action potential recorded from a pacemaker cell in the sinoatrial region of the heart. (A) The greatest (negative) transmembrane potential recorded, but this is not stable and rises steadily until the threshold is reached, whereupon the fast depolarization (B—cf. Fig. 11) occurs. The maximum depolarization peak (C) is not as sharp as for a ventricular cell, and the plateau (D—cf. Fig. 11) does not exist. Repolarization takes place (E), following which the sequence repeats.

are triggered by those with the faster periodicity, thus giving the whole organ its synchronous contraction.

VI. NERVOUS ELEMENTS

Although the action potential of the heart is determined by muscle cells, not by nerve cells, nervous elements are present, particularly in the nodal regions (Figs. 8 and 9). These perform a regulatory role, secreting agents which speed up the periodicity and the conduction velocity (e.g., adrenaline) or slow it down (e.g., acetylcholine). Sensory nerves are also present.

BIBLIOGRAPHY

Berne, R. M., and Levy, M. N. (1981). "Cardiovascular Physiology." Mosby, St. Louis, MO.

Berne, R. M., Sperelakis, N., and Geiger, S. R. (1979). "Handbook of Physiology. Section 2: The Cardiovascular System." Am. Physiol. Soc., Bethesda, MD.

Bers, D. M. (1991). "Excitation–Contraction Coupling and Cardiac Contractile Force." Kluwer Academic Publishers, Dordrecht/Boston/London.

Fozzard, H. A., Haber, E., Jennings, R. B., Katz, A. M., and Morgan, H. E. (1992). "The Heart and Cardiovascular System, Scientific Foundations," 2nd Ed., Vols. I and II. Raven, New York.

Katz, A. M. (1992). "Physiology of the Heart," 2nd Ed. Raven Press, New York.

Langer, G. A. (1990). "Calcium and the Heart." Raven Press, New York.

Nathan, R. D. (1986). "Cardiac Muscle: The Regulation of Excitation and Contraction." Academic Press, Orlando, FL.

Noble, D., and Powell, T. (1987). "Electrophysiology of Single Cardiac Cells." Academic Press, Orlando, FL.

Sperelakis, N. (1995). "Cell Physiology Sourcebook." Academic Press, San Diego.

TerKeurs, H. E. D. J., and Tyberg, J. (1987). "Mechanics of the Circulation." Nijhoff, Dordrecht, The Netherlands.

Cardiogenic Reflexes

ROGER HAINSWORTH
DAVID MARY
University of Leeds

GLOSSARY

Afferent nerves Group of nerve fibers that connect the receptors to the central nervous system. They carry impulses that represent depolarization (i.e., a difference in electrical potential along the surface of the fiber). The number of impulses per unit of time is determined by the intensity of the stimulus to the receptor

Effector organs Organs that form the target of reflex action via efferent nerves

Efferent pathways One or more pathways connecting the central nervous system to organs affected by the reflex. The pathways subserving the connections can involve humoral, neural, or neurohumoral mechanisms

Myelinated nerves Larger-diameter nerves individually covered by a myelinated sheath. The velocity of conduction in these nerves is more than 2.5 m/sec. Nonmyelinated nerves conduct more slowly

Receptor Small sensory organ with a defined structure and connection via an afferent nerve fiber to the nervous system; in some cases a structure has not been defined, and the receptor is presumed by neural activity in its afferent fiber

Reflex Term that describes reception of a signal by the nervous system and the consequent trigger or modification of organ function; reflexes involve receptors, afferent nerves, central connections, efferent nerves, and effector organs

Threshold Least stimulus that gives rise to activity in the afferent nerve fibers or to a measured response of an effector organ

THE TERM "CARDIOGENIC REFLEXES" REFERS TO THE responses, involving nervous pathways to and from the brain, which result from the stimulation of nerves ending in the heart. Because of difficulties in localization of the stimuli in people, most of our knowledge of these reflexes has stemmed from experiments involving anesthetized animals. The nervous receptors responsible for these reflexes have been demonstrated in both atria, the ventricles (mainly left), and the coronary arteries. Activation of the cardiac receptors occurs in response to physiological events, including changes in the volumes or pressures within the cardiac chambers. Many of the afferent fibers exhibit tonic activity, that is, they are active at normal levels of cardiac filling and pressures. Some only become active in response to chemical stimulation, and this could result from the administration of exogenous pharmacological agents or from chemical changes occurring in some pathological states. The efferent pathway of cardiogenic reflexes controls the rate and force of contraction of the heart itself, the degree of constriction of blood vessels, and the secretion of several hormones. Cardiogenic reflexes thus have an important role in the control of the cardiovascular system.

I. INTRODUCTION

In the study of any reflex, certain components must be defined. These are the stimulus (the event that

ENCYCLOPEDIA OF HUMAN BIOLOGY, Second Edition, VOLUME 2. Copyright © 1997 by Academic Press. All rights of reproduction in any form reserved.

triggers the response), the receptor (the specialized transducer in the body that detects the stimulus and transmits it to a nerve), the afferent nervous pathway (which conveys the information to the brain or the spinal cord), the pathways through the central nervous system, the efferent nervous pathway (which conveys information from the brain or the spinal cord), the effector organ (the heart, blood vessel, etc., which changes as the result of the reflex), and the response. To define these components, complex preparations are required, and these are difficult even in anesthetized animal experiments, and impossible in humans. The study of reflexes in the cardiovascular system is particularly difficult, and the study of reflexes arising from the heart in intact humans is nearly impossible. For this reason almost all of our knowledge of cardiogenic reflexes emanates from studies of anesthetized animals, and therefore most of what is contained in this contribution is based on results obtained from animal experiments. [*See* Cardiovascular System, Anatomy; Cardiovascular System, Physiology and Biochemistry.]

II. AFFERENT NERVES FROM THE HEART

A. Nerve Endings

The heart, like most organs, is innervated by the autonomic nervous system, and afferent impulses run in the vagal and the sympathetic nerves. Although much is known concerning the afferent nerves from the heart, much less is known concerning the receptor organs attached to these nerves. Two particular structures have been described in the heart: complex unencapsulated endings that are seen in the atria and a fine network of nerve fibers seen throughout the endocardium (Fig. 1).

Complex unencapsulated endings are the discrete branching ends of nerves. They are attached to larger nerve fibers that have a myelin sheath and run in the vagus nerves, and they are located in the cardiac atria on the endocardial (i.e., inner) surface of the heart. The greatest concentrations of these receptors are at the junctional regions between the venae cavae and the pulmonary veins, with the atria. These endings are the only ones that have definitely been shown to be receptors.

A fine network of branching beaded nerve fibers extends over the entire endocardial surface of both the atria and the ventricles. The function of this network is unclear. It might be partially formed from branches of the complex unencapsulated endings. Another possibility is that it is attached to the many fine nonmyelinated nerves known to innervate the heart. It has also been suggested that the nerve net might be concerned with the motor innervation of the heart.

B. Afferent Discharges in Nerves from the Atria

Myelinated vagal afferent nerves have been shown to be attached to the atrial receptors (complex unencapsulated endings). The activity in these nerves has been studied by electrical recording of the discharges traveling toward the central nervous system. The activity in the fibers has been classified as follows: type A, which shows a burst of activity coinciding with contraction of the atria, identified by the *a* wave of the atrial pressure; and type B, activity coinciding mainly with atrial filling, the *v* wave of pressure. Also present is an intermediate discharge pattern that has the features of both types A and B. There has been much debate concerning the significance of the various discharge patterns. However, it is likely that there is no fundamental difference in the atrial receptors, and the different discharge patterns result from different locations of the receptors. The type of discharge of a single receptor can be changed following hemorrhage, infusions, and changes in heart rate (Fig. 2)

In general, the activity of atrial receptors is related to atrial volume and pressure. They signal the degree of atrial filling, which is dependent on, among other things, blood volume, and hence these receptors have sometimes been referred to as volume receptors.

There are many more small nonmyelinated nerve fibers (C fibers) ending in the atria, compared with the myelinated nerves. The activity of these fibers is also stimulated by increases in atrial pressure and volume. However, in general, they have a higher threshold and require higher pressures for their excitation. Many are silent at normal pressures and become active only at abnormally high levels of atrial pressure.

In addition to vagal nerve endings, there are nerve fibers that run in the sympathetic rami. These afferent sympathetic nerves seem to be stimulated by events similar to those affecting the vagal afferents. One interesting feature of these nerves is that frequently they are attached to more than one terminal. Sometimes a single fiber can be stimulated by events in both an atrium and a ventricle.

Some sympathetic afferent nerves can be stimulated by mechanical events, whereas others are excited only by the application of chemicals. Some of the excitant chemicals might be released during injury of the heart

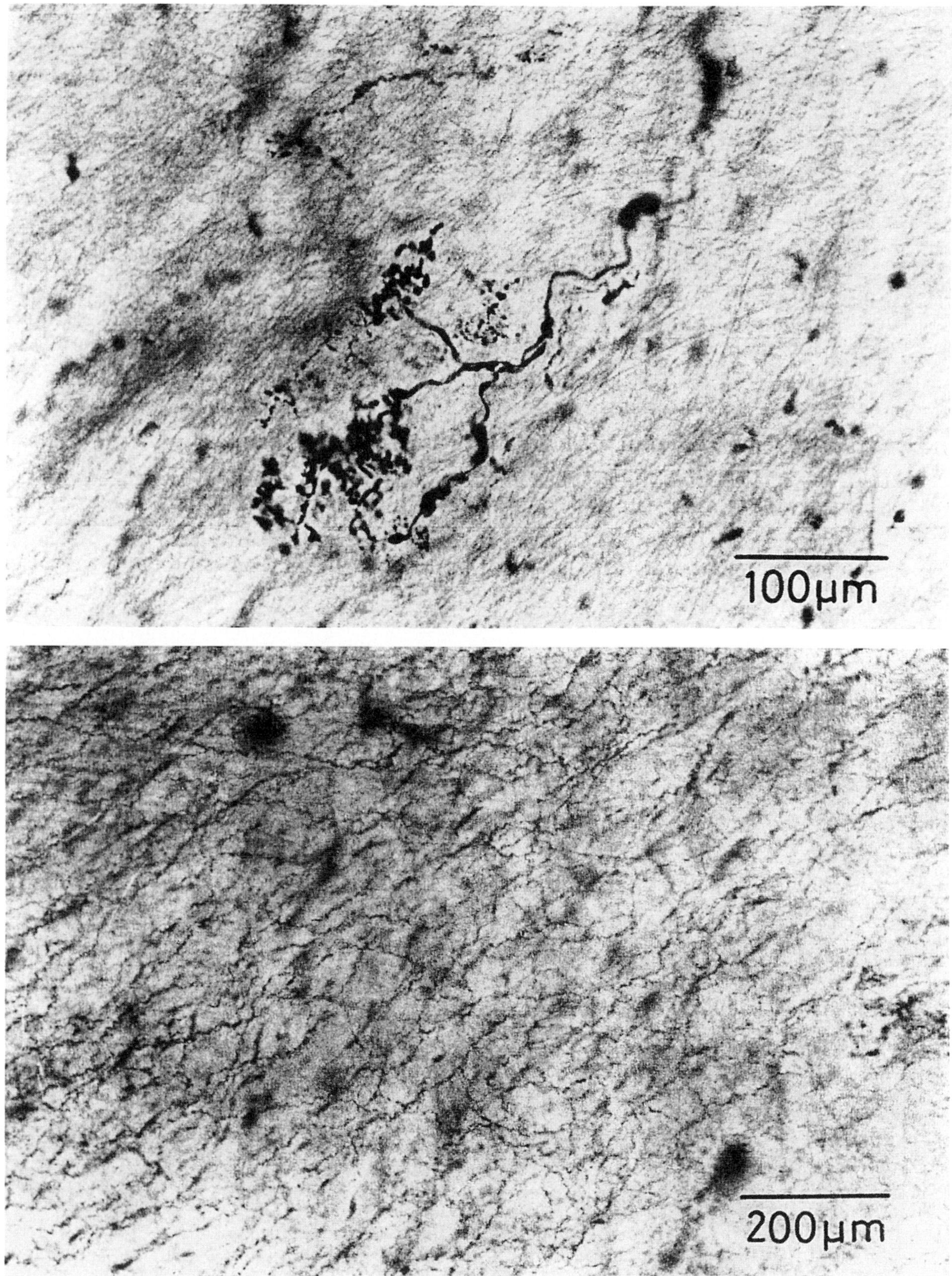

FIGURE 1 Example of two types of nervous structures in the dog's atrium. The two frames are whole-thickness preparations of endocardium obtained from the pulmonary venoatrial junction and supravitally stained with methylene blue. (Top) A complex unencapsulated ending believed to be an atrial receptor attached to an afferent nerve. (Bottom) Nerve network.

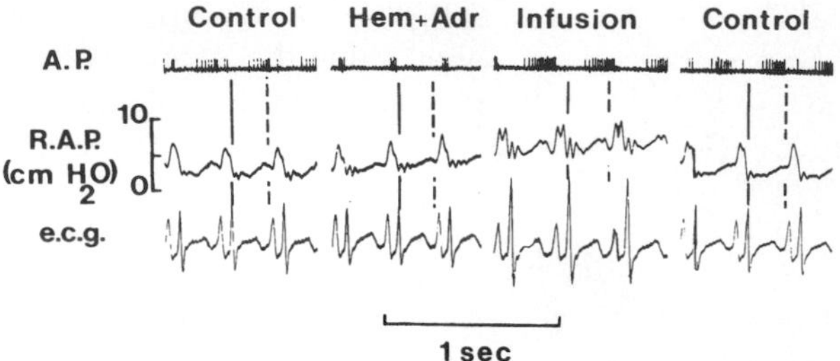

FIGURE 2 Parts of experimental recordings showing changes in the pattern of discharge of an atrial receptor in a cat. Shown are action potentials (A.P.) in a single afferent vagal nerve fiber, right atrial pressure pulse (R.A.P.), and the electrocardiogram (e.c.g.). The vertical lines are drawn to relate temporal events in atria and the afferent nerve: the solid line indicates the end of the *a* wave of atrial systole, and the dashed line indicates the peak of the *v* wave of atrial filling. Control (left): the pattern of activity is of the intermediate type; Hem + Adr: the activity has changed to type A pattern after bleeding and administration of adrenaline; Infusion; the activity has changed to type B pattern after infusion of dextran; Control (right): the receptor has regained its intermediate pattern of discharge.

(e.g., when blood supply is inadequate in coronary artery disease), and it has been suggested that they might serve as pain receptors.

C. Activity in Nerves from the Ventricles

In contrast to the atria, the ventricles are supplied almost exclusively by nonmyelinated vagal afferents. Some of these nerves clearly respond to changes in ventricular pressure, although whether they respond mainly to ventricular contraction or to ventricular filling is unclear. Some have a phasic discharge related to ventricular pressure, whereas in others the discharge is apparently random.

Some ventricular nerve fibers are not readily activated by mechanical events, but they are strongly stimulated by toxic and irritant chemical agents applied directly to the receptor site or injected into the coronary circulation. As discussed in Section IV, chemosensitive ventricular afferents can cause powerful reflex responses, but what constitutes their normal stimulus or what is their physiological role is not known.

The ventricles also receive an afferent sympathetic innervation, sometimes coterminal with atrial fibers. These might also be stimulated by mechanical and chemical events.

D. Activity in Nerves from the Coronary Arteries

It is now known that there are nerves that respond to changes in pressure in the coronary arteries as opposed to those that respond to ventricular events (Fig. 3). Unlike ventricular mechanoreceptors, which are attached to nonmyelinated nerves and require high pressure to change their activity, coronary receptors are attached to myelinated nerves and respond to physiological pressure changes. Coronary receptors have been localized to the left main coronary artery and the proximal portions of the anterior descending and circumflex arteries.

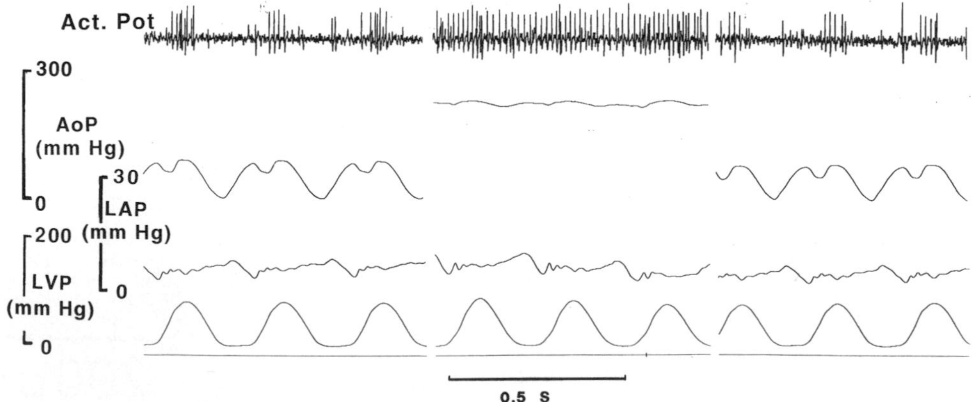

FIGURE 3 Afferent discharges in vagus neve from a coronary arterial mechanoreceptor. Traces of action potentials; aortic root pressure (AoP = coronary arterial pressure); left atrial pressure (LAP) and left ventricular pressure (LVP). An increase in coronary pressure results in an increase in afferent nervous activity. [From M. J. Drinkhill *et al.* (1993). Afferent discharges from coronary arterial and ventricular mechanoreceptors in anaesthetized dogs. *J. Physiol.* **472**, 785–799.]

III. REFLEXES ARISING IN THE ATRIA

A. Methods for Stimulation of Atrial Receptors

Electrophysiological studies outlined earlier have shown that the discharge from atrial receptors is increased by procedures that increase atrial filling. These include fluid infusion, immersion in water, and changing body position from upright to supine or head down. The disadvantage of these procedures is that they do not provide a localized stimulus to atrial receptors, and the resulting responses would be due to changes in the activity from many cardiovascular reflexogenic areas.

The function of atrial receptors has been studied in experimental animal preparations in which attempts have been made to localize the stimulus to the reflexogenic areas of the atria. Experiments have been performed in which the blood flow into the atria was increased or the outflow from the atria was obstructed. However, the disadvantage of these techniques is that the changes in pressure might affect other regions of the heart.

Since most of the myelinated atrial afferents are situated at the great venoatrial junctions, localized distension of these regions by balloon has provided a useful method for discretely stimulating atrial receptors and studying their function.

B. Responses of the Heart

Discrete stimulation of either right or left atrial receptors results in a reflex increase in heart rate (Fig. 4). The afferent pathway for this reflex has been investigated by graded cooling of the vagal nerves. The response of heart rate to stimulation of atrial receptors

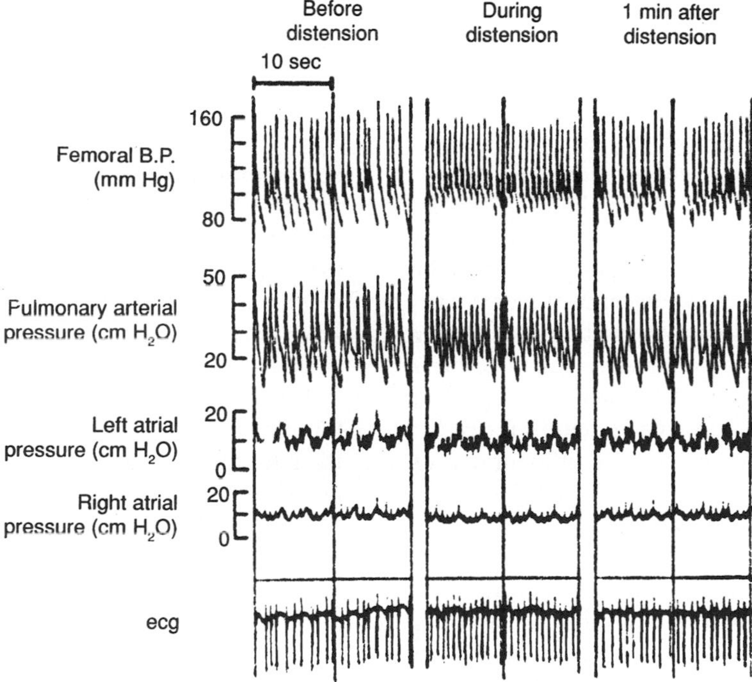

FIGURE 4 Example of the reflex response of an increase in heart rate to discrete stimulation of left atrial receptors in a dog. Shown are femoral arterial pressure (B.P.), pulmonary arterial pressure, left atrial pressure, right atrial pressure, datum line, and the electrocardiogram (ecg). The middle column shows recordings during stimulation of atrial receptors, by distending small balloons at the pulmonary venoatrial junctions, and is to be compared with the control recording (left column) and recovery after the release of balloon distention (right column). During stimulation of atrial receptors, the heart rate increased without significant changes in cardiovascular pressures, indicating the discrete nature of the stimulus. [From J. R. Ledsome and R. J. Linden (1964). A reflex increase in heart rate from distension of the pulmonary-vein–atrial junctions. *J. Physiol.* **170**, 456–473.]

is slightly reduced at 16°C, and at 12°C it is nearly abolished. This is the range of temperatures that blocks the increase in the frequency of nerve impulses in the myelinated vagal afferents, but not in nonmyelinated nerves. The afferent pathway of the reflex, therefore, lies in myelinated vagal afferent nerves.

The reflex increase in heart rate can also be abolished by cutting the sympathetic nerves to the heart, by blocking the release of norepinephrine from the nerve terminals, or by giving β-adrenoceptor-blocking drugs to prevent their action. The efferent pathway of the reflex, therefore, lies in the cardiac sympathetic nerves.

The efferent limb of the heart rate reflex has several peculiarities. The nerves engaged in the atrial reflex seem to affect only the sinoatrial node, because there is no effect on the force of contraction of the heart that would occur if sympathetic nerves to the myocardium were excited. Also, it has been shown that the particular nerve fibers involved in the atrial receptor reflex are not affected by several other reflexogenic stimuli, including the stimulation of baroreceptors, chemoreceptors, visceral afferents, and cutaneous nerves. Other cardiac sympathetic nerves are affected by all of the other stimuli, but not by atrial receptors. Atrial receptors thus seem to have their own discrete pathway.

There have been several reports that stimulation of atrial receptors results in reflex bradycardia. However, most of the studies on which these reports are based can be critized owing to inadequate localization of the stimulus. Nevertheless, the possibility remains that greater degrees of atrial distension, particularly when the atrial receptors attached to myelinated nerves are for some reason ineffective, might result in reflexes mediated by the nonmyelinated nerves. Thus, bradycardia might occur in some circumstances as a result of excitation of nonmyelinated afferent nerves.

C. Responses of Blood Vessels

The reflex effects of discrete stimulation of atrial receptors have been studied in several vascular beds. Electrophysiological studies of the efferent activity in sympathetic nerves supplying a number of areas have yielded interesting results. These showed that, whereas sympathetic efferent activity increased in cardiac fibers, it decreased in nerves supplying the kidney (Fig. 5). There was no change in activity in the lumbar and splenic nerves. The effects on the resistance to blood flow reflect this pattern of activity.

Stimulation of atrial receptors has been shown to have no significant effect, in the steady state, on resistance to blood flow in the limb circulation. A transient vasodilatation immediately following distension has sometimes been reported, which is probably due to excitation of nonmyelinated afferent nerves.

As would be expected from the electrophysiological studies of activity in renal nerves, atrial receptor stimulation results in dilatation of renal resistance vessels, which normally would be expected to lead to an increase in renal blood flow. Unlike the variable transient response in the limb, the renal vasodilatation persists for as long as the stimulus is maintained.

The only other vascular bed shown to be affected by the stimulation of atrial receptors is the coronary circulation (i.e., the blood vessels supplying the heart muscle). Recent studies have shown that stimulation of atrial receptors results in the constriction of coronary vessels, leading to a decrease in their blood flow. It should be appreciated in this context that coronary blood flow is largely dependent on the work of the heart and that factors that increase heart work, including increases in heart rate, greatly increase coronary blood flow. Therefore, the reduction in flow due to atrial receptor stimulation can only be seen if heart rate and heart force are controlled.

D. Effects on Urine Flow

Stimulation of atrial receptors results in an increase in urine flow (Fig. 6). Experiments to demonstrate this have involved discrete distension of the venoatrial junctions, obstruction of the mitral valve by balloons, increases in atrial pressure by volume infusions, or increases in venous return to the heart by procedures such as water immersion.

The mechanism responsible for the diuresis is uncertain. It has been shown to occur in animals in which the hypothalamus has been destroyed; therefore, it is not dependent on changes in antidiuretic hormone (ADH), although there is a decrease in the level of ADH, and this could contribute to the response. Renal nerves might also contribute, although an increase in urine flow occurs even in denervated kidneys perfused with blood from a donor animal in which the atria are distended.

The increase in urine flow is sometimes accompanied by an increase in sodium excretion. Whereas the diuretic response seems to be mainly humorally mediated, the natriuresis is dependent on the concomitant changes in heart rate and changes in renal nerve activity.

A humoral agent that has received much attention in recent years is atrial natriuretic peptide. This pep-

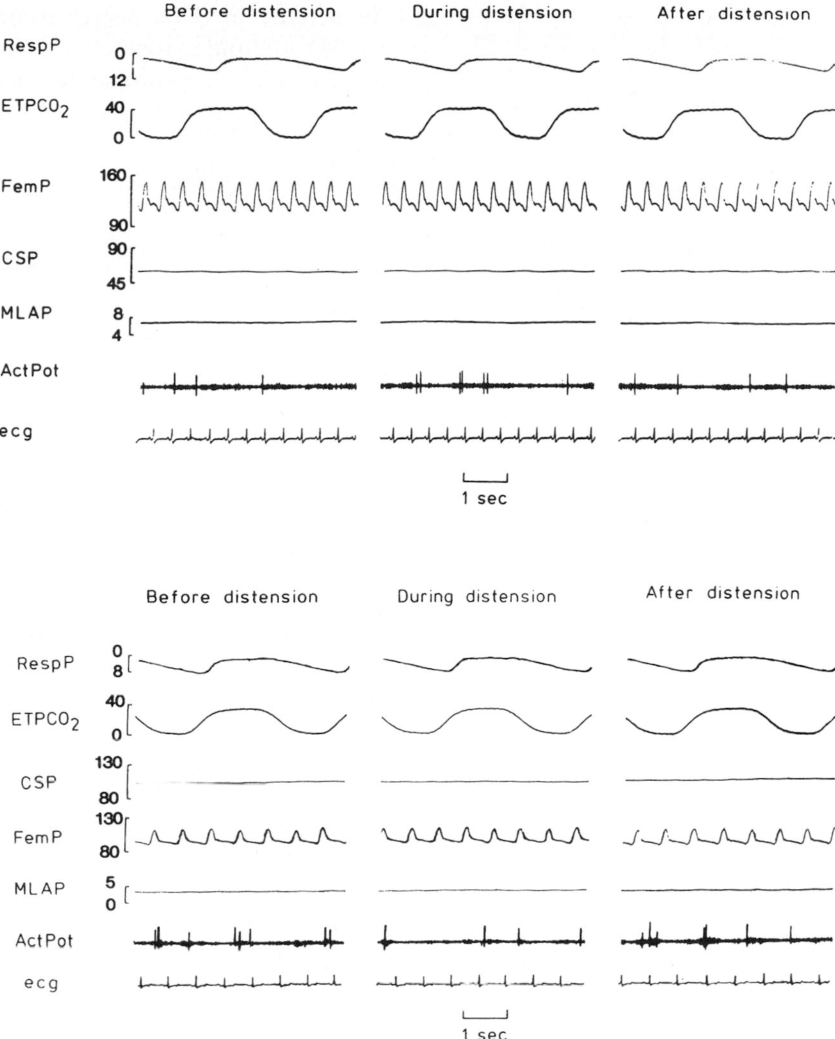

FIGURE 5 Examples of a reflex increase in activity of an efferent cardiac sympathetic nerve fiber (top) and a decrease in activity in an efferent renal sympathetic nerve fiber (bottom). Both frames show endotracheal or respiratory pressure (RespP), end-tidal carbon dioxide partial pressure ($ETPCO_2$), femoral arterial blood pressure (FemP), carotid sinus perfusion pressure (CSP), mean left atrial pressure (MLAP), action potentials (ActPot), and the electrocardiogram (ecg). The units of pressure are millimeters of mercury. As in Fig. 4, the middle column contains the recording taken during discrete stimulation of atrial receptors and is sandwiched by two columns of initial control and recovery period recordings.

tide is released from atrial myocytes in response to stretch, and might be involved in the response to increases in atrial volume, in addition to the responses to stimulation of afferent nerves. However, it is not essential for the diuretic response to atrial distension. It is also possible that an undiscovered hormone(s), might contribute to the response.

In summary, therefore, discrete stimulation of atrial receptors results in increases in urine flow and small and inconsistent increases in sodium excretion. The diuresis is mediated largely by a humoral agent(s). The natriuresis appears to be dependent on changes in renal nerve activity and concomitant hemodynamic changes.

E. Effect on Hormones

Vasopressin or ADH is released by the posterior part of the pituitary gland in response to an increase in the osmolality of blood. Blood osmolality is mainly

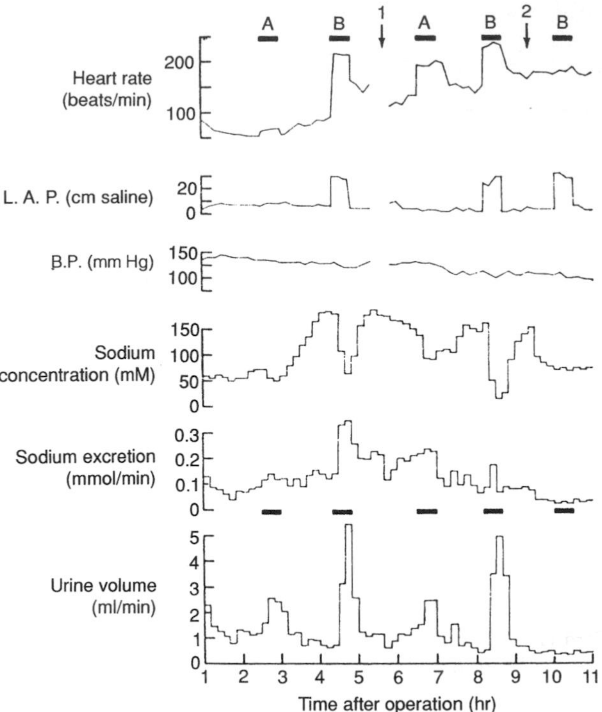

FIGURE 6 Example of the reflex increase in urine flow and the urinary rate of sodium excretion in a dog in response to atrial distension. (A) Small balloons distended in the pulmonary venoatrial junctions to stimulate atrial receptors with no obstruction to blood flow in the heart. (B) Balloon distended in the left atrium to cause mitral obstruction. There is a break in the record at 1, when the atrial balloon was replaced because it leaked. (2) The right vagus nerve as cut in the neck, and the left vagus nerve was cut at the level of the upper border of the aorta. [From J. R. Ledsome and R. J. Linden (1968). The role of left atrial receptors in the diuretic response to left atrial distension. *J. Physiol.* **198**, 487–503.]

dependent on the concentration of sodium. There are believed to be osmoreceptors that control the release of this hormone. As mentioned earlier, there is evidence that the release of ADH into the blood is partly under the control of atrial receptors. An increase in plasma volume results in an increased stimulation of atrial receptors, and this leads to a reduction in the release of ADH and a consequent water diuresis.

Renin is a hormone that is released by the kidney in response to either a decrease in blood pressure to the kidney or activation of the renal nerves. Renin has no direct cardiovascular effects, but it is responsible for the formation of angiotensin II, a powerful vasoconstrictor agent. The stimulation of atrial receptors results in a decrease in the plasma level of renin. The mechanism of this response has been shown to involve renal nerves. The precise mechanism by which the activity in renal nerves controls renin secretion is not fully understood. It was generally believed to involve solely a β-adrenergic mechanism. However, β-adrenoceptor-blocking drugs do not completely prevent this response. The control of renin release involves an additional pathway that is neither adrenergic nor cholinergic.

Plasma cortisol is controlled by another pituitary hormone, adrenocorticotropic hormone. The stimulation of atrial receptors results in a decrease in this hormone and a consequent decrease in the plasma level of cortisol.

IV. REFLEXES FROM VENTRICULAR RECEPTORS

Nonmyelinated ventricular afferent nerves may be excited by mechanical stimulation, chemical stimulation, or both. The chemosensitive afferents differ from arterial chemoreceptors in that they do not respond to changes in blood levels of oxygen, carbon dioxide, or acidity, but only to extraneous chemicals such as veratridine or capsaicin, or chemicals released as a consequence of myocardial hypoxia or ischemia, such as bradykinin and prostaglandins. Stimulation of chemosensitive afferents by injection of the chemicals into the coronary circulation results in a profound slowing of the heart and dilatation of blood vessels, causing a steep fall in blood pressure (Fig. 7). This response is known as the Bezold–Jarisch reflex.

The role of ventricular mechanoreceptors is uncertain. Although some are active at normal ventricular pressures, they do not respond in a predictable way to increases in pressure. They may be stimulated by abnormally high ventricular pressures or when the ventricle is abnormally distended. It has also been suggested that they may respond to increases in the force of ventricular contraction. The responses to stimulation of ventricular mechanoreceptors are probably decreases in heart rate and dilatation of blood vessels, although adequate physiological studies are lacking.

V. REFLEXES FROM CORONARY RECEPTORS

Increases in the pressure distending the coronary arteries result in reflex dilatation of resistance blood vessels (Fig. 8). Responses occur when the pressure is

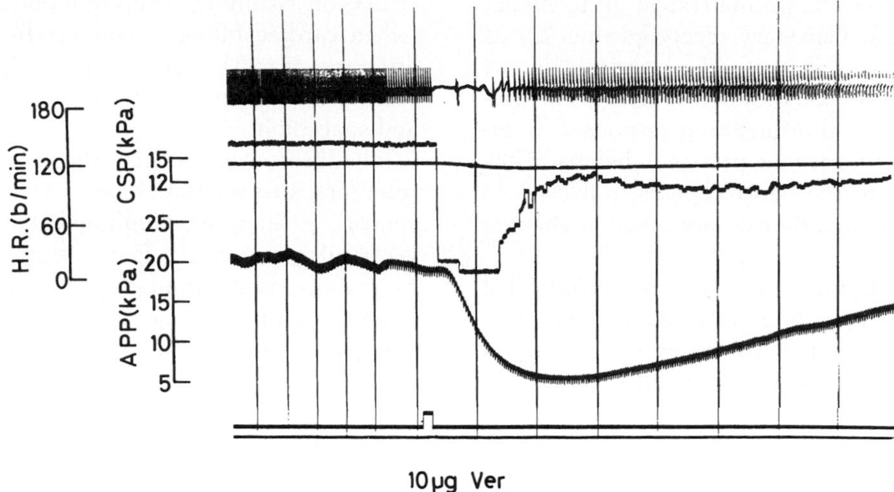

FIGURE 7 Reflex responses to coronary chemoreflex. Traces of electrocardiogram, heart rate, carotid sinus pressure (CSP), and perfusion pressure to vascularly isolated limb perfused at constant flow (APP). Injection of veratridine into coronary circulation resulted in profound bradycardia (slowing of heart) and vasodilation (fall in arterial perfusion pressure). 1 kPa = 7.5 mm Hg. [From K. H. McGregor, R. Hainsworth, and R. Ford (1986). Hind limb vacular responses in anaesthetized dogs to aortic root injections of veratridine. *Quart. J. Exp. Physiol.* **71**, 577–587.]

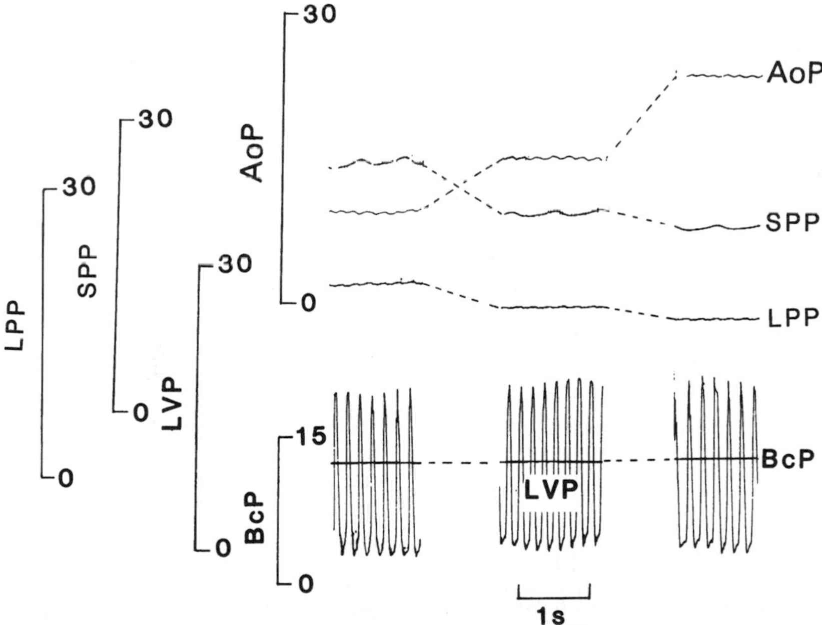

FIGURE 8 Vasodilatation in response to increase in coronary arterial pressure. Traces of aortic root pressure (AoP = coronary arterial pressure), systemic arterial and limb perfusion pressures (SPP and LPP), brachiocephalic arterial pressure (BcP; controls carotid and aortic baroreceptors, held constant), and left ventricular pressure (LVP). Pressures are in kPa, 1 kPa = 7.5 mm Hg. An increase in coronary arterial pressure results in dilatation in systemic and limb vascular beds. [From J. K. A. Al-Timman *et al.* (1993). Reflex responses to stimulation of mechanoreceptors in the left ventricle and coronary arteries in anaesthetized dogs. *J. Physiol.* **472**, 769–783.]

changed from below the normal range up to higher than normal levels. Coronary receptors thus act as arterial baroreceptors in that, like receptors in the aorta and carotid sinuses, they detect changes in arterial blood pressure and bring about responses (negative feedback) that buffer the pressure changes. They seem to differ from the baroreceptors, however, in that, at least in the dog, they do not result in changes in the heart rate.

It is now apparent that blood pressure is controlled not only by receptors in the carotid arteries and aortic arch but also by receptors in the coronary arteries and probably also in still undiscovered regions.

VI. REFLEXES MEDIATED BY SYMPATHETIC AFFERENT NERVES

Afferent nerves running in the sympathetic nervous system, like vagal afferents, are activated by chemical and mechanical stimuli. Stimulation of these afferent nerves has been shown to result in spinal reflexes (not involving the brain) that lead mainly to increases in activity in efferent sympathetic nerves. This causes increases in the heart rate and the constriction of blood vessels, and therefore increases in blood pressure. Supraspinal mechanisms are also involved, as seen, for example, by inhibition of efferent vagal activity to the heart.

The reflex effects of stimulation of cardiac sympathetic afferent nerves are thus opposite those from stimulation of vagal afferent nerves. Whereas the vagal reflexes function as a "negative feedback system" and consequently inhibit the cardiovascular system, the sympathetic afferents appear to excite it. The role of this mechanism is far from clear, and, at least in anesthetized animals, the vagal inhibitory mechanisms seem to be dominant. However, it is possible that sympathetic reflexes have a modulating role in cardiovascular control.

VII. PHYSIOLOGICAL AND PATHOPHYSIOLOGICAL ROLE OF CARDIOGENIC REFLEXES

A. Atrial Reflexes

Any event that alters cardiac filling is likely to result in a change in the activity of atrial receptors. Therefore, they are likely to contribute to the reflex changes occurring during postural changes (cardiac filling decreases on assuming the upright position) and exercise (when cardiac filling is enhanced). The stimulus to atrial receptors is likely to change during many other events, including water immersion, acceleration, and spaceflight.

It has been suggested that the principal role of atrial receptors is to regulate heart size and to ensure that ventricular filling is maintained at a level for optimal ventricular function. Thus, increased heart filling causes increased stimulation of atrial receptors that, by causing an increase in heart rate as well as by decreasing blood volume through the urinary response, acts to reduce the extent to which the atria, and the ventricles, are distended.

There is evidence that atrial receptors might be destroyed and lose their function in congestive heart failure, thus possibly contributing to the changes that occur in this condition. This functional "deactivation" of the atrial receptors might contribute to the water and salt retention, and eventually to the clinical condition of edema, in which extra fluid is obvious in dependent parts of the body. Deactivation of atrial receptors causes decreases in urine flow and in the rate of urinary sodium excretion, increases in the activity of the renin–angiotension system and vasopressin, and vasoconstriction of the renal vessels.

It is known that a higher incidence of renal failure is associated with shock resulting from extensive bleeding than with acute heart failure. This difference has been suggested to be due in part to deactivation of atrial receptors during bleeding, with a resulting renal vasoconstriction that is greater than that occurring in heart failure. Following bleeding the deactivation of atrial receptors could lead to an increase in the plasma cortisol level, which would protect body tissues and enhance the vasoconstrictive effects of sympathetic nerves on blood vessels to maintain normal arterial blood pressure levels. The increase in the activity of the renin–angiotension system and in the blood level of vasopressin and the decrease in urine flow would result in fluid retention, to compensate for lost blood, and vasoconstriction, to maintain normal arterial blood pressure.

Paroxysmal tachycardia is a condition in which the heart has bouts in which it beats abnormally fast. This condition is sometimes associated with an increase in urine flow. During the rapid heart rate the atria cannot empty normally into the ventricles, and they become distended. This increases the stimulus to atrial receptors, leading to an increase in urine flow.

It is possible to suggest a role for atrial receptor reflexes in the pathophysiological control of the coronary circulation. The stimulation of atrial receptors results in a reflex increase in heart rate, and this, in turn, increases coronary flow. However, the direct reflex effect of the stimulation of atrial receptors is to reduce coronary flow by sympathetic vasoconstriction. This effect might help to maintain uniform perfusion to all layers of the ventricular wall, which is important since the limitation to blood flow to the inner layers is believed to constitute a mechanism for myocardial ischemia during increases in heart rate, particularly in the presence of coronary narrowing. The clinical condition relating to these events is labeled "angina" and is caused by pathological narrowing of the coronary arteries. In such patients the atrial receptor reflex might have a protective role.

The role of nonmyelinated afferent nerves from the atria and the sympathetic afferents are unknown.

B. Ventricular and Coronary Receptors

Ventricular mechanoreceptors, which are attached to nonmyelinated afferent nerves, are probably of little importance in normal cardiovascular control because of the high pressures required to change their activity and the erratic nature of the responses. It has been suggested that, because some receptors respond to increased force of contraction, particularly when the ventricle is nearly empty, they may initiate the vasovagal response (this is the abrupt dilatation of blood vessels and cardiac slowing that cause people to faint in response to low cardiac filling or emotional stimuli). It is now thought unlikely that this is a major cause of the vasovagal response, partly because experiments in animals in which ventricular filling was reduced and the cardiac sympathetic nerves were stimulated to increase the force of contraction did not induce the response. Furthermore, a similar response occurs in people who have had a heart transplant and therefore have ventricles without a nerve supply.

It has been suggested that, following an obstruction to a coronary artery (coronary thrombosis), the damaged region of the heart becomes distended and this causes stimulation of ventricular mechanoreceptors with consequent decreases in heart rate and blood pressure.

Ventricular chemosensitive afferents may play a role in the responses occurring during coronary artery narrowing. When the blood supply to the heart muscle is inadequate, particularly when its work is increased during exercise, various chemicals accumulate and these stimulate chemosensitive afferent nerves. When sympathetic afferent nerves are stimulated, this mediates cardiac pain (angina pectoris) and may also lead to increases in heart rate and blood pressure. Stimulation of chemosensitive nerves running in the vagus nerves, on the other hand, results in slowing of the heart and a fall in blood pressure. The resultant response seems to depend on the region of the heart that is affected; reduction in flow to the inferolateral wall usually decreases blood pressure, whereas when the anterior wall is affected the pressure is likely to increase and there may be dangerous abnormal heart rhythms.

Coronary arterial mechanoreceptors probably function as arterial baroreceptors in that they detect changes in arterial blood pressure and bring about responses that tend to restore the pressure toward its previous level. Not only is this likely to be an important homeostatic control system but it is also of relevance to many physiological studies in which it is incorrectly assumed that if carotid and aortic baroreceptors are controlled, all high-pressure regulatory mechanisms are abolished. Responses from stimulation of coronary receptors may also explain what many investigators previously had thought to be ventricular reflexes.

BIBLIOGRAPHY

Al-Timman, J. K. A., Drinkhill, M. J., and Hainsworth , R. (1993). Reflex responses to stimulation of mechanoreceptors in the left ventricle and coronary arteries in anaesthetized dogs. *J. Physiol.* **472**, 769–783.

Drinkhill, M. J., Moore, J., and Hainsworth, R. (1993). Afferent discharges from coronary arterial and ventricular mechanoreceptors in anaesthetized dogs. *J. Physiol.* **472**, 785–799.

Hainsworth, R. (1990). Atrial receptors. *In* "Reflex Control of the Circulation" (I. H. Zucker and J. P. Gilmore, eds.). Telford, Caldwell, NJ.

Hainsworth, R. (1991). Reflexes from the heart, *Physiol. Rev.* **71**, 617–658.

Hainsworth, R., McWilliam, P.N., and Mary, D. A. S. G. (eds.) (1987). "Cardiogenic Reflexes." Oxford Univ. Press, Oxford, England.

Linden, R. J., and Kappagoda, C. T. (eds.) (1982). Atrial receptors. *Monogr. Physiol. Soc.* **39.**

Malliani, A. (1982). Cardiovascular sympathetic afferent fibres. *Rev. Physiol. Biochem. Pharmacol.* **95**, 11–74.

Mary, D. A. S. G. (1992). Reflex effects on the coronary circulation. *Exp. Physiol.* 77, 243–270.

Cardiovascular Hormones

ROGER HAINSWORTH
DAVID MARY
University of Leeds

GLOSSARY

Endocrine Glands Organs that release substances (hormones) directly into the bloodstream. These include the pituitary gland (also known as hypophysis), which is found at the base of the brain and is made of two parts: the posterior pituitary, which has neural structures (neurohypophysis), and the anterior pituitary, which has a glandular structure (adenohypophysis). The adrenal glands are located at the top of the kidneys and are really two distinct organs: the adrenal cortex (outside shell), which is made of glandular cells rich in cholesterol for the manufacture of steroid hormones, and the adrenal medulla, which contains granular cells grouped around blood vessels. Other endocrine glands are mainly single organs and include the thyroid gland in the neck with the parathyroid glands adjacent to it, the thymus in the upper part of the chest, and the sex glands.

Hormones Chemical substances that are released from endocrine glands directly into the bloodstream. The substances are transported by blood to exert specific effects on cells and organs that may be remote from the site of origin. The hormones have profound effects on regulation of body functions in a manner that is slower than that mediated by nerve fibers and is not localized to specific regions.

"CARDIOVASCULAR HORMONES" REFERS TO HORmones that are intimately connected with the function of the heart and blood vessels (cardiovascular system). The conventional definition of a hormone is that it is a chemical agent that is released from specialized organs (endocrine glands) into the bloodstream, where it is transported to target organs to trigger or modify their functions. The intimate relationship of cardiovascular hormones to the function of the cardiovascular system implies that the hormone has a significant physiological effect on the function of the heart or blood vessels under normal circumstances. Also, changes in cardiovascular function, for example, changes in arterial or venous blood pressure, are involved in the regulation of the rate of release of the hormone. [*See* Cardiovascular System, Anatomy; Cardiovascular System, Physiology and Biochemistry.]

I. INTRODUCTION

Hormones are regulatory substances that are released directly from endocrine glands into the circulation and are transported by the circulation to various target organs. So, in the sense that all hormones are carried by the bloodstream, all could be regarded as cardiovascular. However, in this article the term cardiovascular hormone is used to describe only those hormones that are secreted by organs that are controlled in a major way by cardiovascular events and that have among their target organs the heart or blood vessels. This definition is intended to exclude hormones that have only indirect effects on the circulation. For example, the hormones secreted by the thyroid gland have the effect of accelerating metabolic activity and this indirectly increases blood flow to the active tissues. But the thyroid hormones are not primarily concerned with cardiovascular control and

ENCYCLOPEDIA OF HUMAN BIOLOGY, Second Edition, VOLUME 2. Copyright © 1997 by Academic Press. All rights of reproduction in any form reserved.

are therefore not considered as cardiovascular hormones. The extent to which hormones, which can be classified as cardiovascular, are actually involved in cardiovascular control varies considerably. At one extreme are the adrenal medullary hormones, epinephrine (adrenaline) and norepinephrine (noradrenaline), which are really an extension of the sympathetic nervous system and are intimately concerned with cardiovascular control. At the other extreme, and less obviously cardiovascular hormones, are the adrenal cortical secretions. These steroid hormones are influenced to an extent by cardiovascular events and among other effects do play a role, largely permissive, in circulatory control. Other endocrine systems lie between these extremes in the extent of their cardiovascular involvement. [*See* Endocrine System.]

In this short review we have categorized cardiovascular hormones into (1) those secreted by organs that have as their major function the secretion of hormones, that is, the established endocrine glands; (2) organs that are not primarily endocrine glands but nevertheless do secrete substances into the circulation that have cardiovascular actions; and (3) "local hormones" that are mainly formed in blood endothelial (lining) cells and have only local effects. The main cardiovascular sites of action are shown in Table I.

II. ENDOCRINE GLANDS WITH MAJOR INVOLVEMENT IN CARDIOVASCULAR CONTROL

A. The Adrenal Medulla

The adrenal glands are paired organs that lie superior to the kidneys. They have two major components, the cortex or outer part of the gland, which is controlled by the anterior part of the pituitary gland (see the following), and the medulla or central part of the gland, which is controlled by preganglionic sympathetic nerves. Sympathetic nerves are a part of the nervous system that is concerned with bodily functions (autonomic nervous system). They leave the spinal cord in the thoracic and lumbar regions and stimulate a second nerve in a node called a ganglion. The second nerve then courses to its target organ, where norepinephrine is released and its effect is directly on its target, for example, a blood vessel. In the case of the adrenal medulla, the first nerve from the spinal cord, the preganglionic nerve, terminates on the chromaffin cell (secretory cells) in the gland and these release hormones into the circulation. Unlike the postganglionic nerves, however, which release norepinephrine at their target, the adrenal medulla releases both norepinephrine and a related substance, epinephrine, in proportions of about 30:70, into the

TABLE I

Main Sites of Action of Cardiovascular Hormones[a]

Hormone	Heart	Arteries (resistance)	Kidney: Diuresis[b]	Kidney: Natriuresis[c]
Epinephrine	+	+/−[d]		
Norepinephrine	+	+		
Angiotensin-II		+		−[e]
Vasopressin		+	−	
Cortisol		+[f]	−[g]	−[g]
Cardiac peptides		−	+	+
Endothelium-derived relaxing factor	−			

[a]+ indicates excitatory response, for example, constriction of arteries or increase in urine flow; − indicates inhibition of these effects.
[b]Increase in urine flow.
[c]Increase in urinary sodium excretion.
[d]Vessels in skeletal muscles dilate, whereas others constrict.
[e]Effect is through aldosterone.
[f]Effect is complementary and essential to action of norepinephrine.
[g]A related hormone, aldosterone, is much more effective.

circulation. The effects of the adrenal medullary hormones are similar to those of the sympathetic system, but are much more diffuse. [*See* Adrenal Gland.]

1. Control of Adrenal Medullary Secretions

Epinephrine and norepinephrine are released into the blood by the adrenal glands as part of the "fright and flight" reaction. Under conditions of stress, particularly when it necessitates aggressive or evasive responses, the sympathetic nervous system becomes strongly excited and this excitation includes the nerves supplying the adrenal medulla.

In addition to stimulation by "higher centers" in response to emotional events, adrenal medullary secretion is influenced by reflexes that are involved in circulatory control. These include baroreceptors and chemoreceptors, as well as changes in levels of pressure and blood gases in the blood perfusing the brain.

Arterial baroreceptors are nerve endings found in some arteries, particularly the carotid sinus (a dilatation at the origin of the internal carotid arteries) and the arch of the aorta. They are also found in the walls of other arteries, including subclavian and coronary arteries. They are stimulated by both stretch and rate of stretch, and thus they respond not only to increases in arterial blood pressure but also to increases in pulse pressure that result from larger volumes of blood being pumped by the heart. The responses to stimulation of baroreceptors have the effect of decreasing blood pressure and the main mechanism for achieving this is to decrease the activity of sympathetic nerves and to inhibit secretion of the adrenal glands. Thus one effect of a decrease in blood pressure is to cause an increase in the stimulation of the adrenal medulla.

Chemoreceptors are situated in the carotid and aortic bodies. These are highly vascular structures, situated near the corresponding baroreceptors, and they respond to decreases in the level of oxygen and increases in carbon dioxide and acidity in arterial blood. Stimulation of chemoreceptors causes increased activity in sympathetic nerves to blood vessels and increased adrenal medullary secretion as well as increased pulmonary ventilation. Interestingly, stimulation of carotid chemoreceptors inhibits the cardiac sympathetic nerves.

A reduction in the flow of arterial blood supplying the brain can also stimulate sympathetic and adrenal activity. The same response also occurs when the oxygen level decreases or the carbon dioxide and acidity increase outside their physiological ranges.

2. Effects of Epinephrine and Norepinephrine

The responses to increased levels of the adrenal medullary catecholamines (epinephrine and norepinephrine) are in some ways similar to the effects of sympathetic nerve stimulation. However, there are important differences. Because the nerves release only norepinephrine, the effects of sympathetic stimulation are predominantly excitatory. The adrenal medulla, however, releases mainly epinephrine, which also has inhibitory effects. Furthermore, sympathetic effects can often be localized; for example, atrial receptor stimulation increases sympathetic activity to the sinoatrial node and decreases it to the kidney, and has no effects on other regions. Catecholamines secreted into the circulation reach all regions that have a blood supply. Thus, epinephrine dilates the smooth muscle in the airways (facilitating increased breathing in exercise), dilates blood vessels supplying skeletal muscle and thereby increases muscle blood flow, and constricts blood vessels elsewhere. This is in contrast to the sympathetic nerves, which decrease blood flow generally.

The effects of catecholamines have been divided into the α or excitatory actions and the β_1 effects, which are also excitatory to the heart and cause an increase in rate and force of contraction, and β_2 effects, which dilate airway muscle and blood vessels to skeletal muscle.

B. The Renin–Angiotensin–Aldosterone System

The renin–angiotensin–aldosterone system is a series of enzymatic reactions (Fig. 1) that is initiated by the release of the proteolytic enzyme renin. Renin is synthesized and stored in the juxtaglomerular cells of the kidney. These cells line the arteries supplying the renal glomeruli (the site at which blood is filtered at the first stage of formation of urine) (Fig. 2). The only established role of renin is to break down a protein in the blood, angiotensinogen, to form a decapeptide named angiotensin-I. Angiotensin-I is almost inactive, but is cleaved by another enzyme, angiotensin-converting enzyme (ACE), mainly as the blood flows through the lung. The resulting octapeptide, angiotensin-II, is a very potent cardiovascular hormone, but one that has a half-life of only about 30 sec. Angiotensin-II is further degraded to angiotensin-III, a heptapeptide, and angiotensin-II and -III promote the release of

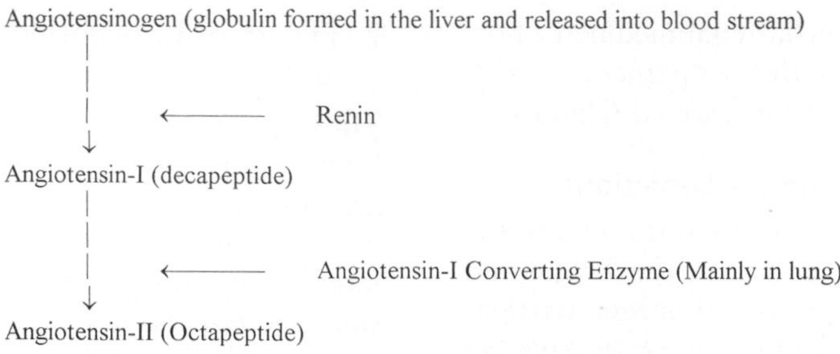

FIGURE 1 The renin–angiotensin system.

aldosterone, which is a mineralocorticoid that enhances sodium retention by the kidney.

1. Control of Renin Release

Renin release is controlled by pressure receptors and chemical receptors in the kidney, and by neural and hormonal influences.

The juxtaglomerular cells of the kidney function as intrarenal baroreceptors in that they respond to decreases in local blood pressure to stimulate the release of renin. Thus the rate of renin release is inversely related to intrarenal blood pressure.

The second mechanism involves the cells of the kidney tubules, actually the initial part of the distal

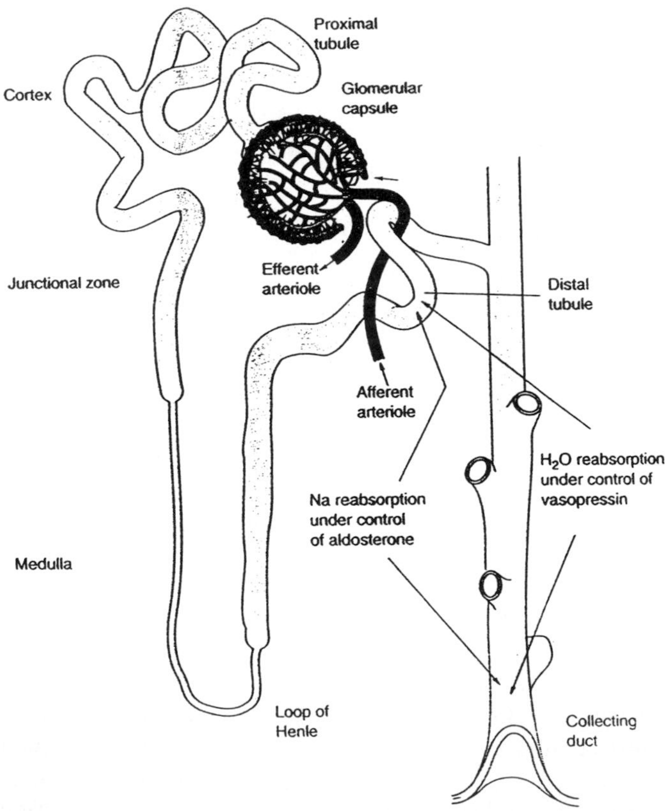

FIGURE 2 Diagram illustrating the sites of action of aldosterone and antidiuretic (vasopressin) hormones. The part of the kidney responsible for the formation of urine in the glomerulus and its transport through tubules to the collecting duct is shown (collectively known as the nephron). The layers of the kidney are indicated on the left: cortex (outer shell) and medulla (inner part) and the junctional zone in between.

convoluted tubule, known as the macula densa. The rate of release of renin is inversely related to the amount of sodium passing into the tubules and its rate of transport into the tubular cells. For instance, a decrease in sodium excretion, resulting from body salt depletion or low blood pressure, results in an increase in renin release.

Activity in sympathetic nerves supplying the kidneys also enhances renin release. This effect is mediated mainly by β-adrenoceptors. The sympathetic discharge to the kidney is influenced by several reflex mechanisms, including baroreceptors, chemoreceptors, and atrial receptors.

Renin release is also enhanced by adrenal medullary catecholamines. It is inhibited by atrial natriuretic peptides, cortisol, and vasopressin. It is also inhibited by increased levels of angiotensin and aldosterone, which thus function as a negative feedback mechanism.

2. Effects of Angiotensin-II

Angiotensin-II is an extremely potent vasoconstrictor. It causes a powerful constriction of smooth muscle, particularly in small arteries and arterioles, and thus increases arterial blood pressure. The renin–angiotensin system is powerfully excited in circumstances in which blood pressure or blood volume is abnormally low. Its role in normal control of blood pressure remains to be established, although drugs that inhibit the formation of angiotensin-II by inhibiting angiotensin-I-converting enzyme are frequently used to lower blood pressure in hypertensive patients.

The other action of angiotensin, the stimulation of aldosterone, has the effect of causing sodium retention (Fig. 2) and thereby to increase extracellular, including plasma, volume. This effect may also contribute to hypertension.

C. Antidiuretic Hormone (Vasopressin)

Antidiuretic hormone (ADH) is synthesized in the supraoptic nucleus of the hypothalamus and passes actually along nerve fibers, the hypothalamo–hypophyseal tract, to the posterior part of the pituitary. In response to stimulation it is released from the nerves in the posterior pituitary into the bloodstream. Although it is often called vasopressin, to describe its action on blood vessels, its primary target is the kidney.

1. Control of ADH Secretion

Osmoreceptors in the anterior part of the hypothalamus respond to increases in osmotic pressure in the extracellular fluid. Decreases in osmotic pressure, associated with low plasma levels of sodium, result in low levels of plasma ADH. Although under most circumstances the plasma osmotic pressure is the main factor determining ADH release, other factors may also be of importance. Very high levels of ADH occur during stress, such as anesthesia and surgical trauma. [*See* Anesthesia.]

The release of ADH is also under the control of cardiovascular reflexes, in particular the atrial receptor reflex. Atrial receptors are nerve endings attached to stretch receptors in the wall of the atria and the adjoining great veins. They are excited by increases in atrial volume and pressure. Thus increases in the filling of the heart, caused among other things by increases in blood volume, result in increases in the afferent nervous activity in the vagus nerves from the atrial receptors. Apart from the reflex increase in heart rate and the decreased activity in the renal sympathetic nerves, the other major effect of stimulation of atrial receptors is to inhibit ADH release.

2. Effects of Antidiuretic Hormone

The principal target of ADH is on the distal tubules and collecting ducts of the kidney (Fig. 2) and the effect is to increase their permeability to water and thereby to enhance water reabsorption and increase urinary concentration. An increase in blood volume, which increases the stimulation of atrial receptors, causes a decreased secretion of ADH and a water diuresis. Because atrial receptors are excited by increases in atrial and blood volumes, they are sometimes referred to as volume receptors.

Antidiuretic hormone, or vasopressin, also has the effect of increasing the contraction of the smooth muscles in small arteries and thereby to increase arterial blood pressure. This effect seems to be minor under most physiological conditions. The concentrations of vasopressin required to increase blood pressure are much higher than those needed to cause maximal antidiuresis. However, part of the reason why the effect of ADH on blood pressure is small may be due to its central effect, which is to enhance cardiovascular reflexes, particularly the baroreceptor reflex. Therefore the direct vasoconstrictor action is opposed by an enhanced reflex inhibition of efferent sympathetic vasoconstrictor activity.

D. Adrenal Cortical Hormones

The adrenal cortex releases three groups of steroid hormones with overlapping actions. These are glucocorticoids (that enhance glucose formation), mineralocorticoids (that act on minerals in the body), and sex hormones. The first two are also known collectively as corticosteroids. [*See* Steroid Hormones; Steroid Hormone Synthesis.]

Cortisol is a steroid hormone (one of the glucocorticoid hormones) that is released in the blood and is mainly transported in a form bound to plasma proteins. The release of cortisol is mainly controlled by the blood level of adrenocorticotrophic hormone (ACTH), which is itself a hormone released by the anterior pituitary gland. Also, the concentration of cortisol in the blood can inhibit the release ACTH and therefore limit the further release of cortisol.

The rate of release of cortisol is controlled by the baroreceptor reflexes and the atrial receptor reflex; stimulation of these reflexes decreases the rate of release of cortisol, and unloading of these receptors reflexly increases the rate of cortisol release. It is also believed that angiotensin-II and vasopressin may promote the release of cortisol; these hormones interact to enhance sodium conservation by the kidney.

Cortisol has several well-established actions in the body, including regulation of metabolism, anti-inflammatory, and sodium-retaining effects. The sodium-retaining effect is less potent than that of aldosterone, but it can also lead to conservation of sodium ions and increases in plasma volume. Another important action of cortisol is to maintain the integrity of sympathetic vasoconstrictor actions on peripheral blood vessels. These actions of cortisol have constituted the concept that cortisol is released during stresses. Unlike the adrenal medulla, complete loss of adrenal cortical function is incompatible with life.

III. CARDIAC PEPTIDES

The link between atrial distension and increase in flow of urine has been known for 40 years. It is established that stimulation of atrial receptors induces an increase in urine flow partly because of a reflex inhibition of antidiuretic hormone secretion and other mechanisms, including a decrease in renal nerve activity. More recently it has been shown that atrial myocytes secrete a substance that is released into the circulation and may lead to a diuresis, natriuresis (increased sodium excretion), and hypotension. This substance is now known as atrial natriuretic peptide (ANP). Another related substance has been initially isolated from extracts of brain tissue, but is now known to be stored in the ventricles of the heart; this is referred to as brain natriuretic peptide (BNP). A third type has been found in the vascular endothelium and is denoted C natriuretic peptide.

A. Factors Influencing Release of ANP

Despite an enormous amount of research, the physiological role of ANP is still not firmly established. Release into the circulation occurs in response to distension of the heart, particularly the atria. Thus increases in blood concentration have been shown to occur in situations in which the atria may be distended, for example, when changing from upright to lying position. Changes in blood volume also influence ANP release. It is also known that plasma levels of ANP are raised in heart failure, possibly because of distension of the heart. Increases in heart rate and plasma osmolarity are also associated with elevated levels of plasma ANP.

B. Effects of ANP

High blood levels of ANP or BNP cause increased urinary excretion of sodium and water. There is vasodilatation within the kidney and elsewhere and the release of renin and aldosterone is suppressed. Thus the effect of ANP is to decrease plasma and extracellular fluid volumes, to lower arterial and central venous blood pressures and to promote the movement of fluid out of the vascular compartment.

It should be emphasized that the effects of ANP are the direct consequence of cardiac distension and so do not involve reflexes. It is unclear whether ANP actually has a major physiological role. Atrial distension leads to a diuresis through reflex mechanisms. ANP release may also contribute to the diuresis, however, physiological degrees of atrial distension may not result in plasma levels of ANP, which are sufficiently high to evoke appreciable diuresis. The physiological importance of ANP thus still remains to be established.

IV. LOCAL FACTORS

In recent years it has become apparent that the endothelium, or lining cells, of blood vessels has a major influence on the state of constriction or dilatation of the blood vessels themselves. The most prominent of these "local hormones" is endothelium-derived relaxation factor (EDRF), now known to be nitric oxide. Many other vasoactive agents are released by the endothelial cells, including prostaglandins, thromboxane, and endothelins. Prostaglandins are known to dilate blood vessels, whereas thromboxane and endothelins cause vasoconstriction. These agents are also involved in inflammatory conditions and in promoting or inhibiting blood clotting.

Nitric oxide (NO) is closer to the definition of cardiovascular hormones than other local hormones, being released from the endothelium into the bloodstream. NO is continuously released under basal conditions, but the rate of release can be modified by several influences. Most prominent among the stimuli acting to enhance NO release is shear stress on the vessel by the flowing blood. An increase in blood flow enhances the release of NO. Release can also be enhanced by a number of agents, including bradykinin and acetylcholine.

The effects of NO are normally localized because NO is very rapidly inactivated by hemoglobin in the blood. The principal effect is to cause relaxation of the smooth muscle in blood vessel walls. The responses of the blood vessels to several agents are dependent on intact endothelium. Substances that normally dilate blood vessels, including 5-hydroxytryptamine, bradykinin, and acetylcholine, have the opposite effect in the absence of endothelium. Another action of NO at the smooth muscle cell is to block the release of the vasoconstrictor transmitter norepinephrine and thus to enhance further the dilatation of the blood vessels.

BIBLIOGRAPHY

Hainsworth, R. (1991). Reflexes from the heart. *Physiol. Rev.* **71**, 617–658.

Hainsworth, R., and Drinkhill, M. J. (1995). Regulation of blood volume. *In* "Studies in Physiology, Cardiovascular Regulation" (D. Jordan and J. Marshall, eds.), pp. 77–91. Portland Press, London.

Keeton, T. K., and Campbell, W. B. (1981). The pharmacologic alteration of renin release. *Pharmacol. Rev.* **31**, 81–227.

Lang, C. C., and Struthers, A. D. (1992). Atrial and brain natriuretic peptide—A dual natriuretic peptide system potentially involved in circulatory homeostasis. *Clin. Sci.* **83**, 519–527.

Moncada, S., Palmer, R. M. J., and Higgs, E. A. (1991). Nitric oxide: Physiology, pathophysiology and pharmacology. *Pharmacol Rev.* **43**, 109–142.

Cardiovascular Responses to Weightlessness

DWAIN L. ECKBERG
Hunter Holmes McGuire Department of Veterans Affairs Medical Center and Medical College of Virginia

GLOSSARY

Autonomic nervous system Extensive array of sensory and motor nerves that modulates internal organ function automatically. Norepinephrine and acetycholine are the principal cardiovascular autonomic neurotransmitters

Baroreflex Neural pressure-regulating reflex that opposes changes of blood pressure. During standing, heart rate increases and resistance vessels constrict, in response to changes of traffic carried over sympathetic and vagus autonomic nerves

Cardiovascular deconditioning Syndrome that occurs in astronauts after space travel, comprising reduced exercise capacity and, with standing, unusually reduced mean blood pressure, increased heart rate, and, in extreme cases, loss of consciousness

Orthostatic hypotension Unusually large reduction of blood pressure in the standing position

Syncope Loss of consciousness. This extreme manifestation of post-spaceflight cardiovascular abnormalities may occur with standing, because of reduction of blood flow to the brain. "Presyncope" signifies lightheadedness and feelings of impending loss of consciousness

Vasovagal reaction Vasovagal reactions are simple faints that occur under a variety of circumstances, including standing. They are due to reflex augmentation of vagus nerve traffic to the heart and reduction of sympathetic nerve traffic to vessels, which lead to abrupt heart rate slowing and reduced blood pressure and blood flow to the brain

HUMAN CARDIOVASCULAR RESPONSES TO MICROgravity are a source of fascination, in part because they cannot be studied on earth. Cardiovascular function is normal in space (unless untoward responses are provoked experimentally), and the abnormalities that occur because of space travel are seen only after astronauts (and cosmonauts) return to the earth's gravitational field. Most astronauts experience unusually large reductions of arterial pressure and increases of heart rate when they stand after sojourns in space. Symptoms with standing range from none, to slight lightheadedness, to frank loss of consciousness. These postflight abnormalities are thought to result, in as yet unexplained ways, from a cascade of cardiovascular changes that begins in the first minutes and hours of weightlessness. According to this construct, loss of normal gravitational forces shifts body fluids from the lower to the upper body, and translocation of blood stretches cardiac chambers and sequentially provokes release of endogenous diuretic hormones, diuresis, and blood volume reduction. Postflight cardiovascular changes disappear spontaneously within days to weeks and, as far as is known, leave no residual abnormalities.

ENCYCLOPEDIA OF HUMAN BIOLOGY, Second Edition, VOLUME 2. Copyright © 1997 by Academic Press. All rights of reproduction in any form reserved.

I. HISTORY

The earliest animals in space—dogs, rats, mice, and flies—were launched amid predictions of massive organ failure, including particularly cardiovascular collapse. The first human in space, 27-year-old Yuri Gagarin, placed in earth orbit by the Soviet Union on April 12, 1961, returned from his 89-minute flight as proof that exposure to weightlessness does not lead to cardiovascular collapse. However, the ninth man in space, the American Wally Schirra, returned from a 9-hour, 13-minute orbital mission in 1962 and became the first astronaut to report cardiovascular symptoms after a space mission. The symptoms he described have been labeled "cardiovascular deconditioning" and comprise orthostatic hypotension, narrowing of pulse pressure (the difference between systolic and diastolic arterial pressures), lightheadedness or frank syncope, tachycardia, and reduced exercise tolerance. Some of these changes are depicted in Fig. 1. Orthostatic hypotension occurs in most astronauts and seems to be independent of mission duration. Symptoms are highly variable. In some instances, they are of trivial importance—transient lightheadedness after prolonged standing. The Soviet astronaut Anatoly Levchenko flew an airplane immediately after he returned to earth from an 8-day mission. In other instances, symptoms may be disabling; however, precise data on symptoms of astronauts, their severity, and their duration are not available.

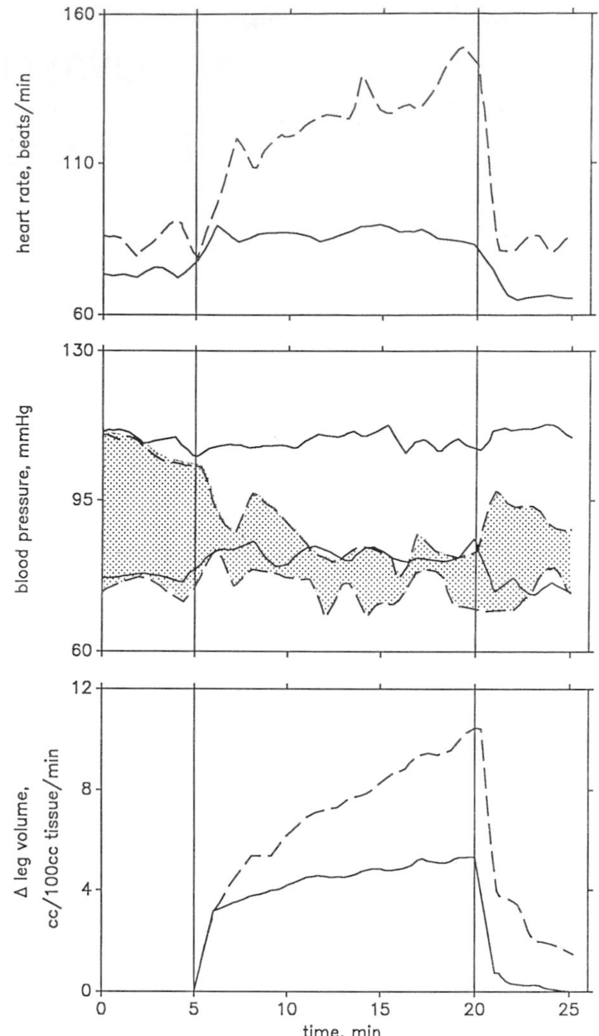

FIGURE 1 Responses of one astronaut to passive 70° tilt before and after a *Gemini* mission in 1965. Postflight responses are indicated by dashed lines.

II. THE DATA BASE

Humans have been in space for over a quarter century, and a wealth of scientific information has been obtained during human space missions. However, the process of scientific knowing is different for physiological research done on humans in space than on earth, and what understanding has been derived from space research must be interpreted in light of the considerable constraints imposed by this unusual type of research. This data base differs from the usual scientific data bases in important ways.

First, when astronauts escape the laws of gravity, scientists who study their physiology do not simultaneously escape the laws of statistics. An abiding problem for human space research has been small sample sizes. Typically, four or fewer astronauts are studied during a single mission, and data obtained are published in terms of responses of individual astronauts, who, curiously, may be referred to by name (or by function—"the lunar module pilot"). Given vagaries of human responses to experimental interventions, these sample sizes may be too small for statistical analysis. Indeed, in publications that report human physiological data from space, indications of mean and variance are uncommon, and indications of statistical significance are rare. Moreover, it is virtually unheard of for the same experiment to be repeated in space with the same astronaut as subject. Therefore, the simple requirement that the reproducibility of scientific data be documented may be impossible to fulfill in space. It even is difficult to do the same experiment on different subjects during serial space missions; mission durations are variable, the conditions of missions vary, and other scientists are waiting to obtain their own unique data from space.

Second, the types of experiments that can be done with astronauts as subjects are much more limited in space than on earth. Above all, the experiment must be safe. Many types of measurements that are routine in laboratories on earth, such as direct measurement of pressure in the heart or an artery, may be very difficult or impossible to obtain in space. However, research in space is becoming increasingly sophisticated. Several astronauts have been launched into space with catheters whose tips lie at the junction of the superior vena cava and right atrium. The American Space Shuttle and the Russian Mir Space Station have been equipped with freezers for storage of blood samples, mass spectrometers for analysis of respiratory gases, and echocardiographs for analysis of heart chamber size and function. Astonishingly, plans are being laid to record human nerve traffic directly, with fine electrodes inserted through the skin, during the Space Shuttle mission, Neurolab.

Also, as on earth, it is important that the experiments not alter the physiology being studied. Although it may be of great interest to know how plasma hormone and neurotransmitter concentrations change in space, withdrawal of too much blood (by the dozens of scientists who are interested in a large variety of plasma constituents) may reduce blood volume and distort the physiological measurements being obtained.

Related to this issue is the probability that flight surgeons may "intervene" in the interest of crew safety. One example of this is the treatment of space adaptation syndrome; studies of autonomic function after administration of scopolamine, amphetamine, metaclopromide, or promethazine may be of uncertain value. Another is the practice of giving crew oral salt and fluids immediately before they reenter the earth's gravitational field. This practice is understandable, since the combination of a blood volume deficit (see Section V) and increased gravitational force along the long axis of the body during reentry is undesirable. However, this practice may compromise studies of body fluid volumes and autonomic function in the hours immediately following return to earth.

Third, one of the most scarce commodities in space is astronauts' time. Studies tend to be constrained by an inflexible "time-line," and experiments most likely to succeed are those that can be done quickly. An extension of this constraint is that the best research is that which can be done by the astronaut on himself or herself. If an experiment must be "tended" by another astronaut, the astronaut time required is doubled.

Fourth, as on earth, the kind of research that can be undertaken is constrained by the devices available, and equipment that may be readily available in most laboratories may not be available for use in space. For example, during the first quarter century of space travel, it was not possible to freeze plasma samples in space for subsequent analyses of hormones and neurotransmitters; this capability came late.

Fifth, information on what *is* known from physiological research in space may be difficult or impossible to gain access to. Results may not be published in rigorously peer-reviewed journals; as a result, usual standards for scientific reporting may not be followed, and it may be difficult to determine what measurements were made and what the results were. Also, large amounts of in-flight data are buried in technical reports and government documents that may not be readily available. (Much of the literature that has not been peer-reviewed, or "gray" literature, has been included in the new National Library of Medicine–National Aeronautics and Space Administration data base, Spaceline.) Most of the data regarding human physiology during long-term space missions was published by Soviet (including Russian) scientists in Russian. Even if translations are available, Russian nomenclature may not be familiar to Western readers.

Sixth, although increasing numbers of women are serving as astronauts, there are no published data on how responses of women to space travel differ from those of men.

Notwithstanding the manifold impediments to development of a sound, statistically robust data base on physiological changes in humans during weightlessness, published data may be sufficient to support reasoned conclusions. Moreover, the probability is that many critical missing measurements in the data base will be obtained soon. In the discussion that follows, an effort is made to indicate how solid the scientific data are.

III. HEADWORD FLUID SHIFT

Some of the earliest observations made during space missions indicated that astronauts in space look and feel different than they do on earth.They have puffy, ruddy, edematous faces, speak with adenoidal twangs, have difficulty breathing through their noses, and have distended neck veins and small legs ("bird legs" according to actual measurements of leg volumes). Soviet infrared photographs of astronauts in space document dilation of face and neck veins, and these changes persist for months in space. All can be explained by headword migration of body fluids during weightlessness. This concept is central to current understanding of cardiovascular changes occurring in space (but see the following).

Postflight cardiovascular changes are thought to result from a chain of events that is initiated in the earliest minutes of spaceflight, or even before spaceflight begins. Weightlessness causes blood and fluid that normally would be pulled toward the feet during most (about two-thirds) of the day to be shifted toward the head. This shift of fluids is believed to trigger a cascade of events that results ultimately in postflight "cardiovascular deconditioning." A corollary of this notion, and one that is probably no longer tenable, is that such headword shifts are *sustained* during spaceflight.

The simplest way to test the fluid shift hypothesis is to measure volume in heart chambers. Data from Russian, French, and American experiments indicate that left ventricular end-diastolic volume (i.e., the volume just before systolic contraction) is increased during the first mission days, is reduced to subnormal levels after about the fourth day, and remains subnormal thereafter, for the duration of the mission. Figure 2 depicts average heart rate and echocardiogram measurements from four American crew members obtained during a Space Shuttle mission in 1985. (These limited data also show that during the first day, stroke volume is elevated. Since heart rate is elevated, cardiac output is transiently supranormal.) Similar conclusions have been drawn by Russian and German scientists from measurements of electrical thoracic impedance (called "rheography").

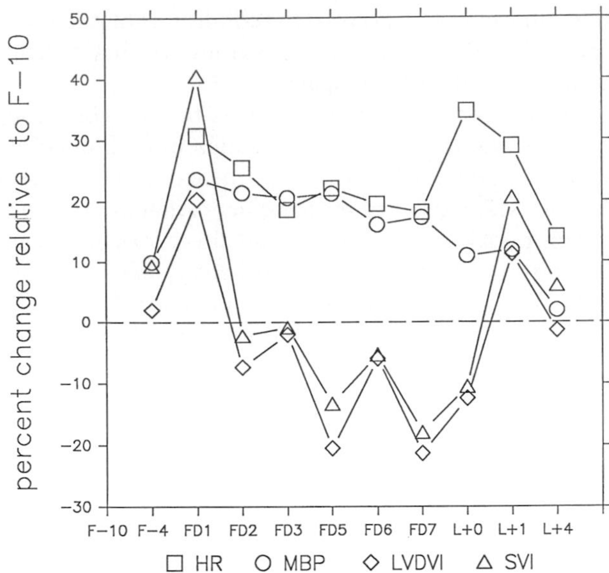

FIGURE 2 Average data obtained from four astronauts during a 1985 Space Shuttle mission. F, flight; FD, flight day; L, landing; HR, heart rate; MBP, mean blood pressure; LVDVI, left ventricular end-diastolic volume index (volume divided by body surface area); SVI, stroke volume index. [Adapted with permission from M. W. Bungo (1989). Echocardiographic changes related to space travel. *In* "Two-Dimensional Echocardiography and Cardiac Doppler" (J. N. Shapira and J. G. Harold, eds.), 2nd Ed. Williams & Wilkins, Baltimore.]

IV. INTRATHORACIC PRESSURE CHANGES

A less satisfactory alternative to actual *volume* measurements is measurement of *pressures* in intrathoracic veins. Use of pressure measurements to obtain indexes of volumes is based on the assumption that the relation between pressure and volume—compliance—is constant in space. Venous compliance is determined by a complex interplay among viscoelastic and myogenic properties, blood volume, and neurohumoral venoconstrictor and venodilator influences. There is good evidence that leg vein compliance increases in space. Indirect evidence suggests that cardiac compliance increases early during space missions: central venous pressure (not quite cardiac pressure, see the following) is reduced, at times when cardiac chamber volumes are increased.

Central venous pressure has been measured with saline-filled catheters, or with a catheter with a pressure transducer mounted on its tip, in four astronauts during three Space Shuttle missions. For such measurements, the catheter tip is inserted into an arm vein and advanced until its tip is in the superior vena cava, near its junction with the right atrium. Pressure recording begins before launch, and continues during the astronauts' first hours in space. The first recording of central venous pressure during weightlessness was made by Jay C. Buckey and colleagues on June 5, 1991, during the Spacelab Life Sciences-I (SLS-1), the first Space Shuttle mission dedicated to life sciences research. A recording from this mission is depicted in Fig. 3. This astronaut, and other astronauts on American Space Shuttle missions, was launched with his legs positioned at a higher level than his heart. As a result, central venous pressure before launch was already higher than it was when prelaunch measure-

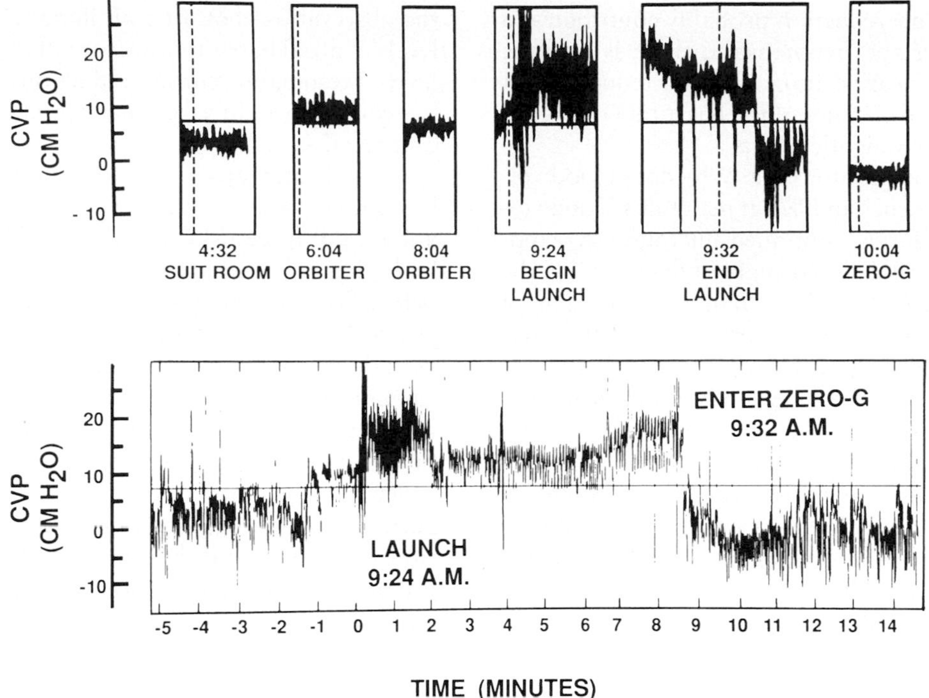

FIGURE 3 Central venous pressure (CVP) measured in one astronaut before launch and during the earliest minutes of the Spacelab Life Sciences-1 Space Shuttle mission in 1991. [Reproduced with permission from J. C. Buckey, F. A. Gaffney, L. D. Lane, B. D. Levine, D. E. Watenpaugh, and C. G. Blomqvist. (1993). *N. Engl. J. Med.* 328, 1853–1854.]

ments were made in the suit room (first two upper panels). By the time the launch began (lower panel), central venous pressure had returned to the level recorded in the suit room, before the astronaut had entered the Space Shuttle. During the launch, central venous pressure was substantially elevated. However, within one minute after the Space Shuttle engines were powered down, central venous pressure had fallen to *below* prelaunch levels.

In summary, the available evidence provides qualified support for the notion that weightlessness leads to headword migration of body fluids. These data point toward a transient increase of central venous volume, followed by a later reduction to *below* prelaunch supine levels. Subnormal central venous pressure levels occurring at times when left ventricular (and presumably right atrial) volumes are increased (Fig. 1) suggest that the compliance of cardiac chambers is increased during the earliest hours of astronauts' exposure to weightlessness. Existing data do not explain why facial edema persists despite reduction of central venous pressure. Moreover, none of the data discussed explain how this brief shift of fluid volume can cause persistent reduction of blood volume or postflight orthostatic hypotension. Apparently, however, this temporary increase in the degree of heart chamber stretch leads to an early diuresis. A brief elevation of plasma atrial natriuretic factor was observed early in one Space Shuttle mission. Presumably, release of this hormone sets in motion changes of fluid balance that lead to the reductions of blood volume, which are discussed next.

V. BLOOD VOLUME CHANGES

A central component of the cardiovascular deconditioning of weightlessness is the reduction of blood volume that occurs in all missions, including those as brief as 8 hours. Plasma volume falls early in missions and stabilizes at about 12% less than preflight levels, after 30 to 60 days of weightlessness. (A parallel 10% reduction of red blood cell mass occurs also. This is unexplained; it appears to be due to impaired red cell production by bone marrow, since there is no evidence for excessive destruction of red cells or hemorrhage.)

This blood volume reduction probably contributes to postflight orthostatic hypotension; there is a loose, but significant linear relation between reductions of blood volume and changes of the heart rate responses to standing after weightlessness.

Two important questions are: Why does blood volume reduction occur? and What maintains blood volume reduction during continued microgravity exposure? Blood volume is strongly influenced by the degree of stretch of cardiac chambers and that of baroreceptors in the thoracic aorta and carotid arteries. [*See* Vascular Capacitance.] The usual (i.e., for about two-thirds of the time) body position for humans is sitting or standing. It is likely that during daily activities, blood volume is adjusted at levels adequate to compensate for downward shifts in the upright position. Although astronauts stand, they do not experience the usual footward pull of blood volume that occurs on earth. Therefore, in space, blood volume may be reduced, because usual postural reductions of blood volume in the upper half of the body no longer occur.

There are at least two mechanisms that adjust blood volume according to the degree of filling of cardiac chambers and vessels in the upper part of the body. First, baroreceptors sense the degree of stretch and alter autonomic nerve activity. Increases of baroreceptor input to the central nervous system (such as might occur with headward blood volume shifts at the beginning of space flight) lead to reductions of sympathetic nerve activity to the kidney and, consequently, to reductions of plasma renin activity and the circulating peptides angiotensin-II and aldosterone. Reductions of angiotensin-II and aldosterone reduce thirst and increase kidney salt and water excretion.

Second, increased atrial and ventricular stretch leads to increased levels of atrial natriuretic factor and increased excretion of salt and water. This mechanism appears to be invoked early during space missions. Nitza M. Cintron and her colleagues at Johnson Space Center documented a brief elevation of plasma atrial natriuretic factor early in one Space Shuttle mission. This peptide (and possibly other diuretic hormones) is a good candidate to set in motion changes of fluid balance that lead to early reductions of blood volume during space flight.

The question of what *maintains* blood volume reduction is harder to answer. The most likely explanation is that after an early diuresis, blood volume stabilizes at levels less than those on earth, because blood is no longer pulled footward as it is during two-thirds of the time on earth. This possibility is supported by the observation that after an initial increase, plasma levels of atrial natriuretic factor fall to below preflight levels. Presumably, chronic reduction of blood volume is mediated by reduced cardiac chamber size and levels of arterial stretch.

A simple, perhaps not very important question that has not yet been answered is, Where does the blood volume go? It seems certain that fluid leaves the lower body; leg volumes are consistently less during than before flight. (Limited data suggest that *arm* volumes are unchanged.) Presumably, although it has not yet been documented, there is increased urine production early in space missions. Since astronauts are launched in the Space Shuttle on their backs, increased renal fluid excretion may begin *before* American missions.

Several types of evidence suggest that a diuresis occurs. The first urine samples obtained during missions are dilute, and this may reflect an excess water excretion. Also, hemoconcentration occurs; this is consistent with early plasma volume reduction. During space missions, crew members routinely lose weight (on average about 3 kg). Most weight loss occurs during the first 2 days of missions. The picture regarding fluid balance early in missions is clouded by another factor: space adaptation syndrome occurs in as many as half of astronauts and may contribute to fluid loss through vomiting and reduction of fluid intake. Moreover, astronauts who do not have space adaptation syndrome may not drink as much fluid in space as on earth.

However, it seems unlikely that cardiovascular deconditioning is due simply to reduction of blood volume (which amounts to about 600 ml). Although a minority (about 5%) of healthy people have orthostatic hypotension and faint after they donate blood, most do not. Most astronauts have orthostatic hypotension after exposure to weightlessness. Indeed, many astronauts experience orthostatic hypotension, even though they take salt and water prior to return to earth. In people studied after head-down bed rest, there seems to be no correlation between the amount of blood volume reduction and the fall of standing blood pressure. Also, if such people are given intravenous saline infusions to replace lost blood volume, or a mineralocorticoid to prevent blood volume reduction, they still have orthostatic hypotension.

VI. CARDIAC FUNCTION

The name "cardiovascular deconditioning" may be unfortunate because it implies that weightlessness impairs cardiac performance. Evidence for impairment of cardiac function has been sought exhaustively and not found. This is not to say that cardiac function is unchanged; changes occur, but they probably can be explained fully by changes of blood volume and cardiac loading conditions.

There is solid evidence that exposure to weightlessness reduces heart size. Postflight chest X rays document reductions of heart size from preflight levels. Postflight echocardiograms obtained by American, Soviet, and French physicians show that left ventricular end-diastolic volume declines by an average of about 25% and stroke volume declines by about 19%. There is no correlation between reductions of cardiac dimensions and flight duration. This provides indirect evidence that blood volume reductions occur during the earliest days of missions and stabilize at some lower level.

There also is strong evidence that cardiac function in space is normal. The absolute amounts of blood in the left ventricle before and after contraction are reduced (probably because of reduced blood volume), and the amount of blood ejected with each heart beat ("stroke volume") is correspondingly reduced. However, the percentage of the left ventricular volume ejected with each beat (the "ejection fraction") is normal in space; this suggests that myocardial contractile properties are normal. At rest, there is little or no change of cardiac output (the amount of blood pumped by the heart per minute), in part because heart rate may be faster.

Pre- and postflight echocardiograms indicate that left ventricular wall thickness does not change during weightlessness. Since heart size diminishes, this indicates that total left ventricular mass diminishes. These measurements provide some evidence that the edema of the head and neck observed in all astronauts does not extend to the thorax. Calculated left ventricular mass returns to preflight values very rapidly, and therefore reductions of left ventricular mass may be due simply to reductions of myocardial interstitial or intracellular fluid.

Many people voiced early concerns that cardiac electrical activity or *rhythm* might be abnormal in weightlessness. There is solid evidence from long-term ("Holter") electrocardiogram recordings that the incidence of cardiac arrhythmias *is* greater in space than on earth, particularly during (or, as on earth, after) exercise. The cause of cardiac arrhythmias in space is unknown. It is not known if circumstances leading to arrhythmias in space are reproduced exactly during Holter monitoring on earth; conceivably, astronauts have more arrhythmias in space because of the work they perform, or because the circumstances under which they perform it are more stressful than on earth. In two American astronauts who had frequent ventricular premature beats during activities on the moon, cardiac rhythm may have been abnormal because they trained for their mission in the heat and humidity of a Florida summer and probably were launched with body potassium deficiencies.

VII. AUTONOMIC OUTFLOW

It is not entirely certain that resting heart rate and blood pressure change in space. However, some evidence suggests that heart rate and blood pressure increase in space by about 10–20%. Since cardiac output tends to be normal (reflecting lower stroke volumes but higher heart rates), increases of blood pressure reflect increases of peripheral resistance. These hemodynamic changes probably result from changes of traffic carried over autonomic nerves to the heart and blood vessels. The stimulus for presumed autonomic changes is not known with certainty. It seems likely that autonomic nerve traffic changes in response to reductions of blood volume (or more likely blood pressure). This explanation is not totally satisfactory, however, because on earth, baroreflex responses to arterial hypotension should restore blood pressure toward *normal* levels, but not to *above normal* levels. Thus, autonomic adjustments that lead to increased heart rate and blood pressure may betray *resetting* of the usual relation that obtains between blood pressure and volume and autonomic outflow. Such resetting may be mediated by chronically reduced levels of neural input from cardiac chambers and aortic and carotid baroreceptors. This may in turn modify central processing of baroreceptor information (a process known as "plasticity").

The nature of postulated changes of autonomic nerve output from the central nervous system is unknown, and there even is uncertainty that changes occur at all. It is not known if blood pressure elevation and heart rate speeding are due to increased sympathetic activity, reduced vagal activity, or a combination of these two. [*See* Hypertension.]

The evidence that autonomic nerve traffic changes in space is indirect and inconclusive. During experimentally induced changes of blood pressure on earth, norepinephrine varies as a linear function of sympathetic traffic carried to the important skeletal muscle vascular bed. Therefore, if in space the blood pressure reductions increase sympathetic nerve activity, this should be reflected in increased levels of the principal sympathetic neurotransmitter, norepinephrine. There are very limited data regarding sympathetic neurotransmitters in space. Urinary concentrations of norepinephrine and its metabolites were measured during the American Skylab program in the 1960s and during a joint Russian–French mission, Antares–Mir '92. There was no major change in excretion (and, therefore, integrated neural release) of norepinephrine. B. Kvetnansky from the Slovak Academy of Sciences in former Czechoslovakia and his Soviet coworkers used a system for withdrawing blood from Soviet astronauts and freezing it at −30°C until it could be analyzed on earth. They analyzed blood samples drawn from three astronauts on Days 217 to 219 of a 237-day mission and found that in-flight plasma norepinephrine and epinephrine levels increase in space, but to levels that do not exceed the rather high upper limit of normal. Most recently, Peter Norsk and colleagues reported that average plasma norepinephrine levels in four astronauts studied during the 10-day second German Space Shuttle mission, D2, were increased above levels measured before the mission during supine rest. These results do not explain blood pressure elevations that occur in space; they provide some very preliminary evidence (from very small samples sizes) that sympathetic nervous system activity increases, but to levels that do not exceed the normal range.

Concentrations of both plasma and urinary sympathetic neurotransmitters are supranormal after astronauts return to earth. This suggests that both sympathetic nerve release of norepinephrine and adrenal medullary release of epinephrine are increased. These changes can be explained on the basis of hypotension during most of the day (when astronauts are sitting or standing) and, perhaps, stress associated with return to earth.

The relative tachycardia (and possibly some of the blood pressure elevation) that occurs in space may also be explained on the basis of reduced vagus nerve restraint of heart rate. Vagal–cardiac nerve traffic has not been measured directly in humans; therefore in space as on earth, it must be quantified with indirect measures. Of these, some measure of the variance of heart rate (or more properly, its reciprocal, R-R interval) and mean R-R interval are used. Faster heart rates during and after weightlessness provide some evidence that there is less vagal–cardiac nervous outflow to the heart. A study by Janice M. Fritsch-Yelle and coworkers indicates that R-R interval fluctuations at the breathing frequency are normal on landing day.

Thus, sympathetic outflow appears to be slightly increased, and vagal outflow appears to be normal during and after exposure to weightlessness. Existing data do not identify the autonomic sensors that signal the need for persistent changes of autonomic nerve output. Evidence that weightlessness impairs cardiovascular reflex function is discussed next.

As mentioned, sustained weightlessness cannot be produced and studied on earth. A corollary of this is that standing cannot be produced and studied in space. Although astronauts can and do assume head-up and feet-down positions in space, such standing differs substantially from standing on earth. With standing on earth, blood is drawn toward the feet, intravascular pressures in the upper body fall, and a pressure gradient develops between baroreceptors in the aorta in the chest and the carotid arteries in the neck (pressure is lower in the carotid arteries than in the aorta). There are also pressure on the soles of the feet and changes of nerve traffic from otoliths and semicircular canals to the central nervous system. None of these changes occur with the assumption of upright posture in space.

However, there is an experimental analog that is used to simulate standing in space. Astronauts enter an airtight cylindrical chamber that encloses the lower half of their bodies, below the iliac crest. Pressure measured in this chamber is lowered stepwise by some vacuum device and is measured by a gauge. Lower body negative pressure (or LBNP) draws blood away from the thorax into the pelvis and legs. The amount of blood drawn into the lower body by similar levels of suction is greater in space than on earth. Therefore, the capacity of the venous system is increased. Increased venous compliance may be secondary to reduced neurohumoral constriction of veins, or reduced support of leg veins by atrophic antigravity muscles. Presumably, because of greater sequestration of blood caused by the same amount of suction, heart rate and vascular resistance increase more, and left ventricular stroke volume and cardiac output fall more in space than on earth.

Not surprisingly, astronauts are more likely to experience hypotension and presyncope or syncope during lower body suction in space than on earth. Such hemodynamic changes and symptoms are known as "vasovagal" reactions ("vaso-" because sympathetic nerve traffic to resistance blood vessels decreases and vessels dilate, and "-vagal" because vagal nerve traffic to the heart increases and heart rate slows). Astronauts who have the greatest reduction of blood pressure during lower body suction are likely to have the greatest reduction of blood pressure with standing postflight.

As indicated, vasovagal reactions are due to reflex changes of autonomic nervous activity. However, vasovagal reactions, including syncope or presyncope during lower body suction, occur on earth as well as in space. It is not clear that lower body suction in space is comparable to lower body suction on earth. On earth, lower body suction usually is not performed in subjects from whom blood has been withdrawn; when it is, subjects experience syncope at lower intensities of suction than before blood withdrawal. Therefore, the apparent increased susceptibility of astronauts to presyncope or syncope during lower body suction may result simply because during and immediately after space missions, they begin and end lower body suction at lower blood volumes than they had prior to missions.

One hypothesis to explain the orthostatic hypotension that occurs in astronauts after spaceflight is that their baroreflexes do not respond appropriately to standing. This possibility is underscored by research in dogs in whom the nerves that sense arterial pressure changes (baroreceptors) have been cut and in whom the brain no longer receives accurate information regarding arterial pressure. Such dogs experience huge swings of blood pressure during their daily experiences and large reductions of pressure when they stand.

Baroreceptor reflex function has been measured in 13 astronauts during space missions. There is strong evidence that spaceflight impairs reflex circulatory control. Baroreceptor function has also been studied in astronauts before and after Space Shuttle missions, and in healthy people on earth subjected to prolonged (in one experiment, 30 days) 6° head-down bed rest. This research employed pressure changes delivered to a neck chamber worn by subjects. Human arterial baroreceptors are located primarily at the bifurcation of the common carotid artery in the neck and the thoracic aorta. Suction applied to the neck lowers tissue pressure outside the carotid artery. When this happens, the pressure inside the artery is suddenly pushing against a vacuum, and the artery expands. Carotid artery expansion increases firing rates of baroreceptor nerves in the wall of the carotid artery, increases vagus nerve traffic to the heart and slows the heart rate, and reduces sympathetic nervous traffic and relaxes blood vessels. Positive pressure applied to the neck exerts opposite effects. Figure 4 shows an astronaut studying her own baroreflex function in space, during the Spacelab Life Sciences-1 Space Shuttle mission in 1991. The inset shows a typical sigmoid baroreflex relation, telemetered from space.

Results of experiments employing neck chambers show that both spaceflight and simulated weightlessness with head-down bed rest lead to impairment of baroreflex function. In one bed rest study, Victor A. Convertino and his coworkers at the Kennedy and Ames Space Centers showed that volunteers who developed the greatest impairment of baroreflex function during bed rest were most likely to faint when they stood erect at the end of the bed rest period. Scientists at the Medical College of Virginia in Richmond and Johnson Space Center have found that the baroreflex function of astronauts is impaired during and after brief Space Shuttle missions. It is not yet clear what contribution the baroreflex malfunction that develops during weightlessness makes to postmission orthostatic hypotension.

VIII. EXERCISE

During comparable levels of exercise in space (if such can be achieved without forces of gravity), heart rate and blood pressure are higher and stroke volume is lower than on earth. Although stroke volume declines, cardiac output (stroke volume × heart rate) may remain normal by virtue of more rapid heart rates. Contraction of the left ventricle, as judged by echocardiography, actually may be increased from preflight levels. All of this suggests that changes measured in space are probably due to blood volume changes, and that cardiac muscle function is normal.

In space, oxygen uptake is slightly less during and slightly more after exercise than before flight. Increases of oxygen uptake after exercise suggest that, in space, astronauts rely more on anaerobic mechanisms than they do on earth. Oxygen uptake during maximal exercise was measured in six astronauts during recent space missions and found to be normal. This stands in contrast with the modest postflight reductions of maximal oxygen consumption that have been reported.

Work performed outside the spacecraft (extravehicular activity) constitutes a special case. This kind of work involves primarily arm and upper body muscle groups and may be exceedingly stressful for astro-

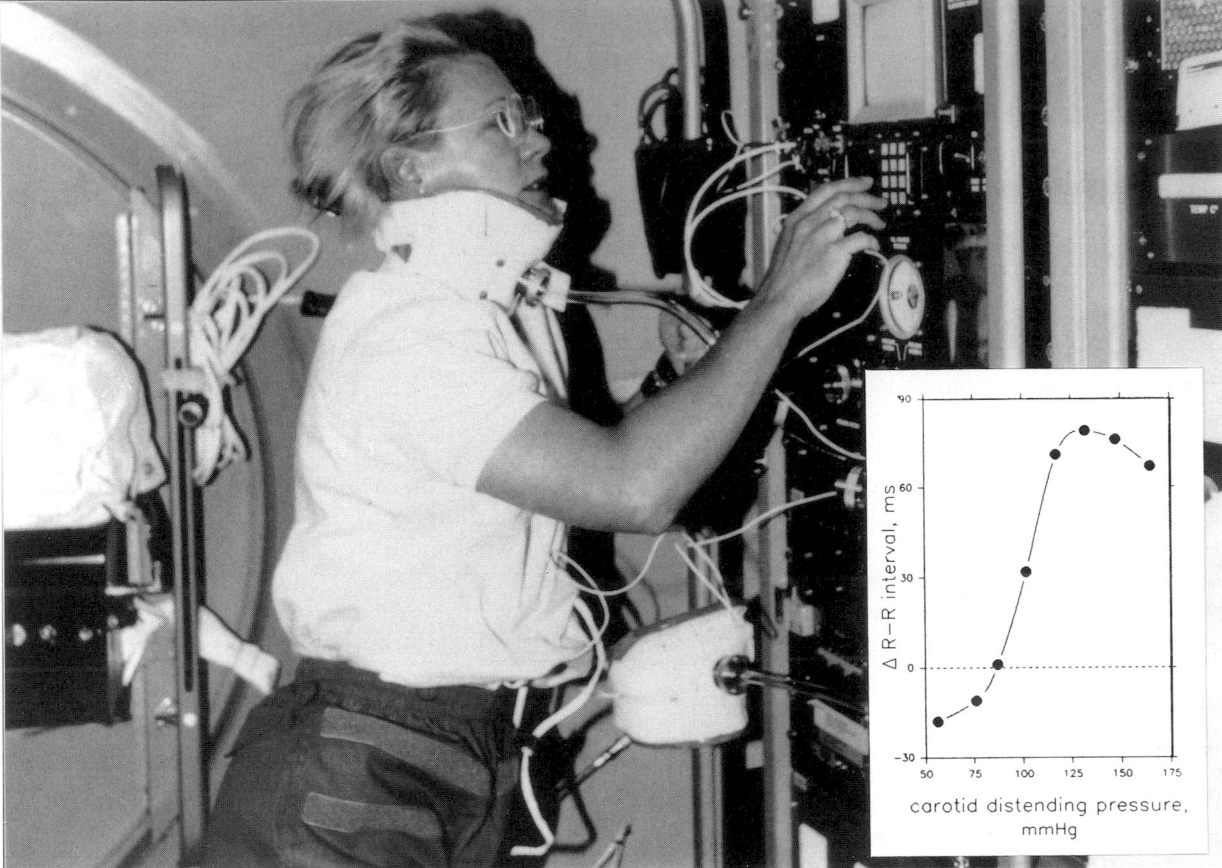

FIGURE 4 Neck chamber used to study baroreflex function during the Spacelab Life Sciences Space Shuttle mission in 1991. The inset shows a typical sigmoid carotid baroreceptor–cardiac reflex relation, telemetered from space. [Reproduced with permission from D. L. Eckberg and J. M. Fritsch. (1993). How should human baroreflexes be tested? *News Physiol. Sci.* 8, 7–12.]

nauts; they may have higher heart and respiratory rates and sweat more, and their body heat production may exceed the capacity of space suit cooling systems to dissipate heat.

Exercise capacity and oxygen consumption clearly are reduced after exposure to weightlessness. Maximal oxygen consumption has not been measured after spaceflight. Reduced aerobic capacity is likey to be multifactorial. First, antigravity muscles atrophy from disuse; this has been documented directly by microscopic examination of muscles of three Soviet astronauts who died tragically because of spacecraft decompression during reentry, and in American astronauts indirectly by postflight magnetic resonance images. Second, muscle efficiency is reduced. Third, blood volume and red blood cell mass, and consequently oxygen carrying capacity, are smaller after spaceflight than before. Fourth, left ventricular stroke volume during exercise is reduced by about 20%; however, because heart rate during exercise is greater, cardiac output is reduced by only about 10%. Diastolic blood pressure is normal, but systolic pressure is elevated. The combination of increased heart rate and systolic pressure indicates that left ventricular myocardial oxygen consumption is increased during exercise after spaceflight. [*See* Exercise and Cardiovascular Function.]

IX. COUNTERMEASURES

For many years, "countermeasures" have been used in space to try to prevent or ameliorate postweightlessness cardiovascular problems. Many American and Russian astronauts take oral salt tablets and drink fluids before reentry. Strong evidence from postflight stand tests indicates that this countermeasure does not prevent postflight orthostatic hypotension. Many crew also perform bicycle or other aerobic exercise during missions. Although some evidence suggests that in-flight exercise prevents postflight cardiovascular abnormalities, this has not been proven. Russian scientists simulate the footward blood volume shift that occurs with standing with a special suit ("Chibis") that applies lower body suction in space during usual activities.

X. SUMMARY

Weightlessness of even brief duration leads to postflight cardiovascular abnormalities, including reduced exercise tolerance and inordinate heart rate speeding and blood pressure reduction with standing. In most astronauts, these changes constitute a nuisance—lightheadedness forces them to sit down. In a few astronauts, cardiovascular changes are disabling—they cannot stand without losing consciousness. All of these changes disappear spontaneously within days to weeks after return to earth, and appear to leave no residual abnormalities.

Data on human cardiovascular function during and after weightlessness must be pieced together from disparate sources. In many instances, observations are too few to permit proper determination of statistical significance. Nevertheless, extant data suggest that the following changes occur when humans are exposed to a weightlessness environment: early during space missions, body fluids shift toward the head, stretch cardiac chambers, provoke release of diuretic hormones, and cause diuresis and blood volume contraction; later during space missions, heart size is reduced to below prelaunch dimensions. In poorly understood ways, as a result of chronic blood volume reductions, cardiovascular reflexes reset and alter the usual relation between blood volume and pressure and autonomic outflow that obtains on earth. After space missions, blood volume deficiency and probably impaired baroreceptor reflex function lead to the cardiovascular changes with standing described earlier.

Certain critical questions have not been answered: If reduced blood volume and pressure lead to reflex heart rate speeding and increases of blood pressure, what maintains heart rate and blood pressure at supranormal levels after blood pressure has returned to normal? Why do heart rate and blood pressure rise above preflight levels? Ambitious American, European, and Russian plans for future human physiological research during space missions are likely to provide answers to these challenging questions and, probably, to questions that have not yet been framed.

BIBLIOGRAPHY

Berry, C. A. (1974). Medical legacy of *Apollo*. *Aerospace Med.* **45**,1046–1057.

Berry, C. A., and Catterson, A. D. (1967). Pre-*Gemini* medical predictions versus *Gemini* flight results. *In* "*Gemini* Summary Conference," pp. 197–218. National Aeronautics and Space Administration, Washington, D.C.

Blomqvist, C. G., and Stone, H. L. (1983). Cardiovascular adjustments to gravitational stress. *In* "Handbook of Physiology: The Cardiovascular System, Peripheral Circulation and Organ Blood Flow" (J. T. Shepherd and F. M. Abboud, eds.), Section 2, Part 2, pp. 1025–1063. American Physiological Society, Bethesda, MD.

Bungo, M. W. (1989). The cardiopulmonary system. *In* "Space Physiology and Medicine" (A. E. Nicogossian, C. L. Huntoon, and S. L. Pool, eds.), pp. 179–201. Lea & Febiger, Philadelphia.

Bungo, M. W. (1990). Echocardiographic changes related to space travel. *In* "Two-Dimensional Echocardiography and Cardiac Doppler" (J. N. Schapira and J. G. Harold, eds.), 2nd Ed., pp. 524–529. Williams & Wilkins, Baltimore.

Catterson, A. D., McCutcheon, E. P., Minners, H. A., and Pollard, R. A. (1963). Aeromedical observations. *In* "*Mercury* Project Summary Including Results of the Fourth Manned Orbital Flight May 15 and 16, 1963," pp. 299–326. National Aeronautics and Space Administration, Washington, D.C.

Convertino, V. A. (1990). Physiological adaptations to weightlessness: Effects on exercise and work performance. 1990. *Exercise Sport Sci. Rev.* **18**, 119–166.

Convertino, V. A., Doerr, D. F., Eckberg, D. L., Fritsch, J. M., and Vernikos-Danellis, J. (1990). Head-down bedrest impairs vagal baroreflex responses and provokes orthostatic hypotension. *J. Appl. Physiol.* **68**, 1458–1464.

Fritsch, J. M., Charles, J. B., Bennett, B. S., Jones, M. M., and Eckberg, D. L. (1992). Short-duration spaceflight impairs human carotid baroreceptor–cardiac reflex responses. *J. Appl. Physiol.* **73**, 664–671.

Fritsch-Yelle, J. M., Charles, J. B., Jones, M. M., Beightol, L. A., and Eckberg, D. L. (1994). Spaceflight alters autonomic regulation of arterial pressure in humans. *J. Appl. Physiol.* **77**, 1776–1783.

Hoffler, G. W., Wolthuis, R. A., and Johnson, R. L. (1974). *Apollo* space crew cardiovascular evaluations. *Aerospace Med.* **45**, 807–820.

Johnson, P. C., Driscoll, T. B., and LeBlanc, A. D. (1977). Blood volume changes. *In* "Biomedical Results from *Skylab*" (R. S. Johnston and L. F. Dietlein, eds.), pp. 235–241. National Aeronautics and Space Administration, Washington, D.C.

Levy, M. N., and Talbot, J. M. (1983). Cardiovascular deconditioning of space flight. *Physiologist* **26**, 297–303.

Norsk, P., Drummer, C., Röcker, L., Strollo, F., Christensen, N. J., Warberg, J., Bie, P., Stadeager, C., Johansen, L. B., Heer, M., Gunga, H.-C., and Gerzer, R. (1995). Renal and endocrine responses in humans to isotonic saline infusion during microgravity. *J. Appl. Physiol.* **78**, 2253–2259.

Sandler, H. (1980). Effects of bedrest and weightlessness on the heart. *In* "Hearts and Heart-like Organs" (G. H. Bourne, ed.), Vol. 2, pp. 435–524. Academic Press, New York.

Cardiovascular System, Anatomy

ANTHONY J. GAUDIN
California State University, Northridge

GLOSSARY

Endothelium A thin membrane of flattened cells that lines the cavity of an organ or blood vessel

Pulmonary Pertaining to the lungs and external respiration

Semilunar Shaped like a half-moon

Serous A thin cellular membrane that lacks glands and secretes a watery product similar to tissue fluid

Systemic Pertaining to the entire body

THE CARDIOVASCULAR SYSTEM CONSISTS OF THE heart and blood vessels. The heart is a muscular pump that moves blood through the vessels, which carry the blood to the tissues and return it to the heart.

I. ANATOMY OF THE HEART

The heart lies within the thoracic cavity, just below the sternum. It is cone shaped, with the apex pointing down and to the left edge of the diaphragm (see Color Plate 1). Large blood vessels join the heart at the base of the cone, which is just behind the sternum. The adult heart is usually about the size of a clenched fist. Its dimensions average 13 cm long, 9 cm at the base, and about 6 cm at its thickest anterior–posterior dimension. It weighs about 350 g. These figures vary from person to person, but in general the heart is slightly larger in men than in women. Disease, a prolonged program of physical exercise, or pregnancy can cause an increase in the size of the heart.

A. Pericardium

The heart is surrounded by a double-layered membrane, the pericardium, or pericardial sac. The inner layer of the sac, the visceral pericardium, or epicardium, is a thin serous (i.e., secretory) membrane closely attached to the surface of the heart. (It is also considered to be the outermost of the three tissue layers of the heart, described in Section I,B.) The outer layer, the parietal pericardium, consists of two subdivisions—a thin serous pericardium and a thick fibrous pericardium—that provides strength. The pericardial layers are separated by the pericardial cavity. This thin space contains a lubricant called pericardial fluid that is secreted by the serous membranes. Thus, the two membranes slide over the heart during heartbeats, with a minimum of friction. Sometimes these membranes become infected or irritated and inflamed, resulting in pericarditis.

B. Muscular Walls of the Heart

The wall of the heart has three layers of tissue: an internal endocardium, a middle myocardium, and the external epicardium, which is the same as the inner pericardial layer just described (Fig. 1). [*See* Cardiac Muscle.]

The endocardium is a thin layer that lines the interior of the heart, covers all internal structures (such as valves), and extends to the blood vessels that attach to the heart. It consists of two layers, including a thin endothelial layer that forms the free surface of the interior of the heart and an underlying fibrous layer that supports the surface layer. The fibrous layer contains blood vessels, elastic and strengthening fibers, and specialized muscle cells (Purkinje fibers) that transmit nerve impulses to coordinate contractions of the heart.

The myocardium, the thickest of the layers, is composed entirely of muscle tissue. The myocardial layer

ENCYCLOPEDIA OF HUMAN BIOLOGY, Second Edition, VOLUME 2. Copyright © 1997 by Academic Press. All rights of reproduction in any form reserved.

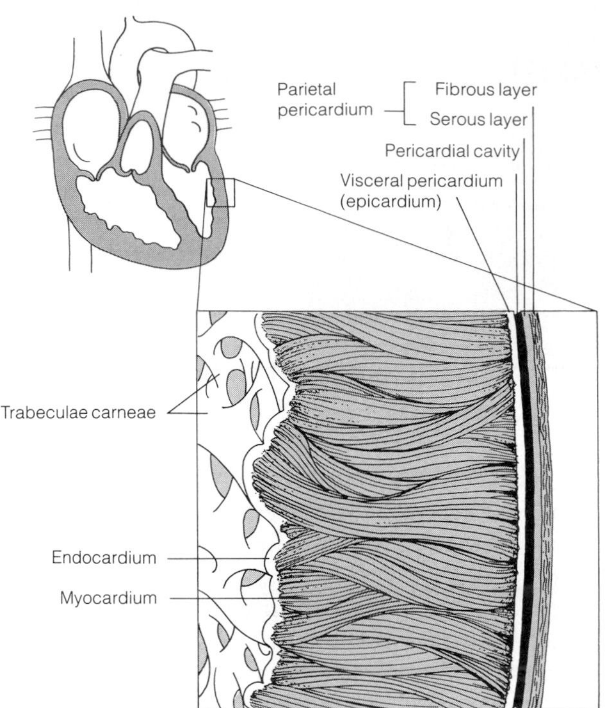

FIGURE 1 The pericardium. [Source: Gaudin, A. J., and Jones, K. C. (1989). "Human Anatomy and Physiology." Harcourt Brace Jovanovich, San Diego, p. 561. Reproduced with permission.]

of the ventricles (Fig. 1) is much thicker than that of the atria; its contraction has sufficient force to propel blood through the lungs and the body. In the ventricles, the muscle fibers are arranged in spiraling bands rather than extending straight from the base to the apex, as in the atria. This arrangement produces contractions that squeeze the blood from the ventricles. The myocardial contractions pump the blood from the ventricles through the circulatory system.

The external surface of the myocardium is relatively smooth, but the internal surface in the ventricles is irregular. In these areas are many folds, columns, and ridges, called trabeculae carneae (i.e., fleshy sticks), and conical projections, called papillary muscles. The trabeculae carneae provide additional strength to the myocardium, while adding a minimum of weight. Papillary muscles prevent the backward flow of blood through heart valves. (Both are discussed in detail in Section I,C,2.)

C. Chambers and Valves of the Heart

The heart has four chambers: right and left atria specialized to receive blood from veins, and right and left ventricles specialized to pump blood into the arteries (see Color Plate 2). The atria have thinner walls than the ventricles and lie at the base of the heart, whereas the ventricles make up the apex and bulk of the heart. The atria and ventricles are separated by muscular walls referred to as septa. The interatrial septum lies between the left and right atria, whereas the interventricular septum separates the two ventricles. In addition to muscle, the septa also contain the fibrous skeleton of the heart, an internal network of fibrous connective tissue (see Color Plate 3). This skeleton has two functions: it provides a place of attachment for cardiac muscle, and it helps support the valves that separate the atria from the ventricles. These valves are the atrioventricular valves and are attached to a portion of the fibrous skeleton known as the coronary trigone. The fibrous tissue of this structure provides support for the valves as they operate during normal cardiac movements while pumping blood.

The chambers in the two sides of the heart form a natural division of the twofold function of the circulatory system. Blood entering the right side of the heart has traveled through the body, where its oxygen supply has been lowered and its carbon dioxide level has been raised. This blood is pumped through the pulmonary circulation of the heart to the lungs. Here, the blood exchanges carbon dioxide for oxygen. Pulmonary circulation restores a high concentration of blood oxygen and lowers the concentration of carbon dioxide. This blood moves from the lungs and enters the left atrium of the heart. The left side of the heart pumps blood through the systemic circulation, the vast system of vessels that carries blood to all tissues of the body. [*See* Cardiovascular System, Physiology and Biochemistry.]

1. Right Atrium

Blood from the body first enters the heart in the right atrium from three major veins: the superior vena cava, inferior vena cava, and coronary sinus (see Color Plate 2). The superior vena cava receives blood from the upper part of the body, whereas the inferior vena cava drains all tissues below the diaphragm. The coronary sinus receives blood from the muscles of the heart itself. The opening of the coronary sinus is an enlarged vein on the posterior surface of the atrium that collects blood coming from the veins of the cardiac walls.

The auricles are ear-shaped extensions of both atria which increase their volume under certain circumstances. During resting conditions, an atrium does not open to its full capacity, expanding only enough to hold the moderate amount of blood entering. During physical activity and exercise, however, the amount of blood entering the heart may double or triple, and at these times the auricles function as a volume reservoir, expanding to their full extent. The auricles thus increase the volume of both atria and the amount of blood they can accommodate. The interior wall of the right atrium is relatively smooth, whereas the auricle contains numerous ridges of cardiac muscle, called the musculi pectinati (i.e., comb-like muscles). Musculi pectinati function in much the same way as trabeculae carneae, providing the heart with additional strength with a minimum of weight.

The opening between the superior vena cava and the right atrium is unobstructed, but a flap-like extension from the auricle wall at the base of the inferior vena cava appears to be the remnant of an important functional structure in the fetus. This flap-like structure directs blood through an opening in the fetal heart called the foramen ovale (i.e., the oval hole). This structure has no apparent adult function. In the adult heart, the fossa ovalis, an oval depression, is visible in the interatrial septum. This is a remnant of the former position of the fetal foramen ovale.

2. Right Ventricle

Blood then passes from the right atrium into the right ventricle. The right ventricle is a triangular chamber occupying the right apex of the heart and resting on the diaphragm. The right atrioventricular (tricuspid) valve is a specialized structure that separates the right atrium and ventricle (see Color Plate 4) and consists of two portions: cusps and cords. The three triangular cone-shaped cusps extend across the opening between the chambers. These flexible cusps have their peripheral edges attached to the fibrous skeleton, and the central edges project into the cavity of the right ventricle. The interior surface of the ventricle is not smooth and flat. Instead, it consists of trabeculae carneae and several finger-like papillary muscles that project into the cavity of the ventricle (Fig. 1 and Color Plate 4). The papillary muscles are attached to the free edges of the cusps of the atrioventricular valve by elongate tendinous cords called chordae tendineae.

As the right atrium fills with blood, it contracts, and the blood enters the right ventricle. The force of blood passing through the tricuspid valve forces the cusps down into the right ventricle, providing a clear pathway into the chamber. After the right ventricle fills with blood, it contracts, squeezing the blood out of the chamber toward the lungs. This action also puts pressure on the undersides of the three atrioventricular cusps and causes them to close by pushing them upward. The simultaneous contraction of the papillary muscles pulls on the chordae tendineae, which prevents the cusps from prolapsing into the right atrium. As a result, the opening is sealed and blood does not leak back into the right atrium.

Blood leaves the right ventricle through the pulmonary trunk. The base of this large artery lies in the upper medial front region of the right ventricle. Trabeculae carneae are absent here. The entrance to this artery contains a three-part pulmonary semilunar valve composed of pocket-like folds (cusps) of endothelium (see Color Plates 4 and 6). During ventricular contraction, blood forces the cusps open, but they close rapidly when the ventricle relaxes. Closure of the semilunar valve prevents backward blood flow.

3. Left Atrium

The left atrium resembles the right atrium, but its myocardial layer is slightly thicker. Its auricle projects anteriorly and superiorly, and partially covers the pulmonary trunk. Blood from four pulmonary veins enters the right atrium. Their openings are unobstructed with valves, and relaxation of the atrial walls allows the chamber to fill with oxygenated blood. The musculi pectinati and the left side of the fossa ovalis are also in the left atrium. Contraction of the left atrium sends blood through the atrioventricular opening into the left ventricle (see Color Plate 5).

4. Left Ventricle

The left ventricle occupies the left apical position of the heart. Its myocardium is approximately three times as thick as that of the right ventricle. This chamber propels blood throughout the entire body when it contracts. Its lining resembles that of the right ventricle, with numerous trabeculae carneae and several papillary muscles.

The separation between the left atrium and ventricle is maintained by the left atrioventricular valve, also called the mitral, or bicuspid, valve. It has two, instead of three, cusps connected through numerous chordae tendineae to papillary muscles that project from the wall of the ventricle. The bicuspid valve functions in precisely the same way as the tricuspid valve: Both prevent backflow of blood.

Blood leaves the left ventricle through the aorta, the largest blood vessel in the body. The base of the aorta lies at the superior medial edge of the left ventricle. Its orifice is closed by an aortic semilunar valve (see Color Plate 6), similar in function and structure to the pulmonary semilunar valve. It prevents the backward flow of blood from the aorta.

D. Cardiac Intrinsic Blood Supply

An adult heart contracts approximately 75 times each minute, every day of our lives. This effort results in an extensive amount of work by cardiac muscle cells, and they require a constant supply of oxygen and nutrients. They also produce large amounts of waste materials that must be removed. The myocardium has "first call" on the blood leaving the left ventricle. The right and left coronary arteries originate just above the cusps of the aortic semilunar valve (see Color Plates 2 and 6). They form the coronary circulation, an intrinsic supply of freshly oxygenated blood to the heart. The left and right vessels mainly supply corresponding sides of the heart, but extensive branching, along with numerous interconnections, called anastomoses, between their branches, results in a complex circulatory pattern as the arteries extend from the base of the ventricles to the cardiac apex. One advantage of extensive branching and anastomoses is that blood arriving at any spot in the myocardium may originate in any one of several large branches. Alternative sources of blood for a specific area are called collateral circulation, and if blockage occurs in one branch leading to a specific area, increased flow through collateral circulation may be sufficient to satisfy the needs of the myocardial cells in that area.

Venous blood in the myocardium is collected by several cardiac veins that anastomose extensively with each other. Almost all of the blood that passes through the myocardium empties into the coronary sinus, an enlarged vein that drains into the right atrium. Only a few small veins empty into the right atrium or into the ventricles directly.

Coronary circulation is essential to the function of cells in the heart. Blood inside the chambers does not supply heart muscle directly; this function is provided instead by the coronary circulation. A partial blockage of one coronary artery can result in a decrease in the critical blood supply to the myocardium distal to the blockage and, when severe enough, may cause severe chest pain, known as angina pectoris. Cessation of blood flow consequent to complete blockage may result in a myocardial infarction, commonly called a heart attack. An infarction causes the death of the part of the heart cut off from nutrients and oxygen, where the dead muscle is then replaced with fibrous connective tissue. Because this scar tissue lacks the ability to contract, the heart loses some of its function. Massive infarctions affecting large areas of cardiac tissue can result in death.

II. BLOOD VESSELS

The groups of blood vessels are arteries, veins, and capillaries (see Color Plate 7). Arteries conduct large quantities of blood from the heart. Distal to the heart, they branch into smaller arteries, called arterioles. These branch into still smaller vessels, which eventually become microscopic capillaries. Capillaries connect arteries and veins. They distribute blood to the tissues and bring it into proximity with the cells. Capillaries form dense networks in practically all body tissues. At the efferent end of a network, the capillaries fuse into slightly larger vessels, called venules, which in turn fuse into veins. Veins collect blood from the tissues and return it to the heart. For the most part, blood is confined within the blood vessels, making the human cardiovascular system a "closed" circulatory system. Some of the blood's fluid component does escape from the capillaries and bathes the body tissues in their immediate vicinity. Most of this fluid reenters capillaries in the same network and is returned to the veins. The rest is collected by the lymphatic system, which in turn returns it to veins.

A. Arteries

Arteries are blood vessels with thick walls (see Color Plate 7), ranging from about 2.5 cm to approximately 0.5 mm in diameter. The wall of an artery has layers, or tunicae. The innermost layer, the tunica intima, lines the cavity or lumen of the artery. Also called the tunica interna, it is composed of a thin layer of cells (i.e., the endothelium), whose free border is in contact with the blood. The tunica intima is continuous with the cardiac endocardium and extends throughout the cardiovascular system as a lining only one cell layer thick. The tunica media forms a cylinder of smooth muscle and elastic connective tissue external to the tunica intima. The tunica externa is nearly as thick as the tunica media, but consists of fibrous connective tissue, with only small numbers of smooth muscle fibers.

An arteriole is arbitrarily defined as any artery less than 0.5 mm in diameter.

B. Capillaries

The capillaries are the smallest of blood vessels (see Color Plate 7). They are microscopic in size, approximately 0.01 mm in diameter, nearly the diameter of a red blood cell. Some capillaries are smaller than the red blood cells they carry, forcing the cells to bend into a "C" shape as they squeeze through the vessel. This increases the surface of the blood cell in contact with the wall of the capillary. The wall of a capillary consists of a single endothelial layer of cells resembling the pieces of a jigsaw puzzle. Capillaries are usually between 0.5 and 1 mm long. In active tissues, such as skeletal muscles, liver, lungs, nervous system, and kidneys, capillaries are so numerous that few tissue cells are more than one or two cells away from a capillary. In relatively inactive tissues, such as tendons and ligaments, capillaries are not nearly so numerous. Capillaries are absent from certain tissues, including cartilage, the cornea of the eye, the epidermis of the skin, and other epithelial (i.e., covering and lining) tissues.

The capillaries provide the only place where materials can enter and leave an unruptured vessel. Some materials move through the wall of the capillary by means of diffusion or active transport. Serum also can leave the capillaries under hydrostatic pressure, through junctions between the endothelial cells. Certain types of white blood cells also possess the ability to squeeze between the lining cells, leave the capillary, and wander through body tissues. This ability is called diapedesis and usually belongs to white blood cells that engulf and destroy dead body cells and microbial invaders.

C. Veins

The smallest veins, called venules, are formed when several capillaries joint together. They are similar to capillaries, but slightly larger in diameter. Farther from the capillary bed, a thin tunica media, consisting of only a few muscle fibers and some fibrous connective tissue, appears in the venule wall. At the point at which the tunica externa appears, the venules become veins.

Veins conduct blood from the venules back to the heart. The smallest veins have a thin wall consisting of the three tunicae—interna, media, and externa—found in arteries (see Color Plate 7). The major structural difference between arteries and veins, however, lies in the relative thickness of the tunicae. In a vein, the tunica media and tunica externa are much thinner than in an artery, and, consequently, the walls of veins tend to bulge more easily under pressure than those of arteries.

Another distinctive feature of veins is the presence of valves, particularly in the veins of the arms and legs (Fig. 2). These valves are folds of the tunica intima, resembling the semilunar valves of the aorta and the pulmonary artery. As shown in Fig. 2, the valves allow

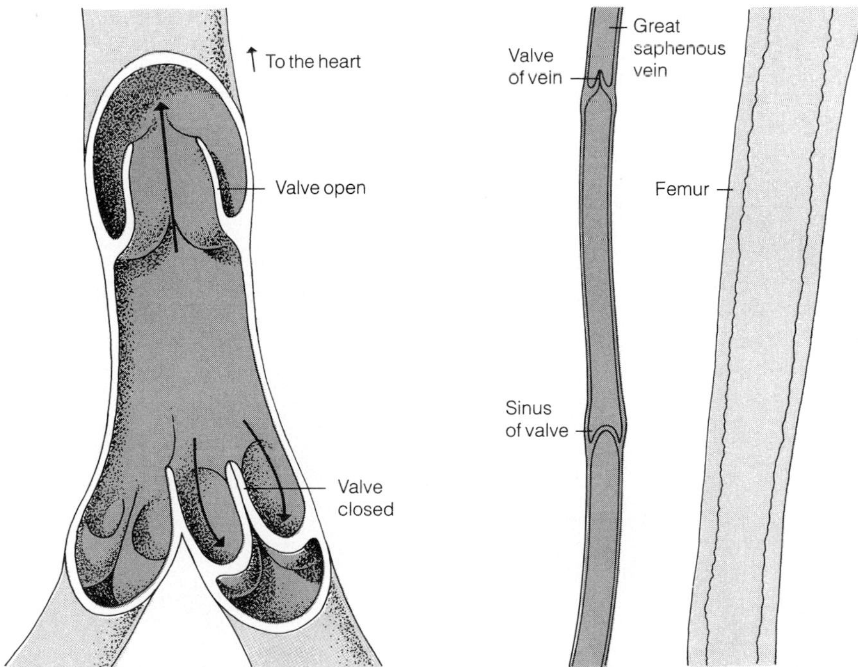

FIGURE 2 Valves in a vein. [Source: Gaudin, A. J., and Jones, K. C. (1989). "Human Anatomy and Physiology." Harcourt Brace Jovanovich, San Diego, p. 587. Reproduced with permission.]

the blood to flow in one direction only: toward the heart. A decrease in pressure in the vein on the side of the valve distal to the heart causes the valve to fill and close. On a warm summer day, when the veins just beneath the skin are dilated, their valves may stand out as small lumps on the legs.

III. ANATOMY OF THE CIRCULATORY PATHWAYS

The routes that supply and drain the lungs form the pulmonary circulation, whereas those that supply and drain other tissues are the systemic circulation.

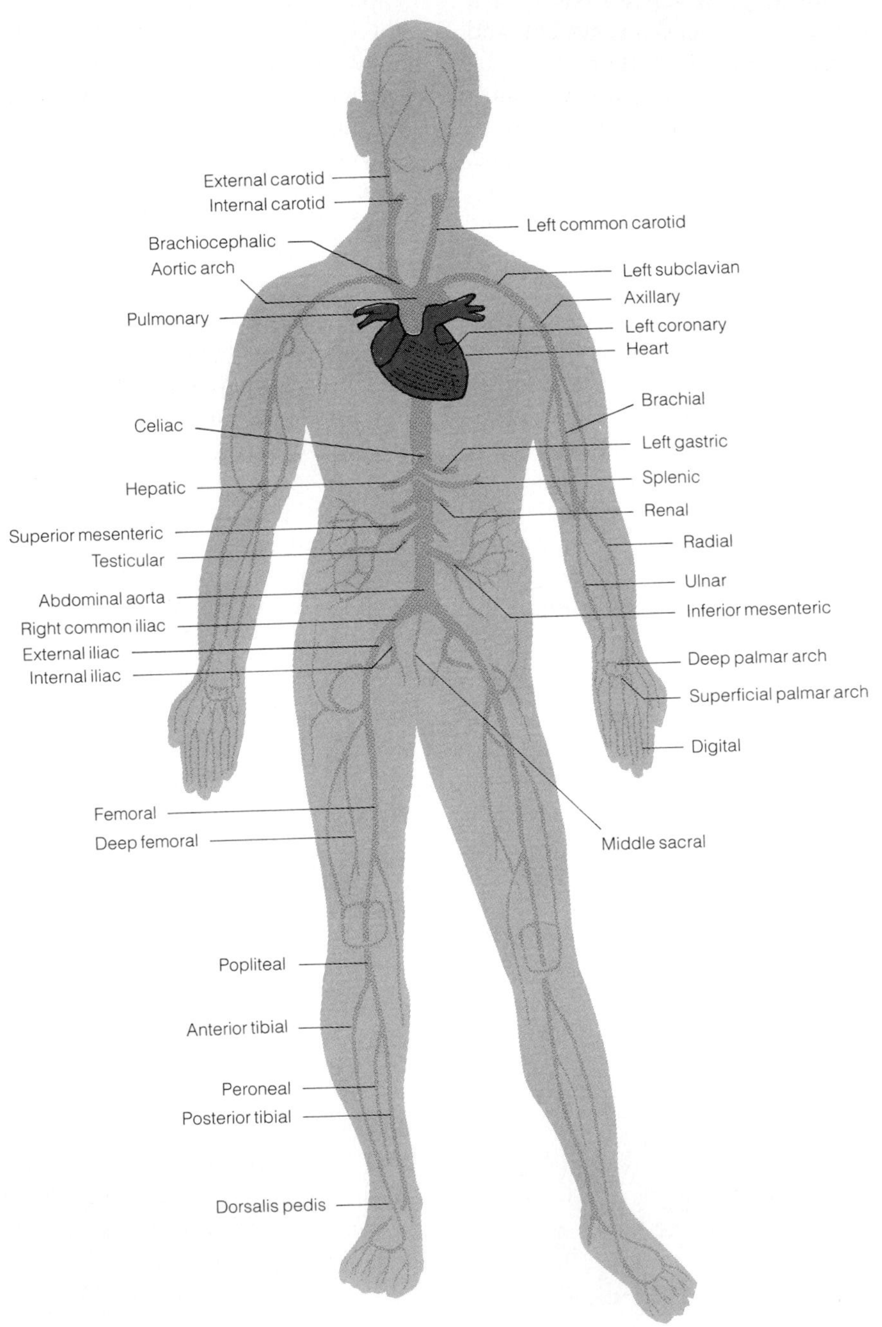

FIGURE 3 Systemic arteries. [Source: Gaudin, A. J., and Jones, K. C. (1989). "Human Anatomy and Physiology." Harcourt Brace Jovanovich, San Diego, p. 590. Reproduced with permission.]

A. Pulmonary Circulation

Blood in the right ventricle of the heart has come from the tissues. It has surrendered to them much of its oxygen and taken from them carbon dioxide produced by the metabolic activities of the cells. Contraction of the right ventricle sends this blood past the pulmonary semilunar valve into the pulmonary trunk (see Color Plate 8). The short pulmonary trunk branches into right and left pulmonary arteries that lead toward the lungs. Shortly before entering the lungs, each pulmonary artery branches into lobar arteries leading to lobes of the lungs, two on the left and three on the right. The lobar arteries undergo additional branching into smaller arteries and arterioles until they finally branch into capillaries closely associated with the alveoli of the lungs. Their close proximity to the lumina of the alveoli facilitates diffusion of carbon dioxide from the blood and into the alveoli and oxygen from the lungs into the capillaries. [*See* Respiratory System, Anatomy.]

After passing through the capillary beds of the lungs, the blood collects in venules, which fuse to form numerous veins. These continue to fuse, eventually forming four large pulmonary veins, two from each lung, which lead to the left atrium. Note that the pulmonary arteries carry deoxygenated blood, while the pulmonary veins carry oxygenated blood. This situation is the reverse of that present in all the other arteries and veins in the adult body.

The pulmonary circuit is responsible for transporting blood from the heart for the purpose of replenishing its oxygen supply and removing its carbon dioxide. This blood is not used to nourish the tissues of the lungs or respiratory passages, nor is it used as a source of oxygen by these tissues. For these purposes, the lungs have their own set of systemic arteries.

B. Systemic Circulation

Systemic circulation is that part of the vascular system that carries blood from the left ventricle to the tissues and returns it to the right atrium. This blood supplies oxygen and nutrients to the body, while simultaneously removing waste materials and, in some cases, important cellular products, such as hormones. It is formed by the systemic arteries and systemic veins.

1. Systemic Arteries

The aorta is the major artery of the systemic circuit. It begins at the superior medial border of the left ventricle, proceeds upward for a short distance, then executes a sharp U-turn (see Color Plate 8) and extends down the length of the thorax and abdomen until it reaches the pelvic girdle, where it splits into two branches (Fig. 3).

Table I lists the major branches of the aorta illustrated in Fig. 3, along with the organs they supply.

TABLE I
Major Systemic Arteries

Artery	Body areas supplied
Axillary	Shoulder and axilla
Brachial	Upper arm
Brachiocephalic	Head, neck, and arm
Celiac	Divides into left gastric, splenic, and hepatic arteries
Common carotid	Neck
Common iliac	Divides into external and internal iliac arteries
Coronary	Heart
Deep femoral	Thigh
Digital	Fingers
Dorsalis pedis	Foot
External carotid	Neck and external head regions
External iliac	Femoral artery
Femoral	Thigh
Gastric	Stomach
Hepatic	Liver, gallbladder, pancreas, and duodenum
Inferior mesenteric	Descending colon, rectum, and pelvic wall
Internal carotid	Neck and internal head regions
Internal iliac	Rectum, urinary bladder, external genitalia, buttocks muscles, uterus, and vagina
Left gastric	Esophagus and stomach
Middle sacral	Sacrum
Ovarian	Ovaries
Palmar arch	Hand
Peroneal	Calf
Popliteal	Knee
Posterior tibial	Calf
Pulmonary	Lungs
Radial	Forearm
Renal	Kidney
Splenic	Stomach, pancreas, and spleen
Subclavian	Shoulder
Superior mesenteric	Pancreas, small intestine, ascending and transverse colon
Testicular	Testes
Ulnar	Forearm

2. Systemic Veins

The systemic veins collect blood from the tissues and return it to the right atrium (Fig. 4). With few exceptions, capillary blood is collected in venules that aggregate into larger veins. In some places, venous blood collects in sinuses, as in the brain (dural sinuses), in the heart (coronary sinus), and in the liver, spleen, and adrenal glands (where they are called sinusoids). Sinuses and sinusoids are lined with endothelial cells that are continuous with those of capillaries and veins, but lack the muscular tunica media of veins. Instead, their walls are chiefly fibrous connective tissue. Even-

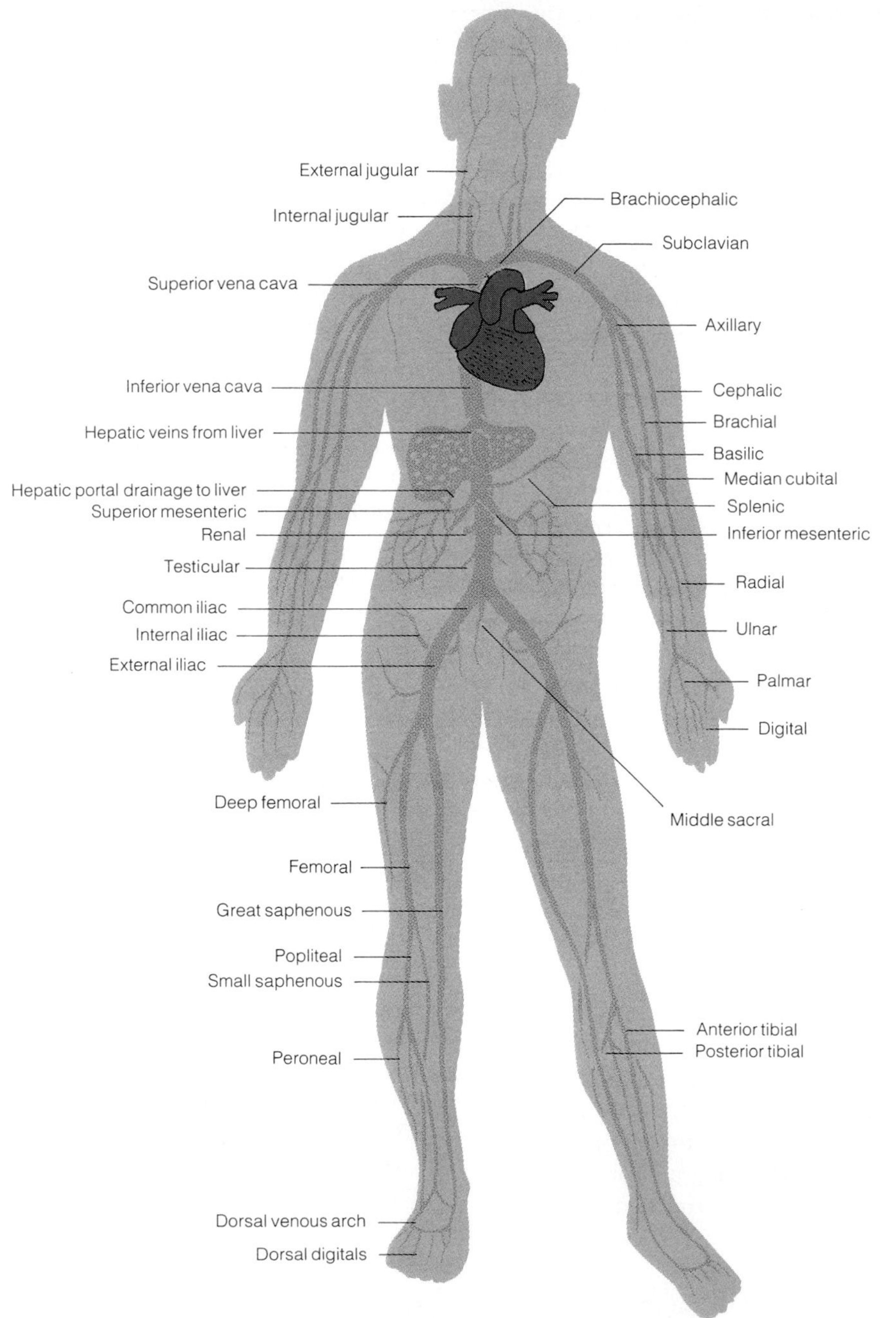

FIGURE 4 Systemic veins. [Source: Gaudin, A. J., and Jones, K. C. (1989). "Human Anatomy and Physiology." Harcourt Brace Jovanovich, San Diego, p. 598. Reproduced with permission.]

tually, all veins drain into the superior vena cava, inferior vena cava, or coronary sinus, which in turn drain into the right atrium.

Superficial (i.e., cutaneous) veins lie just beneath the skin in superficial fascia. They receive blood from superficial tissues, yet freely communicate with deeper veins. Superficial veins are usually visible on the surface of the body, especially in the arms and legs. Deep veins tend to parallel arteries and usually bear the same names.

Table II lists the major veins illustrated in Fig. 4, along with the organs they drain.

TABLE II
Major Systemic Veins

Vein	Body areas drained
Anterior tibial	Shin
Axillary	Shoulder and axilla
Basilic	Superficial regions of upper arm
Brachial	Deep regions of upper arm
Brachiocephalic	Head, neck, and arm
Cephalic	Superficial regions of upper arm
Common iliac	External and internal iliac veins
Deep femoral	Thigh
External iliac	Leg
External jugular	Superficial regions of face, scalp, and neck
Femoral	Thigh and leg
Great saphenous	Thigh and leg
Hepatic	Liver
Inferior vena cava	Tissues and organs below diaphragm
Internal iliac	Urinary bladder and reproductive organs
Internal jugular	Brain and deep regions of face
Internal thoracic	Thoracic wall
Median cubital	Forearm
Ovarian	Ovary
Peroneal	Leg
Plantar arches	Feet
Popliteal	Knee and calf
Posterior tibial	Calf
Radial	Forearm
Renal	Kidney
Small saphenous	Thigh and leg
Subclavian	Shoulder and arm
Superior mesenteric	Pancreas, small intestine, ascending and transverse colon
Superior vena cava	Tissues and organs above diaphragm
Testicular	Testes
Ulnar	Forearm

C. The Hepatic Portal System

"Portal" systems are circulatory routes that begin and end in capillaries. In the hepatic portal system (see Color Plate 9), blood collected from the capillaries of the intestine, spleen, and pancreas flows into veins that reach the liver, where they branch into the sinusoid system of that organ. Two major veins contribute to the hepatic system, namely, the superior mesenteric vein, which collects blood from the small intestine, and the splenic vein, which drains the spleen and pancreas and, through the inferior mesenteric vein, the large intestine. These two veins merge into the large hepatic portal vein (not to be confused with the hepatic vein), which leads into the liver. The portal vein, after collecting several smaller veins that service other digestive organs, enters the liver and ramifies into smaller veins and venules, eventually terminating in sinusoids. Blood from the sinusoids collects in the hepatic veins and eventually empties into the inferior vena cava.

BIBLIOGRAPHY

Birnholz, J. C., and Farrell, E. E. (1984). Ultrasound images of human fetal development. *Am. Sci.* **72**, 608–613.

DeVries, W. C., and Joyce, L. D. (1983). The artificial heart. *Clin. Symp.* **35**, no. 2.

Klocke, F. J., and Ellis, A. K. (1980). Control of coronary blood flow. *Annu. Rev. Med.* **31**, 489–508.

McMinn, R. M. H., and Hutchings, R. T. (1988). "Color Atlas of Human Anatomy," 2nd Ed. New York Med. Publ., Chicago.

Melloni, J. L., *et al.* (1988). "Melloni's Illustrated Review of Human Anatomy." Lippincott, Philadelphia.

Rigotti, N. A., Thomas, G. S., and Leaf, A. (1983). Exercise and coronary heart disease. *Annu. Rev. Med.* **34**, 391–412.

Rushmer, R. F. (1976). "Structure and Function of the Cardiovascular System," 2nd Ed. Saunders, Philadelphia.

Cardiovascular System, Pharmacology

M. GABRIEL KHAN
University of Ottawa

GLOSSARY

Angina pectoris Chest pain caused by severe but temporary lack of blood and oxygen to a part of the heart muscle (see Fig. 1)

Atheroma/atherosclerosis Hardened plaque in the wall of an artery; the plaque is filled with cholesterol, calcium, and other substances. The plaque of atheroma hardens the artery, hence the term "atherosclerosis" (sclerosis-hardening) (see Fig. 2)

Coronary heart disease (CHD) See coronary thrombosis and ischemia (Fig. 1)

Coronary thrombosis Blood clot, thrombosis, in a coronary artery blocking flow to a part of the heart muscle; also called a heart attack or myocardial infarction

Hypercholesterolemia High blood cholesterol

Hypertrophy Enlargement caused by an increase in size of cells; the muscle cells of the left ventricle enlarge under the influence of extra work imposed by prolonged hypertension

Ischemia Temporary lack of blood and oxygen to an area of cells (e.g., heart muscle) usually caused by severe obstruction of the artery supplying blood to this area of cells. Thus the term "ischemic heart disease" is synonymous with coronary artery disease or CHD

Myocardial infarction (infarct) Death of an area of myocardium caused by blockage of a coronary artery by blood clot and atheroma; medical term for a heart attack (see Fig. 1)

Platelets Small disc-like particles that circulate in the blood and initiate the formation of blood clots. Platelets clump and form little plugs, which arrest bleeding

CORONARY HEART DISEASE (CHD) IS PREVALENT worldwide. The disease manifests itself as recurrent chest pain, heart attacks, sudden death, and heart failure (Fig. 1). Appropriate treatment of CHD entails the correction or modification of the underlying pathophysiology (Fig. 2). Hypertension is high blood pressure that affects more than 80 million North Americans, causing stroke, kidney failure, left ventricular hypertrophy, heart failure, and CHD. These complications can be prevented by treatment with antihypertensive agents.

Pharmacological agents used in the management of heart and vascular disease will be discussed with emphasis on their mode of action and pharmacokinetics, as well as their ability to prevent the disease process, ameliorate symptoms, and prevent death.

I. CHOLESTEROL-LOWERING DRUGS

Cholesterol reaches the blood from food eaten and from biosynthesis in the liver. Cholesterol is insoluble in blood and must attach to a carrier protein that transports it as a lipoprotein molecule. Most of the cholesterol is carried by a particle that has a low density and is thus termed *low-density lipoprotein* (LDL) cholesterol. LDL cholesterol is the main culprit in the formation of atheromatous plaques in arteries (Fig. 2). In addition, a small amount of cholesterol is carried by a *high-density lipoprotein* called HDL cholesterol. The latter is termed "good" cholesterol

ENCYCLOPEDIA OF HUMAN BIOLOGY, Second Edition, VOLUME 2. Copyright © 1997 by Academic Press. All rights of reproduction in any form reserved.

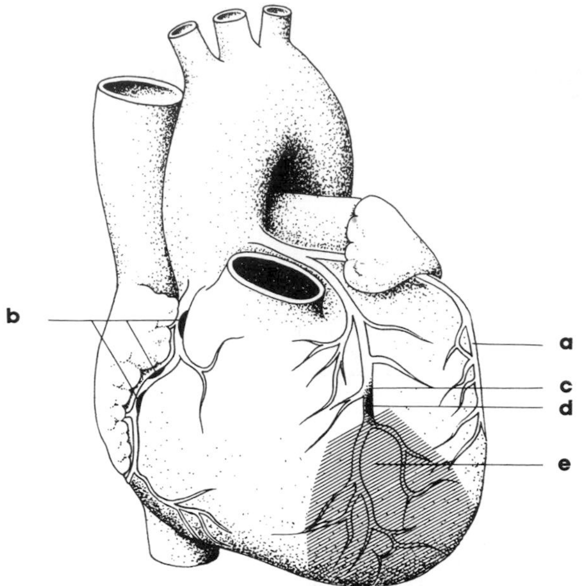

FIGURE 1 Coronary heart (artery) disease. (a) Normal coronary artery; (b) obstruction of a coronary artery by atherosclerosis, causing less blood to reach the heart muscle, producing chest pain, called *angina;* (c) blood clot; (d) complete obstruction of a coronary artery by atherosclerosis and blood clot (coronary thrombosis), causing (e); (e) damage and death of heart muscle cells (i.e., a heart attack, myocardial infarction). [From M. Gabriel Khan (1996). "Heart Trouble Encyclopedia," Stoddart, Toronto.]

as a high blood level is associated with a lowered incidence of CHD. An effective drug should cause at least a 20–40% decrease in blood LDL cholesterol. Drugs that are capable of increasing HDL cholesterol 20–40% are potentially useful. [*See* Cholesterol.]

A. Bile Acid Sequestrants

Bile acid sequestrants are resins that are insoluble in water; they bind bile acids to form insoluble complexes that are not absorbed in the intestine, resulting in increased excretion of bile acids in the feces. Because LDL cholesterol is degraded in the liver to bile acids, the increased excretion of bile stimulates LDL cholesterol breakdown to bile acids, thus lowering LDL cholesterol in the blood. This therapy also increases the activity of LDL receptors on the membrane of liver cells; the LDL bind to the receptors and are removed from the blood. A severe reduction in LDL-receptor density is present in patients with genetic, familial hypercholesterolemia. In these patients, LDL cholesterol cannot be incorporated into cells and remain at high levels in the blood. Bile acid sequestrants are capable of causing a 20–30% reduction in LDL cholesterol, but in practice, only 6–15% reduction is obtained because the unpalatable taste and gastric side effects cause poor patient compliance. A 0–10% increase in HDL cholesterol may occur. Commonly used bile acid binding resins are cholestyramine, colestipol, and divistyramine. Bile acid sequestrants have been shown to decrease obstruction of arteries in animals due to atheromas, and this finding has been confirmed by angiographic studies in humans. Cholestyramine has been shown to reduce slightly the incidence of fatal and nonfatal heart attacks, but a decrease in total mortality rate has not been demonstrated. [*See* Bile Acids.]

B. HMG-CoA Reductase Inhibitors

These agents are inhibitors of 3-hydroxy-3-methylglutaryl-coenzyme A (HMG-CoA) reductase, the key enzyme that operates early in the biosynthetic pathway that leads to the formation of cholesterol. This enzyme performs the rate-limiting step in the pathway by catalyzing the conversion of HMG-CoA to mevalonic acid. When the enzymatic step is blocked by a drug, HMG-CoA does not accumulate, because it is water soluble and is readily broken down to innocuous metabolites. These drugs are very effective: modest oral doses cause a 20–25% reduction in LDL; high doses can reduce levels by about 35% in susceptible individuals. The drugs used are the fungal product lovastatin and simvastatin, which is a semisynthetic compound derived from lovastatin. Pravastatin is produced by microbial transformation of mevastatin. Fluvastatin is a fully synthetic HMG-CoA reductase inhibitor.

Clinical trials indicate that these agents slow the progression of obstruction in coronary arteries caused by plaques of atheroma and decrease the incidence of fatal and nonfatal heart attacks.

C. Fibrates

Fibric acid derivatives activate plasma lipoprotein lipase, which breaks down the lipid component of *very low-density lipoproteins* (VLDL), termed *triglycerides.* Their effect is a decrease in hepatic production of VLDL and an inhibition of the excretion of VLDL, resulting in a reduction in serum triglycerides. However, high blood triglycerides are not a key factor in the production of atheroma and CHD. Fibrates decrease triglycerides by 30–40%, LDL cholesterol by 5–12%, and cause a 5–15% elevation of HDL

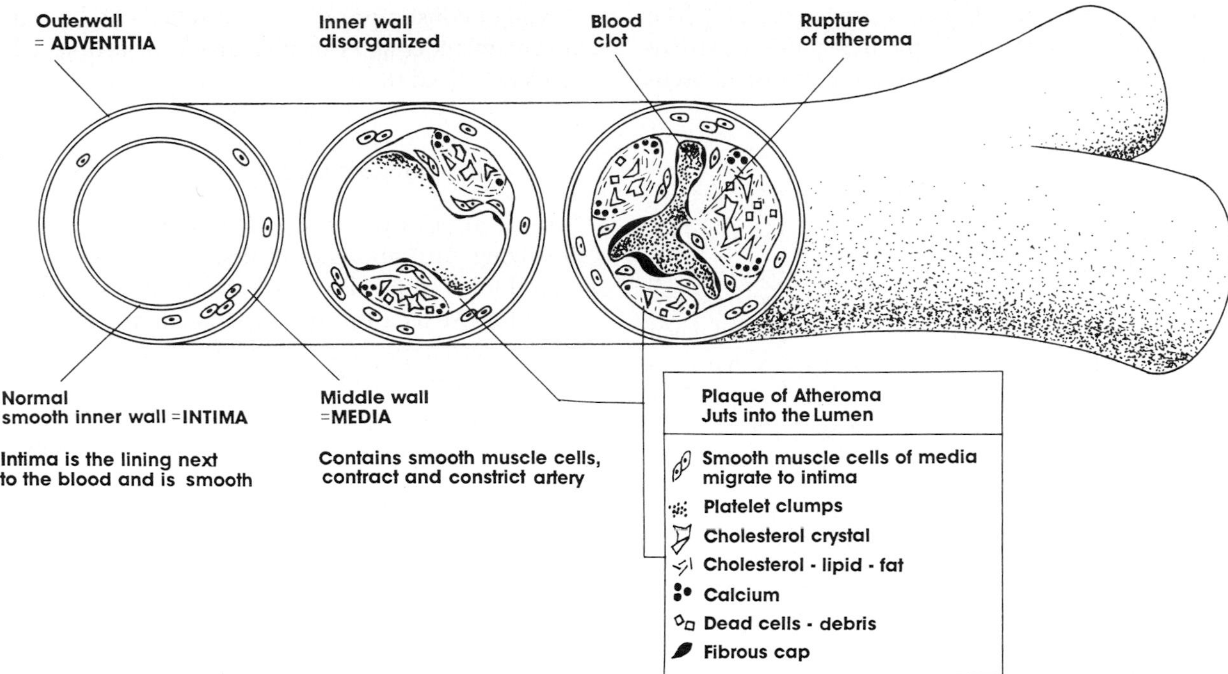

FIGURE 2 Atherosclerosis of artery. [From M. Gabriel Khan (1996). "Heart Trouble Encyclopedia," Stoddart, Toronto.]

cholesterol. Available drugs include gemfibrozil, bezafibrate, fenofibrate, and etofibrate. Large clinical trials using gemfibrozil for a 5-year study period showed a small decrease in the incidence of nonfatal heart attacks, but no significant reduction in fatal heart attacks or death rate.

D. Nicotinic Acid (Niacin)

Nicotinic acid inhibits secretion of lipoproteins from the liver, causing a 20–40% reduction in blood triglycerides and about a 12% fall in LDL cholesterol. The drug causes a 5–20% increase in HDL cholesterol. In a large study with patients followed for 15 years, nicotinic acid produced approximately 11% fewer total deaths and 12% fewer cardiac deaths than placebo. The salutary effect is so small that in practice the drug is hardly justifiable in view of the large doses required and the bothersome adverse effects it produces.

II. ANTIPLATELET AGENTS

When circulating blood platelets come in contact with damaged surfaces, they release a powerful platelet-clumping substance, thromboxane A_2, that causes platelets at the site to clump and stimulate other factors in the blood to form a clot (Fig. 2). Antiplatelet agents block the action of thromboxane A_2, thus preventing the formation of platelet clots. Antiplatelet agents are useful in the prevention of clots in arteries but are less effective in preventing clots in veins.

A. Aspirin

Platelets contain arachidonic acid that is converted to thromboxane A_2 by an enzyme, cyclooxygenase. Acetylsalicyclic acid (ASA, aspirin) irreversibly acetylates cyclooxygenase and thus prevents the formation of an endoperoxide intermediate, which is the precursor of prostaglandins and thromboxanes. A dose of 80–160 mg aspirin completely destroys cyclooxygenase activity for the 4- to 7-day life of the exposed platelets. Thus, the action of aspirin on platelets and on bleeding time lasts several days. Aspirin at the dose of 160 mg given to patients at the onset of a heart attack or to patients with severe angina reduces the incidence of myocardial infarction and death.

Aspirin administered to patients after a heart attack and continued for several years reduces the incidence of recurrent heart attack and improves survival. The drug has proven effective in the prevention of stroke and death in patients with cerebral transient ischemic attacks.

B. Dipyridamole

Dipyridamole is a weak inhibitor of platelet aggregation but is more effective than ASA in preventing platelet adhesion to foreign material such as implanted heart valves. The drug effect is increased when combined with ASA. Dipyridamole used alone does not reduce the incidence of heart attacks and does not prolong life in any subgroup of patients with heart disease.

C. Ticlopidine

Ticlopidine is a potent inhibitor of platelet aggregation. This agent alters the platelet membrane directly and does not inhibit cyclooxygenase, thus, unlike aspirin, the drug does not alter thromboxane production and does not cause gastric ulcers. Clinical trials have documented the role of ticlopidine in preventing stroke and death in patients with cerebral transient ischemic attacks or poststroke; the drug is recommended when aspirin in contraindicated. A dose of 250 mg twice daily is effective. Ticlopidine may cause a fall in the white blood count in some patients and caution is necessary.

D. Omega-3 Fatty Acids

Omega-3 fatty acids, eicosapentaenoic (EPA) and docosahexaenoic (DHA) acids, increase the production of a vasodilator prostaglandin, prostacyclin, in the endothelial lining of blood vessels. The vasodilator effect of prostacyclin counterbalances the constrictor action of thromboxane A_2. EPA and DHA inhibit the formation of thromboxane A_2 by reducing levels of its precursor, arachidonic acid, as the substrate for the enzyme cyclooxygenase, which generates thromboxane A_2. This twofold action in preventing platelet aggregation and maintaining arterial dilatation prevents blood clotting and inhibits atheroma formation. These fatty acids also increase levels of activators [e.g., tissue plasminogen activator (tPA)] that dissolve clots. EPA and DHA concentrates produce about 10–40% lowering of blood triglycerides but do not significantly affect levels of LDL and HDL cholesterol. Omega-3 fatty acids are plentiful in ocean fish, especially mackerel, herring, cod, salmon, and tuna.

A high consumption of EPA and DHA by native Greenlanders confers on this group of people a low incidence of CHD.

III. THROMBOLYTIC AGENTS

Normal arteries produce minute amounts of a clot-dissolving thrombolytic agents, tPA, that dissolves some blood clots. tPA has been produced commercially for use as a thrombolytic agent and is intravenously given to patients with heart attacks. All thrombolytic agents produce their salutary effects by actively converting a circulating blood protein, plasminogen, to plasmin, a powerful fibrin-dissolving, lytic substance. Available thrombolytic agents include streptokinase and tPA.

A. Streptokinase

When introduced in the body by intravenous injection, streptokinase forms an activator complex with circulating plasminogen that converts free plasminogen to plasmin, which then causes dissolution of clots. Streptokinase has a short half-life of about 80 min because it degrades to smaller fragments, resulting in loss of activity. Streptokinase is an effective thrombolytic agent, especially if given within 4 hr of the onset of chest pain that heralds a heart attack. Streptokinase, given by intravenous infusion within 6 hr of onset of a heart atack, causes a significant reduction in cardiac death rate. The combination of streptokinase and aspirin has been shown to be superior to the use of either drug alone.

B. Tissue Plasminogen Activator

tPA binds specifically to the fibrin component of a blood clot. tPA interacts with plasminogen through a cyclic fibrin bridge, resulting in conversion of plasminogen to plasmin, which then causes lysis of fibrin. tPA has a short half-life, and about 80% of the drug is removed from the blood within 10 min. The drug is cleared mainly by the liver. tPA given by intravenous infusion to heart attack patients within the first 6 hr of onset has effects that are slightly superior to those observed with streptokinase.

C. Anisoylated Plasminogen Streptokinase Activator Complex

Anisoylated plasminogen streptokinase activator complex (APSAC) is a 1-to-1 molecular combination of streptokinase and plasminogen, with a catalytic center protected by a chemical group. In the blood, the chemical group is removed, and APSAC becomes active. APSAC has a delayed onset of action because it takes time to remove the protecting group. One advantage of APSAC is its effectiveness when given intravenously as a quick bolus injection.

IV. ANTICOAGULANTS

Anticoagulants prevent thrombus (clot) formation. Clot formation is initiated when several clotting factors enter a cascade of enzymatic reactions that finally converts prothrombin to thrombin. Thrombin converts the circulating protein fibrinogen to strands of fibrin, which polymerizes to form a firm clot. The two commonly used anticoagulants are heparin, used intravenously, and warfarin, used orally. Recently, direct thrombin inhibitors, argatroban, hirudin, and hirulog, have been submitted to clinical trials in patients.

A. Thrombin Inhibitors (Argatroban, Hirudin, Hirulog)

These specific thrombin inhibitors are more effective than heparin, oral anticoagulants, and aspirin in preventing clot formation in coronary arteries. In small clinical trials, the intravenous use of these agents has proven superior to heparin. Like all global inhibitors of thrombin, they have the potential for severe bleeding and as yet no effective antidote is available.

B. Heparin

Heparin is inactivated by gastric acidity and must be given intravenously or, for some situations, subcutaneously. In the circulation, heparin is bound to a plasma protein, antithrombin III, and the complex inhibits the action of thrombin. Thus, heparin neutralizes the action of thrombin and other clotting factors, preventing conversion of fibrinogen to a fibrin clot. The drug takes effect within minutes, with an elimination half-life of 60–90 min.

C. Oral Anticoagulants (Coumarins, e.g., Warfarin)

Prothrombin is a vitamin K-dependent clotting factor manufactured in the liver. Warfarin competitively inhibits the effects of vitamin K, thus reducing the availability of prothrombin for conversion to thrombin in the final stages of the clotting cascade. In the absence of thrombin, fibrinogen cannot be converted to a fibrin clot. These agents are well absorbed orally and have a half-life of 24–36 hr and are given as tablets once daily.

Oral anticoagulants do not prevent heart attacks. They are more effective in preventing the formation and propagation of fibrin thrombi that occur in veins. In the case of venous thrombosis in the legs, oral anticoagulants are given for a few months after a course of intravenous heparin therapy.

V. ANTIANGINAL DRUGS

The major determinant of oxygen supply to the heart muscle is coronary blood flow. In individuals with atheroma, blood oxygen supply is fixed because of the narrowing of the coronary arteries, and anginal pain occurs when there is an increase in one or more of the three factors that determine myocardial oxygen demand: the heart rate, the velocity of contraction of the heart muscle, and the product of heart rate and blood pressure. Therefore, angina occurs when there is an imbalance between oxygen supply and demand. The pharmacological agents used in the management of angina are directed at correcting or improving this imbalance. Oxygen supply can be improved only by dilatation of the narrow artery at the site of obstruction. Pharmacological agents taken orally are not able to dilate continuously segments of diseased and narrowed arteries for days or weeks. Thus, the emphasis is placed on decreasing myocardial oxygen demand. Of the groups of antianginal drugs to be discussed, the beta-blockers are superior in this regard. [*See* Atherosclerosis.]

A. Beta-blockers

The catecholamines, epinephrine and norepinephrine, are released from sympathetic nerve endings and as hormones from the adrenal glands. Catecholamines effect their major action at receptor sites called *beta receptors* that are present on the surface of cells in various organs and tissues, in particular the heart and arteries. Catecholamines are a group of stimulants that cause an increase in the force and velocity of contraction of the heart and an increase in heart rate and blood pressure. Beta-blockers resemble the catecholamines, but block their effects at the beta-receptor sites. They decrease both heart rate and blood pressure, and consequently the oxygen requirement. By causing a decrease in the force and velocity of contraction of the heart muscle they further decrease the oxygen requirement. An important effect derives from the fact that the coronary arteries are perfused with blood during the short moment of diastole when the heart is not contracting. Beta-blockers, by slowing the heart rate, increase the diastolic period and thus increase blood flow through the coronary arteries. [*See* Adrenal Gland; Catecholamines and Behavior.]

Beta-blockers are easily absorbed from the gut; some are metabolized by the liver and others are eliminated by the kidneys. Available preparations include atenolol, acebutolol, metoprolol, nadolol, propanolol, sotalol, and timolol.

Beta-blockers decrease the incidence of heart attacks and sudden death in patients with ischemic heart disease. They are also effective antianginal agents; about 80% of patients treated are expected to get from 50–75% relief of their anginal pain. Also, bothersome palpitations, abnormal heart rhythms, are abolished. Beta-blockers prolong life in patients who are given the drug immediately after an acute heart attack, continuing for a period of 2 years.

B. Nitrates

Nitroglycerin tablets placed under the tongue quickly relieves the pain of angina. This drug, as well as orally administered nitrate preparations, dilate veins throughout the body that return blood to the heart. This rapid pooling of blood in the peripheral circulation reduces the work of the heart, thus less oxygen is required and chest pain is relieved. Nitrates bind to "nitrate receptors" in the vascular smooth muscle wall of arteries, causing its relaxation and dilatation of the veins. Orally administered nitrates are rapidly metabolized in the liver, limiting their availability to vascular receptors. Sublingual, transdermal, and intravenous preparations partially overcome this problem. The effectiveness of nitrate tolerance commonly decreases after several days of continuous use because of a relative depletion of sulfhydryl groups in vascular smooth muscle cells. A daily 10-hr nitrate-free interval is necessary to allow generation of an adequate supply of sulfhydryl groups and to restore vascular responsiveness. Available preparations include sublingual nitroglycerin, glyceryl trinitrate, oral nitroglycerin tablets, isosorbide dinitrate, and cutaneous preparations; an intravenous preparation is used in severe cases of angina. The antianginal effect of oral nitrates is about half that observed with beta-blockers. Oral or cutaneous nitrate preparations have not been shown to prevent death.

C. Calcium Antagonists

Calcium is transported from the exterior to the interior or cells via a system of tubules called *slow calcium channels*. Calcium reaches the interior of the muscle cell, binds to a regulatory protein (troponin) that removes the inhibitory action of another protein (tropomyosin), and using energy supplied by adenosine triphosphate (ATP) allows the interaction between the muscle filaments, containing myosin and actin, respectively, with consequent contraction of the muscle cell. The slow calcium channels are selectively blocked by a class of agents known as calcium channel blockers or calcium antagonists. Their action prevents calcium influx into the cell, thereby causing relaxation of muscle in arterial walls and consequently dilation of arteries. This effect causes a reduction in blood pressure, and less work is imposed on the heart. These drugs cause modest dilatation of the coronary arteries in some patients, but this may be minimal at the site of atheromatous obstruction.

One group of calcium antagonists, the dihydropyridines of which amlodipine, felodipine, isradipine, nifedipine, and nitrendipine are examples, has their actions mainly on arteries and has no effect on the electrical conduction system of the heart. In contrast, calcium antagonists of the phenylalkylamine structure, verapamil, or the benzothiazepine, diltiazem, have actions on the arteries as well as on the heart. Consequently, verapamil and diltiazem cause slowing of the heart rate and a variable decrease in the force of contraction of the heart muscle. Calcium antagonists are easily absorbed when taken by mouth and are mainly metabolized in the liver. The agents reduce chest pain of angina to about the same degree as beta-blockers but do not prevent heart attacks or prolong life.

VI. ANTIARRHYTHMIC DRUGS

Disturbances of the heart beat, or arrhythmias, may arise from the ventricles (i.e., ventricular arrhythmias) or in the atria above the ventricles (i.e., "supra" ventricular arrhythmias). Cardiac cells have an inherent pacemaker potential that is normally suppressed by the powerful, natural generator or pacemaker that resides in the heart's sinus node and produces the heart beat in each living human. The normal impulses (i.e., action potential) generated in the pacemaker reach the ventricles through an intricate network of fibers in which they are conducted at different velocities. Arrhythmias or abnormal heart rhythms are produced by either the generation of abnormal impulses or enhanced automaticity of impulses and/or disturbances of impulse conduction. Antiarrhythmic drugs suppress these mechanisms and stop the abnormal rhythm. Of particular importance are the abnormal rhythms that occur during the onset of myocardial infarction (MI). Such abnormal rhythms, termed *ventricular ectopics* or *ventricular premature beats,* may occur in runs and suppress the normal heart beat. Multiple ectopic beats occurring consecutively are designated ventricular tachycardia and can worsen, producing ventricular fibrillation (VF), a condition in which the ventricle quivers instead of contracting. During VF there is cardiac arrest, no blood is expelled from the heart, the brain is deprived of blood, and the individual loses consciousness.

The electrocardiogram (ECG) picks up the heart's electrical impulses that are transmitted through the skin of the chest. Figure 3 illustrates how myocardial cells generate an action potential in phase 0 through a fast influx of sodium ions (Na^+) into cells, which increases the resting potential (voltage) of the cell (depolarization). Later (phase 3), the cell returns to its resting potential with an efflux of potassium ions (K^+). Most antiarrhythmic agents produce their effects by decreasing the rate at which Na^+ enters the myocardial cell in phase 0 (Fig. 3). Thus, generation of the action potential of an abnormal impulse is dampened and does not reach sufficient magnitude to produce abnormal beats. Drugs in this category include quinidine, disopyramide, procainamide, flecainide, and propafenone. Beta-blockers, lidocaine, and the majority of antiarrhythmic agents decrease the rate of automaticity of abnormal rhythms by depressing phase 4 of the action potential as indicated by the arrow in Fig. 3. Lidocaine is often given intravenously during the first few hours of MI to abolish and prevent serious arrhythmias. The action of lidocaine is immediate, but of short half-life (about 10 min) because the drug is quickly metabolized in the liver. Oral derivatives of lidocaine have not proven to be a major success: they do not abolish serious arrhythmias or prolong life and have adverse effects. Other agents (i.e., amiodarone and a unique beta-blocker, sotalol) cause a prolongation of the action potential (phase 2) and thus retard the generation of an abnormal impulse. Also, as indicated in Fig. 3, an increase in the absolute refractory period (phases 1 and 2) protects from dangerous impulse stimuli. During a 20- to 30-msec vulnerable period in phase 3 (in Fig. 3), a strong electrical stimulus or ventricular ectopic beat can readily trigger ventricular tachycardia and VF.

Several of the mentioned antiarrhythmic agents prevent ventricular tachycardia and have the potential to save life. Beta-blockers suppress some serious arrhythmias that occur during an MI; when given to patients after a heart attack and continued for as long as 2 years, beta-blockers have been shown to save lives and reduce the incidence of sudden deaths. Other antiarrhythmics suppress serious arrhythmias but have not been shown to prolong life. Amiodarone appears to increase survival in patients who have survived a cardiac arrest. Because of side effects, the drug is used as a last resort. In a significant number of patients with concomitant heart failure and a sick heart, some antiarrhythmic agents may actually increase the incidence of dangerous arrhythmias and/ or worsen heart failure.

The atrioventricular (AV) node slightly decreases the rate of conduction of impulses from atria to ventricles, which represents a protective effect. In some individuals, abnormal ectopic beats at the frequency of 150–200/min arise in the atria. They are conducted through the AV node to the ventricles, which then beat at 150–200. This condition is called *paroxysmal atrial tachycardia* (PAT) and is not uncommon in healthy young adults with normal hearts. The fast heart beats can be reduced by drugs such as digitalis, beta-blockers, or a calcium antagonist, verapamil, that slow impulse traffic through the AV node. In a similar condition, *atrial fibrillation,* instead of regular beats, the atria quivers or fibrillates. Impulses are generated at 400–500/min and bombard the "tollgate" AV node, reaching the ventricles irregularly. The heart beat becomes irregular. Agents (e.g., digitalis) that reduce impulse traffic through the AV node slow the ventricular rate to a normal level of 70–90 beats/min. Beta-blockers and some calcium antagonists also slow the heart rate in patients with atrial fibrillation.

VII. HEART FAILURE

Heart failure is usually the result of a diseased heart. The commonest cause is a weak heart muscle as a result of one or more heart attacks, severe valve disease, hypertension, and rare diseases that affect the heart muscle. The weakened heart muscle fails to pump sufficient blood from the left ventricle into the

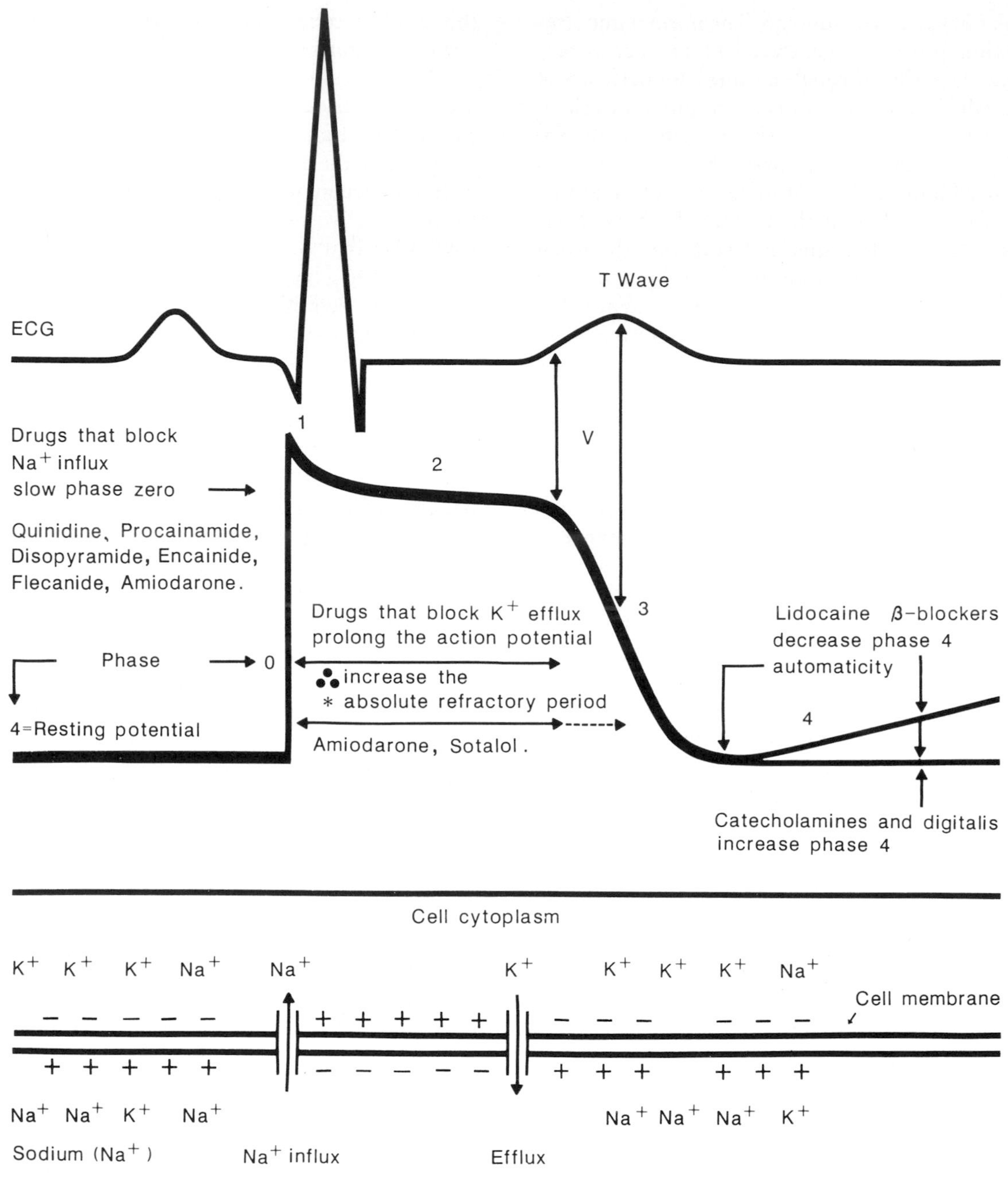

FIGURE 3 Antiarrhythmic drug action. Asterisk indicates absolute refractory period; during phases 1 and 2, a stimulus evokes no response: an arrhythmia cannot be triggered. V, vulnerable period.

arteries to supply organs and tissues adequately. The proportion of blood that cannot be ejected from the left ventricle backs up into the lungs, causing congestion of veins. Fluid containing sodium and water leaks into the sponge work and air sacs of the lungs, causing severe shortness of breath. [*See* Cardiac Muscle.]

Figure 4 shows the mechanisms that operate during heart failure. The cardiac output falls, less blood reaches the kidney, and sensors initiate the secretion of renin in the kidney. Renin is an enzyme that converts circulating angiotensinogen to angiotensin I. A "converting enzyme" converts angiotensin I to angio-

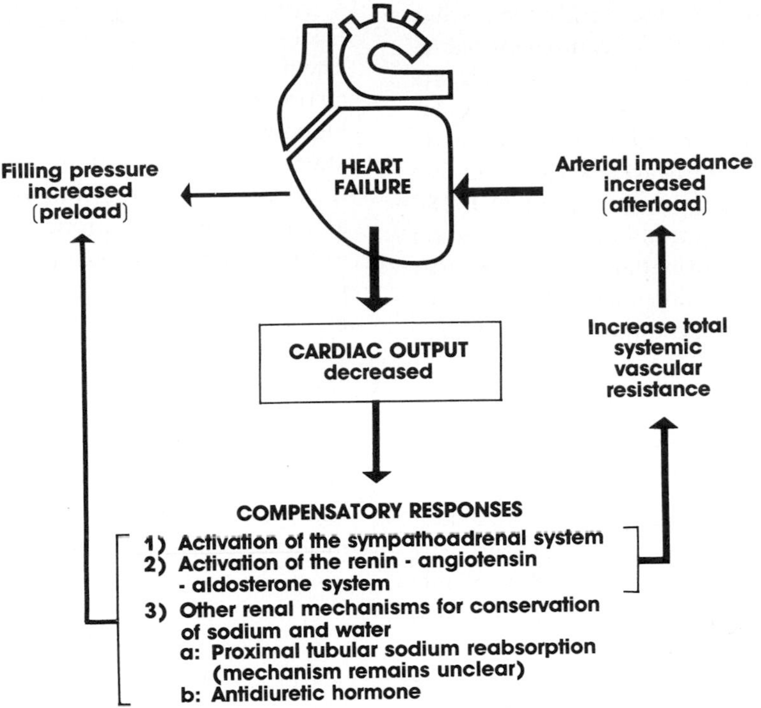

FIGURE 4 Pathophysiology of heart failure. [From M. Gabriel Khan (1995). "Cardiac Drug Therapy," 4th Ed.). Saunders, London.]

tensin II. The latter compound constricts arteries, thereby increasing systemic vascular resistance (SVR), normally a compensatory mechanism when blood volume is reduced, as in blood loss. Constriction of arteries, however, increases arterial impedance (afterload), thereby increasing the work of the failing heart. This and other compensatory mechanisms therefore become counterproductive. A major medical maneuver in the treatment of patients with heart failure is to block angiotensin "converting enzyme" by drugs called *angiotensin converting enzyme (ACE) inhibitors,* thereby reducing the work of the heart. Another normal compensatory mechanism is reabsorption of sodium and water by the kidney under the effect of the renin–angiotensin–aldosterone system and other systems. Sodium and water are returned to the circulating blood, imposing further strain on the failing heart. This is seen as increase in preload in Fig. 4. Congestion in the lungs occurs; sodium and water leak out into the alveoli and into other areas of the body such as the legs, forming edema. In edema, the legs are brine-logged, not just water-logged.

The three main treatments that assist the failing heart are diuretics, which reduce preload; the inotropic agent digoxin, which increases cardiac output; and vasodilators, in particular ACE inhibitors, which reduce the arterial impedance, or afterload.

A. Diuretic Drugs

Diuretics cause the kidneys to excrete the several liters of sodium and water that were retained in the body. Available diuretics include furosemide and thiazides. Diuretics do not repair the damaged heart muscle and have no effect on the disease process. They are effective, however, in relieving severe shortness of breath and edema. [*See* Diuretics.]

B. Inotropic Agents

Digitalis and digoxin increase the force and velocity of myocardial contraction. These compounds inhibit the function of the sodium pump located in the membrane of heart muscle cells. This pump, using ATP energy, pumps Na^+ out of the cell and K^+ into the cell and is mainly responsible for maintaining the resting potential of the cardiac muscle cells. The action of the drugs results in an increase in intracellular sodium accompanied by an increase in intracellular calcium. Calcium within the muscle cell stimulates the force of myocardial contraction. Consequently the cardiac output increases, and some of the compensatory responses that are counterproductive to the body are halted. Occasionally, HF is due to a fast irregular heart beat, called atrial fibrillation, a condition in which the top chamber of the heart beats at about 300 beats/min and the ventricles beat at about 150–180 beats/min. In patients with atrial fibrillation, digoxin reduces the heart rate to normal and improves heart failure. Clinical trials have documented the efficacy of digoxin in improving the symptoms of heart failure and quality of life.

C. Vasodilators

ACE inhibitors dilate arteries and cause some dilatation of veins so that afterload and preload to the heart are reduced (Fig. 4). Available ACE inhibitors include captopril, enalapril, lisinopril, benazepril, quinapril, ramipril, spirapril, and trandolapril. They prevent further hypertrophy of the heart, are effective in reducing symptoms such as shortness of breath, and improve the exercise capacity of individuals.

D. Nitrates

The actions of nitrates are discussed in Section V. They dilate veins, pooling blood in the periphery. Because less blood is returned to the overworked heart, they reduce the heart's preload. Available preparations include isosorbide dinitrate and nitroglycerin.

Diuretics, digoxin, and ACE inhibitors individually do not improve survival, but their combination has been shown to prolong life in patients with heart failure.

VIII. ANTIHYPERTENSIVE AGENTS

In most cases, hypertension (i.e., elevated blood pressure) is due to an increase in the state of contraction (tone) of arterial muscles, which produces an increase in systemic vascular resistance. With the exception of beta-blockers, the final action of all antihypertensive agents is to produce mild to moderate dilatation of arteries, thus lowering the SVR against which the heart must pump. The various groups of drugs achieve this goal by different mechanisms. [*See* Hypertension.]

A. Diuretics

The major action is on the renal tubules, causing excretion of sodium and water and thus a decrease in blood volume and a fall in blood pressure. However, after several months of use, the body compensates for the slight reduction in blood volume, and the major action of chronic diuretic therapy is limited to the loss of sodium, which causes mild dilatation of arteries and a reduction in SVR. Diuretics also cause potassium excretion in the urine. Potassium loss must be prevented because it contributes to abnormal heart rhythms and to a slight increase in cardiovascular deaths. Available preparations include thiazides (e.g., hydrochlorothiazide and bendrofluazide). Diuretics do not prevent heart enlargement or hypertrophy, which carries a risk of sudden death. They reduce the incidence of stroke to the same extent as other agents.

B. Beta-Blockers

The exact mechanism by which beta-blockers reduce blood pressure is not completely understood. Their action on the heart is given in Section V. Reduction in heart rate and in the force of cardiac contraction result in a mild decrease in cardiac output, which lowers blood pressure. These agents decrease renin production by the kidneys and decrease the effects of the sympathetic nervous system on the heart and arteries. They decrease the speed and velocity at which blood is ejected from the heart into the arteries. Thus, there is less hydraulic stress on the walls of arteries, especially at branching points, resulting in less wear and tear. The effect is apparent in the management of aneurysms of the aorta, a condition in which the weakened artery wall dilates and finally ruptures (dissection). In such patients, although blood pressure is usually low, beta-blockers are given intravenously to reduce the ejection velocity of blood and to prevent further dissection and weakening of the arterial wall. Beta-blockers prevent left ventricular hypertrophy and reduce the incidence of stroke and heart attacks. Although the effect is modest, they do prolong life.

C. ACE Inhibitors

The action of ACE inhibitors is given in Section VII. They can prevent the production of angiotensin II, resulting in dilatation of arteries, decrease in SVR, and a fall in blood pressure. ACE inhibitors are useful agents for the management of both heart failure and high blood pressure.

ACE inhibitors do not cause an increase in heart rate as seen with other vasodilators. The blocking of angiotensin II causes a decrease in the secretion of a hormone, aldosterone, by the adrenal glands, which results in a favorable excretion of sodium and retention or conservation of potassium. ACE inhibitors prevent left ventricular hypertrophy, but there is no indication that they are superior to diuretics or beta-blockers in prolonging life in hypertensive patients.

D. Calcium Antagonists

The actions of calcium antagonists are given in Section V. They cause peripheral dilatation of arteries, resulting in a decrease in SVR and a fall in blood pressure. Available preparations include amlodipine, felodipine, nifedipine, nitrendipine, verapamil, and diltiazem.

E. Vasodilators

Calcium antagonists and ACE inhibitors are special groups of vasodilators. Other vasodilators include hydralazine, which has a direct dilating effect on arteries. Prazosin, indoramin, and trimazosin are alpha$_1$-adrenergic receptor blockers that act directly on nerve endings in the walls of arteries. These drugs cause dilatation of arteries, a fall in blood pressure, with a compensatory increase in heart rate as well as an increase in the ejection velocity of the heart. Consequently, they do not prevent left ventricular hypertrophy and are contraindicated in patients with aneurysms. Sodium and water are retained by the kidney, so that the antihypertensive effect usually decreases with prolonged use. The addition of a diuretic is often needed to excrete retained sodium and water, thus potentiating the antihypertensive effect.

F. Centrally Acting Drugs

These drugs modulate or inhibit sympathetic outflow from the central nervous system, which increases tone and constriction of arteries. Available drugs are methyldopa, clonidine, guanfacine, guanabenz, and reserpine. The final action of these drugs results in dilatation of arteries, with consequent retention of sodium and water by the kidney. Often they require the addition of a diuretic to complement the blood pressure lowering the effect by removing sodium and water. Centrally acting drugs do not consistently prevent left ventricular hypertrophy.

G. Conclusion

The majority of patients with hypertension can be controlled with a beta-blocker, a diuretic, and/or both. Beta-blockers are as effective as other groups of drugs in lowering blood pressure. All antihypertensive agents, by reducing blood pressure, prevent the occurrence of fatal and nonfatal strokes and heart failure. Only beta-blocking agents decrease the incidence of fatal and nonfatal heart attacks. Cigarette smoking increases hepatic degradation of propranolol and other hepatic metabolized beta-blockers and can prevent beneficial effects on mortality. Timolol, a partially metabolized drug, effectively decreases the death rate in smokers and nonsmokers with CHD. Pharmacological agents should be evaluated not only in terms of their effects on amelioration of symptoms, but also on their ability to prolong life.

BIBLIOGRAPHY

ISIS-2 (Second International Study of Infarct Survival) Collaborative Group. (1988). Randomised trial of intravenous streptokinase, oral aspirin, both, or neither among 17,187 cases of suspected acute myocardial infarction: ISIS-2. *Lancet* **2**, 349–360.

Khan, M. Gabriel (1996). "Heart Trouble Encyclopedia," Stoddart, Toronto.

Khan, M. Gabriel (1995). "Cardiac Drug Therapy," 4th Ed. Saunders, London.

Cardiovascular System, Physiology and Biochemistry

MARC VERSTRAETE
University of Leuven

GLOSSARY

Afterload Impedance to ventricular emptying during systole or the sum of all of the loads against which the myocardial fibers must shorten during ventricular contraction

Baroreceptors Receptors or sensors for stretch in the arterial wall

Cardiac index Cardiac output per minute per square meter of body surface

Cardiac output Product of the volume of blood ejected during each heart beat (left ventricular stroke volume) multiplied by the heart rate

Chemoreceptors Sensors in the arterial wall for chemical changes, such as oxygen or carbon dioxide

Inotropic state Vigor of contraction of heart muscle, reflected in the speed and capacity of shortening of the myocardial fibers at a given load

Preload Passive load that establishes the initial length of the myocardial fibers prior to contraction, reflected by the end-diastolic volume of the ventricle (Starling's law of the heart)

Ventricular ejection fraction Left ventricular stroke volume divided by the left ventricular end-diastolic volume

IN SINGLE-CELL ORGANISMS THERE IS NO NEED FOR a respiratory or cardiovascular system, and oxygen can diffuse readily through the cell wall. In small multicelled animals this vital gas can still diffuse through the intermediate tissues in sufficient quantity. When vertebrates struggled on land and became too big to rely on small air-carrying tubes that ramify throughout the body, they required a better method of distributing oxygen. In all higher vertebrates the advanced system consists basically of one air pump, the lungs, and two liquid pumps: the two ventricles of the heart. The system in humans is only a sophisticated extension of the simple diffusion mechanism acceptable to the smallest creatures. The diffusion process is similar, but the distance from the external air into the final destination is greater; blood is the transport agency that compensates for the distance. Blood has to be brought near to the surface (the lungs can be considered an interiorized external surface), so that the oxygen of the inspired air can diffuse into the red blood cells, which are then moved through the pump function of the heart near to all tissues, so that oxygen can then diffuse into them.

I. HEART

The heart lies in the center of the chest, protected by the sternum and flanked on each side by the lungs, and weighs less than one pound in the adult. Starting its pumping system 8 months before birth and continuing to beat thereafter 4000 million times over the course of a lifetime, it is probably the only organ of the human body which never comes to even a temporary rest.

The right and left ventricle beat simultaneously and have the same output, albeit under remarkably different pressure regimens. The blood returns from the body via two large veins (the venae cavae) to the right

ENCYCLOPEDIA OF HUMAN BIOLOGY, Second Edition, VOLUME 2. Copyright © 1997 by Academic Press. All rights of reproduction in any form reserved.

atrium. Another vein, the coronary sinus, drains the blood present in the heart muscle itself, also to the right atrium. From this compartment the blood flows through an opening and closing device called a valve, made of three leaflets (i.e., the tricuspid valve), and flows to the right ventricle. During contraction of the ventricle, called systole (a Greek word for "contraction"), the blood is pumped at a pressure of approximately 20–30 mm Hg into a short pulmonary circuit ending in the lungs.

There, it flows through capillaries surrounding the lung alveoli, which unite to eventually form four pulmonary veins (two for each lung), which bring the blood back to the heart and enter the left atrium. This blood, oxygenated during its passage through the pulmonary circuit, is passed into the left ventricle through the mitral valve connecting the two compartments of the left heart. The muscularly well-developed left ventricle pumps the blood in the aorta, from which branches distribute into the whole body, forming the systemic circulatory circuit (Fig. 1). The output of both ventricles is about 5 liters per minute, which increases sixfold during severe exercise; however, the left ventricle develops a pressure of about 140 mm Hg, approximately five times higher than the peak pressure generated by the right ventricle, which supplies the much shorter pulmonary circuit. [*See* Cardiovascular System, Anatomy.]

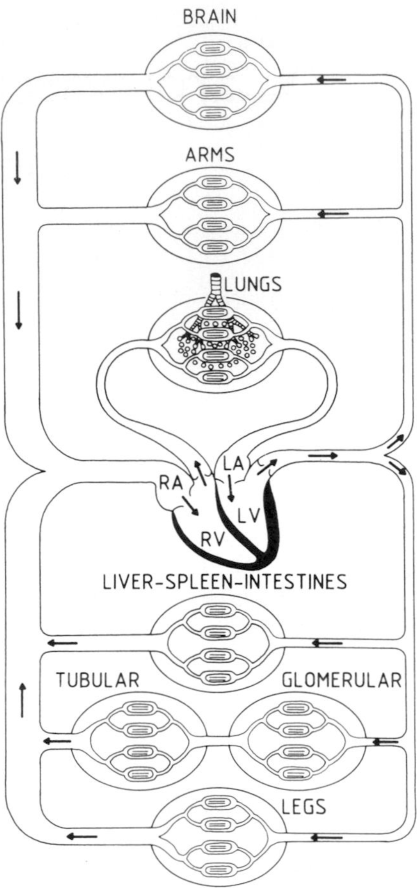

FIGURE 1 Arrangements of the parallel routes by which the circulation passes from the aorta to the venae cavae. For the systemic arterial circulation the arterial pressure is determined by the product of cardiac output and the resistance to flow offered by the various vascular beds arranged in parallel.

A. Cardiac Muscle and the Contractile Process

The heart is essentially a muscular organ. Considering its untiring function, the cardiac muscle, termed the myocardium, is a special kind of muscle. In both its structure and function it is intermediate between striated muscle and smooth muscle. Like the latter, it is able to contract and relax rapidly; like skeletal muscle, it shows cross-striations. The necessity of continuous functioning of the heart demands that the myocardium also resembles the smooth muscle in its ability to carry out long-lasting activities. Microscopically, there is evidence that cardiac muscle is made up of cells with a single nucleus, as is true for smooth muscle. The demand for sustained activity is greater in the case of the heart than for any other organ; however, its activity is not strictly continuous. Although the heart beats without ceasing for over a half-century, there is a very short pause between each contraction of the ventricles, called diastole, during which these compartments refill with blood. [*See* Cardiac Muscle.]

The innermost layer of cardiac muscles in the two ventricles is arranged in a circular manner. The adult left ventricle is roughly cylindrical and is made of a thick muscular layer (8–10 mm); the right ventricle is triangular in the frontal projection and its wall is much thinner (3–4 mm). Outside there are two oblique or spiral layers, arranged at almost right angles to one another and tending on contraction to pull the ventricular chamber toward the valve rings or vice versa.

When a ventricle contracts, it does so concentrically, shortening its diameter, and also longitudinally, reducing its length by pulling on its attachment to the valve rings. The total silhouette of the heart appears surprisingly immobile while it is contracting, despite the volume of blood being ejected when the two ventricles contract. This is because of the reciprocal volume changes in the atria and the ventricles; when the latter contract, the right and left atria fill with blood coming from the systemic or pulmonary circulation,

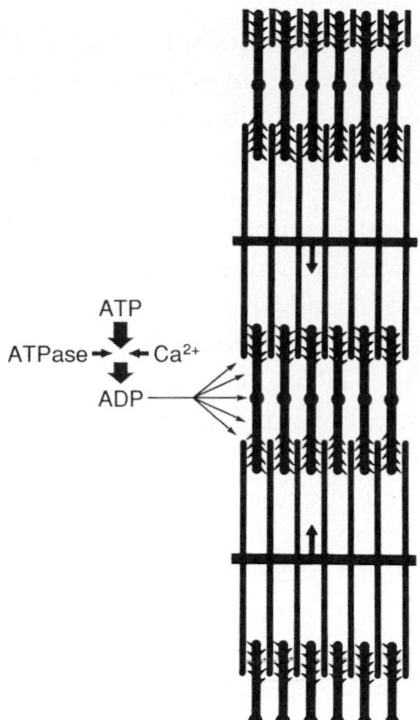

FIGURE 2 The "sliding filament" model of myocardial contraction.

respectively. The left ventricle has an ellipsoid shape, shortens more in its short axis than in its long axis, and normally empties about two-thirds of its content during ejection.

The cardiac muscle consists of columns of large cylindrical muscle fibers that branch to form a network. Each fiber consists of a membrane, the sarcolemma, surrounding bundles of small fibers, the myofibrils. Within the myofibrils the contractile proteins are arranged longitudinally into repeating units, the sarcomeres. These are composed of myofilaments, which are macromolecular complexes of contractile proteins. One myofilament is thick and is composed of the protein myosin; the other is thin and is composed of the protein actin. Both slide past each other to produce shortening of the myofibrils. Filaments made up of actin are drawn further and further between the thick myosin filaments, so that the sarcomere shortens. The thin and thick filaments remain constant in length during the process, but interact by virtue of cross-bridges formed between them (i.e., the actomyosin interaction (Fig. 2). The resulting contraction is rapid, as each muscle fiber shortens at a speed equal to several times its length in 1 second.

B. Energy and Metabolism of the Heart Muscle

As just noted, the "sliding filament" model relates structural changes to functional events during contraction of the myocardium. The movement of the thin filaments (actin) can be accounted for by the cross-bridges, which are the projecting parts of the myosin filaments to specific sites on the actin filaments. As force develops between the myofilaments, the actin is pulled toward the center of the sarcomere. To shorten further, cross-bridges must detach and reattach to new binding sites. The velocity of muscle-shortening depends on the speed at which the cross-bridge can attach, pull, detach, and resume the original position. The force development by the muscle depends on the number of cross-bridges attached at a given time.

Cardiac muscle can govern its contractility. The force and velocity of myocardial contraction are regulated to a large extent by the amount of free calcium ions released from the sarcoplasmic reticulum. In the relaxed state the actin molecule is not accessible for the cross-bridges because it is embraced by other protein complexes, troponin and tropomyosin, which are located periodically along the actin filaments. In the absence of free calcium ions, troponin works through tropomyosin, which courses along the actin filament to prevent actin from interacting with myosin (Fig. 3). With the arrival of the action potential at the myocyte, calcium is released and binds to the troponin–tropomyosin complex, the actin molecules become accessible for the cross-bridges, and the actomyosin interaction starts.

The sequence of events leading to muscle contraction can be schematized as follows: The electrical impulse traveling along the membrane of a muscle fiber reaches the sarcoplasmic reticulum and releases calcium ions. These combine with the troponin component of the thin filament. Troponin acts as a latch, blocking the formation of an active complex between myosin and actin. Calcium binding unlocks the blocked state, allowing ATP-charged myosin to interact with actin. When this occurs, ATP is hydrolyzed by the ATPase contained in myosin (actomyosin ATPase) for the cycling of the cross-bridges and thus for the propulsive force: swiveling of myosin and pulling of thin filaments toward the center of the sarcomere. Successive cycles (binding of ATP to myosin, detachment of myosin from actin, and reattachment in a new position) result in the continued sliding of the thin element (Fig. 2). Relaxation of the actin–myosin myofibrils occurs as the result of an inhibition produced by the troponin–tropomyosin complex in the presence of a low intracellular calcium concentration. Many drugs (including digitalis, sympathomimetic amines, calcium channel antagonists, and phosphodiesterase inhibitors) have an influence on myocardial

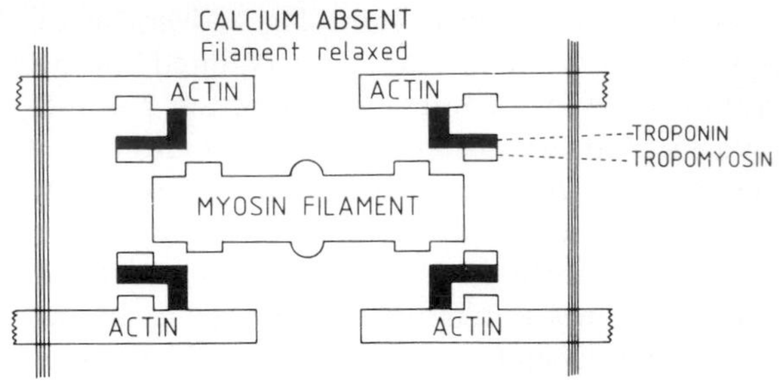

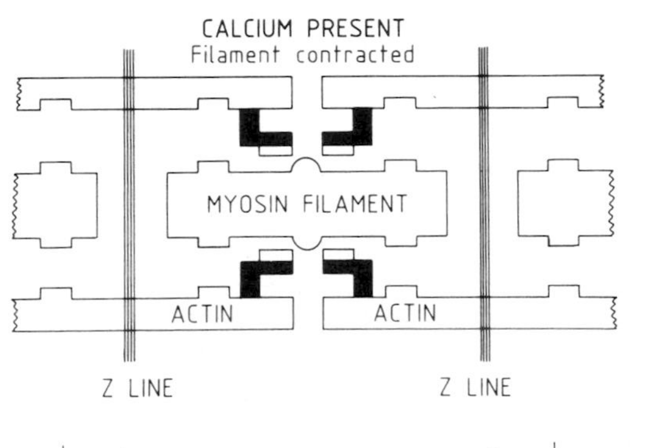

FIGURE 3 In the absence of calcium, the interaction between actin and myosin is prevented by the troponin–tropomyosin complex. In the presence of calcium, the troponin–tropomyosin complex no longer blocks the site at which interaction between actin and myosin takes place, and the contractile event can occur.

contractility through their effects on available intracellular calcium ions.

To perform its essential function, the myocardium, like all other tissues of the body, must be lavishly supplied with oxygen and Na, K, Ca, glucose, and other nutrients. Exchange from the blood flowing through the heart cavities contributes to the nourishment of the inner heart layer, the endocardium, but is inadequate to ensure nutrition to the large mass of the myocardium, owing to the rapid transit of the blood. Indeed, although many liters of blood circulate through the heart, this is of no avail to the deeper layers of the cardiac wall. To this end, a separate and complex circulatory system subserves the myocardium and the pericardium, the latter being the outer structure that encloses the heart. This system is called the coronary circuit (from the Latin word *corona,* which means "crown," as the coronary vessels encircle the heart like a crown).

The principal arteries of this system, the right and left coronary arteries, branch off the aorta near its base, lie on the surface of the heart underneath the epicardium, and supply all of the blood to the myocardium and the epicardium. In keeping with its almost continuing activity, the myocardium is richly supplied with arterioles and capillaries. An intercommunicating network of veins drains 85% of the myocardial blood supply to a collecting venous sinus, the coronary sinus, which empties in the right atrium, just above the tricuspid valve; smaller veins collect the remaining blood and also enter the right atrium.

The energy of the contractile process of the heart is supplied by ATP and creatine phosphate produced by aerobic metabolism. The total energy expenditure of the normal heart can be equated by its oxygen consumption. In the normal heart, anaerobic metabolism (not utilizing oxygen) might account for about 5% of the energy utilized. This proportion might increase to 20% or more during exercise in patients with heart failure. Thus, to maintain contractile activity, the cells must regenerate ATP. The muscles contain phosphocreatine, which, in the presence of the enzyme creatine phosphotransferase, can rapidly regenerate ATP according to the equation phosphocreatine + ADP $\rightarrow$ creatine + ATP. Since the amount of ATP is less than that of phosphocreatine and since creatine phosphotransferase is abundant, phosphocreatine plays a major role in the instantaneous regeneration of ATP. Although phosphocreatine provides an immediate reserve of energy, it is not sufficient to sustain prolonged activity. [*See* Adenosine Triphosphate (ATP).]

Ultimately, regeneration of energy-rich phosphate compounds involves the breakdown of metabolic substrates such as fatty acids, glucose, and lactate. This can occur aerobically or anaerobically. The former results not only in the production of more ATP per molecule of substrate but also in the formation of carbon dioxide and water as end products, which diffuse easily from the contracting cell. By contrast, the anaerobic consumption of substrates terminates with the production of fixed acids. When the concentration of the latter rises, intracellular acidosis becomes unavoidable, since these acids are not freely diffusible. The various end products of metabolism exert a positive feedback on the intracellular enzymatic processes as well as on the oxygen delivery to the working cells, which continuously adjust energy production to its utilization. When the supply of blood to the heart is cut off or reduced, metabolism (but not contraction) can continue at a lower level by anaerobic pathways.

The heart is a lipid-consuming organ, and glucose is used relatively less than in skeletal muscle. Free fatty acids are the chief source of energy. They are derived from plasma-free fatty acids and triglyceride and, in exercise, from myocardial stores of triglyceride. Keto acids and lactate are additional important substrates that supply energy for ATP formation.

C. Functional Characteristics

The heart is inherently rhythmic; this explains why a heart of a mammal will continue to beat for some time when cut off completely from its nerve supply. The initiation of the heart beat is to be found in a specialized neuromuscular tissue, called the sinoatrial (SA) node, which is a small strip of tissue, about 1 × 2 cm in length, and 3 mm at its widest point. This SA node is located in the posterior wall of the right atrium near its junction with the superior vena cava. The SA node sets the heart rate, because its cells have the greatest ability to spontaneously initiate an action potential. This property of being a pacemaker is due to an inherent instability of the cell membrane that is caused by the interplay between three currents: an outward potassium current that is deactivated and two inward currents that are activated, that is, the calcium current and the typical "pacemaker" current, called i_h, because it is activated on hyperpolarization.

As an impulse is generated, it immediately spreads in all directions through the atrial muscle in a ripple pattern, similar to that of waves generated when a stone is thrown in water. The impulses reach a specialized tissue called the atrioventricular node, situated at the base of the interatrial system. To permit sufficient time for complete atrial contraction, the conduction of the impulse in the atrioventricular node is slightly delayed, allowing atrial depolarization to precede ventricular depolarization by approximately 140 msec. From the atrioventricular junction the electrical impulse activates atrioventricular bundles of specialized cells, called the bundle of His, which contain typical cardiac cells. Conduction of the impulse occurs through these cells (nerve fibers are present and may eventually modulate activity, as in any other part of the heart). These bundles pass into the interventricular septum, and its right and left bundle branches project downward and around the tip of the ventricles; their end branches form the network of Purkinje, which terminates in the ventricular muscle. The ventricles contract slightly out of phase. The left ventricle begins to contract about 50 msec after the right one.

To adjust, the heart is connected to the sympathetic and parasympathetic autonomic nervous systems. These two systems provide a reciprocal neural control of the heart. Stimulation of the sympathetic system

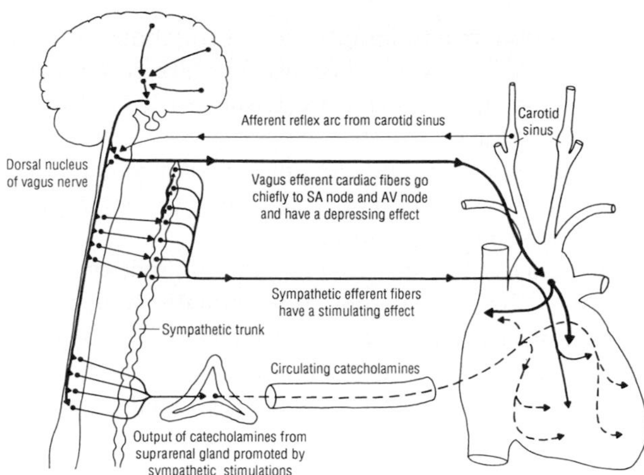

FIGURE 4 Neural and humoral regulation of cardiac function.

increases the release of norepinephrine by the nerve endings that are richly distributed throughout the atrium and the ventricles, allowing reflex regulation of the contractility of the myocardium. The sympathetic nerves also heavily innervate the sinoatrial node and the atrioventricular junction, where increases in sympathetic tone accelerate the heart rate and improve conduction velocity. In contrast, stimulation of the parasympathetic vagus nerve is associated with the release of acetylcholine by the numerous nerve endings in the two atria, the sinoatrial node, and the atrioventricular junction, but not beyond the latter or in the ventricles. Activation of the parasympathetic system slows the heart rate (Fig. 4). [*See* Autonomic Nervous System.]

II. ARTERIAL AND CAPILLARY SYSTEM

A. Structure

The inner layer of the heart, arteries, and veins is lined by endothelial cells. This endothelial lining, the endothelium, provides a smooth, low-friction inner surface in contact with the blood. Arteries form a tubing system, taking the blood from the left ventricle and distributing it throughout the body. The much shorter pulmonary artery system similarly distributes the blood from the right ventricle to the lungs. As the left ventricle pumps the blood under high pressure in a rhythmic manner, the receiving large arteries must be elastic and able to recoil during systole, which results in pulsating waves; the recoil imparts energy to the blood, maintaining its flow throughout the diastole. This is possible because the arteries possess strong elastic walls. Elastic tissue provides for the storage of energy by stretching and relaxing; fibrous elements serve to limit the amount of stretch and to prevent overdilatation of the wall or even rupture under a surge of pressure.

As arteries become smaller, the number of elastic fibers decreases and they are gradually replaced by layers of circular or spiral smooth muscle fibers. By reducing in size, the arterioles offer more resistance to blood flow. However, because of their well-developed muscular layers, they can actively vary their wall tension and diameter, and thus the resistance they offer. The constriction or dilatation of arterioles in specific areas of the body (e.g., skeletal muscle, the skin, and the abdominal region) not only regulates blood pressure, but also varies the distribution of blood in various parts of the body.

Metarteries are still smaller arterial structures connecting the arterioles and the capillaries; they are "thoroughfare channels," because major capillary networks arise from them. The capillary bed is more like a fishnet of interconnecting vessels than a simple parallel arrangement. The flow of blood from the metarteries into the small but numerous capillaries is regulated by precapillary sphincters (ring-shaped muscles surrounding the vessels). Opening and closing of these sphincters alternatively irrigate different capillary networks. It is in the capillaries that the ultimate goal of the cardiovascular system is fulfilled. In these tiny vessels, oxygen and other essential substances can leave the bloodstream mainly by diffusion to the tissues, from which metabolites can drain away in the reverse direction. The flow in capillaries is not steady or even uniform in direction. Instead, there is vasomotion, in which individual capillaries are seen to close completely and then reopen. As a result the flow is intermittent, and it often proceeds in the reverse direction when the vessel reopens. This to-and-fro intermittent capillary flow serves as a periodic "rinsing" of the blood. The contents of a capillary are immobilized for a few seconds, while diffusion and exchange of gases and materials can proceed.

B. Function

The low pressure in the pulmonary artery (about one-fifth of the pressure in the systemic circulation) is, in part, accounted for by the large number of open capillary beds in the respiratory surfaces of the lungs. In addition, pulmonary arterioles have thinner muscular coats and larger diameters than do arterioles in the systemic circulation.

There is vigorous pulsatile outflow from the left ventricle at each beat into the aorta. The rise in pressure at each systole causes an expansion of the aorta and the large arteries, because of their well-developed elastic layer and because blood enters the arterial tree faster than it leaves through the small-bore arterioles. The large elastic arteries store blood under pressure during systole and release it again during diastole, so that the run-off of blood in the peripheral circulation is continuous. The effect is to conserve energy in accelerating the blood column and to reduce the pressure that the smallest vessels must withstand. The energy loss from friction as the blood passes through numerous arteries with decreasing diameter is considerable, and a pulsatile square-wave pattern of flow becomes gradually damped and converted in a continuous flow.

The general organization of the systemic circulation is such that the arterial bed serves as a pressure reservoir from which the circulations to the various organs operate in parallel. Each organ takes the blood supply that it requires by regulating its local vascular resistance primarily on the basis of the metabolic needs, whereas the total peripheral vascular resistance is primarily controlled by cardiovascular reflexes and maintains the pressure in the arterial system. For example, during exercise the vascular resistance in exercising skeletal muscles decreases markedly, and, despite the increase in cardiac output, the blood pressure might decrease were it not for a reflex increase in vascular resistance in nonexercising muscles and certain other areas.

III. VENOUS NETWORK

A. Structure

The small vessels collecting blood from the capillaries are venules. They are simple endothelial tubes supported by some collagenous tissue and, in a larger venule, by a few smooth muscle fibers as well. As venules continue to increase in size, their walls begin to show some characteristics of the arterial wall structure, but are considerably thinner in proportion to the diameter of the vessel.

Smaller veins begin to possess a helical–circular coat of smooth muscle cells located in the middle coat, which contains less muscle and elastic tissue than is found in the arterial wall. Owing to the structure of their walls, veins are much more distensible, can be compressed more easily, and are less well adapted to holding high pressure within them than arteries are. The larger veins eventually collect in one of the two venae cavae, which drain in the right atrium. The venous blood flow is slow because of the greater cross-sectional area, but more continuous than in arteries. Veins are low-resistance channels and only a small pressure gradient exists along the major ones. At intervals in the long veins, particularly in the limbs, the endothelial lining is pulled out to form cup-shaped valve cusps, similar in design to the semilunar cup-shaped valves of the heart. These valves permit unidirectional flow only (i.e., toward the heart).

The systemic veins are not simply a series of passive tubes for the transport of blood back to the heart. They are a reservoir of variable capacity whose compliance is about 30 times greater than that of the arterial tree. The capacity of the peripheral veins can be modulated by contraction or relaxation of the smooth muscle cells in their walls.

B. Function of Veins

The greater part (i.e., 60%) of the blood volume is contained in the low-pressure side of the circulation, which includes the postcapillary systemic veins, the right heart, and the pulmonary circulation. The total blood volume (one-twelfth of the body weight) is maintained relatively constant by appropriate reflex adjustment of renal hemodynamics and hormonal effects on the renal tubules that preserve or excrete sodium and water. As a consequence, mainly of hydrostatic forces, shifts of fluid do occur between the different parts of the cardiovascular system, especially the cardiopulmonary vessels and the systemic veins. To compensate for these shifts, the veins are capable of active and passive changes in capacity that serve to modulate the filling pressure of the heart by adjusting the central blood volume. The active expulsion of blood is due to contraction of the venous smooth muscle cells, which are under autonomic sympathetic nervous control; the passive change is due to a decrease in venous distending pressure, resulting from constriction of the precapillary resistance vessels. These changes are brought about mainly by alterations in sympathetic nerve signals.

The effective venous distending pressure (i.e., transmural pressure) is determined by the pressure of the blood within the vein minus the counterpressure exerted by the surrounding tissues. The venous blood pressure depends on the arterial pressure and the diameter of the arteriole; when the latter dilate, the venous pressure is augmented, leading to an increase in venous capacity. The venous blood pressure also depends on the hydrostatic load, which increases by the gravitational forces on standing, but to the same extent as the arterial pressure, nullifying changes in the driving force of flow in the venous and arterial systems. The increased venous pressures on standing, however, increases capillary pressure and thus filtration and will also augment the pooling of blood at the venous side. Peripheral venous pooling can be counteracted by the massage function of the leg muscles during their contraction. Because of the presence of valves, the direction of flow is unilateral, and the resulting translocation of blood from the legs to the large-capacitance veins improves the filling of the heart and augments the stroke volume. Also because of the increase of intraabdominal pressure during inspiration, blood tends to be drawn into the thorax by virtue of the increased pressure gradient; conversely, on expiration the intrathoracic pressure rises and blood flow stops. Forced expiration (e.g., to raise the intrathoracic pressure) might considerably arrest the return of blood to the thorax temporarily. The venous return to the right heart is further enhanced, since the venous pressure inside the thorax decreases below atmospheric pressure during respiration.

IV. HEMODYNAMICS

A harmonious circulatory system, under a relatively stable perfusion pressure, has to allow for a wide variation in flow, in both the cardiac output and its distribution to different organs and tissues in response to varying demands. To this end, three main features of the circulation have to be kept under control.

A. Regulation of Arterial Blood Pressure

Arterial blood pressure is principally controlled by the activity of the vasomotor and cardiac centers, located in the medulla.

The vasomotor center functions largely as a regulator of the tone of smooth muscles of arterioles by acting through the sympathetic division of the autonomic nervous system. An increase or decrease in tone is brought about by an increase or decrease in the release of norepinephrine at sympathetic nerve endings. Spontaneous activity of the vasomotor center maintains arteriolar tone, which can be increased or decreased by input to the center from peripheral receptors, other brain centers, or substances in the bloodstream. By stimulating specific areas of the vasomotor center, an excitatory (pressor) or inhibitory (dilator) portion can be identified.

The cardiac center governs the heart rate via sympathetic and parasympathetic (vagus nerve) innervation of the sinoatrial node. Increased sympathetic stimulation (which increases the release of epinephrine) and decreased parasympathetic stimulation (which decreases the release of acetylcholine) accelerate the heart rate; the reverse actions slow it. An increase or decrease in the force of ventricular contraction is brought about largely by increasing or decreasing the sympathetic stimulation of cardiac muscle. The best-known and most important regulatory mechanism mediated by these medullary centers is the baroreceptor reflex, adaptive changes in blood pressure initiated by arterial pressor receptors stimulated by tension in the wall of the vessel and thus by pressure within the vessel.

B. Distribution among the Various Local Circulations

The amount of blood flowing to an individual organ is determined by the difference between the arterial and venous pressures in the vessels supplying the organ and by the vascular resistance of the organ. Although the arterial and venous pressures change in situations such as exercise, emotional stress, and eating, most of the alterations in the distribution of blood flow are the result of changes in vascular resistances in the organ (Fig. 1). There are two resistances: the arteriolar and precapillary. The former is controlled by the contraction of smooth muscle in the walls of the small arterioles and is mainly under neural control.

Alterations in the diameter of arterioles affect not only the flow, but also the pressure within capillaries and veins. The precapillary resistance is due to smooth muscle fibers controlling the mouths of the capillary channels and, by governing the number of open capillaries, they control the capillary exchange area. If many sphincters are closed, the blood flows through the main capillary channels, few in a number and relatively large. If the sphincters relax, blood is diverted through a much denser network of capillaries and runs slowly in close contact with the tissues. Precapillary sphincters are influenced mainly by local vasodilator metabolites—produced, for example, during exercise—that readily diffuse through the tissues and cause the sphincters to relax.

C. Cardiac Output

At each level of activity, in health or in disease, a complicated interplay automatically adjusts the extent of shortening of myocardial fibers and, consequently, the stroke volume and the cardiac output. There are four principal determinants of the stroke volume: preload (end-diastolic volume or passive load), which establishes the initial muscle length of the cardiac fibers prior to contraction; afterload (impedance to ventricular empyting during systole or the sum of all the loads against which the myocardial fibers must shorten during systole, including the aortic impedance, the peripheral vascular resistance and the viscous mass of the blood); the contractility or inotropic state of the heart (reflected in the speed and capacity of shortening of the myocardium); and the coordinated pattern of contraction. A fifth determinant, the heart rate, sets the cardiac output, which is the product of stroke volume and heart rate.

All of these determinants interrelate, but their relationships are not fixed; they play greater or lesser roles depending on the functional state of the heart. Thus, when the inherent contractility (i.e., inotropic state) is impaired, stroke volume and cardiac output might be maintained by ventricular dilatation (i.e., the Frank–Starling mechanism). Indeed, according to the Frank–Starling law, there is a direct relationship between end-diastolic fiber length and ventricular work, as the energy of contraction is a function of the length of the muscle fibers.

V. Lymphatic System

A. Structure

In humans the exchange of fluid across the capillaries throughout the body into the interstitial fluid amounts to about 20 liters per day. This moves in both directions, and the great bulk is returned to the systemic circulation through the walls of the capillaries and the venules. Approximately 10–20%, however, is returned to the systemic circulation by a circuitous route, the lymphatic system. The lymph capillaries start with microscopic blind ends and consist of a single layer of overlapping endothelial cells attached by anchoring filaments to the surrounding tissue.

With muscular contraction these fine strands might distort the lymphatic vessel to open spaces between the endothelial cells and permit the entrance of protein, large particles, and cells present in the interstitial fluid. The small lymphatic vessels form a network in almost all tissues and have the same relationship to tissue spaces as have blood capillaries. Small lymphatic vessels converge to become larger and unite in lymphatic trunks, which have layered, but still thin, walls. These larger vessels contain valves and join finally to form the two large lymphatic ducts that empty into the left and right subclavian veins. The content of the lymph (Latin for "water") is similar to that of plasma, except that the lymph has only one-half as much protein. Because of the thinness of their basement membranes, the lymphatic capillaries are much more permeable to macromolecules than are the blood capillaries, so that protein molecules that escape from the latter are taken up into the lymph channels.

There is no pump to move the lymph; its movements are principally due to the squeezing action of the adjacent muscular tissues. Lymphatic circulation is consequently slow and uncertain.

B. Function

Phylogenetically, the need for a lymphatic system arose as soon as a closed cardiovascular system developed. In mammals the necessity for providing a high-pressure system to ensure an adequate supply of oxygen to the tissues created a situation favoring the transudation of fluid from capillaries. The role of this mechanism was minimized by the increase in plasma proteins, which exerted a considerable oncotic pressure. However, albumin that leaves the vascular compartment cannot be returned to the blood, since back-diffusion into the capillaries cannot occur against the larger albumin concentration gradient. Were the protein not removed by the lymph vessels, it would accumulate in the interstitial fluid and act as an oncotic force to draw more fluid from the blood capillaries to produce increasingly severe edema. In addition, there still remains the problem of clearing the tissue spaces of substances that had leaked out or that were not absorbed by the blood. As 50% or more of the total circulating protein escapes from the blood vessels, this extravascular protein cannot all be resorbed into the blood capillaries and hence must return by way of the lymphatic vessels, which act as an overflow drainage system. In 24 hours a volume of fluid equivalent to the total plasma volume (approximately 4 liters) and over 50% of the total circulating fluid passes in the thoracic duct alone, emphasizing the importance of the lymphatic system.

BIBLIOGRAPHY

Abramson, D. I. (ed.) (1962). "Blood Vessels and Lymphatics." Academic Press, New York.

Berne, R. M., and Levy, M. (eds.) (1972). "Cardiovascular Physiology." Mosby, St. Louis.

Braunwald, E. (ed.) (1996). "Heart Disease. A Textbook of Cardiovascular Disease." Saunders, Philadelphia.

Burton, A. C. (ed.) (1965). "Physiology and Biophysics of the Circulation. An Introductory Text." Yearbook, Chicago.

Forrester, J. M. (ed.) (1985). "A Companion to Medical Students. Anatomy, Biochemistry and Physiology." Blackwell, Oxford, England.

Hardin, G. (ed.) (1961). "Biology. Its Principles and Implications." Freeman, San Francisco.

Jacob, S. W., Francone, C. A., and Lassow, W. J. (eds.) (1978). "Structure and Function in Man." Saunders, Philadelphia.

Schlant, R. C., Alexander, R. W., O'Rourke, R. A., Roberts, R., and Sonnenblick, E. H. (eds.) (1994). "The Heart." McGraw–Hill, New York.

Shepherd, J. T., and Vanhoutte, P. (eds.) (1975). "Veins and Their Control." Saunders, Philadelphia.

Shepherd, J. T., and Vanhoutte, P. (eds.) (1979). "The Human Cardiovascular System. Facts and Concepts." Raven, New York.

Smith, A. (ed.) (1968). "The Body." Allen & Unwin, London.

Wyngaarden, J. B., and Smith, L. H. (eds.) (1988). "Cecil's Textbook of Medicine." Saunders, Philadelphia.

Carnivory

JOHN D. SPETH
University of Michigan

GLOSSARY

Australopithecines Extinct hominids of the Pilocene and Lower Pleistocene (ca. 6 to 1 million years ago) in Africa that walked erect (bipedally) and had human-like hands and teeth, but much smaller cranial capacities and more ape-like jaws and skulls. Several species have been recognized (e.g., *Australopithecus afarensis, A. africanus, A. robustus,* and *A. boisei*)

Hominid Primates belonging to the family Hominidae, which includes modern humans (*Homo sapiens sapiens*) and all premodern members of the genus *Homo,* as well as closely related ancestral forms known as Australopithecines

Homo erectus Extinct human species found throughout Africa and southern Eurasia during the Lower and Middle Pleistocene, dating between about 1.8 million and 300,000 years ago; had a cranial capacity intermediate between the Australopithecines and anatomically fully modern humans, and was probably the first hominid to use fire

Neanderthal Last extinct human form (*Homo sapiens neanderthalensis*) prior to the appearance of anatomically fully modern humans (*H. sapiens sapiens*), dating to the Upper Pleistocene between about 150,000 and 35,000–40,000 years ago

Pleistocene Geological period, known popularly as the Ice Age, that lasted from about 1.75 million to 10,000 years ago; generally divided into three periods: Lower (ca. 1.75 million to 700,000 years ago), Middle (ca. 700,000 to 130,000 years ago), and Upper (ca. 130,000 to 10,000 years ago)

Pliocene Geological period immediately preceding the Pleistocene that lasted from about 6 million to 1.75 million years ago

Primate Mammalian order that includes prosimians (e.g., lemur and loris), New and Old World monkeys (e.g., howler, spider, macaque, and baboon), apes (e.g., gibbon, orangutan, chimpanzee, and gorilla), and hominids (e.g., Australopithecines, *Homo erectus,* Neanderthal, and modern human)

Taphonomy Field that studies the processes (e.g., natural decay, trampling, transport by flowing water, and transport and destruction by carnivores) that affect an assemblage of bones from the time an animal dies until its bones have become incorporated into the fossil record

THE FREQUENT CONSUMPTION OF MEAT BY MODern humans, often in prodigious quantities, has attracted the attention of scholars since Darwin. Because humans are believed to have evolved from a primate ancestor, and since most living primates—including chimpanzees, our closest living relatives—subsist largely on vegetal foods (e.g., fruits, leaves, and shoots), the development of the human proclivity to kill and consume the flesh of animals, particularly large and dangerous ones, has come to play a central role in most theories of hominid origins and evolution. These theories, often referred to collectively as the hunting hypothesis, are now being challenged by new insights from living hunters and gatherers, human nutrition, chimpanzee behavior in the wild, archaeology, and taphonomy.

I. HUNTING HYPOTHESIS

That humans are descended from a primate ancestor and that we very likely share a common ancestor with

ENCYCLOPEDIA OF HUMAN BIOLOGY, Second Edition, VOLUME 2. Copyright © 1997 by Academic Press. All rights of reproduction in any form reserved.

the chimpanzee, a great ape found in the tropical forests and woodlands of sub-Saharan Africa, are widely accepted today by most scholars concerned with hominid origins and evolution. Based on genetic, paleontological, geological, and other evidence, there is also a growing consensus that the divergence between the human and ape lines probably occurred around 6 million years ago, plus or minus about 1 million years. Every year, as new hominid fossils are found and previous finds are reanalyzed and more securely dated, as detailed paleoenvironmental and taphonomic studies reveal the context in which these early hominids lived, died, and became incorporated into the fossil record, and as our understanding of the fundamental adaptive and evolutionary processes that give rise to behavioral and genetic differentiation are refined, the when, where, and how of hominid origins and evolution are becoming increasingly clear.

Anthropologists and biologists are also approaching this fascinating problem from the other end of the time scale. Some are studying the behavioral ecology and diet of chimpanzees and other living primates, whereas others are focusing on the adaptations and subsistence practices of contemporary hunter–gatherers or foragers (living peoples whose life-styles are thought to most closely approximate the way our ancestors lived for several million years prior to the emergence of farming and animal husbandry, a series of dramatic economic and social changes that occurred only within the last 10,000 years). These behavioral and comparative studies of living populations help pinpoint important similarities and differences between humans and our closest living nonhuman relatives, and in the process shed light on the very processes that gave rise to the human line. [*See* Primate Behavioral Ecology.]

One of the most striking contrasts between contemporary hunter–gatherers and most living primates, a difference that was clearly recognized over a century ago by Darwin, is that humans are highly successful predators, even of large and dangerous prey. And though humans are basically omnivores, they are capable of eating (and often do eat) prodigious amounts of meat. Nowhere is this more strikingly apparent than among traditional Eskimos living in the Arctic; these hunters subsist almost entirely on a diet of meat obtained by killing large and often dangerous sea mammals, including seals, walruses, and even whales, as well as caribou, polar bears, and a variety of other terrestrial mammals. Most primates, on the other hand, are largely vegetarian, deriving the bulk of their diet from fruits, leaves, shoots, nuts, resin, and other plant parts.

Darwin, in fact, believed this contrast was so fundamental that he used it as the basis for a theory of human origins that, somewhat modified, remains persuasive today. In its modern form the theory—often known as the hunting hypothesis—holds that during the Miocene in sub-Saharan Africa—roughly 25 to 6 million years ago—climatic conditions gradually deteriorated, becoming drier and more seasonal. These changes favored the expansion of open woodlands and savanna grasslands at the expense of dense tropical rain forests. Many species of great ape were unable to cope with these dramatic habitat changes and became extinct, but some populations of one of these species, thought by most to have been the common ancestor of both chimpanzees and humans, adapted to them by turning to the one obvious new resource that these expanding grasslands offered: large herds of antelopes and other herbivores. But to do this effectively, these "protohominids," known as Australopithecines, would have exposed themselves to new dangers, especially from the many other predators that also preyed on the herbivores. These included lions, leopards, several species of hyena, and many other large dangerous carnivores. To protect themselves and to more effectively kill large herbivores and cut through their thick tough hides, the hominids began to rely on hand-held sharp-edged tools, probably first of wood, bone, or perhaps even shell, then of flaked stone.

Gradually, a positive-feedback relationship emerged in which increasing tool use by the hominids favored greater reliance on a bipedal stance in order to free their hands, as well as greater intelligence (and hence larger brain size) as they came to rely more and more on tools and other forms of learned or culturally mediated behavior. These changes were also accompanied by a gradual reduction in the size of their initially large and intimidating canines as the function of these teeth for both defense and gaining access to tough or inaccessible vegetable foods was supplanted by tools.

More recent variants of this basic hunting hypothesis attribute other characteristically human traits to human carnivory as well. For example, some anthropologists have argued that the bonanza of meat provided by a single large animal carcass favored the emergence of food-sharing, not just between mother and offspring or among siblings, as is common in primates, but among a much broader network of less closely related kin (and perhaps even among nonkin). The hunting of larger and more dangerous prey might also have contributed to the development of the division of labor that typifies most contemporary foraging societies, in which males do the hunting and then transport meat back to a central place to provision females with dependent young and to share with other members of the group. A few anthropologists have even argued that the skills and the knowledge needed to successfully locate and kill large dangerous game favored the development of new and more effective means of communication, a process that culminated in the emergence of true language. These factors are also thought to have favored increased levels of cooperation among larger numbers of hunters, enhancing the importance of group integration beyond the biological family and ultimately giving rise to new and more complex forms of social organization. Hunting and meat-eating, therefore, have often been viewed as the locus of the pivotal selective forces that gave rise to protohominids and gradually transformed these remote ancestors into what we are today. [*See* Evolving Hominid Strategies.]

Despite its elegance and compelling simplicity, various lines of evidence are now beginning to raise serious doubts about the validity of the hunting hypothesis. Although no equally comprehensive and compelling new theory has emerged to replace it, our views about the role of carnivory in the evolutionary history of the human line have begun to shift, in some respects dramatically, and they likely will continue to shift. The new evidence and insights are coming from many different sources, including ethnographic studies of modern hunter–gatherers, field studies of wild free-ranging primates (especially chimpanzees), human nutrition, archaeology, taphonomy, and a host of others. In subsequent sections the more important of these new insights are considered and the role that human carnivory might have played in hominid origins and in the subsequent evolution of our species is reevaluated.

II. PROTEIN IN HUNTER–GATHERER DIET

Hunter–gatherers (or forages) are people who live entirely or largely without the benefits of agriculture or animal husbandry. Prior to the end of the Pleistocene, which occurred about 10,000 years ago, humans everywhere subsisted exclusively by hunting (including fishing) and gathering, and even after the appearance of farming communities, many populations continued to live as foragers, some up to the present. Even as recently as the period of European colonial expansion in the 15th and 16th centuries, there were still hundreds of hunting–gathering groups throughout the world. For example, in the Old World these included Chukchi in Siberia; Ainu in Japan; Agta and Batak in the Philippines; Australian Aborigines; Andaman Islanders; Vedda in Sri Lanka; and Pygmies, Hadza, and Bushmen (San) in sub-Saharan Africa. In the New World there were also many foraging groups, including Eskimos, Dogrib, and Mistassini Cree in the Arctic and Subarctic; Shoshoni, Paiute, Washo, Chumash, and many others in the Great Basin and California in the western United States; Seri in northern Mexico; and Ache, Hiwi, Ona, Yaghan, and numerous others in South America.

All of these historically or ethnographically known foragers do (or did) a great deal of hunting and attribute a lot of social importance to it, although the amount of time they devote to hunting and the actual contribution of animal foods to their diet vary greatly from group to group. Despite this variability, there is a broad positive correlation between geographic latitude and the proportion of meat in the diet. Not unexpectedly, in the Arctic, plant foods contribute only minimally to the diet, whereas hunting of both land and sea mammals, as well as fishing, constitutes the bulk of the foods consumed, both by weight and in terms of total calories. Outside the Arctic, however (especially as one approaches the equator), the contribution of plant foods to the larder increases sharply, attaining levels in excess of 60–80% by weight of the total food intake in many (although not all) tropical groups.

This observation—that, outside the Arctic, plant foods form a substantial, and often the dominant, component of the foragers' diet—is not new. It was already clearly recognized over a century ago by Friedrich Engels, but for some reason was brushed aside by most anthropologists until the 1960s, when intensive fieldwork among the Kalahari San (Bushmen) in southern Africa and similar work among Australian Aborigines made it impossible for anthropologists to continue to ignore the importance of plant foods in the diet of most hunter–gatherers. In fact, the change in perspective was so dramatic that some anthropologists now insist that we refer to foraging populations as gatherer–hunters, rather than hunter–gatherers.

This change in perspective has also led anthropologists to question some of the basic assumptions of the classic hunting hypothesis. If plant foods constitute the bulk of the forager diet in many areas of the tropics today, perhaps this was also true of the early hominid diet in the distant past, since our earliest ancestors, the Australopithecines, evolved in the tropics and archaeological evidence indicates that hominids probably remained there for several million years, not expanding into the more temperate environments of Eurasia until about 1 million years ago or even more recently, long after the emergence of a more advanced form, *Homo erectus*. This observation, of course, in no way implies that carnivory played *no* role in the origins or evolution of early hominids. It does suggest, however, that the hunting hypothesis might be overly simplistic in downplaying or ignoring the role of plant foods (and, of course, nonsubsistence factors) in this process.

Contemporary hunter–gatherers have also provided other insights that are gradually transforming our views about the nature and evolutionary significance of human carnivory. For example, for many years anthropologists, like their colleagues in the medical and nutritional fields, felt that meat was important first and foremost as a source of high-quality protein. After all, meat is one of the richest natural sources of protein, and the protein is nutritionally ideal, because it possesses essential amino acids that the human body is incapable of synthesizing on its own and because the amino acids are present in proportions that are optimal for use by the body. In comparison, many plant foods are impoverished in protein, the protein might be difficult to digest, and one or more critical amino acids might be underrepresented or missing. [*See* Proteins (Nutrition).]

From this perspective hunting was seen primarily as an efficient strategy for acquiring high-quality protein and protein-derived calories, and much of human evolution was thought to have been driven by technological and social developments that enhanced the success with which foragers were able to capture this critical and presumably limiting macronutrient. Fat—which generally accompanies meat as marbling within the muscle tissue; as subcutaneous layers along the back, neck, chest, and belly; as deposits around the internal organs; and as deposits within the marrow cavities of the limb bones—though an obvious source of calories, was seldom given adequate attention by scholars of human evolution, and (as discussed in the next section) has often received extremely "bad press" from nutritionists because of its probable links to the so-called diseases of civilization: obesity, atherosclerosis, high blood pressure, adult-onset diabetes, and heart disease.

There is no question that protein is an extremely important macronutrient, both as a source of amino acids necessary for normal growth and maintenance and as a source of calories. However, recent studies of living foragers, combined with new insights from medicine and nutrition, are beginning to suggest that we might have placed too much emphasis on the acquisition of high-quality protein while underestimating the importance of fat (and carbohydrate) to peoples whose sustenance came entirely from wild resources. As a consequence it has also become necessary to reexamine some of the basic assumptions of the traditional hunting hypothesis.

The tremendous importance of fat to foragers is immediately evident in their hunting and butchering decisions. For example, they frequently seek out and kill only the fattest animals, ignoring the lean ones, and they often then select just those cuts of meat and marrow bones that have the highest fat content, abandoning the remaining parts or feeding them to their dogs. Sometimes this behavior merely reflects an excess of meat from a particularly successful hunt that permits the participants the luxury of choice. However, foragers also abandon entire animals, if, upon butchering, they prove to be fat depleted, even when the hunters themselves are short of food, and despite the fact that they might already have invested considerable effort to locate and kill the animals. In other words, foragers often seem to be hunting for calories, not protein, and specifically for calories provided by a nonprotein source (i.e., fat). This behavior is evident year-round, but paradoxically it becomes most pronounced not when food is abundant, but during periods of food shortage, when the animals are leanest, the hunters' supply of vegetable foods has dwindled, and the hunters themselves are losing weight. Why don't the hunters eat all of the meat made available by a kill, regardless of its fat content, using the excess protein for energy? The nutritional reasons that underlie this seemingly counterintuitive behavior are providing valuable insights into the nature and evolution of human carnivory.

III. PROTEIN AND HUMAN NUTRITION

Travel accounts and ethnographies worldwide are filled with graphic descriptions of the almost glutonous attitude of foragers when it comes to fat; they often devour the fattiest organs and marrow immediately at the kill, including hunks of fat sliced from the back fat, and whenever possible they consume rendered animal fat or fatty marrow mixed in with other foods at almost every meal. Among many foragers, animal fat (including fish oil) is so highly prized that it is widely traded, and it is frequently used in rituals.

The intense interest in fat shown by hunter–gatherers seems to stand in stark contrast to the dire warnings of nutritionists and medical researchers, who see a close connection between high-fat diets and many of the diseases of civilization. Their views about the role of fat in the human diet are beginning to change, however, and hunter–gatherers have played an important part in this change. For example, traditional Eskimos consume prodigious amounts of fat, amounts far in excess of what most people in modern industrial nations eat, and yet they display virtually no signs of these diseases. One reason, of course, is that Eskimos are normally far more active than we are and burn off the calories that less active people would store as fat. They also burn off more calories just to keep warm.

But there is another reason that Eskimos and other foragers can consume so much fat without apparent health problems. Nutritionists have found that the fatty acid composition of fat from domestic animals is different from that of fat from wild animals, a reflection of the "unnatural" diet fed to the domestic forms, and it is the "domestic" fat, not fat per se, that appears to be most closely linked to the diseases of civilization. The concern we now show about the composition of fat in our diet is reflected, for example, by the fact that most of us eat margarine, which is high in polyunsaturated fats, rather than butter, which contains more saturated fats. The growing importance of fish oils rich in so-called ω-3 fatty acids reflects a similar concern. [*See* Fats and Oils (Nutrition).]

Nutritionists are also concerned, though, about the quantity of fat in the diet, not just its composition. How does one account then for the fact that all hunter–gatherers, not just Eskimos, display such an avid interest in fat? One reason, of course, is that fat makes foods taste good, since many flavor-enhancing substances are soluble only in fat, and it also conveys a feeling of satiety when one has eaten. Fat is also a critical source of both fat-soluble vitamins and essential fatty acids. Although foragers are almost certainly unaware of the existence of either nutrient, they are well aware that the inclusion of adequate fat in their diet contributes to an overall feeling of well-being and good health. Moreover, fat is a highly concentrated source of energy, supplying more than twice the calories per gram than either protein or carbohydrate, and fat is more efficiently metabolized than protein.

Unlike the situation in modern industrialized nations, however, where fatty foods are abundant and readily available year-round at the neighborhood supermarket, in the world of foragers living outside of the Arctic fat is actually quite scarce. Most wild animals, even in prime condition, are generally much leaner than domestic ones, and during the late winter and spring (or late dry season and early rainy season in tropical latitudes), as forage abundance and quality decline, wild animals can become very lean. This is especially true for pregnant or nursing female animals supporting a developing fetus or newborn calf during periods when their own food intake is restricted. It is also true for calves, which often have difficulty getting enough forage during stressful seasons, and for older animals, whose teeth have begun to wear out, reducing their ability to masticate food.

Thus, fat is usually abundant only during seasons when animals are in prime condition, and at such times it is consumed by foragers with great zeal. During ensuing seasons of shortage, however, both animals and foragers become lean again. As a consequence, obesity and other diseases of civilization that are linked to excessive fat intake are not a problem to foragers as long as they are living under traditional conditions in which food intake follows an annual "boom–bust," or feast–famine, cycle. Only when foragers become settled and acquire the "benefits" of a Western diet do they begin to show the same diseases of civilization that we do.

There is another important reason that animal fat is critical to hunter–gatherers, particularly during times of stress, when plant foods become scarce or unavailable. There appears to be an upper limit to the total amount of protein that one can safely consume on a regular basis. This limit—best expressed as the total number of grams of protein per unit of lean body mass that the body can safely handle—is about 300 g, or roughly 50% of one's total calories under normal nonstressful conditions. Protein intakes above this threshold, especially if they fluctuate sharply from day to day, can exceed the rate at which the liver can metabolize amino acids and the body can synthesize and excrete urea, leading ultimately to hypertrophy and functional overload of the liver and the kidneys, elevated (even toxic) levels of ammonia in the blood, dehydration and electrolyte imbalances, severe calcium loss, micronutrient deficiencies, and lean tissue loss.

Perhaps of greater significance for the success and long-term viability of foraging groups, however, is the recent suggestion by nutritionists that the safe upper limit to total protein intake for pregnant women might actually be considerably lower than 300 g. Several studies have shown that supplementation of maternal diets with protein in excess of about 20% of total calories usually leads to declines, not gains, in infant birth weight, and often to increases in perinatal mortality as well. Infants who are born prematurely appear to be most severely affected by such high maternal protein supplements. Declines in infant birth weight might be seen in situations in which maternal weight increases in response to the dietary supplementation, and the decline becomes more pronounced the greater the mother's protein intake is above the 20% threshold. Birth weight also declines when the mother's total calorie intake is restricted, but is most extreme when the diet is both low in energy and very high in protein.

Thus, adult foragers must normally get more than one-half of their total daily calories from nonprotein sources in order to remain healthy, and pregnant women must get considerably more. When plant foods are abundant, this poses little problem (as long as the plant foods themselves are not excessively rich in protein), but during seasons or interannual periods of food shortage, when vegetable resources dwindle and wild animals become very lean, finding adequate nonprotein energy sources becomes increasingly difficult.

One obvious way that modern foragers cope with this problem is to become extremely selective in the animals they hunt and in the body parts they consume, discarding those parts that are too lean. They also invest considerable effort to extract the grease that is dispersed within the spongy bones, by smashing up joints and vertebrae and boiling out the precious lipids. Despite these efforts, by the end of the dry season, foragers like the Kalahari San might be forced to consume almost 2 kg of lean meat per person per day (roughly 420 g of protein), and yet their body weight declines. This astounding meat intake, not uncommon among contemporary foragers, means that protein at times can contribute up to 70% or more of their daily caloric intake, a level far in excess of the 300 g suggested earlier as the safe upper limit, and therefore almost certainly a sign of stress, not affluence, as many anthropologists once thought. And for pregnant women, such high protein intakes could have devastating reproductive consequences.

Modern hunter–gatherers have the benefit of stone or steel axes to smash up bones and clay or iron pots to boil them in to extract the grease. These seemingly mundane items of material culture, absolutely critical to modern foragers during periods of resource stress, are actually comparatively recent additions to their technological inventory. Stone axes and ceramic vessels did not make their appearance until the very end of the Pleistocene at the earliest, and of course metal items are very recent introductions. The use of stone-boiling to extract the grease, a much more labor-intensive process in which stones are heated in a fire and then transferred to a watertight hide, gut, or basketry container, probably also dates to the latter part of the Pleistocene or later. Thus, prior to the Upper Pleistocene the grease in the spongy bones might have been largely, if not entirely, inaccessible to foragers, and as a consequence their reliance on hunting and meat-eating might have been quite different from the pattern we see among modern hunter–gatherers.

IV. CHIMPANZEE CARNIVORY

If the chimpanzee is our closest living relative, as most scholars now believe, then patterns of behavior shared by both chimpanzees and modern foragers should provide invaluable clues about the behavior of protohominids. Until the early 1960s, however, when Jane Goodall began her pioneering work in the Gombe National Park—a strip of rugged forested terrain along the eastern shore of Lake Tanganyika in Tanzania—little was actually known about free-ranging chimpanzees. Over the three decades since her studies began, our understanding of chimpanzees and other primates under natural conditions has improved immeasurably, and, not surprisingly, our views about the origins and evolutionary significance of human carnivory have been altered in the process.

In Darwin's time, and in fact right up to the 1960s, chimpanzees were thought to be vegetarians, actually frugivores, animals whose diet includes large quantities of fruits. As field studies proceeded, however, it became clear that chimpanzees also ate insects, especially termites, which they collected from nests using simple "fishing" tools made from grass stems. Perhaps the most dramatic change in our views of the chimpanzee diet came when researchers first began to observe these supposedly peaceful vegetarians killing and eating mammals, especially the young of several species of monkeys and small antelopes. Chimpanzees were clearly omnivores, not vegetarians.

Until the chimpanzees became habituated to the presence of their human observers, however, researchers had great difficulty following them closely enough to get reliable information on the frequency of meat-eating or the conditions that triggered the behavior. Thus, in the early stages of the Gombe research, the chimpanzees were provisioned with bananas at a central feeding station so that they could be observed more easily. Unfortunately, the concentrated food resource provided by the bananas also attracted baboons, a situation that often led to intense competition and conflict between the two primate species. Many of the cases of meat-eating seen in the early stages of research at Gombe occurred when chimpanzees killed and consumed immature baboons near the provisioning site. Most researchers therefore tended to attribute chimpanzee meat-eating to the competitive environment artificially created by food provisioning.

Over the years, however, more and more incidences of killing and meat-eating were observed, and they continued to be reported long after provisioning had ceased. Moreover, cases of killing and meat-eating began to be reported from other chimpanzee study sites as well, in both Central and West Africa. As field studies progressed it became evident that chimpanzee predation was neither just an occasional event nor an artifact of provisioning, but occurred repeatedly and with design. It is now clear that chimpanzees actually hunt deliberately, often cooperatively, and share meat with other group members. In fact, recent work in the Ivory Coast has shown that West African chimpanzees hunt, on average, almost once every 3 days, far more frequently than do their counterparts in Gombe. They also hunt in larger groups, focus more on adult prey, share the kill more actively and widely, and in at least one case consumed an estimated 1.4 kg of meat and bone per individual. And though chimpanzees do not use tools to kill their prey, the Ivory Coast chimpanzees occasionally use sticks to extract marrow from the limb bones.

These ongoing studies of chimpanzees in the wild clearly demonstrate that our closest living relatives hunt and share meat, behaviors that until quite recently anthropologists had reserved solely for humans. In fact, it now seems likely that the common ancestor of both chimpanzees and humans already hunted, making it improbable that protohominids were the first primates to "discover" carnivory, a discovery that supposedly then transformed them from apes into hominids. The insights from chimpanzee studies have in no way made the hunting hypothesis obsolete, however. Instead, many anthropologists now feel that the explanation for hominid origins lies not in the discovery of hunting per se, but in the quantity of meat that early hominids were able to procure; in other words, in their success at locating and killing large game.

V. HUNTING AND SCAVENGING IN PREHISTORY

Since the pioneering work by Mary and Louis Leakey in the 1950s and 1960s at Olduvai Gorge in Tanzania, our understanding of the archaeological record of early hominids has improved dramatically. Many well-preserved sites have been painstakingly excavated, mapped, and analyzed, providing tantalizing "snapshots" of the behavior of our earliest ancestors as much as 2 million years ago. In case after case the sites revealed the close juxtaposition of sharp-edged stone flakes and heavy-duty chopping or pounding tools with the bones of many different species of animal, ranging in size from rodents, birds, and turtles to huge dangerous ones such as buffaloes, hippopotamuses, and relatives of the modern elephant. The repeated association of tools and bones seemed to be incontrovertible proof that early hominids were avid and highly successful hunters and provided one of the principal cornerstones of the traditional hunting hypothesis.

While archaeology was uncovering the hard evidence of past human activities, another field—taphonomy—was beginning to explore other issues that would ultimately force archaeologists to reexamine many of their basic assumptions. Taphonomy is the study of the processes that can alter and distort an assemblage of bones from the time an animal dies until its bones become incorporated into the fossil record. Thus, for example, grassland-dwelling animals (e.g., wildebeest) might drown while crossing a stream, becoming fossilized in fluvial deposits alongside the bones of crocodiles, hippos, and other water-loving species with which wildebeest, in life, are obviously unassociated; softer, less dense animal bones, such as the proximal ends (i.e., those closest to the body) of femurs, tibias, and humeri, might decay more readily than more durable ones, such as distal humeri or teeth; smaller bones (e.g., those of rodents) are more likely than larger ones (e.g., from antelopes or elephants) to be trampled into the substrate and preserved; lighter, less dense bones are more likely than compact ones to be winnowed away by a flowing stream; hyenas might selectively destroy some of the marrow bones and remove others from a hominid campsite after the hominids have abandoned the locality; and so forth.

Taphonomic studies began to make it clear that the mere juxtaposition of stone tools and bones did not demonstrate their functional association. Flowing water could have brought the two together in a channel deposit, or hyenas and humans could each have taken advantage of the same shade tree, but at slightly different times, producing a fortuitous association of bones and tools. Because of these basic taphonomic questions, archaeologists found themselves almost having to start from scratch, demonstrating through painstaking analyses what they had previously merely assumed. Though tedious, this process of reevaluation, which is still going on, has proven to be extremely fruitful.

For example, we now know that some, but by no means all, of the early sites where stone tools and bones occur together do, in fact, represent places where hominids butchered animal carcasses. The best evidence is provided by unambiguous cut marks on many of the bones, produced by sharp-edged flakes, and by use–wear studies of some of the stone tools themselves, which reveal distinct polishes on their edges, shown by experimental work to be the product of meat-cutting. Other, somewhat less direct and hence more controversial, evidence includes the fact that bones of many species occur together in a single place, a pattern different from that normally found at hyena or lion kills. The proportions of different skeletal elements in some of these sites make it unlikely that the bones were transported there by hyenas or other carnivores.

The taphonomic studies of these sites have raised a different issue, however—one that has far-reaching implications for the hunting hypothesis. Many of the bones at these sites have gnaw marks, punctures, and other clear evidence of carnivore damage. Moreover, even in sites where it can be shown convincingly that flowing water and differential decay played no significant role in altering the composition of the bone assemblages, less dense limb elements with lots of spongy tissue (e.g., the proximal ends of femurs, tibias, and humeri) are, nevertheless, conspicuously underrepresented. Detailed studies of the feeding behavior of many different carnivores suggest that these bones have most likely been destroyed or carried off by hyenas.

These observations have raised an even more fundamental question, one that had not been seriously considered until quite recently. Did humans kill the animals, and hyenas scavenge the remains that littered the campsite once its human occupants had left? This scenario, of course, is entirely compatible with the traditional archaeological view. Or did lions or hyenas kill the animals, and humans merely scavenge the carcasses for edible scraps of meat and marrow after the carnivores had finished feeding or were driven off by the humans? Many scholars now believe the latter is more likely.

If early hominids were basically scavengers, not hunters, what role did scavenging play in hominid origins and evolution? The answer to this question remains far from resolved, but certain facets of the problem are beginning to become clear. One facet concerns the kinds of scavenging opportunities that would have been available to early hominids as they began to exploit the expanding woodlands and open grasslands of sub-Saharan Africa. Ongoing studies of hyena scavenging in the wild are providing some interesting insights.

Hyenas are voracious carnivores, hunting as well as scavenging from carcasses killed by other predators, and in a matter of minutes they can consume the entire carcass, bones and all, of a small-sized antelope, such as a gazelle. Not unexpectedly, however, the larger the carcass, the more of it the hyenas are likely to leave behind, other things being equal (e.g., the level of competition among hyenas, or between hyenas and other predators, at the site of the carcass). Thus, hyenas might consume most of the carcass of medium-sized animals, such as wildebeest and zebras (live weights between about 100 kg and 350 kg), but abandon the vertebrae, skull, and lower limbs partially or largely intact. From these elements early hominids would have been able to scavenge scraps of edible muscle tissue, marrow, and, of course, brain. Sharp flakes would have made it possible for them to open the skin around the limb bones in order to get at the marrow within the shaft; a stone or limb bone could have been used as a hammer to break open the shaft or to open the skull to get at the brain. Without appropriate boiling technology, however, grease in the spongy tissue of the limb bones and vertebrae probably would have been largely inaccessible.

Hyenas often abandon the largest carcasses—those weighing in excess of about 350 kg (e.g., the buffalo, rhinoceros, giraffe, and elephant)—more or less intact. These would have provided early hominids with a bonanza of meat and marrow, but kills or natural deaths of these megafauna are infrequent events.

These observations suggest that early hominid scavengers would have been most likely to encounter the partial remains of medium-sized animals, from which they could have gleaned scraps of meat and marrow, particularly from the lower limbs, and perhaps brain. If they transported the edible parts back to a central place to process them in comparative safety (and perhaps to share with other members of the group), these are the skeletal elements that we can expect to find in greatest abundance in Pliocene–Pleistocene archaeological sites.

An important issue that remains unresolved, however, is whether early hominids were passive or active scavengers, that is, whether they had to wait until hyenas and other predators had finished feeding on a carcass or instead were able to drive them away and thereby get earlier access to it. If they were active scavengers, they would have been able to transport back to their home base many more of the less marginal meaty parts, such as the upper limbs and the rump.

The difficulty in resolving whether early hominids were primarily active or passive scavengers is largely methodological at this point. Understandably, faunal analysts generally focus on those bone fragments that can be identified realiably to species. Unfortunately, usually only the joints (i.e., epiphyses) of the limbs are suitable; shaft fragments are virtually impossible to identify to species. The proximal epiphysis of the humerus and the tibia and both epiphyses of the femur, however, contain lots of spongy tissue that is relished by carnivores and easily destroyed. This means that even if early hominids regularly transported these meaty elements back to their camp, hyenas would almost certainly carry them off as soon as the camp was abandoned. The result is that, to the faunal analyst, the archaeological bone assemblage would appear to be devoid of meaty upper limbs.

To get around this problem, archaeologists are now beginning the painstaking task of identifying the thousands of shaft fragments, obviously not to species, but to approximate carcass size. The initial results have been intriguing but controversial. They suggest that, at least on some early hominid sites, upper limbs originally might have been fairly well represented, but only shaft fragments now remain. This indicates that early hominids might have been active scavengers capable of driving away predators from carcasses before the meatiest portions had been totally devoured. On the basis of these shaft fragment studies, some archaeologists have even suggested that early hominids might have hunted these medium-sized animals, a position that, if correct, brings us full circle to where we began when taphonomic insights first challenged the hunting hypothesis.

Given the considerable evidence now available for successful and regular hunting by chimpanzees, most notably by those in the Ivory Coast, it seems extremely likely that our Australopithecine ancestors were, in fact, successful hunters, at least of the small- and medium-sized classes, although it remains a matter of conjecture whether these small hominids were capable of intimidating hyenas and lions and driving them away from freshly killed carcasses.

Scavenging studies have revealed another interesting facet of hyena behavior. During the rainy season many herbivores move away from permanent water sources into the open grasslands, where they rely on temporary pools of water. In these open habitats hyenas are extremely aggressive scavengers (and hunters) and often leave little remaining of a carcass that could then be scavenged by hominids. Groups of hyenas might even drive lions away from a carcass before they have finished feeding. In contrast, during the dry season, when many herbivores stay much closer to permanent water sources, hyenas often leave lion kills untouched or only partially devoured. The reason for this seasonal difference in hyena feeding behavior appears to be related to their fear of lions, which are much more likely to ambush them in the dense thickets and woodlands near water courses than out in the open. This suggests that the opportunities for hominid scavenging are likely to have been greater during the dry season than during the rainy season and close to water courses (assuming, of course, that early hominids were less intimidated in vegetated areas by lions than modern hyenas seem to be). In addition, the dry season is also the time when very young and very old or sickly animals are most likely to die of hunger, disease, or other natural causes, providing additional potentially scavengable carcasses.

The dry season is also the time when both hominids and herbivores are most likely to be suffering from resource shortages and hence are relatively lean. This would also be the time, therefore, when hominids would have to be most selective in their scavenging, targeting especially the few parts remaining on a carcass that retain fat—the brain, the mandible, and the marrow bones of the lower limbs, the last elements in the legs of a severely stressed animal to become fat depleted.

Thus, answers to these critical and interesting questions—whether early hominids were hunters or scavengers and, if the latter, whether they were active or passive scavengers; also, whether they procured meat year-round or focused primarily on scraps of fatty tissue gleaned from carcasses during periods of food stress—hinge ultimately on the ability of archaeologists to find ways to determine, among other things, whether upper limbs were actually present in substantial numbers at early hominid sites and on their success in establishing the seasonality of these ancient occupations. Both are seemingly straightforward methodological tasks, but both so far have proven to be exceedingly difficult.

The many profound changes in our ideas about early hominid carnivory are also affecting our views about hunting and meat-eating in later stages of human evolution. Until quite recently, the traditional evolutionary scenario went more or less like this: Once early hominids had learned to hunt animals, much of subsequent human evolution revolved around improvements in the effectiveness with which they were able to kill more and larger game, a positive-feedback relationship that involved increasing intelligence, better forms of communication (and ultimately language), more effective intragroup cooperation, and, of course, better weaponry. Already by 500,000 years ago, or even earlier, hominids—by this time *Homo erectus*—were believed to have been highly successful big-game hunters, killing animals as large and dangerous as buffaloes, rhinoceroses, and even elephants. One site of this time period in Spain, known as Torralba, was even thought to have been the location of a communal elephant drive, where many of these huge beasts were supposedly stampeded into a bog and killed there by hunters.

In the past few years, however, many of the sites on which this evolutionary scenario was based have been reexamined from a taphonomic perspective, and, perhaps not surprisingly, the evidence for big-game hunting is being called into question. The consensus that has emerged seems to be that even *Homo erectus* was primarily a scavenger rather than a hunter, at least of larger animals, although this conclusion remains far from proven.

Debate now centers more on the hunting prowess of Neanderthals (*Homo sapiens neanderthalensis*), the last occupants of Europe and the Near East prior to the appearance of anatomically fully modern humans, some 35,000–40,000 years ago. Sites dating to the period of the Neanderthals—referred to by archaeologists as the Middle Paleolithic (or middle Old Stone Age)—often contain the remains of large mammals, including wild cattle, horses, bison, mammoths, and giant cave bears. Some of these sites, in fact, contain dozens or even hundreds of individuals of a single species, suggesting that Neanderthals not only could kill these large animals, but perhaps did so in large communal drives.

At the moment, however, opinion is divided concerning the proper interpretation of the faunal remains. Some archaeologists feel that many of the large mammal remains, particularly those found in cave sites, have little or nothing to do with humans, but were instead dragged into the cave by hyenas that made their dens there after the human occupants had left these sites. These archaeologists argue that Neanderthals probably hunted only the smaller, more docile species (e.g., deer), and scavenged from the carcasses of the larger ones. True big-game hunting, in this view, emerged late in human evolution, probably not until the appearance of anatomically fully modern humans, ca. 35,000 years ago. Others feel that the transition to full-fledged hunting economies occurred somewhat earlier, perhaps midway through the Middle Paleolithic, roughly 70,000–90,000 years ago.

At stake in this debate is not just the issue of whether anatomically premodern humans, perhaps as recently as the Upper Pleistocene, were scavengers rather than hunters, but whether they actually possessed the necessary cognitive (and, of course, technological) sophistication to successfully plan, coordinate, and carry out a communal hunt of extremely large dangerous prey. Answers to questions such as these will bring us much closer to understanding how a small-brained quadrupedal ape was transformed into the unique creature that we are today.

BIBLIOGRAPHY

Binford, L. R. (1981). "Bones: Ancient Men and Modern Myths." Academic Press, New York.

Blumenschine, R. J. (1987). Characteristics of an early hominid scavenging niche. *Curr. Anthropol.* **28**, 383.

Bunn, H. T. (1986). Patterns of skeletal representation and hominid subsistence activities at Olduvai Gorge, Tanzania, and Koobi Fora, Kenya. *J. Hum. Evol.* **15**, 673.

Eaton, S. B., Shostak, M., and Konner, M. (1988). "The Paleolithic Prescription." Harper & Row, New York.

Goodall, J. (1986). "The Chimpanzees of Gombe." Belknap, Cambridge, MA.

Gordon, K. D. (1987). Evolutionary perspectives on human diet. *In* "Nutritional Anthropology" (F. E. Johnston, ed.), pp. 3–39. Liss, New York.

Lee, R. B., and DeVore, I. (eds.) (1968). "Man the Hunter." Aldine, Chicago.

O'Connell, J. F., Hawkes, K., and Jones, N. B. (1988). Hadza scavenging: Implications for Plio/Pleistocene hominid subsistence. *Curr. Anthropol.* **29,** 356.

Speth, J. D. (1987). Early hominid subsistence strategies in seasonal habitats. *J. Archaeol. Sci.* **14,** 13.

Speth, J. D. (1989). Early hominid hunting and scavenging: The role of meat as an energy source. *J. Hum. Evol.* **18,** 329.

Cartilage

JOSEPH A. BUCKWALTER
University of Iowa

GLOSSARY

Appositional growth Enlargement of cartilage by the addition of new cells and matrix to the surface of the tissue

Collagens A family of 14 or more proteins that consist, at least in part, of helical amino acid chains. Some assume the form of fibrils and give cartilages their form and tensile strength; others do not form fibrils and might help organize and stabilize the meshwork of fibrillar collagens

Elastic cartilage Type of cartilage distinguished by a high concentration of elastin in the matrix; forms parts of the external ear, epiglottis, and laryngeal and bronchiolar cartilages

Elastin Protein that forms fibrils and sheets that can be deformed, without rupturing or tearing, and then return to their original shape and size

Enchondral ossification Replacement of cartilage by bone through cartilage mineralization, cartilage resorption, and bone formation

Fibrous cartilage Type of cartilage distinguished by a high concentration of type I collagen fibrils in the extracellular matrix; forms parts of the intervertebral disk, pubic symphysis, tendon and ligament insertions, and intraarticular menisci

Glycoproteins and noncollagenous proteins Molecules consisting of protein and a small number of monosaccharides or oligosaccharides; some might help organize and maintain the macromolecular framework of the cartilage matrix and help chondrocytes adhere to the matrix framework

Glycosaminoglycans Polysaccharide chains formed from repeating disaccharide units containing a derivative of either glucosamine or galactosamine and at least one negatively charged carboxylate or sulfate group; those found in cartilage include hyaluronic acid, chondroitin 4-sulfate, chondroitin 6-sulfate, dermatan sulfate, and keratan sulfate

Hyaline cartilage Type of cartilage distinguished by high concentrations of type II collagen fibrils, large aggregating proteoglycans, and water. In the fetus it forms most of the skeleton; during skeletal growth it forms the cartilaginous growth plates, or physes; in adults it persists as the nasal, laryngeal, bronchial, articular, and costal cartilages; and at any age it participates in healing some types of bone and cartilage injuries

Interstitial growth Enlargement of cartilage by the addition of new cells and matrix within the substance of the tissue

Link proteins Small noncollagenous proteins that stabilize proteoglycan aggregates, increase the degree of aggregation, increase the size of aggregates, and influence the spacing of proteoglycan monomers in aggregates

Perichondrium Layer of cells and matrix that covers some cartilage surfaces

Permeability Ease with which water can flow through the cartilage matrix; measured by the force required to cause water to flow through cartilage

Proteoglycan aggregates Molecules formed by the noncovalent association of multiple proteoglycan monomers, multiple link proteins, and a hyaluronic acid filament

Proteoglycan monomers Molecules consisting of glycosaminoglyans covalently bound to protein. Cartilage proteoglycan monomers exist in the form of large aggregating monomers (aggrecans), large nonaggregating monomers, and small nonaggregating monomers

Viscoelasticity State of having both viscous and elastic properties. Viscous fluids tend to flow slowly because of friction between component molecules as they slide past one another. Elastic solids tend to regain their original

ENCYCLOPEDIA OF HUMAN BIOLOGY, Second Edition, VOLUME 2. Copyright © 1997 by Academic Press. All rights of reproduction in any form reserved.

shape after being compressed. The response of a viscoelastic material to loading can be modeled as the combined responses of a viscous fluid and an elastic solid. When subjected to constant load or constant deformation, the response of the material varies with time

BEGINNING WITH FETAL LIFE, CARTILAGE, A CONnective tissue consisting of an abundant extracellular matrix and specialized cartilage cells (chondrocytes), demonstrates remarkable capacities for interstitial and appositional growth, changing shape as it grows. In some locations the development of cartilage, followed by cartilage mineralization and resorption, makes possible the formation of bone. Throughout life the unique material properties of cartilage—including durability, stiffness, resiliency, and viscoelasticity—make possible the normal function of the musculoskeletal, auditory, and respiratory systems.

Cartilage or precartilaginous tissues first appear in the fifth week of human prenatal life, form the initial wholly cartilaginous skeleton, grow rapidly, and remodel as they grow. At about the eighth week of prenatal life, the cartilage begins to be replaced by bone in many areas. During fetal life, infancy, and adolescence, growth plate or physeal cartilage produces longitudinal bone growth and is replaced by bone. When skeletal growth ceases, the growth plates disappear, but other kinds of cartilage remain as essential structural components of the external ear, the respiratory system (including the nasal, laryngeal, tracheal, and bronchial cartilages) and the skeletal system (including intervertebral disks, menisci, costal cartilages, parts of the insertions of tendon and ligaments into bone, and the remarkably durable, almost frictionless, surfaces of synovial joints). At any age cartilage participates in the healing of bone and cartilage injuries.

Like other dense organized connective tissues, including tendon, ligament, and bone, cartilage consists of a sparse population of mesenchymal cells embedded within an abundant extracellular matrix. In most mature cartilage the cells contribute about 5% to the total tissue volume, and the matrix contributes about 95%. The roughly spherical shape of the cartilage cells, or chondrocytes, and the unique composition of the matrices they synthesize and assemble distinguish cartilage from the other connective tissues. Although cartilage lacks nerves, lymph vessels, and blood vessels, chondrocytes are metabolically active and respond to changes in hormonal balance, availability of nutrients, oxygen tension, and mechanical loads. With the exception of the exposed surfaces found in synovial joints and the junctions of cartilage and bone, a thin layer of fibrous tissue, or perichondrium, covers cartilage and separates it from other tissues.

The matrix component molecules—including collagens, elastin, proteoglycans, and noncollagenous proteins—and the organization of these matrix macromolecules give the tissue its material properties. Differences in matrix composition and organization, with resulting differences in appearance and material properties, distinguish three types of human cartilage: hyaline, fibrous, and elastic. Hyaline cartilage, the most abundant type, has been more extensively studied than the others, so most concepts of cartilage are based on this type. As a result, many authors refer to hyaline cartilage simply as "cartilage," obscuring the differences among the three types of cartilage.

I. FORMATION, GROWTH, AND MATURATION

Cartilage forms from groups of mesenchymal cells that cluster together. Each cell then assumes a spherical shape and begins to synthesize and secrete macromolecules that form a cartilaginous matrix. Subsequent promotion and maintenance of chondrogenesis depend on interactions among the cells, and between the cells and the molecules of extracellular matrix. Matrix accumulation separates the cells, and a perichondrium usually forms, marking the boundary between the cartilage and the surrounding tissues. The outer layer of perichondrium consists of fibroblastlike cells and a fibrous matrix; the inner layer contains more spherical cells, with the appearance of immature chondrocytes (chondroblasts). [*See* Extracellular Matrix.]

Cartilage volume can increase either by appositional growth (i.e., the addition of new cells and matrix at the tissue surface from the perichondrium) or by interstitial growth (i.e., growth within the tissue). Interstitial growth occurs by cellular proliferation, synthesis of new matrix, cell enlargement due to synthesis of new cell organelles and cytoplasm, and cell, and possibly matrix, swelling. The plasticity of the cartilage matrix makes interstitial growth possible. Tissues with rigid matrices (e.g., bone or mineralized cartilage) can increase their volume only by appositional growth and therefore usually cannot grow as rapidly or change shape as easily.

During cartilage formation and growth, the cell density is high, as the chondrocytes proliferate rapidly and synthesize large volumes of matrix. With maturation, matrix synthesis slows, cell density declines, and, once growth ceases, chondrocytes rarely divide. In mature tissue the cells decrease their synthetic activity to the level required to maintain the matrix. With aging, cell death further decreases cell density and the composition and structure of at least some matrix macromolecules change. [*See* Cell Death in Human Development.]

II. COMPOSITION

Cartilage consists of three components: chondrocytes, extracellular water, and macromolecules that form the framework of the extracellular matrix.

A. Chondrocytes

Unlike tissues such as liver, muscle, or kidney, the primary functions of cartilage depend on the matrix, not the cells, and the cells form only a small part of the tissue. However, the cells form, maintain, and modify the matrix. Like most mesenchymal cells, chondrocytes surround themselves with the matrix they synthesize and assemble, and they do not form contacts with other cells. They contain the organelles responsible for matrix synthesis, including endoplasmic reticulum and Golgi membranes, and also frequently contain intracytoplasmic filaments and glycogen. At least some chondrocytes have a cilium that extends into the matrix and might sense changes in the matrix.

A close association exists between chondrocyte shape and cell function. With the exception of the flattened ellipsoid cells found immediately deep to the perichondrium, in the superficial zone of articular cartilage and in the proliferative zone of growth plate, chondrocytes have a spherical shape. During chondrogenesis, when undifferentiated mesenchymal cells assume a spherical shape, they generally begin to synthesize the cartilage-specific collagens and proteoglycans. Experimentally, when they lose their spherical shape, they stop making cartilage-specific molecules.

The relationship between chondrocytes and their matrix does not end with the synthesis and assembly of the matrix molecules. Maintenance of normal composition, structure, and function of the tissue depends on continual complex interactions between the chondrocytes and their matrix. Normal degradation of matrix macromolecules, especially proteoglycans, forces the cells to synthesize new matrix components to preserve the tissue. Chondrocytes sense the loss of matrix molecules caused by disease or injury and attempt to replace them. If they fail, the matrix deteriorates and the chondrocytes could die. The matrix limits the types and concentrations of nutrients, hormones, or drugs that can reach the chondrocytes and protects the cells from injury due to normal mechanical loading.

In addition, the matrix acts as a mechanical signal transducer for the chondrocytes. Deformation of the matrix generates signals that are transmitted through the matrix and cause cells to alter their synthetic activity. The signals might be purely mechanical (e.g., pressure or tension on the cell membrane), but deformation of the matrix could generate other types of signals, including changes in electrical potential or ion fluxes. Through these mechanisms, or possibly others, the synthetic function of chondrocytes and, thus, maintenance of the normal matrix composition depend on at least some loading of the tissue.

B. Tissue Fluid

The tissue fluid forms the largest component of cartilage; depending on the type of cartilage and its age, water can contribute up to 80% of the wet weight of cartilage. It contains dissolved gases, small proteins, and metabolites. Because the cartilage water is not contained by cell membranes, its volume, concentration, organization, and behavior depend on its interaction with the structural macromolecules. The nutrients and metabolites essential for chondrocyte function must pass through matrix fluid, and, if the macromolecular framework did not maintain and organize the tissue water and impede its flow through the matrix, cartilage would lose its resiliency and ability to resist compression.

C. Structural Macromolecules

The structural macromolecules that contribute 20–40% of the wet weight of cartilage include fibrillar molecules (e.g., collagens and elastin) and nonfibrillar, or ground substance, molecules (e.g., proteoglycans and noncollagenous proteins). The cells synthesize these macromolecules from amino acids and sugars; available evidence suggests that all chondrocytes can synthesize each type of molecule. [*See* Collagen, Structure and Function; Elastin.]

1. Collagens

Collagens, a family of at least 14 (types I–XIV) or more distinct protein molecules, consist, in part, of helical amino acid chains (Fig. 1). In addition to classification by differences in amino acid composition, collagens can be classified into fibrillar interstitial collagens and nonfibrillar collagens by their form and location. In most tissues, including cartilage, fibrillar collagens form a much larger proportion of the matrix than do nonfibrillar collagens, and for this reason the nonfibrillar collagens have been referred to as "quantitatively minor collagens."

Fibrillar collagens appear as cross-banded rope-like fibrils when examined by electron microscopy (Figs. 1 and 2). They give cartilage its tensile strength and form and create an extracellular meshwork for cell attachment and binding or entanglement of the other matrix macromolecules. The recognized fibrillar collagens include types I, II, III, V, and XI. Other collagens, including types IX and XII, associate with the cross-striated fibrils formed by fibrillar collagens. The two most common fibrillar collagens—types I and II—have different distributions within the tissues. Type I collagen forms the fibrillar matrix meshwork of skin, tendon, ligament, bone, meniscus, fibrous cartilage, and annulus fibrosis. Type II collagen fibril forms the collagen meshwork of hyaline cartilage, nucleus pulposus of the intervertebral disk, and vitreous body of the eye and has a higher hydroxylysine content. In general, tissues such as hyaline cartilage, containing primarily type II collagen, have a relatively high concentration of water and ground substance molecules, which gives them a glassy, translucent, or clear appearance. Tissues such as fibrous cartilage, containing primarily type I collagen, tend to have a

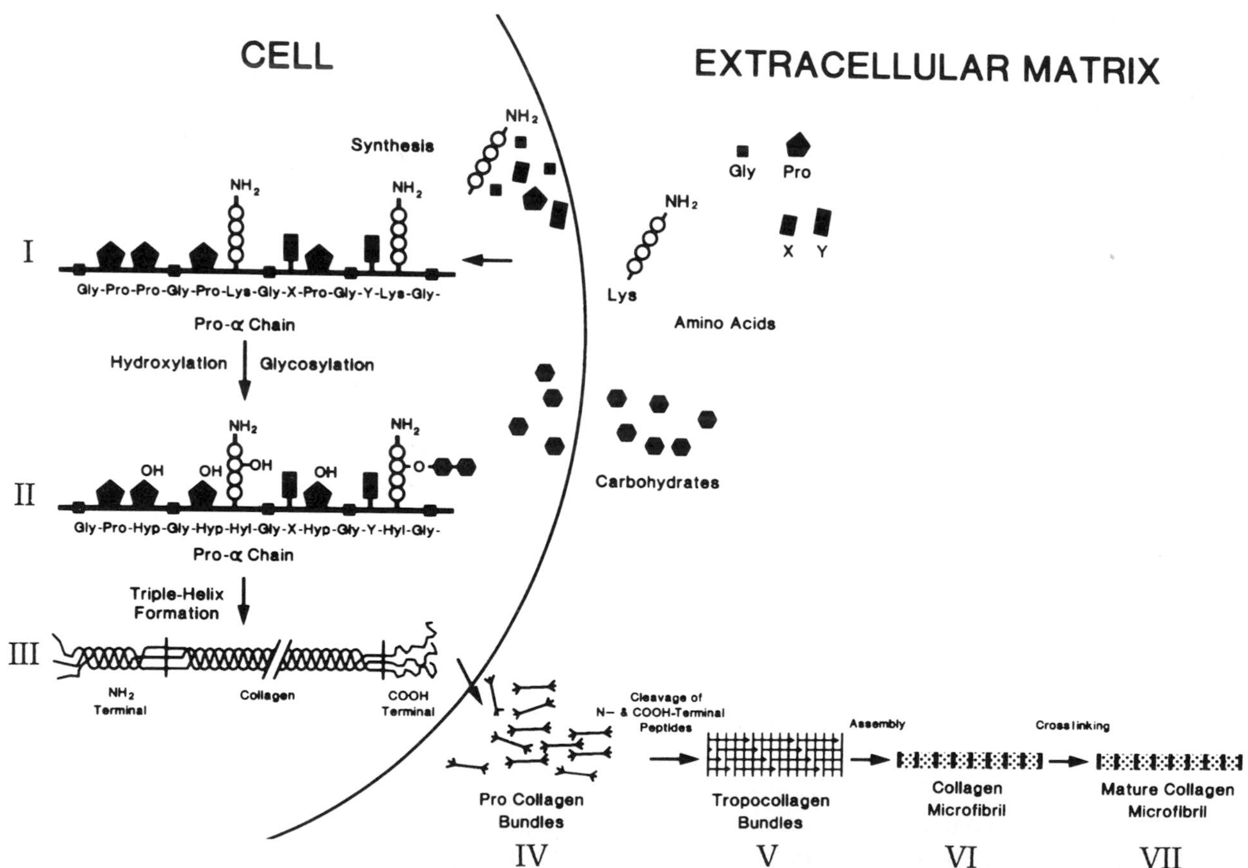

FIGURE 1 The synthesis, secretion, and matrix assembly of fibrillar collagen. (I) The cell links amino acids into Pro-α amino acid chains and then (II) adds carbohydrates. (III) Three Pro-α chains wind around each other to form the collagen triple helix, and (IV) the cell secretes the procollagen into the matrix. Here (V), procollagen peptidases cleave the amino (N)- and carboxy (COOH)-terminal peptides to form the collagen or tropocollagen molecule. (VI) Tropocollagen molecules then align themselves in a staggered arrangement to create collagen microfibrils. (VII) Cross-linking within and among collagen molecules strengthens the microfibril. Microfibrils aggregate to form the collagen fibrils seen by electron microscopy. [Reproduced from J. A. Buckwalter and R. R. Cooper (1987). The cells and matrices of skeletal connective tissues. *In* "The Scientific Basics of Orthopaedics" (J. A. Albright and R. A. Brand, eds.), p. 19. Appleton & Lange, East Norwalk, CT.]

FIGURE 2 Electron micrograph of a human infant annulus fibrosus showing the rope-like cross-banded collagen fibrils paralleling an immature elastic fiber consisting of amorphous elastin and glycoprotein microfibrils. Notice that patches of amorphous elastic (*) have accumulated within the more darkly stained microfibrils. [Reproduced from J. A. Buckwalter and R. R. Cooper (1987). The cells and matrices of skeletal connective tissues. *In* "The Scientific Basics of Orthopaedics" (J. A. Albright and R. A. Brand, eds.), p. 21. Appleton & Lange, East Norwalk, CT.]

lower concentration of water and ground substance molecules, which makes them more opaque.

The nonfibrillar collagens might lie close to cells, form part of basement membranes, or associate with other matrix macromolecules, including fibrillar collagens. When examined by electron microscopy, they can appear as fine threads or filaments that can form networks or mat-like structures. Some types of nonfibrillar collagens can be found directly adjacent to or in the region of the chondrocyte cell membranes, where they form part of the pericellular matrix. Others help organize and stabilize the extracellular matrix, including the fibrillar collagen meshwork.

2. Elastin

Although elastin, like collagen, can form protein fibrils (Figs. 2 and 3), the fibrils lack the cross-banding pattern and differ in amino acid composition, conformation of the amino acid chains, and material properties. Elastin can also form sheet-like structures. In any shape, elastin can undergo some deformation, without rupturing or tearing, and when unloaded it can return to its original shape and size. Like collagens, elastin consists primarily of a protein with a small carbohydrate component. However, elastin contains little hydroxyproline, no hydroxylysine, and two unique amino acids: desmosine and isodesmosine. Unlike the precise ordered arrangement of the amino acid chains in helical collagen molecules, the elastin amino acid chains assume a variety of cross-linked random coil conformations that allow the molecule to stretch and recoil without rupturing. Elastin usually first appears within aggregates of glycoprotein microfibrils (Fig. 3). Elastin then accumulates until the enlarging elastic fiber appears to consist of a central region containing only elastin surrounded by a thin layer of microfibrils.

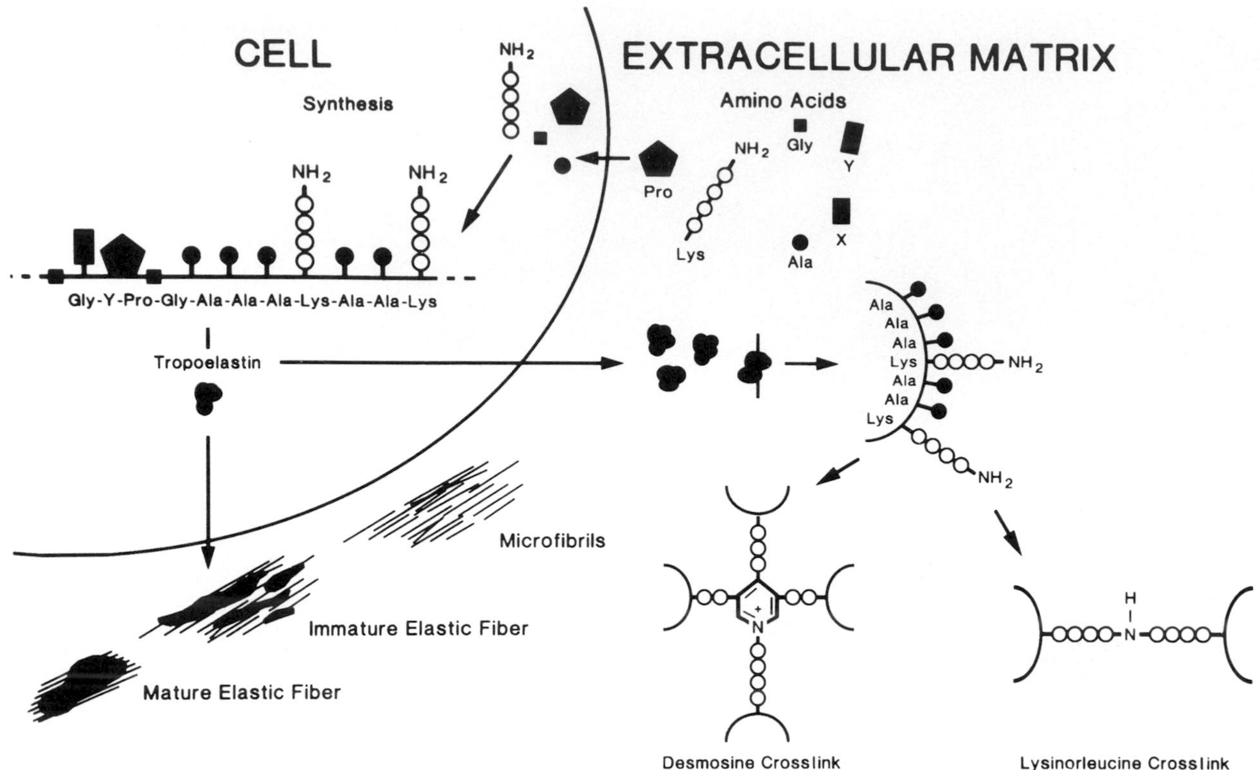

FIGURE 3 The synthesis, secretion, and matrix assembly of elastic fibers. Immature elastic fibers consist almost entirely of glycoprotein microfibrils. Amorphous elastin progressively accumulates in the region of the microfibrils until the mature elastic fiber consists almost entirely of elastin, with a few peripheral microfibrils. The cross-linking of the amino acid chains help give elastin its special material properties. [Reproduced from J. A. Buckwalter and R. R. Cooper (1987). The cells and matrices of skeletal connective tissues. *In* "The Scientific Basics of Orthopaedics" (J. A. Albright and R. A. Brand, eds.), p. 21. Appleton & Lange, East Norwalk, CT.]

Only in elastic cartilage, large blood vessels, and some ligaments (e.g., the nuchal ligament and ligamentum flavum of the spine) does elastin significantly contribute to the structure and material properties of human tissue.

3. Proteoglycans

Proteoglycan monomers, molecules consisting of polysaccharide chains bound to protein, are the major macromolecule of the cartilage ground substance. They contain little protein (about 5%) and consist primarily of special polysaccharide chains called glycosaminoglycans. These, in turn, consist of repeating disaccharide units containing a derivative of either glucosamine or galactosamine. Each disaccharide unit also contains at least one negatively charged carboxylate or sulfate group, so that the glycosaminoglycans form long strings of negative charges. Cartilage glycosaminoglycans include hyaluronic acid, chondroitin 4-sulfate, chondroitin 6-sulfate, dermatan sulfate, and keratan sulfate. Proteoglycan monomers have multiple forms (Fig. 4), including large nonaggregating proteoglycans, small nonaggregating proteoglycans (aggrecans), and large aggregating proteoglycan monomers. The latter molecules can exist as individual monomers or as aggregates consisting of multiple monomers, hyaluronic acid, and small proteins called link proteins.

The source and function of the large nonaggregating proteoglycans (Fig. 4) remain uncertain. They might derive from the breakdown of aggregating proteoglycans or they might represent a distinct population of proteoglycans that have a function similar to that of the aggregating proteoglycans. The small nonaggregating proteoglycans (Fig. 4) might contain chondroitin sulfate, and at least some small proteogly-

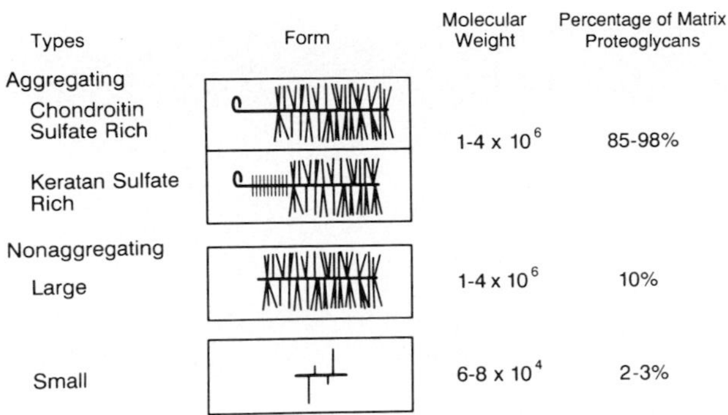

FIGURE 4 The known types of cartilage proteoglycan monomers. The percentage of matrix proteoglycans given in the last column refers to hyaline cartilage. Each monomer type consists of a protein core (the central line of each drawing) and covalently bound glycosaminoglycan chains (the projecting side arms). The longer side arms of the aggregating monomers represent chondroitin sulfate; the shorter side arms, keratan sulfate. In hyaline cartilage, aggregating monomers rich in chondroitin sulfate and aggregating monomers rich in keratan sulfate have been identified. The aggregating monomers are shown with a folded region of their protein core to represent the region that binds to hyaluronic acid.

cans contain another glycosaminoglycan, dermatan sulfate. The dermatan sulfate proteoglycans might form specific associations with collagen fibrils.

Aggregating monomers consist of protein core filaments with multiple covalently bound oligosaccharides and longer chondroitin and keratin sulfate chains (Figs. 4 and 5). Because each of the glycosaminoglycan chains creates a string of negative charges that bind water and cations, proteoglycans in solution fill a large volume. Because water molecules are bipolar, with a negative charge at the central oxygen atom and positive charges at the hydrogen atoms, the ordered array of negative charges interacts with large volumes of water. Since adjacent negatively charged glycosaminoglycan chains repel each other, they tend to maintain the molecule in an extended form and draw water into their molecular domain, creating swelling pressure within the matrix. An intact collagen fibril meshwork limits the swelling of the proteoglycans, but loss or degradation of the collagen fibril meshwork allows the tissue to swell, increasing the water concentration and decreasing the proteoglycan concentration.

In the matrix, most of the aggregating proteoglycan monomers form a noncovalent association with hyaluronic acid and link proteins. The hyaluronic acid filaments form the backbone of aggregates that can reach a length of more than 10,000 nm, with more than 300 monomers (Fig. 6). The link proteins stabilize the association between monomers and hyaluronic acid and might have a role in directing the assembly of aggregates in the matrix. Aggregates might help anchor monomers within the matrix, preventing their displacement, and organize and stabilize the macromolecular framework of the matrix. They also might help control the flow of water through the matrix and help maintain water within it. Because of their ability to interact with water, proteoglycans help give cartilage both its stiffness to compression and its resilience (Fig. 7) and might contribute to its durability.

4. Noncollagenous Proteins

Less is known about the noncollagenous proteins and glycoproteins than about collagens, elastin, or proteoglycans. Only a few noncollagenous proteins of cartilage have been identified, and although their functions have not been defined, they form a significant part of the macromolecular framework of cartilages. These proteins, which might have a small number of attached monosaccharides or oligosaccharides, might help organize and maintain the macromolecular structure of the matrix and the relationship between chondrocytes and the matrix.

Link proteins help organize and stabilize the matrix through their effects on proteoglycan aggregation. Other noncollagenous proteins, including chondronectin, fibronectin, and anchorin CII, might influence the behavior of the cartilage cells. Chondronectin is thought to mediate the adhesion of chondrocytes to the matrix and help stabilize the phenotype of hyaline cartilage chondrocytes, whereas fibronectin, under experimental conditions, can cause chondrocyte-like cells to assume the form and function of fibroblast-like cells. Anchorin CII might have functions similar to those of chondronectin.

III. MATERIAL PROPERTIES

Articular cartilage provides load bearing, a low-friction surface, and resilience and distributes loads across synovial joints. Although only a few millimeters thick, it usually performs these functions for 80 years or more without significant deterioration. Intervertebral disks withstand the large loads caused by bending and lifting, usually without rupturing. When loaded, or when the spine twists, flexes, or extends, they deform and then regain their original size and shape. Tendon and ligament insertions have great tensile strength, yet they remain pliable and rarely tear from repetitive bending and loading. The cartilage of

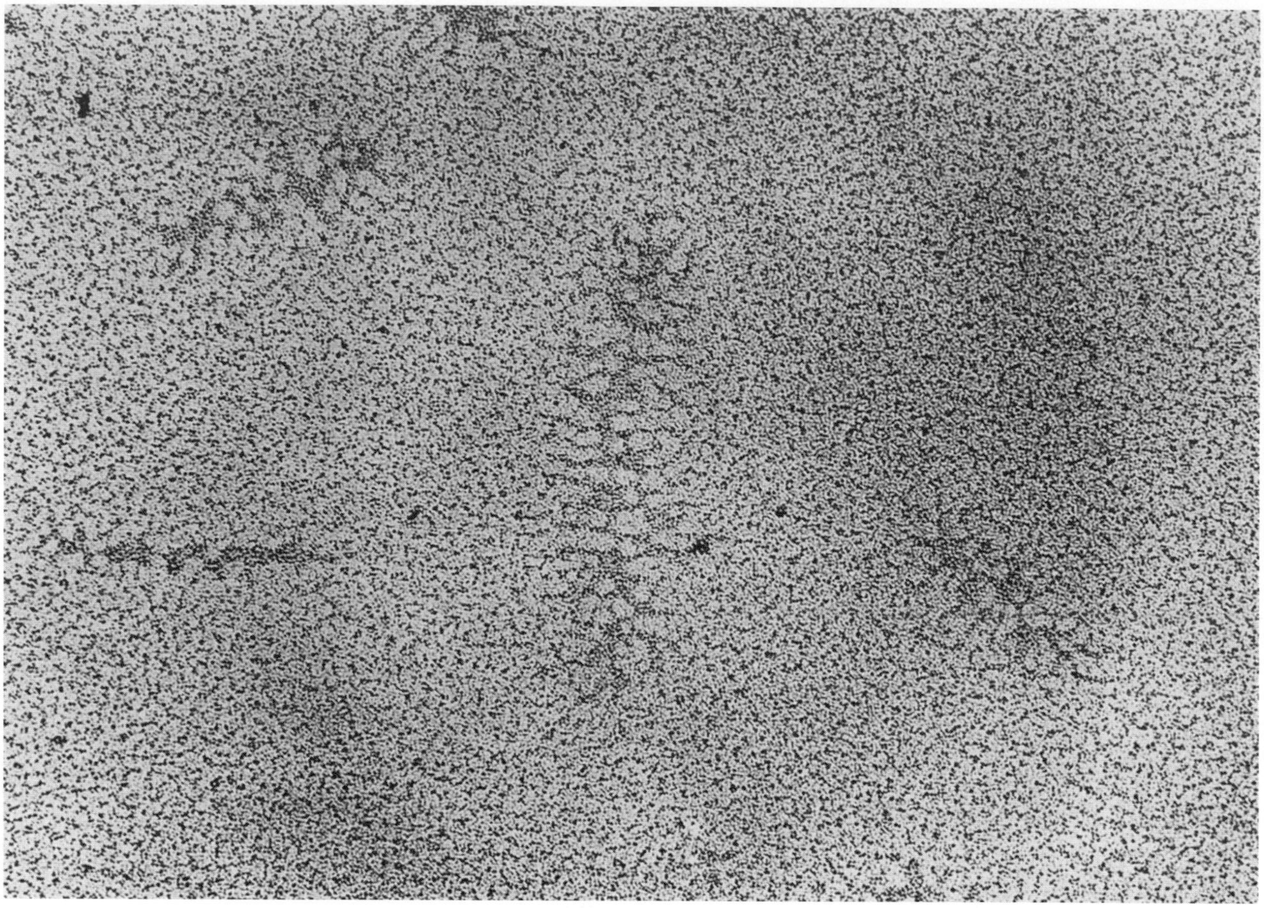

FIGURE 5 Electron micrograph of a large proteoglycan monomer. It is not possible to determine from this micrograph whether this monomer can aggregate. The densely packed glycosaminoglycan chains attached to the central protein filament obscure the filament in many areas. [Reproduced from J. A. Buckwalter (1983). Articular cartilage. *Am. Acad. Orthop. Surg., Instr. Course Lect.* **32**, 355.]

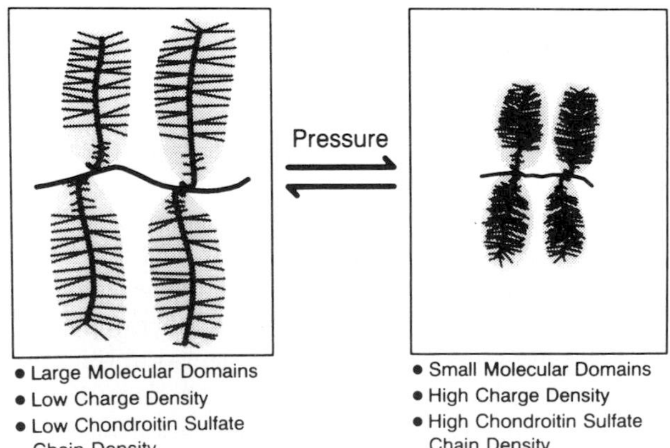

FIGURE 6 The reversible expansion of aggregated proteoglycan monomers in solution. Forcing the negatively charged glycosaminoglycan chains closer together drives water from the molecular domain and increases the resistance to further compression. Release of the compression allows water to return to the molecular domain.

the nose, respiratory passages, ears, larynx, and ribs preserves the shape of these structures, resists deformation, yet remains flexible.

A. Matrix Composition, Structure, and Material Properties

The behavior of cartilage when loaded depends on the composition and organization of the matrix, that is, the concentration, properties, and organization of the matrix macromolecules, the water content, and the physical and electrical interactions between water and the macromolecular framework. Since cartilage matrix consists of a macromolecular framework filled with water, the tissue can be considered a fluid-filled porous solid, and porosity (i.e., the ratio of the pore volume within the tissue to the total volume of the tissue) can be estimated by the water content. Since some cartilages have a water content of 60–80%,

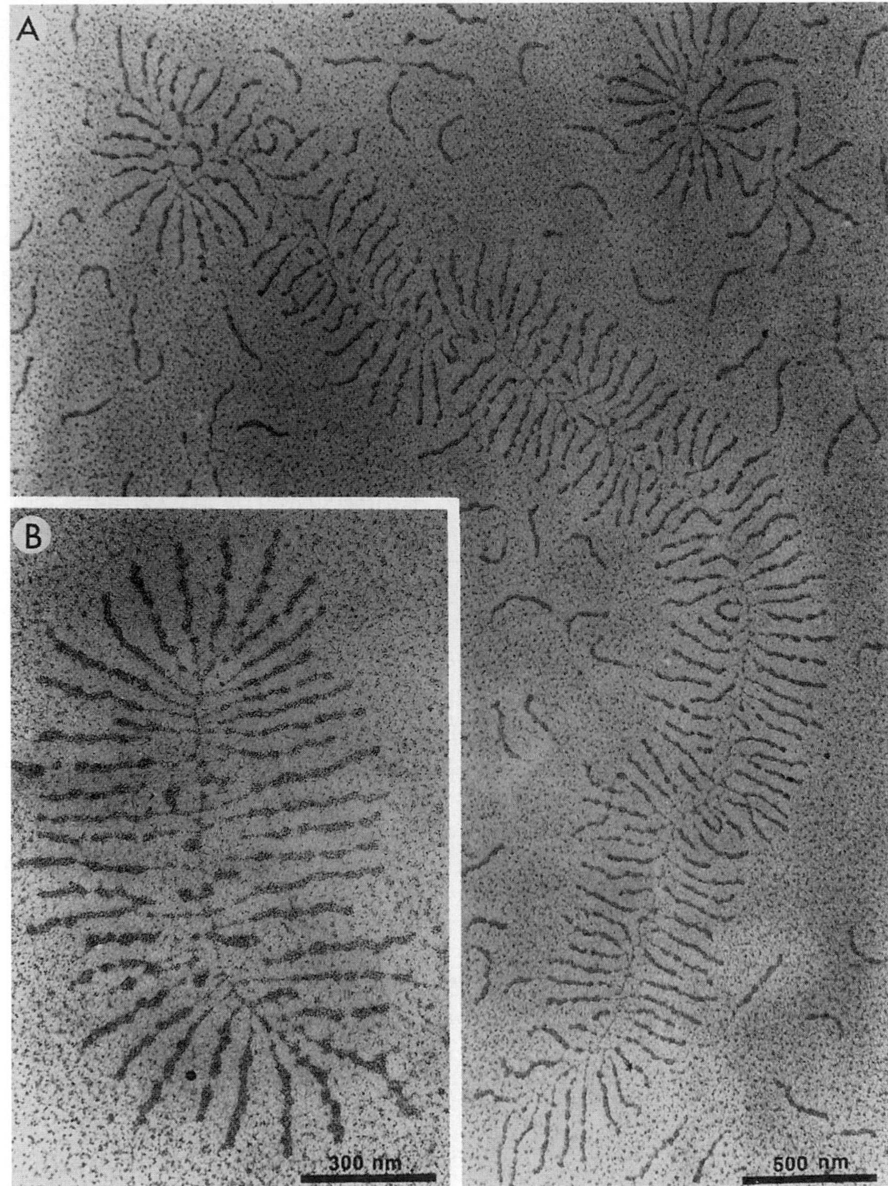

FIGURE 7 Electron micrographs of proteoglycan aggregates. The central filaments are hyaluronic acid. The projecting side chains are proteoglycan monomers. In this preparation the glycosaminoglycan chains are collapsed around the monomer protein cores. (A) A large proteoglycan aggregate. (B) A moderate size proteoglycan aggregate. [Reprinted by permission of VCH Publishers, Inc., 220 East 23rd St., New York, NY 10010. From "Collagen and Related Research," Vol. e489–504 (1983), Fig. 1, p. 492.]

their porosity is relatively high. Interactions between the water and the pore walls of the macromolecular framework of the matrix generate resistance to the flow of water and determine the permeability of cartilage (i.e., the ease with which water can flow through the matrix).

Because cartilage matrix consists of a solid macromolecular framework filled with water, it behaves as a viscoelastic material; that is, its response to loading combines viscosity, a property characteristic of fluids, with elasticity, a property characteristic of solids. The response of viscoelastic materials to constant load or a constant deformation varies with time. When subjected to a constant load, a viscoelastic material responds with an initial deformation, followed by slow progressive deformation, until it reaches an equilibrium state; this behavior is called creep. When subjected to constant deformation, a viscoelastic material responds with high initial stress, followed by a slow progressive decrease in the stress required to maintain the constant deformation; this behavior is called stress relaxation. Creep and stress relaxation might be caused by fluid flow through the matrix or by deformation or movement of the matrix macromolecules. To a point cartilage can restore its original form after deformation by reversing fluid flow and restoration of the macromolecular framework (Fig. 6).

B. Effect of Differences in Matrix Composition

Differences in matrix composition cause differences in the material properties of cartilage. Collagen fibrils provide tensile strength, but little resistance to compression; elastin provides less resistance to tensile loads, but can be deformed without damage and then regain its size and shape; and the interaction of proteoglycans and water provides resistance to compression, swelling pressure, and resilience, but little tensile strength. Therefore, cartilage with a high concentration of collagen fibrils oriented parallel to an applied tensile load has great tensile strength. A high concentration of elastin creates a flexible tissue, such as the cartilage of the external ear, which can be repetitively stretched, bent, or twisted without permanent damage and almost instantly regains its former shape and size after deformation. A tissue with a high concentration of large aggregated proteoglycans can have great compressive stiffness, swelling pressure, and resilience. In general, a higher proteoglycan concentration in hyaline cartilage is associated with a lower water concentration, decreased permeability and porosity, and increased compressive stiffness, whereas a lower proteoglycan concentration is associated with a higher water concentration, increased permeability and porosity, and decreased compressive stiffness.

C. Effect of Differences in Matrix Organization

The organization of matrix macromolecules is not necessarily uniform throughout the tissue, and therefore, even if there are no differences in composition, differences in matrix organization can cause differences in material properties. For example, the material properties of the same region of articular cartilage differ with the orientation of that region relative to the axis of joint motion. Presumably, differences in collagen fibril organization and orientation cause the differences in tensile strength and stiffness at different orientations. The material properties of articular cartilage also differ with depth from the articular surface. These differences might be related to changes in the collagen fibril orientation, the relationship between collagen fibrils and proteoglycans, matrix composition, or other factors. It is likely that more subtle alterations in matrix organization—such as the degree of proteoglycan aggregation, the size of proteoglycan aggregates, the length of chondroitin sulfate chains, the size of proteoglycan monomers, and the strength and stability of the interactions among collagens, proteoglycans, and other matrix molecules—also affect the material properties of the matrix.

IV. NUTRITION

At all ages chondrocytes have significant nutritional requirements. Their proliferative and synthetic activities in growing cartilage require a steady supply of nutrients; but even chondrocytes in mature cartilage are metabolically active. Although they rarely divide, they must synthesize molecules to compensate for degradation of those in the matrix, particularly proteoglycans. Vascular canals have been reported within cartilage, but most of them probably allow vessels to pass through the cartilage, rather than supply the chondrocytes. Therefore, for chondrocytes to survive and maintain the matrix, nutrients and metabolites must efficiently and rapidly pass through the matrix for significant distances. This might occur by simple diffusion, convection, or a combination of the two. Transport by convection is based on interstitial fluid flow caused by cartilage deformation as a result of cartilage loading; repetitive loading might therefore help maintain chondrocyte nutrition.

V. CARTILAGE TYPES

Adult human cartilages—hyaline, fibrous, and elastic—differ in distribution within the body, matrix composition (Figs. 8A and 8B), material properties, gross and microscopic appearances, and function. Within each type, cartilages vary considerably, and intermediate forms exist, one of which is the repair

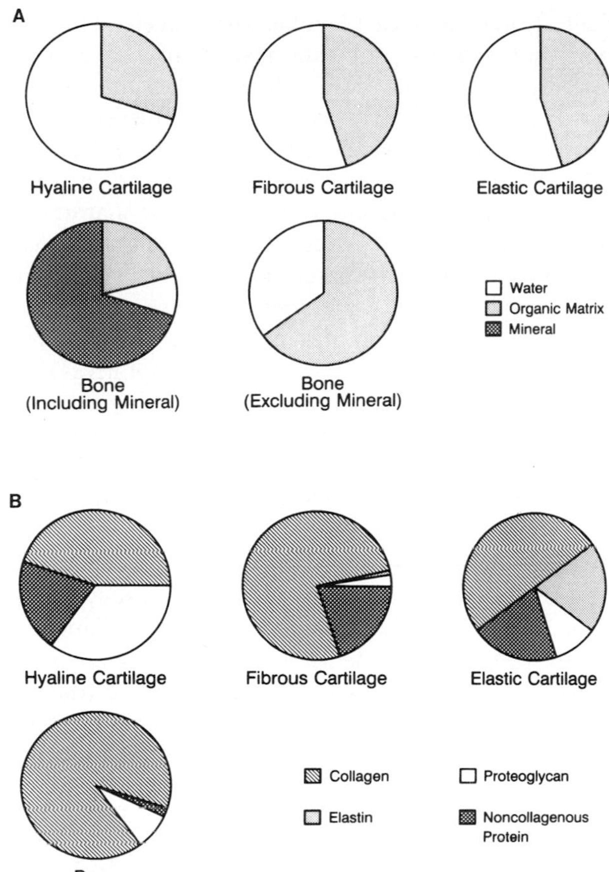

FIGURE 8 (A) Composition of the three cartilage types and bone. Notice that hyaline cartilage has the highest concentration of water. Excluding the mineral component, bone has the lowest concentration of water. (B) The organic matrices of the three cartilage types and bone. Notice that hyaline cartilage has a much greater concentration of proteoglycans and that a high concentration of elastin distinguishes elastic cartilage. The organic matrices of bone and fibrous cartilage have similar compositions: a high concentration of fibrillar type I collagen and a relatively low concentration of proteoglycans.

cartilage that forms during healing of cartilage and bone.

A. Hyaline

Hyaline cartilage, the most widespread and extensively studied human cartilage, received its name because of the clear appearance of the matrix (from the Greek *hylos* meaning "glass"), as seen by light microscopy. It is smooth, slick, and firm to the touch. In the fetus, hyaline cartilage forms most of the skeleton before it is resorbed and replaced by bone through the process of enchondral ossification. It forms the physeal cartilages that produce longitudinal bone growth, until growth ceases and they are resorbed and replaced by bone. In adults, hyaline cartilage persists as the nasal, laryngeal, bronchial, articular, and costal cartilages.

The highly specialized macromolecular framework of hyaline cartilage (Fig. 8B) consists of fibrillar type II collagen, large aggregating proteoglycans, large nonaggregating proteoglycans, small nonaggregating proteoglycans (Fig. 4), and noncollagenous proteins. Hyaline cartilage matrix also contains at least two quantitatively minor collagens—types IX and XI—and might also contain types V and VI. Type IX collagen associates with the surface of type II-containing collagen fibrils and could be involved in the organization and stabilization of the type II collagen meshwork and the interaction of the collagen meshwork with other matrix molecules. Type XI collagen might form part of the type II fibrils and help determine their diameter. In addition, hyaline cartilage has a higher water content than the other cartilage types or bone (Fig. 8A), and it does not contain elastin (Fig. 8B). This specialized matrix gives hyaline cartilage optimal material properties for serving as the articular surface of synovial joints, including stiffness, viscoelasticity, resiliency, and durability.

B. Fibrous

Fibrous cartilage forms important structural parts of the intervertebral disk, pubic symphysis, and tendon and ligament insertions into bone. A specialized form of fibrous cartilage makes up the intraarticular menisci. Unlike hyaline cartilage, it is opaque and densely collagenous; its cut surface is often rough, and its matrix appears fibrillar when examined by light microscopy. Also, unlike hyaline cartilage, the fibrillar collagen of fibrous cartilage is type I, the concentration of fibrillar collagen is higher, and the concentrations of water and proteoglycans are lower (Figs. 8A and 8B). Furthermore, few fibrous cartilages have a high concentration of large aggregating proteoglycans, and at least some fibrous cartilages contain small amounts of elastin. Fibrous cartilage generally has great tensile strength, but does not form smooth, durable, low-friction surfaces, as does hyaline cartilage. Its composition and appearance make it difficult to distinguish it from other dense fibrous tissues, including some regions of tendons and ligaments. For example, in tendon and ligament insertions the transition between the substance of the tendon or ligament and fibrous cartilage is almost imperceptible.

At other locations fibrous cartilage is distinguished by the presence of spherical chondrocytes, rather than the flattened fibroblasts or fibrocytes of dense fibrous tissue. It is likely that the composition, matrix organization, and material properties of fibrous cartilage differ from those of other dense fibrous tissues, but these possible differences have not been extensively examined.

C. Elastic

A high concentration of elastin (Fig. 8B) gives elastic cartilage, the rarest form of human cartilage, a yellowish hue that distinguishes it from the bluish-white often translucent, appearance of hyaline cartilage or the off-white color of fibrous cartilage. Elastic cartilage forms the auricle of the external ear, a major portion of the epiglottis, and some of the laryngeal and bronchiolar cartilages. Bending or twisting the external ear shows that elastic cartilage lacks the stiffness of fibrous or hyaline cartilage, but that it has remarkable flexibility and the ability to be deformed and then regain its original shape.

D. Repair

Bone or cartilage injuries frequently result in the formation of repair cartilage. This tissue usually has a matrix composition and a light-microscopic appearance intermediate between hyaline and fibrous cartilages. It could contain both types I and II collagen and regions with high concentrations of hyaline cartilage like proteoglycans. In healing bone fractures, new bone replaces the repair cartilage. If a bone fracture fails to heal, cartilage might persist at the nonunion site, form a fibrous or cartilaginous union of the bone fragments, or cover the ends of the bone fragments with hyaline-like cartilage to form a false synovial joint, called pseudoarthrosis. The formation of repair cartilage from undifferentiated mesenchymal cells demonstrates that, even in adult humans, these cells have the potential to differentiate into chondrocytes or chondrocyte-like cells and synthesize the components of the cartilage matrix.

VI. CARTILAGE MINERALIZATION AND ENCHONDRAL OSSIFICATION

Mineralization of cartilage (i.e., deposition of relatively insoluble calcium phosphate in the matrix, particularly in hyaline cartilage) is part of skeletal formation and growth and the healing of some bone fractures. Mineralization, which makes the pliable cartilage matrix rigid, precedes the replacement of cartilage by bone during enchondral ossification. Minerlization might also occur with aging and in some diseases, such as pseudogout. By adversely affecting the material properties of the matrix, it might accelerate deterioration of the tissue. The presence of calcium and phosphate ions alone does not cause cartilage to mineralize. In addition, the matrix must be prepared for mineralization. Probably, matrix components that inhibit cartilage mineralization must be altered or removed and, possibly, matrix components that promote mineralization must be added.

The process that converts cartilage into bone, called enchondral or intracartilaginous ossification, includes cartilage mineralization as part of a complex sequence of cell and matrix changes. As the chondrocytes enlarge, the surrounding matrix mineralizes. Capillary sprouts then invade the matrix, some cartilage is removed, and osteoblasts lay seams of osteoid (the organic matrix that mineralizes to form bone) over the mineralized cartilage. Shortly thereafter, the osteoid mineralizes to form primary bone, and osteoblasts add further layers of osteoid. Eventually, chondroclasts and osteoclasts remove the calcified cartilage and primary bone, and osteoblasts form mature lamellar bone in their place.

VII. SPECIALIZED FORMS OF HYALINE CARTILAGE

Although all human cartilages have special features, two forms of hyaline cartilage have particularly high degrees of organization and complex functions.

A. Articular

The function of synovial joints (e.g., the knee or the shoulder) depends on articular cartilage, which forms their bearing surfaces. It distributes loads, minimizing peak stresses on subchondral bone; it can be deformed and regain its original shape; it has remarkable durability; and it provides a load-bearing surface with unequaled low friction. In articular cartilage the chondrocytes organize the matrix components of hyaline cartilage in a unique way to allow it to perform these essential functions. The most apparent organization of the matrix is the changing structure and composi-

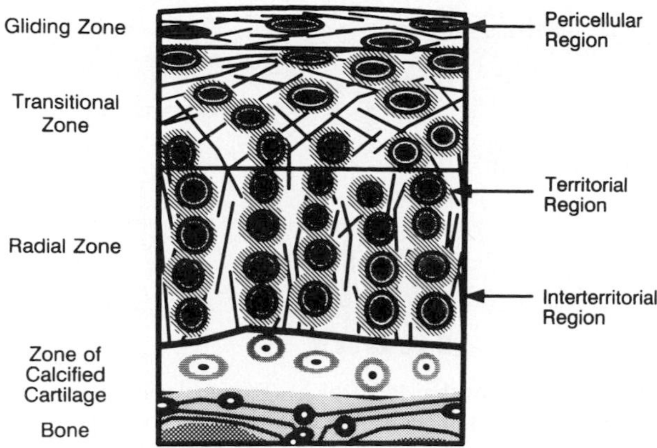

FIGURE 9 Articular cartilage. The tissue can be divided into four zones: the superficial, or gliding, zone; the transitional, or middle, zone; the radial, or deep, zone; and the calcified cartilage zone. In most areas three matrix regions can be identified: pericellular, territorial, and interterritorial. Interterritorial matrix collagen fibrils, represented by lines, tend to lie parallel to the joint surface in the superficial zone and perpendicular to the joint surface in the deep zone.

tion between the joint surface and the subchondral bone. Although the changes are not abrupt, this organization can be described by dividing articular cartilage into four successive zones, beginning at the joint surface: the superficial, or gliding, zone; the middle, or transitional, zone; the deep, or radial, zone; and the zone of calcified cartilage (Fig. 9). Within the zones, distinct matrix regions or compartments can generally be identified: the pericellular matrix, the territorial matrix, and the interterritorial matrix (Fig. 9). [*See* Articular Cartilage and the Intervertebral Disc; Articulations, Joints between Bones.]

1. Cartilage Zones

The thinnest articular cartilage zone—the gliding, or superficial, zone—forms the surface of the joint (Fig. 9). A thin cell-free layer of matrix, consisting primarily of fine fibrils and relatively little polysaccharide, lies directly adjacent to the synovial cavity. Immediately deep to this layer, elongated flattened chondrocytes are arranged with their major axis parallel to the articular surface. These relatively inactive cells contain small volumes of endoplasmic reticulum, Golgi membranes, and mitochondria. Little or no hyaluronic acid is present in this region, and the proteoglycans associated with the collagen resist extraction to a greater degree than those from other zones, suggesting a particularly strong association between collagen and proteoglycan.

The transitional, or middle, zone has several times the volume of the superficial zone. Its more spherical cells contain a greater volume of endoplasmic reticulum, Golgi membranes, mitochondria, and glycogen. They also contain occasional cytoplasmic filaments. The larger interterritorial matrix collagen fibrils are more randomly oriented than are those of the gliding zone.

In the deep, or radial, zone the cells resemble the spherical cells of the transitional zone, but tend to arrange themselves in a columnar pattern perpendicular to the joint surface. This zone has the largest collagen fibrils, the highest proteoglycan content, and the lowest water content, so that, from the superficial zone to the deepest portion of the deep zone, water content decreases and proteoglycan content as well as the diameter of the fibrils of interterritorial matrix collagen increase.

The zone of calcified cartilage separates the softer hyaline cartilage from the stiffer subchondral bone. Collagen fibrils penetrate from the deep zone directly into the calcified cartilage, anchoring the articular cartilage to the bone. The cells are smaller than are those found in other regions and appear relatively inactive.

It seems reasonable to speculate that the differences in matrix composition and organization among articular cartilage zones reflect differences in mechanical function. The superficial zone forms a thin tough layer that might primarily resist shear. The transitional zone might allow a change in the orientation of the collagen fibrils from the superficial zone to the deep zone, and the deep zone might primarily resist compression and distribute compressive loads. The calcified cartilage zone might provide a transition in material properties between hyaline cartilage and bone, as well as anchor the articular surface to the bone.

2. Matrix Regions

The matrix regions (Fig. 9) differ in their proximity to chondrocytes, their collagen content, their collagen fibril diameter, the orientation of their collagen fibrils, and their proteoglycan and noncollagenous protein content and organization.

Chondrocyte cell membranes appear to attach to a thin layer of pericellular matrix that surrounds the cell. This smallest matrix region appears to contain little or no fibrillar collagen, although it could contain some of the quantitatively minor nonfibrillar collagens. The predominant macromolecules of the pericellular matrix appear to be proteoglycans, noncollagenous proteins, and glycoproteins. Chondrocyte cell membranes seem to adhere to the pericellular matrix, thereby indirectly anchoring the cells to the macromolecules of the other matrix compartments. This arrangement might have a role in transmitting mechanical signals from the matrix to the cell.

An envelope of territorial matrix surrounds the pericellular matrix of each chondrocyte and, in some cases, pairs or clusters of chondrocytes and their pericellular matrices. For example, in the radial zone a territorial matrix surrounds each chondrocyte column. The thin collagen fibrils of the territorial matrix nearest the cells appear to bind to the pericellular matrix, but at a distance from the cell they spread and intersect at various angles, forming a fibrillar "basket" around the cells that could provide mechanical protection for the chondrocytes when cartilage is deformed.

The largest matrix compartment of articular cartilage, the interterritorial matrix, is distinguished from the territorial matrix by an increase in collagen fibril diameter and a transition from the interlacing basket-like orientation of fibrils to a more parallel arrangement. The organization and orientation of the collagen fibrils as they pass from the articular surface to the deeper zones are important features of articular cartilage organization. In the superficial zone the fibrils are oriented primarily parallel to the joint surface, in the transition zone they have a more random orientation, and in the deep zone they line up perpendicular to the joint surface (Fig. 9).

B. Physeal

Human bones elongate not by growth of bone tissue, but by growth of the cartilage forming the physes, or growth plates. The complex organization of these structures makes it possible for them to increase their volume so that they cause longitudinal bone growth and then to convert the tissue they produce into bone. Like articular cartilage, they have a layered or zonal organization and a regional organization. The zones differ from those of articular cartilage (Fig. 10), but the matrix regions, like those of articular cartilage, can be identified as pericellular, territorial, and interterritorial.

The layers, or zones, are the reserve, proliferative, and hypertrophic (Fig. 10). Reserve zone cells show relatively little evidence of metabolic activity. Their functions have not been clearly established, but some of them might serve as stem cells for the proliferative zone. In the proliferative zone, cells divide, synthesize extracellular matrix, and assume a highly oriented, flattened, disc-like shape. In rapidly growing bones they create long columns of highly ordered cells that resemble stacks of plates. At the bottom of the proliferative zone, the chondrocytes begin to enlarge, rapidly increasing their volume 5- to 10-fold, and assume a more spherical or polygonal shape. In the lowermost portion of the hypertrophic zone, or the zone of provisional calcification, the longitudinal septae of interterritorial matrix that lie between chondrocyte columns begin to mineralize, the enlarged chondrocytes condense, and metaphyseal capillary sprouts penetrate the unmineralized territorial matrix, invading the cell lacunae. In the metaphysis the mineralized cartilage bars are covered with new bone, resorbed, and eventually replaced by mature bone, duplicating the process of enchondral ossification that converts much of the embryonic cartilaginous skeleton into bone and helps heal some bone fractures.

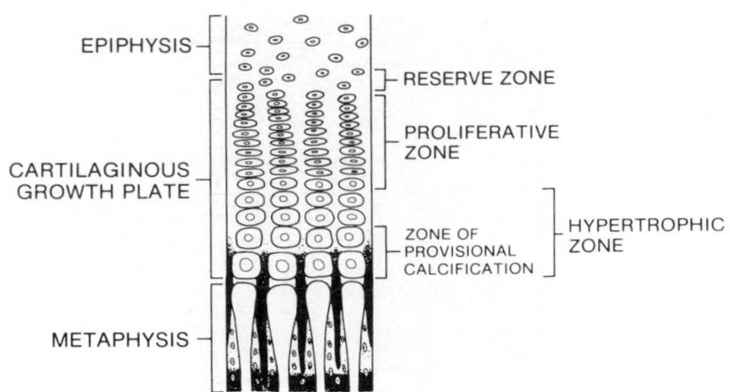

FIGURE 10 Growth plate cartilage. The cartilaginous physis lies between the epiphysis and the metaphysis of the bone and consists of three zones: reserve, proliferative, and hypertrophic. Between the upper proliferative zone and the lower hypertrophic zone, cell volume increases 5- to 10-fold. In the lowermost region of the hypertrophic zone, the zone of provisional calcification, the interterritorial matrix begins to mineralize.

As in articular cartilage, the orientation of the interterritorial matrix collagen fibrils changes among zones. In the reserve and upper portions of the proliferative zone, the collagen fibrils have little apparent orientation, but in the middle and lower regions of the proliferative zone and throughout the hypertrophic zone, the interterritorial matrix collagen fibrils lie parallel to the long axis of the bone.

Growth of long bones requires a directed increase in the volume of the physeal cartilage. The chondrocytes accomplish this by synthesizing new matrix and increasing their volume by the synthesis of cytoplasm and organelles and the accumulation of water. In addition, territorial matrix swelling could contribute to the growth of bone. The swelling of the cells and territorial matrix might be directed to produce longitudinal growth by the physeal perichondrial tissues and the internal organization of the physes, including the transphyseal collagen fibrils of the interterritorial matrix. Through mechanisms that remain poorly understood, the growth plate chondrocytes coordinate these activities to produce symmetrical skeletal growth and prepare the matrix they produce for mineralization.

BIBLIOGRAPHY

Arnoczky, S., Adams, M., DeHaven, K., Erye, D., and Mow, V. (1988). Meniscus. *In* "Injury and Repair of the Musculoskeletal Soft Tissues" (S.-L. Woo and J. A. Buckwalter, eds.), pp. 487–537. American Academy of Orthopaedic Surgeons, Park Ridge, IL.

Buckwalter, J. A., and Rosenberg, L. C. (1988). Electron microscopic studies of cartilage proteoglycans. *Electron Microsc. Rev.* **1,** 87–112.

Buckwalter, J. A., Mower, D., Ungar, R., Schaeffer, J., and Ginsberg, B. (1986). Morphometric analysis of chondrocyte hypertrophy. *J. Bone Joint Surg.* **68A,** 243–255.

Buckwalter, J. A., Hunziker, E., Rosenberg, L., Coutts, R., Adams, M., and Eryre, D. (1988). Articular cartilage: Structure and composition. *In* "Injury and Repair of the Musculoskeletal Soft Tissues" (S. L. Woo and J. A. Buckwalter, eds.), pp. 405–425. American Academy of Orthopaedic Surgeons, Park Ridge, IL.

Caplan, A. I. (1984). Cartilage. *Sci. Am.* **251,** 84–94.

Fawcett, D. W. (1986). Cartilage. *In* "A Textbook of Histology," 11th Ed., pp. 188–198. Saunders, Philadelphia.

Hardingham, T. E., Fosang, A. J., and Dudhia, J. (1992). Aggrecan the chondroitin sulfate/keratan sulfate proteoglycan from cartilage. *In* "Articular Cartilage and Osteoarthritis" (K. E. Kuettner, R. Schleyerbach, J. G. Peyron, and V. C. Hascall, eds.), pp. 5–20. Raven, New York.

Heinegard, D. K., and Pimentel, E. R. (1992). Cartilage matrix proteins. *In* "Articular Cartilage and Osteoarthritis" (K. E. Kuettner, R. Schleyerbach, J. G. Peyron, and V. C. Hascall, eds.), pp. 95–111. Raven, New York.

Kosher, R. A. (1983). The chondroblast and chondrocyte. *In* "Cartilage" (B. K. Hall, ed.), Vol. 1, pp. 59–85. Academic Press, New York.

Mayne, R., and Irwin, M. H. (1986). Collagen types in cartilage. *In* "Articular Cartilage Biochemistry" (K. E. Kuettner, R. Schlegerbach, and U. C. Hascall, eds.), pp. 23–28. Raven, New York.

Mecham, R. P., and Heuser, J. E. (1991). The elastic fiber. *In* "Cell Biology of the Extracellular Matrix" (E. D. Hay, ed.), 2nd Ed., pp. 79–109. Plenum, New York.

Mow, V., and Rossenwasser, M. (1988). Articular cartilage: Biomechanics. *In* "Injury and Repair of the Musculoskeletal Soft Tissues" (S. L. Woo and J. A. Buckwalter, eds.), pp. 427–463. American Academy of Orthopaedic Surgeons, Park Ridge, IL.

Catecholamines and Behavior

ARNOLD J. FRIEDHOFF
New York University Medical Center

RAUL SILVA
St. Luke's Roosevelt Hospital Center

GLOSSARY

Antidepressant drugs Medications used to treat a variety of conditions, such as depression, panic attacks, and obsessive-compulsive disorder. Generally speaking there are three different groups: the tricyclic antidepressants, which include imipramine and the related agents amitriptyline and nortriptyline; the monoamine oxidase inhibitors (MAO inhibitors) such as tranylcypromine and phenelzine; and the newest group, the serotonin reuptake blockers, which include anafranil, fluoxetine, sertraline, and paroxetine

Antipsychotic drugs Medications used for a variety of conditions. The name is derived from the improvement they produce in certain psychotic behaviors such as delusions and hallucinations. The first such agent, chlorpromazine, was synthesized around 1950. Examples of the classic antipsychotic agents are haloperidol and chlorpromazine. New atypical agents include clozapine and resperidol. These medications are also called neuroleptics

Catecholamines Three endogenously produced substances—epinephrine, norepinephrine, and dopamine—that serve as neurotransmitters

Limbic system Group of structures located in the brain that are involved in regulating emotion and its association with behavioral and mental functioning

Neurotransmitters Compounds that are released into interneuronal junctions called synapses. They are released from the axon of a presynaptic neuron and impact on the receptors of the postsynaptic neuron, the nerve cell on the other side of the synapse. This is the chemical means by which the transfer of information occurs in the brain

CATECHOLAMINES ARE POWERFUL CHEMICALS that can be found in neurons throughout the body. The effects of these compounds are responsible for the functioning of the brain even during the early fetal stages of life. They help regulate an endless number of functions, ranging from thinking and mood to motor control. In this article we will review the structure, the anatomical distribution, and the role these substances play in functioning and behavior.

I. NATURE OF THE CATECHOLAMINERGIC SYSTEMS

A. Introduction

Catecholamines are relatively small organic molecules that function in the brain and elsewhere in the body, primarily in a regulatory or modulating role, to keep various systems functioning smoothly in response to demands of the internal and external environment. The most familiar of the three natural catecholamines is adrenaline, or epinephrine (Fig. 1). Its effects have been experienced by all of us, in response to a frightening experience, for example. Its release from the adrenal gland and from nerve cells or neurons regulating heart rate and blood pressure help to put us into a readiness state for fight or flight. Norepinephrine, the closest chemical relative of epinephrine, is more prominently localized in the brain than epinephrine, but is also found in so-called peripheral neurons (those neurons found outside of the brain). In the brain, norepinephrine regulates mood and level of emotional arousal and alertness. Dopamine, the third catecholamine, is prominently involved in regulating motor or movement functions, and also in the coordination of associative thinking and integration of sensory mo-

ENCYCLOPEDIA OF HUMAN BIOLOGY, Second Edition, VOLUME 2. Copyright © 1997 by Academic Press. All rights of reproduction in any form reserved.

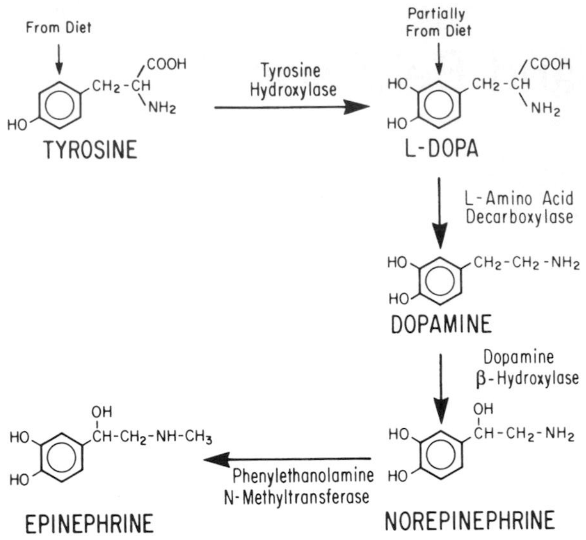

FIGURE 1 Pathway for biosynthesis of the major catecholamines.

tor function. Thus key volitional acts such as movement and thinking are fine-tuned, integrated, and given emotional coloration through the actions of the three catecholamines.

Understanding the role of catecholamines in normal and pathological behavior is important in distinguishing the structural organization and functional aspects of this system in the brain. The relationship of the catecholaminergic system to behavior has been deduced largely by using drugs to alter the function of various components of the neural networks that make up the system.

B. Neurotransmission

Information transfer in the brain is carried on mainly by synaptic transmission, or the passage of a message across synapses or gaps between communicating cells. This occurs through a combination of electrical transmission that takes place within a neuron and the release of a chemical or neurotransmitter that crosses the synaptic gap and then acts on a postsynaptic neuron via specialized detection sites called "receptors"; however, there are some exceptions to this general model. For example, some neurons (not catecholaminergic) relate to each other entirely by change in electrical potential. In many cases involving the catecholaminegic system, other substances are coreleased with the neurotransmitter and modify or modulate its effect. The nature of the effector response can vary depending on the type of receptor, location of the membrane, and nature of the neuromodulators. For example, the stimulation of β_3-adrenergic receptors located in adipose tissues will stimulate the breakdown of fats (lipolysis). This can be contrasted with the stimulation of different α_2-adrenergic receptors, one of which may inhibit the release of certain neurotransmitters at the presynaptic level of adrenergic nerve cells, thereby causing inhibition of norepinephrine release. Meanwhile, stimulation of another α_2-adrenergic receptor located on the membranes of the β cells of the pancreas will cause a decrease in insulin secretion.

The synapse is an important locus for the action of drugs that modify behavior. By blocking reuptake of transmitters, the effect of the transmitter can be enhanced or exaggerated. Conversely, by blocking receptors on postsynaptic cells, transmitter effect can be reduced. A third possibility, which has been exploited pharmacologically, is the modification of the ion exchange involved in electrical transmission. This too can have effects on motor and mental activity.

C. Biosynthesis of Catecholamines

The starting point for the synthesis of all the catecholamines is *l*-tyrosine, which is a nonessential amino acid that can be found in the diet. L-Tyrosine is hydroxylated (gains an OH group) to form dihydroxy-L-phenylalanine, which is also known as levodopa or *l*-dopa. The enzyme responsible for this transformation is tyrosine hydroxylase. In dopaminergic neurons, *l*-dopa is metabolized to dopamine by means of the enzyme dopa decarboxylase. This enzymatic process occurs in the cytoplasmic component of neurons. In noradrenergic nerve cells and in the adrenal medulla, dopamine is transformed to norepinephrine. It has been estimated that approximately 50% of the dopamine synthesized in neuronal cytoplasm of noradrenergic cells is metabolized to norepinephrine. Norepinephrine can then be transformed to epinephrine by the addition of a methyl group (CH_3) to its amino group, through the action of the enzyme phenethanolamine-*N*-methyltransferase. This last step occurs in certain neurons of the brain and in the adrenal medulla (graphic and schematic representations of the biosynthesis and breakdown of catecholamines can be found in many of the references listed in the bibliography). In general, the enzymes described in this section are produced in the neuronal cell bodies and are then transported and stored in nerve endings. Therefore the process of catecholamine biosynthesis takes place within these terminals. The catecholamines synthesized are then taken up and stored in vesicles (chromaffin granules) of the nerve terminals, which are located near the cell membrane. During neural transmission, catecholamines are released from these vesicles into the synaptic cleft. Although certain precursors of catecholamines (such as *l*-dopa) penetrate the blood–brain barrier, the catecholamines do not. Thus all of the catecholamines found in the brain are produced there.

The amount of catecholamines that exist within the adrenal medulla and the sympathetic nervous system is generally constant. Initial changes that occur in the synthesis of these substances, in response to changes, occur in minutes, whereas slower adaptational changes occur over much longer periods, even days in some cases. Catecholamines in the body are maintained at constant levels by a highly efficient process that modulates their biosynthesis, release, and subsequent inactivation.

When an appropriate signal is received by a catecholaminergic neuron, it is transmitted down the axon to the presynaptic terminal, where it initiates the release of quanta of neurotransmitter into the synaptic cleft. The transmitter acts on receptors in follower cells, resulting in the activation or inhibition of these cells.

D. Inactivation of Catecholamines

There are two major means for catecholamine inactivation: reuptake and enzymatic degradation. The reuptake system is fast and highly efficient. It operates through a rapid reuptake of released transmitters back into the presynaptic terminal. The involved transporter reuptake protein has two functions: (1) it rapidly inactivates transmission by removing transmitter from the synapse and (2) it conserves transmitter by re-storing that which is not used in signal transmission. Catecholamines made in the neuron but not stored in terminal vesicles are catabolized by a series of isoenzymes known as monoamine oxidases (MAO), which are located in most living tissues. Another enzyme important in the breakdown of catecholamines released into the synapse is catechol-*o*-methyltransferase. Discussion of all the metabolic steps in degradation of catecholamines is beyond the scope of this article, however, it is important to note that drugs that increase catecholamine levels in the synapse, particularly norepinephrine, are successful antidepressant medications. The concentration of norepinephrine can be altered by two types of drugs: reuptake blockers, which prolong the life of norepinephrine in the synapse by preventing its reentry into the presynaptic neuron, and monoamine oxidase inhibitors, which interfere with breakdown by monoamine oxidase.

From observations of the action of these drugs it has been proposed that depression is the result of low levels of norepinephrine in the brain; however, direct evidence for this proposal has not been found. In support of the proposal, the antihypertensive drug reserpine, which depletes norepinephrine and the other catecholamines, sometimes causes serious depression. Curiously, drugs that increase levels of serotonin, a noncatecholamine found in the brain, are also antidepressants. Norepinephrine and epinephrine also act as hormones when released from the adrenal medulla. Epinephrine is the principal catecholaminergic hormone produced in the medulla. Norepinephrine is the primary neurotransmitter in all postganglionic sympathetic neurons except for those that supply the vasodilator blood vessels of the skeletal muscular system and the sweat glands. The sympathetic nervous system along with the parasympathetic nervous system make up the autonomic nervous system, which helps regulate the visceral functions of the body. The autonomic nervous system has control centers that are located in the spinal cord, hypothalamus, the reticular formation of the medulla oblongata, and other regions of the brain stem. The centers located in the spinal cord and in the brain stem are regulated by the hypothalamus, which also communicates with the pituitary and cerebral cortex. This interconnection enables the complex orchestration of multiple somatic, visceral, and endocrinological functions. [*See* Autonomic Nervous System.]

The noradrenergic system has two major areas of origin in the brain: the locus coeruleus and the lateral tegmental nucleus. The projections of this system extend to all regions of the brain. As explained earlier, dopamine is the precursor in the synthesis of norepinephrine and epinephrine. In addition to this, dopamine has its own complex system and specialized function. The dopamine system is composed of three subdivisions: mesocortical, mesolimbic, and nigrostriatal systems. The mesocortical system extends from the ventral tegmentum to a variety of areas such as the olfactory tubercles, the accumbens, and the prefrontal cortex. The neurons of the mesolimbic system originate in the substantia nigra and the ventral tegmentum and project to the accumbens and amygdala. It is believed that the limbic system is probably more involved in regulating certain mental processes. The nigrostriatal system extends from the substantia nigra to the neostriatal regions. In addition to other functions, the nigrostriatal system is involved in movement. Disturbances of vital structures in this area are related to illnesses such as Parkinsonism.

E. Catecholaminergic Receptor Sites

Catecholamine receptors are proteins embedded in the plasma membrane of a neuron. Activation of these receptors by catecholamines can produce excitatory and/or inhibitory responses. Receptor number, in many cases, is increased or decreased as an adaptive response. For example, blockage of dopamine receptors by antipsychotic drugs, which are dopamine receptor antagonists, often results in a compensatory increase in the number of receptors. A number of types of catecholamine receptors respond to one of the three catecholamines.

1. Dopaminergic Receptors

Five types of dopamine receptors have been identified. They are all called dopamine receptors because they all respond to dopamine and are relatively homologous in structure; however, two types, D1 and D2, can be discriminated pharmacologically by both agonists and antagonists. It is very likely that drugs selective for the other three types will also be found. The ability to selectively activate or inactivate different aspects of the dopaminergic system with drugs that act on one receptor type has made it possible to explore the role that the D1 and D2 dopaminergic system plays in behavior.

a. D1 receptors are found in the caudate nucleus and cortex. There are a variety of extraneural sites where these receptors are located, including the vascular structures of the brain, heart, and renal and mesenteric systems.
b. D2 receptors have been identified in the putamen, caudate nucleus, and striatum, as well as in limbic structures and in low density in the cortex. Two subtypes of D2 receptors have been identified (D2a and D2b), but differences in anatomical location and physiological properties have not been worked out.
c. D3 receptors have been identified in the limbic system.
d. D4 receptors have been recently identified in the frontal cortex, basal ganglia, medulla, midbrain, and amygdala.
e. D5 receptors have also recently been identified in the caudate, putamen, olfactory bulb, and tubercle, as well as in the nucleus accumbens.

2. Adrenergic Receptors

There are two types of adrenergic receptors, with subdivisions within each.

a. α-Adrenergic receptors
   i. α_1-Adrenergic receptors are located on postsynaptic effector cells such as those on the smooth muscles of the vascular, genitourinary, intestinal, and cardiac systems. Additionally, in humans these receptors are located within the liver.
   ii. α_2-Adrenergic receptors inhibit the release of certain neurotransmitters. For example, at the presynaptic level in certain adrenergic nerve cells, these receptors inhibit norepinephrine release, whereas in cholinergic neurons they are responsible for inhibiting acetylcholine release. α_2-Adrenergic receptors are also located in postjunctional sites such as the β cells of the pancreas, in platelets, and in vascular smooth muscle. Although there are at least two subtypes of both α_1- and α_2-adrenergic receptors, the details concerning the actions and localization that would differentiate these particular subtypes have not been worked out.
b. β-Adrenergic receptors
   i. β_1-Adrenergic receptors have been located in the heart, the juxtaglomerular cells of the kidney, and in parathyroid gland.
   ii. β_2-Adrenergic receptors have been identified in the smooth muscles of the vascular, gastrointestinal, genitourinary, and bronchial structures. Additionally, β_2-adrenergic receptors have been located in skeletal muscle and in the liver, as well as on the α cells of the pancreas, which are responsible for glucagon production.
   iii. β_3-Adrenergic receptors are reported to be located in adipose tissue.

F. Plasma Catecholamines

The three catecholamines, when found intact in plasma, do not come from the brain because they cannot cross the blood–brain barrier; however, their metabolites can. Thus, metabolites in plasma originate both in brain and in peripheral tissues. Study of these metabolites has provided certain insights into the role that catecholamines play in behavior. However, direct study of catecholamines in living human brain tissue has not been possible. Fortunately, the new imaging technologies such as positron emission tomography (PET scanning), nuclear magnetic resonance (NMR), and single positron emission computerized tomography (SPECT) open up possibilities for visualizing catecholaminergic function in live conscious human subjects during waking hours. A variety of methods are available for measuring catecholamines in plasma.

II. IMPACT OF CATECHOLAMINES ON BEHAVIOR

Most of the information that is available concerning the functions of catecholamines in regulating human behavior directly results from the use of a group of medications often called psychotropic drugs and antidepressant medications called thymoleptics. Other medications include psychostimulants such as the dextroamphetamines—methylphenidate (most commonly known by its trade name, Ritalin) and *l*-dopa (which has been used to treat Parkinsonism)—as well as a medication that was initially used to treat high blood pressure, reserpine. Most of these drugs impact on more than one system (e.g., dopaminergic, noradrenergic, or serotonergic systems). Catecholamines have been proposed as mediators of most psychiatric illnesses, including schizophrenia, Tourette's syndrome, depression, autism, pervasive developmental disorders, attention deficit–hyperactivity disorder, stereotypic movements, and tremors. Unfortunately, to date no definitive evidence for their eole in any of these has been forthcoming. What is definite, however, is the role catecholamines play in mediating the action of mood-altering, mind-altering, and other types of psychotropic drugs. Antipsychotic drugs that block dopamine receptors reduce the more classic psychotic symptoms (delusions and hallucinations). There is some speculation about which dopamine receptors these agents block to produce improvement, but the prevailing view is that the more traditional agents block D2 receptors whereas the newer atypical agents (such as clozapine) may also block D4 receptors. The fact that agents that block dopamine receptors produce improvement in schizophrenia has led to the proposal that schizophrenia is caused by overactivity of the dopaminergic system. In support of this so-called "dopamine hypothesis," at least one group has reported an increased density of D2 receptors in brains of schizophrenic patients using the relatively new imaging technology called positron emission tomography. Increased density of D2 receptors in postmortem brain tissue from patients with schizophrenia has also been reported. However, most patients have received neuroleptic treatment, which itself can cause these changes. Thus it is not clear whether this increased density is an effect of the pathophysiology or the result of treatment. It is well established that reducing dopaminergic activity with neuroleptics inhibits hallucinatory activity and normalizes delusional or paranoid thinking. It seems probable that the dopaminergic system, particularly the D2 system, has a physiological role in keeping thinking and level of suspiciousness in bounds. Curiously, patients who respond well to antipsychotic medication have a decrease in plasma homovanillic acid (HVA), the principal metabolic of dopamine, during treatment, whereas nonresponders do not. What is odd about these findings is that most plasma HVA does not come from the central nervous system.

Antipsychotics improve certain other symptoms associated with schizophrenia, such as impaired thought processes and attentional problems. Thus it seems that the dopaminergic system may also regulate associative processing and attention. Drugs that improve psychotic symptoms have one more important effect. They produce emotional blunting or so-called "flat affect." Inasmuch as these drugs reduce dopaminergic activity, it seems that dopamine may play a role in affect regulation.

Another illness that may illuminate the role of dopamine in regulation of behavior is Tourette's syndrome. This is an illness with onset usually between the ages of 4 and 8 years of age; however, it can occur at any time. It is characterized by rapid, repetitive movements known as motor tics, which can be as simple as eye blinking or as complex as assuming contorted body positions. In addition to these movements, vocal tics occur—ranging from repetitive coughing and throat clearing to shouting obscene words. These utterances can be a great source of embarrassment to the affected individuals. Both the vocalizations and the motor tics respond to antipsychotic drugs that are, of course, dopamine receptor blockers. This effect on Tourette symptoms occurs even though the patients are not psychotic. Although dopamine is known to play a role in integrating motor movements, there is a distinct possibility that it may also inhibit socially undesirable movements and vocalizations.

It seems that Tourette's syndrome is in some way related to obsessive–compulsive disorder (this latter illness being particularly prevalent in families of Tourette patients). Obsessive–compulsive disorder is often responsive to drugs that increase serotonergic activity. Thus there appears to be a complex interaction between the serotonergic and dopaminergic systems in the regulation of psychomotor activity.

The study of psychological depression and its treatment can also help to illuminate the role of catecholamines in the regulation of behavior. Drugs like the tricyclic antidepressants and the monoamine oxidase inhibitors, both of which increase norepinephrine in the synapse, are useful in treating depressed patients. As a result of those observations, it was first concluded that depression resulted from abnormally low activity of the noradrenergic system. It now appears, however, that increasing norepinephrine levels via drug treatment serves to compensate for unknown pathology in depression. Additionally, all of the drugs useful in treating depression affect other transmitters besides norepinephrine. [*See* Depression, Neurotransmitters and Receptors.]

These observations are, nevertheless, informative. It seems probable that norepinephrine, by regulating its own activity, and in concert with other transmitters, plays a role in the relief and prevention of depression if not in the cause of depression. Norepinephrine may regulate mood, level of emotional arousal, sleep/wakefulness states, and appetite (all of which are often disturbed by depression).

Autism is a serious psychiatric condition that begins in infancy or early childhood. It is characterized by a qualitative impairment in interaction and socialization. Autistic children appear to be oblivious to their surroundings but ironically can react with a temper tantrum if a single toy is moved from its usual location. They are often lacking verbal and nonverbal communication skills. Speech may be limited to repeating a word over and over, and they may not even point to something they want in order to obtain it. Autistic individuals exhibit a restriction of activities and engage in a variety of odd behaviors such as sniffing, twirling and spinning, and inordinate interest in the single function of an object (i.e., staring at a wheel spinning on a toy car for hours). They also sometimes present with violent or self-injurious behavior and temper tantrums. Some of these patients may possess striking talents beyond their apparent cognitive capacity (often referred to as savant-like traits). A few can masterfully play the piano without ever receiving instruction or memorize an entire city's bus routes. [*See* Autism.]

The pervasive developmental disorders are illnesses that may vary in presentation. They may present with only one feature of autism or most of the features (but by definition not all). Though elevated serotonin levels in whole blood seem to be the most consistent finding in autism, there have been reports of increased norepinephrine levels in the plasma of these children when compared to normal control groups. Additionally, the effectiveness of dopamine-blocking neuroleptics on attention and improvement of certain behaviors in autistic children cannot be ignored. One investigation of biological markers in children with pervasive developmental disorder reported that the group that responded to treatment had lower initial plasma levels of HVA.

Attention deficit–hyperactivity disorder (ADHD) is characterized by overactivity, fidgetiness, impulsivity, and distractibility. It is more frequently seen in males and there is usually a family history of the disorder. The illness begins early in life but often is not diagnosed until the child is in school, as its pathology becomes more evident when more controlled behavior is required. There is strong evidence for involvement of the catecholaminergic systems in this illness. Prevailing theories propose a decrease in turnover of both dopamine and norepinephrine. Findings include decreased norepinephrine metabolites in the plasma of these individuals, and treatment involves the use of drugs that have norepinephrine-like effects. Oddly, increases in noradrenergic activity in the activating systems of the brain produce emotional arousal and many of the attendant symptoms of ADHD. In addition, adults given the psychostimulants used to treat ADHD in children have the expected activating effects. Perhaps then the function of the noradrenergic system may be developmentally regulated.

III. CONCLUSIONS

Catecholamines in the brain act at the highest levels of mental function. Although their role in specific mental disorders is not entirely clear, there is little doubt that they modulate, if not mediate, functions like processing of associations, integration of thought processes with movement and speech, emotional tone or affect, mood, appetite, arousal, and sleep/wakefulness state. Most of these functions have not been successfully modeled in nonhuman species, leaving their study to be carried out in living humans. This limitation has made more than inferential conclusions as to behavioral and mental function impossible.

New technological advances in functional brain imaging and in studies of gene expression in accessible human cells have opened new windows into the brain, but definitive studies await further advances.

BIBLIOGRAPHY

Axelrod, J. (1987). Catecholamines. *In* "Encyclopedia of Neuroscience" (G. Adelman, ed.), Vol. I, 1st Ed. Birkhauser, Boston/Basel/Stuttgart.

Davis, K. L., Khan, R. S., Ko, G., and Davidson, M. (1991). Dopamine in schizophrenia: A review and reconceptualization. *Am. J. Psychiatry* **148**, 11, 1474–1486.

Friedhoff, A. J. (ed.) (1975). "Catecholamines and Behavior," Vols. I and II. Plenum, New York.

Friedhoff, A. J. (1991). Catecholamines and behavior. *In* "Encyclopedia of Human Biology" (V. S. Ramachandran, ed.), Vol. II. Academic Press, San Diego.

Gilman, A. G., Rall, T. W., Nies, A. S., and Taylor, P. (eds.) (1990). "Goodman and Gilman's: The Pharmacological Basis of Therapeutics," 8th Ed. Pergamon, New York.

Kaplan, H. I., and Sadock, B. J. (eds.) (1989). "Comprehensive Textbook of Psychiatry," 5th Ed. Williams & Wilkins, Baltimore.

Silva, R. R., and Friedhoff, A. J. (1993). Recent advances in research into Tourette's syndrome. *In* "Handbook of Tourette Syndrome and Related Tic and Behavioral Disorders" (R. Kurlan, ed.). Dekker, New York.

Wilson, J. D., Baaunwald, E., Isselbacher, K. J., Petersdore, R. G., Martin, J. B., Fauchi, A. S., and Rood, R. K. (eds.) (1991). "Principles of Internal Medicine." McGraw–Hill, New York.

CD8 and CD4: Structure, Function, and Molecular Biology

JANE R. PARNES
Stanford University

GLOSSARY

Antigen Foreign substance that is specifically recognized by the immune system

Cytotoxic T lymphocyte T lymphocyte that specifically kills cells expressing the foreign antigen/major histocompatibility complex protein combination for which its T-cell receptor is specific

Helper or inducer T lymphocyte T lymphocyte that provides signals to induce the further differentiation of either B lymphocytes (to secrete antibody) or other T lymphocytes upon recognition of the antigen/major histocompatibility complex protein combination for which its T-cell receptor is specific

Major histocompatibility complex Large genetic region on human chromosome 6 containing genes mediating a variety of immune functions, including the highly polymorphic class I (classic transplantation antigens) and class II (immune response gene products) proteins and the class III (certain complement components) proteins

T-cell receptor Heterodimeric T-lymphocyte surface protein that varies from cell to cell and mediates the specific recognition patterns of T cells

CD8 and CD4 ARE CELL-SURFACE GLYCOPROTEINS expressed primarily on mature T lymphocytes and on their developmental precursors in the thymus. T lymphocytes are responsible for cellular immune responses and are important in protecting the individual against infections by parasites, fungi, and intracellular viruses, as well as against cancer cells and foreign tissues. Although it is the T-cell receptor (TCR) for antigen that is responsible for specific immune recognition, CD8 and CD4 also play important roles in T-cell responses to foreign antigens. These proteins divide mature T lymphocytes into two distinct functional subsets based on their recognition patterns. CD8 and CD4 have been referred to as "accessory molecules" or "coreceptors" because of their ability to interact with the TCR complex to enhance T-cell responses, and as "differentiation antigens" because of their pattern of expression during T-cell development.

I. INTRODUCTION

T-cell precursors that arrive in the thymus from bone marrow initially express very low levels of CD4. This expression is then lost, and the major early population of developing thymocytes (about 5% of thymocytes) is characterized by lack of surface expression of either CD4 or CD8. Such thymocytes are referred to as "double-negatives." The next major developmental stage contains the majority of thymocytes (approximately 80%) and is characterized by simultaneous expression of both CD8 and CD4. On their way to this "double-positive" stage, human thymocytes transiently express CD4 prior to CD8, whereas in most mouse strains CD8 is expressed first. In both cases these small numbers of early cells expressing only CD4 or CD8 can be differentiated from the more mature "single-positive" cells by their lack of surface

ENCYCLOPEDIA OF HUMAN BIOLOGY, Second Edition, VOLUME 2. Copyright © 1997 by Academic Press. All rights of reproduction in any form reserved.

expression of the TCR and CD3, a complex of proteins associated with the TCR and involved in TCR signaling. It is during the $CD4^+$, $CD8^+$ double-positive stage that thymocytes begin to express surface TCR and CD3. Recent studies suggest that it is primarily, though not exclusively, at the double-positive stage that one form of self-tolerance occurs, namely, elimination of thymocytes that are strongly reactive to self-antigens ("negative selection"). It is also thought that a positive selection occurs on double-positive cells (or as they differentiate to single-positive cells) to allow continued differentiation of only those thymocytes that express TCR molecules that will later recognize foreign peptides bound to self-MHC (major histocompatibility complex) proteins. It is not entirely clear how this occurs, since the foreign antigens that may later be encountered by mature T cells are certainly not present. The prevailing view is that thymocytes are positively selected by binding weakly to a complex of self-peptide bound to self-MHC protein (i.e., not strongly enough to be eliminated by negative selection, which operates to remove cells that might later be autoreactive). The mechanisms involved in these selection processes are not fully understood but are the subject of intensive investigation at present. What is clear is that most thymocytes die at the double-positive stage. Those that do differentiate further lose expression of either CD8 or CD4 and become mature "single-positive" CD4 or CD8 cells, with an approximate 2:1 ratio of CD4:CD8. These are the cells that are released into the periphery, and hence mature T cells in peripheral blood consist of approximately 65% CD4 cells and 35% CD8 cells. [*See* Major Histocompatibility Complex (MHC).]

Soon after the identification of CD8 and CD4 through the use of specific antibodies, it was recognized that expression of these proteins on the surface of mature T cells correlates reasonably well with the type of functional activity exhibited by the T cell. In general, CD8 cells are cytotoxic (or possibly suppressor) T cells, whereas CD4 cells are helper or inducer T cells. However, there are some notable exceptions to this generalization, and it is now accepted that the best correlation with CD8 versus CD4 expression is with the recognition properties of the specific TCR expressed by a given T cell. In contrast to antibodies, which recognize soluble forms of foreign proteins, TCRs, which mediate the specific recognition properties of T cells, recognize either foreign peptides bound to self proteins encoded within the major histocompatibility complex or foreign MHC proteins. This requirement for foreign peptide antigen to be bound to self-MHC proteins for T-cell recognition is referred to as MHC restriction of T cells. The MHC proteins involved in this restriction are highly polymorphic from one individual to another and fall into two major categories: class I proteins (HLA-A, -B, and -C), which are the classic transplantation antigens and are responsible for rejection of foreign tissues, and class II proteins (HLA-DP, -DQ, and -DR), which are the products of "immune-response" genes. T cells with receptors that recognize foreign peptides bound to self-class I MHC proteins (or to foreign class I MHC proteins in the absence of foreign peptide antigen) express CD8. Such cells are usually, though not always, cytotoxic (or suppressor) in function. In contrast, T cells with receptors that recognize foreign peptides bound to self-class II MHC proteins (or foreign class II proteins in the absence of foreign peptide antigen) express CD4 and are generally helper or inducer in function. Of note, there are examples of cytotoxic T cells that express CD4, not CD8, and are specific for class II MHC proteins. [*See* T-Cell Receptors.]

Early studies with monoclonal antibodies (mAbs) specific for CD8 and CD4 indicated that these proteins are important in the function of the T cells that bear them, since T-cell function could be blocked by incubation of T cells with such mAbs. These findings, coupled with the knowledge of the correlation between accessory molecule expression and class of MHC protein recognized, suggested the hypothesis that CD8 and CD4 might themselves be receptors for class I and class II MHC proteins, respectively (Fig. 1). However, in contrast to the TCR, which recognizes polymorphic regions on MHC proteins, CD8 and CD4 were hypothesized to recognize invariant, or at least relatively conserved regions on these proteins, since they themselves did not vary in cells of a given individual or from one individual to another. This hypothesis has been verified by binding studies. Cells transfected with genetic constructs encoding CD8 or CD4 and expressing a large amount of these proteins on their surface, or artificial membrane vesicles expressing high levels of these proteins, have been shown to bind to other cells or membrane preparations bearing surface class I or class II MHC proteins, respectively, but not to cells or membranes lacking the appropriate class of MHC protein. These interactions can be inhibited by mAbs specific for either the accessory molecule or its MHC protein li-

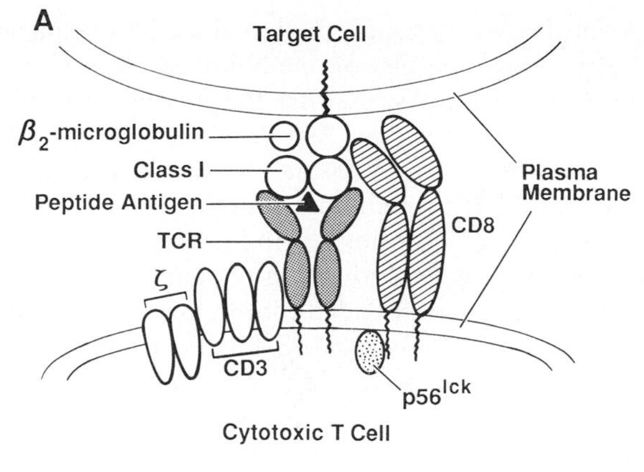

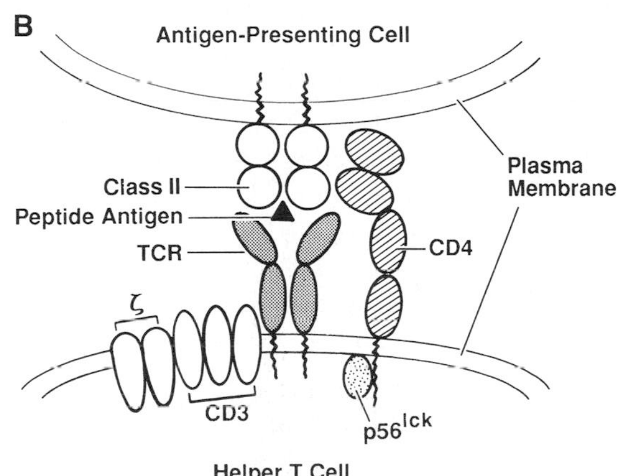

FIGURE 1 Model for interaction between CD4 or CD8 and major histocompatibility complex (MHC) proteins. (A) CD8 on a cytotoxic T cell is shown binding to a nonpolymorphic region of a class I MHC protein on a target cell and interacting with a T-cell receptor (TCR) molecule that is binding both to the same class I MHC protein and to a foreign peptide. CD8 is also shown to be associated with the T-cell-specific tyrosine kinase p56^{lck}. (B) CD4 on a helper T cell is shown binding to a nonpolymorphic region of a class II MHC protein on an antigen-presenting cell and interacting with a TCR molecule that is binding both to the same class II MHC protein and to a foreign peptide. As in the case of CD8, CD4 is also shown to be associated with the T-cell-specific tyrosine kinase p56^{lck}.

gand. The ability of CD8 and CD4 to bind to MHC proteins is thought to aid T-cell responses in two ways: by increasing the avidity of the interaction between the T cell and its target or antigen-presenting cell (i.e., an adhesive role), and/or by facilitating signal transduction to the T cell. It has also been suggested that these molecules may under certain circumstances transmit negative signals to T cells. The functional properties of CD8 and CD4 are described in greater detail in Section IV.

II. CD8 PROTEIN AND GENE STRUCTURE AND EXPRESSION

A. Polypeptide Structure and Subunit Composition

Early biochemical studies of human CD8 led to the conclusion that this molecule consists of disulfide-linked homodimers and higher multimers of a single 34-kDa polypeptide chain on peripheral blood T cells. In the thymus it was noted that the higher multimers contained an additional, larger polypeptide of 46 kDa. The latter has been identified as CD1, a protein recently shown to be involved in selection of and antigen presentation to a distinct subset (NK1$^+$) of T cells and expressed on cortical thymocytes and many antigen-presenting cells. Though CD1 is related to class I MHC proteins and contains a β_2-microglobulin subunit, it is encoded on a different chromosome. The significance of the association of a portion of thymic CD1 with CD8 is not clear. However, CD1 is not associated with CD8 in human mature peripheral T cells (which do not express CD1) or in mouse thymocytes. Recent studies have suggested that a portion of CD8 on human peripheral T cells is disulfide-linked to class I MHC proteins. Again, the significance of this finding is not known, and the topology of such an interaction would have to be quite distinct from that of the noncovalent interaction between CD8 on a T cell and class I MHC proteins of a target cell or antigen-presenting cell.

The initial conclusion that CD8 on human T cells is primarily a homodimer was surprising when one considers that both mouse and rat CD8 had been identified as consisting of heterodimers of two distinct polypeptide chains, CD8α and CD8β. The one human CD8 polypeptide chain that had been identified and studied with mAbs was shown to be the homolog of the mouse CD8α chain. Molecular genetic studies have since demonstrated that the physiological form of human CD8, like its rodent counterparts, is a disulfide-linked heterodimer. The vast majority of CD8$^+$ human peripheral blood T cells express both chains of CD8 (α and β). Those few CD8$^+$ cells that are CD8β^- (1–10%) may be natural killer (NK) cells or another subset of T cells. Most CD8 molecules on the surface of a T cell consist of heterodimers, although a small proportion may consist of CD8α homodimers. Transfection studies into CD8$^-$ cells have demonstrated that CD8α is capable of being expressed on the cell surface as a homodimer, whereas cell-surface expression of CD8β requires heterodimer formation with CD8α. It is likely that the presence of CD8β was missed in human cells because of comigration with the CD8α chain in the initial biochemical studies.

B. CD8α cDNA and Gene Structure

Complementary DNA (cDNA) clones encoding human CD8α were initially isolated by subtractive hybridization techniques using cDNA libraries constructed from mRNA of mouse fibroblasts (L cells) that had been transfected with total human genomic DNA and selected for cell-surface expression of CD8. The predicted amino acid sequence of human CD8α was then determined from the DNA sequence of these clones. The most striking conclusion to be drawn from the protein sequence was that CD8α is a member of what has been called the "immunoglobulin (Ig) gene superfamily," a large family of genes that are evolutionarily related to immunoglobulins. This conclusion is based on the presence in CD8α of an amino-terminal domain that is homologous to Ig variable (V) regions, especially Ig light-chain V regions. This domain of CD8α is of the same size as Ig V and constant (C) regions (approximately 100 amino acids) and contains many of the conserved residues of members of the Ig gene superfamily, including the centrally placed disulfide loop that is so characteristic of Ig "homology units." Computer analyses of the structural characteristics of this V-like domain suggested that this region can fold in a similar manner to Ig domains, and the recently solved crystal structure of CD8α homodimers has confirmed this structure. This Ig V-like domain contains nine β strands forming two β sheets of four and five strands, respectively.

The predicted CD8α protein has a signal peptide of 21 amino acids (cleaved during insertion of the protein into the membrane), followed by a mature protein sequence of 214 amino acids. The external portion of the protein consists of the 96-amino-acid V-like domain and a membrane-proximal, 65-amino-acid hinge-like region or connecting peptide. This is followed by a 24-amino-acid hydrophobic transmembrane segment and a highly basic 29-amino-acid cytoplasmic tail. The cytoplasmic tail contains a Cys-X-Cys motif common to the cytoplasmic tails of both CD8α and CD4 and shown to be critical for interaction with the *src*-family tyrosine kinase p56^{lck} (see Section IV,A). Although the human CD8α sequence suggests the possibility of one N-linked glycosylation site (Asn-X-Ser or Asn-X-Thr), the presence of proline as the variable residue (Asn-Pro-Thr) in the CD8α sequence most likely accounts for the lack of usage of this site. The protein does contain O-linked glycosylation, and this is primarily in the hinge region. The hinge region appears to lack a regular structure and is likely to be in an extended conformation that allows interaction of CD8 with the α3 domain of class I MHC proteins. CD8α is predicted to contain eight cysteine residues: three in the V-like region (two of which form the Ig-like disulfide loop) and two each in the hinge, transmembrane region, and cytoplasmic tail. Biochemical studies of the human protein have demonstrated that the cysteines in the V-like region are not involved in interchain disulfide bridges, and studies in the mouse system indicate that dimerization relies on external cysteine residues. It is therefore likely that either or both of the cysteine residues in the hinge region contribute to dimerization with CD8β, or with another CD8α chain in the absence of CD8β.

The human CD8α gene spans approximately 6.5 kilobases (kb) of DNA and consists of six exons separated by five introns. The gene structure supports the evolutionary relationship between CD8α and Ig genes, since the entire V-like domain of CD8α is encoded within a single exon and because the messenger RNA (mRNA) splicing junctions follow the same rules as for Ig genes (codons are interrupted by introns between the first and second nucleotides, i.e., a 1,2 codon split).

In the mouse system, alternative splicing of CD8α mRNA has been found to yield the production of a shorter form of CD8α (referred to as CD8α') that is missing most of the cytoplasmic tail and does not work as well as CD8α in functional assays. The CD8α' protein is expressed on the cell surface in almost equal amounts to CD8α during most of thymocyte development, but it is almost entirely excluded from the cell surface of mature T cells by a posttranslational control mechanism. An α' form of human CD8α has not been described, and it is likely that CD8α' in the mouse results from an artifact of mRNA splicing. Conversely, an alternatively spliced form of human CD8α mRNA has been observed without a corresponding mouse counterpart: a form of CD8α mRNA that is lacking the sequence encoding the transmembrane region has been identified in human T cells and cell lines. This mRNA results from splicing out of the sequence in exon IV and is predicted to encode a form of protein that would be secreted from the cell. Indeed, such a secreted form of

CD8α has been identified, but its function, if any, remains unknown.

The mouse CD8α gene had been known for many years to map to chromosome 6, closely linked to the Ig κ light-chain locus. The isolation of human CD8α probes allowed the chromosomal mapping of that gene to the short arm of chromosome 2 (2p12) at a location closely linked to the human Ig κ light-chain locus. This close linkage to κ in both mice and humans supports the hypothesis that CD8α and κ derived from a common ancestral precursor. Despite their similarities to Ig genes, the mouse and human CD8α genes are both single copy and do not require rearrangement for expression.

C. CD8β cDNA and Gene Structure

As discussed earlier, human CD8 had been thought to consist of homodimers/homomultimers of a single polypeptide chain on peripheral T cells, and biochemical studies did not reveal the presence of a polypeptide equivalent to mouse and rat CD8β. However, studies with rat and mouse CD8β cDNA probes demonstrated that there is indeed a human gene homologous to rat CD8β and that mRNA corresponding to this gene is expressed in human thymus and peripheral blood T cells. Human thymocyte cDNA clones corresponding to CD8β were then isolated by cross-species hybridization with the rodent CD8β cDNA clones. The majority of these human cDNA clones encode a mature protein of 189 amino acids with 143 residues external to the cell, and transmembrane and cytoplasmic domains of 27 and 19 residues, respectively. The predicted protein is very similar to mouse and rat CD8β, with approximately 56% identical amino acids. Computer analyses of the CD8β sequence have shown that it contains not only an amino-terminal Ig V-like domain, but also a 12-amino-acid sequence with marked similarity to an Ig joining segment (J) immediately thereafter.

Several alternatively spliced forms of human CD8β mRNA have been identified. These vary in their content of sequence encoded by cytoplasmic domain exons and the transmembrane exon. Those lacking sequence encoding the transmembrane exon could potentially yield secreted CD8β protein. However, it is not known to what extent these variants are expressed as protein and what their differential function(s) (if any) might be.

It was recognized from early classic genetic studies that the genes encoding the CD8α and CD8β polypeptide chains are closely linked on mouse chromosome 6. More recent studies have shown that they are separated by only 36 kb of DNA and are in the same transcriptional orientation. In humans, two CD8β genes have been identified. One (CD8β-1) has been shown to be located about 25 kb upstream of human CD8α on chromosome 2 and to be in the same transcriptional orientation. This gene contains nine exons; although most of the coding sequence is clustered within about 20 kb of DNA, the last two (exons 8 and 9) of the four alternatively spliced cytoplasmic exons (which have no known mouse homolog) are located about 45 kb downstream, leading to a total gene locus of about 70 kb. The CD8β-2 gene appears to also be on chromosome 2 but unlinked to CD8β-1 and CD8α. This gene is similar in structure but lacks exons 8 and 9, and hence is far shorter. The 3′ break point of the duplication event occurs 7.9 kb downstream from exon 7, but the upstream site is not known (at least 10 kb upstream of exon 1). The sequences of the two genes are very highly conserved (98.6% identity) in both coding and noncoding regions. Although the CD8β-2 gene appears to be functionally intact, almost all CD8β cDNAs isolated have derived from the CD8β-1 gene. Only a single clone encoding a potentially secreted form of protein derives from CD8β-2. As in the case of CD8α, the CD8β gene organization contains many of the features characteristic of members of the Ig gene superfamily, including the coding of protein domains in separate exons and the 1,2 codon split for splicing junctions. The CD8β genes do not rearrange. Notably, the J-like segment of CD8β is encoded in the same exon as the V-like region, even in the germ line. Like Ig J segment sequences, that of CD8β is followed by an mRNA splicing junction.

As mentioned earlier, transfection experiments demonstrated that the predicted human CD8β protein is expressed on normal human CD8$^+$ T cells as a heterodimer with CD8α. One particular CD8-specific mAb was shown only to bind to the surface of cell lines transfected with both CD8α and CD8β cDNA constructs, and not to the surface of cells transfected with either alone. This mAb, which required the presence of both chains for cell-surface binding, was then shown to bind to the surface of almost all human peripheral blood T cells expressing CD8 (as recognized by other mAbs specific for the α chain).

D. Regulation of CD8 expression

In contrast to CD4 (see Section III), CD8 has a more restricted tissue distribution of expression. To date it has been described on thymocytes and mature T cells, and in some instances on natural killer cells. However, it does not appear to be expressed on other cell or tissue types.

As discussed in the Introduction, CD8 has an interesting pattern of expression during thymocyte development. However, there are currently limited data regarding the mechanisms by which the CD8α and CD8β genes are turned on early in thymocyte development and their expression is later either maintained (in the case of cells bearing TCRs recognizing class I MHC proteins) or turned off (in the case of cells bearing TCRs recognizing class II MHC proteins). It is likely that the major level of control of CD8 expression is transcriptional, but posttranscriptional mechanisms could play some role. Although the expression of the CD8α and CD8β genes is coordinate in most instances, there are some T-cell subpopulations (e.g., CD8 cells in the gut) that express only CD8α and not CD8β. The two genes must therefore also be subject to certain independent regulatory mechanisms. A T-cell-specific enhancer has been demonstrated in the last intron of the human CD8α gene. This enhancer is adjacent to a silencer element and both contain half-Alu repeat sequences that can base-pair to form a cruciform structure that could disrupt the enhancer function. However, the role of these elements in CD8 regulation during T-cell development has not yet been determined.

III. CD4 PROTEIN AND GENE STRUCTURE AND EXPRESSION

A. Polypeptide and cDNA Structure

CD4 is a glycoprotein of approximately 55 kDa, which, in contrast to CD8, immunoprecipitates as a single polypeptide chain. Human cDNA clones encoding CD4 were isolated in much the same way as human CD8 clones. The mature human CD4 polypeptide chain is predicted to be 435 amino acids in length and to be preceded by a 23-amino-acid signal peptide. The protein has two potential N-linked glycosylation sites. There are six cysteine residues external to the cell, and these form three intrachain disulfide loops connecting adjacent cysteines. CD4 has a hydrophobic transmembrane segment of 26 amino acids followed by a basically charged cytoplasmic tail of 38 amino acids. As described earlier for CD8α, the cytoplasmic tail contains a Cys-X-Cys motif that is required for interaction with the tyrosine kinase p56^{lck}.

Comparisons of the predicted protein sequence of CD4 indicate that, like CD8α and CD8β, it too is a member of the Ig gene superfamily. The amino-terminal domain (D1) of CD4 (approximately 100 amino acids) is most strikingly homologous to Ig V regions, and contains the two cysteine residues characteristic of Ig homology units. This V-like region is followed by a short sequence that is similar to Ig J segments, although it is not nearly as closely related as that in CD8β. In contrast to the shorter CD8 structures, there are three additional V-like domains (D2–D4) in CD4, however, they are less closely related than the first. The D2 domain is severely truncated, maintaining only a sequence related to the carboxyl-terminal half of Ig V regions. Although this region contains two cysteine residues, the first of the two is not within the region that is clearly related to Ig V regions and the disulfide loop between the two spans only 28 amino acids. Comparisons to other V regions indicate that this foreshortened domain is most similar to the analogous region of the D1 domain of CD4, suggesting that there may have been a duplication of at least part of the amino-terminal V-like domain. The D2 domain of CD4 is also followed by a J-like sequence. The D3 domain has no cysteine residues, while the D4 domain has a foreshortened disulfide loop (with only 41 amino acids between the two cysteines). These two domains have diverged far more from the IgV structure than the amino-terminal portions of CD4.

The crystal structure of the D1 and D2 domains has confirmed their Ig-like structure. D1 has two β sheets with four and five β strands, respectively, and a conserved intersheet disulfide bond. The smaller D2 domain has only seven β strands and contains an intrasheet rather that the intersheet disulfide bond. D1 and D2 are intimately connected and are likely to form a rigid rod. The crystal structure of the D3D4 fragment has been determined in the rat, and its structural homology to the D1D2 fragment suggests duplication of a two-domain progenitor. Structural predictions suggest that CD4 extends as a rod from the cell surface, enabling it to span the distance required to interact with the $\beta 2$ domain of class II MHC proteins.

B. CD4 Gene Structure

CD4 is encoded by a single gene on human chromosome 12. As is the case for CD8α and CD8β, this gene does not rearrange prior to expression. The gene spans approximately 45 kb and consists of 10 exons separated by 9 introns. The first exon, which contains only noncoding sequence and is located about 10.5

kb upstream from coding sequence, was initially unrecognized, although such an exon was described for the mouse CD4 gene. Another unusual feature of the CD4 gene is the presence of a large intron of about 12 kb dividing the sequence encoding the amino-terminal V-like domain into two exons (exons 2 and 3) approximately halfway through the predicted protein domain. This is an atypical feature for members of the Ig gene superfamily in that each Ig homology unit is usually encoded within a single exon. However, CD4 is not the only example of a divided Ig homology unit. The neural cell adhesion molecule (N-CAM) has five Ig homology units, and the sequence encoding each is split by an intron. The remainder of the intron/exon structure of the CD4 gene correlates well with the predicted protein domains, a feature characteristic of members of the Ig gene superfamily. As is true for other members of the Ig gene superfamily, introns interrupt codons between the first and second nucleotides (1,2 codon split) in all cases except between the exons encoding the evolutionarily unrelated cytoplasmic tail.

C. CD4 Expression

Although CD4 protein appears to be restricted to T-lineage cells in mice, in both rats and humans it has additionally been identified on the surface of macrophage/monocyte cells and the related Langerhans cells. Full-length human CD4 mRNA and protein have also been identified in brain tissue. Macrophages are likely to be the predominant class of $CD4^+$ cells in brain, but the protein also appears to be expressed on at least some neuronal and glial cells. CD4 does not appear to be expressed on normal human B cells, but CD4 mRNA and surface protein have been identified in some Epstein–Barr virus-transformed human B-cell lines. Full-length CD4 transcripts have also been demonstrated in human granulocytes, although these are not known to express cell-surface CD4. The physiological role (if any) of CD4 on non-T cells is unknown.

The regulatory elements controlling CD4 expression are currently being examined. The promoter regions of the human and mouse CD4 genes are highly T cell specific. An enhancer element located 6.5 kb upstream of human CD4 gene has been shown to be required for proper expression of the human CD4 gene in transgenic mice. Two enhancers have been identified in the mouse CD4 gene; one, located 13 kb upstream of the transcription initiation site, appears to be the homolog of the identified human enhancer, whereas the other, located 24 kb upstream of the promoter region, does not have an identified human homolog to date. Both of these seem to be T cell specific, or at least preferential, and the distal enhancer appears to have stronger activity in $CD4^+$ cells. The distal enhancer, however, may be more involved in the regulation of the gene encoding Lag-3, a class II-MHC binding protein, since this enhancer maps upstream of that gene. Although the identified human CD4 enhancer is required for proper expression of human CD4, it is not sufficient. An additional element or elements within or near the gene are required. In the mouse, the additional element has been identified as a silencer within the first intron. This silencer is responsible for the lack of CD4 expression in double-negative thymocytes, as well as turning off expression of the CD4 gene as double-positive thymocytes mature to single-positive CD8 cells. Future studies should provide further documentation of how these various elements, and perhaps others, work together to yield the complicated developmental pattern of CD4 expression.

IV. FUNCTION OF CD8 AND CD4

A. Role of CD8 and CD4 in T-Cell Responses

The initial evidence that CD8 and CD4 play a role in T-cell function came from studies using antisera and subsequently mAbs specific for these proteins. These antibodies were found to block all antigen-driven functions (e.g., cytotoxicity, proliferation, lymphokine release) by T cells that expressed the corresponding molecule and were specific for the appropriate class of MHC protein, although there was clear heterogeneity in the ability of various T-cell clones to be blocked. In the case of cytotoxicity, anti-CD8 mAbs were found to block the formation of conjugates (relatively stable cell–cell contacts) between class I-specific cytotoxic T lymphocytes (CTL) and target cells. Anti-CD4 mAb could block cytotoxicity by class II-specific, $CD4^+$ CTL and induce dissociation of preformed conjugates between such CTL and target cells. These results suggested a role for CD8 and CD4 in the recognition step as opposed to the lytic machinery and led to the hypothesis that the function of CD8 and CD4 is to increase the avidity of the interaction between T cells and antigen-presenting or target cells by binding to class I or class II MHC molecules, respectively. In accord with this idea was the finding that inhibition of cytotoxicity by anti-CD8 or anti-CD4 could be overcome by the approximation of the CTL and target cells by lectins. This hypothesis could also explain the observed heterogeneity in the ability of different T-cell clones to be blocked by anti-CD8 or anti-CD4 mAbs, as well as the finding that bulk cultures of CTL from primary antigen responses could be blocked more easily by anti-CD8 than those from secondary antigen responses. As would be predicted by this model, T cells bearing TCR molecules with apparently low affinity for antigen/MHC were found to be more dependent on CD4 or CD8 interactions (i.e., more easily blocked), whereas T cells with higher-affinity TCRs were less dependent on CD4 or CD8. Finally, as discussed earlier, binding studies have indeed shown that CD8 and CD4 can act as receptors (although with apparently low affinity) for class I and class II MHC proteins, respectively. In the case of CD8α homodimers, binding studies using transfected cells expressing large amounts of surface CD8 have demonstrated that sequence in the membrane-proximal, $\alpha3$ domain of class I MHC proteins is important for binding of these cells to cell lines expressing human class I MHC proteins. Specifically, a highly conserved exposed loop including residues 220–229 and two additional clusters involving amino acids at positions 233 and 235 and positions 245 and 247 have been shown to contribute to this binding in mutagenesis experiments. The potential role of β_2-microglobulin (the light chain of MHC class I proteins) in this binding is unknown, but the $\alpha3$ domain of class I proteins interacts directly with β_2-microglobulin. Similarly, the potential role of CD8β in binding to class I has not yet been elucidated. It is not surprising that the nonpolymorphic CD8 protein would bind to a conserved region on class I. The fact that the TCR engages the more polymorphic $\alpha1$ and $\alpha2$ domains of class I whereas CD8 interacts with the $\alpha3$ domain would theoretically allow a TCR and CD8 to bind to the same class I protein at the same time. Indeed, mutation of key residues of class I important for binding to CD8α has been shown to render cells less competent in activating T-cell responses restricted by the mutant class I proteins, despite the surface expression of other class I molecules with intact $\alpha3$ domains. These findings indicate that simultaneous binding of CD8 and the TCR to the same MHC molecule can be critically important for T-cell activation. Mutational analysis of human CD8α has indicated that sequences in CD8α corresponding to the first and second complementarity-determining regions (CDR) of an antibody combining site are important for interaction with MHC class I. The binding of CD8 to isolated, plate-bound MHC class I proteins has been shown to increase upon activation of the TCR/CD3 complex with soluble anti-TCR mAb. This increased binding is not dependent on interaction of CD8 with the same MHC molecule to which the TCR binds, and it leads to increased signaling for release of cytotoxic granules by CTLs. Such activation-dependent binding requires a tyrosine kinase pathway, as it is blocked by the tyrosine kinase inhibitors. A similar phenomenon has not been described for interaction of CD4 with class II MHC proteins.

Recent mutagenesis studies of class II MHC proteins have identified a region of the $\beta2$ (membrane-proximal) domain of the β chain (residues 137–143) important for binding to CD4. This region of class II MHC proteins is structurally homologous to the CD8α binding site on class I MHC proteins. On the CD4 molecule, mutation within the CDR1- and CDR3-like hoops of the D1 domain and a nearby protruding loop in the D2 domain (all of which are on the same face of the protein) has been shown to diminish interaction with class II MHC proteins. In contrast, the major binding site on CD4 for binding to gp120 of human immunodeficiency virus-1 (HIV-1) (see Section IV,B) is on the opposite face of the CD4 protein within the CDR2-like loop of D1. A variety of studies suggested that CD8 and CD4 are not simply cellular adhesion molecules and that they may additionally be involved in pathways of signal transduction. Although CD8 has been found to play some role in the initial, antigen-independent formation of conjugates between T cells and other cells, other molecules (CD2 and LFA-1) may play a more major role in this process. mAbs specific for CD8 have been shown to block cytotoxicity not only during the phase of conjugate formation, but also during the cytolytic phase. Similarly, mAb specific for CD4 has been shown to inhibit cytotoxicity by class II-specific $CD4^+$ CTL clones at a postbinding step and to have only a small effect on the initial formation of conjugates. Anti-CD4 mAb can also induce the dissociation of preformed conjugates, although this dissociation is much slower and requires higher temperatures than that induced by anti-LFA-1 mAb.

An early clue that CD4 and CD8 might be involved directly or indirectly in the transmission of positive signals to activate T cells came from the finding that both are rapidly phosphorylated at serine and then dephosphorylated upon T-cell activation, but the role of this phosphorylation (if any) is not known. More direct support for a role for CD8 and CD4 in T-cell activation signaling also derived from the demonstration by a variety of different approaches, including coimmunoprecipitation, cocapping, comodulation, and fluorescence resonance energy transfer, that there is a physical association between these proteins and the TCR/CD3 complex. Although this association may be minimal prior to activation, it is increased upon TCR triggering. CD4 has also been shown to cocluster with the TCR to the area of intercellular contact between T cells and antigen-presenting cells during antigen-specific recognition. Furthermore, mAb cross-linking of the TCR/CD3 complex and either CD4 or CD8 via a number of different means was shown to dramatically increase TCR-mediated signaling (e.g., protein tyrosine phosphorylation, Ca^{2+} flux, inositol triphosphate production, lymphokine secretion and proliferation) when suboptimal concentrations of anti-TCR/CD3 mAbs were used.

The role of CD4 and CD8 in signaling during T-cell activation became more clear with the demonstration that both of these proteins are associated with the lymphocyte-specific tyrosine kinase, $p56^{lck}$, which is localized on the inner surface of the plasma membrane (Fig. 1). This tyrosine kinase, which is a member of the *src* family, has been shown to play an important role in TCR/CD3-mediated activation, and mutant cells lacking this kinase cannot be activated through the TCR/CD3 complex. Both CD4 and CD8α associate with $p56^{lck}$ via a Cys-X-Cys sequence in their cytoplasmic tails. A Cys-X-X-Cys motif in the amino-terminal portion of $p56^{lck}$ is critical for this interaction. Cross-linking of CD4 or CD8 results in an increase in associated $p56^{lck}$ kinase activity. Although only the CD8α chain associates with $p56^{lck}$, the presence of the CD8β chain leads to enhanced kinase activity of $p56^{lck}$ upon cross-linking of CD8 or cocrosslinking of CD8 with the TCR/CD3 complex. Additional signaling molecules may also be associated with CD4 and/or CD8. For example, a 32-kDa GTP-binding phosphoprotein has been shown to immunoprecipitate with CD4/$p56^{lck}$ and CD8/$p56^{lck}$ complexes from human T lineage cells.

The most direct demonstration of a physiological role for CD8 and CD4 in T-cell activation has come from gene transfer studies. Gene or cDNA expression vector constructs encoding CD8 or CD4 have been transfected (or infected, in the case of retroviral vectors) into functional T cells, and the effects of cell-surface expression of CD8 or CD4 upon antigen responses have been examined. These studies have shown that CD8 or CD4 can markedly increase antigen responses as long as the appropriate ligand (i.e., class I or class II MHC proteins, respectively) is present on the antigen-presenting or target cell. The requirement for and effects of CD8 or CD4 expression vary depending on the particular T cell and its TCR. In some instances, no response to antigen is present in the absence of one of these coreceptor molecules. In such cases the expression of CD8 or CD4 is absolutely required for a measurable antigen response. In other cases a basal response is present even in the absence of CD8 or CD4 but is increased by its presence. Although homodimers of the CD8α chain appear to be sufficient to enhance T-cell activation, recent studies have shown that heterodimers of the CD8α and CD8β chains function more efficiently to enhance T-cell activation. This may be related at least in part to the increased activity of $p56^{lck}$ in CD8$\alpha\beta$ heterodimers as compared to CD8α homodimers upon CD8/TCR cross-linking, but other/additional mechanisms may be involved.

The ability to assay CD8 and CD4 function by gene transfer into functional T cells has provided a mechanism for establishing the role of different portions of these molecules in enhancing T-cell responses. The external portions of these molecules are clearly required for appropriate ligand binding (i.e., class I or class II MHC proteins, respectively). Gene transfer studies have also demonstrated that CD8 and CD4 function optimally when their cytoplasmic tails are present, although in some instances they can enhance T-cell responses in their absence. This is likely to be at least partially related to the importance of the cytoplasmic tail for the association of CD8 and CD4 with $p56^{lck}$ and possibly other proteins involved in T-cell signaling. The cytoplasmic tail of CD4 has also been shown to be important for modulation of CD4 from the cell surface. Future studies will further dissect the specific functions played by the various domains of CD8 and CD4.

Gene transfer experiments have provided strong support for the notion that CD8 and CD4 can enhance responses by at least two mechanisms: increased adhesion and signal transduction. They have also demonstrated that CD4 and CD8 can have different effects depending on whether they can bind to the same MHC molecule (or at least the same class of MHC molecule) as the TCR (in which case they function as coreceptors) or only at different sites. Studies with mAb blocking had suggested, at least in the mouse system, that CD8 and CD4 could enhance responses in certain but not all instances by binding to MHC proteins to which the specific TCR could not bind. This has been demonstrated in a more direct fashion by gene transfer. For example, gene transfer of human CD8 into a mouse T-cell hybridoma specific for a human class II molecule can stimulate the response to class II as long as class I MHC proteins (the CD8 ligand) are expressed on the antigen-presenting cell. Similarly, a CD4/class II MHC interaction can stimulate the antigen response of a T-cell hybridoma specific for a class I MHC protein. However, this is not always the case, and there are clear instances when only the appropriate coreceptor/MHC protein interaction can stimulate a response. These appear to be cases in which the affinity of the TCR for its ligand is low, and hence the cell is more dependent on a coreceptor. It appears likely (though definitive proof is lacking) that enhanced signal transduction, that is, coreceptor function, mediated by CD8 or CD4 will occur only when one of these molecules and the TCR bind to the same MHC protein, and/or are functionally associated with the TCR, and that responses are stimulated only by an adhesion component when binding of CD4 or CD8 to its ligand is independent of the TCR. The latter may not be sufficient to allow detectable responses in the case of low-affinity TCRs, but may provide clear enhancement of responses mediated by TCRs with somewhat higher affinity.

Regardless of whether there is a clear division of function based on binding to the same or different MHC molecules, the bulk of evidence indicates that, depending on the specific TCR involved (particularly its affinity for a given antigen/MHC) and the antigen/MHC density (or concentration), some responses appear to require binding of CD4 or CD8 to the same MHC molecule as the TCR (coreceptor function), some need binding only at sites apart from the TCR/CD3 complex, and some are totally independent of CD4 or CD8 interactions. What remains unclear is the mechanism(s) by which these molecules enhance function. The stimulation from their adhesion function appears to be a result of increasing the avidity of the interaction between the T cell and target or antigen-presenting cells. The mechanisms involved in enhancing responses when the TCR and CD4 or CD8 bind to the same MHC protein molecule are still not entirely clear. It appears likely that the association of CD4 and CD8 with the tyrosine kinase p56^{lck} allows increased signal transduction when CD4 and CD8 are able to associate with the TCR/CD3 complex. However, the function of p56^{lck} bound to CD4 does not always appear to be totally dependent on its tyrosine kinase activity, at least in the setting of the transfected cells examined in this regard using mutant forms of transfected p56^{lck}. It is possible that protein/protein interactions mediated by the SH2 and/or SH3 domain of p56^{lck} may be of major importance in the enhancement of T-cell activation mediated by CD4 and/or CD8. Future studies should sort out the molecular mechanisms involved in activation enhancement mediated by p56^{lck} bound to CD4 or CD8. It is also possible that an association between CD4 or CD8 and the TCR/CD3 complex alters the conformation of the latter either to increase its affinity for antigen/MHC or to facilitate signal transduction, by approximating p56^{lck} to the TCR/CD3 signaling complex and/or in other ways. It has not been excluded that CD4 or CD8 might directly transmit positive signals to the T cell (i.e., in a manner independent of the TCR/CD3 complex). Another suggestion has been that modulation of CD4 or CD8 from the cell surface in response to activating stimuli might lower the threshold for T-cell triggering. However, it has been shown that modulation of neither CD4 nor the TCR is an absolute requirement for T-cell activation, at least when activation is induced by mAbs.

A variety of studies have shown that mAbs specific for CD8 or CD4 can inhibit T-cell activation induced by lectins or by mAbs specific for CD3 or the TCR, despite the absence of the appropriate ligand for CD8 or CD4 in these systems. These findings led to the hypothesis that CD8 and CD4 might function to deliver a negative signal to the T cell, thereby inhibiting activation. In support of this hypothesis, anti-CD4 mAbs have been shown to block the lectin- or antigen-induced rise in cytoplasmic free Ca^{2+} in a $CD4^+$ T-cell clone or hybridoma, and the anti-CD3 mediated mobilization of cytoplasmic free Ca^{2+} in human peripheral blood T cells. However, there is still no direct evidence for transmission of a negative signal via CD8 or CD4, and gene transfer studies support the concept that the primary physiological function of these molecules is to enhance rather than to block T-cell function. The studies supporting the negative signal model involve mAb blocking as opposed to physiological conditions, and hence steric hindrance must be considered among the possible explanations for the observed results. The evidence for a physical association between CD4 or CD8 and the TCR/CD3 complex increases the likelihood that mAbs binding to CD8 or CD4 could interfere sterically with T-cell activation. Recent studies have indicated that cross-linking of the TCR/CD3 complex without co-cross-linking of CD4 leads to decreased T-cell activation as long as intact CD4 capable of binding p56^{lck} is present. This diminution of signaling does not occur in the absence of CD4 or in the presence of CD4 mutated such that it cannot bind $p56^{lck}$. These results indicate that what has been called negative signaling in at least some cases reflects sequestration of $p56^{lck}$ away from the TCR/CD3 complex, leading to decreased positive signaling. However, there may be specific conditions under which negative signals can be transmitted through CD4 and CD8. [*See* Lymphocyte Responses to Retinoids.]

B. Role of CD4 as the Receptor for HIV-1

In addition to its critical role in T-cell function, CD4 has been shown to serve another important, albeit nonphysiological function as the cell-surface receptor for HIV-1, the retrovirus that is responsible for causing acquired immune deficiency syndrome (AIDS). *In vitro* infection of $CD4^+$ cells by HIV-1 can be inhibited by mAbs directed against CD4, but not by mAbs specific for other cell-surface molecules. Expression of CD4 introduced by gene transfer confers susceptibility to HIV-1 infection upon human cells that otherwise lack CD4 expression and hence are resistant to HIV-1. Cell-surface expression of CD4 is required not only for viral infection, but also for cell fusion (syncytium formation) mediated by HIV-1. CD4 binds to gp120, the exterior envelope glycoprotein of HIV-1, and, as mentioned earlier, the CDR2-like loop in D1 of CD4 is the main binding site. The binding site(s) for CD4 on gp120 is located within the carboxyl-terminal half of gp120; amino acids 410–421 have been shown to be critical for binding, but other sequences and tertiary structure may also be important. Notably, mouse CD4 does not bind HIV-1 despite its homology to human CD4, and mouse cells expressing human CD4 bind HIV-1 but are not susceptible to infection. Numerous studies have shown that genetically engineered soluble forms of human CD4 block infection by HIV-1 *in vitro*, although studies with such forms as therapeutic agents have not been promising *in vivo*. Binding of HIV-1 to cell-surface CD4 has been found to induce a rapid increase in the phosphorylation of CD4, similar to that seen with antigen activation. However, binding of isolated gp120 is not sufficient to induce CD4 phosphorylation, and elimination of the phosphorylation sites in the CD4 cytoplasmic tail (or even most of the cytoplasmic tail) does not block viral entry.

HIV-1 infection has been shown to result in a loss or decrease of cell-surface expression of CD4. One study suggested that the loss of surface CD4 was the result of decreased steady-state levels of CD4 mRNA. However, others have demonstrated that the reduction of CD4 expression is rapid and can occur without a concomitant decrease in CD4 mRNA. Loss of surface CD4 correlates with the presence of intracellular complexes between the envelope glycoproteins (gp120 and gp160) and CD4, suggesting that the reduction of surface CD4 may be a consequence of altered processing and localization of this protein in cells infected with HIV-1.

HIV-1 infection may disrupt the function of $CD4^+$ cells in a variety of ways, most notably by the major decrease in the absolute number of $CD4^+$ cells. This topic is discussed in more detail elsewhere in this volume.

C. Role of CD8 and CD4 during Thymocyte Development

In addition to their function(s) on mature T cells, CD8 and CD4 have also been shown to play an important role during thymocyte development, particularly during the positive and negative selection stages discussed in the Introduction. Disruption of the interaction of CD8 with class I MHC or CD4 with class II MHC proteins via a variety of methods leads to a block in positive selection and hence the development of mature CD8 or CD4 T cells, respectively. These interactions are also required for the elimination (negative selection) of potentially autoreactive CD8 or CD4 T cells, respectively, during thymocyte development. The mechanism(s) of lineage commitment by which developing thymocytes bearing TCRs recognizing class I MHC proteins become mature CD8 cytotoxic T cells and those bearing TCRs recognizing class II MHC proteins become mature CD4 helper T cells have not been fully elucidated. A variety of models have been put forth concerning this question. The instructive model posits that double-positive thymocytes bearing a class I MHC-specific TCR receive a signal to maintain CD8 and turn off CD4 upon interaction of the TCR and CD8 with class I MHC, whereas double-positive thymocytes bearing a class II MHC-specific TCR receive a signal to maintain CD4 and turn off CD8 upon interaction of the TCR and CD4 with class II MHC. The stochastic model postulates that CD4 or CD8 randomly turns off at the double-positive stage after a TCR and coreceptor interaction with MHC, and only those cells that have a TCR and coreceptor with the same MHC class specificity will then get a signal to survive by interaction of these proteins with the selecting MHC protein. Recently, an asymmetric model has been proposed in which an active signal is required only for CD8 cell development, whereas CD4 cell development results from a default pathway. Though these various models have generated a great deal of discussion, it would seem fair to say that the molecular mechanisms of CD8 and CD4 T-cell lineage commitment are not yet fully elucidated.

V. SUMMARY

CD8 and CD4 have been shown to be critical cell-surface proteins on T lymphocytes and not simply molecular markers for two different subsets of T cells. Through their interaction with class I and class II MHC proteins, respectively, they play key roles both in thymocyte development and in the function of mature T cells. During the complex processes of T-cell recognition and activation, CD8 and CD4 play a minimum of two roles, that is, in both adhesion and T-cell triggering. It is likely that the latter function involves an association between CD8 or CD4 and the TCR/CD3 complex. It is also clear that with some but not all T cells, CD8 and CD4 can enhance T-cell responses whether they bind to the same or different MHC molecules as the TCR. However, the molecular mechanisms involved may be different in the two cases, that is, signal transduction (plus adhesion) or only adhesion, respectively. In addition to its physiological function as a receptor for class II MHC proteins, CD4 also serves as the receptor for the AIDS virus HIV-1, thereby playing a role in a devastating pathologic process. Although progress on the understanding of CD4 and CD8 function has proceeded at an accelerating rate in recent years, there are still many open questions regarding how these molecules work, with respect to both their ability to enhance T-cell activation and their role in thymocyte development and lineage commitment. It is likely that the combination of molecular genetics, biochemistry, and cell biology, together with the more recently developed technologies for the genetic manipulation of mice, will fill in many of the current gaps in our knowledge of CD8 and CD4 function.

BIBLIOGRAPHY

Janeway, C. A., Jr. (1993). The T cell receptor as a multicomponent signalling machine: CD4/CD8 coreceptors and CD45 in T cell activation. *Ann. Rev. Immunol.* **10**, 645.

Killeen, N., Davis, C. B., Chu, K., Crooks, M. E., Sawada, S., Scarborough, J. D., Boyd, K. A., Stuart, S. G., Xu, H., and Littman, D. R. (1993). CD4 function in thymocyte differentiation and T cell activation. *Philos. Trans. Roy. Soc. London Ser. B: Biol. Sci.* **342**, 25.

Leahy, D. J. (1995). A structural view of CD4 and CD8. *FASEB J.* **9**, 17.

Miceli, M. C., and Parnes, J. R. (1993). Role of CD4 and CD8 in T cell activation and differentiation. *Adv. Immunol.* **53**, 59.

Parnes, J. R. (1989). Molecular biology and function of CD4 and CD8. *Adv. Immunol.* **44**, 265.

Cell

HENRY TEDESCHI
State University of New York at Albany

GLOSSARY

Axon Elongated process of nerve cells; nerve fibers of most vertebrates contain may axons forming a bundle

Cell culture Growth of cells in a tissue-culture dish or flask in a partially defined medium

Cytoskeleton Complement of fibers and tubules forming a network in the cytoplasm

DNA Deoxyribose nucleic acid, a substance carrying the genetic information of cells

Energy transduction Transformation of energy from one form to another, for example, from a chemical reaction (adenosine triphosphate hydrolysis) to a mechanical event (muscle contraction)

Gamete Sperm or ovum

Messenger RNA Ribose nucleic acid carrying the transcript of the genetic information

Polysomes Complex of messenger RNA and several ribosomes in which each ribosome is attached to a nascent peptide chain

Ribosomes RNA–protein particles involved in protein synthesis

Transcription Synthesis of RNA complementary to the DNA strand, which serves as its template

Translation Synthesis of polypeptide using the instructions contained in the nucleotide sequence of messenger RNA

A CELL IS A COMPARTMENT ENCLOSED BY A membrane and capable, at least under circumscribed conditions, of existing independently. All living organisms are composed of cells—either one or many. Unicellular organisms are those in which a single cell manages the metabolic processes involved in nutrition, growth, maintenance, and reproduction. In multicellular organisms, cells generally carry out specialized functions, that is, they are differentiated. Because their genetic makeup is usually identical (the notable exception being B lymphocytes of the immunological system, where hypermutation is a mechanism allowing for the production of antibody diversity), cells of different groups are said to differ in gene expression. Most cell types express only a tiny fraction of their genes.

Maintenance of the living cell requires the precise orchestration of many complicated chemical reaction sequences. The level of regulation and specificity of the reactions can be achieved only in a carefully designed environment and on intricately ordered catalytic structures. These two requirements can be fulfilled only by cells that are, as often stated, the basic units of life. The plasma membrane enables cells to maintain an internal environment suitable for those reactions that they need to build, repair, and replicate themselves. Without exception, this environment is different from the outside environment. Nevertheless, cells are not isolated from their environment, but rather they continuously communicate with it, exchanging materials and energy with it. Cells also communicate with each other in multicellular organisms, and this communication is continuous among the different cells, just as there is communication among the various intracellular compartments.

I. CELL SYSTEMS

Organisms are classified as prokaryotes or eukaryotes on the basis of their general cellular organization. In contrast to eukaryotes, prokaryotes, which include

ENCYCLOPEDIA OF HUMAN BIOLOGY, Second Edition, VOLUME 2. Copyright © 1997 by Academic Press. All rights of reproduction in any form reserved.

bacteria and cyanobacteria, are most frequently unicellular and have no organelles or well-defined nucleus. Prokaryotes lack a cytoskeleton, and their DNA is a circular structure that is attached to the plasma membrane. Some prokaryotes can function metabolically either aerobically or anaerobically, and others only in one of the two modes. A variety of bacteria can carry out photosynthesis. Some specialized bacteria are chemolithotrophs, that is, they function metabolically by oxidizing inorganic compounds.

Eukaryotes include protists, fungi, plants, and animals. Their cells, most frequently enlongated, are generally larger than those of prokaryotes, ranging in length from a few micrometers to several centimeters, or even long in the cases of special processes such as the axons of neurons. Eukaryotes have a cytoskeleton that is used to give structure and polarity to the cytoplasm. The reactions and structures are present in separate compartments or organelles surrounded by selective membranes. The DNA is present in the nucleus in very long linear molecules that are generally bound to histones and other proteins. During cell division, the DNA condenses to form chromosomes, which are pulled apart by the spindle apparatus (mainly made up of microtubules) to form two daughter cells of identical genetic complement (see Section VII).

Most of our knowledge about human cells has been derived from cells in culture. Much recent information about eukaryotes has been provided by studies carried out with yeast, either *Saccharomyces cerevisiae* or *Schizosaccharomyces pombe*. Yeast have provided the flexibility of microorganisms and the complexity of eukaryotes, and their asexual and sexual reproductive cycles have facilitated genetic analysis. Furthermore, the availability of stable vectors permits the application of recombinant DNA techniques. The universality of function in evolutionarily separate organisms has allowed the application of many of the findings to mammalian systems. Even simpler than yeast, prokaryotic cells are easier to deal with in research and have permitted rapid advances in molecular biology and biochemistry. The simplicity of these organisms results in very short generation times (20 min in some species) and allows them to be grown in large quantities in chemically defined media. These advantages have allowed rapid genetic and molecular studies, which would have been intractable in more complex systems.

II. EUKARYOTIC CELLS

Membranes are responsible for much of the compartmentalization of eukaryotic cells and allow the precise regulation of materials in and out of the compartments. The membranes are composed of phospholipid bilayers and associated proteins. The nucleus, mitochondria, and, in plant cells, chloroplasts are surrounded by double membranes. Many other organelles and vacuoles have a single membrane. A variety of processes take place in these compartments in isolation from the rest of the cytoplasm. Furthermore, some of the membranes are the site of conversion of energy into a chemically useful form.

III. THE MEMBRANES

A. The Plasma Membrane

In addition to integral proteins that are present in the membranes by virtue of their hydrophobicity, other proteins are covalently attached to lipid components, either fatty acids or glycosylphosphatidylinositol. Therefore, their function is likely to be closely associated with interactions with the bilayers. Some of the plasma membrane lipids and proteins are glycosylated (i.e., conjugated to carbohydate components, which project into the extracellular space). The position of the integral membrane proteins across the phospholipid bilayers of the membranes allows them to play a fundamental role in linking the surface of the cell to the cytoskeletal network of the cytoplasm, thus integrating the cell surface and extracellular components with the entire cytoplasmic volume. Furthermore, they function in communication among cells, between cells and the extracellular environment (either the extracellular fluid or extracellular matrix), and in response to chemical signals. Some hormones and most growth factors and cytokines bind to receptors, integral proteins that when activated trigger a cascade of biochemical events leading to activation of transcriptional events or enzyme activities.

Channels made up of integral membrane proteins permit the passage of specific solutes (such as ions) and water through the membrane. In specialized junctions between cells—the gap junctions—channels actually connect the cells by traversing the two plasma membranes, allowing intimate communication between the cytoplasms. Transport proteins, also known as carriers (or transporters), are integral proteins that transport solutes from one side of the membrane to the other. They have a major role in creating a distinct cytoplasmic environment, which differs from the extracellular medium. Either channels or transporters when present in cell sheets tightly sealed by specialized junctions (the tight junctions) are responsible for the directional passage of solute and water in specialized organs such as the kidney.

The transport of solutes against their electrochemical gradients (i.e., both concentration and electrical potential) is an active transport. Many of the transport proteins act as pumps, notably the Na^+,K^+-adenosine triphosphatase (ATPase) of the plasma membrane and the Ca^{2+}-ATPase of the sarcoplasmic reticulum of muscle cells. The transport ATPases couple the uphill transfer of ions against an electrochemical gradient to ATP hydrolysis. In the test tube under special unphysiological conditions, all of these pumps can be run on reverse (i.e., in the direction of the electrochemical gradient) to synthesize ATP from adenosine diphosphate (ADP) and inorganic phosphate (P_i). Other pumps do not depend on the hydrolysis of ATP to provide energy for transport against a gradient. The passage of a solute is coupled to the flux of another solute downhill (frequently Na^+) in the direction of its electrochemical gradient [i.e., the concentration gradient and the gradient imposed by the membrane potential (negative inside)]. [*See* Adenosine Triphosphate (ATP); Cell Junctions; Cell Membrane Transport.]

The active transport of ions coupled to ATP hydrolyis is not completely understood but apparently involves (1) the binding of the ion to high-affinity sites of the transporter molecule on the loading side of the membrane; (2) a phosphorylation of the transporter; (3) a conformational change in the transporter molecule leading to a state in which the binding sites (and hence the transported ions) are first not available to either the outside or the inside of the cell (known as occlusion), and then become available to the discharge site with a decrease in their binding affinity for the ion; and (4) hydrolysis of the phosphorylated transporter. A model summarizing some of these events is shown in Fig. 1. The structure in the middle of the phospholipid bilayer corresponds to the transporter molecule. The solid circles represent the high-affinity groups binding to the transported ion. Figure 1A shows the availability of the binding groups (high affinity) on the leading side of the membrane. A conformational change of the transporter is represented in Fig. 1B. This conformational change has a number of consequences. First, the affinity of the binding sites has been changed to low affinity and, second, they are now available to the discharge side of the membrane. The diagram does not show the phosphorylation of the transporter and the hydrolysis of the phosphate that accompanies these events.

The Na^+,K^+-ATPase is responsible for pumping the Na^+ out of cells and K^+ into cells, so that the intracellular concentration inside cells is high for K^+ and low for Na^+. The tendency for K^+ to diffuse out of cells through K^+ channels is responsible for the resting potential (inside negative) of cells. In excitable cells the influx of Na^+ through the Na^+ channels depolarizes the cell and reverses the potential so that the inside becomes positive, a phenomenon known as the action potential. Recovery to the original resting potential occurs when the K^+ influx compensates for the

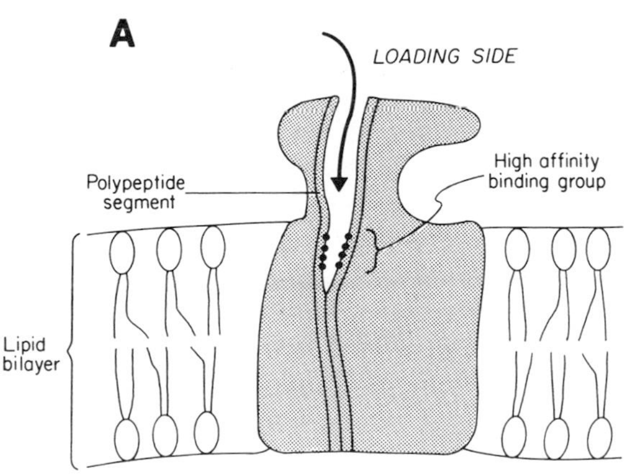

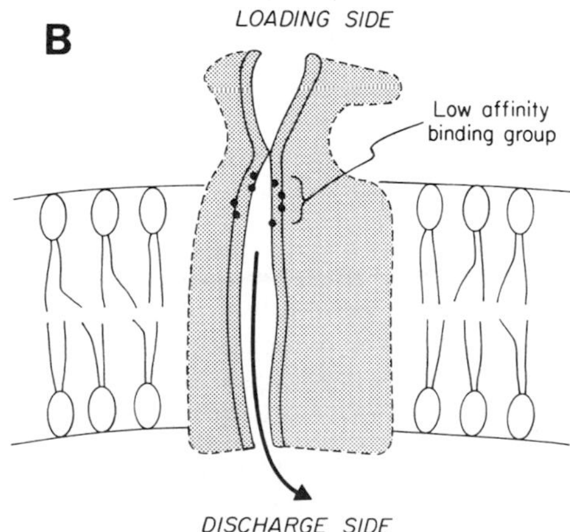

FIGURE 1 Model of a transport protein. The phospholipids in the bilayer are represented by heads (the hydrophilic part of the molecule) and tails (the hydrophobic portions). The structure in the middle represents the protein traversing the membrane, and the double lines represent two of the component polypeptides involved directly in transport. The solid circles represent the sites that bind the ions. The arrows represent the movement of the ion being transported. (A) Conformation of the transport protein in which high-affinity binding groups are exposed to the loading side of the membrane. (B) Conformation of the transport protein where the binding groups (now of low affinity) are exposed to the discharge side.

Na^+ entry. In elongated excitable cells such as nerve cells, or neurons, a depolarization wave corresponding to the action potential travels the length of the axon in the transmission of the electrical signal. The Ca^{2+}-ATPase of the sarcoplasmic reticulum (S.R.) of striated muscle pumps Ca^{2+} out of the cytoplasm into the S.R. to maintain a low cytoplasmic concentration of the ion (<1 μM). When released through Ca^{2+} channels, Ca^{2+} triggers muscle contraction and other metabolic alterations.

Many growth factors and hormones are peptides or chemicals unable to pass through the plasma membrane. They act by binding to membrane receptors, which are integral membrane proteins. In most cases the response to surface-bound hormones involves an amplifying sequence of events referred to as a biochemical cascade. In some cases the receptor activates an enzyme such as adenyl cyclase or phospholipase C, which produce specific compounds that act as intracellular signals or second messengers. Other receptors are ion channels. Still others are tyrosine protein autokinases or activate a protein tyrosine kinase. The initial phosphorylation is followed by the activation of serine/threonine protein kinases and subsequently by a protein kinase cascade. The biochemical cascade eventually results in the activation of targets, frequently proteins, such as enzymes or transcription factors, responsible for the biological effect of the ligand. The activation often involves a phosphorylation of the target. Receptor and ligand are generally taken up by endocytosis (see Section IV,A). In most cases the physiological function of the receptor–ligand complex is independent of this uptake, which is thought to have a regulatory role. There are, however, some notable exceptions (e.g., nerve growth factor) in which the factors are active on intracellular targets.

B. The Nuclear Envelope

The nucleus is enclosed by a specialized double membrane—the nuclear envelope. [*See* Cell Nucleus]. The nuclear envelope is interrupted in many sites by the nucleopore complexes, which are specialized protein complexes that form intricate passageways linking the cytoplasm with the interior of the nucleus. The envelope allows the exchange of solutes with the cytoplasm by two distinct mechanisms. In one mechanism, the passage of molecules is passive and unselective. However, macromolecules above a critical size are excluded. These results suggest an involvement of channels of the nucleopore complexes, approximately 10 nm in diameter. Highly specific exchanges also occur through the nucleopore complexes and involve a complex machinery.

Most proteins are synthesized in the cytoplasm by a process that follows a genetic blueprint provided by messenger RNA (mRNA). mRNA and the RNA of the protein-synthesizing machinery of the cytoplasm, such as ribosomal or transfer RNA, are transcribed from the DNA of the nucleus; therefore continuous traffic takes place between the nucleus and the cytoplasm through the nucleopore complexes. In addition, proteins have been shown to shuttle in an out of the nucleus by a similar mechanism. These proteins are thought to play a regulatory role during development. [*See* Nuclear Pore, Structure and Function.]

C. Energy-Transducing Membranes

Specialized portions of the plasma membrane of bacteria, the inner mitochondrial membrane, or the *thylakoid vesicles* inside chloroplasts, house protein complexes (made up mostly of integral proteins) that have the unique function of transducing the energy released by chemical oxidation–reduction reactions to the synthesis of ATP from adenosine diphosphate (ADP) and P_i. In photosynthetic cells, special membranes collect light and funnel energy into bacteria and chloroplasts to sites that power the movement of electrons through protein complexes in the membranes—the electron transport chain (Fig. 2)—which in turn translocates H^+ (i.e., protons) uphill across the membranes to form a proton electrochemical gradient. Similarly, in other bacteria and mitochondria, the oxidation of substrates powers the movement of electrons and the translocation of protons. The passage of protons downhill in the opposite direction, the direction of the electrochemical gradient, through the ATP–synthase complex (F_0F_1) produces ATP from ADP and P_i. In contrast to the transport ATPases, the ATP–synthases synthesize ATP under physiological conditions. The F_o portion of the ATP–synthases is embedded in the membrane, whereas the F_1 portion is present in the water phase. The components of the F_1 complex of most living cells have been highly conserved, indicating a common evolutionary origin.

IV. THE INTRACELLULAR TRANSPORT SYSTEM

The cytoplasmic system of membranes is very complex and in a dynamic state in which new materials

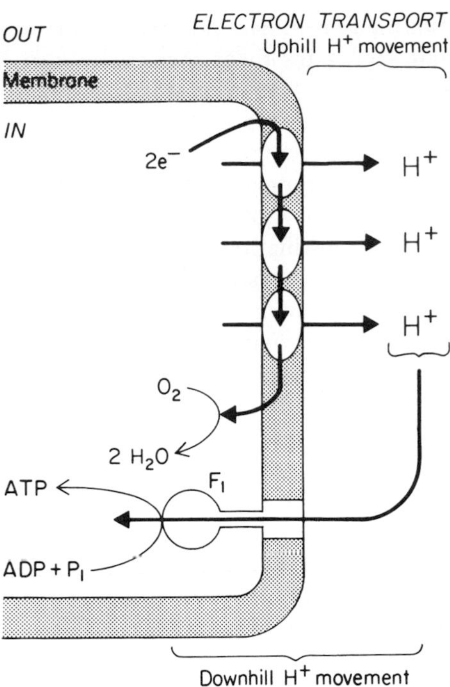

FIGURE 2 Process taking place in transducing membranes. The process represented in this diagram corresponds to the events taking place in mitochondria. The arrows in the membrane represent the passage of electrons in the electron transport chain. Their passage is coupled to the uphill efflux of protons (powered by the electron transport). The downhill return of the protons to the mitochondrial interior (thinner arrow) through the ATP–synthase is coupled to the synthesis of ATP from ADP and P_i.

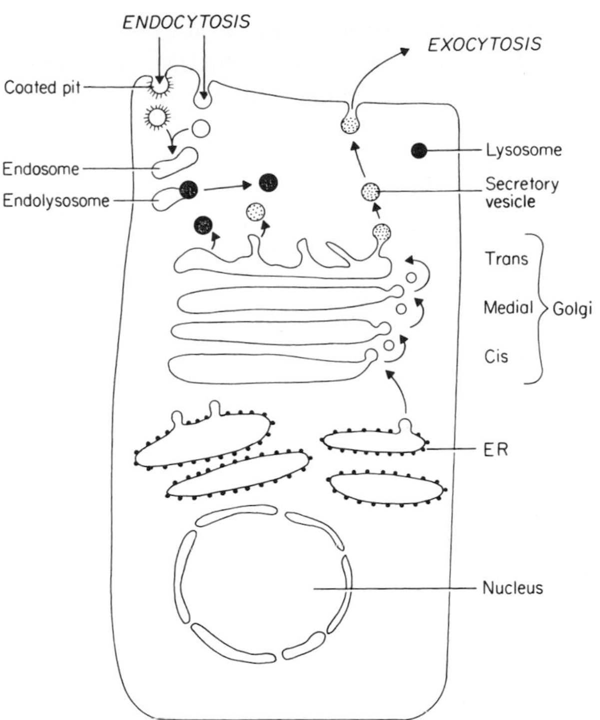

FIGURE 3 Diagram of the intracellular transport system of the cell. Materials continuously enter the cell by endocytosis (shown at top left). Materials synthesized by the endoplasmic reticulum (ER) are present in either the lumen or the ER membrane. They are exported to the Golgi system, and they are moved and processed biochemically from the cis to the trans end. The various components are segregated into vesicles such as lysosomes or secretory vesicles. The secretory vesicles are discharged by exocytosis (shown at top right). All movement of materials takes place in small transport vesicles.

are continuously produced and broken down, taken up by cells, and released to the surroundings. Material is taken up in endocytosis by invagination of the plasma membrane. In contrast, a variety of proteins synthesized in the endoplsmic reticulum are transferred to different sites on the cell or secreted. Secretion occurs by a process called exocytosis, in which the membrane of the vesicles becomes continuous with the plasma membrane and they discharge their contents to the outside. Endocytosis, exocytosis, and intracellular transport are illustrated in Fig. 3.

A. Endocytosis

During endocytosis, a portion of the cell membrane first invaginates and then pinches off to form vesicles, referred to as endosomes. Receptor-mediated endocytosis has been intensely studied. In this form of endocytosis, a ligand, such as a hormone or growth factor, is first bound to a specific receptor. When they involve the coat protein, clathrin, either the receptors are present at the cell surface, in specialized structures called coated pits, or they move to the coated pits after binding the ligand. The coated pits are small specialized indentations at the surface. The vesicles formed by invagination from these pits, the coated vesicles, are surrounded by clathrin. The clathrin is arranged in a characteristic polyhedral basket structure. The coated vesicles lose their coat and the contents are processed to recycle either the receptor or the ligand or both depending on the system. Alternatively, either ligand, receptor, or both are digested together with material taken up nonspecifically in endocytosis. Digestion occurs when the contents are transferred to lysosomes, organelles that contain hydrolytic enzymes.

Smooth invaginations and vesicles have also been shown to be involved in receptor-mediated endocytosis. Despite their smooth appearance, in some cases special electron microscope techniques were able to demonstrate a coat distinct from clathrin consisting of delicate filaments arranged as striations. The major protein component of the coat has been shown to be caveolin. Unlike clathrin, which provides a framework external to the vesicle, caveolin is an integral protein, spanning the membrane. Although smooth and clathrin-coated vesicles at first follow a separate pathway, the final destination of their contents, the lysosomes, is eventually the same.

Phagocytosis is a form of endocytosis in which larger particles (e.g., bacteria) are engulfed by cells (e.g., macrophages) to form large vacuoles.

B. Transport of Newly Synthesized Proteins

Polypeptides destined for intravesicular or plasma membrane sites are inserted through the membrane of the endoplasmic reticulum (ER) while being translated by polysomes. They remain in the membrane when they are to become integral proteins. Alternatively, they are delivered to the interior of the ER vesicle. In either case, the polypeptide is guided through the membrane by a special amino acid sequence of approximately 20 amino acids at the amino end, the signal or leader sequence, which is cleaved almost immediately to produce a mature form of the protein. Oligosaccharides are attached by *N*-glycosylation (i.e., to an aspargine residue of the polypeptide) in the ER.

Mitochondrial or chloroplast polypeptides, which are synthesized in the cytoplasm, are subsequently transferred into the organelles. Some are guided by special signal sequences, but others are thought to be already in their mature form and to be guided by targeting sequence in the interior of the molecule.

The Golgi complex (also called the Golgi) (see Fig. 3) consists of stacks of cisternae (as many as six) distributed apically in the cells. The cis face of the Golgi complex is closely associated with elements of the ER, whereas the trans side is generally toward the periphery of the apical portion of the cell. The oligosaccharide portions of the peptides are processed in the Golgi complex by the enzymatic removal of some of the carbohydrate in a process referred to as trimming. Other oligosaccharides are added in the Golgi apparatus. The processing reactions are sequential in both time and location. Processing starts at the cis portion of the complex and progresses to the medial portion and the processed material leaves at the trans cisternae. From here it moves to a reticular network that segregates them specifically into separate vesicles for delivery to their final destination. The process through the Golgi complex corresponds to highly localized, sequential enzymatic steps characteristic of each Golgi compartment. The transfer through the Golgi (ER to cis Golgi, to medial Golgi, to trans Golgi) to final destinations (i.e., secretory granules, plasma membrane, lysosomes) occurs exclusively through the movement of vesicles. The process of marking the various materials for delivery to specific fate is referred to as targeting.

Different kinds of vesicles have been recognized in intracellular transport. Clathrin-coated vesicles are associated with proteins targeted for the lysosomes and the storage secretory granules. The transport between Golgi cisternae is associated with coatomer-coated vesicles (coated with the coating proteins known as COP) and uncoated vesicles.

The nature of the targeting is not always clear. Lysosomal enzymes are selected to be packaged into lysosomes to their mannose-6-phosphate's (M6P's), which have been added to the N-linked oligosaccharides in the cis portion of the Golgi system. The selection is carried out in a process in which M6P's attached to the lysosomal enzymes are bound to M6P receptors clustered in the inner surface of the membrane of vesicles budding off from the network of tubular vesicles connected to the Golgi apparatus. Proteins destined to the plasma membrane or continuously secreted (in a process known as constitutive secretion) apparently do not need a special signal. However, in polarized cells such as epithelia, which have apical and basolateral membrane domains, sorting signals have been found in basolateral targeted proteins. Furthermore, in the case of apical targeting, proteins attached to glycosylphosphatidylinositol are suspected to be targeted to glycosphingolipid domains. In exocytosis, which discharges the contents of the vesicles to the outside, the membrane of the vesicle is incorporated into the plasma membrane. In this process, the internal portion of the vesicle membrane becomes external in relation to the plasma membrane. For example, the carbohydrate portion of integral glycoproteins, which originally faced the interior of the ER, becomes external. Some secretory proteins are stored to be released following an appropriate physiological message, such as the presence of a hormone (in a process known as regulated secretion). In regulated secretion, some kind of sorting signal is needed to segregate the proteins in the appropriate secretory vesicle. The nature of the signal, presumed to be in the amino acid sequence of the protein, is still unknown. It is unlikely to correspond to an amino acid sequence at the amino terminus of the protein, as shown by substitution experiments. Furthermore, it must be common to several proteins because several secretory products can be stored in the same vesicle.

V. METABOLISM OF MAMMALIAN CELLS

Virtually all chemical reactions in cells are catalyzed by enzymes. The energy generated by metabolism is primarily coupled to the formation of ATP from ADP and P_i. Conversely, most reactions requiring energy are powered by ATP hydrolysis.

A. Catabolic Reactions

The oxidative breakdown of glucose, fatty acids, and amino acids provide the energy needed by cells. The energy reserves are primarily in the form of glycogen (a storage polymer of glucose) and neutral fat deposits. Glycogen serves only as a temporary source of energy, whereas fats can provide energy for longer periods provided some carbohydrate is available. Glycogen is present in the liver and skeletal muscle, fats primarily in adipose tissue. The breakdown of glucose (glycolysis) in the absence of respiratory reactions provides some of the energy, two ATPs per glucose molucule. In contrast, 36–38 ATPs are generated by oxidation of glucose to CO_2. The intermediate common to all reactions involving oxidative metabolism, other than glycolysis, is acetyl-CoA. Its oxidation to CO_2 proceeds via the tricarboxylic acid (TCA) cycle, which also generates reduced nicotinamide adenine nucleotide and reduced flavine adenine nucleotides, which are oxidized by the electron transport chain. The TCA cycle corresponds to a cyclic set of enzyme-catalyzed reactions that take place in the mitochondrial interior. [*See* Glycogen; Glycolysis.]

The forward and backward steps of a pathway are generally catalyzed by different enzymes. This difference allows for separate regulation.

B. Regulation of Metabolism

Regulation can take place through changes in the specific activity of enzymes (i.e., the activity per mole of enzyme). Changes in specific activity can result from allosteric interactions where the binding of certain metabolites to the enzyme changes its activity by modifying its conformation. Feedback inhibition is an allosteric regulation of a metabolic pathway where a metabolic end product decreases the activity of the initial reaction of the pathway.

Enzymes can also be covalently modified so that they are converted from an inactive to an active form. In many cases this takes place in a mechanism involving the phosphorylation of the regulated enzyme. Phosphorylation can be the end result of a complicated cascade of events. For example, as discussed earlier, epinephrine can trigger the production of a second messenger such as cAMP, which in turn facilitates a number of subsequent enzymatic events culminating in the activation of phosphorylase *b* by phosphorylation. The active form of the enzyme (phosphorylase *a*) catalyzes the breakdown of glycogen to produce glucose-1-phosphate.

In other mechanisms, the level of the enzyme can be regulated in response to hormonal or dietary factors by changing the rate of either degradation or synthesis of an enzyme. These effects may result from the alteration of the transcription rate of the corresponding gene, which in turn increases or decreases the production of the enzyme. A variety of regulative events have also been found to occur at the translational level, where the rate of synthesis of certain enzymes is changed. [*See* Metabolic Regulation.]

C. Metabolic Specialization of Cells

Cell of various organs and tissues are metabolically specialized. For example, liver and adipose tissues serve as processors of substrate used as fuel by other cells. Because most of the circulation from the intestine passes directly through the liver, the liver receives the substances taken up by the intestinal absorption and processes them. The liver can take up large amounts of glucose and convert it into glycogen. Adipose cells contain the fat reserves of the body in the form of triacylglycerides. When mobilized, the triglycerides are hydrolyzed by a lipase to form glycerol and fatty acids, and the fatty acids are bound to albumin in the bloodstream. Fatty acids can be metabolized directly by several tissues. The fatty acids and glycerol can also be processed by the liver, and fatty acids released by the liver are contained in low-density lipoprotein. When fatty acids are used in metabolism they form acetyl-CoA, which can be oxidized via the TCA cycle. When acetyl-CoA is present in excess in the liver, acetoacetate and 3-hydroxybutyrate (the ketone bodies) are produced. Ketone bodies are metabolized primarily in the heart and the renal cortex, which prefer it to glucose. Liver cells also metabolize keto acids produced from the breakdown of proteins. [*See* Fatty Acids.]

Under most conditions, brain cells use exclusively glucose, although ketone bodies can replace glucose during starvation. In contrast, muscle can use glucose, fatty acids, and ketone bodies. During a burst of activity, muscle functions primarily by metabolizing the glucose-1-phosphate generated from glycogen. In contrast, fatty acids are the substrate of preferences in resting muscle. Heart muscle generally favors ketone bodies over glucose.

VI. CELL MOVEMENT

The movements of cells follows two basic prototypes: the actin–myosin (usually called actomyosin) system and the tubulin–dynein system. The actomyosin system is epitomized by striated muscle. The system works by the cyclic formation and release of cross-bridges between fibers of actin and myosin. One of the models proposed for actomyosin-based movement involves tilting the myosin heads while attached to the actin fibers so that the actin moves in each cycle. The repetition of the model summated for many fibers would correspond to muscle contraction. The model is illustrated in Fig. 4. The fiber made up of globular subunits is actin. A myosin bundle is shown with a projection containing a double-headed myosin head (Fig. 4), which attaches to actin to a cross-bridge. The movement of the myosin head (the tilt in Fig. 4) produces the movement of the actin. The myosin head (no longer tilted) detaches and then binds to the next cross-bridging site, restarting the cycle. Each cycle hydrolyzes one ATP. When the actin is attached to a structure (in muscle the Z-line), the movement can be converted into work. The head portion of the myosin is responsible for the movement, the cross-bridges, and the energy expenditure, whereas the actin molecule essentially plays a passive role. Molecules such as myosin, which are directly responsible for the transduction of chemical energy into mechanical work, have been referred to as motors.

Similar to the actin–myosin system, in the tubulin–dynein system the tubules formed by tubulin have a passive role, whereas dynein, which in cilia and flagella is attached to a tubule, acts as a motor. When the microtubules are attached (in cilia and flagella to the basal bodies), the dynein–microtubule complexes slide in relation to each other, so that the cilia or flagella, which are made up of highly organized tubule bundles, are forced to bend (Fig. 5).

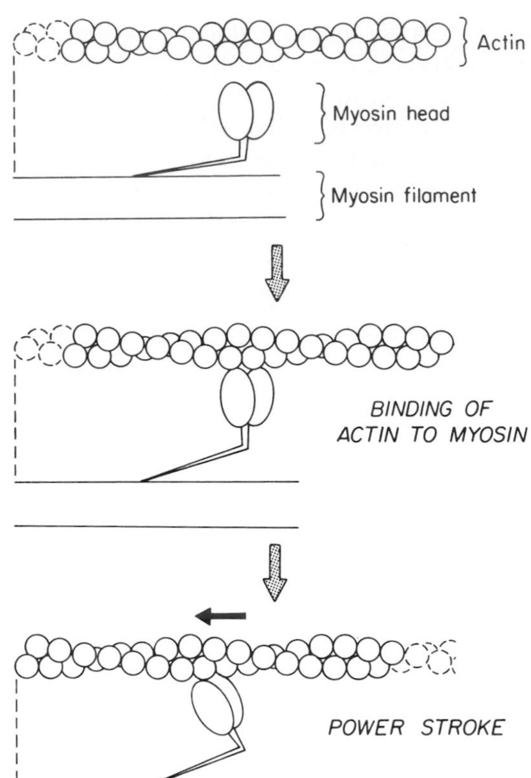

FIGURE 4 The actin–myosin system: the interaction of myosin with actin in skeletal muscle. In the first panel, the myosin heads are not attached to the actin. In the second panel, the myosin head has attached to the actin filament. In the third panel, the myosin head tilts, producing a motion in the actin filament (indicated by the black arrow). The cycle starts again when the myosin head reattaches to the actin filament in the next cross-bridge site.

Analogous systems are present in the cytoplasm, where microtubules (Fig. 6) or actin fibers (Fig. 7) serve as conduits for the movement of the motors, either the myosins or the microtubule-associated mechanochemical proteins such as cytoplasmic dynein or kinesin. Vesicles or other structures attached to the motors are moved along either actin or microtubular linear elements. In microtubular movement, different motors are responsible for movement in different directions along a microtubule.

In muscle, myosin is two-headed and tailed (now called myosin II). Myosin II is also involved in cytokinesis (the actual separation of two daughter cells by contractile events during cell division) and is found in the tail region of ameboid cells, suggesting a role

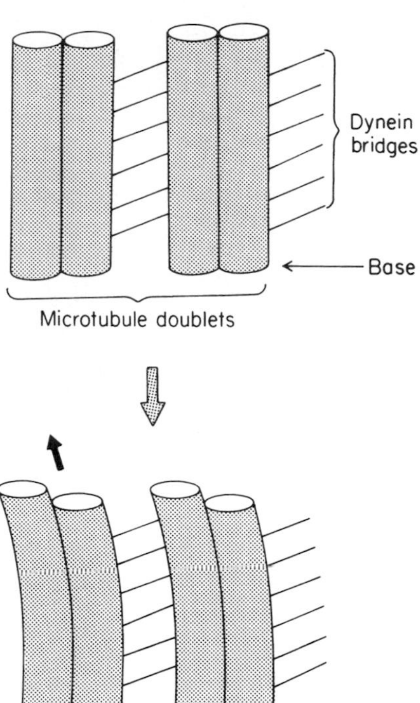

FIGURE 5 The tubulin–dynein system, showing bending of the microtubules that form part of the cilium or flagellum. The double filament on the left is sliding in relation to that on the right, forcing the whole unit to bend. The sliding filament is thought to resemble, in principle, that shown in Fig. 4 for actomyosin.

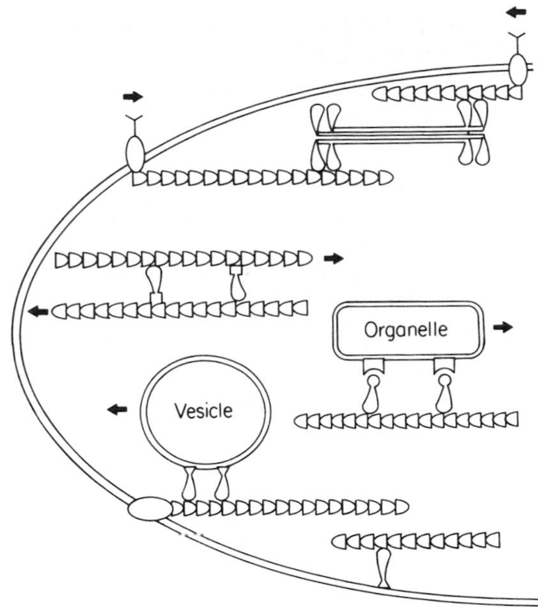

FIGURE 7 Speculative schematic diagram of various actin-based motors. The actin filaments are represented as polar structures. Conventional myosin (myosin II) forms bipolar filaments, which can pull two points in the cell together. When binding two actin filaments, myosin II motors can cause shear (shown in the middle). Alternatively, myosin I can attach to organelles or vesicles directly or indirectly or to actin molecules, thus allowing movement. [Reproduced by permission from J. A. Spudich (1989). In pursuit of myosin function—Minireview. *Cell Reg.* **1**, 1–11.]

in the retraction during ameboid movement. Other myosin are involved in some forms of cell movement. Myosin I, a single-headed variety with a shorter tail, has a role in interactions with the cell surface and movement of vesicles. In ameboid cells, it has been found in the leading edge of moving cells and at sites of phagocytosis. In vertebrate cells, it has been found associated with intestinal microvilli. A general role of this myosin I or myosin I-like molecules is also suggested by their presence in the microvilli of the eye of the fruit fly *Drosophila*. Various possible roles of actin and myosin in movement are illustrated in Fig. 7.

Together with intermediate filaments, actin and microtubules also have a significant role in the organization of the cytoplasm. Their capacity to polymerize and depolymerize undoubtedly play a significant role in determining cytoplasmic structure and it dynamics.

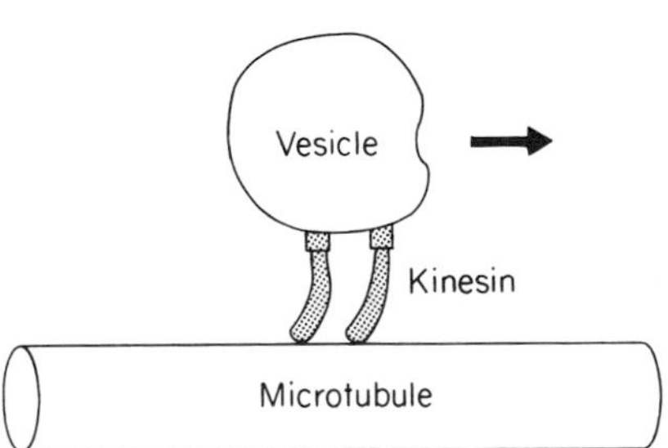

FIGURE 6 Movement of a vesicle along a cytoplasmic microtubule by means of a dynein-like motor.

VII. THE CELL CYCLE

In animal somatic cell division (i.e., mitosis), a tetraploid cell, produced by duplication of the chromosomes, divides to produce two diploid daughter cells. In this process, the nuclear envelope is disrupted and the chromosomes condense into an inactive "transport" form. They are partitioned equally to daughter cells by the mitotic spindle. The daughter cells are separated by a contractile event, cytokinesis, where actin and myosin in the contractile ring pinch the dividing cell into two. These events have been labeled the M phase (M for mitosis). The period between M phases is the interphase.

In cells that are periodically dividing, the interphase nucleus is far from inactive, although there is a pause, the G_1 phase (G for gap). DNA is synthesized during the S phase, followed by another pause, G_2, which precedes the M phase. These events ($M \rightarrow G_1 \rightarrow S \rightarrow G_2$) are referred to as the cell cycle. Although it varies with cell type and physiological conditions, typically the cycle lasts from 16 to 24 hr; the M phase represents 1–2 hr, whereas the S phase may last as long as 8 hr.

Cells that stop dividing are said to be in G_0, and they can stay in that phase indefinitely. A mammalian cell leaves G_0 by the action of external growth factors that bind to specific receptor proteins and activate, producing a cascade of biochemical events that trigger the various steps of cell division. The activation of key entry steps, for example $G_1 \rightarrow S$ or $G_2 \rightarrow M$ transition points, is mediated by cyclins. Generally cyclins are regulated transcriptionally. Cyclins and cyclin-dependent protein kinases (cdk's) are present as complexes that phosphorylate specific targets. These may be transcription factors activating the expression of certain genes or products of genes that suppress cell division directly or indirectly (e.g., RB or p53 genes, where RB stands for retinoblastoma, which is the most common form of cancer produced by mutations of the RB gene, and p53 refers to the molecular mass in kDa of the protein).

Normal cells undergo a limited number of divisions. For example, fibroblasts from a human fetus will undergo approximately 50 divisions. Cells of older individuals are limited to fewer divisions. Transformed cells (e.g., cells that have been made tumor-like by infection with some viruses) can divide indefinitely. This release from inhibition apparently results by activation or malfunction of oncogenes. Oncogenes are present in all cells and are thought to play a role in the regulation of normal growth. When the gene is altered, growth may become uncontrolled. [*See* Oncogene Amplification in Human Cancer.]

Mitosis in animal cells is generally considered to proceed in five M stages and the final step of cytokinesis (Fig. 8). The M stages are prophase, prometaphase, metaphase, anaphase, and telophase. Interphase corresponds to the period between M phases and, therefore, corresponds to $G_1 + S + G_2$. In interphase, the centrosome, or cell center, located on one side of the nucleus, contains a pair of self-replicating centrioles, structures that are arranged at right angles from each other. The centrosome material serves as a microtubule-organizing center and is attached to microtubules. Mouse oocytes and plants lack centrioles. In interphase, arrays of microtubules radiate from the centrosome. The centriole pair replicates during S phase. In prophase the centrosome splits, and the two resulting centrosomes move to opposite sides of the cell, where they serve as focal points for polymerization of the microtubules to form an ester. At prometaphase, the nuclear membrane breaks down, and the microtubules attach to the chromosomes to form a spindle, where the centrosomes have become the spindle poles. Chromatids, the two identical components formed by the replication of each chromosome, remain closely attached. The spindles of higher animals and plants contain several thousand microtubules, whereas in some fungi there may be as few as 50. [*See* Mitosis.]

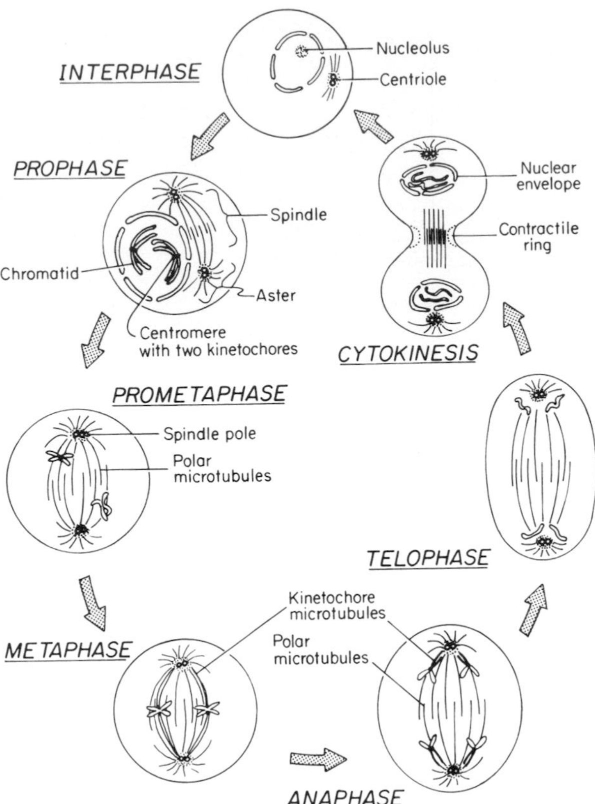

FIGURE 8 Diagrammatic representation of mitosis and cytokinesis. For simplicity, only two chromosomes are shown at the beginning of mitosis. There are 46 chromosomes in human cells, that is, 23 pairs.

The two chromatids are attached to each other near their centromere. The centromere corresponds to specific DNA sequences needed for chromosome separation. A specialized structure, the kinetochore, develops at each centromere. By prometaphase, some of the microtubules (typically 20 to 40), the kinetochore microtubules, have connected to the kinetochores so that the two sister chromatids are attached to opposite spindle poles. The spindle microtubules, which are not attached to the kinetochores, are known as the polar microtubules. Those that are not in the spindle are called astral microtubules. The chromosomes appear to be pushed to the equator, which they reach at metaphase. At anaphase, the sister chromatids are separated, each with one kinetochore. Two simultaneous processes seem to take place. At anaphase A, the chromatids are moved toward their respective pole in a process in which the kinetochore microtubules are shortened. In anaphase B, the poles are separated farther as the spindle elongates by an assembly of microtubules at their distal ends. This polymerization, accompanied by sliding of the astral microtubules, is thought to push the poles apart. At telophase, the chromosomes arrive at the poles, the kinetochore microtubules disappear, the daughter cells separate (cytokinesis), the condensed chromatin unravels, and the nuclear envelope reforms in the two cells.

VIII. MEIOSIS

With the exception of the sex-determining chromosomes, cells are generally diploid (i.e., they have two versions of each chromosome—homologous pairs—each derived from one parent). In contrast, the gametes are haploid, and they contain only one chromosome of each homologous pair. Meiosis is the special cell division that produces gametes, and it differs significantly from mitosis. First, each homologous pair replicates to produce sister chromatids just as in mitosis, however, the four chromatids are physically attached before lining up in the spindle. The first division (division I of meiosis) separates out the homologous pairs to produce two diploid cells. A second division (division II of meiosis) occurs without DNA replication and separates out the sister chromatids to form a haploid gamete. At prophase, the close association between homologous pairs preceding division I allows for recombination. In the process, equivalent parts of the homologous chromosome pair are exchanged, and event known as crossing-over. [*See* Meiosis.]

BIBLIOGRAPHY

Alberts, B., Bray, D., Lewis, J., Raff, M., Roberts, K., and Watson, J. D. (1994). "Molecular Biology of the Cell," 3rd Ed., Chap. 1, 2, 8, and 10–18. Garland, New York/London.

Bray, D. (1992). "Cell Movements," Garland, New York/London.

Cramer, W. A., and Knaff, D. B. (1990). "Energy Transduction in Biological Membranes," Chaps. 3–5.

Gennis, R. B. (1989). "Biomembranes, Molecular Structure and Function," Chaps. 1, 6, 8, and 9. Springer-Verlag, New York/Berlin.

Lodish, H., Baltimore, D., Berk, A., Zipursky, S. L., and Darnell, J. (1995). "Molecular Cell Biology," Chaps. 1, 5, 10, 11, 14–17, 22, 23, 25. W. H. Freeman and Company, New York.

Murray, A., and Hunt, T. (1993). "The Cell Cycle. An Introduction," Chaps. 1 and 3–8, Freeman, New York.

Tedeschi, H. (1993). "Cell Physiology," 2nd Ed., Chaps. 2–5, 10, 12, 13–15, and 17–18. Brown, Dubuque, IA.

Cell Adhesion Molecules

JUDITH A. VARNER
University of California, San Diego

GLOSSARY

Cadherins Family of structurally related intercellular adhesion molecules that mediate the adhesion of like cells to one another by interacting with identical molecules on opposing cell surfaces

Endothelium Cells that line the blood vessel walls and are exposed to the flow of blood

Extracellular matrix Acellular structure filling the spaces between cells that is often composed of collagen, proteoglycans, and fibrillar proteins, such as fibronectin, vitronectin, and laminin

Heterotypic adhesion Intercellular adhesion of usually nonidentical cells through the interactions of two distinct types of molecules on opposing cell surfaces

Homotypic adhesion Intercellular adhesion of usually identical cell types via identical molecules on opposing cell surfaces

Immunoglobulin family Family of adhesion proteins that mediate the attachment of like cells (N-CAM) or unlike cells (V-CAM, I-CAM) that contain structural repeats that are very similar to those found in immunoglobulins

Integrins Family of over 20 structurally related molecules composed of an α and a β subunit that mediate the attachment of cells to surrounding extracellular matrix and to certain proteins circulating in plasma

Intercellular adhesion Attachment of cells directly to other like or unlike cells through the interactions of adhesion molecules on the surface of cells

Leukocytes White blood cells

Selectins Family of adhesion proteins that mediate cellular attachment of unlike cells by recognizing specific carbohydrate sequences (sugars) on the proteins or lipids of opposing cells

CELL ADHESION CONSTITUTES THE ATTACHMENT of cells to other like or unlike cells or to proteins found in the extracellular spaces in either a stable or transient manner.

I. INTRODUCTION

Cellular adhesion is the critical feature that allows the existence of multicellular organisms. It is responsible for an enormous variety of physiological events, which include normal tissue homeostasis, fertilization and embryo implantation, wound healing, cellular movement during development and during wound healing, the movement of lymphocytes from the bloodstream into tissues, and the movement of cells during the metastasis of cancerous cells. Cells use a number of complex and highly specific mechanisms to interact with one another and with proteins in the surrounding extracellular spaces and to contribute to these physiological events. At least four distinct families of cell-surface proteins effect cellular attachment to other cells and extracellular matrices. These are the integrin family, the cadherin family, the selectin family, and the immunoglobulin family. Although these molecules were initially characterized as molecules that mediate the attachment of cells to other cells or to extracellular matrix proteins, it is now clear

ENCYCLOPEDIA OF HUMAN BIOLOGY, Second Edition, VOLUME 2. Copyright © 1997 by Academic Press. All rights of reproduction in any form reserved.

that some of these play dynamic and vital roles in sending information about the local environment to the nucleus of the cells and in determining the fate of the cell. [*See* Extracellular Matrix.]

II. THE INTEGRIN FAMILY OF EXTRACELLULAR MATRIX ADHESION PROTEINS

The integrin family of adhesion proteins are primarily receptors for extracellular matrix proteins that mediate cellular attachment to molecules found in the spaces surrounding cells. The first integrin to be described was the platelet fibrinogen receptor, glycoprotein IIb/IIIa, which plays critical roles in wound healing and thrombosis. Integrins were first described as a family of structurally related molecules just over 10 years ago. Since that time, more than twenty integrins with unique specificities for diverse extracellular matrix proteins and/or cell-surface proteins have been identified (Table I). The mechanisms by which integrins direct the attachment of cells to proteins in the surrounding extracellular spaces and occasionally to other cells have also been characterized. Though originally thought to be a family of cell-surface proteins primarily responsible for anchoring cells to the extracellular matrix, integrins have recently been shown to impact dynamic processes in normal and tumor cells, such as cell migration, proliferation, differentiation, and survival.

The integrin molecule is composed of two structurally distinct transmembrane subunits called α and β subunits. Most α subunits have a molecular mass of approximately 140 kDa (g/mole), whereas β chains generally are smaller in size, usually 95 kDa. α chains contain structural repeats that bind the metals calcium, magnesium, or manganese. β chains are rich in the amino acid cysteine. Both subunits are required for extracellular matrix protein recognition. At least fifteen α and eight β subunits have been identified, and these combine to form more than twenty different $\alpha\beta$ heterodimeric combinations, each with the capability of interacting with a unique set of ligands. Integrins bind to extracellular matrix proteins or cell-surface immunoglobulin family molecules by recognizing and forming close molecular interactions with short peptide sequences present in these molecules. Although some integrins selectively recognize a single extracellular matrix protein ligand, others bind two or more ligands. For example, the integrin $\alpha5\beta1$ recognizes only the matrix protein fibronectin whereas the integrin $\alpha v\beta3$ recognizes fibronectin, vitronectin, laminin, collagen, fibrinogen, osteopontin, and others. Some integrins recognize the short tripeptide sequence Arg-Gly-Asp (RGD) present in several extracellular matrix proteins and in cell-surface molecules, whereas others recognize alternative peptide sequences. Specificity is determined by the unique amino acid combinations that immediately flank these short tripeptide sequences in individual extracellular matrix proteins. In addition, integrins require calcium, magnesium, or manganese ions as cofactors in order to attach to their ligands.

TABLE I
The Integrin Family and Their Ligands

Integrin	Ligand
$\alpha1\beta1$	Collagen, laminin
$\alpha2\beta1$	Collagen, laminin
$\alpha3\beta1$	Fibronectin, laminin, collagen
$\alpha4\beta1$	Fibronectin (CS-1), V-CAM
$\alpha5\beta1$	Fibronectin
$\alpha6\beta1$	Laminin
$\alpha7\beta1$	Laminin
$\alpha8\beta1$	Tenascin, fibronectin, vitronectin
$\alpha9\beta1$	Tenascin
$\alpha v\beta1$	Vitronectin, fibronectin
$\alpha v\beta3$	Vitronectin, fibrinogen, fibronectin, osteopontin, laminin, collagen, von Willibrand factor, thrombospondin
$\alpha v\beta5$	Vitronectin
$\alpha v\beta6$	Fibronectin
$\alpha v\beta8$	Laminin, collagen, fibronectin
$\alpha IIb\beta3$	Fibrinogen, vitronectin, fibronectin, von Willibrand factor, thrombospondin
$\alpha6\beta4$	Laminin
$\alpha4\beta7$	Fibronectin, V-CAM, MAd-CAM
$\alpha^L\beta2$	I-CAM-1, I-CAM-2
$\alpha^M\beta2$	C3b component of complement, fibrinogen, factor X, I-CAM-1
$\alpha^X\beta2$	Fibrinogen, C3b component of complement

Combinations of different integrins on cell surfaces allow cells to recognize and respond to a variety of different extracellular matrix proteins. Though many integrins are found on diverse cell types, a few are selectively expressed on specific cell types. The subfamily of integrin molecules that use the $\beta2$ integrin subunit is found only on immune cells such as T and B lymphocytes, macrophages, monocytes, and eosinophils. The integrin $\alpha6\beta4$ is found primarily in epithelial cells that are attached to a basement membrane. The integrin known as glycoprotein IIb/IIIa is found exclusively on platelets and on the precursor cell to platelets, the megakaryocyte.

Integrins on cells such as epithelial, endothelial, and muscle cells are continuously active; that is, they are able to mediate cellular adhesion without any form of prior stimulation by soluble factors. Integrins on cells that circulate in the bloodstream, however, are present in inactive forms that require prior activation for cells to be enabled to attach via these integrins. This activation event requires an "inside out signal" or a signal from within the cell that is usually generated in response to another external stimulus.

Integrins mediate cellular adhesion to and migration on the extracellular matrix proteins found in intercellular spaces and basement membranes, but they also regulate cellular proliferation. Interaction of integrins with their extracellular matrix protein ligands induces a cascade of events that include the attachment of several components of the actin cytoskeleton to the membrane via the integrin β subunit's cytoplasmic tail, thus reinforcing the strength of the cellular adhesion. In addition, integrin-mediated adhesion leads to the generation of an intracellular signaling cascade that includes activation of a network of kinases. This signaling cascade leads to the synthesis of new proteins that are required for cell growth, also known as immediate early gene products. In contrast, prevention of cell attachment by blocking the integrin–ligand interaction suppresses cellular growth and induces cell death by a process known as apoptosis. In apoptosis, DNA-degrading enzymes cleave DNA into small fragments and thus induce cell death.

The loss of certain integrins from cell surfaces is responsible for some now well-characterized human diseases. Loss of either subunit of the integrin IIb/IIIa induces a serious bleeding disorder, known as Glanzman's thrombasthenia. Loss of the IIIa subunit may also lead to infertility due to a failure of the embryo to implant in afflicted women. Loss of the $\beta2$ subunit induces the serious disorder lymphocyte adhesion deficiency, in which lymphocytes are unable to leave the bloodstream and fight infections in tissues. Loss of or mutations in the genes for the β subunits of the epithelial integrin $\alpha6\beta4$ leads to a skin blistering disease known as epidermolysis bullosa.

Thus, integrins play roles in a number of cellular processes that impact normal tissue homeostasis, wound healing, the immune response, and the development of tumors. These events require integrin roles in the regulation of proliferation and cell survival, as well as cellular motility and invasion.

III. THE CADHERIN FAMILY OF INTRACELLULAR CELL ADHESION PROTEINS

The attachment of like cells to one another is a key feature of solid tissues such as epithelia, which is found in the digestive tract, the breast, the reproductive organs, and the skin. This type of intercellular adhesion is mediated by the cadherin family of adhesion proteins. Cadherins are a family of transmembrane, calcium-dependent cell adhesion proteins that mediate intercellular adhesion in a strictly homotypic fashion (Table II). Unlike integrins, a cadherin molecule is composed of a single subunit. The extracellular regions of cadherins contain five repeating homologous structural units (domains). Each repeat contains calcium binding sites that are unique in structure to the cadherin family. In the presence of calcium, the cadherin molecule adopts a rod-like structure that collapses after its removal. Two cadherin rod-like molecules on the surface of one cell form a tight unit through the interaction of their amino-terminal domains. This unit interacts through its amino-terminal

TABLE II
The Cadherin Family and Their Tissue Distribution

Cadherin	Tissue distribution
E-cadherin	Epithelium
P-cadherin	Placenta
N-cadherin	Nervous system
B-cadherin	Nervous system
T-cadherin	Nervous system
R-cadherin	Nervous system
M-cadherin	Mesenchyme
L-CAM	Liver
Desmogleins	Epithelium
fat	Fruit fly imaginal discs
dachsous	Fruit fly imaginal discs

repeats with other pairs of cadherins on other cells in a zipper-like fashion, thus creating a zipper-like network of interacting cell adhesion molecules.

The original cadherins were first described over 10 years ago. These include unique cadherins from epithelia (E-cadherin), placenta (P-cadherin), neuronal cells (N-cadherin), and liver tissue (L-CAM). The epithelial cadherin, E-cadherin, plays a critical role in maintaining the integrity of the epithelium. It is expressed early in embryogenesis on many cell types but later is restricted mainly to epithelia. Removal of the gene for E-cadherin by a molecular procedure known as a "knockout" leads to early embryonic death, prior to implantation, indicating that this molecule plays a very critical role in development. E-cadherin is often lost from epithelial cells as they become cancerous; this loss is thought to permit increased cell invasiveness and thus metastasis by reducing the intercellular contacts that prevent cellular migration. N-cadherin is present continuously during development of the neural tube. Additional brain cadherins, such as R-, B-, and T-cadherins are present at late stages of brain development. Studies indicate that these cadherins play crucial, deciding roles in the determination of cell fate during embryogenesis. Cells that express E-cadherin become epithelial cells, whereas cells that express N-cadherin become neural or cartilage cells. Additional members of the cadherin family include the more distantly related products of the fruit fly *Drosophila melanogaster, genes, fat* and *dachsous.* These proteins have arrays of 34 and 27 tandom cadherin repeats, respectively. Desmogleins and desmocollins are cadherins found in desmosomes. Desmosomes are structures within epithelia and cardiac muscle that link the cell surface to a dense network of cytoskeletal filamentous proteins, giving the tissue a large degree of tensile strength.

The cadherins have highly conserved cytoplasmic domain sequences that interact with the α, β, and γ (plakoglobin) catenins. Catenins are intracellular proteins that link cadherins to the actin cytoskeleton and that may have the ability to transmit information to the cell nucleus. α-Catenin links cadherins to the actin cytoskeleton and is similar in structure to vinculin, one of the cytoskeletal components that binds directly to integrins and mediates integrin attachment to the actin cytoskeleton. β-Catenin is required for the linkage of α-catenin to cadherin cytoplasmic domain. It binds to the cytoplasmic tail of nondesmosomal cadherins, whereas plakoglobin binds to the tail of both desmosomal and nondesmosomal cadherins. In addition, β-catenin and plakoglobin are found both complexed with cadherins and in a soluble cytoplasmic form. This cytoplasmic form may respond to extracellular information by binding to transcription factors and participating in their movement to the nucleus. At present, changes in the ratio of cytoplasmic to cadherin-bound β-catenin and plakoglobin appear to play significant roles in regulation of morphogenesis and cell fate.

Diseases in which cadherins play a major role include at least two autoimmune skin blistering disorders. Pemphigus vulgaris is a result of inhibition of cadherin molecule homophilic interactions by autoantibodies directed against the cadherin desmoglein-3, leading to loss of cell-to-cell adhesion in the skin and skin blistering. Autoantibodies directed against desmoglein-1 impair cell-to-cell adhesion in more superficial layers of the skin and lead to pemphigus foliaceus.

The cadherins are a large family of homophilic adhesion proteins that play important roles in the determination of cell fate during development and possibly in adulthood. Unlike most of the other adhesion molecules so far described, cadherins recognize identical molecules on neighboring cells that are usually of the same cell type. They are closely associated with a family of intracellular molecules known as catenins, which may play roles in transmitting information from the outside of the cell to the nucleus, thus impacting cell fate.

IV. THE SELECTIN FAMILY OF CELL ADHESION PROTEINS

The selectin family of adhesion proteins is a group of structurally related transmembrane proteins that primarily promote the attachment of leukocytes (white blood cells) to endothelium (blood vessel wall cells). They are lectins, which are proteins that specifically recognize oligosaccharide structures attached to proteins or to lipids. Three selectins have been described: E-selectin (also called CD62L), which is found on endothelial cells; P-selectin (also called CD62P), which is found on platelets and endothelial cells; and L-selectin (also called CD62L), which is found on leukocytes (Table III). Although leukocytes normally are propelled by the flow of blood, selectins permit them to attach loosely to the endothelium and to "roll" along the surface of the blood vessel. By rolling on the endothelium, leukocytes are in position to monitor information from the surrounding tissues indicating a need for these cells to leave the blood-

TABLE III
The Selectin Family and Their Ligands

Selectin	Ligand
L-selectin	Gly-CAM, mucosal adressin cell adhesion molecule (MAd-CAM), CD34
E-selectin	E-selectin ligand-1 (ESL-1)
P-Selectin	P-selectin ligand-1 (PSGL-1)

stream and enter tissue to fight an infection or participate in inflammation. In response to information indicating infection or inflammation, these leukocytes then activate other adhesion proteins in the integrin family, which permits stronger attachments to the endothelium as well as movement between endothelial cells to penetrate into tissues.

Selectins belong to a family of carbohydrate binding proteins known as the C-type lectin family. Structurally, they are similar to other C-type lectins, such as the rate mannose binding protein. The carbohydrate recognition structural element is found in the amino terminus of the protein. The recognition of carbohydrates is dependent on the binding of a single molecule of calcium to this region of the lectin. An adjacent structural region is also thought to play a role in carbohydrate recognition because removal of this domain by a process called deletional analysis inactivates the receptor. All three selectins recognize variations of a tetrasaccharide known as sialyl Lewis x (sLe^x). It contains one molecule each of four different sugars: neuraminic acid (sialic acid), *N*-acetylglucosamine, fucose, and sulfated galactose. This tetrasaccharide can be modified by the absence of sulfate or fucose groups. Even though the three selectins recognize the purified sugar, all bind with greater affinity to proteins or lipids that contain these sugars on their surfaces. Such "gylco"proteins and "glyco"lipids are found primarily on the cell membrane.

E-selectin is found transiently on endothelial cells that have been exposed to cytokines secreted by leukocytes, such as during inflammation. E-selectin recognizes carbohydrate structures found on the surface of leukocytes. The presence of E-selectin on endothelium allows myeloid cells (such as macrophages and neutrophils) and lymphocytes to adhere and roll on the blood vessel wall. P-selectin is stored in the secretory granules of both platelets and endothelial cells and is secreted in response to thrombin, which plays a role in blood clotting during wound healing. P-selectin also binds to carbohydrates found on proteins on the surface of leukocytes, binds to myeloid and lymphocyte cells, and is the major endothelial receptor for a special subset of lymphocytes that are found mainly in skin, called $\gamma\delta$ T cells. Both lymphocytes and myeloid cells can attach and roll on purified E- and P-selectin, indicating that these proteins alone can mediate leukocyte rolling. L-selectin is found only on leukocytes. It is found on all leukocytes with the exception of a subpopulation of lymphocytes known as memory T cells. L-selectin recognizes carbohydrates on proteins on the endothelial cell surface and is shed from the cell surface by proteolytic cleavage after leukocytes are activated by signals from sites of inflammation. The ligands for L-selectin are continuously present on the endothelium in the lymph node and Peyer's patches in the gut and are transiently induced on other endothelia as a result of inflammatory indicators. [*See* Lymphocytes.]

The ligands for the selectins have been identified. L-selectin of leukocytes recognizes oligosaccharides present on at least three different proteins of the endothelium: CD34, MAd-CAM, and Gly-CAM-1. CD34 and MAd-CAM are transmembrane proteins, whereas Gly-CAM-1 is a secreted protein. P-selectin recognizes a mucin-like glycoprotein on leukocytes called P-selectin-dependent ligand (PSGL-1). It also binds to a heat-stable antigen known as CD24. E-selectin recognizes E-selectin ligand (ESL-1), a 150-kDa fucosylated protein found only on myeloid cells. Whereas most selectins recognize glycans that are linked to proteins via oxygen groups present on the amino acid side chains of threonine and serine (O-linked glycans), E-selectin recognizes sugars on ESL-1 that are linked via a nitrogen group on the amino acid side chain of asparagine (N-linked glycans). Unlike E-selectin, P- and L-selectins can bind to sulfated glycans that lack the sugars sialic acid and/or fucose, such as subsets of a polymeric glycan found on proteins and in plasma known as heparin.

Studies of mice lacking the selectins indicate that such mice have impaired disease-fighting abilities, although they develop normally. Leukocytes in such mice are unable to move out of the bloodstream into host tissues to fight infections despite having adequate levels of the higher-affinity integrin receptors for endothelium. L-selectin-deficient mice have fewer lymphocytes binding to peripheral lymph node endothelium and thus less efficient homing to the primary lymphatic tissues. These mice also exhibit deficient neutrophil emigration into the peritoneum. P-selectin-deficient mice are unable to roll on mesenteric vessels and exhibit delayed migration of neutrophils into the peritoneum. They are also defective in the penetration of monocytes into tissues during chronic inflammation. E-selectin-deficient mice have abnormalities in leukocyte rolling and extravasation. When mice are lacking both P- and E-selectin, the ability of leukocytes to roll on endothelium is reduced by 46-fold, suggesting that loss of just one selectin can be somewhat ameliorated by the other selectins.

The selectins are a small family of adhesion proteins with a unique mechanism of action. Unlike the other adhesion protein families, the selectins are carbohydrate binding molecules. They mediate weak attachment of immune cells to cells of the blood vessel wall, but are critical for effective immune system function.

V. THE IMMUNOGLOBULIN FAMILY OF CELL ADHESION PROTEINS

The immunoglobulin superfamily of cell adhesion proteins are heterophilic transmembrane proteins that play roles in the adhesion of like or unlike cells to one another (Table IV). These adhesion proteins are composed of structural units that resemble the basic structural unit found in immunoglobulins, the immunoglobulin fold. These proteins are generally composed of four or more such structural regions and may contain other structural entities as well. The immunoglobulin fold is a structural unit in which most of the amino acids serve as a scaffolding in order to present a few amino acids in an extended loop to interact with other proteins. The first member of this family of proteins to be described was found on neuronal cells and was called N-CAM, or neural cell adhesion molecule. Since then a number of immunoglobulin-like transmembrane adhesion proteins have been identified and described to play adhesive roles in the development of the nervous system, in the proper movement of leukocytes from the bloodstream into tissues to sites of infection or inflammation, and possibly in the formation of new blood vessels during inflammation and tumor growth.

TABLE IV
The Immunoglobulin Family Members and Their Ligands

Immunoglobulin family member	Ligand
N-CAM	N-CAM
Ng-CAM	F11, axonin
Nr-CAM	Axonin
Contactin/F11	Ng-CAM
Fasciclin II	Fasciclin II
Neuroglian	Neuroglian
Axonin	Ng-CAM, Nr-CAM
I-CAM-1	Integrin $\alpha^L\beta2$, $\alpha^M\beta2$
I-CAM-2	Integrin $\alpha^L\beta2$, $\alpha^M\beta2$
V-CAM	Integrin $\alpha4\beta1$, $\alpha4\beta7$
L1	Integrin $\alpha v\beta3$

The original member of this family is the only known vertebrate member that binds in a homophilic fashion to like N-CAM molecules on adjacent cells. This molecule plays a role in axonal growth during development. N-CAM is highly glycosylated with linear polymers of a unique form of sugar known as sialic acid (containing an α 2-to-8 linkage). The glycans (sugar polymers) are attached to the fifth immunoglobulin domain and the amount of the polysialic acid changes during development. As the embryo ages, the amount of glycosylation declines although the level of N-CAM does not. A decrease in the level of polysialic acid correlates with a decrease in plasticity of the neuron. In adults, the only tissue with highly glycosylated N-CAM is the olfactory system, which is capable of synaptic plasticity.

Other neuronal immunoglobulin-like adhesion molecules include Ng-CAM, or neuron glia cell adhesion molecule, Nr-CAM (Ng-CAM-related), and fascilin II and neuroglian, two immunoglobulin-like homophilic adhesion molecules found in the fruit fly *Drosophila melanogaster*. Some immunoglobulin-like adhesion molecules are not transmembrane proteins but contain a short lipid tail that implants into the outer layer of the cell membrane. These include molecules known as F3/F11/contactin, axonin/TAG-1, and BIG1 and BIG2. All of these are neuronal cell adhesion molecules. Unlike N-CAM, these molecules are not homophilic but rather bind other immunoglobulin-like adhesion proteins on opposing cell surfaces in a heterophilic fashion. For example, the immunoglobulin-like adhesion molecule axonin binds the immunoglobulin-like molecules Ng-CAM and Nr-CAM. F11 also binds Ng-CAM. These interactions guide the movement of axons in the developing spinal cord and appear to be required for the appropriate linking of neurons in the spinal cord. The four N-terminal immunoglobulin domains (out of six) of axonin are responsible for the interaction with Ng-CAM.

In contrast to those members of the family that recognize other immunoglobulin family adhesion proteins, immunoglobulin adhesion proteins found on the endothelium interact with members of the integrin family of adhesion proteins. For example, V-CAM and I-CAM are two endothelial cell immunoglobulin-like adhesion molecules that bind to integrins on leukocytes. V-CAM, or vascular cell adhesion molecule, promotes the attachment of leukocytes via their integrin receptor known as $\alpha 4\beta 1$. I-CAM, or inflammatory cell adhesion molecule, promotes the attachment of leukocytes to endothelium via their integrin receptors known as $\alpha^{L}\beta 2$ and $\alpha^{M}\beta 2$. Both of these proteins are found on endothelium near sites of inflammation but not on normal "resting" endothelium.

Another immunoglobulin-like molecule, L1, binds the integrin $\alpha v\beta 3$, which is found on endothelial cells that are in the process of sprouting new blood vessels. In this instance, L1 on one endothelial cell recognizes integrin $\alpha v\beta 3$ on another. Inhibition of this event with reagents that disrupt this interaction prevent the growth of new blood vessels. L1 is now known to bind to integrin $\alpha v\beta 3$ using an RGD tripeptide sequence present in the sixthmost membrane distal immunoglobulin-like domain.

Members of the immunoglobulin family of adhesion proteins are related to one another on the basis of similar structural regions that interact with other adhesion molecules. Most of these family members recognize nonidentical members of the same family, although one recognizes identical molecules. A few of these recognize members of the integrin family of adhesion proteins.

VI. CONCLUSIONS

Cell adhesion proteins come in diverse shapes and sizes. Although there are at least four classes of adhesion proteins that display highly specialized functions and distributions, all adhesion molecules permit the close contact of cells with adjacent cells or specialized protein layers known as extracellular matrices. Cell adhesion proteins are dynamic molecules that promote the movement of cells into and out of tissues and the bloodstream and that often actively determine cell fate. The acquisition of cell adhesion proteins by single-celled animals enabled multicellular organisms to arise and permitted the orderly embryonic development of all multicellular organisms.

BIBLIOGRAPHY

Cheresh, D. A. (1993). Integrins: Structure, function, and biological properties. *Adv. Mol. Cell Biol.* **6**, 225–252.

Gumbiner, B. M. (1996). Cell adhesion: The molecular basis of tissue architecture and morphogenesis. *Cell* **84**, 345–357.

Huber, H., Bierkamp, C., and Kemler, R. (1996). Cadherins and catenins in development. *Curr. Opin. Cell Biol.* **8**, 685–691.

Hynes, R. O. (1992). Integrins: Versatility, modulation, and signaling in cell adhesion. *Cell* **69**, 11–25.

Hynes, R. O., and Lander, A. D. (1992). Contact and adhesive specificities in the associations, migrations, and targeting of cells and axons. *Cell* **68**, 303–322.

Klymkowsky, M. W., and Parr, B. (1995). The body language of cells: The intimate connection between cell adhesion and behavior. *Cell* **83**, 5–8.

McEver, R. P., Moore, K. L., and Cummings, R. L. (1996). Leukocyte trafficking mediated by selectin–carbohydrate interactions. *J. Biol. Chem.* **270**, 11025–11028.

Cell Cycle

WEI JIANG
The Salk Institute for Biological Studies

GLOSSARY

Cell cycle Sequence of events responsible for cell duplication, consisting of four phases: gap1 (G1), DNA synthesis (S), gap2 (G2), and mitosis (M)

Checkpoints Feedback mechanisms that ensure that the next phase of the cell cycle is not initiated before the preceding step has been completed successfully

Cyclin-dependent kinase inhibitors (CDIs) Group of low-molecular-weight proteins that bind to CDKs or cyclin/CDK complexes and inhibit their kinase activities

Cyclin-dependent kinases (CDKs) Family of serine/threonine kinases that are activated by binding to specific cyclins and phosphorylate the unique set of substrates that are essential for sequencial cell-cycle events

Cyclins Family of proteins that bind to and activate CDKs

CELL PROLIFERATION IS THE FUNDAMENTAL PROPerty of life reproduction. The important events of the eukaryotic cell cycle are that cells must duplicate their genetic materials accurately and subsequently segregate them into two daughter cells. Recent studies have deepened our understanding of the role played by cell-cycle components in regulating these events. Perturbation of cell-cycle progression, mainly by deregulated expression of the cell-cycle control genes, can cause abnormal cell growth and thus enhance oncogenesis.

I. INTRODUCTION

The cell cycle consists of four phases: (1) G1 phase, an interval between the end of the cell division and the beginning of DNA synthesis. During G1 phase of the cell cycle, cells undergo protein and RNA synthesis, increase size, and prepare for DNA synthesis. (2) S phase, a period of DNA synthesis. (3) G2 phase, a second interval between the end of DNA synthesis and the beginning of cell division. During G2 phase of the cell cycle, cells undergo another period of growth and prepare for cell division. (4) M (mitosis) phase, a period of cell division. In mitosis, cells undergo chromosome condensation, nuclear envelope breakdown, spindle formation, attachment of chromosomes to the mitotic spindles, and separation of sister chromatids into two daughter cells. The G1, S, and G2 together are known as interphase. After nuclear division (mitosis), the cell cycle is completed by cytoplasmic division (cytokinesis). The nondividing cells exit the cell cycle at G1 into either a quiescent (G0) state, in which cells can reenter the cell cycle after receiving new growth signals, or a terminally differentiated state, in which cells lose their growth ability and thus are unable to reenter the cell cycle. The cell cycle can also be arrested at G1/S and G2/M checkpoints upon various internal and external signals.

Cells proceed through the cell cycle such as interphase, mitosis, and cytokinesis in a strictly orderly fashion. The transitions between these different cell-cycle states are regulated at checkpoints by a group of protein complexes that are activated sequentially during the cell cycle. These protein complexes are composed of a family of serine/threonine protein kinases called cyclin-dependent kinases (CDKs), their positive regulators called cyclins, and their negative regulators called cyclin-dependent kinase inhibitors

ENCYCLOPEDIA OF HUMAN BIOLOGY, Second Edition, VOLUME 2. Copyright © 1997 by Academic Press. All rights of reproduction in any form reserved.

(CDIs). These cyclin/CDK complexes phosphorylate critical substrates that are essential for cell-cycle progression and cell division.

II. CYCLINS, CYCLIN-DEPENDENT KINASES, AND CDK INHIBITORS

A. Cyclin-Dependent Kinases

CDKs are a family of serine/threonine protein kinases that share the sequence homology with their yeast archetypal member CDC28/cdc2 (CDC28 in *Saccharomyces cerevisiae* and cdc2 in *Schizosaccharomyces pombe,* respectively). Genetic and biochemical studies in yeast demonstrate that CDC28/cdc2 is absolutely essential for the cell division cycle (CDC) of these unicellular organisms. This had led to the idea that the homologs of CDC28/cdc2 exist in multicellular organisms but with a more complex fashion. Through various approaches, eight CDK genes have been identified in mammalian cells thus far. They are CDK1 (the homolog of CDC28/cdc2) and CDK2 to CDK8. The kinase activities of CDKs are primarily controlled posttranslationally by binding to their activators and inhibitors since the expression level of most CDKs remains in constant excess during the cell cycle. In addition, CDKs are regulated by other protein kinases and phosphatases.

The best-understood CDK was CDK1 (CDC2). The monomeric CDK1 in cells is inactive and unphosphorylated. To become active, CDK1 must bind to cyclin B and cyclin A and be phosphorylated on a threonine 161 residue (T161). Phosphorylation on the site homologous to T161 appears to be universally required for all CDK activation. The kinase that phosphorylates this site is the CDK-activating kinase (CAK). It is intriguing that CAK itself is composed of a CDK (CDK7), a cyclin (cyclin H), and a third subunit MAT1 that activates and stabilizes the cyclin H/CDK7 complex. The activation of CDK1 is also regulated negatively by phosphorylation. Phosphorylation of CDK1 on tyrosine 15 (Y15) and threonine 14 (T14) residues that overlap with an ATP-binding site inactivates CDK1 enzyme activity. The protein kinases that phosphorylate Y15 and Y15/T14 sites are Wee1/Mik1 and MYT1. To activate cyclin B/CDK1, both phosphorylated Y15 and T14 must be dephosphorylated by Cdc25C phosphatase. See Fig. 1 for a model of CDK activation.

B. Cyclins

Cyclin was originally discovered in marine invertebrates as the protein that accumulated at high levels following the fertilization of eggs and underwent abrupt destruction in mitosis. Since then, cyclins have been isolated from a variety of eukaryotes, from yeast to human. More than a dozen distinct cyclin genes have been found in the human genome based on their conserved sequence motifs with cyclins of other species, patterns of their expression, and their functional roles during cell cycle. They are cyclins A, B1 and B2, C, D1, D2, and D3, E, F, G1 and G2, H, and I. The highest sequence similarity region among the cyclin family is defined as the "cyclin box," which is essential for cyclin and CDK interaction. Although multiple cyclins and CDKs exist, only certain combinations have been detected in mammalian cells.

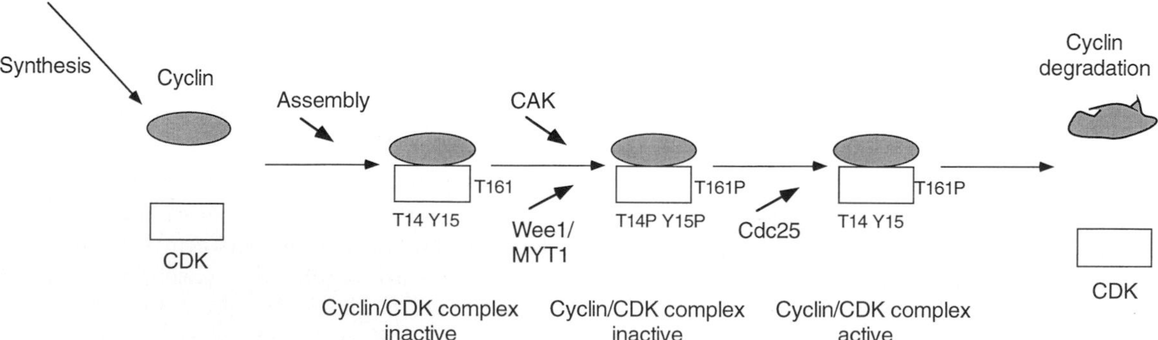

FIGURE 1 Positive and negative regulation of CDK activation. When cyclin is synthesized, CDK and cyclin are assembled to form a complex, which is phosphorylated on T161 and dephosphorylated on T14 and Y15 to become active. The phosphorylation on T161 is mediated by CAK. The phosphorylations on T14 and Y15 are mediated by Wee1 and MYT1 kinases, and subsequently dephosphorylated by Cdc25 phosphatase. Finally, degradation of cyclin leads to CDK inactivation.

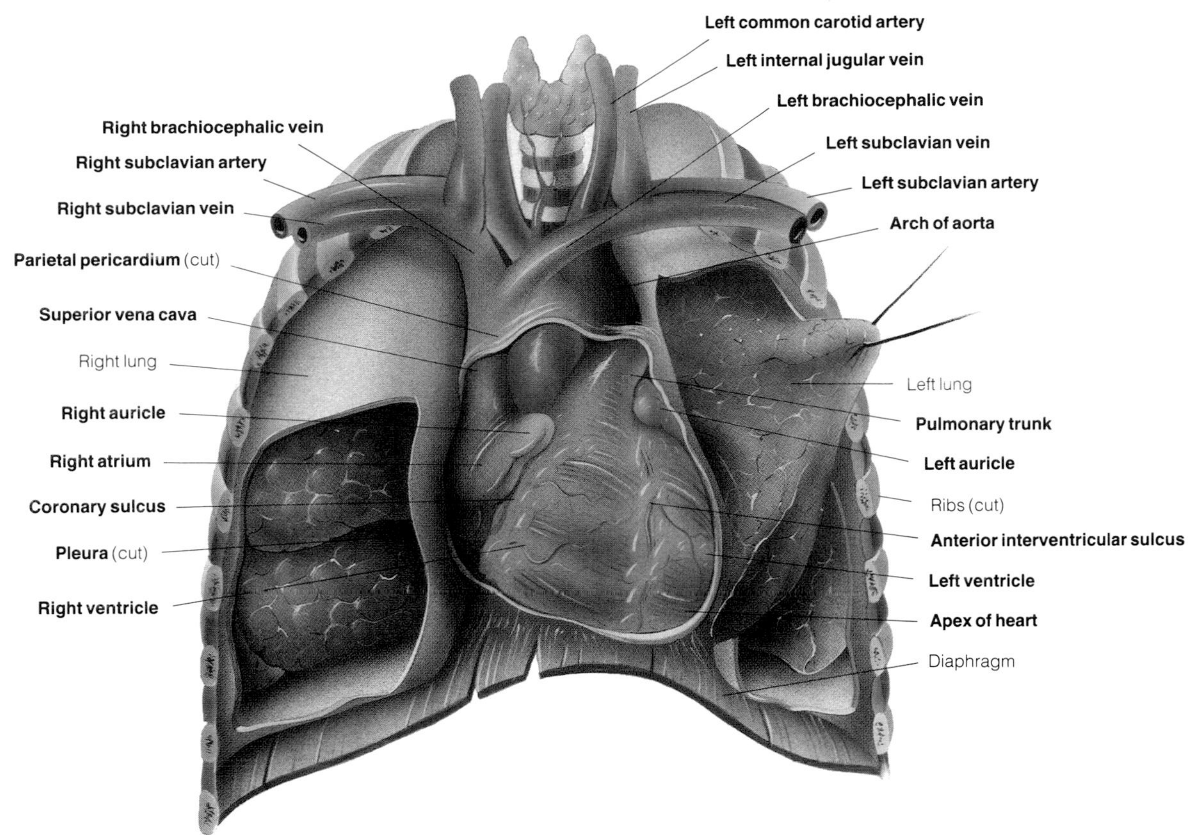

COLOR PLATE 1 The heart in relation to surrounding structures. [Source: Gaudin, A. J., and Jones, K. C. (1989). "Human Anatomy and Physiology." Harcourt Brace Jovanovich, San Diego, p. 559. Reproduced with permission.] [*See* Cardiovascular System, Anatomy.]

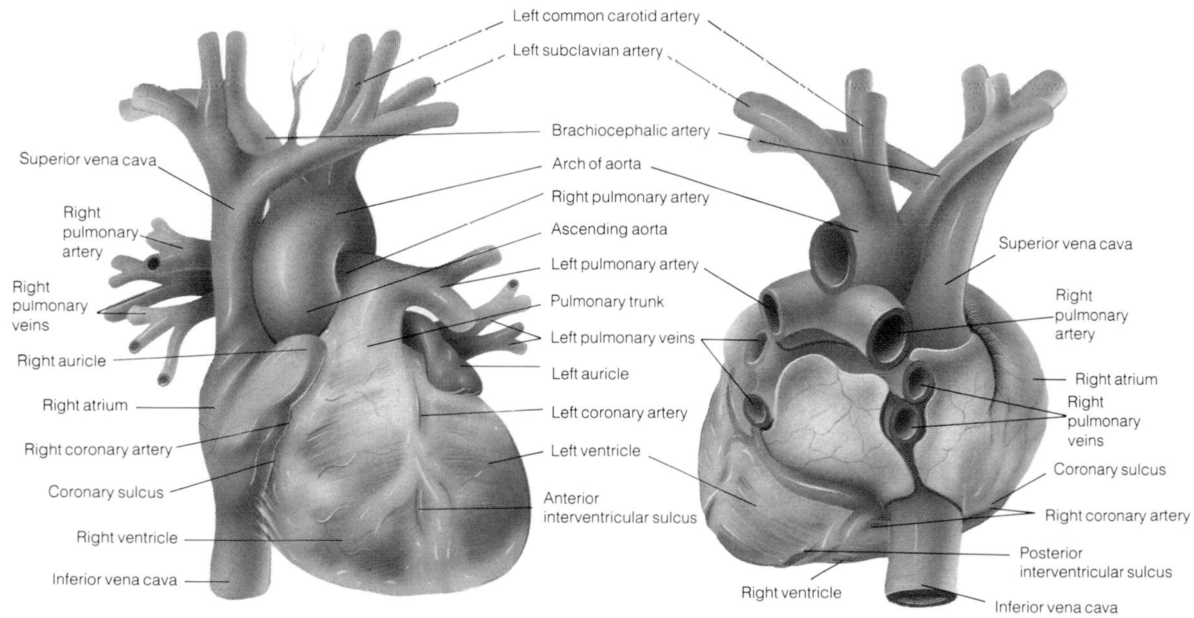

COLOR PLATE 2 The heart in detail. [Source: Gaudin, A. J., and Jones, K. C. (1989). "Human Anatomy and Physiology." Harcourt Brace Jovanovich, San Diego, p. 562. Reproduced with permission.] [*See* Cardiovascular System, Anatomy.]

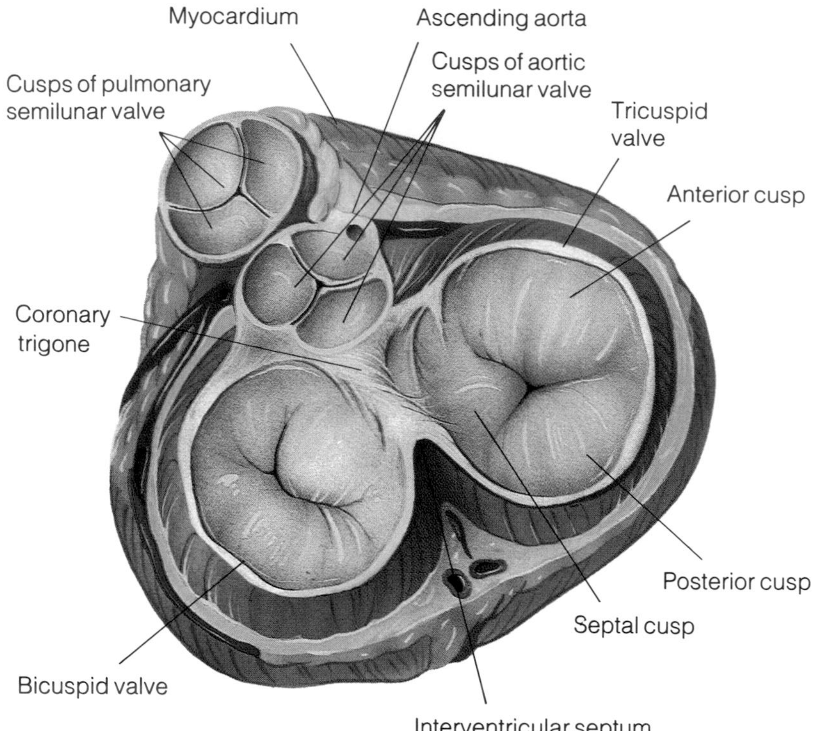

COLOR PLATE 3 The fibrous skeleton of the heart. [Source: Gaudin, A. J., and Jones, K. C. (1989). "Human Anatomy and Physiology." Harcourt Brace Jovanovich, San Diego, p. 563. Reproduced with permission.] [*See* Cardiovascular System, Anatomy.]

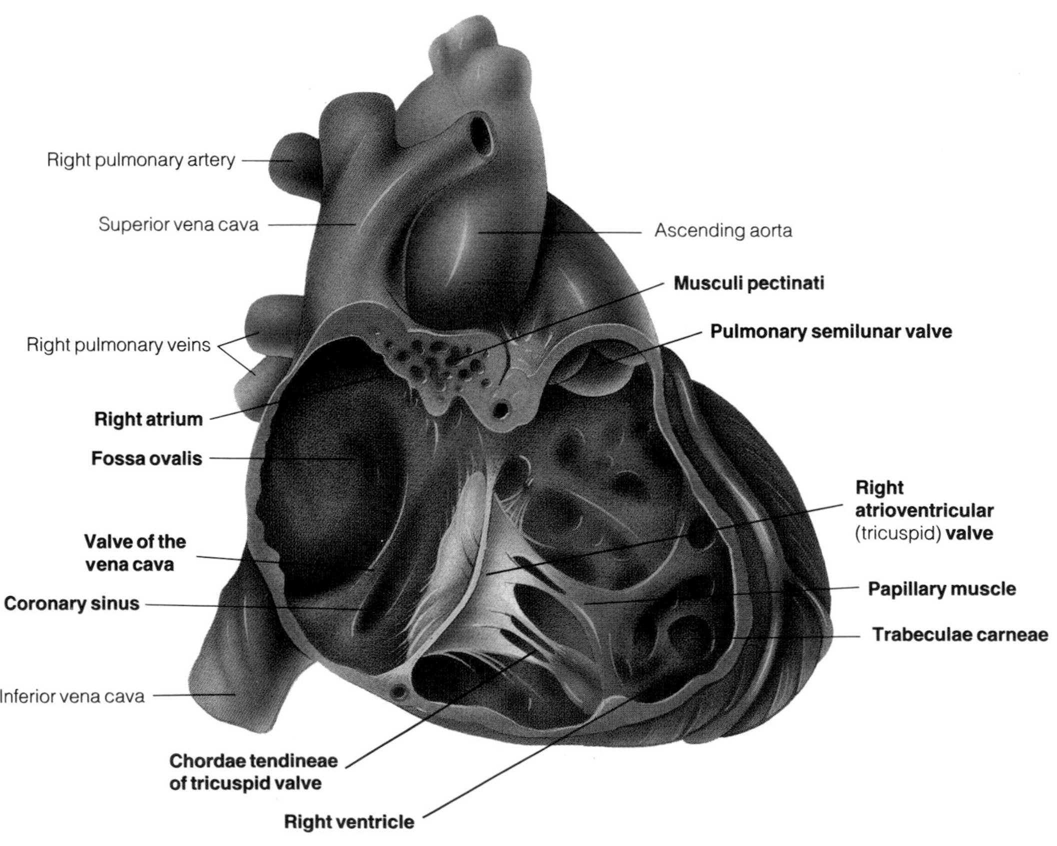

COLOR PLATE 4 A frontal view of the interior of the right atrium and right ventricle. [Source: Gaudin, A. J., and Jones, K. C. (1989). "Human Anatomy and Physiology." Harcourt Brace Jovanovich, San Diego, p. 565. Reproduced with permission.] [*See* Cardiovascular System, Anatomy.]

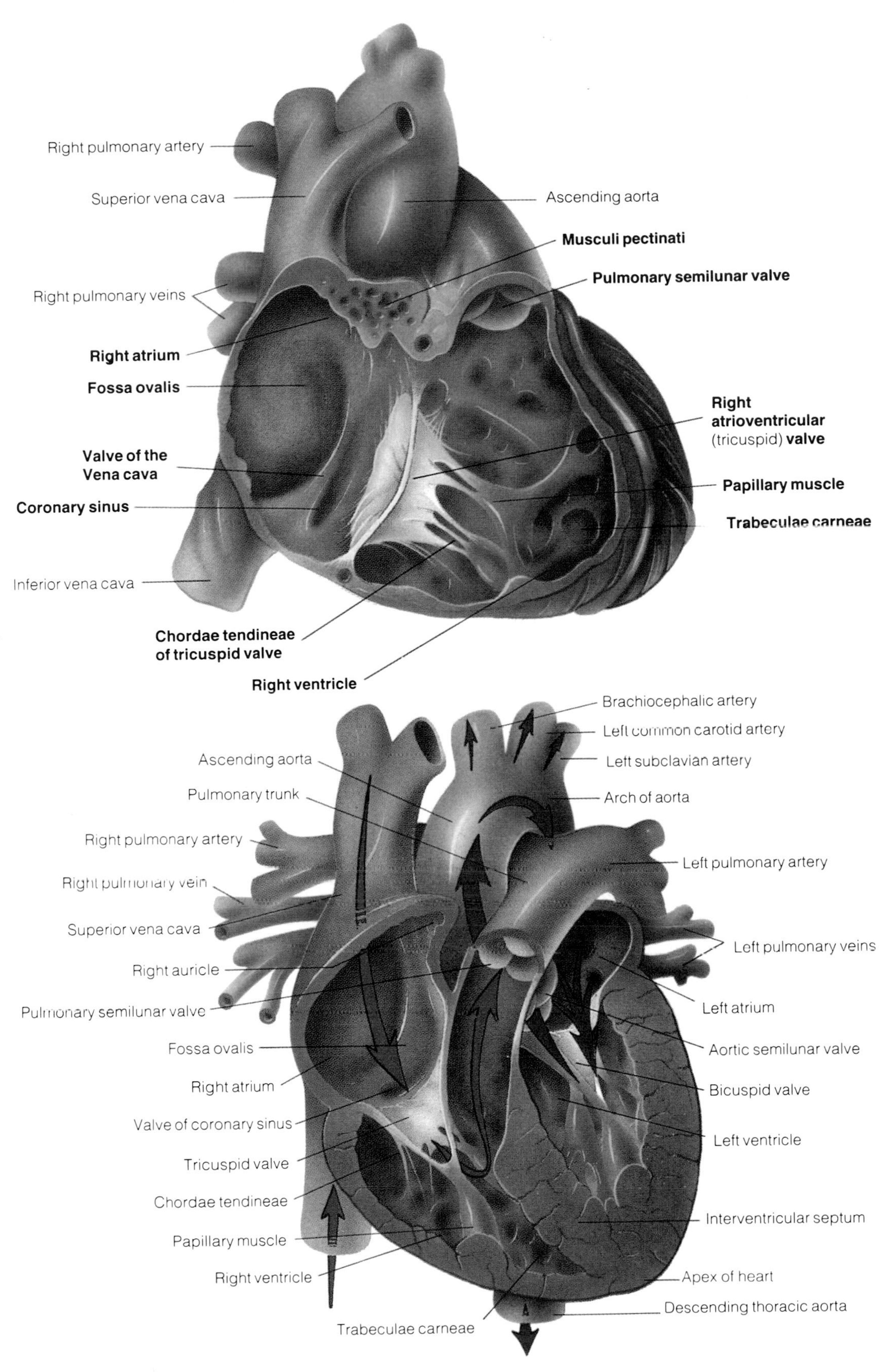

COLOR PLATE 5 The internal anatomy of the heart. [Source: Gaudin, A. J., and Jones, K. C. (1989). "Human Anatomy and Physiology." Harcourt Brace Jovanovich, San Diego, p. 565. Reproduced with permission.] [*See* Cardiovascular System, Anatomy.]

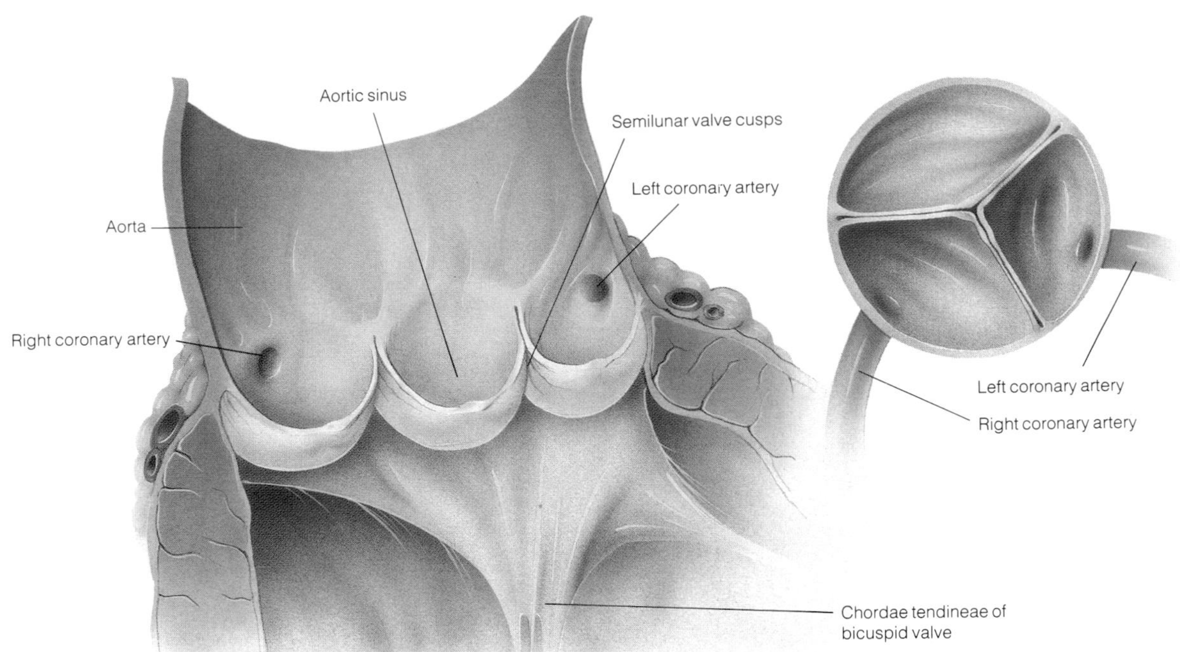

COLOR PLATE 6 The aortic semilunar valve. [Source: Gaudin, A. J., and Jones, K. C. (1989). "Human Anatomy and Physiology." Harcourt Brace Jovanovich, San Diego, p. 566. Reproduced with permission.] [*See* Cardiovascular System, Anatomy.]

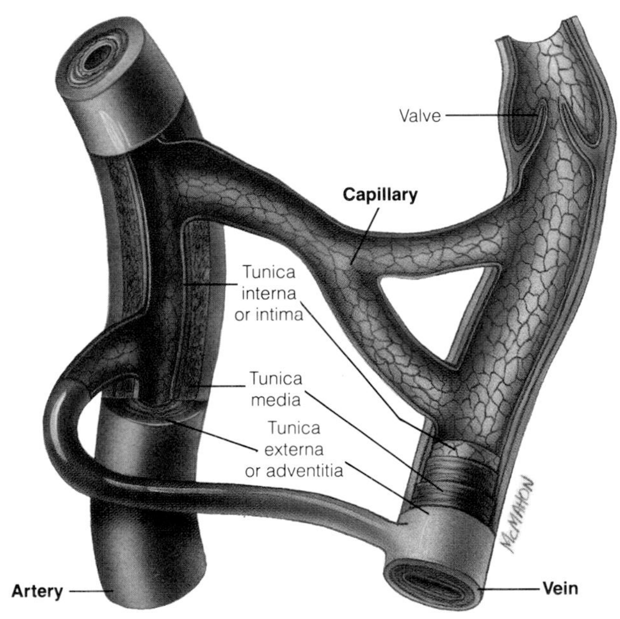

COLOR PLATE 7 Structure of blood vessels. [Source: Gaudin, A. J., and Jones, K. C. (1989). "Human Anatomy and Physiology." Harcourt Brace Jovanovich, San Diego, p. 585. Reproduced with permission.] [*See* Cardiovascular System, Anatomy.]

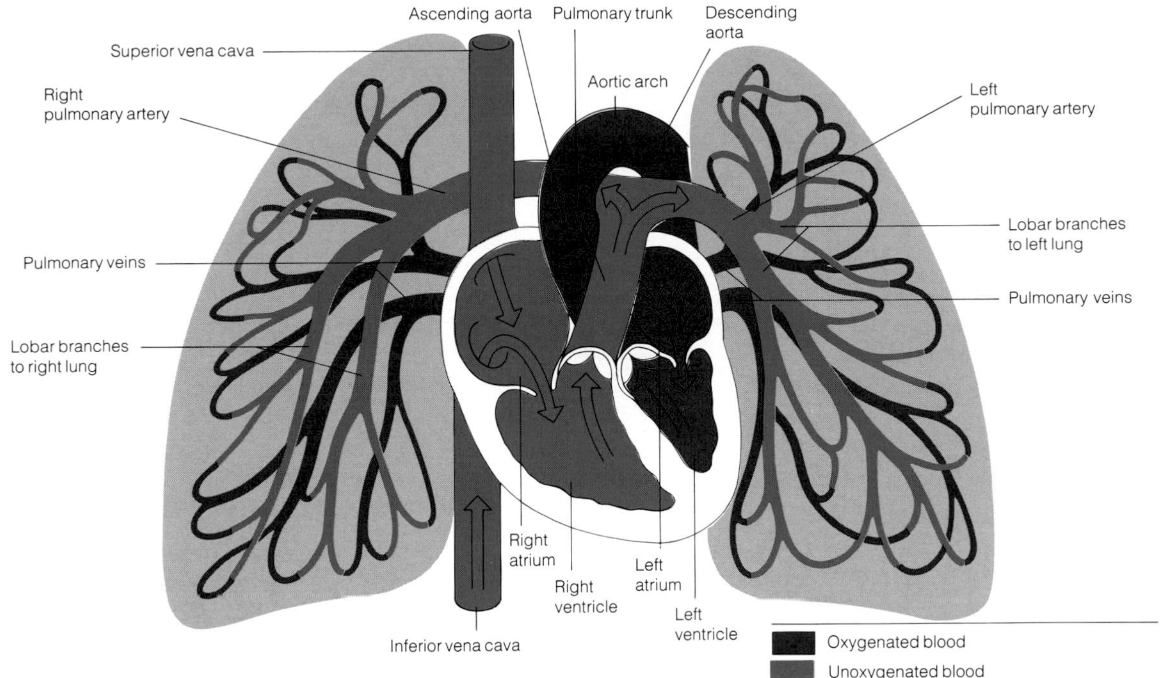

COLOR PLATE 8 Pulmonary circulation. [Source: Gaudin, A. J., and Jones, K. C. (1989). "Human Anatomy and Physiology."] Harcourt Brace Jovanovich, San Diego, p. 589. Reproduced with permission.] [*See* Cardiovascular System, Anatomy.]

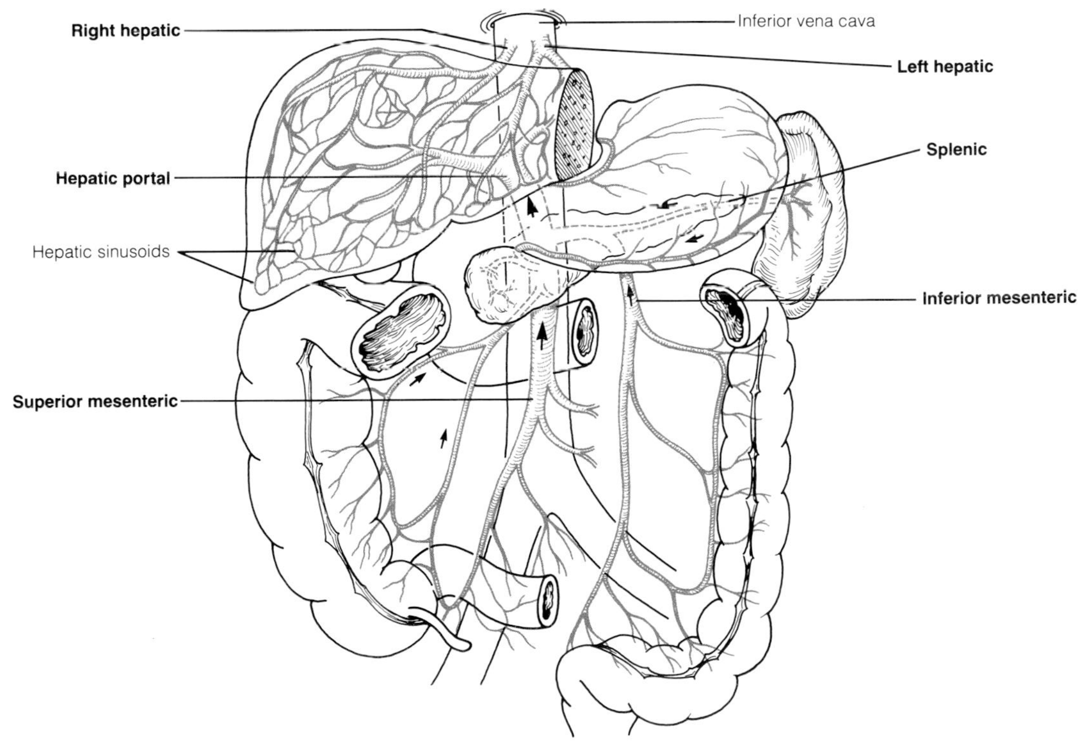

COLOR PLATE 9 Veins of the hepatic portal system. [Source: Gaudin, A. J., and Jones, K. C. (1989). "Human Anatomy and Physiology." Harcourt Brace Jovanovich, San Diego, p. 602. Reproduced with permission.] [*See* Cardiovascular System, Anatomy.]

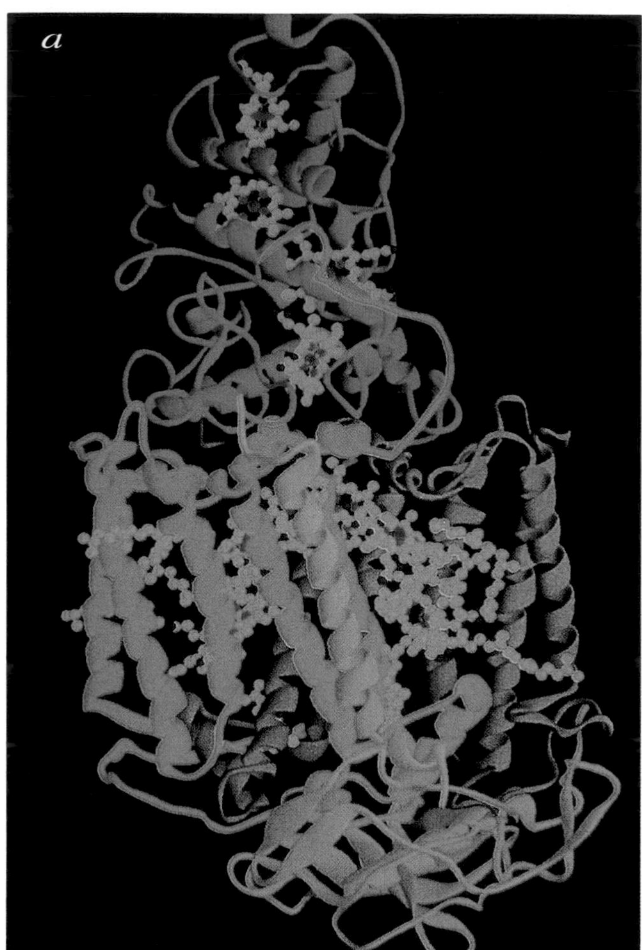

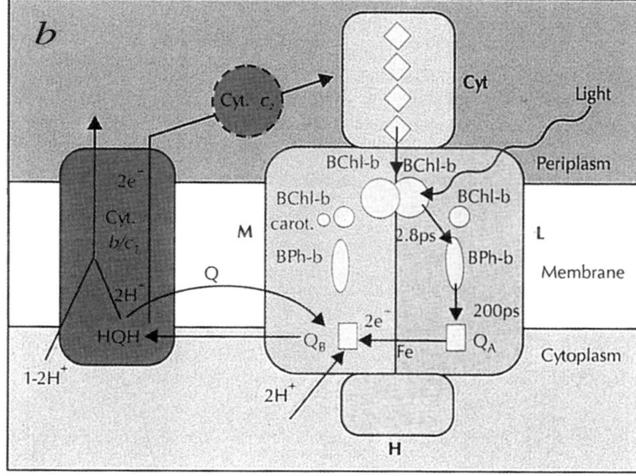

COLOR PLATE 10 The photosynthetic reaction center (RC) from *Rhodopseudomonas viridis.* (a) Overall view of the RC showing the protein subunits as ribbons and the prosthetic groups as ball-and-stick models. Colors: cytochrome, green; L-subunit, brown; M-subunit, blue; H-subunit, purple. Prosthetic groups in atom: carbons, grey; nitrogens, blue; oxygens, red; metals, purple. The middle part of the RC spans the membrane. (b) Schematic view of the reaction center showing the light-driven cyclic electron flow. Absorption of a photon by the primary electron donor, the special pair of BChl-b molecules (linked circles) in the RC, leads to electron transfer to BPh-b (ellipse) in the A-branch within ~3 psec and further to Q_A within ~200 psec. From Q_A the electron is transferred to Q_B within ~25 μsec. Q_B dissociates from the RC, after having received two electrons and two protons. It is replaced by a quinone from a pool in the membrane. The photooxidized primary donor is reduced within 120 nsec (at room temperature) by the periplasmically bound cytochrome. The electrons return to the RC via the cytochrome b/c_1 complex and cytochrome c_2. [Used with permission from J. Deisenhofer (1993). *Structure* **1**, XVIII–XIX.] [*See* Cell Membrane Transport.]

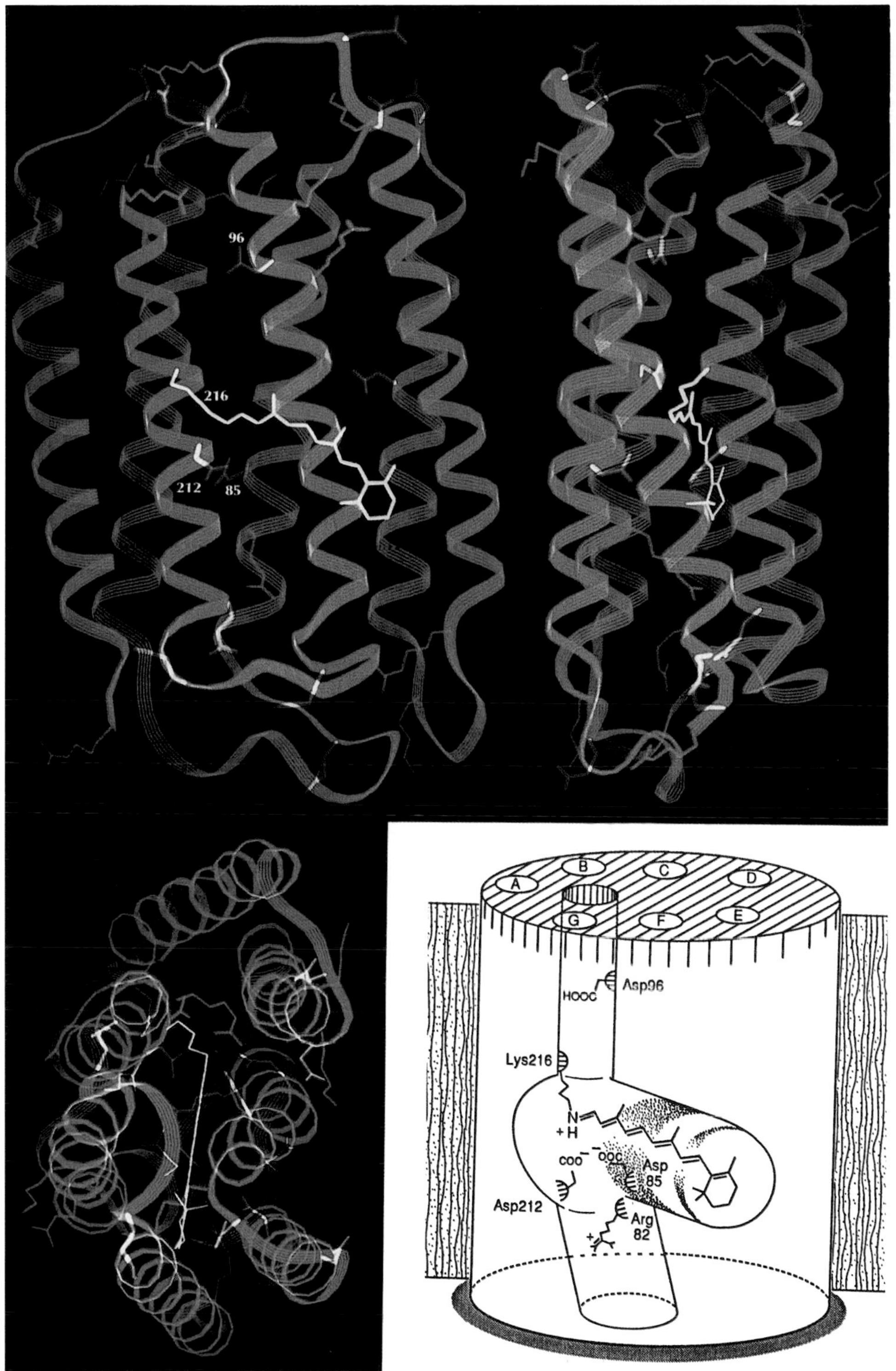

COLOR PLATE 11 Views of the structure of bacteriorhodopsin. (top left and right) Views parallel to the membrane plane with the intracellular surface on top. (bottom left) Projection view perpendicular to the membrane plane viewed from the intracellular surface. The backbone is blue, the retinal and lysine 216 are yellow, aspartate and glutamate are red, and lysine and arginine are blue. The two aspartate residues that are involved in proton transfer are aspartate 96 above and aspartate 85 below the retinal in the top views. Aspartate 212 near aspartate 85 is always unprotonated. The conformations of the loop regions connecting the helices are tentative. [Used with permission from J. M. Baldwin, E. Beckman, T. A. Ceska, K. H. Downing, R. Henderson, and F. Zemlin (1993). *Structure* **1**, XX–XX1.] (bottom right) Artistic impression of the relationship of proton channel to retinal and other key residues involved in proton transport, pictured within the overall boundary of the ground-state bacteriorhodopsin. [Used with permission from R. Henderson, J. M. Baldwin, T. A. Ceska, F. Zemlin, E. Beckman, and K. H. Downing (1990). *J. Mol. Biol.* **213**, 899–929.] [*See* Cell Membrane Transport.]

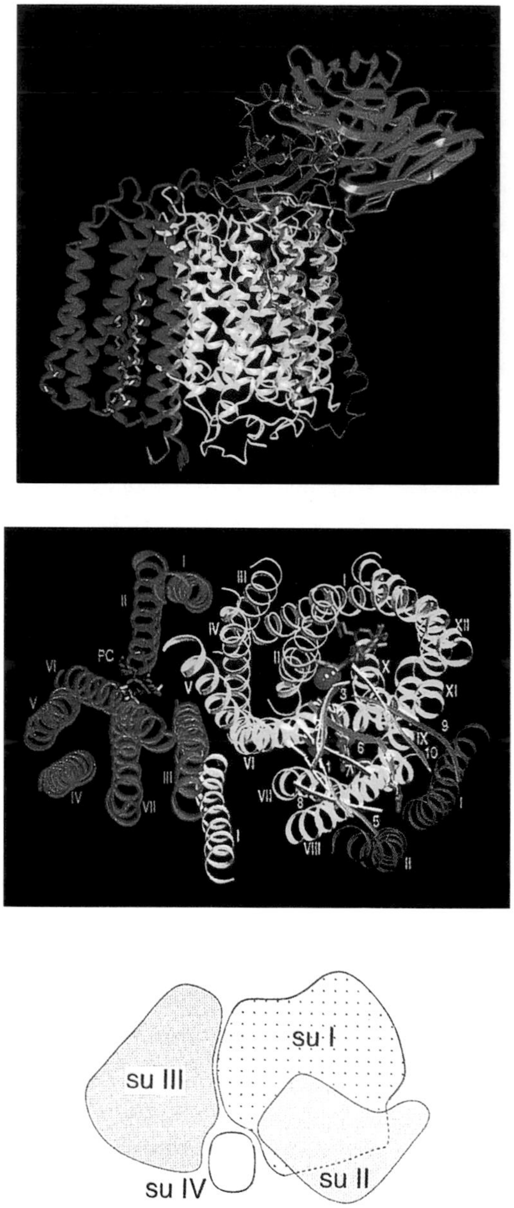

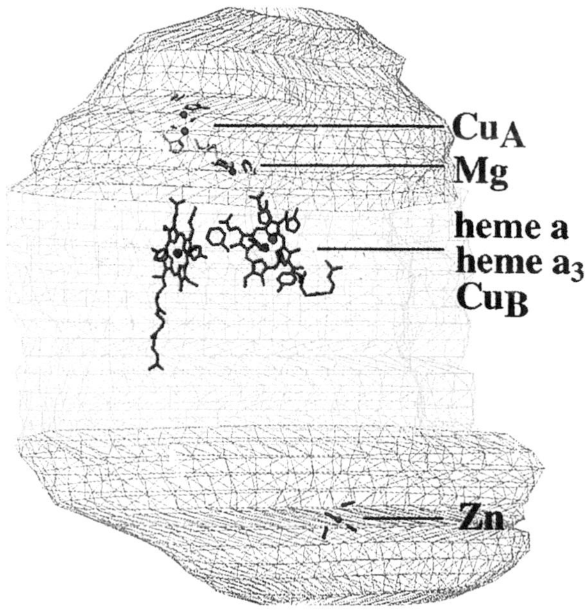

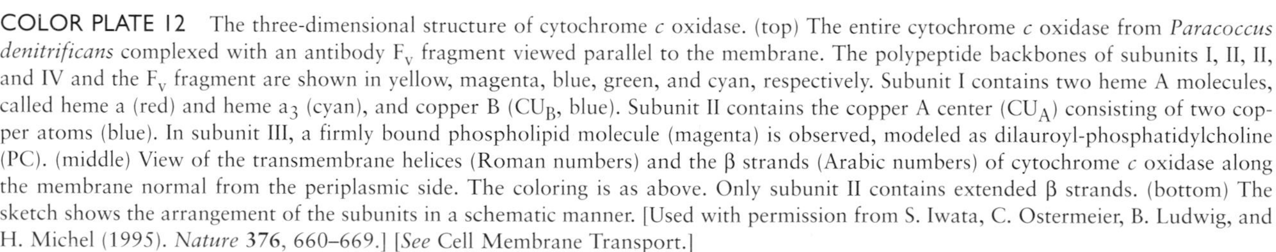

COLOR PLATE 12 The three-dimensional structure of cytochrome *c* oxidase. (top) The entire cytochrome *c* oxidase from *Paracoccus denitrificans* complexed with an antibody F_V fragment viewed parallel to the membrane. The polypeptide backbones of subunits I, II, II, and IV and the F_V fragment are shown in yellow, magenta, blue, green, and cyan, respectively. Subunit I contains two heme A molecules, called heme a (red) and heme a_3 (cyan), and copper B (CU_B, blue). Subunit II contains the copper A center (CU_A) consisting of two copper atoms (blue). In subunit III, a firmly bound phospholipid molecule (magenta) is observed, modeled as dilauroyl-phosphatidylcholine (PC). (middle) View of the transmembrane helices (Roman numbers) and the β strands (Arabic numbers) of cytochrome *c* oxidase along the membrane normal from the periplasmic side. The coloring is as above. Only subunit II contains extended β strands. (bottom) The sketch shows the arrangement of the subunits in a schematic manner. [Used with permission from S. Iwata, C. Ostermeier, B. Ludwig, and H. Michel (1995). *Nature* **376**, 660–669.] [*See* Cell Membrane Transport.]

COLOR PLATE 13 A schematic representation of metal site locations in beef heart cytochrome *c* oxidase. Molecular surface defined with the electron density map at 5 Å resolution is shown by the cage. The yellow represents the transmembrane region with 48 Å thickness. The upper and the lower light blue represent the hydrophilic domanis protruding 37 Å on the cytosolic side and 32 Å on the matrix space, respectively. The three metal centers, Fe_a, Fe_{a3}, and CU_B are located at the same level, 13 Å below the membrane surface on the cytosolic side. Another metal center, Cu_A, is 8 Å above the membrane surface while the magnesium ion is at the membrane surface level. The distances of Cu_A–Fe_a, Cu_A–Fe_{a3}, CU_A–Mg, Fe_{a3}–Fe_a, and Fe_{a3}–Mg are 19, 22, 9, 14, and 15 Å, respectively. [Used with permission from T. Tsukihara, H. Aoyama, E. Yamashita, T. Tomizaki, H. Yamaguchi, K. Shinzawa-Itoh, R. Nakashima, R. Yaono, and S. Yoshikawa (1995). *Science* **269**, 1069–1074.] [*See* Cell Membrane Transport.]

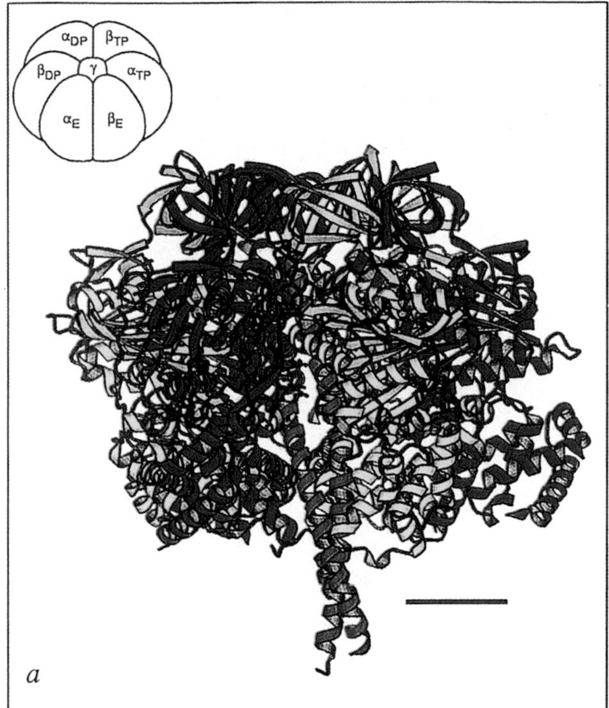

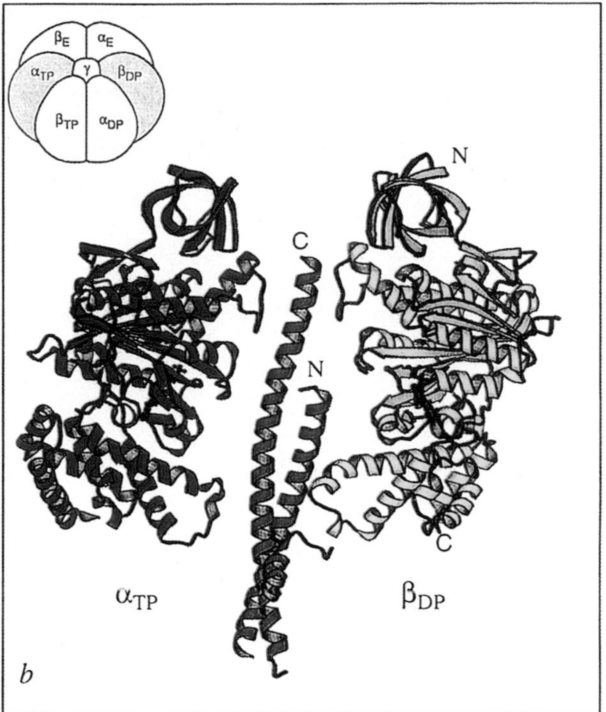

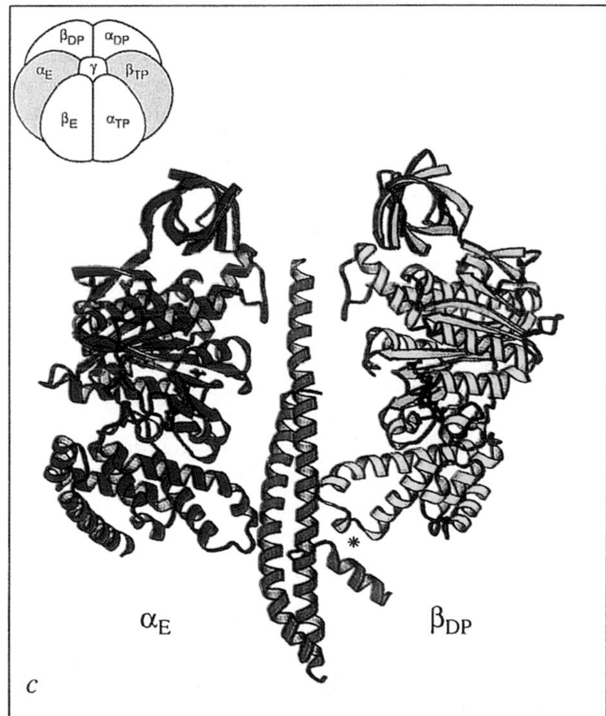

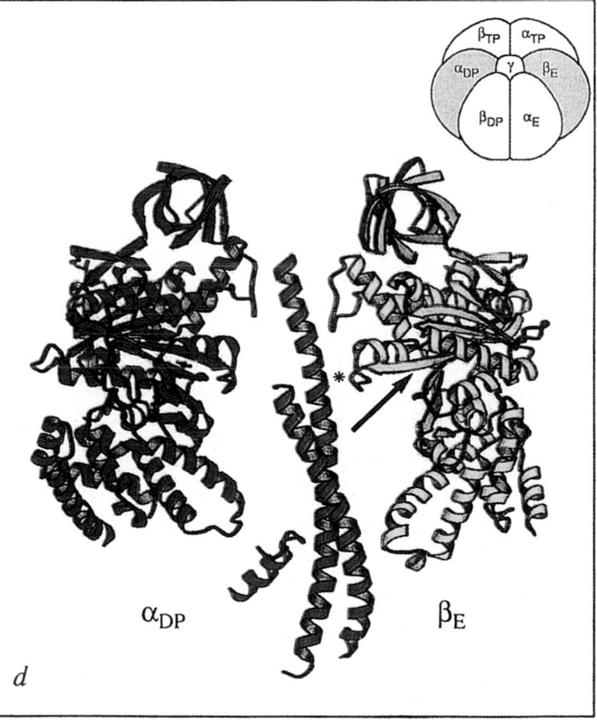

COLOR PLATE 14 The structure of F_1-ATPase from bovine heart mitochondria at 2.8 Å resolution. The α, β, and γ subunits are red, yellow, and blue, respectively; nucleotides are black, in ball and stick representation. AMP-PNP is bound to the three α subunits and to the β subunit defined as B_{TP}. Subunit β_{DP} has bound ADP; subunit β_E has no associated nucleotide. Therefore, the structure probably represents the ADP-inhibited state of the enzyme. (a) A view of the entire F_1 particle in which subunits α_E and β_E point toward the viewer, revealing the antiparallel coiled-coil of the amino- and carboxy-terminal helices of the γ subunit through the open interface between them. The bar is 20 Å long. (b) Subunits α_{TP}, γ, and β_{DP} from a similar viewpoint to that of (a), but rotated 180° about the axis of pseudosymmetry. MgADP, but no phosphate, is bound to subunit β_{DP}. (c) Subunits α_E, γ, and β_{TP} from a similar viewpoint to (a), but rotated by −60°. The asterisk denotes the loop containing the DELSEED sequence with which the ATPase inhibitor protein is thought to interact. (d) Subunits α_{DP}, γ, and β_E from a similar viewpoint to that of (a), but rotated by 60°. The arrow indicates that the β sheet is disrupted in the nucleotide binding domain of subunit β_E, preventing it from binding nucleotide. [Used with permission from J. E. Walker (1994). *Curr. Opin. Struct. Biol.* 4, 912–918.] [*See* Cell Membrane Transport.]

1
Engagement, Descent, Flexion

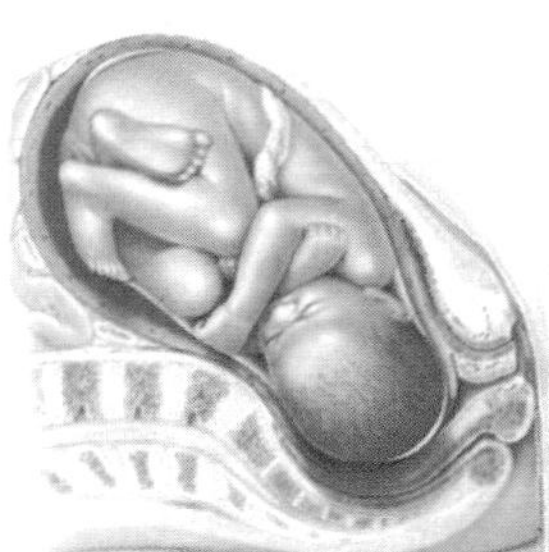

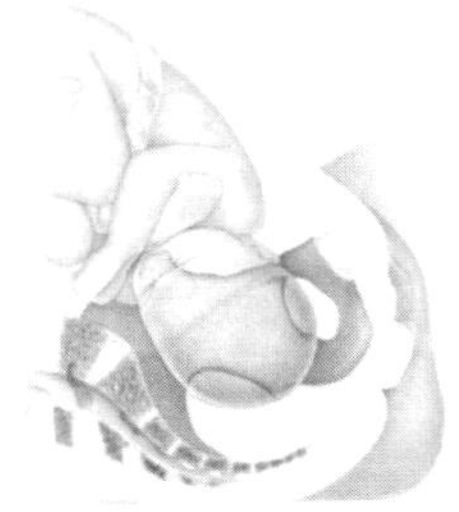

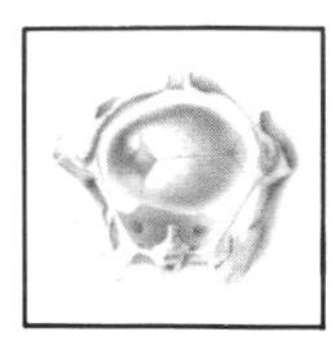

2

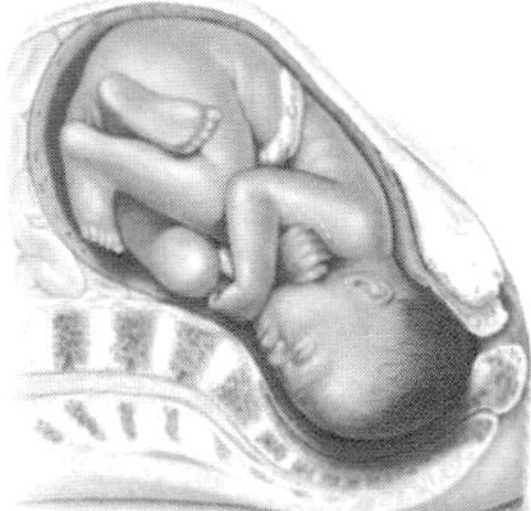

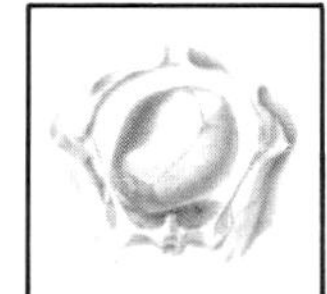

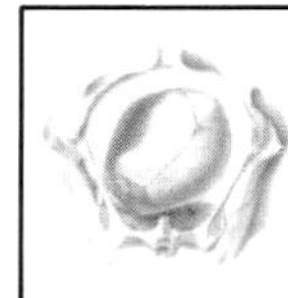

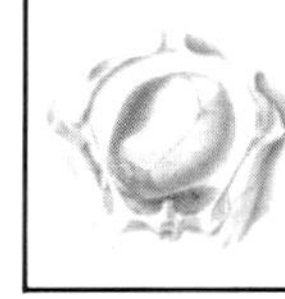

Internal Rotation

3

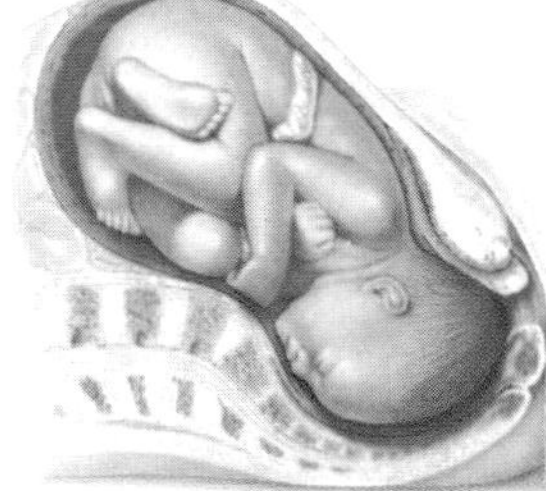

Extension Beginning (Rotation Complete)

4

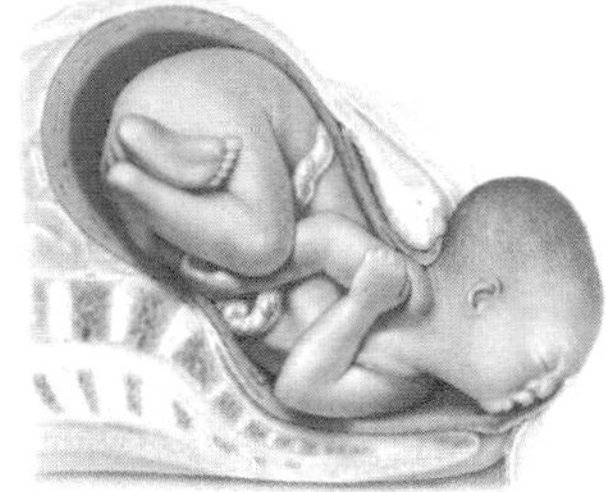

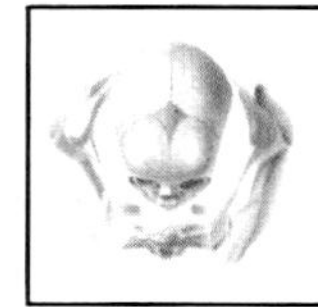

Extension Complete

5

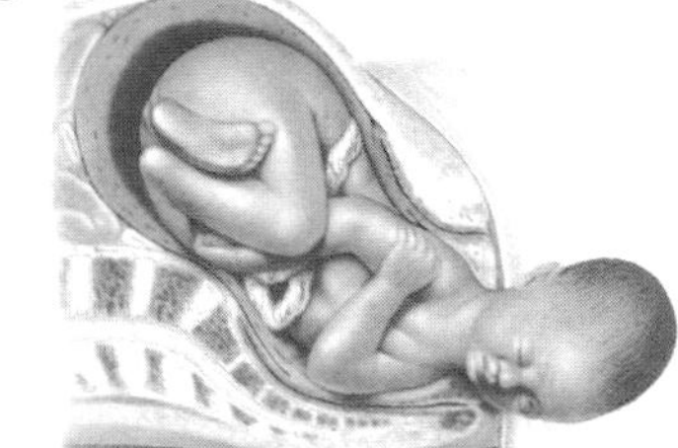

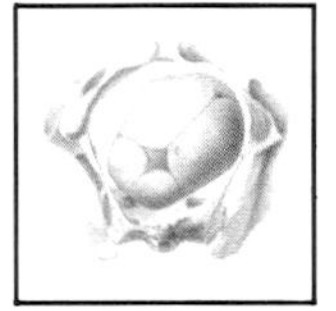

External Rotation (Restitution)

6

External Rotation (Shoulder Rotation)

7

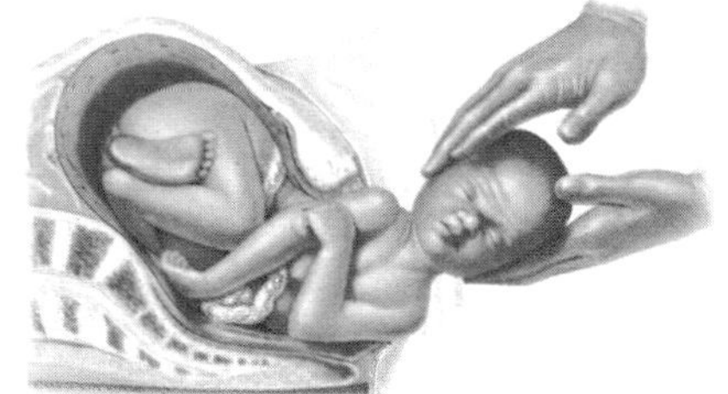

Expulsion

COLOR PLATE 15 Mechanisms of labor. [Reprinted with permission from Ross Laboratories, Columbus, OH 43216, from Clinical Educ. Aid #13 © 1964 Ross Laboratories.] [*See* Childbirth.]

COLOR PLATE 16 Depiction of a dramatic demonstration of color mixture and the law of trichromacy. Circles of light of three wavelengths (e.g., 460, 530, and 650 nm) are partially overlapped on the screen. The three outer crescents show the perceived colors of each of the three original wavelengths (blue, green, and red). The three small triangles show pair-wise mixtures of the three wavelengths, and the central triangle shows the mixture of all three. By varying the intensities of the three light beams, the central triangle can be set to look white. By intuition, further variations of intensities can yield any desired color in the central triangle (e.g., gradually turning down the intensity of the 460 nm light will create colors graduating from white to yellow). The demonstration can be set up with three slide projectors (preferably with variable voltage inputs) and three narrow-wavelength-band filters from a photography store. [Reproduced by permission from Anthony W. Young.] [*See* Color Vision.]

The first identified cyclins were cyclin A and B-type cyclins, which are found in most eukaryotes, including human. There are two distinct B-type cyclin genes, cyclin B1 and B2, in mammalian cells. B-type cyclins accumulate in mid-G2 phase of the cell cycle and predominantly localize in the cytoplasm. After cyclin B binds to B-type cyclin-dependent kinase, CDK1, the cyclin B/CDK1 complex moves to the nucleus at the time of nuclear membrane breakdown. Although both cyclins B1 and B2 form complexes with CDK1 and regulate mitotic events during the cell cycle, it is unclear whether cyclins B1 and B2 are interchangeable. In mammalian cells, cyclin A is expressed during the S phase earlier than B-type cyclins. In contrast to B-type cyclins, cyclin A associates not only with CDK1 but also with CDK2. These complexes accumulate in the nucleus and are active during the S and early G2 phases, indicating that they regulate the progression of these phases of the cell cycle. After cells progress through G2 to M, destruction of cyclins A and B is necessary for cells to exit mitosis. Indeed, cyclins A and B contain a region that is required for protein degradation. Destruction of cyclin A and B proteins is mediated by the ubiquitin pathway, which has been implicated in the turnover of several important cellular regulators.

Cyclins C, D, and E were cloned through their ability to complement a *S. cerevisiae* strain that was mutant in three known G1 cyclins. Whereas the cyclin C mRNA level remains the same through the cell cycle, the cyclin E transcript accumulates periodically and peaks at the G1/S transition. Cyclin C complexes with CDK8 and the complex is found to be a part of the RNA polymerase II holoenzyme. The precise function of cyclin C/CDK8 is unknown, and it may be involved in transcription control during the cell cycle. Cyclin E primarily binds to CDK2 and cyclin E/CDK2 controls the G1/S transition. There are three D-type cyclins (D1, D2, and D3) whose expression is induced in G1 phase by growth factors. There is evidence demonstrating that D-type cyclins associate with several CDKs, including CDK1, 2, 4, 5, and 6, but thus far only the complexes of cyclin D/CDK4 and cyclin D/CDK6 display enzymatic activities. The kinetics and functional properties of these two complexes seem to be similar. The enzymatic activity of cyclin D/CDK4 (or cyclin D/CDK6) peaks at G1 phase, supporting that this type of cyclin regulates G1 to S phase transition.

Several other mammalian cyclins have also been identified within the past few years, including cyclins F, G1 and G2, H, and I. As mentioned earlier, cyclin H/CDK7 is a CAK that can phosphorylate a conserved threonine residue of CDKs (161, 160, and 172 for CDK1, CDK2, and CDK4, respectively) *in vitro*. It is of interest that cyclin H/CDK7/MAT1 complexes are also found in the transcription factor TFIIH and phosphorylate the carboxy-terminal repeat domain (CTD) of RNA polymerase II. Therefore, like cyclin C/CDK8, CAK may also be involved in transcription control. The function and the CDK partner(s) of cyclins F, G1 and G2, and I are currently unknown. Nevertheless, the expression of G-type cyclins can be induced by tumor suppressor p53, suggesting that they may play roles at G1/S and/or G2/M "checkpoints."

C. CDK Inhibitors

It has been demonstrated that in normal human fibroblasts, CDK4 forms either quaternary complexes with a D-type cyclin, a proliferating-cell nuclear antigen (PCNA), and a protein of M_r 21,000 (p21) or binary complexes with a protein of M_r 16,000 (p16). Molecular cloning and characterization of p21 and p16 provided significant insights into possible roles of these low-molecular-weight cyclin/CDKs interacting proteins in the cell-cycle control. Biochemical studies demonstrated that p21 and p16 can inhibit cyclin D/CDK4 kinase activity *in vitro,* therefore they were subsequently defined as CDK inhibitors.

p21 and p16 represent two distinct CDI families. The p21 CDI family includes p21 (also named Cip1, Cap20, Pic1, Sdi1, and Waf1), p27 (Kip1), and p57 (Kip2), which share sequence homology at their amino-terminal region. p21 CDIs can bind to and inhibit the kinase activity of various cyclin/CDK complexes, including cyclin D/CDK4 and 6, cyclin E/CDK2, cyclin A/CDK2, and to a lesser extent cyclin B/CDK1. Overexpression of p21 CDIs in mammalian cells revealed that they cause cell-cycle arrest. The expression of p21 is induced by p53 after DNA damage, indicating that the p21 gene is one of the downstream effectors of p53 that mediates growth arrest in response to DNA damage. Consistently, the p21 −/− cells from homozygous deletion of the p21 gene transgenic mice display a significant defect in their ability to arrest in G1 after DNA damage. However, p21 can be expressed at high levels in p53-negative cells, indicating that the p21 gene must also be regulated in a p53-independent manner. During mouse development, p21 mRNA is localized in a highly selective manner in postmitotic differentiated cells, suggesting that its expression correlates with cell differentiation. p21 is also found at high levels in senescent fibroblast cells. In addition, p21 can inhibit subcellular DNA replication by binding to PCNA through its carboxy-terminal region, which does not show any sequence similarity with p27 or p57 and is not required for the inhibition of cyclin/CDKs kinase activities. These findings indicate that p21 also functions in DNA replication control. [*See* Tumor Suppressor Genes, p53.]

The expression of p27 is increased after cells are treated with transforming growth factor-β (TGF-β), deprived of growth factors, or in contact-inhibited stage, suggesting that it may be involved in antimitogenic signals leading to growth arrest. Transgenic mice that are homozygous for deletion of the p27 gene display an overall increase in body size, hyperplasia of multiple organs, retinal dysplasia, and pituitary tumors, indicating that p27 normally inhibits growth. Expression of p57 mRNA is more tissue-restricted than that for p21 and p27 and is observed in the cell types that are not proliferating actively, suggesting that p57 may also play some role in inhibiting growth.

The p16 CDI family contains p16 (also named Ink4A), p15 (Ink4B), p18 (Ink4C), and p19 (Ink4D). They are primarily composed of several tandemly repeated ankyrin motifs and show no sequence similarity with the p21 CDI family. In contrast to p21 CDIs, the CDIs of the p16 family specifically bind to CDK4 and 6 but not other known CDKs or cyclins and inhibit cyclin D/CDK4 and cyclin D/CDK6 kinase activities. It is postulated that they compete with D-type cyclins for binding cyclin D-dependent kinases. Overexpression of p16 CDIs in cells that contain a functional retinoblastoma protein (Rb), which functions as a growth inhibitor in late G1 phase of the cell cycle, induces G1 phase arrest, whereas the effect cannot be observed in cells that lack a functional Rb. These results are consistent with Rb being a specific substrate for cyclin D-dependent kinases *in vitro,* suggesting that Rb might be the most critical target of these kinases *in vivo* (see Section III).

The structure of p15 (Ink4B) is very similar to that of p16. However, p15 is distinguished from p16 by its 30-fold induction in cells in response to TGF-β, suggesting that this may be one of the mechanisms of TGF-β-mediated cell growth arrest. The p15 and p16 genes are located on the same region of human chromosome 9p21 where the chromosomal deletions have been found in many human cancers (see Section IV). It is of interest that the p16 gene can express an alternate transcript that contains a new first exon (called exon β1) spliced onto the remaining p16 exons 2 and 3. This transcript reveals an alternative reading fame and encodes a protein p19^{ARF} that shows no sequence similarity with known proteins including p16 and p21 CDIs. Although overexpression of p19^{ARF} in cells can cause cell-cycle arrest, there is no evidence demonstrating that this protein binds to and inhibits any cyclin/CDK. Therefore, how p19^{ARF} induces cell-cycle arrest remains to be determined. The p18 (Ink4C) and p19 (Ink4D) genes are located to human chromosomes 1p32 and 19p13, respectively. Although the biochemical properties of p18 and p19 are indistinguishable from those of p16, there is no evidence thus far indicating that either p18 or p19 is involved in human cancer.

III. CELL-CYCLE CONTROL

The sequential activation of a series of cyclin/CDK complexes mediates the eukaryotic cell-cycle progression. Therefore, the active individual cyclin/CDK complex must phosphorylate the unique set of substrates that are essential for the given cell-cycle event. For instance, G1 phase cyclin/CDK complexes phosphorylate and regulate proteins necessary for G1 to S phase transition; S phase cyclin/CDK complexes phosphorylate and regulate proteins necessary for DNA replication; and G2 and M phase cyclin/CDK complexes phosphorylate and regulate proteins necessary for G to M phase transition, mitosis, and cytokinesis.

In mammalian cells, the cyclin/CDK complexes most closely linked to the G1 phase of the cell cycle are cyclin D/CDK4/6 and cyclin E/CDK2. As a cell enters the cell cycle from quiescence (G0), D-type cyclins and cyclin E are synthesized during G1 phase. D-type cyclins are expressed in cell lineage-specific fashion and most cells express cyclin D3 and either D1 and D2. The expression of the D-type cyclin genes is induced as part of the delayed early response to mitogenic stimulation. The D-type cyclins have a very short half-life ($t_{1/2} < 25$ min), therefore, the synthesis of this type of cyclin can be ceased immediately after growth factor withdrawal. It has been suggested that the D-type cyclins may act as growth factor sensors. The assembly of D-type cyclins with CDK4/6 is also regulated posttranslationally by mitogens, although the mechanisms remain to be determined. After cyclin D/CDK4 and 6 complexes are formed, the cyclin D-bound CDKs undergo phosphorylation on the threonine residue (T172 in CDK4) by a CAK to acquire catalytic activity.

The activities of cyclin D-dependent CDK4/6 are detected in early G1 phase and peak at middle to late G1 phase of the cell cycle. The holoenzymes of cyclin D/CDK4/6 have a distinct substrate preference for Rb other than histone (H1), which is the common *in vitro* substrate for other CDKs. Several lines of evidence suggest that Rb is a key physiological substrate of CDK4/6, although other CDKs may contribute to Rb phosphorylation in other phases of the cycle. The D-type cyclins are able to bind to Rb through their amino-terminal LXCXE motif, which is also found in Rb binding oncoproteins E1A, large T antigen, and E7. Overexpression of cyclin D1 under an inducible promoter in G1 phase causes Rb phosphorylation earlier than in the normal cell cycle and shortens the G1 phase of the cell cycle. Conversely, microinjection of anticyclin D1 antibodies in early to middle G1 phase of the cell cycle results in cell-cycle arrest before S phase in cells that contain a functional Rb protein, but not in cells lacking Rb protein. Overexpression of p16 CDIs in cells blocks G1 cell-cycle progression in a Rb-dependent manner. Thus, in middle to late G1, cyclin D/CDK4 and 6 phosphorylate Rb and then the phosphorylated Rb protein loses its growth inhibition ability and releases the transcription factors, such as E2F, whose activities promote S phase entry.

Cyclin E is expressed periodically at maximum levels in G1/S phase transition. After assembling with cyclin E-dependent kinase CDK2, the holoenzyme also undergoes phosphorylation at threonine residue T160 in CDK2 by a CAK to acquire catalytic activity. Cyclin E/CDK2 kinase activity peaks at the G1/S transition, suggesting that it controls the ability of cells to enter the S phase. Experiments demonstrate that cyclin E/CDK2 regulates a G1/S phase transition somewhat differently from that promoted by D-type cyclin-dependent kinases. Although overexpression of cyclin E in fibroblasts also accelerates the G1 phase progression, it does not trigger rapid Rb phosphorylation. Cooverexpression of cyclin D1 and cyclin E in fibroblasts causes additive acceleration of G1 progression. Microinjection of antibodies against cyclin E prevents S phase initiation even in the cells lacking a functional Rb. Together, these results indicate that cyclin E/CDK2 likely phosphorylates other unknown key substrates that are perhaps required to trigger the actual onset of DNA replication.

Once cells enter the S phase, cyclin E is degraded and CDK2 forms complexes with cyclin A. The kinase activity of cyclin A/CDK2 appears shortly after cyclin E/CDK2, concomitant with the onset of measurable DNA synthesis. Microinjection of cyclin A antisense expression plasmids or antibodies against cyclin A into mammalian fibroblasts blocks entry into S phase. Cyclin A colocalizes with sites of DNA replication in S phase nuclei, suggesting that it may directly participate in the assembly, activation, and regulation of DNA replication. In addition, overexpression of cyclin A in normal rat kidney (NRK) cells is sufficient to overcome the S phase block and allow repeated cell cycles in suspension. Since *in vitro* studies using SV40 DNA virus as a model system demonstrate that CDKs are required for assembling initiation complexes at the replication origins, it would not be surprising that the replication origin recognition complex (ORC) proteins, minichromosome maintenance (MCM) proteins, and CDC6-like protein(s) that are involved in DNA replication are the substrates of cyclin-dependent CDK2 kinases.

The G2 to M phases and the mitosis of the cell cycle are governed mainly by cyclinA/CDK1 and cyclin B/CDK1. The best-understood cyclin-dependent kinase activity involved in this part of cell-cycle transition is that of cyclin B1/CDK1. In G2, cyclin B1 and CDK1 together form mitosis-promoting factor (MPF) and initiate the G2/M transition, in which MPF phosphorylates a number of molecules, including histone H1, lamins, nucleolin, caldesmon, myosin light chain, and pp60^{src} involved in mitosis. In M phase, the MPF activates the ubiquitin-dependent proteolytic system, causing both cyclin B destruction and initiation of anaphase. After mitosis, the destruction machinery for cyclin B is switched off and the daughter cells progress into the next cell cycle. See Fig. 2 for a model of cell-cycle control.

IV. CELL CYCLE AND CANCER

Since the cell-cycle genes normally control cell-cycle progression, aberrations in these genes might cause cell loss of cell-cycle control and thus enhance oncogenesis. Indeed, large bodies of evidence indicate the possible roles of these cell-cycle regulatory proteins in multistage carcinogenesis. The first direct evidence came from cloning of human cyclin A. The human cyclin A gene was isolated from human hepatitis B (HBV) viral insertional mutagenesis in a single human hepatocellular carcinoma. This integration event resulted in the accumulation of a stabilized HBV pre-S-cyclin A chimera protein. Overexpression of this chimera protein is thought to contribute to liver cell transformation. Furthermore, cyclin A associates with adenovirus E1A in virally infected cells. The associa-

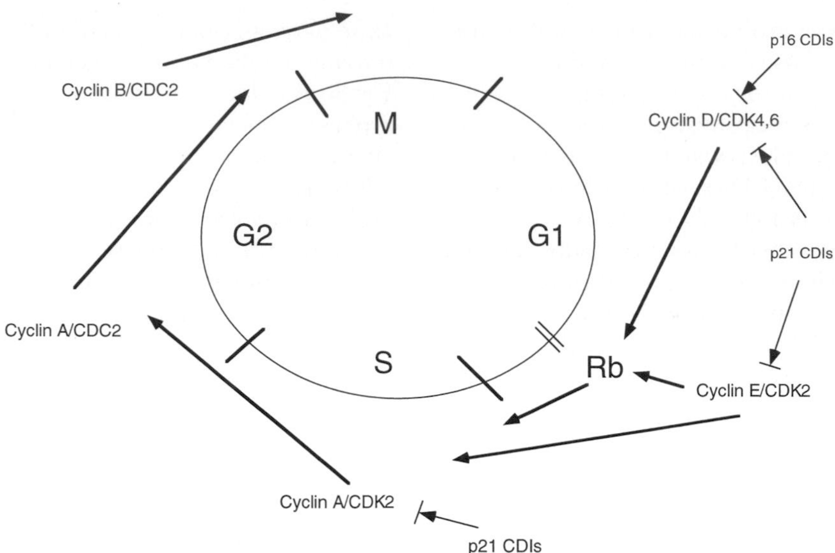

FIGURE 2 Simplified model of cell-cycle progression controlled by cyclins, CDKs, and CDIs (for details see text).

tion of cyclin A with E1A, together with the demonstration of multimeric protein complexes, including cyclin A, CDK2, Rb-related protein p107, and E2F, suggests that cyclin A might be involved in transformation induced by adenovirus. Overexpression of the human cyclin E gene was also detected in several types of human cancer. In addition, deregulating cyclin A and cyclin E synthesis could be involved in the anchorage-independent growth properties of transformed cells since fibroblasts expressing ectopic cyclin A and cyclin E are able to grow in agar suspension. [*See* Adenoviruses.]

In fact, the strongest connection between cyclin and oncogenesis was from studies on cyclin D1. Several lines of evidence indicated that cyclin D1 was involved in human tumorigenesis. Cyclin D1 was originally isolated as a gene that was rearranged on the human chromosome 11q13 region in a subset of parathyroid adenomas (referred to as *prad1*). This rearrangement was due to a chromosomal inversion inv(11) (p15;q13), resulting in the cyclin D1 gene being under the control of the parathyroid hormone (PTH) promoter. Thus, deregulated overexpression of cyclin D1 was detected in this type of tumor. The human chromosome 11q13 region was also frequently found as the site of reciprocal translocation t(11 : 14) (q13 : q32) in centrocytic B-cell lymphoma and leukemia. The rearrangement region, termed *bcl*-1, was assumed to contain a protooncogene that could be activated by this translocation. Genomic mapping studies from the *bcl*-1 breakpoint identified that cyclin D1 was the *bcl*-1 protooncogene. Overexpression of the cyclin D1 gene was detected in those primary tumors and cell lines carrying the t(11 : 14) translocation. In addition, amplification and overexpression of the cyclin D1 gene were detected in several types of human somatic tumors, including esophagus, breast, lung, liver, head, and neck. [*See* Oncogenes and Protooncogenes.]

Consistent with these results, cyclin D1 has been shown to substitute or partially substitute for certain oncogenes in cellular transformation assays. Overexpression of cyclin D1 in rat fibroblasts accelerates G1 phase progression and induces tumors in nude mice. In addition, cyclin D1 overexpression cooperates with an activated *ras* oncogene to transform primary baby rat kidney cells or rat embryo fibroblasts and with *myc* oncogene to induce B-cell lymphomas in transgenic mice. Furthermore, overexpression of cyclin D1 alone under the control of the mouse mammary tumor virus long terminal repeat in transgenic mice results in mammary hyperplasia and carcinoma in females in a pattern that correlates with age, level of expression, and number of pregnancies. Conversely, expression of an antisense cyclin D1 construct in a human esophageal carcinoma cell line, in which the cyclin D1 gene is amplified and overexpressed, causes inhibition of cell growth and loss of tumorigenicity in nude mice. Thus, overexpression of cyclin D1 plays a role not only in establishing but also in maintaining the transformation of cells.

Amplification and overexpression of the CDK4 gene were also found in human gliomas and sarcomas. A point mutation of the CDK4 gene was detected in some sporadic and familial melanomas. The mutated CDK4 creates a tumor-specific antigen and can disrupt cell-cycle regulation exerted by CDI p16. Given the cases of cyclin D1 and CDK4 in oncogenesis, it would not be surprising that deregulation of CDIs may be involved in tumor development. The notable one is p16. The p16 gene is located on human chromosome 9p21 region, which is perhaps the most common site of chromosomal alterations in human cancers. Indeed, the p16 gene (also called MTS1 for multiple tumor suppressor 1) is found to be rearranged, deleted, or mutated in many tumor cell lines and some types of human primary cancer, including pancreatic adenocarcinomas, esophageal squamous cell carcinomas, glioblastomas, leukemias, and lung and bladder carcinomas. Point mutations in the p16 gene were subsequently identified in familial melanoma (MLM) patients, supporting the view that p16 is the MLM tumor suppressor gene. Consistent with these findings, mice carrying a targeted deletion of p16 develop spontaneous tumors at an early age. Furthermore, inactivation of the p16 gene expression by upstream "CpG island" methylation was also detected in human tumors in which the p16 gene was not deleted or mutated. Together, since overexpression of cyclin D1 and CDK4, loss of p16, and loss of Rb have similar effects on G1 progression, they do represent a common pathway that is disrupted in most human tumors.

The p15 gene (also called MTS2) is adjacent to the p16 gene on the human chromosome 9p21. Although mutations in the p15 gene were found to be rare, the gene was commonly deleted in the tumors that also harbored the p16 gene deletion, suggesting that it is a putative tumor suppressor gene. Genetic alterations of other cell-cycle genes, including p21 CDIs, have not been detected in human cancers, although they play important roles in cell proliferation, differentiation, and cell-cycle control, suggesting that they might contribute indirectly to tumorigenesis. The link between p21 and p53, which is commonly mutated in human cancers, suggests that p21 can be an important mediator of p53-dependent tumor suppression. p53 is thought to be activated when DNA is damaged in G1, which would increase the amount of p21 and lead to inhibition of G1 CDK activities, thus preventing cells from entering the S phase. Conversely, in cells in which the p53 gene is mutated, the failure to induce p21 after DNA damage might allow the replication of damaged DNA and could contribute to the increased incidence of chromosomal abnormalities in transformed cells. The link between cyclin D1 and its dependent kinases and p21 and p27 suggests that p21 and p27 can be the important mediator of the cyclin D1/CDK4/p16/RB oncogenic pathway. In growing cells, p27 may be sequestered in cyclin D/CDK4 and 6 complexes. When cells are treated with TGF-β, increased expression of p15 can cause displacement of p27 in cyclin D/CDK4 and 6 complexes and release of p27, which would then be able to inhibit cyclin E/CDK2, resulting in an arrest at the G1/S transition. Transformed cells often display impaired contact inhibition of growth and fail to arrest in response to treatment, suggesting that p27 function may be altered during carcinogenesis. Furthermore, overexpression of Cdc25B has been found in some human breast tumors and tumor-derived cell lines. There are three Cdc25 genes (Cdc25A, B, and C) in the human. As discussed earlier, Cdc25C regulates the G/M transition by dephosphorylating and activating CDK1. Recent evidence indicates that Cdc25A and Cdc25B regulate G1 progression, perhaps by dephosphorylating and activating G1 CDKs. Therefore, increased expression of Cdc25B or A may lead to premature activation of G1 CDKs, thus causing abnormal G1 progression and oncogenesis.

BIBLIOGRAPHY

Draetta, G. F. (1994). Mammalian G1 cyclins. *Curr. Opin. Cell Biol.* **6**, 842–846.

Hartwell, L. H., and Kastan, M. B. (1994). Cell cycle control and cancer. *Science* **266**, 1821–1828.

Hunter, T., and Pines, J. (1994). Cyclins and cancer. II. Cyclin D and CDK inhibitors come of age. *Cell* **79**, 573–582.

Morgan, D. O. (1995). Principles of CDK regulation. *Nature* **374**, 131–134.

Sherr, C. J., and Roberts, J. M. (1995). Inhibitors of mammalian G1 cyclin-dependent kinases. *Genes Dev.* **9**, 1149–1163.

Cell Junctions

ALAN F. LAU
University of Hawaii at Manoa

GLOSSARY

Actin cytoskeleton Intracellular filament structure composed of actin protein that helps create the ultrastructure of a cell

Basement membrane Fibrous layer containing laminin-5 which underlies the epithelium and separates it from the connective tissue

Desmosome Cell-to-cell adhesion structure that is located in the lateral plasma membrane of epithelial cells and interconnects their intermediate filaments

Gap junction Major communicating junction that creates aqueous channels directly interconnecting the cytoplasms of adjacent cells

Hemidesmosome Cell-to-basement membrane adhesion structure that is localized to the basal surfaces of epithelial cells and is associated with intermediate filaments

Intermediate filament Intracellular filament structure composed of keratin or desmin that contributes to the cytoskeletal network of the cell

Tight junction Major occlusive junction that creates a barrier separating the contents of the lumen from the extracellular fluids; it is located in the lateral membrane of the epithelial cell near the lumen and is associated with the actin cytoskeleton

Transmembrane protein Protein that passes completely through a lipid membrane one or more times

Zonula adherans Cell-to-cell adhesion structure that is located in the lateral plasma membrane of epithelial cells and is attached to the actin cytoskeleton

IN ANIMALS, EPITHELIA ARE SHEETS OF CELLS THAT line all body cavities and cover the free surfaces of the body. Epithelial tissues perform an important function of creating barriers to the movement of solutes, water, and cells from one body organ to another. The ability of epithelial cells to perform this barrier function and to maintain the structural integrity of the tissue are dependent upon the formation of cell junctions which are specialized structures in the plasma membrane of the cell that establish close physical connections between neighboring epithelial cells or an epithelial cell and its underlying basement membrane. In addition, nearly all metazoan cells in an organ or tissue are able to share small ions and other metabolites through another type of membrane cell junction. Cell junctions can be divided into three major functional groups: (1) occlusive junctions, which prevent the passage of water and solutes from one side of the epithelial sheet to the other; (2) adhesive junctions, which cement cells to each other or to the underlying basement membrane; and (3) communicating junctions, which establish electrical and chemical connections in cells of a tissue or organ by allowing the passage of ions and other small molecules directly between the cytoplasms of adjacent cells.

I. OCCLUSIVE JUNCTIONS

One common function of epithelia is to separate bodily fluids with different chemical compositions from each other. For example, the mammalian small intestinal epithelium separates contents of the gut lumen from fluids in the extracellular space and connective tissue on the opposite side of the epithelial cell sheet. Selected chemicals and nutrients in the intestine are actively pumped into the epithelial cell across its

ENCYCLOPEDIA OF HUMAN BIOLOGY, Second Edition, VOLUME 2. Copyright © 1997 by Academic Press. All rights of reproduction in any form reserved.

apical membrane surface (facing the lumen). Nutrients then diffuse out of the epithelial cell across basolateral membranes (toward the basement membrane) into the extracellular space, and ultimately into the surrounding connective tissue and the bloodstream. The apical and basolateral surfaces of the epithelial cell plasma membrane contain specific and distinct proteins that are involved in these transport processes. In intestinal epithelial cells, the tight junction is the barrier which prevents influx of intestinal fluid into the extracellular space and connective tissues, diffusion of transported nutrients back into the intestinal lumen, and the mixing of the distinct transport proteins located in the apical and basolateral plasma membrane surfaces.

A. Tight Junctions

Tight junctions (zonula occludens) are usually formed in the lateral plasma membrane, just below the apical surface of the polarized epithelial cell (Fig. 1). Tight junctions encircle the entire cell and are connected to similar structures in adjacent surrounding cells. In conventional electron microscopy, tight junctions appear as very close contact points between adjacent membrane regions with a corresponding loss of the extracellular space. In freeze fracture electron microscopy, tight junctions appear as a belt of web-like strands that encircle the apical region of the cell. The strands appear to be made by two rows of proteins, with each row contributed by an adjacent cell. The proteins bind to one another which seals the intervening extracellular space and limits the exchange of most water-soluble molecules between the lumen of the organ and its interstitial space. It is also thought that the web of crisscrossing protein fibers prevents the mixing of transport proteins that are differentially localized to either the apical or the basolateral surfaces of the cell's plasma membrane.

A number of proteins appear to make up the tight junction (Fig. 2 and Table I). Occludin is an integral transmembrane protein of approximately 66 kDa containing four hydrophobic domains which probably cause it to span the plasma membrane four times. Tight junctions likely form by the extracellular adhesion of occludin molecules in the membranes of two apposed cells, thereby closing off the intervening extracellular space. Other proteins have been associated with the cytoplasmic aspect of tight junctions. Zonula occludens-1 (ZO-1) is a 220-kDa peripheral protein located closest to the plasma membrane in tight junctions. Because ZO-1 interacts with the cytoplasmically located C-terminal domain of occludin and binds to spectrin tetramers, it may play a role in the architecture of tight junctions. In addition, the actin cytoskeleton may be involved in the maintenance of tight junction structure because portions of the actin cytoskeleton closely underlie the cytoplasmic aspect of tight junctions and actin depolymerizing agents, such as cytochalasins, modulate tight junction structure. Other proteins, such as cingulin and ZO-2, have also been found in tight junctions, but their role in tight junction structure or function is currently unclear.

II. ADHESIVE JUNCTIONS

Adhesive junctions are found in many cell types, including epithelial cells, but they are particularly evident in cells that undergo strong mechanical stresses, such as intestinal and cardiac cells and the skin epithelium. These specialized membrane structures link together the intracellular cytoskeletal elements of neighboring cells. This architecture creates tissues and organs with sufficient mechanical strength and stability to resist the mechanical stresses they encounter. There are two major classes of adhesive junctions that differ in form and biochemical makeup: the zonula adherans, which anchor together the actin cytoskeletons of adjacent cells through adhesion of the classical transmembrane cadherin molecules, and the desmosomes, which link the keratin intermediate filament networks of neighboring cells through adhesion of transmembrane desmosomal cadherin molecules. [*See* Cell Adhesion Molecules.]

A. Zonula Adherans

The zonula adherans is most pronounced in epithelium where strong contractile or mechanical forces are at work such as in the small intestine. The zonula adherans encircles the apical region of the cell's lateral membrane, just below the tight junction, forming a continuous zone of adhesion between cells (Fig. 1). For this reason, the zonula adherans is sometimes referred to as the "adhesion belt." Intercellular adhesion is accomplished through the interactions of transmembrane glycoproteins, called cadherins, which are Ca^{2+}-dependent homophilic adhesion receptors (Table I). In the presence of calcium, cadherin molecules located in the membrane of one cell will bind to cad-

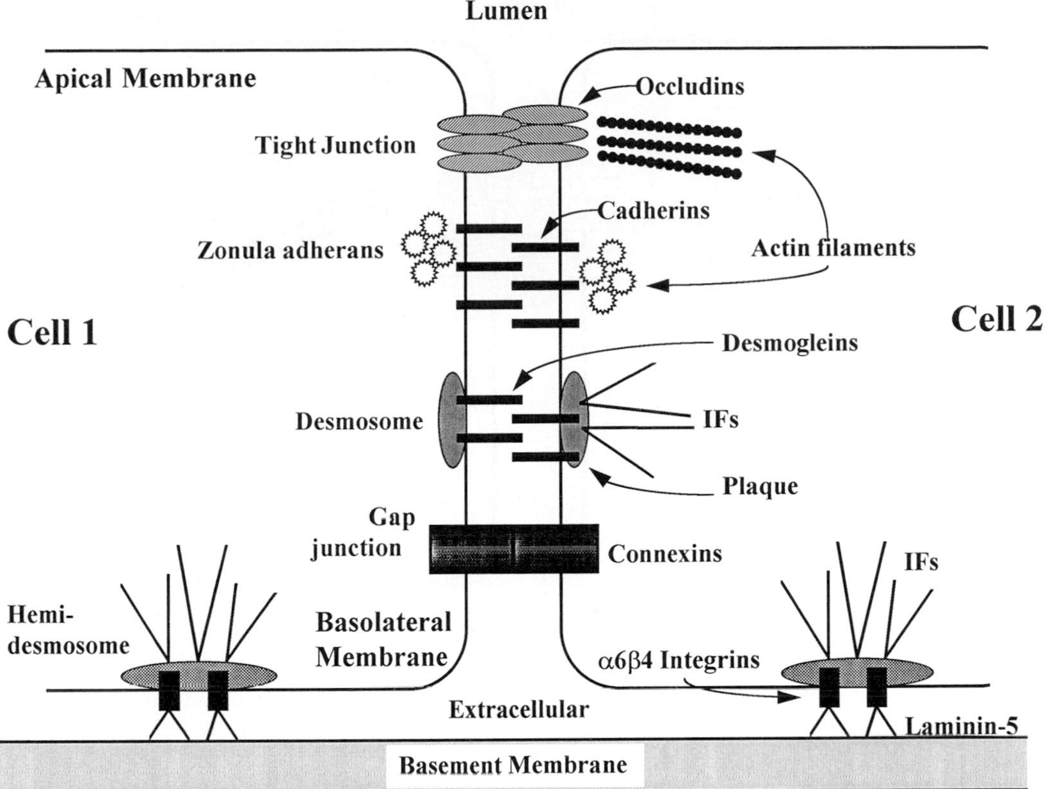

FIGURE 1 Major cell junctions in epithelial cells. The major cell junctions (labeled in Cell 1) are depicted as they are arranged in the plasma membranes of two adjacent epithelial cells. The transmembrane adhesive proteins and the associated intracellular cytoskeletal networks made of actin or intermediate filaments (IFs) are shown (labeled in Cell 2). For clarity, the intracellular adapter proteins are not shown. Locations of the apical, basolateral plasma membranes, and the basement membrane containing laminin-5 are shown.

herins in the membrane of an adjacent cell. One recent model of this interaction depicts the formation of cadherin dimers in one epithelial cell resulting in their orientation away from the cell's membrane surface (Fig. 3). This arrangement might permit the interdigitation of cadherin's N termini which contain homophilic-binding domains, thus creating a linear "zipper-like" array between two adjacent cells. The zonula adherans is closely associated over its entire length with a prominent bundle of actin filaments that run parallel to it and just inside of the plasma membrane. Intracellular attachment proteins, such as α-catenin and β-catenin, are required for cadherin-based cell adhesion. Because α-catenin has actin-binding activity, it is probably involved in connecting the transmembrane cadherin molecules to the actin cytoskeleton. In this manner, the zonula adherans effectively establishes a continuous actin cytoskeletal network across the entire epithelial cell sheet.

B. Desmosomes and Hemidesmosomes

Desmosomes and hemidesmosomes are specialized membrane structures responsible for cell–cell and cell–substratum adhesion, respectively. They are found primarily in epithelial cell layers (skin epidermis, bladder and mammary gland epithelia), but are also present in cardiac muscle and in cells covering the surface of the brain. Although their function is similar to the zonula adherans, desmosomes and hemidesmosomes differ in form because they are tied not to the actin cytoskeleton, but to the intracellular intermediate filament (IF) network. Desmosomes and hemidesmosomes in epithelial cells bind to the keratin IF

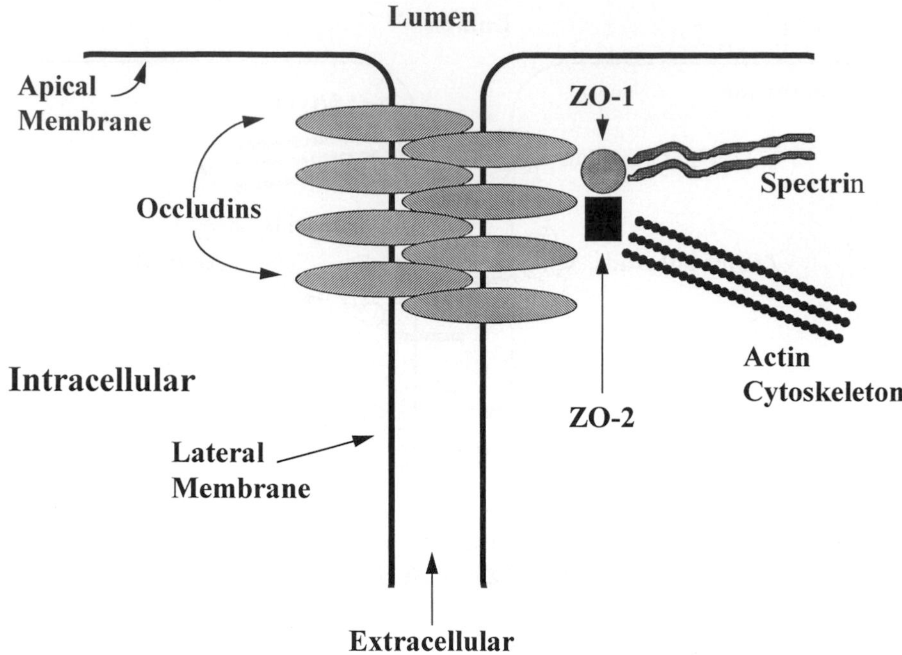

FIGURE 2 Structure and composition of tight junctions. Formation of tight junctions occurs in adjacent cells by the interaction of occludin molecules in the lateral plasma membrane just below the apical membrane surface. The four-pass transmembrane occludins are associated with ZO-1 and ZO-2 adapter proteins, spectrin, and the actin cytoskeleton.

system whereas in cardiac muscle they bind to desmin-containing IFs. The intermediate filaments appear to insert into the desmosomes and hemidesmosomes in a near perpendicular manner. By creating a network of IFs that is distributed across all cells in a tissue as well as being anchored to the basement membrane, desmosomes and hemidesmosomes contribute to the tensile strength of the tissues comprising organs, re-

TABLE I
Molecular Components of Major Cell Junctions

Cell junction	Transmembrane component	Cytoplasmic adapters	Cytoskeletal system
Tight junction	Occludin (four passes)[a]	ZO-1 ZO-2 Cingulin Spectrin	Actin
Zonula adherans	Cadherin (single pass)	α-Catenin β-Catenin Vinculin Plakoglobin	Actin
Desmosome	Desmoglein Desmocollin (single pass)	Desmoplakin Plakoglobin	Intermediate filament (keratin, desmin)
Hemidesmosome	$\alpha6\beta4$ integrin[b] (single pass) BP180 (PBAG2)[c]	BP230 (BPAG1)	Intermediate filament (keratin, desmin)
Gap junction	Connexin (four passes)	?	?

[a]The number of times the protein crosses through the plasma membrane.
[b]The $\alpha6\beta4$ integrin receptor binds to laminin-5 located in the basement membrane. All other transmembrane proteins bind the same molecule located in the membrane of the neighboring cell.
[c]BP, bullous pemphigoid; BPAG, bullous pemphigoid antigen.

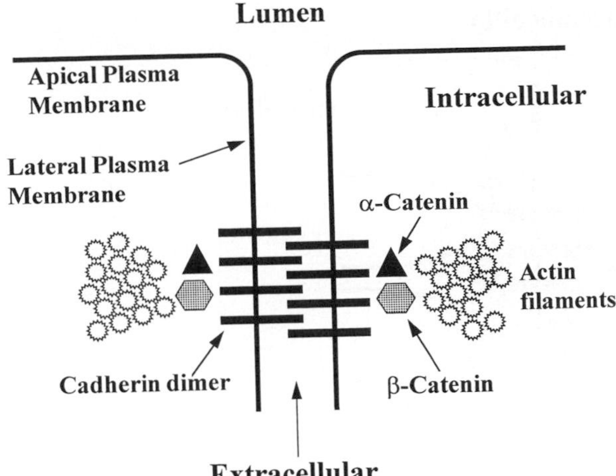

FIGURE 3 Structure and composition of zonula adherans. Zonula adherans is formed between cells by the extracellular association of Ca^{2+}-dependent homophilic binding of transmembrane cadherin molecules across the cell's lateral plasma membrane. The zonula adherans generally lies immediately below the tight junction and forms a contiguous belt of adhesion around the entire cell. The α-catenin and β-catenin proteins are associated with the zonula adherans as are bundles of actin filaments which course in parallel fashion just beneath the plasma membrane.

sisting forces that would act to destroy cell-to-cell or cell-to-basement membrane associations.

Desmosomes are most pronounced in epithelial cells where they appear as "spot welds" located at the lateral membrane surfaces between adjacent cells (Fig. 1). These "spot welds" represent points of cell–cell adhesion that are generated by relatively new members of the cadherin family of adhesion proteins, known as "desmosomal cadherins." This subdivision of cadherins includes desmogleins and desmocollins which, like the classical cadherins, are single-pass transmembrane glycoproteins (Table I). As for the classical cadherins, the desmosomal cadherins also exhibit calcium-binding and adhesion properties localized to their extracellular domains and conserved regions in their cytoplasmic domains which are likely involved in binding to intracellular adapter proteins. It is believed but not yet proven that the desmosomal cadherins function as cellular adhesion proteins in a manner similar to that for the classical cadherins comprising the zonula adherans.

Electron micrographs of desmosomes reveal a pair of electron-dense plaques (one in each cell), sandwiching a plasma membrane region with numerous fibers reaching outward from the plaques into the cell interior (Fig. 4). Each dense cytoplasmic plaque is thought to contain the cytoplasmic domains of the transmembrane desmoglein adhesion molecules and the major desmosomal plaque molecules, plakoglobin and desmoplakin. Plakoglobin is a member of the gene family that contains β-catenin and the tumor suppressor gene adenomatous polyposis coli (APC) that is mutated in certain familial forms of colon cancer. Because plakoglobin binds to the cytoplasmic domains of both desmoglein and desmocollin, it may serve as an adapter linking the transmembrane adhesion proteins to the IF network. Desmoplakin is the most abundant component of desmosomal plaques. It is a modular dumbbell-shaped protein with two globular ends corresponding to its amino and carboxyl termini and a central α-helical region. Studies have demonstrated that the carboxyl terminus of desmoplakin mediates binding to keratin IFs which radiate outward from the desmosomal plaques into the cytoplasm. The amino terminus of desmoplakin associates with the desmosomal plaque itself. These results suggest that desmoplakin functions as a molecular link anchoring IFs in the desmosomal plaque.

Hemidesmosomes are one of the major cell junctions involved in the adhesion of cells to the basement membrane. It is abundant in basal membranes of epidermal basal cells and serves to anchor these cells to the underlying basement membrane (Fig. 1). Because of this feature, the hemidesmosome exhibits only one electron-dense plaque structure located just inside of the plasma membrane. IFs appear to radiate up and away from the plaque into the cell's cytoplasm. The transmembrane component of the hemidesmosome that binds to the basement membrane is the heterodimeric $\alpha6\beta4$ integrin receptor (Fig. 5 and Table I). The $\beta4$ subunit is unique among β integrins because it possesses a very long cytoplasmic tail (over 1000 amino acids) that may be critical in the binding of the $\beta4$ subunit to keratin IFs. The $\alpha6$ subunit lacks a similar long cytoplasmic tail; however, its extracellular region is involved in binding to the $\beta4$ integrin subunit itself and the BP180 transmembrane protein. BP180 is one of the reactive antigens in the autoimmune blistering disease, bullous pemphigoid (see Section IV,A). The $\alpha6\beta4$ integrin receptor binds to the basement membrane protein, laminin-5, also known as epiligrin. In stratified squamous epithelial tissues, laminin-5 has been found to be concentrated in the basement membrane immediately underneath the hemidesmosome. This interaction helps cement the basal cell to the basement membrane and promotes the integrity of the epidermis. The intracellular hemidesmosome plaque contains the 230-kDa protein BP230, which is another antigen for autoantibodies

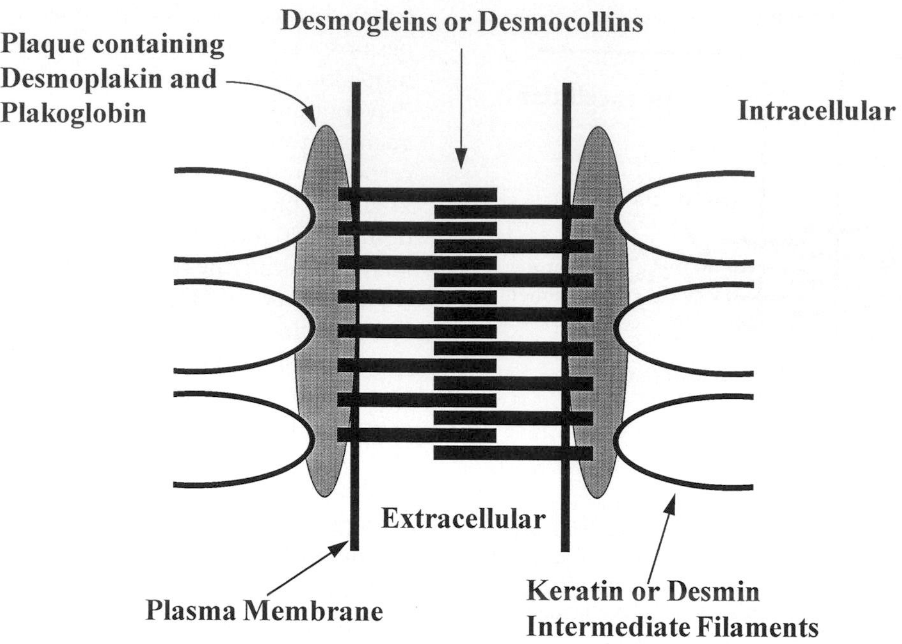

FIGURE 4 Structure and composition of desmosomes. The desmosome is formed between adjacent cells by calcium-dependent association of the "desmosomal cadherins," desmoglein or desmocollin, through their extracellular amino termini. These molecules are associated with the cytoplasmic plaque which also contains desmoplakin and plakoglobin. Keratin- or desmin-containing intermediate filaments insert into the desmosomal plaque of each cell and radiate outward into the cell's cytoplasm.

found in some bullous pemphigoid patients. The BP230 plaque protein belongs to a family of proteins that includes desmoplakin, the major component of desmosomes. The BP230 resembles desmoplakin with globular ends and a rod-like central portion. Like desmoplakin, BP230 also contains an IF-binding site in its carboxyl terminus. The architecture of the hemidesmosome is dependent on BP230 because its loss

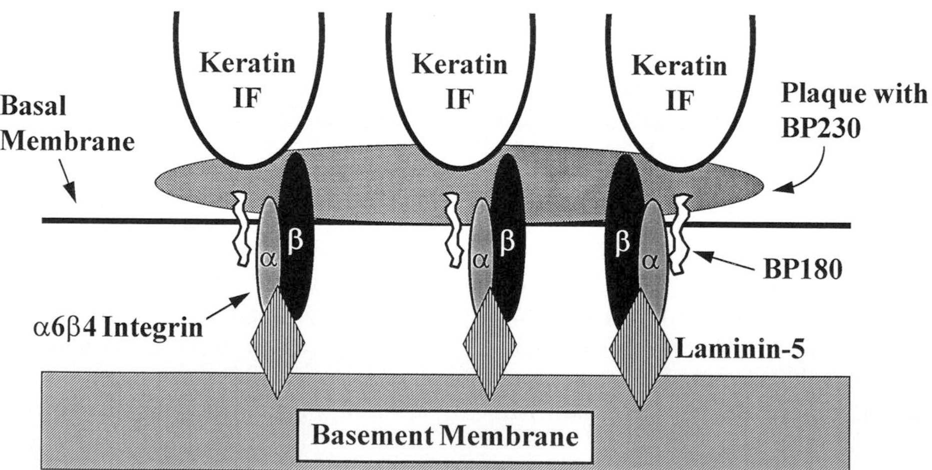

FIGURE 5 Structure and composition of hemidesmosomes. Hemidesmosomes form at the basal membrane surface of epithelial cells and the underlying basement membrane. The transmembrane $\alpha6\beta4$ integrin receptor binds laminin-5 in the basal lamina (not depicted) of the basement membrane. In turn, the $\alpha6\beta4$ integrin receptor is anchored in the cytoplasmic plaque which also contains the transmembrane BP180 and intracellular BP230 pemphigoid antigens. The hemidesmosome is also connected to the cell's intermediate filament (IF) network.

from cells results in a lack of the innermost portion of the hemidesmosomal plaque and the normally associated IFs.

III. COMMUNICATING JUNCTIONS

In metazoan cells, communicating junctions have a distinctly different function from the previously discussed junctions that create fluid barriers or cell adhesion sites. Communicating junctions are engaged in the direct exchange of small molecules under approximately 1000 Da between the cytoplasms of adjacent cells. The major communicating cell junction in metazoans, including humans, is the gap junction.

A. Gap Junctions

Gap junctions exist in almost all fully differentiated metazoan cells, with the notable exceptions of skeletal muscle cells, circulating blood cells, and spermatozoa. Gap junctions are thought to have a wide array of functions, including maintenance of synchronous contraction of cardiac and smooth muscle cells (intestinal and uterine), transmission of signals in neuronal electrotonic synapses, metabolic cooperation in avascular organs (lens), regulation of secretion by the pancreas, pattern formation during development, and the control of cell growth and oncogenic transformation. Gap junctions are thought to achieve these diverse effects because they establish a unique form of intercellular communication that allows the direct passage of ions, small metabolites, and regulatory molecules such as calcium, inositol triphosphate, and cAMP between adjacent cells. For example, the synchronous contraction of the heart muscle is dependent on the electrical signal created by the flow of ions through gap junctions between adjacent cardiac muscle cells.

Gap junctions in the lateral plasma membrane actually represent a large array of intercellular membrane channels that span the extracellular gap between cells (Fig. 1). Each intercellular channel is constructed by the docking of two hemichannels, or connexons, across the extracellular space (Fig. 6A). Each connexon is formed separately and contributed by one of the two adjacent cells. The intercellular channel creates an aqueous pore of approximately 2–4 nm in diameter that traverses the two plasma membranes, connects the cell's cytoplasms directly, and seals them from the extracellular milieu.

The connexon is formed by a hexamer of proteins called connexins; the complete intercellular channel

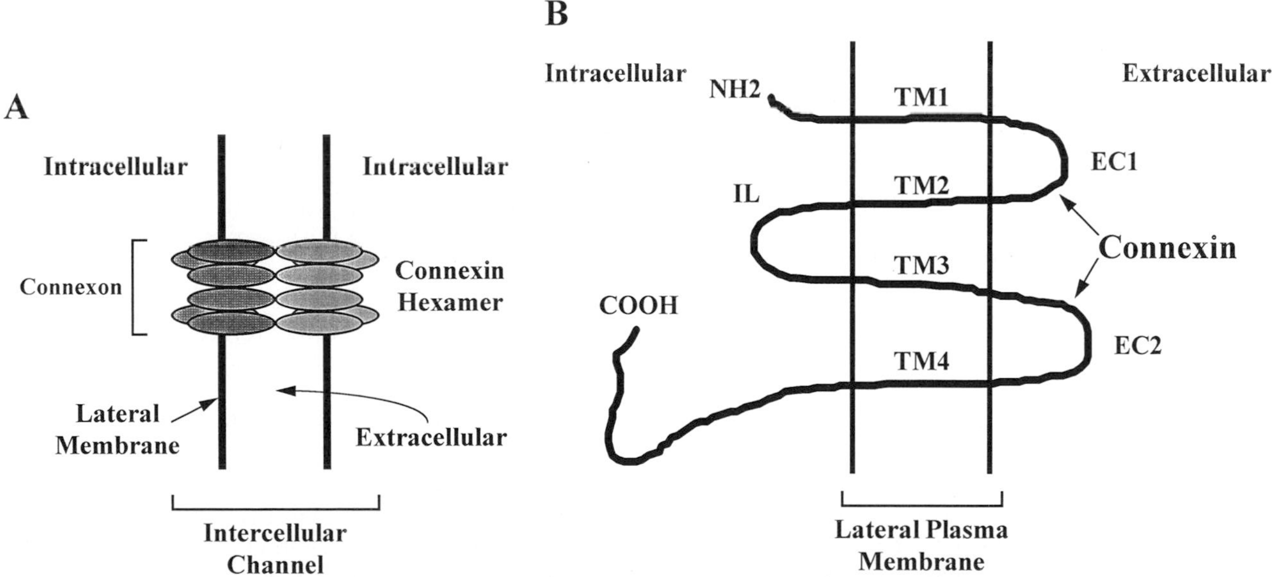

FIGURE 6 Structure and composition of gap junctions. (A) The gap junction intercellular channel is formed across the lateral plasma membranes of two adjacent cells by the extracellular association of two connexons. Each connexon, or hemi-channel, is formed by six connexin proteins. The complete intercellular channel is formed by two connexons containing a total of 12 connexins. (B) The orientation of a single connexin molecule in the lateral plasma membrane of one cell is shown. Connexin passes through the membrane four times creating four transmembrane regions (TM1-4), two extracellular loops (EC1, EC2), one intracellular loop (IL), and cytoplasmically located amino and carboxyl termini.

consists of 12 connexin molecules (Fig. 6A). Each individual connexin in the connexon is oriented in the plasma membrane such that its hydrophobic regions pass through the lipid bilayer four times (TM1-4), creating two extracellular regions (EC1 and EC2) and three cytoplasmically located domains consisting of the amino and carboxyl termini and an intracellular loop (IL) bordered by the second and third transmembrane regions (Fig. 6B). The two extracellular domains (EC1 and EC2) are involved in docking with the corresponding regions of the connexins in the apposed cell to create the intact intercellular channel. Each extracellular domain contains three conserved cysteine amino acids which form intramolecular disulfide bonds. Unlike the homophilic binding of cadherins, the molecular basis for binding of connexins from adjacent cells is not presently clear.

Evidence indicates that connexins are synthesized in the endoplasmic reticulum of cells, followed by their assembly into connexons in the trans-Golgi network. Connexons are transported to the plasma membrane and subsequently cluster in gap junction plaques, dock with apposed connexons, and form open intercellular channels. The formation of functional gap junction channels has been demonstrated to depend on cell adhesion, mediated by molecules such as E-cadherin. Conversely, signals passed through gap junctions may alter cell–cell adhesion. Thus, there appears to be a complex interplay between cell adhesion and signals passed through gap junction channels.

Connexins belong to a growing multigene family currently containing 13 distinct members. Sequence analysis has indicated that the amino terminus and the transmembrane regions of connexins are relatively conserved, whereas the cytoplasmic portion of connexin containing the carboxyl terminus tends to exhibit the greatest size difference and sequence diversity. Different connexins are named according to the species of origin together with the abbreviation of the gene name (Cx) and their molecular mass predicted from the complementary DNA. Thus, rCx43 refers to the connexin protein of 43,000 Da that was originally discovered in rat myocardium.

Because different cells can express different connexins, the possibility of selective gap junctional communication between different cells exists. A cogent example of this condition is the Purkinje fiber conduction system in the heart. Purkinje fibers, which conduct the electrical signal to contract throughout the heart, express mainly Cx40 whereas heart muscle cells express primarily Cx43. Because these connexins do not dock with one another to create patent gap junctions, an electrical signal can be propagated throughout the length of the Purkinje fibers without the unwanted occurrence of premature stimulation of heart muscle. However, some of the cells within the terminal branches of the Purkinje fibers that contact heart muscle cells do express Cx43. This arrangement permits the creation of gap junctions between terminal Purkinje cells and muscle cells and the transfer of the electrical impulse to the surrounding heart muscle cells resulting in the synchronous contraction of the myocardium.

Gap junctions can be regulated by a number of mechanisms, including changes in intracellular calcium and pH and by protein kinase phosphorylation. For example, a decrease in intracellular pH that may occur in ischemic heart muscle cells may produce a closure of gap junctions. Cyclic AMP-dependent protein kinase A generally stimulates phosphorylation of Cx32 and upregulation of gap junctional communication. However, stimulation of protein kinase C by the phorbol ester tumor promoter TPA results in Cx43 phosphorylation and the disruption of gap junction function. Tyrosine protein kinases such as v-Src and the epidermal growth factor receptor also stimulate Cx43 phosphorylation and reduce gap junctional communication.

IV. CELL JUNCTIONS IN HUMAN DISEASE

The fact that cell junctions play numerous important roles in the establishment and maintenance of cellular architecture and homeostasis is reflected in their involvement in several human diseases.

A. Desmosomes and Hemidesmosomes in Human Disease

The human autoimmune skin disease pemphigus is characterized by intradermal blistering and defective keratinocyte adhesion. In pemphigus foliaceous (PF), autoantibodies are directed against the desmosomal transmembrane protein desmoglein1 (dsg1). The lesion in this syndrome is in the upper epidermis where splitting of the cell layers occurs due to insufficient intercellular adhesion. Symptoms include a scaly crusted epidermis, but no oral membrane erosion. Because dsg1 is expressed in the same upper epidermal cell layers, it is believed that the PF autoantibody disrupts dsg1 homophilic binding and consequently abrogates its adhesive function (Fig. 4 and Table I). This notion has been substantiated by the replication of PF in mice by the injection of autoimmune antibody against dsg1. In the case of pemphigus vulgaris (PV), large skin blisters and oral mucosa erosions are evident in patients. The autoimmune antibody in PV has been found to react against desmoglein3, which is localized to the cells of the lower epidermis, including the basal cells. A loss of cells in the basal layer disrupts the barrier function of the epithelium and leads to leakage of interstitial fluid and blister formation.

Bullous pemphigoid (BP) is also an autoimmune disease that is characterized by the presence of skin blisters and oral mucosa lesions in a substantial number of patients. Circulating autoantibodies in these patients react with the BP180 (BPAG2) antigen which is one of the transmembrane proteins localized to the hemidesmosome (Fig. 5 and Table I). Electron microscopy of a patient's epidermis reveals either a complete lack of hemidesmosomes or the presence of morphologically defective ones. Mice injected with antibodies against BP180 develop the same histological lesions observed in the epidermis of patients.

B. Gap Junctions in Human Disease

In Charcot-Marie-Tooth (CMT) disease, an inherited condition characterized by peripheral neuropathy due to demyelination, at least 42 distinct mutations have been discovered throughout the Cx32 gene (point mutations, deletions, and frame shifts). Cx32 has been reported to be expressed in Schwann cells, not at cell–cell contact sites, but uniquely at paranodal locations and Schmidt–Lantermann incisures. This distribution suggests that Cx32 channels may form between myelin turns of the same cell and not between different Schwann cells. Perhaps this arrangement may permit a more rapid diffusion of ions, nutrients, and other small molecules radially through the myelin wrap and sustain the neuron–glial interactions that are necessary for proper maintenance of the myelin sheath.

Elimination of the Cx43 gene in mice by homologous recombination resulted in the homozygous animals surviving to term but dying shortly after birth with labored breathing and cyanosis. The major defect in these animals was enlargement of the right ventricle of the heart where the pulmonary artery exists to the lungs and obstruction of the outflow tract by excess tissue. This defect blocked the pulmonary artery and prevented adequate blood flow to the lungs of these animals. These results suggested that Cx43 was required for the normal development of the mouse heart. Cx43 may also be essential for normal development of the human heart. Patients with visceroatrial heterotaxia, a familial condition characterized by a spectrum of developmental abnormalities of the heart, contain mutations in Cx43 where serine amino acids are substituted by prolines. The response of the mutant Cx43 to protein kinase activation was altered in cells expressing the mutant gene, suggesting that the function of the mutant Cx43 discovered in visceroatrial heterotaxia patients may also be altered.

C. Cell Junctions and Cancer

Another intriguing possible involvement of cell junctions in human disease is in the genesis of human cancer. The cell–cell adhesion molecules ZO-1, ZO-2, and E-cadherin and the gap junction proteins Cx26 and Cx43 are suspected tumor suppressor gene products. The ZO-1 and ZO-2 intracellular proteins of tight junctions have significant sequence and functional similarities to the lethal(1)discs large-1 (*dlg*) tumor suppressor gene of *Drosophila*. The *dlg* gene product is localized to *Drosophila* septate junctions which are thought to be equivalent to vertebrate tight junctions. Mutations of the E-cadherin gene have been detected in about 50% of gastric carcinomas and, in general, a reduction of cadherin-dependent adhesion has been associated with increased invasiveness and metastatic behavior of human tumors. The APC tumor suppressor gene, which is involved in an inherited form of colon carcinoma, familial adenomatous polyposis, is a member of the plakoglobin/β-catenin family. Finally, although connexin gene mutations have not been discovered in human tumors, expression of Cx26 and Cx43 has been found to be deficient in human mammary carcinoma cell lines, suggesting their possible tumor suppressive role. In addition, induction and maintenance of the neoplastically transformed cell phenotype by diverse stimuli (chemical carcinogens, tumor promoters, oncogenes, growth factors) have generally been associated with the disruption of gap junctional communication. [*See* Tumor Suppressor Genes.]

BIBLIOGRAPHY

Alberts, B., Bray, D., Lewis, J., Raff, M., Roberts, K., and Watson, J. D. (1994). "Molecular Biology of the Cell," 3rd Ed. Garland Publishing, New York.

Anderson, J. M., Balda, M. S., and Fanning, A. S. (1993). The structure and regulation of tight junctions. *Curr. Opin. Cell Biol.* **5**, 772.

Bruzzone, R., White, T. W., and Paul, D. L. (1996). Connections with connexins: The molecular basis of direct intercellular signaling. *Eur. J. Biochem.* **238**, 1.

Citi, S. (1993). The molecular organization of tight junctions. *J. Cell Biol.* **121**, 485.

Garrod, D. R. (1993). Desmosomes and hemidesmosomes. *Curr. Opin. Cell Biol.* **5**, 30.

Goodenough, D. A., Goliger, J. A., and Paul, D. L. (1996). Connexins, connexons, and intercellular communication. *Annu. Rev. Biochem.* **65**, 475.

Green, K. J., and Jones, J. C. R. (1996). Desmosomes and hemidesmosomes: Structure and function of molecular components. *FASEB J.* **10**, 871.

Gumbiner, B. M. (1996). Cell adhesion: The molecular basis of tissue architecture and morphogenesis. *Cell* **84**, 345.

Lodish, H., Baltimore, D., Berk, A., Zipursky, S. J., Matsudaira, P., and Darnell, J. (1995). "Molecular Cell Biology," 3rd Ed. Freeman, New York.

Stanley, J. R. (1993). Cell adhesion molecules as targets of autoantibodies in pemphigus and pemphigoid, bullous diseases due to defective epidermal cell adhesion. *Adv. Immunol.* **53**, 291.

Yamasaki, H., and Naus, C. C. G. (1996). Role of connexin genes in growth control. *Carcinogenesis* **17**, 1199.

Cell Membrane

ALAN F. HORWITZ
University of Illinois at Urbana–Champaign

GLOSSARY

Amphipathic Term used to describe molecules that have both polar and apolar domains

Diffusion Motion of a molecule arising from thermal energy (motions) and random encounters with neighbors. The result is described by a statistical quantity: the mean squared distance traveled in a given time

Hydropathy plot Used to predict membrane-spanning regions of integral membrane proteins from their primary sequence. A measure of the hydrophobicity of each amino acid is plotted against the position in the amino acid sequence

Hydrophilic Term used to describe polar molecules, that is, those that seek out and are soluble in aqueous, or polar, environments

Hydrophobic Term used to describe apolar molecules, that is, those that avoid and are insoluble in aqueous, or polar, environments

Hydrophobic effect Tendency of hydrophobic molecules to avoid contact with water and to interact with other hydrophobic molecules

Lipid bilayer Structure formed by phospholipids and other closely related molecules. Two monolayers of lipid face each other with the polar regions facing the aqueous phase and the apolar, fatty acyl chains facing each other to create a hydrophobic interior

Membrane fluidity Refers to the dynamic nature of cell membranes. In a fluid membrane, constituents execute rapid lateral and rotational diffusion

WHY MEMBRANES? CELL MEMBRANES FUNCTION primarily as barriers that separate either the cell interior from its external environment or a compartment within the cell from the surrounding cytosol. To understand membranes, one must first appreciate that the cell must concentrate and organize molecules that promote growth, survival, and its general well-being and eliminate those that do not. Thus the cell and its interior compartments are not in equilibrium; rather they are containers filled with segregated molecular components. In this light, cell membranes play a key role in organizing the cellular interior. They serve as a permeability barrier that facilitates the cell's task to concentrate and segregate molecules that support growth and survival, segregate and eliminate those that are toxic, and facilitate cellular processes.

The barrier function of the plasma membrane, the limiting membrane that surrounds the entire cell and separates it from the extracellular environment, dictates many of the activities associated with it. The cell requires mechanisms to regulate the entry and exit of components, both large and small. Thus transport proteins that mediate the accumulation of nutrients and ions, for example, are among membrane constituents. Membrane vesicles pinch off to allow entry of larger components (endocytosis). Vesicles inside the cell fuse to the cell surface to allow exit of components (exocytosis) and delivery of new membrane. Membranes also serve as conduits for "social" communication between the cell's exterior and interior. Thus hormone and growth factor receptors transmit signals from agents that reside outside of the cell into characteristic cellular responses that include proliferation, gene expression, cytoskeletal organization and migration, and differentiation. Many cells are also equipped for direct communication with neighboring cells. Examples include gap junctions, which allow small molecules to flow between cells, and the localized release of small molecules like neural transmitters, which

ENCYCLOPEDIA OF HUMAN BIOLOGY, Second Edition, VOLUME 2. Copyright © 1997 by Academic Press. All rights of reproduction in any form reserved.

propagate electrical activity between cells like nerve and muscle. [*See* Cell Membrane Transport.]

Membranes also play a structural role by resisting the forces to which cells are exposed, thus keeping the cell intact. In this role it is often assisted by membrane-associated elements including the cell wall and cytoskeleton.

In addition to the plasma membrane, which encompasses the entire cell and separates the cytoplasm from the surrounding, the interior of eukaryotic cells also has a variety of different membranes that function to compartmentalize the cytoplasm. This compartmentalization greatly enhances the efficiency of many cellular processes and facilitates trafficking to their appropriate target residences within the cell. For example, enzymes that degrade proteins are found in lysosomes, whereas much of the protein synthetic apparatus is found on the endoplasmic reticulum.

I. FLUID MOSAIC MODEL AND ITS ORIGINS

An early description of a plasma membrane as a structure enclosing cells appears early in the nineteenth century. Microscopic observations on contracted muscle cells show a residual structure at the ends of the cell, where the contractile apparatus was no longer present. In the early twentieth century, measurements of the surface area occupied by the lipids extracted from the red cell membrane suggested that there was sufficient lipid to encircle the cell twice. From this observation, the notion of a lipid bilayer emerged. Near the middle of this century, electron microscopic observations revealed a common image, which suggested similar structural features among various membranes. Biochemists found that membrane proteins required detergents for extraction and were insoluble in aqueous media. This pointed to some membrane proteins, at least, as different from soluble proteins and interacting with the hydrophobic milieu of the bilayer rather than binding to the bilayer surface via electrostatic association. In the late 1960s, physical measurements on lipid bilayers showed that membrane lipids were in a fluid (melted) state, that is, they display considerable dynamics in the membrane plane. Membrane proteins were also shown to diffuse and rotate rapidly. And finally, studies on the primary structure (sequence) of membrane proteins showed that they are modular with a distinct membrane-spanning, hydrophobic segment and cytoplasmic and extracellular domains.

All of this was synthesized in the fluid mosaic model of membrane structure, which appeared in the early 1970s. The key tenets of the model are that the fluid lipid bilayer is the key structural feature of membranes and that membrane proteins are free to diffuse in this bilayer. An important concept from the model was the shift from viewing membranes as a dynamic rather than static structure. It also provided a physical-chemical rational for membrane structure based on the hydrophobic effect, that is, the tendency of hydrophobic regions to segregate and associate with each other and for hydrophilic regions to be exposed to water. This model remains the most useful conceptualization of membrane structure. Though it anticipated that some membrane components might not move, it is now clear that many, if not most, membrane protein components show constrained mobility.

II. MEMBRANE COMPONENTS

What are the major components of membranes and do they vary among different membranes? Answers to these questions require methods for the purification of different membranes. This is relatively easy for most bacteria and the red cell, for example, since they have only a single membrane, the plasma membrane. Purification of membranes from higher eukaryotic cells presents a more daunting problem since they also have internal membranes (e.g., the endoplasmic reticulum, Golgi apparatus, mitochondria). However, owing to their differing functions, these membranes tend to have different components that produce membranes with differing densities. Thus centrifugation through a medium of varying density (e.g., a sucrose density gradient) often affects a reasonable purification of different membranes. Antibodies that identify molecules unique to specific membranes are also used, with good success, to purify different cellular membranes.

Proteins and lipids (including their glycosylated derivatives) are the major constituents of essentially all membranes (Fig. 1). As indicated earlier, the ratio of lipid to protein varies considerably among different organisms and membranes. This reflects their differing functions. Mitochondria, for example, efficiently produce energy and contain many different proteins; they have a relatively high protein-to-lipid ratio. Myelin, on the other hand, functions primarily to insulate nerves and is not highly metabolically active. It has a low protein-to-lipid ratio.

The lipid composition of different membranes varies considerably. Although cholesterol and phos-

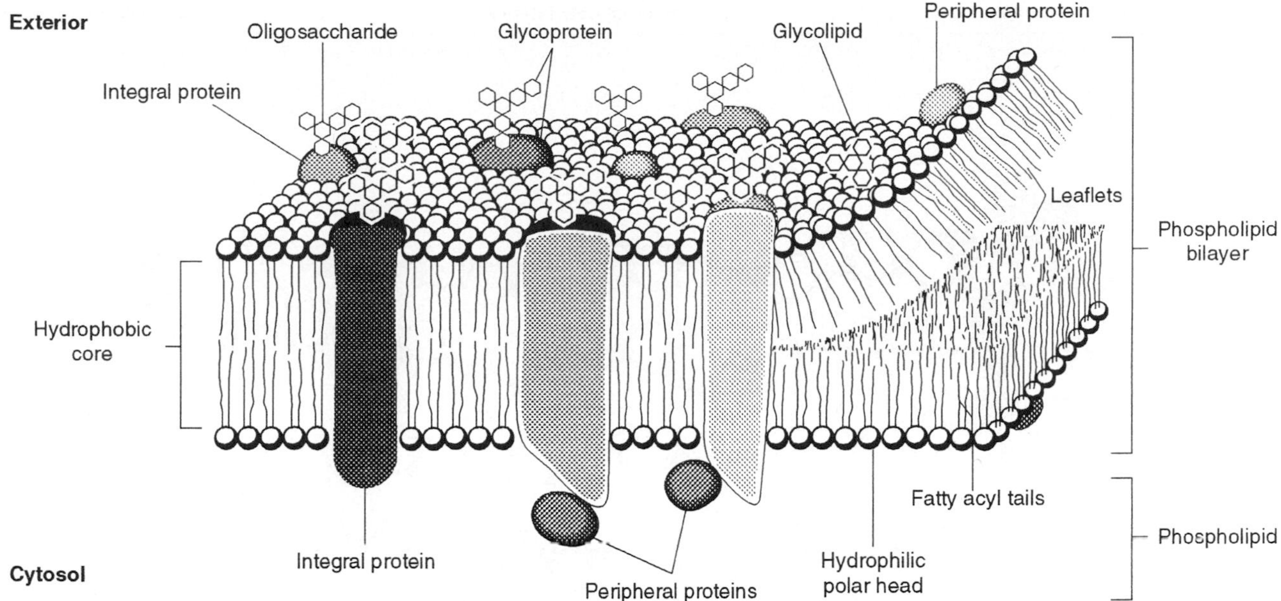

FIGURE 1 Schematic diagram of a typical biological membrane. The phospholipid bilayer, the basic structure of all cellular membranes, consists of two leaflets of phospholipid molecules whose fatty acyl tails form the hydrophobic interior of the bilayer; their polar, hydrophilic head groups line both surfaces. Integral proteins have one or more regions embedded in the lipid bilayer; most span the bilayer as shown. Peripheral proteins are primarily associated with the membrane by specific protein–protein interactions. Oligosaccharides bind mainly to membrane proteins; however, some bind to lipids, forming glycolipids. [From H. Lodish *et al.* (1995), Fig. 14-1.]

phatidylcholine are prominent lipids in animal cells, they are absent in bacteria. Plant cells tend to have relatively large amounts of galactolipids, which are not abundant in animal cells. Even within a given cell, the different membranes have varying compositions. Mitochondria, for example, are highly enriched in diphosphatidylglycerol when compared to the plasma membrane.

Membrane protein composition also varies considerably. This follows from the function-determining role of proteins. For example, the major function of the sarcoplasmic reticulum in muscle is the accumulation and release of calcium. Therefore it has relatively few proteins, and its major components function in calcium fluxes. The bacterial membrane has many activities that reflect its diverse functions and environments. Hence, its protein composition is relatively diverse.

III. LIPID BILAYERS

The lipid bilayer is the major structural feature of membranes (see Fig. 1). It is the matrix in and on which other membrane components reside. It is also the primary permeability barrier. Phospholipids, and some structurally related lipids, spontaneously form bilayers in aqueous solutions. This property of phospholipids follows directly from their modular, amphipathic nature and the hydrophobic effect. These lipids have structurally separated apolar (hydrophobic) and polar (hydrophilic) domains (Fig. 2). The polar region includes the phosphoryl moiety and the polar headgroup to which it is esterified. The fatty acyl chains comprise the apolar region. In bilayers, the polar groups face the water and the apolar fatty acyl chains face each other in the membrane interior. This structure minimizes contact of the apolar, or hydrophobic, regions with water. It is this tendency of apolar regions to avoid water and associate with one another that provides the driving force for bilayer structure (the hydrophobic effect). The intermolecular interactions among the fatty acyl chains in the hydrophobic interior are weak. Thus, the hydrophobic interior is fluid, that is, dynamic. The lipid molecules in bilayers diffuse; that is they exchange places (translate) and rotate rapidly.

Despite their dynamic nature, there is increasing evidence that membrane lipids are not distributed randomly on the cell surface. Instead, some reside in domains, or patches, where they likely associate with some activity that either requires or is activated by

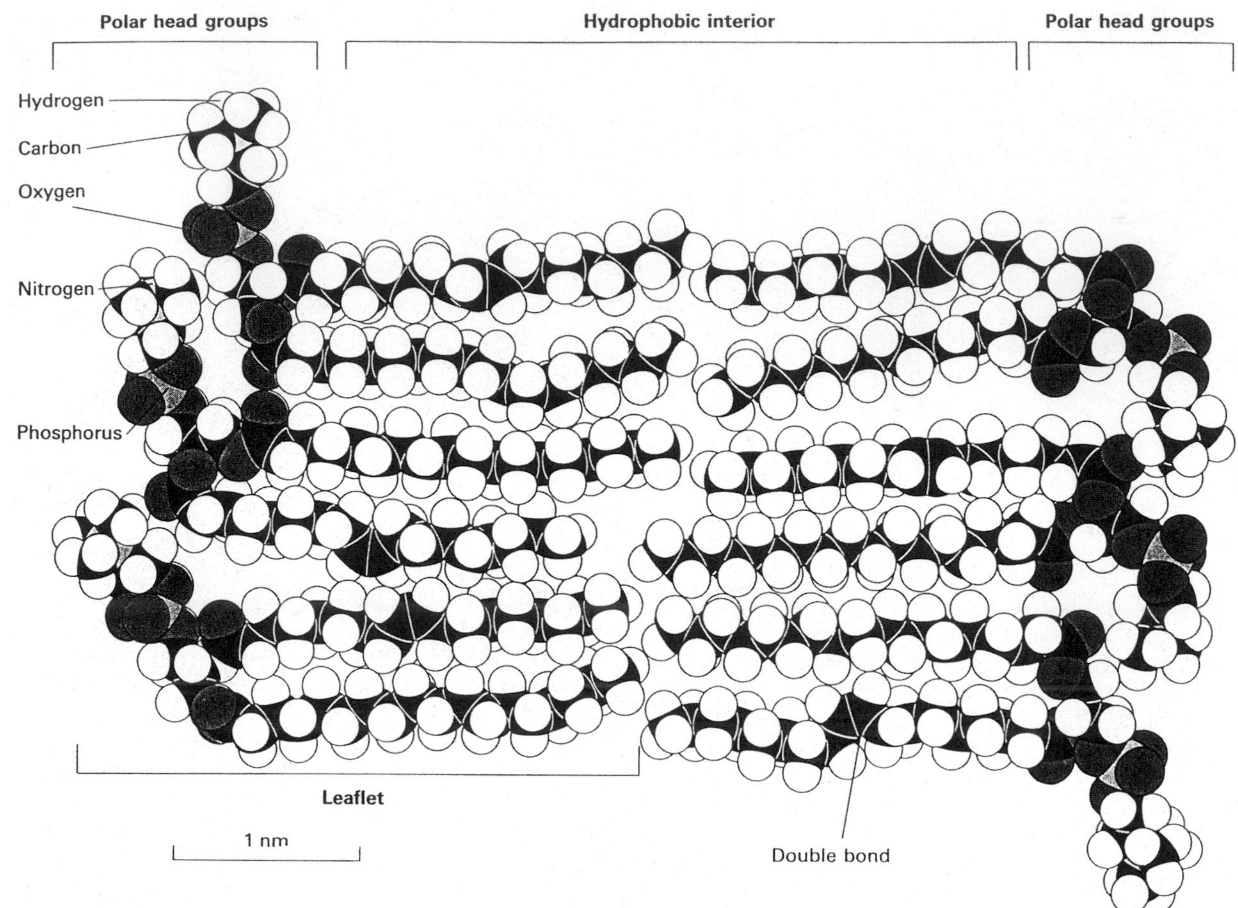

FIGURE 2 A space-filling model of a typical phospholipid bilayer. The hydrophobic interior is generated by the fatty acyl side chains. Some of these chains have bends caused by double bonds. The different polar headgroups all lie on the outer, aqueous surface of the bilayer. [From L. Stryer (1988). "Biochemistry," 3rd Ed., p. 289. W. H. Freeman and Company. Courtesy of L. Stryer.]

them. On some cells the patches are small, whereas on others, like epithelia or sperm, they are extensive and occupy an entire membrane region, for example, the epithelial basal lateral surface.

In contrast to their rapid diffusive motion, lipids seldom flip from one side of the bilayer to the other since this would require that the polar headgroup pass through the hydrophobic interior of the bilayer—an energetically unfavorable event. This allows the bilayer halves to be populated by different lipids. Such asymmetries are commonly observed. In the red cell, for example, the majority of the phosphatidylcholine and sphingomyelin reside on the extracellular face of the bilayer, whereas phosphatidylethanolamine and phosphatidylserine face the cytoplasm. In addition, glycolipids, which are present in most plasma membranes, reside exclusively on the extracellular leaflet.

It is not clear how these lipid asymmetries are generated: however, flipases, enzymes that catalyze the transbilayer migration of lipids, appear to be involved. The targeting of glycolipids to the outer leaflet is dictated by the addition of carbohydrates in the lumen of the Golgi apparatus. Though the existence of lipid asymmetry is well documented, its function is less clear. It does seem likely, however, that asymmetric bilayers serve to position certain lipid classes more appropriate to the activities and functions with which they associate (see the following).

If the only role of lipids were to form a bilayer structure, one would not expect the large variety of different lipids that are observed in nature. Although some of the diversity undoubtedly arises from evolutionary redundancy, there are other reasons as well. The lipid bilayer of membranes needs to be fluid at physiologic temperatures. It is well known, for example, that shorter chains and more unsaturated fatty acids produce bilayers that are fluid at lower temperatures. In bacteria, the growth temperature can vary considerably; the lipid composition changes as the growth temperature changes to produce a fluid bilayer at the growth temperature. In higher eukaryotic organisms, like mammals, the fatty acid composition is dictated in part by dietary fatty acids. One role that cholesterol plays in these organisms is to buffer the melting transition of bilayer lipids and thus keep them in a fluid state despite wide variations in their fatty acid composition. Some lipids play regulatory, or signaling, roles. Phosphatidylinositol, for example, can be metabolized to form important mediators that function in intracellular regulation of many key intracellular processes. Certain long-chain, polyunsaturated fatty acids can be metabolized to form eicosanoids, like the prostaglandins, which function as hormones. Some phospholipids, like phosphatidylserine, interact with specific enzymes (e.g., protein kinase C) that enhance their enzymatic activity. And finally, glycolipids appear to facilitate adhesion, among other less well characterized activities. [*See* Lipids.]

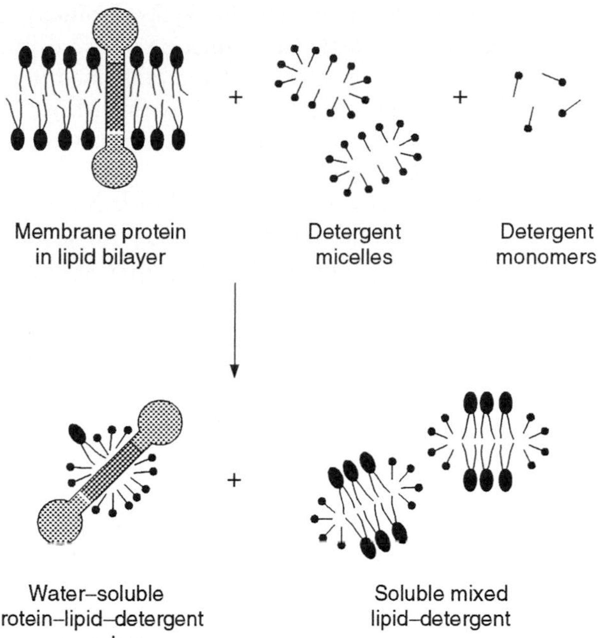

FIGURE 3 Solubilizing membrane proteins with a mild detergent. The detergent disrupts the lipid bilayer and brings the proteins into solution as protein–lipid–detergent complexes. The phospholipids in the membrane are also solubilized by the detergent. [From B. Alberts *et al.* (1994), Fig. 10-19.]

IV. MEMBRANE PROTEINS

Membrane proteins are typically characterized by the method used to extract them from membranes. Using this criterion, they fall into two general classes: peripheral and integral. Peripheral membrane proteins are only weakly associated, via ionic interactions, with the bilayer surface or another membrane protein (see Fig. 1). They can be extracted using techniques commonly employed for the isolation of soluble proteins. These include high salt concentrations (ionic strength), extremes of pH, and chelators of divalent metal ions like EDTA. Integral membrane proteins, in contrast, have domains that reside within the hydrophobic core of the bilayer and require detergents for solubilization. Detergents are also amphipathic molecules, that is, they have two regions: polar and apolar. However, their geometry differs from that of phospholipids. Therefore they form micelles, small, closed monolayer vesicles, rather than bilayer vesicles. They break up the bilayer structure and solubilize membrane proteins by replacing neighboring lipids with detergent molecules (Fig. 3). The result is a soluble, mixed micelle containing detergent, lipid, and protein. The detergent keeps the membrane proteins soluble by interacting with their hydrophobic, membrane-spanning regions. Nonionic detergents, like Triton X-100, are among a class generally classified as mild detergents; they retain most biological activities. Stronger detergents, like sodium dodecylsulfate (SDS), not only solubilize but also denature proteins by unfolding the hydrophobic core.

Integral membrane proteins usually have one or more 20- to 24-amino-acid stretches consisting almost exclusively of hydrophobic amino acids. This is the number of amino acids required to form a bilayer-spanning α helix. The α-helical structure presents the hydrophobic amino acid side chains to the hydrophobic interior of the bilayer while keeping the polar (amide) part of the amino acid away from it. Many membrane proteins span the membrane several times; this is a common feature of transport and other proteins that communicate between the outside and inside of the membrane (membrane proteins that span only once usually mediate such transmembrane communication by forming dimers). Proteins that do span more than once often form what is termed a helical hairpin. They pass through the membrane, make a tight turn, and then pass through the membrane again. These membrane-spanning sequences are usually evident in hydropathy plots, in which the hydrophobicity of each amino acid is plotted as a function of its position in the primary sequence. Thus, like lipids, membrane proteins have a modular structure with distinct hydrophobic domains as well as domains that have hydrophilic surfaces (Fig. 4).

The extracellular domain of membrane proteins is usually glycosylated on asparagine or serine residues (see Fig. 1). These carbohydrate chains vary somewhat in composition. They are added posttranslationally in the endoplasmic reticulum or the Golgi apparatus. The function of carbohydrates is not entirely clear. Since they are very hydrophilic and often charged, it is likely that they protect the protein and keep it from interacting nonspecifically. Some carbohydrate sequences mediate cell–cell adhesion, which is particularly important in the inflammatory response in which leukocytes interact with the endothelium via a class of adhesion molecules called selectins.

The cytoplasmic regions of membrane proteins tend to be highly conserved among different species. Some have enzymatic activity, often a protein kinase, which mediates signal transduction phenomena. Others contain sequences that serve as ligands that interact with intracellular molecules or serve to direct the protein to a particular compartment or membrane region. The latter is seen on epithelial cells, for example, where a specific amino acid sequence(s) targets certain proteins to the apical membrane domain.

There is a class of membrane proteins that are more usefully classified by their molecular properties rather than their extractability. These proteins are anchored to the membrane by covalently attached lipids, lipid anchors, rather than sequences of membrane-spanning amino acids (see Fig. 4). Three kinds of lipid anchors have been described thus far: fatty acids, like myristate and palmitate, phosphatidylinositol, and prenyl groups. Proteins anchored by phosphatidylinositol reside on the extracellular surface of the bilayer, whereas the other two, which comprise many regulatory (signal transduction) molecules and oncogenes, reside on the cytoplasmic surface. Regulatory molecules and oncogenes require this membrane anchorage to be active.

Few high-resolution structures of membrane proteins have been reported. The challenge, in large part, is due to their insolubility in aqueous solutions. The detergents required for their solubility interfere with crystal formation, which is required for the X-ray diffraction techniques used to determine protein structures. However, low-resolution structures derived from electron diffraction analyses of two-dimensional membrane protein arrays (e.g., gap junctions or the purple membrane from *Halobacterium halobium*) have provided important information on the orientation of membrane-spanning domains. Recently, novel detergents have been designed that are compatible with crystal formation. Using these detergents, some

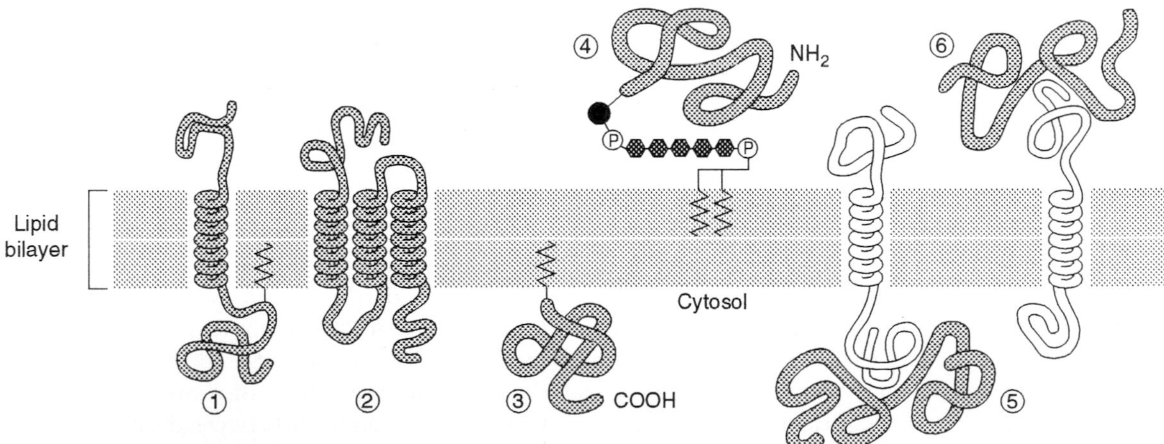

FIGURE 4 Six ways in which membrane proteins associate with the lipid bilayer. Most transmembrane proteins are thought to extend across the bilayer as a single α helix (1) or as multiple α helices (2); some of these "single-pass" and "multipass" proteins have a covalently attached fatty acid chain inserted in the cytoplasmic monolayer (1). Other membrane proteins are attached to the bilayer solely by a covalently attached lipid—either a fatty acid chain or prenyl group—in the cytoplasmic monolayer (3) or, less often, via an oligosaccharide, to a minor phospholipid, phosphatidylinositol, in the noncytoplasmic monolayer (4). Finally, many proteins are attached to the membrane only by noncovalent interactions with other membrane proteins (5) and (6). [From B. Alberts *et al.* (1994), Fig. 10-13.]

high-resolution membrane protein structures have been determined. Finally, crystal structures of the active regions (domains) from membrane proteins, which lack the bilayer-spanning regions, are being reported more often.

A general structural feature of membrane proteins is that the membrane-spanning helices tend to be parallel to each other but perpendicular to the membrane plane (Fig. 5). For molecules that function as channels, several helices bundle together to form a pore, which resides in the center of the bundle. The pore is lined with some polar residues to facilitate the transport of polar molecules. Relatively subtle changes in the relative orientation of the membrane-spanning helical bundles appear to regulate channel function. [*See* Membranes, Biological; Protein Targeting, Basic Concepts.]

V. MEMBRANE DYNAMICS

The diffusion of many membrane proteins has been measured in a variety of ways. The evidence is overwhelming that they execute both rotational and translational diffusion. The proteins (and lipids) sense an effective lipid bilayer viscosity that is about 100 times that of water. It is similar to that of a light machine oil, like that used for bicycles. The rates of diffusive motion in the membrane plane can be described by the equation $\langle y^2 \rangle = 4Dt$, where $\langle y^2 \rangle$ is the mean squared distance diffused, D is the diffusion coefficient, and t is the time. Note that the distance diffused has a square root dependence on time. A protein like rhodopsin, a visual protein, moves laterally about 1 μm in the membrane plane in a second. It rotates roughly 20 times in a microsecond.

As outlined earlier, the rapid diffusive motion follows from the relatively weak intermolecular interactions between the hydrophobic molecules in the membrane bilayer. Recall that it is the avoidance of water rather than the strong attractive forces between hydrophobic residues that drives bilayer structure. Thus, the intermolecular interactions between the fatty acyl residues on one lipid and the exposed hydrophobic residues of adjacent membrane lipids and proteins are weak. This results in the dynamic nature of both membrane lipids and proteins.

Although some membrane proteins do execute rapid diffusion, most show anomalous diffusion rates.

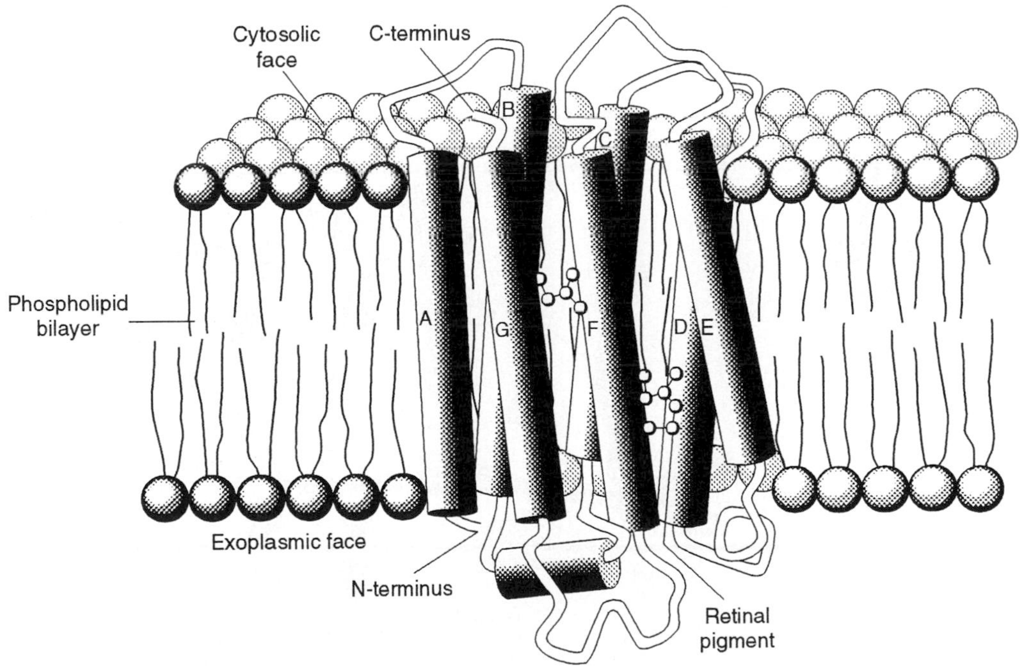

FIGURE 5 Overall structure of bacteriorhodopsin as deduced from electron diffraction analyses of two-dimensional crystals of the protein in the bacterial membrane. The seven membrane-spanning α helices are labeled A–G. The retinal pigment is covalently attached to lysine 216 in helix G. The approximate position of the protein in the phospholipid bilayer is indicated. [Adapted from R. Henderson *et al.* (1990). *J. Mol. Biol.* **213,** 899.]

That is, for a given protein, a substantial fraction do not diffuse, and those that do diffuse exhibit rates that are much slower than expected. This anomaly is not due to intrinsic properties, like size, of different membrane proteins. Theory predicts that the size of the membrane protein contributes only very modestly to the diffusion rate. Thus other explanations are more likely. For example, the energy-transducing, purple membrane of the bacterium *Halobacterium halobium* is a highly aggregated patch of bacteriorhodopsin, an integral membrane protein. Coated pits, which are involved in internalization of some receptors, are also large clusters of membrane proteins. Linkage to a cytoskeletal component like actin serves to tether proteins to an immobile skeleton (Fig. 6). Many membrane proteins are so anchored, including the anion transport molecule in the red cell. Finally, some membrane proteins are anchored to extracellular structures. These include other cells, as seen for gap junctions, which mediate intercellular communication, or connective tissues (extracellular matrices), as seen with cell-surface receptors for extracellular molecules like fibronectin.

Why do those membrane proteins that do diffuse exhibit reduced diffusion rates? There appear to be several reasons. One is that many membrane proteins reside in cytoskeletal corrals, or fences, with dimensions in the range of a fraction of a micron. These cytoskeletal domains are dynamic and confine the proteins transiently (about 20 sec). Other membrane proteins associate transiently with cytoskeletal elements. Finally, some proteins associate with motor proteins and are transported quickly to specific cellular loci like the cell edge.

The fluid nature of membranes is required for many membrane functions. Growth factor-initiated signal transduction often requires the ligand-induced clustering of receptors for their activation, which occurs via cross-phosphorylation. Likewise the clusters of receptors found at sites of adhesion and other cellular junctions appear to concentrate from a diffusion-trap mechanism. Finally, a fluid bilayer is required for processes like endocytosis and membrane fusion, as well as for properties like deformability.

VI. MEMBRANE LINKAGES

The cytoskeletal and extracellular linkages with the plasma membrane mentioned earlier are often elabo-

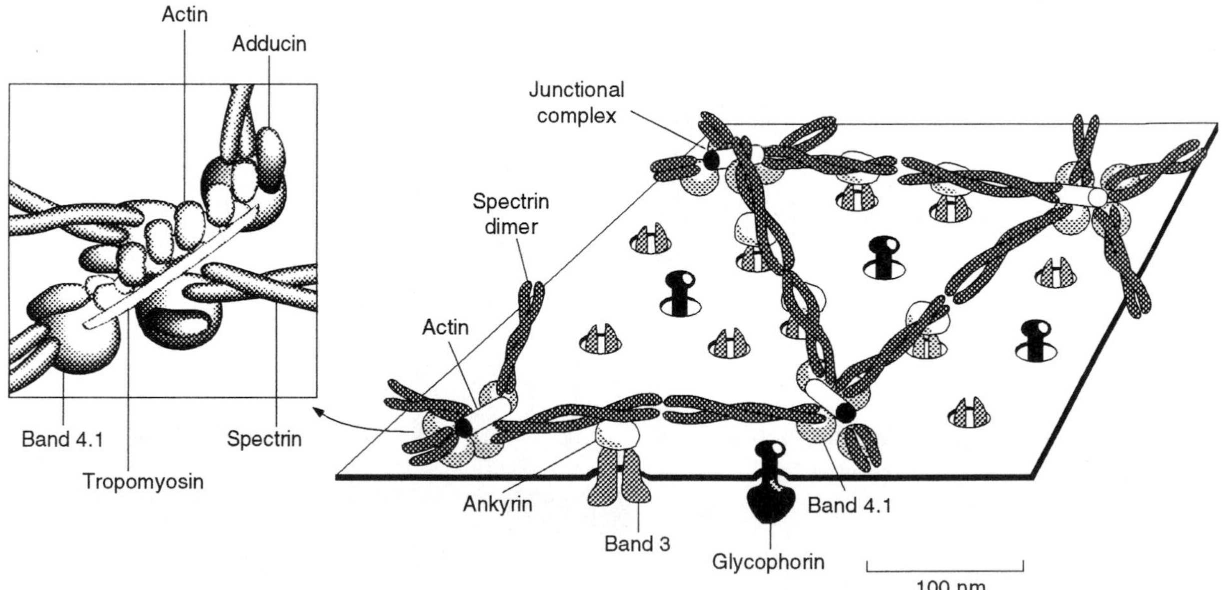

FIGURE 6 The spectrin-based cytoskeleton on the cytoplasmic surface of the human red blood cell membrane. Spectrin dimers associate head-to-head to form tetramers that are linked together into a net-like meshwork by junctional complexes composed of short actin filaments, tropomyosin, band 4.1, and adducin (enlarged in the box on the left). The cytoskeleton is linked to the membrane by the indirect binding of spectrin tetramers to some band 3 proteins via ankyrin molecules. [Right image courtesy of T. Byers and Branton (1985). *Proc. Natl. Acad. Sci. U.S.A.* **82,** 6153–6157.]

rate and serve important structural and regulatory roles. The molecular interactions of several have been described. The linkage between the red cell anion channel, band 3, and the cytoskeleton was the first linkage described in molecular detail and is among the best understood (see Fig. 6). The anion channel binds, via its cytoplasmic domain, to ankyrin, which in turn interacts with spectrin, which is a part of the actin cytoskeleton in red cells. Mutants that lack molecules in the red cell cytoskeleton show dramatic phenotypes—the cells lose their biconcave shape. In contrast to normal red cells, all of the anion channels diffuse at the expected rate for unrestrained diffusion. Though described initially for red cells, homologues of both ankyrin and spectrin are present in other cells and appear to serve analogous linkage functions.

Another linkage is that between the cell exterior and the cytoskeleton. Some members of the integrin family of cell-surface receptors mediate such a linkage. These integral membrane proteins bind to components of the extracellular matrix (connective tissues) like fibronectin, collagen, and laminin. These extracellular molecules are part of a complex, highly organized fibrillar matrix. The integrin cytoplasmic domains bind to cytoskeletally associated molecules. Thus they form structural linkages between the cell exterior and interior. These linkages are highly organized. In addition to structural molecules, they also contain many regulatory molecules that mediate proliferation, differentiation, gene expression, cytoskeletal organization, and so on. Therefore, these linkages serve as signaling centers as well as structural supports. Connections between cells (cell–cell adhesions) are also highly organized. Epithelial cells, for example, have many different junctions (Fig. 7). The tight junction serves to separate the cell surface into two distinct domains—the apical and basolateral—and renders the connections between the cell impermeable. Other junctions, like the desmosome, are largely structural. The junction between nerve and muscle, the

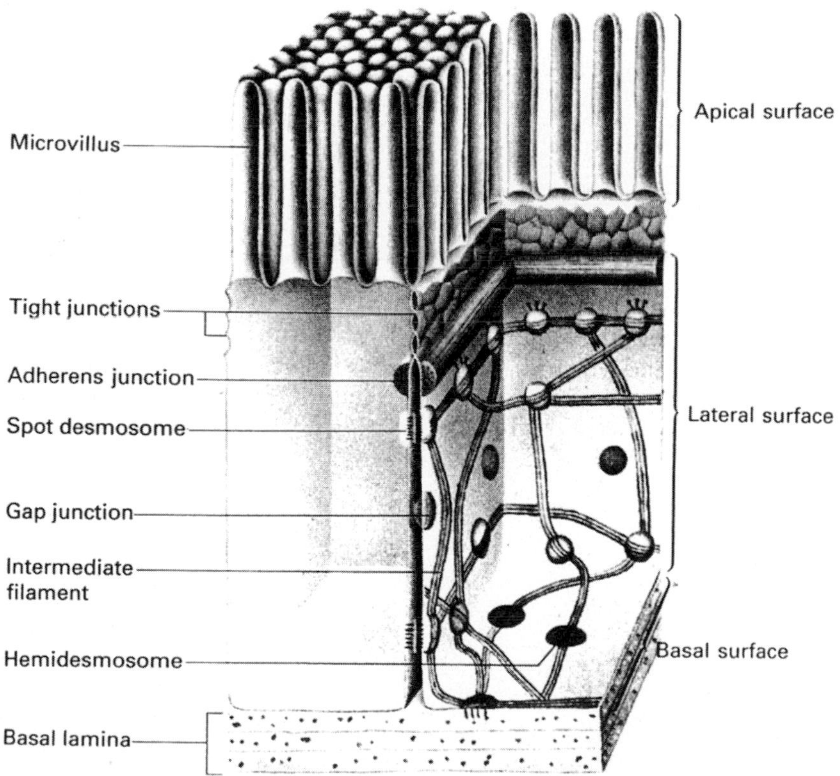

FIGURE 7 Schematic diagram of intestinal epithelial cells and principal types of cell junctions that connect them. The basal surface of the cells rests on a fibrous network of collagen and proteoglycans (basal lamina), which supports the epithelial cell layer. The apical surface faces the intestinal lumen. [From B. Alberts *et al.* (1994), Fig. 14-42.]

neuromuscular junction, is also highly organized with many different molecules, including some specific to that junction. In addition to holding the nerve and the muscle together, it mediates the transmission of chemical stimuli from the nerve to the muscle.

BIBLIOGRAPHY

Alberts, B., Bray, D., Lewis, J., Raff, M., Roberts, K., and Watson, J. D. (1994). "Molecular Biology of the Cell," 3rd Ed., Chap. 10, pp. 476–506. Garland Press, New York/London.

Bennet, V., and Gilligan, D. M. (1993). The spectrin based membrane skeleton and micro scale organization of the plasma membrane. *Annu. Rev. Cell Biol.* **9**, 27–66.

Chow, M., Der, C. J., and Buss, J. E. (1992). Structure and biological effects of lipid modifications on proteins. *Curr. Opin. Cell Biol.* **4**, 629–636.

Devaux, P. F. (1993). Lipid transmembrane asymmetry and flip-flop in biological membranes and in lipid bilayers. *Curr. Opin. Cell Biol.* **3**, 489–494.

Glaser, M. (1993). Lipid domains in biological membranes. *Curr. Biol.* **3**, 475–481.

Kusumi, A., and Sako, Y. (1996). Cell surface organization by the membrane skeleton. *Curr. Opin. Cell Biol.* **8**, 566–574.

Lodish, H., Baltimore, D., Berk, A., Zipursky, S. L., Matsudaira, P., and Darnell, J. (1995). "Molecular Cell Biology," 3rd Ed., Chap. 14, pp. 595–631. Scientific American Books, New York.

Nelson, W. J. (1992). Regulation of cell surface polarity from bacteria to mammals. *Science* **258**, 948–955.

Sheets, E. D., Simson, R., and Jacobson, K. (1995). New insights into membrane dynamics from the analysis of cell surface interactions by physical methods. *Curr. Opin. Cell Biol* **7**, 707–714.

Silvius, J. R. (1992). Solubilization and functional reconstitution of biomembrane components. *Annu. Rev. Biophys. Biomol. Struct.* **21**, 323–348.

Singer, S. J. (1990). The structure and insertion of integral membrane proteins. *Annu. Rev. Cell Biol.* **6**, 1–39.

Yeagle, Y. P. (1993). "The Membranes of Cells," 2nd Ed. Academic Press, San Diego.

Cell Membrane Transport

ANTHONY N. MARTONOSI
State University of New York, Health Science Center at Syracuse

HIROSHI NAKAMURA
Osaka Prefectural College of Health Sciences

GLOSSARY

Active transport Carrier-mediated transport of an ion or molecule against its electrochemical gradient driven by an energy source, usually ATP

Carrier Transmembrane protein capable of interaction with the transported ligand and with an appropriate energy source facilitating its transport across the membrane

Channel Ligand or voltage-gated selective pore in the membrane that permits the translocation of ions or molecules down their electrochemical gradients

Passive transport or facilitated diffusion Carrier-mediated transport of an ion or molecule down its electrochemical gradient

Proton-motive force Electrochemical potential of protons consisting of the H^+ gradient (ΔpH) and the transmembrane electrical potential ($\Delta\psi$)

Secondary active transport Carrier-mediated transport of an ion or molecule against its electrochemical gradient utilizing the downhill transport of an ion or molecule as the source of energy

BIOLOGICAL MEMBRANES SEPARATE CELLS FROM their environment and subdivide the cell interior into metabolic compartments (nuclei, mitochondria, endoplasmic reticulum, Golgi apparatus, lysosomes, peroxisomes, secretory vesicles, etc.) that perform unique but interrelated functions. The communication between cells, and between the metabolic compartments within the cells, is achieved by exchange of ions, metabolites, nucleic acids, and proteins through transport systems of varying complexity and specificity located in the surface and intracellular membranes.

I. GENERAL PROPERTIES OF MEMBRANE TRANSPORT SYSTEMS

The membrane transport systems consist of membrane-spanning proteins that are anchored in the lipid bilayer phase of the membrane by one or more α-helical stretches consisting of 20–25 hydrophobic amino acids; the rest of the molecule is exposed to the water phase on both sides of the membrane. The membrane-spanning segments usually form a transmembrane channel for the passage of specific ions or molecules, whereas the exposed hydrophilic domains contain the catalytic or regulatory sites that determine

ENCYCLOPEDIA OF HUMAN BIOLOGY, Second Edition, VOLUME 2. Copyright © 1997 by Academic Press. All rights of reproduction in any form reserved.

the rate, direction, and specificity of the transport process. Crystallographic and spectroscopic techniques are beginning to yield detailed information about the three-dimensional structure of several transport proteins that aid the formulation of the molecular mechanism of the transport process.

The simplest case is a *hydrophilic pore or channel* across the membrane that opens in response to the binding of a ligand or a change in membrane potential, allowing the movement of ions or molecules down their electrochemical gradient. The flux through the channels is determined by the electrochemical gradient of the transported ion or molecule and by the dimensions and internal structure of the channel. The "gap junctions" between cells are wide pores that permit free passage of a variety of molecules up to ~1000 molecular weight from one cell into the other, whereas the channels for Na^+, K^+, and Ca^{2+} are tailored for a specific ion.

The transporters involved in *passive facilitated diffusion*, or in *active transport,* interact with the transported ion or molecule on one side of the membrane in one conformation (E_1) followed by a change in conformation (E_2) releasing the transported species on the other side. The transporter may return to the E_1 state unloaded (net transport) or carrying a second substrate (exchange). In passive transport the driving force for the movement is the electrochemical gradient of the transported species and no additional energy source is required. Examples of such passive transport are the glucose and anion transporters of erythrocytes.

In active transport the movement occurs against an electrochemical gradient and therefore requires some energy source. In the ion-motive ATPases involved in Ca^{2+}, Na^+, K^+, and H^+ transport, the energy source is the hydrolysis of ATP, whereas the *Escherichia coli* phosphotransferase system uses phosphoenolpyruvate for active sugar transport. A number of transporters use ion gradients as energy sources. For example, the gradient of Na^+ ions across the cell surface provides energy for the active transport of amino acids and sugars in the intestine and for the active transport of Ca^{2+} from the heart muscle into the extracellular medium.

Complex proton pumping assemblies in the chloroplast membranes and in some bacteria use light energy to generate proton-motive force that serves as the energy source for the synthesis of ATP. An analogous process in the mitochondria converts the energy gained from oxidation of fuel molecules into proton-motive force and finally into ATP. Recent structural studies provide a fascinating insight into the molecular mechanism of these processes.

In this article we survey the role of transport processes from the capture of light energy in photosynthesis through the production and utilization of ATP, to the communication between cells and subcellular compartments by a complex network of passive and active transport systems.

II. CONVERSION OF SOLAR ENERGY INTO "PROTON-MOTIVE FORCE"

A. Photosynthesis

Energy consumed in biological processes originates from solar energy trapped by photosynthesis. The first step in this process is the absorption of light by chlorophyll. Photons absorbed by several chlorophyll molecules are channeled into the photosynthetic reaction center, where the energy of excited electrons is transformed into chemical energy. In green plants this transformation is mediated by two kinds of photosystems. Photosystem I generates reducing power in the form of reduced nicotinamide adenine dinucleotide phosphate (NADPH), whereas photosystem II transfers electrons from water and produces O_2. Electron flow within each system and from photosystem II to I generates a transmembrane proton gradient that drives the synthesis of ATP through an ATP synthase, coupling proton flux to ATP formation.

A landmark achievement was the solution of the structure of the photosynthetic reaction center from the purple bacterium (*Rhodopseudomonas viridis*) by X-ray crystallography at 2.3 Å resolution (Color Plate 10). The reaction center (molecular weight of 150,000) consists of four polypeptide chains containing 12 prosthetic groups and one iron atom. Three of the polypeptide chains are anchored in the membrane by α helices. The structure permits the tracing of the path of light-driven cyclic electron flow from the primary electron donor to the cytochromes in agreement with earlier biochemical studies. Analysis of the structures of photosystem I and II and of the light-harvesting chlorophyll a/b complex is also in progress.

B. Bacteriorhodopsin

Bacteriorhodopsin is a membrane protein (MW 25,000) in the purple membrane of the salt-loving bacterium *Halobacterium halobium* that converts light energy into a proton gradient. The protein forms a 1 : 1 complex with the retinal chromophore that serves as a light-absorbing group and gives the molecule its purple color.

Electron microscopy at several tilt-angles followed by three-dimensional reconstruction yielded a structure at 3.5 Å resolution that shows seven transmembrane α helices surrounding a proton channel. The channel is blocked by retinal near the middle of the bilayer (Color Plate 11). The retinal is joined to the polypeptide by a protonated Schiff-base through a lysine residue. Upon absorption of light, photoisomerization of the all-*trans* retinal into 13-*cis* retinal initiates a chain reaction of H^+ transfers from the Schiff-base nitrogen to aspartate 85 and from arginine 82 to the external medium. The Schiff-base is reprotonated from Asp 96, which in turn accepts a proton from the cytosol. The net result is the transfer of a H^+ from the cytosol to the outside of the cell during each photocycle. The proton gradient generated in this process serves as an energy source for the synthesis of ATP through an ATP synthase.

III. GENERATION OF H^+ GRADIENTS BY OXIDATION OF FUEL MOLECULES

In animal cells the source of high-energy electrons is oxidation of fuel molecules (sugars, fats, and amino acids) through dehydrogenase enzymes that produce NADH and $FADH_2$. The flow of electrons from NADH and $FADH_2$ to O_2 through the electron transport chain of the mitochondrial inner membrane is associated with pumping of protons out of the mitochondrial matrix. The proton-motive force generated in this process consists of a proton gradient across the mitochondrial inner membrane (ΔpH) and a transmembrane electrical potential ($\Delta\psi$).

Important insight into the mechanism of this process was gained with the determination of the structure of cytochrome oxidase by X-ray crystallography at 2.8 Å resolution. Cytochrome oxidase is the terminal enzyme of the electron transport chain that catalyzes the transfer of electron from cytochrome *c* to O_2. The cytochrome oxidase isolated from a soil bacterium (*Paracoccus denitrificans*) contains 4 subunits (Color Plate 12). The bovine mitochondrial cytochrome oxidase consists of 13 subunits (Color Plate 13), 3 of which are coded by mitochondrial genes. Both cytochrome oxidases contain two heme A molecules (a and a_3) and two copper centers (CuA and CuB). The electrons from cytochrome *c* are first transferred to the CuA center, then to heme a, and finally to the binuclear center formed by heme a_3 and CuB, where O_2 is reduced to $2H_2O$. Four protons are translocated to the cytosolic side of the membrane for each pair of electrons flowing through the oxidase. ATP is synthesized when protons are allowed to flow back to the mitochondrial matrix through the proton channel of the F_1–F_0 ATP synthase located in the mitochondrial inner membrane.

IV. COUPLING OF PROTON FLUXES TO THE SYNTHESIS OF ATP

During a single day a resting adult turns over about 50 kg of ATP and intense muscle activity may further increase this energy utilization by five- or sixfold. The synthesis and hydrolysis of ATP are tightly coupled and subject to precise regulation, which assures the maintenance of nearly constant cellular ATP levels even during intense activity.

The key enzyme of ATP synthesis is the F_1–F_0 ATP synthase of mitochondria, chloroplasts, and bacterial membranes, which couples transmembrane proton fluxes to the synthesis of ATP. The energy from sunlight or from oxidation of food molecules generates a gradient of protons (H^+) across the membrane. The F_1F_0 complex provides a channel for these protons to flow back across the membrane, coupling the proton flux to the synthesis of ATP. The process occurs without an identifiable phosphorylated enzyme intermediate.

A. The Structure and Function of ATP Synthase

The ATP synthase of bovine mitochondria is a multisubunit complex consisting of a globular F_1 domain, which contains the catalytic sites for ADP and inorganic phosphate, and a membrane-embedded F_0 domain that forms the proton channel across the mitochondrial inner membrane. The F_1 and F_0 domains are connected by a stalk of ~45 Å length. Energy furnished by proton flux through the F_0 channel is relayed to the catalytic F_1 domain by conformational changes transmitted through the stalk, resulting in the synthesis of ATP and its release from the active site of the enzyme.

The F_1-ATPase can be separated from the membrane domain after disruption of the stalk. It has a molecular weight of 370,000 and consists of five distinct subunits with a stoichiometry of $3\alpha:3\beta:1\gamma:1\delta:1\varepsilon$. The isolated F_1 catalyzes only ATP hydrolysis; it is incapable of ATP synthesis because of the absence of F_0, which contains the H^+ channel. However, the isolated F_1-ATPase can be coupled to the F_0 domain by coassembly with other subunits (OSCP, F_6, subunits b and d) and this restores the proton flux-dependent ATP synthesis. The catalytic nucleotide binding sites of the F_1ATPase are in the 3β subunits. The homologous α subunits also bind nucleotides, but these nucleotides do not exchange during catalysis and their role is still a mystery.

X-ray crystallography of F_1-ATPase crystals grown in D_2O medium in the presence of ADP and the ATP analogue adenylylimido-diphosphate (AMPPNP) established the structure of the enzyme at 2.8 Å resolution (Color Plate 14). The three α and three β subunits are arranged in an alternating fashion to form a ring that is held together by a crown of β barrels. The central cavity of this ring accommodates an α helix from the C terminus of the γ subunit, together with portions of a second helix from the N terminus. The sleeve surrounding the γ subunit is hydrophobic and may act as a molecular bearing permitting the rotation of the $\alpha_3\beta_3$ ring relative to the γ subunit. The overall structure is asymmetric. AMPPNP was bound to all three α subunits and to one of the β subunits (β_{TP}). Another β subunit bound ADP (β_{DP}), while the third had no bound nucleotide (β_E). The γ subunit may prevent the β_E subunit from adopting the nucleotide binding configuration. Rotation of the $\alpha_3\beta_3$ ring around the γ subunit is expected to alter cyclically the structure of nucleotide binding sites, permitting the sequential binding of ADP and inorganic phosphate (Pi), the formation of ATP, and its release from the active site at each of the β subunits.

The structure derived from X-ray crystallography is consistent with the binding change hypothesis of ATP synthesis based on kinetic, equilibrium binding, and chemical modification studies (Fig. 1). According to this hypothesis, the three catalytic sites cyclically alternate between three distinct states: an open state, a loose state occupied by ADP and Pi, and a tight state occupied by ATP. The synthesis of ATP would occur spontaneously, as it is energetically favored by the hydrophobic environment of the catalytic site, and the energy supplied by the proton flow through F_0 is assumed to promote the release of ATP from the enzyme. The basic mechanism is ancient, as F_1–F_0-type ATP synthases are present even in archaebacteria.

The transition between the three states is assumed to be driven by the rotation of the $\alpha_3\beta_3$ ring relative to the γ subunit propelled by the proton-motive force. Based on the maximum rate of ATP synthesis (~400/sec), the rate of rotation is expected to be about 130/sec. A similar proton flux-driven rotational mechanism operates in the bacterial flagellar motor, where the energy of H^+ flux is used for mechanical work, propelling the bacteria either forward or in

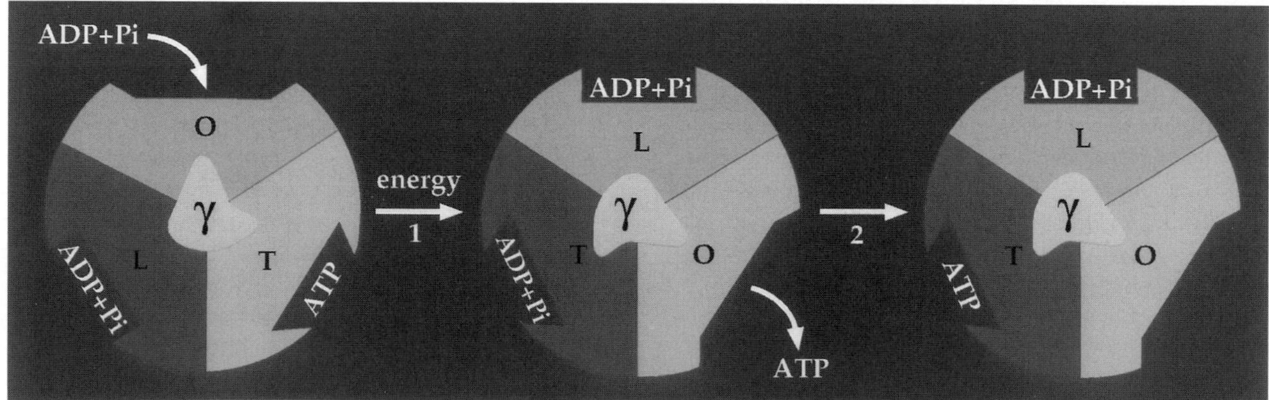

FIGURE 1 The binding-change mechanism for ATP synthesis. The central asymmetric mass representing the $\gamma\delta\varepsilon b_2$ subunits rotates relative to the three $\alpha\beta$ pairs, which in this illustration remain stationary. This rotation forces the three catalytic sites to undergo conformational changes associated with substrate binding and product release. T, L, and O stand for tight, loose, and open conformations and refer to the affinities of catalytic sites for ligand. In step 2, ATP forms spontaneously from tightly bound ADP and Pi. [From T. M. Duncan, V. V. Bulygin, Y. Zhou, M. Hutcheon, and R. L. Cross (1995). *Proc. Natl. Acad. Sci. U.S.A.* **92**, 10964–10968.]

reverse at various rates. The angular velocity of rotation is proportional to the proton-motive force and the maximum velocity is about 100–300/sec. The torque is generated by multiple force-generating units (8–16) and about 1000 protons move during each revolution of a flagellum.

B. The ATP–ADP Translocase

As ATP and ADP do not diffuse freely across the mitochondrial inner membrane, an ATP–ADP translocase catalyzes the entry of ADP into and the exit of ATP out of the mitochondrial matrix:

$$ADP^{3-}_{cytoplasm} + ATP^{4-}_{matrix} \rightleftarrows ADP^{3-}_{matrix} + ATP^{4-}_{cytoplasm}$$

The ATP–ADP exchanger is a homodimer of identical 30-kDa subunits. Each contains a single nucleotide binding site, where Mg-free ATP or ADP are bound with similar affinity. The translocase moves only when nucleotide is bound and the rate of movement is faster with ATP than with ADP because of the extra negative charge of ATP. Therefore, the rate of exchange is regulated by the rate of ATP synthesis. The movement depends on a membrane potential (negative in the matrix) that is decreased by the exchange of ATP^{4-} for ADP^{3-} and must be constantly restored by electron flow through the respiratory chain. [*See* Adenosine Triphosphate (ATP).]

V. ION-MOTIVE ATPases: THE GENERATION OF CATION GRADIENTS

The ion-motive ATPases use the free energy of ATP hydrolysis to catalyze the translocation of H^+, Na^+, K^+, and Ca^{2+} across various cellular membranes against an electrochemical gradient. Two major classes of ion-motive ATPases have been identified.

The vacuolar (V-type) H^+-transporting ATPases are found in the clathryn-coated vesicles, endosomes, lysosomes, secretory granules, and Golgi vesicles of animal cells, in the vacuoles of fungi and yeast, and in the tonoplasts of plants. They are structurally and mechanistically related to the F_1-ATPase, but operate physiologically in the direction of ATP hydrolysis coupled to H^+ movement.

The P-type ATPases of surface membranes and sarco/endoplasmic reticulum belong to a structurally distinct family. The translocation of ions by these ATPases is stoichiometrically coupled to the phosphorylation of the enzyme by ATP at a specially reactive aspartyl carboxyl group, with the formation of an aspartyl–phosphate enzyme intermediate. The family of P-type ATPases includes the Ca^{2+}-ATPases of plasma membranes and endoplasmic reticulum, the Na^+-K^+-ATPase and H^+-K^+-ATPase of surface membranes, the heavy metal-transporting ATPases, and several other cation-transporting enzymes of bacteria.

Recent crystallographic, spectroscopic, kinetic, and molecular genetic approaches yielded interesting glimpses into the structure and mechanism of both classes of enzymes. Even though the molecular mechanism of ion translocation remains a mystery, these new observations define the problem and prepare the ground for their solution.

A. The Vacuolar (V-Type) ATPases

The V-type ATPases are ATP-dependent proton pumps that are responsible for the acidification of a variety of intracellular compartments. The acidification in the endosomes causes dissociation of ligands from receptors, facilitating receptor recycling. In the lysosomes, the acid pH is necessary for optimal activation of hydrolases. The low pH within secretory vesicles is important in the uptake of neurotransmitters and other small molecules.

The V-type ATPases are structurally related to the F_1–F_0 class and share common evolutionary origin (Fig. 2). Whereas the F_1–F_0 ATP synthase utilizes the proton-motive force for ATP synthesis, the V-ATPases operate in reverse and use the free energy of ATP to produce proton gradients across intracellular membranes and on the cell surface. Like the F_1–F_0 ATP synthases, the V-ATPases contain a V_1 domain, which is responsible for ATP hydrolysis, and a V_0 domain, which forms the proton channel. The subunit composition of the V_1 domain is A_3B_3CDE for a molecular weight of about 500,000. The A and B subunits of the V-ATPase are homologous structurally and functionally to the β and α subunits of the F_1-ATPase. The catalytic sites of ATP hydrolysis are located on the A subunits, which also contain the critical glycine-rich loop, the latter required for ATP hydrolysis in the F_1-ATPase β subunits. The B subunit lacks the glycine-rich loop and appears to be noncatalytic, although it binds nucleotide analogues. Nevertheless, the B subunits are expected to contribute to the catalytic nucleotide binding sites since modification of

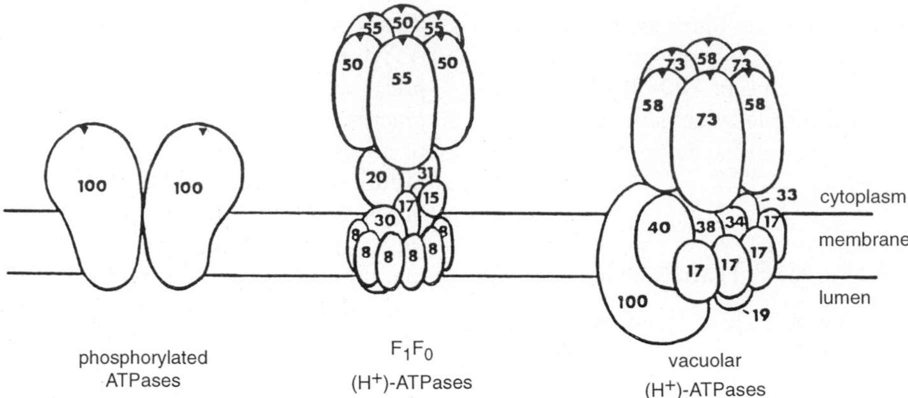

FIGURE 2 Structure of three major classes of ATP-driven cation pumps. Shown as representatives of each class are the Ca^{2+}-ATPase of sarcoplasmic reticulum, (phosphorylated class), *E. coli* H^+-ATPase (F_1F_0 class), and the H^+-ATPase of clathrin-coated vesicles (vacuolar class). Vacuolar and F_1F_0 classes share a number of important structural properties, including complexity of subunit composition, the presence of 3 copies each of 2 nucleotide binding subunits (v, nucleotide binding sites), and the presence of 6–12 copies of a subunit that forms part of the dicyclohexylcarbodiiuride (DCCD)-inhibitable proton channel (17,000 MW in vacuolar H^+-ATPase and 8000 MW in F_1F_0 H^+-ATPase). These and other structural similarities suggest that vacuolar and F_1F_0 classes of proton pumps are derived from a common evolutionary ancestor. [From M. Forgac (1989). *Physiol. Rev.* **69**, 765–796.]

critical residues of the B subunit located at the presumed A–B interface dramatically decreased enzymatic activity.

The integral V_0 domain is a 250-kDa complex of the following composition: 100, 38, 19, and 16_6 kDa (proteolipid). The 16-kDa proteolipid of eukaryotic V-ATPases may have evolved by gene duplication from an ancestral proteolipid of 8 kDa that is present in all F-ATPases and in archaebacterial V-ATPases. This gene duplication produced far-reaching mechanistic consequences. All F and V-ATPases that have short (8 kDa) proteolipid are capable of ATP synthesis and the H-flux : ATP ratio is usually near 4 : 1. Eukaryotic V-ATPases that have large proteolipid (16 kDa) catalyze only ATP hydrolysis with an H : ATP ratio of only 2 : 1. Dissociation of F_1 from F_0 yields an active F_1-ATPase and generates proton leak through the F_0 channel. By contrast, the separated V_1 domain of eukaryotic V-ATPases has no ATPase activity and there is no proton leak from the residual V_0 domain. The archaebacterial V-ATPase that contains short (8 kDa) proteolipid readily yields active V_1 domain after separation from V_0, and there is enhanced proton leak from V_0 after removal of the V_1 sector, similarly to the F_1-ATPase. These observations show that the region linking the catalytic domain to the membrane domain contributes to the differences between F_1F_0 and the V-ATPases.

V-ATPases reside at high densities in the plasma membranes of specialized insect and vertebrate cells, such as the proton-secreting cells of frog skin and urinary bladder, the proton-transporting cells of kidney, the osteoclasts, macrophages and neutrophils, the cells of Malpighian tubules, and the K^+-secreting goblet cells of insect gut. In some of these cells the density of V-ATPases in the cell membrane may be as high as 15,000 ATPase molecules per μm^2 surface area and may form two-dimensional crystalline arrays. The V-ATPase distribution in these cells is usually polarized. The mechanisms involved in the selective accumulation of V-ATPases in specific regions of certain cell types are unknown.

In vertebrate cells, plasma membrane V-ATPases are primarily used for H^+ transport, whereas in insect cells they generate membrane potential. In a lipid bilayer in the absence of other ion channels, a V-ATPase would generate large membrane potential with relatively small changes in H^+ concentration due to rapid approach to electrochemical equilibrium. However, if anion channels are also present, the V-ATPase can drive substantial acid flux, and if cation/H^+ antiporters or symporters are present it can produce alkalinization. Therefore, the physiological consequences of V-ATPase function depend on a complex interplay between several ion-transporting systems. This may explain why in certain cell membranes V-ATPases, whereas in others P-type H^+ transport ATPases are used to produce H^+ fluxes.

B. The P-Type Ion-Motive ATPases

The P-type ion-motive ATPase family includes:

1. The Na^+-K^+-ATPase that is responsible for the maintenance of Na^+ and K^+ gradient across the cell-surface membranes of animal cells.
2. The Ca^{2+}-ATPases of plasma membranes and endoplasmic reticulum that regulate the Ca^{2+} content of cells and the distribution of Ca^{2+} between the cytoplasm and the endoplasmic reticulum.
3. The H^+-K^+-ATPase of gastric mucosa that is responsible for the secretion of hydrochloric acid in the stomach.
4. The plasma membrane H^+-ATPases of plants, fungi, and yeast.
5. The CPX class of heavy metal-transporting ATPases that translocate both essential and toxic metals across bacterial and animal membranes.
6. The bacterial Mg^{2+} transport ATPases.

All of these enzymes share a fundamentally similar reaction cycle that involves transitions between two major conformations ($E_1 \longleftrightarrow E_2$) linked to the phosphorylation of a reactive aspartyl group of the active site by ATP. The enzymes bind with high-affinity metal ions on the cis side of the membrane in the E_1 state. This is followed by phosphorylation ($E_1 \sim P$), conversion into the $E_2 \sim P$ state, and translocation of the bound cation across the membrane. During this process the affinity of the enzyme for the translocated cation is reduced and the ion is released on the trans side of the membrane. The binding site is now able to bind one or more counterions from the trans side and, following hydrolysis of the phosphoenzyme intermediate, the carrier returns to the E_1 state carrying the counterion(s). During each of these cycles one ATP is hydrolyzed, coupled to the translocation of cations with a well-defined stoichiometry (Fig. 3).

There is considerable sequence homology between the various P-type transport ATPases, suggesting that they are all derived from a common early ancestor. Their molecular weight is in the range of 100,000–130,000 and based on electron crystallography and prediction from amino acid sequences they each contain similar structural domains (Figs. 2, 4, 5, and 6). The large pear-shaped cytoplasmic domain represents about two-thirds of the mass of the protein and based on mutagenesis and chemical modification studies contains the catalytic sites for ATP hydrolysis. The cytoplasmic domain is linked by a stalk to the membrane domain that is assumed to contain 8–10 transmembrane helices that form the channel for the translocation of cations. High-resolution structures are not available for any of the P-type transporters and the finer structural details are speculative. As the estimated distances between the ATP binding and cation transport sites are on the order of 30–40 Å, it is assumed that the coupling of ATP hydrolysis to cation transport involves long-range interactions between the cytoplasmic and the membrane domains. The structural basis of these interactions is unknown.

In spite of their basic similarity, each P-type transport ATPase has unique structural and mechanistic features and fulfills a distinct physiological role. A brief discussion of these distinctive properties follows.

1. The Na^+-K^+-ATPase

During each cycle of ATP hydrolysis the Na^+-K^+-ATPase catalyzes the transfer of three Na^+ ions from the cytoplasm to the extracellular medium and two K^+ ions from the extracellular medium into the cytoplasm (Figs. 3 and 4). The 3 Na^+/2 K^+ stoichiometry implies that the transport is electrogenic, maintaining negative potential on the cytoplasmic side of the membrane. Both Na^+ and K^+ are transported actively against their electrochemical gradients at the expense of the free energy of ATP hydrolysis, maintaining the high K^+ (140 mM) and low Na^+ (11 mM) concentration in the cytoplasm relative to the 140 mM Na^+ and 5 mM K^+ concentration in the extracellular medium and blood plasma. In addition to being optimal for the functioning of cellular enzymes, the gradient of Na^+ and K^+ ions represents a battery of energy for coupled transport of Ca^{2+}, sugars, amino acids, and neurotransmitters, maintains membrane potential that is the basis of nerve and muscle function, and contributes to the regulation of cell volume.

The Na-K pump contains, in addition to the catalytic α subunit of 113 kDa with binding sites for ATP and cations, a smaller (55 kDa) highly glycosylated β subunit (see Fig. 4). Tissue-specific isoforms exist both for the α (α_1, α_2, α_3) and for the β (β_1, β_2, β_3) subunits. The β subunit appears to be involved in the maturation of the enzyme and in its localization on the plasma membrane, but its role in the catalytic mechanism is a mystery.

A unique property of the Na^+-K^+-ATPase is its tight interaction with digitalis glycosides (ouabain, di-

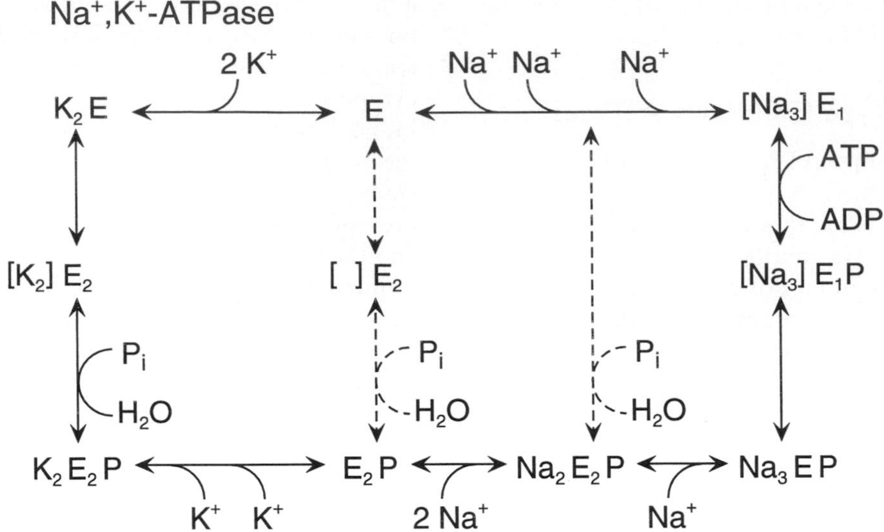

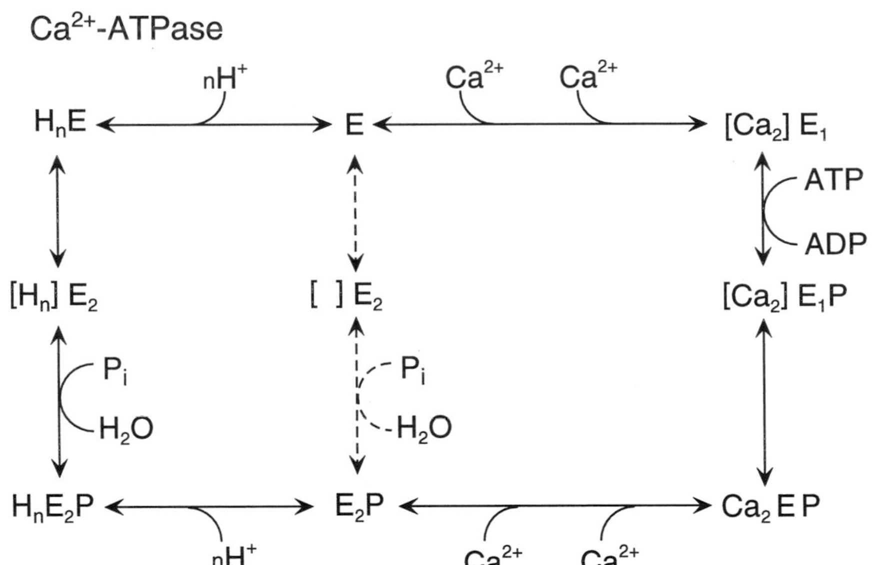

FIGURE 3 Simplified reaction cycles of the Ca^{2+}- and Na^+-K^+-ATPases showing binding and dissociation steps for the cations and the phosphoryl group, as well as transitions between different conformational states of the enzymes. The cytoplasmic side of the membrane is upward and the extracytoplasmic side is downward. Brackets indicate occluded states of the cation binding sites. E_1 indicates the form with chemical specificity for reaction with ATP/ADP. E_2 indicates the form with chemical specificity for reaction with Pi/H_2O. Unlabeled forms are those that are difficult to stabilize and define experimentally. A tentative H^+-countertransport limb is shown for the Ca^{2+}-ATPase according to recent findings. Dashed lines indicate alternative reaction pathways that may be used under certain conditions. [From J. P. Andersen and B. Vilsen (1995). *FEBS Lett.* **359,** 101–106.]

goxin) at multiple sites in both the extracellular and membrane domains with inhibition of Na-K transport. The resulting increase in cytoplasmic Na^+ concentration increases the inward flux of Ca^{2+} through the $Na^+:Ca^{2+}$ exchanger with secondary increase in cytoplasmic Ca^{2+} concentrations that in turn strengthens the contraction of heart muscle. This mechanism is the basis of the widespread use of digitalis glycosides in the therapy of heart failure.

2. The Ca^{2+} Transport ATPases

The cytoplasmic free Ca^{2+} concentration in a resting cell is on the order of 0.02–0.03 μM, at an extracellular Ca^{2+} concentration of ~3 mM. The low cytoplasmic Ca^{2+} concentration is vital for proper cell function and it is maintained by continual pumping of Ca^{2+} out of the cell through the plasma membrane Ca^{2+}-ATPase (PMCA), and by sequestering of cytoplasmic Ca^{2+} in the endoplasmic reticulum lumen through the sarco/endoplasmic reticulum Ca^{2+}-ATPase (SERCA). In some cells, a Na^+–Ca^{2+} exchanger also contributes to the extrusion of calcium from the cell.

Both PMCA- and SERCA-type ATPases operate by the $E_1 \rightarrow E_2$ mechanism, coupling ATP hydrolysis to Ca^{2+} translocation with the transient formation of a phosphoenzyme intermediate (see Fig. 3). However, they are the products of different genes and there are significant differences in their structure and regulation.

a. The Ca^{2+}-ATPase of Plasma Membranes

The plasma membrane Ca^{2+}-ATPase occurs in at least four distinct isoforms (PMCA 1–4) and several splicing variants with calculated masses in the range of 127,300–138,800. The Ca^{2+} pump is organized in the membrane with a topology similar to that of other P-type ATPases except for the extended C-terminal domain (Fig. 5). This contains the calmodulin binding site flanked by two acidic α-helical stretches and sites of phosphorylation by protein kinase A and protein kinase C. The calmodulin binding domain functions as a repressor of pump activity by folding over the molecule and limiting the access of substrate to the catalytic site. Binding of Ca–calmodulin to the calmodulin binding site relieves this inhibition and activates the enzyme whenever the Ca^{2+} concentration of the cytoplasm is high, indicating a need for Ca^{2+} extrusion. The enzyme can also be activated by proteolysis that removes the C-terminal domain and by acidic phospholipids that bind to a well-defined site in the cytoplasmic domain of the Ca^{2+}-ATPase. The Ca^{2+} affinity (K_m) of the Ca^{2+}-ATPase is 10 μM in the resting state and 0.5 μM in the activated state, and the dissociation constant (K_D) of Ca–calmodulin binding is ~1 nM. Therefore, PMCA is expected to be fully active when the Ca^{2+} concentration in the cytoplasm rises to micromolar levels.

b. The Ca^{2+} Pump of Sarco/endoplasmic Reticulum

The SERCA family of Ca^{2+}-ATPases is encoded by three distinct genes (SERCA 1–3) with specific expression in several spliced isoforms in different tissues ranging in molecular weight from ~109,000 to 115,000. The SERCA 1a and 1b isoforms are characteristic for the sarcoplasmic reticulum of fast-twitch skeletal muscle, the SERCA 2a isoform for slow-twitch skeletal and cardiac muscles, and the SERCA 2b and 3 isoforms are present in smooth muscles and in nonmuscle cells. The expression of the various isoforms is differentially regulated. In adult skeletal muscle the density of the Ca^{2+}-ATPase in the sarcoplasmic reticulum is close to 30,000 ATPase molecules per μm^2 membrane area, representing ~80% of the protein content of the sarcoplasmic reticulum.

Electron crystallography of Ca^{2+}-ATPase in several conformations yielded a low-resolution structure of the enzyme. The large pear-shaped cytoplasmic domain contains the catalytic site for ATP hydrolysis whereas the Ca^{2+} channel is presumed to be formed by the transmembrane α helices (Fig. 6). The nature of the structural changes associated with ATP hydrolysis and Ca^{2+} translocation is unknown, but may involve a rearrangement of helices as shown in Fig. 6.

In resting muscle cells the sarcoplasmic reticulum lumen contains much of the Ca^{2+} content of the muscle cell, reaching concentrations of several millimoles. During muscle activation the depolarization of surface membranes activates the voltage-gated dihydropyridine receptors of the T tubules. These in turn trigger Ca^{2+} release from the sarcoplasmic reticulum by opening Ca^{2+} channels located at specialized junctional regions with the T tubules. Within a few milliseconds the Ca^{2+} concentration in the cytoplasm rises several hundredfold with activation of the contractile apparatus. During muscle relaxation, Ca^{2+} is pumped from the cytoplasm back into the sarcoplasmic reticulum by the Ca^{2+}-ATPase and resting Ca^{2+} levels are restored in the cytoplasm within a fraction of a second. In very fast-acting muscles the frequency of contraction may reach 200–300/sec and a Ca^{2+} release–reabsorption cycle is completed within 3–5 msec. For each ATP hydrolyzed, two Ca^{2+} ions are translocated across the

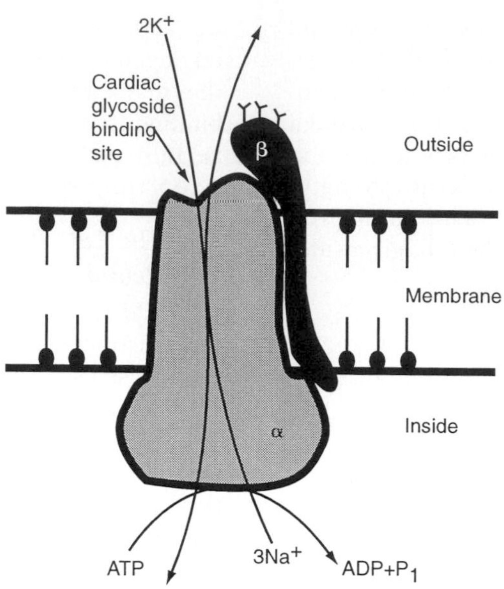

Extracellular

H1 H2 H3 H4 H5 H6 H7 H8 H9 H10

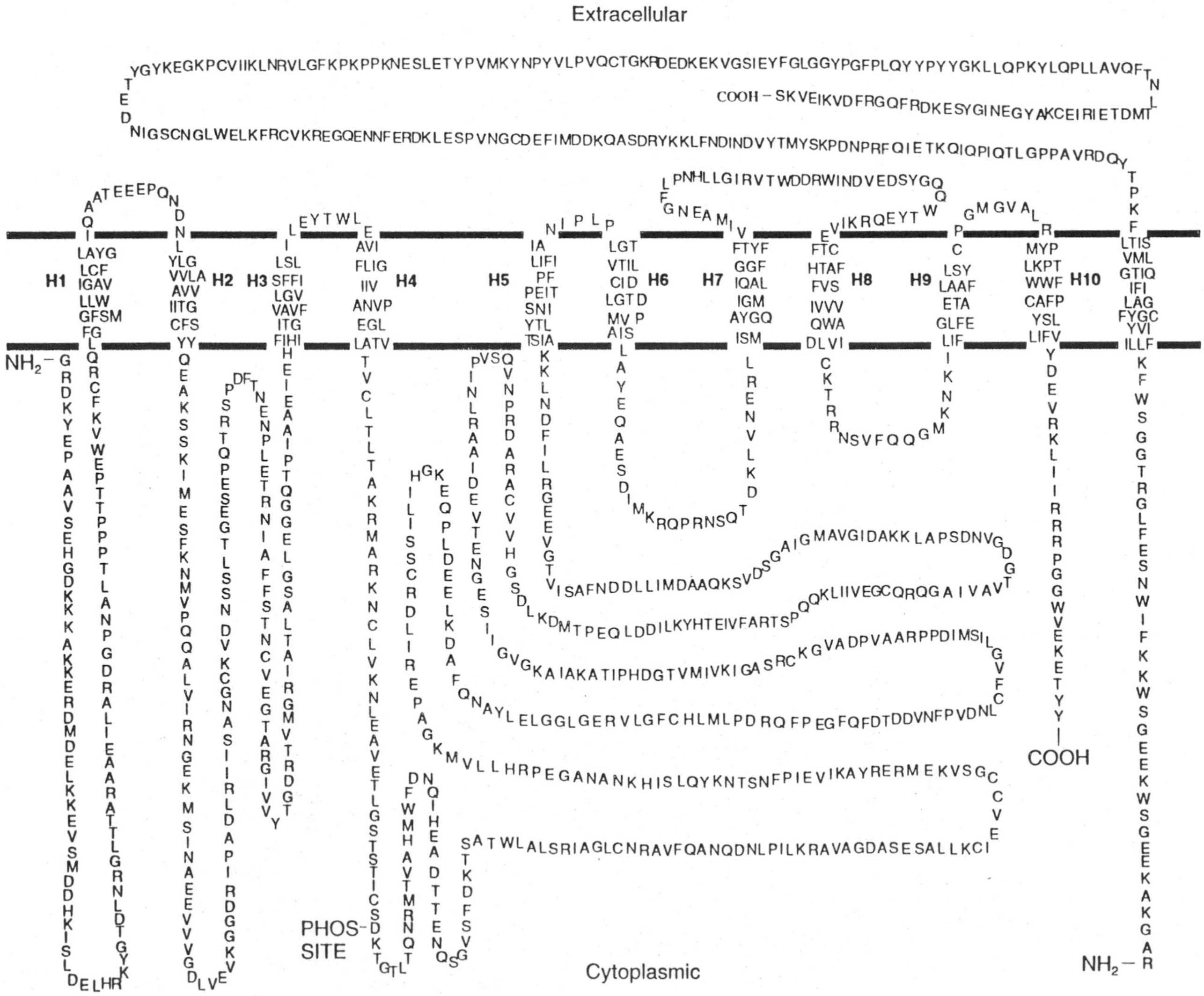

NH2–

COOH

PHOS-SITE

Cytoplasmic

COOH

NH2–

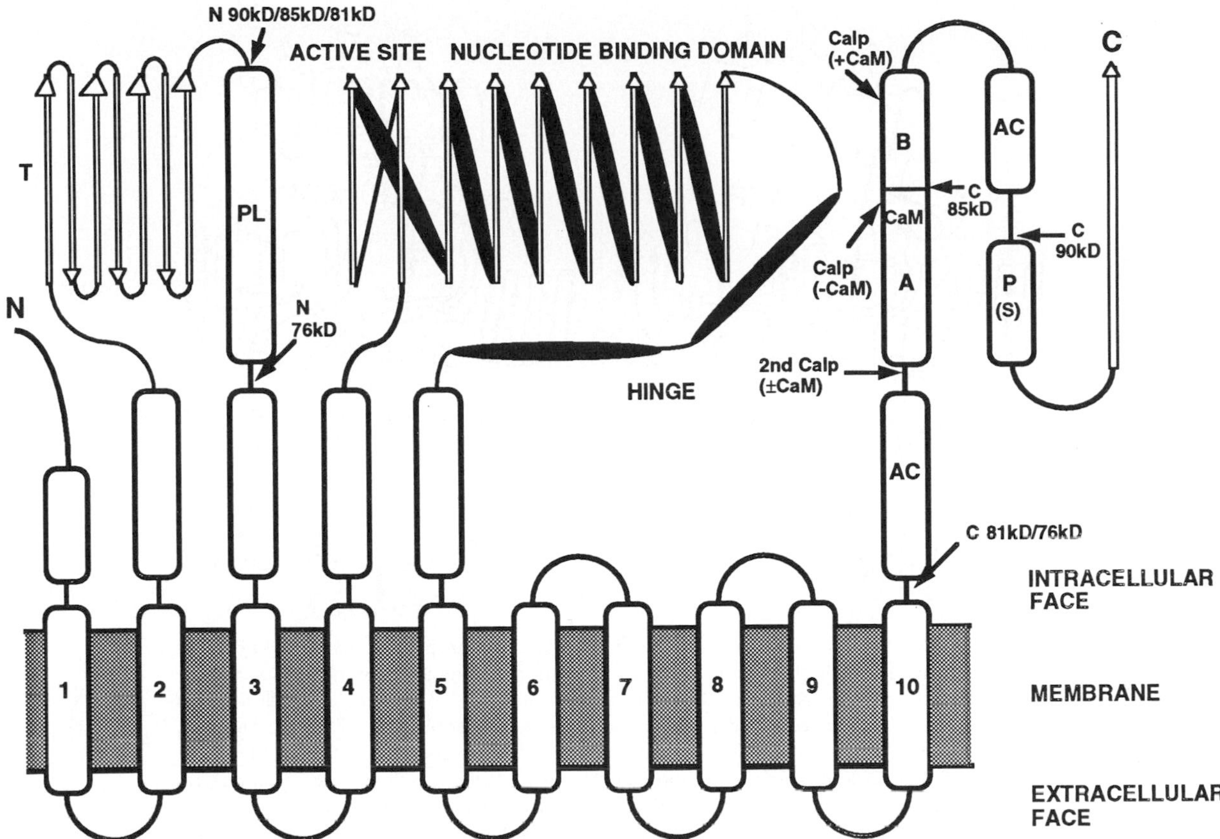

FIGURE 5 Proposed model for the overall topology of plasma membrane Ca^{2+}-ATPase (PMCA), including the putative transmembrane topology (TM1 to TM10) and the assignment of important domains. Open rods and black bars correspond to putative α helices, and arrows denote β sheet elements. The N-terminal (N 90 kDa/85 kDa/81 kDa, N 76 kDa) and C-terminal locations (C 90 kDa, C 85 kDa, C 81 kDa/76 kDa) of the tryptic cleavage sites leading to the production of major proteolytic fragments are indicated, as are the sites of calpain attack in the presence [Calp (+CaM)] and absence [Calp (−CaM)] of calmodulin. The site of secondary, calmodulin-independent, calpain attack is labeled 2nd Calp (±CaM); AC, acidic regions flanking the calmodulin binding domain; C, C terminus; CaM, calmodulin binding domain consisting of subdomains A and B; N, N terminus; T, transduction domain; P(S), region containing the serine residue susceptible to phosphorylation by the cAMP-dependent protein kinase; PL, phospholipid-sensitive region. [From E. E. Strehler (1991). *J. Membrane Biol.* **120**, 1–15.]

membrane (see Fig. 3). Because of the very high concentration of Ca^{2+}-ATPase in the sarcoplasmic reticulum, only one or two cycles of ATP hydrolysis are normally required during relaxation to pump all the activating calcium back into the sarcoplasmic reticulum. In Brody's disease the functional Ca^{2+}-ATPase content of fast-twitch skeletal muscles is sharply reduced, and this is manifested in slowing of the relax-

FIGURE 4 Transmembrane model of Na-K-ATPase. The amino acid sequence is that of sheep α_1 and β_1 subunits. Top: The catalytic site of the α subunit is on the cytoplasmic side of the membrane, whereas the binding site for inhibitory cardiac glycosides is on the extracellular surface. The details of the interaction between α andβ subunits are unknown. The probable stoichiometry of transport is three Na^+ and two K^+ per ATP hydrolyzed. Bottom: The transmembrane disposition of the α and β subunits is shown based on predictions from the amino acid sequence. There are 10 probable transmembrane segments in the α subunit with both N and C termini on the cytoplasmic side. The β subunit has one putative transmembrane domain with its N terminus in the cytoplasm and its C terminus outside. [From J. B. Lingrel, J. V. Huysse, W. O'Brien, E. Jewel-Motz, R. Askew, and P. Schultheis (1994). *Kidney Int.* **45**(Suppl. 44), S32–S39.]

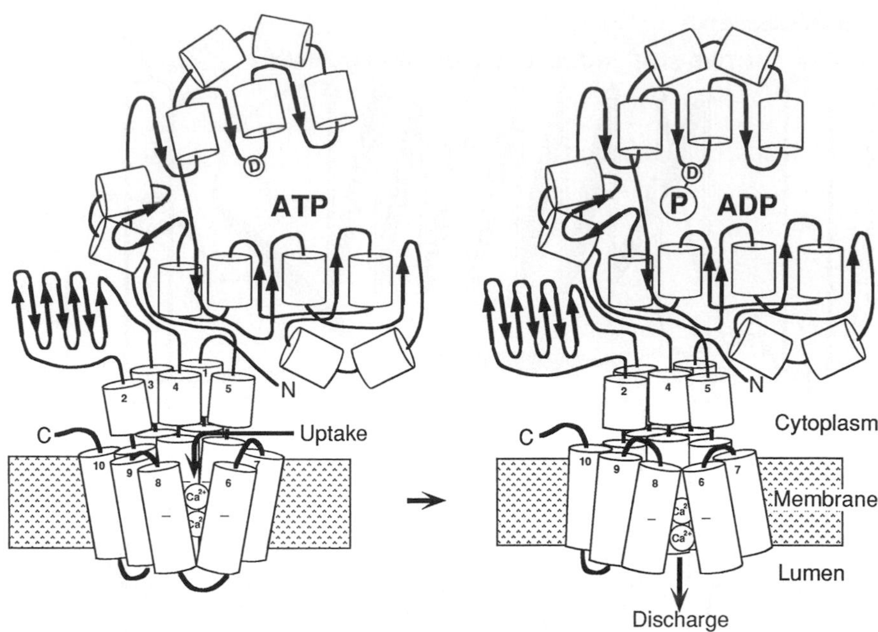

FIGURE 6 Model illustrating a possible mechanism of Ca^{2+} transport by the Ca^{2+}-ATPase. In the E_1 conformation, high-affinity Ca^{2+} binding sites located near the center of the transmembrane domain are accessible to cytoplasmic Ca^{2+}, but not to luminal Ca^{2+}. The sites are made up from amino acid residues located in proposed transmembrane sequences M4, M5, M6, and M8. Conformational changes induced by Ca^{2+}-dependent phosphorylation of Asp 351 by ATP in the cytoplasmic domain lead to the E_2 conformation in which the high-affinity Ca^{2+} binding sites are disrupted, access to the sites by cytoplasmic Ca^{2+} is closed off, and access to the sites by luminal Ca^{2+} is gained. The Ca^{2+} transport cycle thus involves binding of cytoplasmic Ca^{2+} to high-affinity sites in one conformation and release of the same Ca^{2+} to the lumen when the high-affinity sites are disrupted in transition to the second conformation. [From D. H. MacLennan, D. M. Clarke, T. W. Loo, and I. S. Skerjanc (1992). *Acta Physiol. Scand.* **146**, 141–150.]

ation phase of muscle due to the slower rate of reabsorption of calcium from the cytoplasm into the sarcoplasmic reticulum.

The massive fluxes of Ca^{2+} across the sarcoplasmic reticulum membrane during activity would be expected to generate large fluctuations in the membrane potential of sarcoplasmic reticulum. This is prevented by the presence of cation and anion channels in the sarcoplasmic reticulum membrane that conduct compensating ion fluxes. As a result, the membrane potential of sarcoplasmic reticulum remains near zero during all phases of its activity.

The SERCA-type Ca^{2+}-ATPases are regulated primarily by the Ca^{2+} concentrations in the cytoplasm and in the lumen of sarcoplasmic reticulum. With a dissociation constant (K_D) of 0.1 μM for cytoplasmic Ca^{2+} and ~1 mM for lumenal Ca^{2+}, the Ca^{2+}-ATPase is near equilibrium in the resting muscle and maximally activated only during contraction. In contrast to the plasma membrane Ca^{2+}-ATPases, the SERCA ATPases have no binding site for Ca–calmodulin.

In cardiac muscle a regulatory protein, phospholamban, is present that inhibits the Ca^{2+}-ATPase. The inhibition is relieved by phosphorylation of phospholamban through cAMP-dependent protein kinase. This may be one of the mechanisms by which norepinephrine, a well-known activator of protein kinase A, produces stimulation of cardiac muscle. Activation of Ca^{2+}-ATPase by norepinephrine permits faster reloading of sarcoplasmic reticulum Ca^{2+} stores after each contraction, allowing larger Ca^{2+} release at the next stimulus.

The Ca^{2+}-ATPase of sarcoplasmic reticulum is present in the form of oligomers (dimers and tetramers) in the native membrane and the cooperative interactions within these oligomers influence the kinetics of Ca^{2+} transport.

3. The Gastric H^+-K^+-ATPase

The gastric H^+-K^+-ATPase is a heterodimer composed of an α subunit of 1034 amino acids that contains the site for ATP hydrolysis and cation binding and a glycosylated β subunit of 291 amino acids that assists in the expression of the enzyme in the plasma membrane. Like in the ATPases involved in Na^+, K^+, and Ca^{2+} transport, the gastric H^+-K^+-ATPase α subunit has a large cytoplasmic domain connected with a stalk to the membrane domain composed of 10 putative transmembrane helices. The β subunit has a single transmembrane segment. It is suggested that oligomeric interactions between $\alpha\beta$ units may contribute to enzymatic activity.

In the absence of the β subunit, the catalytic subunit is retained in an unstable form in the rough endoplasmic reticulum. When β subunits are present, the $\alpha\beta$ heterodimer is stabilized and moves to the smooth endoplasmic reticulum within the still nonpolarized cell, where it remains enzymatically inactive. Stimulation of the parietal cells causes the relocation of the enzyme from the cytoplasm to the canalicular membrane, where it is exposed to luminal K^+ and becomes active in acid secretion.

The H^+-K^+-ATPase pumps H^+ out of the cytoplasm in exchange for K^+ entering the cells; the process is coupled to ATP hydrolysis in analogy with the Na^+-K^+-ATPase. In the E_1 conformation the ion binding site faces the cytoplasm and binds H^+, whereas in the E_2 conformation it is oriented toward the outside and binds K^+. The E_1–E_2 conversion is associated with the formation of a phosphorylated enzyme intermediate ($E \sim P$). The secreted H_3O first enters into the lumen of canaliculi, where the HCl concentration may rise to 160 m*M* (pH 0.8). The acid reaches the stomach cavity through a narrow pore at the apex of canaliculi. The activation of acid secretion is accompanied by activation of K^+ and Cl^- conductance in the canalicular membrane.

In view of the pathogenic role of acid secretion in duodenal or gastric ulcer and other acid-related diseases, a massive effort is under way to develop proton pump inhibitors. The currently available drugs—omeprazole, lansoprazole, and pantoprazole—are derivatives of 2-pyridyl-methylsulfinyl benzimidazole. They have a pK_a of about 4.0. The drugs accumulate in the secretory canaliculi and, after conversion into cationic sulfenamides, form disulfide bonds with pairs of critical SH groups in the H^+-K^+-ATPase. The reactive cysteines are in the small extracytoplasmic domain of the protein. Omeprazole reacts only with the pumps present in the canaliculi and, since the pumps turn over on the surface membrane with a half-life of ~50 hr, significant amounts of newly synthesized uninhibited enzyme appear on the surface continually, leading to rapid recovery of acid secretion from inhibition.

Another group of drugs acts by competition with K^+ on the extracytoplasmic surface. The substituted imidazopyridine and aryl quinoline derivatives contain protonatable nitrogen. They show surprising selectivity for gastric H^+-K^+-ATPase in spite of the fact that their binding site is in a region analogous to the ouabain binding site of Na^+-K^+-ATPase.

4. The Plasma Membrane H^+-ATPase of Plants and Fungi

The H^+-ATPase of plant and fungal plasma membranes consists of a single polypeptide of ~100 kDa with the usual structural and functional characteristics of P-type ATPases. Besides regulating external and internal pH, the H^+-ATPase builds up H^+ electrochemical potential that drives many secondary transport systems for ions and metabolites. In some plant cells the activity of H^+-ATPase may account for 20–50% of total ATP turnover.

The yeast ATPase is activated *in vivo* by glucose. The activation is greatly reduced by removal of 11 amino acids from the carboxyl terminus. It was suggested that this segment produces autoinhibition by folding over the active site and that phosphorylation of ATPase during glucose metabolism neutralizes this inhibition. A similar mechanism appears to regulate the plant ATPases as well.

5. The Heavy Metal-Transporting CPx-Type ATPases

The heavy metal-transporting ATPases translocate both essential and toxic metals across the membranes of bacterial and animal cells. They represent a newly defined subgroup of P-type ion-motive ATPases characterized structurally by a conserved cystein-proline-x (CPx) sequence near the N terminus that is required for heavy metal binding. Other distinguishing features from common P-ATPases are the unique topology of the transmembrane domain, a conserved histidine–proline sequence in the cytoplasmic domain, and the absence of four transmembrane segments at the C terminus. The currently known transport activities include Ag^+, H^+, Na^+, K^+, Ca^{2+}, Mg^{2+}, Cd^{2+}, and Cu^{2+}.

6. Mg^{2+}-Transporting P-ATPase

Salmonella typhimurium possesses three distinct Mg^{2+} transport systems, one constitutive and two (MgtA and MgtB) expressed only at low to medium Mg^{2+} concentration. Both MgtA and MgtB have the characteristic sequences of P-type ATPases but interestingly they have higher homology to eukaryotic than to prokaryotic enzymes. The transcription of MgtB can be increased 1000-fold by lowering the extracellular Mg^{2+} concentration from 1 m*M* to 1 μM. Curiously, both MgtA and MgtB appear to mediate the transport of Mg^{2+} down its electrochemical gradient, raising questions about the significance of ATP hydrolysis. [*See* Ion Pumps.]

VI. THE ATP BINDING CASSETTE TRANSPORTERS: DRUG RESISTANCE

The ATP binding cassette (ABC) transporters form a superfamily of "traffic ATPases" that mediate the selective movement of a large variety of solutes across biological membranes. The more than 100 ABC transporters identified so far in species ranging from bacteria to humans share a common organization of four core domains that may be expressed individually or may fuse together to form large multidomain proteins. Two predicted transmembrane domains (TM) consisting of 6–12 α helices span the membrane, forming pathways that may determine the selectivity of the transporter. Two ATP binding domains located on the cytosolic surface are assumed to couple ATP hydrolysis to solute translocation and provide the energy for transport against an electrochemical gradient. The ATP binding domains of various ABC transporters share up to 40% sequence identity. The structure of the ATP binding domain is distinct from those of other ATP binding proteins, aiding the identification of new ABC genes.

Whereas the functional units of expressed ABC transporters usually contain two ATP binding and two transmembrane domains, the organization of ABC genes is quite varied both in prokaryotes and in eukaryotes. In prokaryotes the ABC genes are usually contained in operons. The ATP binding and transmembrane (TM) domains are sometimes expressed as separate units that assemble to form the functional transporter as in the oligopeptide permease of *Staphylococcus typhimurium* (Fig. 7). Alternatively, the two ATP binding domains, or the two transmembrane domains, may fuse into single polypeptides as in the ribose transporter or Fe-hydroxamate transporter of *E. coli,* respectively. Several ABC transporters involved in bacterial solute transport also contain associated periplasmic binding proteins.

The organization of ABC genes also varies in eukaryotes (see Fig. 7). In the P-glycoprotein/multidrug resistance (Pgp/MDR) family, the four domains are linked in succession—TM-ATP-TM-ATP—to form a single protein. In the yeast PDR5 the sequence is ATP-TM-ATP-TM. In the cystic fibrosis conductance regulator (CFTR) a regulatory domain (R) is inserted in the middle of the sequence—TM-ATP-R-TM-ATP—which may enhance the regulatory potential of the transporter. The mammalian TAP1–TAP2 peptide transporter, involved in the presentation of the major histocompatibility complex class I antigen, is a homodimer of two fused TM–ATP domains.

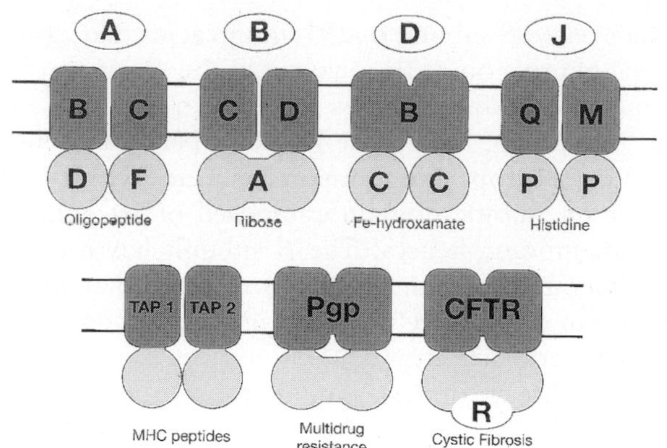

FIGURE 7 Domain organization of ABC transporters. In some transporters, the four core domains are expressed as separate polypeptides (e.g., the oligopeptide permease of *S. typhimurium*). In others, the domains can be fused in a variety of configurations. For example, the two ATP binding domains are fused into a single polypeptide in the ribose transporter of *E. coli,* and the two transmembrane domains are fused in the Fe-hydroxamate transporter of *E. coli*. The mammalian TAP1–TAP2 peptide transporter, associated with major histocompatibility complex (MHC) class I antigen presentation, is a complex of two polypeptides, each comprising a transmembrane domain fused to an ATP binding domain. The human multidrug resistance Pgp and the cystic fibrosis gene product CFTR have all four domains fused into a single multidomain polypeptide. In some transporters, one polypeptide is present as a homodimer in the transport complex to provide the full complement of four domains (e.g., HisP in the histidine permease of *S. typhimurium*). Certain ABC transporters include "extra" domains such as the periplasmic binding proteins of bacterial uptake systems or the regulatory R domain of CFTR, which provide specific, additional functions required by individual transporters. [From C. F. Higgins (1995). *Cell* **82**, 693–696.]

Insight into the molecular basis of the transport function of ABC proteins will require determination of their three-dimensional structure at atomic resolution. The urgency of this task is underscored by the clinical significance of several ABC proteins in human diseases. The human P-glycoprotein (Pgp or MDR) confers multidrug resistance against chemotherapeutic drugs used in cancer treatment, diminishing their effectiveness. The Pgh1 protein of *Plasmodium falciparum* contributes to antimalarial drug resistance by promoting the elimination of chloroquine. Mutations in the sulfonyl urea receptor (SUR) are associated with hyperinsulinemic hypoglycemia as a result of unregulated insulin secretion. Mutations in the CFTR cause a variety of channel defects that constitute the pathophysiology of cystic fibrosis.

A. The Mechanism of Action of P-glycoprotein

Cloning experiments revealed that P-glycoprotein is not a single protein but a small family encoded by two genes in humans (MDR1 and MDR2) and by three genes in mouse (mdr1, mdr2, and mdr3). Only the products of human MDR1 and mouse mdr1 and mdr3 can transfer multidrug resistance.

The P-glycoprotein (MW 170,000) is a component of the plasma membrane that catalyzes the ATP-dependent removal of an enormous variety of chemicals from the cells, including such commonly used drugs in cancer chemotherapy as doxorubicin, daunorubicin, vincristine, vinblastine, etoposide, and taxol. During exposure to these drugs the P-glycoprotein content of plasma membrane increases and multidrug resistance develops. Contrary to earlier suggestions, the outward transport of the drugs is not likely to be mediated by changes in the pH gradient (ΔpH) or in the electrical potential ($\Delta\psi$), since in yeast secretory vesicles expressing mdr3 the accumulation of vinblastine was not affected by membrane potential and was independent of proton gradient or proton movements. Therefore, the transport probably requires direct interaction of the transported ligand with P-glycoprotein followed by an ATP-dependent translocation across one or both leaflets of the bilayer. The mode of interaction of the transporter with drugs of such varied structure remains a mystery. There is no simple relationship between the rate of drug transport and ATP hydrolysis. Though some activators of ATP hydrolysis such as verapamil are transported by Pgp, others such as progesterone are not transported, and not all known transported substrates stimulate ATP hydrolysis.

Highly purified, reconstituted P-glycoprotein by itself is capable of catalyzing the transport of daunorubicin or Hoechst 33342, suggesting no requirement for accessory proteins. Nevertheless, expression of Pgp in several cell types increased the magnitude of cell-swelling-associated Cl^- currents and imposed protein kinase dependence on channel activation. Therefore, in addition to its drug transport activity, the Pgp proteins also regulate heterologous channel proteins either by direct interaction or through intermediate proteins. [*See* Gene Amplification.]

A multidrug resistance related protein (MRP) was identified in a multidrug-resistant cancer cell line that can confer resistance even at normal levels of MDR1. MRP can function as a transporter of glutathione-conjugated molecules and may contribute to resistance against drugs conjugated by glutathione. The structure of MRP is closely related to the cystic fibrosis conductance regulator.

B. The Cystic Fibrosis Conductance Regulator

The CFTR contains a large hydrophilic regulatory (R) sequence between the two TM–ATP domains. The R domain has multiple sites for phosphorylation. CFTR is unique among ABC transporters since it functions as a cAMP-activated nonrectifying Cl^- channel that permits downhill Cl^- flux across the membrane. In addition, the CFTR also regulates several other ion channels that may contribute to the pathophysiology of cystic fibrosis. Among these is an outwardly rectifying Cl^- channel known as ORCC that is activated in normal cells by protein kinase A but fails to be activated by cAMP in cystic fibrosis cells. The normal CFTR imposes cAMP regulation on ORCC in reconstituted systems. The increased Na^+ absorption in cystic fibrosis epithelia is due to increased activity of an amiloride-sensitive Na channel. The Na^+ channel can be inhibited by protein kinase A in a heterologous expression system in the presence of normal CFTR, decreasing the epithelial absorption of Na^+. Finally, CFTR also controls the cAMP-dependent activation of epithelial K^+ channels that provide the driving force for Cl^- secretion.

In summary, the normal function of CFTR assures the proper activation of the Cl^- channels in both CFTR and ORCC, the opening of the K^+ channels, and the closing of the Na^+ channels, thus maintaining normal ion balance in the epithelial tissue of the lung.

C. The Sulfonyl Urea Receptor

The sulfonyl urea receptor is a recently discovered ABC protein that imposes sulfonyl urea sensitivity on the ATP-sensitive K^+ channels (K_{ATP}) of pancreatic β cells. Expression of K_{ATP} in the absence of SUR does not lead to detectable K^+ conductance. Coexpression with SUR reconstitutes a K^+ channel conductance that is similar to that of endogenous K^+ channels. Activation of the K_{ATP} channels by SUR hyperpolarizes the cells, decreases Ca^{2+} entry through voltage-sensitive Ca^{2+} channels, and inhibits insulin secretion. Binding of sulfonyl ureas to the receptor activates insulin release by inhibition of the K_{ATP} channel. Mutations in the SUR gene are associated with persistent hyperinsulinemic hypoglycemias.

VII. COUPLED TRANSPORT PROCESSES (SYMPORT AND ANTIPORT)

The Kedem equation provides a general description of transport coupling in terms of the following parameters:

$$J_S = (1/R_S)(\Delta\mu s + zsF\Delta\psi + R_{SW}J_W + R_{SA}J_A + R_{SB}J_B + R_{SR}J_R)$$

The movement (J) of an ion or molecule (S) across the membrane is subject to the driving forces of concentration gradient ($\Delta\mu s$), electric field ($\Delta\psi$), water flow (J_W), interaction with other ions or molecules (J_A, J_B . . .), and chemical couplings with a transporter or with a metabolic reaction (J_R). R_S is the resistance of the solute to movement, whereas R_{SW}, R_{SA}, R_{SB}, and R_{SR} are other types of resistances to coupled flow related to the various interactions identified by the equation. Direct coupling with a transporter (J_R) is in the center of interest.

The coupled transport processes can be classified into at least two major categories:

1. *Symport or cotransport* is a coupled transport of two solutes moving in the same direction across the membrane, mediated by a single transporter protein. Examples of symport are the Na^+-linked uptake of sugars, amino acids, and neurotransmitters in animal cells, and the analogous H^+-linked uptake of sugars, amino acids, and nucleosides by plants, fungi, and bacteria.

2. *Antiport or exchange* is the coupled transport of two solutes moving in opposite directions across the membrane, mediated by a transporter protein. Examples of antiport are ATP–ADP exchange in mitochondria, $Na^+:Ca^{2+}$ exchange in cardiac muscle, $Na^+:H^+$ exchange in many cell types, and the $Cl^-:HCO_3^-$ exchange in red blood cells. In most of these processes the downhill transport of one of the ligands provides energy for the uphill transport of the other. The mechanism of coupled transport processes will be illustrated by a few selected examples.

A. Na^+–Glucose Cotransport

This classic example of ion-coupled metabolite transport originates from observations made 40 years ago that the intestinal active transport of sugars is dependent on sodium ions in the lumenal bathing solution. In the intervening years, the Na^+–glucose cotransporter (SGLT1) was identified as a member of the large family of 12-helix membrane proteins, which include transporters of glucose, amino acids, nucleosides, inositol, pantothenate, and drugs in bacteria, plants, and animal cells. The active transport of sugar is driven by the Na^+ electrochemical potential across the cell membrane. A possible mechanism of the transport process is outlined in Fig. 8. When the carrier faces the intestinal lumen in the absence of Na^+, its

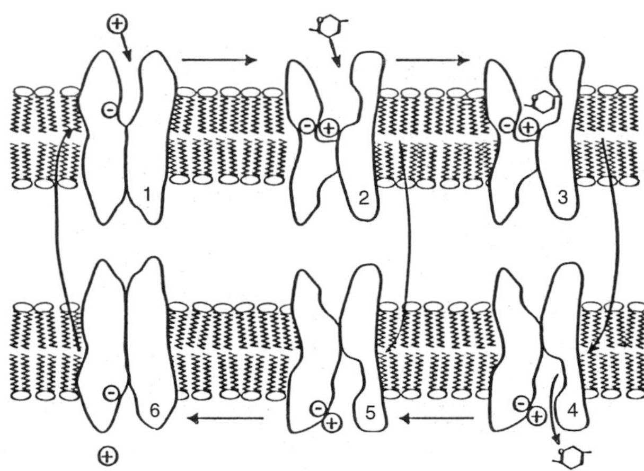

FIGURE 8 A six-state kinetic model for Na^+–glucose cotransport. SGLTl is shown as a charged protein which exists in two major conformations. The transport cycle is shown for Na^+–glucose influx when the external solution contains high Na^+ and sugar concentrations, the intracellular solution is low in Na^+ and sugar, and the membrane potential is −50 mV. Two Na^+ ions bind at the outside surface (in a well 30% across the membrane field), and the change in conformation from 1 to 2 allows glucose to bind (2 to 3). The fully loaded complex undergoes a major conformational change to deliver the ligands to the internal surface (3 to 4), where two Na^+ ions and a sugar molecule are released into the cell. The potential difference facilitates the return of the protein to the outward-facing conformation (6 to 1). Computer simulations of comprehensive kinetic data provide estimates of the associated rate constants: k_{12}, 80,000/mol²/sec; k_{21}, 500/sec; k_{23}, 1×10^5/mol/sec; k_{34}, 50/sec; k_{32}, 20/sec; k_{45}, 800/sec; k_{54}, 1.8×10^7/mol/sec; k_{56}, 10/sec; k_{65}, 50/mol²/sec; k_{61}, 5/sec; k_{16}, 35/sec; k_{25}, 0.3/sec; and k_{52}, 1.4/sec. At voltages different from 0, k_{12} and k_{21} are changed by a factor exp $\pm 0.3\mu$ and k_{16} and k_{61} by exp $\pm 0.7\mu$, where $\mu = VF/RT$, and 0.3 and 0.7 are the fraction of the field sensed by Na binding and SGLT1 conformational changes. [From E. M. Wright, D. D. F. Loo, M. Panayotova-Heiermann, and K. J. Boorer (1994). *Biochem. Soc. Trans.* **22**, 646–650.]

affinity for glucose is low. With increase in Na^+ concentration, two Na^+ ions are bound, inducing a structural change in the carrier that permits high-affinity glucose binding. The fully loaded carrier then undergoes a major conformational change that causes the translocation of Na^+ and glucose across the membrane and their release into the cytoplasm. The low cytoplasmic Na^+ concentration is maintained by the Na^+-K^+-ATPase, permitting the continued accumulation of glucose inside the cell driven by the electrochemical gradient of Na^+ ions. The final step in the transport cycle is the reorientation of the ligand binding sites of the carrier from the cytoplasmic to the outside surface of the membrane facilitated by the membrane potential. The net result is the translocation of two Na^+ and one glucose molecule from the lumen into the cytoplasm in each cycle of the transporter. A similar six-state ordered model was also proposed for the Na–myoinositol, H^+–dipeptide, H^+–glucose, and H^+- or Na^+–amino acid transporters.

The Na-dependent accumulation of glucose ceases in the presence of ouabain, which inhibits the Na^+-K^+-ATPase and allows the cytoplasmic Na^+ concentration to rise. Changes in membrane potential also influence the Na-dependent glucose transport by changing the Na^+ affinity and the conformational equilibrium of the carrier. The estimated dissociation constant (K_D) for Na^+ binding on the external surface is ~80 mM at 0 mV and 25 mM at −150 mV.

The Na–glucose cotransporter has the characteristics of a phlorizin-sensitive water channel. In the presence of glucose a secondary active transport of water accompanies the glucose uptake, which may amount to 5 liters of water taken up through the Na–glucose transporter per day in the human small intestine.

Apart from freeze-fracture studies on the cloned transporter expressed in oocytes, there is no high-resolution structural information on the Na–glucose transporter. Expression of SGLT1 in oocytes was accompanied by increase in the density of 75-Å-diameter freeze-etch particles in the membrane to values exceeding 4000/μm^2 surface area. The particles may represent either monomers or dimers of the transporter. Activation of protein kinases A or C increased the maximum rate of transport by human SGLT1 parallel with the increase in its density on the plasma membrane, suggesting that PKA or PKC regulates transport rate by modulating the concentration of SGLT1 in the plasma membrane through effects on the rates of endocytosis and exocytosis.

In patients with glucose–galactose malabsorption, several types of mutations were identified in the coding region and at the splice sites of the SGLT1 gene. Some of these mutations caused defects in the transport mechanism, such as decreased affinity for sugars, whereas others affected the targeting of the mutant protein to the surface membrane.

B. The $Na^+ : Ca^{2+}$ Exchanger

The $Na^+ : Ca^{2+}$ exchanger is an important regulator of cytoplasmic free Ca^{2+} concentration, transferring one Ca^{2+} ion out of the cell in exchange for three Na^+ ions entering the cytoplasm. The source of energy for the uphill, outward transport of Ca^{2+} is the electrochemical potential of Na^+ that, in turn, is maintained by the Na^+-K^+-ATPase. The $Na^+ : Ca^{2+}$ exchangers expressed in cardiac muscle, kidney, and most other cell types are alternatively spliced products of the NCX1 gene located on human chromosome 2, while in brain and skeletal muscle a second isoform is also expressed, coded by the NCX2 gene located on human chromosome 14. The brain also contains an NCX3 isoform. Little is known about the functional significance of isoform diversity and about the mechanisms that regulate the tissue specificity of their expression.

The predicted structure of NCX1 contains 970 amino acids that are arranged in 11 transmembrane segments (TMS) and a large cytoplasmic domain of 520 amino acids. The NCX2 gene product contains 921 amino acids and its predicted structure is similar to that of NCX1. The homology between various members of the exchanger superfamily is particularly pronounced in the TMS 1 and 2 and in the TMS 8 and 9 domains.

In addition to the Na^+ and Ca^{2+} transport sites, both NCX1 and NCX2 have a high-affinity Ca^{2+} regulatory site in their cytoplasmic domains. Deletion of a large part of the cytoplasmic domain removed the Ca^{2+} regulation, but left the transport activity unaffected, suggesting that the ion translocation function is associated with the transmembrane domain. The Ca regulatory sites have been localized to amino acids 371–508 in the cytoplasmic domain. The Ca affinities of the Ca regulatory sites are 0.3 μM in NCX1 and 15 μM in NCX2. The K_D for Na is close to 30 mM for both isoforms. The cytoplasmic domain contains a Na^+ inactivating site near TMS5.

It was suggested that following depolarization of the membrane at elevated internal Na^+ concentrations, the $Na^+ : Ca^{2+}$ exchanger may operate in reverse and provide Ca^{2+} for the activation of muscle. There is no solid evidence that such reversal occurs physiologically.

C. The Anion Transporter of Red Blood Cells (Band 3 Protein)

The anion transporter of erythrocytes catalyzes rapid $Cl^-:HCO_3^-$ exchange across the cell membrane. The process facilitates CO_2 entry into blood in tissue capillaries and the elimination of CO_2 from the blood in pulmonary capillaries.

The anion exchanger is a 95-kDa glycoprotein (band 3 protein) present in $\sim 10^6$ copies per cell, representing ~25% of the protein content of erythrocytes. The N-terminal half of the molecule (43 kDa) forms a hydrophilic cytoplasmic domain that contains binding sites for other cytoskeletal proteins and contributes to the regulation of cell shape. The C-terminal half (~53 kDa) is highly hydrophobic, consists of 14 putative α-helices, and forms the transmembrane pathways for anions.

The $Cl^-:HCO_3^-$-exchange occurs by the usual double displacement mechanism. The carrier alternates between at least two distinct conformations, facing the inside and outside, respectively. The transitions between the two states are coupled to the transfer of Cl^- in one direction and the HCO_3^- in the other direction. As predicted by this model, an inward-directed Cl^- gradient ($Cl_o^- \gg Cl_i^-$) increases the proportion of inward-facing carriers and vice versa. Due to this shift in conformational equilibrium, the apparent affinity of an anion on the cis side is influenced by the concentration of an anion on the trans side. Interestingly, even with equal Cl^- concentrations (150 m*M*) on the two sides of the membrane, there are more inward-facing states, indicating some intrinsic asymmetry in the carrier. For these reasons it is difficult to determine the intrinsic affinity of the carrier for an anion. Estimates based on competition with the stilbene disulfonate channel inhibitors give an apparent K_D for Cl^- of ~20–80 m*M*.

Based on physical evidence, the carrier protein may exist as a dimer (or perhaps tetramer) in the membrane, although radiation inactivation studies give a target size of only 59 kDa for the binding of 4,4′-dibenzamidostilbene-2,2′-disulfonate, and isolated band 3 monomers appear to be functional in anion transport when reconstituted in lipid vesicles.

At rest the red cell anion exchange is not a vitally important process for survival. Some red cells (lamprey and hagfish) do not possess anion exchangers, and deletion of its gene in mice and cattle did not interfere with survival, although it caused chronic hemolytic anemia. However, during muscle activity or with inadequate ventilation or circulation the process becomes essential and this may explain its conservation during vertebrate evolution.

Mutations in the band 3 gene cause changes in red cell shape, together with altered anion transport. For example, the defective anion transport in hereditary ovalocytosis is due to substitution of Lys 56 with Glu and a deletion of amino acids 400–408 at the boundary between the cytoplasmic and membrane domains. By contrast, the band 3 protein mutations in acanthocytosis are accompanied by increased anion transport activity. The consequences of these mutations are varied due to the interaction of band 3 protein with several cytoskeletal components and because of the participation of other exchangers in anion transport.

The $Cl^-:HCO_3^-$-exchanger is also present in the conducting ducts of kidney medullae and in gastric mucosa, where it provides a pathway for HCO_3^- transport during acid secretion.

VIII. FACILITATED DIFFUSION (UNIPORT)

Several metabolites (glucose, nucleosides, amino acids, etc.) pass through cellular membranes passively, but aided by transporters that interact with them and accelerate their rate of translocation. These transporters also operate by cyclic changes between two conformations, binding the transported solutes on one side of the membrane and releasing them on the other side. In contrast to exchange (antiport), the transport is unidirectional (uniport), and it is assumed that the transporter returns unloaded to the starting conformation. However, since the molecular mechanism of the process is generally unknown, countertransport of some unidentified ligand (water, H^+, etc.) cannot be excluded. The facilitative transporters of sugars and amino acids in mammalian cells provide interesting illustrations of the mechanism of these processes.

A. The Family of Mammalian Facilitative Glucose Transporters

In contrast to the intestinal Na : glucose cotransporter (SGLT1) that utilized the electrochemical Na^+ gradient to drive the active transport of glucose and galactose into the cells, the energy-independent, facilitated diffusion of hexoses is mediated by the GLUT family of glucose carriers. The seven members of the GLUT family (GLUT 1–7) show tissue-specific expression with pronounced differences in their affinities for glucose, galactose, fructose, and other hexoses, but share with each other and with a large superfamily of other membrane transport proteins a unique structure consisting of 12 transmembrane α-helical segments with intracellularly located C and N terminals (Fig. 9). The six helices in the N-terminal and in the C-terminal halves of the molecule are assumed to form two compact domains connected by a long intracellular loop between helices 6 and 7. Mutagenesis of conserved amino acids in helices 7, 10, and 11 and deletion of the intracellular C-terminal segment interfere with glucose transport, suggesting that the C-terminal domain contains the binding sites for glucose and drives the conformational changes associated with glucose translocation. Mutation of glycine residues to aspartate or arginine in helices 1 and 2 of the N-terminal domain interfered with the insertion of the protein into the target membrane, suggesting a role in membrane insertion. GLUT 1–5 are associated with the plasma membranes of various cells, whereas GLUT7 is a component of the endoplasmic reticulum. The targeting to the plasma membrane or to the endoplasmic reticulum involves specific targeting mechanisms.

The inward- and outward-facing glucose binding sites may be structurally separate, as cytochalasin B competitively inhibits sugar efflux by binding at or near the inward-facing site, whereas 3-iodo-4-azido-phenethylamido-7-*O*-succinyl deacetylforskolin (IAPS-forskolin) and 2-*N*-4(1-azi-2,2,2-trifluoroethyl)benzoyl-1,3-(D-mannos-4-yloxy)-2-propylamine (ATP-BMP) bind to the outward-facing binding site and competitively inhibit glucose influx. Truncation of the C terminus of GLUT1 locks the transporter in the inward-facing conformation, which has low affinity for exofacial ligands such as ATP-BMP.

The tissue-specific isoforms of the GLUT family show distinct kinetic and regulatory characteristics with important metabolic implications.

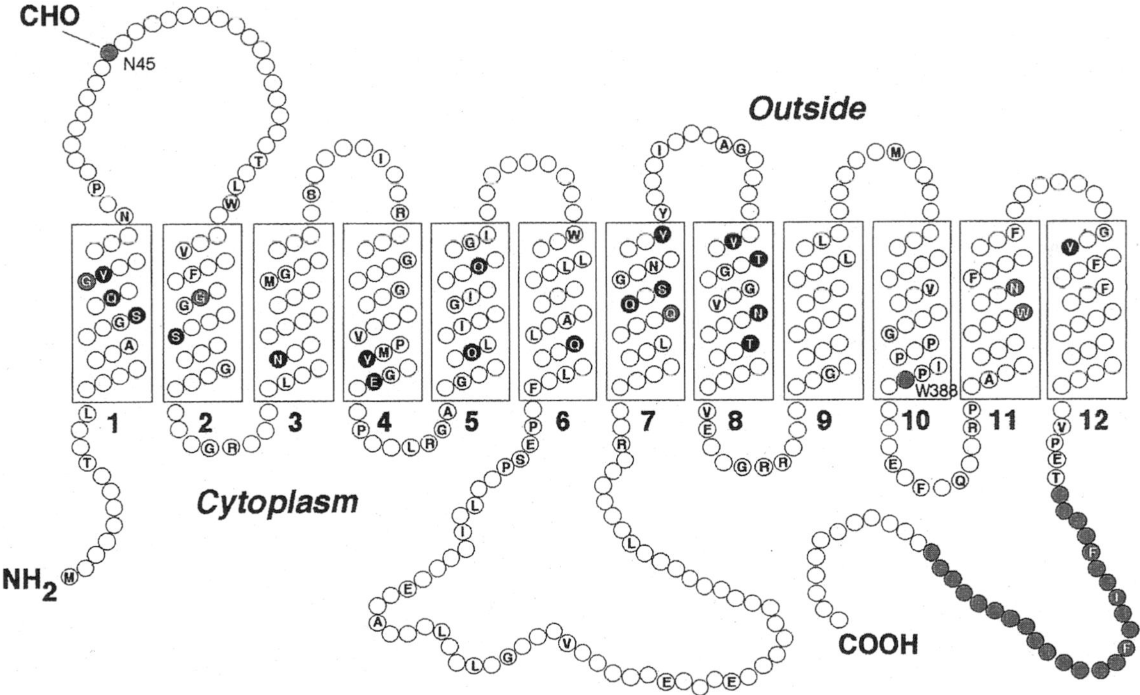

FIGURE 9 Model for the orientation of mammalian facilitative sugar transporters in the membrane. The 12 transmembrane helices are shown as boxes and are numbered 1–12. The potential site for N-glycosylation in the extracellular loop connecting transmembrane segments 1 and 2 is identified. Invariant residues are noted using the single-letter abbreviations. Amino acids with polar residues in the lipid bilayer are shown in solid circles. Mutagenesis or deletion of some of these residues, as well as residues in the C-terminal tail, affects transporter function. [From G. I. Bell, C. F. Burant, J. Takeda, and G. W. Gould (1993). *J. Biol. Chem.* **268**, 19161–19164.]

GLUT1 is the principal glucose transporter of erythrocyte membranes, where it comprises about 3–5% of the membrane proteins. It is also expressed at particularly high levels in cells of the blood/brain barrier, kidney, colon, lactating mammary glands, and fetal tissues and in tissue culture cells. The K_m for net influx of glucose through GLUT1 is relatively low (~1.6 mM) and it functions as a unidirectional transporter when the extracellular glucose concentration is low and the intracellular need for glucose is high.

Although GLUT1 is generally considered to be a constitutively expressed transporter located on the cell surface and responsible for the basal glucose uptake, there are instances of major changes in the glucose transport activity in cells that express only GLUT1. For example, the level of GLUT1 in the mouse mammary gland is relatively low during most of pregnancy, but rapidly increases more than 10-fold with the onset of lactation. Removal of the litters causes a rapid decrease in the level of GLUT1, reaching baseline values within 24 hr. The increase in GLUT1 expression is induced by prolactin, presumably to meet the increased demand for glucose in milk production. Although the effect of prolactin on mammary gland involves increased synthesis of GLUT1, the increased glucose uptake in hemopoietic progenitor cells under the influence of interleukin-3 may be due primarily to translocation of GLUT1 from intracellular sites to the cell surface. Increased expression of GLUT1 is observed in cultured cells following transformation under the influence of mitogens and during glucose starvation.

GLUT2 is unique among the various isoforms because it catalyzes the transport of both glucose and fructose. It is present in liver, in the basolateral membrane of intestine, in proximal tubules of kidney, and in pancreatic β cells. GLUT2 is a low-affinity (K_m = 15–66 mM) but high-capacity transporter for glucose. Therefore, the rate of glucose transport catalyzed by GLUT2 is proportional to glucose concentration over a wide physiological range, providing a rationale for its localization in the liver, where it facilitates rapid glucose efflux during gluconeogenesis, and in the intestine and kidney, where it catalyzes the fast transepithelial fluxes of glucose. In pancreatic β cells, GLUT2 was suggested to serve as a sensor of blood glucose concentration. In the liver, GLUT2 is the primary transporter of fructose.

GLUT3 is the glucose transporter in neurons and is also present in placenta and testis, but is absent from muscle. It has high affinity for glucose (K_m = 2–5 mM) and acts together with GLUT1 to satisfy the high glucose demand of the actively metabolizing brain cells.

GLUT4 is the insulin-stimulated glucose transporter of skeletal muscle, heart, and fat tissue. In adipose cells and in muscle, insulin increases the rate of glucose transport severalfold, without much change in the apparent affinity of the transporter for glucose (K_m = 2–5 mM). Insulin produces this effect by recruitment of GLUT4 from intracellular sites to the plasma membrane. For example, in rat adipocytes, insulin increases the exposure of GLUT4 on the cell surface by 15- to 20-fold, with only lesser change in the exposure of GLUT1. This effect is apparent within 2 min after exposure to insulin and it is explained largely by an increase in the rate of exocytosis, without much change, or perhaps some decrease, in the rate of endocytosis.

In contrast to the virtual absence of GLUT4 on the cell surface in the basal state, a large portion of GLUT1 is maintained on the cell surface even in the absence of insulin. The structural basis of this difference is unclear. Expression of GLUT4 in heterologous cells, which are not normally insulin responsive, results in intracellular sequestration of GLUT4 that is maintained even after insulin treatment. This indicates that expression of GLUT4 on the cell surface requires some mechanism that operates only in insulin-responsive cells. The intracellular pool of GLUT4 is depleted in the adipose tissue of some Type II insulin-resistant diabetics, but in the muscles of these individuals little or no changes were observed in GLUT4 mRNA or protein content.

GLUT5 is the principal fructose transporter of the small intestine, where it is located in the apical brush border on the luminal side of epithelial cells. It is also present in smaller amounts in muscle, brain, and fat tissue but it is absent from the liver. Its K_m for fructose is 6 mM but the affinity for glucose is very low. Neither the localization nor the activity is affected by insulin, consistent with the lack of insulin effect on fructose uptake by muscle or adipose tissue.

GLUT6 is an inactive pseudogene, related to GLUT3.

GLUT7 is localized in liver endoplasmic reticulum and participates together with glucose-6-phosphatase and GLUT2 in the release of glucose from the liver into the circulation. The retention of GLUT7 in the endoplasmic reticulum may be explained by the KKMKND targeting motif in the C terminus of the protein. Otherwise GLUT7 is closely related structurally to GLUT2.

The evolution of the six isoforms of GLUT to serve the diverse metabolic requirements of various tissues is an interesting example of metabolic adaptation.

B. The Mammalian Amino Acid Transporters

Mammalian cells mediate amino acid transport by a complex and overlapping set of transporters that differ in kinetic properties, substrate recognition, and requirement for Na^+ ion cotransport.

The Na-dependent systems A and ASC are expressed in most cells and transport small aliphatic amino acids. System A is inducible by hormones, growth factors, and starvation; it is highly pH sensitive and transports *N*-methylamino-α-isobutyric acid (methyl-AIB). System ASC is not inducible, is pH insensitive, and does not transport methyl-AIB. The widely expressed Na-dependent system X^-_{AG} transports glutamate and aspartate, whereas system β transports β-alanine and taurine.

Other Na-dependent transporters are more tissue specific. The broad specificity B^o, B, and B^{o+} systems are concentrated in kidney, intestine, and fibroblasts. System Gly transports glycine and sarcosine in liver, erythrocytes, and brain, whereas system N is the transporter for glutamine, histidine, and asparagine in the liver.

Among the Na-independent amino acid uniporters, the most widespread are system L, which transports mainly branched chain and aromatic amino acids, and system Y^+, which transports lysine, histidine, and arginine. Other Na-independent transporters of more specific localization are b^{o+} in fibroblasts and blastocysts for the transport of neutral and basic amino acids, and system X_C^- in hepatocytes and fibroblasts for the transport of glutamine and cystine. Several transport systems with overlapping specificities may occur in the same cell membrane.

All amino acid transport systems are stereospecific and favor L-amino acids, but otherwise their specificity is limited.

The nearly ubiquitous Na-independent system Y^+ for the transport of cationic amino acids (arginine, lysine, and histidine) serves as an example to illustrate the mechanism of an amino acid uniport.

1. The Structure and Function of Cationic Amino Acid Transporters

The mouse cationic amino acid transporter (CAT) genes *mCAT1* and *mCAT2* are located on chromosomes 5 and 8 and encode three proteins named CAT1, CAT2, and CAT2A. All cells express the *mCAT1* transcript (CAT1). The *mCAT2* gene encodes two tissue-specific isoforms—CAT2 and CAT2A—that result from alternative splicing. The three CAT isoforms share similar predicted structures consisting of 14 transmembrane helices (TM1–TM14), with both N and C termini exposed on the cytoplasmic surface (Fig. 10). CAT1 and CAT2 share 61% sequence identity whereas CAT2 and CAT2A differ only in a 41-amino-acid segment located between the

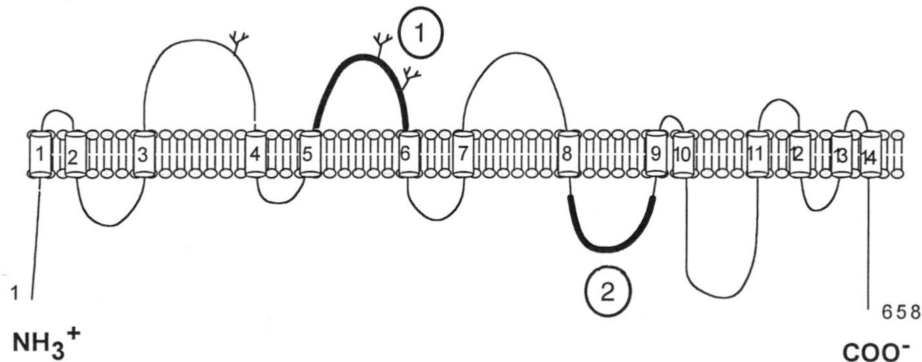

FIGURE 10 Model of mCAT protein. Locations of the virus binding site of mCAT1 ① and the alternatively spliced region of mCAT2/2A ② are shown, assuming there are 14 membrane-spanning (TM) domains. In the alternative 12-TM model, domains 7 and 9 do not cross the membrane, changing the orientation of the alternatively spliced region with respect to the membrane. The predicted glycosylation sites are indicated by branched structures. [From C. L. MacLeod (1996). *Biochem. Soc. Trans.* **24**, 846–852.]

TM8 and TM9 helices. The external loop of CAT1 between TM5 and TM6 contains a receptor site for the mouse ecotropic virus.

The transport properties of CAT1 and CAT2 analyzed after expression of cDNA in *Xenopus* oocytes are nearly identical and characterized by high affinity, relatively low capacity, and stimulation from the trans-side of the membrane. By contrast, the CAT2A isoform has low affinity and high capacity and lacks trans-stimulation. These unique properties of CAT2A are determined by the alternatively spliced 41-amino-acid segment, between helices 8 and 9. Transfer of this region to CAT1 and CAT2 in chimeric proteins confers on them the properties of the donor CAT2A.

Upon addition of basic amino acids to voltage-clamped oocytes expressing a system Y^+ transporter, a concentration-dependent saturable inward current appears that correlates with the uptake of radioactive amino acid, suggesting that one positive charge is carried into the cell per molecule of substrate taken up. The inward current is Na independent and its characteristics are consistent with a facilitated diffusion mechanism involving binding and release of substrates at sites exposed alternately to the extracellular or intracellular medium. At external arginine concentrations of 0.01–1 m*M*, the influx of arginine increased exponentially with hyperpolarization of the membrane (e-fold increase/-59 mV). The efflux of arginine from oocytes preloaded with arginine increased exponentially with depolarization (e-fold increase/-59 mV). Therefore the carrier operates as an electrogenic uniporter in both directions across the membrane. The observations suggest that the charge movement arises from a conformational transition of the unliganded transporter, as it switches from one side of the membrane to the other. Similar results were obtained with lysine and ornithine as transported substrates.

Arginine is the exclusive and immediate precursor of nitric oxide, a signaling molecule with a recognized role in the relaxation of vascular smooth muscle and in brain function. The availability of arginine may limit the rate of agonist-induced NO synthesis. Vasoactive agonists hyperpolarize the membrane parallel with the induction of NO-mediated vasodilation, suggesting that the increased arginine influx induced by hyperpolarization may contribute to the sustained NO synthesis.

The functional significance of the existence of tissue-specific CAT isoforms is unclear. All tissues, except intact liver, express CAT1; all tissues except kidney, intestine, and liver express CAT2, and liver is the only tissue that exclusively expresses CAT2A, the low-affinity isoform. The differential pattern of expression reflects differences in the promoter structures of the *mCAT1* and *mCAT2* genes that impart distinct sensitivity to hormones and growth factors.

IX. CHANNELS AS INSTRUMENTS OF SIGNAL TRANSDUCTION

The plasma membranes of cells act as diffusional barriers and as electrical insulators. Therefore the exchange of signals between cells requires specific receptors and channels that serve as conduits of material exchange, with associated changes in membrane potential.

Most tissue cells communicate with their immediate neighbors through relatively simple gap junctions by direct exchange of ions, small metabolites, growth factors, and second messengers. By contrast, the long-distance communication between the nervous system and the peripheral organs involved in secretory or motor activity occurs through specialized contacts called synapses that control the membrane potential of the postsynaptic (muscle or secretory) cells. In these processes the signal is carried from the nerve cell to the synapses as a wave of potential change in the nerve membrane generated by the sequential opening and closing of channels for Na^+ and K^+ ions. The signal transmission from nerve to the effector organ involves the release of a transmitter substance from the presynaptic cells that acts on the postsynaptic receptors triggering the final response. The chemical nature of the transmitter varies with the type of synapse.

As examples of channel activities we shall discuss the function of gap junctions and the coupling of nerve excitation to contraction in muscle cells.

A. Gap Junctions: The Cellular Internet

Apart from a few terminally differentiated cells, such as skeletal muscle and circulating blood cells, most cells of normal tissues communicate through gap junctions. In the nervous system and in the heart, these channels permit fast signal transfer and synchronization of electrical responses. During embryogenesis the gap junctions coordinate the activities of dividing cells and control pattern formation, while in fully differentiated cells they contribute to homeostatic control.

Malignant transformation is frequently accompanied by loss of junctional communication, suggesting that gap junctions play an important role in growth control.

The basic unit of a gap junction is the connexon, a hexamer of six connexin subunits. The connexons of one cell associate end to end with the connexons of a neighboring cell to form the gap junction channels (Fig. 11). These channels then further associate into gap junction plaques, which are extended quasi-crystalline regions on the cell–cell interface readily visible by electron microscopy.

The connexins belong to a large family of structurally related proteins (Cx) ranging in molecular weight between 26,000 and 50,000. The dominant form in human liver is Cx26, in brain Cx32 and Cx43, whereas in heart five distinct isoforms are present (Cx37, Cx40, Cx43, Cx45, and Cx50). The different isoforms are preferentially distributed in different regions of the heart. The various connexin isoforms share a basic topology, consisting of four transmembrane helices (M1–M4) with the C and N termini exposed on the cytoplasmic surface and two extracellular loops (E_1 and E_2) between the M_1 and M_2 and M_3–M_4 helices. The interactions between connexons of adjacent cells are mediated by the E_1 and E_2 loops. One of the transmembrane domains (M3) is amphipathic, suggesting that it may contribute to the lining of the channel. The cytoplasmic domains are variable among different connexins and are assumed to be involved in the regulation of channel activity.

The gap junctions generally permit the passage of inorganic anions or cations (Na^+, K^+, Ca^{2+}, H^+, Cl^-, PO_4^{2-}, etc.) and small hydrophilic molecules with diameters of 15 Å or less (such as ATP, ADP, cAMP, cGMP, inositol-1,4,5-trisphosphate). Different connexins are endowed with unique properties with respect to unitary conductance, ionic permeability, size cutoff, and channel gating that provide a rationale for their tissue-specific distribution. As sources of additional complexity, connexons may be *homomeric,* containing a single connexin species, or *heteromeric,* containing different connexins, and connexins of the same type may form *homotypic* channels, whereas different connexons may associate into *heterotypic* channels. The second extracellular domain (E_2) is the major determinant of the compatibility between different types of connexons. Developmental stage- and tissue-specific expression of the complexon isoforms is a powerful mechanism for specifically compartmentalized communication between cells. [*See* Cell Junctions.]

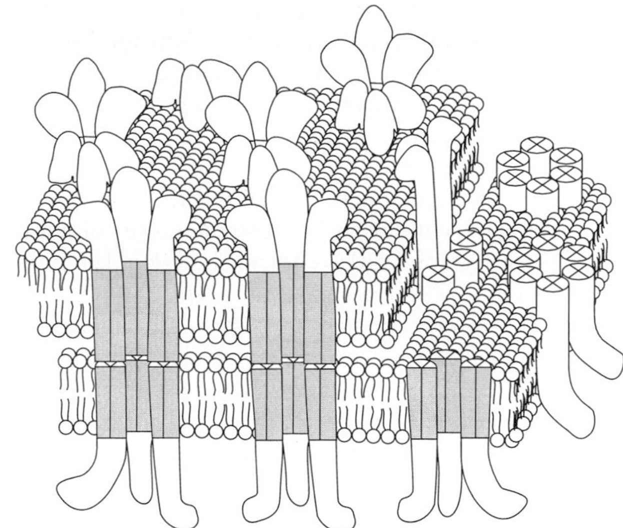

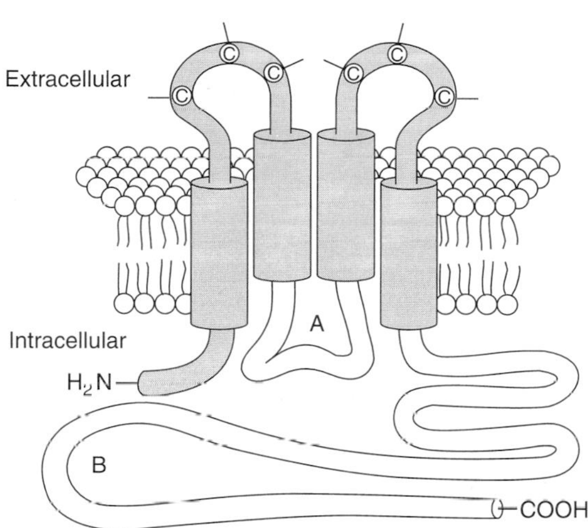

FIGURE 11 Top: Structural model of a gap junction based on electron micrographic and X-ray diffraction studies of isolated liver and heart gap junctions. Bottom: Topological model of connexin orientation within the junctional plasma membrane. Shaded regions represent sequences shared among all connexins. The unshaded predicted cytoplasmic domains (A and B) correspond to unique connexin-specific regions. [From E. C. Beyer and R. D. Veenstra (1994). "Handbook of Membrane Channels" (C. Peracchia, ed.), pp. 379–401. Academic Press, San Diego.]

The synthesis of complexons is regulated at the level of transcription, translation, and intracellular processing, and their functional state is under dynamic control by noncovalent and covalent modification of the channel structure. The conductance of Cx43 channels decreases with decrease in cytoplasmic pH to 6.2 or by increase in cytoplasmic Ca^{2+} above 10 μM; such conditions may arise during cardiac ischemia or other forms of cell death, facilitating the separation of dying cells from healthy tissue and promoting wound healing. Phosphorylation of connexins has a direct effect on the conductance of gap junctions but may also regulate their transit from the site of synthesis in the endoplasmic reticulum to the cell surface. The half-life of connexins is only 1.5–5 hr, and therefore ubiquitin-mediated proteolysis also controls junctional communication.

Aberrations in connexon expression have been implicated in cancer, cardiac ischemia, cardiac hypertrophy, Charcot-Marie-Tooth disease, and viscero-atrial heterotaxia, but the pathogenesis of connexon abnormalities is still not well understood.

B. The Coupling of Nerve Excitation to Muscle Contraction

Nerve excitation triggers muscle contraction by releasing Ca^{2+} from the sarcoplasmic reticulum that activates the contractile proteins of myofibrils. The process requires the participation of voltage-gated Na^+ and K^+ channels in the motor nerves and in the plasma membrane of muscle that generate the action potential wave, the acetylcholine-activated ion channel of the neuromuscular junction that serves as a synapse between the nerve and muscle, the voltage-gated dihydropyridine receptors of the transverse tubules that regulate the conductance of the sarcoplasmic reticulum Ca^{2+} channels, and the Ca^{2+} binding regulatory

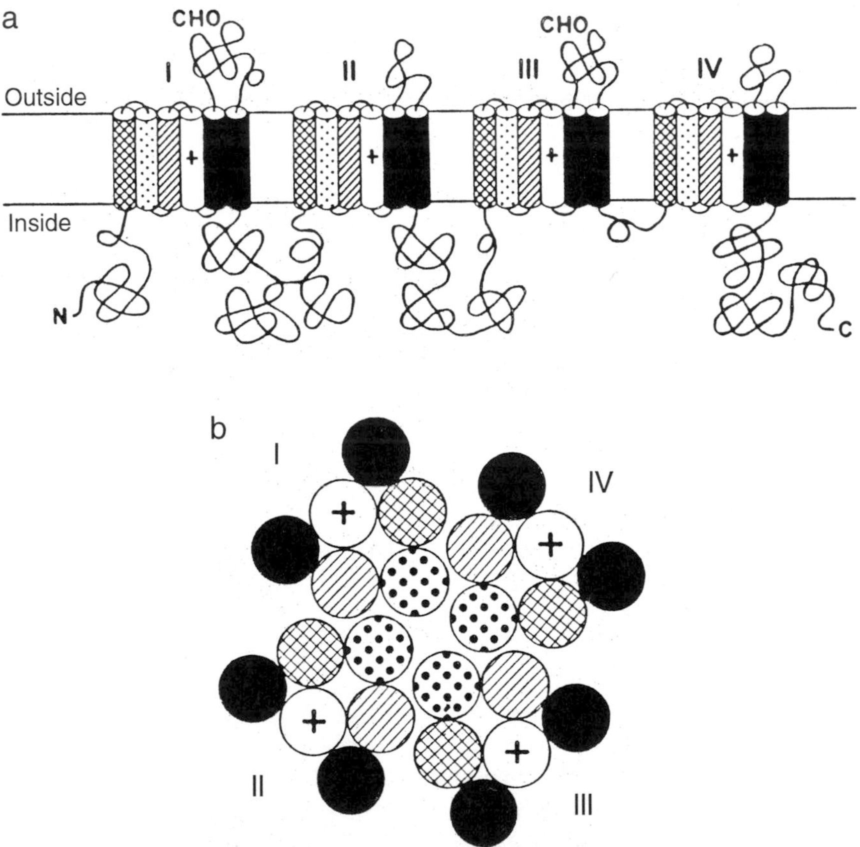

FIGURE 12 Proposed transmembrane topology of the sodium channel (top) and the proposed arrangement of the transmembrane segments viewed in the direction perpendicular to the membrane (bottom). Top: The four homology units spanning the membrane are displayed linearly (I–IV). Segments S1–S6 in each repeat (I–IV) are shown by cylinders. Putative sites of N-glycosylation (CHO) are indicated. Bottom: The ionic channel is represented as a central pore surrounded by the four homology units. Segments S1–S6 in each repeat (I–IV) are shown by circles marked as above. [From M. Noda, T. Ikeda, T. Kayano, H. Suzuki, M. Takeshima, M. Kurasaki, H. Takahashi, and S. Numa (1986). *Nature (London)* **320,** 188–192.]

proteins of the contractile apparatus that impart Ca^{2+} sensitivity on actomyosin. The Ca^{2+} released from the sarcoplasmic reticulum during each contraction is returned during relaxation by the Ca^{2+} transport ATPase, which is the major protein component of sarcoplasmic reticulum membrane. Each of these proteins of the E–C coupling apparatus occur in several isoforms that are coordinately expressed during muscle differentiation, with functional characteristics that match the physiological demands. The elucidation of this complex sequence of reactions is one of the landmark accomplishments of membrane biology that provides deep insight into the physical and chemical basis of channel function.

1. Generation of Action Potential and Its Propagation in Motor Nerves

The nerve cells, like all mammalian cells, have high K^+ and low Na^+ concentrations in their cytoplasm generated by the Na^+-K^+-ATPase. In the resting state the permeability of the surface membrane to Na^+ is low and the membrane potential (~ −60 mV) is close to the equilibrium potential of K^+ ions (~ −75 mV). A slight change in membrane potential during excitation of the nerve cell opens the voltage-sensitive Na^+ channels, permitting the entry of Na^+ ions into the cell; this in turn causes further depolarization by a positive feedback, changing the membrane potential from −60 mV to +30 mV (the Na^+ equilibrium potential) within a millisecond. At this positive potential the Na^+ channels are converted into a closed, inactive state that cannot be activated by depolarization. The K^+ channels open, permitting the outflux of K^+ ions, which returns the membrane potential to −75 mV (the K^+ equilibrium potential) followed by reequilibration to the resting level of −60 mV within a few milliseconds. The wave of depolarization, called action potential, spreads from the motor nerve cell through the motoneuron toward the neuromuscular junction at a rate of 10–25 m/sec.

The voltage-gated Na^+ channels isolated from nerve and muscle cells are large proteins containing close to 2000 amino acids. Analysis of their amino acid sequence revealed four internal repeats (I–IV) with homologous sequences that may have evolved from a single ancestor by internal duplication. Each internal repeat has five hydrophobic (S1, S2, S3, S5, and S6) and one positively charged (S4) segment. The S4 segment contains four to eight arginine and lysine residues located at every third position, with hydrophobic amino acids in between (Fig. 12). The six segments of each repeat are assumed to form transmembrane α helices arranged in a pseudosymmetric fashion, with the positively charged S4 segment serving as voltage sensor. The S4 segments are assumed to move outward in response to depolarization, causing the structural transition through several closed states into the open state.

Tetrodotoxin and saxitoxin block Na^+ channels by binding with high affinity ($K_D = 10^{-9}\,M$) to the extracellular mouth of the pore. Two rings of amino acids contributed by each of the four domains form the putative receptor site of tetrodotoxin. Local anesthetics and several antiarrhythmic drugs block Na^+ channels by entering the pore from the cytoplasmic side in the open state.

The Na^+, K^+, and Ca^{2+} channels isolated from various sources share fundamental structural relationships in the location of residues that contribute to the formation of the channel pore (Fig. 13). Amino acids in the SS1–SS2 region between helices S5 and S6 ap-

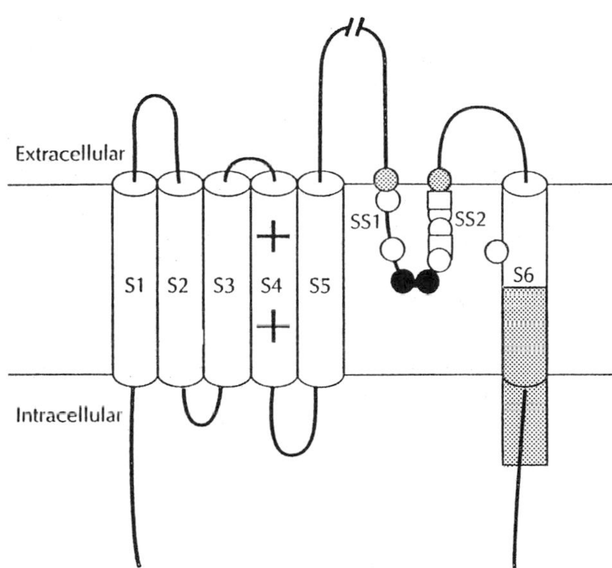

FIGURE 13 Pore-forming regions of the homologous domains of voltage-gated ion channels. The transmembrane folding pattern of a single homologous domain of a voltage-gated ion channel is shown with the short segments SS1 and SS2 and the positions of amino acid residues and segments implicated in pore formation marked. Open circles, residues required for ion conductance and selectivity of K^+ channels; open rectangles, residues required for ion conductance and selectivity of Na^+ and Ca^{2+} channels; lightly shaded circles, residues required for high-affinity binding of pore blockers of K^+ channels; darkly shaded circles, residues required for both ion conductance and binding of pore blockers of K^+ channels; shaded area in S6 is the segment of Ca^{2+} channel that binds phenylalkylamine pore blockers; + indicates the presence of positively charged residues that serve as gating charges. [From W. A. Catterall (1994). *Curr. Opin. Cell Biol.* **6**, 607–615.]

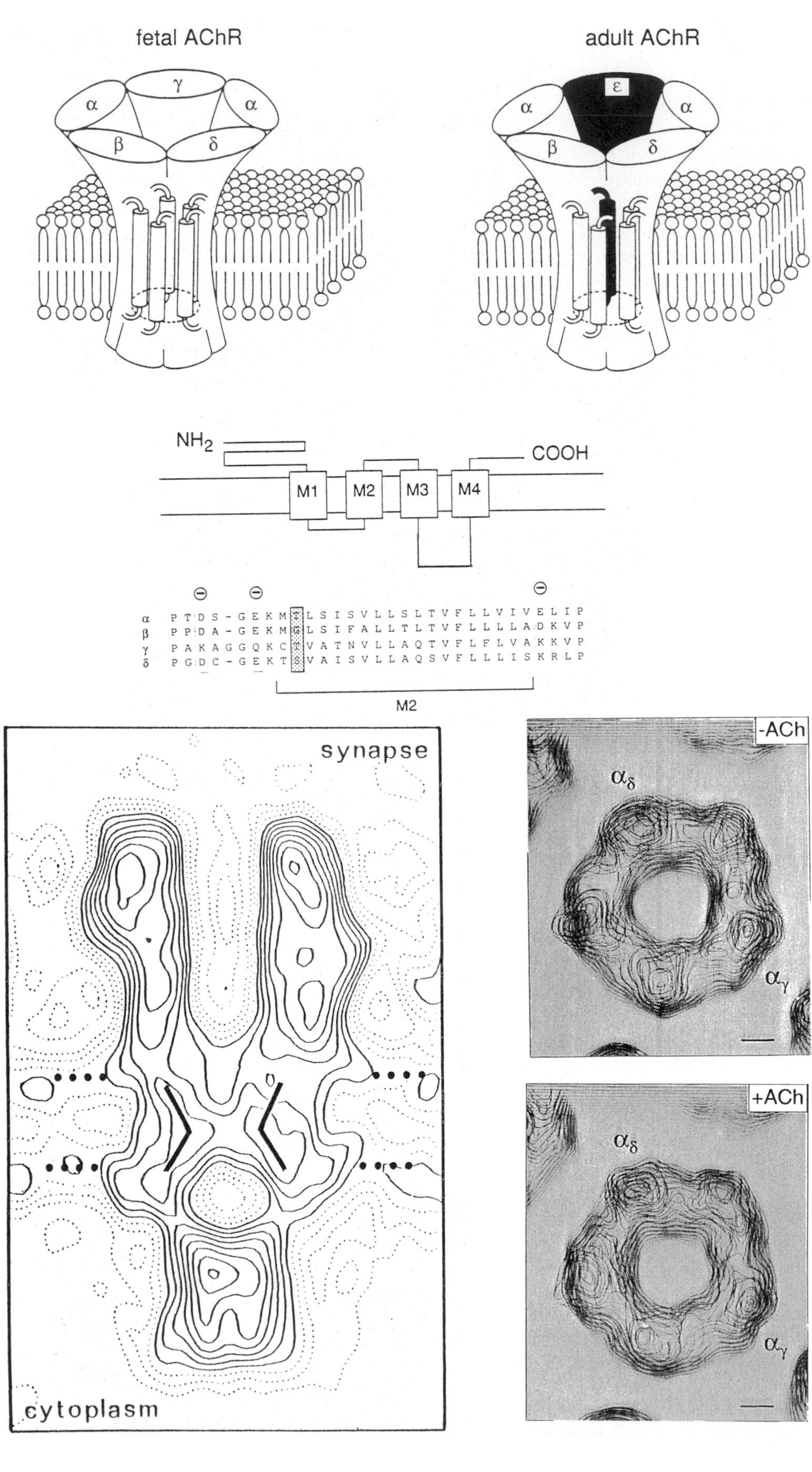
fetal AChR
adult AChR
NH2
COOH
M1
M2
M3
M4
M2
synapse
cytoplasm
-ACh
+ACh

pear to be key determinants of ion selectivity. Mutation of only two of these critical amino acids is sufficient to confer on the Na^+ channel the permeability properties of Ca^{2+} channels.

At positive membrane potential the Na^+ channel enters from the open into a closed inactivated state, in which it does not open upon depolarization. A short cytoplasmic peptide segment between domains III and IV was identified as the inactivation gate, which blocks the intracellular mouth of the pore in the inactivated state. Mutation or deletion of this segment or application of antipeptide antibodies directed against the S3–S4 loop slows the inactivation process.

2. Signal Transmission from the Motor Nerve to Muscle at the Neuromuscular Junction

When the action potential of the motor nerve reaches the neuromuscular junction it releases packets of acetylcholine from the presynaptic vesicles into the synaptic cleft. The acetylcholine binds to the nicotinic acetylcholine receptor of the postsynaptic membrane and increases its conductance to Na^+ and K^+. At a membrane potential of -70 mV, the current flowing through an open acetylcholine receptor channel is ~2.7 pA or 1.7×10^7 ions per second, which corresponds to a conductance of 27 picosiemens (pS).

The acetylcholine receptor (AchR) of the nerve–muscle synapse is a protein complex of ~290 kDa; it consists of a ring of five homologous subunits ($\alpha_2\beta\gamma\delta$) that form a pathway for ions across the membrane (Fig. 14). The γ subunit of fetal AchR is replaced by a structurally similar ε subunit in adult AchR, with increase in the amplitude of elementary currents. The channel pore is narrow (9–10 Å) across the membrane but extends on both sides into 20-Å-wide mouths that project 65 Å into the synaptic cleft and 15 Å into the cytoplasm of the muscle cell. Although only the α subunits bind acetylcholine, the five subunits have similar amino acid sequences, each forming four putative hydrophobic domains (M1–M4) that penetrate across the membrane. The long N-terminal and the short C-terminal tails are in the synaptic clefts, and the short links between M1–M2 and the longer links between M3–M4 are on the cytoplasmic side (see Fig. 14).

Conductance measurements on acetylcholine receptors with point mutations in the M2 transmembrane segments and in the flanking residues identified four amino acid positions that are important for cation transport through the open channel. Negatively charged amino acids bordering the M2 segment form two rings at the channel's extracellular entrance and one ring at the intracellular entrance that accumulate cations and exclude anions by electrostatic effects. The selection between monovalent cations is probably due to sieving by hydroxyamino acids located within the M2 segment at the constricted region of the pore (see Fig. 14). The binding sites for acetylcholine in the α subunits are located about 30 Å above the synaptic surface of the bilayer. The acetylcholine binding sites of the two α subunits are nonequivalent because of the influence of the neighboring γ and δ subunits, and bind acetylcholine with different affinity.

By rapid freezing of samples after acetylcholine addition, the open state of the channel can be trapped and its electron microscopic image compared with the image of the channel in the absence of acetylcholine. Acetylcholine opens the channel by initiating small rotations of the subunits in the extracellular domain that spread through the shaft to the membrane domain, where the M2 helices draw the gate-forming side chains away from the central axis of the channel (see Fig. 14). The open state of the channel is stabilized by side-to-side interaction between the transmembrane helices. Recordings of single-channel activity indicate that the liganded receptor flickers between closed and open states:

FIGURE 14 The structure of the acetylcholine receptor. Top: Schematic diagram of the subunit composition of end plate channel subtypes in fetal (left) and adult (right) skeletal muscle. Second row from top: Schematic drawing of assumed transmembrane folding of AchR subunits as suggested from hydropathy analysis. N- and C-terminal ends are extracellular. Third row from top: Localization of the anion rings and the selectivity filter of the AchR channel determined by single-channel conductance measurements of recombinant AchR channels carrying mutations in the M2 transmembrane domain. [Top three figures are from B. Sakman (1992). *Science* **256,** 503–512.] Bottom left: Channel in profile with positions of the transmembrane rods (dark lines) and the estimated limits of the bilayer (dotted lines) superimposed. Bilayer thickness is 30 Å. [From N. Unwin (1993). *J. Mol. Biol.* **229,** 1101–1124.] Bottom right: Views from the synaptic cleft of the mouth of the channel before and after activation by acetylcholine, made by stacking successive 2-Å-spaced sections on top of one another. [From N. Unwin (1995). *Nature* **373,** 37–43.]

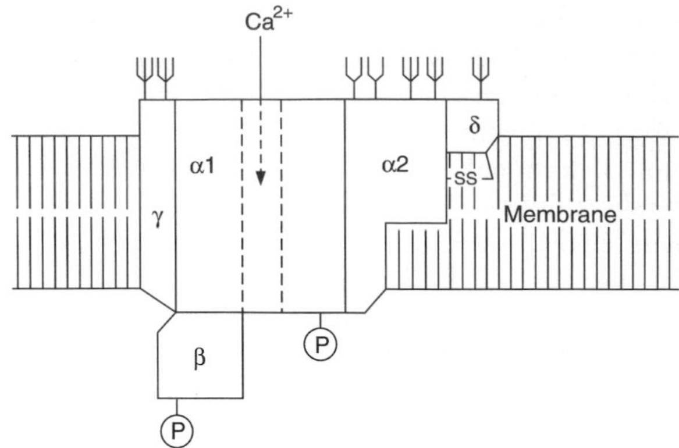

Subunit structure of transverse tubule voltage-sensitive calcium channel

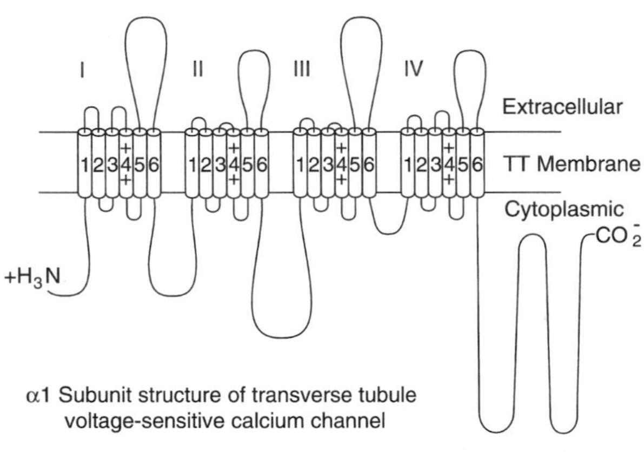

α1 Subunit structure of transverse tubule voltage-sensitive calcium channel

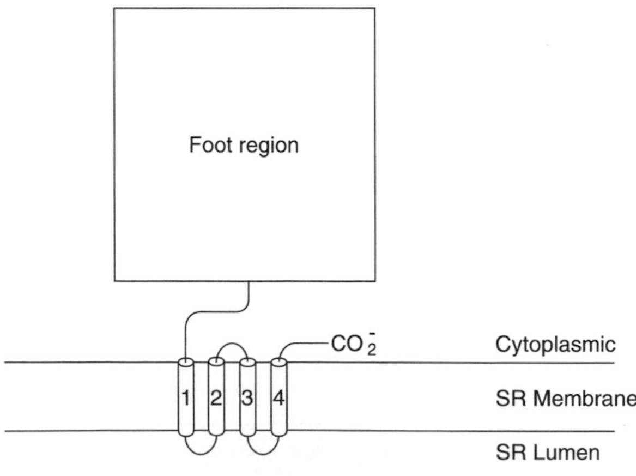

Sarcoplasmic reticulum calcium release channel

FIGURE 15 Channels of excitation–contraction coupling. Top: The subunit composition of the voltage-sensitive Ca^{2+} channel (dihydropyridine receptor) of the transverse tubules of skeletal muscle. The α_1 subunit serves both as voltage sensor and as voltage-gated Ca^{2+} channel in excitation–contraction coupling. Middle: The transmembrane topology of the α_1 subunit of T tubule dihydropyridine receptor. Bottom: The disposition of sarcoplasmic reticulum Ca^{2+} channel (ryanodine receptor) in the junctional face membrane of sarcoplasmic reticulum. The foot region is formed by the N-terminal cytoplasmic domain that links the sarcoplasmic reticulum to the T tubules. The C-terminal region forms the transmembrane domain that contains the Ca^{2+} channel. The precise number of transmembrane helices is unknown. [From W. A. Catterall (1995). *Cell* **64**, 871–874.]

$$\begin{array}{ccccccc} \text{R} & \longleftrightarrow & \text{(Ach)R} & \longleftrightarrow & \text{(Ach)}_2\text{R} & \longleftrightarrow & \text{(Ach)}_2\text{R} \\ \text{closed} & & \text{closed} & & \text{closed} & & \text{open} \\ & & \updownarrow & & \updownarrow & & \\ & & \text{(Ach)R} & \rightleftarrows & \text{(Ach)}_2\text{R} & & \\ & & \text{inactive} & & \text{inactive} & & \end{array}$$

After a brief open period the receptor is reversibly inactivated and will recover its activity only after the acetylcholine concentration is lowered to nanomolar levels by acetylcholinesterase. Since the channel opening is fast (time constant: 20 μsec) while the inactivation is slow (time constant: 50–100 msec), the probability of open state during the initial activation at high acetylcholine concentration is near 100%.

3. Excitation–Contraction (E–C) Coupling in Skeletal Muscle

The depolarization of the postsynaptic membrane initiated by acetylcholine propagates as an action potential wave through the surface membrane and the transverse tubular system (T tubules) into the interior of muscle fiber. The transverse tubules are extensions of the surface membrane for inward conduction of the excitatory stimulus. The junctions between the transverse tubules and the sarcoplasmic reticulum (triads, diads, etc.) contain the two key components of the E–C coupling apparatus. One of these is the dihydropyridine receptor (DHPR) of the T tubule, which serves as voltage sensor in E–C coupling by responding to changes in membrane potential. The other is the sarcoplasmic reticulum Ca^{2+} channel (ryanodine receptor, RyR), which serves as the channel for the release of activating Ca^2 from the sarcoplasmic reticulum into the cytoplasm. The DHPR and RyR are physically linked to each other either directly or through contacts with other proteins.

a. The Structure and Function of Dihydropyridine Receptors

The dihydropyridine receptor is a complex of five subunits ($\alpha_1\alpha_2\beta\gamma\delta$) (Fig. 15) that further associate to form tetrads in the transverse tubule membrane. The

α_1 subunits of the DHPR complex, similarly to those of the Na channel, are large proteins that consist of about 2000 amino acids distributed into four internal repeats (I–IV), with six helical transmembrane segments in each repeat (S1–S6). One of these repeats (S4) is the voltage sensor element (see Fig. 15). The SS1–SS2 region between the S5 and S6 segments may form a hairpin loop in the membrane and is assumed to line the channel pore.

There are distinct isoforms of DHPR in skeletal and cardiac muscles that differ in the kinetics of Ca^{2+} channel activation and in the Ca^{2+} dependence of E–C coupling. By expressing chimeric α_1 subunits in dysgenic mouse skeletal muscles that lack dihydropyridine receptor, the cytoplasmic loop between repeats III and IV was identified as the determinant of skeletal type, Ca^{2+}-insensitive E–C coupling and repeat I as the determinant of the kinetics of channel activation.

b. The Sarcoplasmic Reticulum Ca^{2+} Channel (Ryanodine Receptor)

The ryanodine receptor Ca^{2+} channel is a tetramer of subunits each with a molecular weight of 560,000. It is located in the junctional face membrane of sarcoplasmic reticulum in tight association with the T tubules. It was first seen by electron microscopy as dense structures (feet) extending into the junctional gap of ~120 Å and forming bridges between the T tubules and sarcoplasmic reticulum. Its high affinity for ryanodine, a plant alkaloid, proved useful in its isolation and characterization.

The RyR consists of a large N-terminal cytoplasmic domain with conduction pathways for Ca^{2+} and a channel-forming C-terminal domain, with 4–10 putative transmembrane helices (16–40 per tetramer) (see Fig. 15). It is a cation-selective channel with unusually large conductance for monovalent cations (~750 pS with 250 mM K^+) or Ca^{2+} (150 pS with 50 μM Ca). The Ca^{2+} release through the channel is stimulated by cytoplasmic Ca^{2+} (μM), ATP (mM), or caffeine (mM) and inhibited by Mg^{2+} (mM). Ryanodine at low concentration locks the channel in an open subconductance state, causing persistent Ca^{2+} release and activation of muscle contraction.

The physiological activation of the Ca^{2+} channel is triggered by a voltage-dependent structural change in the dihydropyridine receptor that is transmitted to RyR, opening the Ca^{2+} channel. Surprisingly, not all RyR complexes are in contact with DHPR tetrads. The activator of these unattached ryanodine receptors may be Ca^{2+} released into the junctional gap either from outside through the DHPR Ca^{2+} channels or from the sarcoplasmic reticulum through the attached ryanodine receptors. The opening of sarcoplasmic reticulum Ca^{2+} channels increases the cytoplasmic Ca^{2+} concentration from the resting level of 0.02 μM to 10–20 μM within a few milliseconds following the action potential, causing the activation of muscle contraction and an increase in the rate of catabolic processes required for ATP synthesis. A few milliseconds later the Ca^{2+} channel closes and Ca^{2+} is slowly reabsorbed into the sarcoplasmic reticulum by ATP-dependent active transport through the sarcoplasmic reticulum Ca^{2+}-ATPase.

In malignant hyperthermia, a point mutation in the ryanodine receptor increases its sensitivity to volatile surgical anesthetics (such as halothane), causing persistent Ca^{2+} leak from the sarcoplasmic reticulum. The increase in cytoplasmic Ca^{2+} concentration generates a "catabolic storm" in which body temperatures rise to lethal levels, due to a simultaneous increase in the rates of ATP hydrolysis and ATP synthesis. This vicious cycle can be broken by administration of dantrolene, an inhibitor of Ca^{2+} release.

The delicate mechanism for the regulation of cytoplasmic Ca^{2+} concentration exemplified by the E–C coupling apparatus is not unique to muscle but is also found in brain and in other cells. Besides the RyR, many cells contain a receptor for inositol-1,4,5-trisphosphate as a separate mechanism for the regulation of the Ca^{2+} stores of endoplasmic reticulum.

X. CONCLUSION

During the last decade, interest has shifted from the kinetic description and classification of transport processes to the analysis of the structure of transporters and their molecular mechanisms. This has already provided some insight into the structural basis of inherited diseases affecting metabolite or ion transport and has proved helpful in diagnosis and therapy. In the future, the trend toward structural aspects will extend to the characterization of conformational transitions associated with transport processes aimed at a physical description of the coupling between energy utilization and the interaction of transporters with the transported ions or molecules.

BIBLIOGRAPHY

Abrahams, J. P., Leslie, A. G. W., Lutter, R., and Walker, J. E. (1994). Structure at 2.8 Å resolution of F_1-ATPase from bovine heart mitochondria. *Nature* **370**, 621–628.

Andersen, J. P. (1995). Dissection of the functional domains of the sarcoplasmic reticulum Ca^{2+}-ATPase by site-directed mutagenesis. *BioSci. Rep.* **15**, 243–261.

Bell, G. I., Burant, C. F., Takeda, J., and Gould, G. W. (1993). Structure and function of mammalian facilitative sugar transporters. *J. Biol. Chem.* **268**, 19161–19164.

Beyer, E. C., and Veenstra, R. D. (1994). Molecular biology and electrophysiology of cardiac gap junctions. *In* "Handbook of Membrane Channels" (C. Peracchia, ed.), pp. 379–393. Academic Press, San Diego.

Bruzzone, R., White, T. W., and Goodenough, D. A. (1996). The cellular internet: On-line with connexins. *BioEssays* **18**, 709–718.

Carafoli, E. (1992). The Ca^{2+} pump of the plasma membrane. *J. Biol. Chem.* **267**, 2115–2118.

Catterall, W. A. (1992). Cellular and molecular biology of voltage-gated sodium channels. *Physiol. Rev.* **74**(4), S15–S48.

Catterall, W. A. (1994). Molecular properties of a superfamily of plasma-membrane cation channels. *Curr. Opin. Cell Biol.* **6**, 607–615.

Cross, R. L. (1994). Our primary source of ATP. *Nature* **370**, 594–595.

Deisenhofer, J., and Michel, H. (1989). Nobel lecture. The photosynthetic reaction centre from the purple bacterium *Rhodopseudomonas viridis. EMBO J.* **8**, 2149–2170.

Deisenhofer, J., Epp, O., Sinning, I., and Michel, H. (1995). Crystallographic refinement at 2.3 Å resolution and refined model of the photosynthetic reaction centre from *Rhodopseudomonas viridis. J. Mol. Biol.* **246**, 429–457.

DeMeis, L. (ed.) (1995). Ca^{2+} transport: Pump and channels. *BioSci. Rep.* **15**, 241–408.

Forgac, M. (1989). Structure and function of vacuolar class of ATP-driven proton pumps. *Physiol. Rev.* **69**, 765–796.

Fozzard, H. A., and Hanck, D. A. (1996). Structure and function of voltage-dependent sodium channels: Comparison of brain II and cardiac isoforms. *Physiol. Rev.* **76**, 887–926.

Franzini-Armstrong, C., and Jorgensen, A O. (1994). Structure and development of E–C coupling units in skeletal muscle. *Annu. Rev. Physiol.* **56**, 509–534.

Gottesman, M. M., and Pastan, I. (1993). Biochemistry of multidrug resistance mediated by the multidrug transporter. *Annu. Rev. Biochem.* **62**, 385–427.

Gottesman, M. M., Pastan, I., and Ambudkar, S. V. (1996). P-glycoprotein and multidrug resistance. *Curr. Opin. Genetics Dev.* **6**, 610–617.

Gould, G. W., and Holman, G. D. (1993). The glucose transporter family: Structure, function and tissue-specific expression. *Biochem. J.* **295**, 329–341.

Grigorieff, N., Ceska, T. A., Downing, K. H., Baldwin, J. M., and Henderson, R. (1996). Electron-crystallographic refinement of the structure of bacteriorhodopsin. *J. Mol. Biol.* **259**, 393–421.

Gurnett, C. A., and Campbell, K. P. (1996). Transmembrane auxiliary subunits of voltage-dependent ion channels. *J. Biol. Chem.* **271**, 27975–27978.

Henderson, P. J. F. (1993). The 12-transmembrane helix transporters. *Curr. Opin. Cell Biol.* **5**, 708–721.

Henderson, R., Baldwin, J. M., Ceska, T. A., Zemlin, F., Beckmann, Z. E., and Downing, K. H. (1990). Model for the structure of bacteriorhodopsin based on high-resolution electron cryomicroscopy. *J. Mol. Biol.* **213**, 899–929.

Higgins, C. F. (1995). The ABC of channel regulation. *Cell* **82**, 693–696.

Huber, R. (1989). Nobel lecture. A structural basis of light energy and electron transfer in biology. *EMBO J.* **8**, 2125–2147.

Iwata, S., Ostermeier, C., Ludwig, B., and Michel, H. (1995). Structure at 2.8 Å resolution of cytochrome *c* oxidase from *Paracoccus denitrificans. Nature* **376**, 660–669.

Jencks, W. P. (1989). How does a calcium pump pump calcium? *J. Biol. Chem.* **264**, 18855–18858.

Karlin, A., and Akabas, M. H. (1995). Toward a structural basis for the function of nicotinic acetylcholine receptors and their cousins. *Neuron* **15**, 1231–1244.

Khorana, H. G. (1988). Bacteriorhodopsin, a membrane protein that uses light to translocate protons. *J. Biol. Chem.* **263**, 7439–7442.

Khorana, H. G. (1992). Rhodopsin, photoreceptor of the rod cell. An emerging pattern for structure and function. *J. Biol. Chem.* **267**, 1–4.

Krebs, M. P., and Khorana, H. G. (1993). Mechanism of light-dependent proton translocation by bacteriorhodopsin. *J. Bacteriol.* **175**, 1555–1560.

Kumar, N. M., and Gilula, N. B. (1996). The gap junction communication channel. *Cell* **84**, 381–388.

Lederer, W. J., He, S., Luo, S., DuBell, W., Kofuji, P., Kieval, R., Neubauer, C. F., Ruknudin, A., Cheng, H., Cannell, M. B., Rogers, T. B., and Schulze, D. H. (1996). The molecular biology of the Na^{+}–Ca^{2+} exchanger and its functional roles in heart, smooth muscle cells, neurons, glia, lymphocytes, and nonexcitable cells. *Ann. N.Y. Acad. Sci.* **779**, 7–17.

Lingrel, J. B., and Kuntzweiler, T. (1994). Na^{+},K^{+}-ATPase. *J. Biol. Chem.* **269**, 19659–19662.

Lingrel, J. B., Van Huysse, J., O'Brien, W., Jewell-Motz, E., Askew, R., and Schultheis, P. (1994). Structure–function studies of the Na,K-ATPase. *Kidney Int.* **45**(Suppl. 44), S32–S39.

MacLeod, C. L. (1996). Regulation of cationic amino acid transporter (CAT) gene expression. *Biochem. Soc. Trans.* **24**, 846–852.

Maguire, M. E. (1992). MgtA and MgtB: Prokaryotic P-type ATPases that mediate Mg^{2+} influx. *J. Bioenerg. Biomembr.* **24**, 319–328.

Martonosi, A. N. (1995). The structure and interactions of Ca^{2+}-ATPase. *BioSci. Rep.* **15**, 263–281.

McGivan, J. D., and Pastor-Anglada, M. (1994). Regulatory and molecular aspects of mammalian amino acid transport. *Biochem. J.* **299**, 321–334.

Meissner, G. (1994). Ryanodine receptor/Ca^{2+} release channels and their regulation by endogenous effectors. *Annu. Rev. Physiol.* **56**, 485–508.

Moller, J. V., Juul, B., and le Maire, M. (1996). Structural organization, ion transport, and energy transduction of P-type ATPases. *Biochim. Biophys. Acta* **1286**, 1–51.

Neher, E. (1992). Nobel lecture. Ion channels for communication between and within cells. *Neuron* **8**, 605–612.

Numa, S. (1986). Molecular basis for the function of ionic channels. *Biochem. Soc. Sympos.* **52**, 119–143.

Pedersen, P. L., and Amzel, L. M. (1993). ATP synthases. *J. Biol. Chem.* **268**, 9937–9940.

Pedersen, P. L., and Carafoli, E. (1987). Ion motive ATPases. I. Ubiquity, properties, and significance to cell function. *Trends Biochem. Sci.* **12**, 146–150.

Pedersen, P. L., and Carafoli, E. (1987). Ion motive ATPases. II. Energy coupling and work output. *Trends Biochem. Sci.* **12**, 186–189.

Philipson, K. D., Nicoll, D. A., Matsuoka, S., Hryshko, L. V., Levitsky, D. O., and Weiss, J. N. (1996). Molecular regulation of the Na^+–Ca^{2+} exchanger. *Ann. N.Y. Acad. Sci.* **779,** 20–28.

Reithmeier, R. A. F. (1994). Mammalian exchangers and co-transporters. *Curr. Opin. Cell Biol.* **6,** 583–594.

Rios, E., Pizarro, G., and Stefani, E. (1992). Charge movement and the nature of signal transduction in skeletal muscle excitation–contraction coupling. *Annu. Rev. Physiol.* **54,** 109–133.

Sachs, G., Shin, J. M., Briving, C., Wallmark, B., and Hersey, S. (1995). The pharmacology of the gastric acid pump: The H^+, K^+ ATPase. *Annu. Rev. Pharmacol. Toxicol.* **35,** 277–305.

Sakmann, B. (1992). Nobel lecture. Elementary steps in synaptic transmission revealed by currents through single ion channels. *Science* **256,** 503–512.

Schneider, M. F. (1994). Control of calcium release in functioning skeletal muscle. *Annu. Rev. Physiol.* **56,** 463–484.

Senior, A. E., Al-Shawi, M. K., and Urbatsch, I. L. (1995). The catalytic cycle of P-glycoprotein. *FEBS Lett.* **377,** 285–289.

Shapiro, A. B., and King, V. (1995). Using purified P-glycoprotein to understand multidrug resistance. *J. Bioenerg. Biomembr.* **27,** 7–13.

Solioz, M., and Vulpe, C. (1996). CPx-type ATPases: A class of P-type ATPases that pump heavy metals. *Trends Biochem. Sci.* **21,** 237–241.

Tanabe, T. (1994). Structure and function of skeletal muscle and cardiac dihydropyridine receptors. *In* "Handbook of Membrane Channels" (C. Peracchia, ed.), pp. 177–186. Academic Press, San Diego.

Tanner, M. J. A. (1993). Molecular and cellular biology of the erythrocyte anion exchanger (AE1). *Sem. Hematol.* **30,** 34–57.

Tsukihara, T., Aoyama, H., Yamashita, E., Tomizaki, T., Yamaguchi, H., Shinzawa-Itoh, K., Nakashima, R., Yaono, R., and Yoshikawa, S. (1995). Structure of metal sites of oxidized bovine heart cytochrome *c* oxidase at 2.8 Å. *Science* **269,** 1069–1074.

Tsukihara, T., Aoyama, H., Yamashita, E., Tomizaki, T., Yamaguchi, H., Shinzawa-Itoh, K., Nakashima, R., Yaono, R., and Yoshikawa, S. (1996). The whole structure of the 13-subunit oxidized cytochrome *c* oxidase at 2.8 Å. *Science* **272,** 1136–1144.

Unwin, N. (1989). The structure of ion channels in membranes of excitable cells. *Neuron* **3,** 665–676.

Unwin, N. (1993). Nicotinic acetylcholine receptor at 9 Å resolution. *J. Mol. Biol.* **229,** 1101–1124.

Unwin, N. (1995). Acetylcholine receptor channel imaged in the open state. *Nature* **373,** 37–42.

Walker, J. E., and Collinson, I. R. (1994). The role of the stalk in the coupling mechanism of F_1F_0-ATPases. *FEBS Lett.* **346,** 39–43.

Wright, E. M., Loo, D. D. F., Panayotova-Heiermann, M., and Boorer, K. J. (1994). Mechanisms of Na^+–glucose cotransport. *Biochem. Soc. Trans.* **22,** 646–650.

Wright, E. M., Loo, D. D. F., Turk, E., and Hirayama, B. A. (1996). Sodium cotransporters. *Curr. Opin. Cell Biol.* **8,** 468–473.

Cell Nucleus

HENRY TEDESCHI
State University of New York at Albany

GLOSSARY

Diploid Containing two homologous copies of each chromosome

Gametogenesis Formation of gametes, that is, sperms and oocytes, by meiosis

Interphase Period between cell divisions

Mitosis Cell division in which two diploid daughter cells are produced

rRNA RNA component of the ribosome

Spindle Collection of microtubules forming part of the apparatus needed for cell division

Transcription Synthesis of RNA complementary to the DNA strand, the former serving as template

Translation Synthesis of polypeptides using the instructions contained in the nucleotide sequence of mRNA

THE NUCLEUS IS THE CELLULAR COMPARTMENT OF eukaryotes that contains nearly all the DNA of the cell. Eukaryotic cells have a nuclear envelope that separates two distinct compartments, the cytoplasm and the nucleoplasm. Because the transcription and maturation of mRNA occur in the nucleus and the translation of the mRNA (i.e., the synthesis of polypeptides) occurs in the cytoplasm, the nuclear envelope is important in regulating the transport of macromolecules. The structural organization of the nucleus is complex and comprises a number of structures, many of them associated with the envelope.

I. NUCLEAR ORGANIZATION

The envelope of the nucleus is in contact with a network of filaments in both the cytoplasmic and the nucleoplasmic sides. The filaments on the nucleoplasmic side line the inner face of the membrane and form a compact lining, the nuclear lamina (Fig. 1). The nuclear envelope is interrupted by the nucleopore complex, which juts into the cytoplasm and the nucleoplasm. A nuclear envelope lattice is also thought to provide a network on the nucleoplasmic side, covering both the lamina and the nucleopore complex (see Figs. 6 and 7).

Prominent in the nucleus is the nucleolus, which corresponds to DNA loops formed by several chromosomes (10 in human cells). The nucleolus contains a cluster of rRNA genes known as the nucleolar organizing region.

The centrosome is a cytoplasmic organelle on one side of the nucleus. The centrosome contains a pair of centrioles at right angles to each other. It is a major microtubule organizing center with a special role in cell division and in the separation of chromosomes during mitosis. Each centriole begins dividing during the S phase of the cell cycle, and eventually the two daughter centrioles migrate to opposite sides of the dividing cell. They provide organizing centers for the microtubules, which form the mitotic (or meiotic) spindle [*See* Mitosis.]

Most of the density inside the nucleus when viewed with the electron microscope corresponds to chromatin, the DNA–protein complex of eukaryotic chromo-

ENCYCLOPEDIA OF HUMAN BIOLOGY, Second Edition, VOLUME 2. Copyright © 1997 by Academic Press. All rights of reproduction in any form reserved.

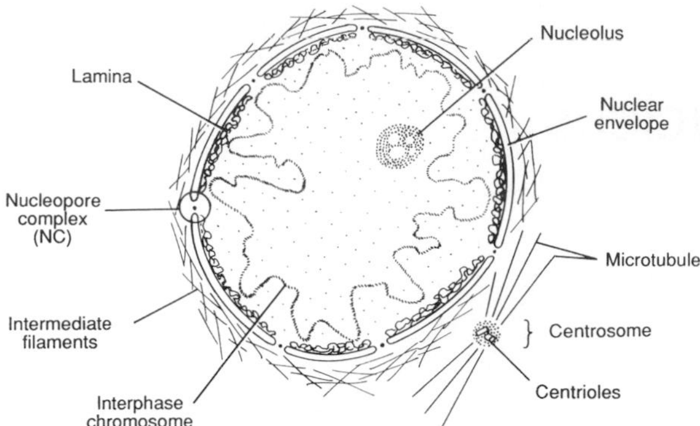

FIGURE 1 Schematic representation of interphase nucleus. Note that the nuclear envelope lattice shown in Figs. 6 and 7 has not been included. This lattice is internal to the lamina and the nucleopore complex.

somes. The denser areas represent heterochromatin, an unusually compact form of chromatin that is transcriptionally inactive.

II. THE CHROMOSOMES

A. Arrangement of the Chromosomal Components

Each human chromosome contains a single linear molecule of DNA estimated to be several centimeters in length. DNA has been shown to occur in three different helical structures, all formed from antiparallel strands, that is, one strand has a 5′ end facing the 3′ end of the other strand. The form present in cells is probably mostly the B-form (B-DNA). The B-DNA is a right-hand helix with the base pair normal to the helix axis. B-DNA corresponds to the DNA model originally proposed by Watson and Crick. The antiparallel arrangement is illustrated in Fig. 2. Part A of the figure illustrates the numbering system used to distinguish the atoms present in the deoxyribose of the deoxynucleotides joined to form the DNA molecule; part B shows the antiparallel arrangement (shown in a straight model rather than a helix). The helices can bend, kink, and unravel. The bends can be rather sharp at sites where special nucleotides are present. [*See* DNA and Gene Transcription.]

The most abundant proteins of the chromatin are the histones. The remaining proteins of the chromatin are the nonhistone proteins. The histones are relatively small proteins containing between 100 and 200 amino acids with a high proportion of the amino acids lysine and arginine. Histones play a major role in the orderly folding of DNA so that molecules, which in fully extended form would be several centimeters long, are packed in structures a few micrometers in length. The DNA–histone complexes have a beaded appearance when extended and viewed with the electron microscope. Each bead is a nucleosome. Each nucleosome is formed by a histone octamer core on which the DNA helix is wound twice (see the individual beads in Fig. 2C). It has been estimated that an average eukaryotic gene contains as many as 50 nucleosomes. Nucleosomes are thought to be absent from areas from which they have been competitively displaced by sequence-specific DNA-binding proteins. [*See* Histones.]

Chromatin is highly condensed so that the nucleosomes are packed on top of each other, producing a thread 30–40 nm in diameter (see Fig. 2C), which is then folded further. The most condensed state is present during cell division, when the chromosomes can be clearly seen with the light microscope. [*See* Chromatin Folding.]

During interphase, the chromosomes of most cells are not visible with the light microscope. Two specialized kinds of chromosomes, which are visible with light microscopy, have been very useful in studying interphase chromosomes. The lamp brush chromosomes of growing oocytes form distinguishable chro-

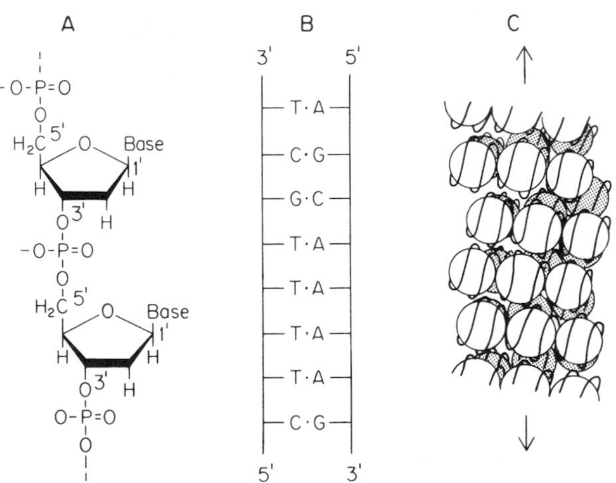

FIGURE 2 Diagram of DNA organization. (A) The backbone of DNA strands and the numbering system used to identify atoms in the deoxyribose component. The numbers also indicate the direction of the DNA strand. (B) Antiparallel arrangement of DNA, which is shown here in extended form. (C) Possible arrangement of DNA (the thread) in relation to the histones (the balls) in chromatin. This unit would then be folded further in condensed chromatin. Each ball with the associated thread corresponds to a nucleosome.

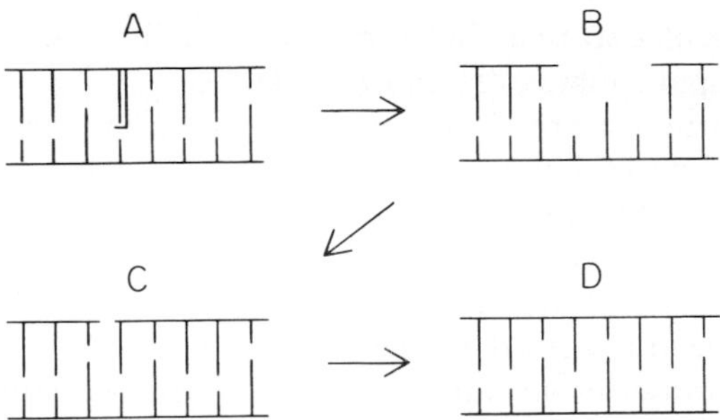

FIGURE 3 Diagram illustrating the DNA repair mechanism. The backbone of the DNA is represented by horizontal lines, whereas the bases are represented by vertical lines. (A) An altered nucleotide indicated by the open rectangle. (B) The damaged piece is removed by repair nuclease. (C) The missing part is resynthesized by DNA polymerase using the complementary strand as a template. (D) The resynthesized piece is reattached to the strand by DNA ligase.

matin loops. The loops apparently correspond to very active transcription sites. The polytene chromosomes of the secretory cells of fly larvae undergo multiple replication cycles without cell division, and the various chromosomal copies, several thousand more than the normal amount, remain side by side. After staining, the chromosomes appear banded and the arrangement shows many landmarks, which have permitted extensive cytogenetic studies with the light microscope. The heavily stained areas correspond to highly condensed chromatin.

B. Replication and Repair

One of the major functions of the nucleus is to maintain chromosomal DNA as intact and unaltered molecules. DNA repair mechanisms mostly depend on repairing a strand from the information contained in its complementary strand. When both are damaged in the same location, the genetic information is lost. The process of DNA repair is represented diagrammatically in Fig. 3. The mechanisms of repair involve recognition and removal of the damaged nucleotides by repair nucleases (Fig. 3B). The DNA polymerase binds to the 3′ OH end of the cut DNA and adds one nucleotide at a time using the complementary strand as a template (Fig. 3C). In a final step, DNA ligase attaches the restored segment to the strand under repair (Fig. 3D). [*See* DNA Repair.]

In addition to repair, the nucleus must provide mechanisms for the replication of chromosomes and the appropriate distribution to the daughter cells during mitosis and meiosis so that the appropriate chromosome complement is present in both daughter cells. The latter functions are associated with particular sequences in the DNA.

The huge eukaryotic DNA molecule is efficiently duplicated in segments—the replicons. The number of separate replicons in mammalian DNA is as high as 20,000–30,000. To replicate, a DNA molecule needs a replication origin, where an initiator protein will bind to catalyze the formation of a replication fork, after which a multienzyme system proceeds to replicate the DNA. In DNA replication, each strand serves as a template for the replication of a new strand. The replication is said to be semiconservative because each daughter cell inherits chromosomes containing an old and a new strand.

Replication involves separation of the two complementary strands. As new strands are being formed, the resulting structure is Y-shaped and is called a DNA replication fork. The replication proceeds asymmetrically at each replicon. Both strands are synthesized from the 5′ to the 3′ direction (Fig. 4). Only one of the two strands has a 3′ end at the initiation site (the strands are antiparallel) so that only one (the leading strand) can serve as a template for a

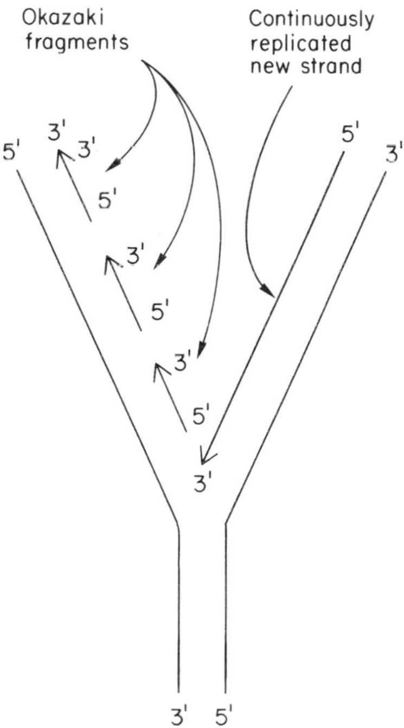

FIGURE 4 DNA replicating fork. The arrows indicate the direction of duplication. The long arrow indicates the continuous duplication of one of the strands from its 5′ end. The smaller arrows show the replication of the other strand discontinuously in small fragments, also from the 5′ ends of the segments.

continuous 5′ to 3′ synthesis. The other strand (the lagging strand) is replicated in pieces 100 to 200 nucleotides long in the 5′ to 3′ direction, the so-called Okazaki fragments. The fragments are then joined by DNA ligase, the same enzyme that functions in repair. The replication is said to be semidiscontinuous.

The DNA polymerase can extend a previously started chain at the 3′ OH end; however, it cannot initiate the replication without a primer. In this case the primer is an RNA strand about 10 bases long, paired to one strand of DNA to provide the free 3′ OH end. Five DNA polymerases have been recognized in mammals. Polymerase α is involved with the replication of the lagging strand, whereas polymerase δ functions in the leading strand. β and ε are repair polymerases of the nucleus and polymerase γ is a mitochondrial polymerase.

The centromeres play an important role in mitosis. They are specific DNA sequences needed for the attachment of the chromatids to the mitotic spindle during cell division.

C. Chromosomes and Transcription

The total DNA in a mammalian nucleus has enough nucleotides to code for as many as 3 million proteins; however, it has been estimated that only a tiny fraction, perhaps as few as 50,000–100,000, are actually produced. This is because only a small portion of the DNA sequences are transcribed and, furthermore, during RNA processing, the nucleus eliminates much of the transcribed RNA.

RNA synthesis is the result of transcription of DNA. One of the two DNA strands acts as a base-pairing template for the nascent RNA strand. Transcription converts genetic information into a form that can be used by the cytoplasmic machinery to produce proteins of the amino acid sequence predetermined by the genes. Transcription occurs very rapidly, much faster than DNA replication.

Transcription is catalyzed by RNA polymerases. RNA polymerase II transcribes protein coding genes, whereas polymerase I synthesizes the large ribosomal RNA molecules and polymerase III synthesizes a variety of small and stable RNAs such as transfer RNA (tRNA) or small nuclear RNA (snRNA) (see the following). Transcription factors (TF) are sequence-specific DNA-binding proteins needed for the initiation of RNA synthesis. The TFs bind specific segments of the DNA and selectively recruit RNA polymerase molecules so that the enzyme repeatedly transcribes the appropriate portion of the DNA.

Portions of the DNA of the chromosomes are specialized to function as regulatory DNA segments, capable of binding the regulatory transcription factors, which either enhance or inhibit transcriptional activity. Many of these segments are upstream in relation to the transcribed segment (i.e., on the 5′ side of the transcription start point), but other locations are also possible.

Initiation involves the binding of RNA polymerase to double-stranded DNA. This requires unwinding the DNA and recognition of a specific DNA sequence, the promoter. The RNA chain grows as the RNA polymerase moves along the DNA, a process known as elongation, transiently forming a DNA–RNA hybrid. Termination involves recognition of a terminator sequence. At this point the transcription ceases, and the DNA–RNA association is disrupted.

A diagrammatic representation of the transcription of DNA and the processes needed to produce mature mRNA is shown in Fig. 5. Polymerase II produces a large transcript, termed heterogeneous nuclear RNA (hnRNA), which eventually forms mRNA. Newly synthesized hnRNA is immediately complexed to proteins, forming particles larger than nucleosomes. The 5′ end of the nascent RNA strand is capped by the addition of methylated guanylate. The cap plays a role in protein synthesis, protecting the RNA from degradation and directing it to the cytoplasm through the nucleopore complex. The 3′ end, the last to be synthesized, is also modified by addition of a poly-A tail, a segment of 100–200 residues of adenylate. The role of the poly-A tail is unknown but it may prevent degradation. Most of the transcribed RNA is degraded; only approximately 5% of the RNA originally in the hnRNA reaches the cytoplasm. The sequences retained are those responsible for coding the amino acid sequence of the proteins (corresponding to exons) and in some cases signal sequences required for transfer of the proteins into membranes or vesicular cytoplasmic compartments. Those degraded are the noncoding segments (corresponding to introns).

The DNA corresponding to a gene is completely transcribed. This transcript is then processed so that the intron sequences are removed, a process known as splicing. The RNA is cleaved at the boundary between intron and exon, and the exon ends are subsequently joined together by components of a large complex termed the spliceosome. This complex contains a number of special snRNA molecules, which are re-

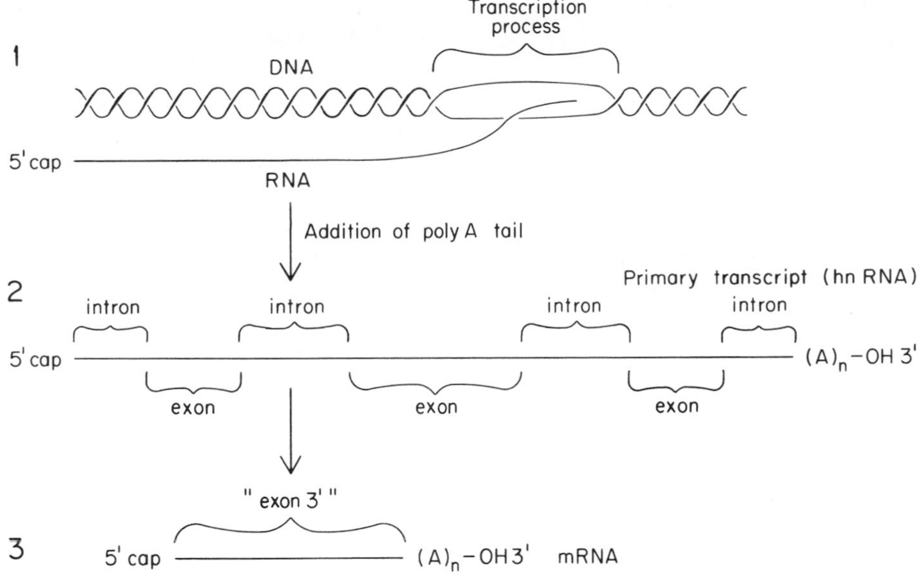

FIGURE 5 Diagrammatic representation of transcription and RNA maturation. (1) The RNA is transcribed from one of the DNA strands and immediately capped. (2) A poly-A tail is attached. (3) The introns are removed from the primary transcript.

sponsible for recognizing the appropriate sequences at which splicing is to take place. The spliceosome is analogous to the ribosomes and has been found to contain as many as 50 proteins and snRNA molecules. Nascent transcripts have been examined with the electron microscope in sites of high transcriptional activity after spreading them on a grid. Nascent hnRNA transcripts can be seen as smooth fibrils 5 nm thick. These have been referred to as perichromatin fibers. The transcripts are progressively longer as transcription proceeds along the DNA coding thread. However, some of the transcripts are shorter than expected from their position, indicating that splicing has occurred. These observations confirm that splicing is generally cotranscriptional. However, the two processes can occur independently. Many primary gene transcripts can be spliced differently to produce different forms of the same protein (this is called alternative splicing).

III. THE NUCLEAR ENVELOPE

The nuclear envelope is double, containing an inner and an outer membrane (see Fig. 1). The envelope is discontinuous at the nucleopore complex, where the two sets of membranes are joined. The outer membrane has been frequently found with attached ribosomes, and the envelope resembles the endoplasmic reticulum, with which it is sometimes continuous. The nuclear envelope is disassembled into vesicles during cell division.

Chromatin is attached to the nuclear envelope. Much of the nuclear heterochromatin, the much condensed form of chromatin present at interphase, is attached to the inner membrane. Although this DNA is not involved in transcription, it may have a role in the overall organization of chromatin. The arrangement of the interphase chromosomes in relation to the nuclear envelope is easiest to follow in polytene chromosomes, for example, in the salivary gland of *Drosophila*. At least in this case, each chromosome occupies a separate spatial domain, and specific loci in the chromosomes are attached to the nuclear envelope. Different kinds of polytene cells from the same organism seem to have a different kind of chromosomal arrangement.

The inner membrane is lined with a fibrous protein meshwork, the nuclear lamina, which has been characterized in some eukaryotic cells. The fibers of the lamina have strong similarities to the cytoplasmic intermediate filaments (e.g., similar amino acid sequences) and serve as a point of attachment of the chromosomes (see Fig. 1). Furthermore, the lamina is attached to the nucleopore complex of the nuclear envelope.

A. The Lamina and the Nuclear Envelope Lattice

The lamina is a ubiquitous component associated with the nuclear envelope of eukaryotes. It has been found in mammals, the African frog *Xenopus*, and even invertebrates such as the clam or the fruit fly *Drosophila*. The laminae are 10–50 nm in thickness. Three protein components, named lamins, have been identified and found to be 60–80 kDa. At least five different lamin species have been recognized in *Xenopus* oocytes. The fibers of the lamina are 10 nm in thickness and have an axial repeat of 25 nm.

In mammalian cells, lamins A and C are identical in their first 566 amino acids; lamin B has not been studied as thoroughly. Lamins A and C both have a 350-amino-acid segment, which is very similar in sequence to a segment of intermediate filaments present in the cytoplasm. The two lamins form dimers from side-to-side association of monomers. The resulting structure is rod-shaped with two globular heads at one end.

Different lamins seem to predominate depending on the developmental stage. In mammals, a lamin (either lamin B or a very similar protein) is present before implantation, whereas A and C make an appearance during organogenesis. The distribution of the lamins is also affected by gametogenesis.

At least in mammals, the lamina is associated with the nuclear envelope through lamin B. Lamin B also remains attached to vesicles formed when the nuclear envelope dissociates during cell division.

More recently, a nuclear envelope lattice has been demonstrated in the nuclear face of isolated nuclear envelopes. This network is made up of a highly regular sheet of fibers. It is internal to the lamina and is attached to the basket structures (Figs. 6 and 7) of the nucleopore complex. The structure of the nuclear envelope shown in Fig. 6 was obtained from a study using nuclei isolated from *Triturus* oocytes.

The isolated nuclear envelope was examined with high-resolution scanning electron microscopy (HRSEM), after fixing and critical point drying and coating with a thin layer of chromium or tantalum.

B. The Nucleopore Complex

The nuclear envelope is traversed by specialized channel structures, the nucleopore complexes (NCs). In vertebrates there are 10–20 NCs per μm^2 or 2000 to 4000 per nucleus. Lamellae resembling the nuclear envelope, the annulate lamellae, are often found in the cytoplasm. Their relationship to the nuclear envelope is not clear.

The NC corresponds to a circular opening in the nuclear envelope. In this region, the inner and outer membranes are joined. The NC has a nuclear and a cytoplasmic ring (Fig. 7), both about 120 nm in diameter and composed of eight subunits each. Eight spokes restrict the opening to about 80 nm. A central structure (central granules or plug, CP in Fig. 7) in the pore opening is thought to be involved in mediated transport of macromolecules. Recent electron microscopy used in conjunction with image reconstruction techniques has revealed the presence of eight channels in the walls of the NC, approximately 10 nm in pore diameter. Cytoplasmic filaments (CF in Fig. 7) are attached to the cytoplasmic ring (CR) through the cytoplasmic particles (P). A basket structure (NB) of the NC faces the nucleoplasmic site (see also Fig. 6A). The filaments are thought to be the first component of the NC to bind proteins actively transported into the nucleus (see the following).

A number of NC proteins, mostly glycoproteins, have been identified. Proof of a localization in the NC requires a preliminary identification followed by a demonstration of its localization using immunoelectronmicroscopy. This proof is needed because it is difficult to isolate NCs without some accompanying contamination. The cDNA sequence of many of these proteins has been determined. One of the striking similarities between these proteins is the presence of repeat motifs of four or five amino acids. Several proteins have been implicated in transport of macromolecules because of their binding to compounds (such as wheat germ agglutinin, WGA) or antibodies that block transport. [*See* Nuclear Pore, Structure and Function.]

Transport takes place by two distinct mechanisms. The NC is the site of mediated active transport. Diffusion of a variety of molecules also occurs passively and is thought to occur through channels in the NC. In this process the channels are thought to act as a sieve, allowing small molecules through and blocking the passage of molecules above a critical size.

In the passive passage through the NC, macromolecules used as probes diffuse at a rate inversely related to their size. Molecules of approximately 20–40 kDa or above (e.g., dextrans or nonnuclear proteins) have been found not to enter in significant numbers. The evidence suggests that the channels have a functional

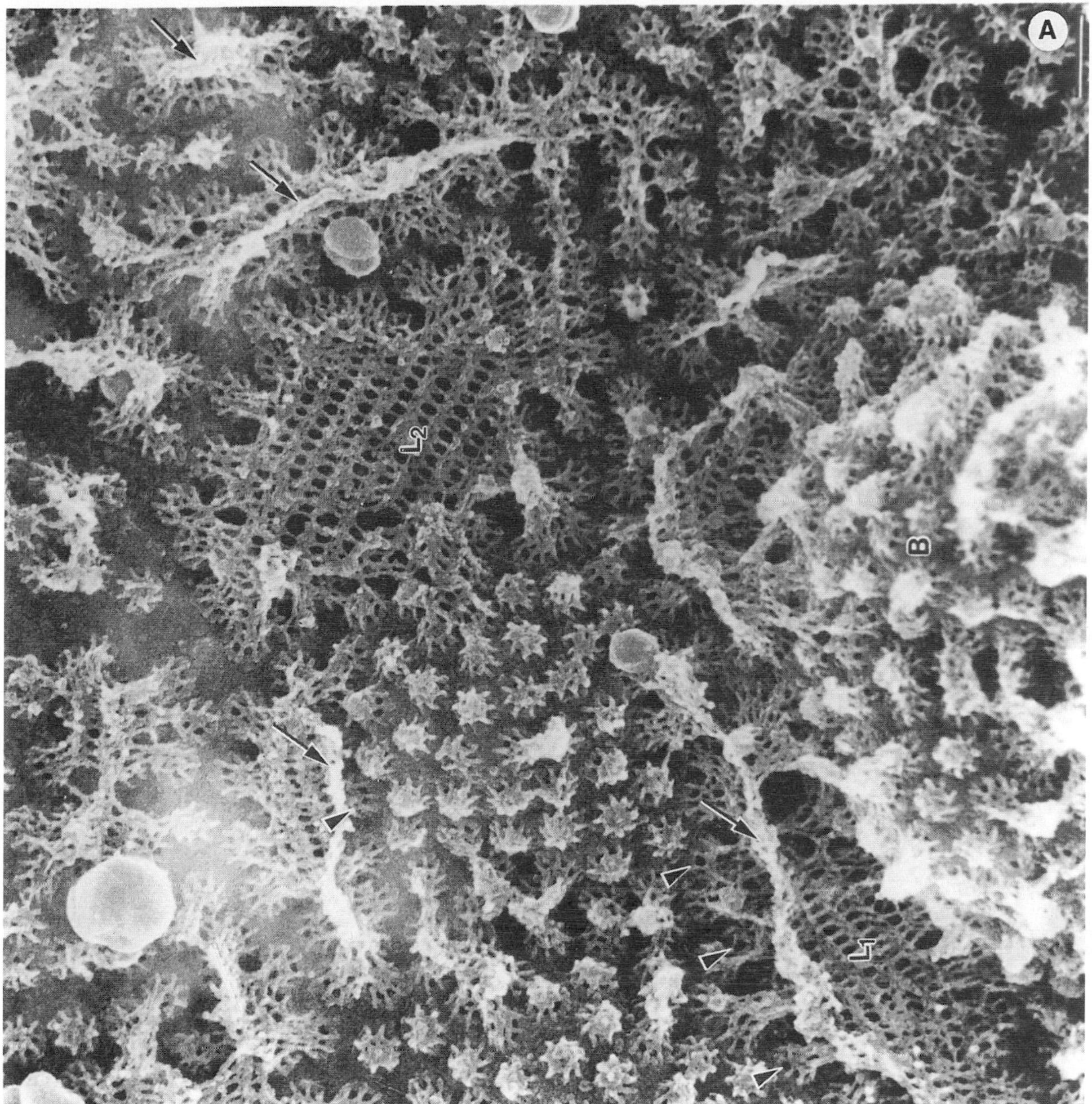

FIGURE 6 (A) Nucleoplasmic face of the nuclear envelope showing the nucleopore complex (NC) baskets and the nuclear envelope lattice (NEL), indicated by L_1 and L_2. Some of the NEL appear "rolled up" (arrows), showing the baskets underneath (arrowheads). "B" in the figure indicates a bulge. Bar = 200 nm. (B) Nucleoplasmic face of the nuclear envelope showing details of the NEL. Part a: The NC is shown below the network (arrow with bar). Arrowheads indicate woven complex elements; medium arrows indicate thin fibers; the large arrow indicates fibers joining adjacent thin fibers. "b" in the figure indicates a nuclear pore complex at the edge of NEL. Bar = 100 nm. Part b: Detail of NEL showing complex woven element, including fibers joining adjacent thin fibers (arrows) and fibers that appear to be branches of the thin fibers over the top (arrowheads) of other branches. The inset shows a reconstruction. Bar = 50 nm. [Reproduced with permission from M. W. Goldberg and T. D. Allen (1992). *J. Cell Biol.* **119**, 1429–1440.]

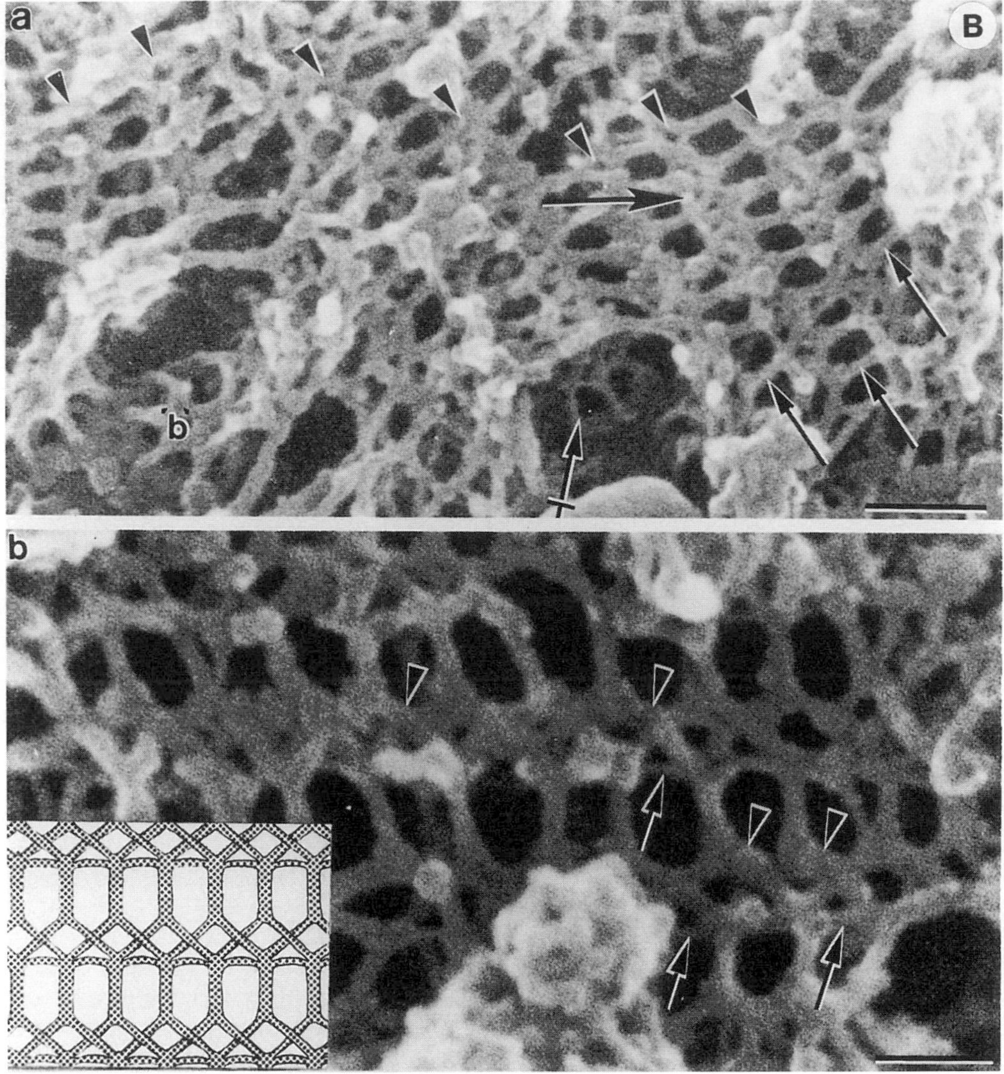

FIGURE 6 *(Continued)*

diameter of about 10 nm. These channels may correspond to those observed in the NC (see the foregoing). In contrast, some proteins (e.g., some proteins extracted from the nucleus) enter rapidly from the cytoplasmic side and can be accumulated by the nucleus even when very large. In fact, large particles such as colloidal gold can be transferred when coated with these proteins.

In at least some nuclei, electrophysiological studies indicate that the permeability to low molecular weight electrolytes is low and there is evidence of ion (e.g., K^+) conducting channel behavior. These findings have led to the suggestion that the channels are part of the NC and that they are involved in the transport of proteins. How the passive or active transport of macromolecules is related to these channels is still not clear.

The import of proteins into the nucleus requires the presence of a special sequence, the nuclear localization sequence (NLS), apparently recognized by a cytoplasmic receptor. A class of NLSs corresponds to a short domain of basic amino acids generally containing four to five amino acids. This class is epitomized by the SV40 large T antigen. Another class contains two basic amino acid domains similar to that of the SV40 large T antigen. In this case, both domains are needed. Despite the two different signals, the same receptors are involved in both systems. Synthetic NLSs attached to nonnuclear proteins allow them to be accumulated in the nucleoplasm. More recently an-

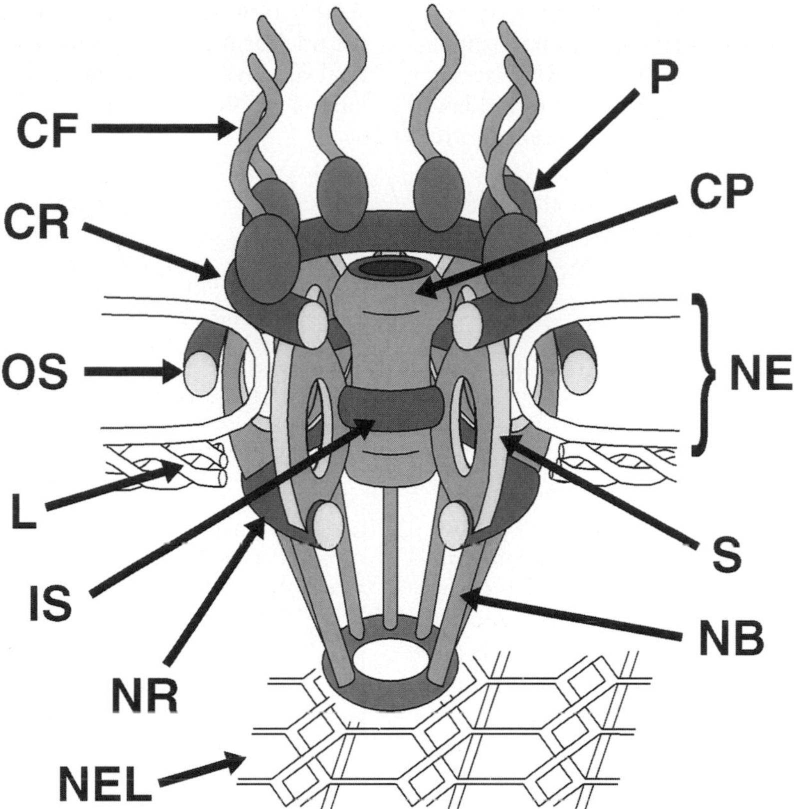

FIGURE 7 Three-dimensional model of a nucleopore complex drawn approximately to scale. The cytoplasmic face is uppermost. Certain facing portions have been omitted for clarity. CF, cytoplasmic filament; P, cytoplasmic particles; CR, cytoplasmic ring; OS, outer spoke ring; IS, inner spoke ring; NR, nuclear ring; S, spoke; CP, central plug or transporter (the buttresses connecting it to the IS have been left out); NB, nuclear basket; NE, nuclear envelope; L, lamina; NEL, nuclear envelope lattice. [Reproduced with permission from M. P. Rout and S. R. Wente (1994). *Trends Cell Biol.* 4, 357–365.]

other nuclear localization sequence of 38 amino acids has been identified (M9). M9 can also function as a nuclear export signal (NES). Other proteins contain shorter sequences with an NES function. In addition, nuclear retention sequences of 80 amino acids have been found in some proteins restricted to the nucleus. The details of protein transport may be much more complex since some proteins are shuttled from the cytoplasm to the nucleus and back to the cytoplasm depending on the developmental stage. The details of the transport have begun to be examined recently *in vitro*. Apparently, the initial attachment of the periphery of the NC, possibly at the cytoplasmic filaments, and the placement over the transporter do not require hydrolysis of ATP, which is required for the translocation that occurs through the pore itself (at the location of the transporter or central plug in Fig. 7). Only four proteins have been found to be needed for reconstituting *in vitro* the nuclear import of proteins.

Because the movement of proteins in and out of the nucleus reflects developmental stages, the transport system is thought to have a role in the regulation of gene expression. The transfer of steroid hormone–receptor complex, which is necessary for hormone action, occurs through this mechanism. Other regulators of gene expression are thought to depend in some manner on the translocation of proteins into the nucleus (e.g., in the case of interferon-α or the cAMP-response element binding protein).

The RNA transcribed in the nucleus undergoes many processing steps as discussed earlier. Following these, the RNA is transferred to the cytoplasm in a very selective manner. A small part of the RNA is transferred; the remainder remains in the nucleus, where it is degraded. These results suggest a role of the NC transport system in posttranscriptional regulation of gene expression. The export of mRNA seems to require the cap region of the mRNA. Cap binding proteins, acting analogously to the NLS-receptors are thought to be involved in RNA export.

The active transfer of either protein or RNA is vectorial, that is, in a single direction. A variety of observations indicate that the same pathways are involved for either proteins or RNA.

C. Assembly and Disassembly of the Nuclear Envelope

During mitosis, the nuclear laminae depolymerizes in response to their phosphorylation at the onset of mitosis. Similarly, the nucleopore complex disassembles. This is followed by breaking down of the nuclear envelope into vesicles. During depolymerization of the lamina, lamin B remains associated with the vesicles while lamins A and C are dissolved in the cytoplasm. When the nuclear envelop reassembles in the final stages of telophase, the lamins are dephosphorylated and repolymerize on the surface of the chromosomes, which may contain receptors for the vesicles. The reassembled lamina binds the vesicles to reform the nuclear envelope and the NCs are reformed. The interdependence of lamins, nuclear pores, and vesicle assembly is not entirely clear. Lamin assembly requires the nuclear membrane. However, the assembly of the nuclear envelope can take place in the absence of lamina or NC assembly.

BIBLIOGRAPHY

Alberts, B., Bray, D., Lewis, J., Raff, M., Roberts, K., and Watson, J. D. (1994). "Molecular Biology of the Cell," 3rd Ed., Chaps. 3, 6, 8, 9, and 12. Garland, New York/London.

Bustamante, J. O. (1994). Nuclear electrophysiology, *J. Membrane Biol.* **138**, 105–112.

Goldberg, M. W., and Allen, T. D. (1992). High resolution scanning electron microscopy of the nuclear envelope: Demonstration of new, regular, fibrous lattice attached to the basket of the nucleoplasmic face of the nuclear pores. *J. Cell Biol.* **119**, 1429–1440.

Görlich, D., and Mattaj, I. W. (1996). Nucleocytoplasmic transport. *Science* **271**, 1513–1518.

Jans, D. A., and Hübner, S. (1996). Regulation of protein transport to nucleus: Central role of phosphorylation. *Physiol. Rev.* **76**, 651–685.

Lewin, B. (1994). "Genes V," Chaps. 19 and 29–31. Oxford Univ. Press, Oxford, England/New York.

Lodish, H., Baltimore, D., Berk, A., Zipursky, S. L., Matsudaira, P., and Darnell, J. (1995). "Molecular Cell Biology," 3rd Ed., Chaps. 9–12. W. H. Freeman, New York.

Rout, M. P., and Wente, S. R. (1994). Pores for thought: Nuclear pore complex proteins. *Trends Cell Biol.* **4**, 357–365.

Spector, D. L. (1993). Macromolecular domains within the cell nucleus. *Annu. Rev. Cell Biol.* **9**, 265–315.

Cell Receptors

NITA J. MAIHLE
Mayo Clinic

GLOSSARY

Cell receptor Typically a protein molecule in either prokaryotic or eukaryotic cells that binds to a specific ligand and responds by transmitting information through the cell, resulting in a characteristic biological response

Coated vesicle Specialized region of cell membrane that carries ligand-occupied receptors into the cell via vesicular transport

Effector systems Downstream mediators that help the receptor relay information provided to the cell through ligand binding

Endocytosis Invagination of the cell membrane and vesiculation; process by which cell-surface receptors enter the cell

Ligand Any of a number of pharmaceutical or naturally occurring molecules that can interact with specific cell receptors

Receptor binding assays Assays using labeled ligand to determine receptor number in a given cell type

Receptor down-modulation Any of a variety of mechanisms used by cells to attenuate signal transduction by ligand; it often involves structural modification of the receptor and/or removal of receptor from the cell surface

Receptor recycling Process by which previously internalized cell-surface receptors are returned to the cell surface

Signal amplification Process by which the initial signal delivered by the ligand is increased during its transmission through the cell

CELLS RECEIVE IMPORTANT INFORMATION FROM the world around them on an ongoing basis. Is it time to move, divide, differentiate, respond to a nearby cell, or respond to a foreign antigen? All of these actions are based on information communicated to the cell from its environment. Chemical messengers of diverse structure, namely, hormones, growth factors, metabolites, and nutrients, are responsible for transmitting this information. The cellular molecules responsible for receiving this information are typically proteins referred to as cell receptors. Cell receptors can be broadly divided into two categories: cell-surface receptors and intracellular receptors.

I. CELL-SURFACE RECEPTORS

Cell-surface receptors enable the cell to respond to the outside world without the need for the chemical messenger to cross the cell membrane. This is important since many chemicals outside the cell could potentially damage proteins and nucleic acids inside the cell. Therefore, the cell is entirely surrounded by a protective hydrophobic barrier—the cell membrane (Fig. 1). The cell membrane is composed of a lipid bilayer studded with membrane proteins, some of which actually span the lipid bilayer. Many of these so-called transmembrane proteins are cell-surface receptors, which can be a single protein or several closely associated proteins that function as a single receptor. The extracellular portion of the receptor is responsible for binding to specific molecules referred to as ligands. The cytoplasmic or intracellular portion

ENCYCLOPEDIA OF HUMAN BIOLOGY, Second Edition, VOLUME 2. Copyright © 1997 by Academic Press. All rights of reproduction in any form reserved.

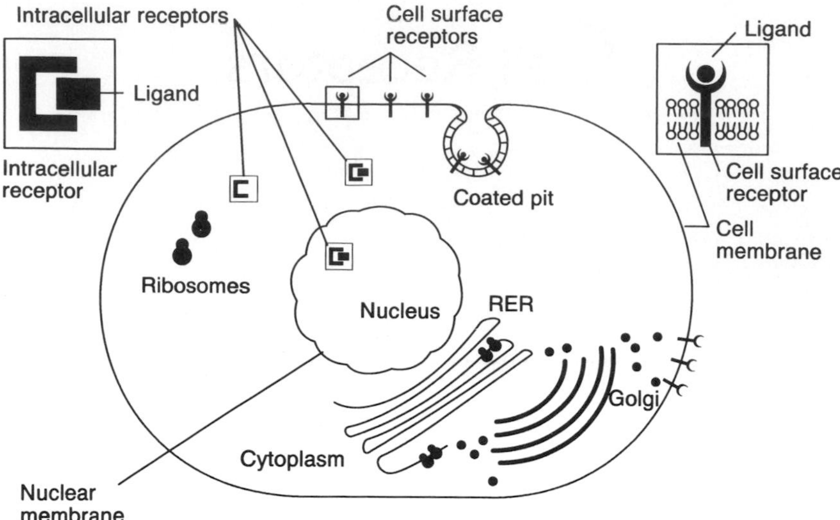

FIGURE 1 Schematic illustration of cell receptors. Cell-surface receptors bind to ligands in the extracellular environment. Ligand-occupied cell-surface receptors are frequently internalized via specialized membrane invaginations called coated pits (see text). Cell-surface receptors are synthesized in the rough endoplasmic reticulum. Intracellular receptors bind to ligands capable of crossing the cell membrane. Intracellular receptors are synthesized on free ribosomes in the cytoplasm and can be localized in either the cytoplasmic or nuclear compartments.

of the receptor may itself have an enzymatic activity that it uses to transmit information to other molecules within the cell, or it may be associated with other signal-transducing molecules such as kinases or guanosine triphosphate (GTP)-binding proteins. [*See* Cell Membrane.]

II. INTRACELLULAR RECEPTORS

Intracellular receptors bind to specific ligands capable of crossing the cell membrane. This usually means that the ligand is sufficiently hydrophobic (e.g., a steroid hormone) to cross the lipid bilayer, either by diffusion or by way of a facilitated mechanism. Other intracellular receptors may respond to energy, such as light, as in the photoreceptor cell of the eye. Ligands that bind to intracellular receptors may be aided in their entry into or transport through the cell by specific binding proteins. However, these binding proteins do not respond to the information carried by the ligand; they simply facilitate the delivery of the ligand to its receptor. Intracellular receptors can be located in either the cytoplasmic or nuclear compartments of the cell. Like cell-surface receptors, intracellular receptors can respond to ligand binding directly, for example, by binding to and directly activating specific DNA sequences responsible for regulating gene expression, or they may respond to ligand binding by forming complexes with other signaling molecules that continue the cascade of signal transduction. In both cases, cell-surface receptors and intracellular receptors are capable of more than binding to a specific ligand, they are also capable of effecting a biological response to the ligand. This important property distinguishes cell receptors from other ligand binding proteins.

III. RECEPTOR RESPONSES AND EFFECTOR SYSTEMS

In all receptor–ligand interactions, information carried by the ligand is transmitted to the receptor as a structural or conformational change in the receptor molecule itself. It is this conformational change that somehow enables the receptor to relay the information carried by the ligand to other compartments of the cell. In some receptor systems, this conformational change may activate the intrinsic enzymatic activity of the receptor molecule, or result in new associations of the receptor with other signaling molecules. Cell receptors that are bound to ligand may also self-associate with other ligand-bound receptors, a process referred to as receptor dimerization.

Alternatively, distinct signal transduction molecules may further modify the structure of the receptor. The most common effector systems involved in cellular growth and differentiation involve phosphorylation by specific kinases or GTP hydrolysis mediated by specific GTP binding proteins. Other receptor systems may effect biological responses by mediating ionic fluxes (e.g., Ca^{2+}) or by directly regulating the transcription of specific genes by binding to DNA regulatory sequences. Often these effector systems are components of a signal transduction cascade pathway that can greatly amplify the initial signal output generated by the initial ligand–receptor interaction.

The biological end points initiated through ligand–receptor interactions vary widely. As outlined earlier, ligand–receptor interactions often result in changes in gene expression. The expression of new gene products may be needed to alter a cell's shape or position. Ligand–receptor interactions may also contribute to the cell's ability to divide or to differentiate into a more highly specialized cell type. Genetic alterations in the genes encoding cell receptors and their associated signaling partners also contribute to a variety of human diseases, including cancer.

IV. RECEPTOR DOWN-MODULATION AND DESENSITIZATION

How is the potent signal initiated by the receptor–ligand interaction attenuated? For cell-surface receptors, ligands may be routed into the cell through a vesiculation and intracellular trafficking mechanism referred to as receptor-mediated endocytosis. Ligand-occupied receptors cluster in the plane of the membrane over specialized or "coated" membrane regions that become invaginated. These invaginations become internalized and are carried into the cytoplasm as coated vesicles. Once inside the cell, both ligand and receptor can be degraded or recycled, depending on the nature of the receptor. Receptors involved in transporting essential cellular nutrients are frequently recycled (e.g., low-density lipoprotein or transferrin receptors). Activated growth factor receptors (e.g., epidermal growth factor receptor) involved in signaling cell division and differentiation are often routed to a cytoplasmic compartment called the lysosome, where they are degraded. In either instance, the number of receptors available on the cell surface for binding to ligand is reduced. In a third category of cell-surface receptor down-modulation, receptors may become phosphorylated by specific kinases, or the ligand-binding domain of the receptor may be proteolytically removed as another means of turning off the signaling capacity of the receptor.

Intracellular receptors such as steroid hormone receptors can also be either degraded (e.g., the progesterone and glucocorticoid receptors) or recycled following ligand binding. Once activated by hormone binding, these receptors may become further structurally modified by phosphorylation or other posttranslation events. These receptors may also become associated with specific binding proteins that may regulate their cellular transport. Unoccupied hormone receptors are frequently sequestered in the cytoplasm by specific chaperone proteins. These proteins are released upon ligand binding to the receptor, facilitating delivery of the hormone-occupied receptor to its target in the nucleus. Other hormone receptors may be transported to the nucleus following synthesis. Like cell-surface receptors, hormone receptors frequently dimerize following ligand binding. However, the precise mechanism of hormone release from these receptors, and the mechanisms involved in receptor recycling, are not yet well understood.

V. RECEPTOR RECYCLING AND BIOSYNTHESIS

In contrast, the recycling of cell-surface receptors has been well characterized. Once ligand-occupied cell-surface receptors are internalized via endocytosis, the intravesicular compartment becomes acidic, causing the ligand to be released from the receptor. This release allows the receptor to be recovered in a ligand-free state through a vesicular sorting process. Recycled cell-surface receptors are thereby replaced on the cell surface by a vesicular transport mechanism.

Alternatively, the cell-surface receptor may be degraded, requiring new receptor biosynthesis. Cell-surface receptors are typically synthesized in the rough endoplasmic reticulum (see Fig. 1) and become further modified through glycosylation in the Golgi apparatus before being transported to the cell surface through a vesicular transport mechanism. [*See* Golgi Apparatus.]

Intracellular receptors, in contrast, are typically soluble cytoplasmic or nuclear proteins that are synthesized on free ribosomes in the cytoplasm (see Fig.

1). Once synthesized, however, these proteins rapidly become associated with specific binding and chaperone proteins that may modify their cellular localization or function. In general, these receptors exhibit specific structural domains that are correlated with specific receptor functions. For example, for steroid hormone receptors a cysteine-rich DNA binding domain can be distinguished from the hormone binding domain. Other regulatory regions of the receptor that can modulate hormone binding, nuclear translocation, and DNA binding may also be part of the steroid hormone receptor's structure.

VI. RECEPTOR ASSAYS

Cell receptors are classically defined by two criteria: their ability to specifically bind to ligand and their ability to exert a biological response following ligand binding. Therefore, most receptor assays are designed to measure either or both of these receptor functions.

A. Ligand Binding and Specificity

One key property of all receptors is their ability to bind to a *specific* class of ligands. In other words, not just any ligand can bind to any receptor. There may, however, be multiple members within a given ligand family, although ligand family members are usually structurally conserved. This ability of the receptor to discriminate between various classes of ligand is usually mediated by a high affinity of the receptor for the ligand, that is, ligand binding occurs readily and dissociation of the bound ligand is rare. The number of binding sites is also typically proportional to the number of receptors, and therefore an estimate of receptor number per cell can be determined using radioisotopically labeled ligand binding to perform ligand binding assays. In general, cell receptors of all classes exhibit finite and saturable binding. [*See* Receptors, Biochemistry.]

B. Measurement of Biological Responses to Ligand Binding

In addition, when a ligand binds to its cognate receptor a characteristic biological response is elicited. As described earlier, these responses vary from activation of downstream signal transducers (e.g., cAMP, ionic fluxes, activation of kinases, GTP binding proteins) to biological end points such as secretion, motility, activation of transcription, cell division, and cell differentiation. Quantitative assessment of one or more of these biological responses can also be used as an accurate measurement of receptor–ligand interactions.

VII. SUMMARY

All cells communicate with their environment by interacting with a variety of extracellular messengers. These messengers, called ligands, have specific structural properties that allow them to interact with high affinity with specific cell receptors. Cell receptors can be broadly divided into two categories: cell-surface receptors and intracellular receptors. Both classes of receptor are typically proteins with specific ligand binding domains. In addition to their ability to recognize specific ligands with high affinity and specificity, both classes are capable of stimulating characteristic biological responses following ligand binding. These responses are mediated by signal transduction pathways characteristic of the ligand–receptor interaction and may result in important biological end points such as changes in gene expression, cell shape and position, cell division, and cell differentiation.

BIBLIOGRAPHY

Kishimoto, T., Taga, T., and Akira, S. (1994). Cytokine signal transduction. *Cell* **76,** 253.

Limbird, L. E. (ed.) (1996). "Cell Surface Receptors: A Short Course on Theory and Methods," 2nd Ed. Kluwer Academic Publishers, Boston.

Mangelsdorf, D., Thummel, C., Beato, M., Herrlich, P., Schutz, G., Umesono, K., Blumber, B., Kastner, P., Mark, M., Chambon, P., and Evans, R. M. (1995). The nuclear receptor superfamily: The second decade. *Cell* **83,** 835.

Pawson, T. (1995). Protein modules and signalling networks. *Nature* **373,** 573.

Cell Signaling

ALEXANDER LEVITZKI
The Hebrew University of Jerusalem

GLOSSARY

Acetylcholine Neurotransmitter of the nervous system

Agonist Compound that activates a receptor; this compound can be a hormone, a neurotransmitter, or a synthetic drug

Antagonist Compound that binds to a receptor but does not activate it; it therefore blocks the action of an agonist on the receptor

Autocrine Form of self-regulation by a cell, which secretes a hormone or a growth factor that affects the same cell through a receptor

Cerevisiae Species of yeast; baker's yeast is *Saccharomyces cerevisiae*

Chemotaxis Movement of a cell or a microorganism up a gradient of a chemical attractant or down the gradient of a repellent

Cytokines Mediators of signal transduction, mainly secreted by immune cells

Growth factor Molecule that stimulates cell growth through a specific receptor

Kinase Enzyme that utilizes adenosine triphosphate or another nucleoside triphosphate to phosphorylate other molecules including proteins

Ligand Molecule that binds to a binding site on a receptor or on an enzyme

Muscarinic Class of acetylcholine receptor to which molecules such as muscarine bind specifically

Paracrine Form of regulation of a cell by a neighboring cell that secretes a hormone or a growth factor

Saccharomyces *Saccharomyces cerevisiae;* see *Cerevisiae*

Second messenger Molecule generated by a biochemical effector system coupled to a receptor; when the receptor binds the agonist, it activates a biochemical system that generates a second messenger, which transmits the message by activating other biochemical systems

Serotonin 5-Hydroxytryptamine, a neurotransmitter

Steroids Class of hormones whose structure is derived from cholesterol

Triiodothyronine Hormone also known as T_3, secreted by the thyroid gland

Tyrosine *p*-Hydroxyphenylanine, an aromatic amino acid

SINGLE CELLS AS WELL AS WHOLE TISSUES RESPOND to environmental signals such as nutrients, hormones, growth factors, chemoattractants, and toxic materials. In almost all cases, the first event involves interaction of the signaling molecule with a cell-surface receptor. This interaction is followed by a transmembrane signaling event triggered by the activated receptor. A biochemical signal is thus transmitted through the cell membrane into the cytoplasm, awakening the cell to produce a characteristic biochemical response. Many signals, such as those produced by growth factors, antigens, or steroids, are further propagated to the nucleus. Signal transduction to the nucleus induces expression of specific sets of genes that in turn induce the cell to divide to differentiate or to produce signaling molecules. In recent years, many of the signaling elements that transmit the signal from the receptor into the cytoplasm and the nucleus have been identified and characterized. The molecular

ENCYCLOPEDIA OF HUMAN BIOLOGY, Second Edition, VOLUME 2. Copyright © 1997 by Academic Press. All rights of reproduction in any form reserved.

mechanisms that underlie cellular signaling have become an important focus in biological research.

I. CELL–CELL COMMUNICATION

At the turn of the 20th century, surgical removal of an organ (i.e., pancreas) from an animal resulted in a severe physiological deficiency. This deficiency could be remedied by implanting the appropriate glandular tissue of the pancreas or by injecting pancreatic extract into the animal from which the pancreas was removed. The extracts were then purified by biochemical procedures to yield pure hormone. These findings led to the understanding that separate (Greek *krinen*) organs within (Greek *endos*) the body control the action of other organs by secreting an active chemical substance into the blood vessels. This, in turn, led to the definition of a new branch of science known as endocrinology, the scientific discipline that explores the mode of action of hormones. The term hormone (from the Greek, meaning "arousing substance") was first introduced to describe the action of a specific chemical substance extracted from intestinal cells that stimulates the pancreas to secrete its enzymatic content; this specific hormone was named secretin. Secretin is carried by the blood from the intestinal cells to the pancreas, where it stimulates the latter to secrete its enzymes into the intestine. In general, hormone refers to a chemical substance secreted by a gland remote from the organ that it affects. The hormone travels through the bloodstream from the secreting organ to the target tissue (Fig. 1). [*See* Endocrine System.]

The discovery of acetylcholine as the chemical transmitter of nerve stimuli was a historical landmark and accelerated the acceptance of the notion that cellular signaling is mediated by chemical transmission.

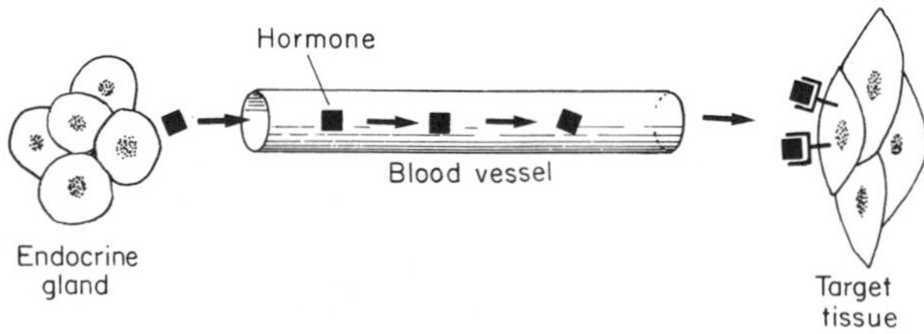

FIGURE 1 Hormones and their action. Hormones are synthesized by glands remote from their target organs and reach their receptors through the bloodstream. The hormone affects the target tissue by interacting with specific receptors. Binding of the hormone to its specific receptor triggers the cell to respond.

Until the discovery of acetylcholine, many investigators believed that nerve stimulation was a purely electrical phenomenon with no involvement of transmitter molecules. Once both hormones and neurotransmitters were recognized, the stage was set for the discovery of many other chemical signals. Indeed, growth factors and local mediators (which are very similar to hormones but can also travel through the tissue fluids, for short distances) were subsequently identified. All classes of chemical signals—neurotransmitters, hormones, growth factors, cytokines, and local mediators—coordinate the activities of all organs in a multicellular organism. [*See* Neuroendocrinology; Soft Tissue Repair and Growth Factor.]

Some signals are mediated through contact signaling, in which one cell touches the next cell so that a membrane-bound signaling molecule interacts directly with a receptor on the neighboring cell. These types of specific interactions govern the specificity with which one cell recognizes a neighboring cell, thus generating a specific tissue. Direct cell–cell signaling can also be mediated through a gap junction; this molecular device is like a tube within the cell membrane, through which signaling molecules can flow from one cell to another. [*See* Cell Junctions.]

Not only mammalian cells respond to chemical signaling, for bacteria, fungi, yeast, and plant cells also communicate with each other through chemical signals. The signaling cell itself can be a target for another signaling cell (Fig. 2). Thus, a complex network of signals allows the cell, the tissue, and the whole organism to function efficiently in its environment. The signaling molecules are named according to the functions they convey, as detilaed in the next section.

II. SIGNALING MOLECULES

All cellular signals are transmitted by chemically defined molecules, which interact with their target cells through specific receptor molecules. In this section, we discuss in some detail the three principal classes of signaling molecules: neurotransmitters, hormones, and growth factors and cytokines, with a brief discussion of other chemical signals, such as chemotactic molecules and pheromones.

A. Neurotransmitters

Neurotransmitters are molecules produced by nerve cells and related at the nerve terminal by specific mech-

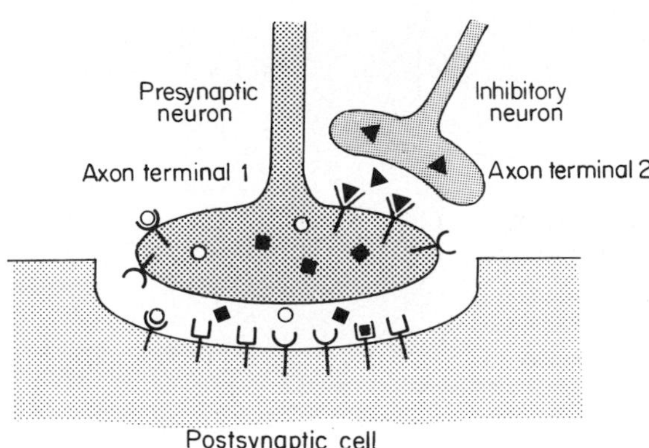

FIGURE 2 The action of neurotransmitters. Axon terminal 1 can release two different neurotransmitters, schematically represented by solid squares (■) and open circles (○). These diffuse across the synaptic cleft to bind to their respective receptors at the postsynaptic cell. The neurotransmitter (○), for example, can bind to specific receptors that reside on the nerve terminal itself and, therefore, are termed autoreceptors. The nerve terminal can be innervated by yet another nerver (e.g., an inhibitory neuron). The axon terminal of the inhibitory neuron, axon terminal 2, releases an inhibitory transmitter (▲) that binds to a specific receptor on axon terminal 1. The binding of the inhibitory neurotransmitter attenuates the action of the affected neuron by inhibiting the release of its neurotransmitters.

anisms, whose molecular details are far from being fully understood. The released neurotransmitters interact with specific receptors on the enervated target tissue and also with receptors located at the nerve terminal itself (see Fig. 2). The neurotransmitters travel by diffusion no more than 500–1000 Å to their target and can thus signal the target cells within milliseconds. Indeed, the nervous system is known for its ability to transmit signals rapidly and precisely.

The role of the receptor at the nerve terminal is to regulate and modulate release of neurotransmitters; therefore, they are referred to as autoreceptors (or presynaptic receptors). One nerve terminal can synthesize and release two chemically different neurotransmitters, as represented in Fig. 2. Neurotransmitters are either low-molecular-weight (MW 100–300) molecules, such as noradrenaline (norepinephrine), serotonine, and glycine, or medium-sized neuropeptides, such as enkiphaline (MW 550), substance P (MW 1200), and β-endorphin (MW 3500). Neurotransmitters travel by diffusion over a very short distance across the synaptic cleft to interact with their receptor on a nearby cell (Fig. 2). The nerve terminal is often innervated by yet another nerve cell. In this case, the cell possesses another set of receptors, which respond to a different neurotransmitter (e.g., an inhibitory neuron). Thus, when nerve cell 1 receives a signal from nerve cell 2 through the inhibitory neurotransmitter, the output of its own signaling neurotransmitters is attenuated. Therefore, the postsynaptic response will be lower when cell 2 is active than when it is quiescent. Clearly, both nerve cell 1 and nerve cell 2 can be innervated by still other nerve cells, generating the network of signaling. Indeed, understanding the network in the brain is essential for understanding how different behavioral patterns emerge. Elucidation of the mechanism of action of each signal is insufficient; one needs to investigate the network of cellular signals to understand the complete signaling pattern of a whole brain area. To some extent, this observation is also true for individual cells, because each cell possesses an array of receptors that respond to different chemical stimuli. Integration of the various responses to the different stimuli, acting through specific receptors on the cell surface, results in a characteristic pattern of response. [*See* Neurotransmitter and Neuropeptide Receptors in the Brain; Receptors, Biochemistry.]

B. Hormones

Hormones are molecules that are released from an endocrine gland into the bloodstream and elicit a response at any cell that expresses a specific receptor to the hormone (see Fig. 1). Tissues, because they are very different from each other, can respond simultaneously to the same hormone. In contrast to neurotransmitters, hormones act at a distance and reach their target cells through the bloodstream. This process is much slower than signaling by neurotransmitters released at nerve terminals. Indeed, the response time to hormones is minutes, or even hours, subsequent to their release into the bloodstream. The chemical variety of hormones is much larger compared with that of neurotransmitters. Hormones range from small hydrophilic molecules, such as (−)noradrenaline (norepinephrine), to very hydrophobic steroid hormones, to medium-sized polypeptide hormones, such as insulin (MW 6000) and the multisubunit human chorionic gonadotrophin (MW 39,000). Many cells possess more than one receptor for different hormones; therefore, the net response of the cell depends on the spectrum of hormones interacting with the cell at any particular time.

Depending on the target cell and its biochemical makeup, the same factor can induce different biochemical responses. Unicellular yeasts such as *Saccharomyces cerevisiae* also respond to specific peptide

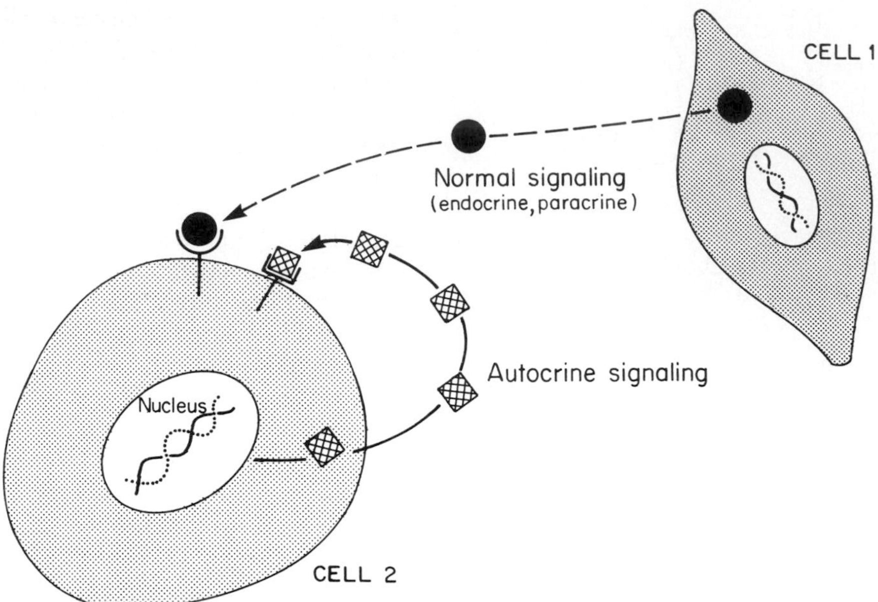

FIGURE 3 Autocrine signaling is compared with endocrine and paracrine signalings, which are more widespread.

pheromones, which induce mating between two mating (sex) types of the yeast. The response through receptors is a regulatory element in unicellular organisms as well as in the cells of multicellular organisms. The molecular mechanisms that underlie signal transduction in molecular organisms are similar to those found in multicellular organisms. [*See* Peptide Hormones and Their Convertases; Steroids.]

C. Growth Factors

Growth factors such as epidermal growth factor (EGF), nerve growth factors (NGF), and platelet-derived growth factor (PDGF), are peptides of molecular weight 6000–30,000. Different cell types require different growth factors for growth, especially at early stages of development. For their growth or differentiation, different cells are now thought to require unique combinations of growth factors, each of which interacts with a specific receptor on the cell surface. Growth factors are often secreted in a tissue close to the target tissue but may also travel like a hormone via the bloodstream to a distant organ. In contrast to hormones and neurotransmitters, growth factors induce cells to proliferate (divide) and also induce characteristic biochemical transformations within the cell (differentiation). Some cells respond to the growth factor that they secrete. This type of regulation is known as autocrine signaling (Fig. 3). There are also factors that inhibit cell growth and, therefore, counteract the action of growth factors. These are frequently called antigrowth factors or growth inhibitors, which probably play a role in growth arrest.

D. Other Chemical Signals

Although neurotransmitters, hormones, and growth factors are the major signaling molecules in multicellular organisms, other types of signaling molecules exist for unicellular organisms. For example, T cells secrete a variety of mediators called lymphokines, interleukins, or cytokines, and these molecule also interact with specific receptors, mainly on immune cells, leading them to respond. Also, simple nutrients such as glucose can be signal molecules in addition to being major metabolites. When the unicellular yeast *S. cerevisiae* is exposed to glucose, it is induced to grow and divide. Thus, aside from being a substrate for energy-producing reactions, glucose signals the yeast to "wake up" by stimulation of proliferative biochemical pathways. Also, bacteria respond to chemical signals such as nutrients and repellents. For example, the amino acid aspartic acid interacts with specific receptors on the surface of the bacterial membrane and induces the bacteria to swim toward the higher concentrations of the nutrient. Thus, nutrients can act as chemotactic signals for the bacteria.

E. Signaling in Plants

Knowledge of the molecular basis of cellular signaling in plants is still in its infancy, but the emerging pattern is similar to that found in the animal kingdom. Plant hormones induce cellular growth and differentiation of specialized tissues. The molecular mechanisms of plant cellular signaling are currently under intensive investigation, indeed, the elements of signal transduction pathways in plants is being rapidly unraveled.

III. RECEPTORS: THE SELECTORS OF SIGNALS

A signal can trigger a cell to respond only if the cell possesses a receptor that can selectively bind the signaling molecule. For each chemical signaling molecule there is at least one type of protein receptor molecule. With the improvement of biochemical techniques and the advent of gene cloning, it is now possible to purify receptor proteins, clone and express receptor genes in various cell systems, and explore in detail their properties and mechanisms of action. Receptors are no longer defined only by pharmacological assays or by their ability to bind radioactively labeled drugs. At present, some of the receptors can be expressed even in large amounts and eventually be crystallized, such that their structure can be analyzed using X-ray methods. Indeed, the enzymatic domain of the insulin receptor protein and of the FGF, receptor (fibroblast growth factor) have been crystallized, allowing a more detailed study of their mechanism of action.

For most signaling molecules one finds a family of receptors, where each member is expressed in different cell types. These different species of receptors, which respond to the same molecule, can be either very different from each other or closely related. Thus, two types of cells can respond to the same signaling molecule, but the signaling molecule triggers different responses in the two different types of receptors. A well-studied example is the β-adrenoceptor and the α_2-adrenoceptor, both of which bind and respond to noradrenaline (norepinephrine) and adrenaline (epinephrine) (Fig. 4). The β-adrenoceptor activates adenylyl cyclase, inducing elevation of intracellular cyclic adenosing monophosphate (cAMP), whereas α_2-adrenoceptor inhibits adenylyl cyclase, causing a decrease of cAMP production. The two receptors are closely related in their gross structural features but differ in the details of the structure.

The β-adrenoceptor *activates* the adenylyl cyclase system by activating guanosine triphosphate binding proteins (G_s), whereas the α_2-adrenoceptor *inhibits* adenylyl cyclase through its inhibitory G protein (G_i). The two receptors also differ in their hormone-binding protein domain, although both bind the same adrenaline or nonadrenaline molecules. The structural differences in the binding sites allow the design of selective drugs for each type of receptor. For example, l-(−)propanolol binds to β-adrenoceptor but not to α_2-adrenoceptors, whereas clonidine and yohimbine bind to α_2-adrenoceptors and not to β-adrenoceptor.

Another example is the existence of two closely related receptors with subtle structural differences: the two types of β-adrenoceptor. β_1- and β_2-adrenoceptor differ in their selective affinity toward adrenaline and noradrenaline, but both activate adenylyl cyclase. It is interesting and of physiological relevance that the heart expresses β_1-adrenoceptors, whereas the lung expresses β_2-adrenoceptors. The difference between the two subtypes of β-adrenoceptors in the heart and lung allows the design of β_1- and β_2-selective drugs, which act selectively on either the heart or lung.

Receptors for growth factor belong to a small number of families. In each family the basic structure of all the receptors within the family is similar. The small but significant differences generate the diversity of growth factors that activate them and the multitude of biochemical outputs.

IV. MECHANISMS OF TRANSMEMBRANE SIGNALING

Most receptors reside on the cell membrane and transmit their signal across the membrane. Some receptors, however, are cytoplasmic or nuclear, and the signaling molecule must penetrate the cell membrane to reach its target receptor. The two classes of receptors function differently (Fig. 5). Membrane receptors are physically coupled to an intracellular biochemical apparatus, which is activated once an agonist binds to the receptor. Cytoplasmic receptors also activate intracellular biochemical machinery when they bind an agonist. The receptors are largely DNA-binding proteins and therefore regulate transcription.

A. Intracellular Receptors

Steroids, triiodothyronine (T_3), retinoids, and vitamin D are hydrophobic hormones that bind to their intra-

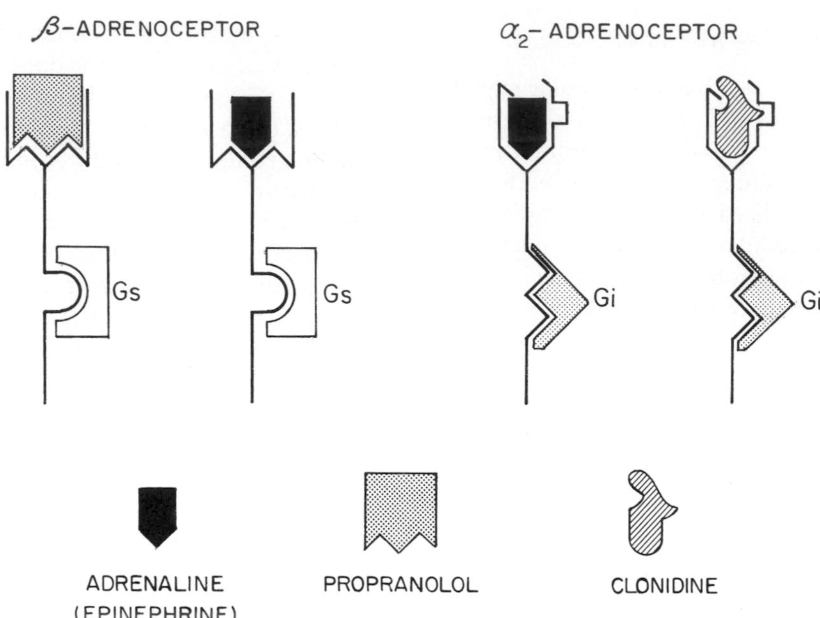

FIGURE 4 Selectivity in the design of receptors. Both the β-adrenoceptor and the α_2-adrenoceptor bind adrenaline. Propanolol, however, is a selective drug for β-adrenoceptor, whereas clonidine is selective to the α_2-adrenoceptor. The schematic representation shows that it is possible to design selective drugs that exploit the difference in the binding domain of the two receptors. The common structural denominator of the β-adrenoceptor and the α_2-adrenoceptor allows adrenaline (and noradrenaline) to bind to both G_s, stimulatory GTP-binding protein which mediates the activation of adenylyl cyclase by stimulatory receptors like the β-adrenergic receptors; G_i, inhibitory GTP-binding protein, which mediates the inhibition of adenylyl cyclase by inhibitory receptors like the α_2-adrenergic receptor.

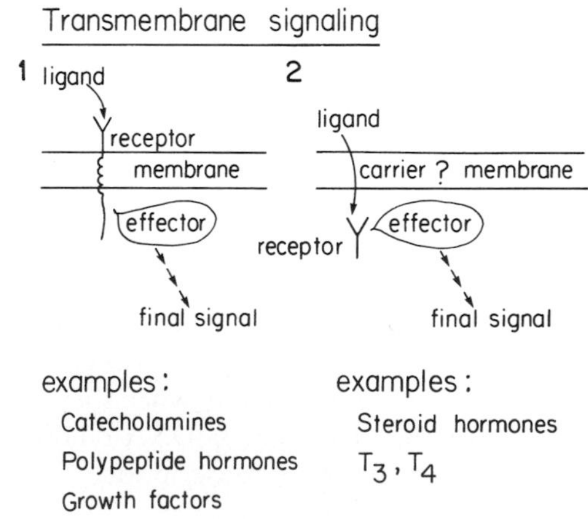

FIGURE 5 Transmembrane signaling, in principle, can occur by two mechanisms. 1. The ligand binds to a membrane-bound receptor, and signal transduction is a transmembrane molecular event. 2. The ligand crosses the membrane to reach an intracellular receptor, which resides in the cytoplasm or the nucleus.

cellular receptors after crossing the cell membrane. Whether these hormones penetrate the cell membrane passively or through a specific membrane protein is still unclear. The intracellular steroid or T_3 receptor changes its conformation upon binding of the hormone and is activated. Following this process, the activated steroid or T_3 receptor binds to specific regulatory DNA regions, leading to the expression of specific genes. Different genes are activated in different cell types because the steroid or T_3-responsive DNA elements are linked to different genes. Similar general principles are applicable to vitamin D_3 receptors and for receptors to retinoic acid. A more detailed discussion on the mode of action of steroid hormones and other ligands for nuclear receptors is given elsewhere.

B. Mechanisms of Signal Transduction by Membrane Receptors

Until the early 1970s, signaling events were defined by characteristic pharmacological responses or biochemical events distal to the initial binding event. For example, the secretion of enzymes from the parotid gland is triggered by acetylcholine, which interacts with a particular class of muscarinic acetylcholine receptors. Twitching of guinea pig ileum induced by serotonin defines a class of serotonin receptors. These signals are distal to the initial binding event. In recent years, more has been learned about the early biochemical events triggered by the binding of signaling molecules and the subsequent events that follow, which bring about the distal, final response.

The biochemical system that is physically activated by the ligand-bound receptor is usually called the effector system. From a biochemical perspective, one can identify a limited number of effector systems. The number of receptors that can be defined pharmacologically, namely, according to their ligand specificity, exceeds by far the number of biochemical effectors. This observation immediately suggests that receptors for different ligands may have common structural features, which allow them to interface and activate identical effector systems. Indeed, distinct receptors that activate adenylyl cyclase have similar effector domains but very different ligand-binding domains. Distinct receptors that inhibit adenylyl cyclase also seem to share similar effector domains (Fig. 6). Similarly, different receptors that all activate phospholipase C seem to have similar effector domains. The design principle extends to all families of receptors. Table I summarizes the types of biochemical effector systems coupled to receptors; a few examples of pharmacologically very different receptors that couple to the same effector system are listed. Table II summarizes the classes of chemicals that activate different effector systems. The table demonstrates that the same effector systems can be activated by different ligands.

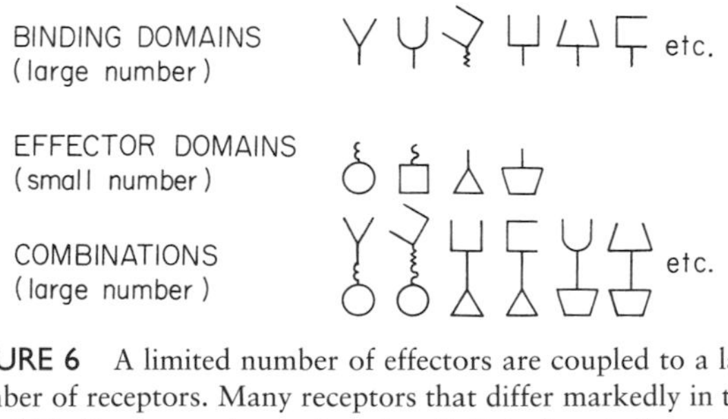

BINDING DOMAINS (large number) etc.

EFFECTOR DOMAINS (small number)

COMBINATIONS (large number) etc.

FIGURE 6 A limited number of effectors are coupled to a large number of receptors. Many receptors that differ markedly in their ligand-binding properties are functionally linked to similar or identical effectors. Therefore, a limited biochemical repertoire may be used to generate the diversity of response found. Different combinations can be expressed in a variety of cells and tissues, thus generating the response characteristic to that cell or tissue.

Different receptors differ in the details of the biochemical reactions characterizing each of the three basic steps. In many cases the cascade of the biochemical reactions, triggered by the activated receptor, involves the recruitment of protein signaling molecules to the activated receptor, which are responsible for an intracellular signal propagation.

C. Basic Features of Cellular Signaling

For all types of cellular signaling there are a number of common underlying principles.

1. The signaling molecule binds to a specific receptor, which is always a protein molecule.
2. Binding of the signaling molecule to the receptor triggers a structural change at the receptor, which in turn triggers a cascade of biochemical reactions, culminating in the final cellular response. In many instances the first biochemical reaction led to the formation of a small molecule such as cAMP, cyclic guanosine monophosphate, diacyl glycerol, and inositol-trisphosphate (IP_3). The ligand that triggers receptor activation is commonly known as the first messenger, whereas the small molecules produced or released inside the cell that transmit the message further are known as second messengers.
3. The activated receptor also triggers with a time lapse, a response that eventually initiates a mechanism that terminates the primary response. This biphasic nature of the response lends its transient nature (Fig. 7), which is the essence of a signaling event.

D. Amplification, Adaptation, and Sensitivity

Triggering of a response by the ligand-bound receptor results from a structural change within the receptor molecule. The conformational change in the receptor converts the receptor to its active form. The active form interacts with the effector, activating it to produce the primary biochemical signal (Table I). Often, a small number of receptors activate a large number of enzyme molecules as an intermediate step in the signaling event, or as an end result of the primary response, or both. This phenomenon is known as amplification. For example, the β-adrenoceptor system has one receptor that activated a dozen adenylyl cyclase molecules, each of which produces about 100–200 molecules of cAMP before the activity of the enzyme decays. Each cAMP molecule subsequently activates the enzyme cAMP-dependent protein kinase,

TABLE I
Types of Chemical Signaling

No.	Primary signaling event	Examples of final response	Types of receptors involved	Mechanisms of signal transduction
1	Activation of adenylyl cyclase (elevation of cAMP)	Enzyme secretion, steroid biosynthesis, glycogen breakdown, activation of ion channels, etc.	β-adrenoceptors, glucagon, ACTH, prostaglandins, serotonin, adenosine (A2), etc.	Inactivation of the receptors by phosphorylation and/or removal from the cell membrane
2	Inhibition of adenylyl cyclase (inhibition of cAMP production)	Attenuation of effects triggered by cAMP	α_2-adrenoceptors, adenosine (A1), somatostatin	As in No. 1
3	Activation of phospholipase C (elevation of inositoltrisphosphate and diacylglycerol)	Neurotransmitter release, platelet aggregation	Muscarinic acetylcholine receptors, α_1-adrenoceptors, glucagon	Probably similar to No. 1
4	Activation of Cl^- channels	Inhibitory nervous signals, sedation	Glycine, GABA	Unknown
5	Opening of Na^+/K^+ channel	Peripheral muscle contraction	Nicotinic (acetylcholine)	Conformational change
6	Opening of Na^+ channels	Excitation	Glutamate	Unknown
7	Opening of Ca^{2+} channels	Isotropic effect in the heart	Adrenaline	As in No. 1
8	Activation of protein tyrosine kinases	Proliferation of cells, glucose uptake into cells and its utilization	EGF, insulin	Internalization of the ligand-bound receptor, dephosphorylation
9	Activation of protein tyrosine phosphatases	Growth arrest(?), entry into mitosis, T-cell response	T-cell receptor	Unknown

ACTH, adrenocorticotropin; GABA, γ-aminobutyric acid.

TABLE II
Types of Signaling Molecules

Types of signaling molecules	Classification	Types of biochemical signal triggered (examples)
Amino acids	Neurotransmitters, hormones	Opening of channels (glycine, GABA, glutamate)
Amino acid derivatives: catecholamines (adrenaline, noradrenaline, dopamine, octopamine), serotonin, histamine	Neurotransmitters, hormones	Activation and inhibition of adenylyl cyclase, activation of phospholipase C
Small peptides (MW 1000)	Neurotransmitters, hormones, growth factors	Activation and inhibition of adenylyl cyclase, activation of the pheromone response in *S. cerevisiae,* chemotaxis of neurophils, activation of phospholipase C
Large peptides and proteins	Hormones, growth factors, lymphokines, cytokines	Activation and inhibition of adenylyl cyclase, activation of phospholipase C, activation of mitogenesis
cAMP	Second messenger in high eukaryotes, primary messenger in *Dictiostelium discoideum*	Activation of protein kinases, aggregation of the slime mold *D. discoidum*
Prostaglandins	Local (paracrine) mediators	Activation of adenylyl cyclase, other unknown effects
Steroids, vitamin D_{31}, retinoic acid	Hormones	DNA binding and transcription of certain genes

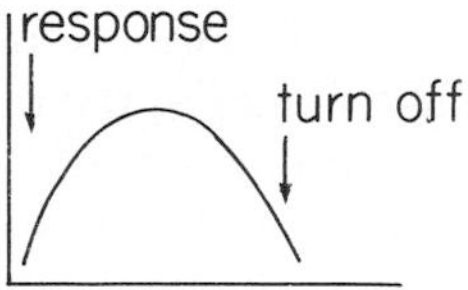

FIGURE 7 The transient nature of signaling. Binding of the ligand to its receptor elicits a response that is turned off after a time delay. The time scale is different for different receptor systems. For some receptors, the response is maximal within milliseconds, and the whole signal is over within well less than a second. For other systems, the time scale is in minutes or hours. The signal must have a transient nature.

which, once activated, phosphorylates at least hundreds of protein molecules. Some of these molecules are themselves enzymes, whose activity is either activated or shut off. The overall amplification of the signal in this system is about 10^6. Amplification is found in almost every system in which a receptor on a cell surface transmits a signal through the membrane into the cell. In ion-gated receptors, many ions pass through a single channel opened upon receptor activation, also producing an amplified signal. The very significant amplification produced by receptors allows nature to economize on the number of receptors and enables the cell membrane to accommodate many kinds of receptors, in addition to other proteins.

Sensitivity of a receptor system is controlled not only by amplification but also by the number of receptors on the cell. Thus, cells with a high number of receptors display enhanced sensitivity to the stimulating ligand, whereas cells with a low number of receptors are less sensitive. Another possible mechanism to increase the sensitivity of a receptor system to its stimulating ligand is cooperativity. In this case, the receptor is composed of a number of protein subunits, with two or more ligand-binding sites. The response is triggered only when all the receptor-binding sites are bound with the stimulating ligand; therefore, the response of the receptor system becomes cooperative. For example, the nicotinic acetylcholine receptor is activated only when two molecules of acetylcholine bind to the receptor. This property allows the nicotinic receptor system to respond fully within a narrow range of acetylcholine concentrations and in a cooperative fashion. This is similar to the cooperative binding of oxygen to hemoglobin, which allows hemoglobin to undergo full oxygenation or deoxygenation over a narrow range of oxygen concentrations.

Adaptation is another hallmark of many receptor systems. Following a time delay, the activated receptor triggers a secondary cascade of biochemical events, which lead to the termination of the signal, thus lending it a transient nature (Fig. 7). For example, the depolarization response of the nicotinic receptor is triggered within 1 msec, whereas the desensitization in muscarinic receptor of the response sets in within 20–100 msec. In the nicotinic receptor, the acetylcholine-occupied receptor is slowly changing its conformation, so that its Na^+/K^+ channel (Table I) is closed. In the β-adrenoceptor system, activation of adenylyl cyclase occurs within seconds, whereas desensitization occurs within 2 min. In this case, the receptor becomes phosphorylated at specific sites, a modification that uncouples the receptor from the adenylyl cyclase system. Tyrosine kinase receptors (Table I) activate intracellular events within 1 min. Activation of the receptor involves the phosphorylation of tyrosine residues on the intracellular domain of the receptor molecule. These phosphorylated sites become docking areas for signaling molecules, which propagate the signal within the cell. Termination of the signal occurs via removal of phosphates from the intracellular domain of the receptor by phosphatases or by internalization of the ligand-bound receptor, thus preventing it from interacting with intracellular signal transducers.

V. ABERRANT CELLULAR SIGNALING

Abnormal cellular signaling can lead to pathological conditions. It can result from a few factors: abnormal secretion of the signaling molecule, its complete absence, mutational alterations in the receptor molecule, a change in the number of receptor molecules, or a combination of these phenomena. For example, diabetes can result from the absence of insulin-secreting cells or blockade of proinsulin processing to insulin. Severe forms of diabetes also result from a mutated insulin receptor or the presence of anti-insulin receptor antibodies, where, in both cases, the receptor cannot respond to the hormone. These forms of diabetes do not respond to insulin and, therefore, are difficult to treat. Amplification of the gene coding for EGF receptor is associated with a number of cancers such as gliomas and squamous cell carcinomas. Psoriasis also represents a good example of aberrant autocrine regulation, namely, the phenomenon in which a cell that secretes a growth factor also responds to it. The psoriatic skin cells overproduce the transforming growth factor α (TGFα), which interacts in an autocrine fashion with EGF receptor of the cell secreting it. This feature seems to cause these cells to hyperproliferate and is believed to play a role in the psoriatic state of the skin. [*See* Transforming Growth Factor, Alpha.] Another component that plays a role in psoriasis is the enhanced signaling of the cytokine IL-6, which probably plays a role in the inflammatory component of the disease.

Autocrine stimulation in pathological situations stems from partially relaxed, underregulated expression of the genes that code for growth factors and results in cells becoming independent of external growth factors. This phenomenon is very characteristic of malignant cells and is most probably partially responsible for their escape from the regulated programmed growth of normal cells.

Aberrant signaling can result not only from the constant exposure of the receptor to its activation ligand but also from mutations that result in an altered, constitutively active receptor. In the normal cell, the growth factor receptor is inactive as long as the growth factor is not bound to it. The abnormal receptor is in its active form all the time; therefore, its growth-promoting activity is constitutive or "on" all the time. Enhanced signaling of a receptor can also result from mutation(s) that lead to its overexpression. Receptor overexpression is also a hallmark of numerous cancers and other proliferative diseases.

VI. CROSS-TALK BETWEEN SIGNALING PATHWAYS

Signaling events are either unidirectional or divergent by activating various sets of intracellular signal transducer elements. It has been found that signaling systems can cross-talk. Steroids acting through stimulation of gene expression by the steroid-bound receptor induce the biosynthesis of β-adrenoceptors, thus leading to enhanced response to catecholamines. Growth factor receptors have been demonstrated to activate certain isotypes of phospholipase C, thus elevating intracellular Ca^{2+}, which is a typical effector system for other receptors too (Table I). Activation of protein kinase C through certain receptor systems can cause phosphorylation of growth factor receptors and attenuate their response to the growth factor.

Cells usually possess not only growth-promoting signaling pathways but also growth-inhibiting pathways. Growth inhibitory signals induce the modulation of growth signals, thus altering cell fate. Thus, TGFβ can attenuate the response of a cell to growth stimulation signals. Similarly, the mating pheromone can transiently arrest the growth of a *S. cerevisiae* cell at G1 and allows it to mate with its opposing mating type.

Because each cell possesses a large number of receptors, they are usually challenged by combinations of chemical signals, sometimes even conflicting in nature, which activate interacting biochemical systems. For example, adrenaline activates both β-adrenoceptors and α_1-adrenoceptors on the surface of liver cells. Therefore, both cAMP and Ca^{2+} levels are elevated within the cell, activating glycogen breakdown through phosphorylase kinase, which is activated by both Ca^{2+} and cAMP-dependent protein kinase. For this response, the effect of the two intracellular messengers is synergistic. However, actions of two signals may also be antagonistic. For example, if a cell harbors a receptor that activates adenylyl cyclase and also a receptor that inhibits it, and if the two receptors are activated simultaneously, the level and pattern of cAMP production depend on the actions of the two opposing receptors. The time constant of the responses of signaling systems may be different, thus enabling the transient signal to be of a characteristic pattern. For example, binding of a β-agonist to its receptor triggers both the activation of cAMP formation but with a short time delay and the activation of desensitization/down-regulation of the receptor. This sequence of events lends to the system its transient nature, a hallmark of a signaling event. Many examples of this nature are known.

The foregoing examples illustrate the complexity of cellular signaling even at the level of a single cell. Clearly then, understanding cellular signaling within a tissue or organ is a formidable task. Investigation of both response pattern of the cell and the molecular details of each signaling event is essential for understanding how a single cell responds to signaling molecules. Because cells are usually in contact with each other and communicate with each other through gap junctions, and cell–cell receptor ligand systems, the response of the whole tissue to its environment is a complex phenomenon. [*See* Cell.]

VII. DIRECT CELL–CELL SIGNALING

In multicellular organisms, cells of the same tissue or of different tissues physically interact with each other. Cells can allow passage of nutrients and metabolites, as well as of signaling molecules such as cAMP, though gap junctions. For example, allowing cAMP to pass between cells enables cells not in direct contact with an adenylyl cyclase-activation hormone to respond to it, if the neighboring cell possesses a receptor system coupled positively with adenylyl cyclase. The oligomeric protein device, which constitutes the gap junction, allows cooperation between the cells and coordination of their biochemical behavior. Since permeability through gap junctions is regulated, these devices become dynamic coordinating elements, which contribute to the response of the tissue as a whole. Another mechanism of direct cell–cell signaling is via cell-adhesion molecules (CAMs). These are transmembrane proteins expressed on one cell that interact with neighboring cells through specific receptors. Certain CAMs provide the signal to both cells to stop dividing and, therefore, are responsible for the well-known phenomenon of contact inhibition. Indeed, it has been recently suggested that the biochemical activity of some of these CAMs may be a protein tyrosine phosphatase, whose biochemical activity opposes the tyrosine phosphorylation associated with growth factor receptors, involved in promoting growth.

VIII. MODIFYING CHEMICAL SIGNALING BY DRUGS

Understanding cellular signaling allows pharmaceutical chemists and clinicians to intervene and modulate the natural pattern of cellular signaling. Understanding the pathophysiology of signaling systems can lead to the design of improved drugs. For example, recognizing that myasthenia gravis is caused by a deficiency of acetylcholine receptors at the neuromuscular junction has enabled a more sophisticated treatment to relieve and even abolish the clinical symptoms of the disease. Recognizing the difference between β_1- (cardiovascular type) and β_2- (lung type) adrenoceptors enables researchers and clinicians to constantly improve drugs for heart and asthma conditions. Similarly, increasing knowledge of Ca^{2+} channels and their mode of action enables researchers to produce improved Ca^{2+} blockers, known for their beneficial effects on heart diseases and other conditions. Even ulcers can be managed or cured by antibiotics and novel drugs rather than the surgeon's scalpel, owing to the understanding of the pathogenesis and the signaling events that lead to the overproduction of acid within the stomach. Refined understanding of how dopamine and γ-aminobutyric acid act in the brain and of their involvement in depression and moods has enabled the development of novel tranquilizers.

IX. SIGNAL TRANSDUCTION THERAPY

The molecular basis of many diseases has been elucidated by the emerging technologies of molecular biology and molecular biochemistry; these discoveries also opened novel modalities for therapy. Thus proliferative diseases such as cancers, atherosclerosis, and psoriasis, as well as inflammatory diseases, are now recognized as diseases of signal transduction. Indeed, antibodies against the HER-2/neu receptor, which is associated with the most malignant forms of cancers of the breast, ovary, and lung, are currently in clinical trials. Low-molecular-weight inhibitors of EGF receptor are also being developed for a number of disorders in which EGF receptor is excessively active. These include papilloma and psoriasis. Similarly, inhibiting the proliferation of vascular smooth muscle cells by blocking proliferative signal transduction pathways is important to block restenosis subsequent to balloon angioplasty or bypass operations. Blocking restenosis in the rat has now been achieved by local delivery to the site of injury of tyrosine kinase blockers (tyrphostins) or by local gene therapy introducing the retinoblastoma protein (Rb), whose expression inhibits cell proliferation. The recent discovery that vascular endothelial growth factor is essential for vascularization of tumors, as well as for development of macular retinopathy, identifies the growth factor and its receptor as targets for therapy. A large number of laboratories and drug companies are already developing angiogenesis inhibitors. The ongoing research on cellular signaling in normal tissues and identification of signaling deficiencies in disease ensure steady progress in generating more selective and therefore less toxic drugs.

BIBLIOGRAPHY

Berridge, M. J. (1985). The molecular basis of communication with the cell. *Sci. Am.* **253**(4), 142–152.

Evans, R. M. (1988). The steroid and thyroid hormone receptor superfamily. *Science* **240**, 889–895.

Levitzki, A. (1984). "Receptors: A Quantitative Approach." Benjamin Cummings, Menlo Park, CA.

Levitzki, A. (1987). Regulation of hormone-sensitive adenylyl cyclase. *Trends Pharmacol. Sci.* **8**, 299–303.

Levitzki, A. (1994). Signal-transduction therapy. *Eur. J. Biochem.* **226**, 1–13.

Levitzki, A., and Gazit, A. (1995). Tyrosine kinase inhibition: An approach to drug development. *Science* **267**, 1782–1788.

Levitzki, A. (1996). Targeting signal transduction for disease therapy. *Curr. Opin. Cell. Biol.* **8**, 239–244.

Cellular Cytoskeleton

CAROL C. GREGORIO
University of Arizona

GLOSSARY

ATP (adenosine 5′-triphosphate) Nucleotide triphosphate composed of adenine, ribose, and three phosphate groups. This is the predominant messenger of chemical energy within cells. The terminal phosphate ester bonds are highly energetic in that their hydrolysis, with perhaps the transfer of phosphate groups to other molecules, occurs with the liberation of free energy

Centrosome Organelle of animal cells that is the major microtubule-organizing center and acts as the spindle pole in mitosis. In most animal cells it accommodates a pair of centrioles

Desmosome Specialized cell–cell junction, created between adjacent cells, characterized by dense plaques of protein. Intermediate filaments interject into this structure

Exocytosis Process by which most molecules are secreted from a eukaryotic cell. The molecules are released extracellularly at the plasma membrane following fusion of the membrane-bound secretory vesicles in which they are packaged

GTP (guanosine 5′-triphosphate) Primary guanine-containing nucleotide triphosphate used in some energy-transfer reactions and in the synthesis of RNA. It has a unique role in microtubule assembly, cell signaling, and protein synthesis

Meiosis Process of cell division in gamete-producing cells in which the number of chromosomes is reduced to one-half by segregation of chromosomal pairs. It comprises two sequential nuclear divisions with just one course of DNA replication

Mitosis Replication and division of the nucleus of a eukaryotic cell resulting in the formation of two new nuclei, each of which has the same number of chromosomes as the parent nucleus

Phagocytosis/endocytosis Cellular processes of binding and internalizing macromolecules and particles from the environment

THE CAPACITY OF EUKARYOTIC CELLS TO UNdergo diverse shape changes, to achieve coordinated and polarized cell motility, and to control the cytoplasmic spatial organization and transport of protein complexes and organelles relies on an intricate network of protein filaments that transverse throughout the cytoplasm, referred to as the cytoskeleton. The cytoskeleton can also furnish structural support, which is particularly important for animal cells, since they lack rigid outer cell walls. The unique activities of the cytoskeleton depend on three types of protein filaments: actin filaments, microtubules, and intermediate filaments. Each type of filament is formed from a different protein subunit: actin for actin filaments, tubulin for microtubules, and a family of related fibrous proteins, such as vimentin, for intermediate filaments. Actin and tubulin are highly conserved evolutionarily and rather ubiquitous throughout eukaryotes, whereas intermediate filaments are much more diverse and are often tissue-specific. The ability of a eukaryotic cell to be spatially organized by cytoskeletal proteins is largely based on their polymerization properties. Each of the three major types of cytoskeletal proteins assemble into linear filaments that can stretch from one side of the cell to the other and serve as tracks for transport between them. The ability of an actin filament or microtubule to participate in distinct cellular functions is intricately dependent on their in-

ENCYCLOPEDIA OF HUMAN BIOLOGY, Second Edition, VOLUME 2. Copyright © 1997 by Academic Press. All rights of reproduction in any form reserved.

teraction with a versatile repertoire of accessory proteins. Some of the binding proteins, for example, are responsible for regulating the assembly of the filaments (by binding either to influence the filamentous form of the molecule or to sequester soluble subunits), whereas others link filaments to one another or to other cellular components. Other proteins, referred to as motor proteins, hydrolyze ATP to move either organelles along the filaments or the filaments themselves. Although actin filaments, microtubules, and intermediate filaments were once regarded as separate entities, increasing evidence suggests that cross talk exists between the three types of filaments and that their functions are coordinated.

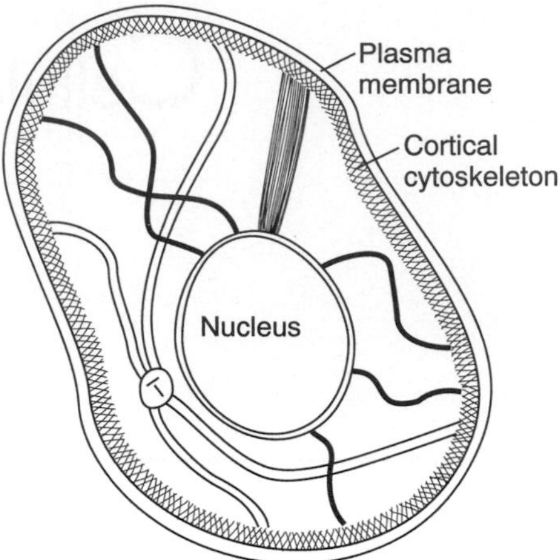

FIGURE 1 A schematic view of the cytoskeleton showing various filamentous elements positioned throughout the cytoplasm in a hypothetical cell. The mechanisms of interaction between extracellular signals, the plasma membrane, cytoskeletal components, and intracellular events is under intense study in numerous laboratories.

I. ACTIN FILAMENTS

Actin filaments, also known as microfilaments [or filamentous actin (f-actin)], are double-stranded helical polymers composed of the protein actin. They appear as flexible structures, with a diameter of 5–9 nm, that are organized into a variety of linear bundles, two-dimensional meshworks, and three-dimensional gels; they infrequently appear as single filaments. Actin is the most abundant protein in many eukaryotic cells, often constituting greater than 5% of the total cellular protein. Actin filaments are polarized. The ends of the filaments are structurally different and consist of a fast-growing (barbed) end that is often oriented at the plasma membrane and a slow-growing (pointed) end that often extends out into the cytoplasm. The "barbed" and "pointed" nomenclature refers to the "arrowhead" motif seen by electron microscopy when actin filaments are decorated with the motor protein myosin.

Yeasts, as well as some other lower eukaryotes, have only one actin gene that encodes a single protein. In contrast, all higher eukaryotes have several isoforms encoded by a family of actin genes, which are separated into three classes based on their isoelectric points. α actins are found in muscle, whereas β and γ actins are the principal constituents of nonmuscle cells.

Actin filaments can form both labile and stable structures in cells. In many cell types, actin filaments are cross-linked into a network by various actin-binding proteins, forming a dense layer in close association with the plasma membrane (Fig. 1). This layer of cytoskeletal components, referred to as the cell cortex, is dynamic and provides mechanical strength to the cells and enables them to perform a varicty of cell-surface movements, such as endocytosis, exocytosis, cytokinesis (cell division), and cell locomotion (e.g., during wound healing or embryonic development). In numerous cell types the cortical actin cytoskeleton is intricately involved in establishing and maintaining cell polarity. It can rearrange rapidly in response to extracellular signals and is therefore considered to be an integral part of the cell's signal transduction pathways. Dynamic surface extensions containing actin filaments are a common feature of animal cells, particularly when the cells are involved in motility or shape changes. For example, the leading edge (i.e., ruffle) of a crawling fibroblast is filled with a dense meshwork of actin filaments. Many cells also extend thin, stiff protrusions called microspikes, which are about 0.1 μm wide and 5 to 10 μm long and contain a loose bundle of about 20 actin filaments. The growing tip (growth cone) of a developing nerve cell axon, called the filopodium, can be up to 50 μm long. All of these are motile structures that can form and retract quickly as a result of regional actin polymerization and depolymerization at the plasma membrane. On the other hand, stable actin filaments are the major component of thin filaments in striated muscle and in the microvilli composing the brush border of the intestinal epithelium. Maintenance of the lengths and polarity of these filaments is crucial for their proper functioning. [*See* Cell Junctions.]

The polymerization of actin is a dynamic process that is regulated by the hydrolysis of a tightly bound nucleotide, adenosine 5′-triphosphate (ATP). Much of what is known about the polymerization properties of actin filaments is derived from *in vitro* assays. Purified actin exists as a monomer in low-ionic-strength solutions and spontaneously polymerizes into actin filaments upon addition of ATP, as well as both monovalent and divalent cations (usually K^+ and Mg^{2+}). *In vitro,* actin filaments engage in treadmilling as a result of the continual addition of actin monomers to the barbed (plus) end and continual loss of actin monomers at the pointed (minus) end, with no net change in filament length. In cells, the polymerization properties of actin filaments are much more complex owing to the influence of actin monomer- as well as actin filament-binding proteins. In many nonmuscle cells, for example, approximately half of the actin is found in cytoplasmic monomeric pools as a result of its interaction with small sequestering proteins such as thymosin β4 or profilin. Moreover, the length of actin filaments is also maintained in part by actin filament capping proteins. These proteins bind to the ends (e.g., gelsolin and capZ at the barbed ends and tropomodulin at the pointed ends) of the filaments and regulate the rate of actin filament elongation and depolymerization.

Other classes of actin filament-binding proteins function to cross-link, sever, bundle, or stabilize the filaments. In addition, others link the filaments to the plasma membrane or forcibly move the filaments relative to one another. For example, gelsolin mediates Ca^{2+}-dependent fragmentation of actin filaments, thereby causing a rapid solation of actin gels. Tropomyosin binds along the length of actin filaments, stabilizing them and altering their affinity for other proteins. Fimbrin and α-actinin cross-link actin filaments into bundles and filamin cross-links actin filaments into loose gels. Spectrin, in conjunction with ankyrin, attaches the sides of actin filaments to the plasma membrane. Various myosin isoforms utilize the energy of ATP hydrolysis to move along actin filaments, carrying adjacent actin filaments against each other in processes such as muscle contraction and cytokinesis. Other myosins function to transport vesicles along the filaments. Furthermore, numerous actin-binding proteins are thought to act in concert to generate movements at the cell surface, including cytokinesis, phagocytosis, and cell locomotion.

Interestingly, while for decades the actins were considered to be a protein family with no surviving relatives, a highly diverse ancient superfamily, referred to as "the actin superfamily," has recently been identified. The members of this family possess the common feature of an actin fold, a tertiary structure that can tolerate enormous sequence diversity. Members of this diverse superfamily include the actins and actin-related proteins (ARPs), as well as noncytoskeletal proteins such as heat-shock proteins, sugar kinases, and several cell-cycle proteins from bacteria. Further study is needed to determine what functional properties are shared by these ancient family members.

II. MICROTUBULES

Microtubules are rigid, hollow cylinders with an outer diameter of 25 nm. They are built from 13 linear protofilaments, each composed of heterodimers of α- and β-tubulin monomers and bundled in parallel to form a cylinder. Like actin filaments, microtubules are polarized, with one end (the plus end) being capable of rapid growth and the other end (the minus end) inclined to lose subunits if not stabilized. Typically, microtubules are oriented with their minus ends attached to a single microtubule-organizing center (MTOC) called a centrosome and their plus ends free in the cytoplasm. Unlike actin filaments, which interact in networks or bundles, microtubules in the cytoplasm function as individual units and are often considered to be the primary organizers of the cytoskeleton. Many of the microtubule arrays in cells are unstable and depend on this lability for their involvement in cellular processes such as cell motility, shape changes, and mitosis, and as tracks for intracellular transport.

Tubulin molecules are diverse. In mammals there are at least six forms of α-tubulin and a similar number of forms of β-tubulin, each encoded by a different gene. In addition, a minor form of tubulin, γ-tubulin, is localized at the centrosome. It is thought that γ-tubulin may interact with α–β tubulin dimers to nucleate the polymerization of nascent microtubules.

As with actin filaments, much of what we know about the dynamic behavior of microtubules is deduced from studying the polymerization of purified tubulin molecules *in vitro*. Pure tubulin will polymerize into microtubules at 37°C as long as Mg^{2+} and guanosine 5′-triphosphate (GTP) are present. Individual microtubules *in vitro* tend to exist in one of two states: (1) steady growth with the plus end elongating at three times the rate of the minus end or (2) rapid, "catastrophic" disassembly. Microtubules in the cell can also exist in these two states; each tubulin molecule will participate in the formation and dismantling of many microtubules in its lifetime. This behavior, called dynamic instability, plays a crucial role in determining the distribution of microtubules in the cytoplasm. The dynamic instability of microtubules requires an input of energy that is derived from the hydrolysis of GTP, shifting the chemical balance between polymerization and depolymerization. Cells modify the dynamic instability of their microtubules for specialized functions. For example, during the M phase of the cell cycle, the rate at which microtubules assemble and disassemble is greatly increased, so that the chromosomes can readily capture growing microtubules and a mitotic spindle can rapidly assemble. Conversely, when a cell differentiates and takes on a defined morphology, the dynamic instability of its microtubules is often suppressed by proteins that bind to the microtubules and stabilize them against depolymerization. The ability to stabilize microtubules in a particular configuration provides an important mechanism by which a cell can organize its cytoplasm.

Slowly growing microtubules are especially labile and prone to catastrophic disassembly, but they can be stabilized by association with other structures that "cap" microtubule ends. Microtubule-organizing centers such as centrosomes protect the minus ends of microtubules and continually nucleate the formation of new microtubules, which grow out in random directions. Any microtubule that encounters a structure that somehow stabilizes or caps its free plus end will be selectively retained, whereas the other microtubules will depolymerize; thus, microtubules originating in the centrosome can be selectively stabilized by events occurring in the cell. It is thought that this selective process largely determines the position of the microtubule arrays in a cell.

After microtubules polymerize, their tubulin subunits can be covalently modified and consequently selectively stabilized by acetylation of particular lysines and by the removal of the tyrosine residue (detyrosination) from the carboxyl terminus of α-tubulin. These alterations are thought to label the microtubule as "mature"; the longer time elapsed after a particular microtubule has polymerized, the higher the percentage of its subunits that will be modified. These alterations also provide sites for the most versatile modification of microtubules, the binding of specific microtubule-associated proteins (MAPs). MAPs function both to stabilize microtubules against disassembly and to mediate their interaction with other cell components, thus creating a functionally differentiated cytoplasm. [*See* Microtubules; Phosphorylation of Microtubule Protein.]

III. INTERMEDIATE FILAMENTS

Intermediate filaments constitute a large and heterogeneous family of tough, rope-like polymers of fibrous polypeptides. They were given this name because when first identified by electron microscopy, it was noted that their apparent diameter (8–10 nm) is between that of the thin actin filaments and the thick myosin filaments. In most animal cells, an extensive network of intermediate filaments surrounds the nucleus and extends across the cytoplasm, providing mechanical and tension-bearing strength to cells and tissues. Unlike microtubules and actin filaments, intermediate filaments appear to be specific to multicellular animals, and even in these organisms they are not represented in every cell type. Intermediate filaments are highly expressed in the cytoplasm of cells that are subject to mechanical stress, such epithelia and muscle cells.

The cytoplasmic intermediate filaments in vertebrate cells are grouped into three classes: (1) keratin filaments, (2) vimentin and vimentin-related filaments, and (3) neurofilaments, each formed by polymerization of their corresponding subunit proteins. By far the most diverse family of subunits is the keratins, which form keratin filaments, primarily in epithelial cells. Based on their amino acid sequence, the keratins are subdivided into two groups composed of type I (acidic) keratins and type II (neutral/basic) keratins. Keratin filaments are heteropolymers of type I and type II keratin polypeptides. In epithelial cells, keratin filaments are anchored at specialized cell junctions referred to as desmosomes, which link adjacent cells together, and to hemidesmosomes, which attach cells to the basal lamina. Since the keratin filaments in individual cells are connected via desmosomes to those of its neighbors, they form an uninterrupted network that spans the entire epithelium. Unlike keratins, vimentin and vimentin-related proteins can assemble into intermediate filaments that are polymers of a single protein species. Vimentin itself is the most widely distributed of the cytoplasmic intermediate filament proteins, being present in many cells of mesodermal origin, including fibroblasts, endothelial cells, and white blood cells. Vimentin-related proteins include desmin, found primarily in muscle cells, and glial fibrillary acidic protein, which forms glial filaments in astrocytes of the central nervous system and in certain Schwann cells in peripheral nerves. These proteins copolymerize with one another, but do not copolymerize with keratins. Nerve cells contain a variety of unique intermediate filaments, which are expressed in different regions of the nervous system, or during specific stages of neural development. The most abundant cytoskeletal components in these cells are the neurofilaments, which extend along the length of an axon. In mammals, three neurofilament proteins have been identified and are classified as NF-L, NF-M, and NF-H for low, middle, and high molecular weight, respectively. Nuclear lamins, which form the fibrous lamina that lies beneath the inner nuclear envelope, are a separate family of intermediate filament proteins. The lamina is usually 10–20 nm thick and is interrupted by nuclear pores, the passageways for macromolecules entering and leaving the nucleus. It is generally assumed that the progenitor of the intermediate filament proteins was a nuclear lamin.

Unlike actin and tubulin, which are globular proteins, the numerous types of intermediate filament protein monomers are all highly elongated fibrous molecules that have an amino-terminal head, a central rod domain, and a carboxyl-terminal tail. Although the monomers of the various types of intermediate filaments differ in amino acid sequence and molecular weights (40,000 to 200,000 daltons), they all contain a homologous central rod domain that mediates the lateral interactions that form the assembled filament. This domain consists of an extended α-helical region containing long tandem repeats of a distinctive amino acid sequence motif referred to as the heptad repeat. This seven-amino-acid motif promotes the formation of an extended coiled-coil structure when the protein dimerizes. In contrast, the globular head and tail domains vary significantly in both size and amino acid sequence without affecting the basic axial structure of the fiber. During assembly, two coiled-coiled dimers associate with each other in an antiparallel manner to form a symmetrical tetramer core. This tetramer then aligns into large overlapping arrays along the axis of the filament to form the nonpolarized intermediate filament. This distinguishes intermediate filaments from actin filaments and microtubules whose functions depend on their polarity. It appears likely that the variable terminal domains that project from the surface of the intermediate filaments enable each filament type to associate with specific components in the cell, so as to correctly target the filaments in a particular cell type. In general, most intermediate filament molecules are assembled into filaments, with few soluble tetramers. Nonetheless, a cell can regulate the assembly of its intermediate filaments and determine their number, length, and location by various mechanisms. One mechanism of control involves the phosphorylation of specific serine residues within the amino-terminal head domain of intermediate filament proteins.

It is apparent that the manner in which intermediate filaments are linked to other cellular components varies greatly among cell types and that the structure is well adapted for mechanical functions since the fibrous subunits associate side by side in overlapping arrays. Thus, in comparison, the filaments can endure larger tension forces than actin filaments and microtubules. [*See* Muscle, Molecular Genetics.]

IV. MOLECULAR MOTORS

Motor proteins move along cytoskeletal protein polymers and generate the majority of the cellular and intracellular movements of living organisms. For example, within minutes, mitochondria and other membrane-bounded organelles change their position by periodic saltatory movements, which are more sustained and directional than the repeated small Brownian movements resulting from random thermal motions. These and other intracellular movements in eukaryotic cells are generated by motor proteins, which bind to either an actin filament or a microtubule and use the energy derived from ATP hydrolysis to move unidirectionally along it. Dozens of different motor proteins have now been identified. They differ in the type of filament to which they attach, the directions along which they traverse, and the "cargo" they transport.

Motor proteins play significant roles in events such as muscle contraction, ciliary beating, cytokinesis, mitosis, and organelle movement. For example, microtubule-dependent motor proteins play an important part in positioning membrane-bounded organelles such as the endoplasmic reticulum and the Golgi apparatus, within a eukaryotic cell. Three superfamilies of motor proteins have been identified. The myosins, which move along actin filaments, are especially abundant in skeletal muscle, where they form a major part of the contractile apparatus. Structurally, myosins have similar motor domains (the part of the protein that is responsible for generating movement), but differ markedly in the domains that are responsible for attaching the myosin molecule to other components of the cell. Other superfamily members are the kinesins, which generally move toward the plus end of the microtubules (away from the centrosome), and the dyneins, which move toward the minus end (toward the centrosome). As with the myosins, each type of microtubule-dependent motor protein carries a distinct cargo as it moves. Cytoplasmic dyneins are involved in organelle transport and mitosis. Kinesins are more diverse than the dyneins, and different family members are involved in organelle transport, mitosis, meiosis, and the transport of synaptic vesicles along axons. Members within each of the superfamilies share a similar motor domain (30–50% amino acid identity) that can function autonomously as a force-generating element. Although the motor domains are specialized for particular types of force-generating activities, many of the unique self-assembly (e.g., filament formation by muscle myosin) or binding interactions (e.g., attachment of motors to membranes) that govern biological function are conferred by the non-motor "tail" domains, which can differ greatly among motors within a given superfamily. The affinity of motors for their protein partners, the cytoskeletal filaments, is dependent on the nucleotide that resides in the active site. The transition from the weak to the strong binding state of the motor–filament complex is thought to elicit the conformational change that allows the motor to produce force and move unidirectionally along the filament.

BIBLIOGRAPHY

Fechheimer, M., and Zigmond, S. H. (1993). Mini-review: Focusing on unpolymerized actin. *J. Cell Biol.* **123,** 1–5.

Klymkowsky, M. W. (1995). Intermediate filaments: New proteins, some answers, more questions. *Curr. Opin. Cell Biol.* **7,** 46–54.

Luna, E. J., and Hitt, A. L. (1992). Cytoskeleton–plasma membrane interactions. *Science* **258,** 955–964.

Mandelkow, E., and Mandelkow, E.-M. (1995). Microtubules and microtubule-associated proteins. *Curr. Opin. Cell Biol.* **7,** 72–81.

Mitchison, T. J. (1992). Compare and contrast actin filaments and microtubules. *Mol. Biol. Cell* **3,** 1309–1315.

Vale, R. D. (1996). Switches, latches, and amplifiers: Common themes of G proteins and molecular motors. *J. Cell Biol.* **135,** 291–302.

Cellular Memory

PATRICK L. HUDDIE
THOMAS J. NELSON
DANIEL L. ALKON
National Institutes of Health

GLOSSARY

Agonist Drug that produces an effect at a receptor site

Antagonist Drug that blocks the effect of an agonist

Associative learning When an organism learns relationships among environmental events; there are two types, classical conditioning and instrumental conditioning

Classical conditioning (Pavlovian conditioning) When two stimuli are presented to an animal in such a way that a response normally evoked by one of the stimuli eventually is elicited by the other stimulus as well. Thus the unconditioned stimulus would normally evoke the unconditioned response, but after associative learning the previously neutral conditioned stimulus will also produce a similar conditioned response. The classic example is the salivation of Pavlov's dogs, where the unconditioned stimulus was food presented to the dog. Pavlov limited his observation to the unconditioned response of salivation. An auditory conditioned stimulus (metronome click) was presented in advance of the unconditioned stimulus and, after several trials, the dog learned to associate the conditioned stimulus with forthcoming food, and salivated before the food was presented (conditioned response)

Conductance Conductance of an ion channel represents the quantity of ions that can pass through the channel in unit time; conductance of a cell membrane represents the sum of all the ion channels in that membrane

Delayed rectifier Type of potassium current that turns on slowly upon depolarization and rectifies because current flows more readily out of the cell than into the cell; hence it is also known as an outward rectifier

Depolarization and hyperpolarization Changes in membrane potential; depolarization is a shift to less negative membrane potentials, whereas hyperpolarization is a shift to more negative membrane potentials

Habituation Occurs when a stimulus elicits a diminished response on successive trials; sensory receptor adaptation or fatigue is not regarded as habituation. An oft-quoted example is the behavior of the polychaete worm *Nereis pelagica,* which contracts back into its burrow when stimulated mechanically or by a shadow. Repeated presentation of one type of stimulus evokes smaller and smaller contractions, but fatigue can be excluded because a novel stimulus produces full contraction after habituation to the original stimulus. Habituation may be a mechanism that limits fatigue; the habituation gradually disappears after a period without the habituating stimulus: this is known as dishabituation

Heterosynaptic Those neuronal interactions in which a third neuron modulates synaptic transmission between two others, by modifying the functions of the presynaptic terminal or the postsynaptic membrane

Instrumental conditioning Experiment that aims to modify the frequency or intensity (or both) of an innate behavior using reinforcement chosen by the experimenter; reinforcement strategies include: rewarding a specific behavior (reward conditioning), punishing a specific behavior (aversive conditioning), and avoidance learning, where the aversive stimulus is delayed if the behavior is displayed

Ion channels Gated protein pores that penetrate lipid bilayers and select which ions may pass through on the basis of size and valence. Some ion channels are opened by changes in membrane potential; others are opened selectively by specific neurotransmitters and second messengers

Iontophoresis Movement of charged drug molecules from a drug-filled electrode to intracellular or extracellular spaces by application of an electric field

Phosphorylation Attachment of inorganic phosphate to a protein; this modifies the conformation of the protein and frequently changes the function of the protein

Phototaxis Movement of an organism toward light

Postsynaptic potentials Caused by a neurotransmitter released from the presynaptic terminal that opens ion channels in the postsynaptic neuronal membrane and causes ionic currents to flow; the currents produce transient depolarizations or hyperpolarizations known as postsynaptic potentials

Quantal release Hypothesis that neurotransmitters are released from the presynaptic terminal in small packets, or quanta, believed to involve the fusion of a single transmitter-containing vesicle with the terminal membrane

Sensitization Gradual increase in the response to a repeated stimulus. Generally, the sensitizing stimulus is noxious or particularly strong

Vestibular organs Organs that detect acceleration; in *Hermissenda* the organ comprises 13 ciliated cells (hence hair cells) surrounding a central fluid-filled space that contains calcium carbonate stones, or statoconia. When accelerations lift the statoconia, the hair cell cilia detect the movement

Voltage clamp Technique used to resolve the difficulty of knowing which membrane currents are involved in electrical activity of excitable cells; many classes of membrane ion channels are voltage dependent, and as the membrane potential changes, the ionic current that passes also changes, and this in turn affects the membrane potential. Thus, it is desirable to hold (clamp) the membrane potential at known voltages and measure the resulting currents in isolation; this is done with electronics that rapidly detect very small voltage changes and respond by passing enough current to oppose the change; this current is equal and opposite to the membrane current involved

CELLULAR MEMORY CAN BE DEFINED AS A CHANGE in neuronal properties consequent upon the association of environmental events, and it is causal, necessary, and sufficient for the expression of the learned response. This is a narrow view, but it is crucial for the study of learning that the chain of events between behavioral modification and molecular changes be made explicit and be causally related.

I. CELLULAR MEMORY AND BEHAVIOR

This article's primary example of cellular memory is that displayed by the marine mollusc *Hermissenda crassicornis;* however, our discussion is informed by the extensive studies of the cellular basis of behavior performed on other animal models, with the restriction that model systems where no discrete cellular component has been implicated in learning-dependent changes have not been considered. We have also considered evidence emerging from the study of other types of cellular transformation, including development and oncogenesis. In addition, we note that many principles of cellular memory are applicable across phyla.

II. A MODEL SYSTEM: ASSOCIATIVE LEARNING IN *HERMISSENDA*

The nudibranch gastropod *Hermissenda* normally exhibits phototaxis. In turbulent ocean conditions, the phototaxis is inhibited; the animal slows or ceases movement, contracts its foot, and adheres more strongly to the substrate. This simple organism, with simple behavior, can be associatively conditioned in a classical conditioning paradigm. The animal can be trained to respond to light (the conditioned stimulus) with inhibition of movement and foot contraction (the conditioned response). [*See* Conditioning.]

The sensory systems involved in these responses are the visual and vestibular organs and are located in the head of *Hermissenda*. The visual system consists of two caudal eyes, each of which comprises five photoreceptors, screening pigments, and a simple lens. The eyes are located above the circumesophageal nervous system and receive inhibitory input from the vestibular organs, the statocysts, located similarly juxtaposed to the circumesophageal nervous system, and in close proximity to the optic ganglion (Fig. 1). The photoreceptors and statocysts interact through direct synaptic connections and via additional synaptic connections through interneurons with the optic ganglion and cerebropleural ganglion.

During training, wild *Hermissenda* are maintained in the laboratory under constant temperature and dim illumination before being placed in the training chamber. A number of animals can be trained in the chamber, which consists of eight seawater-filled tracks, each wide enough to accommodate one animal (Fig. 2). The chamber is attached to a mechanical agitator that moves the plate in the horizontal plane. Above the chamber is a light source with an intervening shutter. The agitator and the shutter are controlled by a computer, so that precise periods of light and turbulent agitation can be provided repeatedly.

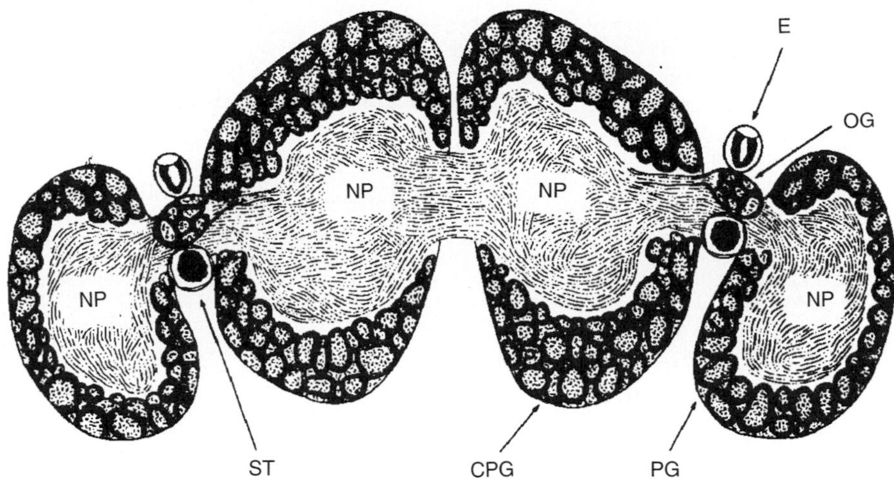

FIGURE 1 *Hermissenda* nervous system. The circumesophageal nervous system has four major ganglia, connected by commissural nerve tracts. Between the cerebropleural ganglion (CPG) and the pedal ganglion (PG) is a triad of structures. The eye (E) is caudal to the small optic ganglion (OG) and the statocyst (ST). Axons from the eye pass through the optic ganglion and are contacted by axons from the statocyst within the CPG (NP = neuropil).

The animals are trained by multiple training trials; after 100 pairings of light and rotation, *Hermissenda* respond to test flashes of light with reduced locomotion and foot shortening. Animals trained with explicitly unpaired stimuli, or naive animals, respond to light with phototaxis and foot lengthening. Thus, the unconditioned response to agitation (foot contraction, etc.) has been transferred to the conditioned response (light) and the animal anticipates the agitation on the basis of exposure to light. Thus, the *Hermissenda* have been conditioned to associate light with forthcoming agitation.

The cellular basis of this behavioral change has been investigated effectively in *Hermissenda* because of the relative simplicity of the sensory and neuronal processing. The elements of this simple nervous system show less specialization than a complex nervous system like our own; that is, each cell (in a simple system) can subserve several functions. This elegant parasimony gives way to neuronal specialization in vertebrate brains. Thus *Hermissenda* photoreceptors transduce light into electrical signals and perform significant processing via their synaptic inputs; with only two Type A photoreceptors and three Type B photoreceptors available in each eye, much of the integration of sensory light occurs at the periphery in the "simple" nervous system.

After training, the membrane resistance of the Type B photoreceptors is increased, which renders the cells more excitable for a constant stimulating current input. This change in the excitability of the B photoreceptor following rotation is a consequence of the properties of the local neuronal network. During rotation, the hair cells increase their output and release more inhibitory transmitter and the B cell is hyperpolarized. After rotation, hair cells undergo an intrinsic poststimulus hyperpolarization and therefore release less inhibitory transmitter onto the B cell and onto an interneuron in the optic ganglion. Both of these actions result in net depolarization of the B cell, since less inhibition comes from the hair cell, and the excitatory input to the B cell from the optic ganglion cell is disinhibited.

After the pairing of light and rotation, the B-cell membrane is depolarized, and calcium-selective voltage-dependent ion channels open. These channels do not inactivate as long as the depolarization is sustained, and while they are open calcium moves into the cell. The total influx of calcium into the B cell is dependent on both the duration of each posttrial depolarization and the number of training trials.

This influx of calcium must be related to the separate effects of light on the B photoreceptor. Light alone causes the release of diacylglycerol from the cleavage of membrane lipid via photosensitive proteins, and phospholipase C. The diacylglycerol is effective in stimulating the membrane-associated activity of a calcium-dependent protein kinase (protein kinase C, or PKC). PKC is stimulated by both diacylglycerol and the increased level of intracellular Ca^{2+},

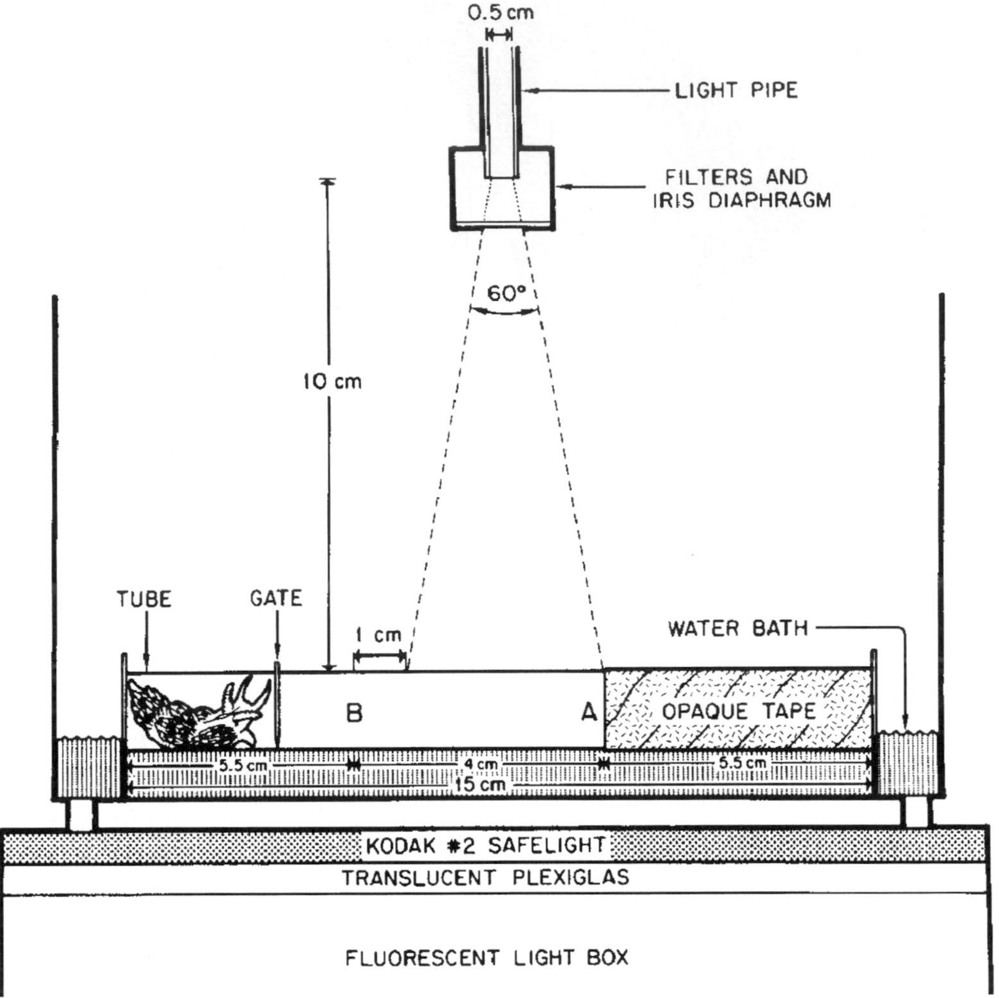

FIGURE 2 Training apparatus for *Hermissenda*. (A) *Hermissenda* held inside the training tracks are illuminated from above when the computer-controlled shutter opens. The training tracks are attached to a mechanical agitator that moves the plate in the horizontal plane; the agitator is also controlled by the computer. (B) Plan view of training chamber. A number of animals can be trained at one time in the chamber, which consists of several seawater-filled tracks, each wide enough to accommodate one animal.

and goes on to cause the increase in membrane resistance. A better understanding of the action of PKC has been gained from studies of voltage-clamped Type B photoreceptors.

The Type B photoreceptor is penetrated with two microelectrodes, one electrode to measure membrane potential and the other to pass current into the cell. The cell is voltage-clamped and the ionic currents that flow during activity are measured.

The dominant outward currents in Type B photoreceptors are potassium currents; K^+ ions are more abundant in the cell than in the extracellular medium and tend to flow down the electrochemical gradient, out of the cell, when K-selective ion channels in the membrane are opened, by changes in membrane voltage or intracellular calcium concentration. Striking reduction of the transient voltage-dependent K^+ current (I_A) and the slower calcium-dependent K^+ current ($I_{K\text{-}Ca}$) can be seen after conditioning (Fig. 3). This change in outward currents is the major cause of the increase in input resistance consequent upon classical conditioning of *Hermissenda*. Further voltage-clamp experiments suggest a role for PKC in the reduction of K^+ currents following conditioning. PKC is activated *in vitro* by a class of lipid-soluble chemicals known as phorbol esters. Application of phorbol ester

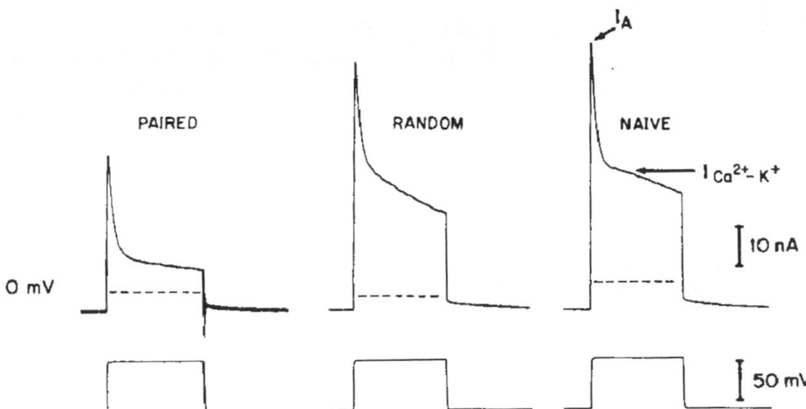

FIGURE 3 Conditioning-induced K^+ current changes in Type B photoreceptors; voltage-clamp records from Type B photoreceptors of *Hermissenda.* The lower trace shows that the photoreceptor membrane potential was held at −60 mV, then stepped to 0 mV for 1 sec. The upper trace shows the resulting currents; the initial spike is the transient voltage-dependent K^+ current (I_A). The slowly declining late component of each current trace is the slower calcium-dependent K^+ current ($I_{Ca^{2+}-K}$). The magnitude of both currents is markedly reduced in animals that received paired light and agitation (record on the left), when compared with naive animals or animals that received explicitly unpaired light and agitation. (The dashed lines indicate the amount of current that was not due to the K^+ current, also called the leak current.)

to the Type B cell, together with elevation of Ca^{2+}, produces reduction of I_A and $I_{K\text{-}Ca}$ analogous to the effect of conditioning.

The enzymatic activity of PKC in the cell is phosphorylation of appropriate substrates. Whether PKC acts directly to phosphorylate K^+ channels or acts on an intermediary is an open question. Recent evidence suggests that an important intermediary may be cp20, a GTP-binding protein of 20 kDa molecular mass, which is phosphorylated after phorbol ester treatment and which inhibits potassium channels when injected into naive Type B cells. [*See* Protein Phosphorylation.]

To understand the consequences of the increased excitability of the Type B cell, we must refer to the known network (Fig. 4). The B photoreceptor inhibits the medial Type A photoreceptor. The medial Type A photoreceptor excites an interneuron (I), which excites a motoneuron that causes locomotion. In addition, conditioning enhances the Type A photoreceptor K^+ currents, and further inhibits the Type A response to light; therefore, after conditioning, the Type B cell inhibits the Type A cell and locomotion is inhibited.

An important question is the causal relationship of these cellular changes to the behavioral changes. We know that for *Hermissenda,* at least, the induction of the cellular change is partially responsible for the predicted behavioral change. Intact *Hermissenda* can be restrained and the B cell impaled with a microelectrode. Pairing of light and depolarizing current causes a significantly slower phototactic response, when compared with control unpaired and sham-operated animals in subsequent behavioral tests. Therefore, training-induced changes at the convergence of the visual and vestibular sensory systems are probably causal for the behavioral change.

Having outlined the salient features of a promising animal model of cellular memory, we would like to consider in more detail the various cellular elements that have been implicated in cellular memory in our own studies and in other systems.

III. MEMBRANE PROCESSES

The cell membrane is the interface between the cell and the world. Input reaches the cell via the membrane, and output is expressed by membrane processes.

A. Ion Channels

1. Ca^{2+} Channels

The slower inward Ca^{2+} current component of the action potential is present to a greater or lesser extent in various types of neurons. The influx of Ca^{2+} is an

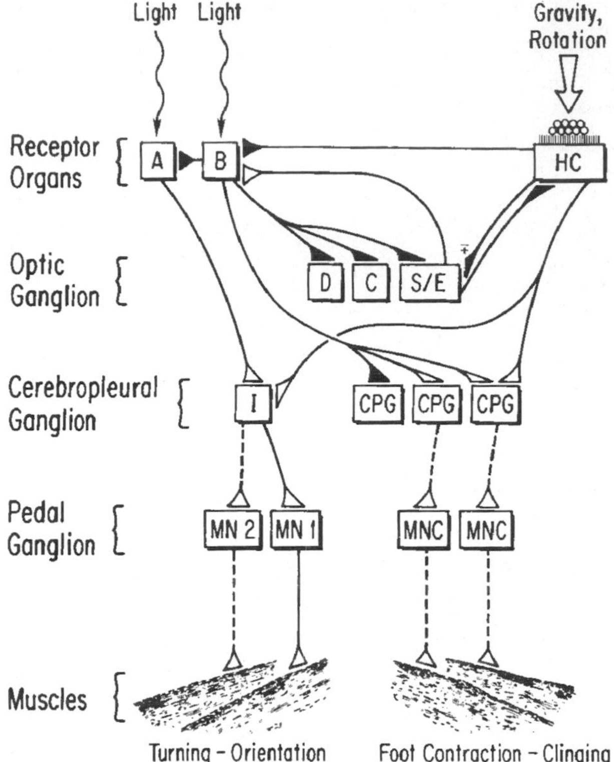

FIGURE 4 Schematic of the *Hermissenda* neural system. This simplified circuit diagram shows the interaction between sensory and motor systems, which underlies associative conditioning. Excitatory synapses are represented by open triangles, and inhibitory synapses by solid triangles. Light stimulates both Type A and Type B photoreceptors. The Type A cell excites an interneuron (I) and, thus, stimulates motoneurons (MN 1, MN 2), which produce muscle contraction and movement toward light. The Type B photoreceptor inhibits the Type A cell. Training makes the B cell more excitable and increases inhibition of the A cell, thus limiting phototaxis. The Type B photoreceptor also sends inhibitory input to neurons D,C, and S/E of the optic ganglion, and both inhibitory and excitatory inputs to neurons of the cerebropleural ganglion (CPG). Excitatory inputs to some of the CPG neurons cause activation of motoneurons (MNC) and contraction of the *Hermissenda* foot.

The hair cells of the statocysts detect acceleration due to gravity or motion and interact with the visual system at several sites. The hair cells directly inhibit the Type B cells and the S/E cell of the optic ganglion. The S/E cell reciprocally inhibits the hair cell but excites the Type B photoreceptor. The hair cells also excite interneurons I and CPB of the cerebropleural ganglion. After hair cell stimulation, the intrinsic postburst hyperpolarization of the hair cell and the slow, depolarizing component of the synaptic potentials at the hair cell to the Type B cell synapse combine, in a pairing-specific, temporally constrained fashion, with increased S/E cell activity and the intracellular sequelae, which follow illumination in the Type B cell to produce an enhancement of Type B photoreceptor excitability specific to conditioning. After conditioning, Type B cells are more excitable and tend to activate the foot contraction side of the network and suppress the phototactic side.

important signal, and the role for Ca^{2+} as an intracellular messenger is discussed in the following. Here we consider the changes in Ca^{2+} channels occurring during learning. In *Hermissenda* after excessive training (e.g., 300 trials), voltage-dependent Ca^{2+} currents are reduced.

Ca^{2+} currents may also be reduced in another model system, reflex habituation of *Aplysia* sensory neurons. *Aplysia* is a gastropod that has been used to study the cellular basis of non-associative learning-related phenomena such as habituation and sensitization.

2. K^+ Channels

Neuronal excitability is primarily regulated by K^+ channels. The K^+ channels determine the duration of the refractory period that follows an action potential. During an action potential, voltage-dependent K^+ channels open and K^+ ions flow out of the neuron down the electrochemical gradient. This outward current is responsible for repolarizing the cell and returning the membrane potential to the resting value after the depolarizing Na^+ and Ca^{2+} currents have terminated. However, immediately following the action potential, the cell is hyperpolarized beyond the normal resting potential, because the dominant membrane conductance is to K^+ ions, and the membrane potential approaches the equilibrium potential for potassium. The postaction potential after hyperpolarization is crucial in determining when the neuron will be able to respond to a new excitatory input or, in the case of spontaneously active neurons, the frequency of action potentials.

Potassium channels are many and diverse. A useful generalization is to divide the K^+ channels into three classes: A currents (I_A), delayed rectifiers, and calcium-dependent K^+ channels. I_A is a fast transient outward current activated by depolarization, which inactivates rapidly even if the depolarization is sustained. The delayed rectifier is a K^+ channel that opens slowly when a threshold depolarization is attained, and remains open while the membrane is depolarized. The Ca-dependent K^+ channels ($I_{K\text{-}Ca}$) are weakly activated by voltage in some but not all neurons, and strongly activated by increased calcium concentration at the intracellular face of the channel. Such an increase in intracellular Ca^{2+} concentration occurs after Ca^{2+} channels open during membrane depolarization, or when intracellular stores of Ca^{2+} are released; thus, the Ca-dependent K^+ channel can be thought to be gated by a second messenger.

Experiments on *Hermissenda* show that both I_A and $I_{K\text{-}Ca}$ are important for the acquisition and expression of cellular memory, and the changes in these currents persist for days following training.

Similar profound changes in K^+ currents are implicated in classical conditioning of rabbits. Nictitating membrane response conditioning of the rabbit is a robust quantifiable behavior modification in which the conditioned stimulus (a tone or light flash) is paired with the unconditioned stimulus (an air puff to the cornea, or a periorbital shock). The unconditioned stimulus elicits eyeball retraction, and the rabbit's third eyelid (the nictitating membrane) closes as a consequence. The animal rapidly learns to blink upon the conditioned stimulus: nictitating membrane closure is used as a convenient measure of the conditioned response.

Analysis of the cellular mechanisms underlying nictitating membrane response conditioning is at an early stage. The hippocampus is not essential but, when present, permits more complex nictitating membrane response conditioning, and under less than optimal conditions. The firing of action potentials by neurons in the CA1 region of the hippocampus (Fig. 5) is increased during training, and the increase in firing precedes the nictitating membrane response by 30–40 msec. The pattern of firing is correlated with the profile of the nictitating membrane response. A cellular mechanism for the increase in excitability of these CA1 neurons may be inhibition of $I_{K\text{-}Ca}$. In hippocampal slices from conditioned animals, the afterhyperpolarization that followed action potentials elicited by depolarizing current injection was reduced when compared with control animals; resting potential and input resistance were unaffected. Single-electrode voltage-clamp studies of CA1 neurons from conditioned and control animals show directly that $I_{K\text{-}Ca}$ is reduced by conditioning. Thus the membrane potential returns more rapidly to the resting potential in conditioned animals, and excitability is enhanced. Additionally, synaptic potentials are enhanced in conditioned animals, also consistent with a reduction in $I_{K\text{-}Ca}$. Thus, the changes seen in *Hermissenda* may be generalized to mammalian brains. [*See* Hippocampal Formation.]

A third system in which K^+ channel modification in learning is suggested is instrumental motor learning in locusts. Both reinforcement and punishment protocols have been used to train locusts to hold the metathoracic leg in particular positions. A simplified preparation was also used, where firing frequency of the anterior adductor coxa motoneuron (AAdC) was considered as the response. The AAdC can be "trained" to increase or decrease its output to the anterior adductor muscle.

In trained AAdC neurons, firing frequency was increased when the preparation was trained to elevate the leg (up-learning) and decreased when the preparation was trained to lower the leg (down-learning). The evidence suggests that up-learning is correlated with a decrease in K^+ conductance of the AAdC mem-

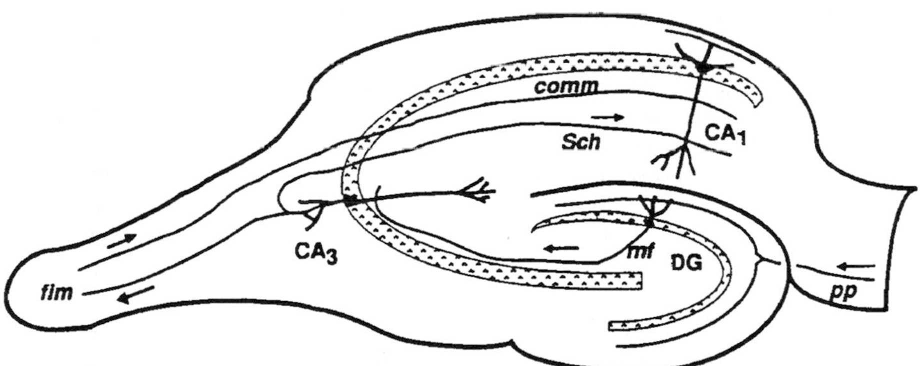

FIGURE 5 Neural pathways and regions of the hippocampus. The hippocampus has a well-defined architecture. When sliced in a particular way, the connections between hippocampal neurons and the input pathways are preserved. Here the excitatory pathways are delineated; the inhibitory pathways are susceptible to damage when the slice is prepared and have not been well characterized yet. The cell bodies (somata) of the CA1 and CA3 neurons are found in the larger speckled area; the somata of the dentate gyrus neurons (DG) are found in the smaller speckled region (which is the dentate gyrus). The excitatory input of the perforant path (pp) synapses upon the dendritic field of DG neurons that project to the dendritic field of CA3 neurons. The output of the CA3 neurons goes out of the hippocampus via the fimbria (fim) and to the CA1 neurons via the Schaeffer collaterals (Sch). The commissural nerve tract (comm) also synapses on CA1 neurons. mf, mossy fibers.

brane and down-learning with increased K^+ conductance. However, the evidence is indirect, and voltage-clamp studies are required to clarify this interesting finding. Moreover, the data also support changes, correlated with learning, in the synaptic inputs to the AAdC neuron. Despite these considerations, the findings suggest that both increases and decreases in K^+ conductance may occur in different types of learning in the same nervous system.

B. Neurotransmitter Receptors

1. Glutamate

Modification of glutamatergic neurotransmission is implicated in two systems used in the study of cellular modifiability: the vestibulo-ocular reflex (VOR) and in the *in vitro* short-term synaptic plasticity model, long-term potentiation (LTP).

The VOR keeps eye position constant as the head moves and depends on head position information generated by the vestibular organs of the inner ear. In the cerebellum, parallel fibers carry vestibular information, and climbing fibers carry visual information via the inferior olive to the dendrites of Purkinje neurons in the flocculus (Fig. 6). The output of the Purkinje cells is inhibitory and projects to the vestibular-ocular relay nuclei. The modifiability of this system is exhibited when reversing prisms are worn. Initially, eye movement is inappropriate for head rotation; the VOR rapidly reverses and again the animal successfully tracks the visual target during rotation of the head. The learned response is retained for times between hours and days. [*See* Eye Movements.]

A cellular locus for this modifiable behavior may be the convergence of the glutamatergic parallel-mossy fibers and climbing fibers on the Purkinje neurons of the flocculus. When glutamate iontophoresis onto Purkinje cells is paired with olivary stimulation (which excites the climbing fibers), the response to glutamate is depressed compared to unpaired controls. Synaptic responses elicited by parallel-mossy fiber stimulation are also depressed by iontophoretic glutamate paired with climbing fiber stimulation. This long-lasting depression is comparable with the long-term depression (LTD) of parallel fiber–Purkinje cell transmission that

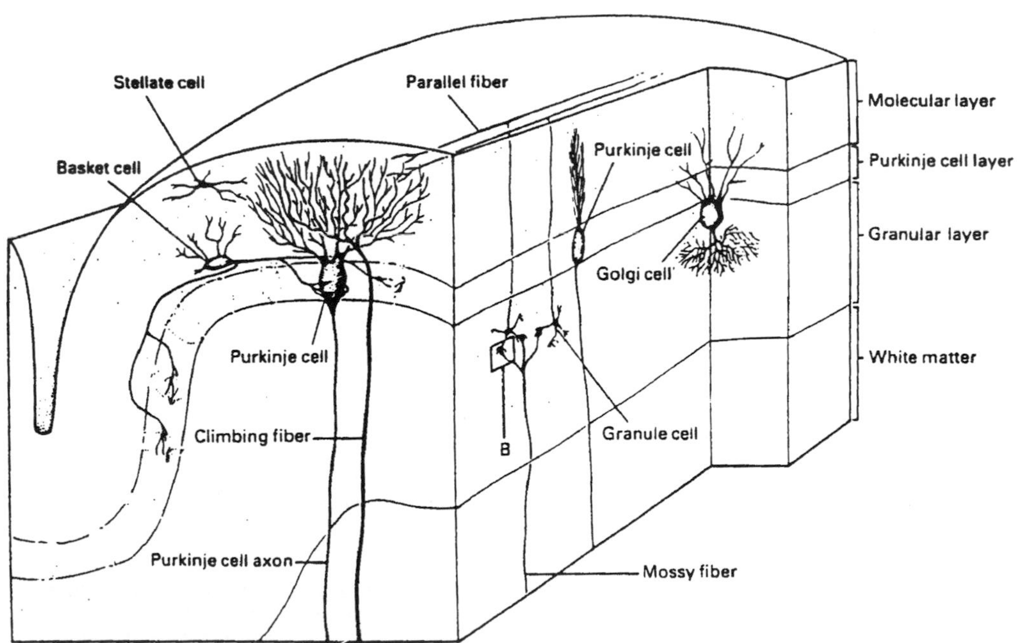

FIGURE 6 Neuronal architecture of the cerebellum. The cerebellar cortex has a regular structure, which aids interpretation of the information flow. Vestibular information comes in via mossy fibers. These synapse upon granule neurons (region marked B). The granule neuron axons bifurcate to become the parallel fibers that synapse on the distal (or upper) dendrites of the Purkinje neurons in the "molecular layer" of the cerebellum. The climbing fibers carry visual information via the inferior olive, to the proximal (or lower) dendrites of Purkinje neurons. The inhibitory output axon of the Purkinje cells goes to the vestibular-ocular relay nuclei. [Reprinted with permission of the publisher, from E. R. Kandel and J. H. Schwartz (1985). The cerebellum. *In* "Principles of Neural Science" (C. Ghez and S. Fahn, eds.), 2nd Ed. Elsevier, New York.]

is caused by pairing parallel fiber and climbing fiber stimulation.

There are three major glutamate receptor subtypes in neuronal tissue, distinguished by agonist specificity: kainate, quisqualate, and *N*-methyl-D-aspartate (NMDA). Normal excitatory transmission in the hippocampus is mediated by quisqualate- and kainate-type receptors.

The glutamate receptors involved in long-term depression appear to be of the quisqualate subtype, because quisqualate, but not kainate or aspartate, can mimic the effects of glutamate. The quisqualate receptor antagonist kynurenic acid also blocks the induction of LTD. Therefore, LTD is thought to be a consequence of quisqualate receptor responses. A role for Ca^{2+} ions in LTD is also suggested by experiments in which the Ca^{2+} chelator EGTA was injected into Purkinje cell dendrites and LTD was abolished. The precise interactions of glutamate, Ca^{2+}, and the climbing fiber input have not yet been elucidated. Ca^{2+} may directly promote glutamate receptor desensitization, by analogy with Ca^{2+}-linked desensitization of nicotinic acetylcholine (ACh) receptors. Ca^{2+} influx into Purkinje cells may be promoted by climbing fiber activity, and then the Ca^{2+} could elevate intracellular cyclic guanosine monophosphate (cGMP) levels by activating guanyl cyclase (a known action of Ca^{2+}). However, cGMP is itself excitatory on Purkinje neurons, and the hypothesis is difficult to test. An unexplored idea is that Ca^{2+} may activate PKC, thus changing the sensitivity of postsynaptic glutamate receptors.

A glutamatergic system widely employed for studying modifiability is LTP in the hippocampal slice *in vitro*. The hippocampus is linked to learning because of clinical and experimental studies of anterograde amnesia. Hippocampal neuroanatomy is highly structured, with laminar organization of neuronal cell bodies and dendritic fields (see Fig. 5). This has facilitated electrophysiological investigation. However, the *in vitro* studies of the physiology of hippocampal synaptic modifiability suffer from lack of a clearly defined behavioral frame of reference. For example, in most cases, the incoming stimuli are electrical shocks to nerve tracts, defined by the experimenter, rather than sensory stimuli appropriate to those nerve tracts. Thus, the wider applicability of these changes in synaptic function remain uncertain for the most part.

LTP was first demonstrated by stimulating dentate gyrus neurons via the perforant path, and the activity of a population of these neurons was recorded with extracellular electrodes; it was found that high-frequency trains of stimuli (up to 100 Hz for 4 sec) would enhance the subsequent population response (the population excitatory post-synaptic potential, EPSP, and population action potential). The potentiation persisted for 30 min to several hours.

Subsequently, LTP has been investigated *in vivo* and *in vitro* and a wealth of information is available. Here we focus on the involvement of glutamate receptor-linked ion channels in LTP.

LTP is induced in the *in vitro* hippocampal slice preparation by brief high-frequency stimulation of an excitatory monosynaptic pathway to both pyramidal and granule neurons in the slice and can be observed for up to 24 hr. LTP requires depolarization of the postsynaptic neuron. LTP of the pyramidal neurons of the CA1 is blocked by the selective NMDA receptor antagonist 2-aminophosphovalerate (APV). In the CA1 and dentate gyrus regions, the NMDA receptor is present and presumably binds glutamate released from presynaptic terminals, but at resting membrane potentials most NMDA receptor-linked ion channels are blocked by Mg^{2+} ions. Depolarization of the neuron expels the Mg^{2+} ion from the NMDA receptor channel and permits the passage of Na^{+}, K^{+}, and Ca^{2+} ions. Under physiological conditions, the fraction of the current carried by Ca^{2+} ions is small. Although the calcium permeability of the NMDA receptor channels of spinal cord neurons exceeds the sodium permeability by about fivefold, the relative physiological extracellular concentrations of Ca^{2+} and Na^{+} mean that the net sodium current exceeds the calcium current. In addition, the voltage-activated calcium channels present in these neurons are activated by depolarization to −30 mV. However, the NMDA receptor-linked channel could provide a significant calcium signal to the intracellular compartment under conditions of LTP, and the NMDA receptor can be considered an important neuromodulatory system for hippocampal LTP in the CA1 and dentate gyrus and in the commissural-associative CA3 pathway. However, LTP is not exclusively linked with NMDA. Kainate appears to serve a similar role in LTP of the mossy fiber to the CA3 neuron pathway. Kainate receptor-linked channels have negligible Ca^{2+} permeability, and their role in LTP may be to overcome endogenous inhibition so that postsynaptic depolarization, and consequent Ca^{2+} influx through voltage-dependent Ca^{2+} channels, can occur. Thus, a general role for NMDA receptors in LTP cannot be assumed. This is particularly the case because NMDA and quisqualate are equally effective in elevating Ca^{2+} in cultured hippocampal neurons, but both are less effective than glutamate or aspartate. In fact, the NMDA antagonist APV is as effective as NMDA itself in elevating intracellular Ca^{2+}.

2. Acetylcholine

Conditioning of the cat eye blink response has been linked with the action of ACh. The response is a short latency eye blink and can be conditioned by pairing an audible "click" with a glabella tap. The sensorimotor cortical neurons that reach the facial nuclei show long-term changes (weeks) in unit activity; their threshold for excitation is lower, and it is suggested that their membrane conductances are altered after conditioning. Pairing depolarization of these neurons with application of ACh increases membrane resistance for approximately 10 min; intracellular injection of cGMP or calcium-calmodulin protein kinase can substitute for ACh. Whether or not the short-term effects of ACh are involved in the conditioning process remains an open question.

3. Dopamine

Long-lasting dopaminergic modulation of neurotransmission has been found in the relatively accessible mammalian sympathetic ganglion preparation. The slow muscarinic EPSP is mimicked by cGMP application and is potentiated heterosynaptically for hours by dopamine acting at dopamine receptors of the D_1 class. This long-term enhancement is mimicked by D_1 agonists and blocked by D_1 antagonists specifically, and appears to involve elevations of the concentration of the intracellular second messenger cyclic adenosine monophosphate (cAMP) in the postsynaptic cell. Thus, dopamine released by adrenergic interneurons within the ganglion facilitates the action of ACh released from cholinergic terminals.

4. Neuropeptides

The gill-siphon withdrawal reflex (GSW) of *Aplysia* is modified by a tetrapeptide transmitter, FMRF-amide (Phe-Met-Arg-Phe-NH_2), released by an interneuron that contacts the presynaptic terminal of the sensory neuron, which innervates the motoneuron responsible for gill withdrawal. Tail shock stimulates the interneuron, hyperpolarizes the sensory neuron, reduces the width of the action potential, and depresses transmitter release. FMRF-amide is found in this interneuron and exogenous FMRF-amide mimics the effects caused by stimulating the interneuron. The behavioral response (inhibition of GSW) is apparent only with weak tail shock; increasing the tail shock causes sensitization. The behavioral paradigm uses a weak tactile stimulation of the siphon as the test stimulus, and tail shock strength is varied. If the frequency of tactile stimulation is increased, the response to weak shock habituates, masking the inhibition. Thus, the inhibition by the FMRF-amide pathway can be masked by habituation. This suggests that FMRF-amide could contribute to the habituation process described earlier. FMRF-amide action involves the second messenger arachidonic acid, which has been shown to activate a potassium channel in the sensory neurons (the S channel).

5. Serotonin

Studies of sensitization of the GSW reflex and the related tail withdrawal in *Aplysia* have led to the suggestion that serotonin, also known as 5-hydroxytryptamine (5HT), is crucial for this behavioral change.

The behavioral paradigm involves two stimuli. The GSW reflex elicited by gentle touch of the mantle or siphon is enhanced by electrical shock stimulation of the head or tail of the animal. In the isolated abdominal ganglion, costimulation of the sensory neuron and the connective bundle from the head facilitates the EPSP observed in the motoneuron during sensory stimulation. The facilitation persists for several minutes, and quantal analysis of the EPSP suggests that the number of quanta per EPSP is increased. Because of intrinsic variability of EPSP amplitude, low EPSP frequency, and polysynaptic and polyneuronal innervation, quantal analysis in this system is not as straightforward as at the neuromuscular junction, for example; therefore, to assert that a reduction in the number of quanta is the only change occurring in this *in vitro* preparation during synaptic facilitation is difficult. Technical difficulties preclude recording from the presynaptic terminals of the sensory neuron, so the electrophysiological analysis of the sensory neuron and the changes induced by connective stimulation has focused on the properties of the neuronal soma, with the assumption that changes at the soma will faithfully reflect presynaptic terminal behavior.

Intracellular recording from the soma of the sensory neuron showed that facilitatory stimulation broadened the action potential (the opposite to the action of the FMRF-amide input to these neurons) by 10–30% under physiological conditions. A probable consequence of spike broadening is increased influx of Ca^{2+}, known to be essential for transmitter release. Under voltage-clamp, short-term facilitation (10–15 min) was found to reduce the steady-state K^+ conductance, and also to reduce a transient K^+ current evoked by depolarization. The consequence of reducing membrane conductance and voltage-dependent K^+ current is to render the neuron more excitable and to slow repolarization following the action potential, which would increase Ca^{2+} influx through voltage-gated Ca^{2+} channels. Exogenous 5HT mimics the effect of connective stimulation, as do agents that block cAMP degradation. A role for elevated intracellular cAMP levels is also supposed.

When long-term potentiation is mimicked in cultured *Aplysia* sensory neurons by 5HT, a "consolidation phase" has been found during which a transcription factor called ApC/EBP (*Aplysia* CCAAT enhancer-binding protein) is induced, even when protein synthesis is pharmacologically blocked. A puzzling aspect of this is that blocking protein synthesis by itself induced ApC/EBP more efficiently than did 5HT, even though blocking protein synthesis also blocks long-term facilitation. Blocking ApC/EBP with antisense RNA or a specific antibody prevented 5HT-induced long-term facilitation. More studies are needed to clarify these results.

Single-channel recordings from sensory neurons show that 5HT can suppress the activity of the S channel, and that cAMP also suppresses this channel, as does the catalytic subunit of the cAMP-dependent protein kinase (protein kinase A, or PKA). However, this is not the only effect of 5HT in these sensory cells. Exogenous 5HT applied to voltage-clamped sensory neurons elevates intracellular Ca^{2+} by a mechanism that is independent of the 5HT effects on the S channel. Despite the attractive simplicity of 5HT-linked facilitation, other studies support a much more complex cellular basis for the sensitization process.

For example, other sets of interneurons are capable of producing the presynaptic facilitation phenomenon. The L28 and L29 group of neurons also have cell bodies located in the abdominal ganglion. The L29 neurons are activated by tail shock and depressed by head stimulation. The L28 neurons are excited by head or tail stimulation. 5HT is not present in any of these neurons, yet stimulation of these neurons in the absence of the cerebral ganglia produces presynaptic facilitation of the sensory neurons. The transmitter at these facilitatory synapses has, therefore, not been identified to date.

Another group of facilitory neurons that synapse upon the sensory neurons uses small cardioactive peptide transmitters, SCP_A and SCP_B; the cell bodies are not localized yet. Therefore, it is probably premature to consider 5HT as the major facilitator in this system.

6. γ-Aminobutyric Acid

γ-Aminobutyric acid (GABA) has recently been implicated in learning in *Hermissenda crassicornis*. As discussed earlier, contiguous stimulation of the Type B photoreceptor by light and the statocyst by agitation causes reduction of photoreceptor K^+ conductance and of phototaxis. Stimulation of the hair cells causes inhibition of action potentials in the Type B cell, and the inhibitory post-synaptic potential recorded in the B cell is blocked by bicuculline, a specific $GABA_A$-receptor antagonist; the $GABA_A$ receptor is commonly linked to chloride ion channels. The hair cells contain GABA, and exogenous GABA also inhibits action potentials in the B cell; however, the situation is more complex. A slowly developing increase in input resistance follows the application of GABA; when a chloride channel blocker is present, the GABA-induced increase in input resistance persists, and a slow depolarization is revealed. The depolarizing response can be blocked by H7, an antagonist of the intracellular enzyme PKC. The joint effect of light and GABAergic hair cell activity is to increase intracellular calcium and increase the inhibition of K^+ channels by PKC. The contiguity of the two stimuli is required to achieve the necessary threshold for a persistent increase in excitability of the B cell. Most significantly, after conditioning, the predominantly inhibitory GABAergic synapse became excitatory for a period of several hours. This phenomenon has also been found to occur during hippocampal long-term potentiation. After paired application of GABA with postsynaptic depolarization, the GABA synapses on CA1 pyramidal cells were temporarily transformed into excitatory synapses. This would result in a radical alteration in the network properties of the hippocampus. [*See* Neurotransmitter and Neuropeptide Receptors in the Brain.]

IV. INTRACELLULAR MESSENGER AND ENZYME SYSTEMS

A. Calcium

The role of calcium as a message critical for synaptic transmission is implicit in the foregoing discussion of the involvement of Ca^{2+} channels. However, Ca^{2+} has a plethora of intracellular effects, and the regulation of intracellular Ca^{2+} concentration ($[Ca^{2+}]$) is an obvious target for mechanisms that affect cellular modifiability. Agents that increase $[Ca^{2+}]$ and are implicated in neuronal modifiability include excitatory amino acids on hippocampal neurons, light and depolarization of *Hermissenda* photoreceptors, and sensitization of *Aplysia* sensory neurons. [*See* Calcium, Biochemistry.]

Calcium is present in the extracellular medium at concentrations between 1 mM (mammals) and 10 mM (marine gastropods). However, the internal free Ca^{2+} is regulated to submicromolar concentrations. Because of this concentration difference, a relatively small change in inward transmembrane Ca^{2+} flux serves as a significant signal. Additionally, cells sequester Ca^{2+} into intracellular compartments, notably the endoplasmic reticulum and mitochondria. Here we consider the evidence for changes in $[Ca^{2+}]$ levels in learning. Later on we examine the cellular targets of Ca^{2+}, which affect changes in neuronal function, including $I_{K\text{-}Ca}$, multifunctional Ca-calmodulin kinase II (CaM-kinase II), calcium-phospholipid-dependent kinase (PKC), calpain, and neurotransmitter release.

Increasing the Ca^{2+} concentration without depolarizing the neuron can be achieved by balanced iontophoresis of Ca^{2+} from intracellular microelectrodes. This has been done for *Hermissenda* and it produces a persistent reduction of K^+ currents I_A and $I_{K\text{-}Ca}$. This is in contrast to the early effect of raising intracellular $[Ca^{2+}]$ (i.e., increase in $I_{K\text{-}Ca}$), and suggests that the long-term effects of Ca^{2+} on K^+ channels are mediated by a third messenger.

It was noted earlier that excitatory amino acids elevate $[Ca^{2+}]$ on hippocampal neurons. The dendritic Ca^{2+} level remains elevated for minutes after repeated applications of glutamate; PKC blockers inhibit the sustained increase in Ca^{2+}, but not the initial transient elevation caused by glutamate. GABA acting at the $GABA_A$ receptor inhibits the rise in $[Ca^{2+}]$; GABA may have other covert effects via the $GABA_B$ receptor. These findings are suggestive in terms of LTP: paired application of glutamate is more effective than continuous single application; PKC activity sustains the Ca^{2+} level, which reciprocally stimulates the activity of PKC; and lastly, GABA apparently is able to reset the Ca^{2+} level to normal, perhaps by closing the Ca^{2+} channels held open by PKC. Ca^{2+} is required for LTP, because removing or increasing extracellular Ca^{2+} can suppress or enhance the development of LTP, and the Ca^{2+} chelator EGTA will block LTP when injected into pyramidal neurons. LTP also increases the uptake of radioactive Ca^{2+} into slices.

B. Cyclic Nucleotides

Evidence suggests that cAMP may be important in facilitation of *Aplysia* sensory neurons and the GSW reflex. Injection of the catalytic subunit of cAMP-dependent protein kinase (PKA) into sensory neurons simulates facilitation, and blockage of adenylate cyclase or PKA blocks facilitation. A cellular action of cAMP and PKA is to close the S channel, thus increasing excitability. In this model, 5HT and the other transmitters implicated in facilitation are postulated to elevate intracellular cAMP levels, causing PKA activation and the closure of S channels.

Learning-deficient *Drosophila* mutants exhibit deficiencies of cAMP metabolism. The short-term memory capacity of the flies was measured in associative conditioning paradigms using olfactory stimuli and electrical shock. The mutant *rut* has low adenylate cyclase levels, and CaM-kinase can no longer stimulate adenylate cyclase. The mutant *dnc* has reduced or absent cAMP phosphodiesterase activity; this enzyme degrades cAMP, and its absence would increase cAMP levels. The fact that both of these mutations show learning deficiencies, but have opposite absolute cAMP levels, shows that the role of cAMP in learning involves a complex interplay between synthesis, the target molecules, and degradation. However, it is not clear from the *Drosophila* studies whether cAMP metabolism is critical for learning or if disrupting cAMP metabolism has general effects that also impinge on learning capacity. Recent genetic experiments have implicated CREB, a cAMP-responsive element binding protein, in learning in both *Drosophila* and mice.

C. Protein Kinases

Protein kinases, which phosphorylate their targets, and whose activity is modulated by a range of second messengers, are linked to memory in several systems. In *Hermissenda,* PKC causes the I_A and $I_{K\text{-}Ca}$ reduction that is causal for the behavioral change. Injection of purified PKC into Type B photoreceptors elevates I_A and $I_{K\text{-}Ca}$; however, pretreatment of the photoreceptors with the PKC activator phorbol ester reveals an opposite effect of PKC injection, in that K^+ currents are reduced. The phorbol esters cause translocation of cytoplasmic PKC to the cell membrane, and it seems that PKC must be in the membrane to cause inhibition of K^+ channels, whereas soluble cytoplasmic PKC is a K^+ channel activator. The translocation step also increases PKC sensitivity to Ca^{2+}.

After nictitating membrane response conditioning, rabbit hippocampal CA1 neurons show reductions in K^+ currents, and phorbol ester treatment mimics the effect of conditioning. Conditioning also causes translocation of PKC to the membrane.

K^+ channels may be only one target for PKC; the growth-associated protein GAP-43 (variously known as F1 or B50) is also phosphorylated by PKC. Phorbol esters are known to sustain synaptic potentiation in the hippocampus and PKC is translocated during LTP; therefore, it is interesting that phosphorylation of GAP-43 is increased in LTP. GAP-43 is found in presynaptic membranes and growth cones and appears to be important for neurite extension. However, because GAP-43 is not present in dendrites, it may not be directly involved in synaptic changes in dendrites.

CaM-kinase II is probably involved in producing LTP, because LTP is blocked by calmidazolium, which is a calmodulin antagonist, and the kinase inhibitor H7 (which also inhibits PKC). One hypothesis is that autophosphorylation of CaM-kinase II is permissive for the proteolytic activity of calpain, a Ca^{2+}-dependent protease, and that calpain proteolyzes cytoskeletal and membrane skeleton proteins such as spectrin, leading to changes in dendritic morphology. Both the number and shape of dendrites may be changed during LTP, but a correlation with calpain action has not been shown.

The cAMP-dependent protein kinase, PKA, was mentioned earlier in connection with facilitation in *Aplysia* sensory neurons. Behavioral sensitization is followed by a decrease in the number of regulatory subunits of PKA in sections of the whole abdominal ganglion, which would presumably lead to unregulated protein phosphorylation by the catalytic subunit of PKA.

In the *Drosophila* learning mutant *tur*, the cAMP-binding affinity of PKA is reduced. As discussed in relation to cAMP, it is difficult to know if this deficiency is specific to mnemonic mechanisms.

Both PKA and CaM-kinase II have been injected into *Hermissenda* Type B photoreceptors and cause reductions of the potassium currents I_A and $I_{K\text{-}Ca}$. However, the changes induced by PKA do not mimic the changes produced by conditioning, and are probably less specific in this system than those of PKC and CaM-kinase II. Interestingly, these experiments illustrate how essential a clear understanding of the biophysical effects caused by training is in isolating the molecular mechanisms underlying behavior; all of these kinases are potent reagents and will almost invariably have nonspecific effects that must be quantitatively distinguished from memory-specific effects.

D. GTP-Binding Proteins

Evidence for involvement of guanosine triphosphate (GTP)-binding proteins in memory processes has emerged. In trained *Hermissenda*, the photoreceptors contain increased levels of a phosphorylated 20-kDa protein, known as cp20. This protein binds GTP and also has GTPase activity, features characteristic of G proteins. Injection of cp20 into the Type B photoreceptors of naive animals produces reductions in I_A and $I_{K\text{-}Ca}$ that mimic the effect of conditioning. cp20 is a specific high-affinity substrate for PKC and undergoes a shift in its subcellular localization after phosphorylation. It is a member of the ARF (adenosine diphosphate ribosylation factor) family of GTP-binding proteins, which are involved in transport of proteins in the Golgi apparatus. Thus, cp20 may also have effects on protein transport. [*See* G Proteins.]

E. Oncogenes and Proto-oncogenes

Both learning and carcinogenesis involve long-term modifications of cellular function, and this has stimulated researchers to seek common mechanisms. Phorbol esters, for example, stimulate PKC, are tumor promoters, and elevate expression of proto-oncogene proteins, which are involved in cell growth and division. Many proto-oncogenes are now known, and several have been suggested to play a part in memory.

Ras proto-oncogenes are GTP-binding proteins with 21 kDa molecular mass. Injection of *ras* into Type B photoreceptors modulates K^+ currents and also inhibits axonal transport; this is comparable to the actions of cp20. Whether *ras* is acting at sites that would normally be targets for cp20, or whether the *ras*-like proteins known to exist in *Hermissenda* have effects similar to cp20, is not known.

The proto-oncogene *c-fos* can be induced by LTP, as well as many other agents, including heat-shock and high-potassium depolarization. The *fos* protein and the related oncogene protein *jun* form a heterodimer that binds to the DNA regulatory site AP1 and induces transcription of various proteins; however, a specific protein product consequent upon LTP is not known yet. Another protein elevated after hippocampal LTP is *egr-1,* which is one of a family of proteins with zinc-finger DNA binding and regulatory activity. The role of changes in transcription and translocation associated with behavior modification and learning is discussed in the next section. [*See* Oncogenes and Proto-oncogenes.]

V. GENE EXPRESSION

Modification of gene expression is an attractive model for sustaining the lifetime of memories. Many early experiments showed that long-term memory was especially impaired by systemic administration of protein synthesis inhibitors. Similar, more recent experiments on molluscs have also shown effects on long-term facilitation or conditioning. However, the physiological insult provided by inhibiting protein synthesis makes it difficult to infer much from these studies.

However, more precise information has been gained from molluscan nervous systems. In *Hermissenda,* at least 21 species of mRNA are elevated in the eyes of conditioned animals, and conditioning also increases ^{32}P incorporation into mRNA immediately after training and for up to 4 days after the end of training.

Polyribosomes located in dendritic spines can translate mRNA at the site of synaptic activity. Selective transport of mRNA in dendrites in culture has also been observed; thus there could be site-specific mRNA translation at synapses. Such selective site-specific mRNA translation has not yet been observed in the context of learning.

Much attention has been given to the nuclear early genes (e.g., *c-myc* and *c-fos*) that are rapidly activated following stimulation and could go on to regulate transcription of other "late" genes that could be responsible for maintenance of memories. Again, this is a tempting idea, but it has not been shown to occur in a behaviorally relevant context. Another potential mechanism could be through protein kinase C, which has been shown to phosphorylate and activate the nuclear enzyme topoisomerase II; this enzyme is implicated in the regulation of DNA and the transcription of specific genes.

Possible changes in gene expression include novel proteins from previously unexpressed genes, novel proteins formed by alternative splicing of mRNA sources from distinct DNA exons, and changes in the proportion of isoforms of extant proteins, in addition to simply changing the level of expression of an extant protein. Alternative splicing has been shown to modify potassium channel function in *Drosophila* nervous systems. Isoforms of regulatory proteins, which show different levels of expression in different tissues, include PKC and PKA; however, isoform differences between particular neurons remain to be demonstrated. One area in which changes in protein levels could be expressed markedly is changes in cellular morphology, and in fact several proteins are linked to morphological changes relevant to behavioral modifiability. The extracellular glycoproteins ependymins are synthesized more after training, in both goldfish and rodents; the ependymins polymerize when the Ca^{2+} concentration falls. It is possible that ependymins would undergo polymerization at active synapses as the flux of Ca^{2+} ions into the cell transiently depleted the local Ca^{2+} concentration. Another protein class linked to modifiability is the S-100 polypeptides, which increase neurite extension in culture. The last example of an extracellular protein implicated in modifiability is the proteoglycan produced by the *per* gene in *Drosophila,* mutations of which impair courtship behavior. Such extracellular proteins could play a role in the changes in dendritic spine shape observed in LTP.

In summary, there is evidence that gene expression changes during learning. Membrane processes do alter the activity of intracellular messengers such as PKC, which have nuclear targets, and changes in the transcription of mRNA and translation of proteins have been seen in learning. Major questions remain: How is the mRNA or its protein product specifically directed to the active region of the neuron? Are common mechanisms available, or do different types of gene expression regulation occur in different neurons within the same animal?

VI. INTEGRATING EXPERIENCE AND OUTPUT

The adaptive advantage of memory is presumably that an organism learns from previous events to respond appropriately and successfully to new environmental challenges. The types of learning most amenable to cellular analysis are relatively simple when compared with human learning, yet may be different only in degree, rather than in kind, at the cellular level.

Generating a model for cellular memory processes assumes that the same events occur in all neurons capable of modification. From the preceding discussions this is patently not the case. However, several general principles do emerge. First, the importance of temporal association of stimuli for achieving a critical threshold of excitation mirrors the behavioral constraint of contiguity.

Second, the spatial dimension is particularly relevant for neuronal function. PKC activity in the plasma membrane is different from PKC activity in the cytosol; however, which synapse out of thousands of synapses acquires translocated PKC is critical for the specificity of the function of that neuron.

Finally, many varieties of incoming signals are transduced to evolutionarily ancient intracellular signals, which are shared by neurons from different phyla; this observation argues for common targets for the intracellular messengers and supports the use of the experimental models to explore the basis of our own biology. The second messengers of interest include Ca^{2+}, cAMP, phospholipid, and perhaps arachidonic acid. Additionally, third messengers are prominent in molecular mnemonics, the protein kinases, and the proto-oncogene proteins in particular, and may be especially important in regulating gene expression and thus the memories that persist for our lifetimes.

BIBLIOGRAPHY

Alkon, D. L. (1987). "Memory Traces in the Brain." Cambridge Univ. Press, Cambridge, England.

Alkon, D. L. (1989). Memory storage and neural systems. *Sci. Am.* **260**(7), 42–50.

Alkon, D. L., and Nelson, T. J. (1990). Specificity of molecular changes in neurons involved in memory storage. *FASEB J.* **4**(6), 1567–1576.

Anderson, B. J., and Steinmetz, J. E. (1994). Cerebellar and brainstem circuits involved in classical eyeblink conditioning. *Rev. Neurosci.* 5(3), 251–273.

Bank, B., LoTurco, J. J., and Alkon, D. L. (1989). Learning induced activation of protein kinase C. *Mol. Neurobiol.* **3**, 55–70.

Davis, R. L. (1996). Physiology and biochemistry of Drosophila learning mutants. *Physiol. Rev.* **76**(2), 299–317.

Dudai, Y. (1989). "The Neurobiology of Memory." Oxford Univ. Press, Oxford, England.

Farley, J., and Alkon D. L. (1985). Cellular mechanisms of learning, memory, and information storage. *Annu. Rev. Psychol.* **36**, 419–494.

Glickstein, M., Yeo, C., and Stein, J. (1987). "Cerebellum and Neuronal Plasticity," Plenum, New York.

Gormezano, I., Prokasy, W. F., and Thompson, R. F. (eds.) (1987). "Classical Conditioning." Lawrence Erlbaum, Hillsdale, NJ.

Ito, M. (1989). Long-term depression. *Annu. Rev. Neurosci.* **12**, 85–102.

Ito, M. (1993). Synaptic plasticity in the cerebellar cortex and its role in motor learning. *Can. J. Neurol. Sci.* **20** (Suppl. 3), S70–74.

Lynch, G. (1986). "Synapses, Circuits, and the Beginnings of Memory." MIT Press, Cambridge, MA.

Lynch, G., Kessler, M., Arai, A., and Larson, J. (1990). The nature and causes of hippocampal long-term potentiation. *Prog. Brain Res.* **83**, 233–250.

Levitan, I. B. (1994). Modulation of ion channels by protein phosphorylation and dephosphorylation. *Annu. Rev. Physiol.* **56**, 193–212.

McPhie, D., Matzel, L., Olds, J., Lester D., Kuzirian A., and Alkon, D. L. (1993). Cell specificity of molecular changes during memoroy storage. *J. of Neurochem.* **60**, 646–651.

Nelson, T. J., and Alkon, D. L. (1989). Specific protein changes during memory acquisition and storage. *BioEssays* **10**, 75–79.

Nelson, T. J., Yoshioka, T., Toyoshima, S., Han, Y.-F., and Alkon, D. L. (1994). Characterization of a GTP-binding protein implicated in both memory storage and interorganelle vesicle transport. *Proc. Nat. Acad. Sci. USA* **91**, 9287–9291.

Rudy, B. (1988). Diversity and ubiquity of K channels. *Neuroscience* **25**, 729–749.

Wickman, K. D., and Clapham, D. E. (1995). G-protein regulation of ion channels. *Curr. Opin. Neurobiol.* 5(3), 278–285.

Cell Volume, Physiological Role

DIETER HÄUSSINGER
Heinrich-Heine University of Düsseldorf

FLORIAN LANG
University of Tübingen

GLOSSARY

Cell hydration Describes the cellular water content. Cell volume changes in the present context reflect osmotic water shifts into or out of the cell; the terms cell hydration and cell volume are used here synonymously

Mitogen-activated protein kinases Members of a protein kinase family that are activated in response to growth factors and mitogens

Organic osmolytes Organic compounds, which are specifically accumulated inside the cell, when extracellular osmolarity increases and which are specifically released from the cells, when ambient osmolarity decreases

Regulatory volume decrease When cells are suddenly exposed to hypoosmotic fluids, they initially swell like osmometers, but then downregulate their volume within minutes

Regulatory volume increase When cells are suddenly exposed to hyperosmotic media, an initial osmometer-like shrinkage is followed by an upregulation of cell volume within minutes

CELL VOLUME HOMEOSTASIS IS OF CRUCIAL importance for overall cellular function. It is therefore not surprising that volume regulatory mechanisms have been found in almost every cell studied so far. However, these volume regulatory mechanisms are apparently not designed to maintain absolute cell volume constancy; they act as dampeners in order to prevent excessive cell volume deviations, which would otherwise result from cumulative substrate uptake or osmotic stresses. These volume regulatory mechanisms can even be activated in the resting state by hormones, resulting in changes in cell volume. On a short-term time scale, cell volume changes are almost exclusively due to alterations of the cellular water content. Cell hydration, i.e., cell volume, is dynamic and can change within minutes under the influence of anisoosmolarity, hormones, nutrients, and oxidative stress. Most importantly, small fluctuations of cell hydration are an independent and potent signal, which regulates cellular metabolism and gene expression. This creates a simple, but elegant way for adaptation of cell function to environmental challenges. Cell swelling and shrinkage lead to certain opposite patterns of cellular metabolic function. Apparently, hormones and amino acids can trigger those patterns by altering cell volume. Thus, cell volume homeostasis does not simply mean volume constancy but rather the integration of events which allow cell hydration to play its physiological role as a regulator of cell function. The interaction between cellular hydration and cell function has been extensively studied in liver cells, but evidence is increasing that regulation of cell function through alterations of cell hydration also occurs in other cell types.

I. CELL VOLUME REGULATION

Regulation of cell volume is primarily achieved by the transport of electrolytes across the cell membrane and can be studied following exposure of cells to anisoos-

ENCYCLOPEDIA OF HUMAN BIOLOGY, Second Edition, VOLUME 2. Copyright © 1997 by Academic Press. All rights of reproduction in any form reserved.

motic fluids. This particular approach should not be seen exclusively as an unphysiological tool for cell volume modification. For example, a hyperosmotic environment is created in the renal medulla during antidiuresis, which makes cell volume regulatory mechanisms mandatory. During intestinal absorption of water or nutrients, portal venous blood may become slightly hypo- or hyperosmotic. Physiologically more important, however, are cell volume changes due to cumulative substrate uptake into the cells and under the influence of hormones. When cells are suddenly exposed to hypoosmotic media, they initially swell like more or less perfect osmometers but within minutes retain almost their original cell volume. This behavior has been labeled regulatory cell volume decrease (RVD). Conversely, shrinkage induced by hyperosmotic exposure is followed by a volume regulatory increase (RVI), which brings back cell volume largely (but not completely) to the starting level. The mechanisms responsible for RVD and RVI are cell type and species dependent, but in general involve alterations of ion transport across the plasma membrane. In addition, some cell types augment RVD and RVI by release or accumulation of nonelectrolytes (socalled "osmolyte strategy"), by metabolic disposal/generation of osmotically active compounds, and by endo/exocytosis. [*See* Cell Volume, Regulation.]

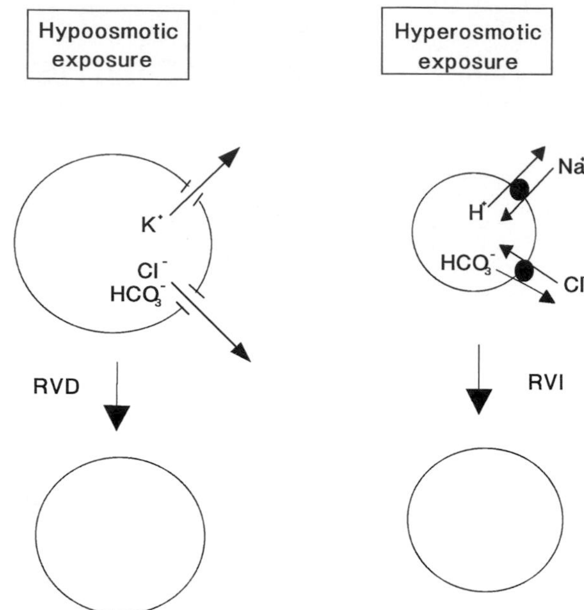

FIGURE 1 Regulatory volume decrease (RVD) and regulatory volume increase (RVI) in hepatocytes following hypoosmotic and hyperosmotic exposure, respectively. RVD is achieved by parallel activation of K^+ and Cl^- channels, whereas RVI involves parallel activation of Na^+/H^+ exchange and Cl^-/HCO_3^- antiport. The volume regulatory ion transport systems can also be activated in the resting state by hormones, resulting in cell swelling or shrinkage, respectively (compare to Fig. 2).

A. Ionic Mechanisms of Cell Volume Regulation

In many mammalian cells, RVD is the result of cellular release of K^+, Cl^-, and HCO_3^-, whereby the primary mechanism leading to the release of these ions is dictated by the cell type. Depending on whether the resting membrane potential is closer to the chloride (e.g., hepatocytes) or to the potassium equilibrium potential (e.g., astrocytes, lymphocytes), RVD caused by activation of K^+ and Cl^- channels is associated with either hyperpolarization or depolarization of the plasma membrane, respectively. In hepatocytes, RVD following hypoosmotic exposure is brought about by an opening of barium and quinidin-sensitive K^+ and anion channels (Fig. 1).

RVI following hyperosmotic exposure of cells is accomplished by an uptake of ions by parallel activation of amiloride-sensitive Na^+/H^+ exchange and Cl^-/HCO_3^- exchange, opening of Na^+ channels, or activation of loop diuretic-sensitive Na^+-K^+-$2Cl^-$ cotransport. In addition, the loss of cellular ions is simultaneously minimized in most cell types by a reduction in membrane conductances. RVI in liver is at least in part achieved by parallel activation of Na^+/H^+ exchange and Cl^-/HCO_3^- exchange (Fig. 1). In contrast to other cell types, Na-K-2Cl cotransport does not appear to appreciably participate in hepatic RVI, even though the carrier probably exists in the hepatic cell membrane and its activation by insulin is followed by an increase of cell volume. [*See* Cell Volume, Regulation.]

B. Osmolytes

In addition to the ionic mechanisms of cell volume regulation, some cell types specifically accumulate or release organic compounds, so-called organic osmolytes, in response to cell shrinkage or cell swelling, respectively. Osmolytes need to be nonperturbing solutes that do not interfere with protein function even when occurring in high intracellular concentrations. Such a prerequisite may explain why only a few classes of organic compounds, viz. polyols, such as inositol and sorbitol (e.g., astrocytes, renal medulla, lens epithelial cells), methylamines such as betaine and α-glycerophosphorylcholine (e.g., renal medulla, liver macrophages), and certain amino acids such as taurine (e.g., Ehrlich ascites tumor cells) have evolved as osmolytes in living cells. Different mechanisms contribute to the intracellular accumulation of osmolytes during hyperosmotic stress: (i) decreased degradation (α-glycerophosphorylcholine), (ii) increased synthesis (induction of aldose reductase), (iii) increased uptake following the induction of specific Na^+-coupled transporters (e.g., for myoinositol, betaine, taurine), and (iv) possibly regulation of osmolyte efflux from the cell via the yet poorly characterized routes. The enhanced synthesis of sorbitol from glucose by aldose reductase under these conditions involves an increased expression of the enzyme due to activation of the encoding gene. Likewise, the Na^+-dependent transporters for inositol and betaine are induced upon hyperosmotic exposure. The genes coding for these transporters have been cloned and a hypertonicity-sensitive element has been identified in the regulatory region of the betaine transporter gene (BGT1). Whereas the ionic mechanisms of cell volume regulation are almost immediate in onset and are completed within 5–20 min after onset of the osmotic challenge, the process of intracellular osmolyte accumulation takes hours or days, but can produce intracellular organic osmolyte concentrations of several hundred millimoles per liter. This is especially important in the renal medulla, as medullary fluid osmolarity can increase up to 3800 mosmol/liter during antidiuresis and decrease to 170 mosmol/liter during diuresis. In the antidiuretic state (high extracellular osmolarity in renal medulla), intracellular osmolarity increases in renal medullary cells as the result of the accumulation of inositol and betaine, which are taken up via concentrative Na^+-dependent transporters and as the result of increased synthesis of sorbitol and α-glycerophosphorylcholine. Conversely, transition from the antidiuretic to the diuretic state and accordingly from a high to a low ambient osmolarity leads within minutes to a dramatic increase in the permeability of medullary cells to organic osmolytes, thus facilitating the rapid efflux of osmolytes from the cells in response to the decline of the extracellular osmolarity. Here, osmolyte-specific transport systems (so-called "permeases") and unspecific ion channels are thought to be involved in this response, but the mechanism underlying their regulation is far from clear.

C. Other Mechanisms

Evidence suggests that cell swelling stimulates exocytosis, which plays a role in RVD in some cell types, such as hepatocytes. Hypoosmotic liver cell swelling transiently increases the plasma membrane surface due to a microtubule-dependent exocytosis at the canalicular and basolateral membrane of the hepatocyte. These exocytotic mechanisms may bring about the insertion of transporter molecules into the plasma membrane, as was shown for bile acid exporter molecules in the canalicular liver membrane. Theoretically, polymerization/depolymerization reactions can decrease/increase the intracellular concentration of osmotically active solutes. Indeed, hypoosmotic cell swelling favors the net conversion of amino acids and glucose into protein and glycogen, respectively, whereas hyperosmotic cell shrinkage triggers opposite net fluxes. The relative contribution of such metabolic changes to overall cell volume regulation remains to be established. At least in liver, glucose conversion into glycogen should have little impact on cell volume homeostasis as the plasma membrane is freely permeable to glucose, and modulation of glycogen/glucose interconversion will have no effect on the intra/extracellular glucose concentration gradient.

II. PHYSIOLOGICAL MODULATORS OF CELL VOLUME

A. Cumulative Substrate Transport

One of the most important challenges for cell volume homeostasis is the cumulative uptake of osmotically active substances, such as amino acids, by specific Na^+-dependent transport systems. These transport systems can create intra/extracellular amino acid concentration gradients across the plasma membrane of up to 20 by utilizing the energy of the transmembrane electrochemical Na^+ gradient. Na^+ cotransported with the amino acid is in turn exchanged for K^+ by the Na^+–K^+–ATPase. The accumulation of amino acids and K^+ inside the cells leads to cell swelling, which in turn triggers a volume regulatory K^+ efflux (Fig. 2). The latter mechanism does not restore liver cell volume; it only prevents cell swelling from becoming excessive as would otherwise be predicted from the continuing accumulation of the amino acid inside the cell. In liver, physiological fluctuations in portal vein amino acid concentration in response to the feeding/starvation cycle are accompanied by parallel alterations of liver cell volume. The extent of amino acid-induced cell swelling seems largely related to the existing steady-state intra/extracellular amino acid concentration gradients. Further, it is modified by hormones and the nutritional state due to (i) regulation of expression of amino acid transport systems in the plasma membrane, (ii) modification of the electro-

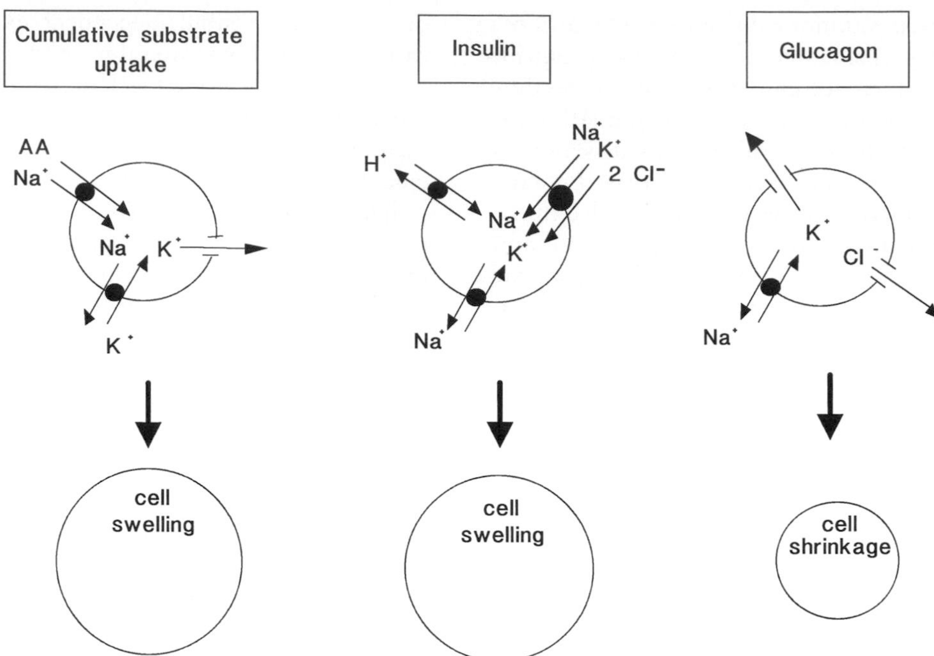

FIGURE 2 Modulators of liver cell hydration. (Left) Cell swelling is induced by amino acids which accumulate inside the cells by Na^+-dependent transport systems. Na^+ is in part exchanged against K^+ by Na^+/K^+-ATPase. The accumulation of the amino acid and a cation inside the cell leads to cell swelling and may activate volume regulatory K^+ efflux. Nonetheless, the cell remains swollen as long as the amino acid is present. (Middle) Insulin activates amiloride-sensitive Na^+/H^+ exchange and bumetanide-sensitive Na-K-2Cl cotransport as well as Na^+/K^+-ATPase, resulting in a cellular accumulation of K^+, Na^+, and Cl^- and cell swelling. (Right) Glucagon depletes cellular K^+ by activating Cl^- channels and quinidine/Ba^{2+}-sensitive K^+ channels, thereby inducing cell shrinkage. The hormone-induced cell volume alterations act like another signal which participates in mediating the amino acid and hormone effects on hepatic metabolism: cell swelling triggers an anabolic, cell shrinkage a catabolic pattern of cellular function.

chemical Na^+ gradient as a driving force for Na^+-coupled transport, and (iii) alteration of intracellular amino acid metabolism.

B. Hormones

Hormones modulate liver cell volume by affecting the activity of volume regulatory ion transport systems. In liver, insulin stimulates amiloride-sensitive Na^+/H^+ exchange, loop-diuretic-sensitive Na-K-2Cl cotransport, and the Na^+/K^+ ATPase, i.e., transport systems that are also turned on for RVI in liver and many other tissues. The concerted activation of these transporters leads to the cellular accumulation of potassium, sodium, and chloride and consequently cell swelling (Fig. 2). Insulin-induced cell swelling and cellular K^+ accumulation are abolished in the presence of bumetanide plus amiloride. Glucagon activates Na^+/K^+ ATPase, but simultaneously depletes cellular K^+, probably due to a simultaneous opening of Ba^{2+}- and quinidine-sensitive K^+ channels. As a result of the cellular Na^+, K^+, and probably Cl^- depletion, hepatocytes shrink (Fig. 2). The physiological relevance is underlined by the finding that half-maximal effects of insulin and glucagon on liver cell hydration are found at hormone concentrations normally present in portal venous blood *in vivo*, i.e., at 10^{-9} and 10^{-10} *M*, respectively. Other hormones also modify hepatocellular hydration; swelling is also induced by insulin-like growth factor-1 (IGF-1), α-adrenergic agonists, and bradykinin, whereas adenosine, extracellular ATP, serotonin, and vasopressin lead to cell shrinkage.

Glucagon and Ca^{2+}-mobilizing hormones were also identified as modulators of mitochondrial matrix volume, but may affect the cytosolic and mitochondrial water spaces in opposite directions. For example, glucagon induces cell shrinkage and simultaneously swells the mitochondria. However, both cell and mitochondrial swelling occur under the influence of phenylephrine. The hormone-induced increase of mitochondrial matrix volume is thought to be due to a pyrophosphate-stimulated K^+ influx into the mitochondria involving adenine nucleotide translocase.

C. Other Effectors on Cell Volume

Oxidative stress exerted by hydroperoxides induces hepatocellular shrinkage due to an opening of Ba^{2+}-sensitive K^+ channels. Cell shrinkage and K^+ channel opening also occur when hydrogen peroxide is generated intracellularly during the oxidation of monoamines. Apparently the balance between intracellular metabolic H_2O_2 generation and its removal by detoxication systems such as catalase and glutathione peroxidase is one determinant for hepatocellular K^+ balance and accordingly cell volume. Oxidative stress may also augment cell shrinkage in other cell types because hydrogen peroxide stimulates a K^+ conductance in pancreatic B cells and oxidative stress inhibits Na-K-2Cl cotransport in vascular endothelial cells. K^+ channel opening under the influence of urea at concentrations found in ureamia also induces cell shrinkage.

Hypoxia reduces the activity of membrane Na^+-K^+-ATPase; this results in the accumulation of Na^+ (and Cl^-) inside the cell and cell swelling. However, the duration of hypoxia required for appreciable cell swelling can be quite variable, whereas swelling itself depends on the rate of Na^+ entry/exit and the extent of compensatory exocytosis. [*See* Cell Volume, Regulation.]

TABLE I
Effects of Hypoosmotic Cell Swelling on Liver Function

Liver cell swelling increases
Protein synthesis
Glycogen synthesis
Lactate uptake
Pentose phosphate shunt
Amino acid uptake
Glutamine breakdown
Glycine oxidation
Ketoisocaproate oxidation
Acetyl-CoA carboxylase
Lipogenesis
Urea synthesis from amino acids
MAP kinase activity
Glutathione (GSH) efflux
Taurocholate excretion into bile
Actin polymerization
Microtubule stability
Exocytosis
pH in vesicular compartments
mRNA levels of c-*jun*, ornithine
Decarboxylase, β-actin, tubulin
Liver cell swelling decreases
Proteolysis
Glycogenolysis
Glucose-6-phosphatase activity
Carnitine palmitoyltransferase I activity
Glutamine synthesis
Urea synthesis from NH_4^+
Biliary glutathione disulfide (oxidized) (GSSG) release
Cytosolic pH
mRNA levels for PEPCK and tyrosine aminotransferase
Viral replication
Cyclooxygenase-2 expression in activated liver macrophages (Kupffer cells)

III. CELL VOLUME CHANGES AS A SIGNAL MODULATING CELL FUNCTION

A. General Considerations

Current evidence suggests that the cellular hydration state is an important determinant of cell function and that hormones, oxidative stress, and nutrients exert in part their effects on metabolism and gene expression by a modification of cell volume. The concept that cellular hydration acts as an independent signal on liver cell function is based on the following observations. (i) Persistent alterations of metabolism occur within minutes in response to anisoosmotic cell volume changes (Table I) and there is a dose–response relationship between the extent of cell hydration change and the metabolic response. (ii) Volume-sensitive signal transduction pathways have been identified that link cell hydration to cell function. (iii) Cell swelling in response to amino acids explains several metabolic effects of amino acids, which cannot be related to their metabolism, such as stimulation of glycogen synthesis or inhibition of proteolysis. (iv) Several metabolic hormone effects can be mimicked by equipotent anisoosmotic cell swelling or shrinkage and some hormone effects disappear when the hormone-induced cell volume changes are prevented.

Thus, cell hydration changes in response to physiological stimuli act as a signal which helps to adapt cellular metabolism to alterations of the environment (substrate, tonicity, hormones). Na^+-dependent amino acid transport systems in the plasma membrane not only play a role in amino acid translocation, they also act as transmembrane signaling systems by altering cellular hydration in response to the substrate supply. Such a signaling role may shed a new light on the long-known heterogeneity of transport systems among different cell types and their different expression during development, rendering specific amino acids a more or less potent signal for cell function. Likewise, transmembrane ion movements under the influence of hormones are an integral part of hormonal signal transduction mechanisms with alterations of cellular hydration acting as another "second messenger" of hormone action. However, the exact place of hormone-induced cell volume changes in the network of known hormone receptor-activated intracellular messenger systems remains to be established.

B. Cellular Hydration and Protein Turnover

Hepatocellular hydration is a major site of proteolysis control in liver: cell swelling inhibits, whereas cell shrinkage stimulates autophagic protein breakdown under conditions when the proteolytic pathway is not already fully activated. A close relationship exists between the proteolytic activity and cell hydration in liver, regardless of whether cell volume is modified by hormones, glutamine, glycine, alanine, bile acids, the K^+ channel blocker Ba^{2+}, or anisotonicity. The effects of glutamine, alanine, glycine, insulin, IGF-1, and glucagon on proteolysis can quantitatively be mimicked by equipotent anisotonic cell volume modulation, leading to the conclusion that the known antiproteolytic effect of insulin and several (but not all) amino acids is transmitted in large part by an agonist-induced cell swelling, whereas stimulation of proteolysis by glucagon is mediated by cell shrinkage. In line with this, the antiproteolytic action of insulin disappears when insulin-induced cell swelling is prevented in the presence of inhibitors of the Na^+/H^+ antiporter and the Na^+-K^+-$2Cl^-$ cotransporter. The nutritional state exerts its control on proteolysis by determining the swelling potencies of hormones and amino acids. For example, in the fed state the antiproteolytic effect of glycine is only about one-third compared to that found after 24 hr of starvation due to an about three-fold higher swelling potency of glycine during starvation, which is explained by an upregulation of the glycine-transporting amino acid transport system A in starvation. It should be emphasized that not all amino acids exert their antiproteolytic effect via changes of cell hydration; for example, leucine and phenylalanine are potent inhibitors of proteolysis, yet they exert little effect on cell volume. Apparently other mechanisms of proteolysis control come into play.

Liver cell swelling not only inhibits proteolysis, but also stimulates protein synthesis. In contrast, cell shrinkage stimulates proteolysis but inhibits protein synthesis. That cell hydration affects both protein degradation and synthesis in opposite directions is a finding that has a direct bearing on the problem of protein catabolic states in disease.

The mechanisms of how cell hydration exerts control on proteolysis are not settled, but it is known that intact microtubules are required and a role of protein phosphorylation has been suggested. The latter could reside at the levels of ribosomal protein S6 phosphorylation, acidification of prelysosomal, endocytotic/autophagic vesicular compartments, and possibly, but not yet proven, at the level of phosphorylation of microtubule-associated proteins.

C. Amino Acid Transport and Metabolism

Amino acid transport not only modifies cell volume, but conversely, cell volume can exert control on amino acid transport. For example, hyperosmolarity induces Na^+-dependent transport systems for neutral amino acids in a variety of cell types. Apparently, neutral amino acids are used here as organic osmolytes in order to counteract cell shrinkage. However, amino acid transport may also be stimulated in response to cell swelling. One example is the hepatic glutamine-transporting system N, and the previously established amino acid-dependent short-term stimulation of amino acid transport is due to amino acid-induced cell swelling. Hypoosmotic cell swelling increases hepatic alanine and glutamine uptake, whose intracellular degradation rate is controlled by transport rather than by metabolism. A swelling-induced hyperpolarization of the cell membrane may not only augment Na^+-coupled substrate transport, but may also explain why sinusoidal glutathione efflux from the liver is increased following hypoosmotic liver cell swelling because glutathione release was shown to be under the control of the membrane potential.

In rat liver, hypoosmotic cell swelling switches hepatic glutamine balance from net release to net uptake. This is due to a stimulation of flux through glutaminase in periportal hepatocytes and a simultaneous inhibition of glutamine synthesis in perivenous hepatocytes. The swelling-induced activation of glutaminase is most likely due to simultaneous mitochondrial swelling, which alters the attachment of the enzyme to the inner mitochondrial membrane. Likewise, swelling stimulates glycine oxidation in perfused rat liver and in isolated mitochondria. These findings suggest that the regulation of amino acid metabolism by anisotonicity can be explained by concomitant alterations of mitochondrial matrix volume. Also, stimulation of glycine and glutamine oxidation by glucagon, cAMP, and Ca^{2+}-mobilizing hormones involves hormone-induced increases of mitochondrial matrix volume, although these agents simultaneously lower whole cell hydration. Thus, mitochondrial pathways, such as glutamine and glycine oxidation and respiration, are stimulated not only by hypoosmolarity, but also by cAMP, glucagon, and Ca^{2+}-mobilizing hormones, despite opposing effects of these effectors on cell volume.

Hypoosmotic cell swelling stimulates urea synthesis and ammonia formation from amino acids, but inhibits when ammonia is used as the sole substrate for urea synthesis, although the swelling of isolated mitochondria was shown to stimulate citrulline synthesis. Inhibition of urea synthesis from ammonia by cell swelling is probably due to a block of the urea cycle at the step of argininosuccinate synthesis due to a cell swelling-induced disturbance of the transfer of reducing equivalents via the malate/aspartate shuttle and an impaired aspartate regeneration during cell swelling.

D. Carbohydrate and Fatty Acid Metabolism

Like protein turnover, carbohydrate metabolism in liver is critically dependent on cell hydration (Table I). Hepatocyte swelling inhibits glycogenolysis, glycolysis, and glucose-6-phosphatase activity, but simultaneously stimulates glycogen synthesis, flux through the pentose phosphate pathway, and lipogenesis. Opposing effects occur in response to cell shrinkage. The stimulatory effect of glutamine and other amino acids on glycogen synthesis and lipogenesis is due to amino acid-induced cell swelling. Glycogen synthesis and lipogenesis are controlled by the activity of glycogen synthase and acetyl-CoA carboxylase; both enzymes are subject to regulation by phosphorylation/dephosphorylation. Swelling of isolated rat hepatocytes activates glycogen synthase in parallel to acetyl-CoA carboxylase and decreases glycogen phosphorylase activity, suggesting an interference of cellular hydration with protein phosphorylation. It was proposed that the activation of glycogen synthase in response to hypoosmotic cell swelling is at least in part due to a lowering of intracellular chloride concentration, which deinhibits glycogen synthase phosphatase. Activation of glycogen synthase by glutamate may contribute to the effects of glutamine-induced but not of hypoosmolarity-induced cell swelling. The swelling-induced stimulation of flux through the pentose phosphate shunt and an increased NADPH provision for glutathione reductase may explain why during oxidative stress the cellular losses of oxidized glutathione are smaller when the hepatocellular hydration state increases.

Hypoosmotic incubation of hepatocytes slightly stimulates lipogenesis from glucose and inhibits carnitine *o*-palmitoyltransferase. Acetyl-CoA carboxylase, a key enzyme in fatty acid synthesis, is activated in response to hypoosmotic and amino acid-induced cell swelling. As with glycogen synthase, activation of acetyl-CoA carboxylase is due to deinhibition of a protein phosphatase, which occurs in response to a lowering of intracellular chloride and/or an increased intracellular concentration of glutamate and aspartate. Lipogenesis is stimulated by glutamine, proline, and alanine, but not by aminoisobutyrate, histidine, and asparagine, and no correlation was detectable between the potency of these amino acids to swell the hepatocytes on the one hand and their potency to stimulate lipogenesis on the other hand. This may suggest that volume changes per se may not play a major role in mediating the lipogenic effect of amino acids. A similar conclusion was derived from studies on ketogenesis. The inhibition of hepatic ketogenesis by glutamine, proline, alanine, and asparagine is probably not due to amino acid-induced cell swelling because no effect on ketogenesis was observed with the nonmetabolizable amino acid analog 2-aminoisobutyrate, despite cell swelling. Furthermore, hypoosmotic exposure stimulates ketogenesis from ketoisocaproate, but not from other ketogenic substrates. However, during ketogenesis, hypoosmotic (hyperosmotic) cell swelling (shrinkage) shifts the mitochondrial $NADH/NAD^+$ system to a more oxidized (reduced) level. This is explained by a stimulation of the respiratory chain, which accompanies mitochondrial swelling.

E. Liver Cell Hydration and Bile Acid Excretion

In the hepatocyte, conjugated bile acids are taken up at the sinusoidal (basolateral) side by a Na^+-dependent carrier and are excreted at the canalicular (apical) membrane by means of a specific transport ATPase.

The canalicular secretion step is rate controlling for overall transcellular bile acid transport. In liver, transcellular taurocholate transport is strongly dependent on cellular hydration: cell swelling stimulates, whereas cell shrinkage inhibits canalicular bile acid secretion, regardless of whether cell volume is modified by anisoosmolarity, amino acids, or insulin. The swelling-induced stimulation of taurocholate excretion into bile is dependent on intact microtubules and due to an increase of transport capacity (V_{max}), which doubles within minutes when hepatocellular hydration increases by about 10%. Evidence has been presented that cell swelling/shrinkage modifies the taurocholate secretion capacity due to a microtubule-dependent insertion/retrieval of canalicular bile acid transporter molecules into/from the canalicular membrane. The swelling-induced stimulation of taurocholate excretion is sensitive to tyrosine kinase inhibitors or G-protein inhibitors. These inhibitors also block the swelling-induced activation of mitogen-activated protein (MAP) kinases in rat hepatocytes, suggesting a causal relationship between swelling-induced MAP kinase activation and bile acid transport.

F. Acidification of Early Endocytotic Vesicles

In liver, cell swelling (shrinkage) leads to a rapid alkalinization (acidification) of intracellular vesicular compartments as revealed in studies on acridine orange fluorescence and the fluorescence of endocytosed fluoresceine isothiocyanate-labeled (FITC-) dextran. The cell volume sensitivity of vesicular pH reflects the response of an early endocytotic compartment (intravesicular pH around 6), but not of more acidic lysosomal compartments (pH around 5). Given the important role of vesicular acidification for receptor–ligand sorting, exocytosis, and protein targeting, interference of cellular hydration with these processes becomes likely. Cell volume also affects receptor-mediated endocytosis: hyperosmotic exposure inhibits galactosyl receptor-mediated, but not fluid phase endocytosis in isolated hepatocytes. The mechanism of how cell volume influences the pH in early endocytotic vesicles is not fully settled. However, it is mediated by microtubule-, G-protein-, and tyrosine-kinase dependent, but Ca^{2+} and cAMP-independent mechanism, which also leads to a swelling-induced activation of MAP kinases. Vesicular acidification requires the presence of a chloride conductance in the vesicular membrane in order to dissipate the membrane potential generated by the H^+ pump and to augment the acidification process. A current working hypothesis suggests that swelling-activated protein kinases mediate the volume sensitivity of vesicular acidification by modulating chloride channel activity.

G. Cellular Hydration and Gene Expression

Cellular hydration also affects cellular metabolism on a long-term time scale by modifying gene expression. This involves not only osmoregulatory genes (whose mRNA levels increase in response to hypertonic stress), such as genes for aldose reductase or for osmolyte transporters such as the Na^+-coupled myoinositol (SMIT) and betaine (BGT1) transporters in renal cells and astrocytes, but also the expression of genes coding for proteins that are not necessarily linked to osmoregulation. Examples for the latter include the hypoosmolarity-induced increases of mRNA levels for β-actin, tubulin, and ornithine decarboxylase, the hyperosmolarity-induced stimulation of cyclooxygenase-2 expression in activated liver macrophages, and the cell volume-dependent expression of tyrosine aminotransferase (TAT) and phosphoenolpyruvate carboxykinase (PEPCK) in liver. TAT and PEPCK mRNA levels markedly increase in response to hyperosmotic cell shrinkage, but decrease in response to cell swelling. Stimulation of proteolysis and induction of enzymes involved in gluconeogenesis (e.g., PEPCK) and amino acid breakdown (e.g., TAT) following cell shrinkage are a paradigm for the coordinated regulation of functionally linked processes by cell volume. Anisoosmotic exposure affects the expression of early immediate genes; examples include the increase of c-*jun* (but not c-*fos*) mRNA levels in response to liver cell swelling and the induction of Egr-1 and c-*fos* mRNA following hyperosmotic treatment of Madin Darby canine kidney (MDCK) cells. Viral replication also depends on host cell hydration. For example, hypoosmotic swelling of duck hepatocytes inhibits replication of duck hepatitis B virus by about 50%, whereas hyperosmotic shrinkage stimulated its replication four- to fivefold.

The mechanisms of how cell volume changes affect gene expression are largely unknown, but changes in the ionic composition, the cytoskeleton, and protein phosphorylation are likely candidates. A hypertonic stress-responsive element has been identified in the 5′-flanking region of the mammalian BGT1 gene (betaine transporter) in MDCK cells; however, the transacting factor(s) remains to be characterized. Hypertonic cell shrinkage in this cell type leads to a protein kinase

C-dependent activation of MAP kinases; however, the role of these protein kinases in inducing the betaine transporter is doubtful. However, the swelling-induced induction of c-*jun* mRNA in hepatoma cells may be due to MAP kinase activation, which occurs within 1 min in these cells in response to hypoosmolarity. Regulation of PEPCK mRNA levels by cellular hydration does not involve protein kinase C activation or changes in cAMP levels, but is sensitive to the protein kinase inhibitor H7. MAP kinases probably play a minor role in the cell volume-dependent regulation of the PEPCK gene.

H. Coupling of Membranes in Epithelial Transport

Transcellular epithelial transport requires the entry of transported substrates across one cell membrane and its extrusion at the other. In face of the usually high transport rates as compared to the respective intracellular pools, mechanisms are required that ascertain the matching of transport across the two opposing cell membranes. In several epithelia, cell volume has been recognized to be an important element in the apical to basal cell membrane coupling, as illustrated by the following two examples.

In intestine and proximal renal tubules, glucose and amino acids are transported by Na^+-coupled uptake across the apical (brush border) cell membrane. Na^+ is extruded in exchange for K^+ by the Na^+/K^+ ATPase at the basolateral cell membrane and K^+ thus accumulated exits the cells through K^+ channels within that cell membrane. Excessive luminal uptake of Na^+ and substrate leads to cell swelling, which then activates the K^+ channels. The activation of the K^+ channels not only serves to limit cell swelling, but helps maintain the electrical driving force for Na^+-coupled transport.

In Cl^- secreting epithelia, activation of K^+ and Cl^- channels during the stimulation of epithelial secretion may lead to cell shrinkage due to cellular KCl loss. The shrinkage then turns on volume regulatory Na-K-2Cl cotransport, which may partially recover cell volume and at the same time supply the cell with further Cl^- for secretion.

I. Excitability

Volume-activated Cl^- currents could depolarize the cell membranes of excitable cells leading to the opening of voltage-gated Ca^{2+} channels, an increase of intracellular Ca^{2+}, and subsequent activation. This sequence of events results in vasoconstriction following the swelling of vascular smooth muscle cells. However, osmotic cell shrinkage leads to vasodilation. Moreover, an enhanced activity of Na^+/H^+ exchanger activity with resulting cell swelling has been implicated in the generation of one type of essential hypertension. On the one hand, cell swelling should enhance the contractility of smooth muscle cells and, on the other hand, enhanced Na^+/H^+ exchange activity should favor cell proliferation and thus hypertrophy of vascular smooth muscle cells.

Cell volume regulatory mechanisms may be similarly important for neuronal excitability. γ-Aminobutyric acid (GABA), for instance, activates K^+ and Cl^- channels. The resulting KCl loss leads to cell shrinkage and to a decrease of intracellular Cl^-. The cell shrinkage activates the Na-K-2Cl cotransport, which not only restores cell volume but prevents a dissipation of the Cl^- gradient. Beyond its potential effect on cell membrane potential and intracellular Ca^{2+} activity, swelling could affect the excitability of neuronal cells by its alkalinizing influence on the pH of secretory granules, which is known to govern transmitter uptake and metabolism. Moreover, the swelling of glial cells results in the release of osmotically active substances, which at the same time serve as neurotransmitters, such as taurine, GABA, glutamate, and aspartate. Possibly as a result of the just-described interactions, increased plasma osmolarity decreases and reduced plasma osmolarity increases the susceptibility to epileptic seizures.

J. Hormone Release

Swelling of pancreatic β cells leads to the activation of unselective ion channels with subsequent depolarization, activation of voltage-gated Ca^{2+} channels, entry of Ca^{2+}, and release of insulin. Furthermore, cell swelling has been shown to trigger the release of prolactin, gonadotropin-releasing hormone, luteinizing hormone, thyrotropin, aldosterone, atrial natriuretic factor, and renin.

In contrast to those hormones, vasopressin is released during cell shrinkage, which apparently leads to disinhibition of a stretch-inactivated cation channel. The activation of this channel leads to depolarization and accelerated action potentials. In addition to vasopressin, the release of nitric oxide (NO) and histamine has been shown to be stimulated by an increase of ambient osmolarity.

K. Migration

The migration of leukocytes is stimulated by chemoattractants such as formylpeptides, which lead to polarization and microfilament reorganization of the cells. *n*-Formyl-methionyl-leucyl-phenylalanine (FMLP) stimulates Na^+/H^+ cotransport in neutrophils, leading to cell swelling. Inhibition of Na^+/H^+ exchange impedes migration. Similarly, inhibition of Na-K-2Cl cotransport with bumetanide inhibited the migration of transformed MDCK cells. In those cells, migration further requires the operation of K^+ channels, which are activated by oscillating intracellular Ca^{2+} activity. Inhibition of these channels similarly prevents migration. It is attractive to postulate that migration involves a volume regulatory decrease at the tail and a volume regulatory increase at the leading edge of a migrating cell.

L. Cell Proliferation

A myriad of mitogenic factors have been shown to activate Na^+/H^+ exchange and/or Na-K-2Cl cotransport, and in a wide variety of cells proliferation has been observed to be paralleled by an increase of cell volume. In fibroblasts, *ras* oncogene expression was similarly paralleled by enhanced Na^+/H^+ exchange and Na-K-2Cl cotransport activity, leading to an increase of cell volume by about 30%. The growth factor-independent proliferation of the *ras* oncogene expressing cells is sensitive to amiloride and furosemide, i.e., to blockers of Na^+/H^+ antiport and Na-K-2Cl cotransport, suggesting a role of cell swelling induced by activation of these transporters for cell proliferation. In lymphocytes, mitogenic signal also activate these transporters and may shift the set point of cell volume regulation to higher resting values, which may be an important prerequisite for cell proliferation.

When exposed to bradykinin, bombesin, or serum, *ras* oncogene expressing cells respond to bradykinin with oscillations of cell membrane potential secondary to oscillations of intracellular Ca^{2+} activity. These oscillations of intracellular Ca^{2+} activity lead to a depolymerization of the actin filaments, which presumably accounts for the set point shift of cell volume regulation. Obviously, cell division at some point requires gain of cell volume. It has not yet been determined whether the increase of cell volume is important for the stimulation of protein synthesis, inhibition of proteolysis, alkalinization of lysosomal vesicles, etc.

IV. CELL VOLUME SENSING

Little is known about the structures sensing the changes in cell hydration. Because cell volume is a physical property of the cell, sensing should occur physically and/or mechanically. Physical volume sensing could involve water shift-dependent concentration changes of one or more intracellular constituents, which may act to regulate volume regulatory transport systems and/or intracellular signaling pathways. One model postulates that the extent of macromolecular crowding, i.e., the cytosolic protein concentration, will determine the tendency of intracellular macromolecules to associate with the plasma membrane and consequently their enzymatic activity. It is conceivable that cellular hydration may in such a way interfere with the activity of protein kinases and phosphatases and that changes in protein phosphorylation may trigger both; volume-regulatory responses and alterations in cellular metabolism and gene expression. In yeast, histidine kinases, which are putative integral membrane proteins, may act as osmosensors with the signal being transduced by autophosphorylation and subsequent phosphate transfer to an aspartate residue in the receiver domain of a cognate response regulator molecule in order to regulate a MAP kinase-like protein kinase cascade. Candidates for mechanical cell volume sensing are the cytoskeleton, ion conductance regulator proteins (e.g., pI_{Cln}), and stretch-regulated ion channels. The molecular mechanisms of stretch activation of these channels are still unclear, but may involve the liberation of fatty acids from the membrane and interactions with the cytoskeleton as initial events.

V. CELL VOLUME SIGNALING

Current understanding of the intracellular signaling events which couple cell hydration to cell function is incomplete and complicated by the fact that cell volume signaling may depend on the cell type under study and the mechanism of how cell swelling is achieved (e.g., hypoosmotic versus amino acid-induced swelling). For example, in jejunal enterocytes the RVD in response to cumulative substrate uptake is sensitive to inhibitors of protein kinase C, whereas the RVD following hypoosmotic exposure is not.

Depending on the cell type under study, various protein kinases and phosphatases may participate in cell volume signaling. Current interest focuses on the regulation of mitogen-activated protein kinases and related protein kinases, such as Jnk by osmotic stress. Hyperosmotic stress activates MAP kinases in yeast and MDCK cells, whereas MAP kinases are activated in response to hypoosmotic cell swelling in rat hepatoma cells, rat hepatocytes, the human intestine 407 cell line, and primary astrocytes. A signal transduction sequence, which is initiated by the osmotic water shift across the plasma membrane and ultimately leads to changes in cell function, has been identified in rat hepatoma and liver cells (Fig. 3). Here, hypoosmotic cell swelling results within 1 min in a pertussis-toxin, cholera toxin-, and genistein-sensitive, but protein kinase C- and Ca^{2+}-independent phosphorylation of the MAP kinases Erk-1 and Erk-2, suggesting that liver cell swelling leads to a G-protein-mediated activation of a yet unidentified tyrosine kinase, which acts to activate a pathway toward MAP kinases. Interruption of this signaling sequence at the level of G-proteins or tyrosine kinase not only prevents hypoosmotic MAP kinase activation, but also the swelling-induced alkalinization of endocytotic vesicles and stimulation of bile acid excretion. This swelling-activated signaling cascade resembles that triggered by growth factor receptor activation and may explain why cell swelling acts like an anabolic signal in liver with respect to protein and carbohydrate metabolism. MAP kinases have multiple protein substrates, such as the microtubule-associated proteins MAP-2 and Tau, and other protein kinases, such as S6 kinase, which could transduce cell volume effects on protein and carbohydrate turnover. The link, however, is not yet established. The swelling-induced activation of MAP kinases is followed by an increased phosphorylation of c-Jun, which may explain—due to autoregulation of the *c-jun* gene—the increase in *c-jun* mRNA levels 30 min after the onset of cell swelling. However, in addition to Erk-1 and Erk-2, other Jun kinases may also be activated by cell swelling. A swelling-induced phosphorylation of transcription factors, such as c-Jun, may explain the influence of cell hydration on gene expression. A new subfamily of protein kinases, the stress-activated protein (SAP) kinases, has been described. SAP kinases are activated by different forms of intra- and extracellular stress and act as c-Jun ki-

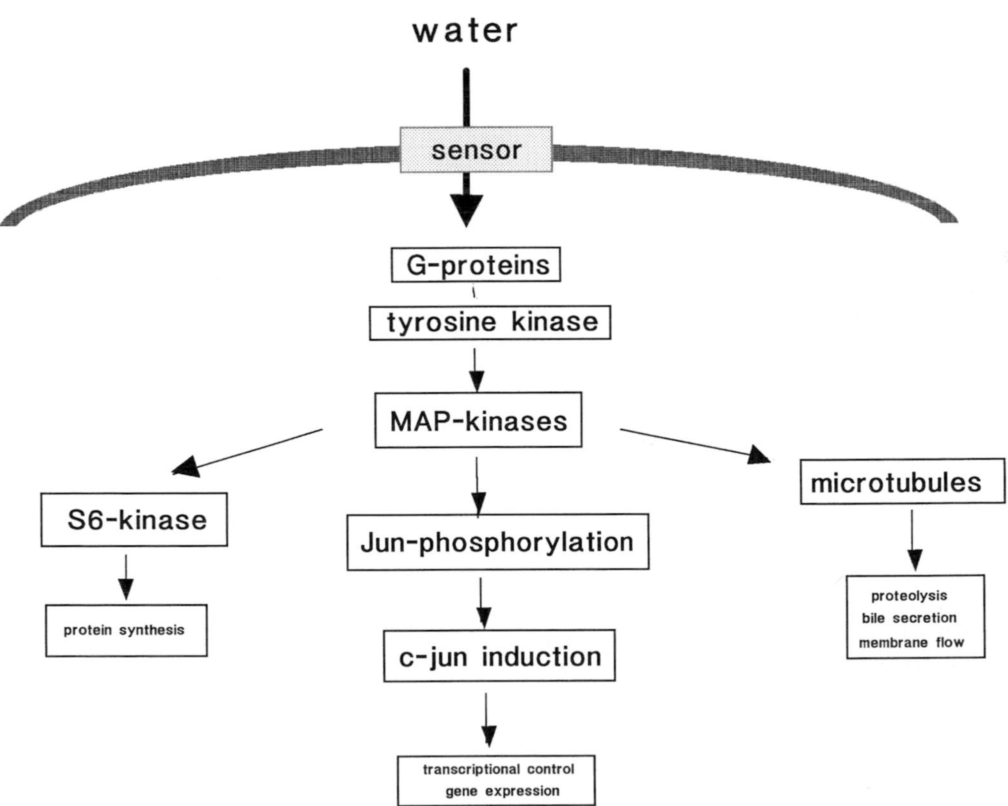

FIGURE 3 Cell volume signaling in the hepatocyte. The hypothetical scheme focuses on the role of MAP kinases, but it should be kept in mind that the signaling sequence depicted is by far incomplete.

nases; they are activated also by hyperosmotic stress. Also, protein phosphatases participate in the regulation of cell function by cell hydration: hypoosmotic hepatocyte swelling lowers the intracellular chloride concentration, thereby leading to a deinhibition of glycogen synthase phosphatase.

Microtubules apparently play an important role in transducing some metabolic alterations in response to changes of cellular hydration as suggested by the colchicine sensitivity of the swelling-induced alkalinization of endocytotic vesicles, the inhibition of proteolysis, and the stimulation of transcellular bile acid transport. It remains to be established to what extent changes in the phosphorylation of microtubule-associated proteins are involved in the microtubule- and MAP kinase-dependent cell volume signaling. However, there are also cell volume-sensitive pathways (e.g., glycine oxidation, pentose phosphate shunt) that are not affected by microtubule disruption. Cell swelling leads within 1 min to an increased polymerization state of β-actin and increases the stability of microtubules; however, the relevance of actin filaments for cell volume signaling in liver is unknown. Whereas hypoosmotic exposure is followed by an increase of intracellular Ca^{2+} in astrocytes and other cell types, this has not been observed consistently in hepatocytes. The significance of other phenomena accompanying cell swelling, such as a decrease of pH_i, transient cell membrane hyperpolarization, and stimulation of inositol-1,4,5-trisphosphate formation, for cell volume signaling on metabolism is not yet clear.

VI. CLINICAL ASPECTS

Only a few aspects are addressed. Contemporary clinical medicine pays careful attention to the hydration state of the extracellular space, but not enough to cellular hydration, probably because of the lack of routinely applicable techniques for the assessment of cell volume in patients. However, it should be kept in mind that cell hydration is determined primarily by the activity of ion and substrate transporting systems in the plasma membrane and to a minor extent by the hydration state of the extracellular space. The role of cell hydration in regulating protein turnover is an important one, partly because it has a direct bearing on the problem of the pathogenesis of protein catabolic states in the severely ill. Indeed, a close relationship between cell hydration in skeletal muscle and whole body nitrogen balance has been established in patients, irrespective of the underlying disease. The working hypothesis put forward is that cell shrinkage in liver and skeletal muscle triggers the protein catabolic state found in various diseases. Although this implies that the degree of cell hydration determines the extent of nitrogen wasting, the pathogenetic mechanisms leading to cell shrinkage may well be multifactorial and could involve disease-specific components.

New evidence shows that hepatic encephalopathy is a primary disorder of astroglia. Ammonia induces glial swelling as the result of glutamine accumulation in astrocytes. Both glial swelling and brain edema are recognized as major events leading to brain dysfunction following acute ammonia intoxication or acute fulminant liver failure. However, astrocyte swelling may also be of pathophysiological significance in chronic hepatic encephalopathy. As demonstrated in humans *in vivo* by proton nuclear magnetic resonance spectroscopy, an early finding in chronic hepatic encephalopathy is the cerebral loss of an osmosensitive myoinositol pool, which is released from astrocytes in response to swelling. It was hypothesized that swelling-induced alterations in glia function may trigger disturbances of glial–neuronal communication, thereby provoking the clinical picture of chronic hepatic encephalopathy.

BIBLIOGRAPHY

Chamberlin, M. E., and Strange, K. (1989). Anisosmotic cell volume regulation: A comparative view. *Am. J. Physiol.* **257,** C159–C173.

Garcia-Perez, A., and Burg, M. B. (1991). *Physiol. Rev.* **71,** 1081–1115.

Halestrap, A. P. (1989). *Biochim. Biophys. Acta* **973,** 355–382.

Häussinger, D., Gerok, W., and Lang, F. (1994). *Am. J. Physiol.* **267,** E343–E355.

Häussinger, D., and Lang, F. (1992). Cell volume and hormone action. *Trends Pharmacol. Sci.* **13,** 371–373.

Lang, F., and Häussinger, D., eds. (1993). "Interaction of Cell Volume and Cell Function." Springer-Verlag, Heidelberg.

Sarkadi, B., and Parker, J. C. (1991). Activation of ion transport pathways by changes in cell volume. *Biochim. Biophys. Acta* **1071,** 407–427.

Strange, K., ed. (1994). "Cellular and Molecular Physiology of Cell Volume Regulation." CRC Press, Boca Raton, FL.

Cell Volume, Regulation

FLORIAN LANG
GILLIAN L. BUSCH
University of Tübingen

DIETER HÄUSSINGER
Heinrich-Heine University of Düsseldorf

GLOSSARY

Compatible osmolytes Organic substances, such as certain polyols, methylamines, and amino acids, that do not significantly perturb protein function over a wide range of concentrations and that are thus utilized by cells to create intracellular osmolarity

Macromolecular crowding High concentrations of macromolecules (macromolecular crowding) influence the kinetics and equilibria of many biochemical reactions. Cellular hydration is thought to modify cellular function in part by its effect on the concentration of cellular macromolecules

Osmotic pressure gradient across a cell membrane Difference of osmotic activity between the intracellular and extracellular fluid creates an osmotic pressure gradient ($\Delta\pi$) defined as: $\Delta\pi = R \cdot T \cdot \Sigma\sigma \cdot \Delta c$, where R is the gas constant, T the temperature in °K, σ the reflection coefficient, and Δc the concentration difference of any substance across the cell membrane. The reflection coefficient describes the relative permeability of the substance and usually varies between 0 (fully permeable) and 1 (fully impermeable)

Regulatory cell volume decrease (RVD) If swollen beyond the volume regulatory set point, cells decrease cellular osmolarity by metabolism and extrusion of ions and organic osmolytes along with osmotically obliged water. As a result, the cells decrease cell volume toward the set point

Regulatory cell volume increase (RVI) If shrunken below the volume regulatory set point, most cells increase cellular osmolarity by metabolism and cellular accumulation of ions and organic osmolytes along with osmotically obliged water. As a result, the cells increase cell volume toward the set point

Stretch-activated ion channel Ion channel that is directly or indirectly activated by cell membrane stretch, for example, during osmotic cell swelling

Stretch-inactivated ion channel Ion channel that is directly or indirectly inactivated by cell membrane stretch and that is activated by decrease of cell membrane tension, for example, during osmotic cell shrinkage

Volume regulatory set point Range of cell volume, where the cell volume regulatory mechanisms are silent or their net effect on cellular osmolarity is zero. If the actual cell volume is above the set point, the cells display RVD, if actual cell volume is below the set point, the cells display RVI

THE CONSTANCY OF CELL VOLUME IS CONTINuously challenged by alterations of extracellular and/or intracellular osmolarities which are followed by respective water fluxes across the cell membrane. Any alteration of cell volume jeopardizes the constancy of the intracellular milieu and cell function by altering the concentrations of all intracellular components. Beyond that, excessive cell swelling eventually leads to disruption of the cell membrane and thus to cell death. Thus, a most obvious prerequisite of cell survival is the avoidance of excessive cell volume alterations. To defend the constancy of their volume, cells have developed a wide variety of cell volume regulatory mechanisms, including ion transport across the

ENCYCLOPEDIA OF HUMAN BIOLOGY, Second Edition, VOLUME 2. Copyright © 1997 by Academic Press. All rights of reproduction in any form reserved.

cell membrane, generation or disposal of organic osmolytes, and regulation of cellular metabolism. These mechanisms are triggered by alterations of cell volume of only a few percent and usually act in concert to maintain cell volume within narrow limits. Nevertheless, cell volume regulation may be incomplete, leaving small deviations from set point cell volume. Moreover, the set points of cell volume regulatory mechanisms may be shifted by hormones and other substances, leading to respective alterations of cell volume. Thus, cells are frequently subjected to minor alterations of cell volume, which in turn affect a myriad of cell volume-sensitive cellular functions. In this article, a synopsis of cell volume regulatory mechanisms and volume-sensitive cellular functions will be followed by a discussion of factors challenging the constancy of cell volume. In a related article, cellular functions modified by cell volume will be reviewed. [*See* Cell Volume, Physiological Role.]

I. CELL VOLUME REGULATORY MECHANISMS

With very few exceptions (e.g., the apical cell membrane of the thick ascending limb of Henle's loop), mammalian cell membranes are highly permeable to water, which follows a hydrostatic and osmotic pressure gradient. Water may permeate by diffusion or through water channels. The rigidity of animal cell membranes is too small to allow the buildup of significant transmembrane hydrostatic pressure gradients. Thus, the crucial determinant of transmembrane water flux and thus of cell volume is an osmotic gradient across the cell membrane. Such a gradient can arise from any alteration of extracellular and intracellular osmolarity. Volume regulatory mechanisms serve to dissipate the osmotic gradient and thus to prevent or reverse changes of cell volume.

A. Volume Regulatory Ion Transport

Most extracellular and intracellular osmolarity is created by ions. Thus, ion movement is the most efficient means to alter osmotic gradients across the cell membrane. Ions are used to counterbalance the osmotic disequilibrium created by cellular accumulation of organic substances, and are thus important to maintain cell volume in steady state. Moreover, ion transport across the cell membrane is utilized to counteract rapid perturbations of intra- and extracellular osmolarity. [*See* Cell Membrane Transport.]

1. Ions in the Maintenance of Steady-State Volume

Ions are required for the maintenance of steady-state volume, since in pursuing their metabolic tasks cells accumulate a number of osmotically active substances, such as amino acids. The excess cellular osmolarity must be counterbalanced by uneven distribution of ions: the Na^+/K^+-ATPase extrudes Na^+ in exchange for K^+. The latter tends to exit the cell through K^+ channels, creating a cell-negative potential across the cell membrane. This cell membrane potential drives anions such as Cl^- out of the cell. As a result, intracellular Cl^- concentration is below that of the extracellular space, the difference amounting to some 40 to 100 m*M* for most cells. The low intracellular Cl^- concentration allows the accumulation of osmotically active organic substances. It should be kept in mind that a cell membrane potential of only −18 mV is required to reduce intracellular Cl^- concentration to half of the extracellular Cl^-, that is, to allow accumulation of 55 mmol/liter of osmotically active organic substances. In most cells, the cell membrane potential is more hyperpolarized than required for maintenance of osmotic equilibrium, and intracellular Cl^- concentration is higher than thermodynamic equilibrium.

In most cells, inhibition of Na^+/K^+-ATPase by ouabain leads eventually to cell swelling, as outlined in the following. As shown for hepatocytes, however, cell volume can be maintained during inhibition of Na^+/K^+-ATPase by HCl sequestration into acidic intracellular vesicles, which are subsequently expelled by exocytosis. The HCl accumulation is thought to be achieved by H^+-ATPase in parallel to Cl^- channels. At least theoretically, H^+-secreting cells, such as the intercalated cells of the renal collecting duct, could similarly maintain their volume with electrogenic H^+ extrusion (H^+-ATPase) in parallel to Cl^- channels or H^+/K^+-ATPase in parallel to K^+ and Cl^- channels.

2. Volume Regulatory Decrease (RVD)

Volume regulatory ion fluxes are the most rapid means to restore osmotic equilibrium across the cell membrane (Figs. 1 and 2). Following cell swelling, cells have to release ions to decrease intracellular osmolarity (Fig. 1). Most cells release K^+ and Cl^- upon swelling through the activation of K^+ channels and/or anion channels. Anion channels allow the passage not only of Cl^- but also of other anions such as HCO_3^- and even organic substances, such as taurine and negatively charged amino acids.

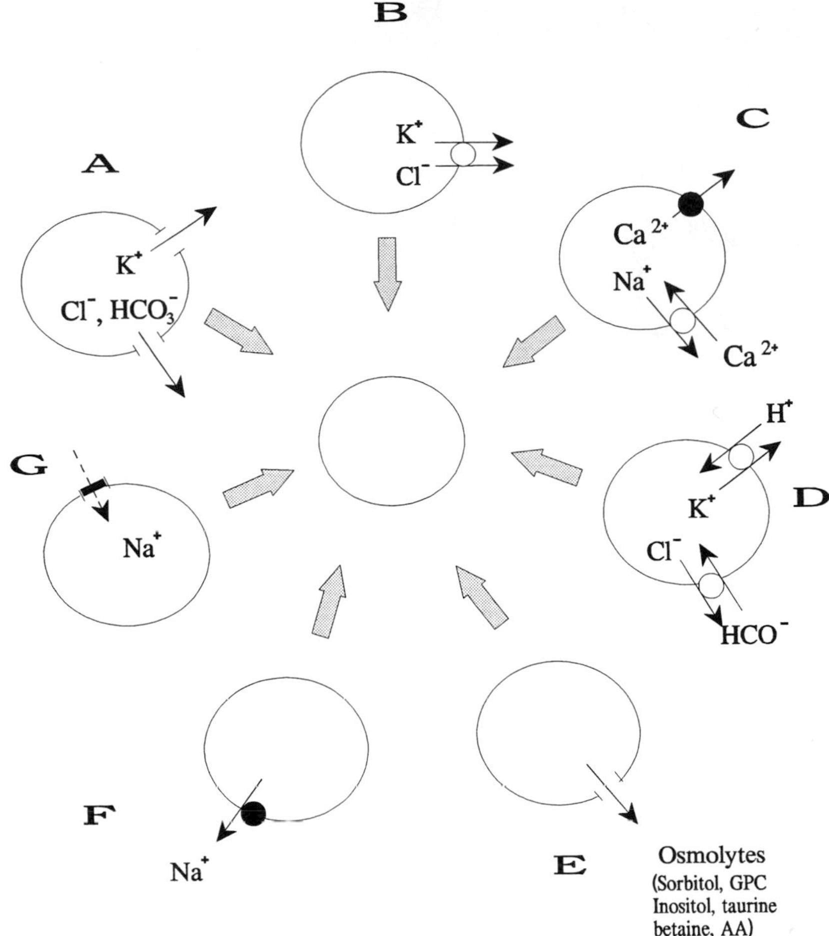

FIGURE 1 Synopsis of some cellular mechanisms allowing for regulatory cell volume decrease. A: Activation of K^+ channels and unselective anion channels (allowing the passage of Cl^- and HCO_3^- among other anions); B: KCl cotransport; C: Ca^{2+}-ATPase in parallel to Na^+/Ca^{2+} exchange; D: K^+/H^+ exchange in parallel to Cl^-/HCO_3^- exchange; E: release of organic osmolytes (sorbitol, glycerophosphorylcholine, inositol, taurine, betaine); F: Na^+-ATPase; G: inhibition of Na^+ channels.

According to conventional electrophysiology and patch-clamp analysis, a wide variety of different K^+ and Cl^- channels are activated. Obviously many different channel proteins from different families are utilized for cell volume regulation. Accordingly, several different proteins have been cloned and are considered to serve as cell volume regulatory ion channels.

Among the cloned K^+ channels invoked to serve cell volume regulation are the Kv1.3 (n-type K^+ channel), the Kv1.5, and the minK channels.

Cloned Cl^- channels originally invoked to serve RVD include the ClC-2 channel, BRI-VDAC, I_{Cln}, and the P-glycoprotein (or MDR protein). Alternatively, P-glycoprotein and I_{Cln} were suggested to regulate the volume regulatory Cl^- channel and expression of P-glycoprotein was shown to be stimulated by hypertonicity. However, the role of P-glycoprotein in cell volume regulation has been questioned. In any case, many of the properties of cell volume regulatory anion channels are not explained by the known cloned channels but rather additional anion channels must be operative.

Some cells release ions through KCl symport or parallel K^+/H^+ exchange and Cl^-/HCO_3^- exchange (leading to KCl loss). In Na^+-rich dog red blood cells, swelling leads to extrusion of Na^+ by reversal of Na^+/Ca^{2+} exchange in parallel to Ca^{2+}-ATPase.

Usually more cations (K^+ and Na^+) are lost from

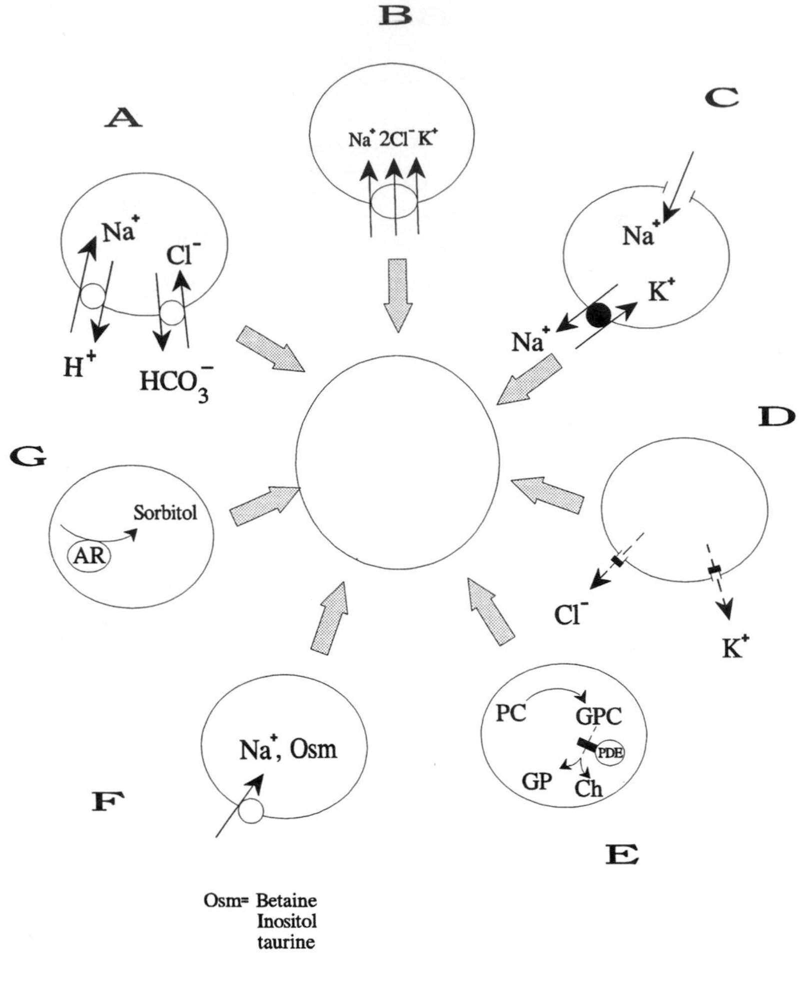

FIGURE 2 Synopsis of some cellular mechanisms allowing for regulatory cell volume increase. A: Parallel activation of Na^+/H^+ exchange; B: $Na^+,K^+,2Cl^-$ cotransport; C: Na^+ channels (in parallel to Na^+/K^+-ATPase); D: inhibition of K^+ channels and Cl^- channels; E: inhibition of degradation of glycerophosphorylcholine (GPC); F: accumulation of the osmolytes betaine, inositol, and taurine by Na^+ coupled transport; G: formation of sorbitol by activation of aldose reductase.

cells than Cl^-. The gap is at least partially due to loss of HCO_3^-. Most HCO_3^- lost is replaced by CO_2, and the H^+ thus generated is bound to intracellular buffers. Thus, the exit of HCO_3^- is limited by the intracellular buffer capacity. The HCO_3^-, which is replaced by CO_2, does not directly contribute to cell volume regulation but allows the cellular loss of K^+.

3. Volume Regulatory Increase (RVI)

Following cell shrinkage, cells accumulate ions through $Na^+,K^+,2Cl^-$ cotransport, parallel operation of Na^+/H^+ exchange and Cl^-/HCO_3^- exchange (leading to gain of NaCl), and opening of Na^+ channels (Fig. 2). The parallel activation of Na^+/K^+-ATPase leads to replacement of thus accumulated Na^+ with K^+. Ion loss through ion channels is minimized by inhibition of K^+ and/or Cl^- channels.

Four members of the Na^+/H^+ exchanger family have been cloned. Among those, NHE-1, NHE-2, and NHE-4 are stimulated, whereas NHE-3 is inhibited by cell shrinkage. The putative volume-sensitive site at the NHE-1 molecule has been identified and is apparently distinct from the sites regulated by Ca^{2+} and growth factors.

Several members of the volume regulatory Na^+,K^+, $2Cl^-$ cotransporters have been cloned.

Some cells do not undergo RVI during exposure to hypertonic extracellular fluid. The same cells, if exposed to hypotonic extracellular fluid, show RVD and, if reexposed to isotonic fluid, first shrink and then display RVI (secondary RVI or RVI on RVD). In these cells, primary RVI may be prevented by increased intracellular Cl^- activity, as detailed in Section II,I.

B. Osmolytes

The cellular accumulation of electrolytes to counterbalance enhanced extracellular osmolarity is limited because of the effect of ion concentration on protein structure and function. Moreover, alterations of ion gradients across the cell membrane would affect the respective transporters: an increase of intracellular Na^+ activity, for instance, might reverse Na^+/Ca^{2+} exchange and thus increase intracellular Ca^{2+} activity, which would in turn affect a multitude of cellular functions.

To circumvent the untoward effects of disturbed ion composition, cells produce "compatible osmolytes," organic molecules specifically designed to create osmolarity without compromising other functions of the cells.

Three groups of osmolytes are used in mammalian cells: polyalcohols, such as sorbitol and inositol; methylamines, such as glycerophosphorylcholine and betaine; and amino acids and amino acid derivatives. The osmolytes are specifically important for cell volume regulation in the renal medulla, where extracellular osmolarity may reach fourfold isotonicity, and in the brain, where cell volume alterations cannot be tolerated owing to the rigid skull and where alterations of ion composition would affect excitability.

Cellular osmolyte accumulation can be achieved by stimulated uptake, enhanced formation, or decreased degradation. Decrease of intracellular osmolyte concentration is accomplished by degradation or release.

1. Glycerophosphorylcholine

Glycerophosphorylcholine (GPC) is formed by deacylation of phosphatidylcholine. The reaction is catalyzed by a phospholipase A_2, which is distinct from the arachidonyl-selective enzyme. GPC is broken down by the GPC:choline phosphodiesterase, which degrades GPC to glycerol-phosphate and choline. Increase of osmolarity by extracellular addition of either NaCl or urea inhibits the phosphodiesterase and thus leads to accumulation of GPC. GPC is released by swollen cells through a mechanism that does not require increase of intracellular Ca^{2+} activity.

2. Sorbitol

Sorbitol is produced from glucose under the catalytic influence of aldose reductase. Cell shrinkage leads to increase of cellular ionic strength, which stimulates the aldose reductase transcription rate, leading to increase of the respective mRNA. With increasing activity of this enzyme, sorbitol formation is enhanced. Osmolarity does not affect mRNA stability or enzyme degradation. Cell swelling stimulates the release of sorbitol through putative channels, which are thought to be inserted into the cell membrane by fusion of vesicles. The fusion is triggered by increase of intracellular Ca^{2+} activity.

3. Inositol, Betaine, and Taurine

Myoinositol (inositol), betaine, and taurine are taken up into cells by Na^+-coupled transport. The transporters for each of these osmolytes have been cloned. Increased cellular ionic strength but not urea stimulates the transcription of the transporters and thus cellular inositol accumulation. Similar to sorbitol, inositol, betaine, and taurine are rapidly released from swollen cells.

4. Amino Acids

Besides betaine and taurine, a wide variety of amino acids are utilized for cell volume regulation, including glutamine, glutamate, (*n*-acetyl)aspartate, γ-aminobutyric acid (GABA), β-alanine, glycine, serine, proline, and threonine. Although the intracellular concentration of most individual amino acids is quite low, the sum of the amino acids significantly contributes to cellular osmolarity in cells exposed to isotonic extracellular fluid. Thus, amino acids are utilized as osmolytes. Cell volume regulates the intracellular concentration of amino acids by modulation of both metabolism and transport.

Cell shrinkage stimulates proteolysis and inhibits protein synthesis and, conversely, cell swelling inhibits proteolysis and stimulates protein synthesis. Furthermore, cell swelling stimulates breakdown of glutamine and glycine as well as cellular release of several amino acids. Accordingly, cellular amino acid concentration increases upon cell shrinkage and decreases upon cell swelling. The amino acids are probably important during adaptation to minor changes of extracellular osmolarity, however, their contribution is negligible for the adaptation to the excessive osmolarities in the kidney medulla.

5. Other Substances

Besides amino acids, numerous organic metabolites contribute to cellular osmolarity. Several metabolic pathways known to be sensitive to cell volume may modify the concentrations of these metabolites and thus contribute to cell volume regulation. Cell swelling increases glycogen synthesis and inhibits glycolysis, thus decreasing the concentrations of glucose-phosphate and its metabolites, for example, of the glycolytic pathway. Furthermore, cell swelling has a stimulatory, albeit weak, effect on lipogenesis. As detailed in the following, cell volume changes interfere with a great number of other metabolic functions, which to some extent may modify cellular osmolarity. The overall impact of these effects on cellular osmolarity is probably modest. However, the influence of cell volume on several metabolic pathways is of paramount importance for regulation of metabolic function.

C. Further Metabolic Pathways Sensitive to Cell Volume

The influence of cell volume is not restricted to the metabolism of macromolecules and of osmolytes, as outlined in the foregoing, but also modifies a variety of other metabolic functions. Cell swelling inhibits glucose 6-phosphatase activity and glycolysis, stimulates flux through the pentose phosphate pathway, and enhances glutathione (GSH) efflux into blood. It inhibits the release of GSSG (oxidized glutathion). Cell swelling stimulates glycine oxidation, glutamine breakdown, and formation of NH_4^+ and urea from amino acids, but inhibits urea synthesis from NH_4^+. Other metabolic functions stimulated by cell swelling include ketoisocaproate oxidation, acetyl-CoA-carboxylase activity, and lipogenesis from glucose.

Most of these metabolic functions are influenced in the opposite direction by cell shrinkage. As mentioned in Section II,M, the expression of several genes is regulated by cell volume.

II. INTRACELLULAR SIGNALING OF CELL VOLUME REGULATION

A. Macromolecular "Crowding"

Cell swelling leads to dilution and cell shrinkage to concentration of cellular proteins. The concentration of intracellular proteins, on the other hand, markedly influences their function. In erythrocytes, the volume regulatory set point could indeed be varied by manipulation of intracellular protein concentration. The set points of both KCl symport and Na^+/H^+ exchange appear to be determined by macromolecular "crowding." It has been suggested that among the enzymes, sensitive to ambient protein concentration, is a kinase that is inactivated by protein dilution during cell swelling and activated by protein "crowding" during cell shrinkage. As detailed in the following, this kinase may inhibit the volume regulatory KCl cotransport. Its inactivation during cell swelling would then reverse the inhibition of volume regulatory KCl efflux.

Because of interaction of proteins with ambient electrolytes, macromolecular crowding is reduced by increasing ionic strength, which indeed shifts the volume regulatory set point to smaller volumes. Similarly, urea decreases the stability of proteins and thus reduces macromolecular "crowding." Indeed, urea does activate erythrocyte KCl transport and Na^+/Ca^{2+} exchange, inhibits the Na^+/H^+ exchanger in erythrocytes and thick ascending limb cells, and activates hepatocyte K^+ channels leading to cell shrinkage. In erythrocytes, the effects of urea were reversed by okadaic acid, pointing to involvement of phosphorylation.

B. Cytoskeleton

1. Actin Filaments

Cell volume modifies actin polymerization and expression of β-actin and tubulin (see the following). Regulatory cell volume decrease is inhibited in several tissues by cytochalasin B, which inhibits actin assembly. Thus, an intact actin filament network is required for activation of at least some of the volume regulatory mechanisms.

One putative target of the actin filament network is the K^+ channels, which cannot be activated in isolated membrane vesicles devoid of cytoskeleton. It has also been speculated that the cytoskeleton may participate in the insertion of channels into the plasma membrane, in the regulation of channels by kinases (see Section II,H) and in the activation of channels by membrane stretch (see Section II,C).

Furthermore, the cytoskeleton is thought to be involved in the activation of the Na^+/H^+ exchanger, which does contain putative cytoskeleton binding sites. Actin depolymerization activates the $Na^+,K^+,2Cl^-$ cotransporter and, in vesicles devoid of cytoskeleton, the $Na^+,K^+,2Cl^-$ cotransporter is permanently active.

2. Microtubules

Cell swelling stabilizes the microtubule network and increases the expression of tubulin. Colchicine, which disrupts the microtubule network, inhibits regulatory cell volume decrease in some, but not all, cells tested. Microtubules may be involved in the activation of Cl^- channels. Beyond that, microtubules are critical for the mediation of the effect of cell volume on acidic cellular compartments, cellular proteolysis, and bile acid transport, as detailed in Section II,L. [*See* Microtubules.]

C. Cell Membrane Stretch

A variety of ion channels are activated by cell membrane stretch (Table I). These channels may be selective for K^+ or less frequently for anions, thus directly serving cell volume regulation. Most of these channels, however, are nonselective cation channels, allowing the passage of K^+, Na^+, and Ca^{2+}. Because of the cell negative membrane potential, the respective electrochemical gradients favor the cellular accumulation of Na^+ and Ca^{2+} rather than cellular loss of K^+. Thus, these channels are not likely to directly serve cell volume regulation. Ca^{2+} entering the cells through these channels, however, is thought to activate Ca^{2+}-sensitive K^+ and/or Cl^- channels.

It is debatable whether or not stretch-activated channels participate in the fine-tuning of cell volume, since considerable stretch is required to activate these channels. These channels may possibly represent a last line of defense but are not involved in the response to moderate changes of cell volume.

In antidiuretic hormone (ADH)-secreting cells, certain unselective cation channels respond to the cell membrane stretch with inactivation (stretch-inactivated channels). The shrinkage of these cells during increase of extracellular osmolarity reverses the inhibition of these channels, leading to depolarization, accelerated action potentials, and release of ADH.

D. Cell Membrane Potential

The influence of cell swelling on cell membrane potential depends on the ion channels preferably activated and on the potential difference prior to cell swelling. Prevailing activation of K^+ channels and a low starting cell membrane potential favor hyperpolarization (e.g., hepatocytes), whereas prevailing activation of anion or nonselective cation channels and a high starting cell membrane potential favor depolarization (e.g., Ehrlich ascites tumor cells, lymphocytes, pancreatic β-cells, astrocytes, opossum kidney cells, and neuroblastoma cells). In some cells, a transient hyperpolarization due to activation of K^+ channels is followed by a more sustained depolarization due to activation of anion channels (e.g., Madin Darby Canine kidney, MDCK, cells).

The alteration of cell membrane potential may influence the activity of further ion channels. A depolarization of the cell membrane may open voltage-sensitive ion channels, such as n-type K^+ channels (in lymphocytes) and voltage-sensitive Ca^{2+} channels. The latter allow entry of Ca^{2+} (see Section II,F).

E. Cytosolic pH

Cell swelling leads to cytosolic acidification, which has been explained by exit of HCO_3^- through the anion channels, by release of H^+ from acidic intracellular compartments, and by enhancement of Cl^-/HCO_3^- exchange due to decreasing cellular Cl^- activity.

Cell shrinkage is frequently observed to alkalinize cells, an effect mainly due to activation of Na^+/H^+ exchange.

The role of cytosolic pH in cell volume regulation has not been explored. It may at least contribute to inhibition of glycolysis and thus to the decreased release of lactic acid during cell swelling.

F. Ca^{2+}

Following cell swelling, Ca^{2+} increases in a variety of cells, whereas it seemingly remains constant in others. Swelling may increase intracellular Ca^{2+} by both activation of Ca^{2+}-permeable channels in the cell membrane and Ca^{2+} release from intracellular stores. Ca^{2+}-permeable channels may be triggered by cell membrane stretch and/or cell membrane depolarization (see the foregoing), and Ca^{2+} release from intracellular stores is presumably stimulated by inositol phosphates.

The increase of intracellular Ca^{2+} may, with or without interaction with calmodulin, activate Ca^{2+}-sensitive K^+ channels and/or Cl^- channels and thus trigger regulatory cell volume decrease. However, in a variety of tissues, increase of intracellular Ca^{2+} was

TABLE I

Factors Altering Cell Volume[a]

Factor	Example cells
Factors leading to cell swelling	
Insulin	Hepatocytes
Mitogens	Variety of cells
Growth hormone	Chondrocytes
Interleukin	Lymphocytes
Bradykinin	Hepatocytes
Antidiuretic hormone (ADH, AVP)	Glial cells
Glucocorticoids	Hepatocytes, fibroblasts
Mineralocorticoids	Leukocytes
Estrogens	Astrocytes, parathyroid glands
Progesterone	Astrocytes, parathyroid glands
Testosterone	Parathyroid glands
Somatostatin	Colon cells
Adenosine	Erythrocytes
α-Adrenergic stimulation	Hepatocytes
β-Adrenergic stimulation	Erythrocytes, salivary glands, sweat glands
Methacholine	Sweat glands
Glutamate	Glial cells, neurons
Kainate	Neurons
NMDA	Brain
Aspartate	Neurons
Deoxyadenosine	Lymphoblastoid cells
cAMP	Sweat glands
cGMP	Barnacle muscle
Arachidonic acid	Glial cells
ras oncogene	Fibroblasts
Phorbol esters	Necturus gallbladder
Genistein	Tumor cells
Superoxide (O_2^-)	Erythrocytes
Amino acid uptake	Hepatocytes, proximal renal tubule, intestine cells
Glucose uptake	Necturus gallbladder, proximal renal tubule, intestine cells, vascular smooth muscle, mesangial cells
Increase of K_0^+	Hepatocytes, gallbladder epithelium, glial cells, retinal Müller cells, neurons, adrenal glomerular cells
Ba^{2+}, quinidine*	Proximal renal tubule, erythrocytes, hepatocytes, A6 cells, MDCK cells
Ouabain*	Necturus gallbladder, collecting duct PC, neurons, platelets
NH_3	Astrocytes, opossum kidney cells
Acidosis	Proximal renal tubule, neurons, glial cells
Butyrate* (SCFFA)	Enterocytes, erythrocytes, leukocytes
Cytochalasin B	Lymphoblast cells
Colchicine	Lymphoblast cells
Vinblastine*	Lymphoblast cells
N-Methylformamide chlorpromazine*	HT29 cells, erythrocytes
Hydroxyurea	Endothelial cells
Ethanol	Hepatocytes, adenohypophysial cells, cardiac cells
Dideoxycytidine	Monoblastoid cells
Mercurials	MDCK cells
Dioxin*	Hepatocytes
Hyperthermia	Chondrocytes, osteoblasts
Photofrin	Tumor cells
Electrical field stimulation	Outer hair cells
Factors leading to cell shrinkage	
ATP	Endothelial cells, hepatocytes
Glucagon	Hepatocytes
VIP	Intestine
ADH	Hepatocytes, MDCK cells
Atriopeptin (ANF)	Glial cells, cardiac myocytes
NO	Heart
α-Adrenergic stimulation	Salivary glands
Isoprenaline	NPCE eye
Acetylcholine*	Salivary glands, sweat glands, enterocytes
Bradykinin	Enterocytes, fibroblasts, endothelial cells
Histamine	Enterocytes
Thrombin	Enterocytes
Serotonin	Hepatocytes, leech glial cells
Adenosine	Hepatocytes, renal collecting duct
fMLP	Granulocytes
Corticostatic peptides	Enterocytes
cAMP	Hepatocytes, necturus gallbladder, MDCK cells, barnacle muscle, pulmonary epithelium, intestine, NPCE eye
cGMP	Heart
A 23187	Pulmonary epithelium, enterocytes, erythrocytes, fibroblasts
Okadaic acid*	Hepatocytes
Ouabain	Neurones, cardiac myocytes
Decrease of K_0^+	Leech glial cells, PCE eye
Removal of Na_0^+	Muscle cells, PCE
Removal of Cl_0^-	Kidney, PCE, toad bladder, amphibian skin
Removal of Ca^{2+}	Muscle cells
Starvation	Hepatocytes
H_2O_2	Hepatocytes
Elastin peptides	Fibroblasts
Urea	Hepatocytes, erythrocytes
Mastoparan	MDCK cells
NDS	Enterocytes
Furosemide	Macula densa cells, MDCK cells
MAG = 3but	HL60 leukemic cells
Ethanol	Prolactin-secreting cells, thyrotropin-secreting cells
Amphotericin B	Cornea epithelium, macrophages
Lead	Erythrocytes
Cisplatin	Renal tubule cells
Noise	Auditory hair cells

[a]Abbreviations: SCFFA, short-chain fatty acids; VIP, vasoactive intestinal peptide; NDS, neutrophil-derived secretagogue; MAG = 3but, monoacetone glucose 3-butyrate; NMDA, *N*-methyl-D-aspartate; PC, principal cell; PCE, pigmented ciliary epithelium; NPCE, nonpigmented ciliary epithelium; *, or similarly acting drugs.

not required for regulatory cell volume decrease. Clearly, Ca^{2+} activation of K^+ and Cl^- channels may contribute to, but is frequently not crucial for, regulatory cell volume decrease.

The increase of intracellular Ca^{2+} activity during cell swelling may enhance contractility (e.g., of vascular smooth muscle cells), stimulate exocytosis, and trigger hormone release [e.g., insulin, prolactin, gonadotropin-releasing hormone, luteinizing hormone, thyrotropin, aldosterone, atriopeptin (ANF), and renin].

G. G-Proteins

Inhibition of regulatory cell volume decrease or of swelling-induced increases of intracellular Ca^{2+} by pertussis toxin suggests the involvement of G-proteins in the intracellular signaling. Furthermore, small G-proteins have been implicated in cell volume regulation. Swelling of enterocytes is following by tyrosine phosphorylation of a $p125^{FAK}$. The inhibition of this phosphorylation by *Clostridium botulinus* C_3 exoenzyme, which depolymerizes the actin filament network by ADP-ribosylation of rho, blunts the volume regulatory anion efflux. G-proteins and/or cytoskeletal elements might be involved in the exocytotic insertion of volume regulatory anion channels into the cell membrane.

H. Protein Phosphorylation

Phosphorylation of a wide variety of proteins is observed during both cell swelling and cell shrinkage. [*See* Protein Phosphorylation.]

1. Cell Swelling

Mechanical stress or cell swelling has been found to stimulate protein kinase C, adenylate cyclase, MAP kinase, and the protein kinase Jnk and to foster tyrosine phosphorylation of several proteins.

Inhibitors of kinases have been found to interfere with regulatory cell volume decrease. How these events link to activation of the various volume regulatory ion transporters is, however, still incompletely understood. Volume regulatory KCl cotransport, for instance, is thought to be activated by dephosphorylation and inactivated by phosphorylation. Swelling was suggested to inhibit a kinase, favoring dephosphorylation. Nothing is known about the properties of this kinase, which appears to be distinct from protein kinases A and C. Some evidence indicates the involvement of the cytoskeleton in the swelling-induced inhibition of the kinase.

2. Cell Shrinkage

Osmotic cell shrinkage has been shown to activate protein kinase C, whereas cAMP formation and cAMP-dependent phosphorylation have been shown to remain unaffected.

As shown in shark rectal gland and duck salt gland, cell shrinkage stimulates serine and threonine phosphorylation of the $Na^+,K^+,2Cl^-$ cotransporter, which appears to be activated by phosphorylation. Phosphorylation of the volume regulatory Na^+/H^+ exchanger, on the other hand, is apparently not affected by cell shrinkage and not required for activation during cell shrinkage.

Most recently, cell shrinkage has been shown to enhance the expression of a putative serine/threonine kinase (h-Sgk).

I. Chloride

Beyond its influence on the driving force of $Na^+,K^+,2Cl^-$ cotransport, a decreased intracellular Cl^- concentration is thought to play a permissive role for the activation of both $Na^+,K^+,2Cl^-$ cotransporter and Na^+/H^+ exchanger. If extracellular osmolarity is made hypertonic by increase of extracellular NaCl concentration, the increase of intracellular Cl^- activity could thus impede regulatory cell volume increase. Accordingly, some cells are unable to regulate their volume during exposure to hypertonic extracellular fluid. If intracellular Cl^- is lowered by prior RVD or by activation of Cl^- channels with cAMP or hormones, the same cells do accomplish RVI.

J. Mg^{2+}

The dilution of intracellular solutes affects the concentration of Mg^{2+}, which in turn has been described to inhibit volume regulatory KCl cotransport and $Na^+,K^+,2Cl^-$ cotransport, but probably plays only a minor role in the volume regulatory activation of the transporters. During cell swelling, the effect of Mg^{2+} only partially accounts for the activation of KCl cotransport. Nevertheless, a decrease of intracellular Mg^{2+} concentration participates in regulatory cell volume decrease. Conversely, an increase of intracellular Mg^{2+} activity stimulates the Na^+/H^+ exchanger and the $Na^+,K^+,2Cl^-$ cotransporter and may thus participate in regulatory cell volume increase.

K. Eicosanoids

Cell swelling has been shown to activate a phospholipase A_2, possibly through decrease of macromolecular "crowding" and/or increase of intracellular Ca^{2+}. The 15-lipoxygenase product hepoxilin A_3 activates volume regulatory K^+ channels in platelets and the 5-lipoxygenase product leukotriene LTD_4 activates volume regulatory K^+ and/or Cl^- channels and/or volume regulatory taurine release in several cells. Accordingly, inhibition of the 5-lipoxygenase by nordihydroguaiaretic acid has been shown to inhibit activation of volume regulatory anion channels and impair regulatory cell volume decrease. However, the inhibitory effect of nordihydroguaiaretic acid was not reversed by addition of LTD_4, indicating that the drug may affect volume regulation by more than inhibition of lipoxygenase.

The enhanced formation of leukotrienes may parallel a decreased formation of prostaglandin E_2 (PGE_2), an effect possibly accounting for inhibition of PGE_2-sensitive Na^+ channels. On the other hand, in ciliary epithelial cells, PGE_2 was thought to mediate the activation of volume regulatory K^+ channels during cell swelling. Ketoconazole, an inhibitor of epoxygenase (cytochrome *P*-450), impedes volume regulatory efflux of osmolytes, such as sorbitol, betaine, myoinositol, or amino acids from several cell types. However, the inhibitory effect was not reversed by addition of hydroxyeicosatetraenoic acid (HETE), indicating that the inhibitory effect on osmolyte flux was not due to inhibition of epoxygenase.

The fatty acid composition of the cell membrane can be modulated by dietary polyunsaturated fatty acids, which lead to enhanced formation of leukotrienes and thus to acceleration of regulatory cell volume decrease.

L. pH in Acidic Cellular Compartments

As evidenced from acridine orange and fluorescein isothiocyanate (FITC) dextran fluorescence, cell swelling leads to alkalinization of acidic cellular compartments, whereas cell shrinkage enhances the acidity in those compartments. The alkalinization of acidic cellular compartments in hepatocytes occurs not only if cell swelling is due to decrease of extracellular osmolarity, but also if cell swelling is caused by inhibition of K^+ channels and by concentrative uptake of amino acids. Since the effect of cell volume on proteolysis and bile acid excretion and on pH in hepatocellular acidic cellular compartments is partially inhibited by colchicine and colcemid, the signal from cell volume to the acidic cellular compartments is likely to somehow involve microtubules. It appears that the influence of cell volume on the pH of acidic cellular compartments is not confined to hepatocytes but involves acidic compartments in a great variety of cells, such as pancreatic B cells, glial cells, neurons, proximal renal tubules, MDCK cells, alveolar cells, and fibroblasts. Accordingly, the functions of these compartments may be modified by alterations of cell volume.

M. Gene Expression

Cell volume influences the expression of a wide variety of genes. The most obvious examples are the hypertonically stimulated expression of enzymes or transporters serving accumulation of osmolytes, such as the aldose reductase, and the Na^+ coupled transporters for betaine, taurine, and inositol (see Section I,B,3).

Other genes shown to be expressed under the stimulatory influence of cell shrinkage include P-glycoprotein, phosphoenolpyruvate carboxykinase (PEPCK), tyrosine aminotransferase, HSP70, Egr-1, c-fos, αB crystallin, and laminin B_2. Genes turned on by cell swelling include β-actin, tubulin, ornithine decarboxylase, cycloxygenase-2, c-jun, and tissue plasminogen activator.

The mRNA for PEPCK is decreased by cell swelling.

In mice, a gene (rol) has been identified that renders erythrocytes resistant to osmotic lysis. The product of this gene is likely to be involved in the regulation of volume regulatory K^+ fluxes. However, the precise function of this gene has remained elusive.

N. ATP

The volume regulatory Cl^- channel in glioma is apparently inhibited by decreased ATP concentration, as it occurs during energy depletion. On the other hand, nonselective cation channels in exocrine pancreas and colonic crypt cells and ATP-sensitive K^+ channels in a variety of cells are inhibited by increase of ATP. As observed in pancreatic B cells, cell swelling activates the ATP-sensitive K^+ channels.

III. CHALLENGES OF CELL VOLUME CONSTANCY

A. Alteration of Extracellular Osmolarity

The extracellular osmolarity of animals is usually under strict regulation and does not undergo dramatic changes. Nevertheless, a variety of clinical conditions could lead to increases or decreases of extracellular osmolarity. These alterations trigger cell volume regulatory mechanisms, leading, for instance, to release of organic osmolytes from the brain. Local alterations of extracellular osmolarity may occur in the gastrointestinal lumen and to a lesser extent in the portal circulation during intestinal absorption. The most profound alterations of extracellular osmolarity are encountered in the kidney medulla. Transition from antidiuresis to diuresis may be paralleled by a decrease of medullary osmolarity from 1400 to 300 mosmol/liter.

B. Alteration of Extracellular Ion Composition

An increase of extracellular K^+ concentration depolarizes the cell membrane, decreases the electrical driving force for Cl^- exit, and thus favors cellular accumulation of Cl^- and cell swelling. Similarly, K^+-induced cell swelling could be accomplished by KCl uptake through KCl cotransport or Na^+,K^+,2Cl cotransport. Conversely, a decrease of extracellular K^+ concentration has been shown to shrink cells. Moreover, as would be expected, isotonic replacement of extracellular Na^+ with impermeant cations has been shown to shrink cells.

Increases of extracellular HCO_3^- concentration impede HCO_3^- exit. If the exit is electrogenic, an increase of HCO_3^- hyperpolarizes the cell membrane, which retains cellular K^+. The cellular accumulation of $KHCO_3$ then leads to cell swelling.

Isotonic replacement of extracellular Cl^- with gluconate favors cell shrinkage owing to cellular loss of Cl^- (and K^+). On the other hand, proprionate, lactate, and acetate may enter the cell in the nonionic moiety, dissociate within the cell, create an intracellular acidosis, and thus stimulate Na^+/H^+ exchange. Eventually the cells swell from accumulation of Na^+ (or K^+) and acid.

During correction of extracellular acidosis in the course of the treatment of diabetic ketoacidosis, the increasing extracellular pH allows the cells to extrude H^+ through the Na^+/H^+ exchanger, leading to cell swelling.

C. Energy Depletion

Energy depletion affects cell volume primarily through impairment of Na^+/K^+-ATPase. The electrogenic entry of Na^+ depolarizes the cell membrane and thus favors the exit of K^+ through K^+ channels. The gradual dissipation of the chemical K^+ gradient decreases the K^+ equilibrium potential, cell membrane potential, and thus the electrical driving force for Cl^- extrusion. Eventually, the cellular accumulation of Cl^- and Na^+ results in cell swelling. The time course of these events may be very slow and critically depends on the rate of Na^+ entry, which can be reduced by cooling or by inhibition of specific transport proteins, such as Na^+ channels (e.g., in neurons), Na^+ coupled transport (e.g., proximal renal tubule), Na^+,K^+,$2Cl^-$ cotransport (e.g., thick ascending limb of Henle's loop), or Na^+/H^+ exchange (many cells).

Paradoxically, some cells shrink transiently following impairment of Na^+/K^+-ATPase. This shrinkage presumably results from activation of K^+ and Cl^- channels by Ca^{2+}, which is accumulated within the cell via the Na^+/Ca^{2+} exchanger because of the increase of intracellular Na^+ activity (Fig. 3).

During ischemia of the brain, additional factors, such as intracellular acidosis, may lead to cell swelling. Furthermore, cell swelling in cerebral ischemia is favored by increase of extracellular K^+ concentration and by extracellular accumulation of glutamate, which stimulates cationic channels through *N*-methyl-D-aspartate (NMDA) receptors and leads to subsequent accumulation of Na^+, depolarization, and uptake of Cl^-.

D. Substrate Transport

In a wide variety of cells, most importantly in specialized epithelia such as intestine and renal proximal tubules, some amino acids and carbohydrates are accumulated by Na^+ coupled transport. Na^+ thus taken up is exchanged for K^+ by the Na^+/K^+-ATPase. The transport eventually leads to cell swelling owing to cellular accumulation of substrates and K^+. Limitation of cell swelling requires the activation of volume regulatory mechanisms, most importantly of K^+ channels.

E. Metabolism

The degradation of macromolecules, such as proteins and glycogen, to the osmotically more active monomers (amino acids and glucosephosphate) increases the cellular osmolarity. The degradation of glucose or amino acids to acids such as lactic acid imposes an acid load on the cell, requiring activation of Na^+/H^+ exchange with additional accumulation of Na^+ (or K^+). Complete degradation to CO_2 or export of amino acids or glucose, on the other hand, decreases intracellular osmolarity. Thus, even beyond

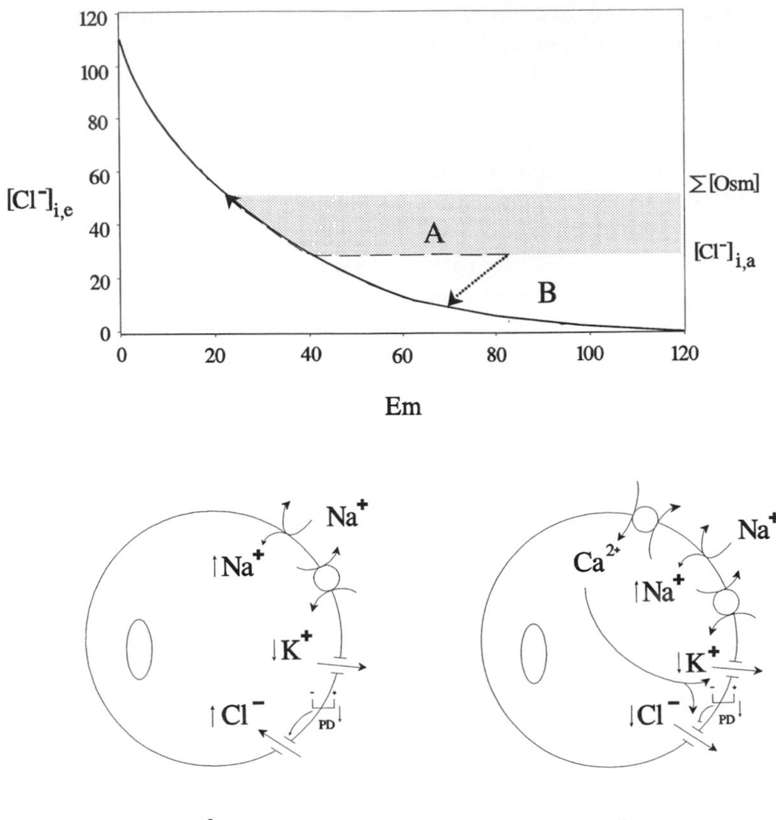

FIGURE 3 Mechanisms of cell shrinkage and cell swelling during inhibition of Na^+/K^+-ATPase. Inhibition of Na^+/K^+-ATPase leads to a gradual depolarization of cell membrane most importantly due to a decrease of the chemical gradient for K^+. This depolarization decreases the electrical driving force for the efflux of Cl^-, thus eventually reversing the electrochemical gradient for Cl^-. The equilibrium intracellular Cl^- activity, $[Cl^-]_{i,e}$, is a curvilinear function (solid line in upper panel) of the potential difference across the cell membrane (E_m). In most intact cells, the actual intracellular chloride activity, $[Cl^-]_{i,a}$, is higher than in equilibrium. The sum of readily releasable osmolytes (Σ[Osm]) is even higher. In either case (A or B), the inhibition of Na^+/K^+-ATPase increases intracellular Na^+ activity, the time course depending on the rate of Na^+ entry. The depolarization due to electrogenic entry of Na^+ favors K^+ exit and Cl^- accumulation. (A) The depolarization due to dissipating K^+ gradient leads to entry of Cl^- and cell swelling, as soon as the $[Cl^-]_{i,e}$ increases above Σ[Osm] (dashed arrow). (B) In cells with Na^+/Ca^{2+} exchange and Ca^{2+}-sensitive K^+ and/or Cl^- channels, the increase of Ca^{2+} could transiently shrink the cells owing to activation of K^+ and Cl^- channels, leading to a decrease of $[Cl^-]_{i,a}$ despite depolarization and increase of $[Cl^-]_{i,e}$ (dotted arrow). Like the cells illustrated in A, these cells will eventually swell from dissipation of the K^+ gradient and depolarization.

the formation or disposal of osmolytes, cell metabolism is a challenge for cell volume constancy. Hence it is not surprising that many metabolic pathways, namely, the metabolism of macromolecules, are sensitive to cell volume (see the preceding).

In diabetes mellitus, the enhanced glucose concentrations increase the formation of sorbitol through aldose reductase and thus disturb osmolyte metabolism. Sorbitol accumulation has been implicated in the generation of several diabetic complications such as neuropathy, retinopathy, microangiopathy, and cataracts. On the other hand, excessive extracellular glucose concentrations may lead to cell shrinkage in diabetes mellitus.

Starvation has been shown to shrink cells, presumably due to reduced availability of amino acids.

F. Hormones

A wide variety of hormones and mediators have been shown to alter cell volume (see Table I). The hormones and mediators may swell cells by activation of Na^+/H^+ exchange and $Na^+,K^+,2Cl^-$ cotransport (e.g., insulin and mitogens), stimulation of cation channels (e.g., glutamate), or inhibition of K^+ and/or Cl^- channels. They may shrink cells by activation of K^+ and/or Cl^- channels (e.g., acetylcholine). The effects of the hormones are mimicked by the respective elements of intracellular transmission (see Table I).

By modifying the cell volume, the hormones trigger cell volume regulatory mechanisms, including ion transport across the cell membrane, enzymes of volume-sensitive metabolic pathways, as well as gene expression. Thus, cell volume becomes an integral part of the intracellular signaling of these hormones.

G. Other Regulators

Besides hormones and mediators, several drugs and toxic substances modify cell volume and thus trigger cell volume regulatory mechanisms (see Table I). As would be expected, inhibition of K^+ channels (Ba^{2+}, quinidine) may lead to cell swelling and inhibition of $Na^+,K^+, 2Cl^-$ cotransport to cell shrinkage. The ability of urea to alter cell volume regulatory set point was discussed earlier. Furthermore, a wide variety of other substances, including ethanol and heavy metals (lead, mercury), have been shown to alter cell volume. The underlying mechanisms are in large part elusive.

The factors altering cell volume may contribute to the physiology and pathophysiology of a wide variety of conditions. For instance, in several stress situations, such as surgical intervention, acute pancreatitis, severe injury, burns, and sepsis, a decrease of muscle intracellular space has been observed leading to reversal of the inhibition of proteolysis and thus to hypercatabolism. The mechanisms underlying muscle cell shrinkage have not yet been explored.

IV. CONCLUSIONS

The constancy of cell volume is continuously challenged by alterations of extracellular osmolarity and, more importantly, by altered intracellular osmolarity. A multitude of cell volume regulatory mechanisms serve to defend the constancy of cell volume. These include ion transport across the cell membrane, such as K^+, anion, and unselective ion channels, KCl symport, Na^+/H^+ exchange, and $Na^+,K^+,2Cl^-$ cotransport; formation or disposal of osmolytes, such as glycerophosphorylcholine, sorbitol, inositol, betaine, taurine, and amino acids; and altered metabolism of macromolecules, such as proteins and glycogen. The number of cell volume regulatory mechanisms is matched by a similarly wide range of intracellular signals altered by cell volume, including macromolecular crowding, the cytoskeleton, Ca^{2+}, H^+, Mg^{2+}, Cl^-, G-proteins, phosphorylation, and eicosanoids. The signals and effectors can be triggered by hormones through alteration of cell volume. Thus, cell volume is an integral part of the machinery regulating cellular function. This machinery is exploited by hormones to exert their effects on target cells, as amplified in a separate article.

BIBLIOGRAPHY

Beyenbach, K. W. (ed.) (1990). Cell volume regulation. *In* "Comparative Physiology," pp. 1–25. Karger-Verlag, Basel.

Boyer, J. L., Graf, J., and Meier, P. J. (1992). Hepatic transport systems regulating pH: Cell volume and bile secretion. *Annu. Rev. Physiol.* **54**, 415–438.

Garcia-Perez, A., and Burg, M. B. (1991). Renal medullary organic osmolytes. *Physiol. Rev.* **71**, 1081–1115.

Garner, M. M., and Burg, M. B. (1994). Macromolecular crowding and confinement in cells exposed to hypertonicity. *Am. J. Physiol.*, C877–C892.

Häussinger, D., and Lang, F. (1991). Cell volume in the regulation of hepatic function: A potent new principle for metabolic control. *Biochim Biophys. Acta* **1071**, 331–350.

Hoffman, E. K., and Dunham, P. B. (1995). Membrane mechanisms and intracellular signalling in cell volume regulation. *Cytology* **161**.

Kinne, R. K. H., Czekay, R.-P., Grunewald, J. M., Mooren, F. C., and Kinne-Saffran, E. (1993). Hypotonicity-evoked release of organic osmolytes from distal renal cells: Systems, signals, and sidedness. *Renal Physiol. Biochem.* **16**, 66–78.

Lang, F., and Häussinger, D. (eds.) (1993). Interaction of cell volume and cell function. *In* "Advances in Comparative and Environmental Physiology," Vol. 14, pp. 249–277. Springer-Verlag, Heidelberg.

Parker, J. C. (1993). In defense of cell volume? *Am. J. Physiol.* **265** (*Cell Physiol.*), C1191–C1200.

Schultz, S. G. (1992). Membrane cross-talk in sodium-absorbing epithelial cells. *In* "The Kidney—Physiology and Pathophysiology," (D. W. Seldin and G. Giebisch, eds.), Vol. 1, pp. 287–299. Raven, New York.

Central Nervous System, Toxicology

STATA NORTON
University of Kansas Medical Center

GLOSSARY

Astrocyte The most common nonneuronal cell of the brain, often called a "nurse cell" for its role in sustaining the metabolic integrity of the neuron

Axon Extension of the neuron through which impulses, in the form of sodium and potassium ion shifts, are transmitted from one neuron to another

Dendrites Processes of the neuron that receive information from other neurons

DNA Deoxyribonucleic acid; carries the genetic information of the cell

Encephalopathy Clinical condition in which the functions of the brain are disorganized; signs vary from mild (drowsiness or confusion) to severe (coma or seizures)

Endoplasmic reticulum Cell organelle involved in the synthesis of some cellular proteins

Microglia Nonneuronal cell that responds to brain injury and aids in removing inflammatory material and dead cells

Mitochondria Cell organelles that contain part of the synthetic mechanism for energy production in the cell using oxygen

Mitosis Process by which a cell doubles and through which cell numbers increase during growth of the organism

Myelin Lipid component of the cellular sheath surrounding some axons; it is produced by oligodendrocytes and insulates axons from each other

Neuron Cell in the nervous system specialized for receiving, storing, and transmitting information

Neuropathy Clinical condition in which the function of the nervous system is abnormal; sensory neuropathies involve increased or decreased sensations of vibration, temperature, pressure, or pain; motor neuropathies include rigidities, spasms, and weakness

Neuropil Process of cells, including neurons, astrocytes, and oligodendrocytes, that fills the tissue of the brain between the cell bodies

Oligodendrocyte Nonneuronal cell in the brain that produces processes that surround the axons of many neurons

Synapse Junction of an axon with another neuron. Specialized chemicals (neurotransmitters) are stored in vesicles on the presynaptic (transmitting) side and are liberated into the synaptic cleft by a nerve impulse. The postsynaptic (receiving) side contains receptors that combine with the transmitter, causing a response in the postsynaptic cell

THROUGHOUT LIFE, THE HUMAN BODY IS EXPOSED to toxic substances from various sources. A toxic substance is a chemical that, in sufficient concentration, causes damage to bodily structure or function. Useful, or even essential, chemicals may be toxic at concentrations greater than those required by the body. The term "toxic agents" includes not only chemicals but also physical agents, such as X rays, which can be harmful at high exposure levels. These definitions contain two important concepts in the study of toxic agents: Dose response is the concept that the intensity of the response of living organisms to a toxic agent is related to the dose; the second concept, duration response, is that the intensity of the response is related to the duration of exposure. Our awareness of the effects of toxic agents on the body is through the nervous system. It has been noted that the two most common adverse effects of drugs are headache and nausea, both of which originate in the central nervous system, regardless of the organ damaged by the toxic

ENCYCLOPEDIA OF HUMAN BIOLOGY, Second Edition, VOLUME 2. Copyright © 1997 by Academic Press. All rights of reproduction in any form reserved.

agent. Furthermore, the nervous system is the direct target for some types of toxic agents, and damage to the nervous system can result in a wide range of effects on an individual. The cells of the central nervous system as targets for toxic agents and the functional consequences are discussed here.

I. HISTORICAL BACKGROUND

The 16th-century physician and alchemist Paracelsus has been credited with first proposing the concept that all chemicals are poisons at some dose. His interest in chemistry and that of his contemporaries marked the beginning of the discovery of thousands of chemicals of benefit to humans. The science of toxicology is thus an outgrowth of the science of chemistry and the growing understanding of the chemicals, both natural and man-made, that can, on one hand, save lives and, on the other, damage cells. In the development of the biological sciences, the central nervous system has been called the last frontier, because the principles governing the relationships between its complex structures and their functions are still being elucidated. The study of toxicology as it relates to the central nervous system depends on the knowledge of nervous system biology, yet it has also contributed to our understanding of how the nervous system functions.

II. EXPOSURE TO TOXIC AGENTS

A. Unintentional

The environment contains many man-made or naturally occurring chemicals to which living organisms are sometimes exposed at levels that result in toxicity. The types of exposure are highly varied, from brushing the skin against a poison ivy leaf to inhaling tetraethyl lead in vaporized gasoline. The most serious exposures in air include lead, carbon monoxide, ozone, and oxides of nitrogen and sulfur. These chemicals are present in the atmosphere of cities primarily as a result of combustion for the production of energy in various forms. Exposure to toxic chemicals also may occur from ingestion of food or water contaminated from agricultural chemicals, industrial effluent, or the disposal of hazardous substances. Concern is often expressed for the increased incidence of types of cancer as a result of environmental exposure. The central nervous system, however, is the target in some exposures that do not result in cancer.

The possibilities of exposure in the workplace are numerous. For chemicals used in the production of many consumer goods, standards for acceptable exposure of workers have been set, which are revised as new information becomes available on the effects of different levels and durations. Good sources for current information are the publications of the American Conference of Governmental Industrial Hygienists, which include threshold limit values and biological indices of exposure for many chemical and physical agents in the workplace.

B. Intentional

The therapeutic use of drugs and the addition of chemicals to food have the same intent, namely, human benefit. Much study has been devoted to ensure that chemicals used as drugs or food additives lack serious toxicity to the nervous system. In the case of drugs, the benefit is sometimes balanced against an acceptable risk. For example, in determining the optimal dose for the treatment of acute leukemia by vincristine and similar drugs, the possibility of development of peripheral neuropathy, which results in weakness in leg or arm movement, must be balanced against the therapeutic effect on the cancer. In recent years, a most notable example of concern over the effects of food additives on the central nervous system has been the proposed link between certain food dyes and hyperactive behavior in children. After considerable controversy, substantial evidence against such a link was obtained by controlled studies of hyperactive children. Currently, there are no major studies implicating food additives with disorders of the central nervous system. However, the controversy did focus on an important consideration, that is, the greater susceptibility of segments of the population to some toxic chemicals. Not only children, in whom the nervous system is still developing, but also aged individuals may be uniquely sensitive.

Intentional exposure to toxic substances also takes place when individuals ingest, inject, or inhale chemicals with effects on the central nervous system, such as ethyl alcohol, cocaine, heroin, toluene, and benzene. Each of these chemicals has the ability to produce acute toxic changes. High doses may cause death through depression of the brain or other effects. Single doses of low levels have reversible effects. Prolonged use may cause irreversible changes.

III. RESPONSES OF THE NERVOUS SYSTEM

Effects of toxic agents on the nervous system can be evaluated in two basic ways: examination of structure and examination of function. Either structure or function or both may be altered by exposure, and the effects may be reversible or irreversible, depending on the agent, dose, and duration of exposure.

A. Structure of the Nervous System

The tissue of the central nervous system of all mammals, including humans, is composed of three major cell types (neurons, astrocytes, and oligodendrocytes), and between these cell bodies, the processes of the cells constitute the neuropil. Blood capillaries and some other specialized cells, such as microglia, complete the tissue of the brain. Each of the three major cell types has been recognized as a selective target for some toxic substances.

Neurons are the cells responsible for transmission, interpretation, and storage of information in the central nervous system. Their characteristic appearance, when the cell body and all of its processes are visualized, shows processes that extend from a central core, called the perikaryon, or cell body (Fig. 1). The large nucleus in the cell body is surrounded by cytoplasm richly filled with the cell organelles necessary for producing the energy to sustain very active metabolism. Energy is required by the neuron for the transport of nutrients down the processes, for the synthesis of chemical transmitters of information, and for maintaining ion homeostasis. Not surprisingly, neurons are at risk of damage whenever the levels of oxygen or glucose, supplied by the blood capillaries, are even temporarily reduced. Oxygen and glucose are the basic chemicals required for the extensive aerobic metabolism that neurons perform.

Astrocytes have been termed the "nurse cells" for the neurons, tending to cluster around large neurons (Fig. 2). Astrocytes are more numerous than neurons and, in some parts of the brain, outnumber them 10 to 1. The processes of the astrocyte are found against the cell body of the neuron and also wrapped around the endothelial cells lining the blood capillaries. [*See* Astrocytes.]

Oligodendrocytes produce spiral wrappings of myelin that insulate axons, neuronal processes that may extend long distances in the nervous system, often in

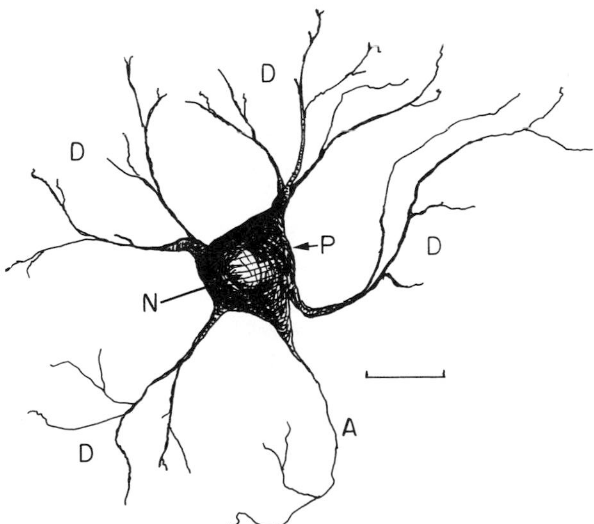

FIGURE 1 A neuron showing the perikaryon (P), nucleus (N), dendrites (D), and axon (A). Bar, 50 μm.

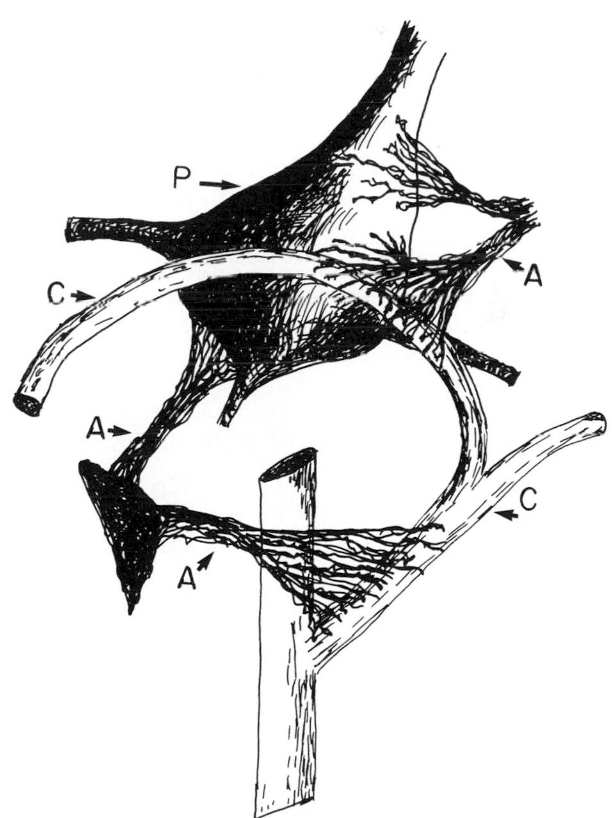

FIGURE 2 A neuron partially enclosed by processes from astrocytes (A). The astrocytes also contact the adjacent capillaries (C). P, perikaryon.

tracts with other axons insulated from each other by myelin.

B. Morphological Response of Neurons

Examination of a neuron with the light microscope or, at higher magnification, with the electron microscope reveals characteristic changes in structure in response to high doses of many toxic chemicals. Any serious interference with the neuronal production of energy from glucose will distort the fluxes of calcium ions, which require energy. Moreover, the pH of the cell will decrease from the accumulation of lactic acid. Both increased intracellular calcium levels and lowered pH activate damaging proteolytic enzymes within the cell. The appearance of the neuron under these conditions is diagrammed in Fig. 3. The nucleus becomes eccentric, the endoplasmic reticulum becomes dispersed, and the cell body and the mitochondria swell, owing to an influx of water. If the process is not rapidly reversed by restoration of an adequate source of energy, the nucleus will become pycnotic (i.e., condensed, with irregular margins), surrounded by a clear area of cytoplasm in the swollen dying cell. Since neurons cannot be replaced by cell division in the adult brain, death of the neurons may result in serious functional disability. However, there appears to be redundancy in carrying out function, so that function is not lost unless large numbers of neurons are affected.

In addition to cell damage by alterations in energy production, there are other targets for toxic chemicals in the neuron. Reversible effects are common in the synaptic area, where two neurons communicate. As noted earlier, the axon, often myelinated, is the process that transmits impulses to neurons in other areas. The axon makes synaptic contacts with the cell bodies, axons, or dendrites of other neurons. At each synapse, packets of special chemicals, called neurotransmitters, are liberated from the terminals of the axon into the synaptic cleft (Fig. 4). These transmitters act on the postsynaptic neuron by combining with specialized proteins called receptors. The synapse is, therefore, a target for chemicals that may alter the synthesis or release of transmitters, interfere with the postsynaptic actions of transmitters, or even mimic the normal transmitter.

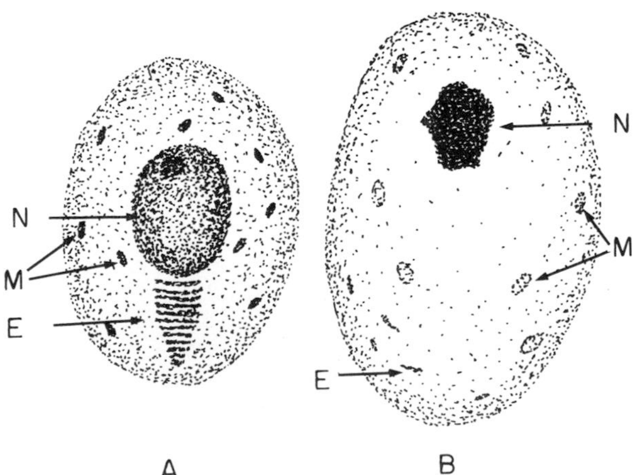

FIGURE 3 The changes in a neuron (A) following exposure to a toxic substance (B) that interferes with energy production. N, nucleus; E, endoplasmic reticulum; M, mitochondria.

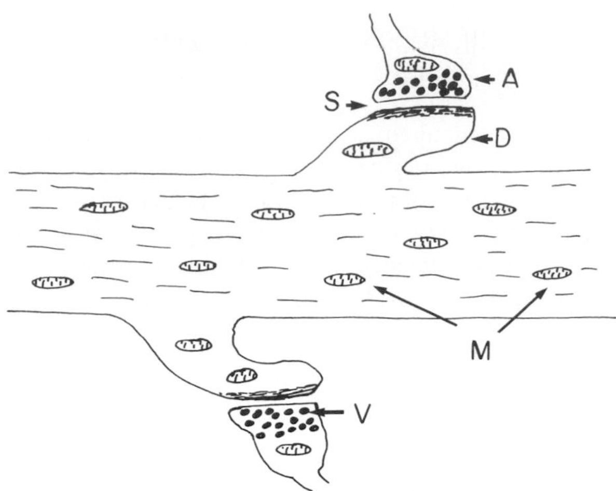

FIGURE 4 Synapse between an axon (A) and a dendrite (D). Vesicles (V) containing neurotransmitter are located in the terminal of the axon. The transmitter is released into the synaptic cleft (S). M, mitochondria.

C. Functional Response

The types of response to exposure to toxic agents are related to the areas of the nervous system that are affected. Their intensity ranges from headache, which disappears when exposure ceases, to permanent damage, with motor weakness or paralysis. Fatal exposures occur when neuronal function is severely compromised. For chemicals whose exposure occurs in the workplace at levels that result in recognizable pathology, the cause is usually correctly identified and measures are taken to reduce further exposure. The Federal government sets accepted occupational exposure levels for these chemicals. Low-level exposures may result in the insidious or slow development of damage, which is difficult to diagnose correctly. There is, however, controversy about "threshold" effects, that is, effect at the level below which no damage occurs following prolonged exposure. For example, although lead exposure from leaded gasoline and lead-containing paint and putty has been studied extensively, the lowest level of lead with a deleterious effect on learning in children is not certain. The complexity of obtaining precise epidemiological data compounds the problem of identifying toxic effects at low levels of exposure.

Table I lists some substances known to cause different functional types of damage, together with the mechanism and site of action in the nervous system. The site of damage varies from the synapse to the axon, to the insulating sheath of myelin, or to the cell body. For some agents, the action is indirect instead of damaging neurons directly (e.g., an agent that alters the oxygen supply to the neuron).

Botulinum toxin is formed by the bacterium *Clostridium botulinum,* which grows anaerobically in improperly canned foods. The toxin reaches the synapse of axons innervating skeletal muscle and blocks the release of acetylcholine, the neurotransmitter for skeletal muscle, resulting in muscular paralysis.

Carbon monoxide is formed during incomplete combustion. When inhaled, the gas combines with hemoglobin in the blood to form carboxyhemoglobin, preventing the formation of oxyhemoglobin, necessary for transporting oxygen to the central nervous system and other tissues. Some neurons die from lack of oxygen at levels that do not kill many other cells.

Methyl bromide is a gas used for fumigation of stored grain. It is fatal to insects and, in high concentrations, damages all organisms. When inhaled, the capillaries in the lungs are first affected. Damage to brain capillaries follows as methyl bromide is transported in the blood, preventing proper oxygen supply to the neurons. Since the damage to the capillaries of the central nervous system is generalized, an acute encephalopathy develops, characterized by a loss of consciousness and repeated seizures.

Methyl mercury is an organic form of mercury used for protection of seed grain from insects and is not retained in the growing plant. Methyl mercury, formed from mercury by bacteria in natural bodies of water, can be picked up by organisms in the food chain and finally by fish, which are consumed by humans. This process resulted in a serious episode of human and animal poisoning by methyl mercury in Minimata Bay in Japan in the 1950s.

Tetrodotoxin is a toxic chemical derived from improperly prepared pufferfish. This fish is considered a delicacy in Japan and is commercially prepared by specially trained workers. Tetrodotoxin is present only in the liver, which must be removed without contaminating the meat. Its effect is to block the sodium channels of the nerve axon, thus preventing transmission of its nerve impulse. This chemical has been of great interest and usefulness for studying the physiological mechanisms surrounding transmission of the nerve impulse.

Triethyltin is one of several organic tin compounds used in industrial processes. Its target is the oligodendrocyte. The rings of myelin around axons are split apart and fluid collects between the rings, resulting in generalized brain edema, which can be serious. However, when the acute phase is passed, the oligodendrocyte can resynthesize the myelin wrapping of the axons.

TABLE I
Examples of Responses to High-Level Exposures

Agent	Functional response	Site of effect	Mechanism of effect
Botulinum toxin	Paralysis	Peripheral synapse of nerve with muscle	Block of the release of transmitter
Carbon monoxide	Coma	Neurons throughout the central nervous system	Depletion of oxygen through formation of carboxyhemoglobin
Methyl bromide	Encephalopathy	Capillaries throughout the central nervous system	Block of blood flow
Methyl mercury	Sensory loss, especially visual	Small neurons of the brain and the spinal cord	Death of small neurons
Tetrodotoxin	Paralysis	Peripheral synapse of nerve with muscle	Block of the transmission of nerve impulse
Triethyltin	Encephalopathy	Myelin throughout the central nervous system	Damage to oligodendrocytes

IV. TOXIC AGENTS

A list of toxic agents causing adverse responses of the nervous system would be lengthy; however, the list of agents causing major problems with irreversible or fatal outcomes is much shorter. Table II encompasses agents important for either their frequency of occurrence or the seriousness of the consequence of exposure. A more complete listing is available in the books given in the bibliography. Each of the agents in Table II illustrates a different type of toxicity, which is discussed briefly here.

Ionizing radiation includes emissions of high-energy subatomic particles from radioactive chemicals, both naturally occurring and man-made, and from cosmic radiation. In the body, high-energy particles give rise to "reactive radicals," reactive chemical species that can damage cells. If the repair mechanisms of the cell are overwhelmed, cell death can occur. The most sensitive cells are those in the process of cell division. Thus, the fetus, with large numbers of dividing cells, is especially at risk.

Lead is a contaminant of the atmosphere from the exhaust of automobiles burning leaded gasoline. It may also be emitted by some industrial processes. In the body, lead, like calcium, can be stored in, or mobilized from, bone. Blood levels of lead are a measure of current exposure, but not of body burden. When high blood levels of lead are reached, the central nervous system becomes seriously involved. The primary target is probably the capillaries. Generalized damage to the vascular system of the brain results in encephalopathy, a condition of brain edema with coma and recurring motor seizures. Children are more likely to develop encephalopathy than adults. Atmospheric lead is insufficient to achieve such high blood levels, which require other sources of exposure, usually the ingestion of lead. A major concern with atmospheric levels is that the continued exposure of children to levels even below those causing encephalopathy may have deleterious effects on brain function.

Methyl mercury selectively damages small neurons in the central nervous system, especially those of the area of the brain involving vision, the visual cortex, and the sensory cell bodies of the spinal cord in the

TABLE II
Toxic Agents Affecting the Central Nervous System

Agent	Type of effect
Environmental contaminants	
Ionizing radiation	Kills dividing cells; adult resistant; fetus sensitive
Lead	Atmospheric lead inhaled; chronic exposure may cause slowed learning in children
Methyl mercury	Chronic or acute exposure to high levels causes sensory loss; fetus and child more sensitive than adults
Nitrates	Converted to nitrites in intestine; high levels can cause methemoglobinemia in babies
Occupational exposures	
Acrylamide	Causes peripheral neuropathy; axons of large sensory and motor nerves degenerate from chronic or acute exposure
Carbon disulfide	Chronic inhalation results in peripheral neuropathy and encephalopathy
Manganese	Chronic inhalation causes emotional disturbances, followed by parkinsonian-like muscle rigidities
Organophosphate insecticides	Acute exposure through skin contact or inhalation results in accumulation of the transmitter acetylcholine
Foods and food additives	
Absinthe	Severe encephalopathy has developed with continued ingestion of wormwood-containing liquor
Lathyrus sativus	Excessive consumption of this pea causes peripheral neuropathy
Methyl alcohol	Contamination of some alcoholic beverages (improperly prepared "moonshine" liquor) has caused blindness or generalized encephalopathy
Triorthocresyl phosphate	Contamination of food has occurred from reuse of containers; peripheral neuropathy results
Drugs and chemicals of abuse	
Alcohol (ethyl alcohol)	Acute intoxication from ingestion; chronic use is addictive; selective brain damage and peripheral neuropathy in adults; fetal alcohol syndrome in offspring of alcoholic mothers
Atropine	Acute overdose reversibly blocks part of the autonomic nervous system and causes hallucinations
Benzene	Chronic inhalation causes neuronal death and affects blood-forming tissues
Vincristine	When used therapeutically as an anticancer drug, may cause neuropathy as a side effect

dorsal root ganglia. The functional consequences are constriction of the visual field, or even blindness, and various abnormalities of sensation, such as numbness or tingling in the arms and legs.

Nitrates are occasionally found in high concentrations in well water, usually as a result of the agricultural use of nitrate-containing fertilizer. Nitrates in drinking water are not toxic to children or adults with normal liver function, but babies, in whom liver function is incompletely developed, may be unable to convert nitrites (which are formed from nitrites by bacteria in the gut) to nontoxic nitrates. Nitrites combine with hemoglobin in the blood to produce methemoglobin, a non-oxygen-carrying form. With severe methemoglobinemia, babies develop a bluish color of the skin and may become comatose from lack of oxygen.

Acrylamide is used industrially in the formation of plastic polymers. Occupational exposure tends to be chronic at low levels. With prolonged exposure, some workers develop a peripheral neuropathy associated with damage to the large axons, and both sensory and motor alterations. There may be some involvement of the autonomic nervous system, since the clinical signs include peripheral vasodilation (reddening of the skin) and sweating in the hands and feet. As with most toxicity from chemical exposure, there is recovery upon withdrawal from exposure, unless damage to the axons is severe.

Carbon disulfide is of historical importance, because the effects of chronic exposure were documented carefully in studies of workers exposed to the chemical in processing rubber. This episode is an example of the development of acceptable standards of exposure to protect workers. It was found that, after exposure for long periods (months or years, depending on levels in the air), workers developed neuropathies involving both sensory and motor fibers, and signs of brain involvement, including fatigue, sleep disturbances, and psychological depression.

Manganese is mined for industrial use in several countries. A central nervous system disorder has been characterized in miners exposed chronically to dust during production. The most specific effect is a type of parkinsonism with muscle stiffness and facial rigidity. Lesions have been found in areas of the brain involved in the control of motor movement. Emotional disturbances may accompany or precede the motor rigidity. [*See* Parkinson's Disease.]

Organophosphate insecticides, such as parathion, are agricultural chemicals that must be used with care to prevent excessive skin contact or inhalation during application. Exposure by either route causes a block of the enzyme cholinesterase that destroys the neurotransmitter acetylcholine. This is the same as the mechanism of toxicity to insects, which also use acetylcholine as a neurotransmitter. When the cholinesterase is blocked, the synapse continues to be activated by the neurotransmitter, and the postsynaptic tissue continues to respond. In the autonomic nervous system, this results in excessive salivation and bronchial secretions, pupillary constriction, and bradycardia. In skeletal muscle, there are repeated small contractions, called fasciculations.

Absinthe is an anise-flavored liquor that, in the 19th century, was flavored with wormwood, an essential oil from the plant *Artemisia absinthium.* Absinthe oil caused neuronal death in the brain and, with repeated exposure, serious psychological disturbances and madness. The oil is no longer used in flavoring.

Lathyrus sativus is a form of pea that is of toxicological importance to countries where this pea is occasionally consumed in large quantities, for instance, during famine. The condition of lathyrism is a spastic paralysis of the legs caused by damage to the large motor neurons controlling the skeletal muscle. The small amounts present in most diets have no effects.

Exposure to methyl alcohol has occurred when it was a contaminant of some improperly prepared alcoholic beverages, such as "moonshine." It is not present in significant concentrations in commercial beverages. Acute exposure to large amounts may cause death from respiratory failure or blindness in those who survive.

Triorthocresyl phosphate ingestion has resulted in several outbreaks of paralysis. Exposure has been the result of unintentional contamination of food or drink, when stored in industrial containers previously used for triorthocresyl phosphate and still containing residual amounts. The chemical causes damage to large axons in the arms and legs, with weakness and sensory disturbances. Triorthocresyl phosphate is an example of a chemical that causes delayed neurotoxicity. A few chemicals, such as this one, do not cause effects in a few hours. A single exposure may cause severe paralysis, but only after a delay of about 2 weeks. This delay is not shortened by larger doses. Recovery is slow and may be incomplete when poisoning is severe.

Alcohol (ethyl alcohol) is the most commonly used drug of abuse. The central nervous system consequences of acute intoxication with alcohol are well known. Chronic ingestion results in peripheral neuropathy, which is partially reversed by dietary thiamine (vitamin B_1); damage to the brain results in Korsakoff's syndrome, in which there is a severe loss of memory. The fetal alcohol syndrome can develop in a baby exposed repeatedly *in utero* to alcohol from maternal ingestion. Since there is no placental barrier to alcohol, the fetal blood level is the same as the mother's. Damage to the baby includes mental and growth retardation. The precise blood levels and length of exposure that are deleterious to the fetus are not known. [*See* Alcohol Toxicology.]

Atropine and related chemicals have many therapeutic uses, but have also been abused for their disorienting effects on the brain. Atropine blocks the action of the neurotransmitter acetylcholine. The effects on the autonomic nervous system are dry mouth, flushing, tachycardia, and pupillary dilation. Hallucinations are common during overdose. Recovery is complete in a few hours.

Benzene and some other solvent gases are abused by inhalation to obtain effects on the central nervous system, which resemble those produced by ethyl alcohol. The effects are reversible, but repeated exposures have been linked with gradual deterioration of personality. Benzene has additional toxic actions on other organs, notably damage to blood-forming tissue. Industrial exposure levels are kept below those causing toxicity.

Vincristine is a drug with valuable therapeutic actions at levels that may cause irreversible toxicity, an example of the need to balance risk and benefit to the individual. Vincristine is effective against some types of cancers at levels apt to cause irreversible peripheral neuropathy. In combination with other anticancer drugs, it can be used at lower doses, which are still effective but have a lower risk of neuropathy.

V. SPECIAL CONSIDERATIONS

A. Immature Nervous System

It has been noted that the immature nervous system is more sensitive to toxic agents than the adult nervous system. The dividing cell is at risk because mitosis puts high metabolic demands on a cell and because DNA is more sensitive to damage during replication. The division of neurons ceases shortly after birth in the human, but maturation proceeds after birth for some time. Neuronal processes grow and synaptic contacts are completed postnatally. Astrocytes continue to divide. The formation of myelin by oligodendrocytes is a prolonged process, paralleling the growth of axons. The entire period of formation and maturation of the nervous system is a time of increased risk from exposure to toxic substances.

B. Geriatric Nervous System

Altered susceptibility to toxic substances with increasing age is an important consideration. Since the liver and the kidney are the major organs of destruction and excretion, respectively, of toxic agents, any deterioration of these organs will result in increased levels of the chemicals during exposure, increasing the risk of a toxic response. In addition, there is some loss of neurons with the aging process, although the rate of loss is uncertain. With depletion of neurons, a toxic response is again more likely to result.

C. Research on Central Nervous System Toxicology

The emphasis in the manifestations of toxicity reported here has been on those agents and exposures causing marked functional and cellular damage; they are the best known. It is important, however, to study the consequences of exposures below levels producing dramatic clinical signs and pathology, because the level of exposure to many toxic agents cannot be reduced to zero and must be controlled to minimize risk.

The detailed study of toxic agents that damage the central nervous system is important in improving understanding of some degenerative diseases of the brain, such as Alzheimer's disease and Huntington's chorea. Scientists are finding that similar processes may be involved in the death of neurons even in different diseases.

BIBLIOGRAPHY

Amdur, M. O., Doull, J., and Klaassen, C. D. (eds.) (1991). "Casarett and Doull's Toxicology," 4th Ed. Pergamon, New York.

American Conference of Governmental Industrial Hygienists (ACGIH) (1987). "Threshold Limit Values and Biological Exposure Indices for 1987–1988." ACGIH, Cincinnati.

Haschek, W. M., and Rousseau, C. G. (eds.) (1991). "Toxicologic Pathology." Academic Press, San Diego.

Hathcock, J. N. (ed.) (1982). "Nutritional Toxicology," Vol. 1. Academic Press, New York.

Mitchell, C. L. (ed.) (1982). "Nervous System Toxicology." Raven, New York.

Narahashi, T. (ed.) (1984). "Cellular and Molecular Neurotoxicology." Raven, New York.

Spencer, P. S., and Schaumburg, H. H. (eds.) (1980). "Experimental and Clinical Neurotoxicology." Williams & Wilkins, Baltimore.

Cerebral Specialization

ROBIN D. MORRIS
WILLIAM D. HOPKINS
DUANE M. RUMBAUGH
Georgia State University

GLOSSARY

Asymmetry Unequal, biased; one side of the brain is better at a given function than another

Cerebral hemispheres Two main sides of the brain that are typically referred to as the left and right hemispheres. Each neocortical hemisphere is divided into four anatomical regions called lobes (temporal, parietal, occipital, and frontal)

Laterality Side of the brain that provides primary control of a given function

CEREBRAL SPECIALIZATION, ALSO REFERRED TO as cerebral dominance, cerebral asymmetry, or functional asymmetry, represents the concept that certain mental abilities or processes are performed more efficiently by one cerebral hemisphere (i.e., one part of the brain) compared to another. Globally, this concept is referred to as cerebral lateralization.

I. HISTORICAL BACKGROUND AND CONCEPTS

Over 100 years ago clinical scientists such as Broca, Dax, and Wernicke described patients who, because of damage to their left cerebral hemispheres caused by strokes or injuries, were unable to talk or understand what others said to them, but at the same time were able to perform other non-language-related activities. As further studies of such patients were undertaken, the classic tenet, called cerebral dominance, was formed, which suggested that the left cerebral hemisphere was crucial for language functions and was the dominant hemisphere. These early ideas suggested that language-related skills were the most important mental abilities and that the left hemisphere controlled the right hemisphere, which came to be known as the nondominant, or minor, hemisphere. In addition, a special link between this left hemisphere dominance for language and right-handedness was suggested.

About 50 years later researchers began to find that damage to the right cerebral hemisphere also created specific types of mental deficits—not in the area of language functions, but in the area of visuospatial abilities. Because of these new findings, it became clear that the right hemisphere also contained important abilities. As evidence for other specialized abilities within each hemisphere began to accumulate, it was realized that each hemisphere was "dominant" for unique skills, and the concept of a dominant hemisphere slowly lost meaning. Currently, if one is speaking of cerebral dominance, the specification of cerebral dominance "for what" is required. Because of this history, it is probably more accurate to discuss such relationships as evidence of cerebral specialization.

In addition to the changing view of cerebral dominance and hemispheric specialization over time, a more complex model of lateralization of such functions has also developed. Numerous studies have shown that the extent of hemispheric differences for most mental functions is not as pure as was once

ENCYCLOPEDIA OF HUMAN BIOLOGY, Second Edition, VOLUME 2. Copyright © 1997 by Academic Press. All rights of reproduction in any form reserved.

suggested. For example, although most studies of right-handed subjects present evidence that the left cerebral hemisphere is primarily involved in language-related functions, it has been shown that the right cerebral hemisphere has the capacity to understand some language and perform some language-related functions, although not with the effectiveness or capacity of the left. Within this framework most mental or cognitive functions can be performed by either hemisphere to some extent, but among individuals one hemisphere is more specialized, dominant, and efficient than the other.

Finally, recent studies have suggested that cerebral specialization is found in species other than humans, including nonhuman primates, birds, and rats. Although these studies are recent and require additional confirmation, they suggest that environmental and genetic factors could influence the development of different patterns of cerebral specialization in different species and that cerebral specialization is not unique to humans.

II. ANATOMICAL ASYMMETRIES

Since evidence for cerebral specialization began to be observed, there has been interest in whether the actual asymmetries in mental functions were related to asymmetries in the underlying brain anatomy. Although such relationships would not prove that the anatomical differences caused the pattern of cerebral specialization, they might lead to further evidence regarding the ontogeny or phylogeny of such patterns. It should be noted, though, that there are competing and interactive theories suggesting that such cerebral specialization is affected more by environmental factors, or an interaction between environmental and genetic factors, than by morphological factors.

A wide range of anatomical brain structures has been examined for left–right asymmetries. These studies have included measurements of the length, width, volume, weight, and angulation of numerous brain structures. Such studies have been performed directly on postmortem brains or via neuroimaging techniques (e.g., computerized tomography, nuclear magnetic resonance imaging, and arteriograms) *in vivo*. [*See* Magnetic Resonance Imaging.]

Anatomical areas that have been shown to be larger or longer in the left cerebral hemisphere compared to the right have included the lateral posterior nucleus, frontal operculum, fusiform gyrus, insula, occipital lobe, occipital horn of the lateral ventricle, parietal operculum, planum temporale (i.e., the posterior and superior surface of the temporal lobe), and sylvian fissure. Right hemisphere structures that have been shown to be larger or longer than those in the left have included the frontal lobe, Heschl's gyrus, and the medial geniculate. Overall, the most consistent and asymmetric region is the temporoparietal area, most right-handers showing larger left hemisphere structures in this area. The right hemisphere, on the other hand, has been described as being larger overall and extending more forward in the skull than the left hemispehre. Besides these anatomical asymmetries, there is also evidence for asymmetries in the underlying neurotransmitter systems (e.g., dopamine). These anatomical asymmetries have been observed in infants and children, and although there is evidence that each hemisphere could develop at different rates, the sequence of developmental neuroanatomical changes has not been clearly identified.

III. FUNCTIONAL ASYMMETRIES

A. Studies of Patients with Neurological Disorders

There have been two main groups of patients with neurological disorders who have provided the mainstay of research on cerebral specialization: patients with lateralized brain lesions or damage and patients who have undergone a commissurotomy (i.e., split-brain surgery) because of uncontrolled seizures. Patients with damage in one cerebral hemisphere due to a stroke, injury, surgery, or some other problem have been studied extensively to identify those mental functions that might be affected following such and injury.

The classic experimental paradigm for this type of research is to show that patients with one damaged hemisphere cannot perform a certain task or function, whereas patients with the other hemisphere damaged can perform the task. Even more specific localization of the effects of the damage can be made by comparing patients with damage in various areas of the same hemisphere. For example, patients with left frontal lobe damage might be compared to patients with left temporal lobe damage on certain language tasks. By this process the components of the system, and their links, can be identified. Unfortunately, such studies do not directly provide evidence for the exact functions of the damaged area, but provide information regarding how the remaining and intact parts of the brain work without that part. Because of this limitation, studies of neurological patients might not always provide accurate information regarding how a normal (i.e., non-neurologically damaged) brain might be specialized and might function.

Split-brain patients are those who have had their corpus callosum, the main bundle of fibers connecting the two cerebral hemispheres, surgically separated because of spreading seizures that cannot be controlled by other means. This surgery results in each of the two cerebral hemispheres being more independent and being unable or only partially able to communicate with each other. By special procedures researchers are able to present information to one hemisphere at a time and to study its ability to process it without the typical enlistment of the contralateral (i.e., opposite) hemisphere. For example, although the right hemisphere might be able to see a picture presented to it, the patient cannot name the picture; but if given a multiple-choice array of pictures that represent the possible answers, the patient can point to the correct answer using his or her left hand. If the same picture is presented to the left hemisphere, the patient can name the picture but cannot point to it with his or her right hand. Such a demonstration shows the different capabilities of the two hemispheres when they cannot pass information between them, as is typical.

B. Studies of Normal Subjects

There are many problems with using brain-damaged patients to study normal brain functioning. Therefore, a number of methods have been developed to study the cerebral specialization of people without brain damage. One of these is the visual half-field technique, which uses the neuroanatomical features of the visual system to present visual information to one hemisphere or the other. If a picture is presented very quickly (usually in less than 200 msec) in one visual field using a special machine called a tachistoscope, (not to one eye, but to either the left or right of the visual midline), the information is directed into the contralateral hemisphere. By comparing how well the left hemisphere performs such a task compared to the right hemisphere, one can identify processing asymmetries. In other words, by comparing how fast or how well one hemisphere performs the presented task compared to the other one can document patterns of cerebral specialization.

Another method is the dichotic listening technique, which uses the anatomical features of the auditory system to assess cerebral asymmetries. This is typically done by presenting different words, sounds, music, or related auditory information to each ear at the same time. Therefore, a person might hear the word "dog" in his or her right ear, but "pig" in the left ear. How fast a person responds, which words are heard and recalled, or how well a specific stimulus is identified are measures used to compare the processing of information given to the right and left ears. Given that a majority of the auditory pathways are contralateral (i.e., crossed) in nature, stimuli presented in the right ear go directly to the left hemisphere, whereas stimuli presented to the left ear go directly to the right hemisphere.

Another method of study, (called dihaptic study), involves a similar paradigm using tactually presented stimuli that are presented bilaterally to each hand. There have also been studies using olfactory stimuli. These laterality paradigms have yielded results consistent with those from studies of brain-injured patients.

IV. DIFFERENCES IN CEREBRAL SPECIALIZATION

One of the more basic questions addressed within the area of cerebral specialization is whether men and women differ in their patterns of abilities. Besides the typically cited findings that females are better in verbal-related functions while males are better in spatial-related functions, most research suggests that women might exhibit more bilateral (i.e., in both hemispheres) representation of various abilities when compared to men, although all of these differences are small. In other words, men appear to be more lateralized, or show greater asymmetry in specialization, than women.

There has also been a great deal of interest in differences in patterns of cerebral specialization between people with right- and left-handed preferences. A relationship between handedness and the lateralization of language has been identified since the earliest studies in this area. A large majority of the human population (i.e., >90%) is right-handed, and almost all of these right-handers (i.e., >98%) show a left cerebral hemisphere specialization for language-related functions. On the other hand, there is great debate about the pattern of cerebral specialization in left-handers. Most studies suggest that over 60% of left-handers have left cerebral hemisphere specialization for language-related functions, while there is great debate about how the remaining 40% of the left-handers are specialized. Most studies suggest that about 20% of the left-handers have right cerebral hemisphere specialization for language-related functions, whereas the remaining 20% are more bilateral in their representation of language abilities. The degree of specialization for non-language-related abilities does not appear to be as strong or specific, although most right-handers (i.e., 70%) show right hemisphere specialization for visuospatial abilities, whereas only about 33% of left-handers show such a relationship.

V. ASYMMETRIES IN OTHER SPECIES

The question of whether animals other than humans manifest lateral asymmetries has been of considerable interest to scientists. Recent evidence suggests that anatomical asymmetries exist in nonhuman primate brains, similar to those found in humans. For example, the length of the left sylvian fissure is longer than that of the right in both monkeys and chimpanzees. Also, patterns indicate a right frontal–left occipital extension in nonhuman primates. Thus, if anatomical asymmetries are correlated with functional asymmetries, then some manifestation of these asymmetries should exist in the behavior of the organisms.

Two issues have been the central focus of studies. The first involves the distribution of lateral asymmetries within a given species. The second involves the homologous or analogous relationship between asymmetries observed in one species relative to another, including humans.

Many individual subjects within a given species show hand, or paw, preferences. However, the overall distribution of hand/paw preferences appears to be equal between left- and right-preference individuals. This is different from the pattern observed in humans, which has prompted many researchers to conclude that asymmetries do not exist in species other than *Homo sapiens*.

Recently, some have suggested that species, particularly nonhuman primates, might show population levels of hand preference, but the direction and degree of asymmetry differ from those observed in humans. Thus, the features of hand preferences in nonhuman species at this time do not seem well defined. Moreover, the pattern of asymmetry might differ among different species. Thus, until further research is conducted, the features and parameters that account for hand preferences in nonhuman animals remain inadequately defined.

Although most studies of hemispheric specialization in nonhuman animals have focused on hand/paw preferences, there is a growing body of literature suggesting that certain cognitive asymmetries exist in other species. The principal technique for evaluating these processes has been to teach an animal to perform a particular task and then lesion an area of the brain. If that particular area is involved in the task being studied, then decrements in performance should be observed. For example, songbirds have been shown to have a left hemisphere dominance in the production of songs learned early in life. If the left hypoglossal nerve is severed in these birds, their capacity to produce songs becomes severely disrupted, whereas if the right nerve is severed, song is not affected. Similar findings have been reported in rodents as well as some primate species in terms of auditory processing of species-specific sounds.

Although one could argue that the processing of species-specific sounds is identical or homologous to that of language processing in humans, this comparison is difficult. The cognitive operations and component analysis involved in language processing might be different in humans relative to other species. Thus, the processing of auditory stimuli presented in a sequential manner might be an underlying asymmetry shared among many species, but the process of extracting meaning, syntax, or intonation from this stream of sounds could involve different aspects of neuropsychological functions not found in other species. Thus, perception of species-specific sounds might be a function analogous to language processing in humans and might have evolved by means of different selection processes.

VI. EVOLUTION OF HEMISPHERIC SPECIALIZATION

In the conceptualization of hemispheric specialization, one must think about its evolutionary significance and what prompted its emergence in animals, including humans. Many theories have been espoused elucidating the evolutionary precursors to hemispheric specialization and the environmental factors that shaped its development. Some have argued that language and cerebral dominance evolved in a parallel manner owing to the close link between manual specialization and language processing. For example, some people argue that the first language used by *Homo sapiens* was a manual, or gestural, language. If a left hemisphere dominance for hand preference already existed in early humans, then assuming they would have utilized their dominant hand for a gestural language, language could have evolved in the left hemisphere. Still others argue that early humans did not use a manual or gestural language, but rather utilized an auditory communication system. They argue that the left hemisphere had a specialization for processing and producing sequential movements. Since auditory signals (e.g., speech) are sequential, the left hemisphere subsequently took control of processing speech sounds. Although both theories can account for current findings in the literature, neither explains how the left hemisphere initially came to have a hemispheric specialization. In other words, it is unknown why the brain operates asymmetrically at all. [*See* Language, Evolution.]

It has been suggested that perhaps the greatest advantage to having asymmetrical processing in the brain is because it nearly doubles the cognitive capacity of an organism. If each half of the brain subsumes separate operations but communicates with the other half, it potentially provides for twice as many capacities. Without complementary specialization, cognitive function is redundant for both halves of the brain. Within this theoretical framework neither language nor other cognitive functions need to be assumed as major selection pressures for the evolution of hemispheric specialization in humans. Instead, at both an individual and a species level, the selection pressure for asymmetrical processing of relevant information within the niche of that species would serve as the basis for differential processing of stimuli within the brain.

Whatever the case, hemispheric specialization has emerged as a most fascinating and controversial issue in the realm of science. Its application and implications for the fields of education, biology, and psychology are only beginning to emerge. One goal of science is to explain and understand phenomena. Scientists have only touched the surface with regard to understanding the evolution, phylogeny, and process of hemispheric specialization in animals and humans.

ACKNOWLEDGMENTS

The work repored here was supported by National Institutes of Health Grants RR0165 and NICHD:06016, and NASA Grant NAG 2-438.

BIBLIOGRAPHY

Beaton, A. (1986). "Left Side, Right Side: A Review of Laterality Research." Yale Univ. Press, New Haven, CT.

Gazzaniga, M. (1985). "The Social Brain: Discovering the Networks of the Mind." Basic Books, New York.

Geschwind, N., and Galaburda, A. M. (1985). Cerebral lateralization: Biological mechanisms, associations, and pathologies. A hypothesis and a program for research. *Arch. Neurol.* **42,** 428–459.

Harnad, S., Doty, R. W., Goldstein, L., Jaynes, J., and Krauthamer, G. (1977). "Lateralization in the Nervous System." Academic Press, New York.

Hopkins, W. D., and de Waal, F. B. D. (1995). Behavioral laterality in captive bonobos (*Pan paniscus*): Replication and extension. *Int. J. Primatol.* **16,** 261–276.

Hopkins, W. D., and Morris, R. (1993). Handedness in great apes: A review of findings. *Int. J. Primatol.* **14**(1), 1–25.

Kolb, B., and Whishaw, I. Q. (1990). "Fundamentals of Human Neuropsychology." Freeman, San Francisco.

Ward, J. P., and Hopkins, W. D. (eds.) (1993). "Primate Laterality: Current Behavioral Evidence of Primate Asymmetries." Springer-Verlag, New York.

Cerebral Ventricular System and Cerebrospinal Fluid

J. E. BRUNI
University of Manitoba

GLOSSARY

Arachnoid villi/granulations Finger-like projections of the arachnoid mater into the dural venous sinuses and/or their venous lacunae that function as one-way valves returning cerebrospinal fluid to the venous blood

Cerebrospinal fluid Watery, clear, colorless liquid produced largely by the choroid plexuses and contained within the cerebral ventricles, central canal of the spinal cord, and the subarachnoid spaces surrounding the central nervous system

Choroid plexus Vascularized epithelium located within each of the ventricles of the brain that produces the cerebrospinal fluid

Hydrocephalus Condition in which there is either overproduction, obstructed circulation, or impaired absorption of cerebrospinal fluid, resulting in accumulation of fluid within the cerebral ventricles or subarachnoid spaces, raised intracranial pressure, and consequent neurological and functional disturbances

Ventricles Cerebrospinal fluid-filled cavities located within each cerebral hemisphere, the diencephalon, and the brain stem

I. CEREBRAL VENTRICLES

The brain and spinal cord are not solid structures. Deep within their substance are cerebrospinal fluid (CSF)-filled cavities or chambers called ventricles. There are four such chambers: the paired lateral ventricles, a midline third ventricle, and a fourth cerebral ventricle. They communicate with each other, with the central canal of the spinal cord, and with the subarachnoid space (SAS) that surrounds the central nervous system. [*See* Brain; Spinal Cord.]

A. Lateral Cerebral Ventricles

The lateral ventricles are paired C-shaped cavities located deep within the white matter of each cerebral hemisphere. Each consists of a central part or body and three horns, the anterior or frontal, the inferior or temporal, and the posterior or occipital horn (Fig. 1). The lateral ventricle in each hemisphere communicates with the vertical midline third ventricle via a narrow channel called the interventricular foramen (of Monro).

The frontal horns of the lateral ventricle project into the frontal lobes of the brain and become confluent with the body of the ventricle at the level of the interventricular foramen (Figs. 1 and 2). The large neocortical commissure, the corpus callosum, forms the roof of the frontal horns of the ventricle. Their lateral walls and floor are formed in part by the caudate nucleus. A thin membranous partition called the septum pellucidum separates the two frontal horns in the midline (Fig. 2).

The body of the lateral ventricle is a triangular cavity that extends from the interventricular foramen to the level of the splenium of the corpus callosum posteriorly. Its roof and floor are formed by the corpus callosum and dorsal surface of thalamus, respectively. The fornix borders the medial wall and floor, whereas

ENCYCLOPEDIA OF HUMAN BIOLOGY, Second Edition, VOLUME 2. Copyright © 1997 by Academic Press. All rights of reproduction in any form reserved.

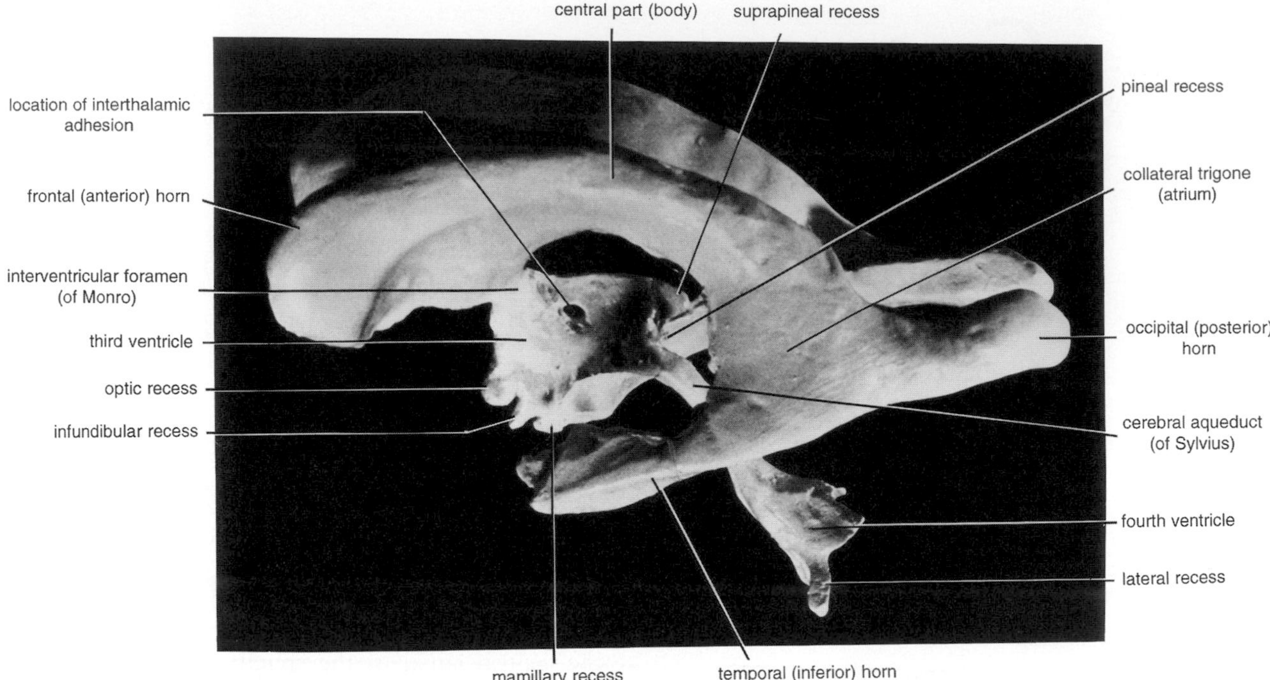

FIGURE 1 Cast of the ventricles of the human brain. [Reproduced with permission from D. G. Montemurro and J. E. Bruni (1988). "The Human Brain in Dissection," 2nd Ed. Oxford Univ. Press, New York.]

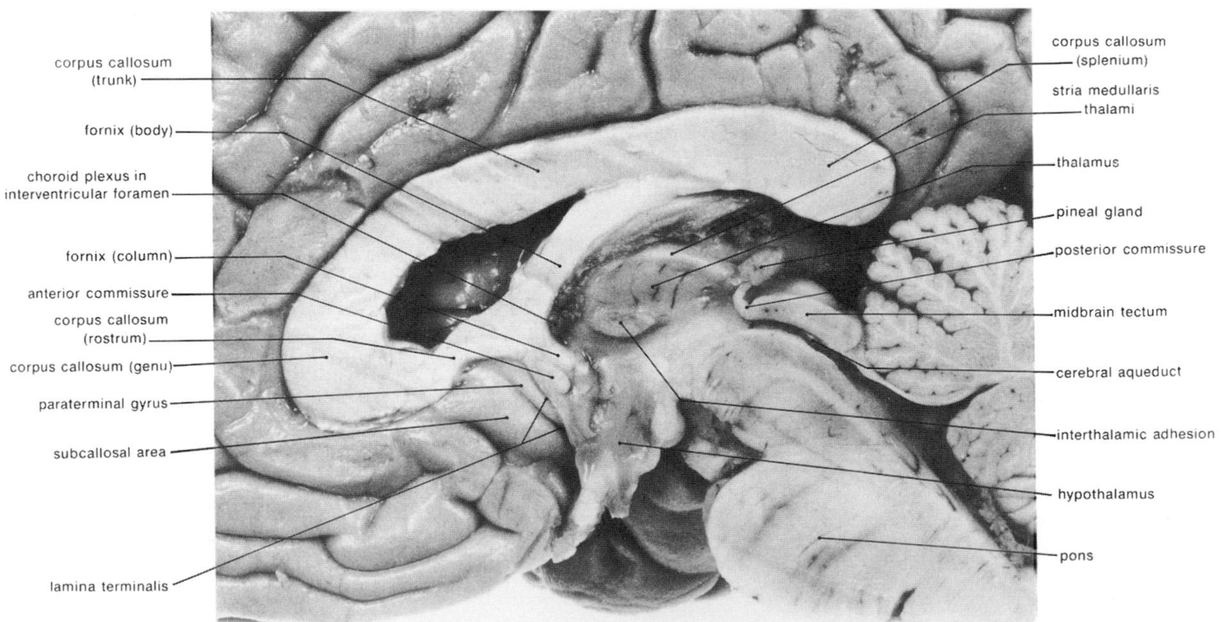

FIGURE 2 Medial view of the right half of the hemisected human brain showing the lateral, third, and fourth ventricles, the interventricular foramen, and cerebral aqueduct *in situ*. The septum pellucidum has been removed along its attachment to the corpus callosum and the fornix to provide a window into the frontal horn of the right lateral ventricle. [Reproduced with permission from D. G. Montemurro and J. E. Bruni (1988). "The Human Brain in Dissection," 2nd Ed. Oxford Univ. Press, New York.]

the tail of the caudate nucleus contributes to the floor and lateral wall.

The temporal horn is an extension of the ventricle into the temporal lobe. In its roof is the tail of the caudate nucleus. The hippocampus is located along its floor and medial wall, whereas the lateral wall consists of white matter, the geniculocalcarine tract, and arcuate fasciculus of the cerebral hemisphere. The choroid plexus is present along the medial wall of the body and temporal horn of the lateral ventricles.

The occipital horn is a similar extension of the ventricle directed caudally into the center of the occipital lobe of the brain. The shape and size of this part of the ventricle is most variable and can be asymmetrical, almost nonexistent, or a significant diverticulum that extends for some distance into the occipital lobe. The occipital horn is surrounded mainly by white matter, including the geniculocalcarine tract, and the radiations of the corpus callosum of the cerebral hemisphere. At the junction of the body with the occipital and temporal horns, the ventricle assumes an expanded triangular shape known as the collateral trigone or atrium.

B. Third Cerebral Ventricle

The third cerebral ventricle is a narrow slit that extends vertically in the midline between the two halves of the forebrain region known as the diencephalon (Figs. 1 and 2). The third ventricle is bounded anteriorly by the lamina terminalis, which extends from the corpus callosum above to the optic chiasma below. The third ventricle communicates with the lateral ventricles anteriorly by way of the interventricular foramen (of Monro).

Dorsally, the lateral wall of the third ventricle is formed by the gray matter of the thalamus, the largest of the four divisions of the diencephalon. The two halves of the thalamus are joined in the midline in ~70% of brains across the cavity of the third ventricle by a mass of thalamic gray matter called the massa intermedia or interthalamic adhesion (Fig. 2).

Posteriorly, the dorsolateral walls of the third ventricle are formed by the smallest division of the diencephalon, known as the epithalamus (Latin *epi* = "above" + thalamus). This region consists of a number of structures that include an endocrine organ known as the pineal gland. In the midline of this region the third ventricle becomes continuous with the narrow cerebral aqueduct (of Sylvius), which traverses the periaqueductal gray matter of the midbrain. The choroid plexus is suspended into the third ventricle all along its narrow roof.

The ventrolateral wall and floor of the third ventricle are formed by the small division of the diencephalon known as the hypothalamus (Latin *hypo* = "below" + thalamus) (Fig. 2). A shallow horizontal groove on the lateral wall of the third ventricle, called the hypothalamic sulcus, separates the hypothalamus below from the thalamus above.

From the cavity of the third ventricle there arise a number of small extensions called recesses (optic, pineal, suprapineal, infundibular recesses), which project into surrounding diencephalic structures (Fig. 1). The most notable of the recesses is the infundibular recess, which is a funnel-shaped projection from the floor of the ventricle into the small stalk (infundibular) that connects the pituitary gland to the basal hypothalamus in the midline (Fig. 1).

C. Fourth Ventricle

The fourth ventricle is a diamond-shaped cavity, situated within the dorsal pons and dorsal anterior or open part of the medulla oblongata (Figs. 1 and 2). This ventricle is also sometimes called the rhomboid fossa because of its four-sided diamond-like configuration. Its roof is tent-shaped and formed largely by the cerebellum. The choroid plexus is present in the roof of the fourth ventricle. Its lateral walls are formed by the cerebellar penduncles, which are three pairs of robust fiber bundles that anchor the cerebellum to each of the three regions of the brain stem. The fourth ventricle is continuous rostrally with the cerebral aqueduct (of Sylvius) in the midbrain and caudally with the central canal of the spinal cord via the closed part of the medulla oblongata.

Posteriorly, the fourth ventricle communicates with the extracranial subarachnoid space, specifically the large subarachnoid cistern known as the cisterna magnum, through three openings or foramina; the midline foramen of Magendie and the paired lateral foramina (of Lushka) (Fig. 1).

Situated immediately beneath the floor of the fourth ventricle are many nuclei that belong to certain of the cranial nerves. [*See* Cranial Nerves.] A chemoreceptive zone capable of eliciting vomiting and also one of seven circumventricular organs (see Section III) known as the area postrema is located on either side of the posterior midline of the fourth ventricle.

II. CENTRAL CANAL OF THE SPINAL CORD

The central canal of the spinal cord is a narrow channel that traverses the midline of the transverse gray commissure (lamina X of Rexed) throughout the length of the spinal cord. It is continuous with the canal of the medulla oblongata and the fourth ventricle and terminates below, in a fusiform expansion (the terminal ventricle) within the conus medullaris. Not unlike the cerebral ventricles, it is also lined by an epithelium consisting for the most part of a single continuous layer of cells known as ependymal cells. Although it is a continuation of the fluid-filled ventricular system and is present throughout the spinal cord, it is usually not patent after the 20th year of life.

III. CIRCUMVENTRICULAR ORGANS

Closely associated with the cerebral ventricles, particularly the third ventricle, are seven specialized-structures known as circumventricular organs. They include the subfornical organ, the organum vasculosum of the lamina terminalis or supraoptic crest, the median eminence, the neurohypophysis, the subcommissural organ (SCO), the pineal body, and the area postrema (AP). All except the AP are unpaired midline structures located within the walls of the third cerebral ventricle. The CVOs all contact the CSF of the ventricular cavity or SAS, they have a high capillary density, and all except for the SCO lack a blood–brain barrier. Although a detailed discussion of these organs is beyond the scope of this article, they are believed to serve as neural transducers responding to substances contained in the blood and/or CSF and are involved in diverse functions such as neuroendocrine regulation, body fluid homeostasis, regulation of circadian rhythms, and vomiting.

IV. CEREBROSPINAL FLUID

A. Formation and Composition

The CSF, a clear, colorless fluid with a specific gravity (1.004–1.007 g/cm^3) slightly greater than that of water, resembles an ultrafiltrate of blood plasma. Under normal conditions, it contains very little protein and has a lower pH and lower concentrations of glucose, potassium, calcium, bicarbonate, and amino acids than does blood plasma. The CSF concentrations of sodium, chloride, and magnesium, however, are larger than those in plasma. Few cells, mainly lymphocytes, are usually present in the CSF. The presence of 0–8 cells/mm^3 in infants and 0–5 cells/mm^3 in adults, however, is usually considered normal. Any changes in the composition of the CSF, such as increased protein concentration, decreased glucose, or changes in its appearance (color or cloudiness), are usually indications of some pathological processes.

Under normal conditions, CSF is formed by the choroid plexus of the four ventricles by a process of active choroidal epithelial cell secretion combined with passive capillary diffusion. The major portion of CSF, however, is formed by the choroid plexus of the lateral ventricles. Although contentious, it has also been suggested that at least 10% of CSF may be formed at sites other than the choroid plexus, particularly under pathological conditions. The ependymal lining of the ventricles and the endothelium of brain capillaries have been considered potential sites of extrachoroidal production. In humans, the total volume of CSF contained within the ventricular system and the subarachnoid space is estimated to be about 140 ml (~10% of brain weight) and the rate of formation is approximately 0.35 ml/min or 500 ml/day. The ventricular system alone is believed to contain from 15 to 40 ml of CSF and ~75 ml surrounds the spinal cord. In humans, a circadian variation in CSF production has recently been demonstrated by Nilsson *et al.* (1994) in which there is a nocturnal increase in production that reaches twice the daytime values at about 2:00 A.M. CSF is produced at a rate sufficient to replace the total volume of CSF approximately two to three times over in 24 hr. In other species, the rate of formation has been determined to be proportional to choroid plexus weight. The normal pressure of CSF as measured by lumbar puncture varies from 25 to 70 mm H_2O in infants and from 65 to 195 mm H_2O in adults while in the recumbent position.

Since the composition of CSF is different from that of an ultrafiltrate of plasma, it must be produced by active choroidal epithelial cell secretion and not just passive capillary diffusion. In the production of CSF, some components like water, enter the CSF by simple diffusion whereas others like sodium ions require active transport mechanisms and expenditure of energy by the choroidal epithelium. Because the secretory process is energy dependent, it can be affected by drugs that act on metabolic processes.

I. Choroid Plexus

The choroid plexus is present within each of the four ventricles of the brain and is principally responsible for the production of CSF. In each lateral ventricle, the choroid plexus extends along the choroidal fissure at the junction of the medial wall and floor from the level of the interventricular foramen to the end of the temporal horn. It is also invaginated into the roof of each of the third and fourth ventricles. It has a characteristic lobulated appearance and consists of a single continuous layer of epithelial cells overlying a highly vascular central core (Fig. 3). The epithelial cells, sometimes referred to as the lamina epithelialis or choroidal epithelium, are derived from the ependymal lining of the ventricles; the blood vessels and connective tissue core derive from a vascular fold of pia mater termed the tela choroidea.

a. Choroidal Epithelium

The choroidal epithelium consists of a single continuous layer of simple cuboidal cells (~10 μm diameter) resting on a basal lamina and forming a regular pattern of domes or bulges on the surface of the choroid plexus. They contain a large centrally placed spherical nucleus surrounded by an abundant lucent cytoplasm (Fig. 3). Their luminal surface is covered with thin cytoplasmic projections known as microvilli, which serve to increase the area available for exchange. Apically, the lateral surfaces of adjoining cells are bound by tight junctions obliterating the intercellular space and forming a barrier to extracellular movement of substances to or from the CSF. The barrier function normally served by brain capillaries is shifted to the epithelium itself in the choroid plexus. The site of the sodium pump involved in sodium secretion by the choroid plexus is located at the microvillus surface. This supports the idea that sodium from the blood enters choroidal cells passively down an electrochemical gradient and then is actively pumped across the apical surface into the CSF. The basolateral surface of the cells is complexly infolded, also increasing the area available for exchange.

The internal structure of these cells reflects their function as producers of CSF. They typically contain a large number of mitochondria especially apically, which are responsible for the high metabolism and the energy for active transport. Cisternae of granular endoplasmic reticulum are distributed throughout the cytoplasm. The Golgi apparatus is ill defined and is represented by numerous stacks of cisternae especially prominent apically and at the sides of the nucleus. Smooth endoplasmic reticulum in the form of tubules and vesicles is extensive, particularly in some cells, and clear vesicles (30–40 nm diameter) are present, particularly in the apical cytoplasm. There is a vesicular transport from the basal to the apical pole of the cells. Vesicles originate by pinocytosis at the basal end of the cell and move to the luminal surface to be discharged into the CSF.

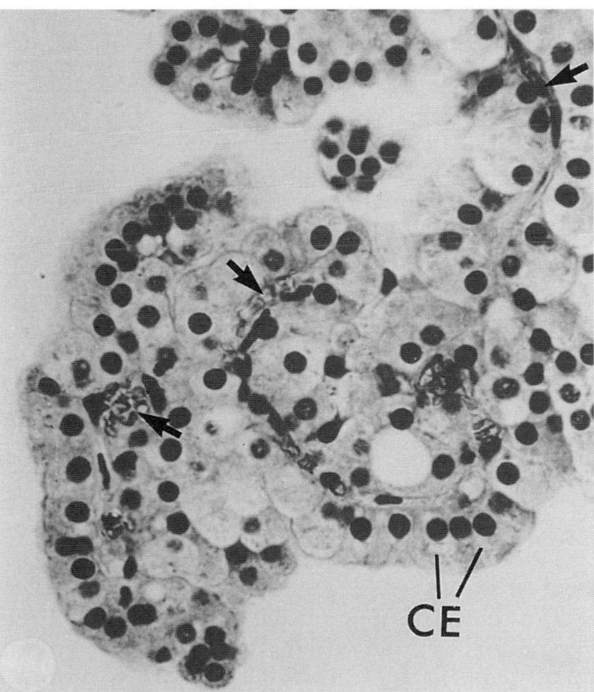

FIGURE 3 Choroid plexus from a rabbit brain as seen with the light microscope. CE, choroidal epithelial cells; arrows, vascular connective tissue core.

b. Connective Tissue Core

Underlying the choroidal epithelial cells and basal lamina on which they rest is the connective tissue core of the choroid plexus. It is composed of a loose, collagenous fiber network surrounding a large complement of small arteries, arterioles, and large venous sinuses (Fig. 3). It is particularly distinguished by the unusually large-caliber capillaries with thin endothelial walls that lack of a blood–brain barrier. Equally unusual is the fact that the capillary endothelium has fenestrations bridged by diaphragms. An extensive array of nerve fibers supply both the walls of blood vessels and the secretory choroidal epithelium itself. This innervation has also been shown to influence the formation of CSF.

c. *Functions of the Choroid Plexus*

The choroid plexus is primarily concerned with the formation of CSF by a combination of active secretion and passive capillary dialysis. It has also been implicated in absorption of substances from the ventricles (i.e., active transport of substances out of the CSF and into the blood against a concentration gradient). The ultimate composition of CSF, therefore, may result from a combination of both secretory and absorptive processes. In addition, arterial pulsations generated by choroidal blood vessels may assist with the circulation of CSF. Greitz *et al.* (1992) has reported that expansion of intracranial arteries in general produces brain expansion, ventricular system compression and may be the main driving force for intraventricular CSF flow.

B. Circulation of CSF

From the lateral ventricles, CSF passes into the third ventricle through the interventricular foramen. Very little if any reflux of CSF into the contralateral lateral ventricle is believed to occur. From the third ventricle, the CSF reaches the fourth ventricle via the cerebral aqueduct. CSF then leaves the ventricular system at the level of the medulla oblongata through three apertures: the midline foramen (of Magendie) and the paired lateral foramina (of Lushka). These apertures communicate with enlargements of the subarachnoid space known as the cisterna magna and the cisterna pontis, respectively.

The subsequent circulation of fluid within the subarachnoid space also follows a prescribed course. From the lateral foramina and cisterna pontis, CSF flows anteriorly along the base of the brain to the interpeduncular and chiasmatic cisterns, ascending slowly to the Sylvian fissure and along the lateral convex and medial surfaces of the hemispheres. From the midline foramen the flow of CSF is directed forward over the cerebellar hemispheres toward the tentorial incisure and superior cistern or downward into the subarachnoid space surrounding the spinal cord.

Only a small amount of CSF is thought to reach the fourth ventricle by ascending the central canal of the spinal cord, which in fact is occluded in most adults after the 20th year of age. Although the amount of CSF occupying the spinal subarachnoid space comprises about one-half of the total volume, the downward flow of CSF around the spinal cord in normal humans has been the subject of debate. Some have argued that there is no true spinal flow other than gravity-assisted diffusion. Others have demonstrated a rapid descent of spinal CSF. The significance of a downward spinal flow of CSF is that it provides an alternate path for CSF resorption and permits sampling of cerebral metabolites by withdrawal of CSF from the lumbar cistern.

The ultimate fate of the CSF in both its intracranial and spinal course is to drain into venous channels that it reaches by way of arachnoid villi (see the following). The circulation of CSF through the ventricular system and subarachnoid space in this manner has been attributed to pressure waves in the fluid generated by arterial pulsations within the choroid plexus and intracranially, the pressure gradients produced by production and absorption of CSF and currents induced by ependymal cilia. The latter more likely contributes more to mixing of CSF and movement of particulate matter within the CSF than to its bulk flow.

C. Cerebrospinal Fluid Absorption

The main route by which CSF gains access to the blood is through the arachnoid villi. These are microscopic tufts of pia-arachnoid mater that project into venous channels at CSF–vascular interfaces (Fig. 4). Aggregations of villi that are visible macroscopically are referred to as arachnoid granulations. Villi are finger-like projections consisting of a cellular and fibrous connective tissue core surrounding fluid-filled spaces that are continuous with the subarachnoid space. Villi function as one-way valves returning CSF from the SAS to the dural venous sinuses because the hydrostatic pressure of the CSF usually exceeds venous pressure. Although the precise nature of transport between the SAS, villus spaces, and the venous channels is still disputed, CSF absorption is thought to involve (1) a system in which there is free intercellular communication between the CSF and venous system via open channels within the villi, (2) vesicular or vacuolar transport across cells or along endothelial lined pores, and/or (3) a closed system in which there is diffusion across a permeable membrane separating the CSF and vascular systems. Each is considered consistent with the active transport of protein and CSF and a passive valve-like action on the part of the arachnoid villi.

Although CSF, particularly in primates and humans, empties mainly into the dural venous channels

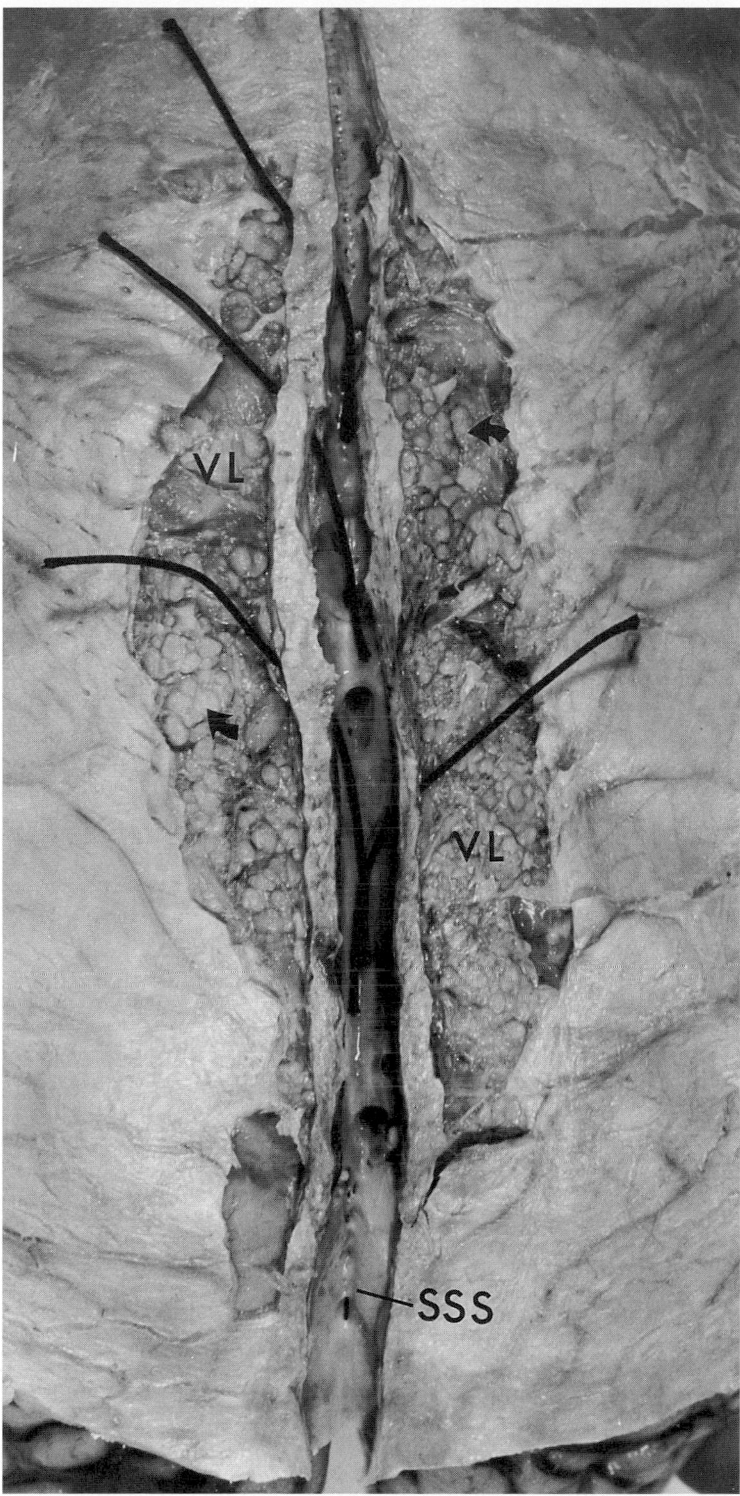

FIGURE 4 Illustration showing the superior sagittal sinus (SSS) and adjoining venous lacunae (VL). The dural roof of the sinus and the lacunae has been removed to show the arachnoid granulations (arrows). The points of communication between the sinus and the venous lacunae are shown by the threads. [Reproduced with permission from D. G. Montemurro and J. E. Bruni (1988). "The Human Brain in Dissection," 2nd Ed. Oxford Univ. Press, New York.]

(sinuses and lacunae) by way of the arachnoid villi, alternate pathways of CSF absorption have also been described. Small amounts of CSF are known to be absorbed across the walls of cerebral capillaries within the brain parenchyma following migration transependymally and along perivascular spaces. Some absorption also occurs via pial vessels and lymphatic channels adjacent to extensions of the subarachnoid space that surround spinal and certain cranial nerves. The presence of arachnoid villi/granulations around spinal nerve roots and veins are consistent with absorption of CSF at spinal cord levels. Involvement of these alternative routes of absorption are thought to be particularly important under pathological conditions such as hydrocephalus (see Section VI). It should also be noted that the primary route of CSF outflow in many mammals is along perineural and paravascular prolongations of the subarachnoid space into lymphatics of the nose and neck.

V. FUNCTIONS OF THE CEREBROSPINAL FLUID

Cerebrospinal fluid performs the following functions:

1. The brain and spinal cord are immersed in a fluid medium that makes them buoyant. A 1.4-kg brain in air weighs only 50 g in CSF affording a measure of physical support and protection against rapid movements and trauma.
2. The CSF serves a nutritive function for both neurons and the supporting glial cells of the nervous system.
3. The CSF acts like a lymphatic system, providing a vehicle for removing waste products of neuronal metabolism.
4. The extracellular space of the brain is in free communication with the CSF and the composition of the fluid in the two compartments is similar. The CSF therefore, plays an important role in maintaining the constancy of the local environment for all cells of the nervous system.
5. The CSF circulates within and over the brain and spinal cord. The presence of a number of biologically active principles (releasing factors, hormones, neurotransmitters, metabolites) within the CSF suggests that it may also function as a transport system.
6. The H^+ and CO_2 concentrations in the CSF (pH) may affect pulmonary ventilation, cerebral blood flow, heart rate, vasomotor, and behavioral activity.
7. By virtue of its composition and changes thereof, the CSF reflects the state of health of the nervous system.

VI. HYDROCEPHALUS

Excess CSF intracranially due to overproduction or impaired absorption or circulation results in a condition known as hydrocephalus (Greek for water + head). Overproduction of CSF is a rare cause of hydrocephalus but may occur with some tumors of the choroid plexus. Hydrocephalus is primarily an affliction of children, occurring as a congenital disease or sequelae of cranial tumors, germinal matrix hemorrhage, or meningitis. In the adult, hydrocephalus is less common, usually arising from subarachnoid hemorrhage, head injury, tumors, meningitis, or unknown causes. [*See* Hydrocephalus.]

Because the brain is encased in a vault of bone, any increase in the volume of CSF must be compensated for by a corresponding reduction in volume of another intracranial compartment or else an increase in intracranial pressure (ICP) ensues. In hydrocephalus there is a progressive enlargement of the ventricles, increased ICP, interstitial edema, diminished cerebral blood flow, and compression of nervous tissue. The extent of the CSF pressure rise depends on the severity of the obstruction and the ability of the brain to compensate. In infants with open cranial sutures for example, compensation is achieved by head enlargement.

The clinical symptoms of hydrocephalus in the adult include headache, nausea/vomiting, disturbances of gait and vision, altered consciousness, impaired memory, papilloedema, and general malaise. If the condition is left untreated, the disturbances of metabolism, neurotransmission, and so on that ensue from the increased ICP lead to progressive intellectual, emotional, and motor dysfunction and eventually death. Although a number of drugs can affect the rate of formation or absorption of CSF, they tend not to be effective clinically. The preferred treatment for hydrocephalus is surgical ventriculostomy, or "shunting" of CSF usually into the peritoneal cavity; procedures which themselves are not without complications that result in some morbidity and mortality.

A. Types of Hydrocephalus

External or communicating hydrocephalus is caused by a blockage somewhere outside the ventricular system proper. The ventricles remain in communication with one another and the subarachnoid space of the brain and spinal cord. Enlargement of the SAS over the cortical surface is usually due to inadequate absorption at the level of the arachnoid villi.

Internal or noncommunicating hydrocephalus is a condition in which obstruction occurs within the ventricular system at or somewhere before the outlet foramina to the SAS, causing the ventricular system alone proximal to the obstruction to enlarge. A particularly vulnerable site for obstruction is the narrow cerebral aqueduct (of Sylvius).

Hydrocephalus *ex vacuo* is a condition characterized by increased CSF volume, ventricular and subarachnoid space size as a result of brain atrophy. In this circumstance intracranial volume remains the same and there is no increase in intracranial pressure.

BIBLIOGRAPHY

Black, P. McL., and Ojemann, R. G. (1990). Hydrocephalus in adults. *In* "Neurological Surgery" (J.R. Youman, ed.), 3rd Ed., Vol. 2. Saunders, Philadelphia.

Cserr, H. F., Harling-Berg, C. J., and Knopf, P. M. (1992). Drainage of brain extracellular fluid into blood and deep cervical lymph and its immunological significance. *Brain Pathol.* **2,** 269–276.

Davson, H. (1972). Dynamic aspects of cerebrospinal fluid. *Dev. Med. Child Neurol.* **14** (Supp 27), 1–16.

Di Chiro, G., Hammock, M. K., and Bleyer, W. A. (1976). Spinal descent of cerebrospinal fluid in man. *Neurology.* **26,** 1–8.

Gomez, D. G., Chambers, A. A., Di Benedetto, A. T., and Potts, D. G. (1974). The spinal cerebrospinal fluid absorptive pathways. *Neuroradiology* **8,** 61–66.

Greitz, D., Wirestam, R., Franck, A., Nordell, B., Thomsen, C., Stahlberg, F. (1992). Pulsatile brain movement and associated hydrodynamics studied by magnetic resonance phase imaging. *Neuroradiology* **34,** 370–380.

Gross, P. M., and Weindl, A. (1987). Peering through the windows of the brain. *J. Cerebr. Blood Flow Metabol.* **7,** 663–672.

Hammock, M. K., and Milhorat, T. H. (1973). Recent studies on the formation of cerebrospinal fluid. *Dev. Med. Child Neurol.* **15** (Supp. 29), 27–34.

Kida, S., Pantazis, A., and Weller, R. O. (1993). CSF drains from the subarachnoid space into nasal lymphatics in the rat. Anatomy, histology and immunological significance. *Neuropathol. Appl. Neurobiol.* **19,** 480–488.

Lindvall, M., and Owman, C. (1984). Sympathetic nervous control of cerebrospinal fluid production in experimental obstructive hydrocephalus. *Exp. Neurol.* **84,** 606–615.

Lowhagen, P., Johansson, B. B., and Nordborg, C. (1994). The nasal route of cerebrospinal fluid drainage in man. A light microscope study. *Neuropathol. Appl. Neurobiol.* **20,** 543–550.

Maillot, C. (1991). The perispinal spaces—constitution, organization and relations with the cerebrospinal fluid. *J. Neuroradiol.* **18,** 18–31.

Montemurro, D. G., and Bruni, J. E. (1988). "The Human Brain in Dissection," 2nd Ed., Oxford Univ. Press, New York.

Nilsson, C., Stahlberg, F., Gideon, P., Thomsen, C., and Hendriksen, O. (1994). The nocturnal increase in human cerebrospinal fluid production is inhibited by a beta 1-receptor antagonist. *Am J. Physiol.* **267,** R1445–1448.

Pollay, M., and Davson, H. (1963). The passage of certain substance out of the cerebrospinal fluid. *Brain* **86,** 137–150.

Rall, D. P. (1964). The structure and function of the cerebrospinal fluid. *In* "Cellular Functions of Membrane Transport" (J. Hoffman, ed.), pp. 269–282. Prentice–Hall, Englewood Cliffs, NJ.

Weller, R. O., Kida, S., and Zhang, E-T. (1992). Pathways of fluid drainage from the brain—morphological aspects and immunological significance in rat and man. *Brain Pathol.* **2,** 277–284.

Cerebrovascular System

TOMIO SASAKI
University of Tokyo

NEAL F. KASSELL
American Association of Neurological Surgeons

GLOSSARY

Anastomosis Natural communication between two blood vessels by collateral channels

Autoregulation Relative constancy of blood flow during alterations in arterial pressure

Innervation Supply of nerves to a body part

CEREBRAL BLOOD VESSELS DIFFER STRUCTURALLY, physiologically, and pathologically from those of extracranial sites. These differences appear to be related to the phenomena responsible for the unique functions of the highly specialized tissues of the central nervous system. From a basic as well as a clinical point of view, currently available information on the features of the cerebrovascular system is provided in this article.

I. DEVELOPMENT OF THE CEREBROVASCULAR SYSTEM

The development of the cerebral circulation system is divided into five stages:

1. The formation of a primordial endothelial vascular plexus from cords of angioblasts.
2. The differentiation into primitive capillaries, arteries, and veins.
3. The stratification of the vasculature into external, dural, and leptomeningeal or pial circulation.
4. The rearrangement of the vascular channels to conform to the marked changes in surrounding head structures.
5. The late histological differentiation into adult vessels.

When the sequential order of this process fails, clinically significant lesions may occur. Faulty formation of the capillary bed during the second stage results in arteriovenous malformations. The changes during stage 3 would then determine whether the arteriovenous malformations would be situated in the scalp, the dura, or the brain. [*See* Brain.]

During the first 8 weeks of fetal life, transitory arteries appear and serve as anastomoses between the primitive internal carotid artery and the bilateral longitudinal neural arteries. Bilateral longitudinal neural arteries will later fuse to form the basilar artery. The trigeminal, otic, and primitive hypoglossal arteries are transitory. These transitory arteries become progressively atrophied (first the otic, then the hypoglossal, and finally the trigeminal) and then disappear completely. Occasionally, these vessels persist into adult life.

II. ANATOMY

The arterial supply of the brain is derived from two pairs of arterial trunks: the internal carotid arteries and the vertebral arteries.

ENCYCLOPEDIA OF HUMAN BIOLOGY, Second Edition, VOLUME 2. Copyright © 1997 by Academic Press. All rights of reproduction in any form reserved.

A. The Carotid System

1. The Internal Carotid Artery

The internal carotid artery may be divided into four segments: cervical, intrapetrosal, intracavernous, and supraclinoid. The fine carotico-tympanic artery arises from the intrapetrosal segment. Numerous small branches arise from the intracavernous segment. There are three main trunks: the meningohypophyseal trunk, the inferior cavernous sinus artery, and the capsular arteries. The meningohypophyseal trunk divides into the inferior hypophyseal artery, the dorsal meningeal artery, and the tentorial artery. The inferior cavernous sinus artery supplies the third, fourth, and sixth nerves, the meninges of the middle fossa, and the gasserian ganglion. Three major branches derive from the supraclinoid portion of the internal carotid artery: the ophthalmic artery, the posterior communicating artery, and the anterior choroidal artery.

The ophthalmic artery generally originates from the medial wall of the internal carotid artery. It enters into the orbit through the optic canal, ventrolaterally to the optic nerve.

The posterior communicating artery courses posteriorly and medically to join the posterior cerebral artery. The thalamoperforating arteries arise from the posterior communicating artery.

The anterior choroidal artery arises from the posterolateral aspect of the internal carotid artery approximately 2 mm proximal to the origin of the anterior cerebral artery (Figs. 1 and 2). From its origin, the anterior choroidal artery courses posteromedially in the chiasmatic cistern. It then follows the optic tract running between the cerebral peduncle and the medial surface of the uncus. It continues laterally and passes through the choroidal fissure to enter the temporal horn of the lateral ventricle. There are connections between the anterior choroidal artery and the lateral posterior choroidal artery in the region of the lateral geniculate body and the choroid plexus of the lateral ventricle. The anterior choroidal artery supplies blood to the posterior two-thirds of the optic

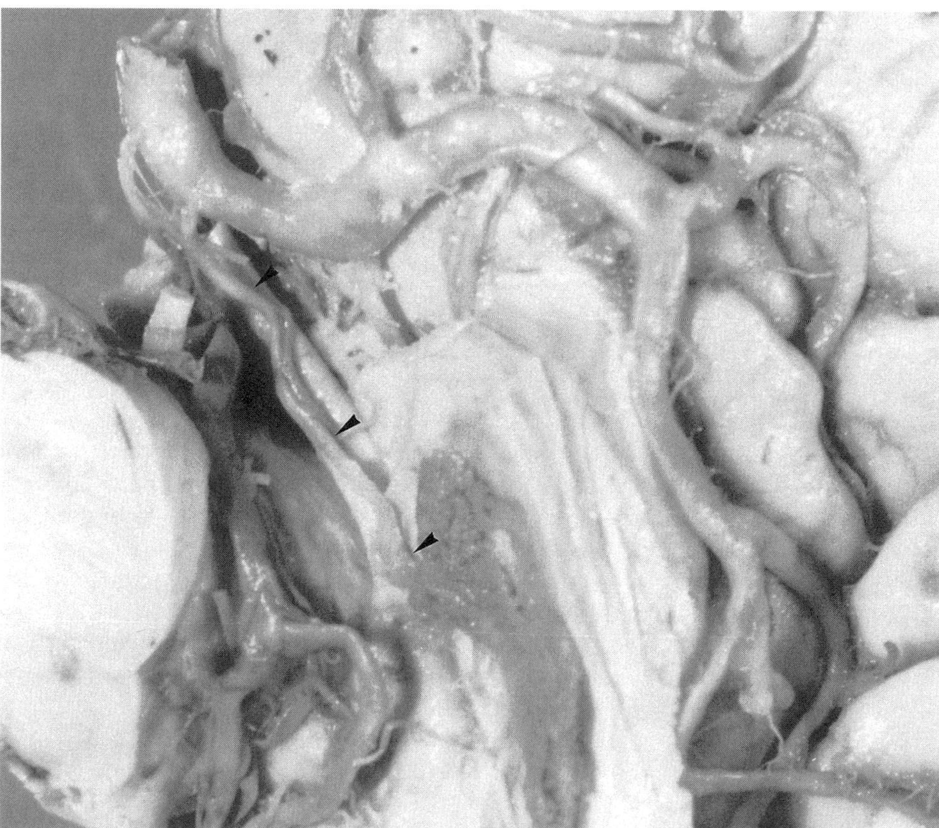

FIGURE 1 Photograph of the base of a brain. The tip of the temporal lobe is removed and the anterior choroidal artery (arrowheads) is exposed. Top is rostral.

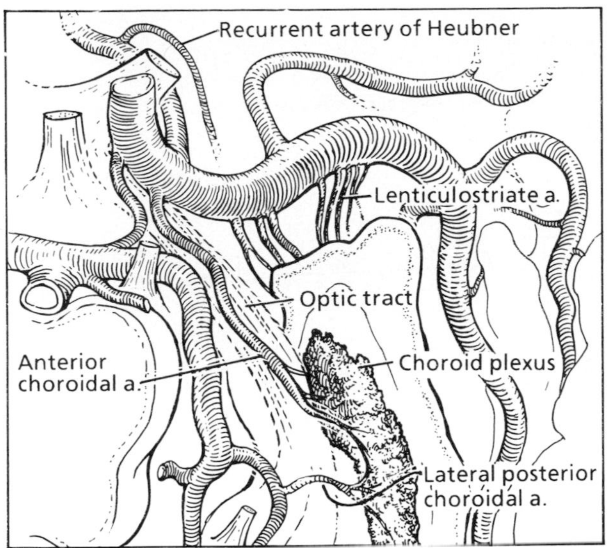

FIGURE 2 Schematic representation of the anterior choroidal artery, the posterior choroidal artery, the recurrent artery of Heubner, and the lenticulostriate arteries (based on Fig. 1).

tract, the external part of the lateral geniculate body, part of the piriform cortex, the uncus of the temporal lobe, the amygdaloid body, the tail of the caudate nucleus, the anterior third of the base of the cerebral peduncle, the substantia nigra, the upper parts of the red nucleus, a lateral portion of the ventral anterior and ventral lateral nuclei of the thalamus, the inferior half of the posterior limb of the internal capsule, and the retrolenticular fibers of the internal capsule.

2. The Anterior Cerebral Artery

From the point of bifurcation, the anterior cerebral artery courses medially and anteriorly above the optic nerve and chiasm. In the interhemispheric fissure, it is joined to the opposite anterior cerebral artery by the anterior communicating artery. Several small branches arise from the anterior communicating artery to supply the infundibulum, optic chiasm, and preoptic area of the hypothalamus. The anterior cerebral artery then ascends in front of the lamina terminalis to the level of the genu of the corpus callosum and passes backward above the corpus callosum in the pericallosal cistern.

Eight cortical arteries derive from either the pericallosal artery or the callosomarginal trunk: the orbitofrontal, frontopolar, anterior internal frontal, middle internal frontal, posterior internal frontal, paracentral, superior internal parietal, and inferior internal parietal arteries (Fig. 3). The anterior cerebral arteries supply the internal part of the orbital surfaces of the frontal lobe, the corpus callosum, the anterior two-thirds of the internal surfaces of the hemispheres, and a parasagittal segment of the external surfaces of the hemispheres.

The recurrent artery of Heubner originates at the precommunical segment of the anterior cerebral artery proximally to the anterior communicating artery (usually immediately before the origin of the anterior communicating artery) and, in a few cases, more distally (Fig. 2). It courses laterally to enter the anterior perforated substance just medial to the lenticulostriate arteries. The recurrent artery of Heubner supplies the head of the caudate nucleus, putamen, and anterior limb of the internal capsule.

3. The Middle Cerebral Artery

The middle cerebral artery passes laterally in the lateral cerebral fissure to the sylvian fissure (Fig. 2). From this horizontal segment derive the lenticulostriate arteries and the anterior temporal arteries. The lenticulostriate arteries enter the anterior perforated substance and supply the substantia innominata, the lateral portion of the anterior commissure, most of the putamen and the lateral segment of the globus pallidus, the superior half of the internal capsule and the adjacent corona radiata, and the body and head of the caudate nucleus (except for the anteroinferior portion).

The anterior temporal branches of the middle cerebral artery supply the temporal pole and the lateral aspect of the temporal lobe. Distal to the horizontal segment, the middle cerebral artery turns upward and posteriorly, and enters the depths of the sylvian fissure to reach the insula. In this region, the middle cerebral artery divides into its major branches (frontal, parietal, and temporal). These branches leave the sylvian fissure, circumnavigate the frontal, parietal, or temporal opercula, and give off 12 cortical branches.

1. Frontal branches: orbitofrontal, prefrontal, precentral, and central arteries.
2. Parietal branches: anterior parietal, posterior parietal, and angular gyrus arteries.
3. Temporal branches: temporo-occipital, posterior temporal, middle temporal, anterior temporal, and temporal polar arteries.

The small perforating branches derive from the insular segment of the middle cerebral artery and supply the claustrum and external capsule.

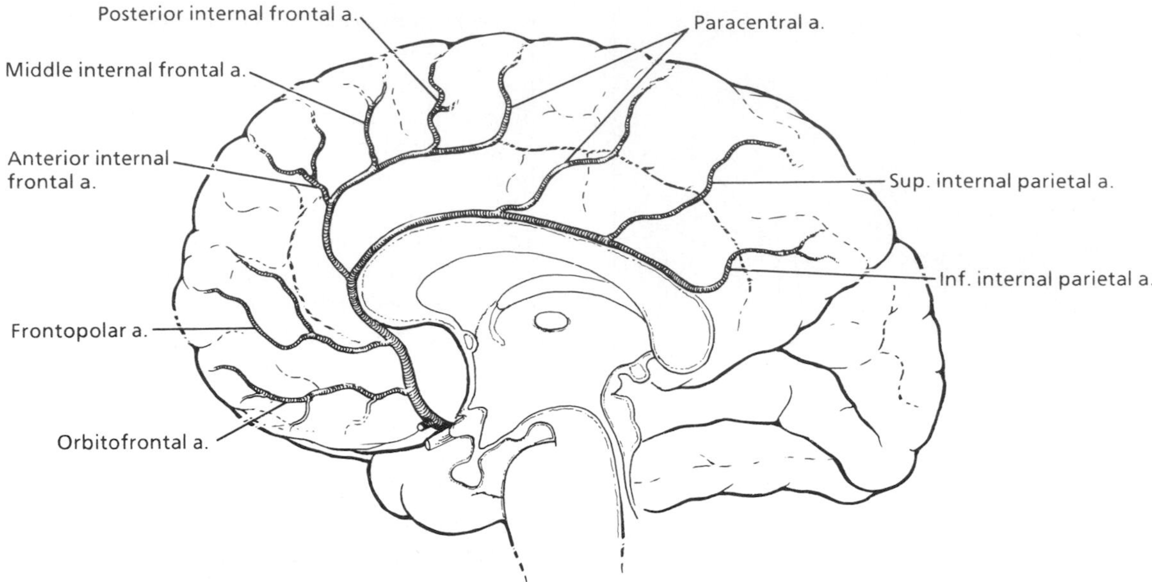

FIGURE 3 Schematic representation of the anterior cerebral artery and its branches.

B. Vertebrobasilar System

1. The Posterior Cerebral Artery

The posterior cerebral arteries derive from the rostral end of the basilar artery and course posteriorly in the perimesencephalic cisterns to encircle the cerebral peduncle. This proximal trunk of the posterior cerebral artery may be divided into peduncular, ambient, and quadrigeminal segments. The posterior communicating artery arises from the midportion of the peduncular segment. The quadrigeminal segment continues posteriorly beneath the splenium of the corpus callosum to terminate in cortical branches. Four main cortical branches (anterior temporal, posterior temporal, parieto-occipital, and calcarine arteries) arise usually from the ambient segment or from the quadrigeminal segment. These cortical arteries supply the inferior and internal surfaces of the temporal lobe, the internal surface of the occipital lobe, and the posterior part of the precuneus. The posterior cerebral artery also provides the perforating branches to the midbrain and the thalamus.

The mesencephalic branches include the interpeduncular perforating branches, the peduncular branches, and the circumflex mesencephalic branches. These mesencephalic branches arise from the peduncular segment of the posterior cerebral artery and supply the cerebral peduncle, the red nucleus, the substantia nigra, and the tegmentum.

The thalamic branches include the thalamoperforating arteries and the thalamogeniculate arteries. The thalamoperforating arteries arise from the posterior communicating artery and the proximal peduncular segment of the posterior cerebral artery. The thalamoperforating arteries supply the posterior chiasm, the optic tract, the posterior hypothalamus, part of the cerebral peduncle, and the anterior and medial portions of the thalamus.

The thalamogeniculate arteries usually derive from the ambient segment of the posterior cerebral artery and supply the lateral geniculate body and the posterior and lateral portions of the thalamus.

One medial posterior choroidal artery and at least two lateral posterior choroidal arteries belong to the group of the posterior choroidal arteries arising from the posterior cerebral artery (Figs. 2 and 4). The medial posterior choroidal artery usually arises from the proximal segments (peduncular or ambient) of the posterior cerebral artery and curves around the midbrain. It approaches the region lateral to the pineal body and then courses forward to the roof of the third ventricle. This vessel supplies the tectum, the choroid plexus of the third ventricle, and the dorsomedial nucleus of the thalamus.

In most instances, the lateral posterior choroidal arteries originate from the ambient segment of the posterior cerebral artery. These vessels initially course laterally to enter the choroid fissure. The anterior branch extends forward to supply the choroid plexus of the temporal horn. The posterior branch courses

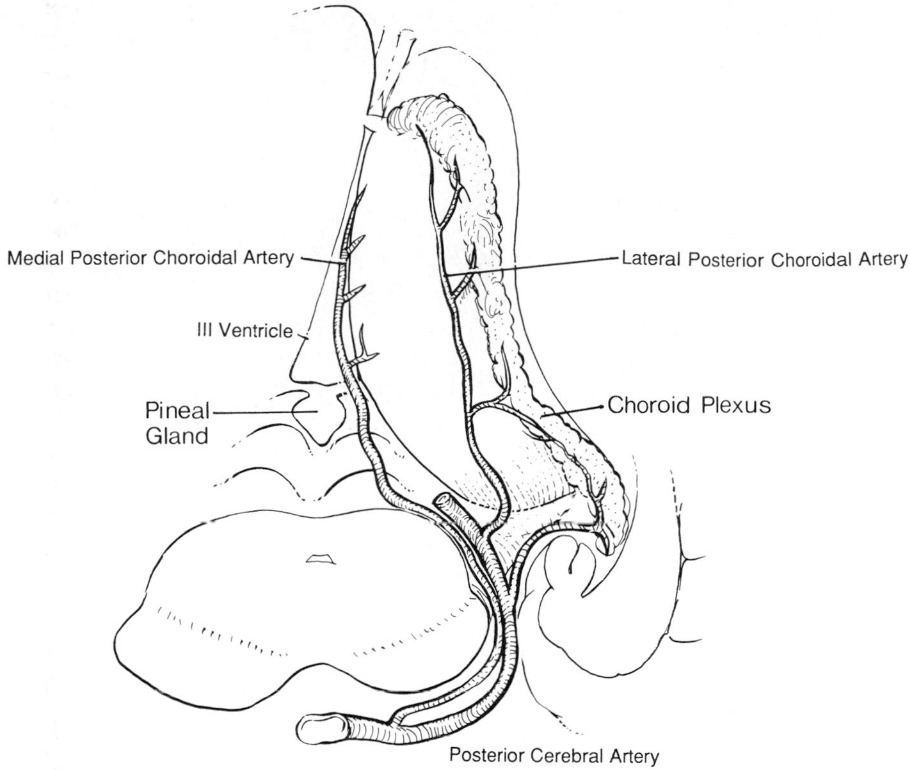

FIGURE 4 Schematic drawing of the medial posterior choroidal artery and the lateral posterior choroidal artery, as viewed from a dorsal and posterior perspective.

posteriorly around the pulvinar, supplying the choroid plexus of the trigone and the body of the lateral ventricle. The posterior branch also supplies the fornix, most of the dorsomedial nucleus of the thalamus, the pulvinar, and part of the lateral geniculate body.

2. The Superior Cerebellar Artery

The superior cerebellar artery originates from the rostral part of the basilar artery. It courses posterolaterally in the perimesencephalic cisterns to encircle the upper pons and lower mesencephalon. On reaching the cerebellum, it divides into several cortical branches. These cortical branches supply the lateral and superior surface of the cerebellar hemispheres, and the superior vermis. From these cortical arteries numerous branches extend deeply into the cerebellum and supply the superior medullary velum, the superior and middle cerebellar peduncles, and the intrinsic cerebellar nuclei, including parts of the dentate nucleus.

3. The Anterior Inferior Cerebellar Artery

The anterior inferior cerebellar artery arises from the inferior third of the basilar trunk and courses laterally and downward. It sends small branches into the pons that supply the lateral aspect of the pons from the junction of the upper and middle third of the pons down to the upper part of the medulla. Within the cerebellopontine angle cistern, the anterior inferior cerebellar artery is usually situated ventral to the acoustic-facial nerve bundle. The internal auditory artery usually originates from this proximal portion of the anterior inferior cerebellar artery. After crossing the seventh and eighth nerves, the anterior inferior cerebellar artery courses medially toward the cerebellopontine angle and then extends to the cerebellar hemisphere.

4. The Posterior Inferior Cerebellar Artery

The posterior inferior cerebellar arteries originate from the vertebral arteries at an average distance of 16 mm below the vertebrobasilar junction. The main trunk of the posterior inferior cerebellar artery may be divided into four segments: the anterior medullary, the lateral medullary, the posterior medullary, and the supratonsillar (Fig. 5).

The anterior medullary segment courses posteriorly

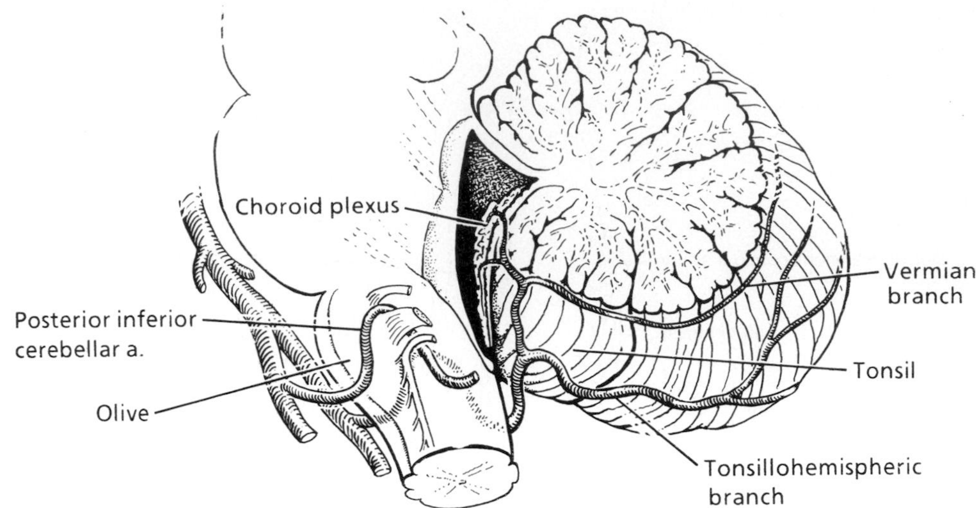

FIGURE 5 Schematic drawing of the posterior inferior cerebellar artery.

within the medullary cistern and winds around the lower end of the olive of the medulla oblongata. It continues posteriorly around the lateral aspect of the medulla as the lateral medullary segment. On reaching the posterior margin of the medulla oblongata, the posterior inferior cerebellar artery (posterior medullary segment) ascends to the anterior aspect of the superior pole of the tonsil behind the posterior medullary velum. The posterior inferior cerebellar artery continues dorsally to the superior pole of the tonsil as the supratonsillar segment. It gives off small branches to the choroid plexus of the fourth ventricle. The supratonsillar segment of the posterior inferior cerebellar artery then bifurcates into two terminal branches: the tonsillohemispheric and vermis branches. The tonsillohemispheric branches descend along the posterior margin of the medial aspect of the tonsil and divide into tonsillar branches and hemispheric branches. In some cases, tonsillohemispheric branches arise adjacent or ventral to the cerebellar tonsil. Vermis branches course below the inferior vermis in the sulcus vallecula between the inferior vermis and the cerebellar hemisphere.

The basilar artery provides three series of perforating branches (paramedian, short circumferential, and long circumferential arteries) to the ventral portion of the pons. The paramedian arteries supply the most medial pontine area, including the pontine nuclei and the corticopontine, corticospinal, and corticobulbar tracts. The short circumferential arteries supply a wedge of tissue along the anterolateral pontine surface. The long circumferential arteries supply most of the tegmentum.

Although the arterial supply to the brain is derived from two carotid and two vertebral arteries, the relative importance of these vessels varies in different species. In primates, the carotid and vertebral circulations participate in approximately equal manner. The carotid system provides a much larger share of the blood supply than the vertebral arteries in horses, and a smaller share in rats. Also, the relative importance of the internal and external carotid arteries varies between species. In dogs, the internal carotid artery is small, and the external carotid artery branches into a complex network of anastomoses, the rete mirabile. Branches of the external carotid artery provide more flow to the brain than the internal carotid artery in that species. In contrast, the internal carotid artery is larger than the external carotid artery and there is no rete in primates. There are numerous collateral pathways in the brain circulation.

5. The Circle of Willis

At the base of the brain, the anatomical pattern of vessels permits collateral blood flow from the internal carotid arteries and the vertebrobasilar system through the circle of Willis. This arterial circle is formed by anterior and posterior communicating arteries and proximal portions of the anterior and posterior cerebral arteries.

6. Leptomeningeal Anastomoses

Anastomoses other than those of the circle of Willis occur over the surfaces of the cerebral hemispheres and cerebellum. Predominantly, leptomeningeal collateral circulation is found in the anastomoses between the anterior cerebral and middle cerebral, and between the middle cerebral and posterior cerebral arteries. On the cerebellar surface, leptomeningeal anastomoses occur between each of the cerebellar arteries and also across the midline to connect with the corresponding arteries of the opposite side. Communications between adjoining branches of the cerebellar arteries are more frequent than in the case of the cerebral arteries.

C. Dural Venous Sinuses

The superior sagittal sinus begins rostrally at the foramen cecum of the frontal bone and proceeds in a caudal direction toward the internal occipital protuberance (Fig. 6). It receives venous outflow from the lateral lacunae and from the ascending superficial cerebral veins. The inferior sagittal sinus courses above the corpus callosum in the free edge of the falx cerebri. This sinus receives small veins that drain the roof of the corpus callosum, the cingulate gyrus, and the adjacent medial hemisphere. The straight sinus receives blood from the inferior sagittal sinus and from the vein of Galen. As it proceeds caudally, this sinus communicates with the superior sagittal sinus and the bilateral transverse sinuses, forming a torcular Herophilis or confluence of sinuses. When the transverse sinuses are poorly developed, the occipital sinus situated in the dorsal aspect of the falx cerebelli and terminating in the confluence of sinuses may persist as a large sinus.

The transverse sinuses course laterally and anteriorly along the attached margin of the tentorium cerebelli and along the groove of the squamous portion of the temporal bone. Near the petrous temporal bone, the transverse sinus receives blood from the superior petrosal sinus, thus communicating with the cavernous sinus. The transverse sinus is continuous with the sigmoid sinus. After traversing the jugular foramen, the sigmoid sinus becomes continuous with the internal jugular vein. The inferior petrosal sinus may enter the most distal segment of the sigmoid sinus, although it usually drains directly into the internal jugular vein. The sphenoparietal sinus receives

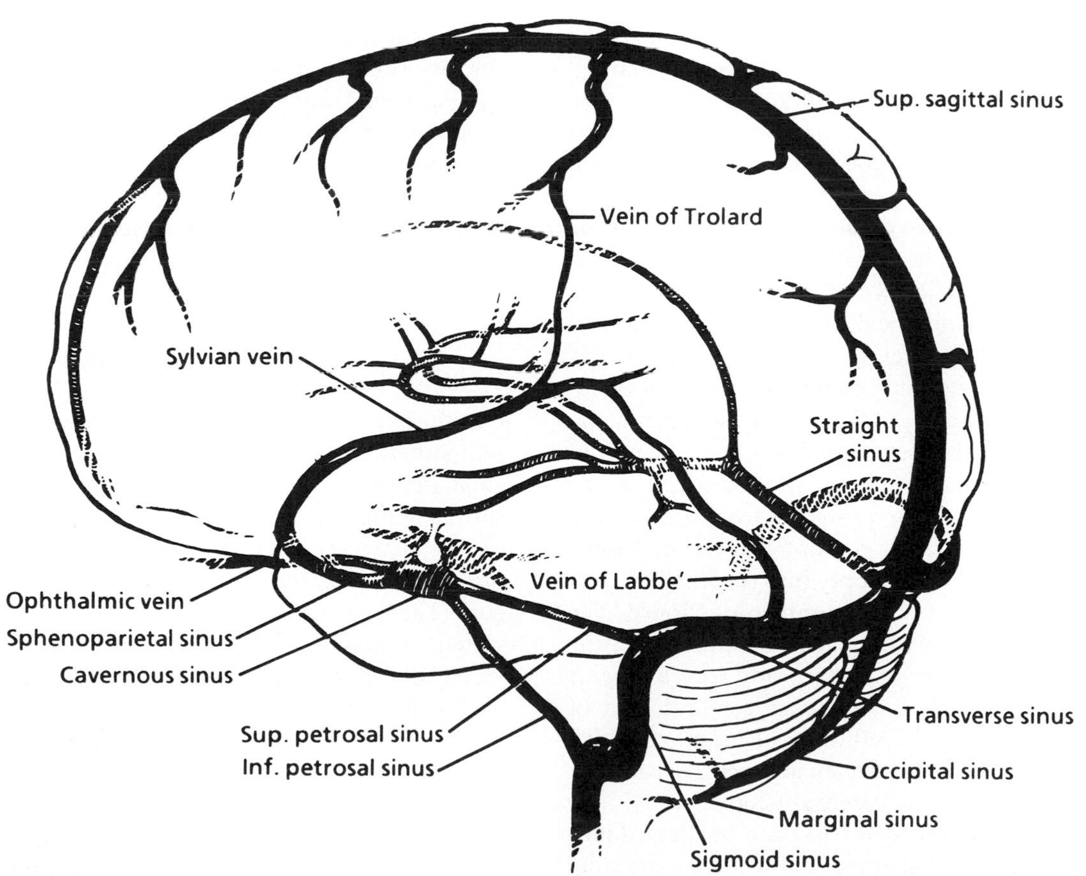

FIGURE 6 Diagram of the intracranial venous sinuses and superficial cerebral veins and deep cerebral veins.

blood from the superficial sylvian vein. It courses in the inferoposterior ridge of the lesser wing of the sphenoid bone and ends in the antero-inferior angle of the cavernous sinus. The cavernous sinus also receives the ophthalmic veins, the inferior and superior petrosal sinuses, and the basilar plexus of veins.

D. Cerebral Veins

The superficial cerebral veins may be divided into three groups: the ascending superficial cerebral veins, the superficial sylvian veins, and the inferior cerebral veins (Fig. 6).

The ascending superficial cerebral veins drain the medial, superolateral, and frontal regions of the cerebral hemisphere. Most of these veins empty into the superior sagittal sinus. The superficial sylvian veins originate near the posterior limb of the sylvian fissure and drain into the sphenoparietal sinus. There are two major anastomotic veins, which connect the superficial sylvian veins with other venous sinuses. The great anastomotic vein of Trolard connects the superficial sylvian veins with the superior sagittal sinus. The lesser anastomotic vein of Labbé connects the superficial sylvian veins with the transverse sinus. The inferior cerebral veins drain the portions of the lateral convexity not drained by the superficial sylvian veins, the ventral surface of the temporal lobe, or the ventral surface of the occipital lobe. These veins drain into the transverse sinus.

The deep cerebral veins drain the blood from the deep white matter, the basal ganglia, and the diencephalon. The medullary veins carry blood in the centripetal direction from the deep structures of the cerebrum toward the lateral ventricles, where they coalesce to form the subependymal veins. The thalamostriate vein is one of the largest subependymal veins. It receives venous return from the caudate nucleus, the lenticular nuclei, the internal capsule, the deep white matter of the posterior frontal lobe, and the deep white matter of the anterior parietal lobe. The subependymal veins empty into the internal cerebral veins at the level of the foramen of Monro. The internal cerebral veins from each side unite under the splenium of the corpus callosum and join the great cerebral vein of Galen. The great cerebral vein of Galen also receives the basal vein of Rosenthal, which originates near the optic chiasm and courses caudally and dorsally around the cerebral peduncle.

The veins of the posterior fossa can be divided into the following three groups: the Galenic draining group, the petrosal draining group, and the tentorial draining group.

The Galenic group of veins receives most of the venous return from the midbrain and from the medial and paravermal regions of the superior cerebellar surface. The major veins of the superficial brainstem include the tectal veins, the lateral mesencephalic vein, the posterior mesencephalic vein, and portions of the longitudinally directed anterior pontomesencephalic vein. The medial and paravermal regions of the superior cerebellar surface are drained by the superior vermian vein and by the precentral cerebellar vein.

The petrosal vein is a major vessel of the petrosal draining group. The veins of the great horizontal fissure, the transverse pontine vein, the brachial vein, the vein of the lateral recess of the fourth ventricle, and the lateral pontine vein converge toward the anterior angle of the cerebellum, where they form a single stem, the petrosal vein. The petrosal vein usually drains into the superior petrosal sinus.

The tentorial draining group includes the inferior vermian vein and the superior and inferior hemispheric veins. The inferior vermian vein is formed by the union of the superior and inferior retrotonsillar tributaries behind the cerebellar tonsil. It proceeds posterosuperiorly through the inferior paravermian sulcus toward the transverse sinus. The superior hemispheric veins run posteriorly and inferiorly on the superior surface of the cerebellar hemisphere and open into the transverse sinus. The inferior hemispheric veins, on the other hand, run superiorly on the inferior aspect of the cerebellar hemisphere and drain into the transverse sinus.

III. INNERVATION

Cerebral vessel receives dual sympathetic and parasympathetic innervation. The adrenergic innervation emanates primarily from the ipsilateral superior cervical ganglion, but arteries in basal and medial areas receive bilateral innervation. The preganglionic nerves originate in the intermediolateral column of the spinal cord. The innervation of arteries in the internal carotid system is more extensive than the innervation in the vertebrobasilar system. A central noradrenergic pathway, originating primarily in the locus ceruleus, also innervates cerebral vessels, particularly smaller arteries and capillaries.

The parasympathetic (cholinergic) innervation emanates mostly from the sphenopalatine ganglion and possibly from the otic ganglion. The preganglionic nerves originate in the superior salivatory nucleus and project via the greater superficial petrosal nerve, a branch of the seventh cranial nerve, to the sphenopalatine ganglion. The lesser superficial petrosal nerve may be the preganglionic projection to the otic ganglion and originates in the inferior salivatory nucleus. Like the sympathetic innervation, cholinergic innervation is more dense in the anterior circulation than in the posterior circulation.

Recent immunohistochemical studies have shown that neurotransmitters within the sympathetic and parasympathetic innervation to the blood vessels coexist with vasoactive peptides, that is, with vasoactive intestinal polypeptides (VIPs) in the cholinergic nerves and with neuropeptide Y in the adrenergic nerves. Nerve fibers immunoreactive for calcitonin gene-related peptide (CGRP) have also been identified in the cerebral arteries. CGRP coexists with substance P. CGRP- and substance P-containing fibers originate in the trigeminal ganglion of the same side. CGRP, substance P, and VIP dilate cerebral arteries, the relative potency being CGRP–substance P–VIP. On the other hand, neuropeptide Y constricts cerebral arteries; however, the functional role of neuropeptides in the control of the cerebral circulation is currently a matter of speculation.

IV. REGULATION OF CEREBRAL BLOOD FLOW

There are several mechanisms by which cerebral blood flow (CBF) may be regulated: autoregulation, neurogenic regulation, chemical regulation, and metabolic regulation.

A. Autoregulation

Autoregulation is defined as the relative constancy of blood flow during alterations in arterial pressure (Fig. 7). CBF is maintained constant over a wide range of blood pressure compared to in other organs. Lower and upper limits of autoregulation are approximately 40 and 160 mm Hg, respectively. Decreases in arterial pressure cause vasodilation. Maximum dilation of cerebral pial arteries has been observed at blood pressures of around 40 mm Hg. On the other hand, increases in arterial pressure cause vasoconstriction within the range of autoregulation. However, a marked increase in arterial pressure (usually >200 mm Hg) induces dilation of pial arterioles and is irreversible for several hours. In such situations, CBF is markedly increased, a phenomenon called breakthrough of autoregulation. In hypertensive patients, the pressure–flow curve is shifted to the right with higher limits of autoregulation (Fig. 7).

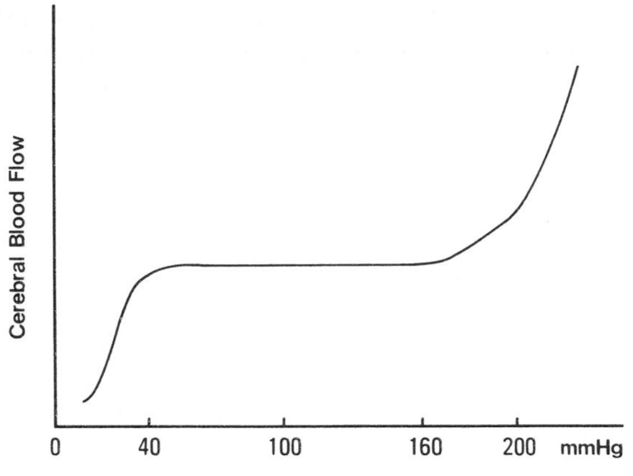

FIGURE 7 Schematic representation of the autoregulation of cerebral blood flow. In normotensive patients, lower and upper limits of autoregulation are approximately 40 and 160 mm Hg, respectively.

B. Neurogenic Control

Cerebral vessels are richly innervated, but the role of sympathetic nerves in regulation of CBF is unclear. The α-adrenoceptors of cerebral vessels are less sensitive to norepinephrine than the α-receptors of other vessels. Transmural electrical stimulation produces much less contraction in cerebral arteries than in peripheral arteries, and electrical stimulation of sympathetic nerves has little or no effect on CBF under normal conditions.

Several studies have shown that acetylcholine produces concentration-dependent vasodilation of cerebral vessels. However, the role of cholinergic nerves in the regulation of CBF has been highly controversial.

C. Chemical Control

Hypercapnia induces pial arteriolar dilatation and increases CBF. Topical application of solutions with low pH on the brain surface produces pial arteriolar dilatation. This vasodilatory response can be obtained when changes in pH are induced either by changes in CO_2 at constant HCO_3^- concentration or when changes in HCO_3^- concentration are made at constant

pCO_2. However, pCO_2 or HCO_3^- concentrations have no effect on pial arteriolar caliber unless the pH is allowed to change. The most important mechanism of action of CO_2 seems to be its local effect on vascular smooth muscle mediated via a change in extracellular fluid pH.

D. Metabolic Regulation

It is well established that there is a tight coupling between brain metabolism and CBF. Alterations in the level of activity of the brain are accompanied by parallel changes in both brain metabolism and CBF. For instance, motor activity in the arm is accompanied by parallel increases in both blood flow and O_2 consumption in the contralateral sensorimotor area of the cortex. The mechanisms by which the brain metabolism influences CBF are not clearly understood. However, it has been assumed that the active neurons release vasodilator metabolites, which presumably reach the blood vessels by diffusion. Among several vasodilator metabolites, adenosine and K^+ have been considered to be promising candidates. Adenosine has a vasodilatory effect on pial arterioles at very low concentrations, and the concentration of adenosine in the brain increases during increased functional activity of the brain. K^+ produces vasodilation of pial arterioles between concentrations of 0 and 10 m*M*. Several studies have shown that K^+ is released into the extracellular fluid during increased functional activity of the brain, to an extent adequate to achieve maximum vasodilation of the pial arterioles.

E. Role of Vascular Endothelium

Recent studies have demonstrated that vascular endothelium continuously synthesizes or releases a large number of vasoactive substances, such as prostacyclin, endothelium-derived relaxing factor (EDRF), endothelium-derived contracting factor (EDCF), and others. These discoveries have led to important new concepts in vascular pathophysiology.

Prostacyclin, synthesized in endothelial cells of cerebral (Fig. 8) as well as peripheral vessels, relaxes vascular smooth muscle cells and inhibits platelet aggregation and adhesion. EDRF is released when muscarinic receptors on endothelial cells are stimulated by acetylcholine (Fig. 9). Many other substances release EDRF from endothelial cells. They include arachidonic acid, bradykinin, histamine, substance P, norepinephrine, 5-hydroxy tryptamine, calcium ionophore A23187, adenine nucleotides, thrombin, vasopressin, endothelins (ETs), and changes in shear stress. Vascular smooth muscle is the target for the biological action of EDRF, which inhibits the contractile process by stimulating guanylate cyclase, leading to the production of cyclic GMP. EDRF also potently inhibits platelet aggregation and adhesion. EDRF has a very short half-life (approximately 6 sec), and it may be simply nitric oxide (NO) **or an NO-containing compound. Nitric oxide is synthesized from the semiessential amino acid L-arginine by the enzyme NO synthase in endothelium. The endothelial NO synthase may be active under basal conditions and can be further stimulated by increases in intracellular calcium that occur in response to receptor-mediated agonists or calcium ionophore.**

EDCF was identified as a vasoconstrictor peptide in the conditioned medium from cultured bovine aortic endothelial cells. It is probably identical to **ET** obtained from the supernatant of cultured endothelia of porcine aorta. **Three ETs (ET-1, ET-2, and ET-3) have been identified, each consisting of 21 amino acids. The vascular endothelial cells produce only ET-1, although ETs are produced on proper stimulation in various cell types. ET-1** produces potent **and long-lasting** constriction of cerebral microvessels as well as major cerebral arteries (Fig. 10). **ETs act on two pharmacologically and molecularly distinct subtypes of receptors (i.e., ET_A and ET_B). The smooth muscle ET_A receptors have been thought to mediate direct vasoconstrictor actions of ET-1, whereas the endothelial ET_B receptors mediate the transient vasodilator response to ET-1 through release of NO. However, the mechanisms of vasoconstrictor action of ET-1 appear to be more complicated. Under certain conditions, ET_B receptors also mediate vasoconstrictor response to ET-1.**

Prostacyclin, EDRF, and EDCF synthesized in cerebral vascular endothelium could contribute to the physiological regulation of cerebral circulation, and the absence or dysfunction of endothelial cells may play a role in pathological vascular events associated with atherosclerosis, thrombogenesis, and cerebral vasospasm.

In addition to the vascular endothelium, neurons and glia can produce NO in response to some stimuli. Therefore, neuronally derived NO may also participate in the regulation of local CBF.

V. THE BLOOD–BRAIN BARRIER

The unique barrier function has been observed not only in the brain capillaries but also in the major cerebral arteries (Fig. 11). Cerebral capillaries differ

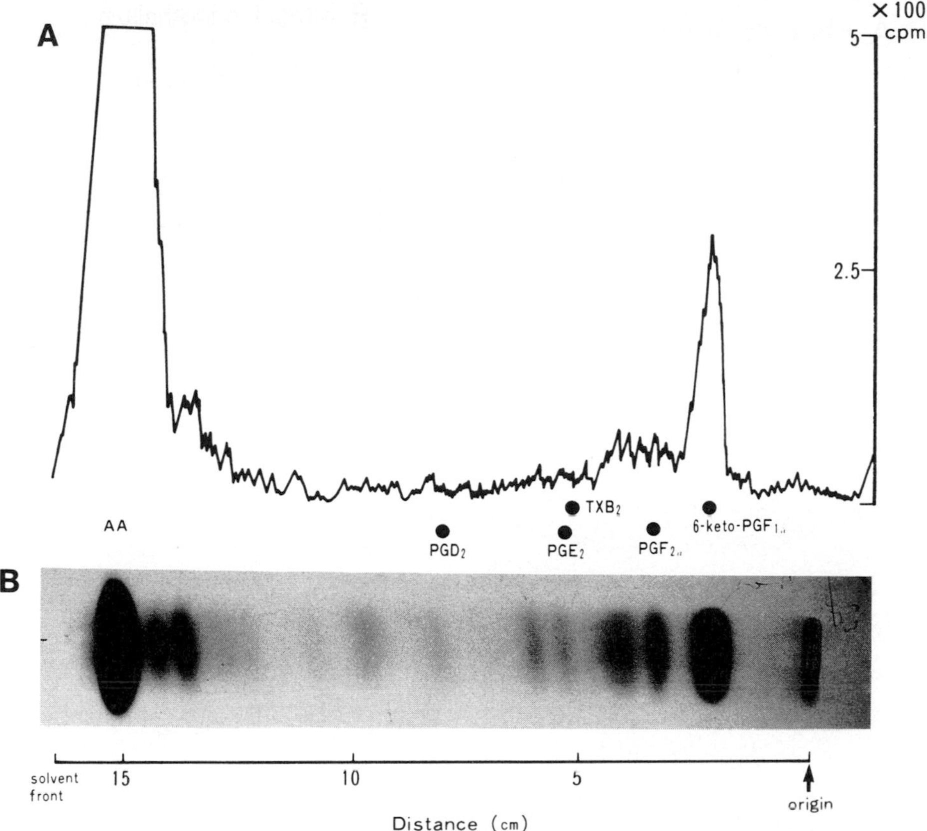

FIGURE 8 Thin-layer radiochromatograms (A) and autoradiograms (B) obtained after incubation of arterial specimens with 0.2 μCi [^{14}C]arachidonic acid (AA) for 30 min at 37°C. [^{14}C]Arachidonic acid was mainly transformed to 6-keto-PGF$_{1\alpha}$, the end product of PGI_2 metabolism, and a very small amount of PGF$_{2\alpha}$ was also noted.

from other capillaries owing to the presence of tight junctions between contiguous endothelial cells and a paucity of pinocytotic vesicles. These morphological features constitute the blood–brain barrier. There is also an enzymatic component in the cerebral vessels; a high activity of monoamine oxidase, which presumably results in effective degradation of the catecholamines that pass the endothelium. The adenosine triphosphatase (ATPase) activity in the brain capillaries is localized to the basal lamina and to the membranes of glial end feet, whereas in nonbarrier capillaries ATPase activity is located in the endothelial pinocytotic vesicles. Such features also may affect the permeability properties and the metabolic function of the blood–brain barrier. It is thought that the development of the blood–brain barrier is induced by contact between cerebral endothelium and astrocytes. In fact, γ-glutamyl transpeptidase, which plays an important role in amino acid transport across the blood–brain barrier, often disappears in cultures of cerebral endothelium, but it can be induced by coculturing the endothelium with glial cells. [*See* Blood–Brain Barrier.]

VI. THE BLOOD–ARTERIAL WALL BARRIER

Endothelial cells in the intradural major cerebral arteries have a barrier function, which is similar to the parenchymal microvessels of the brain. Proximal segments extending 1–4 mm from the origin of the intradural segments of both internal carotid and vertebral arteries have incomplete barrier function. Focal barrier deficiency was also noted at the branching sites of intradural major cerebral arteries and may relate to atherogenesis.

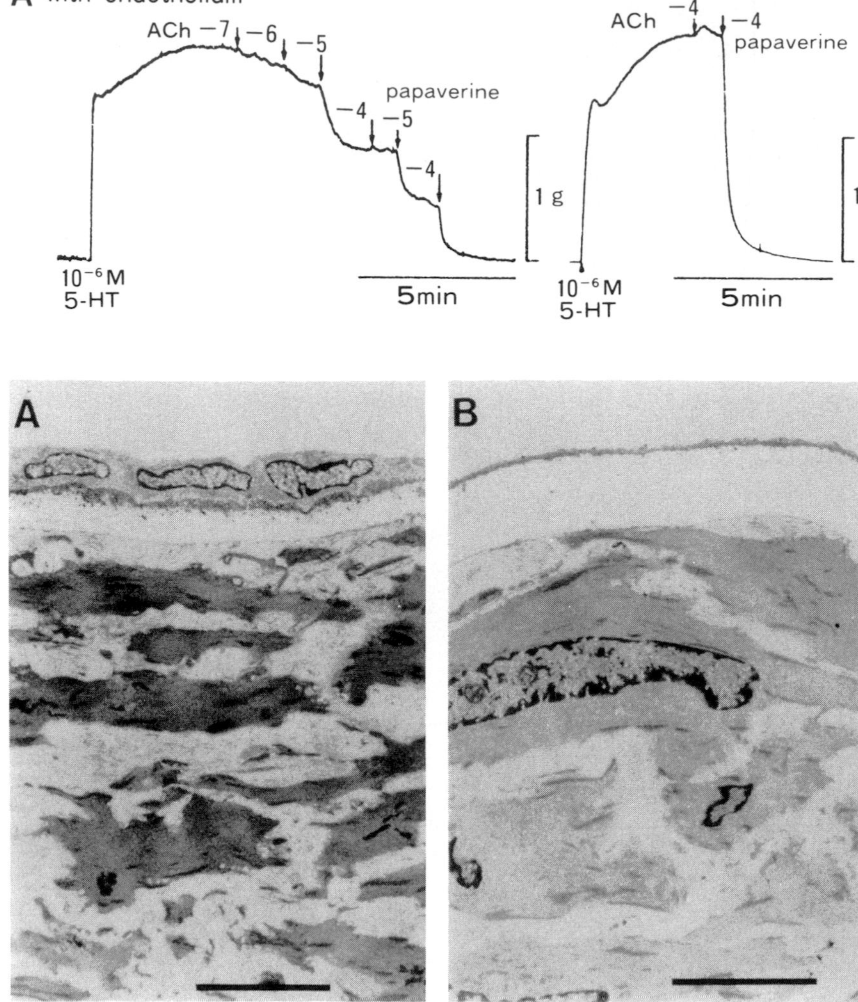

FIGURE 9 Upper: Effect of acetylcholine (ACh) on serotonin (5-HT)-induced contraction of rabbit basilar artery with or without endothelium. (A) Typical pattern of dilator response induced by ACh in basilar artery with intact endothelium. (B) Typical pattern of effect of ACh on basilar artery after removal of endothelium. Ordinate indicates absolute contractile tension of rabbit basilar artery. The number with the arrow indicates the log molar concentrations of ACh or papaverine in the bath. Lower: Electron microscopic views of the rabbit basilar artery with (A) or without (B) endothelium.

VII. PHARMACOLOGICAL CHARACTERISTICS OF CEREBRAL ARTERIES

The responses of cerebral vessels to vasoactive substances differ from those in systemic vessels. Cerebral vessels, although receiving sympathetic innervation equal to that of the systemic vessels, are relatively insensitive to norepinephrine, perhaps because they have fewer α-adrenoceptors. The differences may relate to the different embryonic origin of the two types of vessels, which arise from different primordial cells. The site of fusion of the arteries corresponds to the abrupt change in responsiveness to norepinephrine. The sensitivity of cerebral arteries to norepinephrine differs between species; norepinephrine is a much more potent cerebral vasoconstrictor in the human and monkey than in rabbits, dogs, or cats. It has been recently proposed that the different sensitivity of species differences may be due to differences in

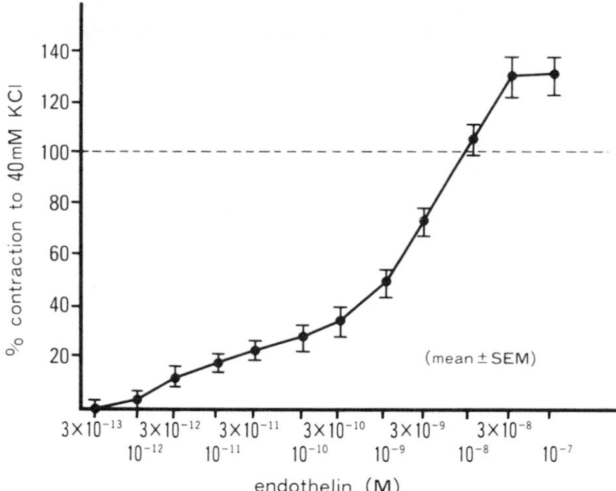

FIGURE 10 Concentration–response curve to endothelin for the canine basilar arteries.

postjunctional α-adrenoceptors. Systemic arterial smooth muscles contain mainly postjunctional α_1-adrenoceptors, whereas systemic venous smooth muscles contain both α_1- and α_2-adrenoceptors. Postjunctional α-adrenoceptors in canine and feline cerebral arteries resemble the α_2-subtype, whereas in monkey and human cerebral arteries, they resemble the α_1-subtype. In pig cerebral arteries, norepinephrine is a potential vasodilator. This is probably due to the dominancy of β-adrenoceptors in pig cerebral arteries.

Species differences in cerebral artery responses to other vasoactive substances have also been reported. Histamine is a prominent vasodilator in human cerebral arteries, whereas it is a strong vasoconstrictor in rabbit cerebral arteries. Human and monkey cerebral arteries are more sensitive to thromboxane A2 than are canine cerebral arteries.

Recent studies using calcium channel-blocking agents have revealed that contractions of cerebral arteries are more dependent on an influx of extracellular calcium than are other systemic arteries. Thus, contractile responses of the basilar artery to serotonin depend almost entirely on the influx of extracellular calcium. In contrast, contraction of the saphenous artery in response to serotonin is produced by release of intracellular calcium as well as by influx of extracellular calcium. These findings suggest that the inhibition of the influx of extracellular calcium into cerebral arterial smooth muscle cells with calcium channel-blocking agents may increase CBF after an ischemic stroke or may prevent the occurrence of cerebral vasospasm following subarachnoid hemorrhage.

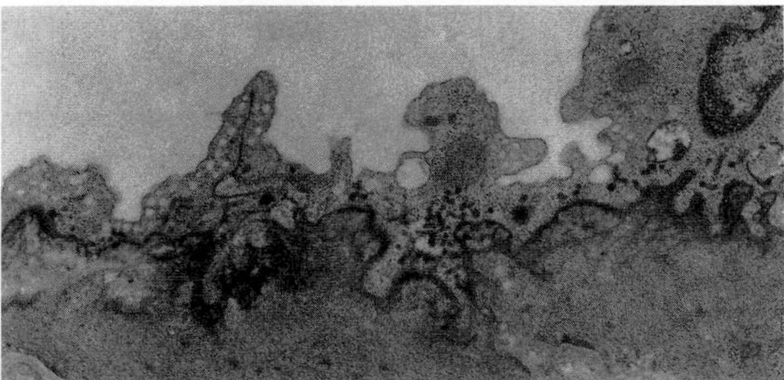

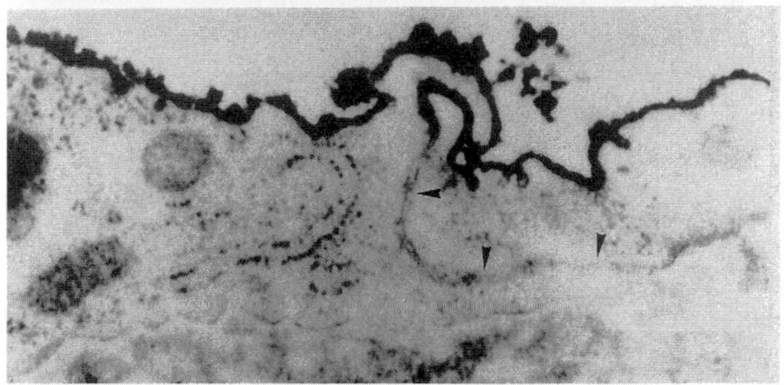

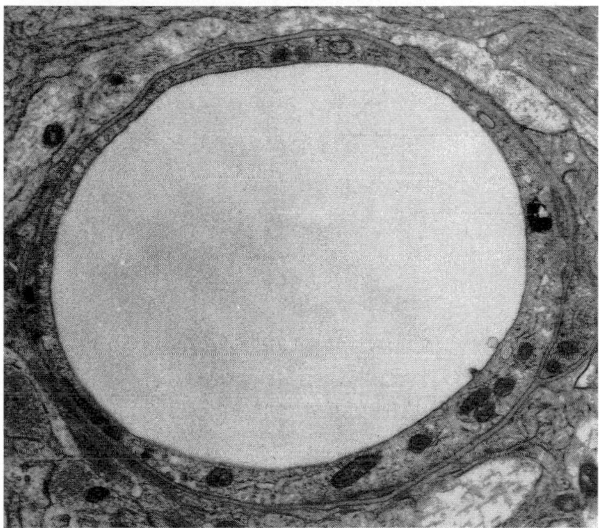

FIGURE 11 Blood–brain barrier and blood–arterial wall barrier. Horseradish peroxidase (HRP) was administered intravenously, and the animal was sacrificed by perfusion-fixation 10 min after the infusion. Upper: The arteriole in the outer layer of small intestine of a rat. HRP permeated into the subendothelial space. Note the HRP reaction products in the interendothelial cleft, plasmalemmal vesicles, and the subendothelial space. Middle: The basilar artery of a rat. Interendothelial cleft (arrowheads) was devoid of HRP reaction products. HRP did not permeate into the subendothelial space. Lower: The brain capillary of a dog. HRP reaction products did not permeate into the basement membrane and into the adjacent extracellular space.

BIBLIOGRAPHY

Bevan, J. A. (1981). A comparison of the contractile response of the rabbit basilar and pulmonary arteries to sympathomimetic agonists: Further evidence for variation in vascular adrenoceptor characteristics. *J. Pharmacol. Exp. Ther.* **216,** 83–89.

De Mey, J. G., and Vanhoutte, P. M. (1981). Uneven distribution of postjunctional alpha$_1$- and alpha$_2$-like adrenoceptors in canine arterial and venous smooth muscle. *Circ. Res.* **48,** 875–884.

Edvinsson, L., Ekman, R., Jansen, I., McCulloch, J., and Uddman, R. (1987). Calcitonin gene-related peptide and cerebral blood vessels: Distribution and vasomotor effects. *J. Cereb. Blood Flow Metabol.* 7, 720–728.

Furchgott, R. F., and Zawadzki, J. V. (1980). The obligatory role of endothelial cells in the relaxation of arterial smooth muscle by acetylcholine. *Nature* **288,** 373–376.

Heistad, D. D., and Kontos, H. A. (1983). Cerebral circulation. *In* "Handbook of Physiology" (J. T. Shepherd, F. M. Abboud, and S. R. Geiger, eds.), pp. 137–182. American Physiological Society, Bethesda, Maryland.

Iadecola, C., Pelligrino, D. A., Moskowitz, M. A., and Lassen, N. A. (1994). Nitric oxide synthase inhibition and cerebrovascular regulation. *J. Cereb. Blood Flow Metabol.* **14,** 175–192.

Parmer, R., Ferrige, A. G., and Moncada, S. (1987). Nitric oxide release accounts for the biological activity of endothelium-derived relaxing factor. *Nature* **327,** 524–526.

Sasaki, T., Kassell, N. F., Torner, J. C., Maixner, W., and Turner, D. (1985). Pharmacological comparison of isolated monkey and dog cerebral arteries. *Stroke* **16,** 482–489.

Streeter, G. L. (1918). The developmental alterations in the vascular system of the brain of the human embryo. *Contrib. Embryol.* **8,** 5–38.

Toda, N. (1983). Alpha adrenergic receptor subtype in human, monkey and dog cerebral arteries. *J. Pharmacol. Exp. Ther.* **226,** 861–868.

Yanagisawa, M., Kurihara, H., Kimura, S., Tomobe, Y., Kobayashi, M., Mitsui, Y., Yazaki, Y., Goto, K., and Masaki, T. (1988). A novel potent vasoconstrictor peptide produced by vascular endothelial cells. *Nature* **332,** 411–415.

Yanagisawa, Y. (1994). The endothelin system. A new target for therapeutic intervention. *Circulation* **89,** 1320–1322.

Chemical Carcinogenesis

R. L. CARTER
University of Surrey and Royal Marsden Hospital

GLOSSARY

Carcinogenesis Origin and development of tumors. The term applies to all forms of tumors and not just to carcinomas (cf. *tumorigenesis* in American usage). *Carcinogenic agents/carcinogens* are causal factors of tumors, and they include exogenous factors (chemicals, physical agents, viruses), endogenous factors (genetic predisposition, hormones), and also general factors such as nutrition and reproductive activities. A distinction may be drawn between *genotoxic* carcinogens, which react directly with and mutate DNA, and *nongenotoxic* carcinogens, which act through other epigenetic mechanisms

Mutation Permanent change in the amount or structure of the genetic material (the *genome*). The alteration may involve a single gene, a block of genes, or a whole chromosome. Effects involving single genes may be a consequence of modifications of single DNA bases (*point mutations*) or of larger changes, including deletions, within the gene. Effects on whole chromosomes may involve changes in number and/or structure. Chemicals that produce such changes are *mutagens*, and mutagenic carcinogens are described as *genotoxic*

Toxicokinetics Mathematical description of rates of absorption, distribution, and elimination of chemicals

Tumor (syn. neoplasm) Mass of abnormal, disorganized tissue characterized by excessive and uncoordinated cell proliferation and by impaired differentiation. *Benign* tumors generally show a close morphological resemblance to their tissue of origin, grow slowly, and do not disseminate; they are rarely fatal. *Malignant* tumors resemble the parent tissue less closely and are composed of increasingly abnormal pleomorphic cells, grow rapidly, and disseminate (metastasize) to distant sites. When untreated, they are fatal. The nomenclature of tumors is based on the tissue of origin and the presence of benign or malignant features. *Cancer* is a general synonym for malignant tumors. [*See* Cell, Genetics, DNA.]

CARCINOGENIC CHEMICALS ARE SUBSTANCES that are causally associated with an increased risk of tumors in humans and/or animals. Their tumor-inducing activity may reside in the parent compound or in associated metabolites. Human exposures to carcinogenic chemicals occur in occupational, environmental, and social or cultural contexts. Most chemical carcinogens exert their effects after prolonged exposure, show a dose–response relationship, and act on a limited number of susceptible target tissues. There is usually a long latent period between the first encounter with a carcinogenic chemical and the appearance of a tumor. The mechanisms of chemical carcinogenesis are complex, with tumors developing in a multistage process in which both genotoxic and nongenotoxic events occur. Chemicals designated as *genotoxic* react with DNA, usually by forming covalent bonds with DNA bases. Various forms of genomic damage follow. If this damage is not lethal to the cell and is imperfectly repaired, it will be transmitted in the form of one or more mutations when the cell divides. Additional mutations occur as the tumors evolve. Nongenotoxic events in chemical carcinogenesis are poorly understood. *Nongenotoxic* chemicals do not bind to DNA and act mainly on cell proliferation as mutagens or cytotoxins or by disturbing cellular and tissue homeostasis. Genotoxic and nongenotoxic mechanisms are linked, and it is through cumulative, *combined* genotoxic and nongenotoxic damage that chemically induced tumors develop and progress.

ENCYCLOPEDIA OF HUMAN BIOLOGY, Second Edition, VOLUME 2. Copyright © 1997 by Academic Press. All rights of reproduction in any form reserved.

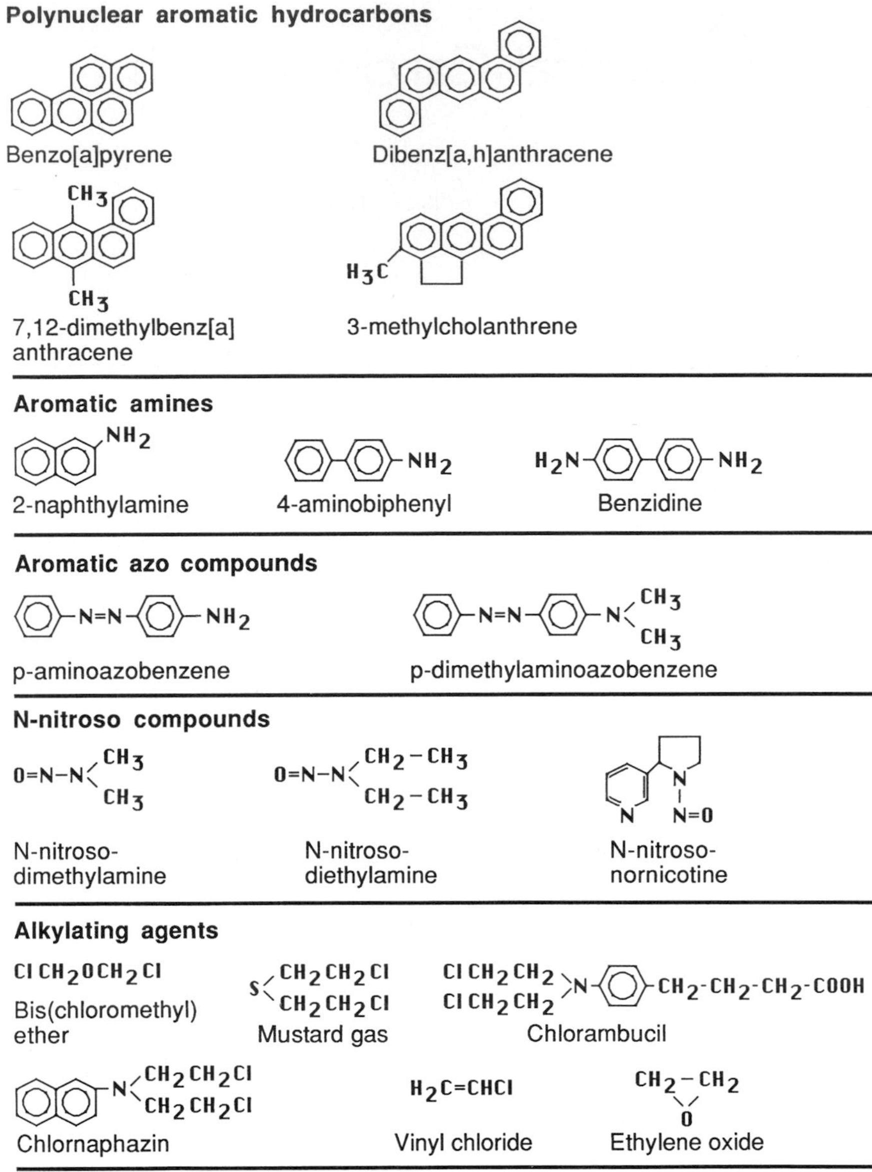

FIGURE 1 Some representative chemical carcinogens.

I. CHEMICALS AS CARCINOGENS

Most human cancers are caused by various extrinsic agents that include chemicals. Such chemicals are structurally diverse and comprise naturally occurring substances and hormones as well as synthetic compounds. Some examples are illustrated in Fig. 1. Several human carcinogens occur as *mixtures* (cigarette smoke and other pyrolysis products, ores) in which it may be difficult or impossible to attribute effects to any one constituent. The mode of action of chemical carcinogens is determined by the initial route(s) of exposure, the processes of absorption and distribution within the body, and (usually) metabolism. For genotoxic carcinogens, chemical structure provides clues to likely carcinogenic activity and (in some instances) to target organ susceptibility. These "structural alerts" cannot, however, be viewed in isolation: a suspicious compound may be innocuous because it is not absorbed or it is rapidly detoxified or excreted. Most genotoxic carcinogens undergo metabolic changes and are converted from inactive *procarcino-*

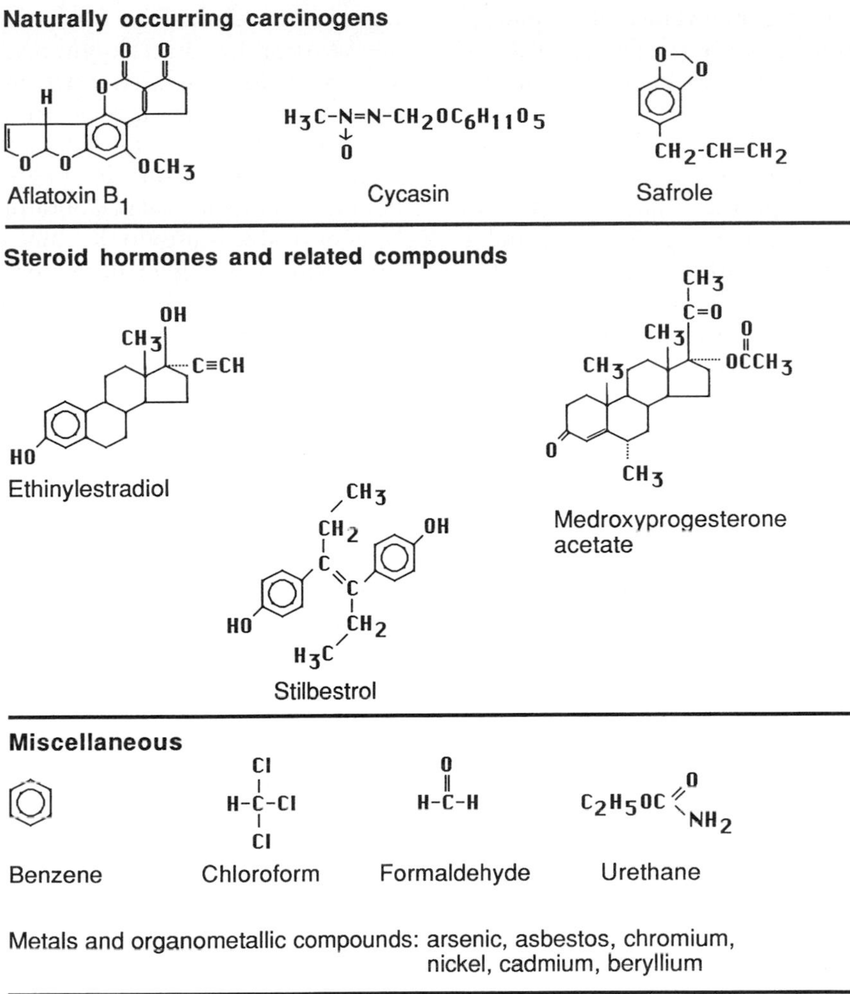

FIGURE 1 *(Continued)*

gens to *ultimate carcinogens* that bind to, and irreversibly alter, DNA in one or more target sites. Some genotoxic carcinogens react directly with DNA without previous metabolic activation. The extent and nature of metabolic activation of most nongenotoxic carcinogens are less clear; such chemicals, by definition, do not react with DNA (see Sections III and IV).

Although this account is concerned with chemical carcinogens, they cannot be viewed as causal agents in isolation. Several different chemicals may interact in an additive or multiplicative fashion (asbestos and cigarette smoke), and chemical carcinogens may operate with other exogenous factors such as *physical agents* (ionizing radiation, UV light) and *oncogenic viruses* (Epstein–Barr, hepatitis B–EBV, HBV). Endogenous or constitutional factors in the host (genetic susceptibility, hormonal status) may also contribute, emphasizing the multifactorial nature of the carcinogenic process.

II. CHEMICALLY INDUCED TUMORS IN HUMANS

Direct evidence that certain chemicals are carcinogenic in humans is provided by epidemiological studies. Data are available for industrial chemicals and processes, and for chemicals encountered in more general cultural and environmental contexts. Much of the information is derived from case–control and cohort studies. Details of methodology fall outside the scope of this account, but certain general points should be stressed. These include the importance of examining sufficiently large numbers of accurately matched test and control subjects, a sufficiently long time scale, an adequate level of exposure in the test group, the use of suitable statistical methods, and the exclusion of bias and confounding factors. Clear evidence for carcinogenic activity will appear as a large relative risk between exposed and unexposed populations and a dose–response relationship with respect to increasing risk and increasing exposure, the latter involving considerations of *duration* of exposure and *concentration* of the suspected agent or agents. Identification of a carcinogenic chemical is aided if the tumors occurring among the exposed group are unusual in terms of their site, morphology, or the clinical setting in which they are encountered, for example, a particular tumor type developing at an uncharacteristically early age.

The case is strengthened by independent confirmation from other, comparable studies. However, epidemiological findings are often more problematic. Circumstances where people come into contact with a single chemical are unusual, and many industrial and environmental exposures involve complex and inconstant mixtures encountered under variable conditions. Duration of exposure and concentration of the agents concerned are often difficult to determine for a whole group, let alone for individuals, although methods are now available for assessing uptake of suspect carcinogens by measuring their levels in body fluids, measuring substances bound as adducts to macromolecules such as DNA and proteins, and—at molecular and genetic levels—measuring mutational spectra (see Sections III and IV). Unrecognized confounding factors, particularly cigarette smoking, will distort the results. A small increase in the incidence of a common tumor may be undetected because of its high prevalence in the unexposed group and among the population as a whole. An inherent limitation of the epidemiological approach is imposed by the long latent period (usually a minimum of 10–15 years and sometimes as long as 30 years) that elapses between first exposure to carcinogenic chemicals and the appearance of a tumor. For example, the current epidemic of lung cancer in developed countries is the consequence of social trends that started in the 1930s and 1940s. The best epidemiological investigations all have inevitable limits of sensitivity, and it is essential that these are recognized.

These comments should be borne in mind when examining Tables I and II, which summarize data drawn from the International Agency for Research on Cancer (IARC) on chemicals and chemical processes regarded as carcinogenic for humans. The chemicals are structurally diverse, although genotoxic substances predominate. They act on a range of target tissues, most of them inducing tumors at or near the points of initial contact—skin, mouth, or respiratory tract. There is reasonable congruence between the findings in humans and in laboratory animals exposed by comparable routes to the same materials, albeit at proportionately higher doses for longer times. The contributions made by the various categories to overall mortality from cancer are difficult to quantitate; but the "best estimates" published by R. Doll and R. Peto in 1981, based on standardized cancer death certification rates in the United States, are still a valuable guide to the relative distribution of several cancers in industrialized societies, where chemicals can be implicated (in whole or in part) as causual agents.

The authors calculate that industrial cancers account for ~4% of all deaths from malignant disease, with occupational bladder and lung cancer making up ~15% and ~20%, respectively, of all deaths from these two tumors. Asbestos is responsible for ~5% of all fatal lung cancers. Among nonoccupational chemical exposures, Doll and Peto conclude that therapeutic drugs are associated with ~0.5% of cancer deaths and alcohol with ~3%. Cigarette smoke is the single most important carcinogenic agent, causually associated with ~30% of all cancer deaths and, worldwide, calculated to be responsible for more than one million deaths from lung cancer each year. The risk ratios for lung cancer associated with passive smoking have been extensively studied since the time of Doll and Peto's review. The topic remains controversial, but most authorities find a small, consistent increase. The contribution made by smokeless tobacco products to cancers of the mouth and pharynx could not be reliably quantitated but, in parts of the world where such tumors are common, they are important etiological agents (see Table II). (Tumors of the mouth and pharynx are infrequent in developed countries, although on a global scale they rank high on the list of major fatal cancers—fourth among males, sixth among females, and sixth overall.)

Approximately 2% of cancer deaths are attributed to pollution of air, water, and food and—very speculatively— ~35% to dietary factors. Both categories are complex. The diet, for example, may contain carcinogenic chemicals in the form of plant products (pyrrolizidine alkaloids, flavenoids), mycotoxins (aflatoxins) and additives, contaminants, and constituents acquired during the processing and preparation of foodstuffs (*N*-nitroso compounds, pyrolysis products). But for people living in developed societies, general factors such as total fat content and excessive calorie intake (overnutrition, obesity) are more important. Certain naturally occurring dietary substances (vitamin C, tocopherols, phenols) may exert *anticarcinogenic* effects by preventing the formation of carcinogens from precursors, and others (vitamin A derivatives, selenium, isothiocyanates, glutathione) may block access of carcinogens to their target tissues. Useful leads come from studying the relatively simple traditional diets eaten by people who live in rural communities in nonindustrial countries. It is likely, for example, that dietary factors are important etiological agents in esophageal cancer. This tumor shows wide differences in incidence over small areas and is particularly common in parts of southern and eastern Africa and in contiguous areas of Iran, Soviet Central Asia, and northwest China, where other risk factors such as alcohol and tobacco are unlikely to be implicated. In Western countries, dietary factors are also implicated in cancers of the stomach, large intestine, pancreas, and breast—all sites that are conspicuously absent from Tables I and II. A problem common to environmental pollutants is the occurrence of known or putative carcinogens—such as benzene, 1,3-butadiene, or nitroarenes in vehicle exhausts or halogenated hydrocarbons in drinking water—that are widely distributed at very low concentrations. Entire populations are likely to be exposed for most or all of their lives. The situation is different from occupational contexts, where a small, more homogeneous group of workers is exposed to higher concentrations during their hours of work, and it is extremely difficult to investigate.

III. CHEMICALLY INDUCED TUMORS IN LABORATORY ANIMALS

The chemical industry in the United States and other developed countries has grown exponentially since the late 1930s, with production of synthetic chemicals virtually doubling in each decade. Increasing numbers of new chemicals—therapeutic drugs, food additives, agrochemicals, veterinary and consumer products, industrial chemicals—are introduced whose potential carcinogenic effects in humans are unknown. Considerable reliance is therefore placed on studies in laboratory animals, and two questions arise: How effectively do animals function as human surrogates in terms of their response to carcinogenic chemicals? And to what extent do studies in animals clarify some of the more fundamental aspects of chemical carcinogenesis that cannot be investigated in people?

The design, conduct, and interpretation of carcinogenicity tests in laboratory animals fall outside the scope of this article. Although protocols are now reasonably standardized, there are several controversial issues. These include the supplementary use of nonrodents in certain contexts, problems posed by some test chemicals (selection of dose levels, routes of administration), classification of certain morphological lesions, evaluation of "spontaneous" tumors occurring (sometimes at high frequency) in untreated control animals, and the correct use of appropriate biostatistics. Extrapolation of results from animals to humans is extremely complicated. Two stages are involved: extrapolation *within* the test species, from the high doses used in the bioassay down to "no adverse effect levels" and on to the low doses proposed for human exposures; and then *across* species from rodents to humans. The first extrapolation involves consideration of dose–response data and the possible existence of a threshold for tumor induction. Dose–response relationships at low concentrations may not be linear (although they are often assumed to be so) and the concept of threshold is not tenable for chemicals that act through genotoxic mechanisms (see Section IV). Extrapolation from animals to humans is even more problematic, particularly with respect to the use of results from long-term tests as a basis for quantitative assessments of human risk. Although such calculations are widely used in radiation biology, they are more difficult to apply to chemicals whose carcinogenic activity is dependent on toxicokinetics and metabolism. Major biological assumptions are frequently made in the calculations, and different mathematical models may generate widely varying estimates of risk based on the same data from a single compound.

There are acknowledged limitations in long-term animal tests but, where the responses of rodents and animals can be directly compared (see Tables I and II), the results are more often similar than divergent. Congruence is particularly close for genotoxic substances. Seven of the chemicals listed in Tables I and II—4-aminobiphenyl, bis(chloromethyl) ether, diethylstilbestrol, melphalan, 8-methoxypsoralen + UV-A, mustard gas, vinyl chloride—were predicted to be human carcinogens on the basis of effects first described in rodents; six of them are genotoxic. (The exception is diethylstilbestrol, an interesting and debatable compound that appears to show both genotoxic and nongenotoxic effects.) Tumors induced in

TABLE I
Industrial Chemicals and Processes Identified as Carcinogenic for Humans

	Routes of exposure[a]	Main target sites for tumors	Reproduced in experimental animals[b]	General comments
I. Industrial Chemicals				
Aromatic amines				*Aromatic amines:* major carcinogens in dyestuff and rubber industries, first suspected at end of 19th century. Early examples of carcinogens acting at site *distant* from initial contact. Pronounced species variation in target sites for tumors—bladder in dogs compared to liver in rats and mice—reflecting different patterns of metabolism (dog unable to acetylate aromatic amines). Potent genotoxic activity.
4-Aminobiphenyl	inh, o, sk	bladder	d, rab	
Benzidine	inh, o, sk	bladder	d	
2-Naphthylamine	inh, o, sk?	bladder	d, ham, mon	
Ethylene oxide	inh	bone marrow, lymphoid system	mo, rat	*Ethylene oxide:* important chemical intermediate; smaller amounts used for sterilization and fumigation but higher levels of exposure occur in these groups. Evidence for human and animal carcinogenicity is not extensive, but the compound is a very potent direct-acting alkylating agent that produces mutagenic and clastogenic effects in many *in vitro* and *in vivo* systems.
Metals and metallic compounds				
Arsenic and certain arsenic compounds	inh, o, sk	skin, lung	ham (lung, limited data)	*Metals and metallic compounds:* occur as complex ores, frequently contaminated with other metallic substances, and as by-products of refining processes. Bioavailability depends on local persistence, solubility, and valency. Animal data for arsenic still very limited. Modes of carcinogenic action unclear. Little evidence for genotoxicity except for hexavalent chromates—CR VI (genotoxic effects may follow reduction in the cell to CR III, which binds to nucleic acids)—and for ionic cadmium. Other metals may damage DNA polymerases, which results in disordered DNA synthesis.
Beryllium and certain beryllium compounds	inh	lung	rat	
Cadmium and certain cadmium compounds	inh	lung	rat	
Chromium compounds, hexavalent	inh	lung	rat	
Nickel and certain nickel compounds	inh, o	nasal cavity, paranasal sinuses, lung	mo, rat (lung)	
Mineral fibers, naturally occurring				*Asbestos:* carcinogenic effects best documented for crocidolite, amosite, chrysotile, and anthophyllite; exposures usually mixed. Tumors of mesothelium (pleural and peritoneal mesotheliomas) very rare in general population and their occurrence serves as valuable marker for probable exposure to asbestos or related fibers. Strong multiplicative effects with smoking for lung cancer but not for mesothelioma. Carcinogenic activity related to physical dimensions of fiber—<2.5 μm thick and 10–80 μm long—rather than chemical structure, Mode of action still unclear, but the fibers damage chromosomes, induce aneuploidy, and are cytotoxic to mesothelial tissues in which they set up repeated cycles of cell death and cell regeneration.
Asbestos	inh, o	lung, mesothelium	rat (lung) rat, ham (mesothelioma)	
Talc-containing asbestos fibers	inh	lung, mesothelium	inadequate data	
Miscellaneous organic chemicals				
BCME and CMME Bis(chloromethyl) ether, Chloromethylmethylether	inh	lung	mo, rat	*Mustard gas:* carcinogenic effects described in production workers but not in troops exposed on the battlefield in World War I. An alkylating agent, predictably genotoxic.
Benzene	inh, o, sk	bone marrow	mo (lymphohemopoietic)	*Vinyl chloride:* associated with very rare form of liver tumor—angiosarcoma. Not pathognomonic of previous exposure to vinyl chloride but a strong pointer, comparable to mesotheliomas in individuals exposed to asbestos. Metabolized to form a genotoxic alkylating agent.
Mustard gas	inh	lung	mo	
Vinyl chloride	inh, sk	liver	mo, rat, ham	

Polynuclear aromatic hydrocarbons				*Polynuclear aromatic hydrocarbons:* human exposure is to complex mixtures and not to single compounds, at least 15 of which are carcinogenic for experimental animals. The earliest recognized occupational carcinogens: scrotal/skin cancers described among "climbing boys" (chimney sweeps) in 18th-century London by Percival Pott and, later, among workers exposed to shale oils, mineral oils, and coal tars. Repeated applications of coal tar shown to produce local skin cancers in rabbits and mice in 1915 and 1918. First pure carcinogen—benzo(*a*)pyrene—isolated from coal tar by E. Kennaway in 1932 at the Institute of Cancer Research, London. Abundant at low levels in the environment, generated by incomplete combustion of organic material.
Coal tar	inh, sk	skin, lung	mo, rab (skin)	
Coal tar pitches	inh, sk	skin, lung	mo (skin)	
Mineral oils, untreated and lightly treated	inh, sk, o	skin	mo	
Shale oils	inh, sk	skin, lung	mo (skin), rat (lung)	
Soots	inh, sk	skin, lung	mo (skin)	
II. Industrial Processes				A group of processes that have been shown, on epidemiological grounds, to carry a clear carcinogenic risk, although the chemicals responsible are not identified. Animal data are, for the most part, unhelpful. Polynuclear aromatic hydrocarbons (particularly coal tar pitch volatiles) may be involved in *aluminum production, coal gasification, coke production,* and *iron and steel founding*. The *manufacture of auramine* and *magenta* involves exposure to other chemicals including aromatic amines. Carcinogenic wood-dusts associated with *furniture and cabinet making* appear to be mostly hardwoods such as oak and beech. The tumors—adenocarcinomas of the nasal cavity and paranasal sinuses—are usually very rare and they may provide important clues for previous exposure: see also mesotheliomas (asbestos) and hepatic angiosarcomas (vinyl chloride). Exposure to *mists and vapors from strong inorganic acids* relates mainly to sulfuric acid, encountered in the manufacture of isopropanol and in the steel industry. *Painting* obviously represents exposure to widely diverse uses. Studies show excess of cancers at several sites with a consistent large excess of lung cancer, too big to be attributed to confounding by smoking. No increased risk for lung or other cancers among paint manufacturers.
Aluminum production	inh	lung		
Auramine production	inh, o, sk?	bladder		
Boot and shoe manufacture and repair	inh	nasal cavity, paranasal sinuses		
Coal gasification	inh, o, sk	skin, bladder, lung		
Coke production	inh	lung		
Furniture and cabinet making	inh	nasal cavity, paranasal sinuses		
Iron and steel founding	inh	lung		
Magenta manufacture	inh, o, sk?	bladder		
Mists and vapors from strong inorganic acids	inh	nasal cavity, paranasal sinuses, larynx, lung		
Painting	inh, o?, sk	lung, other sites		
Rubber industry	inh	bone marrow, lymphoid system, other sites?		

[a]inh, inhalation; o, oral; sk, skin.
[b]mo, mouse; ham, hamster; rab, rabbit; mon, monkey; d, dog.

TABLE II
Other (Nonindustrial) Chemicals and Products Identified as Carcinogenic for Humans

	Routes of exposure[a]	Main target sites for tumors	Reproduced in experimental animals[b]	General comments
I. Medicinal				
Alkylating agents				*Alkylating agents:* nucleophilic substances that alkylate purine and pyrimidine bases. Some (e.g., cyclophosphamide) require metabolic activation. Effects on DNA include single base substitution, single- and double-strand breaks, inter- and intrastrand cross-linkages. Potent genotoxic effects. Predictably carcinogenic, but valuable in chemotherapy of potentially fatal malignant disease. Chlornaphazine no longer in clinical use. Best-documented combined regime is MOPP—nitrogen mustard, vincristine, procarbazine, and prednisone; almost certainly associated with increased risk of acute nonlymphocytic leukemia, non-Hodgkin lymphoma, and perhaps solid tumors (lung, breast). Effects extremely complex, with interaction of several drugs and, often, irradiation.
Chlorambucil	oral and/or parenteral (usually intravenous)	bone marrow	mo, rat (lymphohemopoietic)	
Chlornaphazine	oral and/or parenteral (usually intravenous)	bladder	inadequate data	
Cyclophosphamide	oral and/or parenteral (usually intravenous)	bladder, bone marrow	rat (bladder)	
Melphalan	oral and/or parenteral (usually intravenous)	bone marrow	mo (lymphohemopoietic)	
Methyl-CCNU	oral and/or parenteral (usually intravenous)	bone marrow	inadequate data	
Myleran	oral and/or parenteral (usually intravenous)	bone marrow	mo (lymphohemopoietic)	
Thiotepa	oral and/or parenteral (usually intravenous)	bone marrow	mo, rat (lymphohemopoietic)	
Certain combined chemotherapy regimes	oral and/or parenteral (usually intravenous)	bone marrow, lymphoid system		
Anabolic steroids (oxymethalone)	o	liver	no data	Evidence incriminating *oxymethalone* based on descriptive studies; no analytical studies reported but drug almost certainly carcinogenic.
Immunosuppressants				*Immunosuppressants:* most data from heavily immunosuppressed transplant patients; less information from patients receiving immunosuppression for medical conditions. Associated with oligoclonal lymphoproliferative disorders as well as lymphomas; also Kaposi sarcomas. Often short (<2 yr) latent period. Lesions may regress if immunosuppressant treatment is stopped. Closely associated with EB virus, either reactivated or as a newly acquired infection.
Azathioprine	o, intravenous	lymphoid system	mo, rat	
Cyclosporin	o or intravenous	lymphoid system	limited data	
Estrogens				*Steroidal estrogens and estrogen/progestin combinations:* difficult to test in realistic protocols in laboratory animals. Appear to act mainly as tumor promoters for liver, breast, and other sites. No consistent evidence for genotoxic effect.
Steroidal (estrogen replacement)	o	endometrium	mo, ham	
Nonsteroidal (diethylstilbestrol)	transplacental	cervix, vagina	inadequate data	*Combined oral contraceptives* protect against ovarian and endometrial cancer. Altered relative risks for breast cancer remain controversial; significant trend in risk of breast cancer with total duration of use among young women below age 36 (?).
Estrogen/progestin combinations, oral contraceptives				
Combined	o	liver		
Sequential	o	endometrium	inadequate data	
Miscellaneous				
Analgesic mixtures containing phenacetin	o	urothelium	—	*Phenacetin:* given alone, induces urothelial tumors in rats and mice; the mixtures appear to be negative.
8-Methoxypsoralen+ UV-A	o + sk	skin	mo	*8-Methoxypsoralen + UV-A:* carcinogenic effects in humans and in experimental animals, depending on the combination; 8-methoxypsoralen alone is inactive.
Arsenic	o	skin	—	*Arsenic:* an archaic medicine used at 1% aqueous solution of potassium arsenite (Fowler's Solution).
Coal tar	sk	skin	mo, rab	*Coal tar preparations:* used in chronic eczema and psoriasis.

II. Environmental				
Aflatoxins	o	liver	mo, rat, ham, mon, etc.	*Aflatoxins:* mycotoxins produced by *Aspergillus glavus.* Dose-related association between intake of aflatoxin-contaminated foodstuffs (some cereals, ground nuts) and hepatocellular carcinoma in southern and eastern Africa, China, and Southeast Asia. Problems of confounding with preexisting hepatitis B infection adequately addressed. Aflatoxin B, induces liver tumors in many laboratory animals, although species susceptibility varies (rat exquisitely sensitive; adult mouse, but not neonate, refractory). Strongly genotoxic.
Erionite	inh	mesothelium (pleura)	mo. rat	*Erionite:* a widely distributed silicate, usually associated with other zeolite minerals. Carcinogenic effects so far identified in remote villages in Turkey, where heavy long-term exposure (soil, road dust, building materials) is associated with high incidence of mesothelioma.
Arsenic	o	skin	—	*Arsenic:* high concentrations in drinking water associated with skin cancer in parts of Chile, Argentina, Silicia, and Taiwan.
Asbestos	inh	mesothelium	rat, ham	*Asbestos:* environmental exposure seen in household contacts of asbestos workers and (in some studies) in people living near asbestos factories and mines.
III. Social/Cultural				
Alcoholic beverages				*Alcohol:* no consistent differences between carcinogenic effects of commercial beers, wines, or spirits. No reliable data from animals for pure ethanol, although main metabolite, acetaldehyde, is carcinogenic in rats and hamsters.
Betel quid + tobacco		mouth, pharynx, larynx, esophagus	inadequate data	
		mouth	limited data	
Tobacco smoke		lung; also mouth, pharynx, laryn, esophagus, bladder, pancreas	mo, rat, d (lung) rat (mouth) ham (larynx)	*Tobacco smoke:* the single most important chemical carcinogen, now responsible for ~30% of all cancer deaths in developed countries. Strongly genotoxic.
Smokeless tobacco products		mouth	rat (mouth, limited data)	*Smokeless tobacco products:* long established use in many different parts of the world—particularly India, Pakistan, parts of Southeast Asia, Soviet Central Asia, parts of the Middle East—with consumption up to 5 kg/yr. Wide variations in composition. Some products contain tobacco-specific *N*-nitrosamines at much higher levels (mg/kg) than tobacco smoke.
Chinese-style salted fish		nasopharynx	rat (nasal, paranasal, and mouth, limited data)	*Chinese-style salted fish:* large amounts consumed in southern China, Hong Kong, and Singapore. Some samples contain high levels of dimethylnitrosamine. Nature of interaction is uncertain. Note general difficulties in testing "social/cultural" carcinogens in laboratory animals in a realistis manner.

[a]inh, inhalation; o, oral; sk, skin.
[b]mo, mouse; ham, hamster; rab, rabbit; mon, monkey; d, dog.

laboratory animals by nongenotoxic chemicals are difficult to evaluate (see Section IV), but they still provide acceptable models for studying nongenotoxic mechanisms.

The toxicokinetics and metabolism of carcinogenic chemicals are major factors that determine whether an active or inactive (detoxified) chemical moiety is present in the target tissues. Two main groups of enzymes are involved in metabolizing foreign compounds such as chemical carcinogens: the family of cytochrome P-450-dependent mono-oxygenase isoenzymes and various conjugating enzymes that catalyze the formation of glucuronides, sulfate esters, and mercapturic acids. Most of these metabolizing enzymes are located in the endoplasmic reticulum (which is disrupted by homogenization to form "microsomes," hence the generic term "microsomal enzymes"). The cytochrome P-450 mono-oxygenase isoenzymes are particularly important: distinct P-450 gene families have diverged during evolution from a common ancestor and nearly 80 separate P-450 genes, divided into 14 families, have now been identified. The various isoenzymes may be species-, tissue-, or substance-specific so that extrapolation of carcinogenic effects by compounds that are activated by such enzymes can be difficult. They catalyze a wide variety of reactions such as epoxidation, N-oxidation, and hydroxylation. It is becoming apparent that the metabolism of certain classes of compounds, notably polycyclic aromatic hydrocarbons and aromatic amines, shows *individual*, genetically determind variation (polymorphism) in both humans and animals. The implications for individual sensitivity or resistance to the carcinogenic effects of such substances are of great interest and potential importance.

The metabolic generation of active carcinogenic moieties is best understood in relation to genotoxic chemicals. In most instances, inactive *procarcinogens* are converted into *ultimate carcinogens*—strongly electrophile reactants that bind covalently to nucleophile sites in DNA in target tissues to form carcinogen–DNA adducts. The terms "electrophile" and "nucleophile" refer to agents that acquire or donate electrons, respectively, in the course of chemical reactions. It is the identification of electrophilic, DNA-reactive groups within chemical structures that forms the basis for predictive structure–activity relationships with respect to potential genotoxic and carcinogenic effects. Details of the steps involved in individual activation pathways fall beyond the scope of this account, but examples include formation of *epoxides* (polynuclear aromatic hydrocarbons, aflatoxins, vinyl chloride) and N-hydroxylation and esterification (aromatic amines, azo-dyes, urethane) with generation of *carbonium or nitrenium ions.* Certain alkylating agents function as *direct-acting* ultimate carcinogens without previous metabolic activation. Examples include mustard gas, nitrogen mustard, melphalan, bis(chloromethyl)ether, and ethylene oxide.

The process whereby ultimate carcinogens bind covalently to DNA can only be noted briefly here. All four nitrogenous bases of DNA may act as targets, particularly the reactive (i.e., nucleophilic) purines. Some examples of binding sites are shown in Fig. 2. There are extensive data for alkylating agents, indicating covalent binding to N and O atoms in all four bases. N^7-alkylguanine accounts for at least 70% of bound product, but alkylation at other, subsidiary sites (e.g., O^6-guanine N_2-adenine) may be particularly important in terms of subsequent carcinogenic activity. Binding sites for several chemical carcinogens are still unknown. Other modes of interaction between genotoxic carcinogens and DNA have now been recognized. Some anthracycline drugs intercalate with DNA, a process of physical binding such that they become wedged between the stacked bases of the DNA double helix.

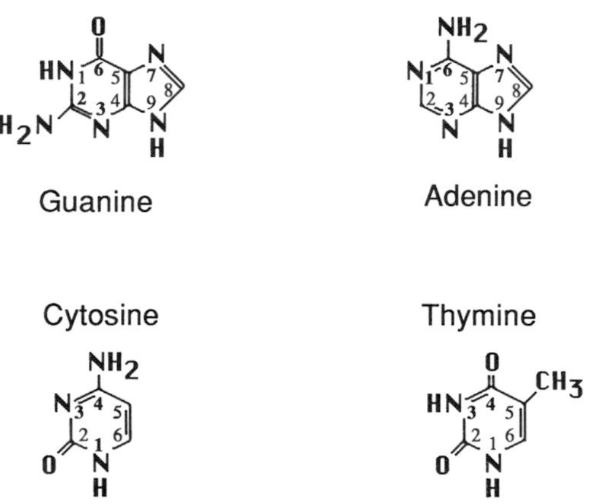

FIGURE 2 Binding sites for ultimate reactive carcinogens in purine and pyrimidine bases.

- PURINES
  Guanine (G) N-7, 0-6, N-3 alkylating agents; NH_2-2 polynuclear aromatic hydrocarbons; C-8 2-acetylaminofluorene, azo-dyes?
  Adenine (A) N-1, N-3, N-7 alkylating agents; NH_2-6 polynuclear aromatic hydrocarbons
- PYRIMIDINES
  Cytidine (C) N-1, N-3 alkylating agents; NH_2-4 polycyclic aromatic hydrocarbons?
  Thymine (T) N-3, N-4 alkylating agents.

Sensitive techniques such as radioimmunoassay and ^{32}P postlabeling procedures are available for measuring carcinogen–DNA adducts that can be used as markers of exposure. Adducts are also formed between carcinogens and proteins such as albumin or hemoglobin, which are more readily accessible.

The immediate consequences of carcinogens binding covalently to DNA include single base substitution, depurination, single- or double-strand breaks, and (in the case of bifunctional alkylating agents such as mustard gas) inter- or intrastrand cross-linkages. The longer-term outcome depends on the extent and nature of the altered DNA. The damage may be extensive and irreversible; for example, a cross-linking adduct may prevent replication of DNA or synthesis of an essential protein and the cell will die. Alternatively, the damage may be restored by error-free DNA repair. Or the cell may survive with imperfectly repaired DNA, which, in the course of subsequent cell and division, will be transmitted to daughter cells in the form of one or more mutations. Somatic mutations are a fundamental feature of tumor development and lead on to a brief consideration of carcinogenic mechanisms. [*See* DNA Repair.]

IV. MECHANISMS OF CHEMICAL CARCINOGENESIS

Carcinogenesis is a multistage process in the course of which there is progressive emergence of the aberrant clonal phenotype that constitutes the cancer cell. A series of separate genotoxic and nongenotoxic events is involved that are closely interdependent: neither genotoxic nor nongenotoxic events, *alone*, are sufficient to induce a tumor. Both are necessary but neither (in themselves) is a sufficient determinant for tumor development.

A. Genotoxic Events

In the present context, genotoxic events—mutations—are induced by genotoxic chemicals. The starting event in carcinogenesis (often called *initiation*) consists of one or more mutations in one or more viable cells; additional mutations occur at various stages during tumor development. Mutations are discussed in detail elsewhere but, in simple terms, they can be considered in three groups: point mutations, chromosomal mutations, and genomic mutations. *Point mutations* are a change in the sequence in one or a few codons that occurs either by base substitution or by deletion or addition of one or more bases. *Chromosomal mutations* are qualitative alterations in chromosomal structure that follow breakage and reunion of chromosomal material during cell division. They include inversions and translocations. *Genomic mutations* involve a change in the total number of chromosomes in the genome. Loss or gain of a single chromosome (aneuploidy) may occur as a result of nondysjunction during mitosis. Many of these mutations can be demonstrated in simple *in vitro* and *in vivo* test systems. A panel of such tests, using well-validated *in vitro* and *in vivo* assays in an ordered sequential manner, helps to identify probable genotoxic carcinogens and, for known carcinogens, gives valuable clues to their mechanism of action. [*See* Mutation Rates.]

It is now feasible to identify the particular genes that are modified as a result of carcinogen-inflicted mutations and, to simplify again, the main areas of interest relating to oncogenes and tumor-suppressor genes.

Oncogenes are activated (mutated) forms of proto-oncogenes that are present in the normal genome. They are highly conserved and they encode proteins that control normal cell growth and differentiation by acting as growth factors, growth factor receptors, signal transducers, protein kinases, and transcriptional activators. Proto-oncogenes are converted into oncogenes in various ways. Qualitative changes may be induced by point mutations and by chromosomal rearrangement, whereas quantitative alterations may result from other sorts of chromosomal rearrangements and from gene amplification. Irrespective of mechanisms, activation of proto-oncogenes is a dominant-like effect such that the mutated cell *gains* gene function that overrides or adds to the gene products that are normally expressed. Although the nature and detailed function of the altered gene products are poorly understood, the net effect is increased and continuous signaling for cell division and a sustained, uncontrolled proliferative state. Some examples of oncogenes in human and rodent tumors are listed in Table III. Their occurrence in human neoplasms is variable both within and between tumor types. More information (particularly for the *ras* oncogenes) has come from chemically induced tumors in rodents, where oncogene levels are generally higher. Certain segments of the *ras* oncogene appear to be particularly susceptible to point mutations inflicted by chemical carcinogens (codons 12, 13, 61) and the mutations produced may be remarkably consistent—the A $\rightarrow$ T

TABLE III
Examples of Oncogenes in Human and Experimental Tumors[a]

A. Human Tumors

Present in one or a few tumor types:

abl	chronic myeloid leukemia	*C-myc*	Burkitt lymphoma, lung
N-myc	neuroblastoma	*neu/erb B2*	breast

Present in several tumor types at varying incidence:
H-, *K-*, *N-ras* genes present in DNA from 10 to 20% of unselected series of fresh tumor biopsies and in tumor cell lines

Distribution in biopsies from specific tumor types:

colon 11/27 *K-ras*	lung 5/39 *N-ras*	liver 3/10 *N-ras*	
26/66 *K-ras*			
bladder 2/38 *H-ras*	breast 0/16	stomach 0/26	cervix uteri 0/30
1/15 *H-ras*			

Increased evidence of *ras* oncogenes in tumor cell lines

B. Experimental Tumors

Carcinogen	Species	Tumor	Oncogene	
7,12-Dimethylbenz(*a*)anthracene (DMBA)	rat	breast	*H-ras*	6/6
7,12-Dimethylbenz(*a*)anthracene + TPA	mouse	skin	*H-ras*	33/33
1,8-Dinitropyrene	rat	soft tissues	*K-ras*	
3-Methylcholanthrene	mouse	soft tissues	*K-ras*	2/2
Methyl(methoxymethyl)nitrosamine	rat	kidney	*K-ras* *N-ras*	
Dimethylnitrosamine	rat	kidney	*K-ras*	10/11
Diethylnitrosamine	mouse	liver	*H-ras*	7/7
N-Nitroso-*N*-methylurea (NMU)	rat mouse	breast thymus	*H-ras* *N-ras* *K-ras*	16/61 5/5 5/5
Vinyl carbamate	mouse	liver	*H-ras*	7/7
N-Hydroxy-2-acetylaminofluorene	mouse	liver	*H-ras*	7/7
Aflatoxin	rat	liver	*K-ras*	7/7
Tetra-aminomethane	rat mouse	lung	*K-ras*	
Transplacental				
N-Nitroso-*N*-methylurea (NMU)	rat	nervous system	*neu*	4/5
N-Nitroso-*N*-ethylurea	rat	nervous system	*neu*	2/3
Spontaneous	mouse	liver	*H-ras*	11/13

[a]Oncogenes are given three-letter codes based on animal or tumor from which they were first derived: *abl*, Abelson mouse leukemia virus; *C-myc*, avian myelocytoma virus; *N-myc*, human neuroblastoma; *neu*, rat neuroglioblastoma; *erb B*, avian erythroblastosis virus; *ras: H-* (Harvey) rat sarcoma virus; *K-*, (Kirsten) rat sarcoma virus; *N-*, human neuroblastoma cell line.

transversion of the second nucleotide of codon 61 of *H-ras* induced by 7, 12-dimethylbenz(*a*)anthracene (DMBA) in mouse skin, and the G → A transversion of the second nucleotide of codon 12 of *H-ras* induced by *N*-nitroso-*N*-methylurea (NMU) in rat breast. [*See* Oncogene Amplification in Human Cancer; Oncogenes and Proto-oncogenes.]

Tumor-suppressor genes differ from oncogenes in that they are involved in carcinogenesis only when, as a result of mutation, they are deleted or inactivated—there is a *loss* rather than a gain of normal gene function. The best-documented examples are the *Retinoblastoma* (*Rb*) *gene*, which is lost or inactivated in children with retinoblastoma, and in several adult tumors, and the *p53-gene*—probably the most commonly mutated gene in all sporadic cancers. The *p53* gene encodes a nuclear phosphoprotein that is concerned with the normal control of the cell cycle. It has been described as a checkpoint control for recognizing DNA damage. It acts by slowing the cell cycle at the

G_1/S phase to allow DNA repair to take place, and by initiating programmed cell death (apoptosis). In brief, *p53* monitors genomic stability, and its loss confers selective advantages for clonal expansion. Mutations sustained by *p53* show several important features. Most of them are concentrated in the sequence covered by three highly conserved exons (5 to 8), and their location and type serve to define distinctive "mutational spectra." The spectra vary from different tumor types and, to some extent, for different carcinogenic chemicals. Thus, *p53* mutations are found in about 60% of all human cancers and in 70% or more of specified histological types, such as various carcinomas of the mouth, lung, and thyroid. Distinctive mutational spectra have recently been described among hepatocellular carcinomas in patients from high-risk areas associated with exposure to aflatoxin B_1 and in tobacco-associated cancers of the lung. The scope for developing what is essentially molecular epidemiology will be obvious. [*See* Tumor Suppressor Genes; Tumor Suppressor Genes, p53; Tumor Suppressor Genes, Retinoblastoma.]

The complexities of this large and rapidly growing area of investigation are daunting. The distinction between oncogenes and tumor-suppressor genes, for example, is not absolute, as some mutant mis-sense *p53* proteins represent a gain of gene function and behave like oncogenes. Ostensibly the same mutations may be induced either early or late in the development of different tumor types. For any one tumor type, the detailed interplay between oncogenes and tumor-suppressor genes is obscure. Colorectal cancers, for example, are known to acquire several different mutations as they develop, including activation of the *K-ras* oncogenes on chromosome 6 and loss of various suppressor genes (*APC, P53, DCC*) on chromosomes 5, 17, and 18.

B. Nongenotoxic Events

Compared to genotoxic substances, the events associated with nongenotoxic agents are poorly understood. Such substances are, by definition, devoid of activity in a range of well-validated, well-conducted genotoxicity tests. They are chemically diverse and lack any "structural alerts" comparable to those described for genotoxic chemicals. No predictive screening tests are available comparable to those used for demonstrating genotoxic activity. Therefore, two important approaches to studying these compounds cannot be used. Certain other general differences between genotoxic and nongenotoxic chemicals should be noted, although exceptions can be set against each of them. Nongenotoxic substances usually exert their effects at large doses over long periods of time, and the effects are initially reversible. Genotoxic compounds are active at low concentrations and, in the case of potent chemicals such as diethylnitrosamine, may induce tumors after a single exposure; the mutations produced are irreversible. Genotoxic carcinogens commonly induce tumors in two or more species, in both sexes, and often at a number of different sites, whereas nongenotoxic agents show more restricted effects with respect to species, strain, and target organ. The types of tumors induced by nongenotoxic agents often overlap, with neoplasms found in matched but untreated control animals, the dose–response is characteristically steep, and threshold and no-effect levels can often be demonstrated—a crucial difference from genotoxic compounds where (theoretically) no threshold operates. Some nongenotoxic agents act through receptors [phorbol esters, 2,3,7,8-tetrachlorodibenzo-*para*-dioxin (TCDD), peroxisome proliferators, hormones], whereas others (saccharin, certain antioxidants, several hepatotoxic and nephrotoxic compounds) do not.

Nongenotoxic events in carcinogenesis are dominated by tissue changes rather than the intrinsic chemical characteristics of the substances that induce them. The most important of these tissue changes are associated with cell proliferation. Many nongenotoxic agents act as mitogens or cytotoxins, or they may disturb normal autocrine, paracrine, or endocrine control processes. The patterns of disordered cell proliferation vary. Nongenotoxic substances such as phenobarbitone or some trophic hormones, which act purely as mitogens, typically induce an early, short-lived, and often intense burst of cell division followed by a long period of sustained proliferation at lower but still elevated levels. Cytotoxic substances such as carbon tetrachloride, on the other hand, cause cell damage and cell death, which evoke reparative proliferation, the sequence of cell death and cell regeneration occurring in cycles. In both instances, the target tissues or organs enlarge as a result of hyperplasia and hypertrophy. The net result is progressive disorganization of physiological homeostasis within a target site, culminating in the development of tumors. For this outcome, it is essential that the stimulus is appropriately intense and that there are adequate numbers of susceptible target cells that are capable of responding. The target tissues in rodents that appear to be most susceptible to nongenotoxic carcinogens are the liver, kidneys, and endocrine glands—

tissues with stable cell populations that, nevertheless, are capable of a high rate of proliferation if appropriately (or inappropriately) stimulated. The basis for such selectivity of target sites is poorly understood but it is likely to include genetic susceptibility, expression of receptors, and (sometimes) species-, strain-, and sex-dependent metabolic pathways.

The processes whereby nongenotoxic agents stimulate cell proliferation are also poorly understood. Cell division may follow direct damage at the target site, but in other circumstances the stimulus for cell proliferation is indirect, for example, following stimulation of peroxisomes and/or microsomes. Some examples are illustrated in Table IV. At a molecular level, evidence is beginning to emerge that some nongenotoxic agents may simulate normal growth factors or growth signals. They may also interact with intact proto-oncogenes or their activated (mutated) forms, modulating gene expression with resulting changes in protein synthesis.

It is important to emphasize that enhanced and disordered cell proliferation alone is insufficient to generate tumors, but it is within the context of deranged cell proliferation that genotoxic and nongenotoxic events converge. The effects of nongenotoxic compounds on cell proliferation serve to expand clonal populations already defined by previous mutations; but enhanced proliferation itself may introduce

TABLE IV

Examples of Nongenotoxic Agents with Carcinogenic Effects[a]

Liver

Phenobarbital, dieldrin, butylated hydroxytoluene (BHT), halogenated biphenyls, some estrogenic steroids

Carcinogenic effects usually more common in mice than rats. Substances stimulate proliferation of smooth endoplasmic reticulum with activation of P-450-dependent mono-oxygenase isoforms. Mechanisms linking activated mono-oxygenases with subsequent cell proliferation are obscure.

Hypolipidemic agents, trichloroethylene, Di(2-ethylhexyl)phthalate (DEHP)

Carcinogenic effects confirmed to rodents. Dependent on proliferation of *peroxisomes*. Receptor-mdiated modulation of gene expression. Increased activity of intercellular enzymes responsible for β-oxidation of long-chain fatty acids, microsomal enzymes, and enzymes involved with cell proliferation. Details of mechanisms linking peroxisome proliferation and stimulation of cell growth are, again, obscure.

Carbon tetrachloride, chloroform, tetrachloro-ethylene, 1,1,2,2-tetrachloroethane

Carcinogenic effects seen mainly in mice. Compounds induce repeated cycles of cell damage and cell death, followed by regenerative hyperplasia.

Kidney

Chloroform

Direct cytotoxic effect on renal tubules inducing tumors in rats (males > females) but not mice.

Tri and perchloroethylene, pentachloroethane, 1,4-dichlorobenzene, unleaded gasoline

Affect male rats through *$\alpha 2\mu$-globulin nephropathy* $\alpha 2\mu$-Globulin is a low-molecular-weight protein produced *only in the liver of male rats*. Filtered at glomeruli, reabsorbed in renal tubules, and broken down by lysosomes. Certain chlorinated hydrocarbons (see above) bind to $\alpha 2\mu$-globulins, forming complexes that are retained in tubular cells. Lysosomes are damaged and renal tubular epithelium dies. Cycles of cell death, followed by regenerative hyperplasia, ensue.

Bladder

Sodium salts of weak acids such as saccharin and ascorbate, *o*-phenylphenate (Ca^{2+} salts, parent acids generally inactive), nitrilotriacetic acid (NTA), calculi-inducing agents

Carcinogenic effects mainly in rats. Large doses (usually ≥1% of diet) for long periods. Various mechanisms involved. Sodium salts listed above induce (physiological) osmotic diuresis with chronic polyuria and polydipsia. Compensatory bladder enlargement follows with mucosal hyperplasia. Other contributory factors include failure of normal urine concentration mechanisms in old rats, associated with progressive, age-related nephropathy. High doses of NTA cause loss of Ca^{2+} and other cations from urothelium. The cells degenerate, and cycles of cell death and regenerative hyperplasia follow.

Thyroid

Chemical goitrogens of various classes, naturally occurring and synthetic

Act mainly by *inhibiting synthesis* of thyroid hormones (T_3, T_4) by thyroid follicular cells, e.g., by inhibiting thyroid peroxidase or interfering with deiodination mechanisms; may also *increase hormone catabolism* and *inhibit $T_4 \rightarrow T_3$ conversion*. Low blood levels of T_3, T_4 provoke sustained increase in circulating TSH from the anterior pituitary gland, which stimulates follicular hyperplasia and development of benign and malignant tumors. TSH-dependent tumors will regress if goitrogenic stimulus is withdrawn.

[a]The table emphasizes the overriding importance of *tissue changes* evoked by nongenotoxic agents. Some of the mechanisms are highly specific with respect to species, strain, and sex. A clear understanding of such processes is essential when regulatory authorities are required to evaluate likely human hazards and risks.

new mutations as well as augmenting existing ones. Chronically proliferating cells are at a greater risk of acquiring mutations in the course of DNA synthesis and, with shortened cell-cycle times, they are less able to correct them by error-free DNA repair.

The larger and longer acting the proliferative stimulus, the more likely it is to be accompanied by genomic damage. Sustained and intense mitogenesis is itself mutagenic. Some nongenotoxic compounds may act as indirect genotoxins. Active oxygen species, for example, are elaborated in the vicinity of inflammatory responses evoked by tissue damage and may possibly modify local DNA bases. *Spontaneous* or *endogenous* mutations may occur as a result of various forms of DNA damage such as depurination or deamination of 5-methylcytosine, or infidelity of DNA polymerases, in the absence of an extrinsic mutagen. Genomic damage of this kind is likely to accumulate as tissues and organs age. (Specific mutations are regularly present in some types of "spontaneous" tumors that develop in old, untreated laboratory rodents.) The notion that populations of initiated cells exist in target sites that have already sustained some mutational damage as a result of either extrinsic or intrinsic mutagens emphasizes the importance of nongenotoxic events in carcinogenesis.

The effects of nongenotoxic agents on cell proliferation are often described as *promotion* and, by extension, the agents themselves are sometimes designated as promoters. Promotion was originally regarded as clonal expansion of initiated cells in strictly defined experimental systems where tumors were induced by one exposure to a subcarcinogenic dose of an initiating agent (a genotoxic carcinogen) followed by prolonged treatment with a promoting agent (a nongenotoxic carcinogen). It is now recognized that initiation is the first but not the only mutational event that imposes clonal restriction on subsequent cell proliferation; and in the case of nongenotoxic agents, considerable nonclonal proliferation may be established before one or more mutations are sustained and a clonal population of genetically altered cells emerges. Furthermore, some nongenotoxic agents act in ways other than by stimulating and perverting normal cell proliferation. Certain phorbol esters, for example, alter cell–cell communication, which is likely to impair the normal processes whereby adjacent cells with transformed or abnormal phenotypes are suppressed. Other nongenotoxic agents are immunosuppressive, facilitating expression of oncogenic viruses or deranging normal immune surveillance.

Finally, it should be noted that some genotoxic carcinogens exert both genotoxic and nongenotoxic effects in the target tissues and may be regarded (using a rather old term) as "complete" carcinogens. Although separate genotoxic and nongenotoxic effects are both required for tumor development, they do not necessarily depend on separate genotoxic and nongenotoxic agents.

BIBLIOGRAPHY

Ames, B. N., and Gold, L. S. (1990). Too many rodent carcinogens: Mitogenesis increases mutagenesis. *Science* **249**, 970–971.

Ashby, J. (1991). Determination of the genotoxic status of a chemical. *Mutat. Res.* **248**, 221–231.

Ashby, J., and Tennant, R. W. (1991). Definitive relationships among chemical structure, carcinogenicity and mutagenicity for 301 chemicals tested by the US NTP. *Mutat. Res.* **257**, 229–306.

Ashby, J., *et al.* (1994). Mechanistically-based human hazard assessment of peroxisome-proliferator-induced hepatocarcinogenesis. *Hum. Exp. Toxicol.* **13**, (Suppl. 2), 1-117.

Balmain, A., and Brown, K. (1988). Oncogenic activation and chemical carcinogenesis. *Adv. Cancer Res.* **51**, 147.

Box, J. L. (1989). *Ras* oncogenes in human cancer: A review. *Cancer Res.* **49**, 4682–4689.

Butterworth, B. E. (1990). Consideration of both genotoxic and non-genotoxic mechanisms in predicting carcinogenic potential. *Mutat. Res.* **239**, 117–132.

Cavenee, W. K., and White, R. L. (1995). The genetic basis of cancer. *Sci. Am.* **272**, 50–58.

Cohen, S. M., and Ellwein, L. B. (1990). Cell proliferation in carcinogenesis. *Science* **249**, 1007–1011.

Doll, R. (1991). Progress against cancer: An epidemiologic assessment. *Am. J. Epidemiol.* **134**, 675–688.

Evans, H. J. (1993). Molecular genetics of human cancers. *Br. J. Cancer* **68**, 1051–1060.

Greenblatt, M. S., Bennett, W. P., Hollstein, M., and Harris, C. C. (1994). Mutations in the p53 tumor suppressor gene: Clues to cancer etiology and molecular pathogenesis. *Cancer Res.* **54**, 4855–4878.

Hildebrand, B., Ashby, J., Grasso, P., Sharratt, M., Bontinck, W. J., and Smith, E. (eds.). (1991). Early indicators of non-genotoxic carcinogenesis. *Mutat. Res.* **248**, 211–376.

International Agency for Research on Cancer. (1972–1996).

"IARC Monographs on the Evaluation of Carcinogenic Risks to Humans," Vols. 1–66 (a continuing series). IARC, Lyon, France.

Weinberg, R. A. (1989). Oncogenes, anti-oncogenes and the molecular basis of multistep carcinogenesis. *Cancer Res.* **49**, 3713–3721.

Chemotherapeutic Synergism, Potentiation and Antagonism

TING-CHAO CHOU
Memorial Sloan-Kettering Cancer Center, Cornell University

DARRYL RIDEOUT
Research Institute of Scripps Clinic

JOSEPH CHOU
University of California, San Francisco

JOSEPH R. BERTINO
Memorial Sloan-Kettering Cancer Center, Cornell University

GLOSSARY

Antagonism Interaction in which the effect is smaller than the expected additive effect

Combination index (CI) Introduced to provide a quantitative measure of the degree of drug interaction for a given end point of effect measurement; CI = 1, <1, and >1 indicate additive, synergistic, and antagonistic effects, respectively

Dose-reduction index Introduced in 1988, provides a measure of how much the dose of each drug in a synergistic combination may be reduced at a given effect level compared with the doses of each drug alone; toxicity toward the host may be avoided or reduced when the dose is reduced

Isobologram Graph representing the equipotent combinations of various concentrations (or doses) of two drugs that produce the same effect; it can be used to identify synergism, additivism, or antagonism

Potentiation, inhibition In drug combinations, a potentiator or inhibitor will augment or depress the effect of other drug(s), whereas by itself it has no effect. The presentation of results is straightforward, usually by a percentage or multiple of changes incurred as a result of the administration of the potentiator or inhibitor

Selectivity index, therapeutic index Selectivity index is the ratio of the median-effect concentrations (or doses) for toxicity to the host in relation to that concentration for the target, such as tumor cells or pathogens (e.g., the ratio of the median-effect toxic dose versus the median-effect therapeutic dose: ID_{50}/ED_{50}); the chemotherapeutic index has a similar meaning but usually refers to a therapeutic situation and may refer to different effect levels (e.g., 10% toxic dose versus 90% therapeutic dose: TD_{10}/ED_{90}); the higher the values of these indices, the safer the drug

Synergism Interaction in which the effect is greater than the expected additive effect

I. WHY COMBINATION CHEMOTHERAPY?

When two or more treatment modalities are used for combination therapy (e.g., chemotherapy, radiation therapy, thermotherapy), the effects can be synergistic, additive, antagonistic, or a mixed effect (e.g., antagonistic at low-dose levels, synergistic at high-

ENCYCLOPEDIA OF HUMAN BIOLOGY, Second Edition, VOLUME 2. Copyright © 1997 by Academic Press. All rights of reproduction in any form reserved.

dose levels). For combination chemotherapy, a multiple-target approach with several drugs can be used, or multiple drugs active with different mechanisms against a single target can be used. For example, in the treatment of cancer, combination chemotherapy attempts to deal with the heterogeneous cell populations or to avoid the development of resistance to make the eradication of the malignant cells feasible.

A. Main Goals of Combination Chemotherapy

1. To increase therapeutic efficacy against the target(s) by synergism, thus lowering the dose of one or both drugs for a given effect.
2. To decrease toxicity or untoward side effects against the host by reducing the dose(s) in synergistic combinations.
3. To minimize or delay the development of drug resistance by use of lower doses and/or multiple targets for attack. The criteria for comparing the development of resistance by a single drug and by multiple drugs are based on a given degree of drug effect and a given duration of exposure to the drugs. Increased therapeutic efficacy decreases the length of therapy for a given degree of therapeutic effect; in turn, the shorter duration of therapy may decrease the risk of developing resistance.
4. To achieve selective synergism against the target (pathogenic microorganisms, cancer cells, or parasites) and/or selective antagonism toward the host, hence increasing the therapeutic index.

Chemotherapy is a double-edged sword: it can harm both the target and the host. Efficacy must be balanced against toxicity. Combination chemotherapy offers options to exploit treatment advantages with respect to efficacy, toxicity, and schedules of treatment.

If the mass-action law is strictly followed, the combined effect of two drugs should be additive. If synergism or antagonism occurs, one or more mechanisms may be involved, which may or may not be understood. A study of dose–effect relationships permits the determination of whether or not synergism exists and, if so, the quantification of the degree of synergism. However, such a study does not provide information with regard to how and why synergism occurs. Therefore, two issues must be addressed separately: the determination of synergism–antagonism and the interpretation of synergism—antagonism.

B. Some Selected Examples of Drug Combinations

Numerous examples of synergism and antagonism in biomedical literature and reviews are available. The following are some selected examples of drug combinations.

(1) Drugs that inhibit two enzymes forming part of a single, vital metabolic pathway may exhibit synergism or antagonism owing to their combined effects on the synthesis of the final metabolic product or products. Because of branching metabolic pathways, feedback inhibition, and potentiation by metabolites, it is usually impossible to predict whether synergism or antagonism will occur in such antimetabolite combinations without detailed knowledge of the pathways and extensive calculations. One clinically useful example of antimetabolite combinations involves sulfamethoxazole and trimethoprim, both of which inhibit reactions required for bacterial DNA synthesis with an end result of strong synergistic effect.

(2) A combination of the prodrugs leucovorin and 5-fluorouracil exhibits a clinically relevant synergistic effect against gastrointestinal and other carcinomas because of simultaneous binding of methylene tetrahydrofolate and 5-fluorodeoxyuridine monophosphate (the active metabolites) to the enzyme thymidylate synthase. The combination, in contrast to either methylene tetrahydrofolate or FdUMP alone, forms a tighter binding ternary complex with the enzyme, thus preventing *de novo* DNA synthesis.

(3) Almost all drugs are metabolized to some extent in the human body, mainly from the activity of enzymes found in normal tissues and/or at the site of the disease. For example, penicillin and related antibacterial agents (β-lactams) can be converted to inactive metabolites by the enzyme β-lactamase found in some resistant strains of bacteria. Hence, although β-lactams alone are not effective for the treatment of infections caused by these bacteria, combinations of β-lactams with β-lactamase inhibitors such as sulbactam are effective. Sulbactam potentiates the antibacterial activity of β-lactams by preventing the breakdown of β-lactams by the bacterial β-lactamase, thereby overcoming drug resistance.

(4) Most chemotherapeutic agents must reach targets inside cells (as opposed to surface targets) to exert their effect. Tumor cells and microbes may be resistant to drugs because of slow rates of drug influx and/or rapid rates of drug efflux. The action of such drugs can be enhanced by adding a second agent that affects the outside of the cell. For example, penicillin inhibits the synthesis of cell-wall components in bacteria such

as *Streptococcus mitis,* thereby enhancing the permeability of streptomycin. When streptomycin and penicillin are used in combination, relatively more streptomycin reaches its intracellular target (the ribosome). As a result, streptomycin and penicillin exhibit therapeutically useful antibacterial synergism. Some tumors possess or develop resistance to vinblastine, doxorubicin, and other drugs because these molecules are actively pumped out of the cell by an efflux pump known as the p-glycoprotein. This resistance can be reversed by clinical agents such as verapamil and quinidine, which inhibit the activity of the pump.

(5) The simplest possible mechanism of synergism involves direct covalent combination between two agents to form a more bioactive molecule. A number of aldehyde–hydrazine derivative combinations exhibit antineoplastic or antibacterial synergy *in vitro* because cytotoxic hydrazones, formed *in situ,* are more cytotoxic than each drug alone.

(6) Combinations of drugs exhibiting synergism can be superior to single drugs in preventing the development of drug resistance in tumors and microbes. If the drugs act on two different macromolecular targets, the probability of a mutation or mutations leading to resistance to both drugs simultaneously is quite small. Two drugs that exhibit collateral sensitivity can also be very useful in this regard: if the target cell or microbe becomes resistant to one agent, it may simultaneously become hypersensitive to the second agent in the combination. For example, tumors that become resistant to 6-mercaptopurine lose their ability to synthesize purines from hypoxanthine and thus become highly susceptible to methotrexate, a folate antagonist that blocks the *de novo* biosynthesis of purines.

(7) 3′-Azido-3′-deoxythymidine (AZT) and recombinant interferon (IFN) have been shown to be strongly synergistic against human immunodeficient virus, type 1 (HIV-1), replication. It is well established that HIV-1 is an etiologic pathogen for acquired immunodeficiency syndrome. AZT inhibits reverse transcriptase, which is required for HIV-1 replication, whereas IFN affects the late-stage, viral particle assembly in the HIV-1 life cycle. Examples of the quantitative determination of synergistic interaction between AZT and IFN against HIV-1 are given in Figs. 1–3.

II. WHAT IS SYNERGISM?

The most common objective for combination chemotherapy is to achieve synergistic drug effects, although in special situations antagonism is exploited. An example involves the use of a high dose of methotrexate

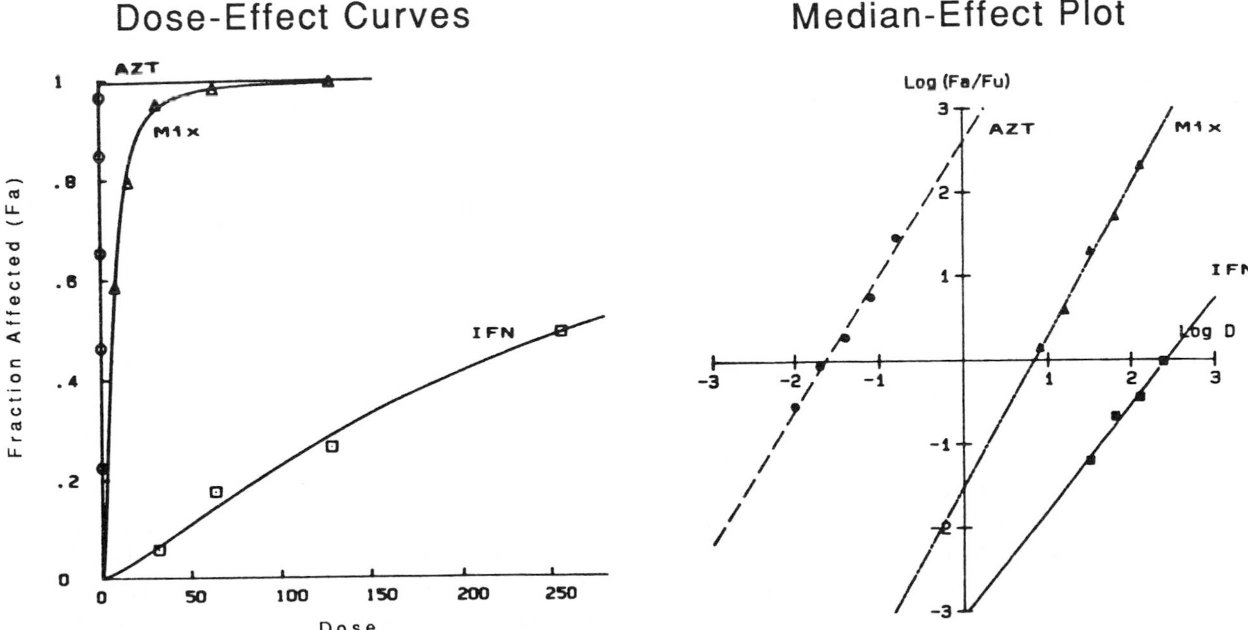

FIGURE 1 Dose–effect curves and the median-effect plot for inhibition of human immunodeficiency virus, type 1, (HIV-1) by recombinant interferon α (IFN) and 3′-azido-3′-deoxythymidine (AZT), and AZT (in μM) and IFN (in U/ml) mixture (1 : 800). Data were obtained using reverse transcriptase assays. The parameters obtained are $(D_m)_{AZT} = 0.0233\ \mu M$, $(m)_{AZT} = 1.593$, $(D_m)_{IFN} = 263.06$ U/ml, $(m)_{IFN} = 1.262$, $(D_m)_{mix} = 6.856$, $(m)_{mix} = 1.803$. m and D_m are obtained from the slope and intercept, respectively.

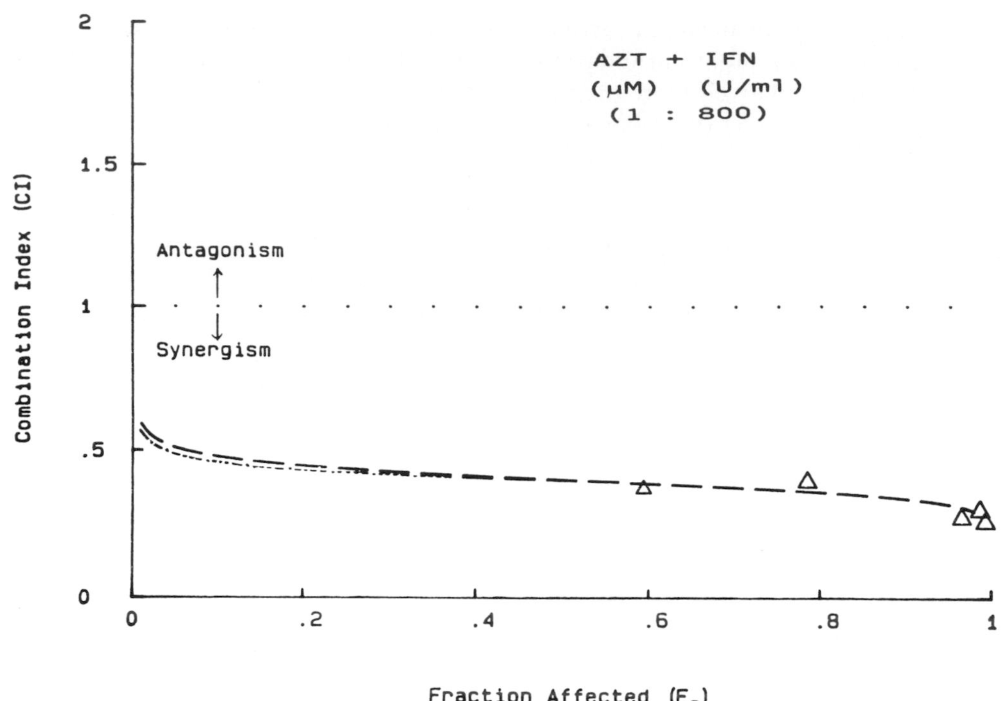

FIGURE 2 The F_a–CI plot showing synergism—antagonism with combination index (CI) as a function of fraction affected (f_a). Parameters shown in Fig. 1 were used for the calculations. CI values for each combination data point are shown with triangles. Dashed and dotted lines are computer-simulated results of CIs assuming mutually exclusive and mutually nonexclusive interactions, respectively, between AZT and IFN. The results show strong synergism between AZT and IFN with CI < 1. In this example, both exclusive and nonexclusive assumptions give very similar results.

to produce drastic therapeutic and toxic effects followed by citrovorum rescue (from toxicity) for cancer treatment.

Though the pursuit of synergy among therapeutic agents extends as far back as the recorded history of medicine, the concept of synergy was not clearly and vigorously defined in mathematical terms until recently. By definition, synergism is more than an additive effect and antagonism is less than an additive effect. Therefore, the crucial question is: What is an additive effect? Unless additivity of drug effects is unambiguously defined, concepts of synergism and antagonism are meaningless.

To simplify the discussion, the combination of two drugs will be considered first. The terms synergy, synergism, and synergistic effect are frequently mentioned in biomedical literature, but synery is not clearly defined. Many factors need to be considered for drug combination studies: If drug 1 has a hyperbolic dose–effect curve and drug 2 has a sigmoidal dose–effect curve, how do we predict the expected additive effect? If two drugs are synergistic at ED_{50} (the dose for 50% effect), they are not necessarily synergistic at ED_{90} (the dose for 90% effect). Among different methods frequently used in drug combination studies, the fractional product method of Webb (1963) for calculation of expected additive effect does not take into account the shapes of dose–effect curves, whereas the classic isobologram of Loewe (1953) assumes that all drugs have similar (mutually exclusive) effects that have theoretical limitations.

III. UNDERLYING THEORY: THE MEDIAN-EFFECT PRINCIPLE OF THE MASS-ACTION LAW

Pharmacological receptor theory and enzyme kinetics are based on the mass-action law. Both have evolved into well-developed disciplines. By using enzyme kinetic systems as models and by carrying out systematic derivation of equations for single substrate–single product reactions and multiple substrates–multiple products reactions with various mechanisms, in the

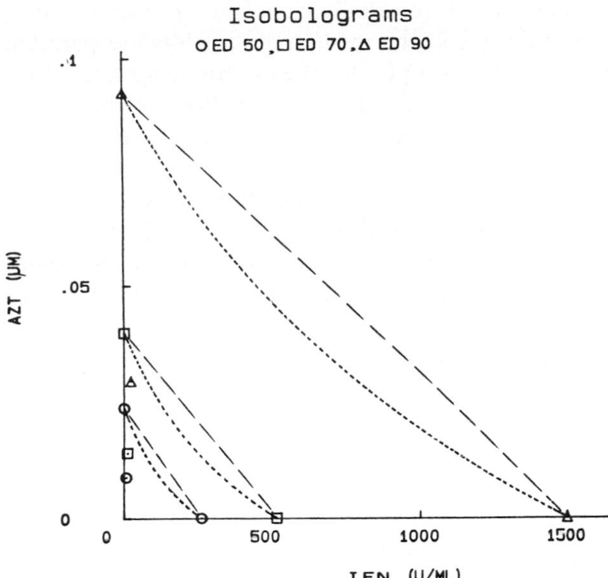

FIGURE 3 The ED_{50}, ED_{70}, and ED_{90} isobolograms. Their construction is based on experimental data, and parameters are shown in Fig. 1. Symbols shown are calculated ED_{50} (○), ED_{70} (□), and ED_{90} (△) for each drug alone (IFN on x-axis and AZT on y-axis) and their equieffective combinations. The points that fall into the lower left indicate synergism, and the points that fall into the upper right antagonism. Dashed and dotted lines depict equieffective simulations for additive effect based on mutually exclusive and mutually nonexclusive assumptions, respectively. The results showed strong synergism between AZT and recombinant interferon α (IFN) against HIV-1 replication.

absence and in the presence of single or multiple inhibitors (competitive, noncompetitive, and uncompetitive) and their permutations, hundreds of equations have been derived and amalgamated into a general median-effect principle (MEP). A few general equations that describe relationships between dose and effect are presented in the following. These equations are no longer restricted to different reaction mechanisms or different mechanisms of inhibition, because the kinetic constants such as K_m, K_i, and V_{max} are canceled out.

A. The Distribution Equation

$$\frac{K_i}{I_{50}} = \frac{E_x}{E_t}, \tag{1}$$

where K_i is the inhibitor ligand–enzyme dissociation constant, I_{50} is the concentration required for 50% inhibition, E_x is the amount or concentration of enzyme receptors or species available for ligand binding at the steady state, and E_t is the amount or concentration for the total enzyme receptors or species. The equation shows that K_i will never be greater than I_{50}. The distribution equation provides the basis for deriving numerous equations that eventually lead to the generalized dose–effect equations (discussed next).

B. The Median-Effect Equation

$$f_a/f_u = (D/D_m)^m, \tag{2}$$

where f_a and f_u are the fractions affected and unaffected, respectively, by a dose (D); D_m is the median-effect dose (ED_{50}); and m is a coefficient signifying the shape of the dose–effect curve. The values of $m = 1$, >1, or <1 indicate a hyperbolic, sigmoidal, or negative sigmoidal dose–effect curve, respectively.

In the median-effect equation, both sides represent dimensionless quantities that relate dose (right side) and effect (left side) in the simplest possible way. Thus, when two basic parameters, m and D_m, are determined, the entire dose–effect curve is described; that is, by rearranging the median-effect equation, the dose required for any given effect may be calculated and the effect for any given dose estimated.

The logarithmic form of Chou's median-effect equation yields

$$\log(f_a/f_u) = m\log(D) - m\log(D_m). \tag{3}$$

Therefore, a plot of dose–effect data with $x = \log(D)$ with respect to $y = \log(f_a/f_u)$ determines m (the slope) and D_m (the antilog of x-intercept) values. This plot is referred to as the median-effect plot. These procedures provide a mathematical basis for simulating dose–effect relationships with a computer.

The MEP, as depicted by its equation, can be readily used for deriving basic equations in biochemistry. Thus, the Michaelis–Menten equation in enzyme kinetics, the Hill equation for higher-order allosteric interactions, the Scatchard equation for receptor binding, and the Henderson–Hasselbalch equation for acid–base equilibrium can all be derived from the MEP in one or two steps. In retrospect, it is not surprising that the fractions afffected and unaffected (f_a and f_u) are equivalent to the fractions saturated and unsaturated, the fractions occupied and and unoccupied, the fractions bound and free, and the fractions ionized and unionized, respectively. The median-effect dose (D_m) for 50% effect is equivalent to half-saturation (K_m), half-occupied (K), half-bound (K_D), and half-ionized (K_a). Thus, four seemingly different equations are based on the same underlying principle.

The Hill equation and the median-effect equation have similar mathematical forms; however, they have elements of basic differences: (1) The Hill equation was derived for a primary ligand such as substrate or agonist, whereas the median-effect equation was derived for a reference ligand, such as inhibitor or antagonist. The equations were obtained from completely different routes. (2) The Hill equation was derived relative to V_{max} as 100%, which determination requires extrapolation and approximation, whereas the median-effect equation relates the observed effect to the uninhibited control value, which can be precisely determined.

C. Multiple-Drug–Effect Equation

The median-effect equation just described is for single drugs. A similar approach can be used for multiple-drug situations by systematic derivation of equations and conversion into fractional inhibitions. Depending on whether the effects of two drugs are mutually exclusive (i.e., same or similar modes of action) or mutually nonexclusive (i.e., different or independent modes of action), two different equations can be combined as shown by Eq. (4):

$$\left[\frac{(f_a)_{1,2}}{(f_u)_{1,2}}\right]^{1/m} = \left[\frac{(f_a)_1}{(f_u)_1}\right]^{1/m} + \left[\frac{(f_a)_2}{(f_u)_2}\right]^{1/m} + \left[\frac{\alpha(f_a)_1(f_a)_2}{(f_u)_1(f_u)_2}\right]^{1/m} = \frac{(D)_1}{(D_m)_1} + \frac{(D)_2}{(D_m)_2} + \frac{\alpha(D)_1(D)_2}{(D_m)_1(D_m)_2}, \quad (4)$$

where subscripts for the parameters f_a, f_u, D, and D_m denote single drugs (1 or 2) or the combination of two drugs (1, 2). When the effects of two drugs are mutually exclusive, the combined effect is the sum of the first two terms (i.e., $\alpha = 0$), and when the effects of two drugs are mutually nonexclusive, the combined effect is the sum of three terms ($\alpha = 1$). When $m = 1$, regardless of whether $\alpha = 0$ or $\alpha = 1$, the equation provides an exact solution for describing dose–effect relationships based on the proposed model. When $m \neq 1$ and $\alpha = 0$ (mutually exclusive effects of two agents), the equation gives a precise solution; however, when $m \neq 1$ and $\alpha = 1$ (mutually nonexclusive effects of two agents), it gives a first-degree approximation in which the most prominent species of ligand–target interaction in the distribution is taken into account.

When $\alpha = 0$ (mutually exclusive), if lines for drug 1 and drug 2 in the median-effect plots are parallel, the median-effect plot for the mixture with serial dilution will give lines parallel with those of the parent compounds. When $\alpha = 1$ (mutually nonexclusive), if the theoretical graphs of the median-effect plots for drug 1 and drug 2 are parallel, the median-effect plot for the mixture of drug 1 and drug 2 with serial dilution will concave upward. In this case, exclusivity of two drugs can be diagnosed. However, the median-effect plots for each drug alone in actual experiments are frequently not parallel, and exclusivity cannot be determined in this way. Therefore, in actual dose–effect analyses with microcomputers, the calculations are carried out both ways in appreciation of the fact that the method (or any other available methods) does not always lead to an unequivocal diagnosis for exclusivity or nonexclusivity of action.

D. The Isobologram Equation

Although the use of isobolograms has a long history in diagnosis of additivity, synergism, and antagonism, there is no explicit derivation of equations from defined models to justify their use. By using the multiple-drug–effect equation described in Eq. (4), the ED_{50} isobologram can be derived simply by setting $(f_a)_{1,2} = (f_u)_{1,2} = 0.5$. Therefore,

$$\left[\frac{(f_a)_1}{(f_u)_1}\right]^{1/m} + \left[\frac{(f_a)_2}{(f_u)_2}\right]^{1/m} + \left[\frac{\alpha(f_a)_1(f_a)_2}{(f_u)_1(f_u)_2}\right]^{1/m} = 1 \quad (5)$$

or

$$\frac{(D)_1}{(D_m)_1} + \frac{(D)_2}{(D_m)_2} + \frac{\alpha(D)_1(D)_2}{(D_m)_1(D_m)_2} = 1. \quad (6)$$

The multiple-drug–effect equation [Eq. (4)], as well as the isobologram equation [Eq. (5)], describes the additive effect of drug combinations. If experimental results deviate from the additive effect, synergism or antagonism is evident. The left side of the isobologram equation [Eq. (5)] has been designated as the drug combination index (CI), wherein CI = 1, <1, and >1 represent additivism, synergism, and antagonism, respectively. Computer software for automated construction of isobolograms has been developed according to Eqs. (5) and (6).

The ED_{50} isobologram equation has been extended to the isobologram for other effect levels. The generalized equation is given by

$$\frac{(D)_1}{(D_x)_1} + \frac{(D)_2}{(D_x)_2} + \frac{\alpha(D)_1(D)_2}{(D_x)_1(D_x)_2} = 1. \quad (7)$$

When $\alpha = 0$ and $\alpha = 1$, the classic isobologram and the conservative isobologram are obtained, respectively.

From Eq. (7) it becomes evident that the conservative isobologram equation (when $\alpha = 1$) always yields slightly less synergism than the classic isobologram equation (when $\alpha = 0$). Thus, if the effects of two drugs are mutually nonexclusive, then the indiscriminate use of the classical isobologram will result in overestimation of synergism.

The sham test, which assumes that two drugs under study have the same identity, provides an internal check of the validity of the classic isobologram but not of the conservative isobologram. If a drug is divided into two test tubes in a double-blind experiment for drug combination studies, the results will be necessarily additive because they are the same compound. Because they are the same compound, they are necessarily mutually exclusive as adopted by the classic isobol concept. In real experimental situations, however, there is little chance that two drugs will have identical effects.

TABLE I
Comparison of Applicability of Different Methods

Methods	Dose–effect curve characteristics			
	Mutually exclusive[a] (similar mode of action)		Mutually nonexclusive[b] (independent mode of action)	
	First order[c]	Higher order[d]	First order[c]	Higher order[d]
Webb's fractional product method[e]	no	no	yes	no
Loewe's isobologram method[f]	yes	yes	no	no
Multiple-drug–effect equation[g]	yes	yes	yes	yes

[a]Mutually exclusive drugs in a mixture give a parallel median-effect plot with respect to the parent compounds.

[b]Mutually nonexclusive drugs in a mixture give an upwardly concave dose–effect curve with respect to the parent compounds.

[c]Hyperbolic dose–effect curve.

[d]Sigmoidal dose–effect curve.

[e]Webb (1963): $i_{1,2} = 1 - [(1-i_1)(1-i_2)]$ or $(f_u)_{1,2} = (f_u)_1(f_u)_2$, where i is fractional inhibition and f_u is the fraction unaffected.

[f]Loewe (1957). *Pharmacol. Rev.* 9, 237–242.

[g]Chou and Talalay (1984). Also known as the median-effect principle and the combination index method.

IV. METHODS OF QUANTITATION OF SYNERGISM—ANTAGONISM

At least three major methods with sound theoretical bases can be used for determining synergism, additivism, or antagonism. These methods are (A) the fractional product method of Webb, (B) the classic isobologram method of Loewe, and (C) the CI method of Chou and Talalay.

Because the equations for A and B can be derived from C, the usefulness and limitations of A and B can be clearly defined (Table I). Although isobolograms have a long history in drug combination studies, there are still empirically modified proposals in this field. Thus, a pocket isobologram method has been proposed by some investigators. This method includes reading off from dose–effect curves, and by adding and subtracting the distances on the graphs, it estimates the displacement, referred to as the additive effect, as a range rather than a specific value or index. This empirical method is similar to 2 + 2 = 3 to 5, which has no obvious theoretical basis. Another proposed isobologram method by some investigators uses the following equation: $P_{ab} = P_a + P_b(1 - P_a)$, where P_a and P_b are the fractions of the organisms responding, respectively, to agents A and B, and P_{ab} is the fraction responding to their combination. However, no evidence indicates that this equation actually describes the classic isobologram of Loewe. In fact, it is identical to the fractional product method of Webb, which although widely used has severe limitations, as shown in Table 1.

V. COMPUTERIZED AUTOMATION OF ANALYSIS

Since the median-effect principle of the mass-action law and its equation and the multiple-drug–effect equation and its isobologram provide theoretical bases, it has been possible to interrelate the experimental scheme of design to equations, the parameters, the graphics, and computerized automation of data analysis. The algorithms and examples of stepwise procedures for the determination of synergism, additivism, and antagonism have been reviewed, and computer software for IBM-PC and Apple II computers has been developed.

For a typical *in vitro* combination of two drugs, the concentrations of each drug alone and their mixture are serially diluted in five to six steps (e.g., twofold series) to give a proper dose range and dose-density spacing. Ideally, the dose range is set between two concentrations below IC_{50} (D_m) and four concentrations above IC_{50}, and the combination ratio for drug 1 and drug 2 is initially kept at their IC_{50} ratio (i.e., equipotency ratio). Because the mixture is serially diluted, the combination ratio remains constant. Mixtures at other ratios can also be made, and they may provide an opportunity to estimate the optimal combination ratio for maximal synergy.

The dose and effect numerical data are first entered into a microcomputer (e.g., IBM-PC) by using software (Chou and Chou, 1987), the median-effect plot (Fig. 1) can be readily constructed, in which $x = \log(D)$ is on the abscissa and $y = \log(f_a/f_u)$ is on the ordinate. The plot is then subjected to linear regression analysis for the determination of the two basic parameters m and D_m and related statistics [e.g., the linear correlation coefficient (r) and mean $\pm$ SE for the slope and intercepts]. By using the m and D_m values for each drug and their combination, the dose (D_x) required for any degree of effect (f_a) can be calculated from the rearrangement of the median-effect equation:

$$D_x = D_m[f_a/(1 - f_a)]^{1/m}. \tag{8}$$

The D_x values are then substituted in the multiple-drug–effect equation [Eq. (7)] for calculating the CI value:

$$\text{CI} = \frac{(D)_1}{(D_x)_1} + \frac{(D)_2}{(D_x)_2} + \frac{\alpha(D)_1(D)_2}{(D_x)_1(D_x)_2}, \tag{9}$$

in which $(D)_1$ and $(D)_2$ (in the numerators) in combination ($D_{1,2}$) give the same x% inhibitory effect (f_a). If $(D)_1$ and $(D)_2$ concentration ratio is $P:Q$, then $(D)_1 = (D)_{1,2} \times P/(P + Q)$ and $(D)_2 = (D)_{1,2} \times Q/(P + Q)$. $(D)_{1,2}$ for x% inhibition (f_a) can be similarly calculated from $m_{1,2}$ and $(D_m)_{1,2}$. In this way, the CI values for any degree of effect (f_a) can be simulated to construct an F_a–CI table or an F_a–CI plot (Fig. 2). The classic isobolograms and the conservative isobolograms are also automatically constructed (Fig. 3). Furthermore, the CI values for the actual experimental combination data points will also be calculated automatically to indicate the degrees of synergism and/or antagonism.

It should be noted that isobologram and the F_a–CI plot should give identical conclusions about synergism, additivism, or antagonism. The isobologram is dose-oriented, whereas the F_a–CI plot is effect-oriented. Isobolograms show synergism–antagonism at several effect levels, whereas F_a–CI plots show synergism–antagonism at all effect levels simultaneously.

Assuming that two drugs are mutually exclusive, the dose reduction index (DRI) can be calculated by:

$$\begin{aligned} \text{CI} &= \frac{(D_x)_{1,2} \times P/(P + Q)}{(D_x)_1} + \frac{(D_x)_{1,2} \times Q/(P + Q)}{(D_x)_2} \\ &= \frac{(D)_1}{(D_x)_1} + \frac{(D)_2}{(D_x)_2} = \frac{1}{(\text{DRI})_1} + \frac{1}{(\text{DRI})_2}, \end{aligned} \tag{10}$$

where $(D)_1:(D)_2 = P:Q$; $(D_x) = D_m \times [F_a/(1 - F_a)]^{1/m}$; $(\text{DRI})_1 = (D_x)_1/(D)_1$; and $(\text{DRI})_2 = (D_x)_2/(D)_2$.

DRI depicts by how many multiples the dose can be reduced because of synergism for a given degree of effect when compared with a single drug alone. When the therapeutic efficacy is maintained, reduction of the dose will result in lower toxicity toward the host and may also reduce or delay the development of drug resistance.

The method for dose–effect analysis for the combination of three drugs has been similarly developed and has also been subjected to computerized automation by the procedures of Chou and Chou.

BIBLIOGRAPHY

Berenbaum, M. C. (1977). Synergy, additivism and antagonism in immunosuppression: A critical review. *Clin. Exp. Immunol.* **28**, 1.

Bertino, J. R., and Mini, E. (1987). Does modulation of 5-fluorouracil by metabolite or metabolites work in the clinics. *In* "New Avenues in Developmental Cancer Chemotherapy," Vol. 8, pp. 163–184. Academic Press, New York.

Chou, J., and Chou, T.-C. (1987). "Dose–Effect Analysis with Microcomputers: Quantitation of ED_{50}, Synergism, Antagonism, Low-Dose Risk, Receptor Ligand Binding and Enzyme Kinetics." Manual and Software Disk for Apple II and IBM-PC Series. Elsevier-Biosoft, Cambridge, England.

Chou, T.-C. (1976). Derivation and properties of Michaelis–Menten type and Hill type equations for reference ligands. *J. Theor. Biol.* **59**, 523.

Chou, T.-C., and Rideout, D. (eds.) (1991). "Synergism and Antagonism in Chemotherapy." Academic Press, San Diego.

Chou, T.-C., and Talalay, P. (1981). Generalized equations for the analysis of multiple inhibitions of Michaelis–Menten and higher order kinetic systems with two or more mutually exclusive and nonexclusive inhibitors. *Eur. J. Biochem.* **115,** 207.

Chou, T.-C., and Talalay, P. (1983). Analysis of combined drug effects: A new look at a very old problem. *Trends Pharmacol. Sci.* **4,** 450.

Chou, T.-C., and Talalay, P. (1984). Quantitative analysis of dose–effect relationships: The combined effects of multiple drugs or enzyme inhibitors. *Adv. Enzyme Regul.* **22,** 27.

Chou, T.-C., and Talalay, P. (1987). Applications of the median-effect principle for the assessment of low-dose risk of carcinogens and for the quantitation of synergism and antagonism of chemotherapeutic agents. *In* "New Avenues in Developmental Cancer Chemotherapy" (K. R. Harrap and T. A. Connors, eds.), pp. 37–64. Bristol-Myers Symposium series. Academic Press, New York.

Chou, T.-C., Motzer, R. J., Tong, Y., and Bosl, G. J. (1994). Computerized quantitation of synergism and antagonism of taxol, topotecan, and cisplatin against teratocarcinoma cell growth: A rational approach to clinical protocol design. *J. Nat'l Cancer Inst.* **86,** 1517.

Goldin, A., and Mantel, N. (1957). The employment of combinations of drugs in the chemotherapy of neoplasia: A review. *Cancer Res.* **17,** 635.

Hitchings, G., and Burchall, J. (1965). Inhibition of folate biosynthesis and function as a basis for chemotherapy. *Adv. Enzymol.* **27,** 417–468.

Loewe, S. (1953). The problem of synergism and antagonism of combined drugs. *Arzneimittel-Forsch.* **3,** 285.

Rideout, D., Calogeropoulou, T., Jaworski, J., and McCarthy, M. (1990). Synergism through direct covalent bonding between agents: A strategy for rational design of chemotherapeutic combinations. *Biopolymers* **29,** 247.

Rothenberg, M., and Ling, V. (1989). Multi-drug resistance: Molecular biology and clinical relevance. *J. Natl. Cancer Inst.* **81,** 907–913.

Steel, G. G., and Peckham, M. J. (1979). Exploitable mechanisms in combined radiotherapy-chemotherapy: The concept of additivity. *Int. J. Radiat. Oncol.* **5,** 85.

Webb, J. L. (1963). Effect of more than one inhibitor. *In* "Enzyme and Metabolic Inhibitors," Vol. 1, pp. 66, 488. Academic Press, New York.

Chemotherapy, Antiparasitic Agents

WILLIAM C. CAMPBELL
Drew University

GLOSSARY

Chemotherapy Treatment of disease by chemical means (i.e., by drugs)

Cyst Thick-walled quiescent stage of a protozoan parasite; in the case of the intestinal amebae, the stage shed in the feces of the host

Trophozoite Feeding and growing stage of a protozoan parasite

PARASITIC DISEASES ARE AMENABLE TO CONTROL by such means as proper sewerage, clean water, and, in the case of some parasites, controlling the intermediate hosts or invertebrate vectors that transmit them. In practice, however, these means are successful only where political and economic resources allow, and that excludes a large portion of the modern world. In the case of a few parasites, immunoprophylaxis may soon become practicable, but at present it is not in routine use for any human parasitic disease. Thus the control of these diseases still devolves largely on the age-old concept of chemotherapy (i.e., the prevention or cure of disease by administration of chemicals known to destroy or inactivate the infectious agent).

Central to the concept of chemotherapy is the concept of "differential toxicity." As Sir William Osler pointed out, it is dosage that determines whether a substance is a medication or a poison. Modern antibacterial medications generally enjoy a wide margin of safety; their toxicity for the microorganism is vastly greater than for humans. The armanentarium of antiparasitic drugs, however, still contains some weapons of marginal safety as well as those that can be deployed with confidence. In the accompanying tables, dosages of particular drugs are given as a matter of reference, but the proper use of drugs requires consideration of the prevailing clinical circumstances as well as an awareness that recommendations change in the light of new information. The dosages in the tables are therefore intended to be representative rather than definitive.

Many parasitic infections are diseases of the developing nations of the tropics. The use of chemotherapy in these countries is often constrained by lack of adequate medical, paramedical, and economic resources. Under these conditions, chemotherapy, like sanitation and vector control, becomes a political as well as scientific problem.

For the present purpose, parasitic diseases will be considered under the headings of diseases caused by protozoa, roundworms, flukes, tapeworms, and arthropods. By convention, bacteria, viruses, and fungi, no matter how parasitic their existence, are not included and may be found elsewhere. To provide greater accessibility for the nonspecialist, individual diseases are listed under the name of the disease rather than the parasite that causes it. This is a significant consideration only for the "older" diseases such as malaria and amebiasis, in which the names of the disease and the parasite are dissimilar.

I. DISEASES CAUSED BY PROTOZOA

A. General Comments

The success of chemotherapy in the treatment of protozoal infections remains highly variable. On the one hand, there are several diseases for which a standard

ENCYCLOPEDIA OF HUMAN BIOLOGY, Second Edition, VOLUME 2. Copyright © 1997 by Academic Press. All rights of reproduction in any form reserved.

TABLE I
Chemotherapy of Diseases Caused by Protozoa[a]

Disease	Representative treatment	Other drugs
Protozoan diseases primarily of the digestive tract		
Amebiasis		
Intestinal	Metronidazole (750 mg tid, 5–10 days), followed by iodoquinol (650 mg tid, 20 days)	Tinidazole, dehydroemetine, paramomycin, diloxanide furoate
Hepatic	Metronidazole (750 mg tid, 10 days)	Dehydroemetine, tinidazole
Cryptosporidiosis	None (see text)	
Giardiasis	Quinacrine HCl (100 mg tid, 5 days)	Metronidazole, tinidazole, furazolidone
Cyclosporosis	Sulfamethoxazole (160 mg) plus trimethropim (800 mg) bid, 3 days	
Protozoan diseases primarily of the urinogenital tract		
Trichomoniasis	Metronidazole (250 mg tid, 7 days, or 2000 mg once)	Tinidazole
Protozoan diseases primarily of the blood		
Malaria (prevention)	Chloroquine phosphate (500 mg, weekly)	Pyrimethamine plus sulfadoxine; proguanil, doxycycline, mefloquine
Malaria (cure)	Chloroquine phosphate (1000 mg, followed 6 hr later by 500 mg; then 500 mg daily for 2 days)	Quinine dihydrochloride (i.v.); quinine sulfate with pyrimethamine, sulfadiazine, or tetracycline; mefloquine
Trypanosomiasis (African)	Eflornithine (see text)	Suramin, pentamidine, melarsoprol, tryparsamide
Protozoan diseases primarily of other tissues		
Trypanosomiasis (American)	Nifurtimox (8–10 mg/kg/day in 3 or 4 divided doses × 60 days)	Benznidazole
Leishmaniasis	Stibogluconate-sodium (20 mg/kg/day i.m. or i.v. × 20 days)	Pentamidine, amphotericin B
Toxoplasmosis	Sulfadiazine (1 g qid × 3–4 weeks) plus pyrimethamine (25 mg daily × 3–4 weeks)	Clindamycin, spiramycin
Pneumocystis carinii pneumonia (PCP)	Pentamidine isethionate (4 mg/kg i.v. over 2 hr, daily × 14 days)	Trimethoprim plus sulfamethoxazole; dapsone, atovaquone

[a]Treatment regimens (oral unless otherwise stated) are given for illustrative, not prescriptive, purposes. bid, twice daily; tid, thrice daily; qid, four times daily; i.v., intravenously; i.m., intramuscularly.

and fairly acceptable treatment has been adopted (Table I). These include amebiasis, vaginal trichomoniasis, and giardiasis. For a second group of protozoal diseases, effective drugs are available, but their practical application has severe limitations. These may be in the form of unacceptable toxicity, as in the case of the drugs currently in use for the treatment of African trypanosomiasis, South American trypanosomiasis (Chagas' disease), and leishmaniasis in its various forms. Alternatively, the limitations may be in the form of drug resistance, as is notably the case in malaria. At the other end of the spectrum are protozoal diseases for which treatment is either totally unavailable or amounts to no more than a few experimental drugs of varying degrees of clinical promise and investigational legitimacy. Diseases in this unfortunate category include cryptosporidiosis, *Pneumocystis carinii* pneumonia (PCP), and amebic meningoencephalitis.

B. Protozoal Diseases Primarily of the Intestinal Tract

1. Amebiasis

The mainstay of treatment for amebiasis is the nitroimidiazole compound, metronidazole:

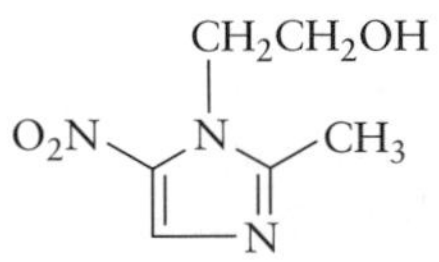

CH_2CH_2OH

O_2N N CH_3

N

This drug is active against the so-called luminal amebae in the colon (those thought to be more-or-less free in the canal) and against the amebae invading the wall of the colon and inducing abscesses in other organs, especially the liver. The side effects of metronidazole are fairly common but not usually severe. Like other nitroimidazoles, it has been associated with carcinogenicity and mutagenicity in animal and microbial test systems. In practice, treatment with metronidazole is generally followed by a course of treatment with one of the "luminal" drugs, such as iodoquinol. When asymptomatic "cyst passers" are treated with either kind of amebacidal drug, cysts disappear from the stool of the patient, but this is believed to represent activity against the trophozoite or growing stage of the parasite (and consequent cessation of cyst production) rather than activity against the cysts themselves. [*See* Amoebiasis: Infection with *Entamoeba histolytica*.]

2. Cryptosporidiosis

There is no known effective treatment for cryptosporidiosis. Spiramycin, azithromycin, paromomycin, and octreotide have been associated with clinical amelioration, but clear-cut evidence of efficacy is not presently on hand.

3. Giardiasis

Giardiasis may be treated with the old antimalarial compound quinacine hydrochloride:

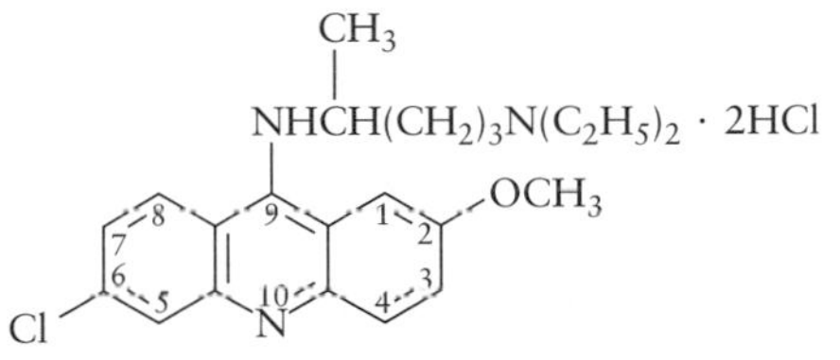

CH_3
$NHCH(CH_2)_3N(C_2H_5)_2 \cdot 2HCl$
OCH_3
Cl N

The nitroimidazole compounds, metronidazole and tinidazole (below), are also effective and are approved for this use in some countries.

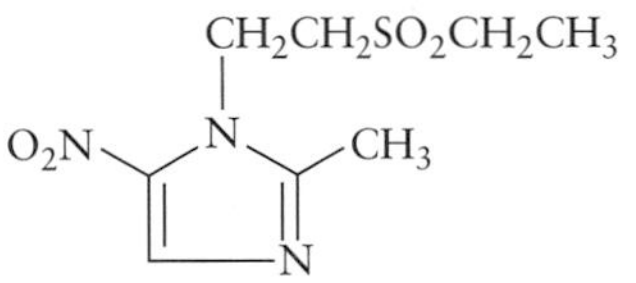

$CH_2CH_2SO_2CH_2CH_3$
O_2N N CH_3
N

C. Protozoan Diseases Primarily of the Urogenital Tract

1. Trichomonal Vaginitis

The mainstay of treatment of trichomoniasis is metronidazole. It is highly effective, but sexual partners of treated patients should also be treated. If this is not done, treatment is likely to be defeated by reinfection.

D. Protozoan Diseases Primarily of the Blood

1. Malaria

Malaria is the chemotherapist's dream—and nightmare. From ancient South American folklore, Western medicine learned that the bark of a certain tree would alleviate the torment of "intermittent fevers," and for more than a century it has been known that a chemical substance named quinine was the active ingredient. Subsequent political events led to a search for alternative remedies, and highly effective man-made antimalarial compounds were eventually developed; but the success of the synthetic chemist has been thwarted in part by the success of the malarial parasite in developing ways of eluding the effects of treatment. Drug resistance is not rare in the world of chemotherapy, but because of the enormity of malaria as a disease of humans, and because its treatment has depended so heavily on one or two drugs, the emergence of drug resistance has brought about an enduring crisis. [*See* Malaria.]

a. Prevention

Until recently, the prevention of malaria has rested almost entirely on the 4-amino-quinoline drug chloroquine:

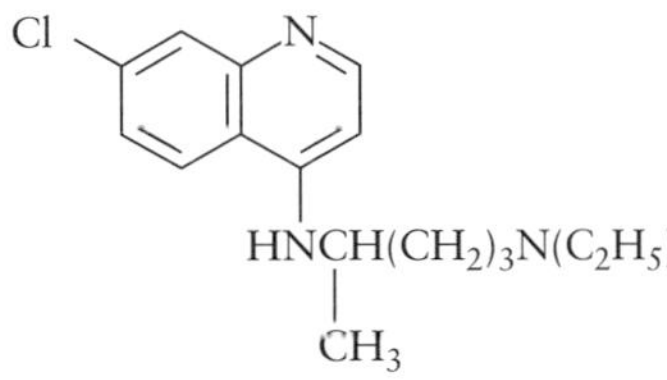

Cl N
$HNCH(CH_2)_3N(C_2H_5)_2$
CH_3

For the most part (i.e., when the situation is not complicated by drug resistance), malaria can be prevented by swallowing a tablet of chloroquine once a week. The drug is not a "causal prophylactic" (i.e., it will not prevent the sporozoites inoculated by mosquitoes from reaching the liver and developing into asexual multiplicative stages). It is a "clinical prophylactic," because it is active against the stages that multiply in the erythrocytes and cause attacks of malaria. It is, moreover, active against the erythrocytic stages of all four major species of human malaria. If chloroquine medication is continued for 6 weeks after departure from a malarious region, any infection of *Plasmodium falciparum* or *P. malariae* that may have been acquired will "burn itself out" [i.e., all the liver stages will have matured and released the next (asexual) generation of parasites into the blood, where they will take up residence, and chloroquine vulnerability, in the erythrocytes]. In the case of *P. vivax* and *P. ovale,* however, some of the hepatic stages will persist for long periods and may become activated long after the traveler has left the endemic region. To prevent this, the 8-amino-quinoline drug primaquine can be taken to kill the hepatic stages and so prevent their "seeding" the erythrocytes and causing attacks of malaria. Because of the potential toxicity of primaquine, it may be more prudent for the erstwhile traveler to eschew primaquine treatment and simply wait to see if fever ensues. If this happens, the erythrocytic phase can be cured with chloroquine, and radical cure can be achieved with primaquine. In that way, primaquine will be used only when the risk of toxic reaction is offset by a definite need for treatment. In regions where chloroquine resistance is known or suspected, the standard chloroquine prophylactic regimen should be supplemented with, or replaced by, other antimalarial drugs. Because of the constantly changing pattern of drug resistance, chemoprophylactic recommendations are becoming increasingly "tailor-made" to suit a given time and place.

b. Cure

Chloroquine (or a related 4-amino-quinoline) is an effective cure for all four species of malaria in humans, but there are qualifications that may be a matter of life and death. If the malaria is due to *P. vivax* or *P. ovale,* chloroquine should be supplemented with primaquine to prevent relapses (the principle involved being the same as that discussed earlier). If the malaria is due to *P. falciparum,* which may rapidly prove fatal, consideration must be given to the question of whether the particular strain of parasite has become resistant to the drug or is likely, on geographical grounds, to have become so. Where resistance is known or suspected, quinine is the drug of choice. It may be given orally (as the sulfate) or parenterally (as the dihydrochloride). In critically ill patients, parenteral administration is often advisable. In many instances, quinine treatment will not prevent recurrence of malarial attacks (in the 6 weeks or so in which *P. falciparum* remains capable of reinvading the erythrocytes) and so the quinine treatment is often supplemented with a pyrimethamine–sulfadiazine combination. Alternatively, tetracycline or clindamycin (which have antimalarial activity in their own right) may be used to supplement quinine in the treatment of chloroquine-resistant cases of *P. falciparum* malaria. Where parasite strains are also resistant to quinine, resort should be made to other drugs, including mefloquine (the hydrochloride salt is shown here) and possibly halofantrine. Unfortunately, resistance to mefloquine has already begun to appear.

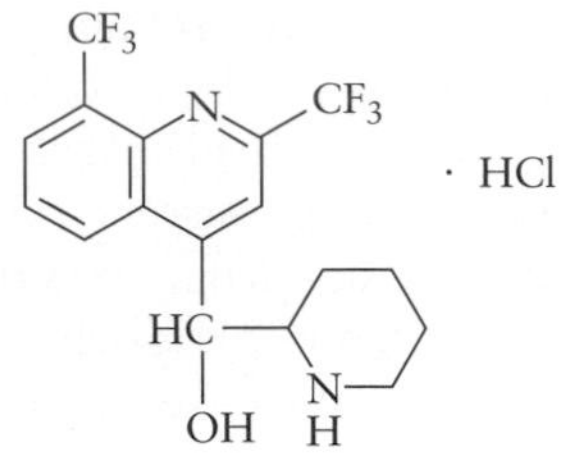

CF_3, N, CF_3, · HCl, HC, OH, N, H

A drug of exceptional promise is artemisinin, the active principle in the Chinese herbal extract quinghaosu. Various derivatives, including artemether, artesunate, and dihydroartemisinin, are being evaluated. Artemether (the methyl ester), which may soon receive regulatory approval, is effective against strains of *P. falciparum* that are resistant to several standard drugs. Prudent use of this life-saving medication will be of the utmost importance, and it should not be used for routine prophylaxis or for therapy in the absence of resistance to other drugs.

Desipramine, a tricyclic psychotropic drug, has been shown to reverse choroquine resistance in experimentally infected monkeys. Unless such stratagems can be exploited clinically, or malaria can be controlled by other means, the discovery and development of new antimalarial drugs remain an important objective.

2. African Trypanosomiasis (Sleeping Sickness)

African trypanosomiasis, caused by subspecies of the *Trypanosoma brucei* complex, may be prevented by intramuscular injections of the diamidine compound pentamidine. Cure of the clinical disease relies on intravenous injections of the naphthylamine sulfonic acid compound suramin or, especially if the disease has progressed to the point of involvement of the central nervous system, intravenous injections of melarsoprol (below left) or tryparsamide (below right).

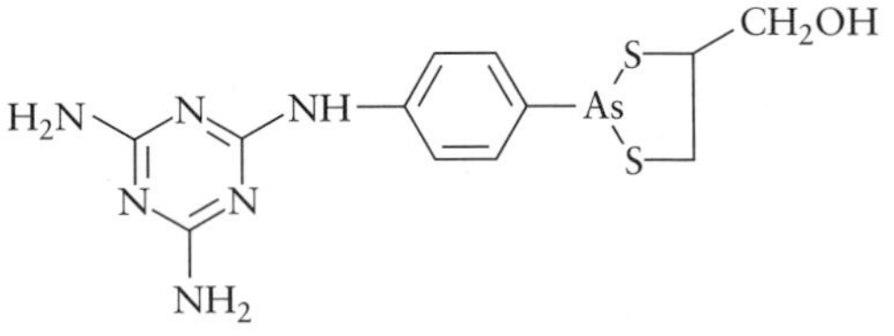

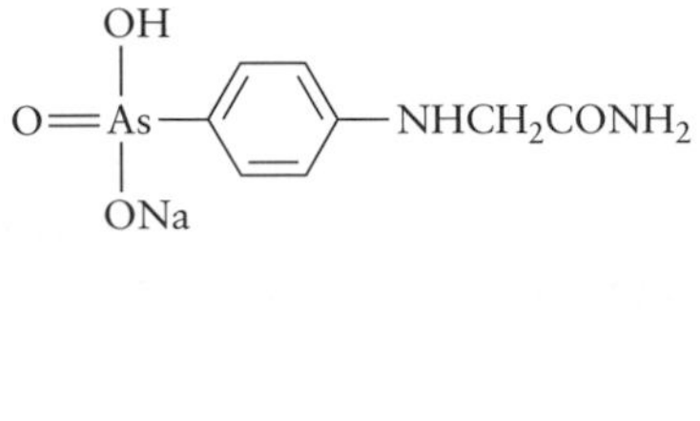

H_2N, N, NH, As, S, CH_2OH, S, N, N, NH_2

OH, O=As, $NHCH_2CONH_2$, ONa

These old drugs are very toxic, especially the last, and much effort has been made to develop safer medications. Among the most promising are eflornithine, allopurinol, bleomycin, and certain nitrofurans. The efficacy of eflornithine is particularly striking; it now appears to be the drug of choice in Gambian sleeping sickness caused by *T. b. gambiense,* especially in cases in which the central nervous system has become involved. Its rapid efficacy in comatose patients has earned it the sobriquet "resurrection drug," and its efficacy and safety have earned it the approval of the U.S. Food and Drug Administration. Unfortunately it is better given intravenously than orally, and the current dosage is 100 mg per kg, given four times daily for 14 days. Attempts are being made to simplify the treatment regimen and to enhance efficacy against infections caused by *T. b. rhodiense*. [*See* Trypanosomiasis.]

E. Protozoan Diseases Primarily of Other Tissues

1. American Trypanosomiasis (Chagas' Disease)

The drugs used for South American trypanosomiasis are quite different from those used for its African counterpart. They are nifurtimox (shown below) and benznidazole, and they are both hazardous and cumbersome to use. Even their efficacy is somewhat debatable, especially in chronic cases.

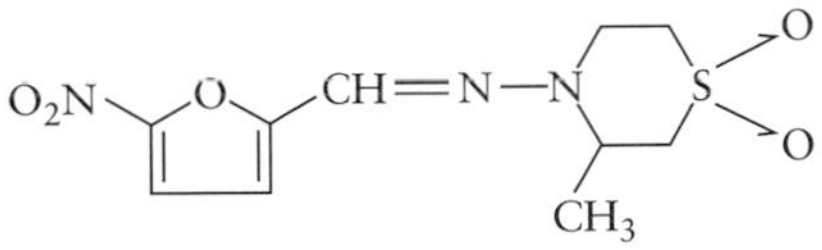

2. Leishmaniasis

The use of pentavalent antimonials has transformed visceral leishmaniasis from a routinely fatal disease to one that is generally curable. This represents a major achievement of 20th-century chemotherapy but must not obscure the fact that these drugs are very toxic and inconvenient to use. Treatment typically consists of parenteral administration of stibogluconate sodium. For certain types of leishmaniasis, pentamidine or amphotericin B may be used. [*See* Leishmaniasis.]

3. Toxoplasmosis

The selection of drugs for the treatment of toxoplasmosis is based largely on laboratory experimentation and uncontrolled clinical trials. Nevertheless, the use of sulfa drugs synergized with dihydrofolate reductase inhibitors (e.g., sulfadiazine plus pyrimethamine or sulfamethoxazole plus trimethoprim) has become standard. Folinic acid should be given to offset the adverse effects of the antifol drugs. In cerebral toxoplasmosis, pyrimethamine with clindamycin has been effective. In preliminary trials, azithromycin has given variable results. In pregnant women the macrolide antibiotic spiramycin is sometimes used to obviate the risk of teratogenicity associated with other treatments.

4. *Pneumocystis carinii* Pneumonia

Pneumocystis carinii pneumonia is essentially a disease of immunosuppressed patients. It has come to the fore because of the emergence of AIDS as an epidemic, and thus the evaluation of drug efficiency has been of short duration and has been fraught with the difficulties of drug evaluation in severely ill patients and complicated clinical situations. Recent molecular studies suggest that *P. carinii* is a fungus rather than a protozoan, so its inclusion in this section is provisional. The distinction is not trivial because awareness of a phylogenetic relationship is important in guiding the direction taken by chemotherapeutic studies.

Two drugs have been found effective in PCP and are in current clinical use: pentamidine (shown here) and a combination of trimethoprim and sulfamethoxazole.

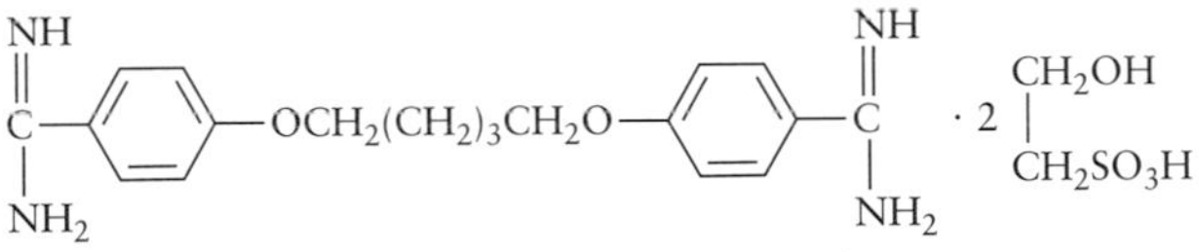

The trimethoprim–sulfamethoxazole combination (given p.o. or i.v.) is generally preferred, but it is not well tolerated in AIDS patients. For reasons that are not understood, some patients do not respond favorably to either drug. In the treatment of various infections, pentamidine is given by intravenous or intramuscular injection; but in cases of PCP, it is now being given by aerosol inhalation to achieve maximum local effect with minimum systemic toxicity. Both forms of treatment (i.e., pentamidine and various combinations of trimethoprim and sulfas or sulfones) may be used to prevent recurrence of pneumonia in treated patients or to prevent the disease in high-risk (immunosuppressed) groups of people. Other drugs with some degree of efficacy in the treatment of PCP include atovaquone and a combination of clindamycin and primaquine.

5. Amebic Meningoencephalitis

There is no standard treatment for invasion of the central nervous system by the amebae *Naegleria* or *Acanthamoeba*. Drugs such as amphotericin B may prove useful, but clinical management will depend on the site of invasion and other factors.

6. Clyclosporosis

Cyclospora, which has been something of a mystery, is now considered to be a protozoan parasite of the coccidial type. The severe diarrheal disease that it causes has been treated successfully by a combination of sulfamethoxazole and trimethoprim.

II. DISEASES CAUSED BY ROUNDWORMS

A. General Comments

For roundworm infections of the gastrointestinal tract, current chemotherapy is generally very satisfactory, but for extraintestinal roundworm infections it is not (Table II). The drugs (anthelmintics) used for gastrointestinal helminthiasis have mostly been developed for the routine control of comparable infections in domestic animals and have reached a high degree of chemotherapeutic sophistication. Similar transfer of technology from animal to human is just beginning to occur in the case of the worms living in extraintestinal sites.

B. Roundworm Infections of the Intestinal Tract

1. Common "Soil-Transmitted" Nematode Infections

Narrow-spectrum anthelmintics, directed at one or two particular species, are sometimes used—a prime example being piperazine, which is inexpensive and widely used to eliminate *Ascaris* in children. In recent times, however, broad-spectrum anthelmintics have become so effective and so safe that they are increasingly used, especially in the tropics, where multiple infections are extremely common. The leading broad-spectrum anthelmintics are pyrantel pamoate, mebendazole, and albendazole (structures shown sequentially below). All of these compounds are effective against the common species of *Ascaris, Ancylostoma, Necator,* and *Trichostrongylus.* Efficacy against the hookworms (*Ancylostoma* and *Necator*) is seldom perfect, and larger or more frequent doses may have to be used. Albendazole enjoys an advantage in that it is generally effective as a single dose, and is becoming increasingly more widely used. Mebendazole and albendazole are also effective against the whipworm, *Trichuris.* All three are inadequately active against *Strongyloides,* and for this helminth the older drug thiabendazole is still used despite its greater frequency of side effects.

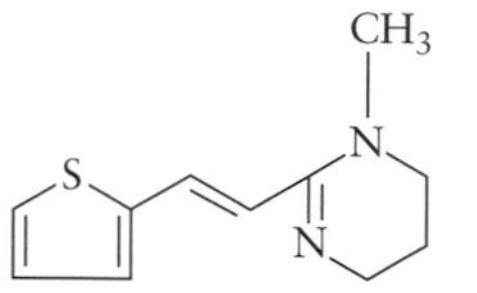

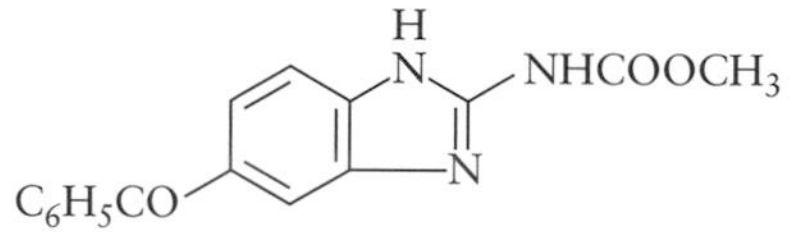

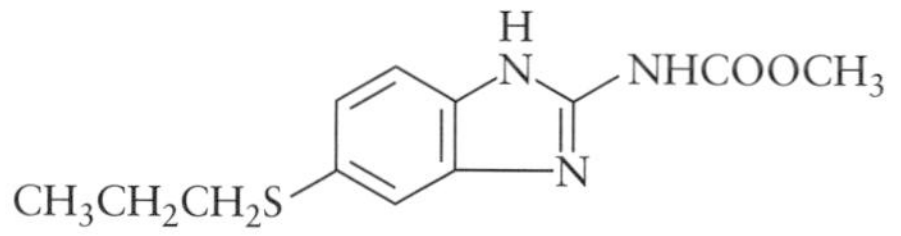

CH_3 S N N

H N $NHCOOCH_3$ N C_6H_5CO

H N $NHCOOCH_3$ N $CH_3CH_2CH_2S$

2. Pinworm Infection (Enterobiasis)

The broad-spectrum anthelmintics pyrantel pamoate, mebendazole, and albendazole are all highly active against the pinworm *Enterobius vermicularis* and are much more convenient to use (and usually much safer) than the drugs used previously.

3. Miscellaneous

Infections caused by *Capillaria philippinensis* may be treated with mebendazole, but a prolonged treatment regimen may be needed. No treatment is known for anisakiasis caused by ingestion of uncooked fish harboring larvae of various anisakine nematodes.

C. Roundworms of the Extraintestinal Tissues

1. Invasive Strongyloidiasis

Although *Strongyloides stercoralis* is a parasite of the small intestine, its most dramatic and life-threatening aspect is its propensity to invade other tissues, especially in immunosuppressed patients. Thiabendazole appears to be less effective in such "invasive strongyloidiasis" than in the intestinal disease, but albendazole has shown much promise in early trials. Ivermectin, although active against the intestinal *Strongyloides* in preliminary trials, has not yet been adequately tested against the invasive condition.

TABLE II
Chemotherapy of Diseases Caused by Helminths[a]

Disease	Representative treatment	Other drugs
Roundworm infections of the intestinal tract		
Common "soil-transmitted" nematode infections	Mebendazole (100 mg bid × 3 days)	Pyrantel, albendazole
Pinworm infection (enterobiasis)	Pyrantel pamoate (1 mg/kg, once; repeat after 2 weeks)	Mebendazole, albendazole
Roundworm infections of the extraintestinal tissues		
Invasive strongyloidiasis	Thiabendazole (25 mg/kg bid × 2 days)	Albendazole
Trichinellosis (trichinosis)	Mebendazole (300 mg tid × 3 days, then 400 mg tid × 10 days)	Albendazole
Onchocerciasis		
Adult worms	Suramin (200 mg i.v. to test tolerance; then 1 g i.v. weekly × 5 weeks)	
Microfilariae	Ivermectin (0.15 mg/kg once; repeat every 6 or 12 months)	Diethylcarbamazine
Filariasis (lymphatic)		
Adult worms	Diethylcarbamazine (50 mg once, increasing over 3 days to 2 mg/kg tid × 21 days)	
Microfilariae	Diethylcarbamazine (see text)	Ivermectin
Loiasis		
Adult worms	Diethylcarbamazine (5 mg/kg daily × 21 days)	
Microfilariae	Diethylcarbamazine (5 mg/kg daily × 21 days)	Ivermectin
Dracunculosis	Metronidazole (250 mg tid × 10 days)	Thiabendazole, mebendazole
Fluke infections		
Schistosomiasis, clonorchiasis, paragonimiasis	Praziquantel (20 mg/kg tid, 1 day)	Oxamniquine (*S. mansoni*)
Fasciolopsiasis, metagonimiasis	Praziquantel (25 mg/kg tid, 1 day)	
Fascioliasis	Bithionol (40 mg/kg every other day for 12 doses)	Dehydroemetine
Tapeworm infections		
Intestinal infections, various species	Praziquantel (20 mg/kg once)	Niclosamide
Extraintestinal larval infections	See text	

[a]Treatment regimens (oral unless otherwise stated) are given for illustrative, not prescriptive, purposes. bid, twice daily; tid, thrice daily; i.v., intravenously.

2. Trichinellosis (Trichinosis)

The treatment of trichinosis, apart from symptomatic treatment, which is beyond the scope of this article, depends on the use of benzimidazoles. Thiabendazole, the first to be introduced, has been superseded by mebendazole as the drug of choice. On the basis of early clinical trials, it would appear that mebendazole in turn will be replaced by albendazole (200 mg bid, taken with a fatty meal to aid absorption).

The benzimidazoles can be used prophylactically in those rare circumstances in which they can be administered within a few hours or days after ingestion of infected meat. If given within a few hours, the drugs can be expected to prevent maturation of the worms in the intestine. If given within a few days, they will suppress the reproduction of the worms and so prevent the shedding of larvae and subsequent invasion of the musculature. When used therapeutically (i.e., when the progeny of the worms have already settled in the patient's muscles), the benzimidazoles provide clinical improvement, although it is not clear whether this is due to direct destruction of the larvae or to inhibition of their metabolic processes with consequent reduction in the response of the host to metabolic products. In extremely severe cases of trichinosis, treatment must naturally be considered in the context of the patient's overall condition and the possibility, at least theoretical, of host reaction to the protein from dead parasites. [*See* Trichinosis.]

3. Onchocerciasis

Once *Onchocerca volvulus* has matured in the human body and the female worms have released their numerous offspring (microfilariae) to populate the skin, destruction of the adult worms may not prevent or cure the skin lesions or ocular damage associated with the infection (because these are caused by the microfilariae). Killing the adults will, however, limit the number of microfilariae produced, and therefore the severity of the lesions. Moreover, if the microfilariae are killed by drug treatment, destruction of the adult worms will prevent reinvasion of the skin and eyes by newly shed microfilariae. Suramin (shown here) is the only drug used in humans for the destruction of the adult worms (macrofilaricidal effect)

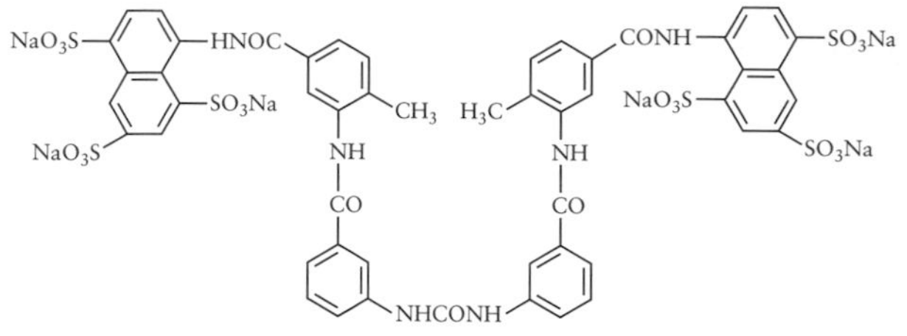

It requires multiple intravenous injections (a serious liability in endemic regions) and is very toxic.

To attack the microfilariae in the dermal and ocular tissues, diethylcarbamazine (DEC) (below left) was used for many years but is now being replaced by ivermectin (below right).

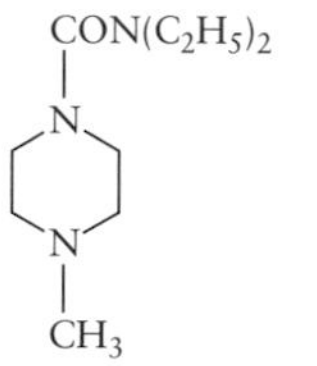

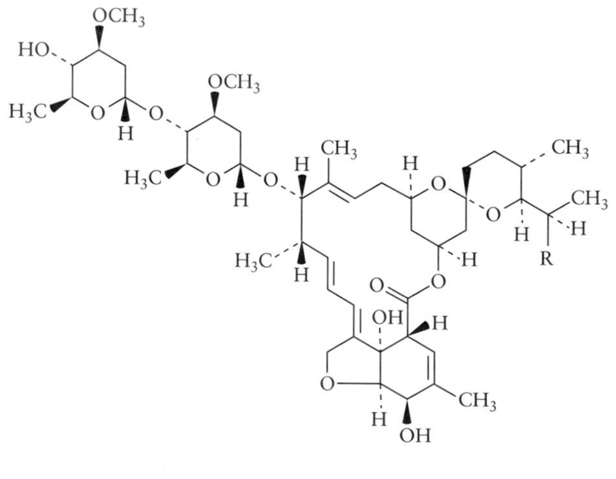

It has long been believed that the adverse reactions associated with DEC therapy (intense itching, etc.) have been caused by host reaction to protein liberated when microfilariae are killed, and thus they would follow the use of any microfilaricidal drug. Nevertheless, reactions after ivermectin therapy, although they do occur and are probably caused by dead microfilariae, are both fewer and less severe than those after DEC. Moreover, ivermectin is given as a single oral dose, whereas DEC requires daily oral doses for a period of about 2 weeks. Reinvasion of the skin by newly shed microfilariae appears to take longer after ivermectin than after DEC. For all of these reasons, ivermectin has been used in community-based trials, and if good results continue to be reported, it will become the drug of choice for both individual patients and community control programs.

4. Lymphatic Filariasis

In lymphatic filariasis (caused by *Wuchereria bancrofti* or *Brugia malayi*), the adult worm is the primary pathogen. Multiple doses of DEC give clinical amelioration and probably cause the death of at least some of the adult worms. DEC is also effective against the microfilariae in the bloodstream and so can be used to prevent transmission from human to mosquito. It has been successfully employed in this way to control the disease in geographically isolated regions. As in the case of onchocerciasis, ivermectin is active against the microfilariae when given as a single oral dose. It now appears that, at least in some endemic areas, DEC is similarly effective as a single oral dose.

5. Other Filariases

Diethylcarbamazine is effective against the microfilariae of *Dipetalonema streptocerca* and *Loa loa* and may induce serious allergic reactions to liberated parasite antigens. However, DEC has little activity, or at least little documented activity, against the microfilariae of *Mansonella perstans* or *M. ozzardi,* and this is in accord with the general lack of hypersensitivity reactions after treatment. Mebendazole or ivermectin may prove to be the microfilaricidal drug of choice for some or all of these four filarial infections. No drug is known to be effective against the adult worms.

6. *Dracunculus medinensis* (Guinea Worm)

Metronidazole and thiabendazole have been found useful in treating guinea worm infection. The adult worms are more easily extracted from the subcutaneous tissue after treatment, but it is not clear to what extent this is due to their anthelmintic efficacy or to their nonspecific anti-inflammatory effect. Use of either drug requires consideration of potential toxicity.

7. Miscellaneous

Visceral larva migrans, caused by migratory larvae of dog ascarids, is not diagnosed frequently enough to allow systematic evaluation of therapy. There have been reports that thiabendazole treatment has resulted in clinical amelioration of the disease. Cutaneous larva migrans, caused by the migratory larvae of dog or cat hookworms, is readily treated with thiabendzole. Oral treatment is effective, but topical treatment is preferable because it is highly effective and obviates the side effects associated with systemic treatment. There is evidence that albendazole is effective against *Gnathostoma,* and animal studies suggest that benzimidazoles may be useful in *Angiostrongylus* infection.

III. DISEASES CAUSED BY FLUKES (TREMATODES)

A. General Comments

The treatment of trematode infections of humans has been revolutionized in recent years by the discovery of praziquantel:

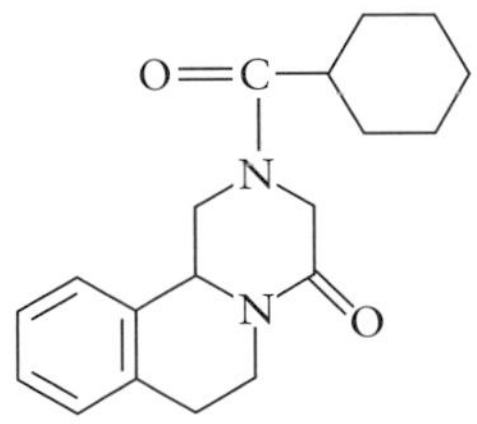

The drugs used previously for such infections were characterized by poor efficacy and potent toxicity. Some of them had to be administered in difficult and dangerous ways. Praziquantel, however, is given orally, and although adverse reactions are by no means uncommon, they are seldom severe.

B. Intestinal Infections

Praziquantel is effective against *Fasciolopsis buski, Heterophyes heterophyes,* and *Metagonimus yokogawai.*

C. Extraintestinal Infections

1. Clonorchiasis, Opisthorchiasis, and Paragonimiasis

Praziquantel is effective against the liver flukes *Clonorchis sinensis* and *Opisthorchis viverrini* and against the lung fluke *Paragonimus westermani.*

2. Fascioliasis

Remarkably, praziquantel does not seem to be effective in the treatment of infections caused by the liver fluke *Fasciola hepatica.* This is especially remarkable because the flukes do take up the drug when exposed to it *in vitro.* Bithional and dehydroemetine are effective, and encouragement may be found in a report that triclabendazole, used for the treatment of fascioliasis in domestic animals, was effective in cases of human fascioliasis.

3. Schistosomiasis

Praziquantel is the drug of choice for infections caused by *Schistosoma haematobium, S. mansoni, S. japonicum,* and *S. mekongi.* For *S. mansoni* infection, oxamiquine (shown here) is an effective and well-tolerated alternative.

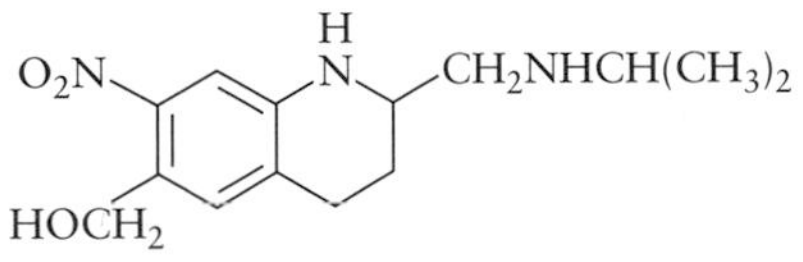

IV. DISEASES CAUSED BY TAPEWORMS (CESTODES)

A. General Comments

The tapeworms share with their platyhelminth relatives, the flukes, a susceptibility to praziquantel, and this has greatly simplified the clinical management of these infections. Unfortunately, however, the most serious tapeworm pathogens are the cystic larvae of certain species, and the cystic infections are extremely difficult to manage in the clinic even with the availability of drugs with some degree of specific activity.

B. Intestinal Infections

Praziquantel is effective against the "adult" or strobilate stage of *Taenia saginata* (beef tapeworm), *T. solium* (pork tapeworm), *Diphyllobothrium latum* (fish tapeworm), and *Dipylidium caninum* (dog tapeworm that occasionally infects humans). The salicylanilide compound niclosamide may be used as an alternative and is well tolerated. Both drugs cause disruption of the tapeworm "segments" or proglottids, and in the case of *T. solium* infection, this is a cause for some concern. The eggs liberated by the breakdown of gravid proglottids may be swept forward into the stomach as a result of retching and vomiting, and the eggs may thus be activated by gastric fluid. As they pass back into the small intestine, such eggs are capable of hatching and releasing larvae that penetrate the wall of the small intestine and become cysticerci in the extraintestinal tissues, as would normally happen in the pig host. It is not clear whether there is a significant probability of this happening, but cysticercosis, if induced by this or other means, is a disease of such clinical significance that even a low probability cannot be dismissed lightly.

C. Extraintestinal (Cystic) Infections

The clinical management of cystic tapeworm infection is difficult, and both surgical and chemotherapeutic interventions must take into account the size and location of the cysts. Drugs capable of destroying the cysticerci of *T. solium* are known (i.e., praziquantel, flubendazole, albendazole, and metrifonate), but the optimum dosage regimen and guidelines for clinical use have yet to be standardized.

Similarly, the hydatid cysts of *Echinococcus granulosus* are known to be susceptible to the action of benzimidazole compounds, and albendazole (400 mg bid for at least a month) is used as an adjunct or alternative to surgical removal. Release of protoscolices from ruptured cysts is always a hazard, but it may be minimized by local application of praziquantel or albendazole. There is evidence from studies in laboratory animals that parenteral administration of benzimidazole compounds may be more effective than oral administration against cystic tapeworms, but this has not been confirmed or extended to human use.

V. DISEASES CAUSED BY ARTHROPODS (ECTOPARASITES)

A. General Comments

The bites of bedbugs, fleas, and flies may require topical treatment to relieve local inflammation, but the control of these insect pests depends on environmental measures rather than human chemotherapy. However, itch mites and lice are more truly parasitic, and human infection may call for specific treatment.

B. Diseases Caused by Mites, Ticks, and Lice

For scabies (infection with *Sarcoptes scabiei*), the standard treatment is topical application of lindane (γ-benzene hexachloride) or the pyrethroid compound permethrin (where registered for such use).

Infections with the follicle mite, *Demodex folliculorum,* rarely need treatment, but some infections have been treated successfully with lindane lotions or sulfur ointments. Severe skin lesions associated with *D. folliculorum* have responded well to treatment of the patient with oral doses of metronidazole (200 mg daily for 6 weeks). This may have been the result of a drug-induced change in the microhabitat (sebum) rather than a direct effect on the mites.

Chigger mites (*Trombicula* sp.) induce local reactions that can be alleviated by local analgesics, and chigger bites can be prevented by application of repellents containing *N*, *N*-diethyl-*m*-toluamide (DEET). Impregnation of clothing with permethrin will protect against chigger bites, not by repellency but by acaricidal action.

Ticks (Ixodidae and Argasidae) are more important as transmitters of disease than as causes of disease. They are commonly removed mechanically, and their bites may be prevented by applying repellents such as DEET or dimethylphthalate (CMP) to skin or clothing. Prevention can also be achieved by impregnating clothing with the pyrethroid compound permethrin, which will kill at least some ticks that crawl onto a treated garment. It cannot be assumed that these repellents and acaricides will protect against all stages of all tick species.

For lousiness (infestation with *Pediculus capitis* or *Phthirus pubis*), the standard topical lindane treatment is being replaced in some quarters by topical application of malathion or pyrethroids (synergized with piperonyl butoxide).

Ivermectin, which is effective against itch mites and biting lice in domestic animals, has not yet been evaluated for efficacy against these ectoparasites in humans.

VI. DISCOVERY OF ANTIPARASITIC DRUGS

The drugs that are in routine clinical use for the treatment of parasitic infections were discovered empirically (i.e., by trial-and-error testing in laboratory systems or by chance observation made in the course of clinical or laboratory investigation). Some of the drugs now entering clinical trial are the products of research into the comparative biochemistry of parasite and host. Both approaches to new drug discovery are thoroughly entrenched, as also, to a lesser extent, is the "semirational" approach in which compounds are empirically screened in an assay designed to detect activity against some particular biochemical function of the parasite.

Regardless of the approach taken to the discovery of compounds with antiparasitic activity, it should be remembered that such discovery is only the first step in a long and difficult process. The probability of finding a new substance with antiparasitic activity is high; but there is only a low probability of carrying it, over innumerable manufacturing, economic, toxicological, and regulatory hurdles, to the point of clinical or commercial success.

VII. MODE OF ACTION OF ANTIPARASITIC DRUGS

A. General Comments

The biochemical mechanism by which a drug destroys a parasite is known in some instances with reasonable certainty. For other drugs, one or more mechanisms have been proposed with some plausibility, whereas for others the mechanism remains almost wholly mysterious. It is important to remember that the clinical usefulness of a drug, although dependent on its biochemical effect on the parasite and relative lack of effect on the host, is in no way dependent on our knowledge of those effects. Such knowledge, however, is important in many ways (e.g., for understanding in its own right, for the discovery of new compounds with similar mode of action but different properties or effects, for developing assays in which compounds are screened empirically for a given kind of biochemical action, and as an aid in obtaining the approval of regulatory agencies).

Another pitfall in the consideration of mode of action is the tendency to conclude that a particular biochemical effect is the mode of action of a drug just because the effect has been discovered. To take just one of many possible examples, the discovery that benzimidazoles (structure of thiabendazole shown below) inhibit the fumarate reductase of certain helminths led to widespread, although not universal, acceptance of that effect as the mode of action.

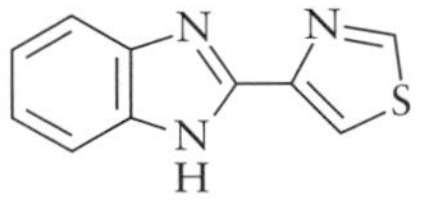

The opinion was bolstered by the observation that the enzyme from a benzimidazole-resistant strain of worm was less easily inhibited than the corresponding enzyme from a sensitive strain. Yet later studies led to the replacement of that hypothesis with an entirely different one (see the following). An observed effect that turns out not to be the mode of action (i.e., not the effect that destroys the parasite) may be unconnected with the actual mode of action or may be secondary to it. A drug may, for example, inhibit the uptake of glucose by a worm parasite, but that might be merely a consequence of the drug's primary action and a mere reflection of the "ill health" of the worm. Then, too, an effect observed *in vitro* may not be relevant to the *in vivo* situation. Similarly, an effect observed *in vitro* or *in vivo* under experimental conditions might have little relevance because the concentrations of drug employed in the experiment might vastly exceed those obtained under clinical conditions.

The effect by which a parasite is destroyed may be one that is directly lethal; but for internal parasites, a nonlethal effect may be just as good (i.e., just as bad for the parasite) as a lethal effect. Intestinal worms will readily be expelled from the host if they are paralyzed and unable to hold their position in the gut; but even extraintestinal worms may die and be resorbed by the host if they are displaced from their normal microhabitat (as schistosomes, e.g., are driven from the mesenteric veins to the liver sinusoids after antimonial therapy) or if they are physically or physiologically altered in such a way as to make them susceptible to the host's immune response.

B. Mode of Action of Some Antiprotozoal Drugs

Quinine, chloroquine, and perhaps mefloquine appear to have a common mode of action. In parasitized erythrocytes, a lytic degradation product of hemoglobin, ferriprotoporphyrin IX, is sequestered in a vacuole within the malarial parasite in a nontoxic form (the "malaria pigment"). When chloroquine is taken up by the malaria parasite, the drug binds to the heme compound, forming a complex that apparently cannot be sequestered and that aggregates into clumps that destroy the parasite. The action of mefloquine may be similar, but it is apparently not identical, because mefloquine is effective against many chloroquine-resistant strains of the malaria parasite. The mode of action of the new antimalarial drug artemisinin is not known, but it has been suggested that the mechanism may involve the production of toxic oxygen intermediates or the inhibition of protein synthesis.

The mechanism of chloroquine resistance is not fully understood. Resistant strains of the parasite accumulate less chloroquine than do ordinary strains, and this appears to depend not on a decrease in the rate at which the parasite takes up the drug, but rather on an increase in the rate at which it pumps the drug out. The phenomenon can be reversed *in vitro* by the calcium blocker verapamil and can be reversed *in vitro* and *in vivo* by the antidepressant drug desipramine, which is a calcium antagonist and has weak antimalarial activity in its own right. Pyrimethamine (shown below) inhibits dihydrofolate reductase and is much more potent against the enzyme of the malarial parasite than against the corresponding mammalian enzyme.

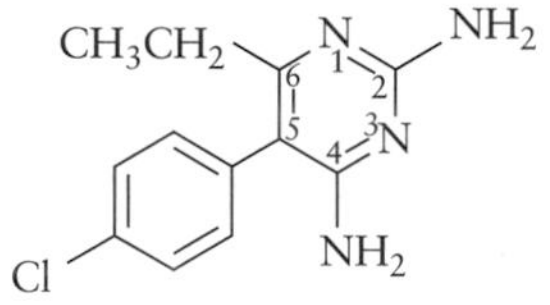

By preventing the reduction of dihydrofolate to tetrahydrofolate (a cofactor essential for the synthesis of thymidylate, purine nucleotides, and certain amino acids), the drug disrupts the multiplication process (shizogony) of the parasite within the host erythrocyte. Resistance to the drug, at least in some strains of *P. falciparum,* is associated with a point mutation in the gene encoding dihydrofolate reductase-thymidylate synthase. The sulfonamides that are used in combination with pyrimethamine are PABA analogues, acting at an earlier stage of folate metabolism.

The mode of action of suramin remains unclear, but it may act by inhibiting the phosphorylation of regulatory proteins in trypanosomes or by inhibiting enzymes involved in glucose metabolism. The drug does not readily penetrate trypanosomes *in vitro,* and its effects occur slowly *in vivo.* The trivalent arsenicals, however, enter trypanosomes quickly, and death (probably caused by the inhibition of enzymes of glucose metabolism) occurs rapidly *in vivo* and *in vitro.* Several compounds have promising activity against trypanosomes but have not yet reached the stage of clinical use. Among these, purine analogues such as allopurinol may kill trypanosomes because those organisms cannot synthesize purines and so must acquire them from exogenous sources.

Eflornithine is α-difluoromethylornithine, and it inhibits the enzyme ornithine decarboxylase, thus preventing the synthesis of the polyamine putrescine. Thc enzyme is involved in both mammalian and trypanosomal cell division, and the eflornithine blocks the enzyme in both cases (indeed it was originally developed to halt the multiplication of cancer cells). Fortunately, however, it is much more potent against the parasite enzyme than against the host enzyme.

C. Mode of Action of Some Antinematodal Drugs

The benzimidazoles, which include several important broad-spectrum antinematodal drugs, appear to act by blocking the assembly of the microtubules essential for maintaining cell structure and function. Most of the other antinematodal drugs act by interfering with neurotransmission. Some (e.g., levamisole, pyrantel, and morantel) depolarize cell membranes by acting on cholinergic receptors.

The organophosphate anthelmintics (not used to a significant extent in human medicine) inhibit the acetylcholinesterase of nematodes more readily than they inhibit the corresponding enzyme in the host and so induce paralysis of the worm at dosages tolerated by the host. Piperazine, which has been used widely for the treatment of ascaris infection, causes a flaccid paralysis because of blockade of GABA-mediated neurotransmission.

Ivermectin destroys nematodes and arthropods by provoking an influx of chloride ions through cell membranes. It is not clear whether the drug acts on receptors in neuronal membranes or in muscle membranes, but in any case paralysis ensues. This induced hyperpolarization of cell membranes was formerly thought to be due to the opening of GABA-gated ion channels but is now thought to result from the opening of ion channels mediated by L-glutamate.

D. Mode of Action of Some Drugs Used Against Flukes and Tapeworms

Praziquantel, which is highly active against both flukes and tapeworms, has multiple and only poorly understood effects. In both flukes and tapeworms, the drug causes a rapid paralysis of muscle and a severe vacuolization of the tegument. The biochemical mechanisms involved, however, are apparently not identical. In the case of flukes, or at least schistosome flukes, calcium and magnesium ions are involved in both the muscle and the tegument effects. It has been suggested that the impaired muscle function and damaged tegument make the worms more vulnerable to the host immune response. In the case of tapeworms, the efficacy of praziquantel depends on the availability of endogenous calcium ions rather than on the uptake of exogenous calcium ions as in the case of schistosomes.

The results of *in vitro* and *in vivo* studies suggest that oxamniquine destroys schistosomes by irreversibly inhibiting the synthesis of nucleic acids and proteins. Niclosamide, a salicylanilide still used for the treatment of tapeworm infections, is probably an uncoupler of oxidative phosphorylation in cestode mitochondria.

E. Mode of Action of Some Drugs Used Against Ectoparasites

The modern drugs used to treat ectoparasitic diseases act on the nervous system of arthropods (unlike the older and more toxic compounds that blocked energy metabolism).

Lindane (γ-benzene hexachloride), like several other insecticides, affects the central nervous system of insects and results in tremors, convulsions, and paralysis. It apparently causes excess release of the neurotransmitter acetylcholine at nerve terminals, and the effects can be blocked by a cholinergic blocking agent. The precise mechanism of action, however, is unclear, and it is likely that calcium ions are involved in the effect on neurotransmitter release. Lindane may, in fact, antagonize the inhibitory neurotransmitter GABA and so lead to increased excitatory action at the synapse, with consequent hyperactivity typical of this and many other insecticides. The mode of action as it applies to human lice and itch mites is conjectural, because the pertinent biochemical studies have been done on other arthropods.

Malathion, like other organophosphates, is thought to inhibit the acetylcholinesterase that is needed to inactivate the neurotransmitter acetylcholine arthropod synapses. Disruption of sodium ion flow across cell membranes is probably involved.

Pyrethroids block neurotransmission by depolarizing the cell membranes, and this in turn is brought about by interfering with the flow of ions across those membranes. The compounds may exert this action directly on the membranes or they may affect the operation of ion pumps. Some pyrethroids may act on the GABA receptor or the ionophore portion of the receptor complex, thus interfering with the operation of chloride ion channels. The action of ivermectin on arthropods also appears to involve chloride ion channels and to be similar to the action seen in nematodes, except that nerve transmission is blocked, at least in some insects, at the neuromuscular junction rather than the interneuron–neuron synapse as in the case of nematodes.

BIBLIOGRAPHY

Anon. (1990). "Drugs Used in Parasitic Diseases." World Health Organization, Geneva.

Anon. (1993). Drugs for AIDS and associated infections. *Medical Lett.* **35,** 79–85.

Anon. (1993). Drugs for parasitic infections. *Medical Lett.* **35,** 111–120.

Campbell, W. C. (1983). Progress and prospects in the chemotherapy of nematode infections of man and other animals. *J. Nematol.* **15,** 608–615.

Campbell, W. C. (1986). The chemotherapy of parasitic infections. *J. Parasitol.* **72,** 45–61.

Campbell, W. C., and Rew, R. S. (eds.) (1986). "The Chemotherapy of Parasitic Diseases." Plenum, New York.

Denham, D. A. (ed.) (1985). "Chemotherapy of Parasites" Symposium of the British Society for Parasitology, Vol. 22. *Parasitology* **90**(4), 613–721.

Gustafsson, L. L., Beerman, B., and Abdi, Y. A. (1987). "Handbook of Drugs for Tropical Parasitic Infections." Taylor & Francis, London.

Gutteridge, W. E. (1985). Existing chemotherapy and its limitations. *Br. Med. J.* **41,** 162–168.

Hooper, M. (ed.) (1987). "Chemotherapy of Tropical Diseases." John Wiley & Sons, Chichester, England.

James, D. M., and Gilles, H. M. (1985). "Human Antiparasitic Drugs: Pharmacology and Usage." John Wiley & Sons, Chichester, England.

Mansfield, J. M. (ed.) (1984). "Parasitic Diseases: The Chemotherapy," Vol. 2. Marcel Dekker, New York.

Peters, W., and Richards, W. H. G. (eds.) (1984). "Antimalarial Drugs," Vols. I and II. "Handbook of Experimental Pharmacology," Vol. 68. Springer-Verlag, New York.

Sturchler, D. (1982). Chemotherapy of human intestinal helminthiases: A review with particular reference to community treatment. *Adv. Pharmacol. Chemother.* **19,** 129–154.

Vanden Bossche, H., Thienpoint, D., and Janssens, P. G. (eds.) (1985). "Chemotherapy of Gastrointestinal Helminths." Springer-Verlag, Berlin.

Chemotherapy, Antiviral Agents

THOMAS W. NORTH
University of Montana

GLOSSARY

Latency State in which viral genome is present in cell but is not expressing the genes necessary for virus replication

Nucleoside analog Compound with a structural modification of the base or sugar of commonly occurring nucleosides

Resistant mutants Mutant virus that can replicate in the presence of a drug that is inhibitory to wild-type virus

Selectivity Degree to which a drug inhibits virus replication more effectively than it inhibits cell growth or function

Target Viral enzyme or component against which the active form of a drug exerts its inhibitory effect

CHEMOTHERAPY HAS BECOME IMPORTANT FOR treatment of viral infections that cannot be prevented with vaccines. Antiviral drugs are available for treatment of herpesvirus and influenza A infections in humans. There is now hope for development of antiviral drugs to combat other viruses that afflict humans, including human immunodeficiency virus (HIV), the causative agent of the acquired immune deficiency syndrome (AIDS). Advances in chemotherapy of herpesvirus infections, most notably with nucleoside analogs, have been of particular importance because they point the way to general strategies that may be used against any virus.

I. ANTIVIRAL STRATEGIES

The most effective strategy yet employed for viral chemotherapy is the inhibition of a virus-encoded enzyme that is essential for its replication. The successful antiherpes drugs are nucleoside analogs that are metabolically activated to their respective nucleotides, and these selectivity inhibit the herpesvirus-encoded DNA polymerase. Each herpesvirus encodes a DNA polymerase that is uniquely required for replication of its DNA and cannot be replaced by a cellular DNA polymerase. Although the herpes and cellular DNA polymerases catalyze similar reactions, they are sufficiently different in physical properties and, most importantly, in their sensitivities to inhibitors. This has enabled development of selective inhibitors that block herpesvirus DNA synthesis without affecting DNA synthesis in uninfected cells. Similarly, the most promising agents for treatment of AIDS are nucleoside analogs, include 3′-azido-3′-deoxythymidine, which are metabolized to nucleotides, and these are selective in their action upon reverse transcriptase. This enzyme, which is necessary for the replication of retroviruses, has no counterpart in uninfected cells.

The selectivity of some of the antiherpes nucleosides can be achieved in another way. This strategy takes advantage of the fact that some herpesviruses encode an enzyme (thymidine kinase) that can phosphorylate certain nucleoside analogs. Several of the antiherpes nucleosides derive at least some of their selectivity from the fact that they are phosphorylated by this enzyme, but not by any cellular enzymes. Thus, the active forms of these nucleosides, their corresponding nucleotides, are formed in virus-infected cells, but not in uninfected cells.

Although the viral enzymes required for DNA synthesis have received much attention, there are numerous other viral targets that might be exploited for

ENCYCLOPEDIA OF HUMAN BIOLOGY, Second Edition, VOLUME 2. Copyright © 1997 by Academic Press. All rights of reproduction in any form reserved.

chemotherapy. Each virus encodes a specific set of proteins that are required for its replication, and each of these is a potential target for chemotherapy. Many of these proteins are unique and have no counterpart in uninfected cells. These include structural proteins as well as enzymes. Viral nucleic acids might also be considered chemotherapeutic targets for "drugs" (such as ribozymes) that will recognize specific viral sequences and inactivate or destroy them.

Attempts have been made to develop antivirals that inhibit cellular enzymes. A rationale for this approach is that certain pathways or enzymes are more critical for virus replication than for cell functions. If such events are critical to viruses in general, it may provide a strategy for development of broad-spectrum antivirals. Of course, this approach is more likely to result in toxicity to host cells and tissues.

II. ANTIHERPES DRUGS

All of the drugs that have been approved or show promise for use in treatment of human herpesvirus infections exert their activity by inhibition of the herpesvirus-encoded DNA polymerase. Most of these are nucleoside analogs that must be phosphorylated to their corresponding triphosphates in order to exert antiviral activity. This metabolism is critically important because the activity of an antiviral nucleoside is dependent on the amount of the triphosphate formed within cells. Also, selectivity can be achieved at the level of activation if a viral-encoded enzyme is required to perform one of the phosphorylations.

The first two drugs that were approved for treatment of herpetic infections in humans were nucleoside analogs, 5-iododeoxyuridine (IUdR, idoxuridine) and 9-β-D-arabinofuranosyladenine (araA, vidarabine) (Fig. 1). These were approved for treatment of disorders caused by herpes simplex virus (HSV). Both of these nucleosides are metabolized to their corresponding 5′-mono, di-, and triphosphates by cellular kinases. It is the 5′-triphosphates (IdUTP and araATP) that are the biologically active forms. IdUTP can substitute for dTTP and be incorporated into DNA by DNA polymerases. However, it is not a selective drug because it can be incorporated into DNA by both

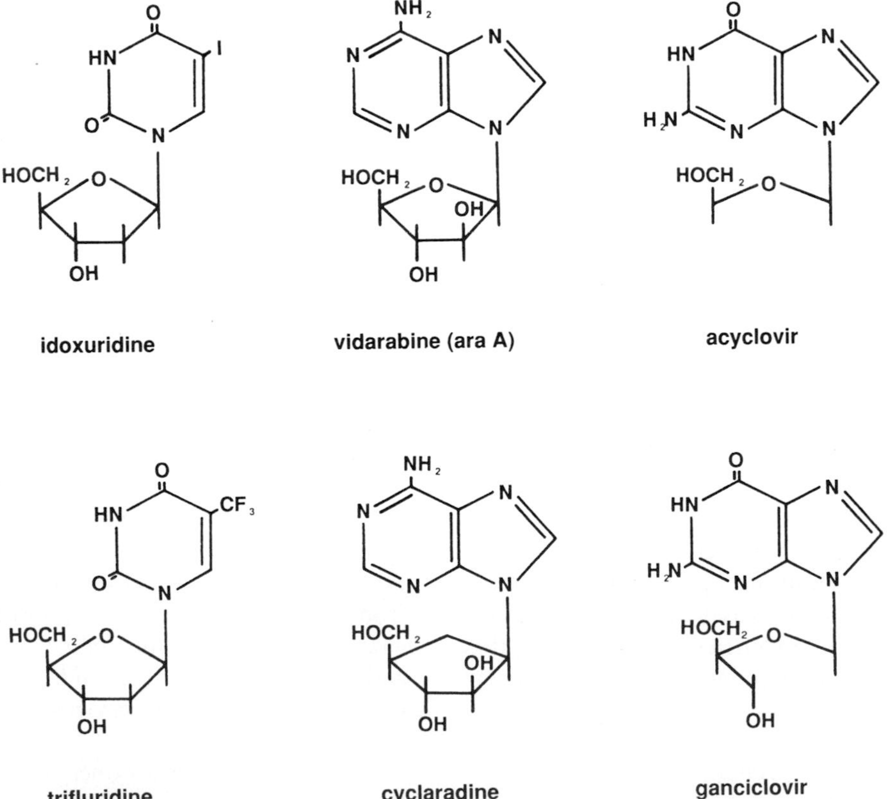

FIGURE 1 Nucleoside analogs with selective antiherpesvirus activities. All of these except cyclaradine have been approved for use in humans.

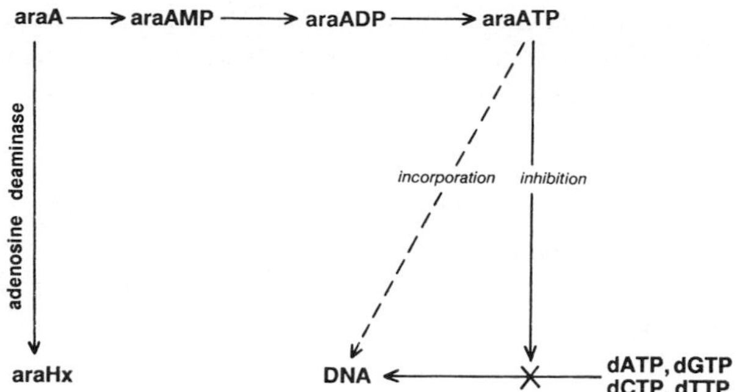

FIGURE 2 Metabolism and actions of 9-β-D-arabinofuranosyladenine (araA).

cellular and herpesvirus DNA polymerases. Accordingly, IUdR has proven too toxic for systemic use, but has been used topically for many years in treatment of herpetic keratitis caused by HSV. Another nucleoside analog, trifluorothymidine or trifluridine (F_3-TdR, Fig. 1), has more recently been approved for treatment of herpes keratitis. Its mechanism and lack of selectivity are similar to that of IUdR.

In contrast to these, araATP is a selective inhibitor of herpesvirus DNA polymerases. It takes substantially higher concentrations of araATP to inhibit cellular polymerases than are necessary to inhibit herpes DNA polymerases. This selectivity enabled araA to become the first drug approved for systemic use in treatment of human herpesvirus infections. The metabolism of araA is shown in Fig. 2. In addition to its activation to araATP, it is metabolically inactivated by adenosine deaminase to 9-β-D-arabinofuranosylhypoxanthine (araHx). This occurs rapidly in humans owing to high levels of adenosine deaminase present in blood and most tissues. This, combined with a low solubility, has limited the therapeutic efficacy of araA. The antiviral activity of araA has been successfully enhanced through combination with inhibitors of adenosine deaminase, such as erythro-9-(2-hydroxy-3-nonyl)adenine or 2′deoxycoformycin, although such combinations have not been approved for clinical use. Nevertheless, this approach has stimulated development for analogs of araA that are not substrates for adenosine deaminase. One of the most promising of these compounds is carbocyclic araA (cyclaradine, Fig. 1). This drug has excellent activity against herpes simplex virus types 1 and 2; the reason it has not been developed for clinical use is unclear.

AraATP exerts its selective antiviral activity through effects on the herpesvirus DNA polymerase. It is both an inhibitor of this enzyme and a substrate that is incorporated into DNA. It is not clear which of these two events is more important for the antiviral activity. Unlike some other antivirals that will be discussed, araA is not a chain terminator. It is incorporated internally into DNA chains.

A major advance in viral chemotherapy occurred with development of an acyclic nucleoside analog, 9-(2-hydroxyethoxymethyl)guanine (acyclovir, Fig. 1), by the Burroughs Wellcome Co. This nucleoside analog is metabolized to the corresponding triphosphate in a manner similar to the nucleosides described earlier. But there is one critical difference in the metabolism of this analog: the first phosphorylation is catalyzed *only* by a virus-encoded enzyme. This viral-specified activation enables a tremendous increase in selectivity over that achieved with analogs that are activated by cellular kinases.

The metabolism of acyclovir is shown in Fig. 3. The first step, formation of acyclovir monophosphate,

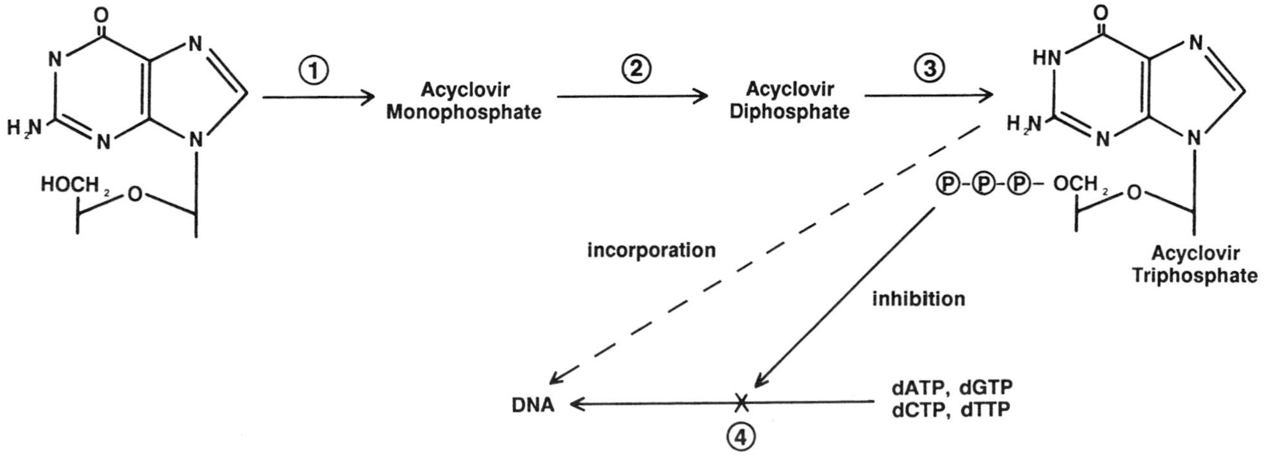

FIGURE 3 Metabolism and action of acyclovir. The enzymes involved are: (1) the herpes-encoded thymidine kinase, (2) cellular GMP kinase, (3) cellular nucleoside diphosphokinase, and (4) herpes-encoded DNA polymerase.

is carried out by the herpesvirus-encoded thymidine kinase. Three of the human herpesviruses (HSV-1, HSV-2, and herpes zoster) encode this enzyme. They have a broad substrate specificity and will phosphorylate thymidine, 2′-deoxycytidine, and many nucleoside analogs, including acyclovir. Cellular thymidine kinases have a much narrower substrate specificity and do not effectively phosphorylate 2′-deoxycytidine, acyclovir, or many other of the nucleoside analogs that are phosphorylated by the viral thymidine kinases. In fact, acyclovir is not effectively phosphorylated to a monophosphate by any cellular enzyme. Thus, acyclovir triphosphate is formed only in cells infected with these herpesviruses, and acyclovir is effective in treatment of infections by these three herpesviruses.

Acyclovir monophosphate is further phosphorylated to the corresponding di- and triphosphate by cellular kinases. The 5′-triphosphate is a competitive inhibitor of herpesvirus DNA polymerases. This inhibition is also selective. Acyclovir triphosphate inhibits the HSV DNA polymerase at a concentration 10- to 50-fold lower than is required to inhibit cellular DNA polymerases. This selectivity, combined with the selective phosphorylation discussed earlier, gives acyclovir a wide margin of safety. Acyclovir is also incorporated into DNA and serves as a chain terminator since the acyclic sugar analog lacks a suitable acceptor for addition of the next nucleotide. Recent evidence indicates that acyclovir can also irreversibly inactivate the HSV DNA polymerse.

Acyclovir is widely used to treat infections by HSV-1, HSV-2, and herpes zoster. It shows little or no activity against viruses that do not encode a thymidine kinase capable of phosphorylating it. Thus, it is much less active against some other members of the herpesvirus family, such as cytomegalovirus (CMV) and Epstein–Barr virus. However, another acyclic nucleoside analog, 9-(1,3,-dihydroxy-2-propoxymethyl)guanine (ganciclovir, DHPG, Fig. 1), has been approved for treatment of CMV infections in humans. This compound must also be phosphorylated to a triphosphate to exert its activity. The enzyme responsible for this activation is not thymidine kinase, but rather a phosphotransferase encoded by the UL97 gene of CMV. Ganciclovir is selectively activated to the triphosphate in CMV-infected cells because uninfected cells cannot effectively phosphorylate it.

Nucleoside analogs are not the only selective antiherpes drugs that are targeted against the herpes DNA polymerase. Several pyrophosphate analogs, including phosphonoacetate and phosphonoformate, are selective inhibitors of herpesvirus DNA polymerases. Clinical use of these pyrophosphate analogs has been limited because of their accumulation in bone, but foscarnet has been used for treatment of CMV infections.

It should be noted that all of these antiherpes drugs are replication inhibitors. However, herpesviruses are able to establish a state of latency in which the viral genome is established within cells and is not replicating. During latency the virus is not susceptible to replication inhibitors. Nevertheless, antiherpes drugs are effective in treatment of disease, but infections can recur when drug is removed. This will be discussed further in Section VI.

The advances that have occurred in development of antiherpes drugs have been facilitated by the availability of target enzymes to study (DNA polymerase and thymidine kinase), and by the availability of animal models to test drugs and therapeutic strategies. The rabbit eye model was particularly important in early studies for treatment of ocular herpes infections because it enabled one eye to serve as a control for drug therapy of the other. For systemic therapy a mouse model in which HSV causes rapid encephalitis has been very important. A guinea pig model for vaginal infections by HSV-2 has been particularly important in development of therapeutic strategies for genital herpes. The coordination of *in vitro* and *in vivo* approaches has been important in development of chemotherapy for herpesvirus infections.

III. CHEMOTHERAPY FOR OTHER VIRUSES

There are relatively few other viruses for which there are approved drugs or rational chemotherapeutic strategies. Amantadine has been approved for prophylaxis and symptomatic management of influenza A infections since 1966. It and a closely related compound, rimantadine, block replication of influenza A viruses. Amantadine inhibits replication of influenza A viruses through effects on the viral M2 protein. This protein forms an ion channel that is believed to be essential for virus uncoating through modulation of the pH of intracellular compartments. Influenza A mutants resistant to amantadine possess an altered M2 protein.

Attempts to develop antiviral drugs for rhinoviruses and other picornaviruses have focused compounds that bind to the virion and prevent uncoating. These attempts have been limited by lack of a suitable animal model for rhinovirus infection. [*See* Influenza Virus Infection.] Promising results have been obtained with oxazolines, such as WIN 54954, which are potent inhibitors of rhinoviruses and other picornaviruses. These compounds bind to a hydrophobic pocket in the viral capsid protein and block either attachment of virus to the cell or the uncoating of virus within cells.

Several thiosemicarbazones have antiviral activity and one of these, 1-methyl-β-isatin-thiosemicarbazone (methisazone), is effective for treatment of smallpox (although eradication of smallpox has eliminated the need for this).

IV. BROAD-SPECTRUM ANTIVIRALS

It is improbable that broad-spectrum antivirals can be directed against viral enzymes because viruses are a diverse group that do not share a common target. Nevertheless, there are compounds that have broad-spectrum antiviral activity. These compounds exert their effects through inhibition of cellular enzymes, and presumably these pathways are more critical to viral replication than to cellular functions. It might be expected that these approaches would not be as selective as targeting of drugs to viral enzymes. Ribavirin (1-β-D-ribofuranosyl-1,2,4-triazole-3-carboxamide), another nucleoside analog, inhibits a broad spectrum of RNA and DNA viruses. Its exact mechanism is not clear, although it is known to inhibit a cellular enzyme, IMP dehydrogenase, and to cause decreases in levels of GTP. Nucleotides of ribavirin may also inhibit specific viral functions. Several inhibitors of *S*-adenosylhomocysteine hydrolase have also been shown to have activity against a broad spectrum of viruses.

Perhaps the most promising approach in the development of broad-spectrum antivirals is through inducers of interferons. Interferons are broad-spectrum, antiviral glycoproteins that are synthesized by the body in response to a variety of stimuli, including virus infections. A large number of different types of compounds can induce interferons. Another promising approach is the development of immunomodulators capable of enhancing the ability of the immune system to protect against virual infections. These strategies will be discussed in other articles. [*See* Interferons.]

V. CHEMOTHERAPY OF AIDS

The rapid spread of HIV and problems in vaccine development have triggered massive efforts to develop drugs and chemotherapeutic strategies for treatment or management of AIDS. Hope has been provided by 3′-azido-3′-deoxythymidine (AZT) and 2′,3′,-dideoxynucleosides (Fig. 4), nucleoside analogs that work in a manner quite analogous to many of the antiherpes drugs. The urgency of the AIDS problem has also stimulated a variety of other approaches and evaluation of all HIV components as potential chemotherapeutic targets. Some of these may provide new general strategies for viral chemotherapy. [*See* Acquired Immunodeficiency Syndrome, Virology.]

Not surprisingly, the most promising target (and the first one exploited) is the HIV reverse transcriptase (RT). This enzyme, which catalyzes synthesis of a DNA copy of the virion RNA, is absolutely essential for replication or expression of the HIV genome. RT is the target for active forms (the corresponding 5′-triphosphates) of several nucleoside analogs that are able to selectively block replication of HIV. These include AZT and the dideoxynucleosides.

AZT was the first drug effective enough in the inhibition of HIV replication to be approved for treatment of AIDS patients. Although its use has been limited somewhat by toxicity to bone marrow, AZT enables marked improvement and increased life expectancy in some patients.

Like other nucleoside analogs, AZT must be activated by phosphorylation to its corresponding 5′-mono-, di-, and triphosphates. This metabolism is accomplished by cellular kinases. The 5′-triphosphate (AZ-TTP) is a competitive inhibitor of the HIV RT. It is also a substrate for this enzyme and is incorporated into DNA. Upon incorporation, AZT is a chain terminator since the 3′-azido group cannot serve as an acceptor for the next incoming nucleotide. The selectivity of AZT is due to the fact that AZ-TTP is not an effective inhibitor or a substrate for cellular DNA polymerases. Thus, treatment with AZT leads to a selective inhibition of viral DNA synthesis. It has not been determined whether it is the inhibition of RT, the chain termination, or the combination that is responsible for the antiviral activity.

Several dideoxynucleosides have been shown to be selective inhibitors of HIV replication, and these work in a manner similar to that of AZT. These are metabolized to their corresponding nucleotides and the 5′-triphosphates are competitive inhibitors of the HIV RT. The dideoxynucleotides are also incorporated into DNA and are well-characterized chain terminators in bacterial systems (as evident from their use in sequencing of DNA). The antiviral activity of these compounds is presumably due to their inhibition of

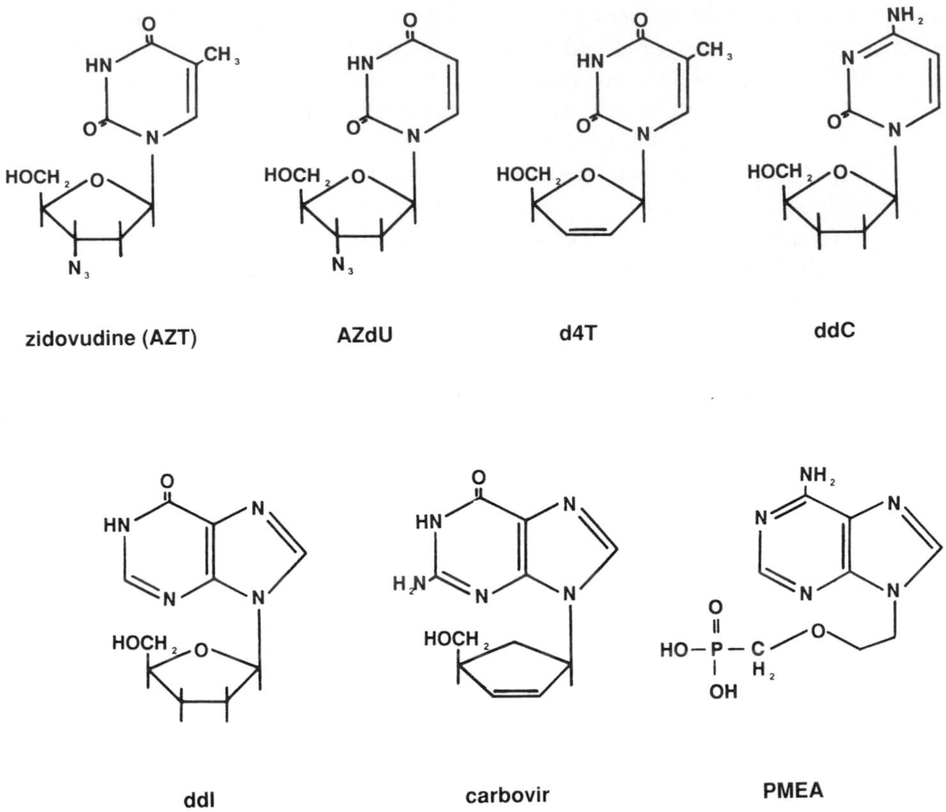

zidovudine (AZT) AZdU d4T ddC

ddI carbovir PMEA

FIGURE 4 Some of the nucleoside analogs that are selective inhibitors of the human immunodeficiency virus.

RT and/or their incorporation into and termination of viral DNA. Two dideoxynucleosides are approved for use in AIDS therapy. These are dideoxycytidine (ddC), which is metabolized to ddCTP, and dideoxyinosine (ddI), which is metabolized to ddATP (Fig. 4). The limiting toxicity of these dideoxynucleosides, which cause peripheral neuritis, is different from that of AZT; neither of them causes suppression of bone marrow. Another nucleoside analog that has been approved for AIDS therapy is 2′,3′-dideoxy-2′,3′-didehydrothymidine (d4T), a thymidine analog that appears to be less toxic than AZT.

Many other nucleoside analogs have shown promise as inhibitors of HIV replication, and some of these are shown in Fig. 4. These include 3′-azido-3′-deoxydeoxyuridine (AZdU), carbocyclic-2′,3′-dideoxy-2′,3′-didehydroguanine (carbovir), (−)-β-L-2′,3′-dideoxy-3′-thiacytidine (3TC), and a group of phosphonyl compounds such as 9-(2-phosphorylmethoxyethyl)adenine (PMEA). Interestingly, some of the phosphonyl derivatives have excellent antiherpes activity. Nonnucleoside inhibitors of HIV RT have also been developed and some of these, such as nevirapine, have excellent anti-HIV activity *in vitro* and in humans.

Other HIV enzymes have been identified and are candidate targets for chemotherapeutic approaches. One of these is RNase H, an activity that is present on the same protein as reverse transcriptase. RNase H is required to degrade the RNA strand of the RNA–DNA hybrid that occurs after RT has made a DNA copy of viral RNA. Removal of the RNA is necessary to enable synthesis of double-stranded DNA. Another enzyme that has received attention as a chemotherapeutic target is integrase. This enzyme is necessary for integration of proviral DNA into the host genome. Although these activities have been well characterized, selective inhibitors have not yet been identified.

HIV also encodes a protease that has received much attention as a target for chemotherapy. The HIV protease is essential for replication of HIV. It cleaves large polyproteins into the individual viral enzymes and structural proteins, and may also be necessary for assembly and/or maturation of the virion. The

HIV enzyme is an aspartyl protease; it has been purified and its crystal structure determined. This information has enabled development of several active site-directed inhibitors of the HIV protease, and several of these have excellent anti-HIV activity *in vitro* and *in vivo*. It is likely that one or more of these will be approved for clinical use in the near future.

Several other targets have been proposed as having potential for chemotherapy of AIDS. One approach is to block attachment of HIV to cells. Attachment involves a specific cellular protein, the CD4 receptor, found on the surface of susceptible lymphocytes and other cells permissive for HIV infection. Attachment of HIV to this receptor can be blocked by providing an excess of soluble CD4 protein, and attempts have been made to develop this into a viable chemotherapeutic strategy. A problem of this approach is that soluble CD4 will not inhibit virus replication once the virus enters the cell, and so the block must be complete (or nearly so) in order to be effective. HIV also encodes several regulatory proteins that control expression of viral genes. These have been suggested as chemotherapeutic targets, but a complication is that some proteins from host cells or from other viruses may be able to substitute for these regulatory elements. Yet another approach being initiated utilizes a viral nucleic acid as a target. This involves development of reagents that will react with and inactivate or destroy specific sequences of viral RNA and DNA. This approach, although potentially very selective, is only in the very early stages of development. [*See* CD8 and CD4: Structure, Function, and Molecular Biology.]

A serious limitation in the development of strategies for chemotherapy of AIDS has been the lack of a suitable animal model. Much work is in progress to develop such a model and some of the approaches look promising. A simian immunodeficiency virus has been identified, and its reverse transcriptase is very similar to that of HIV. This will be useful for some studies, but nonhuman primates are not available in sufficient numbers for large-scale chemotherapeutic studies. A number of rodent models are in development, and some of these should provide useful information. However, the murine retroviruses used in these studies are quite different from HIV, as are their reverse transcriptases and the clinical course of their infections. One promising development is transplantation of the human immune system into immunodeficient mice to produce SCID/hu mice. These mice produce human lymphocytes and are susceptible to infection by HIV. It is not clear whether the mice develop an AIDS-like disease; nevertheless, these mice may be useful for chemotherapeutic studies if they can be generated in sufficient numbers and the properties of the transplanted immune system are reproducible. Yet another model, which the author is working with, is feline immunodeficiency virus (FIV). This virus, like HIV, is a lentivirus, and it causes an immune deficiency in cats that is very similar to AIDS in humans. The FIV reverse transcriptase is very similar to the HIV reverse transcriptase in physical properties and sensitivities to the active forms of several antivirals, including AZ-TTP. This model promises to be useful for coordination of *in vitro* and *in vivo* studies.

If successful chemotherapeutic strategies for AIDS can be devised, it is likely they will point the way for chemotherapy of other human retroviral infections. Successful strategies should also provide insight for general approaches to combat other classes of viruses.

VI. LATENCY, RESISTANCE, AND OTHER CHALLENGES

All of the effective antivirals currently available are replication inhibitors. However, many viruses (including retroviruses and herpesviruses) are able to establish latent states in which the viral genome is present in infected cells, but it is not actively replicating. Latent virus is not susceptible to these replication inhibitors; after the drug is removed the virus can reactivate and undergo replication. It is evident from the situation with herpesviruses that the existence of latency does not preclude success with replication inhibitors. The situation with AIDS awaits further evidence. However, if the goal is to totally eliminate virus, ways must be devised to eliminate latent virus. One possibility is to devise drugs that will reactivate the virus; it might then be eliminated with a replication inhibitor. For this to be successful the activation from latency will have to be complete. It is not known whether this will be possible. Another approach is to kill all virus-infected cells, including those harboring latent virus. If no viral genes are expressed during latency, this approach might require agents directed at the viral genome. If latency cannot be alleviated, alternative approaches might be continuous or periodic use of replication inhibitors, or the development of drugs that block reactivation. Herpes simplex virus requires the viral-encoded thymidine kinase for efficient reactivation, and inhibitors of this enzyme have shown promise in blocking HSV reactivation.

Another major problem in viral chemotherapy is the emergence of drug-resistant mutants. HSV mutants resistant to each of the antiherpes drugs have been isolated *in vitro*. Resistance to acyclovir (or other analogs that require activation via the viral thymidine kinase) can occur from mutations in either the thymidine kinase gene or the DNA polymerase gene. Resistance to araA or phosphonoformate is due only to mutations in the DNA polymerase gene. Resistance to acyclovir occurs clinically and some of these mutants are cross-resistant to other antivirals. Resistance to phosphonoformate also occurs readily. However, clinical resistance of HSV to araA has not been reported and most of the araA-resistant HSV were isolated from their resistance to other drugs. Although araA is much less selective than acyclovir, this slower development of resistant virus might be attractive in some instances.

Drug-resistant mutants of HIV readily emerge during therapy of AIDS patients. Such mutants were first isolated from patients treated 6 months or longer with AZT, and they are prevalent in patients on prolonged AZT therapy. Drug-resistant variants were also isolated from patients following therapy with other approved anti-HIV nucleosides 2′,3′-dideoxyinosine and 2′,3′-dideoxycytidine, and in trials with several other nucleoside analogs or nonnucleoside inhibitors of HIV RT. This problem is not limited to RT-targeted drugs. Mutants resistant to protease inhibitors have been selected *in vitro* and isolated from patients. Because HIV and other retroviruses have very high mutation frequencies, it is expected that drug resistance will be a serious problem with drugs targeted to any HIV protein. Likewise, it is expected that resistance will arise to nucleic acid-targeted agents, such as ribozymes, owing to variation in the sequence of the target site. Strategies to combat resistance will be needed for successful therapy of AIDS.

Resistance of HIV-1 to AZT is usually associated with two or more mutations, commonly in codons 41, 67, 70, 215, and 219 of HIV RT. Similarly, mutations responsible for resistance to other RT or protease inhibitors have been identified. The biochemical basis for resistance has been well defined for mutants resistant to nonnucleoside inhibitors of RT. For example, mutants resistant to nevirapine encode an RT that is highly resistant to the drug. Likewise, mutants resistant to protease inhibitors have a protease that is resistant to these inhibitors. However, the mechanisms responsible for resistance to nucleosides are not clear. RTs from most AZT-resistant mutants of HIV are similar to the wild-type enzyme in their susceptibilities to the 5′-triphosphate of AZT. The failure of RTs from these mutants to correlate with viral drug resistance suggests that factors in addition to RT are involved in resistance to AZT. A better understanding of viral resistance to AZT is needed, and such information should be useful in design of strategies to combat drug resistance.

Drug resistance is encountered in other areas of chemotherapy, such as cancer and tuberculosis. The most successful approach to combat this is combination chemotherapy. The likelihood of multiple resistance occurring during combination chemotherapy is a product of the probabilities of resistance to each drug; combinations of two or three drugs directed at different targets are usually sufficient to eliminate problems of resistance. It is generally recognized that long-term therapy of AIDS will require such combinations of drugs directed at two or more targets.

Despite these problems, there is a great deal of promise for the future of viral chemotherapy. In a relatively short period this field has surpassed the era of random screening of compounds, and the feeling by many that selective antivirals might be impossible because of the reliance of viruses upon host machinery for much of their metabolism. Now there are selective antivirals for certain viruses, and strategies that should enable development of drugs for other classes of viruses. Nucleoside analogs able to selectively inhibit the DNA polymerase of DNA viruses or the reverse transcriptase of retroviruses have been developed. A similar approach might be employed to develop selective inhibitors of the RNA-dependent RNA polymerases of many RNA viruses. The large-scale efforts toward AIDS chemotherapy should provide other general strategies. Many disciplines are contributing to the development of chemotherapeutic strategies and identification of viral targets, and it is likely that many further developments will be facilitated by collaborative, multidisciplinary approaches.

BIBLIOGRAPHY

Elion, G. B. (1984). Acyclovir. *In* "Antiviral Drugs and Interferon: The Molecular Basis for Their Activity" (Y. Becker, ed.), pp. 71–88. Martinus Nijhoff, Boston.

Hirsch, M. S., and D'Aquila, R. T. (1993). Therapy of human immunodeficiency virus infection. *N. Engl. J. Med.* **328**, 1686–1695.

Hirsch, M. S., and Kaplan, J. C. (1987). Antiviral therapy. *Sci. Am.* **256**, 76–85.

Mitsuya, H., Yarchoan, R., Kageyama, S., and Broder, S. (1993). Targeted therapy of human immunodeficiency virus-related disease. *FASEB J.* **5**, 2369–2381.

North, T. W., and Cohen, S. S. (1984). Aranucleosides and aranucleotides in viral chemotherapy. *In* "International Encyclopedia of Pharmacology and Therapeutics, Section III. Viral Chemotherapy, Vol. 1" (D. Shugar, ed.), pp. 303–340. Pergamon, Oxford, England.

Richman, D. D. (1991). Antiviral therapy of HIV infection. *Annu. Rev. Med.* **42**, 69–90.

Richman, D. D. (1993). Resistance of clinical isolates of human immunodeficiency virus to antiretroviral agents. *Antimicrob. Agents Chemother.* **37**, 1207–1213.

Chemotherapy, Cancer

DANIEL E. EPNER
Baylor College of Medicine, Houston Veterans Affairs Medical Center

I. DRUG DISCOVERY

Cancer strikes over a million Americans each year. Fortunately, many cancers that were fatal only a few decades ago are now curable. Cancer treatment is complex and involves three complementary modalities: surgery, radiation, and chemotherapy. Cancer surgery was first performed a century ago by William Halsted, who developed the radical mastectomy for the treatment of breast cancer. At about the same time, Wilhelm Roentgen discovered X-rays and thereby laid the groundwork for modern radiation therapy. Although surgical and radiation techniques have advanced dramatically since the days of Halsted and Roentgen, neither modality has a major role in the curative treatment of cancer that has spread throughout the body, or metastasized.

Early in this century, Paul Ehrlich developed the concept of "chemotherapy," which is the systemic treatment of disease by chemical agents that restore health without significantly harming the patient. Chemotherapy was first used to treat infectious diseases. The dawn of the era of effective antibiotics occurred when penicillin was isolated from a fungal culture in 1929. By World War II, chemotherapy was successfully applied to the treatment of cancer, and since then many highly effective anticancer drugs have been discovered. As a result, several types of metastatic cancer, such as testicular cancer, acute leukemia, and lymphoma, are largely curable with chemotherapy. Chemotherapy also prolongs survival and provides symptomatic relief for patients with many other types of cancer.

Discovery of cancer chemotherapy drugs has resulted from serendipity, empiricism, and rational drug design. Some of the most effective drugs used today were discovered when a clinician or basic scientist astutely observed that a compound unexpectedly inhibited tumor growth in patients or in experimental models. Other drugs were discovered through screening programs at the National Cancer Institute and elsewhere in which thousands of promising compounds were tested for their ability to inhibit cancer growth. Still others were derived by the synthesis of compounds with chemical structures similar to natural metabolites or to drugs with known efficacy. In some cases, these analogs proved to be more effective and less toxic than the compounds from which they were derived. Major conceptual advances in molecular and cellular biology have led to an increased emphasis on rational drug design and a focus on novel therapeutic strategies. These advances have led to the discovery of several new drugs, some of which have already entered clinical trials.

A. Alkylating Agents

The importance of serendipity in the identification of anticancer drugs is dramatically illustrated by the discovery of nitrogen mustard, which was originally studied for its potential as a vesicant in chemical warfare. During World War II, several military personnel were accidentally exposed to nitrogen mustard, which was unexpectedly found to kill normal blood cells that can give rise to lymphoma, a form of cancer. Under the cloak of wartime secrecy, Alfred Gilman and Louis Goodman then studied the usefulness of

ENCYCLOPEDIA OF HUMAN BIOLOGY, Second Edition, VOLUME 2. Copyright © 1997 by Academic Press. All rights of reproduction in any form reserved.

nitrogen mustard for the treatment of lymphoma. They found it to be highly effective for patients who had failed radiation, the only available treatment at that time. [*See* Lymphoma.]

Nitrogen mustard and other alkylating agents work by chemically modifying DNA, thereby altering its structure and function. These compounds exert their cytotoxic effects on cells throughout the cell cycle but have quantitatively greater activity against rapidly proliferating cells, such as cancer cells. The synthesis of analogs of nitrogen mustard and other alkylating agents led to the eventual clinical use of a wide variety of such drugs, listed in Table I.

B. Antimetabolites

Antimetabolites are compounds that bear a close structural resemblance to substances required for normal physiological functioning. They inhibit cell division by interfering with the utilization of these critical metabolites. The earliest use of an antimetabolite as a chemotherapeutic agent resulted from an astute observation by Sidney Farber, who noticed that patients with leukemia experienced an acceleration of their illness when treated with folate. A series of folate antagonists were then provided to Farber and colleagues by medicinal chemists. Although the mechanisms by which antifolates worked were unknown at the time, laboratory studies showed that modified folates clearly inhibited tumor cell growth. Before long, the folate antagonist aminopterin was tested for its activity against advanced acute leukemia in children, and 10 of the first 16 patients demonstrated clinical improvement. Aminopterin has since been replaced by another antifolate, methotrexate, which is commonly used to treat leukemia, breast cancer, head and neck cancer, lymphoma, and primary bone tumors. Methotrexate also cures over half of the women with disseminated choriocarcinoma, as demonstrated by Li and Hertz in the 1950s.

TABLE I
Commonly Used Alkylating Agents

Drug	Major clinical indication
Nitrogen mustard	Hodgkin's lymphoma
Melphalan	Multiple myeloma
Chlorambucil	Chronic lymphocytic leukemia
Busulfan	Chronic myelogenous leukemia
Cyclophosphamide	Breast cancer Non-Hodgkin's lymphoma
Ifosfamide	Sarcoma Testicular cancer
Nitrosoureas (BCNU, CCNU)	Brain tumors

In contrast to the serendipitous discovery of antifolates, three other classes of antimetabolites were discovered largely by rational drug design. 5-Fluorouracil (5-FU), the most commonly used fluoropyrimidine, was synthesized by Charles Heidelberger and colleagues after uracil was shown to be utilized by cancer cells more efficiently than by normal cells. Purine analogs, including 6-mercaptopurine and 6-thioguanine, were synthesized in the early 1950s by Hitchings and Elion. 6-Mercaptopurine is used to treat acute lymphocytic leukemia, whereas 6-thioguanine is used to treat acute myelogenous leukemia. Rational drug design also had a major role in the discovery of adenosine analogs, such as fludarabine, which is used to treat chronic lymphocytic leukemia. Other antimetabolites were discovered by empiricism. For instance, cytosine arabinoside, a drug commonly used to treat acute leukemias, was isolated from the sponge *Cryptothethya crypta.*

C. Platinum Analogs

In 1965, Rosenberg fortuitously observed that an electric current passing through platinum electrodes inhibited bacterial cell division. This finding led to the discovery that platinum analogs inhibited cancer cell division in murine tumor model systems. Initial clinical trials demonstrated antitumor activity in patients with advanced malignancies, but platinum compounds were found to injure normal kidney cells. Once methods were developed to overcome this toxicity, *cis*-platinum assumed a major role in the treatment of testicular cancer, ovarian cancer, squamous cell cancer of the head and neck, and lung cancer. Newer compounds, such as carboplatin, have been developed that do not injure kidney cells.

D. Natural Products: Antitumor Antibiotics and Plant Derivatives

As a result of the great advances in the development of antibacterial antibiotics in the 1940s, potent anticancer drugs were sought from fermentation broths of soil microbes, including bacteria and fungi. Screening of these natural compounds led to the discovery of some of the most effective cancer chemotherapeutic agents used today (see Table II). For instance, daunorubicin, used in the treatment of leukemia, was isolated from a colony of the fungus *Streptomyces* in 1957. Additional research to induce mutant strains of *Streptomyces* resulted in the isolation of doxorubicin, a closely related compound with potent activity against many types of solid tumors.

Plants have also provided effective cancer chemotherapy drugs. One of the earliest plant-derived drugs resulted from a serendipitous observation in the mid-1950s by Noble and colleagues, who studied extracts from the Jamaican periwinkle plant, *Vinca rosea*. Tea made from the leaves of this plant was reported to be of benefit in diabetes, a disorder characterized by elevated blood sugar levels. When tested in animals, however, extracts of *V. rosea* proved to have no effect on blood sugar levels but instead killed the animals within a week. Postmortem examinations of the animals demonstrated that they died of an infection related to bone marrow suppression. Later clinical trials showed that the active compound, a precursor of the anticancer drug vincristine, was effective for the treatment of leukemia and other cancers. Laboratory studies also elucidated its unique mechanism of action: inhibition of microtubule formation. Screening of plant derivatives more recently led to the discovery of Taxol, which was first isolated from the bark of the Pacific yew tree in 1971. In the few years since it was introduced into clinical practice, Taxol has made a major impact on the treatment of several types of cancer, especially ovarian cancer. Efforts are currently underway to identify other plant derivatives with activity against cancer and to explore the potential of compounds isolated from marine organisms, which so far represent a largely untapped resource. [*See* Leukemia.]

TABLE II
Natural Products Commonly Used to Treat Cancer

Drug	Major clinical indication
Antibiotics	
Doxorubicin	Lymphoma
	Breast cancer
	Bladder cancer
Daunorubicin	Acute leukemias
Bleomycin	Testicular cancer
	Lymphoma
Plant derivatives	
Vincristine	Acute lymphocytic leukemia
	Lymphoma
Vinblastine	Bladder cancer
Taxol	Ovarian cancer
VP-16 (Etoposide)	Lung cancer
	Testicular cancer

F. Hormonal Treatments

George Beatson of Glasgow in 1896 described the regression of advanced breast cancer in premenopausal women following removal of the ovaries, which produce estrogen. This was the first demonstration of hormonal dependence of a cancer in humans, made long before the endocrine functions of the ovary were defined. At the same time, J. W. White of Philadelphia showed that surgical removal of the testes, which are the main source of androgen in the body, caused the shrinkage of prostate glands in dogs and benefited men with prostate hypertrophy. Charles Huggins then demonstrated in 1941 that patients with disseminated prostate cancer were dramatically improved by castration or treatment with diethylstilbestrol, a synthetic estrogen that antagonizes androgen action.

These landmark observations laid the foundation for current treatment strategies for breast and prostate cancer. Many women with breast cancer are treated with tamoxifen, a nonsteroidal antiestrogen that competes with estrogen at its receptor, whereas men with metastatic prostate cancer are still treated with (1) surgical castration, (2) diethylstilbestrol, or (3) drugs that inhibit androgen production by the testes combined with drugs that block the binding of the androgen to its receptor. Hormonal treatments are also being evaluated as possible preventive agents for breast and prostate cancer. [*See* Breast Cancer Biology; Prostate Cancer.]

G. Biologic Therapy

Biologic therapy produces antitumor effects primarily through the action of natural host defense mechanisms or by the administration of substances naturally produced in the body. The first use of biologic therapy is generally credited to William Coley, who in 1893 observed regression of lymphoma in a patient who developed a bacterial skin infection. Coley hypothesized that the bacterial infection nonspecifically stimulated the patient's immune system, which then attacked the cancer. He later found that administering a mixture of bacteria to patients with cancer had a beneficial effect in some cases. Coley's work has not been successfully repeated, but it spurred the testing of many bacterial products as possible immunotherapy agents. One such product, BCG, is commonly used today for the treatment of superficial bladder cancer. BCG is administered by direct instillation into the bladder and acts as a nonspecific immune stimulant.

Major advances in molecular and cellular biology have led to the discovery of dozens of cytokines, protein molecules that regulate the immune system and have a myriad of other biologic effects. Cytokines can now be produced in large quantities by recombinant DNA technology and are being tested for their anticancer properties. One such cytokine, interferon α, has revolutionized the treatment of hairy cell leukemia, a rare hematologic malignancy that previously responded poorly to treatment. Eighty to 90% of patients with hairy cell leukemia respond to interferon α; those who do not often respond to either chlorodeoxyadenosine or deoxycoformycin, two new antimetabolites. Interferon is also active against chronic myelogenous leukemia and lymphoma, and may prove useful in the treatment of selected nonhematologic malignancies. Ongoing clinical trials of other biologic agents will undoubtedly lead to major advances in cancer treatment. [*See* Cytokines and the Immune Response; Interferons.]

II. DRUG DEVELOPMENT

Drug discovery is but the first step in the arduous process of developing new cancer treatments. Newly discovered drugs are first tested in animals to identify potential side effects, pharmacokinetic properties, optimal formulation, and the spectrum of antitumor activity. They are then tested in phase I, II, and III human trials. New treatments are not considered "standard" or "established" until they are shown to be superior to other available treatments.

A. Toxicities of Chemotherapy

Phase I trials are designed to determine the maximum tolerated dose of a particular medicine or combination of medicines and to identify side effects. Because most chemotherapy drugs target cancer cell division, they also tend to harm normal cells that divide rapidly, such as those found in the bone marrow, scalp, and gut. As a result, chemotherapy often causes bone marrow suppression, hair loss, nausea, diarrhea, and sore mouth. Certain drugs can also injure the kidneys, heart, lungs, and liver. Bone marrow suppression can cause anemia, infections, and bleeding due to decreased levels of red blood cells, white blood cells, and platelets, respectively. Bone marrow toxicity can now be minimized with the use of colony-stimulating factors, such as G-CSF. These newly discovered natural proteins stimulate bone marrow to recover more quickly following chemotherapy. Nausea can also be controlled with the use of newly discovered antiemetics. Although most side effects of chemotherapy are reversible after the discontinuation of treatment, some drugs can cause permanent organ damage. For instance, anthracyclines, such as doxorubicin and daunorubicin, can cause permanent heart damage, and bleomycin can cause permanent lung damage. Certain chemotherapy drugs have also been linked to the development of secondary malignancies several years after successful treatment of the primary cancer.

B. Efficacy

Phase II and III clinical trials are designed to determine the efficacy of a particular new drug or combination of drugs. Phase II trials identify tumor types for which the treatment appears promising, whereas phase III studies determine whether a new treatment is (1) more effective than a standard therapy or (2) as effective but less toxic than a standard therapy. Well-designed clinical trials focus on a single, major question and have sufficient statistical power to yield a clear answer. This hypothesis-driven, empiric approach to drug development has led to curative treatments for many patients with leukemia, lymphoma, testicular cancer, and choriocarcinoma.

Several studies have shown that a patient's response to chemotherapy is highly dependent on his or her overall physiologic status. Patients who are active and have no significant health problems other than cancer generally respond much better to chemotherapy and experience fewer side effects than patients who are debilitated. In fact, the risks of chemotherapy for severely debilitated patients often outweigh potential benefits.

III. DRUG RESISTANCE

Despite great advances in cancer treatment in recent decades, tumors frequently become resistant to chemotherapy. In 1979, Goldie and Coldman proposed that tumor cells become drug resistant as a result of genomic instability, a cardinal feature of cancer. As DNA in cancer cells accumulates mutations, translocations, and other alterations, it becomes structurally and functionally dynamic, unlike DNA in normal cells, which is stable and tightly regulated. This genomic instability gives rise to tumor cell heterogeneity. Even though cells within a tumor appear similar at the microscopic level, they are actually heterogeneous at the molecular level. As cancer cells evolve, they become resistant to chemotherapy by several mechanisms. Reduced intracellular accumulation of drug is one of the most common mechanisms of antineoplastic drug resistance. For instance, reduced activity of the folate-binding protein or the folate transporter can lead to methotrexate resistance. In tumor cells that accumulate drug normally, resistance may be due to changes in drug activation, inactivation, or cofactors. In some cases, drug resistance is due to alteration of the intracellular drug target. For instance, alterations in dihydrofolate reductase, one of the main targets for methotrexate, can lead to drug resistance. Drug resistance can also result when cancer cells acquire the ability to repair DNA strand breaks caused by chemotherapy.

A. Combination Chemotherapy

One strategy for circumventing drug resistance is the use of multiple drugs in combination. Combination chemotherapy accomplishes three important objectives not possible with single agent treatment: (1) it provides maximal cell kill within the range of toxicity tolerated by the host for each drug, (2) it provides a broader range of coverage of resistant cell lines in a heterogeneous tumor population, and (3) it prevents or slows the development of new resistant lines. The strategy of using combination chemotherapy to overcome drug resistance has largely been successful. Unfortunately, many common cancers, such as prostate, lung, colon, and breast, are resistant even to multiple drugs.

There are several mechanisms for multidrug resistance. One mechanism is the overexpression of an energy-dependent drug efflux pump, P-glycoprotein, encoded by the *MDR1* gene. Another mechanism is decreased topoisomerase activity. Topoisomerases are nuclear enzymes that catalyze the formation of transient DNA strand breaks, facilitate passage of DNA strands through these breaks, and promote rejoining of the DNA strands. The activity of drugs that target topoisomerases, such as etoposide and doxorubicin, is thought to depend on the DNA cleavage activities of these enzymes. Another common mechanism of multidrug resistance is overexpression of enzymes that detoxify chemotherapy drugs, such as glutathione *S*-transferase.

B. Dose Intensification

Drug resistance is a relative concept. Theoretically, even drug-resistant cancer cells can be eliminated by extremely high doses of chemotherapy. This rationale is the basis for bone marrow transplantation. Bone marrow consists of immature blood cells in all stages of development. The least mature bone marrow cells are stem cells, which give rise to mature blood cells. Stem cells are not usually harmed by standard doses of chemotherapy, so marrow suppression is reversible. However, patients who undergo bone marrow transplantation are given chemotherapy in such high doses that even their stem cells are killed and their marrows are irreversibly damaged. They are then given back healthy bone marrow which restores their ability to form blood. Patients who undergo autologous transplantation receive their own marrow harvested prior to chemotherapy, whereas those who undergo allogeneic transplantation receive marrow from a matched donor, usually a sibling. Although most stem cells reside in the bone marrow cavity, some circulate in the bloodstream. Techniques have been developed for harvesting circulating stem cells from patients prior to treatment with high dose chemotherapy and transplanting them instead of complete bone marrow.

IV. INTEGRATING CHEMOTHERAPY WITH SURGERY AND RADIATION

A. Adjuvant Chemotherapy

When Halsted developed the radical mastectomy a century ago for treating women with breast cancer, he could not have known that the operation would later prove to be excessive for some women while inadequate for others. It is now known that much less extensive operations with far fewer side effects are just as effective for eliminating *localized* breast cancer. However, surgery, no matter how extensive, cannot adequately eliminate cancer that has unknowingly metastasized prior to the operation. Such clinically occult metastases eventually grow and cause life-threatening problems. Chemotherapy offers the only hope of completely eliminating metastases, which are theoretically most sensitive to treatment while still undetectable clinically. Based on this rationale, clinical trials involving adjuvant, or postoperative, chemotherapy for women with breast cancer were begun a few decades ago. Those trials showed significant survival advantages for certain subgroups of women treated with chemotherapy. Adjuvant chemotherapy has also been shown to be beneficial in a few other circumstances, such as in treatment of bone sarcoma. Several clinical trials are underway to determine whether adjuvant chemotherapy will be useful in the treatment of other tumor types.

B. Neoadjuvant Chemotherapy

Neoadjuvant chemotherapy, like adjuvant chemotherapy, is used in conjunction with surgery, but it is given prior to the operation. Neoadjuvant chemotherapy has several theoretical advantages. First, it can shrink large tumors and thereby make them easier to remove surgically. Second, it attacks metastatic deposits at the earliest possible time, when they are small and hopefully chemosensitive. Third, it allows clinicians to quickly determine whether a tumor is chemosensitive in the event that further chemotherapy is required later. Neoadjuvant chemotherapy, however, also has theoretical disadvantages. For example, if the tumor proves to be resistant to chemotherapy, definitive surgical treatment is delayed by neoadjuvant therapy, and metastases can form during the delay. Many clinical trials have addressed the role of neoadjuvant chemotherapy for treatment of a variety of tumors. So far, this approach appears to hold promise for treatment of certain types of cancer, such as invasive bladder cancer and esophageal cancer. Future clinical trials will further define the role of neoadjuvant chemotherapy for treating these and other cancers.

C. Combined Treatment with Chemotherapy and Radiation

Chemotherapy is also combined with radiation in selected cases. Certain chemotherapy drugs, such as *cis*-platinum, have been shown to increase the sensitivity of some tumors to radiation. These drugs are therefore often given in relatively low doses at the same time as radiation. Chemotherapy and radiation are also given sequentially in some cases. For instance, many patients with Hodgkin's disease and a large tumor mass within the chest receive both radiation and chemotherapy despite the significant side effects associated with the combined treatment. Chemotherapy and radiation are also used in combination for treating cancers of the lung, head and neck, and other sites. Unfortunately, combining chemotherapy and radiation can lead to severe side effects and, in the case of lung cancer, results in only a small survival advantage.

D. Combined Treatment with Surgery, Radiation, and Chemotherapy

In some cases, such as the management of seminoma, all three modalities are used in conjunction. Seminoma is a highly curable form of testicular cancer that is very sensitive to both radiation and chemotherapy. Unilateral orchiectomy is required for the proper management of men with seminoma and other forms of testicular cancer. Following surgery, patients whose tumors have not spread beyond the pelvis are then treated with radiation, which is curative in most cases. Some patients, however, develop distant metastases at a later time. Those patients are then treated with combination chemotherapy, which is also curative in most cases.

Postoperative radiation treatment is also commonly used if a tumor of any kind is found to be impossible to fully resect at the time of surgery. Despite postoperative radiation treatment, however, such patients often return months to years later with metastatic disease that was undetectable at the time of surgery. Depending on the type of tumor and clinical situation, many patients then benefit from chemotherapy.

V. EMERGING STRATEGIES

Most chemotherapy drugs commonly used today target cell division by a variety of mechanisms. Increased cell division, however, is but one of the many hallmarks of cancer. Other cardinal features of cancer include invasion and metastasis, angiogenesis, abnormal energy metabolism, decreased apoptosis, immortality, abnormal gene expression, and abnormal cell structure. Major conceptual breakthroughs have led to a better understanding of cancer at the molecular level and to the introduction of several novel therapeutic strategies, some of which have already entered clinical trials.

A. Apoptosis

Cell division in normal tissues and in the developing fetus is balanced by programmed cell death, or apoptosis. When this balance is lost as a result of either increased cell division or decreased apoptosis, tumors can form. Many kinds of cancer, such as prostate cancer, have low rates of cell division but also have low rates of apoptosis. These tumors therefore respond poorly to available chemotherapy drugs that target cell division. A great deal of effort has been devoted to clarifying the molecular mechanisms of apoptosis. This effort has resulted in the identification of new compounds which stimulate apoptosis, some of which have shown promising results in animal studies.

B. Angiostasis

Regardless of whether a tumor grows as a result of increased cell division or decreased apoptosis, it cannot grow larger than about 1–2 mm^3 before it requires its own blood supply. New blood vessel growth, or angiogenesis, occurs not only in tumors, but also in fetal tissues and in mature tissues during wound healing. Several investigators, most notably Judah Folkman, have clarified the molecular mechanisms of angiogenesis. They have identified several natural proteins that promote angiogenesis and other factors, such as angiostatin, that inhibit it. Preclinical studies have shown that certain angiostatic factors prevent the growth of metastatic tumors in animals. Tumors within these animals remain microscopic, and therefore harmless, indefinitely. Based on these promising results, clinical trials of angiostatic agents for treatment of patients with a variety of tumors are already underway.

C. Monoclonal Antibodies

Antibodies are proteins produced by B lymphocytes, a type of immune cell, in response to a foreign molecule or invading organism. They bind to the foreign molecule or cell extremely tightly, thereby inactivating it or marking it for destruction by other immune cells. Unfortunately, natural antibodies alone are often not enough to eliminate cancer cells. Nonetheless, scientists have dreamed for years of using antibodies as "magic bullets" to specifically seek out and destroy cancer cells. This dream may soon become a reality. Monoclonal antibodies, which bind to a single molecular target, can now be produced in the laboratory in virtually unlimited quantities. Monoclonal antibodies have been developed that recognize molecules expressed almost exclusively by cancer cells. In some cases, these antibodies are chemically linked to a toxin or to a radioisotope that enhances the anticancer effect. So far, results from animal studies and preliminary clinical trials seem very promising. [*See* Monoclonal Antibody Technology.]

D. Gene Therapy

Genes represent another form of "magic bullet" that can theoretically be used to attack cancer cells. The recent renaissance in the field of molecular genetics has led to the identification of several genes that play key roles in tumor growth. Most cancer cells overexpress one or more oncogenes, which promote cell division. Inhibition of oncogenes with antisense molecules represents one potential cancer treatment strategy. Antisense molecules are nucleic acid sequences that bind specifically to a complementary strand of RNA or DNA, thereby inhibiting production of the relevant gene product. Although laboratory studies appear promising, several technical hurdles need to be overcome before antisense treatments for cancer become feasible. [*See* Gene Therapy in the Treatment of Cancer.]

Many tumors also underexpress one or more tumor suppresser genes, which normally keep cell division in check. *p53* is a tumor suppresser gene that is inactivated in more than 75% of tumors. Reintroduction of a normal *p53* gene into tumor cells that express mutant p53 protein represents another gene therapy strategy. Clinical trials employing this and other gene therapy strategies appear promising but are still probably years away from making a major impact on treatment. [*See* Tumor Suppressor Genes, p53.]

E. Differentiating Agents

People have known for many years that a diet rich in fruits and vegetables can help prevent cancer. However, scientists have only recently discovered that many of the substances in food that protect against cancer are vitamins. Vitamins act by causing cancerous or precancerous cells to differentiate into mature, normal cells. One such substance, all-*trans*-retinoic acid, which is similar to vitamin A, has revolutionized the treatment of acute promyelocytic leukemia. all-*trans*-Retinoic acid induces complete remission in 90% of patients with promyelocytic leukemia and dramatically reduces associated bleeding complications. Waun Ki Hong and colleagues have also shown that differentiating agents can reduce the incidence of head and neck cancer in patients who are at high risk for developing them. Differentiating agents will undoubtedly play a significantly greater role in the treatment and prevention of cancer in the future. [*See* Cancer Prevention.]

F. Application of Circadian Chronobiology to Cancer Chemotherapy

Although most clinicians give little thought to the time of day that a drug is given, several studies by William Hrushesky and others suggest that the therapeutic effect may be maximized and toxicity may be minimized if drugs are administered at carefully selected times of the day. Temporal variability in antitumor effect and toxicity can be explained in part by two important observations. First, the pharmacokinetics of many anticancer drugs show consistent and reproducible circadian temporal variation, depending on the time of their administration. Second, most normal tissues are more or less sensitive to the effects of drugs at specific times of the day. Some tumors also exhibit similar rhythmic susceptibility patterns during the circadian cycle. The availability of portable infusion pumps capable of delivering multiple drugs at particular times of day has made clinical application and testing of these principles possible.

VI. SUMMARY

Cancer strikes over a million Americans each year. While metastatic cancer was once uniformly fatal, the majority of patients with lymphoma, testicular cancer, choriocarcinoma, and acute leukemia are now curable with chemotherapy. Chemotherapy also prolongs survival and provides symptomatic relief for patients with many other types of cancer. Discovery of cancer chemotherapy drugs has resulted from serendipity, empiricism, and rational drug design. Most chemotherapy drugs currently in use target cancer cell division. Unfortunately, they also harm normal cells that divide rapidly and therefore cause several side effects. Patients who are active and have no major medical problems other than cancer generally derive the most benefit from chemotherapy and experience the fewest side effects.

Despite great advances in cancer treatment in recent decades, several major challenges lie ahead. First, we need to develop novel therapeutic strategies that target hallmarks of cancer other than rapid cell division, including invasion and metastasis, angiogenesis, abnormal energy metabolism, decreased apoptosis, immortality, abnormal gene expression, and abnormal cell structure. Second, we need to develop ways of applying conceptual breakthroughs in molecular biology to the diagnosis and treatment of cancer. This process of bridging the gap between the laboratory and the patient will require continued, intensive research. Third, we need to determine how to combine chemotherapy with surgery and radiation optimally, a process that will require constant clinical empiricism. Finally, we need to learn how to prevent cancer through conceptually simple, yet elusive, measures such as smoking cessation, which alone would eliminate the vast majority of lung cancers and greatly reduce the incidence of several other common malignancies.

BIBLIOGRAPHY

Abeloff, M. D., Armitage, J. O., Lichter, A. S., and Niederhuber, J. E. (eds.) (1995). "Clinical Oncology." Churchill Livingstone, New York.

Bertino, J. R. (ed.) (1996). "Encyclopedia of Cancer." Academic Press, San Diego.

Chabner, B. A., and Longo, D. L. (eds.) (1996) "Cancer Chemotherapy and Biotherapy: Principles and Practice," 2nd Ed. Lippincott-Raven, Philadelphia.

DeVita, V. T., Hellman, S., and Rosenberg, S. A. (eds.) (1993). "Cancer: Principles and Practice of Oncology," 4th Ed. Lippincott, Philadelphia.

Haskell, C. M. (ed.) (1995). "Cancer Treatment," 4th Ed. Saunders, Philadelphia.

Peckham, M., Pinedo, H., Veronesi, U. (eds.) (1995). "Oxford Textbook of Oncology." Oxford University Press, Oxford/New York.

Shimkin, M. B. (1977). "Contrary to Nature: Being an Illustrated Commentary on Some Persons and Events of Historical Importance in the Development of Knowledge Concerning Cancer." U.S. Department of Health, Education, and Welfare, Washington, D.C.

Childbirth

VANDA R. LOPS
University of California, San Diego

GLOSSARY

Multigravida Woman who is pregnant for the second, or more, time, also referred to as Gravida 2, 3, 4, etc. (GII, GIII, GIV, etc.)

Primagravida Woman who is pregnant for the first time, also referred to as Gravida 1 (GI)

CHILDBIRTH IS THE END RESULT OF LABOR, WHICH is defined as the process by which the products of conception—fetus, amniotic fluid, placenta, and membranes—are expelled from the uterus through the vagina to the outside world. Variously referred to as accouchment, confinement, and/or parturition, it is the culmination of a gestational period, which can last from 19 days in the mouse to over 1 year in the elephant. The purpose of this article is to present a description of the childbirth process as it occurs in the human gravida after a gestational period of 10 lunar months, 9 calendar months, or 280 days. This description will include a discussion of the theory as to why and how labor begins and the various stages of labor, as well as the factors determining the successful outcome of labor, the powers (uterine contractions), passage (bony pelvis), and the passenger (fetus).

I. LABOR

Labor is divided into three stages. Stage I begins with the first uterine contraction, which causes the cervix to dilate and efface. It continues on to full dilatation, 10 cm or 5 finger breadths in diameter, and complete effacement, during which the cervix shortens and thins, becoming continuous with the lower uterine segment (Fig. 1B).

Stage II begins with full dilatation and ends with the delivery of the infant. The period of time from the delivery of the infant to the delivery of the placenta is referred to as Stage III. Many authorities include a fourth stage during which the mother's physical conditions, vital signs, bleeding, and so on stabilize after the demanding process of birth.

As previously stated, labor usually begins after a gestational period of approximately 10 lunar months or 40 weeks. Labor beginning prior to 37 weeks of gestation is referred to as premature. Eutocia is the term applied to normal labor in which cervical dilatation and descent of the fetus follow accepted time guidelines (Fig. 2). Dystocia or difficult, complicated labor refers to labor in which the dilatation pattern and/or descent of the fetus are abnormally protracted or cease altogether. A possible cause of this is cephalopelvic disproportion, a situation in which either the fetus is too large to fit through the bony pelvis or, conversely, the bony pelvis is too small to accommodate the fetus.

Theories explaining the initiation of labor abound, but the exact mechanism or sequence of events that cause labor to begin are still unclear. It is known, however, that the process is facilitated by the interaction of a number of hormones, including oxytocin, prostaglandins, and the steroid hormones estrogen and progesterone. The most currently accepted theory is that the combination of contractile protein (actinomysin) in the muscle cell and the peaking of stimulatory factors such as prostaglandins, oxytocin, and oxytocin receptors overcome the relaxing action of progesterone on the uterine muscle, thus causing the

ENCYCLOPEDIA OF HUMAN BIOLOGY, Second Edition, VOLUME 2. Copyright © 1997 by Academic Press. All rights of reproduction in any form reserved.

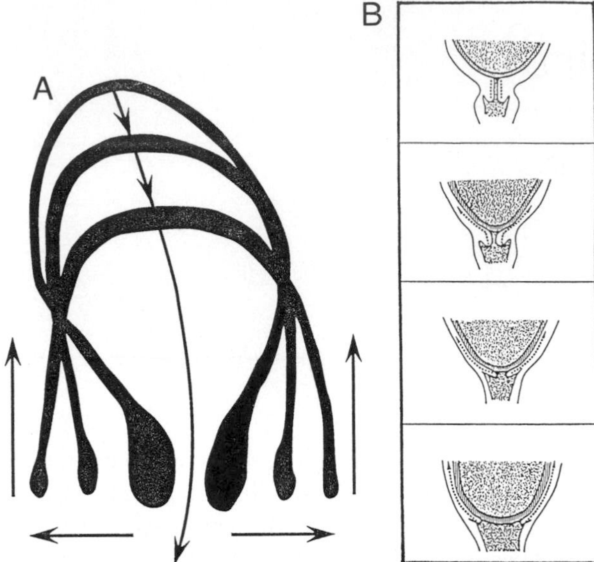

FIGURE 1 Sagittal (A) and anterior–posterior (B) views of dilatation and effacement. [Adapted with permission from J. A. Ingalls and C. N. Salerno (1983). "Maternal and Child Health Nursing." Mosby, St. Louis.]

muscle to begin to contract. [See Uterus and Uterine Response to Estrogen.] A schematic representation of this theory is presented in Fig. 3.

II. STAGES OF LABOR

The first stage of labor can be divided into a latent and active phase. The latent phase, which extends from the beginning of labor to approximately 3 cm of dilatation, is characterized by cervical effacement rather than aggressive dilatation. Most dilatation occurs during the active phase, which lasts from 3 cm to approximately 9–10 cm.

A short deceleration phase follows the active phase, ending with full dilatation, and is characterized by progress in the descent of the presenting part, that is, that part of the fetus coming down through the bony pelvis first. Many believe that this phase is integrated into and is a part of Stage II. It must be remembered that although cervical dilatation and effacement are viewed by many as progress in labor, true progress must also include descent of the presenting part through the bony pelvis.

During pregnancy and labor, the fetus floats in amniotic fluid within the fetal membranes, or bag of waters. These membranes can spontaneously rupture at any time during labor or, in some cases, be artificially ruptured to stimulate labor forces. This closed fetal environment serves not only as a protection against infection but also as a cushion against the stress of uterine contractions during labor. Accepted time guidelines for this stage are shown in Fig. 2 and are one factor utilized in assessing the normalcy of a labor pattern.

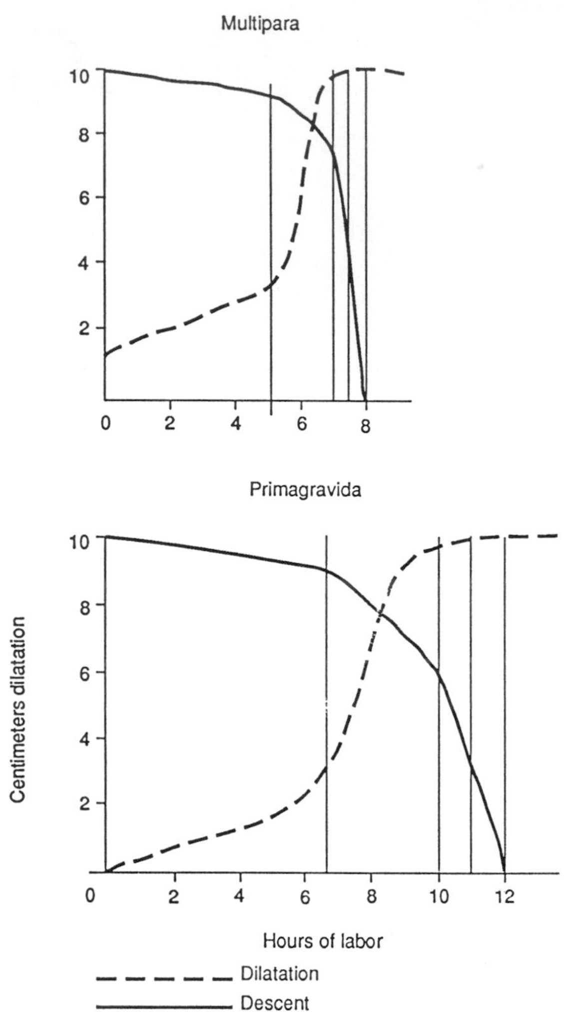

FIGURE 2 Mean labor curves. [Adapted from E. A. Friedman (1978). "Labor: Clinical Evaluation and Management," 2nd Ed. Appleton–Century–Crofts, New York.]

Stage II extends from full dilatation to the delivery of the infant. This is truly the working stage of labor, for the parturient, with each contraction will experience the intense urge to bear down, thus assisting in the descent of the fetus through the birth canal. This

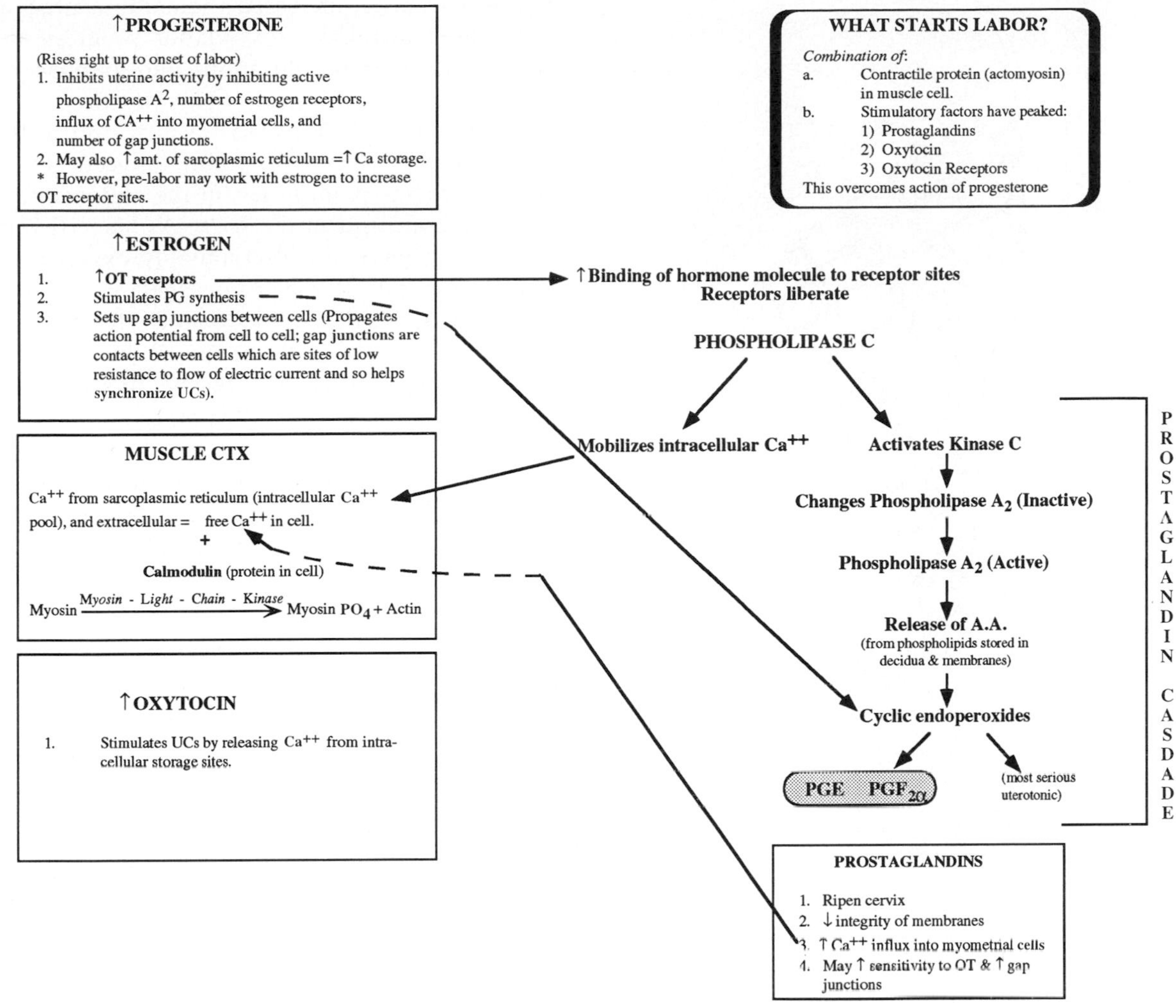

FIGURE 3 The initiation of labor.

stage may last as long as 2–3 hr in the primagravida or 30 min or less in the multigravida.

The last stage of labor, Stage III, begins with the delivery of the infant and ends with the delivery of the placenta and membranes. After delivery of the infant, uterine contractions temporarily cease. When they resume, they are usually painless and directed toward causing the placenta to separate from the uterine wall and be delivered. As the uterus contracts, there is a sudden and dramatic decrease in the size of the placental site. Because the placenta is semirigid, it cannot shrink to compensate for this and, thus, begins to detach from the uterine wall. [See Placenta.]

Once completely separated, the placenta takes the same route as the fetus down through the birth canal to the outside world. This emptying of the uterus allows contractions of interlacing fibers of the myometrium, in turn causing constriction of the large maternal blood vessels, which had provided circulation to the placental site. This results in maternal hemostasis. (Fig. 4) It lasts approximately 5–10 min; however, if the placenta has not been delivered by 30 min, appropriate intervention to assist in the process is warranted.

III. THE POWERS, PASSAGE, AND PASSENGER

A. Powers

During pregnancy, uterine contractions are isometric and irregular in duration, frequency, and intensity. In

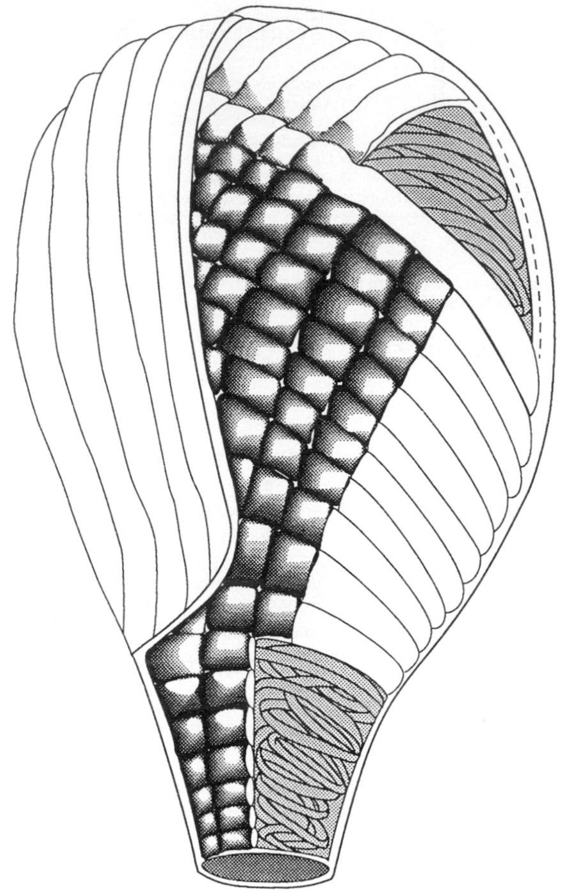

FIGURE 4 Interlacing of the uterine muscle fibers. [Adapted with permission from J. S. Malinowski, C. G. Pedigo, and C. R. Phillips (1989). "Nursing Care during the Labor Process," 3rd Ed. F.A. Davis, Philadelphia.]

labor, these contractions achieve a marvelous sense of coordination and regularity characterized by a contraction and relaxation phase, with the uterus always maintaining a certain degree of resting muscle tone. Relaxation between contractions is essential for continued fetal oxygenation, for resting the uterine muscle, and for maternal rest. An important feature of uterine contractions is that they achieve fundal dominance, that is, they are strongest in the upper portion of the uterus (fundus), and their intensity and duration diminish as the contraction wave moves down the body (corpus) of the uterus.

In early Stage I, the uterine contractions when measured in mm Hg are mild (15–30 mm Hg), lasting anywhere from 15 to 20 sec and occurring every 10–20 min. These contractions follow a regular pattern and can be accompanied by abdominal cramps, backache, increased vaginal discharge, and/or ruptured membranes. As labor progresses they increase in strength, duration, and frequency, so that by late Stage I into Stage II, they can be >50–80 mm Hg in intensity, lasting >80–100 sec, and occurring every $1\frac{1}{2}$–3 min. Subjectively, they are perceived as painful, felt in the abdomen above the pubic bone and in the back. As Stage II begins, the laboring patient will commonly experience the most discomfort. If the membranes have not ruptured previously, they may do so at this point. The contractions may be accompanied by severe low backache, leg cramps due to the pressure of the descending fetus, and a stretching sensation deep in the pelvis, with the beginning of the urge to bear down. Ironically, most women experience relief from the discomfort once they begin, and continue to bear down with each contraction.

B. Passage

Although the passage includes not only the bony pelvis but the vagina and introitus as well, most concern revolves around the shape and size of the former, since both the vagina and introitus are distensible and can be enlarged surgically if necessary. The bony pelvis, however, does not distend or significantly increase its diameters during the labor process.

The bony pelvis consists of the inlet, or entrance, much like a door leading into a room; the cavity, which can be compared to the spatial diameters of the room; and an outlet, or exit, much like the door leading from the room. Each of these components has diameters through which the fetal head must fit. Figure 5 demonstrates the actual numerical value in centimeters of these diameters. It is interesting to note that the largest diameter of the inlet is in the transverse, the average midpelvic diameter is about 10–12 cm, and the pelvic outlet is largest in its anterior posterior plane. Bony pelvic assessment via internal examination is done in early labor when the diameters can be compared to the estimated fetal size. At present, unless dramatically contracted pelvic diameters are found, and regardless of fetal size, all parturients are given a trial of labor.

C. Passenger

The most usual presentation of the fetus to the birth canal is the vertex presentation, in which the

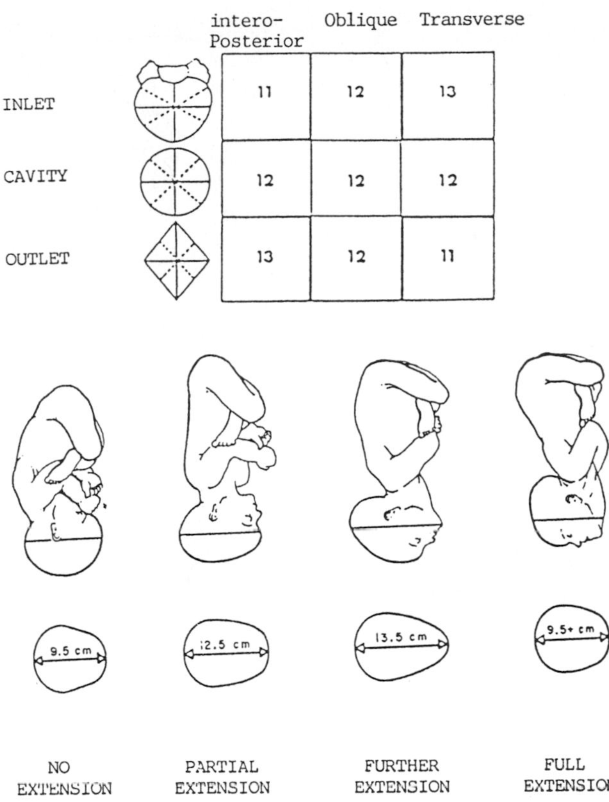

FIGURE 5 Comparison of pelvic diameters and fetal head diameters with degrees of extension. [Adapted from H. Oxorn and W. R. Foote (1986). "Human Labor and Birth." Appleton–Century–Crofts, New York.]

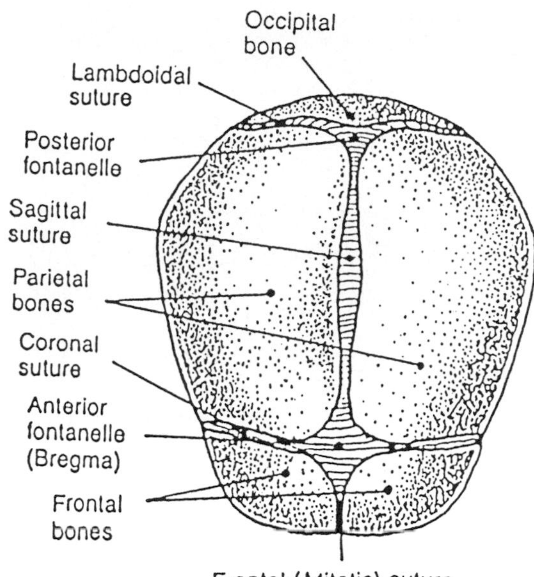

FIGURE 6 Superior view of the fetal skull. [Reproduced with permission from J. A. Ingalls and C. M. Salerno (1983). "Maternal and Child Health Nursing." Mosby, St. Louis.]

head (vertex) or, more specifically, the occipital region of the head enters the bony pelvis inlet first. The fetal head is also described in terms of diameters. It must be remembered that the bones of the fetal head are not fused as in the adult. Figure 6 gives the nomenclature for the various bony processes as well as the dividing lines (sutures) and areas (fontanels) between them. The fontanels, commonly referred to as the "soft spots," are two in number and differ in shape. The diamond-shaped anterior fontanel and the triangular-shaped posterior fontanel provide the practitioner with landmarks, via vaginal examination, to determine the actual position or placement of the vertex in relation to the quadrants of the bony pelvis during labor. In the most usual position, the fetal head is well flexed, chin to chest. This presenting diameter (suboccipital bregmatic) from the occiput to the anterior fontanel is the smallest diameter of the fetal head and, thus, the most favorable to fit through the various pelvic diameters. As the head deflexes, the presenting diameter, which must maneuver through the bony pelvis, changes. Figure 5 allows comparison the possible fetal vertex diameters with the bony pelvic diameters. It also clearly demonstrates that certain positions of the fetal head will result in presenting diameters greater than the pelvic diameters, thus impeding the ability of the head to descend through the bony pelvis, resulting in the earlier mentioned cephalopelvic disproportion.

D. Cardinal Movements of Labor

The raison d'être of the mechanisms of labor is that the smallest fetal diameters will attempt to conform to the largest pelvic diameters, so that the passage through the birth canal is accomplished as easily and efficaciously as possible.

Engagement, the first mechanism, is said to occur when the lowermost bony portion of the fetal vertex has reached the midpoint between the inlet and outlet of the pelvis (Fig. 7). As the vertex continues to descend, flexion of the head to the chest, if not complete, will also continue. As these two mechanisms are occurring, the vertex is in a transverse position (OT) with the fetal face looking out across the pelvis. Upon

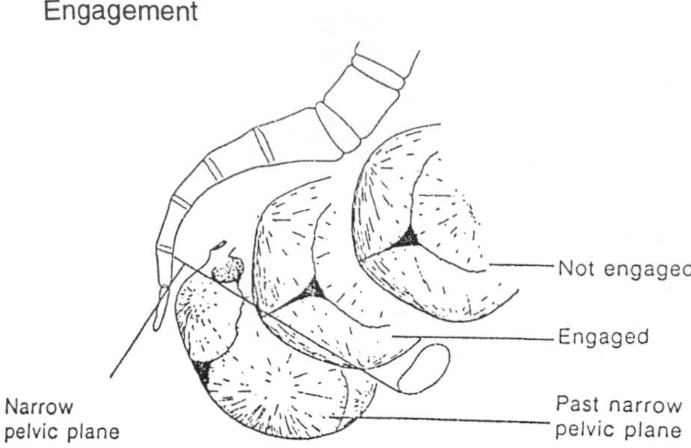

FIGURE 7 The engagement mechanism in labor.

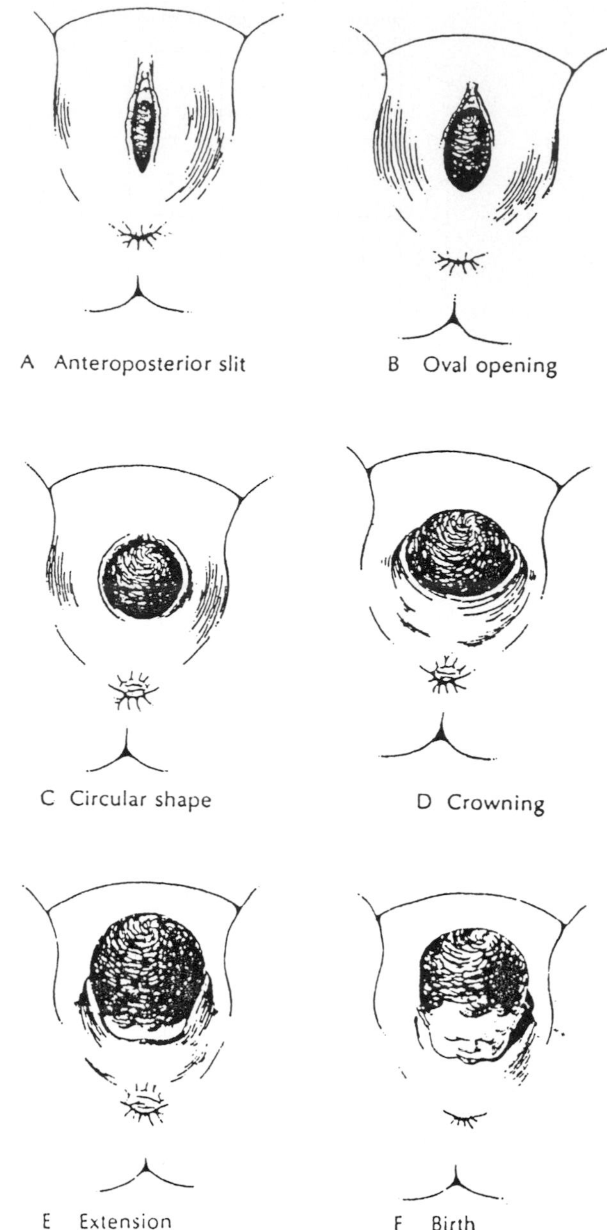

FIGURE 8 Dilatation of the introitus due to extension of the head. [Reprinted from H. Oxorn and W. R. Foote (1986). "Human Labor and Birth." Appleton–Century–Crofts, New York.]

reaching the midpelvis and muscles of the pelvic floor, which offer resistance to further descent, the vertex begins to pivot or rotate away from these obstacles, seeking the path of least resistance. It thus rotates 90° so that the vertex is no longer in the transverse but is now in the anterior–posterior position, most commonly with the occiput or back of the head anterior and the fetal face toward the mother's spine. Because of this internal rotation of the vertex, the shoulders are now in the transverse so that their diameter as they enter the pelvis is passing through the largest diameter of the pelvic inlet.

The fetal head continues to descend through the pelvis under the forces of the uterine contractions and the bearing down efforts of the laboring patient. The next mechanism, extension, occurs when the vertex is at the maternal perineum. At this point, the birth canal slopes upward, so that with each contraction and maternal pushing force, the chin slowly extends away from the chest as if attempting to avoid the resistance offered by the muscles and soft tissue of the maternal pelvic floor. Figure 8 demonstrates how increasingly more of the fetal head becomes visible at the introitus as extension occurs.

Once extension is complete, the head is delivered and will begin an external 90° rotation maneuver that duplicates internal rotation, which will ultimately place it in the same position (OT) in which it first engaged (Fig. 9). In completing this, the shoulders, which had previously engaged in a transverse position, will now, because of this external rotation of the head, internally rotate and assume an anterior–posterior position to pass easily between the ischial spines. They, and the rest of the infant, then deliver spontaneously via the mechanism of expulsion (see Color Plate 15).

This then is the childbirth process. It is the phenomenon by which each generation duplicates itself, beguiling in its simplicity and rationality and overwhelming in its perfection.

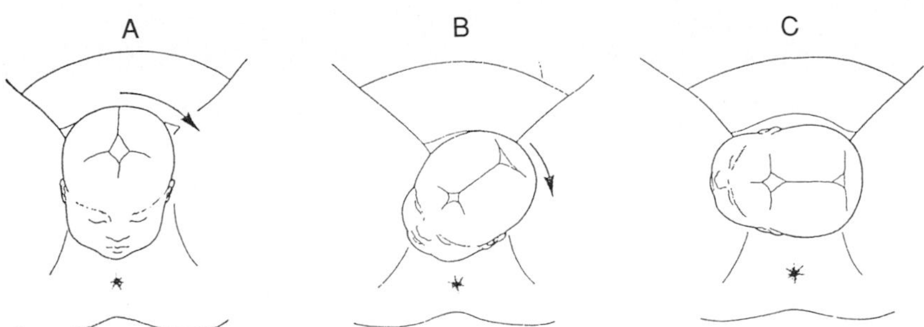

FIGURE 9 Restitution and external rotation from the left occiput anterior (LOA) position. (A) The head is born by extension and restitutes 45° to undo the twist on the neck. (B) The head externally rotates a further 45° in the same direction as restitution, owing to internal rotation of the shoulders through 45°. External rotation, although named for the head, is effected by the shoulders and if prevented by the accoucheur can result in impacted shoulders. (C) The face is directed laterally and the shoulders now lie in the anteroposterior diameter of the pelvic outlet.

BIBLIOGRAPHY

Albers, L. L., Schiff, M., and Gorwoda, J. C. (1996). *Obstet. Gynecol.,* **87**(3), 355–359.

Aldrich, C. J., D'Antona, D., Spencer, J. A., Wyatt, J. S., Peebles, D. M., Deply, D. T., and Reynolds, E. O. (1995). The effect of maternal pushing on fetal cerebral oxygenation and blood volume during the second stage of labor. *Br. J. Obstet. Gynaecol.* **102**(6), 448–453.

Ciray, H. N., Guner, H., Hakansson, H., Tekelioglu, M., Roomans, G. M., and Ulmsten, U. (1995). Morphometric analysis of gap junctions in non-pregnant and term pregnant human myometrium. *Acta Obstet. Gynecol. Scand.* **74**(7), 497–504.

Creasy, R., and Resnik, R. (1989). "Maternal-Fetal Medicine," 2nd Edition. W. B. Saunders Co., Philadelphia.

Cunningham, F. G., MacDonald, P. C., Gant, N. F., Leveno, K. J., Gilstrap III, L. C., Hankins, G. D. V., and Clark, S. L. (1997). "Williams Obstetrics," 20th ed. Appleton & Lange, Stamford.

Flynn, A. M., Kelly, J., Jollins, G., and Lynch, P. F. (1978). Ambulation in labour. *Br. Med. J.* **2**, 591–593.

Friedman, E. (1970). An objective method of evaluating labor. *Hosp. Pract.,* **827**.

Friedman, E. (1970). The functional division of labor. *Am. J. Obstet. Gynecol.* **109**, 274–280.

Fuchs, A. R., Fuchs, F., Hysslcin, P., and Soloff, M. S. (1984). Oxytocin receptors in the human uterus during pregnancy and parturition. *Am. J. Obstet. Gyn.* **150**, 734.

Gomez, R., Ghezzi, F., Romero, R., Munoz, H., Tolosa, J. E., and Rtojas, I. (1995). Premature labor and intraamniotic infection. *Clin. Perinatol.* **22**(2), 281–342.

Gough, G. W., Randall, N. J., Generier, E. S., *et al.* (1990). Head to cervix forces and their relationship to the outcome of labor. *Obstet. Gynecol.* **75**(4), 613–618.

Hendricks, C., and Brenner, W. (1970). Normal cervical dilatation patterns in late pregnancy and labor. *Am. J. Obstet. Gynecol.* **106**, 1105.

Higuchi, T. (1995). Oxytocin: a neurohormone, neuroregulator, paracrine substance. *Jpn. J. Physiol.* **45**(1), 1–21.

Kilpatrick, S. J., and Laros, R. K. (1989). Characteristics of normal labor. *Obstet. Gynecol.* **74**(1), 85–87.

Kredentser, J. V., Embree, J. E., and McCoshen, J. A. (1995). Prostaglandin F2 alpha output by amnio-choriondecidua: relationship with labour and prostaglandin E2 concentration at the amniotic surface. *Am. J. Obstet. Gynecol.* **173**(1), 199–204.

Maesel, A. (1990). Cerebral blood flow during labor in the human fetus. *Acta Obstet. Gynecol. Scand.* **69**(6), 493–495.

Moore, T., Reiter, R. C., Rebar, R. W., and Baker, V. V. (1993). "Gynecology and Obstetrics: A Longitudinal Approach." Churchill Livingstone, New York.

Roberts, J., Mendez-Bauer, C., and Woodell, D. A. (1983). The effects of maternal position on uterine contractility and efficiency. *Birth* **10**(4), 243–249.

Roberts, J., and Wooley, D. (1996). A second look at the second stage of labor. *JOGNN* **25**(5), 415–423.

Romero, R., Brody, D., Oyarzun, E., Mazor, M., Wu, Y., Hobbins, J., and Durum, S. (1989). Infection and labor: III Interleukin-1: A signal for the onset of parturition. *Am. J. Obstet. Gynecol.* **160**, 1117–1123.

Schneider, H., Progler, M., Ziegler, W. H., and Hunch, R. (1990). Biochemical changes in the mother and the fetus during labor and its significance for the management of the second stage. *Int. J. Gynecol. Obstet.* **31**(2), 117–126.

Cholesterol

N. B. MYANT
Royal Postgraduate Medical School, Hammersmith Hospital, London

GLOSSARY

Apolipoprotien B Protein component of low-density lipoprotein (LDL) responsible for the recognition and binding of an LDL particle by an LDL receptor

Chylomicrons Triglyceride rich particles secreted by the small intestine into the blood circulation during the absorption of fat

Foam cells Cells derived from macrophages and containing cytoplasmic droplets filled with esterified cholesterol

High-density lipoprotein Small, dense lipoprotein particle secreted into the plasma in nascent form by the liver and intestine

HMG-CoA reductase Membrane-bound enzyme that catalyzes the reduction of HMG-CoA to mevalonic acid, the rate-limiting step in the synthesis of cholesterol from acetyl-CoA

3-Hydroxy-3-methylglutaryl-CoA (HMG-CoA) Substrate for HMG-CoA reductase

Low-density lipoprotein Lipoprotein that carries most of the cholesterol in human plasma

Low-density lipoprotein receptor Glycoprotein attached to the plasma membrane of most cells and possessing a specific region that binds lipoproteins containing apolipoprotein B or apolipoprotein E in a suitable conformation

Mevalonic acid Branched C_6 compound, formed by the reduction of HMG-CoA, that serves as intermediate in the synthesis of cholesterol and of several essential nonsterol metabolites

Remnant particle Cholesterol-rich particle produced by the partial intravascular degradation of chylomicrons and very low-density lipoproteins

Very low-density lipoprotein Triglyceride-rich lipoprotein secreted into the plasma by the liver

CHOLESTEROL IS AN ESSENTIAL COMPONENT OF the membranes of all animal cells, of the myelin sheaths of nerves, and of the outer shell of plasma lipoprotein particles. In certain specialized cells, cholesterol acts as precursor of bile acids and steroid hormones. Most cells satisfy the bulk of their cholesterol requirements for maintenance, growth, and specialized functions by two regulated processes: (1) the intracellular synthesis of cholesterol from small molecules and (2) the receptor-mediated uptake, from the external medium, of cholesterol-rich particles called low-density lipoprotein(s) (LDL). LDL is the terminal product of the intravascular metabolism of very low-density lipoprotein (VLDL), a lipoprotein secreted by the liver.

The body as a whole acquires exogenous cholesterol by absorption of dietary cholesterol from the lumen of the small intestine. Cholesterol absorbed from the intestine is secreted into the intestinal lymphatics and, thence, into the bloodstream, as a component of large, fat-enriched particles called chylomicrons. As soon as they enter the circulation, chylomicrons are degraded by lipoprotein lipase (an enzyme bound to the luminal surfaces of the blood capillaries) to produce cholesterol-rich "remnant" particles, from which most of the fat has been removed. Chylomicron remnants are taken up by the

ENCYCLOPEDIA OF HUMAN BIOLOGY, Second Edition, VOLUME 2. Copyright © 1997 by Academic Press. All rights of reproduction in any form reserved.

liver and degraded to release free cholesterol within liver cells. Some of this cholesterol is reexcreted into the intestine in the bile and is then reabsorbed after mixing with dietary cholesterol. Thus, biliary and dietary cholesterol participate together in an enterohepatic circulation. Dietary cholesterol, taken up by the liver as chylomicron remnants, inhibits hepatic synthesis of cholesterol. As a result of this regulatory mechanism, and of the inhibition of cholesterol synthesis in nonhepatic tissues by the uptake of LDL, the amount of cholesterol in the whole body of a fully grown individual remains roughly constant from day to day.

I. STRUCTURE AND PROPERTIES

Cholesterol is a member of a class of naturally occurring compounds called sterols. Figure 1 shows the structural formula of cholesterol. The cholesterol molecule has a nucleus of four fused rings common to all sterols, methyl groups at C-10 and C-13, a branched side chain, an OH group at C-3, and a double bond at C-5. The presence of the polar group at C-3 enables cholesterol to interact with other lipids such as fatty acids and phospholipids. Pure cholesterol is a solid at body temperature (melting point 150°C) and is essentially insoluble in water. However, the cholesterol in normal cells and in plasma is prevented from crystallizing by its association with phospholipids, whereas cholesterol in normal bile is held in micellar solution by the detergent action of bile salts and lecithin (see Section IV). The presence of the nuclear double bond makes the ring system flatter than it would be otherwise. This flattening effect facilitates the insertion of cholesterol between the molecules of a biological membrane and may have played a part in the natural selection of cholesterol as the predominant sterol of vertebrates. [*See* Lipids.]

Cholesterol

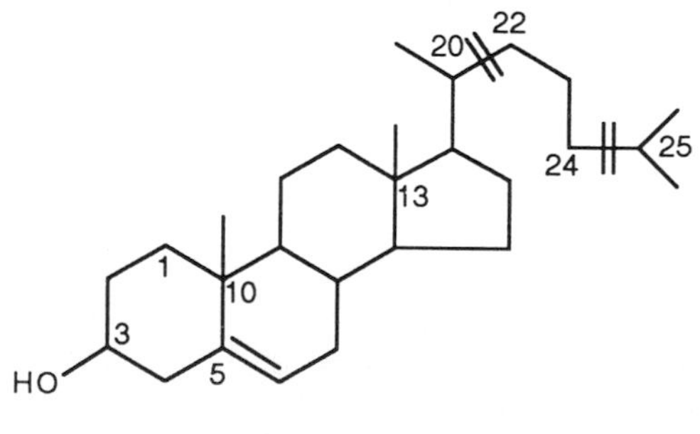

FIGURE 1 Structural formula of cholesterol ($C_{27}H_{45}O$) drawn with the side chain at upper right in folded conformation. Carbon atoms mentioned in the text are numbered. The H atoms attached to carbons are omitted; strokes at C-10, -13, -20, and -25 are methyl groups. Double bars show the positions of cleavage in the formation of bile acids and steroid hormones.

II. FREE AND ESTERIFIED CHOLESTEROL

Most of the cholesterol in the body as a whole is present in free, or unesterified, form; however, in certain tissues and in plasma, a significant proportion of the total is esterified at the 3 position with long-chain fatty acids (those with 12 or more carbon atoms). Cholesterol esterified with fatty acids is less polar than free cholesterol because the 3-OH group is masked by the fatty acyl chain. Esterification of cholesterol within cells is catalyzed by the enzyme acyl-CoA–cholesterol acyltransferase (ACAT), so called because it transfers the fatty acid residue of acyl-CoA to cholesterol. The CoA ester of oleic acid is the preferred acyl substrate for ACAT. Cholesteryl esters are formed in plasma by the transfer of a fatty acid residue from lecithin (a phospholipid) to cholesterol. The enzyme catalyzing this reaction is called lecithin–cholesterol acyltransferase (LCAT). The preferred fatty acid residue for human LCAT is linoleic acid.

III. DISTRIBUTION

A. In the Whole Body

Cholesterol is present in all tissues, but the concentration varies widely among different tissues, from <100 mg/100 g fresh weight in adipose tissue and muscle to >3 g/100 g in brain, nerve, and adrenal glands. An adult human has about 1 g of cholesterol per kg body weight, corresponding to some 60 g in the whole body. About 1 g of cholesterol is lost from the body each day by conversion into bile acids and steroid hormones, and by fecal excretion of unabsorbed biliary cholesterol. In the steady state, this loss is exactly balanced by endogenous synthesis and absorption of dietary cholesterol. Nearly one-fifth of all the cholesterol in the body is present in the myelin sheaths of nerve fibers. Despite its importance in health and disease, the cholesterol in plasma accounts for $<10\%$ of the total in the body. More than 80% of the total cholesterol in the body is unesterified. This reflects the predominant role of cholesterol as a membrane component (see Section IV). In plasma and adrenal glands, about 70% of the total cholesterol is esterified with fatty acids.

B. Within Cells

In most cells, cholesterol is present largely as the free sterol in the plasma membrane and, to a much smaller extent, in the endoplasmic reticulum and the mitochondrial and lysosomal membranes. In the cells of most tissues, small amounts of esterified cholesterol are dispersed throughout the cytoplasm. In contrast, the hormone-producing cells of the adrenal cortex contain large quantities of esterified cholesterol confined within cytoplasmic droplets. When these cells are called upon to secrete steroid hormone, cholesteryl esters in the storage droplets are rapidly hydrolyzed to release free cholesterol (see Section V,B). Cytoplasmic droplets filled with esterified cholesterol also occur in cells of the macrophage class, giving them a foamy appearance. Focal accumulations of foam cells are responsible for the presence of raised fatty streaks in the arterial wall. Fatty streaks may be the precursors of atherosclerotic lesions.

C. Developmental Aspects

The total amount of cholesterol in the human body increases throughout fetal and postnatal development. This increase is due mainly to the laying down of new cell membranes required to keep pace with the multiplication and growth of cells in all tissues. In addition, there is a relative increase in the cholesterol content of certain tissues at particular stages of development, as in the developing brain. Myelination in the white matter of the human brain is most rapid from the seventh month of fetal life to the end of the first few months of postnatal life. During this period, the concentration of cholesterol in whole brain increases twofold and the total amount of brain cholesterol increases severalfold.

IV. CHOLESTEROL AS A COMPONENT OF MEMBRANES

The basic component of all biological membranes is a double layer of phospholipid molecules arranged with their fatty acid chains parallel to one another and facing inward into the hydrophobic interior of the bilayer (Fig. 2).

The physical properties of an artificial phospholipid bilayer containing no sterol are such that it could not function efficiently as a biological membrane. At temperatures below a critical phase-transition temperature (T_c) the membrane would exist in a "crystalline" state in which the fatty acid chains of the phospholipids are immobilized. In this state, permeability to solutes is limited and the viscosity of the bilayer is too high to permit lateral diffusion of proteins and other macromolecules. At temperatures above the T_c, the bilayer exists in an unstable "liquid-crystalline" state in which the fatty acid chains are freely mobile, the viscosity of the bilayer is greatly reduced, and the phospholipids are held together only by noncovalent interaction between their polar head groups.

When free cholesterol is added to a phospholipid bilayer, the cholesterol molecules enter the spaces between the phospholipid molecules, interacting noncovalently with their fatty acid chains. The effect of this is to stabilize the bilayer, over a wide range of temperature, in an intermediate-gel state. In this state, the first 8–10 carbon atoms of the fatty acid chains are immobilized and the remaining carbons are freely mobile (Fig. 2). The cholesterol–phospholipid ratio

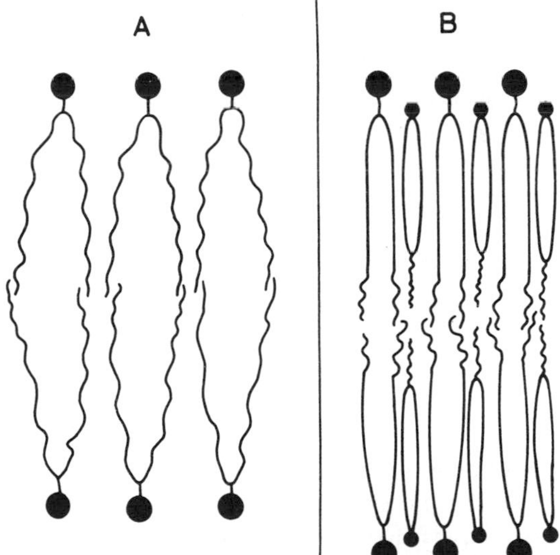

FIGURE 2 Effect of cholesterol on the physical state of a phospholipid bilayer at a temperature above the phase-transition temperature. In the absence of cholesterol (A), the chains are fluid. In the presence of cholesterol (B) (50 moles %), the first 8–10 carbon atoms of the chains are rigid and the bilayer is condensed. The cholesterol molecules are shown with their hydroxyl groups (•) close to the head groups of the phospholipids (●).

required for optimal functioning of cell membranes under physiological conditions varies from one type of membrane to another. In humans, there is about one molecule of cholesterol for every phospholipid molecule in the plasma membrane of the red cell, whereas in mitochondrial membranes the cholesterol–phospholipid molar ratio is <0.1.

In addition to influencing the physical state of the bulk phospholipids of a cell membrane, free cholesterol may also affect the fluidity of a limited zone of membrane in the immediate vicinity of a membrane-bound enzyme. This may alter the conformation of the enzyme and, thus, affect its state of activation. [*See* Membranes, Biological.]

V. CONVERSION INTO BILE ACIDS AND STEROID HORMONES

A. Bile Acids

1. Formation and Secretion of Bile Salts

Bile acids are formed from unesterified cholesterol in the liver by a series of enzymatic reactions culminating in the oxidative cleavage of the side chain between C-24 and C-25 (see Fig. 1). The C_{24} bile acids resulting from these reactions are converted into bile salts by conjugation with glycine or taurine. The bile salts, together with free cholesterol and lecithin, are secreted in the bile. In normal bile, all the free cholesterol is dissolved by the combined detergent actions of bile salts and lecithin. The bile is stored in the gallbladder in the fasting state and is discharged into the duodenum and, thence, into the jejunum, in response to a fatty meal. [*See* Bile Acids.]

2. Enterohepatic Circulation

After participating in the absorption of fat and cholesterol from the jejunum, about 95% of the bile salts secreted by the liver are reabsorbed from the ileum into the blood circulation and are taken up by the liver, to be resecreted with newly synthesized bile salts. The small quantity that is not reabsorbed is excreted in the feces. Bile salts taken up by the liver down-regulate the synthesis of bile acids from cholesterol by feedback inhibition.

To maintain a constant amount of bile salt in the biliary system and small intestine, the liver makes just enough bile acid to replace the daily loss in the feces. In the presence of a normal enterohepatic circulation, this is equal to about 400 mg/day. This rate of synthesis is equivalent to nearly half the total daily turnover of cholesterol (about 1 g/day in humans). If the reabsorption of bile salts is diminished, the conversion of cholesterol into bile acids is partially released from feedback inhibition, so that bile acid synthesis increases to keep pace with increased fecal loss. In the extreme case, when all the bile is diverted from the small intestine by an external fistula, the rate of conversion of hepatic cholesterol into bile acid increases 10-fold; consequently, the liver is partially depleted of cholesterol. This results in an increase in hepatic LDL-receptor activity.

B. Steroid Hormones

Steroid hormones are formed from unesterified cholesterol in the gonads, placenta, and adrenal cortex. The initial step is the cleavage of the cholesterol side chain between C-20 and C-22 (see Fig. 1) to give pregnenolone, the precursor of all steroid hormones. The free cholesterol used as substrate for the cleavage reaction is derived from the intracellular stores of esterified cholesterol (referred to in Section III,B). Release of free cholesterol is mediated by the hydrolytic action of an enzyme (cholesteryl ester hydrolase), whose activity is regulated by adrenocorticotrophic hormone in the adrenal cortex and by gonadotrophic hormones in the gonads. The amount of cholesterol converted into steroid hormones is much less than the amount converted into bile acids. In a normal man, <50 mg of cholesterol is converted into steroid hormones each day. During the luteal phase of the sexual cycle in women, 50–100 mg of cholesterol per day may be used for hormone production. In the adrenal cortex and gonads, the stores of cholesterol used for hormone production are replenished partly by intracellular synthesis of cholesterol, but mainly by receptor-mediated uptake of LDL from the external medium. [*See* Steroids.]

VI. SYNTHESIS

A. HMG-CoA and the Rate-Limiting Step

Cholesterol is synthesized in all animal cells by a pathway in which the primary precursor is acetyl-CoA, the methyl and carboxyl carbons of the acetyl residue supplying all 27 carbon atoms of the cholesterol skeleton. The initial steps in this pathway result in the formation of HMG-CoA, the CoA thioester of 3-hydroxy-3-methylglutaric acid (a branched C_6 acid).

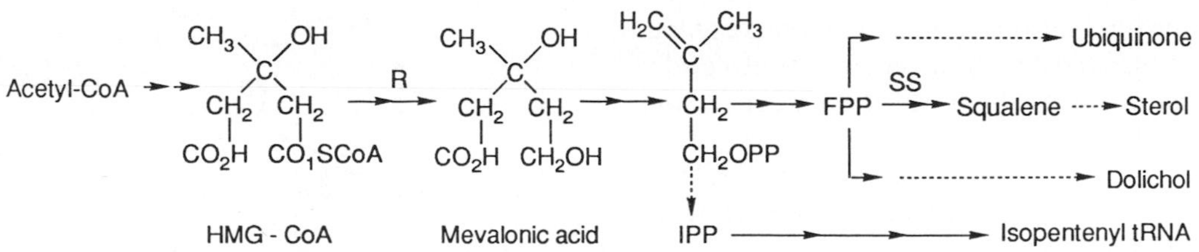

FIGURE 3 Formation of mevalonic acid and the products of its metabolism in animal cells. FPP, farnesyl pyrophosphate; IPP, isopentenyl pyrophosphate; R, reductase; SS, squalene synthetase.

HMG-CoA is reduced by the enzyme HMG-CoA reductase to mevalonic acid. This compound is then decarboxylated to give a branched C_5 isoprenoid unit (Fig. 3). The isoprenoid unit is the building block used in the formation of cholesterol and of various nonsterols, including ubiquinone, dolichols, and tRNAs containing an isopentenyl residue.

The rate at which cholesterol is synthesized in the cell is determined by the rate at which mevalonic acid is generated from HMG-CoA. This, in turn, is determined by the activity of HMG-CoA reductase, the enzyme with the lowest capacity in the biosynthetic chain leading to cholesterol.

B. HMG-CoA Reductase and the Regulation of Cholesterol Synthesis

HMG-CoA reductase is present in all animal and plant tissues. This is not surprising in view of its role in the generation of a precursor required for the formation of sterols and of several nonsterols essential for the growth and multiplication of most eukaryotic cells. The mammalian enzyme is a glycoprotein with its N-terminal portion anchored to the smooth endoplasmic reticulum and its C-terminal portion, which contains the catalytic site, projecting into the cytoplasm.

HMG-CoA reductase is a highly regulated enzyme, its activity varying over a several hundred-fold range under different conditions in the intact cell. Regulation of reductase activity is achieved by modulation of the rates of enzyme synthesis and breakdown and by changes in the state of activation of existing enzyme molecules. Reductase activity undergoes reciprocal changes in response to alterations in the amount of cholesterol present in an intracellular regulatory pool of the unesterified sterol. For example, reductase activity declines when the cholesterol content of the regulatory pool is increased by receptor-mediated uptake of LDL. The inhibitory effect of cholesterol on HMG-CoA reductase is thought to be mediated partly by oxysterols formed within the cell by the enzymatic oxidation of cholesterol. Indirect evidence suggests that oxygenated sterols such as 25-hydroxycholesterol activate a specific protein that, in its activated form, suppresses transcription of the reductase gene by binding to a short segment in the promoter region. Mitogens, growth factors, and hormones such as insulin and thyroxine also influence the rate of synthesis of cholesterol by modulating reductase activity.

The rates of synthesis of cholesterol and nonsterol metabolites of mevalonic acid are regulated independently by multivalent feedback inhibition, a process analogous to the independent regulation, in some bacteria, of the synthesis of several end products derived from a common intermediate.

VII. UPTAKE OF CHOLESTEROL BY CELLS

A. Composition, Origin, and Metabolism of LDL

Most cells in the human body use LDL as their major source of extracellular cholesterol. An LDL particle consists essentially of a hydrophobic core of triglyceride and esterified cholesterol surrounded by a polar shell of phospholipid, free cholesterol, and protein. The phospholipids and free cholesterol are arranged as a monolayer with the phospholipid head groups and the OH groups of cholesterol in contact with the aqueous medium. Each LDL particle has one molecule of a protein called apolipoprotein B (apoB) associated noncovalently with the polar shell. Each molecule of apoB has a single region that binds specifically to an LDL receptor on the surface of a cell (see Section VII,B).

All the LDL in the circulation is derived from remnant particles resulting from the incomplete degradation of VLDL by lipoprotein lipase, an enzyme attached to the luminal surfaces of the blood capillaries. VLDL is secreted by the liver as a triglyceride-rich lipoprotein. During the intravascular conversion of VLDL into LDL, cholesteryl esters, produced by the action of LCAT in the plasma and with linoleate as their predominant fatty acid, are transferred from high-density lipoprotein (HDL) to LDL.

Between one-half and two-thirds of the LDL in plasma is degraded by the LDL-receptor pathway, mainly in the liver (see Section VII,B). Most of the remainder is degraded by other routes not involving specific receptors. However, a small but significant proportion of the plasma LDL is taken up and degraded by cells of the macrophage class after the LDL has been modified by oxidation (see Section VII,C).

B. The LDL-Receptor Pathway

1. Structure and Orientation of the Receptor

The intracellular degradation of LDL by the LDL-receptor pathway begins with the binding of LDL particles to receptors present on the surfaces of most cells in the body. The LDL receptor is a transmembrane glycoprotein with the N terminus outside the cell, a single membrane-spanning segment, and a C-terminal segment projecting into the cytoplasm. The N-terminal segment contains a sequence of seven imperfect repeats that together constitute the binding domain, which binds the apoB molecule in an LDL particle. The LDL receptor also has high binding affinity for another apolipoprotein called apoE. The orientation of the receptor in relation to the plasma membrane and the probable arrangement of the seven repeats are shown in Fig. 4. Most of the LDL receptors on the surface of a cell are clustered in specialized regions called coated pits. The coated pit is a shallow depression on the cell surface with a cytoplasmic coat consisting mainly of a protein (clathrin) arranged in the form of a lattice. LDL receptors are anchored in coated pits by an interaction between the protein coat and the cytoplasmic extension of the receptor.

2. Binding, Internalization, and Degradation of LDL

LDL particles bound to receptors are carried into the interior of the cell by invagination of coated pits, which pinch off from the plasma membrane to form vesicles enclosing bound LDL (Fig. 5). The vesicles fuse with lysosomes, exposing the internalized LDL to digestion by lysosomal enzymes. Before fusion takes

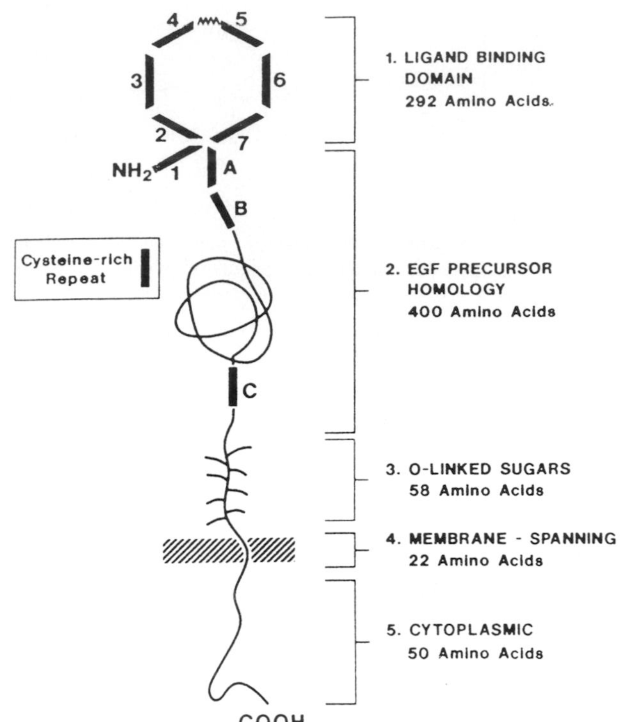

FIGURE 4 Model showing the possible arrangement of elements of the ligand-binding domain of the human LDL receptor and the orientation of the C-terminal and N-terminal portions in relation to the plasma membrane. The essential feature of the binding domain is a hexagonal structure composed of repeats 2, 3, and 4 joined to repeats 5, 6, and 7 by a linker sequence of eight amino acids. EGF, epidermal growth factor. [Reproduced, with permission from Esser *et al.* (1988). *J. Biol. Chem.* **263**, 13282–13290.]

place, the receptors in the vesicles dissociate from their bound LDL and return to the plasma membrane, where they diffuse laterally until they become anchored in coated pits, to begin a new cycle of endocytosis. Lysosomal digestion of LDL results in the complete hydrolysis of its protein component and the hydrolysis of esterified cholesterol to yield free cholesterol. The free cholesterol leaves the lysosomes and becomes available to the cell for membrane formation; conversion into bile acids, steroid hormones, or lipoproteins; or esterification with fatty acids by ACAT in the cytoplasm.

When the amount of cholesterol brought into the cell by the receptor pathway exceeds that required to satisfy its metabolic needs, HMG-CoA reductase activity is suppressed and the synthesis of new LDL receptors declines. Thus, the LDL-receptor pathway fulfills two biological functions. It is the major route for the irreversible removal of LDL from the plasma, and it provides cells with a regulated supply of choles-

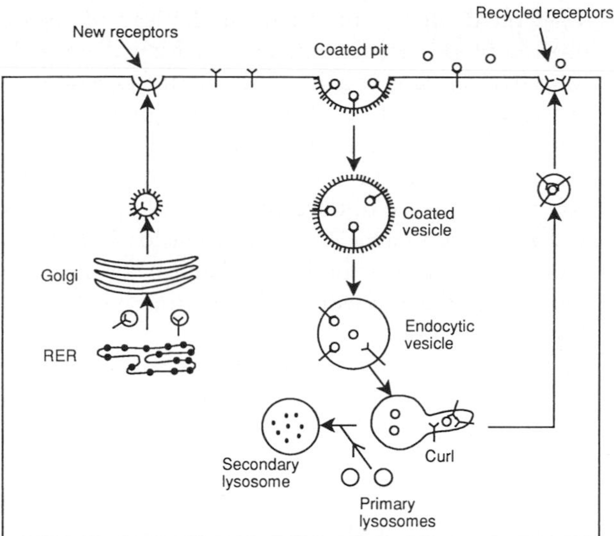

FIGURE 5 Diagram showing the probable sequence of events in the LDL-receptor pathway. The right-hand side shows the endocytosis of LDL (O) bound to LDL receptors (Y) in a coated pit, the recycling of receptors, and the delivery of LDL to a lysosome in which the particles are digested. The left-hand side shows the synthesis of new LDL receptors in the rough endoplasmic reticulum (RER), their transport to Golgi cisternae for processing, the transport of mature new receptors to the plasma membrane, and their lateral diffusion into a coated pit. Endocytosed receptors are thought to return to the plasma membrane in specialized vesicles called compartments of uncoupling of receptor and ligand (CURL). A few internalized LDL particles fail to dissociate from their receptors and are returned to the cell surface by retroendocytosis. [Reproduced, with permission, from N. B. Myant (1990). "Cholesterol Metabolism, LDL, and the LDL Receptor." Academic Press, San Diego.]

terol. In the inherited absence of LDL receptors, LDL accumulates in the plasma at levels that give rise to premature heart disease, but the tissues do not become cholesterol-deficient because their requirement for cholesterol is met by intracellular synthesis.

The quantitative importance of the LDL-receptor pathway varies widely from one tissue to another. In some tissues, such as skeletal muscle, the cell's requirement for cholesterol is satisfied almost entirely by synthesis *in situ*. In others, such as the adrenal cortex in the resting state, >90% of the cholesterol required is supplied by receptor-mediated uptake of LDL. As noted earlier, the liver is responsible for most of the receptor-mediated uptake and catabolism of LDL in the body as a whole. For this reason, liver transplantation has been used successfully in the treatment of patients with an inherited absence of LDL receptors.

C. Receptor-Mediated Uptake of Other Lipoproteins

Some cells obtain extracellular cholesterol by receptor-mediated uptake of lipoproteins other than LDL. Macrophages and other cells of the reticuloendothelial system express receptors (acetyl-LDL receptors) with high affinity for LDL that has been modified by acetylation or peroxidation. Acetyl-LDL is not produced in the body. However, certain cells, including those of the vascular endothelium, generate free radicals capable of oxidizing LDL *in vivo*. Oxidized LDL produced locally may be taken up via acetyl-LDL receptors on macrophages in the arterial wall and may, therefore, be responsible for the formation of foam cells (see Section III,B).

LDL receptors on liver cells, in addition to binding LDL itself, are also responsible for the hepatic uptake of some of the remnants produced by the action of lipoprotein lipase on VLDL. Uptake of VLDL remnants by hepatic LDL receptors is mediated by the presence of apoE in the particles. Decreased activity of hepatic LDL receptors leads to a fall in the rate of removal of VLDL remnants from the circulation, with a consequent increase in the rate of production of LDL (the end product of the intravascular metabolism of VLDL).

The cholesterol-rich remnant particles produced by the partial hydrolysis of chylomicrons (see Section IX) are taken up by the liver by receptors (called apoE receptors) that recognize lipoproteins containing apoE but do not recognize LDL. ApoE receptors mediate the rapid removal of chylomicron remnants that appear in the circulation after a fatty meal.

VIII. REVERSE CHOLESTEROL TRANSPORT

The major route for the removal of cholesterol from the body is by secretion in the bile as bile salts or free cholesterol, followed by excretion in the feces after partial reabsorption from the small intestine. The only tissues outside the liver that are capable of degrading cholesterol are the adrenal cortex and the gonads. Hence, net removal of cholesterol from extrahepatic tissues must largely depend on the transport of cholesterol through the plasma from the periphery to the liver, that is, in a direction opposite to that of the flow of cholesterol into cells by uptake of LDL. This process, known as reverse cholesterol transport, is initiated by the transfer of free cholesterol from cells to small, dense HDL particles present in the interstitial fluid. During the uptake of cholesterol by HDL, the HDL particles are bound to the cell surface by specific high-affinity receptors. The free cholesterol incorporated into HDL is esterified by LCAT, giving rise to cholesteryl ester-rich HDL particles that acquire apoE. These HDL particles deliver their cholesterol to the liver, probably by a combination of receptor-mediated uptake of whole particles (involving LDL and apoE receptors) and direct transfer of cholesterol to hepatocytes without internalization of HDL.

IX. ABSORPTION FROM THE INTESTINE

Cholesterol is absorbed from the jejunum in unesterified form. Dietary cholesteryl esters are hydrolyzed by pancreatic cholesteryl ester hydrolase in the lumen of the jejunum. The free cholesterol so produced, together with that already present in the diet and the bile, is incorporated into mixed micelles containing bile salts, phospholipids, and other lipids. Free cholesterol is then transferred from the micelles to the mucosal cells lining the jejunum. Within the cells, cholesterol is esterified with fatty acids by ACAT, and the esterified cholesterol is incorporated into the triglyceride core of nascent chylomicrons. The chylomicrons are secreted into the intestinal lymphatics and, thence, into the blood circulation. On reaching the circulation, chylomicrons are rapidly converted into remnant particles by the action of lipoprotein lipase (see Section VII,A). Chylomicron remnants have lost nearly all the triglyceride present in the parent particles but have retained most of the esterified cholesterol. These cholesterol-rich particles deliver their load of cholesterol to the liver by uptake mediated by apoE receptors. Thus, the immediate destination of absorbed cholesterol is the liver.

Absorption of cholesterol from the human intestine is incomplete, even when the intake of dietary cholesterol is very low. Under experimental conditions, less than half the cholesterol taken in the diet is absorbed when the daily intake is varied from about 50 mg to >2 g. Because cholesterol is not completely absorbed, the bile provides the body with an exit for cholesterol.

BIBLIOGRAPHY

Brown, M. S., and Goldstein, J. L. (1986). A receptor-mediated pathway for cholesterol homeostasis. *Science* **232**, 34–47.

Gibbons, G. F., Mitropoulos, K. A., and Myant, N. B. (1982). "Biochemistry of Cholesterol." Elsevier, Amsterdam.

Myant, N. B. (1981). "The Biology of Cholesterol and Related Steroids." Heinemann, London.

Myant, N. B. (1990). "Cholesterol Metabolism, LDL, and the LDL Receptor." Academic Press, New York.

Chromatin Folding

JONATHAN WIDOM
University of Illinois at Urbana–Champaign

Revised by
KENNETH W. ADOLPH
University of Minnesota Medical School

GLOSSARY

Chromatin Complex of DNA and histone proteins; the material of which chromosomes are constituted

Chromatosome Nucleosome core particle with an additional 10 base pairs (bp) of DNA at each end (giving 165 bp total) and one molecule of histone H1

Core histones Collective name given to four proteins (H2A, H2B, H3, and H4) that make up the core of the nucleosome

Histone octamer Structured complex containing two each of the four core histones, forming the core of the nucleosome on which DNA is wrapped

Histones Family of small, basic proteins that are primarily responsible for packaging DNA in human and other higher cells

Linker DNA DNA segments that connect the chromatosome to its two immediate neighbors on the chromatin fiber

Nucleosome Subunit of chromosome structure consisting of a stretch of DNA, 165–245 bp in length, folded in a complex with two molecules each of histones H2A, H2B, H3, and H4 and one molecule of histone H1; consists of the chromatosome with parts of two linker DNA segments

Nucleosome core particle Complex containing 145 bp of DNA and an octamer of the core histones

Nucleosome filament Filament of distinct nucleosomes, with chromatosomes separated by visibly extended linker DNA

30-nm filament Intermediate level of chromosome structure produced by folding of the nucleosome filament into a wider fiber having a diameter of approximately 30 nm

HUMAN CELLS CONTAIN TWO SETS OF 23 DISTINCT chromosomes. Each chromosome in a cell contains one molecule of double-stranded DNA that is ~20 Å wide but extremely long; a typical human chromosomal DNA molecule is a few centimeters in length. Such lengths are orders of magnitude greater than the diameter of a cell nucleus. *In vivo,* these DNA molecules are closely associated with a number of proteins that fold the DNA in a hierarchical series of stages and eventually produce a 10,000-fold linear compaction of the DNA preparatory to cell division. Thus, the substrates for important genetic processes such as transcription, replication, recombination, and chromosome division are not bare DNA, but rather are these protein–DNA complexes (chromosomes) in one or another stage of compaction. This article summarizes what is known about the structures and mechanisms of chromosome folding.

I. SUBUNIT STRUCTURE

The lowest level of chromosome structure is based on a repeated motif called a nucleosome. In each nucleosome, a short stretch of DNA, 165–245 base pairs (bp) in length, is locally folded and compacted in a complex with nine proteins: two molecules each of

ENCYCLOPEDIA OF HUMAN BIOLOGY, Second Edition, VOLUME 2. Copyright © 1997 by Academic Press. All rights of reproduction in any form reserved.

histones H2A, H2B, H3, and H4 and one molecule of histone H1. The length of DNA in a single nucleosome is only a small fraction of the overall length of a typical DNA molecule, so this motif is repeated hundreds of thousands of times along the entire DNA length. Thus, nucleosomes can be considered chromosome subunits, except that, owing to the continuity of the DNA, they are connected together in a chain. [*See* DNA in the Nucleosome.]

Stretches at each end of the DNA associated with a single nucleosome in chromatin are accessible to double-stranded DNA endonucleases such as micrococcal nuclease. The earliest stage of digestion of chromatin by such a nuclease produces a size distribution of soluble chromatin fragments (oligonucleosomes). These fragments are the source of chromatin used in the structural studies discussed in subsequent sections of this article. More extensive digestion releases individual nucleosomes, and further digestion degrades the nucleosomes into derivative particles: first into chromatosomes, and ultimately into nucleosome core particles.

A. The Nucleosome Core Particle

The nucleosome core particle is the best understood of these various particles, because it has been crystallized and its structure has been determined by X-ray crystallography. The core particle contains 145 bp of DNA and an octamer of the core histones: two each of the histone proteins H2A, H2B, H3, and H4. [*See* Histones and Histone Genes.]

The particle resembles a disk, with a diameter of 110 Å and a thickness of 57 Å. It is believed to possess dyad symmetry (twofold rotational symmetry), although this is distorted in the crystals owing to packing considerations. The DNA is in the B-form, and is wrapped around the surface of the histone octamer in ~1.8 turns of an approximate flat, left-hand, irregular superhelix. The pitch of this superhelical wrapping varies along the trajectory, with an average of ~28 Å. The location of each of the histones within the octamer core has been deduced from protein–DNA cross-linking data. Each stretch of DNA in the core particle appears to interact predominantly with one histone. The histones appear, from either DNA end, in the sequence H2A, H2B, H4, H3, H3, H4, H2B, and H2A. The central turn of DNA (~80 bp) interacts predominantly with a histone tetramer ($H3_2$–$H4_2$). The remaining ~20 bp of DNA at each end interacts predominantly with one H2A–H2B heterodimer each.

B. The Chromatosome

The chromatosome, from which the core particle is derived, consists of a nucleosome core particle with an additional 10 bp of DNA at each end (giving 165 bp total) and one molecule of histone H1. It is not known where on a nucleosome core particle these additional components are located. No direct structural evidence exists concerning the locations of these extra components, and one must instead rely on indirect biochemical or biophysical studies.

Two of the most likely possible trajectories for the extra 10 bp of DNA at each end include the following: (1) the DNA continues its superhelical wrapping, giving two complete superhelical turns with the 165 bp; or (2) the DNA exits tangentially from the core particle immediately after its 1.8 superhelical turns. Other trajectories are also possible. As discussed in the following, even these two rather similar trajectories lead to significantly distinct predictions for the three-dimensional structure of the nucleosome filament.

Histone H1 is known to sit on the surface of the chromatosome, and several lines of evidence indicate that it is located on a chromatosome in the vicinity of the region where DNA enters and exits the particle. Moreover, H1 has a three-domain structure, consisting of a central, folded, 80-amino-acid region referred to as the globular domain ("GH1"), flanked by extended, highly basic, N- and C-terminal tails, and it is the globular domain that is apparently responsible for determining the ability of H1 to recognize and bind to its site in a chromatosome. The amino acid sequence of the globular domain of H1 is well conserved throughout evolution, and its structure has been determined by two-dimensional nuclear magnetic resonance (NMR) methods.

The two possible DNA trajectories just discussed each lead to a distinct model for the nature of the binding site for GH1. Trajectory 1 leads to a model in which GH1 binds to the surface of the chromatosome in the region where the DNA completes its putative two complete superhelical turns; in this case, the GH1 binding site consists of three coplanar DNA segments. Trajectory 2 creates a pocket on the chromatosome surface in the vicinity of the DNA entry–exit region that is delineated by three DNA segments: one at the back, from the central turn of DNA on the core particle, and one at each side, from the two DNA segments entering and leaving the chromatosome. The expected dimensions for this pocket match the actual diameter of GH1 as measured by small-angle neutron scattering as seen from the two-dimensional NMR structure.

C. The Nucleosome

A nucleosome consists of a chromatosome together with part of each of the two DNA segments, called linker DNA, that connect that chromatosome to its two immediate neighbors on the chromatin fiber. Linker DNA varies in length from one nucleosome to the next about an average value that is characteristic for cells of a given type; this average value itself varies from one cell type to another, within the range ~0–80 bp. The trajectory taken by the two partial linker regions in an isolated nucleosome is not known. The N- and C-terminal domains of histone H1 adopt extended, largely α-helical, structures in solution, and are believed to interact chiefly with linker DNA.

Nucleosomes are heterogeneous in several respects, in addition to the variability of the linker DNA. The sequence of DNA in a nucleosome is of course variable (because essentially all of the genome is packaged in nucleosomes), and this variability seems certain to affect the structure at high resolution. The histones are present in many variants, coded by different genes; moreover, each histone variant is subject to a large number of posttranslational modifications, such as acetylation, methylation, phosphorylation, and the attachment of polyadenosine diphosphate (ADP) ribose and of ubiquitin (a complete small protein). Any of these variations could affect structure.

II. STRUCTURE OF THE NUCLEOSOME FILAMENT

The lowest level of organization of a complete chromosome is the nucleosome filament. The available data suggest a one-dimensional picture of the chromatin filament in which chromatosomes, having a generally conserved composition and structure but differing in detail, are spaced along the DNA at irregular intervals that vary statistically about a well-defined average. At the level of one-dimensional structure, the chromatin filament is ordered but is not regular in the geometric sense.

Chromatin suitable for physical studies is isolated from purified cell nuclei in two steps. First, the extremely long chromosomal DNA molecules are randomly digested *in situ* with an enzyme such as micrococcal nuclease that preferentially attacks the linker DNA. Then, the nuclei are lysed and chromatin fragments (nucleosome oligomers) are released into solution and separated from residual nuclear debris. The chromatin fragments tend to adopt more highly folded states, which are discussed in the following sections of this article. In the absence of any multivalent cations and provided that the concentration of monovalent cations is sufficiently low, these more highly folded states unfold and allow the lowest level of organization to be studied.

A. Nucleosome Filaments versus 100-Å Filaments

Electron microscopic studies led initially to considerable uncertainty regarding the structure of the nucleosome filament. Depending on details of specimen preparation, the chromatin could appear either as a filament of distinct nucleosomes (i.e., with chromatosomes separated by visibly extended linker DNA), now referred to as the nucleosome filament, or as a continuous filament having a width of ~100 Å (presumably, actually 110 Å), referred to as the nucleofilament or the 100-Å filament. It was simply not possible to determine from these studies which, if either, of these two classes of images was representative of the real structure in solution. One could in principle distinguish between two such models by quantitative analysis of hydrodynamic properties—sedimentation coefficients or translational diffusion coefficients—and many such studies were carried out. In practice, these experiments suffered from too many uncertainties to be definitive.

The question of which of the two models—nucleosome filament or 100-Å filament—more accurately represents the species present in solution has now been settled by small-angle X-ray and neutron-scattering experiments. These techniques allow one to measure key model-independent properties of extended structures: the radius of gyration of the cross section (R_c), which is related to the actual diameter or width, and the mass per unit length (m/l), which can be expressed as the number of nucleosomes per unit displacement along the filament. The two different models lead to very different predictions for R_c

and *m/l*, and the experiments are in agreement with the predictions of the nucleosome filament model. It now seems that it was the presence of uranyl acetate, used to enhance contrast, that caused nucleosome filaments to appear in the form of 100-Å filaments.

B. The Nucleosome Filament

The nucleosome filament is the conformation that might be expected on purely physical grounds. Linker DNA segments are, in general, very short compared with the persistence length of naked DNA. (The persistence length is a lengthscale of polymer stiffness; for DNA it is ~500 Å, or ~150 bp.) If linker DNA had the properties of naked DNA, then it would be essentially straight and inflexible and, therefore, might be expected to extend straight from one chromatosome to the next.

Once it was learned that the nucleosome filament was the appropriate structural model, it was possible to give a physical interpretation to a peak observed in the X-ray and neutron-scattering patterns: the diffraction peak arises from (three-dimensional) distance correlations between consecutive nucleosomes along the filament and allows that distance to be measured approximately. The average nucleosome center–center distance measured in this way is found to be in agreement with those expected if the linker DNA is extended (given the known average linker length), and it changes as expected when the average linker DNA length is varied by using chromatin from differing cell types.

Three models for the structure of the nucleosome filament that are consistent with available data are shown in Fig. 1; the structures are shown flattened onto a surface, as occurs when samples are prepared for electron microscopy. The three related structures correspond to three different models for possible trajectories taken by linker DNA (see earlier). Model C corresponds to trajectory 1, in which the DNA on a chromatosome is organized in two full turns and then leaves the chromatosome surface tangentially. Model B corresponds to trajectory 2, in which DNA leaves the chromatosome surface tangentially after only 1.8 turns. Model A corresponds to a trajectory in which the DNA leaves the chromatosome surface tangentially after only 1.5 turns. No experimental evidence supports model A; it is included here only because such a model for the nucleosome filament is implied in certain models for higher levels of folding, discussed in the following.

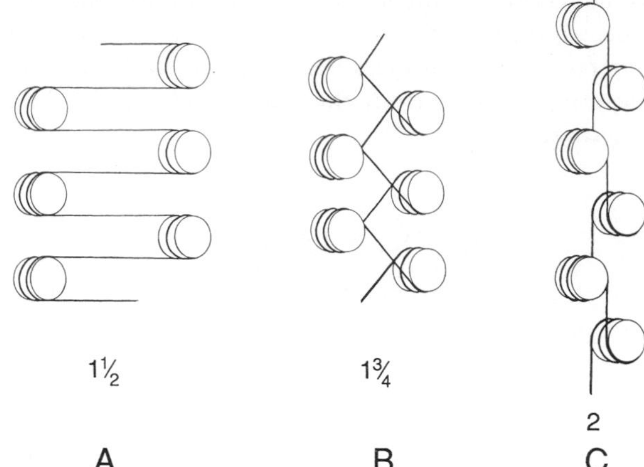

FIGURE 1 Three possible models for the structure of the nucleosome filament. The structures are shown flattened into two dimensions, as happens during electron microscopy. The number underneath each model is the number of superhelical turns made by the DNA as it wraps on the histone surface. The different models lead to different locations in three-dimensional space for the nearest-neighbor nucleosomes and to different numbers and spatial arrangements of DNA segments at the proposed binding site for histone H1. [Drawing by S. Holtz.]

In solution, the natural twist of DNA causes the nucleosome filament to rotate to a variable degree about the axis of each linker, so that the flat faces of the chromatosomes need not be coplanar as illustrated. Such rotations give model B the appearance of a three-dimensional zig-zag. Optical studies of oriented nucleosome filaments in solution suggest that the nucleosomes have their flat faces parallel (±20°) to the axis of the nucleosome filament. Models B and C have important consequences for chromatin types having a very short average linker DNA length. Because of the resistance to torsion and bending of DNA, certain very short linker lengths must be disallowed because they would cause consecutive nucleosomes to overlap in space. This restriction can be alleviated by adding or subtracting ~5 bp of DNA (about half of a helical turn). Many subtly different conformational isomers will exist, even along the same nucleosome filament, depending on the exact length or torsional stress of each linker DNA.

Histone H1 is an important element in the structure of the nucleosome filament. Both electron microscopy and X-ray scattering studies show that when H1 is removed, the distance between consecutive nucleosomes is slightly increased, and the filament is disordered. The characteristic zig-zag appearance in electron micrographs is lost, and the X-ray scattering peak, which arises from distance correlations between consecutive nucleosomes, is reduced in intensity. These results indicate that H1 imposes a locally ordered structure on a nucleosome and its immediate neighbors along the chain.

III. FOLDING OF THE NUCLEOSOME FILAMENT

The next higher level of chromosome structure, referred to as the 30-nm filament, is achieved by intramolecular folding and compaction of the nucleosome filament to produce a shorter, wider fiber having a diameter of approximately 30 nm. The 30-nm filament is a particularly important level of chromosome structure because this is the folded state in which most chromatin is maintained throughout most of the cell cycle. It must be unfolded to allow transcription and replication, and it is further folded to allow meiosis and mitosis. Before considering the structure of the 30-nm filament, it is helpful to consider what is known about the folding process itself.

In vitro, folding of chromatin in the nucleosome filament state into 30-nm filaments is induced by increasing the concentration of cations. The 30-nm filaments remain soluble and can be studied by numerous different physical methods.

An important question is whether or not the 30-nm filaments produced in this way are similar in structure (and not just in appearance) to the 30-nm filaments *in vivo*. Two lines of evidence suggest that they are. First, low-angle X-ray diffraction patterns from 30-nm filaments *in vitro* show the same set of bands characteristic of chromatin in nuclei and in whole living cells (certain cell types have very little cytoplasm, and X-ray diffraction patterns from solutions of these cells turn out to be dominated by diffraction from the chromatin). This is good evidence that the 30-nm filaments produced *in vitro* are similar in internal structure to those produced *in vivo*. Second, chromatin can be isolated from cells without ever exposing it to conditions in which 30-nm filaments would unfold into the nucleosome filament state. Electron microscopic and hydrodynamic studies of such 30-nm filaments have led to the conclusion that there are no significant differences between the 30-nm filaments extracted in that manner and 30-nm filaments formed by refolding from the nucleosome filament state.

One important conclusion from all of these studies is that, at sufficiently high cation concentrations, 30-nm chromatin filaments are no longer soluble and precipitate out of solution. Further addition of multivalent cations causes the aggregated (insoluble) 30-nm filaments to pack very closely together, so that electron density contrast between them is lost.

A. Role of Cations

Essentially, any cation can apparently suffice to cause nucleosome filaments to fold into 30-nm filaments, as judged by electron microscopy, hydrodynamic, optical, and diffraction methods. Inorganic monovalent cations such as Na^+ and K^+, divalent cations such as Ca^{2+}, Mg^{2+}, and Mn^{2+}, trivalent cations such as $Co(NH_3)_6^{3+}$, and organic multivalent cations such as spermidine^{3+} and spermine^{4+} all lead to closely similar 30-nm filament states of chromatin. There remains some debate whether or not any differences do in fact exist.

It is found that the valence of the cation is of primary importance in determining the concentration of cation necessary to induce chromatin folding. Monovalent cations are required at concentrations of 60–80 m*M*, divalent cations at 100 μM–1 m*M*, trivalent cations at 10–100 μM, and tetravalent cations at 1–10 μM. The relatively wide concentration ranges given are due in part to experiments with multivalent cations being done in the presence of a range of monovalent cation concentrations (e.g., buffer cations such as Tris$^+$ or buffer counterions such as Na^+ or K^+) because of cation competition (see the following). When the monovalent concentration is kept constant (and $\leqslant$60 m*M*), cations of the same valence are typically effective within a twofold concentration range. Exceptions to this rule may arise when a cation is particularly bulky or when the cation can bind nonionically to chromatin (e.g., Cu^{2+}).

These data are sufficient to define the role of cations in the stabilization of 30-nm filaments. *A priori,* one can list three possible mechanisms by which cations might act: (1) by binding to a particular site that has chemical selectivity, such as the metal binding site in metalloenzymes; (2) by screening repulsions between small numbers of charges, such as amino acid side chains; or (3) by acting as general DNA counterions, reducing the effective charge per DNA phosphate and screening repulsions between adjacent DNA helices. Model 1 is ruled out by the observed lack of sensitivity to the size or chemical nature of the cation. If model 2 were correct, chromatin folding would be governed by the ionic strength in accord with Debye–Huckel theory. However, the ionic strength necessary to stabilize 30-nm filaments is found to decrease by three orders of magnitude as the valence of the cation is increased from +1 to +4.

The theory governing model 3 is called counterion condensation theory and applies to linear systems having a charge density that exceeds some critical value. In chromatin, roughly one-half of the DNA phosphates (which each carry a formal charge of −1) are neutralized by the positively charged histones; nevertheless, the net linear charge density of the nucleosome filament still exceeds the critical threshold for counterion condensation, and this charge density will increase when nucleosome filaments compact into 30-nm filaments. Counterion condensation theory predicts that the effectiveness of cations as DNA counterions primarily depends on the valence of the cation, with an effect much greater than for the Debye–Huckel theory. This behavior is in qualitative accord with the results obtained for chromatin folding. The theory further predicts that the concentration of a multivalent cation required to induce folding will depend on the monovalent cation concentration because of cation competition. As discussed next, this surprising prediction is verified experimentally. Thus, one concludes that cations induce chromatin folding by reducing repulsions between DNA segments within 30-nm filaments.

B. Cation Competition

An important prediction of counterion condensation theory is that monovalent and multivalent cations compete with each other for binding to chromatin and for stabilizing the 30-nm filament state. This surprising prediction is verified experimentally and leads to complicated behavior. Na^+ (and presumably any other monovalent cation) has dual effects. Sufficiently high concentrations of Na^+ (~60–80 m*M*) cause chromatin to fold into 30-nm filaments even in the absence of Mg^{2+} or other multivalent cations; however, at low concentrations of Na^+ (less than ~45 m*M*) and when Mg^{2+} (or some other multivalent cation) is present and stabilizing the 30-nm filament state, Na^+ has the opposite effect: it competes with the higher-valence cation for binding to chromatin and destabilizes the 30-nm filament state. Cation competition appears to underlie chromatin folding and also the aggregation of 30-nm filaments.

C. Titrations without End Points

Various physical experiments used to monitor the folding of nucleosome filaments into 30-nm filaments have given two very different classes of results. Electron microscopy leads to the conclusion that cation titrations of nucleosome filaments reach a definite end point, in which all of the chromatin is folded into 30-nm filaments that do not change further in appearance on further addition of cations. Neutron-scattering studies show that R_c and m/l reach plateau values at similar cation concentrations, as do optical parameters of oriented chromatin filaments and the rotational relaxation times. A very different view is reached from measurements of sedimentation coefficients, translational diffusion coefficients, light-scattering intensities, and the sharpness of peaks in X-ray solution-scattering patterns. These studies suggest that 30-nm filaments continue to change in structure even after the end point detected by other methods, suggesting that there may be no single 30-nm filament state, and that the only end point of a cation titration of chromatin is the point at which no detectable chromatin remains soluble.

Chromatin can apparently continue to change in structure even past the end points detected in certain experiments; however, such changes must be quite subtle, because some measures do give definite end points, such as the overall appearance in electron micrographs. Perhaps the 30-nm filament should be regarded as a continuum of closely related states.

D. Continuous versus Two-State Folding

If a sufficient concentration of some cation is added to chromatin in the nucleosome filament state, the chromatin refolds into 30-nm filaments. The folding process (as monitored by stopped-flow X-ray scattering) is very fast. In a wide range of physical experiments, it is found that addition of less of the folding cation yields physical parameters (e.g., R_c, m/l, dichroism, sedimentation coefficient) that are intermediate in magnitude between those of nucleosome filaments and those of 30-nm filaments. The question arises: Is the folding of each chromatin filament a gradual (i.e., continuous) process, or is each molecule in one of two limiting states (the nucleosome filament or the 30-nm filament), with the intermediate values of various parameters reflecting cation concentration-dependent fractions of the population of chromatin molecules in each state?

With many physical experiments, it is not possible to settle this question—particularly because a definite 30-nm filament end point may be experimentally inaccessible. However, electron microscopy allows measurements to be made on individual molecules and, therefore, should allow one to distinguish between continuous and two-state folding. Early electron microscopic studies were qualitative but showed that, at intermediate points in titrations, each chromatin filament appeared to be intermediate in width. This has been confirmed by a more recent quantitative study (using scanning transmission electron microscopy) of *m/l* during a Na^+ titration. At intermediate points in the titration, the chromatin was plainly in intermediate states of folding; the two limit states were largely unpopulated. Hence chromatin folding is continuous, not two-state.

E. Structural Consequences

Electron microscopy shows that, in the early stages of a cation titration of nucleosome filaments, the nucleosomes move visibly closer together. Typical results are illustrated schematically in Fig. 2. The open zig-zag appearance (obtained in 1 m*M* Na^+) collapses to a closed zig-zag in which the nucleosomes often appear to be in two closely paired parallel rows (in ~20 m*M* Na^+). As the cation concentration is further increased, the chromatin takes on a range of diverse appearances, implying diverse models for the structure of the 30-nm filaments that will be formed. When the concentration of cations is sufficiently high (~60 m*M* Na^+), the chromatin reaches the 30-nm filament state and becomes more uniform in appearance. Hydrodynamic studies show that the sedimentation coefficient, and (equivalently) the translational diffusion coefficient, continues to increase throughout the folding transition, providing evidence that chromatin folding in solution is accompanied by the packing together of nucleosomes. Similarly, an X-ray scattering peak, which monitors the average internucleosomal distance, is found to move to wider angles, indicating a shortened average internucleosomal distance and, ultimately, to disappear, indicating that most of the nucleosomes are in physical contact with other nucleosomes.

Unfortunately, none of these experiments reveals *which* nucleosomes are packing together in space during chromatin folding. For example, Fig. 2 illustrates that the nearest neighbors (in three dimensions) of nucleosome *i* may or may not be $i - 1$ and $i + 1$. It is simply not possible to answer this question with currently available data. It will be seen in the following section that there are currently several distinct models for the structure of 30-nm filaments that differ chiefly in their connectivity—that is, in their assumptions regarding which nucleosomes along the chain are brought into proximity in the 30-nm filament.

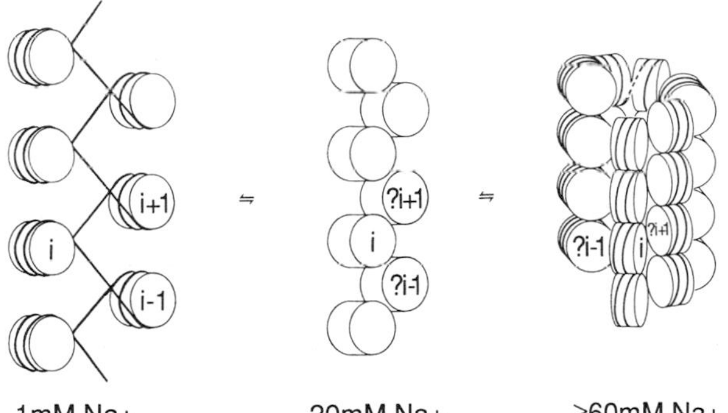

FIGURE 2 Schematic illustrations of the chromatin-folding process as seen by electron microscopy. Typical structures are shown for three different concentrations of Na^+ (as indicated). Increasing concentrations of Na^+ cause the chromatin to fold up. The most compact structure (right side, 60 m*M* Na^+) is approximately 30 nm in diameter, and is called the 30-nm filament. The particular structure shown is the solenoid model; other models are closely similar in appearance (see text). The starting nucleosome filament model (1 m*M* Na^+) is the one thought most likely to be correct, but this is not known. Nucleosome *i* refers to an arbitrary but particular nucleosome in the chain; $i - 1$ and $i + 1$ are its immediate neighbors. The question marks (?) shown in the other structures indicate that those nucleosomes might or might *not* be the one-dimensional neighbors of nucleosome *i*. [Drawing by S. Holz.]

F. Role of Histone H1

Chromatin that is biochemically stripped of histone H1 no longer undergoes cation-dependent folding into 30-nm filaments. Electron micrographs of H1-stripped chromatin in solutions containing sufficient Na^+ or Mg^{2+} for folding of native chromatin into 30-nm filaments show regions of random nucleosome strings interspersed with compact disordered clumps of nucleosomes. The sedimentation coefficient of stripped chromatin does increase during Na^+ or Mg^{2+} titrations, but much less than that of native chromatin. These observations lead to the conclusion that histone H1 thermodynamically stabilizes the 30-nm filament state relative to less compact states. Moreover, H1 appears to confer structure specificity on the compact chromatin, since the nucleosome clumps found with stripped chromatin appear to lack even local order.

Histone H1 has the ability to bind cooperatively to DNA. It now appears that the globular domain of H1 on its own has this same capability. At this time, it is not known with certainty whether or not H1 binds cooperatively to chromatin. H1 exhibits a preference for longer chromatin oligomers, in accord with a cooperative binding mechanism, but simple experiments in which H1-stripped chromatin is titrated with increasing concentrations of H1 fail to reveal any hints of cooperativity. This is a very important issue, because cooperative binding of H1 potentially provides a mechanism for the cooperative folding or unfolding of *regions* of chromosomes, possibly containing several genes. Because RNA and DNA polymerases are large in size compared with nucleosomes, one presumes that 30-nm filaments must be unfolded prior to transcription or replication. Thus H1 could potentially regulate the ability of groups of genes to be transcribed or replicated.

Studies of the binding of GH1 to DNA suggest that cooperativity in binding is mediated by direct contacts between adjacent molecules of GH1. An important question is whether or not such contacts between GH1s exist in chromatin. This is closely related to the question of whether or not 30-nm filaments are cooperatively stabilized.

IV. STRUCTURE OF THE 30-nm FILAMENT

Although a great deal of progress has been made toward elucidation of the structure of the 30-nm filament, several key features of the structure remain unknown. Before considering what is known, it is useful to consider why the problem has been so difficult to study.

First, because nucleosome filaments are heterogeneous in length and in local composition along the filament, 30-nm filaments are too. This greatly reduces the chances for successful crystallization for X-ray crystallography. In electron microscopic studies, structural heterogeneity along the 30-nm filament may sufficiently distort the internal symmetry of the filament as to preclude the application of symmetry-based image reconstruction methods.

The second, and most important, problem comes from the phase diagram for the cation-induced folding of chromatin. The ionic conditions that one wishes to use, to best stabilize the 30-nm filament state, cause the 30-nm filaments to aggregate. Furthermore, electron microscopic studies of 30-nm filaments under conditions where they are properly soluble show that the forces between nucleosomes in 30-nm filaments are not large compared with various distorting forces (e.g., surface effects) that are unavoidably encountered during sample preparation. The aggregation that accompanies the further addition of cations is noncrystallographic; crystallization is not possible in such conditions. When internal symmetry may not exist, structures can still be solved by electron microscopy, using tomographic methods. However, tomographic analysis of the structure of 30-nm filaments in aggregates is extraordinarily difficult, if not impossible. When the concentration of multivalent cations is sufficiently high, electron density contrast between 30-nm filaments in the aggregates is lost altogether.

A. Nucleosome Packing

The problems associated with methods for directly determining the structure of 30-nm filaments have led to an alternative approach. Here, one attempts only to specify the packing of nucleosomes within the 30-nm filament; one can then impose on this packing model the detailed structure of the nucleosome core particle, as determined from crystallographic studies. The location of histone H1 and the path of the linker DNA (i.e., the connectivity) need to be determined separately, because these components are not present in nucleosome core particles.

The packing of nucleosomes within 30-nm filaments has been addressed through a wide range of experiments. Electron microscopy suggested a number of distinct, mutually exclusive possibilities. It has been possible to distinguish between these through X-ray diffraction studies of partially oriented samples of 30-nm filaments. Since the structure of individual nucleosome core particles was known (from crystallographic studies, as discussed earlier), the diffraction patterns from the 30-nm filaments could be deciphered. Other important data were provided by measurements of optical properties such as dichroism or birefringence of 30-nm filaments that were oriented by electric or fluid-flow fields. Hydrodynamic studies and small angle-scattering measurements of R_c and m/l provide additional constraints on possible models.

There is now general agreement that, in 30-nm filaments, nucleosomes are packed edge-to-edge in the direction of the 30-nm filament axis, and radially around it, as illustrated in Fig. 2. The nucleosomes are oriented with their flat faces parallel $\pm 20°$ to the filament axis. The packing is believed to be helical, but key helical parameters such as the handedness and the number of "starts" remain matters of debate. In the best-studied case, there are roughly six nucleosomes per 11 nm (110 Å) translation along the filament axis. The number need not be an integer and, indeed, may vary along the filament: evidence indicates that it depends on the length of linker DNA, which varies from one nucleosome to the next.

Although some aspects of nucleosome packing are known, three key unknown features remain: the rotational setting of each nucleosome in the 30-nm filament about its disk axis, the linker DNA trajectory, and the location within the 30-nm filament of the globular domain (and other domains) of histone H1. These three unknown structural features are all closely related. Geometric considerations imply that, for the case of chromatin having a very short (~0 bp) average linker length, specifying any one of these determines the others implicitly.

Our current inability to specify the rotational setting of nucleosomes has an important consequence. We wish to know which groups on nucleosomes in 30-nm filaments interact to stabilize the structure. As discussed earlier, we know the location within a nucleosome of each of the different proteins. Unfortunately, since the rotational settings are not known, we cannot specify which of these groups are neighbors in 30-nm filaments.

Different possibilities for these three related unknown properties lead to differing models for 30-nm filament structure.

B. Current Models of 30-nm Filament Structure

Three different models of 30-nm filament structure have received the greatest attention in the recent literature and are the most likely to be correct: the solenoid model, the twisted ribbon model, and the crossed linker model. Each of these models is derived by compacting a nucleosome filament in a different but apparently natural manner. All of these models look quite similar, because they are based on the same data for nucleosome packing. In each case, their appearance is similar to that shown in Fig. 2. They differ, however, in their connectivity, their implied location of the globular domain of H1, and their implied rotational settings of nucleosomes. Each model is consistent with much, but not all, of the available data.

In the solenoid model, the nucleosome filament is compacted by first bending the linker DNA and thus bringing consecutive nucleosomes together in space, producing a closely apposed nucleosome filament. This filament is then compacted progressively through a range of one-start helical structures having a constant pitch of 110 Å (nucleosomes packed edge-to-edge with their faces roughly parallel to the direction of the filament axis) and an increasing number of nucleosomes per turn, until the limiting 30-nm filament structure is achieved. The helix could be right- or left-handed. Linker DNA connects laterally neighboring nucleosomes around the filament axis. The lateral neighbors of nucleosome i are $i + 1$ and $i - 1$. This model is the one illustrated in Fig. 2.

In the revised twisted ribbon model, the zig-zag form of the nucleosome filament is considered as a ribbon, forming the 30-nm filament by wrapping helically about an axis, which then becomes the 30-nm filament axis. The fundamental unit of this model is two nucleosomes in length (a zig-zag pair), so the filament that results is two-start. The connectivity is vertical: linker DNA, which must be bent, connects between nucleosomes that are in contact edge-to-edge in the direction of the 30-nm filament axis. The lateral neighbors of nucleosome i are $i + 2$ and $i - 2$.

The crossed linker model allows the linker DNA to remain straight. A zig-zag nucleosome filament is simultaneously twisted about its axis and compressed along the axis to produce the 30-nm filament. The starting nucleosome filament axis and the final 30-nm filament axis are coincident. Depending on how the twisting and compression are coordinated, both one- and two-start helices can result. The handedness is specified. The lateral neighbors of nucleosome i are again $i + 2$ and $i - 2$.

These three different models each make different predictions regarding the locations of GH1, linker DNA, and the rotational settings of nucleosomes. The solenoid model and the crossed linker model imply the same location for GH1 and, therefore, the same rotational setting of nucleosomes: nucleosomes are oriented about their disk axis with the proposed GH1-binding site—the DNA entry–exit region—facing inward toward the center of the 30-nm filament. By contrast, the revised twisted ribbon model implies that nucleosomes are rotated about their disk axis such that the GH1-binding site points up or down. Consecutive nucleosomes are packed vertically; the GH1-binding site of the lower nucleosome faces up and that of the higher nucleosome faces down, so that the two GH1s are in contact.

The only compelling evidence supporting any one of these models and apparently contradicting the others comes from qualitative or quantitative image analysis of selected electron micrographs, within the constraints imposed by the X-ray diffraction data.

C. Tests of Current Models

The solenoid model and the revised twisted ribbon model require linker DNA to bend during chromatin folding. By contrast, the crossed linker models allow linker DNA to remain straight. Can linker DNA bend as required by the solenoid and revised twisted ribbon models? This question has been addressed in the author's laboratory by studying the physical properties of chromatin fragments containing just two nucleosomes, connected by one linker ("dinucleosomes"). It is found that during cation titrations of the dinucleosomes, the two nucleosomes go from an average center–center separation of 150 Å, as expected for fully extended linker DNA, to an average separation of 0 Å: linker DNA can bend to bring *consecutive* nucleosomes in chromatin into contact.

The twisted ribbon model can be tested by determining the extent to which one can make chemical cross-links between the globular domains of histone H1. Simple geometric considerations for this model lead to the conclusion that globular domains of H1 can be in contact only in pairs (on two vertically connected nucleosomes). Therefore, the longest polymer of globular domains that could be produced given this model is of length two. By contrast, the solenoid model and crossed linker models allow for contacts between GH1s along the entire chromatin filament length.

The crossed linker model can be tested by determining the diameter of "30-nm" filaments produced by chromatin having ~0 bp of linker DNA. The maximum possible diameter in this case is 250 Å, and even this wide a fiber can only be produced by unwrapping DNA off the surface of a nucleosome so that there are only $1\frac{1}{2}$ turns on each nucleosome (Fig. 1, model A). If the DNA on each nucleosome is wrapped in $1\frac{3}{4}$ turns or 2 turns (Fig. 1, models B and C), as believed, the maximum filament diameter for the crossed linker model is reduced toward ~220 Å. By contrast, the solenoid and revised twisted ribbon models do not place limits on possible filament widths.

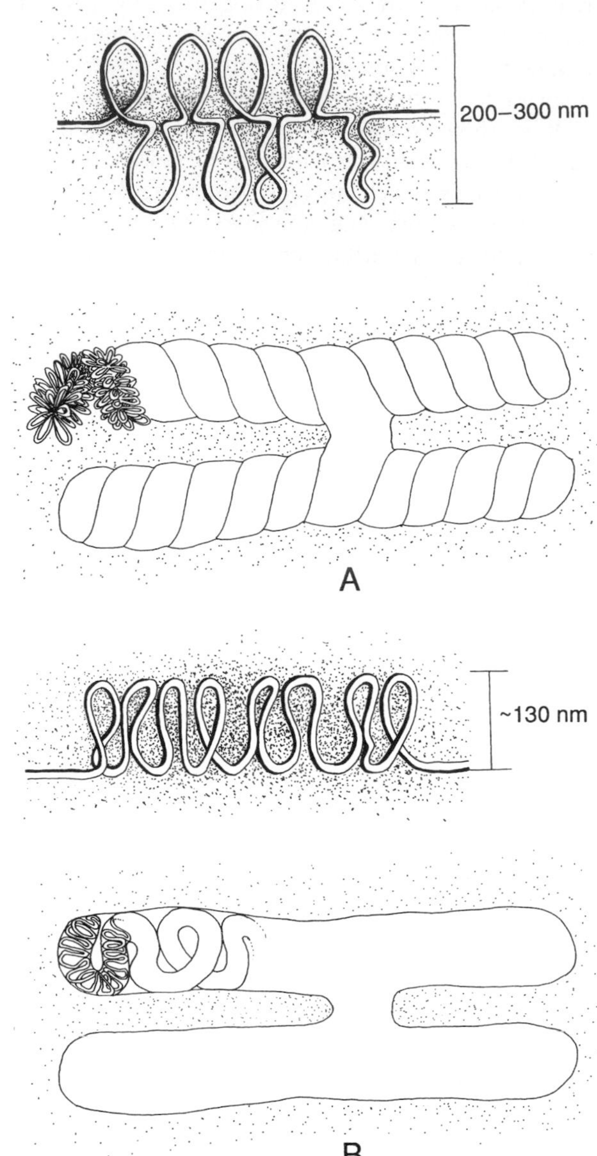

FIGURE 3 Models of higher levels of folding. (A) Top: the 30-nm filament (shown as a featureless thread) is locally organized into loops, giving a shorter fiber having a width or diameter of ~200–300 nm. The loops of the 30-nm filament may twist about themselves. Bottom: the 200- to 300-nm-wide fiber is coiled helically into a mitotic chromatid. The figure shows a pair of chromatids that are connected at the centromeres. These two chromatids will separate to the two daughter cells at cell division. The two chromatids may have opposite handedness for this final stage of helical coiling, as shown. (B) Top: the 30-nm filament is locally folded (not necessarily organized in loops) into a shorter fiber having a width of ~130 nm. Bottom: the 130-nm fiber is locally folded into the mitotic chromatid. This could be accompanied by twisting of the 130-nm fiber about its long axis (not illustrated). Again, two chromatids are shown connected at their centromeres. [(A) is adapted from the work of J. B. Rattner, (B) from A. S. Belmont. Drawings by A. Ellingwood.]

V. HIGHER LEVELS OF STRUCTURE

During meiosis and mitosis, chromosomes are folded into highly compact and specialized structures. It is useful to characterize these by a "packing ratio," which is defined as the ratio of the contour length of a chromosomal DNA molecule relative to the length of that same DNA molecule packed in chromatin, in one or another stage of folding. For meiotic chromosomes in the form of synaptonemal complexes, the packing ratio is typically in the range of 300- to 1000-fold. For mitotic chromosomes, the packing ratio is ~10,000-fold. For comparison, the packing ratio for the 30-nm filament level of structure is ~40-fold. [*See* Meiosis; Mitosis.]

It is believed that both meiotic and mitotic chromosomes are constructed by further folding of chromatin in the 30-nm filament state. For meiotic chromosomes, the evidence is limited and comes from qualitative electron microscopic studies of the ultrastructure of synaptonemal complexes. These show 25- to 30-nm-diameter fibers, which are interpreted as chromatin in the 30-nm filament state, that loop out and then rejoin the body of the synaptonemal complex.

Much more information is available regarding the structure of mitotic chromosomes. Electron microscopic and X-ray diffraction studies have shown convincingly that mitotic chromosomes are produced by further folding of 30-nm filaments; however, light and electron microscopy have led to two different classes of model for the nature of this folding.

Light and electron microscopic studies of mitotic chromosomes that have been treated with standard methanol–acetic acid mixtures or with NaCl–polyglutamic acid have provided evidence that the final level of compaction consists of helical coiling of a fiber that probably has a diameter of roughly 200 nm. The biochemical effects of the methanol–acetic acid treatment are not known; the chief effect of the NaCl–polyglutamic acid treatment is probably to remove some or most of the histone H1 from the chromosome. The manner in which the 30-nm filament is compacted into the 200-nm filament is not clearly established, but some investigators believe that the 30-nm filament is first gathered into loops, at intervals, which are subsequently coiled into a 200-nm filament. Also, evidence suggests that, in two sister chromatids attached at their centromeres, these helical gyres may have opposite handedness, giving the two chromatids mirror symmetry. This model is illustrated in Fig. 3A.

A different view is provided by quantitative three-dimensional tomographic studies of mitotic chromosomes *in situ* or after biochemical isolation. These studies fail to reveal any evidence for a final stage of helical coiling. Evidence of substructure on a 100- to 130-nm scale suggests the existence of filaments having that diameter. There is also some indication that these 100- to 130-nm fibers are produced by repeated folding of the 30-nm filament. This view of mitotic chromosome folding is illustrated in Fig. 3B.

At this time it is not possible to reconcile these two views of mitotic chromosome structure. It is possible that the final stage of compaction really is via helical coiling and that this level of structure is simply obscured by the general close packing of 30-nm filaments within mitotic chromosomes that are maintained in a native-like state. It is equally possible that the helical coiling is not an aspect of native mitotic chromosome structure, but that it arises artifactually after removal of some of the histones (e.g., H1) and a subsequent unfolding of loops or folds of 30-nm filament. This question may be resolved as the tomographic reconstructions are taken to higher resolution. It could presently become possible to trace long stretches of 30-nm filament through three-dimensional space in a chromatid, despite the generally high density of material. Alternatively, it should be possible to find ionic conditions in which 30-nm filaments within mitotic chromosomes are stable, but where subsequent levels of compaction are made artificially less dense (i.e., the chromatids swollen slightly) by reducing the strength of interactions between 30-nm filaments. Promising conditions can be identified from the phase diagram for chromatin folding and aggregation.

BIBLIOGRAPHY

Adolph, K. W. (ed.) (1988). "Chromosomes and Chromatin," Vols. 1–3. CRC Press, Boca Raton, FL.

Adolph, K. W. (ed.) (1990). "Chromosomes: Eukaryotic, Prokaryotic, and Viral," Vols. 1–3. CRC Press, Boca Raton, FL.

Alberts, B., Bray, D., Lewis, J., Raff, M., Roberts, K., and Watson, J. D. (1994). "Molecular Biology of the Cell," 3rd Ed. Garland, New York.

Cold Spring Harbor (1993). "DNA and Chromosomes," Cold Spring Harbor Symposia on Quantitative Biology, Vol. 58. Cold Spring Harbor Laboratory Press, Plainview, NY.

Lodish, H., Baltimore, D., Berk, A., Zipursky, L., Matsudaira, P., and Darnell, J. (1995). "Molecular Cell Biology," 3rd Ed. Freeman, New York.

Widom, J. (1989). Toward a unified model of chromatin folding. *Annu. Rev. Biophys. Biophys. Chem.* **18**, 365–395.

Wolffe, A. P. (1992). "Chromatin Structure and Function." Academic Press, San Diego.

Chromatin Structure and Gene Expression in the Mammalian Brain

IAN R. BROWN
TINA R. IVANOV
University of Toronto

GLOSSARY

Chromatin conformation Folding of nucleosomes into a solenoid chromatin fiber

DNA-binding protein Class of nonhistone nuclear proteins that bind to DNA regulatory sequences, thereby activating gene transcription

DNA regulatory sequence DNA sequence involved in the control of gene transcription

DNase I Enzyme that preferentially digests an open or decondensed chromatin conformation

Northern blot Method for determining the size and tissue distribution of specific RNA species

Nucleosome Organization of DNA and histones into repeating nucleoprotein units

IN THE MAMMALIAN BRAIN UP TO ONE-THIRD OF all single-copy genes (i.e., protein-coding sequences) can be transcribed as nuclear RNA. The proteins encoded by these genes are responsible for the complexity of brain structure and function. In cortical neurons, high levels of gene transcription may be facilitated by modifications in chromatin organization that render genes more accessible to the transcriptional machinery. Gene transcription also requires the interaction of nuclear proteins with DNA regulatory sequences. Approaches to investigating this aspect of gene regulation in the brain include the mapping of DNase I-hypersensitive sites, the introduction of cloned genes into cells grown in tissue culture, gel retardation assays, and DNase I protection "footprinting."

I. TRANSCRIPTION

A. Analysis of Brain RNA by Nucleic Acid Hybridization

The complexity of brain structure and function dictates that many different proteins be synthesized. An estimate of the number of genes transcribed in the mammalian brain can be calculated by first determining the sequence complexity of brain RNA. This value can be obtained by the use of nucleic acid hybridization assays that measure the extent of RNA hybridization either to single-copy DNA (i.e., primarily protein-coding sequences) or to a cDNA population (i.e., reverse transcripts of brain mRNAs). One study, which investigated the complexity of adult rat brain RNA, reported that approximately 20% of single-copy DNA is expressed as cytoplasmic RNA. If one assumes a mammalian genome size of 1.8×10^9 base pairs (bp), the RNA sequence complexity in the brain is 3.6×10^8 bp, which could encode approximately 240,000 different proteins of molecular mass 50,000 daltons. A fraction of these DNA sequences, however, do not encode RNAs. [*See* DNA and Gene Transcription.]

Studies in which a cDNA population derived from brain poly(A)$^+$ RNA is hybridized to RNA from the liver or the kidney indicate that 65% of these transcripts are brain specific. When similar experiments

ENCYCLOPEDIA OF HUMAN BIOLOGY, Second Edition, VOLUME 2. Copyright © 1997 by Academic Press. All rights of reproduction in any form reserved.

are performed using nonneural tissue, significantly lower estimates of sequence complexity are obtained. For example, in the liver and the kidney approximately 57,000 and 38,000 genes, respectively, are transcribed and expressed in a total cytoplasmic RNA fraction. These experiments indicate that the level of gene transcription is at least fourfold higher in the brain than in nonneural tissues and that the majority of the neural transcripts are unique to the brain.

B. Clonal Analysis of Brain RNA

An estimate of RNA complexity in the adult rat brain can also be obtained through the analysis of clones from a cDNA library derived from brain mRNA. One hundred ninety-one clones were randomly selected, the only restriction being that the cDNA be longer than 500 bp. The cDNA clones were classified based on their tissue distribution by Northern blot analysis. Of these clones, 154 detected mRNA species expressed in rat tissues (the remaining clones could represent cloning artifacts).

Class I clones (29 of the 154) hybridize to RNAs that are present in essentially equivalent amounts in the brain, liver, and kidney. These clones could correspond to genes coding for "housekeeping" proteins, which perform common functions in most cell types. Class II clones (41 of the 154) correspond to RNAs that are expressed in the three tissue types, but are differentially regulated. Class III and IV clones (43 and 41 of the 154, respectively), approximately one-half of all clones, correspond to mRNA species that are uniquely expressed in the brain. Those of class IV are of particular interest, in that they are thought to correspond to rare brain-specific mRNAs present at low levels in only a small population of brain cells.

To calculate RNA sequence complexity and obtain an estimate of the number of genes transcribed in the brain, all four classes of clones must be considered. Calculations for class I–III clones indicate that approximately 2640 different mRNAs are present in the brain at an abundance of 0.01% or greater and have a sequence complexity of 8×10^6 nucleotides. Accurate estimates of abundance and mRNA size are difficult to obtain for class IV clones, since the majority of these clones do not detect mRNA species by Northern blot analysis. If a value of 1.4×10^8 is taken for total brain mRNA sequence complexity, class IV clones should account for a complexity of 1.3×10^8. The rarest detectable class III clones detected mRNAs with an average length of 3900 nucleotides. Given the trend that rare brain mRNAs tend to be longer than more abundant species, one can estimate that 30,000 class IV mRNAs ($1.8 \times 10^8/3900$) exist in the brain.

C. Structure of Brain-Specific mRNAs

By analyzing 39 of the brain-specific clones, particularly a subset of 15 clones corresponding to rare mRNA, several unique characteristics of brain-specific mRNAs have been observed. Rare brain-specific transcripts that account for most of the mRNA complexity and mass are, on average, 5000 nucleotides in length, compared to 1600 nucleotides in nonneural transcripts. Structurally, the brain transcripts contain both longer coding and noncoding regions. The longer coding regions could indicate that some brain-specific proteins are synthesized as complex polyproteins (e.g., the proopiomelanocortin gene, which gives rise to many hormones and neuropeptides). This arrangement could allow for the coordinate expression of sets of proteins in the brain. Such a mechanism could be suitable for a highly specialized tissue that has recently undergone extreme expansion in function and has not had sufficient evolutionary time to scatter coding regions throughout the genome and set up coordinate regulation. The presence of longer noncoding regions in brain mRNA species could provide signals for molecular regulatory events that are unique to the brain.

II. ORGANIZATION OF CHROMATIN IN CORTICAL NEURONS

A. Introduction

Hybridization analysis of neuronal and glial nuclear RNA to single-copy DNA suggests that the high sequence complexity of brain RNA is primarily due to neuronal nuclear RNA. To accommodate this complex transcriptional activity, modifications to neuronal chromatin structure might be required. In cortical neurons in the mammalian brain, a temporal correlation has been observed between a rearrangement of chromatin structure and an increase in the RNA template activity of isolated nuclei.

B. Developmental Shortening of the Nucleosomal DNA Repeat Length

Eukaryotic DNA is organized into repeating nucleoprotein units, termed nucleosomes. Each nucleosome is composed of a core particle consisting of two copies of each of the four histones H2A, H2B, H3, and H4, about which are wrapped $1\frac{3}{4}$ turns of double-stranded DNA. Histone H1 is associated with a variable length of linker DNA, which bridges two adjacent core particles. The total length of DNA associated with a nucleosome (i.e., the core particle plus the linker region) is determined by digesting purified nuclei with the enzyme micrococcal nuclease. This enzyme cuts chromatin within the linker region. By electrophoresing the digestion products through a low-percentage polyacrylamide or agarose gel, the nucleosomal DNA repeat length can be determined. For a wide variety of cell types this value is approximately 200 bp. Cortical neurons in the adult mammalian brain demonstrate an atypically short nucleosomal repeat length of 165–170 bp, which reflects a shorter length of the linker DNA. [*See* DNA in the Nucleosome; Histones.]

In mouse, rat, and rabbit, the short nucleosomal repeat length in cortical neurons is not present at birth, but appears during the first week of postnatal development, when cortical neurons are postmitotic, postmigratory, and about to enter a differentiation phase. Observations in guinea pig cortical neurons and rat hypothalamic neurons support this hypothesis. For these two neuronal populations the shift to the short repeat length occurs *in utero*. In guinea pigs the gestation period is three times as long as in rat or mouse, such that the guinea pig brain is more advanced in differentiation at birth. A similar situation occurs for rat hypothalamic neurons, in that the shortening of the repeat length occurs by Day 19 of gestation, which developmentally correlates to the differentiation of the hypothalamus.

A shortening of the nucleosomal repeat length from 200 bp to 167 bp has also been demonstrated for cultured neurons from 16-day fetal rat in response to the thyroid hormone triiodothyronine (T3). The shortening of the nucleosomal repeat length follows a lag period of 15 days. This rearrangement in chromatin structure could result from a series of molecular events mediated by the binding of the T3 hormone–receptor complex to specific chromatin regions. It is of interest that the number of chromatin-associated T3 receptors has been reported to significantly increase at birth.

C. Analysis of Histones in Neurons of the Cerebral Cortex

The shortening of the nucleosomal repeat length in cortical neurons has been investigated in relation to possible changes in histones of the nucleosome core and/or linker region. By polyacrylamide gel electrophoresis, no significant qualitative differences in histones have been detected between cortical neurons, which possess the short repeat length, and glial or kidney nuclei, which possess the typical 200-bp repeat length. The major difference in histone composition is quantitative. Cortical neuronal nuclei have a low H1 content, with approximately 0.5 H1 molecule per nucleosome. This 50% reduction in the amount of linker histone in cortical neurons could affect the second order of chromatin packing, which involves the folding of nucleosomes into a solenoid structure (see Section II,E).

Fluctuations in the relative proportions of H1 subtypes have also been reported during postnatal development. These changes, however, are not unique to cortical neurons and are detected in nondividing cell cultures. A histone H1 variant, $H1^0$, accumulates in cortical neurons during postnatal development following the shift in repeat length (8–18 days postnatally). In nonneural systems $H1^0$ has been implicated in the maintenance of the differentiated cell type.

D. DNase I Sensitivity and Transcriptional Activity of Cortical Neurons

The DNase I digestibility and the RNA template activity of cortical neurons before and after the shift to the atypical short repeat length have been investigated. Pancreatic DNase I recognizes and preferentially digests open or decondensed chromatin. Experiments were performed to compare the digestion kinetics of [^{3}H]thymidine-labeled neuronal nuclei isolated from rat cortex at 1 and 8 days postnatally. Similar digestion kinetics are observed for cortical neuronal nuclei and kidney nuclei isolated 1 day after birth. At 8 days of postnatal development, however, the DNase I digestibility of the neuronal nuclei, but not that of the kidney nuclei, is increased. This suggests that neuronal chromatin is in a less condensed state after the shift in nucleosomal repeat length. Support for this concept comes from studies on yeast that demonstrate a short DNA repeat length and decondensed chromatin. In addition, electron-microscopic studies indicate that adult neuronal nuclei contain little, if any, discernible heterochromatin (i.e., condensed chromatin).

An analysis of the *in vitro* transcriptional activity of isolated rat neuronal nuclei relative to the appearance of the short nucleosomal repeat length has also been performed. These experiments show that neuronal nuclei isolated after the shift (i.e., at 11 days) incorporate higher levels of [^{3}H]UTP into RNA, relative to 1.5-day-old neuronal nuclei. Assays for RNA transcription of isolated kidney nuclei, which do not undergo postnatal shortening of the nucleosomal repeat length, revealed no developmental increases in template activity. These experiments indicate that the postnatal appearance of the short repeat length in cortical neurons is accompanied by a decondensation of chromatin and an elevation of transcriptional activity.

E. Chromatin Conformation of Genes in Cortical Neurons

The DNase I digestion experiment described in the previous section analyzes total neuronal chromatin and does not address whether differences in chromatin conformation exist between genes that are transcribed and genes that are transcriptionally silent in cortical neurons. To determine whether differences exist in chromatin conformation (i.e., the folding of nucleosomes into a solenoid) between these two gene classes, Southern blot analysis of DNase I digests has been performed. In this analysis DNA is purified from DNase I-digested nuclei and cut with a restriction enzyme, and the products are analyzed by agarose gel electrophoresis. The DNA is then transferred to a nylon membrane and hybridized with labeled DNA probes that correspond to genes that are (e.g., the neuron-specific enolase gene) or are not (e.g., the albumin gene) transcribed in cortical neurons.

Neuron-specific genes demonstrate an enhanced sensitivity to DNase I in cortical neurons compared to liver nuclei. This result is in agreement with experiments that show a correlation between gene transcription and increased DNase I sensitivity. The albumin gene, however, is also relatively sensitive to DNase I in cortical neurons. This suggests that both transcribed and nontranscribed genes are relatively sensitive to nuclease digestion in cortical neurons. The short repeat length chromatin in cortical neurons therefore could result in a chromatin solenoid fiber of reduced stability that renders all DNA sequences sensitive to DNase I. This reduced stability might be caused by conformational restraints imposed by the short nucleosomal repeat length and/or the 50% reduction in linker histone H1. A role for H1 in higher-order chromatin packing is suggested by the observation that an H1 polymer lies close to the fiber axis and stabilizes the helical folding of the nucleosomes into the solenoid structure. A partial removal of H1 might be sufficient to disrupt the H1 polymer scaffold and partially unfold the chromatin solenoid.

The functional significance of the relatively unfolded chromatin conformation in cortical neurons is not known. One possibility is that this chromatin conformation allows cortical neurons to regulate gene transcription more efficiently by bypassing an initial step in gene activation—namely, the unfolding of chromatin regions, which renders genes accessible to the transcriptional machinery. This state of transcriptional readiness in cortical neurons could relate to the complexity of gene transcription in the postnatal mammalian brain, the plasticity of postnatal brain development, and the ability of the brain to respond quickly to environmental stimuli.

III. NUCLEAR REGULATORY PROTEINS

A. Introduction

Histones are a relatively simple class of nuclear proteins involved in the coiling of DNA into nucleosomes and the folding of nucleosomes into a solenoid. They are considered to have an inhibitory effect on gene transcription. The positioning of a nucleosome at transcription regulatory sequences can prevent the formation of the transcription initiation complex. Also, considering their involvement in the higher-order folding of nucleosomes into a solenoid, histones might inhibit transcription by packaging a gene within the chromatin solenoid such that it is not accessible to the transcriptional machinery. Nonhistone nuclear proteins, by comparison, are a complex class of proteins involved in a variety of nuclear events, including gene activation.

B. Gel Analysis of Brain Nonhistone Nuclear Proteins

Analyses of brain nonhistone nuclear proteins by one- or two-dimensional gel electrophoresis have demonstrated differences, primarily quantitative, between brain regions and cell types and during brain development and aging. Brain nuclei contain more high-molecular-weight proteins than do other tissue types. Large differences also exist in the amount of nuclear protein within the different classes of brain nuclei. Cortical neuronal nuclei contain almost twice as much nonhistone nuclear protein compared to non-astrocytic glial nuclei. These quantitative differences could reflect differences in template activity among cell types, given the observation that chromatin more actively engaged in RNA synthesis has a higher nonhistone protein content.

Developmentally, differences in nonhistone nuclear proteins have been detected in cerebellar and cortical neurons when neurons have stopped dividing and have begun to differentiate. The significance of these changes is difficult to determine, because the functions of the majority of these proteins are not known. These developmental differences could represent fluctuations in the level of a variety of proteins with diverse enzymatic or regulatory functions. In cortical neurons, for example, fluctuations likely occur in the level of the DNA replicative enzyme, given the cessation of cell division at birth. Also, during early postnatal development, when terminal cortical neuron differentiation initiates, changes in gene expression occur (i.e., an increase in brain RNA sequence complexity). Fluctuations therefore could occur in the synthesis of the enzyme DNA-dependent RNA polymerase II or transcription regulatory proteins.

C. Methods for Identifying Regulatory Sequences Associated with Brain-Specific Genes

Analyzing brain nonhistone nuclear proteins by gel electrophoresis, while providing evidence for cell type and developmental differences, does not allow for determination of their functional significance. This experimental approach is also limited in that only the most abundant proteins are studied. One important group of nuclear proteins that, owing to their low abundance, would be difficult to detect by gel analysis are those involved in the transcriptional regulation of brain-specific genes. Because of their low abundance, these nuclear proteins are frequently characterized by their target DNA binding sites.

One method of identifying DNA regulatory sequences is the mapping of DNase I-hypersensitive sites. These sites encompass short stretches (i.e., 100–200 bp) of chromatin that are very sensitive to low levels of DNase I and are situated primarily, but not exclusively, at 5′ flanking sequences of active and potentially active genes. Structurally, these sites correspond to a discontinuity in the repeating arrangement of nucleosomes along the DNA strand. In nonneural systems, DNase I-hypersensitive sites are located near gene regulatory sequences, where the binding of nuclear proteins could be responsible for the generation of these sites.

Another method for identifying regulatory DNA sequences involves (1) the introduction of a cloned gene (including its flanking regulatory sequences) into cells grown in tissue culture and (2) subsequently assaying for transcriptional activity. Such transfection experiments utilize cultured cells that can transcribe that gene. Modifications must be made to the cloned gene such that its transcription can be distinguished from the cells' endogenous mRNA. Usually, putative regulatory sequences are physically linked to a "reporter" gene, whose activity can be easily monitored (e.g., bacterial genes coding for either chloramphenicol acetyltransferase or β-galactosidase activity). To analyze the role of a short DNA sequence as a transcriptional regulatory element, deletions are performed. The absence of reporter gene activity indicates the involvement of the deleted sequence in transcriptional control.

These gene constructions can also be introduced into animals by injection into fertilized eggs (i.e., transgenic mice). This method, although technically more difficult, is superior to the transfection of cell cultures, since the introduced cloned gene is exposed to a complete set of cell-specific transcriptional signals during embryonic development.

D. Analysis of DNA-Binding Proteins

After identifying putative DNA regulatory regions, experiments can be performed to determine whether they contain protein-binding sequences. The most direct method of investigation is gel retardation assays. These experiments involve the incubation of a protein containing nuclear extract with a short labeled DNA fragment containing the putative regulatory sequence. Following electrophoresis through a low-ionic-strength polyacrylamide gel, DNA complexed to protein is detected as bands with reduced mobility (relative to the free unbound DNA). To identify the protein binding site at the nucleotide level, DNase I protection footprinting experiments are performed. DNA sequences that bind nuclear proteins are protected from digestion by DNase I and, as such, generate a footprint, or gap, within the nucleotide ladder following electrophoresis through a denaturing polyacrylamide gel. Once the specific binding sequence has been identified, the next step is to purify and characterize the regulatory nuclear protein. This can be accomplished by passing a nuclear extract through a chromatography column bearing a matrix to which is bound the target DNA with the binding sequence for that protein.

E. A Role for Silencer Elements in the Tissue Specificity of Neuronal Gene Transcription

The tissue specificity of gene expression in neuronal populations is likely due to the chromatin conformation of the gene, the presence of noncoding DNA regulatory elements that either activate or repress gene transcription, and trans-acting transcription factors that bind to the regulatory elements.

For a number of organs, the tissue specificity of gene transcription is frequently the result of a DNA consensus sequence to which a tissue-specific transcription factor binds to activate gene transcription in the appropriate tissue. Although sequences required for the transcription of a particular gene in a neuronal population have been identified, a neuron-specific enhancer consensus sequence has not been proposed. Comparisons of DNA regions flanking several neuronal genes have revealed short stretches of sequence homology, however, a functional significance still needs to be demonstrated.

In neuronal populations, a specific activator consensus sequence has not been proposed. However, a common silencer element, termed a neural-restrictive silencer element, has recently been described for two neuron-specific genes (the SCG10 gene, which encodes a neuronal-specific membrane-associated protein that accumulates in the axons and growth cones of growing neurons, and the type II sodium channel gene). For these two genes, neuron-specific transcription is due to a combination of positive regulatory elements that direct transcription in both neuronal and nonneuronal tissues and a silencer element that restricts expression in nonneuronal cells. This silencing mechanism appears to be required for these two genes since both are members of multigene families, some members of which are expressed outside the nervous system. These genes may share, with other family members, activator sequences that permit the transcription of the gene in a number of cell types. The tissue-specific expression is mediated by the neural restrictive silencer element and a nuclear factor that binds the silencer element in nonneuronal cell types. Other silencer elements have also been reported to restrict the transcription of a gene to neuronal subsets.

BIBLIOGRAPHY

Brown, I. R. (1983). The organization of DNA in brain cells. *In* "Handbook of Neurochemistry" (A. Lajtha, ed.), Vol. 5. Plenum, New York.

Brown, I. R., and Greenwood, P. D. (1982). Chromosomal components in brain cells. *In* "Molecular Approaches to Neurobiology" (I. R. Brown, ed.). Academic Press, New York.

Cestelli, A., Di Liegro, I., Castiglia, D., Gristina, R., Ferraro, D., Salemi, G., and Savettieri, G. (1987). Triiodothyronine-induced shortening of chromatin repeat length in neurons cultured in a chemically defined medium. *J. Neurochem.* **48**, 1053.

Chaudhari, N., and Hahn, W. E. (1983). Genetic expression in the developing brain. *Science* **220**, 924.

Chikaraishi, D. M. (1979). Complexity of cytoplasmic polyadenylated and nonpolyadenylated rat brain ribonucleic acids. *Biochemistry* **18**, 3249.

Greenwood, P. D., and Brown, I. R. (1982). Developmental changes in DNase I digestability and RNA template activity of neuronal nuclei relative to the postnatal appearance of a sort DNA repeat length. *Neurochem. Res.* **7**, 965.

Greenwood, P. D., Silver, J. C., and Brown, I. R. (1981). Analysis of histones associated with neuronal and glial nuclei exhibiting divergent DNA repeat lengths. *J. Neurochem.* **37**, 498.

Heizmann, C. W., Arnold, E. M., and Kuenzle, C. C. (1980). Fluctuations of non-histone chromosomal proteins in differentiating brain cortex and cerebellar neurons. *J. Biol. Chem.* **25**, 255, 11504.

Hoyle, G. W., Mercer, E. H., Palmiter, R. D., and Brinster, R. L. (1994). Cell-specific expression from the human dopamine β hydroxylase promoter in transgenic mice is controlled via a combination of positive and negative regulatory elements. *J. Neurosci.* **14**, 2455.

Ivanov, T. R., and Brown, I. R. (1983). Developmental changes in the synthesis of nonhistone nuclear proteins relative to the appearance of a short nucleosomal DNA repeat length in cerebral hemisphere neurons. *Neurochem. Res.* **9**, 1323.

Ivanov, T. R., and Brown, I. R. (1989). Genes expressed in cortical neurons—Chromatin conformation and DNase I hypersensitive sites. *Neurochem. Res.* **14**, 129.

Ivanov, T. R., and Brown, I. R. (1992). Interaction of multiple nuclear proteins with the promoter region of the mouse 68-kDa neurofilament gene. *J. Neurosci. Res.* **32**, 149.

Milner, R. J., and Sutcliffe, J. G. (1983). Gene expression in rat brain. *Nucleic Acids Res.* **11**, 5497.

Mori, N., Schoenherr, C., Vandenbergh, D. J., and Anderson, D. J. (1992). A common silencer element in the SCG10 and type II Na^+ channel genes binds a factor present in nonneuronal cells but not in neuronal cells. *Neuron* **9**, 45.

Sutcliffe, J. G. (1988). mRNA in the mammalian central nervous system. *Annu. Rev. Neurosci.* **11**, 157.

Chromosome Anomalies

JEANNE M. MECK
Georgetown University Medical Center

ROBERT C. BAUMILLER
Xavier University

JAN K. BLANCATO
Oncor, Inc.

GLOSSARY

Aneuploidy In human genetics, refers to loss or gain of whole chromosome(s)

Genome Haploid set of chromosomes

Haploid 23 chromosomes (i.e., 1–22 plus X or Y)

Isochromosome Chromosome whose p and q arms are identical genetically

Karyotype Arrangement of the chromosomes of a single cell in pairs in ordered array

p Arm Short arm

q Arm Long arm

-Somy With prefix, indicates the number of copies of a specific chromosome in a cell line (e.g., trisomy 21 or monosomy X)

CHROMOSOME ANOMALIES OCCUR SPONTANEously in all organisms at a low rate. Radiation, certain chemicals, and other environmental stresses can increase the frequency of such changes. The human genome is remarkably free of any major structural chromosomal polymorphisms even in populations that have been breeding isolates for many hundreds of years. This uniformity argues for a unique origin of *Homo sapiens*. This argument is strengthened by the fact that the higher apes have a genome remarkably similar to that of humans in its content but differing by a number of structural alterations, each of which can occur spontaneously but is almost invariably lost from a population.

Chromosome alteration resulting in loss or gain of autosomal material is generally lethal. Alterations that are balanced but produce altered quantities of genetic material in some fraction of the gametes subsequent to normal recombination or to normal meiotic division are rapidly lost from a population, usually persisting for three generations or less.

I. POLYPLOIDY

The human is a diploid species with a relatively brief haploid stage in the male sperm cell and a transient haploid stage in the female, which occurs after sperm penetration and presyngamy.

Triploidy (three genomes, 69,XXX; 69,XXY; 69,XYY) usually occurs because two sperm enter the egg and two male pronuclei join with the haploid female pronucleus. Triploidy may also result if the female product after meiosis I fails to undergo meiosis II successfully and the diploid female nucleus joins with the male pronucleus. Rarely, a fetus can survive to birth, and cases have been reported to have survived for several weeks.

Tetraploidy (92,XXYY; 92,XXXX) occurs because of chromosome doubling without nuclear division after syngamy. The diploid XY or XX zygote becomes

ENCYCLOPEDIA OF HUMAN BIOLOGY, Second Edition, VOLUME 2. Copyright © 1997 by Academic Press. All rights of reproduction in any form reserved.

tetraploid and almost always fails to reach delivery. Other mechanisms leading to tetraploidy can be postulated, but these would necessitate the congeries of two or more exceedingly rare phenomena. Polyploidy is an evolutionary mechanism in plants but is not tolerated in higher organisms.

II. AUTOSOMES

A. Changes in Chromosome Number

There are 23 pairs of chromosomes in the normal human karyotype. The autosomes, or nonsex chromosomes, have been numbered 1 through 22, approximating decreasing size order. The 23rd pair is the sex chromosomes. In the normal male, one X and one Y chromosome are contained in each cell; in the normal female, there are two X chromosomes per cell and no Y chromosomes. [*See* Chromosomes.]

Approximately one of every 200 liveborns has a numerical chromosome abnormality. The most common autosomal trisomy is trisomy 21 (i.e., three copies of the #21 chromosome in each cell). This chromosome abnormality is responsible for Down syndrome and affects approximately one of every 800 liveborns. Other less common autosomal abnormalities include trisomy 13 (Patau syndrome), trisomy 18 (Edward syndrome), and, rarely, trisomy 9 and trisomy 22. No other autosomal trisomies in every cell in the liveborn have been definitively documented. [*See* Down Syndrome, Molecular Genetics.]

By contrast, trisomies for every autosome have been documented in recovered products of conception. Half of all miscarried fetuses have a chromosome abnormality, and the most common of these is the autosomal trisomy. Trisomy 16 is the most common and is found in 32% of chromosomally abnormal spontaneously aborted pregnancies. Autosomal monosomies (i.e., only one copy of a particular chromosome) are believed to occur in early first-trimester spontaneous abortions. Monosomy 21 and 22 are the only exceptions and have been reported in rare instances in liveborns. [*See* Abortion, Spontaneous.]

The human organism tolerates little autosomal aneuploidy. Addition or loss of even a small amount of genetic material is usually not compatible with development. A conceptus with an imbalance is often lost before or in the process of implantation. Even trisomy 21 represents at best only 12% of those aneuploid fetuses that reach 8–12 weeks gestation.

Mitotic and meiotic numerical errors occur by the mechanisms of nondisjunction and anaphase lag. Mitosis is the cell division process that occurs in somatic cells. Before mitosis, replication of the chromosomal material occurs so that each chromosome consists of two strands or chromatids connected at the centromere. During the process of mitosis, the chromatids of each chromosome are separated and pulled to opposite poles. A new nuclear and cytoplasmic membrane then surrounds each set of chromosomes. The resulting daughter cells will have identical genetic material. When an error in the separation of the two strands of the chromosome occurs (nondisjunction), this can lead to the existence of two karyotypically different cell lines. This condition is known as mosaicism. For example, a chromosomally normal embryo may undergo a single nondisjunctional mitotic event during embryogenesis in which a trisomy 21 cell and a monosomy 21 cell are formed amid the normal cells. The monosomic cell line cannot reproduce itself competitively; however, the trisomy 21 cell line can. Many individuals who have a mixture of normal and abnormal cells are believed to be the result of a trisomic fertilized egg that has lost the extra chromosome in one cell, which then proceeds to form a lineage of normal cells among the abnormal cells. An individual with such a mixture of normal cells and trisomy 21 cells is mosaic for Down syndrome. Such individuals may have a higher mental function and less severe physical abnormalities than those who have no normal cells. However, this is not always the case because the distribution of normal to abnormal cells as observed in a lymphocyte chromosome analysis may not be the same as that found in other tissues, which may have a greater effect on phenotype such as the central nervous system. [*See* Mitosis.]

When all cells of an individual are karyotypically abnormal, it is usually the result of a meiotic error. Meiosis is a process of cell division in germ cells, which leads to the formation of the haploid egg or sperm. Meiosis consists of two divisions: MI and MII. Before MI, the oogonia or spermatogonia undergo DNA replication so that there are 46 chromosomes, each consisting of two chromatids. During MI, the homologous chromosomes are paired. Then, the homologues migrate to opposite poles. Nuclear division takes place and results in two daughter nuclei. Each of the two daughter nuclei should contain only 23 chromosomes, each with two chromatids. MII is a mitotic-like division in which there is separation of the chromatids in all 23 chromosomes. The chromatids of each chromosome pair go to opposite poles and different daughter nuclei. [*See* Meiosis.]

A numerical chromosome abnormality occurs as a result of a mistake in segregation in MI or MII in the male or female. However, the products of maternal meiotic errors are found more frequently than those of paternal meiotic errors (95% versus 5% in Down syndrome and trisomy 18). MI errors are more common than MII errors. During nondisjunction, there is failure of the homologues to migrate to opposite poles in MI or the chromatids to do so in MII. This results in one daughter cell gamete becoming disomic and the other becoming nullisomic for a particular chromosome. Nullisomy, or lack of a particular chromosome in a gamete, leads to a monosomic embryo. A disomic gamete, however, when joined with a normal monosomic gamete of the other partner, leads to trisomy.

An unusual phenomenon resulting from abnormal cell division known as uniparental disomy has been shown to be a causal mechanism in genetic disease. In uniparental disomy, both homologues of a chromosome pair in a diploid cell originate from the same parent, whereas the other parent makes no contribution for that chromosome pair. One mechanism for uniparental disomy is the union of a normal monosomic gamete and an abnormal disomic gamete with subsequent loss of the single chromosome contributed by the normal gamete. For example, if an egg had two #15 chromosomes and the sperm contributed one #15 at fertilization, this would lead to trisomy 15. If the paternal #15 was subsequently lost in a mitotic division, then that cell and its progeny would have a normal chromosome number but both #15's would be maternal in origin. If the chromosomes in question are identical to each other, perhaps resulting from an error in the second meiotic division, this is known as "isodisomy." However, if the chromosomes from the same parent are different, perhaps resulting from a first meiotic division error, this is termed "heterodisomy." Uniparental disomy cannot usually be detected by means of routine cytogenetic analysis. In general, molecular techniques must be used to track the parental source of the genes for a particular pair of homologues. In a landmark case study, isodisomy was shown to be the cause for the autosomal recessive disease cystic fibrosis, in a child who had only one parent that carried the cystic fibrosis gene. Both of the child's #7 chromosomes that carried the cystic fibrosis gene were shown to be maternal in origin.

Meiotic chromosome errors have been demonstrated to be more frequent in children of women of advanced maternal age. For example, a 26-year-old woman has an age-related risk of 1/1176 to have a child with trisomy 21 (Down syndrome), whereas a 40-year-old woman has a 1/100 risk. There is no such correlation known with errors leading to mosaicism.

There appears to be no influence of socioeconomic status, race, or parity in this phenomenon. There does not appear to be a substantially increased risk for chromosomally abnormal children associated with advanced paternal age. The reason for the increase in chromosomally abnormal offspring in older women is not known. There are, however, several hypotheses.

B. Changes in Autosome Structure

Alteration in the structure of the chromosomes is not associated with advanced maternal age. Instead, it is more likely the result of exposure to radiation, chemicals, or other mutagens.

1. Structural Rearrangements Involving a Single Chromosome

a. Deletion

If a chromosome is broken in one place and the piece not containing the centromere is lost, this is known as a terminal deletion. Cri du chat (5p-) and Wolf-Hirshhorn syndrome (4p-) are examples of terminal deletions (Fig. 1).

If a chromosome is broken in two places and the middle section is lost, this is known as an interstitial deletion. Prader-Willi syndrome and Angelman syndrome involve a small interstitial deletion of 15q; aniridia-Wilms tumor complex involves an interstitial deletion of 11p.

There are a number of syndromes now known to be the result of small deletions. These "microdeletion syndromes" can be terminal or interstitial in nature. The most common ones are listed in Table I. Some of these syndromes may also result from other mechanisms, such as point mutation or translocation.

A ring chromosome is a specialized type of deletion in which the telomeres of the p and q arms are lost. The result is that the "sticky ends" remaining on either end of the chromosome join together to form a ring. Ring chromosomes are rare because they are usually unstable in mitosis and are easily lost.

Individuals who have deletions have partial monosomy for the portion of chromosome material that is missing.

"Contiguous gene syndrome" is a term that refers to microdeletion syndromes that are known to affect more than one gene and thus combine the phenotypes of the disorders involved. Some of these deletions are large enough to be detectable cytogenetically; others

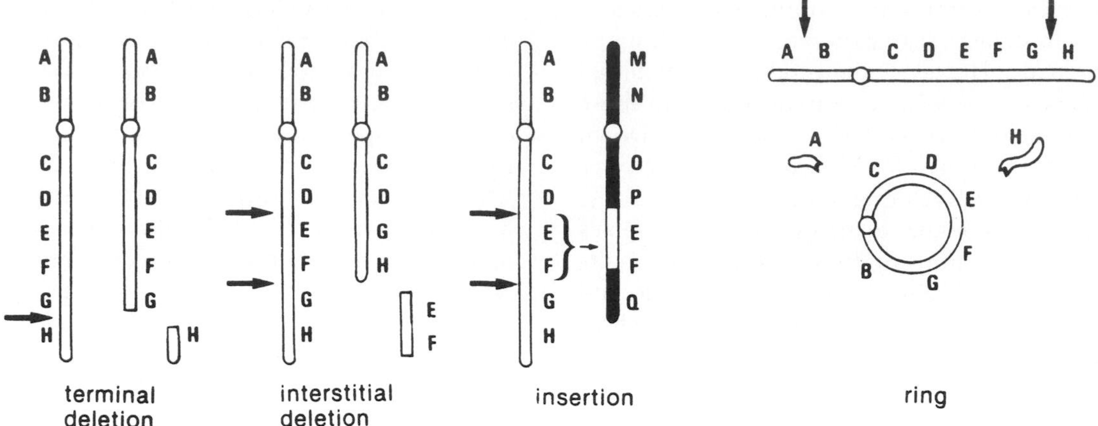

FIGURE 1 Examples of alterations in chromosome structure: terminal deletion, interstitial deletion, insertion, and ring chromosome. [From M. Thompson (1986). "Genetics in Medicine." Saunders, Philadelphia.]

require the use of molecular techniques for detection. Examples of contiguous gene syndromes on the autosomes include the aniridia-Wilms tumor gene complex (11p13) and Langer-Giedion and the trichorhinophalangeal syndromes (8q24).

b. Duplication

Segments of chromosomes are sometimes duplicated by errors in DNA replication or exchange and arranged in a tandem fashion. The duplicated portion can be oriented in the same direction as the original piece of chromosome (direct duplication) or it may be inverted (inverted duplication).

Individuals who have duplications have partial trisomy for a portion of the chromosome.

TABLE I
Microdeletion Syndromes

Syndrome	Chromosome region deleted
Wolf-Hirschhorn	4p16.3
Williams	7q11.23
Langer-Giedion	8q24.11q24.13
Retinoblastoma	13q14
Angelman	15q11q13
Prader-Willi	15q11q13
Smith Magenis	17p11.2
Miller-Dieker	17p13.3
DiGeorge/velocardiofacial	22q11.2

c. Inversion

If there are two breaks in a chromosome and the middle segment is flipped around, this structural change is a chromosomal inversion. There is no extra or missing material, simply a rearrangement. Most inversion carriers are phenotypically normal. Inversions are of two types: pericentric and paracentric. In pericentric inversion, one of the breaks is in the short arm, and the other break is in the long arm. The inverted segment, therefore, contains the centromere. Individuals who have pericentric inversions are at an increased risk of having chromosomally unbalanced offspring, which have duplication of some of the chromosome and deficiency of another portion of the chromosome. The most common pericentric inversion involves the #9 chromosome. It is found in a frequency of 0.13% in the Caucasian population and 1.07% in the Black population. In paracentric inversion, both breaks are in the same arm of the chromosome (i.e., the broken portion is only in the p arm or only in the q arm). The centromere is not contained in the inverted segment.

Individuals who have paracentric inversions are also at increased risk for having chromosomally unbalanced offspring. However, the risk is lower than for pericentric inversion carriers because the resulting abnormalities in the case of a paracentric inversion are dicentric chromosomes (ones with two centromeres) and acentric fragments (no centromeres). Both of these situations are unstable and lead to early embryonic death because of monosomy.

d. Isochromosome

An isochromosome is one in which the two arms of the chromosome are identical (Fig. 2). For example, an isochromosome of the long arm of chromosome 21 consists of two copies of 21q joined by a centromere. It is believed that isochromosome formation can arise by misdivision of the centromere. Instead of dividing longitudinally so that one chromatid goes to each daughter cell, the chromatids divide transversely so that the p and q arms go to different daughter cells.

2. Structural Rearrangements Involving Two or More Chromosomes

a. Translocation

A translocation involves breakage and rearrangement of two or more chromosomes. An individual who has a translocation and maintains the correct amount of chromosome material is known as a balanced carrier. Such individuals are usually phenotypically normal. However, they are at an increased risk of having chromosomally unbalanced (abnormal) offspring. There are two basic types of translocations: (1) robertsonian and (2) reciprocal.

A robertsonian translocation involves breakage at or near the centromeres of two acrocentric chromosomes with fusion of their long arms and loss of their short arms. (An acrocentric chromosome is one in which the centromere is near the end of the chromosome. The p arm of an acrocentric chromosome is not composed of essential structural or regulatory genes.) The acrocentric chromosomes are chromosomes 13, 14, 15, 21, and 22. Any two acrocentrics may join together to form a robertsonian translocation. A balanced carrier of a robertsonian translocation has only 45 separate chromosomes per cell, but one is really two chromosomes joined together. For example, if an individual is a carrier of a robertsonian translocation involving chromosomes 13 and 14, that person will have in addition to one normal chromosome 13 and one normal 14, the translocation chromosome involving the long arms of 13 and 14 joined together. Such individuals can have normal children—some with normal karyotypes and others with balanced carrier karyotypes. They also may conceive fetuses with trisomy 13 or 14 or monosomy 13 or 14. Most of these will end in early spontaneous abortions with only the trisomy 13 conceptuses having a chance to survive gestation.

A reciprocal translocation involves the exchange of chromosome material between any two chromosomes at any point on the chromosome. In this situation, the balanced carrier has 46 chromosomes: however, the two chromosomes involved in the translocation are structurally rearranged (Fig. 3). Like the robertsonian translocation carriers, reciprocal translocation carriers can have normal children—some with normal karyotypes and some with balanced carrier karyotypes. However, they also have an increased risk of having conceptions in which there is partial monosomy of some chromosome material and partial tri-

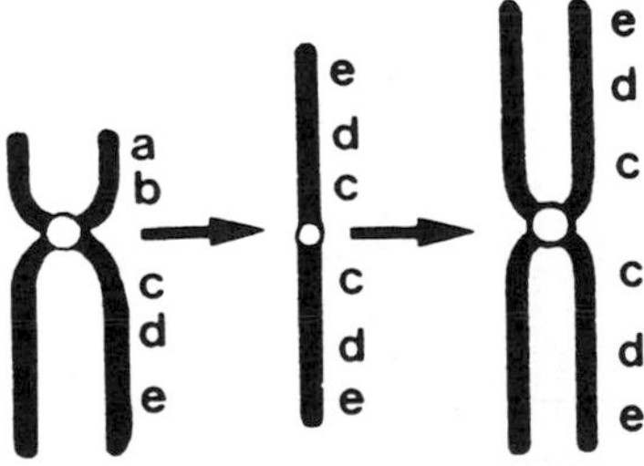

FIGURE 2 Formation of an isochromosome by the mechanism of misdivision of the centromere leading to an isochromosome or duplication of the long arm and complete absence of the short arm. After DNA replication, the isochromosome consists of two chromatids joined together. [From M. Thompson, (1986). "Genetics in Medicine." Saunders, Philadelphia.]

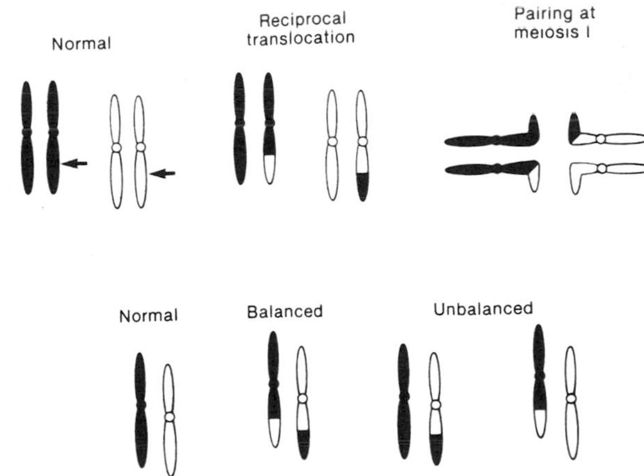

FIGURE 3 Breakage of two nonhomologous chromosomes leading to a reciprocal translocation. Pairing during the first meiotic division results in a cross-shaped figure. Four different types of gametes that may result include: (1) a normal chromosome complement: (2) a balanced chromosome complement: (3) and (4) two different types of chromosome imbalances, each resulting in partial monosomy and partial trisomy. [From M. Thompson, (1986). "Genetics in Medicine." Saunders, Philadelphia.]

somy for other chromosome material. The risk of having chromosomally abnormal offspring varies from one translocation to another and occasionally varies depending on the sex of the carrier parent. In general, the larger the pieces of chromosome material involved in the translocation, the smaller the chance of having a liveborn with a chromosome abnormality. There are also translocations in which three or more chromosomes may be involved. These are known as complex translocations and are rare.

b. Insertions

A piece of chromosome material that is derived from an interstitial deletion may insert itself into another chromosome that has suffered a single break or into another part of the same chromosome from which it came. For example, there may be two breaks in the long arm of a #7 chromosome and one break in the short arm of the #12 chromosome. The portion of 7q can insert itself into the 12p. The terminus of the 7q may join with the remainder of 7, which includes the centromere (Fig. 1).

3. Acquired Versus Constitutional Changes

Chromosome abnormalities associated with birth defects (e.g., Down syndrome) are constitutional abnormalities (i.e., a karyotypic anomaly that occurred before or during embryogenesis). There are also other chromosome abnormalities that occur later in life and are often associated with malignancies. For example, most people who develop chronic myelogenous leukemia (CML) also develop a change in chromosomes 9 and 22 of some of the nucleated cells of their hematopoietic system. This change is a balanced reciprocal translocation between the long arms of chromosomes 9 and 22 and involves a change in the position of the "abl" oncogene normally located on the number 9 chromosome. The derivative #22 chromosome formed from this translocation is known as the "Philadelphia chromosome." The reorientation of the abl oncogene in its new location on chromosome 22 is believed to be directly associated with the development of the malignancy. As the disease progresses, other karyotype anomalies are also noted in individuals affected with CML. These anomalies include: a second Philadelphia chromosome, trisomy 8, trisomy 19, and isochromosome 17q, all in the malignant line only. Some of the acquired chromosome anomalies associated with malignancies are quite specific. In addition to hematologic malignancies, solid tumors have also been shown to undergo karyotypic changes. In contrast to constitutional abnormalities, which usually affect only one or two chromosomes, acquired abnormalities are often quite complex and involve many chromosomes. Frequently, the modal number of chromosomes per cell is different from 46 and may show extreme hypo- or hyperdiploidy. [*See* Chromosome Patterns in Human Cancer and Leukemia.]

III. SEX CHROMOSOMES

A. Changes in Number

The sex chromosomes (X and Y) allow major deletion (monosomy X) and multiple major duplication (XXX, XXY, XYY up to at least XXXXX). The reasons that such gross cytogenetic anomalies are found are twofold: (1) inactivation of all X chromosomes except one and (2) the almost agenic constitution of the Y.

The human Y chromosome has a small portion at the tip of the short arm, which pairs with the tip of the short arm of the X (Fig. 4). The pairing function and subsequent exchange is considered to be necessary to ensure proper disjunction of X and Y at meiosis. That portion of the X that pairs with the Y remains active on every X chromosome. Also on the short arm of the Y is a locus that is necessary for male development initiation, presently termed the "sex determining region of the Y chromosome." Closely linked to the centromere is the H-Y antigen regulator.

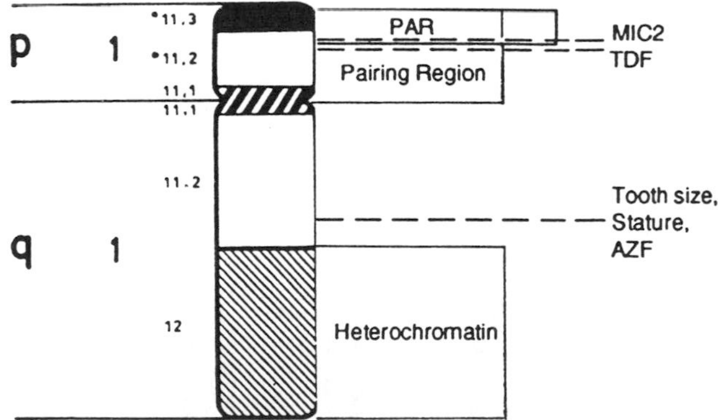

FIGURE 4 Cytogenetic diagram of the human Y chromosome showing the approximate positions of PAR (pseudoautosomal region), pairing region, MIC2 (antigen factor mapped to Xp22.32 and Yp11.3), TDF (testis-determining factor), factors affecting height and tooth size, AZF (factor preventing azoospermia), and heterochromatic region. [From E. Therman (1993). "Human Chromosomes." Springer-Verlag, New York.]

This gene was named from the observation that male to female grafts in isogenic mice did not succeed because of this Y-linked regulator gene(s). Although male specific, neither the product of the H-Y regulatory gene nor its effect on differentiation is known as of this writing. The long arm of the Y has been suspected of containing genes that control spermatogenesis and a gene or genes relating to stature based on height differences between X, XY, and XYY individuals. There are two pseudogenes on the long arm of the Y and a variable-length heterochromatic portion.

The variable length of the heterochromatic region makes the Y polymorphic and thus able to be traced in families and populations at the chromosomal level and on the DNA level. A single Y causes male development regardless of the number of X chromosomes present, but XXY, XXXY, etc., males are sterile. XYY males are not sterile and are taller than their XY sibs on the average. The children of XYY and XY males have a normal sex-determining chromosome complement at similar frequencies (i.e., XX or XY), not XYY or XXY as might be expected. The greater the genetic imbalance (i.e., XYY, XXY, XXYY, XXXXY), the greater the physical and mental impairment. The latter two are severely affected, whereas the XYY individual is only slightly lower in IQ than would have been expected. Socialization of XYY individuals can be normal but is more complicated for XXY, XXYY, and XXXXY individuals because of their sterility, skewed IQ, and sexual ambiguity.

The X chromosome, unlike the Y, is genic, having a gene density similar to that of the autosomes. Genes on the X were mapped before the autosomes because of their phenotypic/genotypic exposure in the male and the ability of knowing about the linkage state in the grandfather and therefore in the mother. Mammalian males and females are amazingly similar. This similarity is ensured by an early gene dosage compensating mechanism, which randomly inactivates the maternal or paternal X chromosome. This activation occurs at the 1000- to 2000-cell stage in the embryo. X inactivation is termed "lyonization" after Mary Lyon, who is credited with first expounding the hypothesis in 1961.

Monosomy X, clinically called "Turner syndrome," is the only monosomy compatible with viable development. X monosomy females are as frequent as one in 2500 females by some reports but rarer by others (one in 10,000), suggesting polymorphic population factors. Data indicate that only 2.5% of monosomy X embryos conceived reach delivery. Thus the relatively agenic Y has a genic constitution that is nonetheless capable of rescuing the X-containing egg. It is thought that the pairing segment of the Y and the always active area of the X are the essential portions.

The Turner female has short stature and gonadal dysgenesis and a wide array of dysmorphic features, each of which occurs with relatively high frequency. The monosomy X female is almost always sterile. The few Turner females who are fertile may have an increased risk of having aneuploid offspring. The condition is usually diagnosed at birth with confirmation by chromosome analysis. The trisomy X individual (XXX) is rarely recognized at birth and often escapes detection throughout life. Menopause is often early. The individual is fertile, and children do not have an increased risk of having offspring with an XXY or XXX chromosome constitution. A lowering of expected IQ occurs in the XXX female as in the XYY male but nevertheless is usually in the normal range. The XXXX and XXXXX females have increasing mental retardation and increasing physical malformations, in both number of systems affected and degree of effect.

A part of the aging process in humans and perhaps contributing to it is the loss of the inactivated X in the female and of the Y in the male. No correlation between any aspect of aging and this loss has been reported.

B. Sex Chromosome Structural Changes

Of peculiar importance for the X chromosome and proper balancing in the genic compensation is the presence of the inactivation center of Xq13. No X chromosome with a deletion of this area has been found. Thus presence of two active X's or portions of two active X's is lethal (the exception being the small, always active region on the short arm of the X, which is capable of pairing with the Y). Therefore any chromosome abnormality resulting in the loss of this area is lethal.

In cases of a balanced X autosome translocation, most cells have the normal X chromosome inactivated, thus leaving one full X active and not influencing the inactivation of part of the autosome attached to an inactivated X. This contrasts with an unbalanced X autosome translocation carrier in whom the normal X usually remains active and the rearranged X becomes inactive. It is believed that the inactivation is random at the time of occurrence but that selection for proper balance is intense, and by birth almost all dividing cells show the appropriate X inactivated.

Various regions of the X are associated with symptoms associated with Turner syndrome. Loss of the Xp2 region results in short stature, whereas loss of the entire p arm results in the full Turner syndrome. Chromosome translocations that are balanced usually have little phenotypic effect except when a breakpoint is in the Xq2 region. This position of the breakpoint leads to gonadal dysgenesis. Such a position effect has been recorded even in the case of a paracentric inversion within the Xq2 region.

Contiguous gene syndromes have also been detected on the X chromosome. Males with these deletions express the abnormal phenotype; females may or may not be affected depending on the percentage of cells in which the X bearing the deletion is active. Whereas structurally abnormal unbalanced X chromosomes are generally selectively inactivated, there is no such selection in females who carry contiguous gene syndromes. Deletions in the region of Xp21 have been associated with a syndrome that includes the phenotypes of Duchenne's muscular dystrophy, glycerol kinase deficiency, adrenal hypoplasia, and mental retardation.

Dicentric chromosomes typically have one active centromere, whereas the other one is inactive or suppressed. In cases in which both centromeres remain active, dicentrics are usually lost in mitosis unless the two centromeres are in close opposition and necessarily travel to the same pole at anaphase. An idic(X)(p22) observed by the author had almost two complete X's jointed together, but only one active centromere. The isodicentric chromosome was inactivated, leaving the normal X as functional.

Because only a small portion of the Y is necessary for normal development, many grossly deleted functional Y chromosomes have been seen in families. Pericentric inversions have been described as well as ring Y's. Translocation of a small portion of the Y to the X or autosome can lead to males with an XX chromosome karyotype. All XX males are thought to be caused by the presence of the testis-determining factor being exchanged to the X by uneven crossing-over or translocational insertion into the X or an autosome.

Y isochromosomes have rarely been described, and only as mosaics. An isochromosome for Y short arms i(Yp) would be expected to have a phenotype of XYY and thus usually escape study. Chromosome techniques in the past were inadequate to distinguish such a chromosome from a partial q arm deletion. Methods for identification of i(Yp) are now available. In the absence of Yp, Yq duplication with X monosomy may not be compatible with successful development.

IV. MOLECULAR CYTOGENETICS

The recently developed technique of fluorescence *in situ* hybridization (FISH) has improved capabilities of detecting structural and numerical chromosomal abnormalities. FISH is based on the same principles as Southern blotting. It employs a DNA probe labeled nonisotopically with a hapten that hybridizes to a target DNA. This specific DNA sequence can then be visualized following binding of the hapten or antibody to a fluorochrome. The specimen for study may be interphase cells, metaphase preparations, or tissue sections that are specially prepared *in situ* on a glass

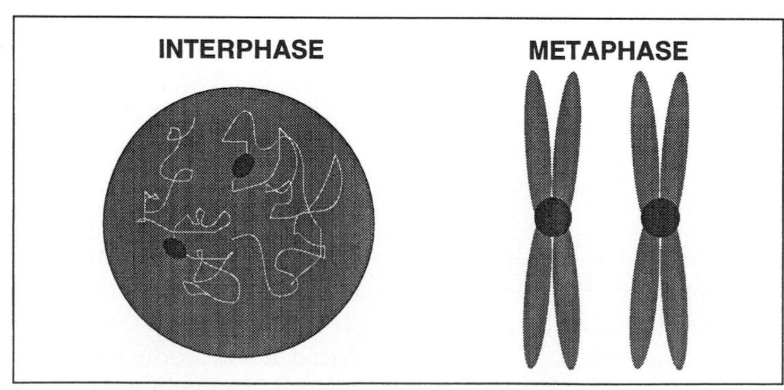

- **Chromosome Identification and Counting**
- **Metaphase or Interphase Chromosomes**

FIGURE 5 Total-chromosome probe applications.

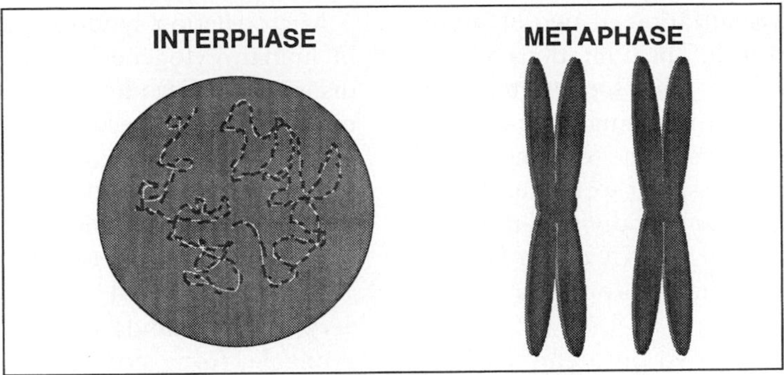

• **Chromosome Identification and Gross Structural Characterization of Metaphase Chromosomes Only**

FIGURE 6 Satellite DNA probe applications.

microscope slide. The DNA in the specimen is denatured and the probe is hybridized to the DNA on the slide *in situ*. The fluorescently labeled target sites are then visualized on the metaphase chromosome or interphase cells through fluorescence microscopy. This new technology is a useful adjunct to classic cytogenetics because it allows for aneuploidy analysis in interphase cells and can assist in resolution of complex chromosome rearrangements.

The ability to detect certain chromosomal abnormalities using FISH has been greatly enhanced by the availability of standardized DNA probes that are chromosome specific. Many of these probes are alpha satellite probes that hybridize to the repetitive DNA at the primary constriction of each chromosome. There are sequence differences in each of the chromosomes with two exceptions (13 and 21; 14 and 22), so that a centromere-specific DNA probe can be used to enumerate the specific chromosomes in a cell. Centromere-specific probes are used for chromosome enumeration or aneuploidy detection. This is particularly valuable for analysis of interphase cells.

The alphoid DNA sequences that are present at the centromere of each chromosome are repetitive targets for chromosome-specific probes. These probes provide bright punctate signals on metaphase chromosomes and interphases (Fig. 5). The ability to analyze interphase cells is a distinct advantage in cases where many cells must be evaluated for statistical studies and in cases where the cells or tissues of interest do not undergo mitosis. Rapid analysis of blood smears, amniocytes, and chorionic villus cells can be performed directly without the cell culture necessary in karyotyping studies. Multiple color labeling of FISH

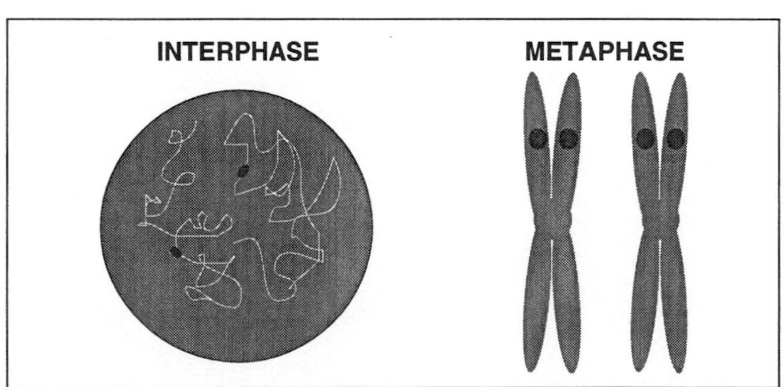

• **Chromosome Identification and Refined Structural Characterization of Both Metaphase and Interphase Chromosomes**

FIGURE 7 Unique sequence probe applications.

probes also allows for visualization of two or more chromosomes in the same cell. The limitations of this technique include specialized microscope filters for multicolor detection. When performing these experiments, the investigator chooses the chromosomes commonly deleted or in excess. For example, if one is studying a blood smear from an individual thought to have Klinefelter syndrome (47,XXY), an X and Y probe labeled with different fluorophores would be used.

Interphase studies can be used on archival tissue sections that have been processed by formalin fixation. Cells can be analyzed within the architecture of the tissue. This is an advantage in cancer pathology because the number and nature of aberrant cells in terms of trisomy or monosomy can be distinguished. A further application is the use of DNA probes for cellular oncogenes, which are commonly amplified many times in tumor tissues. FISH probes for these amplicons can be used to detect these events in tumor tissue sections.

Two types of FISH probes are used primarily in the study of metaphase preparations: whole-chromosome and cosmid contig probes, the latter used for detection of microdeletions. Whole-chromosome probes are made through flow sorting of chromosomes and through microdissection of specific chromosomes from metaphase spreads. They are used to resolve complex chromosome rearrangements and also to identify "markers" or small pieces of additional chromosomal material that cannot be easily identified with standard banding techniques (Fig. 6).

Microdeletion syndrome research is an active area in human cytogenetics. In a number of these rare disorders, a microdeletion occurs on one chromosome of the pair, which results in a characteristic phenotype. FISH studies with cosmid probes from the deleted region can identify the presence or absence of the sequences in cases where high-resolution chromosome banding is not successful (Fig. 7).

The future of FISH as an adjunct to cytogenetics is very bright. Standardized DNA probes are readily available from commercial sources. The use of multicolor detection systems has increased the ability to monitor multiple events within the same cell or tissue, and fluorescent microscopy and computer imaging systems have been developed to assist in the detection of signals and filing of images. The applications are becoming more widespread as the tools in this field accumulate.

BIBLIOGRAPHY

Borganonkor, D. (1994). "Chromosomal Variation in Man," 7th Ed. Wiley–Liss, New York.

Gardner, R., and Sutherland, G. (1989). "Chromosome Abnormalities and Genetic Counseling." Oxford Univ. Press, New York.

Rooney, D. E., and Czepulkowski, B. H. (1992). "Human Cytogenetics: A Practical Approach," Vols. 1 and 2. IRL Press, New York.

Therman, E. (1993). "Human Chromosomes," 3rd Ed. Springer-Verlag, New York.

Chromosome Patterns in Human Cancer and Leukemia

SVERRE HEIM
FELIX MITELMAN
University of Lund

GLOSSARY

Band Chromosomal area distinguishable from adjacent segments by appearing lighter or darker by one or more banding techniques. Bands are numbered consecutively from the centromere outward along each chromosome arm. In designating any particular band, four items of information are therefore required: the chromosome number, the arm symbol, the region number, and the band number within that region. These items are given in consecutive order without spacing or punctuation. For example, 12q13 means chromosome 12, the long arm, region 1, band 3

Centromere Constricted portion of the chromosome, separating it into the long (q) and short (p) arms

Chromosomes Structures in the nucleus, classified according to size, the location of the centromere, and the banding pattern along each arm (Fig. 1). The autosomes are numbered from 1 to 22 in descending order of length; the sex chromosomes are referred to as X and Y. Both the long and short chromosome arms consist of one or more regions

Clone Cell population derived from a single progenitor. It is common practice in tumor cytogenetics to infer a clonal origin when a number of cells are found that have the same karyotypic characteristics. Since subclones may evolve during the development of a neoplasm, clones are not necessarily completely homogeneous

Deletion Abbreviated "del." Loss of a chromosomal segment. del(1)(q23) means loss of all material distal to band q23 on chromosome 1, whereas del(5)(q13q33) means loss of the interstitial segment between bands 13 and 33 on the long arm of chromosome 5

Inversion Abbreviated "inv." A 180° rotation of a chromosomal segment, so inv(16)(p13q22) indicates that the centromeric portion of chromosome 16, between bands p13 and q22, is inverted (Fig. 5)

Karyotype Chromosome complement of cells. The first item to be recorded in describing a karyotype is the total number of chromosomes. The sex chromosome constitution follows next. A normal male karyotype is thus written 46,XY; the normal female complement is 46,XX

Karyotypic alterations Changes that can be either structural, implying that the banding pattern or size of a chromosome is altered, or numerical, which means additional or missing whole chromosomes. A plus or minus sign, when placed before a chromosome number, indicates the gain or loss of that particular chromosome. When placed after a symbol, the sign indicates an increase or decrease in the length of a chromosomal arm. Thus, 47,XY,+21 means a male karyotype with an extra (trisomy) 21, whereas 46,XX,5q− means a female complement that is normal except for loss of chromosomal material from the long arm of one chromosome 5

Landmarks Consistent distinct morphological features of importance in chromosome identification. Landmarks include the ends of chromosome arms (i.e., the telomeres), the centromere, and certain characteristic bands

Region Area consisting of one or more bands and lying between two adjacent landmarks. Regions are numbered consecutively from the centromere outward along each chromosome arm (Fig. 2). Thus, the two regions adjacent

ENCYCLOPEDIA OF HUMAN BIOLOGY, Second Edition, VOLUME 2. Copyright © 1997 by Academic Press. All rights of reproduction in any form reserved.

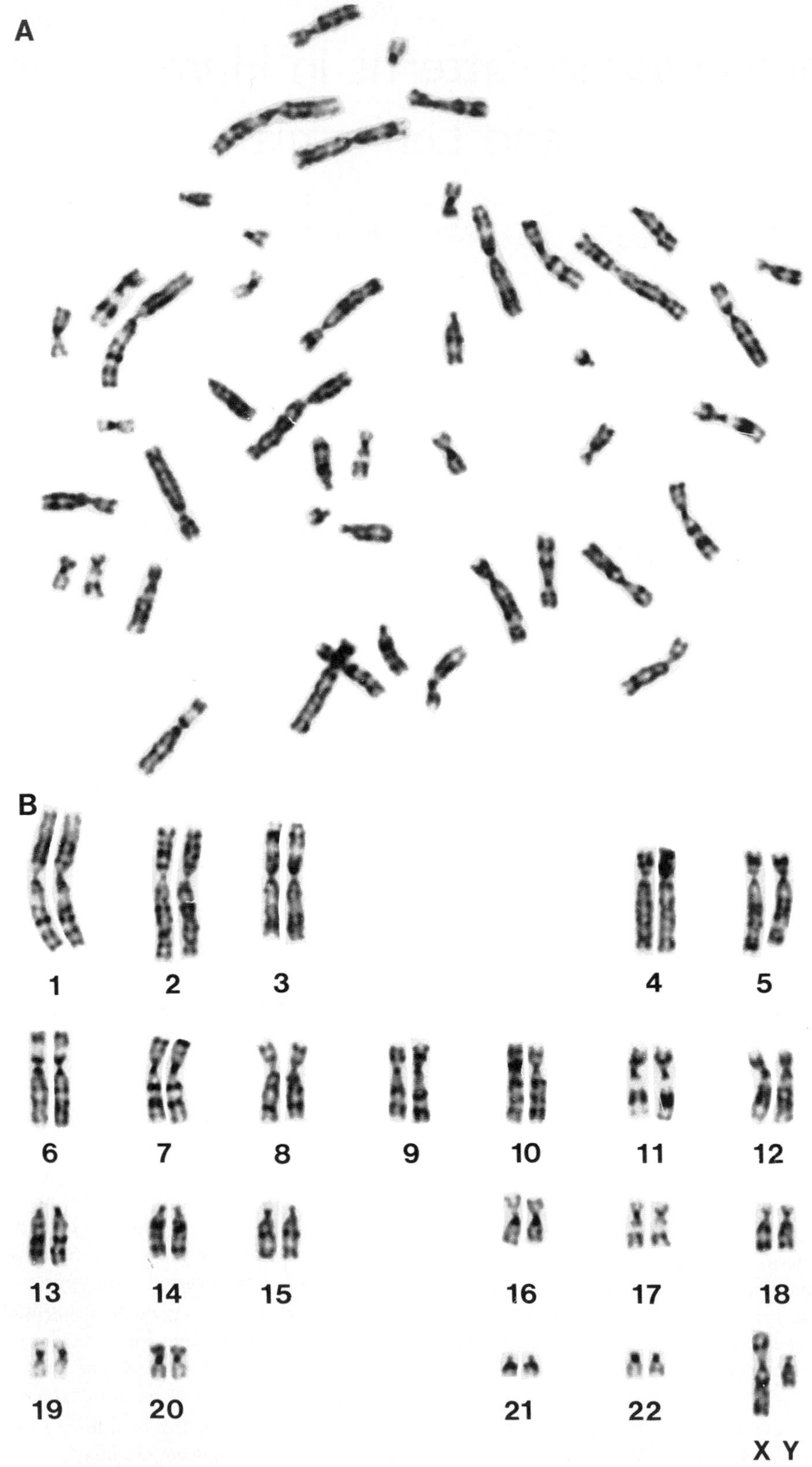

FIGURE 1 Normal metaphase plate (A) and karyotype (B) from a man (46,XY).

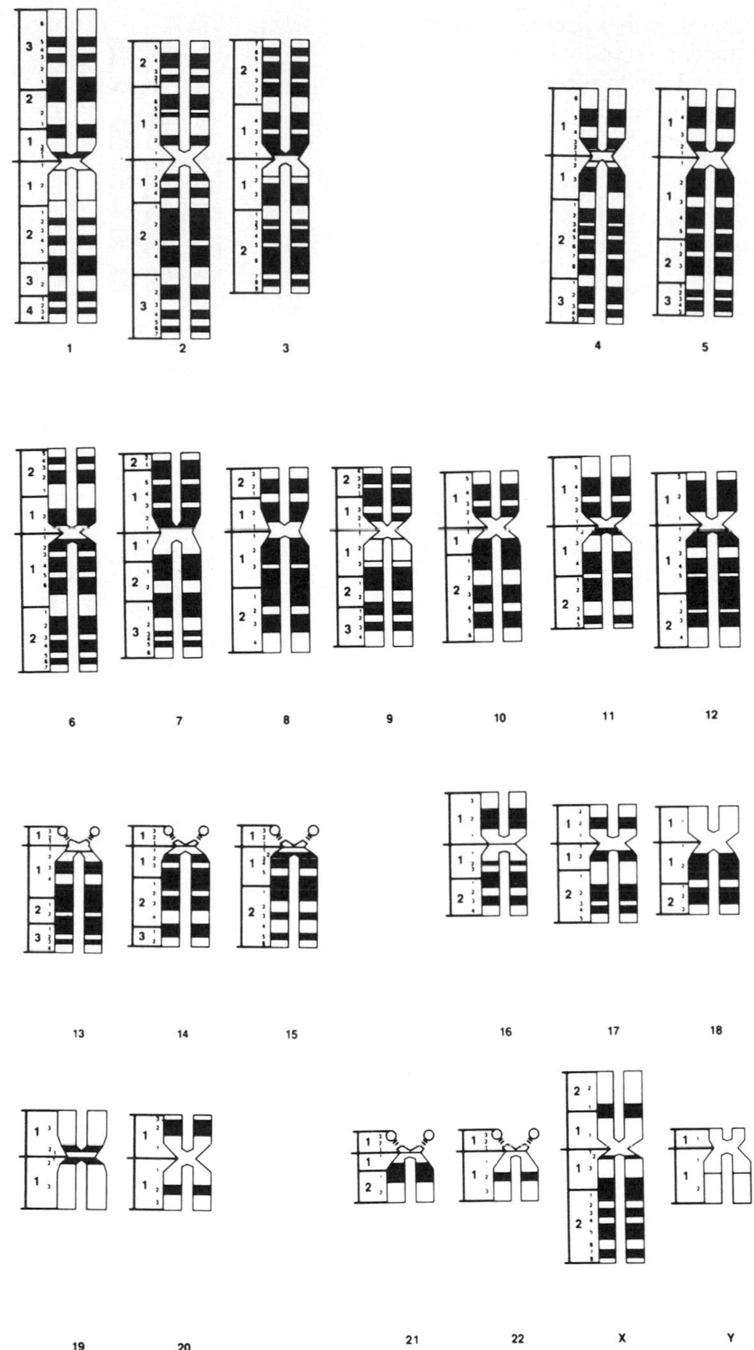

FIGURE 2 Idiotype of the human chromosome complement, illustrating the description of chromosomes as consisting of arms, regions, and bands.

to the centromere are number 1 in each arm; the next, more distal regions are number 2, and so on

Structural aberrations Deviations such as translocations, deletions, and inversions. The chromosome number(s) is specified immediately following the symbol, indicating the type of rearrangement. If two or more chromosomes are altered, a semicolon is used to separate their designations. The breakpoints, given within parentheses, are specified in the same order as the chromosomes involved, and a semicolon is again used to separate the breakpoints

Translocation Abbreviated "t." Movement of a chromosomal segment from one chromosome to another.

46,XX,t(9;22)(q34;q11) thus describes an otherwise normal female karyotype containing a translocation between chromosomes 9 and 22 (Fig. 3), with the breakpoint in chromosome 9 in the long arm, region 3, band 4, and the breakpoint in 22 in the long arm, region 1, band 1.

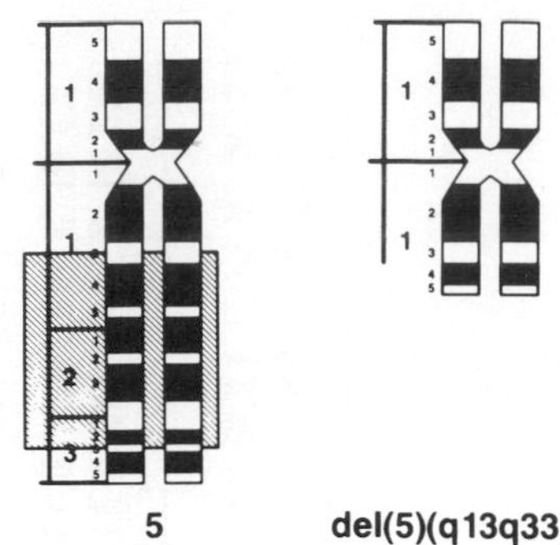

FIGURE 4 del(5)(q13q33), an interstitial deletion of the long arm of chromosome 5, is a common anomaly in myelodysplasia, but it is also seen in acute myeloid leukemia.

CANCER AND LEUKEMIA ARE GENETIC DISEASES not at the organism level, although occasionally this is also the case, but at the level of individual cells. The crucial difference between a neoplastic cell (whether benign or malignant) and a normal one is that the former has undergone DNA change(s) that enable it to escape the organism's normal proliferation control mechanisms. These changes are stable mutations, reproduced with every cell division, and hence confer upon daughter cells the same relative growth advantage that originally established the tumor.

One essential organizational level of the human genome is the chromosome. Acquired cytogenetic changes (chromosomal gains, losses, and relocations) characterize all tumor types that have been extensively studied. The aberrations may be primary, meaning that they occur early in the tumor's life and presumably are essential in its establishment, or secondary, meaning that they occur during later tumor evolution. Their overall distribution is markedly nonrandom, and some tumor or leukemia subtypes are characterized by specific changes. The study of neoplasia-associated chromosomal anomalies teaches us which are the essential genetic changes in different phases of tumorigenesis and in different neoplasms, and also provides some insight into how the pathogenetic effect is achieved. The aberrations also serve as disease markers that help reach a more precise diagnosis, including a better assessment of the prognosis in individual cases.

I. NEOPLASIA AS A GENETIC DISORDER

The essential defect in tumorigenesis is an imbalance between the tendency of neoplastic cells to divide and spread throughout the body and the organism's attempts, through a wide range of control mechanisms, to restrain such growth. Modern medical thinking holds that the reason for the disrupted equilibrium must be sought mainly in the neoplastic cells themselves, not in their regulatory environment. Cancer (and the corresponding malignancies of the hemato- and lymphopoietic organs) is primarily a cellular disease; inherent derangements of the tumor cells, rather than systemic failure to provide appropriate proliferation control, are of the essence.

When tumor cells divide, the tumorigenic quality is faithfully passed on from mother cells to their progeny. The obvious way to achieve this would be via the cells' hereditary material. The first evidence that this is indeed so dates back to the end of the 19th century, when pathologists studying histopathologi-

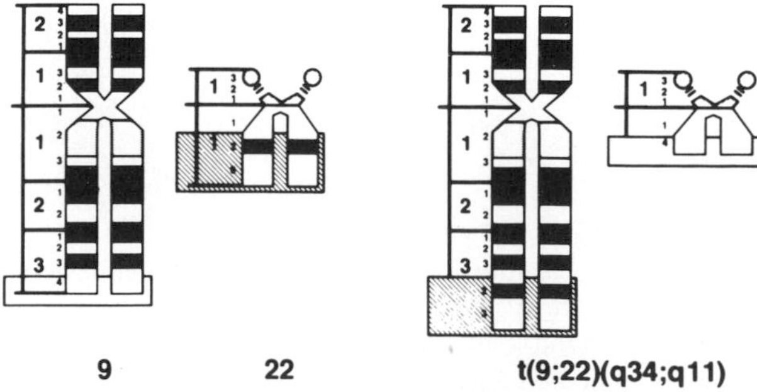

FIGURE 3 t(9;22)(q34;q11) is a characteristic anomaly of chronic myeloid leukemia, but it also occurs in acute leukemia.

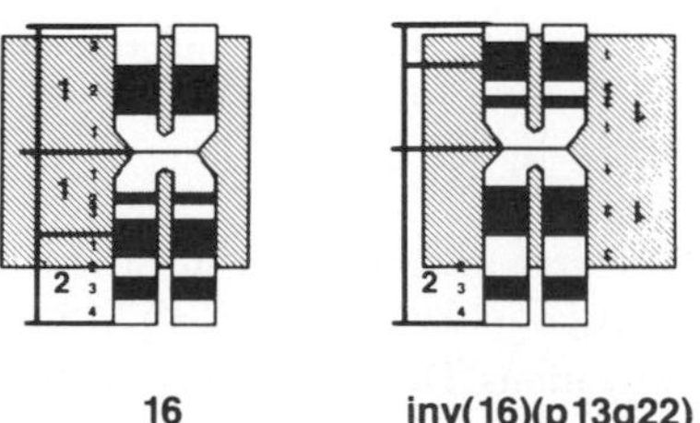

FIGURE 5 inv(16)(p13q22) characterizes a specific myelomonocytic subtype of acute myeloid leukemia.

cal preparations described a multitude of nuclear abnormalities—in fact, chromatin abnormalities—in cancer cells. Numerous methodological improvements during the last four decades—in particular, the introduction of improved tissue and cell culture techniques, the discovery that the drug colchicine causes mitotic arrest in metaphase, the use of hypotonic solutions to separate chromosomes and thereby facilitate their analysis, and the introduction of banding techniques that enable identification of individual chromosomes and chromosomal rearrangements—have revolutionized tumor cytogenetics and confirmed the early suspicions: tumor cells are indeed characterized by chromosomal rearrangements. Furthermore, the regularity of the aberration patterns encountered in the different types of neoplasia indicates that the abnormalities are pathogenetically important, not epiphenomena occurring as by-products during tumorigenesis.

The chromosome is obviously an important organizational level of the human genome, but so is the gene, a much smaller structure. Given that the haploid complement contains about 3×10^9 base pairs of DNA and that, in chromosome preparations of tumor tissue, approximately 300 bands can be discerned, it follows that each band, which roughly corresponds to the unit below which changes will not be cytogenetically detectable, contains, on average, 10^7 base pairs. Since the number of genes in the human genome is probably between 50,000 and 100,000, each band will then harbor something on the order of 200 genes. Extensive DNA-level alterations, not only point mutations but deletions or relocations that may encompass several genes, are thus possible without being cytogenetically visible.

Even if mutations were ubiquitous in neoplasia, cases would therefore probably exist in which the aberrations would not be of a sufficiently large-scale nature to show up as microscopically recognizable chromosomal lesions. As it turns out, perhaps two-thirds of all investigated neoplasms (but with large variations among tumor types) are found to contain acquired chromosome aberrations. This must be interpreted as a minimum estimate of the true frequency; it is highly likely that, in a minority of cases, chromosomal changes really do exist, but the available techniques fail to unravel them.

Granted neoplasia is a cellular disorder, even a genetic one, but does it develop from one or many mother cells? Immunological data and studies of protein and DNA polymorphisms in tumors indicate that, in most instances, the cells derive from a single clone. The cytogenetic evidence bearing on this question is not unequivocal. In hematological and lymphatic malignancies, the chromosomal rearrangements almost always exhibit striking similarities, or may even be identical, in all cells, so the overwhelmingly likely conclusion is that the whole neoplasm resulted from the expansion of a single cellular progenitor. The solid tumor data point in the same direction, with the possible exception of some epithelial tumors, in which several clones with cytogenetically unrelated abnormalities have repeatedly been detected. The most straightforward explanation for the latter findings would be to accept a polyclonal origin. If so, a systematic pathogenetic difference seems to exist between many carcinomas and the majority of other human neoplasms, in which the cytogenetic data strongly favor monoclonal tumorigenesis.

II. PRIMARY AND SECONDARY CHROMOSOMAL ABERRATIONS

Current thinking about carcinogenesis visualizes the process as consisting of three main stages: tumor initiation, promotion, and progression. Although mutational changes are presumably essential only at the initiation stage, genetic alterations are thought to play a role in the other phases of multistep tumorigenesis as well. This theoretical framework is in agreement with observed cytogenetic facts: more often than not, tumor cells contain multiple, not single, genomic aberrations. The important questions then become: Which of the observed alterations are late additions? Which represent the fundamental changes that caused the transition from a normal to a neoplastic cell, or from benign to malignant neoplastic proliferation?

In principle, the acquired aberrations of neoplastic cells may belong to three different categories. Primary aberrations are essential in establishing the neoplasm. They presumably represent rate-limiting steps in tumorigenesis, may occur as solitary cytogenetic changes, and are, as a rule, strongly correlated with tumor type. Secondary aberrations are incurred during later stages of tumor development. They are thus important after the tumor has been established, in tumor progression, and reflect the clonal evolution (see below) during this disease phase. Much evidence indicates that tumor cells are genetically less stable than their normal counterparts, a circumstance that facilitates the emergence of secondary changes. Cytogenetic noise may be a useful term for the massive and bizarre karyotypic rearrangements found in some tumors. In these genomically highly unstable cells, rearrangements are produced at a high rate and, though some or perhaps most are without selective advantage or indeed may be detrimental to the cells that harbor them, their sheer number may obscure the pathogenetically important changes and completely dominate the karyotype.

It is well known that during its lifetime a tumor undergoes a variety of phenotypic changes, usually leading to increasingly aggressive behavior, with infiltration of surrounding tissues and the establishment of distant metastases. The diversification of phenotypic traits is frequently accompanied by genotypic alterations, often involving an overall increase in the genetic complexity of the tumor cells. The modern view on tumor progression holds that additional mutations regularly occur, whereupon the selection pressure confronting the new variant cells determines which will have an evolutionary edge and outgrow their neighbors. Thus, Darwinian selection determines the relative prominence of the various subpopulations of slightly different cells within a tumor: the most fit increase in number, the less fit die away.

The direction of this clonal evolution is determined by the balance between the inherent genomic instability of the tumor cells, which tends to diversify the tumor's genetic constitution, and the selection pressure confronting the cells, which tends to reduce the genetic heterogeneity in favor of maximum adaptation to existing growth conditions. These conflicting tendencies will be operative irrespective of whether the tumor starts out as a monoclonal proliferation or represents a confluence of several simultaneously transformed cells. If genetic instability predominates, then the clonal evolution will lead to increased heterogeneity within the tumor cell population (i.e., genetic divergence), a situation seen in several neoplasms that have been cytogenetically monitored throughout their course. On the other hand, radical alterations in the selection pressure (e.g., the introduction of a new cytotoxic therapy) may well lead to diminished genetic complexity (i.e., genetic convergence) of the tumor. It is important to bear in mind that every cytogenetic investigation of a neoplasm represents only a "snapshot" of the tumor in question, and inferences about its preceding history, let alone about its future characteristics, are necessarily uncertain. This cautionary note also applies to conclusions about the tumor's clonal nature.

III. SPECIFIC CYTOGENETIC ABNORMALITIES IN HUMAN NEOPLASMS

Close to 200 acquired chromosomal aberrations have been consistently associated with human neoplastic disorders. The majority are known in the leukemias and lymphomas, a fact that may be surprising, considering that these diseases—from the points of view of both mortality and morbidity—are quantitatively much less important than solid tumors. The skewed cytogenetic knowledge reflects the technical difficulties encountered; whereas leukemic cells have been easy to sample and are readily brought to divide *in vitro,* solid tumors have been a more problematic material to work with. Only in the last few years have data on the more common human cancers been reported in any number, promising that a clearer picture of solid tumor cytogenetics may soon emerge.

A systematic description of the various specific chromosome aberrations encountered in different tumors would be outside the scope of an encyclopedia of human biology, in particular because rather extensive use of specialized medical terminology would be unavoidable. We restrict ourselves to a few examples of the many cytogenetic–medical associations presently known; for detailed information of this nature, the

TABLE I
Examples of Consistent Karyotypic Anomalies in Hematological Neoplasms

Cytogenetic abnormality	Disease
t(1;19)(q23;p13)	Acute lymphatic leukemia
del(5q)	Myelodysplasia
t(8;14)(q24;q32)	Burkitt's lymphoma
t(8;21)(q22;q22)	Acute myeloid leukemia
t(9;22)(q34;q11)	Chronic myeloid leukemia
+12	Chronic B-cell lymphatic leukemia
inv(14)(q11q32)	Chronic T-cell lymphatic leukemia
t(14;18)(q32;q21)	Malignant lymphoma
t(15;17)(q22;q11)	Acute promyelocytic leukemia
inv(16)(p13q22)	Acute myelomonocytic leukemia

reader may consult the textbooks and articles in the bibliography.

One of the principal lessons learned from the study of chromosomal aberrations in hematological malignancies is that different types of preleukemia, leukemia, and lymphoma are characterized by different anomalies (Table I). Sometimes the specificity of the association is quite high [e.g., between t(15;17) and promyelocytic differentiation of the leukemic cells]; in other instances, it is broader (5q− may be found in a variety of myeloid neoplasms). The normal differentiation pattern of the cell type may be reflected in the acquired aberration: lymphocytic cells of the T lineage have a tendency to have rearrangements of 14q11 (where the T-cell α-chain receptor locus is found), whereas B-lineage lymphocytes regularly have changes affecting the heavy-chain immunoglobulin (*Ig*) locus in 14q32.

In addition to the basic research interest inherent in the mapping of the various aberrations, chromosomal aberrations also convey information of direct clinical importance. First, the finding of an acquired chromosome abnormality proves that a bone marrow or lymph node disorder is neoplastic, which may not have been certain before sampling. Second, the existence of specific or typical anomalies associated with many of the hematological malignancies often enables a more precise disagnosis than might otherwise have been possible. Finally, the presence or absence of certain karyotypic anomalies affects the prognosis of a patient, either adversely or by indicating a more favorable outlook than is generally associated with that particular disorder. Even today, therefore, purely by functioning as disease markers and without taking into account what they reveal about the pathogenetic mechanisms of various cancers and leukemias, the karyotypic aberration patterns may, in given instances, decide the choice of treatment and ultimately the outcome in individual cases.

TABLE II
Examples of Consistent Karyotypic Anomalies in Solid Tumors

Cytogenetic abnormality	Disease
t(X;18)(p11;q11)	Synoviosarcoma
del(1p)	Neuroblastoma
t(2;13)(q35–37;q14)	Rhabdomyosarcoma
t(3;8)(p21;q12)	Mixed tumor of the salivary gland
t(3;12)(q27–28;q13–14)	Lipoma
del(3p)	Carcinomas of kidney and lung
del(11p)	Nephroblastoma (Wilms' tumor)
t(11;22)(q24;q12)	Ewing's sarcoma
t(12;14)(q14–15;q23–24)	Uterine leiomyoma
t(12;16)(q13;p11)	Myxoid liposarcoma
del(13q)	Retinoblastoma
−22	Meningioma

On average, solid tumors (Table II) have more complex karyotypes, with more massive aberrations than have hematological neoplastic disorders. Probably, this only reflects the later sampling of solid tumors, although a more profound biological difference cannot be ruled out. The general lesson from solid-tumor cytogenetics resembles that from the leukemias and lymphomas: different tumor types are characterized by different cytogenetic aberrations. Here, too, the anomalies function as disease markers that make the diagnoses more precise and the assessment of prognosis more reliable; hence, they also indirectly influence the choice of therapy.

IV. PATHOGENETIC CONSEQUENCES OF CHROMOSOME ANOMALIES

The genomic unit currently thought to be the most important is the gene. Everything we know about the functional consequences of genomic rearrangements we explain in terms of gene-level changes; any alterations of larger genomic structures—such as the addition of whole chromosomes or polyploidies—have effects that are presently incomprehensible. Thus, the selective advantage provided by most of the changes we see in cancer cells remains conjectural; we guess that they have somehow made the tumor cells fitter, or we would not have seen them. Correct though this reasoning may be, it offers little new insight into the mechanisms that may be operative.

There are now two groups of genes known to play a direct role in neoplastic proliferation: the oncogenes and the tumor suppressor genes, or antioncogenes. Research on these gene classes is covered elsewhere in this encyclopedia; here, we concentrate only on the cytogenetic aspects of their tumorigenic importance.

The oncogenes may malignantly transform suitable target cells in a dominant fashion (i.e., a single activated oncogene allele is sufficient). Knowledge about them owes much to the study of tumor-producing

RNA viruses (i.e., retroviruses). Later research has shown that the tumor genes—the oncogenes—of the viruses also exist in mammalian, including human, cells, but in a nontumorigenic form, as protooncogenes. The normal function of these genes is presumably in the regulation of cellular proliferation and differentiation, and only when they are inadvertently activated (see the following) do they contribute to tumor growth. The proteins encoded by the oncogenes, of which currently about 10 are known, either are growth factors, have similarities with the cytoplasmic membrane growth factor receptors, interact with the system of secondary messengers that transduce information from the cell's surface to the nucleus, or bind to DNA, usually in complexes with other proteins, and so presumably directly affect transcription. Interference with any of these levels of proliferation control could set off unchecked cell division. [*See* Oncogene Amplification in Human Cancer; Tumor Suppressor Genes.]

It is now known that the loci of protooncogenes within the genome coincide to a remarkably high degree with the breakpoints of cancer-associated chromosome rearrangements. This observation was pivotal in bringing about the understanding that one of the major carcinogenic mechanisms for the activation of protooncogenes is chromosomal change. Thus, although some oncogenes are activated by submicroscopic changes, even point mutations, others normally require chromosome-level alterations to turn them on. In principle, either oncogene activation can involve a qualitative change, meaning that the coding segments of the gene are rearranged, causing the production of an altered protein product, or it may be quantitative, in which case the protein is only insignificantly altered if at all, but is instead produced at a too high rate or is eliminated too slowly. Overproduction can be achieved either by increasing the number of gene copies (i.e., amplification, which may sometimes be cytogenetically surmised when double minute chromosomes or homogeneously staining regions are detected) or by heightened activity on a normal number of loci. Here, we exemplify both qualitative and quantitative oncogene activation by means of cytogenetic rearrangements.

Practically all patients with Burkitt's lymphoma or Burkitt-type acute lymphatic leukemia have in their tumor cells a t(8;14)(q24;q32) or one of the two variant translocations t(2;8)(p12;q24) and t(8;22)(q24;q11) (Fig. 6). The common breakpoint 8q24 harbors the protooncogene *MYC*, which appears to hold the pathogenetic key to the three Burkitt-associated rearrangements. In t(8;14), *MYC* is relocated to the breakpoint region of 14q32, which is within the *Ig* heavy-chain locus. The juxtaposition of *MYC* with constitutively active promoting or enhancing sequences of the *Ig* locus leads to inappropriate *MYC* transcription. In the two variant translocations, the 8q24 breakpoint is slightly more distal (closer to the telomere), so that *MYC* remains on the derivative chromosome 8. The functional results seem to be indistinguishable from the outcome of the more common 8;14 translocation, however. To the immediate vicinity of *MYC* is translocated the *Ig* κ light chain, in the 2;8 translocation, or, in t(8;22), the *Ig* λ light-chain locus. Again, the protooncogene comes under the influence of controlling sequences whose normal function is to maintain a high level of *Ig* locus activity, and this results in untimely or increased production of the *MYC*-encoded protein.

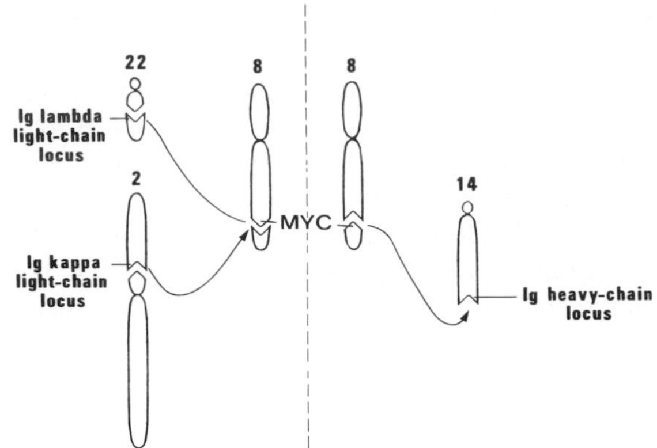

FIGURE 6 t(8;14)(q24;q32) or the variant translocations t(2;8)(p12;q24) and t(8;22)(q24;q11) are characteristic of Burkitt's lymphoma and Burkitt-type acute lymphatic leukemia. The *MYC* protooncogene, which is located in 8q24, is through the translocations brought under the control of regulatory elements whose normal function is to ensure the constitutive activity of the *Ig* genes.

The function of this protein, which binds to the nucleus, is unknown. Presumably, it takes part in the regulation of transcriptional activity, and, through interference with this function, *MYC* activation plays a role in neoplastic transformation. It should be emphasized that, in all three translocations, the coding sequences of the *MYC* protooncogene remain essentially unchanged. The activation therefore exemplifies the quantitative route, that genetic alterations of regulatory sequences may lead to increased or sustained production of an otherwise normal protein and, through this means, exert a major tumorigenic influence.

t(9;22)(q34;q11) (Fig. 3) was the first rearrangement to be specifically associated with a human neoplasm, when the Philadelphia chromosome (named after the city where it was discovered and referring to the derivative chromosome 22 resulting from the 9;22 translocation) was described in patients with chronic myeloid leukemia. Later research has revealed that the breakpoint in 9q34 occurs within an approximately 200-kb region at the 5′ end of the *ABL* protooncogene, and always so that the most 3′ exons (exons 2–11) of the oncogene are translocated to chromosome 22. In 22q11, the breakpoints are restricted to a breakpoint cluster region (bcr) of 5.8 kb, which is part of a much larger *BCR* gene. Through the 9;22 translocation, a novel *ABL–BCR* fusion gene is created (Fig. 7) in which the most 5′ end of *ABL* now consists of the most 5′ *BCR* sequences. Instead of the normal 145-kDa *ABL*-encoded protein, the new fusion gene encodes a 210-kDa protein that is substantially altered in its amino-terminal end and has pronounced tyrosine kinase activity (i.e., it attaches phosphate residues to tyrosine). The cellular target for this new chimeric protein is unknown. The protein was created by the activation of the *ABL* protooncogene by a qualitative change and may represent the first truly malignancy-specific protein known in humans.

The molecular specificities of the Philadelphia chromosome-positive leukemias may be greater than revealed by the cytogenetic findings. A substantial subset of patients with acute lymphatic leukemia and a few patients with acute myeloid leukemia also have t(9;22)(q34;q11). It appears that, in many of these patients, the breakpoints are slightly different from those seen in chronic myeloid leukemia, so that a 190-kDa fusion protein is formed instead of the 210-kDa *ABL* product associated with this disease. How the phenotypic effects of the novel proteins are achieved, especially how they vary in acute and chronic leukemia, is still unknown.

Knowledge about the other major group of cancer-relevant genes—the tumor suppressor genes, or antioncogenes—was originally derived from two sources: cell fusion experiments and the study of hereditary childhood cancers. When cancer cells were fused with normal cells, they were often found to lose the malignant phenotype, indicating that the normal genome contained genes capable of suppressing the genetic changes responsible for malignancy. By sequential loss of chromosomes from such hybrid cells, the malignant behavior often returns, implicating those chromosomes that harbor tumor-suppressing capacity, by inference, tumor suppressor genes. The studies of childhood cancers relevant in this context lead to the finding of small deletions in patients with retinoblastoma or with nephroblastoma (Wilms' tumor) and additional developmental anomalies, consistent with the view that loss of chromosomal material (or, at the submicroscopic level, of genes) may unleash tumor growth. [*See* Retinoblastoma, Molecular Genetics.]

Later examinations of both autosomal dominant childhood cancers and the common sporadic malig-

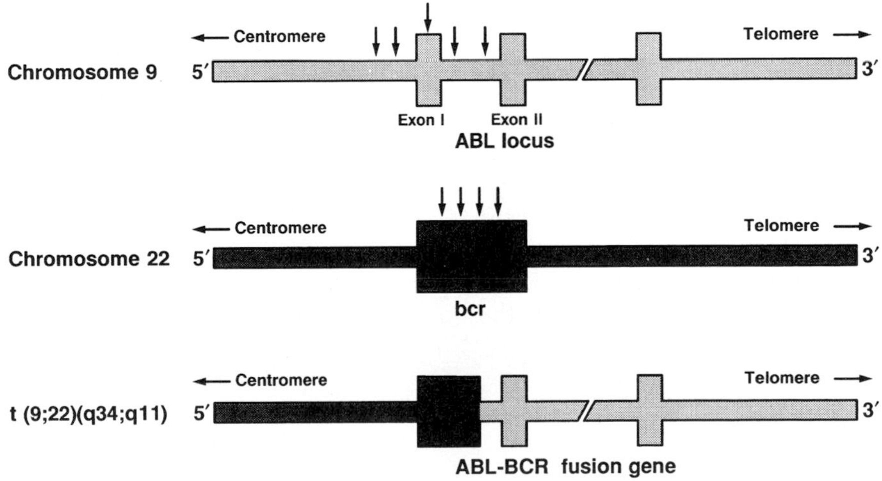

FIGURE 7 The essential molecular consequence of the t(9;22)(q34;q11) in chronic myeloid leukemia is the creation of an *ABL–BCR* fusion gene that, instead of the normal ABL product, encodes a novel 210-kDa protein with strong tyrosine kinase activity.

nancies have confirmed the early hypotheses about the existence of tumor suppressor genes. Whereas the oncogenes seem to bring about neoplastic transformation in a dominant manner (i.e., only one of the protooncogenes of a homolog pair needs to be activated), loss of tumor suppressor genes is a recessive trait at the cellular level. Both loci must be inactivated in order to elicit tumor growth; if one of the alleles is still in the normal wild-type configuration, the phenotype remains nonneoplastic. Inactivation of one of the two suppressor genes in the germ line (thus, all somatic cells have only one functioning allele) generates a dominant predisposition to cancer; tumor development starts only after a chance event inactivates the second allele. The pathogenetic importance of tumor suppressor gene inactivation has been proven for some relatively rare tumor types, such as retinoblastoma (the suppressor locus is in 13q14) and nephroblastoma (a nephroblastoma locus is found in 11p13), and seems likely for several of the more common cancers (e.g., loss of 3p loci in lung and kidney cancers or loss of a 5q locus in large-bowel cancer). The further generality of these and similar findings remains uncertain.

The molecular nature and the precise function of the tumor suppressor genes are unknown. It is also unknown whether they are functionally coupled to the oncogene system of proliferation control, for example, by exerting some kind of regulatory influence on them. The extent to which tumorigenesis through simultaneous oncogene activation and antioncogene inactivation exists, or whether it may indeed be essential in some tumor types, likewise remains wholly conjectural.

In elucidating both classes of cancer genes, the combination of chromosomes analyses and a host of other scientific techniques has been necessary. Prominent among the latter have been molecular genetic investigations, which have added a novel dimension to the cytogenetic analysis of cancer-associated rearrangements. It should be emphasized that the development of the new recombinant DNA techniques by no means obviates the need for continued cytogenetic studies of neoplastic cells, for two principal reasons. First, the cytogenetic and molecular genetic techniques assess two widely different organizational levels of the genome. Functional aspects may exist on both levels that are difficult or even impossible to grasp with too crude or too detailed methods. Today, we know next to nothing about the functional role of higher-order structural units of the genome. The point stressed here is similar to the need, obvious to anyone experienced in histopathology, to use both low and high magnification when tissues or cells are examined under the microscope. Second, only by chromosome analyses can the changes that have occurred in individual cells be examined. This makes cytogenetics uniquely suited to investigation of the clonal nature of tumor cell populations and to analysis of the evolutionary changes they undergo with the passage of time.

BIBLIOGRAPHY

Croce, C. M. (1991). Genetic approaches to the study of the molecular basis of human cancer. *Cancer Res.* **51**, 5015–5018.

Heim, S., and Mitelman, F. (1992). Cytogenetics of solid tumours. *Recent Adv. Histopathol.* **15**, 37–66.

Heim, S., and Mitelman, F. (1995). "Cancer Cytogenetics." 2nd Ed. Wiley–Liss, New York.

Knudson, A. G. (1993). Antioncogenes and human cancer. *Proc. Natl. Acad. Sci. USA* **90**, 10914–10921.

Levine, A. J. (1993). The tumor suppressor genes. *Annu. Rev. Biochem.* **62**, 623–651.

Nowell, P. C. (1992). Biology of disease. Cancer, chromosomes, and genes. *Lab. Invest.* **66**, 407–417.

Rabbitts, T. H. (1994). Chromosomal translocations in human cancer. *Nature* **372**, 143–149.

Raimondi, C. S. (1993). Current status of cytogenetic research in childhood acute lymphoblastic leukemia. *Blood* **81**, 2237–2251.

Rowley, J. D., and Mitelman, F. (1993). Principles of molecular cell biology of cancer: Chromosome abnormalities in human cancer and leukemia. *In* "Cancer: Principles and Practice of Oncology" (V. T. DeVita, S. Hellman, and S. A. Rosenberg, eds.), 4th Ed., pp. 67–91. Lippincott, Philadelphia.

Sandberg, A. A. (1990). "The Chromosomes in Human Cancer and Leukemia," 2nd Ed. Elsevier, New York.

Solomon, E., Borrow, J., and Goddard, A. D. (1991). Chromosome aberrations and cancer. *Science* **254**, 1153–1160.

Stanbridge, E. J. (1992). Functional evidence for human tumour suppressor genes: Chromosome and molecular genetic studies. *Cancer Surv.* **12**, 5–24.

Chromosomes

BARBARA A. HAMKALO
University of California, Irvine

GLOSSARY

Centromere Region of a chromosome involved in segregation; the site of attachment of the spindle apparatus

Chromatin DNA–protein complex that makes up eukaryotic chromosomes

Eukaryotic Possessing a double-membrane nuclear envelope separating the genome from the cytoplasm

Heterochromatin Condensed chromatin

Histones Small, basic, evolutionarily conserved chromosomal proteins

***In situ* hybridization** Technique used to locate specific DNA sequences in cytological preparations

Karyotype Display of mitotic chromosomes in a diploid cell

Kinetochore Specialized, trilamellar, plate-like structures organized on the outer surface of each chromatid that are the sites of attachment of microtubules between the chromatids and the poles of the mitotic apparatus

Nucleosomes Fundamental nucleoprotein subunit of eukaryotic chromosomes

Polymorphism Property of existing in several forms; in molecular genetics, single base changes in the same sequence that are detected via base sequence-specific restriction enzymes

Supercoiling/helicity Coiled coil of double-stranded DNA

Telomere Molecular end of a chromosome

THE GENETIC INFORMATION IN EUKARYOTIC ORganisms is partitioned into units (i.e., chromosomes) that vary in size and number among species. Each chromosome consists of a linear piece of double-stranded DNA complexed with a large number of proteins, some of which are structural, others of which function in DNA duplication, RNA synthesis, or chromosome maintenance. This article deals with the molecular and macromolecular structures of eukaryotic chromosomes.

I. GENERAL FEATURES OF CHROMOSOME STRUCTURE: CHROMOSOME IDENTIFICATION

Chromosomes are visible as distinct entities only transiently, as a consequence of extensive condensation at the time cells are preparing to undergo division. During the remainder of the cell cycle, these structures are sufficiently decondensed to be invisible as defined units, even by high-resolution microscopy. As a result, individual chromosomes can be identified only during mitosis, when cells are dividing and chromosome condensation is maximal. Gross morphological differences, which provide a crude way to distinguish among chromosomes, include their total length and the ratio of the length of the short arm to that of the long arm; convention names the short arm "p" and the long arm "q," relative to the centromere. The only apparent distinction at this level is that centromeres are relatively more condensed (i.e., heterochromatic) than are chromosome arms. This scheme of identification has low discrimination, because many chromosomes look virtually identical by these criteria. However, with a variety of experimental protocols, it is possible to reveal subtle differences in chromosome composition or folding that are reflected in chromosome-specific patterns of alternating densely and lightly stained bands.

ENCYCLOPEDIA OF HUMAN BIOLOGY, Second Edition, VOLUME 2. Copyright © 1997 by Academic Press. All rights of reproduction in any form reserved.

Chromosome banding represents a major advance in the identification of individual chromosomes, because the banding pattern of a given chromosome is essentially a "fingerprint" of that chromosome, which will be identical for any number of individuals, provided that the chromosome is not aberrant. In fact, one significant application of chromosome identification by banding is to reveal chromosomal abnormalities, such as reciprocal translocations and small deletions, which could not be identified by size–arm ratio comparisons. Figure 1A represents a typical human karyotype stained for DNA, and an equivalent karyotype after a banding protocol is shown in Fig. 1B. It is clear that chromosomes that do not look visibly distinguishable in Fig. 1A can be identified by their characteristic and unique banding patterns.

II. GENE MAPPING

It is possible to combine chromosome identification, as just described, with the use of cloned DNA probes to map the position of a probe along the length of a chromosome and, therefore, to correlate specific sequences with specific bands. Such cytological sequence mapping is carried out by *in situ* hybridization. This technique is based on the fact that an appropriately labeled DNA or RNA probe can form a specific hybrid with complementary sequences in a chromosome, provided the chromosomal DNA is denatured. The reaction is formally analogous to hybridization of a probe in solution to purified nucleic acid immobilized on a solid support. Radioactively labeled probes can be mapped relative to chromosome positions after hybridization, emulsion autoradiography, and banding. [*See* Genetic Maps.]

Although *in situ* hybridization with radioactive probes has been invaluable for much early gene mapping, it is slow, because the shorter the target, the longer the exposure time required for a detectable reliable signal. In addition, sites of hybridization using low-abundance probes may be detected in considerably less than half of the relevant chromosomes, making tedious statistical analysis essential.

A major breakthrough was made with the synthesis of nucleotides covalently coupled to biotin. These modified nucleotides are substituted for their radioactive counterparts in standard labeling protocols, and the probes generated are detected rapidly in one of several ways. Detection is possible by bright-field microscopy using either avidin or antibodies against biotin coupled to an enzyme (e.g., horseradish peroxidase or alkaline phosphatase) that generates a colored precipitate as reaction product. Alternatively, if the same ligands are coupled to a fluorescent indicator, hybridized probes can be localized by fluorescence microscopy. Figure 2A is an example of fluorescence detection of a human single-copy DNA probe cloned from chromosome 11. Finally, it is possible to substitute colloidal gold particles for either an enzyme or a fluorochrome to map sequences at high resolution in the electron microscope, as shown in Fig. 2B.

Several modifications of standard *in situ* hybridization methodology provide useful alternatives for gene mapping. In the method named primed *in situ* (PRINS), an unlabeled oligonucleotide is hybridized to denatured chromosomes followed by *in situ* extension in the presence of tagged nucleotides. The small size of the oligonucleotide increases the efficiency of hybridization and the subsequent polymerization results in enhanced incorporation of the tagged nucleotides. A recent improvement in this method incorporates the polymerase chain reaction (PCR) with PRINS, resulting in a tremendous increase in signal strength and sensitivity.

Subsequent to the introduction of biotin-substituted nucleotides, a number of other nonradioactive labeling schemes have been developed that can be used in conjunction with biotin to simultaneously localize several probes. These innovations have proved to be particularly useful in ordering probes as they are generated in conjunction with human genome mapping.

The ability to order probes on metaphase chromosomes is limited by the resolution obtainable in these condensed structures. However, a number of investigators have successfully adapted *in situ* hybridization to localize sequences in interphase nuclei and have shown in several instances that there is a linear relationship between physical distance and intranuclear distance over a range from tens of kilobases to a few hundred kilobases. The resolution can be improved further by lysing interphase nuclei under conditions

FIGURE 1 Representative male human karyotypes. (A) A Giemsa-stained karyotype in which chromosomes of similar dimensions are indistinguishable. (B) A karyotype treated to reveal GTG bands (G bands by trypsin using Giemsa stain). Each chromosome displays a unique and characteristic series of light- and dark-staining bands. Arrows denote Y chromosomes. [Photos courtesy of Lauren Jenkins, University of California, San Francisco.]

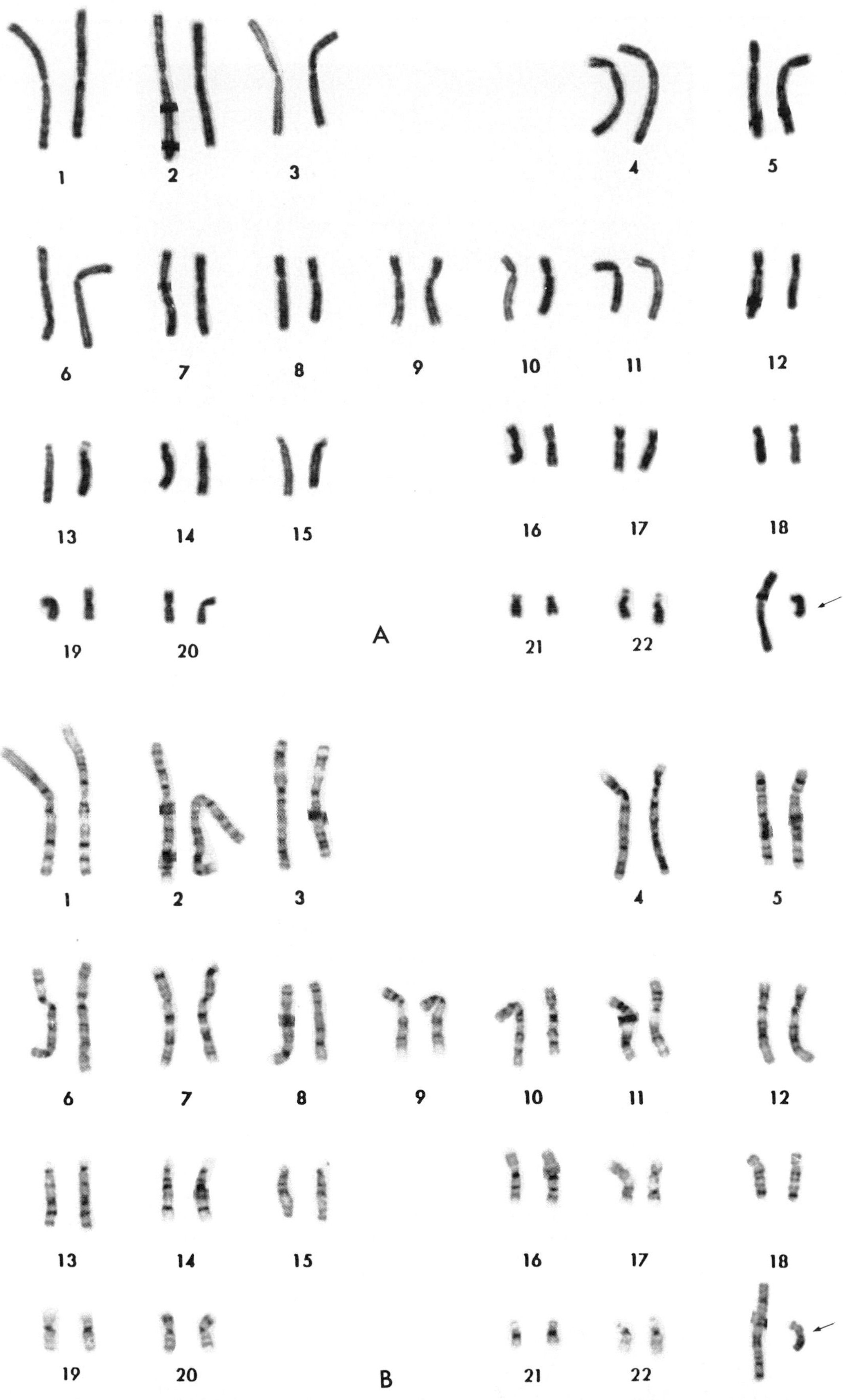

1
2
3
4
5
6
7
8
9
10
11
12
13
14
15
16
17
18
19
20
A
21
22
1
2
3
4
5
6
7
8
9
10
11
12
13
14
15
16
17
18
19
20
B
21
22

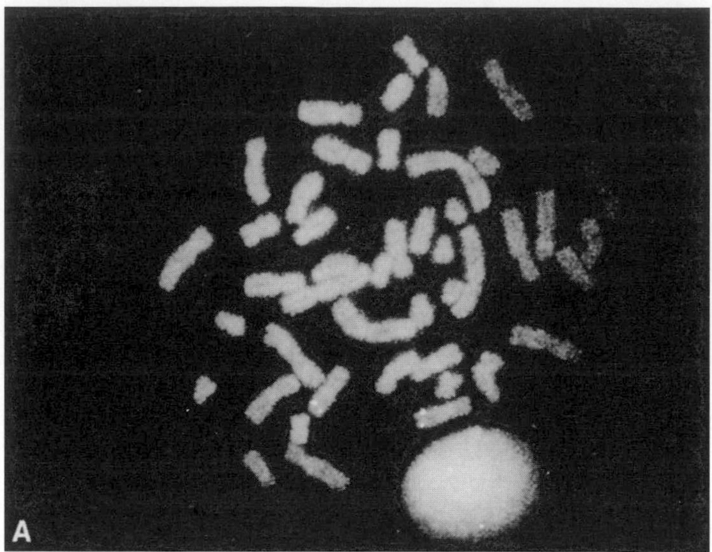

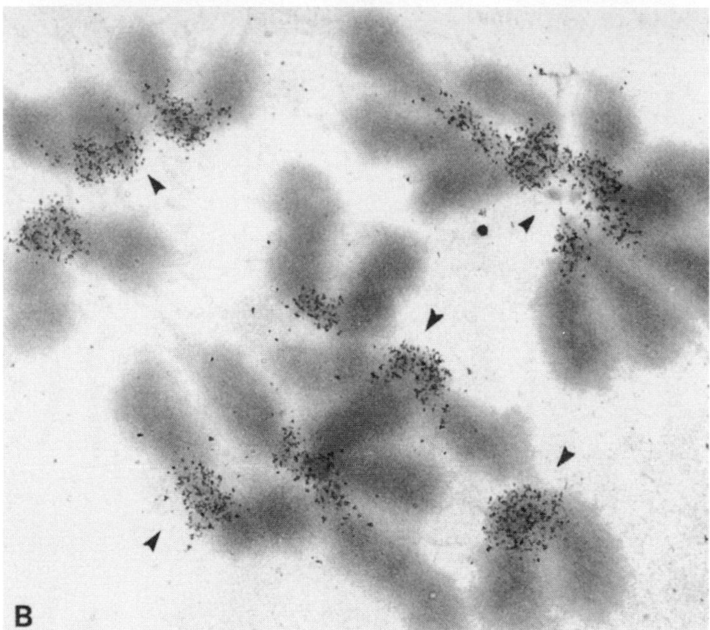

FIGURE 2 (A) A complete human metaphase chromosome spread after chromosomal *in situ* suppression (CISS) hybridization with a biotin-labeled cosmid DNA probe. The probe is detected via fluorescein-labeled avidin (yellowish-green fluorescence), whereas the chromosomes are counterstained with propidium iodide (red fluorescence). The cosmid DNA maps to the long arm of chromosome 11 as seen by the highly specific signals on both chromatids of both chromosome 11 homologs. [Photo courtesy of Peter Lichter, Thomas Cremer, Laura Manuelidis, and David Ward (Yale University) in collaboration with Katherine Call, David Housman (MIT), and Glen Evans (Salk Institute).] (B) Electron micrograph of a group of mouse metaphase chromosomes after *in situ* hybridization with a biotin-labeled probe for centromere-specific DNA sequences and detection via antibiotin antibodies and secondary antibody-coated colloidal gold particles. The centromere regions are labeled with a large number of gold particles (arrowheads). [Photo courtesy of Sandya Narayanswami and Barbara Hamkalo (University of California, Irvine).]

that partially extract chromosomal proteins and extend chromatin fibers. Because of these advances, results from *in situ* hybridization can be directly compared with physical mapping.

It is possible to identify the chromosome to which a genetic marker maps even without a molecular probe by exploiting somatic cell genetics. The fusion of two somatic cells from different species generates interspecific hybrid cells that can be grown in culture. During this growth, chromosomes of one of the parents are preferentially lost. After losing most of the chromosomes of one species, hybrids retaining a single chromosome of that species can be selected, if the retained segment contains a selectable gene.

The chromosome containing the selectable gene can be identified even though each hybrid may retain more than one chromosome, because that would be the only chromosome in common among several independent isolates. The region of interest can be further delineated by combining X rays to cause chromosome breaks in conjunction with continued selection, thus obtaining hybrids with small chromosome fragments containing the selected genes. This approach has been successful in many cases, but the production of chromosomal rearrangements between and within chromosomes resulting from X rays can cause ambiguities.

Another somatic cell genetic approach for chromosome mapping involves treatment of cells to induce the formation of micronuclei, each containing one or a small number of chromosomes. Fusion of small pieces of membrane-bound cytoplasm, each containing a micronucleus (i.e., cytoplasts), with cells from a different species, results in the rapid generation of hybrids containing a small number of chromosomes for use in mapping, as described earlier.

Once a gene is regionally mapped on a chromosome, the obvious next step is to isolate it in cloned form. Many approaches have been used for this purpose, but a particularly useful one is the generation of chromosome-specific recombinant DNA libraries. The most powerful method is to produce a library made from isolated chromosomes. Single-chromosome isolation is now possible by taking advantage of fluorescent dyes that recognize subtle differences in DNA content and base composition to separate the chromosomes by fluorescence-activated cell sorting. [*See* Chromosome-Specific Human Gene Libraries.]

A more direct, but demanding, approach to cloning a region of a chromosome involves microdissection of the region of interest. This technique has been used with considerable success for cloning segments of polytene chromosomes from *Drosophila* (fly) salivary glands, which are amplified as a result of endoreduplication without chromatid separation. Routine microdissection/microcloning of segments of diploid chromosomes is now possible, mainly through the use of PCR to amplify the very small amounts of DNA obtained by microdissection. For example, it is possible to create a minilibrary from as few as ten microdissected chromosome fragments. Although needle microdissection currently is routinely used, there is considerable interest in combining laser fragmentation of chromosomes with laser trapping of fragments prior to cloning.

III. MOLECULAR STRUCTURE OF CHROMOSOMES

The overall molecular composition of chromosomes is rather simple: DNA represents about one-half to one-third of the total mass, the remainder being protein. The amount of DNA per equivalent somatic cell is typically invariant, but the total amount of protein varies, depending in large part on whether the cells are proliferating. This apparent simplicity masks a tremendous complexity when each molecular component is analyzed in detail. The power of molecular biology has resulted in great progress in defining this complexity and in analyzing certain subcomponents in exquisite detail. Nevertheless, many questions remain with respect to the relationships between particular molecular species and their respective functions.

A. DNA

Each chromosome contains one long, linear, double-helical DNA molecule; the longest, if extended, would measure about 7.3 cm. Despite the apparent monotony of the molecule, composed of only four bases, it is possible to molecularly characterize three distinct classes of DNA sequences that coexist in a typical chromosome of higher eukaryotes. They are referred to as unique (i.e., single-copy), moderately repetitive, and highly repetitive sequences. The proportions of the genome represented by each class vary widely among different organisms and, to a certain extent, in the same species. In general, in the haploid genome, a unique sequence exists in one or a few copies, moderately repetitive sequences occur in from a few hundred to a few thousand copies, and highly repetitive sequences appear in tens of thousands to millions of copies.

Additional complexity is apparent within a class, because each is composed of a collection of distinct DNA sequences that are considered together simply for the similarity of their numbers of copies. This heterogeneity shows that these three molecular classes do not correspond to three functional components of the chromosome.

Despite the fact that it is not yet possible to define functions (if they exist) for all types of sequences molecularly identified in the genome, some generalities have emerged. For example, the unique sequence class includes the protein coding sequences, which are distributed throughout most of the length of a given chromosome, interspersed with other DNA classes. This class includes both functional genes and so-called pseudogenes, which are copies of functional genes rendered nonfunctional by mutations. An issue that has yet to be resolved is the distribution of coding sequences relative to the physical map of a chromosome. That is, are genes relatively uniformly spaced on average, or is the arrangement dictated by unidentified rules? The nature and the possible function of the bulk of single-copy DNA also remain enigmas. It certainly contains many unknown genes, but it may also include other types of sequences with different functions.

The middle repetitive sequence class contains both coding and noncoding components. Genes that code for ribosomal RNAs are middle repetitive, as are those coding for histone proteins or immunoglobulin variable regions. To date, however, the majority of moderately repetitive sequences have not been assigned functions. These sequences typically are interspersed in the genome, having varying average periodicities and repeat units of different lengths. Functions suggested for interspersed repeats include cis-acting regulatory sequences involved in the coordinate regulation of unlinked genes and/or origins of replication. In fact, many cis-acting regulatory sequences have now been identified and, based on an ever-expanding data base, it is clear that many genes do share common cis-acting regulatory elements that are, by definition, members of the interspersed repeat class.

The highly repetitive component of the higher eukaryotic genome consists of numerous distinct members of varying lengths and sequences, usually in long tandem arrays interspersed in the genome. The most extensively studied of such sequences tend to be localized at chromosomal centromeres and telomeres. In different organisms, these sequences vary enormously in length and in their number of copies. Since these sequences are not transcribed, it has proved difficult to determine their function, beyond inferences based on their chromosomal positions. For example, although centromeric tandem repeats have been characterized in large numbers of organisms, there is no evidence to support the contention that these sequences actually function to effect chromosome segregation directly; they may play a role in organizing the centromeric chromatin so that it can assemble structures necessary for spindle attachment. [*See* Centromeres, Human.]

Many other functions have been proposed for these sequences, including a role in speciation or in chromosome pairing. The latter possibility is attractive because chromosome-specific versions of at least one centromeric repeat family have been identified for several human chromosomes. The extreme alternate view that these sequences have no function has also been put forward. The sequences located at the ends of chromosomes undoubtedly function to stabilize those sites, as discussed later.

Two human interspersed repeats that appear to be indicators of chromosome organization based on their distribution are the Alu and L1 repeats. The former are short repeats [approximately 300 base pairs (bp) in length] that are relatively (G + C) rich; the others are several kilobases in length and relatively (A + T) rich. Recent *in situ* hybridization data show that both Alu and L1 are nonuniformly distributed along chromosomes with Alu sequences concentrated in regions defined as R bands and L1 enriched in G bands. The bands result from treatments such as those mentioned earlier. This differential distribution in relation to independent chromosomal markers may tell us something about the organization of chromosomes.

Finally, a distinct class of human highly repetitive sequences, with a repeat length between 11 and 60 bp arranged in tandem arrays, has been identified. These repeats exhibit exceedingly high variability in the number of repeats in an array from individual to individual. Their high degree of polymorphism is exploited in human molecular genetics, to map disease loci, and in forensic science, to define a genetic fingerprint unique to each person.

B. Chromosomal Proteins and Chromosome Organization

1. Basic Chromatin Subunits

The proteins associated with chromosomal DNA are divided into two major groups: the histones and the nonhistone chromosomal protein (NHCPs). The former class is defined biochemically as a set of five low-molecular-weight, highly evolutionarily conserved basic proteins, which exist in amounts strictly proportional to DNA content. The primary role of the histones is structural, although there is some evidence that they may also affect gene expression. As implied by its name, the second group is composed of all other proteins associated with the DNA and is, therefore, a large, heterogeneous, and relatively poorly defined collection of diverse proteins with diverse function. Among the NHCPs are enzymes involved in DNA replication, RNA synthesis, protein modification, and hydrolysis; factors required for the positive and negative regulation of gene expression; structural proteins other than the histones; and proteins that function in the maintenance of stable chromosomes (e.g., telomere-binding proteins) and in their faithful segregation (e.g., kinetochore components). With the exception of the proteins with enzymatic activity, only a small proportion of total NHCPs have been isolated and analyzed in detail. In the past few years, however, procedures for isolating DNA-binding proteins have advanced dramatically. These techniques, coupled with biochemical and molecular analyses, allow rapid progress in determining the structure and function of many NHCPs involved in gene expression and chromosome organization.

Histones are the fundamental structural proteins of a chromosome. The core histones (i.e., H2A, H2B, H3, and H4) are found universally in equal stoichiometry, from lower eukaryotes, such as yeast, to mammals. These proteins are remarkably conserved throughout evolution, particularly with respect to their structure in solution and, to a somewhat lesser extent, their primary amino acid sequences. They are rich in basic amino acids, such as lysine and arginine, and can interact with one another in solution with precise stoichiometry to form an octamer containing two molecules of each core histone. The octamer is a compact structure that, in turn, interacts with 146 bp of DNA to form the fundamental structural subunit of all eukaryotic chromosomes, referred to as the nucleosome core particle.

The DNA is wrapped around the outside of the octamer to form a nucleoprotein complex whose structure, based on X-ray diffraction of crystals, approximates a flattened disk. Core particles are arranged along the DNA to give a fiber with the appearance of beads on a string; adjacent beads are connected by thin fibers referred to as linkers. Figure 3A illustrates this configuration as seen in the electron microscope. Adjacent core particles are closely packed *in vivo* and the fifth histone, H1, has been implicated in this and the next higher order of packing.

Linker regions are relatively susceptible to digestion by certain DNases, providing a convenient way to isolate the chromatin subunits. Analysis of digestion products with time reveals a relatively DNase-resistant nucleoprotein structure defined as a monomer nucleosome. Although core particles of virtually identical composition represent the universal chromatin subunit, monomer nucleosomes contain from as little as 165 to over 200 bp of DNA. Differences of DNA length in monomers have been reported between organisms and in different cells of the same organism. The variability in DNA lengths observed in nucleosome preparations is attributed to differences in the linker lengths. Since the fifth histone, H1, is associated with linker DNA, it may have a role in determining the repeat length. Although the significance of variable-length linkers is not yet clear, this variability could be involved in determining how the fundamental nucleosome-containing fiber is organized at the next level of packing. The formation of nucleosomes effects an approximately 7-fold compaction of the DNA relative to its linear length, results in DNA supercoiling, and is the first step in achieving the estimated 10,000-fold packing required to create a visible mitotic chromosome.

Despite the apparent lack of complexity of the histones (i.e., only five proteins), they are, in fact, quite complex, because there are multiple primary amino acid variants of each, and histones can undergo a large number of postsynthetic modifications (e.g., phosphorylation, methylation, acetylation, and ADP ribosylation). If one considers these variations, it becomes clear how complex and mosaic the basic chromatin fiber can be. Although much progress has been made in defining the fundamental chromatin structure, analysis of chromatin fiber mosaicism and its possible relationship to function is a challenge for the future.

2. Higher-Order Structure

Investigations of the higher-order organization of chromosomes have not yet generated a unitary model for the way the fundamental fiber is folded. A number of distinct models have been proposed, based on a variety of types of data, but conflicts among the models have yet to be resolved. One major stumbling block in this area is the inability to isolate higher-order structures under conditions known not to perturb the subtle features of folding. One generalization that is agreed on is that the dimensions of the folded basic fiber are between 20 and 30 nm in width, regardless of the mode of the folding. Figure 3B shows an exam-

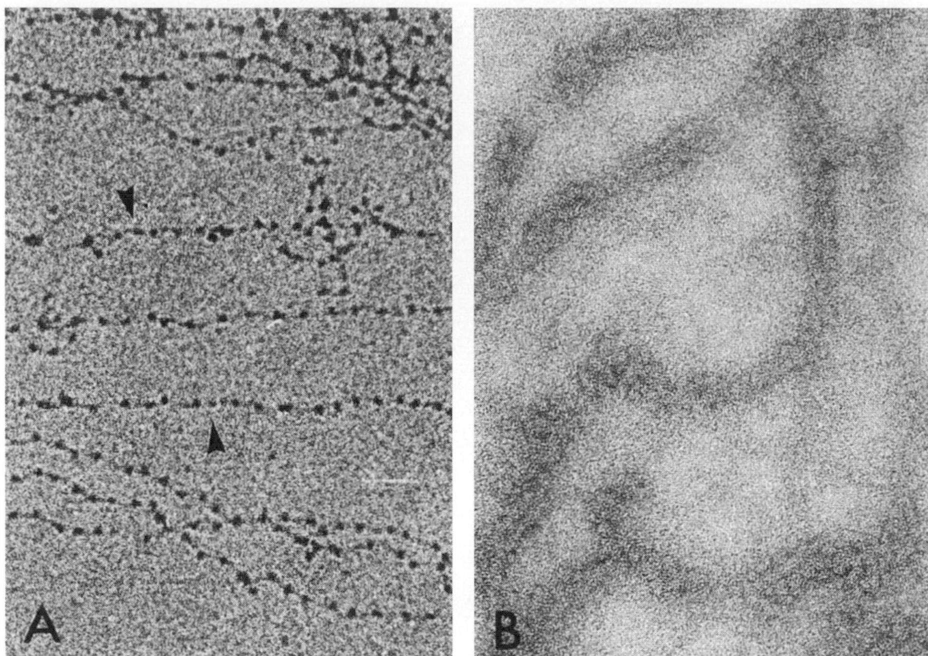

FIGURE 3 Electron micrographs of chromatin fibers released from lysed cultured mammalian cells. (A) The fundamental "beads on a string" (arrowheads) fiber. (B) A specimen prepared under conditions that stabilize 20- to 30-nm higher-order fibers. [Photos courtesy of J. B. Rattner (University of Calgary) and Barbara Hamkalo (University of California, Irvine).]

ple of a folded fiber in which individual nucleosomes can be resolved. It has been estimated that formation of the 20- to 30-nm fiber results in a net condensation of the DNA of about 40-fold, which is not sufficient to create a mitotic chromosome. Thus, several additional levels of folding must occur and, again, there is no general agreement as to exactly how they are effected, although coils and/or loops are frequently proposed as intermediate structures. Some data argue that chromatin folding at these higher levels is mediated by a small number of nonhistone chromosomal proteins, including the enzyme topoisomerase II, which catalyzes reactions that change the superhelicity of DNA.

The organization of chromosomes in interphase nuclei is another unresolved question. The results of molecular and biochemical approaches suggest that the genome is organized or partitioned into domains that can be defined operationally, because of their resistance to attack by specific DNases, presumably as a result of specific protein associations. In some cases, the nuclease-resistant sequences identified in interphase nuclei and in mitotic chromosomes are the same, implying a commonality in the basic organization of the chromosome, rgardless of its state of condensation. The issue of nuclear organization is being studied intensely, so that general rules of organization, if they exist, should emerge in the next few years. Recent studies provide evidence for a model of genomic organization in which chromatin segments defined by nuclease studies (domains) may be insulated from neighboring domains through the existence of boundaries, presumably composed of specialized nucleoprotein structures. On a larger scale, antibodies against proteins involved in splicing identify discrete foci in interphase nuclei, implying that splicing occurs in a compartmentalized fashion. Likewise, sites active in DNA replication are focal. Equally important questions related to higher-order chromosome structure, presently under study, are the structural basis for chromosome banding and whether there are functional correlates to this organization.

3. Position Effects and Chromatin Packaging

When a gene is translocated near or into heterochromatin (e.g., the centromere or telomere), its expression is repressed. This phenomenon, referred to as a position effect, is thought to be the result of packaging of the gene into a heterochromatic structure via association of heterochromatin-associated proteins. A major advance in our understanding of this phenomenon was the identification in *Drosophila* of a heterochromatin protein named HP1. Interestingly, HP1 shares a region of near identity with a protein associated with gene repression, suggesting that this domain can play a structural role as well as a role in the regulation of gene expression. This domain is highly conserved during evolution because it has been identified in organisms as diverse as plants and humans. The next several years promise to provide a detailed molecular dissection of heterochromatin and the relationship between components of heterochromatin (which does not contain genes) and repressed genes, potentially linking the two types of chromatin.

IV. ESSENTIAL FUNCTIONAL ELEMENTS OF A CHROMOSOME

The integrity and stability of eukaryotic chromosomes are dependent on three noncoding genetic elements: sites for the initiation of DNA replication ("origins"), sequences that mediate chromosomal attachment to spindle microtubules (centromeres), and sequences that delineate the molecular ends of chromosomes (telomeres). Progress in identifying these elements has been dramatic in lower eukaryotes, such as yeast. The problem of developing assays for the functions of untranscribed sequences in higher eukaryotes has been a major stumbling block in this regard. It is compounded by the large size of higher eukaryotic genomes, since the elegant experimental approaches that were so successful in lower systems are not readily adapted to larger, more complex, genomes.

A variety of different experimental approaches support the statement that each eukaryotic chromosome possesses a large number of sites at which DNA replication can be initiated. These sites may not be uniformly distributed along the DNA, and the precision with which the bacterial replication machinery recognizes a well-defined origin sequence may not be a feature of eukaryotic initiation. An interesting aspect of eukaryotic origin structure and function is highlighted by the organism's need to regulate DNA replication during development. This point is best illustrated in *Drosophila,* in which, at an early stage in development, cells replicate their DNA in a very short time compared to the time it takes in somatic cells. In these embryonic nuclei, active replication origins are very close together, whereas in somatic cells, many potential origins are not utilized, resulting in distantly spaced replication sites. Cloning and characterization of origin sequences are being attempted by many groups, and successful progress will help define what an origin is, assess the diversity of origin sequences, and define developmental versus somatic origins, if they are distinguishable.

The functional centromere of a higher eukaryotic chromosome has not yet been defined in molecular terms. A centromere is cytologically defined as the primary constriction, a heterochromatic region where sister chromatids remain in close apposition until anaphase, when they begin to move to opposite poles of the mitotic spindle. An essential component of a functional centromere is the kinetochore, a proteinaceous structure that mediates attachment of a sister chromatid to spindle microtubules and orients the direction of chromatid movement. A longstanding hypothesis for the way kinetochores participate in faithful chromosome segregation suggests that they interact specifically, in a currently ill-defined way, with a subset of the DNA sequences found in centromere heterochromatin. Molecular characterization of DNA sequences from yeast, which confer segregation to linked DNA and which, in turn, are associated with distinct chromosomal proteins, supports the general features of this model.

Proof of this model for higher eukaryotic centromeres requires the identification of DNA sequences essential for segregation and a detailed biochemical characterization of the kinetochore, neither of which has been accomplished. The presence of large amounts of tandemly repeated DNA in the centromeric heterochromatin of most eukaryotes led some investigators to propose that these sequences confer the segregation function, but, at the present time, there is no evidence to support this contention. The sera of patients with certain autoimmune diseases contain antibodies that react with various chromosomal regions, including centromeres. Several groups have exploited these antibodies to attempt to define kinetochore polypeptides, but none of the molecules identified has been unambiguously defined as an integral kinetochore component. Thus, the molecular identification of origins of replication and centromeres of higher eukaryotic chromosomes represents fertile areas for future research.

Better understood are the nature and the role of telomeres, or chromosome ends. There are two reasons to suppose that telomeres have unique structural properties. First, it has been known for many years that a broken chromosome end behaves differently than a telomere: telomeres are inert, whereas broken ends are usually "sticky," resulting in the formation of rearranged, often dicentric, unstable chromosomes.

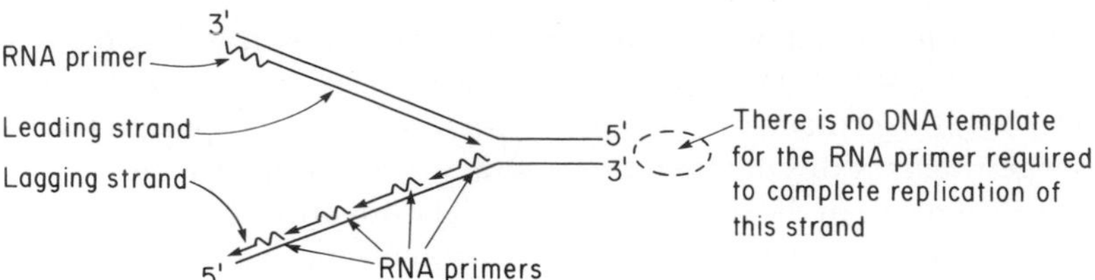

FIGURE 4 Schematic representation of the dilemma faced when replicating the very end of a linear DNA molecule.

Second, if one considers the replication of the end of a linear DNA molecule, it is obvious that DNA polymerase could not replicate it to the very end, for lack of a primer on the discontinuously synthesized strand (Fig. 4). Therefore, telomeres must possess special structural features to allow the complete replication of chromosomes; otherwise, they would be shortened at each replication.

Telomere-functional sequences have now been cloned from a large number of organisms, from yeast to humans. A comparison of the telomere sequences from such evolutionarily divergent organisms reveals striking similarities and remarkable evolutionary conservation, which is explained by their crucial function. They are tandemly repeated (G+C)-rich short sequences, with the 5′ to 3′ polarity of the strand rich in guanine residues, always in the same direction relative to the chromosome end. Chromosomes have different numbers of the simple repeat, and there is evidence for changes in the number of repeats with time. In fact, vertebrate telomeres shorten with age, suggesting a link between telomere loss, chromosome instability, and senescence. The reason for shortening is the apparent lack of or inactivation of telomerase in somatic cells. In contrast, germ cells have very long telomeres, presumably because telomerase is reactivated. The observation that telomerase is also reactivated in a large number of different human tumor cells may provide a novel target for cancer chemotherapy. If telomerase is required for tumor cell proliferation, it would represent a tumor-specific enzyme that could be selectively inhibited without affecting normal somatic cells.

The telomere repeat unit (TTAGGG) identified in human DNA is identical to that found in trypanosomes, an acellular slime mold, and all other vertebrates analyzed, arguing for conservation of an essential function. The identification of human telomeres will be invaluable for the cloning of genes located close to the ends of chromosomes. [*See* Telomeres, Human.]

BIBLIOGRAPHY

Bernardi, G., and Bernardi, G. (1986). *Cold Spring Harbor Symp. Quant. Biol.* **51**, 479.

Blackburn, E. H. (1984). *Annu. Rev. Biochem.* **53**, 163.

de Lange, T. (1994). *Proc. Natl. Acad. Sci. U.S.A.* **91**, 2882.

McKusick, V. A. (1986). *Cold Spring Harbor Symp. Quant. Biol.* **57**, 423.

Paro, R. (1990). *Trends Genet.* **6**, 416.

Van Holde, K. E. (1989). "Chromatin." Springer-Verlag, New York.

Chromosomes, Molecular Studies

CASSANDRA L. SMITH
Boston University

RAFAEL OLIVA
University of Barcelona

SIMON K. LAWRANCE
Oberlin College

DENAN WANG
University of California, San Francisco

GLOSSARY

Allele One of several alternative DNA sequences, usually within a gene

Autosome Chromosome other than a sex chromosome

Consensus sequence DNA sequence containing the most frequently occurring bases, such as from comparing DNA sequences derived from different sources

Contig Ordered overlapping genomic clones or ordered set of sequenced clones

C, T, G, A DNA bases: cytosine, thymine, guanine, and adenine, respectively

DNA hybridization Annealing of an oligonucleotide or polynucleotide to its complementary sequence to form a stable double-stranded structure

Expressed sequence tag (EST) DNA sequence of a gene fragment; usually a portion of a cDNA

Fluorescence *in situ* hybridization (FISH) Hybridization of a fluorescently labeled nucleic acid probe to metaphase chromosomes

Genetic anticipation Increasing severity of the genetic consequences of a disease allele with progressive generations

Genetic imprinting Inherited DNA methylation patterns imposed by parental germline cells

Gradient gel electrophoresis Technique that separates DNA on the basis of thermodynamic stability, thus allowing single-base differences to be detected

Incomplete penetrance Lack of expected phenotype as predicted from a genotype

Inverse polymerase chain reaction Method that allows the region adjacent to a known sequence to be amplified

Karyotype Chromosomal constituents of a genome

Linkage disequilibrium High-frequency cosegregation of linked genes due to suppressed meiotic recombination between the genes

Multiplexing Any methodological approach in which different independent members are treated en masse to save labor

Open reading frame (ORF) DNA sequence that could code for a protein because of the absence of a translational stop codon

P1-artificial chromosome (PAC) P1-bacteriophage artificial chromosome clone fragments up to ~100 kb in size

Phenotype Observable properties of an organism, produced by the genotype in conjunction with the environment

Polymerase chain reaction (PCR) *In vitro* exponential amplification of DNA molecules by multiple cycles of DNA denaturation, primer annealing, and synthesis

Pulsed-field gel electrophoresis (PFG) Electrophoresis technique that fractionates megabase DNA by exposing them to alternating electrical fields

Restriction enzyme site DNA sequence recognized and usually cleaved by its corresponding restriction enzyme

Restriction fragment-length polymorphism (RFLP) Restriction fragment length that is variable in the population

ENCYCLOPEDIA OF HUMAN BIOLOGY, Second Edition, VOLUME 2. Copyright © 1997 by Academic Press. All rights of reproduction in any form reserved.

because of either changes in sequence at restriction sites or insertions/deletions between restriction sites

Sequence tagged repeat polymorphism (STRP) DNA sequence encompassing a simple repeat sequence

Sequence tagged restriction site (STAR) DNA sequence surrounding a restriction site

Sequence tagged site (STS) Random DNA sequence that produces a successful site-specific PCR reaction

Somatic cell Cell of the eukaryotic body other than those destined to become sex cells

Syntenic region Regions of the genome that display conservation of gene location and/or order between two species

Variable number tandem repeat (VNTR) Simple repeating sequence, like a trinucleotide repeat, that differs in length in different individuals

Yeast artificial cloning (YAC) Linear yeast cloning system, usually applicable to clones ranging from 100 kb to more than 1 Mb in size

THE PACE OF MOLECULAR STUDIES ON HUMAN chromosomes parallels the development of an increasing number of techniques that allow dissection of large genomes. Somatic cell genetic techniques were integrated with physical mapping methods that used pulsed-field gel electrophoresis (PFG) and related large-DNA methods, such as large-fragment cloning systems [yeast artificial cloning (YAC), bacteria artificial cloning (BAC), P1-artificial chromosome (PAC)] and linked DNA sequences [restriction fragment-length polymorphism (RFLP), sequence tagged repeat polymorphism (STRP), sequence tagged restriction site (STAR)]. Genetic mapping experiments embraced physical approaches that linked DNA sequences (RFLPs and STRPs) to particular chromosomal locations. These DNA sequence-based genetic markers were used to make chromosomal genetic maps that speeded up the work of gene searchers using reverse genetic or positional cloning approaches. These genetic markers are useful anchors between various types of maps. A summary of human genomic molecular resources available through the Internet are listed in Table I. This article reviews the structure and constituents of human chromosomes and the molecular techniques and approaches currently being used to characterize the human genome.

TABLE I
Useful Human and Mouse Genomic Resources Available Through the Internet

NIH—NCHGR (National Center for Human Genome Research)
www: http://www.nchgr.nih.gov
DOE—OHER (Office of Health and Environmental Research)
www: http://www.er.doe.gov
GDB—chromosomal map and loci data
email: help@gdb.org
ftp: ftp.gdb.org
www: http://www.gdb.org
dbEST/Genbank—DNA sequence data
email: help@ncbi.nlm.nih.gov
ftp: ncbi.nlm.nih.gov
www: http://www.ncbi.nlm.nih.gov/dbEST/index.html
CEPH—YAC physical mapping and human genetic mapping
email: ceph-genethon-map@cephb.fr
ftp: ceph-genethon-map.genethon.fr
www: http://www.cephb.fr/bio/ceph-genethon.map.html
CHLC—genotypes, marker and linage data
email: info-server@chlc.org
ftp: ftp.chlc.org
www: http:///www.chlc.org
Whitehead—map of human and mouse CA repeats
email: genome_database@genome.wi.mit.edu
ftp: genome.wi.mit.edu
www: http://www.genome.wi.mit.edu
Jackson Lab—variety of mouse data
email: mgi-help@informatics.jax.org
ftp: ftp.informatics.jax.org
www: http:///www.jax.org

I. COMPOSITION OF THE HUMAN GENOME

A. Chromosomes

The human genome consists of 23 pairs of nuclear chromosomes and a circular cytoplasmic chromosome located in mitochondria. The nuclear genome consists of 22 autosomal pairs and 1 pair of sex chromosomes designated X and Y. The autosomal chromosomes are numbered on the basis of size: the largest chromosome is number 1; the smallest chromosome is number 21, rather than 22, because chromosome 22 was initially thought to be the smallest.

Each chromosome contains a single linear, double-stranded DNA molecule with a 40% GC (guanine and cytosine) content. The low GC content is due to a deficiency in the dinucleotide sequence 5′ CpG 3′. The rarity of this sequence appears to be caused by two factors. First, *in vivo*, the C residue in this sequence is frequently converted to 5-methyl cytosine (5-MeC). Second, deamination, the most frequent naturally occurring DNA damage, converts C to U (uracil) and 5-MeC to T (thymine). A DNA repair system removes U from DNA, restoring the original C–G base pair. However, repair of a G–T mismatched base pair does not always lead to restoration of the original base pair. Thus, the human genome, like other cytosine methylated genomes, is low in CpG sequences and high in TpG sequences. [*See* Human Genome.]

The state of methylation of a particular region has been correlated with gene expression, carcinogenic changes, and genetic imprinting. Methylation may influence gene expression by promoting the formation of an unusual DNA structure and/or differential protein binding. Carcinogenic changes and genetic imprinting effects are very likely due to changes in gene expression. Maternal and paternal methylation patterns imposed by genetic imprinting in term cells are not entirely equivalent and in some cases lead to differential gene expression in offspring. Genetic imprinting may explain the need for progeny to inherit particular chromosomal regions from a particular parent.

Individual chromosomal DNA molecules are estimated to range in size from 50 to 250 Mb. The entire human haploid genome contains approximately 3000 Mb. This means that a single chromosomal DNA molecule may be as long as 70 mm and that the entire human genome would be nearly 2 m in length if all the chromosomal DNA molecules were laid end-to-end. Compaction of nuclear DNA by proteins allows the entire DNA complement to fit within nuclei that are only 5 μm (5×10^{-6} m) in diameter.

Chromosomes consist of DNA complexed with histone and nonhistone proteins into a structure called chromatin. Histones constitute a group of five low-molecular-weight basic proteins each encoded by multiple genes. Histones H2A, H2B, H3, and H4 package DNA into a structure called the core particle. A second level of packaging is provided by the histone H1, which organizes a 10-nm nucleosomal fiber into a 30-nm fiber with a packaging ratio of 40 : 1. Most DNA is complexed with the same five histone proteins, although some histone genes clearly code for unique subtypes. Some tissue-specific subtypes have been identified, but, in general, the role and deposition of histone subtypes are not clearly understood.

The remaining protein components of chromatin, globally called nonhistone proteins, are heterogeneous. Some are regulatory proteins that control gene expression; others are involved in macromolecular processes such as DNA replication and transcription. Some nonhistone proteins probably contribute to the nuclear scaffold (or nuclear matrix) responsible for higher-order organization of the interphase nuclear chromosomes.

During cell division, chromosomes condense further into highly ordered structures that may be observed by light microscopy (Fig. 1). A constricted portion, called the centromere, serves to correctly partition replicated chromosomes during cell division. The centromere divides the chromosome into small and large arms designated p (for petite) and q (because it follows p in the alphabet), respectively. Giemsa and other staining techniques are used to subdivide the arms into an increasingly finer characteristic set of bands. Most studies divide the genome into about 400 bands, although it is possible to divide the genome into a distinct ~800-band pattern. These bands are numbered on each arm from the centromere to the telomere. For example, the major histocompatibility complex (MHC) is located at 6p21.2 [i.e., the short arm of chromosome 6, band (p)21.2]. Although the precise physical basis for banding is not well understood, bands are useful for identifying and subdividing chromosomes and a number of molecular characteristics have been attributed to them. Giemsa light bands are believed to contain higher amounts of GC base pairs, of genes, of earlier replicating genes, and of *Alu* repetitive elements (see Section I,B).

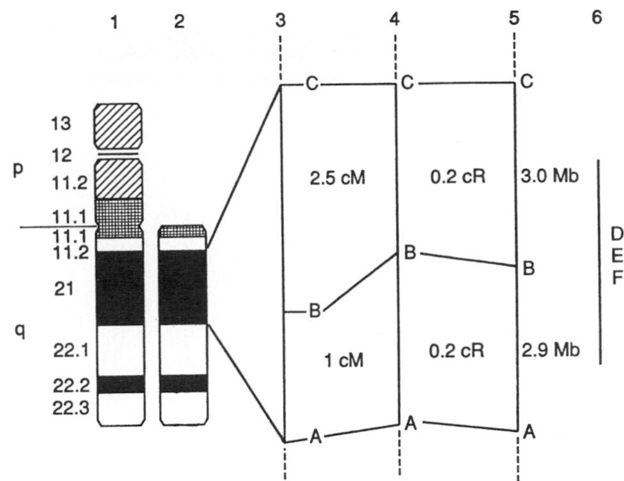

FIGURE 1 Types of chromosomal landmarks that need to be integrated into a genomic data base. Lane 1: Distinct cytogenetic bands of chromosome 21 observed by light microscopy (note the distinct banding pattern of the rDNA repeat 21p12). Lane 2: Fragmented chromosome 21. Lane 3: Genetic linkage map. Lane 4: X-irradiated chromosome map. Lane 5: Physical maps (including restriction maps, STAR maps, clones from overlapping libraries, and DNA sequence). Lane 6: Localization of a marker to a chromosomal region. cM, centimorgans; cR, centirays; Mb, megabase.

Chromosomes contain special structures at their ends called telomeres. These structures are formed from a simple repeat element 5′ TTAGGG 3′ (Table II). Multiple copies of this repeat are added to the 5′ ends of DNA in the absence of DNA template by an

TABLE II
Repetitive DNA Elements in the Human Genome

DNA element	% of genome	Number in the genome	Repeating unit	Type of element	
Alu	5.0	5×10^5	300[a]	SINE	Interspersed
Kpn	3.3	2×10^4	5000[b]	LINE	Interspersed
$(CA)_n$	0.02	5×10^4	~50[b]		Interspersed
stDNA I	0.5	3×10^6		Satellite I	Tandemly repeated
stDNA II	2.0	1×10^7		Satellite II	Tandemly repeated
stDNA III	1.5	9×10^6	~5[c]	Satellite III	Tandemly repeated
stDNA IV	2.0			Satellite IV	Tandemly repeated
Alphoid	0.75	6×10^{-4}	340[d]	Centromeric	Tandemly repeated
$(TTAGGG)_n$	0.01	7×10^4	6[e]	Telomeric	Tandemly repeated

[a]Occurs once every 5 kg.
[b]Occurs once every 50 kb.
[c]A single block of repeat is usually 2 kb in size.
[d]A single block may range from 0.5 to 5.0 Mb.
[e]A single block may be from 5 to 15 kb.

enzyme called telomerase. The added repeat serves as a template for synthesis of the second strand by DNA polymerase. The addition of this repeat and associated histone and nonhistone proteins function to preserve the ends of the chromosome during DNA replication and to prevent their degradation within the nucleus. Telomeres may also help to position chromosomes in interphase (nondividing) nucleus.

B. Repetitive DNA

In prokaryotes, genome size accurately reflects gene number. This is not the case in most eukaryotes, where repeated DNA elements and other noncoding sequences account for substantial portions of the genome. Thus, even though the human genome is large enough to contain a million genes encoding proteins that average 300 amino acids in length, the actual number of genes is probably between 50,000 and 100,000.

A number of human repetitive DNA sequences have already been identified and characterized to one extent or another. However, at this time, there is no coherent picture of the organization and function of DNA repeats. In general, human repetitive sequences may be divided into two types, those that are tandemly repeated and those that are interspersed (see Table II). Tandemly repeated simple sequences, called satellite DNA, are located near centromeres. In humans, there are four families of satellite DNA. Each type consists of thousands of tandem and inverted repeats of short (5–10 bp) sequences. At least 5% of the human genome may consist of satellite sequences. Although not yet demonstrated, these sequences probably play a role in determining chromosome structure and in partitioning chromosomes during mitosis and meiosis. Tandemly repeated centromeric alphoid sequences may be divided into groups, some of which are specific for sets of chromosomes.

Hypervariable minisatellites sequences (also called variable number tandem repeats, or VNTRs) are dispersed, highly polymorphic sequences composed of tandemly repeated sequences about 10–15 bp in length. VNTRs are useful for DNA fingerprinting and for genetic mapping experiments because they are very polymorphic (see the following). A large number of very short VNTRs (2–4 bp in length) are used for polymerase chain reaction (PCR)-based genetic mapping experiments. Particular polymorphisms in tandemly repeating sequences are associated with some human diseases. The polymorphisms within these and other tandem repeated sequences are due to insertions and deletions that arise from recombination or during DNA replication by the close proximity of multiple copies of the same sequence.

The most common interspersed repeat is called *Alu* because it contains a recognition site for the restriction enzyme *Alu* I. Different *Alu* consensus sequences divide the *Alu* repeats into different families. The 300-bp *Alu* repeat is a member of a group of repeats called short interspersed nucleotide repeats (SINEs). *Alu* sequence elements account for as much as 5000 bp of the human genome. Long interspersed nucleotide repeats (LINEs) are characterized by the 5000-bp *Kpn* element, which occurs at one-tenth the frequency of *Alu* elements.

Repeat sequences like *Alu* are dispersed through the genome by retrotransposition, which occurs through the formation of an RNA intermediate as a rare transcription event at particular integration sites. In this process, the enzyme reverse transcriptase forms a complementary DNA, which is then inserted at another location in the genome. Transposition into a gene would cause its inactivation and would very likely be of negative adaptive value. Nonetheless, *Alu* sequences have been tolerated during human evolution, suggesting that their presence in the human genome has not had a negative adaptive value. Some have called retrotransposing sequences parasitic DNA.

Some medium repetitive DNA sequences have unexpected characteristics. For instance, one repeat was found to be inserted in chromosomes and also as an extrachromosomal circular form. The latter form is probably generated by recombination events between adjacent chromosomal elements. Recently, rearrangement of one repeat was associated with differentiation.

Because little or no coding information appears to be contained within repetitive DNA, some researchers refer to repetitive DNA sequences as "junk" DNA. Repetitive DNA, as well as other noncoding "junk" DNA (see Section I,C), has the potential to reveal totally unexpected aspects of the human genome because few (if any) interesting roles have been attributed to them.

C. Genes and Gene Families

The remainder of the genome consists of single- and low-copy DNA sequences. The development of gene cloning and sequencing techniques has provided detailed pictures of the structure of eukaryotic genes. Yet genes are still difficult to recognize from DNA sequence data alone. The presence of a gene may be inferred from the identification of sequences shared with previously identified control elements (e.g., promoters, enhancers, polyadenylation sites, and splice sites) and long coding elements (open reading frames, or ORFs). In addition, putative gene functions sometimes can be predicted from the identification of particular protein motifs with a functional correlate (e.g., adenosine triphosphate binding motif, protease motif, immunoglobulin-like domains, the zinc-finger motif characteristic of DNA-binding proteins).

Eukaryotes have three types of genes defined by the type of RNA polymerase that transcribes them. Each type has distinct structural and regulatory characteristics. RNA polymerase I transcribes the genes for 5.8*S*, 18*S*, and 28*S* ribosomal RNAs (rRNA), which are required for protein synthesis. Genes transcribed by RNA polymerase II include all known protein coding genes as well as some small nuclear RNA genes (snRNA), which are involved in RNA processing (see the following). RNA polymerase III genes synthesize 5*S* ribosomal RNAs as well as the transfer RNAs (tRNA), which are required for translation of RNA polymerase transcripts into proteins.

The cell contains about 100,000 copies of the RNA polymerase II enzyme. This large amount of RNA polymerase II is needed to support generalized transcription. Large amounts of translation components are also required to support a reasonable protein synthetic rate. In fact, the ribosomal RNA and ribosomal protein genes are the most highly expressed genes and the major cytoplasmic components.

The 5.8*S*, 18*S*, and 28*S* rRNA genes are contained within a 45-kb monomer sequence that is tandemly repeated an unknown number of times on chromosomes 13, 14, 15, 21, and 22. Intensive transcription of the 45-kb monomer units containing 5.8*S*, 18*S*, and 28*S* rDNA at specific nuclear locations, called nucleoli, permits their visualization by light microscopy. There are an estimated 100–200 5*S* rRNA genes. Most 5*S* rRNA genes are located in the chromosome 1q telomere region, although other copies might be dispersed over the genome. Genes that code for ribosomal proteins and tRNA are found at multiple locations in the genome. Thus, the genes encoding the various components involved in protein synthesis are not only located on different chromosomes but are also transcribed by different RNA polymerases. It is not clear how synthesis of the various ribosomal components is coordinately regulated.

Genes consist of promoter sequences, regulatory sequences, expressed or coding sequences (exons), and intervening sequences (introns). Promoters are defined by sequences near transcriptional start sites that serve as RNA polymerase recognition sites. Generally, promoters are located at the 5′ end of genes transcribed by RNA polymerase I and II and within genes transcribed by RNA polymerase III. Cis-acting regulatory sequences are bound by transacting factors that regulate the amount, the timing, and the tissue specificity of gene expression.

Regulatory sequences are separated from each other and from promoter sequences: they may be located upstream, downstream, or within a gene. Enhancer sequences promote nearby gene expression, whereas silencer elements decrease gene expression. Enhancer and silencer sequences do not appear to be gene specific and may be located at large distances, 5′ or 3′, from a gene.

DNase I hypersensitivity sites near genes are thought to mark open chromatin configurations at regulatory elements, necessary for RNA polymerase accessibility. For example, a 20-kb region of DNA located more than 40 kb upstream from the human β-globin gene cluster contains multiple DNase I hypersensitivity sites and is required for both tissue specificity and developmentally regulated expression of the β-globin gene cluster. Termed a dominant control region, it may include multiple enhancers. The region may also contain sequences necessary for the attachment of the globin gene complex to the nuclear matrix.

A eukaryotic gene may consist of as little as 10% exons and 90% introns (Fig. 2). Introns are removed from high-molecular-weight nuclear RNA, called hnRNA (heterogeneous nuclear RNA), before translation by the splicing machinery. This means that the average gene coding for a protein of average size (about 30,000 daltons) is considerably larger than the size needed to actually code for the polypeptide product (e.g., 900 bp). Most genes that code for proteins range in size between 10 and 40 kb; however, some genes are very large. The dystrophin gene, whose mutant forms account for Duchenne and Becker muscular dystrophy, is over 2 Mb long, codes for a 400,000-molecular-weight protein, and has about 100 introns. The *BCRA1* gene, responsible for half the cases of early onset hereditary breast cancer, is 10,000 bp in size, whereas the coding region is only 1600 bp in length. In contrast, the protamine gene has a single intron and is only 0.5 kb in size.

The function and evolutionary origin of introns are unresolved issues. Some believe that introns are also genomic "junk." However, introns may provide sev-

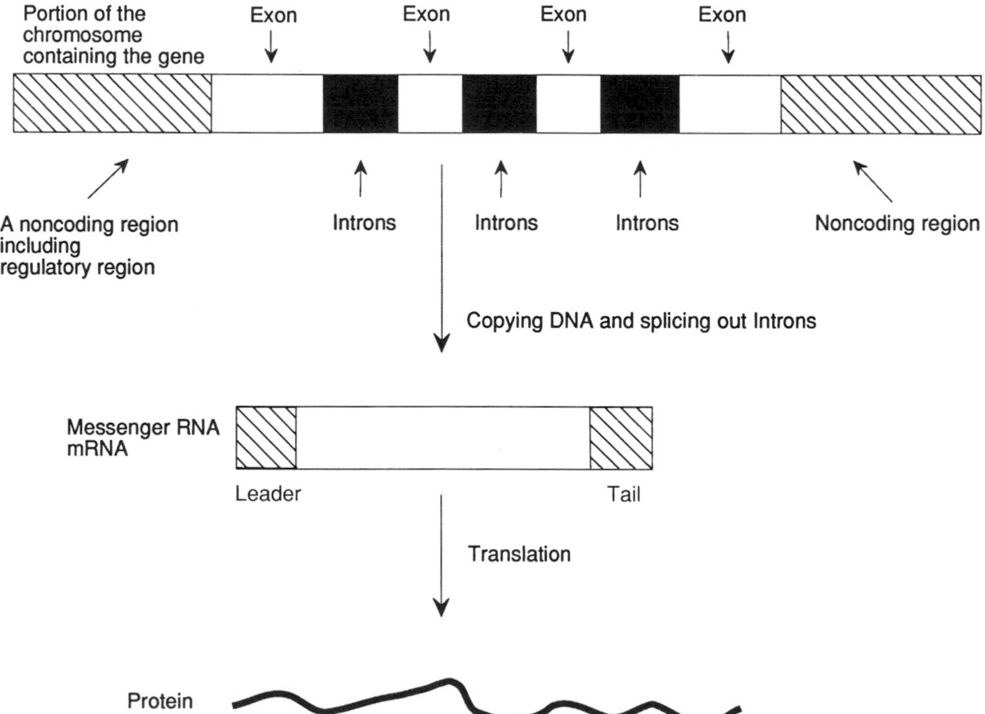

FIGURE 2 Eukaryotic genes are composed of coding and noncoding sequences (introns). Introns are removed from heterogeneous nuclear RNA before translation.

eral functional and evolutionary advantages to the cell. For instance, alternative forms of the same protein can be generated by splicing together different exons; for example, alternative use of an exon encoding a transmembrane domain leads to the formation of either a membrane-bound or secreted forms of immunoglobulin molecules. Another possibility is that novel genes may evolve by "exon shuffling," in which an exon from one gene links to an exon of another gene, generating proteins with new and adaptive properties. If the exon shuffling occurs through a gene duplication, the original exon functions will be preserved. Finally, in some cases ORF within introns may encode small proteins.

Although DNA–RNA renaturation experiments indicate that most genes coding for proteins are present in single copies, many genes share sequence and structural characteristics. For example, the CD8, Thy-1, T-cell receptor, and histocompatibility genes are all members of the immunoglobulin supergene family. Genes within supergene families are believed to have evolved through duplication of a primordial gene that then diverged in structure and function. Members of supergene families are dispersed throughout the genome, often as multigene families. In most cases, close linkage is maintained between multigene family members, possibly reflecting a requirement for coordinate regulation, as in the case of the developmentally regulated hemoglobin gene cluster. In the case of the immunoglobulin and T-cell receptor gene families, linkage must be retained to facilitate the somatic rearrangements necessary for the generation of functional gene products. The significance of linkage is less clear in cases such as the MHC gene complex.

Pseudogenes are inactive copies of functional genes that have accumulated one or more mutations (e.g., splice site defects, deletions, premature termination codons) that preclude their expression. Pseudogenes frequently occur in multigene families. Like parasitic DNA, processed pseudogenes are dispersed in the genome and most likely result from integration of a DNA copy of a processed RNA transcript containing a poly(A) tail. Pseudogenes provide a reservoir of genetic material for evolution.

Genomic structure is conserved through phylogeny. For instance, syntenic regions between mouse and human range from whole chromosome to blocks of DNA ~10 Mb in size. The existence of synteny can aid human genetic studies by helping to identify a homologous gene in an experimental organism such as the mouse, or by identifying candidate genes for human gene searches.

D. Mitochondrial Genome

Mitochondria are cytoplasmic organelles that are the major energy producers of the cell. This organelle contains approximately 10 copies of a small, 16.5-kb, circular chromosome. Unlike nuclear chromosomes, the mitochondrial DNA is inherited only from the mother. The complete sequence of the mitochondrial genome has been determined. The mitochondrial genome encodes its own rRNAs and tRNAs, enzymes of the electron transport system, and a number of as yet unidentified gene products predicted from long ORFs. Mitochondrial genes do not contain introns and contain only small tracts of spacer sequences. Surprisingly, the mitochondrial genetic code is slightly different from the universal genetic code. Several neuromuscular diseases have been associated with mutations found in the mitochondrial genome, perhaps because of the high energy requirements of neurons and muscle cells, which cannot be met when the underlying energy-producing machinery is not functional.

II. GENETIC MAPPING OF HUMAN CHROMOSOMES

Genetic and physical maps of human chromosomes consist of the linear representation of genetic loci. A genetic locus defines a location on a chromosome. The location may be identified by a gene [e.g., expressed sequence tag (EST)], a restriction enzyme site [e.g., RFLP, sequence tagged restriction site (STAR)], a phenotype, a chromosomal band, a rearrangement, a DNA sequence [e.g., sequence tagged site (STS), EST, STAR, STRP], or a molecular clone. The various types of loci are identified by distinct but sometimes overlapping or complementary technologies, including linkage analyses, somatic cell techniques, molecular cloning in bacteriophage, *Escherichia coli,* or yeast, DNA sequencing, chromosome restriction mapping, chromosome walking, chromosome linking, and chromosome jumping. The new field of genomics strives to integrate information obtained from all of these studies into one coherent picture (see Fig. 1).

A. Linkage Analysis

The classic form of a genetic map is derived from studies measuring the frequency of meiotic recombination, which occurs during the formation of germ cells, between chromosomal loci. Two loci that are close together on a chromosome display linkage by cosegregating during meiosis more frequently than loci that are farther apart. In the simplest case, linkage of the two phenotypes, sex and the disease hemophilia, mapped the hemophilia gene to the X chromosome. In the 1930s, the cosegregation frequency of a chromosome X color blindness gene with hemophilia was measured. This allowed an estimate of the distance between the genes and represented the first map of a human chromosome. In practice, phenotypic markers useful for genetic mapping are infrequent in the human genome. Furthermore, many phenotypes are due to multiple genes and some genetic alleles exhibit incomplete penetrance.

About 0.1 to 1% of DNA sequences vary between two individuals. The identification of sequence polymorphisms and the study of their inheritance provide useful genetic markers that are independent of phenotypes. DNA polymorphisms may be identified using restriction enzymes, which cleave DNA at specific DNA sequences and generate DNA fragments of defined sizes. Most RFLPs result when two individuals differ in a sequence defining a restriction enzyme site (i.e., restriction site polymorphism). Some RFLPs arise when two individuals differ in the size of DNA between two restriction sites. The size variation may be due to a polymorphism in a tandemly repeated sequence within the restriction fragment. DNA sequencing around a large number of simple tandem repeats has allowed the development of PCR-based genetic mapping methods. In this case, PCR primers complementary to bases in the unique sequences surrounding the repeated sequences define the ends of the amplified DNA. Both hybridization and/or PCR experiments are used to assess the status of RFLPs and STRPs in human pedigrees. Markers are ordered and organized into genetic linkage groups corresponding to each of the chromosomes. Currently, approximately 1500 RFLPs are cloned and mapped in the human genome and/or about 5000 STRPs. Although this number is impressive, it represents, on average, only one RFLP for every 2 Mb. This means that genetic mapping can usually only narrow the possibilities rather than precisely locate a gene.

Surprisingly, highly polymorphic trinucleotide repeats have been found within exons of normal genes. Besides these normal polymorphisms, an increasing number of neurodegenerative diseases have been associated with increases in particular $(CAG)_n$ repeats coding for glutamine. Some diseases, like fragile X, are associated with a large increase in repeat length (50 → 200), whereas others, like Huntington's disease, are associated with small changes in repeat length (30 → 40). Diseases associated with increases in repeat length often display genetic anticipation, where the age of onset or the severity of the disease correlates with the size of the expanded repeat. Expanded repeats are also associated with changes in DNA methylation. The expanded fragile X repeat can be visualized microscopically, where it appears as a broken or unstained chromosomal region. About 104 fragile sites are distributed throughout the entire genome. Usually fragile sites are observed only after cells have been treated with one of a number of agents that interfere with DNA metabolism. About 20 fragile sites occur rarely and appear to be inherited in a manner similar to that of the fragile X site.

A recent advance in genetic mapping is the application of PCR to analyzing meiotic recombination in sperm. This approach allows highly accurate mapping because each sperm represents a unique meiotic event and large numbers of samples are analyzed. This approach eliminates the need to obtain DNA samples from large numbers of human pedigrees, yet provides recombination frequencies only for male meiosis. Recombination frequencies are different in females. Obviously, the possibility of obtaining the equivalent samples from females (e.g., ovum polar bodies) is quite limited. This method cannot be used to map disease phenotypes.

B. Somatic Cell Genetics

A second category of genetic mapping employs somatic cells rather than examination of inheritance patterns. In its simplest application, loci can be identified in somatic cells by observing abnormal karyotypes in mitotic chromosomes. A familiar example is the correlation of Down's syndrome with trisomy (three copies instead of two) chromosome 21. In some cases, somewhat finer mapping is possible when a phenotype involves a naturally occurring chromosomal rearrangement. For instance, the rare occurrence of X-linked muscular dystrophy in a female was traced to a translocation between X and the ribosomal repeat DNA on chromosome 21. This information was ultimately instrumental in the cloning of the muscular dystrophy gene by chromosomal walking (see the following) from the rDNA.

Genetic loci can also be mapped by *in situ* hybridization of labeled DNA clones to metaphase chromosomes. Recently, highly sensitive methods, employing fluorescence-labeled DNA probes (FISH), have enabled not only the detection of 5-kb sequences but also the ordering of sequences less than 500 kb apart. These methods may also be applied to extended purified genomic and YAC DNAs for high-resolution mapping and have even been used to observe restriction enzyme cleavage of large fragments. DNA sequences may also be localized by electron microscopy using immuno-gold labeling of hybridized probes. Although single-copy sequences have not yet been detected by this technique, electron microscopy has considerable potential for revealing new details of chromosome structure.

A number of different somatic cell variants have been generated for use in genetic mapping experiments. Interspecies somatic cell hybrids or heterokaryons generated by Sendai virus or polyethylene glycol-mediated fusion contain a complete complement of rodent chromosomes plus one or more human chromosomes. The chromosomes may be distinguished by their banding patterns and/or hybridization experiments with chromosome-specific probes. Hybrid cells can be assayed for the presence of a human gene or its expression, thus mapping a gene to the human chromosome present in the somatic cell hybrid. For example, human glucose-6-phosphate dehydrogenase can be detected only in rodent–human hybrid cells containing a copy of human chromosome X. Rodent cell lines harboring mutations in genes such as those involved in DNA repair and purine metabolism have been useful for selecting and identifying chromosomes containing the homologous human gene. Some hybrid cell lines contain fragments of human chromosomes. A panel of cell lines containing different chromosomal subregions may be used to regionally assign genes. For example, a map of the MHC was constructed by correlating karyotypic losses following gamma irradiation with losses in expression of cell-surface proteins encoded by this complex.

An alternative mapping method involves the creation of a panel of clonal cell lines derived from the fusion of the gamma-irradiated monosomic hybrid cells or human cells with unirradiated rodent cells. The panel of cells is isolated that contain different regions of a single human chromosome or regions from multiple human chromosomes, respectively. The panel of cells is then tested for the presence of markers of interest. Genetic linkage in this map is established in a manner similar to mapping by recombination by assuming that the closer two genes are, the more likely they will stay together during X-ray irradiation-induced chromosomal fragmentation. The map distance is in centirays (cR) and roughly correlates with physical distance, since chromosome breaks due to gamma irradiation should be mostly random.

III. PHYSICAL MAPPING

The ultimate physical map of a chromosome is its complete nucleotide sequence. At the moment, this goal remains elusive for complex genomes. Meanwhile, a variety of approaches are being used to construct chromosomal restriction maps. These approaches are the same ones used to construct complete genomic restriction maps for the 5-Mb genome of the bacterium *E. coli*.

Genomic mapping strategies can be broadly classified into "top-down" and "bottom-up" approaches (Fig. 3). Top-down approaches are directed approaches that produce complete maps but that are difficult to automate and do not provide directly cloned DNA sequences. Bottom-up approaches are easy to automate and provide ordered genomic libraries. Bottom-up random approaches rarely produce complete maps unless more directed "endgame" strategies are used, such as conventional chromosome walking (see Section III,A).

A. High-Resolution Physical Maps and Overlapping Libraries

Traditional bottom-up approaches build a map by overlapping cloned sequences. A complete bottom-up map will consist of an ordered set of overlapping DNA clones representing all the DNA in a chromosome or a chromosome subregion. Libraries of DNA fragments from human chromosomal DNA can be isolated in a variety of simple organisms by recombinant DNA technology. These libraries are distinguished by the size of the genomic fragments, vector, and host cell. For instance, libraries constructed in *E. coli* plasmid vectors usually contain short (~4 kb) inserts; lambda-phage libraries contain inserts ranging from 5 to 15 kb in size; cosmid libraries have inserts ranging up to 40 kb in size; PAC and BAC libraries have inserts ranging from 100 to 150 kb in size; and YAC libraries have inserts up to about 1000 kb (Fig. 4). In general, transformation efficiencies decline as insert sizes increase. YAC transformation frequencies are extremely low (i.e., only several hundred clones/μg

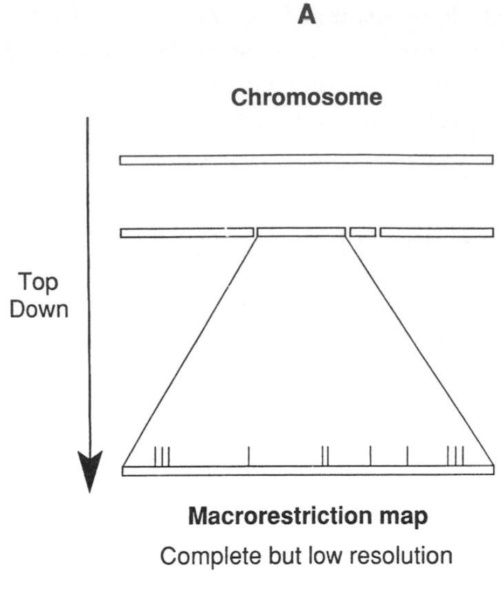

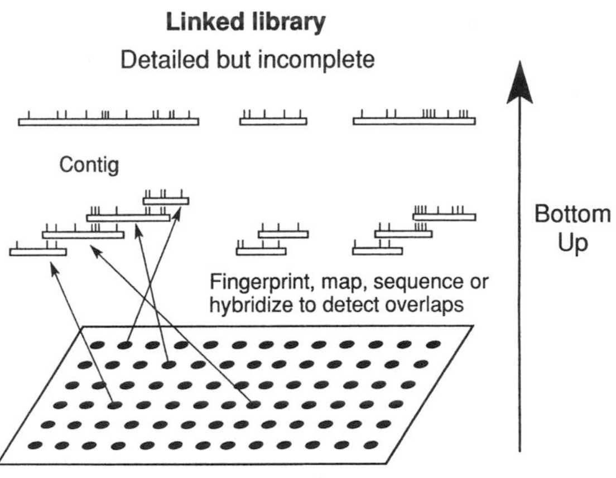

FIGURE 3 Two basic genomic mapping strategies. (A) Top-down approaches produce complete low-resolution restriction maps. (B) Bottom-up approaches produce high-resolution restriction maps and overlapping libraries.

DNA), but fewer clones are required for complete representation of the genome. Any library is potentially incomplete because of unclonable sequences. For instance, some sequences may be unstable as recombinant molecules or may specify toxic products.

Chromosome-specific or region-specific libraries can be constructed using human DNA isolated from flow-sorted chromosomes or from physically dissected chromosome regions. Usually genomic DNA is partially digested with a restriction enzyme or physically sheared at random to the appropriate size for cloning to ensure that the library contains clones that overlap (Fig. 4). Total library size (complexity) is usually five times the size of the genome to ensure, with reasonable certainty (99.9%), that all chromosomal regions are represented. Randomly chosen clones from such a library can then be fingerprinted by one of a variety of methods in order to identify overlaps between clones. Fingerprinting methods include the identification of (1) unordered, shared restriction fragments, (2) ordered restriction sites, (3) complete DNA sequencing, and (4) sequencing around restriction enzyme cleavage sites and the detection of shared sequences by hybridization with (5) specific oligonucleotides or repeats or with (6) the clones themselves. The overlapping clones can then be ordered into contiguous overlapping clone sets called contigs (see Fig. 3B). The final contig spans the entire region of DNA from which the library was constructed.

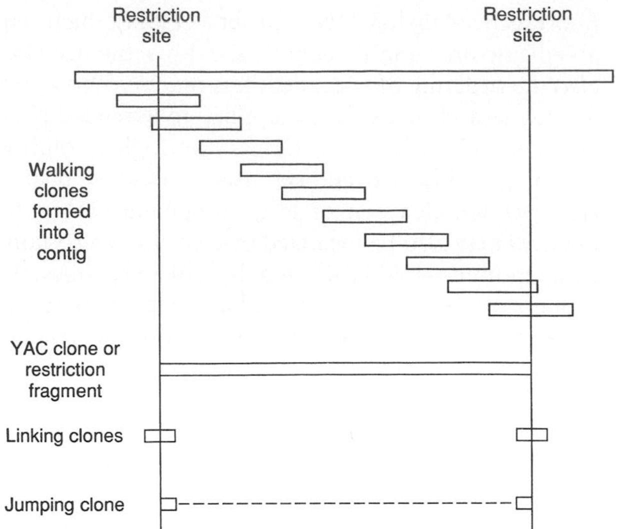

FIGURE 4 Physical characterization and cloning of chromosomes. Originally, overlapping libraries formed by contigs of small DNA pieces were the only methods available for expanding large regions of the genome. Recent advances in PFG and YAC technology greatly facilitate the characterization and isolation of large regions of the genome. Large regions may also be spanned by small linking and jumping clones.

Thus far, only three random methods (methods 1, 2, and 6) have been successful in generating large genomic restriction maps. However, map closure with the first two approaches required chromosome walking. Here, probes from the ends of contigs were used to identify clones to specifically span gaps (see Fig. 4). The number of chromosome walking (discussed next) steps is minimized if single clones span gaps. Hence, large YAC clones are especially useful.

Chromosome walking experiments (method 6) have been used to produce a complete overlapping library of a small 1-Mb prokaryotic genome. The straightforward application of chromosome walking requires repeated cycles of hybridization and probe isolation; consequently, it is extremely laborious and has sometimes been called chromosome crawling. This approach has been eased by the construction of cloning vectors that allow easy end probe generation. The amount of work involved in chromosome walking may be further reduced by multiplexing. Recently, two small bacterial genomes each ≤1.8 Mb in size were completely sequenced using unordered clones. In these cases, the randomly chosen clones were bidirectional sequenced from the cloning sites. Contigs were then created by overlapping the sequence directly. It is not clear how any of these approaches would work on larger genomes.

A number of groups have produced ordered libraries for several human chromosomes (e.g., 16, 19, 21, 22, X), as well as a number of eukaryotic genomes (e.g., *Saccharomyces cereviseae, Schizosaccharomyces pombe, Arabidopis, C. elegans, Drosophila*). About 75% of the human genome is contained in a 33,000-member YAC library. Thus far, a combined top-down and bottom-up approach has proven to be the most efficient method for clone ordering.

B. Low-Resolution Restriction Mapping

The first step in a top-down mapping strategy is the naturally occurring division of the genome into chromosomes. Chromosomal DNA can be further divided by cutting with a restriction enzyme that generates large DNA fragments (Table III). The dinucleotide CpG is contained with the recognition sequence of these restriction enzymes. This dinucleotide is underrepresented in the human genome and is usually methylated (discussed earlier). Since unmethylated CpG-rich islands of DNA have been identified near the 5′ end of many genes, and most enzymes with CpG in their recognition sequence are inhibited by cytosine methylation, most of the restriction enzymes that generate large DNA fragments will preferentially cut at the 5′ end of genes. This means that restriction maps with such enzymes will locate genes (discussed in the following).

The restriction enzymes *Not* I and *Mlu* I have been particularly useful for mapping the human genome because they produce very large restriction fragments. There is some evidence that these enzymes tend to occur in the same CpG islands; thus, they may detect a particular subset of genes. Furthermore, it is estimated that 20% of human genes have a *Not* I site in their associated CpG islands. Although both *Not* I and *Sfi* I have 8-bp recognition sequences, *Sfi* I produces much smaller restriction fragments, probably because its recognition sequence lacks a CpG sequence. Recently,

TABLE III
Restriction Enzymes That Produce Megabase Fragments

Enzyme	Size of recognition site (bp)	Recognition sequence	Average fragment size (kb)	
			In a random sequence	Detected by PFG[a]
Not I	8	GC/GGCCGC	64	1000
Sfi I	8	GGCCNNNN/NGGCC	64	250
Mlu I	6	A/CGCGT	4	1000
Sat I	6	G/TCGAC	4	500
Nru I	6	TCG/CGA	4	300
Pvv I	6	CGAT/CG	4	200
Xho I	6	C/TCGAG	4	200
Nar I	6	GG/CGCC	4	100
Apa I	6	GGGCC/C	4	100
BssH II	6	G/CGCGC	4	100
Sac II	6	CCGC/GG	4	100

[a]Pulsed-field gel electrophoresis.

a number of additional restriction enzymes (e.g., *Ase* I, *Fse* I, and *SgrA* I) and other enzymes involved in site-specific recombination have become available that produce large fragments from the human genomes.

Large restriction fragments can be size-fractionated with PFG. After fractionation, the fragments can be ordered and a map constructed by hybridization experiments using cloned genomic DNA sequences as probes. These probes will hybridize to an unknown location on a large restriction fragment. Hybridization probes may be of several types. For instance, single-copy, genetically mapped probes can serve as anchor points between genetic and physical maps. The more precisely a probe has been genetically mapped, the more precisely it will map a fragment. In the best case, a genetically mapped probe will locate a restriction fragment to within 5–20 Mb on the chromosome. Obviously, the sole use of this approach requires many hybridization experiments with closely spaced probes to ensure that fragments, especially small ones, are not missed.

Telomer-specific clones are especially important because they define the ends of the chromosomes and therefore the ends of the physical maps. In addition, telomeric clones used as hybridization probes to PFG-fractionated DNA partially digested with a restriction enzyme will identify a series of bands that extends in one direction from the end of the chromosome (Fig. 5). Thus, this type of partial restriction enzyme data is very easy to interpret and allows an extensive map construction in a single experiment. On smaller molecules, this type of experiment has been called a Smith-Birnstiel or an indirect-end labeling experiment.

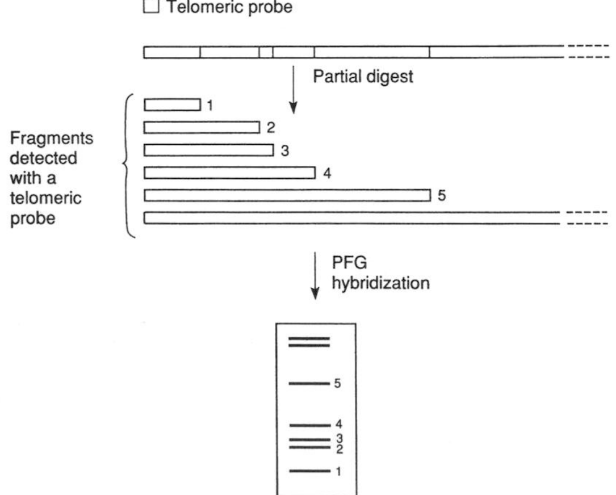

FIGURE 5 Indirect-end labeling experiment with telomeric sequences reveals the restriction sites close to the end of the chromosome. Partial digestion experiments are easy to interpret with telomeric probes because fragments extend in only one direction.

Human telomeres were cloned in yeast by assuming that they would functionally complement half of a YAC. Chimeric YAC–human telomere clones were detected in YAC libraries using an oligonucleotide probe specific for the simple sequence 5′ TTAGGG 3′, which is present at the ends of all higher eukaryotic chromosomes (see Table II). Nearby the simple telomeric repeat is a series of subtelomeric human-specific repeats. These have been used in hybrid cell lines to identify specific human telomeric restriction fragments.

Linking clones allow complete *Not* I restriction maps to be constructed with a relatively small number of hybridization experiments (see Fig. 4). These small clones contain recognition sequences for a particular restriction enzyme. A complete *Not* I linking library of the entire human genome would consist of as few as 3000 clones, whereas a complete *Not* I library for chromosome 21 would consist of only about 50 clones. A *Not* I linking clone used as a hybridization probe identifies two adjacent *Not* I restriction fragments. Linking clones identifying adjacent *Not* I sites will share a common restriction fragment, thereby establishing continuity in the map. The 50 *Not* I restriction fragments expected from chromosome 21 are not likely to overlap in size as long as PFG conditions are adjusted to maximize resolution. Thus, the detection of the same-sized restriction fragment by two chromosome 21 probes usually indicates commonality. However, a PFG fractionation of the total human genome cannot resolve all genomic fragments. Each PFG size class will probably contain several genomic fragments. Therefore, common fragments need to be linked by additional fingerprinting methods.

Megabase restriction fragment-length polymorphism (MRFLP), detected in a series of different cell lines, fingerprints a chromosomal region. The fingerprint is then used to prove continuity between two loci. This method has been termed polymorphism link-up (Fig. 6) and may be used with any type of clone. MRFLPs may be due to either genetic differences or to methylation differences between the cell lines. This approach provides an overview of the diversity of the human genome, because physical maps are made for multiple genomes (see the following). An even more powerful approach involves characterizing restriction enzyme cutting sites by determining the DNA sequence surrounding them (see Section III,C).

Partial digestion strategies can be used to identify neighboring *Not* I fragments and fingerprint chromo-

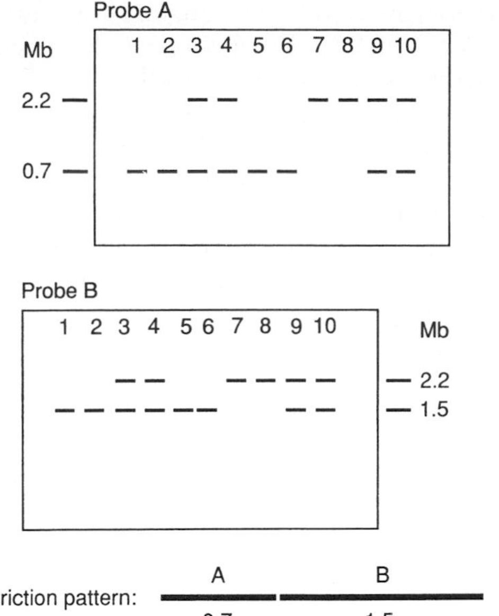

FIGURE 6 Physical map construction using polymorphism link-up. Hybridization of DNA from different cell lines (lanes 1–10) with two putatively linked probes (A and B) leads to the detection of different and common fragment sizes but identical polymorphism patterns.

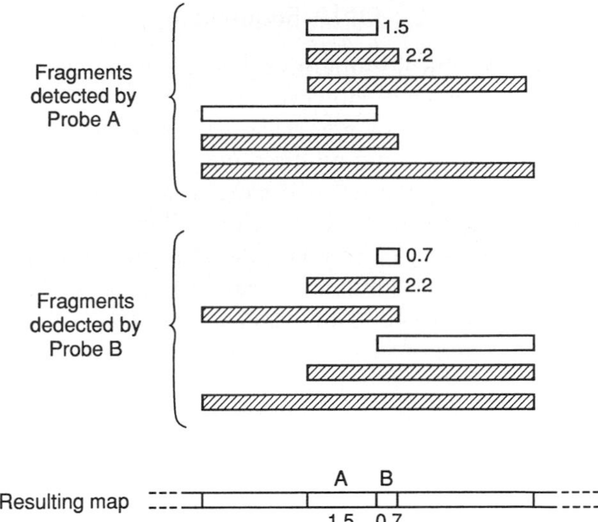

FIGURE 7 Use of partial restriction enzyme digestion to construct physical maps. Single probes used in hybridization experiments to partially digested DNA detect bidirectional partial digestion products. Probes on adjacent fragments detect common (shaded) and unique (unshaded) bands.

some regions (Figs. 4 and 7). Linking clones are particularly useful for partial mapping strategies. Here, each half of a linking clone is used independently as a hybridization probe to partially digested genomic DNA. A series of partial digestion fragments are then identified that expand bidirectionally from the smallest fragment containing the probe. Though more than one restriction map will be compatible with the hybridization data obtained with each half linking clone, the true restriction map must be consistent with the data obtained with both halves of the linking clone (Fig. 7).

Jumping clones are another mapping tool useful for characterizing large genomic regions (see Fig. 4). They contain two short, noncontiguous genomic sequences and are made by deleting the DNA between two distant genomic sequences. Jumping clones are used for rapid chromosome walking over long distances. A *Not* I jumping clone contains DNA sequences from the two ends of a *Not* I fragment (see Fig. 4).

A combination of these approaches has been used to order 60 distinct *Not* I restriction fragments of the human chromosome 21q arm in 9 different cell lines using 80 DNA markers and 11 chromosomal breakpoints. This work showed that there was remarkable large scale conservation of this region of the genome, that is, no large-scale deletions, insertions, or rearrangements were detected in any of the cell line DNA, although restriction site polymorphisms were detected.

It is quite clear that the methods used to create the large-scale restriction map of chromosome 21q cannot be applied to the entire human genome. Hybridization experiments are slow, labor-intensive, costly, and insensitive. The efficiency of these experiments can be increased by using PCR in place of hybridization as the analytical method. PFG lanes containing size-fractionated fragments are sliced and the DNA in each slice is used as a PCR template to test for the presence or absence of a particular DNA locus, as described earlier.

The use of electrophoretically purified *S. pombe* chromosomes and Mb restriction fragments allowed an eightfold increase in clone ordering. These purified DNAs were used as hybridization probes to an arrayed library to sort cosmid clones to different chromosomal regions. Overlaps between clones within a chromosomal region were detected by hybridization experiments that used pools of small genomic restriction fragments. The enhanced efficiency was obtained because hybridization probes from different chromosomes and chromosomal regions were pooled to reduce the total number of experiments.

C. DNA Sequencing

Current technical limitations preclude large-scale genomic sequencing; however, short runs of unique DNA sequences, such as STSs, can be used as clone identifiers and serve multiple purposes. The use of STS eliminates the need to store a large number of clones because the STS of a clone allows the design of PCR primers to reisolate the sequence from any genomic sample. This means that both small and large laboratories can contribute and use an STS data base.

A collection of STS markers around restriction sites (STARs) is helpful in constructing maps (Fig. 8). For instance, sequencing a complete *Not* I linking library for the human genome would require sequencing about 2 Mb of DNA. This is a large, but not unreasonable, amount of sequencing. Sequencing of a complete jumping library of the human genome would require the same amount of work. Simple sequence comparisons between the two libraries will order all the *Not* I sites, but not give the distances between them unless the jumping clones were generated from DNAs of known size. A *Not* I YAC library whose ends are sequenced could be anchored to this map or be used to help create it. Alternatively, end sequences of size-fractionated *Not* I fragments could also provide linkage between two *Not* I sites (see Fig. 8). DNA sequence from the ends of YAC clones or DNA fragments can be obtained by several PCR strategies. The identification and sequencing of *Not* I sites in contigs will link these sites to particular genomic *Not* I STARs.

In practice, *Not* I sites may be good starting points for collecting STAR sequences, but other more frequently occurring STARs must also be collected. The restriction enzyme *Eag* I recognizes the inner 6 bp of the *Not* I site. *Eag* I sites occur three times more frequently than *Not* I sites. Collecting *Eag* I STARs eliminates problems associated with manipulating very large *Not* I fragments, that is, the inability to make jumping clones greater than sequences 500 kb apart and the instability of large YAC clones.

This approach can also be used for finer restriction mapping. For instance, *Not* I–*Eag* I jumping fragments will position sequences adjacent to a *Not* I site next to a nearby *Eag* I site (Fig. 9). Then a type of PCR, called inverse PCR, can be used to obtain the sequence around the *Eag* I site using primers from sequences around the *Not* I site. This approach can be extended by using partial digests as well as more frequently cutting restriction enzymes until, eventu-

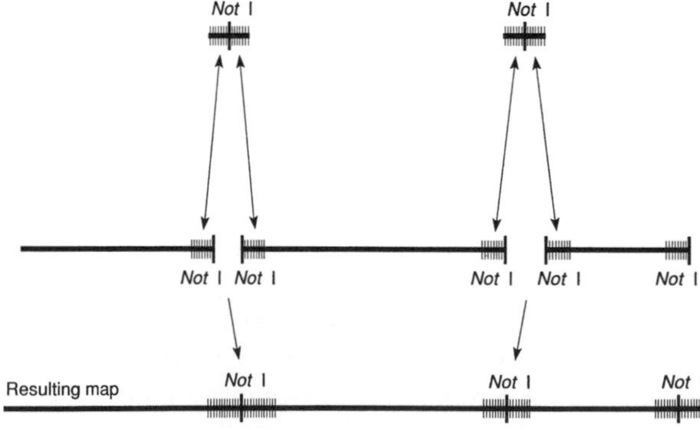

FIGURE 8 Use of STARs to construct physical maps. DNA sequences from a linking clone library are compared (using a computer) with the end sequences of large DNA fragments, YAC clones, or jumping clones. The different DNAs are then linked by sequence similarity to construct a STAR genomic map.

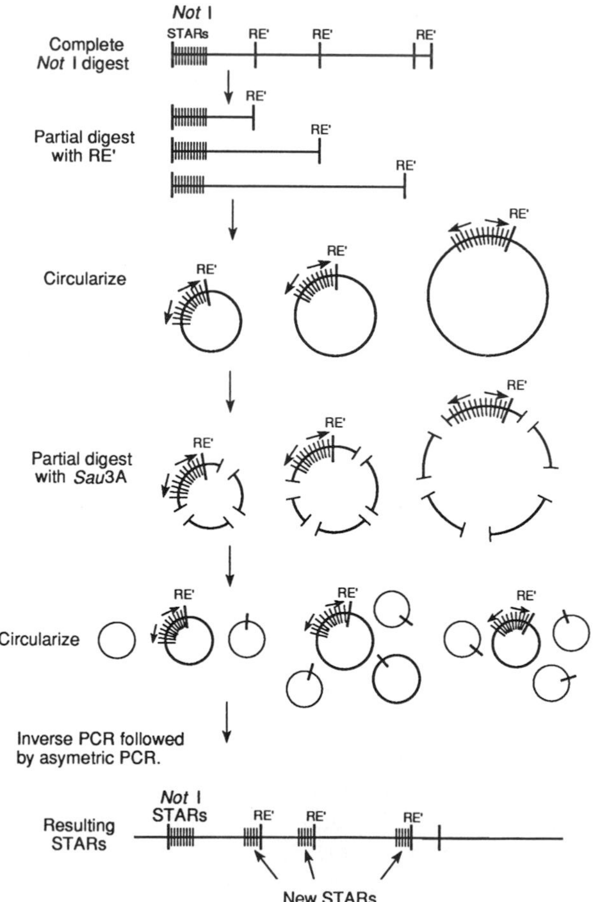

FIGURE 9 Use of existing STARs to define other nearby STARs. Chromosome jumping is used to move sequences around a restriction site to a nearby *Not* I site. Inverse PCR then allows the sequence of the DNA moved close to the *Not* I site to be amplified and subsequently sequenced.

ally, the entire sequence of a genome region is obtained. This approach, unlike many other approaches, provides data that will be continually added to as more sequence data accumulate. There are many more advantages, for instance, the same PCR primers may be used many times. The strategy can also be implemented on genomic DNA and, thus, requires little or no cloning and can be done with raw sequence data with high mistake levels. Lastly, this strategy can be implemented either in a random or directed way or by some combination of approaches.

IV. REVERSE GENETICS/POSITIONAL CLONING

One major interest in studying the human genome involves the identification and isolation of disease genes. Previous studies on disease genes have usually involved the isolation of the protein product. The subsequent purification and protein sequencing of the gene product allowed both antibody and oligonucleotide probes to be produced, which were then used for gene searches in genomic libraries. Now, RFLP mapping allows gene searches to be initiated at the genetic level. This approach has been called reverse genetics and/or positional cloning and begins by identifying a particular RFLP or STRP, which segregates with the disease gene. Subsequently, a variety of methods are employed to identify the gene of interest. For instance, candidate genes may be identified by mapping and cloning of CpG islands in a genetically defined region. The isolation of sequences for these loci allows finer genetic mapping, which may narrow the region if polymorphic loci are cloned. Denaturing gradient gel electrophoresis allows single-base differences between any DNAs to be detected and potentially should allow all markers to be made polymorphic enough for genetic studies. In some cases, such as cystic fibrosis, a more precise genetic location was defined by linkage disequilibrium studies since most patients had the same mutation. However, this is not always the case. Disease phenotypes may originate from one or more mutations at a single locus or from multiple mutations in several loci.

Molecular studies of cloned or PCR-amplified sequences, within candidate gene regions, allow the identification of expressed sequences, conserved through phylogeny (i.e., the assumption is that important and expressed sequences are conserved), and reveal whether or not any physical chromosomal abnormalities are associated with a particular disease phenotype. The identification of molecular abnormalities with disease phenotypes has been instrumental in identifying a number of disease genes. Thus, methods for physically characterizing human chromosomes provide tools for mapping genes.

ACKNOWLEDGMENTS

The authors thank Charles Cantor for discussion. This work was supported by grants (DOA AIBS2154 and DOE DE-FG02-95ER62040.A000) to C. L. S.

BIBLIOGRAPHY

Bickmore, W. A., and Sumner, A. T. (1989). Mammalian chromosome banding—An expression of genome organization. *Trends Genet.* **5**, 144–148.

Bird, A. P. (1987). CpG islands as gene markers in the vertebrate nucleus. *Trends Genet.* **3**, 342–347.

Cantor, C. R., and Smith, C. L. (1989). Large DNA technology and possible applications in gene transfer. *In* "Gene Transfer in Animals: UCLA Symposium on Molecular and Cellular Biology" (I. Verma, R. Mulligan, and A. Beauset, eds.), Vol. 87, new series, pp. 269–281. Alan R. Liss, New York.

Cantor, C. R., Smith, C. L., and Mathew, M. (1988). Pulsed field gel electrophoresis of very large DNA molecules. *Annu. Rev. Biophys. Biophys. Chem.* **17**, 287–304.

Collins, F. S. (1995). Ahead of schedule and under budget: The genome project passes its fifth birthday. *Proc. Natl. Acad. Sci. USA* **92**, 10821–10823.

Condemine, G., and Smith, C. L. (1990). New approaches for physical mapping in small genomes. *J. Bact.* **172**, 1167–1172.

Cox, D. R., and Meyers, R. M. (1992). Bridging the gaps: X-ray breakage of human chromosomes in somatic cell hybrids provides a powerful method for constructing high resolution maps of mammalian chromosomes. *Current Biol.* **2**, 338–339.

Gasser, S. M., and Laemmli, U. K. (1987). A glimpse at chromosomal order. *TIGS* **3**, 16–21.

Orkin, S. H. (1986). Reverse genetics and human disease. *Cell* ? 845–850.

Reik, W. (1989). Genomic imprinting and genetic disorders in man. *Trends Genet.* **5**, 331–336.

Ruddle, F. H. (1981). A new era in mammalian gene mapping: Somatic cell genetics and recombinant DNA methodologies. *Nature* **294**, 115–119.

Smith, C. L., and Cantor, C. R. (1987). Purification, specific fragmentation and separation of large DNA molecules. *In* "Recombinant DNA" (R. Wu, ed.), Methods in Enzymology, Vol. 155, pp. 449–467. Academic Press, New York.

Smith, C. L., and Cantor, C. R. (1997). Consequences of mapping and sequencing the human genome for neurologic diseases. *In* "The Molecular and Genetic Basis of Neurological Disease" (R. N. Rosenberg, S. B. Prusiner, S. DiMauro, and R. L. Barchi, eds., 2nd Ed. Butterworth, Stoneham, Mass. (in press).

Smith, C. L., Lawrance, S.-K., Gillespie, G. A., Cantor, C. R.,

Weissman, S. M., and Collins, F. S. (1987). Strategies for mapping and cloning macro regions of mammalian genomes. *In* "Molecular Genetics of Mammalian Cells" (M. Gottesman, ed.), Methods in Enzymology, Vol. 151, pp. 461–489. Academic Press, New York.

Smith, C. L., Wang, D., and Grothues, D. (1994). Construction of physical maps of chromosomes. *In* "Encyclopedia of Molecular Biology and Biotechnology." VCH Publishers, New York.

Smith, C. L., Wang, D., Broude, N., Bukanov, N., Monastryskaya, G., and Sverdlov, E. (1995). Parallel processing in genome mapping and sequencing. *In* "GenoMethods: A Companion to Methods in Enzymology" (M. Schwab, ed.). Academic Press, San Diego.

Chromosome-Specific Human Gene Libraries

LARRY L. DEAVEN
Los Alamos National Laboratory

GLOSSARY

Cloning Isolation of a single cell and propagating it into a population of cells that are identical to the ancestral cell. In recombinant DNA technology, the procedures used to produce multiple copies of a fragment of DNA

DNA or gene library Set of cloned DNA fragments that represent all or a portion of a genome or the DNA transcribed in a particular tissue

Flow cytometry Method for measuring properties such as DNA content or size of cell populations or subcellular particles and sorting pure fractions from these cells or particles according to those measured properties

Genome One copy of all the DNA in a cell

Vector DNA molecule modified to carry a segment of foreign DNA into a host cell, where it is replicated in large quantities

GENE OR DNA LIBRARIES ARE SETS OF CLONED DNA fragments that represent all or a portion of the genetic information in an organism. The term "library" is somewhat misleading because it implies order, but a gene library is simply a random collection of DNA fragments. Each fragment may or may not contain a complete gene. The fragments are maintained and propagated in a biological cloning vector, usually a bacterial virus or plasmid. If the DNA used to construct a gene library comes from a single human chromosome, the resulting library will be chromosome-enriched or chromosome-specific. Such libraries represent a subset of the genetic information in human cells, and they provide a rich source of DNA fragments from each chromosome for mapping and sequencing genes and for studying genome organization. The most successful approach to making libraries from the DNA in individual human chromosomes has been through the use of chromosomes purified by flow sorting. This article describes the construction process and the application of these libraries to human genetics research, especially the elucidation of molecular detail in the human genome. [*See* DNA and Gene Transcription; Genes.]

I. BACKGROUND

Two major technological developments have made the construction of chromosome-specific gene libraries possible. The first of these, flow sorting, enables the purification of millions of copies of each human chromosome. These sorted chromosomes are required as sources of DNA for library construction. The subject of flow sorting will be discussed in detail in Section II,C.

The second development necessary for library construction was recombinant DNA technology. Although it was possible to construct a library from DNA isolated from the cells of fruit flies (*Drosophila melanogaster*) as early as 1974, application of these methods to larger and more complex genomes had to await the development of more efficient cloning techniques and methods for selecting specific DNA sequences from large numbers of cloned fragments. A series of improvements in these areas was reported over the next three years, and in 1978, the first library

ENCYCLOPEDIA OF HUMAN BIOLOGY, Second Edition, VOLUME 2. Copyright © 1997 by Academic Press. All rights of reproduction in any form reserved.

of human DNA fragments was made. Individual clones containing human γ- and β-globin genes were isolated from this library, clearly establishing a new approach to mapping the human genome and to studies of the structure and function of human genes. [*See* Genetic Maps; Human Genome.]

Since 1978, numerous improvements in recombinant DNA technology have been reported. It is now possible to make a variety of different types of libraries and, therefore, to design the construction of a library to be advantageous for a specific application. For example, by selecting the proper cloning vector, a library can be made that contains small [1–9 kilobases (kb)], medium (15–40 kb), or large (200–1000 kb) pieces of inserted DNA. Some vectors have distinct advantages over others with regard to the quantity of starting material or target DNA required to make a library. Some have features that facilitate the recovery of the inserted DNA fragments and some are better than others at retaining the inserted DNA without deleting or rearranging it. A more complete discussion of the use of different types of libraries is in Section IV. A major constraint in constructing libraries from sorted chromosomes is in the relatively small amount of available target DNA. This limitation requires the use of vectors with high cloning efficiency; this will be discussed in Section II,D.

Although it is possible to make many different types of DNA libraries with some types more useful than others for a given application, it is almost always advantageous to subdivide the DNA of an entire genome before library construction. The resulting set of libraries covers all of the genomic DNA, but each library is less complex than a single library containing all of the cellular DNA. A convenient and possible way to make subsets of human DNA is on a chromosome-by-chromosome basis. To include all of the nuclear DNA in human cells, 24 different libraries are necessary (22 autosomes plus the X and Y chromosomes). Table I lists these human chromosomal types and the amount of DNA in each chromosome. The libraries will vary in size, with the largest (for chromosome 1) being five times as large as the smallest (for chromosome 21).

TABLE I
DNA Content of Human Chromosomes

Chromosome	Mass DNA per chromosome ($\times 10^{-13}$ g)
1	5.14
2	5.03
3	4.19
4	3.97
5	3.81
6	3.56
7	3.31
X	3.17
8	3.02
9	2.86
10	2.82
11	2.81
12	2.77
13	2.26
14	2.14
15	2.08
16	1.93
17	1.76
18	1.68
20	1.38
19	1.31
Y	1.16
22	1.08
21	0.98

II. LIBRARY CONSTRUCTION

A. Sources of Chromosomes

Human chromosomes may be isolated from three different types of cultured cells for sorting on flow cytometers. Two of these, human fibroblast cells and human lymphoblastoid lines, contain only human chromosomes. The third, rodent–human hybrid cells, contains a background of mouse or hamster chromosomes with one or a few human chromosomes in each cell. Each of these cell types has advantages and disadvantages as sources of chromosomes for sorting. These characteristics are briefly described in this section. [*See* Chromosomes.]

Human diploid fibroblasts are cells cultured directly from a small piece of human tissue. The tissue is either cut into fragments or digested into single cell units and placed in a culture flask in growth medium at 37°C. The fibroblast cells begin to divide under these conditions, and the growing cells are propagated by serial transfer into new flasks. The chromosomes in fibroblasts are normal and stable, with little or no tendency to undergo structural rearrangements, provided that the donor of the tissue had a normal karyotype and the tissue was normal. These cells have a finite life span. They grow rapidly for approximately 15 generations, then progressively slow, until they stop dividing entirely at about 50 generations. This is a major disadvantage for isolating large quantities of chromosomes for sorting purposes. Fibroblast cells are useful only during the rapid growth phase when many metaphase cells are present.

Human lymphoblastoid cell cultures are initiated by transforming human lymphocytes with Epstein–Barr virus. They do not undergo cell senescence and continue to divide indefinitely, thus having a distinct advantage over fibroblast cells. The chromosomes in these cell lines are not as stable as those in fibroblasts; however, there are a number of lymphoblastoid lines available that show little or no karyotypic change over long periods of time in culture. They grow as suspension cultures, which is an easy and inexpensive way to propagate cells, but it is difficult to separate mitotic from interphase cells, and isolated chromosomes from these lines contain high concentrations of interphase nuclei. Another potential disadvantage to the use of chromosomes from these lines for library production is the incorporated Epstein–Barr virus. Although the chromosomes appear to maintain stable morphology, the viral incorporation may induce subtle changes in the DNA of the host cells. [*See* Epstein–Barr Virus; Lymphocytes.]

Hybrid cell lines are constructed by inducing two different types of cells to fuse together to become one cell. Hybrids between human cells and mouse cells or Chinese hamster cells are useful for sorting human chromosomes. Initially, the hybrid cell contains two complete sets of chromosomes, one from the human cell and one from the rodent cell. However, when a hybrid cell begins to divide, it loses some of its human chromosomes. Some of the hybrid cells lose all of their human chromosomes, whereas others retain one or more of them for various lengths of time. Single cells can be cloned from a population of hybrid cells, grown as a cloned culture, and analyzed for the specific human chromosomes they contain. Alternative methods for constructing hybrids involve growth in special culture medium that selectively kills cells without the desired chromosome. Through the use of these techniques a series of cell lines are available that contain only 1 to 3 human chromosomes in a rodent background. These lines are necessary to sort those human chromosomes that have very similar DNA content and that flow sorters cannot distinguish from each other (e.g., 9, 10, 11, 12), and are advantageous for sorting many of the other chromosomes. A distinct advantage is that any contaminating DNA present in the sorted chromosomes will be hamster or mouse, which can easily be screened out of a library. On the other hand, these cell lines tend to be unstable, and the desired human chromosome can be lost or rearranged at any time in culture. This undesirable feature requires frequent cytogenetic analysis to make sure the line contains the human chromosome to be sorted in an intact, unrearranged state.

Metaphase chromosomes for sorting are obtained from any of the cell types discussed here by blocking exponentially growing cultures with colcemid. Colcemid holds the dividing cells in metaphase for up to 12 hr, the longest block time used for this work. Hybrids and fibroblasts are grown as flat monolayers of cells attached to 225-cm^2 culture flasks. When the cells enter metaphase they become round and are attached to the surface less securely than the flattened interphase cells. They can be dislodged from the monolayer by gentle treatment with trypsin or by shaking the flask before removing growth medium, leaving the interphase cells still attached to the flask. The mitotic index of the dislodged cells is usually greater than 90%, providing a rich source of chromosomes for isolation and sorting. As mentioned earlier, lymphoblastoid cells grow in suspension and do not attach to culture flasks. They are also blocked with colcemid, but at the end of the colcemid treatment the entire population is collected for *chromosome isolation*. In this case, the percentage of mitotic cells is generally 20 to 60% depending on the growth rate of the cells. After chromosomes are isolated from these cells, interphase nuclei can be removed by gentle centrifugation. This step is important because all sorters operate with low error rates. If even a few interphase nuclei are sorted into the collection tube, they have serious effects on the purity of sorted DNA. By reducing the concentration of nuclei with differential centrifugation, the chances of sorting errors involving nuclei become negligible.

B. Isolation and Staining of Chromosomes

The mitotic cells collected from culture flasks are pelleted by centrifugation and the growth medium is removed. The first step in the chromosome isolation process is to resuspend the cells in a hypotonic solution (usually 75 m*M* KCl). This swells the cells, helps to dissolve the mitotic apparatus, and weakens the cell membrane. After the swelling step, the cells are gently pelleted in a centrifuge, the hypotonic solution is removed, and the cells are resuspended in a chromosome isolation buffer. This suspension is swirled in a vortex apparatus to break open the cell membranes and disperse the chromosomes.

The *chromosome isolation buffer* is probably the most critical element in this process. It must stabilize the chromosomes so that they remain intact during sorting, it must protect the chromosomal DNA from nucleases released when the cells are disrupted, and it must be compatible with the fluorescent stains used for flow cytometric analysis. Several chromosome isolation techniques have been developed involving the use of different kinds of buffers for chromosome stabilization; however, only one of them yields DNA of sufficient molecular weight to construct medium and large insert libraries. This buffer stabilizes chromosomes from many different cell lines for up to 6 months after isolation. It contains 15 m*M* Tris-HCl, 2 m*M* EDTA, 0.5 m*M* EGTA, 80 m*M* KCl, 20 m*M* NaCl, 14 m*M* β-mercaptoethanol, 0.2 m*M* spermine, 0.5 m*M* spermidine, and 0.1% digitonin, at pH 7.2. Although the general procedure outlined here has wide applicability, subtle differences between individual cell lines require optimization of each step to obtain ideal results. Variables that can be adjusted include: colcemid concentration, blocking time, cell concentration during isolation, hypotonic swelling buffer, swelling time, and time of vortex agitation of cells in isolation buffer.

Isolated chromosomes can be stained with a variety of fluorescent dyes for analysis on a flow cytometer. If the chromosomes are stained with a single dye that is specific for DNA, the resolution of single chromosomes is inadequate for sorting at the highest levels of purity. For example, if diploid human chromosomes are stained with propidium iodide, a fluorescent dye that intercalates with DNA, the 24 human chromosomal types will resolve into 17 or 18 peaks based on the total DNA content of each chromosome. However, most of the peaks overlap one another, so an attempt to sort one chromosomal type from a single peak would result in as much as 50% contamination from flanking peaks. This level of purity would be unacceptable for most library applications.

Improvements in resolution can be obtained by staining the chromosomes with two stains, one with an affinity for AT-rich DNA (Hoechst 332558) and one with an affinity for GC-rich DNA (chromomycin A3). When chromosomes from diploid human cells are stained with these dyes and analyzed in a dual-laser flow cytometer, the 24 human chromosomal types are resolved into 21 peaks (chromosomcs 9–12

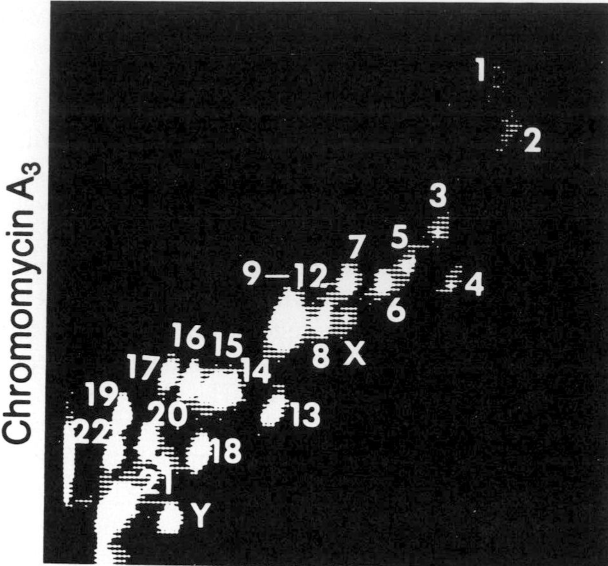

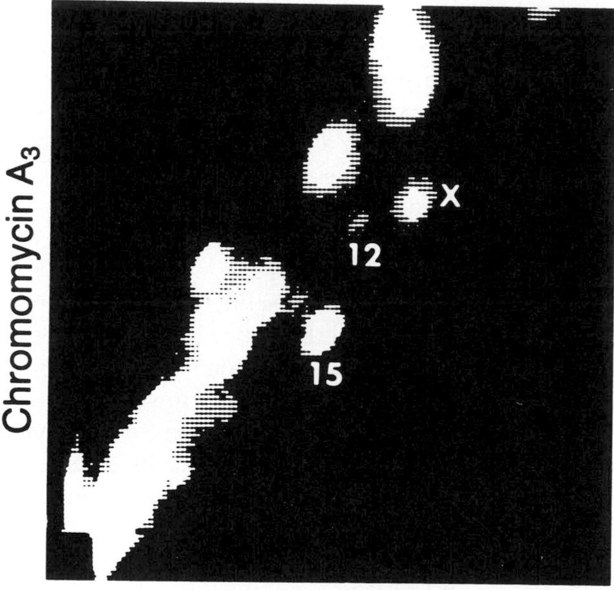

FIGURE 1 (a) A flow histogram of chromosomes isolated from a diploid human fibroblast cell strain (HSF-7). The histogram is composed of the number of fluorescence events versus the fluorescence intensities of the stains chromomycin A3 and Hoechst 33258. Chromosomes 9, 10, 11, and 12 are so similar in total DNA content and in base composition that they form a single, overlapping peak. To sort these chromosomes, hamster–human hybrid cells must be used as chromosome sources. (b) A flow histogram of the chromosomes in a hybrid retaining human chromosomes X, 12, and 15. The hamster chromosomes fall on a separate axis from the human chromosomes due to differences in base composition between the chromosomes of the two species.

overlap in one peak) (Fig. 1a). These stains also differentiate many human chromosomes from the hamster chromosomes in hybrid cells (Fig. 1b). This staining combination differentiates chromosomes on the basis of total DNA content and also on the basis of the relative AT- or GC-richness of each chromosome. Staining is accomplished by adding chromomycin A3 (final concentration 62.5 μM) to the isolated chromosomes for 3 hr and then adding Hoechst 33258 (final concentration 3.7 μM). The stains are usually added to the chromosome suspension one day before sorting.

C. Sorting

Flow cytometers were developed at the Los Alamos National Laboratory during the late 1960s, primarily to measure the DNA content of single cells. Since that time, many improvements have been made, and these instruments can now be used in other applications. The DNA content of organisms as small as bacteria can be detected, and cells or subcellular components can be measured for a number of properties and sorted on the basis of variations in these properties. The basic operating principles of a flow cytometer are illustrated in Fig. 2. The objects to be analyzed and sorted (chromosomes) are carried through the instrument in a fluid stream. The chromosomes enter a flow chamber, where the size and speed of the stream are controlled to permit only one chromosome at a time to cross beams of light from two lasers. The fluorescent dyes, Hoechst 33258 and chromomycin A3, fluoresce as they cross the ultraviolet beam from one laser and the 458-nm beam of the second laser. The resulting fluorescence signals are converted from optical signals to electrical signals by photomultipliers. The signal intensities are proportional to the AT and GC base pair content of each chromosome. The electrical signals are stored in histograms (see Fig. 1) obtained on a pulse–height analyzer; they are also passed to the sorting circuitry of the instrument. After passing through the flow chamber, the fluid stream containing chromosomes is broken up into droplets by a piezoelectric transducer driven by a high-frequency oscillator. If the two fluorescence signals from a chromosome fall within a range defined by a preset sorting window, a changing circuit is activated that applies a short voltage pulse to the portion of the stream containing the chromosome. This happens just before the stream is broken into droplets, so that the droplet containing a desired chromosome is electrically charged. Actually, three droplets are charged to allow for small timing uncertainties. When these droplets pass through a pair of deflection plates, they are deflected into a separate collection vessel, while uncharged droplets containing the unwanted chromosomes go to a waste collection vessel. [*See* Flow Cytometry.]

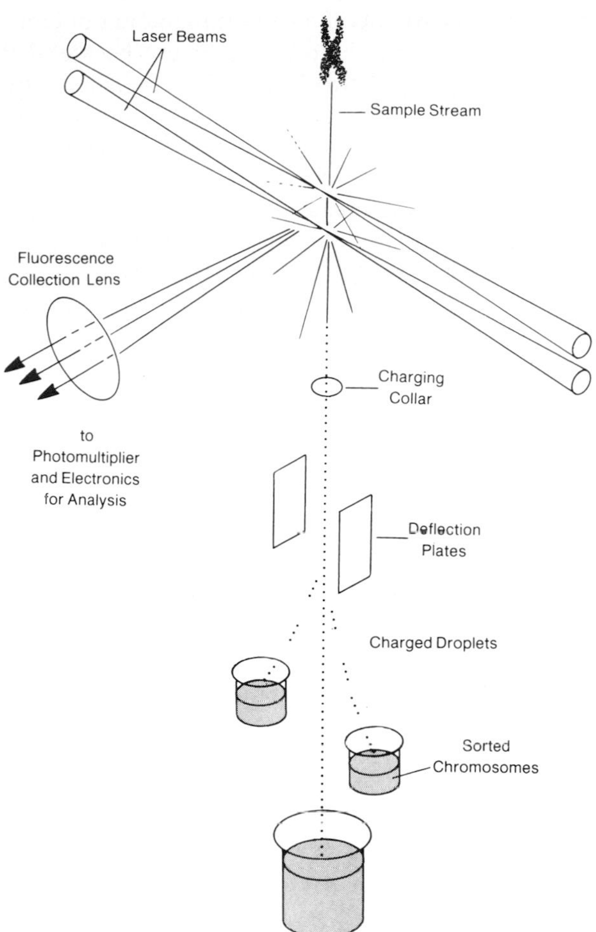

FIGURE 2 A diagram illustrating the operating principles of a dual-laser flow cytometer. A narrow stream of a suspension of stained chromosomes flows across the focused region of two laser beams, where fluorescence properties of each chromosome are measured and stored as electronic signals. If the fluorescence intensities define a desired chromosome, an electric charge is put on the stream, which is then broken into droplets. When the droplets containing the desired chromosome pass through charged deflection plates, they are deflected according to the charge imparted to them. Two chromosomes may be selected for sorting from each suspension.

For library construction, there are two critical aspects to the sorting process. Chromosome sorting must be relatively fast, and for most libraries it must be accomplished with a high level of purity. Library construction would not be practical if it took months of sorting time to obtain a sufficient amount of DNA for starting material, and it would not be useful if the starting material was heavily contaminated with extraneous DNA. Two different types of sorters have been used to purify chromosomes for library construction. They sort at either the conventional rates typical of commercial instruments or at speeds that are approximately an order of magnitude higher. These latter instruments have been developed at the Lawrence Livermore and Los Alamos National Laboratories specifically to aid research projects requiring high numbers of sorted objects. They are called high-speed sorters. Conventional sorters are capable of analyzing 1000–2000 chromosomes/sec and of sorting up to 50 chromosomes of a particular type each second. Optimal chromosome preparations and instrumental performance will yield 1×10^6 copies of a chromosome in 8 hr of operating time. High-speed sorters can analyze 20,000 chromosomes/sec and sort up to 200 chromosomes/sec. Under optimal operating conditions, they will yield 5×10^6 chromosomes in 8 hr of operating time. Thus, high-speed sorters have a distinct advantage over conventional sorters, especially if relatively large amounts of DNA (microgram quantities) are required as starting material for library construction.

The purity of sorted chromosomes is critical to the construction of chromosome-specific DNA libraries, and great effort has gone into purity determinations of sorter outputs. The capability of sorters to differentiate and sort specific objects from a mixed population with high levels of accuracy and precision can be determined with fluorescent microspheres. In test runs, sorters used in this project can sort a single population of microspheres from mixed populations with 95–100% purity at operating speeds. Purity determinations of sorted chromosomes are more difficult and require direct examinations of sorted chromosomes when practical, as well as indirect measurements such as cytological examination of cells used for chromosome sources and library purity determinations.

Indirect examinations that have been employed include:

1. Karyotype analysis (G-banding, Q-banding, G-11 staining, DAPI-Netropsin staining) of metaphase cells from cell strains and lines used as chromosome sources: this analysis confirms that a chromosome of interest is present in the cell culture, the frequency of its presence, and whether it is normal or rearranged.
2. Isozyme analysis of hybrid lines to support karyotype analysis.
3. Measurements of chromosome peak locations and peak volumes in flow histograms and comparisons of that information with cytological karyotype analysis: this type of information on chromosome frequency (peak volume) and chromosome DNA content (peak location) is especially useful when hybrid cells are being used and the human chromosome content is unstable.
4. Examinations of libraries for the presence of Chinese hamster sequences or for extraneous human DNA sequences.

Direct measurements of the purity of sorted chromosomes include:

1. Banding analysis of chromosomes sorted directly onto microscope slides: this method is useful, however, isolated chromosomes are often too condensed to provide high-resolution G-band patterns.
2. DAPI-Netropsin staining of sorted chromosomes: this method is useful only for the human chromosomes with large blocks of centromeric heterochromatin, but it is valuable if isolation methods are used that do not permit G-band analysis.
3. Hybridization of human and hamster genomic DNA to sorted chromosomes: this procedure can be done *in situ* on sorted chromosomes and used to identify hamster chromosomes or human/hamster translocations.
4. Hybridization of human chromosome-specific DNA probes to sorted chromosomes: the major limitations of this approach are the small number of probes available (there are no suitable probes for approximately half of the chromosomes in the human karyotype) and the fact that the chromosomal region covered by these probes is usually limited to the centromere or telomere areas. In chromosomes sorted from hamster/human hybrids, it is possible to miss the detection of hamster DNA in a translocation that does not involve the centromere or telomere area of a human chromosome.

D. Cloning

The final step in the construction of libraries specific for the DNA of a single human chromosome is to insert fragments of the sorted chromosomal DNA into a self-replicating biological *vector*. Although many different types of libraries can be produced, there are four basic steps in the cloning process that are common to all of them: preparation of DNA fragments from DNA isolated from sorted chromosomes and from the vector system; formation of *recombinant DNA molecules* involving both chromosomal and vector fragments; assembly of self-replicating units containing the recombinant DNA molecules; and the amplification of these units into large numbers of copies, perhaps a million times the original starting material. The major challenge in cloning DNA from sorted chromosomes is to maximize the efficiency of each of these steps in order to obtain large libraries from small amounts of target DNA. For conventional cloning, total cellular DNA is easy to obtain and easy to replace if a cloning procedure fails. On the other hand, it may take several weeks to accumulate 1 μg of sorted DNA, so it is essential to utilize it with minimal loss.

The steps involved in constructing a small insert library are illustrated in Fig. 3. In this example, the biological vector is a modified λ *bacteriophage* or bacterial virus called *Charon 21A*. It can accept fragments of human DNA up to 9100 base pairs in length. The first step is to extract DNA molecules from the sorted chromosomes and from the cloning vector. Next the large chromosomal DNA molecules must be fragmented to provide suitable inserts for the vector. This is accomplished by digesting them with a *restriction enzyme* called *Eco*R1. Restriction enzymes recognize specific sequences in the DNA molecules and cut the DNA at each of these sequences. The enzyme *Eco*R1 recognizes a sequence of 6 bases (-G-A-A-T-T-C-) and cuts the phosphodiester bond between the -G-A- bases in this sequence. In addition to the enzymes that recognize six bases, there are also enzymes that recognize four bases or eight bases. Assuming a random distribution of the four bases A, T, G, and C in DNA, a four-base recognition sequence would occur on average every 256 bases, a six-base sequence every 4 kb, and an eight-base sequence every 65 kb. By selecting the proper restriction enzyme, it is possible to cut large DNA molecules into fragments of different average lengths. In our example, in a digest run to completion, *Eco*R1 will reduce the DNA to an average length of 4 kb; however, approximately 33% of the DNA will remain in fragments longer than 9 kb, the upper size in the acceptance range for Charon 21A. This will result in a library that contains only 67% of the chromosomal DNA. To increase the coverage of the chromosomal DNA, a second library would have to be constructed using a different restriction enzyme that recognizes a different six-base sequence.

The linear DNA molecule of Charon 21A is about 42 kb in length, and it contains a single sequence

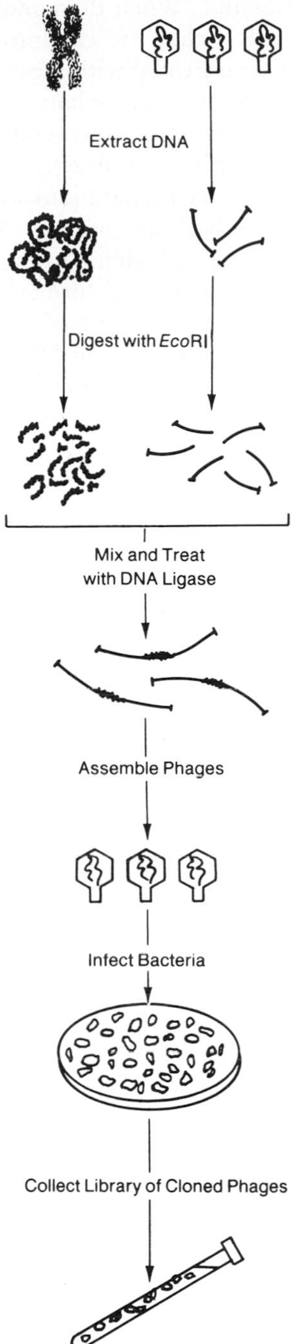

FIGURE 3 A diagram illustrating the major steps involved in the construction of a chromosome-specific DNA library in a λ phage vector. The DNA in the sorted chromosomes and in the vector is isolated and digested with a restriction enzyme called *Eco*R1. The fragments of DNA are ligated together and packaged into infectious page particles. These recombinant phage are then amplified by propagating them on a lawn of host bacterial cells. Aliquots of these amplified phage represent the DNA of the sorted chromosomes.

recognized by *Eco*R1. When these molecules are digested with *Eco*R1, they are cut into two "arms." These arms are then treated with the enzyme alkaline phosphatase to remove a phosphate group. This prevents most of the arms from recombining into intact vector molecules without an insert.

Next the two DNA preparations are mixed, and because they were both cut with *Eco*R1, they both have AATT cohesive ends that become linked by hydrogen bonds. The result is the formation of recombinant DNA molecules consisting of two vector arms linked by a fragment of human DNA. When these recombinant molecules are treated with the enzyme DNA ligase, they are permanently rejoined by the formation of new phosphodiester bonds. To maximize the yield of recombinant molecules, an excess of arms is used in the mixing reaction. Because the arms were treated with alkaline phosphatase, they cannot be rejoined to each other with DNA ligase, and this prevents the formation of large numbers of nonrecombinant contaminates of the library.

The recombinant DNA molecules can now be "packaged" into intact Charon 21A phages by providing the required phage proteins. The resulting infectious phage particles are then seeded onto a monolayer or "lawn" of a suitable host strain of bacteria. Here they inject their DNA into a bacterium, causing the metabolic and replicative apparatus of the host cell to produce more phage protein and DNA. When the cell has produced between 100 and 200 new phage particles, it ruptures and the infection process continues with surrounding cells. The final result is a clear area or plaque on the bacterial lawn. Each plaque contains from 1 to 10 million infective phage particles, and each phage contains a copy of the original recombinant molecule with its human insert. Together they form a library representing the chromosomal DNA used as starting material.

For the construction of libraries with larger inserts, the basic steps described here are repeated, but with different vectors and modified methods of preparing the chromosomal target DNA. Libraries with medium-sized inserts of up to 45 kb are cloned into *cosmid* vectors (Fig. 4). Cosmids are cloning vehicles that incorporate functional properties of λ *phage* and *plasmids.* During the λ phage replication cycle, hundreds of copies of λ DNA form a long chain or concatamer, with each copy of the λ genome joined by a stretch of single-stranded DNA, the cohesive end or "cos" site. The λ packaging enzymes recognize two cos sites 35–45 kb apart in the concatamer, cleave this unit, and package it into a phage head. The cos

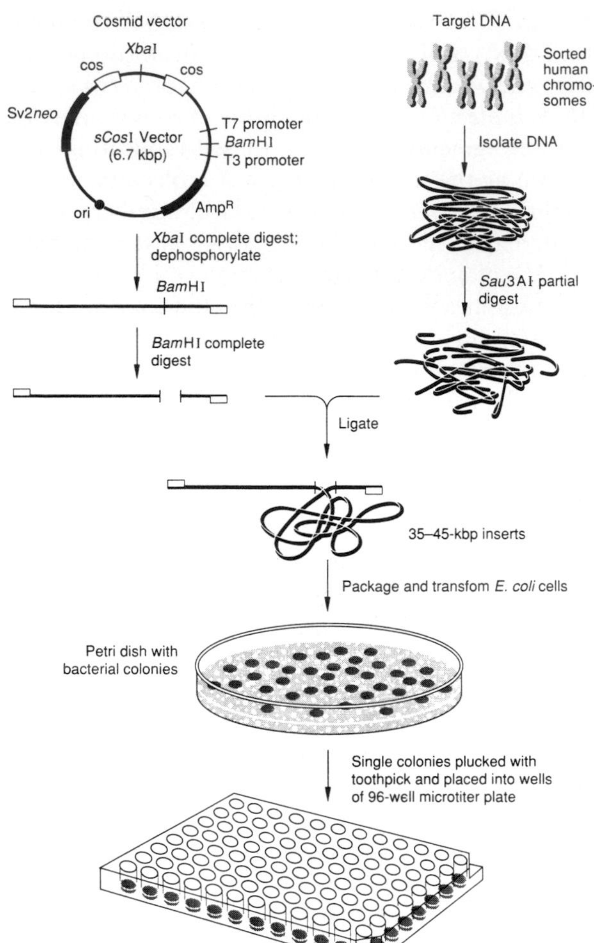

FIGURE 4 Cosmid cloning of DNA from sorted human chromosomes. The cosmid vector (sCos 1) contains two cos sites for rejoining the linear recombinant molecule after transformation. It also contains two selectable markers [resistance to ampicillin (Amp^R) and to neomycin (SV2neo)], a number of restriction sites, a plasmid replicon including an origin of replication (ori), and promoter sequences from the T3 and T7 phages. The T3 and T7 promoters are used to generate end probes to facilitate the assembly of overlapping cosmid clones. The vector molecule is linearized by cutting with the restriction enzyme *Xba*I, then separated into two cloning arms by cutting with *Bam*HI. After fragments between 35 and 45 kbp in length are ligated to the vector arms, the recombinant DNA molecules thus produced are packaged into phage protein coats. The resulting infectious phage particles insert the recombinant molecules into *E. coli* cells, where the molecules cyclize and live as plasmids. To prevent the faster-growing *E. coli* cells from overwhelming the slower ones, each colony is placed in a separate well of a microtiter plate.

sites are the recognition system for reducing concatamers into λ genome units, as part of the packaging process. Plasmids are tiny circular DNA molecules that often contain genes for antibiotic resistance. They reside in bacteria, where they confer antibiotic resistance to the host cells. They are useful as vectors because when a plasmid with an antibiotic resistance gene is used as a vector in plasmid-free host cells that are antibiotic-sensitive, the presence or absence of the plasmid vector can be readily detected by treating the host cells with antibiotic. By incorporating the cos sites of λ phage into plasmids, a new class of cloning vectors (cosmids) was created.

For cosmid cloning, the sorted chromosomal DNA is partially cleaved with a restriction enzyme to yield relatively large pieces of DNA as compared to the complete digestion used for small-insert phage cloning. The cosmid vector is cleaved at the same restriction site to produce cloning arms, each with a cos site at one end. The cosmid arms are then ligated to the partially digested human target DNA fragments. Recombinant molecules consisting of a cloning arm ligated to each end of a fragment of human DNA, where the cos sites are approximately 48 kb apart, can be packaged to produce infectious phage particles. These phages can then be used to introduce their recombinant molecules into a suitable bacterial host. Once inside the bacteria, the recombinant cosmid DNA circularizes and is maintained within the bacteria as a plasmid.

The first cosmid vectors were not as efficient as phage vectors, and the requirement of large amounts of target DNA prevented the construction of cosmid libraries from sorted chromosomes. A number of improvements, especially the addition of a duplicated cos sequence, now make it possible to construct large libraries from submicrogram quantities of DNA. When these new vectors were combined with improved methods for isolating chromosomes with high-molecular-weight DNA, it became possible to construct chromosome-specific libraries in cosmids.

The largest-capacity cloning vectors currently available have an insert acceptance size over 10 times the limit of cosmids. These vectors are called *yeast artificial chromosomes* (YACs). The original vectors contained a cloning site within a gene to facilitate detection of recombinant clones, a yeast centromere, and two sequences that gave rise to functional yeast telomeres. All of these sequences were carried in a single plasmid that can replicate in *Escherichia coli* cells. The basic approach to YAC cloning is similar to that of phage and cosmid cloning. The vector is cleaved at the cloning site and between the telomere sequences. This produces two vector arms that are treated with alkaline phosphatase to prevent religation, and then ligated with large molecules of target DNA. The yeast artificial chromosome is then transformed into yeast cells, where it becomes a stable chromosome. As in the case of cosmid vectors, the initial results suggested that this cloning method would not be practical to combine with chromosomal DNA purified by flow sorting (25 μg of insert DNA was used). However, by making improvements to the chromosome isolation and sorting procedures to provide sorted DNA with high molecular weight, and by utilizing second-generation YAC vectors with high cloning efficiencies, we were able to construct chromosome-specific libraries for several human chromosomes (e.g., 5, 9, 12, 16). These libraries contain inserts that are an average of 200 kb in size, and they have low frequencies of chimeric inserts (inserts that are composed of more than one piece of target DNA). The YAC cloning scheme is shown in Fig. 5.

A new *E. coli*-based vector was described in 1992 that promises to be the most convenient and reliable cloning system currently available. This new vector is based on the *E. coli* F factor and is called a *bacterial artificial chromosome* (BAC). It appears to be quite stable as compared to cosmids or YACs, and it has the potential capacity of 1 megabase; however, inserts on the order of 100 to 200 kb are commonly obtained. The BAC inserts have proven to be an ideal size for both physical chromosome mapping and as starting material for sequencing reactions. Although no libraries have been constructed in BAC vectors from flow-sorted DNA, it is possible to obtain BAC clones from as little as 1 μg of DNA. If improvements can be made in cloning efficiency and in average insert size, it will be possible to make chromosome-specific BAC libraries for at least some human chromosomes.

Two alternative approaches to constructing chromosome-specific BAC libraries have also been attempted or are currently under development and should be mentioned. The use of flow-sorted DNA is desirable but not mandatory for these methods. The first is to assemble a set of chromosome-specific probes and to use these probes to select BACs that represent a chromosome of interest from a total genomic BAC library. One source of chromosome-specific probes is flow-sorted DNA cloned into a small insert vector such as M13. A small amount of sorted DNA can provide thousands of probes, which in turn can be used to identify chromosome-specific BACs in total genomic BAC libraries. This approach was previously used to select clones from YAC libraries, and it proved to be highly successful. A second method involves the use of a new YAC cloning system called *transformation-associated recombination* or TAR cloning. This

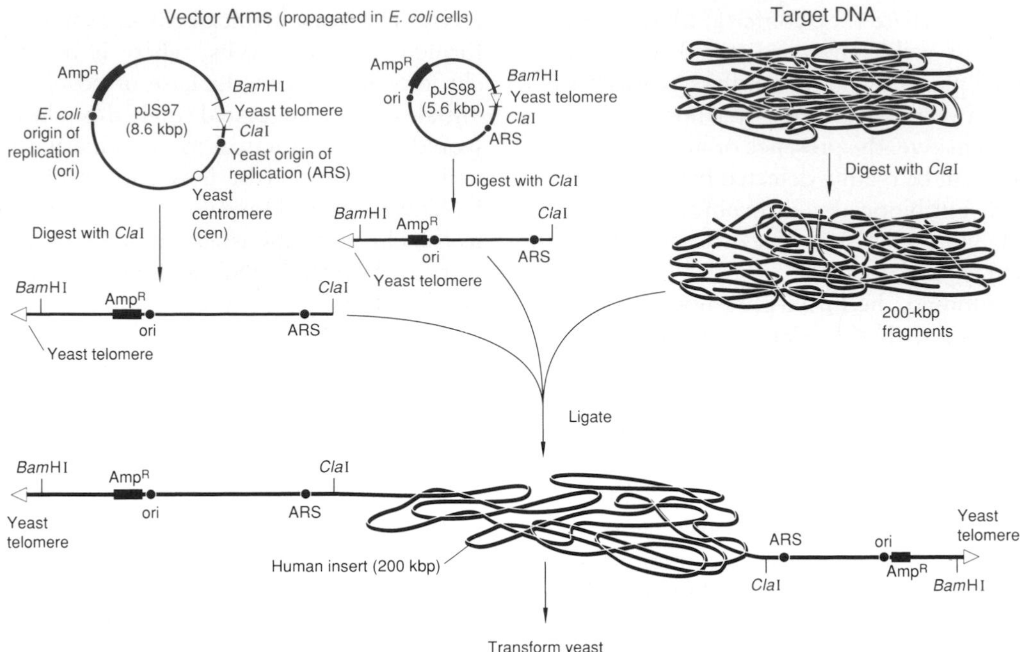

FIGURE 5 YAC cloning. The two arms of the YAC shown are propagated separately in *E. coli* as plasmids pJS97 and pJS98. Each contains an ampicillin-resistance gene, an *E. coli* replicon (including an origin of replication) so that it can propagate as a plasmid, a yeast origin of replication (labeled ARS), a yeast telomere, and several restriction sites, including one for *Cla*I located at the end of the telomere. Only pJS97 contains a yeast centromere and a pigment-suppressor gene that changes the color of yeast colonies. The plasmids are linearized and the target DNA is fragmented, both by cutting with *Cla*I. The vector arms thus produced each have a yeast telomere sequence at one end (arrow) and a *Cla*I tail at the other end. The fragments of target (human) DNA are then ligated to the vector arms to form a YAC that can transform yeast cells.

type of cloning can be used to selectively clone human DNA even if it is mixed with other DNAs. The TAR vectors were constructed with common human repeat sequences such as *Alu* located at the ends of the cloning arms. In this position, they were available to recombine with the *Alu* sequences that are interspersed in large fragments of target human DNA. When the vector and the DNA fragments were transformed into yeast cells, recombination occurred, yielding stable inserts that could be propagated as YACs. When this procedure was used to clone the DNA in a monochromosomal rodent–human hybrid cell line, the TAR cloning enabled the selective isolation of human DNA clones. Only the human DNA was cloned into the vectors because rodent DNA lacks *Alu* sequences. By incorporating the DNA sequences necessary to maintain BAC clones into the TAR vectors, it is possible to transfer the TAR YACs into *E. coli* cells by electroporation, where they can be propagated as YACs. By this means, TAR cloning can be utilized to provide chromosome-specific BACs. The TAR cloning method can also be applied to flow-sorted target DNA because the TAR cloning process eliminates the need for the many *in vitro* enzymatic reactions normally used in cloning schemes and is therefore very efficient.

III. USES OF LIBRARIES

The application of human chromosome-specific DNA libraries to problems in human genetics research has been dependent on two major components: (1) the capability to construct different types of libraries from flow-sorted chromosomes and (2) the parallel development of new techniques in molecular genetics that resulted in novel applications for libraries and new directions in genetics research.

When it became feasible to construct the first chromosome-specific libraries (1983), the most urgent need for them was for human gene mapping and *genetic disease diagnosis*. A basic approach to locating genes involved with disease was to screen a genomic

DNA library for fragments of DNA or probes that were in close proximity to the disease-producing gene. These probes were used in family studies to detect changes in the location of restriction sites in the DNA of closely related individuals who do or do not have the inherited disease. These changes are called restriction fragment-length polymorphisms (RFLPs), and they are inherited as simple genetic traits that obey the Mendelian laws of inheritance. The chromosome-specific libraries helped to overcome the "needle-in-the-haystack" dilemma by providing enriched sources of probes from only one chromosome as opposed to all chromosomes. Human DNA contains families of repetitive sequences that are dispersed throughout the genome. These repeat sequences are not useful as probes because they occur as many as 300,000 times per genome. They must be screened out of a library or masked in order to isolate unique sequence or single-copy probes that map to only one site. Because repeat sequences are dispersed throughout the genome, the larger the insert size in a library, the more likely it is to contain a repeat, hence the decision to make the first libraries from complete digest DNA cloned into a vector that accepts DNA fragments up to 9 kb. The small inserts in these libraries have a relatively high probability of containing only unique sequence DNA. [*See* DNA Methodologies in Disease Diagnosis.]

A limitation of these small-insert libraries is that any fragment greater than 9 kb will not be represented in the library. This deficiency was partially overcome by constructing two libraries for each chromosome. One set of libraries was made using *Eco*R1 as the restriction enzyme and a second using a different six-base cutter, *Hin*dIII. A fragment of DNA excluded from one of these libraries has a high probability of being included in the other.

These small-insert libraries are stored in liquid nitrogen at the American Type Culture Collection in Rockville, Maryland, U.S.A. Aliquots of the original amplification are available to research groups throughout the world. They have been used extensively in the search for unmapped genes, especially those genes responsible for genetic disease. For example, several hundred probes have been isolated and mapped from the chromosome 4 and 7 libraries in the search for the defects responsible for Huntington's disease and cystic fibrosis. Although improved methods now permit the construction of larger-insert libraries, these complete digest libraries continue to be useful. Over 5000 of them have been sent to research laboratories.

A major disadvantage of small-insert libraries is that individual fragments may not contain complete genes. Although many genes are 5–10 kb in length, others are larger (10–50 kb) and still others are huge. The gene for factor VIII, which encodes the blood-clotting factor deficient in humans with hemophilia A, spans at least 190 kb, and the defective gene for Duchenne's muscular dystrophy covers over a million base pairs. In addition to the advantages of having whole genes or even groups of genes in a single cloned insert for molecular studies of gene structure and expression, the initiation of the Human Genome Project made chromosome-specific gene libraries with larger inserts highly desirable. The ultimate aim of the work sponsored by this international project is to determine the sequence of the 3 billion base pairs in the human genome. One way to accomplish this is to organize the fragments of DNA in a library into the linear sequence that exists in an intact chromosome. This ordered library or physical map could then be sequenced one insert at a time until the known sequence extended to the entire length of a chromosome. The construction of physical maps is a formidable task because an average-sized human chromosome requires approximately 3500 cosmid inserts to cover its length. Nevertheless, maps have been constructed for several chromosomes (11, 16, 19) that include extensive coverage in overlapping cosmid clones. The availability of chromosome-specific libraries appreciably enabled and simplified this difficult task.

Because BAC vectors have advantages in insert size and stability as compared to either cosmid or YAC vectors, current chromosomal mapping activity includes the translation of cosmid or YAC maps into maps composed of BAC inserts. Chromosome-specific BAC libraries would make the assembly of BAC-based maps considerably easier than map construction from total genomic libraries.

It should be emphasized that none of the available cloning systems are perfect and each type of library will probably turn out to be incomplete with regard to coverage of all of the DNA in a chromosome. However, by constructing libraries in a variety of vectors, the chances of complete coverage are considerably enhanced. Taken together these sets of libraries provide powerful tools for analyzing and defining the functional properties of the human genome.

ACKNOWLEDGMENTS

This work was supported by the U.S. Department of Energy under Contract W-7405-ENG-36 to the Los Alamos National Laboratory. Figures 1, 2, and 3 are reprinted from *Los Alamos Science,* No. 12, Spring/Summer, 1985, and Figs. 4 and 5 from No. 20, 1992.

I would like to thank the following individuals for scientific contributions or for help in preparing this manuscript: R. Archuleta, N. C. Brown, E. W. Campbell, M. L. Campbell, L. S. Cram, J. J. Fawcett, M. H. Fink, C. E. Hildebrand, J. L. Longmire, L. J. Meincke, P. L. Schor, and M. A. Van Dilla.

BIBLIOGRAPHY

Deaven, L. L., Hildebrand, C. E., Fuscoe, J. C., and Van Dilla, M. A. (1986). Construction of human chromosome specific DNA libraries: The National Laboratory Gene Library Project. *Gene. Eng.* **8**, 317.

Deaven, L. L., Van Dilla, M. A., Bartholdi, M. F., Carrano, A. V., Cram, L. S., Fuscoe, J. C., Gray, J. W., Hildebrand, C. E., Moyzis, R. K., and Perlman, J. (1986). Construction of human chromosome-specific DNA libraries from flow sorted chromosomes. *Cold Spring Harbor Sympos. Quant. Biol.* **51**, 159.

Doggett, N. A., *et al.* (1995). An integrated physical map of human chromosome 16. *Nature* **377**, 335–365.

Kim, U.-J., Shizuya, H., Chen, X.-N., Deaven, L. L., Speicher, S., Solomon, J., Korenberg, J., and Simon, M. I. (1995). Characterization of a human chromosome 22 enriched bacterial artificial chromosome sublibrary. *Genet. Anal.* **12**, 73–79.

Larionov, V., Kouprina, N., Graves, J., Chen, X.-N., Korenberg, J. R., and Resnick, M. A. (1996). Specific cloning of human DNA as yeast artificial chromosomes by transformation-associated recombination. *Proc. Natl. Acad. Sci.* **93**, 491–496.

Longmire, J. L., Brown, N. C., Meincke, L. J., Campbell, M. L., Albright, K. L., Fawcett, J. J., Campbell, E. W., Moyzis, R. K., Hildebrand, C. E., Evans, G. A., and Deaven, L. L. (1993). Construction and characterization of partial digest DNA libraries made form flow-sorted human chromosome 16. *Genet. Anal. Tech. Appl.* **10**(3–4), 69–76.

McCormick, M. K., Buckler, A., Bruno, W., Campbell, E. W., Shera, K., Torney, D., Deaven, L. L., and Moyzis, R. K. (1993). Construction and characterization of a YAC library with a low frequency of chimeric clones from flow-sorted human chromosome 9. *Genomics* **18**, 553–558.

Van Dilla, M. A., *et al.* (1986). Chromosome-specific DNA libraries: Construction and availability. *Biotechnology* **4**, 537.

Van Dilla, M. A., and Deaven, L. L. (1990). Construction of gene libraries for each human chromosome. *Cytometry* **11**, 208–218.

Circadian Rhythms and Periodic Processes

JANET E. JOY
National Research Council

KATHRYN SCARBROUGH
Northwestern University

GLOSSARY

Circadian rhythm Rhythm about a day in length that persists in constant conditions

Entrainment Steady state in which an endogenous rhythm runs synchronously with another rhythm (i.e., *Zeitgeber*) with a constant phase relationship, or phase angle difference

Free-running period Fundamental period of a biological clock when it is not entrained to some forcing oscillation (see *Zeitgeber*); often abbreviated as τ

Pacemaker Localizable entity capable of self-sustaining oscillations and of synchronizing other oscillations

Period Time required for completion of one cycle

Phase Any instantaneous state of a cycle, for example, peak or trough

Phase angle Relative term measuring the relationship between a particular phase in a cycle in relation to some arbitrarily chosen reference point or phase of another cycle. For example, the phase angle between the rhythm of body temperature and the sleep–wake cycle may be described as the difference in hours between the body temperature minimum and the onset of sleep. If the temperature minimum occurs at 10:30 P.M. and sleep onset at 11:30 P.M., then the phase angle difference between these two rhythms is 1 hr

Phase–response curve Plot indicating how the magnitude and direction of a phase shift induced by a single stimulus depends on the time, or phase, at which it is applied

Phase shift Single displacement of an oscillation along the time axis. The means by which a biological clock is reset

Zeitgeber Time giver; any external rhythm that will entrain, or synchronize, an endogenous biological rhythm

BIOLOGICAL RHYTHMS ARE A FUNDAMENTAL element of human physiology. Although we are usually most aware of rhythms associated with eating and sleeping, our lives are also governed by rhythms in hormonal fluctuations, cellular metabolism, and neural activity—to name only a few. Many of these rhythms are driven by internal biological clocks. The study of biological rhythms can be divided into two distinct areas. The first area includes studies of rhythmic processes themselves and addresses questions about the implications of specific patterns of temporal organization. The second area is primarily concerned with the nature of biological clocks and how rhythms are generated. For example, how is the clock itself regulated, where is it located, and what are the elements that make up a biological clock?

I. INTRODUCTION

We live in a periodic universe. The passage of our days, months, and years is dictated by the periodic

ENCYCLOPEDIA OF HUMAN BIOLOGY, Second Edition, VOLUME 2. Copyright © 1997 by Academic Press. All rights of reproduction in any form reserved.

motion of planets. Life on earth evolved in this periodic environment and biological rhythmicity emerged as a basic element of physiological organization. The most prominent rhythms are, of course, the daily cycles that arise from the earth's rotation about its axis, and by far the most widely reported biological rhythms are those that show daily cycles. Daily cycles have been reported for manual dexterity, pain sensitivity, drug metabolism, and cell division—to name only a very few of the many rhythms that rule our lives. Biological rhythms are associated with other geophysical cycles as well. Marine animals living in tidal zones have rhythms that follow the lunar cycle and most animals living away from equatorial regions, where seasonal changes in the physical environment are particularly pronounced, undergo yearly cycles. Biological rhythms also include frequencies that are not associated with known planetary motions. For example, pacemaker cells of the mammalian heart have an intrinsic rhythm of electrical activity with cycles lasting several 100 msec and certain hormones are secreted episodically every few hours.

Human physiology is governed by rhythms and periodic processes that influence every level of organization: from the biochemical to the behavioral. In this article, we focus on the most prominent type of rhythms, namely, those associated with daily fluctuations. Our goal is to outline the role of biological rhythms in human biology, although we illustrate many of the general principles using examples based on animal research, since most of the advances in this field necessarily come from such studies. Sections II–IV discuss the endogenous nature of biological rhythms and how they are regulated. Section V covers the neural basis of biological rhythms. Sections VI–IX explore the development of biological clocks within the individual and the role of biological clocks in selected processes, such as sleep and reproduction. The final section discusses rhythm disorders in human physiology. [*See* Comparative Physiology.]

II. ENDOGENOUS NATURE OF BIOLOGICAL RHYTHMS

A remarkable feature of many biological rhythms is that, although they evolved in response to a periodic environment, they are driven by mechanisms that are independent of periodic environmental input. That is, most biological rhythms are self-sustaining and persist even when all known environmental cycles are eliminated. Such rhythms are said to be endogenous, meaning they are intrinsic to the organism and are not derived from environmental fluctuations. The pattern of wheel-running activity in the golden hamster is a classic example of an endogenous rhythm. A hamster housed in constant darkness and constant temperature will show a very precise rhythm of running activity with a period (τ) that is likely to be about 24.2 hr (Fig. 1). Such a rhythm is said to free-run, because it continues to run free of input from external signals.

The term circadian is derived from Latin for "about" (*circa*) "a day" (*dies*). The observation that the period of the rhythm is close to, but not exactly equal to, 24 hr is typical of circadian rhythms. For example, humans living in temporal isolation with no time cues nonetheless show regularly alternating patterns of sleeping and waking—generally with a period of about 25 hr (Fig. 2). The terms "daily" and "circadian" are not synonymous. A circadian rhythm will be seen as a daily rhythm under natural daylight

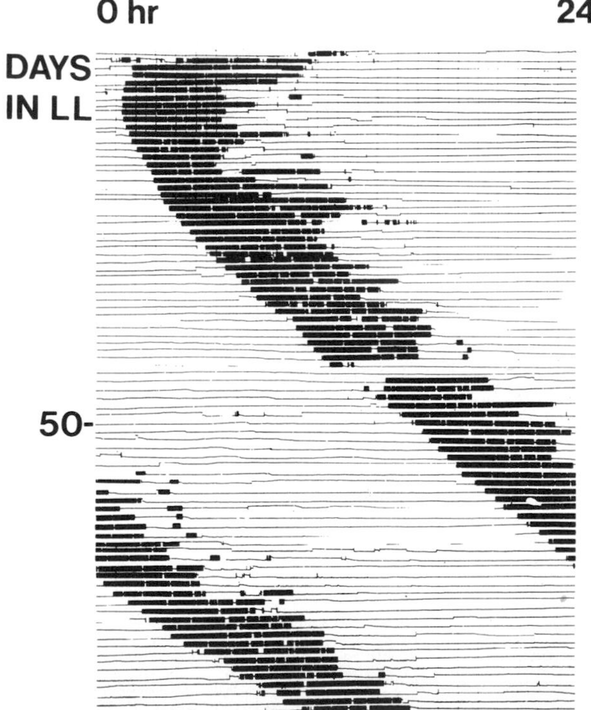

FIGURE 1 Free-running rhythm of wheel-running activity in a golden hamster living in constant light (LL). This record is made by dividing a continuous activity record into 24-hr periods and plotting successive days beneath each other. Activity bouts are indicated by vertical tick marks, which blend together to form a solid band during periods of intense activity. Note the precision of the daily onset of activity. The free-running period (τ) of this animal is 24 hr.

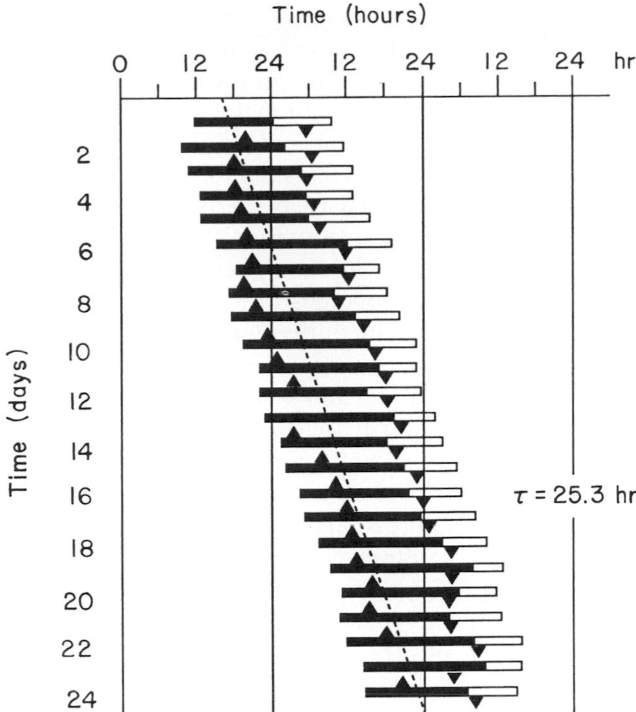

FIGURE 2 Circardian rhythm of wakefulness and sleep (solid and open bars) and of rectal temperature (triangles above bars for maxima, below bars for minima) recorded in a subject living alone in an isolation unit under constant conditions. The free-running period (τ) is 25.3 hr. [From J. Aschoff (1978). *Ergonomics* 39, 739–754.]

cycles; it does not follow that daily rhythms will necessarily show a circadian cycle under constant lighting conditions in the absence of time cues like the transition from light to darkness.

Free-running periods are genetically influenced and show characteristic differences between species as well as between individuals of the same species. The deviations from a strict 24-hr periodicity provide compelling evidence that these cycles are truly endogenous since there are no known environmental cycles with periodicities that could hypothetically drive them.

Many, if not all, daily rhythms are driven by endogenous circadian oscillators. To confirm that a daily rhythm is under the control of an endogenous oscillator (as opposed to being passively driven by daily environmental fluctuations), the rhythm in question must be shown to free-run under constant conditions. Circadian oscillators in animals share certain basic properties: they free-run under constant conditions, they are encoded in the genome, they are generated from neural structures, and they can be entrained, or synchronized, by light.

Circadian rhythms are not unique to higher vertebrates. They are found in plants as well as lower animals, including unicellular organisms. In fact, they have been found in every eukaryotic organism thus far studied. Processes as diverse as spore formation in fungi and color changes in fiddler crabs are known to be regulated by circadian rhythms.

Endogenous rhythms also occur in noncircadian ranges, such as circannual cycles (about a year) or ultradian cycles (less than a day). Hibernation in ground squirrels and migration patterns of garden warblers show circannual cycles that generally have a period of 10–11 months. At the other extreme are the 60-sec ultradian rhythms in the courtship songs of fruit flies. The most prominent ultradian rhythm in human physiology is the 90- to 100-min alternation between the REM and non-REM stages of sleep. (REM is the stage of sleep characterized by rapid eye movements and usually associated with dreaming.) Ultradian rhythms occur in a wide range of frequencies and individual rhythms show considerable cycle-to-cycle variation in length. They are often described as "pseudoperiodic," whereas circadian rhythms, which typically show less than 5% variation in period length, are strictly periodic. As a rule, ultradian rhythms seem to be independently generated as consequences of the particular physiological systems of which they are a part and are not under the control of an endogenous ultradian oscillator. In contrast, circadian rhythms appear to be coupled to an integrated system of circadian oscillators.

III. ENTRAINMENT OF BIOLOGICAL CLOCKS

The circadian oscillator acts as a biological clock. However, since the endogenous period of this clock is typically close to, but not equal to, 24 hr it must be reset each day. It must be synchronized, or "entrained," to the 24-hr day. The most powerful entraining agent, or *Zeitgeber* (German for "time giver"), is the light–dark cycle. Circadian rhythms are exquisitely sensitive to small amounts of light. As little as 1 sec of light per day is sufficient to entrain the circadian period of locomotor activity in the golden hamster. A single light pulse can reset, or phase shift, the rhythm to an earlier phase (phase advance) or to a later phase (phase delay), or cause no change in the rhythm at all (Fig. 3, upper panel). The critical determinant in the ability of a light pulse to reset the rhythm is not the duration of light, but rather the phase of the circadian cycle at which it is applied. The phase–response curve (PRC), in which the size and direction of the phase shift are plotted against the phase at which the pulse is delivered, illustrates the circadian rhythm of sensitivity to light (Fig. 3, lower panel). To measure this rhythm, the PRC must be generated from animals in free-running conditions. The convention of circadian time is used to accommodate individual differences seen in free-running periods. One circadian hour is defined is 1/24 of an individual's free-running period (one circadian hour = $\tau/24$).

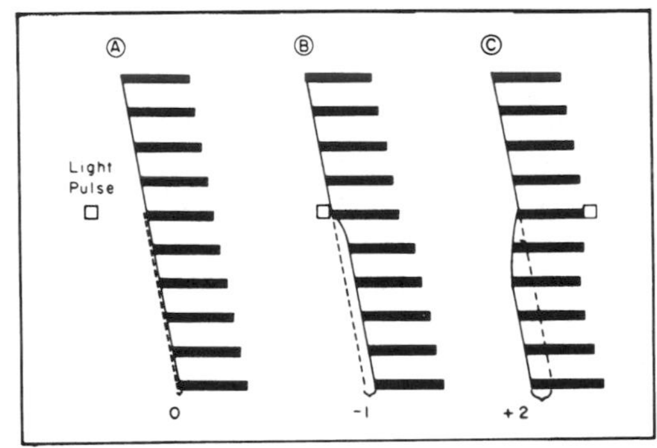

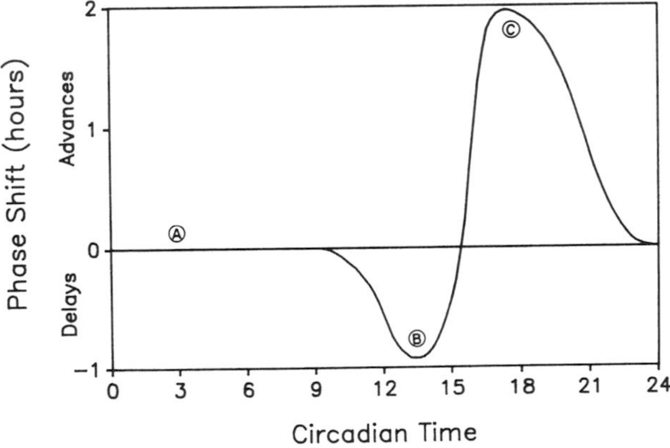

FIGURE 3 Derivation of a phase–response curve (PRC). The top panel shows results from an experiment in which a golden hamster is exposed to a 1-hr light pulse at different circadian times. At all other times, the animal is maintained in constant darkness. Data are plotted as in Figs. 1 and 2. In nocturnal animals, such as the hamster, circadian time 12 is arbitrarily defined as the daily onset of activity. A light pulse given early in the subjective day (**A**) has no effect on the clock, but when given early in the subjective night (**B**) the light pulse causes a 1-hr phase delay, and when given later in the subjective night (**C**) it causes a 2-hr phase advance. The lower panel shows a PRC that summarizes the effect of individual light pulses given at different times spanning a 24 hr period.

The phase–response curve illustrates the behavior of the underlying circadian oscillator, which is often described as the pacemaker of the circadian rhythms it drives. It is important to make the distinction between an oscillator and its output, that is, the rhythm it drives. Circadian rhythms represent the "hands" of the clock, as opposed to being part of the timing mechanism itself. A rhythm that is coupled to a circadian pacemaker provides a marker of the pacemaker's state. Only two properties of a rhythm are reliable indicators of the state of the clock: period length and phase. Amplitude of a rhythm is not a reliable indicator. If a particular rhythm is suppressed or amplified, it does not necessarily follow that the oscillator itself is altered. For example, although cortisol secretion from the human adrenal shows a clear circadian rhythm, the suppression of cortisol by the drug dexamethasone is entirely independent of the circadian pacemaker. In contrast, if the pacemaker is phase-shifted then the cortisol rhythm will also be shifted.

Phase–response curves provide important information about the effect of different stimuli on biological clocks. They are useful in determining whether particular drugs or neurotransmitters (chemicals that transmit signals in the nervous system) are capable of phase-shifting biological clocks. Phase–response curves can also be used to indicate whether a particular stimulus is capable of entraining the biological clock and what the pattern of entrainment will be. Finally, the phase–response curve can be used to predict the phase angle of entrainment. The term phase angle is used to describe the temporal relationship between two separate rhythms, and is defined as the difference between selected phase reference points of any two rhythms (e.g., the entraining rhythms and a biological rhythm). For example, the difference in hours between the time of light offset and the time a hamster begins to run in its activity wheel is a conventional way to measure phase angle of entrainment in a nocturnal animal. This phase angle will vary, depending on the light–dark cycle to which the animal is entrained.

Phase–response curves are typically based on phase shifts that occur after a *single* pulse of light (or any other stimulus that can cause a phase shift). This type of phase–response curve to light for humans is now being established. It appears to compare well with the phase–response curves established for nonhuman animals. That is, phase delays occur in response to light pulses in the evening and phase advances occur in response to early morning light. Humans may require a relatively high-intensity light pulse for phase shifting, whereas maximal phase shifts with 1-hr light pulses of normal room lighting can be obtained in rodents. Comparisons of absolute threshold between nocturnal rodents and humans must, of course, be viewed with caution. In studies on humans, background illumination is generally about 100–400 lux (the intensity of ordinary indoor room light), whereas in rodent studies, light pulses are given against a background of complete darkness.

IV. NONPHOTIC EFFECTS ON BIOLOGICAL RHYTHMS

A variety of nonphotic agents (i.e., unrelated to light), including social factors, feeding schedules, drugs, and numerous neurotransmitter-related substances, have been shown to alter either phase or period of circadian rhythms.

The importance of social factors was first recognized in human studies, in which subjects who were unable to entrain to experimental light–dark schedules were later found to entrain to the same schedules when they were periodically signaled by the sound of a gong to perform routine tasks for the experimenter. The addition of the gong, which was interpreted by the subjects as a social contact, enabled them to entrain to an otherwise weak entraining stimulus. Further support for the hypothesis that social factors can influence human circadian rhythms is the observation that when people are tested in groups, their free-running periods tend to become synchronized with each other.

The role of social cues is particularly, but not uniquely, important in humans. Differences between species most likely reflect differences in social biology. Thus, social interactions do not seem to entrain the relatively asocial squirrel monkey, but will entrain bats that live in large social colonies. The effect of social cues on circadian rhythms may be quite subtle and probably depends on the nature of the social contact.

Periodic feeding schedules have been shown to entrain circadian rhythms of food activity in rats, but this affect appears to be restricted to the entrainment of a rhythm of "food anticipatory" activity that is distinct from other components of activity. Feeding schedules have been reported to entrain circadian rhythms in the squirrel monkey. Although the possible role of meal schedules as a *Zeitgeber* for humans is unknown, such an effect might be expected at least partly owing to the traditional association of mealtimes with social contacts.

Many drugs and endogenous substances have been found to affect the mammalian circadian system. Some of these agents are capable of causing phase shifts, whereas others cause changes in period, but not phase. Chronic treatments with lithium and clorgyline, drugs that are prescribed in affective disorders, as well as ethanol, have been shown to cause slight increases in the length of the free-running period of rodents; whereas the sex hormones estradiol and testosterone have been shown to decrease period length. None of these substances has been shown to cause phase shifts in higher mammals. The benzodiazepines triazolam (Halcion) and diazepam (Valium) cause either phase advances or phase delays, depending on the time of treatment. Numerous neurotransmitters and neurotransmitter-related substances have also been shown to reset rodent clocks, including glutamate (an excitatory amino acid), *N*-methyl-D-aspartate (which mimics the effect of glutamate), serotonin and analogs of this neurotransmitter, muscimol (which simulates the action of the neurotransmitter γ-aminobutyric acid, GABA), carbachol (which simulates acetylcholine), and the neurotransmitter neuropeptide Y.

It is interesting to note that several of these drugs are used to treat sleep and/or mood disorders. Benzodiazepines are widely prescribed as anxiolytics and for insomnia; whereas lithium and clorgyline (a type I monoamine oxidase inhibitor) are prescribed for treatment of depressive disorders. The effects of muscimol and carbachol suggest that neurons containing GABA and acetylcholine, respectively, are important in the regulation of mammalian circadian oscillators. This is discussed further in the next section.

V. NEURAL BASIS OF THE BIOLOGICAL CLOCK

A group of nerve cells, called the suprachiasmatic nucleus (SCN), is thought to be the master pacemaker for circadian rhythms in mammals. The SCN is a bilaterally paired structure in the anterior hypothalamus, a small region at the base of the brain. The two halves of the human SCN lie on the lateral walls of the third ventricle (one of a group of small fluid-filled cavities within the brain) and just above the area where the optic nerves enter the brain. The SCN is a very heterogeneous and complex structure containing extensive connections between nerve cells. Over 25 different types of neurotransmitter-related substances have been found within the SCN region of rats and hamsters. The most abundant neurotransmitter in the brain is γ-aminobutyric acid and the SCN is no exception. Other prevalent neuroactive substances in this region that probably also function as neurotransmitters are vasopressin, vasoactive intestinal peptide, gastrin-releasing peptide, neuropeptide Y, somatostatin, and serotonin.

The SCN is necessary for normal circadian functions and fulfills the basic requirements for an endogenous pacemaker. In rodents, destruction of the SCN abolishes circadian rhythmicity in locomotor activity, sleep–wake cycles, feeding, sexual activity, many hormone rhythms (including pineal melatonin, pituitary prolactin, and gonadotropins), and numerous other functions. Other mechanisms regulating these functions appear to remain intact after SCN lesions; they simply become temporally disorganized. Thus after SCN lesions, animals continue to use running wheels, but instead of restricting most of their activity to a single bout lasting approximately 8 hr, they run for short bursts at irregular intervals. Pathological evidence in humans suggests a similar pacemaker role for the SCN. Tumors that affect the SCN can cause sleep disorders wherein patients repeatedly fall asleep at any time of day, much like SCN-lesioned animals. Sleep in such patients is otherwise normal and is unlike the coma-like sleep that is seen in later stages of such illness.

Three other lines of evidence based on studies in rats and hamsters provide compelling evidence that the SCN contains a self-sustained circadian oscillator. First, the SCN will continue to show circadian rhythms in neuronal activity even after surgical isolation from surrounding brain tissue. Such isolation eliminates rhythmic activity in other brain areas, demonstrating their dependence on neural signals from the SCN. Second, SCN tissue, isolated from the brain and kept *in vitro* under constant lighting conditions, shows circadian rhythms in vasopressin release and neuronal firing rates. Finally, circadian rhythmicity in arrhythmic, SCN-lesioned animals can be restored by neural grafts of SCN tissue from donor animals.

A. Input and Output Pathways

Light entrainment of circadian rhythms in mammals occurs only through the eyes. (Entrainment that occurs independently of the eyes is seen in birds and reptiles, but not in mammals). The primary light entrainment pathway is via a bundle of nerve fibers called the retinohypothalamic tract (RHT), which transmits signals from the retina to the SCN. The RHT pathway has been established in all major mammalian orders, but has not been confirmed in humans, because the neuroanatomical tracing methods, requiring the injection of radioactive substances into the retina shortly before death, cannot be applied.

Entrainment is also mediated through a secondary visual projection, which leads indirectly from the retina to the SCN through intermediary pathways. This projection is called the geniculohypothalamic tract (GHT), which arises from the intergeniculate leaflet (IGL), a subdivision of the lateral geniculate nucleus of the thalamus. The GHT terminates in the SCN, overlapping extensively with the RHT terminals (at least in rat and hamster). Lesions of the IGL or ablation of the GHT do not eliminate entrainment, but do alter the phase–response curve to light and, under certain conditions, also alter the length of the free-running period. The GHT is thought to be important in luminance coding, or mediating the effects of light itensity on the circadian system.

The neuronal pathways that relay signals to the SCN (afferent pathways) and the pathway that relay signals from the SCN to the rest of the brain (efferent pathways) have been characterized in hamsters and rats. As mentioned earlier, the RHT and GHT are the major afferent pathways from which the SCN receives light input. Both the RHT and GHT send signals from the eye on the same side of the body (ipsilateral) as well as from the other side (contralateral) to each side of the SCN. About two-thirds of the fibers go to the contralateral SCN. The SCN also receives afferent connections from many areas of the brain: from the midbrain (raphe nuclei and tegmentum), the hypothalamus (anterior and ventromedial nuclei, arcuate nucleus, and preoptic area), the thalamus (paraventricular nucleus), the septal area, and the hippocampus. In addition, the two halves of the SCN are connected to each other via an extensive system of nerve fibers that traverse the midline in both directions. The efferent pathway project primarily to other areas within the hypothalamus (anterior and ventromedial nuclei, retrochiasmatic area, periventricular area, ventral tuberal area, arcuate nucleus, median eminence, and paraventricular nucleus), as well as to other brain centers: the lateral habenula, midbrain central gray, and septal areas. These widespread connections provide the basis for extensive SCN communication with the rest of the brain and are consistent with the role of the SCN as the temporal coordinator of many diverse rhythms: it is the master pacemaker.

B. Circadian Neural Oscillators Outside the SCN

Evidence pointing to the existence of non-SCN oscillators in mammals comes from experiments with rats that suggest the presence of a "food-entrainable" oscillator lying outside the SCN region. This oscillator is thought to be coupled to a "light-entrained" oscillator, presumably within the SCN. Exposure of rats to daily schedules of food availability will entrain a rhythm of "food anticipatory" activity that is distinct from other locomotor activity rhythm, and that does not appear to be entrained by the light-sensitive oscillator. This rhythm is entrainable by feeding schedules even in SCN-lesioned animals.

The pineal gland (situated just beneath the skull and between the two brain hemispheres) of certain birds and reptiles contains an endogenous self-sustained circadian oscillator. The pineal gland of many birds and reptiles contains both photoreceptive and endocrine cells, which secrete the hormone melatonin. (Although the role of melatonin is not fully understood, it is known to be involved in regulation of the timing of reproduction in many animals, particularly seasonal breeders.) Interestingly, the pineals of many birds and reptiles show circadian rhythms of melatonin release in culture, which can be phase-shifted by light pulses. In contrast, the mammalian pineal contains primarily melatonin-secreting endocrine cells and no photoreceptive cells. The mammalian pineal does not show the circadian rhythms in culture and appears to be under direct control of the SCN. The patterns of variation seen in the pineal of different species suggest that there has been an evolutionary trend away from its role in lower vertebrates as an endogenous circadian oscillator toward a role in higher vertebrates as an organ whose rhythmic output is directly under SCN control. [*See* Melatonin.]

Circadian oscillators have been found in the eyes of many animals, from mollusks to mammals. In the rabbit, a circadian rhythm of electroretinogram (ERG) (a measurement of electrical current generated by retinal nerve cells) of the isolated eye free-runs under constant conditions. In the human eye, ERG measurements, intraocular pressure (which influences diagnostic tests for glaucoma), and the turnover rate (i.e., the rate of renewal) of photoreceptive membranes all show diurnal rhythms. It is very likely that, like other vertebrate retinal rhythms, these rhythms are governed by a circadian oscillator within the eye itself.

VI. DEVELOPMENT OF CIRCADIAN RHYTHMS WITHIN THE INDIVIDUAL

The development of the mammalian circadian system does not depend on exposure to environmental fluctuations. Rat pups that are born and reared under constant lighting conditions will nonetheless show circadian rhythms as adults. Circadian rhythmicity will develop even in pups whose mothers are housed in constant conditions throughout gestation or whose mother's own circadian rhythmicity has been abolished by an SCN lesion.

The biological clock in the SCN begins to oscillate during fetal life. Circadian rhythms can be detected in the fetal rat by the nineteenth gestational day (3 days before birth), at which time the SCN is not yet innervated by the RHT and when it is almost completely lacking in functional synapses (connections at which signals are transmitted between nerve cells). The vast majority of SCN afferents and efferents are formed postnatally and are apparently not necessary for the generation of circadian rhythmicity.

Fetal circadian rhythms are entrained by the mother through unknown mechanisms. Maternal entrainment of postnatal circadian rhythms is less well established. Rat pups are born blind and maternal rhythmicity appears to act as a weak entraining agent for about the first 6 days of life. After this, the pups are able to respond through their own eyes and can be independently entrained by light.

Studies in human infants demonstrate that circadian rhythms in sleeping patterns develop before they are entrained to a 24-hr cycle (Fig. 4). The development of regular 24-hr sleeping patterns in infants is reportedly enhanced by mother–infant interactions, suggesting that maternal factors are also important for the entrainment of circadian rhythms in the human infant.

Daily rhythms appear at different developmental stages, presumably reflecting the maturation of output pathways downstream from the circadian pacemaker. The maturation of a rhythm is typically accompanied by an increase in its amplitude. In the rat, most behavioral and hormonal rhythms are not expressed until 2–3 weeks after birth, when daily rhythms in SCN neuronal firing rates are first detectable. The majority of human circadian rhythms do not become evident until well into the postnatal period. Daily sleep–wake rhythms appear during the second postnatal month, whereas daily rhythms in plasma adrenal hormone (corticosteroid) levels are not established until about 2 years.

VII. GENETICS OF CIRCADIAN RHYTHMS

Genetic analysis of circadian clock mutants from a diverse array of organisms has provided several con-

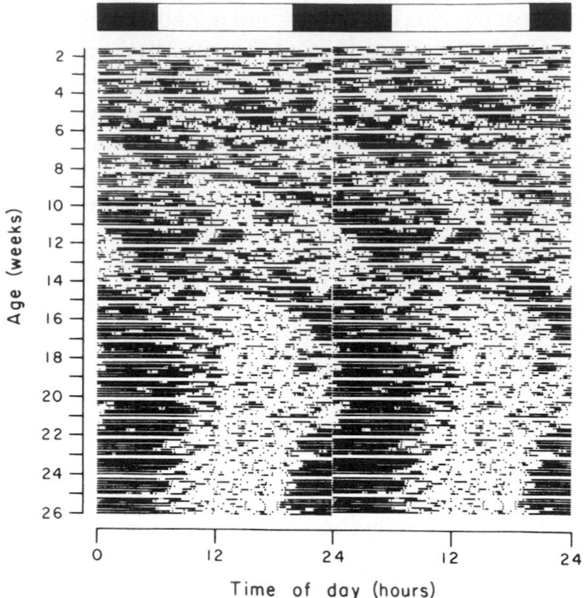

FIGURE 4 The development of sleep–wake rhythmicity in a human infant on a self-demand feeding schedule and a light–dark cycle as indicated at the top of the figure. The black bars in the record indicate time asleep, and the dots represent feedings. The record is repeated and shifted up a day in the right half of the figure to aid visual appreciation of the free-running rhythm. Weekly sets of data are separated by white gaps of no data. [From N. Kleitman and T. G. Eaglemann (1953). *J. Appl. Physiol.* 7, 269. Reproduced with permission.]

ceptual breakthroughs in the field of circadian rhythmicity. It appears that the "clock" is the product of one or very few proteins acting in concert, because single gene mutations can have dramatic effects on circadian timekeeping.

Organisms with circadian clock mutations that are under intensive investigation include *Drosophila* (fruit flies), *Neurospora* (a fungus), cyanobacteria (photosynthetic bacteria), the Syrian hamster, and the laboratory mouse. All of these were induced with mutagens except for the mutation in the Syrian hamster, which was spontaneous and discovered fortuitously. The mutation in hamsters has been dubbed "tau" (for the Greek letter used by convention to denote period), and animals with one copy of the mutant gene (heterozygotes) exhibit a circadian period of about 22 hr, whereas those harboring two copies of the mutant gene (homozygotes) have a circadian period of 20 hr. The mutation in the mouse known as "clock" causes the opposite phenotype. Those animals heterozygous for the mutation have a free-running period of about 25 hr and homozygotes exhibit a very long period of 27 to 28 hr for about 2 weeks under constant conditions and then the animals become arrhythmic.

Many different mutations have been induced in cyanobacteria, some short period and some long period, as well as other mutations that affect the amplitude or persistence of the rhythm. The best-studied circadian mutations in *Neurospora* and *Drosophila* are known as "frequency" and "period." Depending on the exact site of the mutation in these genes, either long circadian periods or short periods or arrhythmicity can be observed. These similarities across phylogeny lead to the hypothesis that the mechanism of the circadian clock may be evolutionarily very old and fairly similar from organism to organism. Interestingly, all of these mutations are semidominant, which means an altered phenotype can be observed in animals harboring only one copy of the mutant gene.

The genes *per* (in *Drosophila*) and *frq* (in *Neurospora*) have been cloned and the messenger RNA and protein encoded by each of these genes is being characterized. Both of these genes appear to encode transcription factors, proteins that control the production of many other proteins by interacting with chromosomal DNA in the nucleus of cells. Moreover, each of these genes seems to control the level of its own protein through the feedback loops. Such feedback systems are probably the mechanism of intracellular timekeeping.

The clock gene in the mouse has been mapped to chromosome 5 and it will be cloned within the next few years. Once the mouse gene is in hand, obtaining the human counterpart will be rapid because the corresponding region in the human genome is already known. These developments hold the promise of a variety of new treatments for sleep and mood disorders, jet lag, and other disruptions of circadian timekeeping.

VIII. SLEEP AND THE BIOLOGICAL CLOCK

The sleep–wake cycle dominates our lives more than any other biological rhythm. Sleep cycles are influenced by ultradian cycles superimposed on circadian rhythms. There are two major components of sleep: patterns of sleep, which refers to the temporal relation between sleep onset and waking, and sleep structure, which refers to the distribution of different sleep stages during the time spent in sleep. Sleep is divided into five stages, generally related to the "depth" of sleep. Stages 1–4 describe increasingly deeper sleep. Stage 5 is typified by rapid eye movements and is called REM sleep. [*See* Sleep.]

Patterns of sleep are timed by the circadian system, although external events, such as daily activities or ambient temperature, may alter either the timing or the structure of sleep. Circadian cycles of fatigue are apparent even during sleep deprivation. This explains why, after struggling to stay awake throughout the night, one is likely to feel less tired during the morning–even though the accumulated hours of lost sleep are greater in the morning.

Three primary variables characterize sleep patterns: (1) sleep latency, that is, the duration of wakefulness preceding sleep onset; (2) sleep length; and (3) timing of sleep onset and termination. Sleep latency is inversely correlated with, and is primarily determined by, the duration of prior wakefulness. Sleep length is only slightly modified by extended periods of wakefulness. Recovery sleep after deprivation rarely matches the amount of sleep lost. For example, a 17-year-old boy chose to undergo 11 days without sleep. When he finally went to bed he slept for only 14 hr, and then resumed his regular 8-hr nighttime sleeping patterns. Further, human subjects living in temporal isolation show no consistent relationship between sleep length and the duration of previous wakefulness. The time of waking is the most regular aspect of sleeping and is thought to be most strongly under circadian control. Sleep onset and termination show consistent relationships with circadian rhythms of body temperature. The most frequently chosen bedtime among experimental subjects is just after the lowest point in the body temperature cycle, and wakeup times typically occur during the rising phase in the temperature cycle. Subjects who went to bed at the low point in the temperature cycle slept for short episodes, whereas subjects who retired close to the peak of the temperature cycle slept much longer.

Ultradian rhythms in sleep structure occur in the 90–100 cycles of alternation between REM and non-REM sleep, with the first REM episode generally occurring 70–90 min after sleep onset. In young adults, REM sleep accounts for 20–25% of total sleep. Unlike circadian rhythms in sleep patterns, the period of the ultradian REM–non-REM cycle does not change when allowed to free-run in constant conditions. In addition, these ultradian rhythms show considerable cycle-to-cycle variation in period length and do not appear to be produced by an endogenous oscillator, but appear to result from interactions between neurons in various areas of the brain stem. In contrast, the proportion of REM sleep within each REM–non-REM cycle increases as the night progresses and appears to be under circadian control.

The secretion of various hormones is regulated by both the timing of sleep and the circadian clock. The relative degree of control by sleep and circadian clock varies with different hormones. The secretion of growth hormone, which shows a clear diurnal rhythm, appears to be primarily dependent on sleep. During sleep deprivation, growth hormone secretion is suppressed, whereas the onset of sleep elicits a large secretory burst of growth hormone whether sleep is delayed, advanced, interrupted, or fragmented.

In contrast, secretion of the pituitary hormone prolactin appears to be controlled jointly by sleep and by the circadian oscillator. Plasma prolactin levels are usually highest during the last 2 hr of a normal sleep period, fall rapidly after sleep ends, and remain relatively low during the day. Increases in prolactin levels are, however, also associated with daytime naps, which has been interpreted as indicating that diurnal rhythms in prolactin are directly controlled by sleep patterns. However, it has since been demonstrated that prolactin secretion is also under separate circadian control. The secretory rhythm of adrenocorticotrophic hormone (ACTH), the pituitary hormone that triggers cortisol release, exemplifies a hormonal rhythm that is under strong circadian control. ACTH and cortisol are secreted in episodes about once every 1–2 hr and the circadian profile appears to be produced by modulating the size of successive secretory pulses. Studies in rodents suggest that, in addition to being primarily dependent on the circadian rhythm of ACTH release, the adrenal cortisol secretion rhythm is amplified by daily variations in the responsiveness of the adrenal to ACTH. Both rhythms persist during sleep deprivation.

IX. RHYTHMS IN REPRODUCTION

Biochemical, neural, hormonal, and behavioral rhythms play a central role in many different aspects of reproduction. Rhythms in reproduction cover a wide spectrum of frequencies: from the 60-sec courtship songs of fruit flies to circannual rhythms of reproductive status in ground squirrels. All the major reproductive hormones are known to fluctuate throughout the day. Hormonal release is predominantly pulsatile, or episodic, with secretory bursts occurring at 1- to

4-hr intervals (Fig. 5). Patterns of hormonal release may be influenced by diurnal modulations in pulse frequency and/or pulse amplitude.

Reproductive processes are regulated by the hormones of the hypothalamic–pituitary–gonadal axis. Many aspects of the reproductive process are regulated through precisely timed changes in pulse frequency and/or amplitude. Pulsatile hormone release is, however, often undetected or underestimated because blood samples are typically taken too infrequently to map rapid fluctuations in serum hormone levels. Hormonal pulsatility is influenced by many factors in addition to circadian signals, particularly, developmental changes and feedback effects from other hormones. For example, testosterone (male sex hormone) has a major effect on the pulsatile pattern of the release of the pituitary luteinizing hormone (LH), which stimulates testosterone release. Castration of both rats and monkeys results in dramatic increases in the frequency and amplitude of LH pulses. In women, the pulse frequency of gonadotropin-releasing hormone (GnRH) is reduced during the luteal phase of the menstrual cycle (the phase after ovulation). This reduction in frequency is thought to be due to the neural actions of progesterone, which is secreted by the corpus luteum of the ovary.

The onset of puberty is characterized by profound changes in diurnal, as well as pulsatile, variations in the pituitary gonadotropic hormones, LH and follicle-stimulating hormone (FSH). In prepubertal children of both sexes, gonadotropins are secreted in low-amplitude pulses during the night. This nocturnal rise is thought to be partially influenced by the circadian system, but is primarily sleep dependent. As a child approaches puberty, the amplitude of the nighttime

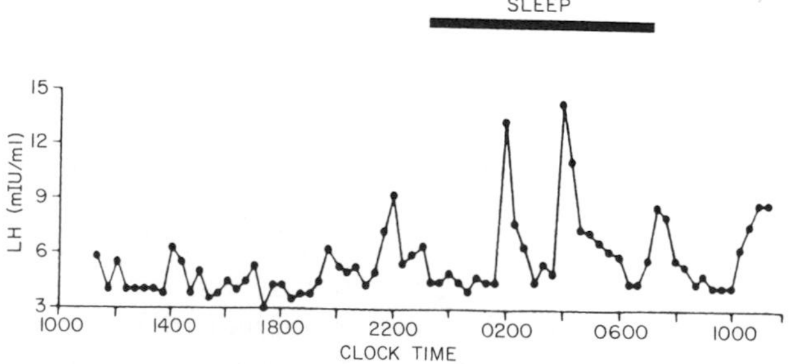

FIGURE 5 24-hr plasma luteinizing hormones (LH) profiles of a 14-year-old boy during a normal sleep–wake cycle. LH release occurs in pulsatile bursts. The nighttime augmentation of LH release is influenced by sleep, as well as by circadian signals that occur independently of sleep. [From S. Kapen *et al.* (1974). *J. Clin. Endocrinol. Metab.* 39, 293.]

pulses increases, thus amplifying the overall circadian variation. In adult men, LH and FSH continue to be released in pulses every few hours, but the diurnal variation is damped and, in most cases, undetectable. Serum testosterone levels fluctuate with about 17–18 pulses per 24-hr period in young adult men, and are modulated by diurnal rhythms with lowest levels generally at night and highest levels in the morning. Since there is little, if any, diurnal variation in serum LH levels (which provide the primary stimulus for testosterone secretion), the diurnal fluctuations in testosterone levels must be due to factors that modify the effectiveness of the LH signal on a daily basis.

In women, the patterns of pituitary LH release are modulated by the menstrual cycle. Ovulation is triggered by a surge in LH release. During the early follicular phase (the phase before ovulation), LH pulses are more frequent during the night. As ovulation approaches, this diurnal modulation of pulsatility disappears and remains absent during the subsequent luteal phase. The timing of the preovulatory LH surge is determined by an interaction between hormonal changes of the menstrual cycle and the circadian system. In normal women, the onset of the LH surge usually occurs in late sleep or early morning. The extent of circadian control has not, however, been precisely defined in humans.

In rodents, which show 4- to 5-day estrous (ovarian) cycles, estrous-related events (e.g., timing of the surge in LH release, ovulation, postovulatory increase in progesterone, and onset of sexual receptivity) are tightly linked to the circadian system and all occur at specific times of day on specific days during the cycle. Interestingly, the neural signal that induces the LH surge occurs in the afternoon every day, but only results in a surge every fourth day. The reasons for this are not completely understood, but we do know that the occurrence of the LH surge requires high levels of estrogen, which occur only on the day before ovulation.

In rodents, the timing of birth is under circadian control. Human births have not been studied in temporal isolation, but are undoubtedly influenced by circadian rhythms. The onset of labor shows a very clear diurnal rhythm, with labor beginning most frequently about 1:30–2:30 A.M. Natural births are more variable in timing than the onset of labor, but also show a diurnal rhythm with births occurring most frequently between 3:00 and 6:00 A.M.

Although the 28-day human menstrual cycle closely matches the 29.5-day lunar cycle, it does not appear to be influenced by the moon's gravitational pull. The correlation is merely coincidental and is not seen in other mammals. Duration of estrous or menstrual cycles in mammals varies widely in length: from the 4-day cycle of the laboratory rat and golden hamster to the 126-day cycle of the Indian elephant.

A number of reproductive disorders are associated with abnormal circadian or ultradian hormonal profiles. Pulsatile patterns of gonadotropin-releasing hormone release from the brain (hypothalamus) are critical for the maintenance of normal profiles of gonadotropin release from the pituitary. In monkeys whose endogenous supply of GnRH has been eliminated, normal patterns of gonadotropin release can be restored with exogenous GnRH, but only under certain treatment schedules. If replacement GnRH is given continuously or at too high a frequency (e.g., once every half hour), gonadotropin levels are suppressed. However, the same overall dose of GnRH, when given in hourly pulses, is sufficient to restore normal gonadotropin profiles.

Anovulatory cycles in women are sometimes traceable to inadequate GnRH pulse frequency, and pulsatile GnRH has been shown to be an effective treatment in hypothalamic amenorrhea. Pulsatile GnRH therapy has also been successfully used for treatment of idiopathic hypogonadotrophic hypogonadism in men. Abnormal patterns in pulsatile release of LH are often associated with abnormal reproductive function in both men and women. In amenorrhea associated with anorexia nervosa, the secretory patterns of LH regress to the pubertal pattern with low daytime pulsatility and increased secretion at night, or to infantile patterns.

Humans are capable of reproducing throughout the year. Most other mammals, however, breed only during restricted seasons. Seasonal cycles in reproduction are generated from both endogenous and exogenous mechanisms. As mentioned earlier, ground squirrels show circannual cycles. They are capable of breeding for only one period of several weeks during each circannual cycle. When housed at constant temperature and photoperiod (i.e., with no seasonal changes in day length), their seasonal breeding cycles will free-run with a period of 10–11 months. A different type of seasonal cycle is seen in the golden hamster, whose reproductive status is determined by photoperiod. In hamsters, exposure to short photoperiods (less than 12 hr of light per 24-hr period) induces gonadal regression. This effect is reversible by transfer to a long photoperiod (more than 12 hr of light per 24-hr period).

Photoperiodic regulation of reproduction relies on a physiological system for the measurement of changing day lengths that involves the circadian clock and integration with hypothalamic centers controlling GnRH. In addition to regulating seasonal cycles, photoperiod also regulates the timing of the onset of puberty in many mammals. The photoperiodic response depends, in part, on the rhythmic pattern of melatonin release. In vertebrates, melatonin release is highest during the night and low during the day. Interestingly, melatonin release is directly suppressed by exposure to light. Although the role of melatonin rhythms in nonphotoperiodic animals remains unclear, there are hints that melatonin may play some role in human reproduction. Puberty in children with pineal tumors is usually disrupted and may be either precocial or delayed.

There are conflicting reports on seasonal fluctuations in human reproductive physiology. The onset of puberty occurs more often during months of longer day lengths. Annual fluctuations in serum testosterone levels have been reported in men, although the basis of such fluctuation is unclear. In human studies it is difficult to independently evaluate separate influences from the physical and social environments. Fluctuations in hormonal levels, as well as sexual activity and birth rate, are likely to be influenced by social and behavioral factors—particularly by the timing of work schedules, annual holidays, and vacations.

X. BIOLOGICAL RHYTHMS AND HUMAN HEALTH

A. Jet Lag

The most abrupt dislocation of the biological clock we are likely to experience is after traveling across several time zones by plane. Depending on the direction of travel, the biological clock must be either phase-advanced or phase-delayed to synchronize with the new time zone. Several days to several weeks may be necessary to become fully resynchronized to the new schedule. During this period one is likely to experience symptoms of jet lag, which include sleep disruption, gastrointestinal disturbances, decreased alertness and attention span, and general feelings of malaise. The process of reentraintment is characterized by internal temporal disorder. Phase relations between different rhythms are disrupted, or internally desynchronized, because some rhythms may take longer than others to become fully reentrained. After a 6-hr phase shift, activity rhythms in human subjects are reentrained within 3 days, whereas it may be a week or more before the rhythms of body temperature or sodium excretion rates are reentrained.

Rates of reentrainment are influenced by the number of time zones crossed, the direction of travel, individual differences, and environmental circumstances. Most people adjust more rapidly to westward travel, which involves delaying the biological clock, than they do to eastward travel. Because the average free-running period of human circadian rhythms is about 24 hr, most people find it easier to sleep an extra hour in the morning than to rise an hour earlier. Differences between individuals in the number of days needed to adjust to new time zones are correlated with the amplitude of their body temperature rhythms. The strength or abundance of entraining agents also influence rates of reentrainment. People who arrive in a new time zone and spend their time indoors will not readjust as quickly as people who go out during the day—thereby exposing themselves to more intense light, social contacts, meal cues, and other reinforcing environmental factors. Despite popular claims, there is no clear evidence that particular diets will minimize jet lag.

B. Shift Work

Though jet travel induces the most abrupt dislocation of the biological clock, the disruptive schedules of shift work affect far more people: 15–25% of the working population in industrialized countries are engaged in shift work. Most shift workers are subjected to rapidly changing schedules that do not allow them to become synchronized. They generally remain at least partially entrained to a normal day–night cycle. Individuals differ greatly in their tolerance for shift work: some report no complaints in adjusting to abnormal shifts, whereas others suffer sleep and digestive disorders. Differences in circadian physiology may account for at least some of these individual differences.

Human performance of many different tasks shows pronounced daily rhythms, particularly in tasks involving repetition and vigilance. It is not surprising then that the incidence of work errors is highest between 3:00 and 5:00 A.M., the low point in daily performance rhythms. This is a problem for worker safety, as well as public safety. We rely on the vigilance of thousands of shift workers who operate nuclear power plants, pilot airplanes, and monitor life-support machines in hospitals, whose shifts cover hours when their performance rhythms are lowest.

A less common type of shift work subjects workers to abnormal period lengths. Nuclear submarine operators in the U.S. Navy work in 6-hr shifts alternating with 12-hr rest periods. These 18-hr cycles are beyond the limits of entrainment for human circadian rhythms and these workers endure long-term disruption of their biological clocks. Experimental subjects exposed to similar cycles have problems with insomnia, emotional disturbances, and impaired coordination. Animal studies have associated such abnormal cycles or frequent phase shift (e.g., weekly shifts of 6 hr or more) with decreased longevity, although it remains unclear whether this also applies to human physiology.

C. Rhythm Pathologies

Most known disorders of biological rhythms are those that interfere with the sleep–wake cycle. The most common complaint is of a difficulty in falling asleep at night. People with insomnia often have no problem once they fall asleep: the problem lies in the timing of sleep. However, many of these people have no problems sleeping when they are able to follow their own schedules, in contrast to the schedule imposed by a typical work day. Certain patients with this "delayed sleep phase" type of disorder can be successfully treated by phase-delaying their sleep cycles until they are in synchrony with "normal" schedules and thereafter maintaining a strict 24-hr sleep–wake cycle.

Early morning waking is a classic symptom of clinical depression. Depressed psychiatric patients also show an earlier onset of REM sleep relative to nondepressed patients. Many rhythms are advanced in depressed patients. These include the daily profiles of prolactin, cortisol, growth hormone, melatonin, and body temperature. These atypical profiles are possibly due to circadian abnormalities (e.g., unusually short cycle lengths) or, alternatively, altered phase relations between the pacemaker and the observed rhythms. Interestingly, a 6-hr advance of the sleep–wake cycle or sleep deprivation for one night can induce a remission lasting several days to two weeks.

Seasonal affective disorder (SAD) is characterized by seasonally recurring episodes of major depression. In the typical pattern, patients become depressed during the fall and winter. The opposite case, in which depression occurs during the summer, has also been reported but is much more rare. Bright-light therapy (i.e., over 5000 lux, where indoor lighting ranges between 100 and 1500 lux) is often effective in alleviating depression in SAD patients within several days, although relapse usually follows within a week or two.

The biological mechanisms underlying the therapeutic effects of bright light are unknown. Proposals that light therapy works by altering melatonin secretion or, alternatively, by causing a phase shift in the circadian oscillator have both been shown to be insufficient to fully explain the effects of light in SAD patients. The causes of SAD are also unclear. Symptoms of SAD may simply represent the extreme end of a continuum of normal seasonal fluctuations. Studies of healthy individuals show clear seasonal variation in mood, with lows occurring during the fall and winter. The timing of seasonal mood cycles for SAD patients and healthy individuals is similar, but the cycles of SAD sufferers show a much greater amplitude. The peaks of both groups are similar, but the lows of the SAD cycles are much lower.

Recent studies show that day length can modulate the duration of melatonin secretion in humans. In addition, healthy women show seasonal rhythms of melatonin secretion but healthy men do not. It may be coincidental that women suffer disproportionately from SAD, but data like these continue to stimulate interest in the possible connections between the dynamics of melatonin secretion and SAD.

D. Medical Implications

Understanding the role of biological rhythms in human physiology has important implications for both diagnosis and treatment of disease. As mentioned in Sections VIII and IX, disruptions in timing are associated with sleep and reproductive disorders. In some cases, the key to a correct diagnosis lies in the temporal aspects of a particular variable. The diagnosis of adrenal dysfunction provides a clear illustration. In Cushing's disease the adrenal is hyperactive, whereas Addison's disease is caused by adrenal failure. In both cases, serum cortisol levels are within the normal daily range, but show a loss in circadian rhythmicity. The critical difference is that in Cushing's disease cortisol levels remain near the normal daily maximum, whereas in Addison's disease cortisol levels remain near the daily minimum. Thus information about the normal range of cortisol variation is insufficient to understand the nature of adrenal function in these cases. It is also essential to understand the temporal structure of cortisol secretion.

Daily rhythms have been shown in the therapeutic effectiveness of many drugs, including anesthetics, toxins, histamines and antihistamines, and drugs used in chemotherapy. Circadian rhythms may influence drug effectiveness in a number of ways: (1) Uptake: The rate of absorption from the intestine into the circulation may show circadian variation. This may also be true for intramuscularly administered drugs. (2) Metabolism: The rate of drug inactivation may show circadian variation. (3) Delivery to the target tissue: Blood volume and extracellular fluid volume show circadian variations. (4) Target tissue response: The target tissue may show circadian variation in the number of affinity of receptor binding sites for the drug.

The influence of circadian rhythms in human physiology is pervasive and temporal organization is a fundamental element of human physiology. Our well-being relies on the biological clocks that maintain this temporal order.

BIBLIOGRAPHY

Aschoff, J. (ed.) (1981). "Handbook of Behavioral Neurobiology," Vol. 4, "Biological Rhythms." Plenum, New York.

Klein, D. C., Moore, R. Y., and Reppert, S. M. (eds.) (1991). "Suprachiasmatic Nucleus. The Mind's Clock." Oxford Univ. Press, New York.

Moore-Ede, M. C., Sulzman, F. M. and Fuller, C. A. (1982). "The Clocks That Time Us." Harvard Univ. Press, Cambridge, MA.

Rosenwasser, A. M., and Adler, N. T. (1986). Structure and function in circadian timing systems: Evidence for multiple coupled circadian oscillators. *Neurosci. Biobehav. Rev.* **10**, 431–448.

Turek, F. W., and Van Cauter, E. (1994). Rhythms in reproduction. *In* "The Physiology of Reproduction" (E. Knobil, *et al.*, eds.), 2nd Ed., pp. 487–540. Raven, New York.

Van Cauter, E., and Turek, F. W. (1995). Endocrine and other biological rhythms. *In* "Textbook of Endocrinology" (L. J. De-Groot, ed.), 3rd Ed. Saunders, Philadelphia.

Wehr, T. A., and Goodwin, F. K. (1983). "Circadian Rhythms in Psychiatry." Boxwood Press, Pacific Grove, CA.

Citrate Cycle

DANIEL E. ATKINSON
University of California, Los Angeles

GLOSSARY

Adenylate energy charge Effective mole fraction of ATP in the adenine nucleotide pool: ([ATP] + 0.5[ADP])/([ADP] + [ADP] + [AMP]); this function is a linear measure of the energy status of the adenylate nucleotides, the main energy-transducing system in the cell

Anabolism Synthesis or building up of cellular constituents

Catabolism Breaking down or degradation of molecules from the diet or from the cell's stores or structure

Flux Rate of flow of material through a metabolic sequence; when the sequence is at steady state, the rate of each reaction is equal to the flux through the sequence as a whole

Modifier Metabolite that modulates the catalytic properties of an enzyme, usually affecting the affinity with which the enzyme binds substrate; a positive modifier causes an increase in affinity for substrate and a negative modifier causes a decrease

THE CITRATE CYCLE (ALSO KNOWN AS THE TRIcarboxylic acid or TCA cycle and as the Krebs cycle) is the central metabolic sequence in most aerobic heterotrophic organisms, including humans and other mammals. Although often thought of as merely the final sequence in the oxidation of glycogen, glucose, and other carbohydrates, the citrate cycle is the most important metabolic crossroads in aerobic metabolism and is involved in both the catabolism (breaking down) and the anabolism (synthesis, or building up) of most of the major classes of metabolites. The cycle or some of its constituent enzymes participate in the oxidation of carbohydrates, proteins, and fats, in the conversion of carbohydrate to amino acids or fats and of proteins and amino acids to carbohydrates or fats, and in the generation of starting materials for many biosynthetic sequences.

I. REACTIONS AND STOICHIOMETRY OF THE CYCLE

In the citrate cycle, oxaloacetate condenses with the acetyl group of acetyl coenzyme A (AcSCoA), forming citrate and liberating free coenzyme A (HSCoA). Citrate is then oxidized in seven steps, with loss of two molecules of CO_2, to oxaloacetate. Thus, in each turn of the cycle, from oxaloacetate back to oxaloacetate, one acetyl group is fully oxidized to carbon dioxide. The reactions of the cycle are diagrammed in Fig. 1. The equations for the reactions and the structures of the intermediates are shown in the Fig. 1 legend.

II. OXIDATIONS

A. Oxidation of Acetate

In the simplest case, the citrate cycle is the route for oxidation of acetate units that are derived from the oxidation of carbohydrates, fats, and some of the amino acids that are liberated in the hydrolysis of proteins. To enter the cycle, acetate units must be activated by formation of a thiol ester with coenzyme A. This activation is built into the degradation sequences; both in glycolysis and in fatty-acid oxida-

ENCYCLOPEDIA OF HUMAN BIOLOGY, Second Edition, VOLUME 2. Copyright © 1997 by Academic Press. All rights of reproduction in any form reserved.

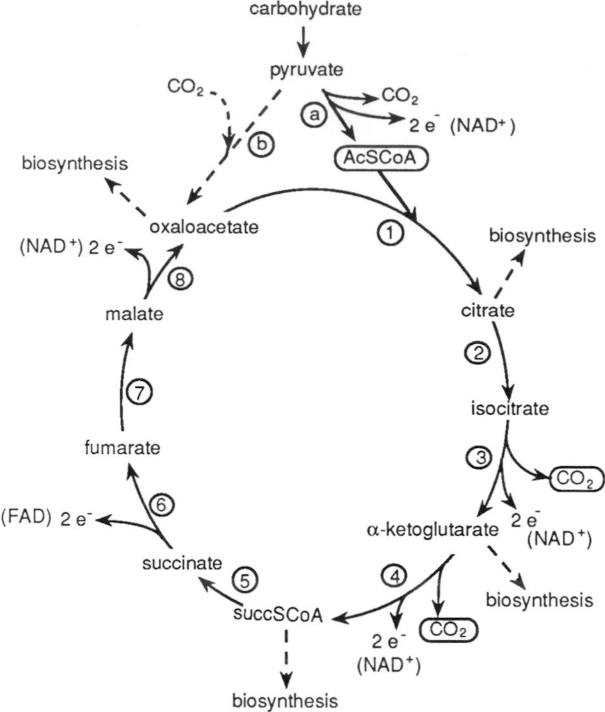

FIGURE 1 The citrate cycle. The input, acetyl coenzyme A, and the output, carbon dioxide, are indicated by circling. The electron transfer agents that accept electrons (e^-) at the oxidative reactions of the cycle are indicated in parentheses. Broken arrows represent reactions that contribute to the generation of biosynthetic intermediates.

The reactions, keyed to the identifying numbers in the figure, and the enzymes that catalyze them:

① citrate synthase:
$AcSCoA + OAA^{2-} + H_2O \longrightarrow citrate^{3-} + H^+ + HSCoA$

② aconitase:
$citrate^{3-} \rightleftharpoons isocitrate^{3-}$

③ isocitrate dehydrogenase:
$isocitrate^{3-} + NAD^+ \longrightarrow \alpha\text{-}ketoglutarate^{2-} + CO_2 + NADH$

④ α-ketoglutarate dehydrogenase:
$\alpha\text{-}ketoglutarate^{2-} + HSCoA + NAD^+ \longrightarrow succSCoA^- + NADH + CO_2$

⑤ succinate thiokinase:
$succSCoA^- + GDP^{3-} + P_i^{2-} + H_2O \longrightarrow succinate^{2-} + GTP^{4-} + HSCoA$

⑥ succinate dehydrogenase:
$succinate^{2-} + FAD \longrightarrow fumarate^{2-} + FADH_2$

⑦ fumarase:
$fumarate^{2-} + H_2O \longrightarrow malate^{2-}$

⑧ malate dehydrogenase:
$malate^{2-} + NAD^+ \longrightarrow OAA^{2-} + NADH + H^+$

Stoichiometry of the cycle (sum of reactions 1–8):

$$AcSCoA + 3NAD^+ + FAD + 3H_2O + GDP + P_i \longrightarrow 2CO_2 + HSCoA + 3NADH + FADH_2 + GTP + 3H^+ \quad (9)$$

Although three protons (H^+) are shown to be generated for each molecule of acetyl coenzyme A oxidized, the citrate cycle does not cause acidification in the living cell. The protons are consumed in the course of the oxidation of NADH:

$$3NADH + 3H^+ + 1.5O_2 \longrightarrow 3NAD^+ + 3H_2O$$

Subsidiary reactions shown in Fig 1 and discussed below are:

ⓐ pyruvate dehydrogenase:
$pyruvate^- + NAD^+ + HSCoA \longrightarrow AcSCoA + NADH + CO_2$

ⓑ pyruvate carboxylase:
$pyruvate^- + CO_2 + ATP^{4-} + H_2O \longrightarrow oxaloacetate^{2-} + ADP^{3-} + P_i^{2-} + 2H^+$

Structures of the intermediates of the cycle:

$citrate^{3-}$:
$H_2C—COO^-$
$HO—C—COO^-$
$H_2C—COO^-$

$isocitrate^{3-}$:
$H_2C—COO^-$
$HC—COO^-$
$HO—CH—COO^-$

$\alpha\text{-}ketoglutarate^{2-}$:
$H_2C—COO^-$
CH_2
$O{=}C—COO^-$

$succSCoA^-$:
$CH_2—CO—S—CoA$
$CH_2—COO^-$

$succinate^{2-}$:
$CH_2—COO^-$
$CH_2—COO^-$

$fumarate^{2-}$:
$HC—COO^-$
$^-OOC—CH$

$malate^{2-}$:
$HO—CH—COO^-$
$H_2C—COO^-$

$oxaloacetate^{2-}$:
$O{=}C—COO^-$
$H_2C—COO^-$

$pyruvate^-$:
$CH_3—CO—COO^-$

tion, the two-carbon unit is generated directly as acetyl coenzyme A. [*See* Coenzymes, Biochemistry.]

Carbohydrates are converted to pyruvate by the reactions of glycolysis. In the process, two molecules of pyruvate are produced from each molecule of glucose, with the transfer of four electrons to the metabolic electron carrier oxidized nicotine adenine dinucleotide (NAD^+). Pyruvate is oxidized to acetyl coenzyme A and CO_2 in a reaction catalyzed by the three-enzyme pyruvate dehydrogenase complex. In this reaction, two electrons are transferred to NAD^+ for each molecule of pyruvate, or four for each starting molecule of glucose; thus, eight electrons are transferred to NAD^+ in the course of conversion of one molecule of glucose, or one glucosyl unit of glycogen, to two molecules of acetyl coenzyme A and two molecules of CO_2. Sixteen additional electrons (eight for each molecule) are lost in the further oxidation of this acetyl coenzyme A to CO_2. This oxidation is carried out through the citrate cycle, which is thus responsible for the removal of two-thirds of the electrons that are lost in the total oxidation of glycogen or hexoses to CO_2. [*See* Glycolysis.]

Storage fats are hydrolyzed to fatty acids and glycerol. The fatty acids are oxidized to acetyl coenzyme A. Because four electrons are lost per two-carbon unit in the conversion of fatty acids to acetyl coenzyme A and eight remain to be removed in the cycle, it can be seen that also in fat degradation the reactions of the citrate cycle are responsible for most of the oxidation. [*See* Fatty Acid Uptake by Cells.]

In the digestion of dietary proteins or the breakdown of proteins of the cell, the first step is hydrolysis. The resulting amino acids are degraded by specific individual catabolic sequences. Among the products of some of these sequences is acetyl coenzyme A, which is oxidized through the citrate cycle.

In the oxidation of each molecule of acetyl coenzyme A, three pairs of electrons are transferred to NAD^+ and one pair to oxidized flavin adenine dinucleotide (FAD). The benefit that the cell derives from oxidation of carbohydrates, fats, and proteins is the provision of these electrons to the oxidative phosphorylation system, which regenerates adenosine triphosphate (ATP) from adenosine diphosphate (ADP) and inorganic phosphate at the expense of the free energy drop that is involved in the transfer of electrons from reduced nicotinamide adenine dinucleotide (NADH) or reduced flavin adenine dinucleotide ($FADH_2$) to molecular oxygen. The reduced carriers NADH and $FADH_2$ that are derived from the oxidative reactions of the citrate cycle are the major source of electrons to drive ATP synthesis in the cells of humans and most other aerobic organisms. [*See* Adenosine Triphosphate (ATP).]

B. Oxidation of Intermediates of the Cycle

Proteins contain 20 kinds of amino acid residues. Each amino acid is catabolized, or degraded, by a specific pathway. Thus, unlike the catabolism of carbohydrates and fats, which in the main follow single pathways and lead to acetyl coenzyme A as the metabolite requiring further oxidation, the degradation of proteins proceeds by many pathways and leads to a number of metabolites. Among these, in addition to acetyl coenzyme A, are α-ketoglutarate, oxaloacetate, and succinyl coenzyme A (succSCoA). Fruits, vegetables, and green parts of plants contain considerable amounts of salts of citrate, isocitrate, succinate, and malate. Thus, a means of oxidation of cycle intermediates is needed.

Reactions 1–8 in Fig. 1 can accomplish only the oxidation of acetyl coenzyme A to CO_2. If any intermediate is supplied from outside the cycle, it will, as far as the cycle proper is concerned, be oxidized only to oxaloacetate. Because every molecule of oxaloacetate that is used in reaction 1 leads to its own replacement by reaction 8, the figure shows no way that a cell could metabolize oxaloacetate itself.

The auxiliary reactions that are called into play to deal with a metabolic or dietary supply of cycle intermediates are diagrammed in Fig. 2. Oxaloacetate is decarboxylated and phosphorylated at the expense of guanosine triphosphate (GTP) (ATP in some species) to produce phosphoenolpyruvate (PEP). Phosphoenolpyruvate is the immediate precursor of pyruvate in glycolysis, and it is converted to pyruvate with generation of a molecule of ATP. Thus, in the conversion of oxaloacetate to pyruvate (equation 10) the net conversion is only the loss of CO_2. One GTP or ATP is used and one ATP is produced, so there is no net effect on the nucleoside triphosphate energy supply.

III. GENERATION OF SYNTHETIC INTERMEDIATES

A. Use of Citrate Cycle Intermediates as Biosynthetic Starting Materials

Humans and other mammals are heterotrophs, that is, they cannot produce carbohydrates and other cell constituents from carbon dioxide, but must depend on autotrophic organisms, mainly green plants, for their nourishment. Therefore, in animal metabolism, the starting materials for all syntheses must be derived from the breakdown of constituents of the diet. All of the many synthetic sequences start from intermediates of the main catabolic pathways: glycolysis, the pentose phosphate pathway, and the citrate cycle. Of

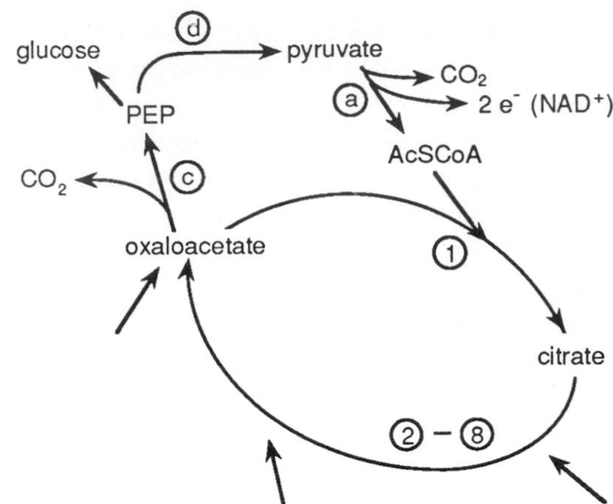

FIGURE 2 Reactions that allow the oxidation of intermediates of the citrate cycle. Two decarboxylations and loss of two electrons (e^-) convert oxaloacetate to acetyl coenzyme A, which enters the cycle and is oxidized in the normal way. Reactions 1–8 correspond to those in Fig. 1. The arrows pointing to the cycle represent entry of cycle intermediates derived from the diet or from catabolism of cell constituents.

Additional reactions:

ⓒ PEP carboxykinase:
oxaloacetate^{2-} + GTP^{4-} + H$^+$ ⟶ phosphoenolpyruvate^{2-} + GDP^{3-} + CO_2

ⓓ pyruvate kinase:
phosphoenolpyruvate^{2-} + ADP^{3-}⟶ pyruvate$^-$ + ATP4^{4-}

Sum of reactions c and d:

$$\text{oxaloacetate}^{2-} + H^+ \longrightarrow \text{pyruvate}^- + CO_2 \qquad (10)$$

Structure of phosphoenolpyruvate:

$$\begin{array}{ccccc} & O^- & & & COO^- \\ & \backslash & & & / \\ HO- & P & -O- & C & \\ & / & & & \backslash\backslash \\ & O^- & & & CH_2 \end{array}$$

With the addition of reactions c and d, the citrate cycle is able to participate in the oxidative degradation of any compound that can be converted to acetyl coenzyme A or to any intermediate of the cycle.

the intermediates of the citrate cycle, citrate, α-ketoglutarate, succinyl coenzyme A, and oxaloacetate serve as starting materials for biosynthetic sequences.

Citrate leaves the mitochondrion (see Section IV) and, by way of the reaction catalyzed by citrate lyase (equation 11), serves as the source of cytosolic acetyl coenzyme A:

$$\text{citrate}^{3-} + \text{ATP}^{4} + \text{HSCoA} \rightarrow \text{AcSCoA} + \text{OAA}^{2-} + \text{ADP}^{3} + \text{P}_i^{2-}. \quad (11)$$

Because acetyl coenzyme A supplies all of the carbon atoms of storage fats, of cholesterol, which is an essential component of membranes, and of the steroid hormones, supplying citrate to support cytosolic production of acetyl coenzyme A is an important function of the citrate cycle. [*See* Cholesterol; Steroids.]

It will be noted that the reaction catalyzed by citrate synthetase (reaction 1 of Fig. 1), the movement of citrate through the mitochondrial membrane, and the reaction catalyzed by citrate lyase, acting together, have the consequence that acetyl coenzyme A and oxaloacetate are moved from the inside of a mitochondrion to the cytosol at the expense of the hydrolysis of a molecule of ATP. As noted, the beneficial consequence is the provision of acetyl coenzyme A for biosynthetic sequences in the cytosol. The oxaloacetate is generally not needed there, and most of it is reduced to malate and taken back into the mitochondria. This sequence of reactions illustrates the generalization that reactions can be made to go in the physiologically useful direction, and materials can be moved to other compartments, by use of ATP. Although it is citrate that actually crosses the mitochondrial membrane, the reactions discussed in this and the preceding paragraph are in effect a pump by which acetyl coenzyme is moved from the mitochondrion to the cytosol, with hydrolysis of ATP supplying the energy.

α-Ketoglutarate is converted by transamination to glutamic acid, an amino acid that is required in the synthesis of proteins. Glutamic acid is also the starting material for synthesis of two other amino acids, glutamine and proline. In addition to its use in protein synthesis, glutamine supplies amino groups in the synthesis of several other types of compounds, including the purines required in nucleic acid synthesis, and is also the primary form in which nitrogen is transported in the blood between tissues and organs in the animal body. [*See* DNA Synthesis.]

Succinyl coenzyme A is required in the synthesis of heme, which is a component of hemoglobin and of cytochromes. [*See* Hemoglobin.]

Oxaloacetate is converted by transamination to aspartic acid, an amino acid required in protein synthesis. Aspartic acid is also the precursor to asparagine, another amino acid constituent of proteins, and of lysine, threonine, methionine, and isoleucine in organisms that can synthesize those amino acids.

B. Replenishment of Cycle Intermediates

In the basic citrate cycle, each molecule of oxaloacetate that is consumed in the synthesis of citrate is replaced by a molecule that is generated when the citrate is oxidized, as shown in Fig. 1. There is no net use of oxaloacetate in that case, as shown by equation 9, but each molecule of citrate, α-ketoglutarate, succinyl coenzyme A, or oxaloacetate that is removed from the cycle for use in biosynthesis prevents the regeneration of one molecule of oxaloacetate. There is thus a loss of oxaloacetate equimolar with the sum of all cycle intermediates used in synthesis. If that loss is not balanced by production of oxaloacetate, the cycle will soon come to a stop because of lack of oxaloacetate for reaction 1.

In mammals, the most important route by which oxaloacetate is replenished is the carboxylation of pyruvate (reaction b in Fig. 1). Thus, pyruvate has two routes of entry into the cycle. When the cycle is operating almost exclusively as an oxidative sequence supplying electrons for regeneration of ATP, the usual situation in muscle, only the entry by way of acetyl coenzyme A (reaction a in Fig. 1) is of importance. However, when a significant amount of synthesis is occurring, as is usually true in liver, kidney, and various other organs, pyruvate becomes an important partition point (see Section VI). Enough pyruvate must be carboxylated to exactly balance the amount of intermediates that are lost from the cycle.

Many kinds of organisms, including plants and many bacteria, replenish oxaloacetate by a different but equivalent reaction. Rather than pyruvate, they carboxylate its immediate precursor in the glycolytic pathway, phosphoenolpyruvate. Because enol phosphates are thermodynamically unstable, this carboxylation is driven by the loss of phosphate, and no ATP is required. Whether pyruvate carboxylase or PEP carboxylase is used, the end result in both cases is that oxaloacetate is generated from a three-carbon metabolite derived from carbohydrate by the glycolytic pathway.

IV. INTRACELLULAR LOCALIZATION OF THE ENZYMES OF THE CYCLE

All of the enzymes of the citrate cycle are found on the inner mitochondrial membrane or in the matrix space inside the membrane. The enzymes of glycolysis are in the cytosol, so pyruvate is produced in the cytosol and must enter the mitochondrion before it can be further oxidized. The entire oxidative phosphorylation system, which converts ADP to ATP at the expense of the energy drop as electrons are transferred from NADH and $FADH_2$ to molecular oxygen, is located on the mitochondrial inner membrane. This location in the interior of the same organelle of both the oxidative phosphorylation system, which is powered by electrons, and the citrate cycle, which supplies most of the electrons (in the form of NADH and $FADH_2$), presumably enhances the efficiency of coupling between the two systems. In addition, the enzymes that convert fatty acids to acetyl coenzyme A are also located in the mitochondrion. Long-chain fatty acids enter the mitochondrion as esters of carnitine, a metabolite that is apparently specialized to participate in the transfer of fatty acids across the inner mitochondrial membrane. Thus, both electrons released in the conversion of long-chain fatty acids to acetyl coenzyme A and those released in the oxidation of acetyl coenzyme A to CO_2 are transferred to NAD^+ or FAD in the immediate vicinity of the system that uses the reduced carriers in the generation of ATP.

V. THE GLYOXYLATE CYCLE

Mammals convert carbohydrates to fat or protein (however, they can make only about half of the amino acids that are needed) and protein to fat or carbohydrate. As we have seen, the reactions of the citrate cycle play important roles in all of these interconversions. Mammals cannot convert fats to carbohydrates or protein, however, because the reactions of the cycle provide no route from acetyl coenzyme A, the product of the first stage of fatty-acid degradation, to anything except carbon dioxide.

The glyoxylate cycle (Fig. 3) is a modification of the citrate cycle that allows acetyl coenzyme A to supply carbon atoms for the production of all of the intermediates of the cycle and, hence, for all of the biosynthetic activities of a cell. Although it does not exist in humans and other mammals, the glyoxylate cycle deserves mention in this treatment because of its great importance in the biosphere generally and its interest as an example of the fact that small changes can allow the same enzymes to participate in sequences that have different functions.

The two enzymes specific to the glyoxylate cycle are isocitrate lyase and malate synthesis (Fig. 3):

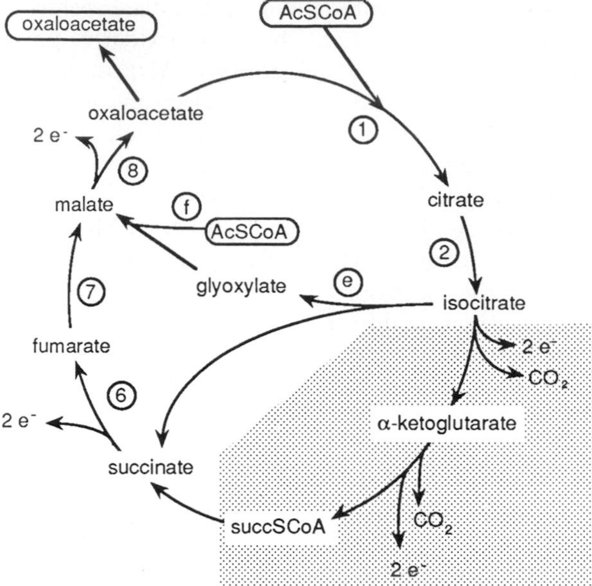

FIGURE 3 The glyoxylate cycle. The decarboxylation reactions of the citrate cycle (stippled area) are bypassed, and a second molecule of acetyl coenzyme A enters at reaction f. The resulting cycle converts two molecules of acetyl coenzyme A to one molecule of oxaloacetate.

ⓔ isocitrate lyase:
isocitrate^{3-} $\longrightarrow$ glyoxylate^{-} + succinate^{2-}

ⓕ malate synthase:
glyoxylate^{-} + AcSCoA + H_2O $\longrightarrow$ malate^{2-} + H^+ + HSCoA

The structure of glyoxylate is O=CH—COO$^-$.

When reactions e and f are combined with reactions 1, 2, and 6–8 of Fig. 1, the overall conversion represented by Fig. 3 is given by reaction 12:

$$\begin{aligned}&2\text{AcSCoA} + \text{FAD} + 2\text{NAD}^+ + 3\text{H}_2\text{O}\\&\quad\rightarrow \text{oxaloacetate}^{2-} + \text{FADH}_2 + 2\text{NADH} \qquad (12)\\&\qquad + 2\text{HSCoA} + 4\text{H}^+.\end{aligned}$$

The citrate cycle oxidizes acetyl coenzyme A to CO_2 and thus serves as the final stage in the oxidation of many metabolites and cell constituents. It seems strange that five of the eight enzymes of the citrate cycle can participate in a metabolic sequence with a chemical direction and physiological function opposite those of the citrate cycle: catalyzing the first stage in the use of acetyl coenzyme A as the starting material for synthesis of all of the constituents of the cell. Figure 3 shows that this reversal of function is achieved by the substitution of reactions e and f for reactions 3–5. Both of the reactions in which CO_2 is released are bypassed (the stippled area of Fig. 3), and a second molecule of acetyl coenzyme A enters at reaction f. Each turn of the glyoxylate cycles starts with one oxaloacetate and generates two, resulting in a net gain. Reaction of the product oxaloacetate with a third molecule of acetyl coenzyme A allows synthesis of citrate and thus of all other cycle intermediates from acetyl coenzyme A. Reactions 3 and 4 are still available, so that a cell that is using acetyl coenyzme A as its sole source of carbon and energy can employ part of its isocitrate in making the α-ketoglutarate and succinyl coenzyme A that it requires for its biosynthetic activities.

Phosphoenolpyruvate, which can be made from oxaloacetate by the action of PEP carboxykinase (Fig. 2), can be converted to glucose and hence to storage glycogen and all of the structural carbohydrates that are required in cell walls and other cell components.

VI. REGULATION

Because the reactions of the citrate cycle have oppositely directed functions of participating in all oxidative degradations and of providing materials for most biosynthetic sequences, regulation of the cycle and its component and associated reactions is of central importance to the economy of the cell. The cycle is related to or affects nearly all of the activities of an organism, so it is not surprising that it should be regulated by many factors that do not impinge on it directly. A complete description of regulatory influences on the cycle would include most of the regulatory and correlative interactions of the cell or organism. We are far from having the information necessary for such a complete description, and only a sampling of what is known can be included here.

Acetyl coenzyme A is the primary fuel of the citrate cycle. Therefore, the flux of material through the cycle must be determined in part by factors that control the rate of generation of acetyl coenzymes A. The metabolism of both carbohydrates and storage fats is regulated by complex systems that respond to many inputs, among them the energy status of the cell, which can be expressed as the adenylate energy charge or the ATP/ADP ratio. As a result of those interactions, the rate at which acetyl coenzyme A is made available by catabolic sequences increases when the energy status of the cell decreases slightly. Thus, the supply of fuel to the cycle increases when an increased rate of ATP production is desirable. Those sequences also react appropriately to organismwide needs by responding to hormonal signals.

At the level of the citrate cycle itself, citrate synthase (reaction 1 of Fig. 1) is regulated by the adenylate system (ATP, ADP, and AMP) in some or all organisms, and by the degree of reduction of the NAD^+/NADH electron-carrier system in some species. The direction of the responses is again such that a greater rate of entry of acetyl coenzyme A into the cycle is favored when the energy status of the cell is slightly below its normal value.

Isocitrate dehydrogenase (reaction 3 of Fig. 1) is sensitively regulated by the adenylate energy-transducing system (ATP, ADP, and AMP). When the energy status is low, this enzyme has high affinity for its substrate isocitrate. Thus, nearly all of the isocitrate that is produced by reactions 1 and 2 will proceed through reaction 3, maximizing the flow through the remainder of the cycle and the rate of donation of electrons to the oxidative phosphorylation system to support regeneration of ATP. When the energy charge is high, the affinity of isocitrate dehydrogenase for its substrate is much reduced. Thus, isocitrate must accumulate to a higher concentration before it can be converted to α-ketoglutarate by action of the dehydrogenase. When the concentration of isocitrate rises, so will that of citrate, with which it is in near-equilibrium. The resulting high level of citrate favors its movement from the mitochondrion to the cytosol, where it is cleaved to supply the acetyl enzyme A that is required for synthesis of storage fats, steroids, and other cell constituents. Citrate is also a strong negative modifier or inhibitor of phosphofructokinase, the most important regulatory enzyme in glycolysis.

To summarize, it appears that if the energy status is poor, nearly all of the isocitrate is channeled along the citrate cycle. The rate of ATP regeneration is maximized. Because of the low concentration of citrate, little of it moves into the cytosol. The rates of biosyntheses that use acetyl coenzyme A are limited by lack of substrate (there are also direct effects of the energy charge on the enzymes of these sequences). This decrease in the rates of syntheses conserves energy (ATP). Because of the low concentration of citrate in the cytosol, the activity of phosphofructokinase is not diminished by citrate binding, and a rapid rate of glycolysis, which supplies acetyl coenzyme A to the citrate cycle, is favored. The result is an increase in the rate of regeneration of ATP.

In contrast, when the energy charge is high, the accumulation of isocitrate and citrate leads to movement of citrate into the cytosol. Under those conditions, energy is available for biosyntheses, and the increased supply of cytosolic acetyl coenzyme A appropriately favors increased rates of synthesis. If the rate of supply of citrate to the cytosol exceeds the level required to meet the cell's synthetic needs, citrate tends to accumulate and inhibit phosphofructokinase, thus decreasing the rate at which glycolysis supplies acetyl coenzyme A to the citrate cycle. Thus, the rate of regeneration of ATP and of citrate itself decreases. The interaction of these and many other related control systems maintains the energy charge, the ATP–ADP ratio, and the concentrations of cell constituents within fairly narrow physiological ranges.

It can be seen from Fig. 1 that pyruvate is an important metabolic branchpoint. If all of the pyruvate that is supplied by glycolysis is oxidatively decarboxylated to form acetyl coenzyme A, the cycle will furnish the maximal number of electrons to the ATP-regenerating machinery, but it will not be able to supply any starting materials for biosynthesis. That condition is approximated closely in a working muscle.

However, under most conditions the cycle must also supply some of its intermediates to biosynthetic sequences, and it is necessary that enough pyruvate be carboxylated to oxaloacetate to exactly match those losses. The responses of pyruvate dehydrogenase and pyruvate carboxylase (reactions a and b in Fig. 1) to variations in the concentration of acetyl coenzyme A are major factors in the partitioning of pyruvate between the two reactions. When intermediates are being withdrawn from the cycle for use in biosynthesis, the concentration of oxaloacetate will decrease. This will limit the rate of reaction 1 (the formation of citrate). Because it is being used more slowly, the concentration of acetyl coenzyme A will rise. Acetyl coenzyme A is a strong negative modifier for pyruvate dehydrogenase (reaction a) and a strong positive modifier for pyruvate carboxylase (reaction b); therefore, when the concentration of acetyl coenzyme A rises slightly, the partitioning of pyruvate changes: more is directed toward oxaloacetate and less toward acetyl coenyzme A. This response will automatically hold the concentration of oxaloacetate at a level that is adequate for citrate synthase, which catalyzes reaction 1. By supplying oxaloacetate at the rate needed to stabilize its concentration, the system is also automatically adjusting the rate of input of oxaloacetate to exactly match the sum of the rates of removal of cycle intermediates.

Most metabolic correlation and control is exerted by adjustment of partitioning ratios at branchpoints. The pyruvate branchpoint involves partitioning only between two alternate routes of entry into the citrate cycle, but it is nevertheless the most centrally important of the metabolic control points in metabolism, because it deals with the fundamental distinction between degradation and biosynthesis. Use of cycle intermediates for biosyntheses, including syntheses of storage carbohydrates and fat, is possible only to the extent that pyruvate is directed toward oxaloacetate, whereas energy for synthesis and other cell activities results in very large part from oxidation of the pyruvate that is directed toward acetyl coenzyme A. Constant adjustment of the balance between the rates of these two reactions that compete for pyruvate allows the citrate cycle to meet the shifting momentary demands of the cell for energy and for biosynthetic starting materials, and to support production of storage glycogen and fat when resources are available in excess of present demand.

BIBLIOGRAPHY

Garrett, R. H., and Grisham, C. M. (1995). *In* "Biochemistry," pp. 598–626. Saunders/Harcourt Brace, New York.

Kornberg, H. L. (1987). Tricarboxylic acid cycles. *BioEssays,* 7, 236–328.

Mathews, C. K., and van Holde, K. E. (1990). *In* "Biochemistry," pp. 467–503. Benjamin/Cummings, Redwood City, CA.

Voet, D., and Voet, J. G. (1990). *In* "Biochemistry," pp. 506–527. Wiley, New York.

Cocaine and Stimulant Addiction

TONG H. LEE
EVERETT H. ELLINWOOD
Duke University Medical Center

JANET L. NEISEWANDER
Arizona State University

GLOSSARY

Anhedonia Loss of ability to obtain pleasure from hobbies and activities

Binge Pattern of stimulant abuse characterized by repetitious administration of high doses so as to sustain stimulant euphoria

Cocainisinus Turn-of-the-century term for the cocaine abuse–addiction syndrome leading to subsequent excessive abnormal behavior

Dysphoria Extreme, uncomfortable, disagreeable feeling

Freebase Free form of cocaine that is produced from cocaine hydrochloride by chemically removing the hydrochloride ions. Unlike the hydrochloride form, the freebase is heat stable and can be smoked

Hypersomnia Intense and prolonged sleep characterized by its excessive duration, often with multiple intense dreaming episodes

Psychomotor Pertaining to combined physical and mental activity, or to movement that is psychically determined

Reinforcement Term used in the learning theories to designate a stimulus, drug, or situation that brings on an increase in incidence of a behavior (e.g., milk that is used to train a hungry rat to bar-press)

Tachyphylaxis Rapid appearance of decreased response to the drug in the body as a result of prior exposure to the drug

STIMULANTS, IN THEIR BROADEST DEFINITION, ARE agents that excite or stimulate various systems in the body. Thus, central stimulants are classified as such for their common ability to stimulate functions mediated by the central nervous system. According to this broad definition, central stimulants may include convulsants, which produce convulsions by excessive stimulation of the brain; however, the term central stimulants is commonly used in a more restrictive sense for those agents that produce alertness, elevated mood and interest, decreased appetite, and increased motor and speech activity. Defined in this fashion, central stimulants include, among others, cocaine, a spectrum of amphetamine-like drugs, caffeine (as well as a group of xanthine drugs sharing a similar chemical structure with caffeine), and some nonamphetamine-type diet pills. Central stimulants have legitimate medical uses; on the other hand, the current cocaine "epidemic" and consequent declaration of a "war on drugs" in the United States have focused on abuse problems associated with certain central stimulants presently in the limelight. The scope of this article is primarily to review selective central stimulants, namely, cocaine and amphetamine, which have specific medical indications but also have been repeatedly abused in history. Because of the limited scope of this article the term stimulant will denote those two drugs; when the term is used in a more general sense, the usage will be clear from the context.

ENCYCLOPEDIA OF HUMAN BIOLOGY, Second Edition, VOLUME 2. Copyright © 1997 by Academic Press. All rights of reproduction in any form reserved.

I. HISTORY

Uses of plants for their central nervous system-stimulating effects date back to prehistoric time. The earliest written record of psychomotor stimulant use is ascribed to the Chinese emperor Shen Nung (c. 3100 BC), who described the medicinal use of the herb Ma Huang, a plant containing the central stimulant ephedrine.

Cocaine also has historical precedents: the native coca plant (*Erythroxylon coca*) has been cultivated in South America since prehistoric time. Archaeological evidence suggests that its leaves were chewed for mental and physical energy, religious or sacramental reasons, and even nutritive sustenance throughout western South America, especially in the high Andes. Bags containing coca leaves and flowers have often been unearthed in burial sites dating back to c. 2500 BC. In the Incan empire, the plant was considered a gift from the gods and served important social and religious functions. The Temples of the Sun were adorned by solid gold models of coca sprigs, and their altars could be approached only by the elite with coca in their mouths. In addition, use of coca was restricted to the aristocracy and other personages designated by royalty. Among those privileged were priests, doctors, and the fabled long-distance runners who relayed messages along the well-developed Incan road system.

Following the Spanish conquest of the Incan empire, use of coca was initially prohibited by the conquistadores as satanic idolatry. However, soon recognizing that the stimulant effect of coca enabled the enslaved natives to endure forced labor, the Spanish reinstated its use; workers were both paid and taxed in coca.

Despite favorable reports from South America about the effects of coca leaves, they were not widely used in Europe until the nineteenth century. Modern investigators have speculated that this lack of popularity was primarily due to the unavailability of fresh coca leaves. Delays during long voyages from South America led to decay of coca leaves, rendering them useless; in addition, the European climate was not conducive to local production of leaves. As expected, isolation of a stable, active compound (cocaine) from coca leaves by Albert Niemann (in 1855) was followed by widespread use of the stimulant.

Perhaps the scientist who did more than any other to elucidate behavioral effects of cocaine was Sigmund Freud. He published several reports in which, from experimentation on himself, he correctly identified cocaine as both a central nervous system stimulant and an euphoriant. At the same time, he recommended its use as an antifatigue formula or aphrodisiac, and for treatment of a wide range of disorders, including asthma, digestive disorders, and alcohol and morphine addictions.

The late nineteenth century can be considered the heyday for cocaine. Positive opinions within the scientific community abounded, and cocaine or cocaine-containing formulations (many of which were patented medicines) were indiscriminately prescribed; sporadic reports of adverse reaction to these products were generally ignored as atypical reaction. Freud himself considered the toxic reactions as a manifestation of character defect. In the public, the high demand for cocaine was captured by entrepreneurs in the form of concoctions. In both Europe and the United States, a tonic mixture of coca extract and wine was used and indeed endorsed by the elite, including United States President William McKinley, Thomas Edison, August Rodin, and Pope Leo XVIII. In 1896, the need for a nonalcoholic substitute for this preparation (alcohol prohibitionist sentiment was high during this period) was exploited by Atlanta entrepreneur John Pemberton, who introduced such a preparation containing extract of coca leaves and caffeine-rich African kola nuts; Coca Cola® was advertised as the "intellectual beverage and temperance drink."

The early twentieth century marked a period in which society finally began to take note of the ever-increasing number of reports on cocaine toxicity. In response to a rise in public concern, the United States federal government began to regulate the manufacture and sale of cocaine-containing preparations. The Pure Food and Drug Act of 1906 required listing cocaine on the labels of all cocaine-containing patent medicines. The Harrison Narcotic Act in 1914 put a further restriction on coca products by forbidding the use of cocaine in proprietary medicines and requiring registration of those involved in coca product trade. With the advent of these restrictive measures, use of cocaine "went underground." Except for a brief resurgence in the 1920s, its use was limited to Bohemian, art, and music subcultures until the early 1970s, when the current epidemic began.

Amphetamine was first synthesized in 1887, and its pharmacology was first studied in 1910. However, because the experiments were performed in anesthetized animals, its central stimulant properties were not discovered until it was independently resynthesized in 1927 as part of a search for synthetic substitutes for ephedrine in the treatment of asthma. In the 1930s, the drug began to be used in nasal inhalers and for treatment of narcolepsy in tablet form; the central stimulant (e.g., euphorogenic) effects of amphetamine became apparent in patients being treated with these preparations. Not surprisingly, the latter discovery was followed by increasing nonmedical uses by the public; it was as if the public had substituted the cheaper, more widely available amphetamine for cocaine, which was by now expensive and tightly regulated. The general public as well as the medical community had forgotten about the previous cocaine epidemic, and major amphetamine epidemics occurred in Japan (1950–1956), Sweden (1964–1968), and the United States (1965–1969) during the following 30-year period.

II. PHARMACOLOGY

A. Catecholamine Effects

Communication between neuronal cells in the brain is mediated by chemicals referred to as neurotransmitters. At the junction between two neurons, there is a small gap referred to as the synaptic cleft. The neuron sending a message releases neurotransmitter into the synaptic cleft. The neurotransmitter diffuses across the cleft and activates receptor sites on the neuron receiving the message. The actions of the neurotransmitters are terminated by, among others, their degradation or reuptake back into the originating neurons for reutilization. Amphetamine and cocaine share the property of potentiating transmission mediated by a particular class of chemicals, namely, catecholamines. Catecholamines, so named because they are amines containing the catechol moiety in their structures, act as mediators of a wide range of neural functions. The two stimulants potentiate responses mediated by catecholamines by increasing their extracellular concentrations; however, the exact mechanisms of action are different. Amphetamine directly causes increased release of catecholamines from nerve terminals, whereas cocaine acts by blocking reuptake of these neurotransmitters. Because reuptake plays a major role in terminating the actions of catecholamines, its blockade leads to results similar to those produced by a direct releaser.

Catecholamines have a wide variety of critical functions in the body. One set of catecholamines, norepinephrine and epinephrine, mediate actions of the sympathetic nervous system, which plays a major role in preparing the body for coping with stress and emergency. Activation of sympathetic activity leads to increased heart rate, dilation of the airways, and diversion of blood from the digestive tract to the muscles. The sympathomimetic effects of the stimulants (i.e., mimicking the action of the sympathetic pathway) are thought to be responsible for many of the complications produced by stimulants.

On the other hand, another catecholamine, dopamine, is involved with mediating incentive motivation (see the following) and modulating a variety of motor and cognitive functions by the brain. Dopamine mechanisms are most often associated with reinforcing effects in incentive-type behaviors. Innate behaviors in lower animals are represented by exploratory stalking, hunting, and foraging behaviors, which appear very early in the young animal's development; these innate behaviors are energized—motivated during this developmental phase without any consumatory reinforcement (e.g., food, specific objects). After interactions with the environment over time, these behaviors become integrated into cascades of object-related behavior. Increasingly, they are integrated during learning into automatic sequences of behavior with attendant emotional responses. High-dose stimulant intoxication in lower animals provokes these exploratory intrinsic behaviors into long trains of stereotyped repetitions of a behavior without object relatedness, which can be triggered by cues associated with injection (e.g., saline injection). If the experimental animal is bar-pressing for small injections of stimulant into a vein or the brain, the drug-administering behavior usually becomes compulsively stereotyped. A drug-associated cue can trigger compulsive drug administration (e.g., monkeys self-administering cocaine when given free access will compulsively dose themselves to death). [*See* Catecholamines and Behavior.]

Cocaine and amphetamine stimulants at low doses induce pleasurable reinforcement, and at high doses intense orgiastic reinforcement. These reinforcement effects have been pinpointed to be secondary to the release of dopamine from terminal areas of very specific dopamine tracts. A variety of studies, including (1) electrical stimulation of these tracts, (2) infusion of dopamine agonists into the terminal area, and (3) blockade of dopamine agonist reinforcement with antagonists, have validated this basis of stimulant-induced reinforcement.

In the early stages of human stimulant use, many everyday behaviors become pleasantly reinforced within an environmental condition. Later, the environmental repertoire becomes increasingly constricted to the stereotyped incentive patterns often focused on the procurement (foraging) and use of the stimulant. Other human stimulant-induced innate patterns of behavior successively evolve from initial exploratory behaviors associated with curiosity to intense stereotyped suspicious behaviors, which then evolve into paranoid stimulant psychoses. The means by which stimulants act on dopamine incentive behavioral systems is currently under intensive scientific study.

B. Local Anesthetic Effect

Local anesthetics block nerve conduction when applied in one area, and although they affect all types of nerves, their clinical use is to block nerves carrying information from pain receptors (e.g., their use by dentists to numb pain). Unlike other stimulants, cocaine also acts as a local anesthetic; in fact, it is the first of its kind to be discovered. Clinical application of cocaine as a local anesthetic was initiated in the late nineteenth century by a Viennese physician who discovered its utility in eye surgeries. However, this practice in ophthalmological surgeries has been abandoned because of its toxic effect on the eye. Current application is limited to topical application in the upper respiratory tract (e.g., nasal mucous membrane, Section II,D).

C. Behavioral Stimulant Effects in Human

Despite their structural and other differences, cocaine and amphetamine are similar in many of their behavioral effects in humans; some experienced abusers may have difficulty distinguishing between the two drugs. Stimulants are among the most powerful euphorogenic drugs known. At even moderate doses, they produce alertness and a sense of well-being and heighten self-esteem as well as increased pleasure derived from various activities. At high doses taken by a rapid route of administration (e.g., "crack" smoking), the euphoria produced by stimulants is often described as consisting of two stages: an intense, "orgiastic-like," short-lasting euphoria (e.g., cocaine "rush"), followed by a more sustained sensation of well-being ("high").

The main difference between cocaine and amphetamine appears to be the duration of their effects. This length of time depends on various factors, including the specific drug's half-life and development of tachyphylaxis and tolerance (see the following). Practically defined, half-life is the time for drug concentration in the blood to decline to one-half of its initial value. Everything else being equal, a stimulant with a longer plasma half-life lasts longer in the body and, consequently, produces longer-lasting euphoria. Thus, based on cocaine's rapid metabolism and plasma half-life of 90 min or less, it can reasonably be predicted that the drug's effects will dissipate much faster than those of amphetamine, with an average half-life of 6–8 hr.

Decreased sensitivity to a drug acquired as a result of prior exposure to the drug is termed drug tolerance; usually, it develops following multiple doses of the drug over a period of time (e.g., days). When it develops more rapidly (e.g., sometimes with the second dosing), it is called tachyphylaxis. Another term, acute tolerance, was also used to describe a similar phenomenon of rapid decrease in sensitivity. Classically, acute tolerance refers to a development of decreased sensitivity while the drug is still present in the body (i.e., the effect of a drug subsides faster than would be predicted by the blood level of the drug). Tachyphylaxis, on the other hand, describes a decreased sensitivity to a *subsequent* dose of a drug (i.e., the same amount of the drug as on the previous occasion produces a lesser effect). Over the years, the distinction between the two terms has become blurred, probably because it is not clear whether or not the two represent *mechanistically* different phenomena.

Much evidence, both clinical and basic, suggests that the central stimulant effects of cocaine and amphetamine are prone to a development of tolerance and tachyphylaxis (acute tolerance). Thus, decreasing sensitivity to cocaine is usually manifested by the user taking multiple, increasing doses over a short period of time (to maintain the cocaine "high"). A similar phenomenon is also observed in amphetamine users (i.e., tachyphylaxis in its classic definition). In addition, because of the longer plasma half-life of amphetamine, a faster decrease in the amphetamine can be readily observed while its blood level falls over the next few hours (acute tolerance). Further reduction in stimulant-induced euphoria over days or weeks (tolerance) is also noted in both legitimate (e.g., treatment of narcolepsy) and illicit uses. It has recently been suggested that incentive motivational processes underlying drug-seeking behavior (i.e., wanting the drug) and drug-induced euphoria (i.e., liking the drug) may be mediated by separate neural systems. Furthermore, incentive motivational processes are not necessarily under conscious control and may develop sensitization, or reverse tolerance, resulting in intense drug craving and compulsive drug use.

D. Medical Application

In medical practice, amphetamines (along with related compounds methamphetamine and methylphenidate) are used for their alertness-producing and appetite-suppressant effects. Clinical practice of utilizing stimulants for sympathomimetic effects (e.g., for airway dilation in asthmatic patients) has been abandoned because of the high-abuse potential and the availability of safer drugs. Stimulants are, at present, prescribed for narcolepsy, which is characterized by a sudden uncontrollable disposition to sleep, and for short-term treatment of obesity, which has been unresponsive to alternative forms of therapy (e.g., repeated dieting and/or other drugs). Interestingly, another major use of these agents is to calm down children with attention-deficit disorder (also called hyperactive child syndrome). The disorder is characterized by impulsive behaviors, low frustration tolerance, distractibility, hyperactivity, and memory difficulties—a set of problems that may be expected to be exacerbated by stimulants at higher doses. It is emphasized that the potential for significant side effects or abuse is high even when stimulants are used under medical supervision; consequently, physicians are warned against imprudent use of stimulants.

Cocaine, which shares the common properties of stimulants, is not used in the foregoing fashion because of, among various factors, its short half-life. The requirement for frequent dosage for the desired therapeutic effects greatly enhances toxic and abuse liability. Currently, cocaine is used as a topical application as a local anesthetic, mostly in the upper respiratory tract. For example, emergency physicians can directly examine a patient with nasal trauma by applying cocaine to the mucous membrane in the nose. The drug reduces pain as well as bleeding (cocaine's vasoconstrictive property constricts capillaries, allowing less blood to flow through the affected area), thus greatly enhancing visualization and treatment of the traumatized area.

III. EPIDEMIOLOGY OF STIMULANT ABUSE

The United States is currently experiencing a cocaine epidemic. In the 1980s, millions of people tried cocaine for the first time, as the positive effects of its use (e.g., alertness, euphoria, increased mental and physical energy) were espoused, indeed proselytized. In 1986, the National Institute on Drug Abuse estimated that more than 3 million people in the United States abused cocaine regularly, more than five times the number of heroin addicts. The spread of cocaine use was associated with a precipitous 15-fold increase in cocaine-related emergency room visits over the preceding 10-year period. In addition to a general upswing in cocaine use, increasing utilization of intravenous injection or freebase ("crack" or "rock") smoking has become a major health policy issue because of the higher propensity of these forms to induce addiction as well as more severe toxicity. The advent of the highly purified form of cocaine freebase known as crack is especially alarming because (1) the form is inexpensive and widely available, (2) it is simple to use, and (3) its rapid absorption and subsequent high blood level is associated with an increased propensity toward compulsive use. Report of a dramatic rise in first-time hospitalization for cocaine abuse (224 in 1984 compared with just 1 in 1980) shortly following the introduction of crack in the Bahamas illustrates the increased risk.

Past experience indicates that several sequential events occur during the establishment of a stimulant epidemic. These factors for amphetamine epidemics could well apply to the 1980s cocaine epidemic.

1. Introduction of the stimulant to the population for recreational purposes, for medical purposes, or for its antifatigue properties.
2. Widespread dissemination of knowledge and, at times, proselytizing of the intensely euphoric stimulant experience.
3. Development of a sufficient core of buying and selling abusers who establish a reliable illegal market for the stimulant.
4. The illegal supply routes providing for immense profits.
5. The proliferation of multiple illegal drug sources compensating for government curbs.
6. Increasing use of rapid onset routes of administration associated with an intensified stimulant effect (e.g., intravenous injection).

One sociological factor aiding in the successful establishment of a stimulant epidemic appears to be the prevailing misguided positive opinions among the general public and scientific community about the particular drug(s). These opinions are often enhanced by positive portrayals from the media, folk heroes, and entertainers, undaunted by lessons of historical experience. For example, as discussed earlier, the cocaine epidemic of the late nineteenth century occurred in the context of highly positive opinions about the drug, in both the general and scientific communities. In addition to being "unequaled as a tonic-stimulant for fatigued or overworked body and brain" (taken from an advertisement for a nostrum containing cocaine), it was espoused as a cure for a variety of ailments, including opium addiction, alcoholism, asthma, tuberculosis, impotence, and digestive disturbances. During this time, reports of adverse reactions to the drug were largely ignored. The subsequent stimulant epidemics of the 1920s, 1950s, and 1960s were again characterized by initial enthusiasm for the virtues of the stimulants untempered by historical perspective.

With respect to the most recent cocaine epidemic, well-documented medical literature from the early twentieth century on adverse effects of cocaine abuse (i.e., cocainisinus) was later dismissed in the 1970s and 1980s as Victorian moralizing. Even recent medical literature was dismissed by the media, which heralded cocaine as the champagne of drugs used by cultural heroes and the elite, and described it as quite safe. For example:

> Cocaine is less harmful than any legal and illegal drugs popular in America. Most of the evidence is that there aren't adverse effects to normal cocaine use. It looks to be much safer than barbiturates and amphetamine and there is no evidence that it has the body effects of cigarettes or alcohol. If I were going out to sell a drug to the public it would probably be cocaine. We might be better off using it as a recreational drug than marihuana. (*San Francisco Chronicle*, Oct. 21, 1976, p. 4)

According to historical experience, the current cocaine epidemic may well abate within the next few years. Society will gradually become wary of the negative aspects of the drug (e.g., its toxicity, the increasingly criminal and violent drug subculture) and will increasingly exert specific legal and medical countermeasures. The restrictive Harrison Narcotic Act in 1914 in the United States was passed largely in response to increasing newspaper publicity about the problems associated with cocaine abuse in the late nineteenth and early twentieth centuries (e.g., aggressive, sex-crazed cocaine "dope fiends"). With respect to the current epidemic, first-time and casual cocaine use has indeed reached a peak and has begun to decline as a result of an intensive campaign against cocaine. Unfortunately, history also warns us that in the future people may again forget about the negative impact of the "cocaine epidemic of the 1980s." Indeed, the abuse of smokable methamphetamine, "ice," is on the rise and may develop into another major stimulant epidemic. Is there a way to prevent this periodic reenactment of this history?

IV. CLINICAL CHARACTERISTICS OF ABUSE

A. Initial, Low-Dose Use

Typically, individuals are initially exposed to single low to moderate doses of stimulants for therapeutic, recreational, and other purposes. The initial recreational form of cocaine use usually takes the form of snorting fine cocaine crystals into the nasal passages, where it is absorbed into the bloodstream by the mucous membrane lining. Taken through this route, cocaine retards the rate of its own absorption because of its powerful vasoconstrictive effect (narrowing of blood vessels). Swallowing pills (e.g., the usual route of administering amphetamine diet pills) is also a relatively slow absorption process. Consequences of delayed absorption are a relatively slow onset of behavioral responses and decreased peak drug concentrations in the blood. The latter results because the body's ability to break down the drug can better keep up with the slow absorption. During the initial low-dose stimulant use, the main factor maintaining the drug-taking behavior appears to be positive responses from others to the user's increased energy and productivity, rather than the stimulant-induced (pharmacological) euphoria per se, which is reduced in its intensity anyway owing to the lower-peak concentrations.

Most stimulant users do not become compulsive stimulant addicts, whose main daily concern is attainment and maintenance of a stimulant "high." The National Institute on Drug Abuse has estimated that 80% of 30 million Americans who have tried cocaine intranasally do not progress to regular users and, thus, either stop or remain periodic recreational users. Predisposing factors to continuing use and then to compulsive stimulant abuse in the remaining 20% are not well known, although desires to reexperience the stimulant high and to avoid the subsequent dysphoria and stimulant craving appear to play a role.

B. Compulsive Abuse

Some low-dose users discover that certain routes of administration intensify euphoria. Thus, they discover that intravenous use or crack smoking can produce extreme pleasure. The individual's daily activities soon become devoted to a search for pharmacological euphoria. Reaction from others, which plays a major role in maintaining the low-dose use, no longer matters, and the individual becomes more socially isolated.

In its most severe form, compulsive stimulant abuse is characterized by binges, in which high doses are repetitively administered in an attempt to "chase" the stimulant high. Binging episodes lasting up to a few days are usually terminated by extreme physical exhaustion and/or exhaustion of the drug supply. The binges may occur one or more times per week and, depending on the available supply, the abuser may consume 10–20 g or more. When money is readily available, as much as $250,000/year may be spent on cocaine.

The propensity of stimulants to produce a compulsive abuse pattern appears to be due to at least three factors: (1) their powerful reinforcing property; (2) the short half-lives of stimulants, which is especially true for cocaine (<90 min); and (3) development of tolerance–tachyphylaxis to the pharmacological euphoria and/or development of sensitization of incentive motivational processes. In addition, high doses and rapid administration routes increase the intensity, leading to a high-dose transition to abuse. [*See* Nonnarcotic Drug Use and Abuse.]

C. Withdrawal

The "crash" phase, the initial phase of stimulant withdrawal, immediately follows a binging episode. Initially marked depressive dysphoria, anxiety, and agitation are noted followed by craving for sleep over the next few hours. Often, the individual uses a wide variety of sedative-anxiolytic drugs such as alcohol to overcome the early agitated state and to induce sleep. Prolonged hypersomnia, often lasting 24–36 or more hr is not unusual during this phase. Notably, addicts report minimal desire for the abused drug during this immediate phase of withdrawal.

As the individual recovers from the crash phase, a period of anhedonia, dysphoria, and decreased mental and physical energy ensues (intermediate withdrawal phase). This phase can last from several days to weeks. Emerging from this state of ennui, or mood and energy dysfunction, a stimulant craving returns, frequently initiating recidivism. Unfortunately, this craving is exquisitely sensitive to various environmental stimuli (often learned during binging episodes through their association with the stimulant high during the binging episodes); for example, individuals in this withdrawal phase can experience a sudden onset of stimulant craving by merely returning to a place that is associated with drug use (e.g., a hotel room used for purchasing crack). With continued drug availability, it is not unusual to observe repetitious cycles of binging with intervening crash and intermediate withdrawal phases over a period of months or even years with a devastating result.

With more sustained abstinence through the intermediate withdrawal phase (e.g., with help from a treatment program), a more "natural" baseline state returns (long-term withdrawal phase). However, although decreased in frequency and intensity, urges to return to stimulant use can recur after months or years of abstinence, most frequently triggered, again, by environmental stimuli. Moreover, a single "taste" of stimulant can induce a full set of behavioral responses, which may have originally taken weeks or months of chronic stimulant use to evolve (termed grease slide return). Thus, the individual may become psychotic (usually a sequela of chronic binging episodes) within minutes to hours of return to stimulant use. Changes observed during the long-term withdrawal phase suggest that chronic stimulant abuse may produce long-lasting (permanent?) residual changes in the brain, a hypothesis that is being actively tested.

V. COCAINE-RELATED MORBIDITY AND MORTALITY

A. Medical Complications

The cocaine-related deaths of famous athletes and entertainers, such as those of Len Bias and John Belushi, have focused attention on several types of stimulant-related toxicity. These examples demonstrate that stimulant-related complications occur even in healthy young individuals and directly contradict various claims of relative safety. Important medical and psychiatric complications that are ascribable to stimulant abuse have been observed for almost a century.

Many of the important stimulants' toxicity is mediated by their actions on the cardiovascular system. Reported cardiovascular toxic pathology associated with stimulant use include spasm of blood vessels (leading to decreased oxygen supply); hypertensive crisis, in which the person's blood pressure suddenly rises to very high levels, causing rupture of large arteries and multiple hemorrhaging of smaller arteries and arterioles; heart attack (acute myocardial infarction); irregular heart rhythms (arrhythmias); inflammation of the heart muscle cells (myocarditis); rupture of large arteries; and multiple hemorrhaging of smaller arteries and of arterioles. Some of these are fatal complications, occurring very rapidly, just minutes after use, precluding treatment; others, if not properly treated, can also lead to death. Importantly, these complications can occur in individuals with no known prior history of medical problems.

Many of these cardiovascular complications have been ascribed to sympathomimetic effects of stimulants (i.e., mimicking the actions of the sympathetic nervous system by enhancing their function; see earlier). For example, because increased heart rate and blood pressure generally increase the heart's oxygen consumption, such changes following stimulants may overwhelm the normal oxygen supply. This oxygen insufficiency, in turn, may ultimately lead to death of heart muscle cells (heart attack). Among other mechanisms that may contribute to cardiac toxicity, the apparent higher propensity of cocaine to produce cardiac arrhythmia (compared with amphetamine or other related compounds) may be secondary to the drug's local anesthetic effect. Local anesthetics in high doses are known to interfere with the heart's conduction of electrical activity, leading to severe arrhythmias.

Stimulant-induced hypertensive episodes have also been associated with bleeding in the brain, which leads to permanent neurological deficit or even death. Exact factors predisposing individuals to this type of complication are not well known. Not infrequently, this complication occurs in individuals with blood vessel abnormalities in the brain such as local dilation (called cerebral aneurysm) secondary to a weakened vessel wall from congenital defect, disease (e.g., chronic high blood pressure), or injury. The sudden hypertension caused by a stimulant becomes too great for the weakened vessel wall to accommodate, leading to rupture and bleeding. Unfortunately, most cases of this structural abnormality are discovered a posteriori during, for example, an autopsy.

In addition to directly causing cerebral bleeding, stimulants can indirectly produce neurological symptoms via their actions on the heart. For example, heart attack or irregular heart beat can reduce blood flow to the brain, depriving the brain of oxygen. Depending on various factors (e.g., length of deprivation or simultaneous occurrence of seizures), this hypoxia (lack of oxygen) can induce a reversible injury or death of brain cells.

Other forms of stimulant-induced toxicity include the induction of seizure and/or uncontrollable high body temperature. In contrast to conditions discussed earlier that can occur at moderately high doses, these two conditions tend to be more associated with very high doses of stimulants. One of the most difficult stimulant overdose cases to treat is the one that occurs in the drug smugglers known as "body packers," who swallow cocaine-filled condoms. One or more of these condoms may rupture, leading to extreme high blood concentrations of cocaine and subsequent induction of fatal convulsions and uncontrollable high body temperatures. Many, if not most, of these cases are fatal despite heroic treatment effort.

In addition to the well-documented cardiovascular and neurological toxicity, other types of medical complications associated with stimulant use include extensive liver or skeletal muscle damages and induction of asthma attacks by crack smoking.

B. Psychiatric Complications

There are significant (more so than for physiological effects) individual differences in psychiatric toxicity of stimulants. In addition, even a single individual's response to the drugs can vary over time; for example, previous history of stimulant abuse is a critical factor determining behavioral responses to a single dose of stimulant. Illustrative of the time dependency is the grease slide phenomenon discussed earlier.

Serious psychiatric complications of stimulant use are not common in occasional, low-dose users; on the other hand, many adverse psychiatric reactions have been described in high-dose users, especially during binging episodes. Thus, an alarming trend toward frequent association of cocaine use with serious psychiatric complications has been shown following the advent of the cheap rapid dose form of crack cocaine.

As described previously, high-dose and/or rapid routes of administration (e.g., crack smoking) are associated with intense euphoria. More dangerously, the exaggerated effect of stimulants is also manifested by impaired judgment, grandiosity, combativeness, and extreme psychomotor stimulation, leading not infrequently to accidents, atypical sexual behavior, or illegal acts.

In contrast to the euphoria associated with single stimulant doses, prolonged binging episodes can be associated with anxiety, irritability, transient panic reaction with terror, and psychosis. Generally, two types of psychotic behaviors are observed: either with or without confusion. Compared with the type without confusion, the one with confusion is more frequently characterized by higher frequency and doses of stimulant and a higher propensity toward violence and hallucinations. The psychosis without confusion tends to last longer than the other type and is characterized by delusions of persecution (false belief that one is being persecuted) and hallucinations in the background of a clear sensorium (e.g., the patient knows who he is, where he is, what the date and year is). This has generated much interest in the scientific community because it can be similar in appearance to paranoid schizophrenia—so similar that at times even experienced psychiatrists have not been able to distinguish between the two. This marked similarity in the two has led to the use of paranoid psychosis as an experimental stimulant model in animals for studying the mechanism(s) involved in human psychoses. [*See* Schizophrenic Disorders.]

VI. TREATMENT FOR COCAINE ABUSE

The primary goal of treatment for cocaine abuse is the initiation and maintenance of abstinence from the drug with subsequent development of personal strategies for relapse prevention. Because of the intense conditioning of cocaine craving that occurs with various environmental stimuli (see earlier), initial avoidance of situations and people associated with stimulant abuse is critical. Avoidance is initially maintained by hospitalization or a move out of the conditioned stimulus-rich environment for a period of time. Subsequently, behavior therapy, involving identification of such risk situations including "mine sweeping the environment for cue triggers" and training in development of specific strategies for avoiding them, is becoming a more prominent feature of treatment. A variety of techniques, including frequent contacts with the patient, peer-support groups, family or couples therapy, urine monitoring, education sessions, and individual psychotherapy, are used to further facilitate successful treatment. In addition to these modes of treatment, supplemental drug therapies are being investigated. There has been some success in treating cocaine dependence with antidepressants. Another approach under investigation is the use of dopamine agonists to alleviate withdrawal symptoms without producing euphoric effects. Similar approaches have been used to treat other types of drug dependence, including methadone maintenance for opiate abuse, benzodiazepines for alcohol detoxification, and the nicotine patch.

VII. SUMMARY

Central stimulants have been used from prehistoric time for their medicinal value as well as their ability to produce mental alertness and a sense of well-being, and to heighten energy, self-esteem, and emotions aroused by interpersonal interactions. Many of these mental effects are thought to be mediated by dopamine; other catecholamines are more responsible for some of the toxicity on the heart. Because of their powerful euphorogenic effects, the stimulants have been abused repeatedly. One specific stimulant dominates each epidemic (e.g., amphetamine in the 1960s, cocaine in the 1980s); cyclic pattern of the abuse is characterized by an initial enthusiasm about the drugs' pharmacological effects despite lessons from the past, followed by widespread abuse and an eventual decline secondary to increasing awareness of their toxicity. Recent data suggest that the current cocaine epidemic may be in its declining stage. One wonders, then, whether or not history will repeat itself in the near future with an epidemic rise in abuse of the stimulant "ice."

BIBLIOGRAPHY

Castellani, S., and Ellinwood, E. H., Jr. (1985). Cocaine: Mechanisms underlying behavioral effects. *In* "Psychopharmacology 2, Part I: Preclinical Psychopharmacology" (D. G. Grahame-Smith, ed.). Elsevier Science Publishers, New York.

Ellinwood, E. H., Jr., and Rockwell, W. J. K. (1988). Central nervous system stimulants and anorectic agents. *In* "Meyler's Side Effects of Drugs" (M. N. G. Dukes, ed.), 11th Ed. Elsevier Biomedical Publishers, New York.

Gawin, F. H., and Ellinwood, E. H., Jr. (1988). Cocaine and other stimulants: Actions, abuse and treatment. *N. Engl. J. Med.* **318,** 1173–1182.

Lago, J. A., and Kosten, T. R. (1994). Stimulant withdrawal. *Addiction* **89,** 1477–1481.

Meyer, R. E. (1992). New pharmacotherapies for cocaine dependence . . . revisited. *Arch. Gen. Psychiatry,* **49,** 900–904.

Miller, N. S., and Gold, M. S. (1994). Dissociation of "conscious desire" (craving) from relapse in alcohol and cocaine dependence. *Ann. Clin. Psychiatry* **6,** 99–106.

O'Brien, C. P., Childress, A. R., McLellan, T., and Ehrman, R. (1990). Integrating systematic cue exposure with standard treatment in recovering drug dependent patients. *Addictive Behav.* **15,** 355–365.

Post, R. M., Weiss, S. R. B., Pert, A., and Uhde, T. W. (1987). Chronic cocaine administration: Sensitization and kindling effects. *In* "Cocaine: Clinical and Biobehavioral Aspects" (S.

Fisher, A. Raskin, and E. H. Uhlenhuth, eds.). Oxford Univ. Press, New York.

Pulvirenti, L., and Koob, G. F. (1994). Dopamine receptor agonists, partial agonists and psychostimulant addiction. *Trends Pharmacol. Sci.* **15**, 374–379.

Robinson, T. E., and Berridge, K. C. (1993). The neural basis of drug craving: An incentive-sensitization theory of addiction. *Brain Res. Rev.* **18**, 247–291.

Weiss, R. D., and Mirin, S. M. (1987). "Cocaine." American Psychiatric Press, Washington, D.C.

Cochlear Chemical Neurotransmission

RICHARD P. BOBBIN
Louisiana State University Medical School

GLOSSARY

Afferent nerve fibers Nerve fibers that conduct action potentials toward the brain

Depolarization Change in the resting membrane potential of a cell in the positive direction, making it less negative

Efferent nerve fibers Nerve fibers that conduct action potentials away from the brain toward the cochlea

Hyperpolarization Change in the resting membrane potential of a cell in the negative direction, making it less positive

Ionotropic Receptor mechanism for a neurotransmitter: the receptor protein is part of an ion channel/receptor protein complex and opens or closes the ion channel directly in response to the presence of a ligand for that receptor

Metabotropic Receptor mechanism for a neurotransmitter: the receptor protein acts via another protein such as a G protein to activate or inhibit enzymes and possibly open, close, or modify ion channels

Muscarinic receptor protein Receptor activated by the neurotransmitter acetylcholine and the drug muscarine, located at the autonomic innervation of glands, cardiac muscle, and smooth muscle, abbreviated M

Neurotransmitter Chemical released from a sensory receptor cell or nerve cell upon depolarization of that cell. Upon release, the chemical diffuses across the gap between the releasing cell and an adjoining cell to act on the adjoining cell to induce a change in the adjoining cell's chemical and/or electrical properties

Nicotinic receptor protein Receptor activated by the neurotransmitter acetylcholine and the drug nicotine, located at the neuromuscular junction of skeletal muscle (abbreviated Nm) and at autonomic ganglia and the central nervous system (abbreviated Nn)

Receptor protein Protein in the membrane of a cell that accepts a chemical messenger (ligand) such as the neurotransmitter, changes its configuration upon accepting the neurotransmitter, and so induces subsequent reactions in the cell, such as opening of an ion channel to allow for diffusion of that ion down its concentration gradient

Synapse Place where neurotransmission occurs between two cells

CHEMICAL NEUROTRANSMISSION IS THE TERM used to describe the process whereby cells use chemicals to transmit information from one nerve cell to another at a synapse. In the cochlea this involves the transfer of information not only between nerve cells, but from the sensory receptor cells to nerve fibers, and from nerve fibers to the sensory receptor cells. In any given cell, there is usually only one primary chemical that is used for this function, the neurotransmitter. Secondary chemicals are also involved, called modulators since they modify the action of the primary chemical.

A chemical must satisfy several criteria before it is considered the proven neurotransmitter or modulator at a synapase. These include: (1) when the candidate is applied to the synapse it must mimic the endogenous compound; (2) the candidate must be present in the presynaptic structure; (3) drugs that antagonize the endogenous compound must also block the exogenously applied candidate; and (4) the endogenous compound must be detected in the extracellular fluid when the synapse is activated. At some synapses these criteria are much more difficult to fulfill than at others. Then the criteria are only guides as to the identity of the transmitter and proof remains an elusive goal.

ENCYCLOPEDIA OF HUMAN BIOLOGY, Second Edition, VOLUME 2. Copyright © 1997 by Academic Press. All rights of reproduction in any form reserved.

I. OVERVIEW OF COCHLEAR NEUROTRANSMISSION

Neurotransmission in the cochlea occurs at four types of synapses (Fig. 1): (1) outer hair cell (OHC) to afferent nerve fiber; (2) inner hair cell (IHC) to afferent nerve fiber; (3) efferent nerve fiber to OHC; and (4) efferent nerve fiber to afferent nerve fiber under the IHCs. The efferent fibers have been subdivided further into medial and lateral. It appears that the OHCs and IHCs utilize one transmitter chemical (i.e., L-glutamate, GLU), and most efferents utilize another transmitter chemical (i.e., acetylcholine, ACH).

The neurotransmitters at all the synapses in the cochlea appear to be released from storage packets called synaptic vesicles as in other neural structures. The synaptic vesicles in the hair cells (HCs) are unusual, however, in that they surround a presynaptic rod-like structure, the function of which is unknown. A depolarization of the presynaptic cell at the synapses induces an opening of voltage-activated Ca^{2+} channels that allows Ca^{2+} to enter the cell. Through a series of chemical reactions the Ca^{2+} triggers the release of the contents of the synaptic vesicles (neurotransmitter) into the synaptic space between the cells. The neurotransmitter then diffuses to the postsynaptic membrane, where it interacts chemically with a receptor protein. This chemical interaction induces a change in the shape of the receptor protein, which in turn changes either the membrane properties of the cell or the concentration of a chemical constituent in the cell. This in turn changes the electrical properties of the postsynaptic cell. The action of neurotransmitters is terminated either by enzymatic destruction or by diffusion and uptake into the cells surrounding the synapse.

II. NEUROTRANSMITTERS OF THE EFFERENT NERVE FIBERS

The medial system of efferent fibers synapse on the OHCs, whereas the lateral system synapses on the afferent nerve endings under the IHCs (Fig. 1). The neurotransmitter at the synapses of these efferents is ACH, the first neurotransmitter identified in the cochlea and in other HC systems. This conclusion was initially based on the localization of cholinesterase (the enzyme that terminates the action of ACH by converting it to acetate and choline) to the efferent

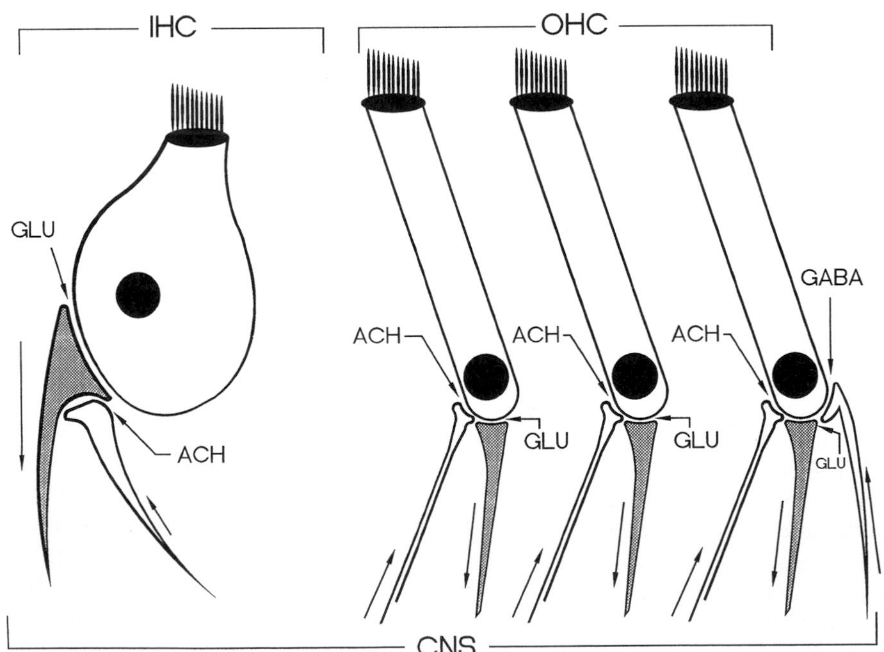

FIGURE 1 Schematic diagram showing the principal neurotransmitters and their location in the cochlea. The arrows indicate the direction of nerve conduction, afferent and efferent. IHC, inner hair cell; OHC, outer hair cells; CNS, central nervous system; ACH, acetylcholine; GLU, glutamic acid; GABA, γ-aminobutyric acid.

nerve fibers in the cochlea. Subsequent studies have strengthened this hypothesis at both the medial and lateral systems of efferent nerve endings in the cochlea. γ-Aminobutyric acid (GABA) appears to be a neurotransmitter at a much smaller population of efferents in both the medial and lateral systems (Fig. 1). Little is known about the chemistry or physiology of the GABA synapse.

The postsynaptic receptor protein for ACH on the OHCs is unusual and has not been definitively categorized. In other systems the ACH receptor is identified as either muscarinic (M) with a metabotropic mechanism of action (e.g., glands, cardiac) or nicotinic (N) with an ionotropic mechanism of action (e.g., neuromuscular junction, Nm; autonomic ganglia, Nn). The receptor in the cochlea appears to be a nicotinic subtype. In the guinea pig, the receptor gives the greatest response to ACH and carbachol with slightly less of a response to suberyldicholine and 1,1-dimethyl-4-phenylpiperazinium (DMPP). On the other hand, nicotine, muscarine, and cytisine give very little if any response. Thus the OHCs have a nicotinic-like receptor that does not respond to nicotine. The basis for classifying the receptor as nicotinic is that the action of ACH at the receptor is more sensitive to antagonism by nicotinic antagonists (e.g., decamethonium, hexamethonium, curare, α-bungarotoxin, κ-bungarotoxin, and trimethaphan) than by muscarinic antagonists (e.g., atropine and pirenzepine). More unusual is the fact that the glycine antagonist strychnine is the most potent antagonist, and bicuculline, a GABA antagonist, blocks though with less potency than the nicotinic antagonists. Thus ACH appears to act on a receptor protein that has a molecular configuration that combines similarities of the glycine receptor, the GABA receptor, and the nicotinic receptor, yet is not very different from the muscarinic receptor. Recent data suggest that the receptor contains an α-9 type of nicotinic subunit.

To date, only the crossed medial efferents have been studied in detail. (All that is known about the lateral efferents is that ACH is the neurotransmitter; the effects are unknown.) The overall effect of activation of the medial efferents is to reduce the output of information from the IHCs to the brain (since the IHCs synapse with approximately 95% of the auditory nerve afferents). This set of efferents synapse predominately on the OHCs, therefore these efferents affect the output of the IHCs indirectly. The OHCs are utilized by the cochlea to transduce sound at low intensities. It is known that sound energy induces a change in the membrane potential of the OHCs, possibly by opening mechanically activated ion channels (i.e., transduction channels) located in the stereocilia (Fig. 2). The transduction channel allows K^+ to enter the OHC. This K^+ then alters the membrane potential of the cell and triggers the opening of L-Ca^{2+} channels, which in turn will result in the entrance of Ca^{2+} and the opening of Ca^{2+}-dependent K^+ channels. The final membrane potential and its subsequent changes depend on the interaction of these ion channels acting in concert with other channels present in OHCs (Fig. 2).

The OHCs will change length in response to the changes in membrane potential. The unidentified molecule or motor responsible for this length change is voltage dependent. Membrane potential changes in a depolarization direction induce a decrease in length

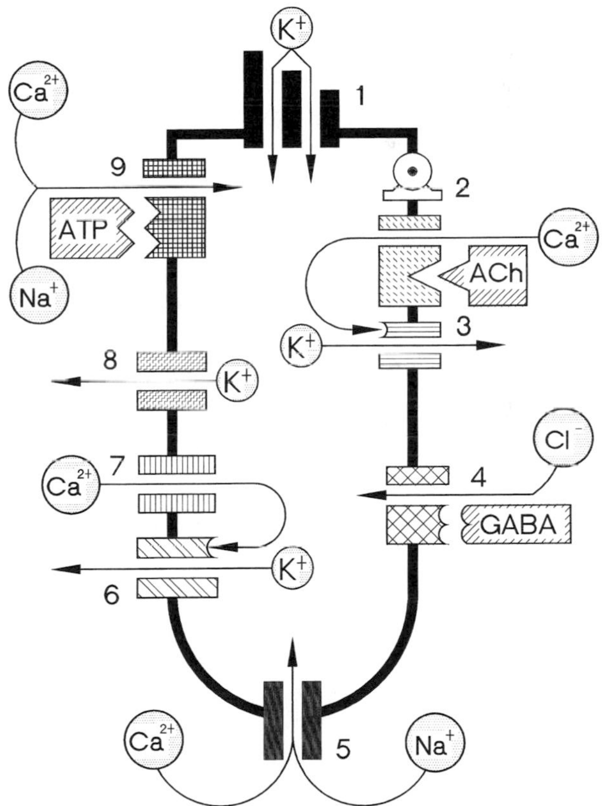

FIGURE 2 Model of an outer hair cell illustrating the various ion channel proteins in the membrane available to alter the membrane potential and so contribute to an alteration in the length of the OHC. 1, transduction channel; 2, voltage-dependent motor molecule; 3, ACH–receptor and ion channel complex; 4, GABA–receptor and ion channel complex; 5, nonspecific cation channel; 6, various calcium-dependent potassium channels (e.g., K_{Ca}); 7, L-type calcium channel; 8, various voltage-dependent K^+ channels (e.g., K_n channel); 9, ATP–receptor and ion channel complex.

and changes in a hyperpolarization direction induce an increase in length. The change in length of the OHCs is thought to affect the stereocilia of the IHCs, probably through a mechanical interaction involving the tectorial membrane.

Stimulation of the efferent nerve fibers electrically or via sound stimulation to the contralateral ear releases ACH from the efferent nerve endings onto the nicotinic-like receptors on OHCs (Fig. 2). Receptor activation probably opens a channel that allows Ca^{2+} to enter the cell, and this Ca^{2+} in turn activates an outward K^+ current that will move the membrane potential toward hyperpolarization (Fig. 2). This will decrease (inhibit) the amplitude of any shortening/lengthening response induced by sound stimulation acting on the OHC stereocilia. In this manner, the efferents physiologically change the functional output of the cochlea by about plus or minus 10 dB. [*See* Ion Pumps.]

Several neuroactive substances have been colocalized to these efferents. The opioid neuropeptides, such as enkephalins, dynorphin B, and α-neoendorphin, are present and distributed within the lateral subdivision of the ACH efferents, being absent from the medial efferents. In addition, dopamine and calcitonin gene-related peptide (CGRP) have been shown in the efferents. These chemicals are most likely released together with ACH or GABA. Since it appears that ACH and GABA open ligand-activated ion channels, then the other colocalized substances may act to modulate this action on the ion channels by possibly regulating levels of second messenger chemicals.

III. NEUROTRANSMITTERS OF THE HAIR CELLS

The other major category of neurotransmitters in the cochlea are those released by the two sets of HCs: the OHCs and the IHCs (Fig. 1). A growing body of evidence demonstrates that GLU is probably the neurotransmitter released by both sets of HCs. There has been difficulty in achieving sufficient evidence for all of the criteria mentioned earlier. For instance, GLU cannot be said to be solely localized to neurons where it is the transmitter because it is ubiquitous, having dominant roles in amino acid, energy, and nitrogen metabolism. Another disturbing facet of the problem is that many endogenous compounds such as L-aspartic acid, L-homocysteic acid, and *N*-acetylaspartylglutamate resemble GLU structurally and so activate GLU receptors. The problem is not unique to the cochlea or to HC systems, but is a problem throughout the nervous system.

As the HC transmitter, GLU may act on several receptor types belonging to the excitatory amino acid category. The receptors have been named after nonendogenous chemicals that activate them. These include both ionotropic receptors, such as α-amino-3-hydroxy-5-methyl-4-isoxazole-propionic acid (AMPA; previously termed quisqualate), kainate, and *N*-methyl-D-aspartate (NMDA), and metabotropic receptors, such as quisqualate (also designated mGluR1 through mGluR7). The AMPA, NMDA, and quisqualate excitatory amino acid receptors appear to have been identified in the cochlea. These receptors are present on the afferent nerve endings that synapse on the IHCs. GLU activation of the AMPA and NMDA receptors opens an ion channel that allows Na^+, K^+, and varying amounts of Ca^{2+} to diffuse down their gradients. In the cochlea this results in depolarization of the postsynaptic, afferent cochlear nerve fiber endings, triggering postsynaptic potentials, and then action potentials at the axon hillock. On the other hand, the type of excitatory amino acid receptor on the afferents synapsing on the OHCs is unidentified. At this synapse, GLU released by the OHCs interacts with afferents that comprise only a very small number (5%) of the total afferent nerve fiber population (95% at the IHCs) in the cochlea. This is such a small population that they are difficult to functionally isolate and study.

GLU has a weak affinity for its receptors. This suggests that a high concentration of GLU (>60 mM) is in the presynaptic storage vesicles, allowing a concentrated solution of GLU to be released onto the receptors. In turn, this low affinity increases the effectiveness of diffusion alone to terminate rapidly the action of GLU. The GLU that diffuses away from the receptors is taken up into neighboring cells through a transport system that is dependent on the presence of Na^+. After being taken up into a cell, the GLU is converted to glutamine, which is then deposited into the extracellular space for uptake into the HCs. In the HCs the glutamine is converted to GLU and repackaged into vesicles for release as HC neurotransmitter.

The action of the neurotransmitter of the IHCs on the afferent nerve endings is antagonized by those chemicals found to block AMPA receptors in other systems. The structural similarity of these antagonists to both GLU and aspartate is shown in Fig. 3. The first blocker tested with selective action in the cochlea was kynurenic acid, which occurs naturally in the body. Up to that time (1987), blockers were either not very active or demonstrated activity on structures

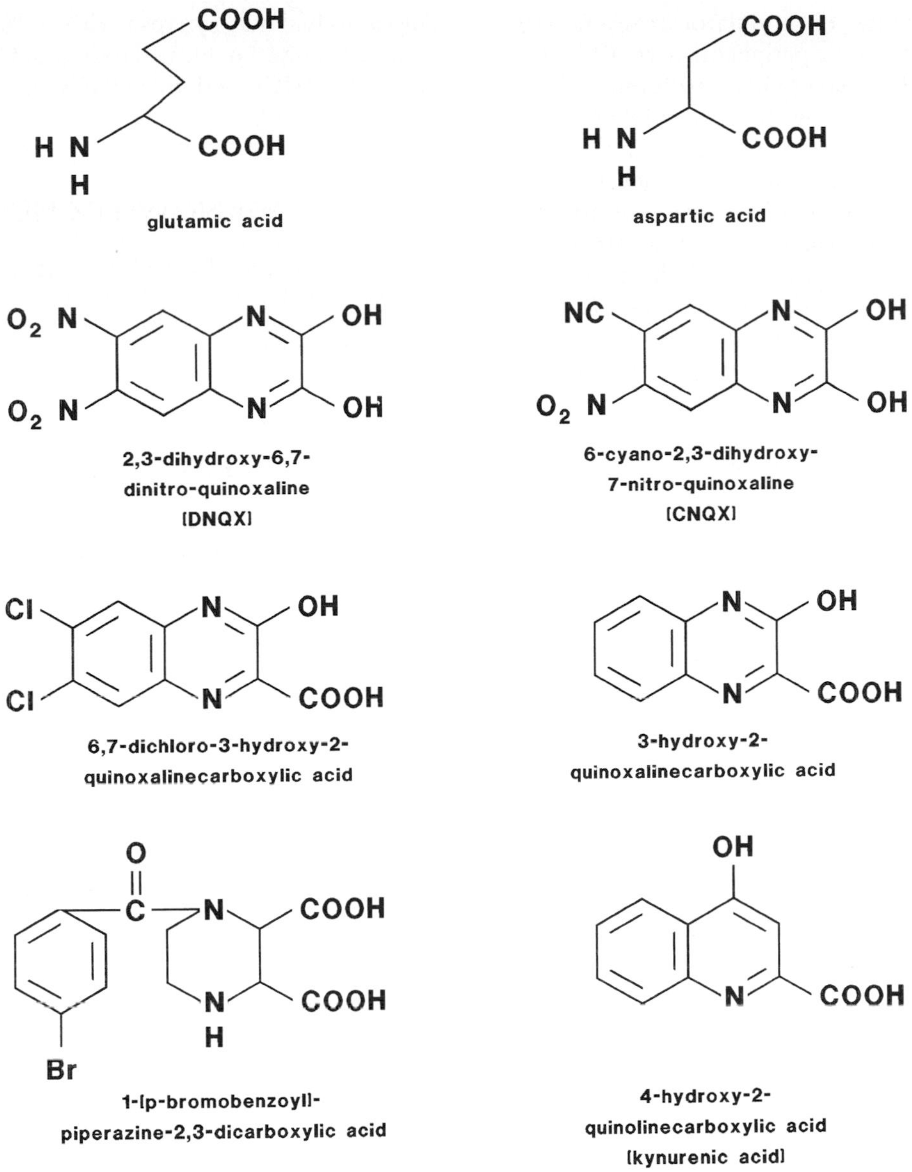

glutamic acid

aspartic acid

2,3-dihydroxy-6,7-dinitro-quinoxaline [DNQX]

6-cyano-2,3-dihydroxy-7-nitro-quinoxaline [CNQX]

6,7-dichloro-3-hydroxy-2-quinoxalinecarboxylic acid

3-hydroxy-2-quinoxalinecarboxylic acid

1-[p-bromobenzoyl]-piperazine-2,3-dicarboxylic acid

4-hydroxy-2-quinolinecarboxylic acid [kynurenic acid]

FIGURE 3 Chemical structures of the neurotransmitter candidates for the hair cells, glutamate and aspartate, and the chemical antagonists of these candidates at the receptors (i.e., AMPA) and at the hair cell-to-afferent nerve fiber junction.

other than the synapse. Other compounds tested and active with about the same potency as kynurenic acid include 1-(*p*-bromobenzoyl)-piperazine-2,3-dicarboxylic acid and 3-hydroxy-2-quinoxalinecarboxylic acid. A 30-fold increase in potency was found with the antagonists 6,7-dichloro-3-hydroxy-2-quinoxalinecarboxylic acid and 6-cyano-7-nitro-quinoxaline-2,3-dione (CNQX). An additional 3-fold increase (100-fold greater than that of kynurenic acid) was found with 6,7-dinitro-quinoxaline-2,3-dione (DNQX). Given the success with these compounds, it appears that in the near future antagonists that are more selective and powerful will certainly become available.

IV. NEUROMODULATORS IN THE COCHLEA

Adenosine triphosphate (ATP), which like GLU plays an integral part in metabolism, appears to play a role as a neurotransmitter or modulator at several structures in the central and peripheral nervous systems. In addition, ATP appears to have profound effects on several types of cells in the cochlea. ATP depolarizes OHCs by opening a cation channel (Fig. 2). In addition, ATP increases intracellular Ca^{2+} levels in IHCs and Deiters' cells. When ATP levels are artificially increased in the fluids surrounding the hair cells and Deiters' cells, they induce profound alterations in the mechanical properties of the cochlea. The role of ATP at these receptors in the physiological functioning of the cochlea is unknown and remains to be determined by future research. Likewise, the cells of origin of the ATP that acts on these receptors on OHCs, IHCs, and Deiters' cells remain to be discovered. [See Adenosine Triphosphate (ATP).]

V. SUMMARY

Our knowledge of neurocommunication between cells in the cochlea is advancing rapidly. ACH and GABA are neurotransmitters of efferent neurons synapsing in the cochlea. Various neuromodulators at the efferent synapses are being identified and their roles defined. ATP appears to have a very important role as a chemical messenger between several cell types in the cochlea. ACH, GABA, and ATP activate ligand-gated ion channels located on the OHCs and together with various voltage-gated ion channels control the voltage of the OHCs and so the length of the cell. The length of the OHCs is important because it affects the output of the IHCs that synapse with 95% of the afferent output of the cochlea. The transmitter released by both the IHCs and the OHCs onto afferent nerve endings is GLU.

ACKNOWLEDGMENTS

Thanks to Maureen Fallon for help in constructing the figures for this manuscript and to Chu Chen, Ph.D., and Anastas Nenov, M.D., for assistance. The author's research cited in this manuscript is supported by grants from NIH, Kam's Fund for Hearing Research, and the Louisiana Lions Eye Foundation.

BIBLIOGRAPHY

Eybalin, M. (1993). Neurotransmitters and neuromodulators of the mammalian cochlea. *Physiol. Rev.* **73,** 309–373.

Erostegui, C., Norris, C. H., and Bobbin, R. P. (1994). *In vitro* pharmacologic characterization of a cholinergic receptor on outer hair cells. *Hear. Res.* **74,** 135–147.

Hoffman, D. W. (1986). Opioid mechanisms in the inner ear. *In* Neurobiology of Hearing: The Cochlea" (R. A. Altschuler, R. P. Bobbin, and D. W. Hoffman, eds.), pp. 371–382. Raven, New York.

Kujawa, S. G., Erostegui, C., Fallon, M., Crist, J., and Bobbin, R. P. (1994). Effects of ATP and related agonists on cochlear function. *Hear. Res.* **76,** 87–100.

Kujawa, S. G., Glattke, T. J., Fallon, M., and Bobbin, R. P. (1994). A nicotinic-like receptor mediates suppression of distortion product otoacoustic emissions by contralateral sound. *Hear. Res.* **74,** 122–134.

Pujol, R., and Lenoir, M. (1986). The four types of synapses in the organ of Corti. *In* "Neurobiology of Hearing: The Cochlea" (R. A. Altschuler, R. P. Bobbin, and D. W. Hoffman, eds.), pp. 161–172. Raven, New York.

Coenzymes, Biochemistry

DONALD B. McCORMICK
Emory University

GLOSSARY

Apoenzyme Protein moiety of an enzyme that requires a coenzyme

Coenzyme Natural, organic molecule that functions in a catalytic, enzyme system

Cofactor Natural reactant, usually either a metal ion or coenzyme, required in an enzyme-catalyzed reaction

Holoenzyme Catalytically active enzyme constituted by coenzyme bound to apoenzyme

Vitamin Essential organic micronutrient that must be supplied exogenously and in many cases is the precursor to a metabolically derived coenzyme

COENZYMES ARE ORGANIC MOLECULES THAT are bound to those enzymes that require their function to catalyze certain biochemical reactions. Though some simple proteins (unconjugated polypeptides) are able to function as enzymes competent to catalyze a modest range of reactions (e.g., hydrolyses), the limited chemical properties of amino acid side chains within proteins do not permit catalyses of many of the numerous essential reactions that operate by diverse mechanisms. Hence, additional reagents more broadly considered as cofactors serve with protein enzymes for reactions that would be difficult or impossible using only simple acid–base catalysis. Among such cofactors are inorganic materials such as metal ions as well as the organic compounds called coenzymes. Coenzymes bind to apoenzymes (proteins) to generate functional holoenzymes. Tightly bound coenzymes are sometimes referred to as prosthetic groups.

Because major portions of some coenzymes cannot be biosynthesized by certain organisms, these precursors, normally made by other organisms, must be supplied exogenously. Many coenzymes are the metabolic result of converting an ingested vitamin, especially those of the B complex, to a form suitable for binding and function in an enzyme system. Of the 13 vitamins presently known to be required in the diet of humans, at least eight (thiamin, riboflavin, niacin, vitamin B_6, vitamin B_{12}, folacin, biotin, and pantothenate) are simply the essential precursors for coenzymatic forms made in our bodies. Because coenzymes are indispensable coreactants in many enzyme-catalyzed reactions involved in the formation, metabolism, and degradation of almost all body components, it follows that the vitaminic precursors for most coenzymes are essential for normal growth and function. [*See* Vitamin A.]

I. THIAMIN PYROPHOSPHATE

A. Chemistry

Thiamin pyrophosphate (TPP) is the principal if not sole coenzyme derived from thiamin (vitamin B_1). The structure of TPP (Fig. 1) indicates that the physiologically active form is the ionized pyrophosphate ester formed at the β-hydroxyethyl substituent of the thia-

ENCYCLOPEDIA OF HUMAN BIOLOGY, Second Edition, VOLUME 2. Copyright © 1997 by Academic Press. All rights of reproduction in any form reserved.

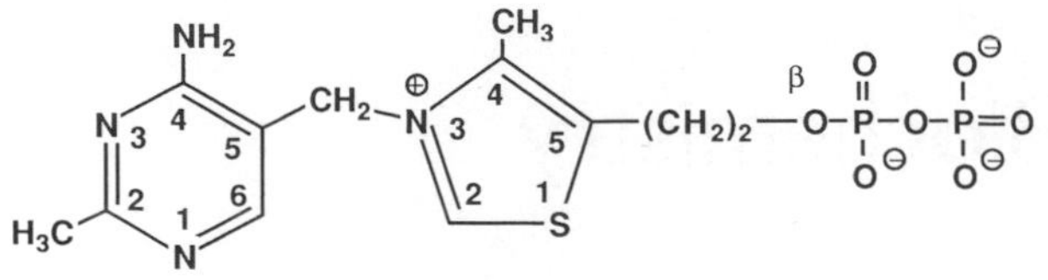

FIGURE 1 Structure with numbering for thiamin pyrophosphate.

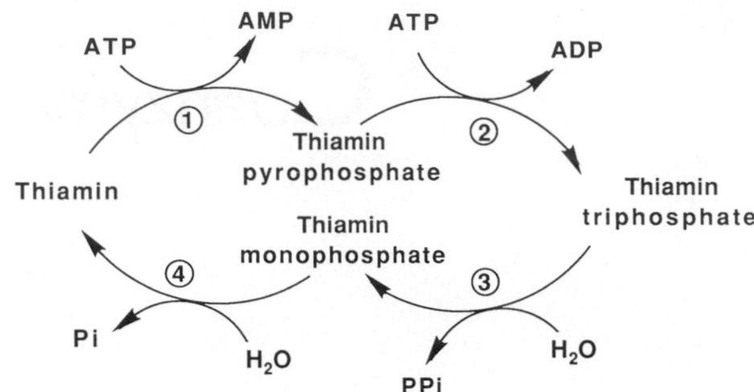

FIGURE 2 Interconversions of thiamin and its phosphates catalyzed by (1) thiaminokinase, (2) TPP-ATP phosphoryl transferase, (3) thiamin triphosphatase, and (4) thiamine monophosphatase. Pi, inorganic phosphate; PPi, inorganic pyrophosphase.

zole moiety of thiamin, which can be systematically named as 3-(2′-methyl-4′-amino-5′-pyrimidylmethyl)-4-methyl-5-(β-hydroxyethyl) thiazole.

TPP, as is the case with the parent vitamin, is relatively stable in acidic aqueous solutions, but sulfite causes cleavage at the methylene bridge to release the substituted pyrimidylsulfonate and the thiazole pyrophosphate. TPP is unstable in alkaline medium because the thiazole portion is subject to base attack at carbon 2. With hydroxyl ion, this leads to a pseudobase, thiazole ring opening and some disulfide under mild oxidizing conditions. Also, the 4-amino on the pyrimidyl portion can attack as an intramolecular base to form the tricyclic amino adduct, which can be oxidized with ferricyanide to yield thiochrome pyrophosphate. The thiochrome-level compound is fluorescent (λ excit = 385 nm; λ emit = 440 nm) and easily detected.

B. Metabolism

TPP is formed in a number of tissues from thiamin, and a fraction is subsequently converted, especially in brain, to the triphosphate. Hydrolysis of the latter to the monophosphate and some further release of free vitamin complete the interconversions shown in Fig. 2. Among the forms involved, TPP predominates in cells. Approximately 30 mg of thiamin-level compounds are found in an adult with 80% as pyrophosphate, 10% as triphosphate, and the rest as thiamin and its monophosphate. About half of the body stores are found in skeletal muscles with much of the remainder in heart, liver, kidneys, and nervous tissue, including brain, which contains much of the triphosphate. Concerning enzymes catalyzing the interconversions, thiaminokinase is widespread, but the phosphoryl transferase and membrane-associated triphosphatase are mainly in nervous tissue.

C. Functions

There are two general types of reactions where TPP functions as the Mg^{2+}-coordinated coenzyme for so-called active aldehyde transfers. First, in decarboxylation of α-keto acids, the condensation of the thiazole moiety of TPP with the α-carbonyl carbon on the acid leads to loss of CO_2 and production of a resonance-stabilized carbanion (Fig. 3). Protonation and release

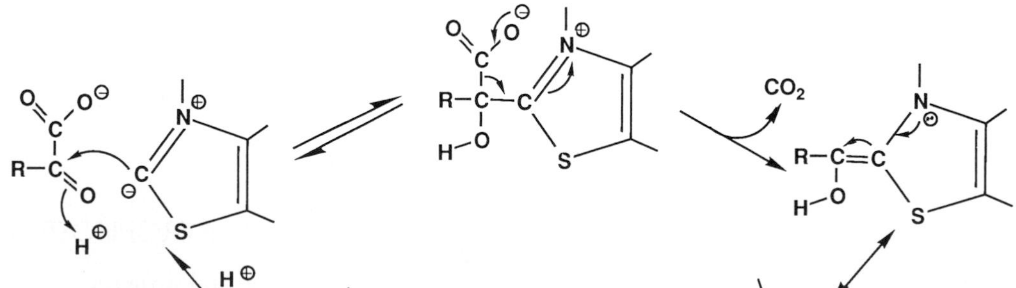

FIGURE 3 Function of the thiazole moiety of thiamine pyrophosphate in α-keto acid decarboxylations.

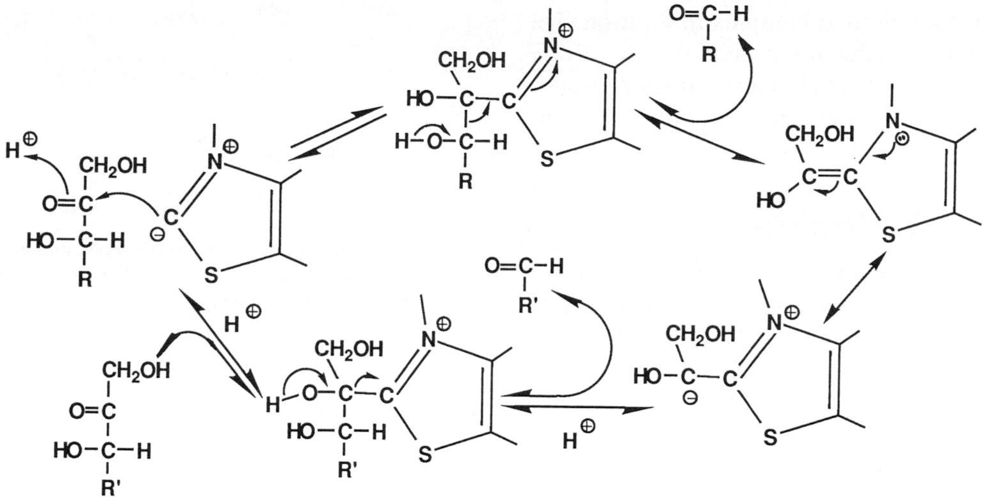

FIGURE 4 Function of the thiazole moiety of thiamine pyrophosphate in transketolations.

of aldehyde occur in fermentative organisms such as yeast, which have only the TPP-dependent dicarboxylase, but reaction of the α-hydroxyalkyl-TPP with lipoyl residues and ultimate conversion to acyl-CoA occur in higher eukaryotes including humans with multienzymatic dehydrogenase complexes, as described in Section IX. The other general reaction involving TPP is the transformation of α-ketols (ketose phosphates). Although specialized phosphoketolases in certain bacteria and higher plants can split ketose phosphates to simpler, released products, the reaction of importance to humans and most animals is a transketolation, as mechanistically illustrated in Fig. 4. Transketolase is a TPP-dependent enzyme found in the cytosol of many tissues, especially liver and blood cells, where principal carbohydrate pathways exist. This enzyme catalyzes the reversible transfer of a glycoaldehyde moiety (α,β-dihydroxyethyl-TPP) from the first two carbons of a donor ketose phosphate to the aldehyde carbon of an aldose phosphate of the pentose phosphate pathway, which also supplies nicotinamide adenine dinucleotide phosphate, reduced (NADPH), needed for biosynthetic reactions.

II. FLAVOCOENZYMES

A. Chemistry

Flavin adenine dinucleotide (FAD) and its immediate precursor flavin mononucleotide (FMN) are two commonly encountered flavocoenzymes. The structure of these (Fig. 5) indicates a common tricyclic isoalloxazine nucleus with a D-ribityl side chain, as found in riboflavin [7,8-dimethyl-10-(1′-D-ribityl)-isoalloxazine]. FMN is riboflavin 5′-phosphate, whereas FAD is extended to include a pyrophosphoryl-linked 5′-adenosine monophosphate (AMP) moiety. In some less commonly encountered, but essential, forms of flavocoenzymes, there is covalent attachment of a peptidyl residue through an electronegative atom, usually in the 8α position.

Flavocoenzymes are relatively more stable in acid than base and are photodecomposed, largely to lumichrome and lumiflavin, by cleavages at the ribityl side chain. In the natural oxidized (quinoid) forms, FMN and FAD are fluorescent (λ excit = 450 nm; λ emit = 530 nm) and readily detected when unbound by protein; however, FAD is significantly quenched

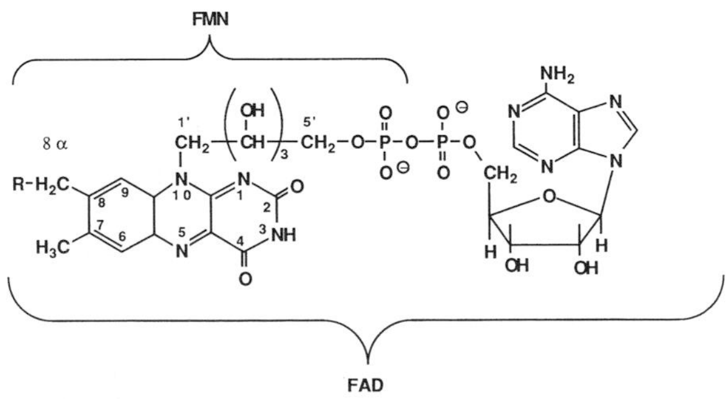

FIGURE 5 Structure with numbering for principal flavocoenzymes, where R = H for FAD or an electronegative atom (N, O, S) with peptide for covalently attached species.

(80%) by its intramolecular complex in solution. The observed oxidation–reduction potential near −0.2 V poises these coenzymes for electron transport usually after the more negative pyridine nucleotides and before cytochromes.

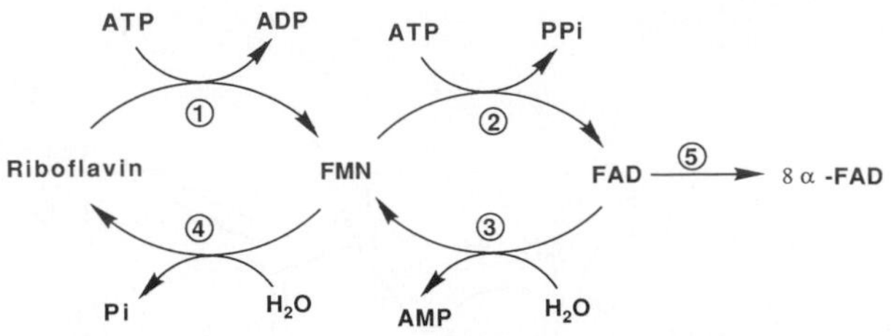

B. Metabolism

Biosynthesis of flavocoenzymes occurs within the cellular cytoplasm of most tissues, but particularly in the small intestine, liver, heart, and kidney. The obligatory first step is the adenosine triphosphate (ATP)-dependent phosphorylation of riboflavin catalyzed by Zn^{2+}-preferring flavokinase. The FMN product can be complexes with specific apoenzymes to form several functional flavoproteins, but the larger quantity is further converted to FAD in a second ATP-dependent reaction catalyzed by Mg^{2+}-preferring FAD synthetase. These coenzyme-forming steps (Fig. 6) are tightly regulated. FAD is the predominant flavocoenzyme present in tissues, where it is mainly complexed with numerous flavoprotein dehydrogenases and oxidases. Less than 10% of the FAD can also become covalently attached to specific amino acid residues of a few important apoenzymes. Examples include the 8α-*N*(3)-histidyl-FAD within succinate dehydrogenase and 8α-*S*-cysteinyl-FAD within monoamine oxidase, both of mitochondrial localization. Turnover of covalently attached flavocoenzymes requires intracellular proteolysis, and further degradation of the coenzymes involves a pyrophosphatase cleavage of FAD to FMN and action by nonspecific phosphatases on the latter (Fig. 6).

ATP ADP ATP PPi
① ②
Riboflavin FMN FAD ⑤ 8α-FAD
④ ③
Pi H_2O AMP H_2O

FIGURE 6 Interconversions of riboflavin and its coenzymes catalyzed by (1) flavokinase, (2) FAD synthetase, (3) FAD pyrophosphatase, (4) FMN phosphatase, and (5) posttranslational modification.

C. Functions

Flavocoenzymes participate in oxidation–reduction reactions in numerous metabolic pathways and in energy production via the respiratory chain. The redox functions of a flavocoenzyme (Fig. 7) include one-electron transfers during which the biologically encountered, neutral, oxidized quinone level of flavin is

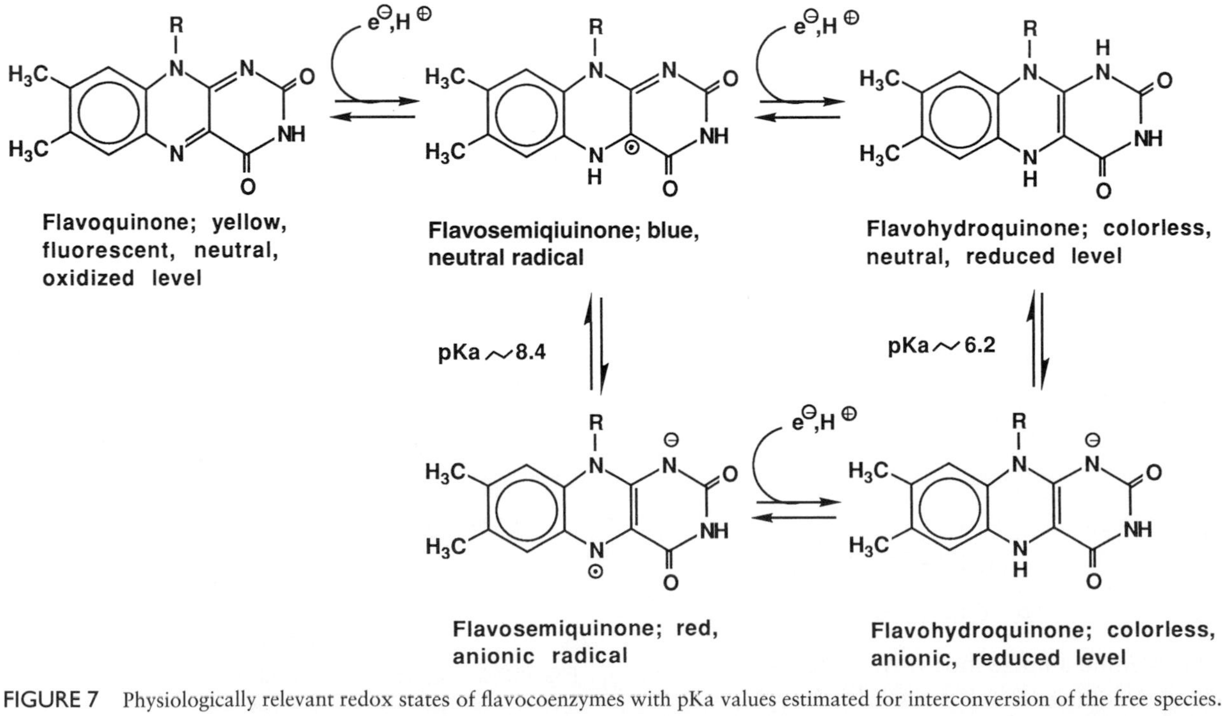

Flavoquinone; yellow, fluorescent, neutral, oxidized level

$e^{\ominus}, H^{\oplus}$

Flavosemiqiuinone; blue, neutral radical

$e^{\ominus}, H^{\oplus}$

Flavohydroquinone; colorless, neutral, reduced level

pKa ~8.4

pKa ~6.2

$e^{\ominus}, H^{\oplus}$

Flavosemiquinone; red, anionic radical

Flavohydroquinone; colorless, anionic, reduced level

FIGURE 7 Physiologically relevant redox states of flavocoenzymes with pKa values estimated for interconversion of the free species.

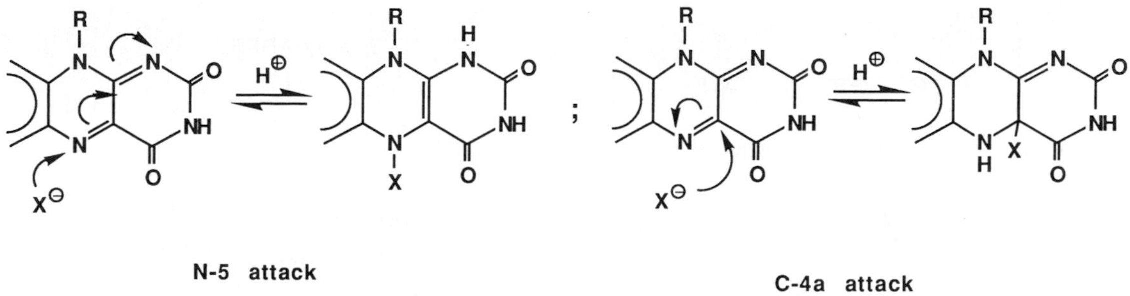

N-5 attack

C-4a attack

FIGURE 8 Reaction types encountered with flavoquinone coenzymes and natural nucleophiles (X^-).

half-reduced to the radical semiquinone, which can exist within natural pH ranges as neutral or anionic species. A further electron can lead to a fully reduced hydroquinone. Additionally, a single-step, two-electron transfer from substrate to flavin can occur (Fig. 8). Such cases as hydride ion transfer from reduced pyridine nucleotide or the carbanion generated by base abstraction of a substrate proton may lead to attack at the flavin N-5 position; some nucleophiles such as the hydrogen peroxide anion add at the C-4a position.

There are flavoprotein-catalyzed dehydrogenations that are both pyridine nucleotide dependent and independent, reactions with sulfur-containing compounds, hydroxylations, oxidative decarboxylations, dioxygenations, and reduction of O_2 to hydrogen peroxide. The intrinsic abilities of flavins to be varyingly potentiated as redox carriers upon differential binding to proteins, to participate in both one- and two-electron transfers, and in reduced (1,5-dihydro) form to react rapidly with oxygen permit wide scope in their operation.

III. PYRIDINE NUCLEOTIDE COENZYMES

A. Chemistry

Nicotinamide adenine dinucleotide (NAD) and its phosphate (NADP) (Fig. 9) are the two natural coenzymes derived from niacin. Both contain an N-1-substituted pyridine 3-carboxamide that is essential to function in redox reactions with a potential near −0.32 V. The oxidized coenzymes are labile to alkali, including nucleophilic addition at the para (4) position, whereas the reduced (1,4-dihydro) coenzymes are labile to acid. Nicotinamide adenine dinucleotide, reduced (NADH) and NADPH (but not NAD and NADP) characteristically absorb light in the near ultraviolet (340 nm).

B. Metabolism

Converging pathways lead to the formation of NAD (Fig. 10), a fraction of which is phosphorylated to NADP. Vitaminic precursors are converted to the coenzymes in blood cells, kidney, brain, and liver. Nicotinate and nicotinamide react with 5-phosphoribosyl-1α-pyrophosphate (PRPP) and nicotinic acid mononucleotide (NaMN) or nicotinamide mononucleotide (NMN), respectively. Additionally in liver, quinolinate from catabolism of tryptophan is similarly converted with concomitant decarboxylation to NaMN. Subsequent reactions with ATP to incorporate the 5′-adenylate portion yield NAD from NMN and deamido-NAD from NaMN. The deamido compound reacts with glutamine in a cytosolic ATP-dependent step to yield NAD, glutamate, AMP, and pyrophosphate. In breakdown, NAD is hydrolyzed to NMN and the latter to nicotinamide, which, in turn, can be converted to nicotinate by a rather widespread microsomal deamidase. An NAD glycohydrolase (NADase) catalyzes hydrolysis of NAD to nicotin-

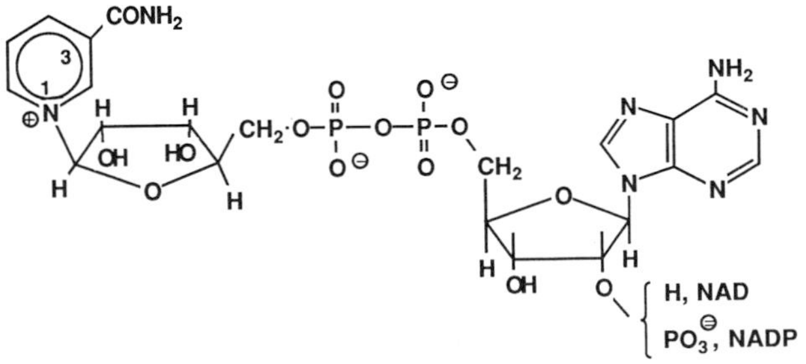

H, NAD
$PO_3^{\ominus}$, NADP

FIGURE 9 Structures with numbering for pyridine nucleotide coenzymes.

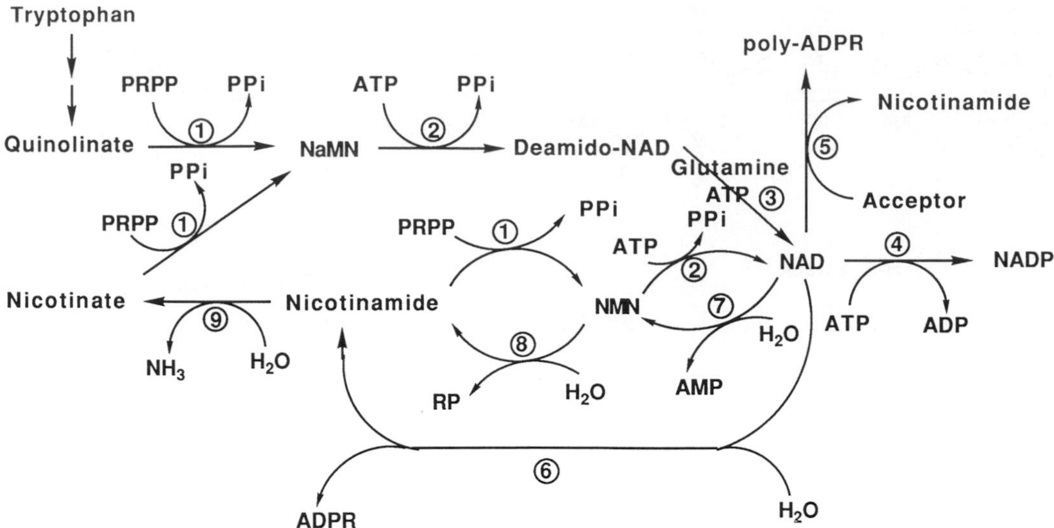

FIGURE 10 Interconversions of niacin-level precursors and the coenzymes catalyzed by (1) phosphoribosyl-transferases, (2) adenylyl-transferases, (3) synthetase, (4) kinase, (5) poly-ADPR synthetase (polymerase), (6) NAD glycohydrolase, (7) NAD pyrophosphatase, (8) NMN hydrolase, and (9) nicotinamide deamidase. RP is ribose 5-phosphate.

amide plus adenosine 5′-pyrophospho-5-ribose (ADPR). Some NAD glycohydrolases have the ability to transglycosidate, that is, transfer the ADPR moiety of NAD to acceptor macromolecules.

C. Functions

Numerous enzymes require the nicotinamide moiety within either NAD or NADP. Most of these oxidoreductases function as dehydrogenases and catalyze such diverse reactions as the conversion of alcohols (often sugars and polyols) to aldehydes or ketones, hemiacetals to lactones, aldehydes to acids, and certain amino acids to keto acids. The common mechanism of operation (generalized in Fig. 11) involves the stereospecific abstraction of a hydride ion from substrate, with para addition to one or the other side of carbon 4 in the pyridine ring of the nucleotide coenzyme. The second hydrogen of the substrate group oxidized is concomitantly removed as a proton and ultimately exchanges as a hydronium ion.

Most dehydrogenases utilizing NAD or NADP function reversibly. Glutamate dehydrogenase, for example, favors the oxidative direction, whereas others, such as glutathione reductase, catalyze preferential reduction. A further generality is that most NAD-dependent enzymes are involved in catabolic reactions, whereas NADP systems are more common to biosynthetic reactions. The additional function of NAD as a substrate for providing the ADPR moiety to modify macromolecules has been recently more appreciated. ADP-ribosyl transferase catalyzes such a modification of the prokaryote elongation factor 2, thereby blocking translocation on ribosomes. Poly-(ADPR) synthetases (polymerases) in eukaryotes catalyze a multiple addition of ADPR from NAD to form

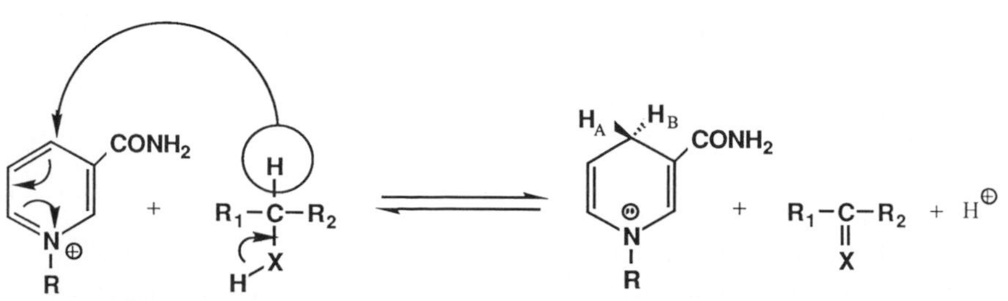

FIGURE 11 The hydride ion transfer for operation of pyridine nucleotide coenzymes.

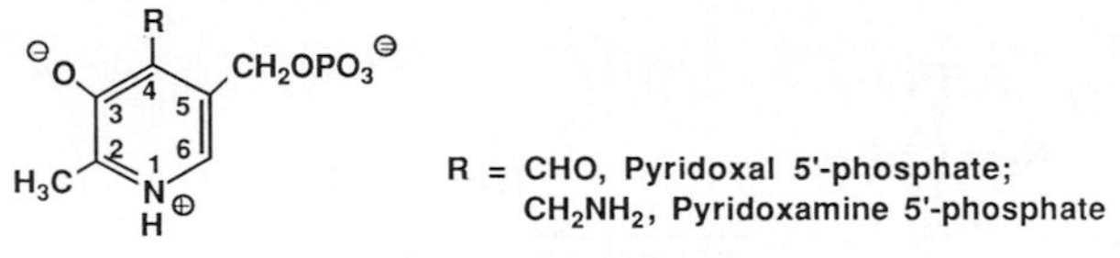

R = CHO, Pyridoxal 5'-phosphate;
CH_2NH_2, Pyridoxamine 5'-phosphate

FIGURE 12 Structure with numbering for coenzyme forms of vitamin B_6.

(ADPR)n-acceptor plus nicotinamide and hydrogen ion. This activity is found in mitochondria and bound to microsomes as well as in nuclei, where it affects operation of DNA. This nonredox function of NAD probably accounts for the rapid turnover of NAD in human cells.

IV. PYRIDOXAL 5′-PHOSPHATE AND PYRIDOXAMINE 5′-PHOSPHATE

A. Chemistry

Two of the three natural forms of vitamin B_6 (pyridoxine, pyridoxal, and pyridoxamine) can be phosphorylated to directly yield functional coenzymes, that is, pyridoxal 5′-phosphate (PLP) and pyridoxamine 5′-phosphate (PMP). The structures of these 4-substituted 2-methyl-3-hydroxy-5-hydroxymethylpyridine 5′-phosphates are shown in Fig. 12. PLP is the predominant and more diversely functional coenzymatic form, although PMP interconverts as coenzyme during transaminations.

At physiological pH, the dianionic phosphates of these coenzymes exist as zwitterionic *meta*-phenolate pyridinium compounds. They are very water-soluble, absorb light in the ultraviolet region, exhibit fluorescence, and, in general, are sensitive to light, particularly at alkaline pH. Both natural and synthetic carbonyl reagents (e.g., hydrazines and hydroxylamines) form Schiff bases with the 4-formyl function of PLP (and pyridoxal), thereby removing the coenzyme and inhibiting PLP-dependent reactions.

B. Metabolism

The metabolic interconversions of vitamin and coenzymatic forms of B_6 are shown in Fig. 13. Each of the three vitamin-level compounds is phosphorylated in the cytosol by ATP-utilizing pyridoxal kinase, which, in mammalian tissues, prefers Zn^{2+}. Most cells of facultative and aerobic organisms contain a cytosolic FMN-dependent pyridoxine (pyridoxamine) 5′-phosphate oxidase responsible for catalyzing the O_2-dependent conversion of pyridoxine 5′-phosphate (PNP) and PMP to PLP. During aminotransferase (transminase)-catalyzed reactions, PLP and PMP interconvert with amino and keto functions of substrate–product participants. Release of free vitamin, mainly pyridoxal when physiological nonsaturating levels of vitamin are absorbed, occurs when the phosphates are hydrolyzed by nonspecific alkaline phosphatase located on the plasma membrane of cells.

C. Functions

PLP functions in numerous reactions that embrace the metabolism of proteins, carbohydrates, and lipids. Especially diverse are PLP-dependent enzymes that are involved in amino acid metabolism. By virtue of the ability of PLP to condense its 4-formyl substituent with an amine, usually the α-amino group of an amino

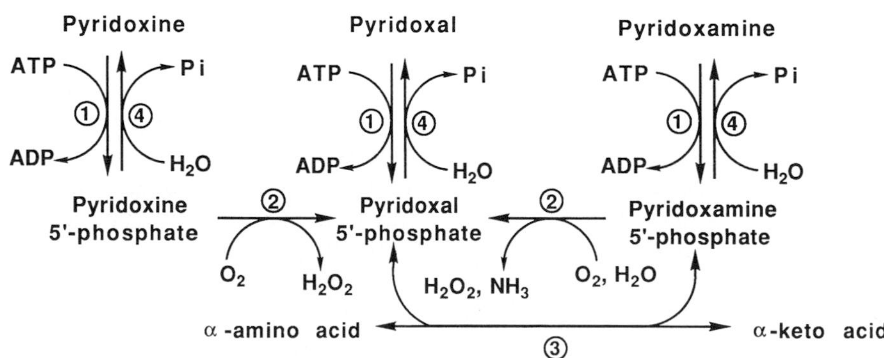

FIGURE 13 Interconversions of the vitamin B_6 group with coenzymes catalyzed by (1) pyridoxal kinase, (2) pyridoxine (pyridoxamine) 5′-phosphate oxidase, (3) aminotransferases, and (4) phosphatases.

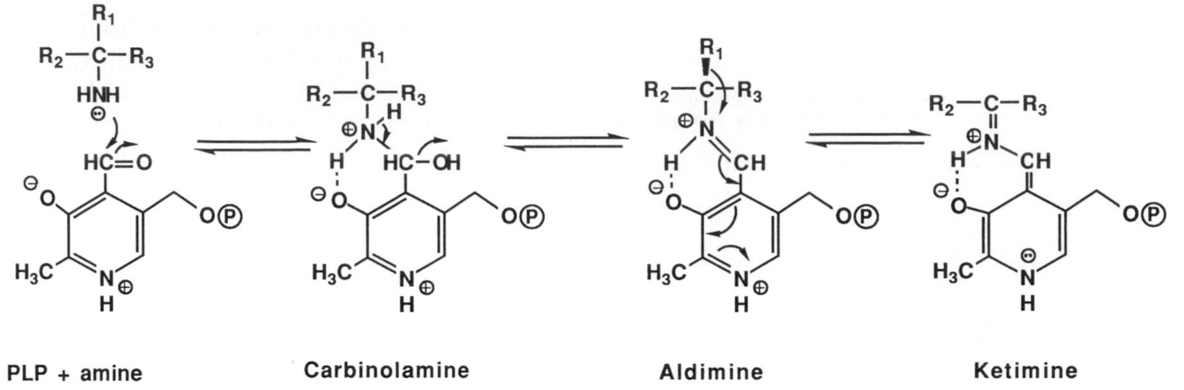

FIGURE 14 Operation of PLP with a generalized amine.

acid, to form an azomethine (Schiff base) linkage, a conjugated double-bond system extending from the α-carbon of the amine (amino acid) to the pyridinium nitrogen in PLP results in reduced electron density around the α-carbon. This potentially weakens each of the bonds from the amine (amino acid) carbon to the adjoined functions (hydrogen, carboxyl, or side chain). A given apoenzyme then locks in a particular configuration of the coenzyme–substrate compound such that maximal overlap of the bond to be broken will occur with the resonant, coplanar, electron-withdrawing system of the coenzyme complex. These events are depicted in Fig. 14.

Aminotransferases effect rupture of the α-hydrogen bond of an amino acid with ultimate formation of an α-keto acid and PMP; this reversible reaction provides an interface between amino acid metabolism and that for ketogenic and glucogenic reactions. Amino acid decarboxylases catalyze breakage of the α-carboxyl bond and lead to irreversible formation of amines, including several that are functional in nervous tissue (e.g., epinephrine, norepinephrine, serotonin, and γ-aminobutyrate). The biosynthesis of heme depends on the early formation of δ-aminolevulinate from PLP-dependent condensation of glycine and succinyl-CoA followed by decarboxylation. There are many examples of enzymes, such as cysteine desulfhydrase and serine hydroxymethyltransferase, that affect the loss or transfer of amino acid side chains. PLP is the essential coenzyme for phosphorylase that catalyzes phosphorolysis of the α-1,4 linkages of glycogen. An important role in lipid metabolism is the PLP-dependent condensation of L-serine with palmitoyl-CoA to form 3-dehydrosphinganine, a precursor of sphingolipids. [*See* Sphingolipid Metabolism and Biology.]

A simpler, but less frequently encountered, variation on the way in which PLP functions is provided by the pyruvoyl terminus of some enzymes. In these electrophilic centers [CH_3-C(β)O-C(α)O-NH-R′], the amino function condenses with the β-carbonyl, while the α-carbonyl enhances electron withdrawal from the resulting ketimine. For example, 5-adenosylmethionine decarboxylase from mammals as well as *Escherichia coli* uses such a system to form spermidine from putrescine and methionine.

V. PTERIN COENZYMES

A. Chemistry

Among natural compounds with a pteridine nucleus, those most commonly encountered are derivatives of 2-amino-4-hydroxypteridines, which are trivially named pterins. Although a number of pterins when reduced to the 5,6,7,8-tetrahydro level function as coenzymes, the most generally utilized are poly-γ-glutamates of tetrahydrofolate (THF) (Fig. 15). The natural derivatives of tetrahydropteroylglutamates responsible for vectoring 1-carbon units in different enzymatic reactions are also shown in abbreviated form in Fig. 15. All of these bear the substituent for transfer at nitrogen 5 or 10 or are bridged between these basic centers. The number of glutamate residues varies, usually from one to seven, but a few to several glutamyls optimize binding of tetrahydrofolyl coenzymes to most enzymes requiring their function.

Less commonly encountered, but essential for some coenzymatic roles of pterins, are those compounds shown in Fig. 16. Tetrahydrobiopterin cycles with its quinoid 7,8-dihydro form during O_2-dependent hydroxylation of such aromatic amino acids as in the

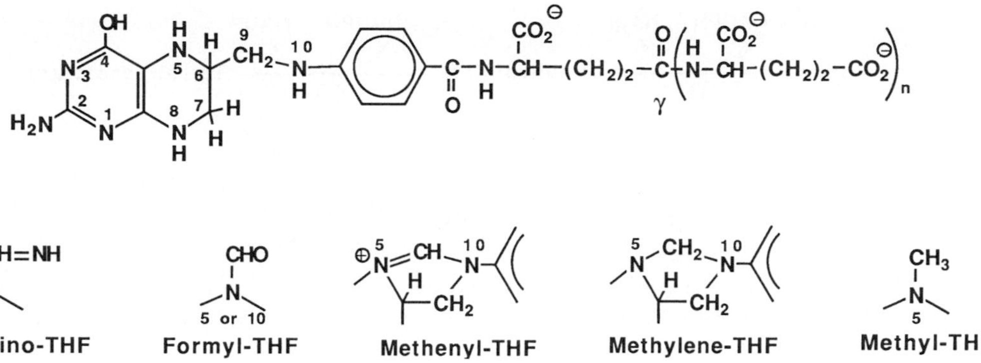

Formimino-THF Formyl-THF Methenyl-THF Methylene-THF Methyl-THF

FIGURE 15 Structures with numbering for tetrahydropteroyl-L-glutamates including the formimino, formyl, methenyl, methylene, and methyl derivatives.

conversion of phenylalanine to tyrosine. The most recently elucidated pterin cofactor in some Mo/Fe flavoproteins (e.g., xanthine dehydrogenase) is given in the right-hand side of Fig. 16.

Pterin coenzymes, most at the tetrahydro level, are sensitive to oxidation and have characteristic absorbance in ultraviolet light. Upon heating in aqueous media below pH 4, the pterin portion of the folyl-type coenzyme tends to split from the *para*-aminobenzoyl glutamate portion. The xanthine dehydrogenase pterin during isolation loses hydrogen from the side chain to generate a double bond between sulfur-bearing carbons.

B. Metabolism

The interconversions of folate with the initial coenzymatic relatives, the tetrahydrofolyl polyglutamates, are shown in Fig. 17. The dihydrofolate reductase necessary for reducing the vitamin-level compound through 7,8-dihydro to 5,6,7,8-tetrahydro levels is the target of inhibitory drugs such as aminopterin and amethopterin (methotrexate). A similar dihydropterin reductase catalyzes reduction of dihydro- to tetrahydrobiopterin. Tetrahydrofolate is intracellularly trapped and extended to polyglutamate forms that operate with THF-dependent systems. In some cases (e.g., thymidylate synthetase), there is a redox change in tetrahydro to dihydro coenzyme, which is recycled by the NADPH-dependent reductase. Turnover of coenzyme to folate at the monoglutamate level requires hydrolytic cleavage of the extra glutamyl residues. Cells of the small intestinal mucosa are especially rich in the γ-glutamyl peptidase ("conjugase") that degrades ingested natural folyl polyglutamates to the more readily absorbed folate.

C. Functions

An overview of some of the major interconnections among the 1-carbon-bearing THF coenzymes and their metabolic origins and roles are given in Fig. 18. Reactions include (1) generation and utilization of formate; (2) *de novo* purine biosynthesis wherein glycinamide ribonucleotide and 5-amino-4-imidazole carboxamide ribonucleotide are transformylated by 5,10-methenyl-THF and 10-formyl-THF, respectively; (3) pyrimidine nucleotide biosynthesis, wherein deoxyuridylate and 5,10-methylene-THF form thymidylate and dihydrofolyl coenzyme; and (4) conver-

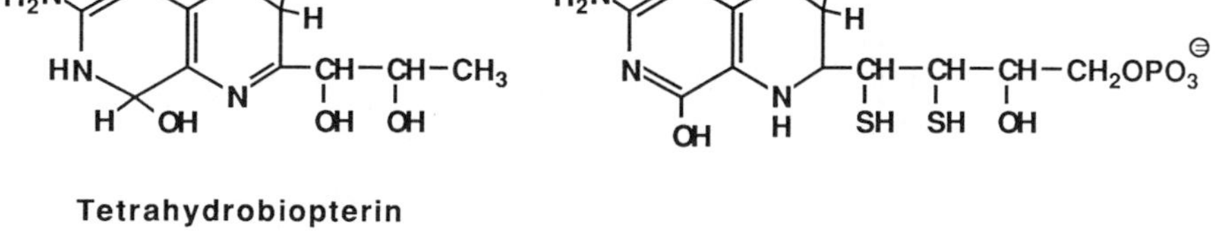

Tetrahydrobiopterin

FIGURE 16 Other pterin coenzymes: tetrahydrobiopterin of phenylalanine hydroxylase and the pterin cofactor of certain Mo/Fe flavoproteins.

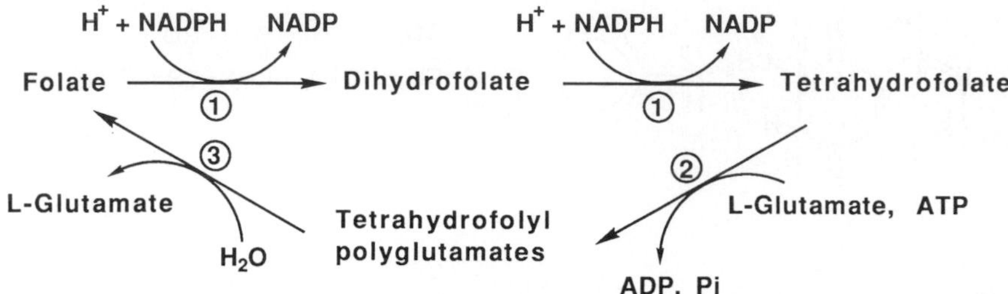

FIGURE 17 Interconversions of folate and its tetrahydro, polyglutamate coenzymes involving (1) dihydrofolate reductase, (2) folylpolyglutamate synthetase, and (3) pteroylpolyglutamate hydrolase.

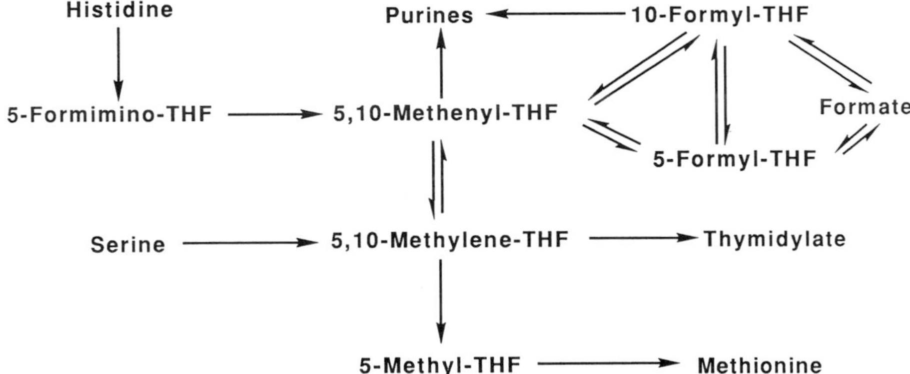

FIGURE 18 Origins, interconversions, and functions of tetrahydrofolyl coenzymes in 1-carbon transfers.

sions of some amino acids, namely, *N*-formimino-L-glutamate (from histidine catabolism) with THF to L-glutamate and 5,10-methenyl-THF (via 5-formimino-THF), L-serine with THF to glycine and 5,10-methylene-THF, and L-homocysteine with 5-methyl-THF to L-methionine and regenerated THF.

VI. B_{12} COENZYMES

A. Chemistry

The coenzyme B_{12} (CoB_{12}) known to function in most organisms, including humans, is 5′-deoxyadenosylcobalamin (Fig. 19). A second important coenzyme form is methylcobalamin (methyl-B_{12}), in which the methyl group replaces the deoxyadenosyl moiety of CoB_{12}. Some prokaryotes utilize other bases (e.g., adenine) in this position originally occupied by the cobalt-coordinated cyanide anion in cyanocobalamin, the initially isolated form of vitamin B_{12}.

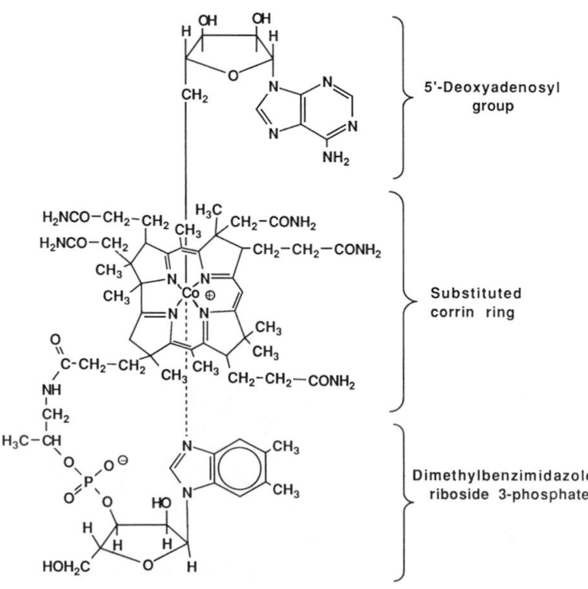

FIGURE 19 Structure with identification of principal components for coenzyme B_{12}.

Coenzyme forms of B_{12} are light-absorbing, photolabile compounds that readily undergo photolysis to yield aquocobalamin (B_{12b}), in which H_2O is coordinated to cobalt in the corrin ring. Acid hydrolysis of CoB_{12} yields hydroxocobalamin (B_{12a}) with a coordinated hydroxyl ion.

B. Metabolism

The metabolic interconversions of vitamin B_{12} as the naturally occurring hydroxocobalamin (B_{12a}) with other vitamin- and coenzyme-level forms and the two B_{12} coenzyme-dependent systems in mammals are given in Fig. 20. As outlined, B_{12a} is sequentially reduced to the paramagnetic or radical B_{12r} and further to the very reactive B_{12s}. The latter reacts in enzyme-catalyzed nucleophilic displacements of tripolyphosphate from ATP to generate CoB_{12}, or of THF from 5-methyl-THF to generate methyl-B_{12}.

C. Functions

Seemingly all CoB_{12}-dependent reactions react through a radical mechanism, and all but one (*Lactobacillus leichmanii* ribonucleotide reductase) involve a rearrangement of a vicinal group (X) and a hydrogen atom. This general mechanism is illustrated in Fig. 21. For the CoB_{12}-dependent mammalian enzyme, L-methylmalonyl-CoA mutase, X is the CoA-S-CO- group, which moves with retention of configuration from the carboxyl-bearing carbon of L(*R*)-methylmalonyl-CoA to the carbon β to the carboxyl group in succinyl-CoA. This reaction is essential for funneling propionate to the tricarboxylic acid cycle. Without CoB_{12} (from vitamin B_{12}), more methylmalonate is excreted, but also the CoA ester competes with malonyl-CoA in normal fatty acid elongation to form instead abnormal, branched-chain fatty acids. As indicated earlier (cf. Fig. 20), methyl-B_{12} is necessary in the transmethylase-catalyzed formation of L-methionine and regeneration of THF. Without this role, there would not only be no biosynthesis of the essential amino acid, but increased exogenous supply of folate would be necessary to replenish THF, which would not otherwise be recovered from its 5-methyl derivative.

VII. BIOTINYL FUNCTIONS

A. Chemistry

The coenzymatic form of biotin only occurs naturally as the vitamin (*cis*-tetrahydro-2-oxothieno[3,4-*d*]-imidazoline-4-valeric acid), which has become amide

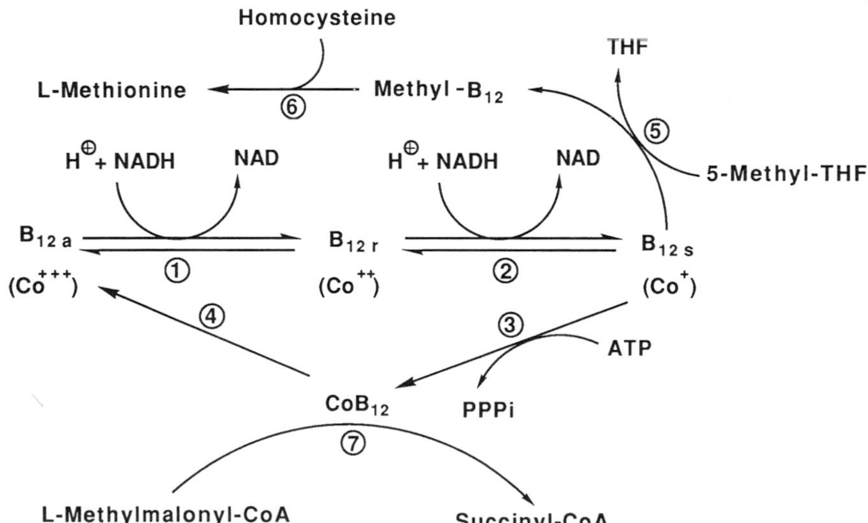

FIGURE 20 Interconversions of B_{12}, methyl-B_{12}, and Co-B_{12} in principal metabolic reactions include (1, 2) B_{12} reductases, (3) deoxyadenosyl transferase, (4) oxidations, (5, 6) methyltetrahydrofolate-homocysteine transmethylase system, and (7) L-methylmalonyl-CoA mutase.

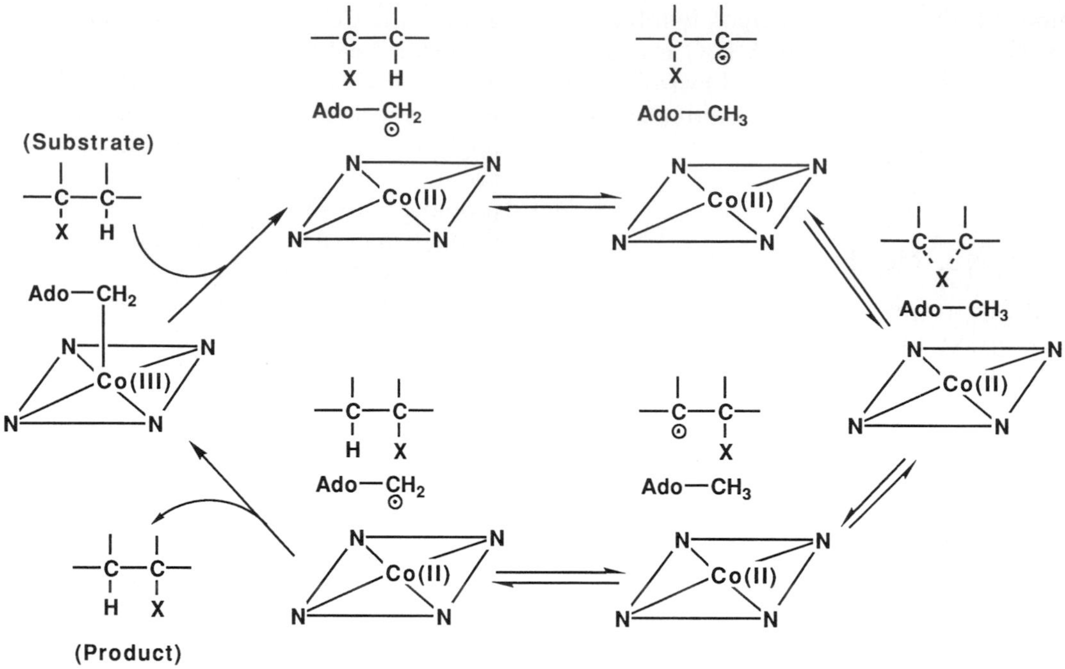

FIGURE 21 Radical intermediates in CoB_{12} enzyme reactions.

linked to the ε-amino group of specific lysyl residues in carboxylases and transcarboxylase (Fig. 22).

Though the two ureido nitrogens are essentially isoelectronic in the biotinyl moiety, the steric crowding of the thiolane side chain near N-3′ essentially prevents chemical additions, which therefore occur at the N-1′ position.

B. Metabolism

Biotin is inserted into enzymes dependent on its operation by a holoenzyme synthetase that forms biotinyl 5′-adenylate as an intermediate from the vitamin and ATP. Usual proteolytic turnover of biotinylated enzymes releases biocytin (ε-*N*-biotinyl-L-lysine), which is further hydrolyzed to release the vitamin and amino acid in a reaction catalyzed by biotinidase (biocytinase, biotin amidohydrolase).

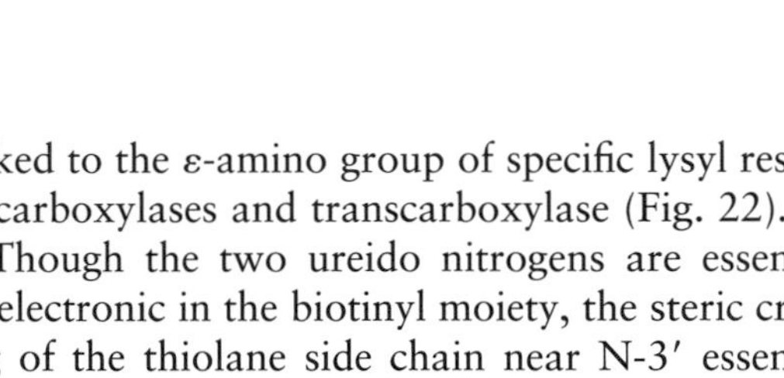

FIGURE 22 Structure with numbering for *d*-biotinyl-L-lysyl coenzyme moiety.

C. Functions

There are nine known biotin-dependent enzymes: six carboxylases, two decarboxylases, and a transcarboxylase. Of these, only four carboxylases are found in tissues of humans and other mammals. These carboxylases, named for their substrates, are (1) a cytosolic enzyme that converts acetyl-CoA to malonyl-CoA for fatty acid biosynthesis, (2) a mitochondrial enzyme that converts pyruvate to oxaloacetate for citrate formation, (3) the enzyme for converting propionyl-CoA to D-methylmalonyl-CoA, and (4) the enzyme that carboxylates β-methylcrotonyl-CoA from L-leucine catabolism to form β-methylglutaconyl-CoA.

Biotin-dependent carboxylases operate by a common mechanism (Fig. 23). This involves phosphorylation of bicarbonate by ATP to form carbonyl phosphate, followed by transfer of the carboxyl group from this electrophilic mixed-acid anhydride to the sterically less hindered and nucleophilically enhanced N-1′ of the biotinyl moiety. The resulting N(1′)-carboxybiotinyl enzyme can then exchange the carboxylate function with a reactive center in a sub-

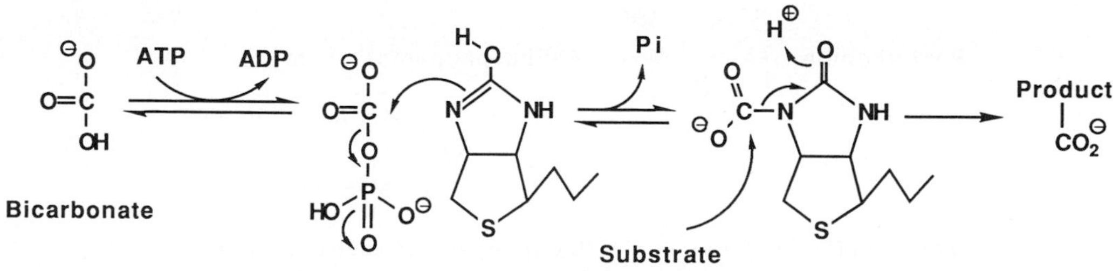

FIGURE 23 Mechanism of carboxylations by biotin-dependent enzymes with the intermediacy of the putative carbonyl phosphate.

strate, typically at a carbon with incipient carbanion character.

VIII. PHOSPHOPANTETHEINYL COENZYMES

A. Chemistry

The 4′-phosphopantetheinyl moiety, derived from the vitamin pantothenate [D-N-(2,4-dihydroxy-3,3-dimethyl-butyryl)-β-alanine] and β-mercaptoethylamine, serves as a functional component within the structure of coenzyme A (CoA) (Fig. 24) and as a prosthetic group covalently attached to a seryl residue of acyl carrier protein (ACP).

Because of the thiol terminus with a pKa near 9, phosphopantetheine and its coenzymatic forms are readily oxidized to the catalytically inactive disulfides.

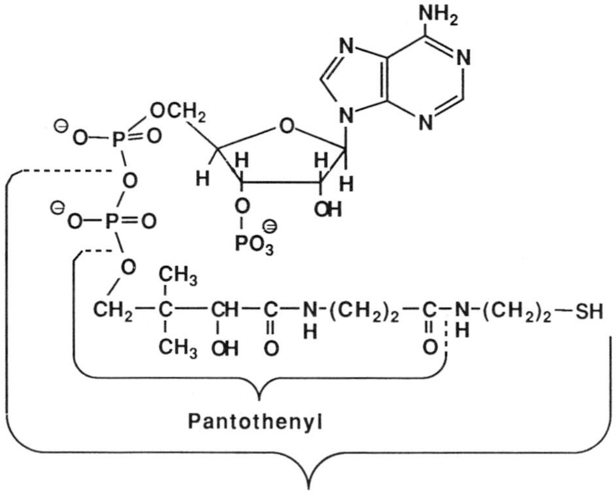

FIGURE 24 Structure for coenzyme A with pantothenyl and 4′-phosphopantetheine as components.

CoA has a strong absorption maximum at 260 nm attributable to the adenine moiety.

B. Metabolism

Within cells, synthesis of CoA and the 4′-phosphopantetheinyl moiety of ACP occurs via successive enzyme-catalyzed conversions (Fig. 25). Following the ATP-driven steps in biosynthesis of CoA, formation of ACP occurs by transfer of the 4′-phosphopantetheinyl moiety of CoA, which binds via a phosphodiester link to apo-ACP in a reaction catalyzed by ACP holoprotein synthase. About 80% of pantothenate in animal tissues is in CoA form, much of it as thioesters. The rest exists mainly as phosphopantetheine and phosphopantothenate. Cleavage enzymes catalyzing hydrolysis of the phosphate moieties (CoA → dephospho-CoA → 4′-phosphopantetheine → pantetheine) and release of β-mercaptoethylamine (cysteamine) and pantothenate from pantetheine operate during turnover and release of the vitamin, some of which is excreted in urine.

C. Functions

The myriad acyl thioesters of CoA are central to the metabolism of numerous compounds, especially lipids, and the ultimate catabolic disposition of carbohydrates and ketogenic amino acids. The chemical properties of the thioester, which has a high group-transfer potential, permit facile acylations and hydrolysis; the ready formation of enolate ions and the carbanion-like character of the carbon α to the carbonyl facilitate condensation reactions. For example, acetyl-CoA, which is formed during metabolism of carbohydrates, fats, and some amino acids, can acetylate compounds such as choline and hexosamines to produce essential biochemicals; it can also condense with other metabolites such as oxaloacetate to supply citrate, and it can

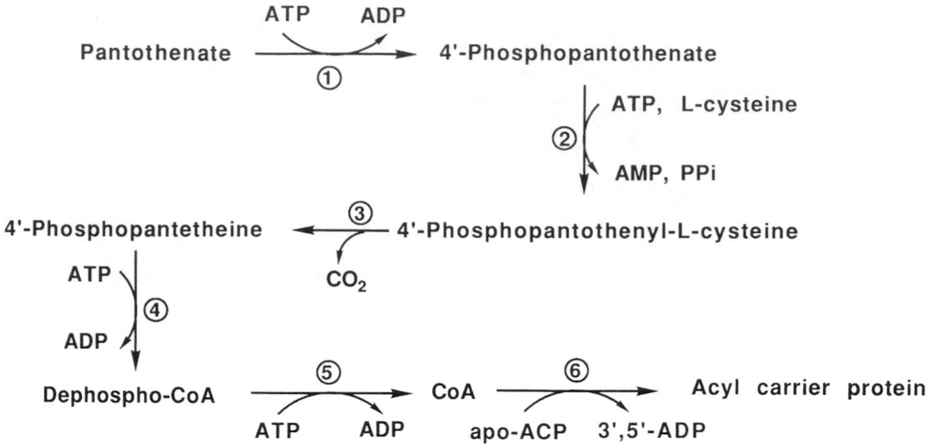

Pantothenate —(1) ATP → ADP→ 4'-Phosphopantothenate

4'-Phosphopantothenate —(2) ATP, L-cysteine → AMP, PPi→ 4'-Phosphopantothenyl-L-cysteine

4'-Phosphopantothenyl-L-cysteine —(3) → CO_2→ 4'-Phosphopantetheine

4'-Phosphopantetheine —(4) ATP → ADP→ Dephospho-CoA

Dephospho-CoA —(5) ATP → ADP→ CoA —(6) apo-ACP → 3',5'-ADP→ Acyl carrier protein

FIGURE 25 Biosynthesis of CoA and the 4′-phosphopantetheine of ACP from pantothenate involving (1) pantothenate kinase, (2, 3) synthase and decarboxylase for 4′-phosphopantothenyl-L-cysteine, (4) a so-called pyrophosphorylase, (5) dephospho-CoA kinase, and (6) ACP synthase.

lead to cholesterol. The reactive sulfhydryl termini of ACP provide exchange points for acetyl-CoA and malonyl-CoA. The ACP-S-malonyl thioester can chain-elongate during fatty acid biosynthesis in a synthase complex.

IX. LIPOYL FUNCTIONS

A. Chemistry

The coenzyme form of α-lipoic acid (thioctic acid) (Fig. 26) occurs in amide linkage to the ε-amino group of lysyl residues within transacylases. Hence, the functional dithiolane ring is on an extended, flexible arm.

Natural *d*-lipoic acid (1,2-dithiolane-3-pentanoic acid) is more soluble in organic than aqueous media and has a relatively weak absorption maximum at 333 nm. Because of the considerable ring strain, the sulfurs are especially subject to chemical and photochemical attack. The reduction potential E_0' at pH 7, 25°C) is -0.3 V and near that of pyridine nucleotide coenzymes.

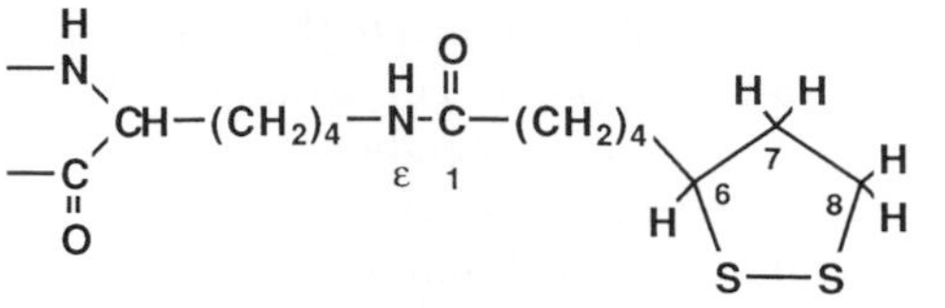

FIGURE 26 Structure with numbering for *d*-lipoyl-L-lysyl coenzyme moiety.

B. Metabolism

Lipoic acid is attached to appropriate lysyl residues of acceptor apoproteins via the 5′-adenylate in an ATP-dependent reaction catalyzed by a holoenzyme synthetase. Proteolytic turnover of lipoyl enzymes involves release of the ε-N-amide of lysine and further hydrolysis to release lipoate.

C. Functions

Operation of lipoyl residues within transacylase subunits of multienzyme complexes that function as α-keto acid dehydrogenases for facultative and aerobic organisms is shown in Fig. 27. The lipoyl function mediates the transfer of electrons and activated acyl groups from the thiazole-attached α-hydroxyalkyl-TPP (cf. Section I). In this process, the disulfide bond is broken and a dihydrolipoyl residue is transiently generated; reoxidation requires coupling to a flavoprotein system. All α-keto acid dehydrogenases are composed of varying numbers of the three basic subunits, that is, TPP-containing α-keto acid decarboxylases, lipoyl-containing transacylases, and the FAD-containing dihydrolipoyl dehydrogenase. In humans and other animals, the three important and distinct mitochondrial α-keto acid dehydrogenases are for pyruvate to generate acetyl-CoA utilizing a transacetylase, α-ketoglutarate that forms succinyl-CoA using a *trans*-succinylase, and branched-chain keto acids from certain amino acids (valine, leucine, isoleucine) on which yet another transacylase is at work to yield the acyl-CoA products.

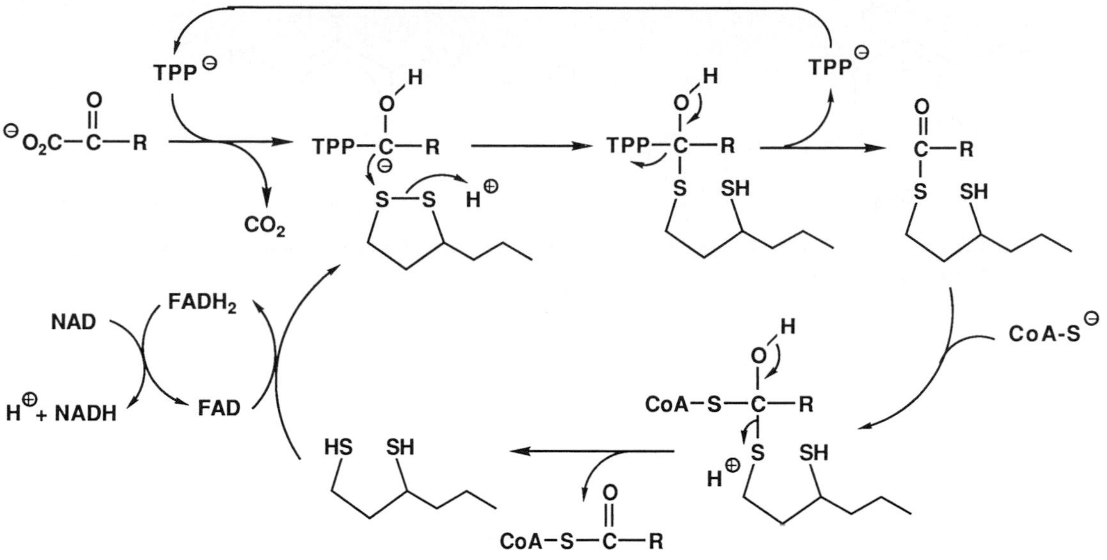

FIGURE 27 Function of the lipoyl-dependent enzymes involved in transacylations following α-keto acid decarboxylations (see Fig. 3). The transfer of an acyl moiety from an α-hydroxyalkyl-TPP to the lipoyl group and thence to CoA results in formation of the dihydrolipoyl group, which is cyclically reoxidized by dihydrolipoyl dehydrogenase.

X. PYRROLOQUINOLINE QUINONE

A. Chemistry

Pyrroloquinoline quinone (PQQ) (Fig. 28) is a more recently studied cofactor that was originally isolated from bacteria possessing a methanol dehydrogenase. It is now recognized to be a functional coenzyme for many quinoproteins that occur in many organisms, including humans, and catalyze a range of oxidation–reduction reactions that replace or supplement other longer-known redox coenzymes. PQQ is usually tightly bound to its apoprotein partners; some of these associations probably involve esterification of the carboxyl groups to amino acid residues in the proteins.

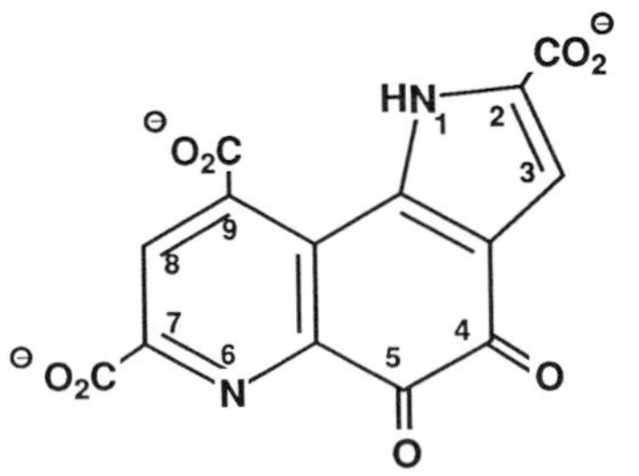

FIGURE 28 Structure with numbering for pyrroloquinoline quinone (PQQ; methoxatin; 4,5-dihydro-4,5-dioxo-1*H*-[2,3-*f*]-quinoline-2,7,9-tricarboxylate).

PQQ is a reddish compound (λ max = 249, 325, and 475 nm at pH 7), which is fluorescent. Although fairly resistant to acidic and basic conditions, the quinone carbonyls are readily attacked by nucleophiles. PQQ is easily reduced to the nonfluorescent quinol with absorbance at 302 nm (pH 7). The redox potential is fairly high at an E_0' (pH 7) near +0.08 V.

B. Metabolism

Stringent experiments with rats suggest that PQQ may join the list of vitamins required by mammals. Microflora in the gastrointestinal tract, however, may symbiotically supply some of this cofactor to the host. Two amino acids have been implicated as the biosynthetic precursors in a *Methylobacterium*. That portion of the structure including ring atoms 6, 7, 8, and 9 and the two carboxyl groups at 7′ and 9′ is derived intact from glutamate; the remainder is from a symmetric product of the shikimate pathway, probably the L-dopaquinone produced from tyrosine in which the pyrrolic N-1 of PQQ was the α-amino group.

C. Functions

PQQ is a redox component of holoenzymes in which the coenzyme can be reduced by successive one-elec-

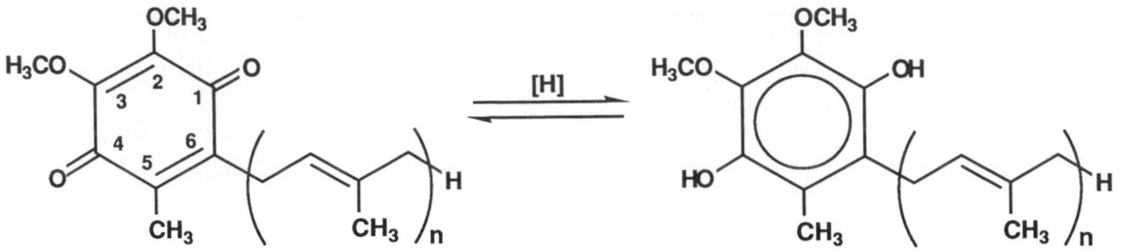

FIGURE 29 Structure with numbering of oxidized and reduced forms of ubiquinones (coenzyme Q) (n = 6–10 isoprenoid units).

tron steps to the radical PQQH˙ and further to $PQQH_2$. In some enzymes, however, no radical intermediate is detectable, so two-electron processes seem to operate in which addition reactions may accompany hydrogen abstractions.

An important mammalian enzyme that may contain a PQQ-like cofactor is lysyl oxidase needed for collagen cross-linking. Dopamine-β-hydroxylase, which is essential for production of the neurotransmitter norepinephrine, contains a somewhat related trihydroxyphenylamine derivative as a cofactor.

XI. UBIQUINONES

A. Chemistry

The ubiquinones (coenzyme Q's) are a group of "ubiquitous" 2,3-dimethoxy-5-methylbenzoquinones substituted at position 6 with variable-length terpenoid chains. The reduction of these to the 1,4-hydroquinone forms is illustrated in Fig. 29.

The ubiquinones are strongly hydrophobic and soluble only in organic solvents. The quinone form has an absorption band at 275 nm, which disappears upon reduction to the quinol (dihydroquinone). As with similar quinones, reduction can proceed by a two-electron process or by two single-electron transfers through a semiquinone intermediate, which can be neutral (QH˙) or anionic ($Q^{\bar{\cdot}}$) depending on pH. The redox potential is +0.11 V.

B. Metabolism

The shikimate pathway is involved in biosynthesis of ubiquinones as well as plastoquinones, tocopherols (vitamin E), and nathoquinones (vitamin K). Because this pathway, also used in formation of phenylalanine and tyrosine, does not occur in mammals, the basic ring structures derive from bacteria and plants. Addition of the side chains, however, can be accomplished by microsomal prenylation in most organisms, including humans.

C. Functions

In eukaryotes, ubiquinones are found mainly in the mitochondrial inner membrane, where they function to accept electrons from several different dehydrogenases and relay them to the cytochrome system. Among substrates serving as electron donors are succinate and glycerol 3-phosphate, which are initially oxidized by flavoprotein systems, and malate, α-ketoglutarate, and β-hydroxybutyrate, which transfer their electrons initially to NAD. Hence, the positioning of ubiquinones in the electron–oxygen transfer process is near cytochrome b, after flavoproteins that couple in some cases with pyridine nucleotide-dependent enzymes, and before cytochromes c_1, c, a, and a_3. The relatively fluid movement of most of the ubiquinone within the phospholipid bilayer of the membrane serves well to provide a more mobile single-electron carrier between the rather large and rigid complexes that are constituted by other redox participants. In this connection, the relatively higher concentration of ubiquinone facilitates contact and the efficiency of electron flow. In heart mitochondria, for example, there is about seven times more ubiquinone than cytochrome a_3.

XII. METHANOGEN COENZYMES

A. Chemistry

The diverse coenzymes required to convert CO_2 to CH_4 are structurally elaborated and given in Fig. 30. These rather novel compounds vary from a complex nickel-porphyrin-like F_{430} to the phenyl ether of methanofuran, the unusual pterin of tetrahydromethanopterin, the deazaflavin of F_{420}, to the somewhat simpler sulfur-containing mercaptoheptanoylthreonine and coenzyme M.

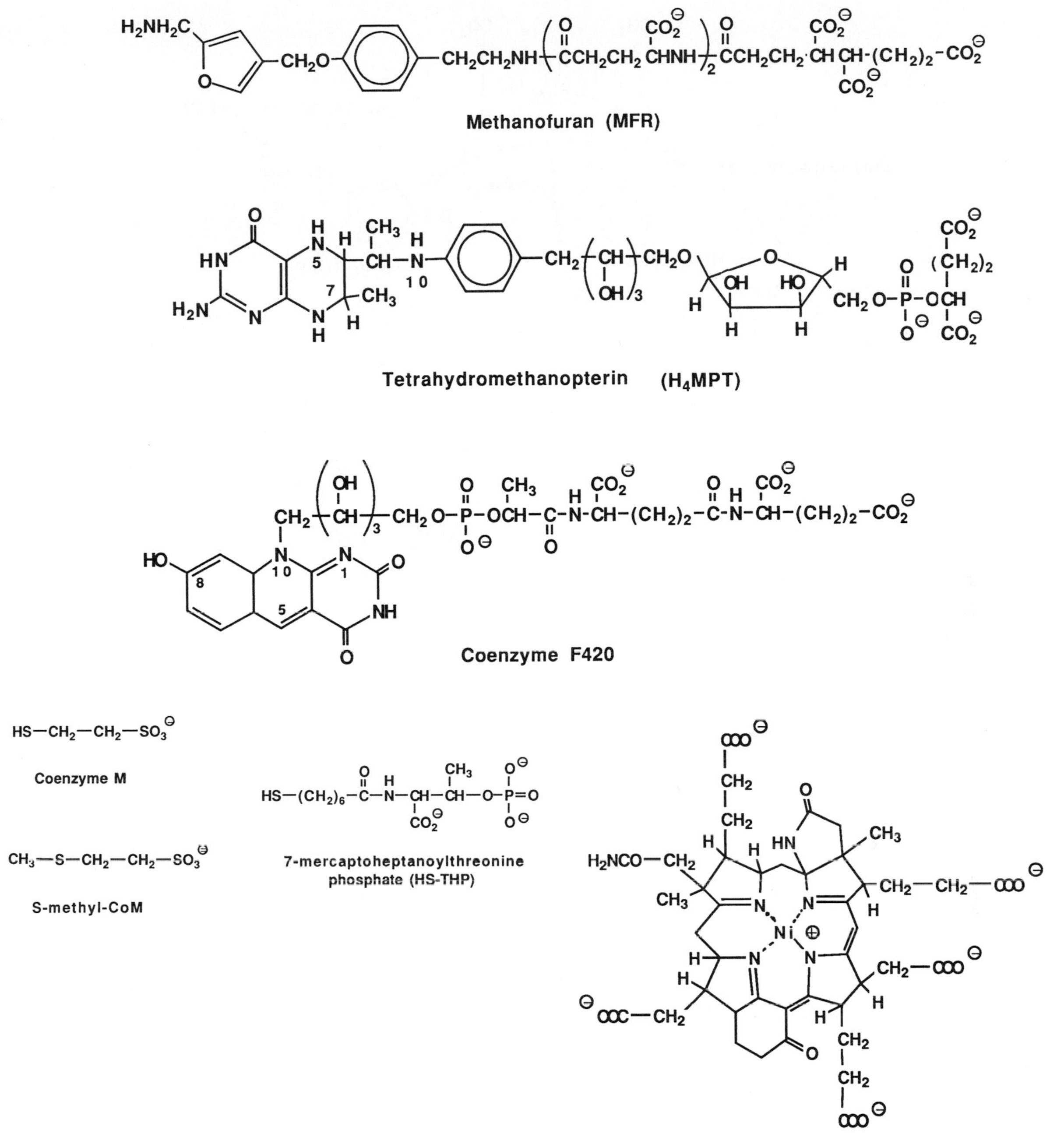

Methanofuran (MFR)

Tetrahydromethanopterin (H_4MPT)

Coenzyme F420

Coenzyme M

S-methyl-CoM

7-mercaptoheptanoylthreonine phosphate (HS-THP)

FIGURE 30 Structures of coenzymes that participate in methanogenesis by certain archaebacteria.

B. Function

The proposed cycle in which the various coenzymes function in the methyl-coenzyme M methylreductase system is given in Fig. 31. The highly specialized archaebacteria (known as methanogens) that can accomplish such reduction of carbon dioxide through the formyl, methenyl, methylene, and methyl stages to methane are widespread in nature. The importance of the coenzymes that participate can be appreciated by the quantitative and global aspects of carbon cycling through methane-forming systems, which are not only in free-living microbes, but also in microbes in the intestinal tract of the human.

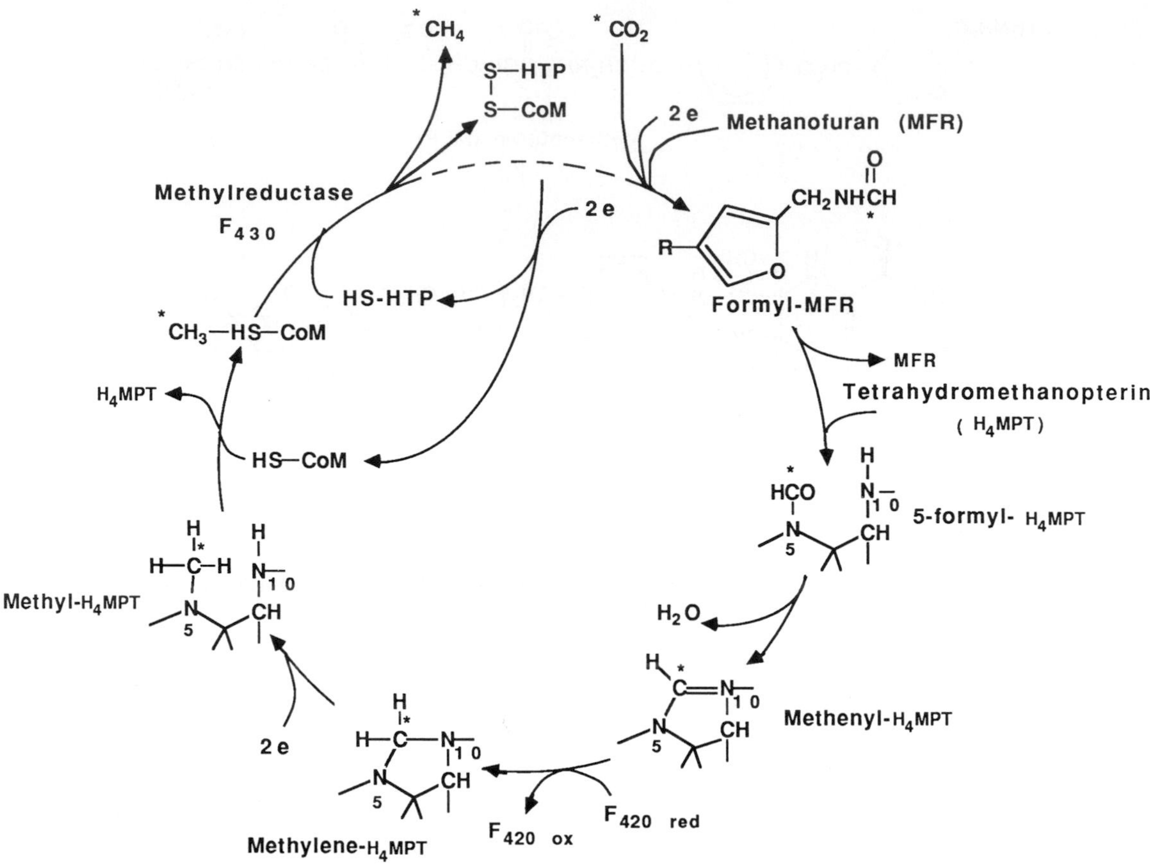

FIGURE 31 Proposed cycle for the reduction of carbon dioxide to methane. The carbon originating from CO_2 is tagged with an asterisk to follow its sequential fate in attachment to coenzymes during ultimate conversion to CH_4.

BIBLIOGRAPHY

Bender, D. A. (ed.) (1992). "Nutritional Biochemistry of the Vitamins." Cambridge Univ. Press, Cambridge, England.

Chytil, F., and McCormick, D. B. (eds.) (1986). Vitamins and coenzymes, Parts G and H. *In* "Methods in Enzymology," Vols. 122 and 123. Academic Press, New York.

Duine, J. A., Frank Jzn, J., and Jongejan, J. A. (1987). Enzymology of quinoproteins. *In* "Advances in Enzymology" (A. Meister, ed.), Vol. 59, pp. 170–212. John Wiley & Sons, New York.

Edmondson, D. E., and McCormick, D. B. (eds.) (1987). "Flavins and Flavoproteins." W. de Gruyter, New York.

Frey, P. A. (1988). Structure and function of coenzymes. *In* "Biochemistry" (G. Zubay, coord. au.). Macmillan, New York.

Korpela, T., and Christen, P. (eds.) (1987). "Biochemistry of Vitamin B_6." Birkhauser Verlag, Boston.

Rouvière, P. E., and Wolfe, R. E. (1988). Novel biochemistry of methanogenesis (minireview). *J. Biol. Chem.* **263,** 7913–7916.

Voet, D., and Voet, J. G. (1995). "Biochemistry," Second edition. John Wiley & Sons, New York.

Cognitive Representation in the Brain

MARK H. BICKHARD
Lehigh University

GLOSSARY

Connectionism Approach to representation as being distributed in the patterns of activations of nodes within a network. Inspired in part by the high interconnectivity of the nervous system

Encodingism View that representation is fundamentally constituted as elements or structures that are known correspondences with what they represent

Ensemble Population of active elements in which the statistical properties of single elements over time are equal to the statistical properties of the population of elements at a single time

Incoherence problem Impossibility of specifying what an encoding is supposed to represent except in terms of some other already available representation, and the incoherence that results when that regress is supposed to halt with foundational encodings

Information processing View of cognition as consisting of the processing—the manipulation, combination, and generation—of symbolic encodings

Interactivism Functional and emergent approach to representation. Representation emerges as interactive differentiations of environments and consequent indications of possible further system activity in the service of goal-directed interactions

Skepticism Argument that it is impossible to check the accuracy of our representations of the world because it is impossible to know anything about the world except in terms of those representations themselves. Any purported check, then, is checking the representations against themselves—it is circular.

Transduction Transformation of form of energy. Also, a supposed generation of sensory encodings from encounters with environmental energy

STUDIES OF THE COGNITIVE ASPECTS OF BRAIN functioning cannot proceed without assumptions concerning the nature of representation. Since the demise of classical associationism, those assumptions, both in cognitive neuroscience and in cognitive psychology in general, have been dominated by the computer-inspired information processing model. In this model, representations are taken as being constituted as symbolic encodings, which are generated, processed, and transmitted by the central nervous system (CNS). This model has dominated so long and so thoroughly that at times it has seemed to attain the level of unquestionable common sense—the way things obviously must be.

More recently, however, several competing alternatives to this standard position, and criticisms of these standard assumptions, have arisen. As a result, future studies will increasingly be forced to take into explicit account their conceptual assumptions concerning representation as well as the neurophysiological and psychological results against which their models are tested.

One important alternative to information processing views is that of connectionism or parallel distributed processing (PDP). Another position has recently emerged in robotics, but has not yet had much impact in brain studies. This approach attempts to eschew representation altogether in favor of systems

ENCYCLOPEDIA OF HUMAN BIOLOGY, Second Edition, VOLUME 2. Copyright © 1997 by Academic Press. All rights of reproduction in any form reserved.

without data structures that can nevertheless accomplish their goals.

All three of these positions, however, information processing and its two alternatives, *share* one basic assumption concerning the nature of representation, in spite of their differences in how that assumption is developed. The assumption in common among them is that representation is constituted as encodings, whether or not these are taken to be symbolic. That assumption is itself subject to severe criticism, which in turn leads to a fourth alternative to all three positions: an interactive conception of representation.

This article is primarily a review of the four positions concerning the nature of representation and some of the arguments among them. With respect to the information processing position especially, this will also involve some illustrative examples of how that approach can be applied to brain functioning. I will be arguing in favor of the interactivist position.

I. INFORMATION PROCESSING

The backbone of the information processing perspective is the presumed flow of information from environment to perception to cognition to language. Information originates in the environment and is processed through the senses into the brain or mind, where further cognitive processing occurs, and where new encodings of resultant mental contents can be generated and transmitted as language utterances. Those utterances, in turn, will be received by an audience and decoded in accordance with their semantics into cognitive contents for that audience. The basic flow, then, is from the environment, through the senses and cognition, and into the environment again as language—from which it will in general reenter the nervous system via the perception and understanding of language.

The various steps of this sequence are of three general kinds: the transduction of new encoding elements in the primary perceptual organs in response to encounters with environmental information, the generation of new encodings on the basis of already present encodings in the various processing steps—this is a form of heuristic and perhaps implicit inference of new encodings on the "premise" of extant encodings—and the emission of encodings via language. All three of these kinds of processes are being investigated, and knowledge of differentiations and specializations within the central nervous system by sensory modality and form of cognition is growing rapidly.

The sensory nervous system is generally considered to provide only two possible forms of basic encoding. The transduction process must result in signals being transmitted along various *axons* carrying some *frequency* of impulse. Basic sensory encodings, then, must be implemented in some combination of line (axon or spatial) and frequency (temporal) encoding.

For example, human color vision is based on three different types of receptors, each attuned to transduce a differing range of electromagnetic wavelengths. The transduction sensitivities of these types of receptors are maximal in, respectively, blue, green, and red ranges of color. This gives rise to a *line encoding* of *color,* which undergoes several staages of further processing in the retina, the visual pathways, and the visual cortex. Both color receptors (cones) and more general light intensity receptors (rods) are distributed spatially over the retina (with cones concentrated in the fovea) and this gives rise to a *line encoding* of relative *spatial position* of light reception. *Light intensity* itself receives a *frequency encoding.* The topology of these spatial relationships tends to be maintained through the several layers of further processing—thus, the spatial encoding of relative position of reception tends to be maintained. The auditory system, for another example, yields primarily a *line encoding* of *frequency,* although lower frequencies seem to involve some degree of *frequency encoding.* [*See* Color Vision; Ears and Hearing.]

These relatively simple correspondences between properties of the stimulus, on the one hand, and lines or frequencies of neural activity, on the other, become more complex and less well understood with progressive steps of processing. Some of the more complicated and well-known examples are the apparent motion and "feature" detectors of the visual system. The feature detectors seem to be sensitive to features such as edges and orientations—important properties of visual boundaries. [*See* Visual System.]

Such models of perceptual encoding are based on single neurons and their activities as the elements of the presumed encodings. However much insight these models may provide for perceptual representation, there are strong reasons to think that single neurons are not the locus of representation in the central nervous system. One consideration is simply that the limiting case of such single-neuron representation yields single-neuron "encodings" of all of our concepts and representations—a common *reductio ad absurdum* of this is the infamous "grandmother neuron" that represents our grandmother. Since there is a continual loss of neurons to cell death, we would experience random and total losses of representation of whatever those neurons represented, including, potentially, our grandmothers. Such unitized representational losses are not found with neural death, thus single neurons cannot be the microanatomical locus of concept representation. Furthermore, activities of single cortical neurons in response to repeated stimulus instances are found in general to be highly unreliable, also not giving a foundation for a single-neuron locus of representational encoding.

One design solution to this unreliability problem (cell death is itself a version of unreliability) is redundancy: if many neurons are serving the same representational function redundantly, then the loss of one or more, or the unreliability of all of them, can in principle be compensated for by the activity of the whole redundant set. A more powerful hypothesis, however, is that the functional unit is not the single neuron at all, but rather local populations of neurons that function as statistical ensembles. The relevant properties of such ensembles would be the temporally and perhaps spatially organized patterns of oscillations within the ensembles, which would modulate the similar activities of other ensembles. Cognitive activity would consist of such modulatory processes disseminating and interacting throughout the brain.

This is a very different notion from the classical "switchboard" model of brain activity in which impulses are generated at perceptual surfaces and then switched to various output neurons within the CNS. In this long outdated switchboard model, even the notion of frequency encoding is distorted in that the switchboard metaphor emphasizes on and off relationships, not frequencies. The ensemble model builds on the spatial and frequency characteristics of neural functioning, provides a redundancy with respect to single-level neurons in that the ensembles will exhibit reliable statistical properties of their oscillations in spite of single-neuron unreliability (this is in effect a reduction of noise or error variance via a larger sample), and, for the first time among the models discussed, acknowledges the endogenous activity of the central nervous system.

This later point is potentially quite important. Neurons in general are not quiescent until stimulated by synaptic transmissions or sensory input. Neurons exhibit baseline frequencies of axonal impulses, varying from neural type to neural type, and varying from zero to high frequency. This intrinsic neural activity is continuous. Sensory inputs do not switch this on or off so much as they modulate the frequencies and patterns in which this ongoing activity takes place. The ensemble notion is a population-level version of this basic point concerning even single neurons.

Perception as a modulation of internal endogenous activity is a very different notion than is suggested by the simple information processing flow from environment to perception to cognition. Modulation is not the same relationship as simple encoded input. Even the information processing models, however, acknowledge the necessity of contributions to perception and cognition from previously learned or innate sources—the sensory information is not adequate to cognition, nor, most models hold, even to perception. Memories, for example, might be postulated as involved in the inferences from basic sensory reception to full perceptions of objects located and moving in space and time. Modulation of ongoing activity, then, is not at such deep variance with such versions of the information processing approach. [*See* Perception.]

Even these forms of information processing models, however, retain the presupposition that all cognition is ultimately input from the environment (or perhaps provided innately). If not present in current sensory input, relevant information must have been provided in earlier experiences and be available in memory. That is, such models, except for the "out" of innatism, force an empiricist epistemology, in which the senses are the source of all knowledge. Empiricist epistemologies have not fared well in epistemology or the philosophy of science. Understanding the necessity of mathematical relationships, such as $1 + 1 = 2$, for example, has been a classical counter to empiricisms—it might be conceivable that we could learn *that* $1 + 1 = 2$ simply from experience, but no amount of experience will ever provide knowledge that this relationship is logically necessary. Mathematics would remain on par with, say, astronomy, in which the number of planets also remains consistent no matter how many times you look, but that number is *not* necessary. For that matter, there does not seem to be any perceptual realm at all for mathematics—we can see pebbles, but not numbers. Many other cognitions, such as of virtues and vices, are not presentable in sensory form. Such considerations suggest that the endogenous activity of the CNS is not simply bringing to bear previously perceived information, but that it is involved in some way in *emergent* representational phenomena.

A convergent consideration for the notion of endogenously active ensembles as units of functional activity is the acknowledgment that the organism is fundamentally engaged in physical activity in the environment. Such interaction with the world requires in most cases *correct timing* of the organism's side of the interactions. Correct timing means neither too fast nor too slow nor in the wrong phase. Walking, for example, is not so much a matter of pushing the legs back and forth as it is a matter of exciting and modulating an intrinsic oscillation in the spinal cord, and of the skeletal–musculatory system itself. Most activities require such timing—driving a car, catching a ball, running, and so on—and timing is fundamentally based on oscillatory phenomena. We would expect, therefore, that oscillations and modulations would be fundamental to the operating design principles of the nervous system. Ensembles not only provide an answer to the problem of unreliability, and pose the problem and the promise of intrinsic endogenous activity, but they are also endowed with the basic solution to the problem of timing in action and interaction.

The information processing approach has no intrinsic place for timing. The sequence of operations on symbolic encodings has all the same *formal* properties no matter what the timing may be of the steps involved in that sequence. It is clear that results may be obtained too late to be of any good and, thus, that *speed* is necessary, but in this view timing is irrelevant to the nature of cognitive activity per se. Cognition abstracts away from the timing considerations that are essential to action, according to this view, but even so we could expect the timing properties of oscillatory phenomena to dominate the functioning of the CNS. I will argue later that timing is in fact *not* irrelevant to cognition in general.

II. CONNECTIONISM

Consideration of the highly parallel manner of functioning of neurons and neural circuitry, and of the enormously complex interconnectedness of neural circuitry, contributed to the inspiration of one major alternative to standard information processing approaches—connectionism or parallel distributed processing. The underlying metaphor for the information processing approach is the von Neumann computer, which has only one locus of processing. Parallel processing in computers can be introduced by multiplying the number of simultaneously active processing units, but something more than this seemed to be taking place in the brain.

One highly persuasive consideration is that the brain accomplishes many tasks, such as various forms of pattern recognition, in a very short amount of time and, therefore, in a very small number of strictly sequential "steps" of processing. One potential solution was to posit that the many steps that seemed to be logically required were carried out simultaneously and in parallel in multiple processing units. This solution, however, retained the basic assumptions of the information processing approach and simply introduced multiple processors, and it was a conceivable approach only when the basic task did not involve internal dependencies that required sequential processing steps—only when the processing could in fact be broken down into multiple parallel streams.

Another strong consideration was that standard information processing approaches had failed miserably at modeling the phenomena of learning. Systems could be designed that succeeded in "learning" things that were very close to what they had been designed to solve, but any significant generalization beyond the problem for which they were designed seemed unattainable. One perspective on this failure comes from noting that the information processing approach construes all processing in terms of the generation and elimination of *instances* or various *types* of encoding elements, but the basic types of encodings themselves must be designed in from the beginning—there is no way to generate new types of symbolic encoding representations.

The major excitement of the connectionist or PDP approaches is that they seem to solve this problem of learning, of the generation of new representations. PDP systems can undergo training with respect to sets of problems, and generate solutions to them that then generalize beyond the training set. At times, the form of the generalizations found seems tantalizingly similar to human solutions to those same problems.

A PDP system engages in two levels of dynamic activity. The primary level is that activity by which a categorization, the representation, of an input array is settled upon. The secondary or metalevel of activity is that by which the system "learns" to correctly categorize such input patterns.

A PDP system is a set of functional nodes, each of which is capable of varying levles of "activation," interconnected by a fixed topology of paths. Each connection between nodes has a weight, which can be positive or negative. The nodes and interconnections are frequently organized into layers, perhaps with feedback among layers. Some subset of nodes are connected directly to the environment, from which they receive "activation," and they in turn activate the nodes to which they are connected in accordance with the weights of the connection paths. Those nodes activate still further nodes in accordance with their connectivities and weights, and so on. The system eventually settles down into a fixed pattern of levels of activation in the nodes, or in some selected subset of the nodes, that is specific to that particular pattern of inputs. The first key to the power and appeal of PDP models is that that pattern of resultant activations can be taken as a classification of the input pattern: it will classify *together* all such input patterns that yield that same (or similar) resultant activation pattern, and will classify as *different* all input patterns that result in some different final activation pattern. The class of possible final activation patterns, then, forms the class of classification *categories* for the possible input patterns. Note that the processes of the settling of the node activations is massively parallel among all the nodes and their weighted interconnectivities.

The second, and most important, key to the appeal of PDP models is that various adjustment rules can be used to adjust the weights of the connections among the nodes, in accordance with various training "experiences," so as to "learn" input pattern classifications. The organization of connections usually remains fixed in such training, but the changes in the *weights* of those connections can change the entire first-level—classifying—dynamics of the overall system. In particular, it can result in differing resultant classifications of the input patterns. In other words, the system can adjust to, can be trained to, new and desired classification schemes. Insofar as the input patterns classifying activation patterns are taken to be *representations* of those categories of input patterns, the system can be construed as generating new representations of novel input categories, something that is impossible in the standard information processing approach.

With proper design and appropriate shifts in interpretation, PDP systems can manifest still other characteristics that are simultaneously powerful, exciting, and reminiscent of the way in which the brain functions. One important example derives from the possibility of the *input* activation pattern being any of several *subpatterns* of the overall *resultant* activation pattern—so that any piece of the overall pattern as input results in the activation of the whole pattern—in which case we have a model of content addressable memory. Content addressable memory is a form of memory that permits memory representations to be accessed directly in terms of their representational *contents,* rather than just in terms of their *location* in the memory organization. Human memory, in particular, manifests this phenomenon.

A different shift in interpretation considers the input patterns themselves to be whole patterns, but the resultant activation pattern to be a composite of the permitted input patterns. Under this interpretation, the system manifests an *association* between the various input patterns—an associative memory, again manifested in human memory.

Connectionist approaches, then, capture a parallelism at least reminiscent of the functioning of the brain, model the emergence of new categorizations, and model a form of content addressable memory, associative memory, and other properties of human memory. They are an exciting alternative to information processing approaches for these and additional reasons, and are being pursued eagerly.

Connectionism is not without its critics, however. One of the most powerful criticisms of the potentialities of PDP approaches turns on what from another perspective is one of their greatest strengths—the singularity and lack of internal structure of the categorizing patterns of activation. This is an aspect of their greatest strength in that emergent, novel representations would be expected to be singular and without internal representational structure. To simply put together already available representations in some new structure is what information processing approaches already do, and does not constitute emergent novel representation. On the other hand, it is precisely the ability to construct new *structures* of already available representations, and, thus, to implicitly capture not only the resultant representation but also its internal representational structure, that is the *forte* of symbolic encoding information processing approaches. It is argued that this componentiality of representation is necessary for both language and cognition alike, and is not provided by PDP approaches. For example, any genuine cognitive system, so the argument goes, that is capable of thinking "John loves Mary" is also capable of thinking "Mary loves John." This generalization of ability is natural in a symbolic encoding framework, but not in the more holistic PDP framework—which is not to conclude that connectionist approaches are incapable of modeling such phenomena. Needless to say, the arguments and explorations continue.

One obvious notion, for example, is the possiblity of hybrid systems in which a basic PDP-type layer provides the representational categories that can then be operated on and processed in a more conventional information processing manner. The potentialities of hybrid systems are still being explored.

Connectionism boasts a natural manifestation of several inherent properties of CNS functioning: parallelism, emergentism, content addressability and associativity, and so on. Nevertheless, there are a number of inadequacies, or at least disanalogies, of connectionist approaches with respect to this comparison. For example, PDP networks "represent" by virtue of static patterns of activation of the nodes, once settled into, whereas the CNS is engaged in continuous ongoing activity. It is at least plausible, and even likely, that cognition in the brain is a function of that activity, and cannot be captured in such static models. In this respect, among others, PDP networks are *not* akin to neural ensembles. A similar observation is that the CNS is engaged in interaction with an environment (internal or external) whereas a standard categorizing PDP network has no comparable outputs at all. Furthermore, although PDP networks do manifest a kind of emergence of categorization abilities, the "learning" rules by which this is accomplished are relatively inefficient and are, in general, *not* plausible as models of learning in the brain.

III. ANTIREPRESENTATIONALIST ROBOTICS

The information processing approach encounters many problems of interpretation. One important example is what is known as the empty symbol problem. The basic notion underlying this problem involves the sense in which encoded symbols in the information processing approach are supposed to represent various events and objects and facts in the environment by virtue of being in correspondence with those events and objects and facts. Transduction, for example, is fundamentally a change in form of energy from some environmental form to some form of neural activity. There is nothing epistemic or representational in this energy-change level of consideration—such changes in form of energy or activity occur ubiquitously in the physical world, without being confused with representation. Transduction in sensory systems, however, is taken to not only be constituted by such energy changes and their resulting correspondences, but those correspondences, in turn, are taken as representing whatever those correspondences are with, whatever was in fact transduced. The empty symbol problem arises, among other ways, from consideration of how those factual correspondences could represent what has been transduced, or some other relevant aspect of the correspondence. Specifically, how does the system know what the correspondences are with, or, among the multitude of things that are in fact in correspondence—light patterns, quantum electron processes in the surfaces of objects, chemical reactions in the retina, and so on—how does the system know which are being represented? The scientist-observer can analyze these correspondences, and analyze as well which of those correspondences seem to be ecologically relevant, but this establishes only which correspondences do occur and which of those would be ecologically desirable to represent. There does not seem to be any way for the system itself to have epistemic contact with whatever it is in causal contact with, so as to obtain transduced *encodings,* not just transduced *energies.* The internal symbols, in other words, seem doomed to be empty. They differ in shape or size or some other formal properties that allow them to be differentiated and operated upon, but they carry no representational content, or at least it is not understood how they can carry any representational content *for the system itself.*

Because of this and related difficulties in the information processing approach, some researchers in robotics have proposed that representation be avoided altogether and, furthermore, that robots can function quite well without any representation at all. Insofar as that is correct, it may be that symbols and data structures are superfluous for robotics, that representation is the wrong level or the wrong sort of abstraction for robots.

In place of notions of the communication and processing of representations, design is in terms of minimally coordinated multiple-activity systems, each of which is competent to its own perceptions and actions. An energy transducer, for example, does not have to establish an encoding of a wall, so long as it successfully controls the locomotion of the system so as to avoid bumping into walls.

In important ways, this is a return to the roots of cybernetics and control theory out of which computer information models evolved. The control of effective interaction with an environment has largely been lost from information processing approaches and considered to be of little interest outside of robotic engineering. In particular, it is not generally understood to have any basic relevance to foundational issues of representation or cognitive science. Roboticists clearly have not been able to ignore such concerns quite so readily.

This antirepresentationalism of some roboticists emphasizes the interactive and hierarchically organized control structure aspects of the nervous system—aspects that are absent in both the information processing and the PDP approaches. I will be arguing that these aspects are not only of practical design relevance, but are intrinsic to the nature of cognition and representation as well.

IV. THE ENCODINGISM COMMONALITY

There is a common notion of representation among these three approaches, namely, that representation is constituted as encodings of what is to be represented. In the information processing approach, this is a direct assumption, with the basic atomic encoding types designed directly into the system. In the PDP approach, these basic encodings are presumed to emerge in the learning process of the PDP network, but what is learned is still a correspondence between patterns that is presumed to constitute a representation—an encoding. The antirepresentationalism position accepts the same notions of encoding representation, but concludes, not that they are wrong, but that robotics can proceed without them. I will argue that all three approaches are in error in this common assumption.

V. WHAT'S WRONG WITH ENCODINGISM?

A prototypic encoding is a representational stand-in. It is some element that is specialized to serve a representational function, to carry a representational content, that is determined by some *other* representation—which may also be an encoding—thus it "stands-in" for that other representation. In Morse code, for example, ". . ." stands-in for "S," whereas, bit patterns serve the stand-in function in computers. In this sense, encodings most certainly exist, and are quite powerful and useful. They specialize and differentiate representational functions, and change the form of representational elements, in such ways as to allow processing to occur that would otherwise be difficult or impossible: ". . ." can be sent over a telegraph wire, whereas "S" cannot.

The term "encoding," however, is also used in a variety of derivative ways that do not comport with these paradigmatic cases. Genes, for example, are often called encodings of the proteins whose construction they control, but they are not *representations* in any legitimate sense: instead, DNA base pair triples and strings of such are elements of a complex control organization that builds proteins. The selectivity of those DNA triples for certain amino acids in the control process is what motivates their being deemed encodings—there is a correspondence involved. This generalization of the paradigmatic encoding notion makes quite clear the seductive power of the correspondence notion, even though, in this case, there is no epistemic or knowing or representing agent at all. It is only the stand-in notion of encoding that is a representation, and therefore it is only this notion that I will analyze in terms of its sufficiency for the general notion of representation.

The existence and importance of encodings are not at issue. What is at issue is the assumption that representation is fully characterized by encodings. There is a complex of related criticisms and arguments against strictly encodingist conceptions of representations, some of which are of ancient provenance. Several of the core components of this complex of arguments will be summarized. I submit that these arguments render any simple encodingism logically incoherent: strict encodingisms cannot make logical sense, and certainly cannot ground models of cognition, at the neural level or more abstractly. That is, strict encodingism is not merely factually false, it logically cannot be true.

The first argument is that of skepticism. The basic skeptical argument first notes that in order to check whether or not our representations are correct we would have to compare those encodings against the world that they are taken to represent. But since we can know the world that is supposedly being represented only via those representations themselves, we cannot ever get independent epistemic access to the world to check our encoded representations of it. Any such check ends up being simply a circular check of our representations against themselves. The conclusion, then, is the classical skeptical contention that we cannot have genuine knowledge of the world since we cannot ever check the accuracy of our representations of it.

This inability to detect representational error has serious consequences. From a programmatic perspective, if representational *error* cannot be naturalistically accounted for, then representation itself cannot be naturalistically accounted for. From a design or modeling perspective, if representational error is not detectable by the system itself, then there can be no guidance to error-guided processes, such as goal-directedness and learning. A great deal of effort over the last years has attempted to account for the very possibility of representational error from within the encoding perspective. The basic problem is that, if representation is constituted as some kind of correspondence with what is represented, then, if the correspondence exists, the representation exists *and* it is correct, whereas if the correspondence does not exist, then the representation does not exist and, therefore, cannot be incorrect. Incorrect representation seems to be impossible. The efforts to solve this problem, however, have been from the perspective of observers or analyzers or designers of the supposed representational systems, and they could not, even if acceptable on their own terms, account for representational error that is detectable by the system itself. That is, they could not account for error-guided processes, such as goal-directedness and learning.

The second argument asks not about the accuracy of our representations, but about the construction of them in the first place. The point is that in order to construct representational elements that are in correspondence with the world, we must already know what those correspondences are to be with, but that cannot occur until the correspondent encodings are constructed. Therefore, there is no way to get started; no way to know what encodings to construct. We must already know the world before we can construct a copy of it; we must already represent the world before we can construct our representations of it.

A closely related consideration is to note that the *factual* correspondences found in sensory transduction between neural activity and environmental events do *not* establish *epistemic* correspondences. The problem concerning how to know which encodings to construct leads in this context to the question of how the system, the CNS, could possibly know what those sensory correspondences are with and, therefore, what those sensory elements are taken to be encodings of. In other words, how can the system turn nonrepresentational energy transductions into representational encodings? It would have to know what the internal neural activity was in correspondence with in the world in order to know what the representational content of that neural activity should be taken to be. It would have to already have its representation of the world in order to construct its representation of the world.

A quick apparent answer to these questions is to mention evolution, and assume that they have been solved there. But the problem is logical, and evolution has no more power to escape them than does maturation, learning, or development. Note that the power of evolution to construct active systems that successfully interact with their environments, along the lines of the antirepresentational robotics position, is *not* questioned by these arguments. What *is* put into question is the ability to construe those transductions as more than useful control signals, to construe them as encodings—where does their representational power, their representational content, come from, and how does it come into being? The factual correspondences that obtain between the environment and neural activity help explain how and why that neural activity is in fact useful, but that useful functioning does not require any encoded representations. How does, or could, evolution, maturation, learning, development, human design, or any other constructive process construct representations out of control organizations, or out of anything else, that is not representational already?

An additional level of consideration derives not from the question of the accuracy of encoded representations, nor from the question of which ones to construct, but from the question of how the system could possibly know what any encodings in a strict encodingist system were even *supposed* to represent, prior to such questions of accuracy or rational construction. For actual encodings, the answer to this question is provided by the representation for which the encoding is a stand-in: the encoding represents the same thing as that for which it stands-in. But this introduces a regress into the origin of the representational contents involved: they might be provided for encoding X in terms of encoding Y, and for Y in terms of Z, and so on, but this regress must end in a finite number of steps. If we consider a supposed grounding level of encodings, which are not stand-ins for any other representations and which are logically independent encoding representations, we find that there is no way to provide the necessary representational content. For some purported grounding level encoding X, we might attempt to define it in terms of some other representations, in which case it would not be at the ground level, contrary to hypothesis. But this leaves us at best with "X represents whatever it is that X represents," and this does not establish X as a representation of anything. The requirement for logically independent encodings in order to ground any strict encodingist model encounters a logical incoherence: logically independent encodings cannot exist.

The underlying reason for all of these problems with encodingisms is that the notion of encodings focuses on, and is enormously powerful for, change of form of representations *that already exist,* and for combinations of representations *that already exist.* There is nothing in the notion of encodings that can explain the *origin* of representation, the *emergence* of representation out of a ground that is itself not already representational. This is so whether that emergence is evolutionary, maturational, learning, developmental, or by design. Encodingisms model things that can be done with representations that are already available, so to assume that representation in the broad sense can be fully captured in a strictly encodingist model is intrinsically incoherent. It encounters a requirement that encodings cannot serve—the requirement to explain representational emergence.

The closest attempt at such explanation within the encodingist framework is that of transduction, or its closely related notion of induction—a temporally extended transduction. But, as we have seen, these do not work. In both cases, the knower must already have the representation—the cognitive category or sensory encoding content—before it can notice or detect or transduce or postulate that "corresponded-with" element or category for its environment. Something more is needed, something that can account for representational emergence.

VI. INTERACTIVISM

Consider a system, or subsystem, in interaction with its environment. The internal course of that interaction will, in general, depend both on the internal organization of the system and on the particulars of the environment being interacted with. At the end of the interaction, the system will be left in some internal state, say state A or state B. If A and B are the only two possible final states of this system or subsystem, then those internal states will serve a function of *differentiating* possible environments into two classes—those environments that leave the system in A, and those that leave the system in B. A simple version of this differentiation involves systems or subsystems that have no outputs, they passively arrive at final internal states, for example, energy transductions.

Note that at this point, the typical encodingist move would be to note that such differentiation establishes correspondences with the differentiated environments and, therefore, A and B *encode* their respective correspondent categories of environments. The first step in this move is correct: the differentiations do establish factual correspondences with whatever it is that is differentiated. The second step is invalid: those factual correspondences do not in themselves constitute encoded representations of what the correspondences are with, of what the differentiations are differentiations of. The differentiations are more primitive than encodings, yet they do involve factual correspondences. It should be clear that the correspondences noted in actual sensory systems are in fact useful as differentiations, and are epistemically nothing more than differentiations—all the system has functional access to is that state A is different from state B, and that it is currently in state A. They are not encodings.

At this point, we are roughly in the position of the antirepresentationalist roboticists—potentially useful signals in a control structure—but with the recognition that something more is necessary. In particular, the antirepresentationalism of the roboticist position *accepts* the basic encodingist notions of representation, and there is independent reason to conclude that those notions are false and incoherent. The emergence of representation must itself be accounted for, and encodingism cannot do that.

We already have the emergence of a representational sort of function, differentiation, out of system organization that is not itself representational. What is yet missing is the emergence of representational content. Purely differentiating states A and B are truly empty; they differentiate, but, in standard senses, they represent nothing. The next task is to account for the emergence of their having representational content.

Suppose that the system with final states A and B is a subsystem of a larger goal-directed system. In this larger system, in general, various alternative strategies and heuristics will be available for attempting to reach various posisble goals. In such a case, given some particular goal at a particular time, some selection among possible strategies and heuristics must be made. It may be that the system makes a choice of strategies or heuristics in part on the basis of whether the differentiating subsystem has reached final state A or B. If so—if, when attempting to reach goal G72 and final state A obtains, try strategy S17, while if final state B obtains, try strategy S46—then such functional connections constitute implicit predications concerning the environments differentiated by A and B. In particular, the predications involved are "state A environments are appropriate to strategy S17; state B environments are appropriate to strategy S46." Such implicit predications are *about* the environments, and can be true or false about those environments. They constitute *representations* of supposed environmental properties. They constitute *representational contents* attached to final states A and B.

Most importantly, they constitute *emergent* representational contents: there is nothing in the system organizations involved that is itself a representation nor that is representational in a more general functional sense. The function of representation emerges in the further selection of system activity on the basis of initial environmental differentiations.

These considerations suffice to show: (1) that the interactive model of representation suffices to account for at least one form, a functional and implicit form, of representation; (2) that this form is capable of emergence from nonrepresentational ground; and (3) that this form is not itself an encoding form. There remain, of course, many questions concerning the interactive approach to representation. Among the most important are those concerned with the adequacy of interactive representation to the many representational phenomena. One important version of the adequacy question focuses on abstract knowledge, such as mathematics; another focuses on language. Essentially, the adequacy questions lead into the basic programmatic adequacy of interactivism in general, and the answers constitute a general model of cognition, perception, representation, and language. These programmatic issues will not be pursued here. What is critical for current purposes is that we have found a conception of representation that is not an encoding, not subject to the many logical incoherencies of encodingism.

The fundamental new property is that interactive representation is functionally emergent in organizations of interactive systems. That is, it constitutes a model of the emergence of representation out of action. As such, it does not fall to the incoherence problem because the representational content is emergent in the strategies and heuristics that are selected, and they do not require any prior representations in order to come into being or to be used. It does not fall to the origins problem because the differentiation into A or B does not require the prior knowledge of what A or B environments are, nor the prior knowledge of which sort of environment the system is currently in. It does not fall to the skeptical problem because the functional information that the system is in, say, an A-type environment is tautologically certain, whereas knowledge of *properties* of A-type environments in the strategies and heuristics are defeasible, and can in fact be tested, checked, by *using* those strategies and heuristics to actually engage the environment, and checking to see if they work, if they succeed. The fact that representation is *emergent* from action has as one critically important consequence that representation can be checked *via* action without encountering the circularities of skepticism. Without that emergence, checking representation via action gets nowhere because there are no determinate representational interpretations of the actions or of their outcomes: there is no determinate crossing from action back to representation.

Further, interactive representation can serve as the ground for encodings: stand-ins for indicator states like A and B can be constructed and processed, and can be useful for exactly the reasons encodings are useful. A simple differentiator might function strictly passively, though that is intrinsically of reduced power relative to interactive versions; one form of such a passive differentiator would be a simple sensory transducer, another would be a PDP network. In this perspective, both the potential power of the connectionist approach and the limitations from their intrinsic passivity and non-goal-directedness are evident: connectionist systems are passive differentiators, and as such they cannot differentiate what would require *interaction* to differentiate, and they have no representational content—their activation patterns constitute empty "symbols." Interactive representation intrinsically and necessarily involves *open, interactive, goal-directed systems* of exactly the sort discussed by the antirepresentational roboticists: such robots, in the interactive view, *do* in fact involve representation—representation in its most fundamental form as a functional aspect of successful goal-directed interaction—and therefore there is a natural bridge to more standard encoding representations *in those circumstances in which encodings would be useful.* The general claim, clearly, is that the interactive model of representation captures the strengths of all three alternative approaches, without encountering their limitations and logical incoherences.

VII. SOME CONNECTIONS AND IMPLICATIONS

Interactivism accommodates the correspondences involved in sensory neural activity, but with a distinctly nonencoding interpretation: those correspondences are created in differentiations that are useful to the further functioning of the overall system in many and various ways, but they do not in any legitimate sense constitute representations of the light patterns, and so on, that they in fact differentiate. The level of analysis concerned with what is in fact differentiated is important to understanding *how* those sensory differentiations manage to be useful to the organism, but such analyses do not constitute analyses of what the organism knows or represents.

Intermodulations of neural ensemble oscillatory activity *are* control relationships. They are precisely what an interactive control system necessarily involves: a *control* system because that is the level at which notions of interactive system and goal-directed system are constituted, and an *oscillatory and modulatory* control system because successful action—thus successful representation in most cases—requires correct timing at all levels. Interactive representation is emergent not from abstractly sequenced action, but from correctly timed *interaction.*

In the interactive view, representation emerges in the organization of the ongoing oscillatory and modulatory activities of the CNS—differentiations of environments, and subsequent differentiations, selections, of further activity. There is no need—in fact, it is *in general* quite inappropriate—to attempt to interpret those oscillations and modulations as *themselves* constituting representations (with the caveat of derivative, secondary encodings specialized for and based on that emergent representational function).

In particular, there is no need to attempt to understand language activity in terms of various encodings being transmitted from homunculus to homunculus in the brain for various processings and understandings. In the context of the currently dominant encoding understanding of the nature of language, it has been difficult to avoid this form of analysis of language phenomena at the level of neural functioning. In fact, language *cannot* be fundamentally an encoding phenomena for exactly the same reasons that perception cannot be: encodings cannot provide representational content that is not already there, including knowledge of what an utterance of a word is supposed to represent. Thus, it would be impossible either to learn or to understand utterances if language were in fact merely encodings. Wittgenstein, among others, made essentially this point some time ago, but, in the absence of alternative conceptions and with the dominance of the information processing approach in general, it has had limited impact. [*See* Language.]

VIII. SUMMARY

Studies of cognitive phenomena in the brain have tended to maintain the same presuppositions concerning the fundamental nature of cognition and representation as has cognitive psychology in general. For some decades, those presuppositions were massively dominated by the information processing view, so much so that it began to take on a sense of taken-for-granted obviousness. More recently, several alternatives to the information processing view, and critiques of that view, have emerged. An unexamined taken-for-grantedness concerning cognition and representation, thus, no longer suffices.

I have reviewed four of these views, and argued that three of them—information processing, connectionism, and a version of robotics—although all interestingly and importantly different from each other, nevertheless share an underlying assumption concerning the nature of representation—an encodingist assumption. Furthermore, this encodingist assumption is wrong and logically incoherent in its foundations.

An alternative interactive model of representation is outlined, and it is argued that it captures the strengths of each of the other three approaches and avoids their limitations and logical weaknesses. This approach introduces new understandings of the nature and significance of sensory and CNS activity, and gives rise to novel questions concerning, and approaches to, phenomena such as language.

BIBLIOGRAPHY

Baars, B. J. (1986). "The Cognitive Revolution in Psychology." Guilford, New York.

Beer, R. D. (1990). "Intelligence as Adaptive Behavior." Academic Press, San Diego.

Bickhard, M. H. (1980). "Cognition, Convention, and Communication." Praeger, New York.

Bickhard, M. H. (1992). How does the environment affect the person? *In* "Children's Development within Social Contexts: Metatheory and Theory" (L. T. Winegar and J. Valsiner, eds.), pp. 63–92. Erlbaum, Hillsdale, NJ.

Bickhard, M. H. (1993). Representational content in humans and machines. *J. Exp. Theor. Artificial Intelligence* 5, 285–333.

Bickhard, M. H., and Campbell, R. L. (1992). Some foundational questions concerning language studies. *J. Pragmatics* **17**(5/6), 401–433.

Bickhard, M. H., and Richie, D. M. (1983). "On the Nature of Representation: A Case Study of James J. Gibson's Theory of Perception." Praeger, New York.

Bickhard, M. H., and Terveen, L. (1995). "Foundational Issues in Artificial Intelligence and Cognitive Science: Impasse and Solution." Elsevier Scientific, Amsterdam.

Brooks, R. A. (1991a). Intelligence without representation. *Artificial Intelligence* **47**(1–3), 139–159.

Brooks, R. A. (1991b). New approaches to robotics. *Science* **253**(5025), 1227–1232.

Brooks, R. A. (1991c). How to build complete creatures rather than isolated cognitive simulators. *In* "Architectures for Intelligence" (K. VanLehn, ed.), pp. 225–239. Erlbaum, Hillsdale, NJ.

Bullock, T. H. (1981). Spikeless neurones: Where do we go from here? *In* "Neurones without Impulses" (A. Roberts and B. M. H. Bush, eds.), pp. 269–284. Cambridge Univ. Press, New York.

Burnyeat, M. (1983). "The Skeptical Tradition." Univ. of California Press, Berkeley.

Campbell, R. L., and Bickhard, M. H. (1986). "Knowing Levels and Developmental Stages." Karger, Basel.

Carlson, N. R. (1986). "Physiology of Behavior." Allyn & Bacon, Boston.

Chapman, D., and Agre, P. (1986). Abstract reasoning as emergent from concrete activity. *In* "Reasoning about Actions and Plans, Proceedings of the 1986 Workshop" (M. P. Georgeff and A. L. Lansky, eds.), pp. 411–424. Morgan Kaufman, San Francisco, CA.

Churchland, P. S., and Sejnowski, T. J. (1992). "The Computational Brain." MIT Press, Cambridge, MA.

Clark, A. (1989). "Microcognition." MIT Press, Cambridge, MA.

Clark, A. (1993). "Associative Engines." MIT Press, Cambridge, MA.

Fodor, J. A. (1990). "A Theory of Content." MIT Press, Cambridge, MA.

Freeman, W. J., and Skarda, C. A. (1990). Representations: Who needs them? *In* "Brain Organization and Memory" (J. L. McGaugh, N. M. Weinberger, and G. Lynch, eds.), pp. 375–380. Oxford Univ. Press, New York.

Gardner, H. (1987). "The Mind's New Science." Basic, New York.

Glass, A. L., Holyoak, K. J., and Santa, J. L. (1979). "Cognition." Addison–Wesley, Reading, MA.

Goschke, T., and Koppelberg, D. (1991). The concept of representation and the representation of concepts in connectionist models. *In* "Philosophy and Connectionist Theory" (W. Ramsey, S. P. Stich, and D. E. Rumelhart, eds.), pp. 129–161. Erlbaum, Hillsdale, NJ.

Hanson, P. P. (1990). "Information, Language, and Cognition." Univ. of British Columbia Press, Vancouver.

Kenny, A. (1973). "Wittgenstein." Harvard Univ. Press, Cambridge, MA.

Koch, C., and Poggio, T. (1987). Biophysics of computation. *In* "Synaptic Function" (G. M. Edelman, W. E. Gall, and W. M. Cowan, eds.), pp. 637–697. John Wiley & Sons, New York.

Maes, P. (1990). "Designing Autonomous Agents." MIT Press, Cambridge, MA.

McClelland, J. L., and Rumelhart, D. E. (1986). "Parallel Distributed Processing: Explorations in the Microstructure of Cognition," Vol. 2, "Psychological and Biological Models." MIT Press, Cambridge, MA.

Nesser, U. (1967). "Cognitive Psychology." Appleton–Century–Crofts, New York.

Palmer, S. E. (1978). Fundamental aspects of cognitive representation. *In* "Cognition and Categorization" (E. Rosch and B. B. Lloyd, eds.), pp. 259–303. Erlbaum, Hillsdale, NJ.

Pellionisz, A. J. (1991). Geometry of massively parallel neural interconnectedness in wired networks and wireless volume transmission. *In* "Volume Transmission in the Brain: Novel Mechanisms for Neural Transmission" (K. Fuxe and L. F. Agnati, eds.), pp. 557–568. Raven, New York.

Pinker, S., and Mehler, J. (1988). "Connections and Symbols." MIT Press, Cambridge, MA.

Port, R., and van Gelder, T. J. (1995). "Mind as Motion: Dynamics, Behavior, and Cognition." MIT Press, Cambridge, MA.

Rumelhart, D., and McClelland, J. (1986). "Parallel Distributed Processing: Explorations in the Microstructure of Cognition," Vol. 1, "Foundations." MIT Press, Cambridge, MA.

Smolensky, P. (1988). On the proper treatment of connectionism. *Behav. Brain Sci.* **11**, 1–74.

Thatcher, R. W., and John, E. R. (1977). "Functional Neuroscience," Vol. 1, "Foundations of Cognitive Processes." Erlbaum, Hillsdale, NJ.

van Gelder, T. (1990). Compositionality: A connectionist variation on a classical theme. *Cognitive Sci.* **14**(3), 355–384.

van Gelder, T. (1991). What is the "D" in "PDP"? A survey of the concept of distribution. *In* "Philosophy and Connectionist Theory" (W. Ramsey, S. P. Stich, and D. E. Rummelhart, eds.), pp. 33–59. Erlbaum, Hillsdale, NJ.

Varela, F. J., and Bourgine, P. (1992). "Toward A Practice of Autonomous Systems." MIT Press, Cambridge, MA.

Collagen, Structure and Function

MARCEL E. NIMNI
Childrens Hospital Los Angeles

GLOSSARY

α Chains Individual polypeptides that coil up around each other to form a collagen molecule

Collagen fibers Bundles of fibrils that can be visualized with the scanning electron microscope

Collagen molecule Triple-helical molecule that assembles spontaneously into fibrils, which are visible with the electron microscope and show a characteristic periodicity

Procollagen Intracellular precursor of collagen

COLLAGEN IS THE SINGLE MOST ABUNDANT ANImal protein in mammals, accounting for up to 30% of all proteins. The collagen molecules, after being secreted by the cells, assemble into characteristic fibers responsible for the functional integrity of tissues such as bone, cartilage, skin, and tendon (Fig. 1). They contribute a structural framework to other tissues, such as blood vessels and most organs. Cross-links between adjacent molecules are a prerequisite for the collagen fibers to withstand the physical stresses to which they are exposed. A variety of human conditions, normal and pathological, involve the ability of tissues to repair and regenerate their collagenous framework. In human tissues, 13 collagen types have been identified.

I. COLLAGEN MOLECULE

The arrangement of amino acids in the collagen molecule is shown schematically in Fig. 2. Every third amino acid is glycine. Proline and OH-proline follow each other relatively frequently, and the Gly–Pro–Hyp sequence makes up about 10% of the molecule. This triple-helical structure generates a symmetrical pattern of three left-handed helical chains that are, in turn, slightly displaced to the right, superimposing an additional "supercoil" with a pitch of approximately 8.6 nm.

These chains are known as α chains, and for the interstitial collagens (types I–III, see Fig. 11) they show a molecular mass of approximately 100 kDa and contain approximately 1000 amino acids. The amino acids within each chain are displaced by a distance $h = 0.201$ nm with a relative twist of $-110°$, making the number of residues per turn 3.27 and the distance between each third glycine 0.87 nm. The individual residues are nearly fully extended in the collagen structure, since the maximum displacement within a fully stretched chain would be approximately 0.36 nm. This separation is such that it will not allow intrachain bonds to form (as does occur in the α helix), and only interchain hydrogen bonds are possible. The exact number of hydrogen bonds that stabilize the triple-helical structure has not been determined. One model describes two hydrogen bonds for every three amino acids, whereas another assumes one.

In addition to these intramolecular conformational patterns, there seems to exist a supermolecular coiling.

ENCYCLOPEDIA OF HUMAN BIOLOGY, Second Edition, VOLUME 2. Copyright © 1997 by Academic Press. All rights of reproduction in any form reserved.

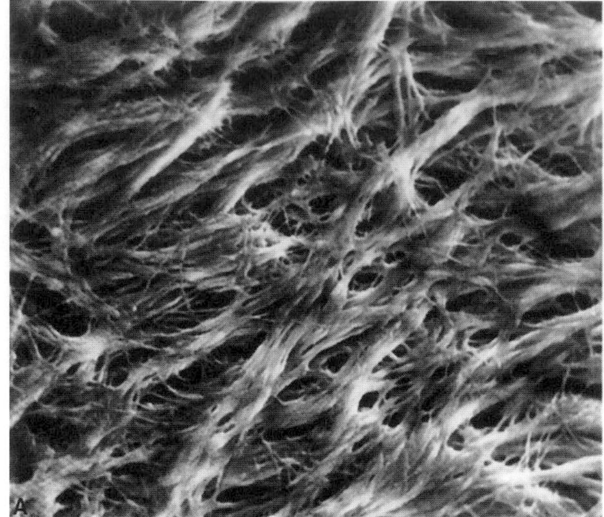

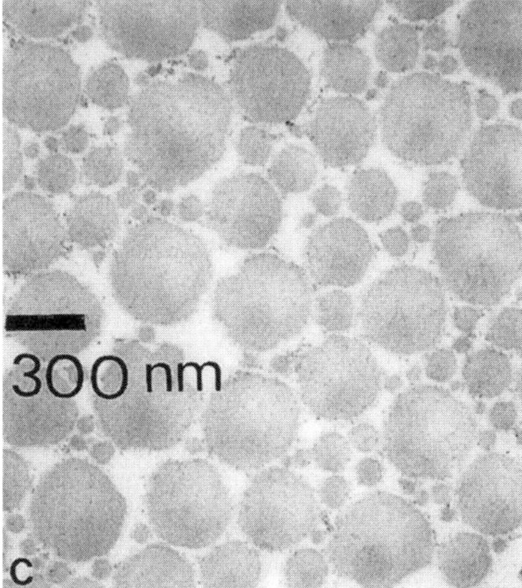

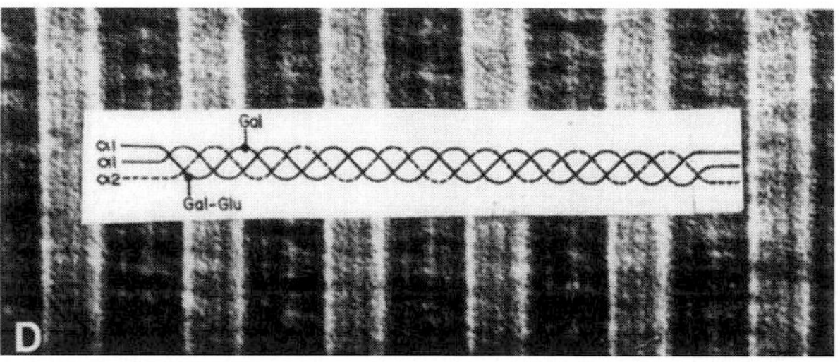

FIGURE 1 (A) Bundles of collagen fibers in the osteoid (i.e., noncalcified bone matrix) lining of a haversian canal (i.e., a vascular tunnel) in a human long bone. The specimen was prepared by cleaning in a solution containing enzyme detergent for 6 hr and was then observed under a scanning electron microscope. Original magnification, ×5400. (Courtesy of Dr. A. Boyde.) (B) Tendon containing collagen fibrils aligned parallel to each other. (C) Cross section of tendon showing fibrils of various diameters (bar = 300 nm). (D) Portion of a native collagen fibril stained with phosphotungstic acid displaying the characteristic 68-nm periodicity, overlaid by a diagram of a collagen molecule (type I) measuring approximately 300 nm.

Microfibrils, possibly representing intermediate stages of packing, have been described.

The process of self-assembly that causes the collagen molecules to organize into fibers is shown schematically in Fig. 3. The thermodynamics of such a system involve changes in the state of the water molecules, many of which are associated with nonpolar regions of the collagen molecule.

II. BIOSYNTHESIS OF PROCOLLAGEN

In order for the organism to develop an extracellular network of collagen fibers, the cells involved in the biosynthetic process must first synthesize a precursor known as procollagen (Fig. 4). This molecule is later enzymatically trimmed of its nonhelical ends, giving rise to a collagen molecule that spontaneously assembles into fibers in the extracellular space. Procollagen molecules have been identified as precursors of the three interstitial collagens (i.e., types I–III). Several of the amino- and carboxy-terminal peptides (i.e., propeptides) have been characterized and their primary sequences have been determined.

The carboxy-terminal propeptides of both pro $\alpha1$ and pro $\alpha2$ chains have molecular masses of 30–35 kDa and globular conformations without any

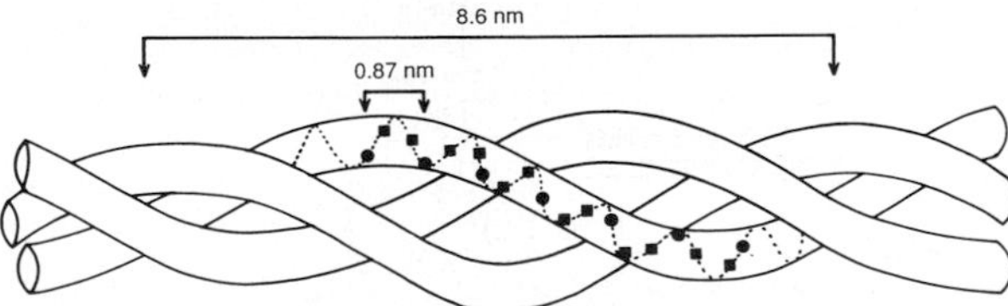

FIGURE 2 The collagen triple helix. The individual α chains are left-handed helices with approximately three residues per turn. The chains are in turn coiled around each other following a right-handed twist. The hydrogen bonds that stabilize the triple helix (not shown) form between opposing residues in different chains (interpeptide hydrogen bonding) and are therefore quite different from α helices, which occur between amino acids located within the same polypeptide. ●, glycine; ■, predominantly imino acids.

collagen-like domain. These peptides contain asparagine-linked oligosaccharide units composed of *N*-acetylglucosamine and mannose. Once the molecule is completed and translocated to the cell surface, the extensions are enzymatically removed from those collagens that form fibrils. Enzymes that selectively remove these extensions can be found in a variety of connective tissues and in the culture media derived from collagen-secreting cells.

A. Gene Expression

Since the discovery about 20 years ago of a distinct form of collagen in cartilage, now known as type II collagen, many other unique molecular species have been observed. Types I–III, V, and XI collagens are categorized as fiber-forming collagens. They all exhibit lengthy uninterrupted collagenous domains and are first synthesized as biosynthetic precursors (i.e., procollagens). Gene cloning experiments have demonstrated that the type I collagen genes are evolutionarily related, for they share a common ancestral gene structure. Human chromosome number 17 contains the coding information for the α1 chain of type I collagen, whereas chromosome 7 codes for its complementary α2 chain. A comparison of the five fibrillar collagens described shows that, with one exception [types III and α2(V) located on chromosome 2], all other genes are located on different chromosomes.

The collagen genes for fiber-forming collagens are large, about 10 times the size of the functional mRNA. Many of the exons (i.e., coding sequences) are 54 base pairs (bp) in length and are separated from each other by large intervening sequences (i.e., introns) that range in size from about 80 to 2000 bp. The gene itself contains 38,000 bp and is complex. The finding that most exons of these genes have identical lengths suggests that the ancestral gene for collagen was assembled by multiple duplications of single genetic units containing an exon of 54 bp (Fig. 5). It is likely that a primordial exon of this size could have encoded for a Gly–Pro–Pro tripeptide repeated six times (3 × 3 × 6). Such a polypeptide of 18 amino acids probably

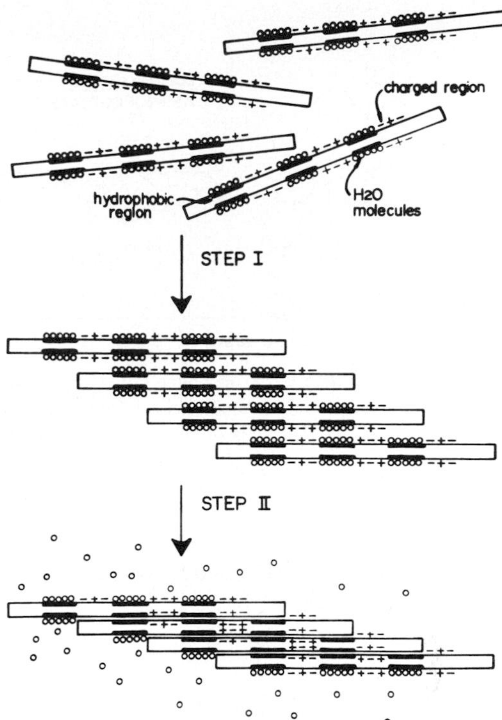

FIGURE 3 The nature of the forces associated with the early interactions of the helical regions of collagen via a mechanism of endothermic polymerization. This occurs *in vivo* and *in vitro* when the molecules still retain remnants of their telopeptides (i.e., nonhelical extensions), but can also occur *in vitro* with molecules devoid of such extensions. Further insight into the role of such telopeptides and modalities of assembly is provided in Figs. 4 and 9. Soluble collagen can be extracted from most tissues by cold neutral salt solutions. If these solutions are warmed to 37°C, the collagen molecules reassemble into native fibers. The upper part of the drawing represents molecules that aligned in a quarter-staggered overlap. The alignment is primarily due to interactions of opposing electrostatic charges, as depicted by + and −. As the temperature approaches 37°C, the hydrogen-bonded water molecules (○), clustered around the hydrophobic regions of collagen, "melt" and expose these nonpolar surfaces. Exclusion of water allows these surfaces to interact with each other, giving rise to hydrophobic bonds that greatly enhance the stability of the fiber. The driving force results from an increase of entropy of the system, since the release of "organized" water from initial sites and its transformation into "random" water increase the disorder of the system. Part of the aging process that leads to a gradual insolubilization of collagen can be associated with this phenomenon, which continues to operate through the life span of an individual.

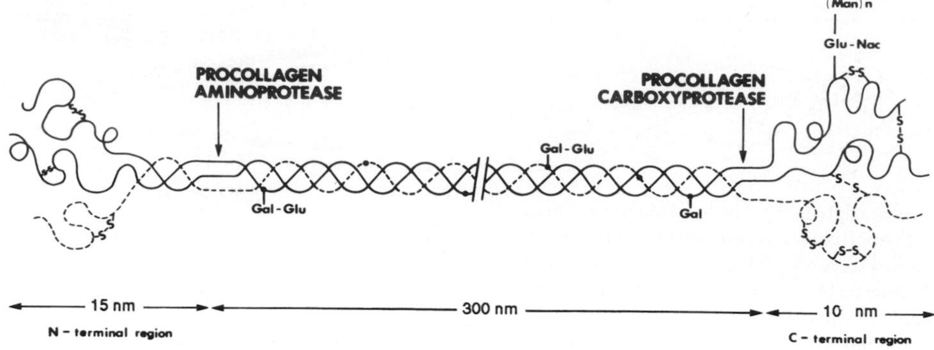

FIGURE 4 Procollagen molecule showing the nonhelical terminal extensions. The amino (N) terminus contains a small helical domain, and the carboxy (C) terminus is stabilized by interchain disulfide bonds. The sites of cleavage by procollagen peptidases are indicated by arrows.

had the minimum length needed to form a stable triple-helical structure.

B. Translational, Cotranslational, and Early Posttranslational Events

After the gene is transcribed, it is spliced to remove introns and to yield a functional mRNA that contains about 3000 bases. Specific mRNAs for each chain and collagen type are translocated to the cytoplasm and translated into proteins in the rough endoplasmic reticulum on membrane-bound polysomes (Fig. 6). As the collagen polypeptide is synthesized in this region, it is modified in important ways. Two major constituents of collagen are the modified amino acids hydroxyproline and hydroxylysine, but neither of them can be directly incorporated into proteins. Instead, proline and lysine are incorporated and then modified by two hydroxylating enzymes, prolyl and lysyl hydroxylases. These enzymes require for their activity ferrous iron, ascorbate, and α-ketoglutarate. The degree of hydroxylation differs from tissue to tissue and probably with the availability of substrate,

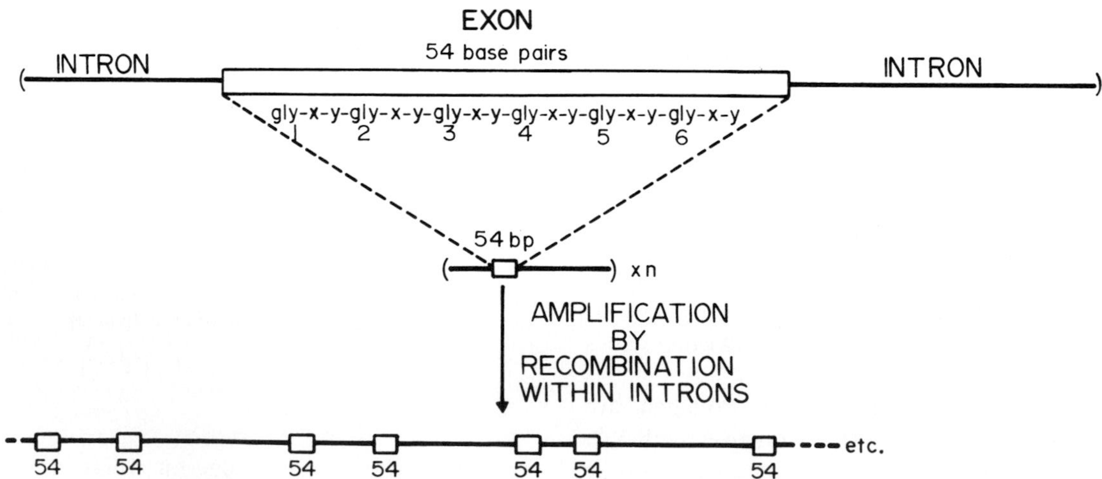

FIGURE 5 The collagen gene is made up of multiple units containing 54 base pairs, each of which corresponds to sequences of 18 amino acids. The conservation of this minimum sequence and the fact that it is repeated in such an exacting fashion provide valuable information to investigators interested in the process of evolution of proteins. (Redrawn from Dr. DeCrombrugghe.)

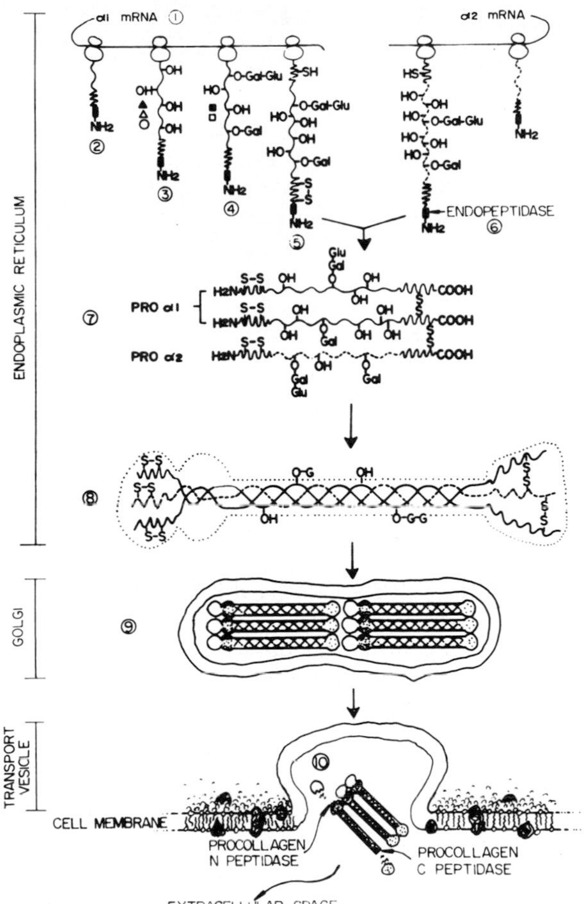

FIGURE 6 Sequence of events in the biosynthesis of collagen. (1) Synthesis of specific mRNAs for the different procollagen chains. (2) Translation of the message on polysomes of the rough endoplasmic reticulum. (3) Hydroxylation of specific proline residues by 3-proline hydroxylase (▲) and 4-proline hydroxylase (△) and of lysine by lysyl hydroxylase (○). (4) Glycosylation of hydroxylysine by galactosyltransferase (■) and addition of glucose by a glucosyltransferase (□). (5) Removal of the amino (N)-terminal signal peptide. (6) Release of completed α chains from ribosomes. (7) Recognition of three α chains through the carboxy (C)-terminal prepeptide and formation of disulfide cross-links. (8) Folding of the molecule and formation of a triple helix. (9) Intracellular translocation of the procollagen molecules and packaging into vesicles. (10) Fusion of vesicles with the cell membrane and extrusion of the molecule accompanied by the removal of the carboxy-terminal nonhelical extensions and part of the amino-terminal nonhelical extensions by specific peptidases.

rates of synthesis, and turnover, as well as the time during which the molecule remains in the presence of the hydroxylating enzymes. The time required for the synthesis of a complete pro α chain is about 6.7 minutes.

As lysyl residues in the newly synthesized pro α chains are hydroxylated, sugar residues are added to the resulting hydroxylysyl groups. Glycosylation is catalyzed by two specific enzymes: a galactosyltransferase and a glucosyltransferase. Once the translation, modifications, and additions are completed, the individual pro α chains become properly aligned for the triple helix to form.

C. Intracellular Translocation of Procollagen and Extrusion into the Extracellular Space

The procollagen molecule, now detached from the ribosome, emerges from the endoplasmic reticulum and moves toward the Golgi apparatus through the microsomal lumen. In the Golgi apparatus the carboxy-terminal mannose-rich carbohydrate extension is remodeled, and the molecules are packaged in vesicles and carried toward the cellular membrane (Fig. 7). [*See* Golgi Apparatus.]

The small aggregates of oriented procollagen molecules are probably trimmed of their nonhelical amino and carboxyl extensions by specific peptidases when they reach the extracellular space. In the case of type I collagen, the first peptidase to act seems to be the aminoprotease; this is followed by a carboxyprotease. In type III collagen the sequence of removal might be reversed.

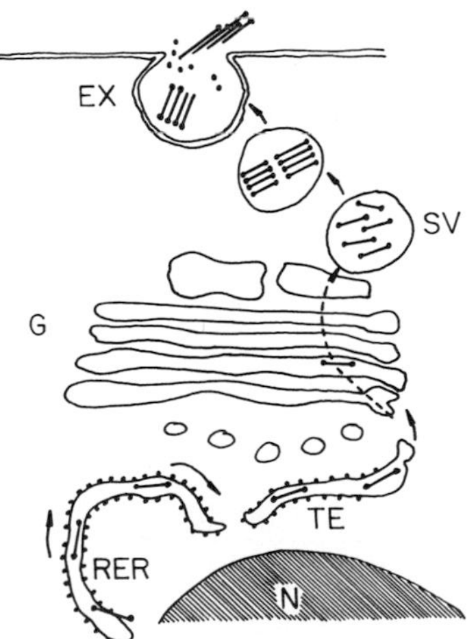

FIGURE 7 Movement of procollagen through the cisternae of the rough endoplasmic reticulum (RER) and through a transitional endoplasm (TE) to the Golgi apparatus (G), where it is packaged into secretory vesicle (SV) prior to extrusion (EX) by exocytosis. N, nucleus.

D. Lysyl Oxidase

Recently formed microfibrils seem to be recognized by the enzyme lysyl oxidase, which converts certain peptide-bound lysines and hydroxylysines to aldehydes (Fig. 8). The enzyme is an extracellular amine oxidase, which has been purified from a variety of connective tissues. It requires Cu^{2+} and probably pyridoxal as cofactors, and molecular oxygen seems to be the cosubstrate and hydrogen acceptor. It is irreversibly inhibited by the lathyrogen β-amino propionitrile (βAPN), a substance found in the flowering sweet pea, *Lathyrus odoratus*. This enzyme exhibits maximal activity when acting on collagen fibrils rather than on monomeric collagen.

E. Fibrillogenesis

The tendency of collagen molecules to form macromolecular aggregates is well known. This tendency is common with most fibrous proteins that form filaments with helical symmetry and that occupy equivalent or quasi-equivalent positions.

The exact mode in which the collagen molecules pack into microfibrils (the precursor of the larger fibrils) still remains a subject for speculation. A five-stranded microfibril was first suggested to account for such a substructure, which would satisfy the condition that adjacent molecules were equivalently related by a quarter stagger (overlaps staggered at intervals equal to approximately one-quarter the length of the molecule) (Fig. 9).

When monomeric collagen is heated to 37°C, it progressively polymerizes, generating a turbidity curve that reflects the presence of intermediate aggregates. The lag phase (persistence of monomers), the nucleation and appearance of turbidity (microfibrils), and the rapid increase in turbidity (fiber formation) have been equated to the way in which the cell might handle this process (Fig. 10).

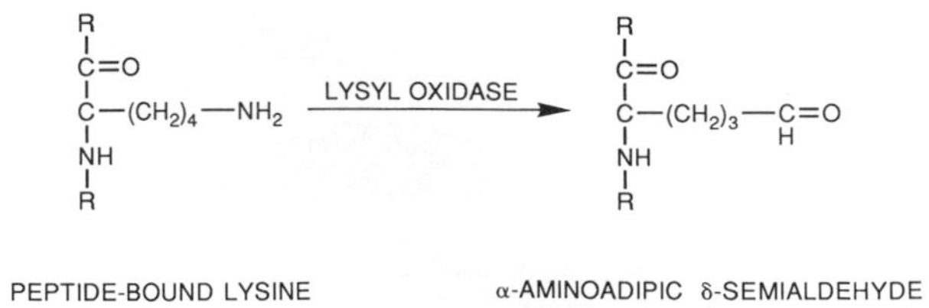

$$\begin{array}{c} R \\ | \\ C{=}O \\ | \\ C{-}(CH_2)_4{-}NH_2 \\ | \\ NH \\ | \\ R \end{array} \xrightarrow{\text{LYSYL OXIDASE}} \begin{array}{c} R \\ | \\ C{=}O \\ | \\ C{-}(CH_2)_3{-}\underset{H}{C}{=}O \\ | \\ NH \\ | \\ R \end{array}$$

PEPTIDE-BOUND LYSINE α-AMINOADIPIC δ-SEMIALDEHYDE

FIGURE 8 The oxidative deamination of peptide-bound lysine by the enzyme lysyl oxidase generates the aldehydes associated with the collagen molecule.

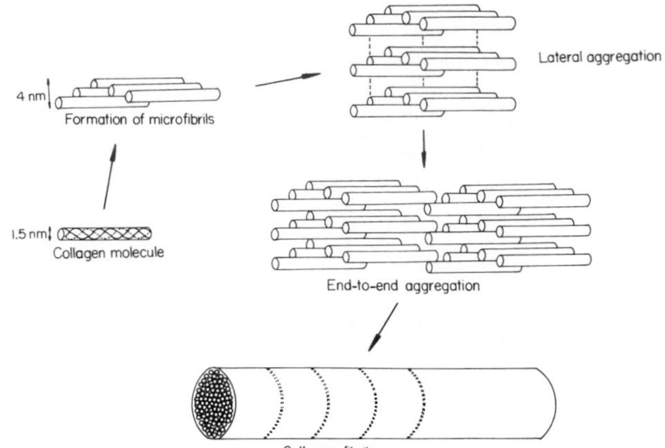

FIGURE 9 Formation of the five-membered microfibril and its potential for lateral and end-to-end aggregation to form fibrils.

III. TYPES OF COLLAGEN

A quarter of a century has passed since we first realized that all collagen fibers within a particular organism are not made up of identical molecules. The different collagen types are usually identified using Roman numerals, assigned as they are purified and characterized. Figure 11 summarizes the molecular characteristics of most of the collagens identified to date.

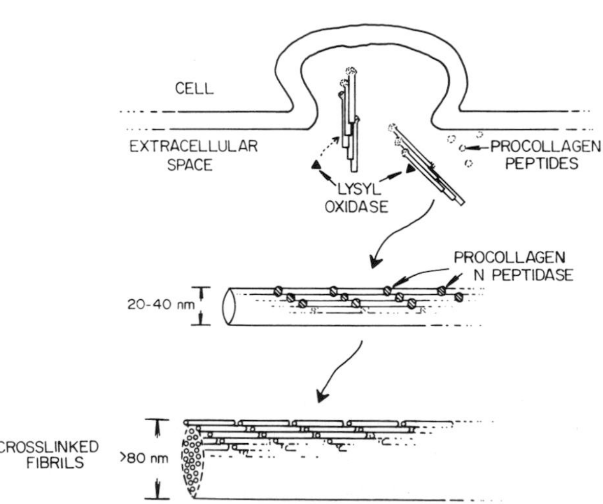

FIGURE 10 Fibrillogenesis: microfibrils in a quarter-staggered configuration have lost their carboxy-terminal nonhelical extensions and part of their amino (N)-terminal extensions. In this form they seem to organize readily into small-diameter fibrils that retain part of the amino-terminal nonhelical extensions. After being relieved of these peptides by a procollagen peptidase, fibrils are able to grow in diameter by apposition of microfibrils or by merger with other small-diameter fibrils.

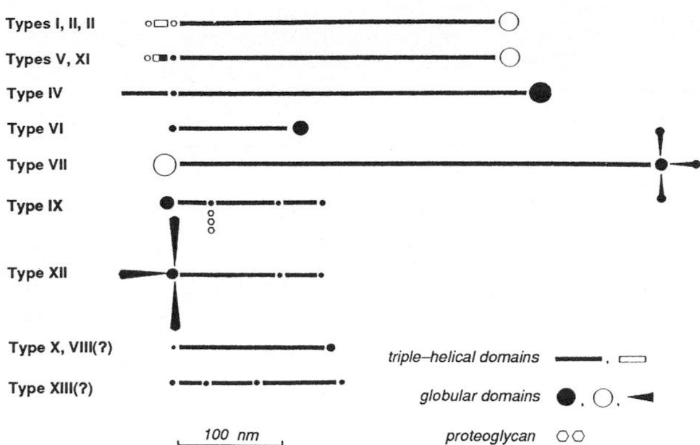

FIGURE 11 Collagens contain both triple-helical (solid and open rods) and globular domains (open and solid circles). Portions of the initially synthesized molecules are removed prior to their incorporation into insoluble matrices (open rods and circles) while the rest of the molecule remains intact in the matrix (solid circles and rods). The domains and their distributions are drawn approximately to scale. [From R. E. Burgeson and M. E. Nimni (1992). *Clin. Ortho. Rel. Res.* **282**, 250.]

A. Type I

Before 1969, type I collagen was the only mammalian collagen known. It is composed of three chains, two identical $\alpha 1$ chains and one $\alpha 2$ chain. Type I collagen is most abundant in skin, tendon, ligament, bone, and cornea, where it makes up 80–99% of the total collagen. Bone matrix is essentially all type I collagen. The most common technique used to isolate this molecule and distinguish it both qualitatively and quantitatively from the other collagens involves the use of solvents of different ionic strengths and pH levels, followed by differential "salting out."

The amino acid compositions of some of the better-characterized and more abundant human collagen types are shown in Table I. The amino acids that are present in a quantity significantly different from that of type I collagen are indicated in bold numbers. On many occasions these differences have been used to identify collagen types or their mixtures and to suspect the presence of new or abnormal collagen species. A simplified diagram illustrating type I collagen and its relationship to the other interstitial collagens, types II and III, is shown in Fig. 12.

The critical importance of type I collagen to normal human physiology is evident from studies of mutations in the $\alpha 1$ and $\alpha 2$ genes. The effect of insertions and deletions, as well as single amino acid substitutions, has been observed. The majority of these defects result in phenotypic changes in bone development characteristic of osteogenesis imperfecta (OI). Substitutions of any other amino acid for glycine in the first position of the repeating tripeptide Gly–X–Y of the triple-helical domain results in varying degrees of OI severity, depending on the specific location of the substitution along the α chain. In general, the gene defects underlying OI appear to diminish the amount of type I collagen accumulated within the bone matrix, because of either intracellular degradation or failure of defective collagen molecules to become correctly incorporated into growing fibers.

B. Type I-Trimer

In addition to the genetically distinct types of interstitial collagens described, another molecular form of collagen, $\alpha 1$(I)-trimer or type I-trimer, has been demonstrated. The $\alpha 1$(I)-trimer consists of three polypeptides that are genetically identical to the $\alpha 1$ chains of type I collagen. It was originally isolated from a virus-induced tumor, and its synthesis was observed in cultures of polyoma virus-induced mouse tumors. Human skin has also been shown to contain $\alpha 1$(I)-trimer. Its amino acid composition is, in general, similar to that of $\alpha 1$(I) from type I collagen, except that the amount of hydroxylysine is increased and lysine is decreased. The content of 3-hydroxyproline is also increased, suggesting that both chains are derived from the same gene. Since the synthesis of the type I-trimer is an early event in the dedifferentiation of chondrocytes in cultures, and accompanies a variety of unusual circumstances associated with cultured cells, its potential for monitoring abnormalities in the collagen biosynthetic pathway may be of significant interest.

C. Type II

Isolation and analysis of collagens from a variety of cartilaginous structures show that type II collagen is made up primarily of molecules containing three identical α chains. The most significant features of cartilage collagen are its high content of hydroxylysine and glycosidically bound carbohydrates. Type II collagen is also present in the intervertebral disk, which contains a central gel-like region, the nucleus pulposus surrounded by concentric rings of highly ordered dense collagen fibers known as the annulus fibrosus. Another tissue that contains appreciable amounts of type II collagen is the vitreous body of the eye. Comparisons of collagens extracted from cartilage and vitreous by pepsin treatment show differ-

TABLE I
Amino Acid Composition of the Human Collagen Chains[a]

Amino acid	α1(I)	α2(I)	α1(II)	α1(III)	α1(IV)	α2(V)	α1(V)	α3(V)	α1(XI)	α2(XI)[b]
3-Hydroxyproline	1	1	2	0	7	3	5	1	· · ·	· · ·
4-Hydroxyproline	108	93	97	125	**133**	106	110	91	98	93
Aspartic acid	42	44	43	42	51	50	49	42	46	50
Threonine	16	19	23	13	20	29	21	19	17	25
Serine	34	30	25	39	37	34	23	34	25	28
Glutamic acid	73	68	89	71	79	89	100	97	107	98
Proline	124	113	120	107	**65**	107	130	98	109	119
Glycine	133	338	333	350	328	331	332	330	334	327
Alanine	115	102	103	96	**37**	**54**	**39**	**49**	**54**	**49**
Half-cystine	0	0	0	**2**	0–1	0	0	1	0	0
Valine	21	35	18	**14**	28	27	17	29	28	18
Methionine	7	5	10	8	13	11	9	8	10	9
Isoleucine	6	14	9	13	**29**	15	17	20	15	16
Leucine	19	30	26	22	**52**	37	36	**56**	35	39
Tyrosine	1	4	2	3	2	2	4	2	2	3
Phenylalanine	12	12	13	8	**29**	11	12	9	11	11
Hydroxylysine	9	12	**20**	5	**49**	**23**	**36**	**43**	**38**	**40**
Lysine	26	18	15	30	9	13	14	15	19	15
Histidine	3	12	2	6	6	10	6	14	6	11
Arginine	50	50	50	46	**26**	48	40	42	45	48
Gal-Hydroxylysine	1	1	4	· · ·	2	3	5	7	· · ·	· · ·
Glc-Gal-Hydroxylysine	1	2	12	· · ·	30	5	29	17	28	34

[a]Amounts are expressed as residues per 1000 total residues. Boldface numbers represent quantities significantly different from those for type I. Ellipses (· · ·) indicate either trace amounts or nondetectable.

[b]α3(XI) appears to be almost identical to α1(II), either a similar gene product or a close genetic variant that has been altered posttranslationally.

ences in amino acid composition, carbohydrate content, and the presence of additional α chains that electrophoretically migrate more slowly than α1(II).

Type II collagen is synthesized during the chondrogenic stages of development of the mesoderm. During amphibian limb regeneration, the tissues undergo three distinct stages of development during the transition from mesenchyme to cartilage and bone. The synthesis and presence of type II collagen correlate with chondrogenic activity. A similar developmental pattern has been observed in the bone matrix induced endochondral calcification system. Genetic mutations of type II collagen genes have been detected in several types of chondrodysplasia. [*See* Articular Cartilage and the Intervertebral Disc; Cartilage.]

D. Type III

When human dermis is digested with pepsin under conditions in which the collagen molecules retain their helical conformation, type I molecules can be separated from type III molecules by differential salt precipitation at pH 7.5. The type III molecules are composed of three identical chains. Characteristic of this collagen is the presence of intramolecular disulfide bonds involving two cysteine residues close to the carboxy-terminal region of the triple helix. Because the ratio of type I to type III collagen changes with age, type III being predominant in fetal skin, this type of collagen is many times referred to as fetal, or embryonic, collagen. Formation of intermolecular disulfide bridges by type III collagen could be of great advantage during early development and wound healing, when collagen is deposited at a rapid rate in order to fill a gap.

Recent evidence suggests that type III collagen may be predominantly located at the fibril surface. It is possible that the collagen type available at the surface may provide a unique interactive edge for other molecules, thus mediating the interactions between fibrils

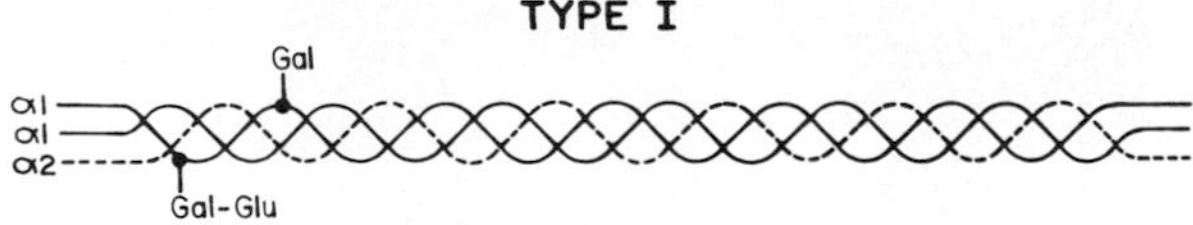

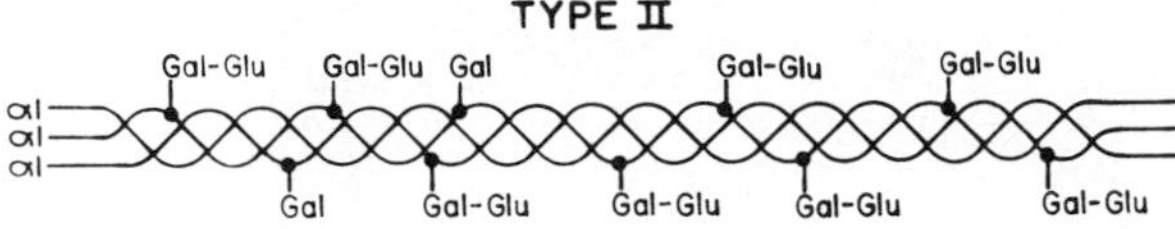

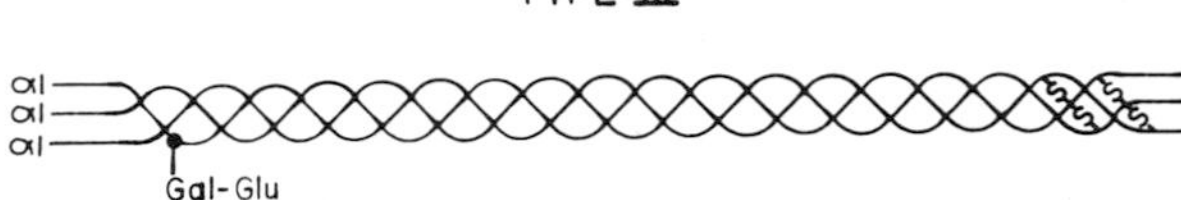

FIGURE 12 Diagram of the three interstitial types of collagen. Type I is present in skin, bone, tendon, etc.; type II is present in cartilage; and type III is present in blood vessels and developing tissues and as a minor component in skin and other tissues. There are differences in the chain composition and degrees of glycosylation. Disulfide cross-links are seen only in type III collagen.

within a fiber or between fibrils and other matrix macromolecules. These complex interactions are likely to underlie the mechanical properties of the tissue.

Normal bone matrix might be the only tissue containing type I collagen that lacks type III. Blood vessels are particularly rich in type III collagen.

E. Type IV

Type IV collagen is the major component of basement membranes and is generally regarded as the most characteristic of a large number of macromolecules present within these specialized connective tissue structures. These macromolecules include, in addition to collagen, glycoproteins such as laminin, fibronectin, and proteoglycans. Basement membranes underlie epithelia and endothelia; are involved in cell differentiation and orientation, membrane polarization, and selective permeability to macromolecules; and are a target for a large number of diseases (Fig. 13).

Type IV collagen differs from the interstitial collagens in its amino acid composition (Table I) and by the fact that it does not organize into a fibrillar structure. This molecule consists of a large triple-helical domain and noncollagenous extensions that make it resemble procollagen. Carbohydrate accounts for 10% of the mass.

It seems that type IV procollagen (with a molecular mass of approximately 180 kDa per chain) is incorporated as such into basement membranes without processing. The globular extensions at the end of the molecule associate to form a network. Thus, association of four molecules at their amino termini gives rise to tetramers that can aggregate further into a regular tetragonal or irregular polygonal meshwork. The pore size of these networks could differ in such a way as to account for the specific functions of basement membranes in different tissues.

F. Type V

Type V collagen was discovered in pepsin digests of placental membranes and other tissues. Three chains—$\alpha1(V)$, $\alpha2(V)$, and $\alpha3(V)$—were initially isolated from this collagen. They are similar in size to the α chains of interstitial collagens, except that the $\alpha1(V)$ chains are larger. Two related collagen chains were isolated from cartilage.

Molecules containing two $\alpha1$ chains and one $\alpha2$ chain are the most common form of type V collagen. In chick cornea, this molecule is present in banded collagen fibers containing type I collagen. A variety of observations support the concept that the type V collagen is found in the interior, but not the exterior, of the fibril. This in turn suggests that type V may be involved in initial events of fibrillogenesis.

Type V collagen is particularly abundant in vascular tissues, where it appears to be synthesized by smooth muscle cells, although it is also present in relatively large amounts within the corneal stroma, which lacks blood vessels. Type V collagen seems to be a unique form of collagen that contributes to the shape of cells by localizing on their surface and participates in the formation of an exocytoskeleton and binding to other connective tissue components.

G. Type VI

The type VI collagen molecule is a heterotrimeric assembly of three genetically distinct chains: $\alpha1(VI)$, $\alpha2(VI)$, and $\alpha3(VI)$. This is one of the first collagens found to contain a triple-helical domain less than 300 nm in length. Electron microscopic visualization of rotary shadowed molecules shows a characteristic dumbbell shape.

Antibodies to type VI collagen indicate that the molecule has a ubiquitous distribution throughout connective tissues. It is shared between cartilage and noncartilaginous tissues. In skin, type VI filaments are highly concentrated around endothelial basement membranes, forming a loose sheath around blood ves-

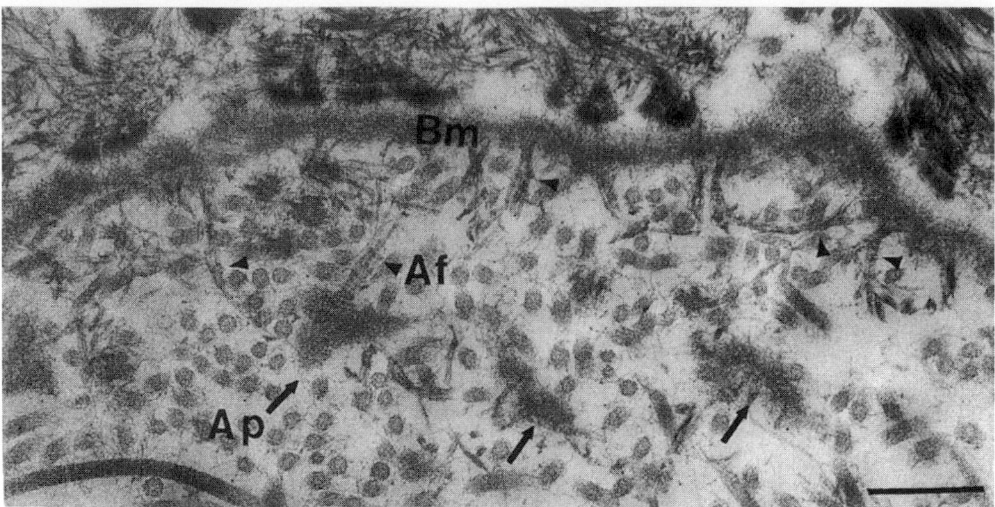

FIGURE 13 Electron micrograph of the basement membrane zone of the dermal–epidermal junction of human skin. The basement membrane containing type IV collagen is indicated (Bm). Anchoring fibrils (Af), containing type VII collagen, are indicated by arrowheads; anchoring plaques (Ap), containing the carboxy-terminal domains of type VII collagen and type IV collagen, are indicated by arrows. Bar = 200 nm. (Courtesy of Douglas Keene, Portland Shriners Hospital, Oregon.)

sels, nerves, and fat cells, separating these dermal and subdermal elements from the surrounding fibrillar network.

The function of type VI collagen is not known, but the ultrastructure of the network suggests that it is an independent fibrous system, perhaps important to the development and maintenance of the spatial separation of distinct tissue components from large, banded fiber bundles, as well as providing structural unity to the entire connective tissue. Type VI collagen has been reported to promote fibroblast attachment and spreading. As the type VI network appears to be readily available to cells that migrate through connective tissues, recognition of this network by cells may be important.

H. Type VII

Type VII collagen, also known as the anchoring fibril network, is the largest collagen thus far described, with a total molecular mass of approximately 1050 kDa for the precursor form of the molecule. The molecule is a homotrimeric assembly of $\alpha 1$(VII) chains.

The tissue distribution of type VII exactly correlates with that of anchoring fibrils. Anchoring fibrils are specialized fibrous structures found within the subbasal lamina of the basement membranes, which secure the lamina densa to the underlying connective tissue matrix by physically trapping collagen fibrils between the lamina densa and the anchoring fibril (Fig. 13).

Individuals lacking anchoring fibrils suffer extensive cutaneous and mucosal blistering, resulting from separation of the dermis from the epidermis along the subbasal lamina.

I. Type VIII

The peptides now known as type VIII collagen were first observed as cell culture products of bovine endothelial cells and rabbit corneal endothelial cells. Current models of the molecular structure predict that type VIII contains little nonhelical structure.

J. Type IX

The first fragment of type IX collagen was isolated from pig hyaline cartilage after pepsin digestion. The intact molecule is unusual in possessing three collagenous domains, each of which is interspersed by a short, noncollagenous, highly flexible domain.

One unusual characteristic of type IX collagen is the presence of a covalently attached chondroitin sulfate on the $\alpha 2$ chain. In chick cartilage, type IX is associated with type II on the surface of the banded fibrils.

K. Type X

A number of unique characteristics distinguish type X collagen from the other collagens. The molecule is 138 nm long, approximately one-half the length of the interstitial collagens. It has a restricted tissue distribution and is synthesized predominantly by hypertrophic chondrocytes during the process of endochondral ossification (i.e., bone formation preceded by cartilage). In the fetal and adolescent skeletons the molecule is a transient intermediate in cartilage, which is later replaced by bone. In the adult skeleton, type X collagen persists in the zone of calcified cartilage that separates hyaline cartilage from the subchondral bone.

L. Type XI

Type XI collagen was first discovered during differential salt fractionation of pepsin digests of hyaline cartilage. The protein is best recognized by the presence of three distinct α chains, smaller than $\alpha1(II)$ chains. This collagen resembles type V collagen in some of its properties. It might regulate the diameter of type II collagen fibrils, act as a specific building block of the basket of fibril surrounding chondrocytes, and mediate collagen–proteoglycan interactions.

M. Type XII

The gene for type XII collagen was recently cloned from tendon fibroblasts. The amino acid sequence predicted from its base sequence is homologous with, but not identical to, that of type IX collagen.

N. Type XIII

This is one of the most recently identified collagenous molecules. It was originally identified from a cDNA clone encoding a short triple-helical sequence. Four different mRNA species have been identified that encode this chain, indicating that the molecules arise from extensive alternative splicing. The encoded proteins are 62,000–67,000 kDa. The mRNAs for this collagen have been detected in the epidermis, hair follicle, and nail roots in skin; the mesenchymal cells of the bone marrow and endomysium; the cartilage growth plate; and smooth muscle. The function of this collagen is unknown.

Based on their molecular conformation and patterns of assembly, collagens are sometimes grouped into classes. As mentioned earlier, class 1 collagens are known as fiber-forming collagens and include types I, I-trimer, II, III, V, and XI. Class 2 collagens are structurally distinct from the class I molecule, but share common features among themselves. All appear to be related, have a short triple-helical domain containing interruptions. Sequence homologies between the class II collagens suggest that they may share common functions. They include types IX and XII collagens. Class 3 collagens include a group of molecules that serve unique functions, such as types IV, VI, VII, and X. Class 4 does not truly represent a unique class but is rather a grouping of molecules that cannot be correctly categorized because of gaps in scientific knowledge. It includes types VIII and XIII.

A schematic diagram illustrating the relative dimensions and fundamental features of most of the collagens described is shown in Fig. 11.

IV. COLLAGEN METABOLISM

Collagen is the most abundant of all body proteins. Tissues (e.g., bone) that are involved in active remodeling are responsible for the major turnover, whereas other less dynamic tissues in the full-grown individual (e.g., skin and tendons) might exhibit slow, almost negligible, turnover. The collagen-synthesizing activity of cells is usually assessed by their ability to synthesize hydroxyproline or by the activities of specific enzymes such as the proline and lysyl hydroxylases.

A. Degradation by Bacterial and Mammalian Collagenases

Because of its triple-helical structure stabilized by hydrogen bonds, the collagen molecules are quite resistant to enzymatic degradation in their native configuration. They can be degraded by collagenases. The first molecules to be isolated were of bacterial origin, specifically *Clostridium histolyticum*. These enzymes are quite specific for collagen, but will also degrade gelatin, which is denatured collagen. They are inhibited by cysteine and other sulfhydryl compounds and by ethylenediaminetetraacetic acid (EDTA), a chelator for divalent cations. These enzymes are specific for peptide bonds involving glycine in a collagen helix conformation. Because of the abundance of this amino acid in collagen (i.e., every third residue), this enzyme generates a large number of small peptides.

The first enzyme derived from animal tissue capable of degrading collagen at neutral pH was isolated from the culture fluid of tadpole tissue. It cleaves the native molecule into two fragments in a highly specific fashion at a temperature below that of denaturation of the substrate. Since this discovery, collagenolytic enzymes have been obtained from a wide range of tissues from animal species. In general, these enzymes have a number of fundamental properties in common: they all have a neutral pH optimum, and they are not stored within the cell, but rather are secreted either in an inactive form or bound to inhibitors. Figure 14 schematically summarizes the fundamental aspects of these enzymes and their modes of action. They appear to be zinc metalloenzymes requiring calcium, and are not inhibited by agents that block serine or sulfhydryl-type proteinases. Nearly all of the collagenases studied so far have a molecular mass that ranges from 25 to 60 kDa. Mammalian collagenases display a great deal of specificity, cleaving the bond Gly–Leu or Gly–Ile. There are slight differences in the amino acid sequences surrounding the scission site; these could account for the differences in the rates at which various collagens are degraded.

The enzyme interacts tightly with the collagen fiber and appears to remain bound to the macromolecular aggregate during the degradation process. Approximately 10% of the collagen molecules in reconstituted collagen fibrils appear to be accessible for binding, in close agreement with the theoretical number of molecules estimated to be present near the surface of the fiber. The *in vitro* data obtained seem to indicate that digestion proceeds until completion, with hopping from one molecule to another without returning to the solution. Collagen from individuals of increasing age becomes more resistant to enzymatic digestion.

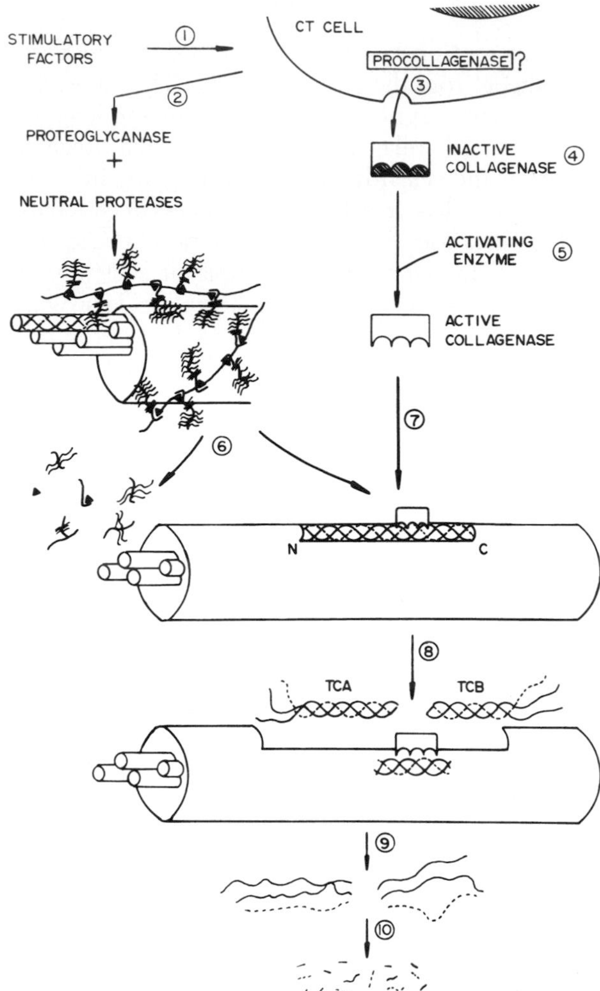

FIGURE 14 Sequence of events leading to the degradation of collagen fibers by the enzyme collagenase. (1) A variety of factors stimulate connective tissue (CT) cells to synthesize collagenase, glycosidases, and neutral proteases. (2) The proteoglycan-degrading enzymes remove the mucopolysaccharides that surround collagen fibers and expose them to collagenase. (3) Inactive collagenase is secreted. (4) The enzyme is usually found in the extracellular space bound to an inhibitor. (5) An activating enzyme removes the inhibitor. (6) Glycosidases complete the degradation of the proteoglycans. (7) The active collagenase binds to fibrillar collagen. (8) Collagenase splits the first collagen molecule into two fragments (TCA and TCB), which denature and begin to unfold at body temperature. The enzyme now moves on to an adjacent molecule. (9) The denatured collagen fragments are now susceptible to other proteases. (10) Nonspecific neutral proteases degrade the collagen polypeptides.

B. Urinary Excretion of Collagen Degradation Products

Because of the relatively large amounts of hydroxyproline (i.e., 12–14%) present in collagen, the assay of this amino acid in body fluids has raised considerable interest. More than 95% of the hydroxyproline is excreted in a peptide-bound form.

Altered urinary hydroxyproline values are found in conditions affecting collagen metabolism systemically, such as endocrine disorders, particularly those conditions accompanied by relatively extensive involvement of bone. The glycosides galactosylhydroxylysine and glucosylgalactosylhydroxylysine, characteristic moieties of the collagen molecule, have been identified in urine and are also used as criteria of collagen degradation.

A significant amount of the collagen synthesized by cells can be degraded intracellularly before secretion, and such degradation products also appear in the urine. It has been estimated that between 10% and 60% of the newly synthesized collagen can be degraded by this route.

V. CROSS-LINKING

A. Intramolecular and Intermolecular Cross-links

Cross-linking renders the collagen fibers stable and provides them with a degree of tensile strength and viscoelasticity adequate to perform their structural role. The degree of cross-linking, the number and density of the fibers in a particular tissue, and their orientation and diameter combine to provide this function. Cross-linking begins with the oxidative de amination of the ε carbon of lysin or hydroxylysine to yield the corresponding semialdehydes and is mediated by the enzyme lysyl oxidase (Fig. 8). Enzymatic activity is inhibited by β-aminoproprionitrile, chelating agents, isonicotinic acid hydrazide, and other carbonyl reagents. Lysyl oxidase exhibits particular affinity for the lysines and hydroxylysines present in the nonhelical extensions of collagen, but can, at a slower pace, also alter residues located in the helical region of the molecule (Fig. 15).

In general, lysine-derived cross-links seem to predominate in soft connective tissues such as skin and tendon, whereas hydroxylysine-derived cross-links are prevalent in the harder connective tissues, such as bone, cartilage, and dentine, which are less prone to yield soluble collagens.

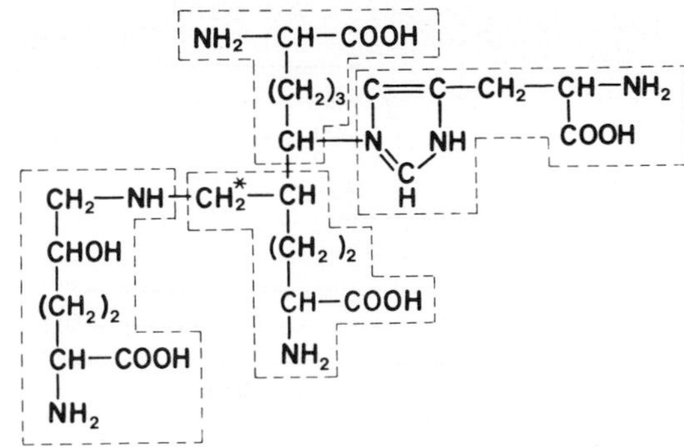

FIGURE 16 Histadinohydroxymerodesmosine: a more complex tetrafunctional covalent cross-link that bridges three different molecules. Two of the residues are part of an aldol condensation product (intramolecular cross-link) and therefore are associated with one single molecule.

Several other cross-links have been identified and their locations have been established. These more complex polyfunctional cross-links can contain histidine (Fig. 16) or can result in the formation of naturally fluorescent pyridinium ring structures (Fig. 17).

The study of collagen cross-linking has advanced steadily, and even though hindered by the difficulty in dealing with an insoluble three-dimensional matrix composed of quarter-staggered molecules, many cross-linking regions, primarily those involving the nonhelical extension peptides, have been identified (Fig. 18). It is interesting in this connection that covalent cross-links between types I and III molecules have

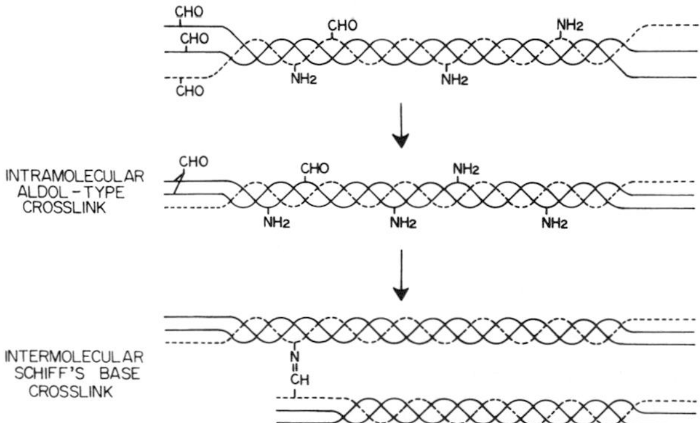

FIGURE 15 Formation of intramolecular and intermolecular cross-links in type I collagen. Intramolecular cross-links occur in the nonhelical regions and involve a condensation reaction between lysine- or hydroxylysine-derived aldehydes within a single molecule. Intermolecular cross-links, on the other hand, involve aldehydes and ε-amino groups of lysine present in different molecules.

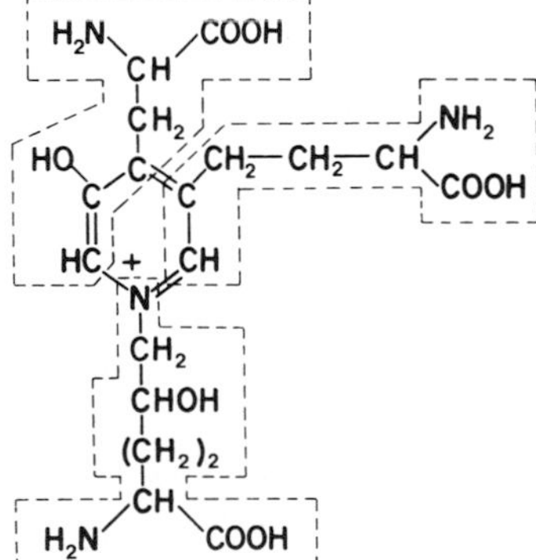

FIGURE 17 Pyridinoline: this trifunctional pyridinium cross-link, which joins three adjacent collagen molecules, can be generated by one hydroxylysine residue and two hydroxylysine-derived aldehydes or by spontaneous interaction of two hydroxylysine-5-keto-norleucine residues formed from a hydroxy-lysine and a hydroxylysine aldehyde.

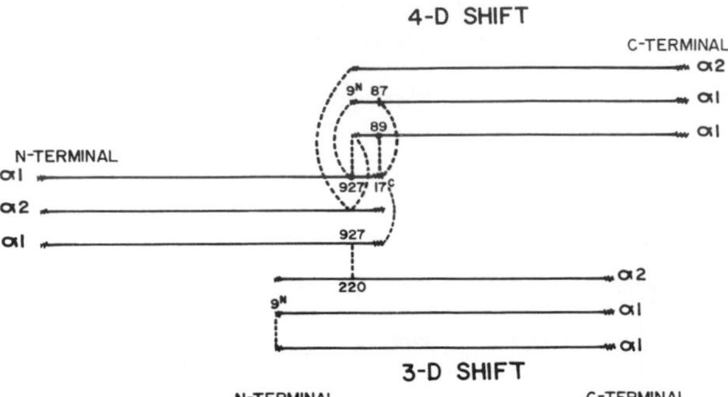

FIGURE 18 Schematic representation of type I collagen molecules aligned in three-dimensional (3-D) and four-dimensional (4-D) staggered positions. The known cross-linking sites are indicated by dashed lines. The intermolecular cross-link formed between hydroxylysine residues 9^N from the amino (N)-terminal region of α1 chains is among the first cross-linking sites to be recognized. The carboxy (C)-terminal hydroxylysyl residue 927 of α1 chain can cross-link to one or two α1 chains through residue 9^N. Hydroxylysyl residue 927 cross-links to 9^N of an α1 chain or to 9^N of an α2 chain. The carboxy-terminal 17^C hydroxylysyl residue of the α1 chain cross-links to hydroxylysyl residue 87 from an α1 chain of another molecule. The residue 17^C might also form an aldol-type intramolecular cross-link with a similar residue of the other α1 chain. Histidine 89 from the carboxy-terminal region of an adjacent molecule adds to this cross-link via a Michael addition. Recently, an intermolecular cross-link between hydroxylysyl residues 927 and 220 in the helical region was found in dentin and bone, supporting the observation that aldehydes do form in the collagen helix.

been recently identified, reflecting the presence of heterogeneous fibrils.

B. Collagen and Aging

The physical as well as chemical properties of collagen change with age. These changes in physicochemical properties have been attributed to the formation of both covalent and noncovalent cross-links. Neutral salt-soluble collagen, which has a low concentration of β components, will generate intramolecular bonds if gelled at 37°C. These intramolecular bonds seem to precede the formation of stable intermolecular cross-links, since these gels can redissolve when cooled to yield a soluble collagen with a higher content of β components (i.e., dimers of α chains) of intramolecular origin.

Although it is attractive, the cross-linking theory of aging awaits further experimental support. Obviously, stabilization of collagen fibers might occur by means other than the formation of new covalent cross-links. An increase in the number of weaker forces that stabilize macromolecules and their aggregates (e.g., van der Waals bonds, ionic interactions, hydrophobic bonds, and combinations of such forces) could account for changes in the physicochemical properties of the collagen fibers. For instance, a slow time-dependent exclusion of intermolecular water could not only lead to an increase in hydrophobic contacts, but could strengthen existing ionic bonds by placing them in an environment of decreased dielectric constant (Fig. 3).

C. Inhibition of Collagen Cross-linking: Aminonitriles and D-Penicillamine

Lathyrism is a connective tissue disorder associated with the ingestion or injection of βAPN and its chemical analogs or extracts of the sweet pea or other members of the Lathyrus family, usually consumed during famine. The skeletal changes observed differ from species to species and vary with age, being much more pronounced in younger animals. The epiphyseal plate, the growth zone of the ends of long bone, is a prime target.

The connective tissue abnormalities are associated with cross-linking defects in collagen and elastin. They are revealed by an increased solubility in hypertonic neutral salt solutions, owing to an inhibition of lysyl oxidase activity. Since copper deficiency also inhibits this enzymatic activity, the similarities of the defects induced by these two mechanisms are readily explainable.

Administration of D-penicillamine to animals and humans also causes an accumulation of collagen soluble in neutral salt in skin and various soft tissues. Two of the more characteristic properties of D-penicillamine—namely, the ability to trap carbonyl compounds and to chelate heavy metals—are of primary significance in impairing collagen cross-linking. The former property manifests itself in all effective dose ranges, whereas the latter occurs only at high doses, far beyond those administered to humans.

The collagen extracted from tissues of animals treated with D-penicillamine is able to form stable fibers *in vitro* and is not deficient in aldehydes, as is that from βAPN-treated animals. In fact, its aldehyde content is even higher than normal, suggesting that the mechanisms of action of βAPN and D-penicillamine are different (Fig. 19).

D. Wound Healing and Vitamin C

Collagen synthesis increases in wounds. There is an increase in the synthesis of type III collagen 10 hours

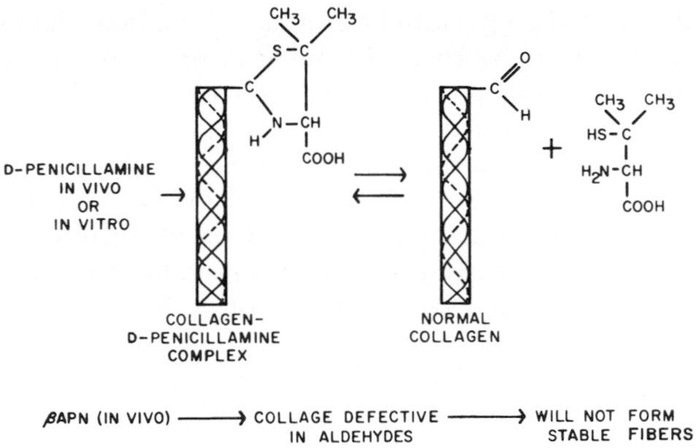

D-PENICILLAMINE IN VIVO OR IN VITRO

COLLAGEN-D-PENICILLAMINE COMPLEX

NORMAL COLLAGEN

βAPN (IN VIVO) ⟶ COLLAGE DEFECTIVE IN ALDEHYDES ⟶ WILL NOT FORM STABLE FIBERS

FIGURE 19 The modes of action of D-penicillamine and aminonitriles, inhibitors of collagen cross-links. D-Penicillamine *in vivo*, as well as *in vitro*, interacts with aldehydes on collagen, preventing them from subsequently participating in the formation of intramolecular and intermolecular cross-links. The reversibility of the collagen defect seen when D-penicillamine therapy is discontinued can be explained by the instability of the thiazolidine complex formed between D-penicillamine and the peptide-bound aldehydes. On the other hand, βAPN, a lathyrogen, inhibits aldehyde formation by irreversibly binding to the enzyme lysyl oxidase.

after wounding, but by 24 hours its synthesis returns to a normal value. The early type III collagen deposited could be important in establishing the initial wound structure and providing a basic lattice for subsequent healing events.

Well-healed mature scars in humans can reopen because of ascorbic acid deficiency. This phenomenon was recognized in ancient times, when sailors on extended sea voyages living on diets devoid of fresh fruits and vegetables noticed a breakdown in skin scars that had been healed for years. This is because metabolic activity is present in a scar, even after the healing process is completed, and ascorbic acid (i.e., vitamin C) is required for the hydroxylation of peptide-bound proline.

VI. COLLAGEN SYNTHESIS BY CELLS IN CULTURE

Various cell types have been shown to synthesize collagen. Human fibroblasts in culture synthesize both types I and III collagen, type I accounting for 70–90% of the total. Cell density in the culture and nutrient availability can alter the ratio of collagen types synthesized: human lung fibroblasts synthesize 26–68% more type III collagen at confluency than at low cell density.

Chondrocytes, which are highly differentiated mesenchymal cells, synthesize a unique blend of collagen (mostly type II) and proteoglycans. The normal pattern of proteoglycan and collagen syntheses can be altered by a variety of factors; these changes occur concomitantly with changes in cell shape. Although chondrocytes in culture are metabolically active, in adult cartilage their metabolic activity is relatively low and mitosis is rarely seen. When degenerative or trauma-related changes occur in the cartilage from joints, chondrocytes recover their ability to divide. The state of differentiation of chondrocytes can be monitored in a quantitative fashion by determining the collagen types they produce.

VII. BIOMECHANICAL PROPERTIES OF CONNECTIVE TISSUES

The biomechanical properties of collagen fibers have been studied primarily in tissues where these fibers are oriented in a parallel fashion (e.g., tendons and ligaments). The response to stretching is, however, difficult to interpret because many variables significantly affect behavior. The toe region (Fig. 20a) of the stress–strain curve reflects the slack associated with elimination of the crimping of the fibers; during this phase the elastic modulus increases steadily. The linear part of the curve (Fig. 20b) is where stress increases rapidly. This is the area from which the modulus is usually calculated and reflects the actual contribution of collagen molecules aligned parallel to each other and stretching of the crystalline network. When the point of maximum stress (Fig. 20c) is reached, the tissue breaks; this is the failure point, which can occur suddenly in bone or can be preceded

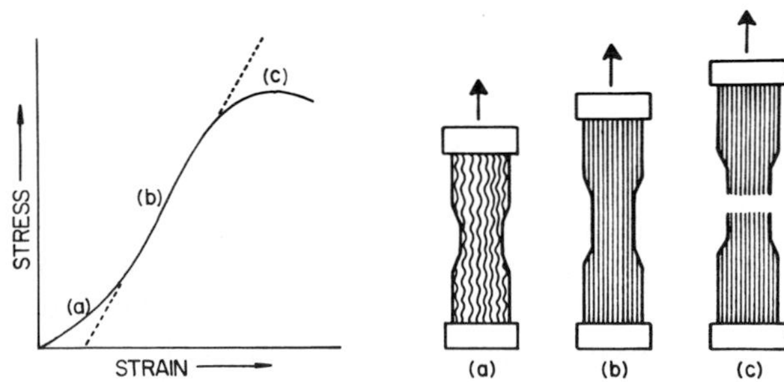

FIGURE 20 Changes in the internal structure of collagenous tissues during the stretching of a tendon or a ligament. (a) Elimination of crimpling. (b) Resistance displayed by the aligned fibers. (c) Failure point.

by a leveling off toward the strain axis, as in skin, ligaments, and tendons.

Denaturation of collagen by heat or chemicals can produce a collapse of the helical structure and a shrinkage in the direction of the longitudinal axis of the fiber. Whereas the collagen molecule (i.e., monomer) in a disperse solution melts at around 37°C, when assembled into fibrils it does not do so until it reaches a temperature of around 60°C. The tension developed during thermal shrinkage and the swelling that accompanies this event have been used to evaluate the physical properties of the collagen networks. Such an approach, particularly measurements of hydrothermal swelling, allow us to quantitate differences in the degrees of cross-linking of collagen that occur in skin with aging.

The cartilage of joints, a hard tissue characterized by a high degree of hydration, owes many of its properties to its ability to shift water molecules from one domain to another. Water is not evenly distributed through the extracellular space; rather, the upper 25% of that cartilage contains approximately 85% water, but as the depth increases the degree of hydration decreases in a linear manner to about 70%. It is hypothesized that one of the major contributors to the destruction of cartilage in osteoarthritis is hydration. The rupture of the collagenous framework seems to occur as a result of osmotic pressure imbalance within the tissue.

When cartilaginous tissue is in contact with a saline solution or a physiological fluid, the proteoglycans within the matrix exert an osmotic pressure that depends on their concentration. This osmotic pressure is able to counteract an externally applied force and enables cartilage to remain hydrated under load. Healthy cartilage is in thermodynamic equilibrium with the synovial fluid and is therefore balanced by a combination of tensile stresses exerted by the collagen network of the tissue and those coming from outside of the network, or applied pressure.

Whereas the molecules in type I collagen are separated by a distance of 14 Å, type II collagen molecules are separated by a distance of 16–17 Å. Measurements made on wet rat tail tendon (mostly type I collagen) show that 55% of the volume of a fibril is occupied by collagen molecules and 45% by water. From these etimates it can be calculated that a wet fibril of type II collagen could contain 50–100% more water than does a fiber of type I collagen that has the same number of molecules. One can easily envision how additional water in type II fibers can be advantageous to tissues such as cartilage and the nucleus pulposus of the intervertebral disks, whose main functions are to absorb or distribute compressive loads, in contrast to type I fibers present in tendon and skin, which transmit tension.

The normal mechanical stresses that act on the connective tissues seem to be essential for maintenance of the cellular activities required for the synthesis, turnover, and organization of macromolecules of the extracellular matrix. Reduction of such stresses by inactivity or immobilization that results in a loss of muscle mass (atrophy of disuse) can also cause a loss of bone (i.e., osteoporosis) and cartilage. Immobility on the periarticular connective tissues often leads to rigid contractures, a major complication of fracture treatment using casts. A large number of cross-links could cause stiffness and be responsible for the contracture.

VIII. COLLAGEN–PROTEOGLYCAN INTERACTIONS

To understand the physical properties of connective tissues such as cartilage and intervertebral disks, it is important to have an understanding of the salient features of the proteoglycan molecules (Fig. 21). There are essentially two types of glycosaminoglycans: those with weak negative changes (e.g., hyaluronic acid) and those with strong negative charges (e.g., the chondroitin sulfates, heparins, and dermatan sulfate, the latter comprising the largest bulk of the proteoglycans). Their distribution and physiochemical characteristics, which contribute to distinct functions, are also unique. Hyaluronic acid, with weak negative charges associated with the carboxylic acid residues present in glucuronic acid, has a tendency to form hydrated gels and can therefore contribute significantly to the viscoelastic fluidity of synovial fluid and to the turgency of the skin of an infant. [*See* Proteoglycans.]

On the other hand, the negatively charged polysaccharides that contain sulfonic acid residues are able to develop strong ionic bonds with the positively charged amino acids on the surface of the collagen fibers, particularly lysine, hydroxylysine, and arginine. Such tissues are more compact, resilient, and less hydrated and exhibit the viscoelastic behavior typified by hyaline articular cartilage. They are also more collagenous than their fluid counterparts. Synovial fluid has no collagen, the vitreous body of the eye has only small amounts of type II collagen, and the skin of a newborn rabbit has less than 2% collagen; this in

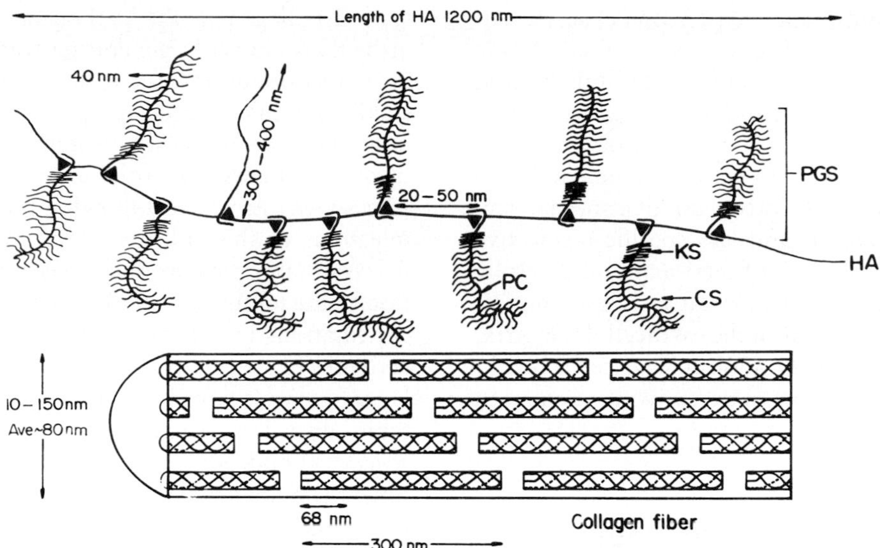

FIGURE 21 Collagen fibers do not exist in a vacuum. They are usually closely associated with the proteoglycans of the ground substance. A collagen molecule is depicted in cartilage, adjacent to a proteoglycan aggregate containing hyaluronic acid (HA), proteoglycan subunits (PGS), and link proteins (▲) that help to stabilize the structure. The PGSs consist of a protein core (PC) from which the negatively charged glycosaminoglycan chains of chondroitin sulfate (CS) and keratan sulfate (KS) radiate.

contrast to a three-month-old rabbit, which has more than 15% collagen. Fetal skin has so little collagen that its wounds can heal without generating scars.

The physiochemical properties of the various types of connective tissues, their viscoelastic properties, the diffusion of macromolecules and of small ions through their midst, and the exclusion of molecules of various molecular masses (e.g., immunoglobulins) are, understandably, different.

IX. MINERALIZATION OF COLLAGEN

A major proportion of human collagen is found in bone. Demineralized bone is almost exclusively type I collagen, with small amounts of characteristic noncollagenous proteins. Animal bones are a primary source of commercial gelatin in which, through a process of boiling, the collagen has been denatured.

The mechanism by which collagen mineralizes is a complex one. Essentially, it involves the nucleation of calcium and phosphate ions around functional groups in collagen and the subsequent formation of calcium phosphates, which mature into highly insoluble hydroxyapatites. This occurs on the surface as well as in the interstices of the fibril. It has been estimated that approximately one-half of the mineral is on the surface of the fibril and the remainder is inside, to a great extent in spaces left during the quarter-stagger packing of the collagen molecules (i.e., holes). Why some collagens mineralize and others do not is not clearly understood. It could be related to the intermolecular distances that result from the lateral packing, which in some cases could restrict the access of Ca^{2+} and PO_4, or from chemical modifications that occur subsequent to collagen deposition in the extracellular space (e.g., cross-linking and phosphorylation).

In addition to bone collagen, occasionally other tissue collagens calcify, usually as a result of pathological events (e.g., diseased heart valves, sclerotic blood vessels, and collagen implants). The mechanisms underlying these two forms of calcification are quite different: whereas bone formation is a cell-mediated event, dystrophic calcification is not.

X. FIBROSIS AND TISSUE REPAIR

Accumulation of collagen in excessive amounts is a major pathological event that underlies several clinical conditions, including pulmonary fibrosis, liver cirrhosis, and retrocorneal fibrous membrane formation, as well as various forms of dermal fibrosis, such as scleroderma, keloids, and hypertrophic scars. Although in many of these diseases the terminal fibrotic lesion is considered to be the sequela of cellular injury, the cell populations injured and the endogenous mediators responsible for the postinjury fibrotic response vary from organ to organ. In many instances we seem to be dealing with an uncontrolled repair mechanism, in which less organized and less specific connective tissue replaces a previously functional and carefully constructed matrix. In other instances we see an imbalance in the homeostasis of the extracellular matrix, in which synthesis of macromolecules exceeds breakdown, the end result being an excessive accumulation of collagen.

XI. COLLAGEN DISEASES

As noted, healthy tissues contain optimum amounts of a well-organized extracellular matrix, of which collagen is a major component. The collagen types present seem to be closely associated with particular structures. Fibril diameter, orientation density, and mode of packing are optimal for the function of diverse tissues such as tendons, cornea, liver parenchyma, bone matrix, and glomeruli. The ability of mesenchymal cells to generate and maintain such structures is paramount to health. An imbalance can severely damage the ability of a tissue or organ to function and can lead to some of the problems briefly mentioned, causing many degenerative or fibrotic diseases to occur.

Wear and tear of the connective tissues, particularly those exposed to harsh biomechanical environments such as encountered in the joint structures (cartilage, tendons, ligaments), can lead to severe disabilities.

Although collagen is generally perceived as a poor antigen, several autoimmune diseases seem to be associated with a loss of tolerance to specific collagen types, particularly IV, VII, and II. Goodpasture syndrome, for example, is a form of glomerulonephritis associated with the production of antibodies to type IV collagen, a major component of the glomerular basement membrane. Searches for the antigenic determinants in this disease, as well as in a heritable form of nephritis, and diabetic nephropathy associated with the nonenzymatic glycosylation of basement membrane collagen led to the identification of new α chains as minor components of type IV collagen.

Type VII collagen appears to be the target for an autoimmune process that damages anchoring fibrils (Fig. 13) and results in blistering due to the disadherence of the epidermis from the dermis. The role of type II collagen in the pathogenesis of rheumatoid arthritis is currently generating tremendous interest, as induction of tolerance, by administering orally small amounts of chicken type II collagen to affected individuals, seems to significantly improve the status of these patients. The role of type II collagen in this disease was suspected following the observation that injections of this collagen, combined with Freund's Adjuvant, into rats produced a generalized inflammation of the joints very similar to rheumatoid arthritis.

Mutations in collagen genes or deficiencies in the activities of specific posttranslational enzymes of collagen synthesis have been characterized in many heritable disorders, such as osteogenesis imperfecta, several chondrodysplasias, several subtypes of the Ehlers-Danlos syndrome (a disease associated with lax joints and weak connective tissue structures), and the renal and blistering diseases already mentioned. In addition, collagen mutations have been found in certain common diseases, namely, osteoporosis, osteoarthritis, and aortic aneurysms, and it is now evident that subsets of patients with these diseases have defects in types I, II, or III collagen, respectively, as predisposing factors. Mutations have so far been identified in only 6 of the more than 30 collagen genes, and thus research into collagen defects is only in its early stages.

Many other diseases of the connective tissues that involve collagen seem to be associated with a decline in the structural integrity of the extracellular matrix, many times associated with the process of aging, yet sometimes independent of it. Some of the vascular changes seen in atherosclerotic blood vessels, the decline of calcifiable matrix (osteoid) that leads to osteoporosis, the thinning of the skin caused by a loss of collagen, and structural changes in our ocular system and of type II collagen in cartilage that may make the molecules more susceptible to proteolysis can result from alterations in the quality or density of the collagen framework of the tissues involved.

Further understanding of the process of collagen biosynthesis at all levels, the nature and structure of the many types of collagen present in the human body, their morphology, physicochemical characteristics, and mechanisms of degradation will greatly contribute to unraveling the mysteries of a fundamental aspect of our structural anatomy.

BIBLIOGRAPHY

Burgeson, R. E., and Nimni, M. E. (1992). Collagen types: Molecular structure and tissue distribution. *Clin. Orthopaedics* **282**, 250–272.

Cunningham, L. W., and Frederiksen, D. W. (eds.) (1982). "Methods in Enzymology," Vol. 82. Academic Press, New York.

Fleischmajer, R., Olsen, B. R., and Kuhn, K. (eds.) (1990). "Structure, Molecular Biology and Pathology of Collagen." The New York Academy of Science, New York.

Glimcher, M. E., and Lian, K. B. (eds.) (1989). "The Chemistry and Biology of Mineralized Tissues." Gordon & Breach, New York.

Kang, A., and Nimni, M. E. (eds.) (1992). "Collagen: Pathobiochemistry," Vol. V. CRC Press, Boca Raton, FL.

Kivirikko, K. I. (1993). Collagens and their abnormalities in a wide spectrum of diseases. *Ann. Med.* **25**, 113–126.

Mecham, R. P. (ed.) (1986). "Regulation of Matrix Accumulation." Academic Press, Orlando, FL.

Nimni, M. E. (ed.) (1988). "Collagen: Biochemistry, Biomechanics and Biotechnology," Vols. I–III. CRC Press, Boca Raton, FL.

Olsen, B. R., and Nimni, M. E. (eds.) (1989). "Collagen: Molecular Biology," Vol. IV. CRC Press, Boca Raton, FL.

Colon Cancer Biology

JOHN M. CARETHERS
University of California, San Diego

GLOSSARY

Adenoma (adenomatous polyp) Benign colonic epithelial growth with dysplastic glandular features

Allelotype Determination of loss of chromosomal segments in tumors

Clonal expansion More rapid growth of the progeny of one cell compared to that of the surrounding cells

Dominant negative effect Oligomerization of a mutated protein with a wild-type protein such that the function of the wild-type protein is attenuated or lost

Familial colon cancer Colon cancer that has developed in persons with germline mutations of genes that put the person at high risk for development of colon cancer, usually by the fourth decade of life

Loss of heterozygosity Loss of a normal allele from the genome

Metastasis Tumor spread to a location distant from the site of origin

Microsatellite instability Excessive expansion or contraction of the length of a repetitive DNA sequence in tumor tissue

Neoplasia Tissue that has lost the normal restraints on its growth, which is capable of clonal expansion

Oncogenes Mutated versions of human cellular genes that encode for proteins typically involved in signal transduction or the regulation of gene expression that can transform normal cells

Sporadic colon cancer Colon cancer that has developed without apparent familial risk, usually in the seventh or eight decade of life

Tumor suppressor gene Gene that encodes a protein that functions to inhibit cell proliferation and tumorigenicity; only one copy of the gene is required for its function

Wild-type allele Normal, nonmutated gene that is the most common sequence found in nature

COLORECTAL CANCER DEVELOPS THROUGH A SEries of genetic events that sequentially permits progressively more aggressive growth characteristics. Transformation of a colon cell to a malignant cell occurs through a multistep process in which certain individual cells gain a growth advantage because of a key genetic alteration, allowing the cell's progeny to undergo clonal expansion.

I. INTRODUCTION

Multiple genetic events must occur for a normal colon cell to evolve into a cancer. In Western societies, a high-fat, low-fiber diet has been epidemiologically linked to the occurrence of colon cancer. The exposure of a colon cell to these and accrual of other factors likely initiate colorectal neoplasia, and subsequent genetic events as a consequence of an advantaged growth process propel the colon cell toward malignancy. In familial colon cancer, initiation and progression of colon cells from normal to malignancy are enhanced by the inheritance of a germline mutation of a tumor suppressor gene that inactivates one of the two copies and increases the likeilhood that cancer will develop. Although one normal allele is enough to prevent the cell from unregulated growth, this allele is the sole blueprint for a key growth regula-

ENCYCLOPEDIA OF HUMAN BIOLOGY, Second Edition, VOLUME 2. Copyright © 1997 by Academic Press. All rights of reproduction in any form reserved.

tor (compared to two normal alleles in nonfamilial cancer). Thus in cells in which a germline mutation exists, an additional mutation or "hit" to the other allele is sufficient to remove the regulatory effects of the gene.

The relative ease of obtainable tissue from the colon by colonoscopy has made colon cancer the best-studied human malignancy. Over the past decade, the adenoma has been the subject of intense scrutiny, and some important genetic clues from this investigation have been elucidated. Adenomatous polyps are the precursors to colorectal adenocarcinoma. As the adenoma grows, a change in its architecture from tubular to villous often occurs. As the adenoma enlarges, the chance of malignancy from within the polyp rises. Carcinoma develops as a clonal outgrowth of some adenomas, and if not removed the carcinoma spreads locally in some cases and systemically to lymph nodes and other organs in other cases. The growth of tissue, from normal epithelium to adenoma to carcinoma, is paralleled by inactivating mutations and loss of alleles with the tumor suppressor genes *APC, DCC,* and *p53* and activating mutations of the K-*ras* oncogene. These findings have led to the proposal of a sequential genetic scheme, termed the adenoma-to-carcinoma sequence.

II. HISTOLOGICAL PROGRESSION OF COLON CANCER

The earliest identifiable lesion of colonic neoplasia is the adenomatous polyp, which over time can develop lethal potential. There is some evidence that dysplastic aberrant crypt foci, colonic lesions seen with magnification after methylene blue staining of grossly normal colonic mucosa, may harbor the same genetic defect as the adenoma, thus making them possible precursor lesions to adenomas. The prevalence of adenomas rises in direct proportion with the frequency of colorectal cancer. Less commonly, colon cancer may develop in flat mucosa, representing the rare instance of malignant conversion that has appeared early in the natural history of colorectal tumorigenesis.

In normal colonic epithelium, a constant turnover of the surface epithelium occurs approximately every 6 days due to the proliferation and differentiation of anchored crypt cells. The lower third of a colonic crypt is the site of the proliferating compartment where mitoses occur; these cells mature as they migrate up the crypt to the surface, where they lose their capacity to divide. Eventually, the cells die and are shed into the lumen of the colon. In the adenomatous polyp, the ordinary crypt compartmentalization is altered. There is continued cell division and lack of differentiation of cells so that the proliferating compartment might consume the entire crypt. The protracted cell division in conjunction with hampered cell maturation and extrusion results in increased surface cell replication. This results in a downward infolding of epithelial cells, which interpolate themselves and branch between normal crypt elements. This process results in the typical branching glandular pattern seen with tubular adenomas. As the adenoma grows, there is an increased growth of the underlying mesenchyme. Where mesenchyme growth matches epithelial growth, no impedance to epithelial proliferation exists, which results in long, finger-like projections of the glandular components. This histological picture is the villous adenoma. Often, an adenoma may display details of both tubular and villous types. Villous adenomas are generally larger than tubular adenomas, in agreement with the concept that the villous subtype is associated with enhanced growth characteristics.

Adenomas are benign neoplasms. The growth is autonomous, but the neoplasm is incapable of invasion or metastasis, which defines malignancy. Although it is clear that colon cancer arises from colon adenomas, it is also certain that most adenomas do not develop into carcinoma. It is unknown what percentage of adenomas progress to cancer, but it must be very small. It is also not certain the time course by which an adenoma may develop into cancer. An estimate from patients with *familial adenomatous polyposis* (FAP), an autosomal dominant disease in which patients develop thousands of adenomas because of germline mutation in the *APC* gene, suggests that a minimal of one to two decades must pass before carcinoma occurs in an adenoma. A shorter time course for the adenoma-to-carcinoma sequence may occur in *hereditary nonpolyposis colon cancer* (HNPCC), an autosomal dominant condition in which patients develop colon cancer at a young age due to mutation in one of four genes involved in DNA mismatch repair (MMR). As the adenoma enlarges, carcinoma develops within the polyp, which identifies the polyp as malignant (Fig. 1). The cancer within the polyp has a growth advantage over the adenomatous portion, and subsequently replaces it. Unfortunately, the accelerated growth by the cancer cells is not contained by the polyp; by mechanisms that are poorly understood, the tumor gains the capacity to extend into and through the colon wall, and eventually spreads to other organs (metastases).

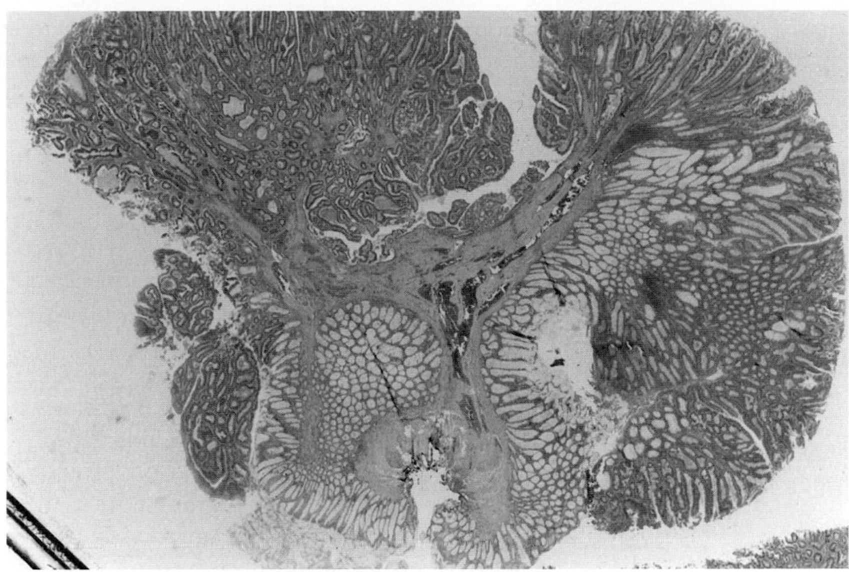

FIGURE 1 This large, malignant, bilobar polyp has a central stalk, predominant villous histology in the right lobe, and carcinoma in the upper left portion of the left lobe. Note the long, dysplastic, finger-like glands in the villous glandular architecture (nonmalignant) and the poorly organized glandular architecture of the cancerous cells.

Each advantage gained in growth, beginning with the adenoma and subsequently progressing to carcinoma, occurs by successive waves of clonal expansion (Fig. 2). Each clone gains its advantage by a genetic alteration within the cell that allows more successful proliferation over predecessors with lesser genetic variations. Other modifications that are acquired during clonal expansion include loss of cell-to-cell adherence, enhanced cell mobility, the elaboration of enzymes that digest the epithelial basement membrane,

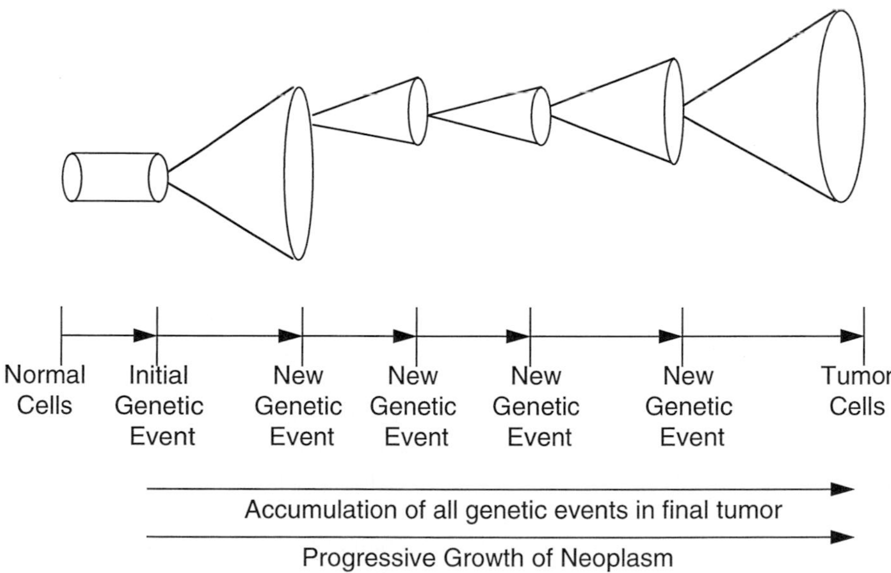

FIGURE 2 A diagram illustrating clonal expansion. After some initial genetic event, one cell acquires a growth advantage over neighboring cells. Subsequently, a new genetic event occurs in one cell, allowing accelerated growth over remaining cells. Each successive cellular expansion of clones occurs after a key genetic event, with the accumulation of the genetic events within the cell culminating in the development of cancerous cells.

and the ability to survive in remote locations. Some of these biological behaviors have been identified as caused by changes in the genetic makeup of the cell.

III. GENETIC MAKEUP OF COLON CANCER

For a colon cell to acquire a growth advantage from a defective tumor suppressor gene, it must lose both copies of that gene because tumor suppressor genes act in a recessive fashion. Typically, one allele becomes mutated (or is inherited as mutated), and the other allele is lost during a flawed mitosis. The protection from uncontrolled cellular proliferation disappers, allowing the cell to undergo clonal expansion of its cellular lineage. These loss of heterozygosity (LOH) events uncover latent heterozygous mutations, and the consequences are the development of malignancy, invasion, and metastases.

Allelotyping has led to the localization of tumor suppressor genes involved in colorectal cancer. Losses of the small arm of chromosome 17 (17p) and the long arm of chromosome 18 (18q) were identified as the most frequent, and these locations were later found to harbor the *p53* and *DCC* genes, respectively. In addition, frequent allelic loss was noted at chromosome 5q, the location of the *APC* gene. Other allelic losses were noted, but the potential tumor suppressor genes at these locales have yet to be identified. Of the tumor suppressor genes identified, the *APC, p53,* and *DCC* genes have been best demonstrated to be involved in colorectal cancer.

In contrast, oncogenes act in a dominant fashion; that is, mutation of one allele is sufficient to uncouple normal growth regulation, regardless of the presence of a normal allele. The mutations in oncogenes are activating mutations and allow the gene product to increase its activity in an unregulated manner. The excessive stimulus contributes to uncontrolled cellular proliferation. In colon cancer, the K-*ras* oncogene has been demonstrated to play a role in colon tumorigenesis and has been particularly associated with the larger and more villous tubular adenomas. Activation of K-*ras* is not necessary for malignant transformation. [*See* Oncogenes and Proto-oncogenes.]

A. *APC* Gene

The adenomatous polyposis coli (*APC*) gene, located on chromosome 5q, is the genetic locus of FAP and Gardner's syndromes (characterized by the development of thousands of colonic adenomas by the mid-20's), as well as some patients with Turcot's syndrome (colonic adenomas and associated brain tumor). In sporadic colon cancer (i.e., in which there is no germline passage of a mutation), the *APC* gene plays a critical role in tumorigenesis. A mutated *APC* gene is found in the majority (>80%) of adenomas and most (>80%) colorectal carcinomas. The mutated *APC* gene is found only in the adenoma and never in the surrounding normal tissue, an indication that the mutation is a somatic (not germline) event. Because *APC* is considered a tumor suppressor gene, inactivation of the second allele must occur for the cell to lose the growth restriction provided by the *APC* gene. Inactivation of the second allele usually occurs by LOH. In some adenomas, the second *APC* allele appeared normal. Loss of *APC* function in those cases may occur by a dominant-negative effect in which the mutated protein heterodimerizes with the full-length *APC* protein, thus inactivating it.

In sporadic colon cancers, half of the mutations found in the *APC* gene occur in a 722-base-pair region between codons 1281 and 1554 (Fig. 3). This area of the *APC* gene coincides with mutations in FAP in which the "profuse" variety of polyposis develops. In this variety of FAP, there are many more polyps (>5000 typically), and carcinoma occurs at an earlier age. As is the case in FAP, the majority of mutations in sporadic colon cancer cause truncation of the *APC* protein. These mutations include insertions and deletions that lead to reading frame shifts that create a stop codon and nonsense point mutations that result in premature stop codons.

Most evidence indicates that *APC* mutations occur early and might be the initial event in sporadic colorectal tumorigenesis. *APC* mutations are found even in small adenomas (<1 cm), and the prevalence of mutation is not higher in carcinomas than adenomas, suggesting no advantage later in the carcinogenesis sequence. LOH of 5q occurs abruptly at the transition of normal colonic epithelium to the adenoma. Simultaneous analysis of K-*ras* revealed mutations present in only 20% of adenomas with *APC* mutations, consistent with the hypothesis that *APC* mutations precede K-*ras* mutations. The presence of a mutated *APC* allele and absence of *APC* gene function indicates that an acquired event has occurred in a single cell, which has given rise to the polyp by clonal expansion. Thus, the *APC* gene is considered the "gatekeeper" for colonic neoplasia. If a colonic crypt cell loses *APC* function, neoplasia may be initiated.

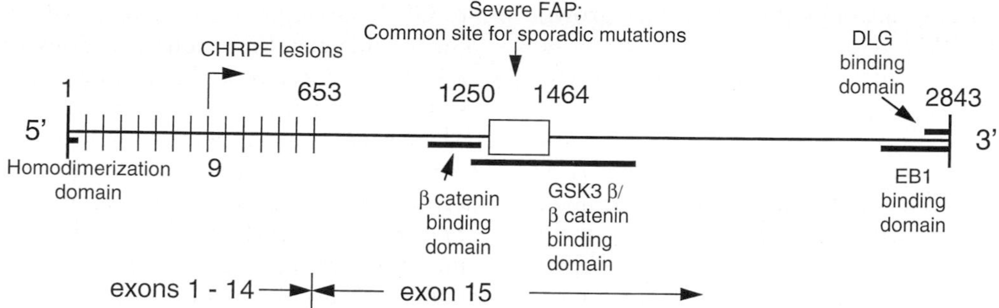

FIGURE 3 A map of the *APC* gene. A homodimerization domain at the amino terminus of the *APC* protein allows one *APC* molecule to form a complex with a second *APC* molecule. β-Catenin, involved in linking *APC* to the zona adherens complex, binds *APC* between codons 1014 and 1210. GSK3β (glycogen synthase kinase 3β), a molecule that phosphorylates *APC* and is involved in the turnover of β-catenin, binds the middle region of *APC* (with β-catenin). The protein EB1, perhaps involved in conveying part of the tumor suppressor function, binds *APC* between codons 2560 and 2843; likewise, DLG (the human homologue of the *Drosophila melanogaster* disc large tumor suppressor protein) binds *APC* between codons 2771 and 2843 and may relay cell polarity and prevent neoplastic proliferation. In sporadic colorectal cancers, a common site for mutation is between codons 1281 and 1554. Similarly, mutations between codons 1250 and 1464 give a virulent phenotype of profuse polyposis in FAP. CHRPE (congenital hypertrophy of the retinal pigment epithelium) lesions in FAP have been associated with mutations between exons 9 and 15.

Analysis of the *APC* protein has identified some clues to its function. The *APC* gene product is expressed in the cytoplasm of the colonocyte and is increased in the upper portion of the colonic crypts in concert with cell maturation. *APC* protein associates with the cell's microtubular skeleton, binds to proteins involved in the zona adherens (a complex that may regulate contact inhibition with neighboring cells), and binds a novel protein, EB1, which may relay part of *APC*'s tumor suppressor function to the nucleus (see Fig. 3). At least part of *APC*'s function appears to activate programmed cell death (apoptosis), thus preventing unregulated growth of senescent colonic epithelial cells.

B. 18q and the *DCC* Gene

The deleted in colorectal carcinoma (*DCC*) gene was found by searching in an area of deletion on chromosome 18q common to a large sample of colorectal tumors. One allele of *DCC* is lost in 71% of colorectal cancers, which matches the observation that LOH of 18q markers occurs in 73% of colorectal cancers. Since it is rare to find a homozygous deletion of *DCC*, it may be that a 50% reduction in gene dose may be important in tumorigenesis. Mutations in the residual allele have not been reported, perhaps due to the large size of this gene, which has hampered full mutational analysis. In adenomas, 18q allelic loss is relatively rare in small adenomas but approaches 50% in large adenomas. Loss of *DCC* generally occurs in adenomas after LOH of the *APC* gene and mutation of the K-*ras* oncogene.

DCC encodes a unique protein with homology to neural cell adhesion molecules and other related cell-surface glycoproteins. The similarity has led to the hypothesis that the *DCC* gene product is involved in cell-to-cell adhesion and cell matrix interactions, which may be important in preventing tumor growth, invasion, and metastases. In patients with colorectal cancer metastases, over 80% have LOH of 18q. In addition, LOH of 18q may identify patients who likely have micrometastases from those who do not.

There are additional tumor suppressor genes in the region of *DCC* on chromosome 18q that may be responsible in part for the biological behavior of colon cancer attributable to LOH of 18q. These include the *DPC4* (deleted in pancreatic carcinoma 4) gene and the *MADR2* (MAD-related 2) gene, but their full role in colorectal tumorigenesis is incompletely understood. In the case of the *MADR2* gene, its protein is involved in intracellular signaling pathways after activation by transforming growth factor β (TGF-β), a growth factor that serves to inhibit colonic epithelial growth. Loss of *MADR2* function would theoretically uncouple the inhibitory growth effects of TGF-β.

C. *p53* Gene

The *p53* gene (so named because its protein is 53 kilodaltons in size) is the most important determinant of malignancy during colorectal tumorigenesis. Its location on chromosome 17p is lost in 75% of colon cancers, which is never observed in benign adenomas, making LOH at the *p53* locus a late event in colorectal tumorigenesis. Loss of 17p occurs at a lower frequency in larger adenomas compared with LOH at 18q, implying that *p53* loss usually occurs after that of 18q. In addition, mutation of *p53* (most are missense mutations) coupled with LOH of the remaining allele coincides with the appearance of cancer within a polyp. Furthermore, LOH of 17p and *p53* overexpression (due to the longer half-life of the mutated *p53* protein) prognosticate a decreased 5-year survival in colorectal cancer patients.

Abnormalities in the *p53* gene can be found in more than half of all human malignancies, making it a component in pathways central to the development of human cancer. *p53* is an important negative regulator of normal cell growth and division. Its gene product is a transcription factor because of its ability to activate the expression of other genes. These include genes that prohibit progression of the cell cycle (*WAF1/CIP1*), activate DNA synthesis after removal of damaged DNA (*GADD45*), and stimulate programmed cell death (including the response to chemotherapy with 5-fluorouracil in cancer patients). The functions attributed to *p53* suggest that the normal protein has the ability to regulate the inherent mutability of human DNA in somatic cells. This has led to the description that *p53* is the "guardian of the genome." [*See* Tumor Suppressor Genes, p53.]

D. K-*ras* Gene

The Kirsten-*ras*-2 (K-*ras*) oncogene, located on chromosome 12p, is one of three members of the ras gene family that encode proteins involved in hydrolysis of guanosine triphosphate (GTP) to guanosine diphosphate (GDP) during intracellular signal transduction. Normally, an inactive K-*ras* protein exchanges GDP for GTP, causing a conformational change. The conformation change allows the active molecule to interact with intracellular effector molecules, conveying a growth signal from the plasma membrane to the nucleus. Subsequently, GTP is intrinsically hydrolyzed to GDP, making K-*ras* protein inactive. In colorectal cancer, missense mutations at codons 12, 13, and 61 allow stabilization of the K-*ras* protein in its active state, which permits continuous signal transduction. This leads to unregulated growth of the colon cell.

In sporadic colorectal cancer, mutations at codons 12, 13, and 61 of K-*ras* have been detected in 47% of carcinomas and 50% of large adenomas. K-*ras* mutations are found in more advanced adenoma stages, and occur after *APC* gene alterations. The role of K-*ras* may be one of a growth facilitator, enabling small adenomas to grow into larger ones. This possibility has been strengthened by lack of mutations of K-*ras* in flat carcinomas, implying that mutated K-*ras* does indeed permit growth into a larger polyp.

IV. MODEL OF COLORECTAL CARCINOGENESIS

A multistep model of genetic progression of colorectal cancer was originally proposed by E. R. Fearon and B. Vogelstein in 1990 after systematic identification of allelic loss in colorectal cancer. For an adenoma to develop, mutation of one *APC* allele with subsequent loss of the second *APC* allele is required. The lack of a normal *APC* protein prohibits its normal growth inhibitory signal and allows the cell to expand clonally into an early adenoma. By chance, one cell in the adenoma acquires a mutation in its K-*ras* oncogene. This mutation allows that cell to clonally expand, overgrowing the cells that have only *APC* mutations, and permits the adenoma to grow larger. During this accelerated wave of growth and increased mitotic activity, gradual accumulations of other mutations occur as subclones develop within the mutant *APC*/K-*ras* clones. These defects may include mutation of the *DCC* gene and other 18q genes followed by LOH of the other 18q allele. Mutation of the *p53* gene occurs during this period, and this may be silent. When LOH occurs at the remaining normal *p53* allele, the malignant phenotype emerges and overgrows the benign cells and has the ability to invade (Fig. 4).

V. MICROSATELLITE INSTABILITY AND COLONIC TUMOR PROGRESSION

In a subset of colorectal tumors—those from patients with HNPCC and those from a subset of sporadic

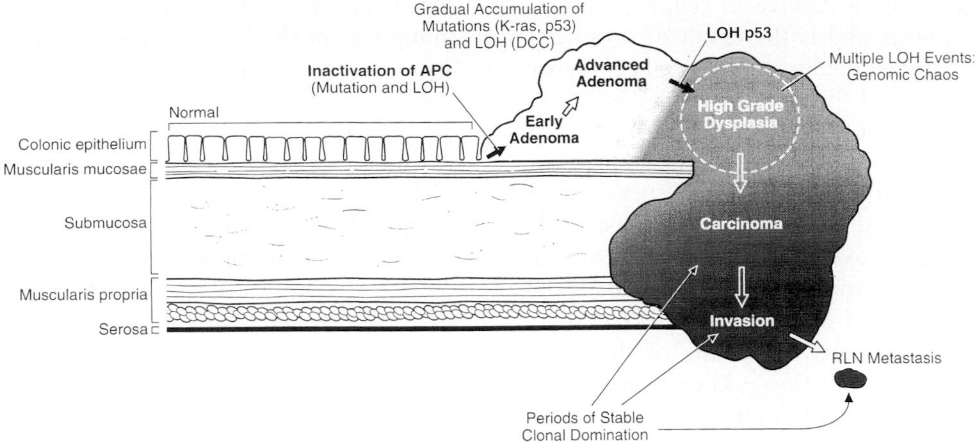

FIGURE 4 A model of genetic alterations mediating neoplastic progression in colon cancer tumorigenesis. *APC* is inactivated by mutation and loss of heterozygosity at the transition between normal colonic epithelium and adenomatous epithelium. Thereafter, successive rounds of clonal expansion occur with accumulation of mutations (i.e., K-*ras, DCC, p53*) and LOH (i.e., *DCC*) that allow acceleration of growth and dominance over lesser-advantaged clones. *p53* inactivation by LOH of 17p first appears in high-grade dysplasia and defines the transition between benign and malignant neoplasia. Successive genetic events as well as prior genetic changes enable the colon cancer to invade and metastasize to regional lymph nodes (RLN) and ultimately distant organs such as the liver. [From C. R. Boland, J. Sato, H. D. Appleman, R. S. Bresalier, and A. P. Feinberg (1995). Microallelotyping defines the sequence and tempo of allelic losses at tumour suppressor gene loci during colorectal cancer progression. *Nature Med.* **1**, 902–909, with permission.]

colorectal cancer—few or no LOH events occur. This finding has led to the identification of an alternative pathway for colorectal tumorigenesis. These tumors are hypermutable due to a defective mismatch repair system. The MMR system is a complex of proteins that edits and directs repair of polymerase errors made during DNA replication. MMR genes recognize and direct repair of nucleotide mispairs and misalignment at short repetitive sequences of DNA (microsatellites) whose length was not accurately copied during DNA replication. Base mispairing causes nucleotide transitions and transversions, introducing point mutations and altering the authentic genetic sequence, perhaps within certain growth regulatory genes. However, typical mutations in *APC, p53,* and K-*ras* appear less often than in tumors lacking microsatellite instability (MIN), consistent with the concept of an alternative pathway of tumorigenesis. Occasionally, microsatellite sequences are present in the coding region of growth regulatory genes. The transforming growth factor β receptor II (TGF-β RII), when bound by TGF-β, is inhibitory toward colonic cellular proliferation. With defective MMR, the microsatellite within TGF-β RII's coding region changes length, rendering the receptor inactive. This removes the normal growth brake provided by TGF-β and allows the cell to undergo clonal expansion.

In sporadic colorectal cancer, MIN is present in approximately 15% of tumors (compared to 95% of HNPCC tumors). In adenomas from HNPCC patients, the mutational spectrum suggests that DNA MMR inactivation might occur before an *APC* mutation. In sporadic adenomas, MIN is rare. This suggests that inactivation of some component of the MMR system might occur after an adenoma has already formed, and such inactivation may drive the polyp toward malignancy. Sporadic polyps that exhibit MIN may progress to carcinoma at an accelerated rate once the *APC* gene is inactivated. The hypothesis proposing a path independent of tumor suppressor gene inactivation in sporadic cancers with MIN is currently being evaluated.

VI. COLON CANCER METASTASES

It has been observed that circulating tumor cells may be found in the mesenteric and systemic circulation of colon cancer patients, but curiously this has no prognostic significance. More importantly, the ability of colon cancer cells to survive in remote locations determines prognosis and is the result of select cells with features that include: loss of cell–cell adherence, proteolytic digestion of the epithelial basement membrane, enhanced cell mobility, entrance into the circulation, adherence to a target tissue's endothelial surface, penetration through the vessel into the target organ, adaptability and growth in a new substrate by clonal expansion, and evasion of the immune system. The most common location for colorectal cancer metastases is the liver, and this "homing" capability is in addition to its downstream locale of the portal blood flow from the intestines. Alterations in cell-surface glycoproteins are common during carcinogenesis and have a major influence on the homotypic and heterotypic cell–cell and cell–substratum interactions that mediate metastases. Primary colon cancer differs from metastatic colon cancer in that metastatic colon cells express sialylated mucin-associated carbohydrate structures that may play a role in adhesion to target endothelial glycoproteins and the target cell basement membrane. Other carbohydrate structures found on colon cancer cells have been demonstrated to bind to molecules such as laminin, a normal component of the basement membrane. Thus, only select colon cancer cells with altered surface membrane structures, among other tactics of cell survival such as neoangiogenesis, have the ability to grow as metastases. [*See* Metastasis.]

VII. SUMMARY

Colorectal cancer develops as a consequence of accumulations of genetic alterations and progressive waves of clonal expansion of cells that have a growth advantage over their progenitors. These alterations include mutations in the K-*ras* oncogene, and a model consisting of a concomitant mutation in one allele together with the loss of the other allele in tumor suppressor genes such as *APC, p53,* and *DCC.* At least two of these genetic changes, those found with *APC* and *p53,* correlate with major histological transitions into, and out of, the adenomatous polyp. A second process occurs in the 15% of sporadic colorectal cancers that exhibit MIN, that is, inactivation of the MMR system, which allows accumulation of mutations in genes and errors at microsatellite loci. The uncontrolled growth and malignant transformation provided by each process may enhance spread of the tumor when certain cells develop metastatic potential, in part by expressing sialylated glycoproteins on its cell surface. The complex nature of colorectal tumorigenesis continues to be studied so that our understanding may lead to an amelioration and prevention of this disease in the future.

BIBLIOGRAPHY

Boland, C. R. (1995). Malignant tumors of the colon. *In* "Textbook of Gastroenterology" (T. Yamada, D. H. Alpers, C. Owyang, D. W. Powell, and F. E. Silverstein, eds.), 2nd Ed., pp. 1967–2026. Lippincott, Philadelphia.

Boland, C. R., Sato, J., Appelman, H. D., Bresalier, R. S., and Feinberg, A. P. (1995). Microallelotyping defines the sequence and tempo of allelic losses at tumour suppressor gene loci during colorectal cancer progression. *Nature Med.* **1**, 902–909.

Fearon, E. R., and Vogelstein, B. (1990). A genetic model for colorectal tumorigenesis. *Cell* **61**, 759–767.

Kinzler, K. W., and Vogelstein, B. (1996). Lessons from hereditary colorectal cancer. *Cell* **87**, 159–170.

Marra, G., and Boland, C. R. (1995). Hereditary nonpolyposis colon cancer (HNPCC): The syndrome, the genes, and a historical prospective. *J. Natl. Cancer Inst.* **87**, 1114–1125.

Color Vision

DAVIDA Y. TELLER
University of Washington

GLOSSARY

Brightness One of the three dimensions of perceived color; perceptual quality that ordinarily varies with the intensity of light; lights of all colors can vary in brightness; also used to refer to white–grey–black variations of the colors of objects

Cones Photoreceptors that serve daytime (photopic) vision and make color vision possible; there are three kinds of cones: short-, mid-, and long-wavelength-sensitive, with maximal sensitivity at approximately 435, 535, and 565 nm, respectively; these terms replace the older nomenclature of blue, green, and red cones, respectively

Hue One of the three dimensions of perceived color (e.g., blue, green, yellow, orange, red, purple)

Isoluminant stimulus Pattern made up of purely chromatic variations, without variations in luminance (e.g., a set of red and green stripes, in which the red and green are matched in luminance)

Luminance Scientifically refined measure for specifying the efficiency (effective intensity) of lights of different wavelengths for human vision; lights that match in luminance will, in general, differ somewhat in (perceived) brightness; luminances of lights of different wavelengths are additive

Metamers Lights of different wavelength composition that are identical in appearance

Photopic vision Vision at daylight illumination levels; roughly, cone-mediated vision, including color vision

Rods Photoreceptors that serve nighttime (scotopic) vision

Saturation One of the three dimensions of perceived color; the variations white–pink–red and white–light blue–deep blue are variations of saturation

Scotopic vision Night vision; vision at dim illumination levels with the eyes adjusted to darkness; scotopic vision is colorless

IN ORDINARY LANGUAGE, THE TERM COLOR vision refers to one's capacity to see colors, or to tell objects or lights apart on the basis of color differences. The term color refers to a particular subset of the variations we perceive in the qualities of lights or objects. We use color terms (e.g., red, yellow, blue, green, purple, pink, lime green) to refer to this group of perceived variations.

It is not immediately obvious why color variations group themselves together as a natural perceptual category. In addition, the category boundary is ambivalently defined. Interestingly, in ordinary language brightness or lightness variations—the perceptual qualities of white, grey, and black—are sometimes included and sometimes excluded from the list of colors. For example, the following two exchanges are both acceptable usage: What color is your dress?—It's black and white. Is that a color TV?—No, it's black and white.

As a scientific discipline, color science encompasses two main domains. The first is psychophysics: the quantitative description of perceptions and their relations to the physical stimuli that give rise to them. In describing color perception quantitatively, color scientists ask questions such as: Can human subjects arrange colors consistently into groups or series on the basis of perceived similarity? In what ways and along how many dimensions do color perceptions vary? How can these and other color data be best represented in geometrical or mathematical terms? And in relating perceived colors to physical stimuli, we ask: What is the relationship between lights, objects, and perceived colors? What factors other than

ENCYCLOPEDIA OF HUMAN BIOLOGY, Second Edition, VOLUME 2. Copyright © 1997 by Academic Press. All rights of reproduction in any form reserved.

the wavelength of light influence perceived colors? These questions are treated in Section I of the present article. [*See* Vision, Psychophysics.]

The second main domain of color science is the attempt to explain color vision, on the basis of neural and mathematical models of information processing in the eye and brain. In this domain, we ask questions such as: How does information about color reach the eye? In what ways do the initial information encoding processes in the eye constrain the colors we see? How is information about color recoded and processed by the eye, and by the brain? Is it eventually recoded into a form that bears a recognizable correspondence to the characteristics of perceived colors? Questions of this kind are treated in Section II.

In Section III we consider the genetic basis of color vision and several kinds of naturally occurring variations in color vision. These include color vision deficiencies or color "blindness," the development of color vision in infants, and the color vision of different species of animals. We close with a comment (Section IV) on the intellectual value of the study of color vision and a list of recommended readings.

I. PERCEPTUAL ASPECTS OF COLOR

A. Light

Electromagnetic radiation is one of the basic forms of physical energy. The range of wavelengths of electromagnetic radiation that is visible to the human eye is called light; these wavelengths range from about 400 to about 700 nm (1 nm = 10^{-9} m).

When a beam of white light (e.g., from a tungsten bulb) passes through a prism, it is spread out to reveal its different wavelengths. Under such conditions, a rainbow, or spectrum, of colors can be displayed. By testing human subjects with light of each wavelength in turn, one can examine the sensitivity of the human eye to each wavelength, and the variation of perceived color with wavelength.

B. Spectral Sensitivity

The human eye is differentially sensitive to lights of different wavelengths. The exact shape of the spectral sensitivity (luminous efficiency) curve and the wavelength to which the eye is maximally sensitive vary somewhat with light levels and methods of measurement. Under scotopic conditions—when the eye is adjusted to dim illumination levels, and dim lights are used for testing—sensitivity is maximal at about 500 nm, whereas under photopic conditions—when lights are at normal room illumination or higher—sensitivity is maximal at about 550 nm. Standard scotopic and photopic curves are shown in Fig. 1A. This shift in spectral sensitivity can sometimes be observed in daily life by noticing that red objects, which send mostly longer wavelengths of light to the eye, become quite suddenly relatively darker compared with blue or green objects as twilight deepens.

C. Spectral and Extraspectral Hues

1. Scotopic Conditions

Under scotopic conditions, all lights, regardless of wavelength composition, look whitish in color, and spots of light of all wavelengths can be made identical in appearance by varying their intensities to set them equal in brightness. This striking absence of color perception at dim illumination levels can be demonstrated by laying out a set of differently colored articles of clothing in the evening. When one wakes up at night in a very dimly illuminated room, the clothing appears colorless and varies only in shades of grey.

2. Photopic Conditions: The Spectral Hues

When higher light levels are used, the expected colors reappear. As wavelength increases from 400 to 700 nm, colors ranging from violets, through blues, greens, yellows, and oranges, to reds are seen, as shown in Fig. 1B.

Interestingly, at the perceptual level, the linear ordering of colors in Fig. 1B does not capture all of the facts of color similarity. In particular, the colors seen at the two spectral extremes—reds and violets—share a reddish component. To be true to an ordering of colors by perceptual similarity, the linear array would have to be bent into a segment of a circle, so that the two spectral extremes approach each other, as shown in Fig. 2A.

3. The Extraspectral Purples

The color purple is extraspectral, that is, there is no wavelength of light that looks purple. Extraspectral colors are produced by superimposing lights of different wavelengths. In particular, if a very short wavelength (violet-appearing) and very long wavelength (red-appearing) light are superimposed, and their relative intensities varied, a continuous series of hues varying from violet through mid-purple, to reddish purple and red can be produced.

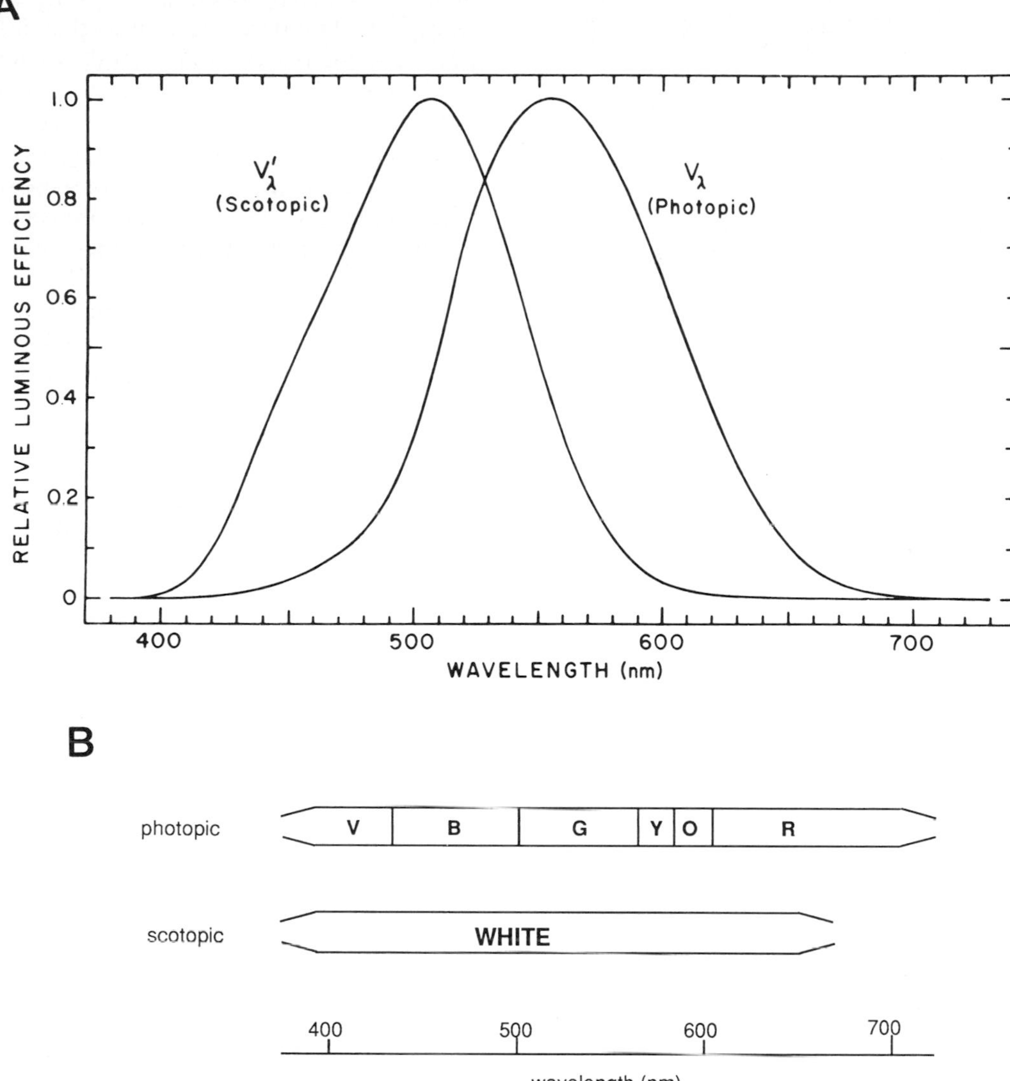

FIGURE 1 Light and color. (A) Variations in the sensitivity (luminous efficiency) of the human eye with variations in the wavelength of light. Under scotopic conditions (very dim illumination levels) the sensitivity maximum occurs at about 500 nm. Under photopic (daylight) conditions, the sensitivity maximum occurs at about 550 nm. The curves V_λ and V'_λ are internationally adopted standard photopic and scotopic curves respectively. [Reprinted, with permission, from J. Pokorny et al. (eds.), 1979, "Congenital and Acquired Color Vision Defects," Grune & Stratton, New York.] (B) The approximate colors perceived by a human subject viewing various wavelengths of light. Under photopic conditions, colors are seen; under scotopic conditions, all wavelengths of light give rise to the perception of whiteness. V, violets; B, blues; G, greens; Y, yellows; O, oranges; R, reds.

D. The Three Dimensions of Perceived Color

1. The Hue Circle

When ordered by perceptual similarity, brightness-matched patches of spectral and purple lights form a complete, continuous circle, called the hue circle. An example of a hue circle is shown in Fig. 2A.

2. Unique and Mutually Exclusive (Opponent) Hues

Beyond the characteristics already discussed, the hues in the hue circle differ from each other in important qualitative respects. The key is that some colors appear to be analyzable into perceptual combinations of other colors, whereas others are not. The unitary, or *unique*, hues—red, yellow, blue, and green—are

A

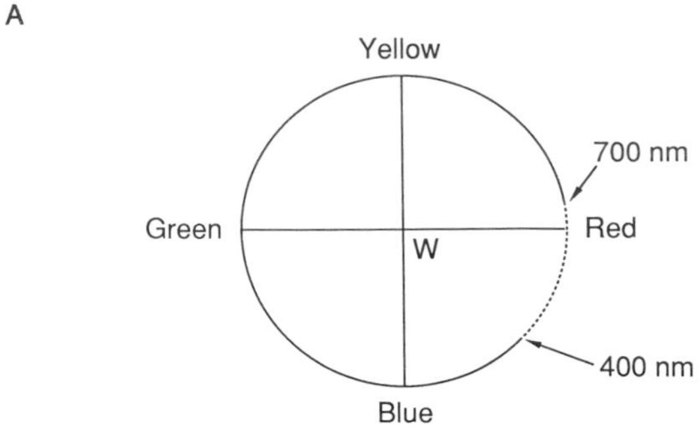

B

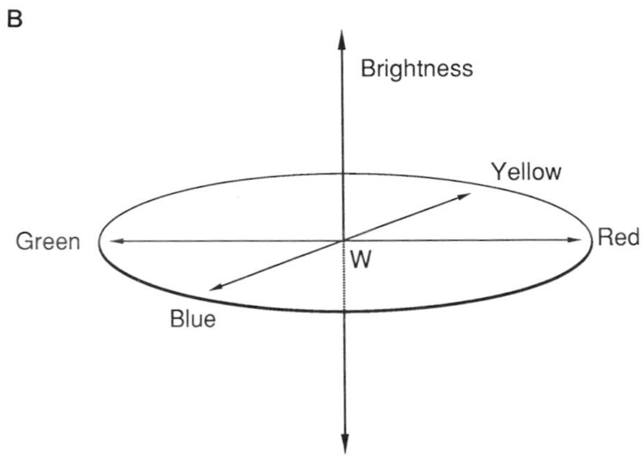

FIGURE 2 Perceptual color spaces. (A) Hue circle. The solid line represents the physical (wavelength) spectrum. The cardinal axes are chosen to correspond to the two pairs of unique and mutually exclusive hues; red–green and yellow–blue. The dotted line represents the purples. White is located at the center, with variations of saturation (e.g., white–pink–red) represented on radii of the circle. A two-dimensional figure is sufficient to represent all of the perceived variations of the colors (hues and saturations) of lights, ordered by similarity. (B) Three-dimensional color space, with brightness represented on the vertical axis. A three-dimensional figure is required to represent all of the perceived variations of the colors of lights when brightness is included.

so-called because subjects report that these colors cannot be perceptually analyzed into subparts. In contrast, *binary* hues, such as purples, oranges, yellow-greens, and blue-greens, can be so analyzed. For example, most subjects agree that orange can be described as a reddish yellow, whereas red cannot be described as an orangish purple. Furthermore, of the unitary hues, there are two perceptually mutually exclusive, or *opponent,* pairs: red versus green and blue versus yellow, so-called because they cannot coexist as perceptual components (most people draw a blank when asked to imagine a reddish-green or a yellowish-blue).

Under ordinary viewing conditions, unique blue, green, and yellow occur at approximately 470, 520, and 575 nm, respectively; these values change somewhat with illumination level and other viewing conditions. Unique red is extraspectral, occurring when a small amount of short-wavelength light is added to a patch of predominantly long-wavelength light. It must be emphasized that there is nothing physically unusual about electromagnetic radiation of these wavelengths. The uniqueness and mutual exclusiveness of particular hues are perceptually, rather than physically, based.

In geometrical terms, the unique hue pairs are often used to define two cardinal axes—red–green and yellow–blue—for the hue circle; this rule was applied in construction of Fig. 2A.

3. Saturation

Besides the purples, other wavelength mixtures yield other extraspectral colors. The white-appearing light from the sun contains all wavelengths in relatively equal amounts. Mixture of this white with increasing proportions of a spectral color (such as red) again yields a continuous series of color variations, from white through pale pink and deeper pinks to red. Such color variations are called variations in saturation. Ordered by perceptual similarity, these colors fit within the hue circle, with white at the center and increasing saturations of each hue arranged on a line from white to the corresponding hue on the circle.

4. Brightness

The perceptual qualities of lights also vary along a third perceptual dimension, that of brightness. In general, variations of the intensity of light, without variations in wavelength, yield variations in perceived brightness. Addition of the brightness dimension to the two-dimensional hue circle yields a three-dimensional color solid, as shown in Fig. 2B. As noted in our original definition of the term color, the brightness dimension can be included or excluded from color terms depending on context.

5. The Color Solid: A Geometrical Representation of Color Vision

In summary, one of the most fundamental and theoretically enticing characteristics of color vision is its three-dimensionality: in mathematical terms, a closed two-dimensional surface, or plane, is both necessary and sufficient to display all of the hues and saturations that can be produced by spots of light, and a third dimension is required for brightness.

E. Wavelength Mixture

1. Complementary Wavelengths

The appearance of whiteness is not constrained to an equal energy mixture of all wavelengths. There is an indefinitely large number of pairs of wavelengths that, mixed together (i.e., superimposed) in the appropriate ratio, produce the perception of white. These pairs of lights are called complementary wavelengths. The placing of complementary wavelengths on the opposite ends of axes through white produces another major constraint on the geometry of the hue solid. Because the originally perceived colors of the two wavelengths are utterly lost from the perceived white of the mixture, the phenomenon of complementarity is perhaps the most dramatically counterintuitive aspect of color perception.

2. Metamers

Metamers are stimuli that differ in wavelength composition but are identical in hue, in saturation, and in brightness, that is, identical in appearance and thus indiscriminable from each other. Perceptual whiteness provides a good example, because unlimited numbers of complementary pairs exist, and beyond these, there are an infinite number of mixtures of three, four, or any number of wavelengths, including the mixture of all wavelengths in equal proportions, that yield the perception of white. These lights can all be adjusted in intensity to form a set of indiscriminable lights, that is, a metamer set.

Moreover, white is not the only color for which metamers exist. The same phenomenon occurs for any other color, although the metamer sets will in general be larger the less saturated the color. Because the members of a metamer set differ in wavelength composition but cannot be told apart visually, the phenomenon of metamerism provides an important example of loss of information by the visual system.

3. Trichromacy

Metamer sets are not haphazard. The form and extent of metamerism is summarized by a general rule called the law of trichromacy. In inexact but intuitively useful form, the law of trichromacy is illustrated in Color Plate 16 and in Fig. 3.

Let A and B be two patches of light. Let A be composed of three superimposed lights of broadly separated wavelengths: λ_1, λ_2, and λ_3 (e.g., 430, 530, and 650 nm respectively). Let B be composed of any other light, of any intensity and wavelength composition (i.e., of any hue, saturation, and brightness). Then it is possible, simply by variation of the intensities of the three wavelengths in patch A, to make the two patches of light match each other exactly:

$$A \equiv B,$$

where the symbol $\equiv$ is used to denote a metameric match. Therefore, roughly speaking, lights of three well-chosen wavelengths, mixed in differing combinations, are sufficient to generate a patch of light of *any* perceived hue, saturation, and brightness.

Two modifications are required to state the law of trichromacy rigorously. First, very saturated lights in patch B cannot be matched by leaving all three wavelengths in patch A; however, if one of the three is moved and combined with the saturated light in patch B, the two spots of light can always be made to match. And second, the wavelengths of the lights in patch A need not be those specified, nor need they be composed of only single wavelengths. Any set of four lights will do, provided that no two of them can be mixed to match a third. In its most general form, the law of trichromacy states that any set of four different lights can be arranged into two patches, such that a metameric match can be produced by variation of the intensities of three of them.

Thus, we arrive by wavelength mixture experiments at the same number we found in the perceptual ordering of colors. Color–brightness vision is a system with three and only three perceptual dimensions, and three and only three degrees of freedom.

F. Luminance

1. The Distinction between Brightness and Luminance

Patches of light that vary in hue and saturation can be equated in brightness by several different methods, but a problem arises in that different techniques and instructions yield systematically different results; therefore, color scientists distinguish between two concepts: brightness (with a new and more precise definition) and luminance.

In formal measurements of brightness, subjects are instructed to match the brightness of each spectral light to the brightness of a white standard light (direct heterochromatic brightness matching), or to the brightness of immediately neighboring wavelengths (step-by-step brightness matching). The results of such an experiment are shown in Fig. 4. Unfortunately, brightness values defined by such techniques suffer

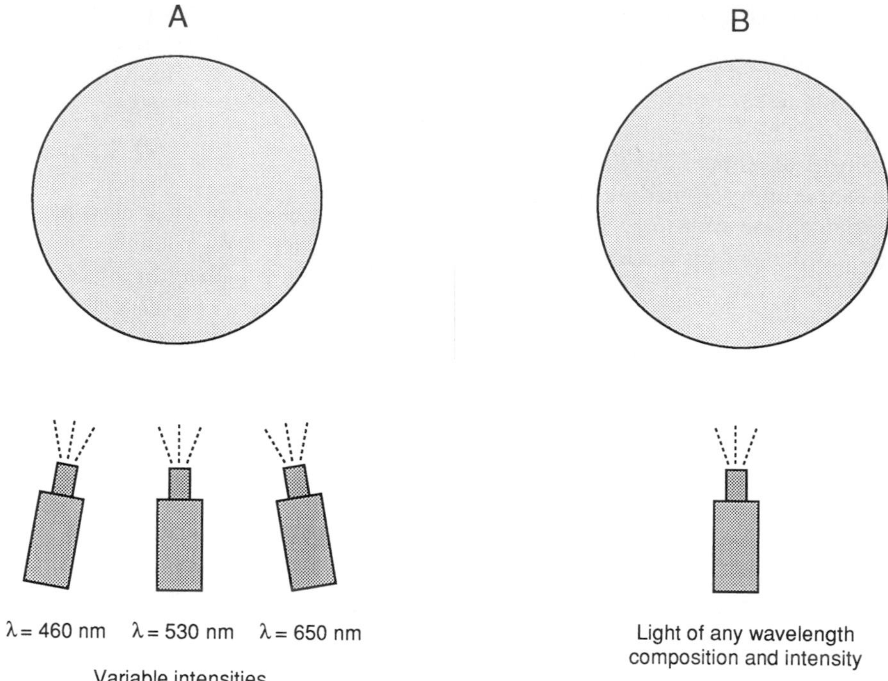

FIGURE 3 Trichromatic matching experiment. Two patches of light—A and B—are set up. Patch A is illuminated by lights of three wavelengths (e.g., 460, 530, and 650 nm); their intensities can be varied. Patch B is a light of any chosen wavelength composition and intensity. The law of trichromacy states that by varying the intensities of λ_1, λ_2, and λ_3, patches A and B can be made to match exactly. Lights of different wavelength composition that match exactly are called metamers.

from the difficulty that they are not strictly additive; that is, if two spectral lights are each matched in brightness to the same white standard, and then each is reduced to half intensity and the two are superimposed, the combined patch is judged to be brighter than the original white light.

The inelegance of a nonadditive metric led color scientists to invent another, more satisfactory metric, called luminance. In flicker photometry, each spectral test light is alternated (flickered) with a standard white light at a rate of 10–15 cycles per second, and the subject's task is to vary the intensity of the spectral light to minimize the perceived flicker. In the minimally distinct border technique, test and standard lights are presented in two exactly contiguous patches, and the subject's task is to vary the intensity of each spectral light to make the edge, or border, between them appear minimally distinct. Although these measures have no particular face validity as quantifications of the brightness dimension, both measures yield highly similar spectral sensitivity curves, and the resulting values are additive in the sense defined earlier. For this reason, luminance rather than brightness is used to specify the visual effectiveness of lights for most scientific purposes. A world standard luminance curve, V_λ, has been established; this is the function shown in Fig. 1.

2. Vision with Isoluminant Stimuli

Isoluminant stimuli are patterns composed purely of color variations without variations in luminance (e.g., a field of luminance-matched red and green stripes). Though black and white striped patterns as fine as 60 cycles per degree (c/d) of visual angle can be resolved, spatial resolution is much more limited (about 10 c/d) for isoluminant chromatic patterns.

In addition to the resolution of spatial detail, many other visual functions are poorly sustained when isoluminant stimulus patterns are used. If stimuli are composed of spatial variations in hue and/or saturation without variations in luminance, the borders separating the colors become perceptually fuzzy and indistinct. There are losses of precision in stereoscopic vision (the ability to perceive depth and distance on

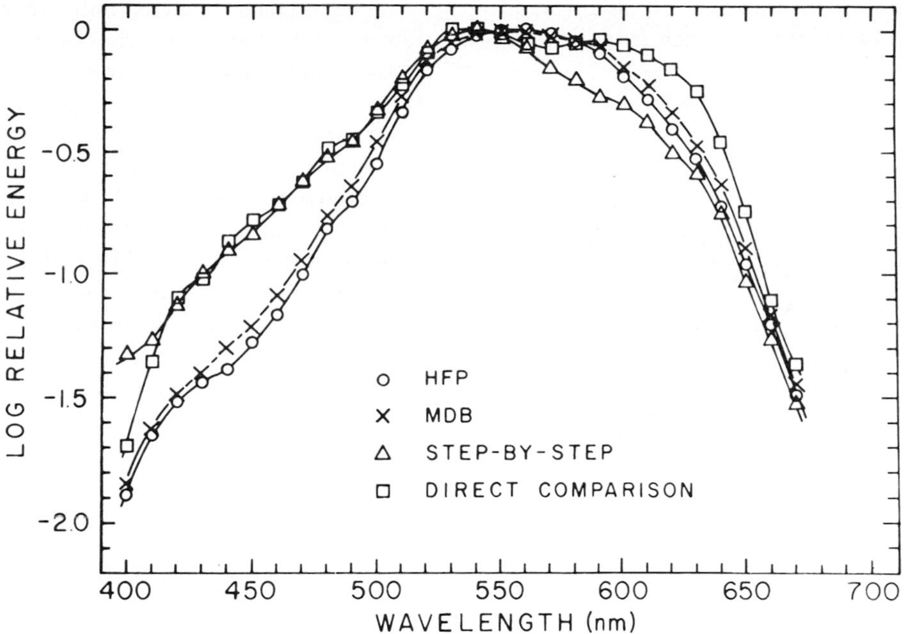

FIGURE 4 Brightness versus luminance. Different methods of equating the "brightness" of lights of different wavelengths lead to different spectral sensitivity curves. Color scientists distinguish between brightness, defined by direct comparisons of brightness between spots of light of different wavelength composition, and luminance, defined largely by a technique called heterochromatic flicker photometry (HFP). Minimally distinct border (MDB) matches agree with HFP matches, while step-by-step comparisons agree with direct brightness matches. [Reprinted, with permission, from J. Pokorny *et al.* (eds.), 1979, "Congenital and Acquired Color Vision Defects," Grune & Stratton, New York.]

the basis of combining inputs from the two eyes), accommodation (the ability to focus the lens of the eye), vernier acuity (the ability to see small spatial offsets), perception of the speed and direction of motion, and other visual functions. These visual losses occur when colors are matched in luminance (not brightness), and this fact, together with the characteristic of addivity, suggests that luminance measures might be revealing the more fundamental property of visual coding.

G. Alternative Axes for Color Space

Finally, we return to the question of the choice of cardinal axes for describing the three-dimensional nature of color perception. The axes originally shown in Fig. 2 were based on directly observable, qualitative perceptual dimensions. But other cardinal axes, based on other perceptual characteristics, are also useful in representing other, more sophisticated characteristics of color vision.

1. The Luminance Axis and the Isoluminant Plane

The additivity property of luminance suggests that luminance be substituted for brightness as the vertical axis of the color solid. The two-dimensional color plane, perpendicular to the luminance axis, then becomes a plane in which luminance does not vary (i.e., an isoluminant plane). Spatial patterns made up of stimuli selected from this plane will show the perceptual losses described earlier.

2. The Tritan Axis

When isoluminant patterns are used, there are in some cases more dramatic losses of vision for some color axes than for others. In particular, there is an axis called the tritan axis, which runs from yellow-green to mid-purple in Fig. 2A. This axis is defined in more detail in Sections II,C,3 and III,A,3. Isoluminant chromatic stimuli composed of lights selected from along the tritan axis show particularly large perceptual losses. Borders, although indistinct on all chromatic axes, "melt" into invisibility for such tritan stimuli.

3. The Independence of Red–Green and Tritan Axes

In addition, certain perceptual phenomena transfer among most chromatic axes but do not transfer between red–green and tritan stimuli. For example, a subject is exposed to a light alternating in time (flickering) between isoluminant red and green. After adapting to this light for several seconds, the subject is asked to detect flicker along the other chromatic axes. He or she will be found to have lost sensitivity for flicker on most axes, including the yellow–blue axis, but little if any loss of sensitivity is found along the tritan axis. Similarly, adaptation to flicker along the tritan axis leads to a minimal sensitivity loss on the red–green axis. This and similar phenomena suggest that these two axes represent stimuli that are processed very nearly independently of each other. To capture the fact of this independence, red–green and tritan axes can be used as the two cardinal chromatic dimensions in color space.

4. An Alternative Color Space

A color space with luminance, red–green, and tritan axes is shown in Fig. 6B. This version of color space departs somewhat from direct description of the perceptual aspects of color (Fig. 2B), particularly in that the tritan axis departs considerably from the old yellow–blue axis. The two color spaces represent different constellations of facts about color vision. We will suggest in the following that a series of recodings of color–brightness information occurs within the visual system, and that different color phenomena reveal the marks left by different stages of neural processing.

H. Contrast Effects

If a disk-shaped test field consisting of a fixed intensity of white light is surrounded by a white ring, or annulus, the test disk can be made to appear any shade of grey or black by variation of the luminance of the annulus. This perceptual phenomenon is called simultaneous contrast. A ratio of about 20 : 1 between annulus and disk yields the perception of black. Similarly, test disks of various wavelengths will appear increasingly darkened by the annulus; for example, a 475-nm light will change from blue through dark blue through navy blue to black, and a 600-nm disk will change from orange to brown to black.

Greys, blacks, browns, and other dark colors do not occur with single patches of light; they are added to the realm of color perception by simultaneous contrast. Color spaces intending to represent the dark colors often use the negative half of the brightness (or "lightness") axis to represent "darkness."

Annuli of different wavelengths will induce approximately complementary hues. For example, a red annulus surrounding a white disk induces a greenish hue in the disk. In combination with luminance variations, chromatic annuli can yield a surprisingly large range of color variations in the disk. Similar phenomena (called successive contrast) occur in the time domain; for example, a white test field following a green one will appear reddish.

I. The Colors of Objects

1. The Complexity of Object Colors

Up to this point, we have discussed mainly the perceived colors of lights. The colors of objects are more complex, for several reasons. First, most objects do not emit light; instead, they are visible because they reflect some of the light that falls upon them. The spectral reflectance function of an object describes the fraction of the incident light of each wavelength reflected by the object. Most objects have broad spectral reflectance functions. Different objects reflect light best in different spectral regions, and the perceived color of an object will depend on the particular mixture of wavelengths which that object sends to the eye.

Second, the spectral characteristics of the illumination that falls on an object, as well as the object's reflectance function, will influence the spectral mixture, and hence one should expect that the illumination under which an object is viewed will influence its perceived color. And third, objects are usually surrounded by other objects, and seen after other objects, so that simultaneous and successive contrast effects also routinely influence the perceived colors of objects.

2. Color Constancy

The concept of color constancy refers to the tendency of the perceived color of an object to remain constant across variations of the spectral composition of the incident light. Despite the difficulties discussed earlier, a fair degree of color constancy does occur in complex natural scenes, across changes among broad-band illuminants, such as direct versus indirect sunlight, and lights produced by tungsten versus fluorescent bulbs.

However, even with broad-band illuminants, color

constancy can be imperfect when one is interested in exact color matches. Changes from daylight to tungsten light can upset the metameric match between one's tie and one's suit, or one's blouse and one's lipstick. Major failures of constancy occur, and major changes of the perceived hues of objects are seen, if the illuminant changes too much or becomes too narrow-band. For example, the low-pressure sodium lamps sometimes used for street lighting provide narrow-band illumination of about 590 nm and yield odd color perceptions.

Because the source of information is not obvious for separating the spectral reflectance functions of objects from the spectral characteristics of the incident illumination, the perceived constancy of object colors has been historically difficult to explain. Computational approaches to this problem are discussed in the following.

J. Summary

Much is known about the perceptual properties of color vision, and their dependencies on the properties of physical stimuli. But there is much about color perception that is specifically not directly predictable from the properties of light. Nothing about the physics of light leads us to expect the closed color circle, the three-dimensionality of color perceptions, the uniqueness and mutual exclusiveness of certain hues, the occurrence of complementary wavelengths, the trichromacy of color mixture, the reductions of visual function at isoluminance, or the fact that a surrounding annulus can completely change the perceptual qualities of a disk of light. We know a lot about which stimuli map to which perceptions, but the mappings are complex and not predictable simply from the physics of light. Their causes lie in the details of information processing within the eye and brain.

II. PHYSIOLOGICAL ASPECTS OF COLOR

A. A Brief Sketch of the Visual System

Light from a visual stimulus is imaged by the optics of the eye onto the retina, a thin layer of neural tissue that lines the back of the eyeball. There it is absorbed by the photoreceptors, neurons (neural cells) specialized to absorb light. The resulting signals are processed through several layers of neurons, before exiting the eye via the neural processes (parts of neurons) that make up the optic nerve. The optic nerve ends at a way station, the lateral geniculate nucleus, which in turn sends neural processes to the primary visual cortex. From there, the visual signals spread in two or more processing "streams" to other more distant parts of the cortex for further processing. [*See* Visual System.]

Modern neuroscientific techniques make it possible to use microelectrodes to record the electrical activity of individual neurons within many different parts of the visual system and brain. Thus, one can select a neuron in, for instance, the lateral geniculate nucleus and listen to its response as lights of different wavelengths and intensities are shone upon the retina. Other modern techniques make it possible to stain the cells from which one records and to trace the pathways that run between the different areas and subareas of the brain. Much of what we know about color processing comes from experiments of this kind, carried out on primates or other animals.

B. Photoreceptors and Photopigments

1. Light and Photopigments

A photopigment is a substance that absorbs light in some portion of the visible spectrum. The normal human eye contains four kinds of photoreceptors, each containing its own unique photopigment. The four photoreceptor types are the long-wavelength-sensitive (LWS), mid-wavelength-sensitive (MWS), and short-wavelength-sensitive (SWS) cones, which have maximal sensitivities at about 435, 535, and 565 nm, respectively, and the rods, which have maximal sensitivity at about 500 nm. As shown in Fig. 5, the four sensitivity ranges overlap, and together cover (and define) the visible spectrum. The properties of photopigments and photoreceptors provide well-accepted explanations for spectral sensitivity, metamerism, and trichromacy.

To discuss the absorption of light by photopigments, we shift to the description of light in terms of packets of energy, called quanta. Lights of different wavelengths have different amounts of energy per quantum, and different probabilities of being absorbed by any given photopigment. It is the differential probability of quantal absorption that gives rise to the spectral sensitivity curves of the four photoreceptors.

The absorption of a quantum of light results in a particular change in shape—an isomerization—of the photopigment molecule that absorbs the quantum. This change of shape, when it occurs, is identical

A

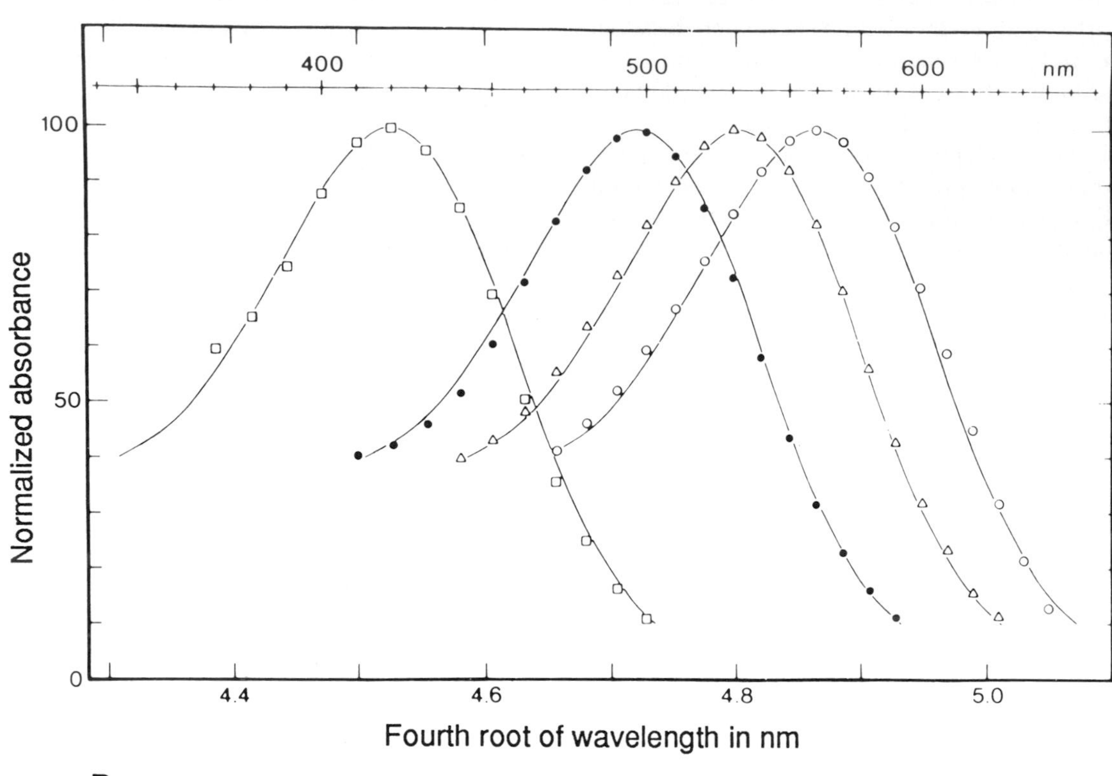

400
500
600
nm
100
50
0
Normalized absorbance
4.4
4.6
4.8
5.0
Fourth root of wavelength in nm

B

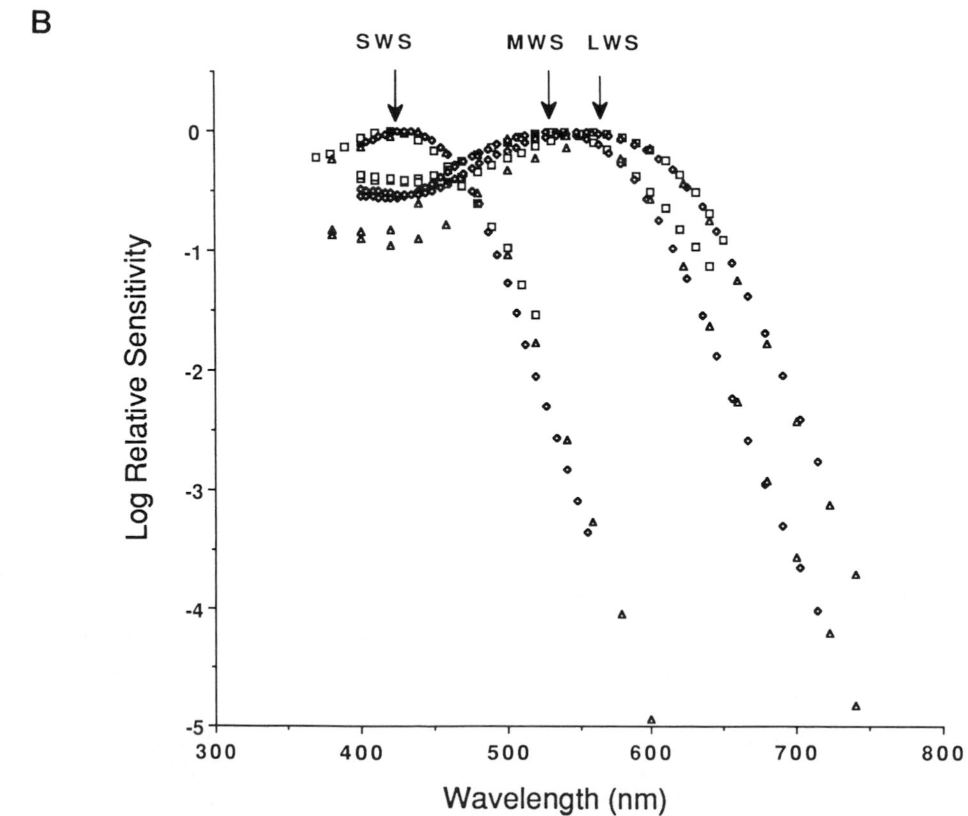

SWS
MWS
LWS
0
-1
-2
-3
-4
-5
Log Relative Sensitivity
300
400
500
600
700
800
Wavelength (nm)

regardless of the wavelength of the quantum. Thus, wavelength information is lost at the instant of quantal absorption. This photochemical fact has profound implications for vision, because it means that individual photoreceptors can signal only the number of quanta they absorb. They cannot signal the wavelength of light, and a matrix of photoreceptors containing only a single pigment would have no means of preserving wavelength information.

2. Properties of Scotopic Vision

At very low illumination levels, rods initiate detectable signals whereas cones do not. The scotopic spectral sensitivity curve, V'_λ in Fig. 1, follows the spectral sensitivity curve of the rod photopigment rhodopsin (modified to allow for the absorption of light in the lens and other optical elements of the eye). Moreover, test patches of all wavelengths, weighted by this absorption spectrum to match them in brightness, are perceptually indistinguishable. The colorlessness of scotopic vision comes about because only a single photoreceptor type is active, and no single photoreceptor type can preserve wavelength information on its own.

3. Photopic Spectral Sensitivity and Luminance

The standard photopic spectral sensitivity curve, V_λ in Fig. 1, can be well fitted by a weighted sum of inputs from LWS and MWS cones. For this and other reasons it is believed that SWS cones contribute little if at all to luminance. They do contribute to perceived brightness, and their signals help to account for the difference between brightness and luminance in the short-wavelength region of the spectrum (Fig. 4).

FIGURE 5 The spectral sensitivity curves of the human photoreceptors. (A) The spectral sensitivities of the rods (●) and the short-wavelength-sensitive (SWS), mid-wavelength-sensitive (MWS), and long-wavelength-sensitive (LWS) cones plotted on a linear ordinate normalized to the maximum of each curve. These data were obtained by direct measurements (microspectrophotometry) of single photoreceptors. [Reprinted, with permission, from Mollon and Sharpe (1983). (B) Spectral sensitivities for the three cone types, obtained by means of three very different techniques—psychophysics (◆), microspectrophotometry (□), and electrophysiological recordings from individual cones (▲). The logarithmic plot illustrates the overlap of the different cone spectra. This overlap is required for the encoding of wavelength information. Different wavelengths of light cause different ratios of quantal absorptions among the three cone types. [Figure provided by Peter Lennie; Adapted from P. Lennie and M. D'Zmura, 1988, Mechanisms of color vision. *CRC Crit. Rev. Neurobiol.* 3, 333–402.]

4. Trichromacy

At photopic illumination levels, cones are functional, whereas rods reach the upper limit of their signaling range and cease to contribute meaningful signals. Logic insists that light of any particular wavelength composition and intensity can do nothing more than to produce a set of three quantum catches: L, M, and S, in the LWS, MWS, and SWS cone types, respectively. If two patches of light A and B produce differing quantum catches in any one or more of the three cone types, then patches A and B are potentially discriminable from each other. But if patches A and B were to produce identical quantum catches in all three cone types, they must be indistinguishable, that is, metamers: if $L_A = L_B$ *and* $M_A = M_B$ *and* $S_A = S_B$, then $A \equiv B$.

The total quantum catch in any photoreceptor is simply the sum of the quantum catches at each wavelength. So, for patch A, composed of the three wavelengths λ_1, λ_2, and λ_3, the quantum catches L_A, M_A, and S_A in the LWS, MWS, and SWS cones, respectively, are

$$L_A = Q_1 l_1 + Q_2 l_2 + Q_3 l_3,$$

$$M_A = Q_1 m_1 + Q_2 m_2 + Q_3 m_3,$$

$$S_A = Q_1 s_1 + Q_2 s_2 + Q_3 s_3,$$

where Q_1, Q_2, and Q_3 are the incident numbers of quanta of wavelengths λ_1, λ_2, and λ_3, respectively; l_1, l_2, and l_3 are the probabilities of quantal catch at λ_1, λ_2, and λ_3, respectively (i.e., the spectral sensitivity of L at each respective wavelength); and the m's and s's are similarly defined. Thus the situation is described by a set of three equations in three unknowns, the unknowns being Q_1, Q_2, and Q_3, the intensities of the three wavelengths of light in patch A.

Light of any wavelength composition in patch B produces a characteristic set of quantum catch values L_B, M_B, and S_B in the LWS, MWS, and SWS cones, respectively. These values can be plugged into the foregoing equations, and, because three equations in three unknowns are guaranteed solution, we know that the equations can be solved for the three light intensity values. The solution yields the intensity values of λ_1, λ_2, and λ_3 needed to make the metameric match.

It follows from these considerations, regardless of the wavelength composition of patch B, that patch A can always be made metameric to patch B by variation of the radiances of its three component lights alone. A negative value of Q corresponds to the physical operation of moving the corresponding light from patch A to patch B. The reader may readily generalize from the special case to the general one of matching any four lights by means of three intensity adjustments.

Thus, the behaviorally described law of trichromacy, as schematized in Color Plate 16 and Fig. 3, is explained exactly in terms of the properties of light and the properties of the photopigments, at the very first stage of retinal processing, the absorption of quanta by photopigments. The particular photopigments we have determine our particular metamer sets; if one or more of the pigments were shifted along the wavelength axis, the metamer sets would all change, but the property of trichromacy would remain. The spectral sensitivities of our photopigments, and the fact that there are three of them, leave irreversible marks on our perception of colors.

From a computational standpoint, information about wavelength composition, to the degree that it is encoded, is available in the relative quantum catches in the three photoreceptors, as these will be invariant across intensity variations. Information concerning the intensity of light is, in principle, available in the absolute levels of quantal catches in the three cone types.

5. A Three-Dimensional Photoreceptor Space

The information encoded by the three cone types can be represented quantitatively in the simplest of all color spaces, with the quantum catches of the three cone types represented on the three axes, as shown in Fig. 6A. Any individual wavelength of light (or wavelength mixture) of fixed intensity is represented as a point in this space. Variations of intensity of a light of fixed wavelength composition occupy a ray extending outward from the origin; variations in intensity produce variations along the ray.

This three-dimensional space, therefore, represents the information about wavelength composition and intensity that is available for visual processing, in the form in which it is available after the absorption of quanta by the SWS, MWS, and LWS cones. Because metamers plot to the same point, this space elegantly represents the phenomena of color mixture and the reasons for metamerism and trichromacy. But aside from its three-dimensionality, it does not look much like the perceptually derived color spaces of Fig. 2. To find the causes of other perceptual aspects of color,

A

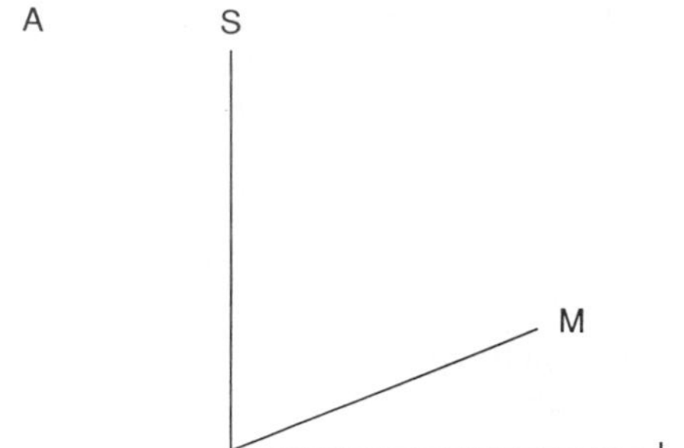

B

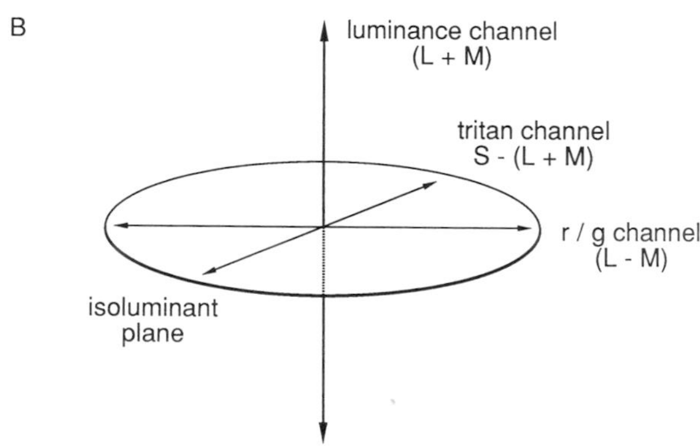

FIGURE 6 Physiologically based three-dimensional color spaces. A. Photoreceptor space. The quantum catches L, M, and S of the LWS, MWS, and SWS cones are plotted on the three axes. Any given light will cause a particular set of quantum catches in the three photoreceptors and, thus, can be represented as a point in this space. All members of a metamer set cause the same pattern of quantum catches, and thereby map to the same point. This space represents the reduced color–brightness information available after the absorption of quanta in the three kinds of cones. B. Three-dimensional space representing the alternate opponent model described in the text. The model suggests that the photoreceptor signals L, M, and S combine in particular sums and differences to make signals in three postreceptional channels. These are a luminance channel (L + M), an r/g channel (L − M), and a tritan channel [S − (L + M)]. The plane perpendicular to the luminance axis is the isoluminant plane. A recoding of this general nature, and perhaps of this specific form, occurs in the early processing stages of the human visual system.

one must therefore look to recodings of these cone-generated signals at later levels of visual processing.

C. Recodings at the Level of Single Neurons

1. The Classic Color-Opponent Model

It has been argued by opponent process theorists for more than a century that the properties of the necessary neural codes for color and brightness can be deduced directly from the characteristics of color perception. The marked perceptual tridimensionality of brightness–color perceptions (Fig. 2) has been taken to imply that brightness–color information must be carried by signals in three separate and relatively independent cell types, or channels, corresponding to and signaling the three perceptual dimensions. These are a putative brightness (or white–black) channel and two putative chromatic channels, the latter having the specific characteristics required by the unique and mutually exclusive hues—red–green and yellow–blue.

The mutual exclusiveness of red and green, and yellow and blue, has been taken to reveal the presence of *opponent coding* in the two chromatic channels. That is, it is argued that the signals in these channels can deviate in either of two opposite and mutually exclusive directions (such as hyperpolarizations and depolarizations of the cell membrane) from a neutral state, and that the two members of a mutually exclusive hue pair are mutually exclusive precisely because they are coded by mutually exclusive deviations of the corresponding neural unit, in its two opposite directions from the neutral value. Unique hues (e.g., yellow–blue) are taken to occur when only the relevant chromatic channel is active, and the other chromatic channel (e.g., red–green) is in its neutral state. Binary hues occur when both chromatic channels are active. Perceived saturation depends on the relative strengths of signals in the brightness versus chromatic channels, and neutral colors (whites, greys, and blacks) occur when both chromatic channels are at their neutral values. In summary, this classic model suggests that three precisely defined classes of cells exist in the visual system: one that responds to luminance variations, one to blue–yellow variations, and one to red–green variations, with the latter two in opponent codes, with neutral values precisely predicted by the perceptually unique hues.

Of course, one need not accept this argument at face value, for two reasons. First, as a theory of visual processing it is incomplete. The early levels of the visual system must preserve and recode sufficient information to allow all aspects of vision, not just color vision, to occur. In particular, the visual system must encode and process information about spatial patterns if we are to recognize objects, and one should expect a code that begins both analyses, probably in some intertwined form. And second, even at more central levels, the requirement of a neural code with such simplistic correspondences to color perception may be more a matter of our own cognitive convenience than of logical necessity.

2. Horizontal Cells

Despite the preceding reservations, direct physiological evidence indicates that an immediate opponent recoding of the outputs from the different cone types does occur. Horizontal cells are retinal neurons that receive direct input from the photoreceptors. In fishes and other nonmammalian species, some classes of horizontal cells indeed exhibit one response (hyperpolarization) to lights of some wavelengths, the opposite response (depolarization) to lights of other wavelengths, and no response (a neutral value) at the transitional wavelength. Thus, the qualitative characteristics required for an opponent chromatic code are found at the very earliest stages of neural processing in some species.

However, for primates, neither the location nor the detailed form of the earliest retinal recodings is yet known. For the reasons described earlier, several stages of recoding will probably occur, and one would expect that a code corresponding to the perceptual dimensions would occur late rather than early in neural processing. To describe the earliest stages of neural recoding, we here adopt a single, relatively simple and current model, which we call the alternative opponent model.

3. An Alternative Opponent Model

The alternative opponent model suggests that three neural channels, which differ in detail from those of the classic opponent model, occur in the human visual system. The alternative model suggests that LWS and MWS cone signals are summed to produce the signal in a luminance channel (L + M). It also suggests two chromatic channels, a r/g channel constituted from the difference signal between LWS and MWS cones (L − M), and a tritan channel, constituted from the difference signal between SWS cones and the sum of LWS and MWS cones [S − (L + M)]. The tritan axis referred to briefly in Section I,F corresponds to the tritan axis in this theoretical coding scheme. The largest difference between this model and the classic model is the substitution of a tritan for a blue–yellow axis. A three-dimensional stimulus space with the putative signals in these three channels represented on the three axes is shown in Fig. 6B. This model suggests, then, that three classes of cells, responsive to luminance, tritan, and red–green stimulus variations, respectively, should be found.

But visual neurons must analyze spatial as well as

spectral information, and we here digress to consider the question of spatial coding.

D. Spatial Aspects of Neural Coding

1. Receptive Fields

Suppose one were to shine a tiny spot of light on the retina and record the response from a single neuron, somewhere within the visual system, as the location of the light is varied. The receptive field of a neuron is that retinal region, or set of photoreceptors, that when illuminated causes a response in the neuron. Most visual neurons receive inputs from many photoreceptors, not just one; that is, they have extended receptive fields. Moreover, visual system neurons exhibit spatial opponency; that is, they respond with an increase of activity to light in part of the receptive field, and with a decrease of activity to light in the rest of the receptive field. These two parts tend to be concentrically arranged, in a so-called center-surround configuration, as shown in the lower part of Fig. 7. Because the maximal response of the neuron will occur when a particular spatial pattern of light—for instance, light covering all of the center, but none of the surround—falls on its receptive field, this receptive field structure begins the analysis of spatial pattern.

2. Combined Spatial and Spectral Opponency

Many neurons in the early stages of visual processing are both spatially and spectrally opponent, with the center of the receptive field receiving inputs predominantly from cones of one type and the surround predominantly from cones of the second (or second and third) type. Examples of such spatially and spectrally opponent neurons, combining all possible combinations of inputs from LWS and MWS cones, are shown schematically in Fig. 7.

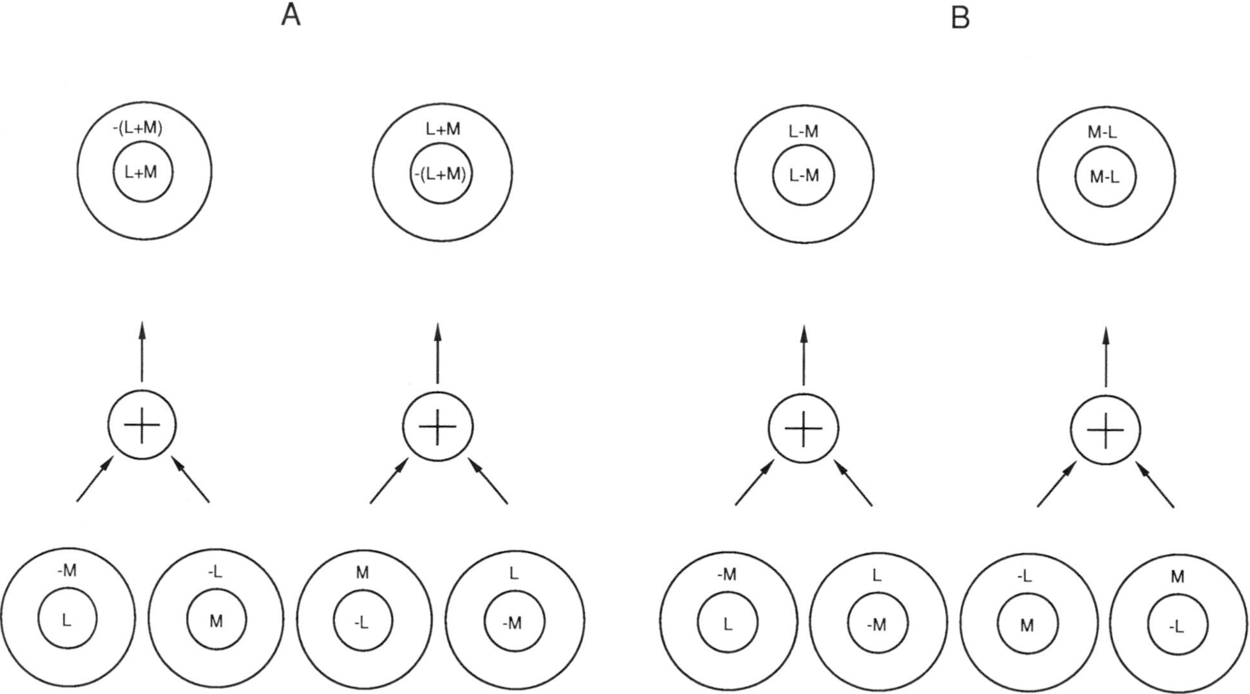

FIGURE 7 Spatial and spectral antagonism between L and M cone inputs. The bottom line of the diagram schematizes the receptive fields of individual postreceptoral visual neurons. These neurons exhibit center-surround antagonism, with one cone type (L or M) providing the predominant input (either positive or negative) to the receptive field center, and the other cone type (M or L) providing the predominant input (either negative or positive) to the surround. Many neurons of this kind occur in the retina and lateral geniculate nucleus of primate visual systems. The upper part of the diagram shows a scheme by which pairs of such spatially and spectrally opponent cells could be combined to yield A, spatially but not spectrally opponent luminance channels, with (L + M) in the center and −(L + M) in the surround, or vice versa; and B, spectrally but not spatially opponent r/g channels, with either (L − M) or (M − L) throughout both center and surround. A recoding of this general nature, and perhaps of this specific form, may occur in the early stages of cortical processing. [Adapted from P. Lennie and M. D'Zmura (1988) Mechanisms of color vision. *CRC Crit. Rev. Neurobiol.* 3, 333–402.]

Cells with this kind of spectral–spatial coding have been seen by many researchers in the early processing stages of the visual system. There is little current consensus as to which combinations of cone types occur, particularly with regard to neurons carrying inputs from SWS cones. In the absence of consensus we again adopt the alternative opponent model and suggest that two basic kinds of opponent cells occur, namely, r/g, with opposed L and M inputs, and tritan, with opposed S and (L + M) inputs. Each basic type can occur in several spatial configurations, such as those shown in Fig. 7, and with different weightings of cone inputs. Neurons of these two types have been seen clearly in one of the most recent studies of cells in the lateral geniculate nucleus (lgn).

Neurons such as these, with both spatially and spectrally opponent receptive fields, obviously carry both spatial and spectral information. Because they respond to both luminance patterns and chromatic patterns, they do not provide a clean separation of luminance from chromatic signals, and the notion that separate luminance and chromatic channels exist at the earliest stages of neural processing may have to be discarded.

3. Emergence of a Luminance Channel

One advantage of the alternate opponent model is that simple recombinations of pairs of such cells at a later stage could, in principle, allow the separation of red–green from luminance channels, as shown in the upper part of Fig. 7. Thus, a recombination scheme of this kind could yield cells corresponding closely to the three channels of the alternative opponent model. Although again there is no consensus on this point, some recent evidence indicates that a recoding of this type occurs in the early stages of cortical processing.

4. Psychophysical Considerations

Because it is easy to imagine that information carried in different cells can be processed independently, the existence of a neural processing stage of this kind provides a ready mechanism for modeling psychophysical observations such as the "melting" borders seen with tritan stimuli and the independence of adaptation effects between r/g and tritan axes, as discussed in Section I,G. If such a model is adopted, it also suggests that important aspects of border perception and flicker adaptation occur at a physiological level at which the chromatic code remains in this particular form.

To return to our starting point: The earliest retinal recoding schemes for human color vision are not yet known. The best currently available evidence suggests that at the level of the lgn and in early cortical processing, the neural color–brightness code corresponds more closely to the alternate color space of Fig. 6B than to the more perceptually derived color space of Fig. 2B. If the uniqueness and mutual exclusiveness of red, green, yellow, and blue are taken to imply the existence of corresponding cells or physiological channels, one must expect further shifts of the chromatic code in individual cells at higher levels of visual processing. The number and kind of later transformations of chromatic axes, and the existence or nonexistence of a stage corresponding to that suggested by the unique and mutually exclusive hues, remain open questions at this time.

E. Parallel Processing Streams

We turn now from consideration of the coding of information in single neurons to the question of parallel processing by larger populations of neurons. It is generally agreed that there are at least two major subpathways in early visual processing. These subpathways originate in the retina and recombine in the cortex to form two or more parallel and largely independent information processing "streams." Different aspects of visual processing occur rather separately and in parallel in these separate streams. Virtually all current theorists agree on the principle of parallel processing, but the exact numbers and specific functions of the different streams remain matters of controversy and speculation.

Parallel processing schemes provide a ready, if perhaps oversimplified, explanation for losses of visual function at isoluminance. If isoluminant stimuli create neural signals confined to a chromatic stream, and if this stream fails to access the neural machinery used for other specific aspects of visual processing—such as border perception, stereopsis, vernier acuity, and motion—then isoluminant stimuli would be incapable of supporting these aspects of visual function.

F. Algorithms for Color Constancy

Recent progress in models of color constancy has taken the form of computational schemes, or algorithms. That is, it has been shown that both the reflectance functions of natural objects and the spectral characteristics of natural illuminants conform to certain simplifying rules (technically, can be approximated by small numbers of basis functions). In principle, a system that had these rules stored in memory, when confronted with a complex visual scene, could factor out the spectral reflectance functions of objects from the illumination spectrum with a fair degree of accuracy, and thus provide approximate color constancy, at least over a limited range of conditions. With this view, constancy would fail when either the illuminant or the reflectance function could not be approximated by the basis functions stored in memory.

Some evidence indicates that a particular area in visual cortex, called cortical area V4, contains cells whose response properties correlate more closely with perceived hue than with wavelength composition. However, there is no broad consensus on a neural model or a neural locus for the carrying out of a color constancy algorithm at this time.

G. Summary

Much is known about the processing of color and brightness information within the primate visual system. The facts of color mixture, including metamerism and trichromacy, can be attributed with confidence to the characteristics of photopigments, whose spectral sensitivity curves are now well established. Beyond the photoreceptors, a series of opponent coding stages is believed to occur. The initial stages, up to and including early levels of cortical processing, appear to form a code resembling, if not corresponding exactly to, the alternate opponent code. The processing of luminance and chromatic information in parallel streams provides a possible explanation for the dramatic losses of visual function seen with purely chromatic stimuli. Transformations of the chromatic code within the chromatic stream have not yet been studied in detail, and the specific neural bases for the more perceptual characteristics of color, including the uniqueness and mutual exclusiveness of specific hues and the partial color constancy exhibited by human subjects, remain elusive at the present time.

III. VARIATIONS OF COLOR VISION

A. Color Vision Deficiencies (Color Blindness)

There are many different forms of color vision loss, both genetic and acquired. The simplest and most common are the "red–green" dichromacies and anomalies. These color deficiencies are surprisingly common (about 8% of Caucasian males are red–green color deficient). The red–green deficiencies and one other category, tritanopia, are discussed here because they provide interesting examples of reduced and altered forms of the normal trichromatic system and test one's understanding of that system. Other types of color vision deficiencies, although equally interesting, are beyond the scope of this chapter.

1. Red–Green Dichromacies

There are two types of red–green dichromacy: protanopia and deuteranopia, involving, respectively, the functional loss of LWS or MWS cones. With the loss of one cone type, color vision is reduced from a three-dimensional (trichromatic) to a two-dimensional (dichromatic) system. If, for example, the LWS cones were lost, the L axis in Fig. 6A would collapse onto the M, S plane, and all stimuli that were originally represented along any line parallel to the LWS cone axis (i.e., discriminable by means of the L signal alone) would become metamers. A dimension would also be lost from Fig. 6B, because without an L signal the luminance (L + M) and r/g (L − M) channels would both carry only the M signal and would be completely redundant. The equation for L would be dropped from the color mixture equations. Mixtures of only two wavelengths in patch A (Fig. 3) would be sufficient to match light of any wavelength composition in patch B. Metamer sets are larger for the dichromat, fewer color discriminations can be made, and two socks that obviously differ in color for the color-normal may match for the dichromat.

2. Red–Green Anomalies

There are also milder red–green deficiencies, called color anomalies. The anomalies probably stem from the replacement of either the LWS pigment (in protanomaly) or the MWS pigment (in deuteranomaly) by a different pigment whose spectral sensitivity (Fig. 5) is shifted along the wavelength axis. If, for instance, the MWS pigment were shifted, the values of m_1, m_2, and m_3 in the color mixture equations would change. Consequently, the intensities Q_1, Q_2, and Q_3 required to match patch A to patch B (Fig. 3) would change. An anomalous subject is still a trichromat, but he or she has different metamer sets than does the color-normal subject. A suit and tie that match for the color-normal may not match for the anomalous subject, and vice versa.

3. Tritanopia

There is a third kind of dichromacy, called tritanopia. Tritanopes lack functional SWS cones. Like protanopes and deuteranopes, the color vision of tritanopes is reduced to two dimensions. The SWS cone axis (Fig. 6A) is lost. The tritan axis (Fig. 6B) is also lost, because the tritan channel [S − (L + M)] is the only channel that carries SWS cone signals; if they are eliminated, this channel becomes redundant with the luminance (L + M) channel. In fact, the tritan axis derives its name from its dependence on the SWS cone signal, and from the fact that it is effectively lost in the reduced vision of tritanopic observers.

B. Genetics

The red–green color deficiencies are transmitted as X-linked recessives, that is, expressed in a male, carried without expression in all of his daughters, inherited by half of her children, and expressed in the males who inherit it. Other forms of genetic color vision deficiencies have other forms of inheritance.

Recently, it has become possible to locate the genetic structures responsible for the rod and cone photopigments in the human genome. The rod photopigment gene is located on chromosome 3, and the SWS cone pigment gene is on chromosome 7. The LWS and MWS cone pigment genes are located in tandem on the X chromosome, as expected from the X-linked recessive mode of inheritance.

Surprisingly, neither the genetic structures for the LWS and MWS pigments nor the genotype–phenotype relationships are as simple as might have been expected. Color-normal individuals often have two or more MWS pigment genes, and a variety of hybrid genes (genes having parts of both the LWS and the MWS pigment genes) also occur. The genotypes of dichromats and anomalous trichromats often show the expected gene deletions and variations; however, genotypes vary even within color vision categories, similar genotypes sometimes occur in subjects with different color vision deficiencies, and fusion genes are often seen in red–green color-normal subjects. The further explication of phenotype–genotype relationships in red–green color vision is an area of intense research effort at present.

C. Development

Newborn infants can certainly see, because they will stare fixedly at bold black and white patterns. They have an adult-like scotopic spectral sensitivity curve and broad, variable photopic curves indicative of the presence of multiple cone types; however, they respond poorly if at all to isoluminant chromatic stimuli. The earliest color discrimination made is probably between red and other colors; this ability is present by 1–2 months postnatal. Discriminations among many stimuli displaying large color differences have been well documented by 2–3 months. Little is known about the progress of color vision from these primitive beginnings to its adult form.

Strong evidence indicates that young infants have functional rods and LWS, MWS, and SWS cones. Their failures to demonstrate full-fledged color vision may indicate a specific immaturity of postreceptoral chromatic pathways, or may simply be one additional manifestation of a broad and general immaturity of both achromatic and chromatic processing.

D. Other Species

Both rods and cones are present in the retinas of most vertebrates. Behavioral studies have provided evidence of at least some color vision in such disparate species as frogs, goldfish, pigeons, cats, dogs, ground squirrels, and many species of primates. The color vision of ground squirrels is dichromatic. The color vision of macaque (Old World) monkeys is trichromatic and highly similar to that of humans. New World monkeys show color vision that is variable, both among species and, in the case of squirrel monkeys, within individuals of the same species.

IV. EPILOGUE

In conclusion, the study of color vision provides a textbook example of a problem in modern systems neurobiology. We have learned much about the ways in which sensory information loss and sensory recodings leave their marks on our perceptual world. Color vision also provides an example of the profitable interplay of concepts from behavioral, mathematical, physiological, genetic, and anatomical sciences, in the accumulation of human knowledge.

BIBLIOGRAPHY

Cornsweet, T. N. (1970). "Visual Perception." Academic Press, New York.

Jacobs, G. H. (1981). "Comparative Color Vision." Academic Press, New York.

Kaiser, P. K., and Boynton, R. M. (1996). "Human Color Vision," 2nd edition. Optical Society of America, Washington, DC.

Lennie, P., and D'Zmura, M. (1988). Mechanisms of color vision. *CRC Crit. Rev. Neurobiol.* **3**, 333–402.

Mollon, J. D., and Sharpe, L. T. (eds.) (1983). "Colour Vision: Physiology and Psychophysics." Academic Press, London.

Piantanida, T. (1988). The molecular genetics of color vision and color blindness. *Trends Genet.* **4**, 319–323.

Pokorny, J., Smith, V. C., Verriest, G., and Pinckers, J. L. G. (eds.) (1979). "Congenital and Acquired Color Vision Defects." Grune & Stratton, New York.

Teller, D. Y., and Bornstein, M. H. (1987). Infant color vision and color perception. *In* "Handbook of Infant Perception. I. From Sensation to Perception" (P. Salapatek and L. Cohen, eds.). Academic Press, Orlando, FL.

Teller, D. Y., and Pugh, E. N., Jr. (1983). Linking propositions in color vision. *In* "Colour Vision" (J. D. Mollon and L. T. Sharpe, eds.). Academic Press, London.

Wyszecki, G., and Stiles, W. S. (1982). "Color Science: Concepts and Methods, Quantitative Data and Formulae," 2nd Ed. John Wiley & Sons, New York.

Comparative Anatomy

FRIDERUN ANKEL-SIMONS
Duke University Primate Center

GLOSSARY

Dermatoglyphics Epidermal ridges with openings of sweat glands, arranged parallel in curved lines

Evolution Study of the historical development of the diversity of life on earth

Locomotion Power of moving from one place to another

Morphology Science of form and structure of animals and plants

Omnivore Organism that eats everything

Pentadactyl Five-fingered

Primates Order of mammals that contains lemurs, lorises, bushbabies, tarsiers, monkeys of the Old and New Worlds, lesser and great apes, and humans

Saimiri Genus name for the New World squirrel monkey

ANATOMY STUDIES ALL ORGANISMS, THEIR structure inside and out, and their similarity to each other. Thus, anatomy to a great extent is comparative. Comparative anatomy is the study of homologous structures in various organisms, their similarities or dissimilarities, ultimately the tool of taxomony, taxonomy being the attempt to classify all organisms into groups according to likeness, to define their relationship to each other, and even to address their possible historical ties with each other—when history is also called evolution.

The comparative anatomy of humans distinguishes us from the rest of the animal kingdom. Within the vertebrate class mammalia and the order primates, including lemurs, lorises, bushbabies, tarsiers, monkeys, apes, and humans, we are taxonomically placed alone in our own family: Hominidae. We are physically different in some ways from the nonhuman primates, but are we really different enough to be in our own taxonomic family? The answer to this question must be left open.

It is the comparison of *Homo sapiens sapiens,* such as you and me, with our closest relatives like the gorilla, the chimpanzee, even the macaque and all the nonhuman primates in general that physically defines our place in nature.

I. INTRODUCTION

For human biology, comparative anatomy provides the tool to understand what is truly human structurally. In the endeavor to answer the ever present human inquiry about what makes us different from all other animals—and how to define the place of humankind within the animate world as far as our anatomy is concerned—we must use comparison. Human biologists regard humans as primates. Primates are the group of mammals that are most similar to us and thus most closely related to us. [*See* Primates.]

In this article, human structure is contrasted with that of one relative, our primate cousin the macaque. The macaque has been chosen for this purpose because it is less advanced structurally and less specialized than, for example, the apes or many other monkeys. The macaque appears, therefore, to be well suited for this comparison.

A. Posture

Among all primates only humans are habitually bipedal. All the other primates are quadrupedal or ex-

ENCYCLOPEDIA OF HUMAN BIOLOGY, Second Edition, VOLUME 2. Copyright © 1997 by Academic Press. All rights of reproduction in any form reserved.

hibit variations of quadrupedality. Bipedality is one of the pivotal human features, and in the following comparison we will start at the feet, going upward through the human body and forward in the body of the macaque, culminating with the head.

B. Variability

Biochemically, all mammal tissues are alike. Even though we discuss "the human" and "the macaque," it must always be remembered that all primates and certainly humans exhibit an incredible degree of variability in all their physical features. This is equally true for monkeys. We must, however, ignore these variations in our brief attempt to comparatively define human structure. "The human" and "the macaque" are abstractions and simplifications.

II. COMPARATIVE ANATOMY

A. Feet

The feet of the macaque are nimble grasping feet (Fig. 1). As such, they are more similar to human hands than to human feet. The great toe is long and robust in both. Human toes 2 through 5, however, are short, whereas these toes are long and nimble in the macaque foot. The entire macaque foot is loose and flexible like the human hand. The human foot, in contrast, is tightly bound by tendons and arched on the inward plantar aspect. Human toes are aligned parallel to each other and have only restricted mobility. Indeed, the big toe of humans appears to be large and strong but only toes 2 through 5 are shortened compared to those of the macaque. The big toe in unison with the second toe of the human is, in fact, the strong counterbalancing axis of the relatively short striding foot that has to provide support for the comparatively enormous length of the upright body (10 in. balancing about 55 in., a ratio of approximately 1 : 5.5, while in the quadrupedal macaque the ratio is only about 1 : 2.7). The stability of the quadrupedally based body is doubtless much greater than that of the bipedal body, with the trunk erect and the entire weight based on only two feet. Even though most primates show a tendency to occasionally adopt an upright posture, they do not usually do so for any length of time. The bony elements of human and macaque feet are essentially the same, except for the weight-bearing talus and calcaneus; these are much more robust and strongly developed in the human foot than in the macaque foot. The phalanges of the human toes 2 through 5 are reduced in length. Also the heel (posterior extension of the calcaneus) is much larger and longer in human feet than in those of macaques, a fact closely related to the strong development of the calcaneus tendon (Achilles tendon) in humans, the tendon of the gastrocnemius muscle that is not nearly as well developed in macaques.

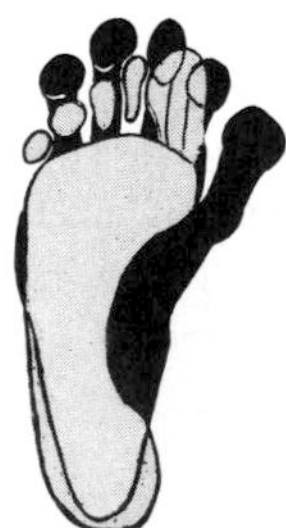

FIGURE 1 Human (shaded) and macaque (black) footprints (brought to approximately the same length). Human footprint superimposed over macaque footprint.

In combination with the alignment of the hallux (big toe) more or less in line with the other toes in humans, the abductor hallucis muscle is less strongly developed in humans than in the monkeys. The same is true for the other intrinsic foot muscles that provide mobility for the toes.

The plantar aspect of primate feet is covered by connective tissue pads and a friction skin that has characteristic dermatoglyphic patterns. All toes in both hand and foot have nails. The primitive toe length pattern as found in macaques is $3 > 4 > 2 > 5 > 1$, with the third toes being the longest. The third toe also represents the functional axis in the grasping foot. The toe length formula in humans is commonly $2 > 1 > 3 > 4 > 5$ or $1 > 2 > 3 > 4 > 5$, the functional axis thus either passing through the second or the first toe (hallux).

The nervous and vascular supply of the foot and leg is basically the same in both primates. Nerves and blood vessels, however, do vary individually to some extent in all primates.

B. Legs

The tibia and fibula are more robust in humans than in macaques, with the muscles of the lower leg bulkier and the muscles themselves comparatively shorter in humans than the equivalent in macaques. These bulky muscles of humans extend into large, strongly developed tendons that are much less powerful in the quadrupedal monkey. Some minor differences in the way the muscles originate and insert in the two primates are closely related to the functional differences of the two different foot types: stability in the human foot and grasping mobility in that of the macaques. The tibia, being the major weight-bearing member of the lower leg, is especially adapted to this task by being comparatively robust in bipedal humans. The upper member of the leg, the femur, is also more robust in humans than in macaques, whereas that of macaques is more defined in its relief. The femoral head and neck are proportionally larger in humans than in macaques. The latter fact is again closely related to the weight-bearing demands of the femur in bipeds. A significant positional and size rearrangement of the hip musculature in the human biped, compared with quadrupedal macaques, results in a unique human pattern of this musculature: these rearrangements are intrinsically tied to the profound differences in the pelvic morphology.

Also, the human legs are proportionally much longer than the legs of the macaque. The great length of the human legs allows a much more efficient stride than short legs could.

C. Pelvis

The pelvis of the macaque has a long and narrow iliac blade that is bent outward. The distal and upper end of the macaque pelvis (the ischium) is covered by cartilaginous padding. These enlarged ends are known as "ischial callosities." When sitting, monkeys rest on these callosities rather than on muscles, as we do (Fig. 2).

Our pelvis, in contrast, is short and wide, and the iliac blades are bent inward in such a manner that the human pelvis is bowl shaped. The large hip muscle, gluteus maximus, is an abductor of the thigh, positioned at the side of the body in macaques, whereas in humans it is very bulky and positioned at the back of the pelvis rather than at its side. Thus, it functions as a powerful extender of the leg in the biped. Humans also sit on this muscle and on the muscles of the back of the upper thigh, the so-called "hamstrings." While the muscles, blood vessels, and nerves of the leg are indeed similar in both macaque and human, the morphology of the pelvis and the concomitant differences of the musculature are closely related to their different ways of locomotion. Thus, for example, the human "hamstrings," gluteus medius and minimus, are involved in the stabilization of the bipedal body when standing. They also abduct and medially rotate the thigh in humans. The gluteus maximus provides power when striding and climbing in the biped and, together with the posterior portion of the medius, rotates the human thigh laterally. The gluteus maximus of macaques covers the side of the femoral articulation of the hip and covers the thick gluteus medius that occupies most of the pelvic fossa. The gluteus minimus also is positioned beneath the medius in the quadrupedal macaque, and these muscles act together in strong retraction of the leg and are major abductors.

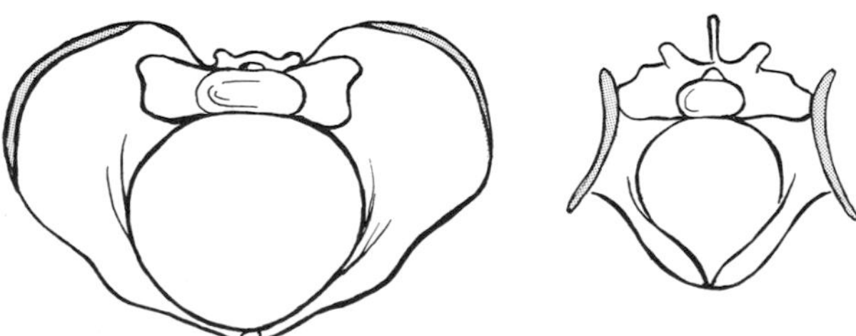

FIGURE 2 Human (left) and macaque (right) pelvis (brought to about the same size) seen from above (human) and the front (macaque). The iliac rim is shaded to show the striking difference in the way they are curved.

The pelvis is connected to the skeleton of the upper body by an element of the vertebral column, where several vertebrae are solidly joined to each other forming the sacrum. The sacrum is combined of three vertebrae in monkeys and of five or six in humans. It is the ilium of the pelvis that articulates with the sacrum. Together the two hip bones and the sacrum shape the pelvic outlet. The macaque pelvis is long and slender, whereas that of humans is wide, short, and shaped like a bowl, and the back of the iliac blade is extended backward and downward. This extension of the iliac blade is the area of articulation with the sacrum (Fig. 3). The sacroiliac articulation is thus situated much closer to and almost above the articulation of the femoral head with the pelvis. The closeness of these two major weight-transmitting articulations within the human pelvis is one of the crucial morphological adaptations within the human body that make true bipedal stance and walking possible. These two articulations transmit the entire weight of the upper body to the legs. In quadrupeds such as the macaque, these two articulations between the vertebral column and the pelvis on the one hand and the pelvis with the head of the femur on the other are positioned behind each other. The articulation of the femur with the pelvis of the macaque is positioned below and behind the sacroiliac connection. The iliac flange of the human pelvis that extends backward and down-

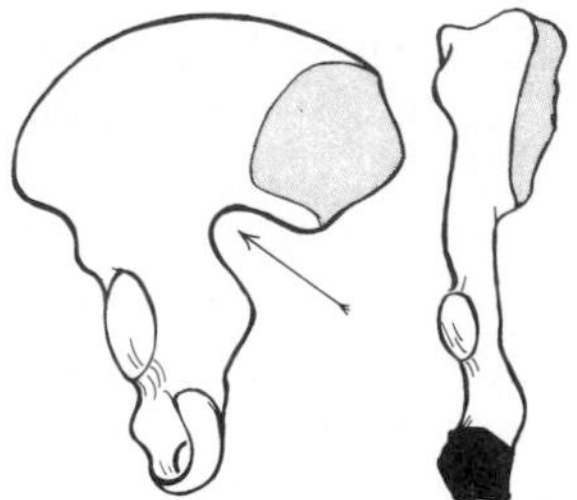

FIGURE 3 Human (left) and macaque (right) hip bones seen from the inside. Ischial callosities in the macaque are shaded dark. The arrow indicates the characteristically human ischiadic notch.

ward encloses an angle with the ischium that is also known as the incisura ischiadica, or the greater ischiadic notch. The greater ischiadic notch is uniquely characteristic of *Homo sapiens* and as such is one of the crucial clues that indicate bipedality.

D. Pelvis as Birth Canal

The pelvis, however, not only functions in locomotion and weight bearing in all primates, it also shapes the canal through which in females the full-term baby passes during birth. The necessity for the fetus to pass through this bottleneck puts a different restraint on pelvic morphology than locomotion. Primates are called primates because their well-developed brains are comparatively large, even in the newborn. Thus, in primates with large brains and usually single, large offspring, the bottleneck of the pelvis can be crucial. A birth problem exists in several primate genera, where the pelvic outlet and the size of the term fetus head are critically close (*Macaca, Saimiri,* and *Homo,* to name only three examples).

In the wild this situation is taken care of by nature's selective forces: if birth is difficult, frequently neither mother nor offspring survive. This used to be true for humans also, but medical intervention (Caesarean section) now counteracts the selective force of birth difficulties caused by the large fetal brain. The sacrum in humans is unusually wide, thus enlarging the birth canal and also providing extensive surfaces for the articulation of the vertebral column with the pelvis and lower limbs. The pelvis contains parts of the urogenital tract. The bottom of the human pelvis is strengthened by muscles that are functionally different from those in macaques because macaques have tails and humans do not. The human sacrum is characteristically bent ventrally, whereas in macaques the last of the three sacral vertebrae is often tilted slightly upward. The width of the primate sacrum is ultimately correlated with the size of the lumbar vertebrae, which are narrow, high, and comparatively long in the macaque, and wide, high, and stout in humans. In humans the lumbar vertebrae arise in a sharp angle upward from the last sacral vertebra. The body of the last sacral vertebra is enlarged ventrally and forms part of a promontory that is caused by the sacrolumbar angulation of the human vertebral column. This promontory is an expression of the upright vertebral column. It is also a point of great stress and frequent injury in humans, an inherent weakness of our upright posture. [*See* Vertebra.]

E. Trunk

The width of the human pelvis is also reflected in the shape of the rib cage above. The lumbar region is short, however, being combined of only five vertebrae in humans, whereas it is long (seven lumbar vertebrae) in macaques. Also, the thorax of macaques is long, deep, and narrow, whereas that of humans is comparatively short, shallow, and wide (Fig. 4). The position of the transverse processes of the thoracic vertebrae is different in the macaque and in humans: they are positioned almost at a right angle to the midsagittal plane of the body (about 75° to 80°) in the former and tilted upward in an angle of about 45° in the latter. As the upper ribs articulate with these processes, the differences of angulation are important. The upper ribs articulate with the thoracic vertebrae in two places: between two adjacent vertebral bodies on the upper ends of the vertebral bodies and on the invertebral disc with the ends of the ribs (capitulum costae) and at the end of the transverse processes of the adjoining thoracic vertebrae with the tuberculum costae. The length of the thoracic vertebral bodies and the angle of rib insertion upon the transverse process both influence the angle of insertion of the ribs. The more dorsal the transverse process of the thoracic vertebrae

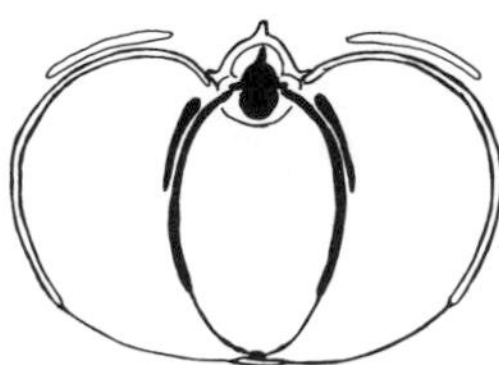

FIGURE 4 Cross sections of the human (white) and macaque (black) rib cage with shoulder blade position (brought to the same dorso-ventral depth).

is angled and the shorter the vertebral bodies of the region are, the steeper is the angulation of the rib. Rib angulation is steeper in humans than in macaques. Also, the way the rib itself is bent at its vertebral end determines the shape of the rib cage. In humans, ribs are bent almost to a quarter circle at the vertebral end, whereas those of macaques are bent only slightly. These arrangements of the rib angulation and position in fact cause the vertebral column to be positioned inside the thorax in humans, whereas it is positioned at the back in macaques. By this means also the center of gravity lies more centrally in the human trunk. A consequence of these features is that the human rib cage is barrel shaped and wider transversely than deep dorsoventrally whereas it is narrow in the monkey. In concordance with this overall shape difference of the thorax, the sternum also is shaped quite differently: it is wide and comparatively short in humans and narrow and long in macaques, where it is also segmented, not fused like the sternum of humans. These differences in the shape of the thorax also have a crucial influence on the position of the shoulder blades and thus the upper arm articulation in the two forms. This articulation is positioned high and somewhat backward on the rib cage in humans and forward and at the side of the thorax in macaques. This, in turn, causes the action radius of the arm to be considerably less restricted in humans than it is in macaques. A rearrangement of the soft tissues accompanies all these differences in the skeletal architecture of the trunk of macaques in accordance with the narrow trunk.

The presacral human vertebral column exhibits a unique and characteristic series of three curves when seen from the side. There is a lordosis (i.e., concave toward the back) in the neck region, a kyphosis (i.e., convex backward) within the thoracic region, and a lordosis in the lumbar region. In fact, the human vertebral column is more similar to an elastic spring than to a true column. Even though it is rigid in specific ways in the different regions, it is also mobile within limits. The possible movements of the vertebrae are channeled by the position of the articulations between single vertebrae that are characteristically different in different regions. All vertebral centers are joined with each other by means of cartilaginous intervertebral discs with the exception of the first and second neck vertebrae (atlas and axis) and the vertebrae of the sacral region that are fused to each other. The characteristic triple curvature of the human presacral vertebral column, the promontory angle between lumbar region and sacrum, and the distinctive ventral curvature of the sacrum itself are all typical only of humans.

F. Shoulder

Whereas the human shoulder blade is triangular with a long medial border, that of macaques is more blade-like with long, almost parallel, fore and hind margins and a short dorsal border. The two differently shaped shoulder blades allow different leverage for the enveloping musculature, which is closely related to the very different use and reach of the forearms. Also, the collar bones are long and angular in humans in conjunction with the barrel-shaped thorax. In macaques, they are comparatively short and straight in accordance with the narrow trunk.

G. Inner Organs

The inner organs of humans and macaques differ mostly in proportion and slightly in position, but they are not characteristic or unique to either one of the two primates compared here.

In the human male, the testicles descend regularly before the infant is born. They do descend at birth in the macaque but are retracted back into the inguinal canal just afterward and only permanently descend into the scrotum at about 6 years of age.

The human penis is comparatively much larger than that of macaques. It is in fact almost the largest among all primates (only the chimpanzee rivals human males in size of the genitals).

It appears that only human females among primates exhibit permanent prominent breasts, which remain large even when not lactating.

H. Arms

One example of the differences of the musculature of the forearm is that the deltoid muscle is a powerful pro- and retractor of the arm in the quadrupedal macaque. It has a larger clavicular insertion in humans than in the monkey and functions as a powerful abductor of the arm and also extends the arm in humans. Also in humans its most ventral portion rotates the arm medially; the most dorsal portion rotates it laterally.

In accordance with the barrel-shaped trunk and lateral position of the shoulder joint, the humeral head in humans is rotated inward dorsomedially (about 45°) compared with the position of the elbow joint of this bone. This torsion of the humeral shaft is also a distinctive human characteristic. The humeral head of macaques faces straight backward and is also proportionally smaller in its diameter than that of humans. Without this rotation of the human humerus, our arms and hands would face outward rather than toward the body in a relaxed position. Also the upper arm of humans is straight rather than curved parallel to the trunk as it is in macaques. In the ulna the olecranon is less prominent in *Homo* than it is in *Macaca,* and at its distal end the styloid process is considerably shorter and rounded in the former instead of pointed as in the latter. Both these differences can be attributed to the difference in use: quadrupedal locomotion rather than free manipulation. Differences in the way the muscles are attached to the two bony elements of the lower arm also reflect the differences in function, which is much more restricted in the macaque. The ulna and radius in humans are positioned parallel to each other when the palm faces up (supination) and then cross over each other (radius over ulna) when the palm faces down (pronation). Macaques are not able to rotate their forearms much, and radius and ulna are positioned close to each other and are tightly bound.

I. Hands

All primate hands are basically built according to the primitive pentadactyl vertebrate plan (Fig. 5).

The human hand is characterized by comparatively greater overall width and a long pollex in contrast to that of the macaque. The hand of macaques is tightly bound and much less mobile and flexible than the human hand that, for instance, can be cupped. As in almost all higher primates, both human and macaque hands have nails on all five fingers. Even though the pollex of the macaque hand is opposable to digits II through V, this ability is much less efficient than the grasping abilities of the human hand. Human hands are not specialized, and they are built according to an old plan—they are structurally primitive. Lined on the palmar surface with a specialized, highly sensitive friction skin with individually characteristic dermatoglyphics, these hands are functionally omnipotent and doubtlessly the executing instruments of human civilization and, ultimately, culture. But this is only possible because these simple hands are doing what a highly evolved brain leads them to do.

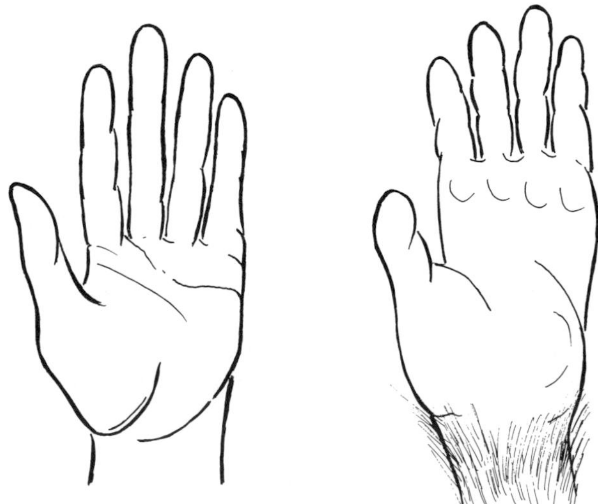

FIGURE 5 Human hand (left) and the hand of a macaque (right) (brought to about the same length).

J. Neck

As we proceed up the neck we get to the pharyngeal region. Even though the anatomy of this region is structurally basically the same in humans and macaques, there is one fundamental difference: humans have language. However, none of the morphological features of the pharyngeal region or the upper airways in humans can be attributed with certainty as the crucial feature that makes language possible. Attempts to identify an anatomical feature that is the key to human language always fail. Even though, for example, the cantilever position of the human epiglottis is a mechanical prerequisite for the possibility of human speech, it can be found in many mammals that cannot talk. Also, the spina mentalis of the human chin is not an indicator for the presence of the ability to have language. Not even the representation of the motor control area of language in Broca's speech area of the brain (inferior frontal gyrus) is anatomical proof of the presence or absence of language. Broca's speech center coordinates the muscles involved in speech. Other cortex areas, however, are necessary for articulate and abstract speech formulation. In essence, the ability to voice abstract ideas cannot be documented anatomically.

K. Head and Face

The head is large and rather spherical in humans due to the dominating size and shape of the brain and braincase. The head of macaques is shaped more like that of a nonprimate mammal, with a long snout and small eyes, each covered above by a frontal torus, no forehead, and small braincase. The vertebral column meets the human head centrally from underneath, but it attaches under the back of the macaque's head. The human foramen magnum is tucked forward under the head and is located almost in the middle of the basicranium (Fig. 6). It is directed backward in the macaque. The facial musculature is highly differentiated in humans compared with that in macaques. The face itself appears to be much larger in humans, which is mainly a function of the missing snout. The monkey face is dominated by this snout and impressive canines, and it is incapable of the varied expressions that are so telling and characteristic of humans. The human maxilla and mandible are tucked under the head. The teeth, even though the same in number and kind as in macaques, are relatively smaller in humans, especially the canines that are incorporated into the rounded arcade of front teeth. Also, the human molars have flattened occlusal reliefs compared with the rather prominent occlusal cusps of macaques. The lower first premolars in macaques are enlarged to an enormous blade-like honing tool for the dagger-shaped upper canine, whereas they are equally small and shaped more or less like incisors in humans, as are the canines.

The human face is also characterized by the delicately chiseled lips—another feature that is uniquely human.

The use of the ear muscles is obsolete in humans; in contrast, the ears of macaques are quite mobile and play a part in social interactions among them. A prominent nose is another feature that is missing in macaques but always found in humans.

The eyes of humans are somewhat closer together than they are in the monkeys and are directed forward.

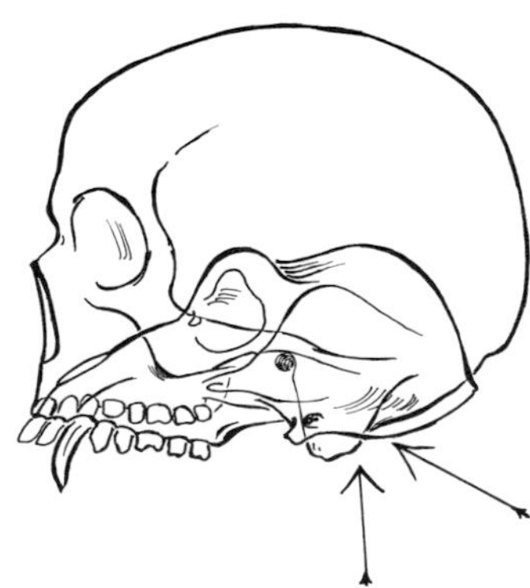

FIGURE 6 Human and macaque skull superimposed (brought to about the same length).

L. Brain and Reproduction

Most striking is the difference in the brain. We can only highlight a few points in our brief comparison. The cortex is absolutely enlarged in humans compared with macaques. The cortex of the human brain is folded into multiple gyri and sulci and also has a higher density of neurons in some areas than does the cortex of the macaque brain. The temporal and frontal lobes especially are enlarged and functionally perfected. It is not the volume of the brain alone that tells about its functional abilities. The sensory–motor cortex is not only larger than in monkeys, but certain areas like that for our hands are enormously enlarged. In humans we find cortex areas that are engaged with the development of skills and that govern foresight and memory as well as language and abstract thinking. A considerable part of our brain is devoted to learning. [*See* Cortex.]

The large brain of humans is also instrumental in important changes concerning human reproduction. It involves strong mother–child dependency. Even though macaque and human females undergo similar reproductive cycles, there are important differences. Macaques have seasonality in their breeding and birthing times; humans do not. Humans are able to breed regardless of seasons. Humans engage in sex without the goal of progeny. Macaque females have color changes that advertise times of highest sexual receptivity to their males: the skin around the head and the anogenital area turns bright red or purple. Humans do not have such signals.

M. Hair

Humans are different from macaques in yet another striking way: humans are naked whereas macaques have fur. Humans usually have only remnants of fur on their scalps, under their arms, in their genital area, over their eyes, and around the mouth of males. Newborn macaques can cling to the mother's fur. Newborn humans are totally helpless, have no fur to cling to, and have to be carried. [*See* Hair.]

III. CONCLUSION: WHAT MAKES US HUMAN?

What do we discover when we now look back at this comparison (Fig. 7)?

Anatomically, humans are in many ways not very different from macaques. Both are primates. What,

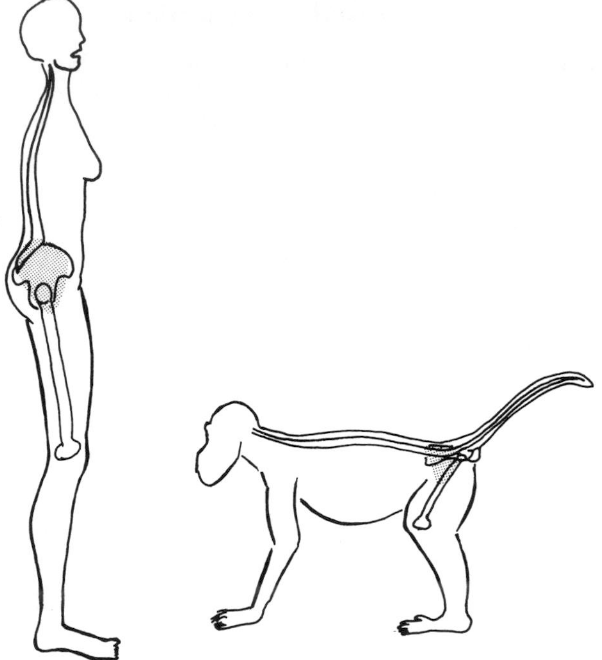

FIGURE 7 Human and macaque outlined, showing vertebral column, hip bone, and upper leg bone. The hip musculature is shaded (not to scale).

then, makes us human and macaques monkeys? What makes humans the most successful primate on earth? Macaques are curious, manipulative, inquisitive, omnivorous, and social. Humans are curious, manipulative, inquisitive, omnivorous, and social. But they are more. Humans are intelligent and stupid. They are compassionate and aloof. They are loving and full of hatred. Humans are imaginative and dull, industrious and lazy, authoritarian and submissive. They are generous and greedy. Humans are immensely creative and unbelievably destructive. How can it be that humans are so overpowering and successful? Is it reproductive prowess? Or is it just that humans are unlimited in being cunning? It seems that these qualities have arisen because of a highly refined large brain, two specialized feet that have taken over locomotion, two able but unspecialized hands, and the ability to communicate through language.

BIBLIOGRAPHY

Aiello, L., and Dean, C. (eds.) (1990). "An Introduction to Human Evolutionary Anatomy." Academic Press, London, San Diego.

Cartmill, M., Hylander, W. L., and Shafland, J. (1987). "Human Structure." Harvard University Press, Cambridge, MA/London, England.

Fleagle, J. G. (1988). "Primate Adaptation and Evolution." Academic Press, San Diego.

Jungers, W. L. (ed.) (1985). "Size and Scaling in Primate Biology." *In* "Advances in Primatology," Vol. VIII. Plenum, New York/London.

Kinzey, W. G. (ed.) (1987). "The Evolution of Human Behavior." State University of New York Press, Albany.

Shipman, P., Walker, A., and Bichell, D. (1985). "The Human Skeleton." Harvard University Press, Cambridge, MA/London, England.

Simons, E. L. (1989). Human evolution, *Science* **245,** 1343.

(This is Duke University Primate Center Publication No. 479)

Comparative Physiology

C. LADD PROSSER
University of Illinois at Urbana–Champaign

GLOSSARY

Hyperosmotic Having a higher osmotic pressure (generally a salt concentration) than that of the surroundings

Poikilotherm An animal whose body temperature varies with the environmental temperature

COMPARATIVE PHYSIOLOGY IS THE DIRECTED discovery of how different organisms solve their life problems, of functional diversity. Comparative physiologists use methods at all levels of biological organization: molecular, organ systems, and intact organisms. This article covers the comparative physiology of animals. Comparative physiology provides explanations of animal distribution (ecology), it gives evidence of evolutionary relationships, and it describes kinds of behavior. The two comparative approaches are metabolic and neural.

Every organism is influenced by multiple environmental factors: water, ions, nutrients, light, mechanical stimuli, gravity, pressure, and other organisms (same and different species) (Fig. 1). Comparative physiology dissects these influences into component mechanisms. Animals display two patterns of response to environmental factors: conformity and regulation. Conformers adjust their internal state to be the same as the environment and are designated by the prefix poikilo: poikiloosmotic (internal osmoconcentration variable) or poikilothermic (cold blooded). Regulators maintain a constant internal state when the environment changes and are designated as homeo: homeoosmotic (internal osmoconcentration constant) or homeothermic (warm blooded). The prefix hetero is used for animals that conform in one body region and regulate in another (e.g., a pig with cold skin and warm liver) or that conform at one time and regulate at another time (nocturnal and diurnal). In general, cells may maintain constancy of a function, e.g., concentration of an ion, different from extracellular fluid, although extracellular fluid is more variable.

I. WATER AND IONS

Living substance (protoplasm) is an aqueous solution and water is essential for life. In multicellular animals the osmotic concentration inside cells is mostly the same as the extracellular osmotic concentration but the ionic composition of the two is very different. For example, intracellular potassium is higher in concentration and sodium is lower than extracellular (Fig. 2). Freshwater animals are more concentrated than their environment; they have evolved mechanisms for excluding excess water (skin, gills) and for excreting water (vacuoles, nephridia, kidneys) (Fig. 3). Many marine invertebrates are at the same osmotic concentration as sea water, some marine/land crabs and cartilaginous fish are hyperosmotic (more concentrated), and bony fish and a few terrestrial crabs are more

ENCYCLOPEDIA OF HUMAN BIOLOGY, Second Edition, VOLUME 2. Copyright © 1997 by Academic Press. All rights of reproduction in any form reserved.

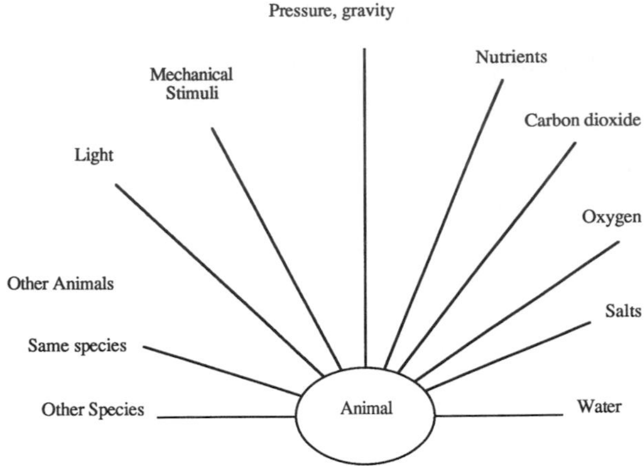

FIGURE 1 Generalized relation between environment and organisms.

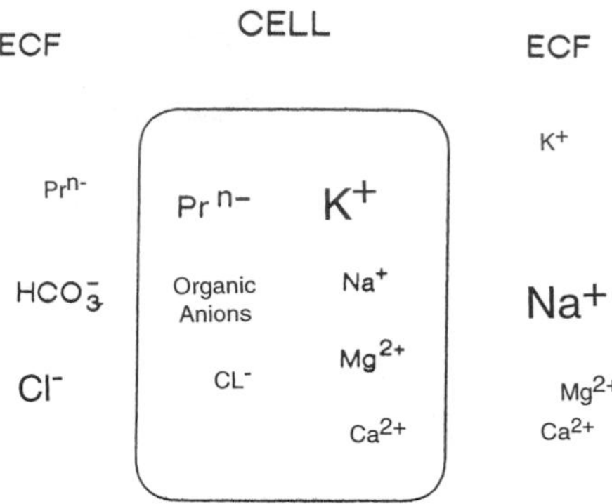

FIGURE 2 Diagram of intracellular (cell) and extracellular (ECF) solutes in animal cells. The relative size of symbols within a compartment indicates relative concentrations within that compartment. Negatively charged ions are Cl^-, HCO_3^- and Protein n^-. Positively charged ions are Na^+, K^+Mg^{2+}, and Ca^{2+}. [From Prosser (1994).]

dilute than sea water (Fig. 4). Terrestrial animals have means of preventing water loss (cuticle, skin) and of obtaining water from scarce sources (desert animals). Mammals such as desert rats and camels can withstand extreme dehydration. All animals take up essential elements (Na, K, Cl, Ca) and exclude less essential ones (Mg, SO_4, Al). Kidneys of vertebrates, both terrestrial and aquatic, have elaborate cellular mechanisms for filtering by glomeruli and for reabsorbing ions and nitrogenous waste products by tubules (Fig. 5). Urine can be hypotonic, isotonic, or hypertonic. Nephridia of annelids, crustaceans, and insects are also highly selective in their excretion. Some fish lack glomeruli and form urine by secretion.

Some transport, across cell membranes and gills, requires metabolic energy, especially transport against a concentration gradient. Most freshwater animals have a combination of passive and active transport (Fig. 6). Active transport is by ion exchange and by ion pumps. In aquatic animals the mechanisms of transport between medium and extracellular fluid (by gills) are similar to those between cells and body fluids, e.g., muscle, nerve, and in kidneys (tubule cells) (Fig. 7). Most primitive animals are osmoconformers (body fluid at same concentration as medium) and ionoregulators; each kind of ion is regulated somewhat independently of others. Some of the water and ion regulation is by digestive tracts. The net effect of many cellular processes is maintenance of osmotic and ionic independence of the environment.

There are special needs for certain elements, e.g., calcium for bone and shells. Nitrogenous waste may be lost by diffusion or by excretory processes. Most aquatic animals excrete ammonia, animals with limited access to water excrete urea, and those with extreme water deprivation excrete uric acid.

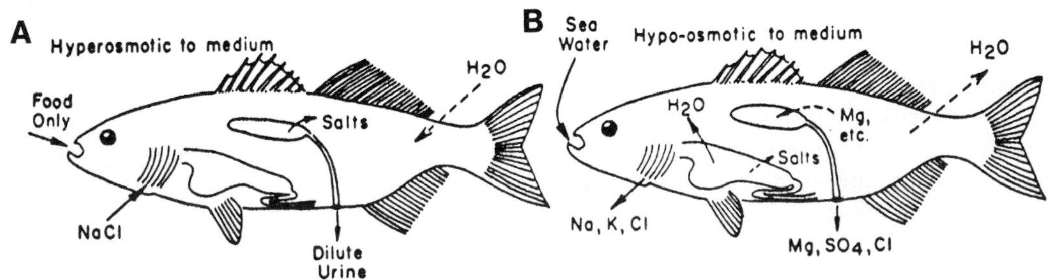

FIGURE 3 Schematic representation of the paths of ion and water movement in osmoregulation of freshwater fish (A) and marine fish (B). Solid arrows indicate active transport and broken arrows indicate passive movement. [From Prosser (1951).]

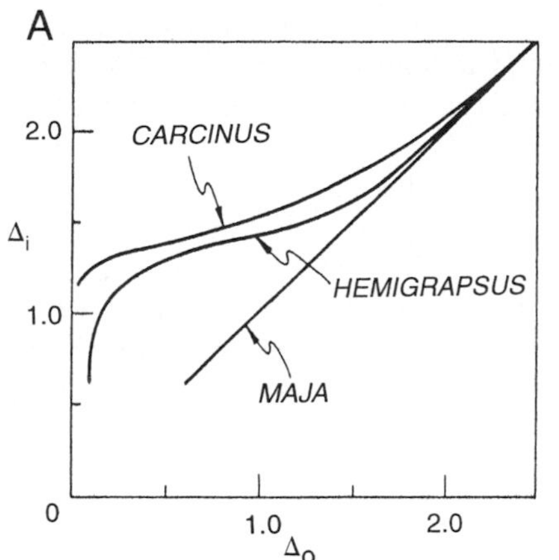

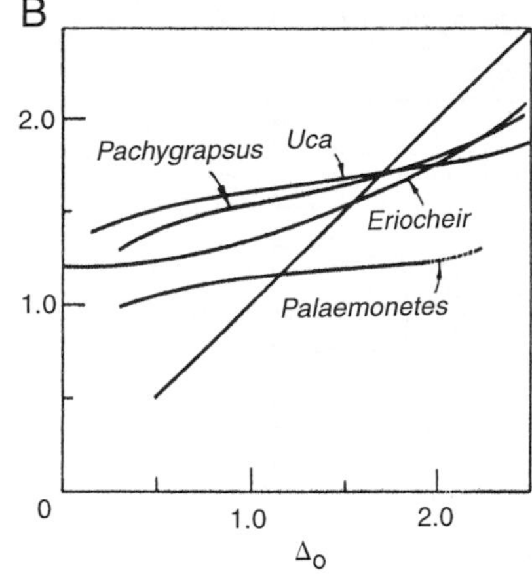

FIGURE 4 Osmotic concentration (Δ_i) in body fluid (hemolymph) and in medium (Δ_o) in several crustaceans. (A) Hyperosmotic regulation in dilute (brackish) water (Carcinus, Hemigrapsus) and osmoconformity in Maia. (B) Hypo- and hyperosmotic regulation in terrestrial crabs (Pachygrapsus, Uca) and eurytolerant crustaceans (Ericheir, Palaemonetes). [From Prosser (1961).]

II. OXYGEN AND CARBON DIOXIDE

All animals are aerobic but all have anaerobic steps in their metabolism and all can survive with reduced oxygen for short periods. External respiration, the uptake of oxygen and the output of carbon dioxide, is by skin or gills in aquatic animals or by lungs or trachaes in air breathers. Frogs use both skin and lungs. Lungfishes use their swim bladders for O_2 supply when their rivers dry up. External respiration is modulated by internal levels of oxygen and carbon dioxide, usually by special receptors, both peripheral and central and by respiratory reflexes. Flow of O_2 and CO_2 is down a concentration gradient; an enzyme, such as carbonic anhydrase, may facilitate CO_2 transport. A few animals rely on dissolved oxygen to supply

FIGURE 5 Cycle of movement of urea and water during water retention in a mammalian kidney unit (nephron). [From Prosser (1994).]

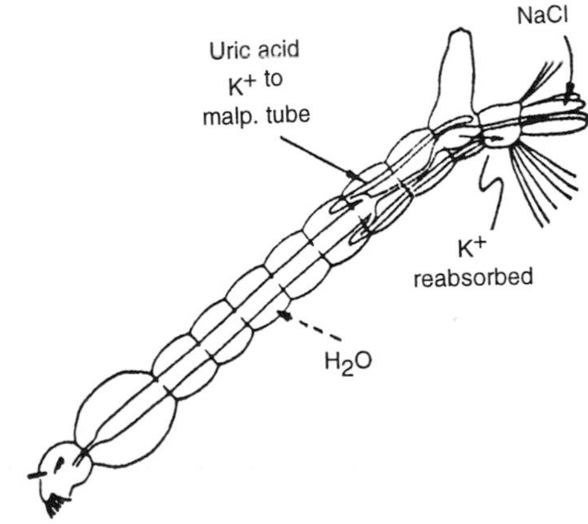

FIGURE 6 Main paths of movement of ions and water in osmoregulation of mosquito larva. Passive entrance of water, active uptake of NaCl from medium, and active reabsorption of potassium from hindgut. [From Prosser (1961).]

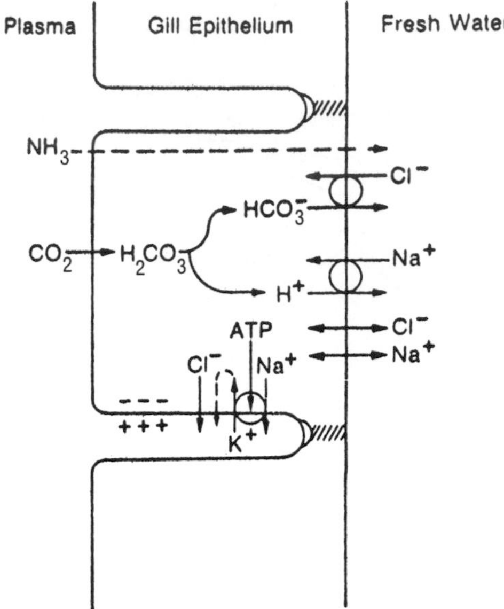

FIGURE 7 Active absorption of sodium and chloride by gill of freshwater fish. Uptake of Na^+ in exchange for H^+ and of Cl^- in exchange for HCO_3^-. The passive inward leak of Na and Cl is according to environmental concentrations. The Na-K pump is indicated between cell and plasma. [From Prosser (1961).]

their tissues but many have a transport pigment in their blood or hemolymph (hemoglobin containing iron in vertebrates, annelids or hemocyanin containing copper in many molluscs and crustaceans and Limulus). The pigment loads oxygen in respiratory organs and unloads it in tissues (Fig. 8). Hemoglobins

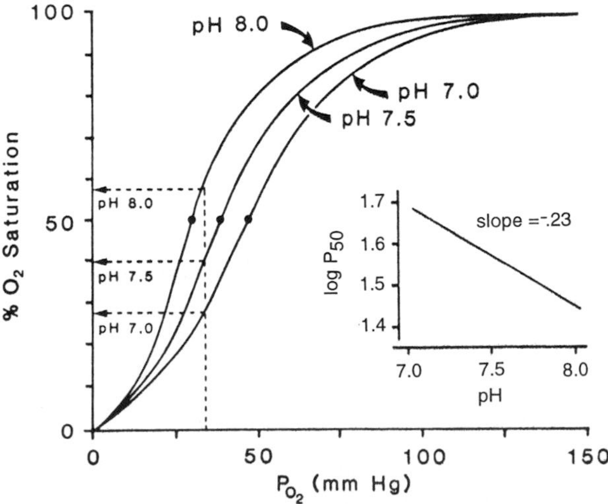

FIGURE 8 Oxygen equilibrium curve of vertebrate blood. The pH change from 8.0 to 7.0 raises the pressure for half-saturation. [From Prosser (1994).]

have evolved several times. Respiratory pigments load and unload oxygen passively along a sigmoid curve as a function of partial pressure of O_2. The loss of CO_2 in lungs or gills also lowers pH; this favors loading of oxygen (Fig. 8). The partial pressure of loading and unloading of O_2 is highly adaptive to environmental oxygen. In fish that live in fast, well-oxygenated water the pressure for 50% loading (P50) is high whereas in fish living in sluggish water the affinity of their Hb for O_2 is low. In insects, air is carried by a system of tubes (tracheas) directly to body cells.

The basic pathways of cellular metabolism are similar in all animals, but significant differences occur between tissues (e.g., muscle and liver). External respiration is modulated by internal levels of oxygen and CO_2, usually by special receptors and central nervous reflexes and hormones. Lungfishes use their swim bladders as an oxygen supply when their rivers dry. The relative importance of different substrates for energy (internal metabolism) is remarkably similar in all animals. Pathways of glycolysis (anaerobic), oxidation (TCA cycle and cytochromes), pentose shunt, and phospholipid enzymes must have evolved very early. [*See* Glycolysis.]

O_2 consumption by active animals may be several times greater than rest or standard metabolism (Fig. 9). O_2 consumption is constant over a wide range of

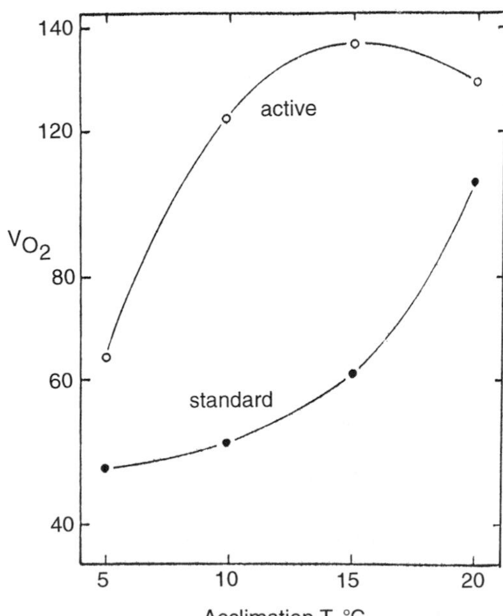

FIGURE 9 O_2 consumption (V_{O_2} as ml O_2/g · hr) at different temperatures of acclimation of spontaneously swimming (active) and resting (standard or inactive) crustaceans. [From Prosser (1994).]

available oxygen; below a critical PO_2, consumption is reduced, especially in aquatic animals. The rate of metabolism (O_2 consumed per gram body weight) increases as body mass decreases such that the metabolic rate of a mouse is high and that of an elephant is low (Fig. 10). Diving air breathers have special adaptations of respiratory organs and transport pigment.

III. TEMPERATURE

Many animals (poikilotherms) are at the same body temperature as their environment; they gain or lose heat from their environment (ectothermy). A few kinds of animals (homeotherms) maintain relative constancy of body temperature and produce their own heat (endothermy). Some animals are heterotherms: their body temperature varies either spatially by body region (heater organs in tuna or thoracic muscles in flying insects) or temporally (night or day) as in hummingbirds and bats. A few kinds of animals can allow their body temperature to fall nearly to freezing (hibernation); some hibernating mammals (e.g., ground squirrels) can arouse periodically whereas other hibernators (a few insects and frogs) tolerate freezing. Some Arctic and Antarctic fishes have antifreeze compounds that permit them to survive in sea water of an osmotic concentration with freezing point slightly below that of the fish blood containing an antifreeze (noncolligative cooling).

All animals are limited in survival at extreme cold and heat (Fig. 11). Brief exposure to above-lethal temperatures may induce synthesis of heat shock proteins that permit survival at subsequent high temperatures.

Direct responses to temperature consist of a rise or a fall in rate functions, a Q_{10} response. Membranes become fluid at high temperatures and solidify in cold; these are changes in lipids. Enzyme activities increase at high temperatures and decrease at low temperatures. Many poikilotherms can alter their temperature tolerance range (both cold and warm) and their enzymes of energy liberation by acclimatization. This process is by changes in cell membrane lipids and in energy-yielding enzymes. Laboratory acclimation is an attempt to model acclimatization in nature. Acclimatory responses are compensatory in raising metabolism in the cold and lowering it in warmth. In acclimation of some fish the size of the liver increases in the cold and decreases in warmth; enzyme activity in total liver is more critical than per milligram of protein (Fig. 12). Mechanisms of acclimation are primarily

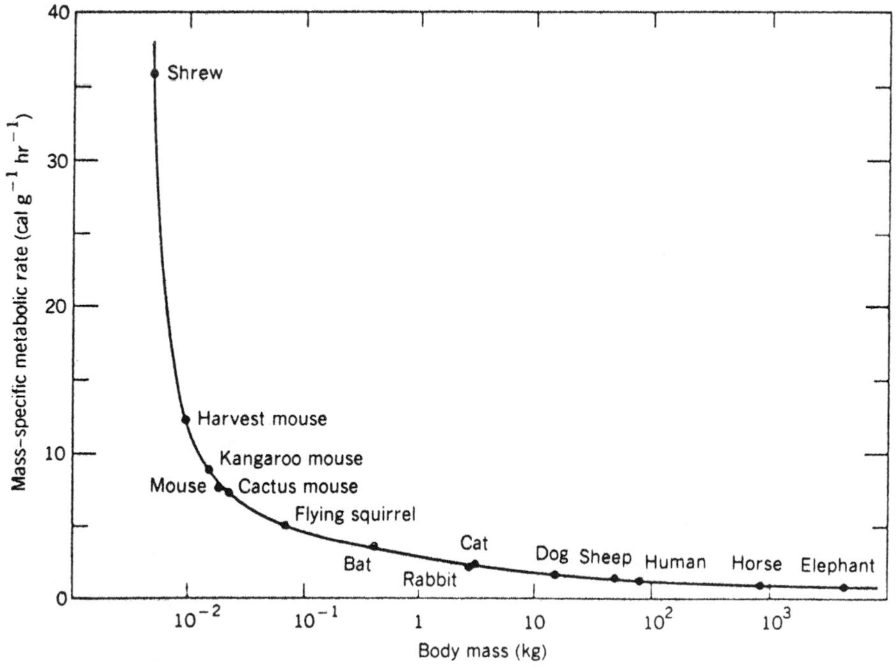

FIGURE 10 Relationship between body mass and metabolic heat production in mammals of widely differing size. [From Prosser (1973).]

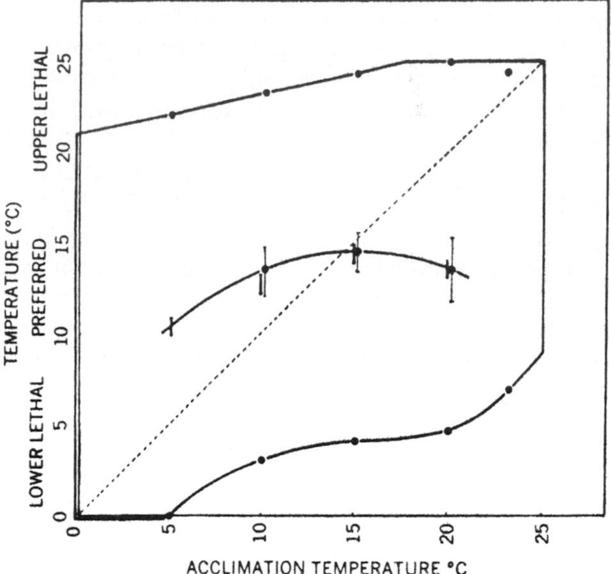

FIGURE 11 Thermal tolerance (high and low lethal) and preferred temperatures of young salmon acclimated to different temperatures. [From Prosser (1973).]

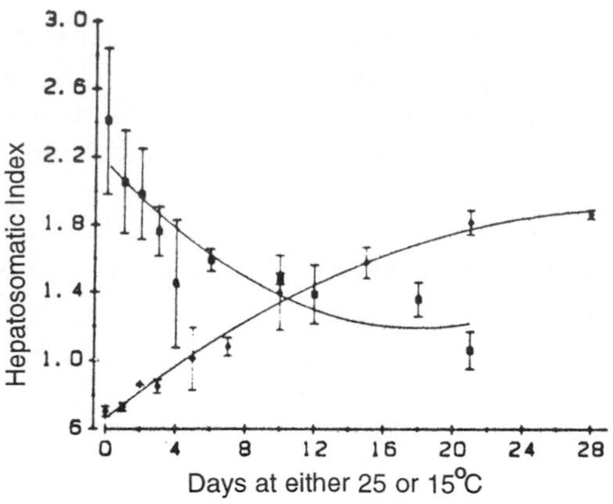

FIGURE 12 Time course of change in size of liver of catfish as the function of days after the transfer from 15 to 25°C (decrease in size) or from 25 to 15°C (rise in size). Each curve is a measure of acclimation. [From Prosser (1994).]

changes in protein synthesis: isozymal selection in some multiform enzymes. Kinetic properties of a given enzyme are genetically determined; for example, the K_m (mid saturation temperature) varies adaptively in Antarctic and tropical fish (Fig. 13). Changes in membrane phospholipid composition may occur more rapidly than changes in metabolic enzymes. The time required for acclimation is determined by the turnover rate of critical proteins. The capacity for acclimation is greater in some animals (e.g., fish) than in others (insects).

Homeotherms regulate body temperature (in both cold and warm environments) by a variety of means: peripheral blood flow, insulating fur or feathers, rate of metabolic heat production, or behavioral selection of a favorable environment. Homeotherms have peripheral receptors for heat and cold and central nervous thermoregulating centers. As ambient temperature falls, heat-conserving mechanisms come into action until a critical low body temperature heat production increases, shivering occurs, and, if this fails, the animal becomes hypothermic (Fig. 14). As the ambient temperature rises, heat loss is increased by peripheral vasodilation, by panting, and by behavioral seeking of shade; at a high critical temperature, regulation fails, body temperature rises, and O_2 consumption and heat shock may result (Figs. 14 and

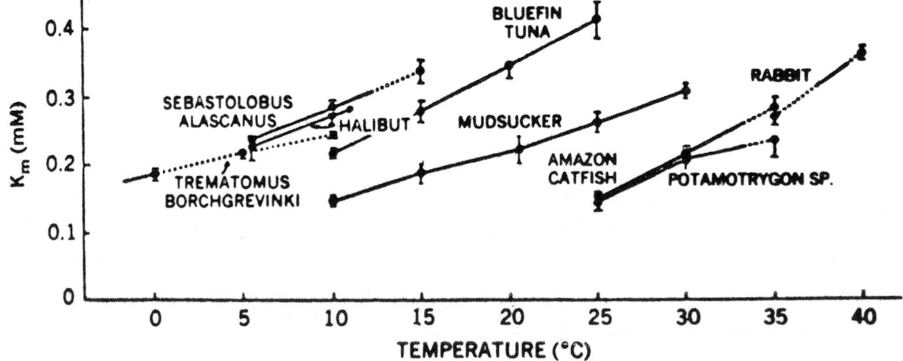

FIGURE 13 Kinetic constants (K_m) for lactate dehydrogenase as a function of temperature of assay for species living in Antarctica (Sebastolobus, Trematomus), temperate (tuna, mudsucker), and tropical (Amazon) water. Differences are genetically determined. [From Prosser (1994).]

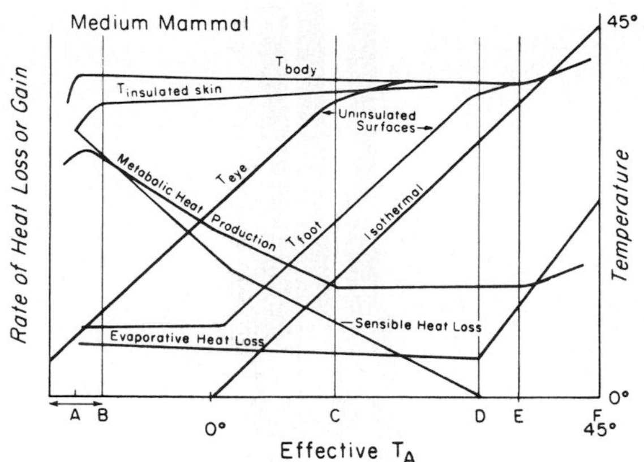

FIGURE 14 Schema of heat loss or gain and of body temperature of a mammal at different environmental temperatures. (A–B) Zone of cold failure, (B) critical low temperature, (B–D) region of maintained body temperature, (D–E) region of evaporative control of body temperature, (D) high critical temperature, and (F) lethal temperature. Different regulatory means are indicated at different ranges of ambient temperature. [From Prosser (1994).]

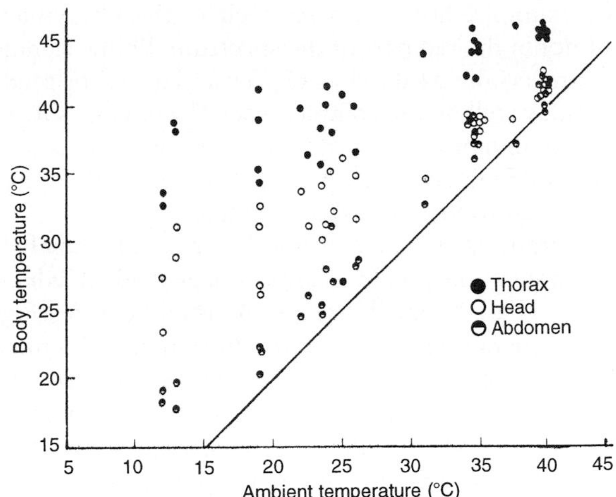

FIGURE 16 Thorax, head, and abdomen temperatures of a bee during flight at different ambient temperatures. Heat production is by muscles of the thorax. [From Prosser (1994).]

15). Most poikilotherms select a preferred temperature and behavioral temperature regulation (Fig. 11). In fish, the brain (hypothalamus) controls behavioral thermoregulation whereas this function is served by circulatory and metabolic means in mammals. Many heterotherms have to raise their body temperature before they can be active: basking in snakes and warm-up of flight muscles and nervous system in flying insects (Fig. 16).

IV. NUTRIENTS

The general requirements for carbohydrates, proteins, and fats are universal and are determined by pathways of intermediary metabolism. Animals differ in their ability to transform basic compounds, e.g., neutral fats and phospholipids into compounds essential for specific reactions. The need for trace substances and vitamins varies greatly; for some animals these are very specific, e.g., carnitine for a few larval insects. Some animals can make essential compounds (vitamins) whereas others must consume them. Some animals have very specific dietary requirements, e.g., beeswax by waxmoths. Selection of diet is by chemical sense: olfaction and taste. Selectivity is extreme in many herbiovores, e.g., some insects feed exclusively on certain plants. Animals can be carnivorous, herbivorous, or omnivorous. Selection by chemical senses and central reflexes (chemoreception) is a subject of much current research.

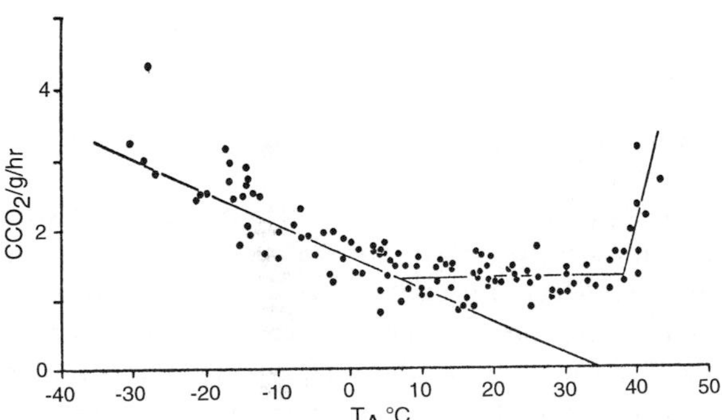

FIGURE 15 Metabolism in birds as a function of ambient temperature. [From Prosser (1973).]

V. LIGHT

All multicellular animals (except for a few cave dwellers and deep sea forms) and most unicellular animals have photoreceptors. These receptors are of several types: simple eyes of molluscs, compound eyes of arthropods, or camera eyes of cephalopods and vertebrates. Remarkably, all eyes use a similar photosensitive pigment: rhodopsin or a related compound. Some receptors are adapted to bright light whereas others are adapted to dim light. A distinction is made between color sensitivity and color vision. Small molecular differences in photopigments permit color sense.

For example, honeybees see well in the ultraviolet, but not in the red part of the spectrum. Photic signals are universally used to locate prey and to escape predators, as well as for complex social behavior. Vision is predominant among the senses in many birds and insects (Hymenoptera). A large fraction of the brain is devoted to vision in most animals.

Compound eyes of arthropods consist of many facets; ommatidia give a mosaic image and in which receptors are coupled directly to visual neurons (Fig. 17). Camera eyes of vertebrates have inverted retinas so that light passes through several neural layers to reach the rods and cones (Fig. 18). An important function of photoreception is the setting of circadian rhythms. The same or different receptors serve both behvior and circadian cycles, but different regions of the central nervous system are used for the two functions. A region of the brain of mammals (suprachiasmatic nucleus) is a seat of a circadian clock which is set by a diurnal light cycle. [*See* Circadian Rhythms and Periodic Processes.]

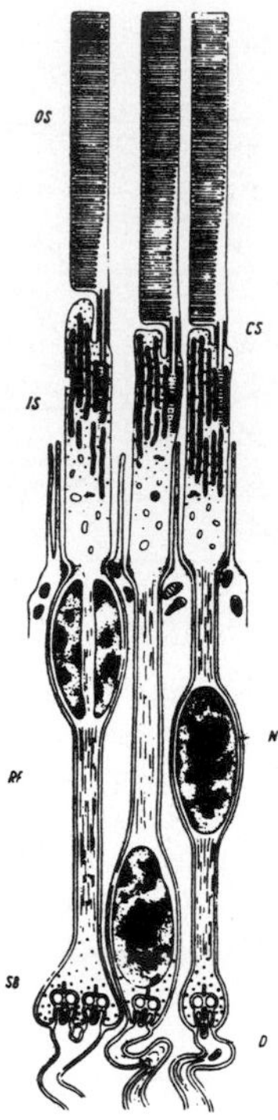

FIGURE 18 Diagrammatic representation of three mammalian rod cells. Outer segment (OS) of membranes that contain photosensitive pigment. Inner segment (IS) with dark-staining nuclei. Contact with bipolar interneurons is at the bottom. [From Prosser (1961).]

VI. HYDROSTATIC AND ATMOSPHERIC PRESSURE

Deep-sea animals are subjected to many atmospheres of pressure, and some of them explode when brought to one atmosphere. Many proteins are denatured at high pressure and deep-sea forms have biochemical adaptations for survival.

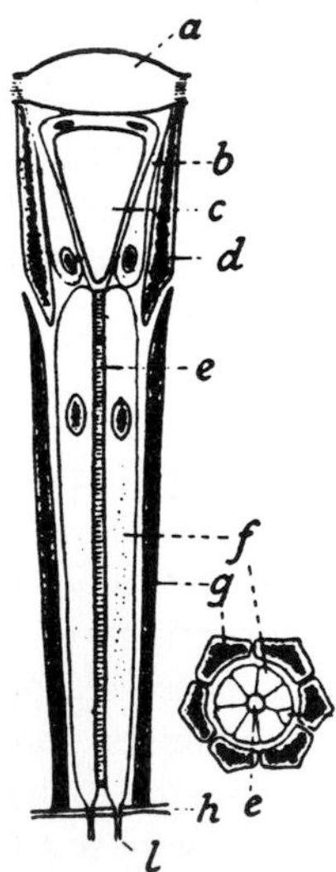

FIGURE 17 Facets or ommatidia of the compound eye. Six retinula cells in cross section. The longitudinal section shows corneal lens (a), crystaline cone (c), iris pigment cells (d), central rhabdome (e), and retinula cells (f) that connect to nerve fibers (l). [From Prosser (1961).]

Reduced atmospheric pressure is experienced by animals living at high altitudes. Molecular adaptations permit O_2 transport from low partial pressure of O_2 into blood pigment. Alpine mammals are uniquely adapted to altitude. Reduced pressure per se has little effect other than low partial pressure of oxygen.

VII. MECHANICAL STIMULI, VIBRATIONS, SOUND, AND GRAVITY

Mechanical properties of substratum, air, and water currents are important for the orientation of many animals. Mechanoreceptors on body surfaces and stretch receptors in appendages are essential for posture and locomotion. Air currents are important for flight in insects and birds whereas water currents are important for swimming fish. Substratum vibrations are behavioral signals for soil-dwelling animals, i.e., fiddler crabs have sensory hairs that are extremely sensitive to motion.

High frequency vibrations (sound) are used for communication within species, e.g., insects, frogs, and aquatic and terrestrial mammals. All phonoreceptors depend on the sensitive membranes of receptor cells. Insect receptors are strands of highly sensitive cells; vertebrate ears are highly organized structures with receptors arranged according to optimum sensitivity to the frequency of stimulation. Sound travels much farther and faster in water than in air, a property used by aquatic mammals for behavioral communication. Recognition of patterned sounds is determined centrally and is much used by frogs, birds, and cetaceans.

Organs of equilibrium are sometimes associated with those of hearing (vertebrates), whereas equilibrium in others is detected by special statocysts (cephalopods).

VIII. STRUCTURES FOR MOVEMENT

Muscles and cilia have several contractile proteins, most importantly dynein in cilia, myosin and actin in muscle. Protoplasmic flow is essential in nerve function chromosome movement in cell division and cell movement in embryonic development. In these the essential protein is kinesin.

Muscles occur in a great diversity of structure; all of them use myosin and actin as the motors. Cleavage of dividing cells makes use of myosin. The theory of contraction, the sliding filament hypothesis, is based on the orientation of myofilament proteins (Fig. 19). Muscles are often classified as striated and smooth; a more functional classification is based on the activation of contractile machinery by cell membrane events. Striations are produced by alternating bands of thick myosin filaments while thin filaments of actin overlap the thick filaments and extend across a so-called I band. In some invertebrate nonstriated muscles the fibers are ribbon shaped with a minimum distance from bottom and top to contractile filaments. In others the striations are diagonal (spiral) and contraction consists of both sliding and shearing. A few postural narrow-fibered muscles of invertebrates contract by a series of folds. Wide-fibered muscles (usually cross-striated) have invaginating tubes connecting the plasma membrane with the intracellular sarcoplasmic reticulum that activates the actin–myosin interaction. In narrow-fibered muscles the activation is direct from plasma membrane to sarcoplasmic reticulum. Many narrow-fibered muscles of invertebrates are postural

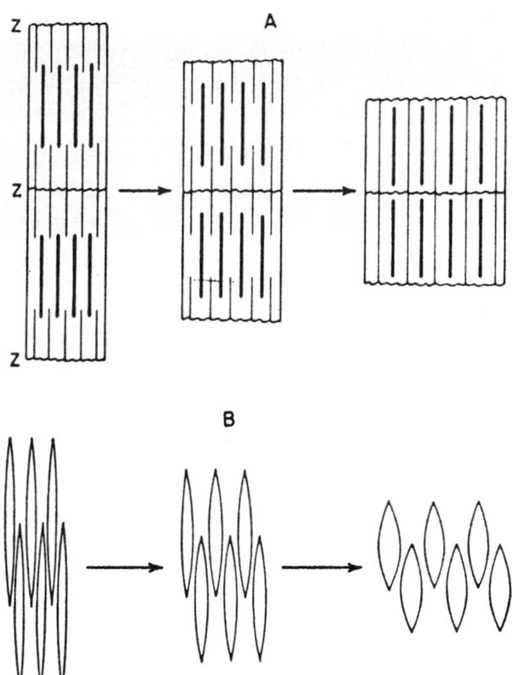

FIGURE 19 Diagrams of the contraction sequence. (A) Events in wide-fibered striated muscle, thick filaments in heavy lines, thin filaments in thin lines, I band crossed by Z-line Sarcomere length from Z to Z. (B) Shortening of narrow-fibered, nonstriated muscle fibers. [From Prosser (1973).]

whereas in vertebrates they are mainly visceral. Calcium is an activating agent in all muscles, and different kinds of muscles use different modes of calcium activation. In vertebrate wide-fibered muscles, calcium is carried by three specific proteins, whereas in narrow fibers the coupling is by a specific chain on the myosin. [*See* Muscle, Physiology, and Biochemistry.]

Some muscles are spontaneously active: heart and gastrointestinal. Postural muscles are activated by motor nerves, some with several excitatory nerves, others with single ones, and some with both excitatory and inhibitory nerves fibers. Insect flight muscles may be oscillatory, several contractions for each motor impulse (Fig. 20B), or they may be synchronous, a contraction for each motor impulse (Fig. 20A). Gradation of contraction in vertebrate postural muscles is controlled by the number and pattern of excitatory impulses. Multineuronal crustacean muscles grade contraction by neuromuscular facilitation and inhibition.

The resting tension of muscle increases with stretching. Active tension increases with length to a maximum and then declines; this is interpreted as due to a change in the overlap of thick and thin filaments

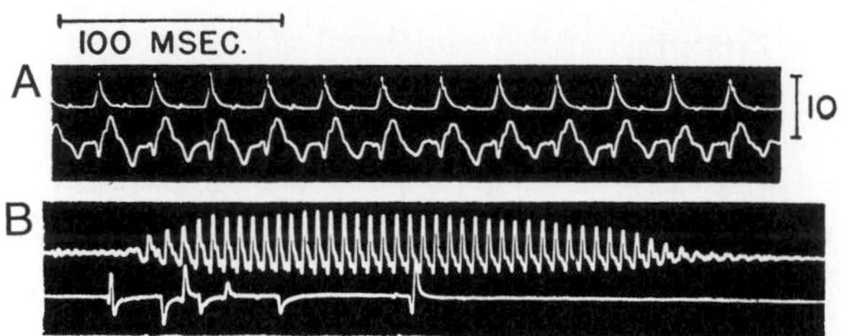

FIGURE 20 (A) Contractions (below) and action potentials (above) in synchronous muscle (moth). (B) Contractions (above) and action potentials (below) in asynchronous muscle (fly). [From Prosser (1961).]

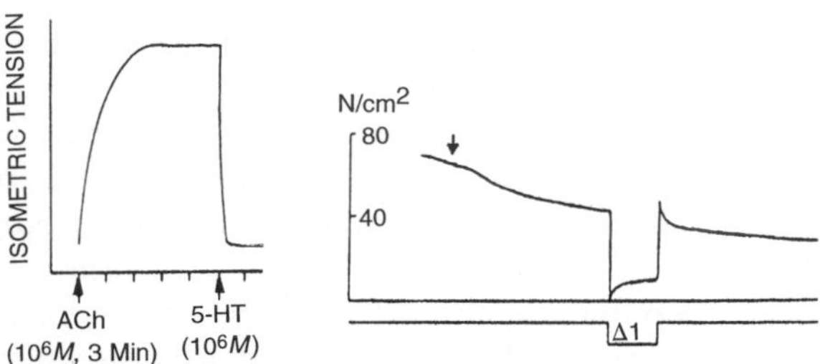

FIGURE 22 Contraction properties of a catch muscle. Stimulation by acetylcholine or by an excitatory nerve. Relaxation by 5HT or by relaxing nerve Δ1. [From Prosser (1961).]

(Fig. 21). "Fast" and "slow" muscles are distinguished partly by different myosins, partly by mechanical properties, and partly by differences in energetics (fast glycolytic and slower oxidative). Muscles differ in speeds of contraction by some 10,000-fold. The fastest muscles are the oscillatory ones of fast-flying insects where wing beats may be as high as 1000 per second; the fastest muscles of mammals are the extra-

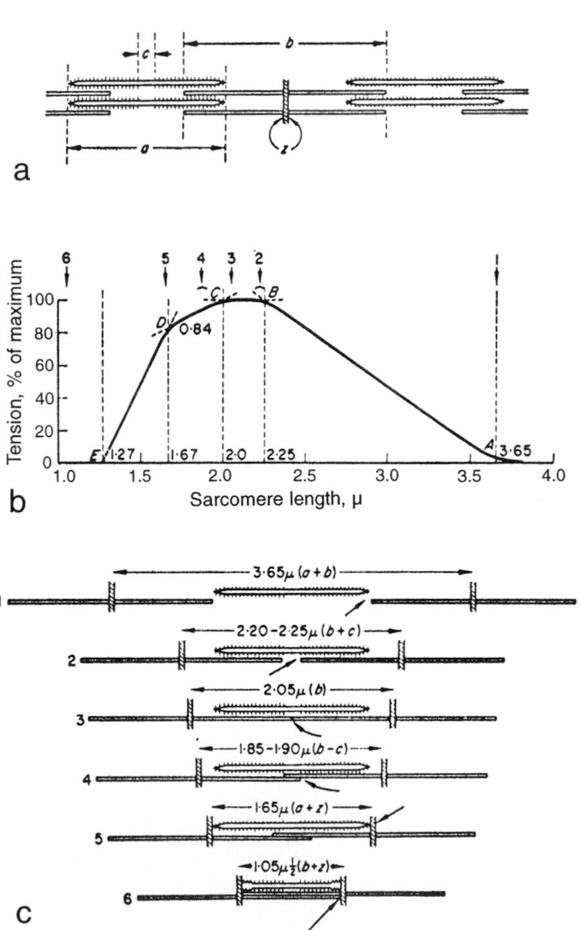

FIGURE 21 Interpretation of tension as the percentage of maximum as a function of sarcomere length. Maximum tension at maximum overlap of thick and thin filaments. Tension declines at both less and more overlap as indicated in diagrams below the tension-length curve. [From Prosser (1973).]

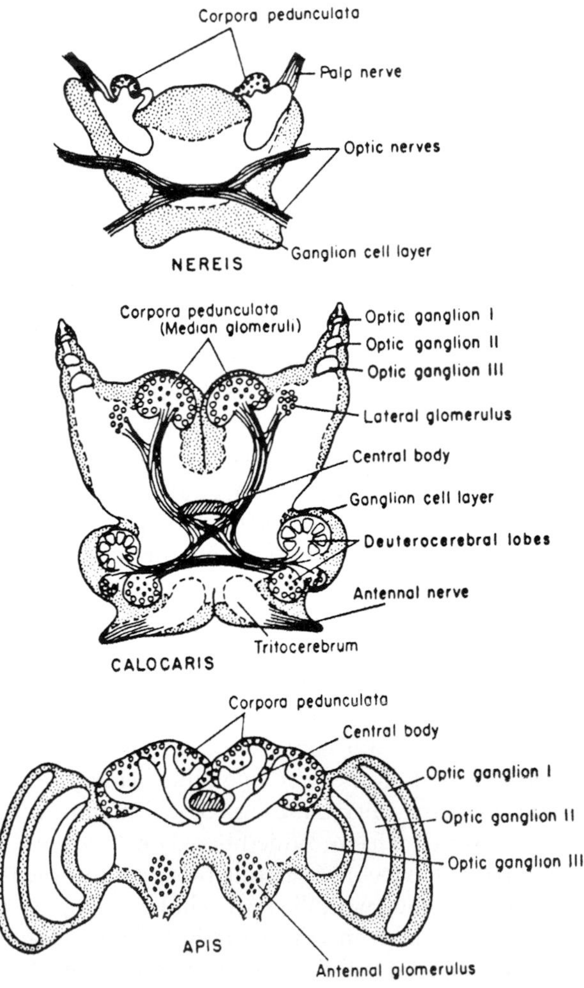

FIGURE 23 Sections of brains of several invertebrates: polychaete worm *Nereis*, crustacean *Calocaris*, and honeybee *Apis*. [From Prosser (1973).]

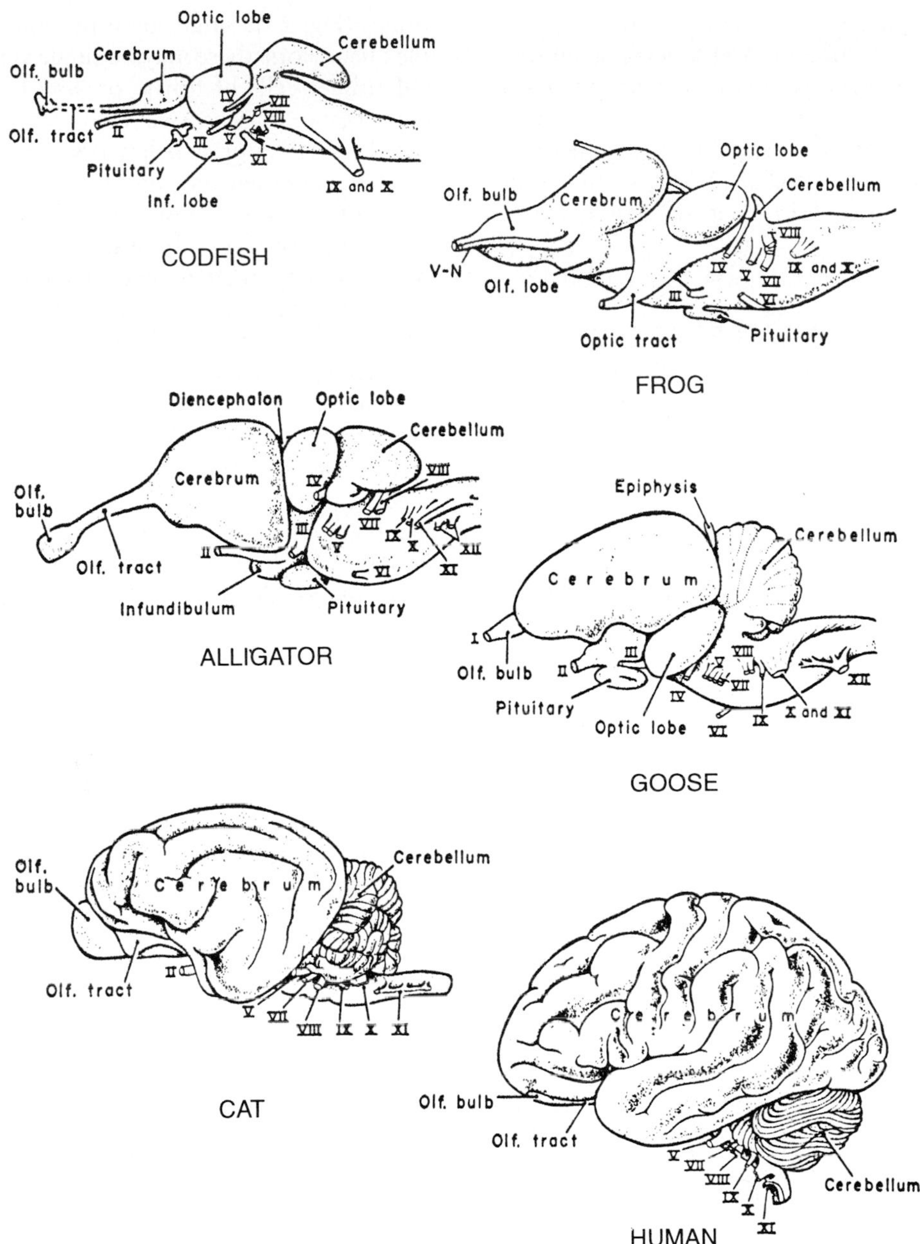

FIGURE 24 Brains of several vertebrates showing relative sizes of various regions. [From Prosser (1973).]

ocular muscles. The slowest muscles are in sea anemones where contractions may take a minute and the slowest in mammals are in visceral organs with a contraction time of seconds. Most locomotor muscles contract in tens of milliseconds. The adductor muscles of bivalve molluscs enter a "catch" state of contraction that is maintained for long periods of time and with very little expenditure of energy (Fig. 22). Muscles are highly diverse in their physiology and each type is adapted for its particular function.

Cilia occur in tubular passages (respiratory tracts and gills); they are used for locomotion in ciliated protozoons, in sperm, and in some free-living larvae. Each cilium has several (usually nine peripheral and two central) tubules of the protein tubulin. Dynein arms project between pairs of peripheral tubules. Protoplasmic flow (e.g., in long neurons) and chromosome movement use tubulins and kinesin as motors.

Nonmuscular effector systems include chromatophores which provide for color change, lightening, and darkening in many animals. These are either neurally or hormonally controlled. Another kind of effector is bioluminescence; deep-sea animals are highly luminescent.

IX. SOCIAL AND PREDATORY INTERACTIONS: ANIMAL BEHAVIOR AND NERVOUS SYSTEMS

The behavior of animals (including humans) is determined by the central nervous system. Nervous systems are organized in several patterns: primitive nerve nets, annelid and arthropod ladder patterns, or tubular arrangements in vertebrates. Most nervous systems are organized in a hierarchical arrangement; anterior regions may be dominant because of the concentration of sense organs anteriorly or anterior regions may be dominant in integrative function (Fig. 23). Some nervous systems have a radial arrangement, as in echinoderms and some coelenterates. All vertebrate brains are organized in a similar plan (Fig. 24). The basic circuitry of brain regions such as cerebellum uses a similar circuit and similar primary neurons in all vertebrates (Fig. 25). Single giant neurons carry out complex behaviors: these may be single very large neurons (Mauthner cells of fishes) or may be formed by the convergence of processes from several neurons (giant squid). The simplest behaviors are reflexes: sensory neurons, interneurons, and motor neurons.

Most nervous systems are capable of some spontaneous activity, i.e., without sensory input. The spontaneous activity of single neurons may be rhythmic, as in brain waves, or irregular, as in arthropod nervous systems. Integration and patterned behavior may result from genetically programmed neural patterns, by interactions in chains of neurons, or by actions of large regions of a brain. Invertebrate animals, especially insects, display complex innate behaviors of feeding, nesting, and reproduction; these behaviors can be modified by conditioning. Vertebrate nervous systems also have innate patterns (e.g., the determinants of specific bird songs) and the capacity for conditioning varies greatly. Whether the modification of simple circuits, e.g., the hippocampal cortex of mammals or the feeding reflexes in gastropod molluscs, is similar to complex learning is debated. Mechanisms of long-term potentiation and inhibition at central synapses are taken as models. Interneuronic transmission is electrical in switchboard-type synapses (giant fibers) but chemical in all integrative synapses. Chemical transmitters are liberated from presynaptic end-

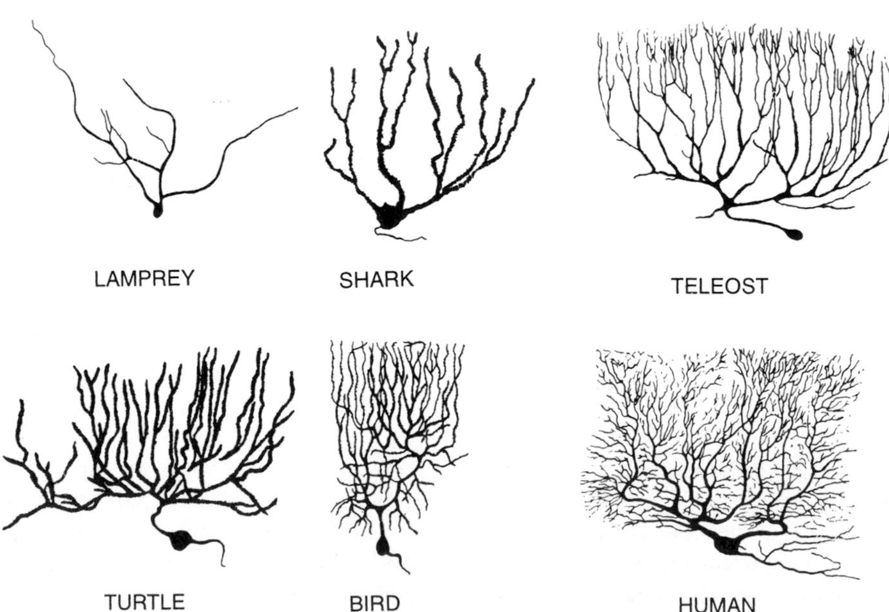

FIGURE 25 Stereograms of five main types of Purkinje cells in cerebellum of six vertebrates. [From Prosser (1973).]

ings and electrical events are initiated postsynaptically (Fig. 26). Presynaptic events may be modulated by converging neurons.

A comparative examination of neurotransmitters and neuromodulators indicates the independent evolution of two classes: (1) amino acids and amines and (2) neuropeptides. Amino acids as chemical signals occur in some prokaryotes and are important in most nervous systems. Glutamic acid and acetylcholine are excitatory in most nervous systems and receptor proteins for them have been isolated. γ-Aminobutyric acid is a widespread inhibitor. Glycine is also inhibitory. By using relatively simple steps, amines are formed from amino acids: by decarboxylation, methylation, hydroxylation, and acetylation. The most widespread transmitter is acetylcholine. A second class are catecholamines (octopamine in arthropods and epinephrine and serotonin in vertebrates). The second class of transmitters–modulators are the peptides that occur in the nervous systems of most animals; these vary in length from 6 to nearly 100 amino acids, many with common amino and carboxyl ends. One group of these, the Phe-Met-Arg-Phe-NH_2 (FMRF) amides, were discovered in molluscs; they occur in great variety in invertebrates, but rarely in vertebrates. The most abundant peptides of vertebrates are both endocrine and neural in origin.

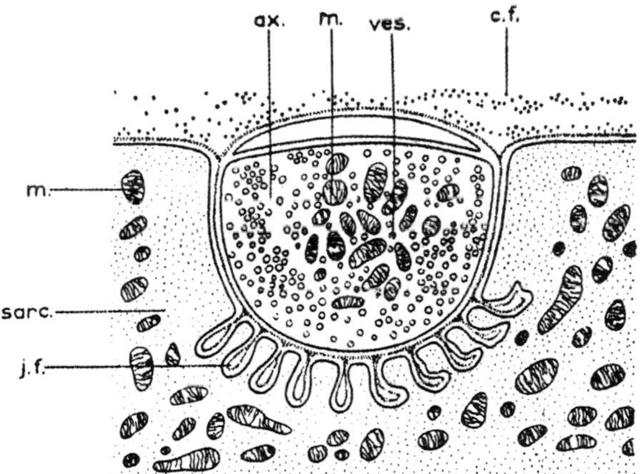

FIGURE 26 A motor end plate of a striated muscle fiber. ves, vesicles where the acetylcholine transmitter is packaged; m, mitochondria; ax, axoplasm; jf, junctional folds in postsynaptic membrane. [From Prosser (1973).]

The majority of behavior consists of patterns of feeding, escaping, courting, and mating. More complex behaviors are social interactions between individuals of the same kind. These include communication by sounds, visual displays, and group organization. The comparative study of behavior and its neural basis is a topic of intense research.

X. APPLICATIONS OF COMPARATIVE PHYSIOLOGY

Comparative physiology or physiological diversity contributes significantly to our understanding of animal distribution. Why some species survive in a stressful environment and others disappear is determined by physiological adaptations to the environment. Comparative studies of nervous systems and behavior, of metabolism, and of water–ion balance provide a basis for extrapolation from one level of biological organization to another and give evidence of evolutionary relations. The comparative approach puts humans in proper biological perspective.

BIBLIOGRAPHY

Hochachka, P., and Somero, G. (1984). "Biochemical Adaptation." Princeton Univ. Press, Princeton, NJ.

Prosser, C. L. (ed.) (1951). "Comparative Animal Physiology." Saunders, Philadelphia.

Prosser, C. L. (ed.) (1961). "Comparative Animal Physiology," 2nd Ed. Saunders, Philadelphia.

Prosser, C. L. (ed.) (1973). "Comparative Animal Physiology," 3rd Ed. Saunders, Philadelphia.

Prosser, C. L. (ed.) (1994). "Comparative Animal Physiology," 4th Ed., Vol. I and II. Wiley-Liss, New York.

Prosser, C. L. (1986). Biological Adaptation; Molecules to Organisms." Wiley-Liss, New York.

Schmidt-Nielsen, K. (1983). "Animal Physiology: Adaptation and Environment," 3rd Ed. Cambridge Univ. Press, Cambridge, UK.

Complement System

MICHAEL K. PANGBURN
University of Texas Health Science Center at Tyler

GORDON D. ROSS
University of Louisville, Kentucky

GLOSSARY

Complement Frequently referring to the entire complement (C) system

C1 inhibitor (C1-INH) Protease inhibitor in plasma that irreversibly inhibits the activity of C1r and C1s, and also the proinflammatory proteases bradykinin and kallikrein

CD59 Membrane protein that protects host cells from their own C by inhibiting the membrane attack complex (MAC)

CR1, CR2, CR3, and CR4 Complement receptor types one, two, three, and four; receptors for different parts of C3 molecules deposited onto activating surfaces

Decay-accelerating factor (DAF) Membrane protein DAF protects host cells from homologous C attack by dissociating C3-convertase enzymes deposited onto host tissue

Mannan-binding protein (MBP) Serum protein MBP is a lectin with specificity for polysaccharides containing mannose; it is responsible for initiating the lectin pathway of C activation

MBP-associated serine protease (MASP) Serum protein MASP is a serine protease that associates with MBP and cleaves C4 and C2 in the lectin pathway of C activation

Membrane attack complex (MAC) C-protein complex of C5b, C6, C7, C8, and polymerized C9

Membrane cofactor protein (MCP) Membrane protein MCP protects host cells from their own C by facilitating the proteolysis of deposited C3b by serum factor I

Opsonization Coating of particles (e.g., bacteria) by serum proteins that facilitates phagocytosis

Paroxysmal nocturnal hemoglobinuria Acquired disorder in which red blood cells are lysed because normal control proteins (e.g., DAF and CD59) are missing from the red cell surface.

Phagocytosis Particle (e.g., bacteria) ingestion by phagocytic white blood cells (neutrophils, monocytes, and macrophages)

Systemic lupus erythematosus Autoimmune disease characterized by a wide spectrum of circulating autoantibodies, some of which form immune complexes and activate C

THE COMPLEMENT SYSTEM IS AN IMPORTANT part of host defense, which functions together with the immune response to provide the effector mechanisms necessary to initiate inflammation, kill bacteria and other pathogens, and facilitate the clearance of bacteria and immune complexes. It is made up of 22 distinct plasma proteins and 12 different membrane proteins. Bacteria or immune complexes trigger activation of complement, resulting in a sequence of reactions in which one component activates another component in a cascade fashion. Along this cascade, inflammation and phagocytosis are initiated, and the terminal event is the generation of cytocidal (cell-killing) activity in the form of membrane-penetrating lesions. Because

ENCYCLOPEDIA OF HUMAN BIOLOGY, Second Edition, VOLUME 2. Copyright © 1997 by Academic Press. All rights of reproduction in any form reserved.

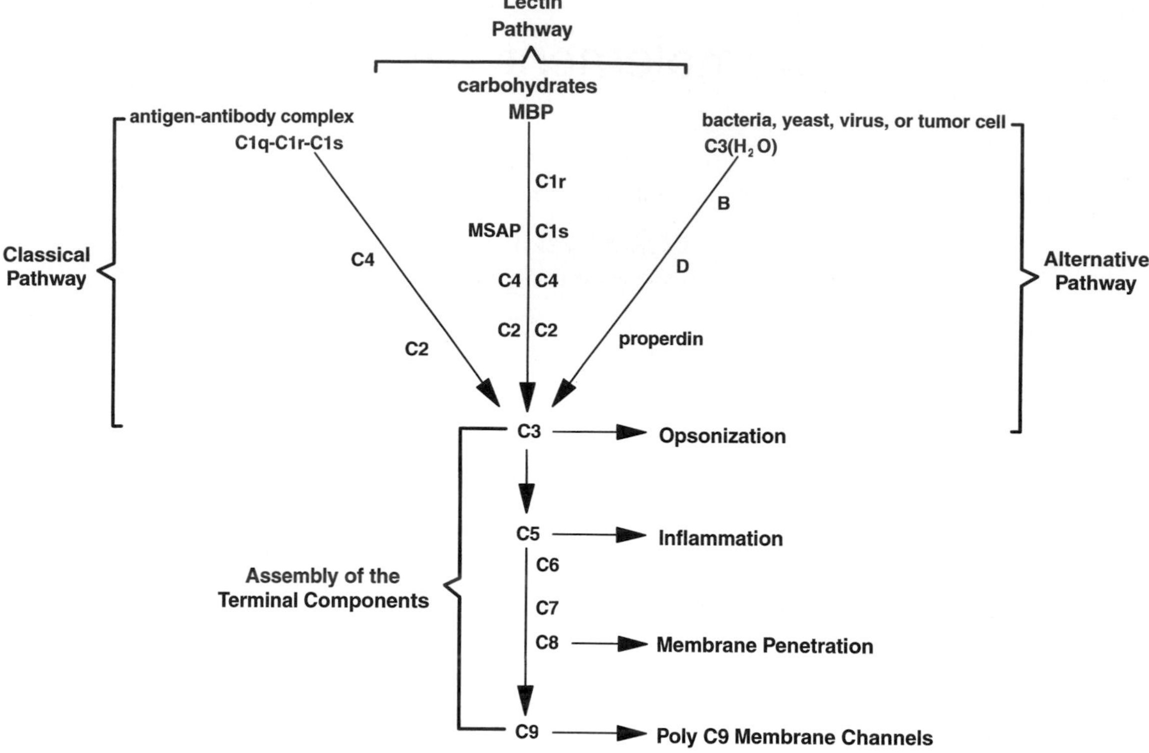

FIGURE 1 The complement system.

of the importance of complement, an inherited or acquired deficiency in any component of the system is frequently associated with either an increased susceptibility to infection or, as will be seen below, a lupus-like syndrome thought to result from the diminished clearance of immune complexes. [*See* Immune System.]

I. INTRODUCTION

The primary function of complement may be defined as the destruction of both foreign organisms and immune complexes. This activity is carried out by two mechanisms. First, coating the particles (e.g., bacteria) with the proteins C3 and C4 results in particle phagocytosis through attachment of the C3/C4 coating to receptors for C3/C4 on macrophages. Second, complement lyses organisms through insertion through cell membranes of hollow tubular structures composed of polymerized C9 molecules. There are three distinct pathways that lead to the deposition of C3: the "classical," "lectin," and "alternative" (Fig. 1). [*See* Macrophages.]

II. CLASSICAL PATHWAY OF COMPLEMENT ACTIVATION

The classical pathway[1] consists of five proteins (i.e., C1, C2, C3, C4, and C5) all present in plasma in inactive form. C3 is also a component of the alternative pathway, and C5 is part of the terminal component pathway that forms the membrane attack complex (MAC). C2, C3, C4, and C5 are present as single molecules, whereas C1 is made up of three noncovalently linked subcomponents (i.e., C1q, C1r, and C1s). Unfortunately, the components were named before their functional properties were elucidated, and it was ultimately found that C4 was misplaced in the sequence of activation, the order being C1, C4, C2, C3, and C5 (Fig. 2). Activation consists of the enzymatic splitting of components. After activation, C1r, C1s,

[1] This section on the classical pathway of complement activation was excerpted with permission from Hughes-Jones, N.C. (1986). The classical pathway. *In* "Immunobiology of the Complement System: An Introduction for Research and Clinical Medicine" (G. D. Ross, ed.), pp. 21–44. Academic Press, Orlando, FL.

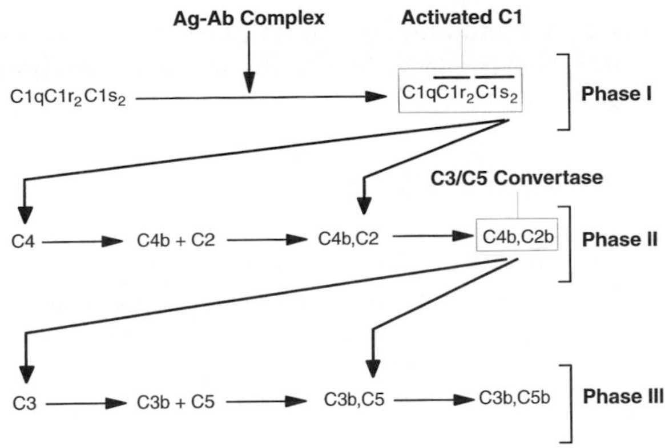

FIGURE 2 Activation of the classical pathway. Two protein-splitting enzymes are generated: activated C1 and C3/C5-convertase. The boxes identify these two enzymes; activated C1 splits both C4 and C2, and C3/C5 convertase splits both C3 and C5.

and C2 become enzymes capable of splitting proteins (proteases), whereas the activated C3 and C4 molecules become capable of binding covalently to immune complexes and cell surfaces.

A. Nomenclature of the Classical Pathway

The nomenclature of the components and of their activated products has evolved as the molecular events were elucidated.

1. The native inactive components are named C1, C2, C3, C4, and C5.
2. Activation of C1 and its subcomponents C1r and C1s is signified by an overline (i.e., $\overline{\text{C1}}$, $\overline{\text{C1r}}$, $\overline{\text{C1s}}$). Activation of C2, C3, C4, and C5 occurs via proteolysis, with the resulting fragments being named by the suffix "a" and "b." In each case, the larger fragment is b (i.e., C2b, C3b, C4b, and C5b); these fragments interact with target membranes.
3. C3b and C4b are also further degraded into "c" and "d" fragments. In each case, the d fragment remains attached to target membranes whereas the c fragment is released into the fluid phase.

B. Overall View of the Molecular Events

The classical cascade is divided into three phases: (1) the formation of activated C1, (2) the formation of the C3/C5 convertase, and (3) the splitting of C3 and C5 to their active forms (Fig. 2). Activated C1 and the C3/C5 convertase are the only enzymes in the classical cascade.

1. Phase I: Formation of Activated C1

The first component, C1, contains three subcomponents: C1q, whose function is the binding of C1 to immune complexes, and C1r and C1s, which are proenzymes. Phase I (Fig. 2) consists of the binding of C1 via C1q to antibody on the target surface. The binding of C1 is followed by the autocatalytic conversion of C1r to an active protease, which then converts C1s to a similar active enzyme. $\overline{\text{C1s}}$ is the active enzyme used in phase II.

Activation of C1 occurs when a surface contains either closely paired IgG antibody doublets or IgM antibody. IgA, IgE, and IgD do not bind C1q and do not activate the classical pathway.

2. Phase II: Formation of the C3/C5 Convertase

The product of phase II is the C3/C5 convertase. The sequence of events is: (1) antibody-bound $\overline{\text{C1s}}$ activated in the first phase cleaves and activates plasma C4; (2) activated C4 molecules (C4b) diffuse to the target surface and become attached close to the $\overline{\text{C1}}$; (3) C2 combines with the bound C4b, and this C2 in turn is also cleaved and activated by the neighboring $\overline{\text{C1s}}$. Phase II thus ends with the formation of a cell-bound C4b,C2b complex (the C3/C5 convertase) which has specificity for splitting both C3 and C5 (Fig. 2).

3. Phase III: The Splitting of C3 and C5

This phase has two functions: (1) attachment of large numbers of C3 molecules to the target surface in order to opsonize the particle for phagocytosis, and (2) cleavage of C5 to initiate the assembly of the membrane attack complex. The sequence is as follows: (a) bound C3/C5 convertase (C4b,C2b) cleaves plasma C3; (b) activated C3 (C3b) attaches ("fixes") close to the C4b,C2b enzyme; and (c) C5 combines with the attached C3b and, as a result of this combination, a modification takes place in the C5 so that it is also susceptible to cleavage by the neighboring C4b,C2b complex. The C4b,C2b enzyme is thus used for the cleavage of both C3 and C5.

Phase III completes the events that generate activated C5 on the target surface. The action of the C3/C5 convertase on C5 is the last enzymatic event; the formation of the membrane attack complex proceeds via the polymerization of C5 to C9 initiated by activated C5.

C. Covalent Fixation of C4 and C3

Cleavage of C4 and C3 at their N-termini results in the disruption of internal thioester bonds, producing C4b and C3b fragments that can covently bind to cell surface sugars or proteins via ester or amide bonds (Figs. 3 and 4).

D. Mechanisms Confining Classical Pathway Activation to Target Membranes

Four mechanisms restrict the classical pathway to target membranes: (1) Activation of C1 requires binding of a specific antibody to an antigen surface. (2) Only C2 that has been modulated by combination with surface-bound C4b can form a C3/C5 convertase after cleavage by $\overline{\text{C1s}}$. Similarly, C5 can only be cleaved by the C3/C5-convertase after it has been modulated by binding to surface-bound C3b. (3) The extremely short life of the C3b asnd C4b combining sites restricts attachment of these molecules to a circular area of 40 nm in radius, centered on the activating C4b,C2b complex. (4) Host cell membrane delay-accelerating factor (DAF) disrupts the C3/C5 convertase deposited onto host tissue.

E. Regulatory Factors

Apart from confining activation to the target, regulatory proteins bring about the rapid destruction of the activated factors at each stage of the classical pathway. These regulatory proteins are required to prevent the complete consumption of plasma C2, C3, and C4 in the fluid phase.

1. Control of C1 Activation

The rapid inactivation of the C1 enzyme to prevent the uncontrolled activation of complement is brought about by C1 inhibitor (C1-INH). C1-INH functions by combining irreversibly with the active sites on both $\overline{\text{C1r}}$ and $\overline{\text{C1s}}$.

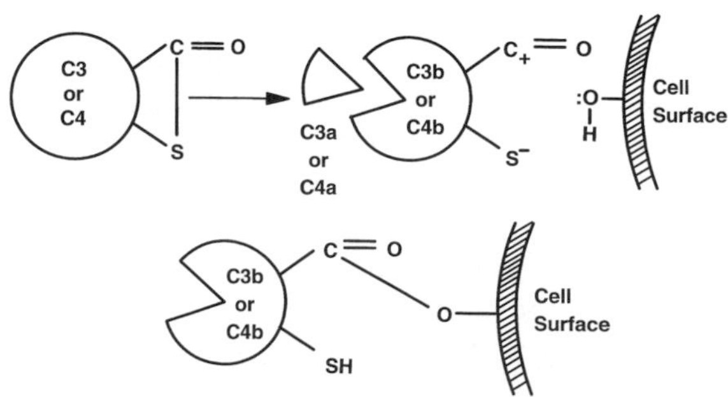

FIGURE 3 Activation and deposition of C3 and C4.

2. Inactivation of C3/C5 Convertase

The C4b,C2b enzyme is inactivated in two stages. First, functional activity is lost by the spontaneous dissociation of C2b from the bound C4b. Two factors promote C2b dissociation [i.e., C4-binding protein (C4BP) and DAF]. C4BP is a plasma protein that combines with C4b preventing further association with C2b. DAF is a protein present within the membranes of many different host cell types which appears to have a similar action to C4BP in bringing about a functional dissociation of C2b from C4b. The second stage of C4b,C2b inactivation is the degradation of the C4b molecule into C4c, which dissociates from the target surface, leaving only the small C4d fragment attached. This degradation is brought about by the enzyme, factor I (I for "inactivator"). Two proteins are known to act as cofactors for factor I proteolysis of C4b:C4BP and the red cell complement receptor CR1.

3. Inactivation of C3b

Factor I is also the normal enzyme that cleaves bound C3b. This inactivation is of the greatest importance in the control of the feedback loop of the alternative pathway (see Section IV,M). It also plays a part in the control of the classical pathway because once C3b is inactivated, it can no longer bind C5 and hence further production of the membrane attack complex is prevented. Cleavage of C3b by factor I can only occur in the presence of a cofactor, and factor H, CR1, or membrane cofactor protein (MCP) can act in this respect. Factor H is a plasma protein that binds to the C3b molecule and acts as a cofactor for the enzyme activity of factor I. Host cell membranes also express MCP, which functions as an additional cofactor to accelerate the breakdown of C3b that has been deposited on host tissue. Factor I splits the α-chain of C3b, inactivating the C3b molecule, so it is termed iC3b. In the case of fixed iC3b exposed to red cells, further cleavage by factor I takes place, but in this case the sole cofactor is red cell CR1. This latter cleavage also occurs in the α-chain with the result that the bulk of the molecule dissociates from the target surface as C3c, leaving the small C3dg fragment attached to the surface. Other proteases (plasmin, trypsin, or elastase) can split off the C3g fragment from C3dg leaving the C3d fragment bound to the surface (Fig. 4).

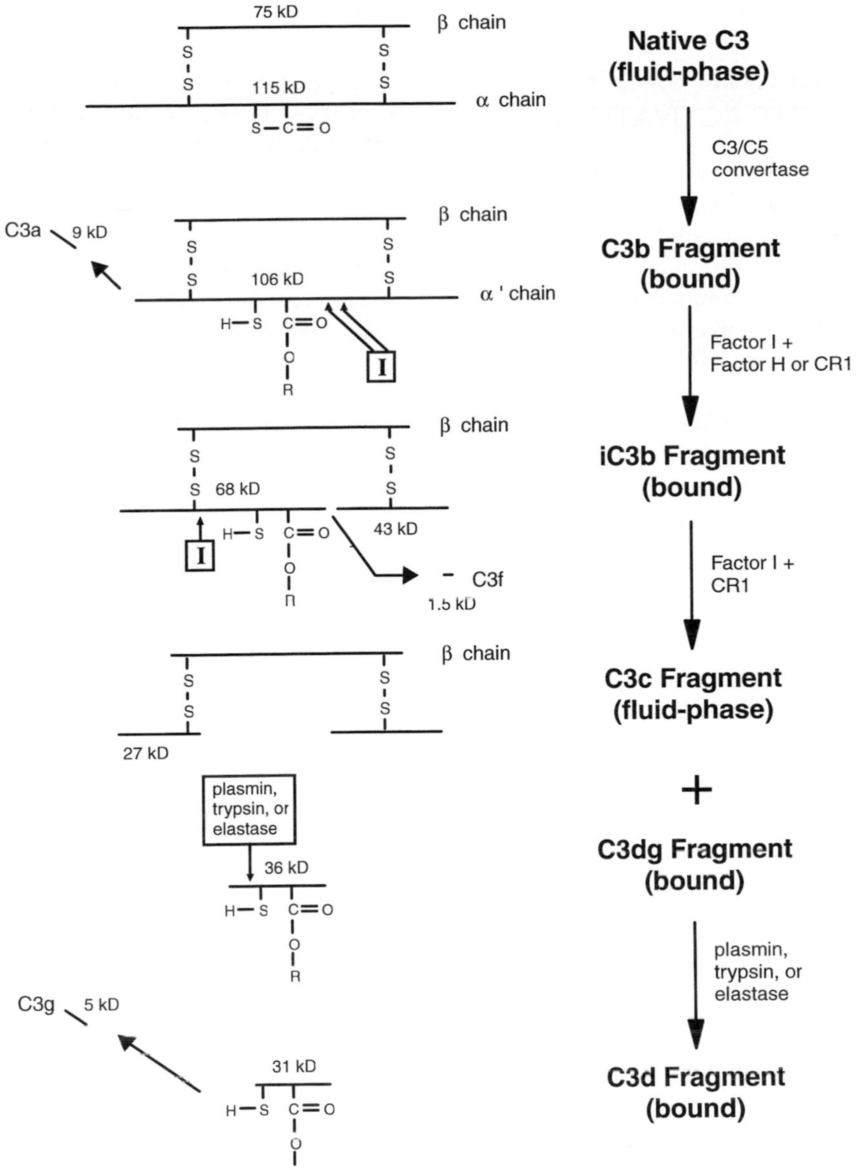

FIGURE 4 Structure of C3 and its physiologic fragments. Native C3 in plasma consists of two disulfide-linked α and β subunits. C3-converse (C4b,C2b or C3b,Bb) proteolyses C3, splitting off the C3a fragment from the N-terminal of the α subunit, causing disruption of an internal thioester bond between a glutamate residue and a cysteine residue on the α subunit (indicated O=C—S). The broken thioester becomes a binding site capable of forming a bond with any nearby hydroxyl group or amino group. The C3b fragment then forms either an ester bond with exposed hydroxyls (such as those of sugar-containing bacterial cell walls) or an amide bond with free protein amino groups. The O=C—O—R indicates this site of ester linkage of C3b to an activating surface. The other end of the broken thioester, the cysteine residue, becomes a free sulfhydryl group (S—H in the figure). The C3b is cleaved rapidly by factor I at two closely spaced sites on the α' subunit, releasing the small C3f peptide and causing a major conformational rearrangement of the C3 molecule that is now called iC3b. The sites of factor I proteolysis are shown with arrows emanting from the boxed letter I. The binding of fixed iC3b to the receptor CR1 exposes a site in the α' subunit where factor I produces a third cleavage, releasing most of the fixed C3 molecule into the fluid phase as the C3c fragment and leaving only the small C3dg piece bound to the surface. The C3dg fragment is not broken down further in normal blood, but is sensitive to a variety of serine proteases that may be present at inflammatory sites (e.g., trypsin, plasmin, or elastase). Proteolysis by one of these enzymes removes the C3g fragment and leaves only the C3d fragment bound to substrates.

III. LECTIN PATHWAY OF COMPLEMENT ACTIVATION

The lectin pathway was not discovered until the early 1990s. Its two unique proteins, mannan-binding protein (MBP) and MBP-associated serine protease (MASP), do not require antibody to identify targets. MBP is a lectin that binds to organisms with mannose-rich cell walls, but not to mammalian cells where mannose is rarely exposed. MBP does not bind to immune complexes. MBP is a member of the collectin family and has structural features similar to C1q, allowing it to bind MASP or C1r-C1s. Once bound to a mannose-rich cell MBP activates its associated MASP or C1r-C1s. The resulting active proteases then activate C4 and C2 of the classical pathway forming a C3/C5 convertase. Subsequent phases of lectin pathway activation are identical to phase II, phase III, and later events in the classical pathway (Fig. 2). Like many other complement proteins, MBP is an acute-phase protein and its concentration in plasma goes up three-fold during infections. Deficiencies of MBP are most serious in young children and result in severe recurrent infections, failure to thrive, and diarrhea.

IV. ALTERNATIVE PATHWAY OF COMPLEMENT ACTIVATION

The alternative pathway[2] provides a natural defense against infectious agents because it is capable of neutralizing a variety of potential pathogens in the absence of specific antibodies. The alternative pathway thus differs from the classical pathway in that it provides an immediately available line of defense that does not require prior immunization. It resembles the classical pathway in that both systems result in the fixation of opsonizing C3 and the initiation of membrane attack via the terminal pathway of complement. The six plasma proteins of the alternative pathway involved in activation perform a continuous surveillance function. They recognize a wide variety of potential pathogens within minutes after such organisms come in contact with plasma. Organisms sensitive to the alternative pathway include certain bacteria and fungi, a number of viruses, virus-infected cells, certain tumor cell lines, parasites such as trypanosomes, and erythrocytes from patients with paroxysmal nocturnal hemoglobinuria (PNH).

Activation of the pathway involves a unique amplification process, which results in the covalent attachment of large numbers of C3b molecules to the surface of the activating particle. The activation process is the result of a dynamic balance between the amplification process and the regulation of this process. Specific regulatory components control this chain reaction-like process, allowing only minimal consumption of the native components.

A. Nomenclature

The five proteins unique to the alternative pathway are factors B, D, H, I, and P (or properdin). C3 is an essential component of all three complement pathways and its numerical designation is retained in the alternative pathway. Two of the proteins (C3 and B) undergo proteolytic cleavage during activation and the fragments are assigned lowercase letters: C3a, C3b, Ba, and Bb. Complexes formed by the noncovalent association of two or more proteins are written using a comma between the symbols (e.g., C3b,Bb).

B. C3

C3 plays a central role in the alternative pathway. Its activated forms (1) participate in initiation of the pathway in the fluid phase, (2) attach covalently in large numbers to biological particles during activation, (3) provide binding sites for C3 receptors on phagocytic cells, and (4) allow activation of C5 which leads to cytolysis. As in the classical pathway, cleavage of C3 produces C3b which can bind to surfaces via the activated thioester site. Simultaneously, other binding sites appear on C3b for factors B, H, I, P, and C5.

Spontaneous low-level hydrolysis of the thioester bond in native C3 gives rise to C3(H_20). The C3(H_2O) behaves as a C3b-like molecule with all properties of C3b except that it lacks a thioester-binding site and thus is only found in the fluid phase (Fig. 5).

C3 and C3b perform a number of functions in the alternative pathway. C3 initiates the pathway through the spontaneous formation of C3(H_20) and C3b is the first protein to attach to the target particle. In contrast, many proteins in the classical pathway (IgG, C1, C4b, C2b) bind to the activating particle surface prior to the attachment of C3b. In the alternative pathway, bound C3b participates in recognizing the

[2] This section on the alternative pathway of complement activation was excerpted with permission from Pangburn, M. K. (1986). The alternative pathway. *In* "Immunobiology of the Complement System: An Introduction for Research and Clinical Medicine" (G. D. Ross, ed.), pp. 45–62. Academic Press, Orlando, FL.

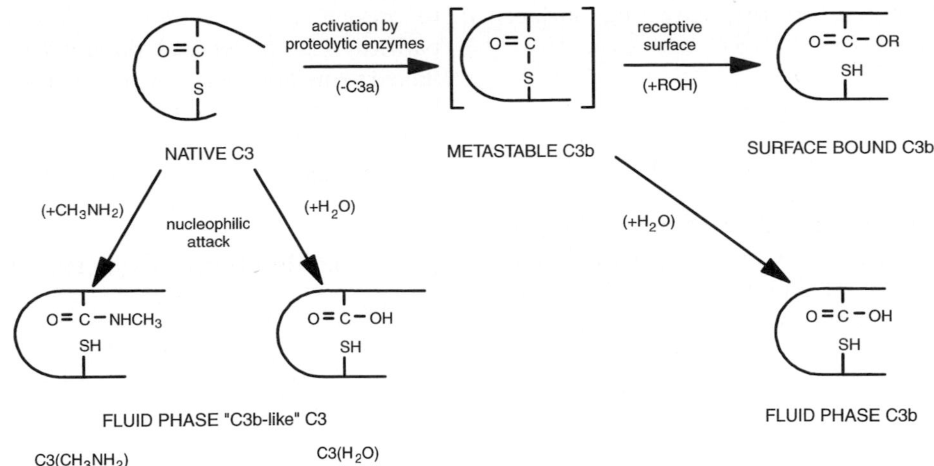

FIGURE 5 C3 activation or denaturation through disruption of its internal thioester bond.

particle as an activator or nonactivator and serves as a subunit of the cell-bound C3/C5-convertase on activators. Activation of C3 by the C3/C5-convertase deposits additional C3b molecules on the surface, simultaneously creating new convertase sites and allowing these enzymes to cleave C5 molecules that become attached to the C3b (Fig. 6).

C. Factor B

Factor B (or B) plays a key role in the alternative pathway. It binds to C3b and forms a proteolytic enzyme that cleaves and activates more C3. Factor B is cleaved only when it is bound to C3b, and this cleavage yields the Bb and Ba fragments. The Bb fragment expresses serine protease activity as long as it remains bound to C3b. Factor B is similar to C2 of the classical pathway. Both proteins form the catalytic subunit of C3-convertases, and both must be bound to be active. C2 is structurally similar to factor B, and the C2 and factor B genes are linked within the major histocompatibility (MHC) locus where they are designated MHC class III genes.

D. Factor D

Factor D (or D) is the enzyme that activates B to form the C3-convertase of the alternative pathway. D is a serine protease that circulates in active form. It is a highly specific enzyme that splits factor B only when B is bound to C3b. The action of D on the complex C3b,B releases the Ba fragment, leaving the C3-con-

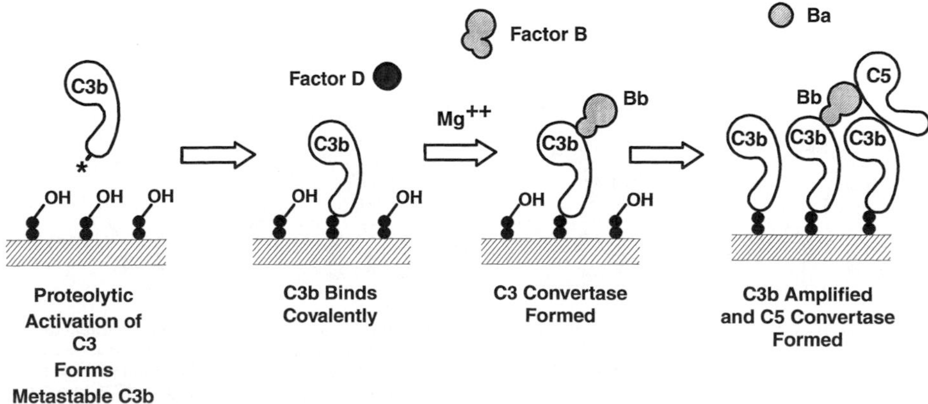

FIGURE 6 Initiation of the alternative pathway of complement activation.

vertase C3b,Bb bound to the activating surface (Fig. 6).

E. Factor H

This protein is a regulator of alternative pathway activation. H binds to C3b and competes with the binding of B. H also binds to the C3b portion of C3-convertases (C3b,Bb) and accelerates the dissociation of Bb from these complexes, thus inactivating them (Fig. 7). As in the classical pathway, H also competes with C5 for binding to C3b. These properties allow factor H to inactivate the enzymatic activity of the C3/C5-convertase and to regulate the use of C5 by the enzymes of both the classical and the alternative pathways.

F. Factor I

Factor I (or I) not only cleaves and inactivates C3b, but also cleaves C3(H_2O). The role of factor I is to prevent formation of the C3-convertase by inactivating C3b permanently (Fig. 7). Failure to block convertase formation leads to consumption of C3 and B through a positive feedback process that is a unique feature of the alternative pathway.

G. Properdin

Properdin (or P) was the first component of the alternative pathway to be identified. Its function is to bind to the C3-convertase (C3b,Bb) and to increase the stability of the complex.

H. Decay-Accelerating Factor

DAF is an important control protein that is an intrinsic component of host cell membranes and functions to prevent assembly of the classical or alternative pathway C3-convertases on normal tissue. DAF is linked to membranes via a phosphatidylinositol glycolipid, and its absence from red blood cells (along with CD59 and HRF, see Section IV,B,2) has been shown to be a major causitive factor in the disease PNH.

I. Membrane Cofactor Protein

MCP is a host cell protein that functions similarly to factor H in reducing the activity of C3b deposited onto host tissue. Unlike red cell CR1, host cell MCP functions only on the cell on which it resides and it cannot exert its cofactor function with factor I on C3b deposited onto adjacent cells that lack MCP.

J. Activation Process

Activation of the alternative pathway involves both reversible and irreversible interactions, and can be divided into four phases: initiation, deposition of C3b, recognition, and amplification (Fig. 8). Activation begins in plasma with the formation of enzymes that cleave C3 and generate C3b. Through its metastable thioester site, C3b may attach randomly to nearby particles, to other proteins, or react with water. All of these forms of C3b are rapidly inactivated unless the particle to which they are attached is recognized as an activator. Recognition involves bound C3b and factor H, as well as DAF. The exact process by which H distinguishes bound C3b on host tissue is not yet understood, but it results in H having a considerably higher affinity for C3b bound to host tissue as compared to C3b bound to bacteria. On the surface of activators, an amplification process rapidly deposits large numbers of C3b around the initial C3b (Figs. 6

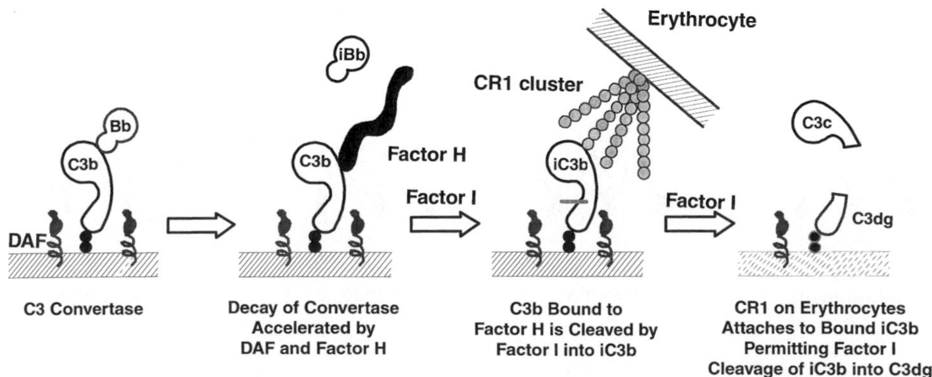

FIGURE 7 Regulation of the alternative pathway on normal tissue surfaces.

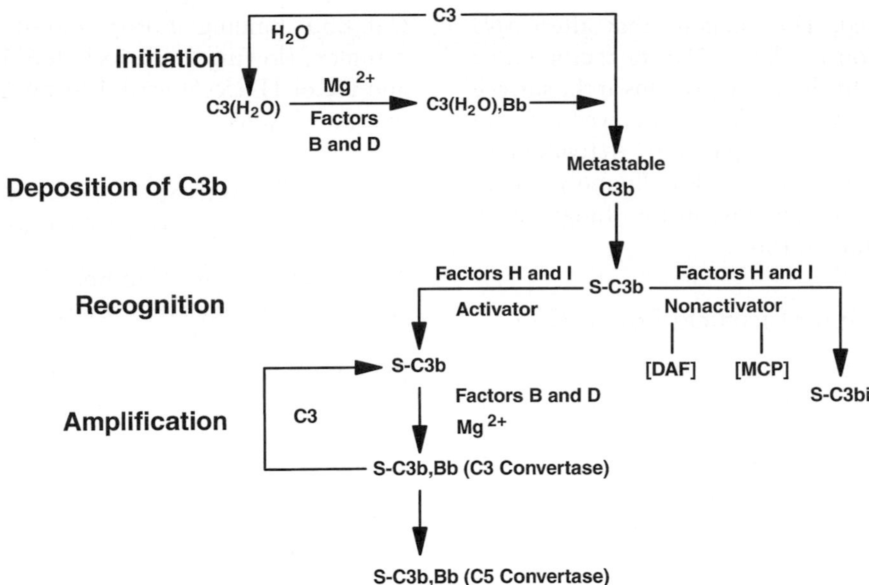

FIGURE 8 Summary of the activation and regulation steps of the alternative pathway of complement activation.

and 8). This is followed by the activation of C5 and the membranolytic proteins.

K. Initiation

Initiation (Fig. 8) involves the spontaneous formation of the chemically and conformationally altered form of C3 [i.e., C3(H_2O)]. C3 altered by hydrolysis of the thioester bond, C3(H_2O), is formed continuously at a very slow rate (0.005%/min) in aqueous solutions. This molecule has all of the functional properties of C3b except that the molecule is resistant to inactivation by H and I for a brief period after formation. C3(H_2O) forms a complex with factor B in the presence of Mg^{2+}. As illustrated in Fig. 8, the C3(H_2O),B complex is activated by factor D forming a fluid-phase C3 convertase. C3(H_2O),Bb is the first enzyme of the pathway capable of generating C3b. The enzyme itself is confined to the fluid phase, but metastable C3b can diffuse up to 300 Å to find and attach to nearby particle surfaces (abbreviated 'S' in Fig. 8). It should be noted that this initiating process does not rely on specific initiators but is spontaneous.

L. Deposition of the First C3b

The ability of the activated thioester in C3b to react with a wide variety of carbohydrates enables the pathway to deposit C3b onto a broad spectrum of organisms. This is important since C3b may be the first host molecule to encounter and initiate a challenge to an invading pathogen in the absence of specific antibodies. A unique feature of alternative pathway activation is that initial C3b attachment appears to be continuous and indiscriminate, occurring on host cells as well as on foreign particles.

M. Recognition

Discrimination between activators and nonactivators occurs soon after initial deposition of C3b. Discrimination is manifested by a reduction in the effectiveness of regulatory factors to control the amplification process when the initial C3b molecules are bound to activator surfaces (Fig. 8). Fluid-phase C3b and C3b bound to host particles are rapidly inactivated by factors H and I. In contrast, when bound to activating particles, both C3b and the C3 convertase are relatively protected from destruction by the fluid-phase regulatory proteins. This appears to be determined by how effectively factor H can interact with activator-bound C3b, as well as by membrane DAF and MCP that function at the host cell surface. C3b bound to activators exhibits a reduced affinity for factor H, while the binding of factors B, I, and properdin to C3b is unaffected by the type of particle to which the C3b is attached. This suggests that alternative pathway recognition resides in C3b or factor H, or is expressed jointly by these two proteins at the surface of the particle. It is not yet clear what structures are recognized by the alternative pathway. Activators include many pure polysaccharides, lipopolysaccharides, certain immunoglobulins, viruses, fungi, bacteria, tumor cells, and parasites.

N. Amplification of Particle-Bound C3b

The C3b-dependent positive feedback process is a unique feature of the alternative pathway. C3b with B, D, and Mg^{2+} forms a C3 convertase capable of generating more C3b (Fig. 9). In the presence of B and D, each subsequent C3b has the potential of repeating this process. The initial C3b is thus amplified in number. H and I limit this process both in the fluid phase and on nonactivating particles. C3b deposited on an activator is relatively resistant to H and I, and the binding of B to C3b and its cleavage by D is unaffected by the nature of the surface to which the C3b is bound. The enzyme responsible for amplification is C3b,Bb or the properdin-stabilized form of this enzyme (see Section IV,O). The enzyme C3b,Bb is labile, and spontaneous dissociation of Bb from the complex results in an irreversible loss of enzymatic activity. Factor H accelerates the dissociation of the C3b,Bb complex, resulting in a very short half-life.

O. Role of Properdin

Properdin binds to the cell-bound alternative pathway C3-convertase, forming the trimolecular complex C3b,Bb,P. Binding of properdin stabilizes the C3b,Bb complex, slowing its dissociation. Both spontaneous and factor H-accelerated dissociation of the Bb subunit are slowed 5- to 10-fold.

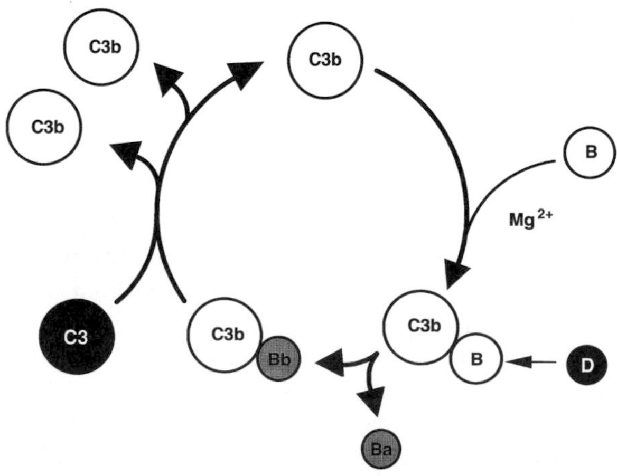

FIGURE 9 Positive feedback loop of the alternative pathway of complement activation.

P. Activation of C5 and the Terminal Pathway

The C3-convertase, C3b,Bb, can function as a C5-convertase provided that additional bound C3b molecules are present. The role of these C3b molecules is the same as in the classical pathway (i.e., to bind C5 and present it to the C3-convertase enzyme).

V. COMPLEMENT CYTOTOXIC ACTIVITY AND TERMINAL COMPONENTS OF COMPLEMENT

Terminal complement protein activity[3] is directed toward the destruction of invading organisms. This occurs via the assembly on target membranes of five proteins (C5b, C6, C7, C8, C9) into the membrane attack complex. The MAC forms transmembrane channels, displaces lipids and other membrane constituents, and causes reorganization of lipids in the phospholipid bilayer. The MAC is only capable of binding to lipid membranes and cannot attach to protein immune complexes or the polysaccharide cell walls of yeast or gram-positive bacteria that lack outer lipid membranes. [*See* Lipids.]

A. Assembly of the MAC

1. Activation of C5

Cleavage of C5 by the classical, lectin, or alternative pathway C3/C5-convertase liberates C5a, leaving the C5b fragment bound to the C3b unit of the C5-convertase. For a limited time, this C3b,C5b complex serves as the acceptor for C6.

2. Formation of the C5b,6 Complex

C5b binds stoichiometric amounts of C6. C5b,6 remains bound to the C3b subunit of the C5-convertase and serves as an acceptor for C7 (Fig. 10).

[3] This section on complement cytotoxic activity and the terminal components of complement was excerpted with permission from Podack, E. R. (1986). Assembly and functions of the terminal components. *In* "Immunobiology of the Complement System: An Introduction for Research and Clinical Medicine" (G. D. Ross, ed.), pp. 115–137. Academic Press, Orlando, FL.

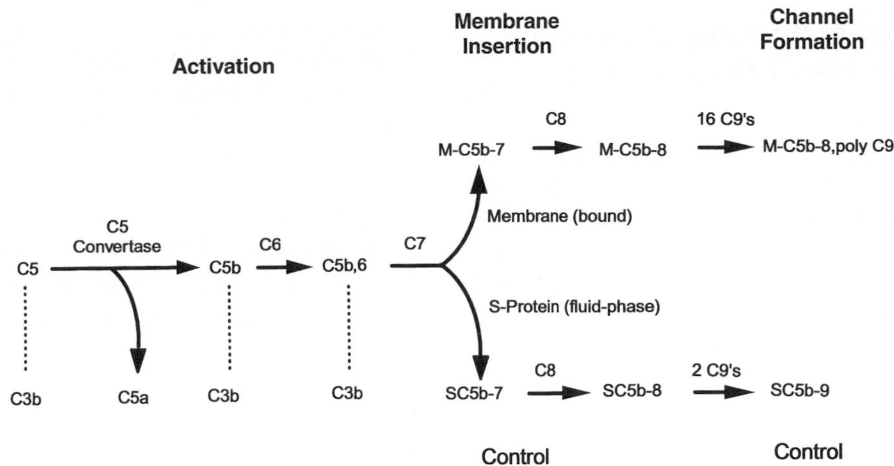

FIGURE 10 Activation of the terminal components and formation of the membrane attack complex.

3. Formation and Membrane Insertion of the C5b-7 Complex

The reactions described so far occur on the hydrophilic surface of membranes or particles, and the proteins involved retain their hydrophilic properties even after complex formation. Binding of C7 to C5b,6 causes an irreversible transition of the hydrophilic proteins to the amphiphilic (i.e., both hydrophobic and lipophilic) C5b-7 complex. In the complex of these three proteins (C5b,C6,C7), a site is exposed that is capable of binding to membranes. Insertion of C5b-7 into membranes bearing C3b is highly efficient and approaches 100%. If, on the other hand, the activating surface is not a phospholipid membrane, such as the protein surface of immune complexes or the carbohydrate of yeast cell walls, the C5b-7 has no substrate for hydrophobic insertion and the complex is released into the fluid phase.

4. Binding of C8 and Formation of Small Transmembrane Channels

On binding of C8 to C5b-7, the C8 portion of the complex inserts itself into the hydrocarbon core of the membrane. Functionally, C5b-8 creates small membrane pores with an effective diameter of approximately 10 Å.

5. Binding of C9, C9 Polymerization, and Formation of Membrane Lesions

Interaction of C9 with C5b-8 complexes causes polymerization of C9, forming "poly C9." Poly C9 is a hollow tubule formed by 12 to 18 molecules of C9 (Fig. 11). This hollow tubular structure (tMAC in Fig. 11) is responsible for poly C9 cytolytic action. Incomplete C9 tubular structures (nontubular MAC or nt-MAC in Fig. 11) are also observed in electron microscopy of MAC-lysed cells, and appear to consist of only 4–8 C9 molecules.

C5b-8 facilitates insertion of polymerizing C9 into membranes. The ultrastructure of the MAC corresponds to poly C9 with the exception that the C5b-8 complex in MAC is detectable as a long appendage attached to the torus of poly C9 (Figs. 10 and 11). The poly C9 is the channel through the membrane, and the subunits of C5b and C8 form the long appendage on MAC-poly C9 torus.

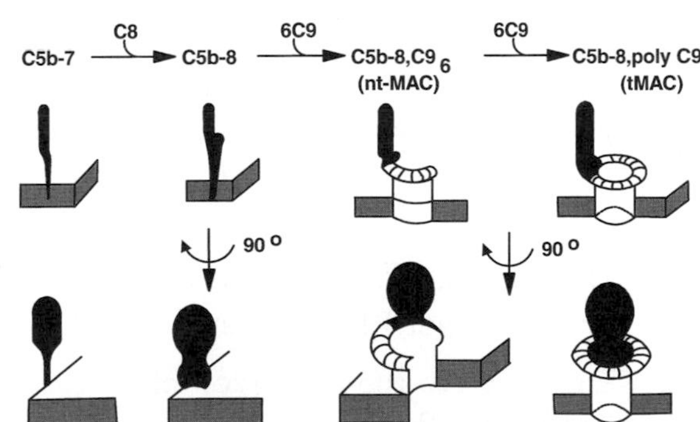

FIGURE 11 Appearance of the membrane attack complex at each stage of formation. These drawings were derived from photographs taken with an electron microscope. The components that make up each stage of the developing MAC complex are shown in the top row. In the bottom row, the appearance of the forming MAC complex is shown from a different angle produced by a 90° clockwise rotation of the MAC shown in the middle row.

B. Regulation of the MAC in the Fluid Phase and Host Cell Membranes

1. Control of the MAC by S Protein

Transfer of the amphiphilic C5b-7 complex between cells is prevented by the plasma S protein and lipoproteins that form a complex with the fluid phase C5b-7. Two to three molecules of S protein bind to released, but not membrane-inserted, C5b-7, giving rise to the SC5b-7 complex. This complex is fully water soluble and hence has lost its capability for membrane insertion. The SC5b-7 complex still reacts with C8 and C9; C9 polymerization, however, is inhibited and only two to three C9 molecules are incorporated into SC5b-9 (Fig. 10). S protein has been shown to be the same as vitronectin.

2. Protection of Host Cells by the Homologous Restriction Factor and CD59

The homologous restriction factor (HRF; also referred to as C8-binding protein) and CD59 are normal components of host cells that inhibit membrane insertion of the MAC. HRF shows homologous species specificity, such that HRF only protects human cells from human MAC; the HRF of most nonhuman cells does not protect them from human MAC. Human cells are lysed efficiently by rabbit MAC, whereas rabbit cells are lysed efficiently by human MAC. HRF acts at the stage of C8 insertion into membranes, whereas CD59 acts somewhat earlier at the stage of insertion of the C5b-7 complex. Similar to DAF, both the HRF and CD59 are attached to membranes via a phosphatidylinositol glycolipid and are deficient on red cells from patients with PNH. Red cells from patients with type III PNH are missing DAF, CD59, and HRF, leading to spontaneous hemolysis via the alternative pathway which occurs after approximately 6 days in circulation. Normal red cells circulate up to 90 days.

C. Functional Effects of the MAC on Target Membranes

1. Cell Death due to the Foramtion of Transmembrane Pores

The MAC forms a pore through membranes of cells that may penetrate through the entire membrane bilayer. MAC complexes that do penetrate the lipid bilayer cause leakage of small salt ions and the rapid uptake of water by the cell in an attempt to balance the higher osmotic pressure of the cellular cytoplasmic constituents that are too large to pass through the MAC pore. With red blood cells, this results in a rapid swelling of cells and hemolysis. Bacteria and nucleated cells may also be killed by the leakage of cellular salt ions without lysis.

2. Functional Effects of the MAC Independent of Pore Formation

The reorganization of lipid bilayers by MAC may adversely affect the stability of bilayer membranes. A second effect of MAC is caused by the displacement of membrane constituents by the insertion of large numbers of MAC. Because MAC occupies a relatively large area in the membrane, the insertion, e.g., into bacterial membranes, increases the total membrane surface area by more than twofold, causing loss of structural integrity. This displacement of membrane constituents and the consequent physical alteration and surface expansion of attacked membranes may cause cell death independent of the effects caused by the formation of pores.

3. Secondary Effects Contributing to Cell Death by the MAC

Pores created by the MAC in membranes of bacteria permit access to and degradation of the peptidoglycan layer of the bacterial cell wall by the lytic enzyme lysozyme. Membrane pores also allow the entry of Ca^{2+} into the intracellular space, triggering a variety of cellular reactions. Pore formation is accompanied by the breakdown of the voltage difference at the two sides of the membrane (membrane potential) and by an efflux of K^+ and entry of Na^+. Compensatory ion pumping along with cell activation by Ca^{2+} entry may be responsible for the rapid depletion of ATP and high energy phosphates observed in target cells. These effects may contribute to cell death. [*See* Ion Pumps.]

VI. INFLAMMATORY FUNCTION OF COMPLEMENT: ANAPHYLATOXINS

Anaphylatoxins[4] are small fragments derived from the complement components C3, C4, and C5 during com-

[4] This section on the inflammatory function of complement and the anaphylatoxins was excerpted with permission from Chenoweth, D. E. (1986). Complement mediators of Inflammation. *In* "Immunobiology of the Complement System: An Introduction for Research and Clinical Medicine" (G. D. Ross, ed.), pp. 63–86. Academic Press, Orlando, FL.

plement activation. As the larger "b" fragments are generated, the smaller "a" fragments formed simultaneously from the amino termini of each component's α-chain (C3a, C4a, and C5a) have the function of causing inflammation. These small polypeptides share a common biological activity termed *spasmogenicity* (i.e., they promote smooth muscle contraction and induce increased vascular permeability).

One particular anaphylatoxin, C5a, plays a unique physiologic role as a mediator of the inflammatory responses. This glycopolypeptide differs from C3a and C4a in three important ways. First, C5a is a considerably more potent biological effector than C3a or C4a. Second, C5a, but not C3a or C4a, retains significant biological activity in serum. Third, C5a exerts a series of unique effects on human granulocytes (blood white cells) and thus promotes their participation in the inflammatory process. [*See* Inflammation.]

A. Analysis of Anaphylatoxin Production as a Measure of Complement Activation

The extent of complement activation as well as the pathway responsible for the activation phenomena may be defined by quantitating anaphylatoxin production. Specific radioimmunoassay (RIA) procedures have been developed that permit selective quantitation of each anaphylatoxin (C3a, C4a, and C5a). Employing these assays, it has been shown that the detection of elevated levels of C4a can be considered as diagnostic of classical pathway activation events. In contrast, the appearance of increased levels of either C3a or C5a, without evidence of increased C4a, is evidence for activation of the alternative pathway without activation of the classical pathway. Initial studies have suggested that the monitoring of C3a levels in patients with autoimmune disease or rheumatoid arthritis is one way of assessing disease activity status. [*See* Radioimmunoassays.]

B. Control Mechanisms

1. Enzymatic Inactivation of the Anaphylatoxins

Once anaphylatoxins are formed, their spasmogenic activities are rapidly abrogated by a normal plasma enzyme [i.e., serum carboxypeptidase N (SCPN)] that acts as an anaphylatoxin inactivator. This enzyme removes the COOH-terminal arginine from each of the anaphylatoxin molecules and converts them to their "des-Arg" derivatives (Fig. 12). SCPN rapidly destroys all of the activity of the C3a and C4a anaphylatoxins. C5a is less affected because it is converted to its des-Arg derivative at a slower rate; in addition, ~5% of the C5a formed during complement activation is resistant to cleavage by SCPN. Moreover, while enzymatically degraded C3a and C4a are biologically inert, the degraded form of C5a (des-Arg-C5a) retains considerable activity as an inflammatory mediator.

2. Breakdown of C5a

A second type of unique control mechanism exists for the regulation of C5a activity. Both C5a and des-Arg-C5a bind to specific receptors on peripheral blood granulocytes. Once bound, the C5a is rapidly internalized by these cells and is completely degraded with loss of activity. Thus, the cells that are activated by C5a during the inflammatory response play the major role in inactivating the molecule.

C. Biologic Activities of the Anaphylatoxins

1. Spasmogenic Properties

The spasmogenic activities normally ascribed to the anaphylatoxins include the ability to induce smooth muscle contraction, promote increased vascular permeability, and cause the release of histamine from mast cells and basophils. Des-Arg-C5a also exhibits a lower level of spasmogenic activity.

The ability to induce smooth muscle contraction is usually defined by measuirng the contraction of

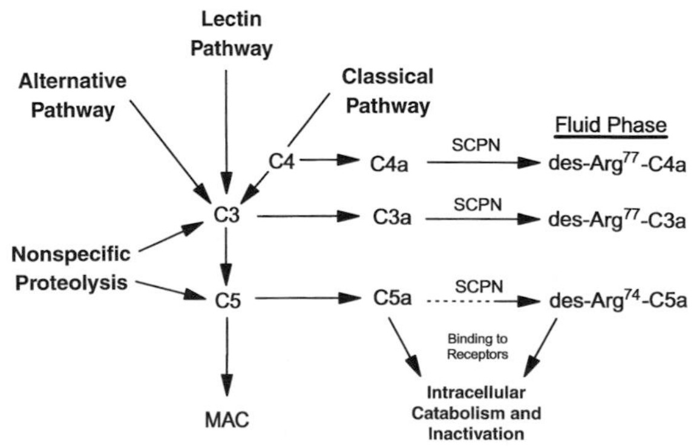

FIGURE 12 Generation and control of anaphylatoxins from the complement system.

guinea pig ileal or uterine tissues which are rich in smooth muscle. Increased vascular permeability may be assessed after intradermal injection of the anaphylatoxins into guinea pigs previously perfused with Evan's blue dye which produces skin bluing at the site of injection or into humans where it causes a wheal and flair response. [*See* Smooth Muscle.]

The spasmogenic properties of the anaphylatoxins may be manifest during the initial phases of the acute inflammatory response, when increased vascular permeability and tissue edema are readily apparent.

2. Granulocyte-Related Activities of Human C5a

C5a is the most potent complement-derived mediator of granulocyte responses thought to be critical to the inflammatory process. The main cellular responses include (1) chemotactic migration; (2) augmented adherence to cells; (3) degranulation, i.e., the release of internal granules full of lytic enzymes; and (4) production of toxic oxygen derivatives. These properties of C5a, rather than spasmogenic activity, account for this molecule's importance as an inflammatory mediator. In addition, both C5a and des-Arg-C5a promote the chemotactic migration of neutrophils (a type of granulocyte) and monocytes (another type of white blood cell). With neutrophils, C5a expresses measurable chemotactic activity at very low concentrations (10^{-10} *M*). Des-Arg-C5a is 10- 50-fold less active than C5a.

C5a and des-Arg-C5a also augment adherence and/or aggregation of neutrophils and monocytes. These phenomena are manifest as a profound transient loss of granulocytes from blood and accumulation in the lung vessels.

Conceptually, the production of even extremely low quantities (picomolar) of C5a at a localized site could promote the adherence of granulocytes to the endothelium, induce their chemotactic migration into the site, and prime them to destroy the eliciting agent. All of these events are observed in inflammatory foci and are important for host defense. However, both C5a and des-Arg-C5a may also act throughout the body and trigger similar types of granulocyte responses in blood or distant organs. When this occurs, C5a-activated granulocytes release cytotoxic substances that destroy normal tissues. In this case, normal host defense mechanisms may actually contribute to the causation of specific diseases (e.g., the adult respiratory distress syndrome or rheumatoid arthritis).

VII. OPSONIZATION AND MEMBRANE COMPLEMENT RECEPTORS

Opsonization[5] is the process by which particles are made readily ingestible by phagocytic cells. Serum proteins (opsonins) coat particles and cause them to bind avidly to phagocytes and trigger ingestion. The complement system plays a major role in opsonization by coating particles such as bacteria with C3. The bacteria then bind to the C3 receptors at the phagocyte surface, with clearance of the bacteria. Viruses, soluble antigen–antibody complexes, and tumor cells are opsonized and removed by a similar mechanism. Other serum proteins, particularly IgG antibacterial antibody and fibronectin, also opsonize bacteria. For each type of opsonin, phagocytes have an opsonin-specific membrane receptor responsible for binding particles coated with that opsonin. In the blood, C3-coated particles and immune complexes are bound first to red blood cells, which transport the bound complexes to macrophages in the liver. The complexes are stripped from the surface of the red cells, which then return to the circulation. At sites of infection, C activation generates C5a, which attracts phagocytic cells via C5a receptors. Once at the site of C activation, phagocytes utilize receptors for fragments of C3 to bind to bacteria or soluble complexes. This facilitates phagocytosis and the release of bactericidal substances and other mediators of inflammation. [*See* Antibody–Antigen Complexes: Biological Consequences.]

Nine types of C receptors are believed to exist, and structural data are now available on all except one.

A. Opsonization By Complement

Fixed C3 and IgG antibodies are the most important opsonins. On particles that activate either classical or alternative pathways, a major proportion of the fixed C3b is not protected from the control proteins of the alternative pathway (factors H and I) and is rapidly converted into fixed iC3b. This fixed iC3b is an important opsonin recognized by three types of phago-

[5] This section on opsonization and membrane complement receptors was excerpted with permission from Ross, G. D. (1986). Opsonization and membrane complement receptors. *In* "Immunobiology of the Complement System: An Introduction for Research and Clinical Medicine" (G. D. Ross, ed.), pp. 87–114. Academic Press, Orlando, FL.

cyte receptors (CR1, CR3, and CR4). Fixed iC3b that binds to CR1 is subsequently broken down into fixed C3dg by factor I; if plasmin or leukocyte elastase is present, the fixed C3dg may be then cleaved to fixed C3d (see Fig. 4). Fixed C3dg and C3d are poor opsonins *in vitro*, and in patients with cold agglutinin disease, red cells may circulate with as many as 20,000 fixed C3dg molecules per cell without being cleared by the macrophage phagocytic system.

B. Membrane Complement Receptors

Membrane C receptors of phagocytic cells have several important functions in mediating chemotaxis, phagocytosis, and release of cytotoxic and inflammatory substances. The functions of C receptors on nonphagocytic cell types, particularly lymphocytes, are less well defined.

The receptors for fixed C3 fragments have overlapping C3 fragment specificities and have been named according to their order of discovery rather than for their specificity. The reader should refer to the diagram of C3 fragment structure (see Fig. 4). Table I lists the major features of the four types of receptors for fixed C3 fragments.

1. Complement Receptor Type One (CR1)

CR1 (CD35) binds to fixed C3b and with lower affinity to fixed C4b and iC3b. Proteolysis of iC3b into C3dg destroys CR1 activity. Several types of cells express CR1, including red blood cells [erythrocytes (E)], phagocytic cells, lymphocytes (B cells and some T cells), neurons, and kidney podocytes. CR1 along with CR3 and CR4 (see Section VII,B,3,4) are the major opsonin receptors on phagocytic cells that promote the ingestion and killing of microorganisms. The CR1 of erythrocytes (E CR1) has two important functions in the clearance of circulating immune complexes. First, circulating immune complexes and particles that activate C are bound rapidly to E via CR1. Erythrocytes then serve as vehicles that transport the complexes to liver and splenic macrophages. Second, E CR1 serves as the cofactor for factor I cleavage of fixed iC3b (or fixed iC4b) into fixed C3dg (or fixed C4d) and fluid-phase C3c (or C4c). This is an important function in immune complex clearance, as fixed iC3b that is not degraded by this mechanism can attach immune complexes to neutrophils (via CR1 or CR3), triggering the activation of neutrophils and release of cytotoxic enzymes, leukotrienes, and toxic oxygen metabolites. While this is an important mechanism for recognition and destruction of bacteria with fixed iC3b on their surface, it can also be turned against the host and result in neutrophil-mediated tissue damage if the fixed iC3b is present on small immune complexes that become trapped in normal tissues [e.g., it is an important mechanism of autoimmune diseases such as systemic lupus erythematosus (SLE)]. [*See* Neutrophils.]

TABLE I
Receptors for Bound Fragments of C3

Type (WHO name)	Leukocyte Antigen Workshop name	Specificity	Structure	Cell-type expression
CR1	CD35	C3b > C4b > iC3b	Four allotypes: A = 190 kDa B = 220 kDa C = 250 kDa D = 160 kDa	Erythrocytes—very low B cells—high; T cells—low Neutrophils, monocytes—high Macrophages—very low Kidney podocytes—high Neurons—high
CR2	CD21	iC3b = C3dg = C3d ≫ C3b Epstein–Barr virus, CD23	140 kDa	B cells—high Thymocytes—very low Follicular dendritic cells—high Pharyngeal epithelial cells—low
CR3	CD11b (α chain) CD18 (β chain)	iC3b > C3dg > C3d, β-glucan, ICAM-1, fibrogen, factor X, collagen, heparan sulfate	165 kDa α chain 95 kDa β chain	Neutrophils, monocytes—high NK cells, activated T cells—high Macrophages—low
CR4	CD11c (α chain) CD18 (β chain)	iC3b > C3dg, fibrinogen	150 kDa α chain 95 kDa β chain	Neutrophils, monocytes—low NK cells—low Macrophages—high

The function of B-cell CR1 has not been completely defined, but several studies have shown that its triggering can enhance activation of the B cells stimulated by other factors. CR1 also aids in antigen recognition by attaching C3b- or iC3b-bearing antigens onto the B-cell surface. No data are available on the possible functions of T-cell CR1. Only small subsets of both helper and suppressor cells express CR1, and the amount of CR1 expressed per cell is approximately 10% of that expressed by B cells.

CR1 exists in four forms that vary in molecular mass from 160 to 250 kDa. These differences result from the presence of variable numbers of homologous 60–70 amino acid repeating sequence motifs ["short consensus repeats" (SCR)] that form the extracellular domain. Homologous SCR structures are found in seven other C3/C4-binding proteins: CR2, factor B, factor H, C4BP, C2, MCP, and DAF. The genes for factor H, C4BP, CR2, MCP, and DAF have been shown to map to a region on chromosome 1 that has been named the regulator of complement activation (RCA) gene locus.

2. Complement Receptor Type Two (CR2)

CR2 (CD21) is expressed by B lymphocytes, lymph node follicular dendritic cells, pharyngeal epithelial cells, and thymocytes, but is absent from mature T lymphocytes. CR2 is not only a receptor for C3 fragments but also has attachment sites for Epstein–Barr virus (EBV) and the IgE Fc-receptor CD23. CR2 is specific for the C3d portion of iC3b, C3dg, and C3d. EBV or large complexes containing these C3 fragments bind to CR2 and cause the activation of B cells. EBV also uses CR2 to gain entry into B cells, thereby causing either a virus infection (the disease infectious mononucleosis) or a rare form of cancer (Burkitt's lymphoma).

CR2 consists of a single polypeptide chain of 140 kDa whose extracellular domain is made up of 15–16 SCR units with a high degree of homology to the SCR units of CR1. The CR2 gene is linked to the CR1 gene.

CR2 has an important function in facilitating the recognition of antigens by B cells. The blockade of CR2 in mice greatly diminishes humoral immune responses to primary protein antigens. Similar findings of defective humoral immunity were also made with mice in which CR2 expression was prevented by a homologous recombination strategy that destroyed the Crry gene that encodes both CR1 and CR2 in the mouse via alternative splicing of mRNA.

3. Complement Receptor Type Three (CR3)

CR3 (also known as Mac-1, CD11b/CD18, or $\alpha_M\beta_2$-integrin) is a major opsonin receptor on neutrophils and monocyte/macrophages involved in the clearance of bacteria. CR3 has dual functions as a receptor for iC3b and as the β_2-integrin adhesion molecule used by granulocytes to adhere to the extracellular matrix. CR3 binds with high affinity to fixed iC3b and with much lower affinity to fixed C3dg or C3d. CR3 also contains a separate lectin site for the recognition of microbial polysaccharides such as yeast cell wall β-glucans and an inducible high-affinity binding site for the intercellular adhesion molecule ICAM-1. The lectin site has major importance in the granulocyte recognition of bacteria and yeast cell walls, and this site must be occupied by a microbial polysaccharide in order for CR3 to trigger phagocytosis of an iC3b-opsonized bacteria or yeast. In addition, CR3 functions, along with CR4 (see Section VII,B,4), to mediate neutrophil adherence to endothelial cells during inflammatory reactions via exposure of a high-affinity binding site for ICAM-1. CR3 may also express high-affinity binding sites for other extracellular matrix proteins such as collagen, fibrinogen, and heparan sulfate. Neutrophils stimulated by C5a express greatly increased amounts of CR3 and CR4, a proportion of which express the high-affinity binding sites for ICAM-1 and fibrinogen, thus promoting neutrophil attachment to the vascular endothelium and allowing the neutrophils to migrate out of blood vessels into sites of infection. Once at the site of infection, CR3 allows attachment of the phagocytes to bacteria and yeast that bear fixed iC3b and β-glucan-like polysaccharides, promoting phagocytosis and a respiratory burst. Patients with a rare inherited deficiency of CR3 and CR4 (leukocyte adhesion deficiency) have repeated life-threatening bacterial infections because their neutrophils are unable either to migrate into sites of infection or to kill serum-opsonized bacteria.

CR3 consists of two glycoprotein chains known as α (165 kDa, the CD11b antigen) and β (95 kDa, the CD18 antigen) that are noncovalently associated.

4. Complement Receptor Type Four (CR4)

CR4 is closely related to CR3, sharing the same β chain, and with 87% homologous α chain. CR4 appears to have the same C3 fragment specificity as CR3, but is expressed preferentially on tissue macrophages. CR3 and CR4 molecules expressed by phagocytic cells differ in their linkage to the cytoskeleton. CR3 is less restrained than CR4 by cytoskeletal connections and consequently is more mobile than CR4. As a result, relatively small amounts of CR3 can aggregate rapidly at the site of particle contact, promoting particle attachment to the cell. Cytoskeleton-linked CR4 initiates the ingestion of particles that are trapped on the membrane via CR3.

5. C5a Receptor

The C5a receptor (CD88) of neutrophils and monocyte/macrophages, a glycoprotein of 40–47 kDa, is responsible for mediating inflammatory responses to C5a and des-Arg-C5a. Recent studies have shown that the same C5a receptor molecule is also expressed by skin mast cells and hepatocytes.

VIII. HUMAN DISEASES IN WHICH COMPLEMENT IS A SIGNIFICANT FACTOR

A. Autoimmune Diseases

Autoimmune diseases frequently involve the development of autoantibodies to normal tissue or blood cells. The immune complexes generated activate the classical pathway, causing major organ damage from complement-mediated cytotoxicity and complement-mediated recruitment of inflammatory cells, which release damaging cytotoxic substances. The major diseases in this class include autoimmune hemolytic anemia, SLE, and rheumatoid arthritis. [*See* Autoimmune Disease.]

B. Inherited Deficiencies of the Complement System

These diseases are very rare, and many physicians may never see one in their clinical practices (Table II). Patients who have a genetic deficiency of C3 are unable to utilize any complement pathway or function in host defense, and usually have a long history of repeated and life-threatening bacterial infections. In contrast, patients with deficiencies of one of the early classical pathway components (C1q, C4, or C2) have an illness resembling SLE, resulting from a diminished ability to clear circulating antigen–antibody (immune) complexes. In these patients the normal immune complex clearance mechanism is absent because the classical pathway does not progress to the C3 stage at which immune complexes are normally attached to erythrocyte CR1 and transported to the macrophage phagocytic system. Patients with deficiencies of one of the control proteins of the alternative pathway (factor H or factor I) are unable to control normal spontaneous activation of the alternative pathway. The continuous consumption of complement in these patients results in low level C3, factor B, and all of the terminal components. Such patients thus have an acquired C3 deficiency because their C3 is continuously consumed by the uncontrolled alternative pathway. As might be expected, they have the same types and frequencies of infections as patients with a primary genetic deficiency of C3. Patients with a deficiency of properdin, factor D, or one of the terminal components (C5, C6, C7, or C8) have recurrent systemic infections with *Neisseria* bacteria, but do not have problems with other types of bacterial infections. In patients with neisserial infection, apparently the cytotoxic function of complement in host defense cannot be replaced by any other function. Deficiency of serum MBP has also become recognized as a deficiency of the C system. Particularly serious in infancy before the development of anti-bacterial immunity, children with MBP deficiency are unable to opsonize bacteria and yeast that have cell walls rich in mannose.

TABLE II
Genetic Deficiency Diseases of the Complement System

Deficient component	Clinical characteristics
C1q	SLE, nephritis, recurrent bacterial infections, hypogammaglobulinemia, high mortality rate
C1r or C1s	SLE, recurrent bacterial infections, arthritis
C1-INH	Hereditary angioedema
C4 or C2	SLE
C3, factor H, or factor I	Recurrent infections with pyogenic bacteria
Factor D or properdin	Recurrent bacterial infections, disseminated gonococcal or meningococcal (*neisseria* bacteria) infections
C5, C6, C7, or C8	Disseminated gonococcal or meningococcal infections; SLE
C9	No apparent disease
Mannan-binding protein	Recurrent bacterial infections and chronic diarrhea in infancy
CR3 and CR4	Recurrent bacterial infections of the skin and gingiva ("leukocyte adhesion deficiency")

C. Deficiency of C1 Inhibitor; Hereditary Angioedema

The most common genetic deficiency of the complement system among Caucasians is deficiency of the C1-INH, which results in the disease called hereditary angioedema (HAE). Absence of C1-INH results in the consumption of plasma C4 and C2 because any spontaneous activation of C1 in the fluid phase is uncontrolled. Cleavage of C4 and C2 in the fluid phase does not result in C4b2 complex formation and thus does not generate a C3-convertase. As a result, only C4 and C2 are consumed without C3 consumption. A diagnosis of HAE is suggested if patients have low levels of C4 but normal levels of C3 because all other disease processes that consume C4 also consume C3. Because patients with HAE have an acquired deficiency of C4 and C2, many develop a SLE-like illness resembling patients with an inherited deficiency of C4 or C2.

Patients with HAE have occasional attacks of severe swelling in the epiglottis, with danger of asphyxiation, or in the extremities. Attacks may be years apart, with the first one occurring sometimes as late as 30 years of age. Most patients have low levels (20 to 50% of normal) of a functionally active C1-INH, whereas 20% have normal levels of a functionally inactive C1-INH. Both types of patients have been treated successfully with androgens which increase plasma C1-INH levels. The disease is autosomal dominant, so that patients have one normal and one abnormal C1-INH gene, and androgens are thought to increase the synthesis of C1-INH by the normal gene. Treatment with normal C1-INH protein may become possible as a result of the cloning of C1-INH cDNA.

BIBLIOGRAPHY

Colten, H. R., and Rosen, F. S. (1992). Complement deficiencies. *Annu. Rev. Immunol.* **10,** 809–834.

Davies, A., and Lachmann, P. J. (1993). Membrane defence against complement lysis: The structure and biological properties of CD59. *Immunol. Res.* **12,** 258–275.

Esser, A. F. (1994). The membrane attack complex of complement: Assembly, structure and cytotoxic activity. *Toxicology* **87,** 229–247.

Fearon, D. T. (1993). The CD19-CR2-TAPA-1 complex, CD45 and signaling by the antigen receptor of B lymphocytes. *Curr. Opin. Immunol.* **5,** 341–348.

Glovsky, M. M. (1994). Applications of complement determinations in human disease. *Ann. Allergy* **72,** 477–486.

Morgan, B. P. (1995). Complement regulatory molecules: Application to therapy and transplantation. *Immunol. Today* **16,** 257–259.

Pangburn, M. K. (1988). Alternative pathway of complement. *Methods Enzymol.* **162,** 639–652.

Reid, K. B. M., and Turner, M. W. (1994). Mammalian lectins in activation and clearance mechanisms involving the complement system. *Springer Semin. Immunopathol.* **15,** 307–325.

Ross, G. D., and Větvička, V. (1993). CR3 (CD11b,CD18): A phagocyte and NK cell membrane receptor with multiple ligand specificities and functions. *Clin. Exp. Immunol.* **92,** 181–184.

ISBN 0-12-226972-1